COMPREHENSIVE
POLYMER SCIENCE

IN 7 VOLUMES

COMPREHENSIVE POLYMER SCIENCE

The Synthesis, Characterization, Reactions & Applications of Polymers

CHAIRMAN OF THE EDITORIAL BOARD

SIR GEOFFREY ALLEN, FRS

Unilever Research and Engineering, London, UK

DEPUTY CHAIRMAN OF THE EDITORIAL BOARD

JOHN C. BEVINGTON

University of Lancaster, UK

Volume 1

Polymer Characterization

VOLUME EDITORS

COLIN BOOTH & COLIN PRICE

University of Manchester, UK

PERGAMON PRESS

OXFORD · NEW YORK · BEIJING · FRANKFURT
SÃO PAULO · SYDNEY · TOKYO · TORONTO

U.K.	Pergamon Press plc, Headington Hill Hall, Oxford OX3 0BW, England
U.S.A.	Pergamon Press Inc., Maxwell House, Fairview Park, Elmsford, New York 10523, U.S.A.
PEOPLE'S REPUBLIC OF CHINA	Pergamon Press, Room 4037, Qianmen Hotel, Beijing, People's Republic of China
FEDERAL REPUBLIC OF GERMANY	Pergamon Press GmbH, Hammerweg 6, D-6242 Kronberg, Federal Republic of Germany
BRAZIL	Pergamon Editora Ltda, Rua Eça de Queiros, 346, CEP 04011, Paraiso, São Paulo, Brazil
AUSTRALIA	Pergamon Press Australia Pty Ltd., P.O. Box 544, Potts Point, N.S.W. 2011, Australia
JAPAN	Pergamon Press, 5th Floor, Matsuoka Central Building, 1-7-1 Nishishinjuku, Shinjuku-ku, Tokyo 160, Japan
CANADA	Pergamon Press Canada Ltd., Suite No 271, 253 College Street, Toronto, Ontario, Canada M5T 1R5

First edition 1989

Library of Congress Cataloging-in-Publication Data

Comprehensive polymer science.

Includes index.
Contents: v. 1. Polymer characterization/volume editors,
Colin Booth & Colin Price—v. 2. Polymer properties/volume
editors, Colin Booth & Colin Price—[etc.]—v. 7. Specialty
polymers & polymer processing/volume editor,
Sundar L. Aggarwal.
1. Polymers and polymerization. I. Allen,
G. (Geoffrey), 1928–   . II. Bevington, J. C.
QD381.C66 1988   547.7   88–25548

British Library Cataloguing in Publication Data

Comprehensive polymer science.

Vol. 1: Polymer characterization
1. Polymers
I. Allen, Geoffrey   II. Bevington, John C.
III. Booth, Colin   IV. Price, Colin
547.7

ISBN 0-08-036205-2 (vol. 1)
ISBN 0-08-032515-7 (set)

Contents

Preface

It is only 60 years since Staudinger's model of the molecular nature of a polymer was becoming universally accepted and the physical states of rubbers, plastics and fibres understood. Unfortunately, for some time many academic chemists continued not to appreciate the full significance of polymerization reactions and physicists tended to regard polymeric materials as inevitably being of indeterminate composition and unamenable to study by conventional physical methods.

Nevertheless, in the 1930s the foundations were laid for the understanding of the main polymerization mechanisms. An industry based on synthetic rubbers, plastics and fibres was soon established. In World War II it played a major strategic role and afterwards grew to be one of the main elements of the heavy chemicals industry. It became recognized that synthetics may be superior to natural materials in their properties and that they may be used for completely new purposes.

Alongside the production of well-defined materials there grew the ability to characterize the structure of polymer molecules and to understand the relationships between methods of preparation and subsequent treatment, structure and properties, both chemical and physical. As a result, a vast literature of polymer science and technology has been generated and four Nobel prizes awarded specifically for contributions to polymer science. Add to this the fact that many biological molecules, including polypeptides, enzymes, antibodies, carbohydrates and so on, are polymers of varying degrees of complexity, then the universality of polymers in the physical and biological sciences and technologies forms a dominant modern theme.

Comprehensive Polymer Science is a series of volumes designed to set down the structure of this vast subject in such a way that researchers and teachers of polymer science and workers in associated fields can find an authoritative and comprehensive account of the topic of immediate interest. That topic is set out in a framework of related subjects. The text is focused on synthetic polymers with little reference to biological macromolecules *per se* but the science underpins both physical and biological systems.

To ensure that the wide coverage is maintained at an authoritative level, more than 250 authors from 20 countries have been enlisted. Their contributions have been organized into a series of major themes:

Volume 1	Polymer Characterization
Volume 2	Polymer Properties
Volumes 3–5	Polymerization Mechanisms
Volume 6	Polymer Reactions
Volume 7	Specialty Polymers & Polymer Processing

Because of the wide coverage the editors were presented with a particularly difficult decision with regard to symbols and nomenclature. The latter does not follow strictly the recommendations of IUPAC nor are symbols consistent throughout the whole work. However, usage in a particular chapter is consistent with the practice in the current literature. Thus a reader will be able to frame new publications in the context of the information presented in this series of volumes.

We should like to acknowledge the way in which the staff at the publisher, particularly Dr Colin Drayton (who initially proposed the project), Dr Helen McPherson and their editorial team, have supported the editors and authors in their endeavour to produce a text that is both complete and up-to-date and that will appeal to industrial and academic researchers alike. *Comprehensive Polymer Science* is a milestone in the literature of the subject in terms of coverage, clarity and a sustained high level of presentation.

GEOFFREY ALLEN
London

JOHN C. BEVINGTON
Lancaster

Contributors to Volume 1

Professor Sir Geoffrey Allen
Unilever Research & Engineering, PO Box 68, Unilever House, Blackfriars, London EC4P 4BQ, UK

Dr N. S. Allen
Faculty of Science and Engineering, John Dalton Building, Manchester Polytechnic, Chester Street, Manchester M1 5GD, UK

Professor E. D. T. Atkins
H. H. Wills Physics Laboratory, University of Bristol, Tyndall Avenue, Bristol BS8 1TL, UK

Professor D. C. Bassett
Department of Physics, J. J. Thomson Physical Laboratory, University of Reading, Whiteknights, PO Box 220, Reading RG6 2AF, UK

Dr G. S. Beddard
Department of Chemistry, University of Manchester, Manchester M13 9PL, UK

Dr C. Booth
Department of Chemistry, University of Manchester, Manchester M13 9PL, UK

Dr F. A. Bovey
AT&T Bell Laboratories, 600 Mountain Avenue, Murray Hill, NJ 07974, USA

Dr D. Briggs
ICI Plc, Petrochemicals and Plastics Division, PO Box 90, Wilton Centre, Middlesbrough, Cleveland TS6 8JE, UK

Dr P. M. Budd
BP Research Centre, Chertsey Road, Sunbury-on-Thames, Middlesex TW16 7LN, UK

Professor G. G. Cameron
Department of Chemistry, University of Aberdeen, Meston Walk, Old Aberdeen AB9 2HE, UK

Professor B. T. P. Chu
Department of Chemistry and of Materials Science and Engineering, State University of New York, Stony Brook, Long Island, NY 11794-3400, USA

Professor F. Ciardelli
Dipartimento di Chimica e Chimica Industriale, Università di Pisa, 35 Via Risorgimento, I-56100 Pisa, Italy

Dr R. O. Colclough
Department of Chemistry, University of Manchester, Manchester M13 9PL, UK

Dr J. V. Dawkins
Department of Chemistry, Loughborough University of Technology, Loughborough, Leicestershire LE11 3TU, UK

Professor D. L. Dorset
Electron Diffraction Department, Medical Foundation of Buffalo Inc., 73 High Street, Buffalo, NY 14203, USA

Dr H. Geerissen
Johannes-Gutenberg-Universität, Institut für Physikalische Chemie, Jakob-Welder-Weg 13, D-6500, Mainz, Federal Republic of Germany

Professor J. C. Giddings
Department of Chemistry, University of Utah, Salt Lake City, UT 84112, USA

Professor G. Glöckner
Technische Universität Dresden, Sektion Chemie, Wissenschaftsbereich, Hochpolymere und Textilchemie, Mommsenstrasse 13, DDR-8027 Dresden, German Democratic Republic

Dr J. J. Gunderson
Dow Chemical Co., Midland, MI 48667, USA

Dr I. H. Hall
Department of Pure and Applied Physics, University of Manchester Institute of Science and Technology, PO Box 88, Sackville Street, Manchester M60 1QD, UK

Professor A. E. Hamielec
McMaster Institute for Polymer Production Technology, Department of Chemical Engineering, McMaster University, Hamilton, Ontario L8S 4L7, Canada

Dr T. Hammond
Birmingham Polymer Group, Department of Chemistry, University of Birmingham, PO Box 363, Birmingham B15 2TT, UK

Dr F. Heatley
Department of Chemistry, University of Manchester, Manchester M13 9PL, UK

Mr D. A. Hemsley
Institute of Polymer Technology, Loughborough University of Technology, Loughborough, Leicestershire LE11 3TU, UK

Dr D. M. Hindenlang
Analytical Research Division, Allied-Signal Inc., Corporate Technology, PO Box 1021R, Morristown, NJ 07960, USA

Professor S. L. Hsu
Department of Polymer and Engineering Science, University of Massachusetts, Amherst, MA 01003, USA

Professor A. D. Jenkins
School of Chemistry and Molecular Sciences, University of Sussex, Falmer, Brighton, BN1 9QJ, UK

Dr K. Kamide
Fundamental Research Laboratory of Fibers and Fiber-forming Polymers, Asahi Chemical Industry Company Ltd., 11-7 Hacchonawate, Takatsuki, Osaka 569, Japan

Professor I. A. Katime
Departamento Quimica Fisica, Facultad de Ciencias, Universidad del País Vasco, Apartado 644, Bilbao, Spain

Dr T. A. King
Department of Physics, Schuster Laboratory, University of Manchester, Manchester M13 9PL, UK

Dr L. A. Kleintjens
Tossaintstr. 5, 6171 HN Stein, The Netherlands

Dr R. Koningsveld
P Visschersstr. 23, 6174 RB Sweikjuizen, The Netherlands

Dr R. S. Lehrle
Birmingham Polymer Group, Department of Chemistry, University of Birmingham, PO Box 363, Birmingham B15 2TT, UK

Dr K. L. Loening
Chemical Abstracts Service, PO Box 3012, 2540 Olentangy River Road, Columbus, OH 43210, USA

Dr P. A. Lovell
Polymer Science and Technology Group, University of Manchester Institute of Science and Technology, PO Box 88, Sackville Street, Manchester M60 1QD, UK

Professor V. J. McBrierty
Department of Pure and Applied Physics, University of Dublin, Trinity College, Dublin 2, Ireland

Dr J. R. MacCallum
Department of Chemistry, University of St Andrews, St Andrews, Fife KY16 9ST, UK

Dr G. R. Mitchell
Department of Physics, J. J. Thomson Physical Laboratory, University of Reading, Whiteknights, Reading RG6 2AF, UK

Dr J. R. Quintana
Departamento Quimica Fisica, Facultad de Ciencias, Universidad del País Vasco, Apartado 644, Bilbao, Spain

Dr R. W. Richards
Department of Pure and Applied Science, Thomas Graham Building, University of Strathclyde, 295 Cathedral Street, Glasgow G1 1XL, UK

Dr M. J. Richardson
Division of Materials Applications, National Physical Laboratory, Teddington, Middlesex TW11 0LW, UK

Dr D. M. Sadler[†]
H. H. Wills Physics Laboratory, University of Bristol, Tyndall Avenue, Bristol BS8 1TL, UK

Dr N. Schlotter
Bell Communications Research, Navesink Research Center, Room 3X 163, 331 Newman Springs Road, Red Bank, NJ 07701-7020, USA

Dr P. Schützeichel
Johannes-Gutenberg-Universität, Institut für Physikalische Chemie, Jakob-Welder-Weg 13, D-6500, Mainz, Federal Republic of Germany

Dr R. D. Sedgwick
Analytical Research Division, Allied-Signal Inc., Corporate Technology, PO Box 1021R,
Morristown, NJ 07960, USA

Dr M. Styring
McMaster Institute for Polymer Production Technology, Department of Chemical Engineering,
McMaster University, Hamilton, Ontario L8S 4L7, Canada

Dr M. Tsuji
Laboratory of Polymer Crystals, Institute for Chemical Research, Kyoto University, Uji,
Kyoto-fu 611, Japan

Dr B. A. Wolf
Johannes-Gutenberg-Universität, Institut für Physikalische Chemie, Jakob-Welder-Weg 13,
D-6500, Mainz, Federal Republic of Germany

Contents of All Volumes

Volume 4   Chain Polymerization II

Volume 5   Step Polymerization

1
Perspectives

GEOFFREY ALLEN
Unilever Research and Engineering, London, UK

1.1 INTRODUCTION

From the standpoint of western science, the origins of plastics technology lie with the medieval Guilds of Horners. Horn softens at 125 °C and can be shaped into a variety of utensils. Alternatively, it can be rolled into thin translucent sheets originally used in lanterns and windows. The Worshipful Company of Horners was founded in the City of London in the 12th century; it is extant but now many of its members are associated with the modern plastics industry.

The entry of rubber into western civilization is more precisely documented. Natural rubber in the form of balls was brought to Europe from the New World by Columbus returning from his second voyage in 1496. There were various reports of ball games played by the Indians, and in 1585 the chronicler Duran wrote, '. . . the ball is as big as a small ball used in 9-pins . . . It has one property which is that it jumps and rebounds upwards and continues jumping from here to there so that those who run after it become tired before they catch it.'

1.2 THE ORIGINS OF RUBBER TECHNOLOGY

In 1748, Charles-Marie de la Condamine,[1] returning from a voyage of scientific exploration in South America, sent to his sponsors, the Academie de Science in Paris, specimens of natural rubber, called cauchu. He wrote, 'There grows a tree which the natives call Hleve. When the bark is cut a white, milky fluid runs out and hardens in the air.' He described the collection of latex from the *Hevea brasiliensis* tree and how rubber utensils were made by drying it down by smoking over leaf fires.

In 1763, cauchu, dissolved in turpentine, was sold as a rubber cement in France. Seven years later, in the preface to 'A Familiar Introduction to the Theory and Practice of Perspective', Joseph Priestley[2] wrote: 'Since this work was printed off, I have seen a substance excellently adapted to the purpose of wiping from paper the marks of a black-lead pencil. It must therefore be of a singular use to those who practice drawing. It is sold by Mr. Nairne, Mathematical Instrument Maker, opposite the Royal Exchange. He sells a cubial piece of about half an inch for three shillings and he says it will last several years'. Thus, cauchu became 'rubber' in English, whereas in France it remains 'caoutchou'.

The first waterproof clothes were manufactured from natural rubber in the UK in 1791, and in 1823 MacIntosh[3] produced much improved products laminated on both sides with cotton fabric. The world production of natural rubber rose to 100 tonnes per annum. One of the drawbacks of the raw rubber used to waterproof garments was its tendency to crystallize into a rigid form at about 0 °C.

Charles Goodyear's work on the sulfur cure of natural rubber was announced[4] in 1844, and, shortly after, Hancock introduced the term 'vulcanization'. From the point of view of rubber technology, the suppression of flow and crystallization brought about by curing was a major step towards turning natural rubber into an engineering material.

By the end of the century, natural rubber plantations were established in Malaysia. The invention of the internal combustion engine and Dunlop's pneumatic tyre[5] (1888) swelled the increasing demand for natural rubber to 10 000 tonnes per annum worldwide.

This phase of rubber technology culminated with the development of synthetic rubbers. Harries,[6] in 1906, polymerized isoprene using a sodium catalyst to form the first synthetic analogue of natural rubber. Hofmann[7] (1907) of Bayer synthesized natural rubber by heating isoprene in an autoclave. However, because 2,3-dimethylbutadiene was more readily available at lower cost, Hofmann used this monomer to make 'methyl rubber'. The blockade of Germany during World War I revived this process, and production ultimately reached 1500 tonnes per annum.

1.3 THE ORIGINS OF PLASTICS TECHNOLOGY

The first synthetic plastic reported was cellulose nitrate, described by Schonbein[8] in 1846, but his attention quickly turned to its exploitation as gun cotton. However, in 1862 Parkes[9] won a medal for his novel material Parkesine and built a factory at Hackney Wick to manufacture this cellulose derivative. Not surprisingly, the factory burned down in 1868. Shortly after this, the material was patented as Celluloid in 1870 by the Hyatts[10] and they won a prize of $10 000 for a substitute material for billiard balls! Next, cellulose acetate was introduced by Cross and Bevan in 1892[11] and it was later used to make viscose rayon fibre.

About the same period, Baekeland[12] emigrated from Ghent to the USA, *via* the UK. In 1907 he announced the first thermosetting resin, 'Bakelite', a member of the phenol–formaldehyde range of plastics.

Thus, at the outbreak of World War I, there were three industrial plastics: cellulose nitrate, cellulose acetate and 'Bakelite'. They were used principally in electrical and aeronautical applications.

1.4 THE CONCEPT OF A POLYMER MOLECULE

Early attempts to determine the chemical nature of polymeric materials were hampered by three factors which made chemists reluctant to adopt the modern concept.

First, organic chemists in the 19th century strove to find precise formulae and to isolate pure substances. Thus, in 1826, Faraday[13] determined the empirical formula of natural rubber to be C_8H_7. Isoprene was obtained[14] from natural rubber by destructive distillation in 1860 and, although the unit C_5H_8 was firmly established, Harries, in 1905, argued[15] for the association of isoprene dimers (structure 1). Only in 1914 did he concede[16] that larger rings of five or seven isoprene units may be involved. A similar progression of ideas marked the evolution of chemical structures for cellulose, starch and proteins. At the turn of the century most scientists did not distinguish between large covalent structures and aggregates of smaller molecules formed by partial bonding or physical association.

$$\begin{bmatrix} CH_2CH=CMeCH_2 \\ | \qquad\qquad | \\ CH_2CH=CMeCH_2 \end{bmatrix}_n$$

(1)

Thomas Graham's identification[16] of colloid substances in 1861, which drew attention to the very slow diffusion of these substances in solution and their inability to pass through semipermeable membranes, was a second obstacle. Although this led early investigators to the view that natural rubber, cellulose, starch and some proteins comprised large molecules, these physical chemists were also preoccupied with explanations in terms of secondary association of smaller molecular units — which was called 'polymerization'.

Third, the introduction by Raoult[17] in 1881 of the cryoscopic method for the measurement of molecular weight in solution should have clinched the issue. Molecular weights of the order of 36 700 for starch and *ca.* 10 000 for cellulose nitrate were established by 1900. Gladstone and Hibbert[18] found the molecular weight of natural rubber to be 6000 'to at least' 12 000. These results were considered to be uncertain because the measurements were at the limits of experimental capability. Certainly it would have been difficult to extrapolate to dilute solution conditions. Further, a slow rate of dissolution and sensitivity to heat added to the doubts. Again, there was an inclination to categorize them as colloidal solutions in which the solute molecules are aggregated.

Thus, although there were fledgling rubber and plastics industries by 1918, few scientists were prepared to consider the molecular structure of polymers as hundreds of repeat units linked together into very large covalently bound molecules.

The next decade saw a dramatic change. Polanyi[19] and then, more comprehensively, Mark and Meyer[20, 21] made systematic X-ray diffraction studies of crystalline polymers, particularly cellulose, silk fibrosin and natural rubber. Up to 1926, these studies established the unit cell dimensions, and demonstrated that they were of the order of the repeat unit in the structure and certainly not large enough to contain a large molecule. In 1928, in a series of papers,[21] Mark and Meyer concluded that the crystal structure of these materials resulted from the packing of long chain molecules. Ultimately, Mark showed that the unit cell dimensions of natural rubber were consistent with one specific structure (**2**), *i.e. cis*-polyisoprene.

(**2**)

Staudinger was undoubtedly the most potent contributor to the present concept of the macro (polymer) molecule. From 1918 to 1920, whilst working at the ETH in Zurich, he transferred his interests from the organic chemistry of ketenes and aliphatic diazo compounds to the study of high molecular weight compounds. He was told by Wieland to: 'drop the idea of large organic molecules . . . Purify your rubber . . . then it will crystallize!' In fact, most of his colleagues in the early 1920s were still of the opinion that such materials were colloidal aggregates of small molecules.

Staudinger[22] first published his concept that high molecular compounds were composed of long covalently bound molecules in 1919. In the following year he exemplified this thesis by describing polymers (or macromolecules, as he called them) of formaldehyde, styrene, vinyl bromide and isoprene. The free valences at the ends of each polymer chain were argued to have low reactivity because, in a high molecular weight material, they would be present in very low concentration, there being only one free valence for several hundred repeat units in a linear molecule.

Having introduced this concept without experimental support, Staudinger went on to deal with the widespread belief that the colloidal properties of polymers stemmed from the aggregation of smaller units by secondary valence forces or physical interactions. He demonstrated that natural rubber could not be distilled in a high vacuum. No lower hydrocarbons were obtained and, furthermore, the colloidal properties of the product were very similar to those of the original material.

The mainspring of modern polymer science is encapsulated in the next series of experiments carried out by Staudinger. He switched from the study of natural molecules to synthesize a series of model polymers, which included poly(oxymethylene), poly(oxyethylene) and polystyrene. The systematic studies on poly(oxymethylene) oligomers and polymers, reported in 1929,[23] demonstrated that the long chain molecules had normal valence bonds and were terminated by conventional groups. The average chain length was determined by estimating the concentration of end groups. The ability of this polymer to crystallize and to form single crystals was demonstrated. This squashed the notion that very long chain molecules would produce only amorphous materials. The work on polystyrene demonstrated that the samples were polydisperse, and that samples prepared under different conditions had different physical properties; notably, different solution viscosities.

Thus, by the end of the decade there was good evidence for the chain formulae (3) and (4) (polystyrene and poly(oxymethylene), respectively) and the chemical structure of *cis*-polyisoprene had been clearly set out. Staudinger supposed that polymer molecules might contain as many as 100 repeat units.

(3)   (4)

The whole episode was capped by a series of papers by Carothers[24] in which he used condensation reactions to prepare polymers having definitive long chain structures, polyesters and polyamides *etc.* The way was now clear for the systematic development of polymer chemistry.

This period also marked the beginning of the systematic study of polymer properties and their relation to chemical structure. In 1929, Meyer and Mark[25] proposed that the properties of vulcanized natural rubber were due to crosslinks between the essentially linear polyisoprene molecules. Staudinger endorsed the concept of a molecular network. Meyer and Mark attempted to deduce the size of crystallites in cellulose from the breadth of the X-ray diffraction spots, and concluded the dimensions corresponded to chain lengths of 50 to 100 units. Staudinger challenged their view, maintaining there was no relationship between the size of the polymer molecule and crystallite size.

Ultimately, Staudinger showed that the dilute solution viscosity of a polymer was related to average chain lengths. He deduced that the relative viscosity was related to molecular weight through the relationship shown in equation (1), which gives a simple method for estimating molecular weight. Unfortunately, this equation leads to the underestimation of molecular weight by a factor of 10 or so. But there was controversy over the use of dilute solution properties to measure molecular weights of individual chains. Meyer and Mark concluded that, even in dilute solutions, polymer molecules aggregated strongly. Staudinger took the opposite view that the molecules were simply dispersed and that molecular weight could be deduced from dilute solution properties.

$$\left(\frac{\eta_r - 1}{c}\right)_{c \to 0} = KM \tag{1}$$

1.5 FIFTY YEARS OF POLYMER CHEMISTRY

Once it was demonstrated that large macromolecules existed, and that they obeyed the rules of synthesis and structure already familiar in smaller molecules, many research schools in industry and academe began the systematic synthesis of new (and old) polymers.

Staudinger had already proposed one of the major routes to long molecules, namely the successive addition of unsaturated molecules and chain reaction to form long molecules (equation 2). A variation on this theme is the ring-opening mechanism, again using an initiator, as shown in equation (3). Carothers, as we have seen, immediately identified condensation reactions as a major route to polymeric materials (equation 4).

$$R\cdot + CH_2{=}CHX \to RCH_2\dot{C}HX \xrightarrow{CH_2=CHX} \cdots \xrightarrow{CH_2=CHX} R(CH_2CHX)_n CH_2\dot{C}HX \tag{2}$$

$$RO^- + \text{(epoxide)} \longrightarrow ROCH_2CH_2O^- \xrightarrow{} \cdots \xrightarrow{} RO(CH_2CH_2O)_n CH_2CH_2O^- \tag{3}$$

$$2HO(CH_2)_n CO_2H \to HO(CH_2)_n CO_2(CH_2)_n CO_2H + H_2O \xrightarrow{HO(CH_2)_nCO_2H} HO(CH_2)_n CO_2(CH_2)_n CO_2(CH_2)_n CO_2H + H_2O \to \cdots \tag{4}$$

For many years these remained the principal routes to organic polymers. Inorganic polymers were made by specific reactions of functional groups of the condensation type rather than by the initiation of unsaturated monomers. The siloxane polymers, however, provide an excellent example of the use of the ring-opening mechanism.

1.5.1 Condensation Polymerization

The work of Carothers at Du Pont, in the period 1929 to 1938, produced a whole range of industrially important polymers *via* the condensation route. The polyamide family of the nylons was quickly identified, and, by the middle of the decade, the potential of one particular member, nylon 6,6 (**5**), as a fibre-forming polymer was recognized. A variety of polyesters were also examined, and these were to lead eventually to the poly(ethylene terephthalate) series of polymers developed industrially in the next few decades in the UK. Nylon and Terylene became the major synthetic fibres.

$$-NH(CH_2)_6NH \text{---} [CO(CH_2)_4CONH(CH_2)_6NH]_n CO(CH_2)_4CO-$$

(**5**)

Not only was the synthetic chemistry studied but, in the 1930s, the basic understanding of the kinetics of polymerization and molecular weight distributions was developed.

Apart from the condensation of difunctional intermediates, there was in this period a growing understanding of multifunctional intermediates, and the recognition that the thermosetting polymers of the type already introduced into industry by Baekeland *et al.* were essentially three-dimensional chemical network structures produced by the condensation of phenol and formaldehyde. The way was open for a variety of 'thermosetting' products.

1.5.2 Addition Polymerization

The formation of a long chain molecule by the repeated addition of unsaturated monomer units requires an initial activation of one monomer, thereafter the chain propagates through the activated species. For a vinyl monomer and an initiator, the process is as shown in Scheme 1. In the early 1930s it was quickly recognized that free radicals were suitable initiators, though many of the ionic initiators were not fully explored until a decade later.

$$R\cdot + M \rightarrow RM\cdot$$

$$RM\cdot + M \rightarrow RMM\cdot$$

$$RM_n M\cdot + M \rightarrow RM_{n+1}M\cdot$$

Scheme 1

1.5.2.1 Free radical polymerization

Taylor and Bates[27] and, separately, Staudinger[28] suggested free radical initiation; the former to account for the polymerization of ethylene in the gas phase, the latter for liquid-phase polymerization of vinyl monomers. There was a clear recognition of a two-step process, chain initiation followed by chain propagation. Organic peroxides, which readily decompose to release free radicals, became popular initiators (equation 5), but azo compounds and azobisnitriles, which are thermally more stable, were soon added to the list. With this polymerization mechanism established, one could look back in history and account for observations by Hofmann and his co-workers on the conversion of 'styrole into a solid substance' and, later, on evidence for photopolymerization. Indeed, in the mid-19th century Bertholet[29] had noted that unsaturated compounds could be polymerized by the application of heat, by a simultaneous reaction and by the 'influence of the nascent state'.

$$(RCO_2)_2 \rightarrow 2RCO_2\cdot \rightarrow 2R\cdot + 2CO_2 \tag{5}$$

As in the case of condensation polymerization, there followed complementary studies of the kinetics of polymerization and of molecular weight distribution. Staudinger and Frost[30] found that the chain length of polystyrene remains constant with increase in monomer concentration. Shultz[31] interpreted this to mean that both chain growth (propagation) and termination involve an interaction of the same species including the 'activated' chain end, as shown, for example, in Scheme 2.

$$RCH_2\dot{C}HX + CH_2=CHX \rightarrow \ldots \xrightarrow{CH_2=CHX} R(CH_2CHX)_n CH_2\dot{C}HX \xrightarrow{CH_2=CHX} R(CH_2CHX)_{n+1}CH_2\dot{C}HX$$

$$2R(CH_2CHX)_n CH_2\dot{C}HX \rightarrow R(CH_2CHX)_n CH_2CHXCHXCH_2(CH_2CHX)_n R$$

Scheme 2

Hence, the chain length would be independent of the rate of initiation. Schultz showed that the distribution of molecular weight is directly related to the mechanism of polymerization. In conditions where the rates of propagation and termination become independent of chain length, Schultz derived a distribution function for molecular weight which remains a cornerstone of our understanding of free radical polymerization. In 1937, termination by disproportionation (equation 6) and chain transfer of the active site, which leads to branched molecular structures (equation 7) were identified by Flory,[32] essentially to complete the free radical polymerization reaction scheme.

$$R(CH_2CHX)_nCH_2\dot{C}HX + R(CH_2CHX)_mCH_2\dot{C}HX \rightarrow R(CH_2CHX)_nCH=CHX + R(CH_2CHX)_mCH_2CH_2X \quad (6)$$

$$\sim\sim CH_2\dot{C}HX \; + \; \sim\sim CH_2CHX\sim\sim CH_2\dot{C}HX \longrightarrow \sim\sim CH_2CH_2X \; + \; \sim\sim CH_2\dot{C}X\sim\sim CH_2\dot{C}HX \quad (7)$$

Equally important in polymer technology in the 1930s was the range of polymers manufactured by free radical polymerization. They included polystyrene, poly(vinyl chloride), high pressure polyethylene and poly(methyl methacrylate), and, in the early 1940s, the most successful synthetic rubber, GRS, a styrene–butadiene copolymer.

1.5.2.2 *Ionic polymerization*

We have noted that methyl rubber, *i.e.* the product of dimethylbutadiene polymerization in the presence of sodium, was of strategic importance in World War I. Stimulated by this, Ziegler[33] produced a series of papers from 1929 to 1934 which proposed an anionic mechanism for this type of polymerization. The reaction of butadiene with cumyl potassium was shown to be a succession of 1,4-additions of the organometallic compound to the double bonds of the monomer (equation 8). Eventually some 1,2-addition of butadiene was noted.

$$R^-K^+ + CH_2=CHCH=CH_2 \rightarrow RCH_2CH=CHCH_2^- K^+ \xrightarrow{CH_2=CHCH=CH_2} R(CH_2CH=CHCH_2)_nCH_2CH=CHCH_2^- K^+ \quad (8)$$

Although the anionic polymerization mechanism was recognized, it was not until 1952 that the kinetics of polymerization were clearly established. Higginson and Wooding[34] used potassium amide in liquid ammonia to polymerize styrene by an anionic addition mechanism (Scheme 3).

$$KNH_2 \rightarrow K^+ + NH_2^-$$

$$NH_2^- + CH_2=\bar{C}HX \rightarrow NH_2CH_2\bar{C}HX$$

$$NH_2(CH_2CHX)_nCH_2\bar{C}HX + CH_2=CHX \rightarrow NH_2(CH_2CHX)_{n+1}CH_2\bar{C}HX$$

$$NH_2(CH_2CHX)_nCH_2\bar{C}HX + NH_3 \rightarrow NH_2(CH_2CHX)_nCH_2CH_2X + NH_2^-$$

Scheme 3

Finally, in 1956, Szwarc[35] used sodium naphthalene in tetrahydrofuran to demonstrate the formation of polystyrene with a very narrow molecular weight arising from the rapid initiation of a number of chains and the absence of a termination mechanism. Flory[36] had shown in 1940 that in such circumstances the polymer should have a Poisson distribution of molecular weight.

The initiation of addition polymerization in vinyl monomers by cationic mechanisms was not clearly recognized until the late 1940s. Polanyi and Evans[37] and Plesch[38,39] *et al.* showed that Friedel–Crafts catalysts, such as BF_3, $AlCl_3$, $TlCl_4$, $SnCl_4$ and, in some circumstances, H_2SO_4, were initiators for styrene, isobutene, butadiene, *etc.* The reactions were very rapid at very low temperatures, and molecular weights were found to be inversely related to the temperature of polymerization. All the catalysts were very strong electron acceptors and the role of water has been a source of great controversy. The chain propagation step is almost certainly the addition of monomer to a carbonium ion (equation 9). However, cationic polymerization proved difficult to study, not only because of the role of adventitious water, but also because of chain transfer and the propensity of carbonium ions to isomerize.

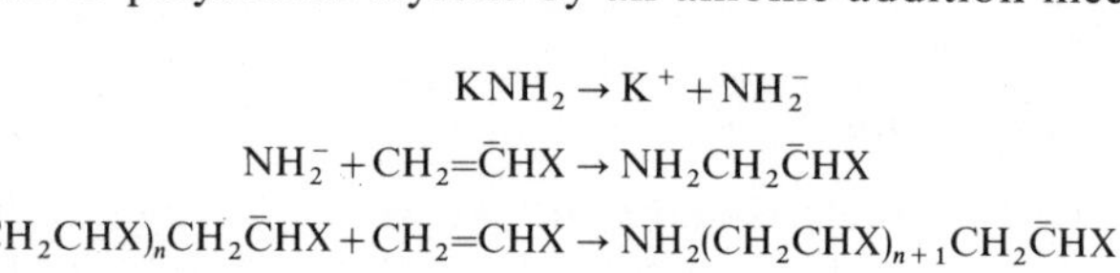

$$\sim\sim CH_2\overset{X}{\underset{Y}{\overset{|}{\underset{|}{C^+}}}} \; + \; CH_2=CXY \longrightarrow \sim\sim CH_2\overset{X}{\underset{Y}{\overset{|}{\underset{|}{C}}}}CH_2\overset{X}{\underset{Y}{\overset{|}{\underset{|}{C^+}}}} \quad (9)$$

1.5.3 Coordination Catalysts

In 1950, Ziegler and Gellert reported[40] that ethylene could add to alkyllithium compounds to extend the alkyl chain. They went on to show that LiAlH$_4$ and triethylaluminum each incorporated ethylene even faster than tetraethyllithium aluminum compounds. These experiments were carried out under pressure in autoclaves, and it was discovered that nickel which had been leached from the stainless steel walls was an inhibitor of polymerization. At that time Ziegler was attempting to produce butane by this type of reaction, and so other metals were investigated. As a result, the crucial discovery was made that triethylaluminum in the presence of transition metal compounds could catalyze the polymerization of ethylene even at atmospheric pressure. Furthermore, the product was different from that obtained from high pressure free radical polymerization. The new material had a much higher melting point (130–140 °C) and density and fewer methyl end groups per molecule and was soon recognized to be 'linear' polyethylene.

Natta[41] was stimulated by these discoveries to use similar catalysts to polymerize propylene. He expected to obtain a rubbery polymer but instead obtained a crystalline material of high molecular weight, melting at 170 °C. The crystal structure showed that there were three monomer units per unit cell and Natta deduced that in long segments of the chain there were runs of asymmetric carbons of the same configuration. Natta also announced the synthesis of linear crystalline polystyrene and poly(1-butene). Thus, stereoregular polymers were discovered and α-alkene polymers, which were previously only known in amorphous states, were now available in crystalline variants. Isotactic and syndiotactic stereoregularity was identified.

Although the use of polymerization conditions to control stereoregularity was discussed by Huggins[42] in 1944, and Schildknecht[43] had noted that heterogeneous catalysts could lead to the formation of a crystalline product, the work of Ziegler and Natta marked the introduction of sterically controlled polymerization. Industry quickly took up this new family of catalysts to make series of polyethylenes, polypropylenes, synthetic *cis*-1,4-polyisoprenes, and so on — all with controlled stereoregularity. The discoveries coincided with the growth of the petrochemical industry and thus the impact was compounded.

1.6 POLYMER SCIENCE

Major developments in our understanding of the properties of individual polymer molecules and assemblies of polymer molecules also followed Staudinger's work of the 1920s, and was bound in with the evolution of addition and condensation polymer chemistry.

Arguments concerning the shape of polymer molecules did not derive from their chemical formulae, as organic chemists were usually content to write down a two-dimensional representation; structure (6), for example. Curiosity about the molecular structure of rubbers and fibres, and particularly about the origin of rubber elasticity, led to the development of statistical models of polymer chains. As early as 1802 Gough[44] had shown that natural rubber became warm when stretched, whereas other solids cooled. In 1859 Joule[45] actually measured the temperature rise as a function of extension, and showed that a rubber strip held under tension retracted as its temperature was raised. He noted differences in behaviour between raw and vulcanized natural rubber. In 1926 Woehlich[46] drew an analogy between the contraction of a tendon on heating and the thermoelastic behaviour of natural rubber, and concluded that the retractive force of a stretched rubber is due to the thermal motion of rigid rod-like particles opposing the alignment induced by stretching. Meyer[47] dismissed this explanation on the grounds that randomly oriented rods in the unstretched form would occupy a much larger volume than the rods arranged in parallel in the oriented state. Joule had in fact noted a slight increase in volume in the stretched specimens.

$$-\overset{\overset{\displaystyle H}{|}}{\underset{\underset{\displaystyle H}{|}}{C}}-\overset{\overset{\displaystyle H}{|}}{\underset{\underset{\displaystyle R}{|}}{C}}-\overset{\overset{\displaystyle H}{|}}{\underset{\underset{\displaystyle H}{|}}{C}}-\overset{\overset{\displaystyle H}{|}}{\underset{\underset{\displaystyle R}{|}}{C}}-\overset{\overset{\displaystyle H}{|}}{\underset{\underset{\displaystyle H}{|}}{C}}-\overset{\overset{\displaystyle H}{|}}{\underset{\underset{\displaystyle R}{|}}{C}}-\overset{\overset{\displaystyle H}{|}}{\underset{\underset{\displaystyle H}{|}}{C}}-\overset{\overset{\displaystyle H}{|}}{\underset{\underset{\displaystyle R}{|}}{C}}-$$

(6)

The concept of a change in shape of the polymer molecule was introduced. Mark *et al.* suggested[48] that in the relaxed state the rubber molecule adopted a spiral conformation, and that on stretching there was a shift from predominantly intramolecular attraction of the double bonds in the chain to intermolecular interaction between chains. Staudinger pointed out[49] that other polymers which did not have strongly interacting bonds on the backbone were also rubbers at high temperatures. Meyer, v. Susick and Valko, in 1932, finally identified[50] the absorption of heat on retraction of a rubber strip as the key to the origin of rubber elasticity, *i.e.* the retractive force arises from a transformation from a state of higher entropy to an elongated state of lower entropy. In the same year Busse added an important molecular concept[51] to this thermodynamic observation, by proposing that internal rotations about single bonds produced 'kinky shapes' in long molecules. To him, the fact that an elastic body could be deformed several hundred per cent and recover implied that there were structural units at the molecular level which also have this property. Thus, the stage was set for the introduction of a statistical model of a polymer molecule and also for the kinetic theory of rubber elasticity.

1.6.1   The Polymer Molecule

Kuhn introduced the representation of a polymer molecule as a random coil[52] in 1934. The simplest model is one in which the molecule is defined by n links each of length l, the links being completely freely jointed to each other and having no cross section. Thus the chain can pass through itself — a phantom chain! Although this model does not correspond to a real polymer chain, which has fixed bond angles, hindered internal rotation and a finite thickness, the statistical behaviour of a random (phantom) coil was already known. It is mathematically identical to the diffusion of a random flight in three dimensions of n equal steps of length l, already solved by Rayleigh[53] in 1919. Thus, for the random coil, its fully extended contour length is nl, but the vector displacement distance r between the chain ends in its random flight conformations has a mean square value of r^2 given by equation (10). Thus the root mean square end-to-end vector distance r is proportional to $n^{\frac{1}{2}}$. Comparison of $\langle r^2 \rangle^{\frac{1}{2}}$ with the contour length nl which represents the limit of extension of the polymer chain shows that the extensibility of a random phantom coil is $n^{\frac{1}{2}}$. Since n is usually > 100 for a polymer chain, this simple model demonstrates the origin of long range elasticity in a polymer molecule.

$$\langle r^2 \rangle = nl^2 \tag{10}$$

Kuhn realized that fixed bond angles would expand the statistical dimensions beyond the freely jointed random coil value. Eyring[54] demonstrated that, for a chain with a fixed bond angle between the links, $\langle r^2 \rangle$ was given by equation (11) at large values of n. Thus, for a polyethylene chain where $\theta = 109° \ 21'$, $\langle r^2 \rangle = 2.0 \ nl^2$. A decade later[55] the effect of hindered internal rotation was analyzed. If ϕ is the internal angle of rotation of a third bond in the chain with respect to the two preceding bonds, defining the transconformation as $\phi = 0$, then equation (12) is obtained. Again the result holds for large values of n and values of $\cos \phi$ not too near unity. In most chains different rotational isomers have different energy values, so that $\overline{\cos} \, \phi$ is in fact temperature dependent, and hence so is $\langle r^2 \rangle$. This rotational isomeric model, which takes into account bond lengths, bond angles and hindered internal rotation in the polymer chain, is now used for detailed statistical analyses of real polymers.

$$\langle r^2 \rangle = nl^2 \left(\frac{1 - \cos \theta}{1 + \cos \theta} \right) \tag{11}$$

$$\langle r^2 \rangle = nl^2 \left(\frac{1 - \cos \theta}{1 + \cos \theta} \right) \left(\frac{1 + \overline{\cos} \, \phi}{1 - \overline{\cos} \, \phi} \right) \tag{12}$$

Kuhn also realized[52] that the finite thickness of the polymer chain would also lead to a chain expansion in an isolated chain. He assumed that this excluded volume effect would cause the mean square end-to-end displacement of the chain to increase with a power of n higher than unity (equation 13).

$$\langle r^2 \rangle = n^{(1+v)} l^2 \tag{13}$$

The excluded volume problem proved to be the most difficult. Not until 1960 was it possible[56] to use Monte Carlo methods to generate large numbers of random steps on a computer. The results were not independent of n, and values for v ranged from 0.17 to 0.22 and were not independent of the type of lattice used. However, in 1964 Edwards[57] demonstrated that, in the limit of an infinitely long chain, with excluded volume, $v = 1/5$. Thus, there now exists a comprehensive theoretical framework for the statistical models of a polymer chain.

1.6.2 Polymer Solutions

Earlier in this chapter we noted that studies of colligative properties of polymer solutions at the turn of the century had delayed the recognition of the molecular characteristics of polymer chains. However, following the development of a statistical model for the polymer chain, Flory[58] and, separately, Huggins[59] in 1942 reported a quantitative statistical mechanical theory of a polymer solution. A lattice model was used to compute the entropy of mixing. Random mixing was assumed, subject only to the constraint that n contiguous cells are required to accommodate one polymer molecule. An energy of mixing was obtained by introducing a simple polymer–solvent interaction parameter. In 1947 Gee and Treloar reported[60] measurements of the vapour pressure of benzene over natural rubber in benzene solution, which supported the Flory–Huggins analysis over a very wide range of concentrations. The theory and experiment showed conclusively that polymer solutions do not obey the laws of ideal solutions because the latter are based on the mole fraction as the composition variable. This supposes that the effect of a polymer molecule on the activity of the solvent would be the same as that of an ordinary solute molecule, which, after all, would be no more than the size of one segment in the polymer chain. The Flory–Huggins theory shows that polymer solutions are better described in terms of the volume fraction as the composition variable.

Flory pointed out,[61] however, that in a poor solvent, cooled to the point of incipient precipitation, the solution would then obey ideal solution laws. At this point the polymer chains would display random flight statistics.

Simultaneously with the understanding of solution thermodynamics, the light-scattering technique was developed by Debye and his co-workers to measure the radius of gyration and hence deduce $\langle r^2 \rangle$ for a polymer in dilute solution. By carrying out the measurements at the temperature at which the solution displays ideal behaviour, it has been possible to measure the random flight (unperturbed) dimensions of polymer chains as a function of molecular weight and structure and thus explore the validity of the statistical models.

1.6.3 Bulk Polymers

Although the elastic properties of natural rubber first drew the attention of scientists and technologists to the properties of polymers in bulk, it was soon apparent, from the stiffening observed in rubber garments exposed for long periods to freezing temperatures and from an experiment carried out by Gough[44] on an extended rubber strip immersed in the cold waters of Lake Ullswater, that even natural rubber could exist in another form. Vulcanization not only reduced the tendency of raw rubber to flow (creep) under a load, it also suppressed the stiffening which occurred on exposure to freezing temperatures. By 1850 it was known that natural rubber existed in two forms.

Staudinger showed conclusively in the 1920s that the high temperature form of each polymer was a rubber and that at lower temperatures some, such as polystyrene, existed as amorphous solids, whilst other polymers, such as poly(methylene oxide), were highly crystalline. Thus, there were at least three states of matter identified for polymeric materials, *viz.* rubber, amorphous solid and partially crystalline solid.

1.6.3.1 *The rubber state*

By 1934, the work of Mark, Meyer and Kuhn had led to the molecular model for a raw rubber as a tangled mass of random coiled molecules each exploring a whole set of random conformations by means of Brownian motion of the repeat units. The conversion of this mass into a network by means of crosslinking at infrequent random intervals along the chains was recognized to be the process of vulcanization.

The kinetic theory of rubber was first developed in its present from[62] by Kuhn in a paper published in 1936. Essentially it was assumed that there were sufficient repeat units between crosslinking points for each chain segment thus defined to act as a random coil. He then attempted to calculate the elastic modulus from the probability distribution of chain extension in undeformed and stretched rubber. This model was developed further in the next few years by Treloar,[63] Flory[64] and by James and Guth.[65] The entropy change of the network was calculated as a function of the extension ratio. The energy change was assumed to be zero and the stress obtained from equation (14), where N is the number of chains in the network and λ is the extension ratio.

$$f=\left(\frac{\delta\Delta F}{\delta l}\right)_{\mathrm{T,V}} \approx \left(\frac{\delta\Delta S}{\delta l}\right)_{\mathrm{T,V}} = NkT(\lambda^2 - \lambda^{-1}) \tag{14}$$

The important conclusions that $f\alpha T$, the absolute temperature, was in accord with several experimental observations, most notably those of Meyer and Ferri[66] on natural rubber. Even more important is the conclusion that the modulus of a rubber is simply dependent on the number of chains in the network, *i.e.* the degree of crosslinking, and that the modulus is independent of the chemical nature of the rubber.

These conclusions formed the basis of rubber technology and the industry developed very rapidly over the next decade. This simple model based on phantom chains is still a useful guide to the science and technology of rubber. Only in the mid-1960s was it proven[67] that there was in fact a small energy component in the stress supported by rubber networks.

We now recognize that the raw rubber is the liquid state of a polymer. Flow occurs slowly because the wriggling chains are entangled and have an inherently low probability of all the segments of chain moving simultaneously in the same direction. Crosslinking maintains all the chains in a permanent network so that flow is suppressed.

1.6.3.2   The glassy state

Meyer and Ferri's careful study[66] on the temperature dependence of the stress in an extended natural rubber held at constant length showed conclusively that the stress was proportional to T. However, at 200 K the slope of the stress–temperature curve suddenly changed sign, and further cooling caused the stress to rise sharply to very high levels. The strip had become inextensible, the 'rubber' elasticity had disappeared. These observations were mirrored in changes in heat capacity in polyisobutene, again at about 200 K, and in changes in the coefficient of thermal expansion in polystyrene at about 370 K. There were no sudden discontinuities in the primary thermodynamic functions F, H, V, *etc.* at this transition. Ultimately, it was recognized that this regime is the amorphous solid form of the polymer, in which the Brownian motion of the chains had been frozen out and they now constituted a tangled rigid mass, *i.e.* the glassy state.

Dielectric and mechanical relaxation measurements eventually showed that the glass transition is associated with the freezing-in of the backbone motion of the polymer chains. Side groups and end groups continue to rotate about the rigid main chain in the glassy state.

1.6.3.3   Crystallization

The first recorded observation of crystallization in a polymer is due to Gough who wrote[44] in 1805: ' . . . If a thong of caoutchoue be extended in water warmer than itself it retains its elasticity. If the experiment be made in water colder than itself it loses part of its retractile power, being unable to recover its former figure, but let the thong be placed in hot water whilst it remains extended for want of spring . . . and it contracts briskly'.

The temperature of the experiment was in the range 0–10 °C, and thus some 75 K above the glass transition, and the phenomenon was certainly crystallization of the stretched rubber followed by melting.

Crystallization was recognized in viscose fibres, cellulose, proteins, *etc.* and natural rubber when the X-ray method was used in the early part of this century to determine molecular structure. We have noted that Staudinger clearly identified poly(oxymethylene) as a crystalline material during the same period.

However, in 1924 Hock[68] quantified Gough's observations, and noted that retraction of the stretched rubber strip did not occur if the sample was kept below 18 °C. Later he observed that, if the stretched unretractable sample was cooled to liquid air temperature and shattered, fibrils were obtained. It was soon realized that these extended specimens were partially crystalline and oriented at the molecular level. The technique of drawing the specimen was to become an essential operation in making synthetic fibres, *e.g.* nylon, poly(ethylene terephthlate), *etc.*

Thus it became clear that polymers could exist in either an amorphous solid state or as partially crystalline material. Bekkedahl's dilatometric study[69] on isotropic natural rubber, published in 1934

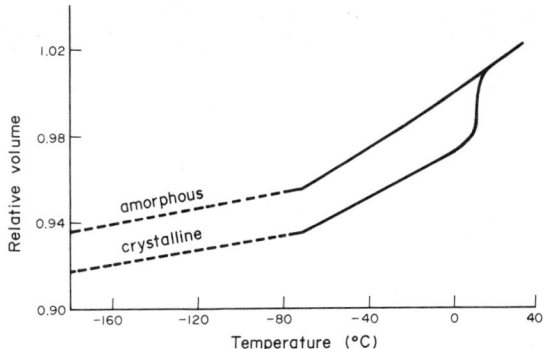

Figure 1 Changes in volume of non-crystalline and partially crystalline natural rubber as a function of temperature

and shown in Figure 1, summarizes neatly the solidification of a crystallizable polymer. If natural rubber is cooled slowly, crystallization occurs in the region of 283 K leading to a fairly sharp contraction in volume. The crystallization process stops at about 273 K and further cooling produces the normal contraction expected of an organic solid. However, at 200 K the rate of contraction diminishes, which is the transformation of the remaining rubbery material into the glass. Thus, above 293 K the specimen is in the rubber state, below 283 K it is partially crystalline and the remaining matrix is rubber. Below 200 K the sample is partially crystalline but now the matrix is glass. On rapid cooling the sample does not have time to crystallize and there is simply a transformation from rubber to glass at 200 K, as shown by the upper curve.

At the time of these experiments some polymers were partially crystallizable and displayed behaviour similar to natural rubber. Others such as polystyrene and poly(methyl methacrylate) were amorphous, simply transforming from rubber to glass on cooling. However, in 1955 several of these amorphous polymers became available with appreciable amounts of stereoregular structures. Such polymers were also crystallizable and behaved similarly to natural rubber.

Thus, the three states of bulk polymers were clearly recognized by the mid-1930s and, with the recognition that some polymer structures were able to generate liquid crystal structures, an adequate foundation was provided for understanding the physical properties of all polymeric materials now available.

1.6.4 Retrospect

This first chapter attempts to give an overview of the growth of polymer science and technology. It is remarkable that the period 1920–1930 transformed our understanding of the nature of polymer molecules. Equally the next 25 years, 1930–1955, saw polymer science comprehensively established. Few basic concepts or new phenomena have been added since then, except that Edwards and de Gennes have spearheaded the redefinition[57] of the theory of polymerized matter.

Despite this rapid evolution of a new science and its importance, not only in physical but also in biological systems, there are still many chemists and physicists who are reluctant to recognize fully its impact. Polymer technology however is difficult to ignore. It now accounts for about 35% of the chemical industry and has strong connections in the industries based on biotechnology.

1.7 REFERENCES

1. C. M. de la Condamine, *Mem. Acad. R. Sci.*, 1751, **17**, 319.
2. J. Priestley, 'A Familiar Introduction to the Theory and Practice of Perspective', Johnson, London, 1770.
3. C. MacIntosh, *Proc. R. Soc. London*, 1843, **5**, 486.
4. C. Goodyear, *US Pat.* 3633 (1844).
5. J. B. Dunlop, *Br. Pat.* 10 607 (1888).
6. C. Harries, *Ber. Dtsch. Chem. Ges.*, **37**, 2708; *Ber. Dtsch. Chem. Ges.*, 1905, **38**, 1195.
7. F. Hofmann, *Chem.-Ztg.*, 1936, **60**, 693.
8. C. F. Schonbein, *Philos. Mag.*, 1847, **31**, 7.
9. M. Kaufman, 'The First Century of Plastics: Celluloid and its Sequel', Plastics and Rubber Institute, London, 1963, chap. 1.
10. J. W. Hyatt and I. J. Hyatt, *US Pat.* 105 338 (1870).

11. C. F. Cross, E. J. Bevan and C. Beadle, *US Pat.* 520 770 (1984).
12. L. H. Baekeland, *US Pat.* 942 699 (1907).
13. M. Faraday, *Q. J. Sci. Arts*, 1826, **21**, 19.
14. G. Williams, *Philos. Trans. R. Soc. London*, 1860, **9**, 254.
15. C. Harries, *Ber. Dtsch. Chem. Ges.*, 1904, **37**, 2708; *Ber. Dtsch. Chem. Ges.*, 1905, **38**, 1195.
16. T. Graham, *Philos. Trans. R. Soc. London*, 1861, **151**, 183.
17. F. M. Raoult, *Ann. Chim. Phys.*, 1888, 6, **15**, 375.
18. J. H. Gladstone and W. Hibbert, *Philos. Mag.*, 1889, 5, **28**, 38.
19. M. Polanyi, *Naturwissenschaften*, 1922, **10**, 411.
20. H. Mark, *Ber. Dtsch. Chem. Ges.*, 1926, **59**, 2982.
21. K. H. Meyer and H. Mark, *Ber. Dtsch. Chem. Ges.*, 1928, **61**, 593, 1932, 1936, 1939.
22. H. Staudinger, *Schweiz. Chem.-Ztg.*, 1919, 1, **28**, 60.
23. H. Staudinger, R. Singer, H. Johner, M. Luthy, W. Kern, D. Russidis and O. Schweitzer, *Justus Liebigs Ann. Chem.*, 1929, **474**, 145.
24. H. W. Carothers, *Chem. Rev.*, 1931, **8**, 353.
25. K. H. Meyer and H. Mark, *Ber. Dtsch. Chem. Ges.*, 1928, **61**, 1939.
26. H. Staudinger, *Z. Angew. Chem.*, 1929, **42**, 67.
27. J. R. Bates and H. S. Taylor, *J. Am. Chem. Soc.*, 1927, **49**, 2438.
28. H. Staudinger, 'Die Hochmolekularen Organischen Verbindungen Kautschuk and Cellulose', Springer, Berlin, 1932, p. 151.
29. M. Betholet, 'Lecons de Chimie professees en 1864 et 1866', Societe Chimique de Paris, Paris, 1866, p. 18.
30. H. Staudinger and W. Frost, *Z. Phys. Chem. B*, 1935, **29**, 235.
31. G. V. Schultz, *Z. Phys. Chem. B*, 1935, **30**, 379.
32. P. J. Flory, *J. Am. Chem. Soc.*, 1937, **59**, 241.
33. K. Ziegler, L. Jakob, H. Wollthan and A. Wenz, *Justus Liebigs Ann. Chem.*, 1934, **511**, 13.
34. W. C. E. Higginson and N. S. Wooding, *J. Chem. Soc.*, 1952, 760, 774.
35. M. Szwarc, M. Levy and R. Milkovich, *J. Am. Chem. Soc.*, 1956, **78**, 2656.
36. P. J. Flory, *J. Am. Chem. Soc.*, 1940, **62**, 1561.
37. A. G. Evans and M. Polanyi, *J. Chem. Soc.*, 1947, 252.
38. P. H. Plesch, *Nature (London)*, 1952, **169**, 828.
39. P. H. Plesch, *J. Chem. Soc.*, 1953, 1659.
40. K. Ziegler and H.-G. Gellert, *Justus Liebigs Ann. Chem.*, 1950, **567**, 179, 185; *Angew. Chem.*, 1952, **64**, 323.
41. G. Natta, P. Pino, P. Corradini, F. Danusso, E. Mantica, G. Mazzanti and G. Moraglio, *J. Am. Chem. Soc.*, 1955, **77**, 1708.
42. M. L. Huggins, *J. Am. Chem. Soc.*, 1944, **66**, 1991.
43. C. E. Schildknecht, S. T. Gross, H. R. Davidson, J. M. Lambert and A. O. Zoss, *Ind. Eng. Chem.*, 1948, **40**, 2104.
44. J. Gough, *Mem. Lit. Philos. Soc. Manchester*, 1805, 2, **1**, 288.
45. J. P. Joule, *Philos. Trans. R. Soc. London*, 1859, **149**, 91.
46. E. Wohlisch, *Verh. Phys.-Med. Ges. Wuzzburg N.F.*, 1926, **51**, 53.
47. K. H. Meyer, *Biochem. Z.*, 1929, **214**, 253.
48. H. Fikentscher and H. Mark, *Kautschuk*, 1930, **6**, 2.
49. H. Staudinger, *Helv. Chim. Acta*, 1930, **13**, 1324.
50. K. H. Meyer, G. v. Susich and E. Valko, *Kolloid-Z.*, 1932, **59**, 208.
51. W. F. Busse, *J. Phys. Chem.*, 1932, **36**, 2862.
52. W. Kuhn, *Kolloid-Z.*, 1934, **68**, 2.
53. Lord Rayleigh, *Philos. Mag.*, 1919, 6, **37**, 321.
54. H. Eyring, *Phys. Rev.*, 1932, **39**, 746.
55. S. Oka, *Proc. Phys.-Math. Soc. Jpn.*, 1942, **24**, 657.
56. F. T. Wall and J. J. Erpenbeck, *J. Chem. Phys.*, 1959, **30**, 634.
57. M. Doi and S. F. Edwards, 'The Theory of Polymer Dynamics', Clarendon Press, Oxford, 1986, chap. 1.
58. P. J. Flory, *J. Chem. Phys.*, 1942, **10**, 51.
59. M. L. Huggins, *J. Phys. Chem.*, 1942, **46**, 151; *J. Am. Chem. Soc.*, 1942, **64**, 1712.
60. G. Gee and L. R. G. Treloar, *Trans. Faraday Soc.*, 1942, **38**, 147.
61. P. J. Flory and W. R. Krigbaum, *J. Chem. Phys.*, 1950, **18**, 1086.
62. W. Kuhn, *Kolloid-Z.*, 1936, **76**, 258.
63. L. R. G. Treloar, *Trans. Faraday Soc.*, 1943, **39**, 36.
64. P. J. Flory and J. Rehner, Jr., *J. Chem. Phys.*, 1943, **11**, 512.
65. H. M. James and E. Guth, *J. Chem. Phys.*, 1947, **15**, 669.
66. K. H. Meyer and C. Ferri, *Helv. Chim. Acta*, 1935, **18**, 570.
67. G. Allen, U. Bianchi and C. Price, *Trans. Faraday Soc.*, 1963, **59**, 2493.
68. L. Hock-Giessen, *Z. Elektrochem. Angew. Phys. Chem.*, 1925, **31**, 404.
69. N. Bekkedahl, *J. Res. Natl. Bur. Stand. (U.S.)*, 1934, **13**, 411.

2

Nomenclature

AUBREY D. JENKINS
University of Sussex, Brighton, UK
and
KURT L. LOENING
Chemical Abstracts Service, Columbus, OH, USA

2.1 INTRODUCTION

2.1.1 Scope and History

The current set of recommendations for naming polymers has been shaped by the emergence and maturing of polymer science combined with work of the International Union of Pure and Applied Chemistry (IUPAC) and national chemical societies on the development and documentation of chemical nomenclature systems. While it is not possible to include the existing IUPAC nomenclature recommendations in total here, the basic principles are excerpted, along with many examples, and included in the sections that follow. For greater detail the original documents should be consulted.[1-11]

Nomenclature rules and definitions for low-molecular-weight compounds were proposed, approved and used long before chemists realized the importance of macromolecular compounds and began treating them as chemical substances. The Geneva Conference, famous for its contributions to organic nomenclature,[12] took place in 1892. Established in 1911, the International Association of Chemical Societies appointed commissions on inorganic nomenclature, organic nomenclature and physical constants. Unfortunately, the activities of these commissions were interrupted by the First World War. After the war, in 1919, the International Union of Pure and Applied Chemistry was founded. It also established commissions on the nomenclature of organic, inorganic and biological chemistry. Out of these commissions came 'The Definitive Report on the Reform of the Nomenclature of Organic Chemistry', adopted in Liège in 1930.[13] The first report of the corresponding inorganic commission was written in 1926 by Delépine.[14] This was followed in 1940 by a comprehensive set of rules.[15] The work of the Commission on Nomenclature of Biological Chemistry was concerned with carbohydrates, proteins, enzymes, fats, *etc.*[16]

After the Second World War, activities in all areas of chemical nomenclature resumed and multiplied with increasing speed. As macromolecules became more important chemically, it became evident that the rules and definitions devised for substances of low molecular weight could, to a considerable extent, also be applied to those of high molecular weight. Some modifications and extensions were, however, necessary.

In 1947 a new IUPAC Commission on Macromolecular Chemistry was officially formed, and the establishment of a nomenclature system for high-molecular-weight polymers was adopted as a main objective. An IUPAC subcommittee on nomenclature was appointed to consider, together with the corresponding organic commission, the adoption of a report referred to as the 'Huggins Report', originated by a subcommittee on nomenclature of macromolecules of the US National Research Council. After considerable discussion, the report was put into final form and unanimously accepted by the Commission in 1951. It was published in English, French and German in 1952.[17]

This report dealt with two types of problems: (i) ideas, concepts and definitions peculiar to the macromolecular field; and (ii) a system for representing linear polymers, by name and formula, on the basis of their composition and structure, rather than on the basis of the monomers from which they were, or might be, derived. In the first category were included special terms and definitions needed because high polymers and other macromolecules are usually mixtures of many different molecular species, differing with regard to molecular weight, composition and structure. Also included were terms related to methods commonly used for polymer characterization, for example methods for obtaining molecular weight averages by measurements of solution viscosities.

In dealing with the second category, a system of naming polymers and representing them by formulae on the basis of the name and formula of the biradical that is the smallest repeating unit (disregarding conformational differences) was devised and recommended for use.

To avoid confusion between the structure-based names and the usual source-based names, it was proposed that the structure-based names always have the termination 'amer'. Thus, the recommended structure-based name for a linear poly(ethylene) was 'polymethamer' and that for poly(oxymethylene) was 'polymethoxamer'.

The 1952 recommendations were extended by a 'Report on Nomenclature Dealing with Steric Regularity in High Polymers', approved by the Commission on Macromolecules in 1963 after a one year trial period of the report published in 1962;[18] but publication of the final report was delayed until 1966.[19] Italian, German, French and Japanese translations were also published.[20] This report on steric regularity was concerned especially with head-to-tail linear polymers and the definition of various types of tactic polymers. It was based to a considerable extent on proposals of Natta and his co-workers.[21,22] Natta proposed that the polymers which he obtained from α-substituted alkenes and which showed a high degree of regularity be called isotactic polymers.

Most of the recommendations in these reports received general acceptance; however, the structure-based system of naming linear polymers was not widely used. Also, with the development of polymer chemistry, the Nomenclature Committee realized that some of the previous definitions required modification and that some new terms and definitions were needed. When the 1952 recommendations were approved, its rules were adequate for most needs; indeed, most polymers could at that time be reasonably named on the basis of the substance used in producing the polymer. In the intervening years, however, the rapid growth of the polymer field dictated a need for modification and expansion of these rules. Thus, the 'Nomenclature of Regular Single-Strand Organic Polymers' was tentatively approved in 1973,[23] and made definitive in 1975[2] as an updating of the structure-based part of the 1952 recommendations. The new recommendations were based on a proposal of the Nomenclature Committee of the Polymer Division of the American Chemical Society.[24] The major changes were the elimination of the '-amer' endings in the names, and changes in the names of the unit biradicals to make them conform to contemporary IUPAC organic nomenclature.

In 1949,[25] the Organic Nomenclature Commission approved changes and additions to the 1930 Liège rules. These new rules involved primarily additions and changes in the names of radicals or substituent prefixes, and rules for the nomenclature of organosilicon compounds were also presented. These rules were followed by the first edition in 1958 of Sections A and B of the present organic recommendations[26] covering hydrocarbons and fundamental heterocyclic systems. In 1965 these Sections were joined with Section C involving characteristic groups. Two other significant developments in organic nomenclature were: (i) the tentative rules for stereochemistry (Section E) issued in 1968; and (ii) Section D, dealing with organometallic compounds, coordination compounds, and organic derivatives of phosphorus, arsenic, antimony, bismuth, silicon and boron, first issued as tentative recommendations in 1973.

With the increased emphasis on structure-based concepts in the growing field of polymer science, many of the basic definitions of the 1952 Recommendations required refinement. Thus, 'Basic Definitions of Terms Relating to Polymers' was issued by the Commission on Macromolecular Nomenclature, first tentatively in 1971 and in final form in 1974,[1] to supersede portions of the 1952 and later recommendations. Two broad sets of definitions were presented. One of these, based on the structure of polymer molecules, was called 'structure-based', and the other, based on processes by which polymeric substances come into being, 'process-based'.

At the same time the Commission on Macromolecular Nomenclature issued a 'List of Standard Abbreviations (Symbols) for Synthetic Polymer Materials'.[11]

Since the development of the 'Nomenclature Dealing with Steric Regularity in High Polymers', issued in tentative form in 1962 and in definitive form in 1966, the increasingly sophisticated techniques for structure determination have greatly enlarged the field of polymer stereochemistry. Thus, the Commission issued 'Stereochemical Definitions and Notations Relating to Polymers'.[5] It employed all the definitions prescribed in the basic definitions published in 1974, and it also introduced new concepts dealing with the microstructure of polymer chains. In addition, it proposed a set of definitions and notations for the description of the conformations of polymer molecules. Consistency with the prior documents, Section E on stereochemistry, and with the IUPAC–IUB 'Abbreviations and Symbols for the Description of the Conformation of Polypeptide Chains'[27] was maintained as far as possible.

Recommendations for the nomenclature of biopolymers have traditionally been issued by a Joint Commission on Biochemical Nomenclature of IUPAC and the International Union of Biochemistry (IUB). These include recommendations for naming polysaccharides,[28] including symbols for specifying the conformation of polysaccharide chains,[29] and for naming peptides,[30] synthetic polypeptides[31] and polynucleotides.[32]

2.1.2 Future Developments

The international authority responsible for making recommendations for polymer nomenclature is Commission IV.1, the Commission on Macromolecular Nomenclature, of IUPAC. Many of its initiatives are based on the preliminary work of individual members of the commission, some develop through cooperation with the Nomenclature Committee of the Polymer Division of the American Chemical Society.

Following work on copolymers, a logical extension would embrace networks, including interpenetrating networks, and a dialog along these lines has taken place with the ACS group over several years: it is likely that recommendations will be forthcoming in about two years' time. Ladder polymers may also be considered in the near future.

The Macromolecular Division of IUPAC decided in 1985 to set up a Working Party on Liquid Crystal Polymers and this group had its first meeting in August 1986. In view of the rapid development of this field, the early emergence of internationally agreed terminology would be most welcome and may be expected.

Some criticism has been expressed with regard to the Basic Terms.[1] With such sophisticated substances as polymers, it is a matter of great difficulty to construct definitions for fundamental concepts that are concise enough to be useful without being so imprecise as to leave scope for confusion and uncertainty. The time may soon come when the IUPAC Commission will choose to review the Basic Terms but sweeping changes could undermine the entire nomenclature system for polymers that has been developed over the last 20 years or so: in the view of the present authors, a complete overhaul of a system which is working reasonably well is too high a price to pay for some relatively minor improvements to the Basic Terms, however intellectually satisfying the changes might be.

2.1.3   Note on Nonideality

. In order to present clear concepts it is necessary that idealized definitions be adopted even though the realities of polymer science are complex. Deviations from ideality occur in polymers at both molecular and bulk levels in ways that have no parallel with the small molecules of organic or inorganic chemistry. Although such deviations may have to be ignored in constructing useful nomenclature, the terminology recommended can usefully be applied to the predominant structural features of real polymer molecules, if necessary accompanied by self-explanatory (if imprecise) qualifications, such as 'almost completely isotactic' or 'highly syndiotactic'. Although such expressions lack the rigor beloved by the purist, every experienced polymer scientist will know that effective communication in this discipline is impossible without them.

2.2   STRUCTURE- AND SOURCE-BASED DEFINITIONS AND NOMENCLATURE

2.2.1   Basic Definitions of Terms Relating to Polymers[1]

Two broad sets of definitions are presented. One of these is based on the structure of polymer molecules and the other on the processes by which polymeric substances come into being. The first type of definition is termed 'structure-based' and the second 'process-based' or 'source-based'. The primary definition of *polymer* is structure-based. The process-based set of definitions is linked to the primary definition of *polymer* through the definitions of the terms *polymerization* and *monomer*. Thus, definitions of categories of polymers (*e.g.* alternating copolymers) or of polymerization processes (*e.g.* alternating copolymerization) may be referred to as process-based, but names for specific polymers which derive from the name(s) of the parent monomer(s) are described as source-based. Although IUPAC has, from 1952 onwards,[17] devoted most of its effort to the development of structure-based names, source-based names may be allowed[23] and the current IUPAC nomenclature for copolymers[4] is source-based. The source-based names have developed by general usage, the convention being that the name of a homopolymer is the name of the parent monomer, prefixed by 'poly' and contained in parentheses if the monomer name consists of more than one word. (See examples in Section 2.2.3.) A detailed list of terms and definitions is presented in Table 1.

2.2.2   Structure-based Nomenclature of Regular Single-strand Organic Polymers[2]

2.2.2.1   Introduction

To the extent that the organic polymer structure can be portrayed as a chain of regularly repeating structural or constitutional repeating units (the terms are synonymous), the structure can be named by the rules outlined below; in addition, provision has been made for including end groups in the name. There is no objection to source-based names where these names are clear and unambiguous; but the use of the structure-based nomenclature detailed below is preferred.

2.2.2.2   Fundamental principles

This nomenclature system rests upon the selection of a preferred *constitutional repeating unit* (CRU) of which the polymer is a multiple; the name of the polymer is simply the name of this

Table 1 Basic Definitions of Terms Relating to Polymers

Term	Definition
1 Primary Definitions	
1.1 Polymer	A substance composed of molecules characterized by the multiple repetition of one or more species of atoms or groups of atoms (constitutional units, see definition 1.3) linked to each other in amounts sufficient to provide a set of properties that do not vary markedly with the addition or removal of one or a few of the constitutional units[a]
1.2 Oligomer	A substance composed of molecules containing a few of one or more species of atoms or groups of atoms (constitutional units) repetitively linked to each other. The physical properties of an oligomer vary with the addition or removal of one or a few of the constitutional units from its molecules[a]
1.3 Constitutional unit	A species of atom or group of atoms present in a chain of a polymer or oligomer molecule[a]
2 Secondary Definitions	
2.1 Monomer	A compound consisting of molecules each of which can provide one or more constitutional units[b]
2.2 Polymerization	The process of converting a monomer or a mixture of monomers into a polymer[b]
2.3 Oligomerization	The process of converting a monomer or a mixture of monomers into an oligomer[b]
3 Derived Definitions	
3.1 Regular polymer	A polymer whose molecules can be described by only one species of constitutional unit in a single sequential arrangement[a]
3.2 Irregular polymer	A polymer whose molecules cannot be described by only one species of constitutional unit in a single sequential arrangement[a]
3.3 Constitutional repeating unit	The smallest constitutional unit whose repetition describes a regular polymer[a]
3.4 Configurational unit	A constitutional unit having one or more sites of defined stereoisomerism[a]
3.5 Configurational base unit	A constitutional repeating unit whose configuration is defined at least at one site of stereoisomerism in the main chain of a polymer molecule. In a regular polymer, a configurational base unit corresponds to the constitutional repeating unit. Two configurational base units are called enantiomeric when they are mirror images at the plane containing the main-chain bonds[a]
3.6 Configurational repeating unit	The smallest set of one, two or more successive configurational base units that prescribes configurational repetition at one or more sites of stereoisomerism in the main chain of a polymer molecule[a]
3.7 Stereorepeating unit	A configurational repeating unit having defined configuration at all sites of stereoisomerism in the main chain of a polymer molecule[a]
3.8 Tactic polymer	A regular polymer whose molecules can be described by only one species of configurational repeating unit in a single sequential arrangement[a]
3.9 Tacticity	The orderliness of the succession of configurational repeating units in the main chain of a polymer molecule[a]
3.10 Isotactic polymer	A regular polymer whose molecules can be described by only one species of configurational base unit (having chiral or prochiral atoms in the main chain) in a single sequential arrangement. In an isotactic polymer molecule the configurational repeating unit is identical with the configurational base unit[a]
3.11 Syndiotactic polymer	A regular polymer whose molecules can be described by alternation of configurational base units that are enantiomeric. In a syndiotactic polymer the configurational repeating unit consists of two configurational base units that are enantiomeric[a]
3.12 Stereoregular polymer	A regular polymer whose molecules can be described by only one species of stereorepeating unit in a single sequential arrangement[a]
3.13 Atactic polymer	A regular polymer whose molecules have a random distribution of equal numbers of the possible configurational base units[a]
3.14 Block	A portion of a polymer molecule comprising many constitutional units, that has at least one constitutional or configurational feature not present in the adjacent portions. The definitions that relate to polymer can also be applied to block[a]
3.15 Regular block	A block that can be described by only one species of constitutional repeating unit in a single sequential arrangement[a]
3.16 Irregular block	A block that cannot be described by only one species of constitutional repeating unit in a single sequential arrangement[a]
3.17 Tactic block	A regular block that can be described by only one species of configurational repeating unit in a single sequential arrangement[a]
3.18 Atactic block	A regular block that has a random distribution of equal numbers of the possible configurational base units[a]
3.19 Stereoblock	A regular block that can be described by one species of stereorepeating unit in a single sequential arrangement[a]
3.20 Block polymer	A polymer whose molecules consist of blocks connected linearly. The blocks are connected directly or through a constitutional unit that is not part of the blocks[a]
3.21 Block polymerization	Polymerization in which a block polymer is formed[b]
3.22 Tactic block polymer	A polymer whose molecules consist of tactic blocks connected linearly[a]

Table 1   (*Continued*)

Term	Definition
3.23 Stereoblock polymer	A polymer whose molecules consist of stereoblocks connected linearly[a]
3.24 Graft polymer	A polymer whose molecules have one or more species of block connected to the main chain as side chains, these side chains having constitutional or configurational features different from the constitutional units comprising the main chain, exclusive of junction points[a]
3.25 Graft polymerization	Polymerization in which a graft polymer is formed[b]
3.26 Monomeric unit Mer	The largest constitutional unit contributed by a single monomer molecule in a polymerization process[b]
3.27 Degree of polymerization of a molecule of a polymer	The number of monomeric units in a molecule of a polymer[b]
3.28 Degree of polymerization of a polymer	The average value of the degree of polymerization of the molecules of a polymer. The method of averaging must be stated: for example, number-average degree of polymerization[b]
3.29 Addition polymerization	Polymerization by a repeated addition process[b]
3.30 Condensation polymerization; polycondensation	Polymerization by a repeated condensation process (*i.e.* with elimination of simple molecules)[b]
3.31 Homopolymer	A polymer derived from one species of monomer[b]
3.32 Copolymer	A polymer derived from more than one species of monomer[b]
3.33 Alternating copolymer	A copolymer in whose molecules two species of monomeric units are distributed in alternating sequence[b]
3.34 Random copolymer	A copolymer in whose molecules two or more species of monomeric units are distributed in random sequence[b]
3.35 Block copolymer	A block polymer derived from more than one species of monomer[b]
3.36 Graft copolymer	A graft polymer derived from more than one species of monomer[b]
3.37 Bipolymer	A polymer derived from two species of monomer[b]
3.38 Terpolymer	A polymer derived from three species of monomer[b]
3.39 Quaterpolymer	A polymer derived from four species of monomer[b]
3.40 Homopolymerization	Polymerization in which a homopolymer is formed[b]
3.41 Copolymerization	Polymerization in which a copolymer is formed[b]
3.42 Alternating copolymerization	Polymerization in which an alternating copolymer is formed[b]
3.43 Random copolymerization	Polymerization in which a random copolymer is formed[b]
3.44 Block copolymerization	Polymerization in which a block copolymer is formed[b]
3.45 Graft copolymerization	Polymerization in which a graft copolymer is formed[b]
3.46 Stereospecific polymerization	Polymerization in which a tactic polymer is formed[b]
3.47 Stereoselective polymerization	Polymerization in which a polymer molecule is formed from a mixture of stereoisomeric monomer molecules by incorporation of only one stereoisomeric species[b]

[a] Structure-based definition.
[b] Process-based definition.

repeating unit prefixed by 'poly'. The unit itself is named wherever possible according to the Definitive Rules for the Nomenclature of Organic Chemistry.[26] For regular single-strand polymers, this unit is usually a bivalent group.

In using this nomenclature, the steps to be followed in sequence are (i) *identify* the CRU; (ii) *orient* the CRU; and (iii) *name* the CRU. Identification and orientation must always precede the selection of the name of the polymer.

(i)   Identification of the constitutional repeating unit

There are many ways to write the CRU for most polymer structures. In simple cases, these units are readily identified (Figure 1). In more complex cases, it is often necessary to draw a large segment of the chain and from it choose all of the possible CRUs (Figure 2).

To allow construction of a unique name, a single CRU must be selected. The following rules have been designed to specify both *seniority* among subunits, *i.e.* the point at which to begin writing the CRU, and the *direction* along the chain in which to continue to the end of the CRU. The preferred constitutional repeating unit will be one beginning with the subunit of highest seniority (see Rule 2).

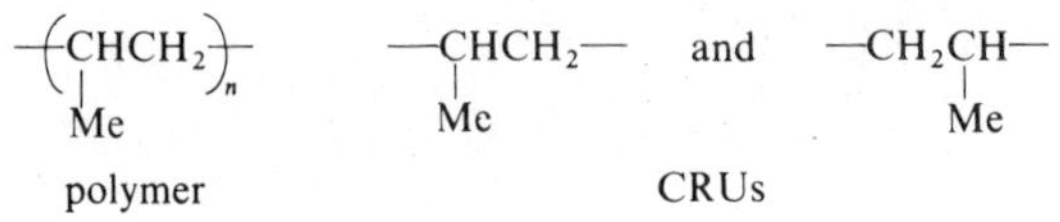

$$\begin{array}{ccc} \left(\text{CHCH}_2\right)_n & \text{—CHCH}_2\text{—} \quad \text{and} \quad \text{—CH}_2\text{CH—} \\ \mid & \mid \qquad\qquad\qquad\qquad \mid \\ \text{Me} & \text{Me} \qquad\qquad\qquad\quad\ \text{Me} \\ \text{polymer} & \text{CRUs} \end{array}$$

Figure 1

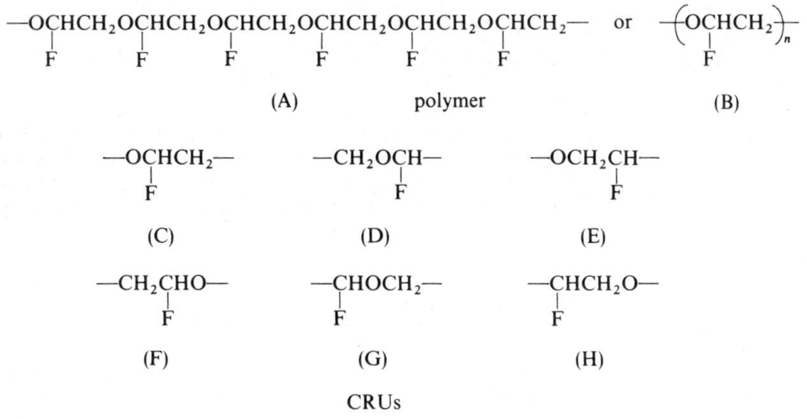

$$-\text{OCHCH}_2\text{OCHCH}_2\text{OCHCH}_2\text{OCHCH}_2\text{OCHCH}_2\text{OCHCH}_2- \quad \text{or} \quad \left(\text{OCHCH}_2\right)_n$$

(with F substituents shown below each CH)

(A)   polymer   (B)

$$-\text{OCHCH}_2- \quad\quad -\text{CH}_2\text{OCH}- \quad\quad -\text{OCH}_2\text{CH}-$$
(with F below)

(C)   (D)   (E)

$$-\text{CH}_2\text{CHO}- \quad\quad -\text{CHOCH}_2- \quad\quad -\text{CHCH}_2\text{O}-$$
(with F below)

(F)   (G)   (H)

CRUs

Figure 2

From this subunit, one proceeds toward the subunit next in seniority. In the preceding example, the subunit of highest seniority is an oxygen atom and the subunit next in seniority is a substituted $-\text{CH}_2\text{CH}_2-$ unit. The parent CRU will therefore be either $-\text{OCH}_2\text{CH}_2-$ or $-\text{CH}_2\text{CH}_2\text{O}-$. Further choice in this case is based on the lowest locant for substitution, so that the CRU is (C) rather than (E), or (F) rather than (H) (see Figure 2).

(ii) Orientation of the constitutional repeating unit

The CRU is written to read from left to right. In the above example, the preferred CRU is therefore (C; Figure 2).

(iii) Naming the constitutional repeating unit

The name of the CRU is formed by citing, in order, the names of the largest subunits within the CRU (Rule 1.21). In the example shown in Figure 2, the oxygen atom is called oxy and the $-\text{CH}_2\text{CH}_2-$ (preferred to $-\text{CH}_2-$ because it is larger and can be named as a unit) is called ethylene; the latter unit substituted with one fluorine atom is called 1-fluoroethylene. The CRU in question is therefore named oxy(1-fluoroethylene), and the corresponding polymer is poly[oxy(1-fluoroethylene)] (B; Figure 2).

2.2.2.3 Nomenclature rules

The rules which follow are essentially directions for the selection of the CRU in a given polymer.

Rule 1 The Constitutional Repeating Unit

Regular single-strand polymer chains can usually be represented as multiples of a bivalent repeating unit which can itself be named. The name of the polymer is *poly(constitutional repeating unit)*. In those cases in which a choice is possible between a bivalent and a higher-valent CRU, the number of free valences (taken to mean classical free valence throughout this chapter) is minimized only after all other orders of seniority have been observed, for example (A) is preferred to (B), but (C) is preferred to (D) (Figure 3; see also Rule 2.12).

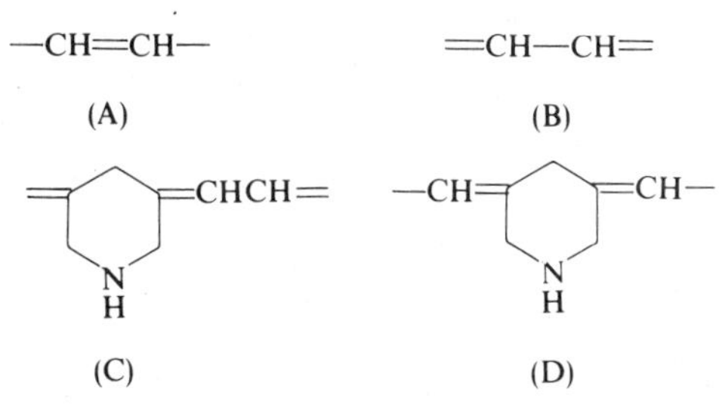

$$-\text{CH}=\text{CH}- \quad\quad\quad\quad =\text{CH}-\text{CH}=$$

(A)   (B)

(C)   (D)

Figure 3

Rule 1.1   The generic name

Linear polymers of unspecified chain length will be named by prefixing 'poly' to the name, placed in parentheses or brackets, of the structural repeating unit of the polymer, *i.e.* the smallest unit of which the polymer is a multiple. If the name of the repeating unit is 'ABC', the corresponding polymer, $+ABC)_n$, is named poly(ABC). Where it is desired to specify chain length, the appropriate Greek prefix (deca, docosa, *etc.*) may be used in place of poly. For a single-strand polymer, the CRU is named within the restriction of directional citation by the IUPAC organic nomenclature rules.[26]

Rule 1.2   Simple constitutional repeating units

1.21. The CRU may contain one or more subunits. Among the possible subunits or combinations of adjacent subunits, the largest possible group, based on main chain atoms and rings only, is to be named (see also Rule 2.11). When the largest group includes the entire CRU, its name, prefixed by 'poly', is the name of the polymer, for example $+CH_2)_n$ is poly(methylene) and $+OCH_2CH_2)_n$ is poly(oxyethylene).

The name of a CRU or any subunit has no relationship to the manner in which the unit was prepared; the name is simply that of the largest identifiable unit and any locants for unsaturation, substituents, *etc.* are dictated by the *structure* of the unit, *e.g.* $+CH=CHCH_2CH_2)_n$ is poly-(1-butenylene) not poly(2-butenylene), which gives a higher locant to the double bond, nor poly(vinyleneethylene), which identifies less than the largest unit in the CRU.

1.22. Identification of the preferred CRU rests on (a) the kinds of atoms or rings in the main chain or (b) on the location of substituents when there is only one kind of main chain atom or ring. Orientation of the CRU in case (a) is determined by the rules of seniority given in Rule 2; in case (b), lowest locants (except when fixed numbering applies; see Rule 1.24) are given to substituents in alphabetical order (Rule 2.42; see Figure 4).

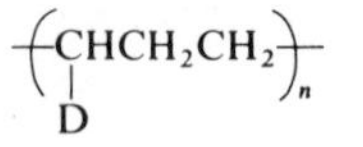

poly(trimethylene-1-*d*)

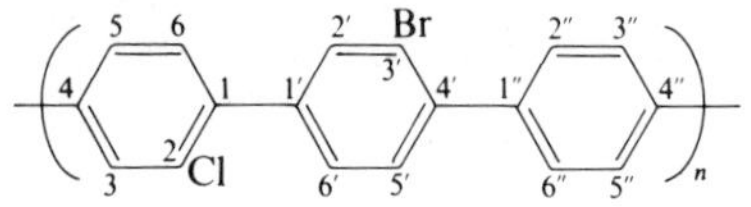

poly(3'-bromo-2-chloro-*p*-terphenyl-4,4''-ylene)

Figure 4

After the CRU and its orientation, reading left to right, have been established, the CRU or its constituent subunits are named to include as many as possible, in order, of (a) the main chain atoms or rings and (b) the substituents within a single name (see also Rule 3.1); for example (**1**) is poly(ethylidene), not poly(methylmethylene); (**2**) is poly(1-phenylethylene), not poly(benzylidene-methylene) or poly(1-phenyldimethylene); (**3**) is poly(1,2-dioxotetramethylene), not poly(succinyl), since substituent positions 1,2 are preferred to 1,4, and identification and orientation of the CRU precede formation of the name; (**4**) is poly(1,3-dioxohexamethylene), not poly(malonyltri-methylene), because the six-carbon chain is the largest unit that can itself be named; and (**5**) is poly(vinylene), not $\neq CH–CH \neq_n$, poly(ethanediylidene).

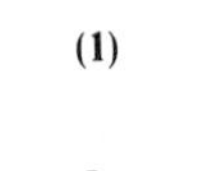

(1)

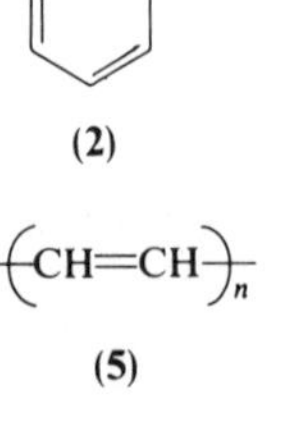

(2)

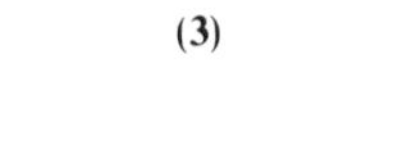

(3)

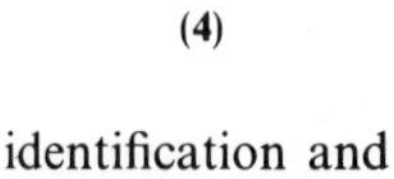

(4)

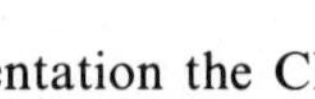

(5)

1.23. If after identification and orientation the CRU is found to contain one or more acyclic bivalent groups having more than two heteroatoms in the main chain, these groups may often be advantageously named by replacement nomenclature.[26] The main chain of the group is named and numbered as though the entire chain were an acyclic hydrocarbon and the heteroatoms are named by means of prefixes 'aza', 'oxa', *etc.* with locants to fix their positions, for example for (**6**) the replacement name is poly(1-oxa-6-thia-4,9-diazanonamethylene-1,3-cyclohexylene, and the systematic name is poly(oxyethyleneiminomethylenethioethyleneimino-1,3-cyclohexylene).

See Rules 2.14 and 2.32 for other examples of the use of replacement nomenclature.

1.24. Groups having fixed numbering retain that numbering in naming the CRU (see also Rules 2.22 and 2.41), for example (**7**) is poly(2,4-pyridinediyl).

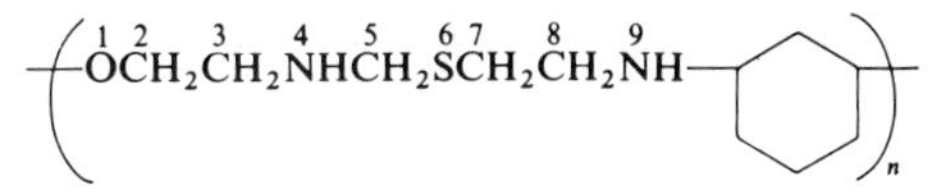

(**6**)

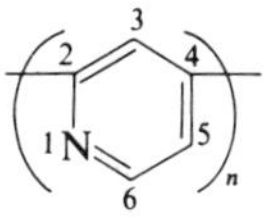

(**7**)

For most acyclic and monocarbocyclic groups, preference in lowest numbers is given to the carbon atoms having the free valences. In other families of compounds, notably the polycyclic hydrocarbons, bridged hydrocarbons, spiro hydrocarbons and heterocyclic ring systems, numbering is fixed for the ring system. Free valences in groups are numbered as low as possible, consistent with the fixed numbering. Since direction through the bivalent group is a requisite parameter in naming polymers, the same fixed numbering is retained for either direction of progress through the group in generating the polymer name, for example (**8**) is poly(2,7-naphthylene), not poly(7,2-naphthylene) or poly(3,6-naphthylene), and (**9**) is poly(tricyclo[2.2.1.0^{2,6}]hept-3,5-ylene).

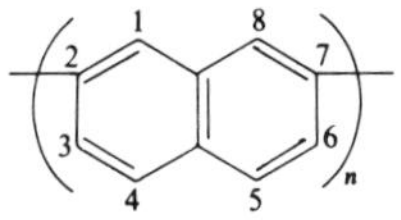

(**8**)

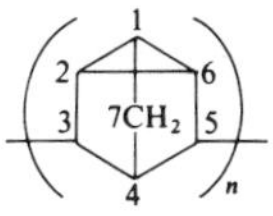

(**9**)

Rule 2 Constitutional Repeating Units Having Two or More Subunits

Many regular single-strand polymers can be represented as multiples of repeating units, such as –ABC–, which consist of a series of smaller subunits, –A–, –B– and –C–. The prototype name of the polymer is poly(ABC), where (ABC) stands for the names of A, B and C, taken in that order. This rule is concerned with the seniority of subunits in identifying the preferred CRU for a given polymer structure.

Rule 2.1   Seniority of subunits and direction of citation

2.11. Polymers having CRUs containing two or more subunits are named with the prefix 'poly' followed in parentheses or brackets by the names of the largest possible subunits cited in order from left to right as they appear in the CRU. The CRU is written from left to right beginning with the subunit of highest seniority and proceeding in a direction defined by the shorter path to the subunit next in seniority, for example (**10**) is poly(oxyterephthaloylhydrazoterephthaloyl), not poly(oxycarbonyl-1,4-phenylenebicarbamoyl-1,4-phenylenecarbonyl).

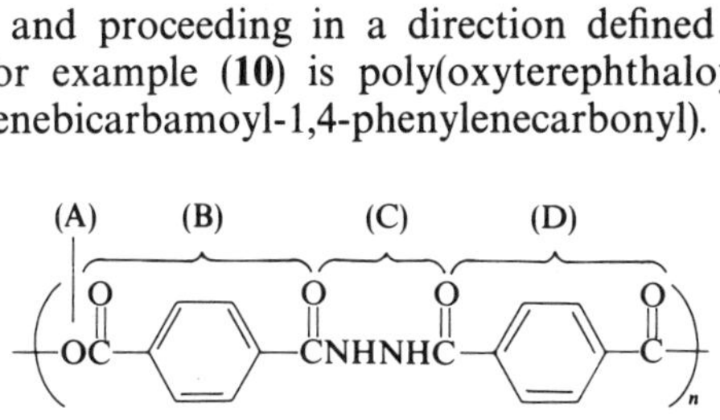

(**10**)

2.12. Usually the CRU in a regular single-strand polymer should be a bivalent group. The starting point for the unit is at the subunit of highest seniority, and citation will be in the direction of the shorter path toward that subunit or subunit combinations of highest seniority.

For citation of the first subunit, the order of seniority among the types of bivalent groups is (1) *heterocyclic rings* (see Rule 2.2), followed by (2) *chains containing heteroatoms* (see Rule 2.3), (3) *carbocyclic rings* (see Rule 2.4), and (4) *chains containing only carbon*, in that order. This order is unaffected by the presence of rings, atoms, or groups that are not part of the main chain, even though

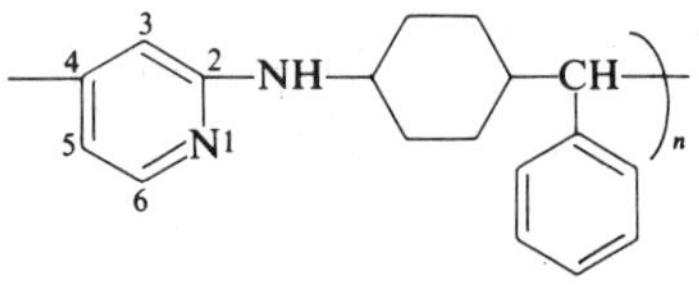

(11)

such substituents could be expressed as part of a trivial name for a bivalent group, for example **(11)** is named poly(4,2-pyridinediylimino-1,4-cyclohexylenebenzylidene).

2.13. Choice of direction along the main chain of the CRU is determined by the shorter path, counting ring and chain atoms individually, from the subunit of highest seniority to the subunit next in seniority, for example **(12)** is poly(4,2-pyridinediylbenzylideneimino-1,4-cyclohexylene).

The possible paths between subunits of first and second seniority necessarily involve subunits of lesser seniority. Except in cases where two paths are of equal length (see Rule 2.14), the number rather than the nature of the atoms involved is the determining factor, so **(13)** is poly(3,5-pyridinediylmethylenepyrrole-3,4-diyloxymethylene), not poly(3,5-pyridinediylmethyleneoxypyrrole-3,4-diylmethylene), in which the longer $-CH_2O-$ path between rings is followed. Where a ring constitutes all or part of a path, the shortest continuous chain of atoms in the ring is selected, for example **(14)** is poly(3,5-pyridinediyl-3,8-acenaphthylenylenepyrrole-3,4-diyl-3,7-acenaphthylenylene) (heavy line denotes path followed).

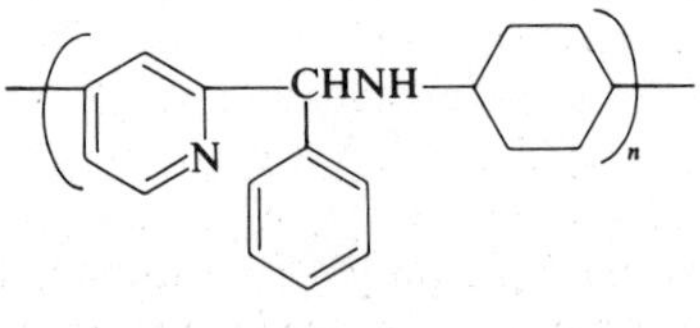

(12)

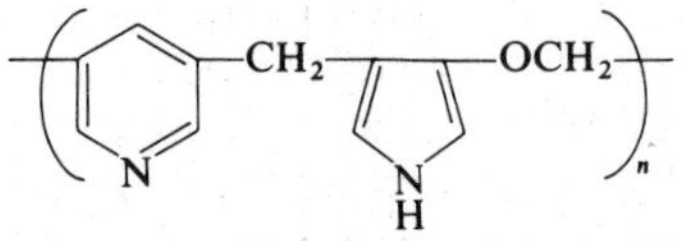

(13)

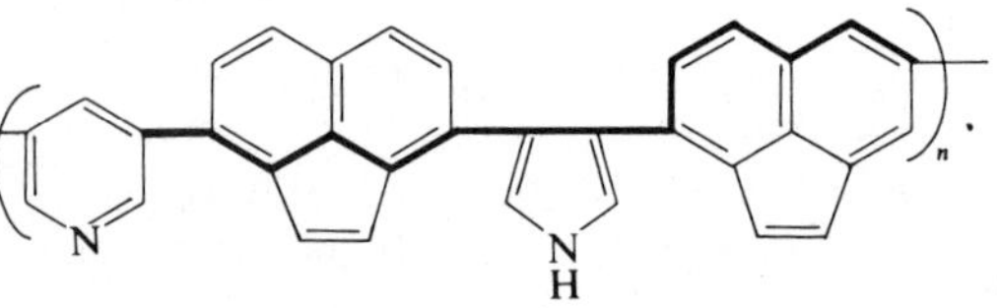

(14)

2.14. When the choice of path determining direction of citation involves paths of equal length to subunits of equal seniority in the normal order of precedence, the choice of path depends upon the kind of subunits in the paths themselves. This condition applies to chains typified by the following generalized structures, where A, B and C are subunits in that order of decreasing seniority, separated by paths of differing lengths x and y which contain units of lower seniority than C: –C–y–B–x–A–x–B–y–C–, –B–y–A–x–A–y–B– or –B–y–A–x–A–x–A–y–B–. The choice of direction is from a subunit A to the nearest part of a path x having highest seniority, or, if two paths x are identical in every respect, to the nearest part of path y having highest seniority, *etc.* until some point of difference is encountered. (See also Rules 2.24 and 2.32.)

Examples of direction of citation based on the constituent parts of paths of equal length are **(15)**, poly(3,5-pyridinediyl-1,4-phenylenemethyleneoxymethyleneiminomethyleneoxy-1,4-phenylenemethylene) (choice of path from heterocyclic ring to O determined by position of phenylene); $\{OCH_2OCH_2NHCH_2CH_2SCH_2NHCH_2CH_2\}_n$ **(16)**, poly(oxymethyleneoxymethyleneiminoethylenethiomethyleneiminoethylene) or poly(1,3-dioxa-8-thia-5,10-diazadodecamethylene) (choice of path from O to S determined by position of NH); and **(17)**, poly(3,5-pyridinediyl-1,3-cyclohexyleneoxytrimethylene) (a portion of a cyclic structure is senior to carbon chain of equal length).

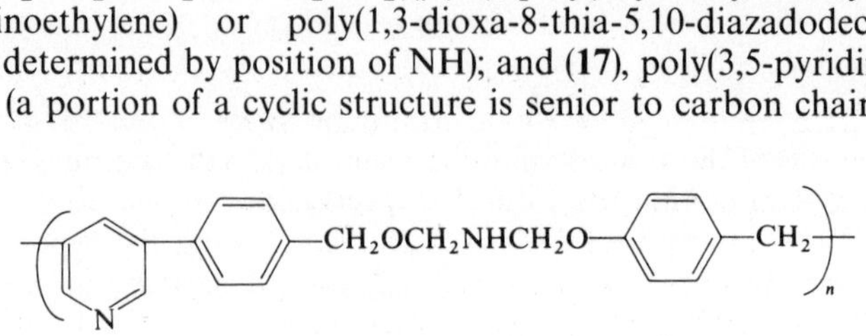

(15)

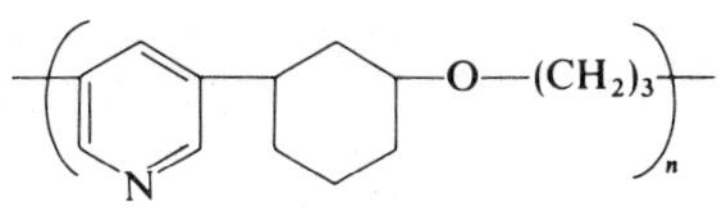

(17)

Where substituents control the choice of CRU, the order of seniority is that given in Rule 2.42, for example **(18)** is poly[thio(2-chlorotrimethylene)thiotrimethylene] and **(19)** is poly(thio-(1-iodoethylene)thio(5-bromo-3-chloropentamethylene)] (direction determined by the lower locant in the first cited subunit after beginning the CRU).

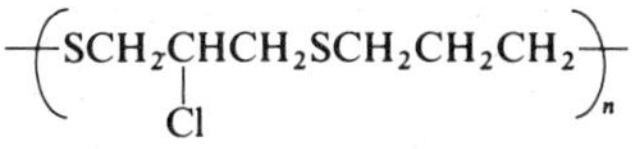

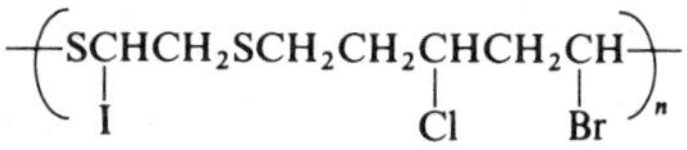

(18)

Rule 2.2   Heterocyclic rings

2.21. A CRU having two or more subunits that include a heterocyclic ring system in the main chain is named by citing first the heterocyclic ring group of highest seniority and proceeding by the shorter path in descending order of preference to (a) another of the same heterocyclic ring (see Rule 2.24), for example poly[3,5-pyridinediylmethylene-3,5-pyridinediyl(tetrahydro-2*H*-pyran-3,5-diyl)] **(20)**; (b) the heterocyclic ring next in seniority (see Rule 2.23), for example poly(2,6-morpholinediyl-3,5-pyridinediyl-2,8-thianthrenediyl) **(21)**; (c) the senior acyclic group containing a heteroatom in the main chain (see Rule 2.31), for example poly(3,5-pyridinediylmethyleneoxy-1,4-phenylene) **(22)**; (d) the senior carbocyclic ring system (see Rule 2.41), for example poly(3,5-pyridinediyl-1,4-phenylene-1,2-cyclopentylene) **(23)**; and (e) the senior acyclic group containing only carbon in the main chain (see Rule 2.42), for example poly(3,5-pyridinediylcarbonyloxymethylene) **(24)**.

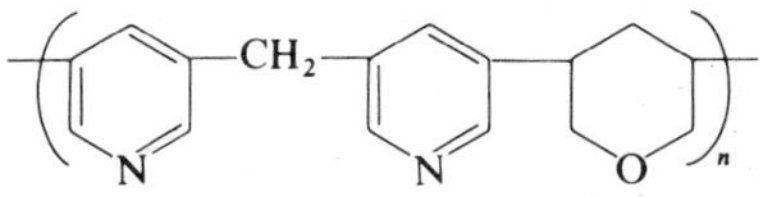

(20)

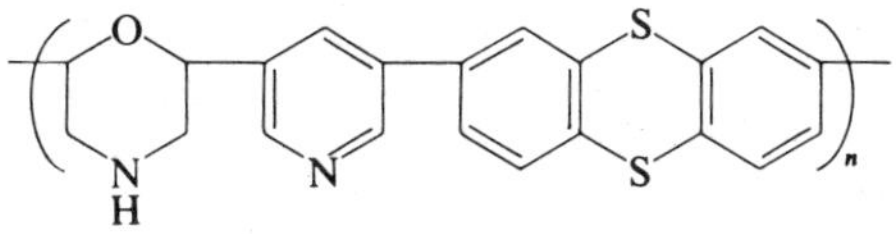

(21)

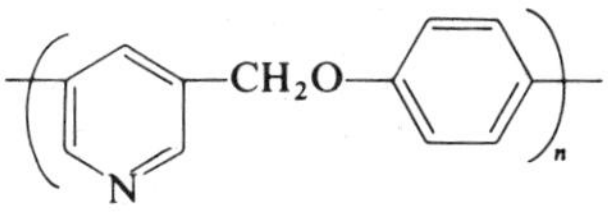

(22)

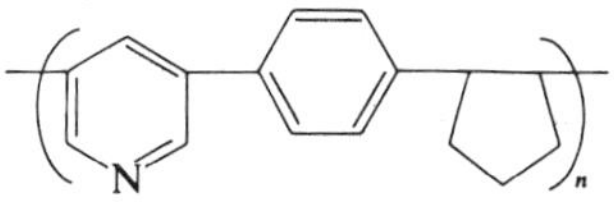

(23)

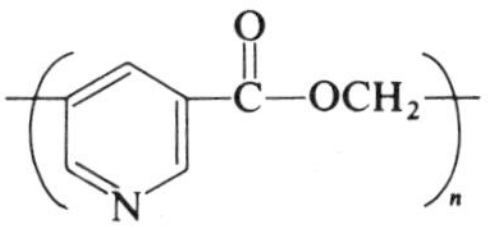

(24)

2.22. Consistent with the fixed numbering of heterocyclic rings, the points of attachment to the main chain of the CRU should have the lowest permissible locants, as shown in poly(2,4-piperidinediyloxymethylene) **(25)** and poly(4,2-piperidinediyloxymethylene) **(26)**. Where there is a choice, the point of attachment at the left side of the ring should have the lowest permissible number, for example poly(4*H*-1,2,4-triazole-3,5-diylmethylene) **(27)**.

2.23. Among heterocyclic ring systems, the descending order of seniority is (a) a ring system with nitrogen in the ring; (b) a ring system containing a heteroatom other than nitrogen as high as possible in the order given in Rule 2.31; (c) a ring system containing the greatest number of rings;

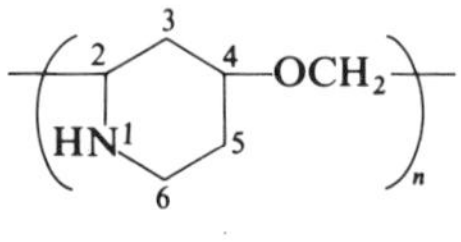

(25) 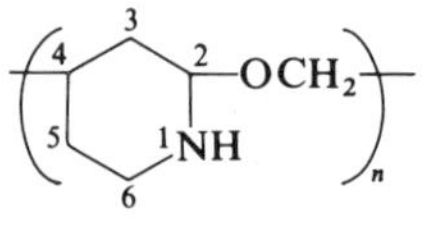   (26)

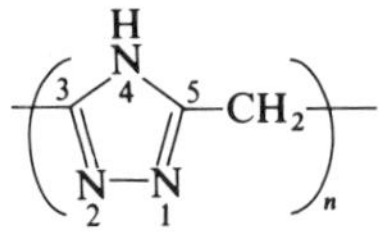

(27) 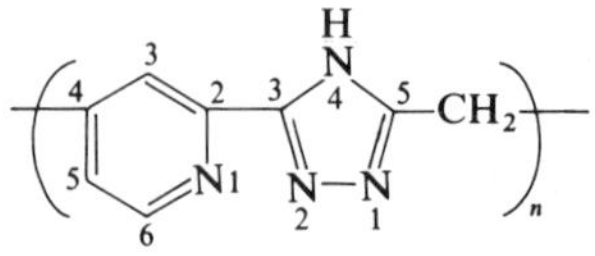   (28)

(d) a ring system having the largest individual ring; (e) a ring system having the largest number of heteroatoms; (f) a ring system containing the greatest variety of heteroatoms; and (g) the ring system having the greatest number of heteroatoms highest in the order given in Rule 2.31. This order is that followed in Rule B-2 of the IUPAC Rules,[26] and is illustrated by poly(4,2-pyridinediyl-4H-1,2,4-triazole-3,5-diylmethylene) (28).

When two heterocyclic subunits differ only in degree of unsaturation, the senior subunit is that having the least hydrogenated ring system, thus (29) is poly(3,5-pyridinediyl-2,4-piperidinediyl).

Among assemblies of identical heterocyclic rings, the ring of highest seniority is that having lowest numbers for the points of attachment between the rings within the assembly consistent with the fixed numbering of the parent ring, for example (30) is poly[(3,3'-biquinoline)-6,6'-diyl], not poly[(6,6'-biquinoline)-3,3'-diyl], and (31) is poly[(2,3-bipyridine)-4,5'-diyl] not poly[(2',3-bipyridine)-4',5-diyl] or poly[(3',4-bipyridine)-2,5'-diyl]. Further choice is based on the number and kind of substituting groups (see Rule 2.42), for example (32) is poly[(4-chloro-3,3'-bipyridine-5,5'-diyl)methylene].

2.24. When the CRU contains two identical rings of highest seniority or more than two such rings of highest seniority separated by identical paths, the direction of citation is determined by the shorter path to the subunit of second seniority. Further choice is based on the shorter path from that subunit to the subunit of third seniority, *etc.*, as indicated in the order of seniority in Rule 2.21 and exemplified by (33), poly(4,2-piperidinediylmethylene-4,2-piperidinediyl-1,2-cyclopentyleneethylene-1,2-cyclopentylenemethylene) (numbers in rings show order of preference in formation of name).

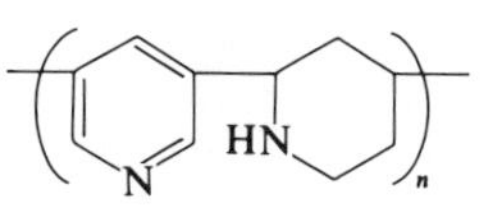

(29) 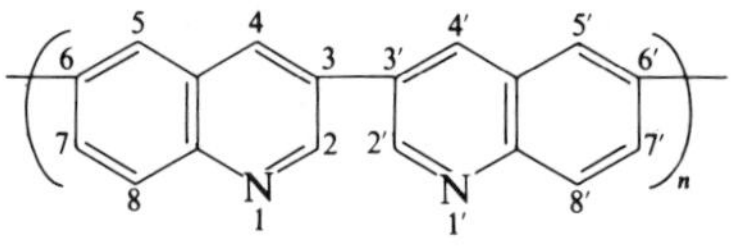   (30)

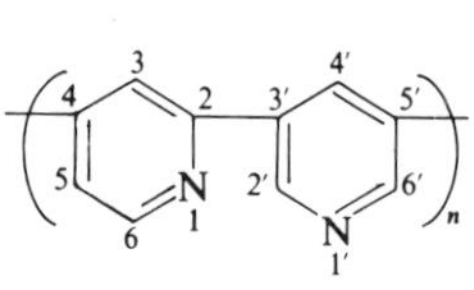

(31) 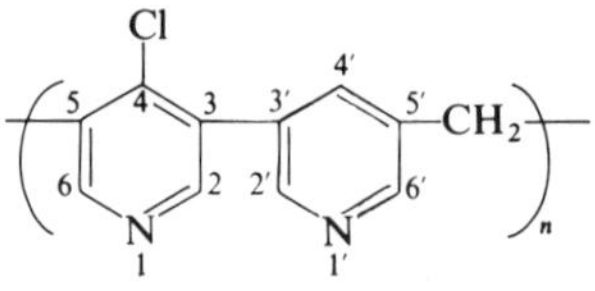   (32)

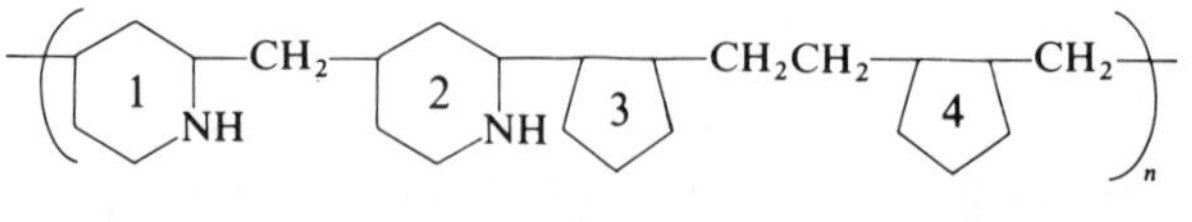

(33)

Rule 2.3   *Heteroatoms in chains*

2.31. Complex CRUs in which the senior subunit is a heteroatom or an acyclic chain with a heteroatom in the main chain are named by citing first the heteroatom of highest seniority and proceeding by the shorter path in descending order of seniority to (a) another heteroatom of the same kind; (b) the heteroatom next in seniority; (c) the senior carbocyclic ring system (see Rule 2.41); and (d) the senior acyclic group containing only carbon in the main chain (see Rule 2.42). For the most common heteroatoms the descending order of seniority is O, S, Se, Te, N, P, As, Sb, Bi, Si, Ge, Sn, Pb, B and Hg; other heteroatoms may be placed within this order as indicated by their positions in the periodic table. Examples include poly[imino(1-oxoethylene)silylenetrimethylene] (**34**); poly(oxymethyleneiminocarbonylthio-1,3-phenyleneethylene) (**35**); and $\{ONHCH_2NHNHCH_2\}_n$, poly(oxyiminomethylenehydrazomethylene) (**36**).

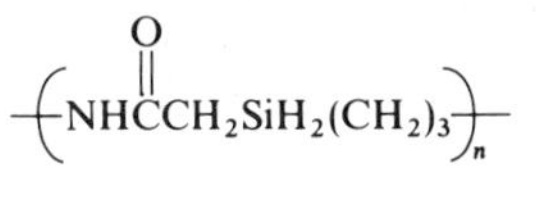

(**34**)

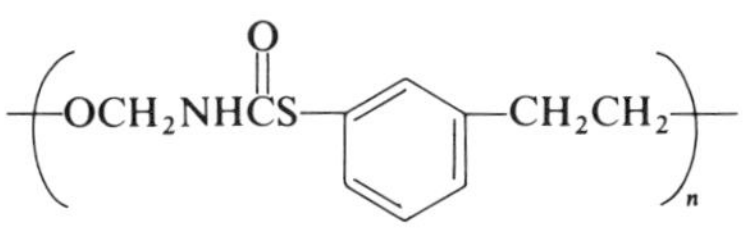

(**35**)

Parentheses must be used in some cases to prevent ambiguity, for example in poly[thio(carbonyl)] (**37**), 'carbonyl' is enclosed in parentheses to differentiate the structure from poly(thiocarbonyl) (**38**).

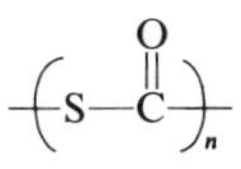

(**37**)    (**38**)

The direction of bonding in unsymmetrical single-atom radicals (*e.g.* =N– or –N= for nitrilo) is indicated by the endings of the names of the adjacent subunits in the CRU, thus (**39**) is poly(nitrilo-1,4-phenylenenitrilo-2-propen-3-yl-1-ylidene-1,4-phenylene-1-propen-1-yl-3-ylidene) and (**40**) is poly(oxycarbonylnitrilo-1,3-propanediylidenenitrilocarbonyl). Direction in a group such as azoxy, *i.e.* –N(O)=N– or –N=N(O)–, is indicated by the prefixes *ONN* or *NNO*, respectively, in that order seniority, for example poly[oxymethylene-*ONN*-azoxy(chloromethylene)] (**41**).

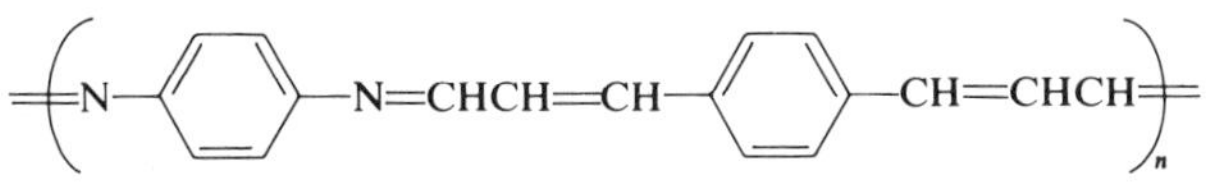

(**39**)

(**40**) 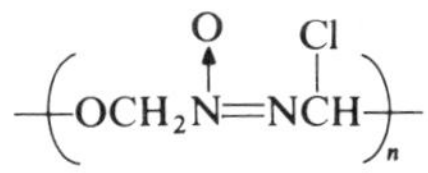   (**41**)

The unsymmetrical group –N=N–NH–, designated 'diazoamino' under the IUPAC Organic Rules,[26] in the present directional nomenclature for polymers is called 'azoimino'.

Among heteroatoms of the same kind, the shortest path and direction between the heteroatoms is chosen first. For a choice between equal paths, the heteroatom of highest seniority is the one most highly substituted, with the order of substituent seniority being that given in Rule 2.42, for example poly(sulfinylmethylenethiotrimethylenesulfonyl-1,4-phenylene) (**42**) and poly[(methylimino)-methyleneimino-1,3-phenylene] (**43**).

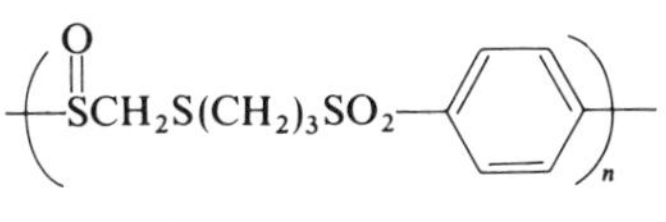

(**42**) 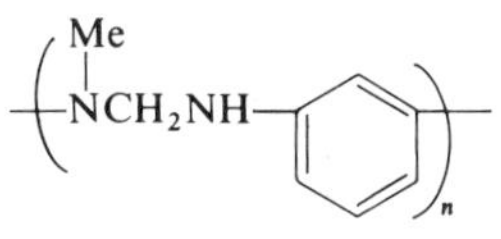   (**43**)

2.32. If the CRU contains two or more heteroatoms of highest seniority or more than two such heteroatoms separated by identical paths, the direction of citation is determined by the shorter path to the subunit of second seniority. Further choice is based on the shorter path from that subunit to the subunit of third seniority, *etc.* as indicated in the order of seniority in Rule 2.31, for example (**44**) is poly(oxymethyleneoxymethyleneoxymethyleneimino-1,3-phenylenemethyleneiminomethylene) or poly(1,3,5-trioxa-7-azaheptamethylene-1,3-phenylene-2-azatrimethylene) (the shorter path X between the NH group and the ring has been taken).

Rule 2.4 *Carbocyclic rings and carbon chains*

2.41. Constitutional repeating units in which the senior subunit is a carbocyclic ring system are named by citing first the carbocyclic ring of highest seniority and proceeding by the shorter path, in descending order of seniority, to (a) another of the same carbocycle; (b) the carbocyclic system next in seniority; and (c) the acyclic group appearing earliest in the alphabet. Carbocyclic ring system seniority is based on complexity, with the ring system of highest seniority being that containing the largest number of rings. Further order of seniority is based on (a) the largest individual ring at the first point of difference; (b) the largest number of atoms common to the rings; (c) the lowest locant numbers at the first point of difference for ring junctions; and (d) the least hydrogenated ring. The basis for further choice is found in Rule C-14.1 of the IUPAC Rules.[26] The direction of citation in CRUs having two or more carbocycles of highest seniority is determined in a manner analogous to that of Rule 2.32. An example is provided by poly(2,7-naphthylene-1,4-phenylene-1,3-cyclohexylene) (**45**).

More than one numbering method may be in use in certain ring systems, such as the spiro hydrocarbons. Generally, in a specific ring system, a ring with unprimed locants is senior to one with primed locants. Points of attachment to the main chain of the CRU receive lowest permissible numbers; examples are poly(spiro[4.5]dec-2,8-yleneethylene) (**46**) (repeating unit named by IUPAC Rule A-41), poly(oxyspiro[3.5]nona-2,5-dien-7,1-ylene-4-cyclohexen-1,3-ylene) (**47**), and poly[(3-chloro-4,4′-biphenylylene)methylene(3-chloro-1,4-phenylene)methylene] (**48**) (not poly[(3′-chloro-4,4′-biphenylylene)methylene(2-chloro-1,4-phenylene)methylene]; the substituent in the preferred ring determines the direction).

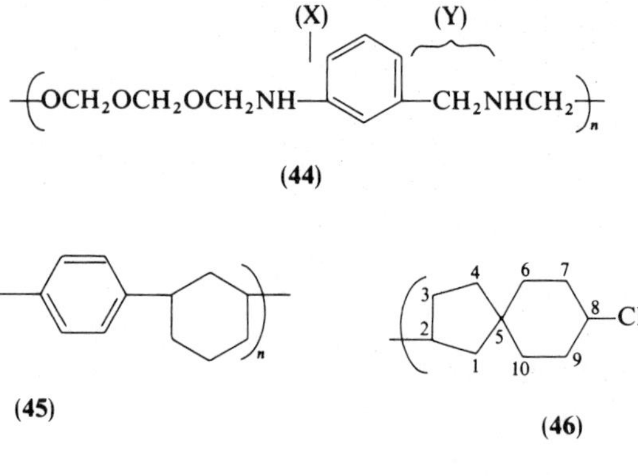

(**44**)

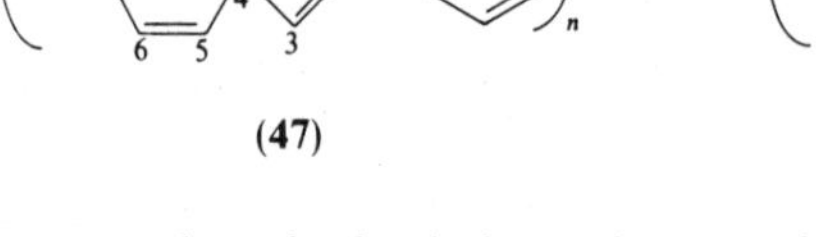

(**45**)

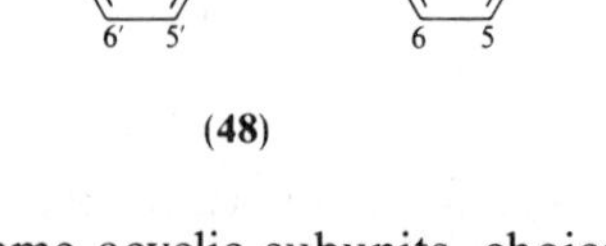

(**46**)

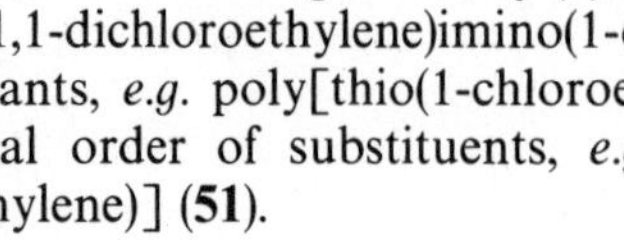

(**47**)

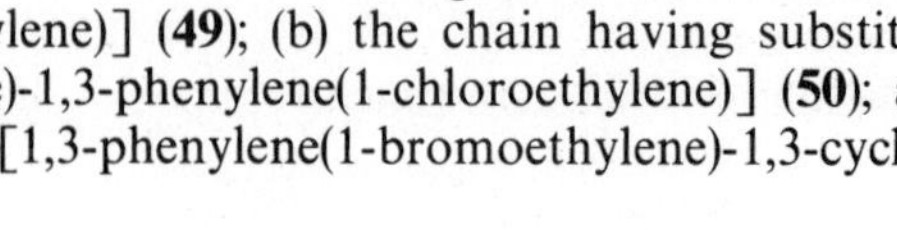

(**48**)

2.42. When equal paths lead through two of the same acyclic subunits, choice of direction is determined, in descending order, by (a) the acyclic chain with the largest number of substituents, *e.g.* poly[oxy(1,1-dichloroethylene)imino(1-oxoethylene)] (**49**); (b) the chain having substituents with lowest locants, *e.g.* poly[thio(1-chloroethylene)-1,3-phenylene(1-chloroethylene)] (**50**); and (c) the alphabetical order of substituents, *e.g.* poly[1,3-phenylene(1-bromoethylene)-1,3-cyclohexylene-(2-butylethylene)] (**51**).

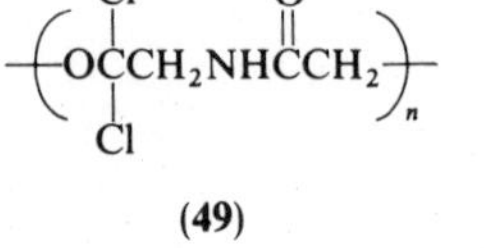

(**49**)

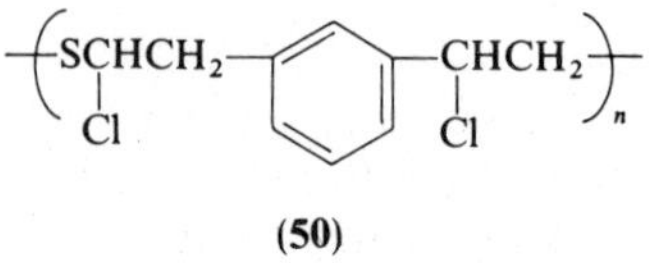

(**50**)

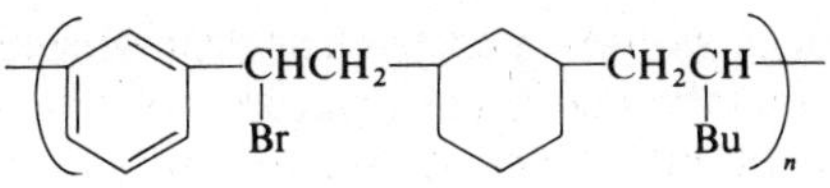

(51)

Rule 3 Substituents

3.1. Substituents to acyclic or cyclic subunits in the main chain of the CRU are included within the trivial name of the subunit wherever such name is approved by the IUPAC Organic Rules[26] (see also Rule 1.22), *e.g.* poly(oxybenzylidene) (52) and poly(oxyoxalyl) (53).

3.2. Substituents along the main chain other than those included in the name of a subunit are denoted by means of prefixes appended to the name of the subunit to which they are bound. In groups not having numbering fixed by other criteria, lowest locants appear at the left side of the group as written in the CRU (see also Rule 2.14), thus (54) is poly[(6-chloro-1-cyclohexen-1,3-ylene)(1-bromoethylene)], not poly[(6-chloro-1-cyclohexen-3,1-ylene)(2-bromoethylene)].

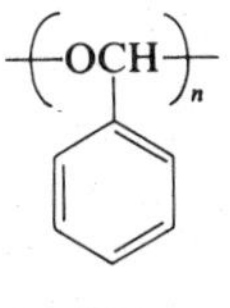

(52) 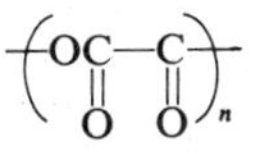   (53) 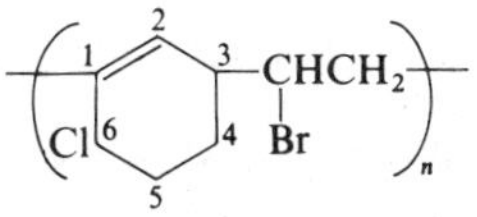   (54)

Functional derivatives that are clearly a part of the CRU are named as substituents to the appropriate subunit by the use of prefixes, *e.g.* poly[oxy[2-(methoxycarbonyl)ethylidene]] (55) and poly[iminomethyleneiminocarbonyl[2-[(2,4-dinitrophenyl)hydrazono]-1,3-cyclopentylene]carbonyl] (56).

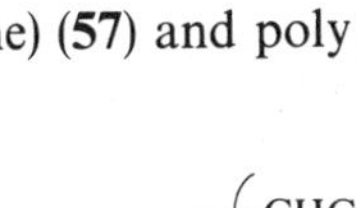

(55)

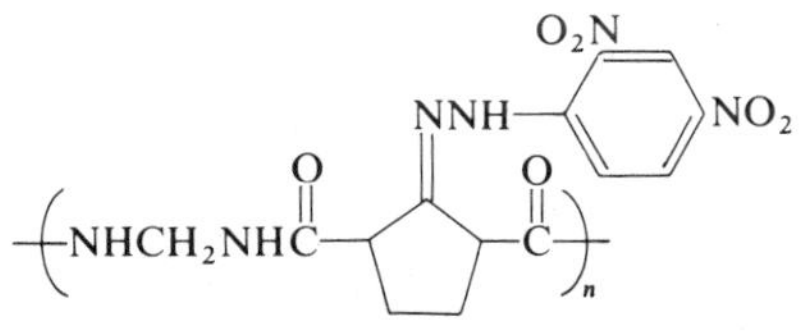

(56)

3.3. Salts and onium compounds of polymers are named by placing the appropriate prefix or suffix together with the name of the CRU in the enclosed part of the polymer name, *e.g.* poly(sodium 1-carboxylatoethylene) (57) and poly[(dimethyliminio)ethylene bromide] (58).

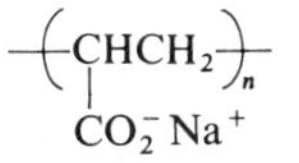

(57) 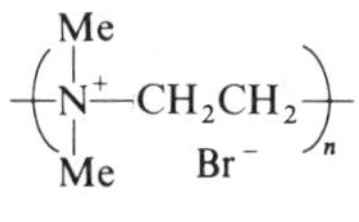   (58)

Certain substituents are frequently expressed as part of a trivial name. The subunit thus named can itself be further substituted without altering the original trivial name, *e.g.* poly[oxy(2-chlorosuccinyl)] (59). The same (in this case, doubly-bonded oxygen) substituents not expressed in a trivial name have no special seniority, *e.g.* poly[imino[1-oxo-2-(phenylthio)ethylene]] (60) and poly[imino(1-chloro-2-oxoethylene)(4-nitro-1,3-phenylene)(3-bromotrimethylene)] (61).

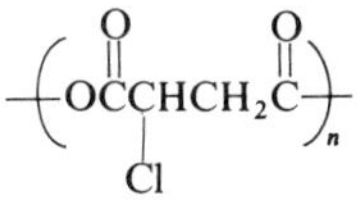

(59) 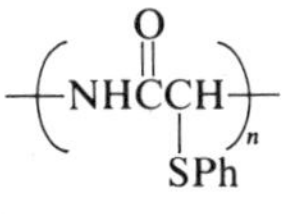   (60) 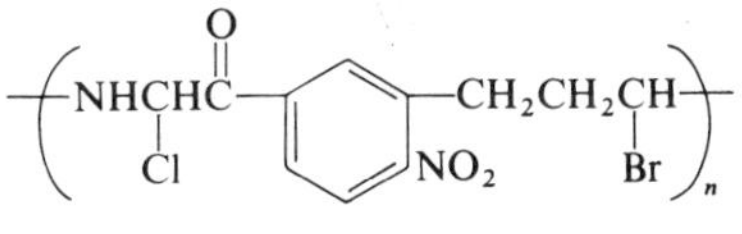   (61)

Substituents in unknown positions in specific subunits are named in the usual way but either without locants or with an *x* locant, *e.g.* poly(*x*-imino-1,2-cyclopentylene) (62), the *x* being required to differentiate this structure from poly(imino-1,2-cyclopentylene) (**63**); and poly[oxycarbonyloxy-(methylethylene)] (**64**) (position of the methyl not stated).

3.4. End groups may be specified by prefixes placed ahead of the name of the polymer. The end group designated by α is that attached to the left side of the CRU written as described in the preceding rules; the other end group is designated by ω, *e.g.* α-(trichloromethyl)-ω-chloropoly-(1,4-phenylenemethylene) (**65**).

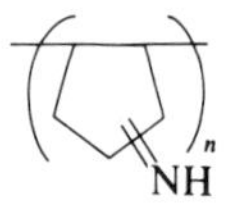

| (62) | (63) | (64) | (65) |

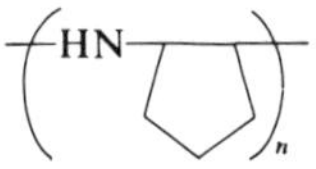

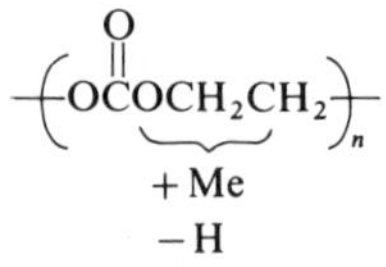

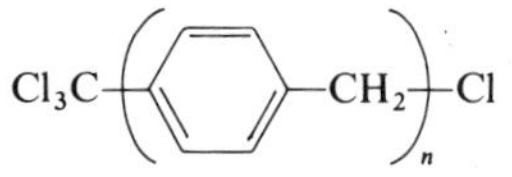

2.2.3 Source-based Nomenclature

The Commission recognized that a number of common polymers have semisystematic or trivial names that are well established by usage; it is not intended that they be immediately supplanted by the structure-based names. Nevertheless, it is hoped that for scientific communication the use of semisystematic or trivial names for polymers will be kept to a minimum.

Table 2 Semisystematic or Trivial Names for Idealized Polymer Structures

Structure	Name	Structure	Name
$+CH_2CH_2\}_n$	Polyethylene; poly(methylene)	$+CHCH_2\}_n$ | O_2CMe	Poly(vinyl acetate); poly(1-acetoxyethylene)
$+CHCH_2\}_n$ | Me	Polypropylene; poly(propylene)	$+CHCH_2\}_n$ | Cl	Poly(vinyl chloride); poly(1-chloroethylene)
Me | $+C—CH_2\}_n$ | Me	Polyisobutylene; poly(1,1-dimethylethylene)	F | $+CCH_2\}_n$ | F	Poly(vinylidene difluoride); poly(1,1-difluoroethylene)
$+CH=CHCH_2CH_2\}_n$	Polybutadiene; poly(1-butenylene)	$+CF_2CF_2\}_n$	Poly(tetrafluoroethylene); poly(difluoromethylene)
$+C=CHCH_2CH_2\}_n$ | Me	Polyisoprene; poly(1-methyl-1-butenylene)	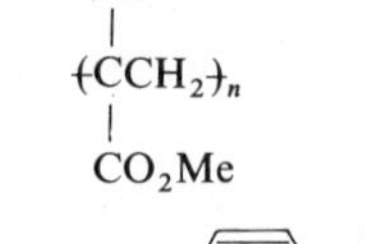 (structure: 2-propyl-1,3-dioxane ring with CH_2)	Poly(vinyl butyral); poly[(2-propyl-1,3-dioxane-4,6-diyl)-methylene]
$+CHCH_2\}_n$ | (phenyl ring)	Polystyrene; poly(1-phenylethylene)	$+CHCH_2\}_n$ | CO_2Me	Poly(methyl acrylate); poly[1-(methoxycarbonyl)ethylene]
$+OCH_2\}_n$	Polyformaldehyde; poly(oxymethylene)	Me | $+CCH_2\}_n$ | CO_2Me	Poly(methyl methacrylate); poly[1-(methoxycarbonyl)-1-methylethylene]
$+OCH_2CH_2\}_n$	Poly(ethylene oxide); poly(oxyethylene)	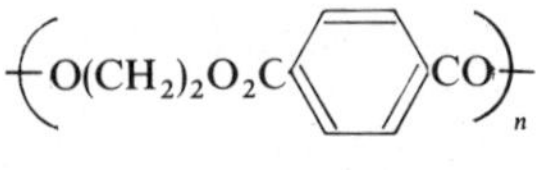 $+O(CH_2)_2O_2C\text{(ring)}CO\}_n$	Poly(ethylene terephthalate); poly(oxyethyleneoxyterephthaloyl)
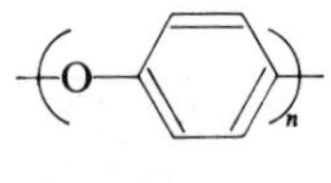 $+O\text{(ring)}\}_n$	Poly(phenylene oxide); poly(oxy-1,4-phenylene)	$+NHCO(CH_2)_4—CONH(CH_2)_6\}_n$	Poly(hexamethylene adipamide); poly[imino(1,6-dioxohexamethylene)iminohexamethylene]; poly(iminoadipoyliminohexamethylene)
$+CHCH_2\}_n$ | CN	Polyacrylonitrile; poly(1-cyanoethylene)	$+NHCO(CH_2)_5\}_n$	Poly(ε-caprolactam); poly[imino-(1-oxohexamethylene)]
$+CHCH_2\}_n$ | OH	Poly(vinyl alcohol); poly(1-hydroxyethylene)		

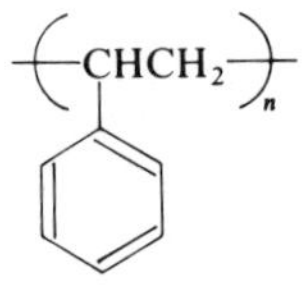

For the idealized structural representations shown in Table 2, the semisystematic or trivial names given are approved for use in scientific work; the corresponding structure-based names are given as alternative names. Equivalent names for close analogs of these polymers, *e.g.* other alkyl ester analogs of poly(methyl methacrylate), are also acceptable. Where the semisystematic name is an obvious source-based name, the polymer referred to is that derived from the indicated source.

2.2.4 List of Standard Abbreviations (Symbols) for Synthetic Polymers and Polymer Materials

The purpose of these abbreviations is to provide uniform contractions of names used for synthetic polymers and polymer materials. Abbreviated terms have evolved through widespread common usage. This compilation of abbreviations has been prepared to promote one rather than several abbreviations for a given material and to avoid the use of the same abbreviation for more than one material.

The abbreviations listed in Table 3 may be used in the scientific and industrial literature of polymer science without further definition. All others, even those appearing in former lists compiled by IUPAC and the ISO[33] should be defined at least once in any document in which they are used. The American Chemical Society has recently compiled a master list of all known abbreviations in the polymer field.[34]

This short list of abbreviations is based on common usage and it is not intended as a model for forming other abbreviations. These abbreviations should be written or printed in capital letters.

Table 3 Abbreviations Used Substantively

Abbreviation	Polymer name	Abbreviation	Polymer name
PAN	Polyacrylonitrile	PS	Polystyrene
PCTFE	Poly(chlorotrifluoroethylene)	PTFE	Poly(tetrafluoroethylene)
PEO	Poly(ethylene oxide)	PVAC	Poly(vinyl acetate)
PETP	Poly(ethylene terephthalate)	PVAL	Poly(vinyl alcohol)
PE	Polyethylene	PVC	Poly(vinyl chloride)
PIB	Polyisobutylene	PVDC	Poly(vinylidene dichloride)
PMMA	Poly(methyl methacrylate)	PVDF	Poly(vinylidene difluoride)
POM	Poly(oxymethylene); polyformaldehyde	PVF	Poly(vinyl fluoride)
PP	Polypropylene		

2.2.5 Nomenclature for Regular Single-strand and Quasi-single-strand Inorganic and Coordination Polymers[3]

2.2.5.1 Introduction

The system for naming polymers in terms of structure previously published[2, 24] dealt with linear organic polymers, primarily those defined[2]* as regular single-strand polymers, and followed as closely as possible established principles of organic nomenclature.[26] Accordingly, constituent subunits of the smallest repeating structural unit, named as bivalent radicals, are combined additively to form the name of the constitutional repeating unit. Extension of this method to linear inorganic and/or coordination polymers is seriously limited by the general lack of a system for naming bivalent radicals, because of the basic difference in philosophy between inorganic nomenclature and organic nomenclature. Furthermore, the constituent units of the constitutional repeating unit in most inorganic and coordination polymers are not bivalent radicals in the usual sense.

The present system is designed to name, uniquely and unambiguously, regular inorganic and/or coordination linear polymers, the constituent subunits of which can be formulated according to usual chemical principles of covalent and/or coordinate covalent bonding and the structures of which can be described by a constitutional repeating unit with at least one terminal constituent subunit that is connected through only one atom to other identical constitutional repeating units, or to an end group. 'Ladder' structures are thus excluded.

A *regular linear polymer* that can be described by a preferred constitutional repeating unit in which *both* terminal constituent subunits are connected through single atoms to the other identical constitutional repeating units or to an end group is called a *regular single-strand polymer* (Figure 5).[2]

---

* Section 2.2.2 of this chapter covers the information in ref. 2.

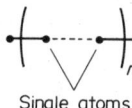

Single atoms

Figure 5

A *regular linear polymer* that can be described by a preferred constitutional repeating unit in which *only one* terminal constituent subunit is connected through a single atom to the other identical constitutional repeating units or to an end group is a *quasi-single-strand polymer*, *i.e.* it does not fit the definition of regular single-strand polymers,[2] but can be named in the same manner (Figure 6).

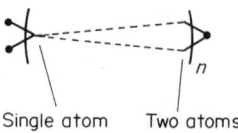

Single atom   Two atoms

Figure 6

Established principles of inorganic and coordination nomenclature[35] are used as far as is consistent with the definitions[1] and basic principles[2,24] of polymer nomenclature already published. As in the nomenclature of organic polymers, these rules apply to structures, not substances, that may be idealized representations of complex systems.

A few polymers with inorganic backbones in which the bonding is primarily covalent have trivial or semisystematic names of long standing, *e.g.* poly(dimethylsiloxane) for $\{SiMe_2O\}_n$,* and poly(dichlorophosphazene) for $\{PCl_2=N\}_n$†; some of these polymers can also be named by the principles for naming organic polymers, *i.e.* by using bivalent radicals,[2,24] *e.g.* poly[oxy(dimethylsilylene)] for $\{SiMe_2O\}_n$, and poly[nitrilo(dichlorophosphoranylidyne)] for $\{PCl_2=N\}_n$. There is no objection to the use of trivial or semisystematic names as long as they are clear and unambiguous, nor is there objection to the use of names based on the principles for naming organic polymers, if names for the bivalent radicals in the structure are clearly established; however, for some structures, the use of the rules given below provides unambiguous names with much less artificiality.

2.2.5.2   *Fundamental principles*

The system of nomenclature for regular single-strand and quasi-single-strand inorganic and coordination polymers is governed by the same fundamental principles of polymer nomenclature developed for single-strand organic polymers.[2] It is based on the selection and naming of a constitutional repeating unit (CRU), defined[2] as the smallest structural unit the repetition of which describes the polymer structure. The name of the polymer is the name of this repeating unit prefixed by the terms 'poly', '*catena*', or other structural indicator, and designations for end groups, if desired.

The name of the constitutional repeating unit (CRU) is formed by citing, in order of appearance along the chain of the CRU, the names of its constituent subunits, which are the *longest structural fragments* of the CRU that can be named by the established principles of inorganic and/or coordination nomenclature.‡ Accordingly, bridging ligands are not broken into subunits smaller than those named by principles of coordination nomenclature for ligands.[2]

---

* Trivial names of 'siloxane' polymers are based on a $-SiH_2-O-$ repeating unit named siloxane which, together with substituents, is enclosed in parentheses or brackets and preceded by the prefix 'poly'. Names of low-molecular-weight siloxane polymers can be named by using the prefix 'oligo' or a numerical prefix in place of 'poly' as described in the organic polymers rules.[2,24] On the other hand, names of specific low-molecular-weight acyclic siloxanes with the general formula $H_3Si\{OSiH_2\}_nOSiH_3$ are formed according to Rule D-6.22.[24] This results in similar, yet not identical names; for example, when $n = 2$ in the general formula above, the name would be tetrasiloxane. In these names, substituents are cited as prefixes to such parent names, for example decamethyltetrasiloxane.

† This name is a hybrid of additive and substitutive nomenclature that does not satisfy fully the rules of either system.

‡ In the rules for naming organic polymers the constituent subunits of the CRU are the largest structural fragments that can be named as bivalent or multivalent radicals according to the established principles of organic nomenclature.[24] Single central atoms, mononuclear coordination centers, and bridging ligands are constituent subunits of CRUs in regular single-strand and quasi-single-strand inorganic and coordination polymers. Polynuclear coordination centers are used as subunits only under certain conditions.

(i) Identification of the constitutional repeating unit (CRU)

In many cases, the polymer structure is simple enough that the CRU and its constituent subunits can readily be identified; for example, in the polymer (**66**; Figure 7), the repeating units are (A) and (B). However, in more complex cases and occasionally in some simple cases it may be necessary to draw out a fairly long segment of the polymer chain in order to identify the possible constitutional repeating units.

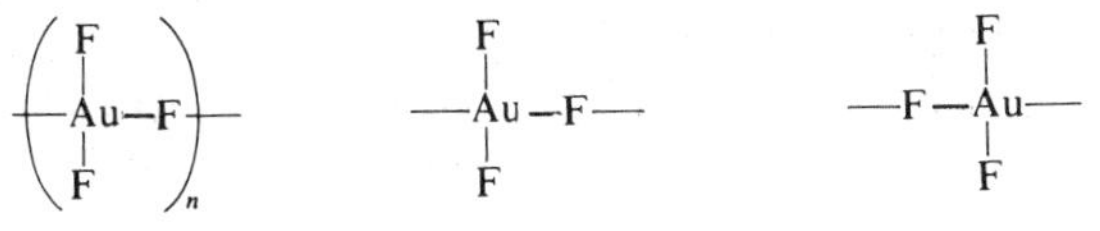

Figure 7

(ii) Orientation of the constitutional repeating unit (CRU)

The constituent subunit of the CRU at which the citation of the CRU begins is the central atom (or coordination center) of highest seniority, *i.e.* the most preferred central atom according to the set of hierarchical rules given in Rule IP-2.0.[5] This center is normally written as the *left* terminal subunit of the CRU. Further refinements to these general principles are given under Rule IP-2.0 of the full document.

(iii) Naming the constitutional repeating unit (CRU)

The name of a CRU of single-strand and quasi-single-strand inorganic or coordination polymers is based on a backbone consisting of central atoms and bridging ligands where present. All inorganic or coordination polymers have one or more central atoms, but may or may not have bridging ligands. Homoatomic inorganic polymers are considered to consist of central atoms only. Coordination centers, mononuclear or polynuclear, and their associated ligands, except for ligands between central atoms in the backbone, if any, are named by the usual principles of coordination nomenclature. Bridging ligands are named as ligands prefixed by the Greek letter μ.

Selection of the largest structural fragments in the backbone that can be assigned multivalent radical names as subunits of a CRU is a fundamental principle in naming linear organic polymers. For naming inorganic and coordination polymers, this principle is applied to the selection of bridging ligands in the CRU. When there is a choice, the largest group that can be named by the accepted methods for naming polydentate ligands is chosen. For example, in the polymer (**67**) the

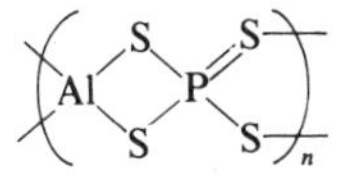

(**67**)

CRU could be considered as two central atoms connected by sulfur ligands. However, the principle of 'largest bridging ligand' requires the bridging ligand to be phosphorotetrathioato(3−). Strict application of this principle to inorganic or coordination polymers would lead to the selection of polynuclear coordination centers as the 'largest' structural fragment in the backbone. Since, at the present time, there are no officially accepted rules for uniquely naming and/or numbering certain types of polynuclear coordination centers, it is not yet convenient in some cases to use polynuclear coordination centers as subunits of a CRU in inorganic and coordination polymers. Hence, the principle of largest subunit is not always applied to coordination centers of a CRU and, in this set of rules, polynuclear coordination centers are used as subunits of CRUs only when it is *not* convenient to express such structural units in terms of their mononuclear coordination centers (see Rule IP-5.0[3]). However, for illustrative purposes, names using polynuclear subunits are given as alternatives for some of the examples in the rules that follow.

Once the names of the constituent subunits of the CRU are determined, the CRU name is formed by citing the name of the senior subunit followed by the names of the other constituent subunits as they occur in the preferred direction along the polymer chain. Detailed rules and many examples are given in the complete document.[3] A few examples are given in Table 4.

Table 4   Names of Constitutional Repeating Units (CRUs)

Regular single-strand inorganic and coordination polymers
Constitutional repeating units with homoatomic backbones:

$$\left(\!\!\begin{array}{cc} F & Me \\ | & | \\ -Si\!-\!Si\!- \\ | & | \\ F & Me \end{array}\!\!\right)_{\!\!n}$$

catena-poly[(difluorosilicon)(dimethylsilicon)] (the subunit with the alphabetically earliest side group is the senior subunit)[a]

Constitutional repeating units with backbones consisting of a mononuclear central atom and one bridging ligand:

$$\left(\!\!\begin{array}{c} NH_3 \\ | \\ -Zn\!-\!Cl- \\ | \\ Cl \end{array}\!\!\right)_{\!\!n}$$

catena-poly[(amminechlorozinc)-μ-chloro]

$$\left(\!\!\begin{array}{c} Ph \\ | \\ -Si\!-\!O- \\ | \\ Ph \end{array}\!\!\right)_{\!\!n}$$

catena-poly[(diphenylsilicon)-μ-oxo][b]

Constitutional repeating units consisting of more than one central atom and no more than one bridging ligand between each central atom of the polymer:

$$\left(\!\!\begin{array}{cc} CN & NH_3 \\ | & | \\ -Ni\!-\!CN\!-\!Cu\!-\!NC- \\ | & | \\ CN & NH_3 \end{array}\!\!\right)_{\!\!n}$$

catena-poly[[bis(cyano-*C*)nickel]-μ-(cyano-*C*:*N*)-(diamminecopper)-μ-(cyano-*N*:*C*)]

Regular quasi-single-strand coordination polymers
Constitutional repeating units with backbones consisting of one mononuclear central atom and two or more bridging ligands, alike or different, or a chelating ligand:

$$\left(\!\!\begin{array}{c} Cl \\ \diagdown \\ Pd \\ \diagup \\ Cl \end{array}\!\!\right)_{\!\!n}$$

catena-poly[palladium-di-μ-chloro] (*catena*-di-μ-chloro-palladium in the 1970 IUPAC Inorganic Rules[35])

Constitutional repeating units containing more than one central atom:

catena-poly[[oxovanadium-di-μ-hydroxo-(oxovanadium)]-[μ-(8-quinolinolato-*N*:*O*)-μ-(8-quinolinolato-*O*:*N*)]]

Single-strand and quasi-single-strand coordination polymers with polynuclear coordination centers

$$(\!+\!octahedro\text{-}W_6(\mu\text{-}Br_8)\,(2,3,4,5\text{-}Br_4)\!+\!(6\!:\!1)\!-\!(Br_4)\!+\!)_{\!\!n}$$

catena-poly[(octa-μ-bromo-2,3,4,5-tetrabromo-*octahedro*-hexatungsten)-6:1-μ-[tetrabromido(2–)]]

Regular single-strand and quasi-single-strand inorganic and coordination polymers with ionic CRUs[c]

$$\left(\!\!\begin{array}{c} Cs^+ \\ \left[\begin{array}{c} Cl \\ | \\ -Cu\!-\!Cl \\ | \\ Cl \end{array}\right]^- \end{array}\!\!\right)_{\!\!n}$$

catena-poly[cesium [cuprate-tri-μ-chloro] (1–)] or *catena*-poly[cesium [cuprate(II)-tri-μ-chloro]]

Designation of stereochemical configuration of a constitutional repeating unit consisting of a mononuclear central atom and one bridging ligand

$$\left(\!\!\begin{array}{c} F \quad F \\ \diagdown \diagup \\ Au \\ | \\ F \end{array}\!\!\right)_{\!\!n}$$

cis-catena-poly[(difluorogold)-μ-fluoro]

Table 4 *(Continued)*

Specification of end groups of linear inorganic or coordination polymers

Cl$\left(\text{S}\right)_n$H   α-chloro-ω-hydro-*catena*-poly[sulfur]

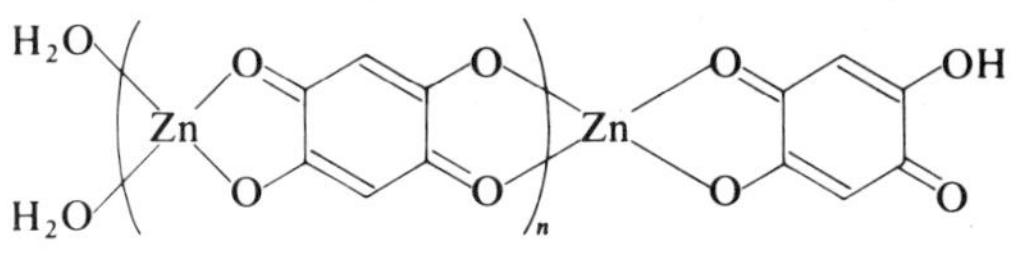

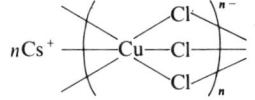

α,α-diaqua-ω-[[2,5-dihydroxy-*p*-benzoquinonato(1−)-O^1,O^2]-zinc]-*catena*-poly[zinc-μ-[2,5-dihydroxy-*p*-benzoquinon-ato(2−)-O^1,O^2:O^4,O^5]]

End groups that may be considered ionic

[O$\left(\text{MoO}_2\text{–O}\right)_nMoO_3$]$^{2-}$ [α-oxo-ω-(trioxomolybdate)-*catena*-poly[(dioxomolyb-denum)-μ-oxo]](2−)

---

[a] According to the rules for naming linear organic polymers this inorganic homoatomic polymer could be named poly(1,1-difluoro-2,2-dimethyldisilane-1,2-diyl).
[b] According to the rules for linear organic polymers this polymer would be oriented and named poly[oxy(diphenylsilylene)].
[c] From an inorganic viewpoint, it might be better to consider these polymers to be salts of polymeric ions as illustrated below:

nCs$^+$ [Cu—Cl, Cl, Cl]$_n^{n-}$

cesium *catena*-poly[[cuprate-tri-μ-chloro](1−)]
cesium *catena*-poly[cuprate(II)-tri-μ-chloro]

2.2.6 Source-based Nomenclature for Copolymers[4]

2.2.6.1 *Introduction*

In principle, a comprehensive structure-based system of naming copolymers would be desirable. However, such a system presupposes a knowledge of the structural identity of all the constitutional units as well as their sequential arrangements within the polymer molecules; this information is rarely available for the synthetic polymers encountered in practice. For this reason an essentially source-based nomenclature system has been developed. Sections 2.2.6.2 and 2.2.6.3 present the principles and mode of application of the system to copolymers of all types in a way that is capable of logical extension to molecules of considerable complexity. For relatively simple copolymers, an alternative system, which sometimes leads to more concise names, is available and this is outlined in Section 2.2.6.4.

Application of this system should not discourage the use of structure-based nomenclature whenever the copolymer structure is fully known and is amenable to treatment by the rules for single-strand polymers.[2,3] Further, an attempt has been made to maintain consistency, as far as possible, with the abbreviated nomenclature of synthetic polypeptides published by the IUPAC–IUB Commission on Biochemical Nomenclature.[31] It is intended that the present nomenclature system[4] supersede the previous recommendations published in 1952.[17]

2.2.6.2 *Basic concept*

By definition, copolymers are polymers that are derived from more than one species of monomer. Various classes of copolymers are distinguished on the basis of the characteristic sequence arrangements of the monomeric units within the copolymer molecules. Generally, the names of monomers are used to specify monomeric units; the latter can be named using the trivial, semisystematic, or systematic form. The classes of copolymers are as follows: unspecified (Rule 1); statistical (Rule 2.1); random (Rule 2.2); alternating (Rule 3); periodic (Rule 4.1); block (Rule 5.1); and graft (Rule 6.1). In those cases where copolymer molecules can be described by only one species of constitutional unit in a single sequential arrangement, copolymers are regular polymers[1] and can, therefore, be named on a structure basis.[2,3] Polymers having monomeric units differing in constitutional or configurational features, but derived from a single monomer, are not regarded as copolymers, in accordance with the basic definitions.[1] Examples of such polymers, which are not copolymers, are: (a) polybutadiene with mixed sequences of 1,2 and 1,4 units; (b) poly(methyloxirane), also known as poly(propylene oxide), obtained through polymerization of a mixture of the two enantiomers, (*R*) and (*S*), and containing both (*R*) and (*S*) units.

The nomenclature system presented here can, however, also be applied to such pseudo-copolymers. Polymers having monomeric units differing in constitutional features, but derived from a homopolymer by chemical modification, can be named in the same way, for example partially hydrolyzed poly(vinyl acetate) containing both ester and alcohol units.

A closely-related alternative system of nomenclature, which may be preferred in some circumstances, is described in Section 2.2.6.4.

2.2.6.3 *Classification and definition of copolymers*

A systematic source-based nomenclature for copolymers must identify the constituent monomers and provide a description of the sequence arrangement of the different types of monomeric units present. These objectives can be achieved by citing the names of the constituent monomers after the prefix 'poly', and by placing between the names of each pair of monomers an italicized connective to denote the kind of arrangement by which those two types of monomeric units are related in the structure. Seven types of sequence arrangement are listed in Table 5 together with the corresponding connectives and examples, in which A, B, and C represent the names of monomers. (The citation of A, B, and C is not intended to reflect an order of seniority, unless such seniority is specified in the rules. As a result, more than one name is often possible.) Each of these types of copolymer is considered in more detail below. When the chemical nature of the end groups is to be specified, the name of the copolymer (as described above) is preceded by the systematic names of the terminal units. The prefix α or ω refers to the terminal unit attached to the left or right, respectively, of the structure, as written, *e.g.* α-X-ω-Y-poly(A-*co*-B).

Table 5   Sequence Arrangements of Copolymers

Type	*Connective*	*Example*
Unspecified	-*co*-	Poly(A-*co*-B)
Statistical	-*stat*-	Poly(A-*stat*-B)
Random	-*ran*-	Poly(A-*ran*-B)
Alternating	-*alt*-	Poly(A-*alt*-B)
Periodic	-*per*-	Poly(A-*per*-B-*per*-C)
Block	-*block*-	PolyA-*block*-polyB
Graft	-*graft*-	PolyA-*graft*-polyB

1.   Copolymers with an unspecified arrangement of monomeric units

Rule 1.   An unspecified sequence arrangement of monomeric units is represented by (A-*co*-B) and the corresponding copolymer has the name poly(A-*co*-B); for example an unspecified copolymer of styrene and methyl methacrylate is named poly[styrene-*co*-(methyl methacrylate)].

2.   Statistical copolymers

Statistical copolymers are copolymers in which the sequential distribution of the monomeric units obeys known statistical laws; for example the monomer sequence distribution may follow Markovian statistics of zeroth (Bernoullian), first, second, or a higher order. Kinetically, the elementary processes leading to the formation of a statistical sequence of monomeric units do not necessarily proceed with equal *a priori* probability. These processes can lead to various types of sequence distribution comprising those in which the arrangement of monomeric units tends toward alternation, tends toward clustering of like units, or exhibits no ordering tendency at all.[36] In simple binary copolymerization, the nature of this sequence distribution can be indicated by the numerical values of a function either of the reactivity ratios or of the related run number.[36, 37]

The term 'statistical copolymer' is intended to embrace a large proportion of those copolymers that are prepared by simultaneous polymerization of two or more monomers in admixture. Such copolymers are often described in the literature as 'random copolymers', but this is almost always an improper use of the term random and the practice should be abandoned.

Rule 2.1.   A statistical sequence arrangement of monomeric units is represented by (A-*stat*-B), (A-*stat*-B-*stat*-C), *etc.*, where -*stat*- indicates that the statistical sequence distribution with regard to A, B, C, *etc.*, units is considered to be known. Statistical copolymers are named poly(A-*stat*-B), poly(A-*stat*-B-*stat*-C), *etc.*, *e.g.* poly(styrene-*stat*-butadiene), poly(styrene-*stat*-acrylonitrile-*stat*-butadiene).

Random copolymers. A random copolymer is a special case of a statistical copolymer. It is a statistical copolymer in which the probability of finding a given monomeric unit at any given site in the chain is independent of the nature of the neighboring units at that position (Bernoullian distribution). In other words, for such a copolymer the probability of finding a sequence . . . ABC . . . of monomeric units A, B, C . . . , *i.e.* P[. . . ABC . . .], is given by

$$P[\ldots ABC \ldots] = P[A] \cdot P[B] \cdot P[C] \ldots = \prod_{i=A,B,C\ldots} P[i] \tag{1}$$

where P[A], P[B], P[C], *etc.* are the unconditional probabilities of the occurrence of various monomeric units. As already noted above, the term 'random' should *not* be used for statistical copolymers except in this narrow sense.

Some authors use the term 'random' to denote the Bernoullian case further restricted by the condition that the monomeric units be present in exactly equal amounts.[38]

Rule 2.2. A random sequence arrangement of monomeric units is represented by (A-*ran*-B), (A-*ran*-B-*ran*-C), *etc.*, where -*ran*- indicates a random sequence distribution with regard to A, B, C, *etc.* units. Random copolymers are named poly(A-*ran*-B), poly(A-*ran*-B-*ran*-C), *etc.*, *e.g.* poly[ethylene-*ran*-(vinyl acetate)].

3. Alternating copolymers

An alternating copolymer is a copolymer comprising two species of monomeric units distributed in alternating sequence. The arrangement –ABABABAB– or (AB)$_n$ thus represents an alternating copolymer.

Rule 3. An alternating sequence arrangement of monomeric units is represented by (A-*alt*-B) and the corresponding alternating copolymer is named poly(A-*alt*-B), *e.g.* poly[styrene-*alt*-(maleic anhydride)].

Alternating sequence arrangements can form constitutionally regular structures and may, in those cases, also be named utilizing the structure-based nomenclature for regular single-strand organic polymers. The example above would then be named poly[(2,5-dioxotetrahydrofuran-3,4-diyl)(1-phenylethylene)].

4. Other types of periodic copolymers

In addition to alternating polymers, other structures are known in which the monomeric units appear in an ordered sequence. Examples are –ABCABCABC– or (ABC)$_n$; –ABBABBABB– or (ABB)$_n$; –AABBAABBAABB– or (AABB)$_n$; and –ABACABACABAC– or (ABAC)$_n$.

Rule 4.1. A periodic sequence arrangement of monomeric units is represented by (A-*per*-B-*per*-C), (A-*per*-B-*per*-B), (A-*per*-A-*per*-B-*per*-B), (A-*per*-B-*per*-A-*per*-C), *etc.*, and the corresponding periodic copolymers are named poly(A-*per*-B-*per*-C), poly(A-*per*-B-*per*-B), poly(A-*per*-A-*per*-B-*per*-B), poly(A-*per*-B-*per*-A-*per*-C), *etc.*

If these polymers are regular, they can also be named according to the structure-based nomenclature for regular single-strand organic polymers.[2] The binary monomer mixture consisting of formaldehyde and ethylene oxide might yield the periodically sequenced copolymer $\{CH_2OCH_2CH_2OCH_2CH_2O\}_n$, which is named poly[formaldehyde-*per*-(ethylene oxide)-*per*-(ethylene oxide)] or poly[formaldehyde-*alt*-bis(ethylene oxide)] or, alternatively, poly(oxymethyleneoxyethyleneoxyethylene).

Rule 4.2. If copolymer structures comprise several types of periodic sites, only some of which are always occupied by particular species of monomeric units (A, B . . .), and sites of the other types are occupied by two or more types of monomeric unit (U, V . . .) in irregular arrangement, the names of the monomers in the latter sites are embraced by parentheses and are separated by semicolon(s), for example the copolymer with the sequence arrangement –AUAVAVAUAVAUAU– is named poly-[A-*alt*-(U; V)].

5. Block copolymers

A block polymer is a polymer comprising molecules in which there is a linear arrangement of blocks, a block being defined as a portion of a polymer molecule in which the monomeric units have at least one constitutional or configurational feature absent from the adjacent portions.[1] In a block copolymer, the distinguishing feature is constitutional, *i.e.* each of the blocks comprises units derived from a characteristic species of monomer. In the sequence arrangements –AAAAAAAA–BBBBBBBBBBBB–, –AAAAAAAA–BBBBBBBBBBBB–AAAAAAAA– and –AABABAAABB–AAAAAAAA–BBBBBBBBBBBB– the sequences –AAAAAAAA–, –BBBBBBBBBBBB– and –AABABAAABB– are blocks.

Rule 5.1. A block sequence arrangement is represented by A_k-*block*-B_m, A_k-*block*-(A-*stat*-B), *etc.*, and the corresponding polymers are named polyA-*block*-polyB, polyA-*block*-poly(A-*stat*-B), *etc.*, respectively. If no ambiguity arises, a long dash may be used to designate block connections, as follows: polyA–polyB. For complex cases, use of -*block*- rather than the long dash is always encouraged. The order of citation of the block names corresponds to the order of succession of the blocks in the chain as written from left to right.

In the following examples, the subscripts $k, m, \ldots$ may be indeterminate or specific (see Rule 5.3). In each case, the first line gives a representation of the block sequence arrangement, the second the corresponding name, and the third an illustration of a specific case.

A_k-*block*-B_m

polyA-*block*-polyB

polystyrene-*block*-polybutadiene

A_k-*block*-B_n-*block*-C_m

polyA-*block*-polyB-*block*-polyC

polystyrene-*block*-polybutadiene-*block*-poly(methyl methacrylate).

Rule 5.2. Where a succession of blocks, such as $-A_k B_n C_m-$ is repeated, the appropriate multiplying prefix is used, *e.g.* $(A_k$-*block*-B_n-*block*-$C_m)_3$ is represented as tris(polyA-*block*-polyB-*block*-polyC).

Rule 5.3. When it is possible to specify the chain length of a block, the appropriate Greek prefix (*e.g.* hecta for 100) may be used rather than poly. Although short sequence lengths are not strictly embraced within the definition of 'block', the same device may usefully be employed by using the general prefix 'oligo' or the appropriate specific prefix (*e.g.* tri), *e.g.* $(A_c$-*block*-B_k-*block*-$C_3)_n$ is poly(oligoA-*block*-polyB-*block*-triC), where c is a small integer corresponding to the degree of polymerization of the oligomeric sequence.

Rule 5.4. Those block copolymers, derived from more than two monomers, that also exhibit statistical block sequence arrangements are named according to the principles of Rule 2.1; for example the statistical sequence arrangement $-A_k$-*block*-B_m-*block*-C_n-*block*-B_m-*block*-A_k-*block*-C_n- is named poly(polyA-*stat*-polyB-*stat*-polyC).

Rule 5.5. In the name of block copolymers with blocks connected by way of junction units X that are not part of the blocks, the name of the junction unit is inserted in the appropriate place. The connective, -*block*-, may be omitted. Thus, A_k-*block*-X-*block*-C_m or A_k–X–C_m is named polyA-*block*-X-*block*-polyC or polyA–X–polyC. The same designations can be applied to block polymers; for example polystyrene-*block*-dimethylsilylene-*block*-polybutadiene or polystyrene–dimethylsilylene–polybutadiene.

Rule 5.6. A block copolymer wherein A_k and B_m blocks, connected through junctions X, are distributed in statistical manner in the polymer molecules, as in $-A_k$-*block*-X-*block*-B_m-*block*-X-*block*-B_m-*block*-X-*block*-A_k-, is named poly[(polyA-*block*-X)-*stat*-(polyB-*block*-X)]. A block copolymer wherein A_k and B_m blocks and junction units X are all distributed in statistical manner, as in $-A_k$-*block*-X-*block*-B_m-*block*-A_k-*block*-B_m-*block*-X-*block*-B_m-*block*-X-*block*-A_k-, is named poly-(polyA-*stat*-X-*stat*-polyB).

6. *Graft copolymers*

A graft polymer is a polymer comprising molecules with one or more species of block connected to the main chain as side chains, these side chains having constitutional or configurational features that differ from those in the main chain.[1] In a graft copolymer the distinguishing feature of the side chains is constitutional, *i.e.* the side chains comprise units derived from at least one species of monomer different from those which supply the units of the main chain.

Rule 6.1. The simplest case of a graft copolymer can be represented by A_k-*graft*-B_m or the arrangement shown in Figure 8; the corresponding name is polyA-*graft*-polyB, where the monomer named first (A in this case) is that which supplied the backbone (main chain) units, while that named second (B) is in the side chain(s). Each of the following examples presents, in order, a

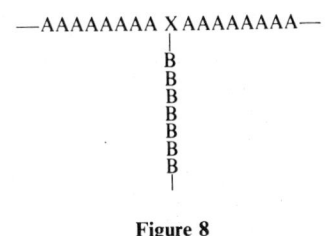

Figure 8

representation of the graft sequence arrangement, the corresponding name, and an illustration of a specific case.

(a) A_k-*graft*-B_m, polyA-*graft*-polyB, polybutadiene-*graft*-polystyrene (polystyrene grafted to polybutadiene); (b) (A_k-*block*-B_m)-*graft*-C_n, (polyA-*block*-polyB)-*graft*-polyC, (polybutadiene-*block*-polystyrene)-*graft*-polyacrylonitrile (polyacrylonitrile grafted to a polybutadiene–polystyrene block copolymer at unspecified sites); and (c) (A-*stat*-B)-*graft*-C_n, poly(A-*stat*-B)-*graft*-polyC, poly(butadiene-*stat*-styrene)-*graft*-polyacrylonitrile (polyacrylonitrile grafted to a statistical butadiene–styrene copolymer at unspecified sites).

Rule 6.2. If more than one type of graft chain is attached to the backbone, semicolons are used to separate the names of the grafts or their symbolic representations, *e.g.* A_k-*graft*-(B_m;C_n), polyA-*graft*-(polyB;polyC), polybutadiene-*graft*-[polystyrene;poly(methyl methacrylate)] (polystyrene and poly(methyl methacrylate) chains grafted to a polybutadiene backbone).

Rule 6.3. Graft copolymers with known numbers of graft chains are named using numeric prefixes (mono, bis, tris, *etc.*), *e.g.* A_k(-*graft*-B_m)$_3$, polyA-tris(-*graft*-polyB), polybutadiene-tris(-*graft*-polystyrene) (three polystyrene grafts per polybutadiene backbone). If the precise site of grafting is known, it can be specified, *e.g.* A_{10}-*block*-(X-*graft*-B_m)-*block*-A_{15}, decaA-*block*-(X-*graft*-polyB)-*block*-pentadecaA, decabutadiéne-*block*-(methylsilanetriyl-*graft*-polystyrene)-*block*-pentadecabutadiene.

The system of naming graft copolymers is also applicable, in principle, to 'star copolymers', where chains having different constitutional or configurational features are linked through a central moiety, *e.g.* A_k-*block*-[X-(*graft*-B_m)$_2$]-*block*-A_k or B_m-*block*-[X-(*graft*-A_k)$_2$]-*block*-B_m, polyA-*block*-[X-bis(-*graft*-polyB)]-*block*-polyA or polyB-*block*-[X-bis(-*graft*-polyA)]-*block*-polyB; polystyrene-*block*-[silanetetrayl-bis(-*graft*-polybutadiene)]-*block*-polystyrene or polybutadiene-*block*-[silanetetrayl-bis(-*graft*-polystyrene)]-*block*-polybutadiene (two polystyrene and two polybutadiene chains attached to a central Si atom).

7. Polymers made by polycondensation or related polymerization

The nomenclature system for copolymers is also applicable to polymers made by condensation polymerization of more than one monomeric species, or, more generally, by polymerization of more than one monomeric species where molecules of all sizes (*i.e.* monomers, oligomers, polymers) can react with each other. One can distinguish the case of polymers made by polycondensation of homopolymerizable monomers from that of polymers made by polycondensation of complementary ingredients that do not usually separately homopolymerize.

Rigorous application of the source-based definition of a copolymer[1] embraces polymers such as poly(ethylene terephthalate) or poly(hexamethylene adipamide) (which are commonly regarded as homopolymers) because two ingredients are, in each case, the usual starting materials of polymerization. If polymers of this type have constitutionally regular structures and are regular polymers, the nomenclature for regular single-strand organic polymers can also be used.[2] This applies, for example, to the polymer derived from terephthalic acid and ethylene glycol, which by source-based copolymer nomenclature would be named as poly(ethylene glycol-*alt*-terephthalic acid) if in fact the polymer has been prepared by a condensation polymerization starting with terephthalic acid and ethylene glycol. However, if the starting material is the partial ester $HOCH_2CH_2OOCC_6H_4CO_2H$, the appropriate source-based name is that of a homopolymer, whereas use of the starting material bis(hydroxyethyl) terephthalate, $HOCH_2CH_2$–$OOCC_6H_4CO_2CH_2CH_2OH$ (extensively employed industrially), would suggest the name poly-[bis(hydroxyethyl) terephthalate]. Regardless of the starting materials used, the structure-based name is poly(oxyethyleneoxyterephthaloyl). The trivial name poly(ethylene terephthalate) is also permitted, because it is so well established in the literature.

For all such polymers made by condensation polymerization of two complementary difunctional ingredients (or 'monomers'), which can readily be visualized as reacting on a 1:1 basis to give an 'implicit monomer', the homopolymerization of which would give the actual product, the single-strand structure-based nomenclature may be suitable insofar as such a polymer is regular and can be represented as possessing a single constitutional repeating unit. It is to be noted that this is applicable only to cases where the mole ratio of the ingredients is 1:1 and the ingredients are exclusively difunctional.

The introduction of a third component into the reaction system necessitates the use of copolymer nomenclature which can logically be developed from the foregoing rules, as the examples below illustrate. The copolymer derived from reaction of ethylene glycol with a mixture of terephthalic and isophthalic acids would be named poly[(ethylene glycol-*alt*-terephthalic acid)-*co*-(ethylene glycol-*alt*-isophthalic acid)], poly(ethylene terephthalate-*co*-ethylene isophthalate), or poly[(ethylene glycol)-*alt*-(terephthalic acid; isophthalic acid)].

A copolymer formed from oligo(adipic acid-*alt*-1,4-butanediol) and oligo(2,4-toluenediisocyanate-*co*-trimethylolpropane) in the presence of trimethylolpropane is named poly[oligo(adipic acid-*alt*-1,4-butanediol)-*co*-oligo(2,4-toluenediisocyanate-*co*-trimethylolpropane)-*co*-trimethylolpropane].

A polymer derived from the condensation polymerization of a single actual monomer, the molecules of which terminate in two different complementary functional groups (*e.g.* 6-aminohexanoic acid) is, by definition, a (regular) homopolymer. When two different monomers of this type react together, the product is a copolymer that can be named in appropriate fashion. For example, if 6-aminohexanoic acid is copolycondensed with 7-aminoheptanoic acid, leading to a statistical distribution of monomeric units, the product is named poly[(6-aminohexanoic acid)-*stat*-(7-aminoheptanoic acid)].

8.   Specification with regard to mass fractions, mole fractions, molar masses, and degrees of polymerization

Whereas subscripts placed immediately after the name of the monomer or the block designate the degree of polymerization or repetition, mass and mole fractions and molar masses (which in most cases are average quantities) may be expressed by placing corresponding figures after the complete name or symbol of the copolymer. The order of citation is the same as for the monomer species in the name or symbol of the copolymer. Unknown quantities can be designated by a, b, *etc.*

Although this scheme can be extended to complicated cases, it is recommended that its application be restricted to simple cases; any treatment of more complicated systems should be explained in the text.

Rule 8.1.   Mass fractions, or mass percentages, for the monomeric units are placed in parentheses after the copolymer name, followed by the symbol '*w*', or the phrase 'mass %', respectively. The order of citation in the parentheses is the same as in the name, for example (a) polybutadiene-*graft*-polystyrene (0.75:0.25 *w*) or polybutadiene-*graft*-polystyrene (75:25 mass %), a graft copolymer containing 75 mass % polybutadiene and 25 mass % grafted polystyrene; and (b) polybutadiene-*graft*-poly(styrene-*stat*-acrylonitrile) (0.75:a:b *w*) or polybutadiene-*graft*-poly(styrene-*stat*-acrylonitrile) (75:a:b mass %), a graft copolymer containing 75 mass % of butadiene units as backbone and unknown quantities in statistical arrangement of styrene and acrylonitrile units in the grafted chains.

Rule 8.2.   Mole fractions, or mole percentages, for the monomeric units are placed in parentheses after the copolymer name, followed by the symbol '*x*', or the phrase 'mol %', respectively. The order of citation in the parentheses is the same as in the name, for example polybutadiene-*graft*-polystyrene (0.85:0.15 *x*) or polybutadiene-*graft*-polystyrene (85:15 mol %), a graft copolymer containing 85 mol % of butadiene units and 15 mol % of styrene units.

Rule 8.3.   The molar mass, relative molecular mass, or degree of polymerization may be included in the scheme of Rules 8.1 and 8.2 by adding the corresponding figures, followed by the symbol M, M_r, or DP, respectively; for example (a) polybutadiene-*graft*-polystyrene (75:25 mass %; 90 000:30 000 M_r, a graft copolymer consisting of 75 mass % of butadiene units with a relative molecular mass of 90 000 as the backbone, and 25 mass % of styrene units in grafted chains with a relative molecular mass of 30 000; and (b) polybutadiene-*graft*-polystyrene (1700:290 DP), a graft copolymer consisting of a polybutadiene backbone with a degree of polymerization of 1700 to which polystyrene with a degree of polymerization of 290 is grafted.

2.2.6.4 *Alternative nomenclature for copolymers*

The nomenclature system for copolymers presented in Sections 2.2.6.2 and 2.2.6.3 is designed to meet the requirement of providing a systematic name for any copolymer, however complex the structure; as a result, the systematic names may be undesirably long in some cases. However, many copolymers reported in the literature have relatively simple structures, which do not necessitate this elaborate system. For these simpler cases, a different, more concise, nomenclature system is recommended

(i) Fundamental principles

The alternative nomenclature system is based on the following principles.

(1) A copolymer is described by the prefix 'copoly' followed by citation of the names of the monomers used (source-based nomenclature). The prefix is used only once, just as the prefix 'poly' is employed only once in naming regular single-strand polymers (called regular polymers hereafter)[2] or the simple copolymers described in the main text, *e.g.* copoly(styrene/butadiene).

(2) The specification of the type of structure in a copolymer (the connectives *block*, *alt*, etc. of the main nomenclature system) is shown by an italicized prefix preceding 'copoly' *e.g.* *block*-copoly-(styrene/butadiene).

(3) Only the names of the monomers are included in the main part of the copolymer name; the terminal units are specified before the main name (using the prefixes α and ω), whereas the junction units between blocks are specified after the main name (using the symbol μ), *e.g.* *block*-copoly-(styrene/butadiene)-μ-dimethylsilylene.

(4) The mass fraction, mole fraction, molar mass, or degree of polymerization of monomeric units in copolymers is specified separately in parentheses after the name.

(5) In addition to the principles stated in (1)–(4) above, some further conventions are used for the more complex copolymers, for example where the structure cannot be classified in a unique fashion.

In general, no seniority rule is provided for the order of citation of the monomer names. In block or periodic copolymers, the order of citation of the names of monomers corresponds to the sequence in which the monomeric units occur in the molecules. In graft copolymers, the initially cited name is that of the backbone (main chain).

(ii) Applications

A1. Application to simple copolymers

In the examples to be cited below, the name derived from the system elaborated in Sections 2.2.6.2 and 2.2.6.3 is given after the alternative name proposed in Section 2.2.6.4(i).

Rule A1.1.   The name of a copolymer consists of the prefix 'copoly' followed by parentheses in which the names of the monomers used are enumerated, separated by an oblique stroke, as in copoly(A/B), copoly(A/B/C), *etc.*, where A, B, and C represent the names of the monomers employed, *e.g.* copoly(styrene/methyl methacrylate), poly(styrene-*co*-methyl methacrylate).

Rule A1.2.   The arrangement of monomeric units in a copolymer, if known, is specified by one of the following italicized prefixes: *stat*- (statistical), *ran*- (random), *alt*- (alternating), *per*- (periodic), *block*-, or *graft*-, as in Section 2.2.6.3, *e.g.* *stat*-copoly(styrene/butadiene), poly-(styrene-*stat*-butadiene); *ran*-copoly(ethylene/vinyl acetate), poly[ethylene-*ran*-(vinyl acetate)]; *alt*-copoly(styrene/maleic anhydride), poly[styrene-*alt*-(maleic anhydride)]; *block*-copoly(styrene/butadiene/methyl methacrylate), polystyrene-*block*-polybutadiene-*block*-poly(methyl methacrylate); and *graft*-copoly(butadiene/styrene), polybutadiene-*graft*-polystyrene.

Rule A1.3.   When monomeric units of one particular kind occur in groups in a periodic copolymer, this can be indicated by a 'polykis' prefix. The repetition of a set of blocks in a block copolymer for a known or unknown number of times can be represented similarly, *e.g.* *per*-copoly(A/B/B/B) = *per*-copoly(A/trisB) = *alt*-copoly(A/trisB), poly(A-*per*-trisB); *block*-copoly-(A/B/C/A/B/C/A/B/C) = *block*-copoly[tris(A/B/C)], tris(polyA-*block*-polyB-*block*-polyC); and *per*-copoly[formaldehyde/bis(ethylene oxide)], poly[formaldehyde-*per*-bis(ethylene oxide)].

Rule A1.4.   When one type of site in a basically alternating or periodic copolymer can be occupied by units derived from two or more monomers, the names of the copolymers are based on the principles elaborated in Sections 2.2.6.2 and 2.2.6.3 and the rules in Section 2.2.6.4(i), *e.g.* *per*-copoly[A/B/(C;D)], poly[A-*per*-B-*per*(C;D)]; and *alt*-copoly[methyl methacrylate/(styrene;1-vinylnaphthalene)], poly[methyl methacrylate-*alt*-(styrene;1-vinylnaphthalene)].

The same procedure can be used to name graft copolymers with two or more different types of branch (B, C, *etc.*) grafted onto a backbone (A), *e.g. graft*-copoly[A/(B;C)], polyA-*graft*-(polyB;polyC); and *graft*-copoly[butadiene/(styrene;methyl methacrylate)], polybutadiene-*graft*-[polystyrene; poly(methyl methacrylate)].

Rule A1.5. Terminal units (preceded by the prefixes α and ω) are specified before the main copolymer name, but junction units between blocks (preceded by the prefix μ) are specified after the main copolymer name. If one type of junction unit occurs in the structure more than once, the appropriate multiplying prefix (bis, tris, *etc.*) may be used; if more than one type of junction unit occurs in the structure, they may be designated μ_1, μ_2, *etc.*, *e.g.* α-butyl-ω-carboxy-*block*-copoly-(styrene/butadiene), α-butyl-ω-carboxy-polystyrene-*block*-polybutadiene; *block*-copoly(styrene/butadiene)-μ-dimethylsilylene, polystyrene-*block*-dimethylsilylene-*block*-polybutadiene; and *graft*-copoly(butadiene/styrene)-polykis(μ-methylsilanetriyl).

Rule A1.6. Specification with regard to mass fractions, mole fractions, molar masses, relative molecular masses, and degrees of polymerization is treated as in Rules 8.1–8.3 of Section 2.2.6.3.

A2.   *Application to more complex copolymers*

The alternative nomenclature can be extended to copolymers of some complexity. The following rules and examples deal with a few such cases. In general, the nomenclature system described in Sections 2.2.6.2 and 2.2.6.3 will be found preferable for complex structures.

Rule A2. When a graft copolymer or block copolymer contains a constituent block which is itself a copolymer, the block is named co(B/C) with a descriptive prefix, if necessary (see Rule A1.2), *e.g. block*-copoly[*stat*-co(styrene/butadiene)/styrene/butadiene], poly(styrene-*stat*-butadiene)-*block*-polystyrene-*block*-polybutadiene; and *graft*-copoly[*stat*-co(butadiene/styrene)/acrylonitrile], poly-(butadiene-*stat*-styrene)-*graft*-polyacrylonitrile.

A3.   *Application to polymers made by polycondensation or related polymerization*

The principles of Sections 2.2.6.2 and 2.2.6.3 and the rules of Section 2.2.6.4 can be applied to polymers made by polycondensation or related polymerization. Examples of alternative names are copoly(6-aminohexanoic acid/7-aminoheptanoic acid), *alt*-copoly(ethylene glycol/terephthalic acid) and *alt*-copoly[ethylene glycol/(terephthalic acid; isophthalic acid)].

2.3   STEREOCHEMISTRY

2.3.1   Stereochemical Definitions and Notations[5]

The first recommendations on nomenclature for polymer stereochemistry were published in 1962[18, 19] but the enormous developments in polymer synthesis and characterization of the last 25 years have rendered this obsolete. Nomenclature relating to the constitution and configuration of macromolecules has been refined (using structure-based concepts)[1, 2] and definitive IUPAC[26] and IUPAC–IUB[27] documents relating, respectively, to organic molecules and to polypeptide chains have appeared. Stereochemical definitions and notations relating to polymers were published in 1981;[5] their principal features are summarized here.

It is convenient if stereochemical formulae for polymer chains are shown as Fischer projections rotated through 90°, *i.e.* displayed horizontally rather than vertically, or as hypothetical extended zigzag chains; the latter occasionally give a clearer indication of the three-dimensional arrangement. It is preferred that the hypothetical extended zigzag chains be consistently drawn with the backbone bond on the extreme left of the formula presented rising from left to right and with the interrupted line, on any given backbone carbon atom, drawn to the left of the full line.

The use of rotated Fischer projections corresponds to the common practice of using horizontal lines to denote polymer backbone bonds, but it is most important to note that this does not give an immediately visual impression of the zigzag chain. In the projections as used in this chapter, *at each individual backbone carbon atom* the horizontal lines represent bonds directed below the plane of the paper from the carbon atom, while the vertical lines project above the plane of the paper from the carbon atom. Thus, the rotated Fischer projection (**68a**) corresponds to (**68b**) and hence to zigzag chain (**68c**), while (**69a**) corresponds to (**69b**) and hence to (**69c**).

Unless otherwise stated, the drawings of configurational base units, configurational repeating units, stereorepeating units, *etc.*, provide information regarding *relative* configurations.

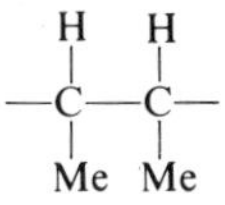

(68a) 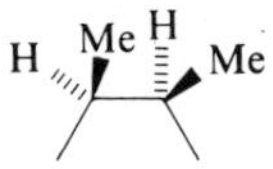 (68b) 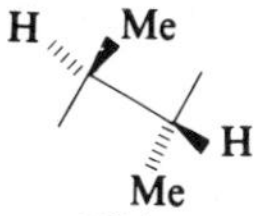 (68c)

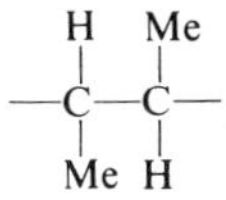

(69a)  (69b) 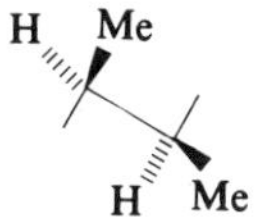 (69c)

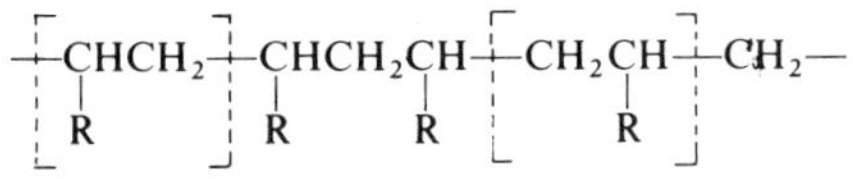

(70)

In a polymer molecule, the two portions of the main chain attached to any constitutional unit are, in general, nonidentical; consequently, a backbone carbon atom that also bears two different side-groups is considered to be a chiral center.

The absence from a formula of any one of the horizontal or vertical lines at a chiral or prochiral carbon atom, or of *cis* or *trans* designations at double bonds, indicates that the configuration of that stereoisomeric center is not known. Also, the convention of orienting polymer structures (and the corresponding constitutional and configurational units) from left to right is used. Thus, the two bracketted constitutional units in (70) are regarded as different, even though the repetition of either one of them would give the same regular polymer.

2.3.1.1 *Basic definitions*

1. *Configurational unit.* A constitutional unit having one or more sites of defined stereoisomerism.
2. *Configurational base unit.* A constitutional repeating unit, the configuration of which is defined at one or more sites of stereoisomerism in the main chain of a polymer molecule. In a regular polymer, a configurational base unit corresponds to the constitutional repeating unit.
3. *Configurational repeating unit.* The smallest set of one, two or more successive configurational base units that prescribes configurational repetition at one or more sites of stereoisomerism in the main chain of a polymer molecule.
4. *Stereorepeating unit.* A configurational repeating unit having defined configuration at all sites of stereoisomerism in the main chain of a polymer molecule.

Note on enantiomeric and diastereoisomeric units. Two configurational units (definitions 1, 2, 3, 4 above) that correspond to the same constitutional unit are considered to be *enantiomeric* if they are nonsuperimposable mirror images. Two nonsuperimposable configurational units that correspond to the same constitutional unit are considered to be *diastereoisomeric* if they are *not* mirror images; for example in the regular polymer molecule $-[CH(Me)CH_2]_n-$, poly(propylene), the constitutional repeating unit is $-CH(Me)CH_2-$ and the corresponding configurational base units are (71) and (72). The configurational base units (71) and (72) are enantiomeric, while the configurational units (71) and (73) cannot be enantiomeric because the constitutional units are different species, according to this nomenclature. The simplest possible stereorepeating units are (71), (74) and (75), and the corresponding stereoregular polymers are (71a), (74a) and (75a).

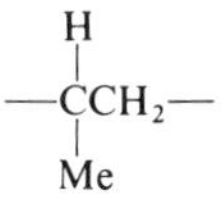

(71) 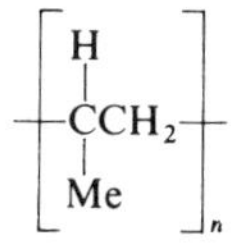 (71a) (72) (73)

an isotactic
polymer (see
definition 7)

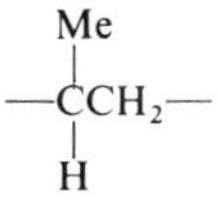

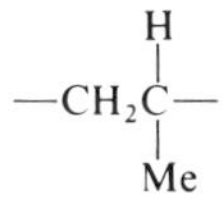

$$Me \quad H$$
$$-\overset{|}{C}CH_2\overset{|}{C}CH_2-$$
$$\underset{|}{H} \quad \underset{|}{Me}$$

(74)

$$\left[\begin{array}{c} Me \quad H \\ +\overset{|}{C}CH_2\overset{|}{C}CH_2+ \\ \underset{|}{H} \quad \underset{|}{Me} \end{array} \right]_n$$

(74a)

a syndiotactic
polymer (see
definition 8)

$$Me \quad Me \quad H \quad H$$
$$-\overset{|}{C}CH_2\overset{|}{C}CH_2\overset{|}{C}CH_2\overset{|}{C}CH_2-$$
$$\underset{|}{H} \quad \underset{|}{H} \quad \underset{|}{Me} \quad \underset{|}{Me}$$

(75)

$$\left[\begin{array}{c} Me \quad Me \quad H \quad H \\ +\overset{|}{C}CH_2\overset{|}{C}CH_2\overset{|}{C}CH_2\overset{|}{C}CH_2+ \\ \underset{|}{H} \quad \underset{|}{H} \quad \underset{|}{Me} \quad \underset{|}{Me} \end{array} \right]_n$$

(75a)

a hypothetical heterotactic
polymer (see Section 2.3.1.2(ii))

5. *Tactic polymer.* A regular polymer, the molecules of which can be described in terms of only one species of configurational repeating unit in a single sequential arrangement.

6. *Tacticity.* The orderliness of the succession of configurational repeating units in the main chain of a polymer molecule.

7. *Isotactic polymer.* A regular polymer, the molecules of which can be described in terms of only one species of configurational base unit (having chiral or prochiral atoms in the main chain) in a single sequential arrangement.

Note. In an isotactic polymer, the configurational repeating unit is identical with the configurational base unit.

8. *Syndiotactic polymer.* A regular polymer, the molecules of which can be described in terms of alternation of configurational base units that are enantiomeric.

Note. In a syndiotactic polymer, the configurational repeating unit consists of two configurational base units that are enantiomeric.

9. *Stereoregular polymer.* A regular polymer, the molecules of which can be described in terms of only one species of stereorepeating unit in a single sequential arrangement.

10. *Atactic polymer.* A regular polymer, the molecules of which have equal numbers of the possible configurational base units in a random sequence distribution.

Some examples of tactic polymers are shown below. For the polymer $\{CH(CO_2R)CH(Me)\}_n$, if only the ester-bearing main-chain site in each constitutional repeating unit has defined stereochemistry, the configurational repeating unit is **(76)** and the corresponding isotactic polymer is **(77)**. In the corresponding syndiotactic case, the configurational repeating unit is **(78)** and the syndiotactic polymer is **(79)**. As the definition of a stereoregular polymer (see definitions 4 and 9) requires that the configuration be defined at *all* sites of stereoisomerism, structures **(77)** and **(79)** do *not* represent stereoregular polymers. The same is true of **(80)** and **(81)**, which differ from **(77)** and **(79)** in that the

$$H$$
$$-\overset{|}{C}CH(Me)-$$
$$\underset{|}{C}O_2R$$

(76)

$$\left[\begin{array}{c} H \\ +\overset{|}{C}CH(Me)+ \\ \underset{|}{C}O_2R \end{array} \right]_n$$

(77)

$$H \qquad CO_2R$$
$$-\overset{|}{C}CH(Me)\overset{|}{C}CH(Me)-$$
$$\underset{|}{C}O_2R \quad \underset{|}{H}$$

(78)

$$\left[\begin{array}{c} H \qquad CO_2R \\ +\overset{|}{C}CH(Me)\overset{|}{C}CH(Me)+ \\ \underset{|}{C}O_2R \quad \underset{|}{H} \end{array} \right]_n$$

(79)

$$\left[\begin{array}{c} H \\ +CH(CO_2R)\overset{|}{C}+ \\ \underset{|}{Me} \end{array} \right]_n$$

(80)

$$\left[\begin{array}{c} H \qquad\qquad Me \\ +CH(CO_2R)\overset{|}{C}CH(CO_2R)\overset{|}{C}+ \\ \underset{|}{Me} \qquad\quad \underset{|}{H} \end{array} \right]_n$$

(81)

sites of specified and unspecified configuration have been interchanged. Examples (**71a**), (**74a**), (**75a**), (**77**), (**79**), (**80**) and (**81**) are tactic polymers. A stereoregular polymer is always a tactic polymer, but a tactic polymer is not always stereoregular because a tactic polymer need not have *all* sites of stereoisomerism defined.

Note. Structure-based names of tactic polymers are formed before the application of adjectives designating tacticity; thus, 'syndiotactic poly(ethylidene)' is preferred to 'syndiotactic poly(dimethylethylene)' because a shorter repeating unit is identified, in conformity with the rules for regular single-strand organic polymers.[2]

Note on atactic polymers. As the definition above indicates, a regular polymer, the configurational base units of which contain one site of stereoisomerism only, is atactic if it has equal numbers of the possible types of configurational base units arranged in a random distribution. If the constitutional repeating unit contains more than one site of stereoisomerism, the polymer may be atactic with respect to only one type of site if there are equal numbers of the possible configurations of that site arranged in a random distribution.

A polymer such as $\pm$CH=CHCH(Me)CH$_2$ $\pm_n$, which has two main-chain sites of stereoisomerism, may be atactic with respect to the double bond only, with respect to the chiral atom only, or with respect to both centers of stereoisomerism. If there is a random distribution of equal numbers of units in which the double bond is *cis* and *trans*, the polymer is atactic with respect to the double bond, and if there is a random distribution of equal numbers of units containing the chiral atom in the two possible configurations, the polymer is atactic with respect to the chiral atom.* The polymer is completely atactic when it contains, in a random distribution, equal numbers of the four possible configurational base units which have defined stereochemistry at both sites of stereoisomerism.

In addition to isotactic, syndiotactic and atactic polymers (and other well-defined types of tactic polymers), there exists the whole range of possible arrangements between the completely ordered and the completely random distributions of configurational base units, and it is necessary to employ the concept of degree of tacticity to describe such systems.

11. *Stereospecific polymerization.* Polymerization in which a tactic polymer is formed; however, polymerization in which stereoisomerism present in the monomer is merely retained in the polymer is not to be regarded as stereospecific. For example, the polymerization of a chiral monomer, *e.g.* D-propylene oxide (D-methyloxirane), with retention of configuration is not considered to be a stereospecific reaction; however, selective polymerization, with retention, of one of the enantiomers present in a mixture of D- and L-propylene oxide molecules is so classified.

12. *Ditactic polymer.* A tactic polymer that contains two sites of defined stereoisomerism in the main chain of the configurational base unit, for example (**82**) and (**83**) are both ditactic.

(**82**)

(**83**)

(**84**)

13. *Tritactic polymer.* A tactic polymer that contains three sites of defined stereoisomerism in the main chain of the configurational base unit, *e.g.* poly[3-(methoxycarbonyl)-4-methyl-*cis*-1-butenylene] (**84**).

14. *Diisotactic polymer.* An isotactic polymer that contains two chiral or prochiral atoms with defined stereochemistry in the main chain of the configurational base unit.

15. *Disyndiotactic polymer.* A syndiotactic polymer that contains two chiral or prochiral atoms with defined stereochemistry in the main chain of the configurational base unit.

---

* With regard to isomerism about double bonds, it is recommended that the *E* and *Z* designations[26] be used, where appropriate, in describing side-chain configurations and in the names of monomers used in source-based polymer nomenclature. In structure-based polymer names and in descriptions of configuration about double bonds in polymer main chains, the use of *cis* and *trans* is preferred.

The following examples illustrate diiso- and disyndio-tacticity: (85) and (86) are diisotactic, (87) is disyndiotactic.* A polymer with the repeating unit (88) is ditactic and may be described as syndiotactic (see definition 8), but it is *not* disyndiotactic.

(85)   (86)

(87)   (88)

The relative configuration of adjacent, constitutionally nonequivalent, carbon atoms can be specified as 'erythro' or 'threo', as appropriate, by adding the required prefix to the terms 'diisotactic' and 'disyndiotactic', as necessary (see Section 2.3.1.2(ii)).

16. *Cistactic polymer.* A tactic polymer in which the main-chain double bonds of the configurational base units are entirely in the *cis* arrangement.

17. *Transtactic polymer.* A tactic polymer in which the main-chain double bonds of the configurational base units are entirely in the *trans* arrangement.

Terms referring to the tacticity of polymers (tactic, ditactic, tritactic, isotactic, cistactic, *etc.*) can also be applied with similar significance to chains, sequences, blocks, *etc.*

Note. Terms defining stereochemical arrangements are to be italicized only when they form part of the name of a polymer; the use of such terms as adjectives, even when immediately preceding names, does not require italics. This practice is illustrated in the following example: isotactic poly(3-methyl-*trans*-1-butenylene), transisotactic poly(3-methyl-1-butenylene) describes (89) and/or (90).

(89)   (90)

18. *Block.* A portion of a polymer molecule, comprising many constitutional units, that has at least one constitutional or configurational feature which is not present in the adjacent portions.

19. *Tactic block.* A regular block that can be described by only one species of configurational repeating unit in a single sequential arrangement.

20. *Atactic block.* A regular block that has equal numbers of the possible configurational base units in a random sequence distribution.

21. *Stereoblock.* A regular block that can be described by one species of stereorepeating unit in a single sequential arrangement.

22. *Tactic block polymer.* A polymer, the molecules of which consist of tactic blocks connected linearly (see example in Figure 9).

23. *Stereoblock polymer.* A polymer, the molecules of which consist of stereoblocks connected linearly (see example shown in Figure 10).

An example of a tactic block polymer is $-A_kB_lA_mB_n-$ (Figure 9). In this case the blocks are stereoblocks but the block polymer is not a block copolymer because all the units derive from a single monomer.

does *not* represent a different disyndiotactic polymer.

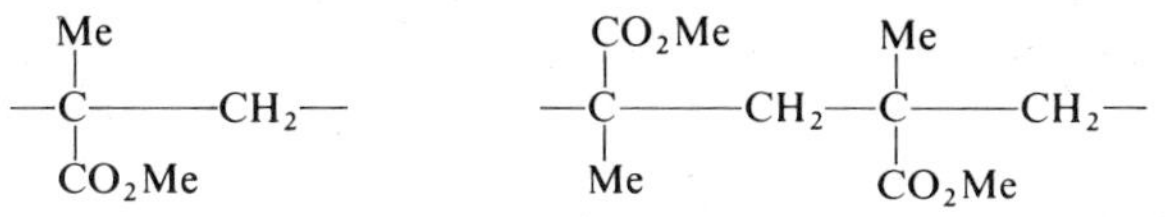

Figure 9

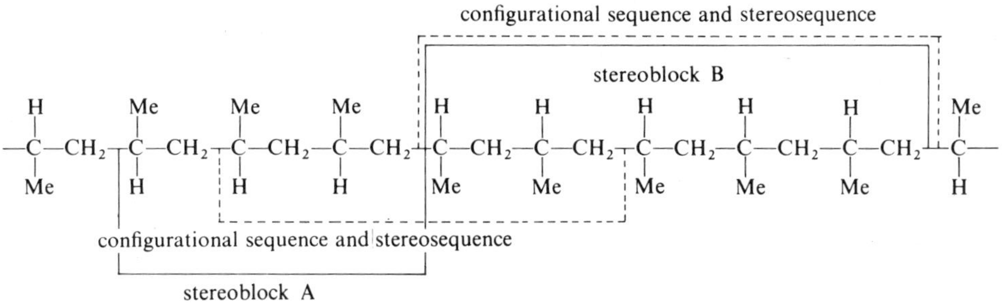

Figure 10

In the example shown in Figure 10 of a regular poly(propylene) chain, the stereoblocks are denoted by ⌐————⌐ . Here, the sequence of identical relative configurations of adjacent units that characterizes the stereoblock is terminated at each end of the block. Note that ⌐——————⌐ represents a configurational sequence, which may or may not be identical with a stereoblock (see definitions 21 and 26). The configurational sequence and stereosequence coincide in this particular case because there is only one site of stereoisomerism in each constitutional repeating unit (compare definitions 26 and 27).

2.3.1.2   Sequences

(i)   Constitutional and configurational sequences

The description of polymer structure revealed by studies of physical properties focuses attention on the distribution of local arrangements present in the molecules, and terms useful in this context are defined below.* (The terms defined here in relation to complete polymer molecules can also be applied to sequences and to blocks, as in Table 1, definition 3.14.)

24. *Constitutional sequence.* A defined portion of a polymer molecule comprising constitutional units of one or more species , *e.g.* $-CH_2CH_2CH_2CH(Me)-$, $-CH_2CH(Me)CH_2CH(Me)-$.

25. *Constitutional homosequence.* A constitutional sequence which contains constitutional units of only one species and in one sequential arrangement, *e.g.* $-CH(Me)CH_2CH(Me)CH_2-$, $\{CH(Me)CH_2\}$. In these two cases, the constitutional unit $-CH(Me)CH_2-$ can be called the constitutional repeating unit of the homosequence.

26. *Configurational sequence.* A constitutional sequence in which the relative or absolute configuration is defined at one or more sites of stereoisomerism in each constitutional unit in the main chain of a polymer molecule. (See Figure 10.)

27. *Stereosequence.* A configurational sequence in which the relative or absolute configuration is defined at all sites of stereoisomerism in the main chain of a polymer molecule.

28. *Configurational homosequence.* A constitutional homosequence in which the relative or absolute configuration is defined at one or more sites of stereoisomerism in each constitutional unit in the main chain of a polymer molecule.

29. *Stereohomosequence.* A configurational homosequence in which the relative or absolute configuration is defined at all sites of stereoisomersim in the main chain of a polymer molecule.

---

* The numbering of definitions in this section continues from that in Section 2.3.1.1.

(ii)   Description of relative configurations

(a) Erythro *and* threo *structures.* The relative configuration at two contiguous carbon atoms in the main chain bearing, respectively, substituents a and b (a ≠ b), is designated by the prefix '*erythro*' or '*threo*', as appropriate (Figure 11), by analogy with the terminology for carbohydrate systems in which the substituents are –OH. See Rule 4.11 of ref. 26. Similar systems in which a higher level of substitution exists may be treated analogously if the *erythro* or *threo* designation is employed to denote the relative placements of those two substituents, one for each backbone carbon atom, which rank highest according to the sequence rule.

erythro   *threo*

Figure 11

(b)   Meso *and* racemo *structures.* Relative configurations of consecutive, but not necessarily contiguous, constitutionally equivalent carbon atoms that have a symmetrically constituted connecting group (if any) are designated as '*meso*' or '*racemo*', as appropriate (Figure 12). The structures

(a)   (b)

Figure 12   (a) *meso*, abbreviation *m*; (b) *racemo*, abbreviation *r*; the symbol —⋁⋁— represents a symmetrically constituted connecting group, such as $-CH_2-$, $-CH_2CH_2-$ or $-CR_2CH_2CR_2-$

(**91**) and (**92**) both have the *meso* relative configuration but the boldly printed carbon atoms in each of the formulae below (**93** and **94**) cannot be considered as in a *meso* arrangement because the connecting group lacks the necessary symmetry. The term '*racemo*' is introduced here as the logical prefix for the designation of an arrangement that is analogous to racemic, in the sense defined above. It is unfortunate that the meaning of the term '*racemic*' current in organic chemistry is not directly applicable to polymers, but the use of the prefix '*racemo*' proposed here should not cause confusion because of the special context. To achieve a full configurational description, it may be necessary to preface the name of a polymer with a compound adjective that combines a term such as '*erythro*', '*threo*', '*meso*', or '*racemo*' with a term such as 'diisotactic' or 'disyndiotactic'. Examples are given in Table 6.

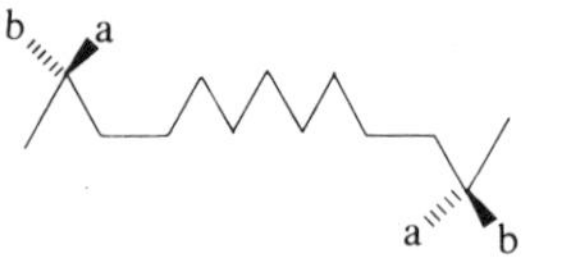

(**91**)    (**92**)

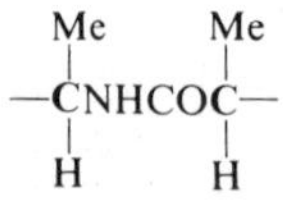

(**93**) 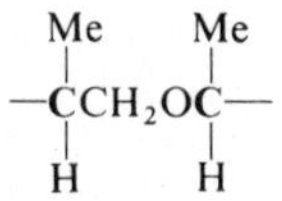   (**94**)

(c)   Stereosequences.   Stereosequences terminating in tetrahedral stereoisomeric centers at both ends, and which comprise two, three, four, five, *etc.* consecutive centers of that type, may be called diads, triads, tetrads, pentads, *etc.*, respectively. Typical diads are (**95**) and (**96**). When it is necessary

Table 6 Examples of Diisotactic and Disyndiotactic Polymers

Erythrodiisotactic polymer

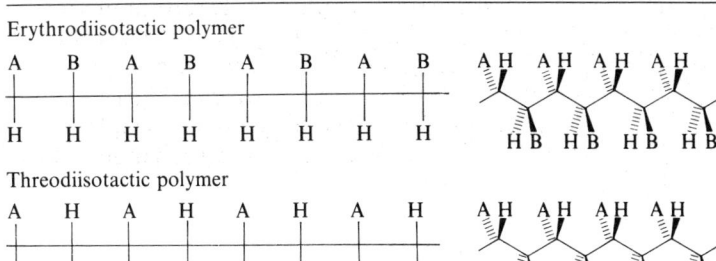

Threodiisotactic polymer

Disyndiotactic polymer[a]

Polymers with chiral centers arising from rings linking adjacent main-chain carbon atoms can be included in this nomenclature:

Erythrodiisotactic polymer

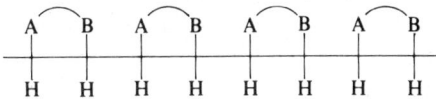

Threodiisotactic polymer

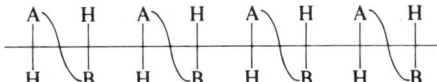

In the last two cases, the chiralities of the asymmetric centers should be designated (*R*) or (*S*) if known.

Erythrodisyndiotactic polymer

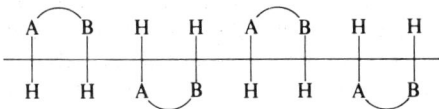

Threodisyndiotactic polymer

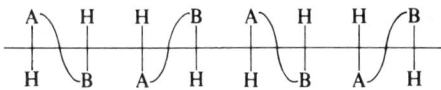

If the rings are symmetrical:

Mesodiisotactic   Racemodiisostatic

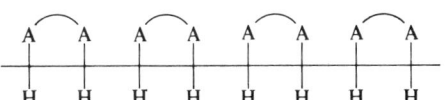   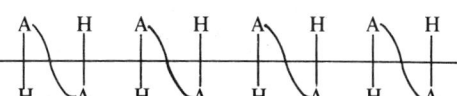

---

[a] This polymer cannot be expressed as erythrodisyndiotactic nor as threodisyndiotactic. Instead:

Erythrodisyndiotactic   Threodisyndiotactic

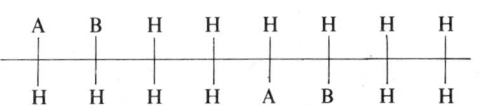   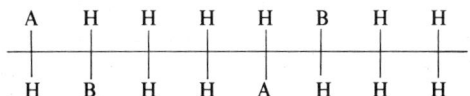

to specify the internal stereochemistry of the group, a prefix is required. In vinyl polymers there are *meso* (*m*) and *racemo* (*r*) diads and *mm*, *mr*, *rr* triads. The latter may be called isotactic, heterotactic and syndiotactic triads, respectively. Stereoregular vinyl polymers can be defined in terms of the regular sequences of diads; thus an isotactic vinyl polymer consists entirely of *m* diads, *i.e.* it corresponds to the following succession of relative configuration –*mmmmmm*–, whereas a syndio-tactic vinyl polymer consists entirely of *r* diads, corresponding to the sequence –*rrrrrr*–. Similarly, a vinyl polymer consisting entirely of *mr*(=*rm*) triads is called a heterotactic polymer.

$$
\begin{array}{cc}
\underset{\text{(95)}}{\overset{\displaystyle \text{Me}\quad\text{Me}}{-\text{CCH}_2\text{C}-}} & \underset{\text{(96)}}{\overset{\displaystyle \text{Me}\quad\quad\text{Me}\,\cdot}{-\text{CCH}_2\text{CH}_2\text{C}-}}
\end{array}
$$

2.4 TERMINOLOGY FOR POLYMERIZATIONS INVOLVING CHIRAL MONOMERS OR RESULTING IN OPTICALLY ACTIVE POLYMERS[7]

The nomenclature in this section is concerned with polymerizations of mixtures of stereoisomers and diastereoisomers, each molecule of which is optically active (chiral monomers), and polym-erizations of prochiral monomer molecules giving optically active polymers. Particular attention is drawn to some of the terms defined here because they are intended to replace current usage, and have received widespread support.

2.4.1 Polymerizations of Chiral Monomers: Introduction

There may be two different types of chiral monomers and polymers depending on the position of the chiral centers. The latter may either become a part of the main chain (as is the case of cyclic monomers bearing one or two asymmetric carbon atoms in the ring), or be a part of a side chain in both the monomer and the resulting polymer (for some cyclic compounds and for vinyl monomers with a chiral substituent).

In the first case, the polymerization will give optically active macromolecules (each containing only one type of configurational repeating unit in the case of a perfect process) only if the reaction is stereospecific. In the second case, stereospecificity is not required, a discrimination of the monomer based on the side chain being sufficient to give an optically active polymer.

First we shall consider polymerizations of stereoisomeric mixtures of monomer molecules, containing one or two asymmetric centers, for which an enantiomer discrimination is achieved. (The case of pure enantiomers is excluded since polymerization of a chiral monomer with retention of configuration is not considered to be a stereospecific reaction;[5] see Section 2.3.1.1 definition.) The polymer sample as a whole formed from such mixtures may or may not be optically active, but the macromolecules are generally optically active.

When the monomer molecules contain only one chiral center, the stereoregular macromolecules are optically active, for example methyloxirane (propylene oxide). When the monomer molecules contain two chiral centers, the macromolecules may be (*e.g.* from *cis*-2,3-dimethylthiirane, *cis*-1,2-dimethylethylene sulfide) or may not be optically active (*e.g.* from *trans*-2,3-dimethylthiirane, *trans*-1,2-dimethylethylene sulfide).

The case of monomers containing one or two chiral centers and consisting of a mixture of enantiomers is considered first, and followed by the case of mixtures of diastereoisomers. Monomers with two chiral centers but having a plane of symmetry will be examined in Section 2.4.2 since they may be considered prochiral.

2.4.1.1 *Polymerizations of enantiomer mixtures*

According to the basic definitions of terms relating to polymers (Table 1, term 3.47),[1] a *stereoselective polymerization* is a polymerization in which a polymer molecule is formed from a mixture of stereoisomeric monomer molecules by incorporation of only one stereoisomeric species.

This broad definition includes the more specific cases defined below, *i.e.* enantioasymmetric, enantiosymmetric, diastereosymmetric and diastereoasymmetric polymerizations.

(i)   Enantioasymmetric polymerization (formerly called 'stereoselective' or 'asymmetric selective' polymerization)

This describes a polymerization in which, starting from a mixture of enantiomers, only one of the enantiomers is polymerized. For example, polymerization of racemic propylene sulfide (methylthiirane), using some optically active initiators, may proceed with incorporation of only one of the two enantiomers. This will result in the production of stereoregular, optically active polymers composed of only one type of configurational repeating unit, if there is either complete retention or complete inversion of configuration. Note that (a) the residual monomer is enriched in the other type of enantiomer, and would ideally become the pure enantiomer at 50% conversion; and (b) polymers formed by enantioasymmetric polymerization are not necessarily stereoregular. For example, the polymerization of an alkenic mixture (with the chiral centers in the side chain) may give a polymer that is optically active without being stereoregular.

Polymerization of racemic *trans*-2,3-dimethylthiirane (racemic *trans*-1,2-dimethylethylene sulfide) using some optically active initiators, but giving optically inactive polymer molecules by incorporation of only one of the two enantiomers, provides an example of enantioasymmetric polymerization.

(ii)   Enantiosymmetric polymerization

A polymerization in which, starting from a mixture of enantiomers, the polymer formed consists of a mixture of two types of chains, each one containing only units derived from one of the enantiomers.

The polymerization of various types of racemic oxiranes (epoxides) or thiiranes (episulfides) can give polymer molecules that are both stereoregular and optically active, the sample being optically inactive through intermolecular compensation. The residual monomer is also inactive by compensation.

Statistical polymerization of two enantiomers should not be confused with either enantiosymmetric or enantioasymmetric polymerization.

2.4.1.2   *Polymerizations of diastereoisomer mixtures*

(i)   Diastereoasymmetric polymerization

A polymerization in which, starting from a mixture of diastereoisomers, only one of the diastereoisomers is polymerized, for example stereospecific polymerization of (*S*)-*sec*-butyl-(*R,S*)-thiirane [(*S*)-*sec*-butyl-(*R,S*)-ethylene sulfide] using optically active initiators.

(ii)   Diastereosymmetric polymerization

A polymerization in which, starting from a mixture of diastereoisomers, the polymer formed consists of a mixture of chains, each one containing only units derived from one of the diastereoisomers, for example stereospecific polymerization of (*S*)-*sec*-butyl-(*R,S*)-oxirane [(*S*)-*sec*-butyl-(*R,S*)-ethylene oxide] using nonoptically active initiators.

2.4.2   Polymerizations of Prochiral Monomers

Polymerizations of monomers that are prochiral molecules may give, depending on the initiator used, stereoirregular or stereoregular polymers, which may be optically active.

These prochiral monomers may be cyclic molecules with a plane of symmetry, or ethylenic molecules in which the double bond is prochiral. In the former case, inversion of configuration generally occurs on ring-opening, giving polymer units with two identical successive chiral centers, and the polymer may or may not be optically active. In the latter case (opening of a multiple bond) one can distinguish two cases: (1) monomers, such as α-alkenes, in which, if the chains formed have identical end groups, the chiral centers formed in the configurational unit have an absolute configuration changing from one end of the polymer chain to the other; (2) that of prochiral monomers, such as benzofuran (coumarone) or indene, which produce chiral centers the configuration of which is not dependent on their position in the polymer chain.

For alkenes, the polymer chains are optically inactive by internal compensation. For benzofuran (coumarone), and similar compounds, individual macromolecules may contain the same configurational units but the sample as a whole may be optically inactive by external compensation, provided the stereospecific initiator is nondiscriminating. Polymerizations giving these various types of optically inactive stereoregular polymers have been considered in the IUPAC document on stereochemical definitions and notations relating to polymers[5] (see Section 2.3.1). On the other hand, with the use of chiral initiators, it is possible to have an 'enantioface discriminating polymerization'[39] giving an optically active polymer, the chains of which contain an excess of one type of configurational unit. The ideal case, for which the configurational units in the various polymer chains are identical, is called an 'enantiogenic polymerization' (or 'asymmetric polymerization'[40]).

(i)   Enantiogenic polymerization (formerly called 'asymmetric polymerization')

The polymerization of prochiral molecules giving a polymer consisting of polymer chains containing only one type of configurational repeating unit and which keeps the same absolute configuration along the polymer chain is known as enantiogenic polymerization. Examples include (a) polymerization of *cis*-2,3-dimethylthiirane (*cis*-1,2-dimethylethylene sulfide) (*meso* compound with *R* and *S* asymmetric carbons), giving an optically active polymer containing only *R,R* or *S,S* configurational repeating units; (b) polymerization of benzofuran (coumarone) giving an optically active polymer containing only one type of configurational repeating unit (*R* or *S*); (c) polymerization of methyl 2,4-hexadienoate (methyl sorbate) with 1,4 addition, to give an optically active polymer containing only one type of configurational repeating units; and (d) polymerization of 1,4-pentadiene with 1,4 addition, giving an optically active polymer containing only one type of configurational repeating unit.

2.5   PROPERTIES

2.5.1   Crystalline Polymers

Recommendations for terminology relating to crystalline polymers have recently been formulated.[8] The paper includes definitions of terms basic to the study of crystallinity in general, and it provides a commentary on points which have a specific implication when applied to polymers. It also contains terminology which has become widely used to describe certain types of polymer crystals (lamellar crystal, shish-kebab) as well as the kinetics of polymer crystallization. Table 7 lists the types of structural disorder typically found in crystalline polymers.

2.5.2   Individual Macromolecules, their Assemblies and Dilute Solutions

Terminology relating to individual macromolecules, their assemblies and dilute polymer solutions has very recently been the subject of recommendations from the IUPAC Commission on Macromolecular Nomenclature.[9] Although many of the terms defined in the new document have been in use for decades, it has not been an easy matter to reach agreement upon the precise shades of meaning which attach to some of these terms, and it will prevent further confusion from developing in the literature if their future use follows the IUPAC recommendations.

The topics covered in relation to individual macromolecules and their assemblies are: molar mass, molar mass distributions and averages; molecular dimensions, coiling, branching and networks; and models of polymer chains. For dilute solutions, the recommended terminology covers: thermodynamics, theory, solvent quality, transport properties, radiation (light) scattering, and separation.

Particular attention is drawn to the terms 'uniform' and 'nonuniform', which are proposed to replace 'monodisperse' and 'polydisperse'. (The former is contradictory, the latter tautologous.) Many other experiments have been proposed to describe polymer samples which either do or do not comprise molecules all having the same size: the IUPAC recommendation is both explicit and simple; it is to be hoped that it will win universal acceptance.

2.6   A CLASSIFICATION OF LINEAR SINGLE-STRAND POLYMERS[10]

The classification scheme presented here facilitates logical indexing. This scheme is limited to linear single-strand organic and inorganic polymers. It is consistent with previous IUPAC recommendations.[1,2,3,5,17]

Table 7 Examples of Structural Disorder Occurring in Crystalline Polymers

Type of structural disorder	Definition	Examples
(i) Lattice distortion	Structural disorder resulting from misalignment of the unit cells within the crystallites	As in usual crystallization, *i.e.* mechanical strain, lattice dislocation, impurities, *etc.*
(ii) Chain orientation disorder	Structural disorder resulting from the statistical coexistence within the crystallites of identical chains with opposite orientations. A typical example is provided by the up–down statistical coexistence of isomorphous, anticlined chains in the same crystal structure	Isotactic polypropylene, isotactic polystyrene, poly(vinylidene difluoride), form II
(iii) Configurational disorder	Structural disorder resulting from the statistical cocrystallization of different configurational repeating units	Atactic polymers capable of crystallization: poly(vinyl alcohol), poly(vinyl fluoride)
(iv) Conformational disorder	Structural disorder resulting from the statistical coexistence within the crystallites of identical configurational units with different conformations	The high-temperature polymorph of *trans*-1,4-poly(butadiene), *cis*-poly(isoprene)
(v) Macromolecular isomorphism	Structural disorder resulting from the statistical cocrystallization of different constitutional repeating units, which may either belong to the same copolymer chains (copolymer isomorphism) or appear separately in different homopolymer chains (homopolymer isomorphism)	
(1) Copolymer isomorphism		Poly(acetaldehyde-*co*-propionaldehyde), poly[bis(3-aminopropyl) ether-adipate-*co*-heptamethylene-diamine-adipate], isotactic poly-(4-methyl-1-pentene-*co*-4-methyl-1-hexene), isotactic poly(1-butene-*co*-3-methylbutene), isotactic poly(styrene-*co*-p-fluoro-styrene)
(2) Homopolymer isomorphism		Mixtures of poly(vinyl fluoride) and poly(vinylidene difluoride); mixture of isotactic poly(4-methylpentene) and isotactic poly(4-methylhexene)

2.6.1 General Principles

The classification consists of a hierarchical scheme for naming polymers according to the chemical constitution of the repeating units in the main chain (backbone). It can be applied to homopolymers, alternating copolymers and other macromolecular substances in which the constitutional repeating units can be identified. All existing linear single-strand polymers are embraced by this classification, which has been designed so as to be capable of extension to include any new structures of this type.

Four hierarchical levels are used; in order of decreasing importance, they are; classes, subclasses, groups and individual polymers.

2.6.1.1 Classes

Polymers are divided into two principal classes on the basis of the constitution of the main chain.

(i) Homochain polymers

Homochain polymers are those in which the main chains are constructed from atoms of a single element.

Rule 1. Homochain polymers are named by placing the name or symbol of the element in the main chain immediately before the expression '-chain polymer', for example carbon-chain polymer or C-chain polymer, sulfur-chain polymer or S-chain polymer (see also Section 2.6.1.2).

Introduction

(ii) Heterochain polymers

Heterochain polymers are those in which the main chains are constructed from atoms of two or more elements.

Rule 2.1.   Heterochain polymers are named by placing the names or symbols of all the elements in the main chain, in parentheses, immediately before the expression '-chain polymer', for example (oxygen,carbon)-chain polymer or (O,C)-chain polymer, (oxygen,nitrogen,carbon)-chain polymer or (O,N,C)-chain polymer (see also Section 2.6.1.2).

Rule 2.2.   The order of citation of the elements in heterochain polymers is that conventionally used in inorganic nomenclature.[3] For the common elements, the order of citation is as follows: O, S, N, P, C, Si, B.

Note that in those cases where the main chain has bonds in common with cyclic structures, *all* atoms in the rings must be considered for classification purposes. Thus (I) is a homochain polymer, whereas (II) is a heterochain polymer (Figure 13). On the other hand, a polymer such as (III; Figure 13) in which the main chain has no bonds in common with the cyclic structure, is classified as a homochain polymer.

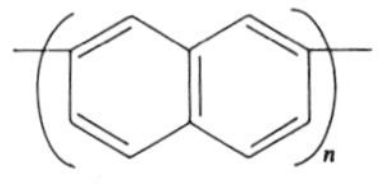

(I) 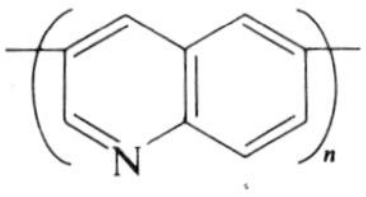   (II) 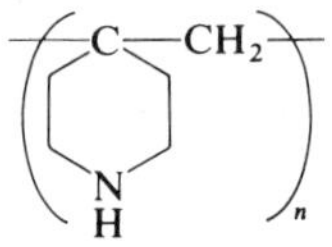   (III)

Figure 13

Rule 3.1.   The presence of specific side groups, or specific elements in side groups, is indicated by placing the name of the side group, or of the element in the side group, immediately before the expression '-sidegroup polymer'.

Rule 3.2.   Where more than one element, or more than one side group, is to be specified, the names of the elements or side groups are placed in parentheses, for example oxygen-sidegroup polymer, hydroxyl-sidegroup polymer [for poly(1-hydroxyethylene), *etc.*], ether-sidegroup polymer [for poly(1-methoxyethylene), *etc.*], (O,N,P)-sidegroup polymer, (ether,amide)-sidegroup polymer.

2.6.1.2 Subclasses

Each class of polymers can be divided into various subclasses, according to the nature of the elements in the main chain. Examples are given in Rules 1 and 2.1 above (Section 2.6.1.1).

2.6.1.3 Groups

Each subclass can be further divided into groups having similar chemical structures. Examples of carbon-chain polymers are polyalkylenes and polyarylenes. Examples of (oxygen,carbon)-chain polymers are polyethers and polycarbonates.

It is convenient to retain such widely accepted terminology as polyacetals, polycarbonates, polyamides, polyesters, nucleic acids, *etc.*, for naming the groups of polymers.

2.6.1.4 Individual polymers

The lowest hierarchical ranking is given to the individual polymers, which are named in accordance with accepted IUPAC nomenclature practice. The reader is referred to refs. 2 and 3 for recommendations for naming organic and inorganic single-strand polymers, respectively.

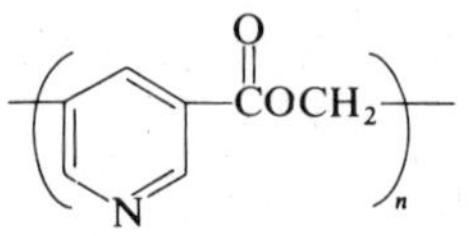

(97)

It is to be noted that, for polymers of complex structure, assignment to more than one subclass and/or group is sometimes possible. For example, the polymer (97) may be classified as follows: class: heterochain; subclass: (O,N,C)-chain polymer; group: polycarboxylate and/or pyridine-chain polymer.

2.7 REFERENCES

1. IUPAC, 'Basic Definitions of Terms Relating to Polymers', *Pure Appl. Chem.*, 1974, **40**, 478.
2. IUPAC, 'Nomenclature of Regular Single-Strand Organic Polymers' (Rules Approved 1975), *Pure Appl. Chem.*, 1976, **48**, 373.
3. IUPAC, 'Nomenclature for Regular Single-Strand and Quasi-Single-Strand Inorganic and Coordination Polymers' (Recommendations 1984), *Pure Appl. Chem.*, 1985, **57**, 149.
4. IUPAC, 'Source-Based Nomenclature for Copolymers', *Pure Appl. Chem.*, 1985, **57**, 1427.
5. IUPAC, 'Stereochemical Definitions and Notations Relating to Polymers' (Recommendations 1980), *Pure Appl. Chem.*, 1981, **53**, 733.
6. IUPAC, 'Stereochemical Definitions and Notations Relating to Polymers' (Provisional Recommendations), *Pure Appl. Chem.*, 1979, **51**, 1101.
7. IUPAC, 'Terminology for Polymerizations Involving Chiral Monomers or Resulting in Optically Active Polymers', *Pure Appl. Chem.*, in course of publication.
8. IUPAC, 'Terminology Relating to Crystalline Polymers' (Recommendations 1984), *Pure Appl. Chem.*, in course of publication.
9. IUPAC, 'Definitions of Basic Terms Relating to Individual Macromolecules, Their Assemblies and Dilute Solutions' (Recommendations 1986), in course of publication.
10. IUPAC, 'A Classification of Linear Single-Strand Polymers' (Provisional).
11. IUPAC, 'List of Standard Abbreviations (Symbols) for Synthetic Polymers and Polymer Materials', *Pure Appl. Chem.*, 1974, **40**, 474; IUPAC, 'Use of Abbreviations for Names of Polymeric Substances' (Recommendations 1986), *Pure Appl. Chem.*, 1987, **59**, 691.
12. H. E. Armstrong, *Proc. R. Soc. London*, 1892, **8** (114), 127.
13. International Union of Chemistry. Commission on the Reform of the Nomenclature of Organic Chemistry, 'Definitive Report, Liège Meeting, 1930'. (Translation with comments by A. M. Patterson, *J. Am. Chem. Soc.*, 1933, **55**, 3905. Appeared in English also in *J. Chem. Soc.*, 1931, 1607.)
14. M. Delépine, 'Revision of Inorganic Chemical Nomenclature, Report of the IUPAC Committee for the Reform of the Nomenclature of Inorganic Chemistry', *Chem. Weekbl.*, 1926, **23**, 86.
15. W. P. Jorissen, H. Bassett, A. Damiens, F. Fichter, and H. Remy, 'Rules for Naming Inorganic Compounds. Report of the Committee of the International Union of Chemistry for the Reform of Inorganic Chemical Nomenclature, 1940', *J. Am. Chem. Soc.*, 1941, **63**, 889.
16. J. E. Courtois, 'Work of Commission on Nomenclature of Biological Chemistry', in 'Chemical Nomenclature', American Chemical Society, Washington, DC, 1953, p. 83 (*Adv. Chem. Ser.*, No. 8).
17. IUPAC, 'Report on Nomenclature in the Field of Macromolecules', *J. Polym. Sci.*, 1952, **8**, 257; *Makromol. Chem.*, 1960, **38**, 1. The French version was published as a supplement to *Bull. Soc. Chim. Fr.* in 1952.
18. M. L. Huggins, G. Natta, V. Desreux and H. Mark, 'Report on Nomenclature Dealing with Steric Regularity in High Polymers', *J. Polym. Sci.*, 1962, **56**, 153.
19. IUPAC, 'Report on Nomenclature Dealing with Steric Regularity in High Polymers', *Pure Appl. Chem.*, 1966, **12** (1–4), 643.
20. M. L. Huggins, G. Natta, V. Desreux and H. Mark, *Chim. Ind. (Milan)*, 1964, **46**, 536; *Makromol. Chem.* 1965, **82**, 1; *Bull. Soc. Chim. Fr.*, 1965, 2127; *Kobunshi Kagaku*, 1965, **14**, 708.
21. G. Natta and F. Danusso, *J. Polym. Sci.*, 1959, **34**, 3.
22. G. Natta, M. Farina and M. Peraldo, *J. Polym. Sci.*, 1960, **38**, 13.
23. IUPAC, 'Nomenclature of Regular Single-Strand Organic Polymers' (Provisional), *Macromolecules*, 1973, **6**, 149.
24. American Chemical Society, 'A Structure-Based Nomenclature for Linear Polymers', *Macromolecules*, 1968, **1**, 193.
25. IUPAC, 'Changes and Additions to the Definitive Report [of the Liège Meeting]', *Compt. Rend. de la Quinzième Conf.*, Amsterdam, 1949, 132.
26. IUPAC, 'Nomenclature of Organic Chemistry, Sections, A, B, C, D, E, F, and H', Pergamon, Oxford, 1979.
27. IUPAC–IUB, 'Abbreviations and Symbols for the Description of the Conformation of Polypeptide Chains' (Tentative), *Biochemistry*, 1970, **9**, 3471; (Provisional), *Pure Appl. Chem.*, 1974, **40**, 291.
28. IUPAC–IUB, 'Polysaccharide Nomenclature' (Provisional), *Pure Appl. Chem.*, 1982, **54**, 1523. Published as 'Recommendations 1980', in *J. Biol. Chem.*, 1982, **257**, 3352.
29. IUPAC–IUB, 'Symbols for Specifying the Conformation of Polysaccharide Chains' (Provisional), *Pure Appl. Chem.*, 1983, **55**, 1269; and as 'Recommendations 1981', *Eur. J. Biochem.*, 1983, **131**, 5.
30. 'Nomenclature and Symbolism for Amino Acids and Peptides' (Recommendations 1983), *Pure Appl. Chem.*, 1984, **56**, 595.
31. IUPAC–IUB, 'Abbreviated Nomenclature of Synthetic Polypeptides (Polymerized Amino Acids)' (Revised Rules), *Pure Appl. Chem.*, 1973, **33**, 437. Published previously as revised rules in *Biochemistry*, 1972, **11**, 942.
32. IUPAC–IUB, 'Abbreviations and Symbols for Nucleic Acids, Polynucleotides and their Constituents' (Rules Approved 1974), *Pure Appl. Chem.*, 1974, **40**, 277. Published previously as revised rules in *Biochemistry*, 1970, **9**, 4022.
33. International Standards Organization (ISO), 'Plastics, Symbols and Codes. Part 1: Symbols for Basic Polymers and their Modifications, and for Plasticizers', *International Standard ISO 1043*, 1984.
34. American Chemical Society, 'Abbreviations for Thermoplastics, Thermosets, Fibers, Elastomers, and Additives', *Polym. News*, 1983, **9**, 101; Part 2, *Polym. News*, 1985, **10**, 169.
35. IUPAC, 'Nomenclature of Inorganic Chemistry', 2nd edn., Butterworths, London, 1971. Also published in *Pure Appl. Chem.*, 1971, **28**, 1.

36. G. E. Ham in 'Encyclopedia of Polymer Science and Technology', ed. H. F. Mark, N. G. Gaylord, and N. M. Bikales, Wiley-Interscience, New York, 1966, vol. 4, p. 165.
37. H. J. Harwood and W. M. Ritchey, *J. Polym. Sci., Polym. Lett. Ed.*, 1964, **2**, 601.
38. H. J. Harwood, *Rubber Chem. Technol.*, 1982, **55**, 769.
39. Y. Izumi and A. Tai, in 'Stereodifferentiating Reactions', ed. Kodansha, Academic Press, New York, 1977, chap. 6.
40. G. Natta and M. Farina, 'Stereochemistry', Longman, Harlow, Essex, 1972, p. 228.
41. IUPAC, 'Recommendations for Abbreviations of Terms Relating to Plastics and Elastomers', *Pure Appl. Chem.*, 1969, **18**, 583.

3
Averages and Distributions

COLIN BOOTH and RICHARD O. COLCLOUGH
University of Manchester, UK

3.1 INTRODUCTION

All synthetic polymers show a distribution of molecular weights, which may be averaged in several different ways. Any physical or performance property of a polymer may be related to one or more average molecular weights, the type of average being determined by the physical averaging process inherent in the method used to measure the property. Many polymers have distributions of other molecular parameters such as chemical composition, stereoregularity, chain branching, *etc.* For example, a linear copolymer has interdependent distributions of chain length, chemical composition and sequence length, and averages of all these parameters determine its properties.

Averages and distributions are well treated in many textbooks, and those of Elias[1] and Odian[2] as well as the classic text of Flory[3] come to mind. Other sources are the comprehensive reviews of averages by Elias and co-workers[4, 5] and the detailed monographs on molecular weight distributions by Tang[6] and by Peebles.[7] The aim here is to summarize the basic principles, including essential definitions and examples of applications. Detailed descriptions and derivations of averages and distributions for particular polymeric systems will be found elsewhere in these volumes.

3.2 DISTRIBUTIONS AND STATISTICAL WEIGHTS

For a synthetic polymer the distribution of a property (Q) is discontinuous. For a high polymer the intervals in Q become vanishingly small compared to the total range and the discontinuous distribution can be replaced by a continuous one. For a low polymer (oligomer) this simplification is not valid.

The discontinuous g distribution of a property Q is the set of statistical weights g_i each corresponding to species i with value of the property equal to Q_i. Common statistical weights are number (n_i), weight (w_i, actually mass) and z ($z_i = w_i M_i$, where M_i = molar mass). These amounts may be normalized (i.e. $\Sigma_i n_i = 1$; $\Sigma_i w_i = 1$; $\Sigma_i z_i = 1$) in which case they are number (or mole) fraction (n_i), weight fraction (w_i) and z fraction (z_i). Such fractions are usually preferred for distributions. (The use of the same symbol for amount and fraction causes no difficulties in context. Similarly the symbol M is used both for molar mass and for molecular weight.)

The corresponding continuous g distribution of Q is $g(Q)\,dQ$ *vs.* Q. When normalized, *i.e.* $\int_0^\infty g(Q)\,dQ = 1$, $g(Q)\,dQ$ is the g fraction of species with property Q in the range Q to $Q + dQ$. The description of the distribution by $g(Q)$ alone implies that $dQ = 1$.

The cumulative forms are respectively $G_i = \sum_{j=1}^{j=i} g_j$ and $G(Q) = \int_0^Q g(Q')\,dQ'$, the latter being properly called the distribution function.[8] Discontinuous and continuous distributions are illustrated in Figure 1.

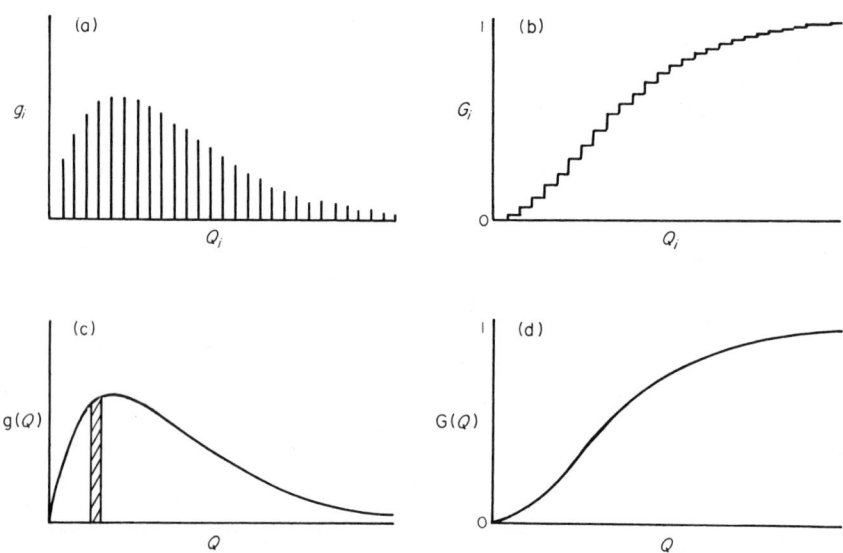

Figure 1 Representations of g fraction distributions of a property Q: (a, b) discontinuous and (c, d) continuous representations of (a, c) frequency and (b, d) cumulative distributions. The shaded area in (c) is the g fraction $g(Q)\,dQ$. Note that g_i and $g(Q)\,dQ$ coincide only if dQ is appropriately chosen

Statistical weights of species i which are uniform in molar mass M_i are related by

$$z_i = w_i M_i, \quad w_i = n_i M_i \tag{1a}$$

or, when normalized, by

$$z_i = w_i M_i / \textstyle\sum_i w_i M_i, \quad w_i = n_i M_i / \textstyle\sum_i n_i M_i \tag{1b}$$

or conversely,

$$w_i = (z_i / M_i) / \textstyle\sum_i (z_i / M_i), \quad n_i = (w_i / M_i) / \textstyle\sum_i (w_i / M_i). \tag{1c}$$

These relationships can be extended to include further statistical weights, *e.g.* $(z+1)_i = w_i M_i^2$, $(n-1)_i = n_i / M_i$. If the species i are non-uniform in molar mass (*e.g.* when i refers to polymer fractions rather than to polymer species) then M_i must be replaced by an appropriate average, *e.g.*

$$z_i = w_i(\bar{M}_w)_i, \quad w_i = n_i(\bar{M}_n)_i \tag{1d}$$

where $(\bar{M}_w)_i$ and $(\bar{M}_n)_i$ denote the weight-average and number-average molar masses of the species i; see Section 3.3.1.

Shape-based statistical weights (*e.g.* length, area, volume) are related to n, w and z by expressions which depend on the geometry of the species.[9] For example the surface area (a_i), which is an important statistical weight in the measurement of the diffusion coefficient, of a spherical species i of density ρ which is homogeneous in radius (R_i), is related to n, w and z by

$$a_i = n_i 4\pi R_i^2 = 3w_i/\rho R_i = 9z_i/4\pi\rho^2 R_i^4$$

3.3 AVERAGES

An average is defined in terms of the property being averaged (Q) and its statistical weight (g_i) [or its continuous distribution $g(Q)$] so that it always has the same physical units as the property in question. If the property and statistical weight are dimensioned, this requirement provides a useful check of the formula used to calculate the average.

The following sections give the formulae for discontinuous distributions. The necessary modifications for continuous distributions are noted in Section 3.3.4.

3.3.1 Simple Averages

The simplest average is the arithmetic average, *i.e.*

$$\bar{Q}_g = \sum_i g_i Q_i \Big/ \sum_i g_i \tag{2}$$

where the sums are over all species i. If the statistical weight is normalized, then $\bar{Q}_g = \sum_i g_i Q_i$ is equivalent to the first moment of the g-fraction distribution of Q relative to the origin. For this reason the simple average is called a one-moment average.

Commonly encountered simple averages include the average molecular weights set out in Table 1. Note that they are defined by equation (2) and that the name is derived from the statistical weight used. Alternative forms are also listed, in accordance with equation (1a). These forms can be useful for calculation, but are not used as definitions since they involve ratios of two moments: in this case ratios of the form $\sum_i g_i Q_i^\beta \Big/ \sum_i g_i Q_i^{\beta-1}$, where β is a small integer.

A similar set of average radii of spheres of uniform density is given in Table 2. Again they are defined by equation (3) and named accordingly. However the alternative two-moment forms differ from those of Table 1, since in this case $R_i^3 \propto M_i$.

Of course the physical property measured for spheres may lead directly to an average of R^3 rather than R. A well-known example is the average cubed hydrodynamic radius of an assembly of spheres

Table 1 Average Molecular Weights

Average	Statistical weight	Definition	Alternative form
Number, $\bar{M}_n$	n_i	$\sum_i n_i M_i \Big/ \sum_i n_i$	$\sum_i w_i \Big/ \sum_i (w_i/M_i)$
Weight, $\bar{M}_w$	w_i	$\sum_i w_i M_i \Big/ \sum_i w_i$	$\sum_i n_i M_i^2 \Big/ \sum_i n_i M_i$
z, $\bar{M}_z$	z_i	$\sum_i z_i M_i \Big/ \sum_i z_i$	$\sum_i w_i M_i^2 \Big/ \sum_i w_i M_i$
			$\sum_i n_i M_i^3 \Big/ \sum_i n_i M_i^2$

Table 2   Average Radii of Spheres

Average	Statistical weight	Definition	Alternative form
Number, $\bar{R}_n$	n_i	$\sum_i n_i R_i \Big/ \sum_i n_i$	
Surface, $\bar{R}_a$	a_i	$\sum_i a_i R_i \Big/ \sum_i a_i$	$\sum_i n_i R_i^3 \Big/ \sum_i n_i R_i^2$
Weight, $\bar{R}_w$	w_i	$\sum_i w_i R_i \Big/ \sum_i w_i$	$\sum_i n_i R_i^4 \Big/ \sum_i n_i R_i^3$
z, $\bar{R}_z$	z_i	$\sum_i z_i R_i \Big/ \sum_i z_i$	$\sum_i w_i M_i R_i \Big/ \sum_i w_i R_i$
			$\sum_i w_i R_i^4 \Big/ \sum_i w_i R_i^3$
			$\sum_i n_i R_i^7 \Big/ \sum_i n_i R_i^6$

(latex particles or microgels) determined by applying the Einstein equation

$$[\eta] = (10\pi N_a/3)\, R^3/M = K R^3/M$$

to the measured intrinsic viscosity of a non-uniform sample,

$$[\eta] = \sum_i w_i [\eta]_i \Big/ \sum_i w_i$$

Substitution gives

$$[\eta] = K \sum_i (w_i/M_i) R_i^3 \Big/ \sum_i w_i = K \sum_i n_i R_i^3 \Big/ \sum_i w_i = K \left(\sum_i n_i R_i^3 \Big/ \sum_i n_i \right)\left(\sum_i n_i \Big/ \sum_i w_i \right)$$

$$\text{or} \quad (\bar{R}^3)_n = [\eta] \bar{M}_n / K$$

Provided that the number-average molar mass is combined with the intrinsic viscosity, the average cubed radius obtained is $(\bar{R}^3)_n = \left[\sum_i n_i R_i^3 \Big/ \sum_i n_i \right]$, which is a number average.

3.3.2 Exponent Averages

A more general definition of a one-moment average is

$$\bar{Q}_g = \left[\sum_i g_i Q_i^\alpha \Big/ \sum_i g_i \right]^{1/\alpha} \tag{3}$$

Equation (2) is the special case when $\alpha = 1$. The exponent average ($\alpha \neq 1$) may be named from the method of measurement of the property.

A well-known example is the viscosity-average molecular weight, $\bar{M}_v = \left[\sum_i w_i M_i^{a_\eta} \Big/ \sum_i w_i \right]^{1/a_\eta}$ which is derived from the measured intrinsic viscosity $[\eta]$ (a weight-averaged quantity $[\bar{\eta}]_w$) *via* the Mark–Houwink equation, $[\eta] = K_\eta M^{a_\eta}$. Similar weight-average exponent averages are the sedimentation average derived from the measured sedimentation coefficient ($\bar{s}_w$) *via* $s = K_s M^{a_s}$, and the diffusion average derived from the measured diffusion coefficient ($\bar{D}_w$) *via* $D = K_D M^{a_D}$.

3.3.3 Two-moment Averages

A useful average is defined as the ratio of moments of the same statistical weight but differing in exponent by unity, *i.e.* by the equation

$$\bar{Q}_g = \sum_i g_i Q_i^\beta \Big/ \sum_i g_i Q_i^{\beta-1} \tag{4}$$

where β takes any value. Equation (2) is the special case $\beta = 1$. A more general definition due to Samarin,[10]

$$\bar{Q}_g = \left[\sum_i g_i Q_i^{\alpha + \beta - 1} \bigg/ \sum_i g_i Q_i^{\beta - 1} \right]^{1/\alpha} \tag{5}$$

gives the exponent average when $\beta = 1$ and the simple average when $\alpha = \beta = 1$.

A well-known example of equation (4) is the average molecular weight obtained from measurements of the weight-average sedimentation and diffusion coefficients, *i.e.*

$$\bar{s}_w = \sum_i w_i s_i \bigg/ \sum_i w_i = K_s \sum_i w_i M_i^{a_s} \bigg/ \sum_i w_i \quad \text{and} \quad \bar{D}_w = \sum_i w_i D_i \bigg/ \sum_i w_i = K_D \sum_i w_i M_i^{a_D} \bigg/ \sum_i w_i$$

by combining them in the Svedberg equation $\bar{M}_{sD} = (K_D/K_s)(\bar{s}_w/\bar{D}_w)$

to obtain $\bar{M}_{sD} = \sum_i w_i M_i^{a_s} \bigg/ \sum_i w_i M_i^{a_D}$

The requirement that the average has the same units as the quantity measured means that the defining equation must be such that the sum of exponents is always unity, as in equation (4). Here

$$a_s - a_D = 1 \quad \text{and} \quad \bar{M}_{sD} = \sum_i w_i M_i^{a_s} \bigg/ \sum_i w_i M_i^{a_s - 1}$$

This type of analysis has been shown to be most useful, not only in defining the average resulting from a combination of techniques but also for interrelating the various averages. Useful listings of two-moment averages and further discussion can be found in the reviews of Elias and co-workers.[4, 5]

3.3.4 Other Forms

Only averages commonly met in polymer science are described above. Other averages may be encountered: see refs. 4 and 5 for an extensive discussion.

For a continuous distribution the general definition (equation 5) takes the form[10]

$$\bar{Q}_g = \left[\int_0^\infty g(Q) Q^{\alpha + \beta + 1} \, dQ \bigg/ \int_0^\infty g(Q) Q^{\beta + 1} \, dQ \right]^{1/\alpha} \tag{6a}$$

and the special cases are as before, *i.e.* $\alpha = \beta = 1$ for the simple average and $\alpha \neq 1$, $\beta = 1$ for the exponent average. Thus the number-average molecular weight is given by

$$\bar{M}_n = \int_0^\infty n(M) M \, dM \bigg/ \int_0^\infty n(M) \, dM \tag{6b}$$

or, for a number-fraction distribution, simply by

$$\bar{M}_n = \int_0^\infty n(M) M \, dM \tag{6c}$$

3.4 DISTRIBUTIONS AND AVERAGES FOR LINEAR POLYMERS

Molecular weight distributions and averages for linear polymers merit particular consideration. Mechanisms of polymerization lead directly to expressions for the number-fraction distribution of chain length (x chain units = degree of polymerization) and the number-average chain length ($\bar{x}_n$). Insofar as x is directly proportional to molecular weight (M), the weight-fraction distribution and the weight and z averages follow, together with the molecular-weight distributions and averages. For high-molecular-weight homopolymers this proportionality can be assumed, the effects of incorporating non-monomeric residues into the molecule being negligible. However for oligomers, copolymers and non-linear polymers, composition is not necessarily independent of x, and this possibility should always be borne in mind. The discussion in this section is confined to chain length distributions and averages for high-molecular-weight homopolymers.

In addition to the polymerization mechanism (initiation, termination, *etc.*, propagation is usually random) the subsequent treatment of the polymer (fractionation, degradation, *etc.*) may also affect the chain-length distribution. Consequently many distributions have been described, some following directly from the predictions of model reaction schemes and others adopted by virtue of their versatility and/or ease of application. The distributions described here are commonly encountered: see ref. 7 for a thorough account.

3.4.1 Exponential Distribution

Schulz[11] derived a chain-length distribution to describe the product of a steady state linear radical-chain polymerization in which the bimolecular reaction of radical and monomer results either in addition or in termination: *viz.*,

$$n(x) = (-\ln p)p^x; \quad w(x) = (-\ln p)^2 x p^x \tag{7a;b}$$

where $n(x)$ is the number-fraction and $w(x)$ the weight-fraction distribution, and where p is the probability that reaction of a chain radical with monomer results in addition rather than termination, so that $p \to 1$ for a high polymer. The two distributions are related by $w(x) = n(x)(x/\bar{x}_n)$ (see equation (1b), $x \propto M$) with $\bar{x}_n = 1/(-\ln p)$ (see equation 6c).

Flory[12] independently derived the same distribution for a linear condensation polymerization in which all chain ends have equal probability (p) of step growth and in which ring formation is excluded, *i.e.*

$$n(x) = (1-p)p^{x-1}; \quad w(x) = (1-p)^2 x p^{x-1} \tag{8a;b}$$

In this case p is identified with the fraction of end groups reacted. The equivalence of equations (7) and (8) for high polymer ($p \to 1$) can be seen by writing equation (7) as $n(x) = p(-\ln p)p^{x-1}$ and substituting $(-\ln p) \approx (1-p)$ and $p \approx 1$. The distribution is often seen in the form of equation (8), and is called the Schulz–Flory or the 'most probable' distribution. The simple averages are

$$\bar{x}_n = 1/(1-p); \quad \bar{x}_w = (1+p)/(1-p); \quad \bar{x}_z = (1+4p+p^2)/(1-p^2)$$

A useful modification for high polymer is the exponential form, obtained by substituting $b = (-\ln p)$ as follows:

$$n(x) = b \exp(-bx); \quad w(x) = b^2 x \exp(-bx) \tag{9a;b}$$

where $b = 1/\bar{x}_n$. These distributions are illustrated in Figure 2.

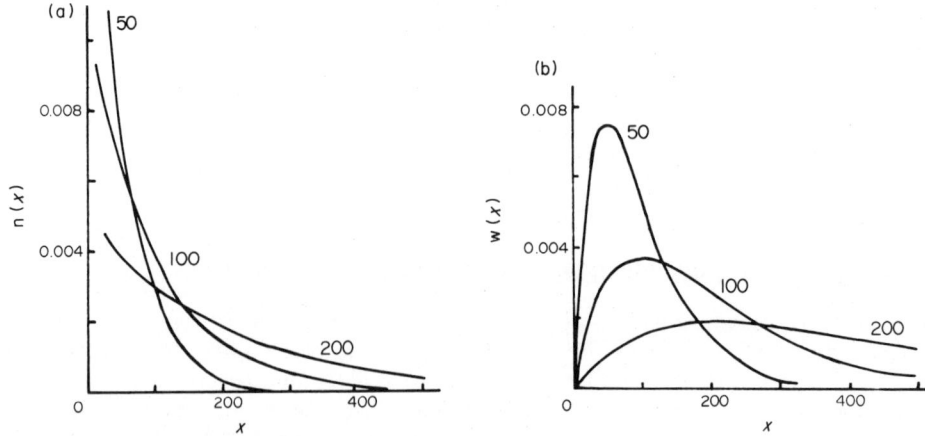

Figure 2 Schulz–Flory chain-length distributions. (a) Number-fraction distribution [$n(x)$ *vs.* x, equation (9a)] and (b) weight-fraction distribution [$w(x)$ *vs.* x, equation (9b)]. The curves are for number-average chain lengths $\bar{x}_n = 50, 100, 200$, as indicated. The maxima in $w(x)$ are at $\bar{x}_n$ (Adapted from ref. 12 with permission of *J. Am. Chem. Soc.*)

The exponential distribution has been generalized (see Schulz[13] and Zimm[14]) to cover a wider set of reaction conditions; *i.e.*

$$n(x) = [b^a/(a-1)!] \, x^{a-1} \exp(-bx) \tag{10a}$$

$$w(x) = [b^{a+1}/a!] \, x^a \exp(-bx) \tag{10b}$$

where $b = a/\bar{x}_n$. The second parameter a allows for a variety of termination conditions in chain polymerization or for incorporation of multifunctional components in step polymerization. Effectively a is the degree of coupling, *i.e.* the number of growing chains which combine to form a dead chain. For example, in a steady-state radical-chain polymerization without chain transfer and with termination *via* bimolecular reaction of two growing chains, $a = 1$ corresponds to formation of two dead chains by disproportionation and $a = 2$ to formation of one dead chain by coupling. Note that $a = 1$ is the Schulz–Flory distribution.

Schulz–Zimm number-fraction and weight-fraction distributions are illustrated in Figure 3. At constant $\bar{x}_n$ the distribution becomes progressively narrower as a is increased. The simple averages are

$$\bar{x}_n = a/b; \quad \bar{x}_w = (a+1)/b; \quad \bar{x}_z = (a+2)/b$$

so that $\bar{x}_w/\bar{x}_n = (a+1)/a$ and $\bar{x}_z/\bar{x}_w = (a+2)/(a+1)$. The viscosity average is $\bar{x}_v = (1/b)[(a+a_\eta)!/a!]^{1/a_\eta}$ where a_η is the Mark–Houwink exponent.

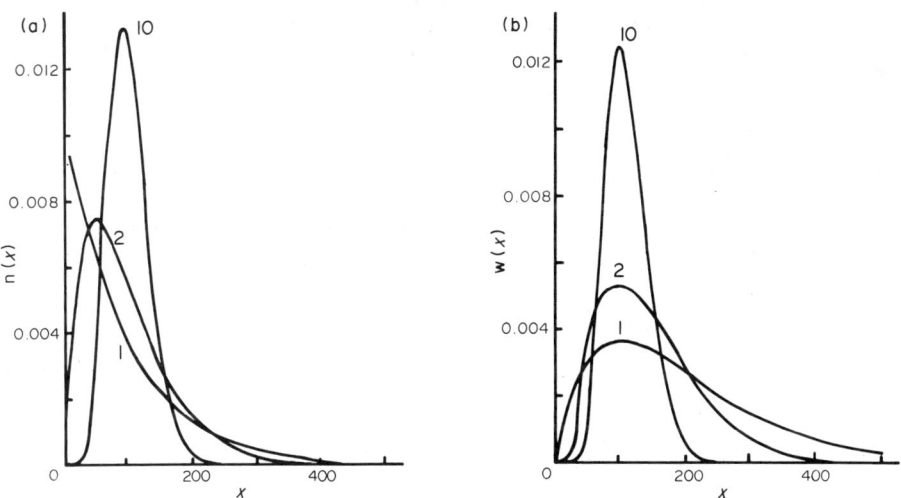

Figure 3 Schulz–Zimm chain-length distributions. (a) Number-fraction distribution [$n(x)$ *vs.* x, equation (10a)] and (b) weight-fraction distribution [$w(x)$ *vs.* x, equation (10b)]. The curves are for number-average chain length $\bar{x}_n = 100$ and parameter $a = 1, 2, 10$ as indicated (*i.e.* $\bar{x}_w = 200, 150, 110$). The maxima in $n(x)$ are at $\bar{x}_n\{1-(1/a)\}$ and those in $w(x)$ are at $\bar{x}_n$ (Adapted from ref. 13)

More generally a can have non-integral values, in which case the factorial is replaced by its gamma function [$a! = \Gamma(a+1)$] and

$$w(x) = [b^{a+1}/\Gamma(a+1)]x^a \exp(-bx)$$

In this form the exponential distribution has been used to represent fairly wide distributions (but not wider than the Schulz–Flory distribution) with the parameters defined empirically by

$$a = \bar{x}_n/(\bar{x}_w - \bar{x}_n); \quad b = a/\bar{x}_n.$$

A more general form of the exponential distribution[15] includes a third empirical parameter, *i.e.*

$$n(x) = [cb^{a/c}/\Gamma\{a/c\}] \, x^{a-1} \exp(-bx^c) \quad \text{and} \quad w(x) = [cb^{((a+1)/c)}/\Gamma\{(a+1)/c\}] \, x^a \exp(-bx^c)$$

The simple averages are $\bar{x}_n = [\Gamma\{(a+1)/c\}/\Gamma\{a/c\}]b^{1/c}$, $\bar{x}_w = [\Gamma\{(a+2)/c\}/\Gamma\{(a+1)/c\}]b^{1/c}$, *etc.*

The special cases discussed above are $a=1$, $b=1/\bar{x}_n$, $c=1$ for the Schulz–Flory distribution and $a>0$, $b=a/\bar{x}_n$, $c=1$ for the Schulz–Zimm distribution. The special case[16] $a=c-1$ and $c>1$, $b=[\bar{x}_n\Gamma(1-1/c)^{-c}$, *i.e.*

$$w(x)=cbx^{c-1}\exp(-bx^c),$$

was first applied to polymers by Tung.[17] This Weibull–Tung distribution can be integrated analytically, *i.e.*

$$\int_0^x w(x)\,dx=1-\exp(-bx^c)$$

which is an advantage for fitting experimental data.[14,15] By contrast the Schulz–Zimm distribution can only be integrated numerically.

3.4.2 Poisson Distribution

Flory,[3,19] in considering the base-catalysed polymerization of ethylene oxide, pointed out that the conditions of instantaneous initiation and no termination lead to a very narrow chain-length distribution equivalent to a Poisson number-fraction distribution, *i.e.*

$$n(x)=[v^{x-1}/(x-1)!]\exp(-v) \tag{11a}$$

where $v=\bar{x}_n-1$. The corresponding weight-fraction distribution is

$$w(x)=(x/\bar{x}_n)[v^{x-1}/(x-1)!]\exp(-v) \tag{11b}$$

which, since $x\approx\bar{x}_n$ for a very narrow distribution, is little different from $n(x)$. The simple averages are

$$\bar{x}_n=1+v, \quad \bar{x}_w=1+v+[v/(v+1)]\approx 2+v$$

The Poisson distribution is compared with the Schulz–Flory distribution in Figure 4.

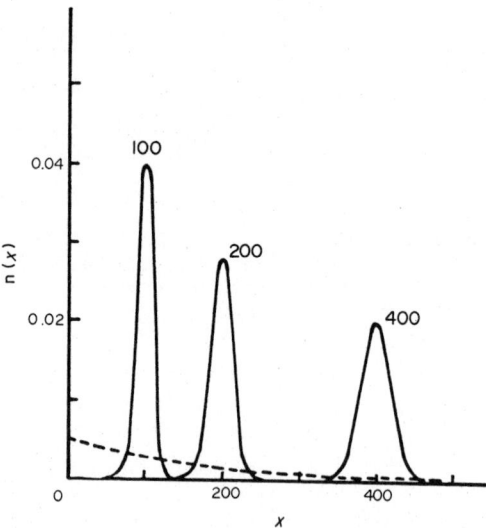

Figure 4 Poisson number-fraction chain-length distribution [$n(x)$ *vs.* x, equation (11a)]. The curves are for number-average chain lengths $\bar{x}_n=100, 200, 400$, as indicated. The dashed curve is the Schulz–Flory number-fraction distribution for $\bar{x}_n=200$. Note that the ratios $\bar{x}_w/\bar{x}_n$ for the Poisson distributions are 1.01 ($\bar{x}_n=100$), 1.005 ($\bar{x}_n=200$) and 1.0025 ($\bar{x}_n=400$); *i.e.* the relative widths of the distributions decrease as their absolute widths increase (see Section 3.5)

3.4.3 Normal Distribution

The normal or Gaussian distribution represents the law of errors about a mean, and is the familiar symmetrical 'bell-shaped' curve. The distribution occupies the range $-\infty$ to $+\infty$ and so, since negative chain lengths have no meaning, has only limited use in the description of chain-length distributions. It may be applied to narrow distributions instead of the Poisson distribution, in which capacity it has the advantage of greater flexibility, due to its two parameters.

A narrow normal weight-fraction chain-length distribution,[9] which overlaps zero to an insignificant extent (see Figure 5a) can be written

$$w(x) = C_w \exp[-(x - \bar{x}_w)^2/2\sigma_w^2] \tag{12a}$$

where $C_w = 1/\{(2\pi)^{1/2}\sigma_w\}$ is the normalizing factor. The parameters are the weight-average chain length ($\bar{x}_w$), which is equivalent to the median value of the weight-fraction distribution (x_{mw}, the chain length at the maximum), and the standard deviation of the distribution (σ_w, the half-width at 0.607 of the maximum height) which is given by[9]

$$\sigma_w^2 = \int_{-\infty}^{+\infty} w(x)(x - \bar{x}_w)^2 \, dx = \bar{x}_z\bar{x}_w - (\bar{x}_w)^2 \tag{13}$$

The number-fraction distribution corresponding to equation (12a) is not a normal distribution but is given by

$$n(x) = (\bar{x}_n/x)w(x) \tag{14}$$

The number-fraction distribution is less symmetrical than the 'bell-shaped' distribution $w(x)$, having an relatively larger low-x fraction, as illustrated in Figure 5(a). If required, a normal number-fraction distribution must be redefined in terms of $\bar{x}_n$ and σ_n, i.e.

$$n(x) = C_n \exp[-(x - \bar{x}_n)^2/2\sigma_n^2]$$

$$\text{where } C_n = 1/\{(2\pi)^{1/2}\sigma_n\} \text{ and } \sigma_n^2 = \bar{x}_w\bar{x}_n - (\bar{x}_n)^2.$$

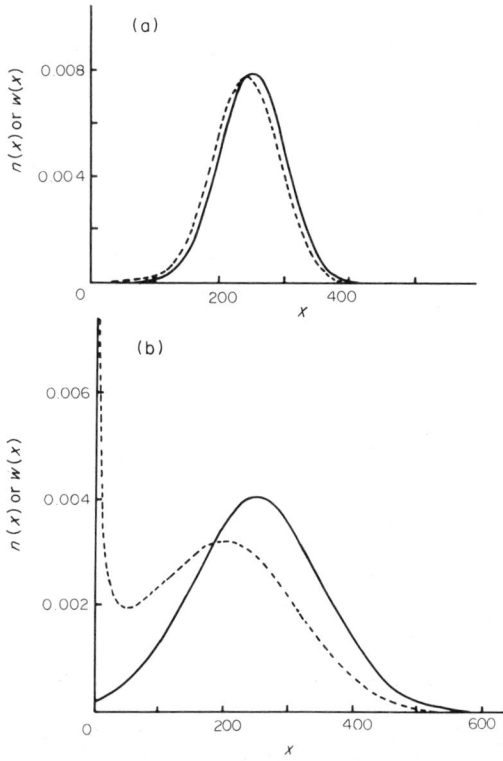

Figure 5 (a) (——) Normal weight-fraction chain-length distribution [$w(x)$ *vs.* x, equation (12a)] for weight-average chain length $\bar{x}_w = 250$ and standard deviation $\sigma_w = 50$; and (----) the corresponding number-fraction distribution [$n(x)$ *vs.* x, equation (14)]. For this distribution $\bar{x}_z/\bar{x}_w = 1.04$ and $\bar{x}_w/\bar{x}_n \approx 1.05$. (b) (——) Normal weight-fraction chain-length distribution [$w(x)$ *vs.* x, equations (12b and 12c)] for weight-average chain length $\bar{x}_w^* - 250$ and standard deviation $\sigma_w^* = 100$; and (----) the corresponding number-fraction distribution [$n(x)$ *vs.* x, equation (14)]. For this distribution $\bar{x}_z/\bar{x}_w \approx 1.16$ and $\bar{x}_w/\bar{x}_n \approx 1.39$

If the weight-fraction distribution is wider and overlaps $x=0$ (even slightly, see Figure 5b) the above equations are not directly applicable. The weight-fraction distribution is defined for the complete distribution from $-\infty$ to $+\infty$ by

$$w^*(x)=C_w^* \exp[-(x-\bar{x}_w^*)^2/2\sigma_w^{*2}] \tag{12b}$$

where $C_w^*=1/\{(2\pi)^{1/2}\sigma_w^*\}$. The actual weight-fraction distribution of the polymer is

$$w(x)=w^*(x)\Big/\int_0^\infty w^*(x)\,dx \tag{12c}$$

In this case the number-fraction distribution is very unsymmetrical (see Figure 5b).

[Note that, as usual, $\bar{x}_w^*$ is equal to the chain length at the maximum (x_{mw}) of $w^*(x)$, and σ_w^* is equal to the half-width at 0.607 of the maximum height of $w^*(x)$.]

3.4.4 Logarithmic-normal Distribution

The problem which restricts the use of the normal distribution can be avoided by using the logarithm of chain length. The log-normal distribution was first used for polymers by Lansing and Kraemer[20] when investigating the wide molecular-weight distribution of gelatin, and later Wesslau[21] found it useful for describing the distribution of Ziegler polyethylene. Consequently it may be referred to as the Kraemer and Lansing distribution or the Wesslau distribution.

The 'bell-shaped' normal weight-fraction distribution applied to the logarithm of chain length is

$$w(\ln x)=C \exp[-\{\ln x-\overline{(\ln x)}_w\}^2/2\sigma^2] \tag{15}$$

where $w(\ln x)\,d\ln x$ is the fraction of chains with values of $\ln x$ in the range $\ln x$ to $(\ln x+d\ln x)$. Here $C=1/\{(2\pi)^{1/2}\sigma\}$ is the normalizing factor, $\overline{(\ln x)}_w$ is the weight-average logarithmic chain length, and σ is the standard deviation of the logarithmic distribution, *i.e.*

$$\sigma=\int_0^\infty w(x)[\ln x-\overline{(\ln x)}_w]^2\,dx$$

This form of the distribution corresponds directly to a symmetrical GPC curve of $w(V)$ against V, where V is the elution volume in a GPC system with a linear calibration of $\ln x$ against V.

The weight fraction of polymer is given either by $w(\ln x)\,d\ln x$ or, for the corresponding weight-fraction chain-length distribution, $w(x)\,dx$. Equating the two weight fractions gives $w(x)=w(\ln x)\,(d\ln x/dx)=w(\ln x)(1/x)$, and hence

$$w(x)=(C/x)\exp[-\{\ln x-\overline{(\ln x)}_w\}^2/2\sigma^2] \tag{16a}$$

The weight-average logarithmic chain length $\overline{(\ln x)}_w$ is equivalent to $\ln x_{mw}$, where x_{mw} is the median value of the weight-fraction distribution. Substitution gives the more familiar form

$$w(x)=(C/x)\exp[-\{\ln (x/x_{mw})\}^2/2\sigma^2] \tag{16b}$$

Due to the log-to-linear conversion $w(x)$ *vs.* x is not bell-shaped (see Figure 6) and the parameter σ cannot be interpreted in the 'normal' way (as it can for equations 12 and 15). Indeed σ and x_m are best regarded as empirical parameters, with values given by equation (17) below, since this allows the log-normal distribution to be used to represent a variety of wide chain-length distributions.

The simple averages are

$$\bar{x}_n=x_{mw}\exp(-\sigma^2/2), \quad \bar{x}_w=x_{mw}\exp(+\sigma^2/2), \quad \bar{x}_z=x_{mw}\exp(+3\sigma^2/2) \tag{17a}$$

$$\text{so that } \bar{x}_w/\bar{x}_n=\bar{x}_z/\bar{x}_w=\exp(\sigma^2) \text{ and } x_{mw}=(\bar{x}_w\bar{x}_n)^{1/2} \tag{17b}$$

The number-fraction chain-length distribution of a polymer with a log-normal weight-fraction distribution is also log-normal, with the same σ but a different median value x_{mn}: *i.e.*

$$n(x)=(C/x)\exp[-\{\ln (x/x_{mn})\}^2/2\sigma^2] \tag{18}$$

where $\ln x_{mn}=\overline{(\ln x)}_n=\ln x_{mw}-\sigma^2$.

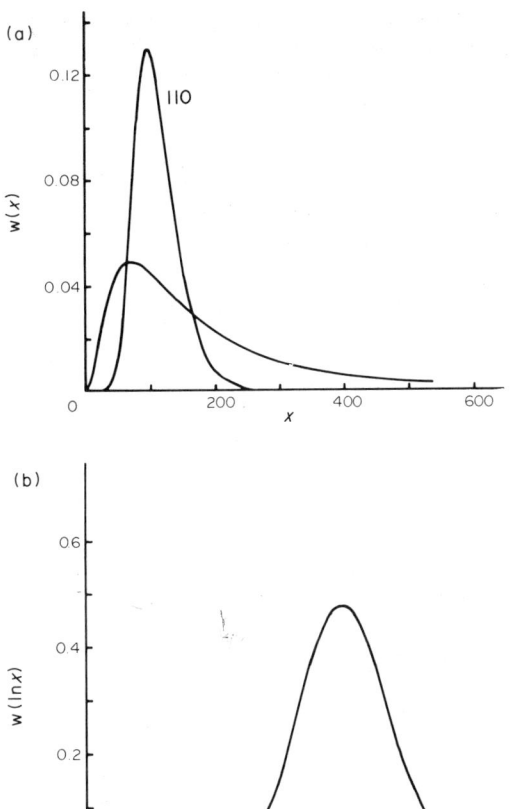

Figure 6 (a) Logarithmic-normal weight-fraction chain-length distribution [$w(x)$ *vs.* x, equation (16a) or (16b)] for number-average chain length $\bar{x}_n = 100$ and weight-average chain lengths $\bar{x}_w = 110$ or 200 as indicated (*i.e.* standard deviation $\sigma = 0.309$ or 0.833, median chain length $x_{mw} = (x_w x_n)^{1/2} = 105$ or 141). The maxima are at $\bar{x}_n/(\bar{x}_w/\bar{x}_n)^{1/2}$. (b) Normal weight-fraction distribution of logarithmic chain length [$w(\ln x)$ *vs.* ln x, equation (15)] for $\bar{x}_n = 100$ and $\bar{x}_w = 200$ (*i.e.* $\sigma = 0.833$, $x_{mw} = 141$. The maximum is at ln $x_{mw} = 4.95$

A generalized form[22] of the log-normal distribution has been shown[23] to be equivalent to using a median x_{ms} where

$$\ln x_{ms} = \ln x_{mw} - (s+1)\sigma^2$$

so that $s = -1$ corresponds to equation (16) and $s = 0$ to equation (18).

3.5 DISTRIBUTION WIDTHS

A familiar measure of distribution width is the ratio of molecular weights $\bar{M}_w/\bar{M}_n$, $\bar{M}_w/\bar{M}_n \to 1$ being narrow and $\bar{M}_w/\bar{M}_n > 2$ being wide. However for a given polymerization mechanism the random nature of the growth process has the consequence that the absolute width of the molecular-weight distribution increases as the average molecular weight increases. This is illustrated in Figures 2 and 4. At the same time, as the average molecular weight increases, the relative width of the molecular-weight distribution, as measured by $\bar{M}_w/\bar{M}_n$, may be nearly constant (Figure 2) or decrease (Figure 4).

A satisfactory measure of the absolute width of a distribution is the standard deviation σ, although it should be noted that the well-known relationships between σ and the normal distribution generally do not apply. For a number-fraction distribution of molecular weight σ_n is defined (see equation 13) by

$$\sigma_n = [\bar{M}_n \bar{M}_w - \bar{M}_n^2]^{1/2} = \bar{M}_n[(\bar{M}_w/\bar{M}_n) - 1]^{1/2}$$

Thus for a given value of the ratio $\bar{M}_w/\bar{M}_n$ the width of the number-fraction distribution, defined as σ_n, increases directly as $\bar{M}_n$ increases.

The width of the number-fraction distribution relative to the number-average molecular weight is

$$\Delta = \sigma_n/\bar{M}_n = [(\bar{M}_w/\bar{M}_n) - 1]^{1/2}$$

as used by Lowry.[24] However the most commonly used measure of the relative width of the number-fraction molecular-weight distribution is the ratio of average molecular weights $\bar{M}_w/\bar{M}_n$,

$$\bar{M}_w/\bar{M}_n = (\sigma_n^2/\bar{M}_n^2) + 1,$$

although Schulz[25] preferred to use $U = (\bar{M}_w/\bar{M}_n) - 1$.

The corresponding quantities for the weight-fraction distribution of molecular weight are

$$\sigma_w = [\bar{M}_w\bar{M}_z - \bar{M}_w^2]^{1/2} = \bar{M}_w[(\bar{M}_z/\bar{M}_w) - 1]^{1/2}$$

for the absolute width, and

$$g = \sigma_w/\bar{M}_w = [(\bar{M}_z/\bar{M}_w) - 1]^{1/2}$$

as suggested by Hosemann and Schramek,[26] for the relative width. However the most commonly used measure of the relative width is the ratio

$$\bar{M}_z/\bar{M}_w = (\sigma_w^2/\bar{M}_w^2) + 1.$$

Both the absolute and the relative measures of distribution width can be useful depending on circumstances. Also it is preferable to use the most direct measure of the width, *e.g.* $\bar{M}_z/\bar{M}_w$ for the relative width of the weight-fraction molecular-weight distribution. The two examples below serve to illustrate these points. It is possible that only $\bar{M}_w$ and $\bar{M}_n$ are available: Pyun[27] has given some detailed guidance on the use of the ratio $\bar{M}_w/\bar{M}_n$ under such circumstances.

(a) Consider a non-uniform oligomer (such as polyethylene glycol 2000) with a Poisson number-fraction distribution of molecular weights. The number-average chain length is $\bar{x}_n = 45$. The polymer crystallizes into lamellae in which the chains are unfolded; *i.e.* the lamellar spacing (*via* small-angle X-ray scattering) is equal to the number-average chain length. The width of the number-fraction distribution of chain lengths is described either relatively by $\bar{x}_w/\bar{x}_n = 1.02$ or absolutely by $\sigma_n = \bar{x}_n[(\bar{x}_w/\bar{x}_n) - 1]^{1/2} = 6.4$. The problem of packing the chains into a lamellar crystal is illustrated in Figure 7. The ratio $\bar{x}_w/\bar{x}_n$ is not very meaningful in this context. The parameter σ_n is a better description: an assignment of (on average) $2\sigma_n = 12.8$ chain units to the disordered lamellar surface layer leaves 32.2 chain units (*i.e.* 72%) for the crystalline layer. This is within the range of crystallinities observed for polyethylene glycol 2000, *e.g.* 65–85% depending on the interpretation of the data.[28,29]

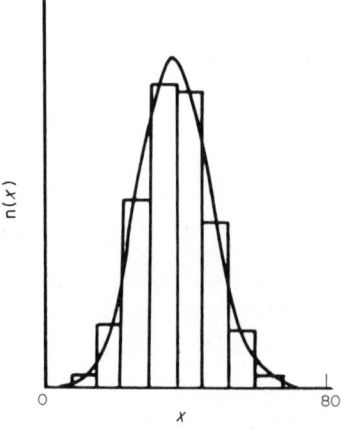
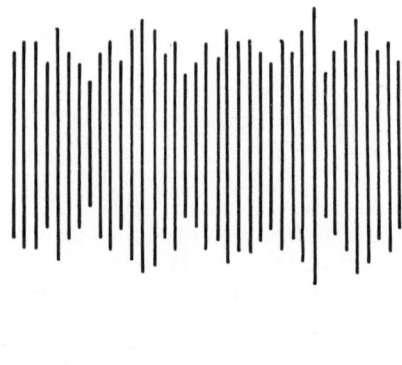

Figure 7 A Poisson number-fraction distribution of chain length [see equation (11a)] for number-average chain length $\bar{x}_n = 45$ represented by (left) a histogram of eight species or (right) a 2-dimensional array. The latter illustrates packing in an unfolded-chain lamellar crystal

(b) As a second example consider an equimolar mixture of two polymeric species with very different molecular weights

$$M_1 = 10^4; \, n_1 = 0.5; \, w_1 = 0.0099;$$

$$M_2 = 10^6; \, n_2 = 0.5; \, w_2 = 0.9901.$$

The average molecular weights (*via* equation (3) or Table 1) are

$$\bar{M}_n = 5.05 \times 10^5; \, \bar{M}_w = 9.902 \times 10^5; \, \bar{M}_z = 9.999 \times 10^5;$$

and the width parameters are

$$\sigma_n = 4.95 \times 10^5; \, \bar{M}_w/\bar{M}_n = 1.96; \, \sigma_w = 0.95 \times 10^5; \, \bar{M}_z/\bar{M}_w = 1.01.$$

The number-fraction distribution is fairly wide, since there are equal numbers of species widely separated in molecular weight, and is well described by σ_n or by $\bar{M}_w/\bar{M}_n$. The weight distribution is narrow, since a single species predominates, and is well described by σ_w or by $\bar{M}_z/\bar{M}_w$. In this example, which models a high polymer, the molecular-weight ratios are the more useful measures.

3.6 DISTRIBUTIONS AND AVERAGES FOR COPOLYMERS

3.6.1 Statistical Copolymers

The description of statistical copolymers requires distributions and averages of chain length, sequence length and composition. Sequence length distributions and averages are accessible *via* NMR. The experimental determination of composition distributions can be difficult since the chain-length and composition distributions are seldom independent of one another.

Here it is convenient to consider a linear statistical copolymerization of two monomers (A and B) in which the reactivity of the chain end is not influenced by the penultimate group. Under these circumstances the sequence-length distribution and averages are analogous to the Schulz–Flory equations (see Section 3.4.1). For a sequence length x_A of A units the number-fraction distribution is

$$n(x_A) = (1 - p_{AA})p_{AA}^{x_A - 1},$$

and the number-average sequence length is $(\bar{x}_A)_n = 1/(1 - p_{AA})$, with corresponding equations for B units and weight distributions. The parameter p_{AA} is the probability of adding an A unit to an A end, rather than a B unit to an A end, and is given by $p_{AA} = r_A[A]/(r_A[A] + [B])$ where r_A is the reactivity ratio (k_{AA}/k_{AB}) and [A] and [B] are molar concentrations of monomer in the copolymerizing mixture. Since the concentrations usually change during copolymerization, the equations in this section refer to copolymer produced instantaneously at given values of [A] and [B].

To define the chain-length/composition distribution[30] a deviation function y is introduced such that $y = \bar{q}_n - q$, where q is the number fraction of A units in a copolymer molecule and $\bar{q}_n$ is its number-average value, given by

$$\bar{q}_n = [A]\{r_A[A] + [B]\}/\{r_A[A]^2 + 2[A][B] + r_B[B]^2\} \qquad (r_B = k_{BB}/k_{AB})$$

The number-fraction distribution for copolymers of chain length x and composition y derived by Stockmayer[30] is

$$n(x, y) = (1 - p)p^{x-1}(x/\pi\phi)^{1/2} \exp[1 + (\bar{x}_n y^2/\phi)] \qquad (19a)$$

where p has its usual meaning of the probability of the chain propagating (*via* an A or a B unit) rather than terminating. The parameter ϕ is given by

$$\phi = 2\bar{q}_n(1 - \bar{q}_n)[1 - 4\bar{q}_n(1 - \bar{q}_n)(1 - r_A r_B)]^{1/2}$$

Equation (19a) is best written in the exponential form, *i.e.*

$$n(x, y) = b^2 \exp(-bx)(x/\pi\phi)^{1/2} \exp[1 + (y^2/b\phi)] \qquad (19b)$$

where $b = 1/\bar{x}_n$. Integration of equation (19b) from $-\infty$ to $+\infty$ with respect to y gives the familiar Schulz–Flory chain-length distribution in the form of equation (9) (Section 3.4.1), and integration from 0 to $+\infty$ with respect to x gives the composition distribution

$$n(y) = 3(1/b\phi)^{1/2}/x[1+(y^2/b\phi)]^{5/2}. \qquad (20)$$

The composition distribution, $n(y)$ *vs.* $q = \bar{q}_n - y$, is illustrated for two values of $\bar{x}_n = 1/b$ in Figure 8. For high molecular weight copolymer the composition distribution is narrow, and the spread of composition is generally insignificant compared with the change in average composition as copolymerization proceeds.

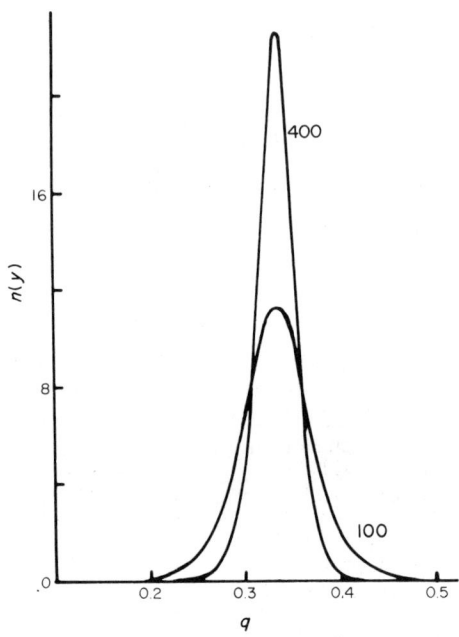

Figure 8 Number-fraction composition distribution [$n(y)$ *vs.* q, equation (20)] for a statistical copolymer. ($y = \bar{q}_n - q$, q = number fraction of A in a copolymer molecule, $\bar{q}_n$ = number-average value of q). The copolymer is formed when monomer concentrations are equal ([A] = [B]) and reactivity ratios are $r_A = 0.5$ and $r_B = 2.0$; hence $\bar{q}_n = 1/3$. The curves are for number-average chain length (A or B units) $\bar{x}_n = 100$, 400, as indicated (Adapted from ref. 7 with permission of Wiley–Interscience)

3.6.2 Block Copolymers

The molecular structure of a block copolymer is relatively simple compared with that of a statistical copolymer. Nevertheless there are inherent difficulties in dealing with a copolymer rather than a homopolymer, which complicate even this case. The example below deals with the determination of average block length.

Consider a diblock copolymer prepared by a sequential process, so that the number and weight-average molecular weights of the first (A) block are known [$(\bar{M}_A)_n, (\bar{M}_A)_w$] together with the average molecular weights and average composition of the copolymer ($\bar{M}_n$, $\bar{M}_w$ and the weight fraction ($\bar{q}_w$) of A units in the copolymer).

Since $\sum_i n_i = \sum_i n_{Ai} = \sum_i n_{Bi}$ and $\sum_i w_i = \left(\sum_i w_{Ai} + \sum_i w_{Bi}\right)$, the number-average molecular weight of the B block is readily calculated from

$$\bar{M}_n = \sum_i n_i M_i \bigg/ \sum_i n_i = \sum_i w_i \bigg/ \sum_i n_i = \sum_i w_{Ai} \bigg/ \sum_i n_{Ai} + \sum_i w_{Bi} \bigg/ \sum_i n_{Bi}$$

i.e. $\bar{M}_n = (\bar{M}_A)_n + (\bar{M}_B)_n$. $\qquad (21)$

However the determination of the weight-average molecular weight of the B block is less straightforward. The weight-average molecular weight of the copolymer is given by

$$\bar{M}_w = \sum_i w_i M_i \bigg/ \sum_i w_i = \sum_i w_i \{q_i + (1-q_i)\} M_i \{q_i + (1-q_i)\} \bigg/ \sum_i w_i,$$

$$\text{or, } \bar{M}_w = \sum_i (w_i q_i)(M_i q_i) \bigg/ \sum_i w_i + \sum_i \{w_i(1-q_i)\}\{M_i(1-q_i)\} \bigg/ \sum_i w_i + 2\sum_i w_i M_i q_i (1-q_i)/\sum_i w_i$$

$$= \sum_i w_{Ai} M_{Ai} \bigg/ \sum_i w_i + \sum_i w_{Bi} M_{Bi} \bigg/ \sum_i w_i + 2\sum_i w_i M_i q_i (1-q_1)/\sum_i w_i$$

Hence, since $\sum_i w_{Ai} = \bar{q}_w \sum_i w_i$ and $\sum_i w_{Bi} = (1-\bar{q}_w)\sum_i w_i$,

$$\bar{M}_w = \bar{q}_w(\bar{M}_A)_w + (1-\bar{q}_w)(\bar{M}_B)_w + 2\sum_i w_i M_i q_i(1-q_i)/\sum_i w_i.$$

or, on introducing a deviation function $y_i = \bar{q}_w - q_i$,

$$\bar{M}_w = [\bar{q}_w(\bar{M}_A)_w + (1-\bar{q}_w)(\bar{M}_B)_w - (2\bar{q}_w - 1)P - Q]/[1 - 2\bar{q}_w(1-\bar{q}_w)]. \tag{22}$$

The new parameters $P = \sum_i w_i M_i y_i \bigg/ \sum_i w_i$ and $Q = \sum_i w_i M_i y_i^2 \bigg/ \sum_i w_i$ depend on the widths of the molecular weight and composition distributions: in particular $P = 0$ when the molecular weight is uniform and $P = Q = 0$ when the composition is uniform.

Equations (21) and (22) are not restricted to block copolymers but apply generally provided that $(\bar{M}_A)_n$ and $(\bar{M}_A)_w$ are identified with the A units in all the sequences in the molecule, and correspondingly $(\bar{M}_B)_n$ and $(\bar{M}_B)_w$ with all the B units. Indeed equations of the form of equation (22) were derived initially with statistical copolymers in mind.[31]

3.7 DISTRIBUTIONS AND AVERAGES FOR NON-LINEAR POLYMERS

The distributions described in Section 3.4 are for linear polymers, or for non-linear polymers with just one branch point per molecule (*i.e.* equation (10), $a = 3, 4, 5$ *etc.*). More complex descriptions are required for non-linear polymers with chains connecting branch points, *i.e.* crosslinks. The basic relations were developed by Flory[32] and Stockmayer.[33] However practical difficulties in the experimental determination of molecular-weight distributions of non-linear polymers, due both to the complexity of their separation by GPC or by solubility and to problems of interpretation, mean that distributions are little used. Average molecular weights can be measured and are of more interest.

3.7.1 Single Multifunctional Monomer

A simple example[33] is the step polymerization of a multifunctional monomer (A_f) of functionality f with all functional groups equally reactive. Prior to gelation and neglecting ring formation, the degree-of-polymerization distributions are

$$n(x) = \Omega[f/(1-pf/2)] \, p^{x-1}(1-p)^{(fx-2x+2)} \tag{23a}$$

and

$$w(x) = \Omega fx \, p^{x-1}(1-p)^{(fx-2x+2)} \tag{23b}$$

where, in this case, $p = p_A =$ fraction of functional groups reacted, and $\Omega = [(fx-x)!/\{x!(fx-2x+2)!\}]$. The Schulz–Flory distribution (see Section 3.4.1) is the special case $f = 2$.

The simple molecular weight averages are[32-34]

$$\bar{M}_n = M_A/(1-pf/2) \tag{24a}$$

$$\bar{M}_w = M_A(1-p)/[1-p(f-1)] \tag{24b}$$

$$\bar{M}_z = M_A[(1+p)^3 - fp^2(3+p)]/[(1+p)\{1-p(f-1)\}] \tag{24c}$$

where M_A is the molecular weight of the chain unit. A polymerizing system gels at a critical extent of reaction given by $p_c = 1/(f-1)$; hence $\bar{M}_w \to \infty$ and $\bar{M}_z \to \infty$ as $p \to p_c$.

Weight-fraction degree-of-polymerization distributions for this case are illustrated in Figure 9. Considered at constant extent of reaction p the distribution is wider for the higher values of the functionality f: see Figure 9(a). This is in sharp contrast to the case of one branch point per molecule (degree of coupling $a = f$) which is illustrated in Figure 3. For a given functionality f the distribution increases in width as the extent of reaction p is increased: see Figure 9(b).

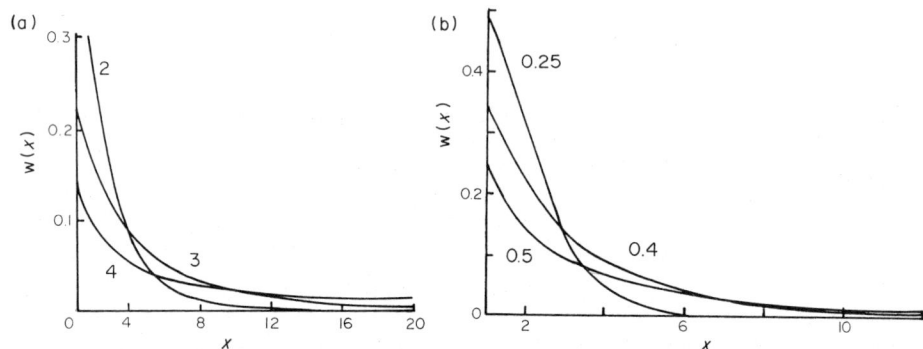

Figure 9 Weight-fraction degree-of-polymerization distributions [$w(x)$ *vs.* x, equation (23)] for the step polymerization of multifunctional monomers (A_f). (a) The effect of functionality, $f = 2, 3, 4$ as indicated, at constant extent of reaction (p = fraction of functional groups reacted). (b) The effect of extent of reaction, $p = 0.25, 0.4, 0.5$ as indicated, when $f = 3$ (Adapted from ref. 32 with permission of *J. Am. Chem. Soc.*)

A related distribution for the step polymerization of a mixture of a multifunctional monomer A_f and a bifunctional monomer A_2 with all functional groups equally reactive has been derived by Stockmayer.[33]

3.7.2   Mixture of Monomers ($A_f + B_g$)

Most technologically important applications involve systems in which the functional groups of a monomer A_f react only with those of a second monomer B_g. The overall degree-of-polymerization distributions and averages are of the form of equations (23) and (24) with p appropriately defined.[2, 3, 32] However more detailed formulations are available.[33-35]

Consider a mixture of a_f mol of a monomer (A_f) of functionality f and b_g mol of a monomer (B_g) of functionality g. The number-fraction distribution for molecules of overall degree of polymerization x (A or B units) prior to gelation, neglecting ring formation, is given by equation (23a) with $p = p_A$. The corresponding distribution for molecules with x A units and y B units is

$$n(x, y) = \Omega[f/\{1 + (f/gr) - pf\}] p^y (1-p)^{(fx-x-y+1)} (pr)^{(x-1)} (1-pr)^{(gy-x-y+1)} \tag{25}$$

where $p = p_A$ = fraction of A functional groups reacted, r = molar ratio of functional groups ($f a_f / g b_g$), and

$$\Omega = [(fx-x)! (gy-y)!]/[x! y! (fx-x-y+1)! (gy-x-y+1)!]$$

The molecular weight averages are

$$\bar{M}_n = [M_A(gr/f) + M_B]/[1 + (gr/f) - grp] \tag{26a}$$

$$\text{and } \bar{M}_w = \frac{M_A^2(gr/f)\{1 + rp^2(g-1)\} + M_B^2\{1 + rp^2(f-1)\} + 2grp M_A M_B}{\{M_A(gr/f) + M_B\}\{1 - rp^2(f-1)(g-1)\}} \tag{26b}$$

where M_A and M_B are the molecular weights of the A and B units. The gel point is given by $rp_c^2 = 1/[(f-1)(g-1)]$.

3.7.3 Mixture of Monomers $\left(\sum_i A_i + \sum_j B_j\right)$

A number distribution for the case of a mixture of A monomers with one or more functional groups $(A_1, A_2, \ldots A_i, \ldots)$ which react only with the functional groups of a similar mixture of B monomers $(B_1, B_2, \ldots B_j, \ldots)$ without ring formation has been published by Stockmayer.[35] The lengthy equation is available in the Polymer Handbook:[36] equation (25) is the special case of a binary mixture. The equations for the average molecular weights are of more interest.

$\bar{M}_n$ is calculated *via* the stoichiometry as the total weight divided by the total number of polymer molecules:

$$\bar{M}_n = \left(\sum_i a_i M_{Ai} + \sum_j b_j M_{Bj}\right) \bigg/ \left(\sum_i a_i + \sum_j b_j - p\sum_j b_i f_i\right) \tag{27a}$$

where $a_1, a_2, \ldots a_i, \ldots$ are the numbers of A monomers and $b_1, b_2, \ldots b_j, \ldots$ the numbers of B monomers, and $p = p_A =$ fraction of A functional groups reacted.

The equation for $\bar{M}_w$ is more complex and it is useful to define:[34]

$$m_a = \sum_i a_i M_{Ai} \bigg/ \sum_i a_i f_i, \quad m'_a = \sum_i a_i M_{Ai}^2 \bigg/ \sum_i a_i f_i, \quad M_a = \sum_i a_i M_{Ai} f_i \bigg/ \sum_i a_i f_i, \quad f_e = \sum_i a_i f_i^2 \bigg/ \sum_i a_i f_i,$$

with analogous quantities $(m_b, m'_b, M_b$ and $g_e)$ for the B unit, and the molar ratio of functional groups $r = \sum_i a_i f_i \bigg/ \sum_j b_j g_j$. In these terms $\bar{M}_w$ is given by

$$\bar{M}_w = \frac{rm'_a + m'_b}{rm_a + m_b} + \frac{rp[p(f_e - 1)M_b^2 + rp(g_e - 1)M_a^2 + 2M_a M_b]}{(rm_a + m_b)[1 - rp^2(f_e - 1)(g_e - 1)]} \tag{27b}$$

Equation (26b) is obtained by simplification of equation (27b). The gel point is given by

$$rp_c^2 = 1/[(f_e - 1)(g_e - 1)] \tag{28}$$

The dependence of weight-average molecular weight on extent of reaction p is illustrated in Figure 10 for commonly encountered cases for which $r = 1$.

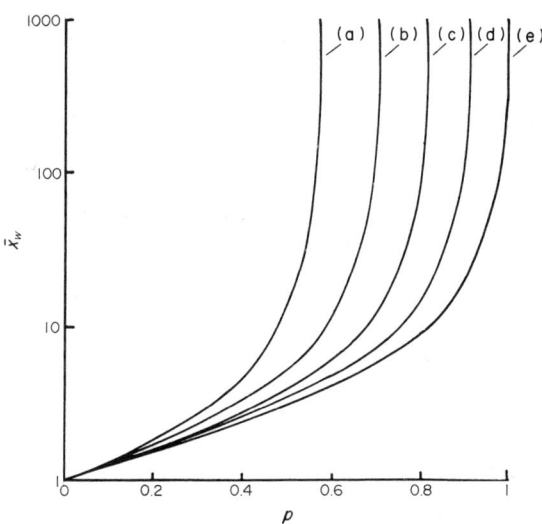

Figure 10 Weight-average degree of polymerization $[\bar{x}_w = \bar{M}_w(p)/\bar{M}_w(p=0)$, logarithmic scale] *vs.* extent of reaction (p = fraction of A functional groups reacted) for step polymerization of mixtures of multifunctional monomers: (a) $A_4 + 2B_2$; (b) $2A_3 + 3B_2$; (c) $2A_3 + 3A_2 + 6B_2$; (d) $2A_3 + 12A_2 + 15B_2$; (e) $A_2 + B_2$. The mole ratio of functional groups A and B in the mixtures (r) is 1. The effective functionality (f_e) of the A monomers in the mixtures is: (a) 4.0; (b) 3.0; (c) 2.5; (d) 2.2; (e) 2.0. The gel point at which $\bar{x}_w \to \infty$, is $p_c = (f_e - 1)^{-1/2}$ when $r = 1$ and the functionality of the B monomer is 2 (see equation 28) (Adapted from ref. 34 with permission of *Macromolecules*)

3.8 EFFECTS OF RING FORMATION

The distributions and averages described in the preceding sections were formulated neglecting the possibility of ring formation. Cyclization in chain-growth polymerizations is generally not important, though it may be in specific systems. By contrast cyclization is a general feature of condensation polymerization (both linear and non-linear) and can be particularly important in dilute systems. Details of cyclic polymers and polymer-forming reactions can be found in a compilation of reviews edited by Semlyen.[37]

For the reversible step-polymerization of a bifunctional monomer, the fractions of rings and chains of a given size at equilibrium may be assumed[38] to depend only on the probability that two ends of the same molecule are adjacent compared with two ends from different molecules. The former (leading to rings) depends on the statistics of chain conformation, the latter (leading to chains) depends on concentration. Results have been obtained for Gaussian[38] and other[37] chain statistics. For a bifunctional monomer and for Gaussian statistics, the weight-fraction distribution for chains is[38]

$$w_c(x) = (1 - w_r)(1 - p')^2 x(p')^{x-1} \tag{29}$$

where w_r is the total weight fraction of rings and p' is an extent of reaction defined by $(1 - p') = (1 - p)/(1 - w_r)$, where p is the fraction of end groups reacted. Equation (29) replaces equation (8b) (Section 3.4.1); w_r can be calculated as described elsewhere.[7,38] The corresponding weight-fraction distribution for rings is[37]

$$w_r(x) = (BM_0/c)(p')^x x^{-3/2} \tag{30}$$

where B is a quantity of statistical origin, c is the concentration (g cm^{-3}) and M_0 is the molar mass of the constitutive unit. Note that $w_r(x)$ increases as c decreases, and that $w_r(x)$ is a monotonically decreasing function of x, as distinct from $w_c(x)$ which approximates the behaviour of $w(x)$ of equation (8b) (see Figure 2).

The fractions of rings and chains resulting from an irreversible (kinetically controlled) polycondensation reaction can be calculated using similar assumptions. The reaction scheme for a bifunctional monomer has been set out by Gordon and Temple[39] and the infinite set of simultaneous differential rate equations so generated can be solved. Calculations[40,41] based on Gaussian statistics show that $w_r(x)$ generally varies as $x^{-3/2}$, as in the equilibrium case. Of course the total weight fraction of rings formed depends on the rate constants assigned to the formation reactions for rings and chains.

3.9 REFERENCES

1. H.-G. Elias, 'Macromolecules', 2nd edn., Plenum, New York, 1984.
2. G. Odian, 'Principles of Polymerization', 2nd edn., Wiley–Interscience, New York, 1981.
3. P. J. Flory, 'Principles of Polymer Chemistry', Cornell, Ithaca, NY, 1953.
4. H.-G. Elias, R. Bareiss and J. G. Watterson, *Adv. Polym. Sci.*, 1973, **11**, 111.
5. H.-G. Elias, *Pure Appl. Chem.*, 1975, **43**, 115.
6. A.-Q. Tang, 'Statistical Theory of Polymer Reactions', Academic Press, Beijing, 1985.
7. L. H. Peebles, Jr., 'Molecular Weight Distributions in Polymers', Wiley–Interscience, New York, 1971.
8. See, for example, M. L. Boas, 'Mathematical Methods in the Physical Sciences', 1st edn., Wiley, New York, 1966, p. 698.
9. G. Herdan, 'Small Particle Statistics', 2nd edn., Butterworths, London, 1960.
10. A. F. Samarin, *Polym. Sci. USSR (Engl. Transl.)*, 1971, **13**, 1153.
11. G. V. Schulz, *Z. Phys. Chem., Abt. B*, 1935, **30**, 379.
12. P. J. Flory, *J. Am. Chem. Soc.*, 1936, **58**, 1877.
13. G. V. Schulz, *Z. Phys. Chem., Abt. B*, 1939, **43**, 25.
14. B. H. Zimm, *J. Chem. Phys.*, 1948, **16**, 1099.
15. Attributed to L. T. Muus and W. H. Stockmayer in reference 7.
16. W. Weibull, *J. Appl. Mech.*, 1951, **18**, 293.
17. L. H. Tung, *J. Polym. Sci.*, 1956, **20**, 495.
18. L. H. Tung, in 'Polymer Fractionation', ed. M. J. R. Cantow, Academic Press, New York, 1967, p. 379.
19. P. J. Flory, *J. Am. Chem. Soc.*, 1940, **62**, 1561.
20. W. D. Lansing and E. O. Kraemer, *J. Am. Chem. Soc.*, 1935, **57**, 1369.
21. H. Wesslau, *Makromol. Chem.*, 1956, **20**, 111.
22. W. F. Espenscheid, M. Kerker and E. Matijevic, *J. Phys. Chem.*, 1964, **68**, 3093.
23. E. P. Honig, *J. Phys. Chem.*, 1965, **69**, 4418.
24. G. G. Lowry, *J. Polym. Sci., Part B*, 1963, **1**, 489.
25. G. V. Schulz, *Z. Phys. Chem., Abt. B*, 1940, **47**, 155.
26. R. Hosemann and W. Schramek, *J. Polym. Sci.*, 1962, **59**, 29.
27. C. W. Pyun, *J. Polym. Sci., Polym. Phys. Ed.*, 1986, **24**, 229.

28. J. P. Arlie, P. A. Spegt and A. E. Skoulios, *Makromol. Chem.*, 1966, **99**, 160.
29. P. C. Ashman and C. Booth, *Polymer*, 1975, **16**, 889; R. C. Domszy, R. H. Mobbs, Y.-K. Leung, F. Heatley and C. Booth, *Polymer*, 1979, **20**, 1204; 1980, **21**, 588.
30. W. H. Stockmayer, *J. Chem. Phys.*, 1945, **13**, 199.
31. W. Bushuk and H. Benoit, *Can. J. Chem.*, 1958, **36**, 1616.
32. P. J. Flory, *J. Am. Chem. Soc.*, 1941, **63**, 3083, 3091, 3096.
33. W. H. Stockmayer, *J. Chem. Phys.*, 1943, **11**, 45.
34. C. W. Macosko and D. R. Miller, *Macromolecules*, 1976, **9**, 199.
35. W. H. Stockmayer, *J. Polym. Sci.*, 1952, **9**, 69; 1953, **11**, 424.
36. L. H. Peebles, Jr., in 'Polymer Handbook', ed. J. Brandrup and E. H. Immergut, 2nd edn., Wiley–Interscience, New York, p. II–405.
37. J. A. Semlyen (ed.), 'Cyclic Polymers', Elsevier, London, 1986.
38. H. Jacobson and W. H. Stockmayer, *J. Chem. Phys.*, 1950, **18**, 1600.
39. M. Gordon and W. B. Temple, *Makromol. Chem.*, 1972, **160**, 263.
40. C. Samoria, E. Vallés and D. R. Miller, *Makromol. Chem., Makromol Symp.*, 1986, **2**, 69.
41. X.-F. Yuan, A. J. Masters, C. V. Nicholas and C. Booth, *Makromol. Chem.*, in press.

4

Colligative Properties

KENJI KAMIDE

Asahi Chemical Industry Company Ltd., Takatsuki, Osaka, Japan

4.1 INTRODUCTION

The colligative properties of a solution are defined as the properties which are determined by the number or the mole fraction of components (solutes and solvent) in the solution and are independent of the nature of the solutes and their molecular weights. If any physical quantities belonging to the colligative property (*i.e.* the colligative quantity) can be measured under thermodynamic equilibrium and also if the total weight of solute dissolved in a given solution is known, the molecular weight of the nonvolatile solute (or in the case of a nonuniform solute, the average molecular weight) can be determined. Unfortunately, both the osmotic pressure of the real solution and the vapor pressure of the vapor phase in equilibrium with the real solution are significantly influenced, in the finite concentration range of the solute, by the solute–solvent interaction and are not unique functions of the number of solute molecules existing in the solution. In general the activity of the solvent can be directly evaluated from the osmotic pressure or the partial vapor pressure. In fact, an extraordinarily large discrepancy from a simple proportionality of the osmotic pressure or the vapor pressure to the solute mole fraction was observed in the late 1920s to early 1930s in polymer solutions and these experimental facts motivated theoretical study by Flory, Huggins *et al.*, based on the lattice model, of the thermodynamics of polymer solutions. Conversely, measurement of the extent of the deviation of polymer solutions from the proportionality relations between osmotic pressure or vapor pressure and solute mole fraction enables us to elucidate the thermodynamic interaction between a polymer chain and a solvent molecule (see equations 50 and 51) and to study the dissolved state of polymer molecules. The osmotic pressure and vapor pressure of the solution can be approximately regarded as colligative in the extremely low polymer concentration range for polymer–good solvent systems, and over the whole range of concentration when the polymer is dissolved in a solvent in which the second and the higher virial coefficients (equations 57–58) are zero. In the above cases, we can determine unambiguously the molecular weight of the polymer from its colligative properties. In fact, in the 1930s the determination of molecular weight of polymers using the colligative properties of the solutions was widely attempted, mainly by Staudinger and his collaborators,[1] for cellulose derivatives and synthetic polymer solutions, giving a direct verification of 'macromolecular compounds'. The importance of the use of colligative properties to determine the molecular weights of polymers and to estimate the polymer–solvent interactions remains unchanged to the present day.

4.2 SIMPLE MOLECULE SOLUTIONS

4.2.1 Ideal Solutions

Let us consider a mixture consisting of a large number of simple molecule components and consider a change in Gibbs' free energy G accompanying the addition of one mole of component i at constant temperature T under constant pressure P to the mixture. This is called the partial molar Gibbs free energy or simply the chemical potential, μ_i. If it is given, over an entire range of the molar fraction of the component i, by

$$\mu_i \ (\equiv (\partial G/\partial n_i)_{P,T,N_j}) = \mu_i^\circ(T,P) + RT \ln n_i \tag{1}$$

where $\mu_i^\circ(T,P)$ is the chemical potential of pure liquid component i, which is a function of temperature T and pressure P, R is the gas constant and n_i is the relative proportion of the i component in the solution (known as the mole fraction of component i), the solution is called an ideal solution, which is the simplest physical model of a solution. Here, the subscript N_j is the number of moles of the j component in the system and the partial differentiation is made under constant T and P at constant composition except for component i.

In general, the partial molar entropy, volume and enthalpy of the component i in the solution, S_i, V_i and H_i, are given in the forms:

$$S_i = -(\partial \mu_i/\partial T)_{P,N_j} \tag{2}$$

$$V_i = (\partial \mu_i/\partial P)_{T,N_j} \tag{3}$$

$$H_i = \{\partial(\mu_i/T)/\partial(1/T)\}_{P,N_j} \tag{4}$$

Here, all the partial differentiations are carried out at constant P (or at constant T) and constant composition except for component i. Note that the equation $\mu_i = H_i - TS_i$ holds. The differences in the values of S_i, V_i and H_i between an ideal solution and pure component i (S_i°, V_i° and H_i°) are

$$\Delta S_i^{id} \equiv \Delta S_i (\equiv S_i - S_i^\circ) = -R\ln n_i \tag{5}$$

$$\Delta V_i^{id} \equiv \Delta V_i (\equiv V_i - V_i^\circ) = 0 \tag{6}$$

$$\Delta H_i^{id} \equiv \Delta H_i (\equiv H_i - H_i^\circ) = 0 \tag{7}$$

In an ideal solution the entropy change of mixing (*i.e.* the mixing entropy) depends only on the mole fraction n_i (equation 5) and neither a volumetric change nor a thermal change occurs on mixing (equations 6 and 7). In other words, an ideal solution, in which all components are randomly mixed, is absolutely athermal.

4.2.1.1 *Vapor–liquid phase equilibrium*

Consider, next, that the solution is in equilibrium with the vapor phase (*i.e.* the vapor–liquid phase equilibrium) as shown in Figure 1(a). Assuming the vapor to be a perfect gas, the chemical potential of the component i in the vapor phase μ_i' is given by

$$\mu_i' = \mu_i^*(T) + RT \ln P_i \tag{8}$$

where P_i is the partial pressure of the component i in the vapor phase in equilibrium with the solution at temperature T and $\mu_i^*(T)$ is the chemical potential of the vapor of pure component i. Since the vapor phase is in equilibrium with the solution, μ_i is equal to μ_i', that is,

$$P_i = n_i \exp[\{\mu_i^\circ(T,P) - \mu_i^*(T)\}/RT] \tag{9}$$

If the vapor pressure of pure liquid component i at the same temperature is P_i°, then from equation (9) we obtain

$$P_i^\circ \equiv \exp[\{\mu_i^\circ(T,P) - \mu_i^*(T)\}/RT]$$

or

$$P_i^\circ = \exp[\{\mu_i^\circ(T,0) - \mu_i^*(T)\}/RT]\,\exp(PV_i^\circ/RT) \qquad (9)'$$

where V_i° is the molar volume of pure liquid component i.

Combination of equations (9) and (9)' yields

$$P_i/P_i^\circ = n_i \qquad (10)$$

provided that $\exp(P_iV_i^\circ/RT) \simeq 1$. Equation (10) is known as Raoult's law.[2] Some solutions, consisting of chemically similar components, obey approximately equation (10). Recent advances in the accuracy and precision of the measurement of ΔV_i, ΔH_i and P_i reveal that almost ideal solutions cannot simply be represented by equation (10). Figure 2 illustrates the compositional dependence of total pressure $P(\equiv P_0 + P_1)$ of the methanol–ethanol system at 24.95 °C,[3] which follows, within the precision of the experiment, Raoult's law. This system is, however, not an ideal solution due to an extremely small, but significant, nonzero heat of mixing.[4]

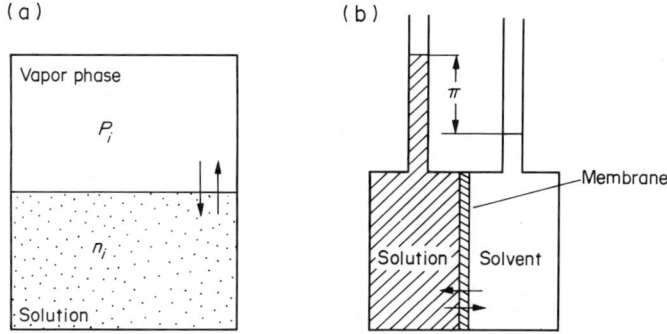

Figure 1 Schematic representation of (a) vapor pressure and (b) membrane osmotic pressure

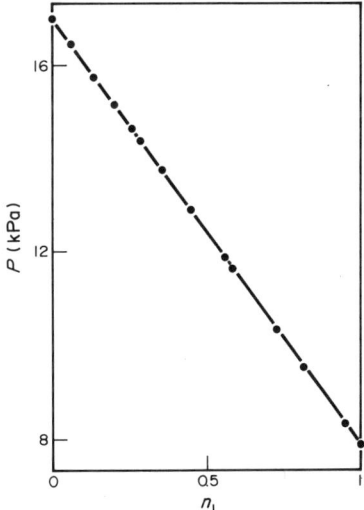

Figure 2 Relationship between the total vapor pressure P and mole fraction n_1 of ethanol for the methanol–ethanol system at 24.95 °C[3]

4.2.1.2 Solution–membrane–solvent phase equilibrium (osmometry)

When a solution and a solvent are separated by a membrane permeable only to the solvent but not to nonvolatile solutes (this is referred to as a semipermeable membrane), the solute molecules are unable to pass through the membrane, resulting in a pressure difference, π, between the two liquid phases, which is defined as the osmotic pressure (Figure 1b). In this case the chemical potential of the pure solvent at T and P, $\mu_0^\circ(T, P)$ (here, the suffix 0 denotes solvent) is the same as that of the solvent component in the solution at T and $P+\pi$, $\mu_0(T, P+\pi)$. This is the condition of osmotic equilibrium or solution–membrane–solvent phase equilibrium. Then $\mu_0^\circ(T,P)$ can readily be represented for an ideal solution by

$$\mu_0^\circ(T, P) = \mu_0(T, P+\pi) = \mu_0^\circ(T, P+\pi) + RT\ln n_0 \tag{11}$$

where $\mu_0^\circ(T, P+\pi)$ is μ_0° at T and $P+\pi$. Note that only solvent that permeates the membrane will have equal chemical potential on both sides.

If $\pi \ll P$, we can use a Taylor series to expand with respect to π the first term of the right-hand side of equation (11) to obtain an approximate equaton:

$$\pi = -(RT\ln n_0)/(\partial\mu_0^\circ(T, P)/\partial P)_T. \tag{12}$$

Combination of equations (3) and (12) yields

$$\pi = -RT\ln n_0/V_0^\circ = -\Delta\mu_0/V_0^\circ \tag{13}$$

$$\text{where } \Delta\mu_0 = \mu_0(T, P) - \mu_0^\circ(T, P) \tag{14}$$

The osmotic pressure of ideal solutions is determined, if $\pi \ll P$, by the mole fraction of the solvent n_0. Equation (13) indicates that the osmotic pressure is a colligative property, because π depends only on n_0.

When the mole fraction of the solute n_1 is sufficiently small as compared with n_0 (i.e. $n_1 \ll n_0$, as in the case of a dilute solution), the total volume of the solution V can be regarded as $V_0^\circ N_0$ (N_0 is the number of the solvent moles in V) and equation (13) can be simplified to:

$$\pi = (RT/V)(n_1 + n_1^2/2 + \ldots) \tag{15}$$

which resembles the gas law. Equation (15) can also be, after some rearrangement, rewritten as a power series in the solute concentration c (i.e. the summation of mass of all solute components present in unit volume of the solution) in the form,

$$\pi = RT(c/M + V_0 c^2/2N_0 M^2 + \ldots)$$
$$= RT(c/M + A_{2,0}^{\text{id}} c^2 + A_{3,0}^{\text{id}} c^3 + \ldots) \tag{16}$$

with

$$A_{2,0}^{\text{id}} = V_0/2M^2 \tag{17a}$$

$$A_{3,0}^{\text{id}} = V_0^2/3M^3 \tag{17b}$$

where M is the solute molecular weight (i.e. the relative molar mass), and $A_{2,0}^{\text{id}}$ and $A_{3,0}^{\text{id}}$ are the second and the third virial coefficients of osmotic pressure for the ideal solution. Note that although $c/M \gg A_{2,0}^{\text{id}} c^2$, even in an ideal solution $A_{2,0}$ does not vanish in the strict sense. In the extremely dilute solution range, equation (16) reduces to the well-known van't Hoff's equation:[5]

$$\pi = (RT/V)n_1 = RT(c/M) \tag{18}$$

Note that equation (18) can apply only to dilute ideal solutions.

4.2.1.3 *Solution–vapor solvent phase equilibrium (vapor pressure osmometry)*

Suppose that at a certain moment t a drop of pure solvent at temperature T_0 and a drop of solution containing $N_0(t)$ moles of solvent and N_1 ($=$ constant) moles of absolutely nonvolatile solute at temperature T are in contact with saturated vapor, whose vapor pressure is $P_0(T_0)$ over the solvent at the temperature T_0. This is the condition of solution–vapor solvent phase equilibrium (Figure 3a). By taking into account the heat balance achieved by the solution drop, the change in the temperature of the drop with time dT/dt can be written in the form:[6]

$$dT/dt = (1/C_p V_d \rho)\{(dN_0(t)/dt)\Delta H - (k_1 A_1' + k_2 A_2')(T - T_0)\} \tag{19}$$

where C_p represents the specific heat under a constant pressure; V_d the volume of the solution drop; ρ the density of the solution drop; $dN_0(t)/dt$ (mol s^{-1}) the condensation rate of the solvent vapor into the solution drop; ΔH the molar heat of condensation; k_1 and k_2 the coefficients of the surface heat transfer (J cm^{-2} s^{-1} K^{-1}); A'_1 the surface area of the solution drop; A'_2 the area of contact between the solution drop and the thermistor or thermocouple including lead wire. As A'_2 is constant, dN_0/dt in equation (19) can readily be shown to be related to the difference between the saturated vapor pressure $P_0(T_0)$ of the ambient pure solvent and the vapor pressure $P(T)$ of the solvent in the solution drop through the formula:

$$dN_0(T)/dt = k_3 A'_1 \{P_0(T_0) - P(T)\} \tag{20}$$

where k_3 (mol g^{-1} s^{-1}) denotes the mass transfer coefficient which varies depending on the detailed mechanism of diffusion.

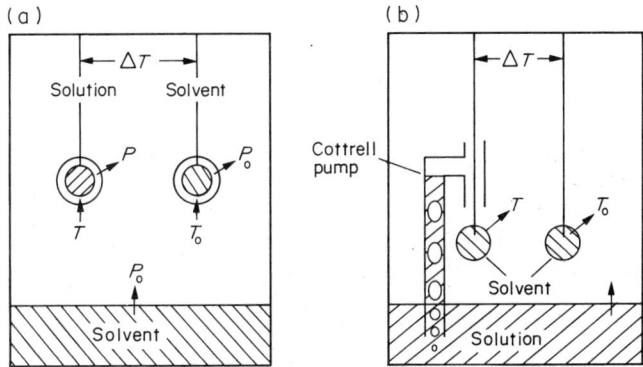

Figure 3 Schematic diagrams of apparatus for (a) vapor pressure osmometry[7] and (b) ebulliometry

Assuming that the temperature of the solution drop T is very close to that of the solvent T_0, that the vapor is an ideal gas and that the solution is an ideal solution, the vapor pressure of the solution drop at T, $P(T)$, can be expressed in the form:[6, 7]

$$P(T) = a_0 P_0(T_0)\{1 + [\Delta H/(RT_0^2)](T-T_0) + 1/2[\Delta H/(RT_0^2)][\Delta H/(RT_0^2) - 2/T_0](T-T_0)^2 + \ldots\} \tag{21}$$

where a_0 is an activity (see equation 32).

In deriving equation (21) the Clausius–Clapeyron relation is employed. Finally equation (19) can be rearranged through use of equations (20) and (21) as follows:

$$dT/dt = (1/C_p V_d \rho)\{k_3 A'_1 P_0(T_0)\Delta H\{1 - a_0[1 + \Delta H(T-T_0)/(RT_0^2) + \Delta H/(2RT_0^2)[\Delta H/(RT_0^2) - 2/T_0](T-T_0)^2]\} - (k_1 A'_1 + k_2 A'_2)(T-T_0)\} \tag{22}$$

Equation (22) is a generalized equation expressing the change in T of the solution drop with time.

When changes in concentration, volume and density of the solution drop due to condensation of solvent vapor are all negligibly small, $dT/dt = 0$. In other words, the temperature difference between the two drops $(T-T_0)(\equiv \Delta T)$ becomes constant. Note that this is a kind of 'steady state' and not an equilibrium state. Careful distinction of these two states should be made. Hereafter, the suffixes s and e mean steady and equilibrium states, respectively.

The temperature difference in the steady state, $(T-T_0)_s$ $(\equiv \Delta T_s)$, can be expressed as a power series of the concentration, from equation (22) after rather tedious calculations:[6, 7]

$$(T-T_0)_s = K_s(c/M + A_{2,v}^{id} c^2 + A_{3,v}^{id} c^3 + \ldots) \tag{23}$$

where $A_{2,v}^{id}$ and $A_{3,v}^{id}$ are second and third virial coefficients, and

$$K_s = K_e/\{1 + RT_0^2(k_1 A'_1 + k_2 A'_2)/(k_3 A_1 P_0(T_0) \Delta H^2)\} \tag{24}$$

$$K_e = R T_0^2 V_0^\circ/\Delta H \tag{25}$$

where K_s is a calibration parameter (cm^3 K mol^{-1}), depending on the combination of solute and solvent and the dimensions of the apparatus. In the hypothetical case of $k_1 = k_2 = 0$, K_s reduces to K_e

and accordingly, $(T-T_0)_s$ coincides with $(T-T_0)_e$. Unfortunately this can never be realized because $k_1 \neq 0$ and $k_2 \neq 0$. The ratio K_s/K_e is a measure of the thermal insulation of the apparatus; in other words, this ratio represents the efficiency of the apparatus. Therefore, vapor pressure osmometry (VPO) is not, in the strict sense, a colligative method. But, many experiments show that in actual cases, K_s can reasonably be regarded as constant independent of the solute nature.[8] Kamide *et al.*[6,7,8] have pointed out that the ambient vapor phase of the solvent chamber in a commercial VPO apparatus is usually unsaturated, and that an observed temperature difference of zero between two pure solvent drops does not afford direct evidence that the ambient vapor is completely saturated, but fortunately the relationship between $\Delta T_s/c$ and c reduces to the corresponding equation (see equation 23), derived previously by assuming that the ambient vapor phase is saturated. The theory indicates that the effect of unsaturation is negligibly small. The VPO theory of binary solutions (equations 23–25) can be generalized to the case of a ternary mixture.[10]

Equations (23–25) give a satisfactory theoretical basis of vapor pressure osmometry (*i.e.* the relative lowering of the vapor pressure of the solvent). Since usually $T-T_0 \ll T_0$, the term containing $(T-T_0)^2$ in equation (21) can be neglected, and one obtains

$$(T-T_0)_s = n_1/\{\Delta H/RT_0^2 + (k_1 A_1' + k_2 A_2')/k_3 A_1' P_0(T_0)\Delta H\} \tag{26a}$$

$$= (K_s/V_0^\circ)n_1 \tag{26b}$$

$$= K_s c/M \tag{26c}$$

The right-hand side of equation (26c) is the first term of the right-hand side of equation (23), which was derived in a more rigorous manner. Equation (26) indicates that, if $T-T_0 \ll T_0$ (*i.e.* in the very dilute concentration range) then $(T-T_0)_s$ can be, to a good approximation, regarded as a colligative property.

4.2.1.4 *Boiling point elevation, freezing point depression*

The temperature at which the equilibrium vapor over a binary solution containing nonvolatile solute equals the external pressure (usually atmospheric pressure), is higher than that of the pure solvent (*i.e.* boiling point elevation or ebulliometry) and similarly, the freezing temperature of a solution is lower than that of the pure solvent (*i.e.* freezing point depression or cryoscopy). When the solution is in equilibrium with a vapor or a solid phase at T and P, the chemical potential of the solvent component in both phases, $\mu_0(T, P)$ and $\mu_0'(T, P)$, is the same: *i.e.*

$$\mu_0(T, P) = \mu_0'(T, P) \tag{27}$$

For an ideal solution $\mu_0(T, P)$ is, of course, given by equation (1). On the other hand, if the pure solvent is in equilibrium with the vapor or solid phase at T_0 and P, equation (28) holds:

$$\mu_0'(T_0, P) = \mu_0^\circ(T_0, P) \tag{28}$$

Combination of equations (1), (27) and (28) leads to

$$\mu_0'(T, P) - \mu_0'(T_0, P) = \mu_0^\circ(T, P) - \mu_0^\circ(T_0, P) + RT\ln n_0 \tag{29}$$

Considering $T - T_0 = \Delta T \ll T$ and using equation (2), equation (29) can be rearranged to:

$$\Delta T/T = -(RT/L_0)\ln n_0 \tag{30}$$

where L_0 is the heat of vaporization (in case of boiling point elevation) or $-L_0$ is the heat of crystallization (in case of freezing point depression). Note that only when the solution and the solvent are placed in two isolated vessels[11] is the difference in the boiling temperature ΔT expressed explicitly by equation (30). In other words in this case thermodynamic equilibrium is attained. But when the solution and the solvent are placed in a single cell[12] (*i.e.* the vapor phase is in equilibrium with the solution and with solvent concurrently as in Figure 3b), ΔT can not be simply expressed by equation (30), because heat transfer from the boiling solution to the vapor phase and to the thermistor can not be ignored and, as in the case of vapor pressure osmometry, only the steady state is set up.

4.2.2 Nonideal Solutions

Mixtures of low molecular weight components with similar chemical structure are expected to behave like ideal solutions, but this is a very rare case, as discussed before. Solutions whose components greatly differ in chemical structure and in polarity exhibit remarkable deviations from ideal solution theory. The chemical potential of the component i in a nonideal solution is generally given by

$$\mu_i(T, P, n_0, \ldots, n_j, \ldots) = \mu_i^\circ(T,P) + RT\ln a_i \tag{31}$$

$$\text{with } a_i = \gamma_i n_i \tag{32}$$

Here a_i and γ_i are the activity and the activity coefficient of the component i, respectively. γ_i represents the extent of deviation from ideality and is a complicated function of n_0 and n_j ($j \neq i$) and the interaction between solutes and solvent. For nonideal solutions, the relations derived for the ideal solution are modified as follows:

$$\Delta S_i = -R\ln a_i \neq \Delta S_i^{id} \tag{5'}$$

$$\Delta V_i \neq 0 \tag{6'}$$

$$\Delta H_i \neq 0 \tag{7'}$$

$$\ln P_i/P_i^\circ = \Delta\mu_i/RT = \ln a_i \tag{10'}$$

$$\pi = (RT\ln a_0)/V_0^\circ \tag{13'}$$

$$(T-T_0)_s = (K_s/V_0)(1-a_0) \tag{26'}$$

γ_0 can be expressed as a polynomial in n_0 and after reducing n_0 to $c, (p_0/P_0^\circ)$, π and $(T-T_0)_s$ are finally given in their virial expansion forms as

$$\ln(p_0/p_0^\circ) = \Delta\mu_0/RT \tag{33a}$$

$$= V_0^\circ(c/M + A_{2,0}c^2 + A_{3,0}c^3 + \ldots) \tag{33b}$$

$$\pi = -\Delta\mu_0/V_0^\circ \tag{34a}$$

$$= RT(c/M + A_{2,0}c^2 + A_{3,0}c^3 + \ldots) \tag{34b}$$

and

$$(T-T_0)_s \simeq (K_s/V_0)(1-a_0) \tag{35a}$$

$$= K_s(c/M + A_{2,v}c^2 + A_{3,v}c^3 + \ldots) \tag{35b}$$

where $A_{2,0}$ and $A_{2,v}$ are the second virial coefficients in membrane osmometry and vapor pressure osmometry respectively, and $A_{3,0}$ and $A_{3,v}$, the corresponding third virial coefficients. $A_{2,v}$ and $A_{3,v}$ are related to $A_{2,0}$ and $A_{3,0}$ through the following relations:[11]

$$A_{2,v} = A_{2,0} + (V_0/M^2)\{(RT_0/\Delta H) - 1/2)(K_s/K_e)^2 + K_s/K_e - 1/2\} \tag{36}$$

and

$$A_{3,v} = A_{3,0} + 2A_{2,0}(V_0/M)\{(RT_0/\Delta H) - 1/2)(K_s/K_e)^2 + K_s/K_e - 1/2\} + (V_0^2/M^3)[(RT_0/\Delta H - 1/2\{2(RT_0/\Delta H$$
$$- 1/2)(K_s/K_e)^2 - K_s/K_e - 2\}(K_s/K_e)^2 - 2(K_s/K_e)^3 + (K_s/K_e)^2 - K_s/K_e + 1/6] \tag{37}$$

Equation (36) is a rigorous equation, but equation (37) is an approximate equation.

Figure 4 shows the activity a_0 of benzene in the *m*-terphenyl–benzene system, determined using equations (33) and (35) by vapor pressure osmometry[7] and vapor pressure measurements.[13] Both methods give approximately the same value of a_0. Equations (33–35) indicate that in the finite concentration range and when $A_{2,0} \neq 0$ and $A_{3,0} \neq 0$ (or $A_{2,v} \neq 0$ and $A_{3,v} \neq 0$), P/P_0, π and $(T-T_0)_s$ are single functions of a_0, but not single function of n_0. In the vicinity of $c \to 0$, the higher terms containing $c^2, c^3, \ldots$ can be neglected and equation (34) reduces to the van't Hoff's equation.

4.3 HIGH POLYMER SOLUTIONS

It has been widely confirmed with numerous experiments that mixtures of macromolecular compounds with a low molecular weight component (*i.e.* polymer solutions) usually reveal re-

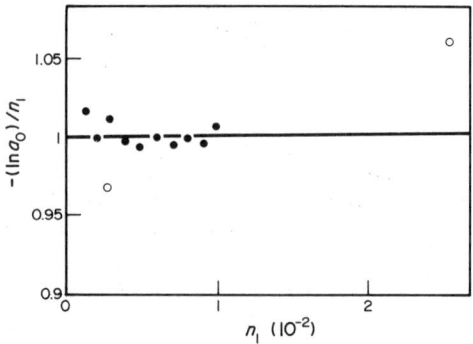

Figure 4   Plot of the ratio of minus logarithm of activity a_0 of benzene and mole fraction n_1 of *m*-terphenyl as a function of n_1.[9] ($\bullet$) vapor pressure osmometry (VPO) at 37 °C (Mechrolab model 302); dilution effect was taken into account. ($\bigcirc$) vapor pressure measurements obtained by Fujishiro *et al.*[13] at 30 °C. Solid line indicates the experimental equation $\log P = 2.076970 - 0.434 n_1 - 0.06569 n_1^2$, obtained from vapor pressure measurements at higher concentration by Fujishiro *et al.*[13]

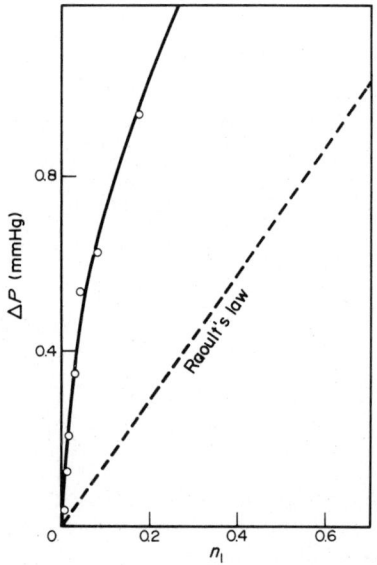

Figure 5   Vapor pressure lowering ΔP plotted as a function of the mole fraction of n_1 of atactic polystyrene (number-average molecular weight $\bar{M}_n = 2.59 \times 10^4$) in cyclohexane at 34 °C. Unfilled circles are experimental results obtained by Krigbaum and Geymer.[14] The broken line represents Raoult's law (equation 10)

markable deviation from the behavior of an ideal solution. Figure 5 illustrates the relations between the depression of the vapor pressure of cyclohexane $\Delta P (= P_0 - P_0^{\circ})$ and the mole fraction of polystyrene n_1, observed by Krigbaum and Geymer[14] for the polystyrene–cyclohexane system at 24.95 °C. Here, the volume fraction ϕ_1 in the original literature has been converted to n_1 by use of Fujiwara's data[15] on the specific volume at 35 °C of the same system. Even in the range $n_1 \sim 0.05$ a significant deviation from ideality of ΔP is observed. In order to explain this 'abnormality' of vapor pressure and osmotic pressure of polymer solutions, the theory, known now as the Flory–Huggins theory, was proposed independently, based on Meyer's lattice model,[16] by Huggins[17] and Flory.[18]

4.3.1   Flory–Huggins Theory

Consider, first, that N_0 solvent moles and N_1 flexible linear polymer solute moles, each consisting of $r (\equiv V_1^{\circ}/V_0^{\circ}; V_1^{\circ}$, the molar volume of the polymer) segments. Assuming random mixing of the polymer and the solvent molecules, the entropy change of mixing ΔS is expressed by

$$\Delta S = -R[N_0 \ln \phi_0 + N_1 \ln \phi_1] \tag{38}$$

$$\text{with } \phi_0 = N_0/(N_0 + rN_1) \tag{39a}$$

$$\phi_1 = rN_1/(N_0 + rN_1) \tag{39b}$$

where ϕ_0 and ϕ_1 are the volume fractions of the solvent and the solute ($=$ polymer), respectively. Equation (38) reduces straightforwardly, if $r = 1$, to equation (40), which is directly derived from equation (5) for an ideal solution.

$$\Delta S^{id} = -R[N_0 \ln n_0 + N_1 \ln n_1] \tag{40}$$

The assumption of random mixing employed in the Flory–Huggins theory is reasonable over the concentration range in which the segment density can be regarded as uniform.

Departure of the linear flexible polymer solution from ideality can be mostly attributed to the unexpectedly large value of ΔS, which is due to the large asymmetry and the large size of the solute compared to the solvent. For a solution of a linear flexible polymer with a degree of polymerization r of 1×10^4 dissolved at a concentration of 0.01 g cm^{-3} in a solvent with a molecular weight M of 30, $\Delta S/R$, calculated using equation (38), is *ca.* 1000 times larger than that of an ideal solution (equation 40) and *ca.* six times smaller than that of a solution in which separated segments are randomly mixed with the solvent, as given by

$$\Delta S = -R[N_0 \ln \phi_0 + rN_1 \ln \phi_1]. \tag{41}$$

Accordingly, the assumption that a real polymer solution is an ideal solution and application of Raoult's law (equation 10) to vapor pressure data leads to extraordinarily small molecular weight values.

From equation (38), the entropy change of mixing for one solvent mole, ΔS_0, is readily obtained:

$$\Delta S_0 (\equiv \partial \Delta S/\partial N_0) = -R\{\ln(1-\phi_1) + (1-1/r)\phi_1\}. \tag{42}$$

The enthalpy change of dilution, ΔH, can be computed to the first approximation, from the total number of first neighbor solvent–segment contact pairs and the change in energy for the formation of a contact pair. Flory approximated the probability that a given site adjacent to a segment is already occupied by a segment for the case of all segments uniformly distributed over the total lattice, leading for the partial molar heat of dilution ΔH_0 (see equation 4) to a van Laar–Scatchard type expression:[19, 20]

$$\Delta H_0 = RT\kappa\phi_1^2 \tag{43}$$

where κ is a parameter independent of T and ϕ_1 (*i.e.* Flory enthalpy parameter).

In a nonathermal solution (*i.e.* $\Delta H_0 \neq 0$), the entropy of mixing will deviate from that derived by assuming random mixing. Then, equation (42) is immediately modified into

$$\Delta S_0 = -R\{\ln(1-\phi_1) + (1-1/r)\phi_1 + (1/2-\psi)\phi_1^2\} \tag{44}$$

where ψ is a Flory entropy parameter and $(1/2 - \psi)$ denotes the extent of deviation from randomness in mixing the solute and the solvent.

The partial molar free energy of mixing $\Delta\mu_0$ is given by

$$\Delta\mu_0 = \Delta H_0 - T\Delta S_0 \tag{45}$$

Substitution of equations (43) and (44) into (45) leads to

$$\Delta\mu_0 = RT\{\ln(1-\phi_1) + (1-1/r)\phi_1 + \chi\phi_1^2\} \tag{46}$$

with

$$\chi = \kappa + 1/2 - \psi \tag{47}$$

χ is a purely phenomenological dimensionless polymer–solvent interaction parameter.

χ can also be divided into entropic and enthalpic parts as follows:

$$\chi = \chi_s + \chi_h \tag{48}$$

$$\text{Here } \chi_h = \kappa \tag{49a}$$

$$\text{and } \chi_s = 1/2 - \psi \tag{49b}$$

Combining equation (10)' with equation (46) we obtain

$$\log P_0/P_0^\circ = \{ln(1-\phi_1)+(1-1/r)\phi_1+\chi\phi_1^2\} \tag{50}$$

Similarly from equation (13)' and equation (45) one obtains

$$\pi = (RT/V_0^\circ)\{ln(1-\phi_1)+(1-1/r)\phi_1+\chi\phi_1^2\} \tag{51}$$

For multicomponent polymers, all chemical homologs, dissolved in solvent, r in equations (46), (50) and (51) should be replaced with its number average, $\bar{r}_n$. For example,

$$\Delta\mu_0 = RT\{ln(1-\phi_1)+(1-1/\bar{r}_n)\phi_1+\chi\phi_1^2\} \tag{46'}$$

In the above theory χ, κ and ψ are assumed to be constants, independent of ϕ_1 and r. Note that $\bar{r}_n$ or the number-average molecular weight $\bar{M}_n$ can be determined by measuring $\Delta\mu_0$ for polydisperse polymer–solvent systems.

There are various methods for evaluating $\Delta\mu_0$ as demonstrated in Figure 6. And from equation (46) the χ parameter can be calculated.

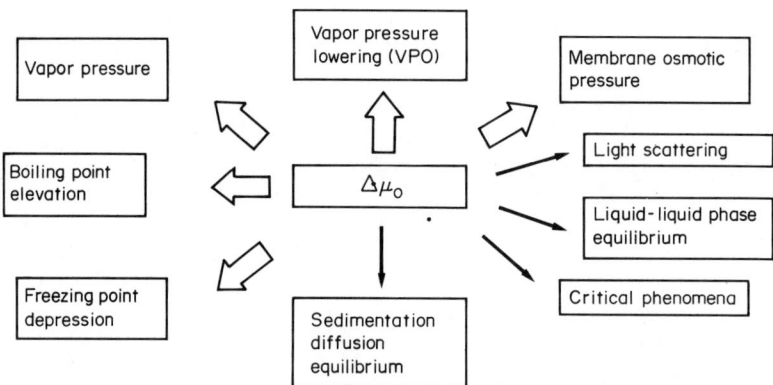

Figure 6   Various thermodynamic properties closely correlated with the chemical potential of the solvent $\Delta\mu_0$ in a polymer solution. Unfilled arrows denote colligative properties; filled arrows denote other properties.

The experimental results indicate that χ is both concentration and molecular weight dependent.[21] For example, χ can be phenomenologically expressed in a power series of the concentration ϕ_1,

$$\chi = \chi_0\left(1+\sum_j p_j\phi_1^j\right) \tag{52}$$

$$\text{with } \chi_0 = \chi_{00}\{1+(k'/\bar{r}_n)\} \tag{53a}$$

$$k' = k_0(1-\theta/T) \tag{53b}$$

Here, χ_0 is a parameter independent of concentration, p_j are the 'concentration-dependence' parameters and θ is the Flory temperature. Equation (53) was first proposed by Kamide *et al.*[21,22] analyzing data on phase separation (by them)[22] and light scattering (by Scholte)[23] for atactic polystyrene in cyclohexane and in methylcyclohexane (Figure 7.)[22] The sign of k' changes in the vicinity of temperature θ.[22]

Finally, the most generalized expressions for $\Delta\mu_0$ and the chemical potential of the polymer with the degree of polymerization r_i, $\Delta\mu_i$, can be written as[21]

$$\Delta\mu_0 = RT\left\{ln(1-\phi_1)+(1-1/\bar{r}_n)\phi_1+\chi_{00}(1+k'/\bar{r}_n)\left(1+\sum_j p_j\phi_1^j\right)\phi_1^2\right\} \tag{54}$$

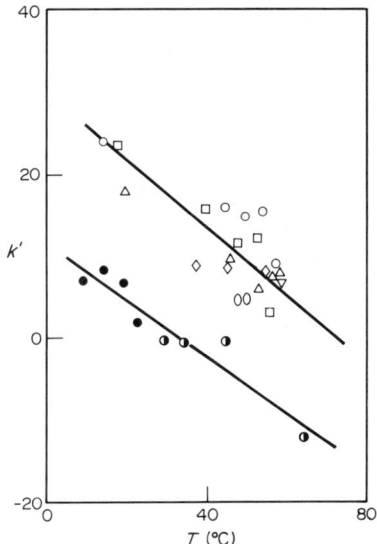

Figure 7 Change in k' parameter with temperature T. Polystyrene (PS)–cyclohexane system;[22] (●) Kamide–Miyazaki by liquid–liquid phase separation;[21] (◑) Scholte by light scattering.[23] PS–methylcyclohexane system (by liquid–liquid phase separation)[22]; volume fraction of PS, (□) $\phi_1 = 5 \times 10^{-3}$; (◇) $\phi_1 = 4.7 \times 10^{-3}$; (○) $\phi_1 = 8.6 \times 10^{-3}$; (0) $\phi_1 = 9.4 \times 10^{-3}$; (△) $\phi_1 = 2.0 \times 10^{-2}$; (▽) $\phi_1 = 1.86 \times 10^{-2}$

$$\Delta\mu_{ri} = RT\left\{\ln\phi_{ri} - (r_i - 1) + r_i(1 - 1/\bar{r}_n)\phi_1 + r_i(1 - \phi_1)^2\chi_{00}\left\{(1 + k'/\bar{r}_n)\times\right.\right.$$

$$\left.\left[1 + \sum_j [p_j/(j+1)]\left(\sum_q (q+1)\phi_1^q\right)\right]\right\} + k'(1/r_i - 1/\bar{r}_n)\left[1/(1-\phi_1) + \sum_j p_j\left(\sum_q \phi_1^q/(1-\phi_1)\right)/(j+1)\right]\right\} \quad (55)$$

where ϕ_{r_i} is the partial molar volume of i component of the polymer and $\phi_1 \equiv \sum_i \phi_{r_i}$.

Equation (55) is derived, using the Gibbs–Duhem relation, from equation (54). If all the parameters in equations (54) and (55) are determined accurately by some adequate method, $\Delta\mu_0$ and $\Delta\mu_{r_i}$, calculated using these parameters, can represent satisfactorily the thermodynamical properties of the polymer solutions over the entire range of T, ϕ_1 and r (or $\bar{r}_n$).[24]

Substitution of equations (52) and (53) into equation (51) and rearrangement lead to

$$\Delta\mu_0 = -RTc\left\{1/\bar{M}_n + (\bar{v}^2/V_0^\circ)(1/2 - \chi_0)c + \sum_j (\bar{v}^{j+2}/V_0^\circ)[1/(j+2) - \chi_0 p_j]c^{j+1}\right\} \quad (56)$$

Here, $\bar{v}$ is the specific volume of the polymer. M_n can be determined from

$$\lim_{c \to 0} (-\Delta\mu_0/RTc).$$

Combination of equation (56) with equation (34) yields

$$A_2 = (\bar{v}^2/V_0^\circ)(1/2 - \chi_0) = (\bar{v}^2/V_0^\circ)\psi_0(1 - \theta/T) \quad (57)$$

$$A_{j+2,0} = (\bar{v}^2/V_0^\circ)(1/(j+1) - \psi_j p_j) \quad (j = 1, 2, \ldots) \quad (58)$$

where ψ_0 and ψ_j are respectively Flory's entropy parameter at infinite dilution and the coefficient of jth power of ϕ_1 in the power series for ψ, see equation (66). $A_{2,0}$ becomes zero at a specific temperature (θ), irrespective of $\bar{M}_n$, since χ_0 is molecular weight independent at θ.

In the case when $A_{2,0} = A_{3,0} = \ldots = 0$, equation (56) reduces straightforwardly to a van't Hoff type equation:

$$\pi = RTc/\bar{M}_n \quad (59)$$

Under the condition of $A_{2,0} = A_{3,0} = \ldots = 0$ or, if not so, in the very low concentration range, the polymer solution behaves like an ideal solution and the temperature at which $A_{2,0}$ is zero is called the Flory θ temperature, and the solvent in which $A_2 = 0$ is realized is called the Flory θ solvent.

The value of $A_{2,0}$ for rigid spheres with radius D dissolved in a solvent can be calculated from the following relation, which is derived by replacing the molar volume in equation (17a) with the molar excluded volume:

$$A_{2,0} = u/(2M^2) \tag{60}$$

For rigid spheres $u = 32\pi D^3 N_A/3$ (N_A; Avogadro's number), and equation (60) can be rewritten as

$$A_{2,0} \text{ (rigid sphere)} = 16\pi N_A D^3/3 M^2 = 4\bar{v}/M \tag{61}$$

where $\bar{v}$ is the specific volume of the spheres. A more rigorous calculation[25] yields the same equation.

Table 1 exemplifies $A_{2,0}$ values for various polymer ($M = 1 \times 10^5$) solutions. Inspection of the Table leads to the conclusion that for a rigid spherical polymer dissolved in a solvent the second and higher terms in parentheses of the right hand side of equation (56) can be satisfactorily ignored as compared with the first term ($1/M$) if M is not larger than 1×10^5–1×10^6. In actual experiments, it was observed that, for aqueous solutions of glycogen and globular proteins including egg albumin and plasma albumin, equation (59) holds over a relatively wide concentraton range.[26] The larger $A_{2,0}$ is a characteristic feature of flexible chain molecules in good solvents.

Table 1 Calculated Values of the Osmotic Second Virial Coefficient $A_{2,0}$ for Various Polymer Solutions ($M = 1 \times 10^5$)

Solution	$A_{2,0}$ (mol cm^3 g^{-2})	Ref.
Ideal solution ($V_0^\circ = 100$ cm^3 mol^{-1})	5×10^{-9}	equation 17
Rigid sphere solution ($\bar{v} = 1$ cm^3 g^{-1})	5×10^{-5}	equation 61
Flexible chain molecule in a good solvent ($\psi = 0.3$, $\theta = 279$ K, $T = 350$ K, $V_0^\circ = 100$ cm^3 mol^{-1}, $\bar{v} = 1$ cm^3 g^{-1})	6.9×10^{-4}	equation 57
Flexible chain molecule in a theta solvent	0	equation 57

4.3.2 Experimental Determination of Parameters

The concentration-dependent coefficients (p_j) have been, heretofore, experimentally determined by various methods including osmometry, isothermal distillation, vapor pressure, critical miscibility, ultracentrifuge and phase separation. Recently, it was demonstrated that p_1 and p_2 can be evaluated concurrently by analyzing either (1) a combination of the phase separation and the cloud point curve,[27] or (2) critical solution temperature and concentration.[28,29] These two methods are based on the fact that the phase separation and the critical phenomena are most greatly influenced by the molecular weight and concentration dependences of the χ parameter. Note that other methods are limited experimentally to a relatively low concentration range and do not enable us to evaluate p_2 accurately. This is the main reason why, in Flory's milestone text book,[30] the values of the χ parameter obtained using vapor pressure and (in part) isothermal distillation equilibrium for poly(dimethylsiloxane), polystyrene and natural rubber were consistent with $p_2 = p_3 = \ldots = 0$. The phase separation method is applicable up to a moderately concentrated solution range, but the experimental accuracy is unfortunately not high enough to estimate p_2 satisfactorily.

Even at the present time, more than 45 years after the first proposal of the Flory–Huggins theory, the atactic polystyrene–cyclohexane system is almost the only system for which all the thermodynamic parameters necessary for describing $\Delta\mu_0$, such as p_1, p_2, θ, ψ_0 and k_0, are determined comprehensively. Figure 8 displays a graphical representation of the variation of the χ parameter with the polymer volume fraction ϕ_1 evaluated for the polystyrene–cyclohexane system by many investigators.[14,27,31–33] Similar results have been obtained[34] for the polystyrene–methylcyclohexane system. In the figure, curves have been calculated from the p_1 and p_2 values evaluated from the critical points. The experimental data points can be reasonably represented by equations (52) and (53) in which the terms higher than ϕ_1^2 are neglected. That is, in the ϕ_1 range of 0–0.65, both p_2 and p_1 are necessary to represent the concentration dependence of χ and, in the comparatively dilute range, there is no sharp distinction in χ between the investigations.

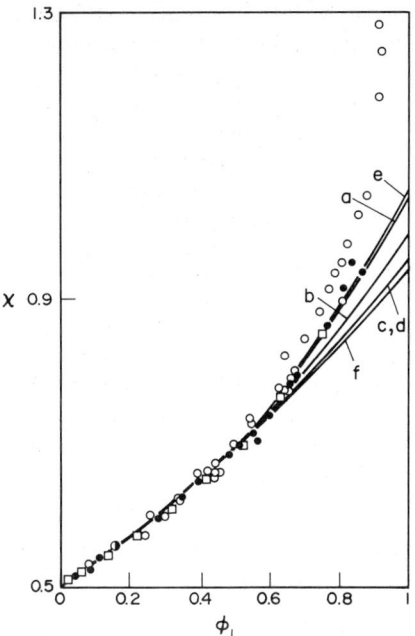

Figure 8 Concentration dependence of the χ parameter for the polystyrene–cyclohexane system:[27] ($\bigcirc$) osmotic pressure or isothermal distillation by Krigbaum and Geymer;[14] ($\bullet$) vapor pressure by Krigbaum and Geymer;[14] ($\square$) ultracentrifuge by Scholte.[23] Lines (a) to (f) are calculated using p_1 and p_2, obtained by experiment; (a) Krigbaum and Geymer;[14] (b) Scholte;[23] (c) Koningsveld et al.;[32] (d) Koningsveld et al.;[44] (e) Kuwahara et al.;[33] (f) Kamide et al.[27]

It is concluded that equation (62), which is a simplified form of equation (52), is an adequate expression of χ from the experimental point of view.

$$\chi = (a' + b'/T + c'/T^2)(1 + p_1\phi_1 + p_2\phi_1^2) \tag{62}$$

Note that in the above methods, p_1 and p_2 are assumed to be temperature independent. In fact this assumption seems to be experimentally acceptable: over a wide temperature range covering both the upper critical solution point (UCSP) and the lower critical solution point (LCSP) of the polystyrene–cyclohexane and polystyrene–methylcyclohexane systems, p_1 was found to be approximately constant 0.6 ± 0.04.[28]

It is interesting to note that if we assume that

$$A_2 = A_3 = A_4 = \ldots = 0 \ at \ T = \theta \tag{63}$$

at the UCSP, the following relations are theoretically expected to hold for χ_0 and the coefficients p_j ($j = 1, 2, \ldots, n$),

$$\chi_0 = 1/2 \tag{64a}$$

$$p_1 = 2/3$$

$$\vdots \tag{64b}$$

$$p_n = 2/(n+2)$$

In the LCSP region, p_1 is not influenced by the free volume in contact with χ_0[35,36] and $\chi_0 p_1 = 1/3$ is also satisfied.

The molecular weight dependence parameter k' can be determined most accurately from the molecular weight dependence of the partition coefficient $\sigma (\equiv (1/r) \log(\phi_{r(2)}/\phi_{r(1)})$, where ϕ_r is the volume fraction of r-mer and the suffixes 1 and 2 refer to the polymer-lean, and polymer-rich phases respectively, both in phase equilibrium). An alternative method is the direct determination of χ or A_2 by light scattering or membrane osmometry, for which, however, we need χ values accurate to four significant digits in order to evaluate k' with two significant figures. Such precision is undoubtedly beyond the accuracy of actual experiments at present.

The molecular weight dependence of the χ parameter also plays an important role in the threshold cloud point curve (CPC) of a polymer solution. Figure 9 shows the CPC of an atactic polystyrene sample in cyclohexane.[26] Full and broken lines are theoretical curves. In the former, only the concentration dependence parameters p_1 and p_2 are considered (*i.e.* $k_0 = 0$ is assumed) and in the latter, in addition to p_1 and p_2, the molecular weight dependence parameter k_0 is taken into account. It can be seen that the experimental cloud point data (open circles) of a polydisperse polymer in a single solvent can be reasonably represented by the theory, provided that the polydispersity of the polymer and the concentration and the molecular weight dependences of the χ parameter are taken into account.

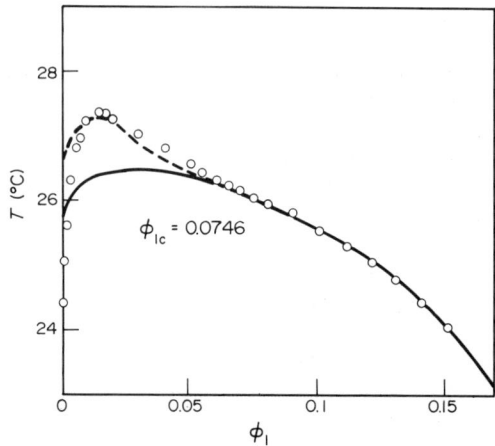

Figure 9 Cloud point curve of polystyrene ($\bar{r}_w = 2117$, $\bar{r}_w/\bar{r}_n = 2.8$) in cyclohexane: ($\bigcirc$) experimental data; (full line) theoretical curve calculated using $p_1 = 0.643$, $p_2 = 0.200$, $k_0 = 0$; (broken line) theoretical curve calculated using $p_1 = 0.643$, $p_2 = 0.200$, $\theta = 34\,°C$ and $k_0 = 108$[27]

Since the χ parameter is a strongly increasing function of the concentration, the κ and ψ parameters are also expected to be concentration dependent, that is

$$\kappa = \kappa_0 + \kappa_1 \phi_1 + \kappa_2 \phi_1^2 + \ldots \tag{65}$$

$$\text{and } \psi = \psi_0 + \psi_1 \phi_1 + \psi_2 \phi_1^2 \ldots \tag{66}$$

That this is so, at least for κ, was calorimetrically confirmed by Fujishiro and his students.[37] The Flory temperature θ and the entropy parameter ψ_0 in equation (66) were experimentally determined from the temperature dependence of A_2, (θ as the temperature at which A_2 becomes zero and ψ_0 from the temperature dependence of A_2 at θ) through use of the relation,

$$\psi_0 = (V_0^\circ/\bar{v}^2)\theta(\partial A_2/\partial T)_\theta \tag{67}$$

which is derived by differentiating equation (57) with respect to T and from the critical point (the temperature T_c and the volume fraction ϕ_{1c}), extrapolating by the Shultz–Flory method[38] to obtain values for a polymer of infinite molecular weight. For all polymer–solvent systems investigated, it was confirmed that the θ temperatures evaluated by the above two procedures agree fairly well with each other, but that ψ_0 values estimated by the second virial method are significantly smaller than those obtained by the critical point method. Stockmayer[39] pointed out the inadequacy of the basic equations (see equations 78' and 80') of the Shultz–Flory theory and proposed an alternative equation (see equations 78 and 80), rigorously derived for polydisperse polymer solutions. Another factor is the concentration dependence of the χ parameter, which was completely neglected in the Shultz–Flory and Stockmayer methods. To solve this problem Koningsveld *et al.*[32] established a method for estimating θ and ψ_0 from data on T_c and ϕ_{1c} for series of solutions differing in average molecular weight. They employed a pair interaction parameter g, expanded as a series function of ϕ_1

in the form*,

$$g \equiv 1/(1-\phi_1) \int_{\phi_1}^{1} \chi \, d\phi_1 = \sum_i g_i \phi_1^i \qquad (68)$$

Here, they assumed that g_0 could be divided into temperature-independent and temperature-dependent components given by

$$g_0 = g_{00} + g_{01}/T \qquad (69)$$

The parameters ψ_0 and θ were calculated from

$$\psi_0 = 1/2 - g_{00} - g_1 \qquad (70)$$

$$\theta = g_{01}/\psi_0 \qquad (71)$$

Note that the concentration and temperature dependences of the g parameter (i.e. g_{00}, g_{01}, g_1 and g_2) were evaluated so that the deviation of the experimental ψ_{0c} and T_c data for all samples from the theory was minimised.

Later, Kamide and Matsuda[28] attempted to clarify whether and to what extent ψ_0 and θ are influenced by the concentration dependence of the χ parameter and by the polymolecularity of the samples and to propose a method for estimating ψ_0 and θ from T_c alone, taking into account the above two factors determined independently by other methods, without assuming a specific temperature dependence of the χ-parameter. Their method is described briefly below.

At the critical point, the following equations can be derived,[27]

$$1/(\bar{r}_w \phi_{1c}) + 1/(1-\phi_{1c}) - \chi_{0c}\{2 + \sum p_j(j+2)\phi_{1c}^j\} = 0 \qquad (72)$$

and

$$1/(1-\phi_{1c}) - \bar{r}_z/(\bar{r}_w \phi_{1c})^2 - \chi_{0c} \sum p_j j(j+2)\psi_c^{(j-1)} = 0 \qquad (73)$$

where χ_{0c} is the critical value of χ_0, $\bar{r}_w$ is the weight-average value of r, and $\bar{r}_z$ is the z-average of r.

Both χ_{0c} and ϕ_{1c} can be obtained simultaneously through application of equations (72) and (73), using a numerical method, to obtain values of $\bar{r}_w$, $\bar{r}_z$ and p_j ($j = 1, 2$). χ_{0c} is related to T_c, ψ_0, θ through the relation[41]

$$1/T_c = \chi_{0c}/(\theta\psi_0) + (1/\theta)[1 - 1/(2\psi_0)] \qquad (74)$$

with

$$\psi_0 \equiv 1/2 - \chi_{0,s} \qquad (75)$$

$$\theta \equiv \chi_{0,h}/(1/2 - \chi_{0,s}) \qquad (76)$$

$\chi_{0,s}$ and $\chi_{0,h}$ are the entropy and enthalpy components of χ_0 respectively (i.e. $\chi_0 = \chi_{0,s} + \chi_{0,h}$). Using χ_{0c}, calculated from equations (72) and (73), and experimental T_c, we can determine θ and ψ_0 from the plot of $1/T_c$ against χ_{0c}. This method is hereafter simply referred to as the Kamide–Matsuda method.[28]

Putting $p_j = 0$ (*for* $j = 1, 2, \ldots$) in equation (72), we obtain

$$\chi_{0c} = (1/2)\{1/\bar{r}_w \phi_{1c}) + 1/(1-\phi_{1c})\} \qquad (77)$$

Substitution of the Stockmayer equation, which was derived rigorously for a multicomponent-polymer–single-solvent system assuming $p_j = 0$ ($j = 1, 2, \ldots$) *i.e.*

$$\phi_{1c} = 1/\{1 + (\bar{r}_w/\bar{r}_z^{1/2}/\bar{r}_z^{1/2})\} \qquad (78)$$

into equation (77) yields

$$\chi_{0c} = (1/2)\{\bar{r}_z^{1/2}/\bar{r}_w + 1/\bar{r}_w\}\{\bar{r}_w/\bar{r}_z^{1/2} + 1\} \qquad (79)$$

---

* χ, χ_0, p_i and p_n are directly related to g, g_i, and g_n through the relations[29]

$$\chi = g - (1-\phi_1)(\partial g/\partial \psi) \quad \text{(a)} \qquad\qquad \chi_0 = g_0 - g_1 \quad \text{(b)}$$

$$p_i = (i+1)(g_i - g_{i+1})/(g_0 - g_1) \quad \text{(c)} \qquad p_n = (n+1)g_n/(g_0 - g_1) \quad \text{(d)}$$

Combining equation (74) with equation (79) we obtain

$$1/T_c = [(1/(2\theta\psi_0)]\{[(1/\bar{r}_w^{1/2}) + (\bar{r}_z/\bar{r}_w)^{1/2}][(1/\bar{r}_w)^{1/2} + (\bar{r}_w/\bar{r}_z)^{1/2}]\} + (1/\theta)[1 - 1/(2\psi_0)]$$ (80)

When it is assumed that $\bar{r}_w = \bar{r}_z$ equations (78) and (79) reduce to the well-known equations derived by Shultz and Flory;[38] *i.e.*

$$\phi_{1c} = 1/\{1 + \bar{r}_w^{1/2}\}$$ (78)′

and

$$1/T_c = [1/(2\theta\psi_0)](1 + 1/\bar{r}_w^{1/2})^2 + (1/\theta)[1 - 1/(2\psi_0)] = [1/(\theta\psi_0)]\{(1/\bar{r}_w)^{1/2} + 1/(2\bar{r}_w)\} + 1/\theta$$ (80)′

From the intercept and slope of the plot of $1/T_c$ vs. $(1/\bar{r}_w^{1/2} + 1/(2\bar{r}_w))$ (Shultz–Flory plot)[38] we can estimate θ and ψ_0. Note that equation (80)′ is strictly valid only for a monodisperse-polymer–single-solvent system, in which the χ parameter is assumed to be independent of polymer molecular weight and concentration.

The Shultz–Flory and Stockmayer theories significantly underestimate $\phi_{1c}[\phi_{1c}(\exp) > \phi_{1c}$ (theo)], but consideration of the parameter p_1 in Kamide–Matsuda's method greatly reduces the disagreement. However, use of the single parameter p_1 is not sufficient to cover a wide range of ϕ_{1c}. As described before, three procedures are available to determine p_1 and p_2 concurrently from the critical solution point (ϕ_{1c} and T_c) (Koningsveld *et al.*[32] and Kamide and Matsuda[28]) or the cloud point curve (Kamide *et al.*).[27] In the UCSP region for a polystyrene–cyclohexane system, Kamide *et al.* determined $p_1 = 0.6$ and $p_2 = 0$ from a phase equilibrium experiment[42] and $p_1 = 0.643$ and $p_2 = 0.200$ from a cloud point experiment[27] (see Figure 9). The same experimental data analyzed by the method of Koningsveld *et al.* and Kamide–Matsuda yield almost the same results for p_1 and p_2: Koningsveld *et al.*[32,44] obtained $p_1 = 0.623$ and $p_2 = 0.290$ by analyzing their critical points data, which were estimated by the phase volume ratio method,[32] and Kamide *et al.*[29] obtained $p_1 = 0.631$ and $p_2 = 0.221$ for the same data by using their own method. Kamide and Matsuda[28] applied Koningsveld's method to literature values of critical points, determined by the threshold cloud point,[40] diameter[33] and phase volume ratio[32] methods, and obtained $p_1 = 0.623$ and $p_2 = 0.308$, which can be compared with values of $p_1 = 0.642$ and $p_2 = 0.190$ estimated by Kamide and Matsuda's method from the same literature data.[32,33,40] Work by Kamide *et al.*[27] on the cloud point curve in the UCSP region for the polystyrene–cyclohexane system yielded results for p_1 and p_2 thoroughly consistent with those by Kamide and Matsuda:[28] $p_1 = 0.64$ and $p_2 = 0.19$. Table 2 collects values of θ and ψ_0 at the UCSP for the polystyrene–cyclohexane system, and includes the p_1 and p_2 values, which are used for calculating θ and ψ_0.

For a given polymer-solvent system the θ values obtained by the Shultz–Flory, Stockmayer and Kamide–Matsuda plots are roughly constant, regardless of the analytical method employed. Detailed investigation shows that at the UCSPs of the polystyrene–cyclohexane and polystyrene–methylcyclohexane systems, where both p_1 and p_2 are positive, the Shultz–Flory and Stockmayer methods overestimate θ by a few degrees; but at the LCSPs of these systems, where $p_1 > 0$ and $p_2 < 0$, these methods underestimate θ slightly. For all the systems, the Koningsveld *et al.* procedure and the Kamide–Matsuda method give almost the same value of θ, within ± 0.1–0.3 K.

Up to now, the Shultz–Flory method has been used mostly for estimating θ, as compiled in the Polymer Handbook, from the critical point data, and θ temperatures estimated thus may have an experimental uncertainty of a few degrees, even if monodisperse polymer samples are used.

As long as the polymolecularity of the polymer samples is small, $(\bar{r}_w/\bar{r}_n < 1.1)$ as in the case of the vast majority of samples used for analysis of the polystyrene–cyclohexane and polystyrene–methylcyclohexane systems, and/or the critical solution points are determined directly as points at which the phase volume ratio is unity, there is no significant difference in θ value, as the theory predicts, between Shultz–Flory and Stockmayer method.

For polystyrene solutions, the absolute magnitude of ψ_0 values estimated by the methods of Koningsveld *et al.* and Kamide and Matsuda are only half to a third of those obtained by the Shultz–Flory and Stockmayer methods. The latter two methods give the same ψ_0 (0.78 ± 0.02). ψ_0 estimated at the UCSP of the polystyrene–cyclohexane system by the second virial coefficient method varies from 0.18–0.36 and averages 0.264, which agrees remarkably well with $\psi_0 = 0.262$ obtained by the Kamide–Matsuda method. For the UCSP of polystyrene–methylcyclohexane, ψ_0 was independently determined from the temperature dependence of the second virial coefficient by Schulz and Baumann[43] as 0.30 and by Kotera *et al.*[47] as 0.16. Their average value (0.23) is near to the value of 0.25 estimated by use of the Kamide–Matsuda method. Hence, as far as a nonpolar polymer

Table 2 Flory θ Temperature and Entropy Parameter ψ_0 at Infinite Dilution at the Upper Critical Solution Point of the Polystyrene–Cyclohexane System

Method	Concentration dependence of χ		θ (K)	ψ_0	
	p_1	p_2			
Critical point					
Shultz–Flory	0	0	306.2 (306.2,[a] 307.0,[b] 307.2,[c] 306.4[d])	0.75 (0.78,[a] 0.79,[b] 1.056[c])	
Stockmayer	0	0	306.5	0.80	
Koningsveld *et al.*					
Critical point (tcp,[b] diameter,[e] pvr[f])	0.623	0.308	305.2	0.29	
Critical point (pvr[f])	0.623[f,g]	0.290[f,g]	305.6[f]	0.30[f]	
Kamide–Matsuda					
Phase separation	0.6[h]	0[h]	306.4	0.35	
Cloud point	0.643[i]	0.200[i]	306.6	0.27	
Critical point (pvr[f])	0.623[f,g]	0.290[f,g]	305.6	0.27	
Critical point (tcp,[b] diameter,[e] pvr[f])	0.642	0.190	305.1	0.27	av. 0.262
Critical point (diameter[e])	0.593	0.551	306.9	0.22	
Critical point (tcp[b])	0.645	0.165	305.1	0.27	
Critical point (pvr[f])	0.631	0.221	305.8	0.27	
			av. 305.9		
Second virial coefficient					
Membrane osmometry	—	—	307.6[j]	0.36[j]	
	—	—	307.6[a]	0.23[a]	
Light scattering	—	—	308.4[k]	0.19[k]	av. 0.264
	—	—	307.0[d]	0.36[l]	
	—	—	307.4[m]	0.18[n]	
			av. 307.6		

[a] Ref. 14. [b] tcp, threshold cloud point; Ref. 40. [c] Ref. 14. [d] Ref. 14. [e] Ref. 38. [d] Ref. 14. [f] pvr, phase volume ratio; Ref. 32. [g] Ref. 44. [h] Ref. 42. [i] Ref. 27. [j] Ref. 45. [k] Ref. 46. [l] Calculated using $\theta(\partial A_2/\partial T)_\theta$ data in Ref. 43. [m] Ref. 47.
[n] Calculated using $\theta(\partial A_2/\partial T)_\theta$ data in Ref. 47.

solution is concerned, ψ_0 values estimated so far by the Shultz–Flory method aré erroneously overestimated owing to the neglect of the concentration dependence of the χ parameter.

From the above, it is clear that the great difference between ψ_0 values obtained from critical points according to the Shultz–Flory procedure and the second virial coefficient is not due to the effect of the polydispersity of the polymer on the critical point, but almost entirely to neglect of the concentration dependence of the χ parameter (p_1 and p_2) in the Shultz–Flory method. Since the second virial coefficient depends only on χ_0, the concentration-independent term of the χ parameter through equation (57), the existence of the concentration dependence of χ has no effect on the results obtained using equation (63), and in this sense the second virial coefficient method is unconditionally preferable from a theoretical point of view.

By analyzing the critical point data for polystyrene in ten solvents and polyethylene in sixteen solvents by the Kamide–Matsuda and Koningsveld *et al.* methods, Kamide *et al.*[29] determined p_1, p_2, θ and ψ_0 for these systems. Except for a few solvents, p_1 values for PS solutions were near to 2/3, which is the value theoretically predicted when $A_2 = A_3 = 0$ at θ (see equation 64b).

Note that the concentration dependence of ψ cannot directly be evaluated, but a significant variation of ψ with ϕ_1 is expected from the fact that both χ and κ have a molecular weight dependence. ψ_0 for polystyrene and polyethylene solutions is negative in the LCSP region and positive in the UCSP region and an athermal solution will be realized, at least for these polymers, in any solvent at a specific temperature between LCSP and UCSP.

κ can be determined in various ways:
(1) From vapor pressures and osmotic pressures measured over a wide range of temperature using the relation

$$\kappa = -(T/R\phi_1^2)[\partial(\mu_0/T)/\partial T]_{p,\phi_1} = T(\partial\chi/\partial T)_{p,\phi_1} \tag{81}$$

(2) From critical phenomena *via* θ and ψ_0 using

$$\kappa_0 = \theta\psi_0/T \tag{82}$$

(3) From the temperature dependence of A_2 determined by membrane osmometry or light scattering in the vicinity of the θ temperature (equation 67) *via* θ and ψ_0
(4) From calorimetry (equation 43)

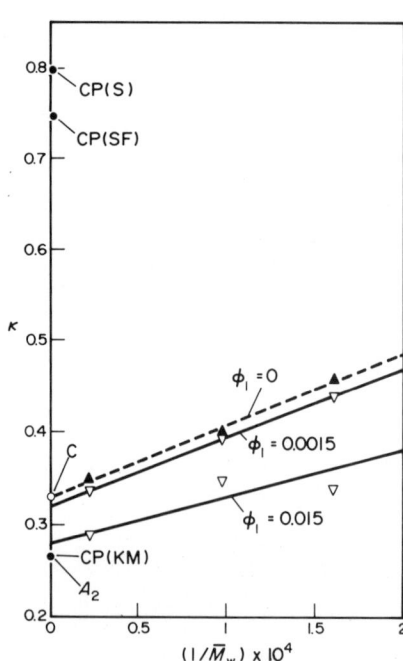

Figure 10 Molecular weight dependence of the enthalpy parameter κ of polystyrene–cyclohexane system at 35 °C. (▲) extrapolated values ($\phi_1 = 0$) from Fujiwara's calorimetrical data (∇);[15] ($\bigcirc$) κ doubly extrapolated from calorimetrical data (C) at $\phi_1 = 0$ to $\bar{M}_w = \infty$; ($\bullet$) value calculated by equation (82) from θ and ψ_0 data (Table 2) obtained from cloud point curve (CP) and by equation (67) from the temperature dependence of the osmotic second coefficient A_2. S, SF and KM in parentheses indicate the use of the Stockmayer, Shultz–Flory and Kamide–Matsuda methods

Values of κ, estimated by calorimetry, decrease with increase in ϕ_1 and with increase in molecular weight: Figure 10 illustrates this for the polystyrene–cyclohexane system at $35\,^\circ\text{C}$.[48] Here, the unfilled triangles are experimental points. The broken line represents values extrapolated to $\phi_1 = 0$ from the experimental data points[15] and denotes the molecular weight dependence of κ_0. The unfilled circle is the value of κ_0 at the limit of $M \to \infty$, obtained by calorimetry. The filled marks are κ_0 values predicted by putting into equation (80) θ and ψ_0 data (Table 2) estimated by applying Kamide–Matsuda and Schultz–Flory methods to literature data at the UCSP. Note that κ_0 calculated by equation (81) corresponds to $M = \infty$. Evidently, κ_0 estimated by the Kamide–Matsuda method coincides within relative error of $\pm 0.5\%$ with that by the second virial coefficient and agrees well within relative error of $\pm 10\%$ with that calorimetrically determined, but κ_0 by the Schultz–Flory method is more than twice the latter. A similar molecular weight dependence of κ_0 has been demonstrated by Kagemoto *et al.*[49,50] for many other polymer solutions.

4.3.3 Summary

From all this work it becomes clear that the following thermodynamic parameters are sufficient to characterize thoroughly a polymer solution: p_1, p_2 (both from critical solution points by the method of Koningsveld *et al.* or that of Kamide and Matsuda), k_0 (from phase separation by the method of Kamide *et al.*), θ and ψ_0 (both from critical solution points by the method of Koningsveld *et al.* or that of Kamide and Matsuda, or from the second virial coefficient), and κ_0 (from θ and ψ_0 or by calorimetry). Once these parameters have been determined, as demonstrated for the atactic polystryene–cyclohexane system, all the thermodynamic phenomena, shown in Figure 6, can be reasonably represented through $\Delta\mu_0$, over the entire range of the concentration. This is one of the final goals in the study of the thermodynamic properties of polymer solutions.

4.4 EXPERIMENTAL DETERMINATION OF NUMBER-AVERAGE MOLECULAR WEIGHT

From a theoretical point of view, all the colligative methods such as vapor pressure lowering, vapor pressure osmometry (VPO), membrane osmometry (MO), ebulliometry and cryoscopy can be used for the determination of $\bar{M}_n$ of macromolecular compounds, because the ratio $\Delta\mu_0/c$ at infinite dilution, *i.e.* $\lim\limits_{c \to 0} \Delta\mu_0/c$, is proportional to $(\bar{M}_n)^{-1}$ (equations 33c, 34c and 35b). The sensitivity of the electrical and/or mechanical detectors needed for measuring these colligative quantities with good accuracy governs the maximum measurable value of $\bar{M}_n$. Table 3 illustrates the maximum $\bar{M}_n$ values measurable by commercially available apparatus for each method. Ebulliometry and cryoscopy, at least with the apparatus commercially available, do not really meet the necessary requirements for determining $\bar{M}_n$ of macromolecules. Attempts to extend the upper limit of $\bar{M}_n$ by cryoscopic and ebullioscopic methods were energetically made, especially in the 1950s. For example, Glover[51] constructed an ebulliometer with a detectable limit of temperature difference of $2 \times 10^{-5}\,^\circ\text{C}$, successfully measuring $\bar{M}_n$ values as high as 1.7×10^5 for polystyrene in toluene. A full account of cryoscopy and ebulliometry is given in reference 52.

In the range $1.0 \times 10^4 \leqslant \bar{M}_n \leqslant 1 \times 10^5$, VPO and MO are the most popular methods for $\bar{M}_n$ determination.

Table 3 The Maximum Number-average Molecular Weight $\bar{M}_n$ Measurable with a High Accuracy Using Commercially Available Apparatus for Polymer Solutions

Method	*Molecular weight* $(\times 10^{-4})$
Membrane osmometry	~ 50
Vapor pressure osmometry	~ 40
Ebulliometry	~ 2
Cryoscopy	~ 2

4.4.1 Vapor Pressure Osmometry

Until the 1950s vapor pressure osmometers were constructed by investigators themselves.[52] Even at present it is not tremendously difficult to make VPO apparatus with higher performance characteristics than those commercially available. Most commercially available VPOs [Hitachi model 115, Mechrolab model 302 (now out of the production) and Knauer (KG Dr. Knauer, DDR)] have a detectable limit of the temperature difference between solution drop and solvent drop of *ca.* 1×10^{-4} °C. Accordingly the measurable upper limit of the molecular weight is of the order of magnitude 10^4. The Hitachi model 117 is based on the principle employed by Kamide *et al.*,[8] who constructed a modified version of the solvent–vapor chamber and thermistor assembly by Dohner *et al.*,[53] and installed them in a Hitachi (molecular weight measurement apparatus) model 115 to obtain a new prototype VPO apparatus with a better sensitivity. With the conventional Wheatstone bridge circuit modified by introducing a matching thermistor, they were able to detect a temperature difference of *ca.* 6×10^{-6} °C. Recently an intermediate version of the Hitachi models 117 and 115, the Hitachi model 114, has come onto the market.

Equation (24) indicates that a large ratio K_s/K_e (*i.e.* a high sensitivity) will be attained under the following conditions: (1) large drop size, (2) solvent with large ΔH, (3) solvent with high vapor pressure, (4) low temperature and (5) reduced total pressure.

Figure 11 shows the dependence of K_s/K_e on the solvent vapor pressure. It can be seen that K_s/K_e increases with increasing vapor pressure of the solvent $P_0(T)$. The classical equilibrium theory fails to explain this result. The full line in the figure is a theoretical curve calculated from equation (24) by assuming that $(k_1 A_1 + k_2 A_2)/(k_3 A_1) = 6 \times 10^7$ cal dyne^{-1} mol^{-1} cm^{-2} K^{-1} (2.5×10^{13} J N^{-1} mol^{-1} cm^{-2} K^{-1}), $\Delta H_0 = 1 \times 10^4$ cal mol^{-1} (4×10^4 J mol^{-1}) and $T = 313$ K. The closed triangles were calculated from equation (24) by assuming that $(k_1 A_1 + k_2 A_2)/(k_3 A_1) = 6 \times 10^7$, and using the experimental values of ΔH and T. The experimental values agree well with the theoretical curve, suggesting that the ratio $(k_1 A_1 + k_2 A_2)/(k_3 A_1)$ is approximately constant regardless of the molecular weight of the solute, the solvent nature (in this case, ΔH) and the temperature when the same VPO apparatus is employed. In this manner it can be demonstrated that the calibration parameter K_s cannot be regarded as an apparatus constant, but that $(k_1 A_1 + k_2 A_2)/(k_3 A_1)$ can be considered as a kind of apparatus constant. This quantity represents the degree of heat transfer from a solution drop to the surroundings, in other words, it is a parameter representing nonideality. Kamide and Sanada[6] have obtained $(k_1 A_1 + k_2 A_2)(k_3 A_1) = 3 \times 10^7$ cal dyne^{-1} mol^{-1} cm^{-2} K^{-1} (1.3×10^{13} J N^{-1} mol^{-1} cm^{-2} K^{-1}) for a Hitachi molecular weight measurement apparatus, model 115 (originally the Hitachi model 115 VPO apparatus). Note that k_3 in equation (24) depends on the mechanism of condensation of the solvent vapor onto the solution drop, but, in any case, k_3 is greatly influenced by the total pressure P_n. This can be understood by considering that the diffusion coefficient of a vapor molecule D_g is approximately proportional to $(P_n)^{-1}$. The effect of P_n on

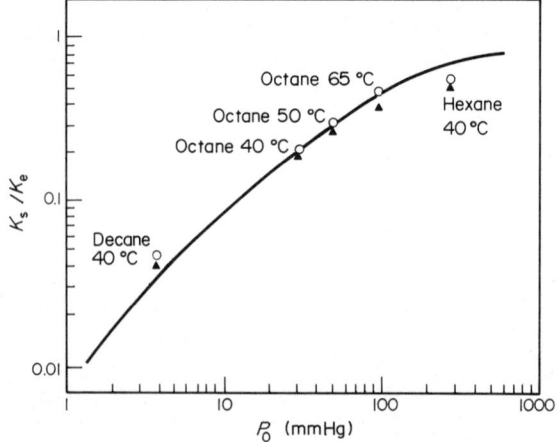

Figure 11   The dependence of the ratio K_s/K_e (equations 24 and 25) on the solvent vapor pressure P_0;[8] ($\bigcirc$), experimental data for unbranched alkanes and tristearin in various solvents; ($\blacktriangle$) calculated points from equation (24) by putting $(k_1 A_1 + k_2 A_2)/(k_3 A_1) = 6 \times 10^7$ cal dyne^{-1} mol^{-1} cm^{-2} K^{-1} (2.5×10^{13} J N^{-1} mol^{-1} cm^{-2} K^{-1}) together with the experimental values of the heat of condensation (ΔH) and temperature (T); (full line) theoretical curve obtained from equation (24) by putting $(k_1 A_1 + k_2 A_2)/(k_3 A_1) = 6 \times 10^7$ cal dyne^{-1} mol^{-1} cm^{-2} K^{-1}, $\Delta H = 10^4$ cal mol^{-1} (4×10^4 J mol^{-1}) and $T = 313$ K

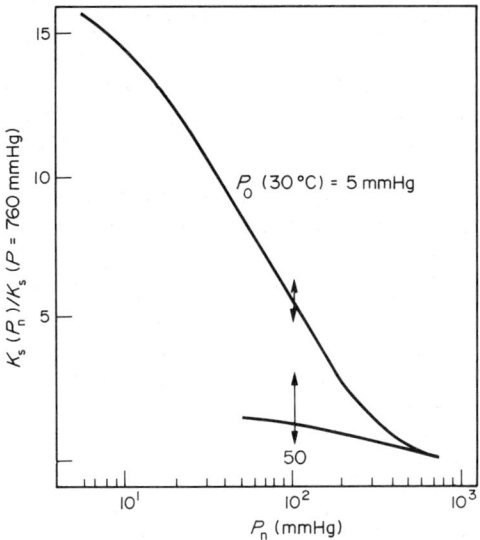

Figure 12 Theoretical relationships between the ratio $K_s(P_n)/K_s$ ($P_n = 760$ mmHg) and total pressure P_n, where K_s ($P_n = 760$ mmHg) is K_s obtained at atmospheric pressure.[6] ($\updownarrow$) experimental data[54] for which $P_n = 100$ mm Hg was assumed: see reference 54.

K_s/K_e becomes greater for a solvent with smaller $P_0(T_0)$. Figure 12 shows the ratio $K_s(P_n)/K_s(P_n = 760$ mmHg) plotted against P_n. Here the full lines are theoretical curves. Experimental data, obtained for $P_n = 1 \times 10^2$ (shown by arrows) and 7.6×10^2 (not shown) mmHg seem to be consistent with the theoretical prediction.

From equation (22) the time necessary to attain the steady state t_s can be calculated:

$$t_s = -(C_p V\rho/c_2) \ln\{c_2\delta T/[c_1 + c_2(T_0 - T)]\} \tag{83}$$

$$\text{with } c_1 = k_3 A_1(1 - a_0)P_0(T_0)\Delta H_0 \tag{84}$$

$$c_2 = k_1 A_1 + k_2 A_2 + \{k_3 A_1 P_0(T_0)\Delta H_0^2 a_0\}/(RT_0^2) \tag{85}$$

Equation (83) can be rewritten as

$$t_s = -(C_p V\rho/c_2) \ln[(1/K_s - 1/K_e)V_0\delta T/(1 - a_0)] \tag{86}$$

Here, δT is the minimum temperature difference detectable by a given apparatus (5.96×10^{-6} °C in Kamide *et al.*'s apparatus). If solvents with small $P_0(T_0)$ (~ 100 mmHg) are employed, t_s is theoretically expected from equation (83) to decrease with an increase in $P_0(T_0)$. This prediction was explicitly confirmed with the experimental data by Higuchi *et al.*,[54] Wilson *et al.*[55] and Kamide–Sanada.[6]

A typical VPO apparatus consists of: (1) two thermistors covered with glass, to which solution and solvent drops are attached; (2) a cell, which is saturated with solvent vapor; (3) a solvent vessel, placed at the lower part of the cell; and (4) an electric circuit, which is often thermostatted. Figure 13 is a schematic sectional view of the Hitachi Model 117.

The calibration parameter K_s is determined from the intercept of the graph of $(T - T_0)_s/c$ against c, for solutions of a standard low molecular weight solute with known M in the given solvent, by using the relation:

$$K_s = \left\{ \lim_{c \to 0} (T - T_0)_s/c \right\}/M \tag{87}$$

Table 4 lists values of K_s for solutions of octadecane, octacosane, hexatriacontane and tristearin in hexane, octane and decane at 40 and 60 °C. It is of interest to point out that, at least in the molecular weight range of 200 to 1000, the parameter K_s appears to be a function of the solvent nature and the temperature and shows no variation with the molecular weight of the solute. Equation (87) is

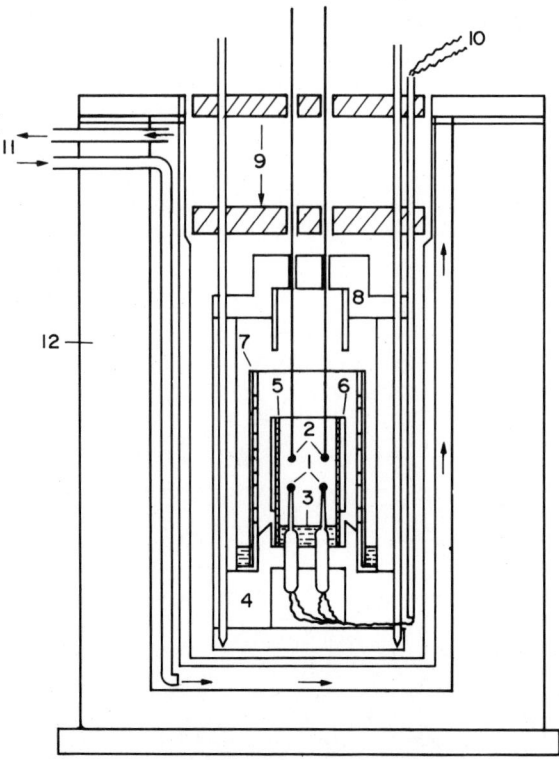

Figure 13   Cutaway view of vapor pressure osmometer Hitachi model 117. (1) thermistor; (2) pipe for dropping in solution and solvent; (3) reservoir for solvent; (4) gasket ring; (5) filter paper; (6) inside cap; (7) cap of filter paper; (8) outside cap; (9) pipe holder; (10) lead wire; (11) pipe for circulating water; (12) insulator.

Table 4   Limiting Value for the Ratio of Steady State Temperature Difference (ΔT_s) and Concentration (c), and the Calibration Parameter (K_s) in Vapor Pressure Osmometry[8] for Solutions of Octadecane, Octacosane, Hexatriacontane and Tristearin in Hexane, Octane and Decane at 40 to 65 °C[8]

Solute	Solvent	Temperature	$\lim\limits_{c \to 0} \Delta T/c$	$K_s \times 10^{-3}$
		(°C)	(K cm³ g⁻¹)	(K cm³ mol⁻¹)
Octadecane	hexane	40	7.80	1.98
	octane	40	2.80	0.71
		50	4.66	1.18
		65	7.93	2.01
Octacosane	decane	40	0.517	0.13
	hexane	40	4.95	1.95
	octane	40	1.93	0.76
Hexatriacontane	hexane	40	3.95	2.00
	octane	40	1.36	0.69
		50	2.22	1.12
		65	4.05	2.05
	decane	40	0.255	0.13
Tristearin	hexane	40	2.32	2.07
	octane	40	0.80	0.71
		50	1.28	1.14
		65	2.28	2.03
	decane	40	0.145	0.13

generalized for the case of a multicomponent polymer solution to yield

$$K_s = \left\{ \lim_{c \to 0} (T - T_0)_s / c \right\} / \bar{M}_n \qquad (87)$$

We can determine K_s for polymers with $\bar{M}_n$ evaluated by other methods in advance. Figure 14 shows plots of $\Delta T_s / c$ vs. c based on data obtained for solutions of some atactic polystyrenes in benzene at 40 °C. Circles and triangles denote duplicate sets of measurements, which gave $\lim_{c \to 0} \Delta T_s / c$ reproducible to $\pm 1\%$ or better. These data, obtained by using Kamide et al.'s apparatus, can be accurately represented by straight lines with positive slopes, drawn by the least squares method, and no deviations from the straight lines were observed even in the low concentration range. In Table 5 the data for K_s and $\lim_{c \to 0} (T - T_0)_s / c$ are collected for solutions of atactic polystyrenes in benzene at 40 °C.

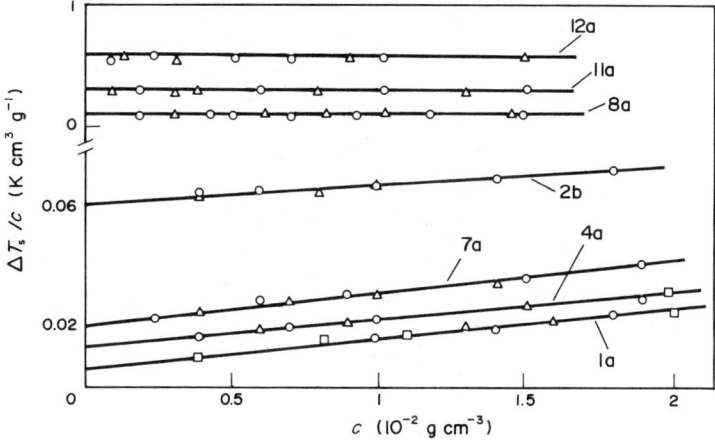

Figure 14 Plot of the ratio of steady state temperature difference (ΔT_s) and concentration (c) against c for solutions of polystyrene standards in benzene.[8] ($\bigcirc$) and ($\triangle$) indicate data obtained in duplicate runs by use of the vapor pressure osmometer Hitachi model 117 at 40 °C. Polymer codes are shown on the curves (see Table 5)

Table 5 Limiting Value for the Ratio of Steady State Temperature Difference (ΔT_s) and Concentration (c), the Calibration Parameter (K_s), and the Number-average Molecular Weight ($\bar{M}_n$), determined by Vapor Pressure Osmometry for Solutions of Atactic Polystyrene (PS) in Benzene at 40 °C

PS sample code	$\lim_{c \to 0} \Delta T_s / c$	$K_s \times 10^{-3}$	$\bar{M}_w \times 10^{-4\,a}$	$\bar{M}_n \times 10^{-4}$		
				as informed[b]	VPO[c]	VPO[d]
	(K cm^3 g^{-1})	(K cm^3 mol^{-1})				
12a	0.576	1.169	0.203	0.205	0.188	0.194
11a	0.310	1.448	0.480	0.440	0.348	0.361 (0.370)[e]
8a	0.110	1.165	1.050	1.030	0.982	1.018
2b	0.059	1.200	2.04	1.930	1.83	1.90
7a	0.0192	0.979	5.10	—	5.63	5.83
4a	0.0124	1.21	9.72	9.17	8.71	9.03
1a	0.0052	0.83	16.00	15.10	20.77	21.50 (19.60)[f]
3a	0.0030	1.23	41.10	38.80	36.00	37.30

[a] By supplier (originally cited by Kamide et al.[8]). [b] $\bar{M}_n$ as informed by the manufacturer. [c] $\bar{M}_n$ was determined by putting $K_s = 1.08 \times 10^3$, which was obtained for benzil. [d] $\bar{M}_n$ was determined by putting $K_s = 0.5$ and $K_e = 2.24 \times 10^3$. [e] Ref. 57. [f] Ref. 56.

The molecular weight dependence of K_s for polystyrene is demonstrated in Figure 15, which also contains data for some standard low molecular weight materials, shown as closed marks. The solid line in the figure is the line of $K_s = 0.5 K_e$. The approximate constancy of K_s over a wide range of $\bar{M}_n$ is verified, but in the strict sense K_s has a tendency to increase gradually as $\bar{M}_n$ of the solute increases. However, to a good approximation K_s is equal to $0.5 K_e$. Furthermore, in previous

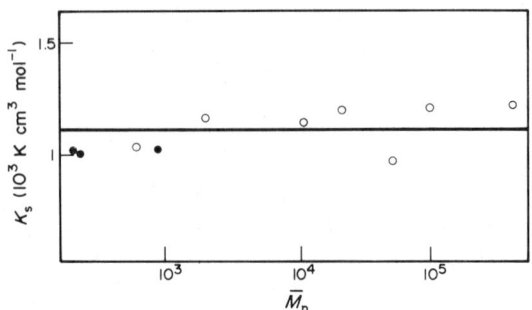

Figure 15   VPO calibration parameter K_s for benzene at 40 °C plotted against number-average molecular weight $\bar{M}_n$:[8] (●)
unbranched alkanes and tristearin; (○) polystyrene. The full line is $K_s = 0.5 K_e$.

experiments the relation $K_s = 0.98 K_e$ had been found for this polymer/solvent system ($600 \leqslant \bar{M}_n \leqslant 1 \times 10^4$) with both a Mechrolab 302 and a Hitachi 115 vapor pressure osmometer. This is consistent with the fact that the quantity $(k_1 A_1 + k_2 A_2)/(k_3 A_1)$ for the better VPO apparatus is also almost half of that for the Mechrolab 302 and the Hitachi 115.

$\bar{M}_n$ for polystyrenes in benzene can be estimated from $\lim_{c \to 0} \Delta T_s/c$ by utilizing either K_s for low molecular weight compounds, such as benzil, or $K_s = 0.5 K_e$. Values of $\bar{M}_n$ obtained in these ways are used in Table 5. Agreement between values of $\bar{M}_n$ given by the supplier and those evaluated by VPO are fairly reasonable (to an accuracy of $\pm 10\%$ or less except for the two samples 11a and 1a), particularly in the case where $K_s = 0.5 K_e$ is employed. From the foregoing discussion it is clear that $\bar{M}_n$ of unknown materials can be accurately determined by VPO over a wide $\bar{M}_n$ range.

Kucharikova[56] noticed that the agreement between the values of $\bar{M}_n$ by VPO and those given by the manufacturers in Table 5 was better than had been first indicated by Kamide *et al.*[8] because the molecular weights used by Kamide *et al.* were not $\bar{M}_n$ but weight-average molecular weights, $\bar{M}_w$. The $\bar{M}_n$ value in the parentheses for sample code 1a in Table 5 was determined by Kucharikova using a Knauer membrane osmometer in toluene at 37 °C, and that for sample code 11a was obtained by Adams *et al.*[57] using gel permeation chromatography (GPC). These values are in good agreement with $\bar{M}_n$ by VPO reported by Kamide *et al.*

In order to estimate the experimental error in the VPO method, Kamide *et al.*[58] repeatedly determined $\bar{M}_n$ of two typical polystyrene and cellulose acetate samples with their apparatus. In this case, the mother solutions, from which the solutions of various concentrations were prepared by adding pure solvent, were freshly prepared for each $\bar{M}_n$ measurement. They obtained as $\bar{M}_n(\text{VPO}) \times 10^{-4}$ 6.7, 7.1, 7.3, 7.5 and 7.5 for a polystyrene sample ASPM 4–8 and 3.6, 3.8 and 4.3 for a cellulose acetate sample EF3–6.[58] From these results $\bar{M}_n(\text{VPO})$ was calculated to be $7.22 \times 10^4 \pm 0.54 \times 10^4$ at the 95% significance level for polystyrene and $3.7 \times 10^4 \pm 0.36 \times 10^4$ at the 95% significance level for cellulose acetate (see Table 6). Evidently the VPO method is reproducible to ± 7–9% or better, which is of the same order of magnitude as that observed in membrane osmometry (MO) or GPC. Of course, the relative experimental error will differ greatly depending on the solvent nature, especially for its vapor pressure and its heat of condensation, and the temperature. However, as explained earlier, the sensitivity of VPO may be improved by using a more volatile solvent.

4.4.2   Membrane Osmometry

Membrane osmometry was first applied to natural polymers, such as gelatine, starch, rubber, protein and some cellulose esters, in the years 1910 to 1930. After a systematic study of natural and synthetic polymers in solution, Schulz[58] established, as early as the 1930s and 1940s, membrane osmometry as the standard method for determination of $\bar{M}_n$ of the polymers. Four types of manual osmometers were in common use in the 1940s: the static osmometer (Schulz design),[59] the dynamic osmometer (Fuoss–Mead design),[60] the static–dynamic osmometer (Zimm–Meyerson design)[61] and the osmotic balance (Jullander design).[62] These osmometers, for which reference should be made to a more detailed and excellent description of their construction and operation,[53] are now out of use because they are tedious and time-consuming in operation.

Table 6 The Number-average Molecular Weights ($\bar{M}_n$) of Fractions of Atactic Polystyrene (PS) and Cellulose Diacetate (CDA) (Total Degree of Substitution DS = 2.46)

| Polymer | Sample code | Membrane osmometry | | VDO[d] | GPC | $\bar{M}_w/\bar{M}_n$ |
| | | Toluene (25°C) Wescan model 231 | THF (25°C) HP[b] model 502 | THF (35.5°C) Kamide et al.[e] | THF (25°C) Shimadzu model 1A | |
				$\bar{M}_n \times 10^{-4}$		
PS	ASPM4-2	—	—	2.3[a]	2.1[a]	1.28
	ASPM4-4	3.6[a]	—	3.5[a]	3.9[a]	1.28
	ASPM4-6	5.1[a]	—	6.0[a]	6.2[a]	1.22
	ASPM4-8	7.1[a]	—	7.2[a]	8.2[a]	1.29
	ASPM4-10	—	—	9.1[a]	10.4[a]	1.24
	ASPM4-12	12.6[a]	—	12.3[a]	12.6[a]	1.19
CDA	EF3-4	—	2.7[a]	2.6[a]	4.1[a]	1.66
	EF3-6	—	4.8[c]	4.1[c,g]	5.0[c]	1.59
	EF3-7	—	5.5[a]	5.4[a]	5.8[a]	1.36
	EF3-10	—	8.7[a]	8.1[a]	8.8[c]	1.25
	EF3-12	—	12.2[c]	11.1[a]	11.9[c]	1.25
	EF3-13	—	12.6[c]	12.9[a]	12.3[c]	1.28

[a] Ref. 58. [b] Hewlett Packard. [c] Ref. 66. [d] Calibration parameter K_s was determined with the aid of benzil and tristearin in THF. [e] Ref. 8. [f] Average value of five determinations. [g] Average of three determinations.

Several automatic membrane osmometers {Mechrolab model 501–503 [Mechrolab Co. (USA)], Shell type (Dohrman Instruments Co., Hallikainen Instruments and J. V. Stavin Co. under the licence of Shell Development Co.), Knauer (KG Dr. Knauer (DBR)] and Wescan model 230, 231 [Wescan Inst. Inc. (USA)]} were popularized in the 1960s and 1970s. Mechrolab's models were the first to appear (initially commercialized by Mechrolab Co., but thereafter by Hewlett–Packard Co.) but now they are out of production. High pressure osmometers were designed[63] for the measurement of osmotic pressures of concentrated solutions.

Figure 16 shows a schematic view of the Wescan osmometer model 230 or 231. Here the upper solution cell is separated from the lower solution cell by the semipermeable membrane. The transfer of solvent from the solvent cell to the solution cell brings about a negative pressure in the solvent cell, which can be measured by a stainless steel diaphragm, attached to a strain gauge transducer. The precision of measurement is 0.5% and the stability is better than 0.02 cm H_2O. The Wescan model 231 can measure osmotic pressures (π) of 0 to 100 cm H_2O over a temperature range of 5 to 130 °C.

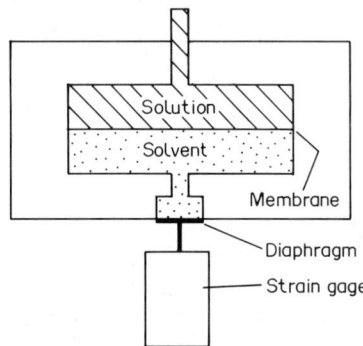

Figure 16   Schematic representation of a membrane osmometer: Wescan model 230 or 231

In the Knauer osmometer the diaphragm is directly connected to an electrode and the displacement of the diaphragm is measured as a change in electrical capacity. In the Shell osmometer, a change in electrical capacity starts a servo-motor by which the liquid level of the solvent storage is readjusted so as to make the displacement of the diaphragm zero.

Many attempts have been made to produce an ideal semipermeable membrane for membrane osmometry.[64] Recent advances in membrane science have improved the performance and quality of the semipermeable membranes for this purpose. At the present time, membranes of cellulose and cellulose acetate are available from Sartorius GmbH (Göttingen, DDR) and Schleicher & Schuell (USA). The former company also produces cellulose nitrate, polytetrafluoroethylene and polyamide membranes. The minimum value of $\bar{M}_n$ measurable by membrane osmometry using these membranes is *ca.* 1×10^4, and the range of $\bar{M}_n$ which can be accurately determined is from 1.5 to 5×10^5.

In actual measurements the following should be carefully confirmed: no leakage of solution and/or solvent, no contamination by air bubbles in either cell and no deformation or swelling of the membrane when installed in the apparatus.

In order to determine $\bar{M}_n$, π is measured for a series of polymer solutions, prepared by changing the polymer concentration c at constant temperature controlled to at least ± 0.01 °C. The intercept at $c \rightarrow 0$ of the plot of π/c *vs.* c yields $(\bar{M}_n)^{-1}$ (equation 34c) and from the slope we can determine A_2 or χ (equation 57). Figure 17 is an example for the polystyrene–toluene system at 25 °C.

In a good solvent the π/c *vs.* c plot often shows upward curvature. In this case the initial slope yields $A_{2,0}$. More accurately the π/c *vs.* c curve can be analyzed according to equation (34) with the aid of a computer to give $\bar{M}_n$, $A_{2,0}$ and $A_{3,0}$.

Table 6 lists $\bar{M}_n$ data for fractions of atactic polystyrene and cellulose diacetate (total degree of substitution DS = 2.46) determined[58] by membrane osmometry, VPO and GPC. The VPO method yields $\bar{M}_n$ values a few percent smaller than those estimated by the MO or the GPC methods. However, the experimental differences between $\bar{M}_n$(VPO) and $\bar{M}_n$(MO) and that between $\bar{M}_n$(VPO) and $\bar{M}_n$(GPC) for any given polymer sample are not significant at the 95% level. By GPC the lowest molecular weight fraction of cellulose diacetate gives a slightly higher $\bar{M}_n$ value compared with the remaining methods. This discrepancy originates from the fact that the reliablity of the GPC

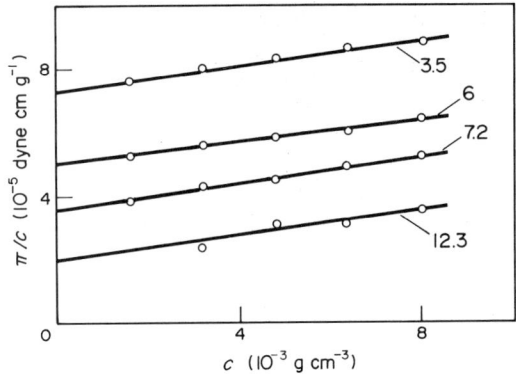

Figure 17 Plot of the ratio of osmotic pressure (π) and concentration (c) against c for polystyrene in toluene at 25 °C.[58] The numbers attached to the lines denote the number-average molecular weights ($\bar{M}_n$) of the samples (see Table 6).

method is comparatively low when applied to a low $\bar{M}_n$ cellulose diacetate, this being associated with the presence of a leading shoulder in the GPC trace.[65] In the last column of Table 6 the ratios of $\bar{M}_w$ to $\bar{M}_n$ for the samples are shown. The molecular weight distribution characteristic of a polymer sample, for which we can determine by VPO an $\bar{M}_n$ almost identical to those obtained by MO and GPC, remains open for further study.

4.5 REFERENCES

1. See, for example, H. Staudinger, 'Arbeitserinnerungen', Hüthig Verlag, Heidelberg, 1961.
2. F. M. Raoult, *Z. Phys. Chem.*, 1888, **2**, 353.
3. S. Takagi, T. Kimura and F. Nishida, unpublished results.
4. A. E. Pope, H. D. Pflug, B. Dacre and G. C. Benson, *Can. J. Chem.*, 1967, **45**, 2665.
5. J. van't Hoff, *Z. Phys. Chem.*, 1887, **1**, 481.
6. K. Kamide and M. Sanada, *Chem. High Polym. (Kobunshi Kagaku)*, 1967, **24**, 662.
7. K. Kamide, K. Sugamiya and C. Nakayama, *Makromol. Chem.*, 1970, **132**, 75.
8. K. Kamide, T. Terakawa and H. Uchiki, *Makromol. Chem.*, 1976, **177**, 1447.
9. K. Kamide, *Chem. High Polym. (Kobunshi Kagaku)*, 1968, **25**, 648.
10. K. Kamide and R. Fujishiro, *Makromol. Chem.*, 1971, **147**, 261.
11. M. Dimbat and F. H. Stross, *Anal. Chem.*, 1957, **29**, 1517; K. G. Schön and G. V. Schultz, *Z. Phys. Chem. (Frankfurt am Main)*, 1954, **2**, 197.
12. N. H. Ray, *Trans. Faraday Soc.*, 1952, **48**, 809; R. S. Lehrle and T. G. Majury, *J. Polym. Sci.*, 1958, **29**, 219; C. A. Glover and R. R. Stanley, *Anal. Chem.*, 1961, **33**, 447.
13. R. Fujishiro, S. Takagi and the late K. Furukawa, *Bull. Chem. Soc. Jpn.*, 1968, **41**, 1313.
14. W. R. Krigbaum and D. O. Geymer, *J. Am. Chem. Soc.*, 1959, **81**, 1859.
15. I. Fujiwara, Ph.D. Dissertation, Osaka City University, 1979.
16. K. H. Meyer, *Z. Phys. Chem., Abt. B*, 1939, **44**, 383.
17. M. L. Huggins, *J. Chem. Phys.*, 1941, **9**, 440; *Ann. N.Y. Acad. Sci.*, 1942, **43**, 1.
18. P. J. Flory, *J. Chem. Phys.*, 1941, **9**, 660.
19. J. J. van Laar, *Z. Phys. Chem., Abt. A*, 1928, **137**, 421.
20. G. Scatchard, *Chem. Rev.*, 1931, **8**, 321.
21. K. Kamide and Y. Miyazaki, *Polym. J.*, 1981, **13**, 325.
22. K. Kamide, T. Abe and Y. Miyazaki, *Polym. J.*, 1982, **14**, 355.
23. Th. G. Scholte, *Eur. Polym. J.*, 1970, **6**, 1063: *J. Polym. Sci., Polym. Phys. Ed.*, 1971, **9**, 1553.
24. See, for example, K. Kamide and S. Matsuda, *Netsu Sokutei*, 1986, **13**, 173.
25. B. H. Zimm, *J. Chem. Phys.*, 1946, **14**, 164.
26. See, for example, I. Sakurada, 'Polymer Chemistry', Kohbunsi Kagaku Kankohkai, Kyoto, 1948, chap. 5.
27. K. Kamide, S. Matsuda, T. Dobashi and M. Kaneko, *Polym. J.*, 1984, **16**, 839.
28. K. Kamide and S. Matsuda, *Polym. J.*, 1984, **16**, 825.
29. K. Kamide, S. Matsuda and M. Saito, *Polym. J.*, 1985, **17**, 1013.
30. P. J. Flory, 'Principles of Polymer Chemistry', Cornell University Press, Ithaca, New York, 1953, chap. XII.
31. Th. G. Scholte, *J. Polym. Sci., Polym. Phys. Ed.*, 1970, **8**, 841.
32. R. Koningsveld, L. A. Kleintjens and A. R. Shultz, *J. Polym. Sci., Polym. Phys. Ed.*, 1970, **8**, 1261.
33. N. Kuwahara, M. Nakata and M. Kaneko, *Polymer*, 1973, **14**, 415.
34. K. Kamide, K. Sugamiya, T. Kawai and Y. Miyazaki, *Polym. J.*, 1980, **12**, 67.
35. D. Patterson and G. Delmas, *Trans. Faraday Soc.*, 1969, **65**, 708.
36. P. J. Flory, *Discuss. Faraday Soc.*, 1970, **49**, 7.
37. See, for example, K. Tamura, S. Murakami and R. Fujishiro, *Polymer*, 1973, **14**, 237; S. Murakami, F. Kimura and R. Fujishiro, *Makromol. Chem.*, 1975, **176**, 3425; Y. Baba, K. Fujimoto, A. Kagemoto and R. Fujishiro, *Makromol. Chem.*, 1977, **178**, 1439, and references cited therein.

38. A. R. Shultz and P. J. Flory, *J. Am. Chem. Soc.*, 1952, **74**, 4760.
39. W. H. Stockmayer, *J. Chem. Phys.*, 1949, **17**, 588.
40. S. Saeki, N. Kuwahara, S. Konno and M. Kaneko, *Macromolecules*, 1973, **6**, 246.
41. See, for example, M. Kurata, 'Thermodynamics of Polymer Solutions', Harwood Academic Publishers, London, 1982, Chapter 2.
42. K. Kamide, Y. Miyazaki and T. Abe, *Makromol. Chem.*, 1976, **177**, 485.
43. G. V. Schulz and H. Baumann, *Makromol. Chem.*, 1963, **60**, 120.
44. R. Koningsveld and L. A. Kleintjens, *Macromolecules*, 1971, **4**, 637.
45. W. R. Krigbaum, *J. Am. Chem. Soc.*, 1954, **76**, 3758.
46. W. R. Krigbaum and D. K. Carpenter, *J. Phys. Chem.*, 1955, **59**, 1166.
47. A. Kotera, T. Saito and N. Fukisaki, *Rep. Prog. Polym. Phys. Jpn.*, 1963, **6**, 9.
48. K. Kamide, S. Matsuda and M. Saito, *Polym. J.*, 1988, **20**, 31.
49. A. Kagemoto, S. Murakami and R. Fujishiro, *Bull. Chem. Soc. Jpn.*, 1967, **40**, 11.
50. A. Kagemoto, S. Murakami and R. Fujishiro, *Makromol. Chem.*, 1967, **105**, 154.
51. C. A. Glover, 'Polymer Molecular Weight Methods', ed. M. Ezrin, Advances in Chemistry Series, 125, American Chemical Society, Washington DC, 1973.
52. K. Kamide, 'Polymer Solutions', ed. A. Nakajima, H. Fujita and H. Inagaki, Experiments in Polymer Science, Series 11, Kyoritsu Publishing Co., Tokyo, 1982, chap. 2.
53. R. E. Dohner, A. H. Wachter and W. Simon, *Helv. Chim. Acta*, 1967, **50**, 2193.
54. W. I. Higuchi, M. A. Schwartz, E. G. Rippie and T. Higuchi, *J. Phys. Chem.*, 1959, **63**, 996.
55. A. Wilson, L. Bini and R. Hofstader, *Anal. Chem.*, 1961, **33**, 135.
56. I. Kucharikova, *J. Appl. Polym. Sci.*, 1979, **23**, 3041.
57. H. E. Adams, E. Ahad, M. S. Chang, D. B. Davis, D. M. French, H. J. Hyer, R. D. Law, R. J. J. Simkins, J. E. Stuchbury and M. Tremblay, *J. Appl. Polym. Sci.*, 1973, **17**, 269.
58. K. Kamide, T. Terakawa and S. Matsuda, *Br. Polym. J.*, 1983, **15**, 91.
59. G. V. Schulz, *Z. Phys. Chem. (Leipzig)*, 1936, **A176**, 317; 1942, **B52**, 1.
60. R. M. Fuoss and D. J. Mead, *J. Phys. Chem.*, 1943, **47**, 59.
61. B. H. Zimm and I. Myerson, *J. Am. Chem. Soc.*, 1946, **68**, 911.
62. I. Jullander, *Ark. Kemi, Mineral. Geol.*, 1945, **21A**, 142.
63. See, for example, N. Kuwahara, T. Okazawa and M. Kaneko, *J. Polym. Sci., Polym. Symp.*, 1968, **23**, 543.
64. See, for example, R. H. Wagner, 'Physical Methods of Organic Chemistry', 1949, vol.1, part I.
65. K. Kamide, T. Terakawa and S. Manabe, *Sen-i Gakkaishi*, 1974, **30**, T-464.
66. K. Kamide, T. Terakawa and Y. Miyazaki, *Polym. J.*, 1979, **11**, 285.

5

Scattering Properties: Light and X-Rays

ISSA A. KATIME and JOSE R. QUINTANA
Universidad del País Vasco, Bilbao, Spain

5.1 INTRODUCTION

Light scattering and small-angle X-ray scattering (SAXS) are valuable methods for the characterization and study of synthetic polymers and biopolymers.

Light scattering from solution allows the determination of the molecular parameters[1-7] (molecular weights, dimensions, shapes, *etc.*) of the scattering particles, and thermodynamic quantities,[8-13] such as virial coefficients, chemical potentials, preferential adsorption coefficients and excess free energies of mixing. To obtain this information it is necessary to know two auxiliary parameters: concentration and refractive index increment. Similarly, SAXS is capable of yielding radii of gyration and, when used on the absolute intensity scale, molecular weights and thermodynamic parameters[14-17] (association constants, degrees of preferential interaction, hydrated volumes, surface to volume ratios, *etc.*). Again, two auxiliary parameters are necessary: concentration and partial specific volume of solute.

The theoretical principles underlying both methods are essentially identical. The differences that exist arise from the difference in the wavelength of the radiation. The dimension range accessible by the methods depends on the wavelength of the radiation used, and the methods are complementary.

5.2   LIGHT SCATTERING

When there is no absorption, the electromagnetic radiation interacts with the atoms or molecules in its path, producing an oscillation in their electronic density. Hence, the molecules act as oscillating electric dipoles that irradiate in all directions.

If the frequency of the scattered radiation is exactly the same as that of the incident radiation (*i.e.* they have the same energy) the scattering is 'elastic'. On the other hand, the scattering process may involve an energy exchange between the radiation and the scattering particles, leading to 'non-elastic' scattering, *e.g.* Raman scattering and Brillouin scattering.[18]

In the case of solutions, concentration fluctuations only contribute to the central elastic part of the scattering spectrum. However, the Brownian movement of the solute molecules creates weak frequency displacements which widen the central peak. This phenomenon is called 'quasielastic scattering' or 'Rayleigh line broadening'.[19-21] In this chapter, we will only deal with elastic scattering.

The elastic scattering produced by independent particles may be classified into three types: Rayleigh scattering, Debye scattering and Mie scattering. Rayleigh scattering is produced by particles with small dimensions compared to the wavelength of the incident radiation ($d < \lambda/20$) and with a refractive index similar to that of the medium. Debye scattering originates from particles with similar dimensions to the wavelength of the incident radiation and, as in the previous case, with a refractive index similar to that of the medium. If the refractive index of the particles differs considerably from that of the medium, and if their dimensions are large enough, they give rise to Mie scattering. In this case, the scattering cannot be considered as the result of a set of induced dipoles fluctuating in the electromagnetic field; the description of Mie scattering is very complicated and the subject of specialized publications.[22,23] Fortunately, scattering behaviour of this type is not common in polymer solutions. We will only deal with Rayleigh and Debye scattering.

Owing to the importance of light scattering, many reviews have been published.[3-7,18-33]

5.2.1   Scattering from a Gas

The theoretical interpretation of light scattering from dilute gases was developed by Lord Rayleigh,[34,35] applying the classical electromagnetic theory. We will consider, first of all, the scattering from a gas, the molecules of which have no thermodynamic interaction, *i.e.* an ideal gas. The system, therefore, consists of a large number of particles moving at random and well separated from each other. The particles have very small dimensions compared to the wavelength of the monochromatic beam used. We also assume that the particles are optically isotropic.

Let us start by considering a plane-polarized electromagnetic wave. Its associated electric field varies sinusoidally in a plane perpendicular to the direction of propagation. In Figure 1, the electromagnetic wave is shown vertically polarized, that is the electric field is parallel to the *z* axis.

The general equation for the electric field of such an electromagnetic wave, propagating through a vacuum, is

$$E = E_0 \exp\left[i(wt - k_0 r)\right] \tag{1}$$

where E_0 is the incident electric vector, *w* is the angular frequency, *t* is the time, k_0 the vectorial wave number and *r* is any position referred to the origin of coordinates. In this case, the electromagnetic radiation follows the *x* direction.

Since we are considering particles that are small compared to the wavelength, we can give each one a unique value *r*, and the local electrical field anywhere within the particle will be the same.

Consider one of these gas molecules located at the origin of coordinates, with its electronic density oscillating in phase with the local electric field. It can be treated as an induced electric dipole, the dipole moment of which **μ**, is proportional to the electric field

$$\mu = \alpha E = \alpha E_0 \exp(iwt) \tag{2}$$

where α represents the polarizability of the molecule.

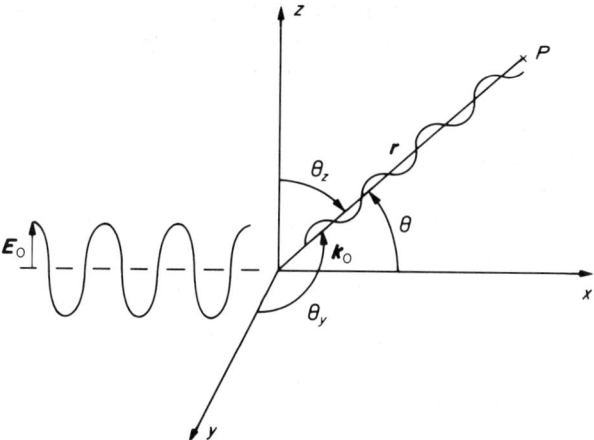

Figure 1 Scattering of vertically polarized radiation by an optically isotropic small particle. E_0 and k_0 are the amplitude of the electric vectors and the vectorial wave number of the incident radiation, r is any position referred to the origin. θ, θ_z and θ_y are the angles between the scattering direction and the x, z and y axes, respectively

This oscillating dipole behaves as a secondary source of electromagnetic radiation, and radiates, in all directions, a plane-polarized electromagnetic wave of identical wavelength to the incident radiation. However, the emission is directional, and the electric field has an amplitude dependent on the dipole-moment component perpendicular to the scattering direction considered. The electrical field of the scattered wave, E_s, at the point P, determined by the position vector r, at such a distance from the dipole that the dipole can be considered as a point, will be perpendicular to r and will be situated on the plane zr (see Figure 1). Its modulus, following the classical electromagnetic theory, is given by

$$E_s = \frac{1}{rc^2} \frac{\partial \mu}{\partial t^2} \sin \theta_z \qquad (3)$$

where c is the velocity of light in a vacuum and θ_z is the angle between the direction of the induced dipole, the z axis in this case, and the scattering direction.

From equations (2) and (3) we obtain

$$E_s = \frac{k_0^2 \alpha E_0}{r} \sin \theta_z \exp [i(wt - k_0 r)] \qquad (4)$$

where $k_0 r$ is the phase displacement of the electric field between the particle and the point P.

Taking into account the fact that the light intensity is defined as the square of the amplitude, the scattered radiation intensity in P is given by

$$I = \frac{16\pi^4 \alpha^2 I_0}{r^2 \lambda_0^4} \sin^2 \theta_z \qquad (5)$$

where $I_0 = E_0^2$ is the intensity of the incident radiation and λ_0 is the wavelength in a vacuum.

Equation (5) shows that the intensity of the scattered radiation is strongly dependent on the wavelength $(\propto \lambda^{-4})$ so that short wavelength radiations are more strongly scattered than longer ones. This is the reason why sunlight makes the sky blue when it is scattered by the atmosphere.

Equation (5) also shows that vertically polarized incident electromagnetic radiation has no scattering intensity in the z direction. From an experimental point of view, it is more interesting to study the distribution of the intensity of scattered radiation in the plane perpendicular to the polarization plane of the incident wave, in this case the xy plane. The scattered intensity in any direction within this plane is

$$I = \frac{16\pi^4 \alpha^2}{r^2 \lambda_0^4} I_0 \qquad (6)$$

Till now, we have analyzed the easiest situation, that is polarized incident radiation. However, unpolarized radiation is often used in experimental work. Any unpolarized radiation can be split up into two polarized waves, perpendicular to each other. These waves have the same intensity but their phases are independent. If I_0 is the unpolarized incident radiation intensity, we can write

$$I_0 = I_{0,\mathrm{V}} + I_{0,\mathrm{H}} \quad (I_{0,\mathrm{V}} = I_{0,\mathrm{H}}) \tag{7}$$

where $I_{0,\mathrm{V}}$ and $I_{0,\mathrm{H}}$ are the vertical and horizontal intensities of the polarized waves. Each of these intensities will be given by an identical equation to (5). The intensity of the scattered radiation in the point P for an unpolarized incident electromagnetic wave can be expressed as

$$I = \frac{8\pi^4 \alpha^2}{r^2 \lambda_0^4} I_0 (\sin^2 \theta_y + \sin^2 \theta_z) \tag{8}$$

where θ_y represents the angle between the y axis and the scattering direction.

We can write the same equation as

$$I = \frac{8\pi^4 \alpha^2}{r^2 \lambda_0^4} I_0 (1 + \cos^2 \theta) \tag{9}$$

where θ is the angle between the propagation direction of the incident wave and the direction of observation.

In Figure 2 are shown the angular distributions of the scattered intensities in the observation plane, xy. They correspond to the intensity from an unpolarized incident wave, $V_\mathrm{V} + H_\mathrm{H}$, and to those from the vertical, V_V, and horizontal, H_H, components.

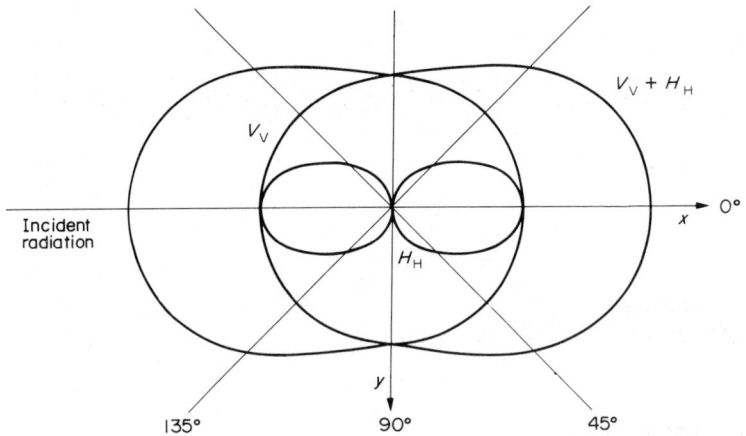

Figure 2 Angular dependence of the intensity of scattered light for an isotropic particle which is small compared to the wavelength of the light: vertically polarized incident light, V_V; horizontally polarized incident light, H_H; and unpolarized incident light, $V_\mathrm{V} + H_\mathrm{H}$

In the case of an unpolarized incident wave, the scattered intensity in the direction of the z axis is not zero, and the wave is polarized horizontally.

$$I_{90°} = \frac{8\pi^4 \alpha^2}{r^2 \lambda_0^4} I_0 \tag{10}$$

The same value is obtained in the direction of the y axis, and, in this case, the scattered wave is vertically polarized.

The maximum value of the scattered intensity is observed in the propagation direction of the incident wave ($\theta = 0°$ or $180°$) and corresponds to twice the value observed in the directions of the z and y axes.

From an experimental point of view, it is more interesting to consider the scattered intensity originating from a unit volume containing N particles. Considering the particles to be independent, the total scattered intensity corresponds to the sum of the scattered intensities of each individual particle.

$$I = \frac{8\pi^4 \alpha^2 N}{r^2 \lambda_0^4} I_0 (1 + \cos^2 \theta) \tag{11}$$

The polarizability α is not easy to obtain experimentally. However, it can be related to the gas refractive index, n, through the dielectric constant, ε,

$$(n^2 - 1) = (\varepsilon - 1) = 4\pi N\alpha \tag{12}$$

In a dilute gas, the refractive index is very close to unity (this being the refractive index of a vacuum). We can therefore express it as a Taylor series, rejecting all terms from the third inclusive, *i.e.*

$$n^2 = 1 + 2\,(dn/dc)c \tag{13}$$

where c is the concentration of the gas, expressed in units of mass/volume, and dn/dc is the refractive index increment of the dilute gas.

Taking into account the fact that the number of molecules per unit volume, N, can be expressed as $N_A c/M$, equations (12) and (13) may be combined with equation (11) to give the intensity of light scattered from unit volume:

$$I = \frac{2\pi^2 M I_0}{r^2 \lambda_0^4 N_A} \left(\frac{dn}{dc}\right)^2 (1 + \cos^2 \theta) \tag{14}$$

where M is the molecular weight of the particles and N_A is Avogadro's constant.

The validity of this equation has been verified experimentally, by light-scattering determinations of Avogadro's constant.[36-38]

To remove the dependence of I on observation distance r and on the incident intensity I_0, the light scattering may be expressed as the 'Rayleigh ratio':

$$R_\theta = \frac{I r^2}{I_0} \tag{15}$$

In this way, equation (14) becomes

$$R_\theta = \frac{2\pi^2 M c}{\lambda_0^4 N_A} \left(\frac{dn}{dc}\right)^2 (1 + \cos^2 \theta) \tag{16}$$

Since I_0 and r can be determined from the optical and mechanical characteristics of the photometer used, the Rayleigh ratio can be obtained experimentally.

5.2.2 Scattering from a Liquid

When considering liquids,[39-41] the molecules are not independent: they are not distributed at random and there is a reduction of intensity of scattered radiation, due to destructive interference. A formal analysis of the interference can lead to an expression for the scattered intensity as a function of the molecular distribution function. This approach is the best one for the study of intermolecular effects and molecular conformations.[26] However, we prefer to use a thermodynamic development that allows us to relate the scattered intensity to macroscopic quantities which can be measured experimentally. This method is particularly useful for the treatment of light scattering from a multicomponent system, in which there is no intramolecular interference.

The thermodynamic treatment of light scattering from liquids and solutions was developed independently by Einstein[42] and Smoluchowski[43,44] and later adapted by Debye[45,46] for macro-molecular solutions.

If the medium is completely ordered and is incompressible, then, for any observation direction different from the propagation direction, we can associate volumes of very small dimensions (compared to the incident wavelength) in pairs in such a way that the distances covered by the scattered rays on their paths to the observation point differ by half a wavelength. The electric fields of the scattered rays will be opposite in phase, and the scattered intensity, therefore, will be zero for any direction other than that of propagation of the incident wave.

Although there is some molecular order in liquids, it is not complete, and, as a consequence, there is a degree of scattering. The density of a liquid in the different volume elements is only constant when averaged over a period of time. If a particular instant of time is considered, the movement of particles, partially at random, leads to density fluctuations at any chosen point, that is to say the number of particles existing in a volume element differs from the corresponding number in another volume element. For this reason, if we consider pairs of elements, as mentioned above, the scattering amplitudes from the different volumes will fluctuate, and a small fraction of them will not be cancelled by interference. The density of a liquid is much greater than the density of a gas, and the fluctuation in the number of existing particles in a volume element is bigger than that exhibited by a similar volume of gas. Therefore, the scattering intensity from a liquid is large compared with that from a gas.

The intensity scattered by a liquid can be easily obtained if we look back at the development made in the previous section. Now, we replace each particle by an elemental volume δV, small enough to be considered as a point when compared to the wavelength of the radiation, but containing enough molecules to allow application of thermodynamic methods. We also consider that the fluctuations in these volumes are independent.

Each elemental volume will have a polarizability α that will fluctuate by a certain amount $\Delta \alpha$ about the average polarizability of the system $\bar{\alpha}$.

$$\alpha^2 = (\bar{\alpha} + \Delta\alpha)^2 = \bar{\alpha}^2 + 2\bar{\alpha}\Delta\alpha + \Delta\alpha^2 \tag{17}$$

Considering the intensity scattered by N elemental volumes within a unit volume, the value $\bar{\alpha}$ is the same for each volume element. Therefore, terms in $\bar{\alpha}$ will cancel for the same reason that the scattered light in an ordered system is zero. Also the average value of $2\bar{\alpha}\Delta\alpha$ will be zero, because $\Delta\alpha$ will take equivalent positive and negative values. Consequently, *via* equation (16), the Rayleigh ratio will be given by

$$R_\theta = \frac{8\pi^4 \overline{\Delta\alpha^2}}{\lambda_0^4 \delta V}(1 + \cos^2 \theta) \tag{18}$$

where $1/\delta V$ is the number of elemental volumes per unit volume.

Taking into account the relation between polarizability and dielectric constant, the polarizability fluctuation can be expressed as

$$\Delta\alpha = \frac{\delta V \Delta\varepsilon}{4\pi} \tag{19}$$

Therefore equation (18) becomes

$$R_\theta = \frac{\pi^2 \delta V \overline{\Delta\varepsilon^2}}{2\lambda_0^4}(1 + \cos^2 \theta) \tag{20}$$

In a pure liquid, the fluctuations in the local value of the dielectric constant are due, in principle, to fluctuations in both temperature and density, but, in general, fluctuations in temperature are negligible and so

$$\overline{\Delta\varepsilon^2} = \left(\frac{\partial\varepsilon}{\partial\rho}\right)^2 \overline{\Delta\rho^2} \tag{21}$$

According to the general theory of fluctuations, a fluctuation in a given variable x with respect to its average value is given by[47]

$$\overline{\Delta x^2} = \frac{kT}{\partial^2 F/\partial x^2} \tag{22}$$

where k is the Boltzmann constant, T is the absolute temperature and F is the Helmholtz free energy of the system. Therefore,

$$\overline{\Delta\rho^2} = \frac{kT}{\partial^2 F/\partial\rho^2} \tag{23}$$

Thermodynamic arguments lead to

$$\left(\frac{\partial^2 F}{\partial\rho^2}\right)_T = \frac{\delta V}{\rho^2}\left(\frac{1}{\beta} - 2P\right) \tag{24}$$

where β is the isothermal compressibility and P the pressure. Under usual experimental conditions $1/\beta \gg 2P$. Rejecting the second term on the right-hand side of equation (24) and combining equations (20)–(24) gives

$$R_\theta = \frac{\pi^2 k T \rho^2 \beta}{2\lambda_0^4}\left(\frac{\partial \varepsilon}{\partial \rho}\right)_T (1+\cos^2 \theta) \tag{25}$$

In the neighbourhood of the gas–liquid critical point, the isothermal compressibility is very large and, therefore, the fluctuations are very important, giving rise to the phenomenon of critical opalescence.[23, 29]

5.2.3 Scattering from a Solution of Small Particles

If we consider a dilute solution,[46, 48, 49] the increment in the dielectric constant $\Delta\varepsilon$ is related both to fluctuations in the number of particles in volume δV (density fluctuations) and to the interchange between solvent and solute molecules, which have different polarizabilities (concentration fluctuations). Therefore,

$$\Delta\varepsilon = \left(\frac{\partial \varepsilon}{\partial \rho}\right)\Delta\rho + \left(\frac{\partial \varepsilon}{\partial c}\right)\Delta c \tag{26}$$

where c is the solute concentration.

If the fluctuations in density and concentration are independent, $\overline{\Delta\rho\Delta c} = 0$ and we have, therefore,

$$\overline{\Delta\varepsilon^2} = \left(\frac{\partial \varepsilon}{\partial \rho}\right)^2 \overline{\Delta\rho^2} + \left(\frac{\partial \varepsilon}{\partial c}\right)^2 \overline{\Delta c^2} \tag{27}$$

If the solution is dilute, the density fluctuations are essentially identical to those existing in the pure solvent. Hence, the excess scattering intensity from the solution compared to the solvent arises only from concentration fluctuations, and the excess Rayleigh ratio can be expressed as

$$\Delta R_\theta = R_{\theta,\,\text{solution}} - R_{\theta,\,\text{solvent}} = \frac{\pi^2 \delta V}{2\lambda_0^4}\left(\frac{\partial \varepsilon}{\partial c}\right)^2 \overline{\Delta c^2} (1+\cos^2 \theta) \tag{28}$$

The refractive index of the dilute solution will have a similar value to that of the solvent (this is a characteristic of Rayleigh scattering) and can be expressed as a power series in concentration. Rejecting all the terms in the series from the third onwards, and taking the square, we obtain

$$n^2 = n_1^2 + 2n_1\left(\frac{dn}{dc}\right)c + \left(\frac{dn}{dc}\right)^2 c^2 \tag{29}$$

where n_1 is the solvent refractive index.

The third term on the right-hand side of equation (29) is negligible compared with the second, and so, following the Maxwell relation $\varepsilon = n^2$,

$$\left(\frac{\partial \varepsilon}{\partial c}\right)^2 = 4n_1^2\left(\frac{dn}{dc}\right)^2 \tag{30}$$

and, applying equation (22),

$$\overline{\Delta c^2} = \frac{kT}{(\partial^2 F/\partial c^2)_T} \tag{31}$$

From thermodynamic arguments

$$\left(\frac{\partial^2 F}{\partial c^2}\right)_T = -\frac{\delta V}{\bar{V}_1 c}\frac{\partial \mu_1}{\partial c} \tag{32}$$

where μ_1 is the chemical potential of the solvent in the solution and $\bar{V}_1$ its partial molar volume.

Combining now equations (28), (30), (31) and (32), the excess Rayleigh ratio is obtained as

$$\Delta R_\theta = -\frac{2\pi^2 n_1^2 kT (dn/dc)^2 \bar{V}_1 c}{\lambda_0^4 (\partial \mu_1/\partial c)_T}(1 + \cos^2\theta). \tag{33}$$

The relation between the chemical potential of the solvent and the osmotic pressure Π

$$\Pi \bar{V}_1 = -(\mu_1 - \mu_1^0) \tag{34}$$

where μ_1^0 is the chemical potential of the pure solvent, leads to

$$\Delta R_\theta = K\frac{RTc}{\partial\Pi/\partial c}(1 + \cos^2\theta) \tag{35}$$

where

$$K = \frac{2\pi^2 n_1^2}{\lambda_0^4 N_A}\left(\frac{dn}{dc}\right)^2 \tag{36}$$

By expressing the osmotic pressure as a power series in concentration,

$$\frac{\Pi}{c} = RT\left(\frac{1}{M} + A_2 c + A_3 c^2 + \ldots\right) \tag{37}$$

we obtain the final expression

$$\frac{Kc(1 + \cos^2\theta)}{\Delta R_\theta} = \frac{1}{M} + 2A_2 c + 3A_3 c^2 + \ldots \tag{38}$$

where A_2, A_3, ..., are the 2nd, 3rd, ..., virial coefficients. This is the general equation for Rayleigh scattering of unpolarized light from a real solution of particles which are much smaller than the wavelength of the light.

In very dilute solutions, the virial coefficients from the third one, A_3, onwards can be neglected. In this way, the above equation allows us to obtain easily the molecular weight of the solute and the second virial coefficient, A_2.

It is interesting to note that equation (38) for an ideal dilute solution, for which all the virial coefficients are zero, could be obtained by considering the solution as a gas formed of particles,[27] with polarizability equal to the excess polarizability of the solution relative to the solvent, *i.e.*

$$\Delta\alpha = (n^2 - n_1^2)/4\pi N \tag{39}$$

where n is the refractive index of the solution at concentration c and n_1 is the refractive index of the solvent.

5.2.4 Scattering from a Solution of Larger Particles

The light-scattering expressions obtained in the above section can only be applied to particles which are much smaller than the wavelength (Rayleigh scattering). For larger particles (Debye scattering) it is observed experimentally that the angular distribution of the intensity is not symmetrical with regard to the zy plane (Figure 1). The intensity of the scattered radiation is reduced by a factor that increases as the observation angle, θ, increases. It is normally considered that this phenomenon takes place when the geometric dimensions of the particles are greater than one-twentieth of the wavelength. Thus, if we consider a wavelength of 632.8 nm in vacuum, which corresponds to a He–Ne laser, the limiting dimension corresponds to a value of about 25 nm. The vast majority of macromolecules (with the exception of the small globular proteins[50]) have larger dimensions than this limiting value.

The reduction in the scattering intensity for a condensed phase, considered in the last section, arises from intermolecular effects. Correspondingly, the lack of symmetry in the angular distribution of the scattering intensity from a macromolecule can be considered as due to intramolecular interference effects. The particle, due to its size, can no longer be considered as a point scatterer, but as a set of point scatterers.

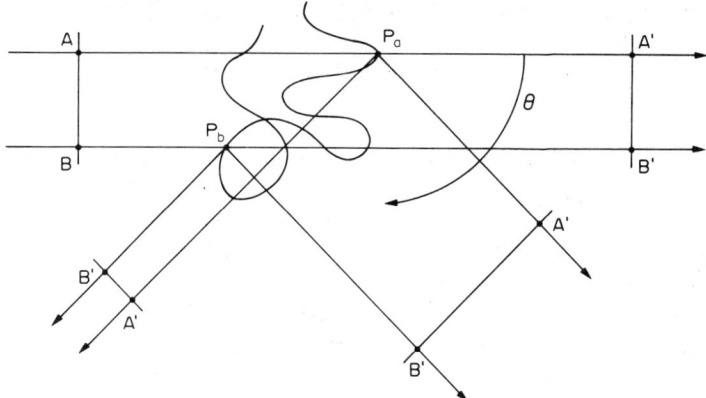

Figure 3 A representation of the phase difference between rays scattered at two points in a molecule more than $\lambda/20$ apart. The difference in phase is greater for rays scattered in the backward direction and therefore the mutual interference is greater

This interference can be easily visualized by considering two scattering centres (Figure 3). By analyzing the difference in the distance travelled by the waves corresponding to each centre, it can be seen that the difference between the path lengths AP_aA' and BP_bB' increases as the angle θ increases, and is zero in the propagation direction of the incident wave, $\theta = 0$. The light-interference effect increases as the angle of observation increases.

This dissymmetry in the scattered intensity caused by the intramolecular interference effect poses a problem, since extrapolation to zero angle (θ) must be made in order to obtain the information outlined in the previous section. However, it is precisely because of this effect that we are able to obtain valuable information about the geometry of the macromolecule.

5.2.4.1 Particle scattering factor, $P(\theta)$

In general, the effect of the angular dissymmetry on the scattering intensity is described in terms of the function $P(\theta)$,[51] which corresponds to the ratio of the real intensity of the scattered light in a given direction to the theoretical intensity which would be observed at the same angle in the absence of intramolecular interference. Since the intensity is proportional to the Rayleigh ratio

$$P(\theta) = \frac{\Delta R_\theta(\text{experimental})}{\Delta R_\theta(\text{without interference})} \tag{40}$$

where ΔR_θ (without interference) is given by equation (38). The function $P(\theta)$ has a value of unity when $\theta = 0$, and it decreases as the observation angle increases. The equations presented in the previous section can be used for large particles if we measure ΔR_θ at several angles and extrapolate to $\theta = 0$.

In order to obtain an expression for the function $P(\theta)$[52] that relates to the geometric parameters of the particle, the electric field of the electromagnetic radiation is assumed to be uniform for the whole molecule. In such a circumstance, the molecule can be represented by a set of dipoles.

If we consider one of these dipoles, at a distance r from the origin of coordinates, the electric field of the electromagnetic radiation at this point is given by

$$E = E_0 \exp[i(wt - kr)] \tag{41}$$

where k is the wave vector in the propagation direction of the incident wave.

The electric field of the scattered radiation at point P (Figure 4), situated at a distance D from the scattering centre, is

$$E_s \propto \frac{\exp(iwt)}{D} \exp[-i(kr - kD)] \tag{42}$$

kr is the phase displacement of the electromagnetic radiation at the scattering point with regard to a wave front in the origin of coordinates. Constants are ignored since we are only interested in the intramolecular interference (that is, expressions for the function $P(\theta)$).

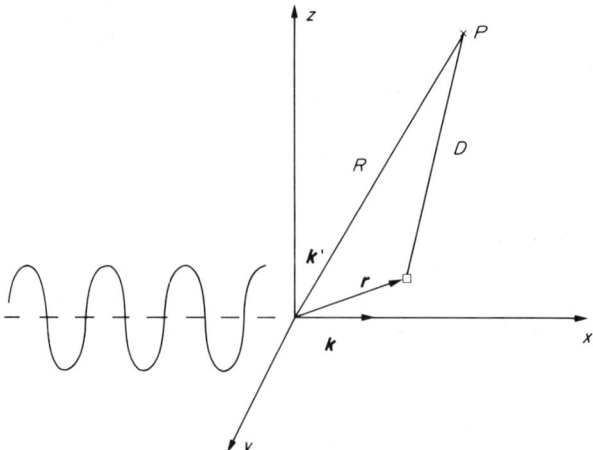

Figure 4 Geometrical relations between vectors associated with incident and scattered light (see text for explanation of symbols)

If $D \gg r$ (Figure 4), then

$$kD = kR - k'r \tag{43}$$

where k' is the wave vector in the direction P. Substituting this expression in equation (42), gives

$$E_s \propto \frac{\exp(iwt)}{R} \exp[-i(kr + kR - k'r)] \tag{44}$$

Definition of the vector $q = k - k'$, the modulus of which is given by

$$q = \frac{4\pi}{\lambda} \sin \frac{\theta}{2} \tag{45}$$

leads to

$$E_s \propto \frac{\exp[i(wt - kR)]}{R} \exp(-iqr) \tag{46}$$

where $\exp[i(wt - kR)]/R$ represents the electromagnetic radiation scattered from the origin and $\exp(-iqr)$ is a phase factor related to the interference effects between the radiation from different dipoles.

If m is the number of scattering centres which constitute the particle, the total scattered radiation is given by

$$E_s \propto \frac{\exp[i(wt - kR)]}{R} \sum_{a=1}^{m} \exp(-iqr_a) \tag{47}$$

or, since the intensity of the scattered radiation is given by the product of the electric field amplitude and its complex conjugate,

$$I \propto \sum_{a=1}^{m} \sum_{b=1}^{m} \exp[-iq(r_a - r_b)] \tag{48}$$

In the absence of destructive interference or for scattering at zero angle θ, all the exponents in the above summation are zero, and

$$I \propto \sum_{a=1}^{m} \sum_{b=1}^{m} 1 = m^2 \tag{49}$$

Therefore, according to the definition of the function $P(\theta)$,

$$P(\theta) = \frac{1}{m^2} \sum_{a=1}^{m} \sum_{b=1}^{m} \exp[-iq(r_a - r_b)] = \frac{1}{m^2} \sum_{a=1}^{m} \sum_{b=1}^{m} \exp(-iqr_{ab}) \tag{50}$$

This expression for $P(\theta)$ corresponds to a molecule fixed in space. However, if the solution is isotropic, the particle will be free to choose any direction, and the measured intensity will be an average over all possible orientations.[28] The expression for $P(\theta)$ corresponding to a molecule with random orientation is[53]

$$P(\theta) = \frac{1}{m^2} \sum_{a=1}^{m} \sum_{b=1}^{m} \frac{\sin q r_{ab}}{q r_{ab}} \tag{51}$$

If the particle is able to change its shape, the distances r_{ab} will be variable, and a second average over time is needed. This is the case for flexible-chain macromolecules.

From the last equation we deduce that $P(\theta)$ will depend on the shape (or average shape) of the molecule. However, this dependence decreases as the magnitude of $q r_{ab}$ approaches zero.

The use of a MacLaurin series for $q r_{ab} \ll 1$ leads to

$$P(\theta) = \frac{1}{m^2} \sum_{a=1}^{m} \sum_{b=1}^{m} \left(1 - \frac{q^2 r_{ab}^2}{3!} + \frac{q^4 r_{ab}^4}{5!} - \dots \right) \tag{52}$$

Considering only the first two terms

$$P(\theta) = 1 - \frac{q^2}{6m^2} \sum_{a=1}^{m} \sum_{b=1}^{m} r_{ab}^2 \tag{53}$$

Independently of the shape of a particle, the square radius of gyration is given by

$$R_g^2 = \frac{1}{2m^2} \sum_{a=1}^{m} \sum_{b=1}^{m} r_{ab}^2 \tag{54}$$

In the case of a flexible macromolecular chain, the mean-square radius of gyration is given by

$$\langle R_g^2 \rangle = \frac{1}{2m^2} \sum_{a=1}^{m} \sum_{b=1}^{m} \langle r_{ab}^2 \rangle \tag{55}$$

Therefore, $P(\theta)$ can be expressed as

$$P(\theta) = 1 - \frac{q^2 \langle R_g^2 \rangle}{3} = 1 - \frac{16\pi^2}{3\lambda^2} \langle R_g^2 \rangle \sin^2 \frac{\theta}{2} \tag{56}$$

This relation between $P(\theta)$ and the mean-square radius of gyration of a macromolecule is a useful general result, since it is independent of molecule shape. The limiting slope at $\theta = 0$ of a plot of $P(\theta)$ against $\sin^2(\theta/2)$ for any macromolecule is $-16\pi^2 \langle R_g^2 \rangle / 3\lambda^2$, which permits determination of $\langle R_g^2 \rangle$. Unfortunately, this slope can only be observed experimentally for values of $R_g/\lambda \gtrsim 0.05$. Because of this we can only use scattering methods to determine particles with dimensions similar to or greater than $\lambda/20$.

5.2.4.2 General equation, Zimm plot

When the particles are large enough to display angular dissymmetry, equation (38) is only valid when $\theta = 0$, since $P(\theta) = 1$ for this angle. Since it is impossible to measure the scattered intensity at zero angle, the experimental scattering intensities must be analyzed as a function of the observation angle, *i.e.*

$$\frac{Kc(1 + \cos^2 \theta)}{\Delta R_\theta} = \frac{1}{M P(\theta)} + 2 A_2 Q(\theta) c \tag{57}$$

The factor $Q(\theta)$, due to intermolecular interference effects at finite concentrations, falls appreciably below unity only for scattering from large particles (high polymers in good solvents) at high values of θ. For $\theta \to 0$ and $c \to 0$, and taking into account the fact that $1/1 - x \simeq 1 + x$ for small values of x, the expression for the scattering intensity for unpolarized light as a function of concentration and observation angle is

$$\frac{Kc(1+\cos^2\theta)}{\Delta R_\theta}\bigg|_{\substack{c\to 0 \\ \theta\to 0}}=\frac{1}{M}\left(1+\frac{16\pi^2}{3\lambda^2}\langle R_g^2\rangle\sin\frac{\theta}{2}\right)+2A_2c \tag{58}$$

This expression is only appropriate for dilute solutions and small values of qr_{ab}.

Equation (58) implies a double dependence of scattering intensities on concentration and observation angle. Therefore, in order to measure the second virial coefficient, the scattering data for each concentration must be extrapolated to zero angle, when

$$\frac{Kc(1+\cos^2\theta)}{\Delta R_\theta}\bigg|_{\theta=0}=\frac{1}{M}+2A_2c \tag{59}$$

Correspondingly, in order to measure the mean-square radius of gyration, the scattering data for each angle must be extrapolated to zero concentration, when

$$\frac{Kc(1+\cos^2\theta)}{\Delta R_\theta}\bigg|_{c=0}=\frac{1}{M}+\frac{16\pi^2}{3\lambda^2M}\langle R_g^2\rangle\sin^2\frac{\theta}{2} \tag{60}$$

The common intercept of the two plots gives the reciprocal of the molecular weight.

Normally, a graphical method, proposed by Zimm,[54,55] is used to do this double extrapolation. It consists of plotting the values of $Kc(1+\cos^2\theta)/\Delta R_\theta$ for each concentration and angle against $\sin^2(\theta/2)+k'c$. The constant k' is arbitrary and is chosen in such a way that the maximum value of $\sin^2(\theta/2)$ is of similar magnitude to $k'c$. This allows a clear separation of all the data points. A practical method, when considering high molecular weights, is to choose a value of k', which, when multiplied by the smallest concentration, gives a value near to unity. For low molecular weights, the highest concentration is used. An example of a Zimm plot is shown in Figure 5.

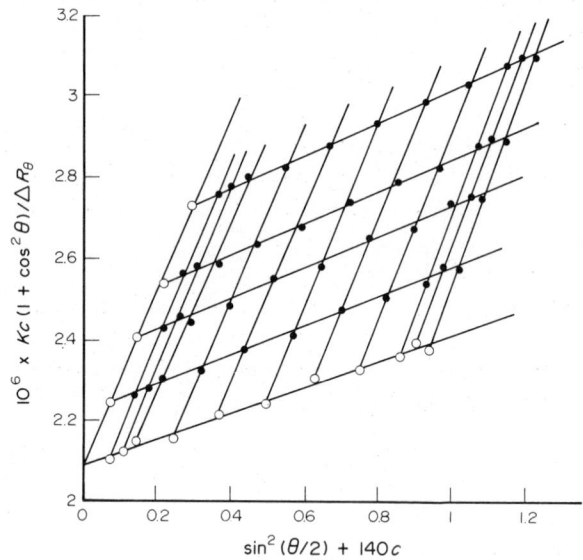

Figure 5   A typical Zimm plot; data obtained from a fraction of poly(methyl methacrylate) ($\bar{M}_w=4.57\times10^5$) in ethyl acetate at 298 K

For very high molecular weights, the ordinate intercept is close to zero, and the values of M and $\langle R_g^2\rangle$ so obtained are uncertain. In this case it is preferable to use the Guinier method,[14] which consists of plotting $\log \Delta R_\theta/Kc(1+\cos^2\theta)$ as a function of $\sin^2(\theta/2)$ for each concentration, and then plotting the slopes and intercepts at $\theta=0$ against concentration.

5.2.4.3   *P(θ) functions for basic particle shapes*

When the dimensions of the particles in solution are very large, *i.e.* the condition $qr_{ab}\ll 1$ does not hold, the first two terms in the series development of the function P(θ) are not enough to represent the accessible experimental data. In this case, $1/P(\theta)$ is no longer a linear function of $q^2 \langle R_g^2 \rangle$, but has a degree of curvature that depends on the shape of the particles. By comparing the experimental curve with those established theoretically for simple geometric shapes,[50] information can be obtained about molecular shape (Figure 6).

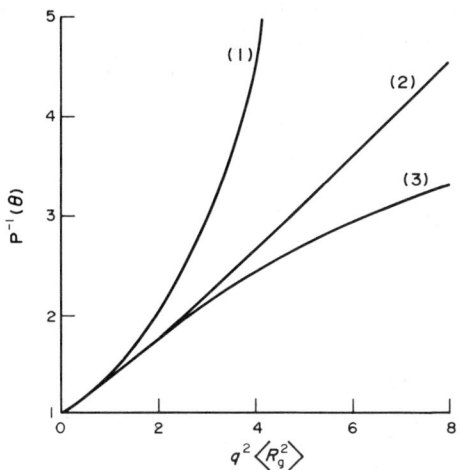

Figure 6 $P^{-1}(\theta)$ as a function of $q^2 \langle R_g^2 \rangle$ for basic particle shapes: (1) spheres, (2) Gaussian coils and (3) rods. $P(\theta)$ is the particle scattering factor, q is equal to $(4\pi/\lambda) \sin(\theta/2)$ and $\langle R_g \rangle$ the radius of gyration of the particle

The first three particle shapes to be studied were an infinitely thin rod,[56] a Gaussian coil[1,50] (corresponding to a flexible macromolecule in a theta solvent) and a homogeneous sphere.[57,58] For these three cases, exact analytical expressions for the function $P(\theta)$ were obtained. Other approximate expressions for particles of more complex shape (*e.g.* chains with excluded volume,[59] chains with persistence length,[60] and discs[61]) have also been obtained.

As a simple example, we derive here the expression for $P(\theta)$ for long thin rods from the general equation (51). The rod consists of $\sigma + 1$ scattering centres in a straight line. The adjacent centres are separated by a distance l, *i.e.* the total length of the rod is $L = l\sigma$. In order to determine the sum in equation (51), we consider that there are $2(\sigma + 1)$ terms for which $r_{ab} = 0$, 2σ terms for which $r_{ab} = l$, $2(\sigma - 1)$ terms for which $r_{ab} = 2l$, and, in general, $2(\sigma + 1 - k)$ terms for which $r_{ab} = kl$, whence

$$P(\theta) = \frac{1}{(\sigma+1)^2} \sum_{k=0}^{\sigma} 2(\sigma+1-k) \frac{\sin qkl}{qkl} \tag{61}$$

For a large number of scattering centres, $\sigma + 1$ can be replaced by σ, and the sum by an integral, *i.e.*

$$P(\theta) = \frac{2}{\sigma} \int_0^\sigma \frac{\sin qkl}{qkl} \, dk - \frac{2}{ql\sigma^2} \int_0^\sigma \sin(qkl) \, dk. \tag{62}$$

Replacing qkl by u, and solving the second integral, by use of the trigonometrical expression $1 - \cos x = 2\sin^2(x/2)$, gives

$$P(\theta) = \frac{2}{qL} \int_0^{qL} \frac{\sin u}{u} \, du - \left[\frac{\sin(qL/2)}{qL/2} \right]^2 \tag{63}$$

The integral in this equation cannot be solved in an analytical way, but its values have been tabulated.[62]

In a similar way, although somewhat more complicated,[47] the expression for $P(\theta)$ for Gaussian coils can be obtained. This expression was first obtained by Debye. For a Gaussian coil with a mean-square radius of gyration $\langle R_g^2 \rangle$

$$P(\theta) = \frac{2}{(q^2 \langle R_g^2 \rangle)^2} [q^2 \langle R_g^2 \rangle - 1 + \exp(-q^2 \langle R_g^2 \rangle)] \tag{64}$$

Similarly, for a sphere of radius R,

$$P(\theta) = \frac{9}{q^6 R^6} (\sin qR - qR \cos qR)^2 \tag{65}$$

For spheres or rods, it is easy[28] to relate the geometrical size (R or L) to the radius of gyration R_g:

$$\text{for a sphere } R_g^2 = 3R^2/5, \quad \text{for a rod } R_g^2 = L^2/12 \tag{66}$$

For the three cases considered above, the function $1/P(\theta)$ has a simple asymptotic behaviour (see Figure 6) at large values of $q^2 \langle R_g^2 \rangle$. For a rod at large qL values,[63]

$$\left. \frac{1}{P(\theta)} \right|_{qL \to \infty} = \frac{2}{\pi^2} + \frac{qL}{\pi} = \frac{2}{\pi} + \frac{4L}{\lambda} \sin \frac{\theta}{2} \tag{67}$$

For a Gaussian chain, at large $q^2 \langle R_g^2 \rangle$ values,[64] the exponential term in the expression for $P(\theta)$ can be omitted, when

$$\left. \frac{1}{P(\theta)} \right|_{q^2 R_g^2 \to \infty} = \frac{q^2 \langle R_g^2 \rangle}{2} \left(1 + \frac{1}{q^2 \langle R_g^2 \rangle} \right) = \frac{1}{2} + \frac{8\pi \langle R_g^2 \rangle}{\lambda^2} \sin^2 \frac{\theta}{2} \tag{68}$$

For a homogeneous sphere with a radius R at large values of qR

$$\left. \frac{1}{P(\theta)} \right|_{qR \to \infty} = \frac{2q^4 R^4}{9} = \frac{512\pi^4 R^4}{9\lambda^4} \sin^4 \frac{\theta}{2} \tag{69}$$

These expressions explain the characteristic curvatures seen in Figure 7. The asymptotic behaviour can be observed experimentally only with particles of very large dimensions ($R_g \gtrsim \lambda/3$).

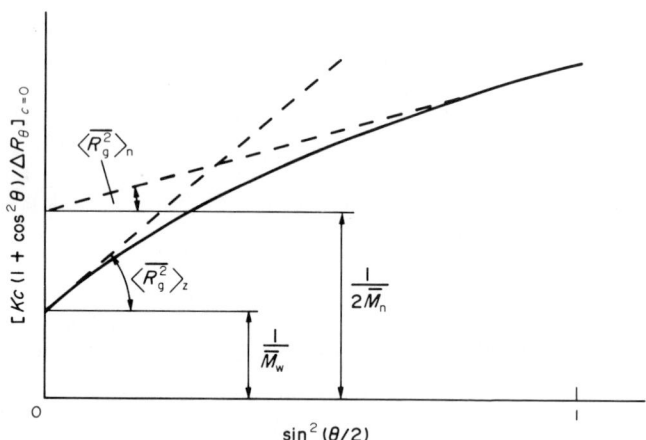

Figure 7   Plot of $[Kc(1 + \cos^2 \theta)/\Delta R_\theta]_{c=0}$ against $\sin^2(\theta/2)$, indicating the information which can be derived from the shape of this plot

5.2.5   Effect of Polydispersity

Most polymeric systems are polydisperse with respect to size and molecular weight. Consequently, the molecular weight and radius of gyration obtained by light scattering are average values.

At infinite dilution, the scattering intensity from an isomolecular species i of small dimensions, compared to the wavelength, is given by

$$\frac{\Delta R_{\theta,i}}{1 + \cos^2 \theta} = Kc_i M_i \tag{70}$$

The intensity scattered by a polydisperse system is the sum of the intensities scattered by the different molecular species

$$\frac{\Delta R_\theta}{1+\cos^2\theta} = K\sum_i c_i M_i \tag{71}$$

The constant K has the same value for all chemically identical species, provided that dn/dc does not depend on the molecular weight. This last condition is not necessarily true for very low molecular weight polymers.

Taking into account the fact that the total concentration is $c=\sum_i c_i$, we write

$$\frac{\Delta R_\theta}{1+\cos^2\theta} = Kc\,\frac{\sum_i c_i M_i}{\sum_i c_i} \tag{72}$$

or, since the expression $\sum_i c_i M_i / \sum_i c_i$ defines the weight-average molecular weight, $\bar M_w$,

$$\frac{\Delta R_\theta}{1+\cos^2\theta} = Kc\,\bar M_w \tag{73}$$

With particles of larger size we have, instead of expression (71), the following

$$\frac{\Delta R_\theta}{1+\cos^2\theta} = K\sum_i c_i M_i P_i(\theta) \tag{74}$$

where $P_i(\theta)$ is the particle scattering factor for the isomolecular species i. This equation can be rewritten as

$$\frac{\Delta R_\theta}{1+\cos^2\theta} = K\sum_i c_i\,\frac{\sum_i c_i M_i}{\sum_i c_i}\,\frac{\sum_i c_i M_i P_i(\theta)}{\sum_i c_i M_i} \tag{75}$$

or as

$$\frac{\Delta R_\theta}{1+\cos^2\theta} = Kc\,\bar M_w\,\overline{P(\theta)_z} \tag{76}$$

where $\overline{P(\theta)_z}$ is a z-average value of the particle scattering factors, given by

$$\overline{P(\theta)_z} = \frac{\sum_i c_i M_i P_i(\theta)}{\sum_i c_i M_i} \tag{77}$$

Substituting for $P_i(\theta)$ from equation (56) gives

$$\overline{P(\theta)_z} = 1 - \frac{q^2}{3}\,\frac{\sum_i c_i M_i \langle R_g^2\rangle_i}{\sum_i c_i M_i} \tag{78}$$

whence

$$\left.\frac{Kc(1+\cos^2\theta)}{\Delta R_\theta}\right|_{\substack{c\to 0\\ \theta\to 0}} = \frac{1}{\bar M_w}\left(1+\frac{16\pi^2}{\lambda^2}\,\overline{\langle R_g^2\rangle_z}\sin^2\frac{\theta}{2}\right) \tag{79}$$

Therefore, all the expressions derived previously are correct for polydisperse systems if we substitute the molecular weight M and the mean square radius of gyration $\langle R_g^2\rangle$ by their average values, $\bar M_w$ and $\langle R_g^2\rangle_z$, respectively.

We must take into account the fact that the interpretation of the average value $\langle R_g^2 \rangle$ depends on the shape of the particles. Therefore, if the variation in the radius of gyration can be expressed by a power relation,

$$\langle R_g^2 \rangle_i = aM_i^x \tag{80}$$

where the exponent x is fixed by the particle shape, *e.g.* for Gaussian coils $x = 1$, for rods $x = 2$ and for spheres $x = 2/3$, then

$$\overline{\langle R_g^2 \rangle_z} = \frac{a\sum\limits_i c_i M_i^{1+x}}{\sum\limits_i c_i M_i} \tag{81}$$

In the case of Gaussian coils, we have

$$a\frac{\sum\limits_i c_i M_i^2}{\sum\limits_i c_i M_i} = \overline{\langle R_g^2 \rangle_z} = \langle R_g^2 \rangle = aM \tag{82}$$

which means that, for polydisperse Gaussian coils, $\overline{\langle R_g^2 \rangle_z}$ is the same as the mean-square radius of gyration of an isomolecular species of the same molecular weight as the z-average molecular weight of the coils. When considering spheres, the molecular weight of the corresponding isomolecular species will be smaller than the z-average value and, for the rods, it will be larger.

If the molecular size is large enough to permit observation of the asymptotic behaviour of the function $P(\theta)$, other average values can be determined. Here, we will only consider the case of Gaussian coils.

As we have already seen, the function $P(\theta)$ for an isomolecular species i, has the following expression for large values of $q^2 \langle R_g^2 \rangle$

$$P_i(\theta) = \frac{2}{q^2 \langle R_g^2 \rangle_i} - \frac{2}{q^4 \langle R_g^4 \rangle_i} \tag{83}$$

Taking into account that, in this case, $\langle R_g^2 \rangle_i = aM_i$, we can write

$$P_i(\theta) = \frac{2}{uM_i} - \frac{2}{u^2 M_i^2} \tag{84}$$

where $u = aq^2$. Substituting this value of $P_i(\theta)$ into equation (77), gives

$$\overline{P(\theta)_z} = \frac{2}{u\sum\limits_i c_i M_i}\left(\sum\limits_i c_i - \sum\limits_i \frac{c_i}{uM_i}\right) \tag{85}$$

Hence

$$\overline{P(\theta)_z} = \frac{z}{u\bar{M}_w}\left(1 - \frac{1}{u\bar{M}_n}\right) \tag{86}$$

For large values of $u\bar{M}_n$

$$\frac{1}{\overline{P(\theta)_z}} = \frac{u\bar{M}_w}{2}\left(1 + \frac{1}{u\bar{M}_n}\right) \tag{87}$$

Combining this equation with the general equation (83) and using the relation $\langle R_g^2 \rangle_i = aM_i$ leads to

$$\left.\frac{Kc(1+\cos^2\theta)}{\Delta R_\theta}\right|_{\substack{c=0 \\ q^2\langle R_g^2\rangle\to\infty}} = \frac{1}{2\bar{M}_n}\left(1 + \frac{16\pi^2}{\lambda^2}\overline{\langle R_g^2 \rangle_n}\sin^2\frac{\theta}{2}\right) \tag{88}$$

Thus, the curve obtained by extrapolation to zero concentration in the Zimm plot shows a linear asymptote, with intercept and slope which lead to the number-average molecular weight and the number-average mean-square radius of gyration. As we have already seen, the first part of the curve allows determination of the weight-average molecular weight and the z-average mean-square radius of gyration (Figure 7).

Benoit *et al.*[65] have demonstrated that, for a wavelength of 325 nm and observation angles between 30° and 150°, the experimental curve obtained by extrapolating to zero concentration corresponds to equation (79) for values of $\langle \overline{R_g^2} \rangle_z^{1/2}$ below approximately 30 nm, but that, for values larger than 120 nm, the curve is better described by equation (88). For intermediate values, the experimental curve has a curvature which increases as the polydispersity increases.

Similar treatments have been made for chains with excluded volume[59] and for rods.[63] For rods, the number-average molecular weight and length can also be obtained.

5.2.6   Effect of Anisotropy

The preceding treatment rests on the assumption that the scattering particles are optically isotropic. In the case of particles which do not show this condition, fluctuations in the orientation of the particles can lead to additional scattering.

An expression for the Rayleigh ratio which accounts for all types of fluctuations is

$$R_\theta^{tot} = R_\theta^{iso} C(\theta) \tag{89}$$

where R_θ^{iso} is the Rayleigh ratio for isotropic particles, and $C(\theta)$ is a factor dependent on the depolarization ratio,[37,66,67] the magnitude of which is experimentally accessible.

The depolarization ratio for vertically polarized incident radiation, ρ_v, is defined as the ratio of the horizontally polarized component, H_V, to the vertically polarized component, V_V, of the scattered radiation

$$\rho_v = \frac{H_V}{V_V} \tag{90}$$

In the same way, the depolarization ratio for unpolarized incident radiation, ρ_u, is defined as

$$\rho_u = \frac{H_u}{V_u} = \frac{H_V + H_H}{V_V + V_H} \tag{91}$$

These ratios can be measured at any observation angle,[26] but generally they are measured at 90°.

For optically isotropic particles, the polarizability is independent of the direction and, therefore, the induced dipole is parallel to the electric field associated with the incident light. Consequently, the scattered light at 90° is plane-polarized perpendicular to the plane of the incident beam and the direction of observation, and the depolarization ratio at this angle is zero. For optically anisotropic particles, the polarizability depends on the direction and is represented by a tensor. It is always possible to choose a system of rectangular coordinates in such a way that the non-diagonal terms cancel. The axes of this system correspond to the three principal polarizabilities of the molecule. If the electric field is parallel to one of these axes, it produces a parallel dipole moment. However, in general, this electric vector is not parallel to any axis and, therefore, the electric vector of the scattered radiation is slightly inclined with regard to that associated with the incident radiation. So, when considering the observation angle $\theta = 90°$ in the horizontal plane, the scattered radiation includes a small horizontal component, and the depolarization ratio, ρ_u, is not zero. In fact, ρ_u attains a maximum value of 0.5 for long rod-shaped particles in which one of the polarizabilities is much larger than the other two.

The Rayleigh ratio of a system of anisotropic particles[37,47] is given by

$$R_{\theta,v}^{tot} = R_{\theta,v}^{iso} \frac{3 + 3\rho_v}{3 - (4 + 7\cos^2\theta)\rho_v} \tag{92}$$

for vertically polarized incident radiation, and

$$R_{\theta,u}^{tot} = \frac{R_{\theta,u}^{iso}}{1 + \cos^2\theta} \frac{6 + 6\rho_u}{6 - 7\rho_u} \left(1 + \frac{1 - \rho_u}{1 + \rho_u} \cos^2\theta \right) \tag{93}$$

when the incident radiation is unpolarized. For the observation angle $\theta = 90°$

$$R_{90,v}^{tot} = R_{90,v}^{iso} \frac{3 + 3\rho_v}{3 - 4\rho_v} \tag{94}$$

and

$$R_{90,u}^{tot} = R_{90,u}^{iso} \frac{6 + 6\rho_u}{6 - 7\rho_u} \tag{95}$$

The correction factor is called the Cabannès factor. It is most important when the particle size is small.

In the case of solutions, when the Rayleigh ratio R_θ is replaced by the excess quantity ΔR_θ, the solvent contribution to both components must be considered in determining the depolarization ratio. The Cabannès factor is close to unity for most macromolecular solutes[68] except for relatively rigid molecules,[61,69,70] as exemplified by a great number of biopolymers. So the effect of optical anisotropy can be neglected for most synthetic polymers, except when very precise molecular weights are required.

5.2.7 Instrumentation

5.2.7.1 Light-scattering photometers

A number of light-scattering instruments have been developed in different laboratories since the technique was adapted for the characterization of high polymers in solutions. Among these are the instruments originally designed by Debye,[46,71] Zimm,[55] Stein and Doty,[72] Bosworth et al.,[73] Cantow and Schulz,[74] Trap and Hermans,[75] Oster,[76] Hadow et al.,[77] Edsall et al.,[78] Stacey and Alexander,[79,80] McIntyre and Doderer,[81] Katime et al.[82-84] and many others.

Several instruments[7,73,85-87] have been developed commercially, for example, by the Phoenix Company (3803 North Fifth Street, Philadelphia 40, PA, USA), the Aminco Instrument Company (Silver Spring, MD, USA), Polymer Consultants Ltd. (Colchester, England), FICA, Shimadzu Seisakusho Ltd. (Kawaramachi-Nijo, Nakagyo, Kyoto, Japan) and Chromatix Inc.

Recently, two new models have been marketed, by Malvern Instruments Ltd. (Spring Lane South, Malvern, Worcestershire, WR14 1AQ, England) and by Amtec (Alpes-Maritimes Technologie, Av. du Dr. J. Lefebvre, 06270 Villeneuve-Loubet, France). These new instruments allow measurements to be made in both the classical (elastic) and the quasielastic mode. Both instruments are similar and use a laser as the light source. The photomultiplier is used as a photon counter, which is connected to a numerical correlator (K7027 (Malvern Instruments Ltd.) and BI-2030 (Brookhaven Instruments Corp., 200 Thirteenth Avenue, Ronkonkoma, NY 11779, USA), respectively). The photodetector is placed on a precision goniometer, so that study of the angular distribution of scattered light is possible.

5.2.7.2 Calibration and correction factors

In practice the light-scattering photometers make relative measurements of scattering intensities. Therefore the scattering intensities from solvent and solution are measured relative to that from a standard. In general the standard used is benzene at a given observation angle (90°). Taking into account the fact that the scattering volume depends on angle it is necessary to introduce a correction factor $(\sin \theta)$ in order to use the standard Rayleigh ratio at angles other than 90°

According to equation (15) the Rayleigh ratio for the standard is given by

$$R_{s,90} = I_s(90°) r^2 / I_0 \tag{96}$$

Thus, the experimental measurements may be expressed as

$$\Delta R_\theta = R_{s,90}(n_0^2/n_s^2)(\Delta I_\theta/I_{s,90}) \sin \theta \tag{97}$$

The ratio n_0^2/n_s^2 is the refractive index correction to the scattering volume due to Hermans and Levinson.[87] Thus equation (58) becomes

$$k'c(I_{s,90}/\alpha\Delta I_\theta)(dn/dc)^2 = (1/M)[1 + (16\pi^2/3\lambda^2)\langle R_g \rangle \sin(\theta/2)] + 2A_2 c \tag{98}$$

where $\alpha = \sin \theta/(1 + \cos^2 \theta)$ and $k' = (2\pi^2 n_s^2/\lambda_0^4 N R_{s,90})$.

Table 1 Light-scattering Data for Benzene at 298 K

Wavelength (nm)	$R_{s, 90} \times 10^6$ (cm^{-1})	*Refractive index*
436	46.3	1.5196
546	16.3	1.5020
578	12.6	1.4987
633	8.96	1.4947

The Rayleigh ratio for benzene has been measured at several wavelengths; the most acceptable values are shown in Table 1.

5.2.7.3 Clarification

Solvents and solutions used for light-scattering measurements should be absolutely dust-free. This is very important because dust particles affect the optical stability of the system and disturb the angular distribution of the scattering intensities appreciably, especially at low angles. The presence of dust is usually shown by marked deviations in the Zimm plot at angles below 45°. The preparation of liquids and solutions for light scattering has been reviewed in detail by Tabor.[88] Organic liquids can be clarified employing several methods, such as distillation, high-speed centrifugation, ultrafiltration, or by a combination of these methods. Filtration and high-speed centrifugation are the most popular. In the former, commercially available cellulose filters, with pore sizes ranging from 0.1 to 0.25 μm in diameter, are used. This method has the disadvantage that species of high molecular weight can be retained on the filter. High-speed centrifugation is probably the most widely used technique. It is necessary to employ centrifugation time of 60–120 min at 20 000–40 000 g. This method is not suitable for use with very polar and dense liquids.

5.2.7.4 Specific refractive index increment

Since the refractive index increment, dn/dc, appears as a squared term in equation (38), it is very important that this quantity is determined accurately and also that measurements are carried out at the same wavelength and temperature as that of the light-scattering measurements. Experimental procedures and tabulated values of dn/dc for many systems have been reported.[3]

This quantity can be measured using either a differential refractometer or an interferometer.[3] The latter method has the better precision. It is necessary to calibrate the differential refractometer or interferometer. The calibration can be carried out using standard aqueous solutions of potassium chloride[90] or sucrose.[91,92]

Scattering intensities can be measured sufficiently accurately when $dn/dc > 0.05$ cm^3 g^{-1}. In order to measure the weight-average molecular weight to $\pm 2\%$ it is necessary to know dn/dc to $\pm 1\%$.

5.2.8 Multicomponent Systems

5.2.8.1 Binary liquid systems

The interpretation of the thermodynamic behaviour of polymers dissolved in mixed solvents requires a knowledge of the thermodynamic properties of a binary mixture of solvents. Activity coefficients and excess Gibbs functions can be adequately determined from light-scattering measurements, following the method developed by Coumou and Mackor[93,94] and later used by others.[9,95,96] This kind of measurement is very useful because it gives information about the interaction parameters, especially in binary liquid mixtures with high boiling points.[95] This is very interesting because the properties of polymers dissolved in a binary solvent mixture depend not only on the interactions between the macromolecules and each of the solvents, but also on the interactions taking place between the solvents themselves.

5.2.8.2 Mixed solvents

When a polymer is dissolved in a binary solvent mixture, one of the solvents usually preferentially solvates the polymer. The solvent excess in the proximity of the macromolecule, compared to the bulk solution, causes interesting phenomena such as cosolvency,[97–100] non-cosolvency,[101,102] preferential solvation[103–108] and the variation of the unperturbed chain dimension, K_θ, of the polymer.[109–111]

With mixed solvents, the analysis of light-scattering data can be affected by the differential sorption of one of the two solvents to the polymer. Two different experimental procedures may be employed to obtain molecular weight; one using the Rayleigh ratio determined at constant solvent composition and the other at constant chemical potential of the solvents.

The determination at constant solvent composition requires the evaluation of refractive index increments at constant chemical potential. Application of molecular or fluctuation theory leads to a general equation which has a supplementary term

$$\Delta R = k'[(dn/dc)_{U_1} + \lambda'(dn/dU_1)]^2 cM = K'(dn/dc)_{\mu_1} cM \tag{99}$$

where $(dn/dc)_{U_1}$ and $(dn/dc)_{\mu_1}$ are the refractive index increments at constant solvent composition and constant chemical potential, respectively, λ' is the sorption coefficient and K' is defined as $K' = 2\pi^2 n^2/\lambda_0^2 N$.

(i) Solvent (1) + solvent (2)

In this kind of mixture the better solvent is always adsorbed by the polymer over the whole composition range.[112–115] In general, low values of the preferential adsorption coefficient are obtained in these systems.[116] If the solvents are polar, the possibility of specific association exists. This can influence the values of the adsorption coefficients, as has been pointed out by Cesteros et al.[117] for the system dimethyl formamide/acetonitrile/poly(methyl methacrylate).

(ii) Solvent (1) + precipitant (2)

For these mixtures, generally, the solvent (1) is adsorbed preferentially by the polymer. The preferential adsorption parameter increases as precipitant composition increases.[118–123] However, there are several mixtures, called 'synergistic solvents', where the addition of the precipitant provokes an increase in the solvent power of the mixture. For such systems at small amounts of precipitant, the preferential adsorption parameter becomes negative because the precipitant is adsorbed preferentially on the macromolecular coil (inverse adsorption).[124–126]

(iii) Precipitant (1) + precipitant (2)

A combination of two precipitants, in both of which the polymer is insoluble, can create a mixed solvent displaying miscibility over a broad range of composition. This effect is called cosolvency.[127–130] For these systems, inversion of the preferential adsorption parameter has been found.[131–136]

There are some important structural properties of the polymer such as its polarity,[115,116,137–139] chemical structure,[140–143] microtacticity,[98,144] molecular weight,[145–149] etc. that are determining factors in preferential adsorption phenomena.

5.2.8.3 Copolymers

Generally, the copolymer chains differ in chemical composition. This phenomenon is called heterogeneity of chemical composition. Thus, solutions of heterogeneous copolymers are in fact multicomponent systems. Owing to the polydispersity in chemical composition, copolymer solutions present a different scattering problem to that of homopolymer solutions.[150–152]

In the case of AB copolymers with polydispersity in molecular weight and chemical composition, but without an appreciable angular dissymmetry, the excess Rayleigh ratio, ΔR_θ, can be expressed as

$$\Delta R_\theta = K'_\theta \sum_i c_i M_i v_i^2 \tag{100}$$

where $K'_\theta = 2\pi^2 n^2(1 + \cos^2\theta)/\lambda_0^4 N$, and c_i is the concentration of the copolymer chains with molecular weight M_i, refractive index increment v_i and chemical composition x_i (x_i is the weight fraction of the constituent A).

The molecular weight calculated by means of the scattered intensity from a copolymer solution using the experimental refractive index increment, v_i, is in fact an apparent molecular weight M^*. Thus, we can write

$$\Delta R_\theta = K'_\theta v^2 c M^* \tag{101}$$

Combining equations (100) and (101) leads to

$$M^* = (1/v^2)\left(\sum_i v_i^2 c_i M_i\right)\Big/\left(\sum_i c_i\right) \tag{102}$$

The refractive index increment of a species i can be expressed as a function of the refractive index increment of its constituents v_A and v_B, assuming the additivity

$$v_i = x_i v_A + (1 - x_i) v_B \tag{103}$$

Substituting this expression in equation (3), the apparent molecular weight becomes

$$M^* = (1/v^2)[x M_w^A v_A^2 + (1 - x) M_w^B v_B^2 + 2 v_A v_B M_w^{AB}] \tag{104}$$

where x is the total weight fraction of the constituent A, M_w^A and M_w^B are the weight-average molecular weights of the constituents A and B, respectively. M_w^{AB} has no direct physical significance, but it can be expressed as a function of M_w^A and M_w^B and the true weight-average molecular weight of the copolymer, M_w

$$M_w^{AB} = (1/2)(M_w - x M_w^A + (1 - x) M_w^B) \tag{105}$$

Introducing the amount $\Delta x_i = x_i - x$, which is the deviation in composition of the species i from the average composition,[2] the apparent molecular weight is given by

$$M^* = M_w + 2P(v_A - v_B/v) + Q(v_A - v_B/v)^2 \tag{106}$$

where

$$P = \sum_i c_i M_i \Delta x_i \Big/ \sum_i c_i, \qquad Q = \sum_i c_i M_i (\Delta x_i)^2 \Big/ \sum_i c_i \tag{107}$$

Parameters P and Q are related to the heterogeneity of chemical composition of the copolymer.

Equation (106) is easy to use experimentally since the plot of M_w^* against $(v_A - v_B)/v$ is a parabola with a vertical axis and, theoretically, three points are enough to determine a curve of such a type. Thus, the true molecular weight of a copolymer, and the parameters P and Q, can be obtained by measuring the apparent molecular weight in several solvents. In the past, in many cases, P and Q have been overestimated; the necessary conditions for a good determination have been analyzed and discussed.[153]

In the case of copolymers with uniform chemical composition, $\Delta x_i = 0$ and therefore M^* does not depend on the refractive index of the solvent and is equal to M_w.

The angular distribution of light scattered by a solution of copolymers has been analyzed in detail by Benoit and Wippler.[154]

5.2.8.4 Polymer mixtures

The growing importance of polymer blends has stimulated developments in the analysis of light scattering from a solution of two polymers.[155-158]

This type of system can be considered as the extreme case of a copolymer solution of just two species.[139,159] Thus, the apparent molecular weight is given by

$$M^* = (1/v^2)[v_A^2 x M_A + v_B^2 (1 - x) M_B] \tag{108}$$

where x is the weight fraction of polymer A.

From the analysis of the scattering intensity as a function of concentration we can deduce the Flory–Huggins polymer–polymer interaction parameter χ_{23}.[159-161] This parameter is a measure of the thermodynamic interaction of the two polymers in solution. In general, the interaction between two different polymers makes only a small contribution to total scattering intensity, and so the determination of χ_{23} carries a large experimental error. However, the information is interesting, as it provides an insight into the compatibility of polymers which is difficult to obtain by other experimental techniques.

5.2.8.5 *Polyelectrolytes*

Light scattering can be used to investigate much more complicated cases, such as polyelectrolytes (including biopolymers). These macromolecules are long chains bearing ionizable groups. Typical examples are the sodium salt of poly(acrylic acid) and quaternized poly(4-vinylpyridine). Linear polyelectrolyte solutions show several characteristic physical properties, the most important of which is the expansion of the polyion coil and its marked change with experimental conditions. Further, the thermodynamic properties are affected by the properties of the counterions. These two factors make the study of the physical behaviour of polyelectrolytes very complicated. Kratochvil[162] and Nagasawa and Takahashi[163] have discussed the problem of light scattering from polyelectrolyte solutions.

The theory of light scattering given above is based on the assumption that the polymer molecules are largely independent of one another, but this is no longer the case when the polymer chains carry ionized groups, because of the electrostatic interaction between charges. The existence of these interactions complicates considerably the interpretation of the results. This complication can be avoided if the range of the electrostatic interactions between macroions is small compared with the size of the volume element. A practical criterion for this is that the Debye–Huckel length $1/\kappa$ should be small compared with the size of the volume element. In general, if $2\sin(\theta/2)/\kappa < 0.05$ the error is less than 1%. For SAXS, however, this condition may be difficult to meet.

Light scattering from a binary system containing non-ionic polymer is described by equations (58) and (59). Light scattering from polyelectrolyte solutions is usually observed in the presence of added salts such as sodium chloride ($0.1\,\mathrm{mol\,l^{-1}}$). Consequently, the above equations developed for binary systems cannot be used without modification for polyelectrolyte/salt/water systems.

The refractive index increment is measured at constant chemical potential, $(\mathrm{d}n/\mathrm{d}c)_\mu$. This is achieved experimentally when the binary solvent (salt/water) and the solution are brought to dialysis equilibrium by use of a membrane permeable to the solvent molecules but not to the solute (Donnan membrane equilibrium). This procedure allows use of the equations obtained for binary mixtures without change. Of course, the radius of gyration of a polyion is affected by the concentration of added salt.[164–170]

5.2.8.6 *Micellar solutions and aggregates*

Under suitable thermodynamic conditions some macromolecules can self-associate to form particles of higher molecular weight.[171–174] These systems can be considered as multicomponent systems due to the great differences in structure between aggregates and free chains. This phenomenon can occur between like or unlike molecules.

A micellar solution is a system containing micelles and free molecules in equilibrium. Such solutions have a critical micelle concentration (cmc); at lower concentrations than the cmc the solution contains only free molecules. Due to this critical behaviour, the dependence of the scattering intensity on concentration shows a sharp change at the cmc.[175,176]

In order to calculate the molecular weight of the micelles, the solution of critical concentration is considered to be the solvent and equation (59) becomes

$$[K(c-c_{\mathrm{c.m.c}})(1+\cos^2\theta)/R_\theta - R_{\mathrm{c.m.c}}]_{\theta=0} = (1/\bar{M}_w) + 2A_2(c-c_{\mathrm{c.m.c}}) + \cdots \qquad (109)$$

where $\bar{M}_w$ is the weight-average molecular weight of the micelles. The specific refractive index is determined from the difference between solution at concentration C and at the cmc.[177] This determination of molecular weight allows the average aggregation number of the micelles to be calculated.

If polymer blends are dissolved in a solvent, it is frequently possible to observe the formation of aggregates.[178,179] Due to the influence of molecular weight on scattering intensity, it is possible to follow the aggregation process with time,[180,181] However, it is very difficult to obtain quantitative results. In some cases it is possible to find aggregates in simple polymer solutions. This phenomenon disturbs the Zimm plot at low angles and leads to a high value of the molecular weight. However, if the aggregate concentration is low, an extrapolation from higher angles can lead to a molecular weight characteristic of the polymer.[182]

5.3 SMALL-ANGLE X-RAY SCATTERING

5.3.1 Theoretical Background

The theory for X-ray scattering (at angles less than 2°) was developed by Guinier and Fournet.[14] Fundamentally, the theoretical outline presented above for light-scattering studies is valid for electromagnetic radiation of all wavelengths. Consequently, it is also valid for small-angle X-ray scattering (SAXS) and neutron scattering (SANS). However, for X-rays, λ is as low as 0.154 nm, and this is much smaller than typical macromolecular dimensions, so structural information over small distances is available from X-ray scattering. The intensity of scattering is a function of the electron density. The form of equation (38) remains the same. Only the expression for the optical constant K requires alteration. In X-ray scattering the term K is given by

$$K_{SAXS} = e^4 (\Delta \rho_e)^2 / m^2 v_e^4 N_A \tag{110}$$

where e and m_e are the charge and mass of the electron, v_e is the velocity of light and $\Delta \rho_e$ is the excess electron density of the solute defined by

$$\Delta \rho_e = (e_e/M) - \bar{v}_2 \rho_e^0 \tag{111}$$

where e_e is the number of electrons carried by a solute molecule, $\bar{v}_2$ the partial specific volume of the solute and ρ_e^0 the number of moles of electrons per unit volume of the solvent medium.

The molecular weight is then obtained from values of the Rayleigh scattering ratio, R_θ, extrapolated to $\theta = 0$ using, for $\lambda = 0.154$ nm, the relation[183,184,255]

$$R_\theta = (4.8 \text{ cm}^{-1}) \bar{M}_w (\Delta \rho_e)^2 c \tag{112}$$

SAXS experiments are fairly difficult to perform, due mainly to the difficulty in accessing the small observation angle but also to the weak scattering. However, the method has provided useful information on polymers with dimensions in the range 1 to 100 nm, and in this respect is complementary to light scattering. The undiffracted X-ray beam is always divergent, due to the design of the target and the X-ray tube. Consequently, since X-rays cannot be focused in the same manner as light, they must be collimated in order to obtain a narrow, nearly parallel beam, as outlined in Figure 8.

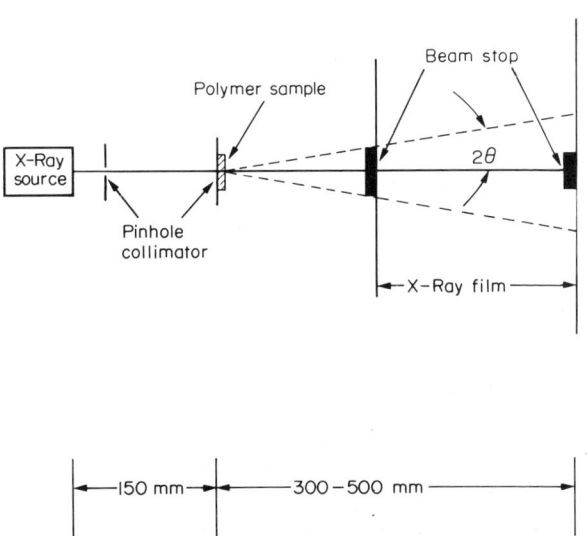

Figure 8 Schematic representation of X-ray diffraction for polymers showing alternative film positions

According to Debye,[185] the intensity scattered by a gas molecule consisting of N atoms with shape factors f_i and mutual distances r_{nm} is given by

$$I_{scat}(s) = I_e \sum_{m=1}^{N} \sum_{n=1}^{N} f_m f_n [(\sin 2\pi s r_{nm})/2\pi s r_{nm}], \quad s = (2/\lambda) \sin \theta \tag{113}$$

where I_e is the scattered intensity from a single electron, λ is wavelength of the radiation used and 2θ is the angle between the incident and scattered beams. From this equation, the angular distribution of the scattering from variously shaped bodies can be calculated by introducing expressions for r_{nm} which are characteristic of the geometry of the molecule. The approximate shape of a scattering particle can be determined by comparing the experimental scattering envelope with calculated envelopes.

Guinier,[186] in 1939, showed that the scattering experiment can yield a characteristic geometric parameter for any particle independently of its shape, namely its radius of gyration $\langle R_g \rangle$. For isotropic particles we can obtain, from expanding equation (113)

$$i_n(s) = i_n(0)[1 - (4\pi^2/3)\langle R_g^2 \rangle s^2 + \ldots] \simeq i_n(0) \exp[(-4\pi^2/3)\langle R_g^2 \rangle s^2] \tag{114}$$

where $i_n(s)$ is the scattered intensity at angle 2θ corresponding to a given value of s, $i_n(0)$ is the normalized intensity extrapolated to zero angle. Thus, at very low angles, if the logarithm of the scattered intensity is plotted against s^2, a straight line with the slope $(4/3)\pi^2 \langle R_g^2 \rangle$ should be obtained (Figure 9). The radius of gyration can be calculated from the slope of this line. The intercept is proportional to the square of the molecular weight.[7] SAXS can be used to determine the radius of gyration of molecules down to a molecular weight of about 300.

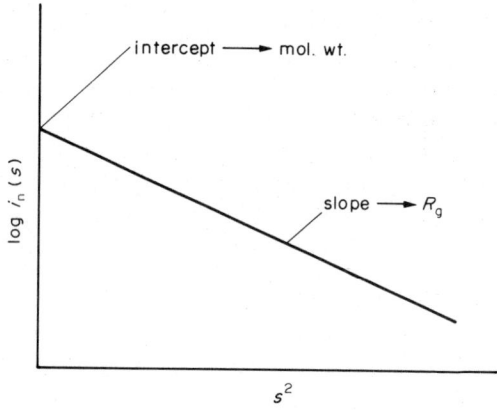

Figure 9   Idealized Guinier plot, where $s = (2/\lambda) \sin \theta$ and $i_n(s)$ is the scattered intensity at angle 2θ corresponding to a given value of s. The radius of gyration is obtained from the slope, and the molecular weight from the intercept

Due to experimental problems, it is rarely possible to use a point source. In general, the geometry of the chosen source is one defined by a narrow slit. In analogy to equation (114), the normalized scattered intensity $j_n(s)$ from an infinite-slit source is given by[14, 256]

$$j_n(s) = j_n(0) \exp(-(4/3)\pi^2 R_a^2 s^2) + \psi(s) \tag{115}$$

where $j_n(0)$ is the value of $j_n(s)$ extrapolated to zero angle, R_a is the apparent radius of gyration and $\psi(s)$ is a residual function expressing the difference between the Gaussian portion of equation (115) and the scattering actually observed.

For an isotropic particle of uniform electron density, at large values of s, and using slit optics[187]

$$\lim_{s \to \infty} s^3 j_n(s) = \lim s^3 j_n(s) + \delta^* s^3 \tag{116}$$

$$j_n^*(s) \equiv j_n(s) - \delta^* \tag{117}$$

where $\lim_{s\to\infty} s^3 j_n(s) = A$ and δ^* are constants and $j_n^*(s)$ is a corrected normalized scattering intensity defined by equation (116).

A plot of $s^3 j_n(s)$ *vs.* s^3 follows the form shown in Figure 10. The intercept of the straight line is A and its slope is δ^*. This last parameter reflects the internal structure of the macromolecule. Knowledge of A and $j_n(s)$ permits the calculation of several other molecular parameters.

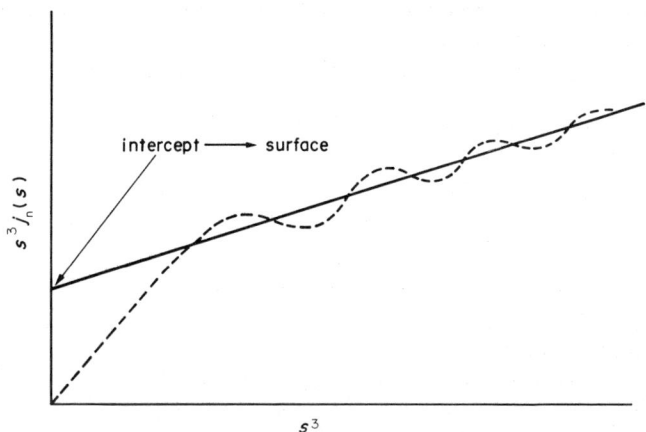

Figure 10 Idealized Soulé–Porod plot, where $s = (2/\lambda)\sin\theta$ and $j_n(s)$ is a corrected normalized scattering intensity

5.3.2 Surface Area Determinations

There are two methods, developed by Porod[188–190] and Debye *et al.*,[12] for evaluating the surface area of a dispersed phase. Porod has shown that for a two-phase system, the surface area per unit mass of dispersed phase S_2 can be calculated from

$$S_2 = K\Phi_1 \lim s^3 j_n(s)/d_2 \tag{118}$$

where the constant K is

$$K = 8\pi/\tilde{Q} \tag{119}$$

for a slit-collimated system, and

$$K = 2\pi^2/Q \tag{120}$$

for a pinhole-collimated system. Q and $\tilde{Q}$ are the total integrals of the scattering intensities (invariant) over the whole angular range defined by

$$Q = \int_0^\infty s^2 i(s)\,ds \quad \text{for pinhole optics} \tag{121}$$

and

$$\tilde{Q} = \int_0^\infty s i(s)\,ds \quad \text{for infinite slits} \tag{122}$$

The invariant Q depends on the volume of particles and is independent of their shape.[188]

In the Debye method, the surface area is evaluated from

$$S_2 = 4\Phi_1/l_p d_2 \tag{123}$$

where l_p is the correlation distance calculated by plotting $1/i^{1/2}$ against $(2\theta)^2$, *i.e.*

$$l_p = (\lambda/2\pi)\,(\text{slope/intercept})^{1/2} \tag{124}$$

The Debye method is experimentally and computationally much more rapid than the Porod method.

In general, X-ray methods give higher specific surface values than other methods, due to the fact that X-rays can access pores inaccessible to other methods.

SAXS can supply valuable information for dilute or dense systems of colloidal particles, swollen polymers, deformation and annealing of polymers and branched polymers.[191-230] SAXS has also been used as a method of investigating deformation and fracture in polymers.[231-238]

5.3.3 Instrumentation

Since X-rays cannot be focused in the same manner as light, they must be collimated in order to obtain a narrow, nearly parallel beam. The main systems used are slits,[239-240] pinholes,[241-248] the Kratky U-bar method,[249-253] Franks curved-mirror X-ray focusing camera[254] and Guinier focusing (single and double) by diffraction. The various methods are illustrated in Figure 11; the simplest is the use of pinholes or slits.

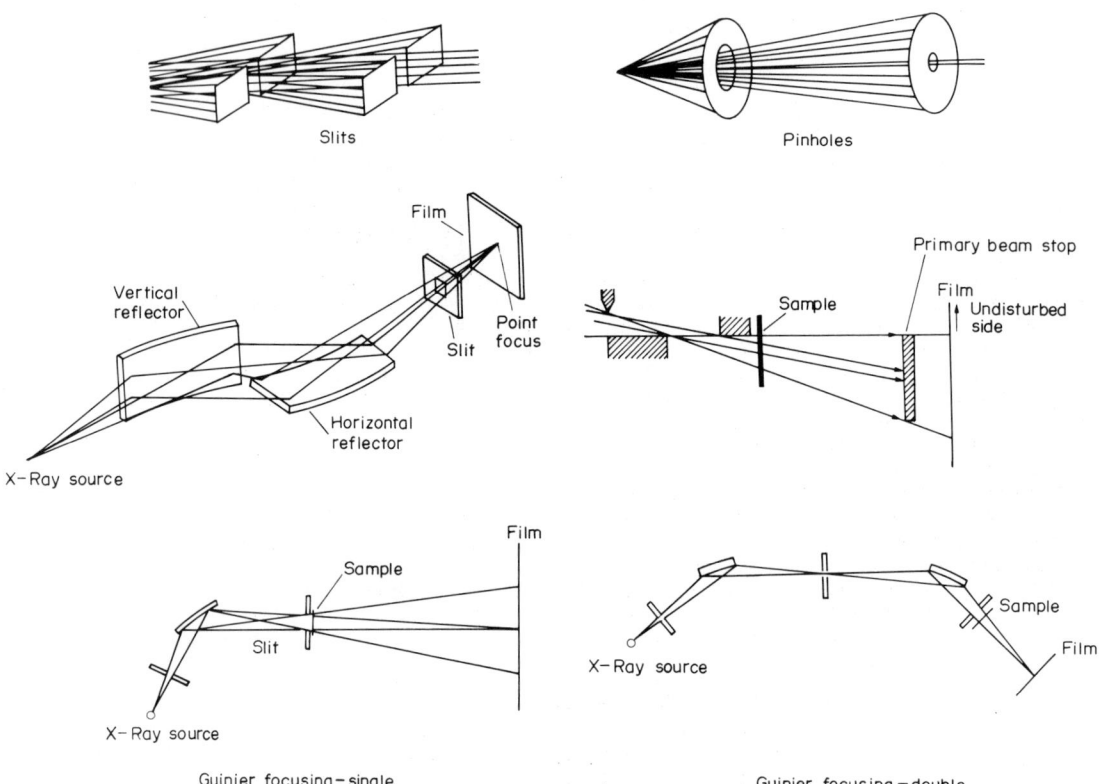

Figure 11 Arrangement of the different collimation systems

The target most often used is copper, with a K_α line at 0.154 nm. The two detection systems used are photographic film and electronic counters. The first method is the oldest and least expensive means of recording the scattered radiation. There are two basic types of counting systems: ionization detectors and solid-state detectors. The last one is the most popular due to its low price.

Samples can be studied both in the solid state and in solution. Films, fibres, rods and sheets can be directly placed in the main beam. In the case of solutions or suspensions it is necessary to determine the background scattering due to the container and liquid employed to dissolve or suspend the particles.

5.4 REFERENCES

1. P. Debye, in 'Light Scattering from Dilute Polymer Solutions', ed. D. McIntyre and F. Gornick, Gordon and Breach, New York, 1964, p. 123; *J. Phys. Colloid Chem.*, 1947, **51**, 18.

2. H. E. Eisenberg, 'Biological Macromolecules and Polyelectrolytes in Solution', Oxford University Press, Oxford, 1976, chap. 4.
3. M. B. Huglin (ed.), 'Light Scattering from Polymer Solutions', Academic Press, London, 1972.
4. N. C. Billingham, 'Molar Mass Measurements in Polymer Science', Kogan Page, London, 1977, chap. 5.
5. M. B. Huglin, *Top. Curr. Chem.*, 1978, **77**, 141.
6. D. McIntyre and F. Gornick (eds.), 'Light Scattering from Dilute Polymer Solutions', Gordon and Breach, New York, 1964.
7. E. P. Pittz, J. C. Lee, B. Bablouzian, R. Townend and S. N. Timasheff, *Methods Enzymol.*, 1973, **27** (D), 209.
8. H. H. Lewis, R. L. Schmidt and H. L. Clever, *J. Phys. Chem.*, 1970, **74**, 4377.
9. B. M. Fechner and C. Strazielle, *Makromol. Chem.*, 1972, **160**, 195.
10. G. V. Schulz and M. Lechner, in 'Light Scattering from Polymer Solutions', ed. M. B. Huglin, Academic Press, London, 1972, chap. 12.
11. Th. G. Scholte, *J. Polym. Sci., Part A-2*, 1971, **9**, 1553.
12. Th. G. Scholte, *Eur. Polym. J.*, 1970, **6**, 1063.
13. G. A. Miller, *J. Phys. Chem.*, 1967, **71**, 2305.
14. A. Guinier and G. Fournet, 'Small Angle Scattering of X-Rays', Wiley, New York, 1955.
15. V. Luzzati, *Acta Crystallogr.*, 1960, **13**, 939.
16. S. N. Timasheff, in 'Molecular Scattering of Light', ed. I. L. Fabelinskii, Plenum Press, New York, 1968, p. 337.
17. S. N. Timasheff, *J. Chem. Educ.*, 1964, **41**, 314
18. G. D. Patterson and J. P. Latham, *Macromol. Rev.*, 1980, **15**, 1.
19. A. J. Hyde, in 'Developments in Polymer Characterization — 1', ed. J. V. Dawkins, Applied Science, London, 1978.
20. B. Chu, 'Laser Light Scattering', Academic Press, New York, 1974.
21. B. J. Berne and R. Pecora, 'Dynamic Light Scattering', Wiley, New York, 1976.
22. K. A. Stacey, 'Light Scattering in Physical Chemistry', Butterworths, London, 1956.
23. I. L. Fabelinskii (ed.), 'Molecular Scattering of Light', Plenum Press, New York, 1968.
24. M. Kerker, 'Electromagnetic Scattering', Pergamon Press, Oxford, 1963.
25. R. L. Rowell and R. S. Stein, 'Electromagnetic Scattering', Gordon and Breach, New York, 1967.
26. E. F. Casassa and G. C. Berry, in 'Techniques and Methods of Polymer Evaluation', ed. P. E. Slade, Dekker, New York, 1975, vol. 4, chap. 5.
27. C. Quivoron, in 'Chimie Macromoléculaire', ed. G. Champetier, R. Buvet, J. Néel and P. Sigwalt, Hermann, Paris, 1970, vol. 2, part 4, chap. 3.
28. C. Tanford, 'Physical Chemistry of Macromolecules', Wiley, New York, 1961, chap. 5.
29. M. Kerker, 'The Scattering of Light and Other Electromagnetic Radiation', Academic Press, New York, 1969.
30. H. Morawetz, 'Macromolecules in Solution', Wiley, New York, 1965, chap. 5.
31. P. J. Flory, 'Principles of Polymer Chemistry', Cornell University Press, Ithaca, NY, 1953, chap. 7.
32. H. G. Elias, 'Macromolecules', Plenum Press, New York, 1984, vol. 1, p. 311.
33. V. N. Tsvetkov, V. E. Eskin and S. Ya. Frenkel, 'Structure of Macromolecules in Solution', Mir, Moscow, 1971.
34. Lord Rayleigh, *Philos. Mag.*, 1871, **41** (4), 447.
35. Lord Rayleigh, *Philos. Mag.*, 1881, **12** (5), 81.
36. J. Cabannes, *Ann. Phys. (Paris)*, 1921, **15**, 5.
37. J. Cabannes and Y. Rocard, 'La Diffusion Moléculaire de la Lumière', Les Presses Universitaires de Paris, Paris, 1929.
38. S. Ewing, *J. Opt. Soc. Am. Rev. Sci. Instrum.*, 1926, **12**, 15.
39. B. H. Zimm, in 'Light Scattering from Dilute Polymer Solutions', ed. D. McIntyre and F. Gornick, Gordon and Breach, New York, 1964, p. 63; *J. Chem. Phys.*, 1945, **13**, 141.
40. C. I. Carr, Jr. and B. H. Zimm, in 'Light Scattering from Dilute Polymer Solutions', ed. D. McIntyre and F. Gornick, Gordon and Breach, New York, 1964, p. 111; *J. Chem. Phys.*, 1950, **18**, 1616.
41. M. Fixman, in 'Light Scattering from Dilute Polymer Solutions', ed. D. McIntyre and F. Gornick, Gordon and Breach, New York, 1964, p. 69; *J. Chem. Phys.*, 1955, **23**, 2074.
42. A. Einstein, *Ann. Phys. (Leipzig)*, 1910, **33**, 1275.
43. M. Smoluchowski, *Ann. Phys. (Leipzig)*, 1908, **25**, 205.
44. M. Smoluchowski, *Philos. Mag.*, 1912, **23** (6), 165.
45. P. Debye, 'The Collected Papers of P. Debye', Wiley–Interscience, New York, 1954.
46. P. Debye, in 'Light Scattering from Dilute Polymer Solutions', ed. D. McIntyre and F. Gornick, Gordon and Breach, New York, 1964, p. 13; *J. Appl. Phys.*, 1944, **15**, 338.
47. G. Oster, *Chem. Rev.*, 1948, **43**, 319.
48. B. H. Zimm, R. S. Stein and P. Doty, in 'Light Scattering from Dilute Polymer Solutions', ed. D. McIntyre and F. Gornick, Gordon and Breach, New York, 1964, p. 7; *Polym. Bull. (Berlin)*, 1945, **1**, 90.
49. W. Heller, in 'Light Scattering from Dilute Polymer Solutions', ed. D. McIntyre and F. Gornick, Gordon and Breach, New York, 1964, p. 37; *Rec. Chem. Prog.*, 1959, **20**, 209.
50. P. Doty and J. T. Edsall, *Adv. Protein Chem.*, 1951, **6**, 35.
51. P. Kratochvil, in 'Light Scattering from Polymer Solutions', ed. M. B. Huglin, Academic Press, London, 1972, chap. 7.
52. P. Debye, in 'Light Scattering from Dilute Polymer Solutions', ed. D. McIntyre and F. Gornick, Gordon and Breach, New York, 1964, p. 139.
53. P. Debye, *Ann. Phys. (Leipzig)*, 1915, **46**, 809.
54. B. H. Zimm, in 'Light Scattering from Dilute Polymer Solutions', ed. D. McIntyre and F. Gornick, Gordon and Breach, New York, 1964, p. 149; *J. Chem. Phys.*, 1948, **16**, 1093.
55. B. H. Zimm, in 'Light Scattering from Dilute Polymer Solutions', ed. D. McIntyre and F. Gornick, Gordon and Breach, New York, 1964, p. 157; *J. Chem. Phys.*, 1948, **16**, 1099.
56. Th. Neugebauer, *Ann. Phys. (Leipzig)*, 1943, **42**, 509.
57. Lord Rayleigh, *Proc. R. Soc. London, Ser. A*, 1911, **84**, 25.
58. R. Gans, *Ann. Phys. (Leipzig)*, 1925, **76**, 29.
59. C. Loucheux, G. Weill and H. Benoit, *J. Chim. Phys. Phys.-Chim. Biol.*, 1958, **55**, 540.
60. A. Peterlin, *J. Polym. Sci.*, 1960, **47**, 403.
61. C. Picot, G. Weill and H. Benoit, *J. Colloid Interface Sci.*, 1968, **27**, 360.

62. M. Abramowitz and J. A. Stegun, 'Handbook of Mathematical Functions', Dover, New York, 1965.
63. A. Holtzer, in 'Light Scattering from Dilute Polymer Solutions', ed. D. McIntyre and F. Gornick, Gordon and Breach, New York, 1964, p. 195; *J. Polym. Sci.*, 1955, **17**, 432.
64. H. Benoit, in 'Light Scattering from Dilute Polymer Solutions', ed. D. McIntyre and F.Gornick, Gordon and Breach New York, 1964, p. 175; *J. Polym. Sci.*, 1953, **11**, 507.
65. H. Benoit, A. M. Holtzer and P. Doty, in 'Light Scattering from Dilute Polymer Solutions', ed. D. McIntyre and F. Gornick, Gordon and Breach, New York, 1964, p. 189; *J. Phys. Chem.*, 1954, **58**, 635.
66. D. S. Subbaramaiya, *Proc.—Indian Acad. Sci., Sect. A*, 1935, **1A**, 709.
67. R. S. Krishnan, *Proc.—Indian Acad. Sci., Sect. A*, 1935, **1A**, 717.
68. G. Vallet and C. Wippler, 'Determination des Masses Moléculaires', Monographies de Chimie Organique, Masson, Paris, 1961, p. 329.
69. P. Horn, *Ann. Phys. (Paris)*, 1955, **10**, 386.
70. H. Utiyama, *J. Phys. Chem.*, 1965, **69**, 4138.
71. P. Debye, *J. Appl. Phys.*, 1946, **17**, 392.
72. R. S. Stein and P. Doty, *J. Am. Chem. Soc.*, 1946, **68**, 159.
73. P. Bosworth, C. R. Masson, H. W. Melville and F. W. Peaker, *J. Polym. Sci.*, 1952, **9**, 565.
74. H. J. Cantow and G. V. Schulz, *Z. Phys. Chem. (Wiesbaden)*, 1954, **2**, 117.
75. H. J. L. Trap and J. J. Hermans, *Recl. Trav. Chim. Pays-Bas*, 1954, **73**, 167.
76. G. Oster, *Anal. Chem.*, 1953, **25**, 1165.
77. H. J. Hadow, H. Sheffer and J. C. Hyde, *Can. J. Res., Sect. B*, 1949, **27**, 791.
78. J. T. Edsall, H. Edelhoch, R. Lontie and P. R. Morrison, *J. Am. Chem. Soc.*, 1950, **72**, 4641.
79. P. Alexander and K. A. Stacey, *Trans. Faraday Soc.*, 1955, **51**, 299.
80. K. A. Stacey and P. Alexander, 'International Symposium on Macromolecular Chemistry', 1954, Milan–Turin.
81. D. McIntyre and G. C. Doderer, *J. Res. Natl. Bur. Stand. (U.S.)*, 1959, **62**, 153.
82. I. Katime, A. Roig, L. M. Leon and S. Montero, *Afinidad*, 1977, **34**, 461.
83. I. Katime, A. Roig, S. Montero and L. M. Leon, *Afinidad*, 1977, **34**, 541.
84. I. Katime, *Afinidad*, 1980, **37**, 387.
85. H. Utiyama, in 'Light Scattering from Polymer Solutions', ed. M. B. Huglin, Academic Press, London, 1972, chap. 3.
86. R. Chiang, in 'Newer Methods of Polymer Characterization', ed. B. Ke, Wiley–Interscience, New York 1964.
87. H. Utiyama, in 'Light Scattering from Polymer Solutions', ed. M. B. Huglin, Academic Press, London, 1972, chap. 4.
88. J. J. Hermans and S. Levinson, *J. Opt. Soc. Am.*, 1951, **41**, 460.
89. B. E. Tabor, in 'Light Scattering from Polymer Solutions', ed. M. B. Huglin, Academic Press, London, 1972, chap. 1.
90. A. Kruis, *Z. Phys. Chem., Abt. B*, 1936, **34**, 13.
91. F. J. Bates, *Natl. Bur. Stand. (U.S.), Circ.*, **C440**, 1942.
92. L. J. Gostling and M. S. Morris, *J. Am. Chem. Soc.*, 1949, **71**, 1998.
93. D. J. Coumou and E. L. Mackor, *Trans. Faraday Soc.*, 1964, **60**, 1726.
94. D. J. Coumou, H. Hijmans and E. L. Mackor, *Trans. Faraday Soc.*, 1960, **60**, 2244.
95. I. Katime and J. R. Ochoa, *Makromol. Chem.*, 1983, **184**, 2143.
96. I. Fernandez–Pierola and A. Horta, *J. Chim. Phys. Phys.-Chim. Biol.*, 1980, **77**, 271.
97. I. Katime and C. Strazielle, *C. R. Hebd. Seances Acad. Sci., Ser. C*, 1974, **278**, 1081.
98. I. Katime, P. Garro and J. M. Teijón, *Eur. Polym. J.*, 1975, **11**, 881.
99. I. Katime and J. R. Ochoa, *Makromol. Chem., Rapid Commun.*, 1982, **3**, 783.
100. I. Katime, L. Gargallo, D. Radić and A. Horta, *Makromol. Chem.*, 1985, **186**, 2125.
101. B. A. Wolf and M. M. Willms, *Makromol. Chem.*, 1978, **179**, 2265.
102. O. Fuchs, *Kunststoffe*, 1953, **43**, 409.
103. I. Katime and C. Strazielle, *Makromol. Chem.*, 1977, **178**, 2295.
104. I. Katime and B. Amo, *Polym. Bull. (Berlin)*, 1980, **2**, 383.
105. I. Katime, R. Valenciano and T. Nuño, *Polym. Bull. (Berlin)*, 1982, **6**, 437.
106. I. Katime, R. Valenciano and J. M. Teijón, *Eur. Polym. J.*, 1979, **15**, 261.
107. A. Dondos, I. Katime and C. Strazielle, *Eur. Polym. J.*, 1983, **19**, 579.
108. I. Katime and J. R. Ochoa, *Eur. Polym. J.*, 1984, **20**, 99.
109. A. Dondos, P. Rempp and H. Benoit, *J. Polym. Sci., Part C*, 1970, **30**, 9.
110. L. Gargallo, *Makromol. Chem.*, 1976, **177**, 233.
111. A. Horta and I. Katime, *Macromolecules*, 1984, **17**, 2734.
112. B. E. Read, *Trans. Faraday Soc.*, 1960, **56**, 382.
113. C. Strazielle and H. Benoit, *J. Chim. Phys. Phys.-Chim. Biol.*, 1961, **58**, 678.
114. H. Lange, *Makromol. Chem.*, 1965, **86**, 192.
115. K. Okita, A. Teramoto, K. Kawahara and H. Fujita, *J. Phys. Chem.*, 1968, **72**, 278.
116. L. C. Cesteros, C. Strazielle and I. Katime, *Eur. Polym. J.*, 1986, **22**, 399.
117. L. C. Cesteros, C. Strazielle and I. Katime, *J. Chem. Soc., Faraday Trans. 1*, 1986, **82**, 1321.
118. I. Katime, R. Valenciano and J. M. Teijón, *Eur. Polym. J.*, 1979, **15**, 261.
119. I. Katime and R. Valenciano, *Polym. Bull. (Berlin)*, 1980, **3**, 431.
120. M. Hert and C. Strazielle, *Makromol. Chem.*, 1974, **175**, 2149.
121. M. Hert, C. Strazielle and H. Benoit, *Makromol. Chem.*, 1973, **172**, 169.
122. C. Strazielle and H. Benoit, *J. Chim. Phys. Phys.-Chim. Biol.*, 1961, **58**, 675.
123. B. Chaufer, B. Sebille and Q. Quivoron, *Eur. Polym. J.*, 1975, **11**, 683.
124. I. Katime, J. Tamarit and J. M. Teijón, *An. Quim.*, 1979, **75**, 7.
125. Z. Tuzar and P. Kratochvil, *Collect. Czech. Chem. Commun.*, 1967, **32**, 3358.
126. H.-G. Elias and O. Etter, *Makromol. Chem.*, 1965, **89**, 228.
127. R. L. Scott, *J. Chem. Phys.*, 1949, **17**, 268.
128. B. A. Wolf, J. W. Breitenbach and H. Senftl, *Monatsh. Chem.*, 1970, **101**, 57.
129. B. A. Wolf, *Adv. Polym. Sci.*, 1972, **10**, 109.
130. B. A. Wolf and R. J. Molinari, *Makromol. Chem.*, 1973, **173**, 241.

131. H. Maillols, L. Bardet and S. Gromb, *Eur. Polym. J.*, 1978, **14**, 1015.
132. J. M. G. Cowie and J. T. McCrindle, *Eur. Polym. J.*, 1972, **8**, 1185.
133. L. Gargallo, D. Radic and I. Katime, *Eur. Polym. J.*, 1981, **17**, 439.
134. L. Gargallo, D. Radic and I. Katime, *Eur. Polym. J.*, 1980, **16**, 383.
135. J. M. G. Cowie and J. T. McCrindle, *Eur. Polym. J.*, 1972, **8**, 1325.
136. J. M. G. Cowie and I. J. McEwen, *Macromolecules*, 1974, **7**, 291.
137. I. Katime and R. Valenciano, *An. Quim.*, 1979, **75**, 258.
138. I. Katime, R. Valenciano and M. Otaduy, *An. Quim.*, 1981, **77**, 405.
139. I. Katime and L. C. Cesteros, *An. Quim.*, 1982, **78**, 161.
140. L. Gargallo, N. Hamidi, I. Katime and D. Radic, *Polym. Bull. (Berlin)*. 1985, **14**, 393.
141. I. Katime, L. Gargallo, D. Radic and A. Horta, *Makromol. Chem.*, 1985, **186**, 2125.
142. L. Gargallo and D. Radic, *Adv. Colloid Interface Sci.*, 1984, **21**, 1.
143. D. Radic and L. Gargallo, *Eur. Polym. J.*, 1982, **18**, 151.
144. I. Katime, J. M. Teijón and J. Tamarit, *An. Quim.*, 1979, **75**, 255.
145. A. Horta and I. Katime, *Macromolecules*, 1984, **17**, 2734.
146. A. Dondos and H. Benoit, in 'Order in Polymer Solutions', ed. K. Šolc, Gordon and Breach, London, 1976.
147. J. Pouchlý, A. Živný and K. Šolc, *J. Polym. Sci., Part C*, 1968, **23**, 245.
148. A. Živný, J. Pouchlý and K. Šolc, *Collect. Czech. Chem. Commun.*, 1967, **32**, 2753.
149. A. Dondos, *Macromolecules*, 1980, **13**, 1023.
150. W. H. Stockmayer, L. D. Moore, Jr., M. Fixman and B. N. Epstein, *J. Polym. Sci.*, 1955, **16**, 517.
151. W. Bushuk and H. Benoit, *Can. J. Chem.*, 1958, **36**, 1616.
152. H. Benoit and D. Froelich, in 'Light Scattering from Polymer Solutions', ed. M. B. Huglin, Academic Press, London, 1972, chap. 11.
153. P. Kratochvil, *J. Polym. Sci., Polym. Symp.*, 1975, **50**, 487.
154. H. Benoit and C. Wippler, *J. Chim. Phys. Phys.-Chim. Biol.*, 1960, **57**, 524.
155. H. J. Karam, in 'Polymer Compatibility and Incompatibility', ed. K. Šolc, Harwood, London, 1982.
156. C. C. Hsu and J. M. Prausnitz, *Macromolecules*, 1974, **7**, 320.
157. R. Koningsveld, L. A. Kleintjens and A. R. Shultz, *J. Polym. Sci., Part A-2*, 1970, **8**, 1261.
158. H. E. Stanley, 'Introduction to Phase and Critical Phenomena', Oxford University Press, Fair Lawn, NJ, 1971.
159. I. Katime, M. Anasagasti, C. Sanz, R. Valenciano and F. Rabagliati, *Mater. Chem. Phys.*, 1987, **16**, 101.
160. M. Anasagasti, I. Katime, R. Valenciano, J. J. del Olmo and C. Sanz, *Mater. Chem. Phys.*, 1987, **16**, 1.
161. R. Kuhn, H. J. Cantow and W. Burchard, *Angew. Makromol. Chem.*, 1968, **2**, 146.
162. P. Kratochvil, *Faserforsch. Textiltech.*, 1973, **24**, 5.
163. N. Nagasawa and A. Takahashi, in 'Light Scattering from Polymer Solutions', ed. M. B. Huglin, Academic Press, London, 1972, chap. 16.
164. A. Takahashi and M. Nagasawa, *J. Am. Chem. Soc.*, 1964, **86**, 543.
165. I. Noda, T. Tsuge and M. Nagasawa, *J. Phys. Chem.*, 1970, **74**, 710.
166. G. C. Berry, *J. Chem. Phys.*, 1966, **44**, 4550.
167. K. Kawahara, T. Norisuye and H. Fujita, *J. Chem. Phys.*, 1968, **49**, 4339.
168. Z. Alexandrowicz, *J. Chem. Phys.*, 1967, **46**, 3789.
169. Z. Alexandrowicz, *J. Chem. Phys.*, 1967, **46**, 3800.
170. Z. Alexandrowicz, *J. Chem. Phys.*, 1967, **47**, 4337.
171. H. G. Elias, in 'Light Scattering from Polymer Solutions', ed. M. B. Huglin, Academic Press, London, 1972, chap. 9.
172. H. G. Elias and R. Bareiss, *Chimia*, 1967, **21**, 53.
173. F. J. Reithel, *Adv. Protein Chem.*, 1963, **18**, 123.
174. C. Price and D. Woods, *Polymer*, 1974, **15**, 389.
175. C. Price, A. L. Hudd and B. Wright, *Polymer*, 1982, **23**, 170.
176. C. Price, N. Briggs, J. R. Quintana, R. B. Stubbersfield and I. Robb, *Polymer Commun.*, 1986, **27**, 292.
177. P. Debye, *J. Phys. Colloid Chem.*, 1949, **53**, 1.
178. V. R. Kuhn and H.-G. Cantow, *Makromol. Chem.*, 1969, **122**, 65.
179. I. Katime, J. R. Quintana, R. Valenciano and C. Strazielle, *Polymer*, 1986, **27**, 742.
180. I. Katime, J. R. Quintana and J. Veguillas, *Eur. Polym. J.*, 1985, **21**, 1075.
181. J. R. Quintana and I. Katime, *J. Chem. Soc., Faraday Trans. 1*, 1986, **82**, 1333.
182. C. Strazielle, *Makromol. Chem.*, 1968, **119**, 50.
183. O. Kratky, *Prog. Biophys. Mol. Biol.*, 1963, **13**, 105.
184. O. Kratky, in 'Small-angle X-ray Scattering', ed. H. Brumberger, Gordon and Breach, New York, 1968.
185. P. Debye, *Ann. Phys. (Leipzig)*, 1915, **46**, 809.
186. A. Guinier, *Ann. Phys. (Paris)*, 1939, **12**, 161.
187. J. L. Soulé, *J. Phys. Radium, Phys. Appl., Suppl.*, 1957, **18**, 90A.
188. G. Porod, *Kolloid-Z.*, 1951, **124**, 83.
189. G. Porod, *Kolloid-Z.*, 1952, **125**, 51.
190. L. Kahovec, G. Porod and H. Ruck, *Kolloid-Z.*, 1953, **133**, 16.
191. P. Debye, H. R. Anderson and H. Brumberger, *J. Appl. Phys.*, 1957, **28**, 679.
192. L. E. Alexander, 'X-ray diffraction methods in Polymer Science', Wiley–Interscience, New York, 1969.
193. R. G. Kirste, in 'Characterization of Macromolecular Structure', National Academy of Sciences of the United States of America, Washington, DC, Publication 1573, 1968.
194. D. Kratky, *Angew. Chem.*, 1960, **72**, 467.
195. W. O. Statton, in 'Newer Methods of Polymer Characterization', ed. B. Ke, Wiley–Interscience, New York, 1964.
196. R. S. Stein and G. L. Wilkes, in 'Structure and Properties of Oriented Polymers', ed. I. M. Ward, Applied Science, London, 1975.
197. T. Asano and Y. Fujiwara, *Polymer*, 1978, **19**, 99.
198. P. C. Ashman and C. Booth, *Polymer*, 1976, **17**, 105.
199. P. C. Ashman and C. Booth, *Polymer*, 1972, **12**, 459.

200. P. C. Ashman and C. Booth, *Polymer*, 1975, **16**, 889.
201. P. C. Ashman, D. R. Cooper and C. Price, *Polymer*, 1975, **16**, 897.
202. F. J. Baltá–Calleja, J. C. G. Ortega and J. M. de Salazar, *Polymer*, 1978, **19**, 1094.
203. B. Chu, M. Pallesen, W. P. Kao, D. E. Andrews and P. W. Schmidt, *J. Chem. Phys.*, 1965, **43**, 2950.
204. J. Clements, R. Jakeways and I. M. Ward, *Polymer*, 1978, **19**, 639.
205. M. Dosière and J. J. Point, *J. Polym. Sci., Polym. Phys. Ed.*, 1977, **15**, 1655.
206. Y. Fujiwara and P. J. Flory, *Macromolecules*, 1970, **3**, 288.
207. S. K. Garg and S. S. Stivala, *J. Polym. Sci., Polym. Phys. Ed.*, 1978, **16**, 1419.
208. A. Garton, R. F. Stepaniak, D. J. Carlsson and D. M. Wiles, *J. Polym. Sci., Polym. Phys. Ed.*, 1978, **16**, 599.
209. T. Hashimoto, H. Kawasaki and H. Kawai, *J. Polym. Sci., Polym. Phys. Ed.*, 1978, **16**, 271.
210. T. Hashimoto, Y. Murakami and H. Kawai, *J. Polym. Sci., Polym. Phys. Ed.*, 1975, **13**, 1613.
211. R. S. Stein and T. Hashimoto, *J. Polym. Sci., Part A-2*, 1970, **8**, 1503.
212. S. Heine, O. Kratky, G. Porod and P. J. Schmitz, *Makromol. Chem.*, 1961, **44/46**, 682.
213. S. Heine, O. Kratky and J. Rappert, *Makromol. Chem.*, 1952, **56**, 150.
214. F. B. Khambatta, F. Werner, T. Russell and R. S. Stein, *J. Polym. Sci., Polym. Phys. Ed.*, 1976, **14**, 1391.
215. H.-G. Kim, *J. Appl. Polym. Sci.*, 1978, **22**, 889.
216. R. G. Kirste and G. Wild, *Makromol. Chem.*, 1969, **121**, 174.
217. R. G. Kirste and W. Wunderlich, *Makromol. Chem.*, 1964, **73**, 240.
218. P. R. Lewis and C. Price, *Polymer*, 1971, **12**, 258.
219. J. Maxfield and L. Mandelkern, *Macromolecules*, 1977, **10**, 1141.
220. M. B. Rhodes and R. S. Stein, *J. Polym. Sci., Part A-2*, 1969, **7**, 1539.
221. I. D. Richardson and I. M. Ward, *J. Polym. Sci., Polym. Phys. Ed.*, 1978, **16**, 667.
222. T. R. Steger and L. E. Nielsen, *J. Polym. Sci., Polym. Phys. Ed.*, 1978, **16**, 613.
223. K. Suehiro, H. Tanizaki and M, Takayanagi, *Polymer*, 1976, **17**, 1059.
224. H. Tanaka and S. Okajima, *J. Polym. Sci., Polym. Lett. Ed.*, 1977, **15**, 349.
225. W. Wenig, *J. Polym. Sci., Polym. Phys. Ed.*, 1978, **16**, 1635.
226. F. P. Warner, W. J. MacKnight and R. S. Stein, *J. Polym. Sci., Polym. Phys. Ed.*, 1977, **15**, 2113.
227. W. W. Yau and R. S. Stein, *J. Polym. Sci., Part A-2*, 1968, **6**, 1.
228. G. S. Y. Yeh, R. Hoseman, J. Loboda-Cačković and H. Čačković, *Polymer*, 1976, **17**, 309.
229. T. Hashimoto, K. Nagatoshi, A. Todo, H. Hasegawa and H. Kawai, *Macromolecules*, 1974, **7**, 364.
230. P. Groisus, Y. Gallot and A. Skoulius, *Makromol. Chem.*, 1969, **127**, 94.
231. P. Grosius, Y. Gallot and A. Skoulios, *Makromol. Chem.*, 1970, **132**, 35.
232. S. N. Zhurkov, V. A. Zakrevskii, V. E. Korsukov and V. S. Kuksenko, *J. Polym. Sci., Part A-2*, 1972, **10**, 1509.
233. S. N. Zhurkov and V. E. Korsukov, *J. Polym. Sci., Polym. Phys. Ed.*, 1974, **12**, 385.
234. S. N. Zhurkov and V. S. Kuksenko, *Int. J. Fract.*, 1975, **11**, 629.
235. S. N. Zhurkov, V. S. Kuksenko and A. I. Slutsker, *Sov. Phys.—Solid State (Engl. Transl.)*, 1969, **11**, 238.
236. S. N. Zhurkov, V. A. Marikhim and A. I. Slutsker, *Sov. Phys.—Solid State (Engl. Transl.)*, 1959, **1**, 1060.
237. S. N. Zhurkov, A. I. Slutsker and V. A. Marikhin, *Sov. Phys.—Solid State (Engl. Transl.)*, 1959, **1**, 1601.
238. S. N. Zhurkov, V. S. Kuksenko and A. I. Slutsker, in 'Fracture, Proceedings of the International Conference on Fracture, 2nd, Brighton, England, April 13–18, 1969', ed. P. L. Pratt, E. H. Andrews, N. E. Frost, R. L. Bell, R. W. Nichols and E. Smith, Chapman and Hall, London, 1969.
239. M. A. Gezalov, V. S. Kuksenko and A. I. Slutsker, *Sov. Phys.—Solid State (Engl. Transl.)*, 1972, **14**, 344.
240. C. R. Worthington, *J. Sci. Instrum.*, 1956, **33**, 66.
241. L. Mandelkern, A. S. Posner, A. F. Diorio and D. E. Roberts, *J. Appl. Phys.*, 1961, **32**, 1509.
242. O. E. A. Bolduan and R. S. Bear, *J. Appl. Phys.*, 1949, **20**, 983.
243. R. S. Bear and O. E. A. Bolduan, *J. Appl. Phys.*, 1951, **22**, 191.
244. W. O. Statton, *J. Polym. Sci.*, 1956, **22**, 385.
245. W. O. Statton and G. M. Godard, *J. Appl. Phys.*, 1957, **28**, 1111.
246. W. O. Statton, *J. Polym. Sci.*, 1959, **41**, 143.
247. W. O. Statton and P. H. Geil, *J. Appl. Polym. Sci.*, 1960, **3**, 357.
248. W. O. Statton, *J. Polym. Sci.*, 1962, **58**, 205.
249. O. Kratky, *Z. Elektrochem.*, 1954, **58**, 49.
250. O. Kratky, *Z. Elektrochem.*, 1958, **62**, 66.
251. O. Kratky, *Kolloid-Z.*, 1955, **144**, 110.
252. O. Kratky, *J. Polym. Sci.*, 1955, **16**, 164; H. Stabinger and O. Kratky, *Makromol. Chem.*, 1978, **179**, 1655.
253. O. Kratky, G. Porod, A. Sekora and B. Paletta, *J. Polym. Sci.*, 1955, **16**, 163.
254. A. Franks, *Proc. Phys. Soc., London, Sect. B*, 1955, **68**, 1054.
255. O. Kratky and H. Leopold, *Makromol. Chem.*, 1964, **75**, 69.
256. V. Luzzati, J. Witz and R. Baro, *Suppl. J. Phys. (Paris)*, 1963, **24**, 141A.

6

Scattering Properties: Neutrons

RANDAL W. RICHARDS

University of Strathclyde, Glasgow, UK

6.1 INTRODUCTION

The properties of neutrons which make their use unique as a probe of condensed matter include: (a) The interaction between neutron and condensed matter is weak, making theoretical treatment relatively simple. (b) Neutron absorption is usually low and thus large sample volumes can be used. (c) The wavelength of neutrons (0.2–2 nm) is appropriate to the study of atomic separation and macromolecular dimensions. (d) Molecular vibrational energies are of the same magnitude as neutron energies (400–0.4 J mol^{-1}) enabling investigation of molecular dynamics. (e) Neutrons possess a magnetic moment which may interact with unpaired electrons and magnetic excitations to provide information on the magnetic structure and dynamics of matter. (f) The neutron–nucleus scattering does not change in a systematic manner as a function of atomic number or atomic mass. It also varies in magnitude from one isotope to another, therefore by simple isotopic substitution the scattering intensity can be radically changed.

Evidently, neutron scattering can be, and is, utilized in many aspects of chemistry and physics and specialist reviews[1-5] are available. However, the fundamental theory is the same for all applications, simplifications and/or extensions being made dependent on the particular aspects investigated. Only a brief outline of the theoretical basis necessary for the investigation of the structure of macromolecules in solution is provided here. More detailed theory can be found in the books by Marshall and Lovesey,[6] Turchin,[7] Squires[8] and Lovesey.[9] A reasonably complete overview of the applications of neutron scattering is available in the text edited by Kostorz[2] whilst periodic reviews of small angle

133

neutron scattering in polymer science have appeared at irregular intervals in the past ten years.[10-15] In the theory section the evaluation of molecular parameters from neutron scattering data will be given and the additional factors required will become apparent. Some emphasis will be placed on concentrated solutions since these have particular relevance to the investigation of scaling concepts in polymer science pioneered by de Gennes.[16] Experimental results will be discussed for selected systems which particularly illustrate the usefulness of neutron scattering in the investigation of macromolecules in solution.

6.2 THEORY

6.2.1 Neutron Cross Section

A neutron of wave vector $k(=2\pi/\lambda)$ has an energy E given by

$$E = \hbar^2 k^2 / 2m \tag{1}$$

where m is the neutron mass and $k = |k|$. Frequently neutron energies are given in meV especially where dynamic processes are being investigated. For structural investigations it is more convenient to quote the neutron wavelength

$$\lambda = (h^2/2mE)^{1/2} = 0.904 E^{-1/2} \text{ nm} \tag{2}$$

The general scattering geometry is shown in Figure 1(a) and what one actually measures is the differential scattering cross section, $(d^2\sigma/d\Omega dE')$, *i.e.* the fraction of neutrons with incident energy E scattered into an element of solid angle $d\Omega \, (= \sin\theta \, d\theta \, d\phi)$ with an energy between E' and $E' + dE'$. If the incident neutron flux is I_0 (number area^{-1} time^{-1}) and we consider only elastic collisions, *i.e.* no energy transfer to or from the specimen, the number of neutrons scattered per unit time into $d\Omega$ is $I_0(d\sigma/d\Omega)d\Omega$. To continue further, consider the case of scattering of neutrons by an isolated bound nucleus. The form of the scattering will depend on the neutron–nucleus interaction potential which is unknown but of very short range. The wavelength of neutrons being much larger means that only s-wave scattering is obtained, *i.e.* it is isotropic and characterized by the scattering length, b. Values

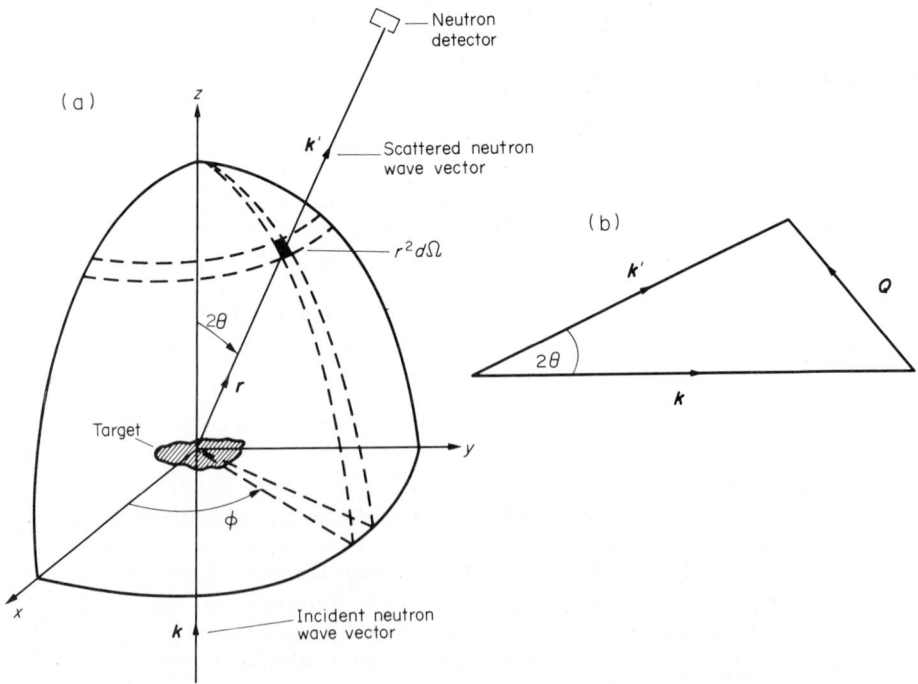

Figure 1   (a) Geometry of neutron scattering. Reproduced by permission of Oxford University Press from ref. 9. (b) Vector diagram of the neutron scattering process

of b vary across the periodic table and also for each isotope at any particular value of z. The neutron–nucleus interaction potential is modelled by the Fermi pseudo-potential for a nucleus at a position R

$$V(\mathbf{r}) = (h^2/2\pi m)b\,\delta(\mathbf{r}-\mathbf{R}) \tag{3}$$

where $\delta(r-R)$ is a delta function of argument $(r-R)$

The differential scattering cross section is given by[9,17]

$$(d\sigma/d\Omega) = \left|(m/2\pi h^2)b\int d\mathbf{r}\exp(-i\mathbf{k}'\cdot\mathbf{r})V(\mathbf{r})\exp(i\mathbf{k}\cdot\mathbf{r})\right|^2$$

Taking the position of the nucleus, R, as the origin, *i.e.* $R=0$, then

$$(d\sigma/d\Omega) = \left|b\int d\mathbf{r}\exp(-i\mathbf{k}\cdot\mathbf{r})\sigma(\mathbf{r})\exp(i\mathbf{k}\cdot\mathbf{r})\right|^2 = |b|^2 \text{ since } \mathbf{k}'=\mathbf{k} \tag{4}$$

Because the scattering is isotropic the *total* cross section is $\sigma = 4\pi|b|^2$.

In all the above we have assumed only one nucleus without spin. Where the nucleus possesses spin the scattering length can take on different values depending on the relative orientation of neutron and nuclear spins. This spin incoherence is combined with isotopic incoherence to give incoherent scattering from the specimen. The source of this becomes clear when the scattering from a rigid assembly of nuclei is considered. For this case and for nuclei at sites i and j with vector positions R_i and R_j then

$$d\sigma/d\Omega = \sum \exp(i\mathbf{Q}\cdot(\mathbf{R}_i-\mathbf{R}_j))\overline{b_ib_j} \tag{5}$$

where Q is the scattering vector defined by the relation $Q=k-k'$. From Figure 1(b), $Q^2=k^2+k'^2-2kk'\cos 2\theta$. Since the scattering is elastic then $k=k'$ and $Q=|Q|=(4\pi/\lambda)\sin\theta$. The value of b_i and b_j depends on the isotope, and its spin, at positions i and j and $\overline{b_ib_j}$ is the average over random nuclear spin orientations and isotope distributions.

For $i\neq j$

$$\overline{b_ib_j} = \overline{b_i}\,\overline{b_j} = |\bar{b}|^2$$

For $i=j$

$$\overline{b_ib_j} = |b|^2$$

Hence

$$\overline{b_ib_j} = |\bar{b}|^2 + (\overline{|b|^2} - |\bar{b}|^2) \tag{6}$$

Replacing equation (6) into equation (5) gives

$$d\sigma/d\Omega = \bar{b}^2\left|\sum_i \exp(i\mathbf{Q}\mathbf{R}_i)\right|^2 + N[|\bar{b}|^2 + (\overline{|b|^2} - |\bar{b}|^2)] \tag{7}$$

$$= (d\sigma/d\Omega)_{\text{coh}} + (d\sigma/d\Omega)_{\text{incoh}} \tag{8}$$

The two contributions to the differential scattering cross section are now evident. Incoherent scattering is isotropic with no phase term and hence no dependence on scattering vector Q. It contains no structural information on the scattering species. The coherent scattering contains all the structural information since it is essentially the Fourier transform of the scattering length correlations in the scattering specimen and describes the interference effects between waves scattered from different nuclei.

6.2.2 Bound Atom Cross Section Values

Several compilations of bound atom coherent scattering lengths, b, and scattering cross sections are available.[18-20] Parameters tabulated include b, the total scattering cross section $\sigma_T(=4\pi|b|^2)$ and the incoherent cross section, σ_I. Coherent scattering cross sections are given by $\sigma_C=4\pi b^2$. Table 1 is a partial listing extracted from the compilation of Sears[18] for nuclei commonly found in macromolecules. From the point of view of using neutron scattering to investigate macromolecular structure, the most significant entries in Table 1 are those pertaining to the isotopes of hydrogen. Not only do the absolute magnitudes differ but also so do the phase of neutrons scattered by protons or deuterons. This has a profound effect on the coherent scattering when deuterium atoms are substituted for hydrogen atoms in polymers or solvents.

Table 1   Neutron Scattering Lengths and Cross Sections

Element	A	b $(10^{12}$ cm)	σ_T $(10^{2\bar{4}}$ cm$^2)$	σ_I $(10^{24}$ cm$^2)$
H	1	−0.374	81.67	79.91
	2	0.667	7.64	2.04
C	12	0.665	5.555	0
N	14	0.937	11.52	0.49
O	16	0.580	4.235	0
F	19	0.565	4.018	0
Si	28	0.411	2.178	0.01
Cl	35	1.166	21.63	5.2

6.2.3   Contrast Factors

Equation (7) is the general equation and when applied to the scattering from a pure material, crystalline or liquid, predicts that coherent scattering would only be observed when strict geometrical conditions are met giving rise to Bragg diffraction peaks for crystalline lattices and structure factors for liquids. For pure amorphous and semi-crystalline polymers these features would also be evident (particularly so for fully deuterated polymer) but they are characteristic of local structure at a length scale of 1 to 2 nm. Interest in this chapter is focussed on the long range structure of macromolecules in solution. Specifically we take a dilute solution of macromolecules, wherein intermolecular interference between coherent scattering from different macromolecules is absent due to the large distances separating them. The *coherent* scattering from a single macromolecule and the solvent is

$$d\sigma/d\Omega = |\textstyle\sum b_i \exp(i\mathbf{Q}\cdot\mathbf{R}_i)|^2 \tag{9}$$

If we consider one solvent molecule then over this length scale and for the values of Q used (*vide infra*) $Q\cdot R_i \ll 1$ and $\exp(iQ\cdot R_i) \approx 1$ and therefore b_i averaged over the solvent molecular volume will give a locally averaged scattering length density, ρ_s, for the solvent. Factorizing out the solvent and macromolecule contributions to equation (9) we have

$$d\sigma/d\Omega = \left|\rho_s\left[\int \exp(i\mathbf{Q}\cdot\mathbf{R})d\mathbf{R} - \int_V \exp(i\mathbf{Q}\cdot\mathbf{R})d\mathbf{R}\right] + \rho(\mathbf{R})\int_V \exp(i\mathbf{Q}\cdot\mathbf{R})d\mathbf{R}\right|^2 \tag{10}$$

The first integral on the right-hand side of equation (10) is over the entire scattering volume and is zero except for $Q=0$ where all scattered waves are in phase. The remaining integrals are over the volume of the macromolecule. Thus,

$$d\sigma/d\Omega = \left|\int_V (\rho(\mathbf{R}) - \rho_s)\exp(i\mathbf{Q}\cdot\mathbf{R})d\mathbf{R}\right|^2 \tag{11}$$

Averaging $\rho(R)$ over the macromolecular segment volume to give a scattering length density ρ, is equivalent to treating the segments as point scatterers and we can write

$$d\sigma/d\Omega = \left|V(\rho - \rho_s)1/V\int \exp(i\mathbf{Q}\cdot\mathbf{R})d\mathbf{R}\right|^2 \tag{12}$$

where $1/V\int_V \exp(iQ\cdot R)dR$ is referred to as the single particle form factor, $F(Q)$, and contains all the structural information of the scattering macromolecule or particle.

The scattering length of a molecule or segment is the sum over the n atoms in the molecule, *i.e.*

$$b = \textstyle\sum_{i=1}^{n} n_i b_i \tag{13}$$

The scattering length density is the scattering length per unit molecular volume:

$$\rho = b/V_m \tag{14}$$

or

$$\rho = N_A b/v\bar{M} \tag{15}$$

where $\bar{M}$ = molecular weight (segment molecular weight for macromolecules), v = partial specific volume and N_A = Avogadro's Number.

Values of scattering length density for some polymers and solvents are given in Table 2. Other values have been tabulated by Ullman.[12]

Table 2 Scattering Length Densities

		Constitutive Unit	
Polymer/Solvent	*Formula*	*Molecular Weight*	$\rho \ (10^{10} \ cm^{-2})$
Polyethylene	C_2H_4	28.05	−0.316
Poly(ethylene-d4)	C_2D_4	32.07	8.24
Polystyrene	C_8H_8	104.15	1.413
Poly(styrene-d8)	C_8D_8	112.19	6.50
Polybutadiene	C_4H_6	54.09	0.467
Poly(butadiene-d6)	C_4D_6	60.12	6.823
Poly(methyl methacrylate)	$C_5H_8O_2$	100.12	1.069
Poly(methyl methacrylate-d8)	$C_5D_8O_2$	108.17	7.03
Cyclohexane	C_6H_{12}	84.16	−0.24
Cyclohexane-d12	C_6D_{12}	96.23	6.01
Toluene	C_7H_8	92.14	0.94
Toluene-d8	C_7D_8	100.19	5.42
Tetrahydrofuran	C_4H_8O	72.11	−0.246
Tetrahydrofuran-d8	C_4D_8O	80.16	5.889
Water	H_2O	18.01	−0.56
Water-d2	D_2O	20.02	6.36

6.2.4 Single Particle Form Factors (Dilute Solutions)

From equation (12) the differential scattering cross section is given by

$$d\sigma/d\Omega = V^2(\rho - \rho_s)^2 F^2(\mathbf{Q}) \tag{16}$$

The form of $F^2(Q)$ depends on the morphology of the scattering particle and there are a multitude of equations available for a wide variety of particle shapes. These can be obtained in the publications of Guinier and Fournet,[21] Glatter and Kratky[22] and particularly useful tabulations have been provided by Burchard.[23,24] Perhaps the most general form of $F^2(Q)$ is the Guinier form (see equation 17) which is valid for any shape of particle.

$$F^2(\mathbf{Q}) = V^2 \exp(-\mathbf{Q}^2 \langle \bar{s}^2 \rangle / 3) \tag{17}$$

Strictly, equation (17) is only applicable in the region of Q where $Q\langle \bar{s}^2 \rangle^{1/2} \leq 1$ with $\langle \bar{s}^2 \rangle^{1/2}$ being the root mean square radius of gyration of the scattering particle.

For polymers, the equation of greatest value is the Debye[25] equation (see equation 18) for point scatterers with a Gaussian distribution about the centre of mass. The correlation with the random walk models of macromolecules is evident,

$$F^2(Q) = 2/u^2 (\exp(-u) - 1 + u) \tag{18}$$

$$u = Q^2 \langle \bar{s}^2 \rangle \tag{19}$$

The vector notation has been omitted from equations (18) and (19) since a random orientation of polymer molecules in solution is assumed. Equation (18) is valid for all length scales of a macromolecule so long as a Gaussian distribution of mass elements about the centre of mass is valid. The form of equation (18) is shown in Figure 2(a) whilst the different regions become more evident in the reduced Kratky plot[26] ($Q^2 \langle \bar{s}^2 \rangle F^2(Q)$ as a function of $Q\langle \bar{s}^2 \rangle^{1/2}$) in Figure 2(b). Initially in Figure 2(a) the intensity falls in an exponential manner with Q^2; thereafter the attenuation is proportional to $1/Q^2$. This latter region becomes evident as the plateau in the Kratky plot of Figure 2(b). All macromolecules with a Gaussian distribution of segments will produce this plateau in a Kratky plot, the asymptotic magnitude of which is 2 when the ordinate is $F(Q)(Q^2 \langle \bar{s} \rangle^2)$. The Q value at which this asymptotic value is reached becomes larger as the radius of gyration decreases.

These regions of different Q dependence of $d\sigma/d\Omega$ can be put on a firmer basis by examining the form of equation (16) [replacing equation (18) for $F^2(Q)$] as the value of Q increases. For

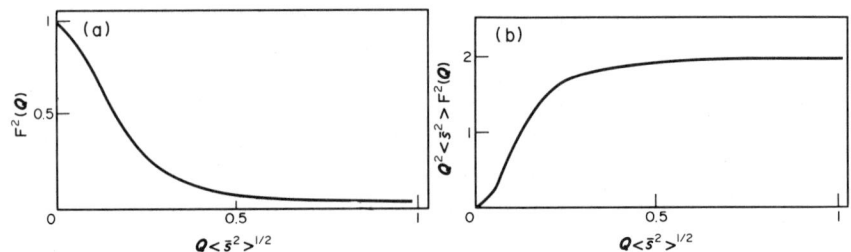

Figure 2 (a) Dependence of the single particle form factor on Q for a Gaussian coil. (b) Kratky plot resulting from the Debye equation

$Q \leq \langle \bar{s} \rangle^{-1/2}$ then by expansion of the exponential

$$d\sigma/d\Omega = V^2(\rho - \rho_s)^2(1 - Q^2\langle \bar{s}^2 \rangle/3) \tag{20}$$

whence

$$(d\sigma/d\Omega)^{-1}V^2(\rho - \rho_s)^2 = 1 + Q^2\langle \bar{s}^2 \rangle/3 \tag{21}$$

This is the Guinier region wherein the mean square radius of gyration may be obtained. For $\langle \bar{s}^2 \rangle^{-1/2} \leq Q \leq a^{-1}$ where a is the persistence length of the macromolecule

$$d\sigma/d\Omega = 2V^2(\rho - \rho_s)^2/Q^2\langle \bar{s}^2 \rangle \tag{22}$$

What is not predicted by the Debye equation is an increase in $F^2(Q)Q^2(\propto Q^{-1})$ for Q values greater than a^{-1} as shown schematically in Figure 3. In this region of scattering vector the scattering pertains to short sequences of the chain behaving as randomly oriented rods. The value of the scattering vector, Q^*, where the change from Q^{-2} to Q^{-1} dependence takes place can be used to obtain the persistence length:

$$a \approx 1.91/Q^* \tag{23}$$

In practice, Kratky plots have been obtained which are much richer in information about the macromolecular configuration than indicated here.[27] Local steric interactions have a great influence on the form of the plateau region and such data is of great value in verifying rotational isomeric state calculations for macromolecular conformation.[28, 29]

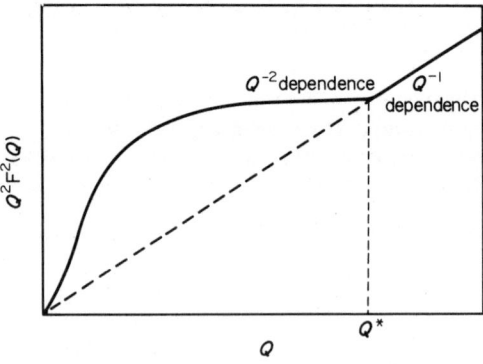

Figure 3 Schematic plot of $Q^2F^2(Q)$ *vs.* Q showing the different regions

6.2.5 Scattering Vector Ranges

We are now in a position to appreciate the scattering vector, Q, values required. From the preceeding section we note that to obtain the radius of gyration of a macromolecule then the Guinier condition $(Q \leq \langle \bar{s}^2 \rangle^{1/2})$ must be fulfilled. Typically, $\langle \bar{s}^2 \rangle^{1/2} \approx 10$ nm whence we must have $Q \leq 0.1$ nm^{-1}. This condition must be *strictly* adhered to for monodisperse polymers but may be relaxed when polydispersity is present.[30] The smallest dimension we would want to examine (without getting into the range of interatomic distances) is that of the statistical step, *i.e.* circa 2.5 nm corresponding to a Q range of *circa* $2\pi/Q$, *i.e.* 2.5 nm^{-1}. From the definition of Q given earlier (following equation 5) and for a typical neutron beam wavelength of 1 nm, these values of Q correspond to angles between $0.5°$ to $20°$ with most of the structural information being obtained at angles of less than $15°$. For these reasons the term small angle neutron scattering is used to describe conformational measurements on macromolecules using neutrons (N.B. the term small angle is inappropriate; the important parameter is Q and light scattering is confined to a much smaller range of Q, this being due to the much larger wavelength of visible light).

6.2.6 Concentrated Solutions

The configuration and thermodynamic properties of macromolecules in semi-dilute and concentrated solutions have been of interest since the mid 1960's.[31-34] Increased interest was engendered by the application of re-normalization group theories[35] and the evaluation of scaling concepts.[16] Fortunately, small angle neutron scattering was developed coincidentally and experimental results could be obtained for the first time on concentrated solutions. For example, the conformational properties of an isolated macromolecule in a concentrated solution were obtained by making up a solution using a mixture of deuterated and hydrogenous polymer such that the deuteropolymer concentration was dilute whilst the overall polymer concentration was high.

In semi-dilute and concentrated solutions macromolecules overlap and excluded volume interactions are attenuated by such intermolecular interactions. The distance between two adjacent intermolecular interactions in one chain is known as the correlation (or screening) length, ξ, because the segments of the chain separated by smaller distances are correlated with each other *via* excluded volume interactions.[16, 32] By Fourier transformation of the pair correlation function for a single macromolecule the single particle form factor discussed earlier is obtained.

There exists another pair correlation function in such concentrated solutions, that between all segments in the solution, referred to as the total pair correlation function. For distances, r, less than ξ this total pair correlation function is dominated by segments from the chain whereon we have chosen our reference segment and moreover excluded volume effects will be evident. At distances greater than ξ the other macromolecules intervene and segments become uncorrelated. The Fourier transform of the total pair correlation function will yield the *scattering law*, $S(Q)^*$, and for $r > \xi$ this is Lorentzian;[32]

$$S(\mathbf{Q}) = f(T,C)/(\mathbf{Q}^2 + \xi^{-2}) \qquad (24)$$

where $f(T,C)$ is an unknown but Q independent function. Hence for regions of $Q < \xi^{-1}$, $(d\sigma/d\Omega)^{-1}$ should have a linear dependence on Q^2. For $r < \xi$ *i.e.* $Q > \xi^{-1}$, the scattering law changes to a $Q^{-5/3}$ dependence commensurate with excluded volume behaviour being evident.

6.3 EXPERIMENTAL

6.3.1 Instrumentation

The majority of small angle neutron diffractometers use a steady state nuclear reactor as the source of neutrons. Whilst particular features may vary from instrument to instrument, they all have several features in common: (a) A neutron wave guide which brings neutrons from the reactor to the diffractometer. This wave guide is constructed so that γ radiation and fast (high energy) neutrons are filtered out. Ideally, as at the Institut Laue-Langevin,[39] the neutron beam is taken from a 'cold' source, which is usually a small volume of liquid deuterium near the reactor core. (b) A monochromator which selects discrete neutron wavelengths from the spectrum of wavelengths available. Monochromators may be crystals of specific substances (*e.g.* beryllium or graphite as at Oak Ridge, Tennessee)[37] or velocity selectors (as at AERE Harwell,[40] and Institut Laue-Langevin).[36] (c) A series of evacuated collimators of adjustable length with the provision for diaphragms of variable size to be mounted at the exit of the collimators. These diaphragms then act as the effective source of the neutron beam. (d) A sample area where specimens can be mounted in the neutron beam under a variety of conditions, *e.g.* temperature, magnetic field, strain, shear stress. (e) An evacuated flight tube in which is mounted a detector which is a two dimensional array of individual detector elements usually 1 cm by 1 cm square or less. Typically there are 4096 of these individual detector elements.

---

*In this terminology the differential scattering cross section is given by $d\sigma/d\Omega = V^2(\rho - \rho_s)^2 S(Q)$

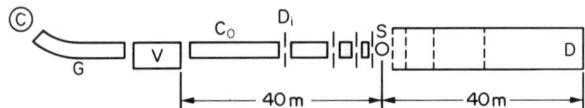

Figure 4 Components of a typical small angle neutron scattering diffractometer: C = reactor core, C_0 = collimator, D = detector, G = neutron guide, D_i = diaphragm, S = sample

(f) A computer for data acquisition and immediate analysis of data and to act as a watchdog and provide on-line control of the diffractometer operation. Figure 4 is a schematic of a small angle neutron diffractometer, whilst full details on individual instruments may be found in the literature[36, 37] or by contacting the separate institutes directly.[38]

A new development is the use of spallation neutron sources, where a pulsed beam of neutrons is produced by the impact of a proton beam on a uranium target. Although the use of pulsed neutron sources is not new, their use for small angle neutron scattering is still very much in the development stage at the time of writing (1986), particularly with regard to data analysis.[41]

6.3.2 Sample and Background Considerations

Specimen thickness should not be greater than that which transmits more than 50% of the incident beam. The transmission, T, is the ratio of the intensity of the neutron beam received on the detector at $Q=0$ after passage through the specimen to the incident unattenuated beam, *i.e.*

$$T = I(0)/I_0 = \exp(-n\sigma t) \tag{25}$$

where t is the sample thickness, σ the total cross section and n the number of scattering nuclei. Typically for specimens where hydrogen atoms form a large component then $t \approx 1$–2 mm. For those systems where deuterium is a major component, (*e.g.* hydrogenous polymer dissolved in a deuterated solvent) then t is of the order of 1 cm. Solutions can be contained in quartz cells, since quartz has almost negligible low angle scattering for neutrons.

Equation (7) shows that the scattering always has an incoherent component which constitutes a scattering background to the coherent scattering and which must be subtracted. Even when the majority of the solution components are deuterated, inspection of Table 1 shows that a small incoherent contribution is still present. Incoherent scattering can be assessed from measurements on a suitable 'blank'. Ideally this blank should contain protons and deuterons in the same proportion as the specimen to obtain the true incoherent scattering. For example, the specimen may be polystyrene-d8 in a hydrogenous solvent, and the blank could be a solution of the monomer in the solvent. The same incoherent scattering will be evident, but since the deuterated monomers are not correlated (*i.e.* linked together) there will be no coherent scattering evident in the region of Q used for the polymer solution.

For concentrated polymer solutions where individual chain dimensions are required, the situation is a little more complicated. From the point of view of neutron scattering we have a three component system (solvent, hydrogenous polymer, deuterated polymer) each with their own scattering length density. Coherent scattering arising from the contrast between hydrogenous polymer and solvent can be avoided by arranging for the polymer and solvent to have the same scattering length density. This can be done if the solvent has hydrogenous and deuterated isomers with scattering length densities of opposite sign. Unfortunately there are few such solvents, cyclohexane, tetrahydrofuran and water being three. Consequently special care has to be taken when subtracting backgrounds where residual contrast exists since the scattering will no longer be completely flat.

6.3.3 Data Analysis and Calibration

For isotropic scattering, the scattered intensity collected on the two dimensional detector can be radially averaged producing a table of Q values and intensities. Background scattering can be subtracted after normalizing background and sample data sets to the same parameter set, *e.g.* thickness, transmission and time or beam monitor count.

If we have a solution of N_p polymer molecules in a solvent and the intensity at a particular value of Q is $I(Q)$ then for a sample–detector separation of r,

$$I(Q) = N_p \frac{d\sigma}{d\Omega} \cdot T \cdot \frac{I_0}{r^2} \varepsilon A \Delta\Omega + \Sigma_I/4\pi \quad . \tag{26}$$

where I_0 is the incident neutron intensity on the sample (neutrons $cm^{-2} s^{-1}$), t the sample thickness with transmission T, ε the detector element efficiency, $\Delta\Omega$ the solid angle subtended by the detector element at the sample and A the cross sectional area illuminated by the beam. The macroscopic incoherent scattering cross section, Σ_I, is obtained from the scattering cross section by the relation $\Sigma_I = \sigma_I \cdot dN_A/\bar{M}$ where d is the density of substance with molecular weight $\bar{M}$.

The cross sectional area is defined by a cadmium diaphragm immediately before the specimen. The incoherent background ($\Sigma_I/4\pi$) can be subtracted out by suitable blank experiments. Apart from known sample parameters N_p, t and T, equation (26) contains instrumental factors which have to be determined using a calibrant of known scattering cross section to obtain $d\sigma/d\Omega$ in absolute units of cm^{-1}. These factors are obtained utilizing scattering data from usually a purely incoherent scatterer of known cross section obtained under the same conditions as the experimental data.[42] Vanadium is an ideal material but regrettably the scattered intensity is low. Light water is predominantly an incoherent scatterer and with much greater intensity than vanadium. Disadvantages of light water as a calibrant are a large fraction of multiple and inelastic scattering and a severe wavelength dependence of the incoherent scattering cross section (Figure 5). Nonetheless, light water is widely used as a calibrant and for a thin, planar water sample then

$$I(Q) = I_0 \Delta\Omega\varepsilon(1 - T_W)Ag/4\pi r^2 \tag{27}$$

where T_W is the neutron transmission of the water and g is a factor to correct for the inelastic scattering which leads to increased forward angle scattering. Thus

$$I_0 \Delta\Omega\varepsilon A/r^2 = 4\pi I(Q)/(g(1 - T_W)) = I(H_2O) \tag{28}$$

Consequently, after subtraction of the background scattering, we have

$$I'(Q) = N_p t T (d\sigma/d\Omega)/I(H_2O) \tag{29}$$

Since $N_p = cN_A/\bar{M}$ and $V = \bar{M}v/N_A$ then

$$I(Q) = (cN_A/\bar{M})(\bar{M}v/N_A)^2(\rho - \rho_s)^2 F^2(Q)t\,T/I(H_2O) = (c\bar{M}v^2/N_A)(\rho - \rho_s)^2 F^2(Q)t\,T/I(H_2O) \tag{30}$$

For an individual macromolecule in solution, and confining ourselves to the Guinier region,

$$I'(Q) = (c\bar{M}v^2/N_A)(\rho - \rho_s)^2 t\,T(1 - Q^2\langle\bar{s}^2\rangle/3)/I(H_2O) \tag{31}$$

$$\text{and} \quad \frac{K^*c}{I'(Q)} = \frac{1}{\bar{M}}(1 + Q^2\langle\bar{s}^2\rangle/3), \text{ where } K^* = N_A I(H_2O)/t\,Tv^2.$$

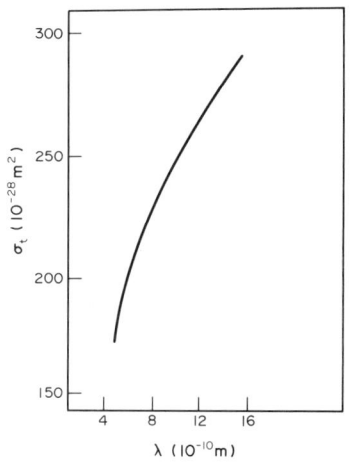

Figure 5 Incoherent scattering cross section for light water as a function of incident neutron wavelength (λ)

Equation (31) is identical in form to the familiar Zimm plot representation used in light scattering.[43] A plot of $K^*c/I'(Q)$ as a function of Q^2 should be linear and the molecular weight, $\bar{M}$, is obtained from the intercept whilst the mean square radius of gyration is given by the slope. Figure 6 shows a plot of $I'(Q)^{-1}$ as a function of the dimensionless group $Q^2\langle\bar{s}^2\rangle$ over the range $0 \le Q^2\langle\bar{s}^2\rangle < 2.0$ for a Gaussian molecule with $\langle\bar{s}^2\rangle^{1/2} = 10$ nm. Although a straight line could be drawn through *all* the data points without too much error, the true mean square radius of gyration will only be obtained for $(Q^2\langle\bar{s}^2\rangle)^{1/2} \le 1$, as the two lines drawn show.

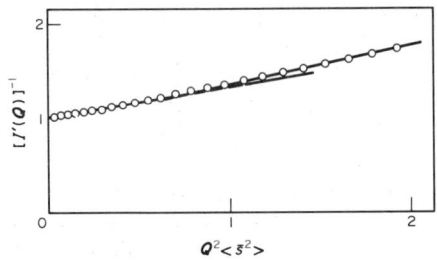

Figure 6   Reciprocal scattering intensity as a function of $Q^2\langle\bar{s}^2\rangle$ calculated from the Debye equation. Note the different slope for $Q^2\langle\bar{s}^2\rangle > 1$

Equation (31) presumes that 'ideal' thermodynamic conditions apply, *i.e.* that there are no long range excluded volume interactions. This is not always the case of course and the more general equation brings in the virial expansion of McMillan and Meyer[44]

$$\frac{K^*c}{I'(Q)} = \frac{1}{\bar{M}}[(1 + Q^2\langle\bar{s}^2\rangle/3) + 2A_2\bar{M}c + 3A_3\bar{M}c^2 \ldots] \tag{32}$$

where A_2 and A_3 are the second and third virial coefficients, the expansion usually being limited to the second virial coefficient. Like all scattering methods, for polydisperse polymers the molecular weight is the weight average whilst the radius of gyration is the *z* average.

Workup and analysis of small angle neutron scattering data at the various institutes is usually carried out by highly flexible suites of computer programmes which enable a variety of form factors and scattering laws to be used.[45]

6.4   SELECTED RESULTS

6.4.1   General

The application of small angle neutron scattering to macromolecules in solution is of most value where selective deuteration is used to the full: for example in the study of concentrated solutions where single chain dimensions are required, or of block copolymers where the conformation of one block is of interest. Studies of macromolecules in *dilute* solution are relatively scarce since light scattering and small angle X-ray scattering are used routinely. However, the wide range of scattering vector available to small angle neutron scattering together with the flexibility in contrast factor makes neutron scattering ideal for the investigation of those molecules whose molecular weights make them too small for the use of light scattering. The range of Q vectors available to the three scattering techniques is shown in Figure 7. Much use of small angle neutron scattering has been made in the investigation of concentrated polymer solutions, particularly with reference to scaling

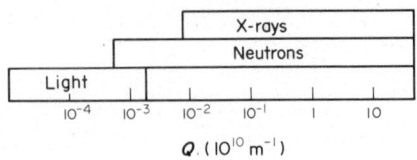

Figure 7   Scattering vector ranges available to light, X-ray and neutron scattering

laws. This has been extended to macromolecular gels (three dimensionally crosslinked networks swollen with solvent) with the aim of studying fundamental aspects of rubber-like elasticity theory. Polyelectrolytes have also been studied and some new insights into their semi-dilute solutions have been obtained.

6.4.2 Dilute Solutions

The configuration of low molecular weight polystyrene under theta conditions (deutero-cyclo-hexane at 38 °C) was investigated by Ballard *et al.*[46] using molecular masses of 600, 2100 and 4000. Only radii of gyration were reported but these were compared with three models of macromolecular configuration: (a) Gaussian coil, (b) rigid rod where $\langle \bar{s}^2 \rangle = L^2/12$, the length of the rod L being given by $L = l_b M_n / M_b$ with l_b and M_b being the length and molar mass of the repeat unit, (c) Kratky–Porod worm-like chain.[47] The latter model interpolated between models (a) and (b) and the experimental data fitted this model best.

Persistence lengths are expected to be larger for rigid macromolecules, and the behaviour of a molecule can be described using the contour length, L_C, and the persistence length, a. Thus when $L_C/a \gg 1$ the Gaussian coil conformation is favoured. Cellulose derivatives are generally considered to be stiff, rod-like molecules and the temperature variation of the persistence length of cellulose tricarbanilate determined by small angle neutron scattering from dioxane solutions has been reported by Gupta *et al.*[48] Values of a ranged from 11.1 nm at 22 °C to 9.1 nm at 80 °C (*cf.* polystyrene $a \approx 1.1$ nm) and agreed well with results from dielectric and stress-optical data.

Partially labelled chains have also been investigated with two aims in mind: (a) to determine the expansion factor of part of a macromolecule and (b) to investigate correlation hole effects. In one investigation[49] the partially labelled chain was an isotopic block copolymer of hydrogenous and deuterated polystyrene. Matsushita *et al.*[49] used three polymers, a fully deuterated polystyrene, a diblock copolymer of deuterated and hydrogenous polystyrene, and a triblock copolymer where the central block was deutero-polystyrene. In all cases the molecular weight of the deuterated blocks was *circa* 2.5×10^4. Measurements were made on carbon disulfide solutions, which have the advantage of being transparent to neutrons since CS_2 has no incoherent scattering cross section. Although differences in expansion factor were small they were outside the experimental error and indicated that the expansion factor of the central portion of the triblock copolymer was greater than that of the end portion but still considerably less than the whole chain expansion factor. A Monte Carlo calculation of the expansion factors gave reasonable agreement with the experimental results.

Duplessix *et al.*[50] also used carbon disulfide as a solvent for a triblock isotopic copolymer of polystyrene with the central portion being deuterated and of molecular weight 13×10^3. In these systems a clear maximum in scattering intensity (I) *vs.* Q was evident for solutions as dilute as 1% wt (Figure 8). This was attributed to the 'correlation hole' effect first described by de Gennes.[16,52]

A cyclic macromolecule is predicted to have a mean square radius of gyration one half of the value for a linear polymer of equivalent molecular weight. Higgins *et al.*[52] studied a series of linear and cyclic poly(dimethyl siloxane)s in deutero-benzene solutions at 292 K. Using full Zimm plots the ratio of the mean square radii of gyration of the linear to cyclic polymers was 1.9 ± 0.2. Furthermore the variation of reciprocal scattered intensity of the cyclic molecules with Q^2 was upwardly convex in agreement with the predictions of Casassa.[53] A fuller analysis of the single particle scattering function for linear and cyclic poly(dimethyl siloxane)s was reported by Edwards *et al.*[54] Monte Carlo calculations of the scattering functions showed that the Kratky plots of cyclic molecules had characteristic maxima not shown by the linear molecules. In particular these calculations showed that a cyclic poly(dimethyl siloxane) with 22 skeletal Si atoms produced the strongest second maximum, this being due to planar ring formation. The Monte Carlo calculation reproduced the experimental scattering data exactly in contrast to the analytical calculations of Casassa[53] and Burchard and Schmidt.[55]

Monte Carlo methods have been used to calculate the single particle scattering functions for model chains and investigate the influence of contrast factor and the offsetting of scattering centres from the main chain backbone.[56] The latter factor produces a considerable reduction in scattered intensity in the plateau region of Kratky plots, due to the increase in the shortest separation of scattering centres in the chain.

An experimental study of the influence of local structure on the scattering function has been carried out by Rawiso *et al.*[57] They used a series of polystyrene samples, a fully deuterated specimen (constitutional unit C_8D_8), a specimen with only the phenyl ring deuterated ($C_8D_5H_3$), and a specimen with a deuterated backbone ($C_8D_3H_5$). Measurements were made at finite concentration

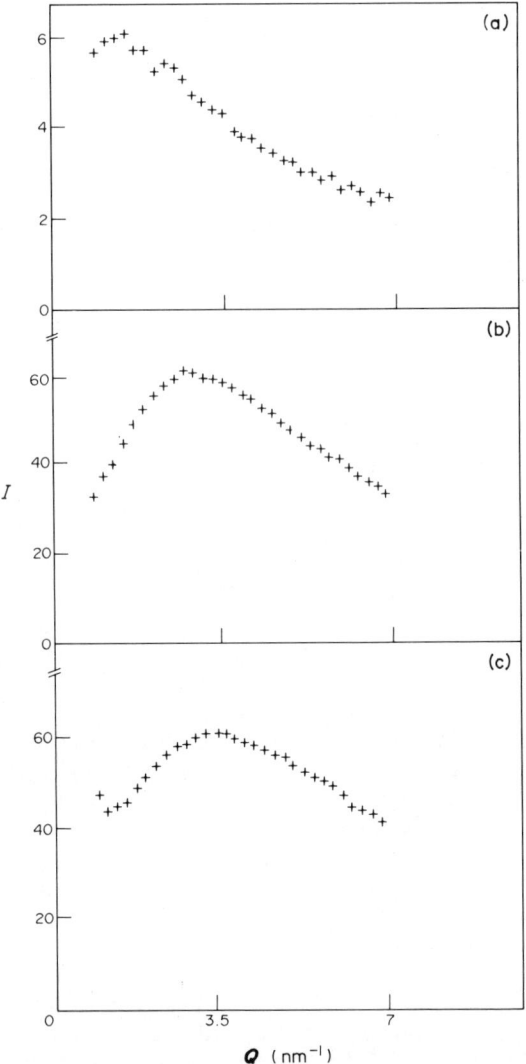

Figure 8 Scattered intensity *I vs. Q* for a partially labelled polystyrene molecule in carbon disulfide solution: (a) 1%, (b) 20%, (c) bulk. Reproduced by permission of IPC Science and Technology Press from *Polymer*, 1979, **20**, 1181

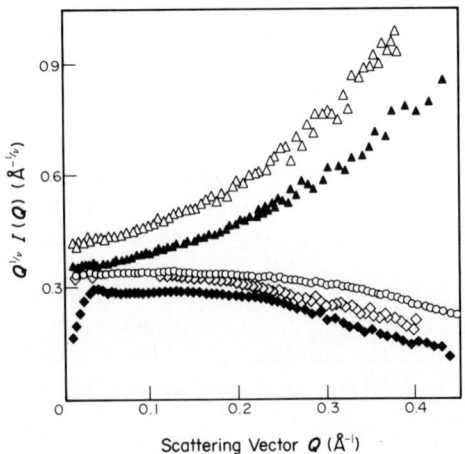

Figure 9 Kratky plots for selectively labelled polystyrene in carbon disulfide solution: $\bigcirc$ perdeuterated polystyrene; $\diamondsuit$, $\blacklozenge$ aromatic ring only deuterated ($\bar{M}_w = 49\,900$, $47\,800$); $\triangle$, $\blacktriangle$ backbone only deuterated ($\bar{M}_w = 1\,630\,000$, $240\,000$). Reproduced by permission of the American Chemical Society from *Macromolecules*, 1987, **20**, 630

in carbon disulfide solution and over a very wide Q range (0.1 nm^{-1} to 6 nm^{-1}). No difference between the various polystyrenes could be detected in the Q range below 0.5 nm^{-1}. Similarly in the intermediate Q range no difference was noted between fully deuterated and phenyl-ring-only deuterated specimens. However, main chain deuterated polymers had a considerably different form in the modified Kratky plot ($Q^{1/v}I(Q)$) as a function of Q^* (see Figure 9). The results were modelled as a thin thread (the backbone) convoluted with a cross section term arising from the finite thickness of the molecule due to the pendant phenyl groups, *i.e.* the macromolecule approximates to a curved cylinder. No attempts were made to use rotational isomeric state type calculations, no doubt because the form of the scattering function at intermediate Q in the presence of excluded volume effects is essentially unknown.[58-61]

6.4.3 Polyelectrolytes

In dilute solution polyelectrolytes produce small angle neutron scattering patterns which do not differ from those for neutral polymers and indeed may be analyzed in the same way. However, the dilute solution regime is at much lower concentration than that for neutral polymers especially when the ionic groups are fully dissociated. In this situation the polyelectrolyte molecule becomes fully extended and the overlap concentration is given by $c^* \simeq \bar{M}_w/N_A L^3$ where N_A is Avogadro's number and L is the length of the molecule which is presumed to be rod-like. At concentrations higher than overlap, the scattering pattern has a maximum and the origin of this maximum has been the focus of interest too.

Early investigations concentrated on examining the rigidity of polyelectrolytes, mainly by determining persistence lengths. Moan and co-workers[62,63] used carboxymethyl cellulose (CMC) in heavy water and poly(methacrylic acid) (PMAA) in heavy water to investigate these effects. For the CMC solutions only the intermediate scattering vector ranges were available. For a degree of neutralization of 0.6 (attained by adding NaOH), the persistence length a decreased linearly with the square root of CMC concentration and was linear with the ionic radius of the fixed charges on the polyelectrolyte molecule. By varying the degree of neutralization, the charge density along the macromolecule could be varied. The persistence length was found to be linear with the effective charge per monomer unit. Taken together, these data confirmed the picture of polyelectrolyte behaviour obtained from other techniques, *i.e.* increased extension of the molecule as the solutions become more dilute (for fixed degree of neutralization) and as the charge density increases (for fixed concentration); this extension being due to the chain being composed of increasingly longer rigid units.

The observation of a peak in the scattering from semi-dilute polyelectrolyte solutions was first attributed to a parallel arrangement of elongated macromolecules in solution.[64] Although there are interactions between the chains, such a model is no longer entertained. Subsequent experiments on salt-free solutions of sodium polystyrene sulfonate (NaPSS) have provided a much greater insight into the structure of polyelectrolyte solutions.[65,66] Typical scattering patterns for deuterated NaPSS in light water are shown in Figure 10; the maximum moves to higher Q values as the polymer concentration increases. Also typical is the very low value of the scattered intensity as Q tends to zero, indicative of a very low value for the osmotic compressibility of the solutions. Addition of simple electrolyte screens the fixed charges from each other and the peak eventually disappears from the scattering which becomes more like that for a Gaussian coil. Further evidence was obtained using semi-dilute solutions of NaPSS in which a few chains were deuterated and the aqueous solvent was a mixture of H_2O and D_2O chosen to match the contrast of the hydrogenous polyelectrolyte chains.[67] The differential scattering cross section is then written as

$$(d\sigma/d\Omega) = (\rho_p - \rho_\omega)^2 N_T (Y_D S_1(\mathbf{Q}) + Y_D^2 S_2(\mathbf{Q})) \qquad (33)$$

where N_T = total number of macromolecules in the solution
Y_D = N_D/N_T the number of fraction of deuterated chains
$S_1(Q)$ = scattering law for one macromolecule $\equiv F^2(Q)$
$S_2(Q)$ = scattering law describing correlations between chains

Rearranging equation (33) gives

$$(d\sigma/d\Omega)/(K_p^2 N_T Y_D) = S_1(\mathbf{Q}) + Y_D S_2(\mathbf{Q}) \qquad (34)$$

Hence plotting of the left-hand side of equation (34) as a function of Y_D (at fixed Q values) gives $S_1(Q)$

as the intercept at $Y_D = 0$ and $S_2(Q)$ as the slope. The variation of the scattering laws so obtained together with the experimentally observed total scattering law ($S_T(Q)$) is shown in Figure 11. $S_1(Q)$ data were analyzed to provide persistence lengths, a_t, which were interpreted as the sum of an intrinsic persistence length of the neutral chain, a_p, and an electrostatic persistence length, a_e, due to

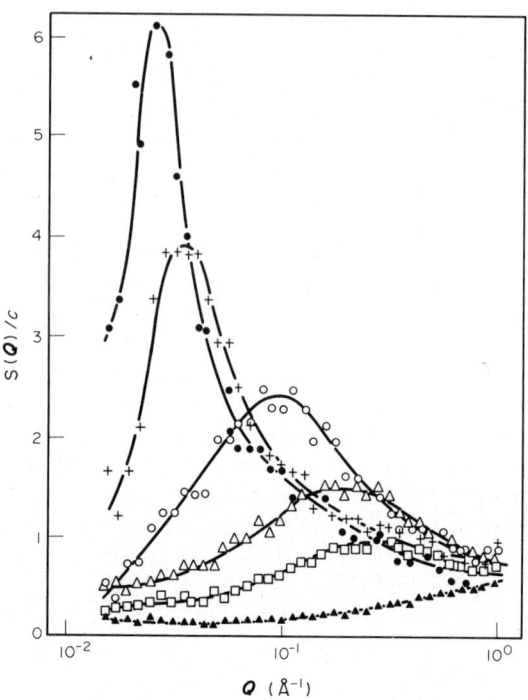

Figure 10 Reduced scattered intensity *vs.* Q for sodium deutero-polystyrene sulfonate in water: ● $c = 10^{-2}\,\mathrm{g\,cm^{-3}}$, + $c = 1.96 \times 10^{-2}\,\mathrm{g\,cm^{-3}}$, ○ $c = 4.76 \times 10^{-2}\,\mathrm{g\,cm^{-3}}$, △ $c = 9.09 \times 10^{-2}\,\mathrm{g\,cm^{-3}}$, □ $c = 13.04 \times 10^{-2}\,\mathrm{g\,cm^{-3}}$, ▲ $c = 23 \times 10^{-2}\,\mathrm{g\,cm^{-3}}$. The ordinate scale is in arbitrary units. Reproduced by permission of Les Editions de Physique from *J. Phys. (Paris)*, 1979, **40**, 70

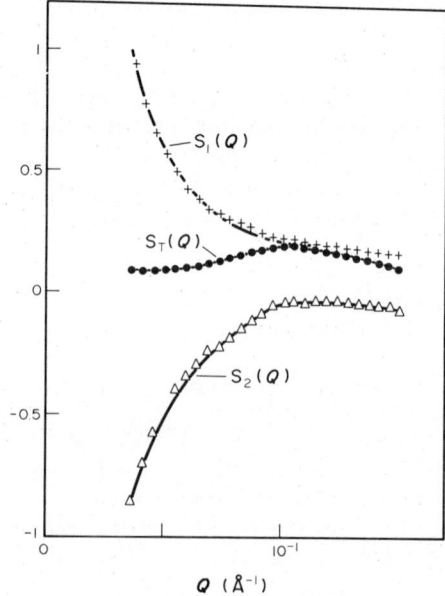

Figure 11 The three scattering laws obtained for deuterated sodium polystyrene sulfonate in light water. The ordinate scale is in arbitrary units. Reproduced by permission of Dr. Dietrich Steinkopff Verlag from *Colloid Polym. Sci.*, 1985, **263**, 955

the fixed charges on the polyion

$$a_t = a_p + a_e \tag{35}$$

A value for a_p was obtained by plotting a_t as a function of (polyelectrolyte concentration)$^{-1/2}$ and extrapolating to a zero value for the abscissa. Values of a_e, obtained as $(a_t - a_p)$ were compared with values calculated by theories of Odijk[68] and Le Bret.[69] Excellent agreement was obtained with the Le Bret calculations and moreover the data did not agree with idea of counterions 'condensing' on the polyion chain. The structure factor $S_2(Q)$, obtained was compared to a correlation hole model introduced by Hayter *et al.*[70] in the discussion of the dynamics of polyelectrolyte solutions and to a modification of Katchalsky's model[71] *i.e.* a parallel packing of stiff rods. The correlation hole model did not agree at all with the data whilst the simple Katchalsky model was also inadequate. Good agreement was obtained when the polyion was modelled not as a rigid rod but as a Kratky–Porod chain with persistence length a_t. This corresponded to the maximum in $S_T(Q)$ arising from alignment of worm-like segments with a separation of $2\pi/Q_{max}$ rather than from a parallel array of rods as suggested by Katchalsky.[71] Since the product $a_t Q_{max} \sim 4.5$, then only one parameter is needed to describe the conformational and structural parameters of polyelectrolyte solutions, since other factors can be obtained by simple arithmetic using the appropriate equations.

6.4.4 Copolymers

There has been considerable interest for a number of years in the dilute solution configuration of block copolymers.[72-75] Two models have been proposed. As a result of the thermodynamic incompatibility between the components a segregated structure is proposed where the two blocks occupy distinct regions of space. The second model is where both blocks interpenetrate each other leading to an expanded conformation. Light scattering has been the major technique used, the method being based on the refractive index dependence of scattered light intensity discussed by Bushuk and Benoit[76] and Leng and Benoit.[77] The light scattering analysis of copolymers requires the use of several solvents with different refractive indices. Regrettably this means that the thermodynamic conditions change and the conformational changes induced as a consequence do not permit unambiguous evaluation of the block copolymer conformation. However, contrast variation methods of this type are ideally suited to small angle neutron scattering. By using deuterated solvents and partly deuterated block copolymers it is possible to adjust the scattering system so that only one selected block has a contrast factor with the solvent.

With modifications for the terminology and symbols, the analysis developed for light scattering from copolymers is equally applicable for small angle neutron scattering. A detailed analysis has been provided by Ionescu *et al.*,[78] the pivotal equation for an AB diblock copolymer being

$$d\sigma/d\Omega = V^2 [K_A^2 F_A^2(Q) w_A^2 + K_B^2 F_B^2(Q)(Q) w_B^2 + 2 K_A K_B w_A w_B F_{AB}^2(Q)]/K_C^2 \tag{37}$$

where subscripts A and B refer to the individual blocks, $F_{AB}^2(Q)$ is an interchain scattering function, and w_A, w_B are the weight fractions of the A and B components.

$$K_i = (\rho_i - \rho_s), \text{ i} = \text{A or B, and } K_C = w_A K_A + w_B K_B,$$

whence

$$d\sigma/d\Omega = V^2 [Y^2 F_A^2(Q) + (1 - Y)^2 F_B^2(Q) + 2 Y(1 - Y) F_{AB}^2(Q)] \tag{38}$$

where $Y = w_A K_A/K_C$. Consequently by adjusting the value of Y it is possible in principle to obtain a differential scattering cross section which corresponds to only one component, *e.g.* when $(1 - Y) = 0$ then $d\sigma/d\Omega = V^2 [Y^2 F_A^2(Q)]$. Therefore by choosing conditions with three widely varying values of Y, all three form factors can be obtained.

Apart from the special conditions mentioned above, the radii of gyration and the molecular weight obtained by small angle neutron scattering from copolymer solutions are all apparent values.

$$\langle \bar{s}^2 \rangle_{app} = Y \langle \bar{s}^2 \rangle_A + (1 - Y) \langle \bar{s}^2 \rangle_B + Y(1 - Y) L^2 \tag{39}$$

where $\langle \bar{s}^2 \rangle_A$ and $\langle \bar{s}^2 \rangle_B$ are the mean square radii of gyration of the A and B blocks and L^2 is the mean square distance between the centres of mass of the two blocks. In a like manner

$$(\bar{M}_w)_{app} = (K_A K_B/K_C^2)(\bar{M}_w)_C + (K_A(K_A - K_B)/K_C^2) w_A(\bar{M}_w)_A + (K_B(K_B - K_A)/K_C^2) w_B(\bar{M}_w)_B \tag{40}$$

where $(\bar{M}_w)_C$ is the true molecular weight of the copolymers.

Evidently equation (39) predicts a convex parabolic dependence of $\langle s^2 \rangle_{app}$ on Y, and such behaviour was indeed found for styrene/isoprene diblock copolymers in toluene and cyclohexane solutions.[79] Variations in contrast were obtained by mixing deuterated and hydrogenous isomers of the two solvents. The parabola obtained by regression on the $\langle \bar{s}^2 \rangle_{app}$ values was in almost exact agreement with the theoretical parabola calculated from the copolymer composition and block mean square radii of gyration (Figure 12). From these data it was ascertained that the radius of gyration of a polystyrene block was little altered from that of homopolystyrene of the same molecular weight, and the authors concluded that there was no evidence for a segregated structure in non-selective solvents.

In one of the Zimm plots of small angle neutron scattering data shown by Ionescu et al.,[79] it is evident that a maximum appears in the scattering pattern. Since the copolymer had deutero-styrene blocks, this maximum could be attributed to a correlation hole effect. A more rigorous analysis of the source of this maximum has been given by Benmouna and Benoit.[80] By using essentially the same equations as Ionescu et al.[78] they showed that a maximum in the scattering of partly-labelled chains can arise from either a contrast effect or a copolymer concentration effect.

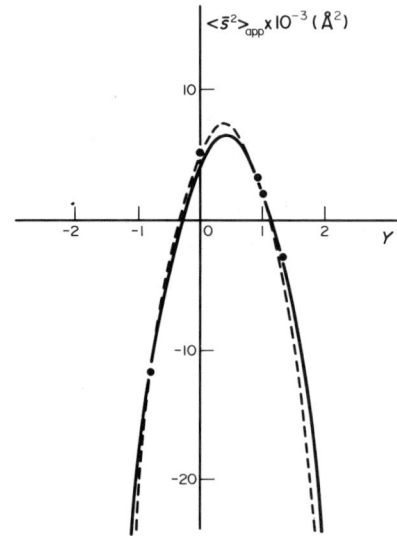

Figure 12   Dependence of the apparent mean square radius of gyration on $Y\,(w_A K_A/K_C)$ for linear deuterostyrene–isoprene block copolymer in cyclohexane at $20\,°C$: ($\bullet$) experimental data; (——) polynomial fit to data; (– – –) theoretical curve. Reproduced by permission of John Wiley Ltd from *J. Polym. Sci., Polym. Phys. Ed.*, 1981, **19**, 1033.

Edwards et al.[81] have also observed maxima in the small angle neutron scattering from a deutero-styrene/methyl methacrylate block copolymer in *p*-xylene at $30\,°C$. Other data also reported[81] showed that this copolymer formed an aggregate in *p*-xylene at this temperature. Furthermore, *p*-xylene and hydrogenous poly(methyl methacrylate) have almost identical scattering length densities and consequently the broad maximum referred to above is seen to be characteristic of the arrangement of the deutero-styrene blocks in the multimer. At $30\,°C$ there was an average of two copolymer molecules in the multimer and the observed scattering was exactly reproduced by a Monte Carlo calculation of the single particle form factor for two rotational isomeric state chains (deutero-polystyrene blocks) attached to an invisible spherical core (methyl methacrylate block).

6.4.5   Concentrated Solutions

Dilute solution behaviour of macromolecules has been thoroughly investigated theoretically and experimentally over the past 40 to 50 years. Significant understanding of their conformational properties has been gained and this is definitively summarized in the monograph by Yamakawa.[82] The conformational properties of amorphous macromolecules in bulk were surmised as being Gaussian in nature, and this supposition was subsequently confirmed using small angle neutron scattering.[83] The question then arose as to the behaviour of chains in between these limits.

Edwards[32-34] has applied mean field theory to both dilute and concentrated solutions. In the latter case the overlap of the macromolecular spatial volumes screens topologically remote sections of an individual chain. As a consequence the excluded volume effect becomes attenuated and the conformation becomes more Gaussian. The characteristic length introduced is ξ_E, the screening length analogous to the Debye–Hückel screening length in simple electrolyte theory. The second approach to concentrated solutions (in actual fact semi-dilute) is based on an analogy drawn between polymer solutions and magnetic phenomena and the use of re-normalization group theory.[16,35] Identical results can be obtained using scaling theory, and a correlation length, ξ, is obtained as the fundamental parameter. In both theories, the macromolecule has self avoiding walk (excluded volume) statistics at length scales less than ξ or ξ_E, but for greater distances the statistics approach Gaussian. The concentration dependence of the radius of gyration and screening length predicted by the two theories differ somewhat.

Daoud and Jannink[84] have extended the concepts embodied in the re-normalization group theory to produce a diagram of polymer solution behaviour in the temperature–concentration plane (Figure 13) and have given scaling relations for the radius of gyration and screening length in the various regions. Scaling relations are simple equations relating properties of the macromolecular solution to the polymer concentration, c, and excluded volume parameter, ω. Perhaps the most widely known is that between radius of gyration and molecular weight $\langle \bar{s}^2 \rangle^{1/2} \sim \bar{M}^\nu$. For semi-dilute solutions in good solvents at a fixed temperature it has been shown that $\xi \sim c^{-m}$ and $\langle \bar{s}^2 \rangle \sim c^n$, where $m = \nu/(3\nu - 1)$ and $n = (1 - 2\nu)/(3\nu - 1)$. Under theta conditions $\nu = 0.5$, whilst at the excluded volume limit $\nu = 0.6$ (see below for a discussion of other values of ν in this latter case). The scaling relations obtained are given in Table 3. Additional regions to this diagram have been proposed by Schaeffer,[85] arising from the interrelation between macromolecular stiffness and thermodynamic quality of the solvent. Here we discuss only the regions proposed on the original diagram because only these have been investigated by small angle neutron scattering. Table 3 also shows the relations obtained by Edwards and co-workers[32-34,86] using mean field theory. Additionally, although Table 3 only shows the exponents, the mean field theories, unlike scaling laws, also include all the front factors in a rigorous form.

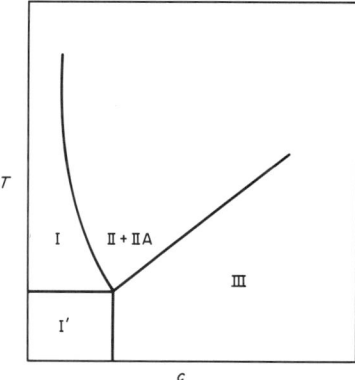

Figure 13   Phase diagram of polymer solution behaviour in the temperature–concentration plane: see text for description of regions

Table 3   Relationships Obtained for Radius of Gyration (*s*) and Correlation Length (ξ) by Scaling Laws and Mean Field Theories

Region	Scaling Laws ξ^2	Scaling Laws $\langle \bar{s}^2 \rangle$	Mean Field Theory ξ^2	Mean Field Theory $\langle \bar{s}^2 \rangle$
I[1]	—	$\bar{M}$	—	$\bar{M}$
I	—	$\bar{M}^{6/5}\omega^{2/5}$	—	$\bar{M}^{6/5}\omega^{2/5}$
II	$c^{-3/2}\omega^{-1/2}$	$\bar{M}c^{-1/5}\omega^{1/5}$	$c^{-6/5}\omega^{-4/5}$	$\bar{M}c^{-1/4}\omega^{1/4}$
IIA	—	—	$\nu^{-1}\omega^{-1}$	$\bar{M}c^{-1/2}\omega^{1/2}$
III	c^2	$\bar{M}$	—	$\bar{M}$

ω = excluded volume parameter $\propto (T - \theta)/T$ where θ is the theta or Flory temperature of the system.

Evidently, the radius of gyration of a macromolecule in a concentrated solution can be obtained by dispersing a few deuterated chains amongst hydrogenous material. Thus the total concentration of polymer can be high whilst that of the labelled scattering macromolecules is dilute. This method was used in all the earlier investigations. Subsequently it has been appreciated that solutions with up to 50% of the total chains deuterated may also be used (*vide infra*).

Screening or correlation lengths may be obtained in several ways. For solutions containing a few marked chains the scattering law ($F^2(Q)$) at low Q will be:

$$F^2(\mathbf{Q}) \propto Q^{-2} \tag{41}$$

since long length scales are probed and the molecule appears to be a Gaussian arrangement of segments of length ξ. As Q increases, shorter length scales are probed until eventually $Q > \xi^{-1}$. At these length scales ($< \xi$) the scattering segments have excluded volume effects and

$$F^2(\mathbf{Q}) \propto Q^{-5/3} \tag{42}$$

The value of Q where this crossover in behaviour takes place, Q^{**}, will give a value for the screening length as $2\pi/Q^{**}$. Although this method has been used,[87] the change from Q^{-2} to $Q^{-5/3}$ behaviour is subtle and, moreover, takes place in regions of Q where the ratio of coherent to incoherent scattering is low.

Rather than make use of the single chain pair correlation function in the manner described above, the total pair correlation function of the system may be used, *i.e.* the correlation function between *all* monomer pairs in the system. For this purpose hydrogenous polymer is dissolved in deuterated solvent at the desired concentration. In this situation, at length scales less than ξ, the total pair correlation function is dominated by units on the same chain and gives $S(Q) \propto Q^{-5/3}$. As the length scale increases, monomer units from other chains intervene and the correlation function becomes attenuated and dies out in a length ξ and produces a scattering law $S(Q) \propto Q^{-2}$. This behaviour can be used in two ways to obtain ξ. Firstly by noting the value of $S(Q^*)$ at which the scattering law changes from Q^{-2} to $Q^{-5/3}$. Although the signal is stronger than the single chain case, the exact value of Q^* where the change takes place is still difficult to ascertain. Secondly, in the region of Q where $Q\xi \leq 1$, the scattering law is given by a Lorentzian:

$$S(\mathbf{Q}) = f(T,C)/(Q^2 + \xi^{-2}) \tag{43}$$

whence the value of ξ can be obtained from the value of slope/intercept of a plot of $S(\mathbf{Q})^{-1} (\propto I(Q)^{-1})$ as a function of Q^2. Such plots are shown in Figure 14 for hydrogenous polystyrene in deuterocyclohexane at different temperatures. The cross over in Q dependence is evident.

The earliest experimental investigations of semi-dilute solutions[88,89] were, in fact, a little in advance of detailed theory. These were measurements of screening lengths as a function of concentration of polystyrene in deutero-benzene, the solutions being semi-dilute *i.e.* in region II of Figure 13. The data were not initially interpreted to obtain scaling exponents but subsequent analysis by King *et al.*[90] produced values of m of 0.72 and 0.79 for the two molecular weights used, in

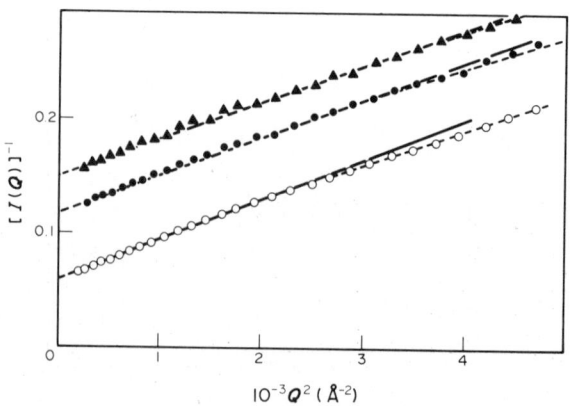

Figure 14 Reciprocal scattered intensity *vs.* Q^2 for a 36% (w/v) solution of polystyrene in deutero-cyclohexane. Note the change in slope at higher Q values where $Q > Q^*$: (○) 31 °C; (●) 46 °C; (▲) 54 °C

excellent agreement with scaling law predictions. These early reports were followed by a detailed study by Cotton *et al.*[91] of polystyrene in carbon disulfide, where values of both radius of gyration and screening length were obtained, the latter from the scattering associated with the total correlation function. The majority of data were for concentrations in the semi-dilute range and the concentration exponent, *m* for the screening length dependence was 0.72. Figure 15 shows the variation in $\langle \bar{s}^2 \rangle / \bar{M}_w$ with concentration, the slope of -0.25 is in excellent agreement with scaling laws. Note that this scaling relation is apparently continuous to the bulk state, somewhat surprising in view of the different exponents predicted by the scaling formulae. In a subsequent paper[92] further regions of the temperature–concentration diagrams were examined for polystyrene in cyclohexane. The combinations of temperatures and concentrations were such that the transition from the theta region (I′) to the semi-dilute good solvent region II was covered. The most striking result was the change in $\log \langle \bar{s}^2 \rangle^{1/2}$ as a function of $\log (T-\theta)$ ($\alpha\omega$, the excluded volume parameter) shown in Figure 16, where the cross over from theta to semi-dilute behaviour is clearly evident. By identifying the temperature at which this cross over occurred, and also cross overs for the screening length in solutions of different concentration (somewhat more tenuously), the full experimental tempera-ture–concentration diagram was constructed.

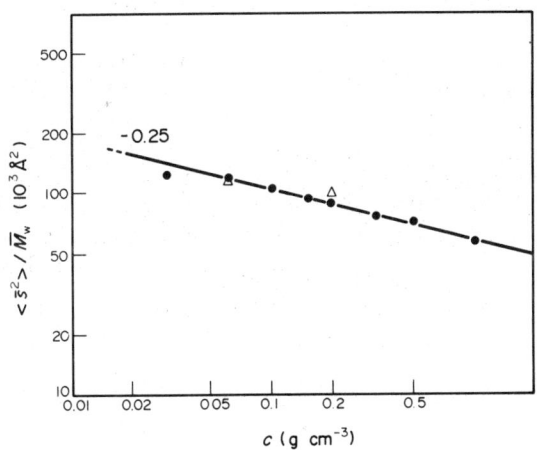

Figure 15 Log–log dependence of mean square radius of gyration on concentration for polystyrene in carbon disulfide solution. Reproduced by permission of the American Chemical Society from *Macromolecules*, 1975, **8**, 804

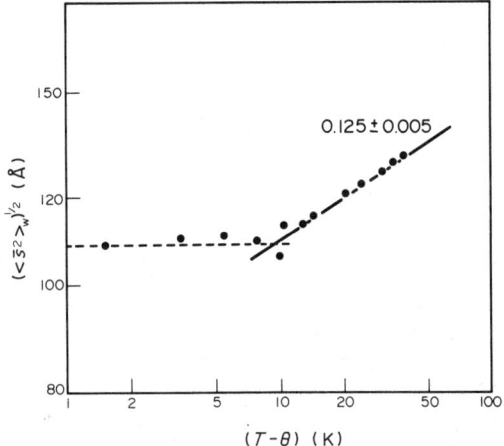

Figure 16 Log–log dependence of the root mean square radius of gyration of polystyrene on temperature for a semi-dilute solution in cyclohexane. Reproduced by permission of the American Institute of Physics from *J. Chem. Phys.*, 1976, **65**, 1101

These results were confirmed by Richards et al.,[93, 94] who also studied polystyrene in cyclohexane solutions. The majority of the data reported in those papers were for a higher concentration region than the semi-dilute good solvent (region IIa in Figure 13), which Richards et al. termed the semi-concentrated region, since it occurred just before the concentrated region (region III). Table 3 shows that exponents differing from the scaling law values should be obtained from mean field theory for the concentration and temperature dependence of $\langle \bar{s}^2 \rangle$ and ξ^2. For solutions of *circa* 37% (w/v) in cyclohexane, the variation of $\langle \bar{s}^2 \rangle$ was linear in $(T-\theta)^{1/2}$, as predicted by mean field theories. Furthermore, the change in screening length with temperature was given by $\xi^{-2} \propto (T-\theta)$, $\langle \bar{s}^2 \rangle$ decreased in proportion to $c^{-1/2}$ over the range $0.4 \le c \,(\text{g cm}^{-3}) \le 0.8$, and, at higher concentrations, the mean square radius of gyration was constant at the theta value. Figure 17 shows these results. Evidently, the temperature–concentration diagram is much more complex than originally set out by Daoud and Jannink.[84] The variation of screening length and mean square radius of gyration with concentration and temperature has also been investigated in the neighbourhood of the lower critical solution temperature ($\approx 220\,^\circ\text{C}$). These results showed that a temperature concentration diagram could be constructed which was a mirror image of that at the upper critical solution temperature originally discussed by Daoud and Jannink.[84]

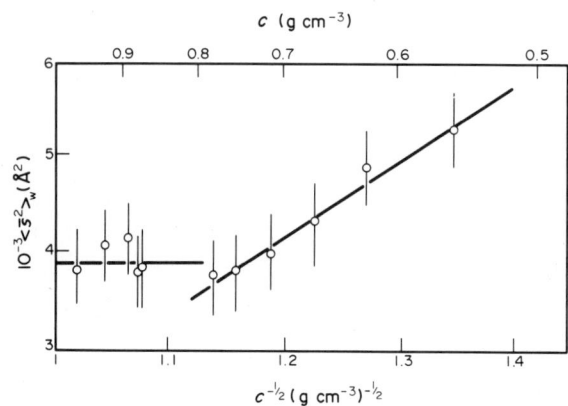

Figure 17   Mean square radius of gyration as a function of root concentration for polystyrene in cyclohexane

Thus far for those systems where we have a proportion of labelled chains dispersed in hydrogen-ous polymer and solvent, the system has been treated as dilute in scattering material, the non-labelled polymer and solvent constituting a background term. However, now that we are aware of the total correlation function and its associated total scattering law, $S_T(Q)$, it is evident that this will always be present and the treatment of scattering from macromolecules in solution (of any kind) has to be revised. At first sight this may be thought to lead to complications in data analysis. However, full discussion[95–97] of this aspect of neutron scattering (and its experimental verification) has shown that this is not so and that higher proportions of labelled material can be used without detriment to the results. Hence for a scattering system composed of labelled polymer, unlabelled polymer and solvent (scattering length densities, ρ_p^D, ρ_p and ρ_s respectively) the scattering law is given by:

$$S(\mathbf{Q}) = (X\rho_p^D + (1-X)(\rho_p - \rho_s)S_T(\mathbf{Q}) + X(1-X)(\rho_p^D - \rho_p)^2 S_s(\mathbf{Q}) \tag{44}$$

where $S_s(Q)$ is the scattering law for a single macromolecule and X is the mole fraction of labelled monomer units in the mixture of labelled and unlabelled polymer. Strictly, this equation is only applicable when the molecular weight and molecular weight distribution of labelled and unlabelled polymer are identical; however, correction terms have been discussed.[98] Values of $S_T(Q)$ and $S_s(Q)$ are extracted from measurements on solutions with different values of X, but the same total concentration, using a weighted subtraction of pairs of data sets.

King et al.[90] have used this method for solutions of hydrogenous and deutrated polystyrene in deutero-toluene with concentrations extending to bulk polymer. Although $S_s(Q)$ was easily obtained and fitted to the Debye equation (equation 18) to give the radius of gyration, $S_T(Q)$ was much smaller. As a consequence, $S_T(Q)$ was only obtained from solutions of hydrogenous polymer alone in the deuterated solvent. For the screening lengths the exponent obtained for the concentration scaling law was 0.70 ± 0.02, in excellent agreement with predictions. Such was not the case for the scaling exponent for the mean square radius of gyration, a value of *circa* 0.16 being obtained, considerably less than that predicted and also observed by others. Attempts were made to rationalize this exponent in terms of the various values of v calculated for the relation $\langle \bar{s}^2 \rangle \sim M^v$ for dilute solutions.[99-101] These data were subsequently re-analysed[102] according to a method suggested by Benmouna and Benoit,[103] which enabled the scaling exponent of the excluded volume interaction parameter to be obtained. Finally it should be noted that both small angle X-ray and small angle neutron scattering results from semi-dilute solutions have been compared,[104] the data being indistinguishable from each other.

6.4.6 Macromolecular Gels

Gels are obtained when a crosslinked network of macromolecules is swollen to equilibrium by a solvent. Scaling laws for gel properties (modulus, screening length, mean square end-to-end distance between crosslinks) are identical to those for semi-dilute solutions when the gel is formed under conditions where the macromolecules are swollen to the excluded volume limit.[16] However, no small angle neutron scattering has been reported on gels prepared under these conditions. Another viewpoint is that a semi-dilute solution is a network with crosslinks of finite lifetime. If the frequency of the probe being used to investigate the solution is smaller than the (lifetime)$^{-1}$ of a transient crosslink, then so far as the probing method is concerned the solution looks like a network.

Davidson *et al.*[105] have used this concept in the examination of the total scattering law from hydrogenous polystyrene networks swollen to equilibrium in deutero-toluene and deutero-cyclohexane at different temperatures. For gels swollen with toluene, as long as the polymer volume fraction was less than *circa* 0.06 the screening lengths were identical with those found for solutions. At higher concentrations an asymptotic value was approached (Figure 18). For gels swollen with cyclohexane only linear behaviour was observed, the concentration exponent being dependent on temperature, a value of approximately 1 being obtained at 35 °C as predicted by both scaling law and mean field theories. At 60 °C however, the exponent was ~ 0.6, much lower than the scaling law prediction and approaching the value predicted by the extrapolation theory of Edwards. Similar behaviour was reported earlier[93,94] for *solutions* in cyclohexane over the same temperature range.

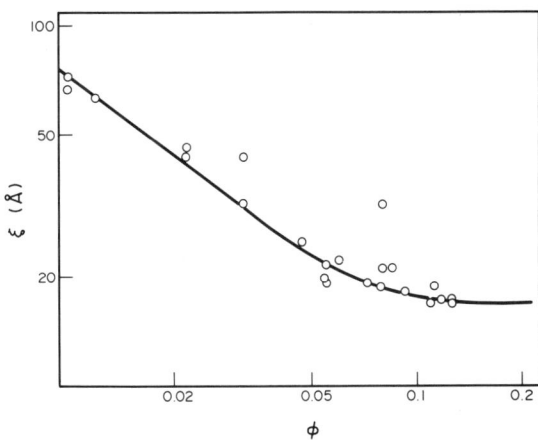

Figure 18 Log–log dependence of the screening length on polymer volume fraction for polystyrene networks swollen in toluene

A major interest in gels is the determination of the radius of gyration of the network molecules and comparison of these values with the theory of rubber-like elasticity. In so far as network chain (*i.e.* a macromolecule between two crosslink points) dimensions are concerned, the two extremes in current theory are affine chain deformation and phantom chain behaviour.[106] In the affine deformation theory the network chain is presumed to deform in exact proportion to the macroscopic dimensions of the network.

$$\langle \bar{s}^2 \rangle / \langle \bar{s}_0^2 \rangle = \lambda^2 \qquad (45)$$

where $\langle \bar{s}_0^2 \rangle$ is the mean square radius of gyration of the undeformed chain. In a phantom network, chains may pass freely through each other and the crosslink points are not fixed but fluctuate about their mean position. For these networks[107] then

$$\langle \bar{s}^2 \rangle / \langle \bar{s}_0^2 \rangle = 1/2[1 + \lambda^2 - 2/f(\lambda^2 - 1)]$$
(46)

where f is the functionality of the network junctions. Rigorously, equations (45) and (46) are only applicable to end linked network chains, however Ullman[108,109] has derived scattering laws for network chains which pass through several crosslink points.

In principle the influence of deformation of conformational properties can be studied by simple extension of a network sample in the neutron beam. In practice, this leads to anisotropic scattering with the intensity of scattering parallel to the stretch direction increasing more rapidly with Q then the scattering perpendicular to the stretch direction. Isotropic deformation of the network can be obtained using solvent swelling. However the concentration of polymer is necessarily decreased, as is the scattered intensity per unit time.

Bastide and co-workers[110] used small angle neutron scattering to determine the radius of gyration in anionicly prepared divinylbenzene-crosslinked polystyrene networks osmotically deswollen by adding polystyrene to the benzene solvent. Figure 19 shows that little change in radius of gyration was observed as the volumetric swelling was decreased. The results were interpreted qualitatively by a de-interdispersion model.[111] In this model two types of crosslinks are noted as being present around any one crosslink chosen as reference. Topological crosslinks are directly connected to the reference crosslink, spatial neighbours are only connected by a long and tortuous path. On swelling the separation between reference and spatial crosslinks increases greatly whilst that between topological neighbours alters but little.

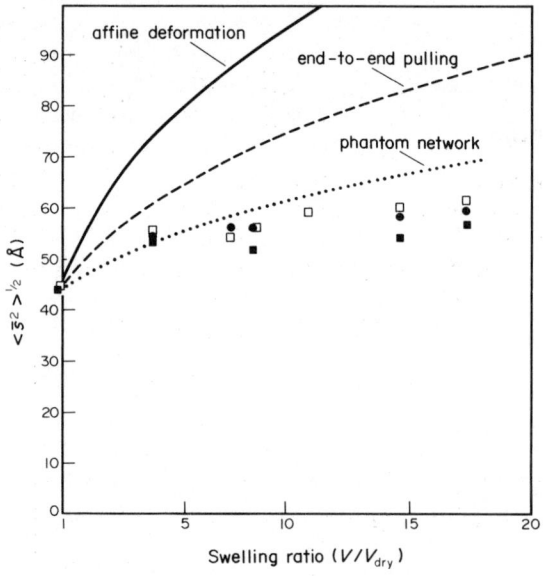

Figure 19 Root mean square radius of gyration as a function of swelling ratio for labelled polystyrene chains in osmotically deswollen networks. The points represent experimental results for different gels and the lines theoretical results as indicated . Reproduced by permission of the Americal Chemical Society from *Macromolecules*, 1984, **17**, 83

Beltzung *et al.*[112] also noted little change in the dimensions of end linked poly(dimethyl siloxane) chains in networks swollen by cyclohexane. Indeed the dimensions were little altered from dilute solution values for isolated molecules.

Multilinked chains in swollen networks have also been studied by Davidson *et al.*[113] using toluene and cyclohexane as swelling agents for networks with differing crosslink densities, and hence differing volumetric swellings. For volumetric swellings of between 8 and 24, no change in the radius of gyration was observed. Similarly, gels which were swollen in cyclohexane at the theta temperature showed no change in labelled chain dimensions from unperturbed values even though the networks were at least threefold swollen. However the most puzzling behaviour observed[113] was the temperature dependence of the labelled chain dimensions in cyclohexane swollen networks: see Figure 20. Since the macroscopic dimensions of the networks were continuously increasing over this temperature range no satisfactory explanation of this phenomenon could be advanced.

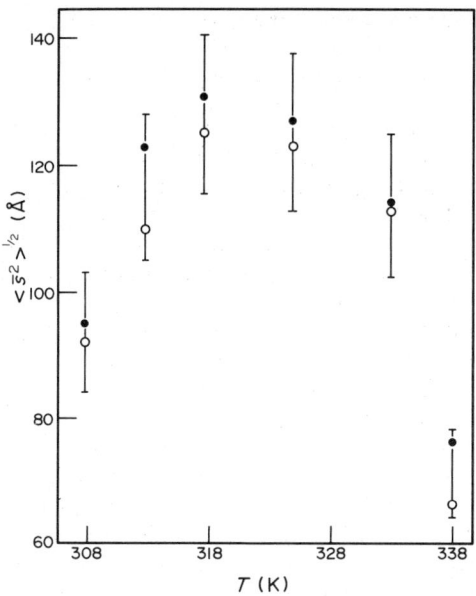

Figure 20 Mean square radius of gyration of multilinked labelled polystyrene chains as a function of temperature in networks swollen in equilibrium in cyclohexane.[113] The points represent the results for different gels

6.4.7 Miscellaneous Systems

The review given above has concentrated on the major areas where small angle neutron scattering has been applied to solutions of macromolecules. Two other areas which promise to yield much interesting information in the future are the study of shear deformation in solution[114] and the segment density profile of macromolecules adsorbed on to a surface.[115,116]

6.5 APPENDIX

Addresses of Institutions with Small Angle Neutron Diffractometers.

(1) Office of the Scientific Secretary
Institut Laue-Langevin
BP156X
38042 Grenoble-Cedex
France

(2) Mr. D. H. C. Harris
Bldg 436
Atomic Energy Research Establishment Harwell
Didcot
Oxon OX11 ORA, UK

(3) Dr. R. J. R. Bennett
Neutron Division
Bldg R3
Rutherford-Appleton Laboratory
Chilton
Didcot
Oxon OX11 OQX, UK

(4) Dr. G. D. Wignall
National Center for Small Angle Scattering Research
Oak Ridge National Laboratory
Post Office Box X
Oak Ridge
Tennessee 37830, USA

(5) Dr. J. R. D. Copley
McMaster Nuclear Reactor
McMaster University
Hamilton
Ontario L85 4K1, Canada

6.6 REFERENCES

1. S. H. Chen and S. Yip (eds.), 'Spectroscopy in Biology and Chemistry, Neutron, X-ray, Laser', Academic Press, New York, 1974.
2. G. Kostorz, 'Treatise on Materials Science and Technology Vol. 15. Neutron Scattering', ed. H. Herman, Academic Press, New York, 1979.
3. J. E. Enderby, *Annu. Rev. Phys. Chem.*, 1983, **34**, 155.
4. B. T. M. Willis, 'Chemical Applications of Thermal Neutron Scattering', Oxford University Press, Oxford, 1973.
5. G. Bacon, 'Neutron Diffraction', 3rd edn., Clarendon Press, Oxford, 1975.
6. W. Marshall and S. W. Lovesey, 'Theory of Thermal Neutron Scattering', Clarendon Press, Oxford, 1971.
7. V. F. Turchin, 'Slow Neutrons', Israel Program for Scientific Translations, Jerusalem, 1965.
8. G. L. Squires, 'Introduction to the Theory of Thermal Neutron Scattering', Cambridge University Press, 1978.
9. S. W. Lovesey, 'Theory of Neutron Scattering from Condensed Matter', Clarendon Press, Oxford, 1984, vols. 1 and 2.
10. R. W. Richards, in 'Developments in Polymer Characterisation-5', ed. J. V. Dawkins, Elsevier Applied Science, London, 1986, p. 1.
11. A. Maconnachie and R. W. Richards, *Polymer*, 1978, **19**, 739.
12. R. Ullman, *Annu. Rev. Mater. Sci.*, 1980, **10**, 261.
13. L. H. Sperling, *Polym. Eng. Sci.*, 1984, **24**, 1.
14. B. Jacrot, *Rep. Prog. Phys.*, 1976, **39**, 911.
15. G. Zaccai and B. Jacrot, *Annu. Rev. Biophys. Bioeng.*, 1983, **12**, 139.
16. P. G. de Gennes, 'Scaling Concepts in Polymer Physics', Cornell University Press, Ithaca, N.Y., 1979.
17. S. W. Lovesey and G. Kostorz, p. 1 in ref. 2.
18. V. F. Sears, Atomic Energy of Canada Limited, Report AECL 8490, Chalk River, Ontario, 1984.
19. L. Koester, H. Rauch, M. Herkens and K. Schroder, KFA Report Jul–1755.
20. S. F. Mughabghab and D. E. Garber, BNL–325 (3rd edn.) vol. 1., Brookhaven National Laboratory, Long Island City, N.Y.
21. A. Guinier and G. Fournet, 'Small Angle Scattering of X-rays', Wiley, New York, 1955.
22. O. Glatter and O. Kratky, 'Small Angle X-ray Scattering', Academic Press, London, 1982.
23. W. Burchard in 'Applied Fibre Science, Volume 1', ed. F. Happey, Academic Press, London, 1978, chap. 10.
24. W. Burchard, *Adv. Polym. Sci.*, 19, **48**, 1.
25. P. Debye, *J. Phys. Colloid Chem.*, 1947, **51**, 98.
26. O. Kratky, *Pure Appl. Chem.*, 1966, **12**, 483.
27. R. Kirste and R. Oberthur, chap. 12 in ref. 22.
28. H. Hayashi, P. J. Flory and G. D. Wignall, *Macromolecules*, 1983, **16**, 1328.
29. C. J. C. Edwards, R. W. Richards and R. F. T. Stepto, *Macromolecules*, 1984, **17**, 2147.
30. R. Ullman, *J. Polym. Sci., Polym. Phys. Ed.*, 1985, **23**, 1477.
31. S. F. Edwards, *Proc. Phys. Soc., London*, 1966, **88**, 265.
32. S. F. Edwards, *J. Phys. A: Math. Gen.*, 1975, **8**, 1670.
33. S. F. Edwards and P. Singh, *J. Chem. Soc., Faraday Trans. 2*, 1979, **75**, 1001.
34. S. F. Edwards and E. F. Jeffers, *J. Chem. Soc., Faraday Trans. 2*, 1979, **75**, 1020.
35. J. des Cloizeaux, *J. Phys. (Paris)*, 1975, **36**, 281.
36. K. Ibel, *J. Appl. Crystallogr.*, 1976, **9**, 296.
37. W. C. Koehler, *Physica B+C (Amsterdam)*, 1986, **137**, 320.
38. See Appendix.
39. 'Neutron Research Facilities at the Institute Laue Langevin High Flux Reactor', Grenoble, 1983.
40. D. I. Page, 'Practical Guide to the PLUTO Small Angle Scattering Spectrometer', HMSO, 1980.
41. ISIS Annual Report, Science and Engineering Research Council, 1986.
42. G. D. Wignall and F. S. Bates, *J. Appl. Crystallogr.*, to be published.
43. B. H. Zimm, *J. Chem. Phys.*, 1948, **16**, 1093.
44. W. G. McMillan, Jr. and J. E. Mayer, *J. Chem. Phys.*, 1945, **13**, 276.
45. R. E. Ghosh, Institut Laue-Langevin Technical Report No. 81GH29T, Grenoble.
46. D. G. H. Ballard, M. G. Rayner and J. Schelten, *Polymer*, 1976, **17**, 349.
47. P. J. Flory, 'Statistical Mechanics of Chain Molecules', Wiley, New York, 1969.
48. A. K. Gupta, J. P. Cotton, E. Marchal, W. Burchard and H. Benoit, *Polymer*, 1976, **17**, 363.
49. Y. Matsushita, I. Noda, M. Nagasawa, T. P. Lodge, E. J. Amis and C. C. Han, *Macromolecules*, 1984, **17**, 1785.
50. R. Duplessix, J. P. Cotton, H. Benoit and C. Picot, *Polymer*, 1979, **20**, 1181.
51. P. G. de Gennes, *J. Phys. (Paris)*, 1970, **31**, 235.
52. J. S. Higgins, K. Dodgson and J. A. Semlyen, *Polymer*, 1979, **20**, 553.
53. E. F. Casassa, *J. Polym. Sci., Part A*, 1965, **3**, 605.
54. C. J. C. Edwards, R. W. Richards, R. F. T. Stepto, K. Dodgson, J. S. Higgins and J. A. Semlyen, *Polymer*, 1984, **25**, 365.
55. W. Burchard and M. Schmidt, *Polymer*, 1980, **21**, 745.
56. C. J. C. Edwards, R. W. Richards and R. F. T. Stepto, *Macromolecules*, 1984, **17**, 2153.
57. M. Rawiso, R. Duplessix and C. Picot, *Macromolecules*, 1987, **20**, 630.
58. T. A. Witten, Jr. and L. Schafer, *J. Chem. Phys.*, 1981, **74**, 2582.
59. S. F. Edwards, *Proc. Phys. Soc. London*, 1965, **85**, 613.
60. P. G. de Gennes, *Rep. Prog. Phys.*, 1969, **32**, 187.
61. K. Shinbo and Y. Miyake, *J. Phys. Soc. Jpn.*, 1980, **48**, 2084.
62. M. Moan and C. Wolff, *Polymer*, 1975, **16**, 776.
63. M. Moan, C. Wolff and R. Ober, *Polymer*, 1975, **10**, 781.
64. J. P. Cotton and M. Moan, *J. Phys. (Paris) Lett.*, 1976, **37**, L75.
65. C. E. Williams, M. Nierlich, J. P. Cotton, G. Jannink, F. Boue, M. Daoud, B. Farnoux, C. Picot, P. G. de Gennes, M. Rinaudo, M. Moan and C. Wolff, *J. Polym. Sci., Polym Lett. Ed.*, 1979, **17**, 379.
66. M. Nierlich, C. E. Williams, F. Boue, J. P. Cotton, M. Daoud, B. Farnoux, G. Jannink, C. Picot, M. Moan, C. Wolff, M. Rinaudo and P. G. de Gennes, *J. Phys. (Paris)*, 1979, **40**, 701

67. M. Nierlich, F. Boue, A. Lapp and R. Oberthur, *Colloid Polym. Sci.*, 1985, **263**, 955.
68. T. Odijk and A. C. Houwaart, *J. Polym. Sci., Polym. Phys. Ed.*, 1978, **16**, 627.
69. M. Le Bret, *J. Chem. Phys.*, 1982, **76**, 6243.
70. F. Nallet, G. Jannink, J. B. Hayter, R. Oberthur and C. Picot, *J. Phys. (Paris)*, 1983, **44**, 87.
71. A. Katchalsky, Z. Alexandrowicz and O. Keden, 'Chemical Physics of Ionic Solutions', ed. B. E. Conway and R. G. Barradas, Wiley, New York, 1966.
72. T. Kotaka, T. Tanaka, H. Ohnuma, Y. Murakami and H. Inagaki, *Polym. J.*, 1970, **1**, 245.
73. T. Tanaka, T. Kotaka and H. Inagaki, *Polym. J.*, 1972, **3**, 338.
74. T. Tanaka, T. Kotaka and H. Inagaki, *Macromolecules*, 1974, **7**, 311.
75. T. Tanaka, T. Kotaka, K. Ban, M. Hattori and H. Inagaki, *Macromolecules*, 1977, **10**, 960.
76. W. Bushuk and H. Benoit, *Can. J. Chem.*, 1958, **36**, 1616.
77. M. Leng and H. Benoit, *J. Chim. Phys. Phys.-Chim. Biol.*, 1961, **58**, 48.
78. L. Ionescu, C. Picot, M. Duval, R. Duplessix, H. Benoit and J. P. Cotton, *J. Polym. Sci., Polym. Phys. Ed.*, 1981, **19**, 1019.
79. L. Ionescu, C. Picot, R. Duplessix, M. Duval, H. Benoit, J. P. Lingelser and Y. Gallot, *J. Polym. Sci., Polym. Phys. Ed.*, 1981, **19**, 1033.
80. M. Benmouna and H. Benoit, *J. Polym. Sci., Polym. Phys. Ed.*, 1983, **21**, 1227.
81. C. J. C. Edwards, R. W. Richards and R. F. T. Stepto, *Polymer*, 1986, **27**, 643.
82. H. Yamakawa, 'Modern Theory of Polymer Solutions', Harper and Row, New York, 1971.
83. J. P. Cotton, D. Decker, H. Benoit, B. Farnoux, J. Higgins, G. Jannink, R. Ober, C. Picot and J. des Cloizeaux, *Macromolecules*, 1974, **7**, 863.
84. M. Daoud and G. Jannink, *J. Phys. (Paris)*, 1975, **37**, 973.
85. D. W. Schaefer, *Polymer*, 1984, **25**, 387.
86. M. Muthukumar and S. F. Edwards, *J. Chem. Phys.*, 1982, **76**, 2720.
87. B. Farnoux, F. Boué, J. P. Cotton, M. Daoud, G. Jannink, M. Nierlich and P. G. de Gennes, *J. Phys. (Paris)*, 1978, **39**, 77.
88. J. P. Cotton, B. Farnoux and G. Jannink, *J. Chem. Phys.*, 1972, **57**, 290.
89. J. P. Cotton, B. Farnoux, G. Jannink and C. Strazielle, *J. Polym. Sci., Polym. Symp.*, 1973, **42**, 981.
90. J. S. King, W. Boyer, G. D. Wignall and R. Ullman, *Macromolecules*, 1985, **18**, 709.
91. M. Daoud, J. P. Cotton, B. Farnoux, G. Jannink, G. Sarma, H. Benoit, R. Duplessix, C. Picot and P. G. de Gennes, *Macromolecules*, 1975, **8**, 804.
92. J. P. Cotton, M. Nierlich, F. Boue, M. Daoud, B. Farnoux, G. Jannink, R. Duplessix and C. Picot, *J. Chem. Phys.*, 1976, **65**, 1101.
93. R. W. Richards, A. Maconnachie and G. Allen, *Polymer*, 1978, **19**, 266.
94. R. W. Richards, A. Maconnachie and G. Allen, *Polymer*, 1981, **22**, 147, 153, 158.
95. S. N. Jahshan and G. C. Summerfield, *J. Polym. Sci., Polym. Phys. Ed.*, 1980, **18**, 1859.
96. A. Z. Akcasu, G. C. Summerfield, S. N. Jahshan, C. C. Han, C. Y. Kim and H. Yu, *J. Polym. Sci., Polym. Phys. Ed.*, 1980, **18**, 863.
97. C. Tangari, G. C. Summerfield, J. S. King, R. Berliner and D. F. R. Mildner, *Macromolecules*, 1980, **13**, 1546.
98. F. Boue, M. Nierlich and L. Leibler, *Polymer*, 1982, **23**, 29.
99. J. C. LeGuillou and J. Zinn-Justin, *Phys. Rev. Lett.*, 1977, **39**, 35.
100. Y. Einaga, Y. Miyaki and H. Fujita, *J. Polym. Sci., Polym. Phys. Ed.*, 1979, **17**, 2103.
101. A. Z. Akcasu and C. C. Han, *Macromolecules*, 1979, **12**, 276.
102. R. Ullman, H. Benoit and J. S. King, *Macromolecules*, 1986, **19**, 183.
103. H. Benoit and M. Benmouna, *Macromolecules*, 1984, **17**, 535.
104. S. Kingusa, H. Hayashi, F. Hamoda, A. Nakajima, K. Kurita, S. Nakajima, M. Furusaka and Y. Ishikawa, *Polym. Commun.*, 1986, **27**, 47.
105. N. S. Davidson, R. W. Richards and A. Maconnachie, *Macromolecules*, 1986, **19**, 434.
106. S. Candau, J. Bastide and M. Delsanti, *Adv. Polym. Sci.*, 1982, **44**, 30.
107. M. Warner and S. F. Edwards, *J. Phys. A: Math. Gen.*, 1978, **11**, 1649.
108. R. Ullman, *J. Chem. Phys.*, 1979, **71**, 436.
109. R. Ullman, *Macromolecules*, 1982, **15**, 582, 1395.
110. J. Bastide, R. Duplessix, C. Picot and S. Candau, *Macromolecules*, 1984, **17**, 83,
111. J. Bastide, C. Picot and S. Candau, *J. Macromol. Sci., Phys.*, 1981, **B19**, 13.
112. M. Beltzung, J. Herz and C. Picot, *Macromolecules*, 1983, **16**, 580.
113. N. S. Davidson and R. W. Richards, *Macromolecules*, 1986, **19**, 2576.
114. P. Lindner and R. C. Oberthur, *Colloid Polym. Sci.*, 1985, **263**, 443.
115. T. Cosgrove, T. L. Crowley and B. Vincent in 'Adsorption from Solution', ed. R. H. Ottewill and C. H. Rochester, Academic Press, London, 1983.
116. K. G. Barnett, T. Cosgrove, T. L. Crowley, T. F. Tadros and B. Vincent, in 'The Effect of Polymers on Dispersion Properties', ed. T. F. Tadros, Academic Press, London, 1981.

7

Photon Correlation Spectroscopy: Technique and Scope

TERENCE A. KING

University of Manchester, UK

Because of factors beyond the editors' control, the submission of the manuscript of this chapter was delayed. In order to minimize any delay in publishing 'Comprehensive Polymer Science' as a whole, the coverage of Photon Correlation Spectroscopy: Technique and Scope appears at the end of this Volume, commencing on page 911.

8

Photon Correlation Spectroscopy: Application to Polymer Solutions

BENJAMIN CHU

State University of New York, Stony Brook, NY, USA

8.1 INTRODUCTION

Laser light scattering (LLS) encompasses angular distribution of absolute integrated scattered intensity, defined as the Rayleigh ratio $R(K)$, and of the time-dependent intensity–intensity correlation function, $G^{(2)}(K,\tau)$ where K and τ are, respectively, the magnitude of the momentum transfer vector and the delay time. In polymer solution characterization, there are four quantities of interest: (i) $\Delta R_{vv}(K)$ representing the excess Rayleigh ratio of vertically polarized incident and scattered light; (ii) $\Delta R_{vh}(K)$ representing the excess Rayleigh ratio of vertically polarized incident and horizontally polarized scattered light, (iii) $G_{vv}^{(2)}(K,\tau)$ denoting the polarized light-scattering spectrum; and (iv) $G_{vh}^{(2)}(K,\tau)$ denoting the depolarized light-scattering spectrum. The symbol Δ is used to denote a difference quantity between the solution and the solvent, *i.e.* $\Delta R = R(\text{solution}) - R(\text{solvent})$. The polarized components, $\Delta R_{vv}(K)$ and $G_{vv}^{(2)}(K,\tau)$, of the absolute integrated scattered intensity and the intensity–intensity time correlation function are more commonly used, since the depolarized components, $\Delta R_{vh}(K)$ and $G_{vh}^{(2)}(K,\tau)$, are applicable mainly for anisotropic polymers. Furthermore, for dilute polymers solutions the depolarized light-scattering intensity is often quite weak, making both $R_{vh}(K)$ and $G_{vh}^{(2)}(K,\tau)$ not easily accessible in practice.

The essential steps in using laser light scattering as an absolute technique to characterize any homopolymer (including alternating copolymers) in solution are: (a) to prepare a dust-free polymer solution; (b) to perform both static ($\Delta R(K)$) and dynamic ($G^{(2)}(K,\tau)$) light-scattering measurements and (c) to analyze the scattering data in order to determine the static and dynamic properties of the polymer in solution.

These three steps are not so easy to realize, since each polymer may present its own specific problems in sample preparation. Laser light-scattering experiments, although having the appearance of being routine, are interactive in nature,[1] and optimization of signal-to-noise ratio in photon correlation spectroscopy (PCS) involves coherence area considerations,[2] which are often not straightforward for the novice. Finally, Laplace inversion of $G^{(2)}(K, \tau)$ from PCS, in order to retrieve information on the characteristic line width distribution function $G(\Gamma)$, is a delicate ill-conditioned problem, not to be treated casually. Thus, the main objective of this chapter is to provide an outline for polymer characterization in the dilute solution regime using laser light scattering. I shall emphasize the absolute calibration of molecular weight using $\Delta R(K)$ and the combination of static $(R(K))$ and dynamic $(G^{(2)}(K,\tau)$ from PCS) light-scattering experiments in order to determine the molecular weight distribution (MWD) and to gain possible additional information on the structure and dynamics of polymers in dilute solution.

The organization of the chapter follows the three steps, *i.e.* preparation, measurement and analysis. In Section 8.2 we briefly describe some accepted procedures for polymer solution clarification. Section 8.3 deals with the practice and the theory of light scattering. Emphasis is on experimental parameters in order to try to optimize the signal-to-noise ratio of $G^{(2)}(K,\tau)$ in PCS and on an outline of the equations used, so that the reader can follow through the light-scattering quantities measured. The Laplace inversion of $G^{(2)}(K,\tau)$ is treated separately in Section 8.4. The treatment in this section is subjective, since it is lengthy to evaluate the specific advantages and disadvantages of different approaches and algorithms. Discussion of specific speciality polymer systems, including linear polyethylene,[3] poly(1,4-phenyleneterephthalamide)[4-6] (PPTA) and an alternating copolymer of ethylene and tetrafluoroethylene[7] (PETFE), starts in Section 8.5. Here, analysis of anisotropic polymers requiring measurements of both $\Delta R_{vv}(K)$ and $\Delta R_{vh}(K)$, as well as considerations of internal motions, in addition to the usual translational motions of center of mass of polydisperse semiflexible chains, such as PPTA, yield additional information related to the polymer structure in terms of the persistence length (ρ). Light-scattering characterizations of speciality polymers are particularly important because, to our knowledge, there are at present no known absolute analytical techniques which could determine the molecular weight and the molecular weight distribution of speciality polymers, such as PPTA and PETFE, except laser light scattering.

In Section 8.6, we discuss the potential application of laser light scattering as a detector following some reasonable separation scheme, such as size exclusion chromatography (SEC). By combining a separation technique with laser light scattering, we shall be able to take advantage of the potentials of both methods and to compensate for their deficiencies. We have shown how such a scheme may be implemented using a prism light-scattering cell.[8]

8.2 POLYMER SOLUTION PREPARATION

A polymer solution suitable for light-scattering measurements has a specific requirement, *i.e.* there should be essentially no other foreign macromolecules or particles, especially those comparable or greater in size than the polymer of interest. In the presence of large dust particles, light-scattering results can often be misleading. Although we can resort to dust discrimination using electronic means, the best way to get reliable light-scattering results is to consider the clarification of the polymer solution as an essential prerequisite. Laser light scattering with well-developed instrumentation and methods of data analysis is an extremely powerful tool for polymer molecular weight and MWD determinations provided that the polymer solutions are properly prepared. The reader is reminded of the significance of this important step.

Dust (or impurity) particles could come from the polymer, the solvent, the light-scattering cell and the air above the polymer solution in the light-scattering cell. It is desirable to establish a routine for clarification of the solvent, the light-scattering cell and the air in contact with the solvent and subsequent polymer solution, although no universal fool-proof procedure can be recommended. Generally, filtered air (or an inert gas) is used throughout the preparation. The light-scattering cell is flushed with a distilled solvent, the best way being to condense the solvent vapor directly in the interior of the light-scattering cell, and the solvent is slowly (and often doubly) distilled and filtered. The polymer to be characterized often contains some dust. Thus, clarification of the polymer solution is an essential step.

One key necessity for the successful preparation of a polymer solution for light-scattering studies is to avoid (or at least to minimize) exposure of clarified polymer solution to the unfiltered atmosphere. After dissolution, the polymer solution should be filtered directly into the (dust-free) light-scattering cell without external exposure. If a one-step filtration process is insufficient to clarify the polymer solution, a closed circulation apparatus,[9] as shown in Figure 1, could be used. However, such a system would usually require large amounts of polymer solution.

For wormlike chains, it is recommended to use high-speed centrifugation for polymer solution clarification. Separate experiments are needed to determine the centrifugal acceleration and duration of centrifugation needed for the clarification procedure. It is important to establish that the clarification process has not altered the polymer characteristics. Normally, dust particles are heavier

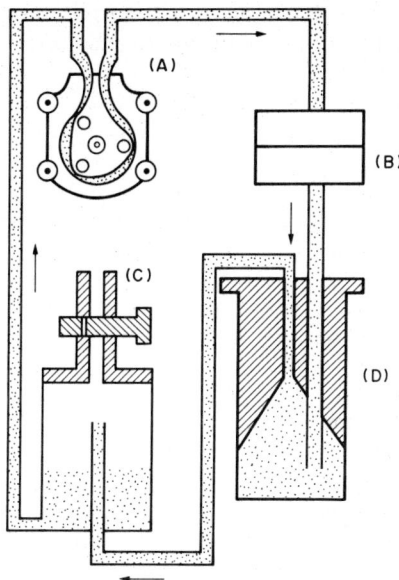

Figure 1 Schematic diagram of the closed-circuit filtration system: (A) silicone-tubing metering pump, (B) filter, (C) air trapping chamber, (D) light-scattering cell (reproduced by permission of Wiley from *J. Polym. Sci., Polym. Phys. Ed.*, 1985, **23**, 2567

and tend to settle in the bottom of the centrifuge tube. On occasion, when the solvent has a high density, such as carbon tetrachloride, certain unwanted impurity or dust particles may gather near the top of the centrifuged solution. Then, it becomes advisable to minimize contact with the upper portion of the centrifuged solution during solution transfer.

8.3 PRACTICE AND THEORY OF LIGHT SCATTERING

8.3.1 Light-scattering Apparatus

The physical principles of laser light scattering are deceptively simple. The uninitiated reader is advised to gain some fundamental knowledge of photon correlation before attempting actual light-scattering experiments. Ford[10] has recently written a good elementary description of the light-scattering apparatus. More details are available in the published literature.[1,11,12]

In optimizing the instrumental parameters used in laser light scattering, there is a conflicting requirement in angular aperture for static and dynamic light scattering. As the particle scattering (or static structure) factor for most dilute polymer solutions has no sharp maxima and minima, it is desirable to use angular apertures ($\Delta\theta$) of a few degrees for the light-scattering detector in order to increase the signal-to-noise ratio, and $\Delta\theta/\theta \approx 0.05$ (or even 0.1) is acceptable for static measurements, with θ being the scattering angle. On the other hand, for dynamic light scattering, the coherence requirement often results in the use of a much smaller angular aperture, *e.g.* $\Delta\theta/\theta \lesssim 0.01$. If both static and dynamic light-scattering experiments are performed using the same optical geometry, a compromise may be achieved based on an empirical observation of getting the maximum scattered intensity without saturating the digital correlator.

The required coherence angle $(\Delta\theta)_{coh}$ is a function of the scattering volume (L_x, L_y, L_z) and the scattering angle θ. If the incident laser beam is propagating in the z direction with polarization in the x direction, the coherence angle in the yz scattering plane[2,13] is $(\Delta\theta)_{coh} = 0.5\lambda/(L_z \sin\theta + L_y \cos\theta)$. If the detection system has an aperture D_A, $L_z = D_A/\sin\theta$. Then

$$(\Delta\theta)_{coh} = 0.5(\lambda/D_A)/[1 + (L_y/D_A)\cos\theta] \tag{1}$$

$(\Delta\theta)_{coh}$ is normally quite small at $\theta < 90°$. For example, if D_A ($\approx L_y$) = 100 μm, $\lambda = 0.366$ μm, $(\Delta\theta)_{coh} \approx 0.1°$ at $\theta = 90°$ and $\approx 0.05°$ at small scattering angles. Thus, for PCS, we need $\Delta\theta/\theta \approx 0.001$

while for static light scattering $\Delta\theta/\theta \approx 0.05$. A compromise covering a factor of 50 may not be feasible under all experimental conditions. Then a change of angular aperture is recommended.

8.3.2 Theory of Light Scattering

8.3.2.1 Intensity of scattered light

The excess Rayleigh ratio $\Delta R_{vv}(K)(\text{cm}^{-1})$ for vertically polarized incident and scattered light from a dilute solution of a monodisperse polymer having a molar mass $M(\text{g mol}^{-1})$ at a finite concentration $c(\text{g cm}^{-3})$ has the form

$$\Delta R_{vv}(K) \approx HMP(K)c(1 - 2A_2 MP(K)c) \tag{2}$$

where $H(\text{mol cm}^2 \text{g}^{-2})$ is equal to $4\pi^2 n^2 (\partial n/\partial c)_T^2/(N_A \lambda_0^4)$ with n, N_A, λ_0 and $(\partial n/\partial c)_T$ being the refractive index, Avogadro's number (mol^{-1}), the wavelength in vacuo (cm), and the refractive index increment, respectively. $A_2(\text{mol cm}^3 \text{g}^{-2})$ is the second virial coefficient and $P(K)$ is the particle scattering factor or the static structure factor. At $KR_g \lesssim 1$, $P^{-1}(K) \simeq 1 + K^2 R_g^2/3$ with $K = (4\pi n/\lambda_0)\sin(\theta/2)$ and R_g being the radius of gyration. For a polydisperse polymer at dilute but finite concentrations

$$\Delta R_{vv}(K) \cong \int HMc_M P_M(K)(1 - 2\bar{A}_2 MP_M(K)c_M + \dots)dM \tag{3}$$

where the subscript M is used to denote the functions for a polymer of molecular weight M, and $\bar{A}_2$ is the average second virial coefficient of the polydisperse polymer. In a strict sense, the second virial coefficient is molecular weight dependent and therefore changes with polydispersity. In the present context, an average A_2 value for the polydisperse polymer is adequate to account for the intermolecular interaction correction provided that the correction is relatively small. If we let $f_w(M) = c_M/c$ be the weight fraction of polymer having molecular weight M, then equation (3) becomes

$$\Delta R_{vv}(K) \cong Hc \int Mf_w(M)P_M(K)[1 - 2\bar{A}_2 P_M(K)Mcf_w(M) + \dots]dM \tag{4}$$

where $c = \int c_M dM$.

Equation (4) represents one formulation of the averaging process. At infinite dilution

$$\lim_{c \to 0} Hc/\Delta R_{vv}(K) \approx \bar{M}_w^{-1}(1 + K^2 \overline{R_{g,z}^2}/3) \tag{5}$$

where $\bar{M}_w$ is the weight average molecular weight and $\overline{R_{g,z}^2}$ is the z average square of the radius of gyration. As $KR_g \to 0$,

$$\lim_{K \to 0} \frac{Hc}{\Delta R_{vv}(K)} \simeq \bar{M}_w^{-1} + 2\bar{A}_2 c \tag{6}$$

and finally as $K \to 0$ and $c \to 0$, $Hc/\Delta R_{vv}(0) = \bar{M}_w^{-1}$.

8.3.2.2 Spectrum of scattered light

The intensity–intensity time correlation function in the self-beating mode for a detector of finite effective photocathode has the form

$$G^{(2)}(\tau, K) = N_s \bar{n}^2 (1 + \beta|g^{(1)}(\tau, K)|^2) \tag{7}$$

where $g^{(1)}(\tau, K)(\equiv \langle E_s^*(K, \tau)E_s(K, 0)\rangle/\langle E_s^*(K, 0)E_s(K, 0)\rangle)$ is the first-order normalized scattered electric field (E_s) time correlation function, $\tau(\equiv N\Delta\tau$, with N and $\Delta\tau$ being the channel number and the delay time increment, respectively) is the delay time, $\bar{n}$ denotes the mean counts per sample time (or delay time increment $\Delta\tau$), N_s is the total number of samples, $A(\equiv N_s \bar{n}^2)$ is the baseline and β is a spatial coherence factor depending upon experimental conditions and is usually taken as an unknown parameter in the data fitting procedure.

As $\tau \to 0$, $g^{(1)}(\tau,K) \to 1$ and $G^{(2)}(0,K) = A(1+\beta)$ with $0 \le \beta \le 1$. The background $A(K)(\equiv N_s \overline{n(K)}^2)$ is related to $\langle I(K,0)I(K,0) \rangle$ (or $\langle |I(K)|^2 \rangle$), where $I(K)$ is the scattered intensity of the *solution*. A more detailed discussion of the quantities measured by PCS has been presented by Pusey and Tough (see Section 4.2 of ref. 14). It is sufficient to summarize here that the scattered electric field amplitude for N identical particles can be written as

$$E_s(K,t) = \sum_{i=1}^{N} b_i(K) \exp[i\boldsymbol{K} \cdot \boldsymbol{r}_i(t)] \tag{8}$$

where $b_i(K)$ is the field amplitude of the light scattered by particle i at scattering wave vector $\boldsymbol{K}$ and $\boldsymbol{r}_i(t)$ is the position of the center-of-mass of particle i at time t. For N non-identical particles,

$$R_{vv}(K) = N \overline{b^2(K)} P(K) \tag{9}$$

where the particle scattering factor is defined by

$$P(K) = [N \overline{b^2(K)}]^{-1} \sum_i \sum_j \langle b_i(K) b_j(K) \exp[i\boldsymbol{K} \cdot (\boldsymbol{r}_i - \boldsymbol{r}_j)] \rangle \tag{10}$$

and similarly the dynamic structure factor $g^{(1)}(K,\tau)$ is given by

$$g^{(1)}(K,\tau) = [N \overline{b^2(K)}]^{-1} \sum_i \sum_j \langle b_i(K) b_j(K) \exp\{i\boldsymbol{K} \cdot [\boldsymbol{r}_i(0) - \boldsymbol{r}_j(\tau)]\} \rangle \tag{11}$$

Equation (11) cannot be simplified in the case of arbitrary polydispersities and interactions.

For intermolecular interactions at finite but dilute concentrations, we write for the diffusion coefficient

$$D = D^0 (1 + \bar{k}_d c) \tag{12}$$

where $\bar{k}_d c$ represents the overall concentration effect of the polydisperse solution with $\bar{k}_d$ and c being the average diffusion second virial coefficient and the total concentration, respectively. The superscript zero denotes infinite dilution.

$$g^{(1)}(K,\tau) = \int_0^\infty G(K,\Gamma) e^{-\Gamma \tau} d\Gamma \tag{13}$$

where $G(K,\Gamma)$ is the normalized characteristic line width distribution. In the presence of internal motions and absence of intermolecular interactions, we can write for each molecular weight M of a polymer molecule

$$\Gamma_M \cong D_M K^2 (1 + f_M R_{g,M}^2 K^2) \tag{14}$$

where Γ_M is an average characteristic linewidth for a particle with molecular weight M in the presence of internal motions and the subscript M denotes a particle having molecular weight M. However, the transformation of $G(K,\Gamma)$ to $f_w(M)$ requires a complex deconvolution procedure since the amplitude contribution at each characteristic linewidth Γ_i could come, not only from the translational motion of one molecule having a translational characteristic time corresponding to Γ_i, but also from the internal motions of other molecules having those possible characteristic times corresponding to Γ_i. Instead, we approximate the corrections due to internal motions and intermolecular interactions as a shift in the measured characteristic linewidth Γ_M having a molecular weight M (to $M + dM$), writing

$$\Gamma_M \cong D_M^0 (1 + \bar{k}_d c) K^2 (1 + \bar{f}_M \langle R_{g,M}^2 \rangle K^2) \tag{15}$$

where $D_M^0 = k_D M^{-\alpha_D}$ and $\bar{f}_M$ denotes an average coefficient for the polymer. We then approximate the amplitude of $G(K,\Gamma)$ to be proportional to $R_{vv}(K)$ as shown in equation (9). Since

$$\ln \Gamma_M = \ln[k_D K^2 (1 + \bar{k}_d c)] - \alpha_D \ln M + \ln(1 + \bar{f}_M R_{g,M}^2 K^2) \tag{16}$$

then

$$d\ln \Gamma_M = \left(-\alpha_D + \frac{\alpha_{R_g} \bar{f}_M R_{g,M}^2 K^2}{1 + \bar{f}_M R_{g,M}^2 K^2} \right) d\ln M \tag{17}$$

where $\bar{f}_M R_{g,M}^2 = k M^{\alpha_{R_g}}$ with $\bar{f}_M$ essentially absorbed by k.

Similarly, with the characteristic linewidth expressed in logarithmic spacing,

$$\int_{\Gamma_{min}}^{\Gamma_{max}} G(\Gamma)d\Gamma = \int_{\Gamma_{min}}^{\Gamma_{max}} \Gamma G(\Gamma)d\ln\Gamma$$

$$= \int_{M_{max}}^{M_{min}} \Gamma G(\Gamma)\left(-\alpha_D + \frac{\alpha_{R_g}\bar{f}_M R_{g,M}^2 K^2}{1+\bar{f}_M R_{g,M}^2 K^2}\right)d\ln M \qquad (18)$$

From equation (14), and in terms of $\ln M$, we get for the *net* signal

$$\Delta R_{vv}(K) = Hc\int M^2 f_w(M) P_M(K)\left[1-2\bar{A}_2 P_M(K)Mcf_w(M)\right]d\ln M$$

$$\propto \int_{M_{min}}^{M_{max}} \Gamma G(\Gamma)\left(\alpha_D - \frac{\alpha_{R_g}f_M R_{g,M}^2 K^2}{1+\bar{f}_M R_{g,M}^2 K^2}\right)d\ln M \qquad (19)$$

Thus, by equating the integrand and in $\ln M$ spacing, we finally have

$$HcM^2 f_w(M)P_M(K)(1-2\bar{A}_2 P_M(K)Mcf_w(M)) = B(K)\Gamma G(\Gamma)\left(\alpha_D - \frac{\alpha_{R_g}\bar{f}_M R_{g,M}^2 K^2}{1+\bar{f}_M R_{g,M}^2 K^2}\right) \qquad (20)$$

For small values of $2\bar{A}_2 P_M(K)Mcf_w(M)$ and $\bar{f}_M R_{g,M}^2 K^2$ (both $\ll 1$), equation (20) is reduced to

$$f_w(M) \cong B^*(K)\Gamma G(\Gamma)/[M^2 P_M(K)] \qquad (21)$$

where $B^*(K)$ is a proportionality constant satisfying $\Sigma f_w(M) = 1$. Equation (21) forms the basis of our molecular weight transform and is valid for a dilute polymer solution in the absence of intermolecular interactions and internal motions. Figure 2 shows schematically the essential steps used in the molecular weight transform.[15] The first-order corrections can be estimated based on equation (20), which is again an approximation, involving only shifts in the x axis without the complex deconvolution on effects of characteristic linewidths due to translation and internal motions on the amplitudes of the characteristic linewidth distribution $G(\Gamma)$ even at infinite dilution. Thus we should consider this approach to be an approximation to the solution by some means of deconvolution which would be model dependent. However, empirically we can show that such a procedure (equation 20) is acceptable even for polydisperse wormlike chains with appreciable internal motions measured at finite concentrations, as shown typically by Figures 3 and 4. The success in being able to scale $g^{(1)}(\Gamma,t)$ using experimentally determined quantities such as the diffusion second virial coefficient $\bar{k}_d$ and the effective radius of gyration $R_{g,app}$ (to be defined in Section 8.5) confirms the validity of this semiempirical correction procedure.

8.4 CORRELATION FUNCTION PROFILE ANALYSIS

Equation (13) is a Fredholm integral equation of the first kind. The associated inversion problem is notoriously ill-conditioned when the data are corrupted by noise and are bandwidth limited because of instrumental limitations. The reader is advised to refer to specific topics in chapters of two recent books[16] dealing with characteristic linewidth distribution analysis for details. The ill-conditioned Laplace inversion problem is likely to be a new topic for the uninitiated reader and can offer pitfalls beyond imagination. Any attempt to extract information buried in noise may lead to physically unreal solutions. In this section we only attempt to provide a direction for the reader to proceed with caution, as the topic requires subtle understanding of the mathematics involved.

Several approaches to the Laplace inversion exist, the most popular ones being those developed by McWhirter and Pike[17,18] and by Provencher.[19] In particular, Provencher provides well-documented programs, CONTIN and DISCRETE, for Laplace inversion of equation (13). The programs have been tested thoroughly by many experts in the field. CONTIN is useful, especially if the reader has an interest in trying to resolve multimodal characteristic linewidth distributions. Unfortunately, the resolution of PCS in this respect is relatively poor even for $|g^{(1)}(\Gamma,t)|$ with noise less than 0.1% and agreement of measured ($\lim_{t\to\infty} G^{(2)}(\Gamma,t)$) and computed ($N_s\bar{n}^2$) baselines to within 0.1%. Thus it becomes important to confirm the conclusions by additional physical measurements. For unimodal characteristic linewidth distributions, simpler and faster algorithms requiring the use

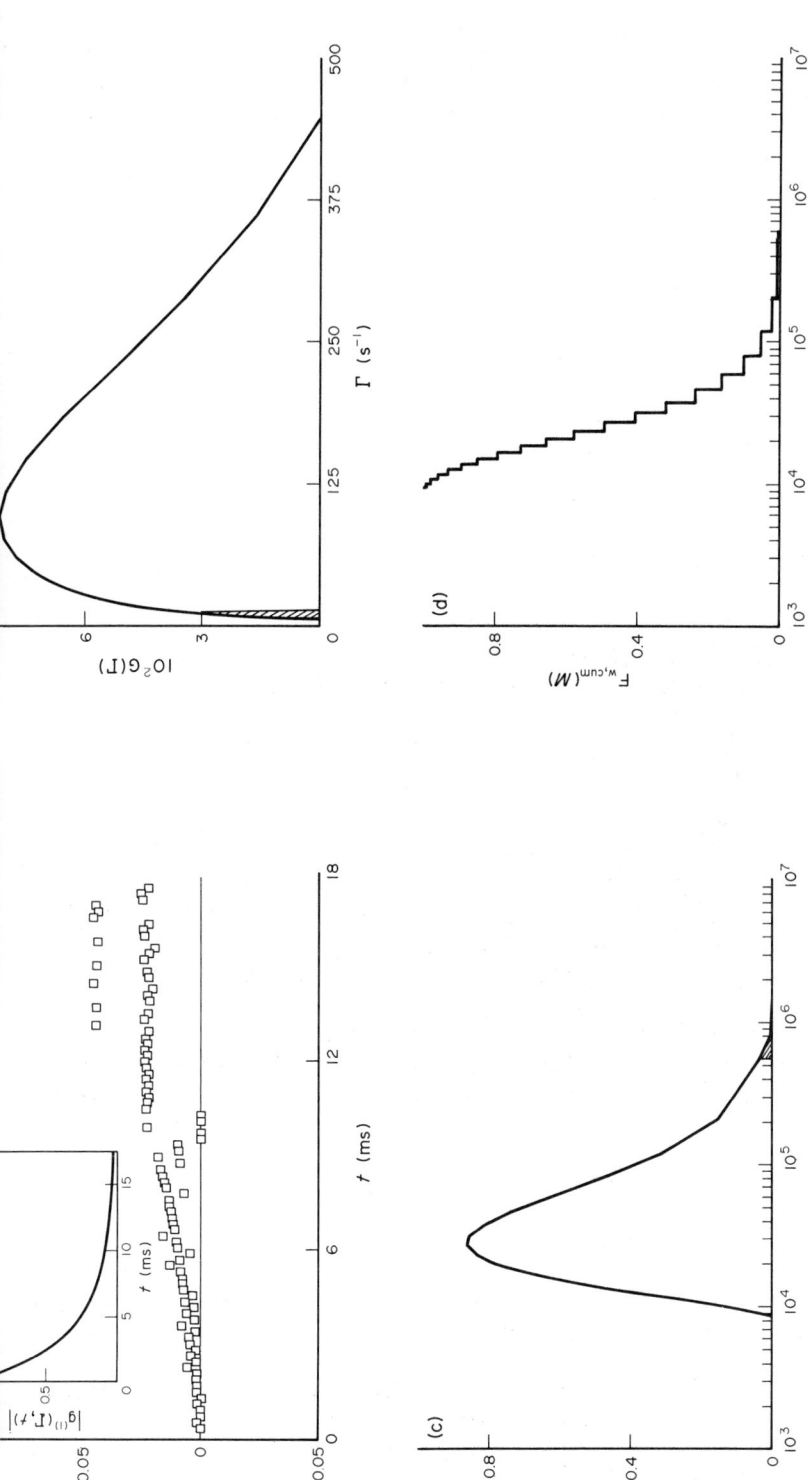

Figure 2 (a) Plot of relative deviation in $|g^{(1)}(t)|$ for the two characteristic line width distributions: $G(\Gamma)$ with and without the 1% shaded section. The shaded contribution of $G(\Gamma)$ corresponds to a baseline adjustment of ~0.05%. A plot of $|g^{(1)}(t)|$ vs. t is inserted to demonstrate the precision required in the time correlation function measurements, i.e. the fittings can always be achieved such that the measured and the computed $|g^{(1)}(t)|$ are indistinguishable within the experimental error limits. Measured and computed baseline must agree within ~0.1%. $t \equiv \tau$ (reproduced by permission of The Society of Polymer Science, Japan, from *Polymer Journal*, 1985, **17**, 225. (b) Plot of $G(\Gamma)$ versus Γ. The shaded area represents 1% of the area of $\int G(\Gamma)d\Gamma$. Results obtained by means of equation (13) using data of Figure 2(a) (reproduced by permission of The Society of Polymer Science, Japan, from *Polymer Journal*, 1985, **17**, 225). (c) Molecular weight distribution of poly(1,4-phenylene terephthalamide) (PPTA) based on $G(\Gamma)$ from Figure 2(b) and equation (15). The shaded area represents the baseline uncertainty in the time correlation function measurement which can be translated into uncertainties in the high molecular weight tail. $F_w(M)$ is the weight distribution (reproduced by permission of The Society of Polymer Science, Japan, from *Polymer Journal*, 1985, **17**, 225) (d) Cumulative molecular weight distribution of PPTA from Figure 2(c). Molecular weight fractions vary from 10^4 to ~5×10^5. The high molecular weight tail corresponding to polymers with $M > 5 \times 10^5$ is seen as within the experimental error limits.

$$F_{w,cum} = \int_{\infty}^{M} F_w(M)MdM \bigg/ \int_{\infty}^{0} F_w(M)dM$$

At $M > 5 \times 10^5 \, \text{g mol}^{-1}$, the amount of polymer in $F_{w,cum}$ has become negligibly small (reproduced by permission of The Society of Polymer Science, Japan, from *Polymer Journal*, 1985, **17** 225)

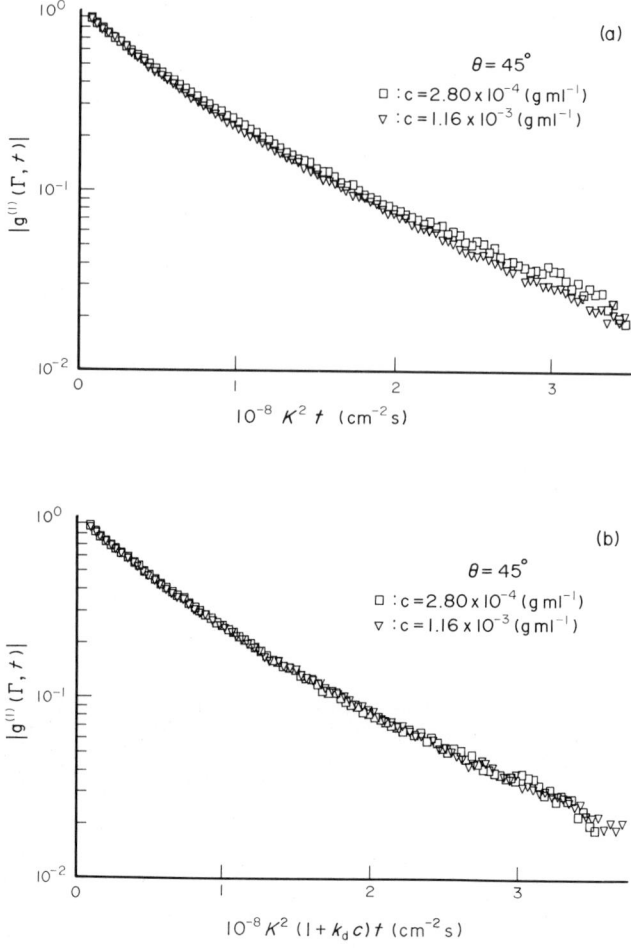

Figure 3 Scaling of $|g^{(1)}(\Gamma,t)|$ due to concentration effect at constant $\theta(=45°)$. $t \equiv \tau$: (a) before scaling; (b) after scaling; $\bar{k}_d = 60\ cm^3\ g^{-1}$ for PPTA ($M_w = 4.3 \times 10^4\ g\ mol^{-1}$) in 96 wt% H_2SO_4 and 0.05 mol dm^{-3} K_2SO_4 at 30 °C (reproduced by permission of Butterworth from *Polymer*, 1985, **26**, 1408)

of only a microprocessor for the Laplace inversion are applicable.[6,20] Fortunately, for most synthetic polymers, the most important polydispersity characteristics are the first few moments of a unimodal size (or molecular weight) distribution function such as the ratios of z average, weight average and number average molecular weights ($\bar{M}_z : \bar{M}_w : \bar{M}_n$) which are within reach by PCS. For relatively narrow characteristic linewidth distributions, the cumulants expansion method[21] becomes applicable

$$\ln|g^{(1)}(\Gamma,t)| = -\bar{\Gamma}t + (1/2)\mu_2 t^2 + \ldots \tag{22}$$

where the average linewidth $\bar{\Gamma} = \int \Gamma G(\Gamma)d\Gamma$ and the second moment $\mu_2 = \int (\Gamma - \bar{\Gamma})^2 G(\Gamma)d\Gamma$. In practice, it is difficult to determine $\bar{\Gamma}$, μ_2, *etc.*, using the cumulants method whenever $\mu_2/\bar{\Gamma}^2 \gtrsim 0.5$. It should also be noted that PCS has a practical limit in determining the polydispersity effect, even in the absence of intermolecular interaction and internal motions, because the amplitude factor in light scattering is proportional to M^2. If we have size ranges expanding beyond the bandwidth limit of the measured correlation function, the motions of those molecules will not be resolved. Similarly, if the amplitude factors of size fractions are buried in the noise, the characteristic linewidth distribution would miss those size fractions. Thus, there are practical upper and lower limits in determining even the variance $\mu_2/\bar{\Gamma}^2$. Normally, $0.01 \lesssim \mu_2/\bar{\Gamma}^2 \lesssim 2$, implying a lower limit for almost monodisperse polymers as well as an upper limit for synthetic polymers, which are very polydisperse.[22]

The discussion above may appear to discourage the uninitiated reader to use PCS for polymer solution characterizations. On the contrary, PCS is a very powerful technique which can be used for routine and rapid characterizations. The emphasis on its limitations is essential so that this newly

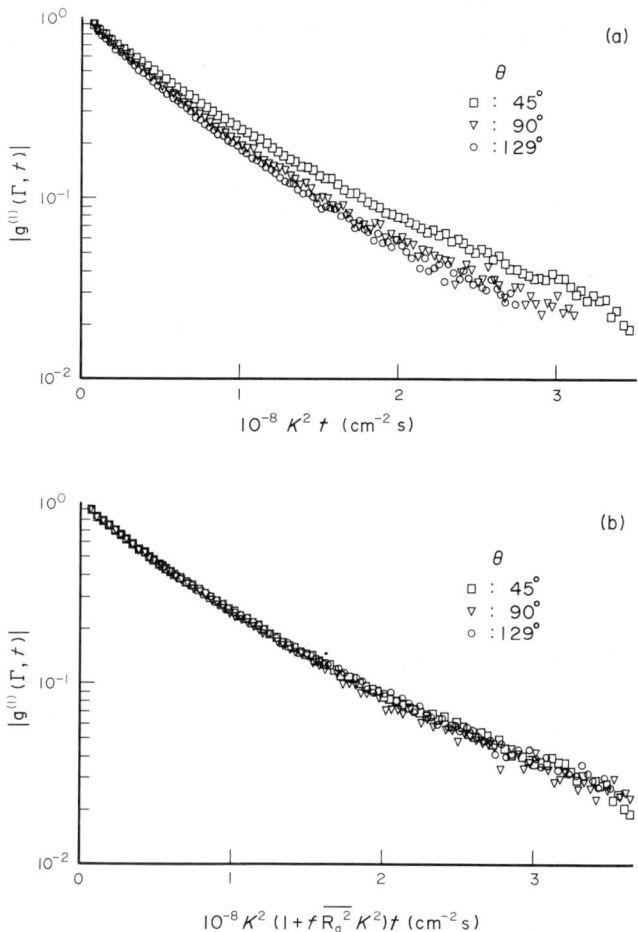

Figure 4 Scaling of $|g^{(1)}(\Gamma,t)|$ due to interference effect at constant concentration ($c = 2.80 \times 10^{-4}$ g cm^{-3}). $t \equiv \tau$: (a) before scaling; (b) after scaling; $f\overline{R_g^2}$ denotes $fR_{g,app}^2$ with f and R_{ga}^2 being 0.2 and 1.2×10^{-11} cm^2, respectively. $\square$: $\theta = 45°$, $K = 1.41 \times 10^5$ cm^{-1}; ∇: $\theta = 90°$, $K = 2.60 \times 10^5$ cm^{-1}; $\bigcirc$: $\theta = 129°$, $K = 3.32 \times 10^5$ cm^{-1}; for the same (as in Figure 2) PPTA ($M_w = 4.3 \times 10^4$ g mol^{-1}) in 96 wt% H$_2$SO$_4$ and 0.05 mol dm^{-3} K$_2$SO$_4$ at 30 °C (reproduced by permission of Butterworth from *Polymer*, 1985, **26**, 1408)

developed technique is not misused. For known polymer systems it is not necessary to use detailed analysis including Laplace inversion. Instead we can measure the difference between two scaled correlation functions, each having its own M_w and MWD, and obtain changes of molecular weight and polydispersity index ($\mu_2/\overline{\Gamma}^2$) directly from the difference curve, as shown in Figure 5, using the basic expression[4]

$$\log|g^{(1)}{}_{M_1}(\Gamma,t^*)| - \log|g^{(1)}{}_{M_2}(\Gamma,t^*)| = \frac{1}{-2.303}[\bar{D}_1^0(1+\bar{k}_d c_1) - \bar{D}_2^0(1+\bar{k}_d c_2)]t^* + \frac{1}{4.606}(\bar{D}_{2,1} - \bar{D}_{2,2})t^{*2} \quad (23)$$

where M_i refers to polymer sample of molecular weight M_i; $\bar{D}_i^0$ is the average translational diffusion coefficient of polymer i at infinite dilution; $\bar{D}_{2,i}$ the corresponding second moment term for the translational diffusion coefficient; and $t^* = tK^2(1 + f\overline{R_g^2}K^2)$ with $f\overline{R_g^2}$ being the experimentally measured average $\overline{f_M R_g^2}$ for the polydisperse polymer. Parameters for the difference curve are easy to compute according to equation (23). The results agree with a detailed analysis based on the Laplace inversion of each individual curve.[4]

8.5 SPECIALITY POLYMER CHARACTERIZATIONS

Laser light scattering offers some unique advantages in characterizing speciality polymers. For example, the experimental difficulties encountered in the light-scattering characterization of linear

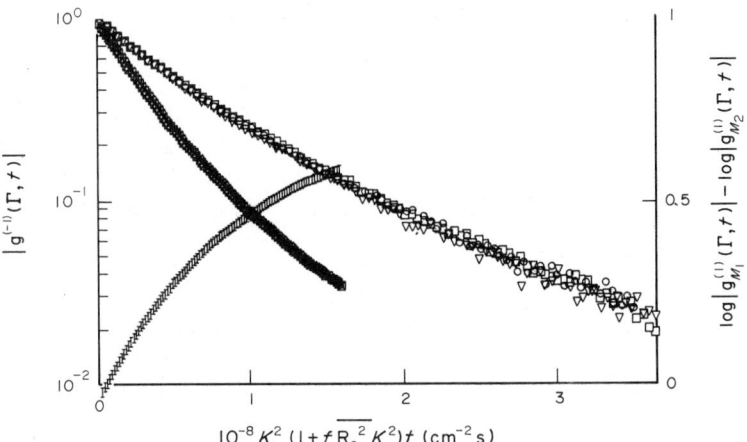

Figure 5 Scaling of $|g^{(1)}(\Gamma,t)|$ due to interference effect at different concentrations and molecular weights of poly(1,4-phenyl-ene terephthalamide) (PPTA)

Fraction	$c(\text{g cm}^{-3})$	$\bar{M}_w(\text{g mol}^{-1})$	f	$R^2_{g,app}(\text{cm}^2)$
PPTA-202	5×10^{-4}	1.35×10^4	0.18	2.1×10^{-12}
PPTA-575	2.8×10^{-4}	4.3×10^4	0.20	1.2×10^{-11}

PPTA in 96 wt% $H_2SO_4 + 0.05$ mol dm^{-3} K_2SO_4 at 30 °C. PPTA-575 ($\square$) $\theta = 45°$; (∇) 90°; ($\bigcirc$) 129°. PPTA-202 ($\times$) 30°; ($\blacksquare$) 120°. Also shown is the difference curve (I, $\log|g^{(1)}(\Gamma,t)| - \log|g^{(1)}(\Gamma,t)|$ for the two PPTA samples. PPTA-575 curve is the same as in Figure 3(b) (reproduced by permission of Academic Press from *J. Colloid Interface Sci.*, 1985, **105**, 473)

polyethylene[3] are quite different from those of poly(1,4-phenyleneterephthalamide)[4,6] (PPTA), commercially known as Kevlar (a registered trademark of DuPont). In the case of polyethylene, we were concerned mainly with the preparation and clarification of polyethylene in 1,2,4-trichloro-benzene at high temperatures. For PPTA, the problems were much more complex[4,5] because PPTA fluoresces at $\lambda_0 = 488$ nm, has a reddish tint denoting absorption in the visible region, is a rod-like anisotropic polyelectrolyte which aggregates easily and requires the addition of electrolytes to form a mixed solvent. Yet we have been able to overcome all such experimental difficulties in order to characterize PPTA.[4-6]

In the presence of molecular anisotropy δ, equations (5) and (6) can be used to measure only apparent values of molecular weight and mean square radius of gyration (M_{app} and $R^2_{g,app}$) such that

$$\bar{M}_w = M_{app}/(1 + 4\delta^2/5) \tag{24}$$

and

$$\overline{R^2_g} = \frac{(1+4\delta^2/5)R^2_{g,app}}{1 - 4\delta/5 + 4\delta^2/7} \tag{25}$$

Depolarized light scattering (ΔR_{vh}) measurements are needed in order to determine the molecular weight of anisotropic polymers in solution. Other procedures remain the same. For wormlike chains the persistence length (ρ) plays an important role. A variety of approaches have been developed to determine ρ.[6,23] The reader is referred to the literature for details.[23]

Recently we succeeded in characterizing an alternating copolymer of ethylene and tetrafluoro-ethylene (PETFE), $\text{-CF}_2\text{CF}_2\text{CH}_2\text{CH}_2\text{-}_x$, by dissolving it in diisobutyl adipate.[7] We were able to clarify the solution, to perform the dilution and to measure static and dynamic light scattering as well as the refractive index increment of PETFE in diisobutyl adipate at 240 °C. Thus, both static and dynamic properties, such as $\bar{A}_2$, $\bar{k}_d$, R^2_g, $\bar{M}_w$ and $\bar{D}$, as well as the MWD, of PETFE in diisobutyl adipate could be studied by laser light scattering.

8.6 SEPARATION TECHNIQUES *VS.* LIGHT SCATTERING

Absolute low-angle light-scattering intensity measurements have been used successfully as a detector for size exclusion chromatography (SEC) and thermal field flow fractionation[24] (TF³).

Measurements of $\lim_{K \to 0} \Delta R_{vv}$ permits determination of M_w at each fraction of solution, resulting in a MWD determination. The lack of resolution in using Laplace inversion based on PCS measurements may be compensated by combining an analytical technique such as SEC and TF[3] with laser light scattering as a detector. Simultaneous measurements of static and dynamic light scattering[8] permit us to determine $\bar{M}_w$, $\bar{R}_g^2$, $\bar{D}$, and the polydispersity index of each fraction. From $\bar{D}$ we can compute an effective hydrodynamic radius $\bar{R}_h$ using the Stokes–Einstein relation. As the ratio $(\bar{R}_g^2)^{1/2}/\bar{R}_h$ is related to polymer structure we can then determine not only MWD but also possible structural changes for each fraction. For SEC the column could act as the prefilter and should simultaneously solve the solution clarification problem which often interferes with reliable light-scattering measurements.

Thus, laser light scattering is complementary to many of the established analytical techniques and has potential for the characterization of speciality polymers. Development of its use as a remote-sensing monitor for polymer characteristic changes in solution is likely to take place in future polymer materials research[25, 26] as the technique is non-invasive. Furthermore, with the use of fiber optics and large scale integrated circuits, the spectrometer can be miniaturized and the sample volume reduced so as to permit a more routine adaptation of LLS as a detector for chromatographic or fractionation methods. New developments in correlator design and Laplace inversion methods[27, 28] continue to improve the PCS capabilities by extending characteristic linewidth measurements over much broader time scales ($\sim 10^7$) and by improving analysis techniques with higher resolving power for the characteristic linewidth distribution.

ACKNOWLEDGEMENT

We gratefully acknowledge support of this research by the National Science Foundation, Polymers Program (DMR 8617820) and the US Army Research Office.

8.7 REFERENCES

1. B. Chu, in 'Proceedings of NATO ASI on the Application of Laser Light Scattering to the Study of Biological Motions', ed. J. C. Earnshaw and M. W. Steer, Plenum Press, New York, 1983, pp. 53–76.
2. B. Chu, 'Laser Light Scattering', Academic Press, New York, 1974.
3. J. W. Pope and B. Chu, *Macromolecules*, 1984, **17**, 2633.
4. B. Chu, C. Wu and J. R. Ford, *J. Colloid Interface Sci.*, 1985, **105**, 473.
5. Q. Ying, B. Chu, R. Qian, J. Bao, J. Zhang and C. Xu, *Polymer*, 1985, **26**, 1401.
6. B. Chu, Q. Ying, C. Wu, J. R. Ford and H. S. Dhadwal, *Polymer*, 1985, **26**, 1408.
7. B. Chu and C. Wu, *Macromolecules*, 1986, **19**, 1285.
8. B. Chu, *US Pat.* 4 565 446 (1986).
9. M. Naoki, I. H. Park, S. L. Wunder and B. Chu, *J. Polym. Sci., Polym. Phys. Ed.*, 1985, **23**, 2567.
10. N. C. Ford, Jr., in 'Dynamic Light Scattering, Applications of Photon Correlation Spectroscopy', ed. R. Pecora, Plenum Press, New York, 1985, chap. 2, pp. 7–58.
11. 'Scattering Techniques Applied to Supramolecular and Nonequlibrium Systems,' ed. S. H. Chen, B. Chu and R. Nossal, Plenum Press, New York, 1981.
12. E. O. Schulz–DuBois (ed.), 'Proceedings of the 5th International Conference on Photon Correlation Techniques in Fluid Mechanics', Springer-Verlag, New York, 1983.
13. H. S. Dhadwal and B. Chu, *J. Colloid Interface Sci.*, 1987, **115**, 561.
14. P. N. Pusey and R. J. A. Tough, in 'Dynamic Light Scattering, Applications of Photon Correlation Spectroscopy', ed. R. Pecora, Plenum Press, New York, 1985, chap. 4, pp. 85–180.
15. B. Chu, *Polym. J.*, 1985, **17**, 225.
16. See for examples, 'Photon Correlation Spectroscopy of Brownian Motion: Polydispersity Analysis and Studies of Particle Dynamics', in ref. 12; 'Essentials of Size Distribution Measurements', in 'Measurements of Suspended Particles by Quasi-Elastic Light Scattering,' ed. B. E. Dahneke, Wiley, New York, 1983.
17. J. G. McWhirter and E. R. Pike, *J. Phys. A: Math. Gen.*, 1978, **11**, 1729.
18. N. Ostrowsky, D. Sornette, P. Parker and E. R. Pike, *Opt. Acta*, 1981, **28**, 1059.
19. See for examples, S. W. Provencher in ref. 12, p. 322, and earlier publications: *Biophys. J.*, 1976, **16**, 27; *J. Chem. Phys.*, 1976, **64**, 2772; *Makromol. Chem.*, 1979, **180**. 201.
20. B. Chu, J. R. Ford and H. S. Dhadwal, *Methods Enzymol.*, 1985, **117**, 256.
21. D. E. Koppel, *J. Chem. Phys.*, 1972, **57**, 4814.
22. B. Chu, M. Onclin and J. R. Ford, *J. Phys. Chem.*, 1984, **88**, 6566.
23. Q.-C. Ying and B. Chu, *Makromol. Chem., Rapid Commun.*, 1984, **5**, 785.
24. J. Janca, J. Chmelik and D. Pribylova, *J. Liq. Chromatogr.*, 1985, **8**, 2343.
25. D.-C. Lee, J. R. Ford, G. Fytas, B. Chu and G. L. Hagnauer, *Macromolecules*, 1986, **19**, 1586.
26. B. Chu and D.-C. Lee, *Macromolecules*, 1986, **19**, 1592.
27. M. Bertero, P. Brianzi, E. R. Pike, G. de Villiers, K. H. Lan and N. Ostrowsky, *J. Chem. Phys.*, 1985, **82**, 1551.
28. A. K. Livesey, P. Licinio and M. Delaye, *J. Chem. Phys.*, 1986, **84**, 5102.

9

Dilute Solution Viscometry

PETER A. LOVELL

University of Manchester Institute of Science and Technology, UK

9.1 INTRODUCTION

One of the most characteristic features of a dilute polymer solution is that its viscosity is considerably higher than that of the pure solvent. This arises because of the large differences in size between polymer and solvent molecules, and can be significant even at very low polymer concentrations, especially for polyelectrolytes and polymers with high molecular weights. Dilute solution viscometry is concerned with accurate quantitative measurement of the increase in viscosity and allows determination of the intrinsic ability of a polymer to increase the viscosity of a particular solvent at a given temperature. This quantity provides a wealth of information relating to the size of the polymer molecule in solution, including the effects upon chain dimensions of polymer structure, molecular shape, degree of polymerization and polymer–solvent interactions. Most commonly, however, it is used to estimate the molecular weight of a polymer. This involves the use of semi-empirical equations which have to be established for each polymer/solvent/temperature system by analysis of polymer samples whose molecular weights are known. Thus the estimates of molecular weight are not absolute. Nevertheless, in comparison to other methods for characterization of polymers in solution (*e.g.* membrane osmometry and light scattering), dilute solution viscometry is simple, fast and inexpensive. It also has the advantage that it is applicable over the complete range of attainable molecular weights. For these reasons, dilute solution viscometry has been the most widely used method of polymer characterization since the birth of polymer science and continues to be used today.

Accounts of dilute solution viscometry appear in many textbooks and review articles.[1-12] The purpose of the current review is to present an up-to-date account of the subject, with particular emphasis upon the measurement of viscosity and the interpretation of the data obtained. Theoretical treatments are introduced only to an extent which is necessary to facilitate an understanding of the material being presented. These aspects are discussed elsewhere.[1-3,5,13-20]

9.2 DEFINITIONS

The viscosity of a fluid is a measure of its resistance to flow when a shearing force is applied, and is most easily defined by considering two parallel layers within a stationary fluid. Each layer is of area A and the layer separation dz is small (Figure 1a). A tangential force F is applied to one of the layers, causing it to move with velocity v. The frictional forces existing between the molecules which constitute the fluid result in some of the energy supplied by the force being consumed in imparting motion to other layers. As a consequence of this viscous drag the second layer also moves, but with velocity $v - dv$. So long as the velocities are below a critical value, the layers will flow uniformly in parallel and the flow is said to be streamline or laminar. The variation in shear strain γ with time t is known as the shear rate $\dot{\gamma}\ (= d\gamma/dt)$ and is equal to the velocity gradient dv/dz when the flow lines are linear. For laminar flow of a Newtonian fluid, $\dot{\gamma}$ is proportional to the shear stress $\tau\ (= F/A)$, the proportionality constant being known as the coefficient of viscosity, dynamic viscosity or, most simply, the viscosity η of the fluid (equation 1).

$$\tau = \eta\dot{\gamma} \tag{1}$$

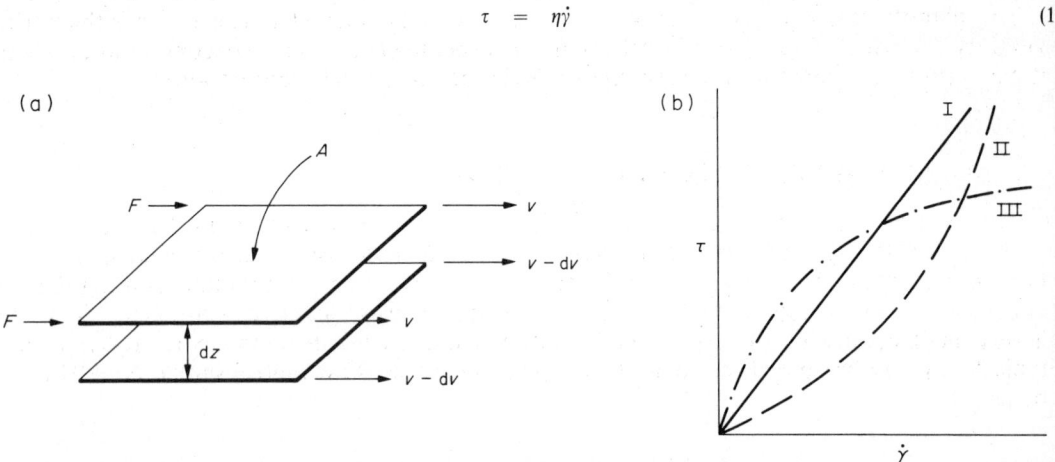

Figure 1 Simple representations of fluid flow behaviour. (a) Velocity gradient resulting from application of a tangential force to one of two adjacent parallel layers in a fluid (see the text for an explanation of the symbols). (b) Characteristic variations of shear stress with shear rate: I, Newtonian fluid (viscosity is constant); II, dilatant fluid (viscosity increases with shear rate); III, pseudoplastic fluid (viscosity decreases with shear rate)

Viscosity has the dimensions $ML^{-1}T^{-1}$ and is usually expressed in units of poise ($1\ P = 1\ dyne\ s\ cm^{-2}$ in the c.g.s. system or $10\ P = 1\ N\ s\ m^{-2}$ in the S.I. system). It is still common practice to use the c.g.s. system for dilute solution viscometry, and this practice is continued here with $g\ cm^{-3}$ being employed as units of concentration rather than the older units of $g\ dl^{-1}$.

Whilst the assumption of Newtonian behaviour is satisfactory for most dilute polymer solutions, it cannot be applied universally. Non-Newtonian behaviour may take the form of either a variation in the apparent viscosity $d\tau/d\dot{\gamma}$ with shear rate (Figure 1b) or a dependence of the apparent viscosity upon time at a given shear rate. For dilute polymer solutions, non-Newtonian behaviour is most commonly observed as a reduction in the apparent viscosity with increasing shear rate, *i.e.* as pseudoplasticity. In order to eliminate this effect it may be necessary to extrapolate the experimental data to zero shear rate.

Absolute measurements of viscosity are not essential in dilute solution viscometry since it is only necessary to determine the viscosity of a polymer solution relative to that of the pure solvent. The

Table 1 Quantities and Terminology used in Dilute Solution Viscometry

Common name	Name proposed by IUPAC	Symbol and definition
Relative viscosity	Viscosity ratio	$\eta_r = \dfrac{\eta}{\eta_0}$
Specific viscosity	—	$\eta_{sp} = \dfrac{\eta - \eta_0}{\eta_0} = \eta_r - 1$
Reduced viscosity	Viscosity number	$\eta_{red} = \dfrac{\eta_{sp}}{c}$
Inherent viscosity	Logarithmic viscosity number	$\eta_{inh} = \dfrac{\ln \eta_r}{c}$
Intrinsic viscosity	Limiting viscosity number	$[\eta] = (\eta_{red})_{c \to 0} = (\eta_{inh})_{c \to 0}$

quantities and terminology used are summarized in Table 1, where η_0 is the viscosity of the pure solvent and η is the viscosity of a polymer solution of concentration c.

The terminology proposed by IUPAC[21] was an attempt to eliminate the inconsistencies associated with the common names, which define as viscosities the quantities that do not have the appropriate dimensions. However, the common names were well established when the IUPAC recommendations were published and the new terminology has largely been ignored.

The quantity of greatest importance for the purposes of polymer characterization is the intrinsic viscosity. It is usually evaluated by extrapolation of experimental data to zero concentration, though it may also be estimated from measurements performed at a single concentration.

9.3 MEASUREMENT OF SOLUTION VISCOSITY

The viscosities of most dilute polymer solutions fall in the range 0.005 to 0.1 P and are normally measured using capillary viscometers. Relative viscosities are usually determined indirectly by performing separate measurements on the pure solvent and the polymer solutions using the same viscometer. The variation in temperature which can be tolerated from one measurement to another is very small, because the viscosities of most solvents decrease rapidly with increasing temperature, typically by 1 to 2% per °C at about 20 °C. Thus the temperature control should be ± 0.01 °C or better.

9.3.1 Capillary Viscometry

Capillary viscometers have many advantages when compared to other types of viscometer. They are of relatively simple construction and inexpensive, they require only small volumes of the solvent/solution whose viscosity is to be measured, typically 2 to 20 cm³, and temperature control is easily achieved by placing the viscometer in a thermostatted water or oil bath. Furthermore, theoretical treatment of flow through a capillary tube is firmly established.[13-17] Thus capillary viscometry has been used almost exclusively for measurements on dilute polymer solutions.

9.3.1.1 Capillary viscometers

Two general classes of capillary viscometer have found use, namely U-tube viscometers and suspended level viscometers. A common feature of these viscometers is that a measuring bulb, with upper and lower etched marks, is attached directly above the capillary tube. Liquid is either drawn or forced into the measuring bulb from a reservoir bulb attached to the bottom of the capillary tube. The time required for the liquid to flow back between the two etched marks is then recorded.

In U-tube viscometers (Figure 2), the pressure head giving rise to flow depends upon the volume of liquid contained in the viscometer. Therefore, it is important to ensure that this volume is the same for each measurement. This is normally achieved after temperature equilibration by adjusting the

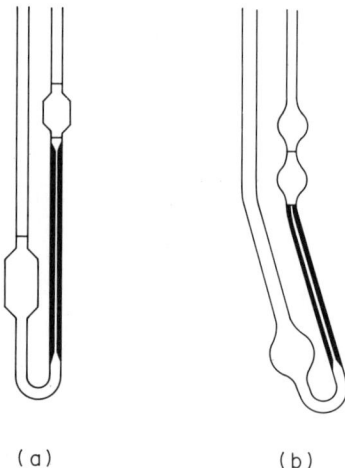

Figure 2   U-tube viscometers: (a) Ostwald, (b) Cannon–Fenske

liquid level to an etched mark just above the reservoir bulb. A further complication in the use of U-tube viscometers is the need for perfect vertical alignment of the viscometer, since slight deviations from the vertical can give rise to significant changes in the pressure head.[15] This problem is essentially eliminated in the Cannon–Fenske viscometer by having the measuring bulb positioned vertically above the reservoir bulb.

Most suspended level viscometers are based upon the design due to Ubbelohde (Figure 3), the most significant feature of which is the additional tube attached just below the capillary tube. This ensures that during measurement the liquid is suspended in the measuring bulb and capillary tube, with atmospheric pressure acting both above and below the flowing column of liquid. These viscometers have a number of advantages in comparison to U-tube viscometers. The variation in pressure head during measurement is smaller, vertical alignment is less critical, and the volume of liquid in the viscometer need not be constant because the position of the suspended liquid level at the bottom of the capillary tube is fixed. The latter feature is particularly useful since it enables solutions to be diluted in the viscometer. Thus, a known volume of the most concentrated solution is placed in the viscometer and its flow time determined. A known volume of solvent is then added. After thorough mixing and temperature equilibration, the flow time of the diluted solution is measured. This procedure is then repeated for several dilutions. The modified Ubbelohde viscometer shown in Figure 3 is especially suitable for this purpose since the minimum volume of liquid required for measurement is smaller than for the original design, thus allowing for more dilutions. Other modifications have been used to allow for control of the atmosphere above the liquid,[22–24] for

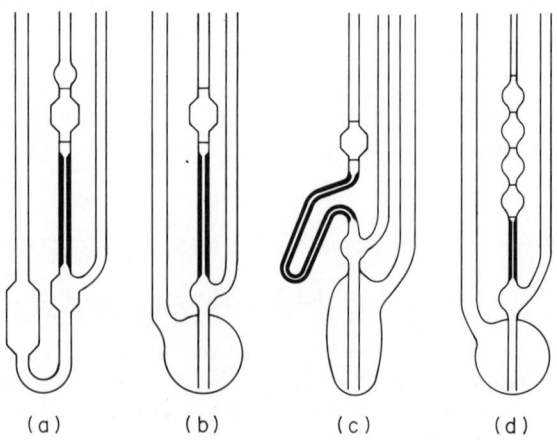

Figure 3   Suspended level viscometers: (a) Ubbelohde; (b) modified Ubbelohde; (c) Desreux–Bischoff type; and (d) Schurz–Immergut type

measurements at high temperatures,[24-26] and for measurements at either low or variable shear rates (Figures 3c and 3d).[27-30] Designs for automation of dilution and flow time measurement have also been described.[31-34]

An important consideration in the use of capillary viscometers is the cleanliness of both the viscometer and the liquids placed in it, since dust particles give rise to incorrect and erratic flow times. Therefore, it is good practice to clean the viscometer in chromic acid prior to use and to filter all solvents and solutions using glass sinters. Several measurements of flow time should be made for each solution and a mean value taken for use in calculations. Flow times should be reproducible to $\pm 0.1\%$ when measured visually using a stopwatch capable of recording time in intervals of 0.1 s. Automatic viscometers enable flow times to be measured with much greater precision, typically $\pm 0.01\%$.

9.3.1.2 Theoretical treatment of flow in a capillary tube

Under conditions of steady laminar Newtonian flow, the volume V of liquid which flows in time t through a capillary tube of length l and radius r is related to both the pressure difference P across the capillary and the viscosity of the liquid by Poiseuille's equation

$$\frac{V}{t} = \frac{\pi r^4 P}{8\eta l} \tag{2}$$

The radial velocity profile corresponding to equation (2) is parabolic, with maximum velocity along the axis of the capillary tube and zero velocity at the wall. Thus the maximum shear rate $\dot{\gamma}_{max}$ is found at the wall

$$\dot{\gamma}_{max} = \frac{4}{\pi r^3}\left(\frac{V}{t}\right) \tag{3}$$

The shear rate is zero at the axis and has an average value $\bar{\gamma}$ given by

$$\bar{\gamma} = \frac{8}{3\pi r^3}\left(\frac{V}{t}\right) \tag{4}$$

During the measurement of flow time, P varies and is normally represented by

$$P = \langle h \rangle \rho g \tag{5}$$

where $\langle h \rangle$ is the average pressure head, ρ is the density of the liquid, and g is the acceleration due to gravity. Hence equation (2) may be rearranged to give

$$\eta = \frac{\pi r^4 \langle h \rangle g \rho t}{8Vl} \tag{6}$$

Poiseuille's equation (2) is not entirely satisfactory because it neglects the energy dissipated in imparting kinetic energy to the liquid and also the energy dissipated by convergence and divergence of the flow lines on entry into and exit from the capillary tube. A kinetic energy correction can be derived theoretically and takes the form of a reduction in P. End effects are normally corrected by introducing an apparent increase in l. When these corrections are included, equation (6) becomes

$$\eta = \frac{\pi r^4 \langle h \rangle g \rho t}{8V(l + nr)} - \frac{mV\rho}{8\pi(l + nr)t} \tag{7}$$

Since n and m are constants for a given viscometer, a convenient form of equation (7) for relative measurements is

$$\eta = A\rho t - \frac{B\rho}{t} \tag{8}$$

where A and B are viscometer constants. These constants are evaluated from measurements performed on liquids for which the viscosities and densities are known accurately.[4,8,17,35-37]

9.3.1.3 Corrections for relative measurements

When relative measurements are made, it is common practice to assume that the solution density ρ is equal to the solvent density ρ_0 and that the kinetic energy correction is negligible. The experimental values for the relative and specific viscosities are then given by the simple expressions (9) and (10), where t and t_0 are the flow times for the solution and solvent respectively.

$$\eta_r^{expt} = \frac{t}{t_0} \tag{9}$$

$$\eta_{sp}^{expt} = \frac{t - t_0}{t_0} \tag{10}$$

In many cases this practice is unacceptable and gives rise to significant errors in evaluation of the intrinsic viscosity. However, the full expressions for η_r and η_{sp} based upon equation (8) are rather cumbersome to use. If it is assumed that ρ is linearly dependent upon the solution concentration c according to the equation

$$\rho = \rho_0 + \left(\frac{d\rho}{dc}\right)c \tag{11}$$

then it is possible to reduce the full expressions to more amenable forms

$$\eta_r = \eta_r^{expt}[1 + \Delta\{1 - (\eta_r^{expt})^{-2}\}][1 + \phi c] \tag{12}$$

$$\eta_{sp} = \eta_{sp}^{expt}[1 + \Delta\{1 - (\eta_r^{expt})^{-1}\}][1 + \phi c] + \phi c \tag{13}$$

for which

$$\Delta = \sum_{n=1}^{\infty} x^n \tag{14}$$

$$x = \frac{B}{At_0^2} \tag{15}$$

$$\phi = \frac{\left(\dfrac{d\rho}{dc}\right)}{\rho_0} \tag{16}$$

The quantity ϕ provides for density correction. If $d\rho/dc$ is unknown, it may be approximated by $(\rho_p - \rho_0)/\rho_p$, where ρ_p is the density of the polymer. The quantity Δ provides for kinetic energy correction and is expressed as a function of the fractional kinetic energy correction for the solvent.

Rearrangement of equation (13) and application of the condition $t \to t_0$ as $c \to 0$ gives an expression for correction of the intrinsic viscosity obtained from use of equation (10)

$$[\eta] = [\eta]^{expt}(1 + 2\Delta) + \phi \tag{17}$$

where

$$[\eta]^{expt} = \left(\frac{t - t_0}{t_0 c}\right)_{c \to 0} \tag{18}$$

The importance of the density correction depends upon the magnitude of $d\rho/dc$, and is often only of significance for values of intrinsic viscosity below $30\ cm^3\ g^{-1}$. In contrast, the kinetic energy correction is significant for many viscometers, as can be seen in Table 2 which gives the kinetic energy correction factors $(1 + 2\Delta)$ for four representative values of t_0 at each of five representative values of the ratio B/A. In order to reduce the kinetic energy correction it is necessary to design the viscometer such that the volumetric flow rate is small, preferably below $0.01\ cm^3\ s^{-1}$. In this respect, small pressure heads and capillary tubes of small diameter are desirable. A modification of the design due to Desreux and Bischoff[30] is particularly suitable for reducing the pressure head (Figure 3c). When selecting a capillary bore, it must be borne in mind that a reduction in the bore size also has the undesirable effects of increasing the susceptibility to clogging and increasing the potential for errors arising from adsorption of polymer on the capillary walls (see Section 9.3.1.5). In practice, a

Table 2 Factors for Kinetic Energy Correction of $[\eta]^{expt}$ for Capillary Viscometers, Calculated according to Equations (14), (15) and (17)

t_0 (s)[a]			$(1 + 2\Delta)$ for $B/A^a =$		
	$10\,s^2$	$100\,s^2$	$250\,s^2$	$500\,s^2$	$750\,s^2$
60	1.006	1.057	1.149	1.323	1.526
120	1.001	1.014	1.035	1.072	1.110
180	1.001	1.006	1.016	1.031	1.047
240	1.000	1.003	1.009	1.018	1.026

[a] Typical values of t_0, B/A and $(1 + 2\Delta)$, for toluene at 25 °C, are: (i) for the standard Ubbelohde viscometer, $t_0 \sim 60$ s, $B/A \sim 140\,s^2$ and $(1 + 2\Delta) \sim 1.1$; (ii) for the modified Desreux-Bischoff viscometer, $t_0 \sim 100$ s, $B/A \sim 10\,s^2$ and $(1 + 2\Delta) \sim 1.002$. Modified viscometers, described in the literature, cover a wide range of values of t_0 and B/A (as set out above).

sensible lower limit for the capillary tube diameter is 0.4 mm. The volume of the measuring bulb should be chosen to give t_0 values in the range 100 to 200 s, which represents a reasonable compromise between the convenience of small t_0 values and the benefits of large t_0 values in reducing the kinetic energy correction.

Corrections for surface tension and drainage effects[28,38] can usually be neglected in relative measurements without introducing significant errors.

9.3.1.4 Non-Newtonian behaviour

The coil dimensions of flexible polymer molecules are sufficiently large to encompass several solution layers moving with different velocities. Thus the molecules are subject to a torque which causes them to rotate and to become deformed, being extended in the direction of flow and compressed perpendicular to this direction. Each region of the coil undergoes periodic extension and compression as the coil rotates. This deformation and alignment of the molecules in the direction of flow reduces their resistance to flow and gives rise to a reduction in the apparent viscosity which is dependent upon the shear rate. The reduction is negligible for polymers with low molecular weights, but its magnitude increases as molecular weight increases due to the increase in coil dimensions. In general, non-Newtonian behaviour can be anticipated for polymers with $[\eta] > 100$ cm^3 g^{-1} and is invariably observed for those with $[\eta] > 400$ cm^3 g^{-1} (Figure 4).[29,39–47] It becomes less prominent

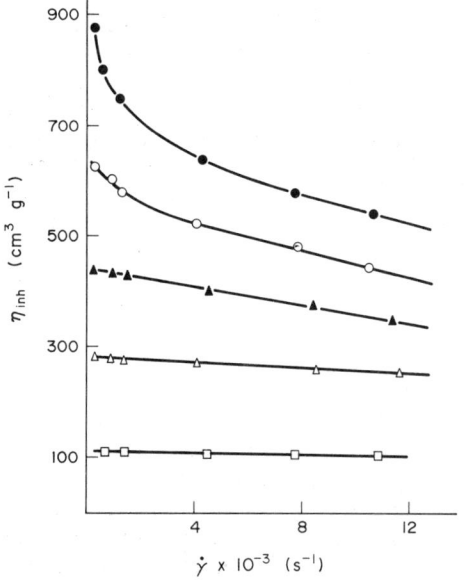

Figure 4 Effect of molecular weight upon the variation of dilute solution viscosity with shear rate. Experimental points, taken from ref. 41, are for solutions of polystyrene fractions in benzene at 25 °C with concentrations in the very dilute range, 3×10^{-4} to 5×10^{-4} g cm^{-3}: ●, $M = 6.5 \times 10^6$; ○, $M = 5.5 \times 10^6$; ▲, $M = 2.22 \times 10^6$; △, $M = 1.24 \times 10^6$; □, $M = 2.88 \times 10^5$

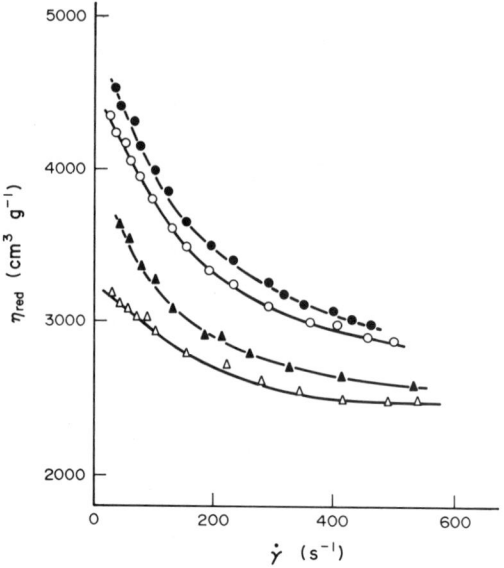

Figure 5 Effect of concentration upon the variation of dilute solution viscosity with shear rate. Experimental points, taken from ref. 60, are for solutions in toluene at 20 °C of a polystyrene sample of molecular weight 35.0×10^6: ●, $c = 5.12 \times 10^{-4}$; ○, $c = 4.08 \times 10^{-4}$; ▲, $c = 2.88 \times 10^{-4}$; △, $c = 1.01 \times 10^{-4}$ g cm^{-3}

as the polymer concentration decreases, but cannot be eliminated by extrapolation to zero concentration (Figure 5). If the shape of the polymer molecules is naturally anisotropic, for example, ellipsoidal or rod-like, then non-Newtonian behaviour is more pronounced and becomes significant at very much lower molecular weights.[42,48-52]

Many attempts have been made to predict theoretically the non-Newtonian behaviour of dilute polymer solutions.[10,19,53-56] However, the models used have been restrictive and agreement with experiment is generally qualitative rather than quantitative. The most commonly quoted expression relating the intrinsic viscosity at finite shear rates to its value $[\eta]_{\dot\gamma=0}$ at zero shear rate is

$$[\eta] = [\eta]_{\dot\gamma=0}(1 - a\beta_0^2 + b\beta_0^4 - \dots) \tag{19}$$

where a and b are constants with values $(a \gg b)$ dependent upon molecular shape and coil dimensions, and β_0 is the reduced shear rate given by

$$\beta_0 = \frac{M[\eta]_{\dot\gamma=0}\eta_0\dot\gamma}{RT} \tag{20}$$

in which M is the molecular weight of the polymer, R is the gas constant and T is the absolute temperature. Peterlin[56] has predicted that $[\eta]$ should pass through a minimum as $\dot\gamma$ increases, though this has only been observed for solutions of very high molecular weight polymers in viscous solvents.[57-59]

If non-Newtonian behaviour is suspected, then measurements must be made over a range of shear rates. Ubbelohde viscometers with several measuring bulbs connected in series vertically above one another[27-29] (Figure 3d) are often used and provide shear rates typically within the range 100 to 4000 s^{-1}. However, the asymptotic tendency towards Newtonian behaviour implies the need for measurements at lower shear rates, and capillary viscometers have been designed for this purpose.[48,60,61] Accurate extrapolation of $[\eta]$ values to zero shear rate is not easily achieved. A simplification of equation (19) is equation (21), for which C is a constant, and this may be used when reliable data are available at low values of $\dot\gamma$, although such data do not always fit well to this equation. Linear extrapolations must be used with caution. A further complication for very high molecular weight polymers and for highly extended polymers is the possibility of shear-induced degradation.[57,62,63]

$$[\eta] = [\eta]_{\dot\gamma=0}(1 - C\dot\gamma^2) \tag{21}$$

9.3.1.5 Effects of polymer adsorption

The observation of anomalous increases in reduced viscosity as the polymer concentration decreases below about 0.001 g cm^{-3} has often been reported for high molecular weight polymers. The origin of this effect was initially the subject of lively debate.[64-69] However, it is now considered to be due to adsorption of polymer at the capillary walls causing a reduction in the effective diameter of the capillary tube, as originally suggested by Öhrn.[66] In fact, the effect is utilized nowadays for the measurement of adsorbed layer thicknesses.[44, 70-72]

Assuming that the solvent flow time is measured in the absence of an adsorbed layer and that the solution flow time is measured in the presence of an adsorbed layer of thickness d, then simple considerations based upon equation (6) with $\rho = \rho_0$ show that the observed (*i.e.* apparent) reduced viscosity η_{sp}^*/c is given by[28,66,69,72]

$$\frac{\eta_{sp}^*}{c} = \frac{\eta_{sp}}{c} + \frac{\eta_r}{c}\left(\frac{4d}{r}\right) \tag{22}$$

Since $\eta_r/c \to \infty$ as $c \to 0$, the apparent increase in the reduced viscosity should become greater as the polymer concentration decreases if d is constant, and should only be significant at very low concentrations if d is in the usual range of 10 to 100 nm. These predictions are in accord with most experimental observations, and in order to eliminate the effects of adsorption, it is normally recommended that intrinsic viscosities be evaluated by extrapolation of data obtained at concentrations above 0.001 g cm^{-3}. Recently, however, the effect has been observed at much higher concentrations and attributed to the formation of much thicker adsorbed layers.[73]. A further complication is that the formation of an adsorbed layer can give rise to a significant reduction in the polymer concentration. Thus it is good practice to allow an equilibrium value of d to be attained with the most concentrated solution and then to measure the flow time with a fresh portion of the solution.[72] A dilution series can then be performed, and finally, the solvent flow time measured. Provided that the value of d remains constant throughout these measurements (*i.e.* desorption is not significant), the effect of the adsorbed layer will be minimized.

9.3.2 Other Methods of Viscometry

Coaxial cylinder viscometers have normally been used when an alternative to capillary viscometry has been required. They are particularly suitable for studying non-Newtonian behaviour and for performing measurements at low shear rates. The use of these viscometers will be briefly described together with that of the capillary bridge differential viscometer. The latter is a recent innovation which enables specific viscosities to be determined directly. Other methods of viscometry are of only limited value for dilute solutions of polymers and will not be described here.

9.3.2.1 Coaxial cylinder viscometers

The simplest form of coaxial cylinder viscometer comprises an outer cylinder which acts as a container for the liquid being studied, and an inner cylinder which is either partially or completely immersed in the liquid and positioned coaxially to the outer cylinder (Figure 6). One cylinder is caused to rotate whilst the other is fixed. For laminar flow of a Newtonian liquid, the angular velocity Ω of the rotating cylinder is related to the torque G applied to it by Margules' equation

$$\Omega = \frac{G(r_2^2 - r_1^2)}{4\pi\eta h r_1^2 r_2^2} \tag{23}$$

where r_1 and r_2 are the radii of the inner and outer cylinders respectively, and h is the depth of immersion of the inner cylinder. Equation (23) is derived assuming that at the fixed cylinder the liquid is stationary, whilst at the rotating cylinder it moves with angular velocity Ω. The shear rate $\dot{\gamma}_r$ at a point in the liquid a radial distance r from the axis of the cylinders is given by

$$\dot{\gamma}_r = \frac{2\Omega r_1^2 r_2^2}{r^2(r_2^2 - r_1^2)} \tag{24}$$

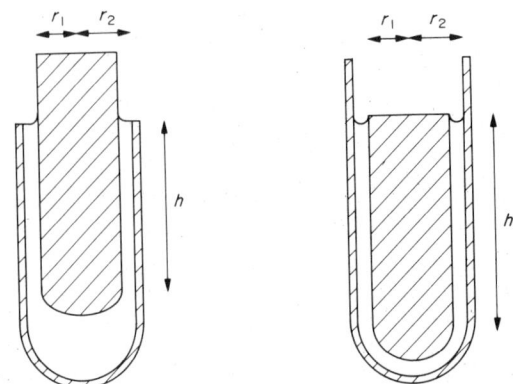

Figure 6   Schematic diagrams showing the recommended positions for the inner cylinder in a Zimm–Crothers viscometer

Provided that the annular gap is small relative to the cylinder radii, the shear rate will not vary greatly across the gap. Also, a wide range of shear rates can be attained by suitable control of Ω.

The design due to Zimm and Crothers[74] has formed the basis for many coaxial cylinder viscometers used in studies of polymer solutions and employs a rotating inner cylinder which floats freely in the liquid. This cylinder is centred by the action of the annular meniscus and attains a depth of immersion which depends upon its mass and the density of the liquid (Figure 6). A constant torque is applied using an external rotating magnet which acts on a piece of steel fixed inside the inner cylinder. The time required for a given number of revolutions of the inner cylinder is measured and yields an average value for its period of revolution t_R. Modification of equation (23) to include the effect of buoyancy on h, and substitution of $\Omega = 2\pi/t_R$, gives

$$\eta = D\rho t_R \tag{25}$$

where D is a constant for a given viscometer operating at a constant torque. Since the kinetic energy correction is zero, equation (25) is entirely analogous to equation (8) in which A is substituted by D and $B = 0$. Thus the errors introduced by assuming $\rho = \rho_0$ are defined by equations (9) to (13) and (16) to (18) in which t and t_0 are substituted by the corresponding t_R and t_{R0} values, and $\Delta = 0$. Corrections for end effects are included in D as an apparent increase in h and, therefore, cancel when relative viscosities are taken. The effect of polymer adsorption at the cylinder walls is negligible because the annular gap is relatively large, typically 1 to 2 mm. Temperature control is usually achieved by fitting a thermostatted jacket around the outer cylinder.

Zimm and Crothers designed their viscometer with low shear rates in mind and reported its use at shear rates down to $0.01\ \text{s}^{-1}$. Thus it is particularly suitable for evaluation of $[\eta]_{\dot{\gamma}=0}$ and for measurements on polymers susceptible to shear-induced degradation. Berry[75] modified the viscometer design in order to improve the stability of the torque when a wide range of temperatures are used. He also used a different method for evaluation of relative viscosities.

A coaxial cylinder viscometer, which allows simultaneous neutron scattering studies to be performed, has recently been reported and used to investigate the effect of shear rate upon molecular shape and dimensions for solutions of polystyrene.[76]

9.3.2.2   *The capillary bridge differential viscometer*

This viscometer was designed by Haney[77–81] and utilizes four matched capillary tubes arranged in a manner analogous to a Wheatstone bridge (Figure 7). Solvent is introduced into the capillary bridge at constant pressure, and calibrated differential pressure transducers measure the pressure P_i at the inlet and also the pressure difference ΔP across the bridge. Two sample reservoirs can be switched simultaneously, either into or out of the flowstreams, by means of two tandem switching valves. Normally, one reservoir contains solvent and the other is filled with the solution to be analyzed. Initially the reservoirs are bypassed and solvent flows through each of the capillary tubes giving rise to a pressure difference ΔP_0 which ideally should be zero. The reservoirs are then switched into the flowstreams and ΔP gradually changes as the solvent is purged from the capillary tube into which the solution is flowing. When the solution completely fills this capillary tube, ΔP attains a new

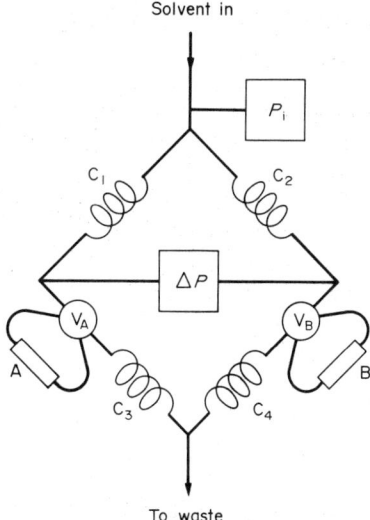

Solvent in

To waste

Figure 7 Schematic diagram of a capillary bridge differential viscometer: C_1, C_2, C_3 and C_4 are matched capillary tubes; V_A and V_B are tandem switching valves; and A and B are sample reservoirs

steady state value ($\Delta P_s + \Delta P_0$), where ΔP_s relates to the difference between the solution and solvent viscosities. The values of P_i and ΔP_s are then used to calculate the specific viscosity of the solution (equation 26).

$$\eta_{sp} = \frac{4\Delta P_s}{P_i - 2\Delta P_s} \tag{26}$$

Equation (26) is easily obtained by application of equation (2) to the flow of liquid through each side of the capillary bridge, assuming that the capillary tubes are of equal flow resistance and that the flow resistance of the other components can be neglected. These assumptions are satisfactory because ΔP_0 is close to zero and very small compared to P_i. Inherent in the use of equation (2) are the further assumptions that conditions of laminar Newtonian flow exist, and also that kinetic energy corrections are negligible. The latter is reasonable since the liquids already possess kinetic energy when they enter the capillary tubes which have a large ratio of length to radius. Assumption of laminar flow is also acceptable. However, the assumption of Newtonian behaviour is not satisfactory for polymers with high values of intrinsic viscosity, particularly as the shear rates are moderately high, typically 1000 to $5000\,\text{s}^{-1}$. Furthermore, this range of shear rates does not allow for accurate evaluation of $[\eta]_{\dot{\gamma}=0}$. The use of very dilute solutions ($c < 0.0001\,\text{g cm}^{-3}$) reduces the magnitude of the effect but does not eliminate it (see Section 9.3.1.4) and creates greater uncertainty in the value of c. Thus the viscometer should be used with caution when non-Newtonian behaviour is suspected. The effects of polymer adsorption may also be present, though it is possible to compensate for them by evaluation of ΔP_0 after the solution has eluted, assuming that the adsorbed polymer is not removed by the flowing solvent.

The sensitivity of the viscometer is an order of magnitude greater than that of a simple capillary viscometer, and it is less sensitive to changes in temperature because η_{sp} is determined directly. A further advantage is that each measurement requires only a few minutes. Haney recommends the use of a single, very dilute solution and assumes that its reduced viscosity is equal to the intrinsic viscosity. He also presents experimental evidence to support this practice, though there are some discrepancies.

There is no doubt that the capillary bridge differential viscometer is a significant development. Nevertheless, its relatively high cost in comparison to simple capillary viscometers is likely to restrict its wide use. A simple modification, however, enables it to be used as an on-line viscosity detector in gel permeation chromatography[82, 83] and it is easy to envisage that this will become a very important area of application.

9.4 EVALUATION OF INTRINSIC VISCOSITY

Viscosity measurements yield data at finite concentrations from which the intrinsic viscosity must be evaluated in order to enable the data to be interpreted properly. The most general relationship between intrinsic viscosity and dilute solution viscosity takes the form of a power series in concentration

$$\frac{\eta_{sp}}{c} = [\eta] + k_1[\eta]^2 c + k_2[\eta]^3 c^2 + k_3[\eta]^4 c^3 + \ldots \ldots \quad (27)$$

where k_1, k_2, k_3, *etc.* are dimensionless constants. Huggins' theoretical analysis of the hydrodynamics of both rigid and flexible polymer molecules[84-86] produced equations which can be reduced to the form

$$\frac{\eta_{sp}}{c} = \frac{[\eta]}{1 - k'[\eta]c} \quad (28)$$

where k' is a dimensionless constant introduced[86] to correct for certain deficiencies in the analysis and is commonly referred to as the Huggins constant. This constant relates to the size and shape of polymer segments, and to hydrodynamic interactions between different segments of the same polymer molecule. Series expansion of equation (28) is permissible provided that $0 < k'[\eta]c < 1$ and gives a convergent series which is equivalent to equation (27) with $k_n = k'^n$ for $n = 1, 2, 3$ *etc.* Thus k_1 corresponds to the Huggins constant, and typically, has values which are essentially independent of molecular weight and which fall in the range 0.3 (for good polymer–solvent pairs) to 0.5 (for poor polymer–solvent pairs). Experimental data for which $k_1 > 0.5$ should be treated with caution since this indicates aggregation of polymer molecules.[10]

Most of the equations and procedures recommended for evaluation of $[\eta]$ can be shown to bear a relationship to equation (27). However, they often yield different values for $[\eta]$ as a consequence of their different levels of approximation. Thus, methods for evaluation of $[\eta]$ should be chosen with due consideration of their validity for the particular set of data to be analyzed. The methods which have been used can be divided into those involving extrapolation of data obtained at a series of concentrations and those involving estimation of $[\eta]$ from a single viscosity measurement.

9.4.1   Methods for Evaluation of $[\eta]$ by Extrapolation of Experimental Data

A large number of equations have been recommended for this purpose. Those most commonly encountered are listed below (equations 29–32), where the subscripts indicate that values for the intrinsic viscosity and the dimensionless slope constants are obtained by application of the corresponding equation to experimental data and may differ from the true values defined by equation (27).

Huggins:[86]
$$\frac{\eta_{sp}}{c} = [\eta]_H + k_H[\eta]_H^2 c \quad (29)$$

Kraemer:[87]
$$\frac{\ln \eta_r}{c} = [\eta]_K - k_K[\eta]_K^2 c \quad (30)$$

Schulz–Blaschke:[88]
$$\frac{\eta_{sp}}{c} = [\eta]_{SB} + k_{SB}[\eta]_{SB}\eta_{sp} \quad (31)$$

Martin:[89]
$$\ln\left(\frac{\eta_{sp}}{c}\right) = \ln[\eta]_M + k_M[\eta]_M c \quad (32)$$

The Huggins equation is simply a truncation of the series expansion of equation (28), and strictly, is only applicable when $[\eta]c \ll 1$. At higher concentrations, experimental data show upward curvature when plotted according to this equation. The Kraemer equation is an approximation of the Huggins equation, from which it may be derived assuming that $\eta_{sp} \ll 1$. Theory predicts that $k_H + k_K = \frac{1}{2}$ when the approximation is satisfactory. The Schulz–Blaschke equation was deduced empirically and is identical to equation (28) with $[\eta] = [\eta]_{SB}$ and $k' = k_{SB}$. Experimental data plotted according to this equation show downward curvature as the concentration increases, though such plots are usually linear to higher concentrations than those obtained by application of the Huggins equation. Similar observations have been made with regard to the Martin equation, which was also deduced empirically. Series expansion of the equation gives an expression which is equivalent to equation (27) with $[\eta] = [\eta]_M$ and $k_n = k_M^n/n!$ for $n = 1, 2, 3$, *etc.* Assuming the solutions are sufficiently dilute to ensure that each of equations (28) to (32) are valid, it is easy to show that

$$[\eta]_H = [\eta]_K = [\eta]_{SB} = [\eta]_M = [\eta] \tag{33}$$

and
$$k_H = k_{SB} = k_M = k' = k_1 \tag{34}$$

Thus, each equation is capable of giving a good estimate of $[\eta]$. Solution concentrations should be chosen such that $\eta_{sp} < 1$, the lower limit depending upon both the precision with which η_{sp} is measured and the magnitude of adsorption effects. Plots of the experimental data must then be carefully inspected for signs of curvature before selecting the data points to be fitted to a straight line for extrapolation. In particular, data points corresponding to $[\eta]c > 1$ should be viewed with suspicion especially for plots according to the Huggins and Kraemer equations. If straight lines are fitted to data which show curvature when plotted according to equations (29) to (32), then the extrapolations are predicted[90] to yield different estimates of $[\eta]$ and k_1

$$[\eta]_{SB} > [\eta]_M > [\eta] > [\eta]_H = [\eta]_K \tag{35}$$

and
$$k_{SB} < k_M < k_1 < k_H \tag{36}$$

These trends are often observed in practice (Figure 8),[91-95] and highlight the possible underestimation of $[\eta]$ when employing the most common method for its evaluation, which involves fitting of the data to both the Huggins and the Kraemer equations (Figure 8a). Sakai[90] has suggested that $[\eta]$ and k_1 are best estimated by taking the arithmetic means of the values obtained from application of

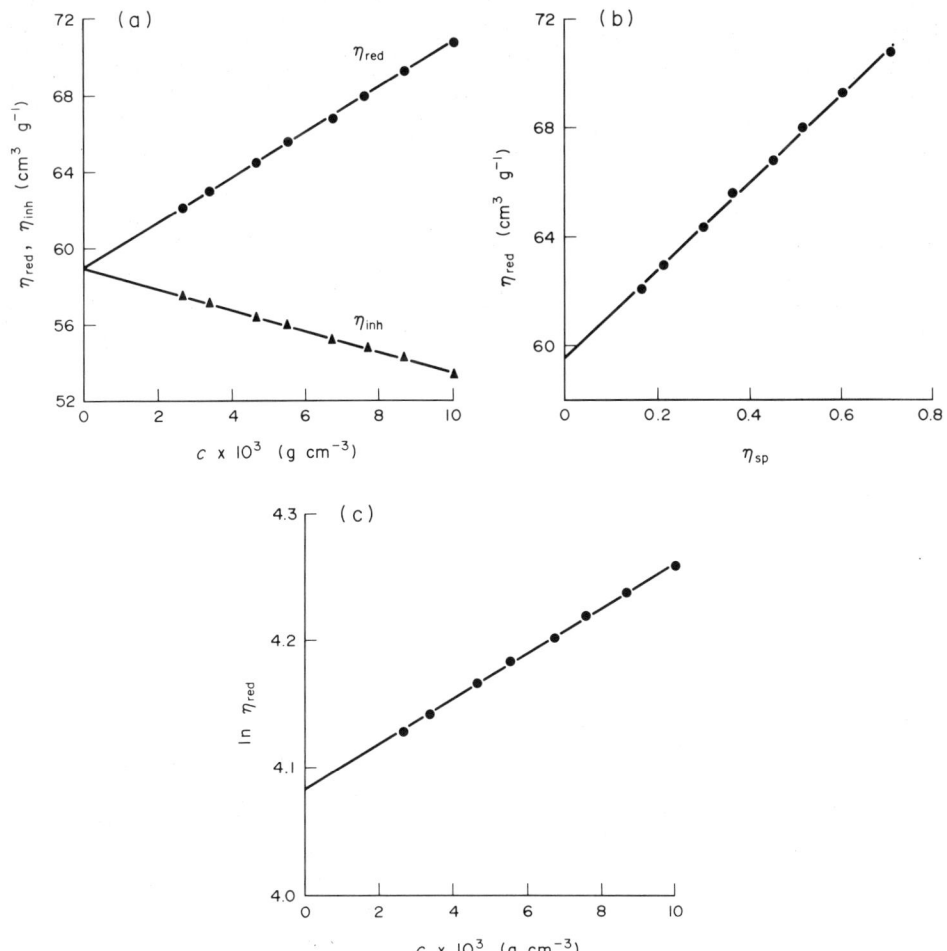

Figure 8 Evaluation of intrinsic viscosity for solutions in toluene of a sample of poly(*t*-butyl acrylate) at 25 °C: (a) dual Huggins–Kraemer plot, which gives $[\eta]_H = [\eta]_K = 59.0 \text{ cm}^3 \text{ g}^{-1}$, $k_H = 0.340$ and $k_K = 0.158$; (b) Schulz–Blaschke plot, which gives $[\eta]_{SB} = 59.6 \text{ cm}^3 \text{ g}^{-1}$ and $k_{SB} = 0.268$; and (c) Martin plot, which gives $[\eta]_M = 59.3 \text{ cm}^3 \text{ g}^{-1}$ and $k_M = 0.300$

the Huggins equation and either the Martin equation (for good polymer–solvent pairs) or the Schulz–Blaschke equation (for poor polymer–solvent pairs). This has some merit when the estimates of $[\eta]$ and/or k_1 are significantly different. Data plotted according to the Martin equation often show linearity at high concentrations but extrapolation from this region must be avoided since it always yields high estimates of $[\eta]$ and low estimates of k_1. Failure to apply density and kinetic energy corrections when they are significant can give rise to curvature in each of the plots and results in poor estimates of $[\eta]$ and k_1. The effects of experimental errors upon evaluation of $[\eta]$ have been investigated, with methods being proposed for their detection and reduction.[96,97]

Heller[98] has derived equations which can be used as alternatives to the Huggins and Kraemer equations. However, these are approximations of the better known equations and have no major advantage in comparison to them, other than greater symmetry in the dual plot. Maron and Reznik[99] have used difference equations obtained by expansion of the Huggins and Kraemer equations to the terms in c^2, followed by subtraction of the latter equation from the former. Their main attribute, besides the higher terms in c, is that the interrelationships between the dimensionless constants are forced to conform with theoretical predictions. For extrapolation of data obtained at higher concentrations, Chee[100] has used a modification of the Schulz–Blaschke equation in which a new parameter is introduced whose value must be determined for each set of experimental data by an iterative procedure. The values of $[\eta]$ obtained are slightly below those resulting from application of the Huggins equation to data obtained at lower concentrations. Lyons and Tobolsky[101] and Fedors[102] have also recommended equations for use over wide ranges of concentration.

9.4.2 Single Point Methods for Evaluation of $[\eta]$

The inconvenience of extrapolation methods for routine analysis has given rise to considerable interest in estimation of $[\eta]$ from a single specific viscosity measurement, particularly when $[\eta]$ need only be evaluated approximately. The majority of single point methods employ equations derived by combination and/or simplification of equations (29) to (31). When using these methods, solution concentrations should be chosen such that $\eta_{sp} \lesssim 0.2$, since higher values can result in poor estimates of $[\eta]$.[103] Thus, η_{sp} must be measured with high precision, for which purpose automatic viscometers are best suited. Besides the method of Haney (see Section 9.3.2.2), the most simple methods for evaluation of $[\eta]$ are based on the following equations

Solomon and Ciută:[104]
$$[\eta] = [2(\eta_{sp} - \ln \eta_r)]^{\frac{1}{2}} c^{-1} \tag{37}$$

Deb and Chatterjee:[105]
$$[\eta] = [3(\ln \eta_r - \eta_{sp} + \tfrac{1}{2}\eta_{sp}^2)]^{\frac{1}{3}} c^{-1} \tag{38}$$

Ram Mohan Rao and Yaseen:[103]
$$[\eta] = [\eta_{sp} + \ln \eta_r](2c)^{-1} \tag{39}$$

Series expansion reveals that equation (37) requires $k_1 = \tfrac{1}{3}$ and that equations (38) and (39) require $k_1 = \tfrac{1}{2}$.[106,107] Hence, these equations are only valid for use with good polymer–solvent pairs, as observed in practice.[103,107–111] Equation (37) has been widely used, and generally, gives good estimates of $[\eta]$ when the above criteria are satisfied. Varma and Sengupta[108] derived additional terms which enable equation (37) to be used for a wider range of systems, including poor polymer–solvent pairs.

Other single point methods[112–117] utilize equations containing adjustable constants which must be evaluated experimentally for each polymer–solvent system. The additional effort required to establish these specific equations is justifiable when a large number of samples of the same polymer are to be analyzed. A nomographic method due to Khan and Bhargava[118] is also widely applicable,[110] but similarly requires establishment of specific nomograms for each polymer–solvent system.

9.5 INTERPRETATION OF INTRINSIC VISCOSITY DATA

Evaluation of $[\eta]$, rather than η_{sp}, simplifies interpretation of experimental data by eliminating the effects of intermolecular polymer–polymer interactions. Thus theories of the solution behaviour of isolated polymer molecules can be applied, and enable $[\eta]$ to be related to molecular weight, molecular dimensions, polymer–solvent interaction parameters and degree of branching.

9.5.1 Theoretical Treatments of Intrinsic Viscosity

The hydrodynamic behaviour of polymer molecules is affected by their chemical and skeletal structure, and by their size and shape as modified by interaction with solvent molecules. Two extremes of hydrodynamic behaviour can be identified, namely free-draining and non-draining. A polymer molecule is said to be free-draining when solvent molecules flow freely past each segment of the chain, and non-draining when solvent molecules within the coil move with the polymer molecule. Therefore theoretical analysis of intrinsic viscosity is complex and most theories are based upon simplifying assumptions, especially with regard to molecular shape and draining behaviour. Some of the simpler theories for linear flexible chains and rod-like molecules are briefly considered in the following sections.

9.5.1.1 Theories for flexible chains

Free-draining models were amongst the first to be considered.[84,85,119-121] Under these conditions each chain segment makes an equal contribution to the frictional coefficient f_0 of the polymer molecule. Therefore, f_0 is directly proportional to the number x of segments in the chain and independent of molecular shape. The dependence of $[\eta]$ upon molecular shape arises from consideration of the rotational motion of the molecule and gives rise to equations of the form

$$[\eta] = \text{constant} \times \langle s^2 \rangle_0 \tag{40}$$

where $\langle s^2 \rangle_0$ is the mean square distance of a chain segment from the centre of mass of the molecule, *i.e.* the mean square radius of gyration. For Gaussian chains with large x, it may be shown that $\langle s^2 \rangle_0 \propto x$. Since the molecular weight M of the polymer molecule is also directly proportional to x, equation (40) may be modified to give

$$[\eta] = F_C M \tag{41}$$

where F_C is a constant for a given polymer/solvent system. This relationship is in accord with Staudinger's empirical hypothesis[122,123] but is unsatisfactory for most systems.

Free-draining models are inadequate because frictional interactions between solvent molecules and chain segments are sufficiently large to perturb the motion of the solvent molecules. Thus the degree of free-draining decreases as the segment density increases, so that solvent molecules near the centre of a Gaussian chain move with it. These perturbations of solvent flow were analyzed by Debye and Bueche[124] who modelled the chain as a sphere of uniform segment density, and by Kirkwood and Riseman[125] who used a Gaussian segment density distribution. The latter treatment is more realistic and yields the following expression for the frictional coefficient of the molecule

$$f_0 = \frac{6\pi\eta_0 \langle s^2 \rangle_0^{\frac{1}{2}}}{(8/3\pi^{\frac{1}{2}}) + (6\pi\eta_0 \langle s^2 \rangle_0^{\frac{1}{2}}/x\zeta)} \tag{42}$$

where ζ is the frictional coefficient of a chain segment. For short chains, the second term in the denominator becomes dominant and $f_0 = x\zeta$, in accord with the theories of free-draining. However, for long chains this term becomes negligible and equation (42) reduces to

$$f_0 = 6\pi\eta_0 \left(\frac{3\pi^{\frac{1}{2}}}{8} \right) \langle s^2 \rangle_0^{\frac{1}{2}} \tag{43}$$

which may be compared to Stokes' law for the frictional coefficient f of an impermeable sphere of radius R

$$f = 6\pi\eta_0 R \tag{44}$$

Thus a high molecular weight Gaussian chain may be modelled as an equivalent impermeable sphere which, with regard to f_0, should be of radius $0.665\langle s^2 \rangle_0^{\frac{1}{2}}$. A similar analysis of $[\eta]$ is possible and Kirkwood–Riseman theory leads to the following equation for $x > 100$[126]

$$[\eta] = 1.259\pi^{\frac{3}{2}} N_A \left(\frac{\langle s^2 \rangle_0^{\frac{3}{2}}}{M} \right) \tag{45}$$

where N_A is the Avogadro constant. This expression may be compared to Einstein's equation[127] for the specific viscosity of a suspension of non-interacting impermeable spheres

$$\eta_{sp} = 2.5\phi_s \tag{46}$$

where ϕ_s is the volume fraction of spheres. Comparison is facilitated by recognizing that $\phi_s = 4\pi N_A c R^3/3M$, where c is the concentration of spheres in units of mass per unit volume. Hence, equation (46) may be written in the form

$$[\eta] = \frac{10\pi N_A R^3}{3M} \tag{47}$$

from which it can be deduced that an equivalent impermeable sphere of radius $0.875\langle s^2\rangle_0^{\frac{1}{2}}$ should be used for modelling $[\eta]$.

A major deficiency of the theories presented so far is their use of unperturbed dimensions for the polymer molecule. In good solvents, polymer molecules in dilute solution expand beyond their unperturbed dimensions due to the effects of volume exclusion and of polymer–solvent interactions, and the radius of gyration is given by

$$\langle s^2\rangle^{\frac{1}{2}} = \alpha\langle s^2\rangle_0^{\frac{1}{2}} \tag{48}$$

where α is the expansion coefficient. Only when polymer–solvent interactions are sufficiently positive to counteract exactly the effects of volume exclusion does $\alpha = 1$; the dilute polymer solution is then said to be under theta conditions.

Flory and Fox[128-130,170] recognized the importance of equation (48) and modified equation (45) accordingly

$$[\eta] = K_\theta \alpha^3 M^{\frac{1}{2}} \tag{49}$$

where

$$K_\theta = \Phi_0^s \left(\frac{\langle s^2\rangle_0}{M}\right)^{\frac{3}{2}} \tag{50}$$

and $\Phi_0^s = 4.22 \times 10^{24}\ \mathrm{mol}^{-1}$ when M has units of mass per mol. They also derived a closed expression for α assuming uniform expansion of a Gaussian chain

$$\alpha^5 - \alpha^3 = 2C_M(\tfrac{1}{2} - \chi)M^{\frac{1}{2}} \tag{51}$$

where χ is the Flory–Huggins polymer–solvent interaction parameter and C_M is given by

$$C_M = (27/2^{\frac{5}{2}}\pi^{\frac{3}{2}})(\bar{v}^2/N_A V_1)(M/\langle r^2\rangle_0)^{\frac{3}{2}} \tag{52}$$

where $\bar{v}$ is the partial specific volume of the polymer, V_1 is the molar volume of the solvent and $\langle r^2\rangle_0$ is the unperturbed mean square distance of separation between the chain ends. For Gaussian chains

$$\langle r^2\rangle_0 = 6\langle s^2\rangle_0 \tag{53}$$

It is instructive to examine the limiting conditions defined by equation (51). Under theta conditions $\chi = \tfrac{1}{2}$, $\alpha = 1$ and equation (49) predicts that $[\eta] = K_\theta M^{0.5}$. For large expansions $\alpha^5 \gg \alpha^3$, so that α^3 is proportional to $M^{0.3}$, and the theory predicts $[\eta] = \mathrm{constant} \times M^{0.8}$. Herein lies the success of Flory–Fox theory since experimentally observed $[\eta]$–M relationships for flexible polymer molecules fall within these theoretical limits.

The Flory–Fox assumption of uniform chain expansion is not satisfactory and individual expansion parameters α_s, α_r and α_η are required for $\langle s^2\rangle_0^{\frac{1}{2}}$ and $\langle r^2\rangle_0^{\frac{1}{2}}$, and in equation (49) respectively. Exact theories express the expansion parameters as power series in the excluded volume parameter z

$$z = (3/2\pi\langle r^2\rangle_0)^{\frac{3}{2}}N^2\beta \tag{54}$$

where N is the number of random flight equivalent chain segments and β is the volume excluded by one chain segment for another (N is not necessarily equal to x since they have different origins, *i.e.* chain statistics and hydrodynamic theory). Short range and long range volume exclusion effects are represented by $\langle r^2 \rangle_0^{\frac{3}{2}}$ and $N^2\beta$ respectively, the latter quantity being the total excluded volume of one polymer coil for another. For small z, the following approximations have been derived[131-133]

$$\alpha_s^2 = 1 + (134/105)z \tag{55}$$

$$\alpha_r^2 = 1 + (4/3)z \tag{56}$$

Hence $\alpha_r > \alpha_s$, revealing that the expanded chain is non-Gaussian and that the relationship between $\langle s^2 \rangle$ and $\langle r^2 \rangle$ does not take the simple form of equation (53). In Flory–Fox theory $\alpha = \alpha_\eta = \alpha_s = \alpha_r$ and β is given by

$$\beta = 2V_1(\tfrac{1}{2} - \chi)/N_A \tag{57}$$

Equations (52), (54) and (57) enable reduction of equation (51) to the following form for small z

$$\alpha^2 = 1 + (3^{\frac{3}{2}}/2)z \tag{58}$$

whereby it can be seen that Flory–Fox theory overestimates the coefficient of z by a factor of about two.

Yamakawa and Kurata[134,135] derived an approximate expression for α_η based upon Kirkwood–Riseman theory in the non-draining limit. For small z this may be written in the closed form

$$\alpha_\eta^3 = 1 + 1.55z \tag{59}$$

Further simplification affords comparison with equations (55) and (56), and gives $\alpha_\eta^2 = 1 + 1.03z$. Hence α_η is considerably smaller than α_s and by combining equations (55) and (59) the relation $\alpha_\eta^3 = \alpha_s^{2.43}$ is obtained.

9.5.1.2 Theories for rod-like molecules

Equation (40) is also applicable to rod-like molecules for which $\langle s^2 \rangle_0 \propto x^2 \propto M^2$. Thus, free-draining theory predicts

$$[\eta]_{\dot\gamma=0} = F_R M^2 \tag{60}$$

where F_R is a constant for a given polymer/solvent system.

By analogy with the treatment of flexible chains, a rod-like molecule may be modelled as an equivalent impermeable prolate ellipsoid. Einstein's theory for impermeable spheres was extended to randomly oriented ellipsoids of revolution by Simha[136] who obtained the expression

$$\eta_{sp} = \omega\phi_e \tag{61}$$

where ϕ_e is the volume fraction of ellipsoids, and for prolate ellipsoids ω is given by

$$\omega = \frac{14}{15} + \frac{p^2}{5}\left\{\left[3\ln(2p) - \frac{9}{2}\right]^{-1} + \left[\ln(2p) - \frac{1}{2}\right]^{-1}\right\} \tag{62}$$

where $p\,(> 10)$ is the ratio of the major to the minor axis of the ellipsoid. For $p > 50$, equation (62) approximates to $\omega = p^2/\ln p$ and can be applied to rod-like molecules of length L and diameter d. The ratio (p_e) for the equivalent prolate ellipsoid is obtained assuming equivalence of length and volume

$$p_e = \left(\frac{2}{3}\right)^{\frac{1}{2}}\left(\frac{L}{d}\right) \tag{63}$$

Since $\phi_e = V_e N_A c/M$, where V_e is the volume of the equivalent prolate ellipsoid, and $V_e \propto L \propto M$, equation (61) can be transformed into

$$[\eta]_{\dot\gamma=0} = K_e\left(\frac{M^2}{\ln M}\right) \tag{64}$$

where K_e is a constant for a given polymer. This equation can be rewritten in the form

$$[\eta]_{\dot\gamma=0} = K_e M^a \tag{65}$$

where $1.72 \lesssim a \lesssim 1.83$ for the corresponding range $10^3 \leq M \leq 10^7$.

Equations (60) and (64) strictly are only applicable to fully extended rigid molecules for which excluded volume effects are absent. Thus, the exponents of M in equations (60) and (65) are independent of solvent. The worm-like chain model[137-139] is more appropriate for stiff chains which are not fully extended and allows for both free-draining and non-draining contributions.[140-144] This yields more complex equations which reduce to the form of equation (64) for fully extended rigid molecules, and to that of equation (49) for flexible Gaussian chains.

9.5.2   Evaluation of Molecular Weight

The following empirical equation was proposed by several workers[145-148] and is known by various combinations of their names, though most commonly as the Mark–Houwink equation

$$[\eta] = K\bar{M}_v^a \tag{66}$$

where $\bar{M}_v$ is the viscosity average molecular weight, and K and a are constants for a given polymer/solvent/temperature system. Generally, $0.5 \leq a \leq 0.8$ for flexible chains, $0.8 \lesssim a \lesssim 1.0$ for inherently stiff molecules (*e.g.* cellulose derivatives, DNA) and $1.0 \lesssim a \lesssim 1.7$ for highly extended chains (*e.g.* polyelectrolytes in solutions of very low ionic strength). The value of K tends to decrease as a increases, and for flexible chains it is typically in the range 10^{-3} to 10^{-1} cm^3 g^{-1}. Rearrangement of equation (66) allows evaluation of $\bar{M}_v$ from $[\eta]$ provided that K and a are known for the system under study.

Since specific viscosities are additive in the limit of infinite dilution, the weight average intrinsic viscosity is obtained for a polydisperse polymer and $\bar{M}_v$ is defined by

$$\bar{M}_v = \left[\frac{\Sigma w_i M_i^a}{\Sigma w_i}\right]^{\frac{1}{a}} \tag{67}$$

where w_i is the total weight of molecules of molecular weight M_i. Thus $\bar{M}_v$ is closer to the weight average $\bar{M}_w$ than to the number average $\bar{M}_n$ molecular weight, and is equal to $\bar{M}_w$ when $a = 1$.

The most widely used method for evaluation of K and a involves measurement of $[\eta]$ for a series of polymer samples with known $\bar{M}_n$ or $\bar{M}_w$. Ideally the samples should have narrow molecular weight distributions (MWDs), but if not, their MWDs should have the same functional form (*e.g.* most probable) and $\bar{M}_w$ should be used in preference to $\bar{M}_n$. The latter procedure yields equations which are similar to equation (66) but in which $\bar{M}_v$ is replaced by either $\bar{M}_w$ or $\bar{M}_n$. It is possible to obtain K and a for equation (66) (*i.e.* in respect of $\bar{M}_v$) from $\bar{M}_w$ and $\bar{M}_n$ provided that the distribution function is known,[169] and this practice is preferable when using calibration samples with broad MWDs.

Generally a plot of $\log[\eta]$ against $\log M$ is fitted to a straight line from which K and a are determined. Theoretically this plot should not be linear over a wide range of M, so that K and a values should not be used for polymers with M outside the range defined by the calibration samples. In practice, such plots are essentially linear over wide ranges of M, though curvature at low M is often observed (Figure 9).[60, 149-153] In the low M region it is common to find $a \approx 0.5$ for flexible chains despite the use of good solvents and in contradiction with the anticipated free-draining behaviour as indicated by equation (41). The failure of equations (41) and (49) to predict behaviour at low M is a consequence of the non-Gaussian character of short flexible chains. Thus, theories for worm-like chains must be employed in order to fully interpret $[\eta]$–M data at low M.[154]

Many methods have been proposed for evaluation of K and a values from gel permeation chromatography measurements.[155-164] They are of particular interest when a range of calibration samples is unavailable, since many of them require only one or two well-characterized broad MWD samples. Other procedures for evaluation of K and a have been proposed.[165-167]

Comprehensive lists of K and a values are available,[149, 168, 169] so that calibration is unnecessary for many polymers. However, when selecting literature values attention must be given to the nature of the calibration samples, in particular their range of M and the form of their MWDs, and also to the method of calibration. In general, K and a show only a slight dependence upon temperature

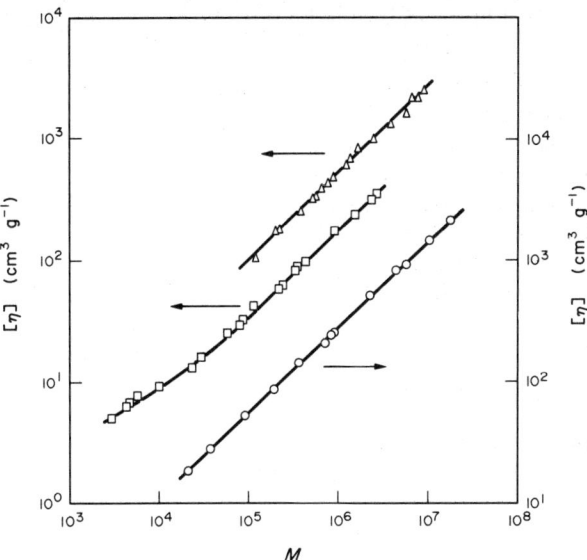

Figure 9 Molecular weight dependence of intrinsic viscosity: $\bigcirc$, experimental points, taken from ref. 60, for solutions of polystyrene in tetrahydrofuran at 25 °C; $\square$, experimental points, taken from ref. 178, for solutions of poly(methyl methacrylate) in benzene at 30 °C; and $\triangle$, experimental points, taken from ref. 209, for solutions of poly(D-β-hydroxy-butyrate) in chloroform at 30 °C

when $a > 0.7$ and may then be used at temperatures slightly different from that at which they were determined.[75,169]

9.5.3 Evaluation of Unperturbed Dimensions and Polymer–Solvent Interaction Parameters

Measurement of $[\eta]$ for a series of samples of known M and narrow MWD also enables evaluation of K_θ and an interaction parameter B which is related to β by

$$B = \beta/C^2 M_0^2 \tag{68}$$

in which

$$C = \frac{\langle r^2 \rangle_0}{nl^2} \quad \text{and} \quad M_0 = \frac{M}{n} \tag{69}$$

where n is the number of real bonds of length l in the chain. The number of random flight equivalent bonds $N = n/C$, so that for Flory–Fox theory

$$B = 2(\tfrac{1}{2} - \chi)\bar{v}^2/V_1 N_A \tag{70}$$

Thus it is possible to evaluate $\langle s^2 \rangle_0^{\frac{1}{2}}$ for any M by assuming a value for Φ_0^s in equation (50), and to evaluate χ from either equation (70) or other similar equations for B.

Most methods for evaluation of K_θ and B are for flexible chains and are based upon equations derived by combining equation (49) with closed expressions for α_η. Flory, Fox and Schaefgen[170] proposed a method based upon equation (51), but this generally yields underestimates of K_θ and is not recommended.[19,168,171] An improved method was proposed by Kurata and Stockmayer,[168] though this requires successive approximations and offers no advantage over the much simpler method of Burchard,[172] and Stockmayer and Fixman.[173] The latter method is based upon equation (59) from which the following expression is derived

$$[\eta]/M^{\frac{1}{2}} = K_\theta + 0.51\Phi_0^r B M^{\frac{1}{2}} \tag{71}$$

where $\Phi_0^r = \Phi_0^s/6^{\frac{3}{2}}$. Thus, $[\eta]/M^{\frac{1}{2}}$ is plotted against $M^{\frac{1}{2}}$ yielding K_θ as the ordinate intercept and B from the slope (Figure 10). Provided that $\alpha_\eta < 1.5$ such plots are linear and give reliable estimates of

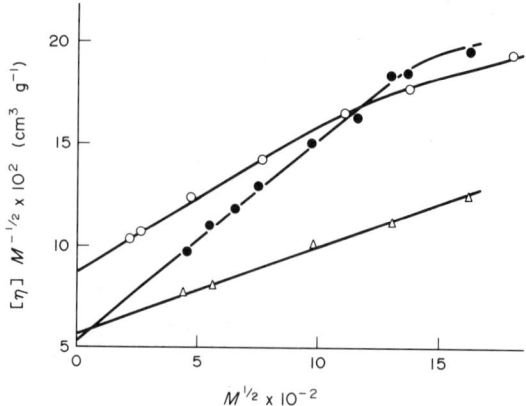

Figure 10 Burchard–Stockmayer–Fixman plots. Experimental points, taken from ref. 171, are for: ●, solutions of poly(ethyl methacrylate) in methyl ethyl ketone at 23 °C; ○, solutions of polystyrene in toluene at 35 °C; and △, solutions of poly(methyl methacrylate) in methyl isobutyrate at 34 °C

K_θ.[19,171] However, deviations from linearity occur at high M for good polymer–solvent pairs, as is also observed for Kurata–Stockmayer plots.[19,171] Such data are best plotted according to the method of Inagaki, Suzuki and Kurata,[174] which is only applicable for $\alpha_\eta > 1.4$.[19,171] When applied correctly, the K–S, B–S–F and I–S–K methods generally yield K_θ values which are in close agreement but give disparate values for B as a consequence of their use of different expressions for α_η. Thus interaction parameters evaluated from $[\eta]$ data must be treated with some degree of caution. Theoretical and experimental evidence[19] suggest that equation (59) is incorrect and that a better expression for $\alpha_\eta < 1.6$ is

$$\alpha_\eta^3 = 1 + 1.05z \qquad (72)$$

Accordingly the numerical constant in equation (71) should be changed from 0.51 to 0.346.

Short flexible chains and stiff chains should not be treated using the methods described above. Methods based upon the theories of worm-like chains are appropriate but difficult to apply. Plots according to these theories have been proposed and may be used to evaluate parameters which characterize worm-like chains.[12,144,154,175,176] A deficiency of these theories, however, is their neglect of excluded volume effects.

A simple empirical equation has been proposed for estimation of K_θ.[177,178] This was shown to be applicable to flexible chain polymers over a wide range of M, and also to worm-like polymers.[177-181] Very simple equations, which appear to give sensible results, have also been proposed for estimation of K_θ, χ and solubility parameters from values of the Mark–Houwink constants K and a.[167,182]

Each of the methods described above requires a value for Φ_0^r so that the necessary calculations can be performed. In Flory–Fox theory Φ_0^r is assumed to be a universal constant equal to 2.87×10^{23} mol^{-1} when M has units of mass per mol. However, other theories give different values for Φ_0^r.[19,183] Furthermore, Φ_0^r applies strictly to non-draining Gaussian chains, so that its value decreases as molecular weight decreases and as chain expansion increases. Equations which give approximate corrections for the latter effect have been proposed.[184,185] However, for good polymer–solvent pairs a suitable experimental value for Φ_0^r is 2.5×10^{23} mol^{-1}.

9.5.4 Branched Polymers

The effect of branching is to increase the segment density within the molecular coil. Thus a branched molecule occupies a smaller volume and has a lower intrinsic viscosity than a similar linear molecule of the same molecular weight. The degree of branching is often characterized in terms of the branching factors in equations (73)–(75), where the subscripts B and L refer to branched and linear polymers of the same molecular weight respectively.

$$g_0 = \frac{\langle s^2 \rangle_{0,B}}{\langle s^2 \rangle_{0,L}} \tag{73}$$

$$g = \frac{\langle s^2 \rangle_B}{\langle s^2 \rangle_L} \tag{74}$$

$$g' = \frac{[\eta]_B}{[\eta]_L} \tag{75}$$

In order to estimate g' it is necessary to measure $[\eta]_B$ and $\bar{M}_{w,B}$ (or $\bar{M}_{n,B}$). Also a Mark–Houwink relationship of the linear polymer must be known for the conditions employed in the measurement of $[\eta]_B$, and is used to calculate $[\eta]_L$ from $\bar{M}_{w,B}$ (or $\bar{M}_{n,B}$), including any necessary corrections for polydispersity. Ideally g_0 should be evaluated, because it corresponds to the effects of branching upon molecular dimensions in the absence of volume exclusion effects, and it is related to the extent of branching by equations derived from simple models.[186] Therefore it is preferable to measure $[\eta]_B$ using poor polymer–solvent pairs under conditions which are close to theta conditions. The simple relationship $g' = g^{\frac{3}{2}} = g_0^{\frac{3}{2}}$ is commonly quoted but is unsatisfactory,[187-191] requiring Φ_0^s and α_η to be independent of branching. The latter assumption is not unreasonable, since the ratio $\alpha_{\eta,B}/\alpha_{\eta,L}$ has been found close to unity for a number of polymers.[191-196] A theoretical study of star-shaped branched polymers by Zimm and Kilb[197] suggested the relation $g' = g_0^{\frac{1}{2}}$ which is often obeyed by polymers with a low degree of long chain branching and also under theta conditions.[188,190,198,199] In general, the relation $g' = g_0^b$ applies where $0.5 \leq b \leq 1.5$ and b increases with the degree of long chain branching. The determination of the value of b represents the major limitation in the use of $[\eta]_B$ for studying the skeletal structures of branched polymers.

9.5.5 Copolymers

The presence of a second type of repeat unit causes the dilute solution behaviour of copolymers to be far more complex than that of homopolymers. Copolymer composition and composition distribution, and the sequence length distributions of the two co-units are the most obvious additional factors which affect the intrinsic viscosities of copolymers. However, interactions between unlike chain segments, and preferential interaction of solvent molecules with one of the co-units, are also of considerable importance. The use of a solvent which is a good solvent for both of the corresponding homopolymers is strongly recommended. This is particularly important for block and graft copolymers since they undergo aggregation/micellization when the solvent is a good solvent for one homopolymer but a poor solvent for the other.[200,201] Mixed solvents have been used but cannot be recommended because they introduce further complications with regard to preferential interactions.

The Mark–Houwink equation is applicable to all classes of linear copolymers provided that the solvent is chosen carefully. However, calibration requires samples with similar average compositions, composition distributions, sequence length distributions and molecular weight distributions. Calibration is further complicated by the difficulties in measuring $\bar{M}_w$ for copolymers (see Chapters 3 and 5 of this volume), so that $\bar{M}_n$ are often used instead. This is satisfactory for copolymers with narrow MWDs but can lead to significant errors when the MWD is broad (see Section 9.5.2). When evaluating $[\eta]$ for block copolymers the extrapolation plot should be inspected for curvature at higher concentrations since this is an indication of aggregation, especially if the slope corresponds to a value for the Huggins constant $k' > 0.5$.[93,202] Use of the Mark–Houwink equation for non-linear copolymers is further complicated by the effects described for branched homopolymers.

The Burchard–Stockmayer–Fixman plot (see Section 9.5.3) has been applied to linear copolymers with some success, provided that the solvent was selected with care. In performing the subsequent calculations, Φ_0^r values for homopolymers can be used without introducing significant errors.[12,200] In addition, the B–S–F plot has been recommended for highlighting anomalies in $[\eta]$–M relationships which are not evident in logarithmic plots of the Mark–Houwink equation.[12]

In view of the complications associated with interpretation of $[\eta]$ data for copolymers, the results obtained must be treated with some degree of caution even when linear plots are observed. Dilute solution viscometry, however, has an additional use for the study of block and graft copolymers, since it provides a simple means of investigating aggregation/micellization phenomena in selective solvents.[200,201,203]

9.5.6 Polyelectrolytes

As a consequence of their poor solubility in organic solvents, polyelectrolytes are most commonly studied as solutions in aqueous media. When completely, or partially, neutralized polyelectrolytes are studied using deionized water as the solvent, a massive increase in specific viscosity is observed in the region of low concentrations (Figure 11). This unique behaviour of polyelectrolytes arises from variations in the degree of dissociation of their ionizable side groups (see Chapter 11 of this volume). Fuoss and Strauss[204,205] found that extrapolation could be performed according to the emipirical equation

$$\left(\frac{\eta_{sp}}{c}\right)^{-1} \;=\; [\eta]^{-1} \;+\; \Upsilon[\eta]^{-1}c^{\frac{1}{2}} \tag{76}$$

where Υ is a constant. The value of $[\eta]$ so obtained corresponds to a rod-like shape for the molecule and is sensitive to the shear rate.

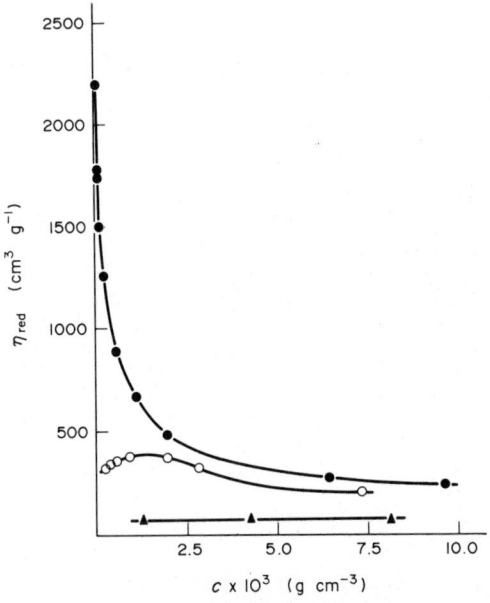

Figure 11 Variation of reduced viscosity with concentration for aqueous solutions of poly(4-vinyl-N-butylpyridinium bromide). Experimental points, taken from ref. 204, are for solutions in: ●, water; ○, 0.001 mol dm^{-3} potassium bromide; △, 0.034 mol dm^{-3} potassium bromide

A more acceptable procedure for evaluation of $[\eta]$ involves the suppression of the effects of ionization by use of an aqueous solution of an inert 1:1 electrolyte as the solvent. The concentration of inert electrolyte is usually maintained constant at 0.1 mol dm^{-3} or above, with theta conditions often being attained at concentrations of the order of 1 mol dm^{-3} for polyelectrolytes with non-polar backbones. In such solvents polyelectrolytes behave as if they were neutral polymers and can be treated accordingly, although, strictly speaking, extrapolation to infinite ionic strength is required. Evaluation of Mark–Houwink constants requires greater care than for simple homopolymers because the effects of ionization must also be suppressed in the measurement of $\bar{M}_w$ or $\bar{M}_n$, and because the constants K and a vary with the degree of neutralization of the polyelectrolyte.[206] Burchard–Stockmayer–Fixman plots are applicable and generally give satisfactory results.[206-208]

9.6 REFERENCES

1. P. J. Flory, 'Principles of Polymer Chemistry', Cornell University Press, Ithaca, New York, 1953.
2. H. Tompa, 'Polymer Solutions', Butterworths, London, 1956.

3. C. Tanford, 'Physical Chemistry of Macromolecules', Wiley, New York, 1961.
4. Ch'ien Jên-yüan, 'Determination of Molecular Weights of High Polymers', Israel Programme for Scientific Translations, Jerusalem, 1963.
5. H. Morawetz, 'Macromolecules in Solution', 2nd edn., Wiley, New York, 1975.
6. H.-G. Elias, 'Macromolecules', Wiley, New York, 1977, vol. 1.
7. N. C. Billingham, 'Molar Mass Measurements in Polymer Science', Kogan Page, London, 1977.
8. P. F. Onyon, in 'Techniques of Polymer Characterization', ed. P. W. Allen, Butterworths, London, 1959, p. 171.
9. S. R. Rafikov, S. A. Pavlova and I. I. Tverdokhlebova, 'Determination of Molecular Weights and Polydispersity of High Polymers', Israel Programme for Scientific Translations, Jerusalem, 1964.
10. W. R. Moore, in 'Progress in Polymer Science', ed. A. D. Jenkins, Pergamon Press, Oxford, 1967, vol. 1, p. 1.
11. J. B. Kinsinger, in 'Encyclopaedia of Polymer Science and Technology', Wiley, New York, 1971, vol. 14, p. 717.
12. M. Bohdanecký and J. Kovár, 'Viscosity of Polymer Solutions', Elsevier, Amsterdam, 1982.
13. Shih-I Pai, 'Viscous Flow Theory I — Laminar Flow', Van Nostrand, New York, 1956.
14. J. R. Van Wazer, J. W. Lyons, K. Y. Kim and R. E. Colwell, 'Viscosity and Flow Measurement', Interscience, New York, 1963.
15. G. Barr, 'A Monograph of Viscometry', Oxford University Press, London, 1931.
16. A. C. Merrington, 'Viscometry', Arnold, London, 1949.
17. A. Dinsdale and F. Moore, 'Viscosity and its Measurement', Chapman and Hall, London, 1962.
18. D. K. Carpenter, in 'Encyclopaedia of Polymer Science and Technology', Wiley, New York, 1970, vol. 12, p. 627.
19. H. Yamakawa, 'Modern Theory of Polymer Solutions', Harper and Row, New York, 1971.
20. See chaps. 3 and 7 of vol. 2 of this work.
21. IUPAC, *J. Polym. Sci.*, 1952, **8**, 257.
22. C. H. Bamford and M. J. S. Dewar, *Proc. R. Soc. London, Ser. A*, 1949, **197**, 356.
23. Z. Priel, M. Sasson and A. Silberberg, *Rev. Sci. Instrum.*, 1973, **44**, 135.
24. A. Nakajima, F. Hamoda and S. Hayashi, *J. Polym. Sci., Polym. Symp.*, 1966, **15**, 285.
25. A. Harness, *J. Polym. Sci.*, 1956, **19**, 591.
26. A. Nasini and C. Mussa, *Makromol. Chem.*, 1957, **22**, 59.
27. J. Schurz and E. H. Immergut, *J. Polym. Sci.*, 1952, **9**, 279.
28. L. H. Cragg and H. van Oene, *Can. J. Chem.*, 1961, **39**, 203.
29. S. M. Shawki and A. E. Hamielec, *J. Appl. Polym. Sci.*, 1979, **23**, 3323.
30. V. Desreux and J. Bischoff, *Bull. Soc. Chim. Belg.*, 1950, **59**, 93.
31. P. Gramain and R. Libeyre, *J. Appl. Polym. Sci.*, 1970, **14**, 383.
32. T. W. Mai, C. E. Foverskov and K. Bak, *Acta Chem. Scand.*, 1970, **24**, 1307.
33. R. B. Simpson, J. S. Smith and H. M. N. H. Irving, *Analyst (London)*, 1971, **96**, 550.
34. J. S. Smith, H. M. N. H. Irving and R. B. Simpson, *Analyst (London)*, 1970, **95**, 743.
35. American Society for Testing Materials, 'ASTM Standards', ASTM D445-53T.
36. Institute of Petroleum, 'Standard Methods for Testing Petroleum and its Products', I. P. Method 71/61.
37. R. C. Weast, 'Handbook of Chemistry and Physics', CRC Press, Cleveland, 1988.
38. G. D. Wedlake, J. H. Vera and G. A. Ratcliff, *Rev. Sci. Instrum.*, 1979, **50**, 93.
39. T. G. Fox, J. C. Fox and P. J. Flory, *J. Am. Chem. Soc.*, 1951, **73**, 1901.
40. W. R. Krigbaum and P. J. Flory, *J. Polym. Sci.*, 1953, **11**, 37.
41. L. J. Sharman, R. H. Sones and L. H. Cragg, *J. Appl. Phys.*, 1953, **24**, 703.
42. F. Bueche, *J. Chem. Phys.*, 1954, **22**, 1570.
43. M. A. Golub, *Can. J. Chem.*, 1957, **35**, 381.
44. C. A. F. Tuijnman and J. J. Hermans, *J. Polym. Sci.*, 1957, **25**, 385.
45. G. M. Guzman and J. M. G. Fatou, *J. Polym. Sci.*, 1959, **35**, 441.
46. E. Passaglia, J. T. Yang and N. J. Wegemer, *J. Polym. Sci.*, 1960, **47**, 333.
47. R. J. Valles, *Polym. Prepr., Am. Chem. Soc., Div. Polym. Chem.*, 1965, **6**, 1041.
48. S. Claesson and U. Lohmander, *Makromol. Chem.*, 1961, **44–46**, 461.
49. S. Newman, L. Loeb and C. M. Conrad, *J. Polym. Sci.*, 1953, **10**, 463.
50. U. Lohmander, *Makromol. Chem.*, 1964, **72**, 159.
51. M. Moan and C. Wolff, *Eur. Polym. J.*, 1973, **9**, 1085.
52. H. H. Neidlinger, G. S. Chen and C. L. McCormick, *J. Appl. Polym. Sci.*, 1984, **29**, 713.
53. W. Kuhn and H. Kuhn, *Helv. Chim. Acta*, 1945, **28**, 1533.
54. M. Čopič, *J. Chim. Phys. Phys.-Chim. Biol.*, 1956, **53**, 440.
55. R. Cerf, *Adv. Polym. Sci.*, 1959, **1**, 382.
56. A. Peterlin, *J. Chem. Phys.*, 1960, **33**, 1799.
57. S. P. Burow, A. Peterlin and D. T. Turner, *Polymer*, 1965, **6**, 35.
58. U. Bianchi and A. Peterlin, *J. Polym. Sci., Polym. Phys. Ed.*, 1968, **6**, 1011.
59. U. Bianchi and A. Peterlin, *Eur. Polym. J.*, 1968, **4**, 515.
60. G. Meyerhoff and B. Appelt, *Macromolecules*, 1979, **12**, 968.
61. E. Unsal, J. L. Duda and E. Klaus, *ACS Symp. Ser.*, 1979, **91**, 141.
62. R. E. Harrington and B. H. Zimm, *J. Phys. Chem.*, 1965, **69**, 161.
63. D. E. Moore and A. G. Parts, *Polymer*, 1968, **9**, 52.
64. D. J. Streeter and R. F. Boyer, *J. Polym. Sci.*, 1954, **14**, 5.
65. S. L. Kapur and S. Gundiah, *J. Colloid Sci.*, 1958, **13**, 170.
66. O. E. Öhrn, *J. Polym. Sci.*, 1955, **17**, 137.
67. M. Takeda and R. Endo, *J. Phys. Chem.*, 1956, **60**, 1202.
68. M. M. Huque, M. Fishman and D. A. I. Goring, *J. Phys. Chem.*, 1959, **63**, 766.
69. H. van Oene and L. H. Cragg, *Nature (London)*, 1961, **191**, 1160.
70. Z. Priel and A. Silberberg, *Pap. Meet. — Am. Chem. Soc., Div. Org. Coat. Plast. Chem.*, 1972, **30**, 556.
71. R. Varoqui and P. Dejardin, *J. Chem. Phys.*, 1977, **66**, 4395.
72. Z. Priel and A. Silberberg, *J. Polym. Sci., Polym. Phys. Ed.*, 1978, **16**, 1917.

73. P. J. Barham and A. Keller, 'On the Validity of the Mark–Houwink Relationship at High Molecular Weights', a paper presented at the meeting on Physical Aspects of Polymer Science, Reading, UK, 9–11 Sept. 1987.
74. B. H. Zimm and D. M. Crothers, *Proc. Natl. Acad. Sci. USA*, 1962, **48**, 905.
75. G. C. Berry, *J. Chem. Phys.*, 1967, **46**, 1338.
76. P. Lindner and R. C. Oberthuer, *Colloid Polym. Sci.*, 1985, **263**, 443.
77. M. A. Haney, *US Pat.* 4 463 598 (1984).
78. Viscotek Differential Viscometer, Viscotek Corp., 1030 Russell Drive, Porter, Texas 77365, USA.
79. M. A. Haney, *Polym. Prepr., Am. Chem. Soc., Div. Polym. Chem.*, 1983, **24**, 455.
80. M. A. Haney, *Am. Lab. (Fairfield, Conn.)*, 1985, **17**(3), 50.
81. M. A. Haney, *J. Appl. Polym. Sci.*, 1985, **30**, 3023.
82. M. A. Haney, *Am. Lab. (Fairfield, Conn.)*, 1985, **17**(4), 116.
83. M. A. Haney, *J. Appl. Polym. Sci.*, 1985, **30**, 3037.
84. M. L. Huggins, *J. Phys. Chem.*, 1938, **42**, 911.
85. M. L. Huggins, *J. Phys. Chem.*, 1939, **43**, 439.
86. M. L. Huggins, *J. Am. Chem. Soc.*, 1942, **64**, 2716.
87. E. O. Kraemer, *Ind. Eng. Chem.*, 1938, **30**, 1200.
88. G. V. Schulz and F. Blaschke, *J. Prakt. Chem.*, 1941, **158**, 130.
89. A. F. Martin, American Chemical Society Meeting, Memphis, April 1942.
90. T. Sakai, *J. Polym. Sci., Polym. Phys. Ed.*, 1968, **6**, 1659.
91. H.-G. Elias and O. Etter, *Makromol. Chem.*, 1963, **66**, 56.
92. F. W. Ibrahim, *J. Polym. Sci., Part A*, 1965, **3**, 469.
93. M. Enyiegbulam and D. J. Hourston, *Polymer*, 1979, **20**, 818.
94. C. I. Simionescu, B. C. Simionescu, I. Neamtu and S. Ioan, *Polymer*, 1987, **28**, 165.
95. B. C. Simionescu, S. Ioan, A. Flondor and C. I. Simionescu, *Angew. Makromol. Chem.*, 1987, **152**, 121.
96. B. Ashar and L. R. Turcotte, *J. Test. Eval.*, 1980, **8**, 235.
97. P. M. Reilly, B. M. E. Van der Hoff and M. Ziogas, *J. Appl. Polym. Sci.*, 1979, **24**, 2087.
98. W. Heller, *J. Colloid Sci.*, 1954, **9**, 547.
99. S. H. Maron and R. B. Reznik, *J. Polym. Sci., Polym. Phys. Ed.*, 1969, **7**, 309.
100. K. K. Chee, *J. Appl. Polym. Sci.*, 1979, **23**, 1639.
101. P. F. Lyons and A. V. Tobolsky, *Polym. Eng. Sci.*, 1970, **10**, 1.
102. R. F. Fedors, *Polymer*, 1979, **20**, 225.
103. M. V. Ram Mohan Rao and M. Yaseen, *J. Appl. Polym. Sci.*, 1986, **31**, 2501.
104. O. F. Solomon and I. Z. Ciută, *J. Appl. Polym. Sci.*, 1962, **6**, 683.
105. P. C. Deb and S. R. Chatterjee, *Indian J. Appl. Chem.*, 1968, **31**, 121.
106. T. Gillespie and M. A. Hulme, *J. Appl. Polym. Sci.*, 1969, **13**, 2031.
107. K. K. Chee, *J. Appl. Polym. Sci.*, 1987, **34**, 891.
108. T. D. Varma and M. Sengupta, *J. Appl. Polym. Sci.*, 1971, **15**, 1599.
109. O. F. Solomon and B. S. Gottesman, *Makromol. Chem.*, 1969, **127**, 153.
110. J. R. Nero and S. K. Sikdar, *J. Appl. Polym. Sci.*, 1982, **27**, 4687.
111. M. H. R. Fanood and M. H. George, *Polymer*, 1987, **28**, 2244.
112. V. E. Hart, *J. Polym. Sci.*, 1955, **17**, 215.
113. S. H. Maron, *J. Appl. Polym. Sci.*, 1961, **5**, 282.
114. J. H. Elliott, K. H. Horowitz and T. Hoodock, *J. Appl. Polym. Sci.*, 1970, **14**, 2947.
115. M. G. Wirick and J. H. Elliott, *J. Appl. Polym. Sci.*, 1973, **17**, 2867.
116. A. Rudin and R. A. Wagner, *J. Appl. Polym. Sci.*, 1975, **19**, 3361.
117. H. U. Khan, V. K. Gupta and G. S. Bhargava, *Polym. Commun.*, 1983, **24**, 191.
118. H. U. Khan and G. S. Bhargava, *J. Polym. Sci., Polym. Lett. Ed.*, 1980, **18**, 803.
119. P. Debye, *J. Chem. Phys.*, 1946, **14**, 636.
120. J. J. Hermans, *Physica (Amsterdam)*, 1943, **10**, 777.
121. H. A. Kramers, *J. Chem. Phys.*, 1946, **14**, 415.
122. H. Staudinger and W. Heuer, *Ber. Dtsch. Chem. Ges. B*, 1930, **63**, 222.
123. H. Staudinger and R. Nodzu, *Ber. Dtsch. Chem. Ges. B*, 1930, **63**, 721.
124. P. Debye and A. M. Bueche, *J. Chem. Phys.*, 1948, **16**, 573.
125. J. G. Kirkwood and J. Riseman, *J. Chem. Phys.*, 1948, **16**, 565.
126. P. L. Auer and C. S. Gardner, *J. Chem. Phys.*, 1955, **23**, 1546.
127. A. Einstein, *Ann. Phys. (Leipzig)*, 1906, **19**, 289; 1911, **34**, 591.
128. P. J. Flory, *J. Chem. Phys.*, 1949, **17**, 303.
129. P. J. Flory and T. G. Fox, *J. Polym. Sci.*, 1950, **5**, 745.
130. P. J. Flory and W. R. Krigbaum, *J. Chem. Phys.*, 1950, **18**, 1086.
131. B. H. Zimm, W. H. Stockmayer and M. Fixman, *J. Chem. Phys.*, 1953, **21**, 1716.
132. M. Fixman, *J. Chem. Phys.*, 1955, **23**, 1656.
133. H. Yamakawa and M. Kurata, *J. Phys. Soc. Jpn.*, 1958, **13**, 78.
134. M. Kurata and H. Yamakawa, *J. Chem. Phys.*, 1958, **29**, 311.
135. H. Yamakawa and M. Kurata, *J. Phys. Soc. Jpn.*, 1958, **13**, 94.
136. R. Simha, *J. Phys. Chem.*, 1940, **44**, 25.
137. O. Kratky and G. Porod, *Recl. Trav. Chim. Pays-Bas*, 1949, **68**, 1106.
138. H. E. Daniels, *Proc. R. Soc. Edinburgh, Sect. A*, 1952, **63**, 290.
139. J. J Hermans and R. Ullman, *Physica (Amsterdam)*, 1952, **18**, 951.
140. J. E. Hearst, *J. Chem. Phys.*, 1963, **38**, 1062.
141. J. E. Hearst, *J. Chem. Phys.*, 1964, **40**, 1506.
142. J. E. Hearst and Y. Tagami, *J. Chem. Phys.*, 1965, **42**, 4149.
143. H. Yamakawa and M. Fujii, *Macromolecules*, 1974, **7**, 128.
144. H. Yamakawa and T. Yoshizaki, *Macromolecules*, 1980, **13**, 633.

145. W. Kuhn, *Kolloid-Z.*, 1934, **68**, 2.
146. H. Mark, in 'Der feste Körper', ed. R. Saenger, Hirzel, Leipzig, 1938, p. 103.
147. I. Sakurada, *Kasen-Koenshu*, 1940, **5**, 33.
148. R. Houwink, *J. Prakt. Chem.*, 1940, **157**, 15.
149. G. Meyerhoff, *Adv. Polym. Sci.*, 1961, **3**, 59.
150. C. Rossi, U. Bianchi and E. Bianchi, *Makromol. Chem.*, 1960, **41**, 31.
151. T. Altares, D. P. Wyman and V. R. Allen, *J. Polym. Sci., Part A*, 1964, **2**, 4533.
152. E. Cohn-Ginsberg, T. G. Fox and H. F. Mason, *Polymer*, 1962, **3**, 97.
153. U. Bianchi and A. Peterlin, *J. Polym. Sci., Polym. Phys. Ed.*, 1968, **6**, 1759. .
154. A. Kaštánek and M. Bohdanecký, *Eur. Polym. J.*, 1985, **21**, 1021.
155. A. R. Weiss and E. Cohn-Ginsberg, *J. Polym. Sci., Polym. Lett. Ed.*, 1969, **7**, 379.
156. C. J. B. Dobbin, A. Rudin and M. F. Tchir, *J. Appl. Polym. Sci.*, 1982, **27**, 1081.
157. C. J. B. Dobbin, A. Rudin and M. F. Tchir, *J. Appl. Polym. Sci.*, 1980, **25**, 2985.
158. X. Zhongde, S. Mingshi, N. Hadjichristidis and L. J. Fetters, *Macromolecules*, 1981, **14**, 1591.
159. V. Grinshpun and A. Rudin, *Makromol. Chem., Rapid Commun.*, 1985, **6**, 219.
160. J. M. Peureux and P. Lochon, *Eur. Polym. J.*, 1983, **19**, 565.
161. H. K. Mahabadi, *J. Appl. Polym. Sci.*, 1985, **30**, 1535.
162. Z. Dobkowski, *J. Appl. Polym. Sci.*, 1984, **29**, 2683.
163. A. Horta, E. Sáiz, J. M. Barrales-Rienda and P. A. Galera Gómez, *Polymer*, 1986, **27**, 139.
164. K. Ito and T. Ukai, *Polym. J.*, 1986, **18**, 593.
165. K. K. Chee, *J. Appl. Polym. Sci.*, 1985, **30**, 1323.
166. D. Jadraque and J. M. Pereña, *Makromol. Chem.*, 1985, **186**, 1263.
167. K. K. Chee, *Polymer*, 1987, **28**, 977.
168. M. Kurata and W. H. Stockmayer, *Adv. Polym. Sci.*, 1963. **3**, 196.
169. J. Brandrup and E. H. Immergut (eds.), 'Polymer Handbook', 2nd edn., Wiley Interscience, New York, 1975, sections IV.1 and IV.3.
170. P. J. Flory and T. G. Fox, *J. Am. Chem. Soc.*, 1951, **73**, 1904.
171. J. M. G. Cowie, *Polymer*, 1966, **7**, 487.
172. W. Burchard, *Makromol. Chem.*, 1961, **50**, 20.
173. W. H. Stockmayer and M. Fixman, *J. Polym. Sci., Polym. Symp.*, 1963, **1**, 137.
174. H. Inagaki, H. Suzuki and M. Kurata, *J. Polym. Sci., Polym. Symp.*, 1966, **15**, 409.
175. M. Bohdanecký, *Macromolecules*, 1983, **16**, 1483.
176. P. P. Shah and H. B. Naik, *Eur. Polym. J.*, 1987, **23**, 73.
177. A. Dondos, *Eur. Polym. J.*, 1977, **13**, 829.
178. A. Dondos and H. Benoit, *Polymer*, 1977, **18**, 1161.
179. A. Dondos and G. Staikos, *Eur. Polym. J.*, 1980, **16**, 1215.
180. G. Staikos and A. Dondos, *Eur. Polym. J.*, 1983, **19**, 555.
181. A. Dondos and V. Skordilis, *J. Polym. Sci., Polym. Phys. Ed.*, 1985, **23**, 615.
182. K. K. Chee, *Polym. Commun.*, 1986, **27**, 135.
183. P. J. Flory, 'Statistical Mechanics of Chain Molecules', Interscience, New York, 1969.
184. O. B. Ptitsyn and Y. E. Eizner, *Sov. Phys.. Tech. Phys. (Engl. Transl.)*, 1960, **4**, 1020.
185. J. F. Douglas and K. F. Freed, *Macromolecules*, 1984, **17**, 1854, 2344.
186. B. H. Zimm and W. H. Stockmayer, *J. Chem. Phys.*, 1949, **17**, 1301.
187. C. D. Thurmond and B. H. Zimm, *J. Polym. Sci.*, 1952, **8**, 477.
188. K. Nagasubramanian, O. Saito and W. W. Graessley, *J. Polym. Sci., Polym. Phys. Ed.*, 1969, **7**, 1955.
189. H. Völker and F. J. Luig, *Angew. Makromol. Chem.*, 1970, **12**, 43.
190. J. Pannell, *Polymer*, 1972, **13**, 2.
191. T. Hama, K. Yamaguchi and T. Suzuki, *Makromol. Chem.*, 1972, **155**, 283.
192. T. A. Orofino and F. Wenger, *J. Phys. Chem.*, 1963, **67**, 566.
193. S. Krozer, *Makromol. Chem.*, 1974, **175**, 1905.
194. J. Herz, M. Hert and C. Strazielle, *Makromol. Chem.*, 1972, **160**, 213.
195. N. Hadjichristidis and J. E. L. Roovers, *J. Polym. Sci., Polym. Phys. Ed.*, 1974, **12**, 2521.
196. J. E. L. Roovers and S. Bywater, *Macromolecules*, 1974, **7**, 443.
197. B. H. Zimm and R. W. Kilb, *J. Polym. Sci.*, 1959, **37**, 19.
198. G. C. Berry, L. M. Hobbs and V. C. Long, *Polymer*, 1964, **5**, 31.
199. J. G. Spiro, D. A. I. Goring and C. A. Winkler, *J. Phys. Chem.*, 1964, **68**, 323.
200. J. V. Dawkins, in 'Block Copolymers', ed. D. C. Allport and W. H. Janes, Applied Science, London, 1973, p. 363.
201. C. Price, in 'Developments in Block Copolymers', ed. I. Goodman, Applied Science, London, 1982, vol. 1, p. 39.
202. S. Krause, *J. Phys. Chem.*, 1964, **68**, 1948.
203. G. E. Molau (ed.), 'Colloidal and Morphological Behaviour of Block and Graft Copolymers', Plenum Press, New York, 1971.
204. R. M. Fuoss and U. P. Strauss, *J. Polym. Sci.*, 1948, **3**, 246, 602.
205. R. M. Fuoss and U. P. Strauss, *Ann. N. Y. Acad. Sci.*, 1949, **51**, 836.
206. I. Noda, T. Tsuge and M. Nagasawa, *J. Phys. Chem.*, 1970, **74**, 710.
207. A. Takáhashi and M. Nagasawa, *J. Am. Chem. Soc.*, 1964, **86**, 543.
208. J. S. Tan and S. P. Gasper, *J. Polym. Sci., Polym. Phys. Ed.*, 1975, **13**, 1705.
209. T. Hirosye, Y. Einaga and H. Fujita, *Polym. J.*, 1979, **11**, 819.

10

Sedimentation and Diffusion

PETER M. BUDD

BP Research Centre, Sunbury-on-Thames, UK

10.1 INTRODUCTION

Sedimentation methods can provide much valuable information about macromolecules in solution. They may be employed to investigate, for example, average molecular weights, molecular weight distribution (MWD), hydrodynamic size, solute shape, buoyant density, association of solute molecules, branching and thermodynamic parameters. There are two main types of sedimentation experiment: (i) sedimentation velocity (SV); and (ii) sedimentation equilibrium (SE).

The study of the diffusion of a macromolecule in solution can also give useful information about the size, shape and mass of the solute.

This chapter provides an introduction to the use of sedimentation methods and of diffusion studies in polymer science.

10.2 INSTRUMENTATION FOR SEDIMENTATION STUDIES

The basic instrument required for most sedimentation studies of macromolecules is an analytical ultracentrifuge — a centrifuge equipped with an optical system to enable the sedimentation behaviour of the solute to be observed and recorded. The first centrifuges with optical systems were developed by Svedberg and co-workers in the 1920s.[1-3] Since then, sedimentation methods have been extensively applied to biological systems, and have also proved useful for the study of synthetic polymers. There have been several reviews of the theory[4-6] and practice[7-11] of analytical ultracentrifugation.

In recent years the most widely used analytical ultracentrifuge has been the Beckman (Spinco) Model E, an instrument which is no longer manufactured, though many are still in operation. There is no purpose-built analytical ultracentrifuge on the market at present, but some modern preparative ultracentrifuges can be equipped with optical systems.

A solution to be centrifuged is contained within a cavity in a cell centrepiece (Figure 1). The cavity is generally sector-shaped so that there is no restriction on sedimentation in a radial direction, and there is usually a second cavity to hold pure solvent as a reference. Various specialized centrepieces have been designed, including some which enable solvent to be layered onto a solution to form a sharp 'synthetic' boundary.[12]

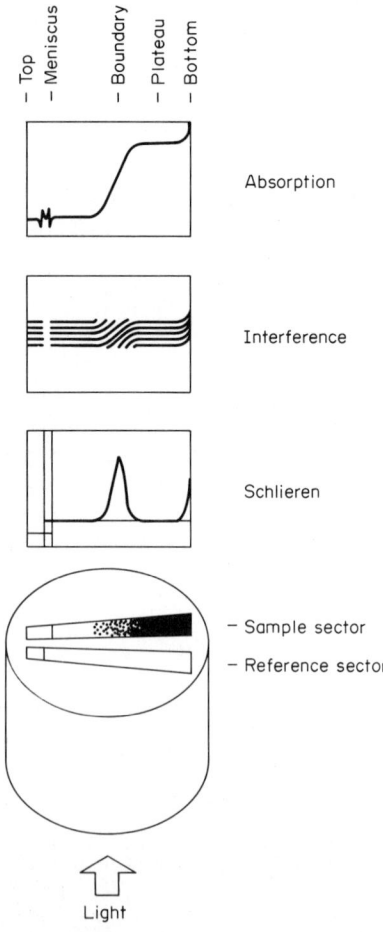

Figure 1   Illustration of a 'double-sector' centrepiece and of results obtained with absorption, interference and Schlieren optical systems; the condition shown is typical of an SV type of experiment

The centrepiece is sandwiched between two transparent windows and, together with various gaskets, held in a cell housing. The whole cell assembly is placed in a rotor such that, when the rotor rotates, the cell crosses the light path of an optical system. A high pressure mercury lamp is often employed as a light source, but some workers have made use of lasers to good effect.[13, 14]

Various types of optical system have been utilized for analytical ultracentrifugation. The absorption, interference and Schlieren systems (Figure 1) are of particular importance.

The absorption system relies on the solute absorbing visible or UV light. Changes in optical density within the cell are recorded photographically or detected by means of a photomultiplier.

The normal, Rayleigh interference system generates a fringe pattern which is related to differences in refractive index between the sample and the reference at each point in the cell. A differential interference system has also been designed.[15]

The Schlieren system gives an image which is proportional to the refractive index gradient at each point in the cell. At low concentrations it is not as precise as the interference system.

In this chapter it is assumed that the solute concentration, c, or the concentration gradient, dc/dr, can be obtained as a function of distance from the centre of rotation, r. The refractometric optical systems give the appropriate information only if every solute component has the same refractive index increment, a condition which usually holds for a polydisperse homopolymer.

10.3 SEDIMENTATION VELOCITY

An SV experiment involves the centrifugation of a solution at a rotor speed sufficiently high for a moving boundary to be seen. Rotor speeds up to 68 000 rev min^{-1} are typical. Schlieren optics are most often employed, in which case a boundary between solute-rich and solute-depleted parts of the cell appears as a peak. For a monodisperse solute a single peak is observed, the width of which depends on the diffusion coefficient of the solute. For a polydisperse solute the shape of the boundary may be related to the MWD. When dealing with mixtures, or if large impurities or aggregate structures are present, multiple peaks may be seen.

10.3.1 Theory

In 1929, Lamm[16] derived, by kinetic reasoning, a differential equation which describes the changes in concentration, c, of a solute with time, t, for a position, r, in a sector-shaped cell. The Lamm equation has since been obtained using irreversible thermodynamics.[5, 6] For a monodisperse, nonelectrolyte solute in a single, incompressible solvent

$$\frac{\partial c}{\partial t} = \frac{1}{r}\frac{\partial}{\partial r}\left[rD\left(\frac{\partial c}{\partial r}\right) - S\omega^2 r^2 c\right] \tag{1}$$

where D is a diffusion coefficient, S is a sedimentation coefficient and ω is the angular velocity. In general S and D are functions of temperature, pressure and concentration. By solving equation (1) for a particular set of conditions the course of an SV experiment may be predicted and procedures for evaluating S and D established. A number of solutions have been given, involving various simplifying assumptions.[5, 6, 17-19] For a multicomponent system an expression similar to equation (1) can be written for each component.

10.3.2 Determination of Sedimentation Coefficient

Svedberg defined S as the rate of movement of a solute in unit field strength

$$S = \frac{dr}{dt}\frac{1}{\omega^2 r} \tag{2}$$

For simple systems S is conveniently determined by plotting the logarithm of the boundary position, $\ln r^*$, against time, t, in which case the slope of the plot is $S\omega^2$. Alternatively S may be evaluated from data obtained at a single experimental time using

$$S = \frac{\ln(r^*/a)}{\omega^2 t} \tag{3}$$

where a is the meniscus position and t is the time from the effective start of the experiment. To account for the period during which the rotor is accelerating, either the effective start is taken as the time the rotor reaches 2/3 final speed, or a 'zero time correction' is employed as an adjustable parameter to fit a set of data.

For a narrow, symmetrical boundary the position of the maximum in the Schlieren image is taken to be $r*$. For broad boundaries it is better, though more time consuming, to use the square root of the second moment, $\bar{r}*$, given by

$$\bar{r}* = \left[\int r^2 \left(\frac{dc}{dr}\right) dr \bigg/ \int \left(\frac{dc}{dr}\right) dr \right]^{1/2} \tag{4}$$

For a polydisperse solute this gives a weight-average S. Sedimentation coefficients are frequently quoted in Svedberg units, one Svedberg being 10^{-13} s.

10.3.3   Concentration Dependence of S

The value of S normally decreases with increase in concentration, and various theories have been suggested to account for this observation.[20-26] For many polymers the variation of S with c may be described by

$$\frac{1}{S} = \frac{1}{S_0}(1 + kc) \tag{5}$$

where S_0 is the limiting sedimentation coefficient and k is a constant which is smaller the poorer the solvent, but is generally finite even for theta conditions.[27,28] For flexible polymers in good solvents the value of k is often close to 1.6 $[\eta]$, where $[\eta]$ is the intrinsic viscosity.[29] The shape of a sedimenting species influences k[30] and shape functions utilizing k have been proposed.[31,32] The influence of solute shape on sedimentation behaviour may be illustrated by a 'hydrodynamically normalized' plot of S_0/S against $c[\eta]$.[33,34] Such a plot shows little sensitivity to molecular weight but the slope is strongly dependent on molecular structure.

An effect of the concentration dependence of S is a sharpening of the boundary, which can be very pronounced for high molecular weight polymers in good solvents.

10.3.4   Limiting Sedimentation Coefficient

A widely used procedure for obtaining S_0 is to extrapolate a plot of $1/S$ against c to $c=0$, in accordance with equation (5). However, it has been suggested that this procedure may give misleading results and that the only justifiable extrapolation is of S against c.[23] Certainly, for strongly concentration-dependent systems great care must be exercized in the choice of extrapolation method, and measurements should be made at the lowest concentrations possible.

The value of S_0 for a nonionic solute is related to its molecular weight, M, its partial specific volume, $\bar{v}$, the density of the solution, ρ, and the limiting molecular translational friction coefficient, f_0, by[35]

$$S_0 = \frac{M(1 - \bar{v}\rho)}{N_A f_0} \tag{6}$$

where N_A is Avogadro's number. The friction coefficient is essentially the same as that which is operative in mutual diffusion[36] and depends on the size and shape of the solute.

10.3.5   Radial Dilution

The solute-rich region of the cell ahead of the boundary in an SV experiment is often referred to as the plateau region. As sedimentation proceeds the concentration in this region decreases, primarily as a result of the sector-shape of the cell cavity. The plateau concentration, $c*$, is related to the initial concentration, c_0, by

$$c* = c_0 \left(\frac{a}{r*}\right)^2 \tag{7}$$

10.3.6 The Johnston–Ogston Effect

On centrifugation of a solution containing two species having different values of S, the apparent concentration of the slow component is higher, and that of the fast component is lower, than when each is investigated alone.[37] This effect was attributed by Johnston and Ogston[38] to the change in S of the slow component at the boundary associated with the fast component. For polydisperse systems the Johnston–Ogston effect leads to a distortion of the boundary.[39]

10.3.7 Pressure Effects

At high rotor speeds large pressure differences, up to several hundred atmospheres, may be generated between the meniscus and the cell bottom. For aqueous systems pressure effects are usually small, but with organic solvents they can lead to S being dependent on position in the cell and on rotor speed.[40] There have been several experimental[41-43] and theoretical[44,45] studies of pressure-dependent sedimentation, and various procedures have been suggested to correct for pressure effects.[46-48] A commonly used practical approach,[49] which is based on a theory for two component systems, relates the experimentally observed S defined by equation (3) to the sedimentation coefficient at atmospheric pressure, S', by

$$S = S'\left\{1 + B\left[\left(\frac{r^*}{a}\right)^2 - 1\right]\right\}$$
(8)

where B includes a pressure-dependent parameter m. A line-fitting procedure is used to evaluate B and S'.

10.3.8 Charge Effects

The sedimentation behaviour of a polyelectrolyte is strongly influenced by charge effects.[50,51] In a solution containing no low molecular weight salt, S is much lower than for an equivalent uncharged polymer. This has been called the 'primary charge effect'[52] and has been attributed to the electric field set up if a charged macromolecule is separated from its counterions. The polyion and its counterions are forced to move together at a rate intermediate between that which each would have alone. Addition of a salt reduces this effect and leads to an increase in S. If the constituent ions of the added salt have different mobilities, a 'secondary charge effect' arises.

Theoretical treatments of charge effects fall into two groups. In the classical approach[52-54] the actual velocity of the polyion is seen as having two components: a contribution from the centrifugal field and a contribution from the internally created electric field. A more rigorous approach[55,56] involves the application of irreversible thermodynamics.

In addition to the 'primary' and 'secondary' charge effects, factors such as chain expansion, the ionic atmosphere and partial drainage may all affect the frictional properties of a polyelectrolyte molecule and hence influence S.[57,58]

10.3.9 Determination of Molecular Weight

Molecular weights may be determined by combining SV and diffusion measurements. For a nonionic, two component system

$$M = \frac{S_0 RT}{D_0(1 - \bar{v}\rho)}$$
(9)

where D_0 is the limiting diffusion coefficient, R is the gas constant and T is the absolute temperature. The treatment of multicomponent systems has been discussed by Eisenberg.[59]

Diffusion coefficients may be determined in the analytical ultracentrifuge,[60] in a classical diffusion cell,[61] or by other techniques such as photon correlation spectroscopy.[62] The sedimentation/diffusion method is capable of high precision.[61] If equation (9) is applied to a polydisperse solute, the average molecular weight obtained depends on the averages of S_0 and D_0 employed.[62-66] For example, combination of a weight-average S_0 and a z-average D_0 gives $\bar{M}_w$.

Combination of S_0 and $[\eta]$ can also yield values for molecular weight. The most common approach utilizes the Mandelkern–Flory relationship[67]

$$M^{2/3} = \frac{S_0 N_A \eta_0 [\eta]^{1/3}}{\Phi^{1/3} P^{-1}(1 - \bar{v}\rho)}$$

(10)

where N_A is Avogadro's number, η_0 is the solvent viscosity and the parameter $\Phi^{1/3} P^{-1}$ generally has a value close to $1.1 \times 10^7 \ \mathrm{mol}^{-1/3}$ (2.5×10^6 if $[\eta]$ is expressed in dl g^{-1}).[68]

In general, the relationship between S_0 and M is of the form

$$S_0 = K M^\alpha$$

(11)

where K and α are constants for a given polymer in a given solvent at a particular temperature over a range of molecular weights. For flexible polymers, α is 0.5 in a theta solvent and less in a good solvent. This relationship has been established for many polymer systems[69] by SV of well-characterized samples. The constants K and α may be evaluated from the corresponding coefficients for diffusion or viscosity.[70]

10.3.10   Determination of Molecular Weight Distribution

For a polydisperse solute, the shape of the boundary depends on the weight distribution of limiting sedimentation coefficients, $g(S_0)$, and is modified by the effects of diffusion, concentration and pressure. Evaluation of $g(S_0)$ enables an MWD to be determined, once the relationship between S_0 and M has been established.

Ignoring, for now, diffusion and concentration effects, but taking account of radial dilution, the distribution of sedimentation coefficients, $g(S)$, is given by

$$g(S) = \frac{1}{c_0}\left(\frac{r}{a}\right)^2 \frac{dc}{dr}\frac{dr}{dS}$$

(12)

In the absence of pressure effects

$$\frac{dr}{dS} = r\omega^2 t$$

(13)

but for a pressure-dependent system it is necessary to use an expression such as[49,71,72]

$$\frac{dr}{dS} = r\omega^2 t\left\{1 - m\left[\left(\frac{r}{a}\right)^2 - 1\right]\right\}$$

(14)

Equation (3) is used to evaluate S.

For a real system one can define an apparent $g(S)$ by equation (12) and then attempt to correct it for diffusion and concentration effects.

If diffusion is significant then a plot of $g(S)$ against S becomes narrower as the experiment proceeds. Diffusional effects on the boundary vary with $t^{1/2}$ whilst effects of polydispersity vary with t. Consequently, extrapolation to infinite time enables a 'diffusion corrected' $g(S)$ to be constructed. At large values of t the apparent $g(S)$ is a linear function of $1/t$.[73-75]

The laboriousness of extrapolation procedures prompted Nekrasov[76-78] to develop a method using model functions to calculate diffusion effects for a single time.

Once a 'diffusion corrected' $g(S)$ has been determined it is still necessary to take account of concentration effects to obtain $g(S_0)$. Usually it is only when working below the theta temperature that the concentration dependence of S is small enough to be ignored.[79] If the relationship between S_0 and the parameter k of equation (5) is known, $g(S_0)$ may be evaluated from data for a single concentration.[77,80] Otherwise, experiments are carried out for several values of c_0 and the 'diffusion corrected' data extrapolated to zero c_0. Several procedures have been proposed for carrying out such an extrapolation,[81-83] none of which has any firm theoretical justification. Empirically, the best method seems to be that of Gralén and Lagermalm,[81] which is to extrapolate to zero c_0 values of S or $1/S$ which correspond to specific values of the integral $\int_0^S g(S)\,dS$.

MWDs obtained by SV generally compare reasonably well with those determined by other techniques.[84-87] However, the procedures mentioned above do not properly correct for the

Johnston–Ogston effect, which leads to a displacement of maxima and a loss of any high molecular weight tail. Nekrasov[88] has proposed a complex computational procedure to take account of this effect.

10.3.11 Branching

Long chain branches on a polymer molecule affect its friction coefficient and hence its sedimentation behaviour. Procedures have been proposed to obtain information on branching by combining studies of SV with those of gel permeation chromatography[89–92] or diffusion.[93] However, for polydisperse samples S can be rather insensitive to degree of branching.[94,95]

10.4 SEDIMENTATION EQUILIBRIUM

A solution subjected to a centrifugal field eventually approaches an equilibrium situation in which the concentration of solute at each point in the cell no longer changes with time. The concentration distribution at SE depends on the MWD of the solute and on thermodynamic interactions with the solvent. An SE experiment generally involves rather lower rotor speeds than those used in SV studies.

10.4.1 Rate of Approach to Equilibrium

Mathematically, true equilibrium is only reached after an infinite time, but a state which is experimentally indistinguishable from equilibrium is reached in a finite time, t_{eq}. Starting with a solution of uniform concentration and operating at constant rotor speed, t_{eq} is inversely proportional to the diffusion coefficient, D, and proportional to $(b-a)^2$, where b and a are the distances from the centre of rotation to the cell bottom and to the meniscus respectively.[96] With a solution column length, $(b-a)$, of around 12 mm, as is commonly used for SV experiments, t_{eq} may be several weeks. For routine use, and in particular for unstable solutes, much shorter experimental times are desirable. Generally solution columns of 3 mm or less are used in SE experiments to shorten t_{eq} to a few days or even a few hours. There are also other ways of reducing t_{eq}. Hexner et al.[97] suggested an overspeeding technique which involves centrifuging at a high speed initially, then reducing the speed when the concentration distribution in the cell is close to that expected for SE at the lower speed. Synthetic-boundary[98] or other specialized centrepieces[99] may also be employed to shorten t_{eq}, provided that the experimental conditions are carefully chosen.[100]

10.4.2 Theory

At first the state of SE was understood in terms of a kinetic picture of sedimentation being balanced by diffusion.[2] In 1953, Goldberg[101] laid the foundations of a thermodynamic theory which provides a rigorous treatment of SE.[5,6] Equations are derived on the basis that, at equilibrium, the sum of the chemical potential and potentials of external forces for each component must be constant throughout the solution.

For a monodisperse, nonelectrolyte macromolecule in a single, incompressible solvent at SE

$$\frac{M(1-\bar{v}\rho)\omega^2 rc}{RT} = \left[1 + c\left(\frac{\partial \ln y}{\partial c}\right)_{T,P}\right]\frac{dc}{dr} \tag{15}$$

where y is an activity coefficient. Equation (15) may be rearranged into various forms suitable for comparison with experimental data.[9,102] For a solute containing many species differing only in molecular weight, such as a typical synthetic homopolymer, expressions similar to equation (15) may be written for each species.[103] On the basis of these expressions, procedures may be established for the evaluation of average molecular weights, MWDs and thermodynamic parameters. The treatment of solutes which are heterogeneous in quantities other than M, such as $\bar{v}$, has been discussed by Kotaka et al.[104] The influence of pressure effects at SE has been considered by Young et al.;[105] however pressure effects are not a serious problem in most SE experiments, as relatively low rotor speeds are employed.

When treating multicomponent systems it is advantageous to utilize a density increment, $(d\rho/dc)$, rather than the buoyancy factor, $(1 - \bar{v}\rho)$.[59,106,107] For a polyelectrolyte in the presence of a low molecular weight salt, or for a polymer in a mixed solvent, the density increment required is that for the condition of constant chemical potential of low molecular weight components, $(d\rho/dc)_\mu$, rather than that of constant concentration, $(d\rho/dc)_{P,m}$. Equilibrium dialysis[59] or isopiestic distillation[108] may be used in evaluating the appropriate increment. When dealing with polyelectrolytes the difference between $(d\rho/dc)_\mu$ and $(d\rho/dc)_{P,m}$ disappears for a salt, such as tetramethylammonium chloride, which itself has zero $(d\rho/dc)$.

10.4.3 Determination of Average Molecular Weights

10.4.3.1 The conventional method

Conventional SE experiments utilize rotor speeds and solution column lengths chosen to give $\lambda \bar{M}_w \approx 1$, where λ is given by

$$\lambda = \frac{(d\rho/dc)(b^2 - a^2)\omega^2}{2RT} \tag{16}$$

An apparent $\bar{M}_w$ may be defined,[110] for a sector-shaped cell, by

$$\bar{M}_w^{app} = \frac{c_b - c_a}{\lambda c_0} \tag{17}$$

where c_b and c_a are the concentrations at the bottom and at the meniscus respectively. With the interference optical system, $(c_b - c_a)$ is the total number of fringes crossed on going from a to b. The value of c_0 is determined in the same optical units by means of a separate, synthetic-boundary experiment or is deduced from a knowledge of the weight concentration, refractive index increment and instrument constants.

For a pseudoideal solution, such as a polymer in a theta solvent, equation (17) gives the true $\bar{M}_w$. For nonideal solutions an extrapolation to zero c_0 is necessary. Fujita[5,6] has shown that

$$\frac{1}{\bar{M}_w^{app}} = \frac{1}{\bar{M}_w} + B(\lambda)c_0 + \ldots \tag{18}$$

Note that the second virial coefficient, $B(\lambda)$, is dependent on the experimental conditions, but the true $\bar{M}_w$ can be determined by extrapolating values of $1/\bar{M}_w^{app}$ obtained for the same λ to zero c_0.[111] If one wishes to extract thermodynamic information from $B(\lambda)$ it is also necessary to extrapolate to zero λ. Expansions for $1/\bar{M}_w^{app}$ in terms of concentration parameters other than c_0 have been proposed.[112,113]

Average molecular weights higher than $\bar{M}_w$ may be determined. For example, $\bar{M}_z$ is obtained from the concentration gradients at the bottom and the meniscus, $(dc/dr)_b$ and $(dc/dr)_a$ respectively, using

$$\bar{M}_z^{app} = \frac{RT}{(d\rho/dc)(c_b - c_a)\omega^2}\left[\frac{1}{b}\left(\frac{dc}{dr}\right)_b - \frac{1}{a}\left(\frac{dc}{dr}\right)_a\right] \tag{19}$$

For nonideal solutions it is necessary to extrapolate $\bar{M}_z^{app}$ to zero concentration to obtain the true $\bar{M}_z$.

10.4.3.2 The short-column method

In a short-column SE experiment[96] the condition $\lambda \bar{M}_w < 1$ is achieved, usually by working with very short solution columns of 1 mm or less. An apparent molecular weight, M_{sc}^{app}, can then be defined by

$$M_{sc}^{app} = \frac{RT}{(d\rho/dc)\omega^2 c_0 r_{mid}}\left(\frac{dc}{dr}\right)_{mid} \tag{20}$$

where $(dc/dr)_{mid}$ is the concentration gradient at the mid-point of the solution column, r_{mid}. For this type of experiment r_{mid} is very close to the hinge-point, the radial position at which the solute concentration at equilibrium equals c_0. A separate evaluation of c_0 is required. If the solution is pseudoideal M_{sc}^{app} is very close to $\bar{M}_w$. For nonideal solutions an extrapolation to zero c_0 is necessary. Advantages of the short-column method are the rapid attainment of equilibrium and, when the Schlieren optical system is employed, the ease of data analysis. This method is not generally as precise as the conventional approach.

10.4.3.3 The meniscus-depletion method

When $\lambda \bar{M}_w \gg 1$ the concentration of solute at the meniscus falls, effectively, to zero. In these circumstances a separate evaluation of c_0 is not necessary as the equivalent information can be obtained by integration of the equilibrium data, so that $\bar{M}_w^{app}$ is given by[114]

$$\bar{M}_w^{app} = \frac{2RT}{(d\rho/dc)\omega^2} \frac{c_b}{\displaystyle\int_a^b c\,d(r^2)} \tag{21}$$

Other average molecular weights may also be obtained, including $\bar{M}_n$. Suzuki has derived virial expansions for $\bar{M}_n$[115] and for higher averages.[116]

Meniscus-depletion experiments have only rarely been applied to polydisperse polymers. Two experimental conditions must be satisfied: (i) zero concentration of all macromolecular species at the meniscus; and (ii) a concentration gradient at the bottom which is not so steep that the optical system cannot resolve concentration changes in that region. With a solution column length of up to 3 mm it is often impossible to satisfy both conditions simultaneously. Longer solution columns starting at a uniform concentration are impractical because of the time required to reach equilibrium; however the use of synthetic-boundary centrepieces[117] to generate long solution columns could make the technique a realistic proposition for many high molecular weight polymers.

10.4.3.4 The variable λ method

The variable λ method involves several experiments conducted at different rotor speeds or with different solution column lengths. The method utilizes the dependence on λ of the equilibrium concentration, or concentration gradient, at one[118] or more[119-121] fixed positions in the solution column. It is possible to determine $\bar{M}_n$, $\bar{M}_w$ and $\bar{M}_z$. The vast amount of experimental work required for the variable λ method prohibits its routine use, but it does provide a theoretically well-grounded route to average molecular weights not easily obtained by other means.

10.4.3.5 Single experiment methods

When dealing with nonideal solutions, the SE methods described above all involve several experiments at various concentrations. If assumptions are made about the thermodynamic behaviour of the solution and about the form of the MWD of the solute, useful information can be extracted from the data of a single experiment. Munk[122-124] has developed single experiment methods for the determination of average molecular weights and of thermodynamic parameters for polymers in good solvents.

10.4.4 Determination of Molecular Weight Distribution

For a pseudoideal, dilute solution the concentration distribution at SE, in a sector-shaped centrepiece, is related to the weight distribution of molecular weights, $g(M)$, by the Fredholm-type integral equation

$$c = c_0 \int_0^\infty \frac{\lambda M g(M)}{\exp(\lambda M) - 1} \exp(\lambda M \xi)\,dM \tag{22}$$

where ξ is given by

$$\xi = \frac{(r^2 - a^2)}{(b^2 - a^2)} \tag{23}$$

There have been a number of attempts to solve equation (22), or similar equations, for g(M). Wales[125] used an expression containing parameters which could be deduced from a polynomial fit to the experimental data. Donnelly[126,127] expressed the observed data as Laplace-transformable functions. Provencher[128] proposed an iterative scheme and Sundelöf[129] suggested a convolution procedure. In 1970 Lee[130] pointed out that the problem of inferring g(M) by inverting equation (22) is 'ill-posed'; that is, minute errors in the experimental data could lead to catastrophically incorrect results. Gehatia and Wiff[131-133] employed regularizing functions to overcome the 'ill-posed' nature of the problem. Jeffcoate and Wragg[134] have proposed two numerical procedures.

The methods mentioned above all attempt to extract g(M) from data for a single λ. Scholte[135,136] has utilized a numerical method employing data for various λ. Adams and co-workers[137,138] have applied the methods of Donnelly and Scholte to nonideal systems and have also derived expressions for non-sector-shaped centrepieces.

There have only been a few examples of the determination of MWDs by SE, and a comprehensive evaluation of the various methods which have been proposed is still awaited.

10.4.5   Sedimentation Equilibrium in a Density Gradient

Centrifugation of a mixed solvent, containing components having different densities, generates a density gradient. At SE a solute collects in a band about the point in the density gradient which corresponds to its buoyant density. The width of the band depends on the molecular weight of the solute, being broader for lower M. The density gradient technique can be employed qualitatively to distinguish between species having different partial specific volumes[139,140] and can be useful for the detection of microgel.[139] Quantitatively, it may be used for the determination of molecular weight[139,141,142] and, for some copolymers, in compositional analysis.[143] Density gradient SE has also been utilized in studies of preferential absorption.[144]

10.4.6   Osmosedimentation

Galembeck and co-workers[145-147] have employed a form of SE experiment in which the solution is separated from pure solvent by a semipermeable membrane. Solvent can flow out of the solution column at the bottom and in at the top of the cell, and equilibrium is attained much more rapidly than is normal in SE experiments. So far, they have used large cells, under gravity or in preparative centrifuges, and have determined the concentration distribution by taking aliquots and measuring the solution density or optical density.

10.5   THE ARCHIBALD METHOD

Archibald[148] found that the ratio S/D can be determined from measurements of c and (dc/dr) at the meniscus, or the bottom, in the early stages of centrifugation. For the meniscus position, a, for example,

$$\frac{S}{D} = \frac{1}{\omega^2 a c_a}\left(\frac{dc}{dr}\right)_a \tag{24}$$

Values of S/D obtained in this way generally have to be extrapolated to $t=0$ and to $c_0=0$ to obtain S_0/D_0.[149,150] Application of equation (9) then enables $\bar{M}_w$ to be calculated.

Archibald's procedure is sometimes called the 'approach to equilibrium' method because the experimental conditions employed are similar to those for SE but the data utilized are for the time before equilibrium is attained. The method underwent a period of popularity because of its relative speed. However, it can be very difficult to obtain the appropriate data with precision, and SE methods are normally to be preferred.

10.6   DIFFUSION

If a concentration gradient is established in a solution, it will tend to relax at a rate governed by a mutual diffusion coefficient, D. Mutual diffusion, and in particular classical methods of determining D, are discussed below. A measure of D is also provided by photon correlation spectroscopy.

The translational mobility of a molecule in solution may be described in terms of a self diffusion coefficient, D_s, which can be determined by tracer methods[151,152] or by nuclear magnetic resonance techniques.[36,153] Generally, D_s equals D only in the limit of infinite dilution.

As well as translational movements, a macromolecule in solution undergoes rotational motions. Techniques which have been employed to study rotational diffusion, mainly of rigid macromolecules, include flow[154] and electric[155] birefringence, electric field light scattering,[156] dielectric relaxation[157] and NMR.[158]

10.6.1 Theory

More than a century ago, Fick[159] deduced that, for a two component system, translational diffusion in a concentration gradient (dc/dx) is described by

$$J = -D\left(\frac{dc}{dx}\right) \tag{25}$$

where J is the flow of particles per second across unit area of a plane perpendicular to the direction of the concentration gradient. Equation (25) is often referred to as 'Fick's first law'. The change in concentration with time, (dc/dt), at a given point x, is given by

$$\left(\frac{dc}{dt}\right) = \frac{d}{dx}\left(D\frac{dc}{dx}\right) \tag{26}$$

which, if D is independent of concentration, reduces to 'Fick's second law'

$$\left(\frac{dc}{dt}\right) = D\frac{d^2c}{dx^2} \tag{27}$$

10.6.2 Classical Determination of D

10.6.2.1 Free diffusion method

In the free diffusion method of determining D, a sharp boundary between solution and solvent is established. The broadening of the boundary is then monitored by an optical or other means. Various types of apparatus have been used.

Means of generating a sharp boundary include: (i) the use of a synthetic-boundary cell centrepiece in an analytical ultracentrifuge,[60,160] although rotor vibrations tend to disturb the boundary and limit the level of precision which can be attained; (ii) the withdrawal of a sliding diaphragm which initially separates solution from solvent;[161] (iii) the sliding together of two parts of a cell, one of which initially contains solution and the other solvent;[162-164] and (iv) the sharpening of an initial, rough boundary by allowing solution and solvent to flow out of a cell through a narrow horizontal slit,[165,166] or by sucking the solvent through a capillary.[166,167]

Optical systems which have been used to monitor boundary broadening include the Schlieren system[164,168,169] as well as Rayleigh,[170] Guoy,[171,172] wedge,[173,174] shearing[175,176] and Moiré[177] interference systems. Another approach uses microfloats to follow changes in density.[178]

These various systems provide a measure of either the solute concentration or its gradient as a function of position and of time, as illustrated in Figure 2. A commonly used way of analyzing gradient data is the height–area method. The height of the gradient curve at the maximum, H, and the area under the curve, A, are measured for various times, t. The slope of a plot of $(A/H)^2$ against t is $4\pi D$. For a polydisperse solute this method provides an average diffusion coefficient D_A given by

$$D_A = \left\{\int_0^\infty D^{-1/2} g(D) dD\right\}^{-2} \tag{28}$$

where $g(D)$ is the distribution of diffusion coefficients. Procedures have been devised for evaluating other averages of D[179-181] and distributions of D.[182]

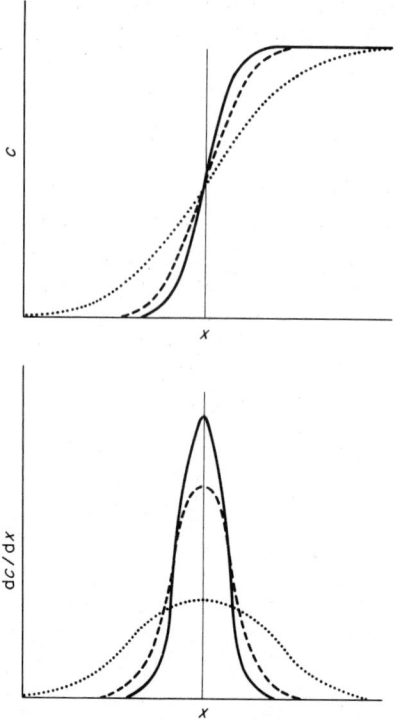

Figure 2 Illustration of the dependence of c and of (dc/dx) on x at three times, t_1 (——) $< t_2$ (----) $< t_3$ (·······), after the start of a free diffusion experiment

10.6.2.2 Membrane method

Another method for determining D utilizes a cell in which two compartments are separated by a porous membrane.[183] Solution is placed in one compartment and solvent in the other; then the change in concentration with time is measured. For D to be evaluated the membrane must be calibrated. This method has been criticized.[151]

10.6.3 Concentration Dependence of D

The mutual diffusion coefficient is related to the translational friction coefficient, f, by

$$D = \frac{RT}{N_A f}\left[1 + c\left(\frac{\partial \ln y}{\partial c}\right)_{T,P}\right] \tag{29}$$

The parameter D depends on concentration for two reasons: (i) there is a hydrodynamic factor of the variation of f with c, which also influences sedimentation velocity; and (ii) there is a thermodynamic factor of the term in brackets in equation (29). For a theta solution D decreases with increase in concentration, but for a polymer in a good solvent D may either decrease or increase.[184]

10.6.4 Limiting Diffusion Coefficient

In the limit of infinite dilution, equation (29) reduces to

$$D_0 = \frac{RT}{N_A f_0} \tag{30}$$

The magnitude of f_0 depends on the size and shape of the diffusing species. For a solid sphere of radius r_h, f_0 is given by Stokes' equation[185]

$$f_0 = 6\pi\eta_0 r_h \tag{31}$$

For a flexible polymer carrying solvent trapped within the polymer domain, r_h is an effective hydrodynamic radius which is related to the radius of gyration, r_g, by

$$r_h = \xi_f r_g \tag{32}$$

where ξ_f is a coefficient whose value is typically in the range 0.55–0.75.[186] Theoretical models have been developed to describe the frictional properties of polymers both with and without drainage from the polymer coil.[187]

Stokes' equation has been extended to nonspherical particles such an ellipsoids.[188] In practice, f_0 is not very sensitive to small deviations from spherical symmetry, but it is possible to distinguish between, for example, rod-like macromolecules and flexible coils, on the basis of measured values of f_0 and r_g.

10.6.5 Charge Effects

For a polyelectrolyte, the charge effects which have been discussed in relation to sedimentation velocity (Section 10.3.8) also influence diffusion. For a solution of a polyelectrolyte containing no low molecular weight salt, D is much higher than for an equivalent uncharged polymer. Addition of a salt leads to a decrease in D. At low concentrations of polyelectrolytes in salt-free solution, discontinuous diffusion fronts have been observed, which arise from an anomolous concentration dependence of D.[170]

10.6.6 Determination of Molecular Weight

As has already been discussed (Section 10.3.9), measurements of D_0 and S_0 can be combined to provide a value of molecular weight. Combination of D_0 and $[\eta]$ can also yield a molecular weight, by making use of the Mandelkern–Flory relationship[67]

$$M^{1/3} = \frac{\Phi^{1/3} P^{-1} RT}{D_0 N_A \eta_0 [\eta]^{1/3}} \tag{33}$$

where the parameters are as previously defined.

Empirically, D_0 is generally found to be related to M by an equation of the form

$$D_0 = K_D M^{-\alpha_D} \tag{34}$$

where K_D and α_D are constants for a given polymer in a given solvent at a particular temperature.[69] For flexible polymers, α_D is 0.5 in a theta solvent and is greater in a good solvent. Application of equation (34) enables a molecular weight distribution to be determined from a distribution of diffusion coefficients.

10.7 REFERENCES

1. T. Svedberg and J. B. Nichols, *J. Am. Chem. Soc.*, 1923, **45**, 2910.
2. T. Svedberg and K. O. Pedersen, 'The Ultracentrifuge', Clarendon, Oxford, 1940.
3. K. O. Pedersen, *Biophys. Chem.*, 1976, **5**, 3.
4. J. W. Williams, K. E. Van Holde, R. L. Baldwin and H. Fujita, *Chem. Rev.*, 1958, **58**, 715.
5. H. Fujita, 'Mathematical Theory of Sedimentation Analysis', Academic, New York, 1962.
6. H. Fujita, 'Foundations of Ultracentrifugal Analysis', Wiley, New York, 1975.
7. H. K. Schachman, 'Ultracentrifugation in Biochemistry', Academic, New York, 1959.
8. R. L. Baldwin and K. E. Van Holde, *Adv. Polym. Sci.*, 1960, **1**, 451.
9. J. M. Creeth and R. H. Pain, *Prog. Biophys. Mol. Biol.*, 1967, **17**, 219.
10. C. H. Chervenka, 'A Manual of Methods for the Analytical Ultracentrifuge', Beckman, Palo Alto, 1973.
11. E. J. Wood, *Appl. Fibre Sci.*, 1979, **2**, 1.
12. H. K. Schachman and W. F. Harrington, *J. Polym. Sci.*, 1954, **12**, 379.
13. D. A. Yphantis, *Methods Enzymol.*, 1979, **61**, 3.
14. W. Mächtle and U. Klodwig, *Makromol. Chem.*, 1979, **180**, 2507.
15. T. Kondo and M. Kawakami, *Anal. Biochem.*, 1981, **117**, 366.
16. O. Lamm, *Z. Phys. Chem., Abt. A*, 1929, **143**, 177.
17. H. Faxén, *Ark. Mat. Astron. Fys.*, 1929, **21B**, 1.
18. W. J. Archibald, *Phys. Rev.*, 1938, **53**, 746.

19. M. Dishon, G. H. Weiss and D. A. Yphantis, *Biopolymers*, 1967, **5**, 697.
20. C. W. Pyun and M. Fixman, *J. Chem. Phys.*, 1964, **41**, 937.
21. P. F. Mijnlieff and W. J. M. Jaspers, *Trans. Faraday Soc.*, 1971, **67**, 1837.
22. H. Elmgren, *J. Polym. Sci., Polym. Lett.*, 1981, **19**, 561.
23. H. Elmgren, *J. Polym. Sci., Polym. Lett.*, 1981, **19**, 567.
24. H. Elmgren, *J. Polym. Sci., Polym. Lett.*, 1982, **20**, 389.
25. C. M. Kok and A. Rudin, *J. Appl. Polym. Sci.*, 1982, **27**, 3357.
26. S. E. Harding and P. Johnson, *Biochem. J.*, 1985, **231**, 543.
27. W. J. Closs, B. R. Jennings and H. G. Jerrard, *Eur. Polym. J.*, 1968, **4**, 639.
28. I. Noda, K. Mizutani and T. Kato, *Macromolecules*, 1977, **10**, 618.
29. M. Wales and K. E. Van Holde, *J. Polym. Sci.*, 1954, **14**, 81.
30. V. S. Skazka and V. M. Yamshchikov, *Vysokomol. Soedin., Ser. A*, 1973, **15**, 213.
31. A. J. Rowe, *Biopolymers*, 1977, **16**, 2595.
32. L. W. Nichol, E. A. Owen and D. J. Winzor, *Arch. Biochem. Biophys.*, 1985, **236**, 338.
33. L.-O. Sundelöf and B. Nyström, *J. Polym. Sci., Polym. Lett.*, 1977, **15**, 377.
34. B. Nyström and J. Roots, *J. Macromol. Sci., Rev. Macromol. Chem.*, 1980, **C19**, 35.
35. H. Elmgren, *J. Polym. Sci, Polym. Lett.*, 1982, **20**, 57.
36. W. Brown, P. Stilbs and R. M. Johnsen, *J. Polym. Sci., Polym. Phys. Ed.*, 1983, **21**, 1029.
37. A. Soda, T. Fujimoto and M. Nagasawa, *J. Phys. Chem.*, 1967, **71**, 4274.
38. J. P. Johnston and A. G. Ogston, *Trans. Faraday Soc.*, 1946, **42**, 789.
39. J. J. Correia, M. L. Johnson, G. H. Weiss and D. A. Yphantis, *Biophys. Chem.*, 1976, **5**, 255.
40. C. Destor and F. Rondelez, *Polymer*, 1981, **22**, 67.
41. W. J. Closs, B. R. Jennings and H. G. Jerrard, *Eur. Polym. J.*, 1968, **4**, 651.
42. B. Nyström and L.-O. Sundelöf, *Chem. Scr.*, 1976, **10**, 16.
43. B. Nyström and J. Roots, *Makromol. Chem.*, 1979, **180**, 2419.
44. M. Dishon, G. H. Weiss and D. A. Yphantis, *J. Polym. Sci., Part A-2*, 1970, **8**, 2163.
45. G. H. Weiss and D. A. Yphantis, *J. Polym. Sci., Part A-2*, 1972, **10**, 339.
46. M. Dishon, M. T. Stroot, G. H. Weiss and D. A. Yphantis, *J. Polym. Sci., Part A-2*, 1971, **9**, 939.
47. P. Vidakovic, C. Allain and F. Rondelez, *Macromolecules*, 1982, **15**, 1571.
48. J. W. A. Van den Berg and P. Le Grand, *Eur. Polym. J.*, 1982, **18**, 51.
49. J. E. Blair and J. W. Williams, *J. Phys. Chem.*, 1964, **68**, 161.
50. M. Nagasawa and Y. Eguchi, *J. Phys. Chem.*, 1967, **71**, 880.
51. P. M. Budd, *Polymer*, 1985, **26**, 1519.
52. K. O. Pedersen, *J. Phys. Chem.*, 1958, **62**, 1282.
53. Z. Alexandrowicz and E. Daniel, *Biopolymers*, 1963, **1**, 447.
54. H. Eisenberg, *Biophys. Chem.*, 1976, **5**, 243.
55. P. F. Mijnlieff, in 'Ultracentrifugal Analysis', ed. J. W. Williams, Academic, New York, 1963, p. 81.
56. R. Varoqui and A. Schmitt, *Biopolymers*, 1972, **11**, 1119.
57. M. Nagasawa, in 'Polyelectrolytes', ed. E. Sélégny, Reidel, Dordrecht, 1974, p. 241.
58. D. Stigter, *Macromolecules*, 1985, **18**, 1619.
59. H. Eisenberg, 'Biological Macromolecules and Polyelectrolytes in Solution', Clarendon, Oxford, 1976.
60. T. M. Aminabhavi and P. Munk, *Macromolecules*, 1979, **12**, 607; 1979, **12**, 1194.
61. V. Petrus, B. Porsch, B. Nyström and L.-O. Sundelöf, *Makromol. Chem.*, 1982, **183**, 1279.
62. P. N. Pusey, in 'Industrial Polymers: Characterisation by Molecular Weight', ed. J. H. S. Green and R. Dietz, Transcripta, London, 1973, p. 26.
63. H. G. Elias, R. E. Bareiss and J. G. Watterson, *Adv. Polym. Sci.*, 1973, **11**, 111.
64. R. E. Bareiss, *Makromol. Chem.*, 1983, **184**, 1509.
65. M. Okabe, *J. Polym. Sci., Polym. Lett.*, 1984, **22**, 477.
66. R. E. Bareiss, *J. Polym. Sci., Polym. Lett.*, 1985, **23**, 549.
67. L. Mandelkern and P. J. Flory, *J. Chem. Phys.*, 1952, **20**, 212.
68. V. Petrus, B. Porsch, B. Nyström and L.-O. Sundelöf, *Makromol. Chem.*, 1983, **184**, 295.
69. P. E. O. Klämer and H. A. Ende, in 'Polymer Handbook', 2nd edn., ed. J. Brandrup and E. H. Immergut, Wiley, New York, 1975, p. IV-61.
70. R. E. Bareiss, *Makromol. Chem.*, 1984, **185**, 1623.
71. I. H. Billick, *J. Polym. Sci.*, 1962, **62**, 167.
72. M. Wales and S. J. Rehfeld, *J. Polym. Sci.*, 1962, **62**, 179.
73. L. J. Gosting, *J. Am. Chem. Soc.*, 1952, **74**, 1548.
74. R. L. Baldwin, *J. Phys. Chem.*, 1954, **58**, 1081.
75. H. Fujita, *Biopolymers*, 1969, **7**, 59.
76. I. K. Nekrasov, *Vysokomol. Soedin., Ser. A*, 1972, **14**, 2252.
77. I. K. Nekrasov, *Vysokomol. Soedin., Ser. A*, 1975, **17**, 439.
78. I. K. Nekrasov and A. M. Kulakova, *Vysokomol. Soedin., Ser. A*, 1977, **19**, 654.
79. C. M. L. Atkinson and R. Dietz, *Br. Polym. J.*, 1974, **6**, 133.
80. T. Kotaka and N. Donkai, *J. Polym. Sci, Part A-2*, 1968, **6**, 1457.
81. N. Gralén and G. Lagermalm, *J. Phys. Chem.*, 1952, **56**, 514.
82. J. W. Williams, W. M. Saunders and J. S. Cicirelli, *J. Phys. Chem.*, 1954, **58**, 774.
83. R. L. Baldwin, L. J. Gosting, J. W. Williams and R. A. Alberty, *Discuss. Faraday Soc.*, 1955, **20**, 13.
84. H. W. McCormick, *J. Polym. Sci., Part A*, 1963, **1**, 103.
85. L. H. Tung and J. R. Runyon, *J. Appl. Polym. Sci.*, 1973, **17**, 1589.
86. K. A. Andrianov, S. S. A. Pavlova, I. P. Tverdokhlebova, N. V. Pertsova, V. A. Temnikovskii and L. N. Pronina, *Vysokomol. Soedin., Ser. A*, 1977, **19**, 466.
87. M. Stickler, *Angew. Makromol. Chem.*, 1984, **123/124**, 85.
88. I. K. Nekrasov, *Usp. Khim.*, 1979, **48**, 1309.

89. L. H. Tung, *J. Polym. Sci., Part A-2*, 1969, **7**, 47.
90. R. Dietz and M. A. Francis, *Polymer*, 1979, **20**, 450.
91. R. Dietz, *J. Appl. Polym. Sci.*, 1980, **25**, 951.
92. L. H. Tung and A. L. Gatzke, *J. Polym. Sci., Polym. Phys. Ed.*, 1983, **21**, 1839.
93. H. Matsuda, I. Yamada, M. Okabe and S. Kuroiwa, *Polym. J.*, 1977, **9**, 527.
94. K. Kamada and H. Sato, *Polym. J.*, 1971, **2**, 489.
95. V. Vošický and M. Bohdanecký, *J. Polym. Sci., Polym. Phys. Ed.*, 1977, **15**, 757.
96. K. E. Van Holde and R. L. Baldwin, *J. Phys. Chem.*, 1958, **62**, 734.
97. P. E. Hexner, L. E. Radford and J. W. Beams, *Proc. Natl. Acad. Sci. USA*, 1961, **47**, 1848.
98. R. A. Pasternak, G. M. Nazarian and J. R. Vinograd, *Nature (London)*, 1957, 179, 92.
99. O. M. Griffith, *Anal. Biochem.*, 1967, **19**, 243.
100. P. A. Charlwood, *Biopolymers*, 1967, **5**, 663.
101. R. J. Goldberg, *J. Phys. Chem.*, 1953, **57**, 194.
102. G. M. Nazarian, *Anal. Chem.*, 1968, **40**, 1766.
103. D. A. Albright and J. W. Williams, *J. Phys. Chem.*, 1967, **71**, 2780.
104. T. Kotaka, N. Donkai and H. Inagaki, *J. Polym. Sci., Part A-2*, 1971, **9**, 1379.
105. T. F. Young, K. A. Kraus and J. S. Johnson, *J. Chem. Phys.*, 1954, **22**, 878.
106. E. F. Casassa and H. Eisenberg, *Adv. Protein Chem.*, 1964, **19**, 287.
107. E. F. Casassa, *Ann. N.Y. Acad. Sci.*, 1969, **164**, 13.
108. D. A. Doughty, *J. Phys. Chem.*, 1979, **83**, 2621.
109. R. Ziccardi and V. Schumaker, *Biopolymers*, 1971, **10**, 1701.
110. W. D. Lansing and E. O. Kraemer, *J. Am. Chem. Soc.*, 1935, **57**, 1369.
111. H. Utiyama, N. Tagata and M. Kurata, *J. Phys. Chem.*, 1969, **73**, 1448.
112. H. Fujita, *J. Phys. Chem.*, 1969, **73**, 1759.
113. K. E. Van Holde and J. W. Williams, *J. Polym. Sci.*, 1953, **11**, 243.
114. D. A. Yphantis, *Biochemistry*, 1964, **3**, 297.
115. H. Suzuki, *Br. Polym. J.*, 1979, **11**, 91.
116. H. Suzuki, *Bull. Inst. Chem. Res., Kyoto Univ.*, 1978, **56**, 89.
117. C. H. Chervenka, *Anal. Biochem.*, 1970, **34**, 24.
118. H. Fujita, *J. Chem. Phys.*, 1960, **32**, 1739.
119. H. W. Osterhoudt and J. W. Williams, *J. Phys. Chem.*, 1965, **69**, 1050.
120. Th. G. Scholte, *J. Polym. Sci., Part A-2*, 1968, **6**, 91.
121. S. Itou, Y. Einaga and H. Fujita, *Polym. J.*, 1976, **8**, 294.
122. P. Munk and M. E. Halbrook, *Macromolecules*, 1976, **9**, 568.
123. P. Munk, T. M. Aminabhavi, P. Williams, D. E. Hoffman and M. Chmelir, *Macromolecules*, 1980, **13**, 871.
124. P. Munk, *Macromolecules*, 1980, **13**, 1215.
125. M. Wales, *J. Phys. Colloid Chem.*, 1948, **52**, 235.
126. T. H. Donnelly, *J. Phys. Chem.*, 1966, **70**, 1862.
127. T. H. Donnelly, *Ann. N.Y. Acad. Sci.*, 1969, **164**, 147.
128. S. W. Provencher, *J. Chem. Phys.*, 1967, **46**, 3229.
129. L.-O. Sundelöf, *Ark. Kemi*, 1968, **29**, 279.
130. D. A. Lee, *J. Polym. Sci., Part A-2*, 1970, **8**, 1039.
131. M. T. Gehatia and D. R. Wiff, *J. Polym. Sci., Part A-2*, 1970, **8**, 2039.
132. M. T. Gehatia and D. R. Wiff, *Eur. Polym. J.*, 1972, **8**, 585.
133. D. R. Wiff and M. T. Gehatia, *Biophys. Chem.*, 1976, **5**, 199.
134. J. M. Jeffcoate and A. Wragg, *J. Polym. Sci., Polym. Phys. Ed.*, 1983, **21**, 123.
135. Th. G. Scholte, *J. Polym. Sci., Part A-2*, 1968, **6**, 111.
136. Th. G. Scholte, *Eur. Polym. J.*, 1970, **6**, 51.
137. E. T. Adams, Jr., W. E. Ferguson, P. J. Wan, J. L. Sarquis and B. M. Escott, *Sep. Sci.*, 1975, **10**, 175.
138. P. J. Wan and E. T. Adams, Jr., *Biophys. Chem.*, 1976, **5**, 207.
139. J. J. Hermans and H. A. Ende, in 'Newer Methods of Polymer Characterization', ed. B. Ke, Interscience, New York, 1964.
140. J. M. G. Cowie and P. M. Toporowski, *Eur. Polym. J.*, 1969, **5**, 493.
141. H. A. Ende, *Chimia*, 1965, **19**, 447.
142. H. Eisenberg, *Biopolymers*, 1967, **5**, 681.
143. C. J. Stacy, *J. Appl. Polym. Sci.*, 1977, **21**, 2231.
144. J. M. G. Cowie, R. Dey and J. T. McCrindle, *Polym. J.*, 1971, **2**, 88.
145. F. Galembeck, P. R. Robilotta, E. A. Pinheiro, I. Joekes and N. Bernardes, *J. Phys. Chem.*, 1980, **84**, 112.
146. A. T. N. Pires, S. P. Nunes and F. Galembeck, *J. Colloid Interface Sci.*, 1984, **98**, 489.
147. S. P. Nunes, F. Galembeck and N. Barelli, *Polymer*, 1986, **27**, 937.
148. W. J. Archibald, *J. Phys. Colloid Chem.*, 1947, **51**, 1204.
149. A. Kotera, H. Matsuda, Y. Miyazawa and E. Joko, *Bull. Chem. Soc. Jpn.*, 1969, **42**, 3093.
150. M. Blazsó and A. Czuppon, *Acta Chim. Acad. Sci. Hung.*, 1973, **76**, 13.
151. R. G. Kitchen, B. N. Preston and J. D. Wells, *J. Polym. Sci., Polym. Symp.*, 1976, **55**, 39.
152. U. Arunyawongsakorn, C. S. Johnson, Jr. and D. A. Gabriel, *Anal. Biochem.*, 1985, **146**, 265.
153. P. T. Callaghan and D. N. Pinder, *Polym. Bull.*, 1981, **5**, 305.
154. R. Cerf and H. A. Scheraga, *Chem. Rev.*, 1952, **51**, 185.
155. I. Tinoco, Jr., *J. Am. Chem. Soc.*, 1955, **77**, 4486.
156. B. R. Jennings, in 'Light Scattering from Polymer Solutions', ed. M. B. Huglin, Academic, London, 1972, chap. 13.
157. A. M. North, *Chem. Soc. Rev.*, 1972, **1**, 49.
158. P. M. Budd, F. Heatley, T. J. Holton and C. Price, *J. Chem. Soc., Faraday Trans. 1*, 1981, **77**, 759.
159. A. Fick, *Ann. Phys. Chem.*, 1855, **94**, 59.
160. M. Adam, M. Delsanti and G. Pouyet, *J. Phys., Lett. (Orsay, Fr.)*, 1979, **40**, 435.

161. O. Lamm, *Nova Acta Regiae Soc. Sci. Ups.*, 1937, **10** (6), 115.
162. H. Neurath, *Chem. Rev.*, 1942,, **30**, 357.
163. S. Claessen, *Nature (London)*, 1946, **158**, 834.
164. B. Chitrangad and H. R. Osmers, *J. Polym. Sci., Polym. Phys. Ed.*, 1980, **18**, 1219.
165. H. Svensson, *Acta Chem. Scand.*, 1949, **3**, 1170.
166. B. Porsch and M. Kubin, *Collect. Czech. Chem. Commun.*, 1971, **36**, 4046.
167. R. Trautman and J. W. Gofman, *J. Phys. Chem.*, 1952, **56**, 464.
168. H. Matsuda, H. Aonuma and S. Kuroiwa, *J. Appl. Polym. Sci.*, 1970, **14**, 335.
169. W. Brown and R. M. Johnsen, *Macromolecules*, 1985, **18**, 379.
170. M. Nagasawa and H. Fujita, *J. Am. Chem. Soc.*, 1964, **86**, 3005.
171. L. J. Gosting, *Adv. Protein Chem.*, 1956, **11**, 476.
172. E. L. Cussler, Jr. and E. N. Lightfoot, *J. Phys. Chem.*, 1965, **69**, 1135.
173. D. R. Paul, *Ind. Eng. Chem., Fundam.*, 1967, **6**, 217.
174. D. R. Paul, V. Mavichak and D. R. Kemp, *J. Appl. Polym. Sci.*, 1971, **15**, 1553.
175. O. Bryngdahl, *Acta Chem. Scand.*, 1957, **11**, 1017.
176. M. Kubin and B. Porsch, *Eur. Polym. J.*, 1970, **6**, 97.
177. V. Bugdahl and B. Rozsondai, *Jena Rev.*, 1967, **12**, 222.
178. M. Gordon and S. Polowinski, *Br. Polym. J.*, 1970, **2**, 182.
179. M. Daune, H. Benoit and Ch. Sadron, *J. Polym. Sci.*, 1955, **16**, 483.
180. E. L. Cussler, Jr., *J. Phys. Chem.*, 1965, **69**, 1144.
181. B. Porsch and M. Kubin, *J. Polym. Sci., Part C*, 1967, **16**, 515.
182. W. Burchard and H.-J. Cantow, in 'Polymer Fractionation', ed. M. J. R. Cantow, Academic, New York, 1967, chap. C3.
183. K. H. Keller, E. R. Canales and S. I. Yum, *J. Phys. Chem.*, 1971, **75**, 379.
184. J. S. Vrentas and J. L. Duda, *J. Polym. Sci., Polym. Phys. Ed.*, 1976, **14**, 101.
185. G. Stokes, *Trans. Cambridge Philos. Soc.*, 1847, **8**, 247; 1851, **9**, 8.
186. C. Tanford, 'Physical Chemistry of Macromolecules', Wiley, New York, 1961, chap. 6.
187. H. Yamakawa, 'Modern Theory of Polymer Solutions', Harper and Row, New York, 1971.
188. F. Perrin, *J. Phys. Radium*, 1936, **7**, 1.

11

Polyelectrolytes

PETER M. BUDD

BP Research Centre, Sunbury-on-Thames, UK

11.1 INTRODUCTION

Polyelectrolytes are polymers possessing many ionizable groups. The combination of polymeric and electrolyte behaviour gives them a number of useful properties, as indicated in Table 1, but also poses problems of characterization. This chapter provides an introduction to the behaviour of polyelectrolytes in solution, discusses the difficulties which this behaviour engenders in the determination of molecular weights and considers means of overcoming these difficulties.

Polyelectrolytes may be biological in origin, such as nucleic acids, or may be synthetic. Some important synthetic polyelectrolytes are listed in Table 2. Polyelectrolytes may conveniently be classified as anionic, cationic or ampholytic, according to whether the ionized polymer carries negative charges, positive charges or both. A number of books,[1-4] book chapters[5-8] and reviews[9-11] have appeared which discuss various aspects of polyelectrolyte behaviour.

Table 1 Examples of the Properties and Applications of Polyelectrolytes

Ability	*Use*
To drastically change the fluid properties of aqueous solutions and suspensions	As thickening agents in food and other products and for enhanced oil recovery
To interact with neutral particles	As flocculants or dispersants, depending on molecular weight and charge density
To interact with small ions	In water softeners
To interact with oppositely charged macromolecules	In selectively permeable membranes

Table 2 Some Important Synthetic Polyelectrolytes

Name	*Repeat unit*	*Name*	*Repeat unit*
Poly(acrylic acid)	$-CH_2CH-$ CO_2H	Poly(phosphoric acid)	$\begin{matrix} O \\ \parallel \\ -O-P- \\ OH \end{matrix}$
Poly(methacrylic acid)	$\begin{matrix} Me \\ \mid \\ -CH_2C- \\ \mid \\ CO_2H \end{matrix}$	Poly(vinylamine)	$-CH_2CH-$ NH_2
Poly(ethylenesulfonic acid)	$-CH_2CH-$ SO_3H	Poly(4-vinyl-*N*-alkylpyridinium chloride)	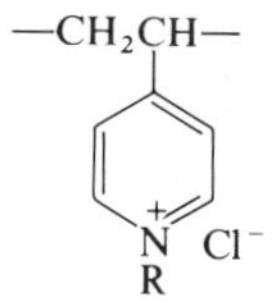 $-CH_2CH-$ (4-substituted pyridinium ring, N^+ Cl^-, R)
Poly(styrenesulfonic acid)	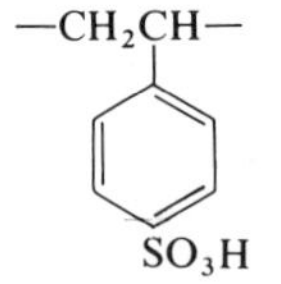 $-CH_2CH-$ (phenyl ring) SO_3H	Poly(ethyleneimine)	$-NHCH_2CH_2-$

11.2 POLYELECTROLYTES IN SOLUTION

In solution, ionization gives rise to a polyion bearing many charges, accompanied by an equivalent number of small counterions. Polyelectrolytes are usually soluble in water and the discussion which follows refers primarily to aqueous solution. Some polyelectrolytes are soluble in solvents other than water and polyelectrolyte effects have been observed in polar organic solvents such as dimethylformamide.[12] If no ionization occurs, the polymer behaves essentially as a nonionic polymer.

11.2.1 Chain Expansion

Repulsion between charged groups on the polymer leads to a flexible polyelectrolyte being more expanded than an equivalent nonionic polymer. At low concentrations in pure water a polyelectrolyte may be almost rod-like. As a simple salt is added to the solution, the charges on the polymer are to some extent shielded from each other and the degree of expansion decreases (Figure 1).

Various other factors influence polyion dimensions. For a weakly acidic or basic polyelectrolyte they can be strongly dependent on the degree of neutralization. Slight changes in chemical structure can profoundly affect solution behaviour; for example, poly(acrylic acid) expands smoothly with increasing charge density, whereas poly(methacrylic acid) undergoes an abrupt transition to a highly expanded form as the charge density increases.[13,14] Ampholytic polymers sometimes behave quite differently from other polyelectrolytes, and may even expand with increasing salt concentration.[15]

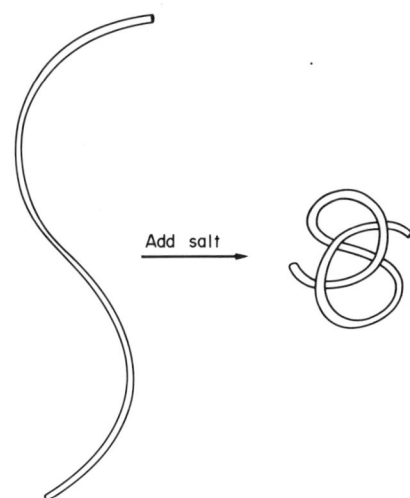

Figure 1 Illustration of the effect of a salt on the degree of expansion of a polyelectrolyte chain

A number of theories of polyelectrolyte expansion have been proposed,[16-26] which have had various degrees of success in explaining experimental observations.

11.2.2 Polyion–Small Ion Interactions

Counterions generated by ionization of a polyelectrolyte in solution are not distributed evenly throughout the solution but tend to remain in the vicinity of the polyion, forming an 'ionic atmosphere'. As a simple salt is added to the solution the thickness of this ionic atmosphere decreases (Figure 2).

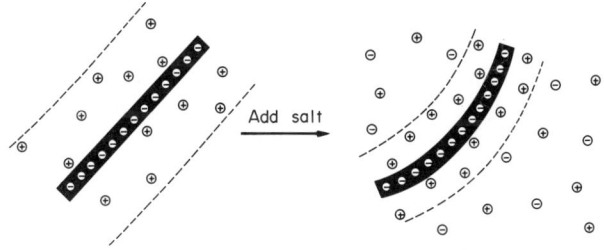

Figure 2 Illustration of the effect of a salt on the thickness of the ionic atmosphere

It is possible that some counterions are bound at specific sites on the polyion. However, over the years there has been much argument regarding this point. Whether or not counterions are specifically bound, the effect of interactions between the polyion and its counterions is to reduce their effective thermodynamic concentration (activity). Thus counterion activity coefficients are very low and there is a profound effect on other thermodynamic quantities. Studies of counterion self-diffusion, by electric conductivity[27] or nuclear magnetic resonance[28,29] can provide an insight into the degree of interaction between a polyion and its counterions.

In order to elucidate the way in which polyions interact with small ions, and to enable thermodynamic properties to be predicted, various approaches have been taken.

11.2.2.1 *Poisson–Boltzmann*

Many workers have taken an approach similar to that of the Debye–Hückel theory of simple electrolytes. A more or less realistic model of a polyion is taken, such as a charged sphere penetrable to small molecules and ions,[16,30] or a rod-like[10,31-37] or coiled[18-21] charged chain, and some form of the Poisson–Boltzmann equation is used to calculate the electrostatic potential around the polyion.

11.2.2.2 Counterion condensation

Some workers, considering a polyion as a charged cylinder[2,38-42] or worm-like chain,[26] have suggested that there is a critical charge density, ξ_{crit}, above which counterions condense on to the surface so as to maintain the net charge density at the critical value. The charge density, ξ, is defined by

$$\xi = e|\beta|/\varepsilon kT, \tag{1}$$

where e is the electronic charge, β is the charge per unit length of the polyion, ε is the relative permittivity, k is the Boltzmann constant and T is the absolute temperature. The value of ξ_{crit} is $|z|^{-1}$ where z is the valence of the counterions. The Debye–Hückel approximation is assumed to apply for interactions with uncondensed counterions.

11.2.2.3 Hypernetted chain

More recently, the hypernetted chain integral equation technique has been applied to polyelectrolytes.[43-45]

11.2.2.4 Computer simulation

Powerful computers have been employed to simulate a portion of a polyion surrounded by small ions, with the distribution of ions computed by allowing each particle to move randomly, then averaging over the configuration generated. Results of such 'Monte Carlo' studies have been compared with predictions of polyelectrolyte theories.[46]

11.2.3 The Donnan Effect

If a polyelectrolyte solution is separated from a solution of a simple salt by a semipermeable (dialysis) membrane, there is at equilibrium a lower concentration of salt on the side of the membrane with polyelectrolyte than on the side without. In other words, salt is excluded from the polyelectrolyte solution to some extent. This effect was first described by Donnan[47] in 1911. He showed that it was a consequence of the requirements (i) that electroneutrality be preserved in the solution; and (ii) that at equilibrium the activities of diffusible components be equal on the two sides of the membrane.

A Donnan membrane distribution parameter, Γ, may be defined, at infinite dilution of polyelectrolyte, by

$$\Gamma = \lim_{c_u \to 0} \left(\frac{C_s' - C_s}{C_u} \right) \tag{2}$$

where C_s' is the molarity of the co-ion (the salt ion with the same sign of charge as the polyion) on the polymer-free side of the membrane, C_s is its molarity at equilibrium on the side with polymer present and C_u is the molarity of the polyelectrolyte per equivalent charged unit. The parameter Γ is related to the mean activity coefficient of mobile ions, $\gamma_\pm$, by

$$\Gamma = 0.5 - \lim_{c_u \to 0} \left\{ \frac{C_s'}{2C_u} \left[\left(\frac{\gamma_\pm'}{\gamma_\pm} \right)^2 - 1 \right] \right\} \tag{3}$$

where the prime refers to the polymer-free side of the membrane. For the 'ideal' case, in which $\gamma_\pm = \gamma_\pm'$, the value of Γ is 0.5. In practice, as mentioned above, the presence of a polyion strongly affects the counterion activity coefficient and hence $\gamma_\pm$. Measured values of Γ are generally much lower than 0.5 at low salt concentrations, increase with increasing salt concentration and may even exceed 0.5 at high salt concentrations. Some experimental values of Γ determined in membrane distribution studies[48,49] are shown in Figure 3.

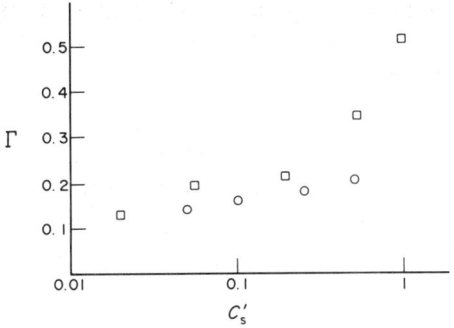

Figure 3   Dependence of Γ on C_s' for poly(ethylenesulfonate)/NH$_4$Cl^{49} ($\bigcirc$) and poly(phosphate)/Me$_4$NBr^{48} ($\square$)

It is often stated that the Donnan effect is suppressed by the addition of salt. This is true in so far as the amount of salt excluded becomes small as a proportion of the total salt concentration, and so has little influence in experiments which depend on the overall salt concentration. However, the actual amount of salt excluded increases with increasing salt concentration, and this has important consequences in the characterization of polyelectrolytes, as will be seen later.

The Donnan effect does not only arise when a membrane is present, but whenever there is a boundary between polyelectrolyte-rich and polyelectrolyte-poor solutions. It thus manifests itself in studies of polyelectrolytes by, for example, size exclusion chromatography (SEC) or sedimentation methods. There are various ways of determining the Donnan parameter Γ, in addition to membrane distribution studies, including SEC,[50, 51] sedimentation velocity,[52] isopiestic distillation[53] and electromotive force measurements using ion-selective electrodes.[54]

11.3   THE DETERMINATION OF MOLECULAR WEIGHTS OF POLYELECTROLYTES

Methods for the determination of molecular weights of polymers can be divided into two groups: primary and secondary. Primary methods, such as colligative properties, static light scattering and sedimentation equilibrium, provide absolute values. Secondary methods, such as viscosity and SEC, are generally more straightforward but require calibration to yield molecular weights. The determination of either a limiting sedimentation coefficient or a limiting diffusion coefficient constitutes a secondary method. However, a combination of the two can give an absolute molecular weight.

In this section, the application of these various techniques to polyelectrolytes is discussed.

11.3.1   General Considerations

All the techniques mentioned below involve preparation of a solution, which in the case of polyelectrolytes is generally an aqueous solution. If 'molecular' information is to be obtained, the solution must be 'molecular' and 'dilute'. Care is required as some polyelectrolytes tend to aggregate in solution. Furthermore, some commercially important polyelectrolytes are of very high molecular weight and the transition from 'dilute' to 'semidilute' solution behaviour can fall within the range of concentrations commonly employed experimentally.

Primary methods of molecular weight determination all require at some point that an accurate weight concentration of the polymer be known. Polyelectrolytes are usually hygroscopic; sodium poly(styrenesulfonate) (NaPSS), for example, typically contains about 10% by weight of moisture. In addition, commercial materials are frequently impure; one 'standard' sample of NaPSS received in the author's laboratory was found to contain 40% inorganic sulfate. Both moisture and impurities need to be taken into account in determining weight concentrations if accurate molecular weights are to be obtained.

As will be seen below, it is generally necessary to study a polyelectrolyte in the presence of a simple salt. However, excessive amounts of salt can lead to separation of a second phase, which may be a liquid, a gel or a solid. Salts containing divalent ions can be particularly effective in promoting phase separation.

Some commercially important polyelectrolytes are copolymers and can possess a distribution of charge density as well as a distribution of molecular weights. This complicates considerably the analysis of characterization data. In the future it is likely that various techniques will be combined to provide details of the two distributions.

11.3.2 Colligative Properties

Colligative properties, such as osmotic pressure and freezing point depression, depend on the number of solute particles in a solution and may be employed in the determination of number-average molecular weight, $\bar{M}_n$ (Volume 1, Chapter 4).

Taking osmotic pressure, Π, as an example, for a nonionic polymer a plot of Π/c against c, where c is the weight concentration of polymer, is linear at low concentrations and $\bar{M}_n$ may be determined from

$$\frac{\Pi}{cRT} = \frac{1}{\bar{M}_n} + A_2 c + \ldots \tag{4}$$

where R is the gas constant and A_2 is a second virial coefficient. For a polyelectrolyte, however, the situation is more complex.

11.3.2.1 Salt-free polyelectrolyte solution

Ideally, Π should be proportional to the total number of particles in solution, but in reality polyion–counterion interactions reduce the number which are osmotically active and Π is much lower than would be expected. The ratio of the actual to the ideal values of Π is termed the osmotic coefficient, ϕ, and has been determined for a number of polyelectrolytes.[55] Generally, ϕ is independent of molecular weight and varies little with concentration, but decreases rapidly with increasing charge density.

If an attempt is made to plot Π/c against c for a polyelectrolyte in salt-free solution, the result is a line which curves upwards with decreasing concentration.[56] Equation (4) cannot be applied and a molecular weight cannot be determined.

11.3.2.2 Polyelectrolyte with added salt

Addition of an excess of simple salt to the polyelectrolyte solution produces a profound change of behaviour. A plot of Π/c against c is then linear, as for a nonionic polymer, and correct molecular weights can be obtained by the application of equation (4). The second virial coefficient is strongly influenced by the Donnan effect.

The question may be asked regarding the calculated molecular weight whether it is that for the polyion alone or whether the counterions are included. The answer is that it depends on how the concentration is measured. An osmotic experiment just counts molecules; molecular *weights* arise because *weight* concentrations are employed and if the counterions are included in the concentration measurement they are included in the molecular weight determination. Both static light scattering and sedimentation equilibrium, which are to be discussed below, may be related to the osmotic work required to produce a change in concentration and so it is true for these techniques too that the molecular weights determined by them are for the components implicitly defined by the concentration measurement.

11.3.3 Static Light Scattering

Static light-scattering experiments provide a measure of weight-average molecular weight, $\bar{M}_w$ (Volume 1, Chapter 5). The light scattered from a solution of a nonionic polymer arises from fluctuations in concentration on a microscopic scale and obeys the relationship

$$\frac{Kc}{\bar{R}_\theta} = \frac{1}{\bar{M}_w P(\theta)} + 2A_2 c + \ldots \tag{5}$$

where $\bar{R}_\theta$ is a measure of the excess of scattering intensity over that of pure solvent at an angle θ, $P(\theta)$ is a function which describes the angular dependence of scattering and K is an optical constant given by

$$K = \frac{2\pi^2 n_0^2}{N_A \lambda^4} \left(\frac{dn}{dc}\right)^2 \qquad (6)$$

where n_0 is the refractive index of the solvent, N_A is Avogadro's number, λ is the wavelength of the light used and (dn/dc) is the refractive index increment. The value of (dn/dc) must be determined by a separate experiment. The limiting value of $Kc/\bar{R}_\theta$, obtained by extrapolation both to zero angle and to zero concentration, is $1/\bar{M}_w$.

11.3.3.1 Salt-free polyelectrolyte solution

For a salt-free polyelectrolyte solution, electrostatic forces are relatively long range and the fluctuation theory on which equation (5) is based breaks down. The intensity of the scattered light is reduced and it is not possible to determine molecular weights.

11.3.3.2 Polyelectrolyte with added salt

Addition of an excess of a simple salt to the system reduces the range of electrostatic interactions and scattering behaviour similar to that of a nonionic polymer is obtained.[57] However, there is still a pitfall; if equation (5) is applied using a refractive index increment $(dn/dc)_{c_s}$ determined on solutions all having the same salt concentration, the resulting molecular weight is incorrect.

Casassa and Eisenberg[4, 58] have shown that, provided the salt concentration is sufficiently high ($> 10^{-3}$ M), equation (5) does hold, to a good approximation. However, the refractive index increment required is not $(dn/dc)_{c_s}$ but $(dn/dc)_{\mu_s}$, the refractive index increment at constant chemical potential of the salt. An increment which is virtually the same as $(dn/dc)_{\mu_s}$ is that at constant chemical potential of all components diffusible through a semipermeable membrane, $(dn/dc)_\mu$. Thus, $\bar{M}_w$ for a polyelectrolyte can be determined by light scattering, as long as the refractive index increment is determined on polyelectrolyte solutions which have been brought into dialysis equilibrium with a salt solution.

A study in the author's laboratory of NaPSS in 0.5 M NaCl showed that if $(dn/dc)_{c_s}$ is used rather than $(dn/dc)_\mu$ a 20% error in molecular weight is incurred. It is not essential for the solutions used in the actual light-scattering experiment to be dialyzed, as dialysis has little effect on the scattered light intensity.

The increment $(dn/dc)_\mu$ differs from $(dn/dc)_{c_s}$ because of the Donnan effect. If the salt has an ion in common with the polyelectrolyte, the two increments are related by

$$\left(\frac{dn}{dc}\right)_\mu \approx \left(\frac{dn}{dc}\right)_{c_s} - \left(\frac{dn}{dc_s}\right)\left\{\frac{M_s}{M_u}\Gamma(1 - c_s \bar{v}_s)\right\} \qquad (7)$$

where (dn/dc_s) is the refractive index increment of the salt at constant concentration of other components, M_s is the molecular weight, c_s the weight concentration and $\bar{v}_s$ the partial specific volume of the salt and M_u is the molecular weight of the polymer per equivalent charged unit. If dialysis is impractical, for example if the sample includes relatively low molecular weight polymer which would be lost through the membrane, it may be possible to calculate $(dn/dc)_\mu$ from $(dn/dc)_{c_s}$ using a value of Γ derived by any of the techniques mentioned earlier. As Γ generally increases with increasing salt concentration, the error incurred if the correct refractive index increment is not used increases with increasing salt concentration.

Vrij and Overbeek[59] have proposed a method for evaluating light-scattering data of charged systems without the need for dialysis or other additional experiments, but their procedure involves assumptions which may not always be valid. They studied half-neutralized poly(methacrylate) in 0.1 M sodium halide solutions. Apparent molecular weights were determined in NaF, NaCl, NaBr and NaI. The square root of apparent molecular weight was plotted against the molar refractive index increment of the salt and the true molecular weight was obtained by extrapolating to zero molar refractive index increment of salt.

11.3.4 Sedimentation Equilibrium

There are various types of sedimentation equilibrium experiment (Volume 1, Chapter 10) which can be used to determine $\bar{M}_n$, $\bar{M}_w$, $\bar{M}_z$ and even, in some cases, complete molecular weight distributions. Analysis of sedimentation equilibrium data requires knowledge of a density increment, $(d\rho/dc)$. A theory of sedimentation equilibrium in polyelectrolyte solutions has been developed by Casassa and Eisenberg.[4,58]

11.3.4.1 Salt-free polyelectrolyte solution

For a salt-free polyelectrolyte solution the attainment of equilibrium is very slow and the system is highly concentration dependent. If data are analyzed as for a nonionic polymer, the quantity $M/(1-Z\phi)$, where Z is the number of charges on the polyion, is determined rather than the true molecular weight.

11.3.4.2 Polyelectrolyte with added salt

In the presence of an excess of a simple salt, a polyelectrolyte behaves much like a nonionic polymer, but problems may still arise.

For example, just as there are several possible refractive index increments which might be used in light-scattering studies, so there are several possible density increments which could be employed in sedimentation studies. Indiscriminate use of the density increment at constant concentration of salt, $(d\rho/dc)_{c_s}$, gives incorrect results. The appropriate density increment for polyelectrolytes is that at constant chemical potential of diffusible components, $(d\rho/dc)_\mu$. This may be determined by density measurements on polyelectrolyte solutions which have been brought into dialysis equilibrium with a salt solution. If the salt has an ion in common with the polyelectrolyte, the two increments are related by

$$\left(\frac{d\rho}{dc}\right)_\mu \approx \left(\frac{d\rho}{dc}\right)_{c_s} - (1-\bar{v}_s\rho)\left\{\frac{M_s}{M_u}\Gamma\right\} \tag{8}$$

where ρ is the density of the solution. Thus $(d\rho/dc)_\mu$ can be calculated from $(d\rho/dc)_{c_s}$ if Γ is known. The increment $(d\rho/dc)_{c_s}$ is equivalent to a buoyancy factor $(1-\bar{v}\rho)$, where $\bar{v}$ is the partial specific volume of the polymer. A study of NaPSS in 0.5 M NaCl[52] showed that if $(d\rho/dc)_{c_s}$ is employed rather than $(d\rho/dc)_\mu$, the resulting molecular weight is 19% lower than the true value.

From equation (8) it can be seen that the difference between $(d\rho/dc)_\mu$ and $(d\rho/dc)_{c_s}$ disappears for a salt for which $(1-\bar{v}_s\rho)$ is zero. Tetramethylammonium chloride is such a salt[60,61] and dialysis is not necessary if, for example, the tetramethylammonium salt of an anionic polyelectrolyte is studied in the presence of tetramethylammonium chloride.[52]

Another problem may arise if a refractometric optical system is used. At sedimentation equilibrium there is a concentration gradient of polyelectrolyte and consequently, because of the Donnan effect, an opposing concentration gradient of salt, which will affect the refractometrically observed distribution. In most cases errors due to this should more or less cancel out, as errors in the same proportion occur in both numerator and denominator in the sedimentation equilibrium equations. However, it is necessary to employ dialyzed solutions and dialysate in synthetic boundary experiments, if they are used to determine initial concentration, although dialyzed solutions are not essential to the actual equilibrium experiments.

11.3.5 Sedimentation Velocity and Diffusion

Both the limiting sedimentation coefficient, S_0, and the limiting diffusion coefficient, D_0, may be empirically related to molecular weight (Volume 1, Chapter 10). For a nonionic polymer S_0 is given by

$$S_0 = \frac{M(1-\bar{v}\rho)}{N_A f} \tag{9}$$

and D_0 by

$$D_0 = \frac{RT}{N_A f} \tag{10}$$

where f is the translational molecular friction coefficient. Combination of equations (9) and (10) gives

$$M = \frac{S_0 RT}{D_0(1 - \bar{v}\rho)} \tag{11}$$

showing that molecular weights can be determined from sedimentation velocity and diffusion measurements.

Mandelkern and Flory[62] have suggested that molecular weights can also be determined by combining either S_0 or D_0 with viscosity data. The method relies on the constancy of the parameter $\Phi^{1/3}P^{-1}$ which is given, when using sedimentation data, by

$$\Phi^{1/3}P^{-1} = \frac{S_0 N_A \eta_0 [\eta]^{1/3}}{M^{2/3}(1 - \bar{v}\rho)} \tag{12}$$

where η_0 is the solvent viscosity and $[\eta]$ is the intrinsic viscosity (limiting viscosity number) of the polymer. For nonionic polymers the value of $\Phi^{1/3}P^{-1}$ is generally close to 1.1×10^7 mol$^{-1/3}$ (2.5×10^6 if $[\eta]$ is expressed in dl g^{-1}).[63]

11.3.5.1 Salt-free polyelectrolyte solution

In both sedimentation and diffusion, the rate of movement of a polyelectrolyte is strongly affected by its charged nature. In the absence of a salt, the sedimentation coefficient, S, for a polyelectrolyte is much lower, and the diffusion coefficient, D, is much higher, than for an equivalent uncharged polymer. This has been called the 'primary charge effect' and has been attributed to the electric field set up if a charged macromolecule is separated from its counterions.[64] The polyion and its counterions are forced to move together at a velocity intermediate between that which each would have alone.

Theoretical treatments of sedimentation and diffusion in polyelectrolyte solutions follow one of two approaches. The classical approach[64-66] sees the actual velocity of the polyion as having two components; a contribution from the centrifugal field (in sedimentation) or chemical potential gradient (in diffusion) and a contribution from the internally created electric field. An alternative and more rigorous approach involves the application of irreversible thermodynamics.[67,68] Essentially similar conclusions are reached.

For a polyelectrolyte PX$_Z$, the classical approach gives, for S_0[4]

$$S_0 = \frac{M(1 - \bar{v}\rho)}{N_A(f + iZf_X)} \tag{13}$$

where f_X is the friction coefficient for the counterion X and i is an effective charge parameter which takes account of interactions of the polyion with its counterions. On average $(1 - i)Z$ counterions are assumed to interact in some way with the polyion, so that it carries an effective charge of iZ. No assumptions are made regarding the nature of the interaction. One may expect the effective charge on the polymer to depend both on the number of any 'condensed' counterions and on the extent of atmospheric screening.

In both sedimentation and diffusion, f for a polyelectrolyte is markedly different from that for an equivalent nonionic polymer, because of the effects of chain expansion and the influence of the ionic atmosphere.

The diffusion coefficient for a polyelectrolyte in the absence of a salt exhibits an unusual concentration dependence[69] which is poorly understood.

11.3.5.2 Polyelectrolyte with added salt

Addition of a simple salt to a polyelectrolyte solution reduces the 'primary charge effect', but can give rise to a 'secondary charge effect' if the constituent ions of the added salt have different mobilities. This secondary effect is, however, small for salts such as NaCl for which both cation and anion have similar friction coefficients.

Taking as an example the sedimentation behaviour of NaPSS in the presence of NaCl,[70] the effect of increasing the salt concentration can be seen in Figure 4, in which $1/S$ is plotted against c for

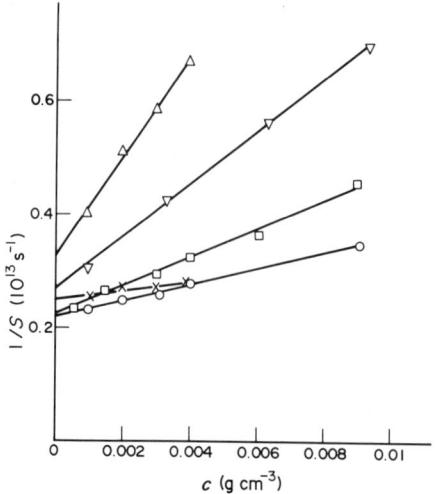

Figure 4   Dependence of reciprocal of sedimentation coefficient on polymer concentration for NaPSS of molecular weight
1.0×10^5 at NaCl concentrations of 0.005 M ($\triangle$), 0.02 M ($\triangledown$), 0.1 M ($\square$), 0.5 M ($\bigcirc$) and 2 M ($\times$)[70]

various salt concentrations. The dependence on polymer concentration decreases and, up to
0.5 M NaCl, S_0 increases.

For a polyelectrolyte PX_Z in the presence of a salt XY, in the absence of secondary charge effects,
S_0 is given by[4]

$$S_0 = \frac{M(1 - \bar{v}\rho) - [0.5iZM_s(1 - \bar{v}_s\rho)]}{N_A f} \tag{14}$$

In experiments involving thermodynamic equilibrium an effective charge parameter may be defined
which is simply related to the Donnan parameter Γ by

$$i = 2\Gamma \tag{15}$$

If it is assumed that equation (15) applies in sedimentation velocity, equation (14) becomes

$$S_0 = \frac{M(d\rho/dc)_\mu}{N_A f} \tag{16}$$

The parameter i, and hence Γ, can be evaluated from sedimentation velocity data.[52,70]

For diffusion of a polyelectrolyte in the presence of an excess of salt equation (10) applies and can
be combined with equation (16) to yield a value for molecular weight, provided all experiments are
performed at the same salt concentration.

It is not so straightforward, however, to use the sedimentation/viscosity or diffusion/viscosity
method for polyelectrolytes. Nagasawa and Eguchi[71] attempted to calculate $\Phi^{1/3} P^{-1}$ using
equation (12) for NaPSS in NaCl at various salt concentrations, C_s. The resulting values were always
lower than those obtained for nonionic polymers and decreased at high salt concentrations, as seen
for two samples in Figure 5.

For a polyelectrolyte the expression for $\Phi^{1/3} P^{-1}$ should be [70,72]

$$\Phi^{1/3} P^{-1} = \frac{S_0 N_A \eta_0 [\eta]^{1/3}}{M^{2/3} \left\{ (1 - \bar{v}\rho) - \dfrac{0.5iM_s(1 - \bar{v}_s\rho)}{M_u} \right\}} \equiv \frac{S_0 N_A \eta_0 [\eta]^{1/3}}{M^{2/3} (d\rho/dc)_\mu} \tag{17}$$

Recalculation of Nagasawa and Eguchi's results using equation (17), which takes account of the
variation of Γ with salt concentration, eliminates the anomolous decrease at high salt concentrations
as seen in Figure 5, but the limiting value of $\Phi^{1/3} P^{-1}$ is still lower than for a nonionic polymer.
Other workers have also obtained low values of $\Phi^{1/3} P^{-1}$ for polyelectrolytes.[73,74]

The low value of $\Phi^{1/3} P^{-1}$ can be attributed to the ionic atmosphere. This influences the
translational friction coefficient, and hence sedimentation and diffusion, but not to the same extent
as the rotational friction coefficient which is operative in viscosity. On decreasing the salt concen-

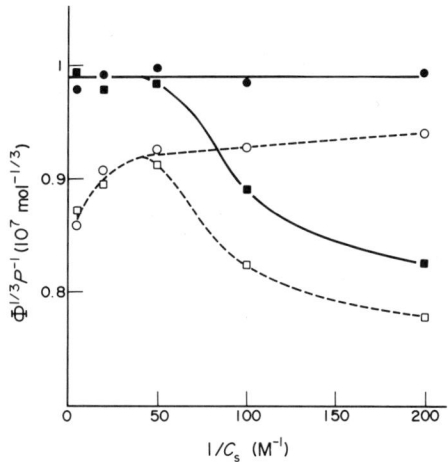

Figure 5 Dependence of $\Phi^{1/3}P^{-1}$ on reciprocal of molar NaCl concentration for NaPSS of molecular weight 3.5×10^5 ($\bigcirc$ $\bullet$) and 1.0×10^6 ($\square$ $\blacksquare$),[69] calculated according to equations (12) (open symbols, dashed lines) and (17) (filled symbols, solid lines)

tration the increasing thickness of the ionic atmosphere causes a decrease in $\Phi^{1/3}P^{-1}$, which may be offset for low molecular weight polyelectrolytes by the effect of increased drainage.

The influence of charge effects on diffusion manifests itself in many techniques. For example, flow field-flow fractionation (Volume 1, Chapter 14) separates macromolecules essentially in terms of diffusion coefficient. Giddings *et al.*[75] have studied potassium poly(styrenesulfonate) by this technique and found that the polymer appeared larger in the presence than in the absence of a salt. This anomolous result arose because the primary charge effect was not taken into account.

11.3.6 Viscosity

The viscosity of a polymer solution depends on the size of the molecule in solution and may be empirically related to molecular weight (Volume 1, Chapter 9). For a nonionic polymer the reduced viscosity, η_{sp}/c, follows the Huggins equation

$$\frac{\eta_{sp}}{c} = [\eta] + k_H [\eta]^2 c \tag{18}$$

where k_H is a constant. Typical behaviour of η_{sp}/c for a polyelectrolyte is illustrated in Figure 6.

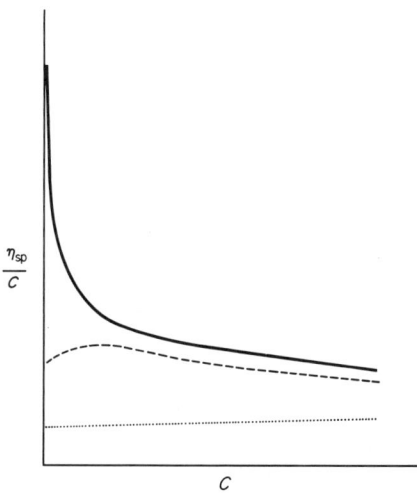

Figure 6 Illustration of typical dependence of reduced viscosity on polymer concentration for a salt-free polyelectrolyte solution (———), a polyelectrolyte with a moderate amount of added salt (-----) and a polyelectrolyte with an excess of salt (··········)

11.3.6.1 *Salt-free polyelectrolyte solution*

For a salt-free polyelectrolyte solution, η_{sp}/c increases markedly with decreasing concentration.[76-79] This increase may be attributed to polyion expansion, combined with increasing interactions between the expanding polyions. Some workers have reported a maximum in η_{sp}/c at low concentrations,[80-84] but this may reflect the presence of ionic impurities.[79]

11.3.6.2 *Polyelectrolyte with added salt*

The addition of even a small quantity of a simple salt greatly reduces η_{sp}/c and a maximum may appear in the η_{sp}/c against c plot. With an excess of salt the behaviour of η_{sp}/c is similar to that for a nonionic polymer and follows equation (18). The value obtained for k_H depends on the way in which the polyelectrolyte is diluted. Dilution procedures which have been employed include: (i) constant molarity of salt, C_s; (ii) constant C_s/C, where C is the molarity of the polyion: (iii) constant $C + xC_s$, where x is typically 0.5 or 1; and (iv) constant activity of salt, achieved by bringing polymer solutions into dialysis equilibrium with a salt solution. Only the last procedure is theoretically justifiable.

The intrinsic viscosity, $[\eta]$, decreases with increasing salt concentration. Experimentally, $[\eta]$ has been found to be proportional to $C_s^{-1/2}$.[85] Early theories of polyelectrolyte expansion predicted a C_s^{-1} dependence, but later theories have gone some way towards explaining the $C_s^{-1/2}$ dependence.[23,25,86,87]

The molecular weight dependence of $[\eta]$ can generally be described quite adequately by the Mark–Houwink equation

$$[\eta] = KM^{\alpha} \tag{19}$$

where K and α are constants for a given polyelectrolyte/salt system at a particular C_s and T. Thus viscosity measurements can be used to provide a measure of molecular weight once the Mark–Houwink constants have been evaluated for a specific system.

A number of theories[88-90] attempt to relate viscosity data to the unperturbed dimensions of a polymer. A frequently used equation is that of Stockmayer and Fixman[88]

$$\frac{[\eta]}{M^{1/2}} = K_{\theta} + 0.51\Phi BM^{1/2} \tag{20}$$

where K_{θ} is related to the unperturbed dimensions of the polymer, Φ is the Flory constant[91] in a theta solvent and B is a long-range interaction parameter. Noda *et al.*,[85] in an investigation of poly(acrylic acid) at various degrees of neutralization, found linear relationships between $[\eta]/M^{1/2}$ and $M^{1/2}$ as predicted by equation (20). However, this approach is only applicable for low molecular weights and fairly low degrees of expansion.

11.3.7 Size Exclusion Chromatography

Size exclusion (or gel permeation) chromatography (SEC) utilizes a column packed with porous beads (Volume 1, Chapter 12). A solution of the polymer is introduced into a stream of solvent flowing through the column and the polymer molecules are separated according to their ability to penetrate the pores. Very large molecules which cannot penetrate the pores elute at a void volume V_0. Very small molecules which penetrate all the pore volume elute later at a total volume V_t. Separation on the basis of hydrodynamic size occurs between V_0 and V_t and calibration of the column enables a molecular weight distribution to be determined. For many nonionic polymers a 'universal' calibration in terms of the hydrodynamic volume parameter $[\eta]M$ applies. If a truly size-based separation is to be obtained there must be no specific interactions between the sample and the column packing material.

A number of column-packing materials have been employed for aqueous SEC,[92,93] including porous silica, frequently coated to reduce interactions with the sample, and gels of hydroxylated polyether copolymers or polyacrylamide. SEC of cationic polyelectrolytes requires specialized packings.[94]

11.3.7.1 Salt-free polyelectrolyte solution

In the absence of a simple salt, polyelectrolytes frequently elute predominately at or near V_0[12,95] and there is little or no separation on the basis of molecular weight. Not only does chain expansion give rise to a large hydrodynamic volume, but also repulsion between the polyion and charged groups on the packing can lead to exclusion of the polyion from the pores.

However, SEC of polyelectrolytes without added salt can provide a test of the purity of the sample, as low molecular weight impurities appear at high elution volume. It has also been suggested that if conductometric and refractometric detectors are employed together, it is possible to obtain information about the distribution of charge density on the polymer. [96,97]

11.3.7.2 Polyelectrolyte with added salt

In general, addition of a salt leads to increasingly high elution volumes for polyelectrolytes, as charges are shielded and the polyion chain contracts. At moderate salt concentrations elution occurs in order of decreasing size, as for a nonionic polymer. If too much salt is added, however, adsorption of the polyion on to the packing can be significant, and there is then no longer a size-based separation. The salt concentration in the eluant thus has to be carefully optimized. Depending on the natures of the polymer and the packing, it may also be desirable to buffer the eluant at a particular pH or to add a surfactant to reduce adsorption.[95]

When a salt is present in the eluant and a detector is used which responds to changes in salt concentration, such as a differential refractometer, a 'Donnan' peak appears at high elution volumes due to salt excluded from the sample region. Provided that the polymer peak is separated from the Donnan peak, it is possible to determine the Donnan parameter Γ.[50,51] Under the conditions of an SEC experiment on a polyelectrolyte PX_Z in the presence of a salt XY[51]

$$\Gamma = \frac{Q_{ex}}{Q_u} \tag{21}$$

where Q_{ex} is the number of moles of salt excluded and Q_u is the number of moles of equivalent units (per charge) of polyelectrolyte.

There have been a number of attempts to apply the 'universal calibration' concept to polyelectrolytes. For some systems it has been reported that, provided the salt concentration is sufficiently high, polyelectrolytes and nonionic polymers obey the same relationship between $[\eta]M$ and elution volume;[97-102] on decreasing the salt concentration, however, there is generally an increasing deviation from the universal calibration plot. For other systems, failure of the universal calibration concept at all salt concentrations has been reported.[103,104]

It is possible to avoid the need for calibration altogether by employing a low-angle laser light-scattering detector, together with a concentration detector, to provide an absolute molecular weight distribution,[105] although there are some problems with this approach in practice.

11.3.8 Electrophoresis

The presence of charged groups on a polymer complicates the application of the various techniques discussed above. However, additional methods of characterization are possible which utilize the behaviour of such a charged species in an electric field.

11.3.8.1 Electrophoresis in solution

If an electric field is established in a solution containing charged particles, the particles will eventually achieve a steady velocity which is proportional to the field strength. The proportionality constant is termed the electrophoretic mobility, m, and depends on the balance between the electrical force acting to accelerate the particle and the frictional forces which oppose its motion.

For an impenetrable sphere with a surface charge Zq, m is given by

$$m = \frac{Zq}{f}. \tag{22}$$

Detailed calculations of m should take account of three effects: (i) the effect of the ionic atmosphere on the frictional behaviour; (ii) the 'retardation' or 'electrophoretic' effect, which reduces the mobility of the polyion because of hydrodynamic interactions with counterions moving in the opposite direction; and (iii) the 'relaxation' effect, which reduces the mobility of the polyion because of electrostatic interactions with the ionic atmosphere, which lags behind the polyion.

The mobility of a flexible polyion depends in a complicated way on the conformation of the polyion, the distribution of charges, the distribution of counterions and the degree of drainage. Both theory[106-108] and experiment[109-111] suggest that the mobility depends on the ratio of the monomeric charge to the monomeric friction coefficient, but is essentially independent of the molecular weight of the polyion.

11.3.8.2 Electrophoresis in gels

Migration of a polyion through a cross-linked gel is impeded to an extent which depends on the degree of cross-linking and on the size of the polyion. Gel electrophoresis thus provides a way of measuring molecular size. The technique has been extensively applied to biological macromolecules[112,113] and has also been employed in the determination of molecular weight distributions of synthetic polyelectrolytes.[114] Reptation theory has been used to describe electrophoresis of polyelectrolytes in gels.[115]

11.4 REFERENCES

1. S. A. Rice and M. Nagasawa, 'Polyelectrolyte Solutions', Academic Press, London, 1961.
2. F. Oosawa, 'Polyelectrolytes', Dekker, New York, 1971.
3. E. Selegny, M. Mandel and U. P. Strauss (eds.), 'Polyelectrolytes', Reidel, Dordrecht, 1974.
4. H. Eisenberg, 'Biological Macromolecules and Polyelectrolytes in Solution', Clarendon, Oxford, 1976.
5. C. Tanford, 'Physical Chemistry of Macromolecules', Wiley, New York, 1961, chap. 7.
6. H. Morawetz, 'Macromolecules in Solution', 2nd edn., Wiley, New York, 1975, chap. 7.
7. E. G. Richards, 'An Introduction to Physical Properties of Large Molecules in Solution', Cambridge University Press, Cambridge, 1980, chap. 9.
8. M. Mandel, in 'Chemistry and Technology of Water-Soluble Polymers', ed. C. A. Finch, Plenum Press, New York, 1983, p. 179.
9. R. W. Armstrong and U. P. Strauss, 'Encyclopedia of Polymer Science and Technology', Wiley–Interscience, New York, 1969, vol. 10, p. 781.
10. A. Katchalsky, *Pure Appl. Chem.*, 1971, **26**, 327.
11. M. Mandel, *Angew. Makromol. Chem.*, 1984, **123/124**, 63.
12. D. R. Scheuing, *J. Appl. Polym. Sci.*, 1984, **29**, 2819.
13. A. Katchalsky and H. Eisenberg, *J. Polym. Sci.*, 1951, **6**, 145.
14. J. L. Koenig, A. C. Angood, J. Semen and J. B. Lando, *J. Am. Chem. Soc.*, 1969, **91**, 7250.
15. J. C. Salamone, I. Ahmed, P. Elayaperumal, M. K. Raheja, A. C. Watterson and A. P. Olson, *ACS Polym. Mater. Sci. Eng.*, 1986, **55**, 269.
16. J. J. Hermans and J. Th. G. Overbeek, *Recl. Trav. Chim. Pays-Bas*, 1948, **67**, 761.
17. W. Kuhn, O. Künzle and A. Katchalsky, *Helv. Chim. Acta*, 1948, **31**, 1994.
18. A. Katchalsky, O. Künzle and W. Kuhn, *J. Polym. Sci.*, 1950, **5**, 283.
19. A. Katchalsky and S. Lifson, *J. Polym. Sci.*, 1953, **11**, 409.
20. F. E. Harris and S. A. Rice, *J. Phys. Chem.*, 1954, **58**, 725.
21. S. A. Rice and F. E. Harris, *J. Phys. Chem.*, 1954, **58**, 733.
22. P. J. Flory, *J. Chem. Phys.*, 1953, **21**, 162.
23. M. Fixman, *J. Chem. Phys.*, 1964, **41**, 3772.
24. Z. Alexandrowicz, *J. Chem. Phys.*, 1967, **47**, 4377.
25. H. W. Chien and A. Isihara, *J. Polym. Sci., Polym. Phys. Ed.*, 1976, **14**, 1015.
26. T. Odijk and A. C. Houwaart, *J. Polym. Sci., Polym. Phys. Ed.*, 1978, **16**, 627.
27. D. Bratko, D. Dolar, A. Godec and J. Špan, *Makromol. Chem., Rapid Commun.*, 1983, **4**, 697.
28. D. Bratko, P. Stilbs and M. Bešter, *Makromol. Chem., Rapid Commun.*, 1985, **6**, 163.
29. R. Rymdén and P. Stilbs, *J. Phys. Chem.*, 1985, **89**, 2425, 3502.
30. F. Oosawa, N. Imai and I. Kagawa, *J. Polym. Sci.*, 1954, **13**, 93.
31. T. Alfrey, Jr., P. W. Berg and H. Morawetz, *J. Polym. Sci.*, 1951, **7**, 543.
32. R. M. Fuoss, A. Katchalsky and S. Lifson, *Proc. Natl. Acad. Sci. USA*, 1951, **37**, 579.
33. S. Lifson and A. Katchalsky, *J. Polym. Sci.*, 1954, **13**, 43.
34. A. Katchalsky, Z. Alexandrowicz and O. Kedem, in 'Chemical Physics of Ionic Solutions', ed. B. E. Conway and R. G. Barradas, Wiley, New York, 1966, p. 295.
35. D. Stigter, *J. Colloid Interface. Sci.*, 1975, **53**, 296.
36. M. Fixman, *J. Chem. Phys.*, 1979, **70**, 4995.
37. G. V. Ramanathan and C. P. Woodbury, Jr., *J. Chem. Phys.*, 1985, **82**, 1482.
38. G. S. Manning, *J. Chem. Phys.*, 1969, **51**, 924, 934, 3249.

39. G. S. Manning, *Annu. Rev. Phys. Chem.*, 1972, **23**, 117.
40. G. S. Manning, *Q. Rev. Biophys.*, 1978, **11**, 179.
41. G. S. Manning, *Acc. Chem. Res.*, 1979, **12**, 443.
42. M. Satoh, J. Komiyama and T. Iijima, *Macromolecules*, 1985, **18**, 1195.
43. R. Bacquet and P. J. Rossky, *J. Phys. Chem.*, 1984, **88**, 2660.
44. E. Gonzales-Tovar, M. Lozada-Cassou and D. Henderson, *J. Chem. Phys.*, 1985, **83**, 361.
45. V. Vlachy and D. A. McQuarrie, *J. Chem. Phys.*, 1985, **83**, 1927.
46. P. Mills, C. F. Anderson and M. T. Record, Jr., *J. Phys. Chem.*, 1985, **89**, 3984; 1986, **90**, 6541.
47. F. G. Donnan, *Z. Elektrochem.*, 1911, **17**, 572.
48. U. P. Strauss and P. Ander, *J. Am. Chem. Soc.*, 1958, **80**, 6494.
49. H. Eisenberg and E. F. Casassa, *J. Polym. Sci.*, 1960, **47**, 29.
50. T. Lindström, A. De Ruvo and C. Söremark, *J. Polym. Sci., Polym. Chem. Ed.*, 1977, **15**, 2029.
51. S. Kadokura, T. Miyamoto and H. Inagaki, *Makromol. Chem.*, 1983, **184**, 2593.
52. P. M. Budd, *Br. Polym. J.*, 1988, **20**, 33.
53. G. Cohen and H. Eisenberg, *Biopolymers*, 1968, **6**, 1077.
54. N. Imai and H. Eisenberg, *J. Chem. Phys.*, 1966, **44**, 130.
55. Z. Alexandrowicz and A. Katchalsky, *J. Polym. Sci., Part A*, 1963, **1**, 3231.
56. U. P. Strauss and R. M. Fuoss, *J. Polym. Sci.*, 1949, **4**, 457.
57. M. Nagasawa and A. Takahashi, in 'Light Scattering from Polymer Solutions', ed. M. B. Huglin, Academic Press, London, 1972, chap. 16.
58. E. F. Casassa and H. Eisenberg, *Adv. Protein Chem.*, 1964, **19**, 287.
59. A. Vrij and J. Th. G. Overbeek, *J. Colloid. Sci.*, 1962, **17**, 570.
60. R. Ziccardi and V. Schumaker, *Biopolymers*, 1971, **10**, 1701.
61. J. P. Kratohvil, *J. Colloid Interface. Sci.*, 1980, **75**, 271.
62. L. Mandelkern and P. J. Flory, *J. Chem. Phys.*, 1952, **20**, 212.
63. V. Petrus, B. Porsch, B. Nyström and L.-O. Sundelöf, *Makromol. Chem.*, 1983, **184**, 295.
64. K. O. Pederson, *J. Phys. Chem.*, 1958, **62**, 1282.
65. Z. Alexandrowicz and E. Daniel, *Biopolymers*, 1963, **1**, 447.
66. H. Eisenberg, *Biophys. Chem.*, 1976, **5**, 243.
67. P. F. Mijnlieff, in 'Ultracentrifugal Analysis', ed. J. W. Williams, Academic Press, New York, 1963, p. 81.
68. R. Varoqui and A. Schmitt, *Biopolymers*, 1972, **11**, 1119.
69. M. Nagasawa and H. Fujita, *J. Am. Chem. Soc.*, 1964, **86**, 3005.
70. P. M. Budd, *Polymer*, 1985, **26**, 1519.
71. M. Nagasawa and Y. Eguchi, *J. Phys. Chem.*, 1967, **71**, 880.
72. H. Eisenberg, *J. Chem. Phys.*, 1962, **36**, 1837.
73. H. Eisenberg and D. Woodside, *J. Chem. Phys.*, 1962, **36**, 1844.
74. M. Stickler, *Angew. Makromol. Chem.*, 1984, **123/124**, 85.
75. J. C. Giddings, G.-C. Lin and M. N. Myers, *J. Liq. Chromatogr.*, 1978, **1**, 1.
76. R. M. Fuoss and U. P. Strauss, *J. Polym. Sci.*, 1948, **3**, 602.
77. A. Oth and P. Doty, *J. Phys. Chem.*, 1952, **56**, 43.
78. T. Alfrey, Jr. and H. Morawetz, *J. Am. Chem. Soc.*, 1952, **74**, 436.
79. R. L. Darskus, D. O. Jordan, T. Kurucsev and M. L. Martin, *J. Polym. Sci., Part A*, 1965, **3**, 1941.
80. H. Fujita, K. Mitshuhasi and T. Homma, *J. Colloid Sci.*, 1954, **9**, 466.
81. H. Eisenberg and J. Pouyet, *J. Polym. Sci.*, 1954, **13**, 85.
82. B. E. Conway, *J. Polym. Sci.*, 1955, **18**, 257.
83. J. A. V. Butler, A. B. Robins and K. V. Shooter, *Proc. R. Soc. London, Ser. A*, 1957, **241**, 299.
84. D. O. Jordan and T. Kurucsev, *Polymer*, 1960, **1**, 185.
85. I. Noda, T. Tsuge and M. Nagasawa, *J. Phys. Chem.*, 1970, **74**, 710.
86. H. W. Chien, C. H. Isihara and A. Isihara, *Polym. J.*, 1976, **8**, 288.
87. N. Imai, *Rep. Prog. Polym. Phys. Jpn.*, 1980, **23**, 95.
88. W. H. Stockmayer and M. Fixman, *J. Polym. Sci., Part C*, 1963, **1**, 137.
89. H. Inagaki, H. Suzuki and M. Kurata, *J. Polym. Sci., Part C*, 1966, **15**, 409.
90. G. C. Berry, *J. Chem. Phys.*, 1967, **46**, 1338.
91. P. J. Flory, 'Principles of Polymer Chemistry', Cornell University, Ithaca, 1953.
92. H. G. Barth, *J. Chromatogr. Sci.*, 1980, **18**, 409.
93. A. R. Cooper, in 'Developments in Polymer Characterisation — 5', ed. J. V. Dawkins, Elsevier, London, 1986, chap. 4.
94. M. Stickler and F. Eisenbeiss, *Eur. Polym. J.*, 1984, **20**, 849.
95. S. N. E. Omorodion, A. E. Hamielec and J. L. Brash, *J. Liq. Chromatogr.*, 1981, **4**, 1903.
96. A. Domard, M. Rinaudo and C. Rochas, *J. Polym. Sci., Polym. Phys. Ed.*, 1979, **17**, 673.
97. C. Rochas, A. Domard and M. Rinaudo, *Eur. Polym. J.*, 1980, **16**, 135.
98. M. Rinaudo, J. Desbrières and C. Rochas, *J. Liq. Chromatogr.*, 1981, **4**, 1297.
99. A. L. Spatorico and G. L. Beyer, *J. Appl. Polym. Sci.*, 1975, **19**, 2933.
100. G. Callec, A. W. Anderson, G. T. Tsao and J. E. Rollings, *J. Polym. Sci., Polym. Chem. Ed.*, 1984, **22**, 287.
101. P. L. Dubin and M. M. Tecklenburg, *Anal. Chem.*, 1985, **57**, 275.
102. P. L. Dubin and C. M. Speck, *ACS Polym. Mater. Sci. Eng.*, 1986, **54**, 194.
103. A. Bose, J. E. Rollings, J. M. Caruthers, M. R. Okos and G. T. Tsao, *J. Appl. Polym. Sci.*, 1982, **27**, 795.
104. A. Hamielec and M. Styring, *Pure Appl. Chem.*, 1985, **57**, 955.
105. J. A. P. P. Van Dijk, F. A. Varkevisser and J. A. M. Smit, *J. Polym. Sci., Polym. Phys. Ed.*, 1987, **25**, 149.
106. J. J. Hermans, *J. Polym. Sci.*, 1955, **18**, 527.
107. J. J. Hermans and H. Fujita, *Koninkl. Ned. Akad. Wetenschap., Proc. Ser. B.*, 1955, **58**, 182.
108. J. Th. G. Overbeek and D. Stigter, *Recl. Trav. Chim. Pays-Bas*, 1956, **75**, 543.
109. E. B. Fitzgerald and R. M. Fuoss, *J. Polym. Sci.*, 1954, **14**, 329.

110. M. Nagasawa, A. Soda and I. Kagawa, *J. Polym. Sci.*, 1958, **31**, 439.
111. I. Noda, M. Nagasawa and M. Ōta, *J. Am. Chem. Soc.*, 1964, **86**, 5075.
112. K. Weber and M. Osborn, *J. Biol. Chem.*, 1969, **244**, 4406.
113. E. M. Southern, *Anal. Biochem.*, 1979, **100**, 304, 319.
114. J.-L. Chen and H. Morawetz, *Macromolecules*, 1982, **15**, 1185.
115. P. Dejardin, O. J. Lumpkin and B. H. Zimm, *J. Polym. Sci, Polym. Symp.*, 1985, **73**, 67.

12

Size Exclusion Chromatography

JOHN V. DAWKINS
Loughborough University of Technology, UK

12.1 INTRODUCTION

Virtually all synthetic polymers have a molecular weight distribution (MWD) which may be represented by the cumulative weight distribution I(M) or differential weight distribution w(M) as a function of molecular weight M.[1] Although the calculation of w(M) for various types of polymerization mechanism was already well documented,[2] an accurate and routine method for the experimental determination of MWD only became possible after 1960 with the development of suitable porous packings for *size exclusion chromatography* (SEC).

The first effective demonstration that polymers may be separated by the size dependence of the degree of solute penetration into a porous packing was reported by Porath and Flodin who employed dextran gels for aqueous *gel filtration chromatography* (GFC).[3] These soft gels are lightly crosslinked polymer networks (*i.e.* prepared with a very small fraction of crosslinking agent), which are highly swollen by water and so have low mechanical stability, which deteriorates as pore size increases. GFC packings in relatively large columns are generally used at low pressures with slow eluent flow rates (< 1 cm^3 min^{-1}) with separations requiring several hours.[4] GFC packings have not been widely used for water soluble synthetic polymers.

The term *gel permeation chromatography* (GPC) was defined by Moore,[5] who developed rigid crosslinked polystyrene gels, covering a wide range of pore sizes, which are suitable for separations of synthetic polymers in organic media. These gels, which are extensively crosslinked, undergo limited or no swelling with the solvent and so have good mechanical stability for separations at high pressures and fast flow rates. GPC packings based on crosslinked polystyrene gels have been widely used for routine polymer characterization and quality control, in particular in determinations of MWD and for characterizing low polymers and small molecules, *e.g.* for prepolymers in resins and for polymer additives.[6,7] The dominant separation mechanism in GFC and GPC is assumed to be *size exclusion*, which requires minimum interactions between solute and gel. Because the crosslinked polymeric packings employed in GFC are often not susceptible to secondary retention mechanisms, efforts have been directed to the development of rigid macroporous packings constituted from hydrophilic polymers for SEC separations of water soluble synthetic polymers.[8-10]

The excellent mechanical stability of inorganic gels led to the development of macroporous silica and glass particles for polymer separations with both aqueous and organic eluents. De Vries and co-workers[11-13] described the use of porous spherical silica beads in GPC and Haller[14-16] developed a controlled method of preparing glass with uniform pores. However, experimental studies indicated that interactions could occur between the polymer and surface sites on the inorganic packing.[17] Work on separations of water soluble polymers has therefore involved inorganic packings deactivated with bonded phases in order to minimize secondary mechanisms arising from interactions.[18]

The resolution of molecules in a complex mixture by a chromatographic method depends on two types of processes. The *separation* process in the stationary phase controls the differential migration of the molecules and includes any secondary retention mechanisms additional to size exclusion which determine the permeation of molecules in the column packing. The *dispersion* process, which generally consists of at least two mechanisms, determines the band broadening of one component, and this broadening is influenced by molecular diffusion and the column packing. If the separation of two components by SEC is considered, giving the chromatogram shown in Figure 1, then the separation process influences the dependence of the retention (or elution) volume V_R on molecular size and the dispersion process determines the base width w of the peak of one component. SEC is widely used for the determination of MWD and average molecular weights of synthetic polymers. An important part of data interpretation is the establishment of a correct calibration relation between M and V_R. An understanding of the separation process is necessary because of the prevalent use of calibration methods which assume that separations of synthetic polymers are controlled solely by a size exclusion mechanism.[19] An understanding of the dispersion process is also required because the experimental chromatogram will always correspond to a MWD which is too broad. Correction methods for chromatogram broadening have been proposed in order to improve the accuracy of the derived data for average molecular weights.[20,21] Following on from the considerable advances in the theoretical understanding and in the experimental practice of liquid chromatography (LC) with microparticulate packings,[22] it has been shown that rigid microspherical packings in short low capacity columns provide high speed separations of polymers and high resolution separations of oligomers and prepolymers by *high performance size exclusion chromatography* (HPSEC). High performance follows from a decreasing contribution of the dispersion process as the particle size of the column packing is reduced. Whilst the decrease in the separation time is obviously desirable, workers involved in polymer characterization regard the precision of the MWD calculated from the experimental chromatogram as very important. If high efficiency columns containing microparticulate packings are used in HPSEC, corrections for chromatogram broadening may be quite small, so that correction procedures may be omitted in the determination of the number average and weight average molecular weights, $\bar{M}_n$ and $\bar{M}_w$, from an experimental chromatogram of a polydisperse polymer.

In this chapter the basic principles of the separation and dispersion processes will be discussed, as will the contribution of non-exclusion secondary mechanisms in some polymer separations. Emphasis will be placed on the use of microspherical packings in HPSEC systems, since HPSEC separations performed with careful attention to instrumentation and procedures give low chromatogram broadening. Consideration is given to achieving reliable and reproducible MWD data for high polymers, together with an assessment of the problems of handling polymers with heterogeneities such as branching and/or copolymer composition. In addition, HPSEC for high speed and high resolution separations of oligomers and prepolymers will be discussed.

12.2 SEPARATION PROCESSES

The volume of eluent V_R required to elute a particular component from a column, as shown in Figure 1, is measured from the injection point to the peak height maximum of the chromatogram.

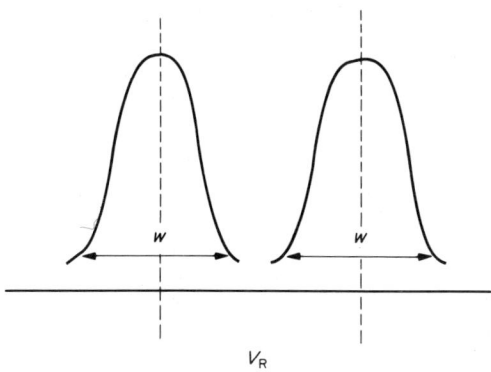

Figure 1 Resolution of two components in size exclusion chromatography; the parameters are defined in the text

The retention time t_R is related to V_R by

$$V_R = t_R f_V \tag{1}$$

where f_V is the volumetric flow rate. In LC[22] the degree of retention is given in terms of the capacity factor k', which is defined by

$$k' = (V_R - V_M)/V_M \tag{2}$$

where V_M is the total volume of solvent in the column. In conventional LC separations an unretained solute will elute at V_M corresponding to $k' = 0$. Therefore, in separations operating by interactive mechanisms in which the solutes are distributed between the stationary phase and the mobile phase (given by V_M), V_R will correspond to $k' > 0$, and several column solvent volumes may be required to elute all solutes.

The basic separation mechanism in SEC assumes that the column packing is 'inert', *i.e.* that interactive mechanisms such as partition and adsorption do not contribute to retention behaviour. The separation is determined by a size exclusion mechanism in which the pore volume in the gel particles accessible to a given molecule is determined by pore size distribution and solute size. Consequently, all molecules will be eluted within one column solvent volume V_M and in a typical separation the values of k' will lie in the range -0.5 to 0. It is necessary therefore to redefine the capacity factor and the volumes of the stationary and mobile phases in SEC. The column solvent volume V_M consists of the interstitial (or void) volume V_o of solvent between porous gel particles defined as the volume of the mobile phase and the volume V_i of solvent within the porous gel particles, defined as the volume of the stationary phase. For SEC the capacity factor k'' by analogy with equation (2) is given by

$$V_R = V_o(1 + k'') \tag{3}$$

The difference between k' and k'' is illustrated in Figure 2.

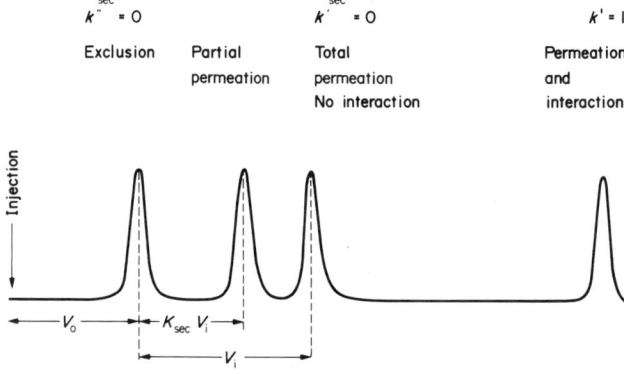

Figure 2 Elution of peaks in gel permeation chromatography; the parameters are defined in the text

12.2.1 Size Exclusion

The separation of a solute of given size in solution is determined by a distribution coefficient K_{SEC} which governs the fraction of V_i that is accessible to this solute. The value of V_R for this solute is given by

$$V_R = V_o + K_{SEC} V_i \tag{4}$$

Very large molecules having a zero value of K_{SEC} will elute at V_o because the sizes of these macromolecules prohibit solute diffusion into the gel pores. Very small molecules, on the other hand, have free access to both stationary and mobile phases, *i.e.* K_{SEC} is unity. As the chromatographic column is washed with solvent, the large molecules are eluted first, followed by solutes of decreasing size, which penetrate an increasing fraction of the solvent within the gel particles. The dependence of V_R on solute is shown in Figure 3 for solutes with a range of sizes, *e.g.* for calibration standards. The shape of the curve in the partial permeation plateau region is related to the pore size distribution within the porous packing.

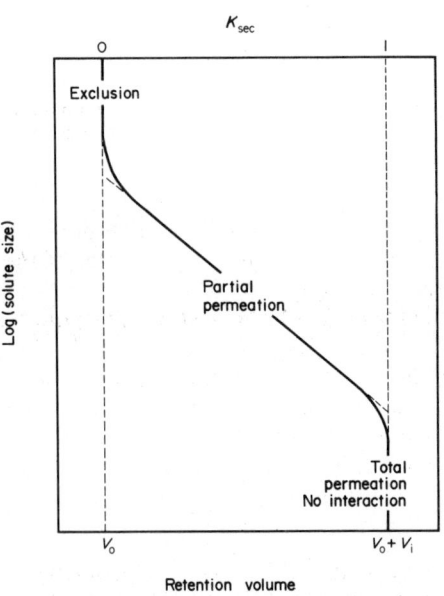

Figure 3 Calibration curve for a size exclusion mechanism; the parameters are defined in the text

Many theoretical models have been proposed for polymer separations by SEC. Whilst flow mechanisms could be important in some experiments, there is considerable evidence indicating that most SEC separations are performed close to equilibrium conditions. For example, at practical flow rates ($f_V \approx 1$ cm^3 min^{-1}), V_R is independent of flow rate and values of K_{SEC} are in good agreement with distribution coefficients determined from static experiments in which a polymer solution is mixed with a porous gel.[23,24] The standard free energy change $\Delta G°$ for the transfer of solute molecules from the mobile phase to the stationary phase at constant temperature T is related to K_{SEC} by

$$\Delta G° = \Delta H° - T\Delta S° = -RT \ln K_{SEC} \tag{5}$$

where R is the gas constant and $\Delta H°$ and $\Delta S°$ are, respectively, the standard enthalpy and entropy differences between solute in the two phases. In a size exclusion model the loss in conformational entropy is considered when a solute molecule is within the pore volume. It is assumed that there are no enthalpy changes when the solute transfers from one phase to the other and when the solute approaches the surface of the porous packing, *i.e.* $\Delta H°$ is zero.

The first theories of size exclusion considered simple geometric models from which the pore volume accessible to a polymer of given size may be calculated.[4] Statistical thermodynamical treatments[23,25] define the distribution coefficient K_{SEC} as the ratio of accessible solute arrangements within the porous packing to those in the mobile phase and from equations (4) and (5) K_{SEC} is given by

$$K_{SEC} = \exp(\Delta S°/R) \tag{6}$$

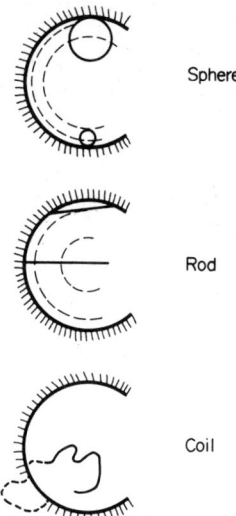

Sphere

Rod

Coil

Figure 4 Size exclusion in a circular pore for hard spheres, rigid rods and random coil chains, showing exclusion zones for centre of sphere and rod and permitted (——) and prohibited (– – –) conformations for a random coil chain

A simplistic representation of this approach is shown in Figure 4 for a solid sphere in a two-dimensional circular pore. The sphere size permits the transfer of the polymer molecule from the mobile phase to a pore within the packing, but the centre of gravity of the sphere cannot approach the pore surface and so is excluded from a volume of solvent around the pore surface. This excluded volume is clearly dependent on the solute size. A similar excluded volume model may be assumed for rigid rod and random coil polymers and may be extended to various models of pore shape.[23,25] For pores such as spheres with radius a, cylinders with radius a and slabs with thickness $2a$ the dependence of K_{SEC} on the size of a polymer chain having n segments, each segment having length b, as K_{SEC} approaches the limiting value of unity is given by

$$K_{SEC} = 1 - 2\lambda\psi(nfb^2/6a^2)^{0.5} \tag{7}$$

in which λ is a numerical factor dependent only on the pore geometry (1, 2 and 3 for the slab, cylinder and sphere respectively), f is the number of arms in a star molecule (1 or 2 for a linear chain), and ψ is a function dependent on f (equal to $\pi^{-0.5}$ for a linear chain). For various pore and solute geometries,[23-25] it can be shown that the expression for K_{SEC} for polymer separations may be generalized in terms of the relation

$$K_{SEC} = \exp(-s\bar{L}/2) \tag{8}$$

where $\bar{L}$ is the mean external length or molecular projection, *e.g.* $\bar{L}$ is equal to the diameter D of a sphere, and s is the surface area per unit pore volume. Casassa[26] has noted that his equation (7) is consistent with equation (8). For separations with K_{SEC} approaching the limiting value of unity he showed that

$$\bar{L} = 4(nfb^2/6\pi)^{0.5}[(3f - 2)/f^2]^{1/6} \tag{9}$$

provided f is not too large. These treatments of the size exclusion mechanism therefore suggest that for a column containing a given packing a characteristic molecular parameter $\bar{L}$ will represent the retention behaviour of all solutes.

From equations (4) and (7)–(9) a size exclusion mechanism predicts a universal calibration relation between solute size and V_R, and experimental results are typically presented as a plot of log (solute size) against V_R as in Figure 3. Benoit and co-workers[27] proposed that the hydrodynamic volume V_h, which is proportional to the product of $[\eta]$ and M, where $[\eta]$ is the intrinsic viscosity of the polymer in the SEC eluent, may be used as the universal calibration parameter. For solid spheres and random coils represented by solid spheres, the product of $[\eta]$ and M is related to V_h of the molecule by the Einstein equation

$$[\eta]M = 2.5N_oV_h = 2.5\pi N_oD^3/6 \tag{10}$$

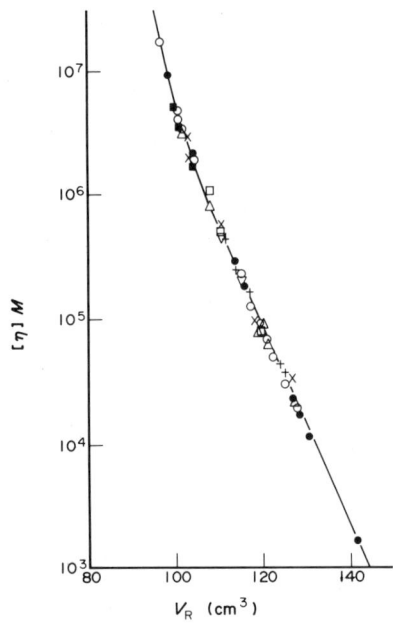

Figure 5 Hydrodynamic volume universal calibration curve for crosslinked polystyrene gels with tetrahydrofuran: ●, linear polystyrene; ○, branched polystyrene (comb type); +, branched polystyrene (star type); △, branched block copolymer of styrene/methyl methacrylate; ×, poly(methyl methacrylate); ○, poly(vinyl chloride); ▽, graft copolymer of styrene/methyl methacrylate; □, polybutadiene (reproduced with permission of John Wiley and Sons, Inc., from ref. 27)

where N_o is the Avogadro constant. Experimental results shown as a plot of $\log[\eta]M$ *vs.* V_R in Figure 5 confirm this universal calibration relation for homopolymers and copolymers separating with crosslinked polystyrene gels and with tetrahydrofuran as eluent. Equations (4) and (6) suggest that K_{SEC} is independent of temperature, a characteristic of a mechanism controlled by entropy changes. The dependence of the relation between solute size (corrected for expansion or contraction) and V_R on temperature may be studied with inorganic packings whose pore dimensions remain constant. Cooper and Bruzzone[28] eluted polystyrene and polyisobutene in trichlorobenzene from porous glass at 25 and 150 °C, concluding that universal calibration behaviour was observed.

Experimental studies with crosslinked polystyrene gels confirm V_h as the universal calibration parameter as long as the eluent, such as tetrahydrofuran, 1,2,4-trichlorobenzene, *o*-dichlorobenzene and chloroform, is a good solvent for linear polystyrene.[19] For separations of polymers in organic media with inorganic packings it is preferred to use an eluent formed from a binary liquid mixture containing one component having considerable affinity for the pore surface so that polymer–substrate interactions are minimized.[29-33] In view of the theoretical justification for equation (8) it is of interest to determine whether the V_h calibration method is valid for other molecular geometries apart from linear chains. The data in Figure 5 include branched polymers and this plot is also applicable to universal calibration of branched (or low density) and linear polyethylene, as shown in Figure 6.[34] The separation of polyethylene containing long chain branching has been studied thoroughly,[35] and almost without exception V_h has been found to be a valid universal calibration parameter. Equation (10) may be extended to macromolecules having a non-spherical geometry. Casassa[26] noted from $\bar{L}$ in equation (8) that V_h could provide universal calibration for random coil and rigid rod polymers for molecular sizes corresponding to a restricted range of axial ratios for the rod-like polymer. This has been confirmed for separations of polystyrene and poly(γ-benzyl L-glutamate) in *N,N*-dimethylacetamide with crosslinked polystyrene gels at 80 °C[36] as shown in Figure 7. For aqueous separations of polyelectrolytes it is recognized that solutions should contain electrolyte in order to reduce electrostatic repulsions along the polyelectrolyte chain. In addition, the ionic strength of the eluent must be sufficiently high to minimize possible interactions between ionic groups on the polyelectrolyte and charges on the surface of the column packing. The plot shown in Figure 8 for sodium poly(styrenesulfonate) and dextran separating on porous glass indicates universal calibration behaviour which has been facilitated by the addition of sodium sulfate to the aqueous eluent.[37] Although Rinaudo and co-workers[38,39] also indicated that universal calibration for separations with porous silica should operate when the ionic strength of the eluent exceeds

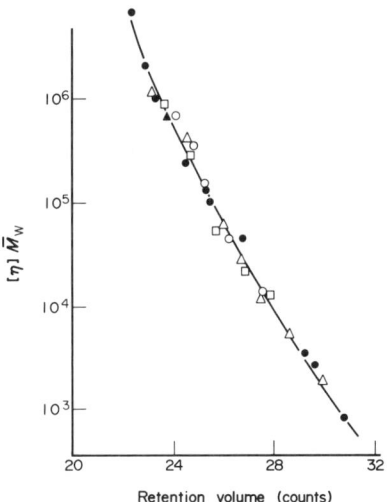

Figure 6 Hydrodynamic volume universal calibration curve for crosslinked polystyrene gels with trichlorobenzene: ●, linear polyethylene fractions; □ △ ○, branched polyethylene fractions (reproduced with permission of John Wiley and Sons, Inc. from ref. 34)

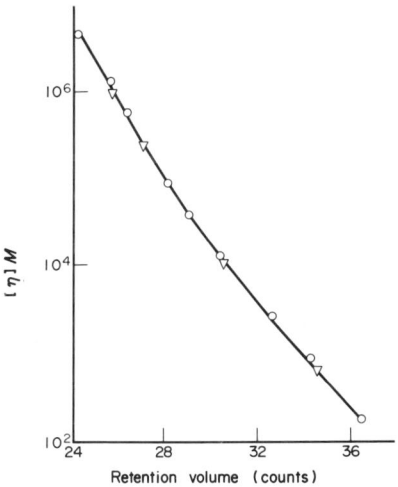

Figure 7 Universal calibration plot for crosslinked polystyrene gels with N,N-dimethylacetamide at 80 °C: ○, polystyrene standards: ▽, poly(γ-benzyl-L-glutamate) (reproduced with permission of Butterworth & Co. Ltd. from ref. 36)

0.05 M, later work on ionic interactions (see discussion of ion exclusion) suggests that polyelectrolytes may not conform to universal calibration.

12.2.2 Secondary Retention Mechanisms

Although size exclusion dominates many polymer separations, some experiments indicate that interactions between polymer and pore surface contribute to retention. A mixed mechanism consisting of size exclusion and a secondary mechanism may be considered as a network-limited separation[40] from which the following retention equation may be derived

$$V_R = V_0 + K_p K_{SEC} V_i \tag{11}$$

where K_p is the distribution coefficient for polymer–gel interaction. Then, equations (4) and (11) are identical with $K_p = 1.0$ for a size exclusion mechanism with no interactions with the stationary phase. Equation (11) may be given a thermodynamic interpretation.[41] Comparison of equations

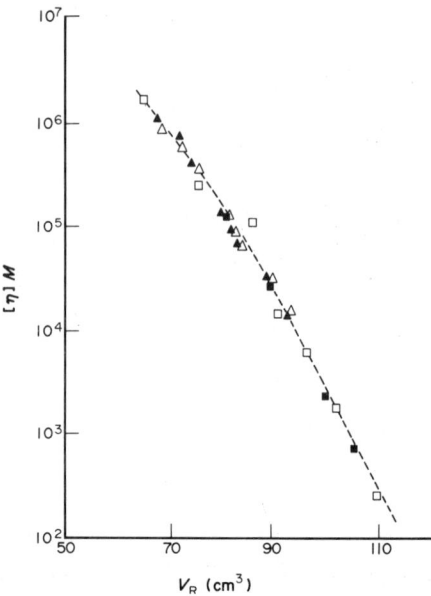

Figure 8 Universal calibration plot for porous glass packing: □ ■, sodium poly(styrenesulfonate); △ ▲, dextran; open symbols, 0.2 M aqueous eluent; closed symbols, 0.8 M aqueous eluent (reproduced with permission of John Wiley and Sons, Inc. from ref. 37)

(4)–(6) and (11) suggests that K_p is given by

$$K_p = \exp(-\Delta H^\circ/RT) \tag{12}$$

i.e. solute–gel interaction effects are determined predominantly by the standard enthalpy change ΔH° on solute transfer to the pore in the gel (as in LC). It follows from equations (11) and (12) that $K_p > 1.0$ corresponds to retardation of polymer in the stationary phase because of attractive interactions and that $K_p < 1.0$ corresponds to early elution of polymer because of repulsive interactions. Equation (11) has been shown to be a reasonable representation of a mixed mechanism for polymers in organic media.[40] Equation (12) clearly suggests that secondary mechanisms should be temperature dependent, and data exhibiting a decrease in K_p as T is raised are discussed elsewhere.[41,42]

Separations with N,N-dimethylformamide (DMF), which is widely used as an eluent for polar synthetic polymers, are influenced by interactions between polymer and stationary phase.[40] Very polar eluents are generally poor solvents for polystyrene. The displacement of calibration curves of V_h for polystyrene in poor and theta solvents to high V_R with respect to other polymers has been observed with separations on crosslinked polystyrene gels.[40,42] A good example[43] of polystyrene separating by a mixed mechanism is shown in Figure 9. The V_h curve for polystyrene in DMF is displaced to high V_R with respect to a curve for the other polymers and copolymers in DMF, which separate solely by size exclusion, whereas universal calibration behaviour is observed with chloroform as eluent. These DMF results have been interpreted in terms of equation (11) with K_p greater than unity.[40] Separations in which V_R is higher than expected from a size exclusion mechanism may arise from secondary *partition* (liquid–liquid) and *adsorption* (liquid–solid) mechanisms. Such interactions between polymer and stationary phase must be weak and reversible so that the polymer is not completely retained in the stationary phase. For some polymer–solvent–gel systems, V_R is even greater than $V_o + V_i$, which is inconsistent with a size exclusion mechanism for which K_{SEC} must lie between zero and unity. For the separations of polar polymers in DMF with crosslinked polystyrene gels, shown in Figure 10, retardation is influenced by interactions between polymer and the stationary phase and is dependent on polymer type.

Secondary mechanisms occurring on crosslinked polystyrene gels may be considered in terms of interaction parameters, permitting a thermodynamic interpretation of K_p.[45–48] A typical expression for the ternary system of solvent (1), polymer (2) and gel (3) is

$$\ln K_p = -(V_2/V_1)\phi_3(1 + g_{23} - g_{12} - g_{13}) \tag{13}$$

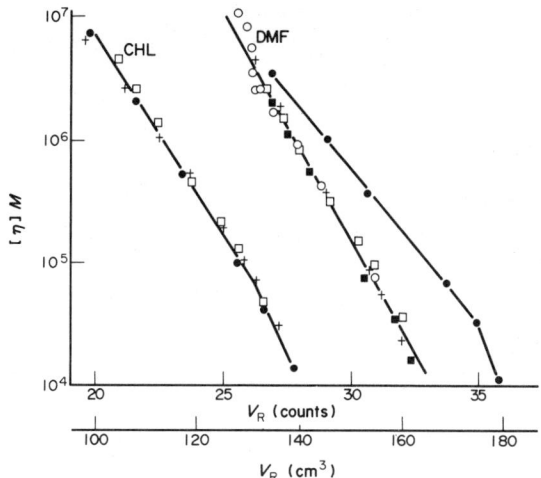

Figure 9 Universal calibration plots for crosslinked polystyrene gels with chloroform (CHL) and *N,N*-dimethylformamide (DMF) at 20°C: ●, polystyrene; □ + ■, copolymers of styrene and acrylonitrile; ○, polyacrylonitrile (reproduced with permission of *Angew. Makromol. Chem.* from ref. 43)

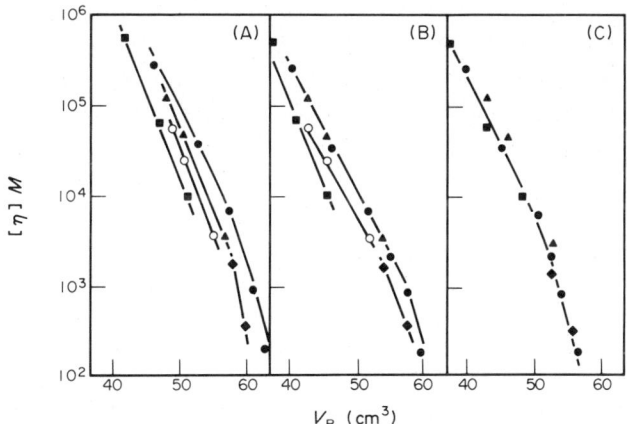

Figure 10 Universal calibration plots for crosslinked polystyrene gels (A), silanized porous glass (B), and porous glass (C), with *N,N*-dimethylformamide: ●, polystyrene; ▲, poly(methyl acrylate); ○, poly(vinylpyrrolidone); ■, poly(*p*-nitrostyrene); ◆, poly(ethylene oxide) (reproduced with permission of John Wiley and Sons, Inc. from ref. 44)

where V_1 and V_2 are molar volumes, ϕ_3 is the volume fraction of gel, and g_{23}, g_{12} and g_{13} are interaction parameters.

Observations of the early elution of some polymers may be explained by a mixed mechanism involving size exclusion and a secondary mechanism consisting of *partial exclusion* due to *polymer incompatibility* with the gel.[45,47-54] A polymer exhibiting repulsive interactions with the gel will generate a positive value of ΔH° in equation (12) leading to values of K_p below unity. Consequently, equation (11) predicts the displacement of a V_h curve to low V_R.[45] Results of K_p below unity have been reported for poly(*N*-3,6-dibromocarbazole), showing that values of K_p below unity decrease as molecular size increases as predicted from a thermodynamic interpretation of polymer incompatibility based on equation (13).[47] Mori and Suzuki[50] have also shown that epoxy prepolymers having values of K_p below unity show the same type of dependence of K_p on molecular size.

Studies of separations of polymers in organic media with inorganic packings have indicated that the plot of log V_h *vs.* V_R may be influenced by the eluent polarity with deviations from the universal calibration plot often being observed.[17] Because of active adsorption sites, polymer retardation and even irreversible retention is always possible with inorganic packings. Attention has been directed to eluents formed from binary liquid mixtures having an alcohol as one component.[29-33,55-57] The

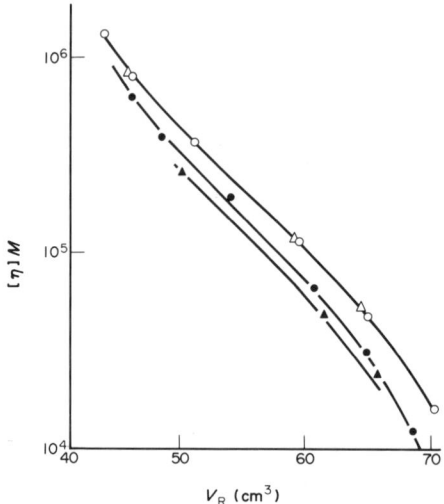

Figure 11 Hydrodynamic volume universal calibration with polystyrene standards for porous silica: ○, benzene; △, chloroform; ●, theta mixture benzene/methanol; ▲, theta mixture chloroform/methanol (reproduced with permission of Huthig and Wepf Verlag from ref. 29)

plots of log V_h for polystyrene in Figure 11 demonstrate that retention data for the binary liquid mixtures are displaced to low V_R with respect to single liquids which are good solvents for polystyrene.[29] The alcohol component, because of its considerable affinity for the surface of the porous silica, will tend to reduce the adsorption of polystyrene. With single solvents such as benzene, chloroform, 1,2-dichloroethane, methyl ethyl ketone and carbon tetrachloride, eluent polarity will be similar to that of polystyrene, which is retarded because it can compete for surface adsorption sites. When the non-solvent component in the binary liquid mixture is less polar than the solvent component, *e.g.* *n*-heptane and methyl ethyl ketone,[30-32] the plot of log V_h is displaced to high V_R with respect to the plot for polystyrene in methyl ethyl ketone. These experimental results demonstrate that reduced eluent polarity will facilitate polymer–substrate interactions. The results may be interpreted in terms of the solvent strength parameter ε° which represents the interaction energy of an eluent with the packing. These observations have been confirmed in the extensive study performed by Campos, Soria and Figueruelo[32] of polystyrene on porous silica with 21 eluents. Their data may be represented by equation (11), and a relation between K_p and ε° has been demonstrated. Audebert[46] has interpreted $K_{SEC}K_p$ in terms of $(\varepsilon_{23} - \varepsilon_{13})$, where ε_{ij} is an interaction energy, showing how the difference in interaction energy determines whether the polymer separates by size exclusion alone, by size exclusion and a repulsive interaction or by size exclusion and an attractive interaction. Experimental studies show how retention behaviour on inorganic packings with organic eluents is determined by polymer polarity.[58,59]

Water soluble synthetic polymers will contain hydrophobic segments with ionic and/or non-ionic hydrophilic groups. Because aqueous separations require the column packing to have a polar structure with or without ionic groups, a range of polymer–substrate interactions may occur, giving rise to various secondary retention mechanisms. It is not always easy therefore to achieve separations dependent only on size exclusion. Inorganic packings generally have anionic groups at the pore surfaces so that electrostatic interactions with polyelectrolytes may generate secondary mechanisms arising from ion exchange, ion exclusion and ion inclusion.[60,61] The magnitude of these mechanisms is extremely dependent on the composition of the eluent. In the *ion exclusion* mechanism the position of the calibration curve for anionic polyelectrolytes small enough in size to enter pores can be shown to depend on the ionic strength of the eluent, a typical example[62] of early elution being shown in Figure 12. Here, sodium poly(styrenesulfonate) in water is prevented from entering pores because of electrostatic repulsion between anionic charges, whereas there is no contribution from ion exclusion when the electrolyte concentration is sufficiently high. Quantitative treatments of a mixed mechanism involving size exclusion and ion exclusion as a function of ionic strength have been proposed and compared with experimental retention data.[63,64] *Ion inclusion* results because a Donnan equilibrium must be achieved even when an excluded high polymer containing charges cannot permeate pores. Consequently, low polymer is forced into the pores to equilibrate with counterions and is retarded in the column. This retardation may be reduced by adding electrolyte,

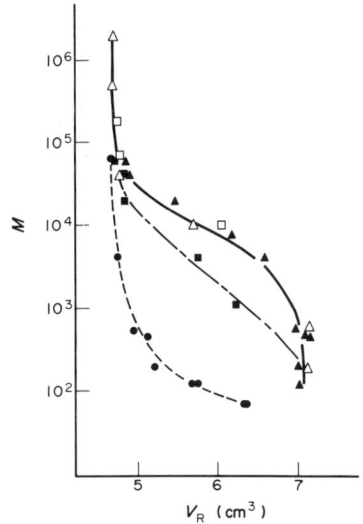

Figure 12 Molecular weight calibrations for aqueous separations of sodium poly(styrenesulfonate) and anionic standards (closed symbols) and dextran and non-ionic standards (open symbols): ● ○, 0.01 M aqueous eluent; ■ □, 0.1 M aqueous eluent; ▲ △, 1.0 M aqueous eluent (reproduced with permission of John Wiley and Sons, Inc. from ref. 62)

but the electrolyte then becomes included, generating 'spurious' peaks at the total permeation volume of the column.[38,65] Chromatographic packings having anionic groups can act as exchange sites with cationic polymers. Attempts to reduce this *ion exchange* mechanism involve lowering the pH and/or raising the ionic strength of the eluent. However, reproducible retention behaviour of cationic polyelectrolytes on non-modified inorganic packings is not easy to attain. Consequently, by analogy with separations of anionic polyelectrolytes with porous glass and porous silica having surface anionic groups, separations of cationic polyelectrolytes have been performed with porous packings to which bonded phases containing quaternary ammonium groups are attached.[66-68] Secondary retention mechanisms in aqueous media may arise not only from ionic effects but also from *adsorption* with non-ionic groups; for example not all silanol groups at a silica surface may dissociate in aqueous media and undissociated silanol groups may hydrogen bond with the polymer. If hydrogen bonding leading to higher V_R is suspected, the addition of urea or guanidine hydrochloride may reduce the interaction.[60] Because of the secondary mechanisms resulting from silanol groups, silicas with bonded phases (*e.g.* resulting from the reaction with γ-glycidoxytrimethoxysilane) have been advocated.[60] However, two problems remain. First, not all silanols are reacted, and, second, adsorption or partition of the polymer with the bonded phase may occur at high ionic strength of the eluent *via* a hydrophobic interaction. A form of hydrophobic binding has been proposed for some separations of polymers in aqueous media with the hydrophilic TSK gel PW.[69] It is emphasized therefore that separations of water soluble polymers by size exclusion with minimum perturbation by secondary mechanisms will depend on a careful choice of eluent composition after considering the nature of possible polymer-packing interactions.

12.3 COLUMN EFFICIENCY

12.3.1 Plate Height and Plate Number

A measure of the efficiency of a chromatography column is the height equivalent to a theoretical plate or plate height H.[70] The plate height for an experimental chromatogram is calculated from the expression

$$H = L/N \tag{14}$$

where L is the column length and N is the plate number. If the chromatograms such as the peaks in Figure 1 are symmetrical, corresponding to a normal error (or Guassian) function, then N may be determined from

$$N = 5.54(V_R/w_{0.5})^2 \tag{15}$$

where $w_{0.5}$ is the width of the chromatogram at half its height. A typical microparticulate packing with particle diameter $\sim 10~\mu m$ will generate a HPSEC column having $N > 20\,000$ plates m^{-1} for a solute eluting at $V_R = V_o + V_i$.

12.3.2 Dispersion Mechanisms

Theoretical interpretations of column efficiency consider the influence of the processes of solute dispersion in the mobile and stationary phases on the plate height H.[70] The basic concepts for general chromatographic separations can be applied to SEC, as proposed by Giddings and Mallik,[71] who examined the dependence of H on the linear flow velocity u of the eluent. Plate height may be thought of as the rate of change of peak (or solute zone) variance (in units of length) relative to the distance migrated L. The variance is the square of the standard deviation σ^2, so that H is defined by

$$H = \sigma^2/L \tag{16}$$

If there are several solute dispersion mechanisms contributing to chromatogram broadening and if these mechanisms are independent of each other, it follows from the laws of statistics that the variance of the chromatogram will be the sum of the variances associated with the individual mechanisms, *i.e.*

$$H = \Sigma\sigma^2/L \tag{17}$$

In the flow of liquid between particles in a packed bed, four major dispersion mechanisms may be identified: (i) dispersion arising from *eddy diffusion* because solute molecules in the mobile phase may move in streamlines between packing particles or have to move in streamlines around particles because the pathway is obstructed; (ii) dispersion arising from *molecular diffusion* in the longitudinal direction in the mobile phase; (iii) dispersion due to *mass transfer in the mobile phase* in which solute molecules move in different streamlines each having different relative velocities; and (iv) dispersion due to *mass transfer to and from the stationary phase* in which some solute molecules are in the stationary phase when others are in the mobile phase. Expressions for σ^2 for all these mechnisms can be derived and, if it is assumed that the mechanisms are *independent*, the dependence of the plate height H on the linear mobile phase velocity u may be written in the form

$$H = A + (B/u) + C_m u + C_s u \tag{18}$$

in which A, B, C_m, and C_s are constants for the dispersion processes (i), (ii), (iii) and (iv) respectively.

Giddings[70] recognized that radial movements of solute molecules will give a rapid interchange of solute molecules between streamlines, so that the molecules will have a range of velocities and will move from an unobstructed streamline between particles to a streamline moving around a particle. It follows that mass transfer in the mobile phase and eddy diffusion are interdependent. This is the basis of Giddings' coupling theory in which the eddy diffusion term is coupled non-additively to a mass transfer term in the mobile phase according to the relation

$$H = (B/u) + C_s u + \Sigma\{1/[(1/A) + (1/C_m u)]\} \tag{19}$$

The dependence of H on u is shown in Figure 13. Equation (19) is the basis of many methods of analyzing experimental plate height data in LC of small molecules.[22]

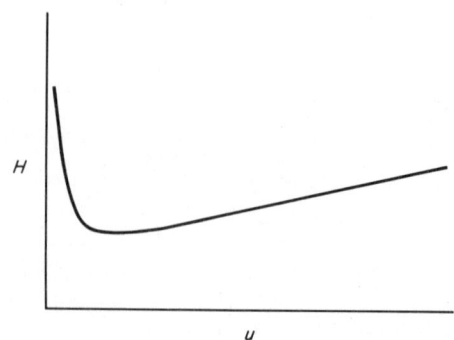

Figure 13 Dependence of plate height (H) on eluent flow rate (u) according to equation (19)

Equation (19) may be simplified for monodisperse high polymers having low diffusion coefficients. Experimental data for H for a solute having a constant retention volume V_R over a range of u may be interpreted in terms of the dispersion mechanisms occurring in the mobile and stationary phases. There is abundant experimental evidence, as reviewed elsewhere,[72] indicating that the first term in equation (19) may be neglected for high polymers at practical flow rates, *e.g.* $u > 1 \text{ mm s}^{-1}$. Experimental plate height data plotted as a function of u in Figure 14 do not display the minimum required by the first term in equation (19). The polystyrene standard PS-1 987 000 (see Figure 14) may be regarded as a non-permeating solute, which is confirmed by the GPC calibration curve.[73] The plate height data for this excluded solute, shown in Figure 14, suggest that for high polymers chromatogram broadening due to solute dispersion in the mobile phase exhibits little or no change with the eluent flow rate u, in agreement with experimental data for non-permeating polymers reviewed elsewhere.[72] Theoretical calculations also suggest that the term (A) is considerably larger than the term $(C_m u)$ for high polymers.[74] Consequently, it is concluded that the eddy diffusion term dominates mobile phase dispersion. It follows that only two dispersion terms, namely eddy diffusion in the mobile phase and mass transfer to and from the stationary phase, have to be considered in the expression for H for a monodisperse high polymer. Then, the dependence of H on u is given by[75]

$$ H = 2\lambda d_p + [R(1 - R)ud_p^2/30D_s] \tag{20} $$

in which λ is a constant close to unity which depends upon the packing, d_p is the packing particle diameter, R is the retention ratio defined by V_o/V_R and D_s is the diffusion coefficient of the solute in the stationary phase.

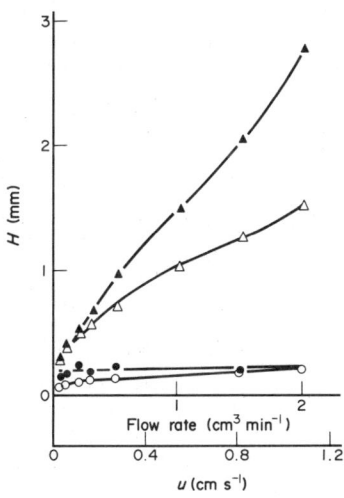

Figure 14 Dependence of plate height (H) on eluent flow rate (u) for polystyrene standards with porous silica: $\triangle$, polystyrene standard $M = 9800$; $\blacktriangle$, polystyrene standard $M = 35\,000$; $\bullet$, polystyrene standard $M = 1\,987\,000$; $\bigcirc$, toluene (reproduced with permission of Elsevier from ref. 73)

Synthetic polymers are not monodisperse, and it is necessary to include a polydispersity contribution to the experimental value of H.[76,77] It follows from equation (17) that equation (20) becomes

$$ H = 2\lambda d_p + [R(1 - R)ud_p^2/30D_s] + \sigma_m^2/L \tag{21} $$

where σ_m (in units of length) is the standard deviation for the true molecular weight distribution. If the permeating polymer has a logarithmic normal distribution,[76,77] then it can be shown that σ_m is related to the true polydispersity $[\bar{M}_w/\bar{M}_n]_T$ by the relation[72-75]

$$ (\sigma_m^2/L) = (L\ln[\bar{M}_w/\bar{M}_n]_T/D_2^2 V_R^2) \tag{22} $$

where D_2 is the slope of the SEC calibration relation between $\ln M$ and V_R. Therefore, equation (21) becomes

$$ H = 2\lambda d_p + [R(1 - R)ud_p^2/30D_s] + (L\ln[\bar{M}_w/\bar{M}_n]_T/D_2^2 V_R^2) \tag{23} $$

The data in Figure 14 for permeating high polymers $(M > 9000)$, and other results also,[74,75] may be represented by equation (23). It can be demonstrated that the mass transfer contribution to plate height in equation (23) increases as the diffusion coefficient decreases and therefore as molecular size increases. Consequently, chromatogram broadening for high polymers will be more extensive at faster eluent flow rates than that for low polymers. It is also evident in equation (23) that the most efficient separations will be obtained with the smallest packing particles because both mobile phase and mass transfer dispersion will be reduced. Procedures for estimating the eddy diffusion term in equation (23) have been discussed,[74, 75] so that it is possible to calculate $[\bar{M}_w/\bar{M}_n]_T$ from the experimental dependence of H on u with equation (23). This method for determining $[\bar{M}_w/\bar{M}_n]_T$ has been used to obtain reasonable values of polydispersity for polystyrene standards[74, 75] and to show that selected proteins may be regarded as monodisperse.[75]

Although equation (23) may be used as a basis for the interpretation of experimental chromatograms for polymers having narrow molecular weight distributions, the practical polymer scientist will generally be concerned with polydisperse samples which will be evaluated in terms of average molecular weights. The experimental polydispersity $[\bar{M}_w/\bar{M}_n]$ may be related to $[\bar{M}_w/\bar{M}_n]_T$ if it is assumed that the chromatogram and the molecular weight distribution are represented approximately by a logarithmic normal function. The experimental value of H is given by the variance σ_E^2 of the experimental chromatogram and σ_m was related to $[\bar{M}_w/\bar{M}_n]_T$ in equation (22). It follows that equation (23) may be transformed to

$$\ln[\bar{M}_w/\bar{M}_n] = \ln[\bar{M}_w/\bar{M}_n]_T + (D_2^2 V_R^2/L)(2\lambda d_p + [R(1-R)ud_p^2/30D_s]) \tag{24}$$

12.4 COLUMN PACKINGS

Essential conditions for the effective fractionation of polymers by SEC are that the pore sizes in the column packing should be comparable to polymer sizes in solution and that the packings should have substantial pore volume, typically $0.5 < V_i/V_o < 1.65$ for macroporous packings. The pore size distribution in a given sample of a porous packing only gives a useful separation range, *e.g.* the plateau region in Figure 3, of about one or two decades in molecular size, as illustrated by the calibration curves for rigid crosslinked polystyrene gels displayed in Figure 15.[78] Consequently, to separate samples over a wide size range, *e.g.* in Figure 5, it is necessary either to use a single column containing several gels having various pore size distributions, see for example the mixed gel calibration in Figure 15, or to have a series arrangement of columns, each covering a different molecular size range. Typical column dimensions for HPSEC are column length = 25–30 cm and internal diameter = 0.7–0.8 cm, and separations are often performed with three or four columns connected in series with short lengths of low volume narrow bore tubing. It is advantageous for the pore size distribution to generate a linear relation between log M and V_R in order to facilitate molecular weight analysis. This may be achieved in the bimodal pore size distribution approach,[79] which involves coupling columns containing only two discrete pore size distributions and approxi-

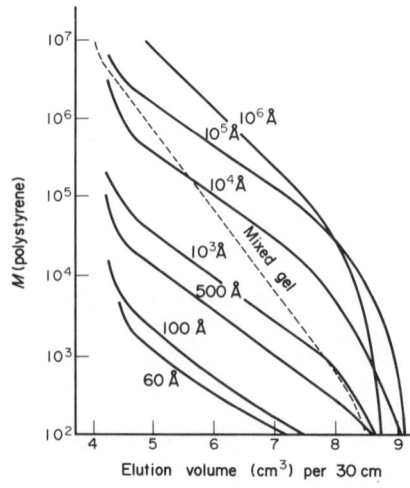

Figure 15 Molecular weight calibrations with polystyrene standards for crosslinked copoly(styrene/divinylbenzene) packings (PLgel) of various porosities as indicated (reproduced with permission from ref. 78)

mately equal pore volumes for each pore size distribution. Wide-range linear calibrations are possible for bimodal column sets with pore sizes differing by about one decade.

Rigid microparticulate packings are generally employed in HPSEC, and full details have been reviewed.[6,18,78,80-83] Packings with particle diameters of ~10 μm are typically used for high polymers, with high resolution separations of low polymers, prepolymers and small molecules being performed with particles having diameters of ~5 μm. Separations of long chain polymers with microparticulate packings must be performed carefully in order to avoid shear degradation during macromolecular diffusion through the column. Macromolecules may experience extensional forces caused by the converging–diverging flow behaviour of the mobile phase as it passes through the packed bed and chain scission may result. Consequently, when separating high polymers with $M > 500\,000$, dilute polymer solutions should be injected, the eluent flow should be reduced to a low velocity, and small packing particle sizes ($d_p < 10$ μm) should be avoided.[84-89] Slurry-packing methods are then preferred for column preparation; experimental packing methods are given elsewhere.[6,83] Many rigid packings are available in prepacked columns. The best SEC performance is obtained with columns containing regular microspheres having a narrow particle size distribution. The permeability of a homogeneous bed formed from such microspheres is optimized and the pressure drop is much lower than for a column packed with irregular particles having a wide size distribution. Typically, the limiting conditions for columns containing rigid packings are pressure drop less than 2000 p.s.i. and linear flow velocity less than 600 cm h^{-1}.

12.4.1 Polymer Packings: Macroporous Gels

These packings are highly crosslinked polymer particles which exhibit a predominantly permanent porosity with little or no swelling. Macroporous polymer gels therefore have high rigidity and retain porosity in the dry state, permitting their use in HPSEC for a wide range of synthetic polymers. Some limited swelling may occur with selected eluents. The crosslinked polystyrene gels in Figure 15 and Table 1 are compatible with a wide range of organic eluents. As explained in Section 12.2 it is preferable to use an eluent with a similar polarity to that of polystyrene, *e.g.* a similar solubility parameter, when adsorption and partition effects are generally absent. Possible disadvantages of rigid organic gels are susceptibility to thermal degradation and a decrease in mechanical stability at the elevated temperatures which are required in separations of polyalkenes and some condensation polymers.

Table 1 Rigid Polystyrene Packings

Fractionation range	Trade name	Supplier
100–10^7 (polystyrenes)	PLgel	Polymer Laboratories
	μ-Styragel	Waters Associates
	Shodex A	Showa Denko
	HSG	Shimadzu
	TSK Type H	Toyo Soda

Polymer packings which are hydrophilic and non-ionic should provide a more versatile column packing for aqueous separations than inorganic packings. Separations of a range of synthetic polymers[8,90,91] have been accomplished with TSK gel type PW (see Table 2 and Figure 16). A comparison of the TSK gel types SW (see Table 4) and PW has been reported.[92] Both these column packings provided acceptable separations for dextran, poly(vinyl alcohol) and poly(vinylpyrrolidone). For polyacrylamide, poly(acrylic acid) and poly(ethyleneimine) it was necessary to add salt to the eluent, but these polymers still displayed adsorption effects with the SW gel, whereas satisfactory separation behaviour was observed with the PW gel.

12.4.2 Inorganic Packings

Rigid inorganic packings have numerous advantages in SEC separations. The packings do not swell and have excellent mechanical and thermal stability. Inorganic packings are much easier to

Table 2 Hydrophilic Polymer Packings: Macroporous Gels

Gel	Fractionation range	Trade name
Hydroxylated polyether copolymer	$1000–7 \times 10^6$ (polysaccharides) $100–8 \times 10^6$ [poly(ethylene oxide)s]	TSK Gel PW Bio-Gel TSK Micropak TSK PW Spherogel PW
Polyacrylamide	$100–10^5$ [poly(ethylene oxide)s]	PLaquagel
Poly(2-hydroxyethyl methacrylate)	$20\,000–5 \times 10^6$ (polysaccharides)	Spheron
Methacrylate glycerol copolymer	$< 400\,000$ (polysaccharides)	Shodex-OH pak
Sulfonated polystyrene	$< 5 \times 10^6$ (polysaccharides)	Shodex Ion-pak
Poly(vinyl alcohol)	$< 5 \times 10^5$	Asahipak GS

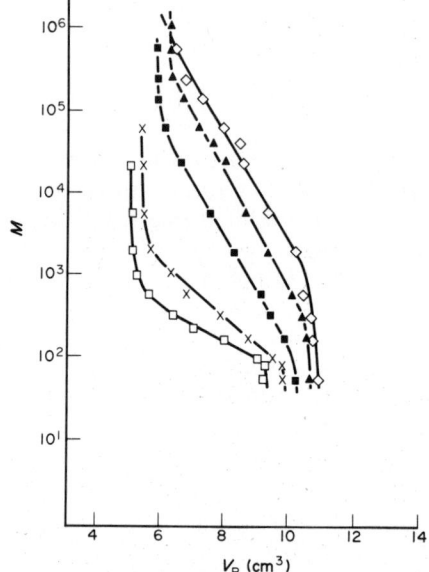

Figure 16 Molecular weight calibrations with poly(ethylene oxide) standards for TSK-GEL PW type columns: □, gel 1000 PW; ×, gel 2000 PW; ■, gel 3000 PW; ▲, gel 4000 PW; ◇, gel 5000 PW (reproduced with permission of Plenum from ref. 92)

handle than crosslinked organic gels and can be regenerated by heating or by treating with acids. The pore size distribution of rigid particles can be determined, *e.g.* by mercury porosimetry, so that theoretical studies of the separation mechanism can be compared with experimental SEC results. Furthermore, the volumes V_i and V_o are not dependent on eluent and temperature, so that the results from a variety of experiments can be compared.

Example of inorganic packings are given in Table 3. Calibration curves reported in ref. 93 for six Porasil porous silicas, with pore diameters in parentheses, are shown in Figure 17. Columns of inorganic packings can be used at high flow rates and high pressures. Inorganic packings are suitable both for aqueous and organic eluents and are particularly suitable for separations at high temperature, *e.g.* polyalkenes. The main deficiency of inorganic packings is the presence of surface sites which may facilitate the interaction of some polymers with the packing as discussed in Section 12.2.2 in the consideration of secondary retention mechanisms.

For polar polymers in both organic and aqueous phases, it is advantageous to coat inorganic packings with a surface-bonded phase in order to minimize polymer–substrate interaction effects. Packings produced by reaction of porous silica or porous glass with γ-glycidoxypropyltrimethoxy-silane have given size exclusion separations for a range of water soluble polymers.[60,61,80,81] Inorganic packings with various bonded phases are listed in Table 4. Calibration curves for aqueous

Table 3 Underivatized Inorganic Packings

Packing	Fractionation range	Trade name
Porous glass	100–4 × 10⁶ (polystyrenes)	CPG
	1000–1.5 × 10⁶ (polysaccharides)	
	3000–8 × 10⁸ (proteins)	
Porous silica	Mean pore sizes from	Porasil
	60 to 25 000 Å	Fractosil
		LiChrospher
		SE Series
		Spherosil
		Zorbax

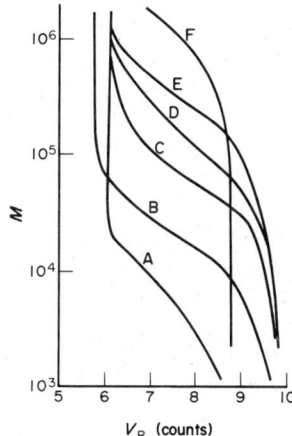

Figure 17 Molecular weight calibrations with polystyrene standards for Porasil silicas: A (< 100 Å), B (100–200 Å), C (200–400 Å) D (400–800 Å), E (800–1500 Å), F (> 1500 Å) (reproduced with permission of Steinkopff Verlag from ref. 93)

Table 4 Inorganic Packings with Bonded Phases

Packing	Bonded phase	Fractionation range	Trade name
Porous silica	—CH₂CHCH₂O— (OH)	100–2 × 10⁵ [poly(ethylene oxide)s] 10³–1 × 10⁶ (proteins)	TSK Gel SW
Porous silica	—(CH₂)₃OCH₂CHCH₂OH (OH)	10³–10⁷ (polysaccharides) 5000–500 000 (proteins)	SynChropak GPC Aquapore LiChrosorb Diol LiChrospher Diol Bio-Sil GFC Aquachrom
Porous silica	Polyether	500–10⁶ (polysaccharides)	μBondagel
Porous silica	Undisclosed	600–5 × 10⁵ (proteins)	Protein column
Porous silica	Polymerized amine	similar to SynChropak GPC	SynChropak CATSEC
Porous glass	—(CH₂)₃OCH₂CHCH₂OH (OH)	Similar to SynChropak	Glycophase CPG Glyceryl CPG

separations of dextrans of poly(ethylene glycol)s with TSK gel type SW[94] are displayed in Figure 18. Silica packings with bonded phases in long term use with eluents having a pH above 8 may exhibit changed retention behaviour because of the dissolution of the silica surface and therefore removal of the bonded phase.

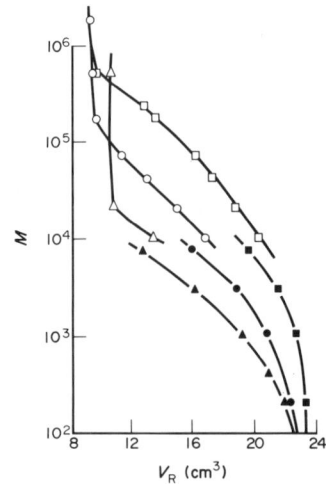

Figure 18 Molecular weight calibrations with dextran standards (open symbols) and polyethylene glycol standards (closed symbols) for TSK-GEL SW type columns: ▲ △, gel G-2000 SW; ○ ●, gel G-3000 SW; □ ■, gel G-4000 SW (reproduced with permission of Elsevier from ref. 94)

12.5 INSTRUMENTATION

12.5.1 Pumping/Injection Systems

The accurate determination of V_R is an essential part of the procedure in the evaluation of molecular weight distributions and average molecular weights of polymers by HPSEC. Short columns require precise measurements of low retention volumes, e.g. $V_i \approx 20$ cm^3 for $L = 120$ cm. Consequently, a constant flow pump which is reproducible and essentially independent of back pressure is preferred. Then, a pulse free liquid flow giving minimum detector noise will be generated and, if the eluent flow is truly constant, V_R may be measured as a time base along a recorder paper, thus avoiding experimental errors in the determination of V_R by siphon or by drop counter. It must be stressed that small variations in flow rate can cause large errors in this procedure for determining molecular weights from the chromatogram of a polymer.[95,96] A useful procedure is to monitor each HPSEC separation in terms of the retention volume of an internal standard.[97]

Injection of polymer solutions into SEC columns is commonly performed with valve loop injectors. The polymer solution must be injected in as narrow a band as possible in order to avoid overloading the short columns and to minimize chromatogram broadening. Therefore, low injection volumes are necessary. Attempts to increase the detector response by raising the polymer concentration should be avoided because of an increase in solution viscosity which may result in additional solute dispersion by a mechanism known as viscous fingering. The latter is most important for high molecular weight polymers, resulting in an increase in retention volume and chromatogram broadening. An examination of the dependence of chromatogram broadening and column efficiency on the injection volume and the concentration of solute has been performed by Mori.[98]

12.5.2 Eluents

The choice of packing for the separation of a given polymer will have to take into consideration the solvent for the polymer in order to assess whether polymer–eluent and eluent–packing interactions may influence secondary retention. Ideally, the eluent should be a good solvent for the polymer, should permit high detector response from the polymer and should wet the packing surface. The most common eluents in SEC are tetrahydrofuran for polymers which dissolve at room temperature, o-dichlorobenzene and trichlorobenzene at 130–150 °C for crystalline polyalkenes, and m-cresol and o-chlorophenol at 90 °C for crystalline condensation polymers such as polyamides and polyesters. For more polar polymers, dimethylformamide and aqueous eluents may be employed, but care is required in avoiding solute–gel interaction effects. Secondary retention mechanisms are always likely to occur when polymer–solvent interactions are not favourable, when polar polymers are separated with less polar eluents, and when packings have active surface sites. Therefore, it may be necessary to select carefully the eluent composition, *e.g.* by the addition of an electrolyte or organic component, after assessing the nature of possible polymer–packing interactions. Even for a non-aqueous eluent such as DMF, the addition of salt has been used to minimize non-exclusion mechanisms.[99–101]

12.5.3 Detectors

In SEC the concentration by weight of polymer in the eluting solvent may be monitored continuously with a detector measuring refractive index, UV absorption or IR absorption. The resulting chromatogram is therefore a weight distribution of the polymer as a function of V_R. When the molecular weight of the polymer in the eluting solvent is measured experimentally with a low angle laser light-scattering detector (LALLS), then the dependence of $w(M)$ on M can be established directly. Alternatively, if universal calibration is valid, as shown in Figures 5 to 8, then at a given V_R the relation

$$\log[\eta]_p M_p = \log[\eta]_{ps} M_{ps} \qquad (25)$$

will apply. Here, subscript p refers to a polymer requiring characterization and subscript ps to polymer standards, which will be polystyrene standards for organic eluents and poly(ethylene oxide) and/or polysaccharide standards for aqueous eluents. Equation (25) permits the determination of the M_p calibration curve when the dependence of $[\eta]_p$ and $[\eta]_{ps}$ on V_R have been established, which may be accomplished by measuring $[\eta]$ with an on-line viscometer detector. This universal calibration approach is an important component of the characterization procedure for branched polymers and copolymers.

The experimental SEC conditions require highly sensitive concentration detectors giving a detector response which is linearly related to polymer concentration. The most common detector for monitoring polymer concentration in the eluent is the differential refractometer. Except for very low polymers, the response of the detector to polymer concentration does not depend on polymer molecular weight. The most sensitive detector is the differential UV photometer, which is appropriate for a polymer with a significant UV absorbance at a convenient wavelength with a non-absorbing eluent. This detector is not affected appreciably by flow pulsations, flow rate changes and temperature fluctuations. In the characterization of copolymers it is necessary to have two detectors in series, *e.g.* a refractometer with either a UV detector or an IR detector. An IR detector is preferred for the detection of polyalkenes at elevated temperatures, because baseline noise and drift are much less than for the refractometer detector.

In LALLS the intensity of scattering from the polymer is expressed in terms of the excess Rayleigh factor R_θ defined as the scattering intensity of the polymer solution minus the scattering intensity of the solvent at a given angle θ normalized with respect to the intensity of the incident beam and the scattering volume. The value of R_θ will be a function of the scattering angle, the polymer concentration and the polymer molecular weight. When measurements are performed at low angles with respect to the incident beam, the equation

$$Kc/R_\theta = (1/\bar{M}_w) + 2A_2 c \qquad (26)$$

is applicable where c is the polymer concentration, A_2 is the second virial coefficient and K is an optical constant defined as

$$K = \frac{2\pi^2 n^2}{\lambda^4 N_o} \left[\frac{dn}{dc}\right]^2 (1 + \cos^2 \theta) \qquad (27)$$

where n is the refractive index of the solvent, λ is the wavelength of the incident beam and the dn/dc is the specific refractive index increment of the polymer solution. In order to apply equations (26) and (27) to an eluting polymer, concentration and LALLS detectors must be attached on-line to the column system, and separate measurements must be performed to establish calibration values of dn/dc and A_2. $\bar{M}_w$, $\bar{M}_n$ and $w(M)$ may then be computed for a polymer. Although light scattering gives good scattering intensities for polymers having high M, there may be little or no LALLS detector sensitivity with $M < 10^4$. For a polydisperse polymer, experimental measurement of M for the chromatogram at high V_R may not be accurate. It follows that when average molecular weights are computed from the distribution $w(M)$ derived from data obtained with concentration and molecular weight detectors, the value of $\bar{M}_w$ is likely to be more reliable than $\bar{M}_n$, which could be substantially in error.[102–105]

Various viscometric detectors have been reported. The polymer concentration must be known in an elution volume increment, and the viscometric and concentration detectors must be attached on-line to the column system in order to establish the intrinsic viscosity $[\eta]$ of the polymer in any increment of eluting solution.[106-110]

12.6 SEPARATIONS

In this section several examples will be presented to illustrate the scope of SEC. The main application of SEC is in the determination of molecular weight distributions of high polymers which is generally straightforward for homopolymers but requires special methods for branched polymers and copolymers. Increasingly, SEC is used for characterizing oligomers, prepolymers and low polymers when high resolution and/or high speed separations are required.

12.6.1 Molecular Weight Distributions

For workers involved in the characterization of high polymers, the precision of the molecular weight distribution $w(M)$ calculated from the experimental chromatogram will be important. The distribution $w(M)$ is calculated from a chromatogram provided the relation between $\log M$ and V_R has been established by calibration or has been determined by LALLS detection. The normalized differential weight distribution is related to a chromatogram by

$$w(M) \;=\; -\frac{dC(V_R)}{dV_R}\,\frac{dV_R}{d\log M}\,\frac{d\log M}{dM} \tag{28}$$

where $d\log M/dM$ is $1/M$ and $C(V_R)$ is the weight fraction of polymer eluted up to retention volume V_R and is related to the ordinate of the chromatogram $F(V_R)$ by

$$F(V_R) \;=\; \frac{dC(V_R)}{dV_R} \tag{29}$$

It follows from equation (24) that the most accurate data for high polymers will be obtained at low eluent flow rates with columns having high efficiencies, *i.e.* well-packed columns containing the smallest particles. A typical molecular weight distribution computed directly from a chromatogram obtained at low eluent flow rate is displayed in Figure 19, where the experimental molecular weight distribution is seen to be broader than a theoretical distribution predicted from the polymerization method.[72] Even when mass transfer dispersion is minimized by performing separations at low u, chromatogram broadening will still be present because of the eddy diffusion term in equation (24).

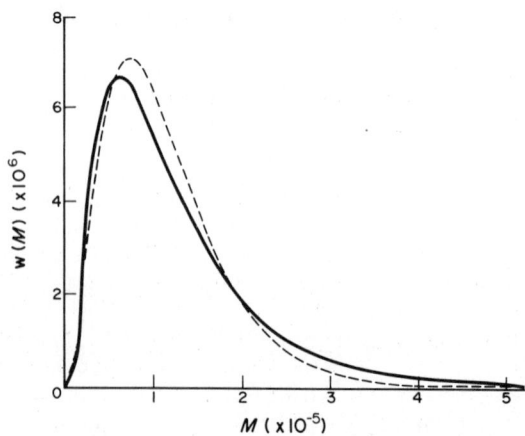

Figure 19 Molecular weight distribution for polydisperse polystyrene: ———, from experimental chromatogram determined by HPSEC with microparticulate crosslinked polystyrene gels ($d_p \approx 10\ \mu m$) at an eluent flow rate of 0.1 cm^3 min^{-1}; ----, theoretical distribution predicted from the polymerization mechanism (reproduced with permission of Butterworth Ltd. from ref. 72)

From the normalized distribution w(M), in equation (28), the number, viscosity, and weight average molecular weights, $\bar{M}_n$, $\bar{M}_v$ and $\bar{M}_w$ respectively, are calculated from the relations

$$\bar{M}_n = 1/\int (1/M)w(M)\,dM \tag{30}$$

$$(\bar{M}_v)^a = \int M^a w(M)\,dM \tag{31}$$

$$\bar{M}_w = \int M w(M)\,dM \tag{32}$$

Here, *a* is the exponent in the Mark–Houwink dilute solution viscosity equation

$$[\eta] = K\bar{M}_v{}^a \tag{33}$$

where *K* is a constant. To demonstrate how the eluent velocity, and therefore mass transfer dispersion, influences the precision of average molecular weights, Dawkins and Yeadon[72] determined the molecular weight distribution of a broad distribution polystyrene as a function of *u* and computed values of $\bar{M}_w$ and $\bar{M}_n$ from the distribution obtained at each flow rate. The fall in $\bar{M}_w/\bar{M}_n$ (of about 5%) as *u* decreases is shown in Figure 20 confirming the prediction of equation (24). Procedures for estimating mobile phase dispersion with equations (23) and (24) have been discussed,[74,75] so that it is possible to calculate $[\bar{M}_w/\bar{M}_n]_T$ from the experimental dependence of *H*, or $[\bar{M}_w/\bar{M}_n]$, on *u*. Values of $[\bar{M}_w/\bar{M}_n]_T$ calculated from data for *H* for polystyrene standards are in reasonable agreement with theoretical values predicted for polystyrene from a 'living' anionic polymerization.[74]

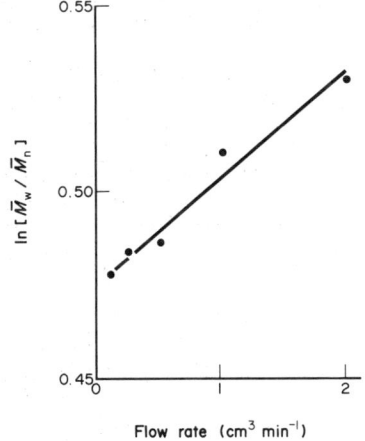

Figure 20 Dependence of measured polydispersity ($\bar{M}_w/\bar{M}_n$) of polystyrene on eluent flow rate (reproduced with permission of Butterworth Ltd. from ref. 72)

Figure 1 indicates that the chromatogram for a monodisperse solute is not a rectangle but a bell-shaped peak, so that a polydisperse polymer having a range of solute sizes generates a chromatogram which is a collection of a large number of overlapping peaks. The distribution w(M) calculated with equation (28) is then broader than the true molecular weight distribution (see Figure 19). Because of dispersion mechanisms, the tails of the chromatogram result from the broadening alone and the solute concentration at a given retention volume depends on the component eluting at V_R and on the broadening contributions from neighbouring components. If the experimental chromatogram is represented by $F(V_R)$ and if w(Y) represents the ideal chromatogram in the absence of broadening, *i.e.* as *H* tends to zero, then these two functions are related by the equation proposed by Tung[111]

$$F(V_R) = \int w(Y)G(V_R, Y)\,dY \tag{34}$$

where $G(V_R, Y)$ is a function describing the broadening contribution for an individual solute component having Y as its retention volume. The use of equation (34) involves choosing an appropriate function for $G(V_R, Y)$ and a numerical technique to solve the equation, as reviewed elsewhere,[20,21,112] but it is not necessarily easy to achieve reliable $w(Y)$ chromatograms with these correction methods. Equations (23) and (24) suggest that reducing dispersion mechanisms will minimize chromatogram broadening, for example by using microparticulate packings (low d_p) and slow eluent flow velocities (low u). Therefore, a chromatogram having a very low broadening contribution may be obtained by selecting the optimum conditions, so that the calculations involved in equation (34) may be omitted.

In view of problems which can be encountered in finding the function $w(Y)$ accurately, an alternative correction method for the average molecular weights may be employed because the moments of the $F(V_R)$ and $w(Y)$ functions are easily related. Thus, a true average molecular weight may be determined from an uncorrected average molecular weight computed from an experimental chromatogram.[20,112] In general, corrections for chromatogram broadening may be ignored if a polymer has a wide molecular weight distribution ($\bar{M}_w/\bar{M}_n > 2$).

12.6.2   Molecular Weight Calibration

In the absence of a LALLS molecular weight detector, the calculation of $w(M)$ in equation (28) from an experimental chromatogram necessitates the determination of the relation

$$\log M = f(V_R) \qquad (35)$$

which is invariably dependent on polymer type and structure. In the simplest situation $\log M$ is linearly related to the function $f(V_R)$ and we can write

$$\log M = D_1 - D_2 V_R \qquad (36)$$

where D_1 and D_2 are constants. Method for establishing the calibration curve defined by equations (35) and (36) can be placed into three categories:[19] the use of calibration standards having narrow molecular weight distributions, the application of procedures involving polydisperse reference materials, and universal calibration requiring a relation between molecular size in solution and molecular weight.

The calibration relation between $\log M$ and V_R for use in equation (28) is readily established with standards of the polymer requiring analysis, provided these standards have narrow molecular weight distributions, i.e. $[\bar{M}_w/\bar{M}_n] \approx 1.1$. The chromatogram of a standard consists of an ordinate $F(V_R)$ which is proportional to the concentration of polymer eluting at retention volume V_R and the molecular weight corresponding to V_R at the peak of $F(V_R)$ is termed M_{peak}. For narrow molecular weight distributions, a plot of log(average molecular weight) *vs.* peak retention volume is satisfactory. Calibration standards for polystyrene, poly(α-methylstyrene), polyisoprene, polybutadiene, polyethylene, poly(methyl methacrylate), polytetrahydrofuran, poly(ethylene oxide) and poly(styrenesulfonate) are available.

Well-characterized standards with narrow molecular weight distribution may not be available, or if they are available, they may not cover a wide range of V_R. When broad distribution reference materials are employed, M_{peak} may not correspond to one of the common average molecular weights, which may lead to errors in positioning the calibration curve. Polydisperse reference materials, *e.g.* dextrans and polyethylenes, generally enable the calibration curve to be constructed over a wide range of V_R by one of the following procedures:[19,113] determination of M_{peak} from the average molecular weights of the reference material; calculation of average retention volumes corresponding to average molecular weights of the reference material; calculation of the calibration curve defined by equation (36). These procedures calculate data for average molecular weights and/or molecular weight distribution from a chromatogram(s) by trial and error methods, in which the calibration curve is adjusted until an acceptable fit of the computed molecular weight data with the experimental information on average molecular weights and/or molecular weight distribution is obtained.

In the absence of well-characterized standards and reference materials of the polymer requiring analysis for part or all of a wide retention volume range, a calibration curve can be established if solute size is controlling separation. The separation mechanism must be size exclusion when the elution behaviour of polymers can be represented by a universal size parameter. The experimental evidence in Figures 5–8 suggests that a plot of $\log [\eta]M$ *vs.* V_R will be the same for all polymers. Thus, a single calibration curve in terms of molecular size established experimentally with standards and/or reference materials then permits the determination of a molecular weight calibration for a given polymer when the relation between the molecular size of that polymer in solution and molecular weight is known. At a given V_R, it follows from equations (10) and (25) that the relation

$$\log M_p \ - \ \log M_{ps} \ = \ \log[\eta]_{ps}/[\eta]_p \tag{37}$$

will apply. Here, subscript p refers to the polymer requiring analysis and subscript ps to an experimental study with standards. It follows from equation (37) that polymers will have different $\log M$ vs. V_R calibrations, unless they follow the same relationship between $[\eta]$ and M. Equation (37) therefore permits the determination of M_p when the dependence of $[\eta]_{ps}$ and $[\eta]_p$ on V_R has been established. Experimentally, the right hand side of equation (37) can be obtained with an automatic viscometer as detector. Alternatively, equation (37) can be converted to a relation between M_p and M_{ps} using the Mark–Houwink constants for standards and the polymer to be analyzed. Substitution of equation (33) into equation (37) and rearrangement gives

$$\log M_p \ - \ \frac{1 \ + \ a_{ps}}{1 \ + \ a_p} \log M_{ps} \ = \ \frac{1}{1 \ + \ a_p} \log \frac{K_{ps}}{K_p} \tag{38}$$

The universal procedure is illustrated in Figure 21 for linear polyethylene in o-dichlorobenzene at 138 °C. The polyethylene fractions have $\bar{M}_w/\bar{M}_n > 1.2$ necessitating the determination of M_{peak} which is not equal to $\bar{M}_n$ or $\bar{M}_w$. Comparison of the polystyrene standards and polyethylene fractions in Figure 21 shows that the two polymers have different $\log M$ calibration curves. Confirmation that the two polymers follow the same plot of $\log[\eta]M$ vs. V_R is given elsewhere.[19] Therefore, equation (38) can be employed to calculate an M_p calibration which in Figure 21 is in satisfactory agreement with the M_{peak} values for the linear polyethylene fractions. Curve AB calculated with equation (38) extrapolates to the experimental behaviour of n-alkanes, *i.e.* curve ABD in Figure 21.

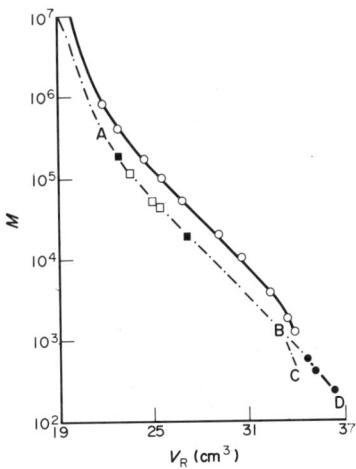

Figure 21 Molecular weight calibration for o-dichlorobenzene at 138 °C: $\bigcirc$, M_{peak} for polystyrene standards; $\bullet$, n-alkanes; $\square$, M_{peak} for linear polyethylene fractions; $\blacksquare$, $\bar{M}_v$ for linear polyethylene fractions; $-\cdot-$ polyethylene calibration calculated with equation (38) (reproduced from ref. 114)

12.6.3 Branched Polymers

When long chain branching (LCB) is present in a polymer, a distribution in branching structure as well as one in M must be considered. Various methods have been developed for the characterization of molecular weight distribution and degree of LCB in branched polymers.[115,116] The total characterization requires an evaluation of both molecular size and molecular weight. If information on w(M) and average molecular weights is required, then on-line LALLS detection in series with a concentration detector will suffice, or the LALLS detector may be replaced by an on-line viscometric detector when the calibration curve M_p in equation (37) may be determined assuming that universal calibration is valid. Because SEC provides a molecular size distribution methods characterizing both w(M) and LCB require an independent measurement of polymer size, either on-line or off-line to the chromatograph, by viscometry.

The principles of methods involving analytical SEC with a concentration detector and the off-line determination of solution viscosity were established by the work of Drott and Mendelson,[117] requiring an experimental chromatogram for the branched polymer, a hydrodynamic volume universal calibration curve, and the intrinsic viscosity of the whole polymer. This treatment depends on two important theoretical assumptions; namely that the branching index λ, defined as the number of branches per unit molecular weight, is constant and that the intrinsic viscosities for linear (l) and branched polymer (br) are related by

$$[\eta]_{br}/[\eta]_l \;=\; g^{0.5} \tag{39}$$

where g depends on the degree of branching. Experimental evidence confirming this approach is shown in Figure 22 where the influence of LCB on molecular weight calibration is clearly evident. However, the assumption of constant λ may not be correct for all branched polyethylenes, and a range of λ values may occur in a given polymer depending on the experimental conditions of polymerization.[118-121] Experimental results for highly branched polystyrenes indicate that the hydrodynamic volume is incorrect for universal calibration,[122-124] and the exponent of 0.5 in equation (39) may not hold for all types of branching.

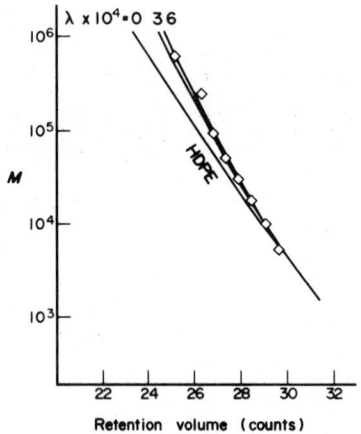

Figure 22 Comparison of experimental calibration data and calculated calibration curves ($\lambda = 3 \times 10^{-4}$ and 6×10^{-4}). HDPE, linear polyethylene fractions; $\diamond$, branched polyethylene fractions (reproduced·with permission of John Wiley and Sons, Inc. from ref. 117)

A variant on the Drott and Mendelson method which uses the same experimental information but does not require theoretical assumptions on branching model and branching index was proposed by Ram and Miltz.[125] From this method a molecular weight calibration for the branched polymer is computed by an iterative procedure, and accurate molecular weight data have been obtained for branched polyethylene[126] and branched poly(vinyl acetate).[127]

12.6.4 Copolymer Composition

Although the refractive index concentration detector is widely used in SEC, the refractive index difference of a solute in a solvent depends on chemical composition, and so a refractometer detector alone cannot discriminate between different chemical groups in copolymers or in a blend of homopolymers. Composition analysis therefore requires on-line multiple detectors. Because of the widespread use of refractometers, a series arrangement of refractive index (RI) and UV detectors is popular,[128-131] though the combination of RI and IR detectors has also been employed.[131] Quantitative determination of copolymer composition distribution by SEC requires the calibration of the detector response defined as the relation between the output signal and solute concentration for each detector. For a copolymer containing styrene units with RI and UV detector responses K_A

and K_s respectively and comonomer units with RI detector response K_B but zero UV detector response, it can be shown that the weight fraction of styrene in the copolymer $W_s(V_R)$ is given by[128,129,132]

$$W_s(V_R) = \frac{K_B F_{UV}(V_R)/F_{RI}(V_R)}{K_s - (K_A - K_B)F_{UV}(V_R)/F_{RI}(V_R)} \qquad (40)$$

where $F_{RI}(V_R)$ and $F_{UV}(V_R)$ are the experimental signals from the two detectors. Data obtained with equation (40) are shown in Figure 23, illustrating that changes in copolymer composition may occur as a function of V_R even though the molecular weight distribution of the diblock copolymer of polystyrene and polytetrahydrofuran may be quite narrow.[132]

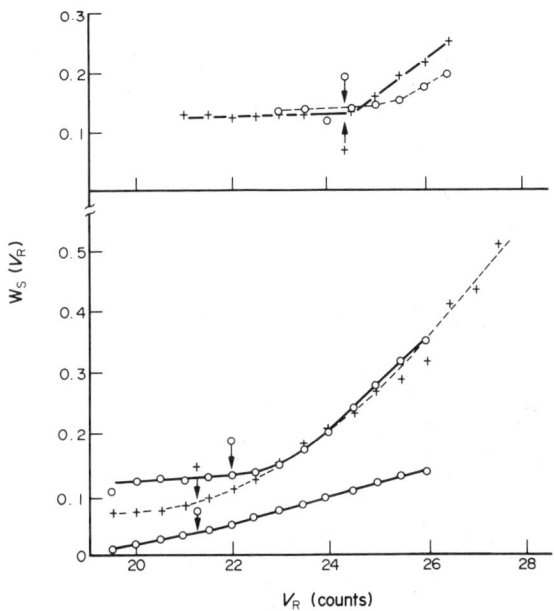

Figure 23 Dependence of copolymer composition $W_s(V_R)$ from equation (40) on retention volume for five diblock copolymers of polystyrene and polytetrahydrofuran (reproduced by permission of Butterworth Ltd. from ref. 132)

When $W_s(V_R)$ varies as a function of V_R, it is clear that $F_{RI}(V_R)$ does not give directly the copolymer concentration. Expressions for the corrected chromatogram $F(V_R)$ in terms of detector signals and responses are given elsewhere.[128,130,132] Methods for composition analysis work best with block copolymers. Caution is advised when applying equation (40) to random (or statistical) copolymers. The spectroscopic absorbance of some functional groups may be influenced by neighbouring comonomer units.[133] The dependence of K_s on copolymer structure will then have to be considered.

With universal calibration given by equation (37) and an on-line viscometer detector, the calibration relation between $\log M$ and V_R may be established for a copolymer.[130] The direct determination of the dependence of $M(V_R)$ on V_R for a heterogeneous copolymer is not straightforward by on-line LALLS detection.[134] Universal calibration equations for block copolymers have been proposed, as reviewed elsewhere.[135]

12.6.5 High Resolution Separations

Separations dominated by a size exclusion mechanism will always have K_{SEC} between zero and unity with all the solutes being eluted within the total solvent volume in the column. This limited range of K_{SEC} determines that high efficiency columns are required for high resolution separations. If a sample contains several species of very different sizes, then peaks for each monodisperse species will be obtained when w is minimized. For the case of two monodisperse solutes 1 and 2 having different sizes, as shown in Figure 1, column resolution R is given by

$$R = 2(V_{R2} - V_{R1})/(w_1 + w_2) \qquad (41)$$

The numerator in equation (41) will depend on separation power which is inversely proportional to the slope of the plot of log(solute size) *vs.* V_R for permeating solutes (see Figure 3), and the denominator will depend on column efficiency according to equation (15). Therefore an increase in separation, *i.e.* the distance between the peak maxima in Figure 1, is dependent on the pore size distribution of the packing, on gel capacity, *i.e.* V_i/V_o, and on the column length. From equations (14), (15) and (20) it is evident that the denominator in equation (41) for column resolution will depend on gel particle size, eluent flow rate, solute size, and the porosity characteristics of the packing which may restrict solute diffusion.

To achieve good resolution, R in equation (41) must be greater than unity, *i.e.* $(V_{R2} - V_{R1}) > w$. The definition of R in equation (41) can be extended by considering molecular weight differences, in terms of the slope D_2 of the linear semi-log calibration defined by equation (36). It follows, as shown by Yau *et al.*,[136] that a general measure of SEC resolution may be developed, and they proposed the specific resolution R_{sp} given by

$$R_{sp} = 0.576/(D_2\sigma) \tag{42}$$

where σ is the standard deviation of a Gaussian peak with $4\sigma = w_1 = w_2$. The number n of components in a sample which can be resolved is related to column efficiency as defined by plate count N. Giddings[137] suggested the following relation

$$n \approx 1 + 0.2N^{0.5} \tag{43}$$

so that $n \approx 21$ for a column with 10 000 plates. The consequence of the restriction of K_{SEC} to values less than 1.0 is that n for SEC separations is considerably less than that for conventional LC (which is about three times larger). A typical example of a low polymer separation is shown in Figure 24 where well-resolved peaks corresponding to the individual components in an epoxy resin are produced.

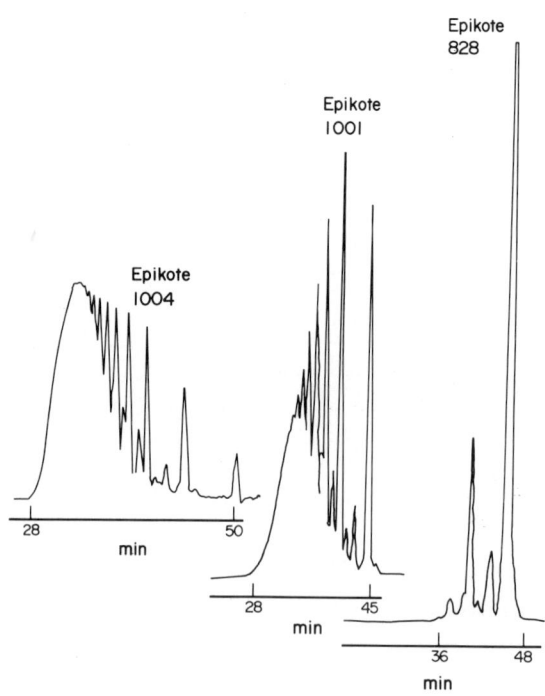

Figure 24 Separations of epoxy resin prepolymers with PLgel packings (5 μm particle diameter) (reproduced with permission from ref. 78)

12.7 REFERENCES

1. L. H. Tung, in 'Polymer Fractionation', ed. M. J. R. Cantow, Academic Press, New York, 1967, chap. E.
2. L. H. Peebles, 'Molecular Weight Distributions in Polymers', Interscience, New York, 1971.
3. J. Porath and P. Flodin, *Nature (London)*, 1959, **183**, 1657.

4. T. Kremmer and L. Boross, 'Gel Chromatography: Theory, Methodology, Applications', Wiley, New York, 1979.
5. J. C. Moore, *J. Polym. Sci., Part A*, 1964, **2**, 835.
6. W. W. Yau, J. J. Kirkland and D. D. Bly, 'Modern Size-Exclusion Liquid Chromatography. Practice of Gel Permeation and Gel Filtration Chromatography', Wiley, New York, 1979.
7. J. Janca, 'Steric Exclusion Liquid Chromatography of Polymers', Dekker, New York, 1984.
8. T. Hashimoto, M. Sasaki, M. Aiura and Y. Kato, *J. Polym. Sci., Polym. Phys. Ed.*, 1978, **16**, 1789.
9. J. V. Dawkins and N. P. Gabbott, *Polymer*, 1981, **22**, 291.
10. H. Hatano, *J. Chromatogr.*, 1985, **332**, 227.
11. A. J. de Vries, M. Le Page, R. Beau and C. L. Guillemin, *Anal. Chem.*, 1967, **39**, 935.
12. M. Le Page, R. Beau and A. J. de Vries, *J. Polym. Sci., Part C*, 1968, **21**, 119.
13. R. Beau, M. Le Page and A. J. de Vries, *Appl. Polym. Symp.*, 1969, **8**, 137.
14. W. Haller, *Nature (London)*, 1965, **206**, 693.
15. W. Haller, *J. Chem. Phys.*, 1965, **42**, 686.
16. W. Haller, *J. Chromatogr.*, 1968, **32**, 676.
17. J. V. Dawkins, in 'Chromatography of Synthetic and Biological Polymers', ed. R. Epton, Ellis Horwood, Chichester, 1978, vol. 1, p. 30.
18. R. E. Majors, *J. Chromatogr. Sci.*, 1977, **15**, 334; 1980, **18**, 488.
19. J. V. Dawkins, in 'Steric Exclusion Liquid Chromatograpy of Polymers', ed. J. Janca, Dekker, New York, 1984, chap. 2.
20. N. Friis and A. Hamielec, *Adv. Chromatogr.*, 1975, **13**, 41.
21. L. H. Tung, 'Fractionation of Synthetic Polymers: Principles and Practices', Dekker, New York, 1977.
22. C. F. Simpson, 'Techniques in Liquid Chromatography', Wiley-Heyden, Chichester, 1982.
23. E. F. Casassa, *J. Phys. Chem.*, 1971, **75**, 3929.
24. M. E. Van Kreveld and N. Van Der Hoed, *J. Chromatogr.*, 1973, **83**, 111.
25. J. C. Giddings, E. Kucera, C. P. Russell and M. N. Myers, *J. Phys. Chem.*, 1968, **72**, 4397.
26. E. F. Casassa, *Macromolecules*, 1976, **9**, 182.
27. Z. Grubisic, P. Rempp and H. Benoit, *J. Polym. Sci., Polym. Lett. Ed.*, 1967, **5**, 753.
28. A. R. Cooper and A. R. Bruzzone, *J. Polym. Sci., Polym. Phys. Ed.*, 1973, **11**, 1423.
29. D. Berek, D. Bakos, T. Bleha and L. Soltes, *Makromol. Chem.*, 1975, **176**, 391.
30. D. Berek, T. Bleha and A. Ozima, *J. Appl. Polym. Sci.*, 1979, **23**, 2233.
31. A. Campos and J. E. Figueruelo, *Makromol. Chem.*, 1977, **178**, 3249.
32. A. Campos, V. Soria and J. E. Figueruelo, *Makromol. Chem.*, 1979, **180**, 1961.
33. T. Spychaj and D. Berek, *Polymer*, 1979, **20**, 1108.
34. L. Wild and R. Guliana, *J. Polym. Sci., Part A-2*, 1967, **5**, 1087.
35. E. E. Drott, in 'Liquid Chromatography of Polymers and Related Materials' (Chromatographic Science Series, ed. J. Cazes, vol. 8), Dekker, New York, 1977.
36. J. V. Dawkins and M. Hemming, *Polymer*, 1975, **16**, 554.
37. A. L. Spatorico and G. L. Beyer, *J. Appl. Polym. Sci.*, 1975, **19**, 2933.
38. C. Rochas, A. Domard and M. Rinaudo, *Eur. Polym. J.*, 1980, **16**, 135.
39. M. Rinaudo, J. Desbrieres and C. Rochas, *J. Liq. Chromatogr.*, 1981, **4**, 1297.
40. J. V. Dawkins and M. Hemming, *Makromol. Chem.*, 1975, **176**, 1795.
41. J. V. Dawkins, *J. Polym. Sci., Polym. Phys. Ed.*, 1976, **14**, 569.
42. J. V. Dawkins and M. Hemming, *Makromol. Chem.*, 1975, **176**, 1815.
43. D. Kranz, U. Pohl and H. Baumann, *Angew. Makromol. Chem.*, 1972, **26**, 67.
44. P. L. Dubin, S. Koontz and K. L. Wright, *J. Polym. Sci., Polym. Chem. Ed.*, 1977, **15**, 2047.
45. J. V. Dawkins, *Polymer*, 1978, **19**, 705.
46. R. Audebert, *Polymer*, 1979, **20**, 1561.
47. J. M. Barrales-Rienda, P. A. Galera Gomez, A. Horta and E. Saiz, *Macromolecules*, 1985, **18**, 2572.
48. B. G. Belenkii and L. Z. Vilenchik, 'Modern Liquid Chromatography of Macromolecules', J. Chromatogr. Library, vol. 25, Elsevier, Amsterdam, 1983, chap. 2.
49. B. G. Belenkii, L. Z. Vilenchik, V. V. Nesterov, V. J. Kolegov and S. Ya. Frenkel, *J. Chromatogr.*, 1975, **109**, 233.
50. S. Mori and T. Suzuki, *Anal. Chem.*, 1980, **52**, 1625.
51. K. H. Altgelt and J. C. Moore, in 'Polymer Fractionation', ed. M. J. R. Cantow, Academic Press, New York, 1967, p. 145.
52. K. H. Altgelt, *Makromol. Chem.*, 1965, **88**, 75.
53. K. H. Altgelt, *Sep. Sci.*, 1970, **5**, 777.
54. A. G. Ogston and P. Silpananta, *Biochem. J.*, 1970, **116**, 171.
55. J. C. Moore and M. C. Arrington, paper presented at the Third International GPC Seminar, Geneva, May 1966.
56. H. A. Swenson, H. M. Kaustinen and K. E. Almin, *J. Polym. Sci., Polym. Lett. Ed.*, 1971, **9**, 261.
57. A. Kotera, K. Furusawa and K. Okamoto, *Rep. Prog. Polym. Phys. Jpn.*, 1973, **16**, 69.
58. K. Nakamura and R. Endo, *J. Appl. Polym. Sci.*, 1981, **26**, 2657.
59. J. E. Figueruelo, V. Soria and A. Campos, *Makromol. Chem.*, 1981, **182**, 1525.
60. H. G. Barth, *J. Chromatogr. Sci.*, 1980, **18**, 409.
61. P. L. Dubin, *Sep. Purif. Methods*, 1981, **10**, 287.
62. A. R. Cooper and D. P. Matzinger, *J. Appl. Polym. Sci.*, 1979, **23**, 419.
63. P. L. Dubin and M. M. Tecklenburg, *Anal. Chem.*, 1985, **57**, 275.
64. M. G. Styring, H. H. Teo, C. Price and C. Booth, *Eur. Polym. J.*, 1988, **24**, 333.
65. A. Domard, M. Rinaudo and C. Rochas, *J. Polym. Sci., Polym. Phys. Ed.*, 1979, **17**, 673.
66. C. P. Talley and L. M. Bowman, *Anal. Chem.*, 1979, **51**, 2239.
67. D. L. Gooding, M. N. Schumuck and K. M. Gooding, *J. Liq. Chromatogr.*, 1982, **5**, 2259.
68. M. Stickler and F. Eisenbeiss, *Eur. Polym. J.*, 1984, **20**, 849.
69. P. L. Dubin, I. J. Levy and R. Oteri, *J. Chromatogr. Sci.*, 1984, **22**, 432.
70. J. C. Giddings, 'Dynamics of Chromatography, Part 1, Principles and Theory', Dekker, New York, 1965.
71. J. C. Giddings and K. L. Mallik, *Anal. Chem.*, 1966, **38**, 997.
72. J. V. Dawkins and G. Yeadon, *Polymer*, 1979, **20**, 981.

73. J. V. Dawkins and G. Yeadon, *J. Chromatogr.*, 1980, **188**, 333.
74. J. V. Dawkins and G. Yeadon, *J. Chromatogr.*, 1981, **206**, 215.
75. J. V. Dawkins and G. Yeadon, *Symp. Faraday Soc.*, 1980, **15**, 127.
76. J. H. Knox and F. McLennan, *Chromatographia*, 1977, **10**, 75.
77. J. H. Knox and F. McLennan, *J. Chromatogr.*, 1979, **185**, 289.
78. F. P. Warner, Polymer Laboratories Limited, personal communication.
79. W. W. Yau, C. R. Ginnard and J. J. Kirkland, *J. Chromatogr.*, 1978, **149**, 465.
80. K. K. Unger and J. N. Kinkel, 'Aqueous Size Exclusion Chromatography', J . Chromatogr. Library, ed. P. L. Dubin, Elsevier, Amsterdam, 1988, vol. 40, in press.
81. K. K. Unger, B. Anspach and H. Giesche, *J. Pharm. Biomed. Anal.*, 1984, **2**, 139.
82. K. K. Unger, *Methods Enzymol.*, 1984, **104**, 154.
83. B. G. Belenkii and L. Z. Vilenchik, 'Modern Liquid Chromatography of Macromolecules', J. Chromatogr. Library, vol. 25, Elsevier, Amsterdam, 1983.
84. H. G. Barth and F. E. Regnier, *ACS Symp. Ser.*, 1984, **245**, 207.
85. H. G. Barth and F. J. Carlin, *J. Liq. Chromatogr.*, 1984, **7**, 1717.
86. J. G. Rooney and G. Verstrate, in 'Liquid Chromatography of Polymers and Related Materials III' (Chromatographic Science Series, vol. 19), Dekker, New York, 1981, p. 207.
87. J. C. Giddings, *Adv. Chromatogr.*, 1982, **20**, 217.
88. D. McIntyre, A. L. Shih, J. Savoca, R. Seeger and A. MacArthur, *ACS Symp. Ser.*, 1984, **245**, 277.
89. D. A. Hoagland, K. A. Larson and R. K. Prud'homme, in 'Modern Methods of Particle Size Analysis', ed. H. G. Barth, Wiley, New York, 1984, chap. 9.
90. Y. Kato, H. Sasaki, M. Aiura and T. Hashimoto, *J. Chromatogr.*, 1978, **153**, 546.
91. Y. Kato, K. Komiya, M. Sasaki and T. Hashimoto, *J. Chromatogr.*, 1980, **193**, 311.
92. T. V. Alfredson, C. T. Wehr and L. Tallman, 'Polymeric Separation Media', ed. A. R. Cooper, Plenum Press, New York, 1982, p. 123.
93. G. D. Wignall, G. W. Longman, M. Hemming and J. V. Dawkins, *Colloid Polym. Sci.*, 1974, **252**, 298.
94. K. Fukano, K. Komiya, M. Sasaki and T. Hashimoto, *J. Chromatogr.*, 1978, **166**, 47.
95. D. D. Bly, H. J. Stoklosa, J. J. Kirkland and W. W. Yau, *Anal. Chem.*, 1975, **47**, 1810.
96. D. D. Bly, W. W. Yau and H. J. Stoklosa, *Anal. Chem.*, 1976, **48**, 1256.
97. E. Kohn and R. W. Ashcraft, in 'Liquid Chromatography of Polymers and Related Materials' (Chromatographic Science Series, vol. 8), ed. J. Cazes, Dekker, New York, 1977, p. 105.
98. S. Mori, *J. Appl. Polym. Sci.*, 1977, **21**, 1921.
99. P. L. Dubin, *J. Liq. Chromatogr.*, 1980, **3**, 623.
100. G. Coppola, P. Fabbri, B. Pallesi and U. Bianchi, *J. Appl. Polym. Sci.*, 1972, **16**, 2829.
101. A. S. Kenyon and E. H. Motthus, *Appl. Polym. Symp.*, 1974, **25**, 57.
102. R. C. Jordan, *J. Liq. Chromatogr.*, 1980, **3**, 439.
103. R. C. Jordan and M. L. McConnell, *ACS Symp. Ser.*, 1980, **138**, 107.
104. G. N. Foster, A. E. Hamielec and T. B. MacRury, *ACS Symp. Ser.*, 1980, **138**, 131.
105. C. Quivoron, in 'Steric Exclusion Liquid Chromatography of Polymers', ed. J. Janca, Dekker, New York, 1984, chap. 5.
106. A. C. Ouano, *J. Polym. Sci., Part A-1*, 1972, **10**, 2169.
107. A. C. Ouano, D. L. Horne and A. R. Gregger, *J. Polym. Sci., Polym. Phys. Ed.*, 1974, **12**, 307.
108. L. Letot, J. Lesec and C. Quivoron, *J. Liq. Chromatogr.*, 1980, **3**, 427.
109. F. B. Maliki, C. Kuo, M. E. Koehler, T. Provder and A. F. Kah, *ACS Symp. Ser.*, 1984, **245**, 281.
110. M. A. Haney, *Polym. Prepr., Am. Chem. Soc., Div. Polym. Chem*, 1983, **24**, 455.
111. L. H. Tung, *J. Appl. Polym. Sci.*, 1966, **10**, 375.
112. A. E. Hamielec, in 'Steric Exclusion Liquid Chromatography of Polymers' (Chromatographic Science Series, vol. 25), ed. J. Janca, Dekker, New York, 1984, p. 117.
113. J. V. Dawkins, *Br. Polym. J.*, 1972, **4**, 87.
114. J. V. Dawkins and J. W. Maddock, *Eur. Polym. J.*, 1971, **7**, 1537.
115. Th. G. Scholte and N. L. J. Meijerink, *Br. Polym. J.*, 1977, **9**, 133.
116. Th. G. Scholte, in 'Developments in Polymer Characterisation — 4', ed. J. V. Dawkins, Applied Science, London, 1983, chap. 1.
117. E. E. Drott and R. A. Mendelson, *J. Polym. Sci., Part A-2*, 1970, **8**, 1361, 1373.
118. A. Cervenka and T. W. Bates, *J. Chromatogr.*, 1970, **53**, 85.
119. E. P. Otocka, R. J. Roe, M. Y. Hellman and P. M. Muglia, *Macromolecules*, 1971, **4**, 507.
120. L. Wild, R. Ranganath and T. Ryle, *J. Polym. Sci., Part A-2*, 1971, **9**, 2137.
121. L. Westerman and J. C. Clark, *J. Polym. Sci., Part A-2*, 1973, **11**, 559.
122. J. Pannell, *Polymer*, 1972, **13**, 277.
123. T. Kato, A. Itsubo, Y. Yamamoto, T. Fujimoto and M. Nagasawa, *Polym. J.*, 1975, **7**, 123.
124. M. R. Ambler and D. McIntyre, *J. Polym. Sci., Polym. Lett. Ed.*, 1975, **13**, 589.
125. A. Ram and J. Miltz, *J. Appl. Polym. Sci.*, 1971, **15**, 2639.
126. L. Wild, R. Ranganath and A. Barlow, *J. Appl. Polym. Sci.*, 1977, **21**, 3331.
127. T. A. Coleman and J. V. Dawkins, *J. Liq. Chromatogr.*, 1986, **9**, 1191.
128. J. R. Runyon, D. E. Barnes, J. F. Rudd and L. H. Tung, *J. Appl. Polym. Sci.*, 1969, **13**, 2359.
129. H. E. Adams, *Sep. Sci.*, 1971, **6**, 259.
130. Z. Grubisic-Gallot, M. Picot, P. R. Gramain and H. Benoit, *J. Appl. Polym. Sci.*, 1972, **6**, 2931.
131. D. J. Harmon and V. L. Folt, *Rubber Chem. Technol.*, 1973, **46**, 448.
132. F. J. Burgess, A. V. Cunliffe, J. V. Dawkins and D. H. Richards, *Polymer*, 1977, **18**, 733.
133. A. Hamielec, *Pure Appl. Chem.*, 1982, **54**, 293.
134. T. Dumelow, S. R. Holding, L. J. Maisey and J. V. Dawkins, *Polymer*, 1986, **27**, 1170.
135. J. V. Dawkins, M. J. Guest and G. M. F. Jeffs, *J. Liq. Chromatogr.*, 1984, **7**, 1739.
136. W. W. Yau, J. J. Kirkland, D. D. Bly and H. J. Stoklosa, *J. Chromatogr.*, 1976, **125**, 219.
137. J. C. Giddings, *Anal. Chem.*, 1967, **39**, 1027.

13

Hydrodynamic Chromatography

MARK G. STYRING and ARCHIE E. HAMIELEC
McMaster University, Hamilton, Ontario, Canada

13.1 INTRODUCTION

As a class of compounds, polymer colloids are of great industrial and technological importance, finding uses directly in, for example, paints, coatings and rubbers. Further, suspension- and emulsion-polymerization procedures, which produce polymers in the form of fine suspensions, account for a very large proportion of synthetic-polymer production. Knowledge of latex particle size and particle-size distribution (PSD) are of paramount importance in both instances. In the first, latex physical properties which largely determine end uses, such as gloss or opacity of films, rheological and electrical properties and latex stability, depend largely on these two factors. In the field of emulsion- and suspension-polymerization, particle size and PSD are key quality-control parameters. In addition, their reliable determination assists in constructing process models for simulation and control of polymerization reactors.

There exist many methods for determining colloid sizes,[1] which vary from approximately 100 Å to tens of microns. HDC covers sizes from approximately 200 too 4000 Å, which spans the range of most synthetic polymeric colloids. For this reason and others, including the techniques' simplicity, accuracy, precision and reliability, it has come to occupy a prominent position in the field of polymer-latex characterization.

HDC shares many features in common with gel-permeation chromatography (GPC), a topic discussed elsewhere in this volume, such as basic instrumentation, data-acquisition and processing techniques, and the fact that chromatographic separation occurs according to size. Indeed, GPC has also been successfully applied to latex characterization and many observations made in this field are relevant to HDC. They will be mentioned wherever appropriate.

Briefly, the HDC technique involves injection of a dilute latex sample into a carrier stream (the aqueous eluant) which is pumped through the system at a constant flow rate. The sample flows through a column(s), packed with non-porous beads, wherein the latex particles are separated according to their size. Large particles exit the column before smaller ones and are detected using a suitable instrument, most commonly a turbidimeter operating in the UV–vis range. This gives an electrical signal with intensity a function of the number, size and chemical composition of the eluting particles which is recorded as a chromatogram. By means of an appropriate calibration in terms of particle size, detector response and instrumental peak broadening, average particle sizes and the PSD may be obtained from the chromatogram.

The first observation of the separation of a compact, spherical macromolecular species on a non-porous substrate (globular proteins on glass beads) was made by Pedersen in 1962.[2] The mechanism which has become most widely accepted for the process, termed separation by flow (SBF), was proposed in 1969 by DiMarzio and Guttman.[3,4] In 1974 the technique was first applied to synthetic-polymer latices by Small[5a] at the Dow Chemical Co., who coined the term HDC. Porous (GPC) packings had, however, been successfully used three years previously by Krebs and Wunderlich[5b] to separate polystyrene latices. A patent for the HDC process was granted[6] and the technology was licensed to the Micromeritics Corp., who now supply instruments commercially. The further development of the technique as applied at Dow has been documented in several publications.[7-12] One of these articles[10a] gives a description of state-of-the-art instrumentation, in which the analysis time per sample is six minutes (compared with approx. 90 minutes for earlier instruments[5]) with a further five minutes required for the calculation of the PSD using a desk-top microcomputer. We would like to point out, however, that although such instrumentation is both impressive and straightforward in operation, it should not be regarded as a 'black box'. Knowledge of basic aspects of chromatography and colloid chemistry are required for properly interpreting the data as will be later clarified.

Academic interest in HDC has also been strong, particularly among workers at the Emulsion Polymers Institute of Lehigh University and their associates.[13-24,47,48] Their contributions have included the development of models successful in accounting for the flow-separation process,[13-16] the design and implementation of methods for improving resolution,[17-20,24] a comparison of HDC with GPC,[17-18] evaluations of the applicability of universal particle-size calibration[21,22] and studies concerning latex recoveries.[23] The field of colloidal chromatography has been reviewed recently[25,26] and the present work may serve as an update.

13.2 EXPERIMENTAL ASPECTS

13.2.1 Apparatus

A typical modern HDC unit is shown schematically in Figure 1. Clearly the instrumentation is exactly as for high-performance GPC (HPGPC), details of which can be found elsewhere.[27] The only major difference is in the packing materials. In Small's original investigation,[5] several packings were used, including beads of polystyrene-*co*-divinylbenzene (mean diameter, $d = 20$ μm), cation-exchange resin beads (Dowex-50 series; $d = 18$, 40 and 58 μm) and spherical glass beads. The 18 μm Dowex packing has subsequently found wide application.[13-15,18-21] In order to reduce analysis times, smaller packing particles are desirable and the high-speed unit described in ref. 10 employs a 15 μm cation-exchange resin. Silicon carbide particles have also been investigated.[28]

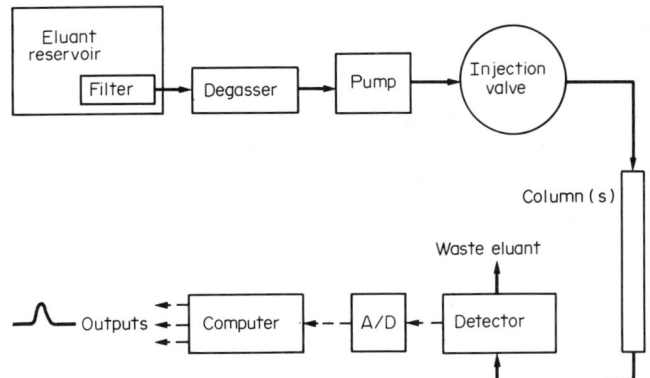

Figure 1 Block diagram of a typical modern HDC apparatus

Eluants for polymeric colloids are based on distilled water containing (usually anionic) surfactant at around the cmc. This introduces steric and electrostatic stabilization to the colloids which, like the packings, are highly hydrophobic by nature, and serves to prevent adsorption of the colloids onto the packing surface.[5,29] Simple electrolytes, *e.g.* sodium nitrate, are usually added to modify the eluant ionic strength, which gives rise to some interesting effects arising from the concomitant changes in the magnitude of the colloidal forces in the system. Bactericides (sodium azide or formaldehyde) are also added.

The remaining hardware can be purchased separately from various chromatographic suppliers if it is desired to custom build one's own apparatus. Modern desk-top microcomputers are sufficiently powerful to handle all the data-acquisition and processing requirements. An integrated unit, the Flow Sizer HDC 5600, is commercially available from Micromeritics Instrument Corp., Norcross, Georgia. An evaluation of this instrument has appeared recently in the literature.[30]

13.2.2 Practice

13.2.2.1 Sample preparation

The experimental procedure is very straightforward. Eluants and suspensions for injection into the instrument are usually filtered before use. Eluant flow rates are typically 0.5 to 2.0 cm³ min⁻¹. Owing to the high sensitivity of UV detectors only very small amounts of sample are required; 0.1 cm³ of sample at 0.005% solids concentration was sufficient according to Small,[5] whereas a tenfold increase in amount was required for the system described by Coll and Fague[29] where a differential refractometer detector was used. Such an increase might lead to earlier plugging of the columns with deposited sample than might otherwise be the case. A UV-absorbing 'marker' species of low molecular weight, usually sodium dichromate at approx. 0.02% concentration, is also incorporated in the sample. This can access the entire interstitial volume in the column and elutes at the end of the chromatographic process. A typical chromatogram obtained for a single colloidal species plus marker is shown in Figure 2.

13.2.2.2 Quantification

By convention the rate of colloid migration through the column is expressed as a dimensionless quantity, the R_F number, which is independent of the flow rate and is simply the ratio of the rate of migration of the colloid to the rate of eluant flow in the interstices. The latter is assumed equal to the rate of migration of the marker and so in terms of the parameters introduced in Figure 2 we have

$$R_F = V_2/V_1 \tag{1}$$

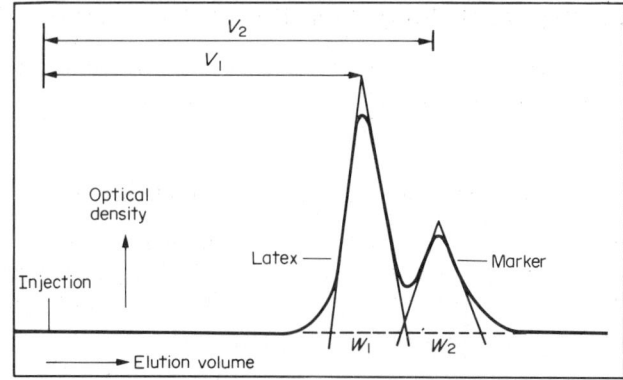

Figure 2 Typical chromatogram obtained by injection of a single colloidal species plus marker. Chromatographic parameters used in evaluating resolution are indicated

13.2.2.3 Colloid deposition and column maintenance

It is well known that colloidal stability is the result of a delicate balance of electrostatic and steric forces present in aqueous solutions which are critically dependent upon such parameters as ionic strength and surface charge.[31] Colloids have a tendency to attach themselves to one another and to other surfaces with which they come in contact, *e.g.* the HDC packing materials. Larger particles tend to deposit more readily than smaller ones. This, together with the small size of the flow channels in packed columns, means that there is an effective maximum size of colloid which can pass through an HDC column. The exact cut-off diameter varies from column to column and with eluant ionic strength, but generally particles larger than approx. 5000 Å will be retained to some extent. This has been amply demonstrated in several particle-recovery studies.[17,23,32] One implication of this phenomenon is that HDC will not yield the true PSD of a polydisperse latex sample which contains particles larger than about 5000 Å. However, there are several ways of circumventing the difficulty. Secchi *et al.*[23] and Rudin and Frick[32] reported the necessity of injecting several samples of 'sacrificial' latex at the beginning of each working day before a constant recovery was obtained. The latter authors also found it useful to keep eluant flowing continuously through the apparatus. These expediencies yielded a size calibration which was reportedly stable for weeks. In addition, the use of larger column packings resulted in diminished loss of larger particles, though at the price of poorer resolution.[23,32] Increasing the surfactant concentration (to around the cmc) also improved recoveries.[23,32]

Another effect of deposition is a gradual deterioration in column performance manifested as a shift in the R_F *vs.* particle-diameter calibration and eventual column plugging. McGowan and Longhorst[10a] reported a typical useful column lifetime of ∼6 months, *i.e.* ∼2000 injections, but that the packing materials could easily be cleaned and re-used.

13.2.2.4 Precision and accuracy

Owing to the reliability of modern chromatographic equipment, particularly the pumps and detectors, the precision of the technique is very good. An uncertainty in peak position for a given latex after 15 injections over two days of ±0.03% has been reported.[10a] Precision in terms of PSD is more difficult to quantify since it depends on having a proper calibration for the instrument and upon the method of data processing. Accuracy, too, depends entirely on the reliability of the calibration of the instrument (for both peak separation and broadening), which is achieved by chromatographing a series of latex standards, usually polystyrenes, having very narrow PSDs. Their diameters need to be previously determined, usually by some absolute technique such as transmission electron microscopy (TEM). At this stage we quote results from ref. 10a in which a standard polystyrene latex was chromatographed and found to have a diameter of 2504 Å by HDC (based on a calibration established using other polystyrene standards), compared with 2423 Å by TEM: a difference of ∼3%. This compares with a figure of 5–7% error in the TEM technique itself as applied to monodisperse latices.[33] We reserve further comments about the all-important calibrations until some further experimental observations and the mechanism and resolution of the technique have been discussed.

13.3 CHROMATOGRAPHIC BEHAVIOUR OF POLYMER LATICES

We intend here to summarize the experimental observations of Small[5] who examined several variables in the process. The trends are applicable to any other HDC system.

13.3.1 Dependence of R_F on Latex Diameter, D, and Packing Diameter, d

Figure 3 shows how R_F varies with d for 2000 Å $\leq D \leq 1$ μm in an unspecified eluant. Clearly for any given packing, R_F increases with D, *i.e.* larger particles elute ahead of smaller ones. Further, as d decreases R_F increases, *i.e.* better separation is achieved with smaller packings.

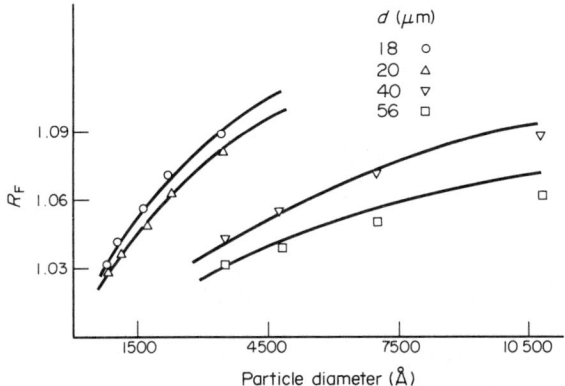

Figure 3 Variation of separation factor with particle and packing diameter. Points indicate actual values;[5] solid lines indicate model predictions.[15] Taken from C. A. Silebi and A. J. McHugh, *AIChE J.*, 1978, **24** (2), 204. Reproduced by permission of the American Institute of Chemical Engineers

13.3.2 Dependence of R_F on Eluant Ionic Strength, I

The same latices were chromatographed in eluants containing varying amounts of NaCl electrolyte to give values of I between 4.25×10^{-4} and 1.76×10^{-1} mol dm^{-3}. Other additives in the eluant were not specified. The results are shown in Figure 4. We see that as I increases, R_F decreases for any given latex. The behaviour observed at high I (curve A) shows a reversal in the expected order of

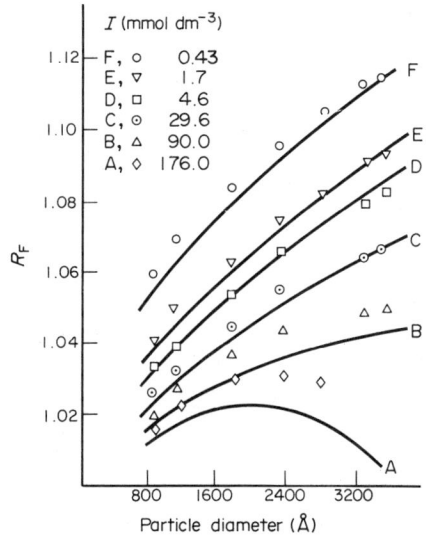

Figure 4 Variation of separation factor with ionic strength. Points indicate actual values;[5] solid lines indicate model predictions. Taken from C. A. Silebi and A. J. McHugh, *AIChE J.*, 1978, **24** (2), 204. Reproduced by permission of the American Institute of Chemical Engineers

elution of latices above approx. 1500 Å diameter. *N.B.* We do not recommend the use of electrolytes containing halide ions in such investigations, due to corrosion of stainless-steel parts in the apparatus. Sodium nitrate is more satisfactory in this respect.[34]

13.3.3 Dependence of Chromatogram Shape on Particle Size

The chromatographic peaks obtained from monodisperse latices are generally symmetrical and of Gaussian form for $D \leq 500$ Å. However, as D increases, the peaks become increasingly skewed, usually towards higher retention volumes, as is shown in ref. 26. Skewing of chromatograms in the forward direction has also been observed.[10b]

13.4 MECHANISM OF HDC

A qualitatively satisfying explanation of the experimental results was forwarded by Small.[5] The interstitial flow channels in the packed columns may be regarded as a series of capillaries. Figure 5a shows one such capillary, across which there exists a parabolic profile of eluant-flow velocities as predicted by Poiseuille's law. A region of maximum velocity exists at the centre of the flow channel, tending towards zero at the packing surface. As a colloidal particle is swept down the capillary it occupies some mean position across the flow profile subject to Brownian diffusion, but is excluded from the packing surface by virtue of its size and hence cannot occupy the slowest streamlines. The larger the particle, the more excluded it is from the surface and so the quicker its passage through the capillary. This explains why R_F increases with D. The marker species is so small that it may sample every streamline, which accounts for its later elution. For small-diameter packings there is a larger surface-area-to-volume ratio and hence a greater effective capillary length so that R_F increases.

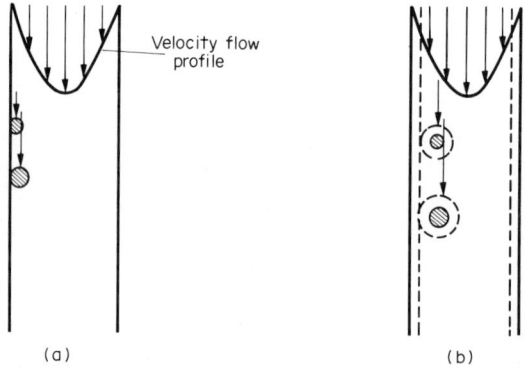

Figure 5   Illustration of migration of colloidal species flowing down a capillary tube (a) in the absence of electrostatic effects and (b) with electrical double layers present

The ionic-strength behaviour is best understood if one considers the interactions between the electrical double-layers[35,36] which exist on both latices and packings. In the case of the latices a charge on the surface, giving rise to a double layer, is a consequence of ionic initiator fragments residing at the surface,[37] together with the presence of adsorbed ionic surfactant,[38] residual either from the emulsion polymerization process and/or from that added to the eluant. It is assumed that this occurs to the packings likewise. Ion-exchange sites on the resin beads contribute further to the charge. The repulsive double-layers between packings and particles have thicknesses which, according to DLVO theory,[35,36] are proportional to $I^{-1/2}$ (increasing I decreases the thickness) and to the charge densities on packing and latex. Now the effect of the double layers on the chromatographic process may be visualized with reference to Figure 5b. They cause the latex to be even more excluded from the packing surface than by virtue of its physical size alone, resulting in an increase in R_F. As more electrolyte is added, the double layers shrink and so R_F diminishes. With regard to curves C, D, E, and F in Figure 4, double-layer repulsion is probably the only influence on colloid transport, other than the hydrodynamic effect. In the instance of curves A and B, however, the latices can approach the packing wall very closely (high I, thin double layers) so that short-range, attractive interactions (van der Waals', Born, London dispersion forces) can make their influence felt, which acts in opposition to the double-layer effect.

Since van der Waals' interactions increase with particle size, one would expect a greater retardation for large latices. This seems to be occurring, particularly in curve A. Moreover, if I is high enough and the latices sufficiently large, the attractive forces become so dominant that latex is deposited on the packing surface and does not elute, as has been observed in practice.[7,34] This phenomenon is analogous to colloidal flocculation in the primary minimum of the potential energy curve as predicted by DLVO theory.[35,36]

Work on quantification of the mechanism took place in the years immediately following Small's original disclosure and a satisfactory model has been forwarded by the Lehigh group.[13-15] Briefly, using a capillary-flow model, the evaluation of the average particle velocity, $\bar{V}_p$, was obtained from the following equation

$$\bar{V}_p = \frac{\int_0^{R_0 - R_p} V_p(r)\exp(-\Phi(r)/kT)r\,dr}{\int_0^{R_0 - R_p} \exp(-\Phi(r)/kT)r\,dr} \tag{2}$$

Here $\Phi(r)$ is the interaction energy between colloid and packing. R_0 and R_p are the radii of the capillary tube and the particle respectively. The marker velocity $\bar{V}_m$ is simply obtained by setting $R_p = 0$. $V_p(r)$ is the particle-streamline velocity and is given by

$$V_p(r) = V_0[1 - (r/R_0)^2] - \gamma V_0(R_p/R_0)^2 \tag{3}$$

where V_0 is the average eluant velocity and γ is the wall-effect parameter calculated from the expressions obtained by Goldman *et al.*[39] $\Phi(r)$ is a summation of contributions from the replusive double-layer and Born forces and the attractive van der Waals' forces, *i.e.*

$$\Phi = \Phi_{DL} + \Phi_B + \Phi_{VW} \tag{4}$$

In order to generate approximations to fit R_F *vs.* D curves at various ionic strengths and R_F *vs.* d curves, certain material parameters, such as A, the Hamaker constant,[40] $\Psi_{0,1}$, the latex surface potential,[41] $\Psi_{0,2}$, the packing surface potential and R_0 the capillary radius[42] were estimated from the literature as indicated by the reference numbers. The good agreement with Small's data is evident in Figures 3 and 4.

13.5 RESOLUTION AND EFFICIENCY

Of primary importance in any chromatographic technique is its resolving power, *i.e.* its ability to discriminate between two solutes quite similar in some property (in this case particle size). In this section we compare several parameters which are indicative of column resolution and efficiency for HDC and the complementary GPC technique.

13.5.1 Column Plate Count, N

With reference to Figure 2, plate count is defined as[27]

$$N_1 = 16(V_1^2/W_1) \tag{5}$$

In a comparative study of an HDC system (Dowex-50, 18 μm packing) and a GPC system (Fractogel silica-glass packings) Nagy *et al.*[20] found values of N of around 20 000 and 600 respectively for similar lengths of column, basing calculations on the dichromate-marker peak in each case. The latter figure is not untypical for porous packings, although improvement by a factor of better than two is easily achievable if controlled-pore glass (CPG-10 series) packings of mesh size 200–400 are used in place of the larger Fractosil packings.[43] The value of N for the high-speed HDC system in ref. 10a was quoted as 63–70 000 plates per metre. The excessive peak spreading in GPC, which makes calculation of PSDs from the experimental chromatograms quite difficult, may be attributed to the extremely low diffusion coefficient[9] ($D \sim 10^{-9}$ cm^2 s^{-1}) of latex particles in aqueous suspension, which means that diffusion into and of the packing pores is very slow.

Separation Methods

13.5.2 Peak Separation, ΔV_p

This is defined simply by

$$\Delta V_p = V_{p,2} - V_{p,1} \qquad (6)$$

Comparison of Figure 6, showing R_F *vs.* D for various porous-packing combinations, with Figure 3 shows that peak separation in GPC is far superior to HDC, although it depends strongly upon the range of packing porosities employed. This is because size separation is achieved by selective exclusion of particles from the pores, as well as by hydrodynamic effects in the interstices.

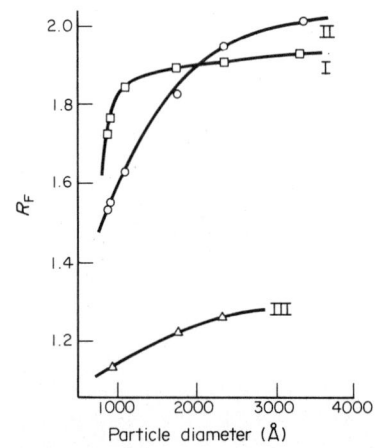

Figure 6 Separation factor *versus* particle diameter for three combinations of porous packings: column set I: porosities 500 Å + 1000 Å + 2000 Å; column set II: porosities 1000 Å + 2000 Å + 3000 Å; and column set III: single porosity 10 000 Å. Taken from D. J. Nagy, C. A. Silebi and A. J. McHugh, *J. Appl. Polym. Sci.*, 1981, **26**, 1567. Copyright© 1981, John Wiley and Sons, Inc. Reprinted by Permission of John Wiley and Sons, Inc.

13.5.3 Specific Resolution, R_s

This quantity is perhaps the best measure of overall column performance. The specific resolution between two particle populations is defined as[27]

$$R_s = 2(V_{p,2} - V_{p,1})/(W_1 + W_2) \qquad (7)$$

As a rule of thumb, when $R_s \geq 1.5$, separation between adjacent populations is almost complete with less than 1% overlap between the two peaks. When $R_s \approx 1$, peak overlap is quite discernible. An R_s factor of better than 0.5 is required for at least partial distinguishability between populations. In Table 1 we present comparative R_s data for GPC and HDC obtained by examining the separation

Table 1 Resolution: Non-porous *vs.* Porous Packings[a]

Particle populations (Å)	R_s, non-porous Three columns	R_s, non-porous One column	R_s, porous One column
880/1760	0.62	0.41	0.14
880/2340	0.91	0.61	0.24
880/3570	1.45	0.94	0.41
1090/2340	0.71	0.49	0.23
380/880	0.33	0.23	0.11
380/1090	0.51	0.36	0.12
380/1760	0.95	0.64	0.26
380/2340	1.23	0.83	0.36
380/3570	1.78	1.17	0.54

[a] Taken from D. J. Nagy, C. A. Silebi and A. J. McHugh, *J. Appl. Polym. Sci.*, 1981, **26**, 1567. Copyright© 1981, John Wiley and Sons, Inc. Reprinted by permission of John Wiley and Sons, Inc.

between populations of various sizes. From these data it would be easy to conclude that HDC is rather superior. However, the GPC packing employed in this study was of a single, relatively large porosity. Much better results are obtained for the small populations, such as 380/880, by switching to smaller-porosity CPG packings.[43]

In deciding which of the two techniques is superior, one must bear in mind the final application. The superior R_s of HDC should better lend it to determining PSDs of samples larger than ~ 1000 Å. Below this size, GPC becomes competitive. The superior ΔV_p of GPC also makes it more suitable for observing subtle shifts in elution volumes of model (monodisperse) colloids resulting from, say, changes in eluant composition[34, 44, 45] or particle chemistry.[46]

13.5.4 Devices for Improving Resolution in HDC

13.5.4.1 Delayed marker injection

This innovation was first described by McGowan and Langhorst in 1982.[10a] It was intended to overcome the problem of overlap of sample and marker peaks when the sample is either small ($D \le 500$ Å) or alternatively has a distribution including small sizes, or contains soluble species of interest, such as surfactant, which would otherwise elute under the marker peak. There are two methods of achieving this. One is to inject the sample and marker separately, using the same valve, with a known time delay. The second is to inject sample and marker simultaneously but at different points on the column, thereby causing the marker to travel an additional length of column, delaying its arrival at the detector.

13.5.4.2 On-column injection

This somewhat more complex method of improving resolution was first reported by Silebi and Viola in 1980[47] and a refinement was recently proposed.[24] The eluant is hereby split into two streams ahead of the injection valve, one passing through the valve, with adjustable flow rate F_i and the second going straight to the column, with adjustable flow rate F_0, where it is distributed around the stream carrying the injected sample. The idea is to minimize sample dispersion in the axial direction in the column. It was found[24] that the best resolution was achieved when $F_i = F_0$, with an improvement of a factor of around 2.5 over a similar instrument not equipped with this device.[48]

13.6 CALIBRATION

In order to obtain accurate average particle sizes and PSDs from the HDC technique, reliable calibrations for particle size and instrumental peak-broadening are essential. In principle these are straightforwardly arrived at by measuring the dependences of R_F and the peak-broadening parameters on D using a series of monodisperse latices whose sizes have been thoroughly characterized by electron microscopy. Kits of well-characterized polystyrene latices are widely available,[49] but the question arises as to whether such calibrations may be applicable to any other colloidal species of interest.

13.6.1 Particle-size Calibration

As demonstrated in Figure 4, R_F for any given latex is strongly dependent on I, and likewise the calibration for a given instrument. This, we recall, is through the effect of I on the double-layer thickness. However, changing the latex chemical type should have some influence on the double-layer thickness by changing $\Psi_{0,1}$. Changing surfactant may also have an influence on $\Psi_{0,1}$, since a molecule of each type of surfactant occupies a different surface area on the latex.[38] Despite this hypothesis, model calculations by Silebi *et al.*[15] indicated that provided one operates in a region where van der Waals' effects on the process are negligible, *i.e.* at low ionic strength, then separation in HDC should be achieved on the basis of size alone, totally independently of particle chemistry. Several practical investigations have been carried out to test the validity of this 'universal' calibration approach. Small *et al.*[7] examined certain variables other than size which it was felt might influence latex elution volumes, such as electrophoretic mobility, amounts of bound carboxylic functionality on the polystyrene surface, eluant pH, the type of surfactant employed in the eluant and latex-particle density. It was stated that all latices eluted in accordance with the polystyrene calibration and with their sizes determined by other methods. However, no characterization data are presented in this publication and details of the other methods are not given.

Some results presented by Coll and Fague[29] also bring into question the applicability of universal calibration. Figure 7 shows their calibration curves established with two different eluants for polystyrene latices, plotted together with data for poly(vinylidene chloride-*co*-acrylonitrile-*co*-acrylic acid) (PVNA) latices. Particle-size data, used in constructing the polystyrene and PVNA curves, do not coincide, with a shift to lower elution volumes (larger effective diameters) for the PVNA latices than predicted by the polystyrene latex calibrations. One possible interpretation is that the acrylic acid groups in the PVNA latices ionize to confer a high surface-charge density on the particle surfaces, thereby setting up a thick double-layer. An alternative explanation is that the PVNA latices swell in the aqueous eluant, a possibility which is discussed in the section on applications.

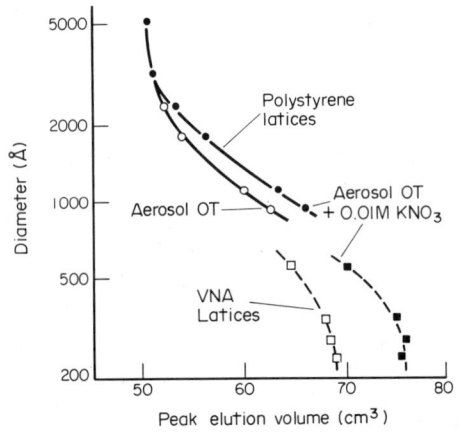

Figure 7 Calibration curves established for two different latex types in two eluants with pH ~ 6.[29] Taken from H. Coll and G. R. Fague, *J. Colloid Interface Sci.*, 1980, **76**, 116. Reproduced by permission of Academic Press, Inc.

Nagy *et al.*[21, 22] have also addressed this problem. In the earlier publication[21] universal calibration was deemed appropriate owing to the fact that elution data for three different types of latex fell on a common curve in the low ionic strength eluant of interest. However, no details concerning the method of characterization of the latices were given. In the subsequent study[22] a calibration was established using monodisperse polystyrene standards. Particle sizes for various other latices were then obtained from their chromatograms through the polystyrene calibration and compared with values obtained by measurements made using a disc centrifuge. The data are presented in Table 2. We see that the values of $\bar{D}_n$ obtained from the disc centrifuge are consistently lower than D from HDC by, on average, 7.5%, whilst the disc-centrifuge $\bar{D}_w$ values are, on average, about a third higher, in one case 105% higher.

The authors of this chapter have recently submitted an investigation whose purpose was to add evidence to this lively debate.[46] Porous packings were chosen owing to their superior peak-separating capabilities which, it was felt, would better show any subtle changes in elution volumes brought about by changes in particle chemistry. Two series of narrow-PSD latices were synthesized and characterized by TEM, one of pure polystyrene and the second being poly(methyl methacrylate) coated on polystyrene in a core-shell morphology. Sizes ranged from 580 to 2070 Å and it was found that the calibration curves for the two species were coincident for all ionic strengths at and below $6.25 \times 10^{-2} \, \mathrm{mol \, dm^{-3}}$, as shown in Figure 8. Changing the surfactant in the eluant and dialysis of the latices had no discernible effect on latex elution volumes. Thus, for the specified conditions, universal calibration was deemed appropriate.

13.6.2 Peak-broadening Calibration

Peak broadening occurs in all packed-column chromatographic processes. It is a manifestation of the fact that not all solute species of the same size elute at the same time due to dispersion effects

Table 2 Comparison of HDC and Disc-centrifuge Particle Size Data for Polymer Latices[a]

Sample no.	Constituent monomers	HDC average particle size (Å)	$\bar{D}_n$ (Å)	Difference (%)	$\bar{D}_w$ (Å)	Difference (%)
1	Vinyl acetate	195	181	−7.2	303	+55
2	Vinyl acetate	200	181	−9.5	334	+67
3	Vinyl acetate	226	194	−14.2	314	+39
4	Vinyl acetate	152	141	−7.2	172	+13
5	Vinyl acetate-co-ethylene	154	138	−10.4	212	+38
6		151	147	−2.6	309	+105
7		230	212	−7.8	254	+10
8	Vinyl acetate-co-	127	120	−5.5	172	+35
9	ethylene-co-vinyl chloride	133	124	−6.8	171	+29
10	Vinyl chloride-co-ethylene	113	101	−10.6	116	+3
11		122	123	−0.8	141	+16
12		114	110	−3.5	130	+14
13	Vinyl acetate-co-butyl acrylate	157	140	−10.8	173	+10

Disc-centrifuge averages span the $\bar{D}_n$, Difference, $\bar{D}_w$, Difference columns.

[a] Taken from D. J. Nagy, *J. Colloid Interface Sci. Lett.*, 1983, **93**, 590. Reproduced by permission of Academic Press, Inc.

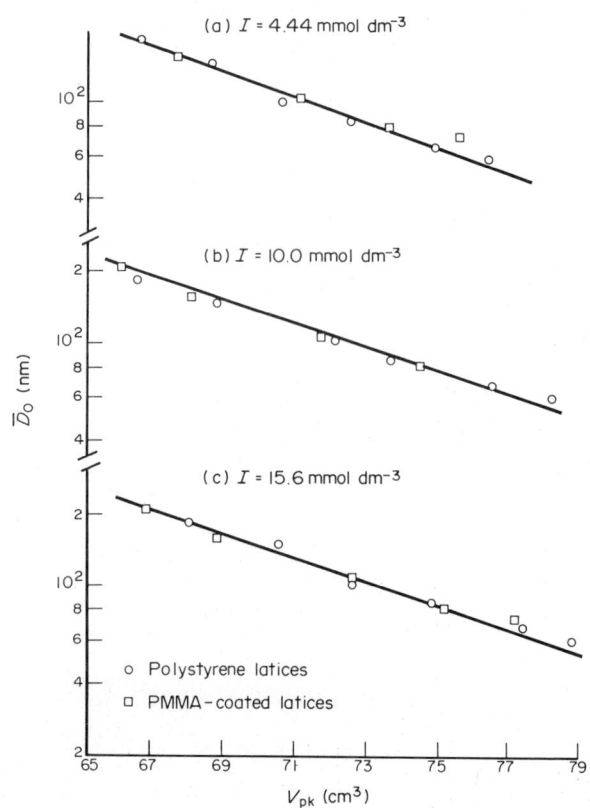

Figure 8 Calibration curves established for polystyrene and poly(methyl methacrylate)-coated latices in eluants at various ionic strengths. [6]

within the column. Important components in broadening are Brownian diffusion across different streamlines, eddying and the large number of interstitial flow paths available to the solutes, each of slightly different length. In the absence of peak-broadening effects, a monodisperse solute would elute as a 'spike' whose width would reflect only the time required to introduce the sample on to the column. However, even in the best systems, monodisperse species give rise to peaks which are approximately Gaussian, or skewed Gaussian in shape. Peaks from polydisperse species have a shape which is characteristic of the instrumental broadening (monodisperse species) together with a contribution from the actual PSD in the sample. In order to obtain an accurate PSD for any species, the raw chromatogram must be corrected for instrumental peak-broadening. Methods for so doing are more fully discussed in the following section. The traditional method of performing such corrections in HDC has been to first obtain chromatograms of a series of monodisperse polystyrene standards, as in particle-size calibration, and to fit the curves with a function of some form (say skewed Gaussian). The monodisperse chromatograms vary in shape with particle size, becoming more skewed with increasing size. This is why a number of different-sized particle standards have to be examined. The parameters obtained for these chromatograms describe the form of the instrumental peak-broadening function. Chromatograms of polydisperse samples are then obtained and mathematically treated to 'subtract' the instrumental effects, yielding chromatograms whose widths depend only on the PSD of the samples.

It has been customary to assume that, like particle-size calibration, peak-broadening calibration is universal. No studies have so far addressed the validity of this assumption, which, if incorrect, would make calculation of PSDs dependent upon having monodisperse standards for each latex type of interest. Work so far has been hampered by the lack of such standards, other than polystyrenes. A study is, however, underway at McMaster[50] on a comparison of peak-broadening calibrations obtained using the same polystyrene and poly(methyl methacrylate)-coated polystyrene latices as described in ref. 46.

13.7 EVALUATION OF PARTICLE-SIZE DISTRIBUTIONS

In this section we set out the theory behind transforming a raw, experimental chromatogram into its corresponding PSD and then illustrate this in some detail with reference to a particular method described in the literature. Some limitations are then discussed.

13.7.1 Theory

Correction of an experimentally-obtained chromatogram, $F(V)$, to yield the peak-broadening-corrected chromatogram whose width is a function only of the PSD, $W(Y)$, reduces to solving the following well-known integral equation[51, 52]

$$F(V) = \int_{V_a}^{V_b} G(V, Y)W(Y)\,dY \tag{8}$$

Here, V is the elution volume and Y the mean elution volume of a single species (referring to particles of a given size). V_a and V_b are the elution volumes where the particle chromatogram respectively departs from and returns to the baseline. $G(V, Y)$ is the peak-broadening function and is the normalized detector response for a monodisperse particle sample with mean elution volume Y. To solve this integral equation for any experimental chromatogram one must first establish the form of $G(V, Y)$. Methods of calibration for peak broadening and for solving the integral equation by both numerical and analytical procedures to obtain $W(Y)$, the average particle diameters and the PSD, are fully documented elsewhere.[19, 51] We now describe the use of $W(Y)$ to calculate the frequency distribution of particle size, $N(D)$.

$$N(D) = \frac{W(Y)D^{-2}(Y)K^{-1}(Y)}{\left[\int_{V_a}^{V_b} W(Y)D^{-2}(Y)K^{-1}(Y)\,dY \right] \left[-\dfrac{dD(Y)}{dY} \right]} \tag{9}$$

Here, $D(Y)$ is the particle-diameter calibration curve and $K(Y)$ is the extinction coefficient for a particle with diameter $D(Y)$. If one is using a UV–vis turbidimetric detector, extinction coefficients of particles in the size range of interest could, in theory, be calculated using equations developed by Mie[53] and appropriate values for the refractive indices of the particles and the eluant. If, on the other hand, a refractometric detector is employed, $K(Y)$ is a function which may be found using the Zimm–Dandliker equations.[54] We would like to point out here that the dependence of chromatogram height on both particle size *and* weight concentration predicted for the UV–vis turbidimeter is quite different from the case in conventional GPC where usually a spectrophotometric detector gives a signal proportional in intensity to weight concentration only.

One can simplify equation (9) and introduce a detector-response factor $\bar{K}(Y)$ as follows

$$N(D) = \frac{W(Y)}{A_T m(Y)\bar{K}(Y)D(Y)} \tag{10}$$

Here A_T is the total area under the chromatogram and $m(Y)$ is the slope of the particle-diameter calibration curve. As is the case with extinction coefficients in equation (9), the detector-response factor may be predicted theoretically (Mie theory or Zimm–Dandliker equations) or measured empirically. The latter is usually the case, since theoretical predictions may not be accurate if such complexities as the presence of residual monomer, surfactant and a hydrated layer on the latex surface are not properly accounted for.[25] This is particularly true when these chemical species are strongly absorbing at the detector wavelength employed.

Resolution in HDC, in terms of dependence of response factor on particle diameter, has been shown to depend on detector type.[55] For a refractometer and a turbidimeter both operating in the Rayleigh regime the respective detector-response dependences are

$$\bar{K} \text{ (refractometer)} \propto nD^3$$

$$\text{and } \bar{K} \text{ (turbidimeter)} \propto nD^6 \tag{11}$$

where n is the number of particles present in the detector cell. This difference in diameter dependence has a remarkable effect on the resolution of smaller particles which may be illustrated by the comparison of correction factors for the number-average particle diameters applicable to these two detector types derived by Hamielec and Singh.[56]

$$\frac{\bar{D}_N(c)}{\bar{D}_N(uc)} \text{ (refractometer)} = \exp\left[\frac{5}{2}(D_2\sigma)^2\right]\left[\frac{1+27\alpha' D_2^3}{1+8\alpha' D_2^3}\right] \tag{12}$$

and

$$\frac{\bar{D}_N(c)}{\bar{D}_N(uc)} \text{ (turbidimeter)} = \exp\left[\frac{11}{2}(D_2\sigma)^2\right]\left[\frac{1+216\alpha' D_2^3}{1+125\alpha' D_2^3}\right] \tag{13}$$

Here $\bar{D}_N(uc)$ and $\bar{D}_N(c)$ are respectively the number-average particle diameters obtained from a chromatogram when uncorrected and corrected for peak broadening, D_2 is the slope of the particle-diameter calibration curve, σ^2 is the variance of the peak-broadening function and α' is a skewing parameter (when $\alpha' = 0$, $G(V, Y)$ is Gaussian).

For small particles ($D \le 500$ Å) we can set $\alpha' = 0$. Comparison of the correction factor for the two detectors shows that the turbidimeter requires a much larger correction than the refractometer, or in other words, the refractometer gives better resolution of small particles. However, were the particles strongly absorbing at the detector wavelength employed, the turbidimeter-detector response would behave as the refractometer (proportional to nD^3) and resolution of the smaller particles would be significantly improved. This has been demonstrated experimentally to be so by Nagy *et al.*[19] Chromatograms were obtained of a bimodal mixture of latices with the detector operating at a wavelength of 254 nm, where polystyrene latices do not absorb significantly, and at a wavelength of 220 nm where the latices do absorb. The improvement in resolution of the smaller particles on switching to the absorbing wavelength is apparent in Figure 9.

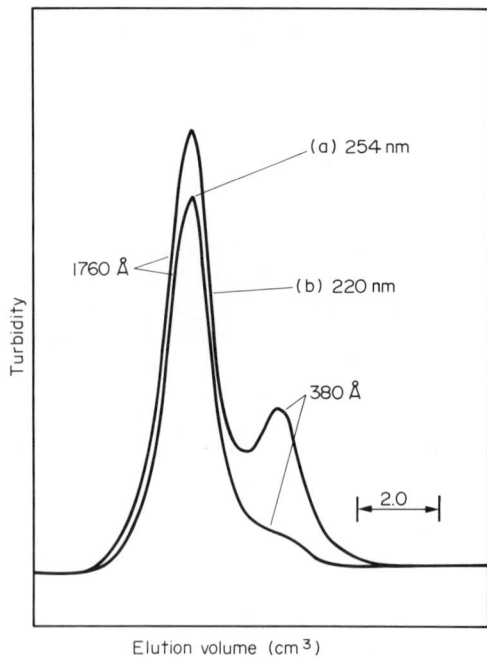

Figure 9 Normalized turbidimetric detector responses from HDC separations of a synthetic bimodal mixture of 880 Å and 1760 Å latices, with the detector operating at wavelengths of (a) 254 nm and (b) 220 nm. The weight ratio of latices is 1 : 1.2. Taken from D. J. Nagy, C. A. Silebi and A. J. McHugh, *J. Appl. Polym. Sci.*, 1980, **26**, 1555. Copyright© 1981, John Wiley and Sons, Inc. Reprinted by permission of John Wiley and Sons, Inc.

A legitimate question at this point concerns the use of a refractometric detector instead of a turbidimeter, to avoid the loss of resolution of small particles, without resorting to changing the wavelength to an absorbing region, which may not always be practicable. One reason why this is not commonly done is that refractometers have much poorer sensitivities than turbidimeters, requiring higher sample concentrations which can lead to shorter column lifetimes through deposition effects. Furthermore conflicting reports as to the exact nature of the refractometric-detector responses have appeared in the literature.[17,29,48,57] The general expression for the refractive index of non-absorbing spheres, derived from Mie theory[53] by Zimm and Dandliker[54] is as follows

$$\frac{d\eta_s}{dc} = \frac{3\eta_m}{2\delta^3 \rho_p} R_e \left[\sum_{n=1}^{\infty} \frac{2n+1}{2n(n+1)} (a_n - b_n) \right] \quad (14)$$

Here η_m and η_s are the refractive indices of medium and dispersion respectively, δ is a dimensionless size parameter ($\delta = \pi D / \lambda$ where D is the particle diameter and λ is the wavelength of the light in the medium), ρ_p is the particle density and a_n and b_n are functions of δ and the particle-to-medium refractive-index ratio, q. In the limit of small δ, this reduces to

$$\frac{d\eta_s}{dc} = \frac{3\eta_m (q^2 - 1)}{2\rho_p (q^2 + 2)} \quad (15)$$

Silebi and McHugh's measurements on polystyrene standards[17] were in accordance with this equation, *i.e.* $\Delta \eta_s$ increased linearly with concentration, independently of particle size. Subsequent data from the same laboratory[48] indicated a steady decrease in $d\eta_s/dc$ with increasing diameter, to the extent that values became negative at high D. Heller and Pugh[57] also observed a decrease when particles larger than 4000 Å were examined. These results, however, bear no resemblance to those of Coll and Fague[29] who observed an increase in detector response by a factor of three when particle diameters increased from 1000 to 5000 Å. It was suggested by Husain *et al.*[25] that these seemingly conflicting data might yet be reconcilable with equation (14) through varying values of q and δ. Whatever the truth of the matter, however, in practical terms the chromatographer must still determine the detector-response function for the latices of interest in order that accurate PSDs may be obtained.

13.7.2 Practice

We refer to McGowan and Langhorst's method of calculating PSDs[10] in order to show how the ideas discussed in the previous section are put into practice. According to this procedure, peak broadening is modelled by a modified Pearson Type VII distribution, which has been used to simulate X-ray diffraction peaks.[58] The chromatogram for a monodisperse latex is described as follows

$$Y_i = [(1 + (V_i - V_0)^2 / M(\sigma n + \varepsilon(V_i - V_0))^2)]^{-M} + VA(1 + L^2 / MK^2)^{-M} \quad (16)$$

Y_i is the height of the peak at elution volume V_i, V_0 is the peak elution volume, A is the peak height at V_0, M is a shape factor, ε is an asymmetry factor, σn is the nominal peak width, SF is a snouting factor, J an integer and V a fraction of SF. Also

$$L = \sigma(J/5 + 2) + V_i - V_0 \quad (17)$$

and

$$K = 3\sigma n J/5 \text{ (or if } K < 6\sigma n/5, K = 4\sigma n/5) \quad (18)$$

Although the mathematics appear complicated, each peak is totally described as a function of six independent variables. Three describe the size and location of the peak (V_0, A and σn) and three others describe its shape (ε, M and SF).

A peak-broadening calibration is obtained by fitting the chromatograms of a series of narrow-PSD latex standards to this Pearson Type VII distribution. These standards are assumed to be strictly monodisperse for the purposes of calculation. Values of ε, M and SF are chosen by the operator and V_0, A and σn are fitted by the computer for each standard. The instrumental broadening is calculated as a broadening factor, BF

$$BF = \sigma n R_F / \sigma m \quad (19)$$

where R_F is the measured R_F of the latex and σm is the width of the fitted marker peak. The width of the chromatograms of the latex standards and hence BF are observed to be a function of particle size. This is shown in Figure 10 where a typical particle-size and instrumental-spreading calibration for a given instrument are illustrated, together with a detector-response curve for the same apparatus.

(a)

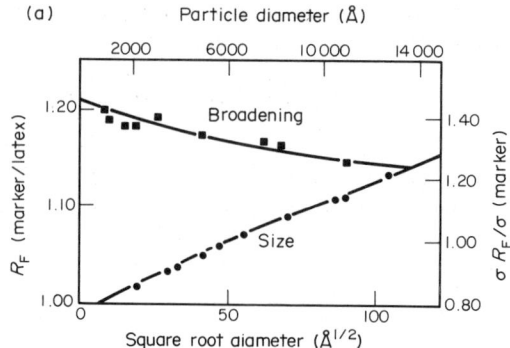

(b)

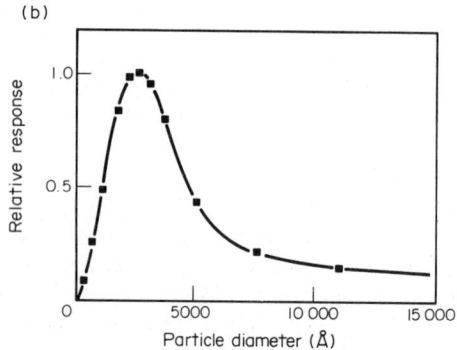

Figure 10 Instrumental calibrations for (a) particle size and peak broadening and (b) detector response at 254 nm.[10] Taken from G. R. McGowan and M. A. Langhorst, *J. Colloid Interface Sci.*, 1982, **89**, 94. Reproduced by permission of Academic Press, Inc.

Having obtained the form of the peak-broadening function and calibrations for both particle size and broadening, we can convert a raw chromatogram into a PSD as follows. First, an initial guess is made at the distribution which is then convoluted with the known peak-broadening function and detector-response curve to calculate a theoretical chromatogram. The original distribution estimate is corrected iteratively, until some convergence criterion is satisfied by using the ratio of the height at each point on the measured chromatogram to the corresponding point on the calculated chromatogram, according the following equations:

$$F_i = \int W_i G_i K_i \, di \tag{20}$$

and

$$W_{i,j+1} = \frac{F_{i,\,\mathrm{calc}}}{F_{i,\,\mathrm{meas}}} W_{i,j} \tag{21}$$

where subscript i denotes points chosen equally spaced in elution volume, j is an iteration counter, W_i is the distribution estimate, G_i is the peak-broadening function, K_i is the detector-response factor and $F_{i,\,\mathrm{calc}}$ and $F_{i,\,\mathrm{meas}}$ are respectively the calculated and measured chromatograms. It has been found expedient in terms of efficiency to use a series of three points on the chromatogram in estimating each $W_{i,j+1}$ as follows

$$W_{i,j+1} = \left[\left(\frac{F_{i-1,\,\mathrm{meas}}}{F_{i-1,\,\mathrm{calc}}} + 2\frac{F_{i,\,\mathrm{meas}}}{F_{i,\,\mathrm{calc}}} + \frac{F_{i+1,\,\mathrm{meas}}}{F_{i+1,\,\mathrm{calc}}} \right) \middle/ 4 \right] B W_{i,j} \tag{22}$$

where B is a 'forcing factor' used to speed convergence.

At each iteration the standard deviation of the error between the calculated and the measured chromatogram, *SD*, is calculated as

$$SD = \frac{100}{F_{i,\,pk}} \left(\sum_i (F_{i,\,calc} - F_{i,\,meas})^2 / i \right) \tag{23}$$

where $F_{i,\,pk}$ is the height of the measured chromatogram at the peak. When *SD* becomes small ($\sim 0.2\%$) or the rate of change of convergence reaches a sufficiently low value ($\sim 1\%$), calculation is terminated and the final estimate of the PSD is obtained.

It is difficult to define the absolute accuracy of a distribution estimate, but one way is to measure the relative amounts of material in known mixtures of monodisperse latices. For example, the mean ratio, calculated from 11 consecutive injections of a known 20/80 mixture of 850 Å/2500 Å monodisperse styrene-*co*-butadiene latices was 19.55/80.45, with a standard deviation of 0.73%.[10]

13.7.3 Limitations

The characterization of co- and ter-polymer latices by HDC has been mentioned in the literature. However, as far as we know, no attention has been paid to the effect of compositional drift within the latex particles on the detector-response factor. It is well known that copolymer characterization by conventional GPC is severely complicated by this,[27] necessitating the use of detectors which can discriminate according to composition, as well as size. The same is presumably true for the detectors employed in HDC and may have a considerable effect on the reliability of data generated by the instruments. In addition, we cannot say with certainty whether or not composition drift, *i.e.* a change in particle chemistry within a sample for both given and varying particle sizes, has an effect on the peak-broadening calibration. This matter, at least, should be further illuminated by publication of the aforementioned study involving peak-broadening parameters of two series of narrow-size-distribution latices.[50]

13.8 APPLICATIONS

We hope already to have demonstrated the utility of HDC in measuring latex particle-size averages and distributions. We now go on to describe a few other areas of application.

13.8.1 Monitoring Particle Growth in Emulsion Polymerization

This application was first mentioned in Smalls' original publication[5] and has been subsequently applied numerous times. Examples are styrene and vinyl acetate emulsion polymerizations studied by Singh and Hamielec,[55] where particle growth in the range 200–6500 Å was successfully followed. Van Gilder and Langhorst[11] have followed the development of both particle size and distribution with conversion in styrene-*co*-butadiene polymerizations.

One advantage of HDC in this application is its speed. In principle the technique could be applied as an on-line monitor of batch or continuous emulsion-production processes, provided sampling problems could be overcome. However, the time delay between sample injection and output of size and distribution data might impede effective control of the reaction conditions. In any event, particle-size data from HDC systems have proven invaluable in the construction of reactor models for simulation and control of industrial-scale processes,[59] as well as for quality control.

13.8.2 Swelling of Latices

The rheological properties of a latex are largely determined by the particle size and distribution of the polymer phase and the volume fraction occupied by that phase. The size of most direct interest is that *in situ*, *i.e.* in the aqueous dispersion medium. HDC can provide such information with far less perturbation of the system than is inherent in some other methods. Electron microscopy, for example, requires removal of the latex from its aqueous environment, frequently accompanied by chemical treatment such as staining, followed by exposure to an intense electron beam.

Since eluant composition can be varied quite widely, Small *et al.*[7] used HDC to examine the change in apparent diameters, D_{app}, of a series of five poly(styrene-*co*-butadiene-*co*-acrylic acid)

(PSBAA) latices of uniform particle size but differing acrylic acid content in eluants of varying pH. As pH was increased, elution volumes were reduced; a shift which was ascribed to increasing neutralization of the carboxylic acid groups which leads to an increased solvation tendency of the ionized sites and hence swelling by uptake of water from the eluant. The effect was larger for a greater acrylic acid content in the latex. Similar experiments were performed on a series of poly(styrene-*co*-butadiene-*co*-vinylic acid) latices, with the acid varying from acrylic to methacrylic to fumaric. Again, D_{app} increased as pH was increased above 7, the effect being larger in the order of the acid groups given above. It must be stressed that the swollen diameters are only apparent as the instrument was calibrated with rigid particles, whilst the terpolymeric latices were deformable and changes in eluant ionic strength on changing pH were not accounted for.

Coll and Fague[29] looked at the pH dependence of elution volumes of VNA latices, as mentioned in Section 13.6.1. Once again increasing pH caused reductions in elution volumes, the effect being greater for the terpolymers having higher acid contents. These workers performed additional light-scattering measurements to demonstrate that the shifts were indeed due to swelling rather than ionic-strength effects. Light-scattering dissymmetry measurements indicated a volume increase of ~10%, whilst inelastic light-scattering data indicated an approximate doubling of latex diameters on changing pH from 3.5 to 9.5, which was more in keeping with the elution-volume shifts.

13.8.3   Agglomeration of Latices

Small[5] has described the utility of HDC in detecting shear-induced agglomeration. Before shearing, the latex in question gave a unimodal chromatographic peak. A shoulder on the low-elution-volume side became apparent upon shearing, which was ascribed to agglomerates. Based on their actual elution position, it was concluded that these agglomerates were doublets and triplets. In a subsequent study,[7] Small *et al.* used HDC to monitor changes in apparent particle size of latices to which had been added certain polymeric thickening agents. It was thus possible to evaluate the thickeners in terms of the concentration required to achieve a certain increase in particle size.

More recently, Van Gilder *et al.*[60] reported the use of HDC as an aid in identifying an optimum binary carboxylated styrene-*co*-butadience latex mixture suitable for paper coatings. The requirement was for a high solids content, ensuring good covering ability, yet a low shear viscosity to facilitate application by a blade coater.

13.8.4   Characterization of Species Other Than Polymeric Colloids

The extension of HDC to inorganic colloids such as silica (Ludox, Du Pont) and carbon blacks was demonstrated by Small.[5] Kirkland[61] later reported on the development of a system which could be used for characterization of colloidal silicas and aluminosilicates in the size range from monomer (silicic acid) to 3000 Å. Semi-rigid porous packings (Sepharose and Sephacryl, Pharmacia) have also recently been found useful for this purpose.[62]

As a class of macromolecular compounds, water-soluble polymers of high molecular weight ($\geq 1 \times 10^6$) have so far proven very difficult to characterize by conventional SEC. However, some such very large molecules have demonstrated susceptibility to being separated according to size on non-porous HDC-type packings. Pedersen's early investigations concerning globular proteins[2] have already been mentioned. More recently, modern HDC methodology has been applied to the separation of xanthan polysaccharides, which are rigid rodlike molecules in aqueous solution, by Prud'homme *et al.*[63] In a subsequent publication[64] it was found that xanthan molecules, which according to other techniques had an approximate length of 1 μm and diameter ~40 Å, eluted at the same position as a polystyrene latex of diameter 0.153 μm. Model calculations were pursued to make inferences about the orientation of such rigid macromolecules under complex two-dimensional flow in microporous media. An extension of this methodology to the field of enhanced oil recovery, where water-soluble macromolecules of high molecular weight are injected into oil wells, forcing remnants of crude oil through porous rocks, was proposed.

Lecourtier and Chauveteau[28] have used HDC to calculate molecular-weight distributions of xanthan samples with good reproducibility.

Prud'homme *et al.* have shown that HDC offers considerable promise for the analysis of other high-molecular-weight, water-soluble polymers of differing conformations besides rigid rods, such as flexible linear-chain coils (partially hydrolyzed polyacrylamides) and flexible branched-chained coils (dextrans). Based on this, Langhorst *et al.*[12] have used the HDC system described in ref. 10a, in conjunction with a low-angle-laser-light-scattering photometer (LALLSP) detector to determine the molecular-weight distributions of non-ionic and partially-hydrolyzed polyacrylamides.

13.9 SUMMARY

Since its inception a relatively short time ago, HDC has become a very widely-used chromatographic technique for determining particle sizes and distributions with considerable ease and rapidity. Its reliability has been well established. However, since the technique does not involve examination of the particles directly, results need to be interpreted in the light of other knowledge about the particle chemistry, for example composition, possible swelling or aggregation effects in the eluant and the likelihood of loss of a certain fraction of the whole distribution by deposition in the columns. The mechanism of the separation is well understood and the evidence gathered to date points to the validity of universal particle-size calibration. This is of considerable value, since calibration of the instrument involves simply chromatographing polystyrene latex standards, which are widely available. It is not yet clear, however, if the same is true for peak-broadening calibration. Further work is also needed to examine the effect of compositional drift in co- and ter-polymer latices as regards the detector-response factor.

Recently the technique has been extended to the characterization of species other than polymeric colloids, most notably water-soluble macromolecules of high molecular weight. We anticipate further publications in this latter field, as well as continued acceptance of HDC as a very important method for sizing sub-micron, synthetic-polymer latices.

ACKNOWLEDGEMENTS

We are grateful to the Natural Science and Engineering Research Council of Canada for financial support throughout the drafting of this chapter. M.G.S. is particularly indebted to the Science and Engineering Research Council of the UK for the award of an Overseas Postdoctoral Fellowship.

13.10 REFERENCES

1. H. G. Barth (ed.), Chemical Analysis Series, 73, 'Modern Methods of Particle Size Analysis', Wiley, New York, 1984.
2. K. O. Pedersen, *Arch. Biochem. Biophys.*, Suppl. 1, 1962, **157**.
3. E. A. DiMarzio and C. M. Guttman, *J. Polym. Sci., Polym. Lett.*, 1969, **7**, 267.
4. E. A. DiMarzio and C. M. Guttman, *Macromolecules*, 1970, **3**, 131.
5. (a) H. Small, *J. Colloid Interface Sci.*, 1974, **48**, 147; (b) K. F. Krebs and W. Wunderlich, *Angew. Makromol. Chem.*, 1971, **20**, 203.
6. H. Small, *US Pat.* 3 865 717 (1975).
7. H. Small, F. L. Saunders and J. Solc, *Adv. Colloid Interface Sci.*, 1976, **6**, 237.
8. H. Small, *Adv. Chromatogr.*, 1977, **15**, 113.
9. H. Small and M. A. Langhorst, *Anal. Chem.*, 1982, **54**, 892A.
10. (a) G. R. McGowan and M. A. Langhorst, *J. Colloid Interface Sci.*, 1982, **89**, 94; (b) M. A. Langhorst, private communication.
11. R. L. Van Gilder and M. A. Langhorst, *Polym. Prepr., Am. Chem. Soc., Div. Polym. Mater. Sci. Eng.*, Sept. 1985, 440.
12. M. A. Langhorst, F. W. Stanely, Jr., S. S. Cutié, J. H. Sugarman, L. R. Wilson, D. A. Hoagland and R. K. Prud'homme, *Polym. Prepr., Am. Chem. Soc., Div. Polym. Mater. Sci. Eng.*, Sept. 1985, 446.
13. R. F. Stoisits, G. W. Poehlein and J. W. Vanderhoff, *J. Colloid Interface Sci.*, 1976, **57**, 337.
14. A. J. McHugh, C. A. Silebi, G. W. Poehlein and J. W. Vanderhoff, in 'Colloid and Surface Science', ed. M. Kerker, Academic Press, New York, 1976, vol. 4, p. 549.
15. C. A. Silebi and A. J. McHugh, *AIChE J.*, 1978, **24**, 204.
16. C. A. Silebi and A. J. McHugh, in 'Emulsions, Latices and Dispersions', ed. P. Becher and M. Yudenfreund, Dekker, New York, 1978, p. 155.
17. C. A. Silebi and A. J. McHugh, *J. Appl. Polym. Sci.*, 1979, **23**, 1699.
18. D. J. Nagy, C. A. Silebi and A. J. McHugh, in 'Polymer Colloids II', ed. R. M. Fitch, Plenum, New York, 1980, p. 121.
19. D. J. Nagy, C. A. Silebi and A. J. McHugh, *J. Appl. Polym. Sci.*, 1981, **26**, 1555.
20. D. J. Nagy, C. A. Silebi and A. J. McHugh, *J. Appl. Polym. Sci.*, 1981, **26**, 1567.
21. D. J. Nagy, C. A. Silebi and A. J. McHugh, *J. Colloid Interface Sci. Notes*, 1981, **79**, 264.
22. D. J. Nagy, *J. Colloid Interface Sci. Lett.*, 1983, **93**, 590.
23. B. M. Secchi, D. L. Visioli and C. A. Silebi, *Polym. Prepr., Am. Chem. Soc., Div. Polym. Mater. Sci. Eng.*, Sept. 1985, 436.
24. J. G. Dos Ramos and C. A. Silebi, *Polym. Prepr., Am. Chem. Soc., Div. Polym. Mater. Sci. Eng.*, Apr. 1986, 268.
25. A. Husain, A. E. Hamielec and J. Vlachopoulos, *J. Liq. Chromatogr.*, 1981, **4**, Suppl. 2, 295.
26. A. Penlidis, A. E. Hamielec and J. F. MacGregor, *J. Liq. Chromatogr.*, 1983, **6**, Suppl. 2, 179.
27. W. W. Yau, J. J. Kirkland and D. D. Bly, 'Modern Size-Exclusion Liquid Chromatography', Wiley, New York, 1979, chap. 5.

28. J. Lecourtier and G. Chauveteau, *Macromolecules*, 1984, **17**, 1340.
29. H. Coll and G. R. Fague, *J. Colloid Interface Sci.*, 1980, **76**, 116.
30. T. W. Thornton and L. B. Gilman, *Polym. Prepr., Am. Chem. Soc., Div. Polym. Mater. Sci. Eng.*, Sept. 1985, 426.
31. D. J. Shaw, 'Introduction to Colloid and Surface Chemistry', 2nd edn., Butterworth, London, 1970.
32. A. Rudin and C. D. Frick, *Polym. Prepr., Am. Chem. Soc., Div. Polym. Mater. Sci. Eng.*, Sept. 1985, 431.
33. J. A. Davidson and H. S. Haller, *J. Colloid Interface Sci.*, 1974, **47**, 459.
34. M. G. Styring, Ph.D. Thesis, University of Manchester (1984).
35. B. V. Derjaguin and L. D. Landau, *Acta Physicochim. URSS*, 1941, **14**, 633.
36. E. J. W. Verwey and J. Th. G. Overbeek, 'Theory of the Stability of Lyophobic Colloids', Elsevier, New York, 1948.
37. J. W. Vanderhoff, H. J. van den Hul, H. J. Tausk and J. Th. G. Overbeek, in 'Clean Surfaces: Their Preparation and Characterization for Interfacial Studies', ed. G. Goldfinger, Dekker, New York, 1969.
38. S. H. Maron, M. E. Elder and I. N. Ulevitch, *J. Colloid Sci.*, 1954, **9**, 89.
39. A. J. Goldman, R. G. Cox and H. Brenner, *Chem. Eng. Sci.*, 1966, **21**, 1151.
40. J. Visser, *Adv. Colloid Interface Sci.*, 1972, **3**, 331.
41. E. J. Schaller and A. E. Humphrey, *J. Colloid Interface Sci.*, 1966, **22**, 573.
42. R. B. Bird, W. E. Stewart and E. N. Lightfoot, 'Transport Phenomena', Wiley, New York, 1960.
43. A. Husain, Ph.D. Thesis, McMaster University (1980).
44. M. G. Styring, C. J. Davison, C. Price and C. Booth, *J. Chem. Soc., Faraday Trans. 1*, 1984, 3051.
45. M. G. Styring, C. Price and C. Booth, *J. Chromatogr.*, 1985, **319**, 115.
46. M. G. Styring, J. A. J. Honig and A. E. Hamielec, *J. Liq. Chromatogr.*, 1986, **9**, 3505.
47. C. A. Silebi and J. P. Viola, *Org. Coat. Plast. Chem.*, 1980, **42**, 151.
48. D. J. Nagy, Ph.D. Thesis, Lehigh University (1979).
49. American Polymer Standards Corp., P.O. Box 901, Mentor, OH, 44061–0901, USA and other suppliers.
50. M. G. Styring and A. E. Hamielec, to be published.
51. A. E. Hamielec, 'Detector Systems for Particle Chromatography', in Chemical Analysis Series, 73, 'Modern Methods of Particle Size Analysis', ed. H. G. Barth, Wiley, New York, 1984.
52. L. H. Tung, *J. Appl. Polym. Sci.*, 1969, **13**, 775.
53. G. Mie, *Ann. Phys.(Leipzig)*, 1908, **25**, 377.
54. B. H. Zimm and W. B. Dandliker, *J. Phys. Chem.*, 1954, **58**, 644.
55. S. Singh and A. E. Hamielec, *J. Appl. Polym. Sci.*, 1978, **22**, 577.
56. A. E. Hamielec and S. Singh, *J. Liq. Chromatogr.*, 1978, **1**, 187.
57. W. Heller and T. L. Pugh, *J. Colloid Sci.*, 1957, **12**, 294.
58. M. M. Hall, Jr., V. G. Veeraraghavan, H. Rubin and P. G. Winchell, *J. Appl. Crystallogr.*, 1977, **10**, 66.
59. C. Kiparissides, J. F. MacGregor and A. E. Hamielec, *Can. J. Chem. Eng.*, 1980, **58**, 56.
60. R. Van Gilder, D. I. Lee, R. Purfeest and J. Allswede, *J. Tech. Assoc. Pulp and Paper Ind. (TAPPI)*, 1983, **66**, 49.
61. J. J. Kirkland, *J. Chomatogr.*, 1979, **185**, 273.
62. D. D. Lasic, *J. Colloid Interface Sci. Notes*, 1986, **110**, 282.
63. R. K. Prud'homme, G. Froiman and D. A. Hoagland, *Carbohydr. Res.*, 1982, **106**, 225.
64. R. K. Prud'homme and D. A. Hoagland, *Sep. Sci. Technol.*, 1983, **18**, 121.
65. D. A. Hoagland, K. A. Larson and R. K. Prud'homme, in Chemical Analysis Series, 73, 'Modern Methods of Particle Size Analysis', ed., H. G. Barth, Wiley, New York, 1984, p. 277.

14

Field-flow Fractionation

JUDY J. GUNDERSON AND J. CALVIN GIDDINGS
University of Utah, Salt Lake City, UT, USA

14.1 INTRODUCTION

Field-flow fractionation (FFF) is a family of high resolution separation techniques especially applicable to macromolecules, colloids and particles.[1-11] The FFF family consists of a number of highly flexible chromatography-like elution techniques that can be adapted to nearly any kind of macromolecular/colloidal separation and characterization problem. The wide applicability of FFF has been hindered by the lack of commercially available instrumentation, a deficiency now being gradually removed.

The concept of field-flow fractionation was first described in 1966.[1] The first successful polymer separation, a crude separation of two polystyrene fractions, was reported in 1967.[12] The methodology developed slowly for a decade as basic experimental and theoretical barriers were breached, mainly at the University of Utah. Since then, FFF techniques have been applied to the separation and characterization of a great number of macromolecules and particulate materials. This work has dealt with species of diverse types and origins, ranging from 10^3 to 10^{18} (essentially a 100 μm particle) in effective molecular weight.[10] The components are separated according to the magnitude of a physical property such as size, diffusion coefficient, electrical charge or molecular weight.[4, 6, 10]

Like chromatography, separation in FFF is achieved by differential migration in a column or channel through which a carrier flows. FFF does not, however, technically fall in the category of chromatographic methods because retention is not caused by the distribution of sample between two phases.[13] Retention in FFF is, instead, induced by the application of an external field or gradient across the channel, perpendicular to the flow axis. The 'field' forces the sample molecules into different flow laminae, having different velocities, within the single flowing phase. FFF can thus be thought of informally as a one-phase chromatographic method.[4, 13]

FFF separations are carried out in a thin (usually 50–500 μm) ribbon-like channel, with an external field or gradient applied across the channel faces as shown in Figure 1a. The FFF channel has an aspect ratio (breadth/thickness) large enough (40–300) that the flow approximates that between infinite parallel plates; the flow of a Newtonian fluid between parallel plates results in a parabolic velocity profile. In such flow, the velocity next to the two major walls of the channel is zero and the maximum velocity occurs in the channel center (see Figure 1b).

The applied external field must be of such a nature as to interact significantly with the sample molecules. By means of the interaction the applied field forces the components to migrate toward one wall, which is referred to as the 'accumulation wall'. The migration of components toward the wall is balanced by diffusion away from the wall, resulting in a very thin (typically 2–20 μm) steady-state layer for each component at the wall. Different levels of interaction with the applied field

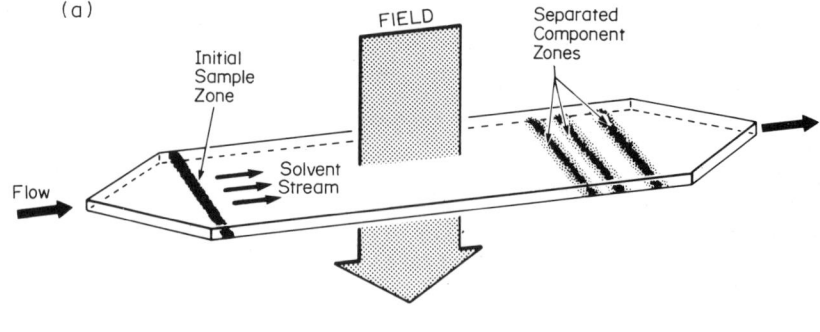

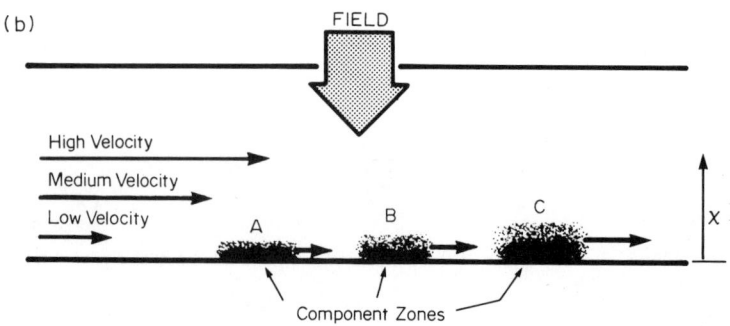

Figure 1 Illustration of a field-flow fractionation channel and the separation of polymer components

for different components result in the formation of layers of different thicknesses near the wall (see Figure 1b).

The parabolic flow carries all the component layers down the channel. Species that form the most compact layers against the wall are displaced more slowly than species forming diffuse layers since the velocity of the carrier is slowest near the wall. This separation process is shown in Figure 1b. Component C has the least interaction with the field and is thus least compressed against the wall, allowing it to sample faster stream filaments than component A which has the strongest interaction with the field. The three species will elute from the column in the following order: C first, followed by B, with A last.

The experimental apparatus for FFF is similar to that for liquid chromatography. It typically consists of a pump to drive the carrier fluid, the separative channel, a detector and a chart recorder to monitor the eluent from the column. A computer may be utilized for data analysis or the control of operating conditions. Samples are injected with a microsyringe or an injection valve at the head of the channel. Only the nature of the separative channel (column) and the mechanism of separation are different in the two methods. The FFF channel is an open ribbon-like conduit, usually formed by clamping a thin spacer, from which the channel volume has been cut out, between two long rectangular plates.

Separation by FFF has several unique advantages.[10, 13] FFF is a very selective separation tool and thus has high resolving power.[5, 14] Since FFF occurs in an open channel that is well defined geometrically with well characterized flow, the basic phenomena (retention and band broadening) underlying component resolution can be rigorously described by theory. This allows first-principle calculations of sample properties from retention data.

FFF is also a highly versatile technique. Experimental conditions may be widely varied to optimize the speed, range and resolution of the separation. Perhaps the most important parameter is the field strength which can be varied accurately between experiments or in the course of one experiment (programming). The programming mode is useful for samples that cover a very broad mass range.[15, 16] In programming, the field strength must decrease during a run to be effective. Following the run, the field may be turned off completely to flush out residual materials.

Another advantage is that FFF occurs in an open channel, minimizing interfacial influences and eliminating clogging. The open channel and uniform flow also result in gentle, nondisruptive migration; FFF operates without abrupt forces, interfacial transport or strong shear gradients. Extensional shear, as found in packed beds, is virtually absent. This makes FFF an ideal tool for the separation of fragile species such as high molecular weight polymers, loosely associated aggregates and delicate biological components.

A large number of different fields may in theory be used in FFF for separation. The choice of field depends on the nature of the samples to be separated.[6] For example, if the sample is a mixture of species with different electrical charges, then an electrical field would be a logical candidate, although other fields might work as well but with separation based on other properties. Four kinds of fields have dominated FFF work. These four fields give four prominent FFF subtechniques: sedimentation FFF, electrical FFF, flow FFF and thermal FFF. Sedimentation FFF has the channel coiled inside a centrifuge and relies on sedimentation forces. This subtechnique separates components on the basis of their mass differences. Flow FFF has a crossflow of carrier fluid which enters and exits the channel *via* semipermeable membrane walls. Flow FFF separates according to differences in diffusion coefficients. Electrical FFF has an electrical field applied across the channel and separates on the basis of differences in electrical charge. Thermal FFF has a steep temperature gradient across the channel and separates according to differences in the thermal diffusion factor. Other possible subtechniques discussed in the literature include concentration gradient FFF,[17] magnetic FFF,[18] shear FFF[19] and dielectrical FFF.[20]

14.2 THEORY OF FIELD-FLOW FRACTIONATION

The theory of FFF has been extensively developed in the years since its inception.[20-25, 27, 28] We will present only the basic elements of the theory necessary to understand the fundamentals and the optimization of FFF systems.

The concentration profile of a component near the accumulation wall is determined from the general mass transport equation. Thus the flux density of a component across a unit area normal to the applied field is described as[29]

$$J_x = -D \frac{dc(x)}{dx} + Uc(x) \tag{1}$$

where x is the distance along the axis parallel to the field axis, D is the diffusion coefficient, $c(x)$ is the component concentration at distance x from the accumulation wall and U is the field-induced drift velocity of the component. The right side of the equation consists of two parts: $-Ddc(x)/dx$, the diffusive flux of the sample away from the wall, and $Uc(x)$, the field-induced flux toward the wall. Once steady state conditions are established (which happens rapidly because of channel thinness), the net flux becomes zero. The steady state concentration profile is found through integration to be[29]

$$c(x) = c_0 e^{-x|U|/D} \tag{2}$$

where c_0 is the concentration at the accumulation wall, x is the distance from the wall, and $|U|$ is the absolute value of U. The drift velocity U is always negative due to our choice of coordinates. Often equation (2) is rewritten as

$$c(x) = c_0 e^{-x/l} \tag{3}$$

where $l = D/|U|$. The parameter l has the dimension of length and is an effective measure of the thickness of the component layer. It is approximately equal to the average distance of the component molecules from the wall.

The parameter l will be examined in more detail because of its fundamental importance in governing experimental behavior. According to the Stokes–Einstein equation, the diffusion coefficient may be written as[29, 30]

$$D = kT/f \tag{4}$$

where k is Boltzmann's constant, T is temperature and f is the friction coefficient. The drift velocity may be related to the force F acting on a component molecule through the equation

$$U = F/f \tag{5}$$

Since l is defined as $D/|U|$, the use of equations (4) and (5) leads to

$$l = kT/F \tag{6}$$

This equation shows that the l value of a component zone is inversely proportional to the force exerted on component molecules by the applied field. It is clear from equation (6) that different levels of interaction (different forces) will lead to different zone thicknesses for different components.

Quantity l is often expressed in the dimensionless form

$$\lambda = \frac{l}{w} = \frac{kT}{Fw} \tag{7}$$

where w is the channel thickness. We term λ the retention parameter.

The migration of components along the channel is driven by flow. The migration is differential because the components are unequally distributed over the nonuniform flow profile. The flow profile of an isoviscous fluid between two infinite parallel plates is given by

$$v(x) = \frac{\Delta p}{2\eta L} x(w-x) = 6\bar{v}\left[\frac{x}{w} - \left(\frac{x}{w}\right)^2\right] \tag{8}$$

where Δp is the pressure drop along the channel, L is the length of the channel, η is the carrier viscosity, $\bar{v}$ is the mean flow velocity and $v(x)$ is the local velocity at a distance x from the wall.

Component molecules generally migrate slower than mean carrier velocity $\bar{v}$ because the components occupy the slow moving wall regions. The extent to which a component is 'retained' relative to the carrier flow is specified by its retention ratio R. This nondimensional parameter is defined as the component velocity divided by $\bar{v}$. R is measured experimentally by dividing the physical (void) volume V^0 of the channel by the volume V_r necessary to elute the sample. Theoretically, R is obtained by formulating a mean component velocity using the concentration distribution of equation (3) and the flow velocity distribution of equation (8). This gives us

$$R = \frac{V^0}{V_r} = 6\lambda\left(\coth\left(\frac{1}{2\lambda} - 2\lambda\right)\right) \tag{9}$$

where $\lambda = l/w$, equation (7). When $\lambda \to 0$, R approaches 6λ.

Equation (9), although valid for a majority of FFF subtechniques, is only correct to a first approximation for thermal FFF. This is a consequence of the applied temperature gradient which causes a variation in the fluid viscosity across the channel. The variation in viscosity leads to a departure from parabolic flow. Equations taking into account this departure have been derived.[31]

If the interaction force F between the sample and the applied field is known, then λ may be calculated from equation (7) and the level of retention predicted from equation (9). Expressions for λ have been determined for the various FFF subtechniques[4] and are summarized in Table 1. This table also shows the sample parameters controlling retention.

Due to the exactness of retention theory, the above calculations can be successfully reversed and the characteristic sample parameters given in Table 1 can be calculated from experimental retention data. This procedure underlies the acquisition of distributions in molecular weight, particle size and other component properties.

Table 1 Expressions for the Retention Parameter λ

Subtechnique	λ Equation	Sample Parameters Controlling Retention
Sedimentation FFF	$\lambda = \dfrac{RT}{GM(1-\rho/\rho_s)w}$	ρ = sample density
		M = molecular weight
Flow FFF	$\lambda = \dfrac{DV^\circ}{\dot{V}_c w^2}$	D = diffusion coefficient
Electrical FFF	$\lambda = \dfrac{D}{\mu E w}$	D = diffusion coefficient
		μ = electrophoretic mobility
Thermal FFF	$\lambda = \dfrac{D}{D_T(\mathrm{d}T/\mathrm{d}x)w}$	D = diffusion coefficient
		D_T = thermal diffusion coefficient

General symbols: w = channel thickness, R = gas constant, T = temperature, ρ = carrier fluid density, G = gravitational acceleration, V° = channel volume, $\dot{V}_c$ = volumetric crossflow rate, E = electrical field strength, $\mathrm{d}T/\mathrm{d}x$ = temperature gradient.

The resolution between components depends upon two factors: the molecular weight (or mass) based selectivity S_M and the degree of band broadening measured by plate height H. The mass-based selectivity can be expressed by[32]

$$S_M = \frac{d \log R}{d \log M} = \frac{d \log V_r}{d \log M} \tag{10}$$

which shows that S_M is simply the percentage change in retention ratio R or retention volume V_r accompanying a 1% change in molecular weight M.

The major contributions to plate height in FFF have been studied in detail.[22, 23] The plate height is theoretically described by

$$H = \frac{2D}{R\bar{v}} + \frac{\chi w^2 \bar{v}}{D} + \Sigma H_i + H_p \tag{11}$$

where D is the diffusion coefficient, R is the retention ratio, $\bar{v}$ is the average linear fluid velocity and χ is the nonequilibrium coefficient. The first term on the right accounts for broadening due to the longitudinal diffusion of component molecules. This term is generally considered negligible due to the sluggish diffusion of macromolecules.

The second (and generally dominant) term on the right is the nonequilibrium (mass transfer) term. This term originates in the nonuniform flow, which causes component molecules at different heights in the zone to be swept forward at different velocities. The nonequilibrium coefficient χ associated with this term is a complicated function of λ.[33, 34] A simple limiting form for χ valid when $\lambda \to 0$ is

$$\chi = 24\lambda^3 \tag{12}$$

It is clear from equations (11) and (12) that as λ decreases, plate height and zone broadening decrease dramatically. Obviously this makes high retention conditions (low R, high V_r) appealing to the experimentalist.

The third term on the right of equation (11) accounts for various contributions due to nonidealities of the system such as imperfect channel design, relaxation processes, width of injection slug, end effects, etc.[35, 36] Many of these terms may be minimized, with proper design, to the point where they are negligible.[35]

When analyzing narrow (nearly monodisperse) polymer samples, the contribution to the plate height due to polydispersity must be considered. A polydisperse sample zone will broaden in the channel due to the fractionation of its individual constituents. This broadening is a separation process and its nonnegligible magnitude, even for samples of low polydispersity, reflects the high resolving power (selectivity) of FFF methods. The polydispersity contribution is given approximately by[37]

$$H_p = L S_M^2 (\mu - 1) \tag{13}$$

where $\mu = \bar{M}_w / \bar{M}_n$, in which $\bar{M}_w$ is the weight average molecular weight and $\bar{M}_n$ is the number average molecular weight.

As noted above, resolution is determined by a combination of selectivity S_M and plate height H. For two components differing by ΔM in molecular weight, the resolution can be approximated by[32]

$$R_s = \frac{S_M N^{1/2}}{4} \frac{\Delta M}{\bar{M}} \tag{14}$$

where $\bar{M}$ is the average of the two molecular weights and N is the number of theoretical plates, $N = L/H$, where L is channel length.

14.3 THERMAL FIELD-FLOW FRACTIONATION

The basic thermal FFF channel design, described elsewhere,[6] is illustrated in Figure 2. It consists of two copper or copper alloy bars with highly polished chrome-plated faces clamped together over a sheet of Mylar from which the channel space has been cut and removed. The channel is rectangular with tapered ends. The two bars are sandwiched between boards of insulating material and the whole assembly is clamped together between aluminum plates. The top bar is heated with cartridge heaters; the heat is removed from the bottom bar by coolant circulation. The temperature difference ΔT established between the bars is generally from 5 to 50 °C, the higher values applicable to lower molecular weight polymers.

The fractionation by programmed thermal FFF (in which ΔT decreases during the run) of polystyrene components over the broad molecular weight range from 2×10^3 to 7.1×10^6 is shown[38]

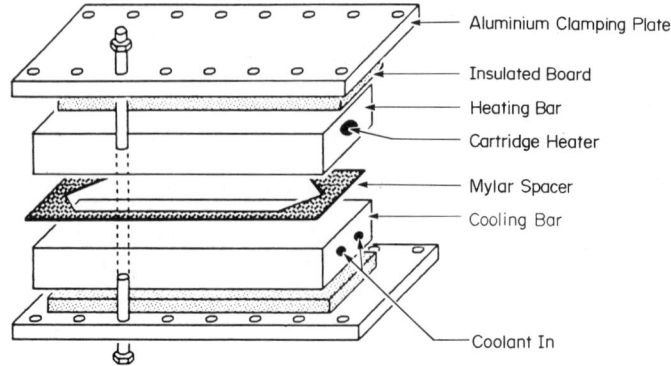

Figure 2   Sandwich structure of a thermal FFF system

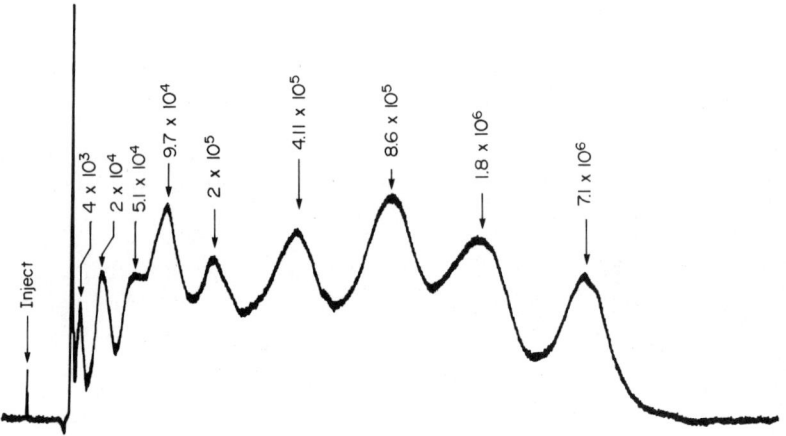

Figure 3   Fractogram illustrating the wide range of applicability of thermal FFF in a single programmed run. The sample consists of a mixture of linear polystyrenes of the designated molecular weights

in Figure 3. Thermal FFF is an excellent tool for the determination of molecular weight distributions due to the wide range of molecular weight that may be covered in one run, particularly with programming. For narrower fractions in which the ratio of the extremes of molecular weight is no more than 100, programming is not necessary but the temperature gradient must be adjusted in each case to best handle the range under consideration. Since the driving force for retention, thermal diffusion, cannot presently be rigorously related to molecular weight, calibration curves are constructed from polymer standards. As in size exclusion chromatography (SEC), these curves are used for determining the molecular weight distribution of the polymer.

The thermal FFF system can be modified to yield high speed polymer analysis by reducing channel thickness w (see last section). However, as in nearly all FFF systems, operating conditions (particularly the flow velocity) can be adjusted to emphasize either speed or resolution. The trade-off between the two is illustrated in Figure 4. Both a five minute and a one minute separation from the same thin channel ($w = 0.05$ mm) system are shown.[39] The former shows superior resolution but reduced speed due to the lower flow velocity employed. The decrease in resolution with increasing flow velocity is a result of the proportionality of the nonequilibrium term of equation (11) to the mean flow velocity $\bar{v}$. The relative importance of speed and resolution must be considered for each separation problem; once decided, the flexibility and predictability of FFF operation gives ready access to the desired result. Details are discussed in the final section.

Since thermal FFF is carried out in an open channel rather than a packed bed, shear degradation is minimized. The gentle nature of FFF makes it an ideal tool for the characterization of ultrahigh molecular weight polymers. Figure 5 is the molecular weight distribution of such a polymer obtained from thermal FFF data.[40] The polymer is a polystyrene sample with a manufacturer-assigned

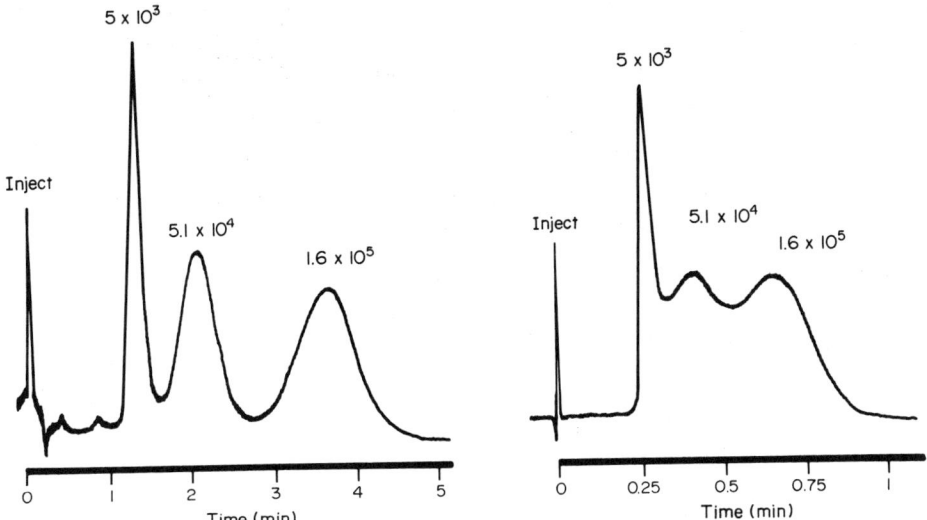

Figure 4 Illustration of the high-speed separation of polystyrene fractions by thermal FFF and the trade-off between resolution and speed. By reducing the flowrate one can gain resolution (left); by increasing flowrate speed is enhanced

molecular weight of 20.6×10^6. If degradation occurred during analysis, it would lead to an underestimation of the polymer's molecular weight. This would be demonstrated by the reinjection of a portion of the eluting peak. Where significant shear degradation has occurred, the sample would elute faster with respect to its first passage through the FFF channel, which would indicate a transformation into smaller species.

A test for shear degradation of the 20.6×10^6 polystyrene sample was performed by reinjection. A portion of the elution peak was reinjected with no significant shift in elution position. This clearly indicates the suitability of thermal FFF for the characterization of ultrahigh molecular weight polymers.[40]

For narrow polymer fractions, such as polymer standards, the polydispersity, $\mu = \bar{M}_w / \bar{M}_n$, is readily obtained from thermal FFF data by measuring band broadening and its velocity dependence.[41, 37] If the first and third terms on the right side of equation (11) are negligible, the flow velocity dependence of the plate height may be written as a simple linear equation

$$H = H_p + C\bar{v} \tag{15}$$

where C is a constant. Figure 6 shows a plot of plate height H vs. flow velocity $\bar{v}$ for a narrow polystyrene standard.[37] The data fall on a straight line as predicted by equation (15). The polydispersity term H_p is the intercept. By means of equation (13) the intercept can be expressed in terms of $\mu - 1$ or simply μ. A $\mu - 1$ scale appropriate to the conditions of the experiments is shown on

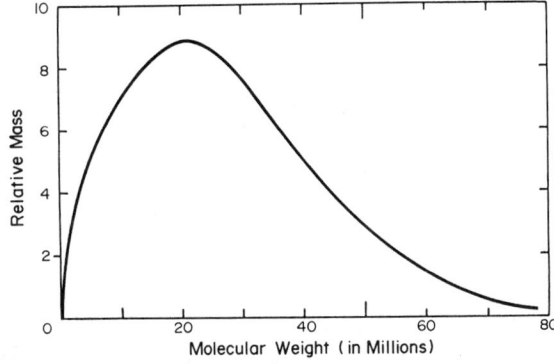

Figure 5 Molecular weight distribution obtained for an ultrahigh molecular weight polystyrene sample by thermal FFF

the righthand side of the figure. The manufacturer-suggested maximum value of μ is 1.06; the value determined by thermal FFF is significantly smaller, 1.003. The figure shows that any μ value significantly above 1.01 is unequivocally ruled out as a polydispersity value. The consistent trend in experimental points suggests that the result, $\mu = 1.003$, is reasonably accurate. This conclusion is further supported by the fact that the slope C of the line shown in Figure 6 is within three percent of its theoretical value.

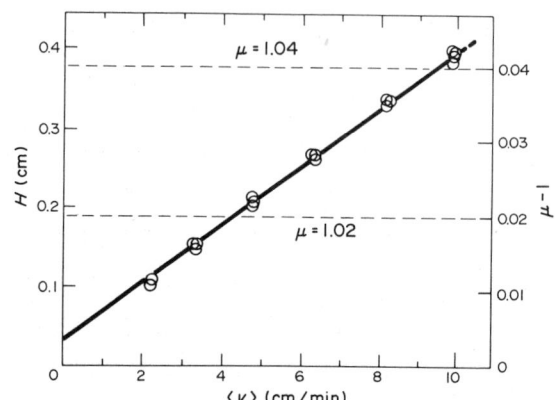

Figure 6   Plate height *vs.* flow velocity plot for a 1.7×10^5 molecular weight polystyrene sample showing a true poly-dispersity value (intercept) of $\mu - 1 = 0.003$

Because the slope of the H *vs.* $\bar{v}$ plot can be calculated in FFF methods, an even simpler method for obtaining polydispersity is available. A single experimental point combined with the theoretical slope can be used to obtain the intercept and thus the polydispersity to a good approximation. The thermal FFF method has an advantage over size exclusion chromatography for obtaining polydispersities because of the simpler band dispersion function and the higher selectivity.[37]

As shown in Table 1, retention in thermal FFF can be broken into two terms: the ordinary (concentration) diffusion coefficient D, which depends almost exclusively on molecular size, and the thermal diffusion coefficient D_T, which depends almost solely on composition. Recent studies in our laboratory have shown D_T to be independent of chain length and configuration. Star-shaped, comb-shaped and linear polystyrene samples of various molecular weights were used to illustrate the lack of dependence of D_T on molecular size or configuration.

Retention in size exclusion chromatography (SEC), by contrast, is based almost exclusively on molecular size. The effective molecular size is generally measured in terms of the hydrodynamic volume, which can be related to the ordinary diffusion coefficient D. Differences in chemical composition are second order effects and usually influence retention in SEC only to the extent that they influence physical size.

Thus, for one polymer type, retention in both thermal FFF and SEC is governed by molecular size. When polymer composition is varied and molecular size held constant, retention in thermal FFF, but not in SEC, changes. This is illustrated in Figure 7. A polystyrene and a poly(methyl methacrylate) fraction cannot be resolved by SEC due to the closeness of their molecular dimensions. The difference in composition, which is reflected in D_T, allows thermal FFF to resolve the two fractions successfully. The additional dependence of retention in thermal FFF on D_T adds a new dimension not influencing retention in SEC,[42] showing some promise for a combination of thermal FFF and SEC to shed light on both compositional and molecular size variations in complex polymeric materials.

We note that despite the common dependence on D, a given increment in D will lead to different fractional shifts in elution volume and different directions of the shift, positive for SEC and negative for thermal FFF. Thus the elution orders are different in the two techniques, as shown by Figures 7 and 3.

In order to better understand thermal diffusion and explore the utility of thermal FFF, a variety of polymers have been investigated. The lipophilic polymers that have been studied along with polystyrene include poly(methyl methacrylate),[42,43] polyisoprene,[42,43] poly(tetrahydrofuran),[43] polypropylene[44] and polyethylene.[44] Thermal diffusion in aqueous systems has also been studied. It

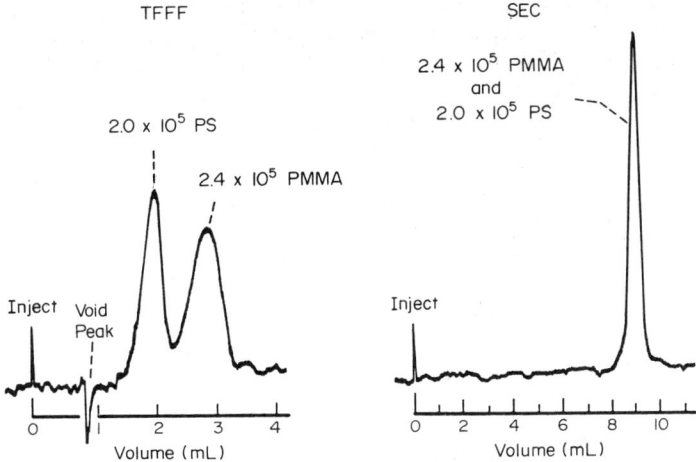

Figure 7 Comparison of fractograms of polystyrene and poly(methyl methacrylate) samples of approximately equal molecular weight obtained by thermal FFF and size exclusion chromatography. The separation achieved by thermal FFF is due to the sensitivity of retention in this method to the chemical composition of the polymer

was established very early that this effect is weak in many aqueous solutions. More recently Kirkland has examined a wider variety of aqueous systems and has determined that some polymer systems show a stronger thermal diffusion effect.[45] Water-soluble materials such as proteins and polysaccharides showed no retention whereas poly(ethylene oxide) and poly(vinyl pyrrolidone) exhibited strong retention, similar to that shown by synthetic polymers in organic solvents. The retention of ionic macromolecules such as sodium polystyrene sulfonate was weak to moderate and depended on the ionic strength of the solvent. Overall, retention in aqueous systems was found to be generally weaker than that exhibited in organic carrier liquids.

14.4 OTHER FIELD-FLOW FRACTIONATION METHODS

Other subtechniques of FFF use other fields or gradients to drive components toward the accumulation wall. Flow FFF employs a crossflow of the carrier as the driving force. The walls of a flow FFF channel are semipermeable membrane–frit combinations with carrier reservoirs on either side. The carrier fluid is pumped across the permeable wall to form a crossflow of fluid for the necessary displacement of components. Retention is a function of the crossflow velocity and the component diffusion coefficient. Since the relationship between the diffusion coefficient and the molecular weight of a component is often well characterized, calibration curves may not be required to determine the molecular weight distribution of a sample. Flow FFF is considered universal in applicability since it requires only that the sample be soluble or dispersable in a carrier liquid and that a semipermeable membrane can be found to retain the sample.

Flow FFF has been shown to be applicable to a wide variety of materials including virus particles,[46] proteins[47] and silica sols,[48] along with hydrophilic[49] and lipophilic[50] polymers. The majority of work has focused on water soluble samples. The ability of flow FFF to fractionate sulfonated polystyrenes[49] is illustrated in Figure 8 for the molecular weight range 4.0×10^4 to 1.3×10^6. In the case of lipophilic polymers, polystyrene samples have been fractionated with ethylbenzene as the carrier. Figure 9 compares the fractograms for nearly monodisperse and polydisperse linear polystyrene samples.[50] The peak broadening clearly reflects the polydispersity, indicating that the observed fractogram represents the molecular weight distribution of the sample.

Electrical FFF employs an electrical field to induce retention. By applying an electrical field, charged macromolecules interact with the field and are separated according to the ratio of electrophoretic mobility to diffusion coefficient (see Table 1). The experimental set-up is very similar to that for flow FFF. Semipermeable membranes are used for the channel walls and a voltage is applied to electrodes that are in electrolytic reservoirs on either side of the channel. Electrical FFF has been shown to be applicable to proteins[51, 52] and has the potential to separate polyelectrolytes.

Sedimentation FFF employs a centrifugal driving force to induce the retention of components. The channel is formed from stainless steel, which may be coated with other material, such as Teflon, for compatibility with the sample or the carrier. The channel is fitted into a centrifuge basket.

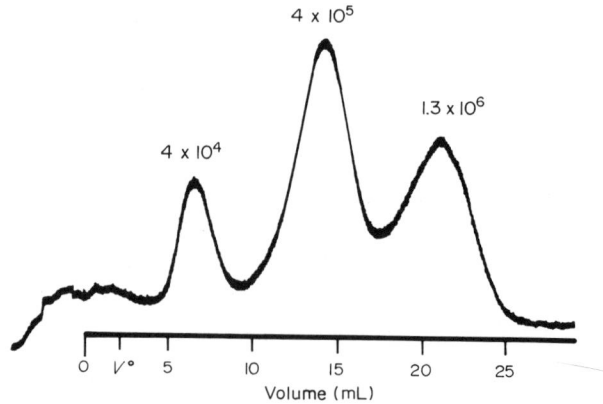

Figure 8   Flow FFF of sulfonated polystyrene fractions in an aqueous medium

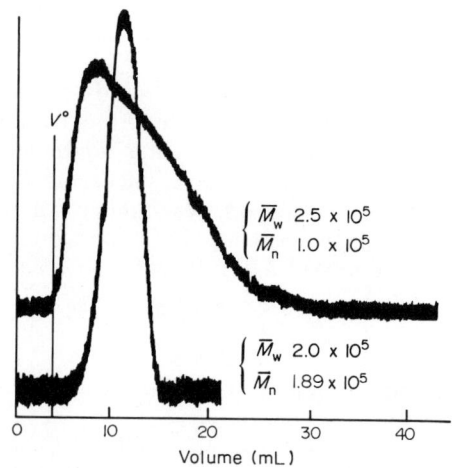

Figure 9   Flow FFF of two polystyrene fractions of different polydispersities in an organic (ethylbenzene) medium

Sedimentation FFF instrumentation is now commercially available.

Although sedimentation FFF has mainly been applied to colloidal particles suspended in aqueous media, it is ideally suited for the molecular weight distribution of ultrahigh molecular weight polymers. This is because retention depends in a calculable way on molecular weight or mass (see Table 1). Thus no calibration is needed, as has been demonstrated with latex particles.[54] Further more, in the limit of high retention, retention volume is a linear function of molecular weight.[13] Thus one not only obtains the molecular weight distribution of the sample directly from the elution diagram, but in a simplified form with a linear molecular mass to time relationship. Programming, however, yields a different mass–time relationship, the form of which varies with the type of programming.

Table 2   Maximum Selectivity Values, $S_M(max)$, for FFF Subtechniques and Size Exclusion Chromatography (SEC)

Method	$S_M(max)$
Sedimentation FFF	1
Flow FFF	0.5–0.55
Thermal FFF	0.5–0.6
SEC	0.1–0.25

Another advantage of sedimentation FFF is that it has a selectivity value, S_M, approaching unity, making sedimentation FFF one of the most selective methods presently available (see Table 2). The combination of these advantageous characteristics will perhaps make sedimentation FFF the method of choice for polymer analysis where it is applicable, namely to higher molecular weight samples.

The lower limit of molecular weight that can be resolved by sedimentation FFF is given very roughly by[3]

$$M \simeq 10^{10}/g \tag{16}$$

where g is the number of gravities. Our work has dealt primarily with high molecular weight colloids employing fields of $\sim 10^3$ gravities, thus restricting our lower molecular weight to $\sim 10^7$. However, the commercially available instrument from DuPont is capable of 3.4×10^4 gravities (1.8×10^4 rpm), which suggests that molecular weights somewhat below 10^6 can be characterized. The instrument is described in the literature as being able to characterize particles in the diameter range 10 nm to 1 μm and macromolecules from 10^6 to 10^{13} in molecular weight.[54] Kirkland *et al.* have studied the retention of water-soluble polymer systems by sedimentation FFF.[54] Included in their work is the determination of the molecular weight distribution of a polyacrylamide sample with a $\bar{M}_w$ of 2.9×10^6.

Two other FFF subtechniques may eventually be available for the characterization of ultrahigh molecular weight polymers. Both techniques are in the early stages of investigation and employ an annular channel design. They are referred to as dielectrical FFF and shear FFF. Dielectrical FFF is based on the dielectrophoresis of neutral particles in the nonuniform electric field of an annular channel. Although no experimental work has been completed, calculations show that the dielectro-phoretic force is strong enough to retain ultrahigh molecular weight polymers dissolved in organic liquids of high dielectric constant.[20] Dielectrical FFF has a selectivity approaching unity, thus matching the selectivity possible in sedimentation FFF. Shear FFF involves a shear-based force which is responsible for migration perpendicular to flow. For shear FFF one of the cylinders making up the annular channel must rotate. The selectivity S_M is predicted to be three or greater, a value considerably higher than that for any other macromolecular separation technique. Although little experimental work has been done, theory suggests that shear FFF would be applicable to polymeric molecules of molecular weight 10^7 and above.[19]

14.5 OPTIMIZATION

We noted earlier that FFF is a very flexible and yet predictable family of techniques for which operating conditions can be varied widely to generate higher resolution, greater speed, wider molecular weight range, *etc.* The best means for exploiting this versatility requires further discussion.

The resolution between two components differing by ΔM in molecular weight was given by equation (14). If we substitute the χ of equation (12) into equation (11), and assume that the nonequilibrium term of equation (11) is the dominant term, then the replacement of N in equation (14) by the resulting L/H term gives

$$R_s = \left(\frac{LD}{384\lambda^3 \bar{v}}\right)^{1/2} \frac{S_M}{w} \frac{\Delta M}{M} \tag{17}$$

This expression shows that a number of factors influence resolution, some of which can be manipulated to increase resolving power. Selectivity S_M depends primarily on the subtechnique (field type) chosen, but to utilize the maximum value of S_M, as shown in Table 2, we need to stay within the range of operation of the subtechnique, roughly defined[32] by $\lambda \leq 0.1$. Equation (17) shows, however, that even smaller values of λ are desirable for enhanced resolution; it is preferable to work in the vicinity of $\lambda \sim 0.1$. Since λ is inversely proportional to the strength of the field or gradient, small λ values are obtained by increasing the field strength to the needed level.

Equation (17) shows that channel thickness w should be minimal. This is why we use thin channels, typically $w = 0.25$ mm, but in some cases, particularly with thermal FFF, down to $w = 0.05$ mm (a value used to produce Figure 4).

We see also that resolution increases with diffusivity D, specifically in proportion to $D^{1/2}$. Since D is inversely proportional to viscosity according to the Stokes–Einstein equation, we can obtain moderate gains by reducing viscosity, either by choosing low viscosity carriers or by increasing the temperature.

Resolution also increases, as shown by equation (17), with $(L/\bar{v})^{1/2}$. We can thus increase channel length L (which is typically ~ 0.5 m) to enhance resolution, but since long channels are awkward to build and operate, we generally prefer to reduce mean flow velocity $\bar{v}$, which has an equivalent effect.

It is important to note that the ratio $L/\bar{v}$ is the passage time t° of a nonretained marker (or void) peak through the channel. The time t_r needed to elute a retained component is proportional to t°, namely

$$t_r = \frac{t^{\circ}}{R} = \frac{L}{\bar{v}R} \tag{18}$$

Thus we see that in maximizing the $L/\bar{v}$ ratio to increase resolution, we are simultaneously increasing the experimental run time. This explains our earlier observation that there is a trade-off between speed and resolution. This is particularly true of the influence of flow velocity, which must be set in accordance with experimental objectives. More details regarding such optimization matters are provided elsewhere.[24,39]

Problems of a different kind may be encountered when attempting to fractionate polymer samples with very broad molecular weight distributions. Earlier peaks are incompletely resolved (because λ exceeds 0.1) and later peaks are slow to appear. Programming techniques will often solve such problems. Major FFF parameters that may be programmed include field strength, flowrate and carrier properties.[15,16] Although field programming is the only programming technique widely implemented, flow programming is in many respects theoretically superior.[56] In flow programming an increasing carrier velocity quickens the elution of later components.

Field strength programming has been studied extensively.[15,38,57-60,16,49] The field strength is usually decreased with time, thus hastening the elution and decreasing the dilution of slower eluting peaks. Some of the field programs that have been studied include linear, parabolic, time-delayed parabolic, exponential and time-delayed exponential. A recent study deals at some length with the optimization of field programming.[61]

Another problem sometimes encountered in FFF is sample overloading. The problem is amplified in FFF by the concentration of the sample in a thin layer near the accumulation wall. This concentration process increases solute–solute interactions, resulting in nonlinear or overloading effects. Polymers are particularly prone to overloading due to chain entanglement and increased viscosity.

While the sample can be diluted to prevent overloading, it may end up too dilute to be detected. The ideal solution to the problem would be to decrease the concentration of sample in the channel but not in the detector. This may be accomplished using a stream splitter.[62] A stream splitter divides the carrier flowstream at the outlet end of the channel such that the concentrated sample layer near the accumulation wall is split away from the bulk of the carrier stream. The concentrated sample stream is then routed to a detector where it provides a greatly enhanced signal. Since the detector signal is increased, the amount of sample injected may be decreased to prevent overloading. Enhancing the detector signal in this way also allows for the analysis of trace materials in the sample. The techniques and uses of stream splitting in FFF are now under active investigation.[49,63,64]

ACKNOWLEDGEMENT

This work was supported by Grant No. CHE-8218503 from the National Science Foundation.

14.6 REFERENCES

1. J. C. Giddings, *Sep. Sci.*, 1966, **1**, 123.
2. E. Grushka, K. D. Caldwell, M. N. Myers and J. C. Giddings, *Sep. Purif. Methods*, 1973, **2**, 127.
3. J. C. Giddings, M. N. Myers, G. C. Lin and M. Martin, *J. Chromatogr.*, 1977, **142**, 23.
4. J. C. Giddings, S. R. Fisher and M. N. Myers, *Am. Lab. (Fairfield Conn.)*, 1978, **10**, 15.
5. J. C. Giddings, *Pure Appl. Chem.*, 1979, **51**, 1459.
6. J. C. Giddings, M. N. Myers, K. D. Caldwell and S. R. Fisher, in 'Methods of Biochemical Analysis', ed. D. Glick, Wiley, New York, 1980, vol. 26, p. 79.
7. J. C. Giddings, *Anal. Chem.*, 1981, **53**, 1170A.
8. J. C. Giddings, M. N. Myers and K. D. Caldwell, *Sep. Sci. Technol.*, 1981, **16**, 549.
9. J. C. Giddings, K. A. Graff, K. D. Caldwell and M. N. Myers, in 'Advances in Chromatography', ed. C. D. Craver, Dekker, New York, 1983, No. 203.
10. J. C. Giddings, *Sep. Sci. Technol.*, 1984, **19**, 831.
11. J. Janča, *Trends Anal. Chem.*, 1983, **2**, 278.

12. G. H. Thompson, M. N. Myers and J. C. Giddings, *Sep. Sci.*, 1967, **2**, 797.
13. J. C. Giddings, *J. Chromatogr.*, 1976, **125**, 3.
14. J. C. Giddings, Y. H. Yoon and M. N. Myers, *Anal. Chem.*, 1975, **47**, 126.
15. F. J. F. Yang, M. N. Myers and J. C. Giddings, *Anal. Chem.*, 1974, **46**, 1924.
16. J. C. Giddings and K. D. Caldwell, *Anal. Chem.*, 1984, **56**, 2093.
17. J. C. Giddings, F. J. Yang and M. N. Myers, *Sep. Sci.*, 1977, **12**, 381.
18. T. C. Schunk, J. Gorse and M. F. Burke, *Sep. Sci. Technol.*, 1984, **19**, 653.
19. J. C. Giddings and S. L. Brantley, *Sep. Sci. Technol.*, 1984, **19**, 631.
20. J. M. Davis and J. C. Giddings, *Sep. Sci. Technol.*, 1986, **21**, 969.
21. J. C. Giddings, *J. Chem. Phys.*, 1968, **49**, 81.
22. M. E. Hovingh, G. H. Thompson and J. C. Giddings, *Anal. Chem.*, 1970, **42**, 195.
23. J. C. Giddings, *J. Chem. Educ.*, 1973, **50**, 667.
24. J. C. Giddings, *Sep. Sci.*, 1973, **8**, 567.
25. M. R. Schure, K. D. Caldwell and J. C. Giddings, *Anal. Chem.*, 1986, **58**, 1509.
26. P. S. Williams, S. B. Giddings and J. C. Giddings, *Anal. Chem.*, 1986, **58**, 2397.
27. E. N. Lightfoot, A. S. Chiang and P. T. Noble, *Annu. Rev. Fluid Mech.*, 1981, **13**, 351.
28. S. Krishnamurthy and R. S. Subramanian, *Sep. Sci.*, 1977, **12**, 347.
29. J. C. Giddings, in 'Treatise on Analytical Chemistry', ed. I. M. Kolthoff and P. J. Elving, Wiley, New York, 1981, part I, vol. 5, chap. 3.
30. P. J. Flory, 'Principles of Polymer Chemistry', Cornell University Press, 1953, chap. 14.
31. S. L. Brimhall, M. N. Myers, K. D. Caldwell and J. C. Giddings, *J. Polym. Sci., Polym. Phys. Ed.*, 1985, **23**, 2445.
32. J. J. Gunderson and J. C. Giddings, *Anal. Chim. Acta*, 1986, **189**, 1.
33. J. C. Giddings, Y. H. Yoon, K. D. Caldwell, M. N. Myers and M. E. Hovingh, *Sep. Sci.*, 1975, **10**, 447.
34. M. Martin and J. C. Giddings, *J. Phys. Chem.*, 1981, **85**, 727.
35. G. Karaiskakis, M. N. Myers, K. D. Caldwell and J. C. Giddings, *Anal. Chem.*, 1981, **53**, 1314.
36. J. C. Giddings, M. R. Schure, M. N. Myers and G. R. Velez, *Anal. Chem.*, 1984, **56**, 2099.
37. M. E. Schimpf, M. N. Myers and J. C. Giddings, *J. Appl. Polym. Sci.*, 1987, **33**, 117.
38. J. C. Giddings, L. K. Smith and M. N. Myers, *Anal. Chem.*, 1976, **48**, 1587.
39. J. C. Giddings, M. Martin and M. N. Myers, *J. Chromatogr.*, 1978, **158**, 419.
40. Y. S. Gao, K. D. Caldwell, M. N. Myers and J. C. Giddings, *Macromolecules*, 1985, **18**, 1272.
41. M. Martin, M. N. Myers and J. C. Giddings, *J. Liq. Chromatogr.*, 1979, **2**, 147.
42. J. J. Gunderson and J. C. Giddings, *Macromolecules*, 1986, **19**, 2618.
43. J. C. Giddings, M. N. Myers and J. Janča, *J. Chromatogr.*, 1979, **186**, 37.
44. S. L. Brimhall, M. N. Myers, K. D. Caldwell and J. C. Giddings, *Sep. Sci. Technol.*, 1981, **16**, 671.
45. J. J. Kirkland and W. W. Yau, *J. Chromatogr.*, 1986, **353**, 95.
46. J. C. Giddings, F. J. Yang and M. N. Myers, *J. Virol.*, 1977, **21**, 131.
47. J. C. Giddings, F. J. Yang and M. N. Myers, *Science*, 1976, **193**, 1244.
48. J. C. Giddings, G. C. Lin and M. N. Myers, *J. Colloid Interface Sci.*, 1978, **65**, 67.
49. K.-G. Wahlund, H. S. Winegarner, K. D. Caldwell and J. C. Giddings, *Anal. Chem.*, 1986, **58**, 573.
50. S. L. Brimhall, M. N. Myers, K. D. Caldwell and J. C. Giddings, *J. Polym. Sci., Polym. Lett. Ed.*, 1984, 22, 339.
51. L. F. Kesner, K. D. Caldwell, M. N. Myers and J. C. Giddings, *Anal. Chem.*, 1976, **48**, 1834.
52. J. C. Giddings, G. C. Lin and M. N. Myers, *Sep. Sci.*, 1976, **11**, 553.
53. J. C. Giddings, G. Karaiskakis, K. D. Caldwell and M. N. Myers, *J. Colloid Interface Sci.*, 1983, **92**, 66.
54. R. L. Blaine, *Res./Dev.*, Sept. 1986, 78.
55. J. J. Kirkland, C. H. Dilks, Jr. and W. W. Yau, *J. Chromatogr.*, 1983, **255**, 255.
56. J. C. Giddings, K. D. Caldwell, J. F. Moellmer, T. H. Dickinson, M. N. Myers and M. Martin, *Anal. Chem.*, 1979, **51**, 30.
57. W. W. Yau and J. J. Kirkland, *Sep. Sci. Technol.*, 1981, **16**, 577.
58. W. W. Yau and J. J. Kirkland, *J. Chromatogr.*, 1981, **218**, 217.
59. J. J. Kirkland and W. W. Yau, *Science*, 1982, **218**, 121.
60. J. J. Kirkland and W. W. Yau, *Macromolecules*, 1985, **18**, 2305.
61. J. C. Giddings, P. S. Williams and R. Beckett, *Anal. Chem.*, 1987, **59**, 28.
62. J. C. Giddings, H. C. Lin, K. D. Caldwell and M. N. Myers, *Sep. Sci. Technol.*, 1983, **18**, 293.
63. J. C. Giddings, K. D. Caldwell and H. K. Jones, in 'Particle Size Distribution Assessment and Characterization', ed. T. Provder, *ACS Symp. Ser.*, 1987, **332**, chap. 15.
64. H. K. Jones, K. Phelan, M. N. Myers and J. C. Giddings, *J. Colloid Interface Sci.*, 1987, **120**, 140.

15

Fractionation

R. KONINGSVELD
Polymer Research Institute, Maastricht, Netherlands,
L. A. KLEINTJENS
DSM Research, Geleen, Netherlands
and
H. GEERISSEN, P. SCHÜTZEICHEL and B. A. WOLF
University of Mainz, FRG

15.1 INTRODUCTION

Synthetic polymers are always inhomogeneous with respect to chain length. The width of the molecular weight distribution may be large or very small, as in products of anionic polymerizations, but strictly one-component polymer samples do not exist. Over the years, many attempts have been made at narrowing chain-length distributions by fractionation. Use can be made of distribution between two phases (two liquids or a solid and a liquid), but the differences in miscibility or solubility between the successive components are so small that isolation of pure components is not feasible and would, if it could be achieved at all, give only vanishingly small yields, since the concentration of components in whole polymers is of the order of 0.01% or less.

Until about 25 years ago, the determination of the molecular weight distribution was the foremost objective of polymer fractionation by phase separation. Since the advent of size-exclusion chromatography, this analytical objective no longer requires the use of time-consuming and essentially inaccurate classical fractionation methods. However, chromatographic techniques are (still) incapable of producing large-size fractions of the order of 100 g and, today, phase-separation methods seem to present the only route to large amounts of narrow-distribution polymers that cannot be obtained by direct synthesis.

The analytical and preparative aspects of classical fractionation have been the subject of extensive theoretical and experimental studies. Fifty years ago, Schulz[1-3] laid foundations that have been used many times but hardly improved. His experimental results not only anticipated the molecular theory of Staverman and Van Santen,[4,5] Huggins[6,7] and Flory[8-10] that was to come a few years later, but also served to question these theories. Schulz's considerations still provide a suitable starting-point, which we shall use here to discuss preparative methods from the standpoint of practical expediency. The majority of these are based on liquid/liquid phase separation, but polymers capable of crystallization can also be fractionated by solid/liquid separation.

Fractionation procedures must be amenable to evaluation, and criteria for the width and location of a molecular weight distribution are called for. Such criteria exist, *e.g.* in the ratios of (a) weight- and number-average molecular weights $\bar{M}_w/\bar{M}_n$; and (b) z- and weight-average molecular weights $\bar{M}_z/\bar{M}_w$. The ratio $\bar{M}_w/\bar{M}_n$ can be shown to be related to the variance, σ_n^2, of the number distribution of chain lengths present in the sample (equation 1).

$$\sigma_n^2/\bar{M}_n^2 \;=\; \bar{M}_w/\bar{M}_n \;-\; 1 \;=\; U_n \tag{1}$$

The quantity U ('Uneinheitlichkeit') was introduced by Schulz.[11] Similarly, the variance of the weight distribution, σ_w^2, is given by

$$\sigma_w^2/\bar{M}_w^2 \;=\; \bar{M}_z/\bar{M}_w \;-\; 1 \;=\; U_w \tag{2}$$

It can be seen from these equations that $U_n \geq 0$; $U_w \geq 0$; $\bar{M}_z \geq \bar{M}_w \geq \bar{M}_n$, the equalities referring to a strictly monodisperse sample containing a single component only.

The quantities U_n and U_w are useful criteria; a successful fractionation will decrease their values and bring them as close to zero as possible. However, U_n and U_w do not reveal how much a fraction's average molecular weight has been shifted relative to that of the parent polymer. For this purpose, Huggins suggested using the quantity f_n, in addition to U_n and U_w.[12,13] For the *k*th fraction we have

$$f_{nk} \;=\; |\bar{M}_{nk} \;-\; \bar{M}_n|/\bar{M}_n \tag{3}$$

where $\bar{M}_n$ and $\bar{M}_{nk}$ are the number-average molecular weights of the parent polymer and the *k*th fraction, respectively. Analogous quantities may be defined for other molecular weight averages.

Preparative fractionation of linear homopolymers being the subject of this chapter, we concentrate on that aspect and do not attempt to review the extensive literature on fractionation in general. In the latter context, references 3 and 14–17 are among those worth consulting.

15.2 FRACTIONATION BY LIQUID/LIQUID PHASE SEPARATION

15.2.1 Fractionation Procedures

There are essentially two procedures used in fractionation by distribution between two liquid phases, *viz.* fractionation by precipitation and by extraction. Other schemes have been described in the literature and represent more or less complicated combinations of precipitation and extraction steps. Here we confine the discussion to the main procedures.

The characteristics of precipitation and extraction fractionation can be clarified with the aid of a ternary system containing a single solvent S and two polymeric homologues P_1 and P_2, differing in molecular weight. An actual molecular weight distribution will contain many more than two components, but can, in a first approximation, be represented by a point X on the P_1P_2 axis in a ternary isothermal diagram. Figure 1 only shows the solvent-rich corner of the composition triangle, because the miscibility gap within the binodal (solid curve) is generally confined to that region. When polymer solutions separate into two phases, one of them usually contains hardly any polymer and the other, though much more concentrated, still contains more solvent than polymer.

A system within the two-phase region with overall polymer concentration $\phi_{X_1} [=(\phi_{P_1}+\phi_{P_2})_1]$ will separate into two phases A_1 and B_1 with overall polymer concentrations $\phi_{X_{1a}}$ and $\phi_{X_{1b}}$. The molecular weight distributions of the fractions in phases A_1 and B_1 are found upon extension of SA_1 and SB_1 to the P_1P_2 axis, where the solvent concentration, ϕ_s, equals zero. Thus we obtain X_{1a} and X_{1b} and see that the phase separation has shifted the compositions of the fractions towards the corners representing the pure components P_1 and P_2. We also note the following:

(i) The ultimate aim (fractions consisting of pure P_1 and P_2) cannot be achieved. No matter how close the binodal comes to the SP_1 axis, or even when at another temperature it intersects that axis, it never assumes a location allowing SA_1 to coincide with SP_1 as long as there is any P_2 in the system. Hence, the fraction X_{1a} will always be contaminated with some P_2.

(ii) The phase-volume ratio $r (= V_a/V_b = a/b)$ is large and, consequently, the relative amount of fraction X_{1b} is small. It separates out as a concentrated solution, which usually settles at the bottom

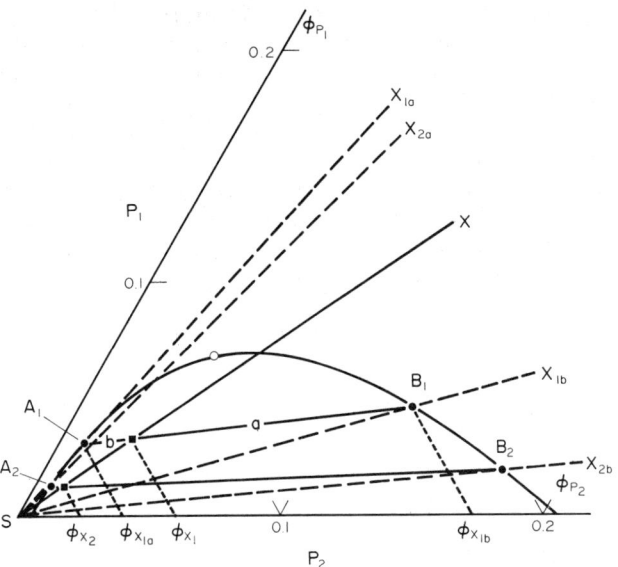

Figure 1 Ternary phase diagram for solutions of a binary polymer (components P_1 and P_2) in a single solvent, calculated with the Flory–Huggins–Staverman model for relative chain lengths $m_1 = 38.2$; $m_2 = 261.8$. The interaction parameter was set equal to 0.5988. Open circle = critical point; ●———●, tie line with coexisting phase compositions; solid curve = binodal; A_1 and A_2, cloud points for binary polymers X_{1a} and X_{2a} at $\chi = 0.5988$; B_1 and B_2, similar for polymers X_{1b} and X_{2b}

of the vessel. For this reason the procedure is called fractionation by precipitation. Most of the polymer remains in the supernatant phase, a, which is a dilute solution containing a polymer with molecular weight distribution X_{1a} that has shifted toward the corner representing the smaller molecules P_1.

(iii) A decrease of the initial polymer concentration from ϕ_{X_1} to ϕ_{X_2} improves the efficiency of the separation for the precipitated fraction X_{2b}. However, the phase-volume ratio is then larger than in the separation at ϕ_{X_1}, so that the yield of fraction X_{2b} is smaller than that of X_{1b}, an undesirable feature in preparative fractionation.

Figure 2 shows what happens if we work at much higher initial concentrations, *e.g.* at ϕ_X. This concentration exceeds the critical composition ϕ_c where the two equilibrium phases ϕ_{X_a} and ϕ_{X_b} are identical. We note that: (a) the fraction in the dilute phase is now the smaller of the two and is contained in the supernatant solution which has the smaller volume (extraction fractionation); (b) the shifts in composition to X_a and X_b are now such that the dilute phase contains the better fraction with respect to U_n, as is seen in Table 1; and (c) the initial concentration can be large, so that the preparative objective is likely to be better served by extraction than by precipitation.

Values of the width and shift criteria are listed in Table 1, which summarizes the above remarks in a quantitative manner. The precipitated fractions X_{1b} and X_{2b} have better U_n and U_w values than the initial polymer X. In the present example extraction (fraction X_a) leads to a smaller U_n but an increased U_w compared to the initial polymer. Other initial distributions, particularly truly multi-component ones, may show different patterns.

Figures 1 and 2 are not merely schematic examples, but were calculated with the Flory–Huggins–Staverman (FHS) theory referred to below. Hence, the situation outlined above is relevant for actual systems, since the FHS model describes the thermodynamic properties of polymer solutions in a qualitatively correct manner.

15.2.2 Molecular Weight Distributions

Actual polymers contain many more than two components. The difference in molecular weight between two successive components is usually small compared to their molecular weights, M, and we follow Schulz[18] in assuming that a molecular weight distribution (MWD) may be represented by a continuous function. In this chapter we use the Γ function, sometimes referred to as exponential or

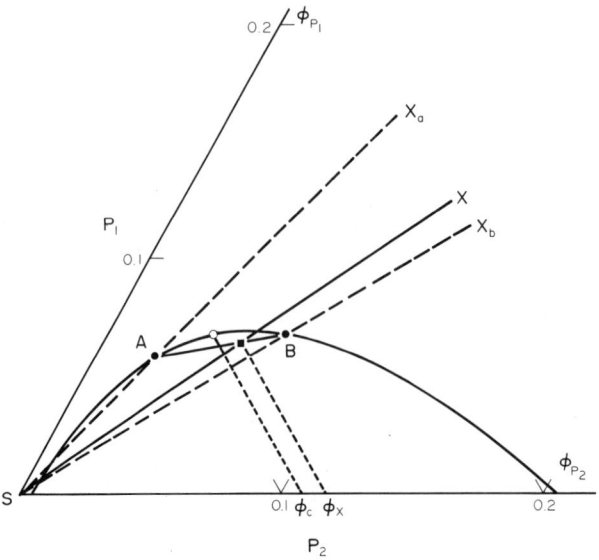

Figure 2 See caption to Figure 1 for definition of symbols; ϕ_c, critical concentration; A and B, cloud points for polymers X_a and X_b.

Table 1 Criteria of Fractionation Efficiency for the Separations in
Figures 1 and 2

Sample	U_n	U_w	f_n
X (initial polymer)	1.24	0.65	0
X_{1b} (fraction)	0.98	0.24	0.65
X_{1a}	0.89	1.10	0.23
X_{2b} (fraction)	0.47	0.08	1.64
X_{2a}	1.00	1.00	0.19
X_b	1.25	0.56	0.08
X_a (fraction)	1.00	1.00	0.19

Schulz–Zimm distribution, to represent mass distributions $w(M)$

$$w(M) \;=\; AM^{\lambda+1}\exp(-\tau M) \tag{4}$$

where $\qquad \lambda \;=\; (1 \;-\; U_n)/U_n \;$ and $\; \tau \;=\; (\lambda \;+\; 1)/\bar{M}_n.$

The exponential distribution is rather symmetrical (Figure 3a) and therefore we also use the asymmetric log-normal function

$$w(M) \;=\; AM\exp[-\beta^2\ln^2(M/M_0)] \tag{5}$$

where $\qquad \beta^2 \;=\; 2\ln(U_n \;+\; 1) \;$ and $\; M_0 \;=\; \bar{M}_n(U_n \;+\; 1)^{-3/2}.$

In both cases the normalization factor A can be used to adapt the function to a desired total mass of polymer. In Figure 3(a) we see that the log-normal distribution assigns more material to the M range smaller than $\bar{M}_n$ than the exponential distribution, but also contains a large amount of very high molecular weight polymer. This limits its applicability in thermodynamic studies of multi-component polymer solutions,[19,20] a feature we do not need to be concerned about in the present context, however.

Sometimes production processes lead to bimodal distributions. In order to check the effect that such initial distibutions may have on fractionation efficiency, we add one example which is obtained by the addition of two suitably chosen log-normal functions (Figure 3b).

Details of the distribution functions used below are given in Table 2.

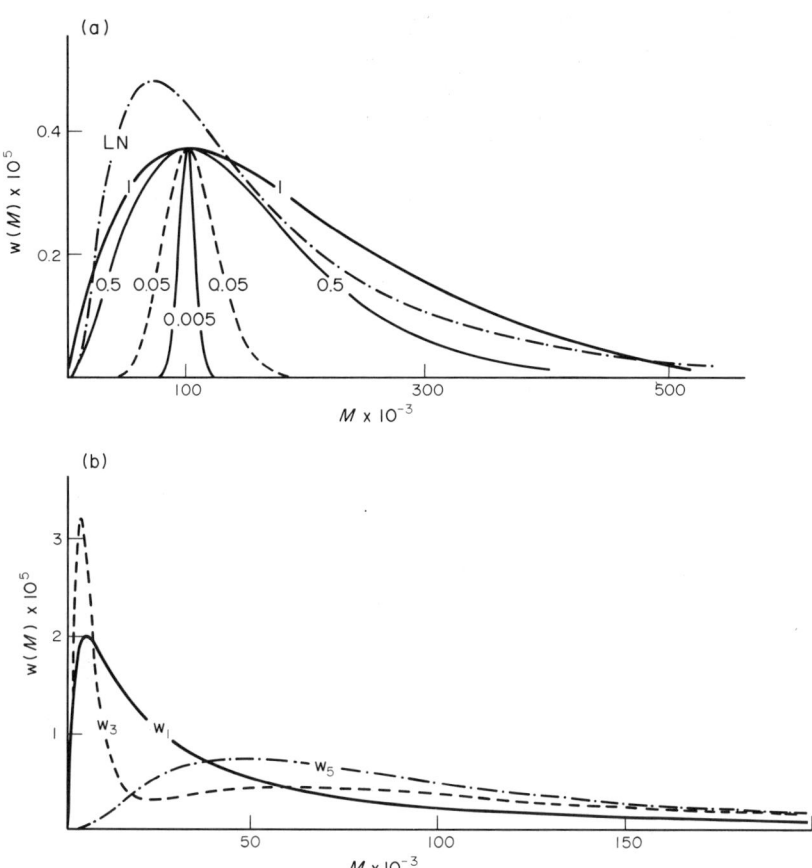

Figure 3 (a) Exponential mass distributions for $\bar{M}_n = 10^5$ and indicated values of U_n; LN, log-normal mass distribution for $\bar{M}_n = 10^5$ and $U_n = 1$; (b) examples of the mass distribution functions used in the calculations; $\bar{M}_w = 1.32 \times 10^5$; w_1, log-normal, $U_n = 9$; w_3, sum of two log-normal functions, $U_n = 9$; w_5, log-normal, $U_n = 1$.

Table 2 The Characteristics of the Initial Distribution Functions Used

Function	$10^{-3} M_w$	U_n	U_w
w_1 (log-normal)	132	9	9
w_3^a	132	9	1.6
w_{31} (log-normal)	7.4	1	1
w_{32} (log-normal)	173.1	1	1
w_5 (log-normal)	132	1	1
w_7 (exponential)	132	1	0.5

^{a} w_3 is a 1:3 mixture of two log-normal functions, w_{31} and w_{32}.

15.2.3 The Schulz Theory of Fractionation

Schulz assumed the distribution of polymer molecules i between two immiscible liquids, a and b, to be governed by their energy difference in the two phases[1-3] and followed Brönsted[21] in setting that energy difference proportional to the degree of polymerization P_i

$$\ln k_i = \ln(c_{ib}/c_{ia}) = kP_i \qquad (6)$$

where c_{ia} and c_{ib} are the concentrations in g cm^{-3} of species i in phases a and b, k_i is the distribution coefficient and k is a proportionality constant depending on temperature. Phase b is assumed to concentrate the higher molecular weight components.

Careful experiments soon led Schulz to reject equation (6) and revise the chain-length dependence of k_i

$$\ln k_i = kP_i^\varepsilon \tag{7}$$

where ε is a system-dependent parameter.[2] His findings have been confirmed in later investigations.[22-28] Some of these indicated that equation (7) needs further improvement if large P ranges are involved (see Figure 4). To date, the improved theory needed for a quantitative treatment of $k_i(P_i)$ has not yet been developed.[29]

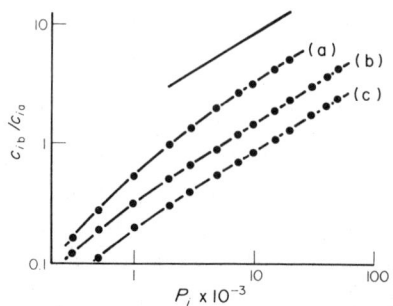

Figure 4 Distribution of linear polyethylene between two liquid phases (solvent, diphenyl ether). Phase separation at three temperatures; (a) 135 °C, (b) 140 °C and (c) 143 °C (the straight line indicates a slope of 0.63)

Assuming the densities of dissolved polymer to be the same in the two phases, we may rewrite equation (6) obtaining

$$\ln k_i = kP_i = \ln(\phi_{ib}/\phi_{ia}) = \sigma m_i \tag{8}$$

where ϕ_{ia} and ϕ_{ib} are the volume fractions of species i in phases a and b. The last equality in equation (8) represents the result obtained from the lattice theory of Flory,[8-10] Huggins[6-7] and Staverman[4-5] (FHS), which includes the parameter m_i, the number of lattice sites occupied by polymer chains i. Obviously, m_i is proportional to P_i and the lattice theory is seen to yield Schulz's expression (6) for k_i which, however, he had already rejected on experimental grounds.

In one respect the FHS approach is superior, because it specifies Schulz's proportionality factor k. We have

$$\sigma = \ln[(1 - \phi_{X_b})/(1 - \phi_{X_a})] + 2\chi(\phi_{X_b} - \phi_{X_a}) \tag{9}$$

where χ is the Flory–Huggins interaction parameter that, to a first approximation, depends on temperature only, albeit in a system-dependent manner.[30]

Having indicated that equation (6) can be given a molecular justification, we proceed to discuss the complete precipitation fractionations which Schulz simulated by calculations based on expression (6) for k_i. He represented the initial MWD by a function arising from kinetic considerations

$$w(P) = -\frac{1}{2}\alpha^P P^2 \ln^3 \alpha \tag{10}$$

where α is a positive number, much smaller than unity.

The MWD of the first fraction in the concentrated phase, b, is given by

$$w_{1b}(P) = w(P)/[1 + r\exp(-kP)] \tag{11}$$

which is derived easily upon replacing c_{ib}/c_{ia} by rw_{ib}/w_{ia} in equation (6), where w_{ia} and w_{ib} are the masses of polymer species i in phases a and b, respectively. The fraction in the dilute phase a has the distribution

$$w_{1a}(P) = w(P)/\{1 + r^{-1}\exp(kP)\} \tag{12}$$

and serves as the initial distribution in the second separation step. The material remaining in solution after a given number of such separations represents the last fraction.

Schulz simulated separations by precipitation into eight fractions, assuming reasonable values for k and r in each separation step. Figures 5 and 6 show some of his results and illustrate that the fraction distributions will overlap considerably, even when each fraction is precipitated twice. Other calculations by Schulz were in accordance with the qualitative conclusions in Section 15.2.1, *i.e.* that the width of the MWD of a precipitated fraction decreases when the amount separated and the initial concentration are decreased.

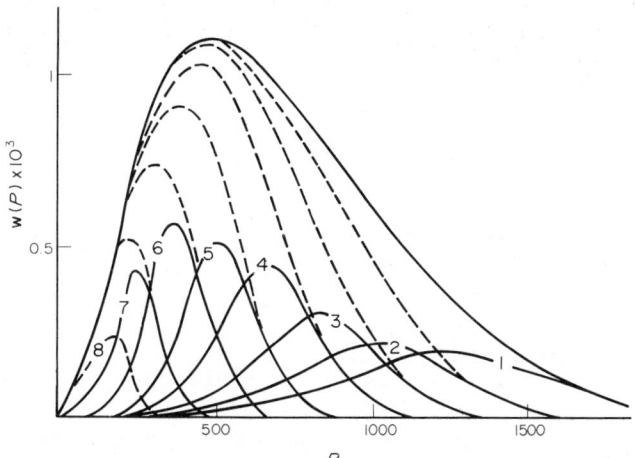

Figure 5   Complete precipitation fraction (7 steps, 8 fractions) calculated by Schulz.[1-3] Dashed curves indicate the fraction remaining in solution after each separation; each fraction precipitated once

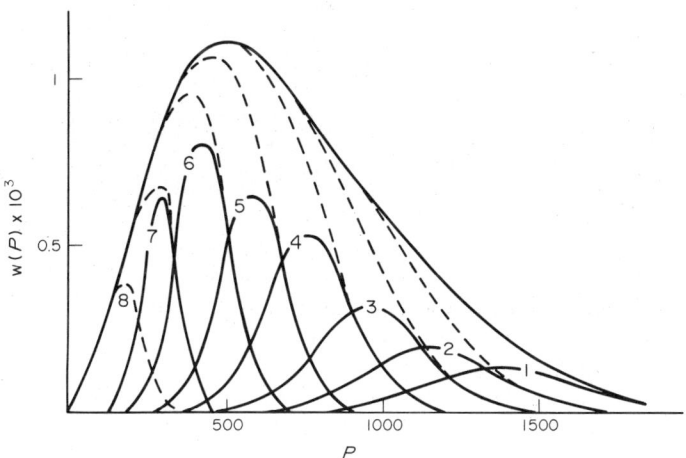

Figure 6   See caption to Figure 5 for definition of symbols; fractions precipitated twice

15.2.4   Later Developments

Since Schulz, many authors have repeated numerical simulations of fractionation procedures, without changing his conclusions.[12,13,19,23,24,31-37] Kamide *et al.*[38] extensively checked and confirmed the results of all these calculations. In the course of this development the calculation methods improved, *i.e.* by the introduction of the FHS model (equations 8 and 9) and of procedures ensuring thermodynamic equilibrium throughout the calculation.

The calculations involve the mass balance

$$\int w_b(M)dM \quad = \quad x \quad = \quad \int w(M)[1 \quad + \quad r\exp(-\sigma m)]^{-1}dM \qquad (13)$$

where x is the mass of the fraction in the concentrated phase. The integrations run from $M=0$ to

$M = \infty$, since we use continuous distribution functions for w(M). Usually, w(M) is normalized to a total mass of initial polymer of 1 g,

$$\int w(M) dM \;=\; 1 \tag{14}$$

and the fraction in the dilute phase then has the mass $(1 - x)$. The lattice model relates the 'relative chain length', m, to molecular weight M by

$$m \;=\; v_p M / V_s \tag{15}$$

where v_p is the specific volume of the polymer in solution (assumed independent of M) and V_s is the molar volume of the solvent.

The initial distributions w(M) are usually chosen so that the integral in equation (14) exists in closed form. The integrals in equation (13), however, will in most cases need numerical treatment. This, in fact, amounts to splitting the distribution up into a number of discrete components and abolishing the continuous functions. Numerical integration of equations like (13) have (under the heading continuous thermodynamics, of which Schulz was one of the earliest exponents) been the subject of thorough studies by Raetzsch *et al.*[39]

The integration is only one of the three fundamental problems involved, the other two being presented by w(M) and k_i. Especially at the low and high M ends of the MWD, the reliability of size-exclusion chromatography in determining w(M) is not very great. Besides, we have the problem of representing the chromatographic curve as closely as possible.[35-38] Probably more important, however, is the uncertainty still existing regarding the distribution coefficient k_i.[29] Evidently, theory has not yet been developed enough to justify better than qualitative considerations of fractionation processes, but this serves the present purpose well enough.

Calculated plots of U_n against x have revealed a pitfall of precipitation fractionation that hitherto seems to have escaped attention.[36,37] If the initial distribution is asymmetric (relatively large U_w compared with U_n) the number distribution of the fraction precipitated under allegedly favourable conditions (small ϕ_x and x) may be wider than that of the initial polymer ($U_{nb} > U_n$), surely an undesirable result. Figure 7 gives some examples that, in view of the qualitatively consistent experimental results of Figure 8, have to be taken seriously. We see that initial distributions of the log-normal type ($U_w = U_n$) may give rise to precipitated fractions (small x) with $U_{nb} > U_n$. If we start with the more symmetrical exponential function ($U_w < U_n$) such behaviour is not encountered.

The reason why a precipitated fraction has a wider number distribution than the initial polymer is to be sought in the shape of w(M) in the lower M range. The denominator in equation (13) increases when M is lowered, and if w(M) is constant in the considered M range or does not rise very steeply, $w_b(M)$ will decrease. However, if w(M) shows a steep rise, as with log-normal distributions, the increase of the denominator may be outweighed and $w_b(M)$ passes through a maximum before it goes to zero at $M = 0$. A two-peak fraction distribution results, comprising a relatively large amount of low molecular weight material and, consequently, a large value of U_n (see Figure 9).

In such cases, extraction fractionation is the preferred procedure. A large amount of the polymer is precipitated (*e.g.* $x = 0.9$) and the fraction is contained in the supernatant solution. Though precipitation will yield more reliable results with more symmetrical initial distributions, extraction is also superior there, which is the reason why Scott[33] and Staverman and Overbeek[40] recommended the extraction procedure outlined above.

Figure 7 gives an example showing that a bimodal initial distribution is also better fractionated by extraction, at least as far as the first fraction is concerned. We are thus led to conclude that the extraction provides the safest route toward a narrow MWD in the fraction. In addition it is the method most amenable to the scaling-up needed for truly preparative fractionation, because the initial concentration does not have a large influence and can be chosen as high as practical operation allows.

15.2.5 Other Expressions for the Distribution Coefficient

Schulz found that the linear relationship between $\ln k_i$ and P_i, postulated by Brönsted and arising from the FHS model, was contradicted by experiment. The improved expression (7) was therefore used by Bohdanecký[41] to check the effect of $\varepsilon \neq 1$ by means of calculations similar to those leading to Figure 7. He found that $\varepsilon \neq 1$ shifted the $U_n(x)$ curves, compared to $\varepsilon = 1$, but the general picture remained unchanged.

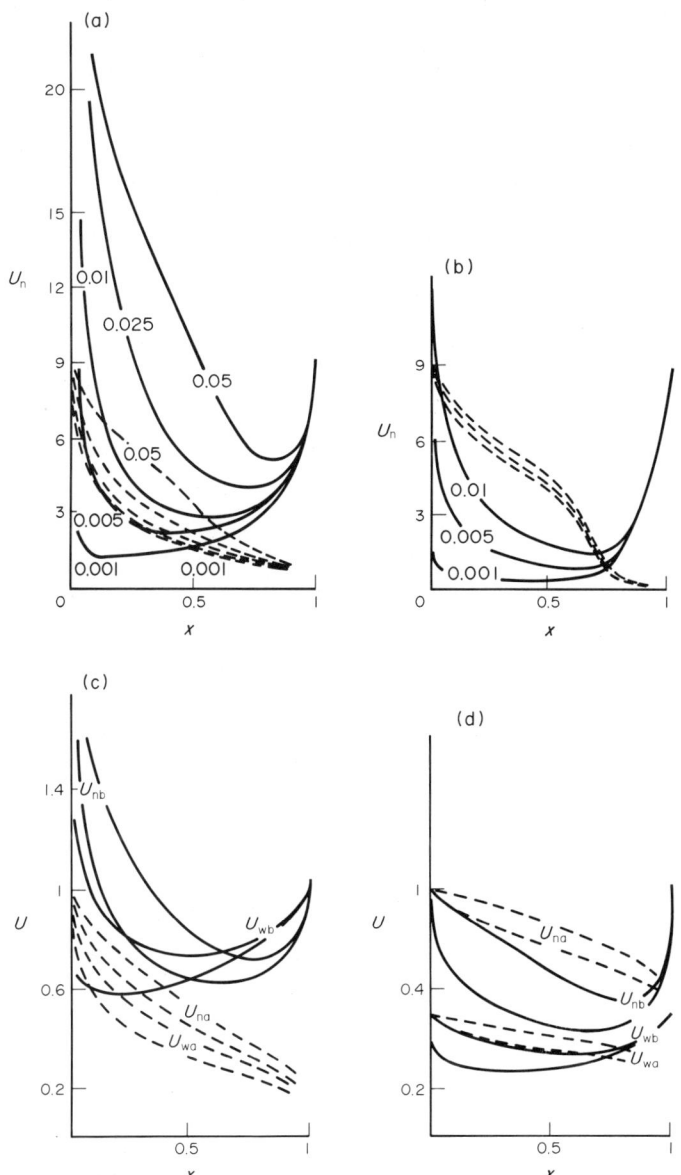

Figure 7 Calculated $U_n(x)$ and $U_w(x)$ curves for various initial distributions, (a) w_1, (b) w_3, (c) w_5 and (d) w_7; and indicated initial concentrations (volume fractions). Solid curves, fraction in concentrated phase; dashed curves, fraction in dilute phase; for w_5 and w_7 the curves represent initial volume fractions of 0.02 and 0.01 (top and bottom, respectively)

Other extensions of the FHS model have been investigated:[37,38] (a) single-solvent systems in which the interaction parameter χ depends on concentration; and (b) mixed-solvent systems in which separation is not effected by a change in temperature (*i.e.* a change in χ in a single-solvent system), but by a variation of the composition of the mixed solvent. The latter consists of a good solvent and a poor one (precipitant), an increasing amount of which will precipitate increasing amounts of the polymer.

In both cases (a) and (b) the model still leads to equation (8), albeit that the separation factor σ assumes a different form. However, the linearity between k_i and m_i is retained and σ, as in equation (8), does not depend on the concentrations of the individual polymer components. Figure 10 shows, in a few examples, that extraction remains the preferable recipe, whatever the type of system.

15.2.6 Large-scale One-step Extraction Fractionation

A decrease of the width of the fraction distribution goes with a decrease of its yield, and sufficient quantities of well-fractionated material will need a considerable amount of initial polymer. It is

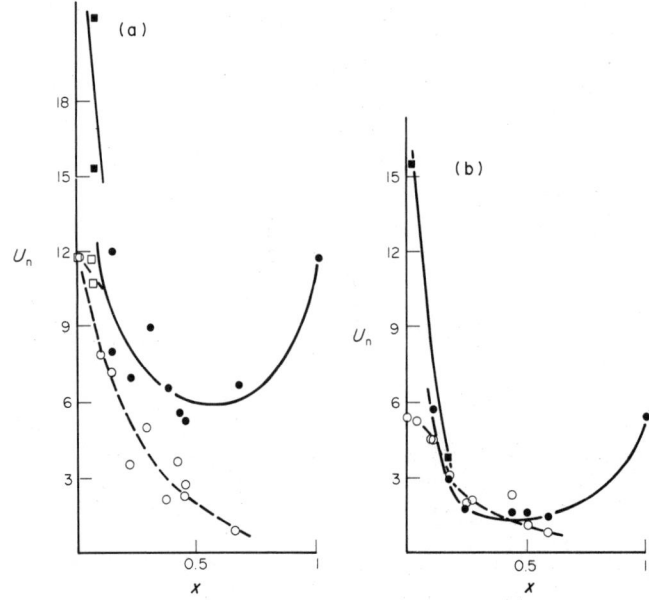

Figure 8 Experimental $U_n(x)$ curves for the system diphenyl ether/polyethylene; solid symbols, concentrated phase; open symbols, dilute phase. Initial concentrations of 1 and 2% by weight (circles and squares, respectively). Initial samples: linear polyethylene; (a) $\bar{M}_w = 1.53 \times 10^5$, $U_n = 12$, $U_w = 5$; and (b) $\bar{M}_w = 5.5 \times 10^4$, $U_n = 5.4$, $U_w = 4.5$

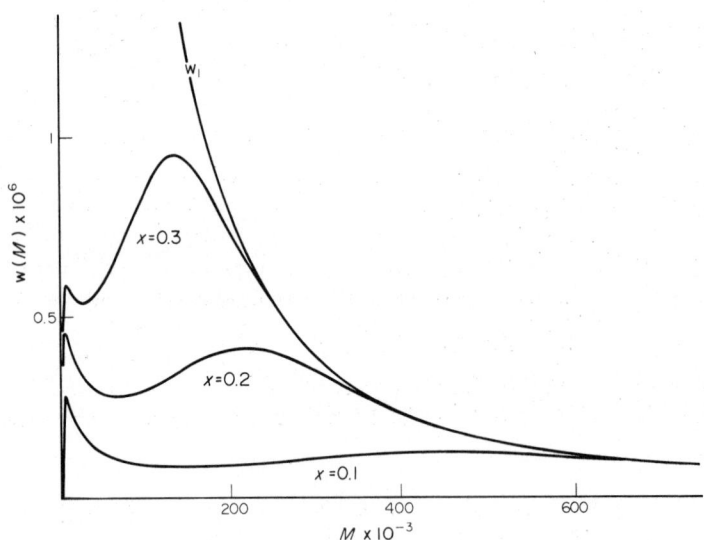

Figure 9 Fraction distributions separated by precipitation from initial distribution w_1, for various values of the mass fraction in the concentrated phase x (initial polymer concentration, $\phi = 0.01$)

shown in Figure 11, calculated for an exponential initial distribution with $\bar{M}_n = 10^5$ and $U_n = 1$, how the necessary amount of starting material rises steeply as the required U_n value for a 100 g fraction is lowered. If several kg of polymer must be handled, *e.g.* in 10% solutions, a pilot-plant scale is soon required to carry out such separations. This has been done[35,42-45] and Table 3 shows some typical results reflecting the effect illustrated in Figure 11.

We note that, though considerable sharpening of the distribution has been obtained, U_n and U_w values close to zero cannot easily be realized. This was already clear from Figure 7. A one-step extraction is effective but calls for many repetitions if very narrow fraction distributions are to be obtained.

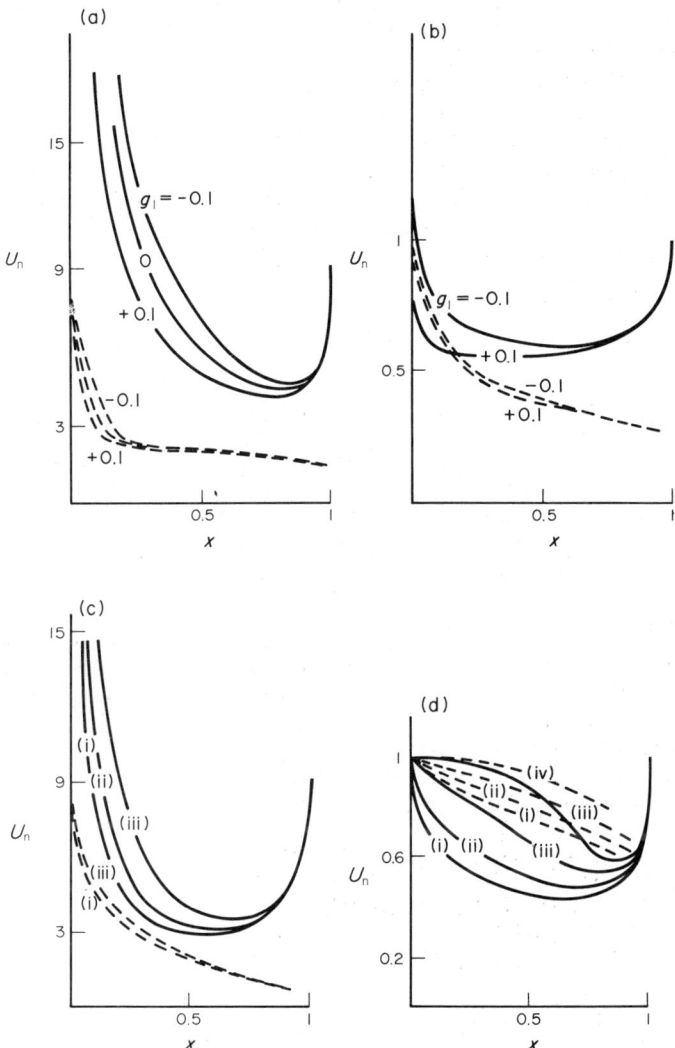

Figure 10 (a) and (b): $U_n(x)$ curves calculated with the FHS model for fractionation in a single solvent with a concentration-dependent interaction parameter, $g = g_0 + g_1\phi$. Initial distribution = sum of two exponential functions: (a) $\bar{M}_w = 10^5$, $U_n = 9$, $U_w = 9$; (b) $\bar{M}_w = 10^4$, $U_n = 1$, $U_w = 1$ (initial concentration, $\phi = 0.01$). (c) and (d): $U_n(x)$ curves calculated with the FHS model for fractionation in a mixed solvent system. Initial distributions: (c) w_1 and (d) w_7. The solvent/polymer, solvent/precipitant and precipitant/polymer interaction parameters are: (i) 0.45/0/0.6; (ii) 0/0/1; (iii) 0/1/1; (iv) 0/1.4/1 (solid curves, fraction in concentrated phase, dashed curves, fraction in dilute phase; initial polymer concentration, $\phi = 0.01$)

Table 3 Large-scale Extraction Fractions of Linear Polyethylene

Sample	$10^{-3}\bar{M}_n$	$10^{-3}\bar{M}_w$	$10^{-3}\bar{M}_z$	U_n	U_w	f_n	Mass (g)
Initial polymer	12	153	900	12	5		
Fraction 1	92	140	330	0.5	1.4	6.7	133
Fraction 2	8.6	55	300	5.4	4.5	0.3	643

15.2.7 Countercurrent Extraction

Continuous extraction provides an obvious practical method to realize repeated extraction steps.[46-48] Englert and Tompa[49] carried out model calculations and showed very convincingly that extremely sharp fraction distributions should be obtainable by countercurrent fractionation. Figure 12,[49] illustrates the potential power of a continuous separation technique. A first run yields the distributions E_1 in the extract and R_1 in the raffinate; note the exceptionally sharp separation. If

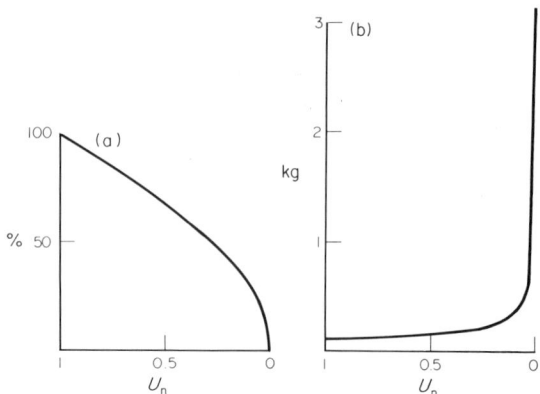

Figure 11 (a) Yield of fraction separated from an exponential distribution ($U_n = 1$; $\bar{M}_n = 10^5$) as a function of the fraction's U_n value; (b) amount of initial polymer needed for a 100 g fraction, as a function of its U_n value

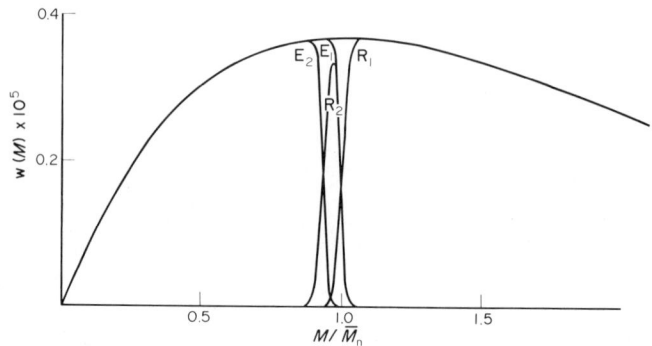

Figure 12 Countercurrent fractionation of an exponential distribution ($U_n = 1$), calculated by Englert and Tompa:[49] E_1, R_1 distributions in extract and raffinate after the first run; E_2, R_2, ditto after the second run (the U_n value of R_2 is 0.003)

the extracted polymer E_1 is subjected to a second countercurrent treatment, extract E_2 results and the fraction obtained is R_2. Its U_n value is 0.003, a very desirable result.

Putting the method into practice met with some problems. The rate of segregation of the phases proved to be too slow.[46] Another problem was to find a partially miscible solvent system in which both phases dissolve enough polymer. Englert and Tompa proposed solving this latter problem by using the incompatibility of most polymers, which often causes solutions of two polymers in a common solvent to be partially miscible. Recently, mixed-solvent systems have been found for a number of polymers that overcome these problems and a breakthrough in this field has been realized, *i.e.* continuous polymer fractionation (CPF),[50-52] described in Section 15.4.

15.3 CRYSTALLIZATION–DISSOLUTION FRACTIONATION (CDF)

An effective method to fractionate crystallizable polymers was developed by Pennings[53-54] some 20 years ago. The procedure is based on the chain-length dependence of the crystallization or dissolution temperature in solution. Usually, this dependence is weak and the success of the method obviously depends on a careful operation.

The polymer is first slowly crystallized from a stirred solution, during which process the chain molecules are sorted, the longest being the first to separate out. The crystals have a fibrillar structure and contain long thin backbones in which the polymer chains are highly extended. The fibrils adhere to the stirrer and form a tenuous network serving as a substrate for shorter chains to crystallize at lower temperatures in a folded chain conformation. When the polymer has crystallized completely it has a structure in which the low M end of the MWD is concentrated in the layers adjacent to the solvent.

In the dissolution procedure that now follows, the polymer is dissolved stepwise at increased temperatures, and the successive fractions usually show a monotonous increase in average molecular weight. The example in Table 4 illustrates that this technique may be very effective.

Table 4 Crystallization–Dissolution Fractions of Linear Polyethylene

Sample	$10^{-3}\bar{M}_n$	$10^{-3}\bar{M}_w$	$10^{-3}\bar{M}_z$	U_n	U_w	f_n	Mass (g)
Initial polymer	8	75	250	8	2.3		
Fraction 1	11	11	12	0	0.1	0.4	33
Fraction 2	8.3	9.3	12.5	0.1	0.3	0	26
Fraction 3	8.4	8.7	11	0	0.3	0	13

The CDF method has been scaled-up to a pilot-plant operation[55] and initial amounts of polymer as large as 5 kg could be split up into 12 fractions. Their $\bar{M}_w$ values, as derived from intrinsic viscosities $[\eta]$, indicated that sizeable shifts in average molecular weight could be obtained (see Table 5).

Table 5 Large-scale Crystallization–Dissolution Fractionation of Linear Polyethylene in Perchloro-ethylene

Sample	$[\eta]$ (dl g^{-1})[a]	$10^{-3}\bar{M}_w^b$	f_w	Mass (g)	Dissolution temp. (°C)
Initial polymer				5000	
Fractions:					
1	0.101	1.6	0.98	80	70
3	0.188	3.9	0.95	140	80
4	0.260	6.0	0.92	300	85
5	0.782	27.6	0.6	468	87
7	1.625	75.6	0	1220	90.5
9	2.22	115	0.5	480	94
11	2.63	147	1.0	177	97.5
12	3.70	235	2.1	599	109

[a] In decalin at 135 °C.
[b] $\bar{M}_w$ calculated from $[\eta] = 4.75 \times 10^{-4}\,\bar{M}_w^{0.725}\,(\bar{M}_w/\bar{M}_n)^{-0.07}$

Since the CDF method is based on differences in crystallization and dissolution temperatures, any structural feature bringing such differences about can be investigated with this method. For instance, the degree of short-chain branching in polyethylene is known to markedly influence the crystallization temperature from solution, and fractions can be prepared with the CDF technique. These fractions will now differ, not only in average chain length, but also in degree of short-chain branching, and additional information is needed to interpret the fractionation results. Size-exclusion chromatography and differential scanning calorimetry have been used for this purpose.[55, 56]

15.4 CONTINUOUS POLYMER FRACTIONATION (CPF)

As demonstrated in Section 15.2.7, model calculations by Englert and Tompa[49] on countercurrent extraction of polymers have shown very convincingly that extremely sharp fraction distributions should be obtainable by that method. Numerous attempts have therefore been made to use it for the purpose of large-scale fractionation. However, in an extensive study of the practical applicability of the partition of polymolecular samples between the phases formed by two incompletely miscible solvents, it was demonstrated that out of 289 such pairs of solvents, only 3 allowed both of the two coexisting phases to contain noteworthy amounts of high-molecular weight material.[57] With all other systems this material was found practically exclusively in one of the phases only.

The thermodynamic reason for this behaviour is obvious. Since solvent 1 and solvent 2 have to exhibit a miscibility gap to be useful for an extraction process, they have to be chemically sufficiently different. Consequently, the χ values of the solvent 1/polymer and solvent 2/polymer systems are normally also quite dissimilar. In this situation, the dependence of polymer solubility on molecular weight is of secondary importance only, so that all polymeric material, irrespective of its molecular weight, is accumulated in one phase.

In the light of the above observations it seemed worthwhile to search for an alternative, with the performance characteristics of countercurrent extraction, but with a very special class of system, characterized by great similarity of the interaction parameters of the polymer with the solvents. Only such systems would permit utilizing the slight variation of polymer solubility with molecular weight for fractionation. Suitable systems have indeed been found, and used for continuous countercurrent extraction.[50-52] The new method for large-scale fractionation, called continuous polymer fractionation (CPF), constitutes an 'intrinsic extraction', in the sense that a sample with a broad MWD is not distributed between two phases also existing in its absence, but between two phases that only come into existence because of the addition of the polymer.

15.4.1 Principle of the CPF Technique

How the CPF method functions can most easily be explained by means of the special case of a quasi-binary system, *i.e.* a solution of a polymer with an MWD in a single theta solvent, although the method is usually operated with quasi-ternary systems (solutions in mixed solvents). At first glance, it would appear impossible to realize a countercurrent extraction with phases composed of only two chemically different species. One has, however, to take into account that the polymeric constituent itself already represents a multicomponent system.

Figure 13 shows schematically how the cloud-point temperature of a quasi-binary system depends on the overall polymer concentration (cloud-point curve). As can be deduced from Figures 1 and 2, the compositions of coexisting phases are not located on the cloud-point curve. A continuous countercurrent extraction can be performed at a temperature T_0, a few degrees below the maximum cloud-point temperature, in a suitable apparatus, *e.g.* a pulsated sieve-bottom column.[51,52] A homogeneous solution of the polymer is used as feed (fd) and the pure theta solvent as extracting agent (ea). The flow rates of these two liquids are chosen in such a manner that the total composition of the mixture within the apparatus corresponds to a point inside the miscibility gap (working point of the CPF). Then, the two phases coexist throughout the process, and the polymer originally contained in the feed spreads into the phase originating from the extracting agent. During this process the polymer species of lower molecular weight are preferentially removed from the feed; because of the countercurrent (chromatographic) nature of the process, the partition of the polymeric species takes place repeatedly, until the feed essentially contains only components of higher molecular weight. This leaves the apparatus as a concentrated solution, denoted as gel (gl), while the extracting agent, having taken up the corresponding material of lower molecular weight, leaves the apparatus as a less concentrated solution, the sol (sl).

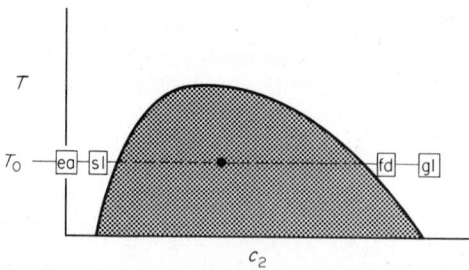

Figure 13 Schematic phase diagram to explain the operation of CPF with a single solvent. At a fixed temperature T_0, a highly concentrated polymer solution (fd) is extracted continuously with pure solvent (ea); the flow rates are so chosen as to give an overall concentration (working point ●) within the (shaded) two-phase region. Under these conditions the feed fd ends up as a gel phase (gl) containing the polymer components of higher molecular weight whereas the *ea* transforms into a moderately concentrated solution (sl) of the components of lower molecular weight.

Since a single partitioning step in CPF corresponds to a normal discontinuous liquid/liquid fractionation experiment (*cf.* Section 15.2), it is obvious that the requirement of comparable amounts of polymer in the two fractions does not constitute a major problem in CPF. With this method, as compared with conventional countercurrent extraction, the alternative for the polymer molecules of different molecular weight is not to be dissolved in either of two solvents, but to be enriched in the more dilute or the more concentrated phase, both of which contain the same solvent.

In order to guarantee satisfactory general conditions for a high degree of reduction of the molecular non-uniformity of the fractions as compared with that of the initial sample, it was found to be necessary to adjust the mass ratio of the fractions in an appropriate fashion. This ratio depends on the initial MWD, as indicated in Section 15.3.2, and must be adjusted in CPF experiments.

Figure 14 gives a schematic representation of the CPF method. The main independent variables are the operating temperature, T_0, the polymer concentration in the feed, c_2^{fd}, and the flow rate $\dot{V}$ of the liquids entering the apparatus (fd and ea). The ratio

$$\dot{q} = \dot{V}^{ea}/\dot{V}^{fd} \tag{16}$$

and c_2^{fd} determine the location of the working point, *cf.* Figure 13 ($c_2^{ea}=0$). The weight ratio of the polymer contained in the sl and gl fractions, respectively, is given by

$$\dot{G} = \dot{m}_2^{sl}/\dot{m}_2^{gl} \tag{17}$$

and can be calculated from the polymer concentrations in the phases leaving the apparatus (c_2^{sl} and c_2^{gl}) and from the corresponding flow rates $\dot{V}^{sl}$ and $\dot{V}^{gl}$. For a steady-state operation of the CPF technique we have

$$\dot{X}^{fd} + \dot{X}^{ea} = \dot{X}^{sl} + \dot{X}^{gl} \tag{18}$$

where $\dot{X}$ stands for the volume and mass fluxes of the entire phases or of any particular component contained in it. Under these conditions the weight-average molecular weight of the original polymer sample and its fractions are related by

$$\bar{M}_w^{fd}\dot{m}_2^{fd} = \bar{M}_w^{gl}\dot{m}_2^{gl} + \bar{M}_w^{sl}\dot{m}_2^{sl} \tag{19}$$

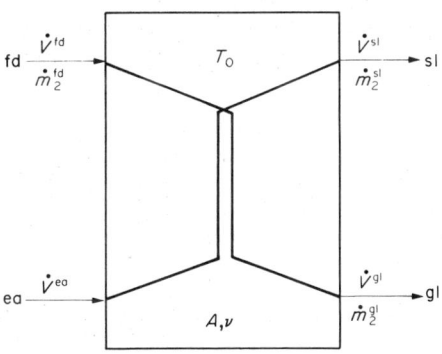

Figure 14 Flow chart of the CPF technique. $\dot{V}$ and $\dot{m}_2$ are volume and polymer mass fluxes (with mixed solvents, index 2 is replaced by 3); A and v are amplitude and frequency of pulsation, respectively (the remaining symbols are defined in the caption to Figure 13)

If the CPF method is operated with quasi-binary systems, there are obviously two main independent variables by which the division of the polymer into two fractions can be influenced: (a) the operating temperature T_0, an increase of which affects $\dot{G}$ when all other factors are kept constant, since the solvent power increases; and (b) the flux ratio $\dot{q}$, an increase of which at constant temperature enlarges the relative amount of the phase extracting the material of low molecular weight and, thereby, $\dot{G}$.

As will be demonstrated in the next section, CPF is, in practice, normally performed under isothermal conditions with mixed solvents. In this case, the variation of temperature can be replaced by the more convenient variation of the composition of the extracting agent, and an increase in T corresponds to an increase in the content of the thermodynamically better solvent. In the following section, this component is indicated by the index 1, the worse solvent by 2 and the polymer by 3.

15.4.2 Present State of the CPF Technique

Continuous polymer fractionation can, in principle, be performed with any apparatus for countercurrent extraction. So far, the experiments have been mainly carried out with a modified commercially available sieve-bottom column with pulsator (Quickfit, Wiesbaden; Otto Fritz GmbH, Hofheim, FRG). A column with a length of 100 cm and an inner diameter of 2.5 cm (total volume 500 cm^3) was used to fractionate a maximum of about 1 g of polymer per minute. The sieve bottoms were approximately 5 cm apart and had a free-space sectional area of about 23%. A plunger pump with a maximum stroke of 60 cm^3 and a frequency range from 0 to 2.5 Hz was used for the pulsation. The transportation of the usually very viscous feed needs an appropriate pump; valveless FMI wobble-plunger pumps have proven to be suitable. In a number of applications it was necessary to thermostat the entire apparatus, including the valve for the regulation of the flux leaving the column at its lower end.

So far, the following polymers have been fractionated by means of CPF: polystyrene (PS), poly(vinyl chloride) (PVC), polyisobutylene (PIB), phenol–formaldehyde resin (novolak) and polyethylene (PE). Each of these substances demonstrates a particular aspect of CPF.

PS was chosen as the first polymer for testing the feasibility of the method and the apparatus, despite the fact that it can easily be synthesized with a narrow MWD. The experiments with PS have shown that CPF can best be performed at room temperature with mixed solvents, with components that are thermodynamically as similar as possible.[50]

In the case of PVC,[51] it turned out to be necessary to change the strategy concerning the selection of the solvent pair, since solutions of this polymer tend to gel thermoreversibly in poor solvents, a behaviour which would naturally render CPF impossible. Under these particular circumstances, the combination of a thermodynamically very good solvent with a very powerful non-solvent, tetrahydrofuran (THF) plus water, proved to be successful in keeping the coexisting phases fluid. This finding can be explained in terms of a pronounced preferential solvation of PVC by THF, which prevents the repeat units in PVC from forming the large number of intersegmental contacts required for the formation of physical gels. It was possible to obtain four fractions with a total mass of *ca.* 400 g and U_n values ranging from 0.15 to 0.2 in two CPF steps, starting from a PVC with $\bar{M}_w = 6.7 \times 10^4$ and $U_n = 1.0$. The sol and gel products obtained in the first run were refractionated, as can be seen in Figure 15.

Experiments with Novolak have shown that it is possible to find a suitable combination of solvents, even if one of them is already fixed due to the production process. Furthermore, these

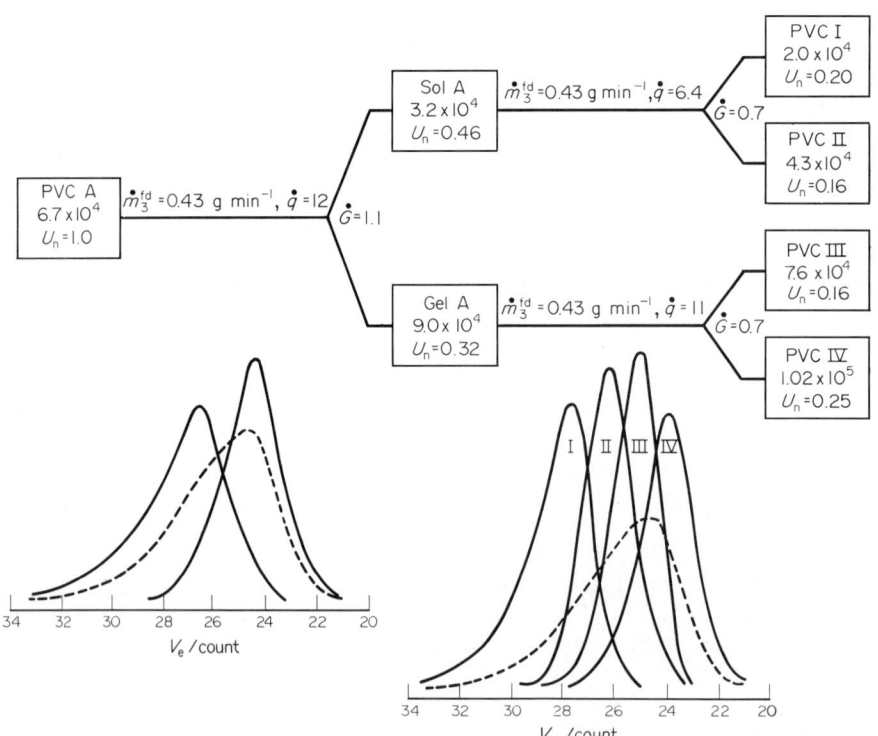

Figure 15 Flow chart of a preparative CPF of PVC. The gels produced in one fractionation step could be used without further treatment as feed in the successive extraction, whereas the corresponding sols had to be precipitated and redissolved for that purpose. Weight-average molecular weights are indicated; curves are size-exclusion chromatography results for initial polymer (dashed) and fractions

experiments demonstrated that the $\dot{G}$ value, *i.e.* the division of the initial material into the fractions, can be tailored in such a fashion that a bimodal initial distribution becomes unimodal in the fractions after only one run.[58]

Ongoing work[59] on the fractionation of practically unbranched PE has demonstrated the necessity for further variations in the application of CPF. With this polymer it is necessary to perform the experiments at temperatures in the region of 130–140 °C, to overcome the crystallization tendency of PE. Further, one is also forced to select a very special single solvent, the density of which is sufficiently different from that of PE that the coexisting phases can be transported through the column by gravitational forces. According to present knowledge, a minimum difference of 0.03 g cm^{-3} is required between the densities of polymer and solvent. Diphenyl ether proved to fulfil this condition; since it is denser than PE, the feed is not introduced at the top of the apparatus but at the bottom.

From the experience gained so far, one may conclude that it should be possible to find suitable solvents or solvent pairs and operating conditions for the CPF method for practically all polymers, although their discovery is not an obvious matter. Naturally, detailed knowledge of the separation mechanisms and means to control them would be of great help in the application of CPF to new substances. For this reason systematic experiments have been performed with PIB.[52] The main conclusions drawn from these experiments will be discussed in the next section.

The question which was asked was the following: 'if the working temperature and composition of the feed are fixed, and if a certain division of the original polymer into fractions, *i.e.* a certain $\dot{G}$ value, is desirable, then what are the best operating conditions to obtain narrow fraction distributions?' Generally speaking there are two options.

In option 1, an extracting agent of comparatively low solvent quality is chosen. This means that large amounts of it are required to transfer the amount of polymer into the sol phase which is required to realize a certain $\dot{G}$ value. In other words, this situation corresponds to large values of $\dot{q}$ and to low polymer concentrations in the column.

Option 2, on the other hand, makes use of thermodynamically better solvent pairs. The procedure leads to smaller $\dot{q}$ values and shifts the working point to larger polymer concentrations.

The situation is illustrated in Figure 16, where the working line on the right represents option 1 and that on the left option 2; the $\dot{q}$ and $\dot{G}$ values corresponding to the individual working points are indicated. It should be noted that, in order to realize a $\dot{G}$ value of 1.6 to 1.7, the solvent requirement in option 1 is about four times as large as that in option 2.

Thermodynamic-equilibrium considerations clearly favour option 1, as follows from the discussions in Section 15.2.4. Though very obvious when precipitation fractionation is concerned, it is further seen in Figures 1, 2, 7 and 10 and in Table 1 that small initial concentrations also improve the

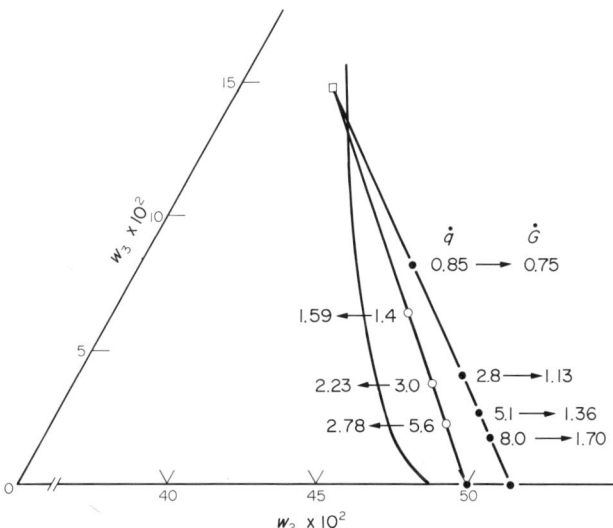

Figure 16 Phase diagram at 25 °C of the system *n*-heptane/methyl ethyl ketone/polyisobutylene ($\bar{M}_w = 9.84 \times 10^4$) and various CPF working points. Flow ratios $\dot{q}$ of extracting agent and feed, and the resulting ratios $\dot{G}$ of polymer mass in sol and gel phase, are indicated. The square outside the miscibility gap represents the feed composition. In all cases the extracting agent consists of polymer-free mixed solvent (w_i = mass fraction of constituent *i*)

efficiency of extraction, albeit to a much lesser extent. Under the non-equilibrium conditions of CPF, however, the situation is more complex, as is demonstrated by the results shown in Figure 17. In this diagram, the weight-average molecular weight of the polymer contained in the sol and gel fractions is plotted as a function of the mass ratio $\dot{G}$ of these fractions for two compositions of the extracting agent corresponding to options 1 and 2. Contrary to the expectation based on equilibrium calculations, the experiments with the better extracting agent turn out to be more efficient, *i.e.* $\bar{M}_w^{sl}$ is lower and $\bar{M}_w^{gl}$ is higher at constant $\dot{G}$, as compared with the worse solvent, despite the approximately fourfold-higher overall polymer concentration.

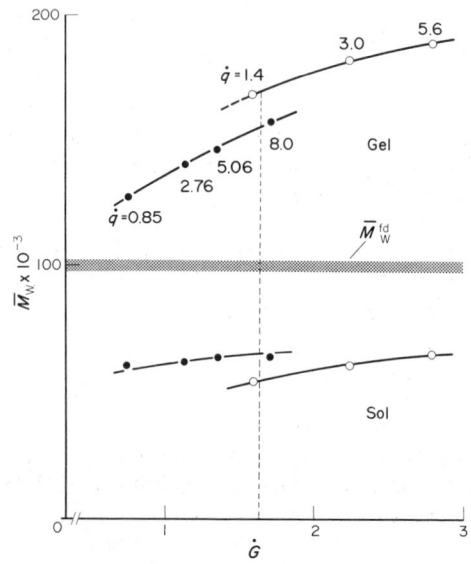

Figure 17 $\bar{M}_w$ values of CPF fractions as a function of the mass ratio $\dot{G}$. The data correspond to the working points shown in Figure 16; $\dot{q}$ values are indicated. Open circles, $w_2^{ea} = 0.50$; filled circles, $w_2^{ea} = 0.52$ (w_2^{ea} = mass fraction of methyl ethyl ketone in the extracting agent)

In order to rationalize this finding, one has to consider kinetic aspects in addition to thermodynamic factors. In CPF, the fractionation efficiency is probably governed by an interplay of thermodynamic driving forces and kinetic limitations of mass transfer between the two phases. As the working point is shifted towards the consolute point of the system, the thermodynamic driving forces decrease, but the corresponding reduction of the interfacial tension allows the rate of mass transfer to increase. This consideration appears to be supported by the present experimental results and allows postulation of the existence of an optimum thermodynamic quality of the extracting agent for each $\dot{G}$ value. If it is too low, fractionation efficiency is poor despite large thermodynamic driving forces, because the transfer of the polymer molecules across the phase boundary is too slow. If, on the other hand, the solvent quality of the extracting agent is too high, the polymer molecules can change phase rapidly but the driving forces become too weak.

For most practical purposes it is far too laborious to exactly determine the optimum solvent quality as a function of $\dot{G}$. However, the general knowledge obtained so far suffices to operate the CPF technique successfully, as is demonstrated by the results in Figure 15. In that particular case, the large-scale fractionation by CPF was performed without any extensive preparatory experimentation, once an appropriate solvent pair was found.

15.4.3 How to Apply CPF to New Polymers

The following procedure is recommended when the solution behaviour of the material to be fractionated deviates from that of the polymers dealt with so far.

First, a suitable solvent pair or single solvent has to be selected. Criteria for the choice are thermodynamic and kinetic aptness plus availability and price. The thermodynamic aspects normally involve the requirement of either a solvent pair of quite similar components or a suitable single solvent. The kinetic aptness of the solvent primarily consists of a sufficient density difference, so that the phases are readily transported through the apparatus by gravitational forces. Further, the viscosity of the solvent should be low so as to aid mass transfer between the phases.

The next important step is the selection of the best working conditions. To this end, one has to first decide, on the basis of the MWD of the polymer to be fractionated, at which molecular weight the cut through the distribution should best be performed, *i.e.* which $\dot{G}$ value appears suitable. Once $\dot{G}$ is fixed, the criteria are that the working point is situated at a reasonably large polymer concentration (to save solvent) and that the difference between the average molecular weights of the gel and sol fractions (f_n and f_w values, see equation 3) is as large as possible. The most important independent variable is the thermodynamic quality of the solvent, which is determined by T in the case of a single solvent and by the composition when a solvent pair is used. The required $\dot{G}$ value is obtained by proper adjustment of the flow rates of feed and extracting agent.

In addition to these more general aspects, some technical features must also be considered; above all, a proper choice of amplitude and frequency of pulsation as far as the apparatus used hitherto is concerned. Only a narrow range of values is available if damming-up of one of the flows entering the column is to be prevented, and a fine dispersion of the phases guaranteed. Finally, it is advisable to select the working conditions, if there is any choice, in such a manner that minor deviations from them have little consequence for the process.

The experience gathered so far with CPF justifies the confident expectation that the method can be applied to practically all polymers, though in some cases it may be necessary to perform extensive preparatory work of the type described in Section 15.4.2.

ACKNOWLEDGEMENT

We thank Professor G. V. Schulz for his kind permission to use redrawings of his classic diagrams in Figures 5 and 6. Thanks are further due to Drs. B. J. Schmitt and P. Wittmer (BASF) for making refs. 22 and 48 available to the authors.

15.5 REFERENCES

1. G. V. Schulz, *Z. Phys. Chem., Abt. A*, 1937, **179**, 321.
2. G. V. Schulz and B. Jirgensons, *Z. Phys. Chem., Abt. B*, 1940, **46**, 105.
3. G. V. Schulz, in 'Die Physik der Hochpolymeren', ed. H. A. Stuart, Springer, Berlin, 1953, vol. II, chap. 17.
4. A. J. Staverman and J. H. Van Santen, *Recl. Trav. Chim. Pays-Bas*, 1941, **60**, 76.
5. A. J. Staverman, *Recl. Trav. Chim. Pays-Bas*, 1941, **60**, 640.
6. M. L. Huggins, *J. Chem. Phys.*, 1941, **9**, 440.
7. M. L. Huggins, *Ann. N. Y. Acad. Sci.*, 1942, **43**, 1.
8. P. J. Flory, *J. Chem. Phys.*, 1941, **9**, 660.
9. P. J. Flory, *J. Chem. Phys.*, 1942, **10**, 51.
10. P. J. Flory, *J. Chem. Phys.*, 1944, **12**, 425.
11. G. V. Schulz, *Z. Phys. Chem., Abt. B*, 1939, **43**, 25.
12. M. L. Huggins, *J. Polym. Sci., Part A-2*, 1967, **5**, 1221.
13. M. L. Huggins and H. Okamoto, in 'Polymer Fractionation', ed. M. J. R. Cantow, Academic Press, New York, 1967.
14. L. H. Cragg and H. Hammerschlag, *Chem. Rev.*, 1946, **39**, 79.
15. G. M. Guzman, in 'Progress in High Polymers', ed. J. C. Robb and F. W. Peaker, Heywood, London, 1961, vol. I, p. 113.
16. M. J. R. Cantow (ed.), 'Polymer Fractionation', Academic Press, New York, 1967.
17. L. H. Tung (ed.), 'Fractionation of Synthetic Polymers', Dekker, New York, 1977.
18. G. V. Schulz, *Z. Phys. Chem., Abt. B*, 1936, **32**, 27.
19. R. Koningsveld, Ph. D. Thesis, University of Leiden, 1967.
20. K. Solc, *Collect. Czech. Chem. Commun*, 1969, **34**, 992.
21. J. N. Brönsted, *Z. Phys. Chem., Bodenstein Festband*, 1931, 79.
22. J. Hengstenberg and F. Käsbauer, 'Festchrift Carl Wurster zum 60. Geburtstag', BASF AG, Ludwigshafen am Rhein, 1960, p. 259.
23. H. Okamoto and K. Sekikawa, *J. Polym. Sci.*, 1961, **55**, 597.
24. H. Okamoto, *J. Polym. Sci., Part A*, 1964, **2**, 3451.
25. J. W. Breitenbach and B. A. Wolf, *Makromol. Chem.*, 1967, **108**, 263.
26. L. A. Kleintjens, R. Koningsveld and W. H. Stockmayer, *Br. Polym. J.*, 1976, **8**, 144.
27. B. A. Wolf, H. F. Bieringer and J. W. Breitenbach, *Br. Polym. J.*, 1978, **10**, 156.
28. H. J. Rüfenacht and P. v. Tavel, *Makromol. Chem.*, 1976, **177**, 2449.
29. E. Nies, Ph. D. Thesis, University of Antwerp, 1983.
30. G. Rehage, *Kunststoffe*, 1963, **53**, 605.
31. V. Desreux and A. Oth, *Chem. Weekbl.*, 1952, **48**, 247.
32. E. V. Sayre, *J. Polym. Sci.*, 1953, **10**, 175.
33. R. L. Scott, *Ind. Eng. Chem.*, 1953, **45**, 2532.

34. L. H. Tung, *J. Polym. Sci.*, 1962, **61**, 449.
35. R. Koningsveld, *Chem. Weekbl.*, 1961, **57**, 129.
36. R. Koningsveld and A. J. Staverman, *J. Polym. Sci., Part A-2*, 1968, **6**, 367, 383.
37. R. Koningsveld, *Adv. Polym. Sci.*, 1970, **7**, 1.
38. K. Kamide *et al.*, in 'Fractionation of Synthetic Polymers', ed. L. H. Tung, Dekker, New York, 1977.
39. See, for example, M. T. Rätzsch and H. Kehlen, *J. Macromol. Sci., Chem.*, 1985, **A22**, 323.
40. A. J. Staverman and J. Th. G. Overbeek, discussion remark to ref. 31.
41. M. Bohdanecký *J. Polym. Sci., Part C*, 1968, **23**, 257.
42. L. A. Kleintjens, Ph. D. Thesis, University of Essex, UK, 1979.
43. J. Van Schooten, H. Van Hoorn and J. Boerma, *Polymer*, 1961, **2**, 161.
44. V. N. Kuznetsov, V. B. Kogan, L. A. Venkstern, S. D. Vogman, T. A. Usatova and V. A. Morozova, *Vysokomol. Soedin., Ser. A*, 1970, **12**, 2768.
45. Z. Roszkowski, *Polimery (Warsaw)*, 1971, **16**, 445.
46. K. E. Almin, *Acta Chem. Scand.*, 1957, **11**, 1541; *Acta Chem. Scand.*, 1959, **13**, 1263, 1274, 1278, 1287, 1293.
47. L. C. Case, *Makromol. Chem.*, 1960, **41**, 61.
48. K. Jäckel, 'Festschrift Carl Wurster zum 60. Geburtstag', BASF AG, Ludwigshafen am Rhein, 1960, p. 269.
49. A. Englert and H. Tompa, *Polymer*, 1970, **11**, 507.
50. B. A. Wolf, H. Geerissen, J. Roos and P. Amareshwar, *Ger. Pat. Appl.* P 32 42 130.3 (1982) (*Chem. Abstr.*, 1984, **101**, 172 154).
51. H. Geerissen, J. Roos and B. A. Wolf, *Makromol. Chem.*, 1985, **186**, 735.
52. H. Geerissen, J. Roos, P. Schützeichel and B. A. Wolf, *J. Appl. Polym. Sci.*, 1987, **34**, 271, 287.
53. A. J. Pennings, *J. Polym. Sci., Part C*, 1967, **16**, 1799.
54. A. J. Pennings, in 'Characterization of Macromolecular Structure', ed. D. McIntyre, NAS Publication 1573, 1968, p. 214.
55. L. A. Kleintjens, unpublished results.
56. V. B. F. Mathot and M. F. J. Pijpers, *Polym. Bull. (Berlin)*, 1984, **11**, 297.
57. A. Dobry, *Makromol. Chem.*, 1956, **18/19**, 317.
58. H. Geerissen and B. A. Wolf, unpublished results.
59. H. Geerissen, P. Schützeichel and B. A. Wolf, in preparation.

16

Chromatographic Cross-fractionation

GOTTFRIED GLÖCKNER
Dresden University of Technology, GDR

16.1 INTRODUCTION

In 1952 a report was given on the influence of the precipitant upon the fractionation of cellulose acetate from solutions in acetone: when *n*-heptane was used, the fraction of highest viscosity had the lowest acetyl value, whereas with water as a precipitant, the fraction of highest viscosity had the highest acetyl content.[1] The authors recognized that the solubility of cellulose acetate is determined both by chain length and composition, and that even the most painstaking fractionation in only one solvent/nonsolvent system could not yield fractions with narrow distributions in chemical composition (CCD) and in molecular weight (MMD). They coined the term 'cross-fractionation' for a scheme that included the fractionation of the raw polymer through the addition of one kind of precipitant (water), and the subfractionation of redissolved fractions through the addition of a different one (*n*-pentane). Rosenthal and White proved the narrowness of the final fractions by repeating the precipitation with water. The authors also suggested cross-fractionation as a general approach in copolymer analysis.

In 1962, Topčiev *et al.* derived equation (1) for the volume fraction $\phi_{DP,x}$ of a copolymer with degree of polymerization DP and composition x in the dilute sol phase and the gel phase, ϕ' and ϕ'', respectively.[2] Teramachi and Nagasawa showed that the values of σ and K are linked to the Huggins' constants of interaction between the copolymer constituents and the solvent components. For a binary A/B copolymer and a binary solvent consisting of the liquids 1 and 2, equation (2) holds. Cross-fractionation by solubility effects requires a solvent/nonsolvent system in which $\chi_A > \chi_B$ and another one in which $\chi_A < \chi_B$ holds.[3]

$$\phi''_{DP,x}/\phi'_{DP,x} = \exp[DP(\sigma + K \cdot x)] \tag{1}$$

$$K = (\phi'_1 - \phi''_1)(\chi_{1A} - \chi_{1B}) + (\phi'_2 - \phi''_2)(\chi_{2A} - \chi_{2B}) \tag{2}$$

The literature dealing with the separation of copolymers has been excellently represented in reviews on polymer fractionation,[4,5] but the list of papers concerned with cross-fractionation is rather short even in a specialized compilation of the literature on fractionation of copolymers.[6]

Irrespective of its importance, cross-fractionation has been performed only seldom. A principal obstacle is often the difficulty of finding the solvent systems which fulfil the above mentioned conditions for χ_A and χ_B. The application of the method for analytical purposes might also be hampered by the effort which is required when classical solubility fractionation is used: about 8–12 weeks of hard labour are necessary for the evaluation of the two-dimensional distribution in size and composition of each copolymer specimen.

The aim of this chapter is a survey of chromatographic (and related) techniques which can facilitate the separations required in cross-fractionation *via* the participation of a stationary phase, and enhance the speed of data acquisition. The contribution is also intended to be a review of the applications of chromatographic cross-fractionation (CCF) known at present.

16.2 BASIC CONCEPTS OF CHROMATOGRAPHIC CROSS-FRACTIONATION

CCF of a binary copolymer requires two chromatographic techniques which must separate the copolymer according to different properties. In the ideal case, one method should yield fractionation by molecular weight and the other by composition. This section deals with suitable combinations of chromatographic techniques.

16.2.1 Orthogonal Size-exclusion Chromatography in Different Solvents

Size-exclusion chromography (SEC) separates polymers by their hydrodynamic volume, given in equation (3), where M is the molecular weight and $[\eta]$ the intrinsic viscosity of a component with the hydrodynamic volume V_h. Substituting the Kuhn–Mark–Houwink equation (4), equation (3') results. K_v and a are constant for a given polymer in a given system. In a different solvent, the same polymer will have different values of the constants K_v and a. The exponent a increases with the thermodynamic quality of the solvent.

$$V_h = [\eta] \cdot M \tag{3}$$

$$[\eta] = K_v \cdot M^a \tag{4}$$

$$V_h = K_v \cdot M^{1+a} \tag{3'}$$

In 1980, coupling two GPC devices and running each with a different mobile phase was suggested.[7] Both devices utilize steric exclusion columns, but the eluent in GPC(2) is thermodynamically poorer than that in GPC(1). Hence, an eluate portion (a 'slice') from GPC(1) will undergo further separation in GPC(2), which will possibly be enhanced by partition or adsorption interactions of the copolymer on the column packing material. The authors proposed the term 'orthogonal chromatography' for the method which they used for the separation of copolymers from styrene and *n*-butyl methacrylate (see Section 16.4.6).

16.2.2 Utilization of Different Modes of Adsorption Chromatography

16.2.2.1 *General remarks on adsorption chromatography and gradient elution*

Adsorption chromatography (AC) is chromatographic separation due to energetic interactions between the solute and the surface of the porous packings. If separation by composition is aimed at, the column should retain one kind of structural unit more strongly than the others. The chance of obtaining these interactions must not depend on the size of the solute if disturbance of the AC separation by SEC effects is to be avoided. Hence, the pores of the packings must be either large enough to give access to all kinds of solute molecules or so small that all of them are excluded. Most of the AC results presented in Section 16.4 have been obtained with small-pore packings.

Chromatographers distinguish between normal-phase and reversed-phase AC. The former utilizes a polar stationary phase which retains the solutes according to their polarity. If gradient elution is performed the polarity of the eluent mixture increases in the course of the experiment. In reversed-phase chromatography (RPC), the stationary phase is less polar than the eluent. An example of a RPC system is a C18 column (*i.e.* packed with porous silica with a bonded layer of octadecyl hydrocarbon chains) in combination with a gradient decreasing in polarity.

According to experience, nonexclusion HPLC of synthetic polymers usually requires gradient elution.[8] Water is a very popular eluent component in low-molecular chromatography. Unfortunately, it is a strong precipitant for most synthetic polymers. Therefore we shall concentrate on the discussion of gradients from nonaqueous solvents.

A normal-phase gradient can be formed, *e.g.* from CCl_4 through the addition of THF or acetonitrile, or a RP gradient from MeOH through the addition of THF or dichloromethane (DCM).

In RPC, separation is primarily governed by solvophobic retention.[9] With polymers, the measures taken in order to make the mobile phase an unpleasant environment for the solute can easily interfere with the boundaries of the narrow solubility window.

16.2.2.2 *Solubility effects*

The dissolution of any material requires a negative change in Gibbs free energy. The increase in entropy, ΔS, is rather large in the dissolution of low-molecular species but much smaller with polymers. Here, the change in enthalpy, ΔH, must be small if positive (or, better, large negative) to ensure solubility. In terms of Hildebrand's concept this means that a polymer will be soluble only in liquids whose solubility parameters are close to that of the polymer under consideration. With low-molecular solutes, the entropy contribution is so large that a broader diversity in the values of the solubility parameters can be tolerated. Hence, the variety of potential eluent components is much greater in low-molecular-liquid chromatography than in polymer chromatography.

In the extreme case, solubility is the predominant factor in polymer chromatography. Examples used for CCF are the temperature-rising elution fractionation of branched polyalkenes[10] or the high-pressure-precipitation-liquid chromatography (HPPLC) of copoly(styrene/acrylonitrile) specimens (see Sections 16.4.1.2 and 16.4.3, respectively).[11,12] Both examples were separations by composition. The supplementary fractionations by molecular weight were performed by SEC.

For solubility-based fractionation by composition a nonsolvent must be chosen with a strong and unambiguous dependence of the cloud point on polymer composition. Alkane hydrocarbons act in this way for copoly(styrene/acrylonitrile)[13] and related copolymers with units of sufficiently differing polarity.[14]

A precipitant for separation by molecular weight should yield cloud points independent of copolymer composition in an adequate range around the mean value. It was, for instance, possible to fractionate copoly(styrene/acrylonitrile) samples with about 17–34 mass-% acrylonitrile according to molecular weight through methanol as a precipitant.[13]

From experience gathered so far it can be inferred that the combination of a solvent with a nonsolvent, of suitable chromatographic strength, seems to be a promising general concept for designing the eluent system in nonexclusion polymer chromatography.

16.2.2.3 *Molecular weight effects in adsorption chromatography*

The adsorption of polymers on a solid surface reaches a plateau value at a rather low concentration of the solute. On alumina, silica or charcoal, the molecular weight dependence of m_a, the amount adsorbed per gram adsorbent, is approximately given by equation (5).

$$m_a = \text{constant} \cdot M^e \tag{5}$$

At very high values of M the exponent e approaches zero. High values ($e \approx 0.3$) have been found for $M < 100\,000$ and thermodynamically poor solvents. In good solvents, the influence of M on adsorption is less pronounced.[15]

A similar effect of solvent quality has been observed in thin-layer chromatography (TLC) investigations. Separations by composition were achieved with good solvents of a suitable polarity whereas separations by molecular weight were reached preferably with thermodynamically poor eluents.[16]

Thus it does not seem impossible to perform both tasks of chromatographic cross-fractionation by the combination of two suitable techniques of AC, especially if solubility effects are utilized.

16.2.3 Prefractionation by Adsorption Chromatography and Subsequent Investigation of the Fractions by Size-exclusion Chromatography

At the present state, SEC is certainly the most versatile method for separation by molecular size. Therefore promising CCF procedures are most likely to include SEC.

It has been suggested that SEC should be used in the second stage after previous separation by composition in order to avoid the disturbance of SEC by chemical heterogeneity *via* changes in a (*cf.* equation 3') and also any possible distortion of the shape of the SEC peak. Using thin-layer chromatography (TLC) for the first separation and SEC for the second one, partially branched polystyrene samples[17] and *triblock*-copoly(methyl methacrylate/styrene/methyl methacrylate)[18] have been investigated. By the same sequence of methods, but with the AC stage performed in a column with dried silica, a *diblock*-copoly(styrene/methyl methacrylate) sample has been analyzed.[19,20]

The definite advantage of using SEC at the last stage is the fact that the elution in SEC is always isocratic. Thus, additional dual detection is possible which provides information on whether the preceeding separation by composition has really met its goal. In the most favourable case this sequence of methods, (i) AC, (ii) dual detection SEC, enables quantitative analyses to be performed even without the need for a set of calibrating copolymers. SEC with Fourier transform IR detection has been successfully employed for the investigation of copoly(styrene/acrylic acid) specimens which had been prefractionated by gradient HPLC on a normal-phase AC column (see Section 16.4.9).[21]

16.2.4 Prefractionation by Size-exclusion Chromatography and Subsequent Investigation of the Fractions by Adsorption Chromatography

The advantages of using SEC for the first separation are: (i) In SEC the distribution constants for all sample components are restricted to the range 0–1. Thus, the chromatograms are relatively short, and even the last eluting portions will not be diluted too much for direct injection during further chromatographic investigations. (ii) SEC is performed isocratically. Hence, all fractions are obtained in the same solvent. The SEC eluent can be a main component of the eluting system in the second separation. (iii) Copolymers, with their tremendous number of constituents differing in molecular weight and composition, are extremely complex mixtures with respect to chromatography. The mutual interference of different species is less dramatic in SEC than in separation mechanisms where the sample constituents compete for active sites on the surface of the column packing.

The drawback of SEC of the complex initial sample is that a fraction of a given hydrodynamic volume may consist of constituents differing in composition and consequently also in molecular weight, as seen in equation (3′). The importance of this effect can be judged by the K_v and a values for the polymers under investigation. From the Kuhn–Mark–Houwink constants given in the literature[22] for THF as a solvent, it can be concluded that in favourable cases the composition effect are only small. This holds true for copolymers of styrene and acrylonitrile (≤ 24 mass-% AN) or methyl methacrylate. (Sometimes, the data for a given system from different sources differ by almost the same degree.) Thus, prefractionation by SEC is a tolerable compromise between the practicability of the experimental procedure and precision.

The subsequent analysis will reveal the chemical composition distribution in each SEC slice. This can be done by utilizing solubility or adsorption effects and employing TLC or HPLC equipment. (The possibility of performing another SEC in a poor solvent was already dealt with in Section 16.2.1).

The prime costs are much lower for TLC than for HPLC, but the latter requires less labour and can be automated. In addition, quantitative evaluation is more difficult with TLC than with HPLC. The combination of SEC and TLC has been used for CCF of *stat*-copoly(styrene/methyl acrylate)[23] and *stat*-copoly(styrene/ethyl methacrylate) specimens.[24]

16.3 OUTLINE OF CHROMATOGRAPHIC CROSS-FRACTIONATION BY HPLC INVESTIGATION OF SIZE-EXCLUSION CHROMATOGRAPHY FRACTIONS

16.3.1 Detection

The detection of the polymer after the second separation is the central question of how to design a CCF procedure. The detection must be sensitive because the two-fold chromatography causes a high dilution, and it must be capable of monitoring the polymer in a multicomponent solvent whose composition varies according to the gradient programme. UV detection is suitable with respect to sensitivity, but what is required are absorbing structural units in the polymers, and eluents which are transparent at the selected wavelength. Copolymers with styrene units can be monitored by UV around 254 nm in eluents composed of alkane hydrocarbons and DCM or THF. Examples are given in Sections 16.4.3, 16.4.4, 16.4.5, 16.4.7 and 16.4.8. For copoly(styrene/acrylonitrile) in 2,2,4-trimethylpentane/THF, the highest sensitivity was observed at 259 nm.

The drawback of UV detection is its restriction to transparent solvents and solutes carrying chromophoric groups. The ideal detection in polymer HPLC should be capable of monitoring any polymer in any solvent without disturbance by the optical properties of the latter and the gradient programme. Evaporating the solvent and measuring the nonvolatile polymer residue is a straightforward approach. It had been put into practice with the moving-wire detector[25-27] and is again utilized in the evaporative light-scattering detector[28-30] and the rotating disc LC/FID detector.[31-33] These instruments have been used successfully for monitoring polymers in gradient elution.[34,35]

16.3.2 The Eluent for Size-exclusion Chromatography Separation

The direct injection of SEC fractions into the HPLC apparatus requires a sample volume of about 100 μl, *i.e.* the HPLC column will be overloaded with respect to the SEC eluent. The first consequence is that the SEC eluent must not have a strong UV absorption if UV detection is used for the HPLC analyses, otherwise, a great part of the HPLC chromatogram may be hidden under the absorption caused by the SEC eluent. The second conclusion is that the disturbance of the chromatographic process in the second column can be minimized if the SEC eluent is, at the same time, a main component of the HPLC eluent system.

THF is a rather common eluent in SEC and fortunately well suited for a direct combination with gradient HPLC. The THF for SEC can even be used with an antioxidant (mostly butylated hydroxytoluene) because this additive is easily separated in the HPLC from the polymer species of interest.

16.3.3 On-line or Off-line Operation

If the two devices are on-line the flow in the first must be stopped during the investigation of every eluate slice by HPLC. The latter takes about half an hour per sample. In practice, about ten SEC fractions should be analyzed. Thus, on-line coupling implies in total a rather long standby time of the SEC. The apparatus can be used more efficiently in off-line configuration where the fractions of interest are collected from an uninterrupted SEC elution. Such a configuration would enable an SEC apparatus to serve several HPLC machines or to perform independent SEC analyses.

Off-line operation with the help of an autoinjection system and an autosampler is just as convenient as on-line operation, but it has the consequence that certain SEC fractions are, at any rate, investigated some hours after being delivered. In this respect, the stabilizer in the SEC eluent has additional importance in preventing the dissolved fractions from being spoiled by reactions with oxygen.

16.3.4   The Eluent for HPLC Separation

With UV detection, the HPLC eluents must be transparent. This condition is especially stringent in gradient elution because the changing concentration of an absorbing solvent would lead to a dramatic shift of the baseline in the course of the programme.

In gradient HPLC with UV detection around 254 nm, the common antioxidants for THF may cause trouble due to their absorption in this region. Nonstabilized THF is required which, however, readily reacts with oxygen forming hazardous peroxides. Therefore, THF for HPLC must be handled with care. The atmosphere in vessels to be filled or the vapour space above the liquid should always be oxygen-free. Purging the THF container in the HPLC apparatus continuously with N_2 or He is indispensible for long-time use of the THF because even the slightest contact with oxygen can yield UV-absorbing compounds or complexes. In line with general recommendations for HPLC, a cover of inert gas should also be applied to the complementary components in the gradient because dissolved oxygen from these eluents can cause an extra UV absorption of THF in the HPLC apparatus. This has been observed with the combination of THF and isooctane.[12]

Apart from these difficulties, gradients based on THF combined with SEC using THF eluent offer the advantage that the solvent of the SEC fractions matches with a main constituent of the gradient HPLC.

The starting eluent must be miscible with the sample solvent and should be a poor solvent of low elution strength. If the starting solvent is too strong the polymer will not be retained properly (see Section 16.3.6). In this case, the solvent power of the starting eluent must be diminished either by increasing the nonsolvent concentration, by choosing a different solvent/nonsolvent combination, or by increasing the difference in polarity between sample and eluent. In normal-phase chromatography, a column of higher activity will also help to gain better retention.

16.3.5   Column Dimensions and Activity

The investigation of SEC fractions without pretreatment has the advantage of saving labour and enabling straightforward automation. The price to pay is the rather large injection volume in gradient HPLC. This volume must be small in comparison with the pore volume of the column.

According to experience gathered so far, the injection volume should not exceed 5% of the available column volume even if the polymer sample is easily retained, as is *e.g. stat*-copoly(styrene/acrylonitrile). *Stat*-copoly(styrene/ethyl methacrylate) specimens are less easily retained; here the injection volume is better restricted to 2% of the available volume.

If the volume ratio is small and the starting eluent poor enough, the polymer injected will be properly retained, and will elute after the required increase in eluent strength.

Another essential factor influencing retention is the activity of the column. In order to restore the initial value a thorough flushing is necessary after each run. This general rule is of special importance in polymer chromatography where traces of the sample may not be completely eluted from the column during the gradient programme. These would lead to drifting retention data, memory effects (*i.e.* elution in subsequent gradient cycles) and reduced column life. In the literature there are numerous reports of such phenomena.[36-38]

It is good practice to use the return gradient for restoring the activity of the column. This requires the use of liquids which are on the one hand good solvents for the polymers investigated and on the other hand chromatographically powerful eluents. THF is quite suitable. Rinsing the column with its ten-fold volume of pure THF has proved successful.[39] Good results have been obtained with cleaning cycles from 100% THF to 100% methanol and back again to 100% THF before returning to the initial eluent composition.

The time required for flushing the column can be reduced by increasing the flow rate. Favourable results have been obtained with 2 ml min^{-1} in the rinsing period and 0.5 ml min^{-1} analytical flow rate on 60×4 mm columns packed with 5 μm particles.

16.3.6 Calibration

Quantitative evaluation of chromatograms requires proper retention and complete elution of the sample. If the starting eluent is too strong, the injection volume too large for a given column, or the column not active enough, separating the sample polymer from the injected solvent will become difficult. Some of the polymer may elute together with the injected solvent. With optical detection, this portion is difficult to quantify and may even elute unnoticed.

Unretained portions are more easily detected if they elute outside the solvent plug. The 'normal' elution of unretained polymer occurs in the interstitial volume of the column. On small-pore packings, a peak appearing at approximately half the elution volume of sample solvent is a distinct signal of insufficient retention. Elution in this position was sometimes observed with *stat*-copoly (styrene/acrylonitrile) samples.

Unretained portions of copolymers with methacrylate units tended to elute in a peak which followed the solvent peak. Occasionally two peaks appeared: i.e. elution of unretained polymer partly in the interstitial volume and partly immediately after the solvent.

From a quantitative point of view, incomplete retention is as bad as incomplete elution during the gradient. On columns with small-pore packings, this is likely to occur with polymers of rather low-molecular weight. Repeated injections of a given sample with comparison of peak areas obtained in such a series may help to detect incompleteness of elution: increasing areas approaching a limit indicate saturation of the column.

Another test involves changing the column activity. This can be done with columns of identical geometry but chemically different packings. In normal-phase chromatography the exchange of a silica column, *e.g.* for a CN column, will cause larger peaks if elution from the silica column is incomplete.

Another and even simpler means is reducing column activity by the addition of a chromato-graphically strong component to the eluent. This can be done as long as retention is complete (*vide supra*). If such an addition, *e.g.* of methanol to a hydrocarbon eluent, causes increasing peak areas the assumption of incomplete elution is verified. With optical detectors, correction must be made for solvatochromic effects.

In the worst case, the system to be investigated and the limitations of chromatographic possibilities will not permit complete retention and complete elution. Then one has to make sure carefully that incompleteness acts upon all sample constituents to the same degree.[71] Only if this is proved may one dare to judge the composition of the whole sample on the basis of results obtained from the properly eluted portion.

The following passages deal with the relations needed for quantitative evaluation of proper chromatograms.

16.3.6.1 *Dependence of size-exclusion chromatography elution volume on molecular weight and composition*

The composition effect can be judged by the values of the Kuhn–Mark–Houwink constants for suitably graded copolymers and the parent homopolymers of the system under investigation. (The latter requirement cannot be met in every case because of insolubility or other basic obstacles.)

16.3.6.2 *Dependence of the detector signal on polymer amount and composition*

This information is required for all detectors employed.[40] The detector response is rather straightforward for the refractive index (RI) detector which is commonly used as a nonspecific mass detector in SEC. The influence of polymer composition on the RI signal is smallest if the refractive indices of the structural units are close together but sufficiently apart from that of the solvent.

The composition dependence of the UV signal is much more complicated because of: (a) the difference in absorptivity for the different constituting units, (b) the effect of solvent composition in gradient elution, and (c) the hypochromic effect of adjacent groups. The latter has been found to be significant with *stat*-copoly(styrene/acrylonitrile)[41,42] and *stat*-copoly(styrene/methyl methacrylate).[43,44]

16.3.6.3 Dependence of the retention in nonexclusion HPLC on molecular weight and composition

The retention in gradient HPLC has been repeatedly claimed to be independent of M. In gradient HPLC with solvent/nonsolvent systems, a molecular-weight effect was always observed. Figures† 6, 7, 11 and 16–18 show decreasing retention with increasing SEC fraction number. Of course, this could be due to a composition shift in the model copolymers but it is improbable that this shift should always be an increase in styrene content with decreasing molecular weight in more than 20 specimens which represent four different copolymer systems and which have been polymerized in three different laboratories — a decrease in styrene irrespective of whether the sample composition is above the azeotropic composition, below or just equal to it. Thus, the faster HPLC elution of the later SEC fractions indicates decreasing retention with decreasing molecular weight.

From the retention time, the gradient programme and the gradient delay the eluent composition at peak position can be calculated. For the systems investigated so far, equation (6) gives the volume fraction of the nonsolvent, ϕ^*, in the eluent portion which delivers a copolymer with molecular weight M.[45-50] Equation (6) had been found previously, relating turbidimetrically measured ϕ^* data and molecular weight.[51]

$$\phi^* = A + B/\sqrt{M} \tag{6}$$

The M dependence is small in comparison with the composition dependence. For copoly(styrene/acrylonitrile), a 1% change in composition has been found to cause a greater effect on retention than a 20% change in M.[45]

The calibration for both M and composition can be performed most efficiently by CCF of model mixtures. Figures 7, 11 and 16–18 contain the respective information for the systems copoly-(styrene/acrylonitrile), copoly(styrene/methyl methacrylate), copoly(styrene/ethyl methacrylate) or copoly(styrene/methoxyethyl methacrylate). For each system, the basic data had been collected within one day.

Repeated checking of the composition calibration by injecting a suitable mixture of model copolymers is recommended because changes in eluent quality will have a pronounced influence on copolymer retention in gradient HPLC. If stock solutions are used one should be aware of incompatibility effects even between specimens of the same system but different composition. Checking for possible precipitation is recommended.

16.3.7 Evaluation of Experimental CCF Results of Binary Copolymers

This section deals with experience in the investigation of copolymers containing one kind of UV-absorbing unit by SEC and gradient HPLC with UV detection. The immediate results of the experiments are (i) the SEC curve with the indication of the cuts between subsequent fractions and (ii) the UV records from the gradient HPLC.

16.3.7.1 Determination of molecular weight and polymer amount in each size-exclusion chromatography fraction

Provided there is knowledge of both the actual SEC calibration and the detector sensitivity, the M value and the portion of each SEC fraction can be estimated from the area and shape of each slice and its elution volume. Summation of these data should yield the average M of the whole sample.

16.3.7.2 Correction of the M effect in HPLC

Investigation of *stat*-copoly(styrene/acrylonitrile) by SEC and gradient HPLC in isooctane/THF yielded the actual values of the constants in equation (6). From the results shown in Figure 7 we

† Figures can be found in Section 16.4

obtained for the slope factor $B = 17.16 + 0.204w$. The quantity w indicates the AN content (mass-%) of the specimen. For the sample in Figure 9 the composition effect on B is within the limits $B = 21.9$ (for 23% AN) and 23.5 (for 31% AN), with $B = 22.55$ for the main portion (26.4% AN). This last value was used in a first approximation of the M correction. We referred all data to $M = 140\,000$ which was near to the number average of that sample, *i.e.* we calculated ϕ_M^* from the individual ϕ_M^* obtained with SEC fractions with a molecular weight M: $\phi_{M,w}^* = \phi_{\text{ref},w}^* + 22.55(140\,000^{-0.5} - M^{-0.5})$.

In a binary gradient with the solvent component B, ϕ_{ref}^* can be found from B% by $\phi_{\text{ref}}^* = 1 - (\text{B\%})/100$. In the S/AN system mentioned we had $(\text{B\%})_{\text{ref}} = 39.015 + 1.178w$, *i.e.* the higher the AN content was, the more THF was required for eluting that sample portion. From this relation, the composition w was obtained through $w = -33.124 + 0.849(\text{B\%})_{\text{ref}}$. The correlation is indicated by $r^2 = 0.9869$. (Unfortunately, in ref. 49, equation (2), the sign was not properly given).

16.3.7.3 Calculation of copolymer composition at any measured HPLC retention time

From the calibration with model mixtures (see Section 16.3.6.3) the composition dependence of the HPLC retention time can be found for M_{ref} by interpolation. With the help of this relation, the composition w can be easily derived from the ϕ_{ref}^* data of each point of the HPLC elution curve.

16.3.7.4 Correction of the UV signal

The height of the UV signal in a HPLC chromatogram, h_w, is determined by the amount of copolymer present and its content in the UV absorbing unit (*e.g.* styrene). Knowing the latter from the previous evaluation and the detector calibration, it is possible to correct the signal h_w which gives the value h representing the actual amount of copolymer present at this point.

16.3.7.5 Evaluation of the two-dimensional distribution of the sample

If there is a strong effect of composition on hydrodynamic volume or on the quantity B in equation (6) the steps 1 to 4 of the previous evaluation must be repeated on the basis of the knowledge gained so far.

Finally, the corrected h data of each HPLC are added up, and also the sums of all fractions. The portion of the starting material contained in a given SEC fraction must be reflected by the ratio of the h sum for that fraction over the total sum. This proves that exact aliquots had been injected into the HPLC apparatus and no concentration change (*e.g.* by solvent evaporation) had occurred. In the case of deviations, the h values must be corrected according to the ratio of the SEC slices to the total area of the SEC curve.

The final h data represent the amount of copolymer having the composition w and the molecular weight M. In order to obtain the contour map of the sample, corresponding values of h must be entered and connected in a plot of composition *vs.* M.

16.4 CROSS-FRACTIONATION OF SYNTHETIC POLYMERS

16.4.1 Partially Branched Polymers

Although branching often occurs as a consequence of chain transfer reactions the final products can be considered as pseudocopolymers of bi- and multi-functional monomers. Linear low-density polyethylenes (LLDPE), also dealt with in this chapter, are actually real copolymers from ethylene and α-alkenes such as butene-1, hexene-1 or octene-1, where the incorporation of the α-alkenes causes short-chain branching.

16.4.1.1 Branched polystyrenes

The investigation of partially branched polystyrene was performed by SEC and TLC.[17] A sample of 2 g was fractionated by SEC on a set of columns packed with polystyrene gels. The 12 fractions obtained were subsequently investigated by TLC on silica with a mixed eluent containing cyclo-hexane, benzene and acetone (12:4:v). The amount of acetone was not specified but from the report on calibrating experiments it can be concluded that v was in the range 0.6–0.8. These eluent mixtures separated the fast running linear polystyrene from the branched portions which were more strongly retained. Figure 1 shows the densitograms of TLC traces from the starting sample and its SEC fractions.

Five species can be distinguished which differ in degree of branching. This was concluded from the hydrodynamic radii derived from the SEC results (see Figure 2) and the effect of the acetone content

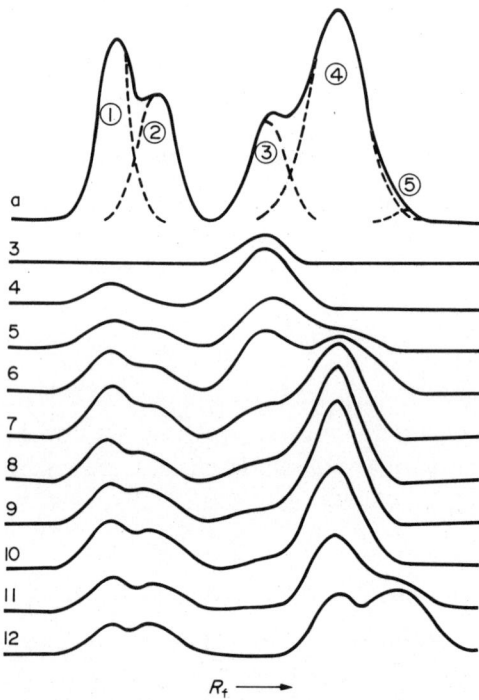

Figure 1 TLC of a partially branched polystyrene and its GPC fractions 3–12: (a) densitogram of the thin-layer chromatogram obtained from the nonfractionated sample with indication of five constituents, *cf.* Table 1, (3–12) densitograms of the fractions, *M* decreasing with increasing number. (Reproduced by permission of Elsevier Scientific Publishing Company, Amsterdam, from *J. Chromatogr.*, 1977, **141**, 72)

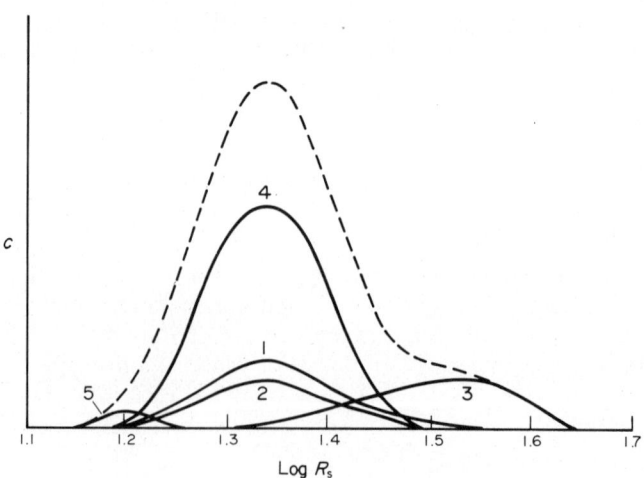

Figure 2 Distribution of hydrodynamic radii R_s of the constituents of a partially branched PS sample, *cf.* Figure 1 (dashed curve: nonfractionated sample). (Reproduced by permission of Elsevier Scientific Publishing Company, Amsterdam, from *J. Chromatogr.* 1977, **141**, 74)

Table 1 Constituents of the Polystyrene Sample shown in Figures 1 and 2

| Component | Molecular weight[a] | | | Relative amount[a] (%) |
	total	backbone	branches	
1	10500	5000	1400	23
2	10500	5000	1900	15
3	19500	(19500)		16
4	10500	10500		42
5	5500	5500		4

[a] B. G. Belenkij and E. S. Gankina, *J. Chromatogr.*, 1977, **141**, 13.

in the TLC eluent mixture on the R_f values of these five components. The curves obtained were compared with those of linear polystyrene standards graded in molecular weight. Results are listed in Table 1.

16.4.1.2 *Branched polyethylenes*

The crystallizability of polyethylene is determined by branching. Temperature-raising elution fractionation (TREF)[10] is a suitable technique for separating polyethylene according to the degree of short-chain branching. In analytical TREF, 50 μg sample material was deposited onto the inert packing material in a thermostated column by slow cooling (1.5 K h^{-1}) of a 0.5% solution in xylene. Subsequently, the polymer was eluted with 1,2,4-trichlorobenzene while the column temperature was steadily increased. The method was calibrated with the help of standards whose –Me content had been measured through IR spectroscopy. Thus, the amount of –Me per 1000 C atoms could be estimated from the elution temperature.

This procedure was scaled up by using 4 g sample material and a 500 × 127 mm TREF column. Thus it was possible to perform CCF by further investigation of the fractions through SEC.[52] The SEC was carried out by using a column set which included a pair of bimodal columns, 10 nm plus 100 nm, and a 400 nm column. The apparatus was equipped with an IR detector and run at a flow rate of 0.7 ml min^{-1}. The CCF method was applied to low-density polyethylene (LDPE) and LLDPE whose short-chain branching is mainly due to inserted α-alkenic comonomers. Figure 3

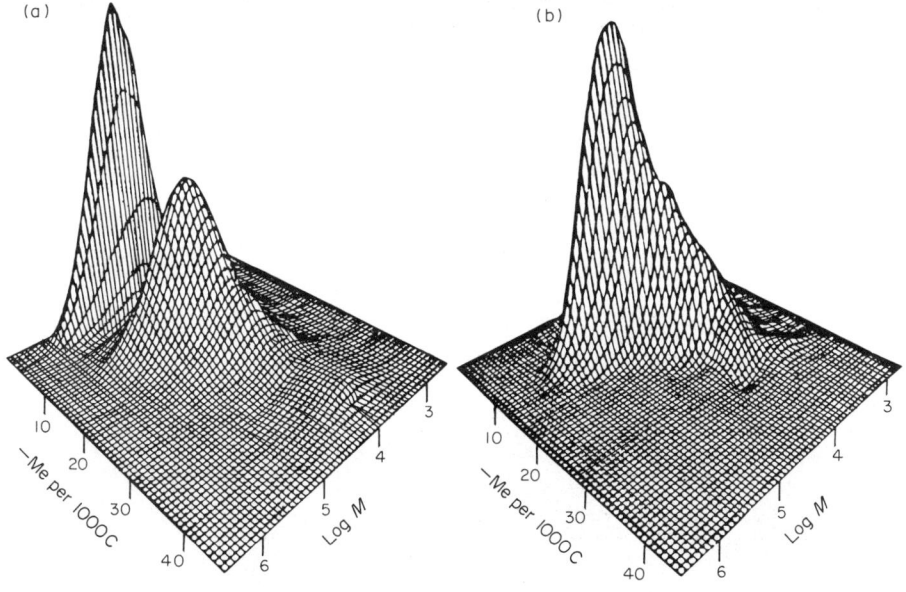

Figure 3 Distribution in short-chain branching and molecular weight of (a) linear low-density polyethylene and (b) PE radically polymerized under high pressure (LDPE). (Reproduced by permission of the American Chemical Society from *Polym. Prepr., Am. Chem. Soc., Div. Polym. Chem.*, 1982, **23**, 134)

shows the frequency distribution for a sample of both LLDPE and LDPE as the function of –Me content and molecular weight.

Nakano and Goto also recognized the potential of TREF for CCF of crystallizable polymers and designed an automated apparatus for the investigation of branched polyalkenes.[53] Stepwise they increased the temperature of the column whose glass-bead packing had been coated with the sample under investigation, and directed the eluate of this column on-line to the SEC apparatus. For this reason the column for the fractionation by crystallizability was only 150×8 mm in size. The void volume of this column was 2.5 ml, which also was the sample volume for the subsequent SEC performed with *o*-dichlorobenzene (*o*-DCB) at 140 °C and a column packed with mixed polystyrene gels. The temperature steps for the TREF column were 2 K in the middle of the operating range (40–140 °C) and larger (10 K) at its edges. The last step was even greater (from 110 to 140 °C).

Figures 4 and 5 show computer-aided drawings of the contour map and the perspective representation of the results from a 1:1 mixture of linear or highly branched polyethylene with 2.0 or 29.5 –Me groups per 1000 C atoms.

16.4.2 Copolymers from Ethylene and Propylene

The separation by composition was first performed in a glass column (1000×80 mm) packed with Celite 545 on which the copolymer sample had been deposited from solution. Fractionation was carried out at (or slightly above) the melting point of the sample under investigation, *i.e.* at a

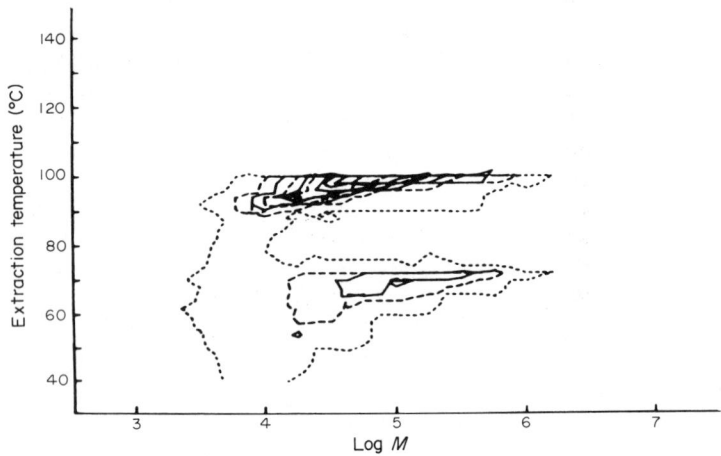

Figure 4 Contour map of the mixture of two polyethylene samples. The sample eluted at about 100 °C contained 2 –Me groups per 1000 C atoms, the other one had 29.5 –Me per 1000 C atoms. (Reproduced by permission of John Wiley & Sons, Inc., New York, from *J. Appl. Polym. Sci.*, 1981, **26**, 4230)

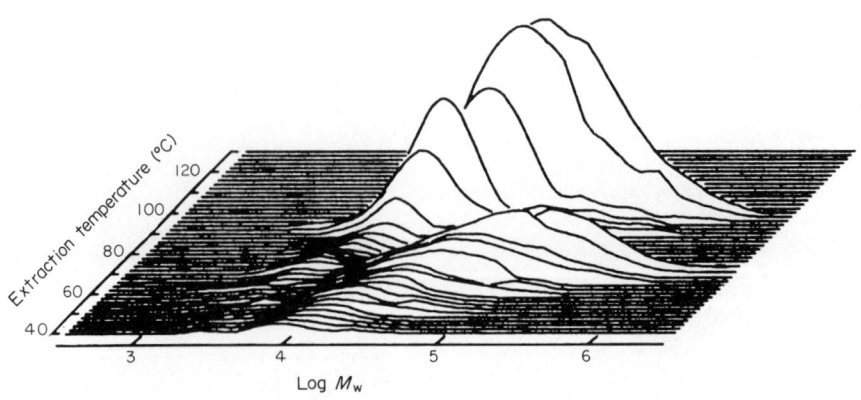

Figure 5 Perspective picture of the two-component mixture: see Figure 4

temperature in the range between between 117 and 123 °C. An exponential gradient was implemented using butyl cellosolve and xylene. Since only the latter is a solvent for the specimens investigated, the gradient was one of thermodynamic quality. SEC was performed in *o*-DCB at 135 °C on a set of four columns packed with PS gels. From the estimated MMD of each fraction and its ethylene content a diagram was constructed which showed the range of chemical composition and molar mass of the starting sample. It indicated the outline of the distribution in molar mass and composition but did not contain the contour lines necessary for full information on MMD and CCD.[54]

16.4.3 Statistical Copolymers from Styrene and Acrylonitrile

Azeotropic copolymers of this system (*stat*-copoly(S/AN); $r_1 = 0.41$; $r_2 = 0.04$) contain 24 mass-% AN or 61.9 mol-% S. Products of about this composition are commercially available. Samples with an AN content below 50 mass-% are soluble in common organic solvents whereas copolymers richer in AN require DMF or DMSO which are solvents even for polyacrylonitrile (PAN).

Ogawa and Sakai reported on the investigation of two model copolymers with 27.8 or 51.3 mass-% AN.[55] About 5–6 g sample material had been loaded from a solution onto a Celite support just before the latter was packed into the column (700 × 27 mm). The first step was separation by composition through elution with a linear gradient (toluene:propanol 1:1)/DMF. The *n*-propanol was added in order to reduce the precipitating power of the nonsolvent toluene. The eluate was collected in 250 ml portions from which the polymer was isolated by precipitation. The AN content of the fractions was estimated from nitrogen analyses.

Fractions of nearly equal composition were used for cloud-point measurements in DMF/ (*n*-propanol:toluene). The data were plotted *vs.* molecular weight, using the AN content as a parameter. The curves showed that the influence of composition exceeded that of *M*. Thus, the first fractionation was mainly a separation by composition.

The size distribution of each fraction was measured by SEC. The results from both sets of experiments were used for the construction of contour maps. The corresponding diagram of the sample with 51.3% AN showed a broad chemical composition distribution with two maxima at about 36 and 68 mass-% AN. The composition distribution of the sample with 27.8% AN was in the limits of 24.5 and 29.6 mass-% with two maxima at about 26 and 29%. A slight increase in AN content with increasing *M* was observed.[55]

Efficient CCF of copoly(S/AN) has been performed by SEC and HPPLC with 2,2,4-trimethylpentane (isooctane)/THF.[45, 47, 49, 50, 56] The isooctane precipitates according to AN content. THF is a solvent. In most experiments it contained about 10 vol-% methanol. This admixture brought the refractive index close to that of isooctane.

A mixture of two S/AN copolymers (I with 16.1 and II with 30 mass-% AN) was fractionated by SEC with THF as an eluent on a set of five *μ*-Styragel columns at a flow rate of 1 ml min⁻¹. After the injection of 0.87 mg sample mixture with 59.3% S/AN I, eluate fractions of 0.5 ml each were collected. These SEC slices were investigated by HPPLC on a C18 column (150 × 4.6 mm; $d_p = 10 \mu m$) without any additional treatment. The multilinear gradient programme started with 10% B (THF + 10% methanol). Within three minutes the B concentration was raised to 60%. The separation was performed by slowly increasing the concentration of B from 60 to 90% within 12 minutes. Figure 6 shows the result of this experiment, which was the first published example of CCF using HPLC equipment for the separation both by size and composition.[56]

Then a mixture of five S/AN copolymers was separated by SEC and HPPLC of the eluate slices. Table 2 gives a compilation of the sample components and their characteristics. A CN bonded-phase column (150 × 4.6 mm; $d_p = 5 \mu m$) allowed almost baseline separation of the copolymers (see Figure 7). The column was operated at 50 °C and the flow rate 1 ml min⁻¹. The gradient was isooctane/THF (50–100% B in 10 min). The CCF required in total about seven hours experimental work. It was performed on 1.04 mg starting material.[49]

Columns packed with RP C18 or bare silica of either small or large pores yielded about the same separating power for S/AN.[11, 57]

Commercial S/AN copolymers have been cross-fractionated by SEC and HPPLC on a reversed-phase C18 column (150 × 4.6 mm; $d_p = 5 \mu m$; $d_o = 6$ nm)[47] and also on the CN column mentioned above.[49] The average AN content of each SEC fraction was calculated from the first moments of the CCD curves measured by HPPLC. For comparison, the average composition of each SEC fraction was independently measured by pyrolysis gas chromatography (PGC) and by dual detection SEC.

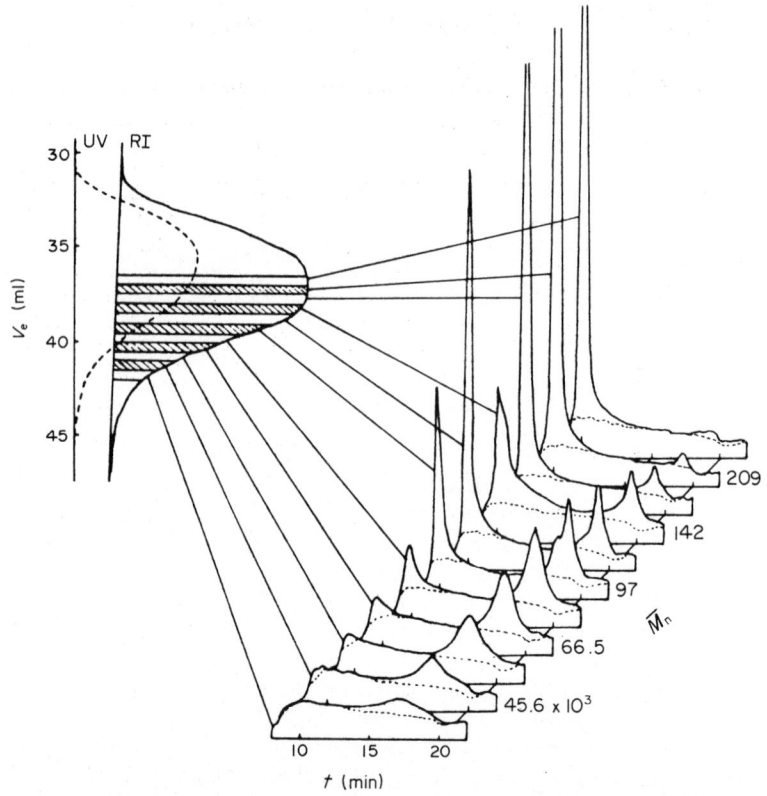

Figure 6 Chromatographic cross-fractionation by HPPLC of SEC fractions. Analysis of the mixture (59.3:40.7) of two *stat*-copoly(styrene/acrylonitrile) samples. The first HPPLC peak is due to the copolymer with 16.1 mass-% AN, $\bar{M}_n$ 325 000, the second to that with 30 mass-% AN, $\bar{M}_n$ 71 000. (Reproduced by permission of Hüthig & Wepf Verlag from *Makromol. Chem., Rapid Commun.*, 1983, **4**, 529)

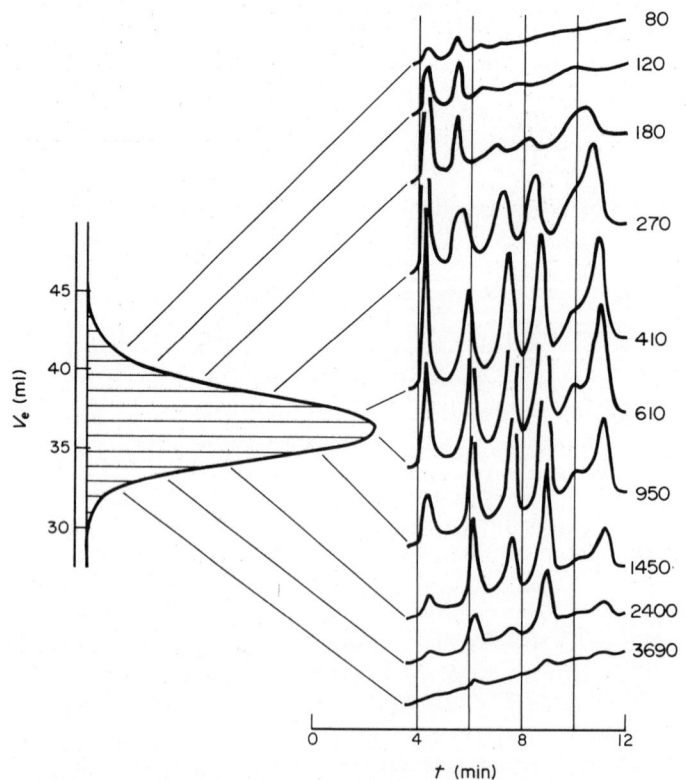

Figure 7 Chromatographic cross-fractionation of *stat*-copoly(S/AN) by HPPLC of SEC fractions. The five model copolymers (*cf.* Table 2) are eluted in a sequence I–V. The molecular weight of the SEC fractions ($\bar{M} \times 10^{-3}$) is indicated at the HPPLC traces. (Reproduced by permission of Elsevier Applied Science Publishers, from 'Integration of Fundamental Polymer Science and Technology', ed. L. A. Kleintjens and P. J. Lemstra, 1986, p. 91)

Table 2 Characteristics of the *stat*-Copoly(styrene/acrylonitrile) Specimens used for the CCF Separation shown in Figure 7

Sample No.	I	II	III	IV	V
Acrylonitrile (mass-%)	16.1	23.0	29.1	36.4	42.7
Styrene (mol-%)	72.6	63.0	55.4	47.1	40.6
Molecular weight $\times 10^{-3}$	325[a]	480[a]	510[a]	380[a]	340[a]
Relative amount in the mixture injected (%)	11.5	22.5	17.7	26.1	22.2

[a] Measured by osmosis.

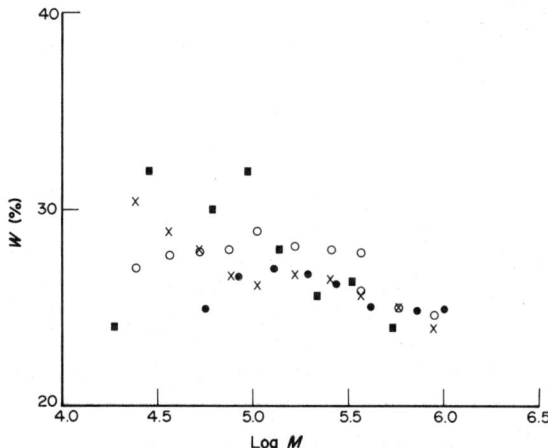

Figure 8 Average AN content of SEC fractions: comparison of CCF results with dual detection SEC (■), pyrolysis gas chromatography (○), CCF on a CN column (*cf.* Figure 9) (×) and CCF on a C18 column (●). (Reproduced by permission of Elsevier Applied Science Publishers, from 'Integration of Fundamental Polymer Science and Technology', ed. L. A. Kleintjens and P. J. Lemstra, 1986, p. 94)

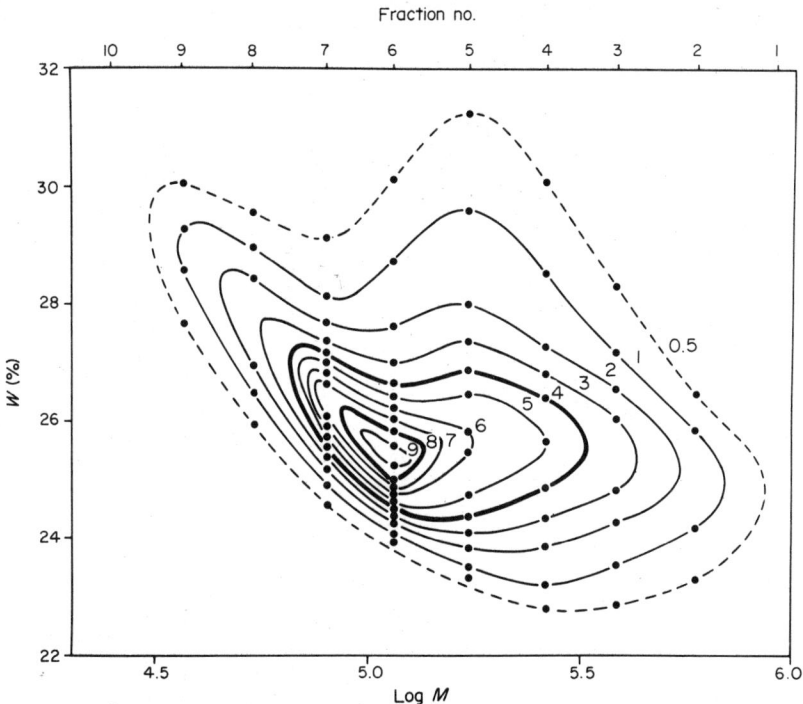

Figure 9 Contour map of a commercial *stat*-copoly(S/AN) sample as obtained through CCF by HPPLC of 10 SEC fractions. HPPLC conditions as in the calibrating experiments shown in Figure 7. (Reproduced by permission of Elsevier Applied Science Publishers, from 'Integration of Fundamental Polymer Science and Technology', ed. L. A. Kleintjens and P. J. Lemstra, 1986, p. 92)

Figure 8 shows results from this investigation, demonstrating reasonable agreement in the high-molecular region.

Whereas dual detection SEC and PGC can only reveal the average composition of an SEC fraction, HPPLC has the additional capacity of evaluating the CCD in each fraction. Figure 9 shows the contour map obtained as the result of CCF of a commercial S/AN copolymer sample.

16.4.4 Copolymers from Styrene and Methyl Methacrylate

16.4.4.1 Block copolymers

Belenkij *et al.* reported in 1975 on the investigation of a *triblock*-copoly(MMA/S/MMA) by preparative SEC of 40 g sample material and subsequent TLC of the fractions using chloroform: methanol mixtures. It was possible to separate poly(methyl methacrylate) homopolymer (PMMA) from block copolymers of the same molecular weight which remained at the start ($R_f = 0$) whereas the PMMA reached R_f values of about 0.5–0.7. The ratio of PMMA to block copolymer in a SEC fraction was estimated from data obtained by pyrolysis gas chromatography of the acetone extracts from the TLC zones of both components.[18]

The complete CCF of a *diblock*-copoly(S/MMA) sample (47:53 mass-%, 739 000 $\bar{M}_n$) was achieved by AC and SEC.[19] The AC yielded separation by composition and was performed in a column (150 × 50 mm) packed with activated dry silica. The chromatographic bed was subdivided by filter paper into nine compartments. The sample (300 mg) was loaded on a pile of filter paper (about 30 sheets) in the lower part of the column shown in Figure 10. The development was performed by soaking an ethyl acetate:benzene mixture (72.5:27.5 vol-%) into the vertically orientated column. It took about five hours until the eluent front reached the last but one compartment at the top of the column. Then the column was immediately dismantled and the polymer extracted from each compartment by THF. The same was done with the paper block where

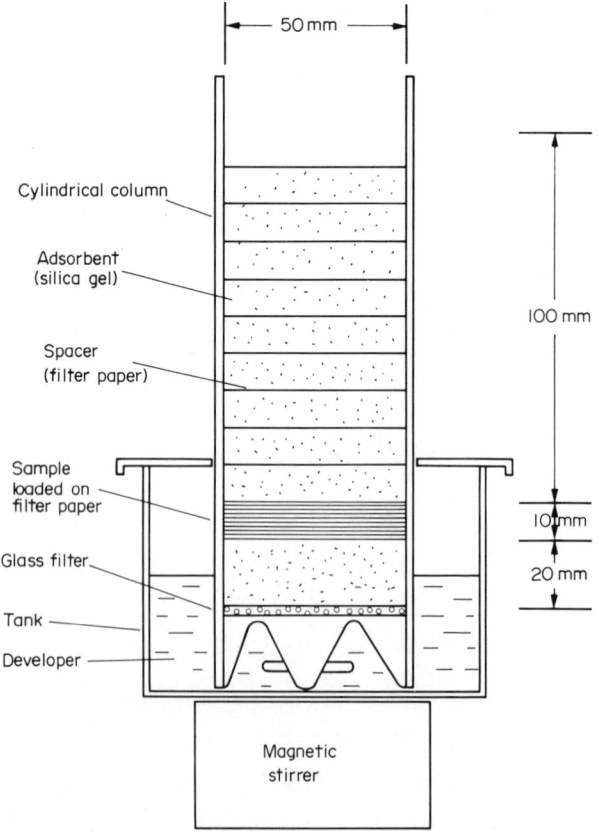

Figure 10 Sketch of the adsorption column used for CCF of *diblock*-copoly(styrene/methyl methacrylate). (Reproduced by permission of Marcel Dekker Inc., from *J. Macromol. Sci., Phys.*, 1980, **B17**, 218)

the immobile remainder of the sample was found. As a whole, about 93% of the initial material was recovered. The MMA content in the fractions ranged from 61.1 mass-% (in the lowest compartment, F8) to 15.1% in F2 at the top of the column. The rest of the sample block contained 79.0% methyl methacrylate.

In 1982, Inagaki et al.[20] showed that the position of the SEC curves of all these fractions was almost the same (see Figure 11). This explained why the SEC investigation of the unfractionated sample by dual detection technique could reveal scarcely any chemical heterogeneity.[19] Every siphon volume contained all sample constituents in a roughly constant ratio. The authors stated that 'no variation of the SEC point-by-point composition does not necessarily guarantee uniformity in composition for a sample copolymer'.[20]

This careful experimental work is an excellent demonstration for the necessity of cross-fractionation.

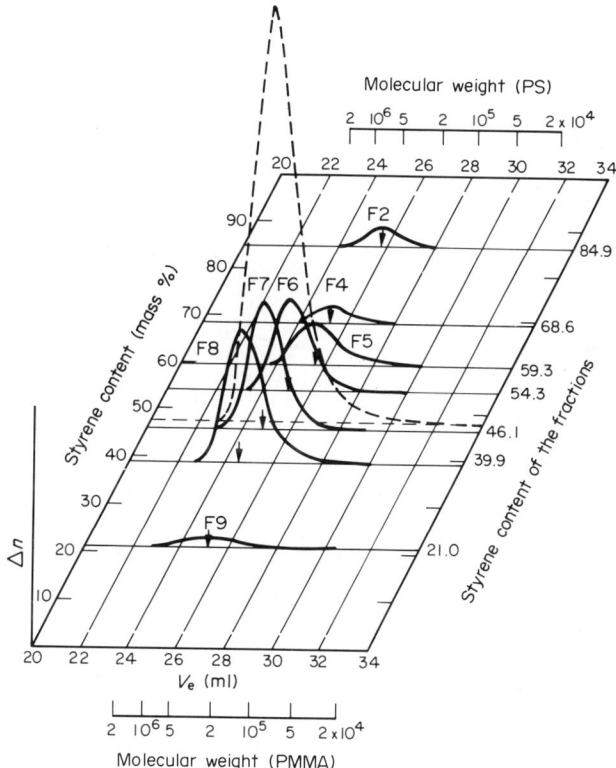

Figure 11 Chromatographic cross-fractionation of *diblock*-copoly(S/MMA): SEC investigation of the fractions obtained by column AC. The dashed curve is from the nonfractionated sample. (Reproduced by permission of the International Union of Pure and Applied Chemistry (IUPAC) from *Pure Appl. Chem.*, 1982, **54**, 317)

16.4.4.2 *Statistical copolymers*

Azeotropic copolymers of this system (*stat*-copoly(S/MMA); $r_1 = 0.53$, $r_2 = 0.49$) contain 47 mass-% MMA or 52.0 mol-% S. The copolymers are soluble in most organic solvents. Alkane hydrocarbons and lower alcohols act as precipitants.

Separation by composition was achieved by gradient HPLC with dichloroethane/THF (exponential from 3 to 20%) on a silica column (250 × 6 mm, $d_o = 6$ nm; $d_p = 9\,\mu$m)[58] as well as with isooctane/(THF + 10% MeOH) on a silica column (150 × 4.6 mm; $d_o = 6$ nm; $d_p = 5\,\mu$m).[59]

CCF has been performed by SEC and gradient HPLC. A sample of about 1 mg was fractionated first by hydrodynamic volume through SEC. The column was a set of two Toyo Soda Mixed Gel Columns GMH6 (2 × 600 × 7.8 mm). The injection volume was $V_o = 0.2$ ml, the sample solvent and the SEC eluent THF at a flow rate of 1 ml min^{-1}. The sample was the mixture of seven model copolymers. Their characteristics and the sample composition are given in Table 3.

The SEC refractive index signal is shown in the left part of Figure 12. The curve does not indicate the components present because the differences in molecular weight are too small.

Table 3 Characteristics of the *stat*-Copoly(styrene/methyl methacrylate) Specimens used for the CCF Separation shown in Figure 12

Sample No.	I	II	III	IV	V	VI	VII
Methyl methacrylate (mass-%)	11.4	23.8	37.0	49.5	64.0	76.2	88.5
Styrene (mol-%)	88.2	75.5	62.1	49.5	35.1	23.1	11.1
Molecular weight $\times 10^{-3}$	160[a]	250[a]	150[a]	185[a]	235[a]	220[a]	220[a]
Relative amount in the mixture injected (%)	12.7	14.0	12.1	12.7	12.7	17.2	18.5

[a] Measured by light scattering.

Eluate portions of 1 ml each were used for subsequent gradient HPLC on a column (60×4 mm; $d_o = 5$ nm; $d_p = 5\,\mu$m) packed with silica Nucleosil 50. The elution was performed at 0.5 ml min^{-1} flow rate with a gradient isooctane/THF (10–80% B in 14 minutes). Figure 12 shows that the seven copolymers had been sufficiently separated in the main SEC fractions 3, 4 and 5. The sample appearing first is the specimen I. The peak sequence is determined by the increase in MMA content. The samples V, VI and VII have comparatively high values of M and, in accordance with this, are scarcely detectable in the low-molecular SEC fractions 6 and 7.

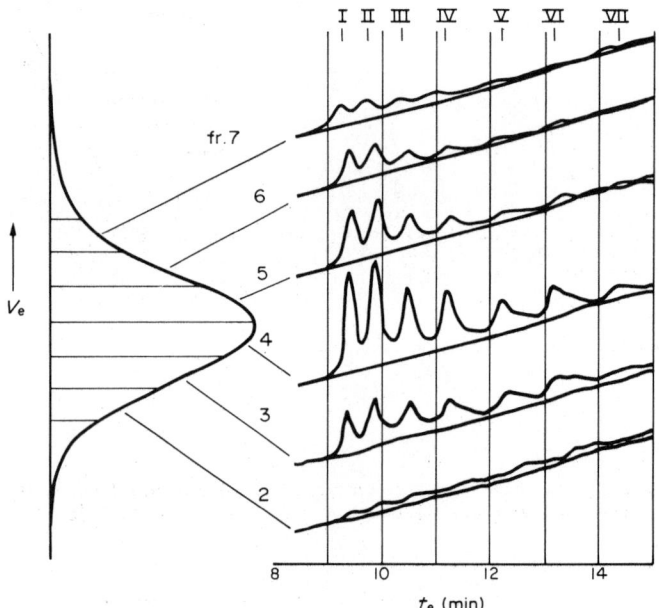

Figure 12 Chromatographic cross-fractionation of *stat*-copoly(S/MMA) by gradient HPLC of SEC fractions having $M \times 10^{-3}$ values 300 (fr. 2), 190 (3), 125 (4), 80 (5), 50 (6) and 30 (fr. 7).

CCF was also attempted by AC and subsequent SEC.[60] About 100 μg *stat*-copoly(S/MMA) were eluted from a silica column (50×4.6 mm; $d_o = 3$ nm; $d_p = 5\,\mu$m) by stepwise increasing the chloroform content in the mixture with dichloroethane. The two main fractions (obtained with DCE:CHCl$_3$ 80:20 or 70:30) from ten repeated elutions were combined and analyzed by SEC.

16.4.5   Statistical Copolymers from Styrene and Methyl Acrylate

Azeotropic copolymers of this system (*stat*-copoly(S/MA); $r_1 = 0.75$; $r_2 = 0.18$) contain 20.1 mass-% MA or 76.6 mol-% S. A high-conversion sample containing 51.7 mass-% MA units was investigated by CCF. Because of the high degree of conversion (92%) and the deviation from the azeotropic point a broad CCD towards higher MA content was expected.[23]

At first, SEC separation according to molecular weight was performed. The column (600×25 mm) was packed with TSK gel and was capable of fractionating 10 mg per injection. The separation into eleven slices was repeated a hundred times. The corresponding ones were united, yielding eleven fractions sufficiently large for further investigation.

The CCD of the fractions was evaluated by TLC on silica through gradient elution with CCl$_4$/methyl acetate (the latter linearly increasing from 6 to 43 vol-%). Standard samples of known composition were used as markers in each TLC run. They enabled the chromatographic traces to be evaluated in terms of composition. Figure 13 shows the result obtained with the help of a TLC scanner, and Figure 14 the contour map of the high-conversion sample. The authors discussed their own work very critically ('In general, the CCD determined by the TLC method may not be sufficiently accurate')[23] but in a thorough examination the results proved sufficiently reliable.

The coupling of high MA contents with high-molecular weight values is understood as the consequence of the gel effect. At a high degree of conversion, the monomer mixture is richer in MA than at the beginning. Hence, the very large macromolecules growing after the start of the Schulz–Trommsdorf effect contain distinctly more MA units. This is clearly seen in Figure 14.

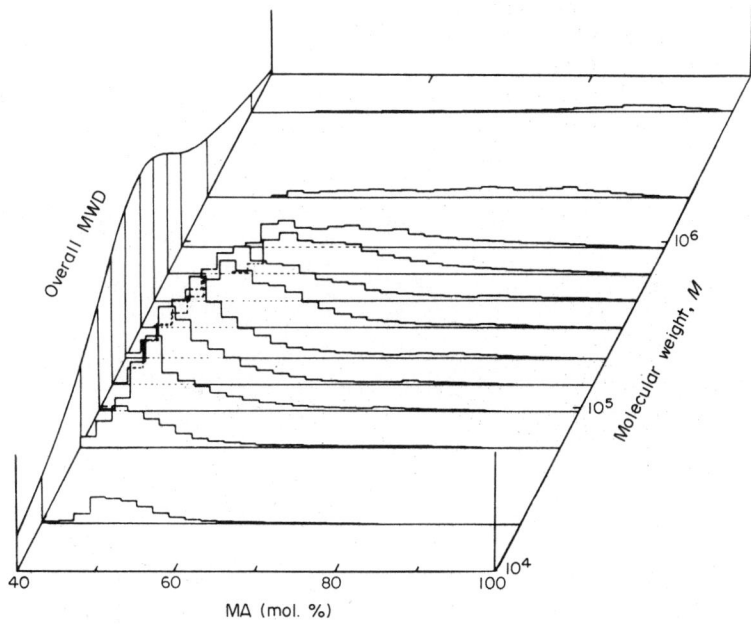

Figure 13 Chromatographic cross-fractionation of *stat*-copoly(S/MA) by TLC of SEC fractions. (Reproduced by permission of the American Chemical Society from *Macromolecules*, 1983, **16**, 544)

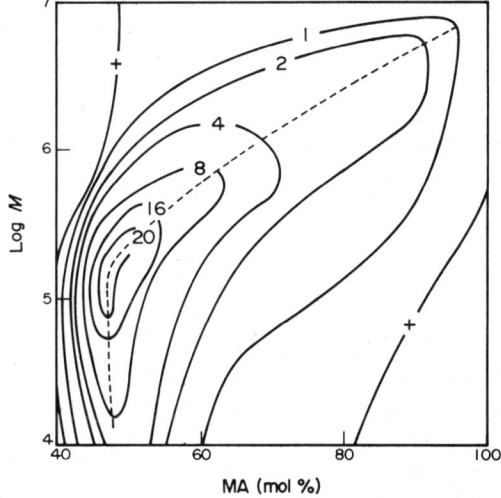

Figure 14 Contour map of a high-conversion *stat*-copoly(S/MA) sample with 56.4 mol-% MA as obtained by CCF through SEC and TLC of the fractions. (Reproduced by permission of the American Chemical Society from *Macromolecules*, 1983, **16**, 544)

16.4.6 Statistical Copolymers from Styrene and *n*-Butyl Methacrylate

Azeotropic copolymers of this system (*stat*-copoly(S/*n*-BMA); $r_1 = 0.63$; $r_2 = 0.64$) contain 58.4 mass-% BMA or 49.3 mol-% S units. Report was given that a copolymer with 33 mol-% S obtained by free-radical polymerization could be separated from admixed PS by 'orthogonal chromatography'.[7] This term refers to coupled-column GPC with different eluents in each device. The column in GPC(1) was packed with *μ*-Styragel whereas in GPC(2) a Bondagel column was used. The separation reported was performed with THF in GPC(1) and a *n*-heptane:THF mixture in GPC(2).

The content of heptane proved decisive. Best results were obtained with 63.8 vol-%. This binary eluent enabled an azeotropic copoly(S/*n*-BMA) to be separated from both parent homopolymers.[61,62] At 60% *n*-heptane, the copolymer peak merged with the peak of poly(*n*-butyl methacrylate)[61] (see Figure 15) and at 65% with the PS peak.[62] The two GPC devices were on-line coupled, but GPC(1) was stopped while GPC(2) was working.

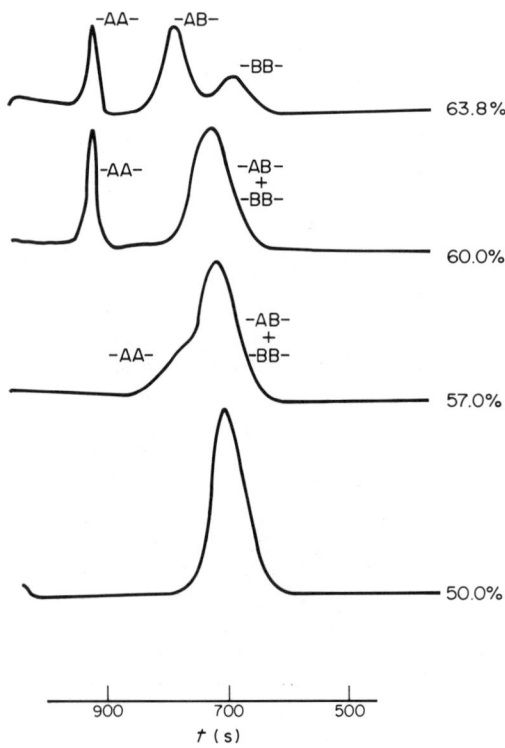

Figure 15 Orthogonal chromatography of the mixture of PS (–AA–), poly(*n*-butyl methacrylate) (–BB–) and azeotropic copoly(S/*n*-BMA) (–AB–). The *n*-heptane content in the eluent of GPC(2) is indicated at the chromatograms. (Reproduced by permission of Marcel Dekker Inc., from *Sep. Purif. Methods*, 1982, **11**, 17)

Most of the results and experimental details were reported in the paper from 1983 but again no example of real cross-fractionation was given. The reason might have been that 'mixed mechanisms in GPC(2) and the coupling of composition with sequence length effects cause calibration referencing retention time to be very difficult'.[63] UV scanning of the peaks from GPC(2) was the main source of information concerning the composition of any constituent.

The advantage of coupling two GPC devices is the possibility of compensating for an adverse influence of molecular weight by adjusting the exclusion potency of the two columns. On the other hand, the separation power for fractionation by composition is comparatively low. To enhance it, one should employ a column with an active packing and perform gradient elution (see Sections 16.2.4 and 16.3).

16.4.7 Statistical Copolymers from Styrene and Ethyl Methacrylate

Azeotropic copolymers of this system ($r_1 = 0.49$; $r_2 = 0.40$) contain 48.2 mass-% EMA or 54.1 mol-% S units. CCF of *stat*-copoly(S/EMA) samples was performed through the combination of

SEC and TLC. The TLC was carried out using 'Chromarods'. These are quartz rods (152 × 0.9 mm) coated with a sinter-fused layer of silica. The samples were spotted near to the lower end and developed along the vertically positioned rods. This was done with a gradient toluene/acetone. After the development, the Chromarods were dried and then slowly moved through the hydrogen flame of an ionization detector. The detector signal provided information on the amount of organic material resting at any given position of the rod. This information and the data from SEC were used for computer-aided plotting of the contour maps of samples polymerized to various degrees of conversion.[24]

Stat-copoly(S/EMA) samples have also been cross-fractionated by SEC and column HPLC with a variety of gradients and stationary phases. The SEC was carried out with 1 mg sample in 200 μl THF. Two Toyo Soda Mixed Gel Columns GMH6 (2 × 600 × 7.8 mm) and THF eluent at a flow rate 1 ml min⁻¹ were used.

Figure 16 shows an example obtained with a silica column and a gradient isooctane/THF (30–70% B in eight minutes). The injections were performed into isooctane: MeOH (99:1). Then the eluent composition was changed gradually to 69:30:1 isooctane:THF:MeOH where the gradient started. 1% MeOH was kept constant during the whole analysis. Such a polar additive is essential for suppressing disturbances in gradient HPLC on silica. The sample was the mixture of five model copolymers whose characteristics are listed in Table 4.

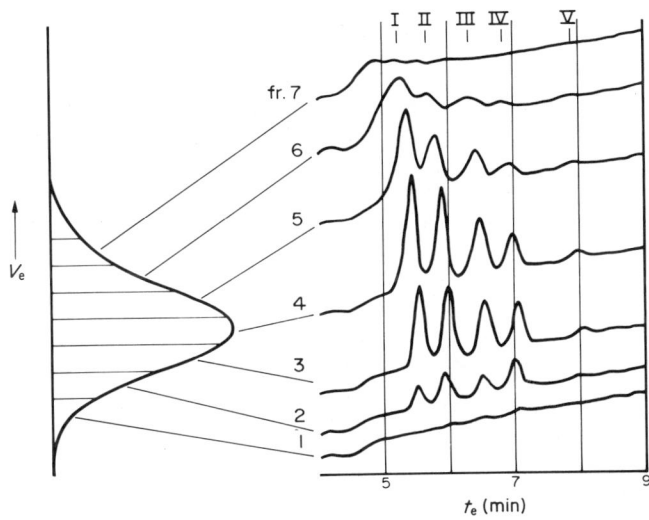

Figure 16 Chromatographic cross-fractionation of *stat*-copoly(S/EMA) by gradient HPLC of SEC fractions on a silica column. The SEC fractions had the following $M \times 10^{-3}$ values: 290 (fr. 1), 135 (2), 85 (3), 55 (4), 35 (5), 22 (6) and 14 (fr. 7)

Table 4 Characteristics of the *stat*-Copoly(styrene/ethyl methacrylate) Specimens used for the CCF Separations shown in Figures 16 and 17

Sample No.	I	II	III	IV	V
Ethyl methacrylate (mass-%)	4.7	32.2	54.6	68.0	92.5
Styrene (mol-%)	95.7	69.8	47.7	34.0	8.2
Molecular weight × 10⁻³	29.5[a]	35.8[a]	40.8[a]	46.0[a]	36.7[a]
Relative amount in the mixture injected (%)	19.6	19.8	20.5	19.6	20.5

[a] By SEC measurements based on a PS calibration.

Figure 17 shows the chromatograms obtained from the same mixture on a CN-bonded-phase column. The HPLC separation had been carried out on the same SEC fractions as in the investigation shown in Figure 16. The gradient and injection conditions were the same in both analyses.

The detector signal at 259 nm reveals the five specimens present in the mixture, whereas the RI curve from the SEC does not give much information. The higher the EMA content of a given sample the more THF was needed for elution, both on the silica and CN column.[64]

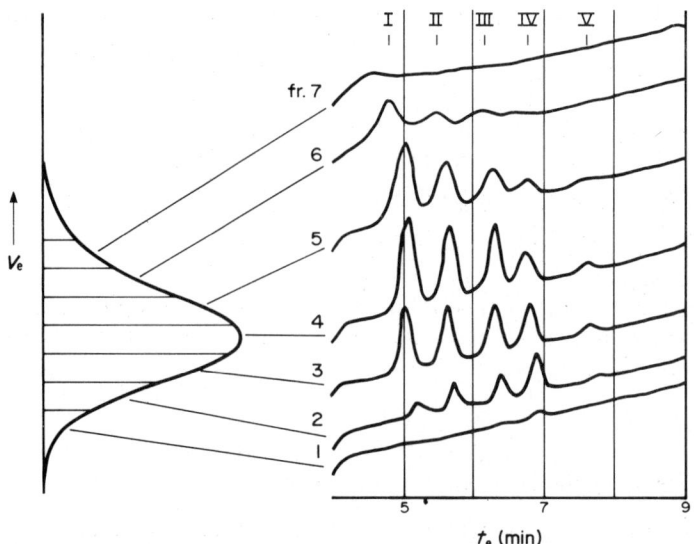

Figure 17 Chromatographic cross-fractionation of *stat*-copoly(S/EMA) by gradient HPLC fractions on a CN bonded-phase column

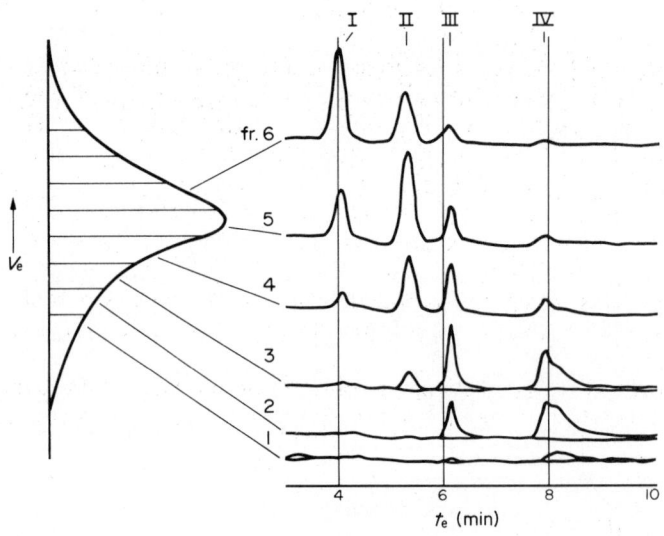

Figure 18 Chromatographic cross-fractionation of *stat*-copoly(S/MEMA) by gradient HPLC of SEC fractions having $M \times 10^{-3}$ values 500 (fr. 1), 200 (2), 150 (3), 100 (4), 65 (5) and 40 (fr. 6)

16.4.8 Statistical Copolymers from Styrene and Methoxyethyl Methacrylate

Azeotropic copolymers of this system (*stat*-copoly(S/MEMA); $r_1 = 0.41$, $r_2 = 0.48$) contain 61.1 mass-% MEMA or 46.85 mol-% S units. Chromatographic cross-fractionation was performed through the combination of SEC and gradient HPLC.[65] Figure 18 shows the results obtained with the mixture of four model copolymers. The characteristics of these specimens are compiled in Table 5.

The SEC was carried out in THF eluent at 1 ml min^{-1} flow rate. A set of two Toyo Soda Mixed Gel Columns GMH6 was used ($2 \times 600 \times 7.8$ mm). The injection volume was 200 μl, the sample amount 1.12 mg.

The gradient HPLC was performed on a CN column (60×4 mm; $d_p = 5$ μm) at 0.5 μl min^{-1} flow rate. The gradient was an increase of the methanol content from 2 to 52% within ten minutes. The other eluent components were isooctane and THF. The elution of the sample constituents occurred in the order of increasing MEMA content. The results correspond with expectations based on the

Table 5 Characteristics of the *stat*-Copoly(styrene/methoxyethyl methacrylate) Specimens used for the CCF Separation shown in Figure 18

Sample No.	I	II	III	IV
Methoxyethyl methacrylate (mass-%)	25.9	49.0	62.4	87.4
Styrene (mol-%)	79.8	59.0	45.5	16.6
Molecular weight × 10⁻³	96	137	173	306
Relative amount in the mixture injected (%)	32.9	26.7	23.0	17.4

molecular weight of the copolymers: IV with the highest M was preferably contained in the high-molecular SEC fractions 1–3, whereas copolymer I with a low-molecular weight was mainly present in the fractions 5 and 6.

16.4.9 Copolymers from Styrene and Acrylic Acid

Azeotropic copolymers of this system (*stat*-copoly(S/AA); $r_1 = 0.15$; $r_2 = 0.25$) contain 44 mass-% AA or 46.9 mol-% S. Samples with AA contents graded from 30 to 70% were investigated by HPPLC and subsequent SEC.[21]

The first stage was gradient elution on a set of two silica columns (each 250 × 4.6 mm; $d_o = 6$ or $d_o = 30$ nm). The gradient was started with cyclohexane:THF:MeOH = 95:2.5:2.5 and reached THF:MeOH = 50:50 within 75 min. Cyclohexane is a nonsolvent for the copolymers under investigation. Thus, the mechanism of the separation was considered to be precipitation of the samples followed by selective redissolution.

The separation by composition was strongly influenced by the molecular weight of the samples which were rather low. This is easily understood by a closer look at equation (6): the $M^{-0.5}$ term is especially effective at low values of M. The difficulties involved with the HPLC separation of nonprefractionated and thus rather complex mixtures were overcome by a strict regime of column operation, and certainly also by the flat gradient.

About ten fractions were investigated by fast SEC with elution times of about three minutes. The high speed of the second separation made on-line operation possible. Thus, the acquisition of the data was accomplished within 75 minutes.

Another advantage of this configuration was that the final isocratic separation enabled UV and FTIR detection. Thus, additional information was obtained concerning the actual composition of the fractions investigated.

The system also saved any trouble with aging solutions or handling a lot of sample vials. In many respects it seems to be very well suited for automated CCF in routine process control.

16.5 CONCLUDING REMARKS

Chromatographic cross-fractionation offers at least two advantages. One of them is the acceleration of data acquisition. With classic fractionation techniques, the investigation of a sample requires about two to three months of skillful labour. CCF is capable of subfractionating 10–12 fractions in about six hours. Since these fractions can be obtained from another apparatus in even shorter time, about four samples per day can be investigated with an autoinjection device and a continuously running apparatus. The equipment mentioned in Section 16.4.9 is even more efficient: with this configuration almost 20 samples could be cross-fractionated within a 24-hour period.

The second advantage is the effect of the stationary phase which facilitates the separation by composition, and the reliability of SEC for the separation by molecular size. Thus, the restricting basic conditions for the interactions between the polymer constituents and the separating solvent systems are partly raised.

In some cases, still another point may be favourable: CCF can be carried out with a tiny amount of material. Several examples mentioned in the previous sections had been performed on 1 mg of starting material. Classic cross-fractionation requires about 10 g.

It can be concluded that CCF is a potential tool in copolymer analysis. It is comparatively new in polymer laboratories, but the basic principle has long been known among chromatographers: in 1978, coupled-column liquid chromatography was applied to the separation of complex mixtures.[66]

It has found application in life sciences for the separation of proteins and other substances of biological importance.[67-70] The mutual exchange of experiences and opinions between polymer scientists and biochemists could accelerate progress in the development of a separating technique which is needed by both groups.

16.6 REFERENCES

1. A. J. Rosenthal and B. B. White, *Ind. Eng. Chem.*, 1952, **44**, 2693.
2. A. V. Topčiev, A. D. Litmanovič and V. Ya. Štern, *Dokl. Akad. Nauk SSSR*, 1962, **147**, 1389.
3. S. Teramachi and M. Nagasawa, *J. Macromol. Sci., Chem.*, 1968, **2**, 1169.
4. G. M. Guzmán, in 'Progress in High Polymers', ed. J. C. Robb and F. W. Peaker, Heywood, London, 1961, p. 113.
5. O. Fuchs and W. Schmieder, in 'Polymer Fractionation', ed. M. J. R. Cantow, Academic Press, New York, 1967, p. 341.
6. G. Riess and P. Callot, in 'Fractionation of Synthetic Polymers', ed. L. H. Tung, Marcel Dekker, New York, 1977, p. 445.
7. S. T. Balke and R. D. Patel, *J. Polym. Sci., Polym. Lett. Ed.*, 1980, **18**, 453.
8. G. Glöckner, in 'Advances in Polymer Science', Springer, Berlin, 1986, vol. 79, p. 159.
9. C. Horváth, W. Melander and I. Molnár, *J. Chromatogr.*, 1976, **125**, 129.
10. L. Wild and T. Ryle, *Polym. Prepr., Am. Chem. Soc., Div. Polym. Chem.*, 1977, **18**, 182.
11. G. Glöckner, H. Kroschwitz, and Ch. Meissner, *Acta Polym.*, 1982, **33**, 614.
12. G. Glöckner and J. H. M. van den Berg, *Chromatographia*, 1984, **19**, 55.
13. G. Glöckner, F. Francuskiewicz, and K.-D. Müller, *Plaste Kautschuk*, 1971, **18**, 654.
14. G. Glöckner, F. Francuskiewicz and H.-U. Reichardt, *Plaste Kautschuk*, 1979, **26**, 431.
15. G. Glöckner, 'Polymercharakterisierung durch Flüssigkeitschromatographie', Deutscher Verlag d. Wissenschaften, Berlin, 1980.
16. H. Inagaki, in 'Advances in Polymer Science', Springer, Berlin, 1977, vol. 24, p. 189.
17. B. G. Belenkii and E. S. Gankina, *J. Chromatogr.*, 1977, **141**, 13.
18. B. G. Belenkii, E. S. Gankina, P. P. Nefedov, M. A. Lazareva, T. S. Savitskaya and M. D. Volchikhina, *J. Chromatogr.*, 1975, **108**, 61.
19. T. Tanaka, M. Omoto, N. Donkai, and H. Inagaki, *J. Macromol. Sci., Phys.*, 1980, **B17**, 211.
20. H. Inagaki and T. Tanaka, *Pure Appl. Chem.*, 1982, **54**, 309.
21. C. S. Weiss, A. M. Dwivedi and J. S. Hazlett, in 'Book of Abstracts, Tenth International Symposium on Column Liquid Chromatography, San Francisco, 1986', ed. R. E. Majors, contr. 605.
22. ASTM Standard D 3593-80, American Society for Testing and Materials, Philadelphia.
23. S. Teramachi, A. Hasegawa and S. Yoshida, *Macromolecules*, 1983, **16**, 542.
24. J. C. J. F. Tacx, Thesis, Eindhoven University of Technology, Eindhoven, 1986.
25. R. J. Maggs, *Chromatographia*, 1968, **1**, 43.
26. R. P. W. Scott and J. G. Lawrence, *J. Chromatogr. Sci.*, 1970, **8**, 65.
27. LCM2 Detector, Pye Unicam Ltd (discontinued).
28. A. Stolyhwo, H. Colin and G. Guiochon, *J. Chromatogr.*, 1983, **265**, 1.
29. A. Stolyhwo, H. Colin, M. Martin and G. Guiochon, *J. Chromatogr.*, 1984, **288**, 253.
30. L. E. Oppenheimer and T. H. Mourey, *J. Chromatogr.*, 1985, **323**, 297.
31. J. B. Dixon (Purdue Research Foundation), *US Pat.* 4 215 090 (1980) (*Chem. Abstr.*, 1980, **93**, 142 383).
32. J. B. Dixon and R. C. Hall (Purdue Res. Found.) *US Pat.* 4 271 022 (1981) (*Chem. Abstr.*, 1981, **95**, 108 167).
33. 945 Universal FID Detector for HPLC, Tracor Instruments, Austin, Texas.
34. W. Scholtan and F. J. Kwoll, *Makromol. Chem.*, 1972, **151**, 33.
35. T. H. Mourey, *J. Chromatogr.*, 1986, **357**, 101.
36. See for example, M. A. Phelan and K. A. Cohen, *J. Chromatogr.*, 1983, **266**, 55.
37. J. W. Crabb and L. M. G. Heilmeyer, Jr., *J. Chromatogr.*, 1984, **296**, 129.
38. R. L. Emanuel, G. H. Williams and R. W. Giese, *J. Chromatogr.*, 1984, **312**, 285.
39. M. Danielewicz and M. Kubin, *J. Appl. Polym. Sci.*, 1981, **26**, 951.
40. S. Teramachi, A. Hasegawa, M. Akatsuka, A. Yamashita and N. Takemoto, *Macromolecules*, 1978, **11**, 1206.
41. R. J. Brüssau and D. J. Stein, *Angew. Makromol. Chem.*, 1970, **12**, 59.
42. L. H. Garcia-Rubio, A. E. Hamielec and J. F. MacGregor, in 'Computer Applications in Applied Polymer Science', ed. Th. Provder, *ACS Symp. Ser.*, 1982, **197**, 151.
43. B. Stützel, Diploma Project, Freiburg University, Freiburg, 1971.
44. B. Stützel, T. Miyamoto and H. J. Cantow, in 'Book of Abstracts, Intern. Symposium on Macromol. Chem., Helsinki, 1972', IUPAC, 337.
45. G. Glöckner, *Pure Appl. Chem.*, 1983, **55**, 1553.
46. G. Glöckner and D. Ilchmann, *Acta Polym.*, 1984, **35**, 680.
47. G. Glöckner, J. H. M. van den Berg, N. L. J. Meijerink, Th. G. Scholte and R. Koningsveld, *J. Chromatogr.*, 1984, **317**, 615.
48. G. Glöckner, V. Albrecht, F. Francuskiewicz and D. Ilchmann, *Angew. Macromol. Chem.*, 1985, **130**, 41.
49. G. Glöckner, J. H. M. van den Berg, N. L. J. Meijerink and Th. G. Scholte, in 'Integration of Fundamental Polymer Science and Technology', ed. L. A. Kleintjens and P. J. Lemstra, Elsevier Applied Science, Barking, UK, 1986, p. 85.
50. G. Glöckner, J. H. M. van den Berg, N. L. J. Meijerink, Th. G. Scholte and R. Koningsveld, *Macromolecules*, 1984, **17**, 962.
51. G. Glöckner, *Z. Phys. Chem. (Leipzig)*, 1965, **229**, 98.
52. L. Wild, T. Ryle and D. Knobeloch, *Polym. Prepr., Am. Chem. Soc., Div. Polym. Chem.*, 1982, **23**, 133.
53. S. Nakano and Y. Goto, *J. Appl. Polym. Sci.*, 1981, **26**, 4217.
54. T. Ogawa, S. Tanaka and T. Inaba, *J. Appl. Polym. Sci.*, 1973, **17**, 319.
55. T. Ogawa and M. Sakai, *J. Polym. Sci., Polym. Phys. Ed.*, 1981, **19**, 1377.
56. G. Glöckner and R. Koningsveld, *Makromol. Chem., Rapid. Commun.*, 1983, **4**, 529.
57. G. Glöckner and J. H. M. van den Berg, in 'Book of Abstracts, Tenth International Symposium on Column Liquid Chromatography, San Francisco, 1986', ed. R. E. Majors, contr. 2902.

58. M. Danielewicz and M. Kubin, *J. Appl. Polym. Sci.*, 1981, **26**, 951.
59. G. Glöckner and J. H. M. van den Berg, *J. Chromatogr.*, 1986, **352**, 511.
60. S. Mori, Y. Uno and M. Suzuki, *Anal. Chem.*, 1986, **58**, 303.
61. S. T. Balke and R. D. Patel, *Polym. Prepr., Am. Chem. Soc., Div. Polym. Chem.*, 1981, **22**, 290.
62. S. T. Balke, *Sep. Purif. Methods*, 1982, **11**, 1.
63. S. T. Balke and R. D. Patel, *Adv. Chem. Ser.*, 1983, **203**, 281.
64. G. Glöckner, M. Stickler and W. Wunderlich, *Fresenius' Z. Anal. Chem.*, 1987, **328**, 76.
65. G. Glöckner, M. Stickler and W. Wunderlich, *J. Appl. Polym. Sci.*, in press.
66. E. L. Johnson, R. Gloor and R. E. Majors, *J. Chromatogr.*, 1978, **149**, 571.
67. See, for example, N. Takahashi, Y. Takahashi, T. L. Ortel, J. N. Lozier, N. Ishioka and F. W. Putnam, *J. Chromatogr.*, 1984, **317**, 11.
68. N. Takahashi, N. Ishioka, Y. Takahashi and F. W. Putnam, *J. Chromatogr.*, 1985, **326**, 407.
69. G. W. Welling, G. Groen, K. Slopsema and S. Welling–Wester, *J. Chromatogr.*, 1985, **326**, 173.
70. C. T. Mant and R. S. Hodges, *J. Chromatogr.*, 1985, **326**, 349.
71. G. Glöckner, *Chromatographia*, 1987, **23**, 517.

17

Structure of Chains by Solution NMR Spectroscopy

FRANK. A. BOVEY

AT & T Bell Laboratories, Murray Hill, NJ, USA

17.1   INTRODUCTION

Nuclear magnetic resonance (NMR) spectroscopy is a most effective and significant method for observing the structure and dynamics of polymer chains both in solution and in the solid state. In the solid state, where the motions of the chains are relatively slow, the resonances are broad owing to the local dipolar field at each observed nucleus. This phenomenon of dipolar broadening (see Section 17.2.3.1) tends to abolish all structural information. In solution, where chain motion is fast (Chapter 18), this effect is averaged nearly to zero and one may obtain detailed structural information. This chapter deals with solution NMR spectroscopy. Corresponding information may be obtained—mainly by ^{13}C spectroscopy—for the solid state by 'magic angle' NMR (Chapter 19), although with substantially decreased resolution.

17.2 BASIC PRINCIPLES

The NMR phenomenon is described in a number of text books and review articles[1-8] which should be consulted for a more detailed treatment.

17.2.1 Magnetic Energy Levels and Transitions

When the spin quantum number of a nucleus is non-zero, it possesses a magnetic moment. The proton is such a nucleus. It has a spin I of 1/2 and when placed in a magnetic field of strength B_0 it will occupy $2I + 1$ quantized magnetic energy states, in this case 2. The relative populations of these states, termed Zeeman levels, are normally given by a Boltzmann distribution. In Table 1, the properties of magnetic nuclei of chief interest are listed. The unit of magnetic field strength is the tesla, T, which is equal to 10^4 gauss (the cgs unit). It should be noted that all nuclei with spin greater than 1/2 have non-spherical distributions of positive charge and therefore also possess electric quadrupole moments.

The separation of Zeeman levels is given by

$$\Delta E = h v_0 = \mu B_0 / I \tag{1}$$

For spin 1/2 nuclei, there are two energy levels for which

$$\Delta E = h v_0 = 2\mu B_0 \tag{2}$$

For spin 1 nuclei, there are three energy levels separated by μB_0. Here, μ is the magnetic moment of the nucleus. These magnetic energy levels are represented in Figure 1. Transitions between energy levels will occur if the system is exposed to a radiofrequency (rf) field B_1 of frequency v_0. In Figure 2 equation (2) is plotted for the proton, the resonance frequency being shown for six magnetic field strengths employed in commercial spectrometers. Fields above *ca.* 2.5 T require superconducting solenoid magnets. The resonance condition for any nucleus may be alternatively expressed as

$$\omega_0 = \gamma B_0 \tag{3}$$

where ω_0 is the frequency expressed in radians s^{-1}, equal to $2\pi v_0$, and γ is the magnetogyric ratio, equal to $2\pi\mu/Ih$. The use of magnetogyric ratios, shown in the fifth column of Table 1, has the advantage that resonant frequencies are always directly proportional to γ, regardless of the spin.

We may also view the resonance phenomenon as arising from the fact that when placed in a magnetic field, a nucleus undergoes Larmor precession (a classical effect) about the direction of B_0 at a rate given by v_0 or ω_0. Transitions between energy levels occur when the rf field frequency equals the Larmor frequency.

17.2.2 The Chemical Shift

Equations (1–3) require a modification which is relatively small but highly significant. Resonance actually occurs, even for nuclei of the same species, at slightly different frequencies depending on the

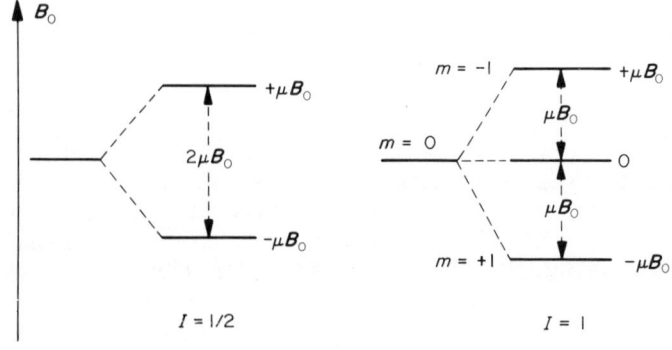

Figure 1   Magnetic energy levels for nuclei of spin 1/2 and spin 1

Table 1 Nuclei of Major Interest for Polymer NMR Spectroscopy

Isotope	Abundance (%)	Resonant frequency in 1 T field	Spin	Magnetogyric ratio γ [a]	Relative sensitivity [b]
^{1}H	99.9844	42.577	1/2	267.43	1.000
^{2}H(D)	0.0156	6.536	1	41.05	0.0096
^{13}C	1.108	10.705	1/2	67.24	0.0159
^{14}N	99.635	3.076	1	19.32	0.0010
^{15}N	0.365	4.315	1/2	−27.102	0.0010
^{19}F	100	40.055	1/2	251.59	0.834
^{29}Si	4.7	8.460	1/2	−53.14	0.0785
^{31}P	100	17.235	1/2	108.25	0.0664

[a] In units of $s^{-1}\,T^{-1} \times 10^{-6}$. [b] For equal numbers of nuclei at constant $\boldsymbol{B}_0$

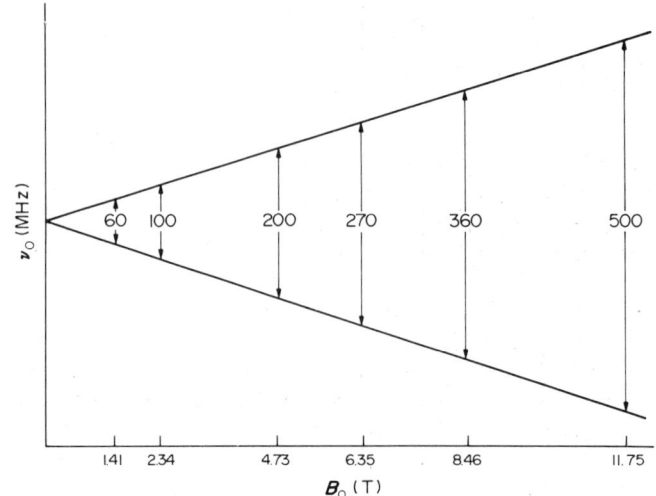

Figure 2 The splitting of magnetic energy levels of protons, expressed as frequency v_0 (in MHz) in varied magnetic field B_0 (in T)

chemical binding and position of the nucleus in the molecule. The cause of this variation is the cloud of electrons about each nucleus, which tends to shield the nucleus against the magnetic field, thus requiring a slightly lower value of v_0 to achieve resonance than for a bare nucleus. Thus, for example, protons attached to or near electronegative groups such as OH, OR, OCOR, CO_2R and halogens experience a lower density of shielding electrons and resonate at higher v_0. Protons removed from such groups, as in hydrocarbon chains, resonate at lower v_0. Similar structural dependence is observed for ^{13}C nuclei. These variations are termed chemical shifts and are commonly expressed in relation to the resonance of tetramethylsilane (TMS) as the zero of reference. The total range of proton chemical shifts in organic compounds is only of the order of 10 p.p.m., *e.g* . *ca.* 1 kHz in a magnetic field of 2.34 T (Figure 2). For ^{13}C nuclei (in common with all other magnetic nuclei), the range is much greater — over 200 p.p.m.; this is the principal reason for the great interest in the study of polymers by ^{13}C NMR.

17.2.3   Nuclear Coupling

17.2.3.1   Direct dipole–dipole coupling

For nuclei so far separated that they do not experience each others' magnetic fields, the local field at each nucleus will be equal to B_0, with a slight correction for electronic screening. But in most solids this condition does not prevail. Each nucleus strongly feels the effect of its magnetic neighbors. Protons are usually the chief contributors to local fields and are sufficiently numerous to have a marked effect not only on proton spectra but on those of other nuclei, of which ^{13}C is the most important. The local field at a ^{13}C nucleus is given by

$$B_{loc} = \pm \frac{h}{4\pi} \gamma_H \frac{(3\cos^2\theta_{C-H} - 1)}{r_{C-H}^3} \tag{4}$$

The $\pm$ sign results from the fact that the local field may add to or subtract from B_0 depending on whether the neighboring proton dipole is aligned with or against the direction of B_0, an almost equal probability. The angle θ_{C-H} is defined in Figure 3; r is the $^{13}C-^1H$ internuclear distance. If r and θ were fixed throughout the sample this interaction would result in a splitting of the ^{13}C resonance into two equal components the separation of which would depend on the orientation of the sample in the magnetic field. In actual polymer samples, which are glassy or microcrystalline, one obtains a summation over many values of θ and r, resulting in a dipolar broadening of many kHz, sufficient to mask all chemical shift information. As the rate of molecular motion begins to exceed the linewidth, the resonance begins to narrow. When the reorienting C–H vectors sample all angles θ_{C-H} in a time short compared to the dipolar coupling (expressed as a frequency) dipolar broadening is reduced to a small value because the time average of the term in parentheses in equation (4) may be replaced by

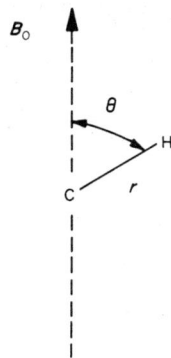

Figure 3 An internuclear C–H vector of length r, making an angle θ with the magnetic field

a space average

$$\int_0^\pi (3\cos^2\theta_{C-H} - 1)\sin\theta_{C-H}\,d\theta_{C-H} = 0 \qquad (5)$$

This condition normally holds in solution. The study and interpretation of dipolar broadened spectra yield important insights concerning molecular motion in solids and are discussed in Chapter 19.

17.2.3.2 Indirect nuclear coupling

Magnetic nuclei may transmit information to each other concerning their spin states not only directly through space but also through the intervening covalent bonds. Rapid tumbling of the molecule does not reduce this interaction to zero. If a nucleus has n sufficiently close, equivalently coupled spin 1/2 neighbors, its resonance will be split into $n+1$ peaks, corresponding to the $n+1$ spin states of the neighboring group of spins. Intensities are given by simple statistical considerations and are therefore proportional to the coefficients of the binomial expansion. Thus, one neighboring spin splits the observed resonance to a doublet, two produce a $1:2:1$ triplet, three a $1:3:3:1$ quartet and so on. The strength of the coupling is denoted by J and is expressed in Hz. The coupling of protons on adjacent saturated carbon atoms, termed vicinal coupling, varies with the dihedral angle ϕ. In the conformations (**1a**) and (**1b**), the *trans* couplings (**1a**), where $\phi = 180°$, are substantially larger than the *gauche* couplings (**1b**), where $\phi = 60°$. J_{gauche} is typically 2–4 Hz and J_{trans} ranges from 8 to 13 Hz. These considerations are of great importance in studying the conformations of polymer chains. The dependence of J on ϕ is generally well described by the so-called Karplus relationship, shown in Figure 4, although the magnitude of the coupling depends somewhat on the nature of the substituents on the bonded carbon atoms.

(**1a**)   (**1b**)

Nuclear coupling between ^{13}C nuclei and directly bonded protons is strong (125–250 Hz). The resulting multiplicity is often helpful in making resonance assignments but does not directly supply information concerning molecular conformation.

17.2.4 Nuclear Relaxation

The separation of Zeeman levels is very small — less than 0.1 J mol^{-1} — and consequently a nuclear spin system is readily disturbed from its equilibrium state. The energy level populations may be equalized by an appropriate pulse of rf energy and may also be inverted. The process of returning

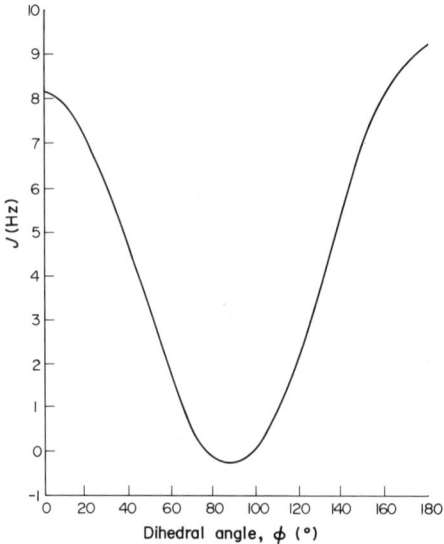

Figure 4   *J* coupling of two vicinal protons, in Hz, as a function of dihedral angle ϕ

from a disturbed state to equilibrium is termed spin–lattice relaxation, characterized by a time T_1. The 'lattice' refers to the other spins in the system and is a term used for both fluids and solids. The thermal link between the observed spin and the lattice is provided by molecular motion. For the relaxation of protons and ^{13}C nuclei in polymers, the only significant mechanism is dipole–dipole interaction with neighboring magnetic nuclei, principally protons. The motions of these neighboring nuclei give rise to fluctuating magnetic fields which have a broad range of frequencies. To the extent that these motions have a component at the resonant frequency ν_0 they will cause spin lattice relaxation because they induce transitions between the energy levels. This is discussed in detail in Chapter 18.

We have considered in Section 17.2.3 the effect of the static neighboring magnetic dipoles arising from protons in broadening the resonances of observed nuclei. Even though the molecules bearing these protons are static or nearly so, the proton magnetic moment is nevertheless precessing about the $\boldsymbol{B}_0$ direction. This constitutes the right sort of field to induce a transition in a neighboring nucleus precessing at the same frequency. By this means a mutual spin exchange or flip-flop occurs which does not change the overall energy of the system but shortens the lifetime of each spin in a given state. This causes an uncertainty broadening which is of similar magnitude to the dipolar broadening. Both effects are included in the characteristic time T_2, which describes spin–spin relaxation and is taken as the inverse of the linewidth:

$$T_2 = \frac{1}{\pi \delta \nu} \tag{6}$$

Spin–spin exchange and dipolar broadening do not always occur together. For dilute nuclei such as ^{13}C, broadening by local proton dipoles is the principal effect because ^{13}C–^{13}C spin flip-flops do not contribute significantly and ^{13}C–^{1}H flip-flops cannot occur owing to the discrepancy in resonant frequencies.

Another contribution to line broadening is inhomogeneity, $\delta \boldsymbol{B}_0$, in the laboratory field. This may be minimized by careful shimming (homogenization of the magnetic field) and is usually included in the experimental value of spin–spin relaxation, commonly termed T_2^*:

$$1/T_2^* = 1/T_2 + \frac{\gamma \delta \boldsymbol{B}_0}{2} \tag{7}$$

The 'true' T_2 may, however, may be separately measured by 'spin echo' methods, which are discussed in Chapter 19. The measurement of T_1 is discussed in Chapter 18.

17.2.5 Experimental Methods

17.2.5.1 CW and pulsed Fourier transform NMR

In NMR spectrometers of older design the rf field was applied continuously ('CW') and nuclei were brought into resonance by sweeping the magnetic field B_0 in strength through the appropriate range of values. This was done by varying the voltage applied to small sweep coils located on the pole faces. Alternatively, one might sweep the frequency of the rf field B_1. These methods have now been entirely displaced by pulse methods, which are much faster and more efficient, particularly for the observation of rare nuclei such as ^{13}C, which has a natural abundance of only 1.1% (Table 1). The rf field is applied as a pulse of resonant energy at a nominal frequency v_0 and of only a few μs duration. Such a pulse is actually of sufficient power to turn the nuclear magnetization vector M_0 90° from the field direction z into the xy plane, where it may be detected. (In the CW method it is only slightly deflected.) This is shown in Figure 5, (a) through to (c). In this vector diagram, the reference frame must be thought of as rotating at the Larmor precession frequency v_0; the magnetization vector M_0 is thus at rest with respect to the z axis (axis designations are primed, to indicate the rotating frame), but is turned into the y' axis by a brief 90° precession about the axis of the B_1 rf field, x'.

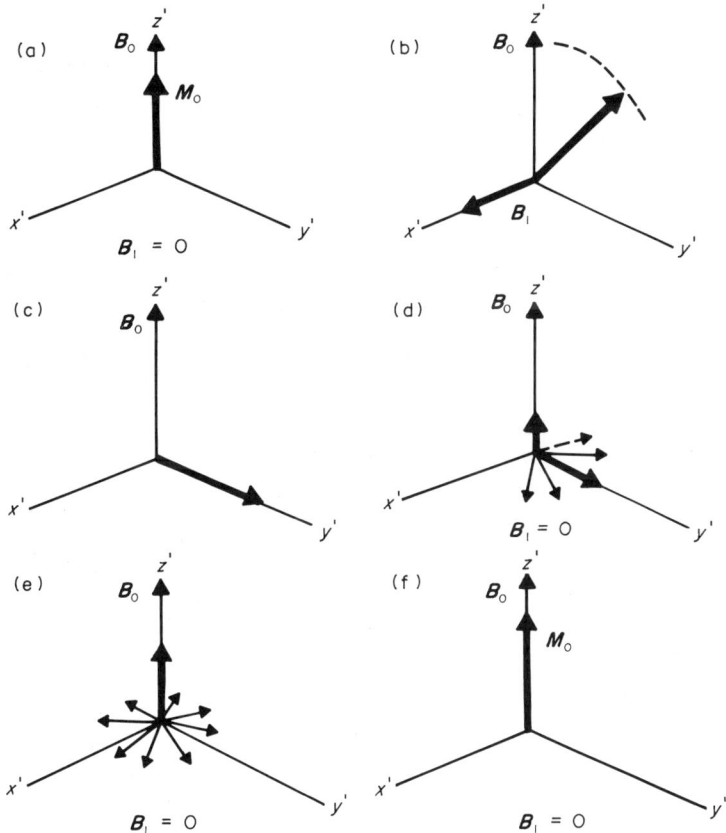

Figure 5 Rotating frame representation of a 90° pulse. Initially the magnetization M_0 is aligned along the magnetic field direction (z'). In (b) and (c) an rf field B_1 is applied along the x' axis, *i.e.* perpendicular to B_0. The duration of the B_1 pulse is sufficient to turn the magnetization vector by 90°; in (d) and (e) the spins begin to dephase in the $x'y'$ plane, corresponding to the spin–spin relaxation process (T_2). The spin–lattice relaxation process (T_1) reestablishes the magnetization along z'. In (f) the equilibrium magnetization has been attained along B_0

After they are turned into the $x'y'$ plane the spin vectors (often termed 'isochromats') relax by the T_1 and T_2 processes we have discussed in Section 17.2.4. The T_2 process is represented as a dephasing of the vectors in the $x'y'$ plane [Figure 5(d) and (e)]; in solids this relaxation is rapid. In solution, the dipole–dipole interactions are averaged out by molecular motion and T_2 may be as long as T_1, or nearly so. Spin–lattice relaxation is shown in Figure 5 as a regrowth of M_0 along the z' axis.

In Figure 6, the pulse is represented in the time domain. Following it, the resonance signal appears as a decaying interference pattern termed a free induction decay or FID. (The envelope of the decay is

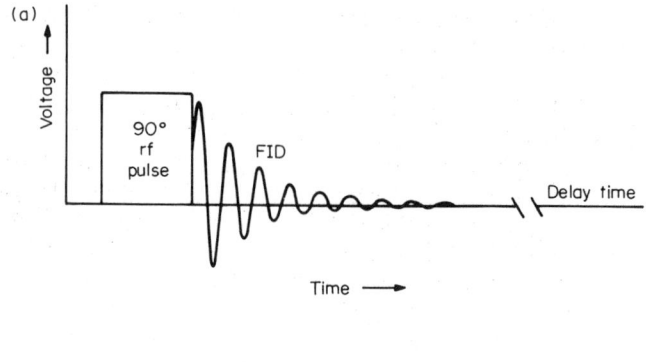

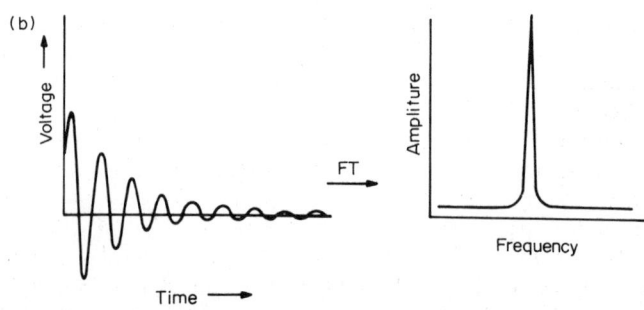

Figure 6   (a) Time domain representation of 90° rf pulse and ensuing free induction decay (FID); (b) Fourier transformation of the time domain FID into the frequency domain NMR signal

a measure of T_2.) This is digitized and stored in a computer memory. For proton signals a single scan may be sufficient, but for ^{13}C and other rare nuclei it may be necessary to accumulate and average several thousand scans, particularly for dilute solutions. The FID, being in the time domain, is difficult to comprehend unless very simple. It is made to undergo a Fourier transformation (FT) to the frequency domain (bottom of Figure 6), a process carried out in the computer of the spectrometer using a stored program.

The rf pulse, although nominally at a single frequency, actually contains a range of frequencies of several thousand Hz, sufficient to cover the chemical shift range of the observed nuclei. Their nuclear moments beat against each other and usually give a much more complex pattern than shown in Figure 6. The frequency range of the pulse depends on its power and duration and is an adjustable parameter.

17.2.5.2   Proton decoupling

We have seen (Section 17.2.3) that ^{13}C nuclei exhibit strong J coupling to directly bonded protons. Unlike proton–proton vicinal coupling, this is not of major structural interest. Spectral simplification may be achieved by removing it and collapsing the multiplets. This is readily done by irradiating the protons with a resonant rf field stronger than that used for proton observation. All carbon spectra shown in the subsequent discussion are obtained with proton decoupling and the carbon resonances therefore appear as singlets.

The decoupling of the protons, termed scalar decoupling, also causes a readjustment of the populations of the ^{13}C energy levels in such a way as to increase the signal strength. The resulting nuclear Overhauser enhancement (NOE) may be as great as three-fold and together with the multiplet collapse results in a striking improvement in signal-to-noise ratio, permitting a corresponding decrease in spectral accumulation time. The quantitative measurement of the NOE is important for chain dynamics studies and is described in Chapter 18.

Irradiating protons in solid polymers with stronger fields than for scalar decoupling can also remove direct dipole–dipole coupling. When combined with magic angle spinning this permits one to obtain fairly well resolved solid state ^{13}C spectra, as we have indicated in the introduction. This is more fully described in Chapter 19.

17.2.5.3 *Advanced techniques*

Certain more advanced NMR methods than those discussed so far have proved useful for the observation of polymer structure in solution. We cannot discuss these in detail. They come under the general heading of multipulse methods.[9,10]

One group of methods is employed for spectral editing, a term referring to the sorting out of resonances in ^{13}C spectra according to whether they have three, two, one or no attached protons. It might seem that this could be most simply done by turning off the proton decoupler and retaining the carbon multiplicities, but in practice this is undesirable. In complex spectra the signals become too weak and the multiplets difficult to discriminate. A better procedure is to apply simultaneous sequences of carefully timed pulses to both the protons and the carbons, designed so that the carbon resonances go to zero, positive or sometimes negative intensities depending on the numbers of attached protons. The resonances become singlets and are enhanced even beyond the normal Overhauser enhancement (except of course for those which are 'edited out' by going to zero). An example is given in Section 17.3.3.3.

Another broader and more important group of methods are those described as two-dimensional or 2D NMR. Here the spectra have two frequency scales at right angles, instead of only one horizontal scale as in 1D NMR. There are two main types of 2D experiments — those that correlate nuclei through interactions based on direct or indirect dipole–dipole interactions and those that resolve chemical shifts and J couplings. All such experiments involve preparation, evolution and detection periods. The direct dipole–dipole correlation experiments also require a mixing time. The mechanics of a 2D experiment are illustrated in Figure 7 for a homonuclear (*e.g.* proton–proton) Overhauser experiment based on direct through-space dipole–dipole interactions.

Using the pulse sequence shown at the upper left in Figure 7, a set of n free induction decays are obtained that differ from one another by equally spaced increments of the evolution time, t_1. It is the systematic incrementing of the evolution time, during which the different spins are labelled by their

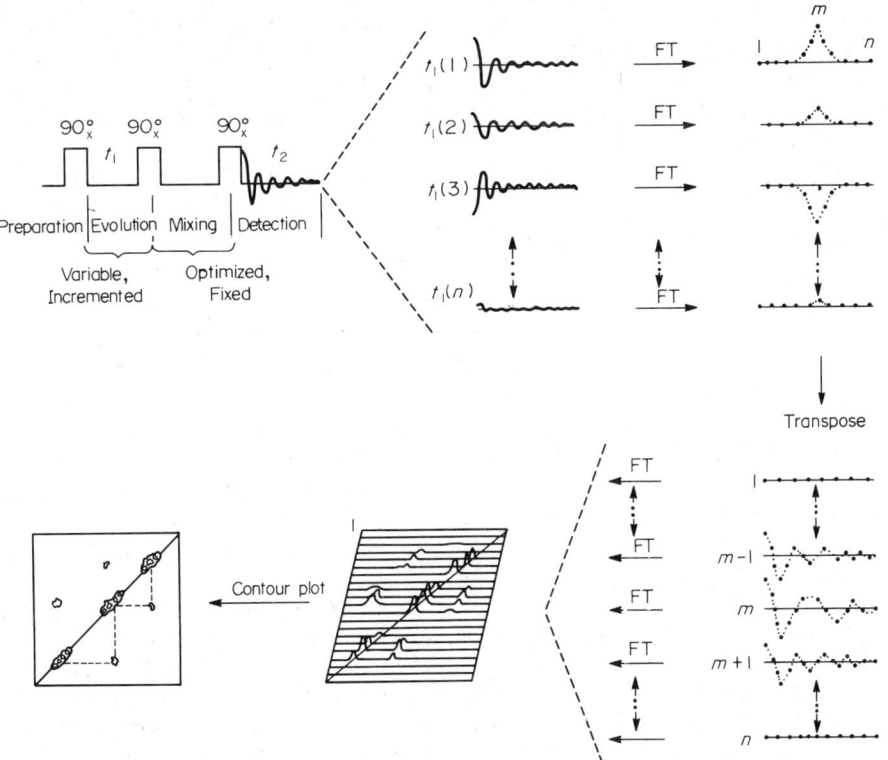

Figure 7 The mechanics of an NOE (nuclear Overhauser enhancement) experiment in the 2D mode

precession frequencies, that gives the experiment its second time dimension and, after Fourier transformation, its second frequency dimension. The transformed spectra are then transposed by constructing new FID's from the columns of points of the original spectra. These new FID's are then themselves Fourier transformed to provide a stacked plot or contour plot (lower left). If a spin were to 'communicate' with its neighbor by exchanging magnetization through dipole–dipole interaction during the mixing time, this would produce an off-diagonal resonance or 'cross peak'. The normal one-dimensional spectrum appears on the diagonal. The data, especially if complex, are most easily visualized on the contour plot, which is in essence a topographical map of the network of spin–spin interactions.

Such methods are of great power and will be illustrated in Section 17.3 for a number of polymers. Useful reviews will be found in refs. 10–12.

17.3 OBSERVATION OF MACROMOLECULAR STRUCTURE

17.3.1 Stereochemical Configuration

NMR is particularly informative concerning stereochemical microstructure, revealing details which can be seen by no other technique. Resolution is in general surprisingly good because linewidth is only weakly dependent on molecular weight or the macroscopic viscosity of the solution.

17.3.1.1 Poly(methyl methacrylate)

In Figure 8 are shown the 500 MHz proton spectra of poly(methyl methacrylate) prepared (a) with a free radical initiator and (b) with fluorenyllithium in toluene, an anionic initiator (see Volume 3, Chapters 1–3). The profound effect of the nature of the initiator is evident in the marked differences between these spectra. To interpret these spectra, let us consider the chain in terms of sequences of two monomer units or dyads. There are two possible types of dyads and they have different symmetry properties. The syndiotactic or racemic (r) dyad has a two-fold axis of symmetry and consequently the two methylene protons are in equivalent environments on a time average over the chain conformations, which are not necessarily in the planar zigzag form shown in the projection formulas (**2a**) and (**2b**). These protons therefore have the same chemical shift and appear as a singlet despite strong two-bond or geminal coupling between them. The isotactic or *meso* (m) dyad has a plane of symmetry but no two-fold axis and so the two protons are non-equivalent and have different chemical shifts. When there is no vicinal coupling to neighboring protons, as is the case in poly(methyl methacrylate), the syndiotactic sequences should exhibit a methylene singlet while the isotactic form should give two doublets, each with a spacing equal to the geminal coupling, *ca.* 15 Hz. We see in Figure 8 that the methylene spectrum of the anionic polymer (b) is almost exclusively a pair of doublets (*ca.* 1.6 and 2.3 p.p.m.); quantitative assessment shows that this polymer is 95% isotactic. The methylene spectrum of the free radical polymer (a) is more complex, but the principal resonance (at *ca.* 1.9 p.p.m.) is a singlet, showing that this polymer is predominantly syndiotactic but more irregular than (b). This is generally the case for vinyl polymers prepared with free radical initiators. (The other resonances are discussed below.)

Planar zigzag conformation

(2a)   (2b)

We see that proton NMR can provide absolute stereochemical information concerning polymer chains without recourse to X-ray or other methods. Somewhat more detailed, but not absolute, information can be gained from the methyl proton resonances near 1.2 p.p.m. (The ester methyl resonances at *ca.* 3.6 p.p.m., which are not shown, are less sensitive to stereochemistry.) In both spectra we note three peaks — or, more correctly, groups of peaks (*vide infra*) — appearing in similar positions but with greatly different intensities. These correspond to the α-methyl groups in the

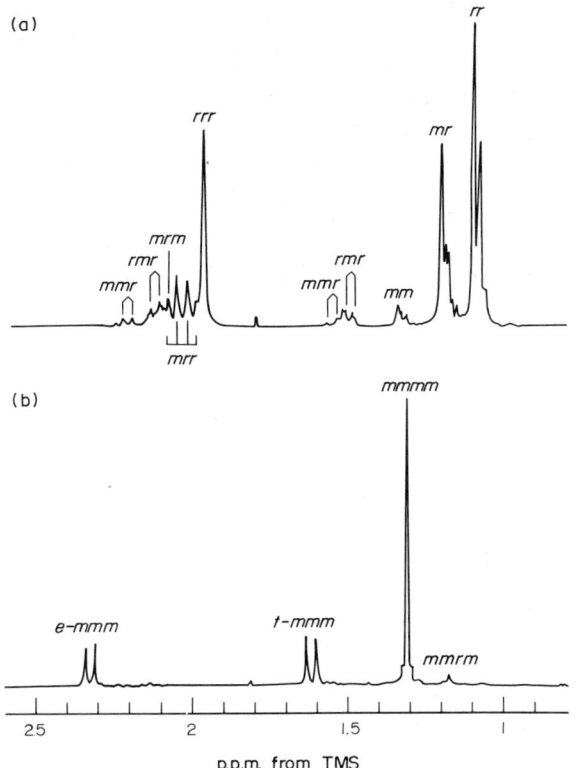

Figure 8 500 MHz proton NMR spectra of (a) predominantly syndiotactic and (b) isotactic poly(methyl methacrylate). The ester methoxyl resonances are not shown[14]

center monomer unit of the three possible triad sequences: isotactic (**3a**), syndiotactic (**3b**) and heterotactic (**3c**). These may be more simply and appropriately designated by the *m* and *r* terminology, as indicated. Measurement of the relative intensities of the *mm*, *mr* and *rr* α-methyl peaks, which appear from left to right in both spectra in this order, gives a valid statistical representation of the structure of each polymer.

(**3a**) Isotactic, *mm*

(**3b**) Syndiotactic, *rr*

(**3c**) Heterotactic, *mr* (or *rm*)

From the triad data one may gain considerable insight into the mechanism of propagation. This is one of the principal uses of such information. Let us designate by P_m the probability that the polymer chain will add a monomer unit to give the same configuration as that of the last unit at its growing end, *i.e.* that an *m* dyad will be generated. We assume that P_m is independent of the stereochemical configuration of the growing chain. The generation of the chain is a Bernoulli-trial process; it is like reaching into a large jar of balls marked *m* and *r* and withdrawing a ball at random. The proportion of *m* balls in the jar is P_m. Since two monomer additions are required to form a triad sequence, it is readily evident that the probabilities of their formation are

$$[mm] = P_m^2 \qquad (8)$$

$$[mr] = 2P_m(1 - P_m) \qquad (9)$$

$$[rr] = (1 - P_m)^2 \qquad (10)$$

The heterotactic sequence must be given double weighting because both directions *mr* and *rm* —
observationally indistinguishable — must be counted. A plot of these relationships is shown in
Figure. 9. It will be noted that the proportion of *mr* (heterotactic) units rises to a maximum when P_m
is 0.5, corresponding to a strictly random or atactic configuration for which $[mm]:[mr]:[rr]$ will be
1:2:1. For any polymer, if Bernoullian, the $[mm]$, $[mr]$ and $[rr]$ sequence intensities will lie on a
single vertical line in Figure 9, corresponding to a single value of P_m. Spectrum (a) in Figure 8
corresponds to these simple statistics, P_m being 0.25 ± 0.01. The polymer corresponding to spectrum
(b) does not. The propagation statistics in this case can be interpreted to indicate that the probability
of isotactic placement is dependent upon the stereochemical configuration of the growing chain and
cannot properly be expressed by a single parameter such as P_m. Free radical and cationic
propagations always give predominantly syndiotactic chains. Anionic initiators may also do so if
strongly complexing ether solvents such as dioxane or glycol dimethyl ether are employed rather
than hydrocarbon solvents as with polymer Figure 8(b).

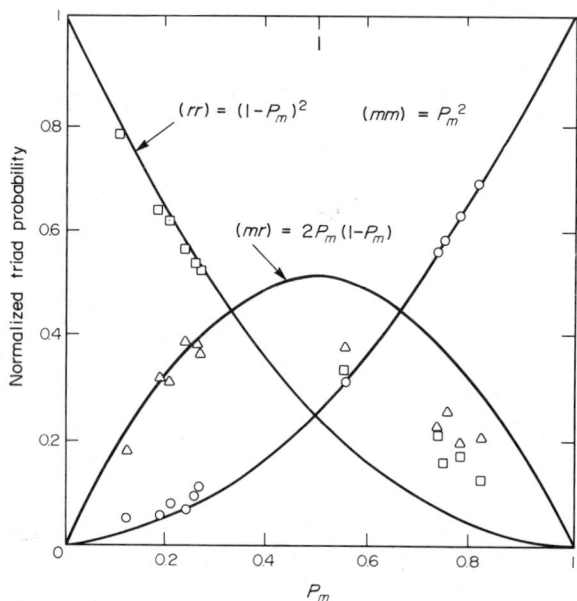

Figure 9 The probabilities of isotactic $[mm]$ heterotactic $[mr]$ and syndiotactic $[rr]$ triads as a function of P_m, the
probability of *m* placement. The points on the left side are for methyl methacrylate polymers prepared with free radical
initiators and those on the right side are for polymers prepared with anionic initiators. For polymer (a) the points for $[rr]$ have
been arbitrarily placed on the $[rr]$ curve and the others fall where they may; for polymer (b), the $[mm]$ points have been placed
on the $[mm]$ curve

It is evident that in the spectra of Figure 8 there is fine structure in both the methylene and methyl
regions which we have not discussed. In spectrum (a) this corresponds principally to residual
resonances of the stereoirregular portions of the chains; in (b) such residual resonances are less
conspicuous. These arise from sensitivity to longer stereochemical sequences than dyad and triad. In
Table 2 are shown planar zigzag projections of such sequences, together with their frequency of
occurrence as a function of P_m, assuming Bernoullian propagation. The tetrads, and all 'even-ads',
refer to observations of β-methylene protons (or β-carbons), while the 'odd-ads' refer to substituents
on the α-carbons (or α-carbons themselves). Resonances for tetrad sequences or higher 'even-ads'
should appear as fine structure in the dyad spectra, while pentad sequences or higher 'odd-ads'
should appear as fine structure on the triad resonances. The assignments to longer sequences as
indicated on the spectra are based on Bernoullian probabilities in spectrum (a); those in spectrum

Table 2 Stereoregular Sequences in Poly(methyl methacrylate)

	Designation	α Substituent Projection	Bernoullian probability		Designation	β-CH₂ Projection	Bernoullian probability
Triad	Isotactic, *mm* (i)		P_m^2	Dyad	*meso, m*		P_m
	Heterotactic, *mr* (h)		$2P_m(1-P_m)$		*racemic, r*		$(1-P_m)$
	Syndiotactic, *rr* (s)		$(1-P_m)^2$	Tetrad	*mmm*		P_m^3
Pentad	*mmmm* (isotactic)		P_m^4		*mmr*		$2P_m^2(1-P_m)$
	mmmr		$2P_m^3(1-P_m)$		*rmr*		$P_m(1-P_m)^2$
	rmmr		$P_m^2(1-P_m)^2$		*mrm*		$P_m^2(1-P_m)$
	mmrm		$2P_m^3(1-P_m)$		*rrm*		$2P_m(1-P_m)^2$
	mmrr		$2P_m^2(1-P_m)^2$		*rrr*		$(1-P_m)^3$
	rmrm (heterotactic)		$2P_m^2(1-P_m)^2$				
	rmrr		$2P_m(1-P_m)^3$				
	mrrm		$P_m^2(1-P_m)^2$				
	rrrm		$2P_m(1-P_m)^3$				
	rrrr (syndiotactic)		$(1-P_m)^4$				

(b) are primarily based on (a). It may be noted that *r*-centered tetrads, *e.g. mrr*, do not necessarily appear as singlets if the sequence as a whole lacks a two-fold axis.

The numbers of observationally distinguishable configurational sequences, or *n*-ads, designated $N(n)$, obey the following relationship

$$
\begin{array}{lcccccccc}
n & 2 & 3 & 4 & 5 & 6 & 7 & 8 & 9 \\
N(n) & 2 & 3 & 6 & 10 & 20 & 36 & 72 & 136
\end{array}
$$

or in general

$$N(n) = 2^{n-2} + 2^{m-1} \tag{11}$$

where $m = n/2$ if n is even and $m = (n-1)/2$ if n is odd. Discrimination of these longer sequences is unlikely to be possible beyond $n = 6$ (hexads) or $n = 7$ (heptads). The observation of such sequences permits rather searching tests of polymerization mechanisms.

Let us now observe isotactic poly(methyl methacrylate) in the 2D proton NMR mode, also at 500 MHz. The spectrum shown in Figure 10[14] is an example of correlated 2D NMR in which the correlation is provided by *J* coupling rather than by through-space dipole–dipole coupling as in the example of Figure 7. In this case, the third 90° pulse is omitted as the scalar interaction is asserted instantaneously without the need for a mixing time. Along the diagonal is the 1D spectrum in contour form. This of itself provides no new information, but the off-diagonal resonances or cross peaks connect *J*-coupled protons. This supplies the whole network of connectivities at a glance. In the present case the cross peaks confirm the geminal coupling in the *meso* methylene dyads. The methylene proton which projects on the same slide of the planar zigzag as the ester groups in the *trans–trans* conformation (4) is designated as *erythro*, the other being *threo*. These doublets are designated as *e* and *t* in Figure 8. Rather unexpectedly, we see a cross peak between the α-methyl resonance and the *erythro* methylene proton but not the *threo* proton. This correlation is brought about by very weak four-bond couplings of the order of 1 Hz and tells us that the preferred conformation of the isotactic chain is *trans–trans*. In this conformation only the *erythro* proton is capable of forming with an α-methyl proton (one-third of the time for each) a 'W'-shaped four-bond path of the kind generally thought to be favorable for such long range coupling. Expressed somewhat differently, the methyl group and the *erythro* proton exhibit in the *trans–trans* conformation the 180° dihedral angle corresponding to maximum coupling according to the

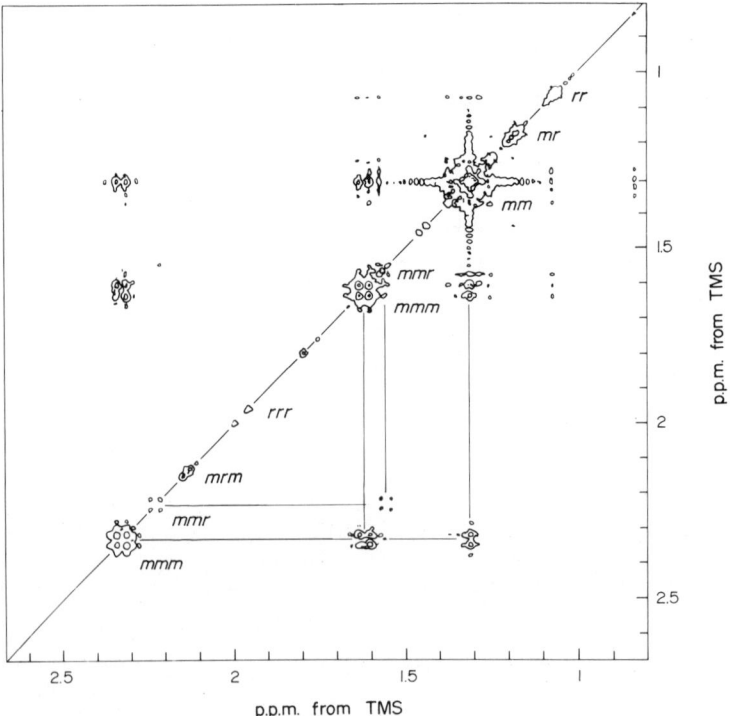

Figure 10   500 MHz proton COSY spectrum of isotactic poly(methyl methacrylate) observed in 10% (w/v) solution in chlorobenzene-d_5 at 100 °C.[14]

correlation of Barfield,[15] whereas the *threo* proton is near a minimum (**5**). This correlation is also observed for methyl methyacrylate oligomers[16] and is in agreement with energy calculations[17-20] indicating a strong *trans* preference.

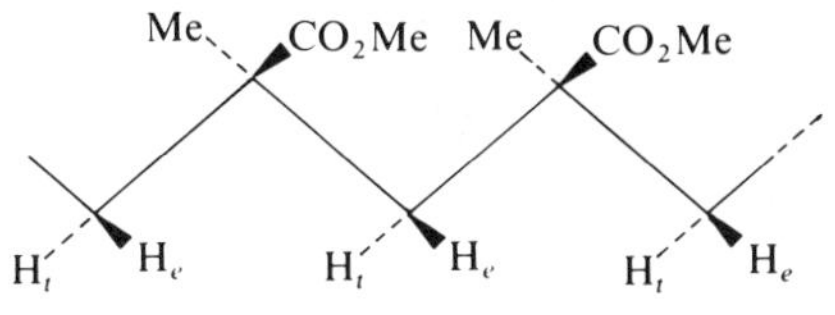

(4)

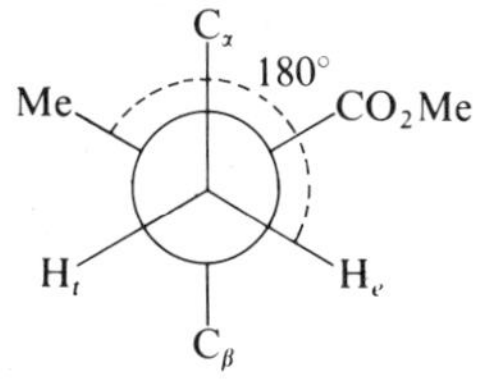

(5)

17.3.1.2 Poly(isopropyl acrylate)

The proton spectrum of poly(methyl methacrylate) is not complicated by vicinal coupling of main chain protons. When such coupling is present the spectra may at first sight appear to be nearly unintelligible. Successful interpretation often requires computer simulation. To do this, one must adopt a spin model that adequately represents a polymer chain without requiring the incorporation of all the protons in a long chain. It has been found that a satisfactory spin model is provided by imagining that the polymer chain is turned back on itself to form a cyclic 'dimer'. This is not to be taken literally but only serves as a spin bookkeeping system. For an isotactic polymer we have the six-spin system shown as (**6a**). For a syndiotactic chain the appropriate model is shown as (**6b**).[21,22]

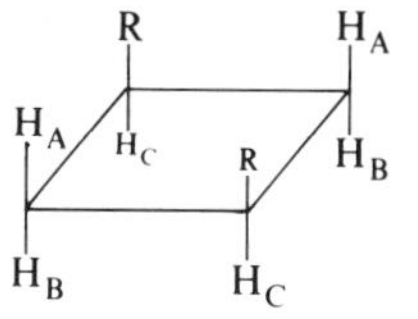

(6a)

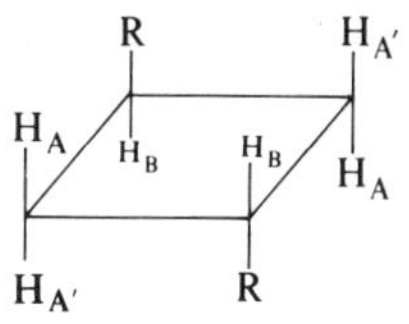

(6b)

In Figure 11(a) is shown the 100 MHz main chain proton spectrum of isotactic poly(isopropyl acrylate) prepared with a Grignard initiator.[23,24] Figure 11(b) shows a computed spectrum matching the observed spectrum and requiring the inclusion of about 5% of syndiotactic sequences (probably *rrm*). The parameters employed are as follows:

Isotactic	Syndiotactic
H_A: 1.69 p.p.m.	H_A: 1.84 p.p.m.
H_B: 2.14 p.p.m.	H_B: 1.58 p.p.m.
H_C: 2.58 p.p.m.	$J_{AB} = -14.0$ Hz
$J_{AB} = -14.0$ Hz	$\frac{1}{2}(J_{AB} + J_{AB'}) = 7.0$ Hz
$J_{AC} = 6.0$ Hz	
$J_{BC} = 7.5$ Hz	

The most significant parameters are the vicinal J_{AC} and J_{BC}; they show the values expected of a *meso* dyad in rapid exchange between $(gt) \leftrightarrows (tg)$ conformers.

17.3.1.3 Polypropylene

The proton spectra of isotactic and syndiotactic polypropylene have been simulated in the way just described for poly(isopropyl acrylate),[25,26] but in general for hydrocarbon polymers ^{13}C NMR is more effective, particularly for the detection of minor but significant structural irregularities. Carbon-13 chemical shifts are very sensitive to chain conformation and therefore to stereochemical configuration because of its strong effect on conformation. We shall expand on this point shortly. First, we see in Figure 12 the 25 MHz ^{13}C spectra of isotactic (top), atactic (middle) and syndiotactic (bottom) polypropylene, observed in 20% 1,2,4-trichlorobenzene solution. The sensitivity to stereochemical configuration is particularly clear in the spectrum of the atactic polymer, in which the methyl resonance is split into peaks corresponding to nine of the ten possible pentad sequences

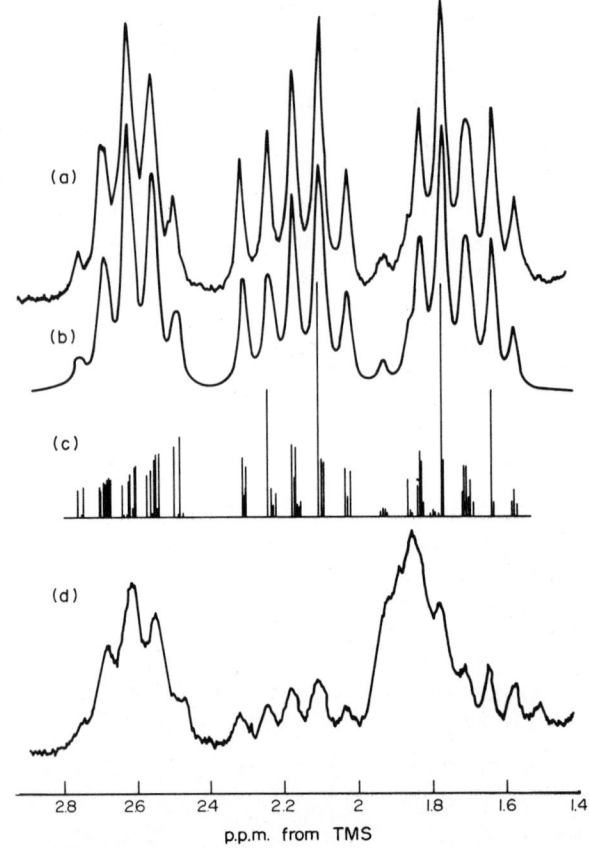

Figure 11 The 100 MHz main chain proton spectrum of predominantly isotactic poly(isopropyl acrylate): (a) observed; (b) calculated with parameters given in the text; (c) 'stick' spectrum corresponding to (b); and (d) observed for 'atactic' polymer. Experimental spectra were obtained using 10% solutions in chlorobenzene at 140 °C.[25]

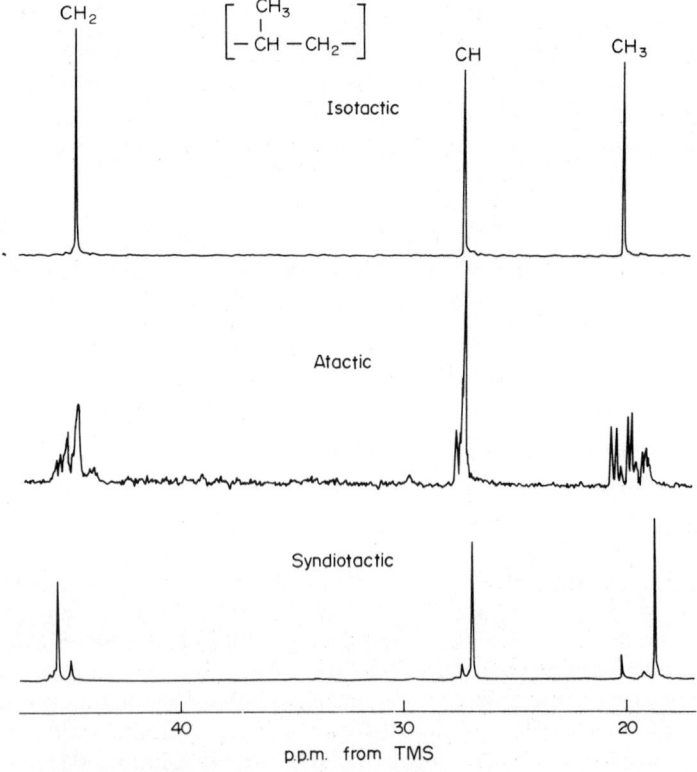

Figure 12 The 25 MHz ^{13}C spectra of three preparations of polypropylene, isotactic, atactic and syndiotactic, observed as 20% (w/v) solutions in 1,2,4-trichlorobenzene at 140 °C (F. C. Schilling, personal communication)

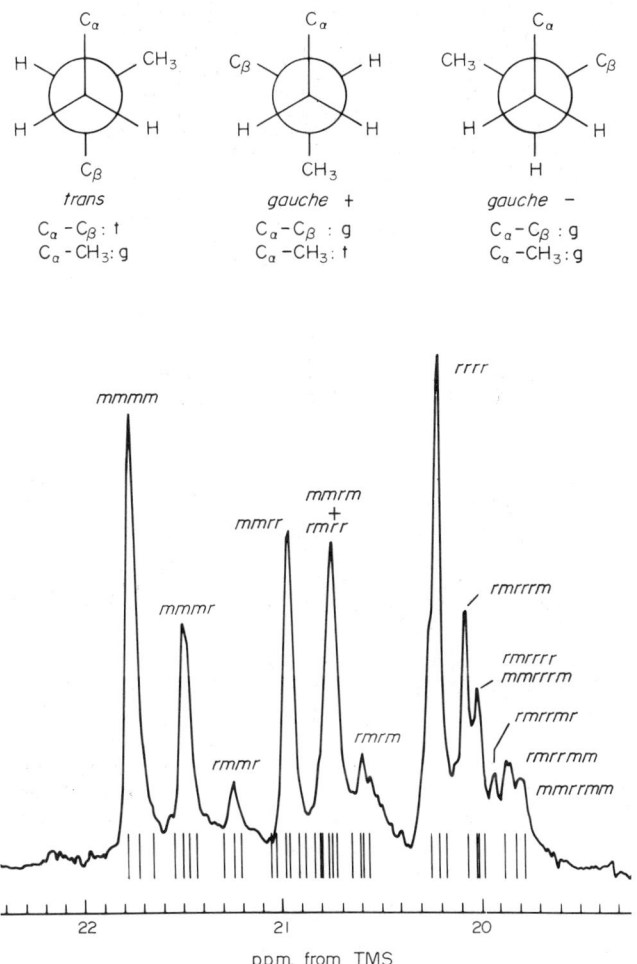

Figure 13 Top: conformations of a fragment of a polypropylene chain. Bottom: 90 MHz ^{13}C NMR spectrum of the methyl region of atactic polypropylene. The 'stick' spectrum shows the RIS-predicted chemical shifts for the 36 methyl heptad sequences, based on the *γ-gauche* effect[27]

(Table 2). It may be observed in Figure 13 that at 90 MHz, heptad chemical shifts can be partially resolved in the methyl region of the atactic polymer spectrum.[27]

It is also noteworthy that the syndiotactic polymer is less stereoregular than the isotactic one and that the configurational statistics of both the atactic and syndiotactic polymers depart markedly from Bernoullian. This is generally the case for chains generated by coordination catalysts (see Volume 4).

Carbon-13 chemical shifts can be effectively rationalized through recognition of a *γ-gauche* shielding effect. This treatment is based on the observation that when two carbons separated by three bonds are *gauche* to each other, they shield each other by *ca.* 5 p.p.m. compared to the chemical shifts of the corresponding *trans* conformation. In Figure 13 (top) we see that in polypropylene the methyl group is *gauche* to the methine carbon (C_a) when the chain is *trans* or *gauche−* but not when it is *gauche+*. Thus, the methyl group experiences differing numbers of *gauche* interactions depending upon the local chain conformation. The methyl carbon chemical shifts can be accurately predicted for all 36 heptad configurational sequences (Figure 13 bottom) from the theoretically calculated populations of *gauche* and *trans* states of the main chain bonds flanking the methine carbon in each sequence.[27] The C_α and C_β chemical shifts may be calculated as a function of stereochemical configuration in the same way. The treatment has been successfully extended to a wide variety of other polymers, and solid state ^{13}C spectra as well (Chapter 19).

17.3.1.4 *Polystyrene*

The earliest reported spectra of a synthetic polymer were of polystyrene.[28-30] At the low frequency then employed (40 MHz) the *meso* methylene protons of the isotactic polymer did not exhibit the expected differentiation of chemical shift. This ambiguity is resolved at higher frequency; at 220 MHz the proton spectrum is consistent only with a differentiation of 0.059 p.p.m.[31] In the ^{13}C spectra a marked dependence on stereochemical configuration is shown by the aromatic C_1 (quaternary) carbon and the main chain β-carbon.[32-35] In Figure 14 are shown the 270 MHz spectra of the main chain protons of isotactic, atactic and syndiotactic polystyrene.[35] The more complex multiplet of the isotactic methylene protons confirms the earlier observation[31] and is in contrast to the triplet resonance of the syndiotactic polymer. (The synthesis of the latter was first reported by Ishihara *et al.*)[35] The atactic polymer exhibits seemingly poorly resolved multiplets because of the many overlapping resonances of tetrad sequences; these have a syndiotactic preference. In Figure 15 are shown the 67.8 MHz spectra of the C_1 carbon of these three polymers. The C_1 carbon of the syndiotactic polymers is shifted upfield by *ca.* 1 p.p.m. from that of the isotactic polymer while the atactic polymer exhibits a band of resonances again consistent with a syndiotactic

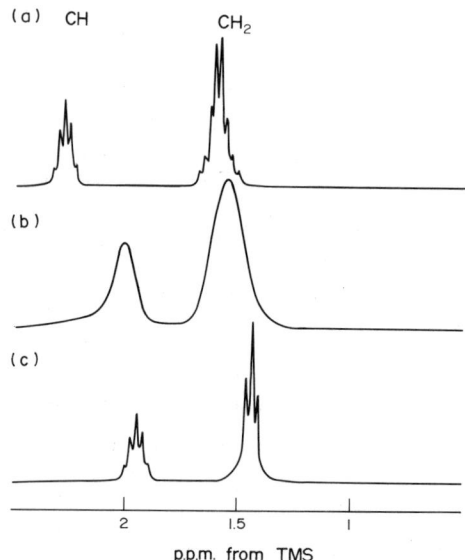

Figure 14 The 270 MHz spectra of main chain protons of three forms of polystyrene: (a) isotactic; (b) atactic; and (c) syndiotactic[35]

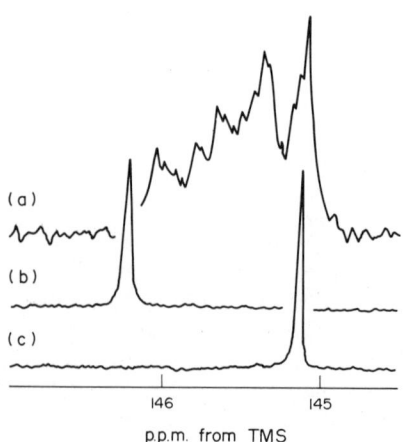

Figure 15 The 67.8 MHz ^{13}C spectra of three forms of polystyrene: (a) atactic; (b) isotactic; and (c) syndiotactic[35]

bias in free radical propagation. This syndiotactic bias characterizes the free radical polymerization of all known vinyl monomers.

17.3.1.5 *Poly(vinyl chloride)*

The proton spectrum of poly(vinyl chloride) presents a fairly complicated interpretive problem[36] similar to those of poly(isopropyl acrylate) and polypropylene but with the added difficulty' that unlike them this polymer cannot be prepared in stereoregular form, although the free radical polymerization becomes increasingly biased toward syndiotactic as the temperature is decreased. The ^{13}C spectrum,[27,37-42] shown in Figure 16, is much more readily interpreted and does not require elaborate computer simulation. Peak assignments are based primarily on expected Bernoullian intensities for a P_m of 0.45. The α-carbon resonance positions are insensitive to the choice of solvent and are well predicted by the *γ-gauche* model discussed in Section 17.3.1.3, whereas the β-carbon chemical shifts show marked solvent dependence, sufficient to reverse some of their relative positions.[27] The reason of this behavior, not exhibited by most polymers, is unclear.

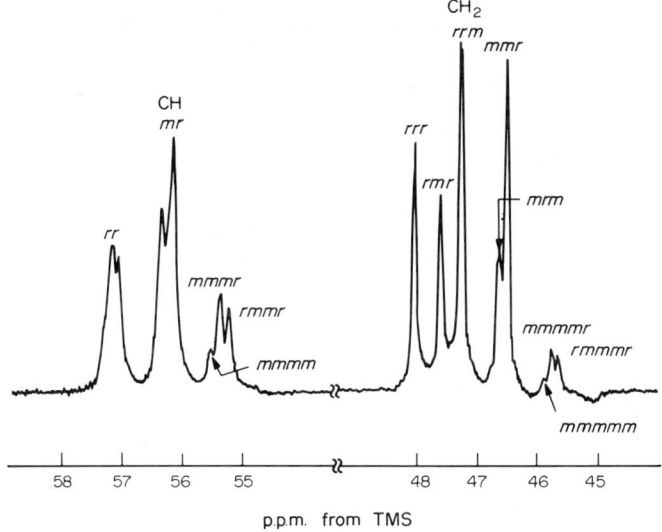

Figure 16 The 25 MHz ^{13}C spectrum of poly(vinyl chloride), observed in 1,2,4-trichlorobenzene at 120 °C.[27]

17.3.2 Geometrical Isomerism

Geometrical isomerism in diene polymers may be readily observed and measured by NMR. In addition to the *cis* ('Z') and *trans* ('E') structures found in natural isoprene polymers, the formation of such polymers by chain propagation of diene monomers may proceed also by incorporation of the monomer through one double bond rather than by 1,4 addition. Thus from butadiene the isomeric chains (**7a–d**) may be produced in pure form by appropriate choice of coordination catalysts.

(**7a**) *trans*-1,4

(**7b**) *cis*-1,4

(**7c**) Isotactic 1,2

(**7d**) Syndiotactic 1,2

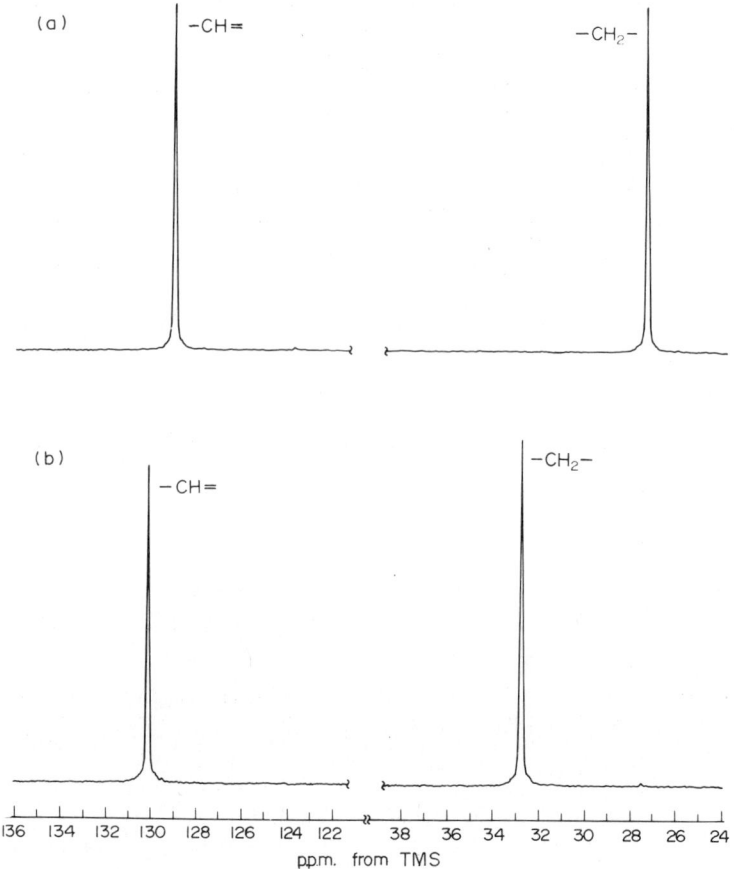

Figure 17   50.3 MHz ^{13}C spectra of (a) *cis*-1,4-polybutadiene; and (b) *trans*-1,4-polybutadiene, observed in CDCl$_3$ at 40 °C
(F. C. Schilling, personal communication)

17.3.2.1 *Polybutadiene*

Proton NMR is moderately effective in the observation of geometrical isomerism but ^{13}C NMR is much more powerful. In Figure 17 are shown the ^{13}C spectra at 50.3 MHz of (a) *cis*-1,4-polybuta-diene and (b) *trans*-1,4-polybutadiene, observed in deuterochloroform solution.[43] The alkenic carbons are only moderately sensitive to geometrical isomerism but the methylene carbons are highly sensitive, being substantially more shielded (by *ca.* 8 p.p.m.) in the *cis* polymer, an effect no doubt closely related to the *γ-gauche* shielding phenomenon (Section 17.3.1.3).

Chains of mixed structure exhibit more complex spectra because of sequence effects. In Figure 18 is shown the 50.3 MHz ^{13}C spectrum of a polybutadiene produced by free radical initiation.[44] At the left (a) is the region of the alkenic carbon resonances, not fully analyzed. The alkenic carbon singlets of the pendant vinyl groups flank those of the 1,4 units. At the right (b) are the aliphatic carbon resonances — mainly 1,4 (and 1,2) methylene groups. This part of the spectrum is shown at two values of gain — 1 × and 5 × — to show the small resonances of the sequences containing 1,2 units. Major peak b corresponds to central methylenes in *cis–cis* units; principal peak d is that of the central 1,4 unit methylene group in *trans–trans* and *trans–cis* units, not discriminated. Peaks a, c, e and m correspond to sequences involving 1,4 units and one 1,2 unit, while the very small resonances f through 1 represent sequences containing two 1,2 units. The overall composition of the polymer is 23% *cis*-1,4, 58% *trans*-1,4 and 19% 1,2. Spectrum (c) is a computer simulation of (b) based on the assumption of a random distribution of units in these proportions. The satisfactory fit shows that free radical propagation in butadiene polymerization is a Bernoullian process with regard to the generation of these isomeric structures.

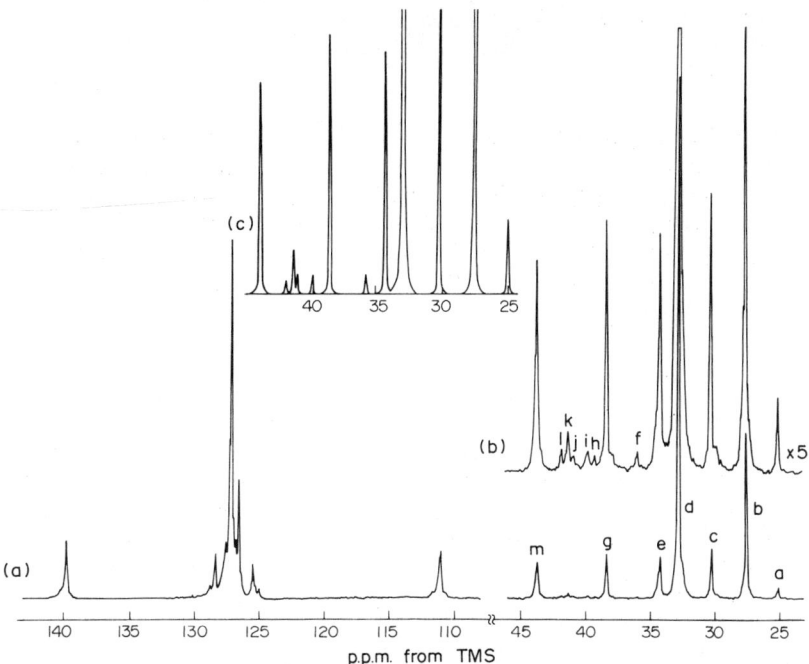

Figure 18 The 50.3 MHz ^{13}C spectrum of free radical polybutadiene, observed at 50 °C in 20% (w/v) solution in CDCl$_3$. (a) Alkenic carbon spectrum; (b) aliphatic carbon spectrum; (c) computer simulation of (b) based on a Bernoullian distribution of *cis*, *trans* and 1,2 units in the proportion of 23:58:19. (L. W. Jelinski, personal communication)

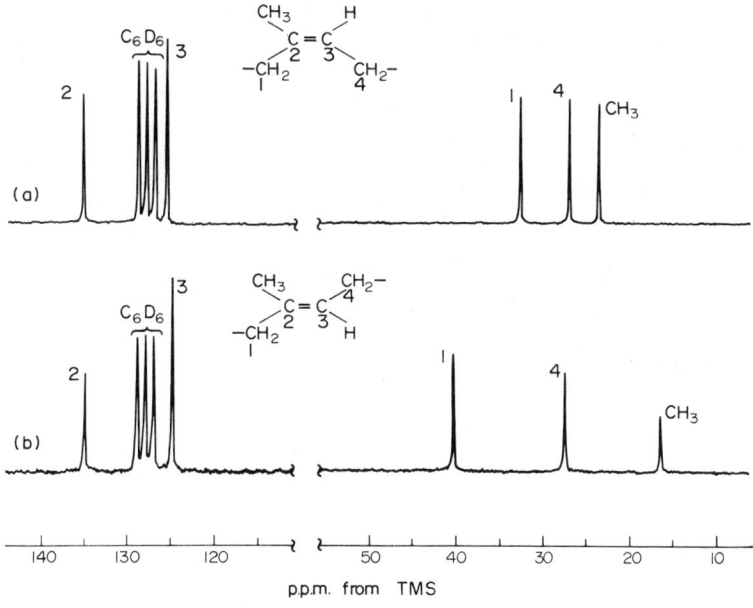

Figure 19 The 50.3 MHz ^{13}C spectra of (a) *cis*- and (b) *trans*-1,4-polyisoprene, observed in 10% (w/v) solution in C$_6$D$_6$ at 60 °C (F. C. Schilling, personal communication)

17.3.2.2 *Polyisoprene*

In Figure 19 are shown 50.3 MHz ^{13}C spectra of *cis*-1,4-polyisoprene and *trans*-1,4-polyisoprene, observed in C$_6$D$_6$ solution.[43] Again, the alkenic carbons are relatively insensitive to isomerism at the double bond but the CH$_3$ and CH$_2$–1 carbons markedly shield each other in a *cis* arrangement compared to the *trans*. The CH$_2$–4 carbon, *cis* to a carbon atom in both isomers, is less affected.

Spectrum (a) is that of natural rubber or *Hevea brasiliensis*. The biochemical pathway to natural rubber is an enzymatic process in which isoprene as such plays no part. The polymer is highly stereoregular, no trace of the *trans* structure being observable. Synthetic *cis*-1,4-polyisoprene is produced commercially using lithium alkyls or Ziegler–Natta catalysts (see Volume 4, Chapters 3 and 6). It contains 2–6% of *trans* units.

17.3.3 Regioregularity

Unsymmetrically substituted vinyl monomers may in principle propagate in either a head-to-tail (**8a**) or head-to-head:tail-to-tail mode (**8b**). In fact, it is observed that (**8a**) is overwhelmingly preferred and the occurrence of mode (**8b**) is reduced to an occasional inverted unit (**8c**). Head-to-head:tail-to-tail inversions may also occur in ring-opening polymerizations; see for example (**8d**).

$$\cdots-CH_2\overset{R}{C}HCH_2\overset{R}{C}HCH_2\overset{R}{C}HCH_2\overset{R}{C}H-\cdots \qquad \cdots-CH_2\overset{R}{C}H\overset{R}{C}HCH_2CH_2\overset{R}{C}H\overset{R}{C}HCH_2-\cdots$$

(**8a**)    (**8b**)

$$\cdots-CH_2\overset{R}{C}HCH_2\overset{R}{C}H\overset{R}{C}HCH_2CH_2\overset{R}{C}HCH_2\overset{R}{C}H-\cdots$$

(**8c**)

$$R-\overset{\diagup}{C}H-CH_2 \longrightarrow \cdots-\overset{R}{C}HCH_2O\overset{R}{C}HCH_2O CH_2\overset{R}{C}HO\overset{R}{C}HCH_2O-\cdots$$
$$\underset{O}{}$$

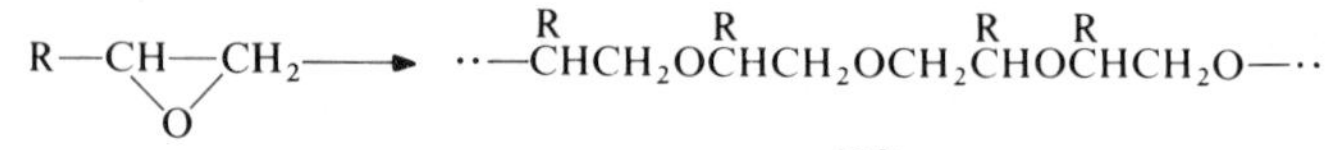

(**8d**)

Fluorine substituted ethylenes are particularly subject to the generation of inverted units because fluorine atoms are relatively undemanding sterically. The physical properties of the polymers may show a marked dependence on the presence of such units, particularly the extent of crystallinity. The presence of ^{19}F offers an effective additional means of detailed study because it is the most readily observed nucleus after the proton (see Table 1) and exhibits a large range of structure–sensitive chemical shifts.

17.3.3.1 Poly(vinyl fluoride)

Poly(vinyl fluoride)—a commercial plastic having high resistance to weathering—has a substantial proportion of head-to-head units.[45,46] The 188 MHz ^{19}F spectrum of a commercial polymer is shown as (a) in Figure 20. The ^{19}F–^1H *J* coupling multiplicity has been removed by proton irradiation just as in ^{13}C spectroscopy. The resonance assignments, designated by capital letters, are shown in (**9**). The stereochemical assignments are also indicated in Figure 20. The *m* and *r* designations that are not underlined (**10a** and **10b**) represent the usual relationships between substituents in 1,3 positions (in planar zigzag projection). The underlined designations (**10c** and **10d**) represent the substituents in 1,2 positions (also in planar zigzag projection). It may be further noted in connection with B resonances that *rm* and *mr* are quite different structures and do not differ merely in direction as with *rm* and *mr*. The fraction of inverted units is *ca.* 11%.

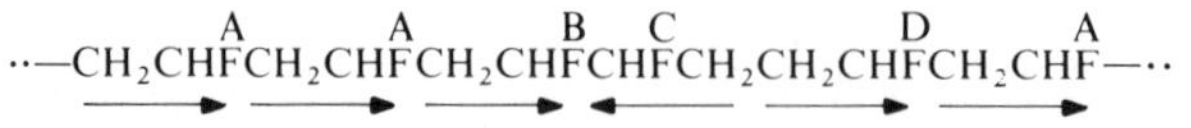

$$\cdots-CH_2\overset{A}{CHF}CH_2\overset{A}{CHF}CH_2\overset{B}{CHF}\overset{C}{CHF}CH_2CH_2\overset{D}{CHF}CH_2\overset{A}{CHF}-\cdots$$

(**9**)

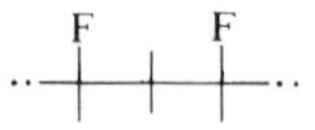

(**10a**) *m* 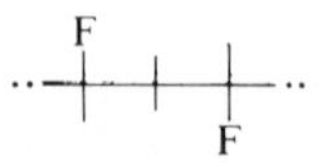   (**10b**) *r* 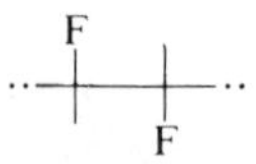   (**10c**) *m̲* 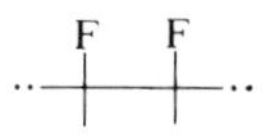   (**10d**) *r̲*

Spectrum (b) is that of a poly(vinyl fluoride) prepared by the following route:[46]

$$CH_2{=}CFCl \xrightarrow[\text{initiator}]{\text{free radical}} (CH_2{-}CFCl)_n \xrightarrow{Bu_3SnH} (CH_2{-}CHF)_n$$

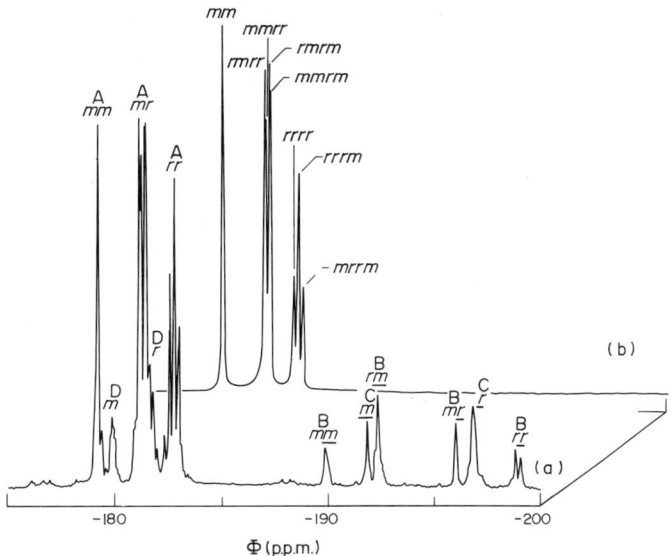

Figure 20 The 188 MHz ^{19}F spectrum of (a) commercial poly(vinyl fluoride); and (b) poly(vinyl fluoride) prepared by reductive dechlorination of poly(1-fluoro-1-chloroethylene); both spectra observed as 8% solutions in *N,N*-dimethylformamide-d_7 at 130 °C.[46]

The steric requirements of the chlorine atom prevent head-to-head addition. The tri-*n*-butyltin hydride removes the chlorine without affecting the fluorine, and the resulting poly(vinyl fluoride) is defect-free. This is shown by spectrum (b) in which all but A resonances have disappeared.

The spectral fine structure which remains is due to stereochemical configuration. Assignments to configurational sequences are not easy to make in this case because there is little or no stereochemical bias in the free radical chain growth ($P_m \simeq 0.48$) and consequently statistical considerations are not helpful. Two-dimensional ^{19}F NMR has enabled detailed assignments to pentad sequences to be made.[47] Correlation of ^{19}F resonances is provided by scalar couplings between fluorines occupying central positions in pentad sequences which share a common hexad. For example, in the hexad sequence (**11**) four-bond ^{19}F–^{19}F coupling between the starred fluorines results in cross peaks connecting the *rmmr* and *mmrm* resonances in the ^{19}F spectrum. This coupling has a magnitude of about 7–8 Hz and is too small to be manifested as splittings. Nevertheless, this coupling and corresponding couplings involving other pentad sequences provide the necessary correlations to assign the whole spectrum.

$$
\begin{array}{ccccccc}
\text{H} & \text{F} & \text{F}^* & \text{F}^* & \text{H} & \text{H} \\
-\text{CCH}_2\text{CCH}_2\text{CCH}_2\text{CCH}_2\text{CCH}_2\text{C}- \\
\text{F} & \text{H} & \text{H} & \text{H} & \text{F} & \text{F}
\end{array}
$$

(**11**) *r* *m* *m* *r* *m*

17.3.3.2 *Poly(vinylidene fluoride)*

Poly(vinylidene fluoride) has long been known but has been discovered only fairly recently to have unusual electrical properties, being both piezoelectric and pyroelectric. The presence of inverted units was observed long ago by Wilson[45,48] and has been more recently confirmed by others.[49-52] It contains 3–6% of the units shown in (**12**).

$$
\begin{array}{ccccc}
& \text{A} & \text{C} \quad \text{D} & \text{B} \\
\cdots-\text{CH}_2\text{CF}_2\text{CH}_2\text{CF}_2\text{CF}_2\text{CH}_2\text{CH}_2\text{CF}_2\text{CH}_2\text{CF}_2-\cdots
\end{array}
$$

(**12**)

The principal defect resonances, shown in the 188 MHz ^{19}F spectrum (Figure 21), are assigned as indicated. Interpretation is somewhat simplified by the absence of asymmetric carbons. In the original work of Wilson[45, 48] the assignments were made on the basis of model compounds. Cais and Kometani[52] have taken a different approach by increasing the number of inversions by the route shown in equation (12).

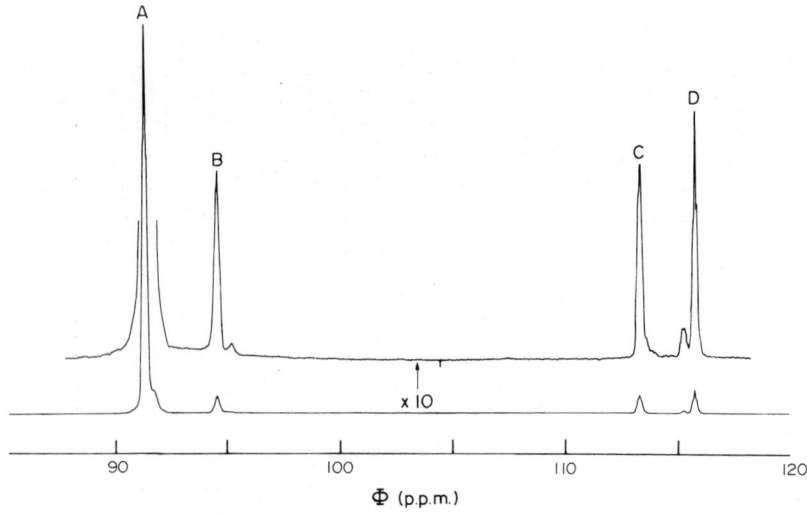

By introducing them in a defined manner their nature is made clear. In Figure 22 are shown the proton-decoupled 470 MHz ^{19}F spectra of polymers containing 3.5 (commercial product, normally prepared), 15 and 23% defects. Because of the higher observing frequency and the increased proportion of inversions, many more resonances can now be resolved. The heptad assignments, established in part by ^{19}F 2D NMR, similar to that employed for poly(vinyl fluoride) (previous section), are expressed in a symbolic form in which 0 stands for CH_2 and 2 stands for CF_2.[52]

Figure 21 The 188 MHz ^{19}F spectrum of commercial poly(vinylidene fluoride) observed in 11% solution in *N,N*-dimethylformamide-d_7 at 27 °C.[51]

17.3.3.3 *Poly(propylene oxide)*

Unlike the vinyl and diene polymers we have considered in previous discussion propylene oxide (13) is truly chiral and its polymers have true asymmetric centres.[53, 54]

(13)

Provided polymerization consistently occurs by cleavage of only one of the C–O bonds, the polymer chain will be of head-to-tail sequence. Unless the monomer is purely *R* or *S*, the chain will in principle contain four types of stereochemical triads; the mirror image structures cannot be discriminated by NMR. It is noteworthy that there are two distinct types of heterotactic sequences (**14a–d**).

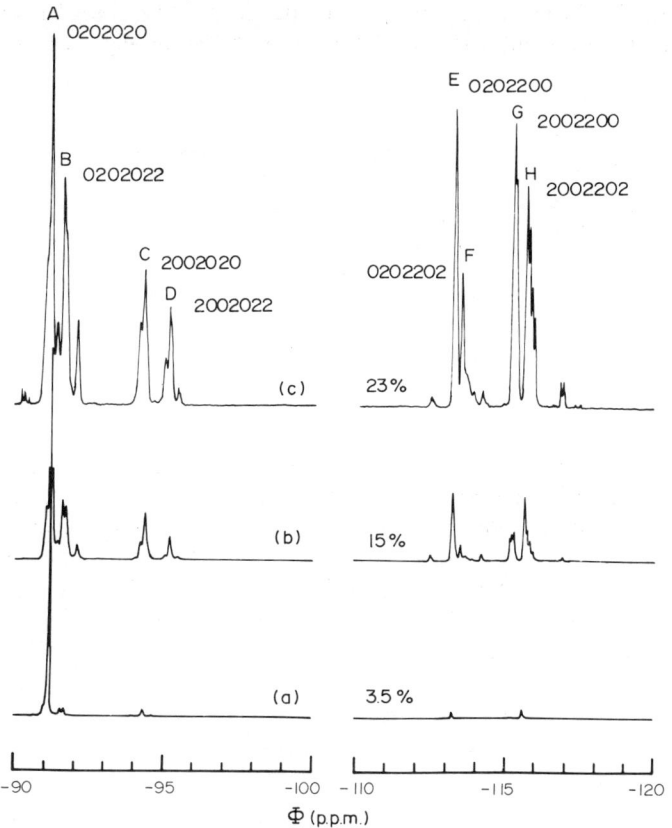

Figure 22 470 MHz ^{19}F spectra of poly(vinylidene fluoride) containing varied proportions of defects, *i.e.* monomer reversals. Spectra observed at 25 °C on 10% solutions in *N,N*-dimethylformamide-d_7. Spectra (a), (b) and (c) are of a sample containing 3.5% (commercial), 15% and 23% defects, respectively. Heptad sequences are designated by a code in which $0 = CH_2$ and $2 = CF_2$. The spectra are vertically scaled so that the peak for sequence A is the same in each.[52]

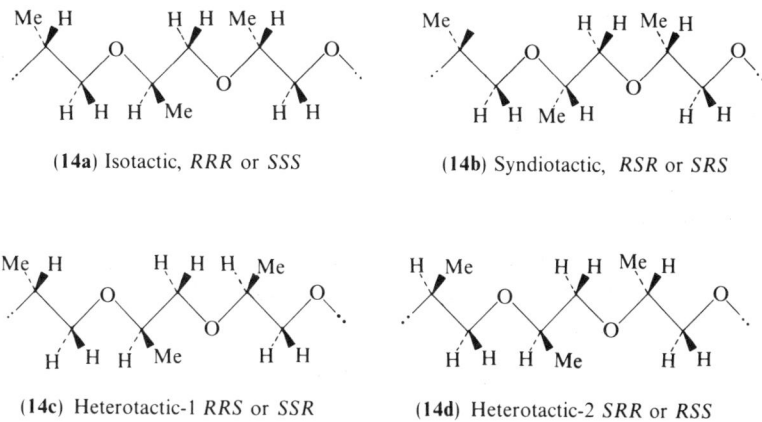

(**14a**) Isotactic, *RRR* or *SSS*   (**14b**) Syndiotactic, *RSR* or *SRS*

(**14c**) Heterotactic-1 *RRS* or *SSR*   (**14d**) Heterotactic-2 *SRR* or *RSS*

If both C–O bonds are cleaved during polymerization, then in addition to the head-to-tail sequences the head-to-head : tail-to-tail triads [represented for *R* chirality in (**15a**)–(**15c**)] will be formed. If both *R* and *S* chiralities may occur, then 12 different such sequences may be produced, or a total of 16 with the head-to-tail triads above. Thus, if propagation is neither stereospecific nor regiospecific the structure of the chain may be quite complex. This is reflected in the NMR spectra.[55–64]

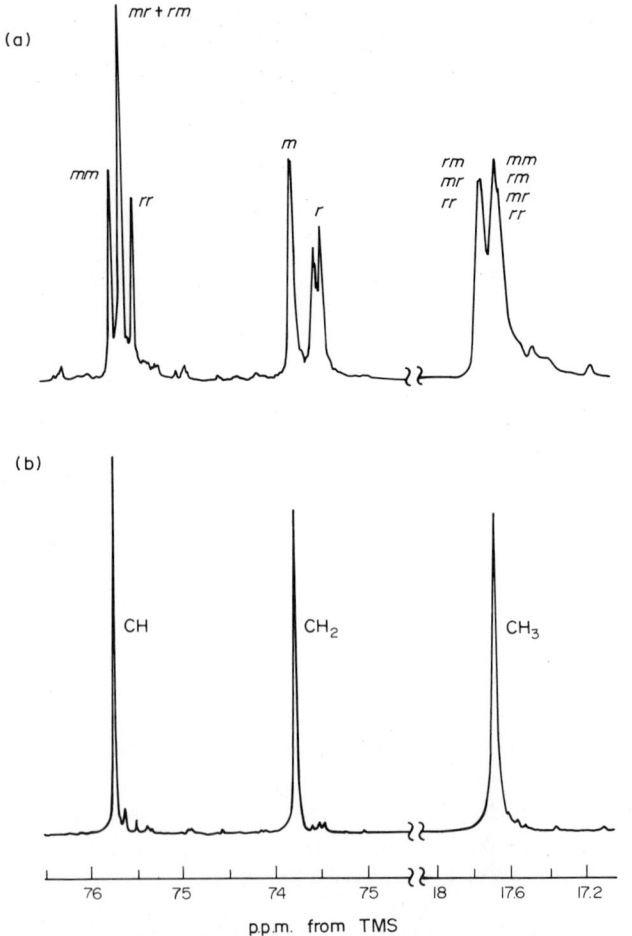

(15a) (T–H:T–T)

(15b) (T–T:H–H)

(15c) (H–H:T–H)

In Figure 23 are shown the 50.3 MHz ^{13}C spectra of atactic (top) and isotactic (bottom) polypropylene oxide.[64] The sensitivity of the ^{13}C chemical shifts to relative chiralities, compared to polypropylene (Figure 12), is markedly reduced owing to the intervening ether oxygen atoms. The two methyl resonances split to four peaks at higher frequency, corresponding to the four head-to-tail triads; the CH and CH$_2$ resonances distinguish only three of these four. Relative intensities are consistent with a random opening of R and S monomers without stereoselectivity by the growing chain end.

Figure 23 50.3 MHz ^{13}C spectra of (a) atactic poly(propylene oxide) of *ca.* 4000 molecular weight; and (b) isotactic poly(propylene oxide). Observed in C$_6$D$_6$ at 23 °C.[64]

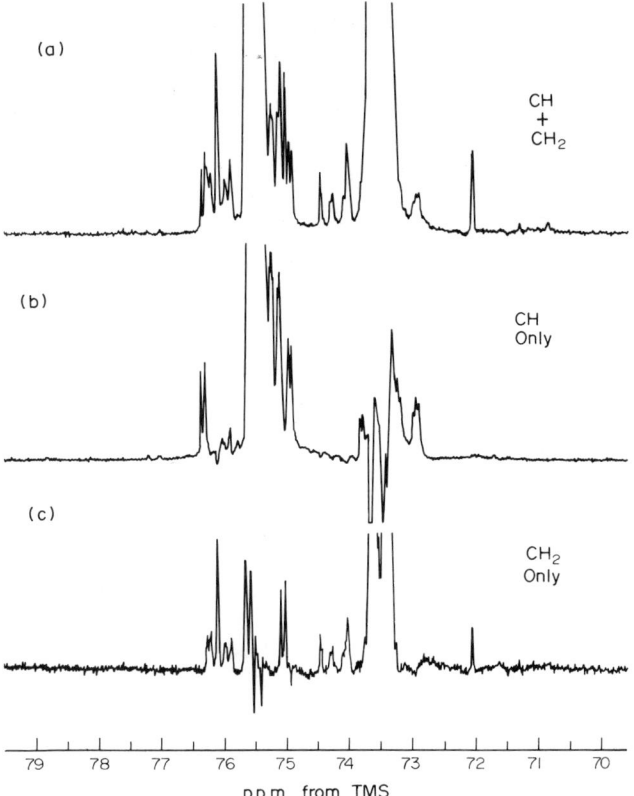

Figure 24 50.3 MHz ^{13}C spectra of atactic poly(propylene oxide) of *ca.* 4000 molecular weight: (a) methine and methylene resonances; (b) DEPT spectrum of methine carbons only; (c) DEPT spectrum of methylene carbons only; observed in C_6D_6 at 23 °C.[64]

    Near the bases of the major resonances may be seen a number of small resonances corresponding to *ca.* 2% of inverted units. (Some end-group resonances are also present.) In Figure 24(a) those minor resonances which are near the major CH and CH_2 peaks of the atactic polymer spectrum are shown at higher gain. They present a complex problem, particularly as head-to-head CH resonances occur in the CH_2 region and tail-to-tail CH_2 resonances occur in the CH region. This fact misled earlier investigators[62] but may be understood in terms of the *γ-gauche* effect (Section 17.3.1.3) and a correct conformational modelling of the poly(propylene oxide) chain.[64,65] Interpretation is also greatly aided by the spectral editing techniques indicated in Section 17.2.5.3. In Figure 24(b) a pulse sequence termed DEPT has been employed[64] with parameters so chosen that only CH resonances appear, CH_2 peaks having zero intensity; in spectrum (c), the parameters are altered so that only CH_2 peaks appear. In each case it will be noted that even the major resonances have been essentially edited out. This has permitted the observation of certain defect resonances otherwise completely obscured. Although complete defect assignments have not been made, comparison of spectra of differing molecular weights permitted the authors to sort out end-group from defect resonances and thereby to demonstrate that the end-groups are **(16)** and **(17)**.

$$\underset{(\textbf{16})}{\cdots-CH_2\overset{\overset{\text{Me}}{|}}{C}HOH} \qquad\qquad \underset{(\textbf{17})}{CH_2=CHCH_2OCH_2\overset{\overset{\text{Me}}{|}}{C}H-\cdots}$$

    This is consistent with a polymerization mechanism in which anionic propagation normally occurs by cleavage of the CH_2–O bond of the monomer, while termination occurs by chain transfer to monomer, as shown in equations (13) and (14).

$$\cdots\!-CH_2CHO^-\cdots M^+ \;+\; CH_2\!-\!\overset{Me}{\underset{\displaystyle O}{\overset{|}{CH}}} \longrightarrow \cdots\!-CH_2\overset{Me}{\underset{}{\overset{|}{C}}HOH \;+\; CH_2\!=\!CHCH_2O^-\cdots M^+ \qquad (13)$$

$$CH_2\!=\!CHCH_2O^-\cdots M^+ \;+\; CH_2\!-\!\overset{Me}{\underset{\displaystyle O}{\overset{|}{CH}}} \longrightarrow CH_2\!=\!CHCH_2OCH_2\!-\!\overset{Me}{\overset{|}{C}}HO\cdots \qquad (14)$$

17.3.4 Branching

Branching in vinyl polymers is a structural variable of substantial importance. It may be introduced deliberately by employing diene or divinyl comonomers, but we are concerned here with branches produced by processes that are under less specific control and involve chain transfer (see Volume 3, Chapter 13) processes of various types. Branches are of particular importance in polyethylene as their presence reduces the melting point and extent of crystallinity. High pressure polyethylene is found by IR spectroscopy (see Chapter 20) to have unusually large numbers of methyl groups, normally expected only at chain ends. When combined with molecular weight measurements these results indicate that there are many more ends than molecules — or in other words that the chains contain branches. Carbon-13 NMR can supply details of the types and distribution of these branches[66-68] because of the sensitivity of ^{13}C chemical shifts to such structural variables, already seen in the discussion of polypropylene (Section 17.3.1.3). In Figure 25 is shown the 50.3 MHz ^{13}C spectrum of a high pressure polyethylene observed in 5% solution in 1,2,4-tri-chlorobenzene at 110 °C. The resonances are labeled according to the scheme inset in the figure, using a large body of information obtained from model hydrocarbons and from ethylene copolymers with small proportions of 1-alkenes. The principal peak, at 30.0 p.p.m., not shown at its full height, corresponds to those methylene groups that are four carbons or more removed from a branch or chain end. It constitutes about 80% of the spectral intensity. The C_1 carbons (*i.e.* methyl groups) and C_2 carbons are the most shielded, branch point carbons the least. Main chain carbons β to the branch are more shielded while those α to the branch are less shielded than unperturbed methylenes. The composition of this polyethylene is shown in Table 3. The predominant branch type is *n*-butyl. Both amyl and butyl branches are believed to be formed by intramolecular chain transfer or 'backbiting', see equation (15). This reaction is evidently most probable when $n=3$ or 4, has a low but finite probability when $n=1$, and zero probability when $n=0$ or 2.

$$\underset{\displaystyle (CH_2)_n}{\overset{\displaystyle -CH_2 \qquad \cdot CH_2}{\diagdown\qquad\diagup}} \longrightarrow \underset{\displaystyle (CH_2)_n}{\overset{\displaystyle -CH\cdot \quad Me}{\diagdown\quad\diagup}} \qquad (15)$$

Table 3   Branching in High-Pressure Polyethylene $\bar{M}_n = 18\,400$, $\bar{M}_w = 129\,000$

Types of branch	Number of branches per 1000 backbone carbons
–Me	0.0
–CH_2Me	1.0
–CH_2CH_2Me	0.0
–$CH_2CH_2CH_2$Me	9.6
–$CH_2CH_2CH_2CH_2$Me	3.6
Hexyl and/or longer	5.6
Total	19.8

The complex appearance of the ethyl branch methyl resonance at *ca.* 11.0 p.p.m. suggests that these branches may occur in groups or with some similar complication. The branches described in Table 3 as 'hexyl and/or longer' are believed to be truly long, as illustrated in the right side of Figure 25. Their frequency of occurrence is estimated from the unique peak labeled L-3-C in Figure 25, corresponding to the third carbon from the chain end. Such branches are too few in number to influence crystallinity but are believed to have a significant effect on melt rheology. They are formed by transfer from the growing chain to a finished polyethylene molecule, resulting in a free radical site on the latter which then adds more monomer.

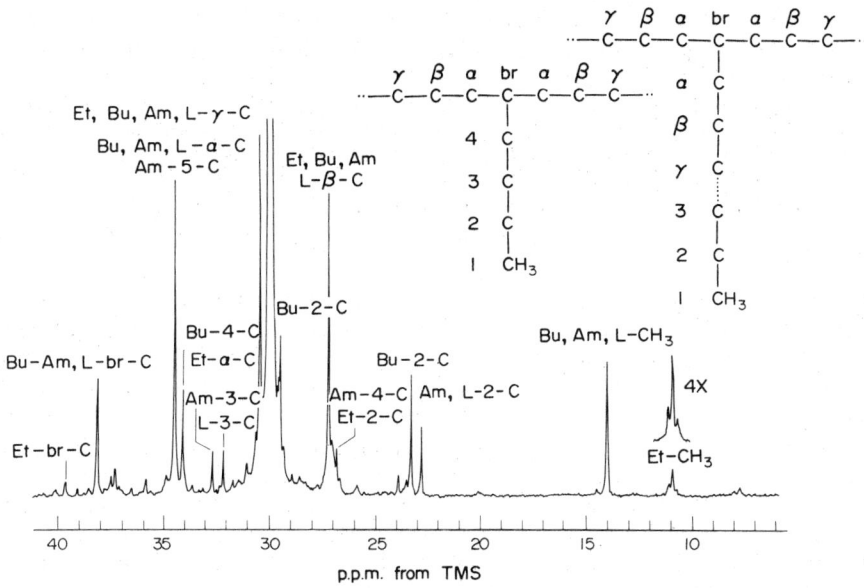

Figure 25 The 50.3 MHz ^{13}C spectrum of high-pressure polyethylene observed in 5% (w/v) solution in 1,2,4-trichlorobenzene at 110 °C (F. C. Schilling, personal communication)

17.3.5 Copolymer Structure

17.3.5.1 Introduction

Copolymers may be broadly divided into three types, as shown in Figure 26: statistical, block and and graft. Block and graft copolymers contain relatively long sequences of one monomer bonded to similar sequences of another. Although they are of major scientific and technological interest, their overall composition is usually known from their method of synthesis and they do not present microstructural problems essentially different from those of homopolymers. Our attention will be confined to the 'statistical' type, in which two or more types of comonomer units are present in each chain (see Volume 3, Chapters 2 and 15). We shall discuss only copolymers of vinyl (or diene) monomers. Copolymers and copolyamides are also significant but their composition is also usually readily predictable from the ratio of comonomers employed.

Figure 26 Basic types of copolymers

The composition of copolymers of vinyl and diene monomers are not in general the same as that of the monomer mixtures from which they are formed and cannot be deduced from the homopolymerization rates of the monomers involved. The relationship between instantaneous copolymer composition and monomer feed composition, *i.e.* the starting ratio of monomers, is given by the following differential equation[69]

$$\frac{d[M_1]}{d[M_2]} = \frac{[M_1]}{[M_2]} \left\{ \frac{r_1[M_1]+[M_2]}{r_2[M_2]+[M_1]} \right\} \tag{16}$$

where M_1 and M_2 represent the two comonomers. The left side of this equation gives the ratio of the rates at which the two monomers enter the copolymer, which in turn must represent the composition of the copolymer being formed at any instant. The ratio $[M_1]/[M_2]$ is the mole ratio of monomers in the feed. The quantities r_1 and r_2 are the reactivity ratios, defined as the ratios of propagation rate constants.

$$r_1 = k_{11}/k_{12} \quad r_2 = k_{22}/k_{21}$$

Here, k_{11} is the rate constant for the addition of monomer 1 to a growing chain ending in a monomer 1 unit; k_{12} is the rate constant for the addition of monomer 2 to the growing chain ending in monomer 1; k_{22} and k_{21} are the corresponding terms for growing chains ending in a monomer 2 unit.

Equation (16) is the copolymer equation in terms of the molar concentrations of the monomers. It is usually more convenient to express this relationship in terms of the mole fraction in both feed and copolymer. The feed mole ratio for monomer 1 is given by

$$f_1 = 1 - f_2 = \frac{[M_1]}{[M_1] + [M_2]} \tag{17}$$

The instantaneous copolymer composition is given by

$$F_1 = 1 - F_2 = \frac{d[M_1]}{d[M_1] + d[M_2]} \tag{18}$$

from which

$$F_1 = \frac{r_1 f_1^2 + f_1 f_2}{r_1 f_1^2 + 2 f_1 f_2 + r_2 f_2^2} \tag{19}$$

(A parallel — but redundant — equation expresses F_2.) We have stated that these relationships deal with the instantaneous composition of the copolymer. Since the comonomers generally do not enter the polymer in the same ratio as in the feed, the latter will drift in composition as copolymerization proceeds, becoming depleted in the more reactive comonomer. As a result, the higher the monomer conversion the more heterogeneous the product. This in no way affects the determination of the overall composition or microstructure of the product but makes it more difficult to interpret in terms of relative reactivities. It is therefore customary in fundamental studies to limit the conversion to about 5% or less, although drifts in composition can be dealt with mathematically. In copolymer production on a practical scale, it is common practice to achieve greater structural regularity by adjusting the monomer input as the reaction proceeds. This usually means withholding the more reactive monomer.

The detailed discussion of various copolymerization cases will not concern us here. Traditionally, the determination of reactivity ratios — which provide important information concerning the behavior of monomers and growing chains — has required the determination of the overall comonomer composition of copolymers prepared from a series of feed ratios. Elemental analysis is most commonly used. A number of computational and graphic methods are employed to do this. The theoretical treatment that predicts overall composition also predicts the frequency of occurrence of comonomer sequences. These can be readily measured by NMR. By NMR it is also possible to observe copolymer stereochemistry and the presence of anomalous units. One can readily detect deviations from the simple model employed here, in which only the terminal residue of a growing chain determines its reactivity; the effect if any, of the penultimate unit, may be clearly observed. It should also be noted that by sequence measurements one can determine reactivity ratios from only a single copolymer provided the feed ratio is known. It may still be desirable to observe a range of composition to assist in resonance assignments but it is not in principle essential.

For a random copolymerization dyad, triad and tetrad sequences may be represented as follows, ignoring stereochemistry

Dyads:	$m_1 m_1$	$m_1 m_2$(or $m_2 m_1$)	$m_2 m_2$
Triads:	$m_1 m_1 m_1$		$m_2 m_2 m_2$
	$m_1 m_1 m_2$(or $m_2 m_1 m_1$)		$m_1 m_2 m_2$(or $m_2 m_2 m_1$)
	$m_2 m_1 m_2$		$m_1 m_2 m_1$
Tetrads:	$m_1 m_1 m_1 m_1$	$m_1 m_1 m_2 m_1$ $(m_1 m_2 m_1 m_1)$	$m_2 m_2 m_2 m_2$
	$m_1 m_1 m_1 m_2$ $(m_2 m_1 m_1 m_1)$	$m_1 m_1 m_2 m_2$ $(m_2 m_2 m_1 m_1)$	$m_2 m_2 m_2 m_1$ $(m_1 m_2 m_2 m_2)$
		$m_2 m_1 m_2 m_1$ $(m_1 m_2 m_1 m_2)$	
	$m_2 m_1 m_1 m_2$	$m_2 m_1 m_2 m_2$ $(m_2 m_2 m_1 m_2)$	$m_1 m_2 m_2 m_1$

The dyad probabilities (*i.e.* frequencies of occurrence) are given by

$$[m_1 m_1] = F_1 P_{11} \tag{20}$$

$$[m_1 m_2] \text{ (or } [m_2 m_1]) = 2F_1 P_{12} = 2F_1(1 - P_{11}) = 2F_2 P_{21} = 2F_1(1 - P_{22}) \tag{21}$$

$$[m_2 m_2] = F_2 P_{22} \tag{22}$$

Here, as we have seen, F_1 and F_2 are the overall mole fractions of m_1 and m_2, respectively. Triad probabilities are given by

$$[m_1 m_1 m_1] = F_1 P_{11}^2 \tag{23}$$

$$[m_1 m_1 m_2] \text{ (or } [m_2 m_1 m_1]) = 2F_1 P_{11}(1 - P_{11}) \tag{24}$$

$$[m_2 m_1 m_2] = F_2(1 - P_{22})(1 - P_{11}) \tag{25}$$

The quantity P_{11} expresses the conditional probability that a chain ending in m_1 will add another m_1 and P_{22} expressses the corresponding probability for m_2. P_{12} is the probability that a chain ending in m_1 will add m_2, equal to the probability that it will not add m_1, *i.e.* $1 - P_{11}$. Corresponding to the four rate constants k_{11}, k_{12}, k_{21} and k_{22} are the four probabilities P_{11}, P_{12}, P_{21} and P_{22}, related by

$$P_{11} + P_{12} = 1 \tag{26}$$

$$P_{21} + P_{22} = 1 \tag{27}$$

since a growing chain has only two choices. We choose to employ P_{11} and P_{22}; it can be shown that they are given in terms of monomer feed mole fractions and reactivity ratios by

$$P_{11} = \frac{r_1 f_1}{1 - f_1(1 - r_1)} \tag{28}$$

$$P_{22} = \frac{r_2 f_2}{1 - f_2(1 - r_2)} \tag{29}$$

from which

$$r_1 = \frac{(1 - f_1)[m_1 m_1]}{f_1(F_1 - [m_1 m_1])} \tag{30}$$

$$r_2 = \frac{(1 - f_2)[m_2 m_2]}{f_2(F_2 - [m_2 m_2])} \tag{31}$$

Entirely analogous relationships apply to triad and tetrad sequences.

17.3.5.2 *Vinylidene chloride/isobutylene system*

The system vinylidene chloride (m_1)/isobutylene (m_2) is appropriate to consider since the copolymer has no asymmetric carbons and no vicinal J coupling. The proton NMR spectrum thus conveys only compositional sequence information.[70,71] In Figure 27 are shown 60 MHz proton NMR spectra of the homopolymers (a) and (b). The homopolymer of vinylidene chloride gives a single resonance for the methylene protons (a); the homopolymer of isobutylene (which can be prepared with cationic but not with free radical initiators) gives singlet resonances of 3:1 intensity for the methyl and methylene protons. The spectrum of a copolymer, prepared with a free radical initiator and containing 70 mol % vinylidene chloride, is shown in (c). The methylene resonances are grouped in three chemical shift ranges: $m_1 m_1$-centred peaks at low field; peaks of methylene protons of $m_1 m_2$-centred units near 3 p.p.m. (18); and CH_2 and CH_3 resonances of m_2-centered sequences (19) at high field. It is evident that tetrad sequences are involved (assignments given in the figure caption). If only dyad sequences were distinguished there would be only three methylene resonances, corresponding to $m_1 m_1$, $m_1 m_2$ (or $m_2 m_1$) and $m_2 m_2$ sequences. The upfield isobutylene peaks show considerable overlap and assignments here are less certain, but these resonances are not required for the analysis.

$$\cdots -\overset{\displaystyle \text{Cl}}{\underset{\displaystyle \text{Cl}}{\text{C}}}\text{CH}_2\overset{\displaystyle \text{Me}}{\underset{\displaystyle \text{Me}}{\text{C}}}- \cdots \qquad\qquad \cdots -\overset{\displaystyle \text{Me}}{\underset{\displaystyle \text{Me}}{\text{C}}}\text{CH}_2\overset{\displaystyle \text{Me}}{\underset{\displaystyle \text{Me}}{\text{C}}}- \cdots$$

(18)   (19)

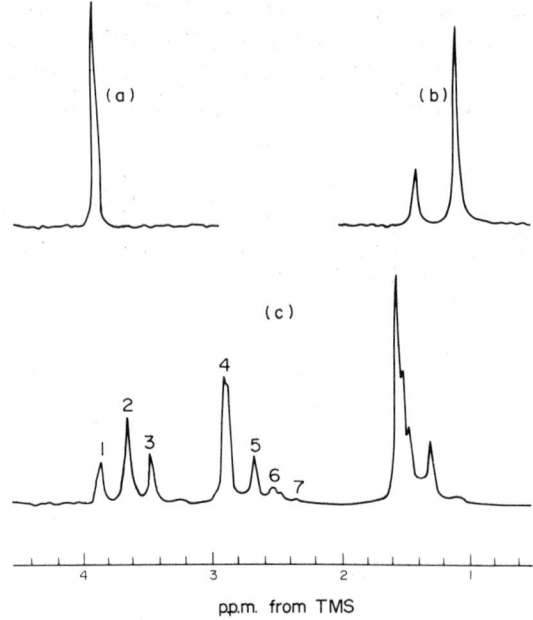

Figure 27   60 MHz proton spectra of (a) poly(vinylidene chloride); (b) polyisobutylene; (c) a vinylidene chloride (m_1)/
isobutylene (m_2) copolymer containing 70 mol % m_1. Peaks are assigned to monomer tetrad sequences as follows:
(1) $m_1 m_1 m_1 m_1$; (2) $m_1 m_1 m_1 m_2$; (3) $m_2 m_1 m_1 m_2$; (4) $m_1 m_1 m_2 m_1$; (5) $m_2 m_1 m_2 m_1$; (6) $m_1 m_1 m_2 m_2$; (7) $m_2 m_1 m_2 m_2$.[72]

From dyad resonances r_1 and r_2 may be calculated. The relative intensity of the $m_1 m_1$ resonances
centered near 3.6 p.p.m. gives $[m_1 m_1]$ as 0.426, normalized over all dyad methylenes, which from the
above relationships gives a value of 3.31 for r_1. Evaluation of the methylene resonance of (**19**) cannot
be readily carried out because of overlaps, even though it would most directly provide a value of r_2.
Instead we note that

$$[m_1 m_2](+[m_2 m_1]) = 2F_2(1 - P_{22}) \tag{32}$$

or

$$[m_1 m_2](+[m_2 m_1]) = 2F_2 - \frac{2F_2 r_2 f_2}{1 - f_2(1 - r_2)} \tag{33}$$

From the group of resonances near 2.8 p.p.m., a value of $[m_1 m_2]$ of 0.56 is obtained, from which a
value of r_2 of 0.04 is calculated. Conventional analysis of this system gives

$$k_{11}/k_{12} = r_1 = 3.3 \tag{34}$$

$$k_{22}/k_{21} = r_2 = 0.05 \tag{35}$$

in agreement within experimental error. These values show that vinylidene chloride radicals prefer
to add vinylidene chloride and that chains terminating in isobutylene units have only a very small
tendency to add another isobutylene.

More searching tests of the propagation mechanism can be obtained from consideration of tetrad
intensities. By such analysis it is found that the relative reactivity of a growing free radical ending in
vinylidene chloride depends upon whether the penultimate unit is another vinylidene chloride unit
or an isobutylene unit; if the latter, the chain is twice as likely to add vinylidene chloride.

17.3.5.3   *Styrene/methyl methacrylate system*

We now consider a case where both monomer sequence and stereochemical configuration must be
taken into account. Copolymer stereochemistry was never considered in early studies of copolymeriz-

ation, or indeed in any work not involving the use of NMR, because there was no way of observing it. In Table 4 are shown the structures, through triads, which must be considered. The representation of longer sequences is omitted, for we find that the number of different sequences $N'(n)$ increases very rapidly with n (compare to equation 11)

$$
\begin{array}{cccccc}
n & 2 & 3 & 4 & 5 & 6 \\
N'(n) & 6 & 20 & 72 & 272 & 1056
\end{array}
$$

or

$$N'(n) = 2^{2(n-1)} + 2^{n-1} \tag{36}$$

The free radical copolymerization of styrene and methyl methacrylate is such a system. It has been intensively studied by NMR.[73-88] It has been found by conventional study that if styrene is taken as monomer 1 and methyl methacrylate as monomer 2, then

$$k_{11}/k_{12} = r_1 = 0.52$$
$$k_{22}/k_{21} = r_2 = 0.46$$
$$r_1 r_2 = 0.24$$

Thus, each growing chain prefers to add the other monomer by a factor of about 2 and the system is of the crossover or alternating type.

Table 4   Configurational Sequences in Copolymers

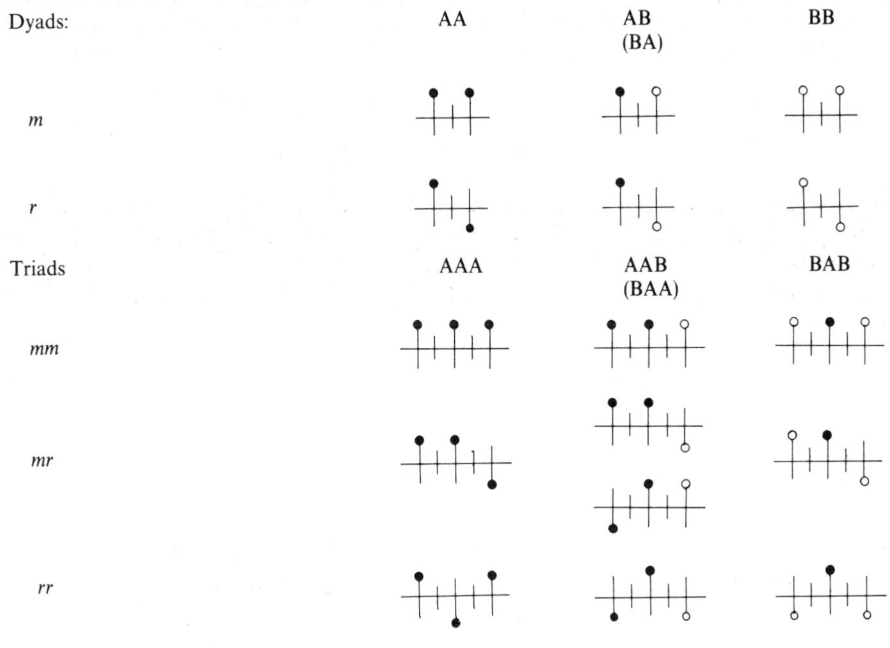

Dyads:	AA	AB (BA)	BB
m			
r			

Triads	AAA	AAB (BAA)	BAB
mm			
mr			
rr			

+ ten others with ● and ○ reversed

In Figure 28(a) is shown the 500 MHz proton spectrum of a random copolymer of styrene and methyl methacrylate containing 50 mol % of each monomer. The resonances appear to be broad, but this is mainly because of a multitude of chemical shifts corresponding to the many possible sequences present; there are too many of these to be individually resolved. When the copolymerization is initiated with diethylaluminum sesquichloride,[84] the monomer sequence becomes rigorously alternating, so that the ten triad sequences shown in Table 4 are reduced to three, which may be termed (arbitrarily), cosyndio, cohetero and coiso, and represented in structures (**20a-20c**) for methyl methacrylate-centred triads.

In Figure 28(b) is shown the 500 MHz proton spectrum of the alternating 1:1 copolymer. The narrowing of the peaks and simplification of the spectrum is evident.[85, 88] The ^{13}C spectrum[88] (not shown) shows clearly that while the monomer sequence is strictly alternating the stereochemistry is

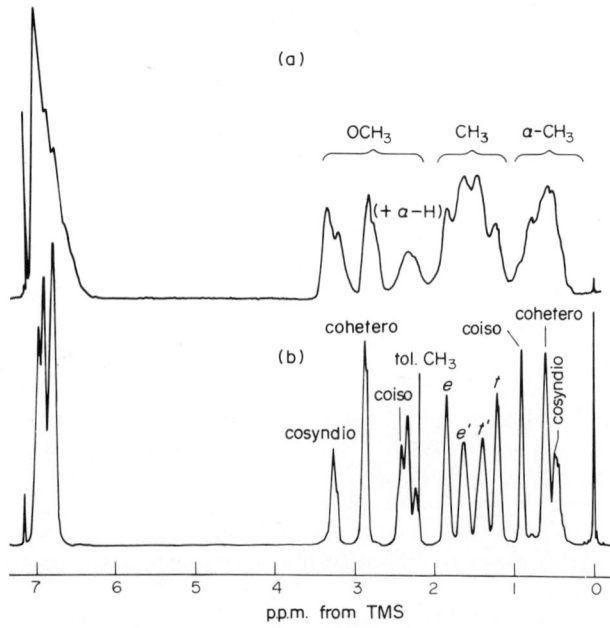

(20a) cosyndio (3.25 p.p.m.)

(20b) cohetero (2.85 p.p.m.)

(20c) coiso (*ca.* 2.20 p.p.m.)

entirely random; thus, the tactic sequences above have a 1:2:1 ratio of probabilities. The region of the ester methyl protons (*ca.* 2.2–3.2 p.p.m.) is split into three distinct groups of resonances in both copolymers, assigned to the configurational sequences as indicated in Figure 28(b) on the basis of phenyl ring current shielding of the methoxyl protons by the adjacent styrene units, an interpretation directly confirmed by the 2D nuclear Overhauser effect proton spectrum.[88] The ring current shielding is maximal when the neighboring phenyl groups are both on the same side of the planar zigzag (the peferred *trans–trans* conformation of the chain) as the ester group. The α-methyl protons

Figure 28   500 MHz proton spectra of (a) the statistical and (b) alternating 1:1 copolymers of styrene and methyl methacrylate, observed in 10% (w/v) solution in hexachlorobutadiene solution at 80 °C.[88]

show an inverse order of shielding for the same reason. Thus NMR is capable of providing information concerning the stereochemistry as well as the monomer sequence in copolymer chains.

17.3.5.4 Ethylene–propylene copolymers

Many other copolymer styrenes have been studied by proton and ^{13}C NMR and a very large literature has accumulated. We briefly consider here a system of much intrinsic interest, which is also the basis of an important class of synthetic elastomers: the copolymerization of ethylene and propylene. This comonomer pair was an early subject for study by ^{13}C NMR;[89-95] the most thorough study is that of Cheng.[95] The system presents complications we have not yet considered for copolymers. First, there is a readily observable content of inverted propylene units, which may be treated as in effect a third comonomer. In addition, we are not always assured that the copolymerization obeys the assumptions inherent in the copolymerization equation (equations 16 and 19). With insoluble coordination catalysts such as those prepared from VCl_3 with aluminum alkyls or chloroaluminum alkyls, there may be two or more catalytic sites producing different polymer chains — varying in propylene content or tacticity or both — and such a system clearly cannot conform to any simple mechanistic picture. However, with soluble vanadium compounds (*e.g.* $VOCl_3$, VCl_4 or vanadium triacetylacetonate) it is believed[95] that the reaction takes place in a homogeneous phase and can be treated as a normal system. There is some difference in reported reactivity ratios depending on the particular catalyst employed, but most of this is probably experimental error.

The reactivity ratios determined by classical means[95] clearly show that ethylene (monomer 1) is much more reactive than propylene: r_1 is in the range 16 ± 2, where r_2 is of the order of 0.05. The reported values of $r_1 r_2$ are close to 1, or somewhat lower, indicating that the nature of the chain end has little influence on the relative rates of addition of the monomers. Figure 29 shows the 25 MHz ^{13}C spectrum of a 50:50 (mole ratio) copolymer, observed in 1,2,4-trichlorobenzene at 120°.[93] The peaks were initially assigned by Crain *et al.*[89] and Zambelli *et al.*[90] on the basis of the Grant and Paul rules[96] for carbon chemical shifts. Later, more detailed assignments were given by Carman *et al.*;[91,92] the most complete assignments have been reported by Cheng.[95] The principal chemical shift influence is that of the methyl-branched carbons, but stereochemical configuration is a substantial perturbation. The methylene carbons are in general the most sensitive and useful measures of the structure. In a sequence of methylene carbons between branch carbons, each CH_2

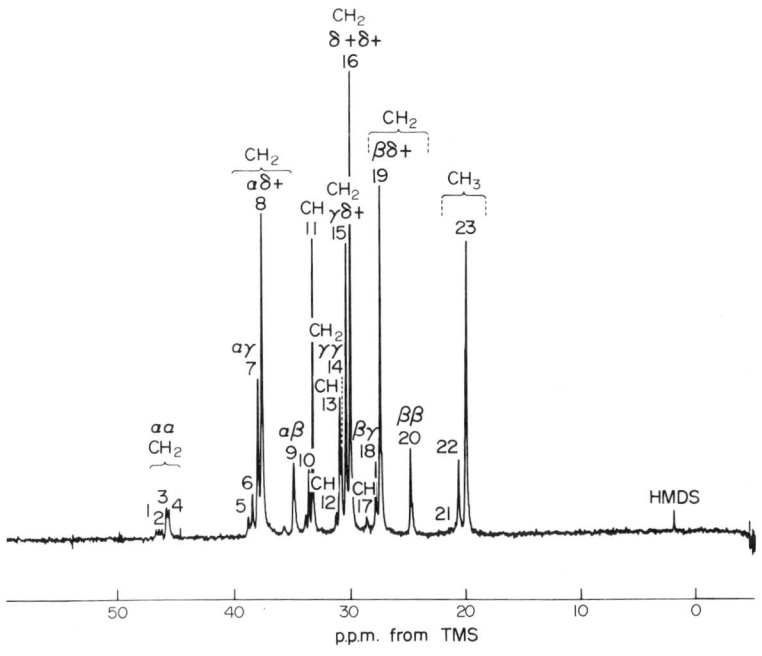

Figure 29 The 25 MHz ^{13}C spectrum of an ethylene–propylene copolymer of 50:50 mol ratio, observed at 120 °C in 1,2,4-trichlorobenzene[93]

45.5–46.5

H Me   H Me

(21a) αα

H Me

34.8

H Me

(21b) αβ, αβ

37.9

H Me   24.9   H Me

(21c) αγ, ββ, αγ

37.5

H Me   27.8

H Me

(21d) αδ⁺, βγ, βγ, αδ⁺

37.5   30.7

H Me   27.4

H Me

(21e) αδ⁺, βδ⁺, γγ, βδ⁺, αδ⁺

37.5   30.4

H Me   27.4

H Me

(21f) αδ⁺, βδ⁺, γδ⁺, γδ⁺, βδ⁺, αδ⁺

37.5   30.4

H Me   27.4   30.0   H Me

(21g) αδ⁺, βδ⁺, γδ⁺, δ⁺δ⁺, γδ⁺, βδ⁺, αδ⁺

gives a distinctive resonance up to a run length of seven. The assignments are shown in **(21a)**–**(21g)**. The Greek letter designations indicate the position of the nearest branch carbon. A plus sign indicates that the branch is at the δ or a more distant position. Methylene carbons that are four or more carbons removed from a branch give the same chemical shift (30.0 p.p.m.) as we have already seen in the case of polyethylene (Section 17.3.4).

17.4 REFERENCES

1. E. D. Becker, 'High Resolution NMR', 2nd edn., Academic Press, New York, 1980.
2. E. Breitmaier and W. Voelter, 'Carbon-13 NMR Spectroscopy', VCH, Weinheim, 1987.
3. T. C. Farrar and E. D. Becker, 'Pulse and Fourier Transform NMR', Academic Press, New York, 1971.
4. C. P. Slichter, 'Principles of Magnetic Resonance', Springer Verlag, Heidelberg, 1978.
5. R. K. Harris, 'Nuclear Magnetic Resonance Spectroscopy', Pitman, London, 1983.
6. F. A. Bovey, 'Chain Structure and Conformation of Macromolecules', Academic Press, New York, 1982.
7. J. C. Randall, 'Polymer Sequence Determination', Academic Press, New York, 1978.
8. L. W. Jelinski, *Chem. Eng. News*, 1985, **62**, 26.
9. C. J. Turner, *Prog. Nucl. Magn. Reson. Spectrosc.* 1984, **16**, 1.
10. R. Benn and H. Günther, *Angew. Chem., Int. Ed. Engl.*, 1983, **22**, 350.
11. R. Freeman and G. A. Morris, *Bull. Magn. Reson.*, 1979, **1**, 5.
12. A. Bax and L. Lerner, *Science*, 1986, **232**, 960.
13. A. Bax, 'Two-Dimensional Nuclear Magnetic Resonance in Liquids', Delft University Press, Holland, 1982.
14. F. C. Schilling, F. A. Bovey, M. D. Bruch and S. A. Kozlowski, *Macromolecules*, 1985, **18**, 1418, and refs. therein.
15. M. Barfield, *J. Chem. Phys.*, 1964, **41**, 3825.
16. P. Cacioli, D. G. Hawthorne, S. R. Johns, D. H. Soloman, E. Rizzardo and R. I. Willing, *J. Chem. Soc., Chem. Commun.*, 1985, 1355.
17. P. R. Sundararajan and P. J. Flory, *J. Am. Chem. Soc.*, 1974, **96**, 5025.
18. P. R. Sundararajan, *J. Polym. Sci., Polym. Lett. Ed.*, 1977, **15**, 699.
19. M. Vacatello and P. J. Flory, *Macromolecules*, 1986, **19**, 405.
20. P. R. Sundararajan, *Macromolecules*, 1986, **19**, 415.
21. R. Chûjô, S. Satoh, T. Ozeki and E. Nagai, *J. Polym. Sci.*, 1962, **61**, S12.
22. W. C. Tincher, *J. Polym. Sci.*, 1962, **62**, S148.
23. C. Schuerch, W. Fowells, A. Yamada, F. A. Bovey, F. P. Hood and E. W. Anderson, *J. Am. Chem. Soc.*, 1964, **86**, 4481.
24. F. Heatley and F. A. Bovey, *Macromolecules*, 1968, **1**, 303.
25. F. Heatley and F. A. Bovey, *Macromolecules*, 1969, **2**, 241.
26. F. Heatley and A. Zambelli, *Macromolecules*, 1969, **2**, 618.
27. A. E. Tonelli and F. C. Schilling, *Acc. Chem. Res.*, 1981, **14**, 233.
28. M. Saunders and A. Wishnia, *Ann. N.Y. Acad. Sci.*, 1958, **70**, 870.
29. A. Odajima, *J. Phys. Soc. Jpn.*, 1959, **14**, 777.
30. F. A. Bovey, G. V. D. Tiers and G. Filipovich, *J. Polym. Sci.*, 1959, **38**, 73.
31. F. Heatley and F. A. Bovey, *Macromolecules*, 1968, **1**, 301.
32. L. F. Johnson, F. Heatley and F. A. Bovey, *Macromolecules*, 1970, **3**, 175.
33. J. C. Randall, 'Polymer Sequence Determination', Academic Press, New York, 1978, pp. 87–92, 116–119.
34. A. E. Tonelli, *Macromolecules*, 1979, **12**, 252.
35. N. Ishihara, T. Seimiya, M. Kuramoto and M. Uoi, *Macromolecules*, 1986, **19**, 2464.
36. F. Heatley and F. A. Bovey, *Macromolecules*, 1969, **2**, 241.
37. C. J. Carman, A. R. Tarpley, Jr. and J. H. Goldstein, *Macromolecules*, 1971, **4**, 445.

38. C. J. Carman, A. R. Tarpley, Jr. and J. H. Goldstein, *J. Am. Chem. Soc.*, 1971, **93**, 2864.
39. C. J. Carman, *Macromolecules*, 1973, **6**, 725.
40. Y. Inoue, I. Ando and A. Nishioka, *Polym. J.*, 1972, **3**, 246.
41. I. Ando, Y. Kato and A. Nishioka, *Makromol. Chem.*, 1976, **177**, 2759.
42. R. E. Cais and W. L. Brown, *Macromolecules*, 1980, **13**, 801.
43. F. C. Schilling, 1983, personal communication. The pioneering observations of diene polymers by ^{13}C NMR were reported by M. W. Duch and D. M. Grant, *Macromolecules*, 1970, **3**, 165. Since then a very large literature has accumulated.
44. L. W. Jelinski, 1982, personal communication. See F. A. Bovey, 'Chain Structure and Conformation of Macromolecules', Academic Press, New York, 1982, pp. 105–110.
45. C. W. Wilson, III and E. R. Santee, Jr., *J. Polym. Sci., Part C*, 1965, **81**, 97.
46. R. E. Cais and J. M. Kometani, in 'NMR and Macromolecules', ed. J. C. Randall, ACS Symposium Series No. 247, 1984, pp. 153–166.
47. M. D. Bruch, F. A. Bovey and R. E. Cais, *Macromolecules*, 1984, **17**, 2547.
48. C. W. Wilson, III, *J. Polym. Sci., Part A-1*, 1963, 1305.
49. R. Liepins, J. R. Surles, N. Morosoff, V. T. Stannett, M. L. Timmons and J. J. Wortman, *J. Polym. Sci., Polym. Chem. Ed.*, 1978, **16**, 3039.
50. R. C. Ferguson and E. G. Brame Jr., *J. Phys. Chem.*, 1979, **83**, 1397.
51. A. E. Tonelli, F. C. Schilling and R. E. Cais, *Macromolecules*, 1982, **15**, 849.
52. R. E. Cais and J. M. Kometani, *Macromolecules*, 1985, **18**, 1354.
53. C. C. Price and M. Osgan, *J. Am. Chem. Soc.*, 1956, **78**, 4787.
54. C. C. Price, R. Spector and A. L. Tumolo, *J. Polym. Sci., Part A-1*, 1967, **5**, 407.
55. K. C. Ramey and N. D. Field, *J. Polym. Sci., Part B*, 1964, **2**, 461.
56. H. Tani, N. Oguni and S. Watanabe, *J. Polym. Sci., Part B*, 1968, **6**, 577.
57. J. Schaefer, *Macromolecules*, 1969, **2**, 533.
58. T. Hirano, P. H. Khanh and T. Tsuruta, *Makromol. Chem.*, 1972, **153**, 331.
59. N. Oguni, K. Lee and H. Tani, *Macromolecules*, 1972, **5**, 819.
60. N. Oguni, S. Watanabe, M. Maki and H. Tani, *Macromolecules*, 1973, **6**, 195.
61. N. Oguni, S. Maeda and H. Tani, *Macromolecules*, 1973, **6**, 459.
62. N. Oguni, S. Shinohara and K. Lee, *Polym. J.*, 1979, **11**, 755.
63. M. D. Bruch, F. A. Bovey, R. E. Cais and J. H. Noggle, *Macromolecules*, 1985, **18**, 1253.
64. F. C. Schilling and A. E. Tonelli, *Macromolecules*, 1986, **19**, 1337.
65. A. Abe, T. Hirano and T. Tsuruta, *Macromolecules*, 1979, **12**, 1092.
66. D. E. Dorman, E. P. Otocka and F. A. Bovey, *Macromolecules*, 1972, **5**, 574.
67. J. C. Randall, *J. Polym. Sci., Polym. Phys. Ed.*, 1973, **11**, 275.
68. F. A. Bovey, F. C. Schilling, F. L. McCrackin and H. L. Wagner, *Macromolecules*, 1976, **9**, 76.
69. F. R. Mayo and F. M. Lewis, *J. Am. Chem. Soc.*, 1944, **66**, 1594.
70. J. B. Kinsinger, T. Fischer and C. W. Wilson, III, *J. Polym. Sci., Part B*, 1966, **4**, 379.
71. J. B. Kinsinger, T. Fischer and C. W. Wilson, III, *J. Polym. Sci., Part B*, 1967, **5**, 285.
72. K. H. Hellwege, U. Johnsen and K. Kolbe, *Kolloid-Z. Z. Polym.*, 1966, **214**, 45.
73. F. A. Bovey, *J. Polym. Sci.*, 1962, **62**, 197.
74. A. Nishioka, Y. Kato and N. Ashikari, *J. Polym. Sci.*, 1962, **62**, S10.
75. Y. Kato, N. Ashikari and A. Nishioka, *Bull. Chem. Soc., Jpn.*, 1964, **37**, 1630.
76. H. J. Harwood, *Angew. Chem., Int. Ed. Engl.*, 1965, **4**, 1051.
77. H. J. Harwood and W. M. Ritchey, *J. Polym. Sci., Part B*, 1965, **3**, 419.
78. K. Ito and Y. Yamashita, *J. Polym. Sci., Part B.*, 1965, **3**, 625, 631.
79. K. Ito, S. Iwase, K. Umehara and Y. Yamashita, *J. Macromol. Sci., Chem.*, 1967, **1**, 891.
80. K. Ito and Y. Yamashita, *J. Polym. Sci., Part B*, 1968, **6**, 227.
81. S. Yabumoto, K. Ishii and K. Arita, *J. Polym. Sci., Part A-1*, 1970, **8**, 295.
82. F. A. Blouin, R. C. Chang, M. H. Quinn and H. J. Harwood, *Polym. Prepr., Am. Chem. Soc., Div. Polym. Chem.*, 1973, **14**, 25.
83. K. Yokota and T. Hirabayashi, *J. Polym. Sci., Polym. Chem. Ed.*, 1976, **14**, 57.
84. H. Hirai, T. Tanabe and H. Koinuma, *J. Polym. Sci., Polym. Chem. Ed.*, 1979, **17**, 843.
85. H. Koinuma, T. Tanabe and H. Hirai, *Makromol. Chem.*, 1980, **181**, 383.
86. A. R. Katritzky, A. Smith and D. E. Weiss, *J. Chem. Soc., Perkin Trans .2*, 1974, 1547.
87. A. R. Katritzky and D. E. Weiss, *Chem. Br.*, 1976, 45.
88. S. A. Heffner, F. A. Bovey, L. A. Verge, P. A. Mirau and A. E. Tonelli, *J. Am. Chem. Soc.*, 1986, **19**, 1628.
89. W. O. Crain, Jr., A. Zambelli and J. D. Roberts, *Macromolecules*, 1971, **4**, 330.
90. A. Zambelli, G. Gatti, C. Sacchi, W. O. Crain, Jr. and J. D. Robert, *Macromolecules*, 1971, **4**, 475.
91. C. J. Carman and C. E. Wilkes, *Rubber Chem. Technol.*, 1971, **44**, 781.
92. C. J. Carman, R. A. Harrington and C. E. Wilkes, *Macromolecules*, 1977, **10**, 536.
93. J. C. Randall, 'Polymer Sequence Determination', Academic Press, New York, 1978, pp. 53–58, 135–138.
94. P. Ammendola, L. Oliva, G. Gianotti and A. Zambelli, *Macromolecules*, 1985, **18**, 1407.
95. H. N. Cheng, *Macromolecules*, 1984, **17**, 1950.
96. G. Crespi, A. Valvasori and U. Flisi, in 'The Stereo Rubbers', ed. W. M. Saltman, Wiley, New York, 1977, p. 365.
97. W. V. Smith, *J. Polym. Sci., Polym. Phys. Ed.*, 1980, **18**, 1573, 1587.
98. D. M. Grant and E. G. Paul, *J. Am. Chem. Soc.*, 1964, **86**, 2984.

18
Dynamics of Chains in Solutions by NMR Spectroscopy

FRANK HEATLEY
University of Manchester, UK

18.1 INTRODUCTION

As described in the previous chapter, polymers in solution normally yield NMR spectra with clearly resolved resonances for structurally distinct nuclei. Since the magnetic relaxation properties of each nucleus are determined by the nature and frequency of the chain motions at that site, one of the most important features of magnetic relaxation in solution in studying polymer dynamics is the ability to provide a detailed insight into molecular motion at the atomic level. For example, backbone and sidegroup motions can be clearly differentiated, or the motions of different components of a copolymer. A further advantage of solution studies compared with solid polymers is that in dilute solution interchain entanglements are alleviated, so allowing the elucidation of the roles played by intramolecular steric conflicts and the solvent viscosity. Polymer motions in solution accessible by NMR cover a very wide frequency spectrum, from reptation in entangled networks at 10^2 Hz or less, through overall tumbling at 10^4 to 10^8 Hz, up to local segmental motions at 10^9 Hz or more. The object of relaxation studies is to determine the nature and rates of those processes controlling relaxation. This chapter principally describes the theory and application of magnetic relaxation studies directed towards this end. The final section considers the use of NMR in studying polymer self-diffusion.

When applied to polymer solutions, the terms 'dilute' and 'semi-dilute' now have a special physical significance.[1] 'Dilute' refers to concentrations below a critical concentration c^* at which chains begin to overlap. In this regime chains interact *via* hydrodynamic interactions with the solvent. c^* is given approximately by $c^* \approx M/N_A R^3$ where M is the molecular weight, N_A is Avogadro's constant and R is the root mean square end-to-end distance. The 'semi-dilute' regime lies between c^* and a concentration c^{**} defined[1] as the concentration at which the number of chain units between interchain entanglements is so small that usual statistics do not apply. c^{**} is independent of M. Most studies of magnetic relaxation have been performed in the dilute and semi-dilute regimes. Where necessary, these two categories will be combined under the term 'moderately dilute'. NMR in more concentrated solutions has special features which are discussed separately (Section 18.5).

The principles of NMR and methods of measuring relaxation times are described in several comprehensive texts.[2]

18.2 PRINCIPLES OF NUCLEAR MAGNETIC RELAXATION

As described in the previous chapter, there are two basic magnetic relaxation processes: (i) relaxation of the nuclear magnetization component parallel to B_0 (longitudinal or spin–lattice relaxation), time constant T_1; and (ii) relaxation of the component perpendicular to B_0 (transverse or spin–spin relaxation), time constant T_2. Both processes arise from fluctuating interactions experienced by the nuclei. In polymers, the most common nuclei studied are protons and protonated ^{13}C nuclei, where the overwhelming relaxation mechanism is dipole–dipole relaxation (DD) arising from the direct mutual interaction of nuclear magnetic moments. For non-protonated ^{13}C nuclei, especially unsaturated nuclei, and also ^{19}F and ^{31}P, the chemical shift anisotropy mechanism may be significant. For nuclei with spin quantum number $I \geqslant 1$, such as ^2H and ^{14}N, the dominant mechanism is quadrupolar relaxation arising from the interaction of the nuclear electric quadrupole moment with electric field gradients. Since all three mechanisms are expressed by a second-order tensor, the theory relating each to molecular motion and the strength of the interaction is analogous. Attention is therefore directed mainly to the DD interaction, with particular reference to ^1H and ^{13}C relaxation.

18.2.1 Correlation and Spectral Density Functions

The DD interaction between two nuclei in a given orientation can be expressed[3] in spherical tensor form

$$H_{DD} = \beta \sum_{q=2}^{-2} (-1)^q A_q Y_2^{-q}(\Omega) \tag{1}$$

where the functions A_q contain spin operators and $Y_2^{-q}(\Omega)$ are standard second-order spherical harmonics involving the polar angles Ω defining the orientation of the internuclear vector relative to the magnetic field. The dipolar interaction constant $\beta = (6\pi/5)^{1/2}(\mu_0/4\pi)\gamma_1\gamma_2 \hbar r^{-3}$ where γ_1 and γ_2 are the magnetogyric ratios and r is the internuclear distance. For fast isotropic motion, H_{DD} averages to zero and does not affect transition frequencies. However in solids and possibly in melts and concentrated solutions, $\langle H_{DD} \rangle$ does not disappear (Section 18.5 and Chapter 19). Fluctuations in H_{DD} due to molecular rotation induce transitions between spin states, thus leading to relaxation. The average time dependence of Y_2^q under the influence of random motion is expressed by an autocorrelation function

$$G(t) = \langle Y_2^q(\Omega, 0) Y_2^{-q}(\Omega, t) \rangle \tag{2}$$

$G(t)$ expresses the average way in which the molecular orientation loses correlation or memory of its previous history under the influence of random molecular movements acting over a time interval t. As t increases $G(t)$ decays monotonically from its initial value $\langle |Y_2^q(\Omega)|^2 \rangle$ to zero. It is convenient at this point to introduce the normalized correlation function $\tilde{G}(t)$ defined as $\tilde{G}(t) = G(t)/\langle |Y_2^q(\Omega)|^2 \rangle$.

From time-dependent perturbation theory, it is found that relaxation depends on $\tilde{G}(t)$ through a spectral density function $J(\omega)$ defined by

$$J(\omega) = \frac{1}{2} \int_{-\infty}^{+\infty} \tilde{G}(t) e^{i\omega t} dt \tag{3}$$

For the case of rotation by isotropic rotational diffusion $\tilde{G}(t)$ is exponential with correlation time τ_c and $J(\omega)$ becomes

$$J(\omega) = \tau_c/(1 + \omega^2 \tau_c^2) \qquad (4)$$

18.2.2 Carbon-13 Relaxation

Only natural abundance protonated ^{13}C nuclei, with relaxation essentially determined by DD interaction with the attached protons, will be considered. Normally, ^{13}C spectra are recorded with broadband proton decoupling in order to simplify the spectrum and under these conditions, T_1 and T_2 of a ^{13}C nucleus in a $^{13}CH_n$ group are given by[4]

$$T_{1C}^{-1} = nQ[J(\omega_H - \omega_C) + 3J(\omega_C) + 6J(\omega_H + \omega_C)] \qquad (5)$$

$$T_{2C}^{-1} = \frac{nQ}{2}[4J(0) + J(\omega_H - \omega_C) + 3J(\omega_C) + 6J(\omega_H) + 6J(\omega_H + \omega_C)] \qquad (6)$$

where $Q = (1/10)(\mu_0/4\pi)^2 \gamma_H^2 \gamma_C^2 \hbar^2 r_{CH}^{-6}$. γ_X is the magnetogyric ratio and ω_X the resonance frequency of nucleus X, and r_{CH} is the C–H bond length.

In addition to T_{1C} and T_{2C}, a third relaxation parameter can be measured in the form of the nuclear Overhauser effect (NOE). The NOE is a change in intensity of the ^{13}C signal resulting from proton decoupling and originates in longitudinal cross-relaxation between ^{13}C and protons. If I_d and I_0 are the ^{13}C intensities with and without proton decoupling respectively, the NOE is expressed in terms of spectral densities by[4]

$$NOE = \frac{I_d}{I_0} = 1 + \frac{\gamma_H}{\gamma_C}\left[\frac{6J(\omega_H + \omega_C) - J(\omega_H - \omega_C)}{J(\omega_H - \omega_C) + 3J(\omega_C) + 6J(\omega_H + \omega_C)}\right] \qquad (7)$$

At this point, it is helpful to consider the dependence of T_{1C}, T_{2C} and the NOE on τ_c for isotropic rotational diffusion (Figure 1). As described in Section 18.4, polymer motion does not usually conform to this ideal, but $J(\omega)$ still resembles equation (4) and the general behaviour of T_{1C}, T_{2C} and the NOE is similar to that depicted.

T_{1C} passes through a minimum when $\omega_C \tau_c \approx 1$. For the condition of extreme narrowing (*i.e.* $\omega_C^2 \tau_c^2 \ll 1$), T_{1C} is independent of ω_C, whereas in the opposite extreme, $T_{1C} \propto \omega_C^2$. In contrast, because of the $J(0)$ term, T_{2C} shows little dependence on ω_C and decreases monotonically with increasing τ_c. For extreme narrowing, $T_{1C} = T_{2C}$, whereas in the opposite regime, $T_{2C} \ll T_{1C}$. The NOE is

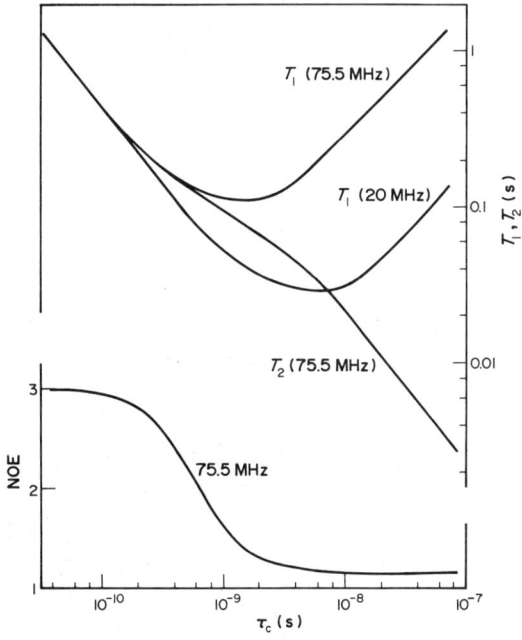

Figure 1 T_{1C} at 20 and 75.5 MHz, T_{2C} at 75.5 MHz and NOE at 75.5 MHz as a function of the isotropic rotational diffusion correlation time for a ^{13}CH group. $r_{CH} = 108$ pm

dependent on τ_c only in the region of the T_{1C} minimum, varying between the asymptotic limits of 2.988 (for $\omega_C^2 \tau_c^2 \ll 1$) and 1.154. Although T_{1C} is much the easiest relaxation parameter to measure accurately, a given T_{1C} corresponds to two possible values of τ_c. The correct one can be selected by measuring additional data such as T_{2C}, NOE or the temperature dependence of T_{1C}.

18.2.3 Proton Relaxation

Proton relaxation is dominated by homonuclear ^{1}H–^{1}H DD interactions, and unlike ^{13}C relaxation, does not normally occur under decoupling conditions. Also there are frequently two or more chemically distinct groups of protons with significant DD interactions within and between groups. For these reasons, proton relaxation is rather more complicated than ^{13}C relaxation.

For a system containing two groups of protons, denoted A and X, for example the methyl and methylene protons in poly(isobutene), spin–lattice relaxation of the longitudinal components M_{zA} and M_{zX} is governed by the coupled equations[5,6]

$$dM_{zA}/dt = -(\rho_{AA} + \rho_{AX})(M_{zA} - M_{zA}^0) - \sigma_{AX}(M_{zX} - M_{zX}^0) \tag{8a}$$

$$dM_{zX}/dt = -(\rho_{XX} + \rho_{XA})(M_{zX} - M_{zX}^0) - \sigma_{XA}(M_{zA} - M_{zA}^0) \tag{8b}$$

$$\rho_{AA} = R[3J(\omega_A) + 12J(2\omega_A)] \sum_{A_j} r_{A_iA_j}^{-6}$$

$$\rho_{AX} = R[J(\omega_A - \omega_X) + 3J(\omega_A) + 6J(\omega_A + \omega_X)] \sum_{X_j} r_{A_iX_j}^{-6}$$

$$\sigma_{AX} = R[6J(\omega_A + \omega_X) - J(\omega_A - \omega_X)] \sum_{X_j} r_{A_iX_j}^{-6}$$

$$R = (1/10)(\mu_0/4\pi)^2 \gamma_H^4 \hbar^2$$

ρ_{XX}, ρ_{XA} and σ_{XA} are as ρ_{AA}, ρ_{AX} and σ_{AX} respectively, but with the subscripts A and X interchanged. The sums such as $\sum_{X_j} r_{A_iX_j}^{-6}$ represent the sum of interactions between one A proton i and all X protons j. The symbols M_{zA}^0 and M_{zX}^0 represent equilibrium values.

In principle equations (8a) and (8b) predict biexponential relaxation but in practice, the recovery of M_{zA} and M_{zX} is found to be closely exponential, with time constants T_{1A} and T_{1X}. Both T_{1A} and T_{1X} pass through a minimum as τ_c varies, but the values of T_{1A} and T_{1X} depend on the correlation time regime. When $(\omega_A + \omega_X)^2 \tau_c^2 \ll 1$, $J(\omega)$ is independent of frequency, and σ_{AX} and σ_{XA} are positive and significantly less than $(\rho_{AA} + \rho_{AX})$ and $(\rho_{XX} + \rho_{XA})$, then T_{1A} and T_{1X} are given to high accuracy by $T_{1A}^{-1} = \rho_{AA} + \rho_{AX} + \sigma_{AX}$ and $T_{1X}^{-1} = \rho_{XX} + \rho_{XA} + \sigma_{XA}$. In general, T_{1A} and T_{1X} are not equal and reflect the different magnitudes of the DD interactions experienced by each nucleus. As τ_c increases however, a regime is entered where conditions are such that $\omega_A^2 \tau_c^2$, $\omega_X^2 \tau_c^2$, $(\omega_A + \omega_X)^2 \tau_c^2 \gg 1$ but $(\omega_A - \omega_X)^2 \tau_c^2 \ll 1$. (This arises because $(\omega_A - \omega_X)$ is a proton chemical shift; *i.e.* ~kHz or less, whereas ω_A, ω_X and $(\omega_A + \omega_X)$ are resonance frequencies *i.e.* ~tens of MHz or more). Thus one finds $J(\omega_A - \omega_X) \gg J(\omega_A)$, $J(\omega_X)$, $J(\omega_A + \omega_X)$. Physically, this relationship indicates that cross-relaxation or spin diffusion between A and X systems, represented by transitions of the type $\alpha\beta \leftrightarrow \beta\alpha$, is much more efficient than relaxation to the lattice, represented by transitions of the type $\alpha\alpha \leftrightarrow \beta\alpha$ and $\alpha\alpha \leftrightarrow \beta\beta$. In these circumstances, the A and X spins maintain a common deviation from equilibrium and relax with the same T_1, which is a weighted average of the individual T_1 values.

$$\left\langle \frac{1}{T_1} \right\rangle = \left(\frac{n_A}{n_A + n_X}\right) \frac{1}{T_{1A}} + \left(\frac{n_X}{n_A + n_X}\right) \frac{1}{T_{1X}} \tag{9}$$

where n_A and n_X are the numbers of A and X nuclei respectively.

On saturating the X protons, the NOE of A is given by

$$\text{NOE}_A\{X\} = 1 + [\sigma_{AX}/(\rho_{AA} + \rho_{AX})](M_{zX}^0/M_{zA}^0) \tag{10}$$

As a function of τ_c, this NOE varies in a similar way to the ^{13}C$\{^1$H$\}$ NOE described above, but the limits are different. At long τ_c when spin diffusion is efficient, $[\sigma_{AX}/(\rho_{AA} + \rho_{AX})](M_{zX}^0/M_{zA}^0) \rightarrow -1$ and $\text{NOE}_A\{X\} \rightarrow 0$. Experimentally this corresponds to disappearance of the A signal by saturation transfer from the irradiated X nuclei. At the opposite extreme of short τ_c, the asymptotic NOE lies between the limits of 1 (when A–X interactions are insignificant compared to A–A) and 1.5 (when A–X interactions are dominant).

Transverse relaxation in multiproton systems is complex.[7] However if T_2 is measured using the Carr–Purcell–Meiboom–Gill spin-echo sequence with a sufficiently rapid pulse repetition rate, spin–spin coupled nuclei all decay with the same T_2, which is the population average of the individual T_2 values given by, *e.g.*

$$T_{2A}^{-1} = \frac{3R}{2}[3J(0) + 5J(\omega_A) + 2J(2\omega_A)]\left[\sum_{A_j} r_{A_iA_j}^{-6} + \sum_{X_j} r_{A_iX_j}^{-6}\right] \tag{11}$$

In the absence of spin–spin coupling, each peak relaxes with its characteristic relaxation time. As a function of τ_c, proton T_2s behave in a similar manner to T_{2C} in Figure 1.

18.2.4 Relaxation in the Rotating Frame

In this technique, the radio-frequency field B_1 is shifted in phase by 90° after a $\pi/2$ pulse, so aligning B_1 with the nuclear magnetization. Transverse relaxation thus effectively occurs as a spin–lattice process in a field B_1 with a time constant denoted $T_{1\rho}$. For proton i interacting with other protons j, $T_{1\rho}$ is given by

$$T_{1\rho}^{-1} = \frac{3R}{2}[3J(2\omega_1) + 5J(\omega_H) + 2J(2\omega_H)]\sum_j r_{ij}^{-6} \tag{12}$$

where $\omega_1 = \gamma_H B_1$. Typically, $\omega_1/2\pi$ lies in the range 10 to 100 kHz, so $T_{1\rho}$ senses the spectral density at low frequencies.

18.2.5 Quadrupole Relaxation

For an axially symmetric electric field gradient, the correlation functions involve the orientation angles of the symmetry axis with respect to the magnetic field. For the most common application in polymers, that of a $C{-}^2H$ unit, this axis is the bond direction. For a nucleus of spin $I = 1$, the quadrupolar T_1 and T_2 are

$$T_{1Q}^{-1} = S[J(\omega_0) + 4J(2\omega_0)] \tag{13a}$$

$$T_{2Q}^{-1} = (S/2)[3J(0) + 5J(\omega_0) + 2J(2\omega_0)] \tag{13b}$$

$S = (3\pi^2/10)(e^2qQ/h)^2$ where (e^2qQ/h) is the quadrupolar coupling constant in Hz. T_{1Q} and T_{2Q} depend on τ_c in a similar manner to T_{1C} and T_{2C} (Figure 1).

18.2.6 Cross-correlation

Rigorous density-matrix analysis of magnetic relaxation in multispin systems[3] shows that cross-correlation between different DD interactions should be taken into account. Experimentally, cross-correlation appears in the form of non-exponential relaxation and different relaxation times for the components of spin–spin multiplets. In practice, the simple expressions given above apply with sufficient accuracy provided that initial relaxation rates are considered and provided that the motional anisotropy is small. One of the few reports of cross-correlation effects in polymers is that of markedly non-exponential methyl ^1H relaxation in poly(2,6-dimethyl-1,4-phenylene oxide)[8] where highly anisotropic motion arises from rapid methyl internal rotation superimposed on slower backbone segmental motions.

18.2.7 Strategies in Characterizing Polymer Motions by Magnetic Relaxation

Unfortunately, the relationship between experimental data and spectral densities is not so direct, nor can data be gathered over such a wide frequency range, that the correlation function may be obtained by a reverse Fourier transformation of equation (3). Instead, some adjustable form is assumed for the correlation function, based on a physically reasonable model, the parameters of which are obtained by searching for the best fit to as wide a range of experimental data as possible. A large data set can be obtained by measuring several parameters at a given resonance frequency (T_1, T_2, NOE) and/or measuring parameters at different frequencies. Although T_2 is rather difficult to measure accurately, it fills an important complementary role to T_1, since T_2 depends on J(0), *i.e.* the total area under the correlation function, whereas T_1 involves components in the resonance frequency region.

In analyzing relaxation data, it is necessary to know the interaction constants. For ^{13}C relaxation, the constant involves the number of protons (n) and the fairly standard C–H distance. Differences in the quantities nT_{1C} or nT_{2C} therefore directly reflect differences in motion. For this reason, the qualitative discussion of polymer dynamics in the following Section concentrates on ^{13}C results. In contrast, ^{1}H relaxation involves numerous ^{1}H–^{1}H internuclear distances which depend strongly on molecular structure and conformation, and the evaluation of the structural sums in equations (8a) and (8b) can be rather uncertain. The principal advantage of ^{1}H relaxation is its very high sensitivity compared with other nuclei, so that very dilute solutions may be studied. In addition, it is somewhat easier to measure T_2 for ^{1}H than for ^{13}C, so giving a wider sample of the spectral density frequency space. This last consideration will assume greater significance in Section 18.4 where the quantitative analysis of dynamic models is considered.

^{2}H relaxation resembles ^{13}C relaxation in that the interaction parameter (in this case the nuclear quadrupole coupling constant) is relatively invariant to chain structure. However, the natural abundance of ^{2}H is only 0.015%, so that practical use of ^{2}H relaxation requires deuteration.

18.3   GENERAL FEATURES OF RELAXATION IN MODERATELY DILUTE SOLUTION

Figure 2 shows ^{1}H T_1 and T_2 data for poly(methyl acrylate) as an illustration of typical relaxation behaviour. As outlined in the previous section, the T_1 values of both methine and methylene protons pass through a minimum, becoming equal at low temperature because of spin diffusion. T_2 in contrast varies monotonically with temperature. In general for random coil chains at normal temperatures, relaxation parameters correspond to correlation times at, or to the low τ_c side of, the T_1 minimum, indicating apparent correlation times of the order of 10^{-9} s or less. Of itself, a value as short as this indicates that the dominant relaxation processes are local segmental motions since the correlation time for overall molecular tumbling is typically 10^{-8} s or greater. Further support for the crucial role of segmental motions stems from the molecular weight and concentration dependence of T_1 and T_2.

18.3.1   Dependence on Molecular Weight

The variation of T_1 (and less frequently T_2) with molecular weight has been studied for a wide variety of chain structures. In all random coil polymers, the principal feature is that in moderately

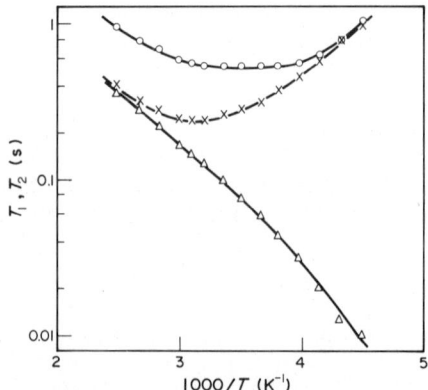

Figure 2   The temperature dependence of ^{1}H relaxation data at 300 MHz for backbone protons of poly(methyl acrylate), 1% (w/w) in toluene-d_8. ○, methine T_1; × methylene T_1; △ methylene T_2 (adapted from F. Heatley and M. K. Cox, *Polymer*, 1981, **22**, 288)

dilute solution, T_1 and T_2 become independent of size above a fairly low critical value. Some of the systems studied, with the approximate critical degree of polymerization (DP) are: poly(oxyethylene)[9] (DP $\approx$ 30), polystyrene[10] (DP $\approx$ 100), poly(isobutene)[11] (DP $\approx$ 50), poly(α-methylstyrene)[12] (DP $\approx$ 90), poly(2,6-dimethyl-1,4-phenylene oxide)[13] (DP $\approx$ 85) and bisphenol A polycarbonate[14] (DP $\approx$ 15). The fact that T_1 and T_2 reach limiting values supports the view that the dominant relaxation processes are local segmental motions independent of the total chain size. Below the critical value, T_1 and T_2 increase with decreasing molecular weight, consistent with an increasing contribution to loss of correlation from overall molecular tumbling. Although a number of synthetic polymers can adopt persistent extended structures in solution,[15] the only well documented study by NMR is the polypeptide poly(γ-benzyl-L-glutamate),[16,17] which forms a molecular solution of regular H-bonded helices under suitable conditions. In this system, the apparent correlation times for backbone motions fulfilled the condition $\omega_C^2 \tau_c^2 \gg 1$, and T_{1C} increased continually with increasing DP up to a value of at least 600, well above typical values for random coils. Here the most important relaxation process is overall molecular tumbling.

The size of segment engaged in those local motions controlling relaxation is not clearly defined. There is experimental data to suggest that up to eight or ten chain bonds may be involved cooperatively. For example, in *alt*-copoly(terephthalic acid/1,6-hexanediol),[18] the longest alkane ^{13}C T_1 at 34 °C is 0.2 s, compared with an estimated 0.6 s for polyethylene at the same temperature (Section 18.3.3). Thus the influence of the ester group is still felt some five bonds or so along the chain. Similarly, in copoly(alkene/styrene)s,[19] the alkylene and phenylethylene unit motions are decoupled when the average sequence length reaches six to eight carbons. In copoly(but-1-ene/propene)s,[20] the ^{13}C T_1 of propylene units is dependent on the sequence structure if a but-1-ene unit is located within two or three units of the propylene unit in question.

18.3.2 Dependence on Concentration and Solvent

Numerous studies have shown that in dilute solutions T_1 and T_2 are essentially constant, but as the concentration increases, T_1 and T_2 vary in a manner reflecting decreased mobility. T_1 studies include poly(oxyethylene),[9] polystyrene,[10,21] poly(methyl methacrylate),[22] poly(dimethylsiloxane)[23] and poly(isobutene).[24,25] T_2 data are available for poly(oxyethylene),[9] poly(dimethylsiloxane),[23] poly(isobutene)[25] and polystyrene.[21,26] Usually the onset of a concentration dependence occurs at a lower concentration for T_2 than for T_1. For ^{13}C relaxation in polystyrene in CDCl$_3$,[21] the concentration at which T_1 and T_2 deviate depends on molecular weight and correlates well with the dilute/semi-dilute boundary c^*. Below c^*, T_1 and T_2 are equal but above c^* they diverge in the order $T_2 < T_1$. The plateau for T_1 and T_2 in dilute solution is perhaps surprising in view of the large changes in macroscopic viscosity. However, although in this regime the effect of changing viscosity *via* concentration is small, the frictional properties of the solvent itself are significant, as shown by investigations of the solvent dependence of relaxation in polystyrene,[27] syndiotactic poly(methyl methacrylate)[28] and poly(oxyethylene).[29,30] The most extensive study is the last of these,[30] covering a solvent viscosity range of 0.4 to 8.4 mPa s using simple halogenated alkanes and aromatic solvents. Empirically it was found that for a series of chemically similar solvents, there is an approximately linear relationship between T_1 and reciprocal solvent viscosity. However, the relaxation times also depend on the solvent chemical structure. There is a particularly large deviation from a correlation with solvent viscosity for θ solvents,[27,28] such solvents apparently behaving with an abnormally high viscosity. A further example is the comparison of poly(oxyethylene) in water and in penta-chloroethane,[29] where the ^{1}H T_1 values are essentially identical yet the viscosity of water is a factor of three less than the latter solvent. Such anomalous solvent effects are probably attributable to changes in chain conformation or strong interactions between polymer and solvent. The fact that in dilute solution T_1 and T_2 are independent of concentration over a wide range yet dependent on the solvent lends support to the view that the effective motions are small-scale local motions dependent on the 'microviscosity' of the immediate chain environment.

One special solvent effect is the phenomenon of micellization of block copolymers when dissolved in a solvent selective for one block. The incompatible component forms the compact micelle core and consequently shows decreased mobility, in some cases attaining solid-like properties even in dilute solution. Investigations of such copolymer systems include styrene *block*-copolybutadiene in organic solvents,[31,32] *block*-copoly[styrene *alt*-co(ethylene/propylene)] in paraffins,[33] alkylene *block*-copoly(oxyethylene) in water [34,35] and oxyethylene *block*-copoly(oxypropylene) in water.[36]

In some instances of apparently homogeneous homopolymer solutions, NMR spectra show evidence of extensive microscopic aggregation. In some solutions of poly(methyl methacrylate),[37,38]

Table 1 ^{13}C Relaxation Times, nT_{1C}, and Apparent Isotropic Correlation Times, τ_c^{app}, for Main-chain Carbons

Polymer	Solvent	Temperature (°C)	Concentration (wt %)	nT_{1C} (s)	τ_c^{app} (ns)	Ref.
Poly(oxymethylene)	HFIP[a]	30	3	0.60	0.082	1
Poly(oxyethylene)	C_6D_6	30	5	2.80	0.018	2
Poly(1,4-epoxycyclohexane)	m-Cresol	32	5	0.065[b]	0.76	3
Poly(styrene oxide)	$CDCl_3$	36	6.5	0.24[b]	0.20	4
Poly(styrene sulfide)	$CDCl_3$	25	15	0.19	0.25	5
Polyethylene	ODCB[c]	100	25	2.70	0.018	6
	ODCB	30[d]	25	1.24	0.04	6
	Decalin-d_{18}	90	5	1.4	0.035	7
Polypropylene	ODCB	100	25	1.13	0.044	6
	ODCB	30[e]	25	0.39	0.13	6
Poly(but-1-ene)	CCl_3CHCl_2	100	10	0.34[f]	0.14	8
Poly(isobutene)	CCl_4	45	5	0.37	0.13	9
Polystyrene	Toluene-d_8	30	15	0.10	0.49	10
Poly(o-methylstyrene)	$CDCl_3$	31	14	0.055	1.2	11
Poly(4-vinyl-N-octylpyridinium bromide)	CD_3OD	27	25	0.069	42	11
Poly(vinyl chloride)	TCB[g]	107	10	0.32	0.15	12
Poly(vinylidene chloride)	HMPA-d_{18}[h]	89	15	0.25[b]	0.20	13
Poly(methyl methacrylate)						
isotactic	$CDCl_3$	38	10	0.126	0.39	14
syndiotactic	$CDCl_3$	38	10	0.080	0.62	14
Poly(methacrylic acid)	$D_2O(pD=8)$	26	4	0.058	1.1	15
Poly(L-lysine)[i]	$D_2O(pD=7)$	30		0.12	0.42	16
Poly(L-glutamic acid)[j]	$D_2O(pD=7)$	30	6	0.080	0.62	17

Poly(1-butene sulfone)	CDCl$_3$	40	25	0.090	23	18
Poly(2-butene sulfone)	CDCl$_3$	30	11	0.112	0.47	19
cis-1,4-Poly(butadiene)	CDCl$_3$	54	20	3.00	0.016	20
trans-1,4-poly(butadiene)	CDCl$_3$	54	20	2.38	0.021	20
Poly(2,6-dimethyl-1,4-phenylene oxide)	CDCl$_3$	30	10	0.093	0.57	21
Bisphenol A polycarbonate	CDCl$_3$	40	10	0.992	0.051	22

[a] Hexafluoro-2-propanol. [b] At 15 MHz. [c] *o*-Dichlorobenzene. [d] Extrapolated from high-temperature data with $E_a = 10.5$ kJ mol^{-1}. [e] As footnote d but $E_a = 14.3$ kJ mol^{-1}. [f] At 22.6 MHz. [g] 1,2,4-Trichlorobenzene. [h] Hexa(methyl-d_3) phosphoramide. [i] In random coil form. [j] All data at 25.14 MHz except as indicated.

1. G. Hermann and G. Weill, *Macromolecules*, 1975, **8**, 171.
2. F. Heatley and I. Walton, *Polymer*, 1976, **17**, 1019.
3. L. L. Chapoy, K. Matsuo and W. H. Stockmayer, *Macromolecules*, 1985, **18**, 188.
4. K. Matsuo, W. H. Stockmayer and S. Mashimo, *Macromolecules*, 1982, **15**, 606.
5. R. E. Cais and F. A. Bovey, *Macromolecules*, 1977, **10**, 752.
6. Y. Inoue, A. Nishioka and R. Chûjô, *Makromol. Chem.*, 1973, **168**, 163.
7. A. A. Jones, G. L. Robinson, F. E. Gerr, M. Bisceglia, S. L. Shostak and R. P. Lubianez, *Macromolecules*, 1980, **13**, 95.
8. F. C. Schilling, R. E. Cais and F. A. Bovey, *Macromolecules*, 1978, **11**, 325.
9. A. A. Jones, R. P. Lubianez, M. A. Hanson and S. L. Shostak, *J. Polym. Sci., Polym. Phys. Ed.*, 1978, **16**, 1685.
10. W. Gronski and N. Murayama, *Makromol. Chem.*, 1978, **179**, 1509.
11. T. Okada, *Polym. J.*, 1979, **11**, 843.
12. F. C. Schilling, *Macromolecules*, 1978, **11**, 1290.
13. K. Matsuo and W. H. Stockmayer, *Macromolecules*, 1981, **14**, 544.
14. J. R. Lyerla, Jr., T. T. Horikawa and D. E. Johnson, *J. Am. Chem. Soc.*, 1977, **99**, 2463.
15. J. D. Cutnell and J. A. Glasel, *Macromolecules*, 1976, **9**, 71.
16. B. Perly, Y. Chevalier and C. Chachaty, *Macromolecules*, 1981, **14**, 969.
17. A. Tsutsumi, B. Perly, A. Forchioni and C. Chachaty, *Macromolecules*, 1978, **11**, 977.
18. R. E. Cais and F. A. Bovey, *Macromolecules*, 1977, **10**, 757.
19. A. H. Fawcett, S. Fee and L. Waring, *Polymer*, 1983, **24**, 1571.
20. W. Gronski and N. Murayama, *Makromol. Chem.*, 1976, **177**, 3017.
21. F. Lauprêtre and L. Monnerie, *Eur. Polym. J.*, 1975, **11**, 845.
22. A. A. Jones and M. Bisceglia, *Macromolecules*, 1979, **12**, 1136.

the integrated high-resolution spectrum intensity was less than expected from the total polymer content. The intensity loss was attributed to a very broad underlying component from aggregates.

For sensitivity reasons, many experimental studies of polymer relaxation have been performed on concentrations considerably in excess of c^* and perhaps even c^{**}. It must therefore be recognized that analyses of the data in terms of segmental motions will incorporate the effects of interchain interactions. As described below, quantitative analysis of relaxation usually involves at least two timescales, a fast time characterizing local motions and a slow time representative of longer range processes. It is the latter which will be most affected by entanglements.

18.3.3 Dependence on Molecular Structure

In order to compare chain dynamics with molecular structure it is necessary to use not only the same sample conditions (molecular weight, solvent, concentration, temperature) but also the same nucleus and resonance frequency. Although there is an unfortunately wide spread of conditions in the literature, the data available are sufficient to provide a useful insight into molecular factors. For this purpose, ^{13}C relaxation data for protonated backbone carbons are convenient since the C–H bond length is reasonably constant. A representative selection of high molecular weight asymptotic values of nT_{1C} for backbone ^{13}C nuclei in CH_n groups is presented in Table 1, together with apparent isotropic correlation times, τ_c^{app}, calculated assuming isotropic rotational diffusion (equation 4). All systems except poly(but-1-ene sulfone) and poly(4-vinyl-N-octylpyridinium bromide) lie on the short τ_c side of the T_1 minimum.

In general, the most mobile systems are simple chains containing heteroatoms [e.g. poly(oxyethylene)] or double bonds (e.g. polybutadiene). This is consistent with the relatively low barriers to rotation[39] about $-CH_2-O-$ and $-CH_2-C=$ bonds (~ 4 kJ mol^{-1}) compared to $-CH_2-CH_2-$ (~ 16 kJ mol^{-1}). It is noteworthy that poly(oxymethylene) (POM) is less mobile than poly(oxyethylene) (POE) though containing more $-CH_2-O-$ bonds. This behaviour can be explained[40] by the fact that POE has several well-populated conformations, whereas POM is substantially restricted to a single conformation, thus inhibiting segmental motions.

As expected, substituents decrease chain mobility. The larger the substituent, the larger is the effect, e.g. in the sequence polyethylene, polypropylene, polystyrene, poly(4-vinyl-N-octylpyridinium chloride). Surprisingly, two geminal substituents, as in poly(isobutene), are not significantly more effective in hindering motion than is one substituent. The explanation, supported by conformational energy calculations,[41] appears to be that in the disubstituted chains steric conflicts are so severe as to raise the energy of all conformations, including transition states, so leaving the barrier unaffected. The isomeric pair poly(but-1-ene sulfone) and poly(but-2-ene sulfone) make an interesting comparison. In spite of having two adjacent substituents, the latter is 50 times more mobile. The difference has been explained[42] in terms of a difference in conformation, the poly(but-1-ene sulfone) being predominantly in a *gauche* conformation with strong ordering between neighbouring $-SO_2-$ dipoles, whereas the conformation of poly(but-2-ene sulfone) is predominantly *trans* with independent $-SO_2-$ groups.

The effect of stereochemistry on relaxation is most marked for the examples quoted *viz.* cis- and trans-1,4-polybutadiene and isotactic and syndiotactic poly(methyl methacrylate). Different stereochemical forms of other vinyl polymers such as polypropylene,[43] poly(vinyl chloride)[44] and polystyrene[45] have also been studied, but differences are slight. Except for polystyrene, isotactic sequences are more mobile.

In general, the order of mobility measured by relaxation in solution approximately parallels the order of glass transition temperatures, so that motions in solution and in the bulk glass appear to be closely related. An interesting study of relaxation in four variants of bisphenol A polycarbonate[46] has shown that good bulk impact resistance correlates with unhindered aromatic ring rotation in solution.

18.3.4 Side-chain Motions

Side-chain rotation superimposed on backbone motions reduces the apparent correlation time by a factor dependent on the relative rates of the motions. Methyl groups in particular exhibit considerable freedom. For example in polypropylene at 100 °C in *ortho*-dichlorobenzene (25 wt %),[47] the CH_3 nT_{1C} of 5.16 s is some five times larger than the backbone value of 1.13 s. In contrast in the *gem*-disubstituted chains poly(isobutene)[24] and poly(methyl methacrylate),[48] nT_{1C} for the α-

CH_3 is only about twice as long due to greater steric interference. In poly(2,6-dimethyl-1,4-phenylene oxide), the CH_3 nT_{1C} is 2.04 s compared with the backbone's 0.093 s, consistent with the very low barrier to rotation for methyl groups bonded to sp^2-hybridized carbons.[39]

For alkyl chains, T_{1C} and hence mobility increases progressively towards the chain ends, as illustrated by the data for poly(4-vinyl-*N*-hexylpyridinium chloride) in Figure 3. The terminal methyl group shows exceptional freedom.

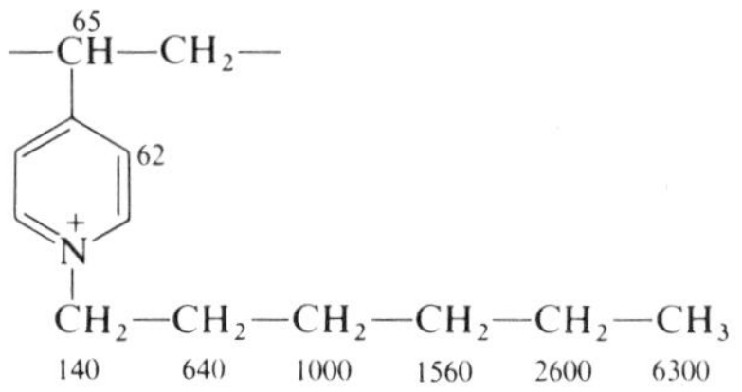

Figure 3 Values of nT_{1C} (in ms) for ^{13}C nuclei in poly(*N*-hexyl-4-vinylpyridinium bromide) at 25.1 MHz, 26 °C, 15% (w/w) in CD_3OD (from T. Yasukawa, D. Ghesquiere and C. Chachaty, *Chem. Phys. Lett.*, 1977, **45**, 279)

In contrast to methyl groups, phenyl ring substituents show low internal mobility. In polystyrene in THF for example,[45] nT_{1C} for the *ortho* ring carbons is only some 30% greater than nT_{1C} for backbone carbons.

18.4 QUANTITATIVE ANALYSIS OF SEGMENTAL MOTIONS

18.4.1 Theoretical Dynamic Models

Compared with Figure 1, experimental values for T_{1C}, T_{2C} and NOE show raised and broadened T_{1C} minima, low T_{2C} values relative to T_{1C} and broadened NOE transitions. In view of the complexity of polymer segmental motions, it is not surprising therefore that the isotropic rotation description is inadequate. The first attempt[49] to interpret ^{13}C relaxation parameters quantitatively invoked a semi-empirical distribution of correlation times, the so-called log-χ^2 distribution. This distribution is unsymmetrical with a tail towards longer correlation times, and was thought to be particularly suitable for polymers. It has been applied successfully to ^{13}C T_1 and NOE data in several polymers.[18,48-51] However, when applied to ^1H relaxation in poly(vinyl acetate),[6] the distribution was unable to simultaneously simulate $^1H–^1H$ cross-relaxation data (dependent on J(ω) at kHz frequencies) and ^1H T_1 values (dependent on J(ω) in the MHz region). Although a suitable distribution function could no doubt be derived to satisfy all the ^1H data, such a solution would not have avoided the general defect of distribution models, namely that they lack detailed insight into the nature of the chain motions. More attractive theories based on a molecular description of segmental motions have been developed.

Segmental motions have been classified[52] into three types, according to the relative positions in the initial and transition states of the 'tails' attached to the mobile segment. In type 1* transitions the tails are unaffected, in type 2* the tails suffer translation only and in type 3* the tails suffer translation and rotation. *A priori*, one would expect type 1* transitions such as those in Figure 4 to be the most favoured since these minimize solvent friction. For such transitions, it has been shown[52] that the diffusion-controlled rate constant for passage (k) over a potential barrier E^* is given by

$$k = \left[\frac{(\alpha_A \alpha_B)^{1/2}}{2\pi \sum_i \xi_i r_i^2} \right] \exp(-E^*/k_B T) \qquad (14)$$

where ξ_i is the friction constant for atom i in the mobile unit situated at r_i from the rotation axis, and α_A and α_B are quadratic force constants in the initial and transition states. If ξ_i is proportional to solvent viscosity η, and α_A, α_B and E^* are independent of solvent, equation (14) yields $k \propto \eta^{-1}$ and hence, since $\tau_c \propto k^{-1}$, $T_1 \propto \eta^{-1}$ in extreme narrowing, as observed experimentally in a homologous solvent series (Section 18.3.2). The type 1* transitions illustrated in Figure 4 form the basis of the most widely used models of polymer motion for relaxation analysis. These transitions occur in a chain distributed on a diamond lattice, and the model is therefore particularly applicable to vinyl polymers and polyethers. The theory was first developed by Valeur *et al.*[53-55] and their treatment

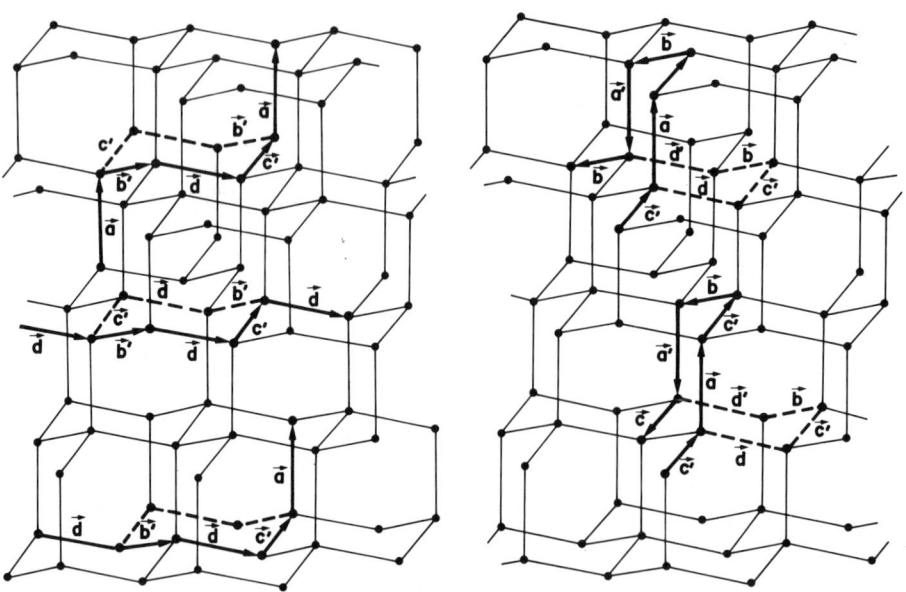

Figure 4   Three-bond (left) and four-bond (right) motions of a chain on a diamond-lattice (reproduced with permission of Wiley from F. Lauprêtre, C. Noël and L. Monnerie, *J. Polym. Sci., Polym. Phys. Ed.*, 1977, **15**, 2127)

will be referred to as the VJGM model. Subsequently modified versions were suggested by Jones and Stockmayer[56] (the JS model) and by Yaris and co-workers.[57-61]

In the VJGM model, both three- and four-bond motions are allowed in an infinite chain with full coupling between bond orientations. Kinetically, the master rate equation contains a diffusive term representing migration of a bond orientation along the chain by three-bond motions and a dissipative term representing the creation and destruction of orientations by four-bond motions. The correlation function for such internal motions is

$$\tilde{G}(t) = \exp(-|t|/\theta)\exp(|t|/\rho)\mathrm{erfc}[(|t|/\rho)^{1/2}] \tag{15}$$

where θ and ρ are correlation times dependent on the transition rates. Other processes may also contribute[55] to θ. The spectral density corresponding to equation (15) has been given by Hunt and Powles.[62]

$$J(\omega) = \frac{\theta\rho(\theta-\rho)}{(\theta-\rho)^2 + \omega^2\theta^2\rho^2}\left\{\left(\frac{\theta}{2\rho}\right)^{1/2}\left[\frac{(1+\omega^2\theta^2)^{1/2}+1}{1+\omega^2\theta^2}\right]^{1/2} + \left(\frac{\theta}{2\rho}\right)^{1/2}\left[\frac{\omega\theta\rho}{\theta-\rho}\right]\left[\frac{(1+\omega^2\theta^2)^{1/2}-1}{1+\omega^2\theta^2}\right]^{1/2} - 1\right\} \tag{16}$$

Note that although this equation was derived from a specific model, it was later pointed out[63] that its significance is more general in that ρ essentially represents cooperative processes and θ may also represent non-cooperative processes such as single-bond rotations.

In the JS modification,[56] only three-bond motions are considered, but conformational coupling is restricted to a finite chain segment thus imposing a sharp cut-off on correlated motions. For a segment of $(2m-1)$ bonds, the correlation function is a discrete distribution of exponential functions

$$\tilde{G}(t) = \sum_{k=1}^{s} B_k\exp(-|t|/\tau_k) \tag{17}$$

where $s = (m+1)/2$. The coefficients B_k and correlation times τ_k depend on m and the conformational transition rate and have been listed[56] for $m = 3, 5$ and 7. The distribution is conveniently characterized by m and the harmonic mean correlation time $\tau_h = \langle\tau_k^{-1}\rangle^{-1}$.

Yaris and colleagues[57-61] have solved the VJGM model for degrees of motional correlation intermediate between the infinite chain VJGM model and the sharp cut-off JS model. They emphasized that diffusive and dissipative motional components are general features of dynamic systems, and that although their theory is based on the actual transitions in Figure 4, the link may be tenuous. It was pointed out[57] that the solution of the diffusive motional component is an integral of wave vectors of plane waves, but physically the wave vectors are limited by a lower bound set by the smallest moveable segment, and an upper bound set by the damping of a conformational displacement by loss processes. If k_A and k_B are wave vectors for the long and short wavelength cut-offs, the correlation function (BY model)[57] is

$$\tilde{G}(t) = \frac{1}{2}\left(\frac{\pi}{t}\right)^{1/2}(\mu_B^{1/2} - \mu_A^{1/2})^{-1}\{\text{erfc}[(\mu_A|t|)^{1/2}] - \text{erfc}[(\mu_B|t|)^{1/2}]\}$$ (18)

where $\mu_{A(B)} = 2k_{A(B)}^2 D/5$ and D is the motional diffusion constant. The corresponding spectral density is

$$J(\omega) = (\mu_B^{1/2} - \mu_A^{1/2})^{-1}\left\{[4(2\omega)^{1/2}]^{-1}\left[\ln\left(\frac{\mu_B - (2\omega\mu_B)^{1/2} + \omega}{\mu_B + (2\omega\mu_B)^{1/2} + \omega}\right) - \ln\left(\frac{\mu_A - (2\omega\mu_A)^{1/2} + \omega}{\mu_A + (2\omega\mu_A)^{1/2} + \omega}\right)\right]\right.$$
$$\left. + [2(2\omega)^{1/2}]^{-1}\left[\tan^{-1}\left(\frac{(2\omega\mu_B)^{1/2}}{\omega - \mu_B}\right) - \tan^{-1}\left(\frac{(2\omega\mu_A)^{1/2}}{\omega - \mu_A}\right)\right]\right\}$$ (19)

Subsequently,[58,59] the long-wavelength cut-off was replaced by a damping term, and the theory was also extended to include the effects of chain–chain interactions[60] and of probes and side-chains.[61] Such additions only modify the rate parameters without affecting the form of the correlation function.

Although type 1* transitions intuitively seem the most plausible mechanism of polymer motion, there is evidence that single-bond rotations (type 3*) may be equally important if not more so. For example excimer fluorescence experiments show little difference between monomeric and polymer systems containing the same chromophore.[64] Further evidence derives from activation energies. Equation (14) implies that activation energies for correlation times can be written as $E_\tau = E^* + E_\eta$, where E_η is the viscosity activation energy. Observed values of E^* are typically about 10–20 kJ mol^{-1}, a range of values which is more consistent with rotation about one bond than two.[39] The existence of feasible one-bond rotations is supported by calculations and computer simulations[65-67] which show that the movement of the attached tails is assisted by torsional distortions in neighbouring bonds. However, the simulations also show that the initial rotation is frequently followed by a second rotation nearby leading effectively to type 1* transitions. Analytically,[68,69] these motions are represented by the correlation function (HH model)[68]

$$\tilde{G}(t) = \exp(-\kappa_0|t|)\exp(-\kappa_1|t|)I_0(\kappa_1|t|)$$ (20)

where I_0 is a modified Bessel function, κ_0 is a rate constant for single-bond rotations and κ_1 a rate constant for cooperative transitions.

18.4.2 Applications of Dynamic Models

The VJGM and JS theories have been extensively applied to experimental data, and have satisfactorily simulated the frequency dependence of T_1, T_2 and NOEs for both ^{1}H and ^{13}C nuclei with only two adjustable parameters, as summarized in Tables 2 and 3. For the VJGM model an extended test was the ^{1}H relaxation study of syndiotactic poly(methyl methacrylate)[70] where the data set included T_1 and T_2 at 300 MHz and T_1 at 80 MHz. Figure 5 shows the excellent agreement between experimental and simulated T_1 and T_2 values. The JS model has been most thoroughly examined for poly(2,6-dimethyl-1,4-phenylene oxide). Dynamic parameters from the original analysis[13] of ^{1}H T_1s at 30 MHz were subsequently used to compare predicted and experimental values of $T_{1\rho}$ at 42.6 kHz[71] and T_1 as a function of resonance frequency[72] with excellent agreement.

From Table 2, the ratio ρ/θ is generally less than unity, indicating preponderant diffusional characteristics of the motion. For poly(vinyl acetate), there is an exceptionally large increase in ρ/θ below $-10\,°$C. This has been attributed[61] to a transition, also seen in dielectric relaxation,[73] from a loose to a compact conformation. In Table 3, the fluorinated polystyrenes have the largest degree of dynamic correlation among the vinyl polymers, as judged by the values of m. This is consistent with the VJGM data in Table 2 where polystyrene in general has the lowest ρ/θ ratio.

The BY model has been applied to ^{1}H relaxation in poly(vinyl acetate)[59] and ^{13}C relaxation in polystyrene,[74] and the HH model to ^{1}H and ^{13}C,[75] and ^{2}H,[76] relaxation in polycarbonates.

Table 2   Applications of the VJGM Model

Polymer[a]	Solvent	Nucleus	Temp (K)	ρ (ns)	ρ/θ	Ref.
PMPO	CDCl$_3$	^{1}H, ^{13}C	323	0.55	0.05	1
PVAC	Toluene-d_8	^{1}H	228	400	100	2
			298	1.5	0.2	
			383	0.03	0.07	
PS	CDCl$_3$	^{1}H	303	0.6	0.08	3
	HCB[b]	^{13}C	317	2.0	0.07	4
	HCB[b]	^{1}H	317	2	1	
PB	PCE[c]	^{13}C	313	0.54	0.33	5
s-PMMA	Toluene-d_8	^{1}H	313	1.5	0.7	6
i-PMMA	Toluene-d_8	^{1}H	313	0.42	0.5	7
PMA	Toluene-d_8	^{1}H	298	0.63	0.4	8

[a] PMPO, poly(2,6-dimethyl-1,4-phenylene oxide); PVAC, poly(vinyl acetate); PS, polystyrene; PB, poly(but-1-ene); s-PMMA, syndiotactic poly(methyl methacrylate); i-PMMA, isotactic poly(methyl methacrylate); PMA, poly(methyl acrylate). [b] Hexachlorobutadiene. [c] Pentachloroethylene.

1. F. Lauprêtre and F. Gény, *Eur. Polym. J.*, 1978, **14**, 401.
2. F. Heatley and M. K. Cox, *Polymer*, 1977, **18**, 225.
3. F. Heatley and B. Wood, *Polymer*, 1978, **19**, 1405.
4. F. Lauprêtre, C. Noël and L. Monnerie, *J. Polym. Sci., Polym. Phys. Ed.*, 1977, **15**, 2127.
5. F. C. Schilling, R. E. Cais and F. A. Bovey, *Macromolecules*, 1978, **11**, 325.
6. F. Heatley and M. K. Cox, *Polymer*, 1980, **21**, 381.
7. F. Heatley and M. K. Cox, *Polymer*, 1981, **22**, 190.
8. F. Heatley and M. K. Cox, *Polymer*, 1981, **22**, 288.

Table 3   Applications of the JS Model

Polymer[a]	Solvent	Nucleus	Temp. (K)	τ_A (ns)	$2m-1$	Ref.
PMFS	CDCl$_3$	^{19}F	298	0.24	9–13	1
PPFS	CDCl$_3$	^{19}F	298	0.33	9–13	1
PMPO	CDCl$_3$	^{1}H	293	3.0	5	2
PIB	CCl$_4$	^{1}H, ^{13}C	318	0.058	5	3
PC	CDCl$_3$	^{1}H, ^{13}C	293	0.14	9	4
PE	Decalin-d_{18}	^{1}H, ^{13}C	363	0.0077	5	5
PVDC	HMPA-d_{18}[b]	^{1}H, ^{13}C	313	0.22	9	6
PMVK	Dioxane-d_8	^{13}C	299	0.54	9	7
PEC	m-Cresol	^{13}C	305	*ca.*0.5	15	8

[a] PMFS and PPFS, poly(m- and p-fluorostyrene); PMPO, poly(2,6-dimethyl-1,4-phenylene oxide); PIB, poly(isobutene); PC, poly(bis-(4-hydroxyphenyl)-2,2-propane carbonate) (bisphenol A polycarbonate); PE, polyethylene; PVDC, poly-(vinylidene chloride); PMVK, poly(methyl vinyl ketone); PEC, poly(1,4-epoxycyclohexane). [b] Hexa(methyl-d_3) phosphoramide.

1. K. Matsuo, K. F. Kuhlmann, H. W.-H. Yang, F. Gény, W. H. Stockmayer and A. A. Jones, *J. Polym. Sci., Polym. Phys. Ed.*, 1977, **15**, 1347.
2. A. A. Jones and R. P. Lubianez, *Macromolecules*, 1978, **11**, 126.
3. A. A. Jones, R. P. Lubianez, M. A. Hanson and S. L. Shostak, *J. Polym. Sci., Polym. Phys. Ed.*, 1978, **16**, 1685.
4. A. A. Jones and M. Bisceglia, *Macromolecules*, 1979, **12**, 1136.
5. A. A. Jones, G. L. Robinson, F. E. Gerr, M. Bisceglia, S. L. Shostak and R. P. Lubianez, *Macromolecules*, 1980, **13**, 95.
6. K. Matsuo and W. H. Stockmayer, *Macromolecules*, 1981, **14**, 544.
7. S. Mashimo, P. Winsor, R. H. Cole, K. Matsuo and W. H. Stockmayer, *Macromolecules*, 1983, **16**, 965.
8. L. L. Chapoy, K. Matsuo and W. H. Stockmayer, *Macromolecules*, 1985, **18**, 188.

18.4.3   Comparison of Models

Although the correlation functions above have apparently widely different forms, they have proved equally effective in practice. The BY model[57] and subsequent modifications[58, 59] for example satisfactorily reproduced ^{1}H relaxation in poly(vinyl acetate) originally analyzed[6] by the VJGM theory, while the VJGM, JS and BY functions proved equally adept in interpreting ^{13}C relaxation in isotactic polystyrene.[74] Likewise, the JS and HH models both gave acceptable interpretation of ^{1}H and ^{13}C relaxation in polycarbonates.[75, 76] Both models gave similar timescales, and both emphasized cooperative transitions over single-bond processes. Numerical comparisons of the VJGM and

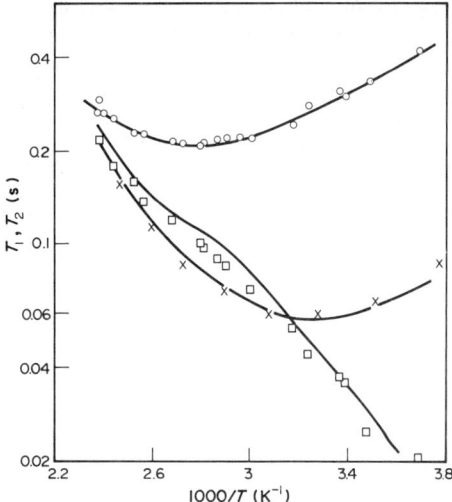

Figure 5 ¹H relaxation data for the backbone CH_2 protons in syndiotactic poly(methyl methacrylate), 1% (w/w) in toluene-d_8. The symbols are experimental data: $\bigcirc$ T_1 at 300 MHz; $\times$ T_1 at 80 MHz; $\square$ T_2 at 300 MHz. The lines are simulations using the VJGM model, equation (16), with $\rho/\theta=0.7$ (adapted from F. Heatley and M. K. Cox, *Polymer*, 1980, **21**, 381)

HH correlation functions[69] and of the JS and HH functions[77] have shown that with suitable choice of parameters, all three can be made to correspond closely. The agreement is due fundamentally to the fact that all are linked by a diffusive mechanism for cooperative motions.[77] A common feature of all model correlation functions is a relatively rapid initial decay followed by a much slower final loss of correlation. This is essentially the same form as a sum of two exponentials, and such a simple expression has received theoretical[78] and practical[21] support in terms of a division of motions into two categories; fast processes meeting the extreme narrowing condition and a single effective slow process. A triexponential expression based on an anisotropic rotational diffusion model has also been successfully employed.[79, 80]

In view of the multiplicity of apparently suitable models, it is not surprising that the ability of NMR to yield an insight into the molecular details of polymer motion has been questioned.[61, 81] Future progress towards this objective appears to lie in two directions. The first is to combine measurements on two or more nuclei in the same system in order to compare correlation functions for internuclear vectors oriented at different angles to the chain. Molecular dynamics simulations have shown that different directions may have quite different correlation functions.[69] An indication of the possible value of this approach is a ¹³C relaxation study of *cis*-1,4-polybutadiene,[82, 83] where it was observed that the nT_{1C} value for the CH carbon is significantly less than that for the CH_2. The difference was rationalized in terms of a specific crankshaft motion in which the CH and CH_2 groups reorient by different angles. The second approach is to obtain information on cross-correlation as well as auto-correlation. Such information is available in its most accessible form in ¹³C relaxation of ¹³CH_2 and ¹³CH_3 groups without decoupling.[3, 84] This technique has considerable potential[85] in characterizing anisotropic motion.

18.5 CONCENTRATED SOLUTIONS

In concentrated samples T_1 and T_2 become strongly dependent on concentration. Increasing concentration has a broadly similar effect to decreasing temperature. For example in the *cis*-polyisoprene/tetrachloroethylene[86] and poly(isobutene)/CS_2[87] systems, T_1 passed through a minimum as a function of concentration at fixed temperature. The effect of high polymer concentration on T_2 is normally a dramatic reduction, leading to such broad lines that chemical shifts are unresolved.[88-91] There are in fact two contributions to the decrease in T_2, one dynamic and the other static. The dynamic contribution arises from the decrease in T_2 due to a decrease in mobility as illustrated in Figure 1. The static effect arises from the presence of very slow motions, variously described in terms of either entanglements[86] or slow evolution of the 'tube' occupied by a chain immersed in the polymer/solvent matrix.[87] If these motions are sufficiently slow, dipole–dipole interactions are not averaged completely. The residual interactions appear as unresolved splittings which contribute to the decay of transverse magnetization, a feature which has been termed 'pseudo-solid' behaviour and which manifests itself experimentally in three ways. Firstly the decay curve in T_2 experiments may be biexponential,[89,92] consisting of a 'fast' component from segments of restricted mobility where residual coupling is largest and a 'slow' component from more mobile segments. The T_1 is usually a single exponential because of spin diffusion. The second consequence of a pseudo-solid component is deviation of the free induction decay (FID) following a 90° pulse from the exponential form occurring in liquid systems. It has been shown theoretically[93] that for a residual dipolar coupling δ_r, the FID is given by

$$F(t) = \exp(-t/T_2)[1 + 3\phi^2 + U(\phi)]^{1/2}/U(\phi)\sqrt{2} \tag{21}$$

where $U(\phi) = [(1 + 3\phi^2)^2 + 4\phi^6]^{1/2}$, $\phi = \delta_r t$ and T_2 is the dynamic contribution to transverse relaxation. The dipolar part of this function resembles a Gaussian but decays more slowly (as $t^{-3/2}$) at long times. Finally the third method of detecting a pseudo-solid component is to rotate the sample at a rate greater than the residual coupling strength.[90,91] If the rotation axis is perpendicular to the magnetic field (the conventional arrangement in iron-cored magnet spectrometers), the residual coupling contribution to the linewidth is halved. The pseudo-solid effect is analogous to that observed in a chemically cross-linked system,[94,95] but with the vital difference that entanglements are transient and are reduced by decreasing concentration or molecular weight, or by increasing temperatures.[87] For rubbers such as 1,4-poly(dienes)[87,90] or poly(isobutene)[91] at ambient temperature, the residual coupling even at 80–90% polymer is only of the order of 100 Hz, about 1% of the rigid lattice broadening, but for the glassy poly(methyl methacrylate) at 60% or more in $CDCl_3$, the residual coupling approaches *ca.* 80% of solid polymer linewidth.[88]

The residual coupling provides a convenient reference frequency for detecting slow motions on this timescale. The criterion of residual coupling has been suggested[96] as an NMR method of defining the onset of polymer coil overlap but the critical concentration defined in these terms, c_v, is greater by an order of magnitude than the ideal overlap concentration c^*. The value of c_v corresponds more closely to the onset of direct polymer–polymer interactions ($c \geq 10/[\eta]$). The concentration c^* has been detected[21] in short-chain polystyrenes by a change in the relative magnitudes of ^{13}C T_1 and T_2 (Section 18.2). The difference between c^* as revealed by the T_1/T_2 comparison and c_v as revealed by residual dipolar coupling is understandable in terms of the development of slow motional components above c^* which affect T_2 more than T_1 but which do not become sufficiently slow to generate residual coupling until a concentration $\sim 10c^*$ is attained.

The concept of free volume has been used successfully to rationalize the temperature and concentration dependence of T_1 and T_2 in *cis*-polyisoprene/tetrachloroethylene[86] and residual dipolar coupling in *cis*-polybutadiene/CS_2 solution,[90] although this approach was not so satisfactory for poly(isobutene).[91] For a solution with polymer volume fraction v_p, the fractional free volume at temperature T is written[86,90]

$$f(T, v_p) = f_s^0 + \alpha_s(T - T_0) + v_p(f_p^0 - f_s^0) + v_p(\alpha_p - \alpha_s)(T - T_0) \tag{22}$$

where the subscripts p and s denote polymer and solvent, f^0 is the fractional free volume at a reference temperature T_0 and α is the coefficient of volume expansion. From the theory of Turnbull and Cohen,[97] the viscosity at given T and v_p relative to some reference state is related to free volume by

$$\eta(T, v_p)/\eta_0 = \exp[B(f(T, v_p)^{-1} - f_0^{-1})] \tag{23}$$

where B is a constant. Here the reference state is taken to be pure polymer at T_0. Making the assumptions that the measured NMR parameter λ (either T_2[86] or δ_r[90]) depends on a correlation time as τ_c^n, and that $\tau_c \propto \eta$, equations (22) and (23) yield a Williams–Landel–Ferry relation[98]

$$\ln \lambda = A' + B'/(T - T_\infty) \tag{24}$$

$$A' = \ln \lambda_0 - B/f_p^0$$

$$B' = nB/[\alpha_s + v_p(\alpha_p - \alpha_s)]$$

$$T_\infty = T_0 - [f_s^0 + v_p(f_p^0 - f_s^0)]/[\alpha_s + v_p(\alpha_p - \alpha_s)]$$

At a given concentration the temperature dependence of T_2 for the polyisoprene system[86] and δ_r for the polybutadiene system[90] obeyed equation (24) very well. The dependence of T_∞ on v_p was linear, consistent with the near equality of α_p and α_s. In both cases, values of f_p^0 derived from the NMR data were close to those obtained from viscoelastic measurements. It thus appears that the motions monitored in these NMR experiments are the same as those determining rheological properties.

18.6 TRANSLATIONAL DIFFUSION

18.6.1 Experimental Technique

Self-diffusion constants are measured by NMR using a variant of the spin-echo technique in which transverse relaxations are compared with and without the application of a field gradient. The preferred method at present is to apply the field gradient, G, as two pulses before and after the π pulse in the standard $[\pi/2-\tau-\pi-\tau-\text{echo}]$ spin-echo sequence.[99] If δ is the width and Δ the separation of the gradient pulses, the magnitude of the echo at time 2τ is attenuated relative to that without a field gradient by the factor $\exp(-\gamma^2 D_s R)$, where $R = \delta^2 G^2(\Delta - \delta/3)$ and D_s is the diffusion constant. The attenuation arises because diffusion during the pulse interval Δ means that the spins do not experience exactly the same field during the second gradient pulse as they did during the first, and so refocussing of the echo is incomplete. The echo amplitude is also attenuated by transverse relaxation after the $\pi/2$ pulse and since a low diffusion constant is usually accompanied by a short T_2, the practical lower limit of D measurable by this method is $\sim 10^{-14}\ \text{m}^2\ \text{s}^{-1}$. The diffusion time Δ ranges from ~ 1 ms to ~ 1 s, corresponding to diffusion distances of the order of μm. For all except the highest molecular weights ($\geq 10^6$) this distance is considerably larger than the coil radius, so that centre of mass diffusion is monitored. For extremely long chains, segmental motions contribute to the measured rate.

In principle, it is possible to detect non-Fickian diffusion by determining D as a function of Δ, and several such cases have apparently been reported.[100,101] However, the reliability of such observations has not yet been established,[102] and clarification of this topic is an important item for the future. Non-Fickian diffusion is revealed by curvature in plots of log(attenuation) vs. R from which D is derived, but it is difficult to separate non-Fickian characteristics from other causes such as polydispersity,[100,103-107] impurities[100,108] and restricted diffusion.[99,109] By Fourier transforming the echo to give a conventional frequency domain NMR spectrum,[110] molecular signals can be clearly differentiated. This not only alleviates the impurity problem, but also opens the door to simultaneous measurements of diffusion constants for different components, such as solute and solvent.[111]

It must be emphasized that NMR measures the self-diffusion coefficient defined[112] in terms of the velocity auto-correlation function of a single particle

$$D_s = \frac{1}{3}\int_0^\infty \langle \mathbf{V}_1(t)\cdot \mathbf{V}_1(0)\rangle\,dt \tag{25}$$

Other techniques, such as photon correlation spectroscopy or flow in a concentration gradient, measure the mutual diffusion coefficient which is associated with relaxation of solute density fluctuations[112]

$$D_m = L_m \frac{1}{c_1}\left(\frac{\partial \pi}{\partial c_1}\right)_{T,\mu_0} \tag{26}$$

where π is the osmotic pressure and the indices 0 and 1 label solvent and solute. L_m is the mutual diffusion Onsager kinetic coefficient which involves solute velocity auto- and cross-correlation functions. D_m is controlled by both hydrodynamic and thermodynamic factors.[113,114] D_s and D_m are necessarily equal only at infinite dilution, but whereas with increasing concentration D_s invariably decreases, D_m may increase in good solvents.[113]

18.6.2 Diffusion in Dilute Solution

In dilute solution ($c < c^*$) standard hydrodynamic theory[115] predicts a variation of D_s with concentration according to

$$D_s^{-1} = D_0^{-1}(1 + k_s c + \ldots) \tag{27}$$

where D_0 is the value of D_s at infinite dilution. In some studies of polystyrene fractions, linear plots of D_s^{-1} vs. c were obtained,[113,116,117] but elsewhere[118] a linear relation between $\log D_s$ and c was observed. The latter relation was also found for poly(oxyethylene)[119,120] and dextran.[121] According to Flory,[122] D_0 varies with molecular weight and RMS end-to-end distance according to

$$D_0 = \frac{kT}{\eta_0 P M^{1/2}} \left[\frac{M}{\langle r_0^2 \rangle} \right]^{1/2} \frac{1}{M^\alpha} \tag{28}$$

where P is a numerical constant, η_0 is the solvent viscosity, and α varies from 0 in a θ solvent to 0.1 in a good solvent. Since the ratio $[M/\langle r_0^2 \rangle]^{1/2}$ is constant for $M \geq 1000$, D_0 should vary as M^n with n lying between -0.5 and -0.6 according to solvent. This relationship is fulfilled for a number of systems, both linear[108,116,117,119,123] and star-branched.[108,123] Values of D_0 for star systems with up to 18 arms were in good agreement with the theoretical prediction[124]

$$D_0(F, M) = D_0(2, M)[2 - F + 2^{1/2}(F - 1)]F^{-1/2} \tag{29}$$

where $D_0(F, M)$ is D_0 for a star with F arms and molecular weight M.

18.6.3 Diffusion in Semi-dilute Solution

For semi-dilute solutions where chain–chain interactions are significant, most attention has been directed towards testing scaling laws for the dynamics of entangled systems,[125] according to which D_s should vary with molecular weight and concentration as $D_s \propto M^{-2} c^{-\gamma}$ where γ varies from 1.75 (good solvent) to 3 (θ solvent). In the first extensive studies of semi-dilute polystyrene solutions,[116,117] the experimental M and c exponents differed significantly from those predicted. The M exponent was found to be -1.4, while γ varied appreciably over the semi-dilute concentration regime. More detailed examination[1] of the concentration dependence in CCl_4 and C_6D_6 showed that γ varied from ~ 1.75 in the lower end of the semi-dilute region to ~ 3 in the upper. Two explanations were offered, one based on a change towards θ conditions as c increases and the other suggesting a change in the statistics of chains between entanglements. As a further complication, a recent study of polystyrene diffusion in THF[126] showed that γ lay very close to 1.75 over a wide range provided that a hitherto unacknowledged correction for free volume changes was applied. Thus although there is a measure of broad agreement with the scaling theory, some of the details require further elucidation.

18.7 REFERENCES

1. P. T. Callaghan and D. N. Pinder, *Macromolecules*, 1984, **17**, 431.
2. (a) D. Shaw, 'Fourier Transform NMR Spectroscopy', Elsevier, Amsterdam, 1976; (b) M. L. Martin, G. J. Martin and J. J. Delpuech, 'Practical NMR Spectroscopy', Heyden, London, 1980; (c) R. K. Harris, 'Nuclear Magnetic Resonance Spectroscopy', Pitman, London, 1983.
3. L. G. Werbelow and D. M. Grant, in 'Advances in Magnetic Resonance', ed. J. S. Waugh, Academic Press, New York, 1977, vol. 9, p. 190.
4. J. R. Lyerla and G. C. Levy, in 'Topics in Carbon-13 NMR Spectroscopy', ed. G. C. Levy, Wiley, New York, 1974, vol. 1, p. 79.
5. J. H. Noggle and R. E. Schirmer, 'The Nuclear Overhauser Effect', Academic Press, New York, 1971.
6. F. Heatley and M. K. Cox, *Polymer*, 1977, **18**, 225.
7. R. L. Vold and R. R. Vold, *J. Chem. Phys.*, 1974, **61**, 2525.
8. R. P. Lubianez and A. A. Jones, *J. Magn. Reson.*, 1980, **38**, 331.
9. K.-J. Liu and R. Ullman, *J. Chem. Phys.*, 1968, **48**, 1158.
10. A. Allerhand and R. K. Hailstone, *J. Chem. Phys.*, 1972, **56**, 3718.
11. Y. Inoue, A. Nishioka and R. Chûjô, *J. Polym. Sci., Polym. Phys. Ed.*, 1973, **11**, 2237.
12. F. Lauprêtre, C. Noël and L. Monnerie, *J. Polym. Sci., Polym. Phys. Ed.*, 1977, **15**, 2143.
13. A. A. Jones and R. P. Lubianez, *Macromolecules*, 1978, **11**, 126.
14. A. A. Jones and M. Bisceglia, *Macromolecules*, 1979, **12**, 1136.
15. G. C. Berry and C. E. Sroog (eds.), *J. Polym. Sci., Polym. Symp.*, 1978, **65**.
16. A. Allerhand and E. Oldfield, *Biochemistry*, 1973, **12**, 3428.
17. P. M. Budd, F. Heatley, T. J. Holton and C. Price, *J. Chem. Soc., Faraday Trans. 1*, 1981, **77**, 759.
18. R. A. Komoroski, *J. Polym. Sci., Polym. Phys. Ed.*, 1979, **17**, 45.
19. A. V. Cunliffe and R. A. Pethrick, *Polymer*, 1980, **21**, 1025.
20. M. Mauzac, J. P. Vairon and F. Lauprêtre, *Polymer*, 1979, **20**, 443.
21. P. Stilbs and M. E. Moseley, *Polymer*, 1981, **22**, 321.
22. Yu. Ya. Gotlib, M. I. Lifshits and V. A. Shevelev, *Polym. Sci. USSR (Engl. Transl.)*, 1975, **17**, 2132.
23. C. Cuniberti, *J. Polym. Sci., Part A-2*, 1970, **8**, 2051.
24. F. Heatley, *Polymer*, 1975, **16**, 493.
25. W. P. Slichter and D. D. Davis, *Macromolecules*, 1968, **1**, 47.

26. F. Lauprêtre, C. Noël and L. Monnerie, *J. Polym. Sci., Polym. Phys. Ed.*, 1977, **15**, 2127.
27. F. Heatley and B. Wood, *Polymer*, 1978, **19**, 1405.
28. F. Heatley and M. K. Cox, *Polymer*, 1980, **21**, 381.
29. K.-J. Liu and J. E. Anderson, *Macromolecules*, 1970, **3**, 163.
30. M. C. Lang, F. Lauprêtre, C. Nöel and L. Monnerie, *J. Chem. Soc., Faraday Trans. 2*, 1979, **75**, 349.
31. F. Heatley and A. Begum, *Makromol. Chem.*, 1977, **178**, 1205.
32. J. Spěváček, *Makromol. Chem., Rapid Commun.*, 1982, **3**, 697.
33. F. Candau, F. Heatley, C. Price and R. B. Stubbersfield, *Eur. Polym. J.*, 1984, **20**, 685.
34. K. Nakamura, R. Endo and M. Takeda, *J. Polym. Sci., Polym. Phys. Ed.*, 1977, **15**, 2095.
35. F. Heatley, H. H. Teo and C. Booth, *J. Chem. Soc., Faraday Trans. 1*, 1984, **80**, 981.
36. J. Rassing, W. P. McKenna, S. Bandyopadhyay and E. M. Eyring, *J. Mol. Liq.*, 1984, **27**, 165.
37. J. Spěváček and B. Schneider, *Makromol. Chem.*, 1975, **176**, 3409.
38. J. Spěváček, *J. Polym. Sci., Polym. Phys. Ed.*, 1978, **16**, 523.
39. D. G. Lister, J. N. Macdonald and N. L. Owen, 'Internal Rotation and Inversion', Academic Press, London, 1978.
40. F. Gény and L. Monnerie, *J. Polym. Sci., Polym. Phys. Ed.*, 1979, **17**, 131, 147.
41. R. H. Boyd and S. M. Breitling, *Macromolecules*, 1972, **5**, 1; P. R. Sundararajan and P. J. Flory, *J. Am. Chem. Soc.*, 1974, **96**, 5025.
42. A. H. Fawcett, S. Fee and L. Waring, *Polymer*, 1983, **24**, 1571.
43. T. Asakura and Y. Doi, *Macromolecules*, 1981, **14**, 72.
44. F. C. Schilling, *Macromolecules*, 1978, **11**, 1290.
45..W. Gronski and N. Murayama, *Makromol. Chem.*, 1978, **179**, 1509.
46. J. F. O'Gara, S. G. Desjardins and A. A. Jones, *Macromolecules*, 1981, **14**, 64.
47. Y. Inoue, A. Nishioka and R. Chûjô, *Makromol. Chem.*, 1973, **168**, 163.
48. J. R. Lyerla Jr., T. T. Horikawa and D. E. Johnson, *J. Am. Chem. Soc.*, 1977, **99**, 2463.
49. J. Schaefer, *Macromolecules*, 1973, **6**, 882.
50. F. Heatley and A. Begum, *Polymer*, 1976, **17**, 399.
51. T. Asakura and Y. Doi, *Macromolecules*, 1983, **16**, 786.
52. E. Helfand, *J. Chem. Phys.*, 1971, **54**, 4651.
53. B. Valeur, J.-P. Jarry, F. Gény and L. Monnerie, *J. Polym. Sci., Polym. Phys. Ed.*, 1975, **13**, 667.
54. B. Valeur, L. Monnerie and J.-P. Jarry, *J. Polym. Sci., Polym. Phys. Ed.*, 1975, **13**, 675.
55. B. Valeur, J.-P. Jarry, F. Gény and L. Monnerie, *J. Polym. Sci., Polym. Phys. Ed.*, 1975, **13**, 2251.
56. A. A. Jones and W. H. Stockmayer, *J. Polym. Sci., Polym. Phys. Ed.*, 1977, **15**, 847.
57. J. T. Bendler and R. Yaris, *Macromolecules*, 1978, **11**, 650 (correction: F. Heatley and J. T. Bendler, *Polymer*, 1979, **20**, 1578).
58. J. Skolnick and R. Yaris, *Macromolecules*, 1982, **15**, 1041 (correction: *ibid*, 1983, **16**, 491).
59. J. Skolnick and R. Yaris, *Macromolecules*, 1982, **15**, 1046 (correction: *ibid*, 1983, **16**, 492).
60. J. Skolnick and R. Yaris, *Macromolecules*, 1983, **16**, 266.
61. B. B. Pant, J. Skolnick and R. Yaris, *Macromolecules*, 1985, **18**, 253.
62. B. I. Hunt and J. G. Powles, *Proc. Phys. Soc., London*, 1966, **88**, 513.
63. P. Tekely, F. Lauprêtre and L. Monnerie, *Macromolecules*, 1983, **16**, 415.
64. T.-P. Liao and H. Morawetz, *Macromolecules*, 1981, **19**, 231.
65. E. Helfand, Z. R. Wasserman and T. A. Weber, *Macromolecules*, 1980, **13**, 526.
66. J. Skolnick and E. Helfand, *J. Chem. Phys.*, 1980, **72**, 5489.
67. E. Helfand and J. Skolnick, *J. Chem. Phys.*, 1982, **77**, 5714.
68. C. K. Hall and E. Helfand, *J. Chem. Phys.*, 1982, **77**, 3275.
69. T. A. Weber and E. Helfand, *J. Phys. Chem.*, 1983, **87**, 2881.
70. F. Heatley and M. K. Cox, *Polymer*, 1981, **22**, 190.
71. A. A. Jones, G. L. Robinson, F. E. Gerr, M. Bisceglia, S. L. Shostak and R. P. Lubianez, *Macromolecules*, 1980, **13**, 95.
72. A. A. Jones and F. P. Shea, *J. Polym. Sci., Polym. Phys. Ed.*, 1982, **20**, 681.
73. S. Mashimo and K. Shinohara, *J. Phys. Soc. Jpn.*, 1973, **34**, 1141.
74. A. A. Jones, G. L. Robinson and F. E. Gerr, *ACS Symp. Ser.*, 1977, **103**, 271.
75. J. J. Connolly, E. Gordon and A. A. Jones, *Macromolecules*, 1984, **17**, 722.
76. J. A. Porco, P. T. Inglefield, A. A. Jones and J. Campbell, *Polym. Prepr. (Am. Chem. Soc., Div. Polym. Chem.)*, 1985, **26**, 172.
77. Y.-Y. Lin, A. A. Jones and W. H. Stockmayer, *J. Polym. Sci., Polym. Phys. Ed.*, 1984, **22**, 2195.
78. G. Lipari and A. Szabo, *J. Am. Chem. Soc.*, 1982, **104**, 4546.
79. G. Hermann and G. Weill, *Macromolecules*, 1975, **8**, 171.
80. C. W. R. Mulder, J. Schriever and J. C. Leyte, *J. Phys. Chem.*, 1985, **89**, 475.
81. P. M. Henrichs, *J. Polym. Sci., Polym. Phys. Ed.*, 1983, **21**, 263.
82. W. Gronski and N. Murayama, *Makromol. Chem.*, 1974, **177**, 3017.
83. W. Gronski, *Makromol. Chem.*, 1977, **178**, 2949.
84. R. R. Vold and R. L. Vold, *J. Chem. Phys.*, 1976, **64**, 320.
85. J. H. Prestegard and D. M. Grant, *J. Am. Chem. Soc.*, 1978, **100**, 4664.
86. A. Charlesby and B. J. Bridges, *Eur. Polym. J.*, 1981, **17**, 645.
87. J. P. Cohen-Addad and A. Guillermo, *J. Polym. Sci., Polym. Phys. Ed.*, 1984, **22**, 931.
88. J. Zajíček, H. Pivcova and B. Schneider, *Makromol. Chem.*, 1981, **182**, 3169.
89. J. Zajíček, H. Pivcova and B. Schneider, *Makromol. Chem.*, 1981, **182**, 3177.
90. J. P. Cohen-Addad and J. P. Faure, *J. Chem. Phys.*, 1974, **61**, 1571.
91. J. P. Cohen-Addad and C. Roby, *J. Chem. Phys.*, 1975, **63**, 3095.
92. J. P. Cohen-Addad, M. Domard and S. Boileau, *J. Chem. Phys.*, 1981, **75**, 4107.
93. J. P. Cohen-Addad and R. Dupeyre, *Polymer*, 1982, **24**, 400.
94. A. Charlesby, *Radiat. Phys. Chem.*, 1979, **14**, 919.
95. A. Charlesby and R. Folland, *Radiat. Phys. Chem.*, 1980, **15**, 393.

96. J. P. Cohen-Addad, *J. Chem. Phys.*, 1979, **71**, 3689.
97. D. Turnbull and M. H. Cohen, *J. Chem. Phys.*, 1961, **34**, 120.
98. J. D. Ferry, 'Viscoelastic Properties of Polymers', 2nd edn., Wiley, New York, 1970, p. 303.
99. E. O. Stejskal and J. E. Tanner, *J. Chem. Phys.*, 1965, **42**, 288.
100. I. Zupancic, G. Lahajnar, R. Blinc, D. H. Reneker and D. L. Vanderhart, *J. Polym. Sci., Polym. Phys. Ed.*, 1985, **23**, 387.
101. A. Peterlin, *Makromol. Chem.*, 1983, **184**, 2377.
102. E. D. von Meerwall, *Rubber Chem. Technol.*, 1985, **58**, 527.
103. D. W. McCall and C. M. Huggins, *Appl. Phys. Lett.*, 1965, **7**, 153.
104. E. D. von Meerwall, *J. Magn. Reson.*, 1982, **50**, 409.
105. E. D. von Meerwall and K. R. Bruno, *J. Magn. Reson.*, 1985, **62**, 417.
106. G. Fleischer, *Makromol. Chem., Rapid Commun.*, 1985, **6**, 463.
107. P. T. Callaghan and D. N. Pinder, *Macromolecules*, 1985, **18**, 373.
108. C. Xuexin, X. Zhongde, E. D. von Meerwall, N. Seung, N. Hadjichristidis and L. J. Fetters, *Macromolecules*, 1984, **17**, 1343.
109. J. E. Tanner, *J. Chem. Phys.*, 1978, **69**, 1748.
110. T. L. James and G. G. McDonald, *J. Magn. Reson.*, 1973, **11**, 58.
111. B. Nyström, M. E. Moseley, P. Stilbs and J. Roots, *Polymer*, 1981, **22**, 218.
112. A. R. Altenberger and M. Tirrell, *J. Polym. Sci., Polym. Phys. Ed.*, 1984, **22**, 909.
113. T. Cosgrove and J. M. Sutherland, *Polymer*, 1983, **24**, 534.
114. W. Brown, P. Stilbs and R. M. Johnsen, *J. Polym. Sci., Polym. Phys. Ed.*, 1983, **21**, 1029.
115. H. Yamakawa, 'Modern Theory of Polymer Solutions', Harper and Row, New York, 1971, chap. 6.
116. P. T. Callaghan and D. N. Pinder, *Macromolecules*, 1980, **13**, 1085.
117. P. T. Callaghan and D. N. Pinder, *Macromolecules*, 1981, **14**, 1334.
118. M. E. Moseley, *Polymer*, 1980, **21**, 1479.
119. J. E. Tanner, K.-J. Liu and J. E. Anderson, *Macromolecules*, 1971, **4**, 586.
120. W. Brown and P. Stilbs, *Polymer*, 1982, **23**, 1780.
121. W. Brown, P. Stilbs and R. M. Johnsen, *J. Polym. Sci., Polym. Phys. Ed.*, 1982, **20**, 1771.
122. P. J. Flory, 'Principles of Polymer Chemistry', Cornell University Press, Ithaca, NY, 1953, chap. 14.
123. E. D. von Meerwall, D. H. Tomich, J. Grigsby, R. W. Pennisi, L. J. Fetters and N. Hadjichristidis, *Macromolecules*, 1983, **16**, 1715.
124. W. H. Stockmayer and M. Fixman, *Ann. N.Y. Acad. Sci.*, 1953, **57**, 334.
125. P. G. de Gennes, 'Scaling Concepts in Polymer Physics', Cornell University Press, Ithaca, New York, 1979.
126. E. D. von Meerwall, E. J. Amis and J. D. Ferry, *Macromolecules*, 1985, **18**, 260.

19

NMR Spectroscopy of Polymers in the Solid State

VINCENT J. McBRIERTY

University of Dublin, Ireland

19.1 INTRODUCTION

The use of NMR to study the structure and dynamics of polymers in solution has been treated elegantly in the two preceding chapters. These presentations give a lucid account of the fundamentals of the method as well as providing vivid illustrations of the power of NMR as a spectroscopic tool. Equally the many challenges encountered in the application of NMR to the solid state have been approached with ingenuity and imagination. These efforts have been richly rewarded particularly in regard to the site specific information contained in high resolution spectra which are now routinely available. Consider first some of the difficulties that arise in the absence of the rapid motional narrowing that typifies the liquid state.

The inherent morphological complexity of solid polymers constitutes a significant impediment to attaining the ultimate goal of relating macroscopic to microscopic events. Characteristically different morphologies can coexist both in semicrystalline homopolymers and in composites of different materials such as blends, segregated block copolymers, filled, plasticized and antiplasticized systems. Indeed, refinements in the sensitivity of NMR have been such that small scale structural and motional heterogeneities are now readily discernable in nominally amorphous polymers such as poly(vinyl chloride) (PVC).[1,2] The difficulties in modelling polymer systems can give rise to significant interpretive problems particularly when the homopolymer in question is a component in a composite.[3-5]

The motion of polymer molecules is complex, involving cooperativity, anisotropy and susceptibility to geometric constraints.[6] Several modes of motion may be active at the same time.[7] The frequency distribution of motions of the backbone chain in particular can be extremely broad and defies description in terms of a single correlation frequency v_c as required in the rudimentary theory of NMR relaxation.[6,8] There is also the related difficulty that the average of the Hamiltonian over the motions encountered in solids is non-zero. Thus, the usual Bloembergen, Purcell and Pound (BPP)[9] or Kubo–Tomita (KT)[10] analyses have to be modified.[11]

Spin diffusion tends to even out magnetization gradients in the perturbed spin system with a concomitant loss of structural and site specific information.[3,6,12,13] These effects are particularly important in solids.

Dipolar interactions generally dominate in the solid in a way that obscures the detailed information contained in the weaker chemical shift and exchange interactions (*cf.* Volume 1, Section 17.2). As a consequence, the interpretation of raw data from low resolution NMR has generally relied, albeit quite successfully, on intuitive notions based on somewhat idealized models for the polymer. This contrasts with the localized molecular information in high resolution NMR when dipolar interactions are small or zero.

In liquids, rapid motion greatly simplifies NMR spectra but there is a concomitant loss of information associated with the additional complexity in solids. In approaching high resolution solid state NMR, these complexities are first removed and then selectively reintroduced to regain the detailed structural and motional information that characterizes the solid state. Recent developments have facilitated the selective manipulation of the Hamiltonian that characterizes the spin system in the solid.[14-18] These experiments are primarily designed to recover, at least in part, the sensitivity of liquid state NMR through suppression of the dipolar contribution by means of cross polarization (CP), dipolar decoupling (DD) and magic angle spinning (MAS) techniques. The result is well illustrated in Table 1 which furnishes typical data for ^{13}C nuclear spin interactions.[19] Note the relative strengths of the various contributions to the total interaction.

The ability to manipulate the spin system through the imposition of external forces rests largely in the fact that the applied static, B_0, and dynamic, $B_1(t)$, magnetic fields interact more strongly with

Table 1   Solution State *vs.* High Resolution Solid State NMR Spectroscopy[a]

Total interaction	=	Zeeman (MHz)	+	Dipolar (kHz)	+	Scalar (Hz)	+	Chemical shift
Solution state		50		0		200		isotropic, single frequency
Solid state		50		50		200		200 p.p.m.-wide chemical shift anisotropy
Solid state NMR technique used to overcome broadening				high power (dipolar) proton decoupling		high power decoupling also removes *J* coupling		magic angle spinning gives the isotropic line

[a] ^{13}C nuclear spin interactions are portrayed for poly(butylene terephthalate) in a field $B_0 = 4.7$ Tesla.[19]

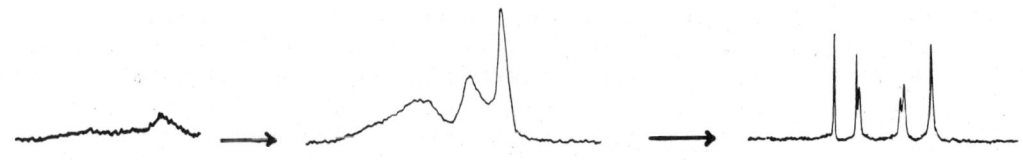

the nuclear dipoles than (a) the dipoles interact with each other, or (b) the dipoles interact with the thermal motions of the host molecules, *i.e.* the lattice. This remarkable situation has led to a wealth of sophisticated multiple pulse sequences which dominate the natural motions of the spins in a controlled way and allow the generation of sharp NMR absorption resonances for example as opposed to the over-damped motions detected in dielectric relaxation for which the dipole motion is tightly coupled to the host lattice.

The theory of coherent averaging, the details of which are beyond the scope of this chapter, permits analysis of many of these experiments.[17, 18] It is in this area that many of the more recent developments of significance have taken place. Notable too, is the way in which the hitherto deleterious effects of spin diffusion have been turned to advantage in the examination of small scale morphology in one-dimensional (1-D)[3, 6, 13, 20] and two-dimensional (2-D) NMR[21, 22] experiments.

In discussing the results of NMR there has been the tendency to adopt two philosophically different approaches. The first exploits Redfield's idea that in many instances the spin system can be described in terms of a spin temperature because of its relative isolation from the host lattice.[23] The second approach applies to those experiments where the effective Hamiltonian describing the spin system is altered, either by mechanically rotating the sample,[24] or by suitably applying radio frequency (rf) fields to reveal more precise and definitive information on the dynamics of the system.[14-18] Numerous examples from the literature will be used here to demonstrate the efficacy of these ideas and will illustrate the wealth of information that can be obtained from solid state NMR. A prerequisite examination of some fundamental concepts and experimental approaches will facilitate subsequent discussion. A prevailing theme throughout will be the way in which this information can be used in conjunction with the results of other experiments to arrive at a more complete understanding of the polymer under examination.

19.2 BASIC CONCEPTS

19.2.1 The Hamiltonian

The properties of the spin system are described in terms of the Hamiltonian

$$\mathcal{H} = \mathcal{H}_Z + \mathcal{H}_{rf} + \mathcal{H}_D + \mathcal{H}_{cs} + \mathcal{H}_Q + \mathcal{H}_J + \mathcal{H}_{SR} \tag{1}$$

The first two contributions are 'external' Hamiltonians which describe the interactions between the nuclear spins and B_0 and the applied rf field $B_1(t)$, respectively. They lead to the resonance condition (Figure 1 of Volume 1, Chapter 17)

$$\Delta E = \hbar \omega_0 = g_N \beta_N B_0$$

$$= \mu B_0 / I$$

$$= \gamma \hbar B_0 \tag{2}$$

where g_N is the nuclear g-factor, β_N is the nuclear magneton, μ is the magnetic moment of the nucleus, I is the spin of the nucleus, γ is the magnetogyric ratio and ω_0 is the resonant frequency. The remaining terms describe the 'internal' response of the spin system, where $\mathcal{H}_D$, $\mathcal{H}_{cs}$ and $\mathcal{H}_Q$ are respectively the dipolar, chemical shift and quadrupolar (for spins with $I > 1/2$ in low symmetry sites) interactions. With few exceptions the indirect electron-coupled interaction $\mathcal{H}_J$ and spin rotation interactions $\mathcal{H}_{SR}$ make an insignificant contribution to $\mathcal{H}$ in solids and may be ignored. It may be noted in passing that $\mathcal{H}_D$ and $\mathcal{H}_J$ couple each spin in the system with all others, whereas the remaining Hamiltonians involve sums of single spin interactions.

The internal Hamiltonians may be treated in a unified way in terms of irreducible tensor notation.[17, 25] It suffices for our purposes merely to list some of the important characteristics illustrating in particular the orientation dependence of each (Table 2).

As mentioned in the introduction, the dipolar Hamiltonian generally dominates and is central to most of the basic theories of NMR in solids. For spins with electric quadrupole moments, $\mathcal{H}_Q$ inevitably controls relaxation. The fact that both $\mathcal{H}_D$ and $\mathcal{H}_Q$, if present, average to zero in non-viscous liquids underpins the remarkable sensitivity of solution NMR to the weaker and richly informative chemical shift and exchange interactions. Clearly the goal in high resolution solid state NMR is either to suppress $\mathcal{H}_D$ and $\mathcal{H}_Q$ to an insignificant level, or to capitalize on the information in these parts of the Hamiltonian.

Table 2 Functional Form of the Various Contributions to the Hamiltonian Appropriate to Solids

Hamiltonian	Spin operator	Dependence on B_0	Isotropic average	Orientational dependence
Dipolar				
homonuclear	$I_1 \cdot I_2 - 3I_{1z}I_{2z}$	None	0	$(3\cos^2\theta - 1)$[a]
heteronuclear	$I_{1z}I_{2z}$	None	0	$(3\cos^2\theta - 1)$
Chemical shift	I_z	Linear	σ	$[3\cos^2\theta - 1 - \eta_\sigma \sin^2\theta \cos 2\phi]$[b]
Quadrupole	$3I_z^2 - I^2 + \eta_Q(I_x^2 - I_y^2)$	None	0	$[3\cos^2\theta - 1 - \eta_Q \sin^2\theta \cos 2\phi]$

[a] θ is the angle between the internuclear vector **r** and B_0 (Figure 1).
[b] (θ, ϕ) are the polar angles of the principal axis system in the laboratory coordinate frame, and η is the asymmetry parameter (see text).

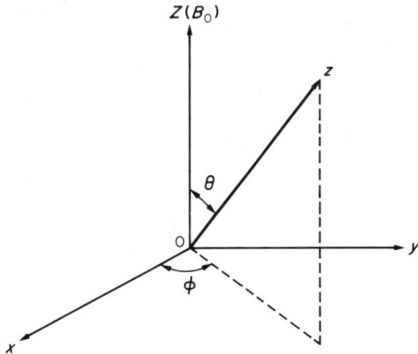

Figure 1 Orientation of a molecular coordinate frame (xyz) relative to the laboratory frame (XYZ) in terms of Euler angles $(\phi\theta\gamma)$. For dipolar interactions, $0z$ is the direction of the internuclear vector r_{12}. For chemical shift and quadrupolar interactions xyz is the principal axis system

19.2.2 The Local Field

The local field conceptually allows one to think in terms of the way in which single nuclear spins behave under the combined influence of B_0, $B_1(t)$ and their local environment. The latter is described in terms of a local field B_{loc} which takes account of the partial, but very useful, magnetic field contribution at the nucleus arising from neighbouring dipole–dipole interactions. In a solid with many such neighbouring contributions, B_{loc} forms a smooth, broad distribution about B_0. In non-viscous liquids, on the other hand, the local field is spatially averaged to zero to give a very narrow distribution about B_0 (*cf.* Volume 1, Section 17.2.3.1). Clearly there are intermediate situations where the onset of specific motions gives rise to partial averaging of the distribution, and consequently the local field is a sensitive probe of the environment in which the resonant nucleus finds itself. In the steady state experiment where the resonant frequency, ω_0, is fixed and B_0 is swept through resonance, the absorption spectrum (or, in practice, its first derivative) for the solid polymer manifests the character of B_{loc}. The linewidth between points of inflection, δB, or the second moment (mean square deviation), M_2, are used to describe the spectrum. Alternatively, the loss of phase coherence arising from the distribution of Larmor frequencies associated with the distribution of B_{loc} can be measured through observation of the decay of transverse magnetization following a burst of rf irradiation. The observed nuclear signal is the free induction decay (FID) described by the spin–spin relaxation time T_2. For a broad distribution in B_{loc}, as in solids, the dephasing time, and therefore T_2, is short (*ca.* 10^{-5} s) whereas in liquids the converse is true with T_2 of the order of tens of milliseconds. Formally, the FID is the Fourier transform of the lineshape of the absorption spectrum and contains identical information.

This rather graphical picture has certain shortcomings in the sense that the local field is viewed as a classical one thereby ignoring the quantum nature of the spin. Nevertheless, this approach accounts for many of the salient features of NMR. As the number of neighbours approaches infinity, the quantum effects become negligible.

Further developing the concept of the local field, it is recognized that even the static (secular) part of B_{loc} has components parallel and perpendicular to B_0. The parallel component principally alters the relative rates of precession of the spins whereas the perpendicular component gives rise to energy exchange between spins as they precess about directions that are not quite aligned with B_0. This exchange, involving energy conserving flip-flop transitions between like spins, has been termed spin diffusion in the introduction.

The time dependent part of the local field, arising from the thermally driven motion of the polymer chains to which the spins are attached, is vital in determining the way in which the perturbed spin system gives up energy to the lattice. Recognizing that spin–lattice relaxation involves transitions between different nuclear magnetic energy levels, it is evident that the process is most efficient for motions at or near the resonant frequency, $\omega_0 = \gamma B_0$. In essence, the role of the rf field is mimicked by the internal time dependent spin interactions and *vice versa*. The spin–lattice relaxation time T_1 characterizes the rate of energy transfer to the lattice.

Relaxation in the resonantly rotating frame involves spin-locking the magnetization M along B_1 using the pulse sequence described in Section 19.7. This leads to a type of frozen equilibrium state since the decay of M along B_1, characterized by the rotating frame relaxation time $T_{1\rho}$, is significantly faster than the return to true equilibrium along B_0 ($T_{1\rho} \ll T_1$). In effect, one is performing a T_1 experiment at the B_1 field while retaining the sensitivity appropriate to the higher static field B_0. Since $B_1 \ll B_0$, $T_{1\rho}$ probes motions of lower frequency which allows more direct comparison with dynamic mechanical and dielectric data (*cf.* Section 19.4.5).

19.2.3 Spin Temperature

One can observe NMR relaxation times on a time scale of microseconds or longer due to the weak coupling of the nuclear spin system to the lattice. The body of spins can retain excess energy for times which are long compared to T_2. ($T_1 \gg T_2$ in solids). It is thus possible to assign a spin temperature T_s to the nearly isolated spin system and to visualize relaxation phenomena as the *thermal* equilibration of the perturbed spins with their surroundings. Recalling that spin diffusion can average out spatial magnetization gradients in a homonuclear spin system at a rate which is of the order of the dipolar couplings, the possibility of a *uniform* spin temperature for the spin system as a whole is thereby allowed (*cf.* Section 19.5).

The applicability of the spin temperature approach requires that there is no oscillatory behaviour. By way of illustration, a pulse of resonant radiation imposes phase coherence on the precessing spins for a period defined by T_2 during which the Boltzmann expression $e^{-\Delta E/kT_s}$ inadequately describes the energy level populations. The validity of the spin temperature approach is reinstated, however, for times which are long compared to T_2.

Thus, for a system of spins at equilibrium, T_s is simply the lattice temperature T. Upon resonant irradiation T_s increases, whereas T remains unchanged. Subsequent spin–lattice relaxation involves the transfer of excess spin energy to the lattice on a timescale defined by T_1. In the rotating frame experiment where the magnetization M is spin-locked along B_1, the spin temperature in this frame may be defined in terms of the Curie expression $M = CB_1/T_s$.[26] Noting that M was initially established in the laboratory frame characterized by a temperature T_L, then $M = CB_0/T_L$ in which case $T_s = T_L(B_1/B_0)$. Thus the spin temperature in the rotating frame is lower than the temperature in the laboratory frame by the factor $B_1/B_0 \ll 1$. The observed decay in the rotating frame therefore stems from a flow of energy from the lattice to the colder spin system.

An interesting situation arises when the amplitude of the spin-locking field B_1 is adiabatically reduced to a value which is less than the dipolar local fields. This has the effect of converting the order associated with the Zeeman polarization with respect to B_1 to ordering of the spins with respect to their local fields. Under these conditions the dipolar energy states are characterized by a temperature which is of the order of $T_L(B_{loc}/B_0)$. This again is a non-equilibrium state characterized by a low spin temperature. The return to the lattice temperature in this case is described in terms of the dipolar spin–lattice relaxation time, T_{1D}.

Another important application of these ideas concerns the system which contains two nuclear spin species such as ^{1}H and ^{13}C. Simultaneous application of resonant radiation to each spin type allows specification of characteristic spin temperatures in the rotating frame. Since the interaction between the two spin systems is weaker than the interaction with the respective rf fields, the spin temperatures are different and are related as follows

$$\frac{T_S(^1\text{H})}{T_S(^{13}\text{C})} = \frac{B_1(^1\text{H})}{B_1(^{13}\text{C})} \tag{3}$$

The rate of energy transfer between the ^{1}H and ^{13}C spin systems is maximized when energy is conserved for pairs of spin flips (as in the case of spin diffusion). This is achieved when

$$\gamma_\text{H} B_1(^1\text{H}) = \gamma_\text{c} B_1(^{13}\text{C}) \tag{4}$$

which is the Hartmann–Hahn condition.[27] Ideally

$$\frac{T_S(^1\text{H})}{T_S(^{13}\text{C})} = \frac{B_1(^1\text{H})}{B_1(^{13}\text{C})} = \frac{\gamma_C}{\gamma_H} \approx \frac{1}{4} \tag{5}$$

The protons are cold relative to the carbon nuclei and under appropriate conditions energy will transfer from the ^{13}C to the ^{1}H spins with a concomitant increase in ^{13}C magnetization and sensitivity (cross polarization). This is fundamental to the routine detection of dilute ^{13}C nuclei.[18,28] Many other interesting and useful applications of the spin temperature concept are described in several excellent texts.[29]

19.3 LINESHAPES AND MOTIONAL AVERAGING

Dipole, chemical shift and quadrupole interactions, to a greater or lesser extent, produce characteristic broadening and/or splitting of NMR spectra. Although these interactions are anisotropic in nature, the continuous spatial distribution of molecular orientations in polycrystalline or powder samples usually results in broad, often featureless spectra with a consequent loss of structural information at a molecular level. The situation is redressed somewhat in studies on oriented polymers (*cf.* Section 19.9.3).[3, 6, 13] Proton spectra from solid polymers are typically 10–100 kHz wide, due largely to homonuclear dipolar couplings, whereas ^{13}C spectral linewidths arising from ^{13}C–^1H interactions are of the order of 1–10 kHz. Quadrupole couplings can lead to extremely broad spectra with linewidths in the range 100 kHz–100 MHz. Chemical shift spectra, however, are narrow, at most of the order of a few hundred p.p.m.

As mentioned in Section 19.2.2, dipolar broadened lineshapes in solids generally are symmetrically disposed about the resonant frequency and often manifest Gaussian character. By comparison, the magnetic shielding of the spin by surrounding electrons and nuclei leads to more interesting lineshapes. The chemical shift shielding tensor is a 3×3 matrix which can be made diagonal by choosing an appropriate molecule-fixed coordinate frame or principal axis system (PAS). The principal components of the shielding tensor are conventionally labelled σ_{11}, σ_{22} and σ_{33}, where σ_{33} corresponds to the direction of greatest shielding. For axial symmetry only two components are required. In isotropic fluids, the observed chemical shift arises from the average of the shielding tensor

$$\sigma = \tfrac{1}{3}(\sigma_{11} + \sigma_{22} + \sigma_{33}) \tag{6}$$

In a powder or polycrystalline sample each inequivalent site gives rise to a characteristic powder lineshape, the details of which reflect the site symmetry and the relative values of the principal components of the shielding tensor. Figure 2 illustrates powder lineshapes for asymmetric and axial situations. Note that the frequencies ω_1, ω_2 and ω_3 are a direct measure of the principal shielding components. That these lineshapes reflect the angular dependence of σ shown in Table 2 is graphically illustrated in Figure 2b where lines of constant chemical shift (or constant frequency) for a typical axial tensor are projected onto the 23 (or 13) plane in the PAS.[17, 30] The lineshape is

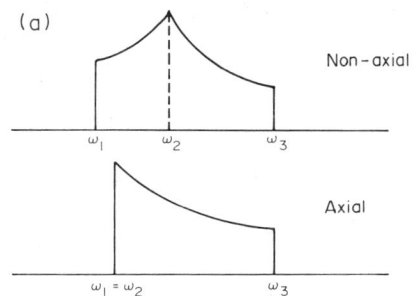

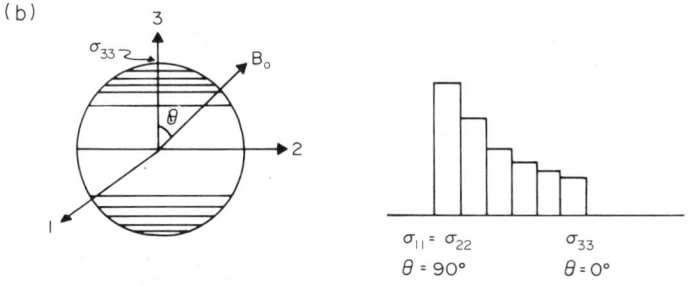

Figure 2 (a) Powder lineshapes for axial and asymmetric chemical shift tensors; (b) lines of constant chemical shift (or frequency) for a typical axial tensor (C_6F_6 at 201 K) projected onto the 23 or 13 plane in the principal axis system (PAS). The histogram (lineshape) is generated from the incremental areas between the line projections (shifted by progressive amounts of 20 p.p.m.)

generated from the incremental intensities or areas as shown. More complex patterns arise in cases where the asymmetry parameter, $\eta_\sigma = (\sigma_{22} - \sigma_{11})/(\sigma_{33} - \sigma)$, is non-zero.

Note also from Table 2 that the angular dependence of quadrupole interactions is identical to that which describes chemical shift effects. Consider the axially symmetric tensor for $I = 1$, as encountered in deuterium NMR. The spectrum comprises the superposition of two axially symmetric patterns, one the mirror image of the other, corresponding to the $0 \leftrightarrow 1$ and $-1 \leftrightarrow 0$ transitions, respectively (Figure 3). The spectrum, first derived by Pake to describe the dipolar interaction between two spins $I = \frac{1}{2}$, is centred on the resonant frequency ω_0.

On the question of motional averaging, it is recalled that the range of molecular motions encountered in solid polymers is wide, bounded at low temperatures by immobility or near immobility and at the other extreme by motions approaching liquid-like proportions. Local motions can attain very high frequencies at normally accessible temperatures.[7] In this context, it is also recalled that spectral lineshapes are sensitive to motions with correlation frequencies ν_c comparable

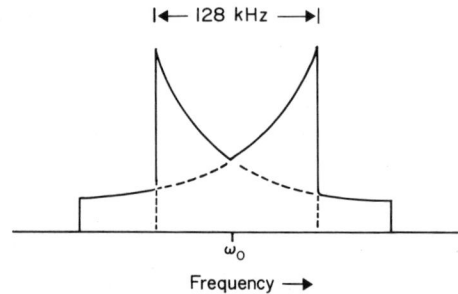

Figure 3 Theoretical ^{2}H lineshape (Pake doublet) for a static C–^{2}H bond. The quadrupole splitting is 128 kHz as indicated

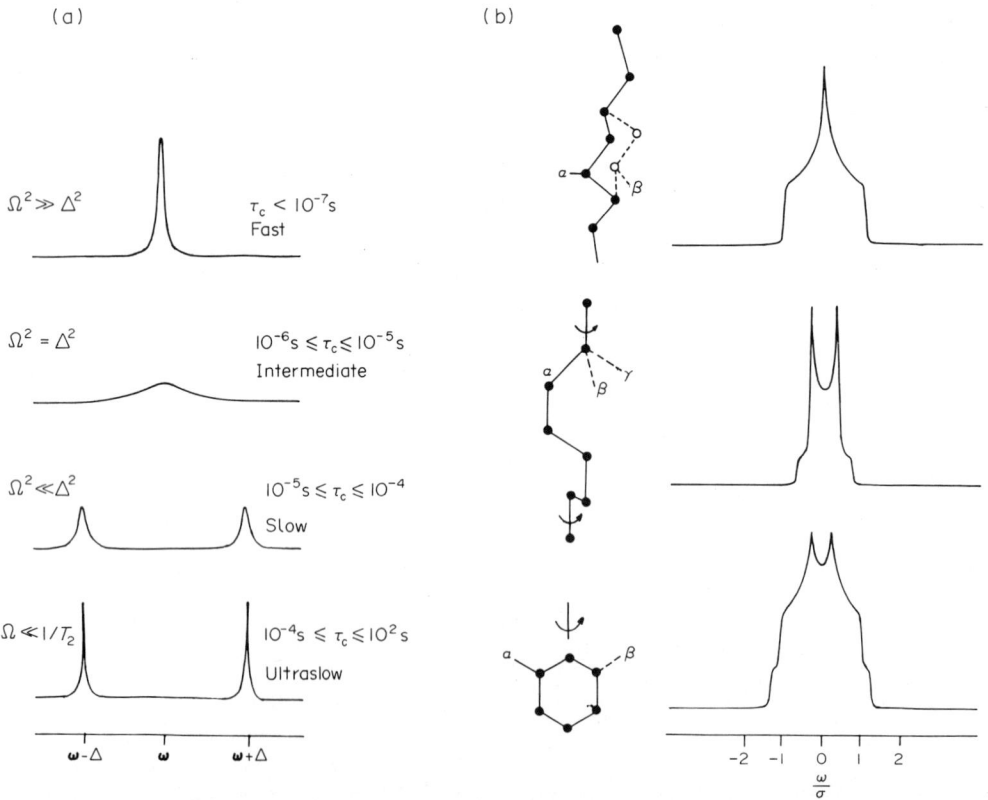

Figure 4 (a) Theoretical lineshapes for deuterons in C–^{2}H bonds resulting from an interchange between two NMR frequencies $\omega \pm \Delta$. The exchange rate is Ω and $\tau_c = (2\Omega)^{-1}$. (b) Theoretical ^{2}H spectra for (i) a kink 3-bond motion; (ii) crankshaft 5-bond motion; and (iii) 180° jump of a phenyl ring. (Reproduced by permission of Springer-Verlag from *Adv. Polym. Sci.*, 1985, **66**, 23)

to or greater than the linewidth. From the range of linewidths encountered in dipolar, chemical shift and quadrupolar spectra, it is clear that information on a wide range of motional frequencies is available.

By way of example, consider the effects of motion on the deuterium lineshape. Spiess[31] has calculated theoretical lineshapes resulting from an interchange between two frequencies $\Omega \pm \Delta$. In the ultraslow exchange limit where the exchange *rate* $\Omega \ll T_2^{-1}$, the lineshape is insensitive to motion; in the slow exchange limit, molecular motion induces extra broadening; for intermediate exchange, the signal encompasses the complete frequency range $\omega \pm \Delta$; and for fast exchange ($\Omega^2 \gg \Delta^2$), a Lorentzian line centred on ω is observed (Figure 4a). Lineshapes for different *types* of motion are illustrated in Figure 4b.[31] Clearly the deuterium lineshapes give convincing graphic evidence of molecular motion.

19.4 NMR RELAXATION: EXPRESSIONS FOR T_1, T_2 AND $T_{1\rho}$

19.4.1 Homonuclear Dipolar Interactions

The Bloembergen, Purcell and Pound (BPP) analysis of NMR relaxation[9] is founded on three basic assumptions: the dominance of dipolar interactions between pairs of spins; the description of the molecular motion responsible for relaxation in terms of a single correlation time τ_c; and the assumption of isotropic motion for which the motional average of the Hamiltonian is zero, as in ideal liquids. The consequent dependence of T_1, T_2, and $T_{1\rho}$ on $v_c = (2\pi\tau_c)^{-1}$ or temperature is portrayed in Figure 5.

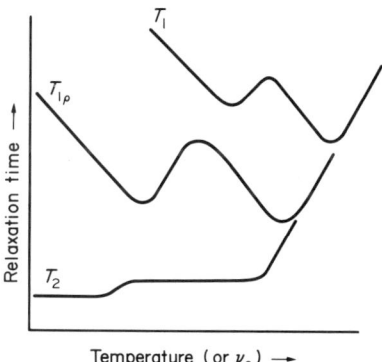

Figure 5 Theoretical dependence of T_1, $T_{1\rho}$ and T_2 on temperature or correlation frequency (v_c). The response depicts a second order transition at lower temperatures and the glass transition at higher temperatures (see text)

Clearly, these assumptions are inappropriate for many of the motions encountered in solid polymers and certain adjustments are necessary. The first modification to the BPP analysis is the construction of a perturbing Hamiltonian whose average is zero and fortunately this can be achieved through a redefinition of the spectral densities describing the motion (*cf.* Volume 1, Section 18.2.1).[11] In the second principal modification, an appropriate distribution of correlation times is incorporated into the expressions for T_1, T_2, $T_{1\rho}$ (and the nuclear Overhauser enhancement factors).[6,8,32]

The revised expressions for like nuclei, undergoing motions that are discrete and well separated and which are characterized by a single correlation time τ_c, may be written[11]

$$T_1^{-1} = 2K \left\{ \frac{\tau_c}{1 + \omega_0^2 \tau_c^2} \sum_k \langle |f_{k1}|^2 \rangle + \frac{\tau_c}{1 + 4\omega_0^2 \tau_c^2} \sum_k \langle |f_{k2}|^2 \rangle \right\} \tag{7}$$

$$T_{1\rho}^{-1} = K/2 \left\{ \frac{\tau_c}{1 + 4\omega_1^2 \tau_c^2} \sum_k \langle |f_{k0}|^2 \rangle \right\} \tag{8}$$

$$T_2^{-2} = K/4 \left\{ \sum_k \langle |f'_{k0}|^2 \rangle \right\} = \frac{\gamma^2 M_2}{2} \tag{9}$$

where $K = 3/2 \left(\dfrac{\mu_0}{4\pi} \right)^2 \gamma^4 \hbar^2 I(I+1)$. The summation over k denotes the usual lattice sum of neighbouring contributions to the reference nucleus and the functions of position coordinates are defined in terms of spherical harmonics[33] as follows

$$\langle |f_{kn}|^2 \rangle = C_n \{ \langle [\![Y_{2n}(\theta_k \phi_k) r_k^{-3}]\!]_{lt}^2 \rangle - \langle [\![Y_{2n}(\theta_k \phi_k) r_k^{-3}]\!]_{ht}^2 \rangle \} \tag{10}$$

$$\langle |f_{k0}'|^2 \rangle = C_0 \{ \langle |Y_{20}(\theta_k \phi_k) r_k^{-3}|^2 \rangle \} \tag{11}$$

where $C_0 = 16\pi/5$, $C_1 = 8\pi/15$ and $C_2 = 32\pi/15$. The brackets $[\![\quad]\!]$ denote, respectively, the motional average over rapid molecular motions at temperatures below (subscripted lt) and above (subscripted ht) the transition.

In deriving equation (9), the exponential correlation function used in the BPP analysis is replaced by a Gaussian one which is more appropriate to the solid; the expression is applied to the plateau regions of T_2 vs. τ_c (or temperature). The situation is somewhat more complex in the vicinity of the transition itself.[9,34] M_2 is the Van Vleck second moment.[35]

The validity of equation (8) requires that $B_1 \gg B_{loc} \approx [M_2/3]^{1/2}$, which can usually be achieved since B_1 is an experimentally adjustable parameter.[36] A special situation arises when $T_1 > \tau_c > T_2$ and $B_1 \approx B_{loc}$. This 'strong collision' domain where almost every molecular motional event results in relaxation is described in terms of the expression[37]

$$T_{1\rho}^{-1} = \frac{2(1-p)}{\tau_c} \left\{ \frac{B_{loc}^2}{B_1^2 + B_{loc}^2} \right\} \approx \frac{1}{\tau_c} \tag{12}$$

The parameter p depends on the details of the motion and is difficult to evaluate, but the approximation $T_{1\rho} \approx \tau_c$ is reasonably good. Note that T_1 and $T_{1\rho}$ in equation (7) and (8) pass through minima when $\omega_0 \tau_c = 0.62$ and $\omega_1 \tau_c = 0.5$ respectively. Changes in T_2 occur when the motional frequencies associated with the transition are of the order of the spectral linewidth. The higher transition in Figure 5 corresponds to the glass transition and ideally, under conditions of extreme narrowing ($\omega_0 \tau_c \ll 1$), the three relaxation times are equal.

It is often convenient in the context of polymers to use the following approximate formulae to estimate the magnitudes of T_1 and $T_{1\rho}$ minima[7]

$$T_{1min} \approx \omega_0 T_{2LT}^2 / \sqrt{2} \tag{13}$$

$$T_{1\rho min} \approx 4\gamma B_1 T_{2LT}^2 \tag{14}$$

where

$$T_{2LT}^{-2} = T_{2lt}^{-2} - T_{2ht}^{-2} \tag{15}$$

If the change in T_2 across the transition is large $T_{2LT} \approx T_{2lt}$.

Expressions derived for the fourth moment, M_4, of the lineshape have been used to test, more rigorously, assumed structural and motional models for the polymer, particularly in oriented form.[38,39] However, computational complexity acts as a major deterrent in the routine use of higher moments.

19.4.2 Heteronuclear Dipolar Interactions

The nature of the dipolar couplings between unlike nuclei not only controls relaxation of the spins involved, but also underpins the experimental methods used to obtain sensitivity enhancement and high resolution. Attention has focussed predominantly upon the ^{13}C–^1H system as discussed by Bovey in Volume 1, Section 17.2.5.2 and Heatley in Volume 1, Section 18.2.2. T_1 and T_2 expressions for proton-decoupled ^{13}C nuclei are given in equations (5) and (6) of Volume 1, Chapter 18, complemented by a third relaxation parameter, the nuclear Overhauser enhancement factor (NOEF), which is a measure of the ^{13}C signal enhancement arising from proton decoupling (equation 7). The NOEF is sensitive to motional correlation frequencies in the region of the $T_1(^{13}C)$ minimum.

The interpretation of ^{13}C rotating frame relaxation ($T_{1\rho}$) is bedevilled by the competing spin–spin contribution involving the equilibration of the spin-locked ^{13}C and ^1H dipolar systems *via* mutual spin flips and the more interesting spin–lattice contributions which probe molecular motions.[40,41] Schaefer *et al.* have shown that $T_{1\rho}(^{13}C)$ does in fact provide information on molecular motion for most glassy and amorphous polymers.[41]

Heteronuclear interactions also contribute to T_2 and to the second moment M_2 of the absorption spectrum (equation 9). The contribution to the lattice sum of interactions between non-resonant neighbouring nuclei (primed notation) is computed in the usual way, but, for systems of spins $I = \frac{1}{2}$, is scaled down by a factor of $4\gamma'^2/9\gamma^2$.[35]

An interesting experiment initially used to explore cross relaxation in anhydrous hydrofluoric acid[42, 43] has been applied to solid poly(vinylidene fluoride) (PVDF).[44, 45] In this transient Overhauser experiment a 180° pulse is resonantly applied to the protons (S spins) and the magnetization of the fluorine nuclei (I spins) is monitored by the subsequent application of a 90° pulse. The observed decay is non-exponential and solution of Solomon's equations[42, 43] yields values for the mean cross relaxation rate as well as the direct relaxation rates of the two spin systems. In the absence of a Hartmann–Hahn type of situation, the energy imbalance in the transfer of energy between the I and S spins is provided by the thermal motions of the molecules at the difference frequency ($\omega_I - \omega_S$). This process has been termed phonon-assisted spin diffusion.[44] The experiment has the added virtue of providing unusually good resolution of T_1 components since the *difference* in two exponentials is observed rather than the sum which is normally the case (*cf.* Section 19.8.1).

19.4.3 Quadrupole Interactions

Relaxation occurs when the interaction between nuclear quadrupole moments and the electric field gradient (EFG) is suitably modulated by molecular motion. The essentials have been treated by Heatley in Volume 1, Section 18.2.5. In the context of polymers, most attention has focussed on the deuterium nucleus $^2H(I = 1)$. For C–^2H bonds, the EFG tensor is normally axially symmetric and is such that the unique principal lies along the C–^2H bond direction. As such there is a characteristic frequency in the static powder pattern (Figure 3) for each orientation, θ, of the C–^2H bond relative to B_0. The spectrum is sensitive to molecular motions of the C–^2H bond vector in the range $\tau_c \approx 10^{-3}$–10^{-7} s.

The sensitive way in which the lineshape reflects local molecular geometry and motion has been exploited in several areas of polymer research, notably by Spiess, Samulski and Jelinski and their co-workers.

19.4.4 Distributions of Correlation Times

A number of analytical distribution functions have been incorporated into the theory of NMR relaxation. The principal step is to introduce a distribution $I(\tau_c)$ into the spectral density functions which characterize the molecular motions

$$J(\omega) = \int_0^\infty \frac{\tau_c I(\tau_c)}{1 + \omega^2 \tau_c^2} \, d\tau_c \tag{16}$$

Lucid accounts of the way in which such distributions affect the NMR response are given elsewhere.[6, 8] Generally, minima are raised and broadened and activation energies for the motion, determined from the limiting slopes on either side of the relaxation minimum, require correction to higher values.

The log-χ^2 distribution introduced by Schaefer[32] is noteworthy in the sense that the skewed distribution has a tail towards longer correlation times which is typical of low frequency cooperative motions, and the logarithmic scale can accommodate the broad range of correlation times required to describe molecular motion in polymers. The Ngai–Williams–Watts fractional exponential correlation function has also been used effectively in the interpretation of polymer relaxation data.[46]

Spiess[31] has examined the effects of a log-Gaussian distribution of correlation times of different widths on ^2H solid echo spectra and distinguishes between *heterogeneous* distributions, where molecular entities move at different rates in different parts of the polymer, and *homogeneous* distributions, where the motion of the entity itself is describable in terms of non-exponential relaxation behaviour.

Although distributions are clearly necessary in the analysis of NMR data for polymers, there are several inherent drawbacks which may be summarized as follows: (a) in many cases, the theoretical fit to the experimental data involves a significant increase in the number of adjustable parameters and therefore the test of assumed molecular motional models is less rigorous; (b) several of the distributions have been routinely carried over from dielectric studies where the angular dependence of the dipole moment vector is described by the first Legendre polynomial and not the second as required by NMR; (c) NMR and dielectric experiments assign different weights to the various molecular motions depending upon the relative concentrations of the electric dipoles and internuclear vectors in the molecule; and (d) no account is taken of the effects of spin diffusion which has no analogue in dielectric relaxation. Resing has shown that a broad distribution of correlation times can lead to an apparent phase transition effect in NMR T_2 data.[47]

19.4.5 Transition Maps

Correlation frequencies (v_c) describing specific molecular motions may be extracted from experimental NMR data according to the methods described in the literature.[7,48,49] The range of frequencies available typically spans 10 decades (Table 3). It is noted that these bounds are dictated by limitations of the experiment rather than by the inherent spectrum of motions as such.

There is remarkably good correlation between NMR relaxation data and the results of diverse other experiments, which include dynamic mechanical, dielectric and differential scanning calorimetric methods. It is conventional to collate data in the form of log v_c *vs.* inverse temperature plots where the family of such plots for a given polymer is the so-called *transition map*. A typical example is shown in Figure 6. Data points tend to lie on well-defined loci indicating that the same molecular motion is responsible for observations from a diverse range of experiments. Loci in the form of straight lines reflect Arrhenius behaviour where the slope of the line provides a direct measure of the activation energy for the motion, ΔE. The curved locus in Figure 6, typical of the glass transition, is usually analyzed in terms of the Williams–Landel–Ferry (WLF) expression.[50] Note that the temperature at which the glass transition is detected is frequency dependent and therefore care must be taken when comparing results from different experiments; T_g, in fact, is taken to be the quasi-static value.

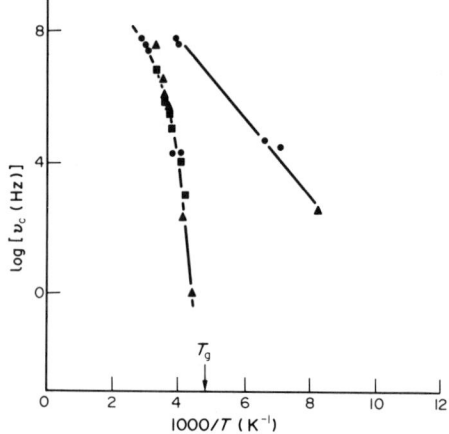

Figure 6 Transition map for poly(isobutylene). The points are shown as NMR data ($\bullet$); dielectric data ($\blacksquare$); and mechanical relaxation data ($\blacktriangle$)

It must be recognized that transition maps take no account of the intensity of a given relaxation (aside from the obvious case where the transition has zero intensity). Thus a point attributable to weak dielectric or mechanical loss may lie on the same locus as a point corresponding to strong NMR relaxation. It is equally difficult to deal with overlapping motions although NMR has a clear advantage over other methods in its ability to detect characteristic motions in different parts of the polymer or composite. Notwithstanding these shortcomings, transition maps have turned out to be an extremely useful starting point in analyzing complex relaxation behaviour in polymers.

19.5 SPIN DIFFUSION

Brief reference was made in Section 19.2.2 to the phenomenon of spin diffusion, whereby excess spin energy may be transferred from one point or region in a homonuclear spin system to another by means of energy conserving flip-flop transitions. Since the rate of energy transfer is of the order of the dipolar coupling and since $T_1 \gg T_2$ in solid polymers, it is clear that excess energy can remain in the spin system for a time which is long compared to T_2 before being transferred to the lattice on the

Table 3 Typical Frequency Ranges Spanned by Different NMR Measurements

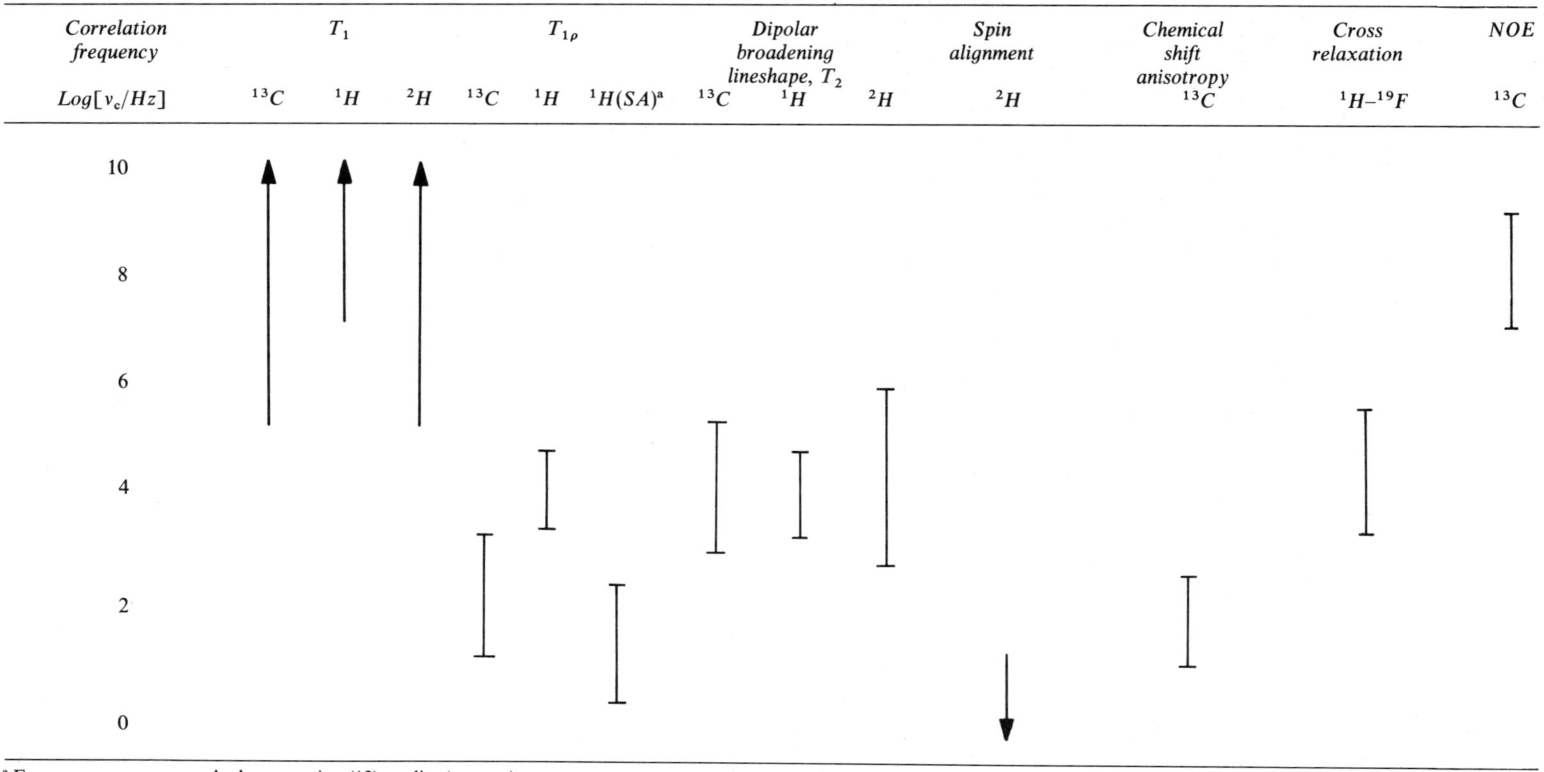

Correlation frequency Log[v_c/Hz]	T_1			$T_{1\rho}$			Dipolar broadening lineshape, T_2			Spin alignment	Chemical shift anisotropy	Cross relaxation	NOE
	^{13}C	^1H	^2H	^{13}C	^1H	$^1H(SA)$[a]	^{13}C	^1H	^2H	^2H	^{13}C	^1H–^{19}F	^{13}C

Correlation frequency axis: 10, 8, 6, 4, 2, 0

[a] Frequency range spanned when equation (12) applies (see text).

timescale of T_1. Spin diffusion tends to average out spatial gradients in the longitudinal magnetization in a way that is somewhat analogous to be behaviour of temperature gradients or to the flow of water in a network of interconnected reservoirs (Figure 7). This leads to a loss of structural detail which has hitherto hampered interpretation of NMR spectra from heterogeneous polymers. More recently, however, these effects have been turned to advantage. A sphere of interaction can be defined around a given relaxing centre[6, 12, 13] and order of magnitude estimates of its spatial dimensions can be obtained from a simplified approach based upon the following expression

$$\langle L^2 \rangle \approx 6D\tau \tag{17}$$

$\langle L^2 \rangle$ is the mean square diffusive path length, D is the diffusion coefficient and τ is the time over which diffusion takes place. T_1 and $T_{1\rho}$ are in the respective time domains of 10^{-1} s and 10^{-3} s and for solid polymers, $D \approx 10^{-11}$–10^{-12} cm^2 s^{-1} in the laboratory frame and one-half this value in the rotating frame.[51] Thus the maximum diffusive path length monitored in a T_1 experiment is of the order of a few tens of nanometres, which implies that an efficiently relaxing entity or region in a polymer can relax neighbouring spins within this spatial dimension but cannot communicate with more remote nuclei. The maximum diffusive path length in the $T_{1\rho}$ experiment is typically an order of magnitude shorter which explains why $T_{1\rho}$ decays are often multicomponent while T_1 decays are not.

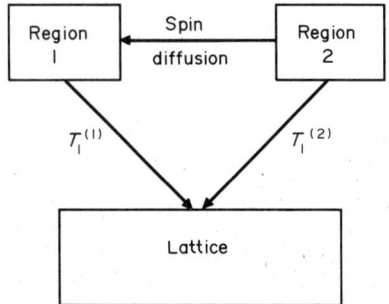

Figure 7   Illustration of the flow of energy *via* spin diffusion between two spin systems in a polymer which are relaxing at different rates. $T_1^{(1)}$ is presumed to be shorter than $T_1^{(2)}$

Whereas spin diffusion and sensitivity is high for proton systems, direct information of a site specific nature is not normally forthcoming, nor is it possible to identify *directly* the available pathways for relaxation. Although ^{13}C NMR does have the prerequisite site specific character, dipolar coupling and therefore spin diffusion between ^{13}C nuclei is significantly weaker. Furthermore, their large internuclear distances in natural abundance incur excessively long diffusion times ($\tau \approx 10$–100 s) for the detection of observable effects.[52] However, in ^{13}C-enriched polymers τ is dramatically reduced to the order of 1 s[53] and exploitation of ^{13}C spin diffusion has now opened up new capabilities to probe very short dimensional scales (<1 nm) in polymers.[22] Recent developments in the use of spin diffusion effects in 2-D NMR to probe structural heterogeneity[21, 22, 52–54] have been alluded to in Volume 1, Section 17.2.5.3 and will be illustrated in Section 19.9.2.

19.6   TOWARDS HIGH RESOLUTION NMR IN SOLID POLYMERS

The generation of high resolution spectra in solid polymers relies upon the imposition of an appropriate time dependence on the normally secular interactions by sample spinning and/or the application of specially designed rf pulse sequences. The object of the exercise is to render $\mathscr{H}_D$ time-dependent with zero average value while retaining contributions of interest such as $\mathscr{H}_{cs}$. Recall that the dipolar Hamiltonian has the form (Table 2)[26]

$$\mathscr{H}_D = \left(\frac{\mu_0}{4\pi}\right) \frac{\gamma_1 \gamma_2 \hbar^2}{2r_{12}^3} [3\cos^2\theta_{12} - 1][I_1 \cdot I_2 - 3I_{1z} I_{2z}] \tag{18}$$

The strategy is either to render $(3\cos^2\theta_{12} - 1)$ zero by appropriate sample spinning or to reorient I_1 and I_2 with suitable rf pulses such that the last term in equation (18) is zero.

Sample spinning can be understood by reference to Figure 8, where rapid rotation of the sample about its cylinder axis results in an average of

$$\overline{[3\cos^2\theta_{12}-1]} = \tfrac{1}{2}(3\cos^2\Delta-1)(3\cos^2\theta'_{12}-1) \tag{19}$$

This follows directly from the addition theorem for spherical harmonics.[33] If Δ, the angle between B_0 and the axis of rotation R, is the so-called magic angle $\cos^{-1}(1/\sqrt{3}) = 54°42'$, then the average, and therefore $\mathscr{H}_D$, is zero. The procedure is known as magic angle spinning (MAS).[24]

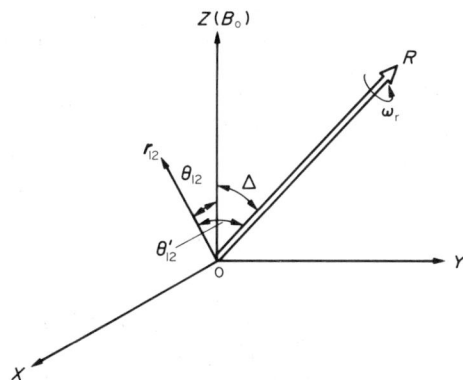

Figure 8 Orientation of the sample spinning axis R relative to B_0: $\mathbf{r}_{12}$ is the internuclear vector in the sample

The second approach recognizes that if $\langle I_i \rangle$ can be induced to spend equal time along the three orthogonal axes of the resonantly rotating frame, this, in effect, is stroboscopically equivalent to rotating the spin about the magic angle and averages the spin part of H_D to zero.[16,28]

In the case of heteronuclear interactions the procedure is more straightforward. Noting that the spin part of $\mathscr{H}_D$ is $I_Z S_Z$ (Table 2), rapid inversion of spins S relative to spins I through strong resonant irradiation will render $\mathscr{H}_D$ zero. This is called dipolar decoupling, as in ^{13}C experiments where the desire is generally to suppress the dominant ^{1}H contributions. Either continuous[55,56] or pulsed[57] irradiation may be used.

Referring again to Table 2, the more complex angular dependence of σ usually generates line broadening in powder samples. MAS reduces σ to $\boldsymbol{\sigma}/\sqrt{3}$ where $\boldsymbol{\sigma}$ is the isotropic average of the chemical shift, as observed in solutions, which is characterized by a single absorption line for each nuclear environment. The price paid for achieving high resolution liquid-like spectra is a loss of information on the anisotropy of the nuclear interactions involved. Note, too, that MAS has the ability to remove only first order quadrupole interactions: $\cos^{-1}(1/\sqrt{3})$ is not a 'magic angle' for second order effects.

The way in which MAS affects $T_{1\rho}$ measurements and chemical shift spectra, particularly in regard to the line broadening which results when molecular motion is on the timescale of MAS rotation speeds, has been examined.[58,59] Procedures have also been developed to obviate complexities arising from spinning sidebands in spectra recorded under MAS.[60,61] The ultimate resolution which can be achieved in multiple pulse experiments is lifetime limited and is generally determined by $T_{1\rho}$.[17]

In general, the imposed rate of averaging must dominate or at least be comparable to the internal motions of the system. Recognizing that the linewidth due to dipolar interactions is typically 10^4–10^5 Hz and that the normal rotor spinning speeds are much lower than this, it is clear that MAS cannot remove strong dipolar couplings. Thus data acquisition often involves a hybrid of experimental approaches. While dipolar decoupling removes the effects of heteronuclear dipole–dipole interactions, the ensuing spectrum still contains broadening due to chemical shift anisotropy; MAS reduces this to its isotropic value. Occasionally, it is advantageous to retain some of the chemical shift anisotropy by spinning at an angle which deviates slightly from the magic angle.

Note finally that $\mathscr{H}_D$ will be weak both for spins that are dilute and remote from each other (as a consequence of the r^{-3} dependence) and for spins with low magnetogyric ratios. These spins are especially amenable to high resolution solid state spectroscopy.

19.7 EXPERIMENTAL PULSE TECHNIQUES

A description of NMR instrumentation can be found in several standard texts.[16-18,62-74] Here attention focusses primarily upon the more important pulse sequences used in routine applications. Those encountered in homonuclear experiments are concisely summarized in Table 4,[30] where the pulse trains may be readily visualized in terms of vector diagrams of which Figure 5 in Volume 1, Chapter 17 is an example. The first seven provide T_1, $T_{1\rho}$ and T_2.[27,36,65-71] The Goldman–Shen[72] and Jeener–Broekhaert[73,74] sequences are central to several important polymer applications to be addressed later. The remaining pulse schemes[75-79] are designed to remove homonuclear dipolar contributions to the linewidth as required in high resolution solid state NMR. It is recalled that this is achieved through modulation of the spin term in $\mathscr{H}_D$ rather than the space term as in earlier experiments using MAS.

The pioneering work in the use of pulse sequences to achieve line narrowing was carried out in Waugh's and Mansfield's laboratories, complemented later by important contributions from Vaughan and co-workers. Typically the WAHUHA four-pulse sequence[75] reduces the linewidth from $\simeq 10^4$ Hz to about 10^2 Hz; eight-pulse sequences lead to an order of magnitude improvement and 24-pulse trains yield another factor of five.[80] Clearly, there is a trade-off between the extent of narrowing achieved and the significant added complexity of the experiment.

There are many variants of these sequences which are designed to facilitate interpretation of spectra from polymers. Typically, in a partially crystalline polymer, it is often the case that the amorphous and crystalline components relax at very different rates at suitably chosen temperatures. Judicious choice of the length of the spin-locking pulse in the $T_{1\rho}$ experiment, for example, suppresses one component and permits direct examination of the other. Such selective discrimination is also an essential element in the Goldman–Shen experiment which differentiates between components on the basis of very different T_2 relaxation times. Following the initial 90° pulse, the component with the shorter T_2 relaxes rapidly, while the component with the longer T_2 is little affected on this timescale. A second 90° pulse realigns the magnetization along B_0 and the ensuing transfer of energy from the long to the short T_2 component by spin diffusion can be subsequently monitored.

Special methods[31,81,82] are required to detect ^{2}H spectra because of the low natural abundance of deuterium, its low magnetic moment and the extreme breadth of the lineshape in solids. It is difficult to generate 90° pulses to achieve uniform excitation of the complete spectrum as required, for example, in the solid echo sequence designed to overcome the effects of recovery time.[66,67] The Jeener–Broekaert sequence has been used to advantage to achieve undistorted lineshapes and in addition has been used by Spiess to measure ultraslow motions:[83] the first 45° pulse not only creates the echo as shown in Figure 9 but also induces quadrupolar order, known as *spin alignment*, which remains for long periods—limited only by T_1—unless molecular reorientation alters the quadrupolar coupling. The second 45° pulse creates an 'alignment echo' which gives an indication of how much molecular reorientation occurred during τ_2. Motions with τ_c as long as 1 s may be studied in this way.

Double resonance techniques are used to study heteronuclear systems, where both spin systems are either simultaneously or sequentially irradiated at their respective resonant frequencies, prior to the detection of one or other of their responses. Attention has focussed on systems containing dilute spins, usually with low magnetogyric ratios and abundant spins with high magnetogyric ratios of which ^{13}C–^{1}H is the most important example. Experimental detection of the rare spins involves

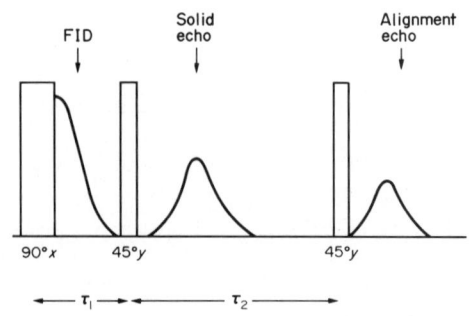

Figure 9 Jeener–Brockaert pulse sequence indicating the formation of the alignment echo

Table 4 Pulse Sequences Commonly Encountered in Solid State NMR[a]

Name	Pulse sequence	Application	Ref.
(1) Free induction decay (FID)	90_x	All interactions contained in spectrum: provides T_2 for solids in the low temperature regime: problems with t_R^b	65
(2) Solid echo	$90_y - \tau - 90_x$	Provides complete FID and T_2	66, 67
(3) Carr–Purcell spin echo	$90_y - (\tau - 180_x)_n$	Removes chemical shift, inhomogeneity and other off-resonance effects	68
(4) Carr–Purcell–Meiboom–Gill	$90_y - (\tau - 180_x - 2\tau - 180_{-x} - \tau)_n$	As in (3) with compensation for phase errors	68, 69
(5) Saturation recovery	$(90_y - \tau)_n$	Yields T_1 and permits discrimination against long T_1	70, 71
(6) Inversion recovery	$180_x - \tau - 90_{-y}$	Yields single and component T_1 values	70, 71
(7) Spin-locking	$90_x\,90_y(\tau)$	Yields single and component $T_{1\rho}$ values	27, 36
(8) Goldman–Shen	$90_x - \tau_1 - 90_{-x} - \tau_2 - 90_x$	Provides information on coupling between spin systems with different T_2 or $T_{1\rho}$ values	72
(9) Jeener–Broekaert	$90_x - \tau_1 - 45_y - \tau_2 - 45_y$	Provides information on direct coupling between dipolar and Zeeman energy baths	73, 74
(10) WAHUHA cycle	$90_x - (\tau - 90_{-x} - \tau - 90_y - 2\tau - 90_{-y} - \tau - 90_x - \tau)_n$	Removes homonuclear coupling to first order	75
(11) Mansfield 6-pulse cycle	$90_{-y} - (\tau - 90_x\,90_{-y} - \tau - 90_{-y} - \tau - 90_{-y} - \tau - 90_y\,90_{-x} - \tau)_n$	Removes homonuclear coupling	76, 77
(12) MREV-8 cycle	$90_x - (\tau - 90_{-x} - \tau - 90_y - 2\tau - 90_{-y} - \tau - 90_{-x} - 2\tau - 90_{-x} - \tau - 90_y - 2\tau - 90_{-y} - \tau - 90_x - \tau)_n$	Homonuclear decoupling: compensation for pulse defects	78, 79

[a] Information extracted from *Surf. Sci. Rep.*, 1981, **1**, 157 (by permission of North Holland Publishing Co.)

[b] t_R is the recovery time of the spectrometer.

dipolar decoupling (DD) and cross polarization (CP), described in Sections 19.2.3 and 19.6 respectively. CP effects polarization transfer from ^1H to ^{13}C nuclei resulting in ^{13}C signal enhancement of up to a factor of four and a concomitant reduction in spectral averaging time by a factor of 16. Note also that the repetition rate for signal averaging is dictated by $T_1(^1H)$ and not by $T_1(^{13}C)$ which is typically an order of magnitude longer.

Progress in this area is continual. For example, a recently reported pulse sequence involving two time reversal echoes (TREV) to generate high resolution spectra of abundant spins is considered to be superior to MREV-8.[84] This sequence has yet to be applied to polymers. More advanced double resonance techniques termed heteronuclear dipolar modulation, separated local field spectroscopy and dipolar rotational spin–echo ^{13}C NMR are described in the original references[85–90] and in a number of excellent review articles.[17, 18, 30] Schaefer and co-workers have used the latter technique to characterize molecular motion in polymers. Briefly, the dipolar coupling between carbons and their attached protons is monitored and the extent to which this coupling is reduced by molecular motion of appropriate frequency is a measure of the *amplitude* of the motion.[91] Undoubtedly, too, multi-quantum and zero-field experiments will have an important role to play in the future.

19.8 MOLECULAR MOTIONS

Primarily for ease of presentation, an arbitrary dichotomy is drawn between molecular dynamics and structure even though the two are initimately related. Before discussing representative examples from the extensive body of literature on the subject it is pertinent to draw some general observations.

Although many of the models used to describe molecular motion in polymers have been developed from fundamentally different standpoints, the fact that they share certain common characteristics[6] is gratifying in the sense that the models must reasonably approximate the behaviour of real systems. Important, too, is the complementary nature of the various experimental approaches both within NMR and with other relaxation experiments. These studies are greatly facilitated by selective isotope enrichment techniques.

19.8.1 Fluorocarbon Polymers

Poly(tetrafluoroethylene) (PTFE), a partially crystalline polymer, has been studied extensively by NMR.[92–99] Aside from understanding its own unique behaviour, there are many related polymer systems of great commercial interest — for example, piezoelectric polymers[100–104] and membranes[105–108] — whose properties can only be understood with prerequisite knowledge of the way in which the $(-CF_2-)_n$ chain behaves.

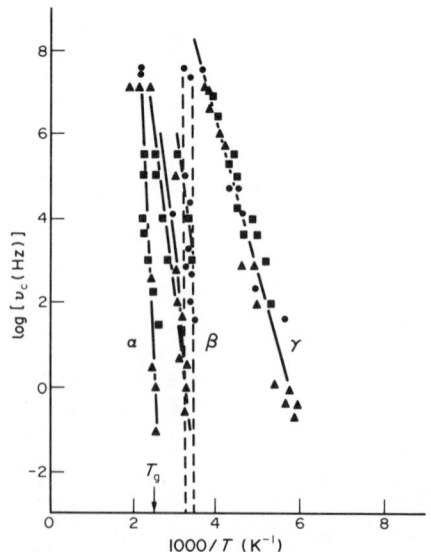

Figure 10 Transition map for PTFE. The dashed lines denote first order transitions: dynamic mechanical data (▲); dielectric data (■); NMR data (●). The α, β and γ relaxations are assigned in the text

The transition map for PTFE, portrayed in Figure 10, is a complex one and features two first order crystal–crystal transitions in the vicinity of room temperature (dashed lines). These transitions almost certainly involve helical interconversion coupled with rotation about the helical axes (β relaxation). The abrupt onset of rotation is clearly demonstrated in low resolution NMR studies on PTFE fibres where relaxation times for the oriented polymer, computed from equations (7)–(11), are in excellent agreement with experimental data.[92, 94] Main chain rotation about the helical axis is also evident in chemical shift data where the crystalline lineshape assumes axial symmetry above the transition temperature (Figure 11).[98] Note that chemical shift spectra are insensitive to translations along the chain as inferred in low resolution linewidth results.[94]

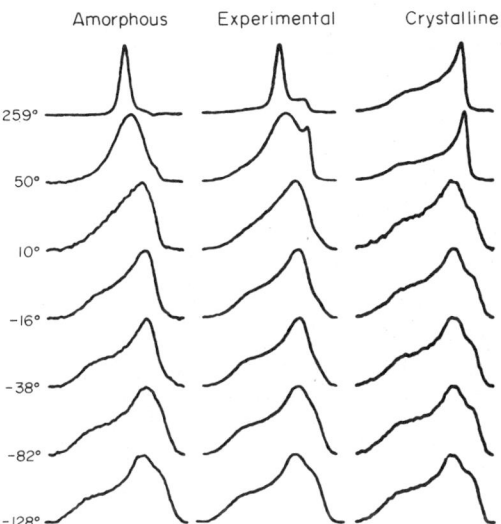

Figure 11 ^{19}F MREV-8 chemical shift lineshapes for PTFE of 68% crystallinity as a function of temperature (°C, as indicated) (reproduced by permission of the American Chemical Society from *Macromolecules*, 1980, **13**, 1635)

The γ relaxation is assigned to rapid reorientation about a local chain axis in the amorphous regions and Vega and English[98] conclude that it is the *angular amplitude* rather than the *rate* which grows with increase in temperature. The amorphous chemical shift lineshape ultimately assumes the symmetry typical of isotropic motions. Separation of the lineshape into crystalline and amorphous parts also yields a satisfactory estimate of the crystallinity. The α relaxation is not detected in NMR measurements.

Fleming and co-workers[99] note the difficulties in obtaining ^{13}C spectra from fluoropolymers because of the sizeable ^{19}F decoupling field strengths required to cope with the large chemical shift anisotropy.

The ability to render poly(vinylidene fluoride) (PVDF), $(-CF_2CH_2-)_n$ pyro- and piezo-electric has attracted much interest. Proton T_1, T_2 and $T_{1\rho}$ data for PVDF are furnished in Figure 12.[100] The results demonstrate clearly the presence of four relaxation processes, three of which are attributable to amorphous polymer and one to crystalline: *I* is the γ relaxation as observed in PTFE; *II* is the β relaxation reflecting the onset of general motions in the amorphous regions; *III* is the β' relaxation assigned to motion of interfacial material; and *IV* is the α relaxation which is manifested in the high temperature $T_{1\rho}$ minimum and the very weak transition in the short T_2 component at 95 °C. The α relaxation is presumed to arise from some form of restricted rotation about the chain axis perhaps with an accompanying translation, in the crystalline regions. The temperature at which rotation sets in correlates well with the conditions under which dipole reorientation can be optimally effected in the poling process (*cf.* Section 19.9.3).

Reference was made in Section 19.4.2 to a transient Overhauser experiment on PVDF which explored cross-relaxation effects between the ^{19}F and ^{1}H spin systems.[44, 45] The fact that cross-relaxation is most efficient when thermal motions have a dominant component at the difference frequency $(\omega_I - \omega_s)$ provides an added sampling frequency for molecular motion (Table 3). Solution to Solomon's equations[42] leads to two relaxation times $T_1 = (\rho + \sigma)^{-1}$ and $D_1 = (\rho - \sigma)^{-1}$ where ρ and σ are respectively the direct- and cross-relaxation rates (Figure 13). Note that two minima are

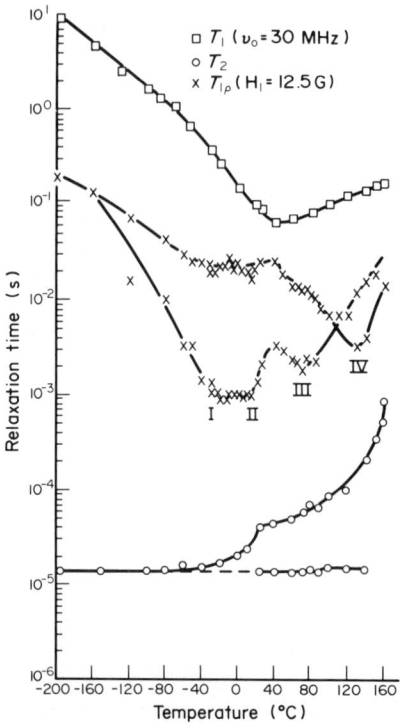

Figure 12 Proton T_1, $T_{1\rho}$ and T_2 data for PVDF recorded as a function of temperature. Two-component relaxation is observed for $T_{1\rho}$ above $-160\,°C$ and for T_2 above $0\,°C$ (reproduced by permission of Wiley from *J. Polym. Sci., Polym. Phys. Ed.*, 1976, **14**, 1271)

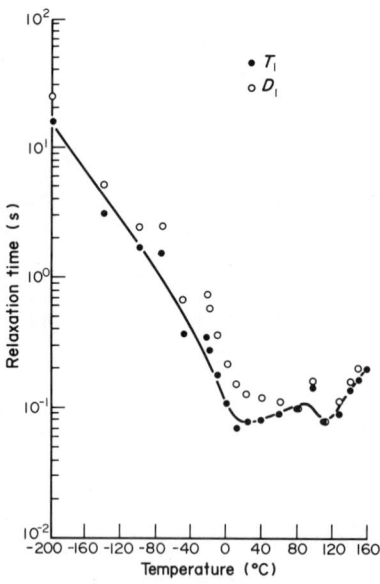

Figure 13 Temperature dependences of the resolved spin lattice times T_1 and D_1 in PVDF (see text) (reproduced by permission of the American Chemical Society from *Macromolecules*, 1977, **10**, 885)

resolved at $+25\,°C$ and $+110\,°C$ respectively; earlier measurements (Figure 12) were unable to resolve the higher temperature minimum which probably manifests the β relaxation.

This section concludes with brief reference to the perfluorosulfonate ionomer membranes. (**1**) developed by du Pont under the 'Nafion' tradename. They are commercially important by virtue of their permiselective properties coupled with an ability to withstand hostile environments. The role of absorbed water and the associated effects of clustering are critical factors in their performance.

$$-(CF_2CF_2)_n-(CFCF_2)_m-$$
$$| $$
$$O-(CF_2CFCF_3-O)-CF_2CF_2SO_3X$$

(1)

NMR has been used to probe selectively (i) the behaviour of the fluorocarbon backbone utilizing ^{19}F resonance, (ii) absorbed water by means of ^1H resonance and (iii) selected cations (for example, $X \equiv {}^{23}Na$) which exhibit magnetic resonance.[105–108] The glass-like character of water at low temperatures, which generally typifies the complex behaviour of water in diverse solid media, is well demonstrated in these systems.[106–107] These attempts to unravel the intricate morphology of the Nafion membranes have been greatly facilitated by an appreciation of the known response of related polymers such as PTFE.

19.8.2 Polycarbonates

Polycarbonates constitute an important class of polymers because of their remarkable impact strength. Bisphenol A polycarbonate (PC) is typical and has the repeat unit (2). ^1H, ^2H and ^{13}C NMR studies clearly identify 180° aromatic ring flips about the C_2 axis as the dominant motion in PC.[91,109–115] This is evident, for example, in the chemical shift anisotropies of carbons on and off the C_2 rotation axis, and is consistent with the deuterium lineshape for PC deuterated selectively in the aromatic ring.[112] Simulation of anisotropic ^{13}C chemical shift patterns[113,114] and dipolar rotational spin-echo ^{13}C spectra[91] yields rates and amplitudes for the motion which encompass a broad range of frequencies centred on *ca.* 300 kHz at room temperature. This is also evident in the shape of the mechanical tan δ and dielectric loss peaks at low temperatures.[116]

(2)

While the various NMR, dielectric and dynamic mechanical measurements on this secondary relaxation in PC are well correlated it is difficult to imagine how 180° flips between symmetric minima of themselves can account for strong mechanical and dielectric loss or indeed to account for the unusually high impact strength of the polymer. It transpires that interpretation of NMR data further requires (i) that motional ring flips are accompanied by oscillations of the order of 30° about the same rotation axes;[115] (ii) that there are other main chain oscillatory or 'wiggling' motions of appreciable amplitude below T_g as evidenced in measurements on PC enriched with ^{13}C at the carbonyl site[113] and through analysis of methyl carbon dipolar patterns;[91] and (iii) that lattice distortions (which are themselves mechanically active) must accompany large amplitude aromatic ring motion in the glass.[115]

The complementary nature of diverse NMR measurements is evident, for example, in the fact that the frequency of the main chain wiggling motion at room temperature exceeds the range to which ^2H spin alignment measurements are sensitive.[112] Schaefer and co-workers[115] conclude that there must be a high degree of *inter-* and *intra*-motional cooperativity which is in keeping with the general notions of energy dissipation and high impact strength.

19.8.3 Poly(butylene terephthalate)

This polymer, denoted PBT, is interesting from a number of points of view: different moieties along the chain exhibit widely different motional and relaxation behaviour (Figure 14);[117] the π-electrons of the aromatic ring of the terephthalate group, unlike those in PC, can participate in conjugation with the carbonyl groups giving rise to resonance stabilization and coplanar C=O and phenyl groups;[118] and the local motions of the alkyl moiety are inherently interesting in the chain dynamic sense.[119]

The motion of the phenyl ring was investigated in (aromatic-d_4)PBT.[120] Solid state ^2H NMR detects three distinct motional regimes. In the crystalline regions the phenyl rings are static on the

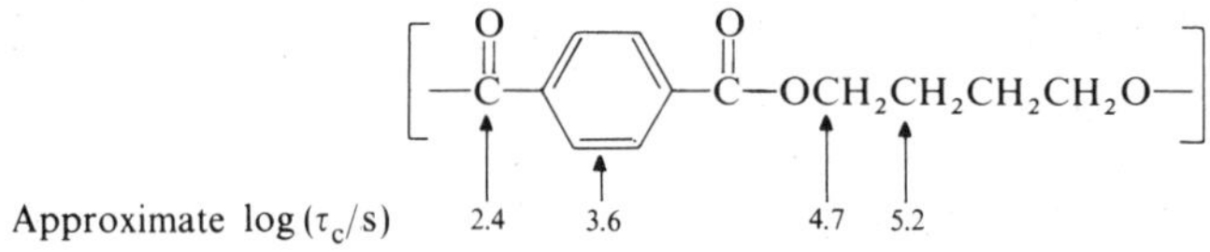

$$\left[\!-\!\overset{\overset{\text{O}}{\|}}{\text{C}}\!-\!\bigcirc\!-\!\overset{\overset{\text{O}}{\|}}{\text{C}}\!-\!OCH_2CH_2CH_2CH_2O\!-\!\right]$$

Approximate $\log(\tau_c/s)$ 2.4   3.6   4.7   5.2

Figure 14 Average correlation times of various carbons along the poly(butylene terephthalate) chain[118]

NMR timescale ($\tau_c > 10^{-3}$ s). As with many other partially crystalline polymers,[6,13,121] there is a region of intermediate mobility in which the rings undergo slow 180° flips, while in the amorphous regions the behaviour is akin to the general pattern of phenyl group motion described earlier, with rapid flips occurring in conjunction with low angle librational motion. Once again conformational considerations are important in the solid.

The extent to which motional heterogeneity prevails in this system is reflected in the spin–lattice recovery of the deuterons in the amorphous component (Figure 15). Note that T_1 is anisotropic across the pattern since the singularities recover at different rates to that of the centre of the spectrum.[120,122] There are also at least two components contributing to the amorphous lineshape at 96 °C. Since T_1 is lower at the higher temperature, phenyl ring motions are clearly on the slow correlation time side of the T_1 minimum.

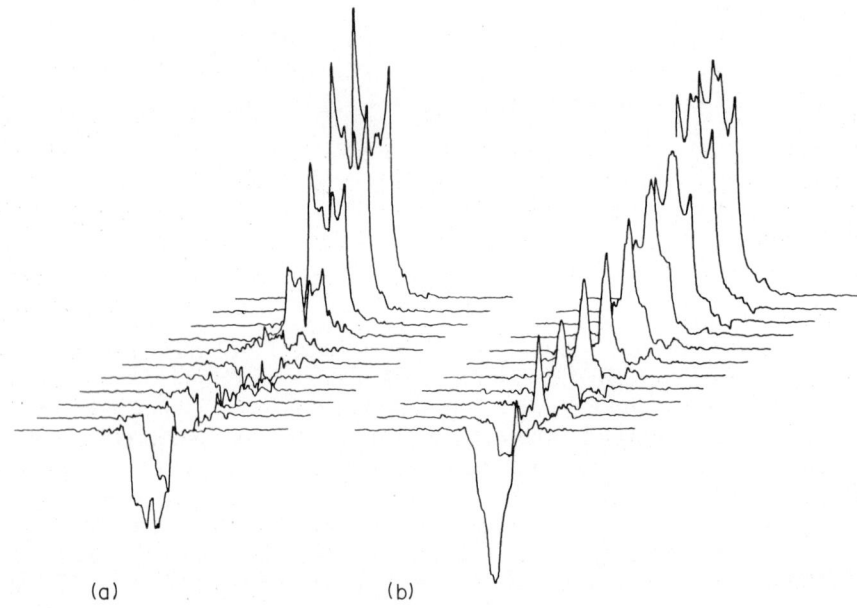

(a)   (b)

Figure 15 Deuterium spin–lattice relaxation behaviour at 55.26 MHz for the amorphous component of PBT at (a) 23 °C and (b) 96 °C. The variable delay in these plots from bottom to top is 1, 10, 20, 30, 40, 50, 100, 200, 400, 800 and 1000 ms (reproduced by permission of the American Chemical Society from *Macromolecules*, 1984, **17**, 2399)

Evidence of three-bond motion has been detected in a PBT sample selectively deuterated in the alkyl moiety. (**3**)[123] Quadrupole echo ^{2}H NMR spectra can be fitted successfully by a two-site model in which the C–^{2}H vector hops between sites separated by the dihedral angle 103°, in keeping with *gauche–trans* conformation transitions.[119] In essence, kink diffusion can occur even when the ends of the participating moiety are pinned, as is essentially the case in PBT (*cf.* Figure 14).

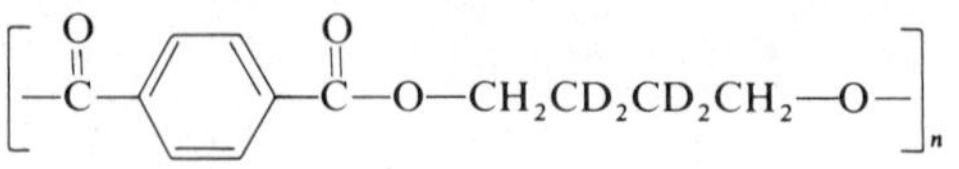

$$\left[\!-\!\overset{\overset{\text{O}}{\|}}{\text{C}}\!-\!\bigcirc\!-\!\overset{\overset{\text{O}}{\|}}{\text{C}}\!-\!O\!-\!CH_2CD_2CD_2CH_2\!-\!O\!-\!\right]_n$$

(**3**)

19.9 STRUCTURAL STUDIES

19.9.1 General Considerations

Implicit in much of the foregoing discussion is the sensitivity of NMR to the environment in which the resonant nucleus finds itself. Studies on polymer morphology, static[6,13,51,124-126] and dynamic[127-130] phase separation, polymer–polymer and polymer–filler interfaces[121] are myriad. Selective deuteration, for example, has given important structural information on the nature of chain re-entry at crystal surfaces. By way of illustration, consider the chemical shift tensor which has site specific character in the sense that detailed information on local geometry and chemical bonding is contained in the chemical shift anisotropy. For polymers, ^{13}C chemical shifts are particularly revealing and there is a growing body of data which facilitates identification of various carbon functionalities. The information listed in Table 5 has been extracted from a recent comprehensive compilation of ^{13}C chemical shielding data.[133]

19.9.2 Heterogeneity in Polymers

Any meaningful discussion of spatial heterogeneity must specify the dimensional scale. A calorimetric experiment, for example, may indicate a homogeneous blend of two component polymers whereas an NMR experiment may not. Indeed, as the discussion in Section 19.5 shows, a system may be deemed to be structurally homogeneous on the basis of a T_1 experiment and heterogeneous in $T_{1\rho}$ measurements. This section examines the utility of NMR in probing spatial heterogeneity on a dimensional scale which is not readily accessible by other means.

Nominally homogeneous poly(vinyl chloride) PVC exhibits distinguishably different types of material when plasticized.[1,2] The addition of 17 wt.% plasticizer (diisodecylphthalate) to PVC [denoted PVC(17)] generates a second component in the FID at 25 °C in addition to the unplasticized PVC signal, which is apparently unchanged (Figure 16a). At 75 °C there are three composite signals in PVC(1), PVC(5) and PVC(17) (Figure 16b).[2] The longest component unambiguously extrapolates to an intensity commensurate with the known amount of plasticizer present, and the shortest component again manifests unplasticized PVC leaving the intermediate

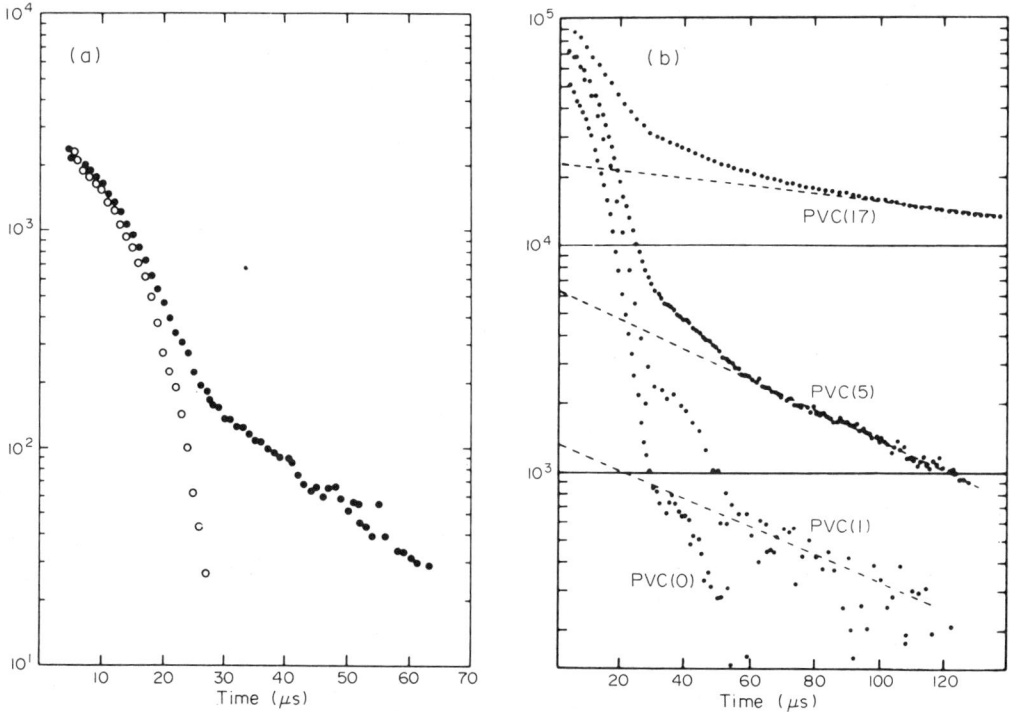

Figure 16 Free induction decays of neat and plasticized PVC. (a) (○) PVC(0) and (●) PVC(17) at 25 °C. (b) PVC(0), PVC(1), PVC(5) and PVC(17) at 75 °C. The number in parenthesis denote the wt.% plasticizer. The ordinates are intensity in arbitrary units (reproduced by permission of the American Chemical Society from *ACS Symp. Ser.* No. 142, 1980, **8**, 147)

Table 5 Chemical Shielding Parameters relative to TMS on the δ scale[a] where liquid $C_6H_6 = 128.7$ ppm and $CS_2 = 192.8$ ppm (upfield is negative).[133]

Functionality	Polymer	Formula	Temperature (K)	Chemical shift parameters				Ref.
				σ_{11}	σ_{22}	σ_{33}	σ	
Alkane carbons	PE	$(-CH_2-)_n$	r.t.	52	39	14	35	134
	PMMA	$(-CH_2CCH_3CO_2CH_3-)_n$	r.t.	50	37	13	33	135
	PVDF	$(-CH_2CF_2-)_n$	297	79	48	29	52	136
Quaternary carbons	PMMA	$(-CH_2CCH_3CO_2CH_3-)_n$	r.t.	56	49	30	45	137
Alkene carbons	cis-1,4 PB	$(-CH_2CH=CHCH_2-)_n$	123	53	43	37	44	136
	trans-PA	$(=CH-)_n$	r.t.	236	115	35	129	137
			r.t	220	142	51	137	138
			r.t	218	138	22	126	139
	cis-PA	$(=CH-)_n$	r.t.	217	143	45	135	140
			r.t.	221	137	20	126	138
			r.t.	219	144	47	137	139
Ethers	PEO	$(-CH_2CH_2O-)_n$	133	91	83	33	69	137
	PET	$(-OCOC_6H_4CO_2CH_2CH_2-)_n$	r.t	80	80	28	63	141
	POM	$(-OCH_2-)_n$	r.t.	114	87	64	88	142
Esters	PET	$(-OCOC_6H_4CO_2CH_2CH_2-)_n$	r.t.	250	122	122	165	141
	PMMA	$(-CH_2CCH_3CO_2CH_3-)_n$	r.t.	268	150	112	177	136
Aromatic carbons	PET	$(-OCOC_6H_4CO_2CH_2CH_2-)_n$	r.t	226	153	15	131	141
Carbon–fluorine	PVDF	$(-CH_2CF_2-)_n$	297	131	120	111	121	137

[a] The relevant carbon atoms are shown in bold.[133]

component attributable to plasticized polymer. Clearly the plasticizer affects certain regions and not others, indicating a heterogeneous system. Use of the Goldman–Shen experiment in conjunction with equation (17) defines the size of the inhomogeneous regions to be no more than a few nanometres.[6]

Consider now a blend of syndiotactic poly(methyl methacrylate) (s-PMMA) and PVC.[143] Knowledge of the degree of miscibility achieved is essential to a full understanding of the macroscopic properties. Figure 17 shows T_1 data for a number of blend compositions where the minimum at *ca.* 0 °C is due to α-CH₃ motion in s-PMMA. At this temperature, T_1 for PVC is an order of magnitude longer implying at most a very limited motion at T_1-sensitive frequencies. As expected, superposition of component relaxation curves is found for a physical mixture of the finely-divided component polymers (Figure 18), whereas a single exponential decay is observed for the blends, indicating that the α-CH₃ groups are communicating with the bulk of protons therein. Thus the two components in the blend are mixed on a dimensional scale of the order of *ca.* 12 nm. Applying a similar reasoning to rotating frame data indicates that the blends are heterogeneous on the shorter dimensional scale probed by $T_{1\rho}$.

Polyurethanes are representative of phase-separated systems comprising hard (usually aromatic or aliphatic diisocyanate) and soft (polyester, polyether or polybutadiene) segments. Assink and Wilkes[144] have studied a polyester soft segment and diphenylmethane diisocyanate hard segment polyurethane. The Goldman–Shen sequence[72] is again used to great effect to determine interdomain spacings which agree with small-angle X-ray scattering estimates. The dependence of soft segment mobility on proximity to the domain interface is also demonstrated.

Jelinski and co-workers[145] used ²H NMR to monitor the hard segment through selective deuteration of the constituent butanediol residue. There are no complications from effects of spin diffusion in these measurements and the amount of interfacial material can be readily quantified.

¹³C spin diffusion measurements on a 75/25 blend of poly(ethylene terephthalate) (PET), doubly ¹³C-enriched at the CH₂ groups, with PC, ¹³C-enriched at the carbonyl group, show that there is intimate mixing on a dimensional scale of less than a nanometre (*cf.* Section 19.5).[22]

The 2-D spectra of blends of atactic polystyrene (PS) and atactic poly(vinyl methyl ether) (PVME), cast from chloroform and toluene respectively, are shown in Figure 19.[54] Cross communication between different moieties generates off-diagonal or cross peaks.[22,146] Such peaks between the aromatic protons in PS and methyl and methine protons in PVME are absent in the blend cast from

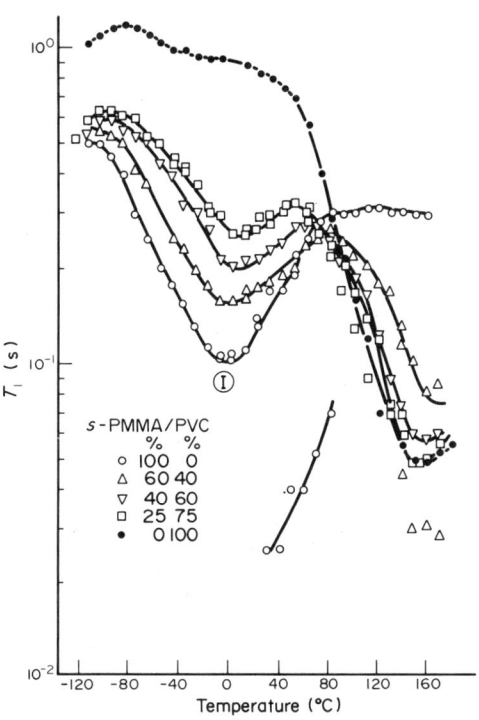

Figure 17 Proton T_1 (40 MHz) data for s-PMMA, PVC and s-PMMA/PVC blends (reproduced by permission of the American Chemical Society from *Macromolecules*, 1985, **18**, 388)

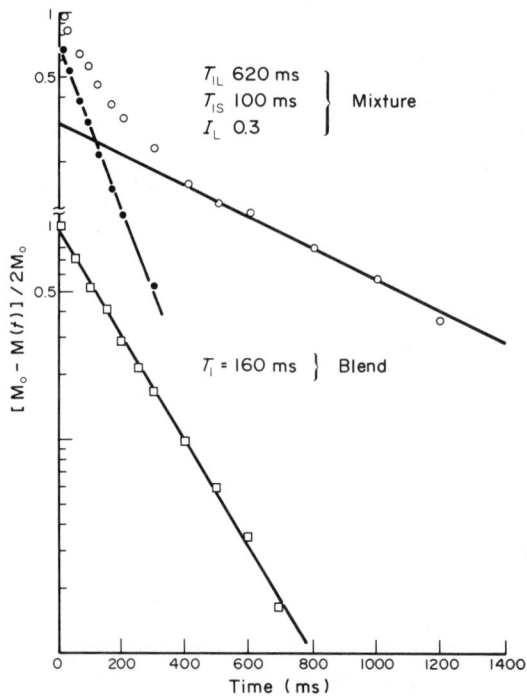

Figure 18 Spin–lattice relaxation curves for a 60/40 physical mixture of *s*-PMMA ($\bigcirc$) and PVC ($\bullet$) and for a 60/40 blend ($\square$) at 0 °C. T_{IL} and T_{IS} are respectively the long and short relaxation time components in the physical mixture. I_L is the fractional intensity of the long component and corresponds to the fraction of PVC protons (0.29) in the mixture. M_0 is the equilibrium and $M(t)$ the partially relaxed longitudinal magnetization (reproduced by permission of the American Chemical Society from *Macromolecules*, 1985, **18**, 388)

chloroform, but present in the blend cast from toluene. Hence component polymer chains are in close proximity only in the toluene cast sample. Ernst and co-workers[54] estimate that the mixed phase contains 60–80% of the total polymer.

The ability to monitor cross relaxation in ^{1}H–^{19}F systems, as described in Section 19.8.1, has been used successfully to obtain information on miscibility in PMMA/PVDF and poly(ethyl methacrylate)/PVDF blends.[45]

19.9.3   Oriented Polymer Systems

Mechanical deformation is often an important step in polymer processing because of the way in which the extrusion, drawing, compression or rolling process beneficially alter the overall properties of the polymer (*cf.* Volume 2, Chapter 13).[147] The development of ultrahigh modulus polymers is a notable example. The anisotropic character conferred on the polymer by mechanical deformation is reflected in the NMR tensors listed in Table 2.

Early studies exploited the dependence of ^{1}H and ^{19}F dipolar linewidths on sample orientation in B_0.[6, 11, 13, 25, 38, 39, 148, 149] The additional information contained in the angular dependence has two important consequences: it facilitates a more rigorous test of the models used to describe relaxation[94] and it permits a quantitative analysis of the statistical distribution of chains in partially ordered systems in terms of moments of the distribution.[25, 149] These moments are an essential ingredient in modelling *any* tensor property of the deformed polymer.[149, 150] Measurements of T_1 and $T_{1\rho}$ have been less informative because of the effects of distributions in τ_c and spin diffusion which tend to average out anisotropy.

There are numerous examples in the literature which demonstrate the central role of distribution moments in predicting properties such as mechanical moduli.[38, 39, 147] An unusually novel application is to monitor the dipole reorientation which occurs when piezoelectric polymers are electrically poled.[102–104] It is fortuitous that the electric dipoles in PVDF are colinear with the crystallographic *b*-axis in the oriented *β* polymorph. Linewidth anisotropy is monitored both as a function of fibre axis alignment and of poling axis alignment (perpendicular to the plane of the film)

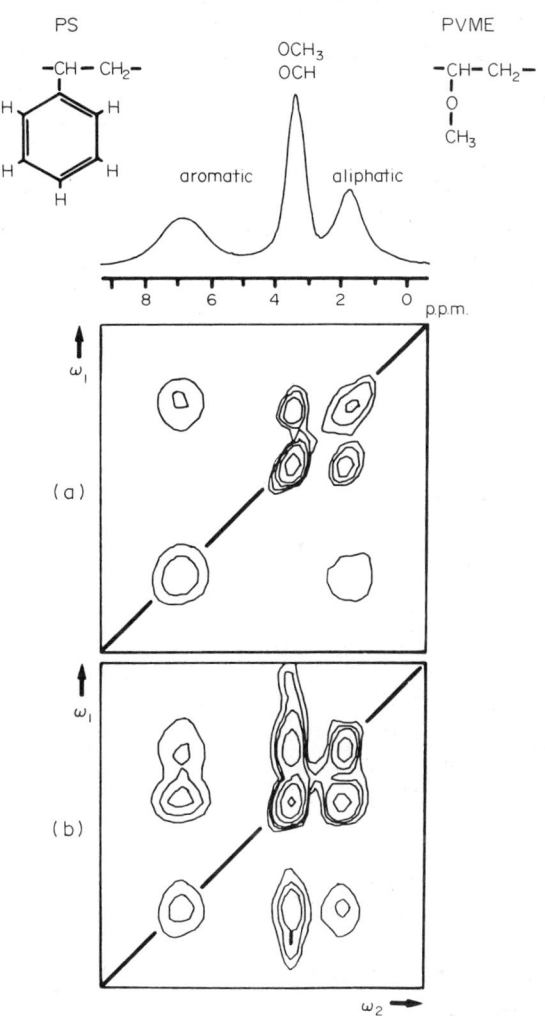

Figure 19 2-D proton spin diffusion spectra of blends of PS and PVME: (a) cast from chloroform and (b) cast from toluene. MAS spinning frequency = 2.8 kHz; $T = 328$ K and mixing time $\tau_m = 100$ ms. The ω_1 and ω_2 frequency dimensions result from Fourier transformation with respect to the evolution and detection periods respectively. (Reproduced by permission of the American Chemical Society from *Macromolecules*, 1985, **18**, 119)

relative to B_0. The tendency for the *b*-axis, and therefore electric dipoles, to reorient towards the poling field is reflected in the NMR linewidth anisotropies. Moments of the electric dipole distribution relative to the poling field are derived and experimental data are used to test various plausible models (Figure 20). The relationship between the second moment of the dipole distribution, $\cos 2\delta$ (where δ is the angle between a typical dipole and the poling direction) and the piezoelectric coefficient d_{31} is established. The results also indicate that electric dipole alignment is far from perfect along the poling direction even in the most strongly poled material. The NMR of other interesting electrically active polymers has been reviewed elsewhere.[3]

The chemical shift tensor also responds sensitively to molecular orientation in deformed polymers. However, its use requires unobscured spectra which typically implies the need to suppress dipolar interactions. Complications can also arise when the principal axis of σ in the molecular coordinate frame is unknown.

A series of papers by Dybowski and co-workers[154-157] builds upon earlier studies[98,152,153] to obtain information on statistical distributions in partially oriented PTFE from ^{19}F chemical shift data. A typical suite of results is shown in Figure 21. Analysis of the first and higher moments of the spectra yield moments of the distribution which in turn generate distribution functions for each draw ratio. In PTFE, it is demonstrated that orientation develops less rapidly than predicted by the pseudo-affine deformation model.[147] Furthermore, the shrinkage observed in oriented PTFE following

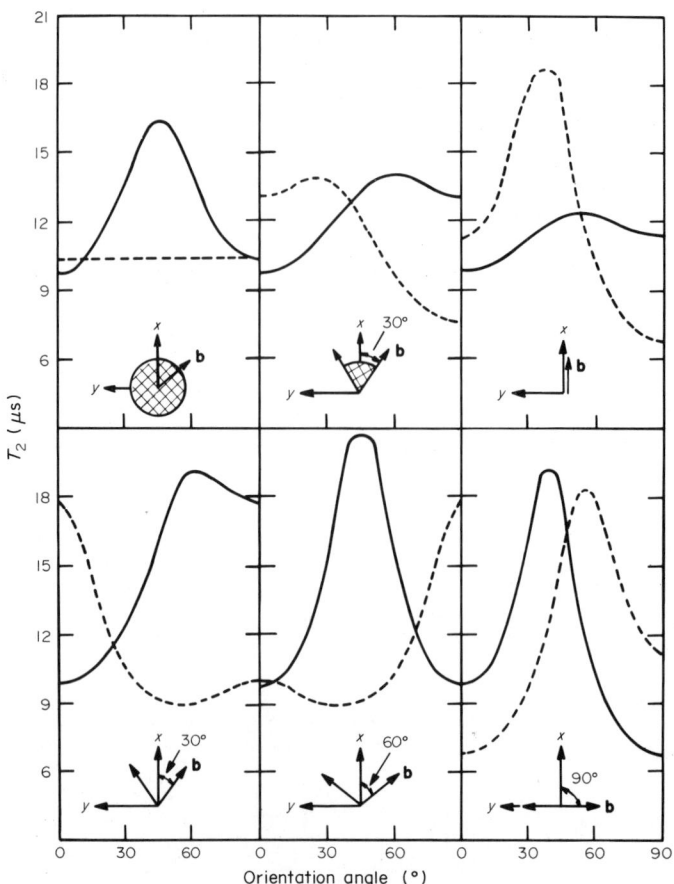

Figure 20 Calculated T_2 behaviour in PVDF for the ideal b-axis distributions as shown. The curves represent T_2 as a function of draw direction (solid line), and poling direction (dashed line) alignment relative to B_0 (reproduced by permission of the American Institute of Physics from *J. Chem. Phys.*, 1982, **17**, 5826)

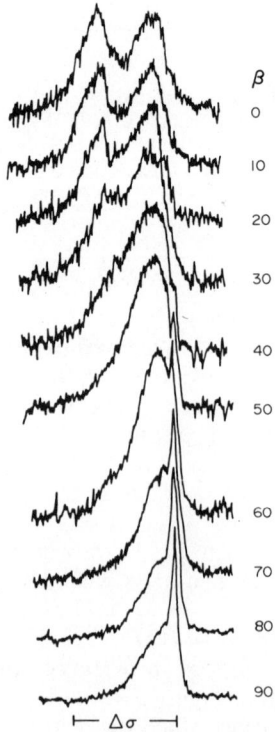

Figure 21 ^{19}F chemical shift spectra of oriented PTFE (140% elongation) as a function of the orientation β relative to B_0. The vertical scaling is not the same throughout. $\Delta\sigma$ is the anisotropy of the effectively axially symmetric shielding tensor (reproduced by permission of Butterworths from *Polymer*, 1982, **23**, 39)

annealing at 300 °C, is accompanied by rapid though incomplete disorientation of crystallites. The residual draw ratio following annealing bears an approximately linear relationship to the second moment of the distribution.[157]

The ^2H lineshape is comparably sensitive to molecular orientation[31] and once again the complete orientation distribution function may be determined. Sillescu and coworkers[25,160] have devised an analytical procedure whereby the ^2H-spectrum is decomposed into subspectra, the relative contributions of which vary according to the sample orientation in B_0. The ability to discriminate between crystalline and amorphous contributions and the near axial symmetry of the EFG about the C–^2H bond for aliphatic deuterons greatly facilitates analysis.

^2H NMR data for uniaxially drawn PE-d_4 (draw ratio, $\lambda=9$) and PE-d_4 single crystal mats vividly illustrate the sensitivity of ^2H lineshapes to local order in deformed polymers.[160] Typical spectra are presented in Figure 22 in which the amorphous component is suppressed. The excellent agreement between experimental and computed lineshapes for different sample orientations in B_0, utilizing only two adjustable parameters, fully justifies the assumed Gaussian distribution of chain axes in the oriented samples.

In an elegant experiment devised by Samulski and co-workers,[161,162] a number of perdeuterated swelling agents are used as probes to investigate molecular order in uniaxially oriented *cis*-1,4-polyisoprene. The ^2H NMR spectra respond to the orientational order conferred on the swelling agent as it progresses through the deformed elastomer matrix. In this way, information is obtained

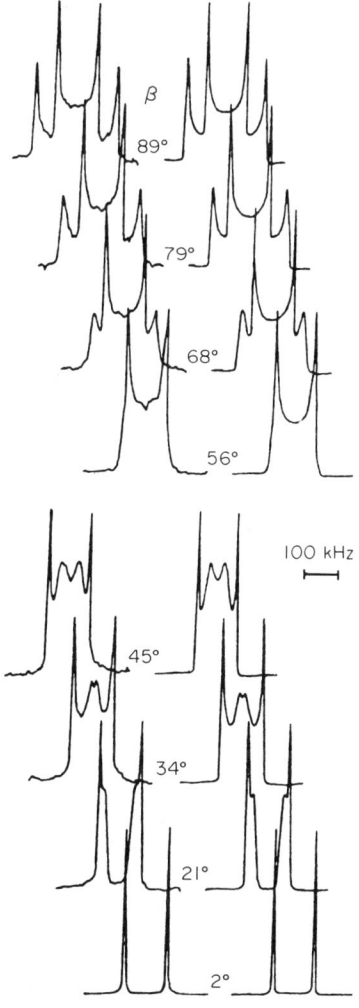

Figure 22 Observed (left) and calculated (right) ^2H spectra of drawn polyethylene ($\lambda=9$) as a function of orientation β relative to B_0. The calculated spectra were convoluted with a Gaussian of variance 0.9 kHz to account for the dipolar coupling between deuterons (reproduced by permission of Butterworths from *Polymer*, 1981, **22**, 1516)

on the short range segmental behaviour of the deformed matrix. An explicit dependence of quadrupolar splitting on draw ratio, temperature and degree of swelling is demonstrated. Guided by the short range nematic-like interactions monitored by ^2H NMR, a lattice model for the swollen network is developed.

19.10 REFERENCES

1. V. J. McBrierty, *Faraday Discuss. Chem. Soc.*, 1979, **68**, 78.
2. D. C. Douglass, *ACS Symp. Ser.*, 1980, **142**, 147.
3. V. J. McBrierty, *Magn. Reson. Rev.*, 1983, **8**, 165.
4. B. Albert, R. Jerôme, Ph. Teyssié, G. Smyth and V. J. McBrierty, *Macromolecules*, 1984, **17**, 2552.
5. V. J. McBrierty, D. C. Douglass and P. J. Barham, *J. Polym. Sci., Polym. Phys. Ed.*, 1980, **18**, 1561.
6. V. J. McBrierty and D. C. Douglass, *Phys. Rep.*, 1980, **63**, 61.
7. D. W. McCall, *NBS Spec. Publ. (US)*, 1969, **301**, 475.
8. T. M. Connor, *Trans. Faraday Soc.*, 1963, **60**, 1574.
9. N. Bloembergen, E. M. Purcell and R. V. Pound, *Phys. Rev.*, 1948, **73**, 679.
10. R. Kubo and K. Tomita, *J. Phys. Soc. Jpn.*, 1954, **9**, 888.
11. V. J. McBrierty and D. C. Douglass, *J. Magn. Reson.*, 1970, **2**, 352.
12. D. C. Douglass and G. P. Jones, *J. Chem. Phys.*, 1966, **45**, 956.
13. V. J. McBrierty and D. C. Douglass, *Macromol. Rev.*, 1981, **16**, 295.
14. P. Mansfield, in 'Progress in NMR Spectroscopy', ed. J. W. Emsley, J. Feeney and L. H. Sutcliffe, Pergamon Press, Oxford, 1972, vol. 8, p. 41.
15. W. K. Rhim, D. D. Elleman and R. W. Vaughan, *J. Chem. Phys.*, 1973, **59**, 3740.
16. C. P. Slichter, 'Principles of Magnetic Resonance', 2nd edn., Springer-Verlag, Berlin, 1978.
17. U. Haeberlen, *Adv. Magn. Reson.*, Suppl. 1, 1976.
18. M. Merhing, 'Principles of High Resolution NMR in Solids', 2nd edn., Springer-Verlag, New York, 1983.
19. L. W. Jelinski, personal communication.
20. A. A. Jones, J. F. O'Gara, P. T. Inglefield, J. T. Bendler, A. F. Yee and K. L. Ngai, *Macromolecules*, 1983, **16**, 658.
21. K. L. Ngai and C. T. White, *Phys. Rev. B: Condens. Matter*, 1979, **20**, 2475.
22. M. Linder, P. M. Henrichs, J. M. Hewitt and D. J. Massa, *J. Chem. Phys.*, 1985, **82**, 1585.
23. A. G. Redfield, *IBM J. Res. Dev.*, 1957, **1**, 19.
24. E. R. Andrew in 'Progress in NMR Spectroscopy', ed. J. W. Emsley, J. Feeney and L. H. Sutcliffe', Pergamon Press, Oxford, 1972, vol. 8, p. 1.
25. R. Hentschel, J. Schlitter, H. Sillescu and H. W. Spiess, *J. Chem. Phys.*, 1978, **68**, 56.
26. A. Abragam, 'The Principles of Nuclear Magnetism', Clarendon, Oxford, 1961.
27. S. R. Hartmann and E. L. Hahn, *Phys. Rev.*, 1962, **128**, 2042.
28. A. Pines, M. G. Gibby and J. S. Waugh, *J. Chem. Phys.*, 1972, **56**, 1776; 1973, **59**, 569.
29. See, for example, M. Goldman, 'Spin Temperature and Nuclear Magnetic Resonance in Solids', Clarendon, Oxford, 1970.
30. T. M. Duncan and C. Dybowski, *Surf. Sci. Rep.*, 1981, **1**, 157.
31. H. W. Spiess, *Adv. Polym. Sci.*, 1985, **66**, 23.
32. J. Schaefer, *Macromolecules*, 1973, **6**, 882.
33. See, for example, A. R. Edmonds, 'Angular Momentum in Quantum Mechanics', Princeton University Press, 1974.
34. E. R. Andrew and J. Lipofsky, *J. Magn. Reson.*, 1972, **8**, 217.
35. J. H. Van Vleck, *Phys. Rev.*, 1948, **74**, 1168.
36. G. P. Jones, *Phys. Rev.*, 1966, **148**, 332.
37. C. P. Slichter and D. Ailion, *Phys. Rev. A.*, 1964, **135**, 1099; 1965, **137**, 235.
38. V. J. McBrierty, I. R. McDonald and I. M. Ward, *J. Phys. D.*, 1971, **4**, 88.
39. V. J. McBrierty, I. R. McDonald, *J. Phys. D.*, 1973, **6**, 131.
40. D. L. VanderHart and A. N. Garroway, *J. Chem. Phys.*, 1979, **71**, 2773.
41. J. Schaefer, E. O. Stejskal, T. R. Steger, M. D. Sefcik and R. A. McKay, *Macromolecules*, 1980, **13**, 1121.
42. I. Solomon, *Phys. Rev.*, 1955, **99**, 559.
43. I. Solomon and N. Bloembergen *J. Chem. Phys.*, 1955, **25**, 261.
44. V. J. McBrierty and D. C. Douglass, *Macromolecules*, 1977, **10**, 855.
45. D. C. Douglass and V. J. McBrierty, *Macromolecules*, 1978, **11**, 766.
46. K. L. Ngai, *Phys. Rev. B: Condens. Matter*, 1980, **22**, 2066.
47. H. A. Resing, *J. Chem. Phys.*, 1965, **43**, 669.
48. F. A. Bovey and L. W. Jelinski, *J. Phys. Chem.*, 1985, **89**, 571.
49. V. J. McBrierty, *Polymer*, 1974, **15**, 503.
50. M. L. Williams, R. F. Landel and J. D. Ferry, *J. Am. Chem. Soc.*, 1955, **77**, 3701.
51. K. J. Packer, J. M. Pope, R. R. Yeung and M. E. A. Cudby, *J. Polym. Sci., Polym. Phys. Ed.*, 1984, **22**, 589.
52. P. Caravatti, J. A. Deli, G. Bodenhausen and R. R. Ernst, *J. Am. Chem. Soc.*, 1982, **104**, 5506.
53. P. M. Henrichs and M. Linder, *J. Magn. Reson.*, 1984, **58**, 458.
54. P. Caravatti, P. Neuenschwander and R. R. Ernst, *Macromolecules*, 1985, **18**, 119.
55. F. Bloch, *Phys. Rev.*, 1958, **111**, 841.
56. L. R. Sarles and R. M. Cotts, *Phys. Rev.*, 1958, **111**, 853.
57. M. Mehring, A. Pines, W. K. Rhim and J. S. Waugh, *J. Chem. Phys.*, 1971, **54**, 3229.
58. D. Suwelack, W. P. Rothwell and J. S. Waugh, *J. Chem. Phys.*, 1980, **73**, 2559.
59. D. L. VanderHart, W. L. Earl and A. N. Garroway, *J. Magn. Reson.*, 1981, **44**, 361.
60. M. Maricq and J. S. Waugh, *Chem. Phys. Lett.*, 1977, **47**, 327.
61. W. T. Dixon, *J. Magn. Reson.*, 1981, **44**, 220.
62. T. C. Farrar and E. D. Becker, 'Pulse and Fourier Transform NMR', Academic Press, New York, 1971.

63. C. Dybowski and R. Lichter (eds.), 'Practical NMR Spectroscopy', Marcel–Dekker, New York, 1984.
64. E. Fukushima and S. B. W. Roeder, 'Experimental Pulse NMR: A Nuts and Bolts Approach', Addison–Wesley, London, 1981.
65. I. J. Lowe and R. E. Norberg, *Phys. Rev.*, 1957, **107**, 46.
66. J. G. Powles and P. Mansfield, *Phys. Lett.*, 1962, **2**, 58.
67. J. G. Powles and J. H. Strange, *Proc. Phys. Soc.*, 1963, **82**, 6.
68. H. Y. Carr and E. M. Purcell, *Phys. Rev.*, 1955, **94**, 630.
69. S. Meiboom and D. Gill, *Rev. Sci. Instrum.*, 1958, **29**, 688.
70. G. H. Weiss, R. K. Gupta, J. A. Ferretti and E. D. Becker, *J. Magn. Reson.*, 1980, **37**, 369.
71. E. D. Becker, J. A. Ferretti, R. K. Gupta and G. H. Weiss, *J. Magn. Reson.*, 1980, **37**, 381.
72. M. Goldman and L. Shen, *Phys. Rev.*, 1966, **144**, 321.
73. J. Jeener and P. Broekaert, *Phys. Rev.*, 1967, **157**, 232.
74. K. J. Packer, *Mol. Phys.*, 1980, **39**, 15.
75. J. S. Waugh, L. M. Huber and U. Haeberlen, *Phys. Rev. Lett.*, 1968, **20**, 180.
76. P. Mansfield and D. Ware, *Phys. Lett.*, 1966, **22**, 133.
77. P. Mansfield, *Phys. Lett. A*, 1970, **32**, 485.
78. W. K. Rhim, D. D. Elleman and R. W. Vaughan, *J. Chem. Phys.*, 1973, **59**, 3740.
79. W. K. Rhim, D. D. Elleman and L. B. Schreiber and R. W. Vaughan, *J. Chem. Phys.*, 1974, **60**, 4595.
80. D. P. Burum and W. K. Rhim, *J. Chem. Phys.*, 1979, **71**, 944.
81. R. Blinc, V. Rutar, J. Seliger, J. Slak and V. Smolej, *Chem. Phys. Lett.*, 1977, **48**, 576.
82. D. E. Wemmer, E. K. Wolff and M. Mehring, *J. Magn. Reson.*, 1981, **42**, 460.
83. H. W. Spiess, *J. Chem. Phys.*, 1980, **72**, 6755.
84. K. Takegoshi and C. A. McDowell, *Chem. Phys. Lett.*, 1985, **116**, 100.
85. M. E. Stoll, *Philos. Trans. R. Soc. London, Ser. A*, 1981, **299**, 565.
86. M. E. Stoll, A. J. Vega and R. W. Vaughan, *J. Chem. Phys.*, 1976, **65**, 4093.
87. R. K. Hester, J. L. Ackerman, V. R. Cross and J. S. Waugh, *Phys. Rev. Lett.*, 1975, **34**, 993.
88. R. K. Hester, J. L. Ackerman, B. L. Neff and J. S. Waugh, *Phys. Rev. Lett.*, 1976, **36**, 1081.
89. J. S. Waugh, *Proc. Natl. Acad. Sci., USA*, 1976, **73**, 1394.
90. J. Schaefer, R. A. McKay, E. O. Stejskal and W. T. Dixon, *J. Magn. Reson.*, 1983, **52**, 123.
91. J. Schaefer, E. O. Stejskal, D. Perchak, J. Skolnick and R. Yaris, *Macromolecules*, 1985, **18**, 368.
92. D. Hyndman and G. F. Origlio, *J. Appl. Phys.*, 1960, **31**, 1849.
93. D. W. McCall, D. C. Douglass and D. R. Falcone, *J. Phys. Chem.*, 1967, **71**, 998.
94. V. J. McBrierty, D. W. McCall, D. C. Douglass and D. R. Falcone, *Macromolecules*, 1971, **4**, 584; *J. Chem. Phys.*, 1970, **52**, 512.
95. M. Mehring, R. G. Griffin and J. S. Waugh, *J. Chem. Phys.*, 1971, **55**, 746.
96. A. N. Garroway, D. C. Stalker and P. Mansfield, *Polymer*, 1975, **16**, 161.
97. J. Schaefer, E. O. Stejskal and R. Buchdahl, *Macromolecules*, 1975, **8**, 291.
98. A. J. Vega and A. D. English, *Macromolecules*, 1980, **13**, 1635.
99. W. W. Fleming, C. A. Fyfe, J. R. Lyerla, H. Vanni and C. S. Yannoni, *Macromolecules*, 1980, **13**, 460.
100. V. J. McBrierty, D. C. Douglass and T. A. Weber, *J. Polym. Sci. Polym. Phys. Ed.*, 1976, **14**, 1271.
101. V. J. McBrierty and D. C. Douglass, *Macromolecules*, 1977, **10**, 855.
102. V. J. McBrierty, D. C. Douglass and T. Furukawa, *Macromolecules*, 1982, **15**, 1063; 1984, **17**, 1136.
103. D. C. Douglass, V. J. McBrierty and T. T. Wang, *J. Chem. Phys.*, 1982, **77**, 5826.
104. V. J. McBrierty, D. C. Douglass and T. T. Wang, *Appl. Phys., Lett.*, 1982, **41**, 1051.
105. K. A. Mauritz, C. J. Hora and A. J. Hopfinger, *Adv. Chem. Ser.*, 1980, **187**, 123.
106. N. G. Boyle, J. M. D. Coey and V. J. McBrierty, *Chem. Phys. Lett.*, 1982, **86**, 16.
107. N. G. Boyle, V. J. McBrierty and D. C. Douglass, *Macromolecules*, 1983, **16**, 75.
108. N. G. Boyle, V. J. McBrierty and A. Eisenberg, *Macromolecules*, 1983, **16**, 80.
109. P. T. Inglefield, A. A. Jones, R. P. Lubianez and J. F. O'Gara, *Macromolecules*, 1981, **14**, 288.
110. P. T. Inglefield, R. M. Amici, J. F. O'Gara, C. C. Hung and A. A. Jones, *Macromolecules*, 1983, **16**, 1552.
111. A. A. Jones, J. F. O'Gara, P. T. Inglefield, J. T. Bendler, A. F. Yee and K. L. Ngai, *Macromolecules*, 1983, **16**, 658.
112. H. W. Spiess, *Colloid Polym. Sci.*, 1983, **261**, 193.
113. P. M. Henrichs, M. Linder, J. M. Hewitt, D. Massa and H. V. Isaacson, *Macromolecules*, 1984, **17**, 2412.
114. J. F. O'Gara, A. A. Jones, C. C. Hung and P. T. Inglefield, *Macromolecules*, 1985, **18**, 1117.
115. J. Schaefer, E. O. Stejskal, R. A. McKay and W. T. Dixon, *Macromolecules*, 1984, **17**, 1479.
116. A. F. Yee and S. A. Smith, *Macromolecules*, 1981, **14**, 54.
117. L. W. Jelinski, J. J. Dumais and A. K. Engel, *Macromolecules*, 1983, **16**, 403.
118. F. A. Bovey and L. W. Jelinski, *J. Phys. Chem.*, 1985, **89**, 571.
119. E. Helfand, *J. Chem. Phys.*, 1971, **54**, 4651.
120. A. L. Cholli, J. J. Dumais, A. K. Engel and L. W. Jelinski, *Macromolecules*, 1984, **17**, 2399.
121. D. C. Douglass and V. J. McBrierty, *Polym. Eng. Sci.*, 1979, **19**, 1054.
122. D. A. Torchia and A. Szabo, *J. Magn. Reson.*, 1982, **49**, 107.
123. L. W. Jelinski, J. J. Dumais and A. K. Engel, *Macromolecules*, 1983, **16**, 492.
124. A. C. Lind, *J. Chem. Phys.*, 1977, **66**, 3482.
125. T. T. P. Cheung and B. C. Gerstein, *J. Appl. Phys.*, 1981, **52**, 5517.
126. J. R. Ebdon, 'Copolymer Characterization by ¹³C NMR', in 'Developments in Polymer Characterization — 2', ed. J. V. Dawkins, Applied Science, Barking, 1980.
127. T. Nishi, T. K. Kwei and T. T. Wang, *J. Appl. Phys.*, 1975, **46**, 4157.
128. T. Nishi, T. T. Wang and T. K. Kwei, *Macromolecules*, 1975, **8**, 227.
129. R. A. Assink, *Macromolecules*, 1978, **11**, 1233.
130. H. Tanaka and T. Nishi, *J. Chem. Phys.*, 1986, **85**, 6197.
131. K. M. Natarajan, E. T. Samulski and R. I. Cukier, *Nature (London)*, 1978, **275**, 527.
132. D. Hentschel, H. Sillescu and H. W. Spiess, *Macromolecules*, 1981, **14**, 1605; *Polymer*, 1984, **25**, 1078.

133. T. M. Duncan, *J. Phys. Chem. Ref. Data*, 1987, **16**, 125.
134. S. J. Opella ad J. S. Waugh, *J. Chem. Phys.*, 1977, **66**, 4919.
135. J. Urbina and J. S. Waugh, *Proc. Natl. Acad. Sci. USA*, 1974, **71**, 5062.
136. H. T. Edzes, *Polymer*, 1983, **24**, 1425.
137. W. W. Fleming, C. A. Fyfe, R. D. Kendrick, J. R. Lyerla, H. Vanni and C. S. Yannoni, *ACS Symp. Ser.*, 1980, **142**, 193.
138. H. A. Resing, D. C. Weber, M. Anderson, G. R. Miller, M. Moran, C. F. Poranski, Jr. and L. Mattix, *Polym. Prepr.*, 1982, **23**, 101.
139. T. Terao, S. Maeda, T. Yamabe, K. Akagi and H. Shirakawa, *Chem. Phys. Lett.*, 1984, **103**, 347; *Solid State Commun.*, 1984, **49**, 829.
140. A. Manenschijn, M. J. Duijvestijn, J. Smidt, R. A. Wind, C. S. Yannoni and T. C. Clarke, *Chem. Phys. Lett.*, 1984, **112**, 99.
141. P. D. Murphy, T. Taki, B. C. Gerstein, P. M. Henrichs and D. J. Massa, *J. Magn. Reson.*, 1982, **49**, 99.
142. W. S. Veeman, *Prog. Nucl. Magn. Reson. Spectrosc.*, 1984, **16**, 193.
143. B. Albert, R. Jérôme, P. Teyssié, G. Smyth, N. G. Boyle and V. J. McBrierty, *Macromolecules*, 1985, **18**, 388.
144. R. A. Assink and G. L. Wilkes, *Polym. Eng. Sci.*, 1977, **17**, 606.
145. J. J. Dumais, L. W. Jelinski, L. M. Leung, I. Gancarz, A. Galambos and J. T. Koberstein, *Macromolecules*, 1985, **18**, 116.
146. N. M. Szeverenyi, M. J. Sullivan and G. E. Maciel, *J. Magn. Reson.*, 1982, **47**, 462.
147. I. M. Ward, 'Structure and Properties of Oriented Polymers', Wiley, New York, 1975.
148. J. O'Brien and V. J. McBrierty, *Proc. R. Ir. Acad.*, 1975, **25**, 331.
149. V. J. McBrierty, *J. Chem. Phys.*, 1974, **61**, 872.
150. V. J. McBrierty, *J. Chem. Phys.*, 1972, **57**, 3287.
151. D. L. Vanderhart, *Macromolecules*, 1979, **12**, 1232.
152. A. N. Garroway, D. C. Stalker and P. Mansfield, *Pulsed Nucl. Magn. Reson. Spin Dyn. Solids, Proc. Spec. Coloq. Ampere*, 1st, 1973.
153. P. Mansfield, M. J. Orchard, D. C. Stalker and K. H. B. Richards, *Phys. Rev. B*, 1973, **7**, 90.
154. A. J. Brandolini, T. M. Apple, C. R. Dybowski and R. G. Pembleton, *Polymer*, 1982, **23**, 39.
155. A. J. Brandolini and C. R. Dybowski, *J. Polym. Sci., Polym. Lett. Ed.*, 1983, **21**, 423.
156. A. J. Brandolini, M. D. Alvey and C. R. Dybowski, *J. Polym. Sci., Polym. Phys. Ed.*, 1984, **21**, 2511.
157. A. J. Brandolini, K. J. Rocco and C. R. Dybowski, *Macromolecules*, 1984, **17**, 1455.
158. G. Hempel and H. Schneider, *Pure Appl. Chem.*, 1982, **54**, 635.
159. H. A. Resing and D. Slotfeldt–Ellingsen, *J. Magn. Reson.*, 1980, **38**, 401.
160. R. Hentschel, H. Sillescu and H. W. Spiess, *Polymer*, 1981, **22**, 1516.
161. B. Deloche and E. T. Samulski, *Macromolecules*, 1981, **14**, 575.
162. H. Toriumi, B. Deloche, J. Herz and E. T. Samulski, *Macromolecules*, 1985, **18**, 304.

20

IR Spectroscopy

SHAW L. HSU

University of Massachusetts, Amherst, MA, USA

20.1 GENERAL INTRODUCTION TO IR SPECTROSCOPY

A substantial amount of literature exists regarding the utility of vibrational spectroscopy for the characterization of polymers. Probably the largest fraction of these reported studies is related to the use of IR spectroscopy to determine the molecular composition of polymers by analyzing the characteristic vibrations of functional groups. The power of vibrational spectroscopy, *i.e.* its selectivity and sensitivity, cannot be overestimated in these applications. Many excellent studies have proven to be of enormous value to polymer scientists in this type of analysis. In this chapter the more recent applications of vibrational spectroscopy are emphasized, *i.e.* the characterization of the microstructures and their relationships to the macroscopic properties of polymers. This is perhaps one of the most often stated goals of polymer scientists. This section provides an introduction to the utility and advantages of vibrational spectroscopy, more specifically IR spectroscopy. More detailed descriptions of applications of spectroscopy to polymer blends, polymer dynamics, and surfaces or interfaces follow.

Generally, sample preparation does not require significant effort, and IR spectra of polymers can be readily obtained. Polymers mixed with potassium bromide and then pressed into pellets, or films prepared from melt or cast from solution, can be easily studied. For bulk samples or powders, or if a concentration profile is needed, the reflectance technique is probably more suitable than transmission. With the recent rapid improvements in instrumentation, quantitative spectroscopic analysis of polymer solutions is not difficult to carry out. The availability of small yet powerful computers attached to the instruments has provided new and powerful analysis routines for structural characterization.

A number of interesting points can be made from the vibrational spectra of the simplest polymer, polyethylene. Often the most interesting or striking feature associated with the vibrational spectrum of a polymer is its simplicity. The IR and Raman spectra obtained for a highly crystalline polyethylene film are shown in Figure 1. A large number of studies have been carried out to elucidate the vibrational features of polyethylene or *n*-alkanes. The earliest representative studies are given in refs. 1–5. For small molecules of N nuclei, there are $3N - 6$ relative vibrations. The six degrees of freedom associated with the overall molecular translations and rotations cannot be considered as vibrations. However, the spectra obtained for polymers are not as complicated as might be imagined considering the extremely high molecular weight of most polymers of interest. In fact, for infinite polymers of well-defined conformation, there are only $3N - 4$ optically active vibrations, where N now refers to the number of nuclei per translationally equivalent structural unit. For polyethylene, this is $-CH_2CH_2-$. For polypropylene N is 27, since there are nine atoms per chemical repeat and three units form a translationally equivalent repeat unit. For highly ordered polymers or oligomers, symmetry analysis or group theory is a powerful tool providing the number of vibrations to be expected and their optical activities.[6-8]

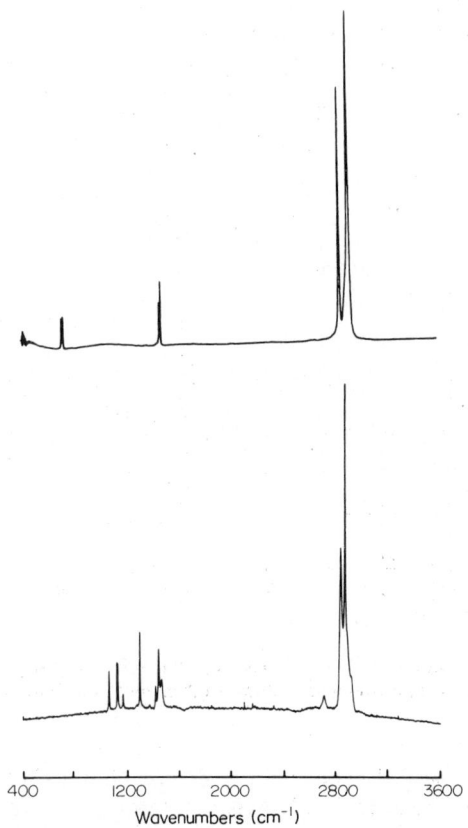

Figure 1 IR and Raman spectra of polyethylene film in the 400–3600 cm^{-1} region: (top) IR spectrum; (bottom) Raman spectrum

Not only do the IR and Raman spectra shown in Figure 1 contain few features, but also it should be noted that none of the bands present in the IR are Raman active. This mutual exclusion simply reflects the fact that the polyethylene chain conformation has an inversion symmetry element. The IR and Raman techniques are complementary and often are both needed for structural characterization.

Most polymer solids contain crystalline sections as well as amorphous regions. Substantial increases in the crystalline order can be measured by thermal annealing. In polyethylene, by measuring a series of samples containing different degrees of crystallinity, well assigned vibrational transitions associated with irregular chain conformations, presumably in the amorphous regions, can be found in the 1300 cm^{-1} region as shown in Figure 2. Since the structures are irregular, the assignments of the transitions cannot be made from symmetry considerations.

If a polymer film is oriented by drawing or extrusion, the polarized IR spectra show tremendous differences in absorption for incident radiation polarized parallel or perpendicular to the drawing direction. Polarized IR spectra of polyethylene are shown in Figure 3. The CH$_2$ rocking vibrations

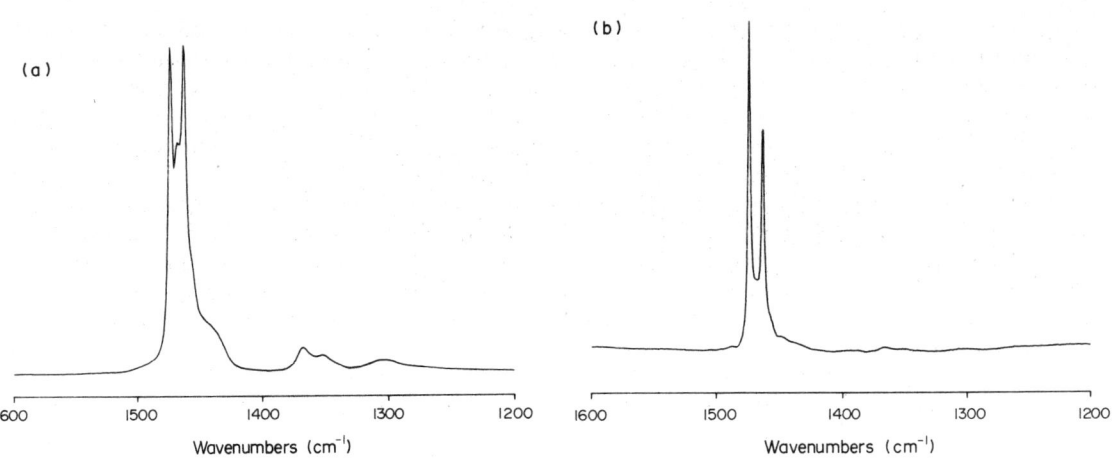

Figure 2 Amorphous bands found for polyethylene: (a) polyethylene film quenched from melt; (b) same film after annealing at 80 °C

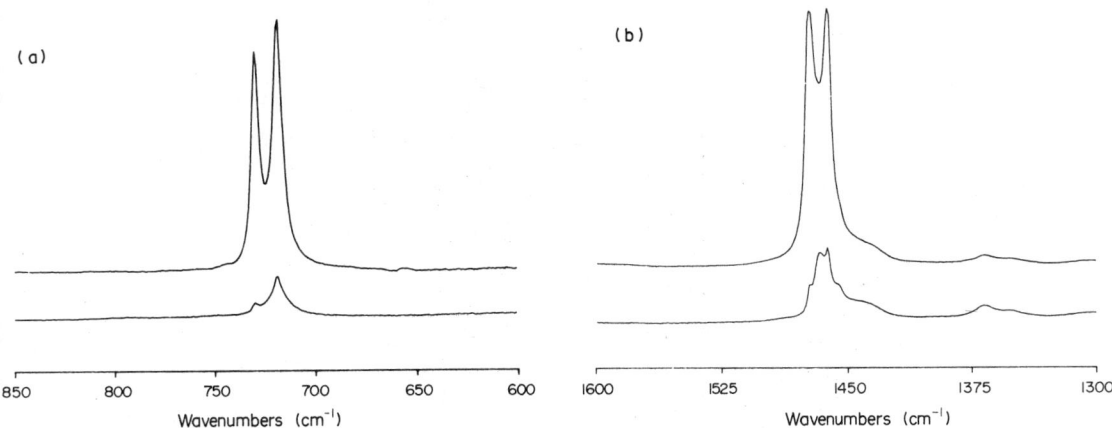

Figure 3 Polarized IR spectra of polyethylene: (a) CH_2 rocking region; (b) CH_2 wagging and bending regions; top is taken with polarization parallel to the draw direction and bottom is taken with polarization perpendicular to the draw direction

in the 700 cm^{-1} region, or the CH_2 wagging or CH_2 bending vibrations in the 1300–1400 cm^{-1} regions exhibit very different polarization behavior. This observation highlights several important features of polarization spectra. First and most importantly, the polarization measured is associated with the direction of the vibration transition moment. The direction of this transition moment relative to the chain segment must be considered in order to determine the overall chain segmental orientation. For some functional groups and extremely well defined polymer structures the transition moment can be estimated. In many cases, however, this is difficult to do. The principal advantage of spectroscopy is that it is a short range technique depending only on local chain conformation and packings. Therefore orientation information obtained from spectroscopy is highly selective and can be used to characterize individual chain segmental orientation in both crystalline and amorphous regions. The relative orientation of the amorphous bands to the crystalline bands can provide a great deal of information regarding the deformation process in forming highly anisotropic fibrous samples.[9-12]

Although not often found in polymers, there are features in the vibrational spectra of oligomers with well-defined chain conformations which are due to the connectivity of the chain of repeat units. These additional spectroscopic features, found in both the IR and Raman spectra of oligomers, have been explained in terms of a coupled oscillator model.[13] Detailed understanding of the vibrations found for oligomers yields a considerable insight into the vibrational spectra obtained for polymers.

The idea that vibrations can couple is a general one, and the coupled oscillator model is very simple. For each of the vibrations with frequency v_0 found for the individual chemical repeat units, a series of vibrations arising from the backbone coupling may be found shifted from the unperturbed value. For a chain with N chemical repeat units or N oscillators, instead of observing a set of N degenerate vibrations of frequency ν_0, a series of vibrations or a progression may exist. Indeed, this has been observed for n-alkanes.[13] Representative IR and Raman spectra for $C_{20}H_{42}$ are shown in Figure 4. The frequency separation of the progression modes is inversely proportional to the unperturbed frequency ν_0. The localized vibrations, such as the CH_2 stretching vibration in the 3000 cm^{-1} region, is hardly perturbed by the interaction along the chain. In contrast, low frequency components such as the CH_2 rocking vibration at 700 cm^{-1}, can be perturbed significantly.

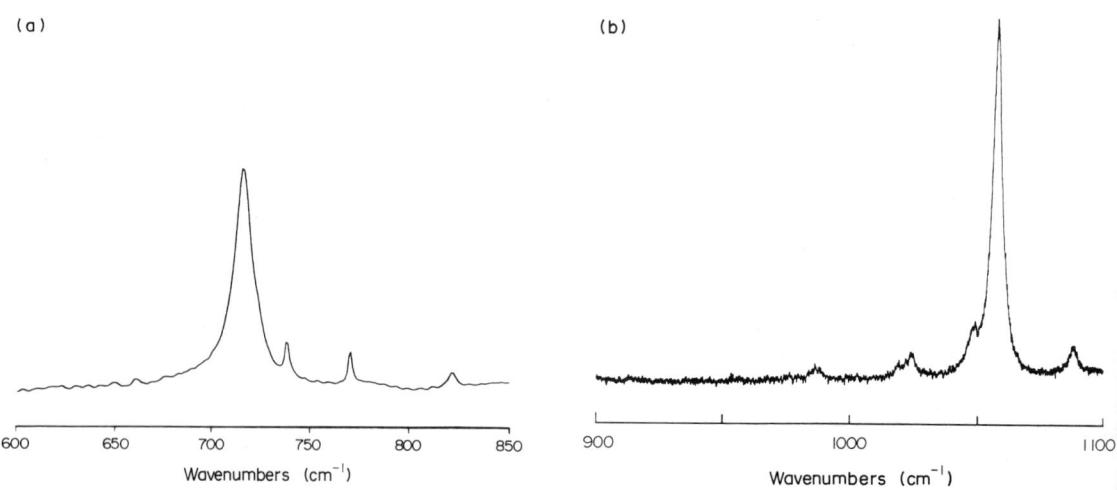

(a)   (b)

Figure 4   Progression bands observed for $C_{20}H_{42}$: (a) IR; (b) Raman

Because the physical origin of the IR and Raman effects are quite different, with one depending on the dipole change and the other on changes in the polarizability tensor, IR spectroscopy is more appropriate for studying chemical composition or side groups and Raman spectroscopy is more appropriate for studying chain conformation, because backbone vibrations of long chain molecules are generally quite intense in Raman. There are few polymer examples of this series of vibrations that have been found and it is difficult to predict the relative intensity of the observed progressions. However, when understood well, knowledge of the changing frequency and character of several vibrations as a function of intrachain coupling has proven to be important in considering the validity of force fields and in developing new techniques to follow conformational changes by selective deuteration experiments.[14]

Another interesting feature observed in the IR spectrum of highly crystalline polyethylene is the doublet in the 700 cm^{-1} region. This doublet arises from crystal field splitting.[15-17] The crystalline structure of polyethylene has been well characterized. There are two chains per crystalline unit cell.[18] Just as there is strong coupling between the vibrations along the polymer chain, the possibility exists that coupling can also occur for equivalent vibrations in the crystalline unit cell. The nonbonded intermolecular coupling between the two chains effectively removes the degeneracy between the equivalent oscillators associated with the chemical units associated with the two individual chains. Instead of one component near 720 cm^{-1}, a doublet at 721 and 730 cm^{-1} is observed (Figure 5). The magnitude of the splitting and the relative intensity of the two components then depends on the specificity and magnitude of the nonbonded van der Waals interactions between the two chains in the unit cell.

Few examples in the literature show such well defined spectroscopic features arising from chain–chain interactions in the three dimensional state as those observed in polyethylene. One must assume that intermolecular interactions are much weaker than intramolecular ones, or that a considerable structural disorder exists making explicit features difficult to observe. There are, however, extremely well defined perturbations due to hydrogen bonding in polypeptides. In this case, there is sufficient three dimensional order, and strong interactions, to make perturbing features easy to observe.[19,20] Coupling between oscillating dipoles has been invoked to explain the extraordinarily large couplings for the amide group vibrations.[20] However, it should be emphasized that there are actually very few vibrational spectra that exhibit features characteristic of three-dimensionally ordered polymer crystals. Detailed characterization of the ones found provide unique opportunities to determine how the polymeric molecules pack.

(a)   (b)

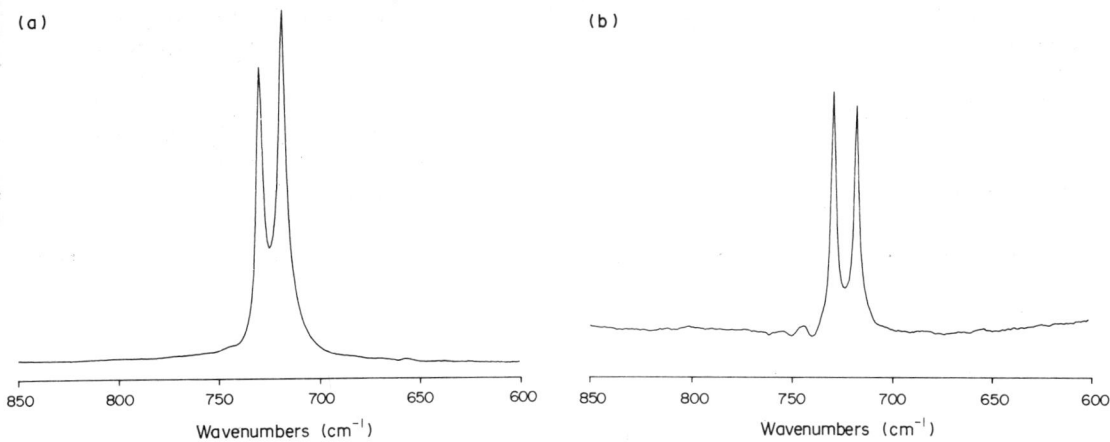

Wavenumbers (cm^{-1})   Wavenumbers (cm^{-1})

Figure 5 Crystalline field splitting observed for polyethylene: (a) film quenched from melt; (b) film annealed at 80 °C

The nature of intermolecular interactions can be directly studied by analyzing the frequency and intensity of the external vibrations, or lattice modes, of the unit cell. These intermolecular vibrations, characteristic of whole chain segment movements, generally reside in the very low frequency region, *i.e.* the far IR region, < 100 cm^{-1}. Fourier transform instruments have improved the signal to noise ratio, making these difficult measurements somewhat less painful. Since intermolecular interactions can be significantly increased by lowering the measurement temperature, the magnitudes of crystal-field splittings may increase substantially and the shape of the lattice modes may narrow with an upward shift in frequency.

One very special case is the temperature-induced structural changes of poly(tetrafluoroethylene). At temperatures below 19 °C, this polymer has a 13/6 helical conformation.[21] Above 19 °C the helical parameter changes to 15 units per seven turns.[21,22] Although a number of scattering studies have been used to characterize this transition, details of the transition have been greatly aided by the

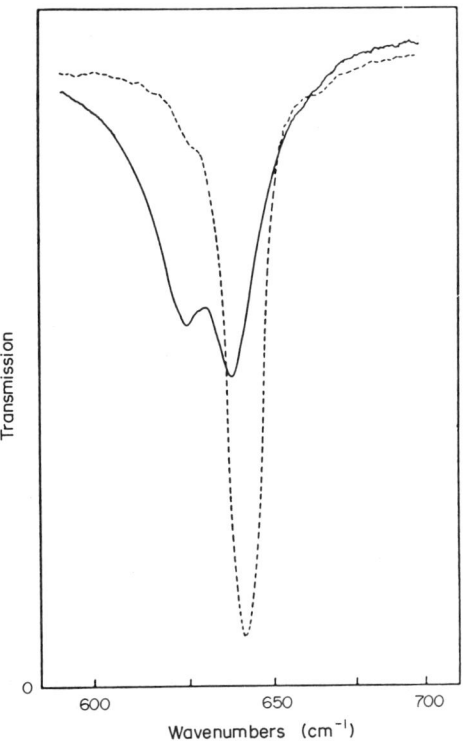

Transmission

Wavenumbers (cm^{-1})

Figure 6 IR spectra near 640 cm^{-1} for PTFE. (------) $T = 100$ K; (———) $T = 302$ K; sample thickness $= 5$ μm. Beckman IR4 spectrometer with KBr prism (from ref. 24)

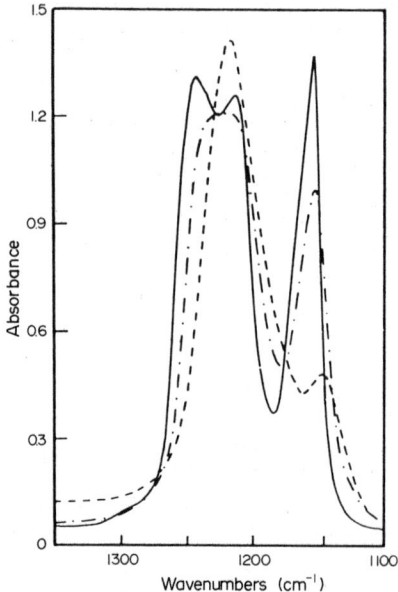

Figure 7 IR absorption spectra in the 1350 to 1100 cm^{-1} region, of powdered PTFE in KBr matrix at different temperatures: (——) room temperature; (–·–·–) 320 °C; (– – –) 395 °C (from ref. 23)

use of IR spectroscopy.[23,24] IR spectra obtained for the two phases at two temperatures below and above the transition temperature are shown in Figures 6 and 7.

In summary, the use of vibrational spectroscopy, more specifically IR spectroscopy, in chemical composition studies is well established. The principal advantage of using vibrational spectroscopy as a morphological tool is that it is extremely selective. Unlike diffraction techniques, which depend on long term order or disorder, vibrational spectroscopy is characteristic of chain conformation or packing at a much more specific or localized level. In the following sections it will be shown that IR spectroscopy can provide unique structural features that are difficult or impossible to obtain with other characterization techniques.

20.2 IR SPECTROSCOPIC STUDIES OF POLYMER BLENDS

The blending of polymers is probably one of the most effective methods available to increase processing efficiency and to enhance macroscopic properties. For example, if one component has a high glass transition temperature and is not easily processable, adding an appropriate second component can lower the glass transition temperature, making the binary mixture much easier to process without significant sacrifices in the mechanical properties associated with the first component. Poly(*p*-phenylene oxide) and polystyrene, which form a thermodynamically stable mixture for all compositions,[25] is such an example. Although polyurethanes, strictly speaking, cannot be treated as polymer mixtures, they represent a case in which very specific phase-separated morphologies can be achieved when segments consist of significantly different chemical structural units. One may also speak of truly phase-separated blends, such as impact-modified polystyrene, in which a dispersed rubber phase is incorporated to improve ductility and impact resistance. The growing recognition of the potential associated with new polymer blends, along with the tremendous range of materials already available, has made research in this area one of the most technologically significant and fundamentally interesting areas of study in polymer science.

In this review, the emphasis will be given to the structural characterization of polymer blends by spectroscopy. Even so, it is an extremely broad subject to discuss. The size or scale of the structural units of interest depend very much on the sensitivity and suitability of the probing technique, and it should not be forgotten that there are significant differences in the physical bases of various characterization techniques. Spectroscopy has made significant contributions in characterizing specific molecular interactions and structural parameters, such as chain conformations and packings in polymer blends.

Due to the large molecular size of polymers, the entropy of mixing is relatively insignificant as a driving force in the mixing of polymers.[26] Therefore the change in the overall free energy on mixing must arise principally from the term related to the heat of mixing, *i.e.* the interaction parameter χ. The value of this term depends on the magnitude and specificity of the interaction between the components. Spectroscopic features assigned to various functional groups of the polymer components may be affected by changes in intermolecular interactions, including permanent dipole interactions, transition dipole interactions and, the strongest, hydrogen-bonding interactions. Interpretation of such spectroscopic features is useful in giving a molecular origin to the thermodynamic parameter.

Many studies presented in the literature show successful utilization of vibrational spectroscopy in this area of study.[27-30] Because of the large magnitude of the forces originating from intramolecular interactions relative to intermolecular ones, the changes in spectroscopic features are generally very small, making their measurement difficult. Changes in hydrogen bonding are somewhat larger and can be observed more easily.

The experimental situation has improved recently, particularly since the use of Fourier transform IR instruments has become relatively common. These new instruments make extremely small differences in band position, shape and intensity easier to detect than in the past. Difference spectra can be obtained by subtracting a spectrum of neat component from those of binary mixtures within the computer associated with the instrument. It is presumed that the spectroscopic differences observed are due to differences in intermolecular interactions between neat and mixed states. However, the IR spectra so obtained may show features not related to the intermolecular interactions in polymer blends, *i.e.* band shape differences, which result from differences in the local index of refraction of the two components, sample thickness and effects due to the optical properties of the sample substrate.[31,32]

In order to use vibrational spectroscopy to follow phase separation behavior, comparison is made between the phase separated state and the homogeneous state. This experiment is difficult to carry out because comparison is actually made between the sum of a rich and poor phase and the homogeneous phase, and the contrast is low. Given the instruments now available, one should not be discouraged. However, great caution should be exercised regarding the interpretation of the spectroscopic features observed.

20.2.1 Spectroscopic Investigation of the Interaction Parameter

Blends of polystyrene (PS) and poly(vinyl methyl ether) (PVME) can be used to illustrate the use of vibrational spectroscopy. A number of studies have shown that solvent, molecular weight, composition and temperature can all affect the compatibility of these two polymers.[33-40] The phase diagram was established several years ago. However, the nature of the intermolecular interaction between the two components has not been firmly established until recently.[30]

In order to search for and characterize spectroscopic features sensitive to changes in phase behavior it should be recognized that, unless one uses the group frequency method, it is difficult to assign vibrational transitions of atactic polymers. Before blends can be studied, those spectroscopic features of the neat polymers which are most sensitive to temperature variations need to be identified. When the temperature of observation is lowered, both the specificity and magnitude of intermolecular interactions may increase sufficiently to enhance spectroscopic features sensitive to interchain interactions.

It is known[29] that vibrations involving the oxygen atom usually exhibit high sensitivity to blend compatibility. In the IR spectrum of PVME a strong doublet at 1085 and 1107 cm^{-1} with a shoulder at 1132 cm^{-1} shows the greatest change in its relative intensity when the sample is cooled or heated, as shown in Figure 8. The 1107 cm^{-1} component dominates at high temperature whereas the 1085 cm^{-1} component dominates at low temperature. In a separate experiment to show the sensitivity of this doublet to structural changes, PVME was quenched from the melt. The IR spectrum, shown in Figure 9 (observed at room temperature), does not show clear evidence of the doublet. Instead, a broad peak is observed in the 1100 cm^{-1} region. The exact assignment of the two components of the 1100 cm^{-1} band is not yet understood, although one can obtain some information from normal vibrational analysis for a series of aliphatic ethers.[41] Therefore, at present one can only conclude that the IR relative intensity is sensitive to structural changes and environment, and may be sensitive to the miscibility of PS/PVME blends.

Previously it has been established that high molecular weight PS/PVME binary mixtures containing ~ 20–80% PS ($M_w \geq 17\,500$) dissolved in chloroform or TCE can be used to cast

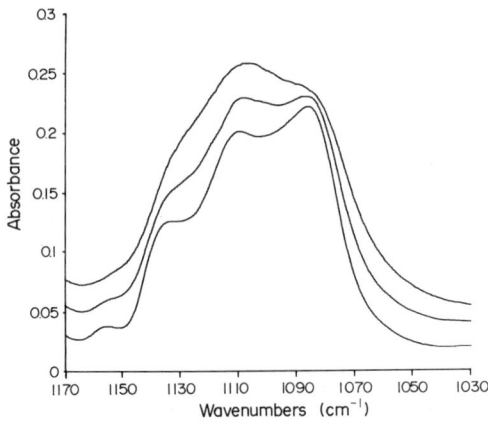

Figure 8   IR spectra near $1100\,cm^{-1}$ of PVME as a function of temperature: (top) 130 °C; (middle) 25 °C; (bottom) −180 °C. The ordinate scale is applicable to −180 °C

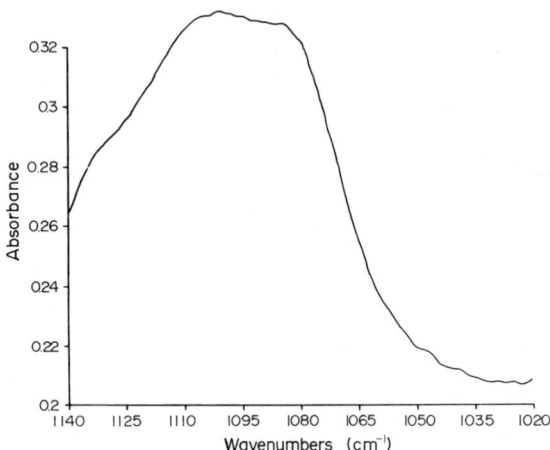

Figure 9   IR spectrum in the $1100\,cm^{-1}$ region for a PVME film quenched from the melt

immiscible samples.[33-35] From the same binary mixtures dissolved in toluene, miscible blends can be prepared. The two types of sample are quite different when examined visually. Totally transparent films are obtained from toluene solution, while the films obtained from TCE or chloroform are grainy in appearance. Pronounced spectroscopic differences are found for the PVME bands in the $1100\,cm^{-1}$ region. Although the frequencies of the doublet do not change, the relative intensities of the two components differ in the two types of samples. For miscible blends, such as the binary mixtures obtained from toluene, the intensity of the $1085\,cm^{-1}$ band is greater than that of the $1107\,cm^{-1}$ component, while for incompatible blends, obtained from chloroform or TCE, the reverse is true (see Figure 10).

A number of IR active bands in PS show small changes in position or shape when PS is blended with PVME. The band most sensitive to phase changes is located near $700\,cm^{-1}$. This band is generally assigned to the aromatic CH out-of-plane bending vibration.[42,43] It is found at $699.5\,cm^{-1}$ for films cast from toluene solutions containing equal molar amounts of PS and PVME. This is to be compared to the peak maximum located at $697.7\,cm^{-1}$ in neat PS. For incompatible blends of PS/PVME, such as films cast from TCE solutions, the peak maximum is usually found at an intermediate position between these two extremes. These different spectra are shown in Figure 11. Similar changes in frequency for this vibration have also been observed for PS blended with poly(2,6-dimethylphenylene oxide).[41]

Additional experiments can be carried out to substantiate that the spectroscopic differences found for PVME or PS are indeed associated with phase behavior. Binary mixtures containing 15% PS are compatible with PVME for all molecular weights and solvents. As can be seen in Figure 12, for

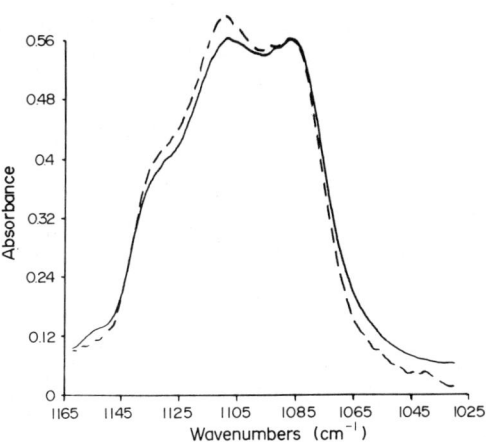

Figure 10 IR spectra in the 1100 cm^{-1} region, of (———) compatible; (– – –) incompatible 50/50 PS/PVME blends

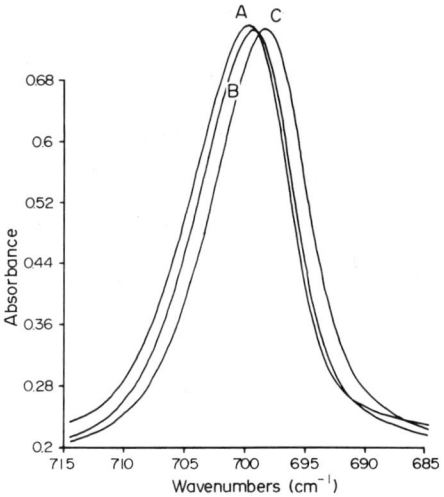

Figure 11 IR spectra in the 700 cm^{-1} region of (A) compatible and (B) incompatible 50/50 PS/PVME blends compared with (C) a spectrum of PS

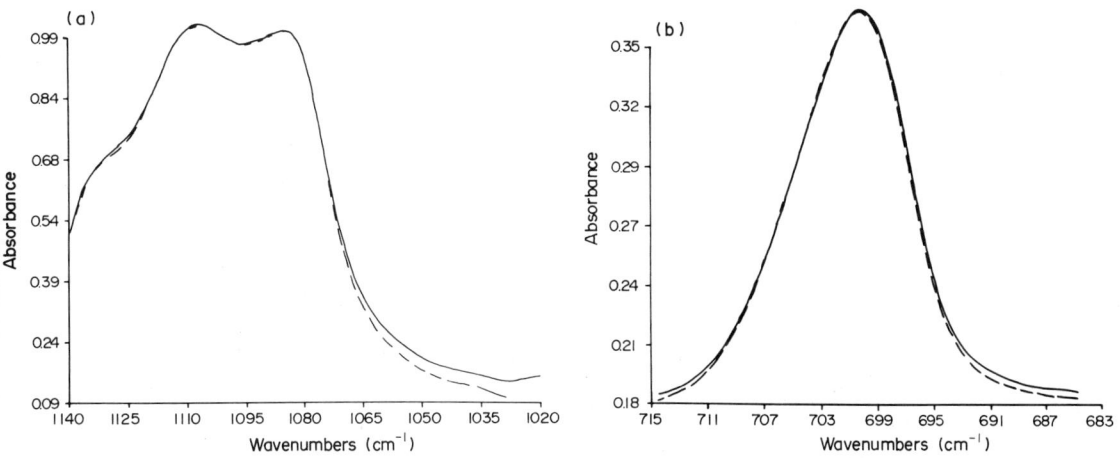

Figure 12 IR spectra in (a) the 1000 cm^{-1} and (b) the 700 cm^{-1} regions obtained for 15/85 PS/PVME incompatible blends: (———) compatible film cast from toluene solution; (– – –) compatible film cast from chloroform solution

binary mixtures containing 15% PS, the PVME and PS bands in the 1100 or 700 cm^{-1} regions are identical for films cast from both toluene and TCE solutions.

PS/PVME blends possess a lower critical solution temperature. Above this critical temperature, a miscible blend can phase separate. For a sample containing equimolar parts of PS and PVME, the initially transparent film will turn slightly bluish at ~130 °C and then optically opalescent at approximately 150 °C. If quenched at this stage, the two phases, one rich in PS and the other rich in PVME, remain separated. However, if the binary mixture is allowed to cool slowly, the two phases return to the homogeneous state. IR spectra have been obtained for binary mixtures containing equal parts of PS or PVME with different thermal histories.[30] The spectroscopic differences observed for either PVME or PS bands in miscible or immiscible samples are entirely consistent with the results presented previously.

At the molecular level, mixing in PS/PVME blends involves dispersion forces from several chemical groups. A negative excess volume of PS and PVME in miscible mixtures suggests that the two polymers can be packed together rather efficiently. As shown previously, the vibrations of PS or PVME which are most sensitive to changes in phase changes are associated with the phenyl ring of PS and the COMe group of PVME. This suggests that the phenyl ring is involved with the OCMe group in the interaction. The same PS band is perturbed in both intensity and frequency when PS is blended with poly(2,6-dimethylphenylene oxide) (2MPPO).[41] These spectral changes have been accounted for by a strong dipole–dipole interaction between the phenyl rings of the PS and the 2MPPO.[41] From molecular models the COMe group in PVME is fairly accessible and may indeed participate in the intermolecular interaction.

Recently, Garcia has confirmed the effect of temperature on the spectra of PS/PVME blends,[44] proposing that the interaction between the components involves partial sharing of the ether lone-pair electrons with the benzene ring of polystyrene. An expected consequence of this interaction is a restriction in the rotational freedom of the ether group in the blend as compared to that in pure PVME.

Monnerie and co-workers have been even more specific in the IR investigation of the interaction in PS/PVME blends. The key result of their experiment is the fact that the lower critical solution temperature of PS/PVME binary mixtures can be perturbed by as much as 40 °C when the polystyrene is deuterated.[45] The polystyrenes used in their experiments were deuterated selectively: either the main chain only, or the ring only, or the whole polymer. Deuteration of the ring affected the LCST nearly as much as did total deuteration.[46] On the other hand, labeling of the main chain had little effect on the phase separation behavior, values of LCST being close to those found for hydrogenous polystyrenes. The results establish that the interaction is indeed associated with the benzene ring.

20.2.2 Spectroscopic Analysis of Chain Conformation in Crystallizable Blends

Among the large number of studies of polymer blends, one very interesting subset consists of binary mixtures with crystallizable components. Familiar examples include blends of chemically dissimilar polymers, *e.g.* poly(vinylidene fluoride) (PVF$_2$) with poly(methyl methacrylate) (PMMA),[47-49] PVF$_2$ with poly(ethyl methacrylate) (PEMA),[50] and poly(ε-caprolactone) (PCL) with poly(vinyl chloride) (PVC);[51-53] or blends of chemically similar polymers, *e.g.* isotactic PMMA with syndiotactic PMMA[54-57] and poly(ethylene oxide) (PEO) with poly(propylene oxide) (PPO)[58] and isotactic polystyrene with copoly(styrene/*p*-methylstyrene).

Crystallization of neat polymers is already a complicated process. Crystallization of polymer blends is even more so. Several factors need to be considered. The nucleation process depends very much on the temperature, and particularly on the difference between the crystallization temperature and the melting temperature. Due to the second component, the suppression of the melting temperature can be significant, thus affecting the nucleation rate.[59] Furthermore, from qualitative considerations, the crystal growth depends on the local concentration of the crystallizable component and on the chain mobility.

An extremely interesting blend studied in our laboratory is the binary mixture of PEO and PMMA. From relative solubility considerations, it has been predicted and verified that PEO and PMMA are miscible in the melt.[60,61] Only a few studies, however, have been carried out to better understand the crystallization kinetics, thermodynamic interaction parameters and microstructures of this specific blend.[61-67] Optical and electron microscopy and X-ray scattering reveal a morphology consisting of PEO crystallites which coexist with an amorphous mixture of PEO and PMMA. More specific information regarding spherulite growth rate, melting point depression and lamellar thickness of PEO crystallites has been obtained by the usual methods.[60-67] Due to experimental constraints, the crystallization kinetics and microstructures of PEO/PMMA blends are most conveniently investigated by scattering methods. Therefore, most studies have given emphasis to samples containing high percentages of PEO, usually above 60% by weight.[61-66] It is to be expected that nucleation and subsequent crystallite growth rate are affected by blend composition, but few attempts have been made to characterize the crystallization behavior and microstructures of blends containing low PEO concentrations, *i.e.* when the degree of crystallinity is low.[68] However, it is interesting to note that the relative ratio of the crystalline forms of PVF_2 (*i.e.* α, β and γ) changes with decreasing volume fraction of PVF_2 in blends with PMMA.[69-71]

Before the crystallization behavior and microstructures of PEO in blends can be analyzed by IR spectroscopy, it is important and informative to examine the vibrational spectra of neat PEO.[73-79] In the crystalline state, the chain conformation of PEO involves internal rotations about the $O-CH_2$, CH_2-CH_2 and CH_2-O bonds forming a *t-g-t* 7/2 helix. Based on normal vibrational analysis, the IR and Raman spectra of the crystalline form of PEO are well understood.[72, 74-79] The vibrational bands in the 1500–800 cm^{-1} region are quite complicated since considerable coupling exists between the local symmetry coordinates of the methylene group, *i.e.* scissor, wagging, twisting and rocking, and the skeletal modes.[73, 75, 77] The bands at 1360 and 1343 cm^{-1} are associated with the CH_2 wagging motion, the 963 and 947 cm^{-1} bands with CH_2 rocking and the band at 843 cm^{-1} with the mixed motion of CH_2 rocking and COC deformation. These bands are sensitive to chain conformational changes. Due to the difficulty in obtaining the spectrum of amorphous PEO at room temperature, it is acceptable to use a spectrum of PEO melt as an approximation. Since the 1360 and 1343 cm^{-1} doublet components are crystalline bands, and since only one component at 1349 cm^{-1} exists in the molten state, the relative intensity of these three bands has been used to characterize the crystallization process.

It is known that when PEO complexes with a mercury halide the chain conformation deviates from the 7/2 helix, forming a $GTG\bar{G}T\bar{G}$ or $T_5GT_5\bar{G}$ conformation,[80, 81] and also that the PEO chain can take up a conformation containing all *trans* sequences when mechanically deformed.[82] Vibrational bands in the 1500–800 cm^{-1} region are associated with these chain conformations.[80, 81, 83] When *trans* sequences are present, bands are found at 1326 cm^{-1} (CH_2 wagging), 862 cm^{-1} (CH_2 rocking) and 1486 cm^{-1} (CH_2 scissor). For example, a rather strong band at 1326 cm^{-1} in the infrared spectrum of the PEO–$HgCl_2$ complex is due to its T_5GT_5G conformation. These same bands for PEO in the melt are characteristically broader than the ones obtained for the crystalline state, reflecting an increase of conformational irregularity.

From thermodynamic considerations for polymer blends, both the nucleation and crystallization growth rates will be reduced by the presence of a noncrystallizable component.[59] This is due to both the reduction in chain mobility and the changes in chemical composition during crystallization. Mandelkern and co-workers first quantitatively considered the reduction in the nucleation rate.[59] When a polymer chain crystallizes in the presence of a noncrystallizable component, it can be expected that the noncrystallizable component outside the crystalline region will result in a systematic depression of its melting point. The interaction between the two components can also contribute to the suppression of melting point, leading to a significant attenuation of the nucleation rate. The magnitude of this effect is greater at lower initial undercoolings and increases rapidly with increasing amounts of the second component.[59] It has been shown that PEO spherulite growth rates are significantly reduced when 40 wt. % of PMMA is present,[62, 63] and the isothermal crystallization rate at 40 °C can be reduced by about three orders of magnitude when the sample contains 50 wt. % PMMA.

The crystallization behavior measured by the spectroscopic technique is consistent with the thermal measurements.[68] In Figure 13 it is clear that the 1360 cm^{-1} and 1343 cm^{-1} doublet is replaced by the 1349 cm^{-1} band with increasing PMMA content, consistent with the interpretation of decreasing PEO crystallinity. It is interesting to observe that for blend samples containing 50% or more PEO, no further spectroscopic changes occur after three days. However, for samples containing 42% or less PEO, the relative intensities of the bands continue to change even after three days, suggesting that crystallization is continuing to take place. As for the sample containing 16% PEO, no spectroscopic evidence of crystallinity is seen even after 23 days.

The morphology of PEO/PMMA blends can be complicated and depends not only on sample composition but also on the crystallization conditions. Each level of the structural sub-units needs to be analyzed by suitable techniques. Small angle and wide angle X-ray scattering, thermal studies and light scattering have all contributed to a better understanding of the overall blend structure. However, one aspect of the crystallizable blends that cannot be studied by techniques other than spectroscopy is the characterization of chain conformation. One of the most interesting results of the

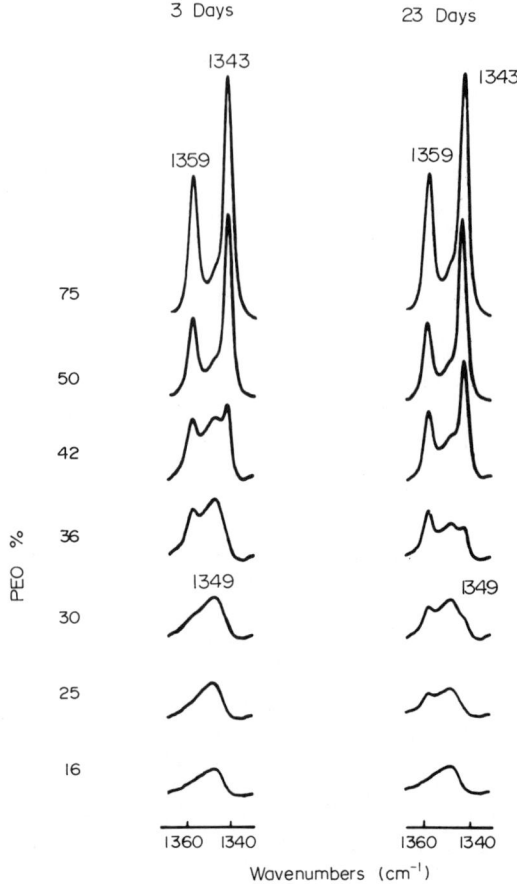

Figure 13 IR spectra in the $1300 \, \text{cm}^{-1}$ region of PEO/PMMA blends at two different crystallization times

PEO/PMMA study is that the intensities of vibrations assignable to *trans* sequences increase as a function of increasing PMMA concentration. For example, in a very clean region of the spectrum, a *trans* band at $1326 \, \text{cm}^{-1}$ is clearly observed in the spectra of the blends (Figure 14). This band is clearly observed even for samples with PEO content as low as 16%. By using the conformation-insensitive CH_2 asymmetric stretching vibration at $2880 \, \text{cm}^{-1}$ as an internal standard for intensity measurements, it is evident that the relative amount of *trans* sequence increases with increasing PMMA content even though the overall crystallinity decreases (Figure 15). This behavior is true for both high and low molecular weight PEO. A comparison of the relative intensities of the $1486 \, \text{cm}^{-1}$ band (*trans* CH_2 scissor motion) and the $1456 \, \text{cm}^{-1}$ (*gauche* CH_2 scissor motion) also shows a transformation from the *gauche* conformation to the *trans* one, since the intensity ratio of these bands changes from 1:7 to 4:5 when PEO content changes from 75% to 16%.

This change in the PEO chain conformation due to the presence of PMMA is interesting. Since we know little regarding either the interaction between the two components or the relative conformational distribution of PEO in the melt, it is hard to predict the specific influence of PMMA in the nucleation and growth of certain forms. The IR spectrum obtained for PEO melt (Figure 16) is typically characterized by broad peaks generally associated with a loss of long range conformational order. Even in this case, it is interesting to note that for neat PEO melt, bands due to the *trans* state, which is a less favorable conformation, can be detected. For example, a broad band centered at $858 \, \text{cm}^{-1}$ is found for PEO melt (Figure 16), intermediate between the $834 \, \text{cm}^{-1}$ of the *gauche* form and the $862 \, \text{cm}^{-1}$ of the *trans* form. Although PEO still crystallizes into crystallites containing the 7/2 helix, the *trans* state becomes the dominant component for noncrystalline blends containing low PEO concentration (see Figure 17). The reason that the *gauche* state is favored for neat PEO is because of the interaction between the oxygen atoms and CH_2 groups.[84] Perhaps a specific interaction with PMMA in the melt alters this interaction, causing the relative change in the microstructures.

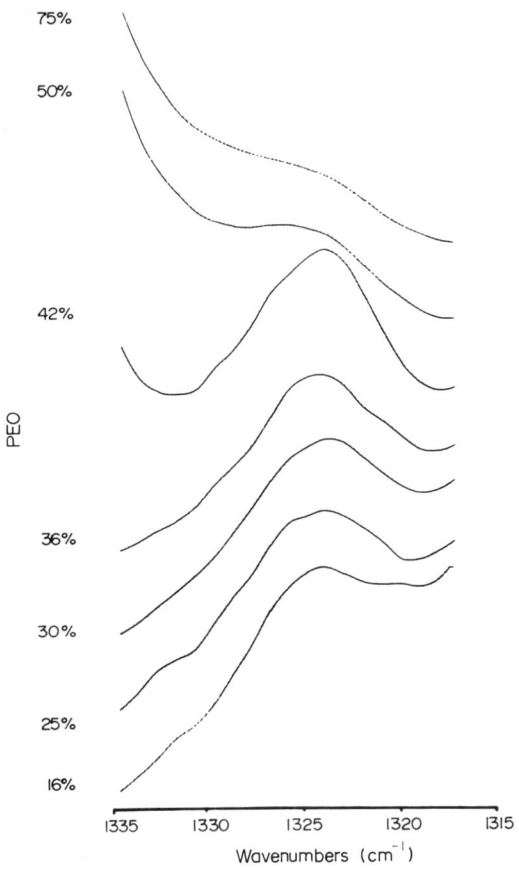

Figure 14 IR spectra in the $1300\,cm^{-1}$ region, of PEO/PMMA blends; $2\,cm^{-1}$ resolution, 300 scans

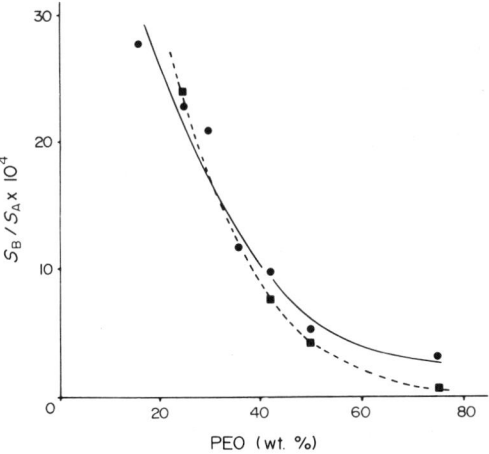

Figure 15 Intensity ratio (S_B/S_A, calculated by area) of the $2880\,cm^{-1}$ (A) and $1326\,cm^{-1}$ (B) bands as a function of PEO concentration: (■) PEO molecular weight 3×10^3, (●) PEO molecular weight 1×10^5

In summary, the spectroscopic data in conjunction with thermal analysis show that the degree of crystallization, melting temperature and crystallization rate and microstructure of PEO in PEO/PMMA blends are all perturbed strongly by the presence of PMMA. There is an increasing number of *trans* sequences in the PEO conformational distribution with increasing PMMA content. A great deal more study is needed in order to understand the specific interaction between PEO and PMMA and thereby to understand the details of the crystallization behavior and microstructure of PEO in blends.

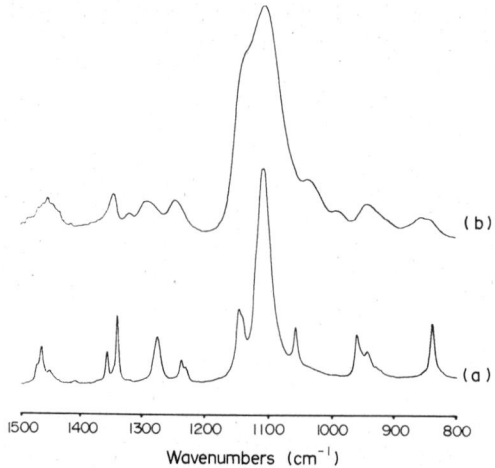

Figure 16   IR spectra of poly(ethylene oxide), resolution $2\,cm^{-1}$, 300 scans: (a) cast from 1,2-dichloroethane solution at room temperature; (b) melt

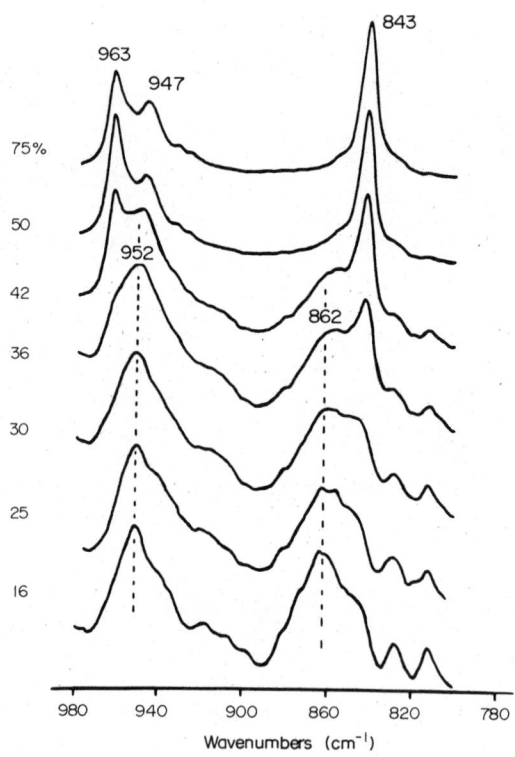

Figure 17 IR difference spectra in the $980–800\,cm^{-1}$ region of PEO/PMMA blends, as a function of PEO concentration; $2\,cm^{-1}$ resolution, 300 scans

Spectroscopy yields information at a localized level. The information obtained in the area of polymer blends complements other characterization techniques. Despite the difficulty in band assignments, spectroscopic studies have increased our confidence in assigning the molecular origins of miscibility in polymer blends.

20.3   TIME-RESOLVED SPECTROSCOPY

The mechanical properties of polymeric materials in bulk are related to their deformational behavior at the microstructural level. When deformed, and depending on stress, strain, and/or thermal history, the structural changes at the micro level may include molecular slippage, reorientation of crystallites and elongation of amorphous segments. Due to the viscoelastic nature of polymers, the stress and the motions at the microstructural level will be out of phase with the strain applied. In order to fully characterize viscoelastic materials, it is desirable to have flexibility in the timescale of measurement while maintaining accuracy and reproducibility at the microscopic and macroscopic levels. Normally, viscoelastic measurements are made at a constant temperature with various frequencies or at a fixed frequency with various temperatures. The usual objective is to construct a curve of the response (mechanical or otherwise) *vs.* log(time) or temperature. The timescale involved is often from seconds to milliseconds. In addition to the usual dynamic–mechanical tests, combinations of dynamic birefringence with X-ray and light scattering have been used to characterize the motions of individual components in multiphase materials.[85] Unlike these techniques, vibrational spectroscopy provides the selectivity needed to directly measure the response of individual molecular segments, but usually lacks the time resolution required for dynamic measurements.

Fourier transform IR spectroscopy (FTIR) has proven to be a powerful tool in polymer characterization. However, it has been applied only to the observation of events that are stationary in time, or at least stationary with respect to the measurement time. The multiplex characteristic (the ability to measure all spectral elements) of the interferometer, together with the high energy throughput, provide FTIR with a substantial gain in signal to noise ratio for a given measurement time, as compared to a dispersive instrument. Hence the use of FTIR to study time-dependent phenomena is feasible, the advantage being that simultaneous measurements of band position, shape and relative intensity are made.

Despite the great improvement in the measurement timescale when FTIR is used, it is impractical, if not impossible, for the interferometer to follow rapidly evolving events with time resolution in the order of milliseconds. The limit is reached when the time period required for one scan of the moving mirror is longer than the time resolution required to describe the physical phenomena. Furthermore, in order to improve the signal to noise ratio, co-adding scans is necessary, a procedure which further degrades the time resolution. Therefore, a number of time-resolved FTIR techniques have been developed to increase the speed of data acquisition to a value suitable for characterizing the dynamics of structural changes in polymers.[86–90]

The principles, representing several different approaches to this type of time-resolved FTIR, have been presented in a number of publications.[86,87,90] Each of these approaches incorporates some form of ordered sampling technique, which generally results in the collection of a large amount of data that must later be sorted to obtain properly time-ordered interferograms. In all cases, spectroscopic data of high temporal resolution is only possible when a definite phase relationship can be established between the time domain of the external event, such as deformation, and the retardation of the interferometer. Unlike all the earlier experiments designed around one-of-a-kind step scanning interferometers,[91,92] most of the recent developments employ commercial fast-scalling interferometers; therefore the phase relationship between the spectrometer and the external event needs to be established.

Several methods have been developed in our laboratory. One scheme has the advantage that accurate phase relationships can be obtained with minimal hardware modification,[88] but the stepwise mechanical deformation used in this particular method gives rise to transient effects and the results cannot be correlated with mechanical information obtained from other techniques, such as rheovibron. Other schemes are more useful, allowing the mechanical deformation event to run continuously,[86,87,89,90] but the external event and the FTIR interferogram scans may have to be initiated independently.[86,87] For experiments requiring high time resolution, significant signal averaging can only be achieved with long measurement times, and some samples are not strong enough to be repetitively deformed for long periods of time. In order to overcome this disadvantage, a separate microprocessor can be incorporated,[87] in addition to the one associated with the FTIR, to control and monitor the deformation event. Since the cycle time of the interferometer is accessible through the signal which starts the FTIR scans, this external microprocessor can record and adjust the relative phase between the external event and the interferometer. Thus the spectroscopic signal corresponding to the same segment of time can be accumulated very efficiently. Since all the programs are written in Pascal, this newly developed scheme also eliminates the need to construct specialized electronic timing circuitry or to write program patches in assembly language.[86]

Alternative methods for obtaining vibrational spectroscopic data with high time resolution show very great potential.[90] Some of the most interesting recent studies are designed around photoelastic modulators in order to obtain extremely accurate polarization changes when samples are mechanically deformed.[90] These new spectroscopic developments have the advantage that the changing absorptions can be directly measured, quite differently from all of the previous studies. By using difference spectra, obtained by successively substracting one from another of a sequence of spectra measured for changing polymer structures, the modulation polarization technique increases significantly the signal to noise ratio.

20.3.1   Studies of Electric Field Induced Changes in Liquid Crystals

The utility of time-resolved IR can be demonstrated by studies of the electric field induced orientation of liquid crystals. Molecular alignment in liquid crystals induced by an external electric field has been extensively investigated, along with its application to electro-optical devices. Details regarding the rate and the degree of molecular orientation in the presence of the applied field are of particular importance.

The orientational distribution functions of *n*-alkylcyanobiphenyl liquid crystals have been invest-igated using a number of techniques.[93-99] Again, the particular advantage of using vibrational spectroscopy is the ability to selectively characterize the local conformation or molecular packing. Furthermore, if the transition moments are well defined, it is possible to interpret the orientations of individual functional groups by measuring the absorption of polarized incident radiation. With the time-resolved technique structural changes occurring within time intervals of milliseconds or less can be measured with confidence.

In one particular experiment,[100] an electric field of 600, 800, 1000 or 2000 V cm^{-1} was period-ically applied to a sample of 4-*n*-pentyl-4'-cyanobiphenyl (5CB) for 0.37 or 0.62 s turned off for an equivalent amount of time, then reapplied. A representative series of spectra of 5CB with time resolution of 10 ms is shown in Figure 18.

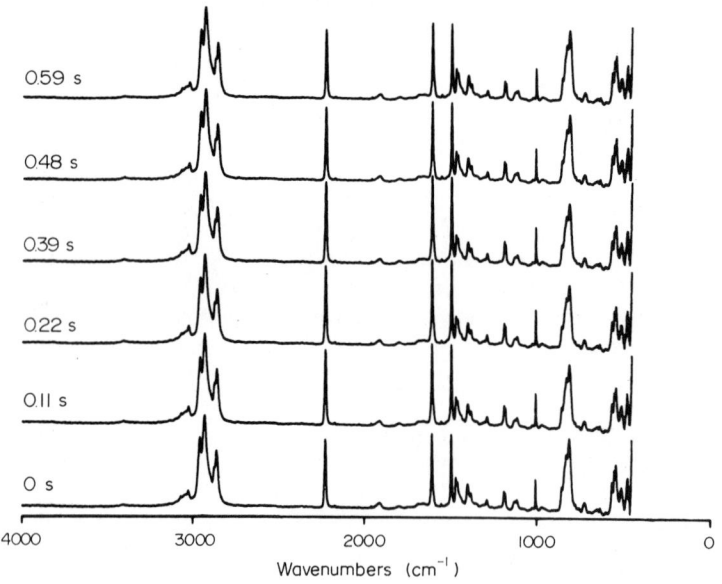

Figure 18   Representative time-resolved spectra of 5CB; polarizer parallel to the electric field, $E = 1000$ V cm^{-1}. Spectra were measured at the time intervals noted

Polarized IR absorption measurements allow the study of the orientation function of individual functional groups in 5CB by measuring the second moment of the orientation function, f, expressed as

$$f = \frac{D-1}{D+2} \frac{D_0+2}{D_0-1} \tag{1}$$

where the dichroic ratio, D, is the relative absorbance for mutually perpendicular polarization of the incident IR radiation. The reference axis for the polarized measurements is defined in the laboratory (electric field direction). For convenience, a reference axis is usually needed in the molecule as well. In polymers, the vector connecting translationally equivalent groups is generally given as this axis. In virtually all experiments of interest, these two axes coincide. The D_0 in the second moment expression is then expressed as follows,

$$D_0 = 2\cot^2\alpha \qquad (2)$$

with α being the angle between the transition moment and defined molecular axis.

The orientation of the methylene sequence in 5CB relative to the rest of the molecule is of considerable interest. The orientation function of this flexible part is expressed as f_{ms}. Since the energy between the *gauche* and the *trans* conformation is not high (nearly 500 cal mol^{-1} (2100 J mol^{-1}), the conformations of the methylene sequences are not expected to be planar zigzag, as observed in the crystalline state, but must be expressed as a statistical distribution. The effective transition moments associated with various localized methylene vibrations also need to be statistically averaged, excluding the extremely high-energy conformation gg' contribution, according to the procedure proposed by Flory.[84] The averaged transition moment direction for each vibration is summarized in Table 1. 4-*n*-Pentyl-4'-cyanobiphenyl (5CB), is shown schematically in Figure 19. It is possible to define the orientation of the methylene sequence, f_{ms}, in terms of an axis, a_2, connecting the α carbon to the terminal carbon, the position of which is evaluated in terms of the statistically averaged conformation discussed above. The orientation of the rigid part, represented by the cyanobiphenyl group, is evaluated with reference to the central axis, a_1, shown in Figure 19. Since the terminal methyl group has well-defined vibrations with respect to the skeletal plane of the methylene sequences, it is possible to define a third orientation function, f_{tm}.

The angle between the long axis of the cyanobiphenyl group and the methylene sequence depends on the conformation of the methylene sequence, and has been determined to be 52° for the statistically averaged conformation.[100] The relative orientation function of the two axes is an

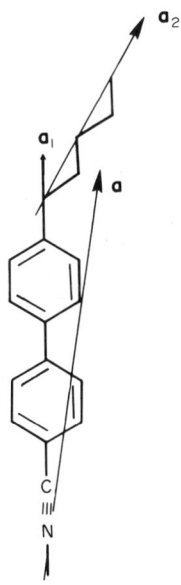

Figure 19 Molecular structure and definition of molecular axes of 5CB

Table 1 Transition Moment Directions in 4-*n*-Pentyl-4'-cyanobiphenyl

Group	*Vibration*	$cos\,^2\alpha^a$
Cyanobiphenyl	C≡N stretching	1.0
Terminal methyl	CH$_3$ asym. stretching	0.0
Methylene sequence	CH$_2$ sym. stretching	0.11436
Methylene sequence	CH$_2$ sym. stretching	0.10960

^{a} α is the angle between transition moment direction and the molecular axis in each group.

important parameter. Thus a third axis, **a**, is defined connecting the nitrogen of the cyanobiphenyl group to the terminal carbon of the methylene sequence, as shown in Figure 19.

The IR absorption spectrum of 5CB in the region of 2750–3150 cm^{-1} is shown in Figure 20. The absorption bands of the C–H stretching vibrations were assigned by reference to the assignments for *n*-alkanes.[2,5] The intensity of each absorption band was obtained by the least-squares curve-fitting method. The combination band overlaps strongly with the methylene asymmetric stretching vibration, and it is difficult to resolve these two bands with accuracy. Therefore, the intensity of the combination band was added to that of the methylene asymmetric stretching vibration after the least-squares curve fitting was carried out. The orientation of the rigid cyanobiphenyl group could be easily measured by following the IR active vibration at 2226 cm^{-1}. This vibration has a well-defined transition moment and is free from any interference from nearby vibrations.

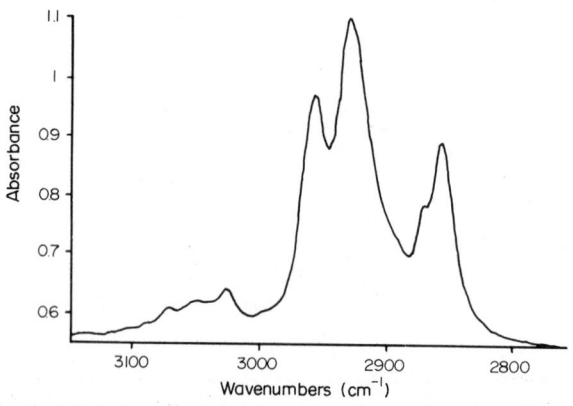

Figure 20   IR absorption spectrum in the 2750–3150 cm^{-1} region of 5CB

The time dependence of the orientation functions, f_{cb}, f_{ms} and f_{tm}, derived from the individual dichroic ratios, are shown in Figures 21–23. In the initial state, the cyanobiphenyl group tends to be oriented perpendicularly to the direction of the electric field, while the methyl and methylene groups show nearly random or slightly parallel orientation. The initial orientation is induced by the wall effect of the sample cell.[101] The orientation function increases with application of the electric field. As shown in Figure 22, the orientation function f_{ms}, calculated from the methylene symmetric stretching vibration, is in good agreement with that obtained from methylene asymmetric stretching vibration, which confirms the validity of the analysis, including the curve resolution of the spectrum, the calculation of the transition moment direction, and the choice of molecular axes. The orientation functions f_{ms} and f_{tm} show analogous behavior in their absolute values, time dependences, and field strength dependences. Conversely, the orientation function f_{cb} responds to the electric field more sensitively than f_{ms} and f_{tm}.

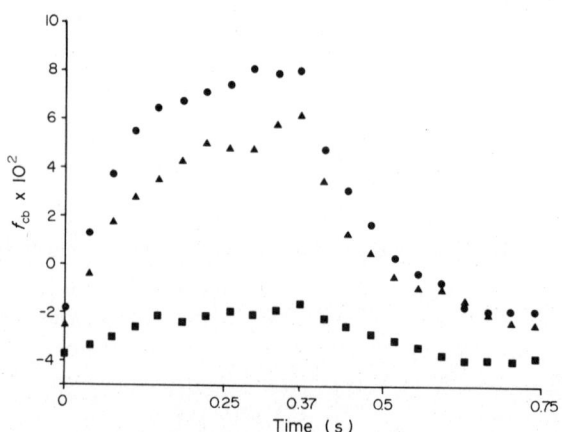

Figure 21   Orientation function of the cyanobiphenyl group of 5CB as a function of time: (●) 1000 V cm^{-1}; (▲) 800 V cm^{-1}; (■) 600 V cm^{-1}

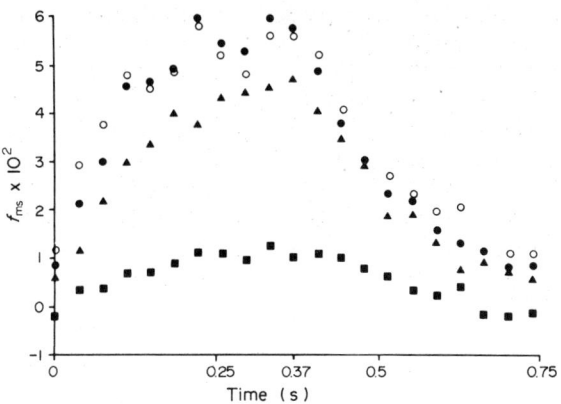

Figure 22 Orientation function of the methylene sequence of 5CB as a function of time: ($\bullet$, $\circ$) 1000 V cm^{-1}, with ($\bullet$) calculated from methylene symmetric stretching vibration and ($\circ$) calculated from methylene asymmetric stretching vibration; ($\blacktriangle$) 800 V cm^{-1}; ($\blacksquare$) 600 V cm^{-1}

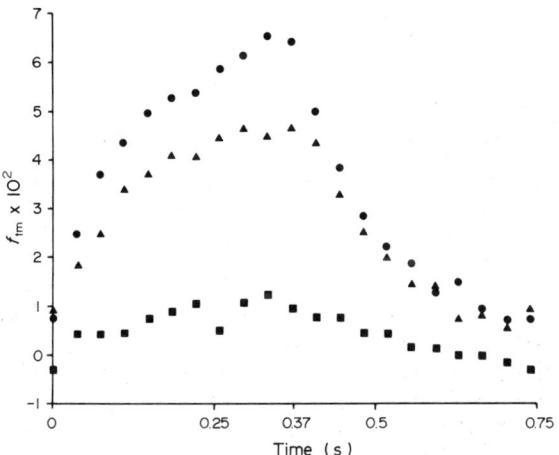

Figure 23 Orientation function of the methyl group of 5CB as a function of time: ($\bullet$) 1000 V cm^{-1}; ($\blacktriangle$) 800 V cm^{-1}; ($\blacksquare$) 600 V cm^{-1}

The electric field induced orientation is dependent on the strength of the electric field. At 600 V cm^{-1}, the electric field induces only very weak orientation. The orientational response to the electric field increases greatly with an increase in field strength to 800–1000 V cm.$^{-1}$

The orientation function, f, of the entire molecule, derived from the individual orientation functions,[100] is shown in Figure 24. This orientation function also increases upon applying an electric field. The rise can be approximated as an exponential function of time

$$f = C[1 - \exp(-t/\tau_r)] \tag{3}$$

as can the decay after the electric field is removed

$$f = C' \exp(-t/\tau_d). \tag{4}$$

In equations (3) and (4), τ_r and τ_d are the relaxation times of the rise and decay processes respectively.

Values of τ_r and τ_d are presented in Table 2 along with the difference between maximum and minimum orientation functions, Δf. The value of the Δf steeply increases with increase in field strength from 600 V cm^{-1} to 800 V cm^{-1}, and saturates at electric field strength higher than 1000 V cm.$^{-1}$

The experimental results can be discussed in terms of the two cases shown in Figure 25. Here, **E** and **L** are the electric field and the optical axis, respectively, and the angle ϕ is the angle of deflection. In case (a) the nematic director is parallel to the surface of the cell, whereas in case (b) the initial

Table 2 Rise Time, Decay Time and Orientation Function
Difference for 4-*n*-Pentyl-4-'cyanobiphenyl

E (V cm^{-1})	τ_r (s)	τ_d (s)	Δf_θ[a]
2000	0.048	<0.1	0.11
1000	0.095	0.126	0.118
800	0.129	0.119	0.103
600	0.134	0.232	0.026

[a] Difference between maximum and minimum orientation functions.

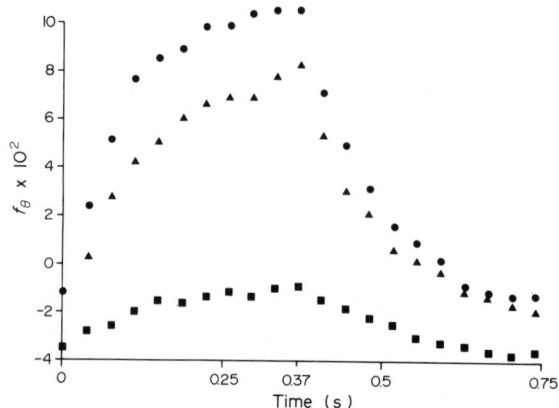

Figure 24 Orientation function of the long axis of the entire 5CB molecule as a function of time: (●) 1000 V cm^{-1}; (▲)
800 V cm^{-1}; (■) 600 V cm^{-1}

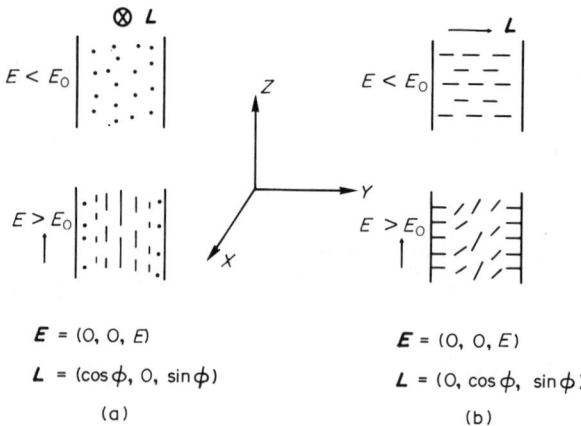

Figure 25 Illustration of molecular alignment above and below threshold electric field; (a) optical axis parallel to wall at
$E < E_0$; (b) optical axis perpendicular to wall at $E < E_0$

direction of the nematic director is perpendicular to the surface of the cell. In both cases, the electric
field is applied perpendicular to the initial direction of the nematic director.

In order to deform nematic liquid crystals, the electric energy is required to overcome the elastic
energy stored in the material. This means that there is a threshold value of the electric field for which
the electric energy is balanced with the elastic energy. The threshold field strengths, $\mathbf{E}_0$, for cases
(a) and (b) are given by equations (5) and (6) respectively.[102,103]

$$\mathbf{E}_0 = \frac{\pi}{x_0}(k_{22}/\varepsilon_0 \Delta\varepsilon)^{1/2} \tag{5}$$

$$\mathbf{E}_0 = \frac{\pi}{x_0}(k_{33}/\varepsilon_0\Delta\varepsilon)^{1/2} \tag{6}$$

where k_{22} and k_{33} are the twist and bend elastic constants respectively, and $\Delta\varepsilon$ is the anisotropy of the dielectric constants. Using acceptable values for $k_{22} = 1.1 - 0.5 \times 10^{-6}$ dyne,[104] $k_{33} = 4.0 - 1.1 \times 10^{-6}$ dyne, (1 dyne $= 10^{-5}$N) and $\Delta\varepsilon = 11.5$,[105] one obtains $\mathbf{E}_0 = 400-600$ V cm^{-1} and $\mathbf{E}_0 = 600-1100$ V cm^{-1}, for cases (a) and (b) respectively. These values agree fairly well with the electric field strengths at which the orientation function starts to increase in the time resolved experiment.

Another theoretical approach to the electric field induced orientation is based on the Boltzmann distribution.[106, 107] When the orientation is caused by a permanent dipole, μ, the orientation function is expressed as

$$f = (3 < \cos^2\theta >_{av} - 1)/2 \tag{7}$$

where θ is the angle between $\mathbf{a}$ and the applied electric field $\mathbf{E}$.

$$\langle\cos^2\theta\rangle_{av} = \frac{\int_0^\pi \cos^2\theta \exp(-\mu E\cos\theta/kT) \sin\theta d\theta}{\int_0^\pi \exp(-\mu E\cos\theta/kT) \sin\theta d\theta} \tag{8}$$

If an induced dipole is the driving force of orientation, the value of $\langle\cos^2\theta\rangle_{av}$ is given by

$$\langle\cos^2\theta\rangle_{av} = \frac{\int_0^\pi \cos^2\theta \exp(\Delta\alpha E \cos^2\theta/kT) \sin\theta d\theta}{\int_0^\pi \exp(\Delta\alpha E \cos^2\theta/kT) \sin\theta d\theta} \tag{9}$$

where $\Delta\alpha$ is the anisotropy of the polarizability tensor. Figure 26 shows plots of the orientation function as a function of μ and $\Delta\alpha$ at a constant electric field of 1000 V cm.$^{-1}$ In order to achieve $f = 0.1$, values of $\mu = 1.6 \times 10^4$ Debye (1 Debye $\equiv 3.336 \times 10^{-30}$ C m) or $\Delta\alpha = 2.6 \times 10^{-15}$ cm^3 are required in the equations. It cannot be determined whether the electric dipole is permanent or induced. However, both these values are much higher than those of a 5CB molecule ($\mu = 4.1$ Debye,[108] $\Delta\alpha = 1.1 - 1.8 \times 10^{-23}$ cm^3).[109]

Thus the experimental results suggest that a mechanism other than a single-molecular process is responsible for the orientation–relaxation process of the 5CB liquid crystal. Generally, liquid crystals form domains in which the molecules are uniaxially aligned. If the molecules are highly oriented in the domain, each domain has a huge permanent dipole and/or a huge anisotropy of polarizability tensor. If an external electric field is applied, the charges accumulated at the boundaries of the domain interact with the external field, causing a torque on the liquid crystal domains. The relaxation time measured is much longer than that of the single-molecular process, because the

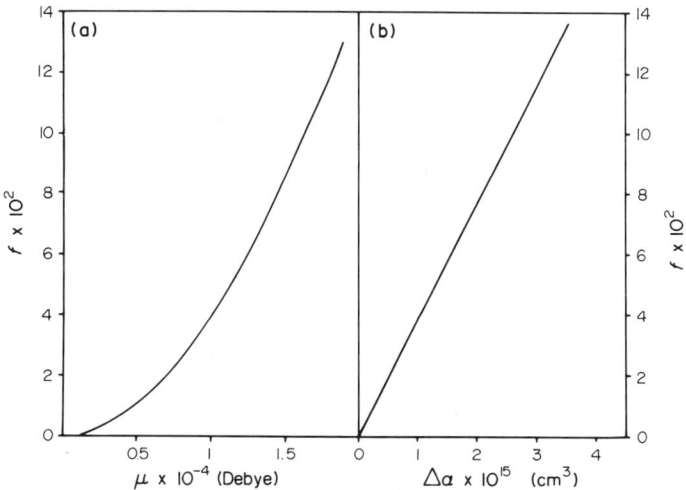

Figure 26 Orientation function calculated on the basis of Boltzman distribution as a function of (a) permanent dipole moment and (b) anisotropy of polarizability tensor. Electric field 1000 V cm^{-1}; temperature 298 K. (1 Debye = 3.336 $\times 10^{-30}$ C m)

mobility of the liquid crystal domains is lower than that of individual molecules. Accordingly, the orientation–relaxation process of the 5CB nematic liquid crystal is interpreted as the electric field induced motion of domains.[100]

20.3.2   Studies of Dynamic–Mechanical Properties

Although not easily carried out, time-resolved studies have proven to be useful in deducing molecular mechanisms giving rise to mechanical properties of polymers.[11,86–90] Systems combining structurally dissimilar polymers are of commercial and academic interest since they offer a convenient route for the modification of properties to meet specific end uses. For example, copolymers of butylene terephthalate and tetramethylene oxide are extremely interesting elastomers. For these copolymers poly(butylene terephthalate) (PBT) crystalline regions have a profound influence on the overall mechanical properties.

It is well known that a stress-induced solid–solid phase transformation occurs in PBT crystals, involving changes in both the chain conformation and packing.[111,112] A number of models based on X-ray studies have been proposed, describing the details of this phase transformation and its relationship to mechanical properties. However, a number of parameters such as time, chemical composition and temperature are still not understood. The vibrational spectra associated with both components in the copolymers, especially the PBT, are well established,[113–115] thus providing a firm basis for the investigation of time-dependent deformation-induced microstructural changes by the time resolved spectroscopic technique.

Two newly developed techniques have been applied.[86,87,89] Several IR active vibrations are highly sensitive to the conformation of the tetramethylene sequence of PBT in the various crystalline phases.[112,114] For example, the 917 cm^{-1} and the 960 cm^{-1} bands (see Figure 27) are assignable to CH_2 rocking in the α and β phases respectively, and the relative intensity of each component can be used to characterize the amount of each phase present when the samples are stressed.

In all successful experiments PBT films were preoriented in order to align the crystals in the stretch direction, thus giving more effective application of the external forces to the molecular chains in the crystal.[86,87] In the time-resolved experiment, the preoriented sample film was stretched an additional 2%. An oscillatory strain was then applied with an amplitude of 3% and a frequency of 1 Hz. For this small strain amplitude, the total absorbance change during deformation was not enough to cause a baseline modulation in the interferogram. Therefore no correction of the interferograms was necessary and, in fact, no spectral artifacts were observed.[87]

When the periodic triangular strain function was applied to the copolymer film, the time-resolved spectroscopic technique revealed changes in the intensities of the two bands characteristic of the two chain conformations. This is shown in Figure 27, where the first spectrum is subtracted from successive spectra obtained 125 ms apart. The subtracted spectra exhibit an increase (above the baseline) and then an eventual decrease of the 960 cm^{-1} band. This change in intensity as a function of strain or time can be used to follow the amount of the β phase present. Similarly, the 917 cm^{-1} band can be used to characterize the amount of the α phase present. The maxima of the change for both the 917 cm^{-1} and 970 cm^{-1} bands correspond to the maximum of the strain value, showing no time lag for the microstructural response, thus indicating that the transition occurs at a rate faster than the temporal resolution of the experiment.

20.4   ORIENTATION MEASUREMENTS

Even though polymers are increasingly accepted as a class of structural materials, in most cases their macroscopic mechanical properties are still much weaker than other solid materials such as metal or ceramics. This is changing rapidly with significant recent advances made in producing high modulus/strength fibers, particularly when the weight is considered. The weak mechanical properties associated with flexible polymers are usually due to segmental disorientation. The amorphous chains, even those in samples of a high overall degree of crystallinity, have ill-defined conformations. Since it takes so little energy to rotate about a single carbon bond, as compared to deformation of valence angles or valence bonds, the compliances associated with amorphous regions are high relative to the crystalline regions. Thus if the noncrystalline regions can be minimized, and if the segmental orientation can be increased, then the overall macroscopic properties can be improved.

Various processing procedures, such as extrusion or drawing, have been used to produce materials from flexible polymers with moduli and strengths close to their theoretical values. An alternative

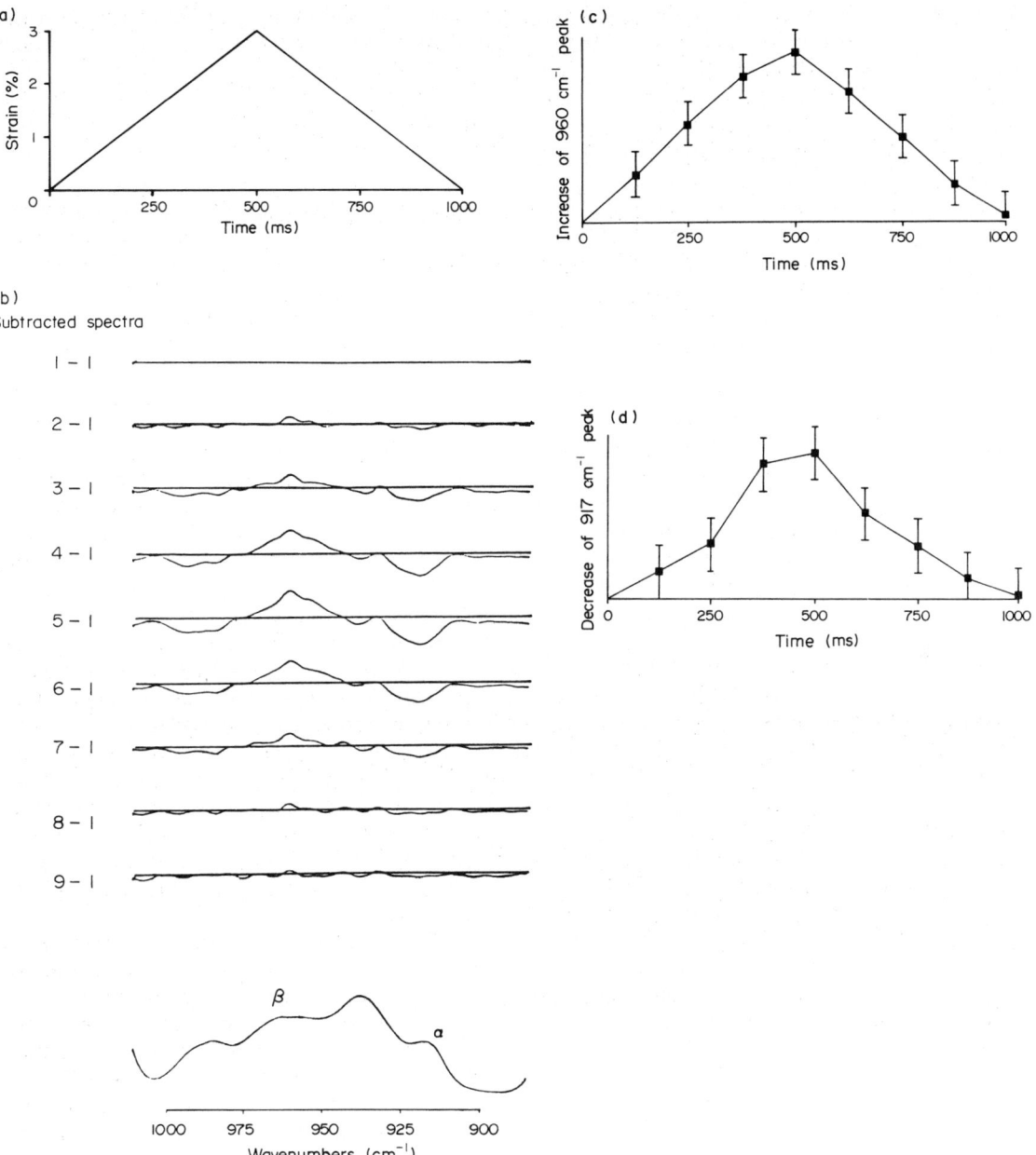

Figure 27 Changes in the conformation sensitive CH_2 rocking region of copoly(BT/TMO): (a) applied strain; (b) difference spectra in the methylene rocking region, showing amorphous and α and β crystalline peaks and their variations as a function of strain—each spectrum is shown after the first spectrum has been subtracted; (c) increase in $960\,cm^{-1}$ band (β form); (d) decrease in $917\,cm^{-1}$ band (α form)

way to obtain high performance materials is to produce fibers or films of rigid polymers from their anisotropic or liquid crystalline states. In both cases, an exact description of segmental orientation is important both for the processing and for the evaluation of the ultimate properties of the polymers under study.

There are various characterization techniques which can yield information on orientation. Many laboratories use wide angle and small angle X-ray scattering, light scattering, birefringence and vibrational spectroscopy. Light-scattering experiments are generally easy to carry out. However, because of the long wavelength (in the order of microns) generally employed, the information obtained is usually only applicable to a description of large scale morphologies, such as the changing shape of the spherulite from a sphere to an ellipsoid. Even though the orientation informtion from X-ray diffraction can be very accurate and complete, it is still limited to the characterization of crystalline regions. The birefringence technique, which depends directly on the anisotropic optical properties of chain segments, has proven to be extremely useful in the characterization of both the amorphous and crystalline regions when used in combination with X-ray measurements. However, orientation measurements with this technique can be influenced by the ill-defined 'form' birefringence.[116]

Vibrational spectroscopy is a useful tool in this aspect of polymer science. Because of the high selectivity of the spectroscopic method, the orientations of individual structural units can be followed with great accuracy. However, it should be noted that the orientation measurements associated with vibrational spectroscopy are inferred from the relative polarizations associated with the changing directions of the transition moments. Therefore the spectroscopic technique is only useful if the bands are well assigned and the transition moments are well defined.

20.4.1 New Developments in Orientation Measurements

Polarized IR absorption experiments are easy to carry out. Highly polarized beams can be obtained. Structural information is inferred from the relative absorption of the two perpendicularly polarized incident radiations. An axis is usually defined in the laboratory as reference. Details of the experiment and possible errors have been described in a number of studies.[117,118]

One of the most interesting recent developments is the use of polarization modulation in the measurement of segmental orientation. Even with the tremendous improvement of signal to noise ratio available with the Fourier transform instrument when compared to the dispersive one, it has not been possible to measure extremely small orientation values. The usual polarization measurements involve two separate measurements, one parallel and one perpendicular, the calculation of their difference in order to obtain the dichroic ratio, and then the calculation of the orientation function.[119] For nearly isotropic materials this type of measurement is limited by the dynamic range of the FTIR instrument. However, with the polarization modulation technique it is possible to measure the differences directly, thus greatly improving the signal to noise ratio. This technique was first used with dispersion instruments to measure deformation-induced orientation changes in polyethylene.[120] The optical experimental setup used in our modulation experiment, shown in Figure 28, is quite similar to those used for vibrational circular dichroism experiments, described in detail elsewhere.[121]

Perhaps the experiment can best be understood from the following series of equations. Additional details can be found in the reference given.[121] The modulated beam is produced by the stress-induced difference in the index of refraction associated with the two perpendicular directions of a photoelastic modulator (PEM), in our case a ZnSe crystal. It is assumed that the crystal is rotated at an angle of α with respect to the laboratory frame of reference. Therefore, the electric field $\mathbf{E}$ seen at the crystal is

$$\mathbf{E} = \begin{pmatrix} \cos\alpha & \sin\alpha \\ -\sin\alpha & \cos\alpha \end{pmatrix} \begin{pmatrix} E_x \\ E_y \end{pmatrix} \qquad (10)$$

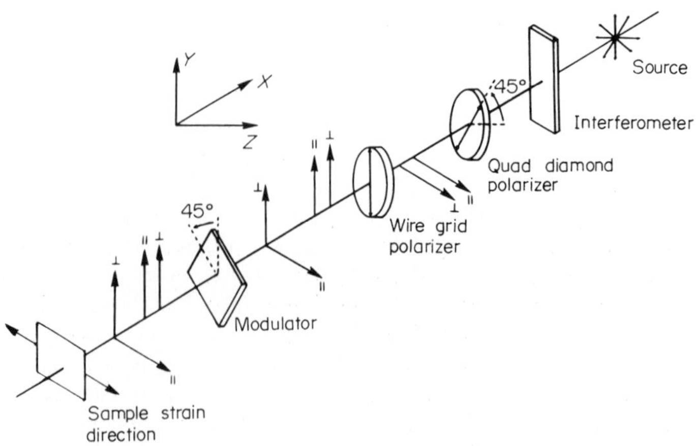

Figure 28 Optical arrangement of elements in IR modulation spectroscopy

For propagation along the z direction and only $E_y \neq 0$, then the PEM sees an electric field of

$$\mathbf{E} = E_y \begin{pmatrix} \sin \alpha \\ \cos \alpha \end{pmatrix} \tag{11}$$

At any instant of time, the velocity of light through the PEM crystal with polarization parallel (p) and perpendicular (s) to the strain differ depending on the refractive indices, thus producing a phase difference, δ, between these two components. This phase difference depends on the wavelength, the thickness of the crystal and the difference in index of refraction.[122]

An additional phase, $e^{i\delta}$, associated with the electric field along the p direction, leads to

$$\mathbf{E} = E_y \begin{pmatrix} e^{i\delta} \sin \alpha \\ \cos \alpha \end{pmatrix} \tag{12}$$

After the crystal, it is necessary to rotate back into the laboratory frame, *i.e.*

$$\mathbf{E} = \begin{pmatrix} \cos \alpha & -\sin \alpha \\ \sin \alpha & \cos \alpha \end{pmatrix} E_y \begin{pmatrix} e^{i\delta} \sin \alpha \\ \cos \alpha \end{pmatrix} \tag{13}$$

$$= E_y \begin{pmatrix} (e^{i\delta} - 1)\cos \alpha \sin \alpha \\ e^{i\delta} \sin^2 \alpha + \cos^2 \alpha \end{pmatrix} \tag{14}$$

Passing through the sample, the extinction coefficients for the incident IR radiation may be quite different for the two axes. The intensity measured at the detector can be approximated by

$$|\mathbf{E}|^2 = |E_y^2|[\cos^2 \alpha \sin^2 \alpha (e^{i\delta} - 1)(e^{-i\delta} - 1)\varepsilon_p^2 + \varepsilon_s^2 (e^{-i\delta} \sin^2 \alpha + \cos^2 \alpha)(e^{i\delta} \sin^2 \alpha + \cos^2 \alpha)] \tag{15}$$

which can be reduced to

$$= E_y^2 [\varepsilon_p (1/4)(2 - 2\cos \delta) + \varepsilon_s (1/4)(2 + 2\cos \delta)] \tag{16}$$

$$= \frac{E_y^2}{2} [\varepsilon_p (1 - \cos \delta) + \varepsilon_s (1 + \cos \delta)] \tag{17}$$

$$= \frac{E_y^2}{2} (\varepsilon_p + \varepsilon_s) + \frac{E_y^2}{2} (\varepsilon_p - \varepsilon_s)\cos \delta \tag{18}$$

where ε_p and ε_s are the relative absorptions between two perpendicular axes in the polymer sample film. The two axes are usually defined as along and perpendicular to the applied strain. Since the crystal is being oscillated continuously at a frequency ω_m, the phase difference δ is given as

$$\delta = \delta(t) = \delta_0 \sin \omega_m t$$

and the intensity at the detector is

$$I = \frac{E_y^2}{2} (\varepsilon_p + \varepsilon_s) + \frac{E_y^2}{2} (\varepsilon_p - \varepsilon_s)\cos(\delta_0 \sin \omega_m t) \tag{19}$$

which is generally written as

$$I = I_{DC} + I_{AC} \cos(\delta_0 \sin \omega_m t). \tag{20}$$

Using the expansion for $\cos(\delta_0 \sin \omega_m t)$ in terms of Bessel functions, the intensity is written as

$$I = J_0(\delta_0) + 2\sum J_n(\delta_0) \cos(n\omega_m t) \tag{21}$$

Thus it is possible to use a lock-in amplifier to select the time-dependent component and eliminate the DC component, so using the entire dynamic range of the IR instrument to measure $\varepsilon_p - \varepsilon_s$ directly.

Extremely accurate polarization measurements can be obtained in this way, even for the amorphous segments in copoly(butylene terephthalate/tetramethylene oxide). These data are shown in Figure 29. It is obvious that accurate orientation measurements can also be achieved at short time intervals. These measurements may eventually be used to deduce the molecular mechanism of deformation.

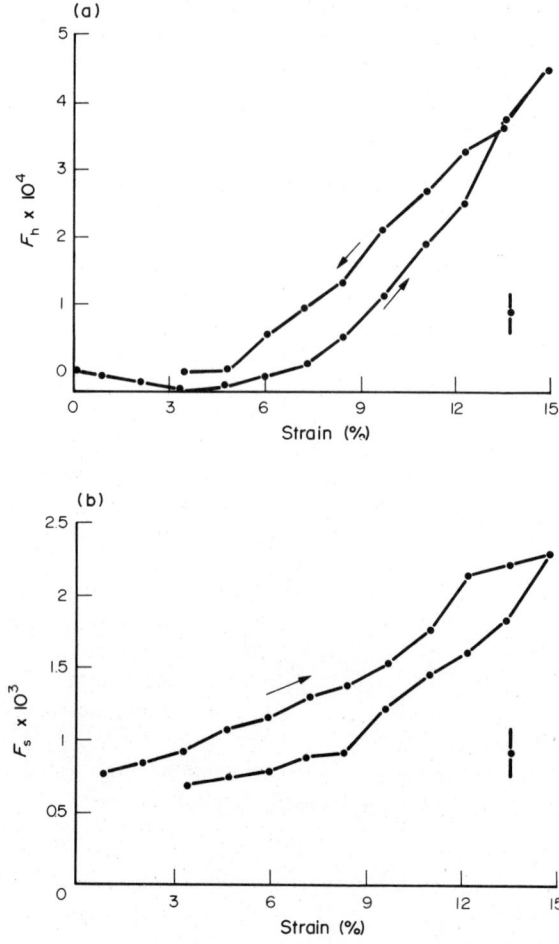

Figure 29 Orientation functions measured for the hard and soft segments in copoly(BT/TMO); 96% hard segments by weight: (a) hard segment orientation F_h; (b) soft segment orientation F_s

20.5 MECHANICAL–VIBRATIONAL SPECTROSCOPY

The macroscopic mechanical properties associated with polymers, such as modulus or strength, are much-studied subjects. The characteristic stress–strain curve associated with polymers is shown in Figure 30. An elastic region is present for small strain values, followed by departure from linearity

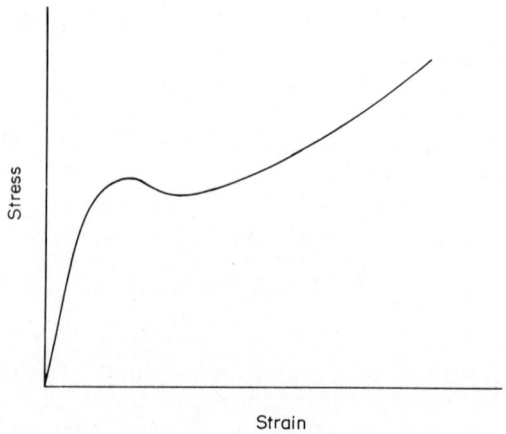

Figure 30 Schematic drawing of a stress–strain curve for a polymer

between stress and strain, and then a yield point. This is followed by a considerable region in which polymers elongate, forming fibrous structures, before finally breaking at a critical stress. In characterizing polymer deformation it is important to evaluate how both the microstructure and the superstructure respond during the deformation. The fundamental question is how to deduce the molecular mechanisms giving rise to the properties observed. If multiple values of stress exist for each value of strain, then there are additional questions regarding the molecular origin of this hysteresis. Given the complexity of polymer morphology, it would be too ambitious to try to form an analytical constitutive equation. Therefore, in most cases, a phenomenological description is the best a polymer scientist can hope for.

When the polymer samples are deformed, a considerable change in the chain segmental orientation can be observed. However, the alignment of polymer chains is never perfect but corresponds to a distribution about a well-defined axis (or axes). The most complete description of this segmental orientation is expressed in terms of spherical harmonics.[123] In the case of IR spectroscopy, all measurements deal with the change in direction associated with transient dipoles, and so only the second moment of this distribution function can be obtained. It should be noted at this point that Raman spectroscopy can reveal a more complete description of the orientation function, as that technique yields both the second moment and the fourth moment of the distribution. The second moment of the orientation distribution function, f, can be expressed as shown in equation (7) or, more commonly, as shown in equation (1). These two expressions can be shown to be equivalent.[124] In fact the major difficulty associated with interpretation of the orientational measurement using IR radiation is the definition of the angle of the transition dipole relative to the chain axis.

Because of the simplicity of its chemical repeat units, and its extremely well-understood vibrational spectrum, polyethylene is undoubtedly one of the most-studied polymers and serves as an ideal example to illustrate the contributions made by IR spectroscopy towards understanding the deformation process. Based on detailed studies of the *n*-alkanes, the vibrational spectra associated with polyethylene are largely understood.[1-5] There are still some points regarding anharmonic contributions which need additional explanation.[125-127] There are many components, such as the multiplet structure in the CH_2 rocking or bending regions, which arise from crystal-field splitting.[15-17] What is not understood well are the features associated with disordered states.

Onogi *et al.* have determined the rotation or the untwisting of the lamellae when the spherulite is elongated at small strains.[128] The orientation distribution functions observed for each of the crystalline axes are shown in Figure 31. These data are complementary and supplementary to the information gathered from birefringence, X-ray and light-scattering experiments.

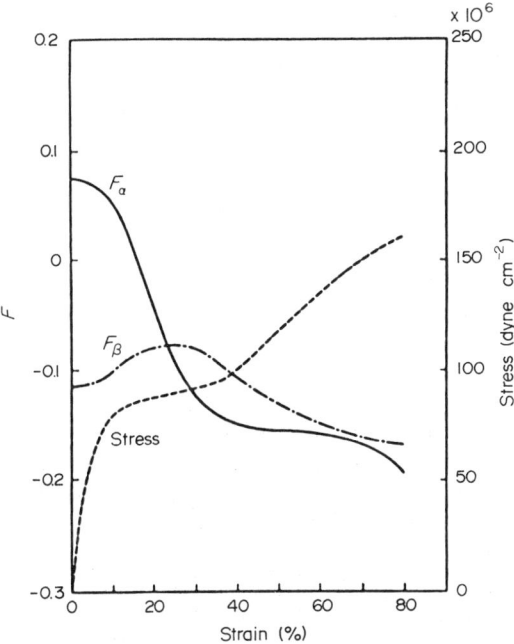

Figure 31 Variation with strain of the orientation functions for the *a*-axis, F_α, and for the *b*-axis, F_β, and also of the stress, for low density polyethylene (1 dyne $\equiv 10^{-5}$ N)

Determination of the orientation distribution has also contributed to a better understanding of the structure–property relationship in the area of polyurethanes. The macroscopic mechanical properties of these copolymers arise from the elastic behavior of the soft segments. The hard segment phase separates into well-defined regions serving as anchoring points. In these materials, vibrations assignable to the hard segments, such as MDI or TDI, or assignable to the soft phase, consisting of chain segments of polyesters, polyethers or polybutadiene, can be easily distinguished.[129-131] If proper care is exercised, one may obtain information about the mixing behavior of hard and soft segments by interpreting spectroscopic features associated with strong intermolecular interactions, such as hydrogen bonding, at the interface between the two components.[132]

A representative spectrum of a sample of MDI-terminated poly(tetramethylene oxide) chain extended with butanediol is shown in Figure 32. The bands in the $1700 \, cm^{-1}$ region, assignable to the amide I vibrations, mainly C=O stretching, usually show three components. The first, at $1732 \, cm^{-1}$, is usually assigned to free C=O groups. The second, at $1700 \, cm^{-1}$ is assignable to the C=O groups participating in strong hydrogen bonding in the ordered region. The third component is associated with isolated hard segments mixed in the soft phase.

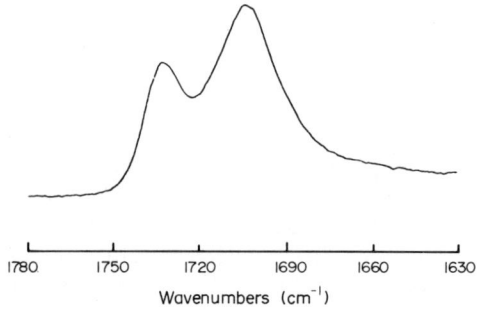

Figure 32   IR spectrum in the $1700 \, cm^{-1}$ region of MDI–butanediol–poly(tetramethylene oxide)–polyurethane

The deformational behavior of the various structural subunits is shown in Figures 33 and 34. The crystalline or highly ordered hard segment regions exhibit negative orientation at low strain values. This behavior is usually associated with the orientation of crystalline lamellae with the long axis parallel to the deformation direction.[133] However, it should be emphasized that this interpretation arises from a quantitative analysis of the deformation behavior of a spherulite[134] with much higher

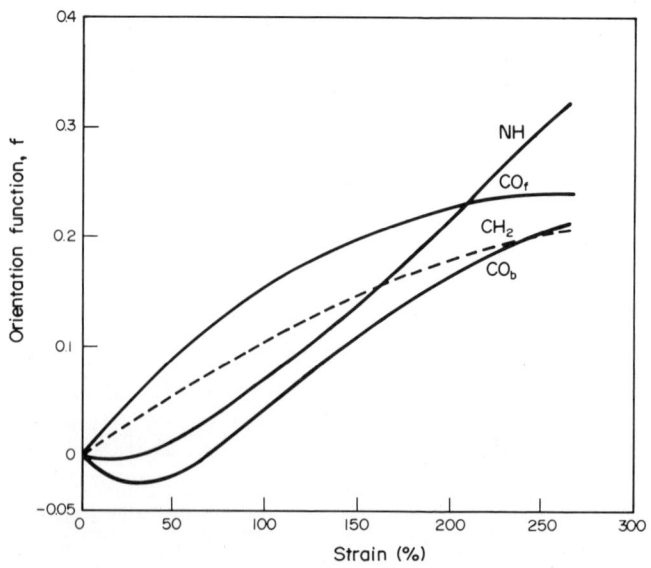

Figure 33   Changes in the orientation functions of various groups *vs.* strain for MDI–butanediol–poly(tetramethylene oxide)–polyurethane (ET-38S) (from S. B. K. S. Lin, S. Hwang, Y. Tray and S. L. Cooper, *Colloid Polym. Sci.*, 1985, **263**, 128)

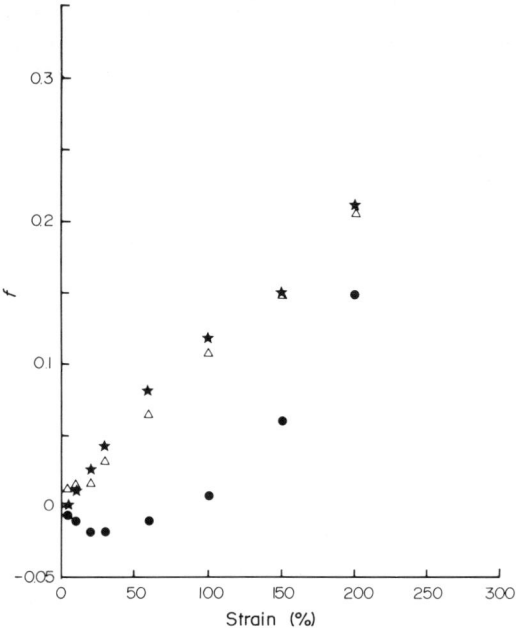

Figure 34 Orientation functions *vs.* strain for MDI–butanediol–poly(tetramethylene oxide)–polyurethane; (●) hydrogen-bonded hard segments; (△) soft segments; (★) isolated hard segments

structural order than that of the polyurethane samples. Since the origins of most absorptions are well understood, both the orientation and the subsequent relaxation behavior of different chain segments can be quite informative, showing the structural component directly responsible for the high modulus of some polyurethanes or the high elastomeric properties of others.[132]

Quantitative analysis of polymer deformation can be based on an analytical function between the orientation function and strain. The relationship is highly model dependent. The often-used pseudo-affine deformation model, introduced by Kratky, treats orientation as occurring *via* the angular rotation of a microstructural element behaving in the same way as a line having end coordinates which separate in the same ratio as the external dimensions of the sample.[10] The segmental orientation function may be interpreted in terms of the orientation function as

$$f = \frac{1}{5N}\left(\lambda^2 - \frac{1}{\lambda}\right) \text{ plus higher order terms} \qquad (22)$$

Using this equation, the number of random links (N) may be calculated from the orientation function and elongation ratio. This relationship is best suited to modeling network deformation at low elongation, *i.e.* $\lambda < N^{1/2}$.[135] Two points need to be considered. (i) The orientation achieved for the end-to-end vector can be high,[135] while that achievable for the individual statistical segment is fairly low.[135] Therefore it is necessary to consider the information obtainable from the characterization technique to be used. For example, a scattering technique such as neutron scattering may yield a considerable amount of long range orientation information. Conversely, the spectroscopic technique yields the orientation of localized chain segments. (ii) Although the model has been used for semicrystalline polymers, the number of segments between 'junction points' is usually so small as to contradict the basic assumption, *i.e.* that the chains can be treated with random statistics.

All of the models available for relating orientation function to strain contain deficiencies. It is difficult to explicitly take into account the molecular weight of the polymers and the measurement temperature. Nor is it possible to formulate a molecular mechanism for chain motion. These general difficulties have impelled us to alter both the experimental method and means of analysis. Instead of constant strain rates, uniaxial step strains were applied to the sample and subsequent decays in chain orientation and stress were measured. It is well known that stress relaxation experiments of this type separate modes of mechanical response according to their respective timescales.[136,137] A molecular model describing segmental motions at the microscopic level during polymer deformation can then be used to analyze the results from mechanical–vibrational spectroscopic experiments. The Doi and Edwards model,[138] in particular, predicts discrete mechanisms of relaxation for strained polymer chains and incorporates the effects of molecular parameters such as friction coefficients and chain molecular weights. If additional mechanisms for chain relaxation are incorporated, as in the treatment of Viovy, Monnerie and Tassin,[139] the model can be applied to multicomponent systems such as miscible blends.

In addition to changes in segmental orientation during deformation, the crystalline unit cell may change as well, affecting the macroscopic properties greatly. For example, the stress hysteresis observed in the stress–strain behavior of poly(butylene terephthalate) (PBT) is believed to be due to a time-dependent reformation of the superstructure, as well as the time-independent crystalline phase transition mentioned earlier.[111,112] A similar mechanism controls the deformation in copoly-(BT/TMO). Because this phase transition can be directly related to mechanical properties, such as the recoverability and toughness in the material, the significant factors which control its character are of practical as well as of fundamental interest.[140]

Although the crystalline phase transition in PBT has been recognized in several X-ray[111,112,141,142] and IR[113–115] studies, significant factors such as the size and perfection of the domains are not yet understood. Due to the identical nature of the crystalline phase in PBT and in copoly(BT/TMO), the same transition is expected to occur in both materials. Because in the copolymers a variable crystallite size can be obtained by changing the concentration of hard segment,[143–145] the effect that size has on the crystalline transition can be examined. In fact, the comonomer concentration dictates the lateral dimensions as well as the thickness of the crystals.

As mentioned earlier, when free from tension PBT crystals exist in the α form, which when stressed transforms reversibly into the β form. Although the difference in unit cell dimensions allows the use of X-ray diffraction to follow the degree of transformation as a function of stress, the sensitivity of this technique to long range order or disorder only makes it rather inefficient. Simultaneous spectroscopic and mechanical measurements, however, not only offer higher time resolution but, based on the conformational difference of the α and β crystalline forms, can be used to study isolated units at all levels of applied stress.

Using the same bands in the 900 cm^{-1} region as in previous analyses,[140] quantitative measurements of the relative change in the α and β contents were obtained by subtracting the first spectrum of the undeformed film from the consecutive spectra collected during deformation. A typical mechanical response over two deformation cycles is shown in Figure 35. Because some plastic deformation of the crystalline superstructure occurs during the first loading, thus affecting the stress, the hysteresis is best measured by the maximum width between the unloading and loading curves.

The orientation of the hard segments as measured by IR dichroism does not increase with additional strain. Therefore the sigmoidal shape of the transformation curves cannot arise from the stress or strain dependent orientation of the crystalline domains. In addition, the hysteresis cycle

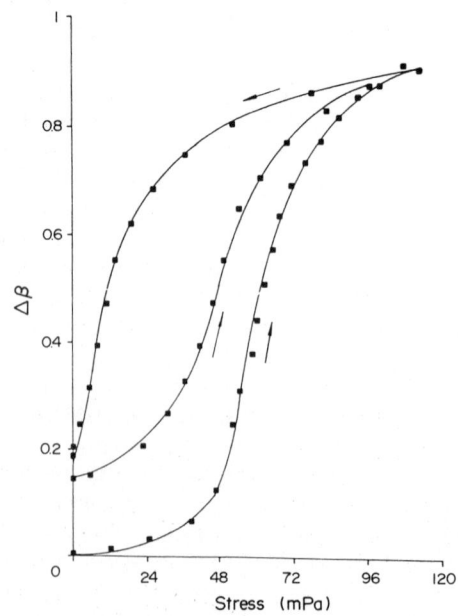

Figure 35 Hysteresis represented by the change in the amount of β phase *vs.* stress for copoly(BT/TMO)

cannot result from any type of hard segment orientation relaxation. Instead, the spectroscopic data suggest that the actual crystallite size determines the cooperativity between and along the crystalline chains, giving rise to the observed hysteresis. The larger the crystal, the greater is the coherence in both types of interactions. Theoretical predictions are in agreement with this picture.[140]

The sigmoidal behavior of the β content as a function of the applied force (Figure 35) and the observed decrease of hysteresis with temperature can be predicted by applying the mean field approximation to a system of units with internal interactions which propagate order from one unit to another. Intermolecular interactions are assumed to occur only between the terephthalate groups belonging to neighboring chains.[146] Different intramolecular interactions arise from the different placements of the terephthalate residues in the α and β crystalline unit cells. In the α form, the plane of the benzene rings is inclined approximately 19° to the c-axis of the unit cell, whereas in the β form, the plane of the benzene rings is nearly parallel to the c-axis. When unlike tetramethylene conformations are adjacent to the same benzene ring, their energies will be perturbed owing to the ill-defined chain packing. Thus dimensional changes in the lattice prompted by the crystalline transition will be hindered, especially in the beginning of the transformation process, and a hysteresis will result.

A Hamiltonian for the whole system can be written, defining the relationship between the force exerted along the chains, the average length of the crystalline segments, the strength of interactions between chains and the relative concentrations of the α and β species.[140] Minimization of the free energy of the system, given as $(H_{MF} - TS)$ where H_{MF} is the Hamiltonian in the mean field approximation and S is the entropy of mixing, yields an explicit solution for the applied force. The transition can be accurately predicted from this model and the expression for the hysteresis ΔF can be derived as

$$\Delta F = 1/b\left[8Kb^2\langle\sigma\rangle - 2k_B T \operatorname{arctanh}\left(\langle\sigma\rangle + \frac{(n-1)^2}{(n+1)nQ}\right)\right] \qquad (23)$$

where K is the interchain force constant, b defines the difference between the c-axis dimensions of the α and β segments, n is the crystallizable chain length, $\langle\sigma\rangle$ is related to the fraction of each crystalline form[140,146] and Q is the mean intramolecular interaction along the segments. It is evident from the equation that the sensitivity of the phase transformation with respect to Q increases as the crystallizable chain length n increases.

20.6 SURFACE CHARACTERIZATION BY IR SPECTROSCOPY

The importance of polymer interfaces and polymer–external surface interactions cannot be overemphasized. To a large extent, interface performance determines the overall mechanical properties of composites and heterophase polymers including polymer blends. How polymers interact with external surfaces is also central to a broad range of applications such as corrosion inhibition, adhesive joint strength, fabrication of electronic components and decorative plating. At present, polymer–polymer interfaces can only be characterized indirectly at best. There are several techniques available including contact angle, X-ray photon electron spectroscopy (XPS) and radioactive labeling, each with particular advantages or disadvantages. Contact angle is probably one of the easiest experiments to carry out, but the information which can be obtained is somewhat limited. XPS can be used to measure the surface composition accurately with high spatial resolution, but the high vacuum requirement may make such experiments difficult to carry out.

Vibrational spectroscopy has proven to be a useful technique in some cases. The attenuated total reflectance technique (ATR) is often used to determine differences between the structures of polymers in surface and bulk phases. Commercial accessories can be purchased which make these spectroscopic experiments easy to perform, although quantitative analysis of the data obtained is still not easily carried out. Examples of ATR applications include chemical composition analysis of polymers,[147] surface orientation resulting from various processing methods[148] and chemical or thermal degradation of polymers.[149,150]

Recently, perhaps because of the dramatic improvements in instrumentation or simply because of changing research perspectives, the study of surfaces has changed considerably. Spatial resolution of a few micrometers is no longer sufficient. Investigation of the mechanical, adhesion and electrical behavior of ultrathin polymer films requires spatial resolution of the order of 10–100 Å. Specular reflectance spectroscopy is a most convenient technique. Furthermore, techniques such as polarization modulation IR spectroscopy have significantly increased the sensitivity,[151,152] and there are the exciting possibilities of photoacoustic spectroscopy.[153-155]

Both development of the techniques and quantitative analysis of the results demand a thorough understanding of the vibrational spectra obtained with the reflection technique. The basic considerations for any analysis must incorporate the real and the imaginary parts of the index of refraction for the polymer films and substrates being studied. The most fundamental relationship is the Kramers–Kronig relationship, usually written as follows[156,157]

$$n(\omega) = 1 + \frac{2}{\pi} P \int_0^\infty \omega' k(\omega')(\omega'^2 - \omega^2)^{-1} d\omega' \tag{24}$$

However, the actual relationship for the index of refraction can be approximated by the damped oscillator model and described as follows[122]

$$n^2 - 1 = 4n \frac{Ne^2}{m} \sum_s \frac{f_s}{\omega'^2 - \omega^2 + i\gamma\omega} \tag{25}$$

where $n(\omega)$ is the complex refraction index, N the number of electrons in unit volume, e the electronic charge, m the mass and f_s the oscillator strength. It is difficult to solve either expression analytically but, with approximations, the calculation of changing optical constants as a function of frequency becomes feasible. Computer algorithms have been written in our laboratory to appraise these optical effects. There are differences in band frequency, shape and intensity between the transmission and reflection IR data arising from optical effects associated with thin films which may have nothing to do with structural differences.

In most analyses, the structural information is obtained from interpreting only a few isolated well-assigned IR bands. Therefore, the exact calculation of the optical effects for thin film or substrates need not be carried out for the entire frequency range, but instead over a limited range of interest.

We have been interested in the surface-induced alignment of liquid crystalline polymers on external surfaces. An example is poly(benzyl L-glutamate), for which representative experimental and calculated curves are shown in Figures 36 and 37.

When compared to transmission spectra, there are substantial differences in the peak frequency and relative intensity. The most dramatic changes are in the doublet observed in the reflectance spectra in the 1650 cm^{-1} region, *i.e.* the amide I (mainly C=O stretching) region. The most obvious interpretation is to assign one of the components to hydrogen-bonded species, and the other one to species free from hydrogen bonding. However the effect can be attributed to optical effects in the film. The film thickness could be calculated from the concentration of the solution used to cast the film and the density of the polymer. (It is also possible to estimate the film thickness from XPS

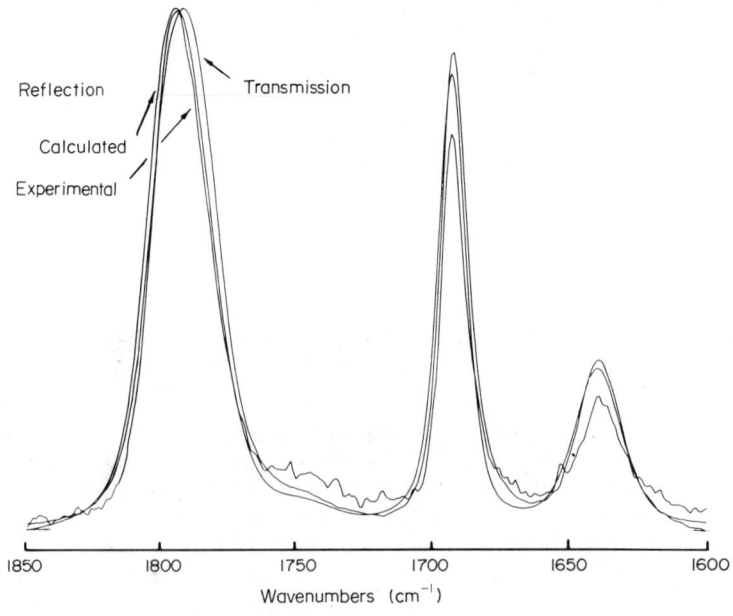

Figure 36   Transmission and reflection (on gold substrate) IR spectra observed for poly(benzyl L-glutamate). The calculated spectrum is obtained as described in the text

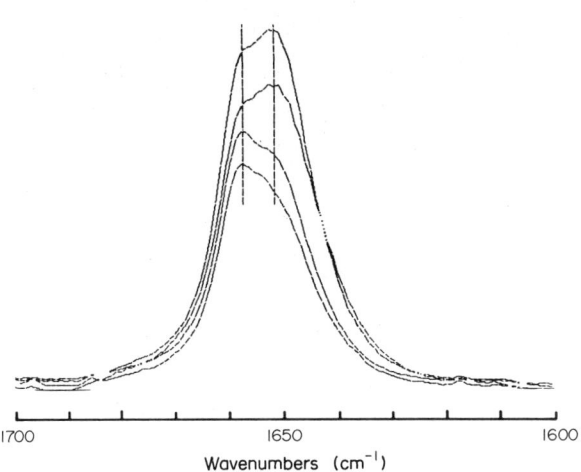

Figure 37 Reflectance IR spectra of PBLG films cast from dilute solution onto a gold surface. Film thicknesses vary between spectra in the range 5000 to 20 000 Å

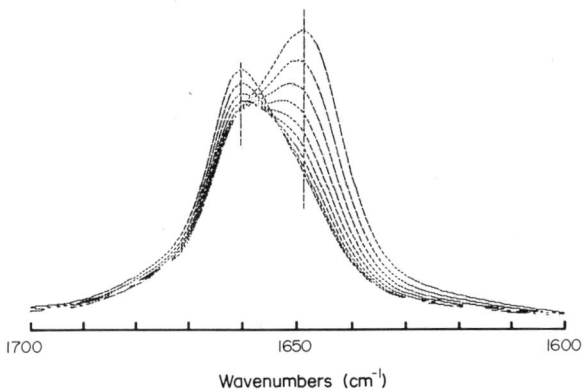

Figure 38 Calculated spectra of PBLG films on a gold substrate as a function of film thickness from 8000 to 16 000 Å

measurements.) Knowing the film thickness, the optical coefficients of the substrate, the angle of incidence and the isotropic transmission spectrum, the spectra to be expected from the reflectance technique were predicted. It was found that the spectra changed quite dramatically with film thickness as shown in Figure 38. Therefore, the two components observed can be completely attributed to optical effects in the thin film on the metallic surface.

20.6.1 Instrumentation

Because of the small amount of sample in the beam, high quality reflectance IR measurements of thin films or polymer surfaces are difficult to obtain. The dynamic range of most instruments is simply not sufficient to measure the absorption with great accuracy. Since most reflectance experiments are carried out using specular reflectance techniques, optical effects complicate the IR spectra obtained, making quantitative analysis quite difficult. A Fourier transform instrument removes many obstacles, making most experiments fairly easy to carry out compared to the opportunities available with dispersive instruments.

If the optical constants for the substrates are well defined, the specular reflectance experiments of thin films can be improved significantly with the use of the double modulation technique described earlier.[151,152] For example, if the substrate used is metallic in nature, the boundary conditions for an incident electromagnetic wave is such that only the electric field with polarization in the plane of scattering (defined as the plane containing the incident and reflected rays) has any energy at the surface. The intensity of the other component polarization is zero near the conducting substrate surface. Hence only one of the two perpendicularly polarized radiations will be absorbed by the thin polymer film. In most cases, the dynamic range of a Fourier transform instrument is more than adequate in differentiating the adsorptions of these two components. However, using the double modulation technique together with phase sensitive detection, a far better signal to noise ratio can be obtained,[152] since the difference between the two components is measured directly.

20.6.2 Applications of the IR Reflection Method

The use of reflectance IR spectroscopy to characterize the composition of chemically or physically absorbed layers on external surfaces can be easily appreciated. However, reflectance spectroscopy can also yield the relative orientation of functional groups or chain segments on external metallic surfaces. As explained earlier the boundary conditions are such that only incident electromagnetic radiation with polarization in the plane of scattering has significant energy at the immediate surface,[122,156,157] since electromagnetic waves with polarization perpendicular to the scattering plane have nodes at the surface and are not absorbed by the thin sample film[122,156,157]. Therefore vibrations with components of the transition moments perpendicular to the metallic surface are expected to appear with enhanced intensity, while those having transition moments nearly parallel to the surface are expected to appear weakly or not at all.

The most definitive studies undoubtedly involve model compounds prepared by the Langmuir–Blodgett technique.[158,159] When the bands are well assigned, and particularly when the directions of the transition moments are well defined, a great deal of information can be derived from the highly polarized reflectance IR spectroscopy of such films on external surfaces.[159] By definition, most high frequency vibrations are well-defined localized vibrations involving few nuclei, thus making transition dipoles easy to define and understand. Examples are the C=O and CH_2 stretching vibrations in hexadecanoic acid. Spectra obtained by Nuzzo and Allara using the reflectance technique are shown in Figures 39 and 40. The coordinate frame used for analysis is defined in Figure 41.

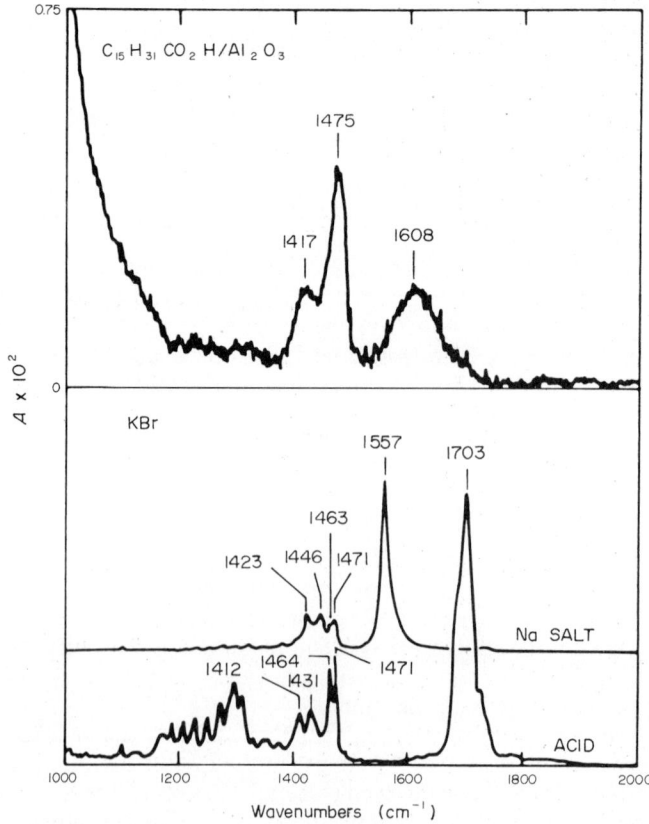

Figure 39   Spectra in the mid-frequency region of n-$C_{15}H_{31}CO_2H$ as a film adsorbed on Al_2O_3 (top) or in a KBr matrix in acid or salt form (bottom). The intensity units for the adsorbate spectrum are adsorbance (from ref. 159)

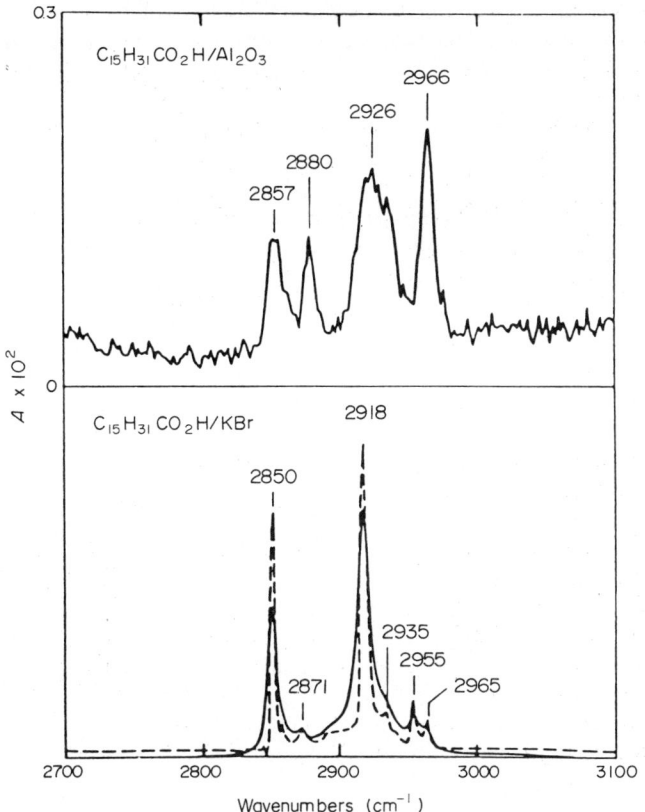

Figure 40 Spectra in the high frequency region of n-$C_{15}H_{31}CO_2H$ as a film adsorbed on Al_2O_3 (top) or in a KBr matrix (bottom). The bulk (KBr) experimental spectrum is accompanied by a deconvoluted spectrum (dashed curve), which represents enhanced resolution (from ref. 159)

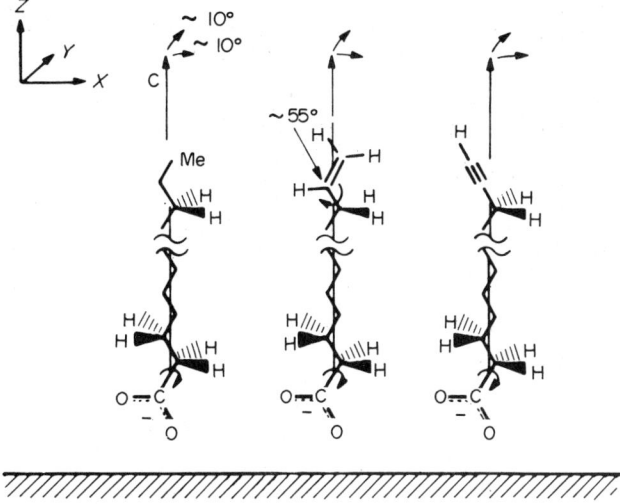

Figure 41 Representations of proposed structures of adsorbate species having fully extended all-*trans* chains. The molecular orientations relative to the surface coordinate system can be arrived at by performing the indicated chain tilting in the X, Y plane ($\sim 10°$) and rotating around the selected C–C bond axes ($\sim 55°$ for the vinyl group) (from ref. 159)

The significant difference between the reflectance and transmission spectra can be accounted for by preferential orientation of the carboxylate groups. Information on the exact orientation of the alkyl tail of the hexadecanoic acid can also be obtained from the high frequency spectra shown in Figure 40. However, some difficulties may be encountered using these vibrations as not all the predicted vibrations can be resolved cleanly in the experimental data. For example, two CH_3

vibrations, asymmetric and symmetric, have perpendicularly oriented transition moments with respect to the plane containing skeletal polymethylene chains.[1–5] When known, all the orientation measurements can be combined to obtain a complete picture regarding the overall orientation of the molecules on the surface.

Optical and X-ray diffraction studies have revealed that the structure of the nematic phases formed by thermotropic liquid crystalline polymers, consisting of a rigid aromatic core and flexible side chains can be characterized as stacks of molecular disks forming columns in a hexagonal array. Consequently they can be modeled by simpler disk-like molecules. Some of the most studied disk-like molecules are homologs of benzenehexa-*n*-alkanoates, denoted as BH*n*, *n* being the number of carbon atoms in a side chain. Vibrational spectroscopy has been used to characterize the confor-mations associated with the methylene sequences and ester groups in various phases. The methylene sequences of BH8 are nearly extended with an all-*trans* conformation in the crystalline phase.[160,161] The degree of this conformational order varies considerably with temperature. Furthermore, conformational isomers are found for the ester groups.

It is of interest to clarify the geometry of these molecules on metal surfaces making use of the unique anisotropy at the interface inherent in the reflectance technique. Due to a lack of suitable vibrations it is difficult to determine the orientation of the central benzene rings of BH*n*. Therefore another series of samples has been prepared, *i.e.* tetrakis(octanoyloxy)*p*-benzoquinone (QTE8), with C=O directly attached to the benzene ring, and the C=O stretching vibration of the quinone ring has been used for the study of orientation of the core.[16]

The carbon–hydrogen stretching region of the transmission spectrum of the isotropic melt and of the reflection spectra of adsorbed films obtained at several temperatures are shown in Figure 42. There are two nearly degenerate CH_3 asymmetric stretching vibrations. The band observed at $2964\,cm^{-1}$ and the component at $2873\,cm^{-1}$ are assigned to the methyl asymmetric and symmetric stretching vibrations respectively. The bands at $2928\,cm^{-1}$ and $2858\,cm^{-1}$ are assigned to the methylene asymmetric and symmetric stretching vibrations respectively. The structural changes corresponding to isotropic–nematic and nematic–crystalline transitions are expected to take place in the temperature range of 56–60 °C. The relative intensities of the bands in this region change significantly as a function of temperature. A distinct shift in the CH_3 asymmetric vibration from $2964\,cm^{-1}$ to $2955\,cm^{-1}$ is observed when the crystalline–nematic transition takes place. From the relative intensity of these bands in the transmission and reflection spectra, it is possible to calculate the relative orientation of individual functional groups on the external surface.[161]

Orientation functions associated with the 'arms' obtained as a function of temperature have been calculated.[161] In the isotropic surface phase at 60 °C the methylene chains are mostly oriented at a slight angle relative to the external surface. The orientation of the skeletal plane is identically 0.0, supporting the view that these arms disorder quite quickly and completely in the isotropic phase. On decreasing the temperature from 60 °C to 56 °C the orientation of the plane containing the

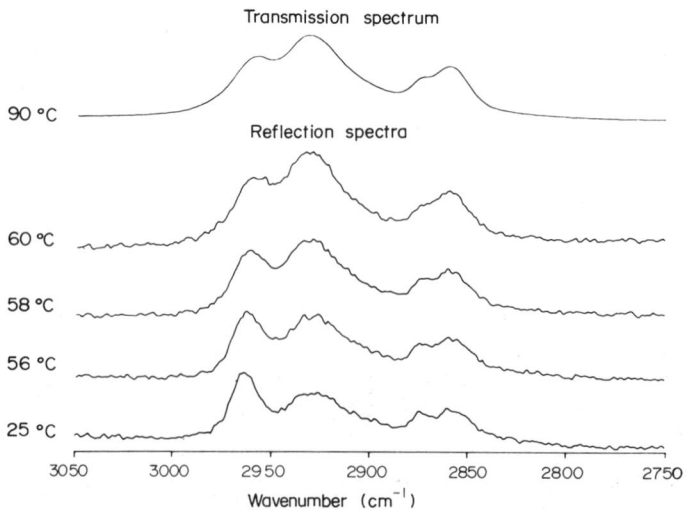

Figure 42 Transmission spectrum of the isotropic melt and reflection spectra in the CH stretching region of adsorbed films of QTE8; as a function of temperature; $2\,cm^{-1}$ resolution, 1000 scans

methylene skeleton increases. Our calculations indicate a substantial increase in this orientation function, to above 0.2, which suggests that most methylene chains are oriented with the skeletal plane perpendicular to the surface.

These theoretical results are supported by the change in the methyl asymmetric stretching vibration. This vibration is observed at 2955 cm^{-1} in the isotropic phase. As mentioned previously, the band is shifted to 2964 cm^{-1} and is sharper in the crystalline state. The two nearly degenerate methyl asymmetric stretching bands are clearly assigned.[3,4] Based on the polarization studies of *n*-alkanes, the transition moment of the higher frequency component is parallel to the skeletal plane and that of the lower frequency component is normal to the skeletal plane. Both components contribute nearly equally to the band seen in the reflection spectrum of the isotropic melt at 60 °C. However, as can be seen in Figure 42, the higher frequency component is much stronger than the lower frequency component in the reflection spectra of the liquid crystalline and crystalline phases temperatures at or below 56 °C, as expected if the C–C–C skeletal plane is perpendicular to the metallic surface in the crystalline phase.

Spectra in the 1900–1600 cm^{-1} region, obtained by both the transmission and reflection techniques, are shown in Figure 43. The IR bands near 1790 cm^{-1} and 1690 cm^{-1} are assigned to the C=O stretching vibrations of the ester groups and the quinone ring respectively. With temperature decreasing from 60 °C to 56 °C, the ester C=O stretching vibration shifts from 1794 cm^{-1} to 1786 cm^{-1}. We have interpreted this frequency shift as being due to conformational change associated with the ester group. The positions of the 1690 cm^{-1} and 1635 cm^{-1} bands show a very small temperature dependence. However, as can be seen in Figure 43, the relative intensities of the bands in this region change significantly as a function of temperature. This intensity of the quinone C=O stretching vibration decreases relative to the remaining two bands with decreasing temperature. Furthermore, the relative intensity of the quinone C=O stretching vibration in the reflection spectrum at room temperature is much smaller than its intensity in the transmission spectrum of the isotropic state, suggesting that the transition moment of the in-plane C=O stretching 1690 cm^{-1} vibration tends to be parallel to the metal surface in the nematic or crystalline phases.

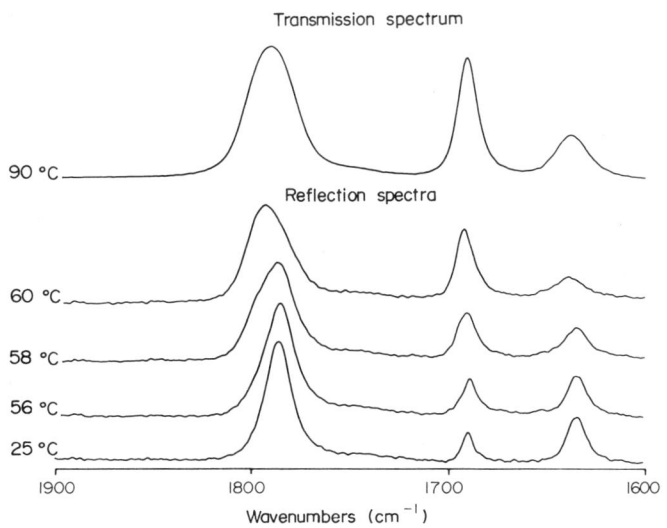

Figure 43 Transmission spectrum of the isotropic melt and reflection spectum in the C=O stretching region of adsorbed films of QTE8 as a function of temperature 2 cm^{-1} resolution, 1000 scans

Semiemperical conformational calculations suggest that the transition moment angle of the ester C=O vibrations relative to the normal of the core plane is expected to be in the range of 31° to 53°. From our intensity data and these theoretical values, the possible range of the orientation function for the aromatic ring can be obtained. It is interesting to note that even in the isotropic state (60–80 °C), the aromatic core shows a slight tendency towards parallel orientation with respect to the metallic surface. This orientation increases with decreasing temperature, particularly in the temperature range of 56–60 °C, reaching values in the range of 0.58 to 0.73 at 25 °C, with perfect orientation being 1. Our data clearly show that the plane of the aromatic core is well aligned with respect to the metallic gold surface.

20.7 CONCLUSION

There has never been any doubt regarding the use of IR spectroscopy as an analytical tool for the determination of chemical composition. Because of the significant improvement in instrumentation in most laboratories, such as the wide use of the Fourier transform technique, and also because of considerably better understanding of features observed, it is now apparent that IR spectroscopy is one of the most powerful characterization techniques available for the study of polymer microstructures.

ACKNOWLEDGEMENT

I deeply appreciate the contributions from my students (Cun Fan, Nick Reynolds, Han Sup Lee) in helping me with the preparation of this manuscript, and particularly David Waldman for his careful preparation of various polyethylene samples and his assistance in obtaining their vibrational spectra. The financial support from National Science Foundation, Polymers Program, Grant number DMR # 8407539; Army Office of Research, Grant number # DAAG 29-84-K-0052 and the support of the Materials Research Laboratory here at the University of Massachusetts over the years have made this research possible. Finally, the preparation of this manuscript would not have been possible without the extraordinary efforts of Sue Rusiecki.

20.8 REFERENCES

1. R. G. Snyder and J. H. Schachtschneider, *Spectrochim. Acta*, 1965, **21**, 169.
2. R. G. Snyder and J. H. Schachtschneider, *Spectrochim. Acta*, 1963, **19**, 85.
3. R. G. Snyder, *J. Mol. Spectrosc.*, 1967, **23**, 224.
4. R. G. Snyder, *J. Chem. Phys.*, 1967, **47**, 1316.
5. J. H. Schachtschneider and R. G. Snyder, *Spectrochim. Acta*, 1963, **19**, 117.
6. S. Bhagavantam and T. Venkatarayudu, 'Theory of Groups and Its Applications to Physical Problems', 3rd edn., Andhra University, Waltair, 1962.
7. L. A. Woodward, 'Introduction to the Theory of Molecular Vibrations and Vibrational Spectroscopy', Clarendon Press, Oxford, 1972.
8. G. Turrell, 'Infrared and Raman Spectra of Crystals', Academic Press, London, 1972.
9. A. Peterlin, *J. Mater. Sci.*, 1971, **6**, 490.
10. O. Von Kratky, *Kolloid-Z.*, 1933, **64**, 213.
11. R. S. Stein, *J. Polym. Sci.*, 1958, **28**, 83.
12. M. E. R. Robinson, D. I. Bower and W. F. Maddams, *J. Polym. Sci., Polym. Phys. Ed.*, 1978, **16**, 2115.
13. R. G. Snyder, *J. Mol. Spectrosc.*, 1960, **4**, 411.
14. R. G. Snyder and M. W. Poore, *Macromolecules*, 1973, **6**, 708.
15. R. S. Stein, *J. Chem. Phys.*, 1955, **23**, 734.
16. M. Tasumi and T. Shimanouchi, *J. Chem. Phys.*, 1965, **43**, 1245.
17. M. Tasumi and S. Krimm, *J. Chem. Phys.*, 1967, **46**, 755.
18. C. W. Bunn and T. C. Alcock, *Trans. Faraday Soc.*, 1945, **41**, 317.
19. T. Miyazawa, *J. Chem. Phys.*, 1960, **32**, 1647.
20. S. Krimm and Y. Abe, *Proc. Natl. Acad. Sci. USA*, 1972, **69**, 2788.
21. C. A. Sperati and H. W. Starkweather, Jr., *Adv. Polym. Sci.*, 1961, **2**, 465.
22. E. S. Clark and L. T. Muus, *Z. Kristallogr., Kristallgeom., Kristallphys., Kristallchen.*, 1962, **117**, 108, 119.
23. G. Masetti, F. Cabassi, G. Morelli and G. Zerbi, *Macromolecules*, 1973, **6**, 700.
24. R. G. Brown, *J. Chem. Phys.*, 1964, **40**, 2900.
25. R. A. Neira–Lemos, Ph.D. Thesis, University of Massachusetts, Amherst, MA, USA, 1974.
26. P. J. Flory, 'Principles of Polymer Chemistry', Cornell University Press, Ithaca, New York, 1953.
27. M. M. Coleman and P. C. Painter, *Appl. Spectrosc. Rev.*, 1984, **20**, 255.
28. M. M. Coleman, D. F. Varnell and J. P. Runt, in 'Polymer Alloys III', ed. D. Klempner and K. C. Frisch, Plenum Press, New York, 1983, p. 59.
29. S. T. Wellinghoff, J. L. Koenig and E. Baer, *J. Polym. Sci., Polym. Phys. Ed.*, 1977, **15**, 1913.
30. F. J. Lu, E. Benedetti and S. L. Hsu, *Macromolecules*, 1983, **16**, 1525.
31. D. L. Allara, *Appl. Spectrosc.*, 1979, **33**, 358.
32. J. P. Hawranek and R. N. Jones, *Spectrochim. Acta, Part A*, 1976, **32**, 99, and refs. therein.
33. T. K. Kwei, T. Nishi and R. F. Roberts, *Macromolecules*, 1974, **7**, 667.
34. D. D. Davis and T. K. Kwei, *J. Polym. Sci., Polym. Phys. Ed.*, 1980, **18**, 2337.
35. T. Nishi, T. T. Wang and T. K. Kwei, *Macromolecules*, 1975, **8**, 227.
36. S. Reich and Y. Cohen, *J. Polym. Sci., Polym. Phys. Ed.*, 1981, **19**, 1255.
37. M. Bank, J. Leffingwell and C. Thies, *Macromolecules*, 1971, **4**, 43.
38. L. P. McMaster, *Macromolecules*, 1973, **6**, 760.
39. L. Zeman and D. Patterson, *Macromolecules*, 1972, **5**, 513.
40. A. Robard, D. Patterson and G. Delmas, *Macromolecules*, 1977, **10**, 706.
41. R. G. Snyder and G. Zerbi, *Spectrochim. Acta, Part A*, 1967, **23**, 391.

42. T. Onishi and S. Krimm, *J. Appl. Phys.*, 1961, **32**, 2320.
43. P. C. Painter, M. M. Coleman and J. L. Koenig, 'The Theory of Vibrational Spectroscopy and Its Application to Polymeric Materials', Wiley, New York, 1982, p. 134.
44. D. Garcia, *J. Polym. Sci., Polym. Phys. Ed.* 1984, **22**, 1773.
45. J. L. Halary, J. M. Ubrich, L. Monnerie, H. Yang and R. S. Stein, *Polym. Commun.*, 1985, **26**, 73.
46. F. Ben Cheikh Larbi, S. Leloup, J. L. Halary and L. Monnerie, *Polym. Commun.*, 1986, **27**, 23.
47. J. S. Noland, N. N.-C. Hsu, R. Saxon, and J. M. Schmitt, *ACS Adv. Chem. Ser.*, 1971, **99**, 15.
48. T. Nishi and T. T. Wang, *Macromolecules*, 1975, **8**, 909.
49. D. R. Paul and J. O. Altamirano, *Polym. Prepr., Am. Chem. Soc., Div. Polym. Chem.*, 1974, **15**, 409; *ACS Adv. Chem. Ser.*, 1975, **142**, 371.
50. J. V. Koleske and R. D. Lundberg, *J. Polym. Sci., Polym. Phys. Ed.*, 1969, **7**, 795.
51. L. M. Robeson, *J. Appl. Polym. Sci.*, 1973, **17**, 3607.
52. L. M. Robeson, *J. Appl. Polym. Sci.*, 1973, **17**, 3607.
53. F. B. Khambatta, F. Warner, T. Russell, and R. S. Stein, *J. Polym. Sci., Polym. Phys. Ed.*, 1976, **14**, 1391.
54. S. Krause and N. Roman, *J. Polym. Sci., Part A*, 1965, **3**, 1631.
55. A. M. Liquori, M. DeSantis Savino and M. D'Alagni, *J. Polym. Sci., Part B*, 1966, **4**, 943.
56. R. G. Bauer and N. C. Bletso, *Polym. Prepr., Am. Chem. Soc., Div. Polym. Chem.*, 1969, **10**, 632.
57. T. Miyamoto and H. Inagaki, *Polym. J.*, 1970, **1**, 46.
58. C. Booth and C. J. Pickles, *J. Polym. Sci., Polym. Phys. Ed.*, 1973, **11**, 595.
59. F. Gornick and L. Mandelkern, *J. Appl. Phys.*, 1962, **33**, 907.
60. I. C. Sanchez, in 'Polymer Blends', ed. D. R. Paul and S. Newman, Academic Press, New York, 1978, p. 135.
61. E. Martuscelli, G. Demma, E. Rossi and A. L. Segre, *Polym. Commun.*, 1983, **24**, 266.
62. E. Calahorra, M. Cortazar and G. M. Guzmán, *Polym. Commun.*, 1983, **24**, 211.
63. E. Martuscelli, M. Canetti, L. Vicini and A. Seves, *Polymer*, 1982, **23**, 331.
64. E. Martuscelli and G. Demma, in 'Polymer Blends: Processing, Morphology and Properties', ed., E. Martuscelli, R. Palumbo and M. Kryszewski, Plenum Press, New York, 1980.
65. E. Calahorra, M. Cortazar and G. M. Guzmán, *Polymer*, 1982, **23**, 1322.
66. M. M. Cortazar, M. E. Calahorra and G. M. Guzmán, *Eur. Polym. J.*, 1982, **18**, 165.
67. D. M. Hoffman, Ph.D. Thesis, University of Massachusetts, Amherst, MA, USA, 1979.
68. X. Li and S. L. Hsu, *J. Polym. Sci., Polym. Phys. Ed.*, 1984, **22**, 1331.
69. B. S. Morra and R. S. Stein, *J. Polym. Sci., Polym. Phys. Ed.*, 1982, **20**, 2243.
70. B. S. Morra and R. S. Stein, *J. Polym. Sci., Polym. Phys. Ed.*, 1982, **20**, 2261.
71. M. M. Coleman, J. Zarian, D. F. Varnell and P. C. Painter, *J. Polym. Sci., Polym. Lett. Ed.*, 1977, **15**, 745.
72. H. Tadokoro, Y. Chatani, T. Yoshihara, S. Tahara and S. Murahashi, *Makromol. Chem:*, 1964, **73**, 109, and refs. therein.
73. J. L. Koenig and A. C. Angood, *J. Polym. Sci., Part A-2*, 1970, **8**, 1787.
74. T. Miyazawa, K. Fukushima and Y. Ideguchi, *J. Chem. Phys.*, 1962, **37**, 2764.
75. T. Yoshihara, H. Tadokoro and S. Murahashi, *J. Chem. Phys.*, 1964, **41**, 2902.
76. Y. Matsui, T. Kubota, H. Tadokoro and T. Yoshihara, *J. Polym. Sci. Part A.*, 1965, **3**, 2275.
77. H. Matsuura and T. Miyazawa, *Bull. Chem. Soc. Jpn.*, 1968, **41**, 1798.
78. M. Yokoyama, H. Ochi, H. Tadokoro and C. C. Price, *Macromolecules*, 1972, **5**, 690.
79. H. Matsuura and H. Murata, *J. Raman Spectrosc.*, 1982, **12**, 144.
80. R. Iwamoto, Y. Saito, H. Ishihara and H. Tadokoro, *J. Polym. Sci., Part A-2*, 1968, **6**, 1509.
81. M. Yokoyama, H. Ishihara, R. Iwamoto and H. Tadokoro, *Macromolecules*, 1969, **2**, 184.
82. Y. Takahashi and H. Tadokoro, *Macromolecules*, 1973, **6**, 672.
83. H. Tadokoro, T. Yoshihara, Y. Chatani and S. Murahashi, *J. Polym. Sci., Polym. Lett.*, 1964, **2**, 363.
84. P. J. Flory, 'Statistical Mechanics of Chain Molecules', Wiley, New York, 1969, p. 166.
85. R. S. Stein, *J. Polym. Sci., Part C*, 1966, **15**, 185.
86. J. E. Lasch, D. J. Burchel, T. Masoaka and S. L. Hsu, *Appl. Spectrosc.*, 1984, **38**, 351.
87. S. E. Molis, W. J. MacKnight and S. L. Hsu, *Appl. Spectrosc.*, 1984, **38**, 529.
88. D. J. Burchell and S. L. Hsu, *ACS Adv. Chem. Ser.*, 1983, **203**, 533.
89. W. G. Fateley and J. L. Koenig, *J. Polym. Sci., Polym. Lett. Ed.*, 1982, **20**, 445.
90. I. Noda, A. E. Dowrey and C. Marcott, *J. Polym. Sci., Polym. Lett. Ed.*, 1983, **21**, 99.
91. R. E. Murphy, F. H. Cook and H. Sakai, *J. Opt. Soc. Am.*, 1975, **65**, 600.
92. H. Sakai and P. Murphy, *Appl. Opt.*, 1978, **17**, 1342.
93. M. S. Sen, R. Brahma, S. K. Roy, D. K. Mukherjee and S. B. Roy, *Mol. Cryst. Liq. Cryst.*, 1983, **100**, 327.
94. N. Kirov, M. Sabeva and H. Ratajczak, *Adv. Mol. Relaxation Interact. Processes*, 1982, **22**, 145.
95. Z. Salamon and A. Skibinski, *Mol. Cryst. Liq. Cryst.*, 1983, **90**, 205.
96. K. Miyano, *Phys. Lett. A*, 1977, **63**, 37.
97. S. Kobinata, Y. Nakajima, H. Yoshida and S. Maeda, *Mol. Cryst. Liq. Cryst.*, 1981, **66**, 67.
98. S. D. Durbin and Y. R. Shen, *Phys. Rev. A*, 1984, **30**, 1419.
99. A. Hatta, *Mol. Cryst. Liq. Cryst.*, 1981, **74**, 195.
100. A. Kaito, Y. Wang and S. L. Hsu, *Anal. Chim. Acta*, 1986, **189**, 27.
101. K. Miyano, *Phys. Rev. Lett.*, 1977, **43**, 51.
102. H. Gruler, T. J. Scheffer and G. Meier, *Z. Naturforsch., Teil A*, 1972, **27**, 966.
103. W. Helfrich, *Mol. Cryst. Liq. Cryst.*, 1973, **21**, 187.
104. H. Hakemi, E. F. Jagodzinski and D. B. DuPré, *J. Chem. Phys.*, 1983, **78**, 1513.
105. P. G. Cummins, D. A. Dunmur and D. A. Laidler, *Mol. Cryst. Liq. Cryst.*, 1975, **30**, 109.
106. C. T. O'Konski, K. Yoshioka and W. H. Orttung, *J. Phys. Chem.*, 1959, **63**, 1558.
107. K. Yoshioka, *J. Chem. Phys.*, 1983, **79**, 3482.
108. H. J. Coles and B. R. Jennings, *Mol. Phys.*, 1978, **36**, 1661.
109. J. P. Parneix and A. Chapoton, *Acta Phys. Pol. A*, 1978, **54**, 667.
110. I. Noda, A. E. Dowrey and C. Marcott, *Bull. Am. Phys. Soc.*, 1986, **31**, 405.
111. R. Jakeways, I. M. Ward, M. A. Wilding, I. H. Hall, J. Desborough and M. G. Pass, *J. Polym. Sci., Polym. Phys. Ed.*, 1975, **13**, 799.

112. M. Yokouchi, Y. Sakakibara, Y. Chatani, H. Tadokoro, T. Tanaka and K. Yoda, *Macromolecules*, 1976, **9**, 266.
113. P. C. Gillette, S. D. Dirlikov, J. L. Koenig and J. B. Lando, *Polymer*, 1982, **23**, 1759.
114. P. C. Gillette, J. B. Lando and J. L. Koenig, *Polymer*, 1985, **26**, 235.
115. K. Holland-Moritz and H. W. Siesler, *Polym. Bull.*, 1981, **4**, 165.
116. R. S. Stein, S. Onogi, K. Sasaguri and D. A. Keedy, *J. Appl. Phys.*, 1963, **34**, 80.
117. R. Zbinden, 'Infrared Spectroscopy of High Polymers', Academic Press, New York, 1964.
118. R. D. B. Fraser, *J. Chem. Phys.*, 1956, **24**, 89.
119. W. Glenz and A. Peterlin, *J. Polym. Sci., Part-A2*, 1971, **9**, 1191.
120. R. S. Stein, *J. Appl. Polym. Sci.*, 1961, **5**, 96.
121. L. A. Nafie and D. W. Vidrine, in 'Fourier Transform Infrared Spectroscopy', ed. J. R. Ferraro and L. J. Basile, Academic Press, 1982, vol. 3, p. 83.
122. R. W. Ditchburn, 'Light', Wiley, New York, 1963.
123. R. J. Roe, *J. Polym. Sci., Part-A2*, 1970, **8**, 1187.
124. R. D. B. Fraser, *J. Chem. Phys.*, 1953, **21**, 1511.
125. R. G. Snyder, S. L. Hsu and S. Krimm, *Spectrochim Acta, Part A*, 1978, **34**, 395.
126. R. G. Snyder and J. R. Scherer, *J. Chem. Phys.*, 1979, **71**, 3221.
127. G. Dellepiane, S. Abbate, P. Bosi and G. Zerbi, *J. Chem. Phys.*, 1980, **73**, 1040.
128. S. Onogi and T. Asada, *J. Polym. Sci., Part C*, 1967, **16**, 1445.
129. R. W. Seymour, G. M. Estes and S. L. Cooper, *Macromolecules*, 1970, **3**, 579.
130. R. W. Seymour, A. E. Allegrezza, Jr. and S. L. Cooper, *Macromolecules*, 1973, **6**, 896.
131. H. Ishihara, I. Kimura, K. Saito and H. Ono, *J. Macromol. Sci., Phys. Ed.*, 1974, **10**, 591.
132. S. B. Lin, K. S. Hwang, S. Y. Tsay and S. L. Cooper, *Colloid Polym. Sci.*, 1985, **263**, 128.
133. S. Onogi and T. Asada, *Prog. Polym. Sci., Jpn.*, 1971, **2**, 261.
134. R. S. Stein, *Acc. Chem. Res.*, 1972, **5**, 121.
135. R. J. Roe and W. R. Krigbaum, *J. Appl. Phys.*, 1964, **35**, 2215.
136. M. Doi, *J. Polym. Sci., Polym. Phys. Ed.*, 1980, **18**, 1005.
137. J. D. Ferry, 'Viscoelastic Properties of Polymers', 3rd edn. Wiley, New York, 1980.
138. M. Doi and S. F. Edwards, *J. Chem. Soc., Faraday Trans. 2*, 1978, **74**, 1789, 1802, 1818.
139. J. L. Viovy, L. Monnerie and J. F. Tassin, *J. Polym. Sci., Polym. Phys. Ed.*, 1983, **21**, 2427.
140. E. Dobrovolny and S. L. Hsu, *Macromolecules*, 1987, **20**, 1022.
141. I. M. Ward and M. A. Wilding, *Polymer*, 1977, **18**, 327.
142. B. Stambaugh, J. B. Lando and J. L. Koenig, *J. Polym. Sci., Polym. Phys. Ed.*, 1979, **17**, 1063.
143. R. J. Cella, *J. Polym. Sci., Polym. Symp.*, 1973, **42**, 727.
144. G. Wegner, T. Fujii, W. Meyer and G. Lieser, *Angew. Makromol. Chem.*, 1978, **74**, 295.
145. R. M. Briber and E. L. Thomas, *Polymer*, 1985, **26**, 8.
146. V. K. Datye and P. L. Taylor, *Macromolecules*, 1985, **18** (4), 671.
147. C. Chang and S. L. Hsu, *J. Polym. Sci., Polym. Phys. Ed.*, 1985, **23**, 2307.
148. J. P. Hobbs, C. S. P. Snug, K. Krishnan and S. Hill, *Macromolecules*, 1983, **16**, 193.
149. S. A. Curran, Ph.D. Thesis, University of Massachusetts, Amherst, MA, 1981.
150. P. C. Lucas, Ph.D. Thesis, University of Massachusetts, Amherst, MA, 1986.
151. W. G. Golden, D. S. Dunn and J. Overend, *J. Catal.*, 1981, **71**, 395.
152. A. E. Dowrey and C. Marcott, *Appl. Spectrosc.*, 1982, **36**, 414.
153. M. W. Urban and J. L. Koenig, *Appl. Spectrosc.*, 1986, **40**, 994.
154. M. W. Urban and J. L. Koenig, *Appl. Spectrosc.*, 1985, **39**, 1051.
155. K. Krishnan, S. Hill, J. P. Hobbs and C. S. P. Sung, *Appl. Spectrosc.*, 1982, **36**, 257.
156. D. W. Berreman, *J. Opt. Soc. Am.*, 1972, **62**, 502.
157. R. G. Greenler, *J. Chem. Phys.*, 1966, **44**, 310; W. N. Hansen, in 'Advances in Electrochemistry and Electrochemical Engineering', ed. P. Delahay and C. W. Tobias, Wiley, New York, 1973, vol. 9, pp. 1–60 and refs. therein.
158. D. L. Allara and R. G. Nuzzo, *Langmuir*, 1985, **1**, 45.
159. D. L. Allara and R. G. Nuzzo, *Langmuir*, 1985, **1**, 52.
160. A. Kaito, M. Kardan and S. L. Hsu, *J. Phys. Chem.*, 1988, in press.
161. M. Kardan, A. Kaito, S. L. Hsu, R. Takur and C. P. Lillya, *J. Phys. Chem.*, 1987, **91**, 1809.

21

Raman Spectroscopy*

N. E. SCHLOTTER
Bell Communications Research, Red Bank, NJ, USA

21.1 INTRODUCTION

Raman spectroscopy is becoming an increasingly important tool in the study of polymeric materials. Improvements in understanding and technique suggest that the potential applications of Raman scattering have barely been initiated. The vibrational information obtained in a Raman spectrum is rich in content about both the chemical and morphological structure of the polymer. Further, the sampling advantages of Raman over many other methods include nondestructive techniques, high spatial resolution, novel sampling geometry, mode selectivity through the use of polarization, and tunable coupling to the electronic structure when desired. Problems such as fluorescence and low signal levels are being overcome with new techniques. As demand grows for high performance polymers, which must exploit extremes of ordering and chemistry, Raman spectroscopy becomes more important as a probe. Raman spectroscopy is being used to study polymers which have high electrical conductivity, optical nonlinearity, strength, and forms of electronic coupling to the environment such as the piezoelectric effect. These important applications and many others indicate that Raman spectroscopy is a tool the polymer scientist will need more frequently in the future. This chapter introduces the basics of Raman spectroscopy as applied to polymers and indicates current directions.

The Raman effect was observed in 1928 by Raman and Krishnan,[1] but wide application was delayed until the development of the laser. Since the advent of the laser, progress has been rapid and what was predominately a laboratory curiosity has become a useful analytical tool. It is possible to

* Reproduced, with modifications, from volume 14 of the 'Encyclopedia of Polymer Science and Engineering', 2nd edn., ed. H. F. Mark, N. Bikales, C. G. Overberger and G. Menges, Wiley, New York, 1988, with permission of the publisher.

analyze Raman spectra in terms of the molecular components, or functional groups, composing the molecule if the vibrational normal modes are reasonably well isolated energetically from one another. As in infrared absorption spectroscopy, there are tables of compiled observations relating the vibrational motions of molecular subunits to frequency ranges.[2, 3]

Since Raman spectroscopy is sensitive to molecular motions, any environmental perturbations or changes in molecular structure that modify these motions can be detected. By careful analysis of the resulting spectra, the source of the modification can be identified. Thus normal coordinate analysis is an integral part of the process used to interpret Raman spectra. Such studies have been very useful for polymers, since it is often more interesting to determine the morphology of the polymer than the molecular constituents of the polymer. For example, usually the repeat unit of the polymer is well known, but the extended conformation and crystal structure are not as well characterized. Raman spectroscopy can determine, in many instances, the most likely crystal structure, lamellar thickness, branch content and processing history.

General reviews of the Raman effect, including descriptions of experimental equipment and technique, may be found in refs. 4–8. Reviews with discussions concerning the application of Raman spectroscopy to polymers are in refs. 9–15. In this chapter only the most recent work is cited to illustrate an example. There are usually many older references to similar work that can be found in the reference lists of the cited article(s). I apologize in advance if I have omitted references that the reader considers important. I have tried to touch all points related to the Raman spectroscopy of polymers, but it is a rapidly growing field and length considerations have forced the adaptation of a survey type of review.

21.2 THEORY OF RAMAN SCATTERING

21.2.1 Isotropic Systems

The Raman effect detects vibrational motions. The vibrational motions that give rise to strong scattering intensity have certain well-defined symmetry properties. By analyzing the polarizations present in the scattered radiation it is possible to determine the symmetry of the vibrational motion giving rise to a particular scattering frequency. Isotropic systems produce Raman scattering in which the modes can be separated into two groups with different symmetry properties corresponding to the rotational invariants of an object in free space. In general, as the ordering of the sample is increased, the number of groups of motions with different symmetry properties that can be spectroscopically separated increases also. For samples that can be fully oriented relative to a laboratory coordinate system, such as molecular single crystals, it is often possible to determine the symmetry of the normal modes exactly. Similar information can be obtained for polymers that have been oriented by processing or crystallization conditions. The effects of orientation on the Raman spectrum will be discussed later, after the basic phenomenon for an isotropic scattering system is presented.

How can information be obtained from a Raman spectrum? First, it is important to realize that Raman scattering is a different effect from IR absorption. The Raman effect is a true scattering effect and as such the light involved interacts with the electronic structure of the molecule that is being modulated by the molecular vibrations, but the light does not undergo any form of energy conversion followed by reconversion and subsequent emission. The interaction of light with the electronic structure of the molecule involves the interaction of the oscillating electric field of the light with the charge distribution or multipole structure of the molecule. In the case of the Raman effect the principal source of interaction is with the induced dipole, which is produced by the oscillations of the electric field of the light wave driving the electron cloud of the molecule. The electron cloud of the molecule is also subject to driving forces from the molecular vibrations, which to a first approximation are linear with the nuclear displacements. This gives rise to a beat frequency between the two driving forces, the electric field of the light wave and the vibrational modes of the molecule. The Raman effect can be considered classically as an inelastic scattering at the beat frequencies; the sum frequency is called the anti-Stokes shift and the difference frequency is called Stokes shift.

To get the correct results for the Raman process it is necessary to take into account the quantum mechanical features of the problem. Fortunately, to describe the features of Raman scattering that are important in this case it is only necessary to bring in the quantum mechanical description of the molecule; the radiation field can be treated classically as a perturbation of the molecular system.

For the Raman effect to occur, the molecule must undergo a transition between an initial and final state. The induced dipole operator P connects these states and forms the transition moment $P(\text{fi})$

$$P(\text{fi}) \quad = \quad \langle f | P | i \rangle \tag{1}$$

where $|i\rangle$ is the initial and $\langle f|$ the final state. The induced dipole moment can be approximated well by the linear term of the induced dipole expansion to describe spontaneous Raman scattering

$$P_0^{(1)} = \boldsymbol{\alpha} \cdot \boldsymbol{E}_0 \quad \text{(at frequency } \omega_0\text{)} \tag{2}$$

where the subscript 0 refers to the maximum amplitude of the vectors. The resulting transition moment amplitude is then given by

$$[P_0^{(1)}](\text{fi}) = \langle f|\boldsymbol{\alpha}|i\rangle \cdot \boldsymbol{E}_0 \tag{3}$$

where $|f\rangle$ and $|i\rangle$ are the time-independent wave functions. The wave functions and the polarizability tensor $\boldsymbol{\alpha}$ are, in general, functions of all the coordinates of the system and $\langle f|\boldsymbol{\alpha}|i\rangle$ is an integral over all coordinate space. Explicitly

$$\langle f|\boldsymbol{\alpha}|i\rangle = \begin{bmatrix} \langle f|\alpha(xx)|i\rangle & \langle f|\alpha(xy)|i\rangle & \langle f|\alpha(xz)|i\rangle \\ \langle f|\alpha(yx)|i\rangle & \langle f|\alpha(yy)|i\rangle & \langle f|\alpha(yz)|i\rangle \\ \langle f|\alpha(zx)|i\rangle & \langle f|\alpha(zy)|i\rangle & \langle f|\alpha(zz)|i\rangle \end{bmatrix} \tag{4}$$

where $\langle f|\alpha(xy)|i\rangle$ is a component of the transition polarizability tensor. As long as the excitation frequency is much larger than the frequency of the vibrational transition frequencies but much smaller than any electronic transition frequency of the system, it is possible to separate the wave functions into a product of components dependent on vibrational coordinates only and on rotational coordinates only. This is done by introducing molecule-fixed axes (X, Y, Z) in addition to the original space-fixed axes (x, y, z). The rotational terms depend on the matrix elements of the direction cosines relating the molecule-fixed to the space-fixed coordinates for the rotational wave functions. The components of the transition polarizability tensor in the space-fixed coordinates can be expressed as

$$\langle f|\alpha(xy)|i\rangle = \sum_{X,Y} [\langle \varphi\{n(f)\}|\alpha(XY)|\varphi\{n(i)\}\rangle \cdot \langle \Phi\{R(f)\}|\cos(xX)\cos(yY)|\Phi\{R(i)\}\rangle] \tag{5}$$

where $n(f)$ and $n(i)$ are the sets of vibrational quantum numbers in the final and initial states; likewise $R(f)$ and $R(i)$ are sets of rotational quantum numbers in the final and initial states. The important feature to recognize is that the vibrational selection rules are determined by the first term of the product. Similarly, the rotational selection rules are independent of the vibrational selection rules in the ground electronic state.

First, consider only isotropic distributions of molecules and note that scattering intensities and polarization properties depend on the products of the vibrational transition polarizability components. Therefore, only the space averages of the products need to be considered:

$$\overline{|\langle f|\alpha(kl)|i\rangle \langle f|\alpha(mn)|i\rangle|} \tag{6}$$

for $(kl, mn) = (xx, xx), (yy, yy), (zz, zz), (xx, yy), (xx, zz), (yy, zz), (xy, xy), (xz, xz), (yz, yz)$. All others are identically zero. For an isotropic system, the symmetric transition polarizability tensor has only two independent terms, which are also rotationally invariant. These are the mean polarizability $\langle \delta \rangle$ and the anisotropy $\langle \beta \rangle$, i.e.

$$\langle \delta \rangle = \tfrac{1}{3}[\langle \alpha(xx) \rangle + \langle \alpha(yy) \rangle + \langle \alpha(zz) \rangle] \tag{7}$$

$$\langle \beta \rangle^2 = \tfrac{1}{2}\{[\langle \alpha(xx) \rangle - \langle \alpha(yy) \rangle]^2 + [\langle \alpha(yy) \rangle - \langle \alpha(zz) \rangle]^2 + [\langle \alpha(zz) \rangle - \langle \alpha(xx) \rangle]^2 + 6[\langle \alpha(xy) \rangle^2 + \langle \alpha(yz) \rangle^2 + \langle \alpha(zx) \rangle^2]\} \tag{8}$$

The terms in equation (6) can be expressed in terms of $\langle \delta \rangle^2$ and $\langle \beta \rangle^2$ as

$$\overline{|\langle \alpha(kk) \rangle|^2} = \frac{45\langle \delta \rangle^2 + 4\langle \beta \rangle^2}{45}, \quad k = x, y, z \tag{9}$$

$$\overline{|\langle \alpha(kl) \rangle|^2} = \frac{\langle \beta \rangle^2}{15}, \quad kl = xy, yz, zx \tag{10}$$

$$\overline{|\langle \alpha(kk) \rangle \langle \alpha(ll) \rangle|} = \frac{45\langle \delta \rangle^2 - 2\langle \beta \rangle^2}{45}, \quad k \neq l, \quad \text{for } x, y, z \tag{11}$$

These results will be used when the depolarization ratios and intensities of an isotropic distribution of molecules are derived.

To understand the vibrational selection rules, the dependence of the transition polarizability components on the vibrational coordinates must be determined. If the first two terms of the series expansion of the components of the polarizability tensor in the normal coordinates $Q(k)$ are substituted into the expression for the transition polarizability tensor components, expressed in molecule-fixed coordinates as in equation (5), the following is obtained

$$\langle \varphi\{n(f)\}|\alpha(XY)|\varphi\{n(i)\}\rangle \;=\; \langle \varphi f|\alpha(XY)|\varphi i\rangle \;=\; (\alpha(XY))_0\langle \varphi f|\varphi i\rangle \;+\; \sum_k \left[\frac{\partial \alpha(XY)}{\partial Q(k)}\right]_0 \langle \varphi f|Q(k)|\varphi i\rangle \quad (12)$$

The zero denotes evaluation at the equilibrium position. If the harmonic oscillator approximation is justified for the system in question the total vibrational wave function will consist of products of harmonic oscillator wave functions of the normal modes. The wave functions $|\varphi f\rangle$ and $|\varphi i\rangle$ can be written as

$$|\varphi i\rangle \;=\; \left|\prod_k \varphi[n(i,k),Q(k)]\right\rangle \;=\; \left|\prod_k \varphi[i,k]\right\rangle, \; |\varphi f\rangle \;=\; \left|\prod_k \varphi[n(f,k),Q(k)]\right\rangle \;=\; \left|\prod_k \varphi[f,k]\right\rangle \quad (13)$$

where $|\varphi[n(i,k),Q(k)]\rangle$ and $|\varphi[n(f,k),Q(k)]\rangle$ are harmonic oscillator wave functions for normal coordinate $Q(k)$ with vibrational quantum numbers $n(i,k)$ and $n(f,k)$ in the initial and final states respectively. On substitution for $|\varphi i\rangle$ and $|\varphi f\rangle$ in equation (12) one obtains

$$\langle \varphi f|\alpha(XY)|\varphi i\rangle \;=\; (\alpha(XY))_0\left\langle \prod_k \varphi[f,k]\middle|\prod_k \varphi[i,k]\right\rangle \;+\; \sum_k\left[\frac{\partial \alpha(XY)}{\partial Q(k)}\right]_0 \left\langle \prod_k \varphi[f,k]\middle|Q(k)\middle|\prod_k \varphi[i,k]\right\rangle \quad (14)$$

The harmonic oscillator wave functions have the properties

$$\langle \varphi[f,k]|\varphi[i,k]\rangle \;=\; \begin{bmatrix} 0 & \text{for} & n(f,k) & \neq & n(i,k) \\ 1 & \text{for} & n(f,k) & = & n(i,k) \end{bmatrix} \quad (15)$$

and

$$\langle \varphi[f,k]|Q(k)|\varphi[i,k]\rangle \;=\; \begin{bmatrix} 0 & \text{for} & n(f,k) = n(i,k) \\ \{n(i,k)+1\}^{\frac{1}{2}}b(k) & \text{for} & n(f,k) = n(i,k)+1 \\ \{n(i,k)\}^{\frac{1}{2}}b(k) & \text{for} & n(f,k) = n(i,k)-1 \end{bmatrix} \quad (16)$$

where

$$b(k)^2 \;=\; h/[8\pi^2 c \tilde{v}(k)] \quad (17)$$

The value of $b(k)$ is for a mass-adjusted normal coordinate $Q(k)$ and $\tilde{v}(k)$ is the frequency (cm^{-1}) of the kth normal mode.

As a result of the orthonormal properties of the harmonic oscillator wave functions the first term of $\langle \varphi f|\alpha(XY)|\varphi i\rangle$ will be zero for Raman scattering. The second term gives nonzero results when only one vibrational quantum number at a time changes by a value of unity. For the kth normal mode if $n(f,k) = n(i,k) \pm 1$ and $n(f,j) = n(i,j)$ for all $j \neq k$, then a nonzero result occurs. If $\Delta n = 1$, the transition moment is associated with Stokes Raman scattering whereas if $\Delta n = -1$, it is associated with anti-Stokes. Concentrating on a single normal mode k and considering only Raman scattering reduces equation (14) to

$$\langle \varphi f(k)|\alpha(XY)|\varphi i(k)\rangle \;=\; \{n(i,k)+1\}^{\frac{1}{2}}b(k)\left[\frac{\partial \alpha(XY)}{\partial Q(k)}\right]_0 \quad \text{if } n(f,k) = n(i,k)+1 \quad \text{and } n(f,j) = n(i,j) \text{ for all other } Q(j)$$

or, equivalently,

$$= \{n(i,k)+1\}^{\frac{1}{2}}b(k)[\alpha'(XY,k)] \quad \text{for Stokes scattering}$$

and

$$= \{n(i,k)\}^{\frac{1}{2}}b(k)[\alpha'(XY,k)], \quad \text{for anti-Stokes} \quad \text{if } n(f,k) = n(i,k)-1 \quad \text{and } n(f,j) = n(i,j)$$
$$\text{for all other } Q(j). \quad (18)$$

Any other set of quantum numbers will reduce the kth term to zero. The assumption of electrical and mechanical harmonicity allows only vibrational fundamentals in the Raman effect. It is further required that at least one element of the derived polarizability tensor be nonzero.

If a single molecule is held with its axes (X, Y, Z) aligned with the space-fixed axes (x, y, z), then the orientational averaging is reduced to

$$\langle \varphi f | \alpha(XY) | \varphi i \rangle = \langle \varphi f | \alpha(xy) | \varphi i \rangle \tag{19}$$

which for Stokes Raman scattering gives

$$[P^{(1)}\{x_0, n(k) + 1, n(k)\}] = (n(i,k) + 1)^{\frac{1}{2}} b(k) \cdot \{[\alpha'(xx,k)] \cdot E_{x_0} + [\alpha'(xy,k)] \cdot E_{y_0} + [\alpha'(xz,k)] \cdot E_{z_0}\} \tag{20}$$

with similar terms for $[P^{(1)}\{y_0, n(k) + 1, n(k)\}]$ and $[P^{(1)}\{z_0, n(k) + 1, n(k)\}]$. Consequently the depolarization ratios, useful for understanding the vibrational symmetry, are identical to the classically derived results for right angle scattering

$$\rho_{\parallel}(\pi/2) = \frac{[\alpha'(xy,k)]^2}{[\alpha'(xz,k)]^2} \tag{21}$$

and

$$\rho_{\perp}(\pi/2) = \frac{[\alpha'(yz,k)]^2}{[\alpha'(yy,k)]^2} \tag{22}$$

The notation $\rho_{\parallel}(\pi/2)$ and $\rho_{\perp}(\pi/2)$ refers to scattering at right angles with the polarization direction parallel to the scattering plane or in-plane, and perpendicular or out-of-plane respectively. For an assembly of freely rotating molecules, the space averages give the same results as obtained classically owing to cancellations in the numerator and denominator.

The principal differences between the classical and the semiclassical pictures occur in the calculation of intensities. For N molecules undergoing a given Stokes shift $n(i,k) + 1 \leftarrow n(i,k)$ the space-averaged squares of the transition moments for Raman scattering depend on $[n(i,k) + 1] \cdot b(k)^2$ resulting in a scattering dependence on $N \cdot [n(i,k) + 1] \cdot f(i,k)$ where $f(i,k)$ is the fraction of molecules in the state $n(i,k)$. Since the harmonic approximation gives identical frequencies for all the single-step transitions (equal energy spacing), the total scattering intensity is proportional to the sum of $N \cdot [n(i,k) + 1] \cdot f(i,k)$ over all states i for a given mode k.

$$I \propto N \sum_i (n(i,k) + 1) \cdot f(i,k) \tag{23}$$

The Boltzmann distribution law gives

$$f(i,k) = \frac{\exp[-(n(i,k) + \frac{1}{2}) hc\bar{v}(k)/kT]}{\sum_i \exp[-(n(i,k) + \frac{1}{2}) hc\bar{v}(k)/kT]} \tag{24}$$

Substituting equation (24) into equation (23) and reducing gives

$$I \propto \frac{N}{1 - \exp[-hc\bar{v}(k)/kT]} \quad \text{(Stokes)} \tag{25}$$

and

$$I \propto N \sum_i [n(i,k) \cdot f(i,k)] = \frac{N}{\exp[+hc\bar{v}(k)/kT] - 1} \quad \text{(Anti-Stokes)} \tag{26}$$

A sample intensity calculation for right angle scattering, as shown in Figure 1, demonstrates the application of the above equations. Here the incident radiation is propagating in the z direction (with y polarization); the scattered radiation is collected in the x direction and analyzed in the y direction. Such an experiment can be equivalently described as a $(\perp, \perp)$, VV, or $z(yy)x$ (in Porto notation[16]) geometry. Right angle scattering is the most common sampling geometry, but appropriate modifications can be made for other scattering geometries, such as backscattering, and the calculations repeated. Continuing, the space-averaged square of the Stokes transition moment amplitude is

$$\overline{[P^{(1)}\{y_0, n(i,k) \quad + \quad 1, n(i,k)\}]^2} \quad = \quad (n(i,k) + 1)b(k)^2 \overline{[\alpha'(yy,k)]^2} \cdot E_{y_0}^2 \tag{27}$$

giving an expression for the scattered intensity as

$$^{\perp}I_{\perp}(\pi/2) \quad = \quad k'(\tilde{v}) \cdot (\tilde{v}(0) \quad - \quad \tilde{v}(k))^4 (n(i,k) + 1) \cdot b(k)^2 \overline{[\alpha'(yy,k)]^2} \cdot E_{y_0}^2 \tag{28}$$

which on substituting for $[\alpha']$ in terms of the derived rotational invariants gives

$$^{\perp}I_{\perp}(\pi/2) \quad = \quad k'(\tilde{v})(\tilde{v}(0) \quad - \quad \tilde{v}(k))^4 \cdot (n(i,k) + 1)b(k)^2 \cdot \frac{45[\delta']^2 \quad + \quad 4[\beta']^2}{45} \cdot E_{y_0}^2 \tag{29}$$

where $k'(\tilde{v}) = \pi^2 c/[2\varepsilon_0]$.

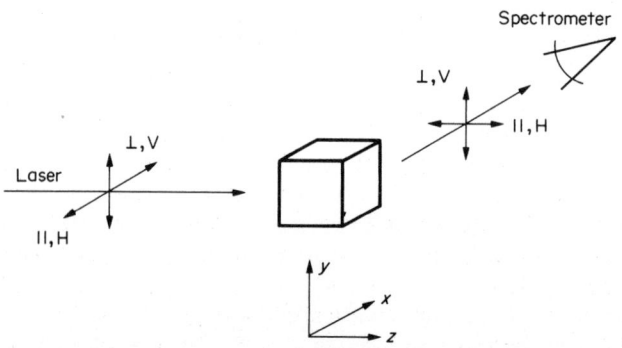

Figure 1 Right-angle scattering geometry. The scattering plane is defined by the incident beam, and the scattering direction by the xz plane. Polarizations perpendicular, or V for vertical, and polarizations parallel, or H for horizontal, are defined relative to the scattering plane

Summing over N molecules on various $n(k)$ levels of the kth vibrational state gives the final results

$$^{\perp}I_{\perp}(\pi/2) \quad = \quad \frac{k(\tilde{v})hN(\tilde{v}(0) \quad - \quad \tilde{v}(k))^4}{8\pi^2 c\tilde{v}(k)\{1 \quad - \quad \exp[-hc\tilde{v}(k)/kT]\}} \cdot \frac{45[\delta']^2 \quad + \quad 4[\beta']^2}{45} \cdot \Phi \tag{30}$$

where $k(\tilde{v}) = \pi^2/\varepsilon_0^2$, $\Phi = \frac{1}{2}\varepsilon_0 c E_0^2$ and $b(k)^2 = h/[8\pi^2 c\tilde{v}(k)]$. Φ is termed the irradiance.

Similarly, the VH or $(\perp, \parallel)$ experiment, or $z(yz)x$ in Porto notation, gives

$$^{\perp}I_{\parallel}(\pi/2) \quad = \quad \frac{k(\tilde{v})hN(\tilde{v}(0) \quad - \quad \tilde{v}(k))^4}{8\pi^2 c\tilde{v}(k)\{1 \quad - \quad \exp[-hc\tilde{v}(k)/kT]\}} \cdot \frac{[\beta']^2}{15} \cdot \Phi \tag{31}$$

As for the isolated molecule above, the depolarization ratio can be calculated from $^{\perp}I_{\parallel}/^{\perp}I_{\perp}$ giving

$$\rho_{\perp}(\pi/2) \quad = \quad \frac{3[\beta']^2}{45[\delta']^2 \quad + \quad 4[\beta']^2} \tag{32}$$

Typically, vibrations with symmetric motions have $\rho_{\perp}(\pi/2) \approx 0$ while vibrations with large non-symmetric motions have $\rho_{\perp}(\pi/2) \approx 3/4$.

The intensity results for anti-Stokes Raman scattering are

$$^{\perp}I_{\perp}(\pi/2) \quad = \quad \frac{k(\tilde{v})hN(\tilde{v}(0) \quad + \quad \tilde{v}(k))^4}{8\pi^2 c\tilde{v}(k)\{\exp[+hc\tilde{v}(k)/kT] \quad - \quad 1\}} \cdot \frac{45[\delta']^2 \quad + \quad 4[\beta']^2}{45} \cdot \Phi \tag{33}$$

and

$$^{\perp}I_{\parallel}(\pi/2) \quad = \quad \frac{k(\tilde{v})hN(\tilde{v}(0) \quad + \quad \tilde{v}(k))^4}{8\pi^2 c\tilde{v}(k)\{\exp[+hc\tilde{v}(k)/kT] \quad - \quad 1\}} \cdot \frac{[\beta']^2}{15} \cdot \Phi \tag{34}$$

The resulting Stokes to anti-Stokes ratio is

$$\left[\frac{\tilde{v}(0) \quad - \quad \tilde{v}(k)}{\tilde{v}(0) \quad + \quad \tilde{v}(k)}\right]^4 \exp\{hc\,\tilde{v}(k)/kT\} \tag{35}$$

The preceding equations (30–35) show the dependence of Raman scattering on both temperature and concentration.

The next section covers modifications to the foregoing results that are needed when the sample being studied is an ordered system. In particular for polymers the ordering is often described as partially ordered. At the same time a brief description of normal modes will be introduced. Detailed application of symmetry and normal mode analysis can be used with the above equations to calculate the intensities of each normal mode. Such detail will not be pursued in this chapter. Theory references cited in the Section 21.1 provide further information.

21.2.2 Ordered Systems

Extensive studies have been made of highly ordered systems, such as molecular crystals.[17] For single crystals with one or more sites, a subset of the polarizability tensor components contribute to the Raman scattering. These components are determined by the relationship of the space-fixed axes to that of the crystal axes. It is convenient to make use of Porto notation[16] where, whether the space-fixed or the crystal-fixed coordinates are chosen, the experiment is described as $A(BC)D$ where A is the incident beam direction, B the direction of polarization of the incident beam, C the direction of polarization analyzed in the scattered beam and D the direction in which the scattering is observed in the selected coordinate system.

For polymers the situation is frequently somewhere between highly ordered and isotropic. A processed crystalline or semicrystalline polymer is partially aligned relative to a set of material axes (*i.e.* draw direction) because of shear forces encountered in drawing, injecting or blowing operations. These conditions of partial alignment can be treated by limiting the space average to a subset of averages in equation (6). For example, averaging about an axis corresponds to a uniaxially drawn fiber. These results have been derived for nearly all orientations.[18,19] Whereas an isotropic system has only two independent terms, the rotational invariants, a uniaxial system has four independent terms and a single crystal has six independent terms. The advantage of ordering a system is that it is possible to extract much more detailed spectroscopic information from the additional polarization experiments available. This leads to detailed considerations of the molecular symmetry and the symmetry of the molecular vibrations.

21.2.3 Molecular Symmetry and Group Theory

Group theory provides a method for correlating the symmetries of the molecular vibrations with the components of the derived polarizability tensor; consequently it is possible to determine for a given Raman experiment which molecular vibrations should be present in the spectrum. Essentially, a given Raman experiment excites a subset of the components of the derived polarizability tensor. The only molecular vibrations which can contribute to the Raman scattering involve the same molecular coordinates as those of the derived polarizability tensor components excited. Numerous sources cover the many aspects of group theory needed to analyze Raman spectra.[4,5,7,8,17,20-25]

The calculation of the normal modes of the molecular system is a separate problem. Typically, a semiempirical approach is used that essentially is based on the concept of masses and springs, but the 'springs' can assume complicated forms. For example, potentials representing stretches, bends, torsions, nonbonded interactions and so forth are used. The force constants are adjusted to give the best fit to experimental data, such as vibrational frequencies and structure, of a test set of molecules. It is then assumed that the parameters derived by fitting the test set are transferable to similar molecules. The resulting normal modes can be sorted by group theory according to their behavior under the operations of the group. The calculated modes can be compared to experimental results as a test of the transferability assumption or used predictively or in further calculations. Calculations based on *ab initio* techniques are becoming practical, but are still limited by computer power.

This process can be understood by an example. In Figure 2 a nonlinear triatomic Y—X—Y is shown with an attached molecular coordinate system. A nonlinear triatomic has $3N$ degrees of freedom, N being the number of atoms. Of these $3N$ degrees of freedom three are rotations and three are center-of-mass translations, leaving $3N - 6$ degrees of freedom, or a total of three normal modes. The normal mode calculation is then done on the molecule to determine the mode structures and symmetries, as illustrated in Figure 3. This molecule belongs to the C_{2v} point group, which has the symmetry operations of the identity, a rotation of 180 degrees about the z axis, a mirror reflection in the xz plane and a mirror reflection in the yz plane. The group table for C_{2v} is given in Table 1. In the C_{2v} group table there are the A_1, A_2, B_1 and B_2 character types. Their behavior under the group operations is given. For example, the character type A_1 looks identical to the original after all groups operations. However, the character type B_1 will be distorted from the original structure when either a C_2 rotation or a σ_{yz} mirror reflection is applied. In a similar manner the normal modes can be sorted according to their behavior under the group operations.

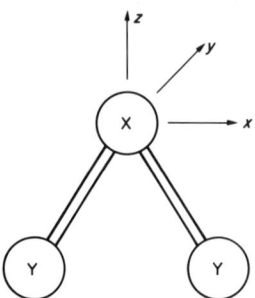

Figure 2 Diagram of a bent Y—X—Y molecule. The molecule is defined to be in the xz plane with the y axis perpendicular to the plane of the molecule

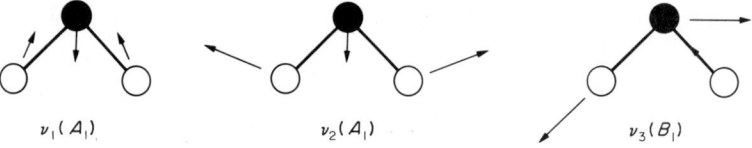

$\nu_1(A_1)$ $\nu_2(A_1)$ $\nu_3(B_1)$

Figure 3 Vibrational modes of a bent Y—X—Y molecule. Modes ν_1, ν_2, and ν_3 are shown with their symmetry types A_1, A_1 and B_1 respectively

Table 1

C_{2v}	E	$C_2(z)$	$\sigma_v(xz)$	$\sigma_v(yz)$	IR	Raman
A_1	1	1	1	1	T_z	$\alpha_{xx}, \alpha_{yy}, \alpha_{zz}$
A_2	1	1	-1	-1	R_z	α_{xy}
B_1	1	-1	1	-1	T_x, R_y	α_{zx}
B_2	1	-1	-1	1	T_y, R_x	α_{yz}

Interpretation of the Raman experiment is now possible. Suppose the molecular coordinates are aligned with the space coordinates. In an experiment with Z-polarized light incident on the molecule and Z-analyzed scattering, only normal modes with induced dipole moments that behave like the zz component of the derived polarizability tensor can be detected. These are the ν_1 and ν_2 modes for the YXY triatomic. However, if an experiment with X-polarized incident light is done and Z-polarized light collected, normal modes that couple these coordinates could contribute. This experiment would detect the ν_3 mode of the nonlinear triatomic.

21.2.4 Line Groups

Line groups are, strictly speaking, applicable only to one-dimensional systems. However, many useful descriptions of polymer spectroscopy refer to the group analysis of the polymer chain in terms of a line group[17] because many isolated polymer chains can be thought of as roughly linear structures. In reality, one is dealing with two subgroups of the three-dimensional space group: the factor group, which is the point group of the repeat unit, and the translational group. The factor group is isomorphic with the crystallographic point group and is consequently very useful because molecular interactions transverse to the chain axis are usually small and can be neglected, leaving one with the internal motions of the repeat unit coupled to the one-dimensional set of translations parallel to the chain. Furthermore, many of the internal motions of the repeat unit are independent of the chain and consequently can be treated as isolated molecular motions. Therefore, the normal modes of the polymer chain can be handled as if the repeat unit is an isolated molecule for many modes, with the remainder of the modes (primarily motions of the chain backbone) approximating to those of a simple line group. Detailed calculations of the phonon structure of polymers must be more precise, but the symmetries of observed vibrational bands can be usefully analyzed with the foregoing simplification.

One application of this approach is to record spectra for sets of oligomers and compare the bands observed as a function of the number of repeat units in the oligomer with calculated dispersion curves for the infinite polymer. This comparison ignores end groups effects in the oligomers but often agrees qualitatively with the shape of the dispersion curve. In effect, it allows one to use vibrational spectroscopy, which is limited to measuring frequencies near zero dispersion, to probe a larger dispersion range. Essentially, each oligomer defines a phase shift on the infinite polymer dispersion curve. A good example of this may be found in Section 21.3.2.1(i) on the longitudinal acoustic mode.

21.2.5 Intensities

In general, intensities have been difficult to calculate because a detailed knowledge of the elements of the polarizability tensor of the molecule as a function of the nuclear coordinates is needed. This would require detailed electronic structure calculations which unfortunately, for most molecules, are still limited by computational restraints. Alternatively, approximate methods are being developed.[26, 27] The electro-optical parametization approach also shows great promise.[27, 28] A general review can be found in ref. 29.

21.3 APPLICATIONS OF RAMAN SCATTERING TO POLYMERS

21.3.1 Chemical Structure and Interactions

Chemical structure refers to local bonding and functional groups. Frequently the Raman spectrum contains bands that can be associated with bonding and functional groups as if these were independent of the overall structure. Although this is not precisely true, since the normal modes of a molecular system include the complete set of atoms in their description, it is often the case that the bulk of the normal mode displacements occur in a limited region of the system because the energy of the given vibration is decoupled from its surroundings by an impedance mismatch. This is equivalent to the behavior of a harmonic oscillator coupled to a system by a very weak spring constant relative to its own spring constant. A typical example is the C—H stretch. Nearly all the amplitude of the normal mode of a C—H stretch is in the motion of the H atom relative to the C atom. Likewise, carbon–carbon double and triple bonds, methylene groups, carboxylate groups, ether linkages, and most other bond arrangements and functional groups have some distinguishing spectral signature.

As a molecular characterization tool, Raman spectroscopy is used in the identification of bonding and functional groups. This is certainly important for new polymers; when combined with other characterization tools, a complete picture of the polymer repeat unit is revealed. Examples are plentiful, such as for polyurethane,[30a] poly(*trans*-alkenylene),[30b] polyacrylamide,[31, 32] poly(oxymethylene),[33] various combinations of tacticity and deuteration of poly(methyl methacrylates),[34] poly(α-methylstyrene),[35] perdeutero isotactic polystyrene,[36] copolymers of isotactic and perdeutero isotactic polystyrene,[37] organosilicon polymer,[38] acrylic polymers,[39] polycarbonate,[40] polysilanes[41, 42] and graphite films.[43] Chemical modification of polyethylene by chlorination has been followed as a function of chlorine content.[44] Chemical change involving the formation of new bonds, such as during a polymerization, can be followed and evaluated by Raman spectroscopy.[43, 45] Figure 4 shows the transformation of poly(*p*-phenylene vinylene) to a graphite structure during pyrolysis.[43] Chemical modification, such as that occurring during ion implantation in polymers, has also been monitored by Raman spectroscopy.[46] Likewise, the modification of polyethylene by methyl and ethyl branches has been studied.[47] The verification of force constants by comparison to vibrational data is another use of the local chemical structure.[48] It is also possible to use Raman spectroscopy as a 'fingerprint' technique for the identification of different materials,[49, 50] without detailing the molecular structures that contribute to the spectra.

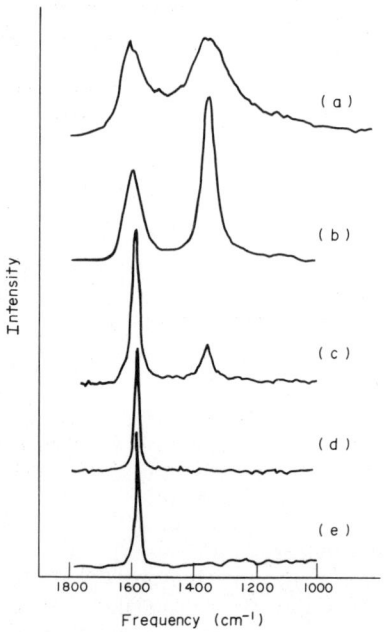

Figure 4 Raman spectra of oriented poly(p-phenylene vinylene) films pyrolyzed at (a) 950 °C, (b) 2000 °C, (c) 2500 °C, (d) 2750 °C, and (e) 3000 °C. As the temperature increases, the film transforms completely to graphitic carbon, as shown by the loss of the 1360 cm^{-1} band and narrowing of the 1580 cm^{-1} band associated with carbon–carbon double bonds (Figure 6 in ref. 43)

Interactions between molecular units and their surroundings can be readily detected by perturbations in the Raman spectrum. Hydrogen bonding, crystal field splitting, chain conformation, chain packing and solvation are all examples of modifications to the surrounding electrical field of the molecular unit that will affect the Raman scattering. Some of these effects modify the electric field of the molecular unit directly whereas others induce changes through external electric field coupling. Hydrogen bonding and changes in the chain conformation directly modify the electronic structure of the molecular unit. Crystal field splitting, chain packing and solvation interact with the molecular unit by coupling external electric fields with the electronic structure of the molecular unit. Both types of perturbations modify the spring constants and effective masses of the molecular unit, leading to shifts in the normal mode frequencies and line shapes.

Examples of chain conformation and packing changes are seen in the melting of poly-(1-methyladenylic acid)[51] and the solvent effects on the sol–gel polymerization used to form silica sols.[52, 53] Similarly, polyacrylamide gels have been studied with varying crosslink concentrations.[54] Likewise, the chain environment of cellulose triacetate[55] and poly(ethylene oxide)[56] have been studied and show differences in their Raman spectra depending on the molecular conformations and packings present. Figure 5 shows the effects of conformational and packing changes in poly(ethylene oxide) in solutions and melt states.[56] Effects of ion clustering or pairing on the Raman spectrum are demonstrated for styrene/sodium p-styrene carboxylate copolymers.[57] Similarly, polymer–salt complexes perturb the chain conformation as shown for poly(ethylene oxide) complexes of alkali metal salts,[58] polyzwitterions[59] and poly(propylene oxide) with sodium thiocyanate.[60]

A variety of external effects can be used to change the environment surrounding the polymer and consequently affect the resulting Raman spectra. A static electric field applied to polystyrene modified the intensities of the modes.[61, 62] Strained chemical bonds due to mechanical stress or surface structures can lead to changes in the Raman frequencies.[63] Detailed studies of the disordering of polyethylene chains indicate that conformational disorder can be directly detected from the Raman spectrum.[64–68] Figure 6 shows the effect of temperature on melt interactions in

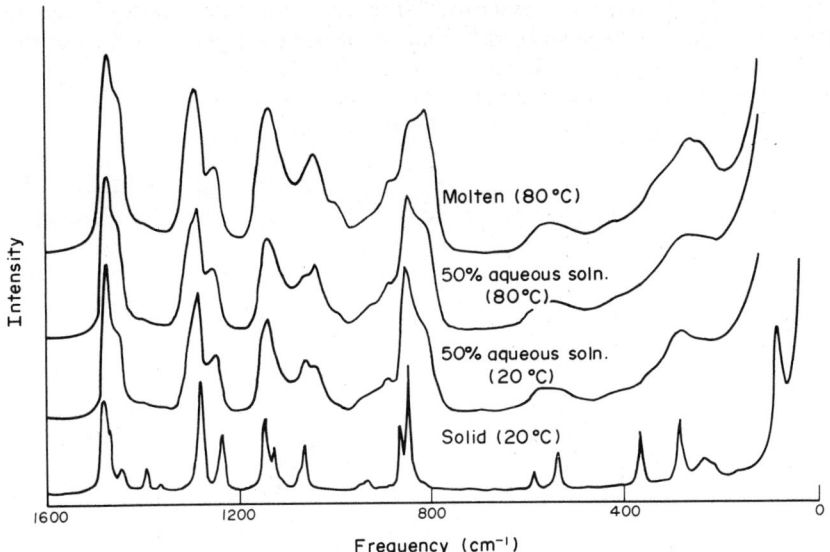

Figure 5 Variations in Raman spectra of poly(ethylene oxide) with average molecular weight 6000. This illustrates the variations in the Raman spectrum as a function of the molecular environment of the polymer chains for molten, solution and solid samples. (a) 80 °C, molten; (b) 80 °C, 50% aqueous solution; (c) 20 °C, 50% aqueous solution; (d) 20 °C, solid (Figure 4 in ref. 56)

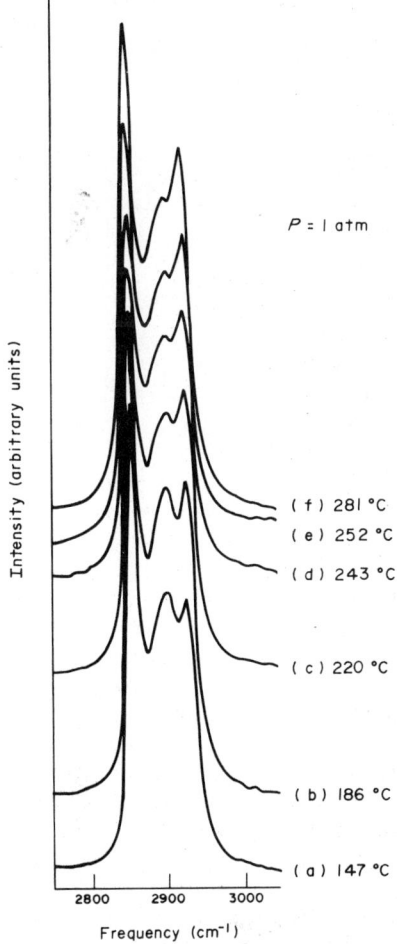

Figure 6 Raman spectra at 1 atm as a function of temperature of molten polyethylene in the C—H stretch region (Figure 17 in ref. 64)

polyethylene in the C—H stretch region of the Raman spectrum.[64] Using a novel instrumental technique it has been possible to derive reorientational correlation times for the phenyl group in polystyrene from high resolution Raman line shapes.[69] Finally, low frequency Raman modes in amorphous poly(vinyl acetate) have been interpreted as phononlike excitations.[70]

Pressure and temperature changes modify molecular interactions. This is very useful because it is then possible to distinguish vibrational structure due to interactions from intrinsic molecular modes by experimentally controlling the temperature or pressure. Phase transitions can be induced, crystal field effects modified and so forth.[71] One example is the use of temperature control to clarify the spectroscopy of polyethylene[72-74] through enhancement of the frequency separation due to crystal field splitting. An example of crystal field splitting in polyethylene is given in Figure 7, which compares room temperature results with those at 10 K.[74] Pressure effects have been used to study changes in the molecular conformation of isotactic polypropylene.[75,76] The Raman spectra of melts of high molecular weight polyethylene indicate the retention of ordered structure in the melt.[77] Both pressure and temperature were used to study the melting of polyethylene and the resulting chain conformations.[66] Similarly, pressure can be used to modify chemical as well as packing interactions, and Raman spectroscopy is well suited to monitoring the formation of new reaction products.[78]

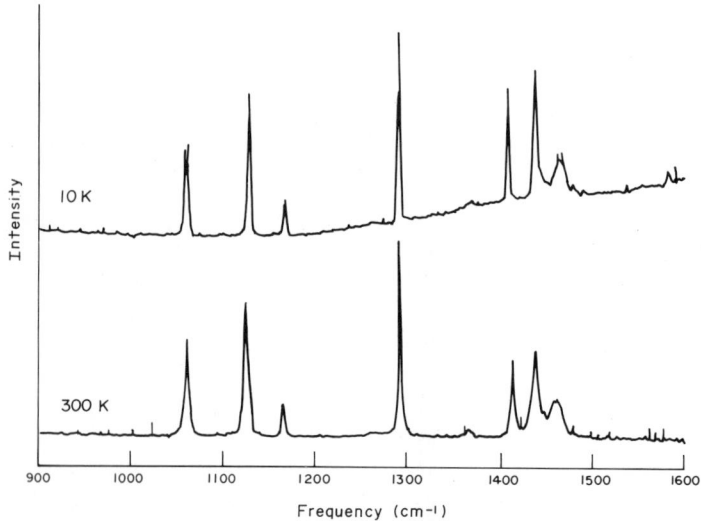

Figure 7 Raman spectra of chain-extended polyethylene recorded at 300 K and 10 K. Splittings of the 1060 and 1295 cm^{-1} bands at low temperature are due to having two chains per unit cell. The appearance of the splittings at low temperature is due, at least in part, to improved resolution from a reduction in thermal motion (Figure 5 in ref. 74)

Another class of Raman studies important to polymer chemistry is that of model compounds. By isolating the repeat unit of the polymer of interest, or of a related structure, it is possible to obtain detailed understanding of the smaller molecular unit that transfers to studies on the related macro-molecule. This is important for several reasons. It is frequently easier to synthesize small molecule variations and isotopically labeled compounds. Normal mode calculations simplify for isolated small molecules. Examples are model compounds of polyethylene,[79,80] polyethers,[81,82] poly(tetra-methylene terephthalate),[83] of disklike liquid crystalline molecules[84] and organosilicon compounds.[38] Figure 8 shows the CH$_2$-bending region of the Raman spectrum for a few poly-ethylene oligomers and distinguishes purely crystalline structure from mixed amorphous/crystalline structure due to the onset of chain folding.[79] Experimentally, small molecules are frequently more tractable to handle than their polymer analogs in terms of dissolution, purification and general sample preparation. Examples of this include the class of polymers known as rigid-rod polymers[85] and many conjugated conducting polymers.

21.3.2 Morphology

In many senses it is in this area that Raman spectroscopy is most useful to the polymer scientist. The Raman effect is sensitive to the molecular environment in many ways. Examples of environ-

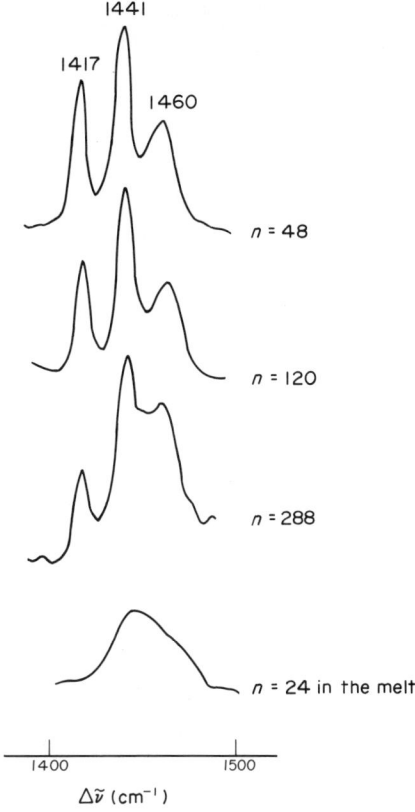

1441

1417

1460

n = 48

n = 120

n = 288

n = 24 in the melt

1400   1500

$\Delta \tilde{\nu}$ (cm^{-1})

Figure 8 CH$_2$-bending region of the Raman spectra of linear alkanes with chain lengths of 24, 48, 120 and 288 carbon atoms. The 24 carbon chain is in the liquid phase whereas the longer chains are recorded in crystalline phases. The 48 and 120 carbon chain samples are crystallized in orthorhombic structures whereas the 288 carbon spectrum indicates a mixture of orthorhombic crystalline structure with amorphous material. This is interpreted as the onset of chain folding giving rise to amorphous material at the fold surface (Figure 6 in ref. 79)

mental effects that perturb the Raman scattering and that can be used to study polymer morphology include crystal packing, orientation effects, processing history, doping and surface interactions. Since the bulk properties of a polymer can be drastically modified by variations in morphology, knowledge of and control over the morphology are technologically important.

Raman spectroscopy has some distinct advantages over other techniques, especially in comparison to infrared methods when morphology is the subject of interest. Raman spectra are frequently obtained with no modification to the sample, which is critical to the preservation of the morphology. Unusual shapes and sizes pose little or no problem since only a surface is needed for sampling. IR measurements frequently require the polymer to be dispersed in salt pellets, and the morphology can be seriously modified in the preparation process. Since Raman excitation is usually from visible lasers, samples can be physically very small (on the order of one micron in diameter) which allows spatial vibrations in morphology to be measured in some cases.

21.3.2.1 *Crystalline structure and morphology*

Crystalline polymers have a number of morphological features that can be studied by Raman spectroscopy.[86] The thickness of the lamellar structure (actually the length of the chain stem that can be at an angle to the surface normal) can be determined by measuring the longitudinal acoustic mode (LAM), which is the accordionlike stretching motion of the chain stem. Likewise, an estimate of the bulk modulus can be obtained from LAM measurements. Oriented polymer samples give Raman spectra that can be analyzed to determine structural details.[87] Examples of such local structure determinations include the separation of planar zigzag from helical chain conformations,[88-90] molecular packings of the chains[91] and the number of chains per unit cell.[92] Figure 9 shows spectral changes as a function of orientation; the observed symmetries and mode intensities

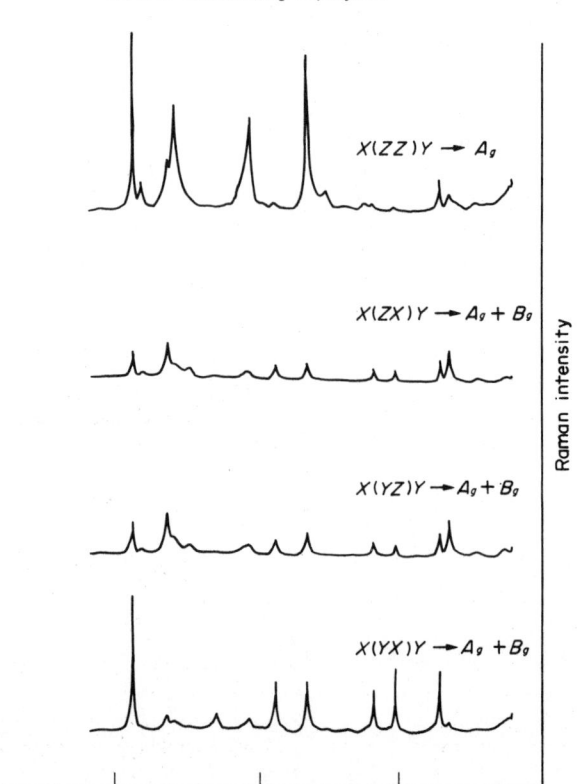

Figure 9 Polarized Raman spectra of ethylene–tetrafluoroethylene (E–TFE) alternating copolymer (Tefzel) filament in a 90° scattering geometry. Symmetry types contributing to each experiment are indicated based on a planar zigzag polymer backbone and give the most consistent interpretation of the data (Figure 3 in ref. 90)

allow the determination that the sample of ethylene–tetrafluoroethylene copolymer has planar zigzag chains rather than helical chains.

Other local structural effects determined include the amounts of remaining amorphous content,[93a, b] amount and type of order induced by processing, defects in structure[94] and in some cases tacticity.[95] Figure 10 shows the effect of tacticity in poly(methyl methacrylate) thin films on Raman spectra. Many phase transitions can be monitored by following the changes in the Raman spectrum associated with the crystalline structure.[91] A review of the structure and properties of crystalline polymers that contains several examples of the use of Raman spectroscopy to clarify polymer structure is given in ref. 96.

Examples of the analysis of chain structure and conformation include *trans*-1,4-polychloroprene,[97] *trans*-1,4-poly(2,3-dichlorobutadiene),[97] *trans*-1,4-poly(2,3-dimethylbutadiene),[98] *trans*-1,4-polyisoprene,[98] isotactic poly(1-butene),[87] isotactic poly(*t*-butyl acrylate),[88] poly(tetrafluoroethylene),[92] poly(1,1,2,2-tetrachlorobutane),[89] vinylidene fluoride–trifluoroethylene random copolymers,[91,99,100] poly(vinylidene fluoride),[101] poly(di-*n*-alkylsilane)s,[102] isotactic and atactic polypropylene,[103] poly(tetramethylene terephthalate),[83] poly(vinylidene fluoride),[104] various polymer blends,[105] polynucleotides,[106-108] poly(ethylene oxide)[109,110] and others. Strain defects and holes formed during crystalline growth in isotactic polypropylene have been detected using Raman spectroscopy.[94]

(i) Longitudinal acoustic mode

The longitudinal acoustic mode (LAM) has become an accepted tool for polymer characterization since many of its features are well understood.[111-119] A detailed review of the Raman active longitudinal acoustic mode can be found in ref. 96. LAM serves as a direct probe of the crystalline lamellar structure, since the morphology, chain defects and fold surface all interact with the

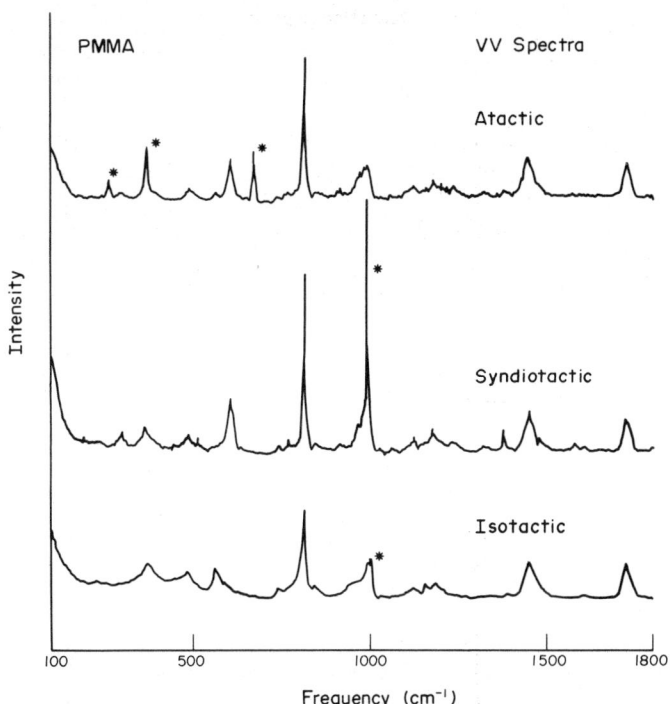

Figure 10 VV polarized Raman spectra of various tacticities of poly(methyl methacrylate). Bands with asterisks indicate residual solvent from the spinning process used to form the films. Sampling was done by waveguide Raman scattering

mode.[120-129] It is insensitive to the amount of amorphous material present[130] and the bandwidth gives an estimate of the distribution of lamellar thickness variations.[131,132]

The LAM mode is the 'breathing' mode of a polymer chain section. It is really a set of Raman active modes that have wavelengths equal to the chain section divided by the odd integers 1, 3, 5, 7, ... and the chain displacements parallel to the chain axis. Figure 11 plots the displacement for LAM modes for $m = 1$–6.[96] Whether observing an oligomer or a stem in the crystalline lamellae, the vibration behaves essentially the same; *i.e.* the frequency varies inversely with the chain length. This indicates that the chains interact only weakly with adjacent chains, the fold surface and the next lamella, leading naturally to the model of the stem as an elastic rod.

The elastic rod model[133] makes use of continuum mechanics to model a set of discrete masses and forces. The uniform elastic rod of length L, density ρ and Young's modulus E has a frequency that can be expressed by

$$v(\text{cm}^{-1}) \quad = \quad \frac{m}{2cL}[E/\rho]^{\frac{1}{2}} \tag{36}$$

where c is the speed of light and $m = 1, 2, \ldots$ For a planar zigzag hydrocarbon system this can be tested by expanding in terms of the number of carbons n in the chain[6] as

$$v(\text{cm}^{-1}) \quad = \quad A(m/n) \quad + \quad B(m/n)^2 \quad + \quad \ldots \tag{37}$$

and fitting to the experimental data. The A term strongly dominates and is approximately 2495 cm^{-1}. Figure 12, which compares observed LAM frequencies of oligomers with the elastic rod model, is similar in shape to the dispersion curve of the acoustic mode of the inifinite chain.[96] The elastic rod model also works with helical polymers; however, the compression of the helical chain depends on the torsional force constant and is correspondingly weaker. This shifts the LAM frequency toward the exciting laser line, often making it difficult to observe. Other continuum models have been proposed to account for small shifts, but the elastic rod model describes all the main features of the LAM mode satisfactorily. There are many model variations.[134] More recent studies include linear aliphatic polyesters,[135,136] hydrogenated polybutadienes,[117,137] isotactic polypropylene,[138] an alternating copolymer of ethylene and chlorotrifluoroethylene[139] and poly-(ethylene oxide),[109,110] as well as polyethylene.[111,117,122]

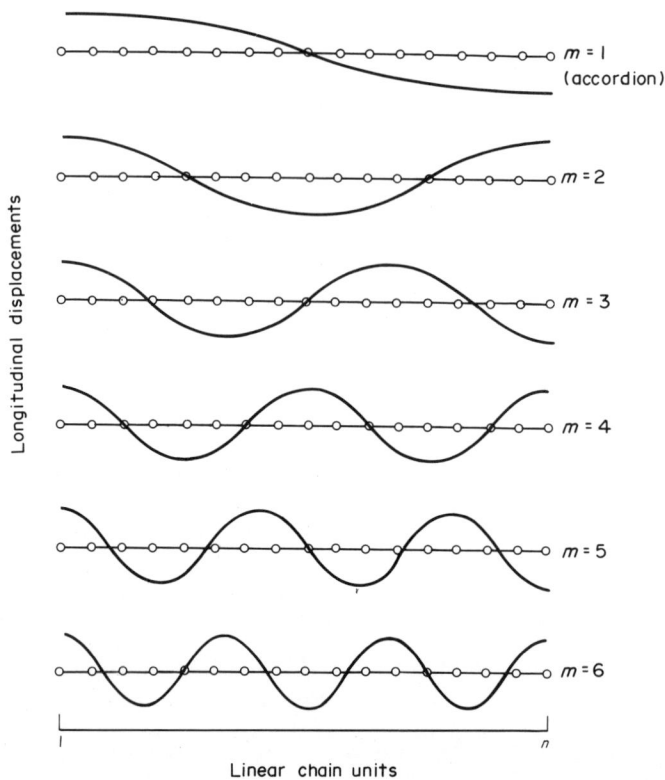

Figure 11   LAM modes of a linear chain drawn to indicate the carbon atom displacements perpendicular to the atom
location in the chain. Actual displacements occur parallel to the linear chain (Figure 4 in ref. 96)

(ii) Local structure

In the crystalline state, local structure refers to the structure of the unit cell and the polymer chains
contained within the cell. When done with polarized excitation and detection, Raman spectroscopy
provides detailed structural relations, or symmetries, that can be determined by matching predicted
spectra for a given structure with experiment. When well-oriented samples are available, it is often
possible to determine the chain conformation (planar zigzag *vs.* helical for example), the number of
chains in the unit cell from the total number of modes and any crystal field splittings, the structure of
the repeat unit (point group of the repeat unit) or unit cell, and packing arrangement of the chains.

Another aspect of local structure in the crystalline state is the incorporation of defects in the
lattice. Incorporated internally, the defects can modify the symmetry of the crystal. At the surface of
single crystals, defects correspond to the fold surface of a single crystal.[140]

21.3.2.2 *Amorphous structure*

A feature analogous to LAM in the crystalline regions has been identified in the amorphous
regions of polymers. It has been named the disordered longitudinal acoustic mode (DLAM) and
seems to obtain its Raman intensity from a statistical distribution of all-*trans* chain segments in the
amorphous region.[141] Earlier studies on liquid alkanes were interpreted relative to liquid crystalline
phases masking the experimental evidence for an amorphous, long range order.[142,143] Figure 13
shows the Raman spectra of liquid alkanes and molten polyethylene; DLAM is the most intense
spectral feature in the low frequency region.[143] Unlike the LAM frequency, which decreases
proportionally to the inverse of the chain length, the DLAM frequency decreases as the inverse of the
square of the chain length plus a constant. If the amorphous regions are the mechanically weak link
between crystalline lamellae, it might be the case that DLAM can be related to actual polymer
strength, not just the maximum bulk modulus as with LAM. This would be of technological
significance if true. Although primarily identified in polyethylene, a variety of other polymers[144,145]
exhibit similar DLAM bands. Raman studies of amorphous contributions to other modes have

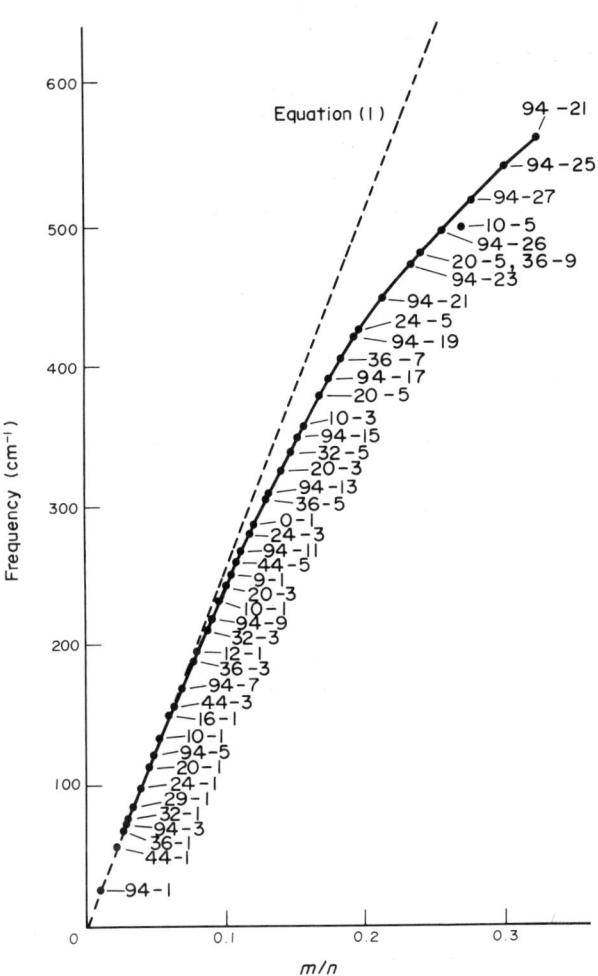

Figure 12 Plot of LAM frequencies as a function of mode/chain length (m/n). Assignments are given as $n - m$. As chain length decreases, the LAM $-$ 1 frequencies show a slight deviation from the elastic rod model indicated by the dashed line. However, higher order modes deviate drastically from the elastic rod model (Figure 5 in ref. 96)

demonstrated that the introduction of *gauche* bonds into the planar zigzag chain of polyethylene are detectable.[146] Similarly, chain distortions in polyethylene due to density fluctuations have been determined from Raman spectra.[147]

21.3.2.3 *Processing effects and orientation*

A typical goal of polymer processing is strength enhancement. One means by which this occurs is through molecular orientation of the chains that can be induced by various techniques of flowing or elongating the polymer. Raman spectroscopy is an excellent tool for studying such orientation effects. Since the Raman scattering tensor can have up to six independent components, corresponding to the linear and cross terms of the induced dipoles that can be accessed, detailed information about orientation in the polymer can be determined. IR, in contrast, can only be used to study the orientation of the permanent dipoles, a linear effect. Thus Raman can probe three dimensional symmetries whereas IR is limited to planar symmetries.

Uniaxial orientation is a common processing effect. Morphology is modified by stretching a filament or extruding the polymer under pressure. The combination of orientation of the polymer and modification of its morphology leads to readily identifiable changes in the polarized Raman spectra. Examples of such processing effects have been studied in polyethylene,[27,148-150] polypropylene,[27] isotactic polypropylene,[151-153] ethylene–tetrafluoroethylene copolymer,[90] poly-(ethylene terephthalate),[154-156] poly(p-phenylene terephthalamide),[157] iodinated nylon-6,[158] poly(vinylidene fluoride),[101] poly(p-phenylene benzobisthiazole),[159] celluloses[160] and others. Figure

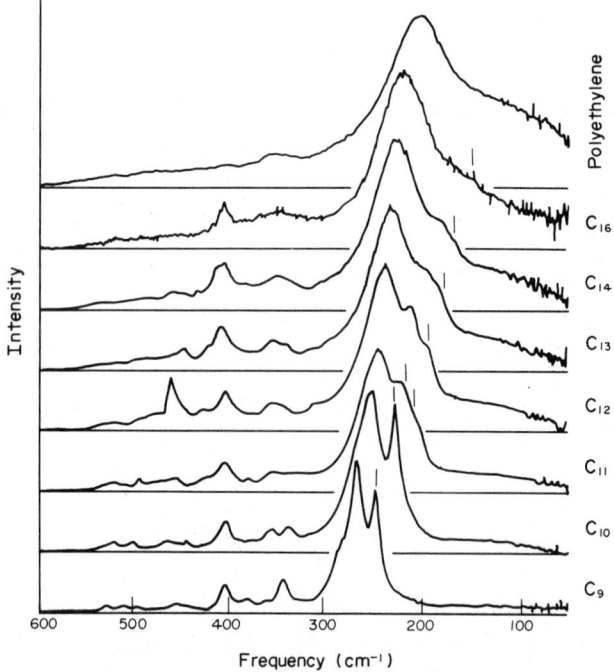

Figure 13 Isotropic Raman spectra of liquid *n*-alkanes (*n* = 9–14, 16) and molten polyethylene. The intense DLAM band shifts with chain length and bandwidths also increase. Intensity is believed to derive from LAM-like motions in the liquid structure (Figure 2 in ref. 143)

14 shows the oriented spectra of poly(vinylidene fluoride), form I, which are indicative of a planar zigzag chain conformation and allow improvement in previous band assignments.[101] A slightly different example of uniaxial orientation is that found in flowing poly(ethylene terephthalate) melts.[161] Other forms of ordering are also possible. Film-processing techniques frequently produce biaxial ordering. Application of the Langmuir–Blodgett film-deposition technique to polymers can produce highly ordered films.[162]

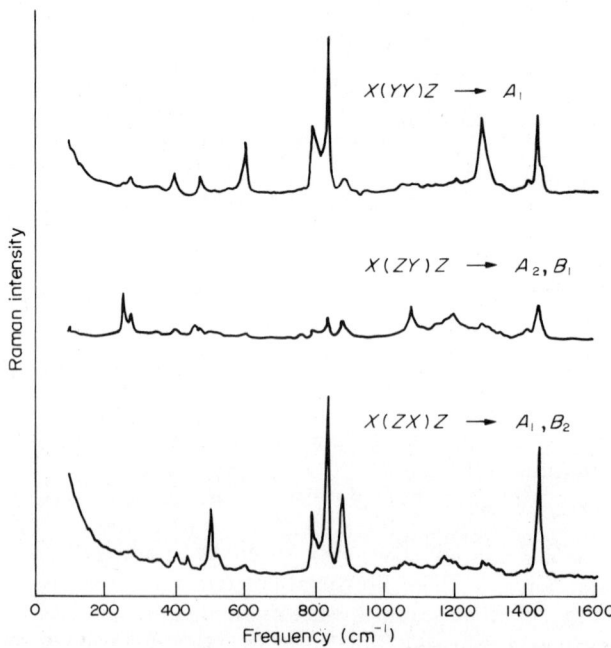

Figure 14 Polarized Raman spectra of a uniaxially oriented poly(vinylidene fluoride) filament. The filament axis is oriented perpendicular to the scattering plane. The spectra have been normalized to the low frequency nitrogen lines. The symmetry types present in each geometry are indicated and correspond to a *trans* planar carbon backbone (Figure 2 in ref. 101)

Raman spectroscopy is also useful in detecting other important orientation effects involving electrical properties of polymers including piezoelectric, pyroelectric, ferroelectric and paraelectric behavior that can be modified by poling techniques.[104] Examples of these phenomena are given in ref. 146. An example of a ferroelectric–paraelectric phase transition is found in vinylidene fluoride–trifluoroethylene random copolymers.[91,99,146]

21.3.3 Conjugated Polymers

Conjugated polymers, which have continuous delocalization along the chain axis, have many unusual properties. Because of their potential technological uses, these properties have attracted a great deal of interest. Properties such as electrical conduction that approaches that of metals and nonlinear optical responses exceeding most inorganics are the main focuses of research.

Conducting polymers form a class of conjugated polymers. The archtypical conducting polymer is of course polyacetylene, $(CH)_x$, with conductivities ranging over 16 orders of magnitude on doping. Such a wide performance range has stimulated interest in lightweight batteries, solar cells and semiconductor applications. Another factor motivating the activity in conducting polymer research has been the theoretical possibility of a novel conduction mechanism.

Because conducting polymers have extended electronic conjugation, which gives rise to visible absorption, resonance Raman spectroscopy (RRS) has been used as a probe of the electronic structure. In the simple particle-in-a-box picture of the electronic structure, the excitation of the laser is tuned to the electron resonance of the conjugated section and motions of the 'box' are enhanced.[163,164] Figure 15 shows the effect of resonance on the Raman spectrum for the $C=C$ stretching mode of polyacetylene with comparison to theory.[163] A review of the RRS of conjugated polymers can be found in ref. 165. Information about the conjugation lengths and the coupling of the backbone to the environment can then be monitored.[164,166–184] Another example of this is seen in the degradation of polyacetylene as a function of electrochemical cycling.[185] Poly(diacetylene)s have been monitored by RRS to determine the effect of severe mechanical deformation on the conjugated backbone[186] and the effect of induced stress at an interface.[187] General studies of poly(diacetylene) crystals, both doped and undoped, have been made.[176a,b] Poly(diacetylene) multilayers have been studied by RRS, and a structural change was detected as the number of layers was increased.[188] Polarized RRS has been used to study the chain orientations in a conducting organic hybrid material.[189] Direct correlation between the conjugation lengths determined by RRS and the conductivity of polythiophenes has also been observed.[166,190]

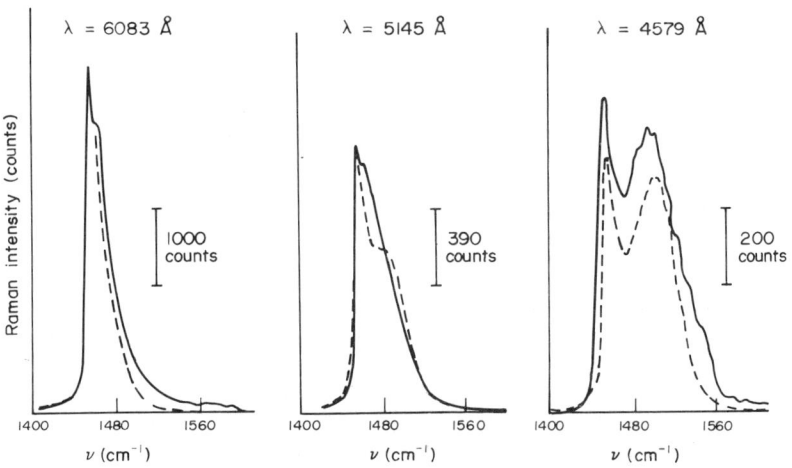

Figure 15 Line shapes of Raman bands for the carbon–carbon double bond stretch mode as measured (———) and as calculated from the interrupted conjugation model (– – –). (a) $\lambda = 608.3$ nm; (b) $\lambda = 514.5$ nm; (c) $\lambda = 457.9$ nm (Figure 1 in ref. 163)

General characterization of conjugated polymers by Raman spectroscopy is also useful. However, because of sample absorption and degradation in many conjugated polymers, this has been applied only to a limited extent. One example is the monitoring of the structural regularity in poly-(perinaphthalene)[191] by Raman band intensities. Following the sample degradation due to photo-oxidation is another example.[192] Interactions of dopants with polyacetylenes and poly(diacetylene)s can be monitored with Raman spectroscopy.[183,193,194] Similarly, Raman spectra of poly(aniline) doped with iodine show structural changes as a function of doping concentration.[195] Structural studies by Raman spectroscopy have been done on poly(p-phenylene) and cis-polyacetylene.[196] Polypyrroles[197] and poly(p-phenylene),[196,198] also conductive polymers, are being studied to determine how their conjugated structures contribute to their conductivity.

The conjugation of the polymer chain backbone is also a major factor in creating novel optical properties. Nonlinear optical response suggests potential uses of these materials in optoelectronic and all-optical devices.[199] Again, RRS can probe the structure of the conjugated chain and provide insight into the optical behavior of these polymers.[199,200] Coherent Raman techniques have also been applied to poly(diacetylene) films to probe their time response and structural modifications due to processing.[201] Resonance Raman studies on poly(diacetylene)-based systems[167,177] show sensitivity to the side groups modifying the electronic structure of the chain.

21.3.4 Polymers on Surfaces

Surface coatings of polymers, ranging from a few angstroms to several microns in thickness, are of interest both scientifically and technologically. Molecular interactions at interfaces, surface wetting, surface-induced orientation and surface chemistry are open areas for research and are critical industrially in adhesives, coatings, surface modification and catalysis. Raman techniques are useful in the study of thin films because of the ability of the laser to probe the surface nondestructively. With care the Raman effect can be measured without disturbing the film or interface. As an example, Raman spectroscopy has been used to study modifications of polyethylene due to surface friction.[134] An overview of polymer surface characterization can be found in ref. 202 which serves as an introduction to a variety of surface spectroscopies applied to polymers, including Raman.

Several approaches have been used to examine small samples. One method is to prepare colloidal materials to increase the amount of surface area, and, for metal colloids, to make use of the surface-enhanced Raman (SERS) effect.[188,203–207] Figure 16 shows SERS spectra for several polymers absorbed on silver colloids.[204] Alternatively, the SERS effect can be used by coating metal diffraction-grating structures with thin films.[208] Similarly, the use of the SERS effect on metal electrodes has allowed surface-adsorbed and surface-reacted polymers to be studied.[209–212] For two-dimensional film structures it has been possible to detect monolayers using RRS and/or waveguide Raman sampling techniques. Array detectors have made it possible to detect monolayers on many other surface configurations and the use of such systems will also be discussed below.

21.4 NEW METHODS AND DIRECTIONS

21.4.1 Fourier Transform Raman Spectroscopy

The Raman spectroscopy of polymeric materials is plagued by the problem of fluorescent impurities, which can be extremely difficult to remove chemically. Fourier Transform Raman (FT-Raman) is one approach to avoiding the fluorescent background that is many orders of magnitude stronger than the Raman signal; the fluorescence is avoided by never being excited. In order to use this approach there are several important experimental considerations. First, a high power continuous wave (CW) near-IR laser is needed to excite the sample. The near-IR radiation is energetically unable to excite fluorescence in most organic materials. Because of the fourth order dependence on frequency of the Raman intensity, a high CW power level is needed to generate a significant quantity of Raman-shifted photons. In order to collect enough of the signal, an interferometer is used to gain the multiplex and throughput advantages. Although it was demonstrated first in 1964[213] it was only marginally feasible at that time. Recent developments in CW Nd:YAG lasers, high rejection laser filters and high sensitivity near-IR detectors have made FT-Raman a viable method.[214–220] Figure 17 shows a comparison of the FT-Raman to conventional scanning Raman. The random copolymer of vinylidene fluoride and trifluoroethylene contained fluorescent impurities that made the conventional spectrum difficult to record compared to the FT-Raman spectrum of the same sample.[220]

In addition to its other advantages, FT-Raman takes advantage of high resolution and spectral accuracy capabilities of the interferometer. Also, most research-grade FTIR interferometers can be adapted for FT-Raman, and the extensive software base used for FTIR can be used for FT-Raman. Potentially it is possible to consider having one instrument for both IR and Raman spectroscopy.

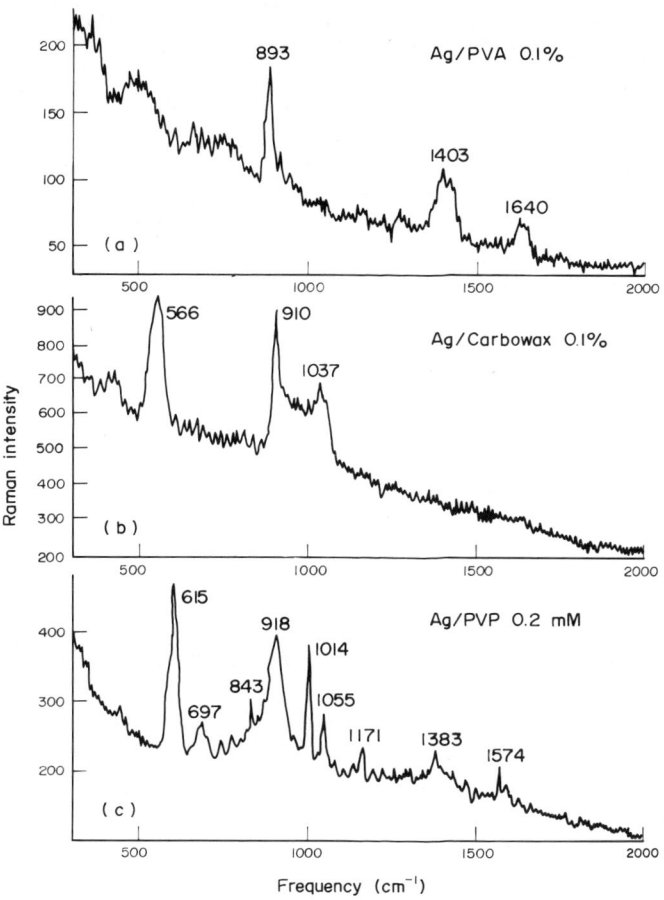

Figure 16 SERS spectra of silver sols stabilized by (top to bottom): (a) poly(vinyl alcohol) (Ag/PVA 0.1%), (b) Carbowax (Ag/Carbowax 0.1%), and (c) poly(vinyl pyridine) (Ag/PVP 0.2 mM). Laser excitation at 647.1 nm (Figure 2 in ref. 204)

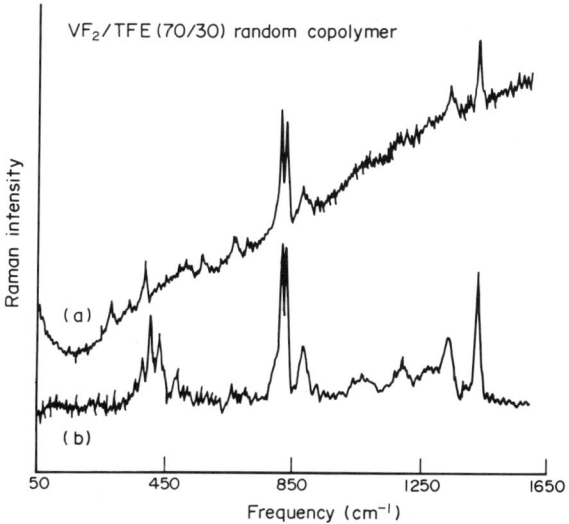

Figure 17 Raman spectra of 70/30 random copolymer of vinylidene fluoride and tetrafluoroethylene. Spectrum (a) is obtained by conventional Raman techniques using 488 nm excitation and photobleaching for 4 h. Spectrum (b) was obtained with the same collection time, but using FT-Raman. Resolution was 4 cm^{-1} with laser powers of 180 mW at 488 nm and 2 W at 1064 nm. Bands at 500 cm^{-1} in (b) are emission lines from the CW Nd:YAG laser. (Figure 6 in ref. 220)

However, FT-Raman has its limitations. Power levels of the laser are high, sensitivity is well behind that achievable with array-based systems, and initial cost is high. An alternative approach which is less sensitive but also eliminates fluorescence has been demonstrated, in which new near-IR detectors are combined with a conventional scanning monochromator and a pulsed Nd:YAG laser to obtain Raman spectra.[221] If near-IR arrays are developed, a great deal of the motivation to do FT-Raman will be lost. In the meantime, if samples are fluorescent, FT-Raman may be the answer.

21.4.2 Array Detectors

Array detectors have greatly enhanced sensitivity. Essentially, the array detector provides the Raman experiment a multiplex advantage and simplifies the mechanical aspects of taking a spectrum. In use, an array detector, which is a position sensitive detector, is coupled to a dispersive unit, that is frequently a triple monochromator, in which the first two stages are run in a subtractive mode and act as a band pass filter. The third stage of the system is the dispersive element that delivers the spectrum to the array. The array detector serves as a digital camera with sensitivities that can far surpass the best film available.

Because of the multiplexing, time is saved when using an array system in comparison to a scanning system. A typical example might be a spectrum from 800 to $1800\,\mathrm{cm}^{-1}$. For a scanning double monochromator this can take more than $1000\,\mathrm{s}$ for steps for $1\,\mathrm{cm}^{-1}\,\mathrm{s}^{-1}$. The array detector obtains the same spectrum in $1\,\mathrm{s}$. Clearly, resolution and the size of the spectral region of interest can modify the amount of time needed by the array to cover the spectral region of interest, but usually this is not a problem. At the other extreme, if sensitivity is needed, the scanning system cannot effectively compete with an array of comparable sensitivity per pixel. In the previous example, if the integration time needed to get a spectrum is $100\,\mathrm{s}$, then the scanning system needs $100\,000\,\mathrm{s}$ to get the same spectrum as the array in $100\,\mathrm{s}$. This is of the order of 30 hours, which is a difficult experiment on a scanning system because of stability problems with the system and sample.

The most commonly available array is the optical multichannel array (OMA) or intensified OMA (I-OMA).[222, 223] This is a silicon diode array with (I-OMA) or without (OMA) a microchannel plate (MCP) intensifier. Several other devices have recently become available to the spectroscopist. The charge-coupled device (CCD) camera is a two-dimensional array system that has improved sensitivity over the I-OMA and is nearly noise free.[224] The other new technology is the intensified resistive anode (IRA) detector that acts as a position sensitive photomultiplier system running in a photon-counting mode.[225, 226] In general, the CCD is apparently the most sensitive detector and the I-OMA the least sensitive. However, operating conditions have not been carefully optimized in all cases and in many cases there may be performance overlap, depending on the sampling method. Considerations of gating speed, readout speed, dynamic range and ease of operation are factors that make careful selection of the array technologies to match the experiment important.

All the array systems described are capable of detecting Raman spectra from a monolayer of organic material under controlled conditions. In the polymer field it should be possible to apply these techniques to the study of surface-adsorbed polymers, interface interactions and kinetics.

21.4.3 Waveguide Raman Sampling

Waveguide Raman sampling (WRS) is a sampling technique for thin film structures.[227, 228] Typically, a guiding film is formed on a substrate in an asymmetric slab waveguide geometry. The film is of higher index material than the layers above and below it and is of good optical quality. By pressing a higher index prism on the surface, it is possible to couple an input laser beam into the guiding layer. A schematic of this is shown in Figure 18.[229] Coupling depends on the thickness of the guiding layer, the relative refractive indices of the layers, the wavelength of the laser, the prism geometry and index, and the angle made by the incoming laser beam relative to a normal to the asymmetric slab waveguide plane. Usually for a one micron thick guiding layer with a visible laser there will be one or two coupling angles that satisfy all the boundary conditions and allow the light to propagate through the guide. The propagating beam can be symmetrically out-coupled to analyze the mode structure. The Raman scattering is collected at right angles to the guide plane.

Since the experimentalist has control over so many parameters in the formation of the waveguide, there are many different geometries that can be used. This fabrication flexibility allows nearly any combination of thin film structures to be studied by WRS. By injecting the laser directly into optically transparent polymer films, Raman spectra of approximately one micron thick films can be

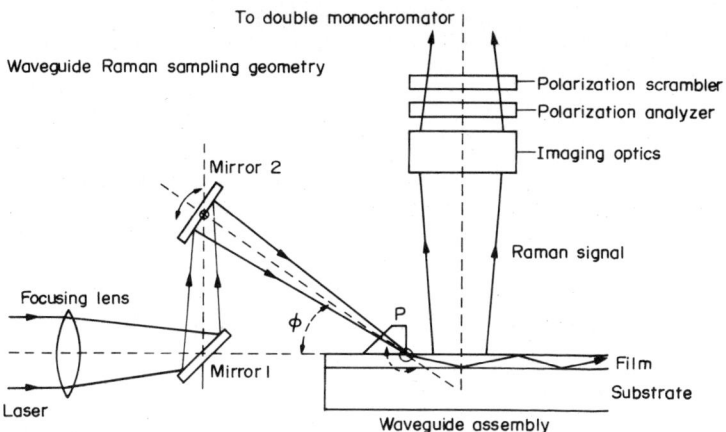

Figure 18 Geometry for waveguide Raman sampling. Coupling angle φ is indicated for prism coupling. The laser is injected into the thin film at the prism corner and propagates perpendicularly to the collection optics when the boundary conditions are met (Figure 1 in ref. 229)

easily obtained. By forming the waveguide from glass and coating organic or polymer films on the surface, Raman spectra can be obtained from 20–200 Å of material.[230-232] Other structures that can be studied include laminates of polymers[233,234] and embedded materials[235] in a polymer guide. Figure 19 shows Raman spectra from a polymer laminate of polystyrene and poly(vinyl alcohol). The spectra correspond closely to mixtures of the pure spectra weighted by the distribution of intensity due to the different guided modes used to excite the spectra.[233] In addition, a flowing gas temperature cell has been developed that works from $-125\,^{\circ}\mathrm{C}$ to $+100\,^{\circ}\mathrm{C}$.[236]

WRS has several major advantages over conventional Raman sampling methods. First, the sampling volume is increased by aligning the propagating beam in the waveguide with the slits of the

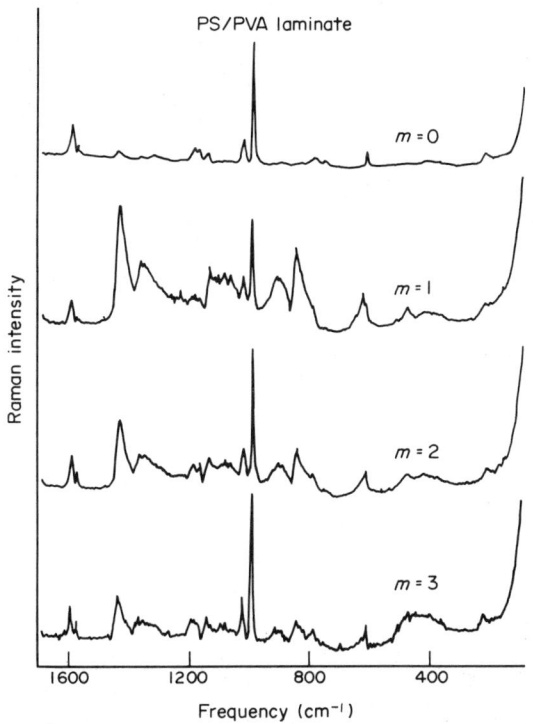

Figure 19 Raman spectra taken from a polystryrene/poly(vinyl alcohol) laminate structure excited by different guided modes in the waveguide. The spectra correspond to mixtures of the isolated polymer spectra weighted by the intensity distribution of the light in the guided mode. The mode for $m = 0$ gives nearly pure polystyrene spectrum. Modes for m greater than zero give mixed spectra, reflecting the composite structure of the guide. (Figure 4 in ref. 233)

spectrometer. Second, the fact the laser travels through (and is confined to) the sample increases the intensity at the sample. In part, the intensity is increased by the one-dimensional focusing in the thin film. The other advantage comes from the use of the propagation direction. To have a similar area illuminated with a conventional line focus would require much higher laser power by factors of ten to one hundred. Finally, it is possible to obtain high quality polarized Raman spectra from WRS.[227,229]

21.4.4 Micro Raman

Micro Raman is the adaptation of an optical microscope as a Raman sampling system. Delhaye and co-workers at the University of Lille are credited with the development of this technique.[237,238] A general review of the micro Raman technique can be found in ref. 239. The system typically uses an optical microscope both to focus the laser excitation on the sample and to collect the scattered Raman signal. The signal collected by the microscope is then optically matched to the double monochromator, and conventional scanning and photon counting are used to collect the signal. By switching mirrors it is also possible to image the sample on the microscope stage on a viewing screen or a video camera to help in positioning the beam focus. Combining the microscope with array detection is a natural extension and promises to minimize beam exposure times on the sample with a consequent reduction in sample damage.[240] There are many advantages to this approach, but it also introduces some additional complexity and ambiguity. The advantages include resolution and sampling areas of the order of a square micron, low laser powers needed owing to the concentration of light by the microscope, imaging of the sample and area discrimination, smaller samples needed, and minimization of fluorescence from some sources.[241] The limitations include the short working distances defined by the microscope objective which limit sampling techniques, loss of polarization information owing to the large solid angle used to collect the Raman signal and high local heating effects. Applications to polymer spectroscopy are most evident for small samples such as single crystal lamella studies, polymer interfaces, phase separations, oriented material, composite systems and thin films.[105,154,241,242] Figure 20 shows oriented Raman spectra from a poly(ethylene terephthalate) fiber 20 μm in diameter.[154] Similarly, the spatial resolution has been used to study the transport of solvents in polymers.[243] Another aspect that is advantageous is the ability to discriminate spatially against fluorescence. Especially in crystalline polymers, where impurities are probably localized to surfaces and amorphous interfaces but still very difficult to remove chemically, using the microprobe it is possible to obtain Raman spectra that are of comparable quality to those obtained using the FT-Raman technique, which eliminates fluorescence interference.[244]

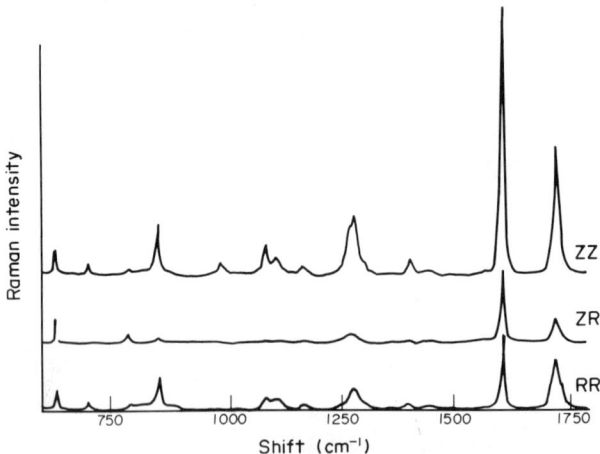

Figure 20   Raman microprobe spectra of a single 20 μm diameter oriented poly(ethylene terephthalate) fiber. The fiber was spin-oriented at 5500 m min^{-1}. Depolarization ratios increase with increasing spin-orientation rates. Laser power was 4 mW at the sample. Z polarization is parallel and R polarization is perpendicular to the fiber. (Figure 6 in ref. 154)

21.5 CONCLUSIONS

Raman spectroscopy has a wide range of applications in polymer science. Some of the methods and techniques in current use have been discussed in this paper, but there are still many variations that are important that have not been given more than a brief description. Examples include methods of varying the temperature and pressure of the polymer, which allow the experimenter to map out portions of the phase diagram, chemical reactions that lead to bonding and structural changes, changes in orientation and morphology as a function of processing, *etc.* Many of these items would require a separate chapter to present adequate detail to do them justice. This is especially true for many of the experimental techniques involving new technologies. Not discussed in this article are any of the nonlinear Raman spectroscopies such as CARS, stimulated Raman, *etc.* Also not discussed is the area of UV resonance Raman, which is potentially applicable to a wide range of polymeric materials, but has only been developed for biological polymers are present. Since intrumentation and experimental methods form a large body of information they have, regretfully, been relegated to the many general reviews cited.

Equally important is the area of interpretation of the Raman spectrum. Normal coordinate analysis is a mature field, but still underdeveloped. Calculations are only adequate for a few classes of molecules, such as hydrocarbons, some conjugated structures and several functional groups. This problem is also shared with those involved in IR spectroscopy. However, there is much less organized spectral data available to the Raman spectroscopist and, consequently, it is not possible to do the same general spectral searches familiar to those in IR spectroscopy. The situation is far from hopeless and mode assignments from small molecules can be used for guidance. In many cases this is all that is needed in polymer studies, since the morphology and not the chemistry is in question.

Raman spectroscopy has many unique sampling techniques that provide vibrational information that would be difficult to obtain in any other manner. However, as with any technique, it must be combined with other analyses to obtain the full picture. In the past there has been a tendency to overlook Raman because of the difficulties in obtaining spectra and interpreting them. Current technologies are solving many of the earlier problems, and it may be possible in the future to obtain Raman spectra as quickly and easily as IR spectra.

21.6 REFERENCES

1. C. V. Raman and K. S. Krishnan, *Nature (London)*, 1928, **121**, 501.
2. R. O. Kagel, 'Raman Spectroscopy', CRC Handbook of Spectroscopy', ed. J. W. Robinson, CRC Press, Boca Raton, FL, 1974, vol. 2, p. 107.
3. G. Varsanyi, 'Vibrational Spectra of Benzene Derivatives', Academic Press, New York, 1969.
4. G. Herzberg, 'Molecular Spectra and Molecular Structure II. Infrared and Raman Spectra of Polyatomic Molecules', Van Nostrand Reinhold, New York, 1945.
5. E. B. Wilson, Jr., J. C. Decius and P. C. Cross, 'Molecular Vibrations: The Theory of Infrared and Raman Vibrational Spectra', McGraw-Hill, New York, 1955.
6. R. F. Schaufele and T. Shimanouti, *J. Chem. Phys.*, 1967, **47**, 3605.
7. H. A. Szymanski, (ed.), 'Raman Spectroscopy, Theory and Practice', Plenum Press, New York, 1970, vol. 2.
8. D. A. Long, 'Raman Spectroscopy', McGraw-Hill, New York, 1977.
9. J. G. Grasselli, M. K. Snavely and B. J. Bulkin, *Phys. Rep.*, 1980, **65**, 231.
10. R. G. Snyder, *Methods Exp. Phys.*, 1980, **16A**, 73.
11. D. B DuPre, *Polym. Test.*, 1983, **3**, 249.
12. G. Zerbi, in 'Advances in Infrared and Raman Spectroscopy', ed. R. J. H. Clark and R. E. Hester, Wiley Heyden, New York, 1984, vol. 11, chap. 6, p. 301.
13. D. L. Gerrard and H. J. Bowley, *Anal. Chem.*, 1986, **58**, 6R.
14. D. L. Gerrard and W. F. Maddams, *Appl. Spectrosc. Rev.*, 1986, **22**, 251.
15. W. F. Maddams, *Am. Lab. (Fairfield. Conn.)*, 1986, **18**, 59.
16. T. C. Damen, S. P. S. Porto and B. Tell, *Phys. Rev.*, 1966, **142**, 570.
17. J. C. Decius and R. M. Hexter, 'Molecular Vibrations in Crystals', McGraw-Hill, New York, 1977.
18. R. G. Snyder, *J. Mol. Spectrosc.*, 1971, **37**, 353.
19. N. E. Schlotter and J. F. Rabolt, *Polymer*, 1984, **25**, 165.
20. M. Hamermesh, 'Group Theory and Its Applications to Physical Problems', Addision-Wesley, Palo Alto, CA, 1962.
21. M. Tinkham, 'Group Theory and Quantum Mechanics', McGraw-Hill, New York, 1964.
22. R. M. Hochstrasser, 'Molecular Aspects of Symmetry', Benjamin, New York, 1966.
23. H. A. Szymanski, (ed.), 'Raman Spectroscopy, Theory and Practice', Plenum Press, New York, 1967, vol. 1.
24. F. A. Cotton, 'Chemical Applications of Group Theory', 2nd edn., Wiley-Interscience, New York, 1971.
25. J. R. Ferraro and J. S. Ziomek, 'Introductory Group Theory and Its Application to Molecular Structure', 2nd edn., Plenum Press, New York, 1975.
26. V. A. Dement'ev, *J. Appl. Spectrosc. (Engl. Transl.)*, 1980, **32**, 266.
27. G. Masetti, S. Abbate, M, Gussoni and G. Zerbi, *J. Chem. Phys.*, 1980, **73**, 4671.
28. M. Gussoni, 'Advances in Infrared and Raman Spectroscopy', ed. R. J. H. Clark and R. E. Hester, Heyden, London, 1979, vol. 6, p. 61.
29. W. B. Person and G. Zerbi, (eds.), 'Vibrational Intensities in Infrared and Raman Spectroscopy', Elsevier, Amsterdam, 1982.

30. (a) L. I. Maklakov, V. L. Furer, V. V. Alekseev and A. L. Furer, *J. Appl. Spectrosc. (Engl. Transl.)*, 1979, **31**, 1285.
    (b) K. Holland-Mortiz and K. van Werden, *J. Polym. Sci., Polym. Phys. Ed.*, 1980, **18**, 1753.
31. M. K. Gupta and R. Bansil, *J. Polym. Sci., Polym. Phys. Ed.*, 1981, **19**, 353.
32. M. M. Gupta and R. Bansil, *J. Polym. Sci., Polym. Lett. Ed.*, 1983, **21**, 969.
33. P. Schmidt, B. Schneider, L. Terlemezyan and M. Mihailov, *Eur. Polym. J.*, 1982, **18**, 25.
34. A. Neppel and I. S. Butler, *J. Raman Spectrosc.*, 1984, **15**, 247.
35. A. Neppel and I. S. Butler, *Spectrochim. Acta, Part A*, 1984, **40**, 1095.
36. P. C. Painter, S. E. Howe and M. M. Coleman, *Appl. Spectrosc.*, 1984, **38**, 184.
37. P. C. Painter, S. E. Howe and M. M. Coleman, *Appl. Spectrosc.*, 1984, **38**, 190.
38. A. L. Smith and D. R. Anderson, *Appl. Spectrosc.*, 1984, **38**, 822.
39. F. Viras and T. A. King, *Polymer*, 1984, **25**, 899.
40. F. Viras and T. A. King, *Polymer*, 1984, **25**, 1411.
41. P. Vora, S. A. Solin, and P. John, *Phys. Rev. B: Condens. Matter*, 1984, **29**, 3423.
42. S. Furukawa, M. Fujino and T. Toriyama, *Solid State Commun.*, 1985, **56**, 363.
43. T. Ohnishi, I. Murase, T. Noguchi and M. Hirooka, *Synth. Met.*, 1986, **14**, 207.
44. J. Stokr, B. Schneider, M. Michailov and S. Stoeva, *Zv. Khim. Bulg. Akad. Nauk (Bulgaria)*, 1982, **15**, 173.
45. U. Ghosh, S. Chattopadhyaya and T. N. Misra, *J. Polym. Sci., Polym. Chem. Ed.*, 1987, **25**, 215.
46. K. Yoshida and M. Iwaki, *Nucl. Instrum. Methods Phys. Res. B*, 1987, **19/20**, 878.
47. D. L. Gerrard, W. F. Maddams and K. P. J. Williams, *Polym. Commun.*, 1984, **25**, 182.
48. K. Tashiro and M. Kobayashi, *Sen'i Gakaishi*, 1987, **43**, 78.
49. P. L. Lang, J. E. Katon, J. F. O'Keefe and D. W. Schiering, *Microchem. J.*, 1986, **34**, 319.
50. B. J. Bulkin, Y. Kwak and I. C. Dea, *Carbohydr. Res.*, 1987, **160**, 95.
51. H. Klump, J. Sturm and W. L. Peticolas, *Ber. Bunsenges. Phys. Chem.*, 1981, **85**, 661.
52. I. Artaki, T. W. Zerda and J. Jonas, *J. Non-Cryst. Solids*, 1986, **81**, 381.
53. T. W. Zerda, I. Artaki and J. Jonas, *J. Non-Cryst. Solids*, 1986, **81**, 365.
54. R. Bansil and M. K. Gupta, *Ferroelectrics*, 1980, **30**, 63.
55. S. P. Firsov and R. G. Zhbankov, *Zh. Prikl. Spektrosk.*, 1982, **37**, 276; *J. Appl. Spectrosc. (Engl. Transl.)*, 1982, **37**, 940.
56. H. Matsuura and K. Fukuhara, *J. Mol. Struct.*, 1985, **126**, 251.
57. A. Neppel, I. S. Butler, N. Brockman and A. Eisenberg, *J. Macromol. Sci., Phys.*, 1981, **19**, 61.
58. B. L. Papke, M. A. Ratner and D. F. Shriver, *J. Phys. Chem. Solids*, 1981, **42**, 493.
59. D. N. Schulz, D. G. Peiffer, P. K. Agarwal, J. Larabee, J. J. Kaladas, L. Soni, B. Handwerker and R. T. Garner, *Polymer*, 1986, **27**, 1734.
60. D. Teeters and R. Frech, *Solid State Ionics*, 1986, **18/19**, 271.
61. F. R. Aussenegg, M. E. Lippitsch and R. Möller, *Opt. Commun.*, 1982, **40**, 263.
62. F. R. Aussenegg and M. E. Lippitsch, *J. Raman Spectrosc.*, 1986, **17**, 45.
63. V. V. Vettegren, *Teubner-Texte Phys. (Prog Polym. Spectrosc.)*, 1986, **9**, 158.
64. S. Abbate, G. Zerbi and S. L. Wunder, *J. Phys. Chem.*, 1982, **86**, 3140.
65. R. G. Snyder, H. L. Strauss and C. A. Elliger, *J. Phys. Chem.*, 1982, **86**, 5145.
66. H. Tanaka and T. Takemura, *Jpn. J. Appl. Phys.*, 1983, **22**, 1001.
67. L. Ricard, S. Abbate and G. Zerbi, *J. Phys. Chem.*, 1985, **89**, 4793.
68. Y. Cho, M. Kobayashi and H. Tadakoro, *J. Chem. Phys.*, 1986, **84**, 4636.
69. D. Samios and T. Dorfmuller, *Chem. Phys. Lett.*, 1985, **117**, 165.
70. F. Viras, K. G. Viras and T. A. King, *J. Polym. Sci., Polym. Phys. Ed.*, 1985, **23**, 609.
71. R. P. Wool, R. S. Bretzlaff, B. Y. Li, C. H. Wang and R. H. Boyd, *J. Polym. Sci., Polym. Phys. Ed.*, 1986, **24**, 1039.
72. I. S. Butler and A. Neppel, *Spectrosc. Lett.*, 1983, **16**, 419.
73. L. B. Smith and R. G. Priest, *Appl. Spectrosc.*, 1984, **38**, 687.
74. N. E. Schlotter and J. F. Rabolt, *Macromolecules*, 1984, **17**, 1581.
75. M. I. Ize-Iyamu, *Mater. Res. Bull.*, 1983, **18**, 225.
76. M. I. Ize-Iyamu, *Spectrosc. Lett.*, 1983, **16**, 29.
77. S. L. Wunder and S. D. Merajver, *J. Polym. Sci., Polym. Phys. Ed.*, 1986, **24**, 99.
78. C. -S. Yoo and M. Nicol, *J. Phys. Chem.*, 1986, **90**, 6726.
79. K. -S. Lee, G. Wegner and S. L. Hsu, *Polymer*, 1987, **28**, 889.
80. P. T. T. Wong, T. E. Chagwedera and H. H. Mantsch, *J. Chem. Phys.*, 1987, **87**, 4487.
81. H. Matsuura and H. Murata, *J. Raman Spectrosc.*, 1982, **12**, 144.
82. G. Sbrana, N. Neto, M. Muniz-Miranda and M. Nocentini, *Spectrochim. Acta, Part A*, 1983, **39**, 295.
83. A. Palmer, S. Poulin-Dandurand and F. Brisse, *Can. J. Chem.*, 1985, **63**, 3079.
84. M. Kardan, B. R. Reinhold, S. L. Hsu, R. Thakur and C. P. Lillya, *Macromolecules*, 1986, **19**, 616.
85. G. M. Venkatesh, D. Y. Shen and S. L. Hsu, *J. Polym. Sci., Polym. Phys. Ed.*, 1981, **19**, 1475.
86. H. Tadokoro, *Polymer*, 1984, **25**, 147.
87. M. Abenoza and A. Armengaud, *Polymer*, 1981, **22**, 1341.
88. N. N. Aylward, *J. Polym. Sci., Polym. Phys. Ed.*, 1981, **19**, 1805.
89. M. M. Coleman, D. F. Varnell, B. A. Brozoski and P. C. Painter, *Polymer*, 1981, **22**, 762.
90. K. Zabel, N. E. Schlotter and J. F. Rabolt, *Macromolecules*, 1983, **16**, 446.
91. K. Tashiro, K. Takano, M. Kobayashi, Y. Chantani and H. Tadakoro, *Polymer*, 1981, **22**, 1312.
92. D. J. Cutler, P. J. Hendra and R. R. Rahalkar, *Polymer*, 1981, **22**, 726.
93. (a) M. Glotin, R. Domszy and L. Mandelkern, *J. Polym. Sci., Polym. Phys. Ed.*, 1983, **21**, 285. (b) R. Alamo, R. Domszy and L. Mandelkern, *J. Phys. Chem.*, 1984, **88**, 6587.
94. A. Galeski and E. Piorkowska, *J. Polym. Sci., Polym. Phys. Ed.*, 1983, **21**, 1313.
95. J. Dybal, J. Stokr and B. Schneider, *Polymer*, 1983, **24**, 971.
96. J. F. Rabolt, *CRC Crit. Rev. Solid State Mater. Sci.*, 1984, **12**, 165.
97. R. J. Petcavich and M. M. Coleman, *J. Macromol. Sci., Phys.*, 1980, **18**, 47.
98. R. J. Petcavich and M. M. Coleman, *J. Polym. Sci., Polym. Phys. Ed.*, 1980, **18**, 2097.
99. K. Tashiro, K. Takano, M. Kobayashi, Y. Chatani and H. Tadokoro, *Polymer*, 1984, **25**, 195.

100. J. S. Green, J. P. Rabe and J. F. Rabolt, *Macromolecules*, 1986, **19**, 1725.
101. L. Lauchlan and J. F. Rabolt, *Macromolecules*, 1986, **19**, 1049.
102. J. F. Rabolt, D. Hofer, R. D. Miller and G. N. Fickes, *Macromolecules*, 1986, **19**, 611.
103. S. D. Merajver, S. L. Wunder and W. Wallace, *J. Polym. Sci., Polym. Phys. Ed.*, 1985, **23**, 2043.
104. K. Tashiro, Y. Itoh, M. Kobayashi and H. Tadokoro, *Macromolecules*, 1985, **18**, 2600.
105. H. Tanaka, T. Ikeda and T. Nishi, *Appl. Phys. Lett.*, 1986, **48**, 393.
106. Y. Nishimura, M. Tsuboi, T. Sato and K. Aoki, *J. Mol. Struct.*, 1986, **146**, 123.
107. G. J. Thomas, Jr., J. M. Benevides and B. Prescott, in 'Biomolecular Stereodynamics IV', ed. R. H. Sarma and M. H. Sarma, Adenine Press, Guilderland, NY, 1986, p. 227.
108. C. Torigoe, Y. Nishimura, M. Tsuboi, J. -I. Matsuzaki, H. Hotoda, M. Sekine and T. Hata, *Spectrochim. Acta, Part A*, 1986, **42**, 1101.
109. K. Viras, F. Viras, C. Campbell, T. A. King and C. Booth, *J. Chem. Soc., Faraday Trans. 2*, 1987, **83**, 917.
110. F. Viras, K. Viras, C. Campbell, T. A. King and C. Booth, *J. Chem. Soc., Faraday Trans. 2*, 1987, **83**, 927.
111. G. Capaccio, M. A. Wilding and I. M. Ward, *J. Polym. Sci., Polym. Phys. Ed.*, 1981, **19**, 1498.
112. R. A. Chivers, P. J. Barham, J. Martinez-Salazar and A. Keller, *J. Polym. Sci., Polym. Phys. Ed.*, 1982, **20**, 1717.
113. G. M. Stack, L. Mandelkern and L. G. Voigt-Martin, *Polym. Bull. (Berlin)*, 1982, **8**, 421.
114. M. Glotin and L. Mandelkern, *J. Polym. Sci., Polym. Lett. Ed.*, 1983, **21**, 807.
115. M. Glotin and L. Mandelkern, *J. Polym. Sci., Polym. Phys. Ed.*, 1983, **21**, 29.
116. R. C. Domszy, R. Alamo, P. J. M. Mathieu and L. Mandelkern, *J. Polym. Sci., Polym. Phys. Ed.*, 1984, **22**, 1727.
117. R. C. Domszy, M. Glotin and L. Mandelkern, *J. Polym. Sci., Polym. Symp.*, 1984, **71**, 151.
118. J. Martinez-Salazar, P. J. Barham and A. Keller, *J. Mater. Sci.*, 1985, **20**, 1616.
119. S. J. Organ and A. Keller, *J. Mater. Sci.*, 1985, **20**, 1602.
120. G. Capaccio, I. M. Ward and M. A. Wilding, *Faraday Discuss. Chem. Soc.*, 1979, **68**, 328.
121. P. H. C. Shu, D. J. Burchell and S. L. Hsu, *J. Polym. Sci., Polym. Phys. Ed.*, 1980, **18**, 1421.
122. A. Peterlin and R. G. Snyder, *J. Polym. Sci., Polym. Phys. Ed.*, 1981, **19**, 1727.
123. J. Runt, I. R. Harrison, W. D. Varnell and J. -I. Wang, *J. Macromol. Sci., Phys.*, 1983, **22**, 197.
124. R. Twieg and J. F. Rabolt, *J. Polym. Sci., Polym. Lett. Ed.*, 1983, **21**, 901.
125. C. Chang and S. Krimm, *J. Polym. Sci., Polym. Phys. Ed.*, 1984, **22**, 1871.
126. D. H. Reneker and J. Mazur, *Polymer*, 1984, **25**, 1549.
127. B. Fanconi and J. F. Rabolt, *J. Polym. Sci., Polym. Phys. Ed.*, 1985, **23**, 1201.
128. C. Chang and S. Krimm, *J. Polym. Sci., Polym. Phys. Ed.*, 1986, **24**, 1373.
129. A. Peterlin, *Croat. Chem. Acta*, 1987, **60**, 103.
130. J. Runt, B. D. Hanrahan and I. R. Harrison, *J. Polym. Sci., Polym. Phys. Ed.*, 1982, **20**, 1687.
131. R. G. Snyder, J. R. Scherer, D. H. Reneker and J. P. Colson, *Polymer*, 1982, **23**, 1286.
132. I. G. Voight-Martin, R. Alamo and L. Mandelkern, *J. Polym. Sci., Polym. Phys. Ed.*, 1986, **24**, 1283.
133. S. Mizushima and T. Shimanouti, *J. Am. Chem. Soc.*, 1949, **71**, 1320.
134. A. Peterlin, *J. Polym. Sci., Polym. Phys. Ed.*, 1982, **20**, 2329.
135. Y. K. Wang, P. H. C. Shu, R. S. Stein and S. L. Hsu, *J. Polym. Sci., Polym. Phys. Ed.*, 1980, **18**, 2287.
136. C. Chang, Y. K. Wang, D. A. Waldman and S. L. Hsu, *J. Polym. Sci., Polym. Phys. Ed.*, 1984, **22**, 2185.
137. D. Dothée, M. Camelot and C. Roques-Carmes, *J. Phys. Colloq. (Orsay, Fr.)*, 1984, **45**, C2-257.
138. P. J. Hendra, J. Vile, H. A. Willis, V. Zichy and M. E. A. Cudby, *Polymer*, 1984, **25**, 785.
139. J. F. Rabolt, *Polymer*, 1981, **22**, 890.
140. G. Zerbi and M. Gussoni, *Polymer*, 1980, **21**, 1129.
141. R. G. Snyder, N. E. Schlotter, R. Alamo and L. Mandelkern, *Macromolecules*, 1986, **19**, 621.
142. E. W. Fischer, G. R. Strobl, M. Dettenmair, M. Stamm and N. Steidle, *Faraday Discuss. Chem. Soc.*, 1979, **68**, 26.
143. J. R. Scherer and R. G. Snyder, *J. Chem. Phys.*, 1980, **72**, 5798.
144. R. G. Snyder and S. L. Wunder, *Macromolecules*, 1986, **19**, 496.
145. L. Mandelkern, R. Alamo, W. L. Mattice and R. G. Snyder, *Macromolecules*, 1986, **19**, 2404.
146. M. Kobayashi, *J. Mol. Struct.*, 1985, **126**, 193.
147. V. I. Vettegren, L. S. Titenkov, A. A. Kusov and Y. V. Zelenev, *Vysokomol. Soedin., Ser. A*, 1985, **27**, 2489.
148. J. -J. Kim and R. T. Bailey, *J. Korean Phys. Soc.*, 1980, **13**, 128.
149. Y. K. Wang, D. A. Waldman, R. S. Stein and S. L. Hsu, *J. Appl. Phys.*, 1982, **53**, 6591.
150. J. C. Merino, J. M. Pastor and J. A. De Saja, *Polymer*, 1985, **26**, 383.
151. C. H. Wang and D. B. Cavanaugh, *J. Appl. Phys.*, 1981, **52**, 6003.
152. L. D. Cambon and L. D. Vinh, *J. Raman Spectrosc.*, 1983, **1**, 291.
153. L. D. Cambon, J. L. Ramonja and D. V. Luu, *J. Raman Spectrosc.*, 1987, **18**, 129.
154. F. Adar and H. Noether, *Polymer*, 1985, **26**, 1935.
155. B. J. Bulkin, M. Lewis and F. J. DeBlase, *Macromolecules*, 1985, **18**, 2587.
156. R. Burzynski, P. N. Prasad and S. Murthy, *J. Polym. Sci., Polym. Phys. Ed.*, 1986, **24**, 133.
157. C. Gallotis, I. M. Robinson, R. J. Young, B. J. E. Smith and D. N. Batchelder, *Polym. Commun.*, 1985, **26**, 354.
158. B. J. Bulkin, F. DeBlase and M. Lewin, *Proc. SPIE, 665, Opt. Tech. Ind. Insp.*, 1986, 234.
159. R. J. Day, I. M. Robinson, M. Zakikhani and R. J. Young, *Polymer*, 1987, **28**, 1833.
160. J. H. Wiley and R. H. Atalla, *ACS Symp. Ser.*, 1987, **340**, 151.
161. P. J. Hendra, D. B. Morris, R. D. Sang and H. A. Willis, *Polymer*, 1982, **23**, 9.
162. R. R. McCaffrey, P. N. Prasad, M. Fornalik and R. Baier, *J. Polym. Sci., Polym. Phys. Ed.*, 1985, **23**, 1523.
163. H. Kuzmany, *J. Phys. Colloq. (Orsay, Fr.)*, 1983, **44**, C3-255.
164. H. Kuzmany and P. Knoll, *J. Raman Spectrosc.*, 1986, **17**, 89.
165. D. N. Batchelder and D. Bloor, in 'Advances in Infrared and Raman Spectroscopy', ed. R. J. H. Clark and R. E. Hester, Wiley Heyden, New York, 1984, vol. 2, p. 133.
166. M. Akimoto, Y. Furukawa, H. Takeuchi, I. Harada, Y. Soma and M. Soma, *Synth. Met.*, 1986, **15**, 353.
167. J. P. Aime, J. L. Fave and M. Schott, *Europhys. Lett.*, 1986, **1**, 505.
168. M. Aldissi, *Synth. Met.*, 1986, **15**, 141.
169. B. S. Elman, M. K. Thakur and R. J. Seymour, *Radiat. Eff.*, 1986, **98**, 139.

170. R. H. Friend, D. D. C. Bradeley, C. M. Pereira, P. D. Townsend, D. C. Bott and K. P. J. Williams, *Synth. Met.*, 1986, **13**, 101.
171. E. Faulques, E. Rzepka, S. LeFant, E. Mulazzi, G. P. Brivio and G. Leising, *Phys. Rev. B: Condens. Mater.*, 1986, **33**, 8622.
172. Z. Iqbal, D. M. Ivory, J. S. Szobota, R. L. Eisenbaumer and R. H. Baughman, *Macromolecules*, 1986, **19**, 2992.
173. B. Jorgensen, R. Liepine and S. Agnew, *Polym. Bull. (Berlin)*, 1986, **16**, 263.
174. G. Masetti, E. Campani, G. Gorini, R. Tubino, P. Piaggio and G. Dellepiane, *Chem. Phys.*, 1986, **108**, 141.
175. D. Schmeltzer, I. Ohana and Y. Yacoby, *J. Phys. C: Solid State Phys.*, 1986, **19**, 2113.
176. (a) M. K. Thakur, B. S. Elman and M. Dagenais, *Chem. Phys. Lett.*, 1986, **125**, 328. (b) Y. Tokura, T. Koda, A. Itsubo, M. Miyabayashi, K. Okuhara and A. Ueda, *J. Chem. Phys.*, 1986, **85**, 99.
177. H. Takeuchi, Y. Furukawa, I. Harada and H. Shirakawa, *J. Chem. Phys.*, 1986, **84**, 2882.
178. G. Youjiang and Y. Lu, *Solid State Commun.*, 1986, **58**, 407.
179. Y. Yacoby and S. Roth, *Synth. Met.*, 1986, **13**, 299.
180. G. You-Jiang and Y. Lu, *Acta Phys. Sinica*, 1986, **35**, 922.
181. R. J. Butera, J. B. Lando and B. Simic-Glavaski, *Macromolecules*, 1987, **20**, 1722.
182. H. Cohen, O. Brafman, E. Ehrenfreund, Z. Vardeny and D. H. Kohn, *Synth. Met.*, 1987, **17**, 389.
183. H. Kuzmany and J. Kurti, *Synth. Met.*, 1987, **21**, 95.
184. C. R. K. Marrian, R. J. Colton, A. Snow and C. J. Taylor, *Mater. Res. Soc. Symp. Proc.*, 1987, **76**, 353.
185. G. Wieners, M. Monkenbusch and G. Wegner, *Ber. Bunsenges. Phys. Chem.*, 1984, **88**, 935.
186. D. N. Batchelder, R. J. Kennedy and D. Bloor, *J. Polym. Sci., Polym. Phys. Ed.*, 1981, **19**, 677.
187. K. Miyano and T. Maeda, *Phys. Rev. B: Condens. Mater.*, 1986, **33**, 4386.
188. Y. J. Chen, S. K. Tripathy, G. M. Carter, B. S. Elman, E. S. Koteles and J. George, Jr., *Solid State Commun.*, 1986, **58**, 97.
189. T. Inabe, W. -B. Liang, J. F. Lomax, S. Nakamura, J. W. Lyding, W. J. McCarthy, S. H. Carr, C. R. Kannewurf and T. J. Marks, *Synth. Met.*, 1986, **13**, 219.
190. C. Taliani, R. Danieli, R. Zamboni, P. Ostoja and W. Porzio, *Synth. Met.*, 1987, **18**, 177.
191. Z. Iqbal, C. Maleysson and R. H. Baughman, *Synth. Met.*, 1986, **15**, 161.
192. N. J. Poole, B. J. E. Smith, D. N. Batchelder, R. T. Read and R. J. Young, *J. Mater. Sci.*, 1986, **21**, 507.
193. F. Ebisawa, T. Kurihara and H. Tabei, *Synth. Met.*, 1987, **18**, 431.
194. J. R. Ferraro, A. Furlani and M. V. Russo, *Appl. Spectrosc.*, 1987, **41**, 830.
195. S. K. Brahma, *Solid State Commun.*, 1986, **57**, 673.
196. D. Rakovic, I. Bozovic, L. A. Gribov, S. A. Stepanyan and V. A. Dementiev, *Synth. Met.*, 1987, **17**, 613.
197. K. M. Cheung, B. J. E. Smith, D. N. Batchelder and D. Bloor, *Synth. Met.*, 1987, **21** 249.
198. S. Krichene, J. P. Buisson and S. Lefrant, *Synth. Met.*, 1987, **17**, 589.
199. P. N. Prasad, *Thin Solid Films*, 1987, **152**, 275.
200. P. N. Prasad, *Proc. SPIE*, 1987, **682**, 120.
201. J. Swiatkiewicz, X. Mi, P. Chopra and P. N. Prasad, *J. Chem. Phys.*, 1987, **87**, 1882.
202. G. Gillberg, *J. Adhes.*, 1987, **21**, 129.
203. S. M. Heard, F. Grieser and C. G. Barraclough, *Chem. Phys. Lett.*, 1983, **95**, 154.
204. P. C. Lee and D. Meisel, *Chem. Phys. Lett.*, 1983, **99**, 262.
205. O. Siiman, A. Lepp and M. Kerker, *Chem. Phys. Lett.*, 1983, **100**, 163.
206. C. J. Sandroff, H. E. King, Jr. and D. R. Herschbach, *J. Phys. Chem.*, 1984, **88**, 5647.
207. D. Fornasiero and F. Grieser, *Chem. Phys. Lett.*, 1987, **139**, 103.
208. K. Metcalfe and R. L. Hester, *Chem. Phys. Lett.*, 1983, **94**, 411.
209. V. Brabec and K. Niki, *Stud. Biophys.*, 1986, **114**, 111.
210. R. Holze, *J. Electroanal. Chem.*, 1987, **224**, 253.
211. J. S. Suh and K. M. Michaelian, *J. Phys. Chem.*, 1987, **91**, 598.
212. J. S. Suh and K. H. Michaelian, *J. Raman Spectrosc.*, 1987, **18**, 409.
213. G. W. Chantry, H. A. Gebbie and C. Helsum, *Nature (London)*, 1964, **203**, 1052.
214. T. Hirschfeld, in 'Laser Raman Gas Diagnostics', ed. M. Lapp, Plenum Press, New York, 1974, p. 379.
215. D. E. Jennings, A. Weber and J. W. Brault, *Appl. Opt.*, 1986, **25**, 284.
216. T. Hirshfeld and D. B. Chase, *Appl. Spectrosc.*, 1986, **40**, 133.
217. D. J. Moffatt, H. Buijs and W. F. Murphy, *Appl. Spectrosc.*, 1986, **40**, 1079.
218. D. B. Chase, *J. Am. Chem. Soc.*, 1986, **108**, 7485.
219. V. M. Hallmark, C. G. Zimba, J. D. Swalen and J. F. Rabolt, *Spectroscopy*, 1987, **2**, 40.
220. C. G. Zimba, V. M. Hallmark, J. D. Swalen and J. F. Rabolt, *Appl. Spectrosc.*, 1987, **41**, 721.
221. M. Fujiwara, H. Hamaguchi and M. Tasumi, *Appl. Spectrosc.*, 1986, **40**, 137.
222. V. M. Hallmark and A. Campion, *Chem. Phys. Lett.*, 1984, **110**, 561.
223. N. E. Schlotter, S. A. Schaertel, S. P. Kelty and R. Howard, *Appl. Spectrosc.*, 1988, **42**, 746.
224. S. B. Dieker, C. A. Murray, J. D. LeGrange and N. E. Schlotter, *Chem. Phys. Lett.*, 1987, **137**, 453.
225. J. C. Tsang, in 'Dynamics on Surfaces', ed. B. Pullman, J. Jortner, A. Nitzan and B. Gerber, Reidel, Norwell, MA, 1984, p. 379.
226. D. K. Veirs, V. K. F. Chia and G. M. Rosenblatt, *Appl. Opt.*, 1987, **26**, 3530.
227. J. F. Rabolt, R. Santo, N. E. Schlotter and J. D. Swalen, *IBM J. Res. Dev.*, 1982, **26**, 209.
228. J. D. Swalen and J. F. Rabolt, *J. Phys. Colloq. (Orsay, Fr.)*, 1983, **44**, 501.
229. N. E. Schlotter and J. F. Rabolt, *J. Phys. Chem.*, 1984, **88**, 2062.
230. J. F. Rabolt, N. E. Schlotter, J. D. Swalen and R. Santo, *J. Polym. Sci., Polym. Phys. Ed.*, 1983, **21**, 1.
231. R. Burzynski, P. N. Prasad, J. Biegajski and D. A. Cadenhead, *Macromolecules*, 1986, **19**, 1059.
232. J. P. Rabe, J. D. Swalen and J. F. Rabolt, *J. Chem. Phys.*, 1987, **86**, 1601.
233. J. D. Swalen, N. E. Schlotter, R. Santo and J. F. Rabolt, *J. Adhes.*, 1981, **13**, 189.
234. D. R. Miller, O. H. Han and P. W. Bohn, *Appl. Spectrosc.*, 1987, **41**, 249.
235. N. E. Schlotter and J. F. Rabolt, *Appl. Spectrosc.*, 1984, **38**, 208.
236. E. Barbaczy, F. Dodge and J. F. Rabolt, *Appl. Spectrosc.*, 1987, **41**, 176.
237. M. Delhaye and P. Dhamelincourt, *J. Raman Spectrosc.*, 1975, **3**, 33.

238. P. Dhamelincourt, F. Wallart, M. Lechlercq, A. T. N'Guyen and D. O. Landon, *Anal. Chem.*, 1979, **51**, 414A.
239. F. Adar, *Sagamore Army Mater. Res. Conf. Proc., 31st*, 1986, **31**, 399.
240. P. Dhamelincourt, M. Delhaye, E. De Silva, J. P. Cornard and B. Lenain, *Proc. Int. Conf. Raman Spectrosc., 10th*, 1986, 20.
241. M. E. Andersen and R. Z. Muggli, *Anal. Chem.*, 1981, **53**, 1772.
242. C. J. Cook, *Energy. Res. Abstr.*, 1987, **12**, 25 335.
243. J. Klier and N. A. Peppas, *Polym. Bull. (Berlin)*, 1986, **16**, 359.
244. Personal communication from F. Adar, Instruments SA, Edison, NJ.

22

Emission Spectroscopy

GODFREY S. BEDDARD
University of Manchester, UK
and
NORMAN S. ALLEN
Manchester·Polytechnic, UK

22.1 GENERAL ASPECTS

In addition to the study of the photophysics of polymers formed from chromophores, or of specifically labelled polymers, fluorescence and phosphorescence techniques can be used in a number of different ways in polymer chemistry. Examples include methods for identifying different polymer samples and their impurities, and techniques used in the study of the inter- and intra-molecular dynamics of polymer chains. Some experiments use the polymer as a substrate, as in hole burning in porphyrin absorption spectra in poly(methyl methacrylate) (PMMA) at low temperatures[1] or as in the ingenious O_2 sensor based on quenching of the excited state of a Ru^{II} complex held in silicone rubber.[2]

Fluorescence and phosphorescence arise from the $\pi\pi^*$ and $n\pi^*$ excited states of aromatics, heterocyclics and aromatic carbonyls (see Figure 1). These excited states are produced by promotion of an electron in the highest occupied molecular orbital to the lowest unoccupied one. In $n\pi^*$ transitions a non-bonding electron, *e.g.* on a carbonyl oxygen, is promoted in the lowest energy transition. In both cases an excited singlet state is produced. In most aromatic molecules the transition moment for allowed transitions ($\pi\pi^*$) lies in the plane of the molecule, thus fluorescence emission from the excited state back to the ground states is in-plane polarized. Under the influence of spin–orbit coupling the singlet excited state may undergo a non-radiative spin-forbidden process (intersystem crossing) to become a metastable triplet state (spin $s = 1$). The spin–orbit coupling is induced by the presence of heavy atoms such as iodine, either chemically attached to the chromophore or in solution (*e.g.* ethyl iodide), as well as by paramagnetic species, such as dissolved O_2.

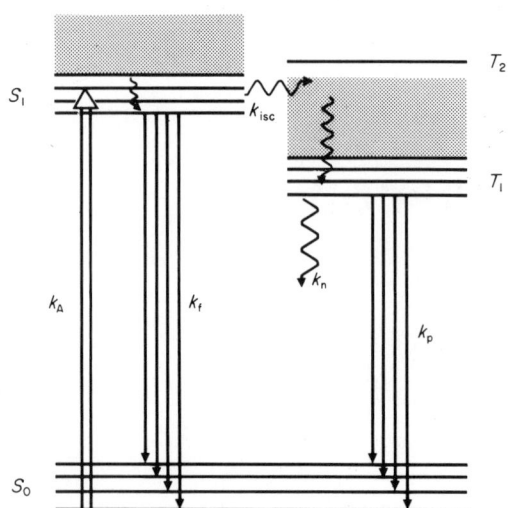

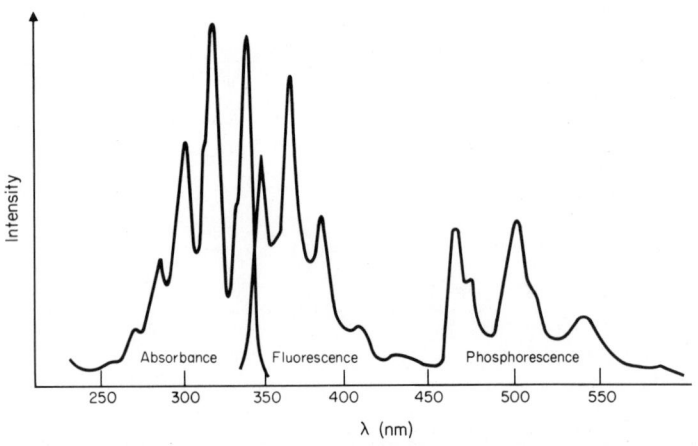

Figure 1   Energy level scheme for $\pi\pi^*$ and $n\pi^*$ transitions. The ground state is S_0, singlet excited states S_1, S_2 and triplet states T_1, T_2. k_A is the rate constant for absorption, k_f for fluorescence from S_1 ($v = 0$) to vibrational levels in S_0. k_p is the rate constant for phosphorescence. Non-radiative processes are shown with wavy lines; k_{isc} is intersystem crossing and k_n is the non-radiative decay of the triplet state. This comprises two parts; isoenergetic transfer to high vibrational levels of S_0 followed by energy loss in collisions with the solvent. Vertical wavy lines in S_1 and T_1 also show energy loss by solvent collisions. The lower spectra show the mirror image relation between absorption and emission in a typical aromatic and also the phosphorescence spectrum

Electrons in non-bonding orbitals also induce spin–orbit coupling, by virtue of their *s* orbital character, rather more than electrons in *p* orbitals. Triplet states convert slowly to the ground state, the transfer being spin forbidden, thus phosphorescence is far longer lived than fluorescence, lasting milliseconds or longer *vs.* nanoseconds for fluorescence. As a consequence bimolecular quenching by dissolved species (such as O_2) often limits the measured phosphorescence lifetime. The spin-forbidden $\pi\pi^*$ phosphorescence is out-of-plane polarized, *i.e.* the transition moment is perpendicular to the molecular plane and, as in intersystem crossing, is induced by spin–orbit coupling. Because the triplet state is lower in energy than the singlet, phosphorescence is observed towards the red of the fluorescence spectrum (see Figure 1).

In molecules with $n\pi^*$ transitions, in contrast to $\pi \to \pi^*$ transitions, the absorption transition moment is out-of-plane polarized, whereas the phosphorescence is in-plane polarized. Heterocyclic molecules often have $n\pi^*$ and $\pi\pi^*$ singlet and triplet states fairly close in energy. From symmetry considerations the intersystem crossings $^1n\pi^* \to {}^3\pi\pi^*$ and $^1\pi\pi^* \to {}^3n\pi^*$ become allowed with spin–orbit coupling, whereas $^1\pi\pi^* \to {}^3\pi\pi^*$ and $^1n\pi^* \to {}^3n\pi^*$ are forbidden, and hence slow, when singlet and triplets have the same symmetry (El-Sayed rules). Thus in benzophenone, where the lowest singlet state is $n\pi^*$, emission is so small as to be unmeasurable, and the triplet $n\pi^*$ yield is unity; but in diphenylanthracene, which has only $\pi\pi^*$ states, the fluorescence yield is almost unity and the triplet yield is almost zero. Clearly most molecules fall between these extreme limits.

Fluorescence and phosphorescence spectra are usually measured by viewing the emission at 90° to the direction of excitation. In a fluorimeter the exciting light is selected in a monochromator and a second monochromator is used to scan the emitted light from near to the excitation wavelength to longer wavelengths. In solution a molecule with a regular (although vibrating) structure such as anthracene has an emission spectrum that is similar to a mirror image of its absorption spectrum, as a result of transitions from S_1 ($v = 0$) to S_0 ($v = 1, 2, 3$) *etc.* (see Figure 1). Because of solvent interactions, the spectrum is broadened and stabilized when compared to the gas phase spectrum. If the excited state also has an appreciable dipole moment, then it can be stabilized by interaction with solvent, when the emission spectrum is red shifted, often by many nanometers, and usually no longer resembles the absorption spectrum. A change in spectral shape also occurs if the excited state undergoes some geometrical change with respect to the ground state. The internuclear separation of the vibrational mode for which the Frank–Condon factors are most favourable in emission may then differ from that in absorption, and the emission spectrum is no longer the expected mirror image of the absorption spectrum.

Phosphorescence is a little more difficult to observe than fluorescence because of the possibility of quenching by impurities before enough time has elapsed for there to be a significant emission probability. The simplest quenching scheme is that due to collisions between an excited molecule and a potential quencher. If A represents the ground state molecule, A* the excited molecule (singlet or triplet), Q the quencher and X another excited state of the molecule or chemical product, then the excitation–quenching sequence is

$$A \xrightarrow{k_A} A^*, \quad A^* \xrightarrow{k_E} A + h\nu, \quad A^* \xrightarrow{k_X} X, \quad A^* + Q \xrightarrow{k_q} A + Q'$$

From an analysis of the kinetics the Stern–Volmer equation

$$1/\tau = 1/\tau_0 + k_q[Q]$$

can be derived, where τ is the excited state lifetime in the presence of quencher, τ_0 ($\tau_0 = (k_E + k_X)^{-1}$) is the value of τ in the absence of quencher, k_E the radiative rate constant and k_q the quenching rate constant. In a fluid solvent, such as cyclohexane, diffusion-controlled quenching rate constants are as high as 3×10^{10} dm^3 mol^{-1} s^{-1}, while for a triplet state of an aromatic $\tau_0 \approx 10^{-1}$ s. Even at a quencher concentration of 10^{-9} mol dm^{-3} the measured τ for the triplet state is more than halved. Fluorescence is less sensitive to quenching than phosphorescence, as lifetimes of $\tau = 10^{-9}$ to 10^{-7} s are typical. Removal of traces of quenchers such as oxygen is almost impossible, considering the low concentrations needed in fluid solutions to avoid triplet quenching. Consequently phosphorescence is usually observed only in solids or in micellar solutions, wherein diffusion of the quencher is slowed. Experimentally phosphorescence can be measured in a fluorimeter with a flash excitation source, or with a continuous lamp chopped with a fast shutter. The detector (photomultiplier) is often gated off electronically while the fluorescence decays and before the longer-lived phosphorescence is detected. Only if the phosphorescence is long enough lived will the decay time be detected in the fluorimeter; if it is shorter lived, or if fluorescence lifetimes are also required, then faster techniques must be used.[3, 4]

The emission yield is defined as the ratio of the rate of emission to the rate of absorption, *i.e.*

$$\phi_0 = \frac{k_E[A^*]}{k_A[A]}$$

At steady state in the absence of quencher the scheme above gives

$$\phi_0 = \frac{k_E}{k_E + k_X}$$

This expression is equivalent to $\phi_0 = k_E \tau_0$, and the Stern–Volmer equations may be recast in terms of yields

$$\phi_0/\phi \;=\; 1 \;+\; \tau_0 k_q [Q]$$

from which k_q is easily calculated if τ_0 is known, either measured directly or estimated from ϕ_0 and k_E, which can sometimes be obtained from the integrated absorption spectrum.[5]

Absolute emission yields may also be calculated from instrumentally corrected spectra if the number of absorbed photons is known. This is a laborious quantity to obtain and is usually done *via* actinometry, *e.g.* with ferrioxalate solutions. Often relative emission yields will suffice, and are measured relative to the known emission yield of a standard compound such as quinine sulfate or 9,10-diphenylanthracene. Care must be taken to correct for the change in the sensitivity of the instrument with wavelength in all emission measurements where spectral shapes differ.

Occasionally excitation spectra are measured in order to identify the origin of an emission. In this experiment the emission wavelength is fixed and the excitation monochromator is scanned over the expected absorption wavelengths. A spectrum related to the absorption is obtained, but modified by the Frank–Condon factors for emission. If the excited state undergoes solvent stabilization or geometrical changes before emission the excitation spectrum can be significantly changed from the absorption spectrum.

The high sensitivity of luminescence methods, which can detect molecules at the femtomole level,[6] makes the detection of trace impurities or emission from minute samples feasible. A disadvantage of the technique is the comparatively broad emission spectra of many aromatic and related chromophores, which often lead to overlapping spectra. Special techniques such as modulation, derivative, synchronous and multiwavelength spectroscopy have been developed to tackle the problems of congested spectra, and allow the analysis of complex spectra.[7]

Simultaneous monitoring with more than one detection wavelength enables the identification of species from compounds having overlapping peaks in liquid chromatography or a related technique. The composite nature of the peak can be resolved as a change in a fluorescence-signal ratio as it passes the detector, since a single compound will give constant fluorescence ratio. Using four detection wavelengths, picogrammes of material have been detected with high dynamic range.[8]

In synchronous fluorescence techniques many spectra are recorded, as both excitation and emission wavelengths are repeatedly scanned at different wavelength separations. A two-dimensional plot is made of excitation against emission wavelengths, in which points of equal fluorescence intensity form the contours. As there is generally some difference in the absorption spectra, species with overlapping emission spectra can often be identified as extra 'hills' in the contour plot.[7] If further clarity is required, the derivatives of the spectra can be plotted. Although no extra information is present in the derivative form of presentation, the eye can often notice subtle differences not obvious in a normal plot.

In the study of the dynamics and photophysics of polymers the steady state methods outlined are of limited use. More information can be obtained by using time-resolved methods: of these fluorescence lifetime measurements have proved to be invaluable. Possibly the most sensitive and direct method is time-correlated photon counting,[4] increasingly used with a picosecond laser source for sub-nanosecond time resolution and increased sensitivity, although for slower time scales fast-gated deuterium lamps are still useful. This technique, described in detail elsewhere,[9] uses a repetitive source and measures the time delay between exciting molecules in a sample and the detection of the first fluorescence photon. These times, when stored as a histogram of number of events at a given time delay *vs.* time delay, give an accurate measure of the excited-state decay. The rates of excited-state quenching, rotational reorientation *via* fluorescence depolarization and energy transfer, as well as time-resolved emission spectra, can be measured.

22.1.1 Fluorescence Probe Analysis

Labelled polymers fall broadly into two classes; those in which isolated chromophores are attached to the polymer as an end group or minor constituent in the chain (type A), and those whose repeat units form the chromophores (type B). Polystyrene is an example of the latter, type B, polymer. Type A labelled polymers have been most used in the study of polymer dynamics, such as end-to-end or end-to-backbone cyclization rates. The photophysics of type B polymers is more complicated because of the variety of different chromophore arrangements and the almost equal probability of light absorption at any site.

Attaching a chromophore to a polymer can influence its photophysical behaviour in ways not experienced in homogeneous solution. The chromophore mobility is strongly influenced by chain dynamics, and inter- and intra-chain interactions. These interactions depend on the polymer configuration, which in turn depends on the type of repeat unit, chain length (average molecular weight), concentration, temperature and solvent. Poor solvents encourage polymer–polymer interactions, reducing mean distances between chromophores and backbone groups: a good solvent removes these interactions and the chromophores are, on average, more isolated from one another.

Probes whose fluorescence is changed in wavelength and/or intensity in media of different polarity have been used to study the effect of different solvents on polymers. For example, dansyl probes (5-dimethylamino-1-naphthalenesulfonyl derivatives) are fluorescent in organic solvents but much less so in water. Dansyl-labelled poly(acrylic acid) exhibits a large increase in fluorescence on aggregation with poly(oxyethylene), indicating the exclusion of water molecules.[10] Similarly for a naphthalenesulfonate-labelled copoly(styrene/divinylbenzene)[11] the response of the probe mirrored the swelling of the polymer gel. In good solvents the probe behaved as if it were surrounded by solvent, but in poor solvents as if it were in pure polymer.

Pyrene is sparingly soluble in water but soluble in organic solvents and polymers such as poly(methacrylate) at low pH. As the pH is increased, an abrupt decrease in pyrene fluorescence intensity and that of its excimers is observed. This occurs at pH 5, indicating an opening of polymer coils at this pH and release of pyrene to the aqueous phase.[12]

22.1.2 Excimers

In concentrated solution, a new excited species (excimer, or exciplex if between dissimilar molecules) containing one excited molecule (M*) and a ground state (M) is formed in equilibrium with the separated species (see Scheme 1a); k_f are the excimer and monomer emission rate constants. The excimer has a broad structureless fluorescence at longer wavelengths than that of the isolated molecular excited state. In solution the excimer emission appears slowly after excitation, due to the formation of the species (MM*) *via* diffusion of M and M*, and subsequently has a long decay as the excited-state equilibrium is established.[13]

Excimers are a common feature in polymers when aromatics such as styrene, naphthalene *etc.* are contained in the repeat unit, but their behaviour is more complex than in fluid solution. Two schemes can be envisaged. In the first two spectrally identical but kinetically distinct monomers (M_1 and M_2) have been proposed.[14] They differ in the extent to which they can undergo energy migration (rate constant k_{ET}) to form excimers. Only one monomer (type M_2) forms the excimer; the

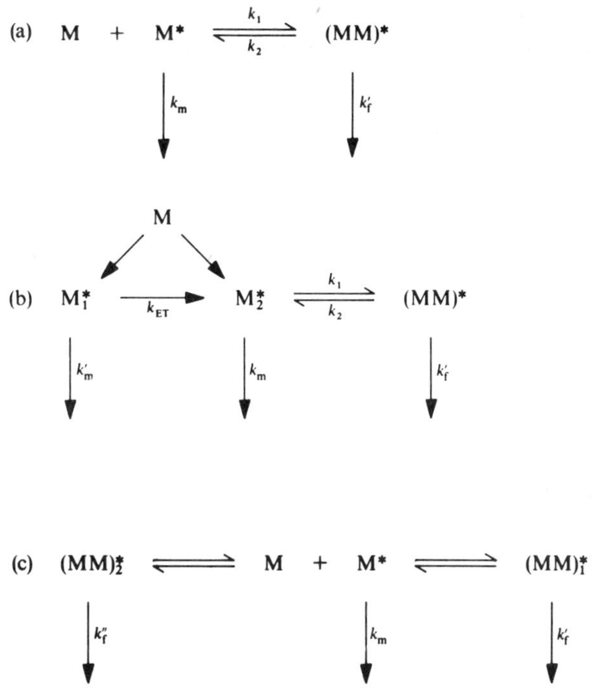

Scheme 1

other has first to transfer its energy to a type M_2 monomer (see Scheme 1b). In the second scheme (see Scheme 1c), which applies to poly(N-vinylcarbazole) (PNVC), one unit is coupled to either of two kinetically distinguishable excimers.[15] The excimer emitting at the normal wavelength of 420 nm has a 28 ns lifetime (at 77 K) but that emitting at 380 nm has a 4 ns lifetime.[16] The PNVC formed by cationic polymerization had more of the 420 nm species and that formed by radical polymerization more of the 380 nm species. These observations are explained by the predominance of syndiotactic sequences in radical-formed polymer, which allow formation of weakly bound excimers emitting at 380 nm, whereas the isotactic sequences in cationic-formed polymer allow formation of more-stable sandwich-type excimers which emit at 420 nm. Confirmation of these assignments for the two excimer bands has come from studies on model compounds.[17,18]

22.1.3 Energy Transfer and Migration

When the emission spectrum of one molecule (the donor) overlaps in wavelength the absorption spectrum of another (the acceptor), non-radiative energy transfer may occur between the two molecules. The acceptor molecule becomes electronically excited and may subsequently fluoresce or phosphoresce and the donor is quenched. The rate of this dipole–dipole or resonance transfer depends on the acceptor absorption and donor emission spectral overlap, the relative orientation of the two molecules, and the inverse sixth power of their separation.[3,5] It is possible therefore for one molecule to quench another while still some distance away (2–5 nm), unlike impurity quenching or excimer formation which occur at very short range (0.5–1 nm), essentially at contact.

Two processes have to be distinguished in energy transfer studies: direct quenching of the excited molecule (donor) by a nearby acceptor in a single step and, secondly, quenching only after many donor-to-donor energy transfer steps, *i.e.* after energy migration. Only in the last step of energy migration is the donor quenched. Energy transfer and migration are very sensitive to coil dimensions: single-step processes become relatively more important as coil dimensions are decreased, because of the decrease in mean donor-to-acceptor separation.

There is growing evidence that migration of energy among chromophores attached to polymer chains can be very efficient. Extensive studies of energy diffusion have been made by Guillet.[19] Several different types of labelled polymers such as copoly(naph-MMA/anth-MMA)were used (naph = naphthalene, anth = anthracene). This copolymer, for example, has efficient (84%) intra-chain singlet–singlet transfer from the naphthalene–PMMA donors to the anthracene–PMMA acceptors.

22.1.4 Fluorescence Polarization

Steady state and time-resolved measurements of fluorescence polarization have proved of great value in determining the local, segmental and overall rotational motion of proteins[20] and synthetic polymers[21] as well as in probing the structures of bilayer membranes and liquid crystals.[22] When polarized light is used to excite a sample, photoselection occurs and the probability of absorption is proportional to $\cos^2\alpha$, where α is the angle between the direction of polarization and the absorption dipole in the molecule (see Figure 2). If emission is viewed through an analyzing polarizer the signal intensity is proportional to $\cos^2\beta$, where β is the angle between the emission dipole and analyzer polarization direction, and the total probability is proportional to $\cos^2\alpha\cos^2\beta$. The emission is commonly observed at right angles to excitation through polarizers set parallel then perpendicular to the excitation polarization (see Figure 2). With assumptions about the distribution and motion of the chromophores the emission probability can be related to the measured intensities.[23] The polarization is often characterized by a quantity called the anisotropy, defined as

$$r = (I_{\parallel} - I_{\perp})/(I_{\parallel} + 2I_{\perp})$$

where the total intensity is $(I_{\parallel} + 2I_{\perp})$ and $I_{\parallel}$ and $I_{\perp}$ are intensities for polarizers set parallel and perpendicular. Occasionally the polarization

$$p = (I_{\parallel} - I_{\perp})/(I_{\parallel} + I_{\perp})$$

may be used, but this quantity is not normalized for the total emitted intensity. Experimentally it is important that a calibration is made, in order to allow for any systematic intensity changes, due to

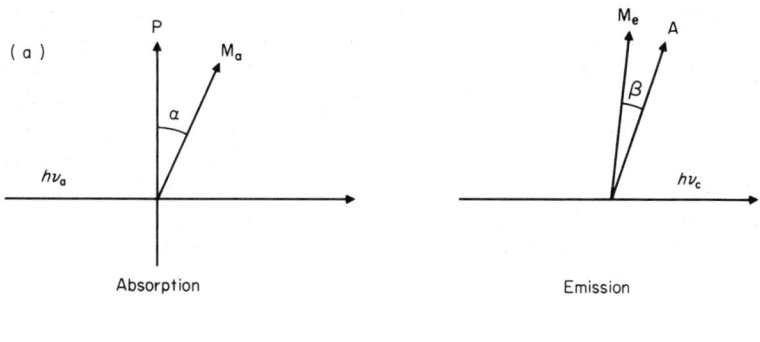

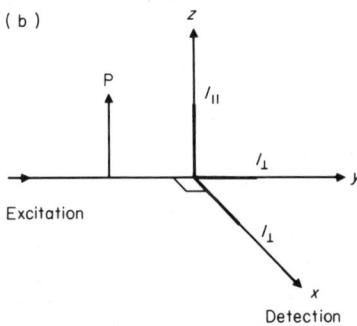

Figure 2 Geometrical arrangement for fluorescence polarization: (a) excitation is polarized along P, the analyzer along A and M is the transition dipole direction; (b) emission is detected along the x direction, excitation is polarized along z and travels along y. Polarizers select emission $I_{\parallel}$ or $I_{\perp}$; the total emission intensity is $I_{\parallel} + 2I_{\perp}$

instrumental effects, between $I_{\parallel}$ and $I_{\perp}$. Both steady state and time-resolved anisotropy can be measured.

The anisotropy is also related to the correlation function for a dipole at zero time and at some time later, *i.e.*

$$r(t) \quad = \quad 2/5 \langle M(0) \cdot M(t) \rangle$$

A model describing the chromophore motion is used to calculate the correlation function; in the simplest case of a sphere of volume V in a solution of viscosity η

$$r(t) \quad = \quad 2/5 \exp(-6Dt)$$

where $D = kT/6V\eta = 1/6\tau(\text{rot})$. An approximate estimation of the rotation time can be made for a spherical polymer: *i.e.* $\tau(\text{rot}) = 3.03 \times 10^{20} \, \eta R^3/T$, where η (cP) is the viscosity of the liquid surrounding the polymer, R(nm) the radius and T(K) the temperature. As r(t) measures the difference in signals, the signal-to-noise ratio worsens as the fluorescence intensity falls. Therefore probes with as long a decay time as possible, compared to the time scale of the longest rotational motion, are preferred. For very slow rotational motions triplet state probes may be used.[24]

If the polymer is in solution, the form of the anisotropy can be loosely described as a fast initial decay due to local motion of the probe and nearby attached groups, followed by a slower decay due to motion of the bulk polymer (or micelle, *etc.*). Probes attached at or near to the end of the chain move much more rapidly than those attached to the interior of the polymer, due to constraints formed by the chain. With most solid polymers only the local motion may be observable, with a large anisotropy persisting to long times, as no bulk rotational motion is possible.

As discussed above, interpretation of the anisotropy relies on a model describing the motions of the probe. The simplest is the Debye–Stokes–Einstein model, which assumes that the rotating body is a rigid ellipsoid moving in an isotropic fluid whose important properties are its temperature and viscosity. The anisotropy comprises five exponential terms, some of which become zero if the absorption and emission dipoles are parallel and the transition dipole lies along one of the symmetry axes of the rotating body.[23] At long times the anisotropy reaches zero as the rotational motion randomizes the initial selection of molecular dipoles. In addition to the whole-body motion of the polymer there is local (or internal) motion of the probe chromophore. This internal motion can be described on the assumption that the probe molecule moves in a restraining potential. The resulting expression for the anisotropy now contains nine exponential expressions involving order parameters P_2 and P_4, diffusion constants parallel and perpendicular ($D_\parallel$ and $D_\perp$) to the chromophore long axis, and angles between the absorption and emission dipoles and the chromophore long axis.[22,23] Choosing chromophores such that these angles are zero simplifies the expression to three exponentials plus a constant. The experimental data does not usually allow analysis of so many terms, usually only two decay times plus any constant anisotropy at long times are measurable. The additional assumption that the diffusion coefficient for perpendicular motion $D_\perp$ is so small as to be zero during the experiment further simplifies the analysis. It is clear, however, that obtaining reliable diffusion constants in polymer systems is very difficult and that, at best, some weighted mean of the diffusion constants is measured.

The fluorescence anisotropy has also to be calculated by considering the various segmental motions (crankshaft, *etc.*) of a polymer chain moving on a tetrahedral lattice.[25,26] Three- and four-bond motions were considered with the three-bond motions dominating, since many four-bond conformations are inhibited in carbon–carbon skeleton polymers. The correlation function has the form of an exponential times an error function

$$\langle M(0) \cdot M(t) \rangle = \exp(t/r)\,\mathrm{erfc}\,(t/r)^{1/2}$$

The relaxation time is r. Two modifications have been made to this model: a polymer will be characterized by a range of relaxation times of which r is a weighted average and, secondly, internal rotation angles and valence angles differ from those of a tetrahedral lattice. This second effect can be allowed for by multiplying the correlation function by $\exp(-t/p)$, where p reflects the deviation from the ideal lattice. This model was found to be reasonably good at describing the anisotropy decay in viscous solutions, but less good at lower viscosities. Polymers with 9,10-diphenylanthracene[26] or anthracene[20] attached either to the backbone of polystyrene or at its ends were used. Motion at the chain ends was found to have a mean rotation time four to five times faster than that of the backbone.

22.2 APPLICATIONS

22.2.1 Cyclization Dynamics

In fluid solution both triplet–triplet annihilation and the formation of excimers depends on the ability of molecules to diffuse freely towards one another. Consequently in a polymer similar groups attached to the chain ends can be used to study the cyclization dynamics. As triplet–triplet annihilation is diffusion controlled, the rate of cyclization can be measured, either from the decay of delayed fluorescence or triplet–triplet absorption. With intense illumination, such as from a ruby laser at 347 nm, a large fraction of the chains in a sample of end-labelled polymer has both anthracenes simultaneously excited to their triplet state.[28] The rapid decay of these triplets (monitored by triplet absorption) measures the cyclization rate when the intrinsic first-order decay of the triplets is taken into account. These rates are in the range 10^3 to $5 \times 10^5\,\mathrm{s}^{-1}$ and vary with chain length N as $K \approx N^{-x}$, where x is 1 in a good solvent (benzene) and $110 < N < 3000$. In a poor solvent such as cyclohexane, $x = 1.5$ for short chains and 1 for $N > 300$.

The kinetics are more complicated when the end groups form excimers. As shown in Scheme 2, the rate of excimer formation k_1 measures the mean rate of cyclization, since the excimer can dissociate

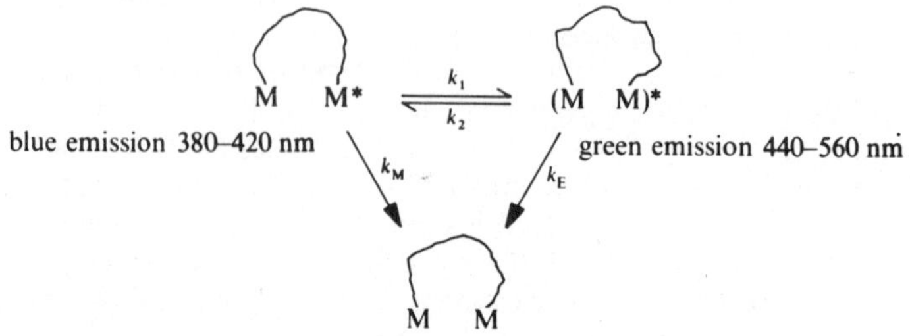

$$
\begin{array}{ccc}
\mathrm{M\quad M^*} & \underset{k_2}{\overset{k_1}{\rightleftharpoons}} & \mathrm{(M\quad M)^*} \\
\text{blue emission 380–420 nm} & & \text{green emission 440–560 nm}
\end{array}
$$

k_M \qquad k_E

$$\mathrm{M\quad M}$$

Scheme 2

either back to an excited end group or to the ground state. The rate constant for decay of the excited end group, k_m, is measured from a singly labelled polymer. With pyrene chromophores as end groups in several different polymer/solvent systems, such as polystyrene (Py–PS–Py) in cyclohexane or toluene and poly(dimethylsiloxane) (Py–PDMS–Py) in toluene, the cyclization rate is in the region of 10^4 s^{-1} and decreases with chain length as $N^{-1.5}$ (see Figure 3).[29,30] Excimer dissociation was found to be more important for short chains, where the fluorescence intensity could be described as the sum of two exponential terms. With longer chains the decay could be described by a single exponential, which is consistent with a single relaxation time of the polymer.

22.2.2 Transitions

In 1962 Oster *et al.*[31] demonstrated that oxygen quenched the phosphorescence of luminescent compounds in polymer films. This observation has since been used as a probe of transitions in many polymers[32] because as the temperature falls, the phosphorescence increases in a discontinuous

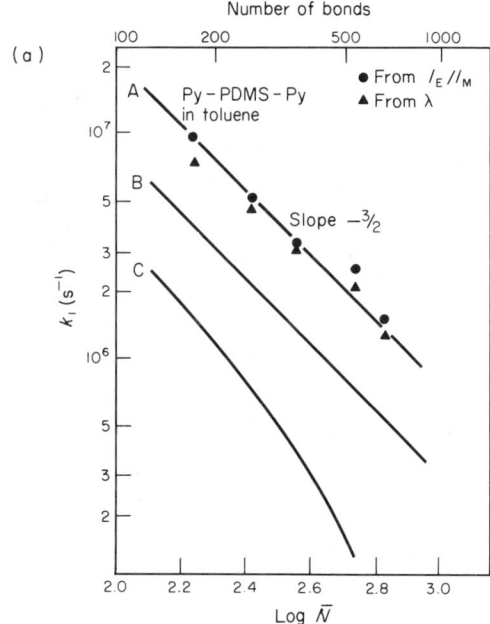

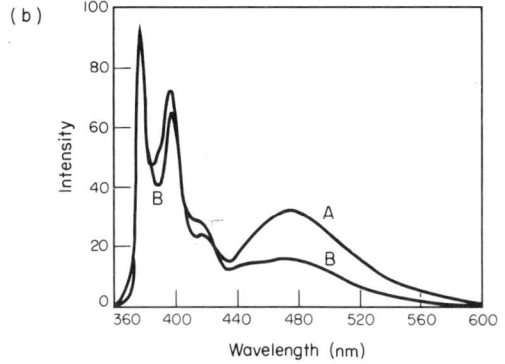

Figure 3 (a) Log–log plot of the mean cyclization rate constant (k_1) *vs.* mean chain length N. Curve A: data from these experiments for Py–PDMS–Py in toluene at 22 °C (● from I_E/I_M; ▲ from fluorescence decay); the line drawn through the points has a slope of −3/2. Curve B: summary of previous results for Py–PS–Py in cyclohexane at 34.5 °C; the experimental slope is −1.52. Curve C: summary of previous results for Py–PS–Py in toluene at 22 °C. (b) Monomer and excimer emission (480 nm) of Py–PDMS–Py (A) and Py–PS–Py (B) in toluene and cyclohexane respectively (after Svirskaya *et al.*, ref. 28).

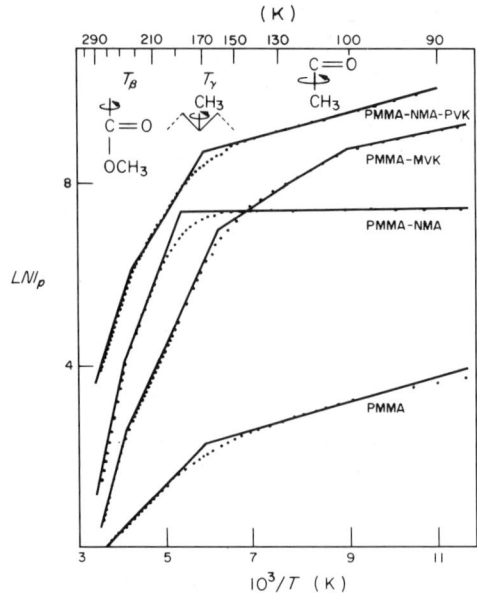

Figure 4   Arrhenius curves for poly(methyl methacrylate) phosphorescence (from Somersall *et al.*, ref. 32).

manner (see Figure 4). The phosphorescence yield is given by the expression

$$\phi_{\mathrm{p}} \;=\; \frac{k_{\mathrm{isc}}k_{\mathrm{p}}}{(k_{\mathrm{f}} + k_{\mathrm{isc}})(k_{\mathrm{p}} + k_{\mathrm{nr}})}$$

where k_{p} is the phosphorescence rate constant and k_{nr} is the sum of all the rate constants of the non-radiative processes depopulating the triplet state (see Figure 1). The phosphorescence intensity is indirectly proportional to the product of the number of photons absorbed and ϕ_{p}. If the temperature of the sample is changed, each of the rate constants (radiative and non-radiative) may vary. It has been observed in studies of aromatics in glasses that the intrinsic change in rate constants is small when compared to the 1000-times increase in phosphorescence observed between 300 and 77 K in polymers such as polyacrylonitrile, polystyrene and copoly(methyl methacrylate/naphthyl methacrylate) (see Figure 4).[32] The low yield at high temperatures is attributable to bimolecular quenching of the triplet state by O_2 molecules. The data was analyzed on the assumption that the O_2 quenching step has an activation energy and that the rate constant takes the form $k_0 = ke^{-\Delta E/kT}$. Incorporating this into the equation for ϕ_{p} and setting $k_{\mathrm{nr}} = (k' + k_0)$ gives

$$1/\phi_{\mathrm{p}} \;=\; A[(k' + k_{\mathrm{p}}) + ke^{-\Delta E/kT}]$$

where $A = (k_{\mathrm{p}} + k_{\mathrm{isc}})/(k_{\mathrm{f}} k_{\mathrm{p}})$. Experimentally the first term $(k' + k_{\mathrm{p}})$ is small compared to k; and a plot of $\ln(\phi_{\mathrm{p}})$ *vs.* $1/kT$ has a slope equal to the activation energy.

At certain temperatures an abrupt change in the slope occurs. This change is attributed to a phase transition in which a particular type of molecular motion begins. These motions involve either large segments of the chain (20–40 units) at the rubber–glass transition temperature (T_{g}), or cooperative but restricted motions of a few units at T_β, or side chain motions involving rotations of small groups at T_γ. The changes in slope of the $\ln(\phi)$ *vs.* $1/T$ plots is attributed to enhanced quenching arising from a greater O_2 diffusion rate above the transitions than below. At sufficiently low temperatures the phosphorescence intensity should become constant as all O_2 motion will be frozen out during the excited-state lifetime. In some polymers a slight increase in phosphorescence with falling temperature is still observed. This could be due to the intrinsic temperature dependences of the rate constants k_{f}, k_{isc}, k_{p}, *etc.* or to the direct influence of residual polymer motions on the excited state. A more detailed discussion of these phosphorescence-quenching results has been given by Guillet.[19, 32]

22.2.3 Miscibility

Both singlet–singlet energy transfer and excimer formation depend upon the proximity of an excited molecule, either to an acceptor or to a similar ground state molecule. If a mixture of two polymers is produced and then rapidly cooled, the early stages of phase separation may be observed by fluorescence methods. The extent of energy transfer and excimer formation, as monitored by the fluorescence of the excited state present, will change with time as interpenetrating chains in the mixture give way to small and growing volumes of the two phases. The change in excimer to chain unit (monomer) fluorescence intensity ratio has been used to study phase separation, *e.g.* spinodal decomposition in the system polystyrene/poly(vinyl methyl ether).[33] The fluorescence detected changes in phase composition long before discernible morphology could be observed by optical microscopy. The growth rate of the dominant concentration fluctuation during the initial stages of spinodal decomposition was found to decrease with increase in the polystyrene molecular weight.

The compatibility of carbazole-labelled PMMA and anthracene-labelled copoly(styrene/acrylonitrile) has been measured[34] from the ratio of the intensity of carbazole–PMAA fluorescence to that of anthracene–SAN (I_C/I_A). Only the carbazole residue was excited at 296 nm and energy transfer to anthracene occurred as the fraction of copolymer in the mixture was increased. A clear decrease in the I_C/I_A ratio occurred at 30–40% copolymer, indicating the limit of compatibility.

Other systems studied by fluorescence methods include blends of poly(2-vinylnaphthalene) with polystyrene and PMMA[35] and anthracene-labelled polystyrene with poly(α-methylstyrene).[36]

22.3 ANALYSIS OF COMMERCIAL POLYMERS

Over the last 25 years luminescence spectroscopy has become an extremely useful analytical tool in polymer science. The technique has provided valuable information on the mechanisms of oxidation and stabilization of polymers[37-39] and is widely used in the identification of polymers and their additives.[37-42]

22.3.1 Type A Polymers

Type A polymers include polyalkenes, synthetic rubbers, aliphatic polyamides, polyurethanes, poly(vinyl halides), aliphatic polyesters, polystyrenes, polyacrylics and polyacetals. Typically the chromophores present in these polymers are carbonyl and α,β-unsaturated carbonyl groups, *e.g.* these chromophores have been found to be responsible for the fluorescence and phosphorescence emissions from polyalkenes,[43-46] aliphatic polyamides,[47-51] synthetic rubbers,[52,53] polyacetals[54,55] and poly(vinyl chloride).[56-58] Low molecular weight aliphatic molecules of this type have very low fluorescence and phosphorescence yields due to spin–orbit-coupling induced transitions between singlet and triplet states. Twisting about unsaturated bonds may also reduce yields. Cyclic α,β-unsaturated carbonyls have also been reported to luminesce.[51] Consequently the origin of the luminescence from this type of polymer has been the subject of some controversy. In the case of polyalkenes, such as low density polyethylene and polypropylene, it has been suggested that the presence of polynuclear aromatics, such as naphthalene, is responsible for the fluorescence emission.[59-62] Some of the fluorescent species could be extracted from the polymer by *n*-hexane, and the fluorescence was subsequently regenerated in air over a period of time.[59-64] However the fluorescence emission spectra of *n*-hexane extracts of four polyalkenes [polypropylene, low and high density polyethylene and poly(4-methylpent-1-ene)] have been found to differ from that of naphthalene recorded under the same conditions.[46] An example of this is shown in Figure 5, where it is evident that the excitation spectrum of naphthalene does not resemble that of the *n*-hexane-extracted species from polypropylene.[46] Furthermore, the results in Figure 6 demonstrate that the fluorescence regeneration process, which follows the extraction of films of polypropylene with *n*-hexane, is an oxidative one, since the regeneration of fluorescence is much faster in pure oxygen than in air.

Aliphatic polyamides, such as nylon 6,6, have more complex spectra. Nylon 6,6 contains at least two major fluorescent species (emitting at 326 and 420 nm), as shown by the spectra in Figure 7.[51] The fluorescence emission at 326 nm is similar to that of the extractable species in polyalkenes, but there is another chromophore which emits at 390 nm and 420 nm which has been associated with ketoimide groups ($-CH_2COCONHCOCH_2-$) introduced into the polymer chain during manufacture and thermal processing.[65] The phosphorescence emission from nylon 6,6 is also complex, which is due to the variety of carbonyl groups present.[51] A recent study[51] combining GC–mass spectrometry and luminescence, has shown that all the fluorescent and phosphorescent species in nylon 6,6 can originate from the aldol condensation reaction of cyclopentanone, the latter being a thermal degradation product of the polymer.

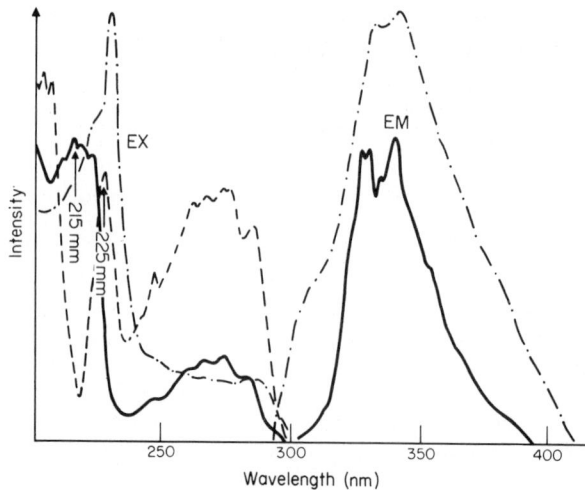

Figure 5   Fluorescence spectra: (———) absorption spectrum of naphthalene in *n*-hexane; (–·–·–) emission spectrum (EM) of an *n*-hexane extract of polypropylene (1 g powder/40 cm^3 *n*-hexane); (– – –) excitation spectrum (EX) of naphthalene in *n*-hexane (reproduced from ref. 46 with permission from Elsevier Applied Science Publishers Ltd.)

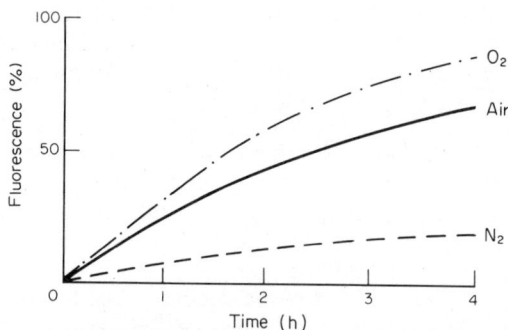

Figure 6   Rate of regeneration of fluorescence emission from *n*-hexane-extracted polypropylene film in: (———) air, (– – –) N_2 and (–·–·) O_2. Excitation at 280 nm; emission at 340 nm (reproduced from ref. 46 with permission from Elsevier Applied Science Publishers Ltd.)

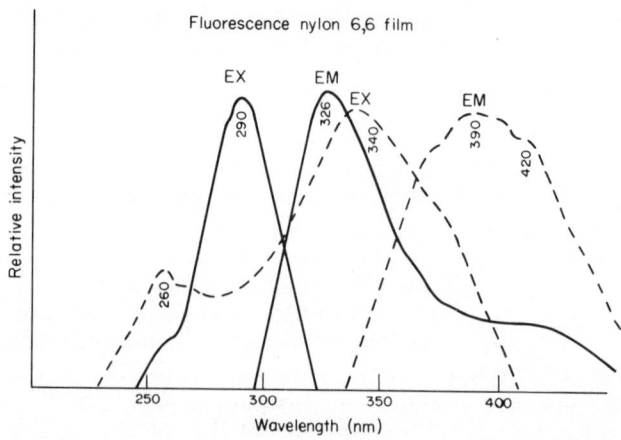

Figure 7   Fluorescence spectra of nylon 6,6 film (100 µm thick). (———) emission spectrum (EM) after excitation at 290 nm and excitation spectrum (EX) produced from species emitting at 326 nm; (– – – –) emission spectrum (EM) after excitation at 340 nm and excitation spectrum (EX) produced from species emitting at 390–420 nm (reproduced from ref. 51)

Unsaturated carbonyl groups are also produced by the thermal oxidation of polybutadiene[66] and PVC.[57] Some early workers attributed the fluorescence of PVC to the presence of polyconjugation,[56] but more recent studies suggest that carbonyl groups are essential for emission to occur.[57,58]

Other commercially important type A polymers are the polyurethanes. In a study of a polyurethane with constitutive unit —$ROCONHC_6H_4CH_2C_6H_4NHCOO$—, where R is an aliphatic polyester or a glycol residue, the phosphorescence emission has been associated[67] with the presence of a benzophenone-type chromophore. Other workers,[68] however, believe that the emission is due to a triplet excimer.

The luminescence spectroscopy of wool has received attention, because of the photodegradation that takes place in sunlight.[69-72] Scoured wool is essentially pure protein which, although the protein has different structures and compositions throughout the fibre, is all referred to as keratin. The fluorescence from wool is primarily that from tryptophan, the excited state of which is influenced by nearby polarizable groups in the protein.[73]

22.3.2 Type B Polymers

Type B polymers include the aromatic polyesters, poly(ether sulfones), aromatic polyamides, some polysulfides, polycarbonates, poly(phenylene oxide), polyimides and many aromatic resins (*e.g.* phenolic resin).

The fluorescence of poly(ethylene terephthalate) has been attributed to an associated ground state complex formed between terephthalate units, whereas the phosphorescence emission originates from the constitutive unit.[74] The emission spectrum of poly(*p*-xylylene) exemplifies the complex nature of the fluorescence emission from highly conjugated aromatic polymers.[75] Its fluorescence emission has three components: arising from the constitutive unit, an excimer and a ground state complex. Similarly the phosphorescence emission from bisphenol A (*i.e.* $HOC_6H_4CMe_2C_6H_4OH$) epoxy resin has also been associated with the constitutive unit, a triplet excimer and a ground state dimer.[76] The phosphorescence of poly(phenylene oxide) [poly(oxy(2,6-dimethyl)-1,4-phenylene)] is due solely to the constitutive unit: in addition to this emission fluorescence also arises from xanthone and quininoid oxidation products.[77]

The fluorescence and phosphorescence spectra of commercial polystyrene are very complex and their interpretation has attracted some controversy in recent years.[59,60,78-83] The main features of the fluorescence spectrum serve to classify this polymer as type B, whereas the presence of impurities leads to features consistent with type A. The fluorescence spectrum is very dependent on the commercial source and purity of the polymer.[81-84] Thin films of 'pure' polymer show only excimer emission (335 nm) on excitation at the polymers absorption maximum ($\lambda_{max} = 250$ nm),[78,83] but all commercial samples also exhibit fluorescence emission at about 280 nm, due to residual styrene monomer.[79,83] In solution emission from both constitutive unit and excimer is observed.[78,80] The intensity of the excimer emission increases with increase in the concentration of the polymer and a red shift in the wavelength maximum also occurs.[80] This has been attributed to intermolecular interaction of adjacent phenyl chromophores.[80] On excitation with light of wavelengths greater than 300 nm, commercial polystyrene also exhibits a fluorescence with emission maxima at 338, 354 and 372 nm,[83] which has been attributed to the presence of *trans*-stilbene linkages. The phosphorescence has been assigned to the presence of acetophenone-type terminal groups.

22.3.3 Identification of Commercial Polymers

Luminescence spectra are used for the identification of polymers. Phosphorescence lifetimes are an additional and useful diagnostic, and are readily determined with modern fluorimeters, by means of a chopped or stroboscopic light source.

The spectroscopic method is rapid and non-destructive and sample preparation is simple. It is highly sensitive, *e.g.* picomolar concentrations of aromatics can be detected by fluorescence spectroscopy. It can best be used to place polymers within a particular class, *e.g.* the nylons. Hence it is best used in conjunction with other analytical techniques, such as IR spectroscopy.

The method is not immediately applicable to polymers which are non-luminescent or only weakly so. However, this problem can sometimes be overcome by thermal oxidation of the material under controlled conditions. The non-volatile oxidation products may be luminescent and are often characteristic of a particular polymer.[37] Some light stabilizers severely reduce the intensity of the polymer emissions and, if present, must be removed by solvent extraction. Other additives, such as

Table 1 Fluorescence and Phosphorescence Properties of Polymers

Polymer	Form	Excitation (nm)	Emission (nm)	Mean lifetime (s)	Chromophore
			Fluorescence		
Poly(ethylene naphthalate)	Chip	320, 344, 357 (s)	370, 389, 405	—	Polymer (dimer)
	Film	344, 357	370, 389, 405	—	Polymer (dimer)
	Fibre	344, 357	370, 389, 405	—	Polymer (dimer)
Poly(ethylene 2,6-naphthalate)	Film	375	435	—	Polymer (excimer)
Polyurethane — MDI-based[3]	Film	372	420	—	Unknown
Nylon 6,6	Chip	357	417	4×10^{-9}	α-Ketoimides
	Fibre	357	417	4×10^{-9}	α-Ketoimides
Nylon 6	Chip	335	390	—	α-Ketoimides
Nylon 6,10	Chip	345, 355	395, 410	—	α-Ketoimides
Nylon 11	Chip	327, 340	375 (s), 395 (s) 385 (s)	—	α-Ketoimides
Nylon 12	Chip	410	450	—	α-Ketoimides
Poly(vinyl fluoride)	Film	290	350	—	α,β-Unsaturated carbonyl (enone or -al)
Poly(vinyl fluoride) (heated)	Film	325	410	—	α,β-Unsaturated carbonyl (dienone or -al)
Poly[oxy(2,6-dimethyl)-1,4-phenylene]	Solution	290	310	—	Polymer
Wool	Film	495, 515	550, 565	—	Xanthanoid (oxidation)
	Fibre	290	345	—	Aromatic tryptophyl residues (N-formylkynurenine)
Poly(m-benzamide)	Film	405	500	—	Polymer
Poly(m-phenyleneisophthalamide)	Film	350, 410 (s)	465	—	Polymer
Poly(m-phenyleneterephthalamide)	Film	375	470	—	Polymer
Bisphenol A-based epoxy resins	Solutions	350	424	—	Impurity in amine
Poly(tetrafluoroethylene)	Film	328	350	—	Unknown
Poly(vinyl alcohol)	Film	258 (s), 295, 330	360, 370 (s)	—	Unknown
Poly(butylene terephthalate)	Solution	255, 290	324	—	Polymer
	Chip	332	400, 420, 450	—	2-Hydroxyterephthalic acid units
Polyethylene (low density)	Powder	230, 265 (s), 300	335 (s), 350	—	Cyclic α,β-unsaturated carbonyl groups of the enone (or -al) type
	Film	230, 273	295 (s), 310, 329 (s), 354 (s), 370 (s)		
Polyethylene (high density)	Film	230, 265 (s), 290	295 (s), 312 (s), 330 (s), 344 (s), 358		
Copoly(ethylene/vinyl acetate)	Film	230, 265 (s), 290	312 (s), 330, 344 (s) 358 (s)		
Polypropylene	Film	230, 285	309 (s), 320		
Poly(4-methyl-pent-l-ene)	Film	230, 285	310, 330		
Polystyrene	Chip	318, 330	336, 354, 368 (s)		trans-Stilbene
Poly(ether sulfone)	Film	320	360	—	Polymer (excimer)
Poly(p-xylylene)	Film	265	296	—	Monomer
		265	365	—	Excimer
Poly(oxymethylene)		307, 327, 339	351 (s), 370, 390 (s)	—	Polymer (dimer)
		210, 285	320	—	α,β-Unsaturated carbonyl groups

Polymer	Form	Excitation (nm)	Phosphorescence (nm)		Assignment
Poly(ethylene terephthalate)	Chip	280, 318, 351	425 (s), 460	0.5	Polymer
	Fibre	284, 310	425 (s), 477	0.7	Polymer
Poly(ethylene 2,6-naphthalate)	Film	375	580	—	Polymer
Poly(urethane — MDI-based	Film	320	423, 455, 489	0.02	Benzophenone-type
Nylon 6,6	Chip	296	400	2.1	
	Fibre	296	430	1.3	
Nylon 6	Chip	282	390 (s), 420, 455 (s)	1.7, 1.6, 1.1	Cyclic α,β-unsaturated carbonyl groups of dienone (or -al) type
Nylon 6,10	Chip	300	430	0.70	
Nylon 11	Chip	296 (s), 273 (s)	423, 450 (s)	1.0, 0.88	
Nylon 12	Chip	268, 286 (s)	363 (s), 410	1.0	
Poly(vinyl chloride)	Film	284	440	0.30	Carbonyl
Poly(tetrafluoroethylene)	Film	260–280[4]	450	0.40	Carbonyl
Poly(vinyl alcohol)	Film	260–280[4]	436	0.40	Unknown
Polyethylene (low density)[5]	Powder	273, 280	367, 381, 391, 405, 416	2.30	Benzoic acid
	Film	278, 280	420	0.60	Benzoic acid
Polyethylene (low density)[6]	Chip	283, 331	370 (s), 435, 455	2.15	
Copoly(ethylene/vinyl acetate)	Film	280, 327	455	0.35	Cyclic α,β-unsaturated carbonyl groups of dienone (or -al) type
Polypropylene	Film	270, 290, 330	420, 445, 480, 510 (s)	0.5–1.2	
Poly(4-methyl-pent-1-ene)	Film	273, 330	430	0.86	
Polyethylene (high density)	Film	275	450	0.35	
Polystyrene	Film	290 (s), 300, 336 (s)	398, 425, 456, 492	0.008	Acetophenone end groups
Poly(ether sulfone)	Film	320	450	0.05	Polymer
Poly(oxymethylene)	Film	290, 320 (s)	415	1.0	α,β-Unsaturated carbonyl species
Wool		320	405, 425, 450	4.76	Tryptophone residues
Bisphenol A epoxy resin	Film	275, 350	460	—	Monomer/dimer and triplet excimer
Poly(oxy(2,6-dimethyl)-1,4-phenylene)		325	460	0.04	Polymer
Poly(butylene terephthalate)	Fibre	305	450	1.2	Polymer
Poly(p-xylylene)		280	402, 430, 447, 461, 478, 493	—	Dimer

[1] Adapted from N. S. Allen, J. Homer and J. F. McKellar, *Analyst*, 1976, **101**, 260 with permission from the Royal Society of Chemistry, London.

[2] (s) = shoulder.

[3] MDI = diphenylmethane 4,4-diisocyanate.

[4] Broad and structureless spectrum.

[5] Prepared using a benzoyl-based catalyst.

[6] Prepared using oxygen or aliphatic peroxide catalyst.

antioxidants and pigments, will almost certainly be present. In many cases these additives will not interfere significantly, but may themselves be identified by luminescence analysis (see Sections 22.3.4 and 22.3.5).

The luminescence characteristics of a number of commercial polymers are shown in Table 1, together with the identities of the chromophores believed to be responsible for the various emissions. All the polymers shown in the table tend to fluoresce in the wavelength range 300–450 nm and phosphoresce in the range 400–600 nm. In certain cases the spectra are highly structured and this can assist in the identification of the polymer. The method of manufacture of a polymer can influence the nature of the luminescence. For example, the phosphorescence emission of polyethylene prepared using a benzoyl-peroxide-type catalyst differs from that of polymer prepared using oxygen as a catalyst.[44] In Table 1 the form of the polymer (chip, film, *etc.*) is noted, as this can give an indication of the processing history of the sample.

22.3.4 Identification of Additives

Many antioxidants and light stabilizers exhibit their own characteristic fluorescence and/or phosphorescence emissions and may therefore be analyzed, perhaps after solvent extraction from the polymer. The fluorescence and phosphorescence properties of a number of antioxidants and light stabilizers have been reviewed elsewhere.[42] The detection limits for additives vary quite markedly and this can be a problem, particularly for commercial materials containing mixtures of additives. Most commercial polymers contain both an antioxidant and a UV stabilizer, and these may have to be extracted and separated, using thin layer chromatography (TLC).

22.3.5 Identification of Pigments

Certain types of pigments, such as dyes, exhibit their own characteristic luminescence spectra. Of the many types of pigments available, white pigments are the most widely used, particularly titanium and zinc oxides. The two crystalline modifications of titanium dioxide, anatase and rutile, may be easily identified by their characteristic emissions.[86-89] At low temperatures anatase exhibits a strong emission spectrum at 540 nm, while rutile exhibits weak emission in the IR region at 815

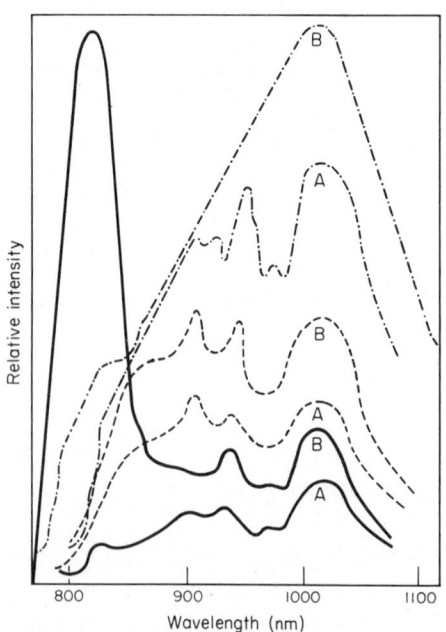

Figure 8 Emission spectra of pigments in polyethylene at room temperature: (———) spectrum of a 'sulfate' processed rutile pigment; (– – – –) spectrum of a 'chloride' processed rutile pigment; (–·–·–·) spectrum of a 'sulfate' processed anatase pigment. The excitation wavelength was 375 nm. Labels A and B indicate sensitivities of × 100 and × 30 respectively (reproduced from ref. 97 with permission)

and 1015 nm. The excitation λ_{max} of the two crystalline forms are 340 and 375 nm respectively. The nature of the surface treatment, often applied to titanium dioxide pigments also affects the intensity of emissions.[87]

The manufacturing history of titanium dioxide pigments can be determined from their characteristic emission spectra in the IR region. For example, at low temperatures rutile pigments manufactured by the 'sulfate' process exhibit emission at 815 nm and 1015 nm, whereas those manufactured by the 'chloride' process exhibit emission at 1015 nm only (Figure 8). Anatase pigments also have an emission at 1015 nm, but of much stronger intensity than that from any of the rutile grades. Zinc oxide has also been reported to luminesce.[90, 91]

22.4 REFERENCES

1. G. Schulte, W. Grond, D. Haaver and R. Silbey, *J. Chem. Phys.*, 1988, **88**, 679.
2. J. R. Bacon and J. N. Demas, *Anal. Chem.*, 1987, **59**, 2780.
3. N. Turro, 'Modern Molecular Photochemistry', Benjamin, Reading, MA, 1978; J. Rabek, 'Experimental Methods in Polymer Chemistry', Wiley, New York, 1980; 'Polymer Photophysics', ed. D. Phillips, Chapman and Hall, London, 1985.
4. C. D. Tran and G. Beddard, *Eur. Biophys. J.*, 1985, **13**, 59.
5. J. B. Birks, 'Photophysics of Aromatic Molecules', Wiley, New York, 1970.
6. T. Nolan and N. Douichi, *Anal. Chem.*, 1987, **59**, 2803.
7. P. John and I. Soutar, *Anal. Chem.*, 1976, **48**, 520.
8. K. Tanabe, M. Glick, B. Smith, E. Voightman and J. Winefordner, *Anal. Chem.*, 1987, **59**, 1125.
9. D. U. Connor and D. Phillips, 'Time Correlated Single Photon Counting', Academic Press, London, 1984.
10. H. L. Chen and H. Morawetz, *Macromolecules*, 1982, **15**, 1445.
11. K. J. Shen, Y. Okahata and T. K. Dougherty, *Macromolecules*, 1984, **17**, 296.
12. T. Chen and J. K. Thomas, *J. Polym. Sci., Polym. Chem. Ed.*, 1979, **17**, 1103.
13. J. B. Birks, *Rep. Prog. Phys.*, 1975, **38**, 903.
14. K. P. Ghiggino, A. J. Roberts and D. Phillips, *Adv. Polym. Sci.*, 1981, **40**, 69.
15. H. Kaufman, W. D. Weielbaumer, J. Bauerbaumer, A. -M. Scholtner and O. F. Olaj, *Macromolecules*, 1985, **18**, 104.
16. A. Itaga, K. Okamoto, H. Masuhara, N. Ikeda, N. Mataga and S. Kusabayashi, *Macromolecules*, 1982, **15**, 1213.
17. F. Evers, K. Kobs, R. Memming and D. R. Terrell, *J. Am. Chem. Soc.*, 1983, **105**, 5988.
18. F. C. DeSchryuer, I. Vandendriessche, S. Toppet, K. Demeyer and N. Boens, *Macromolecules*, 1982, **15**, 406.
19. J. Guillet, 'Polymer Photophysics and Photochemistry', Cambridge University Press, Cambridge, 1985.
20. G. S. Beddard and C. D. Tran, *Eur. Biophys. J.*, 1985, **11**, 243.
21. G. J. Kettle and I. Soutar, *Eur. Polym. J.*, 1978, **14**, 895.
22. L. Best, E. John and F. Jähnig, *Eur. Biophys. J.*, 1987, **15**, 87.
23. T. J. Chuang and K. B. Eisenthal, *J. Chem. Phys.*, 1972, **57**, 5094.
24. R. Cherry, E. Nigg and G. S. Beddard, *Proc. Natl. Acad. Sci. USA*, 1980, **77**, 5899.
25. B. Valeur and L. Monnerie, *Eur. Polym. J.*, 1976, **14**, 1129.
26. N. Hasparyan-Tardiueau, B. Valeur, L. Monnerie and I. Mita, *Polymer*, 1983, **24**, 205.
27. P. Wahl, G. Meyer and J. Parrod, *Eur. Polym. J.*, 1970, **6**, 585.
28. K. Horie, W. Schnabel, I. Mita and H. Ushiki, *Macromolecules*, 1981, **14**, 1422.
29. A. E. C. Redpath and M. A. Winnik, *Polymer*, 1983, **24**, 1286.
30. P. Svirskaya, J. Danhelka, A. E. C. Redpath and M. A. Winnik, *Polymer*, 1983, **24**, 319.
31. G. Oster, N. Geacintov and A. Khan, *Nature (London)*, 1962, **196**, 1089.
32. A. C. Somersall, E. Dan and J. E. Guillet, *Macromolecules*, 1974, **7**, 233.
33. R. Geller and C. W. Frank, *Macromolecules*, 1983, **16**, 1448.
34. H. Morawetz, *Pure Appl. Chem.*, 1980, **52**, 277.
35. S. N. Semerek and C. W. Frank, *Adv. Chem. Ser.*, 1983, **203**, 757.
36. F. Mikes, H. Morawetz and K. S. Dennis, *Macromolecules*, 1984, **17**, 60.
37. N. S. Allen and J. F. McKellar, *Chem. Ind. (London)*, 1978, 907.
38. N. S. Allen, in 'Analysis of Polymer Systems', ed. L. S. Bark and N. S. Allen, Elsevier Applied Science, London, 1982, chap. 4.
39. G. A. George, *Pure Appl. Chem.*, 1985, **57**, 945.
40. S. W. Beavan, J. S. Hargreaves and D. Phillips, *Adv. Photochem.*, 1979, **11**, 207.
41. D. A. Holden and J. E. Guillet, in 'Developments in Polymer Photochemistry — 1', ed. N. S. Allen, Applied Science, London, 1981, chap. 3.
42. G. F. Kirkbright, R. Narayanqswarmy and T. S. West, *Anal. Chim. Acta*, 1970, **52**, 237.
43. N. S. Allen, J. Hamer and J. F. McKellar, *J. Appl. Polym. Sci.*, 1977, **21**, 2261.
44. N. S. Allen, J. Hamer and J. F. McKellar, *J. Appl. Polym. Sci.*, 1977, **21**, 3147.
45. N. S. Allen and J. F McKellar, *J. Appl. Polym. Sci.*, 1978, **22**, 625.
46. N. S. Allen, *Polym. Degradation Stab.*, 1984, **6**, 193.
47. N. S. Allen, J. F. McKellar and D. Wilson, *J. Photochem.*, 1976, **6**, 337.
48. N. S. Allen, J. F. McKellar and D. Wilson, *J. Polym. Sci., Polym. Chem. Ed.*, 1977, **15**, 2793.
49. N. S. Allen, J. F. McKellar and G. O. Phillips, *J. Polym. Sci., Polym. Chem. Ed.*, 1974, **12**, 1233.
50. N. S. Allen and J. F. McKellar, *J. Polym. Sci., Polym. Chem. Ed.*, 1974, **12**, 2623.
51. N. S. Allen and M. J. Harrison, *Eur. Polym. J.*, 1985, **21**, 517.
52. S. W. Beavan and D. Phillips, *J. Photochem.*, 1974, **3**, 349.
53. S. W. Beavan and D. Phillips, *Rubber Chem. Technol.*, 1975, **48**, 692.
54. N. S. Allen and J. F. McKellar, *Polym. Degradation Stab.*, 1979, **1**, 47.

55. O. Nishimura and Z. Osawa, *Polym. Photochem.*, 1981, **1**, 191.
56. E. D. Owen and R. L. Read, *Eur. Polym. J.*, 1979, **15**, 41.
57. N. S. Allen, J. Wooler and K. O. Fatinikun, *Polym. Degradation Stab.*, 1985, **13**, 277.
58. J. F. Rabek, R. Ranby and T. A. Showronski, *Macromolecules*, 1985, **18**, 1810.
59. A. Charlesby and R. H. Partridge, *Proc. R. Soc. London, Ser. A*, 1965, **283**, 312.
60. A. Charlesby and R. H. Partridge, *Proc. R. Soc. London, Ser. A*, 1965, **283**, 329.
61. I. Boustead and A. Charlesby, *Eur. Polym. J.*, 1967, **3**, 459.
62. D. J. Carlsson and D. M. Wiles, *J. Polym. Sci., Polym. Lett. Ed.*, 1973 **11**, 759.
63. Z. Osawa, H. Kuroda and Y. Kobayashi, *J. Appl. Polym. Sci.*, 1984, **29**, 2843.
64. Z. Osawa and H. Kuroda, *J. Polym. Sci., Polym. Lett. Ed.*, 1982, **20**, 577.
65. H. D. Scharf, C. D. Dieris and H. Leismann, *Angew. Markromol. Chem.*, 1979, **79**, 193.
66. S. W. Beavan, P. A. Hackett and D. Phillips, *Eur. Polym. J.*, 1974, **10**, 925.
67. N. S. Allen and J. F. McKellar, *J. Appl. Polym. Sci.*, 1976, **20**, 1441.
68. Z. Osawa and K. Nagashima, *Polym. Degradation Stab.*, 1979, **1**, 311.
69. I. H. Leaver, *Aust. J. Chem.*, 1979, **32**, 1961.
70. I. H. Leaver, *Photochem. Photobiol.*, 1978, **27**, 439.
71. G. J. Smith, *Text Res. J.*, 1976, **46**, 510.
72. K. P. Ghiggino, C. H. Nicholls and M. T. Pailthorpe, *J. Photochem.*, 1975, **4**, 185.
73. G. Smith, M. Thorpe, W. Melhuish and G. S. Beddard, *Photochem. Photobiol.*, 1980, **32**, 715.
74. N. S. Allen and J. F. McKellar, *Makromol. Chem.*, 1978, **179**, 523.
75. Y. Takai, J. H. Calderwood and N. S. Allen, *Makromol. Chem., Rapid Commun.*, 1980, **1**, 17.
76. N. S. Allen, J. P. Binkley, B. J. Parsons, G. O. Phillips and N. H. Tennant, *Polym. Photochem.*, 1982, **2**, 389.
77. N. S. Allen and J. F. McKellar, *Makromol. Chem.*, 1979, **180**, 2875.
78. M. T. Vala, R. Silbey, S. A. Rice and J. Jortner, *J. Chem. Phys.*, 1964, **41**, 2146.
79. L. J. Basille, *J. Chem. Phys.*, 1962, **36**, 2204.
80. T. Nishihara and M. Kaneko, *Makromol. Chem.*, 1969, **84**, 124.
81. G. A. George, *J. Appl. Polym. Sci.*, 1974, **18**, 419.
82. G. A. George and D. K. C. Hodgemann, *Eur. Polym. J.*, 1977, **13**, 63.
83. W. Klopffer, *Eur. Polym. J.*, 1975, **11**, 203.
84. N. S. Allen, J. Homer and J. F. McKellar, *Analyst*, 1976, **101**, 260.
85. N. S. Allen, J. Homer and J. F. McKellar, *Makromol. Chem.*, 1978, **17**, 1575.
86. N. S. Allen, J. F. McKellar, G. O. Phillips and D. G. M. Wood, *J. Polym. Sci., Polym. Lett. Ed.*, 1974, **12**, 241.
87. N. S. Allen, D. J. Bullen and J. F. McKellar, *Chem. Ind. (London)*, 1977, 797.
88. N. S. Allen, D. J. Bullen and J. F. McKellar, *Chem. Ind. (London)*, 1978, 629.
89. N. S. Allen, J. F. McKellar and D. Wilson, *J. Photochem.*, 1977, **7**, 319.
90. K. N. Pandey, P. S. Kanai and V. B. Singh, *Labdev, Part A*, 1971, **9**, 220.
91. G. Winter and R. N. Whittem, *Aust., Def. Res. Lab., Paint Notes*, August 1949, 252.

23

ESR Spectroscopy

G. GORDON CAMERON
University of Aberdeen, UK

23.1 INTRODUCTION

Electron spin resonance (ESR) spectroscopy has found wide application in polymer chemistry and physics in, for example, the identification and study of free radical intermediates formed during degradation, oxidation, irradiation, mechanical fracture, and radical polymerization.[1] In these cases the free radicals are generally transients or relatively short-lived intermediates, and species of this type will be dealt with elsewhere in these volumes. This chapter is concerned with the application of ESR spectroscopy to polymer systems containing stable organic radicals either dispersed in the matrix (spin probes) or covalently bonded at predetermined points to polymer chains (spin labels). The spin-probe and spin-label techniques are used mainly to gain information on dynamics, relaxations and orientations of polymer molecules, and therefore can be compared with allied methods such as mechanical, NMR, dielectric and ultrasonic relaxation measurements, luminescence depolarization and Rayleigh scattering. The ESR methods are sensitive to motions in the frequency range 10^2–10^{11} Hz.

Although a wide range of stable radicals is known, spin labels and probes have been drawn almost exclusively from the very large group of *N,N*-disubstituted NO radicals (nitroxides or nitroxyls) which contain the $>$N$\dot{-}$O moiety. Several factors contribute to this: (i) many easily synthesized nitroxides are remarkably stable, (ii) nitroxides with functional groups that facilitate labelling are available, and (iii) nitroxides have well-defined *g* tensors and hyperfine coupling tensors to the ^{14}N nucleus. The last property means that the shapes, separations and widths of the lines in the ESR spectrum of a nitroxide radical are sensitive to the speed and mode of tumbling, so that analysis of the spectrum can yield dynamic data. This is discussed in Section 23.3.

The spin-probe experiment, which is applied mainly to polymers in bulk, is the simpler of the two but has the disadvantage that probe motion does not reflect directly the motions of the matrix. The spin-label experiment, on the other hand, can yield information on the dynamics of that part of the polymer molecule to which it is attached (inner segments, chain ends, *etc.*), provided that the label does not rotate independently or interfere with the motion of the chain. This valuable feature of spin labelling is shared with other 'reporter group' techniques such as fluorescent labelling, but the ESR technique enjoys the advantage of unusually high sensitivity, which means that in favourable circumstances data may be obtained at spin concentrations of the order 10^{-6} M. Thus dilute solutions of polymers carrying one nitroxide group per chain can be studied. Indeed, in most applications it is necessary to work with relatively low spin concentrations in order to avoid line-broadening due to spin exchange.

In this chapter the basic chemical and theoretical aspects of spin-label and spin-probe investigations are described, with examples chosen to illustrate the scope and limitations of the technique.

23.2 NITROXIDE SPIN-LABEL AND SPIN-PROBE RADICALS

Many of the nitroxide radicals used in ESR studies of synthetic polymers are derivatives of di-*t*-alkylpiperidine and pyrroline in which the bulky side groups hinder decomposition reactions. Nitroxides carrying hydrogen atoms on the carbon adjacent to the nitrogen (shown in **1**) are

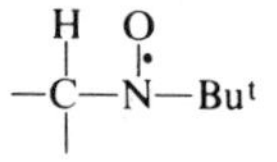

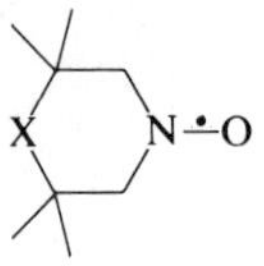

(1)

Table 1 Typical Spin Probe Nitroxide Radicals

2,2,6,6-Tetramethylpiperidine-1-oxy radicals	*2,2,5,5-Tetramethylpyrrolidine-1-oxy radicals*	*Miscellaneous radicals*
X in	Y in	
(2)	**(8)** —NH_2	**(12)**
(3)	**(9)** —OH	**(13)** R^1, R^2 = Me, alkyl
(4)	**(10)** —OMe	
(5)	**(11)**	**(14)**
(6)		**(15)**
(7) R = H, Ph		**(16)**

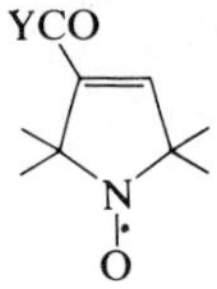

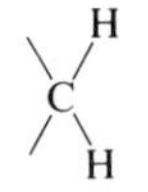

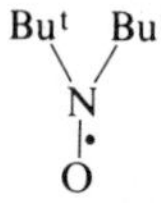

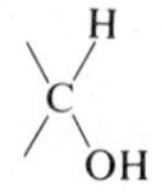

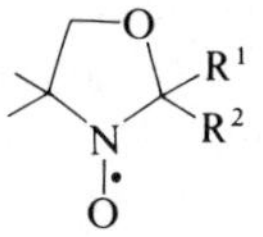

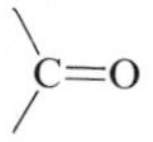

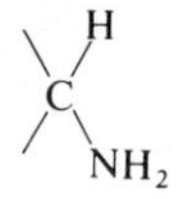

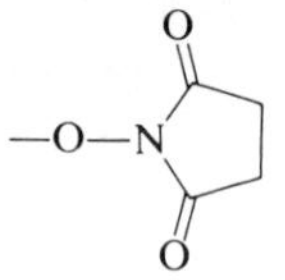

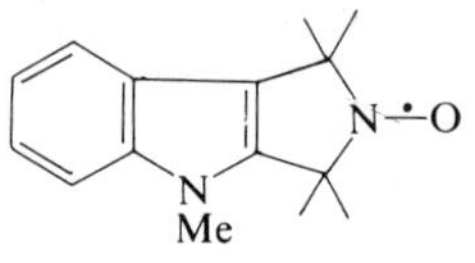

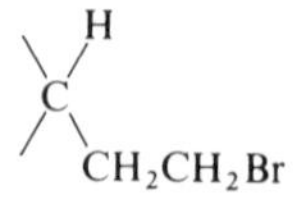

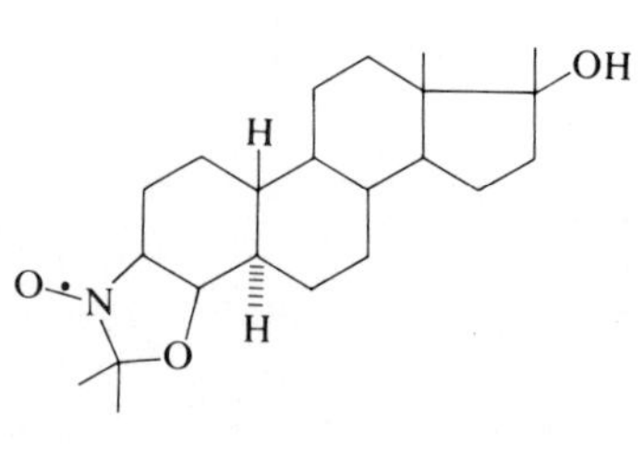

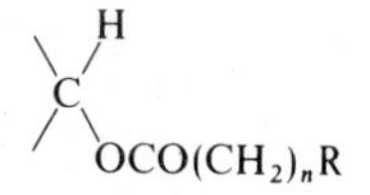

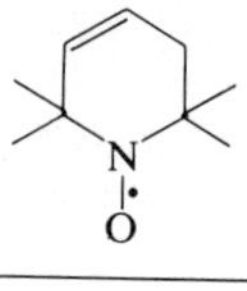

susceptible to disproportionation and other reactions, but at the high dilutions involved in spin-probe and spin-label experiments these decay processes are effectively retarded.

Typical examples of nitroxide spin probes are shown in Table 1. For an exhaustive list the reader is referred to other compilations[2] and to numerous texts and reviews dealing with the chemistry of nitroxides,[3-7] and their applications to polymers.[8-12] The diversity in molecular volume and structure of these probes is an asset to the polymer scientist, for example in providing information on free volume, microheterogeneity and effective segment dimensions in bulk polymers.[13,14]

23.2.1 Synthesis of Spin-labelled Polymers

Table 2 shows a selection of spin-labelled polymers, (28) to (32) being terminally labelled. Of the remainder, which are labelled elsewhere on the chain, (21), (22), (24), (26) and (33) constitute part of the polymer backbone (in-chain).

Methods of synthesizing the spin-labelled polymers can be divided conveniently into two classes: (i) methods which involve some post-polymerization reaction on the polymer, and (ii) those which incorporate the nitroxide (or its immediate precursor) as part of the polymerization process. Examples of labels prepared by both routes are shown in Table 2.

Some of the spin probes listed in Table 1 carry functional groups through which they may be linked, often by a condensation reaction, to an appropriate group on a preformed polymer. Thus, the hydroxyl group on probe (3) can be esterified directly with a carboxyl or an acid chloride group on a polymer to yield label (17) (Table 2).[15] Analogous reactions with the amino-functional nitroxide (5) yield label (18).[16] Labels (19) and (29) may be prepared by reacting free hydroxyl groups on the polymer with suitably functionalized nitroxides,[17] while label (20) is obtained by quaternizing some of the rings on poly(4-vinylpyridine) with the appropriate piperidine nitroxide.[18]

More sophisticated preparative methods are required in other cases. Thus labels (24) and (25) on polyethylene were prepared *via* the Keana synthesis[19] in which a carbonyl group on the polymer is reacted with 2-methyl-2-aminopropan-1-ol according to Scheme 1.[20,21]

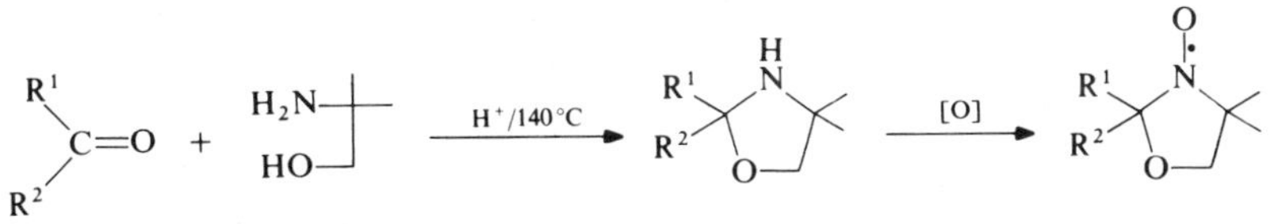

Scheme 1

The same technique can be used to prepare spin probes such as (13) where R^1 and R^2 are alkyl groups. Labels of structure (27) can be synthesized by reacting lightly lithiated polystyrene with 2-methyl-2-nitrosopropane (MNP) or nitrosobenzene as in Scheme 2.

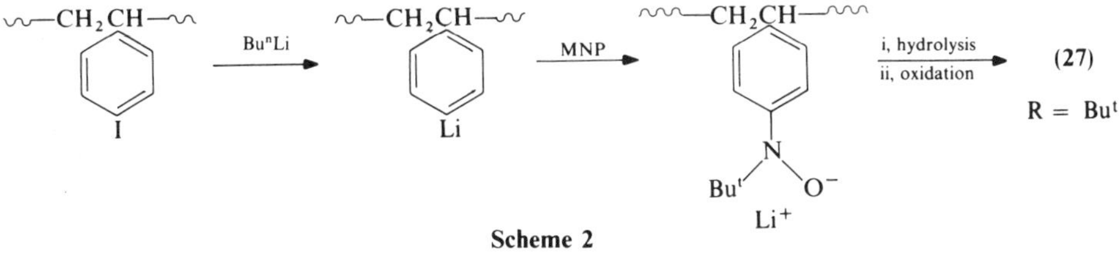

Scheme 2

The structure (23) is labelled a crosslink within an amine-cured epoxy resin resulting from reaction of free epoxy groups with the diamino nitroxide (28a).[22] The end label (28) on the same type of resin is formed by reaction of one free epoxide group with 4-methylamino-2,2,6,6-tetramethylpiperidine-1-oxyl.[22,23]

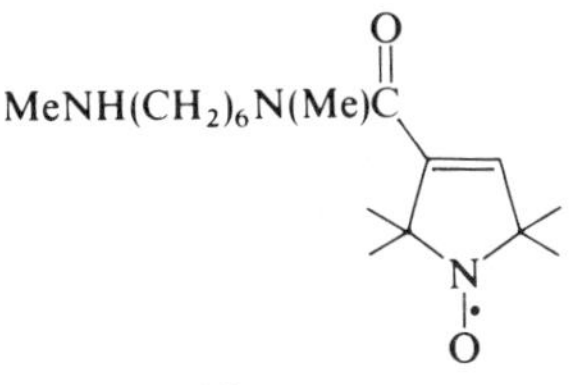

(28a)

Table 2 Representative Examples of Spin-labelled Synthetic Polymers[a]

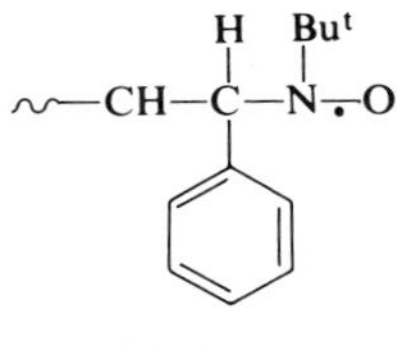

(30)

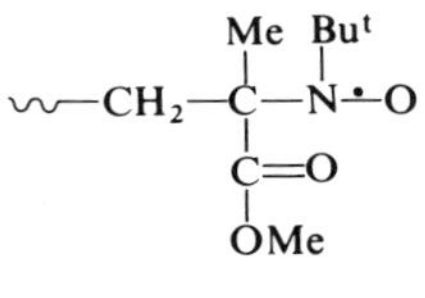

(31)

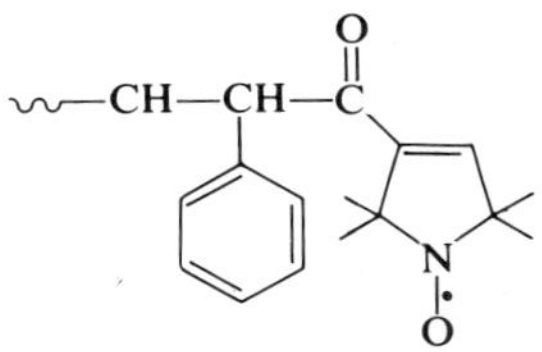

(32)

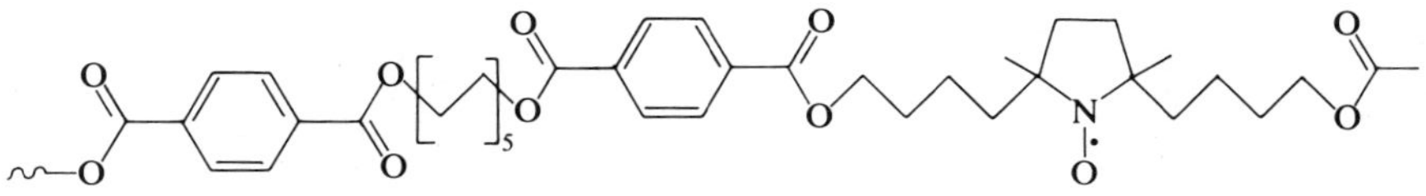

(33)

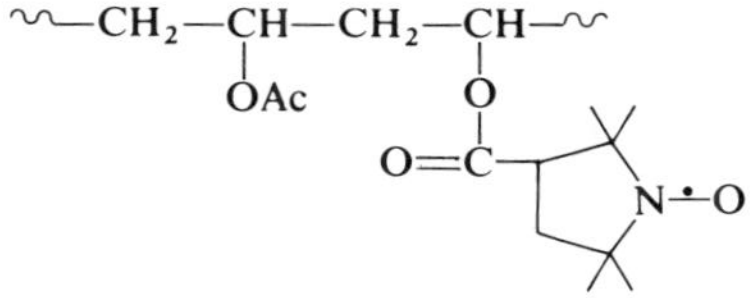

(34)

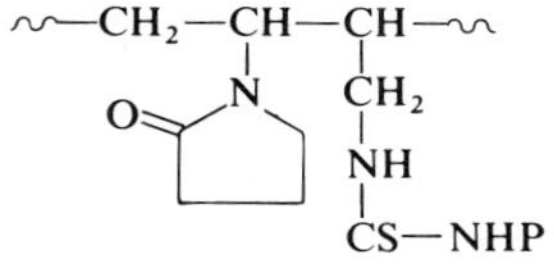

(35)

[a] In structures (17) to (20), (28) and (35) P =

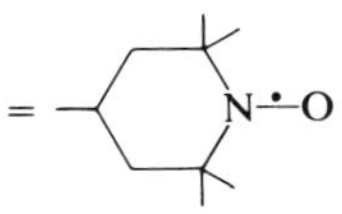

Direct ionic or radical copolymerization of vinyl-substituted nitroxides is not usually feasible because the growing radical or ion reacts preferentially with the nitroxide moiety. It is possible, however, to copolymerize monomers containing hindered cyclic amine or hydroxylamine groups such as the 4-methacryloylpiperidine derivative (36) which can then be oxidized within the resulting copolymer to give the desired nitroxide.[24]

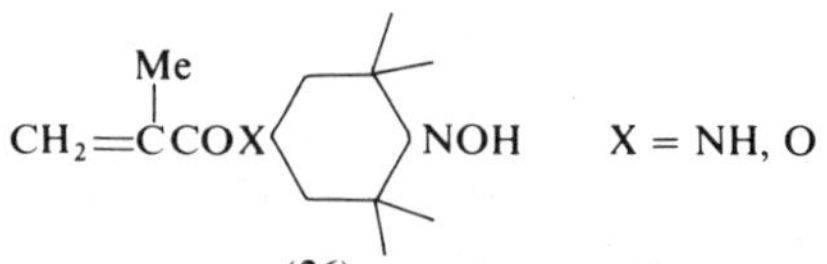

$$CH_2{=}\overset{\overset{\displaystyle Me}{|}}{C}COX\langle\quad\rangle NOH \qquad X = NH, O$$

(36)

The methacrylic derivative (36) therefore provides an alternative route to spin labels (17) or (18). Labels (22) and (26) were incorporated by a similar route, the former by radical copolymerization of 2,2,5,5,-tetramethylpyrroline-3-carboxamide. Such unsaturated cyclics do not copolymerize readily in radical reactions, but enough can be incorporated into polymerizing vinyl acetate to yield a labelled polymer on subsequent oxidation.[25] The in-chain nitroxide (26) was incorporated into polystyrene by copolymerization with 1-acetoxy-2,2,5,5,-tetramethyl-3-vinylpyrroline (Scheme 3).[26] Although the vinylpyrroline can polymerize through the vinyl group alone, the authors found no evidence for this mode of addition.

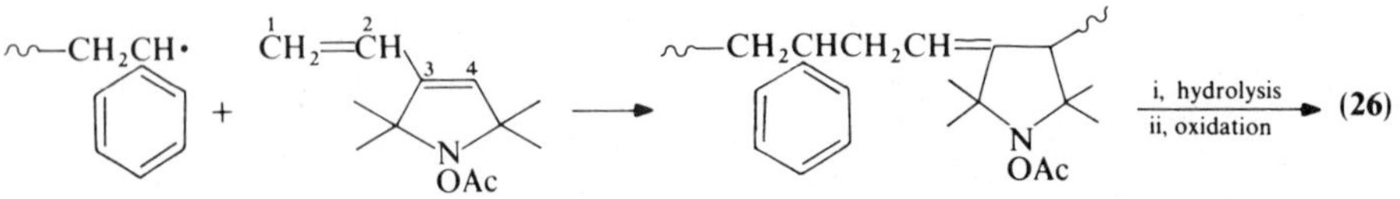

Scheme 3

The 'living' anionic polymerization route can be extremely useful for locating spin labels at desired points in the chain, particularly chain ends. The *t*-butyl nitroxide labels (30) and (31) (*cf.* 1) were attached by terminating living polystyrene and PMMA respectively with MNP.[27,28] The chemistry is essentially the same as in Scheme 2. The in-chain label (21) was prepared by coupling two living polystyrene chains with 2,5-di-*t*-butyl-3,4-diethoxycarbonylpyrrol-1-oxyl (Scheme 4).[29] The mono-ethoxycarbonyl compound would give an end-labelled polymer as does 3-chlorocarbonyl-2,2,5,5-tetramethylpyrroline-1-oxyl which yields (32)[30] with living polystyrene.

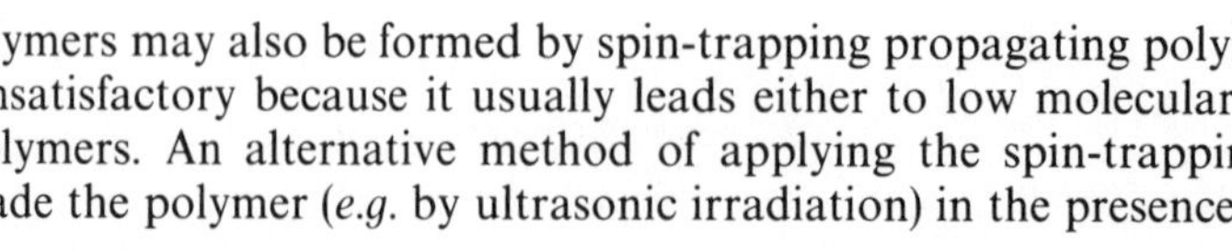

Scheme 4

End-labelled polymers may also be formed by spin-trapping propagating polymer radicals,[31] but this is generally unsatisfactory because it usually leads either to low molecular weight or to very lightly labelled polymers. An alternative method of applying the spin-trapping technique is to mechanically degrade the polymer (*e.g.* by ultrasonic irradiation) in the presence of the spin trap.[32]

23.3 THEORETICAL BACKGROUND

The purpose of this section is to indicate how dynamic data, in the form of the correlation time τ_c, may be derived from ESR spectra of spin-labelled and spin-probed polymers. Since space does not permit a complete, rigorous analysis, frequent reference to the original literature is made. For practical purposes, the relaxation processes are classified into three time regimes: (i) fast $10^{-11}\,s < \tau_c < 10^{-9}\,s$, (ii) slow $10^{-9}\,s < \tau_c < 10^{-7}\,s$, and (iii) very slow $10^{-7}\,s < \tau_c < 10^{-3}\,s$. τ_c may be viewed as the time taken for an axis of the nitroxide group to travel through one radian. Spin-probed polymers at high temperatures and spin-labelled polymers in dilute solution generally give ESR spectra characteristic of the fast regime. In labelled bulk polymers ESR spectra generally reflect relaxation in the slow regime. In each of the three regimes a different approach is required to evaluate τ_c.

23.3.1 Fast Tumbling

An axis system for a nitroxide probe or label is defined in Figure 1. If a nitroxide is trapped in a single crystal of a diamagnetic host in such a way that it is magnetically dilute with the axes defined by Figure 1 in a unique orientation with respect to the crystal axes, then the ESR spectrum is

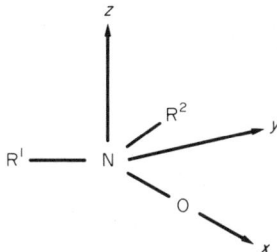

Figure 1 Axis system for a nitroxide group

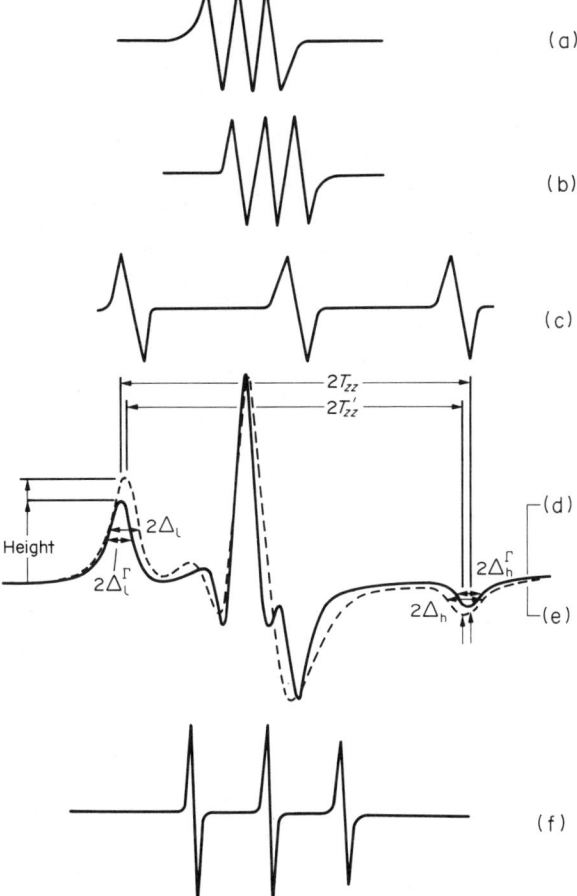

Figure 2 Idealized ESR spectra of nitroxide radicals: (a), (b) and (c) single crystal spectra with the applied magnetic field along the x, y and z principal axes of the g and $\hat{T}$ tensors; (d) rigid limit 'powder average' spectrum; (e) spectrum of nitroxide radicals in slow motion regime; (f) solution spectrum in the fast motion region (spectra (d) and (e) reproduced by permission of the American Chemical Society from ref. 46)

anisotropic, *i.e.* it depends on the angles between the applied magnetic field and the molecular axes of the nitroxide. This is illustrated by Figure 2(a), (b) and (c). The spectrum consists of three lines due to interaction of the unpaired electron with the ^{14}N nucleus ($I = 1$). The splitting of the lines and the position of the centre of the spectrum clearly vary as the direction of the applied field varies with respect to the molecular axes.

When the nitroxide radicals undergo rapid rotational diffusion (for example in dilute solution) the consequences are two-fold. Firstly, the anisotropies are averaged out to give a motionally narrowed spectrum (Figure 2f) whose centre is defined by

$$g_{iso} = \tfrac{1}{3}(g_{xx} + g_{yy} + g_{zz}) \tag{1}$$

and which has a splitting given by

$$T_{iso} = a^N = \tfrac{1}{3}(T_{xx} + T_{yy} + T_{zz}) \tag{2}$$

where g_{xx}, g_{yy}, g_{zz} and T_{xx}, T_{yy}, T_{zz} are the principal values of the g and hyperfine coupling tensors respectively. Secondly, the tumbling modulates the field experienced by the electron and this in turn gives rise to unequal widths among the three lines. If the peak-to-peak width of the first derivative of a Lorentzian line is Δv (Hz), then the line width parameter W is given by $W = (\pi\sqrt{3}\Delta v)$. In general the dependence of W upon m_I, the component of the nuclear spin along the direction of the applied magnetic field, is given by

$$W(m_I) = A + Bm_I + Cm_I^2 \tag{3}$$

For spin-label and spin-probe studies at X-band with isotropic rotational diffusion and near axial symmetry of the coupling tensor $\hat{T}$ ($T_{xx} \simeq T_{yy}$) the coefficients are[33]

$$A = \tau_c \left[\frac{3b^2}{20} + \frac{4}{45}(\Delta\gamma H_0)^2 \right] + \delta \tag{4a}$$

$$B = \tfrac{4}{15}b\Delta\gamma H_0\tau_c \tag{4b}$$

$$C = \tfrac{1}{8}b^2\tau_c \tag{4c}$$

The parameters in equations (4a), (4b) and (4c) are defined as

$$b = (4\pi/3)[T_{zz} - \tfrac{1}{2}(T_{xx} + T_{yy})]$$

$$\Delta\gamma = \frac{|\beta|}{h}[g_{zz} - \tfrac{1}{2}(g_{xx} + g_{yy})]$$

where the principal values of $\hat{T}$ are in Hz. β is the Bohr magneton and H_0 is the applied magnetic field. The coefficient A contains the term δ which represents contributions from sources independent of m_I (*e.g.* spin-exchange). However, the three equations implicit in equation (3) are readily rearranged to eliminate A and hence δ

$$W(m_I)/W(0) = 1 + Bm_I/W(0) + Cm_I^2/W(0) \tag{5}$$

where $W(0)$ represents the peak-to-peak width of the centre line ($m_I = 0$). If $r_\pm$ represents $W(\pm1)/W(0)$ then two values of τ_c may be obtained from

$$\tau_c = \frac{4W_{(0)}}{b^2}(r_+ + r_- - 2) \tag{6a}$$

$$\tau_c = \frac{15W_{(0)}}{8b\Delta\gamma H_0}(r_+ - r_-) \tag{6b}$$

It is usual to obtain accurate values of $r_\pm$ by measuring peak-to-peak intensities Y of the relevant lines, thus

$$r_\pm = W(\pm1)/W(0) = [Y(0)/Y(\pm1)]^{\frac{1}{2}} \tag{7}$$

If the two values of τ_c do not agree, the assumption of isotropic rotation must be suspect, though provided the disparity is not too great an average value is sometimes quoted. It is possible to dispense with the assumption of isotropic rotation but the method of analysis is more difficult to apply.[34, 35]

Finally, it is important to appreciate that the ESR lines from spin-labelled polymers are usually inhomogeneously broadened.[36] That is, they comprise envelopes of unresolved hyperfine lines, usually due to coupling with protons of the label. Use of a deuterated label reduces, but does not completely eliminate, this effect.[37] It is the width of the unresolved line or spin packet which is described by W in equations (3), (5) and (6), and hence in calculating τ_c from inhomogeneously broadened lines a device for calculating W from the observed line width must be applied. One approach is computer simulation to yield a calibration plot of observed *vs.* 'true' line width.[38] Other methods make use of an equation to relate the observed width with the spin packet width.[39,40] Mention should also be made of the spin-echo technique[41] and the method of 'additional broadening'.[42-44] In this last method the natural line width is increased by admitting oxygen until the observed width is essentially the spin packet width. This is equivalent to increasing δ in equation (4a).

23.3.2 Slow Tumbling

The limit of the fast tumbling is reached when the larger of the magnetic anisotropies, in frequency units, is no longer small compared with the frequency of tumbling. This critical point for nitroxides arises when the product $b\tau_c$ exceeds unity; the spectrum then undergoes a marked change in shape and width. The region of change from the fast to the slow regime is fairly sharp and corresponds to a value of $\tau_c \simeq 3 \times 10^{-9}$ s. For correlation times larger than 3×10^{-9} s the typical 'powder average' of a solid is observed. The shape of the slow motion spectrum remains essentially unaltered throughout the range $\infty > \tau_c > 3$ ns, but subtle changes do occur, the most marked of which is the inward shift of the low and high field extrema as the rotational frequency of the radical increases.[45] This and other changes in spectral shape in the slow motion region are shown in Figure 2, which includes one spectrum at the 'rigid limit' (2d; $\tau_c \to \infty$) and one (2e) having a finite correlation time greater than 3 ns.[46]

The lineshape of the ESR spectrum can be computed[47] (as in Figure 2) and this provides the basis of several simplified methods of determining τ_c in the slow region, as described below. The efficiency of the calculation has been improved recently, yielding a substantial saving in computer memory and time.[48] The most rigorous method of determining τ_c is by computer simulation in which the input parameters, correlation time and line width are varied until a good match between the computed and observed spectra is obtained. Anisotropic rotation and various types of rotational diffusion can be accommodated in this method.

For many applications in polymer chemistry simplified methods of estimating τ_c are adequate. All these are based on a relatively straightforward analysis of the outer extrema, usually the inward shift from the rigid limit position with the onset of molecular motion. These methods are reviewed more fully elsewhere[47,49] and only an outline is given here.

Correlation times in the region $10^{-9} < \tau_c < 10^{-6}$ s may be derived from the dependence of τ_c on the parameter κ[50]

$$\kappa = [H(\tau) - H(\tau \to 0)]/[H(\tau \to \infty) - H(\tau \to 0)] \tag{8}$$

In equation (8), $H(\tau)$, $H(\tau \to 0)$ and $H(\tau \to \infty)$ are the magnetic field values corresponding to the maxima of the first derivative absorption line ($m_I = +1$) at the correlation time τ_c, at the 'free' limit ($\tau_c < 10^{-11}$ s) and at the rigid limit ($\tau_c > 10^{-7}$ s), respectively. The relationship between κ and τ_c was derived by theory and was shown to be weakly dependent on the values of the g and $\hat{T}$ tensors, and on the intrinsic line width δ over the range 1–4 G. Although based on an arbitrary jump model for radical reorientation, it has been claimed that the method is also applicable to Brownian diffusion.[43] This method may be useful for estimating correlation times which are just outside the rapid tumbling region but which are too short to be estimated by the methods below.

The extrema separation of a nitroxide spectrum is a monotonic function of τ_c. Within the range $10^{-8} < \tau_c < 10^{-6}$ s the variation of extrema separation has been shown to obey the relationship[51]

$$\tau_c = a(1 - S)^b \tag{9}$$

where S is the ratio T'_{zz}/T_{zz} (see Figure 2), and a and b are constants which depend on the rotational diffusion model and on the intrinsic line width. The value of S has been shown to depend on T_{zz} but simple scaling of results for different values of T_{zz} within the range typical of nitroxides leads to negligible errors. S is insensitive to variations in the g tensor, and in T_{xx} or T_{yy}. Sets of values for a and b for various diffusion models and intrinsic line widths have been published.[47] Equation (9) has proved to be one of the most useful for estimating τ_c in the slow region.

The parameter S becomes insensitive to changes in τ_c when it exceeds *ca.* 10^{-7} s. However, Mason and Freed have shown that a relationship similar to that between τ_c and S exists between τ_c and the dimensionless parameter W_i.[46]

$$\tau_c = a'(W_i - 1)^{-b'} \qquad (10)$$

In equation (10) $W_i = \Delta_i/\Delta_i r$ where $i = 1$ or h corresponding to the low and high field extremum respectively, $\Delta_i = $ the half width at half height of the i extremum (see Figure 2), and $\Delta_i r = \Delta_i$ at the rigid limit. Values of a' and b' have also been tabulated.[46] The widths of the spectral extrema are sensitive to relaxations in the microsecond region and correlation times to $\tau_c < 5 \times 10^{-6}$ s may be measured.

McCalley *et al.*[51] have used a somewhat different theoretical approach to prepare graphs of the inward shifts of extrema *vs.* τ_c. From these curves τ_c can be read off directly. The method is most sensitive in the region $10^{-8} < \tau_c < 10^{-7}$ s. The high field extremum appears to be the more useful because it undergoes a greater shift, remains separate from the central spectral features for shorter values of τ_c, and is insensitive to small variations in the principal values of the g and $\hat{T}$ tensors.

Before using either the shifts or widths of the extrema to calculate τ_c in the slow tumbling region, the appropriate diffusion model must be chosen. This choice may be facilitated by measuring the ratio of the shifts of the high and low field extrema ($\Delta H_h/\Delta H_l$) with change in ΔH_l. This ratio has been shown to be sensitive to the mode of reorientation of nitroxides.[52]

Recently a new method has been developed which may allow determination of correlation times in the slow tumbling region from second derivative spectra.[53,54]

Finally, it should be noted that for many spin-labelled and spin-probed polymers the transition from the slow to the fast tumbling regions occurs over a very narrow temperature range. The phenomenon is analogous to the line width (or second moment) *vs.* temperature transition found in wide line NMR studies. In the motionally narrowed region the extrema separation is around 30 G while in the slow motion region it is 60–70 G. An extrema separation of 50 G is generally taken as characteristic of the transition region, and the temperature associated with this separation, which depends on the particular polymer and probe or label, is referred to as $T_{50\,G}$, after Rabold.[55] This temperature corresponds closely to the condition $b\tau_c \simeq 1$. In as much as the $\hat{T}$ tensor is similar for most nitroxides, $T_{50\,G}$ approximates to an isofrequency point. Applications of $T_{50\,G}$ measurements are discussed later in this chapter.

23.3.3 Very Slow Tumbling

With one exception, all the techniques described so far involve the analysis of the first derivative presentation of the ESR spectrum under linear response conditions. This means that the power of the microwaves inducing transitions is kept low enough for the Boltzmann distribution of populations between the various spin states to remain unperturbed. This distribution is maintained by spin–lattice relaxation processes. At sufficiently high microwave powers the Boltzmann distribution is perturbed and the system is said to be saturated. A competition between two types of process then occurs either to remove the saturation (spin–lattice relaxation) or to transfer it to other parts of the resonance spectrum ('spectral diffusion'). An efficient mechanism for the latter is the modulation of anisotropic magnetic interactions by means of rotational diffusion and this transfers energy between spectral lines corresponding to different molecular orientations.

The exploitation of the observable effects that such variables as microwave power and Zeeman modulation frequency have on a spin system undergoing spin–lattice relaxation and spectral diffusion is known as 'saturation transfer (ST) spectroscopy'.[56,57] The most useful modes of observation developed so far for STESR are the first harmonic of the dispersion and the second harmonic of the absorption signals. The shapes of both are sensitive to correlation times in the region 10^{-3}–10^{-7} s, *i.e* well beyond the region where the conventional ESR spectrum becomes insensitive to any increase in τ_c, though many polymer relaxations in the solid state occur within this time regime. The only satisfactory way in which STESR can be analyzed to yield correlation times is by spectral simulation, for which computer programs are now available.[58]

To date very few applications of STESR have appeared in the literature. Three recent examples, however, deserve mention. In the first, Wasserman[59] measured correlation times of two nitroxide probes in polyethylene over the range 10^{-10}–10^{-5} s using conventional and STESR spectroscopy. In the second example, poly(vinyl acetate) (PVAC) was probed with nitroxides of type (13; Table 1)[60] and from the changes in the STESR spectrum with temperature a characteristic temperature T_R, analogous to $T_{50\,G}$ from conventional ESR spectroscopy, was identified. Both studies reveal that in spin-probed polymers at temperatures where probe motion apparently falls within the frequency range of conventional ESR spectroscopy, a proportion of the probes tumbles sufficiently slowly to fall within the range of the STESR technique. This was interpreted as indicating the existence of two distinct environments for the probes in the polymer matrix; in one type of environment the probes are trapped and almost immobilized, and within the other motion is relatively free. Finally Ohno *et al.*[61] used the STESR technique to calculate correlation times of 10^{-6}–10^{-5} s in PMMA containing spin probes and labels at -80 to $-20\,°C$. These were assigned to side chain motions in the polymer.

23.4 APPLICATIONS OF SPIN LABELS AND SPIN PROBES IN SYNTHETIC POLYMERS

The spin-label technique has been applied to studies of polymers in solution, in bulk and at solid–liquid interfaces. In bulk and plasticized polymers the spin-label and spin-probe techniques are complementary. Illustrative examples of such investigations are given in this section.

23.4.1 Solution Studies

Factors which control the dynamics of a polymer molecule in solution include polymer molecular weight (MW), intramolecular steric and polar interactions, viscous drag exerted by the solvent and interchain interactions. All have been the subject of spin-label experiments.

Polystyrene (PS) labelled with various nitroxides has been the most intensively studied polymer to date. In the most comprehensive of these studies[38,61] PS was lightly labelled in the *p* position to give structure (**27**; R = Bu^t). The solution spectrum in Figure 3 shows coupling to the aromatic protons in addition to the ^{14}N nucleus. After correction for inhomogeneous broadening and overlap of adjacent proton hyperfine lines, correlation times were calculated from the analysis of line widths by equations (4), (5) and (6). The dependence of τ_c on MW for dilute solutions in toluene is shown in Figure 4. The limiting value of τ_c at high MW is characteristic of a 'local' mode of relaxation

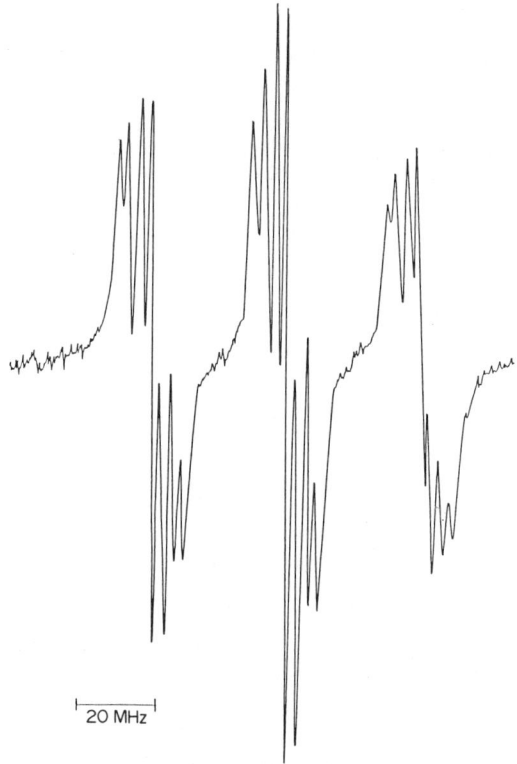

20 MHz

Figure 3 ESR spectrum of a polystyrene fraction ($\bar{M}_n = 1950$) labelled with *p-t*-butyl nitroxide groups (**27**) (1 % in toluene at room temperature) (reproduced by permission of the American Chemical Society from ref. 38)

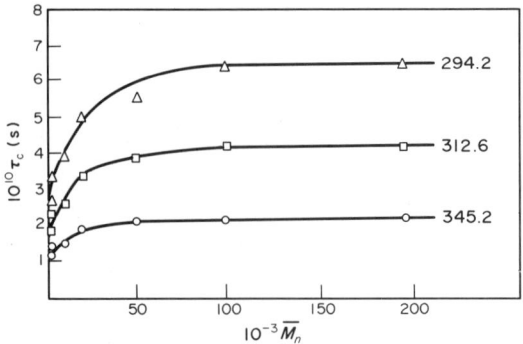

Figure 4   Dependence of τ_c on molecular weight and temperature (K) for 1% solutions of labelled polystyrene in toluene (reproduced by permission of the American Chemical Society from ref. 38)

associated with a correlation time τ_{lm} and attributable to segmental motion. At these values of MW, 'end-over-end' rotation of the whole polymer molecule is very slow in comparison with segmental motion, but as MW decreases whole molecule tumbling begins to contribute significantly to the relaxation and the observed τ_c becomes dependent on MW. If whole molecule tumbling is associated with a correlation time τ_{eoe}, the observed τ_c will be given by

$$\frac{1}{\tau_c} = \frac{1}{\tau_{eoe}} + \frac{1}{\tau_{lm}} \tag{11}$$

Assuming that τ_{lm} is independent of MW, then, at a given temperature, the observed τ_c together with τ_{lm} (the plateau value in Figure 4) enables τ_{eoe} to be determined. For a low molecular weight sample of spin-labelled PS in various solvents values of τ_{eoe} calculated from theory agreed quite well with those derived from equation (11).

That the motion of the label was accurately reflecting that of the polymer in this case was confirmed by the general concordance of data obtained when the *t*-butyl nitroxide label was relocated in the *meta* position of the styryl ring[63] and when the much bulkier *p*-phenyl nitroxide label (**27**; R = Ph) was employed.[38,64] Good agreement between the spin-label results and those from $^{13}C\ T_1$ measurements on the ring and main chain carbon atoms was noted.[38] All these data point to isotropic rotation of the label controlled by motion of the polymer chain.

The studies with the *p*-labelled PS (**27**; R = Bu^t) also revealed the relative effects of solvent and intramolecular interactions on segmental mobility. In the limit of high viscous damping, Helfand showed that the energy barriers to conformational changes in polymers is given by[65]

$$\tau_{lm} = \tau_0 \exp[(E^* + E_\eta)/RT] \tag{12}$$

where E^* is the intrinsic intramolecular barrier and E_η is the activation energy for viscous flow of the solvent. On the plateau region in Figure 4 the temperature dependence of τ_c follows the Arrhenius relationship and $\tau_c \simeq \tau_{lm}$, hence

$$\tau_{lm} = \tau_c = \tau_0 \exp(E_{tot}/RT) \tag{13}$$

and

$$E^* = E_{lm} - E_\eta = E_{tot} - E_\eta \tag{14}$$

Solutions of labelled PS of high MW were examined in three solvents of widely varying viscosity: toluene, α-chloronaphthalene and cyclohexane. The first two are thermodynamically 'good' solvents and cyclohexane is a theta solvent for PS. These characteristics are reflected in the energetics and dynamics of chain motion summarized in Table 3. The relatively high values of E_{tot} (E_{lm}) for α-chloronaphthalene and cyclohexane reflect the viscosity of the solvent in the first case and the tightly coiled configuration of the polymer chain in the second. When E^* is calculated according to equation (14) the values for toluene and α-chloronaphthalene are identical and significantly lower than that associated with cyclohexane. This is a reflection of the extended conformation of the polymer in the first two solvents and the tighter coil in the poor solvent cyclohexane. The correlation times for motion of the label at 298 K also illustrate the contrasting characteristics of the three solvents — toluene a good, highly fluid solvent, α-chloronaphthalene a good but viscous solvent, and

Table 3 Barriers to Rotation for Labelled Polystyrene in Various Solvents[a]

	Solvent	E_η	E_{tot}	E^*	$10^{10}\tau_c$ (s) *at* 298 *K*
Side chain labels[bc]	Toluene	9.00	18.0 ± 0.8	9.0 ± 0.6	5.9 ± 0.5
	α-Chloronaphthalene	17.78	26.4 ± 1.3	8.6 ± 1.3	35.4 ± 3.5[b]
	Cyclohexane	12.55	26.1 ± 0.8	13.6 ± 0.8	10.0 ± 0.3[b]
Chain end labels[d]	Toluene	9.00	14.8 ± 0.8	5.8 ± 0.8	2.4 ± 0.2
	α-Chloronaphthalene	17.78	23.0 ± 0.2	5.2 ± 0.2	11.9 ± 1.5[b]
	Cyclohexane	12.55	21.8 ± 1.4	9.2 ± 0.1	4.9 ± 0.1

[a] All activation energies in kJ mol^{-1}.
[b] Extrapolated for comparison purposes.
[c] From A. T. Bullock, G. G. Cameron and P. M. Smith, *J. Phys. Chem.*, 1973, **77**, 1635.
[d] From A. T. Bullock, G. G. Cameron and N. K. Reddy, *J. Chem. Soc., Faraday Trans. 1*, 1978, **74**, 727.

cyclohexane a poor solvent with fluidity intermediate between the other two. A similar though less comprehensive treatment of polyethylene, carrying labels of structure (**24**), in xylene solution[20] yielded a value of $E^* = 13.9 \pm 1.2$ kJ mol^{-1}. This is reasonably close to the gas phase barrier to rotation in ethane (*ca.* 12 kJ mol^{-1}).

In another study of PS by Friedrich and co-workers, the in-chain label (**21**) was used.[29,66] The polymer had a molecular weight of 62 800 (*i.e.* close to the range where the contribution from whole molecule tumbling is negligible) and a polydispersity of 1.10. The method of labelling ensured that each chain contained one label sited near the centre. The ESR spectrum (Figure 5) shows no proton hyperfine structure, in contrast to that of the polymer with *t*-butyl nitroxide labels on the aromatic

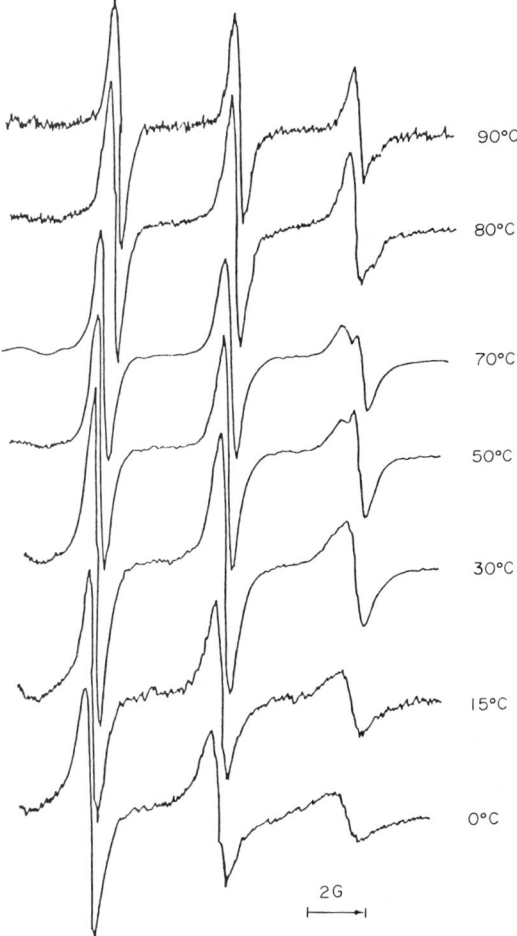

90°C
80°C
70°C
50°C
30°C
15°C
0°C

2G

Figure 5 ESR spectra of solutions of polystyrene labelled with (**21**). (10% in toluene) (reproduced by permission of IPC Science and Technology Press from ref. 29)

ring, but the unresolved couplings to adjacent protons are a source of inhomogeneous broadening. Between 40 and 75 °C the spectrum of the polymer solution appears to be a composite (note the subsplitting of the high field peak above 30 °C and below 80 °C). This could be due to the 'phase transition' in the region of 50 °C.[67] In support of this, the authors successfully simulated the spectra in Figure 5 by summing two spectral components of different correlation times.[66] At 90 °C, where the ESR spectrum is virtually a single component, the label underwent essentially isotropic motion with a correlation time of *ca.* 1.8×10^{-10} s,[29] in good agreement with the data of Bullock *et al.* for side-chain-labelled PS.[38] The value of E^* for the relaxation below 80 °C also agreed well with that obtained by Bullock *et al.* This concordance is pleasing but perhaps a little surprising in view of the relative bulk of label (**21**) and its in-chain position. Friedrich *et al.*[66] also made a critical comparison of data obtained by the ESR and other techniques.

The most recent study of PS involved label (**26**).[26] The ESR spectrum of the polymer in toluene solution was analyzed by the method of Bales[40] over the temperature range -20 to $+93$ °C. In this case the combination of a relatively bulky in-chain label and a double bond effectively slowed down local segmental motion so that the correlation times were increased by an order of magnitude compared with results from other spin-label and ^{13}C NMR studies ($\tau_c = 2.19 \times 10^{-9}$ s at 317 K). However, the value of $E^* = 7.69$ kJ mol^{-1} was not significantly different from those recorded in Table 3 for the side-chain-labelled polymers. From their data Simon *et al.*[26] estimated the effective size of a segment involved in rotational relaxation as two to four monomer units. This is comparable to the figure derived by other workers.[66]

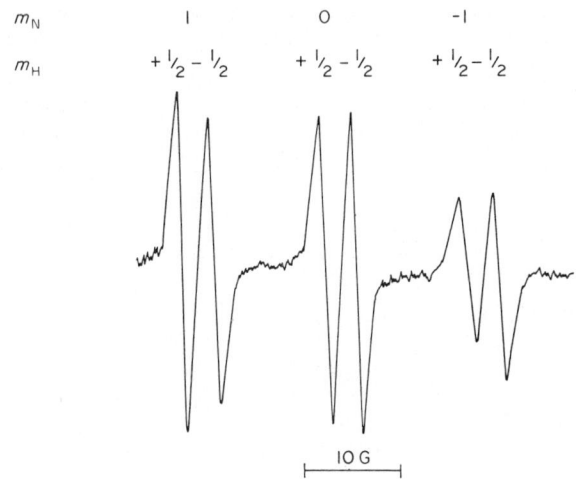

Figure 6   ESR spectrum of end-labelled polystyrene (**30**) (3% solution in toluene at room temperature) (reproduced by permission of the Royal Society of Chemistry from ref. 27)

End-labelled polystyrene (**30**) has also been studied in various solvents.[27] The spectrum is shown in Figure 6 and attention is drawn to the small doublet splitting which arises from the methine proton (*cf.* structure **1**). From the negligible dependence upon temperature and the small magnitude of this splitting (3.32 ± 0.05 G), it can be inferred that rotation about the polymer–label bond does not contribute significantly to the widths of the ESR lines. With a nitroxide label of the more stable di-*t*-alkyl variety this useful additional information would not have been available. The line width analysis, which is discussed fully in ref. 27, is a little more complicated with this type of spectrum. The barriers to rotation of end-labelled PS are compared with the side-chain-labelled polymers in Table 3. The three solvents have a similar effect on relaxations of the labels in both sites but the energy barriers and correlation times are significantly lower for motions of chain ends. This is physically reasonable.

The variance between the results obtained with PS labelled with (**26**) and other relaxation data on this polymer underlines the importance in all labelling experiments of checking that the label accurately reflects the motion of the polymer to which it is bonded and that it does not perturb that motion. In this context it is interesting to note an extensive series of spin-labelling experiments on copolymers of *N*-(2-hydroxypropyl)methacrylamide and 4-nitrophenyl esters of ω-methacryloyl-amino acids of varying chain lengths.[44, 68] The latter provided sites for labelling by an aminolysis reaction with probe (**5**; Table 1), so that the labels were bonded through amide links at the ends of flexible side chains of varying lengths (Scheme 5). Within the series, τ_c in methanol solution decreased monotonically with increasing side chain length from a value characteristic of polymer segments to one approaching that of the free nitroxide (**5**) in methanol. In the last case it would obviously be misleading to associate the motion of the label with that of the polymer chain.

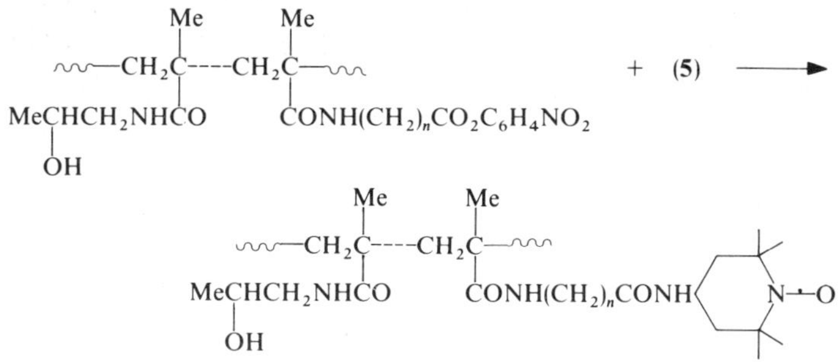

Scheme 5

A similar effect has been noted in dilute solutions of end-labelled poly(benzyl glutamate)[69] which adopts a rigid rod-like α-helix conformation in solution. For polymers of MW above *ca.* 120 000 the motion which dominates the ESR spectrum corresponds to rotation about the helix–label bond. At lower MWs rotation of the rod contributes to the overall motion of the label.

In addition to investigating chain flexibility and segmental motion by the spin-label method, Buchachenko, Wasserman and their co-workers have shown in a series of elegant experiments that the technique is capable of yielding information on the local density of monomer units from host and guest macromolecules in the macromolecular coil, the local translational diffusion coefficients of segments in the coil, and the interpenetrability of macromolecular coils.[70–72] In comprehensive studies of poly(4-vinylpyridine) (PVP) labelled with (**20**) in ethanol solution, these authors worked with labelling densities of 1 label per 100 to 1 label per 5 monomer units. At these densities the ESR line widths are determined mainly by dipolar and electron exchange interactions between neighbouring nitroxide radicals, not by the rotational modulation of the magnetic anisotropies discussed earlier in this chapter, and line broadening from these sources increases with increasing radical concentration. Dipolar interactions are efficient in systems of low molecular mobility and exchange interactions dominate line broadening in highly mobile systems. By studying the dependence of concentration broadening on temperature or viscosity, dipolar and exchange contributions can be separated. In a relatively immobile system the dipolar broadening can be related to the local concentration of labels and hence to the local density of monomer units ρ_{loc} in a relatively straightforward manner. The average monomer unit density $\langle\rho\rangle$, calculated by standard procedures, is compared with ρ_{loc} in Table 4 from which the following conclusions can be drawn: (i) local concentrations of monomer units are only slightly dependent on the molecular weight of the polymer; (ii) in large macromolecules the local densities of monomer units may be five to six times greater than the average density (in accordance with the Gaussian coil model); and (iii) local concentrations of monomer units are almost independent of temperature; *i.e.* both the distribution of monomer units and the coil size are little affected by temperature except at very low temperatures.

At temperatures above 20 °C, where exchange broadening became dominant for labelled PVP, the rate of intramolecular exchange interactions, and hence segment diffusion coefficients, were calculated. The results of this experiment are summarized in Figure 7 where the diffusion coefficient of polymer segments is compared with that of a small nitroxide spin probe. Clearly local diffusion of

Table 4 Average $\langle\rho\rangle$ and Local ρ_{loc} Monomer Unit Densities in Poly-(4-vinylpyridine) (PVP)[a,b]

$M_n \times 10^{-4}$	$\langle\rho\rangle$ (mol l^{-1})	ρ_{loc} (mol l^{-1})	ρ_{loc} at -196 °C (mol l^{-1})
25	0.05	0.30	0.40
5	0.20	0.28	0.36

[a] 1% solution in ethanol at 20 °C.
[b] From A. L. Buchachenko and A. M. Wasserman, *Pure. Appl. Chem.*, 1982, **54**, 507.

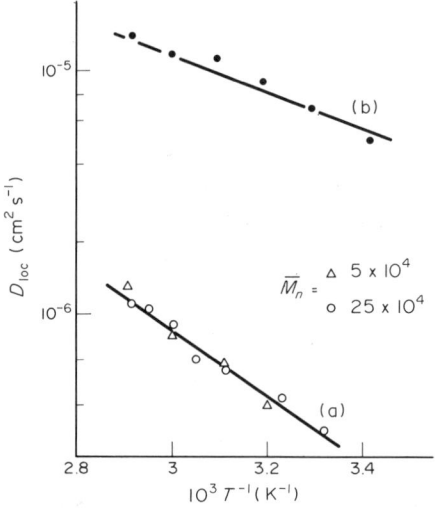

Figure 7 Temperature dependence of local diffusion coefficient of poly(4-vinylpyridine) (PVP) in ethanol solution calculated from exchange broadening: (a) spin label (**20**), (b) spin probe (**6**)[72]

segments is slower by at least an order of magnitude than that of small molecules, and the activation energy for segment diffusion is greater than that for probe diffusion. Also, segment diffusion coefficients are almost independent of polymer molecular weight.

In a separate study, Wasserman and Buchachenko determined from dipolar- and exchange-broadening measurements how monomer unit densities within a reference macromolecule (host) varied as the total polymer concentration in solution was increased.[70–72] This was approached by measuring line widths (of labelled PVP) in concentrated solutions of labelled polymer, and in concentrated solutions of unlabelled polymer containing minor quantities of labelled polymer. In the first case the broadening effect is due to both inter- and intra-molecular interactions of labels, and in the second only to intramolecular interactions. The results are shown in Figure 8. Within the host macromolecule the density of host monomer units is almost independent of total concentration but the density of guest monomer units within the host increases monotonically with concentration. When the polymer concentration reaches about 10 wt. % the contributions of monomer unit density from the reference and guest macromolecules are comparable. At this concentration the macro-molecular coils interpenetrate and overlap to a considerable extent. Bullock, Cameron and Smith[62] observed the effects of this interpenetration in solutions of lightly labelled PS. In this case, the onset of strong intermolecular interactions was evidenced by an exponential rise in τ_c.

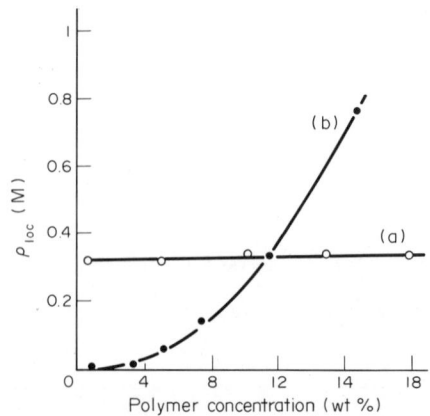

Figure 8 Local concentrations of monomer units ρ_{loc} (mol l^{-1}) of host macromolecules (a) and of guest macromolecules (b) in a macromolecular coil as a function of total polymer concentration in solution (poly(4-vinyl pyridine) (PVP) in ethanol at 20 °C)[71]

All of the results quoted up to this point in this section have concerned polymers with carbon atom backbones. There have, however, been several spin-label studies of polymers with heteroatom chains, chief among them being poly(ethylene oxide) (PEO) end-labelled with (29).[17,73-75] The correlation times, measured by the standard line width analysis, were dependent on solvent viscosity and temperature as before. The energy barrier was *ca.* 6.1 kJ mol^{-1} and it appeared that about one monomer unit was involved in the ESR relaxation process.

Few spin-label investigations of polyesters or polyamides have been reported. In one recent notable example a labelled polyester was prepared by the polycondensation of terephthaloyl chloride and 1,10-decanediol in the presence of the nitroxide diol (33a) to give structure (33; Table 2). The label in this case is within the methylene chain and its relatively high mobility in solution is associated only with that part of the chain and not with chain sections that include terephthaloyl residues.[76]

(33a)

23.4.2 Polymers at Solid–Liquid Interfaces

The spin-label method provides a useful complement to other techniques for studying the adsorption of polymers. When a labelled polymer is adsorbed at a solid–liquid interface the adsorption isotherm may be determined (by using the ESR signal as an analytical device), while information on the polymer–substrate affinity and the polymer conformation at the interface may be inferred from the spectral shape.[25,77-81] When the polymer–surface interaction is weak, the adsorbed molecules are held rather loosely and experience considerable motional freedom in conformations described as 'loops' and 'tails'. The ESR spectra of nitroxide labels in loops or tails are generally of the motionally narrowed type. On the other hand, when the affinity of the polymer for the surface is high, for example when the solvent is thermodynamically poor, the conformation is relatively flat, consisting of 'trains', and motion is greatly restricted. The ESR spectra of labels in trains are of the slow motion variety. These conformational differences are apparent in the ESR spectra of poly(vinyl acetate) (PVAC) labelled with (22) (Figure 9) adsorbed on silica.[25] The spectrum of the polymer adsorbed from a solution in toluene, a relatively poor solvent for PVAC, consists of broad lines characteristic of a slowly tumbling nitroxide, suggesting adsorption in the form of trains. With ethyl acetate, a good solvent, the spectrum is a composite of broad and motionally narrowed lines, indicating that a proportion of segments are in the form of loops or tails. As an interesting comparison, PVAC labelled as in (34) and adsorbed on silica under similar

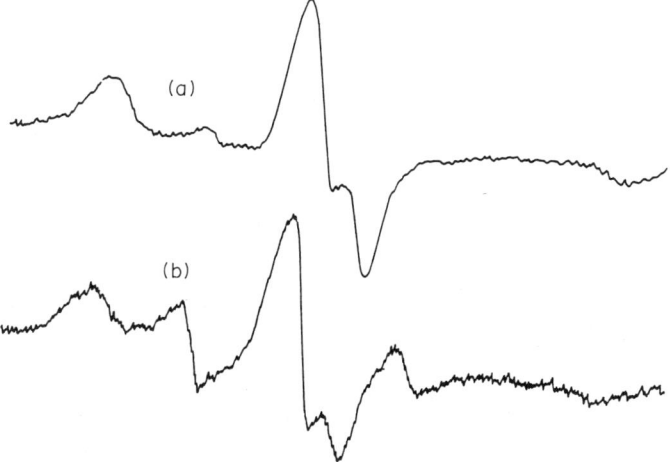

Figure 9 ESR spectra of poly(vinyl acetate) (labelled with **22**) at a silica surface. Adsorbed from (a) toluene solution, (b) ethyl acetate solution at room temperature[25]

conditions, showed a motionally narrowed ESR spectrum.[82] This is almost certainly a consequence of the greater mobility of the pendant label (34). By a process of spectral subtraction on composite spectra like Figure 9(b), Robb and co-workers have succeeded in estimating the relative proportions of loops and trains in poly(vinylpyrrolidone) (label 35) adsorbed on silica.[77] In studies of polymer adsorption it is important to establish that the polymer is not subject to preferential adsorption through the nitroxide label. Despite its simplicity, the spin-label technique has not been widely exploited for studying polymer–surface interactions, though reference should be made here to the work of Miller and co-workers.[69] There have also been a number of studies of grafting to solid surfaces using spin-labelled polymers.[83]

23.4.3 ESR Studies of Polymers in Bulk

In the last two or three years ESR studies of bulk polymers have tended to feature more prominently than solution studies. Spin-probe investigations, in particular, have increased in number.

23.4.3.1 *T_{50G} and the glass transition*

The temperature at which the extrema separation in the ESR spectrum of a spin-probed (less frequently spin-labelled) polymer reaches 50 G is referred to as T_{50G} (see earlier). Values of T_{50G} are obtained conveniently from plots of $2T'_{zz}$ vs. temperature, which are typically sigmoidal in shape (Figure 10).[84] There is an obvious correlation between the T_{50G} values and the glass transition temperatures T_g of the polymers. This occurs because probe motion is usually sensitive to the dynamic state of the host matrix and at T_{50G} the polymer is generally undergoing segmental reorientation. However, the plot of T_{50G} vs. T_g in Figure 11 shows that T_{50G} is greater than T_g.[85] This is explained on the basis that the effective frequency at T_{50G} is about 5×10^7 Hz ($f = (2\pi\tau_c)^{-1}$) whilst the T_g values were measured at ~ 1 Hz. T_{50G} has therefore come to be regarded as a high frequency glass transition temperature. This concept, though useful, has to be applied with discrimination because it is quite conceivable for a probe to tumble with the necessary frequency associated with T_{50G} within a host polymer at a temperature below T_g. In other words, a sigmoidal curve of $2T'_{zz}$ vs. temperature is not immutable evidence of a discontinuity in some property of the surrounding matrix.

In principle an unknown T_g could be read off Figure 11 having measured T_{50G} with probe (7; $n = 0$; R = Ph). Such an approach offers no advantages over conventional methods of determining

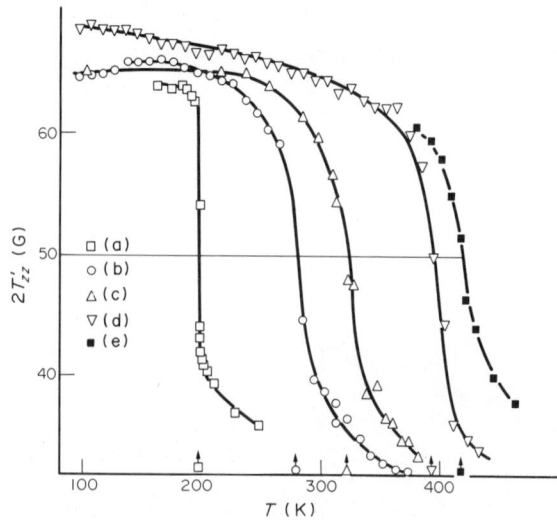

Figure 10 Plots of extrema separation $2T'_{zz}$ vs. temperature of the ESR spectra of probe (7; $n = 0$; R = Ph) in (a) poly(dimethylsiloxane); (b) high density polyethylene; (c) poly(isobutylene); (d) PVC; and (e) polycarbonate. T_{50G} values indicated by arrows (reproduced by permission of Harwood Academic publishers from ref. 84)

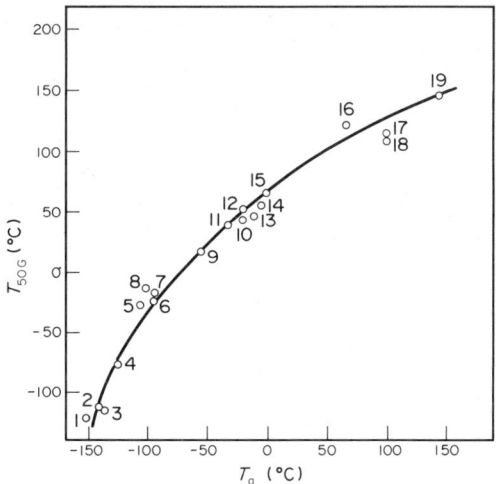

Figure 11 T_{50G} *vs.* T_g for various polymers with probe (**7**; $n = 0$, R = Ph): (1)–(4) poly(dimethylsiloxane)s, (5) Budene, (6) and (7) Diene 55, (8) polypentenamer, (9) ethylene–propylene 50/50 copolymer, (10) polypropylene, (11)–(15) styrene–butadiene copolymers, (16) PVC, (17) and (18) PS, (19) polycarbonate (data from Table 2 of ref. 85)

T_g except in one or two special cases. Thus, Kumler and Boyer[85] employed this technique to estimate T_g of polyethylene.

It is well established that the T_{50G} value for a particular polymer increases as the molecular volume of the probe increases.[86] This is in line with the idea that the mobility of the probe is associated with the free volume in the polymer. It follows, however, that each probe in Table 1 gives a different plot of T_{50G} *vs.* T_g. At temperatures where the probe responds to the glass to rubber relaxation of the host, a free volume model predicts that the correlation time of probe tumbling follows the relationship[13]

$$\ln \tau_c = \ln \tau_\infty + f[2.303 C_{1g} C_{2g}/(T - T_g + C_{2g})]$$ (15)

where τ_∞ = high temperature limit of τ_c, C_{1g}, C_{2g} = constants for a given polymer defined by Williams *et al.*,[87] f = activation volume of probe/activation volume of polymer segment $\simeq$ volume of probe/volume of polymer segment. The formal similarity of equation (15) to the well-known WLF equation is obvious.[87] A plot of $\ln \tau_c$ *vs.* $2.303 C_{1g} C_{2g}/(T - T_g + C_{2g})$ is thus predicted to be linear with a slope of f and intercept $\ln \tau_\infty$. Various spin probes in PVAC followed equation (15),[13] including PVAC doped with (**13**; $R^1 = R^2 = Bu^n$, DBOZ).[60] The latter gave $\tau_\infty \simeq 1.1 \times 10^{-12}$ s which is in good agreement with other published data on PVAC. The value of $f = 0.62$ indicates that the volume of a segment of PVAC at the glass to rubber relaxation is 1.62 times greater than the effective volume of DBOZ. This corresponds to four to seven monomer units per segment, which is in reasonable agreement with data obtained by other methods.

Setting $\tau_\infty = 1.1 \times 10^{-12}$ s and assuming that $\tau_c = 10^{-8}$ s at T_{50G}, equation (15) gives on rearrangement

$$T_{50G} - T_g = C_{2g}[2.303 C_{1g} f/9.1 - 1]$$ (16)

Equation (16) is simpler to apply than the more rigorously based equation (15) and for DBOZ in PVAC gave $f = 0.59$. Values of f for PVAC have also been obtained for the probes (**13**) with $R^1 = R^2$ = Me (TMOZ),[88] $R^1 = R^2 = n$-nonyl (DNOZ)[60] and with PVAC carrying the spin label (**22**).[25] The data are summarized in Table 5. The volumes of the probes vary in the order TMOZ < DBOZ < DNOZ and values of f vary in the same order. The spin-labelled polymer may be viewed in terms of its size and bulk as the ultimate spin probe. In this case, however, it is the labelled segment rather than the whole polymer chain which functions as a spin probe, and on the basis that the labelled and unlabelled segments should have approximately the same volume, f should be close to unity. The values of f in Table 5 are therefore consistent with the underlying theory.

Further evidence in support of the relationship between T_{50G} and T_g comes from the observation that in spin-probed plasticized PVAC T_{50G} varies with composition in precisely the same manner as T_g.[88]

The spin-probe method has also proved to be a sensitive method of measuring the limits of miscibility of the plasticizer dinonyl phthalate and PVAC.[89] The presence of free plasticizer is

Table 5   Values of the Parameter f from $T_{50\,G}$ for Various Spin Probes in Poly(vinyl acetate) (PVAC) and for Spin-labelled PVAC[a]

Spin probe	TMOZ	DBOZ	DNOZ	Spin-labelled PVAC
$T_{50\,G}$(K)	311	370	392	434
f(from equation 16)	0.28	0.59	0.71	0.94

[a] From I. S. Miles, G. G. Cameron and A. T. Bullock, *Polymer*, 1986, **27**, 190.

revealed by the appearance of a motionally narrowed spectrum superimposed on the broad lines arising from plasticized polymer. In this manner it is possible to detect as little as 1% free plasticizer in the system. Minsker *et al.*[90] have recently extended this technique with probe (**2**) to determine the compatibility of poly(vinyl chloride) (PVC) with various plasticizers, in addition to which they were able to study the plasticization kinetics.

It should be noted that much of what has been said above about $T_{50\,G}$ also appears to apply to T_R, an analogous parameter associated with STESR.[60]

All the methods of calculating correlation times mentioned in Section 23.3 assume, and yield, a single value of a correlation time for either spin probe or spin label. However, recent studies of plasticized PVAC doped with spin probes have revealed evidence for the existence of a distribution of correlation times.[91] This was manifest as a double peak in the low field extremum at temperatures just below $T_{50\,G}$, and was explained in terms of a distribution of free volume in the polymer. It was concluded that such a distribution could lead to significant errors in calculating τ_c when probe tumbling is slow, but for motionally narrowed spectra the error in τ_c is likely to be insignificant. Lee and Brown[92] also observed similar spectral features in a spin-probed copolymer of bisphenol A carbonate and dimethylsiloxane. In this case however, the origin of the phenomenon is probably a distribution in local composition rather than free volume.

23.4.3.2   *Relaxations and phase transitions in bulk polymers*

If correlation times for probe or label tumbling are determined, as opposed to simple extrema separation measurements, the potential yield of information is greatly enhanced. In Arrhenius plots of correlation times, transitions are revealed as discontinuities and the slopes of the lines may reveal information on the energetics of relaxation. Some care must be exercised when seeking such information because discontinuities in Arrhenius plots can arise from spurious sources such as inappropriate choice of diffusion model or method of calculating τ_c, variations in degree of anisotropy of probe or label tumbling, *etc.* Typical plots of correlation time *vs.* reciprocal temperature for spin-probed polymers are shown in Figure 12 where the region of change in slope lies close to T_g for each polymer.[92a] At temperatures below T_g the values of correlation times are surprisingly short and activation energies for probe tumbling are low ($3.8 - 11.3$ kJ mol^{-1}). It has been postulated that at $T < T_g$ probe motion is largely independent of macromolecular motion and is

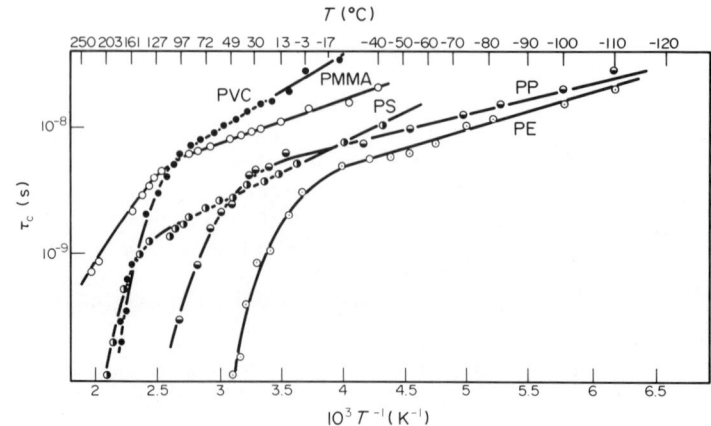

Figure 12   Arrhenius plots of correlation times of probe (**1**) in various polymers (reproduced by permission of Harwood Academic publishers from ref. 9, p. 177)

determined by the static free volume of (or pores within) the polymer. Above T_g, probe motion is determined by dynamic fluctuations due to greatly increased macromolecular motions which are characterized by somewhat higher energies of activation. The existence of pores has also been proposed to explain the motion of probes within polymer gels.[93]

The above comments re-emphasize the point that in the spin-probe experiment it is the motion of the probe and not that of the polymer which is observed. The spin-label experiment is less prone to this drawback and can provide direct information on polymer dynamics, particularly with in-chain labels such as (21) and (24). Unfortunately, there are relatively few examples of spin-label studies which include molten polymers because nitroxides are usually unstable at the relatively high temperatures required. Törmälä, however, succeeded in studying poly(ethylene oxide) (PEO) over a wide temperature range using label (29) and probes (8) and (10), with polymers of molecular weight 200 to 900 000.[10] The collected results of these investigations are plotted as a correlation map in Figure 13.[84] For polymers in the mid-range of molecular weights (1000–22 000) the spin label and probes show four different relaxation regions: (a) the liquid state; (b) the melting region; (c) the solid state; and (d) the 'frozen' state. The frozen state has been divided tentatively into two regions. For comparison purposes Figure 13 includes the low frequency values of the three important transitions T_m, $T_g(u)$ the upper T_g, and $T_g(l)$ the lower T_g. The transition temperatures revealed by the labelled polymers are about 20 °C higher than the corresponding low frequency values. For a given label or probe, τ_c at a specified temperature decreases with increasing molecular weight of the polymer. This is analogous to the change in frequency of the T_g relaxation with variation in molecular weight as observed in dielectric relaxation. The activation energy of both label and probe tumbling in region (c) attains a value of 40 kJ mol^{-1} when $\bar{M}_n > 10\,000$. This is in fair agreement with 'crankshaft' motion of PEO at $T < T_g$, but such agreement may be fortuitous because the polymer is labelled terminally rather than within the chain.

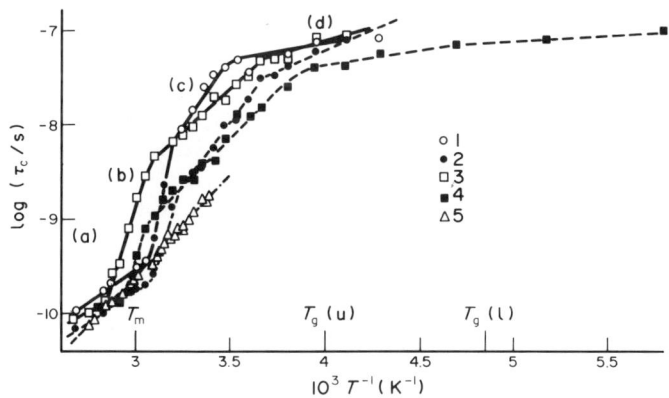

Figure 13 Arrhenius plot for spin-labelled and spin-probed poly(ethylene oxide): (1) label (29) ($\bar{M}_n$ of polymer = 1550). (2) probe (10) ($\bar{M}_n$ = 1550), (3) label (29) ($\bar{M}_n$ = 15 000), (4) probe (10) ($\bar{M}_n$ = 15 000), and (5) probe (8) ($\bar{M}_n$ = 90 000); (a) liquid state, (b) the melting region, (c) the solid state, and (d) the frozen state (reproduced by permission of Harwood Academic publishers from ref. 84)

In an investigation of polyethylene (PE) Bullock *et al.* used the in-chain label (24) which is effectively locked into the backbone.[20] Correlation times covering the range from slow to rapid rotation are shown as Arrhenius plots in Figure 14. Two well-defined transition temperatures are apparent; the upper one corresponds to the melting point T_m and the lower (350 K) to the α transition. Between T_m and T_α the α relaxation is being observed, and below T_α the β relaxation occurs for which the activation energy lies in the range 24–29 kJ mol^{-1}. This is similar to the activation energy derived from proton NMR studies over the same temperature range. Unfortunately, it was not possible to observe the β transition temperature (and hence T_g) because at temperatures below 225 K the conventional ESR spectrum had become too insensitive to yield correlation times. This polymer would be a useful candidate for study by STESR spectroscopy.

Relaxation times of spin probes (4) and (7) (R = Ph; $n = 0$) in poly(oxytetramethylene) have been determined in the temperature range 200–400 K in a recent study along the lines of Törmälä's work on PEO.[94] For both probes slow motion spectra were observed below 270 K and rapid motion spectra above this temperature. Correlation times converted to effective frequencies were plotted on a relaxation map together with dielectric and mechanical relaxation data. It was concluded that in

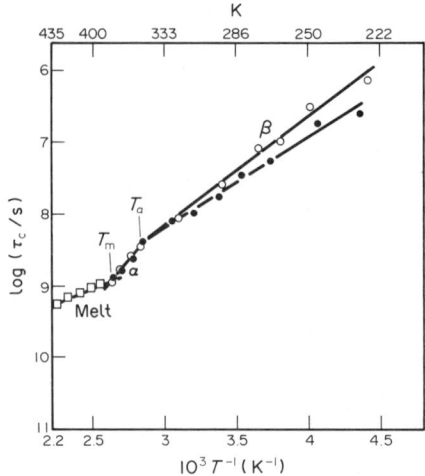

Figure 14 Arrhenius plot for bulk polyethylene labelled with (**24**). Before (○) and after (●) annealing[20]

the slow regime both probes responded to the β relaxation of the polymer and at higher frequencies to merged α and β relaxations. Differences in mobility of the probes were observed, as in previous studies. Isotropic motion of the smaller probe (**4**) was indicated but an improbably high degree of anisotropic motion for the larger probe was suggested. Lembicz and Ulieski have extended these studies to block copoly(ether–ester) elastomers.[95]

Noël *et al.*[96] have recently studied molecular motions and phase transitions in poly(vinylidene fluoride) (PVDF) and related copolymers by fluorescence polarization (FP) and spin-probe techniques [including probes (**3**), (**8**), (**11**) and (**13**)]. Amorphous copolymers containing 30–50 wt. % chlorotrifluoroethylene showed two well-defined relaxations. At low temperatures the local mode γ relaxation with activation energy 27.2 kJ mol^{-1}, was observed. At higher temperatures the glass to rubber β relaxation occurred, and in this region both ESR and FP data showed WLF behaviour. In the case of PVDF two β relaxations were observed. Noël *et al.* note that at $T_{50\,G}$ probe (**3**) in the copolymers did not reflect the glass transition.

23.4.3.3 *Other applications of spin probes and labels*

There are numerous reports of spin-label and spin-probe studies in applications other than those in the preceding two sections. Brown and co-workers[22,23,97] have used the spin probe (**16**) and spin labels (**23**) and (**28**) to investigate the microstructure of amine-cured epoxy resins. The label (**23**) is located at a chain end while (**28**) forms a bridge between two molecules of the epoxy diglycidyl ether of bisphenol A. The ESR spectra of dry samples and those containing up to ~5% solvent (CH_2Cl_2) were of the slow motion type. At solvent contents $\geq 30\%$ the line shape was of the motionally narrowed type. At intermediate solvent concentrations a composite of slow and fast motion spectra was observed for which the characteristic correlation times differed by more than 1.5 orders of magnitude. The slow phase spectrum was identified with nitroxides located in regions of high crosslink density and the fast phase spectrum with regions of lower crosslink density that had been selectively plasticized by solvent. The mobile fraction of spin labels, evaluated from the areas under the fast and slow motion spectra, gave a rough measure of the distribution of crosslink density in the epoxy network.

The spin-probe technique has provided a useful insight into the structure of polyalkenes and other crystalline polymers. The motion of the probe and its distribution within the polymer depend on the degree of crystallinity and the thermomechanical history of the polymer. In general, spin probes tend to concentrate in the amorphous regions of such materials. Meirovitch[98] has employed spin probes of varying dimensions to investigate the effects of cold drawing on polyethylene films. On stretching isotropically doped films there was a dramatic redistribution of probes between oriented and random chain environments, those probes trapped between parallel chains being strongly oriented. As an interesting comparison, a study of the mobility of various spin probes in nylon films showed that only those probes carrying polar amino or amide groups were affected by drawing.[99]

Presumably the orientation of the nylon chains affects the interaction of these probes with the polymer. In the case of polypropylene,[100] drawing between 30 and 110 °C produced a drop in rotational mobility of probe (3) with increasing draw ratio. On annealing, relaxation of the amorphous sections of chains resulted in an increase in the frequency of probe mobility to values greater than those observed in the quenched isotropic polymer. A similar effect of annealing on labelled polyethylene is evident in Figure 14.

Other recent applications of spin probes and labels include investigations of order and molecular mobility in liquid crystalline polymers[101–103] and in thin films of polymers on various substrates,[104] and inhomogeneities in synthetic and natural rubber.[105]

All the spin-probe and spin-label examples quoted in this chapter have involved nitroxides. However, there are one or two examples of polymers carrying other types of label. Hori *et al.*[106,107] and Shimada *et al.*[108,109] have shown that peroxy radicals in polyalkenes and PTFE can also serve as spin labels, providing the same sort of information as nitroxides. These radicals are not in general as stable as nitroxides, nor is it possible to deploy them in the same variety of polymers or applications. Triphenylverdazyl radicals have also been used as spin labels or probes, and in a recent report the verdazyl (36) was attached to a polyamide by condensation of the free amine group with a pendant acid chloride group on the polymer.[110]

(36)

In another example of a less common spin label, polystyrenes containing varying amounts of 2,6-di-*t*-butylphenoxyl units (37) were prepared.[111] Inter- and intra-molecular interactions between the radicals were determined by means of a linewidth analysis which involved simulation of the ESR spectra. There is obviously scope for developing other probes and labels, including transition metal compounds,[112] but it seems unlikely that these will supplant nitroxides.

(37)

23.5 REFERENCES

1. B. Rånby and J. F. Rabek, 'ESR Spectroscopy in Polymer Research', Springer-Verlag, Berlin, 1977.
2. A. R. Forrester, in 'Landolt–Bornstein', vol. 9, 'Magnetic Properties of Free Radicals, Part c1: Nitroxide Radicals', Springer-Verlag, Berlin, 1979, pp. 192–1066.
3. E. G. Rozantsev, 'Free Nitroxyl Radicals', Plenum Press, New York, 1970.
4. A. R. Forrester, J. M. Hay and R. H. Thomson, 'Organic Chemistry of Stable Free Radicals', Academic Press, London, 1968, chap. 5.
5. E. G. Rozantsev and V. D. Sholle, *Synthesis*, 1971, 190, 401.
6. H. G. Aurich and W. Weiss, *Top. Curr. Chem.*, 1975, **59**, 65.
7. H. G. Aurich, in 'The Chemistry of Functional Groups, Supplement F: The Chemistry of Amino, Nitroso, and Nitro Compounds and their Derivatives', ed. Saul Patai, Wiley–Interscience, 1982, part 1, pp. 565–624.
8. G. G. Cameron and A. T. Bullock, in 'Developments in Polymer Characterisation', ed. J. V. Dawkins, Applied Science, London, 1982, vol. 3, chap. 4, p. 107.
9. R. F. Boyer and S. E. Keineth (eds.) 'Molecular Motions in Polymers by ESR', MMI Press Symposium Series, vol. 1, Harwood, Chur, Switzerland, 1980.
10. P. Törmälä, *J. Macromol. Sci., Rev. Macromol. Chem.*, 1979, **17**, 297.
11. L. J. Berliner (ed.), 'Spin Labelling Theory and Applications', Academic Press, New York, 1976.
12. K. J. Ivin (ed.), 'Structural Studies of Macromolecules by Spectroscopic Methods', Wiley, London, 1976. (a) P. Törmälä and J. J. Lindberg, chap. 14; (b) A. T. Bullock and G. G. Cameron, chap. 15; and (c) K. Shimada and M. Szwarc, chap. 16.
13. A. T. Bullock, G. G. Cameron and I. S. Miles, *Polymer*, 1982, **23**, 1536.
14. R. McGregor, T. Iijima, T. Sakai, R. D. Gilbert and K. Hamada, *J. Membr. Sci.*, 1984, **18**, 129.
15. A. T. Bullock, G. G. Cameron and V. Krajewski, *J. Phys. Chem.*, 1976, **80**, 1792.

16. M. C. Cafe and I. D. Robb, *Polymer*, 1979, **20**, 513.
17. P. Törmälä, H. Lattila and J. J. Lindberg, *Polymer*, 1973, **14**, 481.
18. A. M. Wasserman, T. A. Alexandrova and A. L. Buchachenko, *Eur. Polym. J.*, 1976, **12**, 691.
19. J. F. W. Keana, S. B. Keana and D. Beetham, *J. Am. Chem. Soc.*, 1967, **89**, 3055.
20. A. T. Bullock, G. G. Cameron and P. M. Smith, *Eur. Polym. J.*, 1975, **11**, 617.
21. A. T. Bullock, G. G. Cameron and P. M. Smith, *Macromolecules*, 1976, **9**, 650.
22. I. M. Brown and T. C. Sandreczki, *Macromolecules*, 1985, **18**, 2702.
23. I. M. Brown and T. C. Sandreczki, *Chem. Phys. Lett.*, 1979, **64**, 85.
24. T. Kurosaki, K. W. Lee and M. Okawara, *J. Polym. Sci., Polym. Chem. Ed.*, 1972, **10**, 3295.
25. A. T. Bullock, G. G. Cameron, I. More and I. D. Robb, *Eur. Polym. J.*, 1984, **20**, 951.
26. P. Simon, L. Sümegi, A. Rockenbauer, F. Tüdös, J. Csekö and K. Hideg, *Macromolecules*, 1985, **18**, 1137.
27. A. T. Bullock, G. G. Cameron and N. K. Reddy, *J. Chem. Soc., Faraday Trans. 1*, 1978, **74**, 727.
28. A. T. Bullock, G. G. Cameron and J. M. Elsom, *Polymer*, 1974, **15**, 74.
29. C. Friedrich, C. Noël, R. Ramasseul and A. Rassat, *Polymer*, 1980, **21**, 232.
30. G. G. Cameron, E. Ross and I. S. Miles, unpublished work.
31. See, for example, V. A. Lopyrev, L. A. Tatarova, T. I. Vakul'skaya and T. G. Ermakova, *Vysokomol. Soedin., Ser. B*, 1985, **27**, 221.
32. N. Kusumoto and T. Sakai, *Polymer*, 1979, **20**, 1175.
33. A. Hudson and G. R. Luckhurst, *Chem. Rev.*, 1969, **69**, 191.
34. J. H. Freed, *J. Chem. Phys.*, 1964, **41**, 2077.
35. S. A. Goldman, G. V. Bruno, C. F. Polnaszek and J. H. Freed, *J. Chem. Phys.*, 1972, **56**, 716.
36. G. Poggi and C. S. Johnson, Jr., *J. Magn. Reson.*, 1970, **3**, 436.
37. J. S. Hwang, R. P. Mason, L. P. Hwang and J. H. Freed, *J. Phys. Chem.*, 1975, **79**, 489.
38. A. T. Bullock, G. G. Cameron and P. M. Smith, *J. Phys. Chem.*, 1973, **77**, 1635.
39. B. L. Bales, *J. Magn. Reson.*, 1980, **38**, 193.
40. B. L. Bales, *J. Magn. Reson.*, 1982, **48**, 418.
41. I. M. Brown, *J. Chem. Phys.*, 1973, **58**, 4242.
42. A. N. Kuznetsov, A. Y. Volkov, V. A. Livshits and A. T. Mirzoian, *Chem. Phys. Lett.*, 1974, **26**, 369.
43. A. M. Wasserman, T. A. Alexandrova and A. L. Buchachenko, *Eur. Polym. J.*, 1976, **12**, 691.
44. J. Labský, J. Pilǎr and J. Kǎlal, *Macromolecules*, 1977, **10**, 1153.
45. R. G. Gordon and T. Messenger, in 'Electron Spin Relaxation in Liquids', ed. L. T. Muus and P. W. Atkins, Plenum Press, London, 1972.
46. R. P. Mason and J. H. Freed, *J. Phys. Chem.*, 1974, **78**, 1321.
47. J. H. Freed, in ref. 11.
48. G. Moro and J. H. Freed, *J. Chem. Phys.*, 1981, **74**, 3757; 1981, **75**, 3157.
49. G. G. Cameron, in ref. 9, p. 55.
50. A. N. Kuznetsov, A. M. Wasserman, A. U. Volkov and N. N. Korst, *Chem. Phys. Lett.*, 1971, **12**, 103.
51. R. C. McCalley, E. J. Shimshick and H. M. McConnell, *Chem. Phys. Lett.*, 1972, **13**, 115.
52. A. N. Kuznetsov and B. Ebert, *Chem. Phys. Lett.*, 1974, **25**, 342.
53. S. Lee, D. P. Ames and I. M. Brown, *J. Chem. Phys.*, 1982, **76**, 805.
54. D. Kivelson and S. Lee, *J. Chem. Phys.*, 1982, **76**, 5746.
55. G. P. Rabold, *J. Polym. Sci., Part A-1*, 1969, **7**, 1203.
56. L. R. Dalton, B. H. Robinson, L. A. Dalton and P. Coffey, *Adv. Magn. Reson.*, 1976, **8**, 149.
57. D. D. Thomas, L. R. Dalton and J. S. Hyde, *J. Chem. Phys.*, 1976, **65**, 3006.
58. K. Balasubramanian and L. R. Dalton, *J. Magn. Reson.*, 1979, **33**, 245.
59. A. M. Wasserman, *Kem-Kemi.*, 1982, **9**, 56.
60. I. S. Miles, G. G. Cameron and A. T. Bullock, *Polymer*, 1986, **27**, 190.
61. K. Ohno, T. Ishii and J. Sohma, *Jpn. J. Appl. Phys., Part 1*, 1984, **23**, 1385.
62. A. T. Bullock, G. G. Cameron and P. M. Smith, *J. Chem. Soc., Faraday Trans. 2*, 1974, **70**, 1202.
63. A. T. Bullock, G. G. Cameron and P. M. Smith, *Polymer*, 1973, **14**, 525.
64. A. T. Bullock, J. H. Butterworth and G. G. Cameron, *Eur. Polym. J.*, 1971, **7**, 445.
65. E. Helfand, *J. Chem. Phys.*, 1971, **54**, 4651.
66. C. Friedrich, F. Laupêtre, C. Noël and L. Monnerie, *Macromolecules*, 1981, **14**, 1119.
67. Y. Inoue and T. Konno, *Polym. J.*, 1976, **8**, 457.
68. J. Pilǎr, J. Labský, J. Kǎlal and J. H. Freed, *J. Phys. Chem.*, 1979, **83**, 1907.
69. W. G. Miller, W. T. Rudolph, Z. Veksli, D. L. Coon, C. C. Wu and T. M. Liang, in ref. 9, p. 145.
70. A. L. Buchachenko, A. M. Wasserman, T. A. Aleksandrova and A. L. Kovarskii, in ref. 9, p. 33 and refs. therein.
71. A. L. Buchachenko and A. M. Wasserman, *Pure Appl. Chem.*, 1982, **54**, 507.
72. A. M. Wasserman, T. A. Aleksandrova, Yu. E. Kirsch and A. L. Buchachenko, *Eur. Polym. J.*, 1979, **15**, 1051.
73. D. Brown and P. Törmälä, *Makromol. Chem.*, 1978, **179**, 1025.
74. M.-C. Lang, F. Laupêtre, C. Noël and L. Monnerie, *J. Chem. Soc., Faraday Trans. 2*, 1979, **75**, 349.
75. C. Friedrich, F. Laupêtre, C. Noël and L. Monnerie, *Macromolecules*, 1980, **13**, 1625.
76. F. Sundholm, A. M. Wasserman, I. I. Barashkova, V. P. Timofeev and A. L. Buchachenko, *Eur. Polym. J.*, 1984, **20**, 733.
77. K. K. Fox, I. D. Robb and R. Smith, *J. Chem. Soc., Faraday Trans. 1*, 1974, **70**, 1186.
78. I. D. Robb and R. Smith, *Eur. Polym. J.*, 1974, **10**, 1005.
79. I. D. Robb and R. Smith, *Polymer*, 1977, **18**, 500.
80. M. C. Cafe and I. D. Robb, *J. Colloid Interface Sci.*, 1982, **86**, 411.
81. I. D. Robb and M. Sharples, *J. Colloid Interface Sci.*, 1982, **89**, 301.
82. T. M. Liang, P. M. Dickenson and W. G. Miller, *ACS Symp. Ser.*, 1980, **142**, 1.
83. H. Hommel, A. P. Legrand, H. Balard and E. Papirer, *Polymer*, 1984, **25**, 1297.
84. P. Törmälä, G. Weber and J. J. Lindberg, in ref. 9, p. 81.
85. P. L. Kumler and R. F. Boyer, *Macromolecules*, 1976, **9**, 903.
86. N. Kusumoto, in ref. 9, p. 223.

87. M. L. Williams, R. F. Landel and J. D. Ferry, *J. Am. Chem. Soc.*, 1955, **77**, 3701.
88. A. T. Bullock, G. G. Cameron, C. B. Howard and N. K. Reddy, *Polymer*, 1978, **19**, 352.
89. A. T. Bullock, G. G. Cameron and I. S. Miles, *Polym., Commun.*, 1983, **24**, 22.
90. K. S. Minsker, M. I. Abdullin, R. R. Gizatullin and A. M. Wasserman, *Plast. Massy*, 1984, 49.
91. G. G. Cameron, I. S. Miles and A. T. Bullock, *Br. Polym. J.*, 1987, **19**, 129.
92. S. Lee and I. M. Brown, *Macromolecules*, 1979, **12**, 1235.
92a. A. L. Kovarskii, A. M. Wasserman and A. L. Buchachenko, in ref. 9, p. 177.
93. S. G. Starodubtsev, O. K. Boiko, N. R. Pavlova and V. A. Kabanov, *Vysokomol. Soedin., Ser. A*, 1984, **26**, 2205.
94. F. Lembicz, *Makromol. Chem.*, 1985, **186**, 665.
95. F. Lembicz and R. Ulieski, *Makromol. Chem.*, 1985, **186**, 1679.
96. C. Noël, F. Laupêtre, C. Friedrich, C. Léonard, J. L. Halary and L. Monnerie, *Macromolecules*, 1986, **19**, 201.
97. I. M. Brown and T. C. Sandreczki, *Macromolelcules*, 1983, **16**, 1890.
98. E. Meirovitch, *J. Phys. Chem.*, 1984, **88**, 2629.
99. K. Hamada, T. Ijima and R. McGregor, *Macromolecules*, 1986, **19**, 1443.
100. C. L. Choy, W. P. Leung and T. L. Ma, *J. Polym. Sci., Polym. Phys. Ed.*, 1985, **23**, 557.
101. K. Mueller, K. H. Wassmer, R. W. Lenz and G. Kothe, *J. Polym. Sci., Polym. Lett. Ed.*, 1983, **21**, 785.
102. P. Meurisse, C. Friedrich, M. Dvolaitzky, F. Laupêtre, C. Noël and L. Monnerie, *Macromolecules*, 1984, **17**, 72.
103. K. H. Wassmer, E. Ohmes, M. Portugall, H. Ringsdorf and G. Kothe, *J. Am. Chem. Soc.*, 1985, **107**, 1511.
104. V. G. Savkin, V. A. Smurugov and I. O. Delikatnaya, *Dokl. Akad. Nauk SSSR*, 1984, **28**, 340.
105. T. Marinovíc, Z. Veksli, M. Andreis and D. Fleš, *Polym. Bull. (Berlin)*, 1984, **12**, 457.
106. Y. Hori, Y. Makno and H. Kashiwabara, *Polymer*, 1984, **25**, 1436.
107. Y. Hori, S. Shimada and H. Kashiwabara, *J. Polym. Sci., Polym. Phys. Ed.*, 1984, **22**, 1407.
108. S. Shimada, A. Kotaki, Y. Hori and H. Kashiwabara, *Macromolecules*, 1984, **17**, 1104.
109. S. Shimada, Y. Hori and H. Kashiwabara, *Macromolecules*, 1985, **18**, 170.
110. S. A. Belyakov, O. M. Polumbrik, V. T. Dorofeev, V. N. Sokolenko and I. G. Ryoboken, *Vysokomol. Soedin., Ser. B.*, 1984, **26**, 554.
111. D. Braun, P. Törmälä and W. Wittig, *Makromol. Chem.*, 1981, **182**, 2217.
112. R. A. Weiss, J. J. Fitzgerald, H. A. Frank and B. W. Chadwick, *Macromolecules*, 1986, **19**, 2085.

24

Characterization of Surfaces

DAVID BRIGGS
ICI Wilton, Cleveland, UK

24.1   INTRODUCTION

Before 1970 the most surface-sensitive method of polymer characterization was reflection IR spectroscopy — attenuated total reflection (ATR) or multiple internal reflection (MIR) IR (see Chapter 20). Even in its most surface-sensitive configuration the MIR technique samples approximately 0.4 μm (4000 Å) into a polymeric material, the sampling depth being wavelength dependent. Reflection IR also places severe restrictions on sample form since very good contact must be made against a flat crystal surface. The more recently introduced diffuse reflectance and photoacoustic IR spectroscopy methods can overcome some of these sample geometry problems, but the sampling depth remains of the same order as for MIR. Since surface properties of polymeric materials are dominated by structure/composition within the outermost molecular layers, probably no more than tens of angstroms in total thickness, clearly more surface-sensitive probes than IR are essential.

The breakthrough in polymer surface characterization came with the general development of surface spectroscopy methods based on ultrahigh vacuum technology. In 1971 the first X-ray photoelectron spectrum from a polymer was obtained.[1] For several reasons, discussed below, X-ray photoelectron spectroscopy (XPS, also commonly referred to as ESCA) was ideally suited to polymer surface analysis and has subsequently been very widely used. The technique is not without limitations, however, and more recently the considerable potential of secondary ion mass spectrometry (SIMS) has been demonstrated. The first SIMS studies of polymers were carried out as recently as 1980.[2]

This chapter will therefore outline the physical bases of these two techniques and discuss their application in the field of polymer surface characterization.

24.2 X-RAY PHOTOELECTRON SPECTROSCOPY[3]

24.2.1 Physical Basis of XPS

In XPS the sample, inside a high vacuum system (pressure $< 10^{-5}$ Pa), is irradiated with soft X-rays, usually Mg K_α (1253.6 eV) or Al K_α (1486.6 eV). The primary event is photoemission of a core (atomic) electron, but relaxation processes lead also to emission of Auger electrons as shown in Figure 1. The emitted electrons are collected by an electrostatic energy analyzer and detected as a function of kinetic energy (E_K), producing a spectrum such as in Figure 2. The intense, narrow peaks are due to photoelectrons emitted from core orbitals (*e.g.* C 1s). The binding energies (E_B) of these electrons, obtained from the Einstein relation

$$E = h\nu - E_K - \phi \tag{1}$$

($h\nu$ is the X-ray photon energy, ϕ the sample work function) are highly characteristic and allow the identification of all elements except H. The peak intensities are proportional to the number of atoms sampled and, with the aid of appropriate sensitivity factors, atomic compositions can be calculated, with detection limits of typically 0.2 atomic per cent. The much weaker series of peaks at very low binding energy is due to photoelectron emission from the valence band (molecular orbitals). The other, broader peaks are due to Auger electrons whose kinetic energies are given by

$$E_K(ABC) = E_A - E_B - E_C - \phi \tag{2}$$

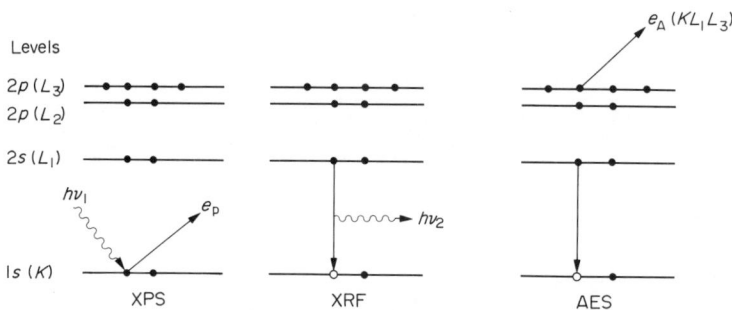

Figure 1 Schematic of the processes involved in X-ray photoelectron emission and subsequent relaxation *via* X-ray fluorescence and Auger electron emission (reproduction permission as in Figure 2)

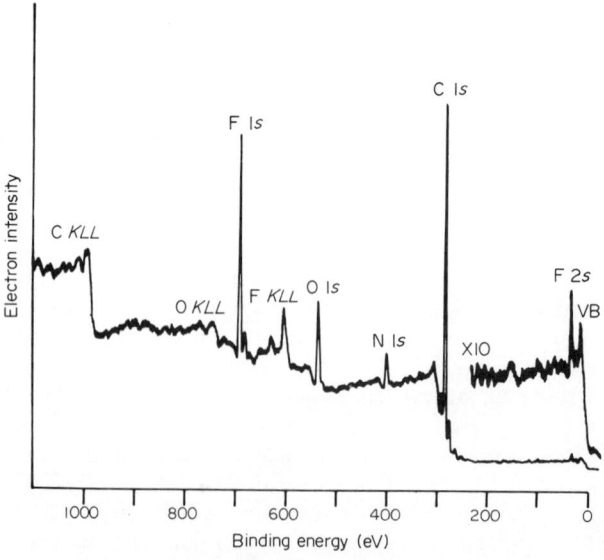

Figure 2 XPS survey scan of the surface of an epoxy adhesive cured in contact with PTFE and then separated by peeling, with some PTFE remaining on the adhesive. Characteristic core level peaks (*e.g.* C 1s), Auger peaks (*e.g.* F KLL) and valence band peaks are all present in this spectrum (reproduced with permission from 'Industrial Adhesion Problems', ed. D. M. Brewis and D. Briggs, Orbital Press, Oxford, 1985).

where A, B and C are respectively the photoionized level, the level of the electron which fills this vacancy and the level of the emitted Auger electron (Figure 1). These kinetic energy values are also element characteristic (but not all elements produce strong Auger signals under typical XPS conditions) and are obviously independent of $h\nu$.

The information from XPS is surface specific because the electrons which give rise to the useful peaks in the spectrum have emerged from the material elastically, *i.e.* without loss of energy. The inelastic mean free path (IMFP) of electrons in a organic solid is given by[4]

$$\lambda = \frac{49}{10^{-3}\rho E^2} + \frac{0.11 E^{0.5}}{10^{-3}\rho} \text{ (nm)} \tag{3}$$

where ρ is the bulk density in kg m^{-3} and E is the electron kinetic energy in eV. Since for many polymers $0.8 < \rho < 1.2$, the first term in equation (3) is negligible compared to the second and $\lambda \propto E^{0.5}$. Experimental determinations of the energy dependence of λ for polymers are somewhat conflicting and subject to rather large errors, but suggest[5] a relationship closer to $\lambda \propto E^{1.2}$. The information depth in XPS is given by

$$d = 3\lambda \sin\theta \tag{4}$$

where d and θ are defined in Figure 3. This equation gives the vertical depth below the surface from which 95% of the observed signal intensity is derived. Thus the depth sampled using Mg K_α for C 1s (963 eV, $\lambda = 12$A)[6] would be ~ 6Å ($\theta = 10°$) and ~ 36 Å (90°).

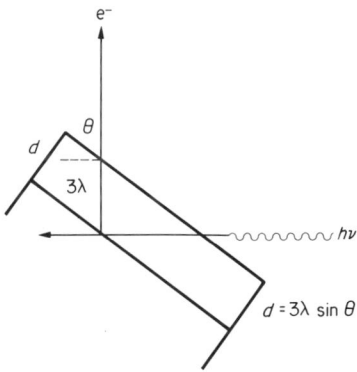

Figure 3 Angular dependance of sampling depth in XPS

24.2.2 Instrumental Aspects of XPS

The photon sources commonly used are Mg $K_{\alpha 1, 2}$ and Al $K_{\alpha 1, 2}$. The doublet nature of the X-ray line is not resolved and the single line widths are 0.7 and 0.85 eV respectively. Thus Mg K_α has an advantage in terms of spectral resolution. However, using a quartz crystal monochromator, Al K_α can be monochromated to give a significant decrease in line width (*e.g.* to ~ 0.4 eV). Some experiments have also been carried out with the higher energy Ti $K_{\alpha 1, 2}$ (4.5 keV).[7] For any given core level the sampling depth will be significantly greater than for the lower energy photons because of the increased kinetic energy of the photoelectrons. However the Ti $K_{\alpha 1, 2}$ line is a broad resolved doublet (causing the photoelectron spectra to be complex) and photoemission cross sections for core levels of interest in polymer studies (C 1s, 0 1s, N 1s) are much reduced.

All insulating surfaces charge up slightly during XPS. This is not normally a problem since the charge build up is small (a few electron volts) and reaches an equilibrium value very rapidly. For various reasons the use of a monochromated Al K_α source exacerbates the effect greatly and an external source of charge compensating electrons has to be provided, usually from an electron 'flood gun'.

There are no particular problems in handling polymeric materials in the vacuum conditions demanded by the technique. Outgassing of volatiles is usually slow, although high additive level rubber formulations are an exception. Radiation damage from the X-ray beam is rarely observed over usual timescales, but sample heating effects due to the proximity of the hot X-ray source may need to be reduced by deliberate cooling.

The most useful instrumental facility is the ability to vary the electron take-off angle (θ in Figure 3) so that changes in composition within the available sampling depth can be studied. This experiment is only appropriate for samples with reasonably flat surfaces.

24.2.3   Information from XPS

24.2.3.1   *Core levels*

As noted above the binding energies (BEs) of the core level peaks are highly element characteristic and all elements except H can be detected. Relative sensitivity factors are such that, excepting Li and Be which have low photoemission cross sections, the most intense core levels show a sensitivity range of roughly one order of magnitude. Peak overlap problems are relatively rare, particularly in polymer surface studies. When photoelectron peaks and Auger peaks overlap (*e.g.* between C 1s and Na *KLL* and Na 1s/Cl *LMM* using Mg K_α) use of an alternative photon source (in this case A1 K_α) moves the photoelectron peaks on the kinetic energy scale. Relative sensitivity factors for the appropriate instrument can either be generated *in situ* by studying homogeneous standards, taken from published tables[8,9] or calculated.[10,11] Peak areas can then be transformed into surface atomic composition data.

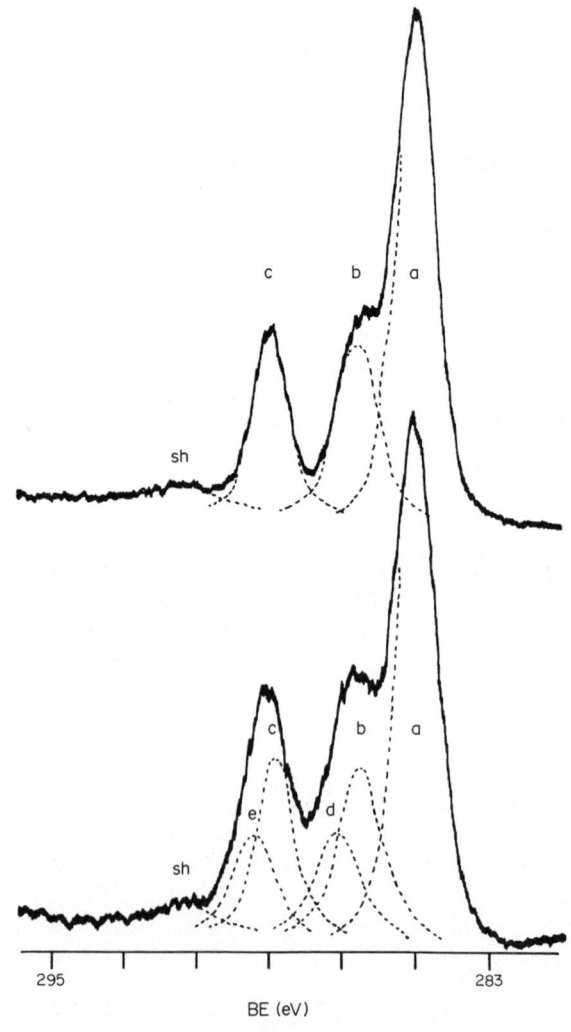

Figure 4 Deconvoluted C 1s profiles from poly(ethylene terephthalate). Upper trace: untreated PET (a) $-\underline{C}_6H_4^-$; (b) $-\underline{C}H_2O^-$; (c) $-\underline{C}O_2CH_2^-$; (sh) shake-up satellite. Lower trace: PET surface after electrical discharge treatment with additional peaks assigned to (d) $\underline{C}$–OH (phenolic) and (d) $-\underline{C}O_2H$ (reproduced with permission from 'Surface Analysis and Pretreatment of Plastics and Metals', ed. D. M. Brewis, Applied Science Publishers, 1982)

Any given core level also shows relatively small variations in BE, depending on the chemical environment, referred to as chemical shifts. This is illustrated by the C 1s peak, obtained at high energy resolution, from poly(ethylene terephthalate) (Figure 4). Clark and other workers[12] have created a large body of literature on this subject through the study of pure polymers and model small molecules coupled with theoretical calculations, much of it reviewed by Dilks.[13] From these data some simple summaries can be made.

(i) C 1s binding energies

(a) Carbon bound to itself and/or hydrogen only, no matter what hybridization, gives $C 1s = 285.0$ eV (often used as a binding energy reference).

(b) Halogens induce shifts to a higher binding energy which can be broken down into a primary substituent effect (*i.e.* on the carbon atom directly attached) and a secondary substituent effect (on the neighbouring carbon atom(s)). These shifts, per substituent, are shown in Table 1.

(c) Oxygen induces shifts to higher binding energy by ~ 1.5 eV per C–O bond (thus O–C–O and $>$C=O give similar C 1s binding energies, *etc*). The secondary effect of X in C–O–X is small (± 0.4 eV) except in the case of $X = NO_2$ (nitrate ester) which produces an additional shift of 0.9 eV.

(d) Nitrogen functionalities have a primary substituent effect which is markedly dependent on the nature of the substituent. Thus C 1s shifts for $-NMe_2$, $-NH_2$, $-NCO$ and $-NO_2$ are 0, 2, 0.6, 1.8 and 1.8 eV, respectively. Both carbon atoms in $-CH_2C\equiv N$ suffer a shift of ~ 1.4 eV. The C 1s shift induced by $-ONO_2$ is ~ 2 eV.

Table 1 Halogen-induced Primary and Secondary Shifts

Halogen	Primary shift (eV)	Secondary shift (eV)
F	2.9	0.7
Cl	1.5	0.3
Br	1.0	<0.2

(ii) O 1s binding energies

O 1s BEs from most functionalities fall within a narrow range of ~ 2 eV around 533 eV. The extremes are seen in carboxyl and carbonate groups in which the singly bound oxygen has the higher BE.

(iii) N 1s binding energies

Many common nitrogen functionalities give N 1s BEs in the narrow region 399–401 eV. These include $-CN$, $-NH_2$, $-OCONH-$ and $-CONH_2$. Quaternization, as in $-\overset{+}{N}H_3$, only increases the BE by ~ 1.5 eV above that of the free amine, largely because of the effect of the counterion.

Oxidized nitrogen functions have much higher N 1s BEs: $-ONO_2$ (~ 408 eV), $-NO_2$ (~ 407 eV) and $-ONO$ (~ 405 eV).

(iv) Other core levels

The only other atoms with variable oxidation states commonly encountered in polymers are sulfur and silicon. The primary effect of sulfur on the C 1s BE is very small (~ 0.4 eV measured in a polysulfone). However, S 2p BEs cover a reasonable range: RSR (~ 164 eV), RSO_2R (~ 167.5 eV), RSO_3H (~ 169 eV).

There have been few reports of Si 2p BEs. Typical silicones having the polysiloxane structure $-OSiR_2O-$ give Si $2p \approx 102$ eV.

It is also worth noting that halide ions can be distinguished from covalently bound halogen. In the case of fluorine the F 1s BE for F^- is ~ 4 eV lower than for C–F (typically 689 eV) whilst the Cl 2p BE for Cl^- is ~ 2 eV lower than C–Cl (typically 201 eV).

The small charging effects mentioned above move all peaks by the same amount. Absolute BEs are established by referencing to a peak of known energy, usually to C 1s $(CH_2) = 285.0$ eV. This structure is present in many polymers anyway, but a 'hydrocarbon' peak is frequently seen from adventitious contamination on surfaces and may even appear *in situ* if the sample is left in the XPS instrument long enough, particularly with the X-ray source operating.

24.2.3.2   *Shake-up satellites*

Shake-up satellites are observed on the low KE side of intense core level peaks. They occur when valence electron reorganization, following photoemission from a core level, results in the excitation of a valence electron to a higher unfilled level ('shake-up'). Conjugated and, especially, aromatic systems show shake-up satellites with intensities of 5–10% of the primary peak and separated by < 10 eV (Figure 4). Early work on hydrocarbon polymers with pendant aromatic groups established[13,14] the importance of two transitions, from the two filled orbitals $b_{1\pi}$ and $a_{2\pi}$ to the unoccupied $b_{2\pi}^{*}$ orbital (*i.e.* $\pi \rightarrow \pi^{*}$). More recently the appearance of shake-up satellites in aromatic polymer spectra has been reviewed[15] and the diagnostic value of these structures in the study of liquid crystal polymer conformation[16] and block copolymer surface composition demonstrated.[17]

The occurrence of shake-up satellites from vinylic unsaturation, at ~ 7 eV from the primary C 1s peak, has been invoked in the determination of the structure of poly(hexafluorobut-2-yne)[18] and to rationalize discrepancies in F/C stoichiometries from plasma polymerized fluoroethylenes.[19] Atoms directly attached to unsaturated centres will also be accompanied by shake-up satellites, and this may have some analytical value.

24.2.3.3   *Valence band spectra*

Photoemission from the molecular orbitals of polymers gives rise to the valence band. The valence band envelope is obviously a sensitive function of polymer structure and provides 'fingerprint' information in situations where core level data is inadequate, although the valence band is of much lower intensity. An example is given in Figure 5 in which purely hydrocarbon polymers are distinguished.[20] The whole subject has been carefully researched both experimentally and theoretically by Pireaux and colleagues and their results have recently been reviewed.[21] These workers

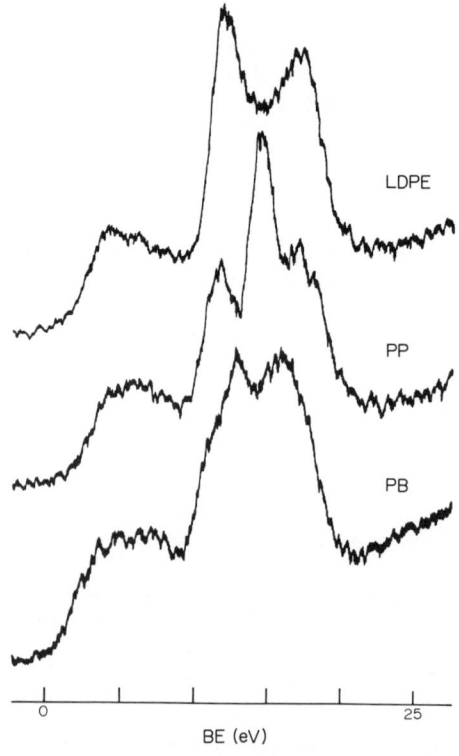

Figure 5   XPS valence band spectra of low density polyethylene (LDPE), polypropylene (PP) and poly(but-1-ene) (PB). These polymers give identical C 1s spectra (reproduction permission as in Figure 4)

have published high-resolution spectra, obtained with an X-ray monochromator, of polymers containing oxygen, fluorine and chlorine as well as hydrocarbon polymers. They have also demonstrated the sensitivity of valence band spectra to various types of isomerization (structural, linkage and stereo-), as well as to tacticity and geometrical conformation.

24.2.3.4 Functional group labelling (derivatization)

It will be realized from Section 24.2.3.1 that core level chemical shifts are frequently ambiguous because of the small dynamic range. Particularly in the case of multifunctional surfaces, often the result of treatments carried out to modify surface properties, major problems of speciation are therefore encountered. Several workers have attempted to overcome these problems by using specific derivatizing agents which react with one functional group and label it with a distinctive element. This procedure may have the additional advantage of increasing the detection sensitivity if the new element has a higher cross section than, for example, carbon, nitrogen or oxygen. Table 2 lists the reactions which have so far been investigated.

Table 2 Functional Group Labelling (Derivatization) for XPS[a, b]

Functional group	Reagent	Product	Ref.
$C=C$	Br_2	$\begin{array}{cc}C-C\\ \mid\ \ \mid \\ Br\ \ Br\end{array}$	22–24
$C=C$	$Hg(CF_3CO_2)_2$ } CCl_3CH_2OH	$\begin{array}{c}C-C\\ \mid\ \ \mid\\ CCl_3CH_2O\ \ Hg(CF_3CO_2)\end{array}$	25
$-CH_2OH$	$(CF_3CO)_2O$	$-CH_2OCOCF_3$	25, 26
$-CH_2OH$	$Ti(acac)_2OPr_2^i$	$-CH_2OTi(acac)OPr^i$	27
$C=C-OH$	$ClCH_2COCl$	$C=C-OCOCH_2Cl$	27
$-CH_2-C\begin{smallmatrix}O\\ \\ \end{smallmatrix}$	Br_2	$-CBr_2-C\begin{smallmatrix}O\\ \\ \end{smallmatrix}$	27, 28
$C=O$	$C_6H_5NH-NH_2$	$C=N-NHC_6H_5$	25, 27, 28
$-CO_2H$	NaOH	$-CO_2^-Na^+$	25–29
$-CO_2H$	$BaCl_2$	$(-CO_2^-)_2Ba^{2+}$	29, 30
$-CO_2H$	CF_3CH_2OH } $C_6H_{11}NCNC_6H_{11}$	$-CO_2CH_2CF_3$	25
$-CO_2H$	(i) KOH; (ii) $C_6F_5CH_2Br$	$-CO_2CH_2C_6F_5$	25
$-CO_2H$	$TlOC_2H_5$	$-CO_2^-Tl^+$	29
$-C-O_2H$	SO_2	$-C-OSO_2OH$	27
$-NH_2$	C_6F_5CHO	$-N=CHC_6F_5$	25
$-NH_2$	$EtS-COCF_3$	$-NH-COCF_3$	32

[a] Reproduced, with permission, from ref. 3. [b] See ref. 54 for a more recent compilation.

The reader is referred to the original literature for reaction conditions and estimates of success. Besides being specific, the derivatizing reaction should proceed rapidly under mild conditions and any necessary solvent should be benign. These last two conditions can be difficult to meet. Reactions which proceed rapidly at room temperature in the solution phase are often sterically hindered in the polymer surface layers. Solvents which permeate into the polymer are likely to aid reaction but may, at the same time, give rise to surface reorganization, *e.g.* functional group migration into the bulk. This aspect of derivatization has been studied in some depth by Everhart and Reilley.[33, 34]

24.2.3.5 Strengths and limitations of XPS

The major strengths of XPS are: ease of sample handling, permitting almost any type of polymeric material to be investigated irrespective of shape, surface morphology *etc.*; minor problems with sample charging; negligible radiation damage in most cases; ease of data quantification. The limitations are: very poor spatial resolution ($\sim 200 \ \mu m$ minimum diameter examined, in most instruments several mm^2 area); relatively poor degree of molecular specificity.

24.3 SECONDARY ION MASS SPECTROMETRY[35]

24.3.1 Physical Basis of SIMS

In SIMS the sample is bombarded by a primary ion beam. About 5% of the material sputtered from the surface consists of positively and negatively charged atoms (elemental ions) or molecular fragments (cluster ions) and these are mass analyzed to produce positive and negative secondary ion mass spectra. In the case of polymer surface characterization[36] it is essential that the experiment is carried out in the 'static' mode, *i.e.* that the accumulated ion dose during spectral acquisition is sufficiently low for the surface to be essentially unperturbed during the measurement. This threshold dose is a function of polymer structure and bombardment conditions, but is as low as 10^{13} ions cm^{-2} for many polymers.[37,38]

Typical bombardment energies are 2–4 keV, Ar^+ being the most widely used primary ion. Under these conditions the primary ion comes to rest ~ 30 Å below the surface. Energy deposition within this depth leads to direct 'knock-on' sputtering around the primary ion track, resulting in much bond breaking and the emission of small fragments. The linear cascade of recoil atoms can lead to the relatively gentle desorption of large molecules or molecular fragments some distance from the primary ion impact site.[39] Desorption of major fragments may also be the result of electronic interactions.[40] At present an exact description of the processes leading to the emission of large secondary ion clusters is unavailable. However, the sputtered fragments can also rearrange and further fragment, leading to the detection of a pattern of stable organic ions, as in conventional organic mass spectrometry, with very high structural information content (Figure 6).

It has been demonstrated,[41] at least for positive secondary ions, that as the energy or mass of the primary ion increases so the overall secondary ion yield increases, together with a relative increase in the yield of larger cluster ions. At the same time the rate of sample damage also increases and so compromises need to be made which also take into account overall detection sensitivity. A 2 keV Xe^+ primary beam of 1 nA cm^{-2} current density is recommended.[41]

Under charged particle bombardment insulating polymers will charge up. Since most of the secondary ions detected have very low energy (< 30 eV) any significant change in surface potential has serious consequences for spectral intensity. Charging is avoided and, therefore, surface potential is largely controlled by simultaneous irradiation with a flood of relatively low energy electrons (discussed in Section 24.3.2).

It is generally assumed in SIMS that the detected secondary ions emerge from a depth equal to about one-third the penetration depth of the primary ion, in which case an information depth for the conditions discussed would be ~ 10 Å. Recent work[42] on a series of model polyurethanes in which variable angle XPS data (described in Section 24.2.2) was compared with SIMS data showed a quantitative correlation only for the XPS results obtained at $\theta = 10°$ (as defined in Figure 3). Under these conditions the XPS sampling depth is indeed of the order of 10 Å, indicating that in static SIMS analysis of polymer surfaces only 1–2 monolayers are probed.

24.3.2 Instrumental Aspects of SIMS

Vacuum conditions for SIMS are rather more stringent than for XPS because of the greater need to avoid surface contamination *in situ*. This is due to (a) the greater surface sensitivity of SIMS and (b) greater confusion of the SIMS spectrum by fragment ions from contamination. Operating pressures of $< 10^{-7}$ Pa (true UHV) are therefore required.

Ion sources can be operated in two ways. The aim is to maintain 'static' conditions, as described above, and the appropriate (very low) ion current density can be obtained either by using a stationary defocussed beam or by using a focussed beam which is scanned raster fashion (TV rate) over a large area. The advantage of the latter approach is that it also allows, in principle, microanalysis and imaging to be carried out. In microanalysis the defined area over which the beam is scanned is much reduced and can be moved from place to place over a heterogeneous sample surface. In imaging, a particular secondary ion of interest is selectively detected by tuning the mass spectrometer to the appropriate mass. As the focussed ion beam scans over the defined area the

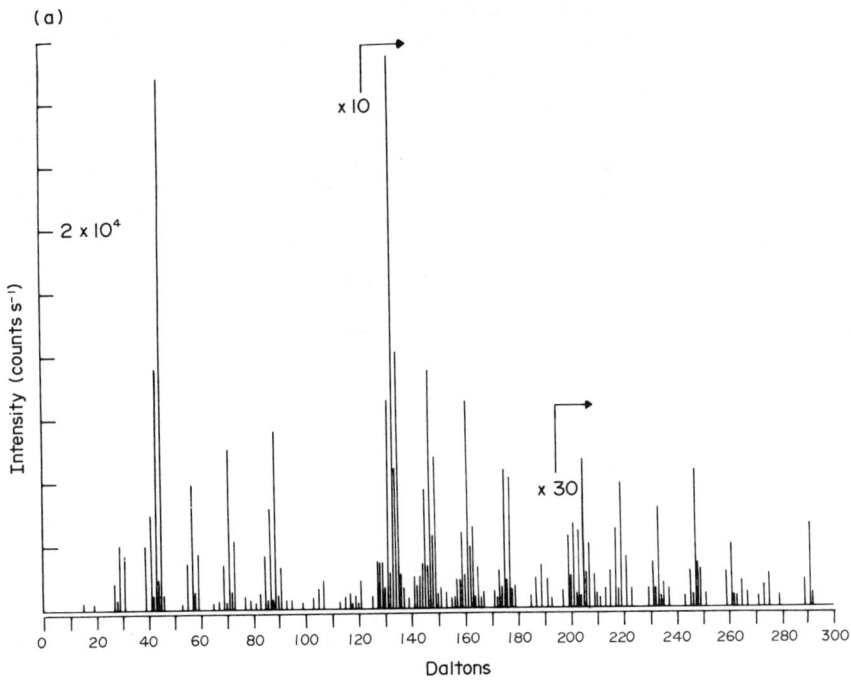

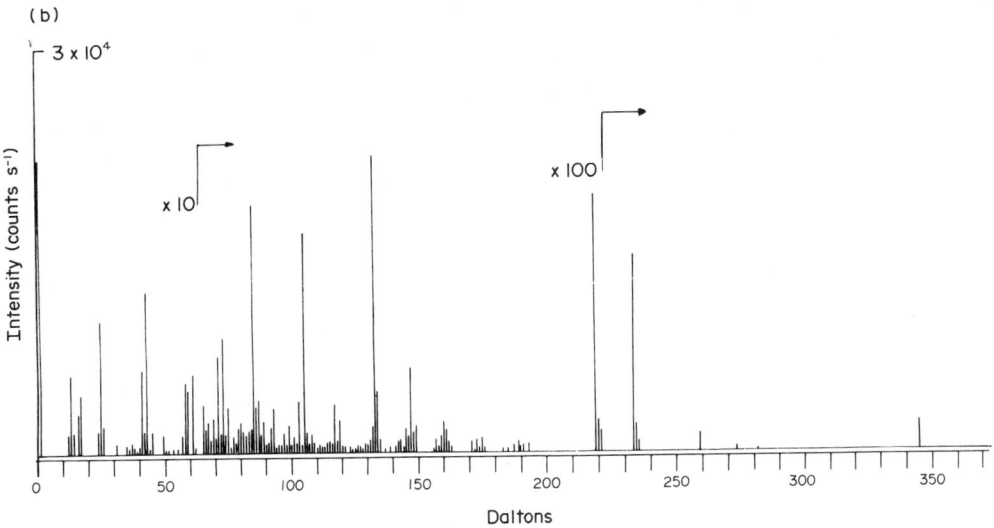

Figure 6   (a) Positive SIMS of poly(ethylene oxide) usng 4 keV Xe$^+$ (1 nA cm^{-2}); (b) negative SIMS of poly(ethylene oxide) using 2 keV Xe$^+$ (1 nA cm^{-2})

secondary ion intensity is fed to an oscilloscope scanning synchronously. This produces an image of the distribution of the chosen species. The spatial resolution is limited to the spot size of the primary ion beam. With conventional noble gas ion sources the minimum probe diameter is $\sim 10\ \mu$m but with liquid-metal ion sources (producing for example Ga$^+$) diameters of 500 Å can be achieved.

Quadrupole mass spectrometers have been used almost exclusively to date for static SIMS experiments. These give adequate mass resolution, have a mass range of several hundred daltons (up to 1000 D) and are compact. However, the collection efficiency is relatively low ($\sim 0.1\%$ of emitted secondary ions detected) and spectrometer transmission falls with increasing mass [$T \propto (1/M)$ $-(1/M^2)$]. Interest is growing in the use of time-of-flight mass spectrometers for several reasons, discussed in Section 24.3.3.4.

As mentioned previously charge neutralization is effected by means of a relatively low energy electron flood. An electron beam is itself capable of producing secondary ion emission (a process often referred to as electron-stimulated desorption), particularly of positive ion clusters.[41,43] This electron-stimulated secondary ion emission (ESIE) must be avoided, especially in imaging experiments where ESIE would provide an unacceptable background intensity over the whole image.[44] Neutralization without ESIE can be achieved by careful control of the electron beam position and intensity. For optimum negative ion detection the surface potential needs to be zero/negative and it appears that this can only be achieved by using increased electron beam current densities.[45,46] Fortunately, ESIE of negative ion clusters appears to be negligible.

An alternative to electron beam charge neutralization which has been successful with inorganic insulators (*e.g.* glasses) is so-called fast atom bombardment (FAB).[47] Here the noble gas ion source is coupled to a charge-exchange cell such that the ion beam is neutralized by a low pressure of gas, *e.g.*

$$\overrightarrow{Ar^+} + Ar \rightarrow \overrightarrow{Ar} + Ar^+$$

Neglible momentum transfer occurs so the diffuse atom beam produced has the same energy as the original ion beam. Although relatively little work on polymers has been carried out, it does not appear to be the case that charging can be completely avoided by this route, even though it may be less severe in many cases. In particular, an electron beam is still required to produce the correct surface potential for negative cluster ion detection.[45]

24.3.3 Information from SIMS

24.3.3.1 *Polymer fingerprint spectra*

As a general rule it is found that SIMS spectra from pure polymers contain peaks due to multiple repeat units (main chain scission) and intact side chains and to fragments of these. Fragmentation pathways have much in common with electron impact mass spectrometry and in several cases a rather close resemblance between SIMS and pyrolysis/EI mass spectra has been observed. Different classes of polymers are therefore readily distinguishable and individual members of a class can usually be identified by careful spectral interpretation. The information content of the combined positive and negative ion spectra is so high that fingerprinting techniques can be used. The strong influence of the repeat unit on the spectra clearly means that copolymer identification is also possible.

Space does not permit the discussion of many examples, but spectra of the polymers of the methacrylate esters illustrate several points. In the positive ion spectra[48] the region above ~ 80 D is very similar in all cases and therefore characteristic of the methacrylate backbone. Below 80 D the spectrum is dominated by fragmentation of the ester side chain. For instance intense peaks at 15, 29, 45 and 57 D characterize, respectively, methyl, ethyl, hydroxyethyl and *t*-butyl groups in the ester side chain. Isomers can also be distinguished. Two of the spectra of the poly(butyl methacrylate)s are shown in Figure 7 and all four isomers can be identified on the basis of the relative intensities of the 29, 43 and 57 D peaks. These spectra are, however, obviously complex and contain many peaks due to $C_xH_y^+$ fragments — these especially dominate the high mass part of the spectra. The negative ion spectra[38] are much simpler as illustrated in Figure 8. Peaks common to the two spectra represent fragments which have lost the alkyl group (R) whereas other peaks represent multiple repeat unit fragments, *e.g.* structures (1)–(4). Simple intense negative ion spectra are especially characteristic of oxygen-containing polymers.

A second example comes from the nylons (polyamides).[49] A typical spectrum from a simple nylon of the type $[-(CH_2)_xCONH-]_n$ is shown in Figure 9. All nylons of this type show $(nM+H)^+$ ions, where M is the repeat unit molecular weight (*e.g.* observed for $n=1$–5 for nylon 6) and $(nM+H-18)^+$ ions probably resulting from the steps outlined in Scheme 1. Similar processes lead to characteristic groups of peaks in the dicarboxylic acid–diamine polyamides which allow the repeat unit to be identified.[49]

$$\text{ununu}\,NH(CH_2)_x[CONH(CH_2)_x]_n\!-\!C\underset{\underset{H}{\displaystyle N}}{\overset{\displaystyle O}{\parallel}}CH(CH_2)_{x-2}CO\text{ununu} \longrightarrow {}^+CH_2(CH_2)_{x-1}[CONH(CH_2)_x]_nCONH_2 \xrightarrow{-H_2O}$$

$$(nM + H)^+$$

$${}^+CH_2(CH_2)_{x-1}[CONH(CH_2)_x]_nCN$$

$$(nM + H - 18)^+$$

Scheme 1

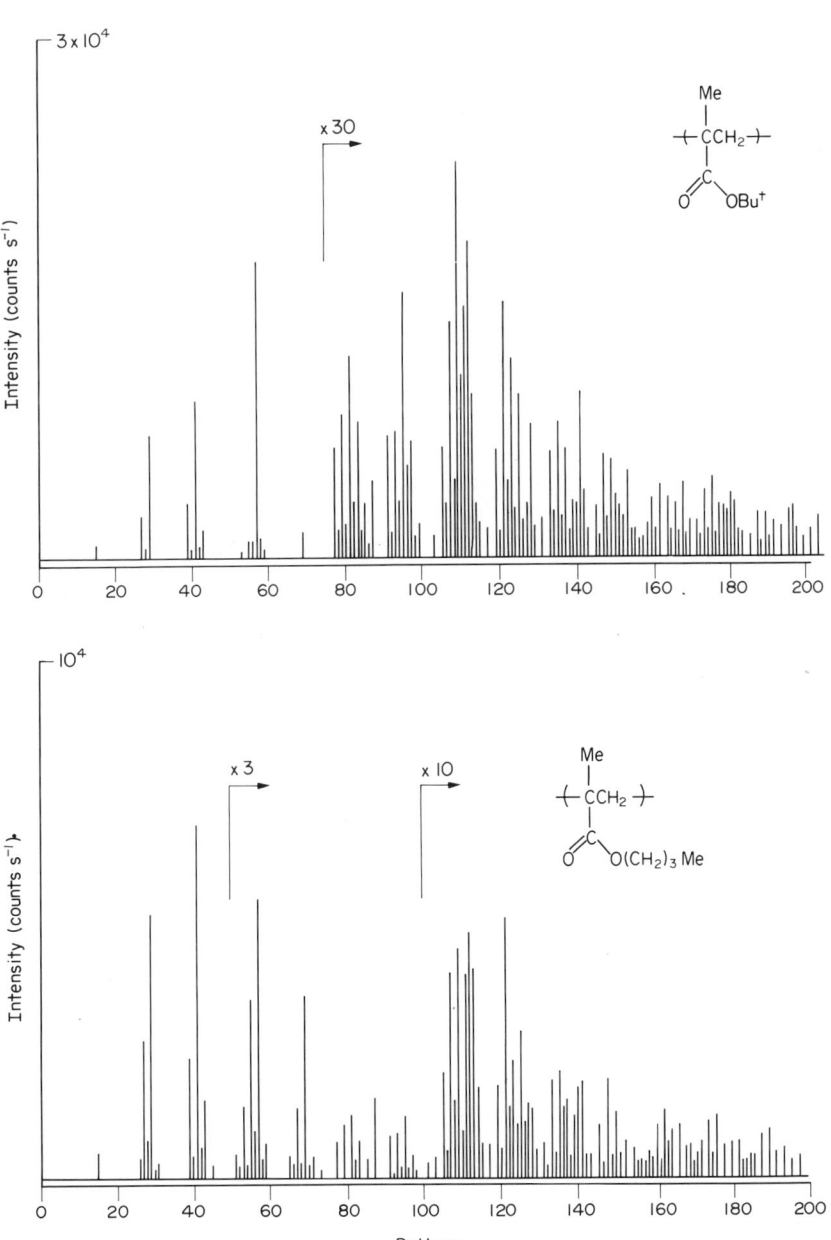

Figure 7 Positive SIMS of poly(*t*-butyl methacrylate) and poly(*n*-butyl methacrylate) using $4\,\mathrm{keV}\,\mathrm{Ar}^+$ ($1\,\mathrm{nA\,cm}^{-2}$) (reproduced from ref. 48 by permission of Wiley)

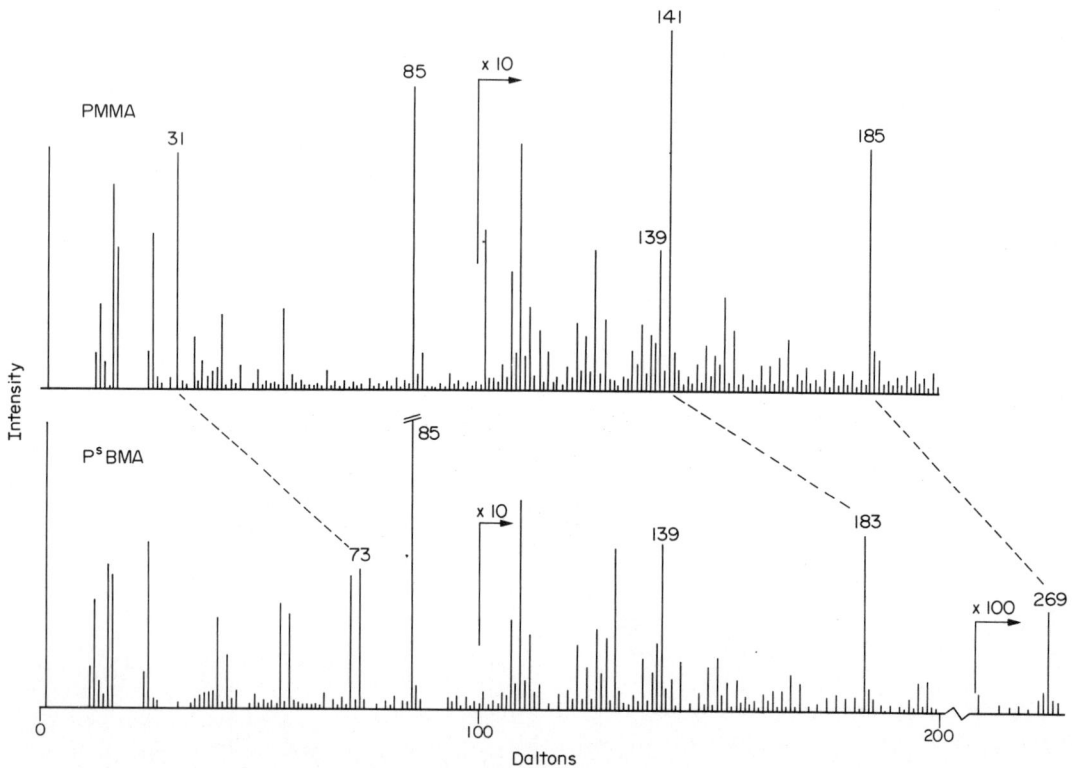

Figure 8 Negative SIMS of poly(methyl methacrylate) (PMMA) and poly(s-butyl methacrylate) (P^sBMA) using 4 keV Xe$^+$ (1 nA cm^{-2})

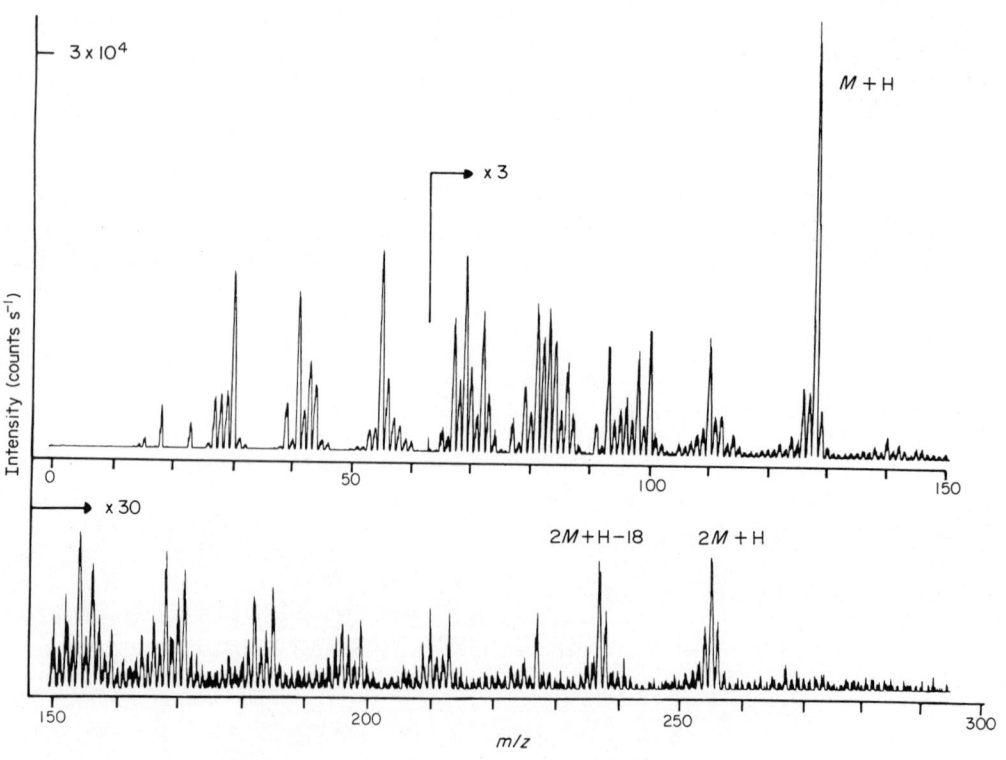

Figure 9 Positive SIMS of nylon 7 using 4 keV Xe$^+$ (3 nA cm^{-2})

24.3.3.2 *Quantitative aspects*

Quantification of SIMS intensity data is well known to be difficult in general because of the strong matrix effects on secondary ion yields. The case of organic surface analysis might be expected to offer even more difficulty. However, it has been found that for closely related structures relative peak intensity data can be used quantitatively. For homogeneous polymers it has been shown[42,48] that relative peak intensities from elemental or quasielemental ions correlate with bulk stoichiometry (atomic ratio) *e.g.* O^-/C^- or $CH^- \propto O/C$ or F^-/C^- or $CH^- \propto F/C$. The intensities of characteristic molecular fragments from the components of copolymers have been correlated with composition in the cases of methacrylate[48] and nylon[49] copolymers. A rather striking use of quantitative data is in the surface analysis of segmented polyurethanes, in which phase segregation in the bulk can often be observed. For a particular series of model polyurethanes based on poly(tetramethylene glycol), the diisocyanate MDI and fluorinated alkanediol chain extenders, relative intensities of peaks characterizing respectively the hard and soft (polyether) segments of the polymer chain have been shown[42] to quantitatively correlate with surface composition (*i.e.* degree of surface segregation of soft segment), as independently assessed[50] by XPS (Figure 10).

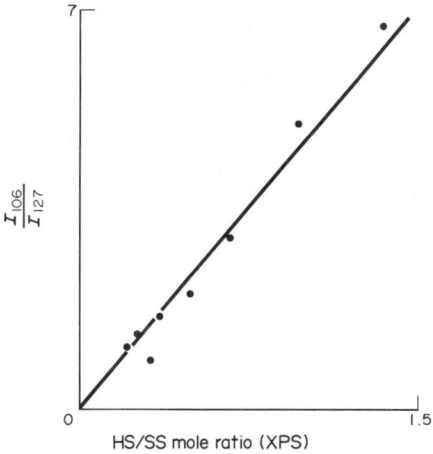

Figure 10 Intensity ratio of two peaks in the positive secondary ion mass spectra of model polyurethanes (see text) which characterize the hard segment (106 D) and soft segment (127 D) respectively, plotted as a function of the hard segment:soft segment mole ratio determined quantitatively for the outer surface using low take-off-angle XPS

24.3.3.3 *Polymer additive/contamination detection*

This is an important requirement in many areas of polymer surface technology. A particular problem with XPS is encountered when small molecules present on a polymer surface do not contain atoms which allow immediate identification *via* specific core level peaks. In many cases the SIMS spectrum contains a quasimolecular ion (*e.g.* MH^+) and together with the fragment ions this allows small molecule identification. An example is given in Figure 11. It has been demonstrated[51] that SIMS can follow the migration of cyclic oligomer (trimer) to the surface of poly(ethylene terephthalate) — in this case the 'small molecule' has the same chemical structure as the polymer but can be identified from its molecular ion.

24.3.3.4 *SIMS imaging and sample damage*

The principle of imaging has been outlined in Section 24.3.2 and an illustration of the result is given in Figure 12. To make full use of the high spatial resolution offered by liquid-metal ion sources, clearly small areas of sample will be irradiated and the rate of sample damage will increase. Figure 13 shows the effect of primary ion damage on PMMA. With a primary ion current of only 50 pA an area of 500 μm × 500 μm will receive an average dose of 10^{13} ions cm^{-2} in 80 s, about the time required to obtain *one* image under fairly ideal conditions, with the instrumentation described.

In order to overcome these problems a new generation of instrumentation is currently being developed, based on pulsed ion sources and time-of-flight (ToF) mass spectrometry. The ToF has

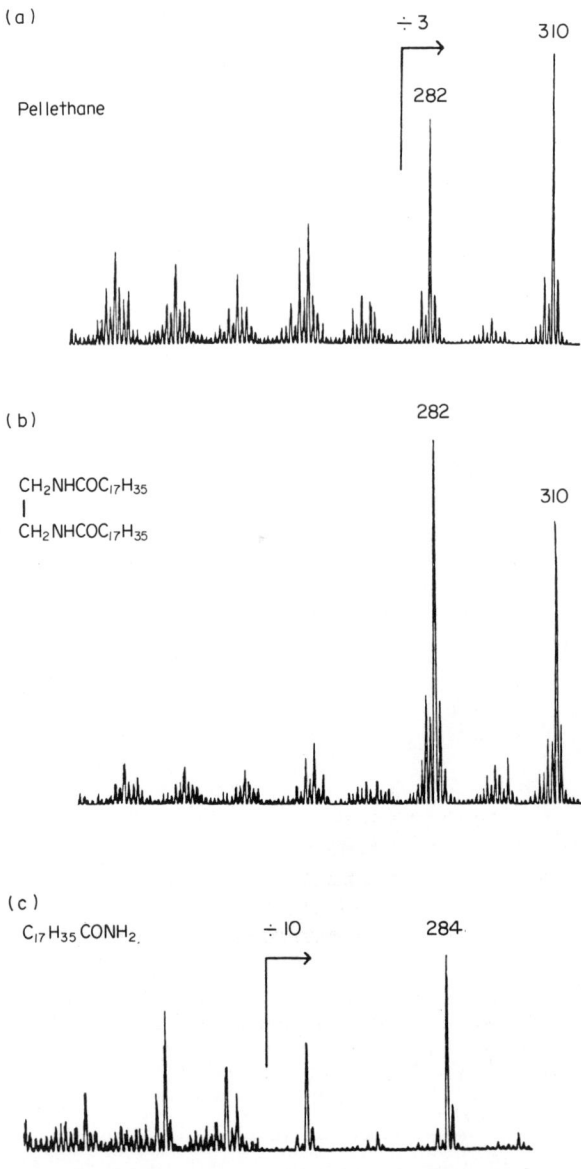

Figure 11 Identification of surface segregant on the surface of a cast film of a commercial polyurethane ('Pellethane') used in biomedical applications. Partial positive SIMS of (a) solution cast 'Pellethane', (b) pure bisethylene stearamide (BES), (c) pure stearamide, all using 4 keV Xe$^+$. Clearly BES, a lubricating agent, is the surface contaminant (reproduced from ref. 36 by permission of Wiley)

superior collection and transmission properties to the quadrupole mass spectrometer, giving several orders of magnitude improvement in detection efficiency (*i.e.* sensitivity). The ToF has, in principle, unlimited mass range, and preliminary work on model polymer systems indicates that exciting developments for polymer surface characterization can be anticipated for this new instrumentation.[52]

24.3.3.5 *Strengths and limitations of SIMS*

The major advantages of SIMS are: high absolute sensitivity and low information depth; molecular specificity; imaging potential. The main disadvantages are: charge compensation and optimization of secondary ion yields; non-straightforward quantification of data.

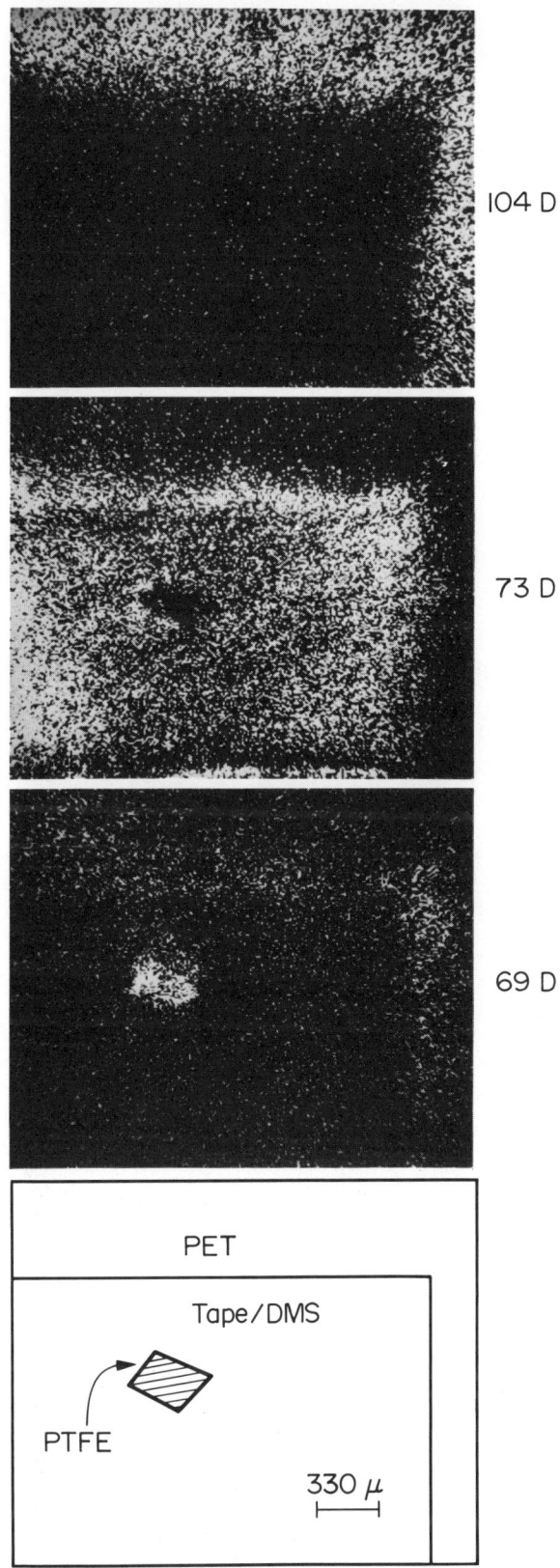

Figure 12 Secondary ion images from a test sample consisting of poly(tetrafluoroethylene) (PTFE) mounted on double-sided tape (with a surface layer of dimethylsilicone (DMS) release agent) in turn mounted on poly(ethylene terephthalate) (PET). The area imaged is shown schematically in the lower part of the figure. Mass 104 D is unique to PET, 73 D to DMS. Peaks at 69 D occur in the positive ion spectra of all three polymers, but most intensely from PTFE (reproduced from D. Briggs and M. J. Hearn, *Spectrochim. Acta*, 1985, **40B**, 707)

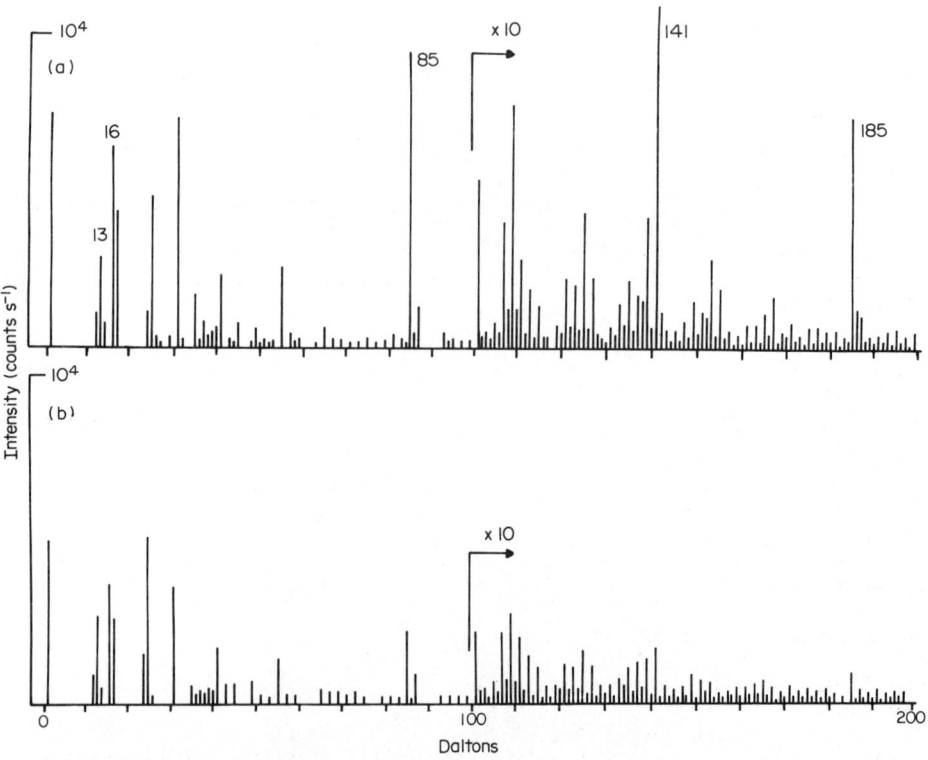

Figure 13 Negative SIMS of PMMA using 4 keV Xe$^+$ (0.5 nA cm^{-2}): (a) acquired using a total dose of 1.5×10^{12} ions cm^{-2}; (b) acquired as in (a) but following irradiation with 4 keV Xe$^+$ (9 nA cm^{-2}) giving a total dose of 5.5×10^{13} ions cm^{-2} (reproduced from ref. 38)

24.4 APPLICATIONS OF XPS AND SIMS

In polymer technology a large number of material properties involve control of surface composition. These include optical properties (*e.g.* haze, gloss, stains), adhesive properties (*e.g.* wettability, printability, bondability, heat sealability and releaseability), electrical properties (*e.g.* static charge-ability), general processing and machine handling properties (*e.g.* friction and 'blocking' of film in reels or stacks), biocompatibility, resistance to weathering and so on. Specific contaminants can also lead to surface crazing and cracking of many plastics.

Additives in plastics and polymeric materials include agents to prevent oxidation (thermal and photochemical), to neutralize acidity, to promote fire retardancy, to aid processing, to lower surface friction, to increase surface conductivity and to prevent 'blocking'. Other additives originating in polymerization processes, *e.g.* stabilizer and emulsifier molecules may be present. The presence on the material surface of these additives may, or may not, be intended, but their identification is necessary in order to establish structure/property relationships and in problem solving.

Surface properties of polymeric materials can also be modified *after* fabrication, thereby optimizing surface and bulk properties of an article semi-independently. Modification methods include chemical (wet) treatment/etching, electrical discharge treatment ('corona', glow discharge), flame treatment and thin film deposition by a variety of methods. Establishing the changes taking place during these modification processes, *e.g.* depth of change, alteration of chemical functionality, is clearly crucial to understanding optimum behavioural properties.

It is in these areas that XPS and SIMS have made a dramatic impact in the last 10 years or so. Since XPS has been used for the longer period of time, the literature is most extensive and has been reviewed.[20] Most of the SIMS literature has already been covered in this article, but its most fruitful area of application to date has been the specific analysis of segregated small molecular additives and very thin films.[53,36] Clearly the techniques complement each other very well and polymer surface characterization benefits enormously from their combined application.[53]

24.5 REFERENCES

1. D. T. Clark and D. Kilcast, *Nature*, 1971, **233**, 77.
2. J. A. Gardella Jr. and D. M. Hercules, *Anal. Chem.*, 1980, **52**, 226.
3. D. Briggs and M. P. Seah (eds.), 'Practical Surface Analysis by Auger and X-Ray Photoelectron Spectroscopy', Wiley, Chichester, 1983.
4. M. P. Seah and W. A. Dench, *Surf. Interface Anal.*, 1979, **1**, 2.
5. R. F. Roberts, D. L. Allara, C. A. Pryde, D. N. E. Buchanan and N. D. Hobbins, *Surf. Interface Anal.*, 1980, **2**, 5.
6. (a) D. T. Clark and H. R. Thomas, *J. Polym. Sci., Polym. Chem. Ed.*, 1977, **15**, 2843; (b) D. T. Clark and D. Shuttleworth, *J. Polym. Sci., Polym. Chem. Ed.*, 1978, **16**, 1093.
7. D. T. Clark, *ACS Symp. Ser.*, 1981, **162**, 255.
8. C. D. Wagner, L. E. Davis, M. V. Zeller, J. A. Taylor, R. H. Raymond and L. H. Gale, *Surf. Interface Anal.*, 1981, **3**, 211.
9. C. D. Wagner, in ref. 3, p. 511.
10. M. P. Seah, in ref. 3, p. 196.
11. M. P. Seah, M. E. Jones and M. T. Anthony, *Surf. Interface Anal.*, 1984, **6**, 242.
12. D. T. Clark and A. Harrison, *J. Polym. Sci., Polym. Chem. Ed.*, 1981, **19**, 1945, and previous papers in this series referred to therein.
13. A. Dilks, in 'Electron Spectroscopy — Theory, Techniques and Applications', ed. C. R. Brundle and A. D. Baker, vol. 4, Academic Press, London, 1981.
14. D. T. Clark and A. Dilks, *J. Polym. Sci., Polym. Chem. Ed.*, 1977, **15**, 15.
15. J. A. Gardella Jr., S. A. Ferguson and R. L. Chin, *Appl. Spectrosc.*, 1986, **40**, 224.
16. L. C. Lopez, D. W. Dwight and M. B. Polk, *Surf. Interface Anal.*, 1986, **9**, 405.
17. R. H. Thomas and J. J. O'Malley, *Macromolecules*, 1981, **14**, 1316.
18. R. D. Chambers, D. T. Clark, D. Kilcast and S. Partington, *J. Polym. Sci., Polym. Chem. Ed.*, 1974, **12**, 1647.
19. A. Dilks, in ref. 7, p. 307.
20. D. Briggs, in ref. 3, p. 359.
21. J. J. Pireaux, J. Riga, R. Caudano and J. Verbist in ref. 7, p. 169.
22. W. M. Riggs and D. W. Dwight, *J. Electron Spectrosc.*, 1974, **5**, 447.
23. D. Briggs, D. M. Brewis and M. B. Konieczko, *J. Mater. Sci.*, 1977, **12**, 429.
24. H. L. Spell and C. P. Christenson, TAPPI, 1978, **1978**, 283.
25. D. S. Everhart and C. N. Reilley, *Anal. Chem.*, 1981, **53**, 665.
26. J. Hammond, J. Holubka, A. Durisin and R. Dickie, 'Abstracts of Colloid and Interfacial Science Section, ACS Miami Meeting, September, 1978', American Chemical Society, Washington.
27. D. Briggs and C. R. Kendall, *Int. J. Adhes. Adhes.*, 1982, **2**, 13.
28. D. Briggs and C. R. Kendall, *Polymer*, 1979, **20**, 1053.
29. C. D. Batich and R. C. Wendt, in ref. 7, p. 221.
30. A. Bradley and M. Czuha, Jr., *Anal. Chem.*, 1975, **47**, 1838.
31. M. Czuha Jr. and W. M. Riggs, *Anal. Chem.*, 1975, **47**, 1836.
32. M. M. Millard and M. S. Masri, *Anal. Chem.*, 1974, **46**, 1820.
33. D. S. Everhart and C. N. Reilley, *Surf. Interface Anal.*, 1981, **3**, 126.
34. D. S. Everhart and C. N. Reilley, *Surf. Interface Anal.*, 1981, **3**, 269.
35. H. W. Werner, in 'Electron and Ion Spectroscopy of Solids', ed. L. Fiermans, J. Vennik and W. Dekeyser, Plenum Press, New York, 1978.
36. D. Briggs, *Surf. Interface Anal.*, 1986, **9**, 391.
37. M. J. Hearn and D. Briggs, *Surf. Interface Anal.*, 1986, **9**, 411.
38. D. Briggs and M. J. Hearn, *Vacuum*, 1986, **11/12**, 1005.
39. C. W. Magee, *Int. J. Mass Spectrom. Ion Phys.*, 1983, **49**, 211.
40. F. R. Kreuger, in 'Desorption Induced by Electronic Transitions, DIET II', ed. W. Brenig and D. Menzel, Springer, Berlin, 1985, p. 271.
41. D. Briggs and M. J. Hearn, *Int. J. Mass Spectrom. Ion Processes*, 1985, **67**, 47.
42. M. J. Hearn, D. Briggs, S. C. Yoon and B. D. Ratner, *Surf. Interface Anal.*, 1987, **10**, 384.
43. D. Briggs and A. B. Wootton, *Surf. Interface Anal.*, 1982, **4**, 109.
44. D. Briggs, *Surf. Interface Anal.*, 1983, **5**, 113.
45. A. Brown and J. C. Vickerman, *Surf. Interface Anal.*, 1986, **8**, 75.
46. D. Briggs and M. J. Hearn, in preparation.
47. D. J. Surman, J. A. van den Berg and J. C. Vickerman, *Surf. Interface Anal.*, 1982, **4**, 160.
48. D. Briggs, M. J. Hearn and B. D. Ratner, *Surf. Interface Anal.*, 1984, **6**, 184.
49. D. Briggs, *Org. Mass Spectrom.*, 1987, **22**, 91.
50. S. C. Yoon and B. D. Ratner, *Macromolecules*, 1986, **19**, 1068.
51. D. Briggs, *Surf. Interface Anal.*, 1986, **8**, 133.
52. D. M. Hercules and L. V. Bletsos, *Polym. Mater. Sci. Eng.*, 1986, **54**, 302.
53. D. Briggs, *Polymer*, 1984, **25**, 1379.
54. J. D. Andrade (ed.), 'Surface and Interfacial Aspects of Biomedical Polymers', Plenum Press, New York, 1985, vol. 1.

25
Optical Activity

FRANCESCO CIARDELLI
University of Pisa, Italy

25.1 CHIRALITY OF MACROMOLECULAR COMPOUNDS

Both in low and high molecular weight compounds optical activity can be observed only in chiral molecules. Analysis of the chirality of a molecule or macromolecule can be made by looking at its symmetry properties. Very simply, a molecule is chiral if all its allowed conformations lack reflection symmetry elements.

In this connection macromolecules differ from conventional low molecular weight molecules as they possess a substantially linear structure along the chain backbone. Accordingly analysis of the symmetry properties has been carried out on the basis of three different models: (i) the infinite length chain; (ii) the finite length chain with equal end groups; and (iii) the finite length chain with different end groups. As recently discussed by Farina[1] point symmetry, valid for molecules having definite and 'discrete' dimensions in all directions, can be used only for the last two models, whereas linear symmetry must be used for the first model which implies an infinite dimension. It is useful to remember that in linear symmetry, in contrast to point symmetry, the new symmetry operation 'translation' and the new symmetry element 'translation axis' are introduced. The analysis for a flexible macromolecule which can assume an extremely large number of conformations is conveniently carried out on the most symmetric of these conformations, which is usually the 'planar zigzag'. The analysis of the derived Fischer projection of an infinite chain indicates that this is chiral when the symmetry plane containing the chain, those perpendicular to the chain and that with translation containing the chain, are all lacking.[1] For finite length chains, chirality is guaranteed by lack of symmetry in the plane containing the chain and the plane perpendicular to the chain at its central point.[1]

By applying these approaches to different polymers derived from a generic vinyl monomer, $CH_2=CHX$, it can be concluded that only atactic macromolecules can be chiral in the first model with infinite length chain. The atactic and the syndiotactic macromolecules with an even number of monomer residues are chiral in the finite chain model with identical end groups, whereas all isotactic, syndiotactic and atactic chains have a chiral structure for the last model with different end groups.[1,2]

In the cases for which the macromolecules are chiral, extremely low optical rotation can be predicted when the molecular weight is sufficiently high, even if a complete separation of the enantiomeric pair were possible. Indeed, in vinyl polymers every stereogenic carbon atom is flanked by two CH_2 groups and its chirality arises only from the different lengths of the two chain sections attached to it. Thus an appreciable contribution to chiroptical properties is conceivable only for asymmetric centers close to the chain ends, the concentration of which decreases with increasing molecular weight. Thus these macromolecules fall more properly under the general classification of cryptochirality.[1,3]

It is well known that many isotactic vinyl polymers assume helical conformations in the crystalline state,[4] but due to the substantial achirality of the macromolecules both screw senses are found in the lattice cells in equal amounts. This is even more true in the melt or in solution, where left-handed and right-handed helical sections alternate even within the same macromolecular chain. Certainly appreciable optical rotation could be observed in the crystalline state provided that crystallization occurred under a chiral field inducing a single screw sense helicity in all chains. Such an optical rotation would be promptly lost on melting or dissolution as an immediate equilibration between the two opposite helical senses would occur due to thermodynamic reasons. While the polymerization of propylene and styrene with a Ziegler–Natta heterogeneous catalyst, based on optically active metalkyls, gave polymers with low optical rotation decreasing with increasing molecular weight,[5] it has been recently claimed that isotactic polypropylene showing appreciable rotatory power only in the solid state has been obtained with optically active homogeneous catalyst systems.[6]

All these observations suggest that macromolecules of the above type are substantially, although sometimes not formally, achiral, and that their symmetry properties are better represented by the infinite length chain. Thus, in order to prepare macromolecules capable of displaying stable and appreciable chiroptical properties in all conditions, the symmetry elements rendering the infinite chain model achiral must be eliminated. There are at least five systems in which this goal has been achieved and which have been actually described.[1,2] These are: 1. polymers of a single enantiomer of a chiral vinyl monomer, for instance in the case of poly[(R)-3-methyl-1-pentene] even the isotactic chain having infinite length lacks the mentioned symmetry planes (1);[7,8] 2. polymers of cyclic monomers such as benzofuran (2);[9,10] 3. polymers containing structural units of the type –A–CHX–B– (with A ≠ B) and derived from a single enantiomer, such as poly[(R)-propylene oxide] (3a)[11–13] and poly[(S)-α-alanine] (3b);[14] 4. copolymers with a well defined sequence distribution and relative stereochemistry of comonomer units;[15,16] these have been obtained by an asymmetric copolymerization using the bifunctional template monomer 3,4-O-cyclohexylidene-D-mannitol-

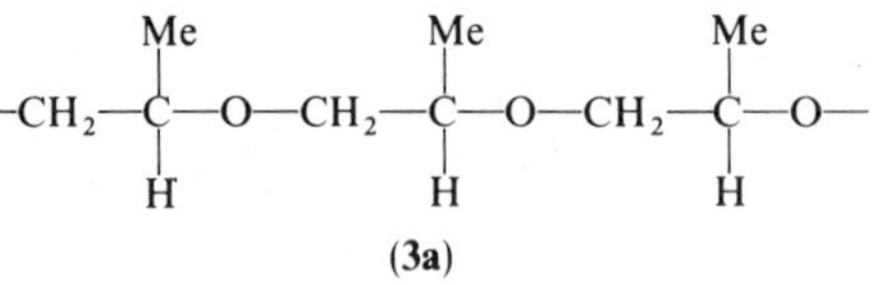

(1)

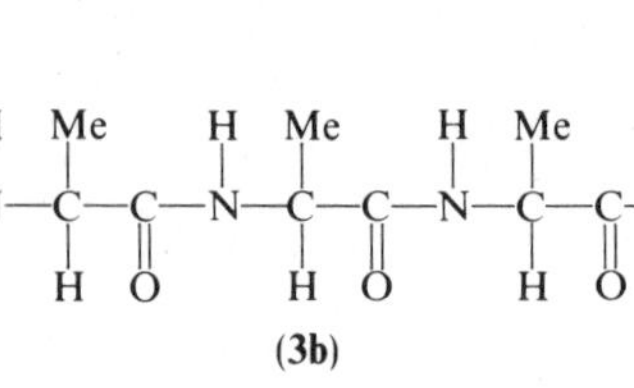

(2)

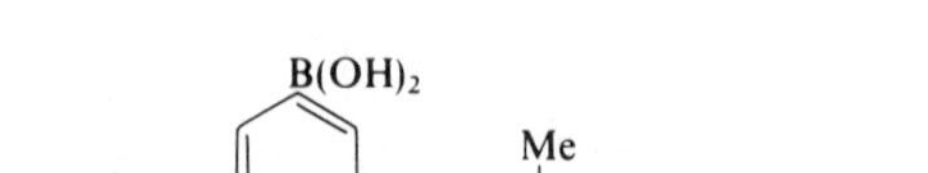

(3a)

(3b)

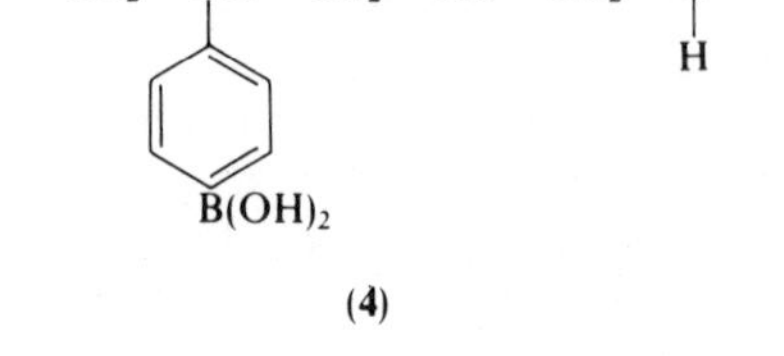

(4)

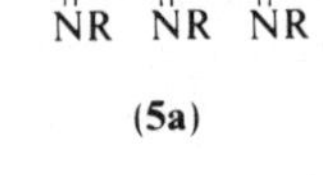

(5a)

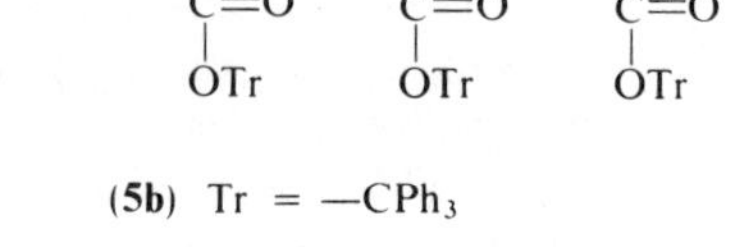

(5b) Tr = —CPh$_3$

1,2-5,6-bis-*O*-4-vinylphenylboronate and a vinyl monomer[17] and after removal of mannitol, macromolecules with structure (4) were obtained; 5. polymers with a very rigid structure and assuming helical conformations of opposite screw sense separated by a high energy barrier not allowing equilibration; typical examples are offered by very rigid structures such as poly(iminomethylene)s (5a)[18] and polymers with very bulky side chains such as poly(trityl methacrylate) (5b).[19]

25.2 MEASUREMENT OF CHIROPTICAL PROPERTIES OF POLYMERS

A linearly polarized light beam passing through an optically active medium generates an elliptically polarized wave. The corresponding ellipse has the major axis rotated by an angle α ('optical rotation') with respect to the original polarization plane of the light beam. The arc-tangent of the minor to major axis of the elliptical vibration gives the 'ellipticity angle' Ψ.[20,21] In order to show non-vanishing optical rotation and ellipticity angle a polymer must consist of chiral macromolecules with a predominant handedness. The measurement of the chiroptical properties α and Ψ is carried out, respectively, by polarimeters and circular dichrographs. These last are substantially the same when the chirality of the medium is due either to chemical or physical reasons, but the treatment of experimental results and the type of information derived are clearly different. Chemists are generally interested in obtaining indications of molecular structure and this is best achieved by measuring chiroptical properties in dilute solutions, where possible artifacts due to orientation, aggregation and supermolecular order, as for instance in liquid crystals and solid crystalline materials, are not observed. Indeed, in the case of polymers most measurements have been performed in solution,[12,22] even if a few examples are also known in the solid state.[23,24]

For solution measurements it is very convenient to use the 'specific optical rotation' and the 'specific ellipticity angle', which are related to optical rotation α and ellipticity angle Ψ as shown in equations (1) and (2), where λ indicates the wavelength of the incident light, T the temperature, l the cell path in dm and c the concentration in $g\,dm^{-3}$. Equations (1) and (2) can be used both for macromolecules and for low molecular weight compounds, whereas some differences arise when computing the corresponding molar properties, according to equations (3) and (4) giving, respectively, the 'molar rotatory power' (or 'molar optical rotation') and the 'molar ellipticity'. When dealing with polymers, M in equations (3) and (4) does not indicate the molecular weight of the macromolecules or the average molecular weight for polydisperse systems, but rather the molecular weight of the constitutional unit. As a consequence, the two quantities are independent of the average number of residues per chain in the polymer, *i.e.* of polymerization degree.

$$|\alpha|_\lambda^T = 100\alpha/lc \tag{1}$$

$$|\Psi|_\lambda^T = 100\Psi/lc \tag{2}$$

$$|\Phi|_\lambda^T = |\alpha|_\lambda^T (M/100) \tag{3}$$

$$|\Theta|_\lambda^T = |\Psi|_\lambda^T (M/100) \tag{4}$$

$$\bar{M} = N_A M_A + N_B M_B \tag{5}$$

In a copolymer of the two monomers A and B, M is replaced by the average value $\bar{M}$ shown in equation (5), where N_A and N_B are the mole fractions and M_A and M_B the molecular weights of the respective monomer residues in the copolymer. Even in the absence of particular interactions between A and B units, $|\Phi|_\lambda^T$ and $|\Theta|_\lambda^T$ in copolymers are composition dependent due to the consequent variation of $\bar{M}$ (see equation 5). A monotonic dependence on composition of copolymer, equal to that expected for the homopolymer mixtures, indicates either absence of major chiral electronic and steric interactions between A and B residues in the copolymer, or no copolymer formation. On the contrary, deviation from linearity demonstrates copolymer formation and can provide specific structural information.[25,26] The situation can be clarified by substituting equation (5) into equations (3) and (4) to obtain equations (6) and (7). To obtain these last two expressions it is assumed that specific quantities of mixtures are given by summation, as for instance $|\alpha| = \Sigma_i w_i |\alpha|_i$ where $|\alpha|_i$ and w_i are the specific rotation and weight fraction of the species i.

$$|\Phi|_\lambda^T = \frac{1}{100} \{(|\alpha|_\lambda^T)_A M_A N_A + (|\alpha|_\lambda^T)_B M_B N_B\} \tag{6}$$

$$|\Theta|_\lambda^T = \frac{1}{100} \{(|\Psi|_\lambda^T)_A M_A N_A + (|\Psi|_\lambda^T)_B M_B N_B\} \tag{7}$$

These last two equations can be converted into equations (8) and (9) by introducing the molar rotation and molar ellipticity of the homopolymers from A and B. Then, by considering also that $N_A + N_B = 1$, two new equations are obtained as shown below.

$$|\Phi|_\lambda^T = (|\Phi|_\lambda^T)_A N_A + (|\Phi|_\lambda^T)_B N_B \tag{8}$$

$$|\Theta|_\lambda^T = (|\Theta|_\lambda^T)_A N_A + (|\Theta|_\lambda^T)_B N_B \tag{9}$$

$$|\Phi|_\lambda^T = N_A\{(|\Phi|_\lambda^T)_A - (|\Phi|_\lambda^T)_B\} + (|\Phi|_\lambda^T)_B \tag{10}$$

$$|\Theta|_\lambda^T = N_A\{(|\Theta|_\lambda^T)_A - (|\Theta|_\lambda^T)_B\} + (|\Theta|_\lambda^T)_B \tag{11}$$

Equations (10) and (11) indicate that by plotting either $|\Phi|_\lambda^T$ or $|\Theta|_\lambda^T$ of the copolymer vs. N_A a straight line is obtained when the two constitutional units have the same chiroptical properties in the copolymer as in the respective homopolymers. This result clearly excludes the possibility that substantial structural changes have occurred in the copolymeric macromolecules. Deviation from linearity, by contrast, can be used for the identification of new structural features. While the optical rotation dispersion cannot be used to separate the individual contributions of A and B, unless one is zero,[26] this is possible for molar ellipticity whenever the chromophores in A and in B absorb in distinct spectral regions. For instance, in the absorbance region of the chromophore A, $(|\Theta|_\lambda^T)_B$ can be zero and then equation (9) is transformed into equation (12). A similar equation can clearly be written for B in its absorption spectral region, provided A does not absorb at the corresponding wavelength.[27-29]

$$|\Theta|_\lambda^T = N_A(|\Theta|_\lambda^T)_A \tag{12}$$

In addition to molecular weight polydispersity, macromolecules are characterized by the presence in solution of an extremely large number of conformations;[30] thus the molar values of chiroptical properties given by the above equations are averages of the many contributions coming from different conformers, the relative amount of which depends on several environmental factors, such as temperature and solvent.[31]

Quite often the molar ellipticity is reported as a function of the difference between extinction coefficients of left (ε_L) and right (ε_R) circularly polarized components of the incident light beam as shown in equation (13). Values of $\Delta\varepsilon$, the 'molar coefficient of dichroic absorption', can be treated in an analogous manner to ellipticity data. Curves obtained by measuring the optical rotation at different wavelengths (optical rotatory dispersion (ORD) curves) are connected to circular dichroism (CD) curves through Kronig–Kramers integral transforms.[20]

$$\Delta\varepsilon = \varepsilon_L - \varepsilon_R = 0.0003|\Theta| \tag{13}$$

25.3 THEORETICAL APPROACHES TO THE CALCULATION OF CHIROPTICAL PROPERTIES IN POLYMERS

A comprehensive review of the theoretical approaches for evaluating optical activity in polymers has been reported in the book by Charney[32] published in 1979. Indeed, these studies are of interest whenever the chiroptical properties of polymers are appreciably different from those of monomeric analogs, the theory being devoted to understanding the origin of these differences. The problem is quite complex due to the lack of an entirely satisfying theory for low molecular weight compounds. In some ways, however, the approach can be more informative as macromolecules can be treated as aggregates of identical chromophores, such as a one-dimensional crystal or a continuous dimer. Clearly such a treatment is possible only if the chain is in an ordered conformation which implies a definite relative geometry of its constitutional units. The most typical examples are poly(α-amino acid)s and proteins, for which Moffitt was able to relate ORD features to the presence of ordered secondary structures in solution.[33,34] Moffitt's theory, successively developed by Tinoco,[35] is based on an approach of perturbative type and is still one of the most useful, at least when groups of electrons do not strongly interact making it necessary to consider exchange strength and charge transfer. The method is rather simple but makes it possible to understand differences in UV absorption and CD between polymers and monomeric analogs, as well as to relate these differences to structure.[36] Indeed, calculations of CD of polypeptides are now possible with good agreement between calculated values and experimental results for the α helix,[37] parallel and β sheets,[38] polyproline[39] and also disordered structures.[40] The treatment of polynucleotides has required a more complex approach due to the presence in each macromolecule of four different aromatic bases,[36] but the results are very satisfying.[41] Similar approaches based on the application of exciton theory to electrostatic interactions among chromophores disposed along a helical chain have allowed the calculation of CD spectra of vinyl polymers with aromatic side chains, with results in excellent agreement with the experimental ones.[42, 43]

Interactions of a single chromophore with the rest of the macromolecule can also be responsible for differences in CD of polymers and monomeric analogs. This behaviour is typically observed when a symmetric chromophore becomes dissymmetric because of insertion in a chiral molecule.[44] Thus in several stereoregular polymers optical activity is strongly affected by typical macro-molecular features such as degree of stereoregularity, conformational arrangement and interaction with solvent, without occurrence of electronic interactions among chromophores, as described through some relevant examples in the following section.

25.4 CHIROPTICAL PROPERTIES AND STRUCTURE

25.4.1 Polymers with Vinyl Structures

In Section 25.1 five ways for introducing chirality into macromolecular structures were indicated for which several actual examples can be found in the literature.

The first method is concerned with the polymers of vinyl monomers having a chiral center in the side group. These polymers are indeed optically active provided that the stereogenic center in the side chain has predominantly a single absolute configuration. In these polymers each constitutional unit contains at least two stereogenic centers, one in the main chain and the other in the side chain. The configurational sequence distribution of the former defines the tacticity of the polymer, while the prevalence of a single configuration in the latter gives the enantiomeric purity. Both clearly affect the absolute values of the chiroptical properties, which increase with increasing enantiomeric purity and degree of isotacticity.[2, 45] This increase is linear when the side chain stereogenic center is far from the backbone and asymptotic when it is in either the α or the β position as found for polyalkenes (6),[7, 8] poly(vinyl ether)s (7)[46-48] and poly(vinyl ketone)s (8).[49, 50]

(6)   (7)   (8)

The cooperative effect anticipated by the asymptotic dependence of chiroptical properties mentioned above has been experimentally confirmed by the study of the chiroptical properties of copolymers between optically active α-alkenes and achiral comonomers which also allows clarification of the meaning of equations (8) and (12). The coisotactic copolymer of (S)-4-methyl-1-hexene (6; $n = 1$; $R^1 = Me$, $R^2 = Et$) with 4-methyl-1-pentene (6; $n = 1$; $R^1 = R^2 = Me$) shows at any composition optical rotation higher than the mixture of the two homopolymers, thus indicating that 4-methyl-1-pentylene (4MP) units in the copolymer contribute to optical rotation, the contribution being of the same sign as that of (S)-4-methyl-1-hexylene (4MH) units. Equation (8) for this particular case assumes the form of equation (14). By assuming that $(|\Phi|_D^{25})_{4MH}$ is the same in the copolymer and in the homopolymer, the values of $(|\Phi|_D^{25})_{4MP}$ can be derived at each composition. This value for the most isotactic fraction is $+160$ against the value of $+240$ calculated for a left-handed helical chain of poly(4-methyl-1-pentene). In order to give such a value, more than 80% of the 4MP units in the coisotactic copolymer must be in the one screw sense helical conformation induced by the chiral comonomer. A similar effect was observed for the coisotactic copolymer of (R)-4-phenyl-1-hexene (6; $n = 1$; $R^1 = Me$, $R^2 = Ph$), with 4MP, for which it was also possible to relate the sign of the dichroic absorption of the phenyl group to the predominant screw sense of the helical chain.[51]

$$|\Phi|_D^{25} = N_{4MH}(|\Phi|_D^{25})_{4MH} + N_{4MP}(|\Phi|_D^{25})_{4MP} \tag{14}$$

Clear evidence of induced optical activity in copolymers of a chiral vinyl monomer with an achiral chromophoric comonomer is offered by the coisotactic copolymerization of optically active α-alkenes with the asymmetric carbon atoms in α position to the double bond, such as (R)-3,7-dimethyl-1-octene (6 $n = 0$; $R^1 = Me$; $R^2 = (CH_2)_3CHMe_2$), with vinylaromatic monomers, such as styrene,[53] methyl-substituted styrenes,[43] 1-[54] and 2-vinylnaphthalenes.[43] In these copolymers the only absorbing moiety at wavelengths over 180 nm is the aromatic chromophore. Accordingly CD investigation in this spectral region can give information about stereochemical arrangement of the aromatic groups without any interference from the aliphatic structure. Indeed, over 180 nm the copolymer CD is given by equation (15), as in this region the ellipticity due to the units from the aliphatic monomer is zero. Equation (15) is clearly derived from equation (9), with $|\Theta|_{Ar}$ and N_{Ar} indicating ellipticity and mole fraction of the aromatic units, and N_{DMO} the mole fraction of the (*R*)-3,7-dimethyl-1-octylene units. Circular dichroism in this region not only indicated that the aromatic units were in a chiral environment but specific features of the dichroic curve could be used to obtain information about chain conformation in solution. Indeed in the region of the strong absorption of the aromatic chromophore, corresponding to the strong allowed $\pi \rightarrow \pi^*$ electronic transition, exciton splitting was observed which could be attributed to the dipole–dipole electrostatic interactions of transition moments of aromatic moieties disposed in a mutual chiral geometry with a predominant handedness. Calculations of the CD curves, performed by means of DeVoe theory,[55] demonstrated that the best fitting between experimental and calculated values was obtained by assuming for aromatic units a single screw sense helix, having the same handedness assumed by the isotactic homopolymers of the corresponding aromatic monomers in the crystalline state.[4,56]

$$|\Theta|_{cop} = N_{Ar}|\Theta|_{Ar} + N_{DMO}0 = N_{Ar}|\Theta|_{Ar} \qquad (15)$$

This concept of induced optical activity in chromophores of achiral monomers when copolymerized with an optically active comonomer has been shown to be valid also for stereoirregular macromolecules. Indeed, free radical copolymers of (−)-menthyl acrylate or methacrylate with vinylaromatic ketones,[57,58] styrene,[59] 4-vinylpyridine,[60] 1-vinylnaphthalene,[61] *N*-vinylcarbazole,[62,63] 4-vinylstilbene,[64] 4-acryloxystilbene,[65] and 4-acryloxyazobenzene[66] all gave evidence of induced optical activity in the electronic transitions of the aromatic or etheroaromatic chromophore. This was in all cases proof of copolymer formation. Moreover, in some cases evidence of exciton splitting was obtained, indicating the presence of a certain conformational order even in these non-stereoregular systems. This type of information can only be obtained at present with the CD technique.[26,67]

The presence of a chromophore in the main chain can be very helpful for demonstrating the dissymmetric arrangement of the backbone. A very simple example is supplied by polymers derived from optically active alkynes which contain a system of alternating double bonds along the chain (9).[68,69] These polymers, in spite of the large number of double bonds in the 1,3 position per macromolecule, absorb at much shorter wavelength than expected, which corresponds to sequences of four to five conjugate double bonds only. This result was explained by considering that the backbone could not assume a planar conformation due to the bulky side chains. The helical structure of the main chain with a predominant screw sense was supported by the strong ellipticity detected by CD, corresponding to a skewed polyene chromophore.[69]

$$-CH=C-CH=C-CH=C-CH=$$

(9)

25.4.2 Polymers with Rigid Helical Structures

The helical structure is *per se* sufficient to provide molecular chirality in polymers but detectable optical activity implies that one screw sense prevails. In the previously mentioned examples the enantiomeric prevalence of chiral centers in the side chains led to the prevalence of one helical screw sense, as the two opposite helical screw senses are in a diastereomeric relation because of their side chains with a single absolute configuration. In polymers not possessing centers of chirality in the side chain, optical activity can be detected only if the two helical senses do not interconvert and can be separated or if they can be produced in a single screw sense under a chiral field.

This type of stereoisomerism due to restricted rotation around single bonds (atropoisomerism) is well documented for low molecular weight organic compounds.[70] The first example in macromolecules was offered rather recently by polymers of isocyanides,[71] also called poly(iminomethylene)s (**5a** and **10**). These polymers have a rigid 4_1 helical structure and poly(*t*-butyliminomethylene)

(10) has been resolved into right- (11) and left-handed (12) helices by column chromatography on an optically active support.[19] The above resolution was taken as proof that polymerization of isocyanides is stereoselective with respect to screw sense, both screw senses being formed in equal amounts when the side chains, R, are achiral as for poly(*t*-butyliminomethylene) (10). When R in (5a) has a chiral center of a single configuration, one screw sense over the other prevails, as has been reported for vinyl polymers.[72] This has been proved by recording the CD spectra of several polyisocyanides with chiral side chains with (S) absolute configuration.[73]

(10)

(11)

(12)

An additional unique case of a polymer which, in the infinite chain model, is not chiral, but which has been obtained in an optically active form is the isotactic polymer of trityl methacrylate (5b). This result does not affect the symmetry considerations and their general validity, as the optical activity of (5b) is actually due to the high chain rigidity which does not allow the originally formed one screw sense helical conformation to 'racemize'. The copolymer of triphenylmethyl methacrylate (trityl methacrylate, TrMA) with optically active (S)-α-methylbenzyl methacrylate (MBMA) prepared in the presence of Bu^{n}Li at −78 °C with THF as the solvent, showed rotatory power dependent on composition, changing from negative to positive with an increase in the content of TrMA units.[74] Coincidentally an increase of the isotacticity was observed by NMR, indicating that the bulky triphenylmethyl side chains of the TrMA units prevented the syndiotactic enchainment. The chiroptical properties of the copolymer were explained by assuming that isotactic sections of TrMA units were in a single screw sense helical conformation. Confirmation came from CD showing that the sign of the dichroic bands associated with ester and aromatic chromophores changes with composition, proving that the related electronic transitions occur in geometrical arrangements of opposite chirality. Further proof was supplied by the homopolymers obtained by polymerization of TrMA in the presence of optically active anionic catalysts. In particular, the polymer obtained at −78 °C in toluene in the presence of the (−)-sparteine/butyllithium complex, showed a very high positive optical rotation of the sodium D-line and strong positive bands were also observed in the CD spectrum between 280 and 208 nm where the electronic transitions of the monomeric unit are located.[19]

25.4.3 Polymers with Stereogenic Centers Attached to Four Different Groups or Atoms

Mirror symmetry elements are lacking in a polymer chain containing in its backbone stereogenic centers (asymmetric carbon atom) directly attached to four different atoms or groups.[75] Indeed these centers maintain their chirality even in the infinite chain model. In these systems, obviously, appreciable optical activity can be observed if one type of chirality of the stereogenic centers prevails in the whole polymer.

Polymers belonging to this group are those obtained by ring opening polymerization of optically active N-carboxy-anhydrides,[76] epoxides,[11-13] episulfides,[77,78] aziridines,[79] lactides,[80,81] and lactones,[82,83] giving constitutional units (13), (3a), (14), (15), (16) and (17), respectively. Similar structures, from the chirality view point, may also be obtained by polycondensation of optically active compounds, such as amino acids, diamines, hydroxydiols and glycols to give optically active aliphatic polyesters (18),[83] polyamides (19)[84] and (20),[85] alicyclic polyamides (21)[86] and (22),[87] aromatic polyamides where chirality may also be due to atropoisomerism (23)[88] and cholesteric

(13)   (14)   (15)   (16)   (17)

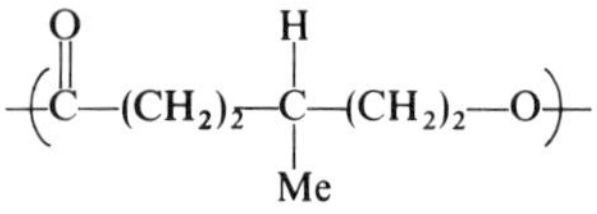

(18) 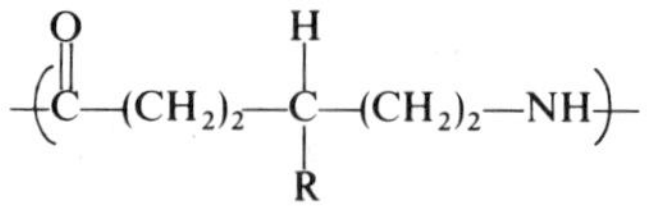   **(19)** 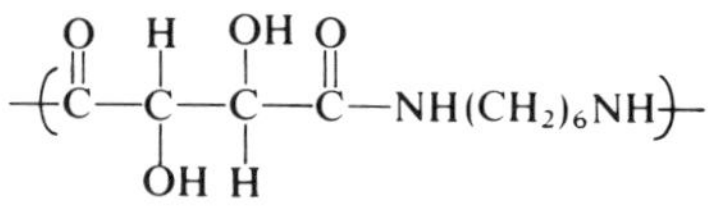   **(20)**

(21) 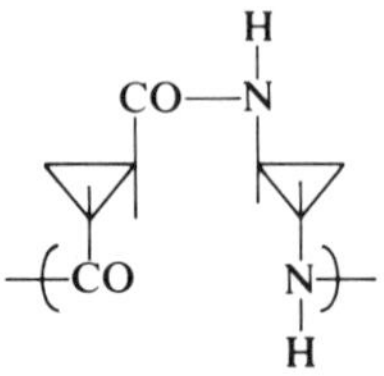   **(22)** 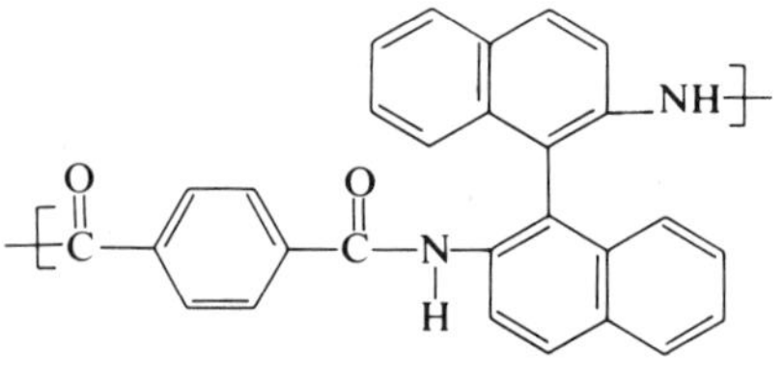   **(23)**

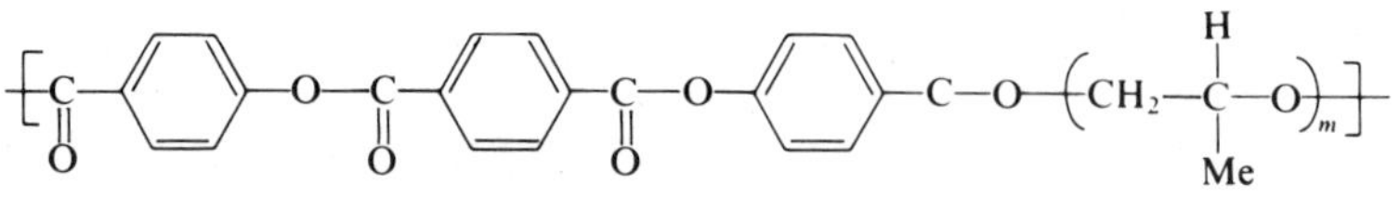

(24) $m = 1, 2, 3$

aromatic polyesters (**24**).[89] Optically active polymers through ring-opening polymerization may also be produced starting with a racemic monomer, and using an optically active catalyst capable of preferentially polymerizing one of the two enantiomers.[75]

In optically active vinyl polymers, where the stereogenic centers responsible for the non-vanishing optical rotation are in the side chains, the stereogenic centers in the main chain do not directly determine chirality, but rather determine stereoregularity and so can only affect the value of the optical rotation. On the contrary, in the polymers of the present type a more direct relationship between optical activity and stereoregularity exists. Indeed in the case, for instance, of poly(propylene oxide), the perfectly isotactic chain is the one formed of units derived from one enantiomer only with head-to-tail enchainment. A polymer consisting entirely of these macromolecules is not only isotactic but also displays the maximum absolute value of the optical rotation. Clearly an equimolar mixture of macromolecules of these types, but derived either from one enantiomer or from its mirror image, is also an isotactic polymer, but does not show any optical activity. In other words the occurrence of optical activity is an indication of some stereoregularity, whereas its absence does exclude stereoregularity.

Some prochiral monomers give rise to polymers with stereogenic centers in the backbone bound to four different groups or atoms. Examples of this class are offered by 1,4- and 1-substituted diene monomers, such as sorbates (**25**)[90,91] and 1,3-pentadiene (**26**),[92-95] as well as by some cyclic monomers, such as benzofuran[9] and naphthofuran.[10] As these monomers do not contain stereogenic centers, which are however formed during the polymerization process, the production of an optically active polymer is due to asymmetric induction by the optically active catalyst or initiator. In the former case the constant control by the catalyst of any incoming monomer molecule gives optically active polymer independently of its molecular weight. By contrast with optically active initiators, which are attached to the chain as end groups, the polymer optical activity rapidly decreases with increasing molecular weight.[1]

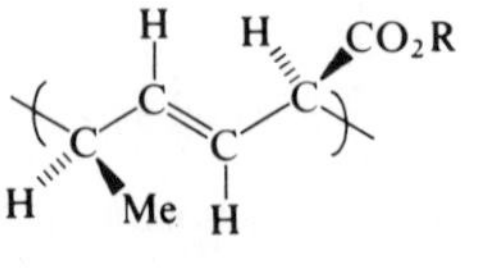

(25) 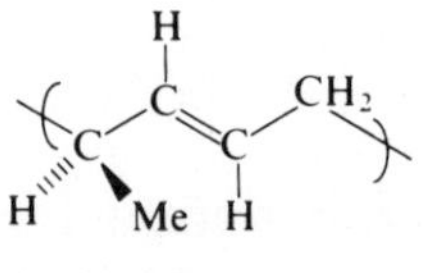   **(26)**

Not very different, at least from the point of view of this chapter, is the situation in alternating copolymers, where every repeating unit contains in the backbone two (**27**), (**28** and three **29**) and four (**30**) stereogenic centers. Again these polymers can display optical activity, provided that one absolute configuration prevails, due to an asymmetric process induced by the optically active

(27)

(28)

(29)

(30)

catalyst or initiator[96,97] or, alternatively, by chiral groups present in the monomer and subsequently removed from the polymer.[98,99] The results in this field are, up to now, relatively modest.

As indicated in Section 25.1, chiral structures can be formed in vinyl copolymers by controlling the sequential distribution. Thus in a copolymer of monomers A and B a sequential distribution of the type –AAB– can give rise to structures with no plane of symmetry even in an infinite chain provided that a proper stereochemistry is achieved. Indeed the polymer depicted in (31) contains asymmetric triads and could in principle display measurable optical activity provided that one configuration prevails.[17] This result has been achieved by copolymerizing optically active 3,4-*O*-cyclohexylidene-D-mannitol-1,2-5,6-bis-*O*-4-vinylphenylboronate[16,17] with styrene, acrylonitrile or methyl methacrylate. In the case of MMA free radical initiation gave copolymers which, after removing D-mannitol, showed negative optical rotation strongly dependent in absolute value on copolymer composition. Copolymers with variable optical rotation have been obtained with several methacrylic derivatives and substituted styrenes.[16] Even if at present the absolute configuration and the amount of structures (4) or (31) are not known, the occurrence of detectable optical rotation confirms the asymmetric structure of the polymers due to the sequence distribution of comonomers in a defined stereochemical arrangement.

(31)

25.5 REFERENCES

1. M. Farina, *Chim. Ind. (Milan)*, 1986, **68**, 62; *Top. Stereochem.*, 1987, **17**, 1.
2. P. Pino, *Adv. Polym. Sci.*, 1965, **4**, 393.
3. K. Mislow and J. Siegel, *J. Am. Chem. Soc.*, 1984, **106**, 3319.
4. G. Natta, *Makromol. Chem.*, 1960, **35**, 94.
5. P. Pino, F. Ciardelli and G. P. Lorenzi, *J. Polym. Sci., Part C*, 1963, **4**, 21.
6. W. Kaminsky, in 'History of Polyolefins', ed. R. B. Seymour and T. Cheng, Reidel, Dordrecht, 1986, p. 257.
7. P. Pino and G. P. Lorenzi, *J. Am. Chem. Soc.*, 1960, **82**, 4745.
8. P. Pino, F. Ciardelli, G. P. Lorenzi and G. Montagnoli, *Makromol. Chem.*, 1963, **61**, 207.
9. G. Natta, M. Farina, M. Peraldo and G. Bressan, *Chim. Ind. (Milan)*, 1961, **43**, 161.
10. G. Bressan, M. Farina and G. Natta, *Makromol. Chem.*, 1966, **93**, 283.
11. M. Osgan and C. C. Price, *J. Polym. Sci.*, 1959, **34**, 153.
12. T. Tsuruta, S. Inoue and I. Tsukuma, *Makromol. Chem.*, 1965, **84**, 298.
13. Y. Kumata, J. Furukawa and T. Fueno, *Bull. Chem. Soc. Jpn.*, 1970, **43**, 3663.
14. M. Szwarc, *Adv. Polym. Sci.*, 1965, **4**, 1.
15. N. Beredjick and C. Schuerch, *J. Am. Chem. Soc.*, 1958, **80**, 1933.
16. G. Wulff, R. Kemmerer, J. Vietmeier and H. G. Poll, *Nouv. J. Chim.*, 1982, **6**, 681.
17. G. Wulff, K. Zabrocki and J. Hohn, *Angew. Chem.*, 1978, **90**, 567.
18. R. J. M. Nolte, A. J. M. van Beijnen and W. Drenth, *J. Am. Chem. Soc.*, 1974, **96**, 5932.
19. Y. Okamoto, K. Suzuki, K. Ohta, K. Hatada and H. Yuki, *J. Am. Chem. Soc.*, 1979, **101**, 4763.
20. A. Moscowitz, *Adv. Chem. Phys.*, 1962, **4**, 67.
21. S. F. Mason, *Q. Rev. Chem. Soc.*, 1963, **17**, 20.
22. F. Ciardelli, 'Optically Active Polymers', in 'Encyclopedia of Polymer Science and Engineering', 2nd edn., Wiley, New York, 1987, vol. 10, pp. 463–493.
23. O. Bonsignori and G. P. Lorenzi, *J. Polym. Sci., Part A-2*, 1970, **8**, 1639.

24. O. Vogl, L. S. Corley, W. J. Harris, G. D. Jaycox and J. Zhang, *Makromol. Chem. Suppl.*, 1985, **13**, 1.
25. P. Pino, F. Ciardelli, G. Montagnoli and O. Pieroni, *J. Polym. Sci., Polym. Lett.*, 1967, **5**, 307.
26. F. Ciardelli, M. Aglietto, C. Carlini, E. Chiellini and R. Solaro, *Pure Appl. Chem.*, 1982, **54**, 521.
27. J. L. Houben, A. Fissi, D. Bacciola, N. Rosato, O. Pieroni and F. Ciardelli, *Int. J. Biol. Macromol.*, 1983, **5**, 94.
28. P. Pino, C. Carlini, E. Chiellini, F. Ciardelli and P. Salvadori, *J. Am. Chem. Soc.*, 1968, **90**, 5025.
29. F. Ciardelli, E. Chiellini, C. Carlini, O. Pieroni, P. Salvadori and R. Menicagli, *J. Polym. Sci., Polym. Symp.*, 1978, **62**, 143.
30. P. Pino, F. Ciardelli and M. Zandomeneghi, *Annu. Rev. Phys. Chem.*, 1970, **21**, 561.
31. F. Ciardelli and P. Salvadori (eds.), 'Fundamental Aspects and Recent Developments in ORD and CD', Heyden, London, 1973.
32. E. Charney, 'The Molecular Basis of Optical Activity', Wiley, New York, 1979, p. 246.
33. W. Moffitt, *J. Chem. Phys.*, 1956, **25**, 467.
34. W. Moffitt and J. T. Yang, *Proc. Natl. Acad. Sci. USA*, 1956, **42**, 596.
35. I. Tinoco, Jr., *Adv. Chem. Phys.*, 1962, **4**, 113.
36. I. Tinoco, Jr., in 'Optically Active Polymers', ed. E. Selegny, Reidel, Dordrecht, 1979, p. 1.
37. R. W. Woody, *J. Chem. Phys.*, 1968, **49**, 4797.
38. E. S. Pysh, *Proc. Natl. Acad. Sci. USA*, 1966, **56**, 825.
39. E. S. Pysh, *Biopolymers*, 1974, **13**, 1563.
40. E. W. Ronish and S. Krimm, *Biopolymers*, 1972, **11**, 1919.
41. I. Tinoco, Jr. and A. L. Williams, Jr., *Annu. Rev. Phys. Chem.*, 1984, **35**, 329.
42. W. Hug, F. Ciardelli and I. Tinoco, Jr., *J. Am. Chem. Soc.*, 1974, **96**, 3407.
43. F. Ciardelli, C. Righini, M. Zandomeneghi and W. Hug, *J. Phys. Chem.*, 1977, **81**, 1948.
44. S. F. Mason, 'Molecular Optical Activity and the Chiral Discriminations', Cambridge University Press, Cambridge, 1982, p. 51.
45. F. Ciardelli, G. Montagnoli, D. Pini, O. Pieroni, C. Carlini and E. Benedetti, *Makromol. Chem.*, 1971, **147**, 53.
46. G. P. Lorenzi, E. Benedetti and E. Chiellini, *Chim. Ind. (Milan)*, 1964, **46**, 1474.
47. P. Pino, P. Salvadori, G. P. Lorenzi, E. Chiellini, L. Lardicci, G. Consiglio, O. Bonsignori and L. Lepri, *Chim. Ind. (Milan)*, 1973, **55**, 182.
48. D. Basagni, A. M. Liquori and B. Pispisa, *J. Polym. Sci., Part B*, 1964, **2**, 241.
49. O. Pieroni, F. Ciardelli, C. Botteghi, L. Lardicci, P. Salvadori and P. Pino, *J. Polym. Sci., Polym. Symp.*, 1969, **22**, 993.
50. Nguyên-tât-Thiên, U. W. Suter and P. Pino, *Makromol. Chem.*, 1983, **184**, 2335.
51. C. Carlini, F. Ciardelli and P. Pino, *Makromol. Chem.*, 1968, **119**, 244.
52. C. Carlini, F. Ciardelli, L. Lardicci and R. Menicagli, *Makromol. Chem.*, 1973, **174**, 27.
53. F. Ciardelli, P. Salvadori, C. Carlini and E. Chiellini, *J. Am. Chem. Soc.*, 1972, **94**, 6536.
54. F. Ciardelli, P. Salvadori, C. Carlini, R. Menicagli and L. Lardicci, *Tetrahedron Lett.*, 1975, 1779.
55. H. De Voe, *J. Chem. Phys.*, 1965, **43**, 3199.
56. V. Petraccone, P. Ganis, P. Corradini and G. Montagnoli, *Eur. Polym. J.*, 1972, **8**, 99.
57. C. Carlini and F. Gurzoni, *Polymer*, 1983, **24**, 101.
58. A. Altomare, C. Carlini, F. Ciardelli and E. M. Pearce, *J. Polym. Sci., Polym. Chem. Ed.*, 1983, **21**, 1693.
59. R. N. Majumdar and C. Carlini, *Makromol. Chem.*, 1980, **181**, 201.
60. R. N. Majumdar, C. Carlini and C. Bertucci, *Makromol. Chem.*, 1982, **183**, 2047.
61. R. N. Majumdar, C. Carlini, N. Rosato and J. L. Houben, *Polymer*, 1980, **21**, 941.
62. E. Chiellini, R. Solaro, A. Ledwith and G. Galli, *Eur. Polym. J.*, 1980, **16**, 875.
63. E. Chiellini, R. Solaro, G. Galli and A. Ledwith, *Adv. Polym. Sci.*, 1984, **62**, 143.
64. A. Altomare, C. Carlini, M. Panattoni and R. Solaro, *Macromolecules*, 1984, **17**, 2207.
65. A. Altomare, C. Carlini and R. Solaro, *Polymer*, 1982, **23**, 1355.
66. A. Altomare, C. Carlini, F. Ciardelli, R. Solaro and N. Rosato, *J. Polym. Sci., Polym. Chem. Ed.*, 1984, **22**, 1267.
67. F. Ciardelli and P. Salvadori, *Pure Appl. Chem.*, 1985, **57**, 931.
68. F. Ciardelli, E. Benedetti and O. Pieroni, *Makromol. Chem.*, 1967, **103**, 1.
69. F. Ciardelli, S. Lanzillo and O. Pieroni, *Macromolecules*, 1974, **7**, 174.
70. E. L. Eliel, 'Stereochemistry of Carbon Compounds', McGraw-Hill, New York, 1962, p. 156.
71. W. Drenth and R. J. M. Nolte, *Acc. Chem. Res.*, 1979, **12**, 30.
72. F. Millich, *Macromol. Rev.*, 1980, **15**, 207.
73. A. J. M. van Beijnen, R. J. M. Nolte, A. J. Naaktgeboren, J. W. Zwikker, W. Drenth and A. M. F. Hezemans, *Macromolecules*, 1983, **16**, 1679.
74. Y. Okamoto, K. Suzuki and H. Yuki, *J. Polym. Sci., Polym. Chem. Ed.*, 1980, **18**, 3043.
75. T. Tsuruta, *J. Polym. Sci., Part D*, 1972, **6**, 179.
76. T. Makino, S. Inoue and T. Tsuruta, *Makromol Chem.*, 1971, **150**, 137.
77. N. Spassky and P. Sigwalt, *Bull. Soc. Chim. Fr.*, 1967, 4617.
78. T. Tsunetsugu, J. Furukawa and T. Fueno, *J. Polym. Sci., Part A-1*, 1971, **9**, 3541.
79. K. Tsuboyama, S. Tsuboyama and M. Yanagita, *Bull. Chem. Soc. Jpn.*, 1967, **40**, 2954.
80. R. C. Shulz and A. Guthmann, *J. Polym. Sci., Polym. Lett.*, 1967, **5**, 1099.
81. H. R. Kricheldorf and J. M. Jonté, *Polym. Bull.*, 1983, **9**, 276.
82. C. G. D'Hont and R. W. Lenz, *J. Polym. Sci., Polym. Chem. Ed.*, 1978, **16**, 261.
83. C. G. Overberger and H. Kaye, *J. Am. Chem. Soc.*, 1967, **89**, 5646.
84. C. G. Overberger and J. H. Kozlowski, *J. Polym. Sci., Part A-1*, 1972, **10**, 2291.
85. Y. Minoura, S. Urayama and Y. Noda, *J. Polym. Sci., Part A-1*, 1967, **5**, 2441.
86. C. G. Overberger, G. Montaudo, Y. Tlishimura, J. Šebenda and R. A. Veneski, *J. Polym. Sci., Part B*, 1969, **7**, 219.
87. C. G. Overberger and T. Nishiyama, *J. Polym. Sci., Polym. Chem. Ed.*, 1981, **19**, 349.
88. R. C. Schulz and R. H. Jung, *Makromol. Chem.*, 1968, **116**, 190.
89. E. Chiellini and G. Galli, *Makromol. Chem. Rapid Commun.*, 1983, **4**, 285.
90. G. Natta, M. Farina, M. Donati and M. Peraldo, *Chim. Ind. (Milan)*, 1960, **42**, 1363.
91. G. Natta, M. Farina and M. Donati, *Makromol. Chem.*, 1961, **43**, 251.
92. G. Natta, L. Porri and S. Valenti, *Makromol. Chem.*, 1963, **67**, 225.

93. G. Costa, P. Locatelli and A. Zambelli, *Macromolecules*, 1973, **6**, 653.
94. M. Farina, G. Audisio and G. Natta, *J. Am. Chem. Soc.*, 1967, **89**, 5071.
95. A. D. Aliev and B. A. Krentsel, *Vysokomol. Soedin., Ser. A*, 1967, **9**, 1464.
96. M. Kurokawa, T. Doiuchi, H. Yamaguchi and Y. Minoura, *J. Polym. Sci., Polym. Chem. Ed.*, 1978, **16**, 129.
97. M. Kurokawa and Y. Minoura, *Makromol. Chem.*, 1980, **181**, 707.
98. E. Chiellini, M. Marchetti, C. Villiers, C. Braud and M. Vert, *Eur. Polym. J.*, 1978, **14**, 251.
99. M. Kurokawa and Y. Minoura, *J. Polym. Sci., Polym. Chem. Ed.*, 1979, **17**, 473.

26

Mass Spectrometry

DAVID M. HINDENLANG and R. DONALD SEDGWICK
Allied-Signal Inc., Morristown, NJ, USA

26.1 INTRODUCTION

A mass spectrometer is used to separate mixtures of gaseous ions according to their masses. The gaseous ions are produced from a neutral sample in an ion source. The oldest established ion sources rely on vaporization of the sample prior to ionization either by electron impact ionization (EI) or chemical ionization (CI).

In EI a gaseous molecule is ionized by collision with an electron of typically 70 eV energy, producing an excited molecular ion, which may dissociate to give structurally related fragment ions. The mixture of the molecular ion and its structurally related fragment ions comprise the mass spectrum.

Modern high resolution mass analyzers can measure relative masses with a precision of better than 1 p.p.m. This can lead to an unambiguous assignment of an atomic composition to a particular mass. This is of immense value when applied to the molecular ion since the relative molecular mass and its atomic composition are found. High resolution mass separation and measurement are made using a double focusing magnetic deflection or a magnetic ion cyclotron resonance (ICR) analyzer. For the majority of applications unit mass resolution is considered to be adequate and can be provided by a quadrupole mass analyzer.

The linking of fragment ions into a stepwise degradation mechanism can be achieved by the use of numerous types of metastable scanning methods using tandem instruments (MS–MS) or sequential techniques using ICR or ion trap mass spectrometry (ITMS). The dissociations of ions can be enhanced by collision-induced dissociation (CID) using collision cells operating at high relative ion energies (keV) for magnetic deflection instruments and lower energies in quadrupole instruments. Using these procedures it is possible to build a detailed fragmentation scheme showing how the fragment ions are interrelated and arise from a common parent molecular ion. The relationships between molecular structures and their mass spectra have been extensively tabulated and an effective mechanistic approach to spectral interpretation exists in the absence of any rigorous theory. Thus spectral interpretation is not precise and structures derived from mass spectral data cannot be treated as definitive in the absence of independent confirmation. Mass spectrometry is, however, capable of giving numerous indications of probable structural features using minute amounts of sample and after a short experimental time.

While EI tends to emphasize the fragmentation of molecules, there exist complementary soft ionization techniques which suppress fragmentation. Thus CI achieves this by converting the gaseous molecule into a stabilized ionic form such as protonated molecule (MH^+), while field ionization (FI) favors molecular ionization to a stable unexcited molecular state.

Most large molecules, including most synthetic and natural polymers, are too involatile to be studied directly by these gas phase ionization methods. They can, however, be used effectively in the following areas of polymer application: (a) to identify and obtain quantitative analysis of volatile additives used in polymer syntheses, polymer processing and stabilization; (b) to characterize polymer structures *via* their volatile products of thermal decomposition; and (c) to characterize high polymer structures by extrapolation from characterization of their volatile low molecular weight oligomers.

More recent developments in methods of ionization have concentrated on techniques of producing gaseous ions from involatile solids including polymers. These are in general means of combining the vaporization and ionization step into a single desorption event. Many involatile species on surfaces can be made to desorb as ions under a variety of bombardment conditions. Thus we have laser desorption, ^{252}Cf and heavy ion plasma desorption, and secondary ion (SI) desorption from solid surfaces and fast atom bombardment (FAB) of liquid surfaces. Field desorption (FD) uses electric fields with thermal assistance to desorb ions from solid surfaces, while desorption chemical ionization (DCI) combines flash vaporization with CI. All these techniques have been applied with varying degrees of success to characterizing oligomeric systems which were too involatile by reason of molecular weight, polarity or ionic state to be studied by traditional gas phase ionization methods. The impetus for the development of these ionization techniques has been in the study of biopolymeric structure with particular emphasis on polypeptides, oligonucleotides, polysaccharides and related antibiotics and compounds of biomedical significance.[1] While this thrust has been decidedly biochemical in nature, the techniques are nevertheless relevant to the study of synthetic polymers and have been used with success in the synthetic polymer field.[2] It would, however, be perverse to discuss these developments outside of the framework in which they developed. We shall therefore discuss applications of these methods to problems in synthetic polymer chemistry against a background of their more publicized achievements in biochemistry.

Finally, there are a series of ionization methods in which charged droplets are desolvated to give gaseous ions. These include electrospray ionization, electrohydrodynamic ionization, thermospray and plasmaspray ionization techniques, all of which have the added potential of providing an interface between liquid chromatography and mass spectrometry.

It is not possible at this time to describe any standard instrumental configuration which is most suitable for studying synthetic polymers. However, certain generalizations can be made as follows.

Small general purpose organic mass spectrometers based on magnetic or quadrupole mass analyzers with gaseous ionization facilities and with GC–MS capability find wide applications in polymer pyrolysis studies[3] and analyses of polymer additives.[4]

SIMS, FAB–MS and FDMS have been widely applied to the analysis of oligomeric mixtures of synthetic and biopolymers in combination with double focusing magnetic mass spectrometers. Developments in these techniques have catalyzed design and production of mass analyzers with mass ranges up to 25 000.

Plasma desorption sources using ^{252}Cf or heavy ion bombardment have been confined to combinations with time-of-flight (TOF) mass analysis methods. These techniques have been consistently successful in the very high mass region, *viz.* $m/z > 20\,000$.

Progress has been made in the last decade in that molecules up to 30 000 molecular weight can now be analyzed by mass spectrometry. However, detailed examination of the references cited will reveal that there is a trade-off of sensitivity with increasing mass, so that the upper mass limits reached in each case represent a limit determined by signal-to-noise ratio. This limit has also been reached using samples of high purity and well-defined molecular mass. A serious problem is the polydispersity introduced by natural isotopic abundance ratios which give a single atomic composition a mass spread of ~ 50 atomic mass units (a.m.u.) at mass 40 000. If we add to this effect the wide dispersion of molecular weights encountered in synthetic polymers, we immediately lower these mass limits by at least an order of magnitude.

Since the range of applicability is essentially determined at the ion source, we have chosen to discuss the applications of mass spectrometry to synthetic polymers in division by method of ionization.

26.2 GAS PHASE IONIZATION TECHNIQUES

26.2.1 Electron Impact Ionization

26.2.1.1 *Thermal degradation*

Without exception, samples of high molecular weight involatile polymers can be analyzed by pyrolysis–mass spectrometry (Py–MS). The polymer is thermally degraded to yield molecular fragments with sufficient stability and volatility to reach an EI ion source coupled with a mass analyzer with sufficient mass range to produce their corresponding mass spectra.

In some of the earliest Py–MS work the structures of natural rubber[5] and polystyrene[6] were proposed from identification of the structures of their monomers and lower oligomers. Refinements of the pyrolysis equipment have followed from the earliest designs of Wall,[7] and Madorsky and Straus[8] through to modern filament and Curie point pyrolyzers.

Three types of polymer behavior have been recognized:[9] (1) polymers which degrade by backbone scission to give high yields of constituent monomers, *e.g.* poly(α-methylstyrene);[10] (2) polymers in which side chain scissions occur leaving the backbone intact — the residues are often very stable unsaturated networks with crosslinking between chains being common, *e.g.* poly(vinyl chloride); and (3) highly crosslinked structures which evolve only minor amounts of volatile products, *e.g.* poly(trivinylbenzene).[11]

It is accepted that identification of the primary products of thermal degradation will be most relevant in determining the identity of the host structure. Thus much early work is complicated by the occurrence of complex secondary reactions and even repolymerization reactions.[9, 12, 13]

Pyrolysis temperature and the sample path length and its temperature between pyrolysis and ion chamber are critical factors which influence the products which are detected. This is a consequence of the competitive nature of the numerous fragmentation processes occurring during and immediately after primary decomposition. Thus Sedgwick and coworkers[14] have described a variable temperature pyrolyzer in which the sample-to-ion-source distance can be varied. Using $(CF_3NOCH_2)_n$ as an example, they showed that at constant temperatures (Figure 1a) oligomers containing up to 16 monomer units at mass 1696 were detected with the pyrolyzer only 1 mm from the source, whereas at 50 cm the highest detected oligomer was the tetramer at 452. Additionally, with a constant path length of variable temperature (Figure 1b), high oligomers were detected at low temperatures but decayed to the lower oligomers as the temperature was raised. A further feature common to this polymer and other linear homopolymers is that the oligomers show a repetitive fragmentation pattern of their EI spectra as illustrated in Figure 2. This is indicative of a regular repeating structure which may frequently be inferred to be carried through to the high polymer.

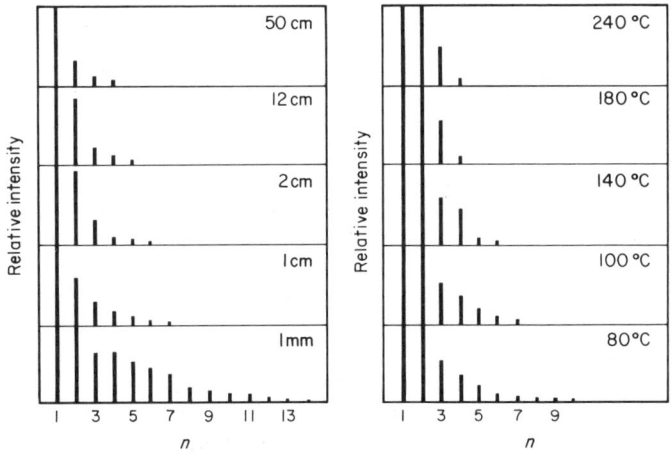

Figure 1 The effect of varying (left) the length of the sample introduction path at constant temperatures (60 °C) and (right) the temperature of the sample introduction path at constant distance (1 cm) on the intensities of successive units of the polymer $(CF_3NOCH_2)_n$ (reproduced with permission from ref. 14)

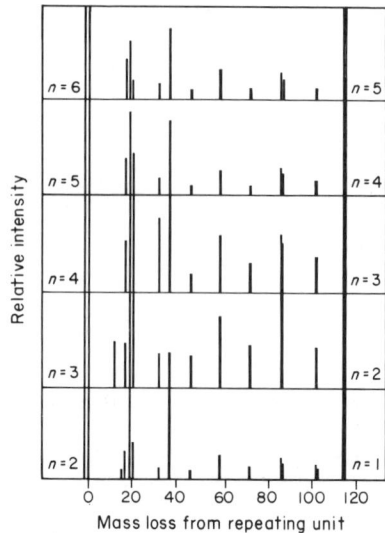

Figure 2 Relative mass spectra of five units of the polymer $(CF_3NOCH_2)_n$. Each interval of 113 mass numbers is shown as a separate spectrum normalized to the peak intensity at $m/z = 113n$, where $n = 2$–6 (reproduced with permission from ref. 14)

Following the work of Kearns,[15] many designs of heated probes are available for the direct insertion of sample into the ion source through a vacuum lock. These provide a convenient means for degrading polymeric samples in association with a general purpose organic mass spectrometer. While this approach is clearly very useful,[16] such systems lack the reproducibility that modern chemometrics have demonstrated to be desirable if maximum information is to be deduced from Py–MS.

Muezelaar and coworkers[17] first showed the extraordinary power of computer-based pattern recognition techniques in classifying very complex biological systems, such as bacteria, by analysis of their Py–MS. They have shown the necessity for reproducibility in either Py–MS or in Py–GC–MS systems and have also recommended operating the EI ion source at low electron beam energy as a means of improving spectral reproducibility. Muezelaar and Lattimer[18] have applied these techniques to a series of synthetic urethane and hydrocarbon polymers. They employed a Curie point pyrolyzer, in which a ferromagnetic wire is inductively heated to a very reproducible temperature using a radiofrequency field. A temperature in the range 500–600 °C is used, dependent on the composition of the wire used. The sample size is kept small to ensure rapid heating of the sample to the Curie point temperature of the wire, thus minimizing competitive degradation reactions and ensuring a high degree of degradation. In a typical example a series of natural rubber vulcanizates[19] were reproducibly pyrolyzed and their mass spectra submitted to principal component (factor) analysis. Mass spectral factors were isolated which correlated with crosslink density as shown in Figure 3.

A widely used alternative to Curie point pyrolysis is to use filament pyrolyzers with temperature regulation by current control. While these can be used to provide very reproducible Py–MS for polymer identification by 'fingerprinting' techniques, they have the added flexibility of being capable of operation in a temperature-programmed mode.[20]

Temperature programming has been advocated as adding an extra dimension to Py–MS and has been used effectively to study the decomposition of several vinyl polymers by Ballisteri *et al.*[21] Total ion current *vs.* temperature profiles are shown in Figure 4. Each polymer shows two peaks, one being at a variable low temperature, and dependent on structure, which corresponds with elimination of HX (X = Br, OH, Cl, OAc) and formation of benzene. All show a peak at ~ 350 °C where the similar unsaturated backbones formed by side chain scission undergo decomposition to a mixture of aromatic and aliphatic hydrocarbons.

While much structural Py–MS is performed entirely *in vacuo*, there is much practical and commercial interest in studying decompositions in inert atmospheres and in air or oxygen. For this type of study a continuous flow tubular furnace is favored.[22] Pyrolysis products are continuously swept from the pyrolysis zone by a gas flow at atmospheric pressure and are admitted to the mass spectrometer using a capillary leak[23] or a jet separator[24] as in conventional GC–MS.

Pyrolysis temperatures may conveniently be determined by thermogravimetric analysis (TGA)

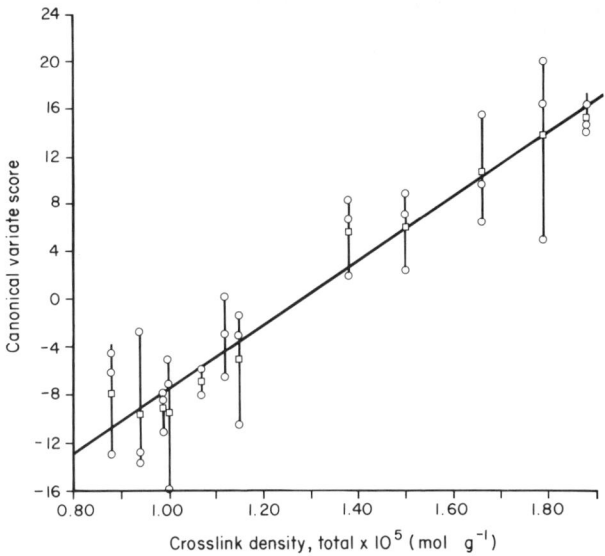

Figure 3 Determination of crosslink density in natural rubber vulcanizates by Py–MS using canonical variate scores obtained from single spectra (○) and the average of triplicate spectra (□) (reproduced with permission from ref. 19)

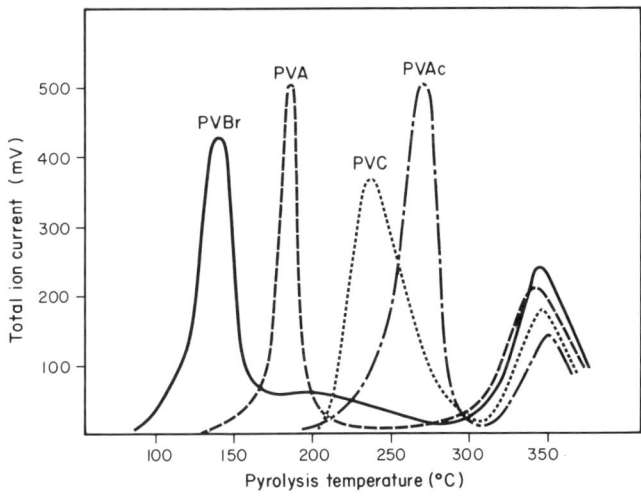

Figure 4 Total ion current *vs.* pyrolysis temperature profiles for a series of vinyl polymers (reproduced with permission from ref. 21)

and this apparatus can be interfaced directly to the mass spectrometer.[25] This allows the sample weight loss to be explained in terms of molecules identified by the mass spectrometer.

Py–MS has been shown to be effective in rapidly distinguishing between similar polymers such as poly(tetrafluoroethylene) (PTFE) and poly(fluoroethylenepropylene) (PFEP).[26] It has also been applied widely to identify unknown polymer samples including copolymer blends.[27]

Many commercial polymer formulations contain numerous additives which act as processing aids or stabilizers against oxidation and photodegradation. These additives can be readily identified since they are frequently distilled out intact in a Py–MS experiment. The identification of additives can be diagnostic of a particular commercial formulation but when present in a formulation where the base polymer is unknown, they represent a hindrance to the successful application of Py–MS. For complex formulations it is usually found that temperature-programmed Py–GC–MS allows separation of additives from polymer pyrolysis products and their independent identification is facilitated. Thus both additives and polymer components were identified in commercial blends of polystyrene with poly(phenylene oxide)[25] and protocols for characterizing intractable rubbers have been published.[27]

Py–MS data are generally easiest to interpret for homopolymers but become exceedingly complex for copolymers where a multiplicity of oligomeric structures exists. Naguya and coworkers[28,29] have tackled this problem and achieved success using Py–GC–MS to first identify GC peaks as monomers, dimers and trimers. The distributions of these oligomers can be related to the copolymer structure by calculation of formation probabilities of all possible diad and triad arrangements. The approach has been applied to methyl methacrylate–styrene copolymers.[30]

Py–GC–MS has also been found to be advantageous in assigning isomeric structures to hydrocarbons formed in the pyrolysis of polyethylenes. Specific branched hydrocarbons can be related to particular types of short branches and can be used to measure their relative abundances in the original polyalkene structures.[31,32] The method has also been applied to infer branching in poly(vinyl chloride) by first reducing it to the corresponding polyalkene.[33]

Pyrolysis MS[34] and GC–MS[35] have been applied to the structure determination of aromatic–aliphatic polyethers.

26.2.1.2 Laser pyrolysis

Lasers have the ability to both pyrolyze and ionize polymeric samples and have been used in both modes. Muezelaar *et al.*[36] have used a 20 W continuous carbon dioxide laser to give rapid and reproducible pyrolysis of a variety of synthetic polymers. The pyrolysates were characterized by a low energy EI source and a quadrupole mass analyzer with a fast ion-counting system. The low energy source produced significant molecular ion relative intensity and was successful in distinguishing polyethylene from polypropylene.[37] They have also used this technology to study butyl rubber, neoprene and nylon. It has also been shown that better reproducibility and hence potential information content of the pyrolysis mass spectra results from scanning the laser beam across the polymer surface.[38] Some complex commercial rubber formulations including additives and fillers were used as test cases to evaluate the method.

26.2.2 Field Ionization

Field ionization (FI) is the ionization of a gaseous molecule by an intense electric field, usually created by a sharp electrode at a high potential.[39] Beckey[40] *et seq.* showed that FI–MS is a soft ionization technique in which molecular ions are the predominating products. It is an ideal technique for simplifying the composite mass spectra of complex volatile mixtures, since each component only contributes its molecular ion to the spectrum.

Lattimer and coworkers[41,42] have combined FI–MS with Py–MS in an elegant variant of the reductive pyrolysis of poly(vinyl chloride). They reduced the PVC in deuterium and used the FI–MS to determine the deuterium distributions in the molecular constituents of the pyrolysates.

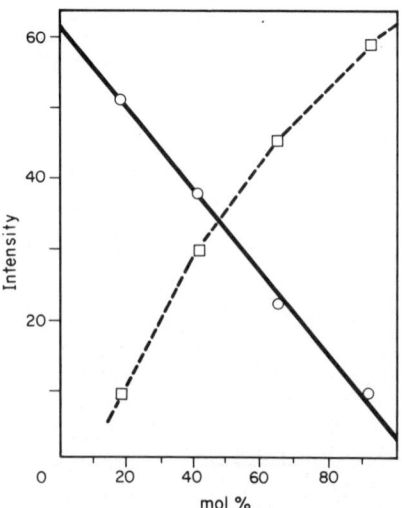

Figure 5 Variation of intensity of the monomer peaks from Py–FI–MS of copolymer of methyl methacrylate ($m/z = 100$, □) and α-methylstyrene ($m/z = 118$, ○) (reproduced with permission from ref. 45)

Hummel and coworkers[43] have also used Py–FI–MS to study a wide range of synthetic polymers including many of commercial significance. They reported almost complete depolymerization of poly(α-methylstyrene)[43] and high monomer yields for polystyrene,[43] polybutadiene[44] and polyacrylonitrile,[45] all of which yielded oligomers up to the pentamer. In contrast polypropylene[46] gave a large variety of products indicating random backbone scissions while poly(vinyl chloride) degraded without depolymerization.[47] Styrene–sulfone copolymers[44] were shown to eliminate sulfur dioxide leaving styrene oligomers which showed a cut-off indicative of the maximum styrene block length. The same workers have also shown some interesting characteristics of a series of copolymers. Copolymers whose corresponding homopolymers degrade to monomer can be analyzed quantitatively for comonomer composition by Py–FI–MS. Thus copolymers of α-methylstyrene with methyl methacrylate[45] pyrolyze to a mixture of monomers which gives the copolymer compositions as shown in Figure 5. Copolymers of a readily depolymerized monomer such as α-methylstyrene with a less readily depolymerized monomer such as acrylonitrile yield Py–FI–MS in which single α-methylstyrene units are attached to short blocks of acrylonitrile.[45] Copolymers of dissimilar monomers such as styrene and vinyl chloride also give interpretable Py–FI–MS, which can be correlated with the degree of blockiness of the samples.[47] A vinyl chloride–propylene copolymer was shown to comprise blocks of poly(vinyl chloride) separated by isolated propylene units.[43]

26.2.3 Sublimation of Oligomers

Using direct heated sample introduction systems at temperatures below the on-set of thermal decomposition, it is possible to distill volatile oligomers from a wide range of polymeric materials. These oligomers have been successfully studied by gas phase ionization methods. Much of the earliest work in this area was aided by the particularly high volatility of oligomers from fluorine-containing polymers.

Bradt et al.[48] reported in 1955 the detection of oligomers of poly(perfluorophenyl) diiodide with masses up to 1755 with $DP_n \leq 11$.

In contrast with decomposing polymers, distilled fractions of stable oligomers tend to increase in molecular weight as the temperature is raised and higher oligomers become volatilized. Since any oligomer is present at low concentration in the polymer, the EI–MS of the species volatile at a constant temperature change with time as the oligomers are sublimed out of the sample. The successive stages of such a distillation are shown in Figure 6.[49] Here an unfractionated sample of the copolymer $(CF_3NOC_2F_4)_n$ prepared from trifluoronitrosomethane and tetrafluoroethylene[50] shows

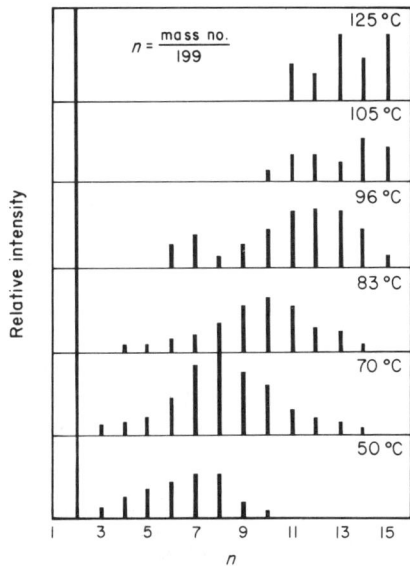

Figure 6 Variation of peak intensity at $119n$ with increasing temperature for the copolymer $(CF_3NOC_2F_4)_n$ (reproduced with permission from ref. 49)

at 50 °C a distribution of oligomers maximizing at $n = 7$ ($m/z = 1393$). The low mass side of this distribution decreases with time as the sample is depleted of the more volatile oligomers. If the temperature is raised, the spectrum is reestablished at a higher molecular weight range. This process is only reversed when the temperature of thermal decomposition is reached, when there is a rapid collapse of the oligomer spectra and only low mass fragments are detected. In this polymer at 220 °C using EI–MS, oligomers of molecular weights up to 6400 have been detected.[49] We conclude that EI–MS can be used to study the lower oligomers found in high polymers and *via* determination of the oligomeric structure it can lead to useful conclusions concerning the polymer. Stable oligomers can be separated from polymers by HPLC or GPC to facilitate MS investigations as for poly(phenylene sulfide)[51] and poly(alkylene sulfide)s.[52, 53]

Examples are known where the high polymers contain low oligomers of a related but different structure, *e.g.* nylon 6 is a linear polyamide which contains cyclic oligomers. Thus while they have undergone different termination reactions, they share a common backbone structure, *ergo* extrapolations from oligomer backbone structure to high polymer backbone structure will be correct.

In polymer syntheses and in the manufacture of industrial polymer it is not uncommon to separate and discard low molecular weight fractions as a step in high polymer purification. These fractions or the crude unfractionated polymer are most amenable to direct MS examination. In some cases it may be possible to modify the polymerization to produce a low molecular weight product specifically for mass spectrometric evaluation. This approach has been used by Ringsdorf and coworkers[54, 55] to characterize a series of substituted aromatic polymers based on an investigation of the structures of some specially synthesized oligomers.[56, 57]

Some oligomeric systems with mass range within the specification of many EI–MS instruments have commercial significance and can be studied with a precision which the established 'classical' methods such as gel permeation chromatography are unable to emulate. This is true for some polyphenyls used as lubricants or hydraulic fluids and polyethers which are used widely in polymer intermediates and in surfactant applications.

Reconsideration of the case of poly(perfluorophenyl) diiodide[48] illustrates that the mass of the molecular ion requires that the polymer chain is terminated at both ends with iodine. This ability to distinguish end groups has been used by Lee and Sedgwick[58] to prove the existence of poly-(perfluorophenyl)s which are terminated by both —H and —F groups, whose existence has mechanistic significance. Similar conclusions were reached for poly(perfluoro-2-butyne), which is effectively terminated by HF addition to the polyene.[59]

In reality, since end groups produce a mass shift in the mass spectrum, cases where end groups are not obvious in oligomer spectra should be rare. The exception is the case where the oligomers are derived only from an exact number of units of the monomer. The structural choices are between a macrocycle or an unsaturated acyclic structure as might arise from a chain transfer step. Unfortunately these two structures are isobaric, eliminating the end group mass shift in the molecular ion. There are, however, end group mass shifts associated with fragment ions involving backbone cleavage and consideration of these processes has been shown to be a viable alternative in fluoro polymers[45] and oligomers from substituted poly(β-alanine)s.[60]

In some cases the converse may be true, *viz.* that the end groups are known but the origin of the fragment ions is ambiguous. This problem was solved in a detailed study of some polyethers by Sedgwick and Lee,[61] using the mass shifts produced by deuteration of one end group and derivatization of the other end group. This detailed definition of the structures of the ionic fragments revealed that certain peaks in the unmodified samples were multiplets, which could be resolved using high resolution techniques. Polyethers with molecular weights up to 1700 a.m.u. were studied and complete assignment of the block structure was made in some samples. Figure 7 shows the distribution of ions of the type $[MeO(EO)_n(PO)_mH - H_2O]^+$ formed by dehydration of the molecular ions of components of this copolymer. The numbers on the diagram represent the block lengths for ethoxy (n) and propoxy (m) units. This distribution of the actual molecular components in the finished copolymer gives $\bar{M}_n$ in agreement with that derived by conventional end group assay. The rather irregular distribution of molecular weights in Figure 7 can be deconvoluted to give the molecular weight distributions of the individual blocks. The method was shown to be capable of clearly distinguishing random and block copolymers and in the latter case the block order was determined.

Direct probe introduction of oligomers of 4,4'-isopropylidenediphenyl carbonate by Wiley[62] was successful in giving spectra up to the mass of the tetramer but the interpretation was frustrated by uncertainties concerning the thermal stability of the system. Ramjit and Sedgwick[63] used EI–MS with a direct introduction system to study the thermally induced ester–ester interchange between blends of poly(ethylene adipate) and poly(trimethylene adipate) to give a random copolymer.

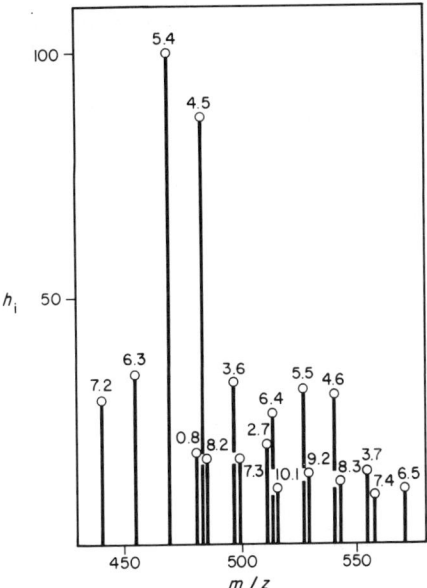

Figure 7 Distribution of ions of the type $[MeO(EO)_n(PO)_mH - H_2O]^+$ formed from a block copolymer. The numbers are the block lengths, *n.m* (reproduced with permission from ref. 61)

26.2.4 Rapid Heating

The thermal decomposition of oligomers during volatilization can be minimized by very rapid heating and ionization. Typically the sample is deposited on a wire in close proximity to the ionizing particle beam. The wire is heated rapidly, at $\sim 1000\,°C\,s^{-1}$ to a high temperature where evaporation is kinetically faster than thermal decomposition.[64] Since the ionization event is very short lived, the spectrometer used must scan very rapidly or a mass spectrographic analyzer must be used. Of the two alternatives the latter is preferable since all ions produced in the thermal spike are detected. This arrangement has been used by Davis and coworkers,[64, 65] using a photographic plate ion detector. Boettger and coworkers[66] have developed an elaborate electro-optical alternative to the photoplate in which the separated ion beams are intercepted by a channel-plate electron multiplier array, which feeds a phosphor connected by fiber optics to an image analyzer. The system is inherently very sensitive and has been used to analyze biopolymers. Developments of this technology have been used to enhance the sensitivity of double focusing magnetic mass spectrometers[67] using the Felgett advantage. Again the main thrust of the applications has been in the biopolymer field but applications to synthetic polymers are clearly not precluded and are eagerly awaited.

Rapid heating of polystyrene using an EI quadrupole system has yielded oligomers up to the undecamer but always in the presence of obvious decomposition products.[68] Changing the ionization mode to CI gave more stable ions as expected and the observed mass range was extended to 3000 a.m.u. The methane CI–MS was used to calculate $\bar{M}_n$ in agreement with that determined classically. Thus while the combination of flash desorption with chemical ionization has been shown to be most promising in a few polymer studies, the technique known generally as desorption chemical ionization (DCI) has been exploited more in the analysis of salts and polar organics used as polymer additives and its main polymer applications have been confined to biopolymers.

26.3 DESORPTION IONIZATION TECHNIQUES

26.3.1 Laser Desorption

Coloff and Vanerborgh,[69, 70] using a continuous laser and a time-of-flight (TOF) mass analyzer, reported that in addition to pyrolytic decomposition the laser also ionized some of the fragments from polystyrene.

A pulsed laser coupled with a TOF–MS is the basis of a commercial system known as LAMMA, laser microprobe mass analysis.[71] Here the high energy laser pulse both degrades and ionizes the sample in a multiphoton process. The spectra are considered to be indicative of a high energy degradation in that molecular ions are not detected. For polyethylene, both positive and negative ion spectra are produced and consist of carbon chains which are dramatically deficient in hydrogen compared with the host structure.

Results from laser desorption[72] (LD–MS) have been improved using short laser pulses to desorb ions from solid samples in an alkali halide matrix. Cation attachment stabilizes the desorbed molecules giving enhanced detection limits. Single pulse operations have been used with appropriate ion detection systems, *viz.* photoplates,[72] electro-optical devices[72] or rapid quadrupole scanning. Longer lasting ion currents were obtained using repetitive laser pulses.[74] While the methodology has been directed to the characterization of materials of biological origin, the application to industrial polymers is not precluded and results from ionic polymers would be of particular interest.

Laser desorption ionization has also been effectively combined with Fourier transform ion cyclotron resonance (FTICR–MS) to study a number of insoluble conducting polymers[75] including a series of heterocyclics.[76]

26.3.2 Field Desorption

Field desorption (FD) uses intense electric fields to ionize molecules deposited on the surface of a specially prepared emitter. The ions desorb from the surface under the influence of the field and solid involatile materials are thus efficiently converted to gaseous ions. The technique was developed by Beckey[40, 77] and has been applied with great success by Schulten[78, 79] to a wide range of involatile materials including biopolymers and synthetic polymers.

The emitter which carries the solid sample is typically a 1–10 μm diameter wire which has been activated by the growth of microscopic dendrites on its surface.[80] Activation is a prolonged process and the resulting emitter is quite fragile and must survive the rigors of sample deposition and the application of high electric fields.[81] Considerable expertise is required for successful operation, yet it is nevertheless one of the most widely used techniques for studying polymers by MS. This wide use is now declining as alternative sputter ion techniques are being shown to be capable of producing comparable data in a more reliable experimental format.

The surface dendrites on the emitter enhance the local electric field to field ionization levels when ionization occurs. Desorption rates and the internal energy of the desorbed ions can be altered by resistive heating of the emitter wire. At low emitter temperatures we have a 'soft' ionization process,[82] while at higher emitter temperatures we observe an increase in ionic fragmentation processes at the expense of the molecular ions. The technique has proven successful with both polar and nonpolar polymers and its usefulness has been maximized by using it in conjunction with higher mass range instrumentation.[83, 84, 85]

The ion emission from the small amount of sample is of high intensity but short lived, consequently there is an advantage in using a mass spectrograph with a focal plane detector. Nevertheless, most work has been carried out using magnetic scanning double focusing mass spectrometers.

Generally, FD–MS shows little evidence for ionic fragmentation with true molecular ions being formed from nonpolar molecules and salts. FD–MS of macromolecules[86] have been reported for oligomers of pivalolactone, caprolactam and isopropylidenediphenyl carbonate,[62, 87] polyesters,[88] polystyrene/poly(propylene glycol)[89] and polystyrene.[90] In this last report Matsuo *et al.* recorded FD–MS spectra of polystyrene molecules up to mass ∼11 000.

FD–MS at high mass resolution is used frequently to determine atomic compositions of molecular species. This technique has been used to characterize[82] an oligomeric phenolic product and a polyurethane extract.

In FD–MS optimization of emitter temperature results in a 'best anode temperature' (BAT). This can be used to optimize formation of ions from particular oligomers and can be programmed to desorb successive oligomers whose summed spectra give a quantitative measure of oligomer distribution. This approach has been used to characterize some epoxy prepolymers[91] and resins,[92] polyglycols and melamine resins.[93] The effect of emitter temperature on the fragmentaion of polymeric ions has been studied for polybutadiene[94] and poly(propylene glycol)[95] with a mass range of several thousands.

The potential to determine accurate molecular weights and detailed molecular weight distribution has already been introduced above. When EI–MS is replaced by FD–MS, this possibility becomes a much more practical proposition to a usefully wide range of materials of interest. The method is obviously limited by the ability of the mass spectrometer to record the highest mass components in the molecular weight distribution of the polymer. Modern magnetic mass spectrometers are capable of operating routinely up to m/z 15 000 and results on biomolecules with molecular weights up to 30 000 have been reported using FAB–MS. While such instrumentation is not yet widely available, the movement of mass spectrometry to higher and higher mass ranges has been a continuous technological development which is not yet exhausted. The trend was initially catalyzed by developments in sputter ion source design and has been fueled by some spectacular successes, particularly in investigations of peptide and protein structure. The full effects of this revolution have not yet worked through to the synthetic polymer field but polymer chemists should be aware that mass spectrometry of molecules up to mass 20 000 is now considered unexceptional. Considering the incisiveness of molecular weight distribution determined by MS compared with the classical methods such as GPC, osmometry, light scattering, viscosity *etc.*, it is surprising how neglected the mass spectrometry of synthetic polymers has been.

A number of careful comparative studies have been carried out,[84, 96, 97] which have clearly demonstrated that FD–MS can be used directly to determine accurate molecular weights, distributions and averages for selected low molecular weight polymers. A typical result is shown in Figure 8 where the distribution of polystyrene molecules with an $\bar{M}_n$ of 1690 is shown in graphic detail.[84] Comparable data have been obtained using FD–MS on polybutadiene, polyisoprene and poly-ethylene;[97] poly(ethylene glycol); poly(propylene glycol) and poly(tetrahydrofuran).[96] For phenol–formaldehyde resins[98] the FD–MS technique gives average molecular weights significantly lower than GPC. FD–MS has been used to identify oligomers separated by liquid chromatography.[99–103]

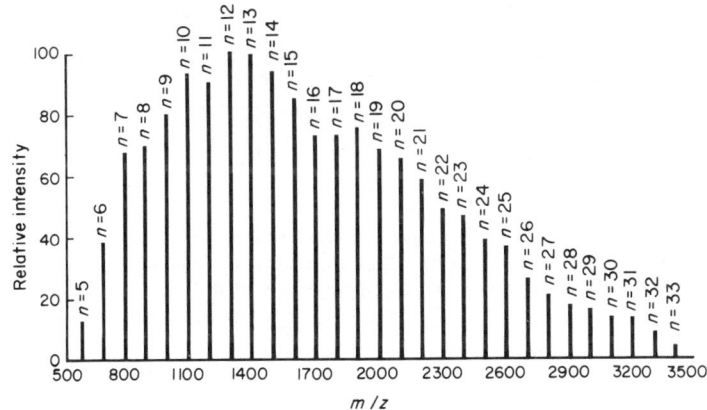

Figure 8 Molecular weight distribution of polystyrene ($M_n = 1690$) determined by FD–MS (reproduced with permission from ref. 84)

FD–MS of a series of synthetic polymers has been used to illustrate the capabilities of a high mass range magnetic mass spectrometer. A mass range up to 12 000 was demonstrated using polystyrene, poly(perfluoropropylene oxide), poly(propylene glycol)s, poly(ethylene glycol)s, nylon 6 and nylon 6,6.[104] A comparison of mass spectrometric average molecular weights, $\bar{M}_n$ and $\bar{M}_w$, with those from classical methods gives tolerable agreement.

26.3.3 Plasma Desorption Mass Spectrometry

Plasma desorption (PD) mass spectrometry was originated by Macfarlane *et al.*,[105] using the high energy, heavy fission fragments from the radioactive decay of the artificial radioisotope ^{252}Cf to both vaporize and ionize involatile molecules. The sample as a thin layer on a metallic foil is bombarded by californium fission fragments which penetrate the foil. Typically molecules are converted to gaseous ions by proton gain and loss giving $(M + H)^+$ and $(M - H)^-$. The ions sputtered from the sample are analyzed by a TOF–MS using a coincidence method of operation triggered by individual fission events. Data is thus accumulated as a series of single events so that data collection times may be long compared with other ionization methods. However, the efficiency of ionization of high molecular weight compounds surpasses that of other ionization techniques.

While developed principally for the study of biomolecules, PD–MS nevertheless has been applied with success to some synthetic polymers. A study of poly(ethylene glycol)s[106] gave molecular weight distributions comparable with other techniques. Sundqvist and coworkers[107,108] have shown that sputter ion efficiency for higher molecular weights can be studied effectively but perhaps inconveniently using heavy ion particle accelerators in place of a radioactive source.

A commercial PD–MS system using ^{252}Cf ionization is available and has been used to study molecules such as interferon.[109] Other biopolymers with molecular weights up to 35 000 have been studied by PD–MS.[108] Applications to industrial polymers have been less frequent but do include reports of good molecular weight averages and distribution for poly(ethyleneimine)s.[110]

26.3.4   Secondary Ion Mass Spectrometry

The use of positive ion beams to sputter ions from surfaces has been a developing branch of mass spectrometry for some years. Used initially to study metallic and modified metallic surfaces, the technique developed as an important means of studying surface reactions. Benninghoven and coworkers[111-114] extended the secondary ion mass spectrometry (SIMS) techniques, with some success, to organic systems. Ions form usually by cation or anion loss[115] and ionization efficiency has been highest for ionic compounds. Generally organic compounds are deposited as a very thin film on a metal target, which is bombarded with inert gas ions (Ar^+ or Xe^+) from a discharge ion source or Cs^+ from a thermal ionization source.[116] Thick films of high molecular weight organic solids, including most polymers, lead to electrical charging of the sample or suffer irreversible surface damage, which has limited the utility of this technique for analytical purposes. The development of time-of-flight SIMS has been shown to be useful for characterization of low molecular weight polymers up to molecular weights of 10 000.[117,118]

26.3.5   Fast Atom Bombardment Mass Spectrometry

A variation of SIMS was introduced by Sedgwick and coworkers,[119] who showed advantages in using neutral particle bombardment and presentation of the solid sample in a liquid matrix. In a fast atom bombardment (FAB) ion source the solid sample is dissolved in a solvent of low volatility and this liquid is deposited on a metallic target. The liquid surface is bombarded with 10 keV xenon atoms and sputtered ions are collected in a conventional mass spectrometer.

FAB–MS was developed initially for the study of highly polar biomolecules which are inaccessible to vapor phase ionization methods.[120] Since the sample is contained in a mobile liquid surface, problems of radiation damage are minimized and long-lived stable mass spectra are obtained. If the sample is itself a liquid, or a solid which can be melted without decomposition to a liquid, it may be studied directly in a FAB ion source.

Ions formed in FAB are typically $(M + H)^+$ and $(M - H)^-$ with structurally significant fragmentation being present. The technique has been used with spectacular success to study polypeptides and small proteins. Thus natural polypeptides with molecular weights up to ~2000 have been fully sequenced using FAB–MS.[121] Peptide fragmentation mechanisms have been tabulated[122] and a standard nomenclature adopted.[123] Protocols have been proposed for the study of protein structure based on enzymatic cleavage followed by sequencing of the fragments by FAB–MS.[124] Similar success has been demonstrated with other biopolymers such as oligonucleotides[125] and polysaccharides.[126]

Since molecular ion species predominate over fragment ions, it is possible to measure accurate molecular weights to much higher masses than fragmentation can be observed. Accurate molecular weights of peptides measured by FAB–MS have progressed from glucagon (3481),[127] insulin (5735)[128] and proinsulin (8680)[129] to trypsinogen (24 000).[130] In the latter case Green and coworkers covered a range of materials traditionally used as calibrants for GPC measurements of peptide molecular weights.

While FAB–MS has been driven by applications to biopolymers, the technique has been clearly demonstrated to be of similar value in the synthetic polymer field.[78,131] Montaudo and coworkers have shown applications to polyesters and polyamides.[132,133]

Following the work of Lattimer,[134] several groups have demonstrated the facility of measuring molecular weight distribution in polyalkoxylates and interest has now spread to investigations of the ion chemistry of these systems using tandem mass spectrometry.[135]

26.4 SPRAY IONIZATION METHODS

The principle of spraying involatile materials as charged droplets into a desolvation chamber designed to reduce them to gaseous ions has received attention with mixed success over many years. Most notable is the work of Dole and coworkers.[136–138] In a project long since abandoned they developed an electrospray system capable of producing a 'macromass' spectrum. For polystyrene solutions molecules of molecular weight 400 000 were believed to be detected as gaseous ions. Recent improvements in ion drift mass spectrometry suggest that this technique is perhaps ripe for reevaluation.[139]

The associated development of an electrohydrodynamic ion source coupled to a double focusing mass spectrometer was introduced by Evans and coworkers.[140] This approach has been refined by Cook and coworkers,[141] who have demonstrated a capability to analyze poly(ethylene glycol)s, but a wide range of analytical ability has not yet been demonstrated.

Of quite recent and already widely demonstrated utility for a wide range of polar organic molecules is the thermospray ionization source of Vestal and coworkers.[142] Here polar analyte compounds in a solution containing a volatile ionic buffer, such as ammonium acetate, are sprayed through a heated nozzle. The heated spray is desolvated in a heated chamber and the resulting ions are sampled into a mass spectrometer, most typically a quadrupole mass analyzer.

Ionization of neutral molecules occurs *via* ion attachment to give $(M + H)^+$, $(M + NH_4)^+$, $(M + MeCO_2)^-$, *etc.* Fragmentation of these ions is minimal and associated ions are not uncommon. This relatively new technique is proving most useful in studying biopolymers up to 3000 molecular weight[143] and extension to synthetic polymer systems seems predictable.

Variants of the thermospray source are being developed further under names such as electrospray[144] and plasmaspray,[145] which claim to use electrical in place of chemical means of ionization. It is premature to judge their utility in polymer chemistry at present but in such a rapidly developing and fruitful area it seems predictable that applications to biopolymers will precede applications to synthetic polymers.

26.5 CONCLUSIONS

Applications of mass spectrometry to high polymers has been traditionally handicapped by an inability to volatilize the sample without decomposition and a lack of mass range in available mass spectrometers. Thus gaseous ion sources were confined to studies of pyrolyzed polymers in which polymer identity and structures were inferred from detailed mass spectrometric analysis of their volatile products. This branch of mass spectrometry is still vitally active and is being advanced to new levels of sophistication and utility using advanced computer-based pattern recognition and artificial intelligence strategies.

The direct analysis of polymers started mainly as series of largely unrelated observations on systems with oligomers of extraordinary volatility and stability, most commonly fluoro polymers. These studies demonstrated the validity of the mass spectrometric approach but these techniques were only exploited effectively after developments in methods for conversion of solid samples into gaseous ions. Of these methods of ionization, field desorption played a major role and now has an established place in studies of synthetic polymers. Developments of sputter ionization methods, particularly PD–MS and FAB–MS, demonstrated the ability to successfully ionize materials with molecular weights measured in tens of thousands. Development of these ion sources has stimulated the production of mass analyzers designed to match their proven mass ranges. While mass spectrometry is still only operating at the lower end of the traditional high polymer molecular weight range, it has nevertheless come of age as a technique for studying high molecular weight materials. While its range of applicability now overlaps with many well-established techniques such as GPC, osmometry, viscometry, light scattering, *etc.* it is clear that mass spectrometry offers potential advantages. Most important is the ability to observe individual molecular characteristics and chemical structure within a distribution without compromising the calculation of the average properties of the distribution, which are traditionally measured.

While mass spectrometry has now moved into the 'kilomass' range, the impetus for the experimental advances has been founded almost exclusively in satisfying the needs of biomedical research. Applications to synthetic and industrial polymer systems have been neglected to a surprising degree, despite the obvious opportunities which exist by simple analogy with the voluminous biomedical literature. It is an opportunity which the present authors hope will be belatedly grasped by mainstream polymer chemists.

26.6 REFERENCES

1. K. Biemann and S. A. Martin, *Mass Spectrom. Rev.*, 1987, **6**, 1.
2. H. R. Schulten and R. P. Lattimer, *Mass Spectrom. Rev.*, 1984, **3**, 231.
3. R. D. Sedgwick, in 'Developments in Polymer Characterization — 1', ed. J. V. Dawkins, Applied Science, London, 1972.
4. R. P. Lattimer and R. E. Harris, *Mass Spectrom. Rev.*, 1985, **4**, 369.
5. T. Midgley and A. L. Henne, *J. Am. Chem. Soc.*, 1929, **51**, 1251.
6. H. Staudinger and A. Steinhofer, *Justus Liebigs Ann. Chem.*, 1935, **35**, 517.
7. L. A. Wall, *J. Res. Natl. Bur. Stand.*, 1948, **41**, 315.
8. S. L. Madorsky and S. Straus, *J. Res. Natl. Bur. Stand.*, 1948, **40**, 417.
9. S. Straus and S. L. Madorsky, *J. Res. Natl. Bur. Stand., Sect. A*, 1958, **63**, 261.
10. S. Straus and S. L. Madorsky, *J. Res. Natl. Bur. Stand.*, 1953, **50**, 165.
11. S. L. Madorsky, 'Thermal Degradation of Organic Polymers', Interscience, New York, 1964.
12. S. L. Madorsky, S. Straus, D. Thompson and L. Williamson, *J. Res. Natl. Bur. Stand.*, 1949, **42**, 499.
13. S. Straus and S. L. Madorsky, *J. Res. Natl. Bur. Stand.*, 1958, **61**, 77.
14. W. T. Flowers, R. N. Haszeldine, E. Henderson and R. D. Sedgwick, *Trans. Faraday Soc.*, 1966, **62**, 1120.
15. G. L. Kearns, *Anal. Chem.*, 1964, **36**, 1402.
16. D. Garozzo, M. Giuffrida and G. Montaudo, *Macromolecules*, 1986, **19**, 1643.
17. H. L. C. Meuzelaar, J. Haverkamp and F. D. Hielman, 'Pyrolysis Mass Spectrometry of Recent and Fossil Biomaterials', Elsevier, New York, 1982.
18. J. M. Richards, W. H. McLennen, H. L. C. Muezelaar, L. C. Henk, J. P. Shockcor and R. P. Lattimer, *J. Appl. Polym. Sci.*, 1987, **34**, 1967.
19. J. L. Savoca, R. P. Lattimer, H. L. C. Muezelaar and J. M. Richards, *33rd Annu. Conf. Mass Spectrom. Allied Top., San Diego*, 1985, 806.
20. I. Lüderwald and H. Ringsdorf, *Angew. Makromol. Chem.*, 1973, **29**, 441.
21. A. Ballisteri, S. Foti, G. Montaudo and E. Scamporrino, *J. Polym. Sci., Polym. Chem. Ed.*, 1980, **18**, 1147.
22. S. Tsuge and T. Takeuchi, *Anal. Chem.*, 1977, **49**, 348.
23. D. S. Chatfield, F. D. Hileman, E. J. Voohees, I. N. Einhorn and J. H. Futrell, in 'Applications of Polymer Spectroscopy', ed. E. G. Brame, Academic Press, New York, 1978.
24. J. G. Moncur, A. B. Campa and P. C. Pinoli, *HRC & CC, J. High Resolut. Chromatogr., Chromatogr. Commun.*, 1982, **5**, 322.
25. G. J. Mol, R. J. Gritter and G. E. Addams, in 'Applications of Polymer Spectroscopy', ed. E. G. Brame, Academic Press, New York, 1978.
26. J. L. Wuepper, *Anal. Chem.*, 1979, **51**, 997.
27. J. B. Pausch, R. P. Lattimer and H. L. C. Muezelaar, *Rubber Chem. Technol.*, 1983, **56**, 1031.
28. T. Nagaya, Y. Suginmura and S. Tsuge, *Macromolecules*, 1980, **13**, 353.
29. S. Tsuge, T. Kobayashi, T. Nagaya and T. Takeuchi, *J. Anal. Appl. Pyrolysis*, 1979, **1**, 133.
30. S. Tsuge, T. Kobayashi, Y. Sugimura, T. Nagaya and T. Takeuchi, *Macromolecules*, 1979, **12**, 988.
31. Y. Sugimura and S. Tsuge, *Macromolecules*, 1979, **12**, 512.
32. S. Tsuge, Y. Sugimura and T. Nagaya, *J. Anal. Appl. Pyrolysis*, 1980, **1**, 221.
33. D. H. Ahlstrom, S. A. Liebman and K. B. Abbas, *J. Polym. Sci., Polym. Chem. Ed.*, 1976, **14**, 2479.
34. G. Montaudo, C. Puglisi, E. Scamporrino and D. Vitalini, *Macromolecules*, 1986, **19**, 870.
35. G. Montaudo, C. Puglisi, E. Scamporrino and D. Vitalini, *Macromolecules*, 1986, **19**, 882.
36. H. L. C. Muezelaar, P. G. Kistemaker and M. A. Posthumus, *Biomed. Mass Spectrom.*, 1974, **1**, 312.
37. P. G. Kistemaker, A. J. H. Boerboom and H. L. C. Muezelaar, *Dyn. Mass Spectrom.*, 1976, **4**, 139.
38. W. H. McLennan, J. M. Richards, H. L. C. Muezelaar, J. B. Pausch and R. P. Lattimer, *34th Annu. Conf. Mass Spectrom. Allied Top., Cincinnatti*, 1986, 817.
39. R. Gomer, 'Field Emission and Field Ionization', Harvard University Press, Harvard, 1961.
40. H. D. Beckey, 'Principles of Field Ionization and Field Desorption Mass Spectrometry', Pergamon Press, Oxford, 1977.
41. R. P. Lattimer and W. J. Kroenke, *J. Appl. Polym. Sci.*, 1980, **25**, 101.
42. R. P. Lattimer and W. J. Kroenke, *J. Appl. Polym. Sci.*, 1981, **26**, 1191.
43. D. O. Hummel, H. D. R. Schüddemage and K. Rübenacker, in 'Polymer Spectroscopy', ed. D. O. Hummel, Verlag Chemie, Weinheim, 1974, p. 355.
44. H. D. R. Schüddemage and D. O. Hummel, *Adv. Mass Spectrom.*, 1968, **4**, 857.
45. D. O. Hummel and H. J. Duessel, *Makromol. Chem.*, 1974, **175**, 655.
46. D. O. Hummel, H. J. Duessel and K. Rubenacker, *Makromol. Chem.*, 1971, **145**, 267.
47. M. Ryska, H. D. R. Schüddemage and D. O. Hummel, *Makromol. Chem.*, 1969, **126**, 32.
48. P. Bradt, V. H. Dibeler and F. L. Mohler, *J. Res. Natl. Bur. Stand.*, 1955, **55**, 323.
49. W. T. Flowers, R. N. Haszeldine, E. Henderson, A. K. Lee and R. D. Sedgwick, *J. Polym. Sci., Polym. Chem. Ed.*, 1972, **10**, 3489.
50. D. A. Barr and R. N. Haszeldine, *J. Chem. Soc.*, 1955, 1881.
51. G. Montaudo, C. Puglisi, E. Scamporrino and D. Vitalini, *Macromolecules*, 1986, **19**, 2157.
52. G. Montaudo, C. Puglisi, E. Scamporrino and D. Vitalini, *Macromolecules*, 1986, **19**, 2689.
53. G. Montaudo, E. Scamporrino, C. Puglisi and D. Vitalini, *J. Polym. Sci., Polym. Chem. Ed.*, 1987, **25**, 1653.
54. R. W. Lenz, I. Lüderwald, G. Montaudo, G. Przbylski and H. Ringsdorf, *Makromol. Chem.*, 1974, **175**, 2441.
55. G. Montaudo, M. Przbylski and H. Ringsdorf, *Makromol. Chem.*, 1975, **176**, 1763.
56. I. Lüderwald, G. Montaudo, M. Przbylski and H. Ringsdorf, *Makromol. Chem.*, 1974, **175**, 2423.
57. G. Montaudo, M. Przbylski and H. Ringsdorf, *Makromol. Chem.*, 1975, **176**, 1753.
58. A. K. Lee, Ph. D. Thesis, University of Manchester, 1971.
59. W. T. Flowers, R. N. Haszeldine, A. Janik, A. K. Lee, P. G. Marshall and R. D. Sedgwick, *J. Polym. Sci., Polym. Chem. Ed.*, 1972, **10**, 3497.
60. I. Lüderwald and H. Ringsdorf, *Angew. Makromol. Chem.*, 1973, **29**, 453.
61. A. K. Lee and R. D. Sedgwick, *J. Polym. Sci., Polym. Chem. Ed.*, 1978, **16**, 685.

62. R. H. Wiley, *J. Polym. Sci., Macromol. Rev.*, 1979, **14**, 379.
63. H. G. Ramjit and R. D. Sedgwick, *J. Macromol. Sci., Chem.*, 1976, **A10**, 815.
64. G. D. Daves, Jr., *Acc. Chem. Res.*, 1979, **12**, 359.
65. W. R. Anderson, Jr., W. Frick and G. D. Daves, Jr., *J. Am. Chem. Soc.*, 1977, **100**, 1974.
66. H. Boettger, C. E. Griffin, D. D. Norris, W. J. Dreyer and K. Kuppermann, *Adv. Mass Spectrom. Biochem. Med.*, 1977, 2513.
67. J. S. Cottrell and S. Evans, *Anal. Chem.*, 1987, **59**, 1990.
68. H. R. Udseth and L. Friedman, *Anal. Chem.*, 1981, **53**, 29.
69. S. G. Coloff and N. E. Vanderborgh, *Anal. Chem.*, 1973, **45**, 1507.
70. S. G. Coloff and N. E. Vanderborgh, *Org. Mass Spectrom.*, 1973, **7**, 1367.
71. J. A. Gardella, Jr., D. M. Hercules and H. J. Heinen, *Spectroscop. Lett.*, 1980, **13**, 347.
72. M. A. Posthumus, P. G. Kristemaker, H. L. C. Meuzelaar and M. C. T. N. de Brauw, *Anal. Chem.*, 1978, **50**, 985.
73. R. Stoll and F. W. Röllgen, *Org. Mass Spectrom.*, 1979, **14**, 642.
74. D. Zakett, A. E. Schoen, R. G. Cooks and P. H. Hemberger, *J. Am. Chem. Soc.*, 1981, **103**, 1295.
75. C. E. Brown, P. Kovacic, C. A. Wilkie, R. B. Cody, R. D. Hein and J. A. Kinsinger, *Synth. Methods*, 1986, **15**, 265.
76. C. E. Brown, P. Kovacic, R. B. Cody, R. E. Hein and J. A. Kinsinger, *J. Polym. Sci., Polym. Lett. Ed.*, 1986, **24**, 519.
77. H. D. Beckey, *Int. J. Mass Spectrom. Ion Phys.*, 1969, **2**, 500.
78. M. Doerr, I. Lüderwald and H. Schulten, *J. Anal. Appl. Pyrolysis*, 1985, **8**, 109.
79. H. R. Schulten, *Int. J. Mass Spectrom. Ion Phys.*, 1979, **32**, 97.
80. H. D. Beckey, M. D. Migahed and F. W. Röllgen, *Adv. Mass Spectrom.*, 1970, **5**, 622.
81. H. D. Beckey, E. Hilt and H. R. Schulten, *J. Phys. E.*, 1973, **6**, 1043.
82. R. P. Lattimer and K. R. Welch, *Rubber Chem. Technol.*, 1980, **53**, 151.
83. H. Matsuda, *At. Masses Fundam. Constants*, 1976, **5**, 185.
84. R. P. Lattimer, D. J. Harmon and G. E. Hansen, *Anal. Chem.*, 1980, **52**, 1808.
85. P. G. Cullis, G. M. Newmann, D. E. Rogers and P. J. Derrick, *Adv. Mass Spectrom.*, 1980, **8**, 1729.
86. H. R. Schulten, *23rd Annu. Conf. Mass Spectrom. Allied Top., Houston*, 1975, 25.
87. P. H. Wiley and J. C. Cook, Jr., *J. Macromol. Sci., Chem.*, 1976, **10**, 811.
88. L. F. Palmer, A. F. Weston and R. A. McDowell, *26th Annu. Conf. Mass Spectrom. Allied Top., St. Louis*, 1978, 181.
89. R. P. Lattimer, K. R. Welch, J. P. Pausch and V. Rapp, *26th Annu. Conf. Mass Spectrom. Allied Top., St. Louis*, 1978, 581.
90. T. Matsuo, H. Matsuda and I. Katakuse, *Anal. Chem.*, 1979, **51**, 1329.
91. J. Saito, S. Toda and S. Tanaka, *Bunseki Kagaku*, 1980, **29**, 462.
92. J. Saito, S. Toda and S. Tanaka, *Netsu Kokasei Jushi*, 1980, **1**, 79.
93. J. Saito, S. Toda and S. Tanaka, *Netsu Kokasei Jushi*, 1980, **1**, 18.
94. A. C. Craig, P. G. Cullis and P. J. Derrick, *Int. J. Mass Spectrom. Ion Phys.*, 1981, **38**, 297.
95. G. M. Neumann, P. G. Cullis and P. J. Derrick, *Z. Naturforsch., Teil A*, 1980, **35**, 1090.
96. R. P. Lattimer and G. E. Hansen, *Macromolecules*, 1981, **14**, 776.
97. R. P. Lattimer and H. R. Schulten, *Int. J. Mass Spectrom. Ion Phys.*, 1983, **52**, 105.
98. R. P. Lattimer, E. R. Hooser, H. E. Diem and C. K. Rhee, *Rubber Chem. Technol.*, 1982, **55**, 442.
99. H. R. Schulten and H. D. Beckey, *J. Chromatogr.*, 1973, **83**, 315.
100. N. E. Evans, D. E. Games, A. H. Jackson and S. A. Mattin, *J. Chromatogr.*, 1975, **115**, 325.
101. H. R. Schulten, *J. Chromatogr.*, 1982, **251**, 105.
102. R. P. Lattimer, D. J. Harmon and K. P. Welch, *Anal. Chem.*, 1979, **51**, 1293.
103. R. P. Lattimer, E. R. Hooser and P. M. Zakriski, *Rubber Chem. Technol.*, 1980, **53**, 346.
104. S. D. Maleknia and C. E. Costello, *34th Annu. Conf. Mass Spectrom. Allied Top., Cincinnatti*, 1986, 825.
105. R. D. Macfarlane and D. F. Torgerson, *Science*, 1976, **191**, 920.
106. B. T. Chait and F. H. Field, *31st Annu. Conf. Mass Spectrom. Allied Top., Boston*, 1983, 619.
107. P. Hakansson, E. Jayasinghe, A. Johansson, I. Kamensky and B. Sundqvist, *Phys. Rev. Lett.*, 1981, **47**, 1227.
108. B. Sunqvist, I. Kamensky, P. Hakansson, J. Kjellberg, M. Salehpour, S. Widdiyasebera, J. Fohlman, P. A. Peterson and P. Roepsdorff, *Biomed. Mass Spectrom.*, 1984, **11**, 242.
109. A. G. Craig, B. Sundquist and I. Kamensky, *35th Annu. Conf. Mass Spectrom. Allied Top., Denver*, 1987, 578.
110. R. Robbins, M. Alai, P. Demirev and R. Cotter, *33rd Annu. Conf. Mass Spectrom. Allied Top., San Diego*, 1985, 403.
111. A. Benninghoven, D. Jaspers and W. Sichtermann, *Appl. Phys.*, 1976, **11**, 35.
112. A. Benninghoven and W. Sichtermann, *Org. Mass Spectrom.*, 1977, **12**, 595.
113. A. Eicke, W. Sichtermann and A. Benninghoven, *Biomed. Mass Spectrom.*, 1980, **15**, 289.
114. A. Benninghoven (ed.), 'Ion Formation from Organic Solids', Springer-Verlag, New York, 1983.
115. R. J. Day, S. E. Unger and R. G. Cooks, *Anal. Chem.*, 1980, **52**, 557A.
116. G. J. Elliott, J. S. Cottrell and S. Evans, *34th Annu. Conf. Mass Spectrom. Allied Top., Cincinnatti*, 1986, 880.
117. I. V. Bletsos, D. M. Hercules, D. Griefendorf and A. Benninghoven, *Anal. Chem.*, 1985, **57**, 2384.
118. I. V. Bletsos, D. M. Hercules, D. van Leyen and A. Benninghoven, *Macromolecules*, 1987, **20**, 407.
119. M. Barber, R. S. Bordoli, R. D. Sedgwick and A. N. Tyler, *J. Chem. Soc., Chem. Commun.*, 1981, 325.
120. M. Barber, R. S. Bordoli, G. J. Elliott, R. D. Sedgwick and A. N. Tyler, *Anal. Chem.*, 1982, **54**, 645A.
121. K. L. Rienhart, L. A. Gaudioso, M. L. Moore, R. C. Pandey, J. C. Cook, M. Barber, R. S. Bordoli, A. N. Tyler, B. N. Green and R. D. Sedgwick, *J. Am. Chem. Soc.*, 1981, **103**, 6517.
122. K. Biemann and S. A. Martin, *Mass Spectrom. Rev.*, 1987, **6**, 1.
123. P. Roepstorff, *Biomed. Mass Spectrom.*, 1984, **11**, 601.
124. H. R. Morris, M. Panico, M. Barber, R. S. Bordoli, R. D. Sedgwick and A. N. Tyler, *Biochem. Biophys. Res. Commun.*, 1981, **101**, 623.
125. T. Matsuo, T. Sakurai, H. Matsuda, M. Matsugi and M. Ikehara, *34th Annu. Conf. Mass Spectrom. Allied Top., Cincinnatti*, 1986, 329.
126. G. R. Her, S. Santikarn, V. N. Reinhold and J. C. William, *35th Annu. Conf. Mass Spectrom. Allied Top., Denver*, 1987, 874.
127. M. Barber, R. S. Bordoli, R. D. Sedgwick, A. N. Tyler, G. V. Garner, D. B. Gordon, L. W. Tetler and R. C. Hider, *Biomed. Mass Spectrom.*, 1982, **9**, 265.

128. M. Barber, R. S. Bordoli, G. J. Elliott, R. D. Sedgwick, A. N. Tyler and B. N. Green, *J. Chem. Soc., Chem. Commun.*, 1982, 936.
129. J. S. Cottrell and B. H. Frank, *Biochem. Biophys. Res. Commun.*, 1985, **127**, 1032.
130. D. Bell and B. N. Green, *35th Annu. Conf. Mass Spectrom. Allied Top., Denver*, 1987, 540.
131. R. L. Cochran, *Appl. Spectroscop. Rev.*, 1986, **22**, 137.
132. A. Ballistreri, D. Garozzo, M. Giuffrida, G. Montaudo, A. Filippi, C. Guaita, P. Manaresi and F. Pilati, *Macromolecules*, 1987, **20**, 1029.
133. A. Ballistreri, D. Garozzo, M. Giuffrida and G. Montaudo, *Anal. Chem.*, 1987, **59**, 2024.
134. R. P. Lattimer, *Int. J. Mass Spectrom. Ion Phys.*, 1983, **55**, 221.
135. J. M. Gilliam and J. L. Occolwitz, *35th Annu. Conf. Mass Spectrom. Allied Top., Denver*, 1987, 624.
136. M. Dole, L. L. Mack, R. L. Hines, R. C. Mobley, L. D. Ferguson and M. B. Alice, *J. Chem. Phys.*, 1968, **49**, 2240.
137. M. Dole, H. L. Cox and J. Giemic, *Adv. Chem. Ser.*, 1973, **125**, 73.
138. K. Nakamae, V. Kumar and M. Dole, *29th Annu. Conf. Mass Spectrom. Allied Top., Minneapolis*, 1981, 517.
139. M. Dole, *33rd Annu. Conf. Mass Spectrom. Allied Top., San Diego*, 1985, 196.
140. B. P. Stimpson, D. S. Simons and C. A. Evans, *J. Phys. Chem.*, 1978, **82**, 660.
141. K. D. Cook, *Mass Spectrom. Rev.*, 1986, **5**, 467.
142. D. Pilosof, H. Y. Kim, D. F. Dyckes and M. L. Vestal, *Anal. Chem.*, 1984, **56**, 1236.
143. P. J. Rudewicz, *Biomed. Environ. Mass Spectrom.*, 1988, **15**, 461.
144. S. F. Wong, C. K. Meng and J. B. Fenn, *35th Annu. Conf. Mass Spectrom. Allied Top., Denver*, 1987, 33.
145. R. H. Bateman, S. T. Krolik, D. J. Jones and H. J. Major, *35th Annu. Conf. Mass Spectrom. Allied Top., Denver*, 1987, 415.

27

Pyrolysis GLC

TIM HAMMOND and ROY S. LEHRLE
University of Birmingham, UK

27.1 INTRODUCTION

The development of gas–liquid chromatography (GLC) by Martin and James[1] in 1952 introduced a technique able to separate and quantify the complex mixture of products arising from a polymer pyrolysis. The earliest work involved a two stage process whereby the polymer was first pyrolyzed to obtain the degradation products; these were then injected into a gas chromatography apparatus.[2]

In 1959, however, single stage pyrolysis gas–liquid chromatography (PGLC) was pioneered almost simultaneously by three sets of workers. They used furnaces,[3] resistive filaments[4] and high intensity radiation[5] to degrade polymer samples mounted at the inlet of a gas chromatography system.

The success of these early workers has led to the widespread application of the technique. Typical applications include the characterization of complex polymer mixtures, the assessment of copolymer composition and microstructure, network and cross-linking assignment, structural analysis of polymers, and the elucidation of degradation mechanisms.

This chapter aims to describe the many techniques available using PGLC and to illustrate the wide range of applications which make it such an important tool for the polymer chemist.

27.2 PYROLYSIS GLC: APPARATUS AND TECHNIQUES

Figure 1 illustrates an example of a modern PGLC system. The polymer sample is degraded in the pyrolysis unit [A] and the degradation products are swept by means of an inert carrier gas stream into the chromatography system [B]. Separation is effected and the products are detected and analyzed at [C]. PGLC systems can therefore be conveniently treated as these three units.

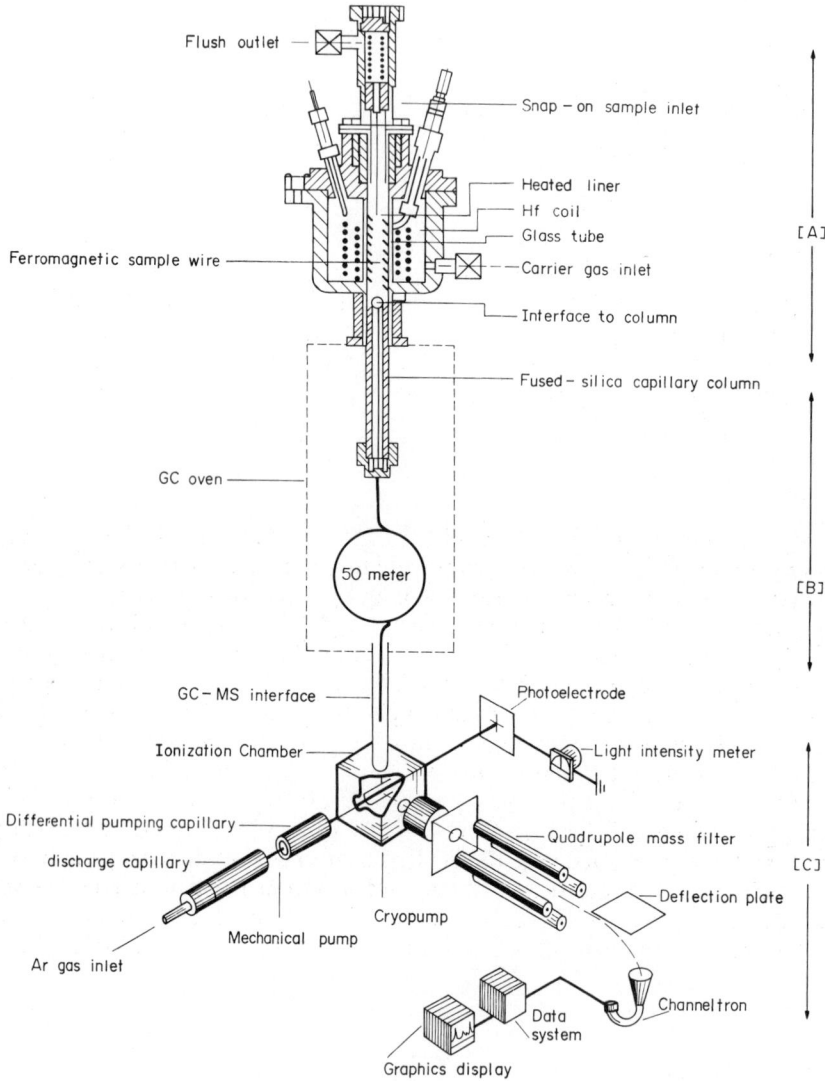

Figure 1 A PGLC system which incorporates mass spectrometric facilities for peak characterization. In this system the pyrolysis unit uses Curie-point heating, and the chromatographic column is a fused-silica capillary which is directly coupled by an interface probe to the source of a photoionization mass spectrometer. Apparatus very much simpler than this is adequate for routine analytical work, but peak characterization facilities are valuable when dealing with systems whose pyrolysis behaviour is totally unknown (reproduced by permission of Elsevier from *J. Anal. Appl. Pyrolysis*, 1985, **8**, 25).

27.2.1 Pyrolysis

An 'ideal' pyrolysis unit must have the following three properties. (i) It must operate at a precise, known temperature. This is essential for mechanistic studies and is required for analytical work if reproducibility is to be achieved. (ii) The pyrolyzer must reach its pyrolysis temperature very rapidly. Figures 2(a) and 2(b) respectively illustrate the ideal temperature profile and the typical temperature profile achieved by simply switching a filament on and off. With long temperature rise-times, the polymer is experiencing a range of degradation temperatures. Such behaviour will lead to anomalous results if mechanistic studies are to be attempted. Short, reproducible temperature rise-times

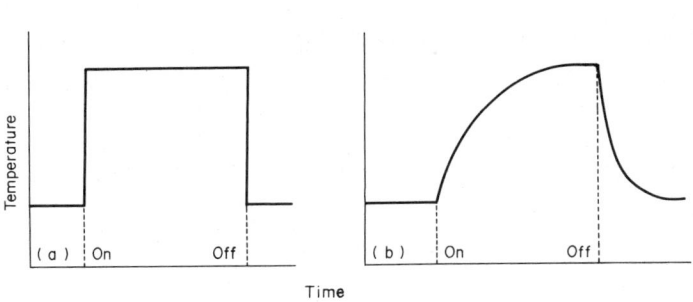

Figure 2 Temperature–time profiles for (a) an ideal pyrolyzer and (b) a more typical unit in which the filament is simply switched on and off

are also required in analytical studies where interlaboratory reproducibility is essential. (iii) The design of chamber and filament must be such that the pyrolysis products are passed rapidly and smoothly into the chromatography system. This reduces the possibility of secondary reactions within hot regions of the pyrolyzer and also improves the chromatographic separation.

Pyrolysis units can be divided into four categories: furnaces, Curie-point pyrolyzers, resistively heated filaments, and lasers and electrical discharge apparatus.

27.2.1.1 *Furnaces*

Furnaces were amongst the first devices used for PGLC.[3] They consist of heated tubes through which the carrier gas stream passes into the gas chromatography inlet. Samples are either introduced into the furnace hot-region by means of gravity[6] or using a push-rod[7] system.

Although furnace pyrolyzers are widely used, they do suffer from several disadvantages compared with other pyrolysis methods. The most important of these is that the sample is not in intimate contact with the heating surface; this leads to poor heat transfer and slow temperature rise-times such that the sample never reaches the furnace temperature during the pyrolysis time. The precise pyrolysis temperature is therefore very poorly defined. The sample temperature is also dependent on the carrier gas flow rate and variations in this lead to poor reproducibility.

Recently, however, there have been some significant improvements in furnace design. Tsuge and Matsubara have built a highly advanced furnace type[8] (Figure 3). This degrades relatively small samples (50 μg) which are suspended over the heating zone. The sample is dropped into the heating zone which is tapered to reduce the 'dead' volume and also to increase the carrier gas velocity. The degradation products are very rapidly transferred into a capillary chromatography system. These authors quote far greater reproducibility and resolution than that obtained for traditional furnace designs.

27.2.1.2 *Curie-point pyrolysis*

This technique seeks to utilize the fact that ferromagnetic metals and alloys cannot exceed a certain limiting temperature (the 'Curie-point') when receiving energy from a radio frequency field.

A typical example of a Curie-point pyrolyzer is given in Figure 4 (ref. 9). The sample is deposited from solution on to a ferromagnetic wire (not shown in the diagram but this is mounted within the quartz tube 2). The induction coil induces an alternating magnetic flux in the wire, resulting in eddy currents which cause a rapid rise in its temperature. When the temperature reaches the Curie-point of the alloy, there is a transition from ferromagnetism to paramagnetism. The energy intake of the conductor falls and the Curie temperature is maintained. Rapid temperature rise-times are achieved with high power units and very good temperature control is claimed. The different pyrolysis temperatures are achieved by selecting ferromagnetic alloys of different composition. Table 1 illustrates the effect of alloy composition on Curie temperature, and lists the rise-times estimated for RF generators of different power outputs.

It can be seen that temperature rise-times are very variable, and depend on the alloy composition and also the RF power. When coiled ferromagnetic wires were used, these times were increased by about an order of magnitude.[10]

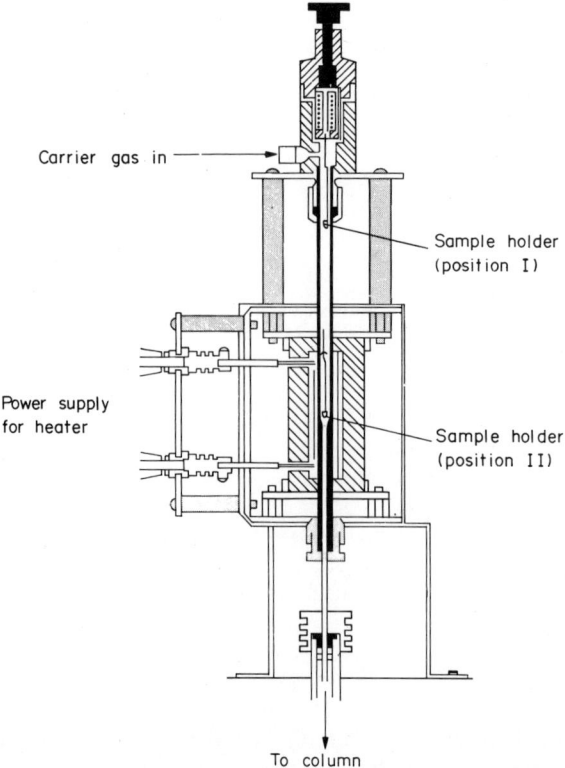

Figure 3 A pyrolyzer design in which the sample at ambient temperature in position I may be dropped to position II in the microfurnace at the chosen pyrolysis temperature. Furnace methods do not rate highly amongst the possible pyrolysis designs; the one illustrated is better than most [reproduced by permission of Elsevier from *J. Anal. Appl. Pyrolysis*, 1985, **8**, 49.]

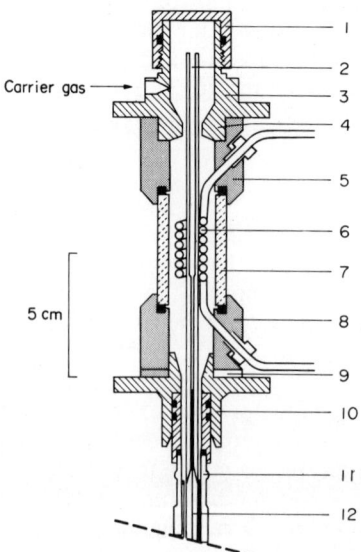

Figure 4 A Curie-point pyrolyzer design. The sample is coated on to a ferromagnetic wire (not shown) which is lowered down the quartz tube (2) into the RF induction coil (6). The latter is a copper tube through which cooling water passes in order to prevent self heating of the coil. After the RF current has pyrolyzed the sample, the products are swept by the carrier gas into the GLC column (12) [reproduced by permission of Pergamon Press, from *Chromatographica*, 1976, **12**, 597.]

Table 1 Curie-point Temperatures for Various Ferromagnetic Alloy Wires, and Rise-times Quoted for Two Types of Curie-point Apparatus[a]

Alloy composition (%)				*Quoted temperature rise-times* (ms)	
Fe	*Ni*	*Co*	*Curie-point* (°C)	*Fischer-Varian* (1500 W)	*Philips* (30 W)
0	100	0	358	300	1300
61.7	0	38.3	400	40	500
50.6	49.4	0	510	150	700
42.0	41.0	16.0	600	70	500
29.2	70.8	0	610	130	1150
33.0	33.0	33.0	700	90	1350
100	0	0	770	110	2100

[a] Reproduced from *Anal. Chem.*, 1972, **44**, 38, with permission from the American Chemical Society.

The use of ferromagnetic tubes as an alternative to wires[9] has been advocated, as these reduce condensation of involatile products on the chamber walls. They do, however, lead to increased temperature rise-times.

One great advantage of Curie-point PGLC is that no electrical contact is required between the power supply and the pyrolysis wire. With this in mind a fully automated system has been described in which coated pyrolysis wires are dropped into the induction coil, pyrolyzed, ejected and replaced by a fresh wire.[11]

The Curie-point method of heating is used extensively, but few workers are aware that it is by no means a 'self thermostatting' method for maintaining filaments at their appropriate Curie-point temperatures. Indeed it has been demonstrated that the filament (and certainly substantial parts of it) may never attain the Curie temperature if the RF generator is of low power,[12] and that the filament may exceed the Curie temperature if the generator is of high enough power to obtain fast temperature rise-times.[13]

27.2.1.3 *Resistively heated filaments*

In its simplest form, a resistive filament apparatus consists of a heating coil or ribbon mounted within a chamber at the head of a gas chromatography system. Power may be controlled by means of an autotransformer (Variac) or a rheostat. These simple systems suffer long temperature rise-times, which means that they can be used only for simple analytical applications.

Improved temperature rise-times have been achieved by boosting the current for the initial second of heating.[14] High voltage pulses have also been used to heat the filament rapidly.[15]

Levy[10] has introduced the boosted capacitance discharge method. A high voltage capacitance is discharged across the filament to cause an initial surge of temperature. A lower current is then supplied to maintain the high temperature. Temperature rise-times of 12 ms are quoted, though some of the results suggest that overshooting of the final temperature occurs during the initial high voltage pulse.

Ericsson[16] has described an apparatus using two half-square-wave voltage pulses. The first high voltage pulse is of variable duration (8–80 ms) and rapidly brings the filament up to its operating temperature. The second pulse maintains this temperature for periods of up to one minute.

A further development of the boosted filament is the Chemical Data Systems 'Pyroprobe'.[17] A schematic representation of the circuit diagram is shown in Figure 5. This uses capacitive discharge for rapid temperature rise-times. The pyrolysis filament (R_1) forms one element of a Wheatstone bridge circuit. When the circuit is 'fired', the bridge is unbalanced and a large current passes through the pyrolysis filament. As the filament heats up, its resistance increases until the bridge is balanced. The reduced current now flowing is then able to maintain the filament temperature. Pyrolysis temperatures are selected by varying resistance R_2. Temperature rise-times of 8 ms to 600 °C and 17 ms to 1000 °C have been quoted.

Recently, however, there has been some criticism of these figures. Wells, Futrell and Voorhees[18] looked at temperature rise-times for several pyroprobes using a photo-transistor for temperature measurement. They found that for very short heating periods (20 ms), the temperature rise-times

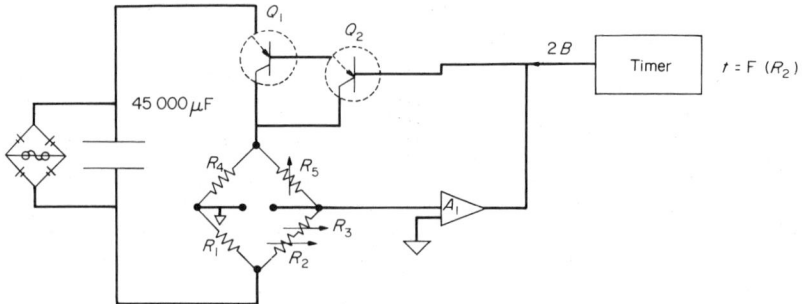

Figure 5 Schematic representation of the 'Pyroprobe' bridge circuit. This seeks to use the discharge of the very large (45 000 µF) condenser to secure fast temperature rise-times [reproduced by permission of the American Chemical Society from *Anal. Chem.*, 1980, **52**, 1783.]

quoted above were applicable. For longer heating periods (20 ms to over 10 s), they found that after the initial rapid temperature rise a second, slower, rise occurred. The difference in temperature between the initial temperature and the final temperature was found to be up to 100 °C. For these longer heating periods, a temperature rise-time of 600 ms was more appropriate.

Probably the most advanced pyrolysis unit to date is that described by Lehrle, Robb and Suggate.[19] This unit uses electronic 'thermocouple feedback control' to give very rapid temperature rise-times and subsequently to control the filament temperature. The unit is illustrated in Figure 6. Very fine thermocouple wires are used to monitor the filament temperature at the point of sample deposition. The thermocouple signal is fed back to the electronic control circuit and is compared with a preset EMF representing the pyrolysis temperature. On 'firing' the filament a very high current (up to 40 A) is supplied which is rapidly reduced as the thermocouple signal approaches the preset EMF. Once the pyrolysis temperature is achieved, the thermocouple continues to monitor and control the filament current and maintain highly accurate temperature control. Temperature rise-times of 50 ms to 800 °C, with temperature control of ±0.5 °C, were displayed on oscillograms.

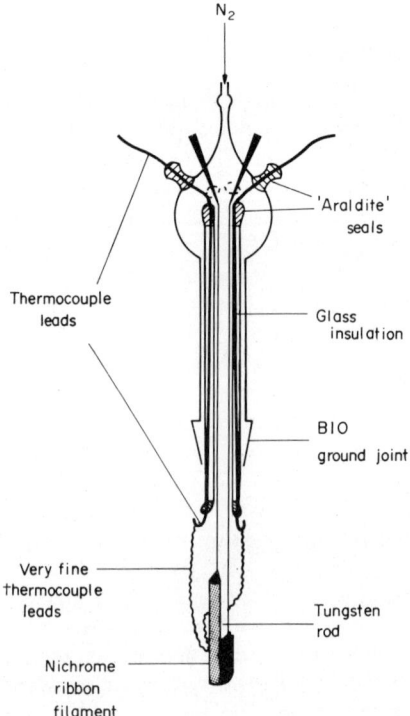

Figure 6 Thermocouple feedback filament. The thermocouple, which is of very low thermal capacity to minimize its response time, is spot welded to that part of the ribbon filament on to which the sample will be deposited [reproduced by permission of Pergamon Press, from *Eur. Polym. J.*, 1982, **18**, 444.]

27.2.1.4 Lasers and other modes of pyrolysis

Lasers have been applied widely for analytical pyrolysis work. A typical example is provided by Fanter, Levy and Wolf[20] who used a pulsed ruby laser to degrade a range of polymers. Temperature rise-times were of the order of microseconds, with mean pyrolysis temperatures in excess of 4000 °C.

One of the major drawbacks of laser pyrolysis is the intense nature of the laser beam. This produces local temperatures in excess of 10 000 °C in a small part of the polymer sample. Such temperatures produce a plasma comprising atoms, electrons, ions and radicals. These plasma products undergo recombination reactions to yield the major product of many laser pyrolyses, acetylene. Products more chracteristic of the polymer structure are produced due to thermal shock caused by collision of the plasma products with the remaining polymer sample. Even further products are formed due to reactions of these products with the acetylene produced in the plasma.

The observed mixture of products is therefore formed by several different fragmentation processes and for this reason high intensity laser pyrolysis can be used only for characterization purposes.

A further problem arises because many materials are transparent to laser radiation. The sample must then be heated by mixing it with an inert absorptive material such as graphite[21] or by coating it on the surface of a cobalt glass rod.[20] However, catalytic effects have been observed, especially when graphite is used.[21]

When defocused lasers are used, more characteristic products are observed. Merrit, Sacher and Petersen[22] used a highly defocused laser to strip the surface of weathered polymers in a study of the surface oxidized region.

Other pyrolytic methods include dielectric breakdown. Barlow, Lehrle and Robb[23] used polymer samples as the dielectric of a high voltage capacitor. The temperatures produced on discharge led to drastic degradation conditions with virtually all polymer types producing large yields of similar degradation products. Only minor yields of products characteristic of the polymer structure were obtained. It was concluded that this technique was of limited utility.

27.2.1.5 A comparison of pyrolysis units

A comparison of pulse mode (filament and Curie-point) and continuous mode (furnace) pyrolyzers has been made by Levy.[24] A summary of his conclusions is given in Table 2.

Table 2 Pulse Mode *vs.* Continuous Mode Pyrolyzers

Pulse mode (*filaments*)	Continuous mode (*furnaces*)
Very small samples can be pyrolyzed	Much larger samples usually employed
Good thermal contact of sample with source of heat. This allows the possibility of bringing the sample rapidly to the required temperature	Slow heat transfer leads to long temperature rise-times. The sample may never reach the temperature of the furnace
Degradation products rapidly swept away from the hot-region	Products may remain in the hot-region for a long time; increased possibility of secondary reactions
The temperature pulse can be accurately reproduced	Very difficult to control the temperature–time profile. The latter may be very sensitive to flow rate

In virtually all respects, furnace methods are inferior to a pulse mode apparatus. However, it should be noted that the apparatus of Tsuge *et al.*[8] (Figure 3) is an example of good furnace design.

The resistively heated filament is the most versatile apparatus since it provides a complete choice of sample temperature. Moreover, advanced units can provide highly accurate monitoring of filament temperature with very rapid temperature rise-times.[19]

Curie-point systems require a different ferromagnetic metal or alloy for each temperature. As the temperature rise-time varies with the composition of the wire and the RF power, it is very difficult to achieve comparable temperature rise-times for a range of temperatures. The position of the Curie-point wire within the induction coil can also lead to anomalous temperature profiles[10, 12] and the temperature attained may not correspond to the Curie-point.[12, 13] These criticisms apart, Curie-point apparatuses are nevertheless widely used.

Lasers and discharge methods are probably the least useful of all the pyrolysis techniques. The lack of temperature control and the extensive fragmentation leads to the technique being restricted mainly to 'fingerprint' comparisons in polymer analysis.

27.2.2 The GLC system

A detailed description of the theory and applications of GLC is beyond the scope of this review. The reader should refer to other sources for more detailed information.[25,26,27]

Early PGLC studies used wide bore, packed chromatography columns for the analysis.[23] However, capillary (open tubular) columns offer far better separation efficiency and resolution (see Figure 7). These are now the preferred column types for most PGLC work.

Capillary columns also have the advantage that they can separate and resolve the very small amounts of material used for kinetic studies (sub-microgram samples). There are however some disadvantages; they can handle *only* very small samples as large product yields from samples over

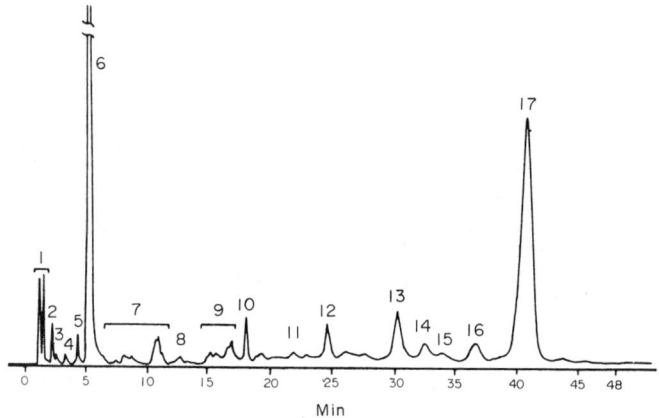

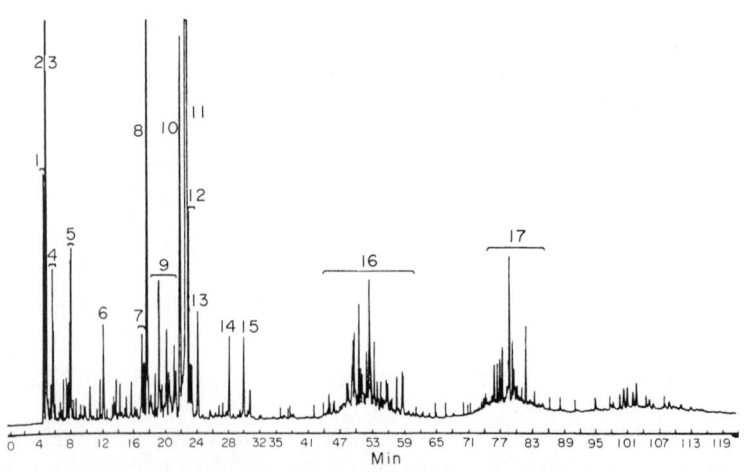

Figure 7 PGLC analyses of a raw natural rubber (SMR 5 CV) obtained using a packed chromatography column (upper chromatogram) and a capillary column (lower chromatogram). These results give an indication of the greater resolving power attainable with capillary columns. (However the comparison is not exact because different stationary liquid phases were used; the capillary column was coated with a silicone gum SE 30, whereas the packed column contained 2.5% of di-2-ethylhexyl sebacate w/w on Chromosorb W supporting phase). The results were obtained by Naveau and Dieu (ref. 65), who assigned the principal peaks 6 and 17 to isoprene monomer and dipentene respectively (upper chromatogram); these compounds appear as peaks 2 and 11 in the lower chromatogram, where the products extend as far as trimers (peaks 16) and tetramers (peaks 17) [reproduced by permission of Elsevier from *J. Anal. Appl. Pyrolysis*, 1980, **2**, pp. 126–127.]

10^{-5} g tend to wash away the column coating. Also, the columns are of very low volume in comparison with the degradation chamber. Since low flow rates are required for optimum separation, this means that it takes long periods to sweep the degradation products from the chamber. This can lead to poor chromatographic resolution.

Both problems are overcome if a *splitting system* is used. Such a system uses high carrier gas flow rates to sweep the degradation products from the chamber. The flow is then split and only a very small proportion of the sample is allowed to enter the chromatography system at much reduced flow rates.

Splitting systems are perfectly adequate for most applications of PGLC. However, they cannot be used for quantitative kinetic studies. This is because such work may require very small samples (10^{-6}–10^{-8} g). It is impossible to detect split products from samples of this size. Also, if repeated degradation experiments are performed, an exactly reproducible split is required each time, and such reproducibility is not attainable with most splitting systems. A further problem is that splitters discriminate in molecular size, preferring to vent the smaller pyrolysis products.

A possible alternative to splitting is to use '*cryogenic focusing*' of the degradation products.[28, 29] A polymer sample is degraded in a rapidly moving carrier gas stream. The degradation products are condensed in a low volume liquid-air cooled trap and the carrier gas is vented. The trap can then be heated up and the degradation products swept directly into the capillary GLC apparatus. Cryogenic focusing provides a technique for studying the *oxidative degradation of polymers*, where oxygen would cause a rapid deterioration in chromatography column performance. Wampler and Levy[30] degraded polyethylene using air as the carrier gas, trapped out the degradation products, vented the air, and then reintroduced the products using an inert carrier gas.

One problem with cryogenic focusing is that the condensing out and reheating of the degradation products leads to the possibility of secondary reactions and the consequent production of new materials uncharacteristic of the pyrolysis process.

It is probably best to use direct (splitless) capillary chromatography using very low volume pyrolysis units. Several successful kinetic studies have been reported using systems of the latter type.

27.2.3 Detection Methods

The detection systems used in PGLC are identical with those used for injected GLC work. Thermal conductivity detectors are typically used for packed column chromatography; flame ionization detectors for capillary GLC.

Peak identification is achieved either by injection of standards, reference to retention indices or by coupling the GLC apparatus to a mass spectrometer. FTIR detectors are also available.[31]

27.3 APPLICATIONS OF PYROLYSIS GLC

27.3.1 Polymer Characterization

The unknown materials are pyrolyzed under specified conditions and the results are compared with those of known samples. There are two principal techniques, outlined below.

27.3.1.1 Single temperature method

The polymer is degraded in a temperature range (usually 500–750 °C) which will give a good variety of characteristic products. The relative retention times and intensities of the peaks are then compared with libraries of characteristic 'fingerprints' of various polymer types. Table 3 lists sources of pyrolysis data for polymer characterization. Typical 'fingerprints' for nitrile rubber degradations are shown in Figure 8 as an example of this approach.

It can be appreciated that with such complex chromatograms, it may often be difficult to distinguish between similar polymeric materials.

Various data handling techniques have been developed to assess the similarity between chromatograms. The subject has been extensively reviewed by Irwin,[32] and a summary of the simpler comparison methods is given in Table 4. More advanced multivariate statistical analysis techniques have also been described[32, 33, 34] but these are usually not needed for polymer comparisons and are more widely used for microbiological analyses.

Table 3   Sources of Characteristic PGLC Fingerprints and Other Data[a]

<table>
<tr><td colspan="4" align="center">*Sources of general PGLC data*</td></tr>
<tr><td>Polymers</td><td>23, 32, 36, 37, 38, 39, 40, 41, 42, 43, 44, 45</td><td>Paints</td><td>52, 53, 54, 55, 56</td></tr>
<tr><td>Rubbers</td><td>35, 46, 47, 48, 49, 50, 51</td><td>Fibres</td><td>57, 58, 59</td></tr>
<tr><td></td><td></td><td>Adhesives</td><td>60</td></tr>
</table>

<table>
<tr><td colspan="4" align="center">*Sources of specific PGLC data*</td></tr>
<tr><td>Polyethylene</td><td>30, 38, 61, 62</td><td>Fluoro polymers</td><td>82</td></tr>
<tr><td>Polypropylene</td><td>63, 64</td><td>Chloro polymers</td><td>83, 84</td></tr>
<tr><td>Polyisoprene</td><td>65, 66, 67</td><td>Polyamides (Nylons)</td><td>85, 86</td></tr>
<tr><td>Polybutadiene</td><td>68, 69, 70, 71, 72</td><td>Aromatic polyesters</td><td>87</td></tr>
<tr><td>PVC/PVA</td><td>73, 74</td><td>Polyethers</td><td>88, 89</td></tr>
<tr><td>Polystyrene/α-methylstyrene</td><td>73, 75, 76, 77, 78</td><td>Polyurethanes</td><td>90, 91, 92</td></tr>
<tr><td>Poly(methylmethacrylate)</td><td>79, 80</td><td>Phenolic resins</td><td>93</td></tr>
<tr><td>Polyacrylonitrile</td><td>19, 81</td><td>Polycarbonates</td><td>94</td></tr>
</table>

[a] Numbers refer to references at end of chapter.

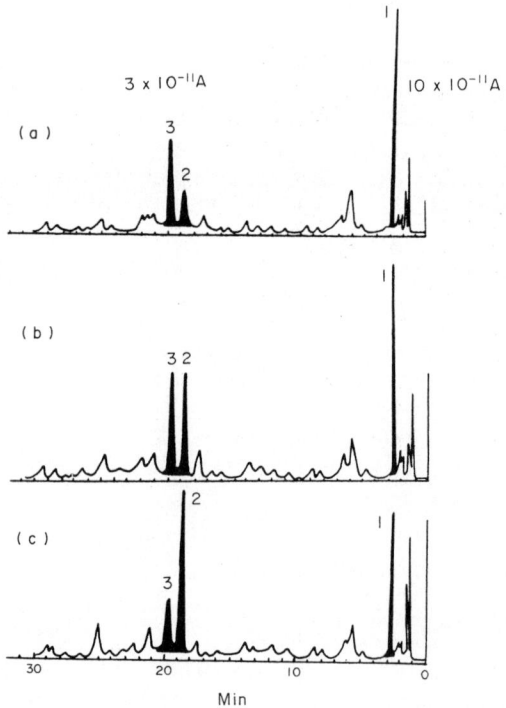

Figure 8   Pyrograms of nitrile rubbers: (a) type SKN-18, (b) type SKN-26 and (c) type SKN-40. Single temperature pyrolyses for 10 seconds at 770 °C. (The column temperature was programmed from 40 °C to 150 °C during each chromatographic analysis.) Peaks 1, 2 and 3 were assigned as butadiene, acrylonitrile and vinylcyclohexene respectively [reproduced by permission of Elsevier from *J. Anal. Appl. Pyrolysis*, 1980, **2**, 32.]

A slightly different technique for polymer blend characterization has been described by Alekseeva.[35] Peaks are selected which are absolutely unique for a particular polymer within a defined range of polymers. These peaks therefore demonstrate the presence of that polymer. This technique has been used to characterize blends of isoprene, butadiene and nitrile rubbers.

27.3.1.2   Sequential degradations (temperature-stepped pyrolysis)

A polymer sample is heated for a few seconds at a sequence of temperatures and the chromatogram recorded at each temperature.[23] This produces a series of chromatograms reflecting the polymer stability and the nature of the degradation products, as shown in Figure 9.

Table 4 Statistical Techniques for Pyrogram Comparison

Definition	Parameters	Notes
Similarity value $S_{i,j}$ Counts the proportion of peaks common to a pair of pyrograms i,j $$S_{i,j} = \frac{100 N_s}{N_s + N_D}$$	N_s = number of peaks common to pyrograms i and j N_D = number of unique peaks in the two pyrograms	Retention time window of 1% allowed Pyrograms identical if $S_{i,j} = 100\%$
Similarity coefficient $S'_{i,j}$ Compares peak intensities in pyrograms i,j $$S'_{i,j} = \sum_{k=1}^{N} \left(\frac{I_i^k}{I_i^k} \right)$$	I_i^k = intensity of kth peak in pyrogram i I_j^k = intensity of kth peak in pyrogram j N = number of peaks in the data set	Quotient arranged so that $I_i^k < I_j^k$ Perfect match if $S'_{i,j} = 1.0$ Values above 0.84 for 13–15 peaks usually mean that chromatograms are identical
Fit factor $F_{i,j}$ For peaks $k = 1$ to pyrograms i and j $$F_{i,j} = 10^3 \left[1 - \frac{\sum_{k=1}^{k=N_2}(I_i^k - I_j^k)^2}{\sum_{k=1}^{k=N}[(I_i^k)^2 + (I_j^k)^2]} \right]$$	As for similarity coefficient	Perfect match: $F_{i,j} = 1000$; typical value for identical pyrograms $F_{i,j} = 975$
Spearman rank order coefficient $r_{i,j}$ $$r_{i,j} = 1 - \frac{6\sum_{k=1}^{N}(R_i^k - R_j^k)^2}{N(N^2 - 1)}$$	The peaks are numbered according to the rank order of their intensities R_i, R_j N = number of peaks in the data set	$r_{i,j} = +1$ pyrograms identical $r_{i,j} = -1$ for maximum deviation

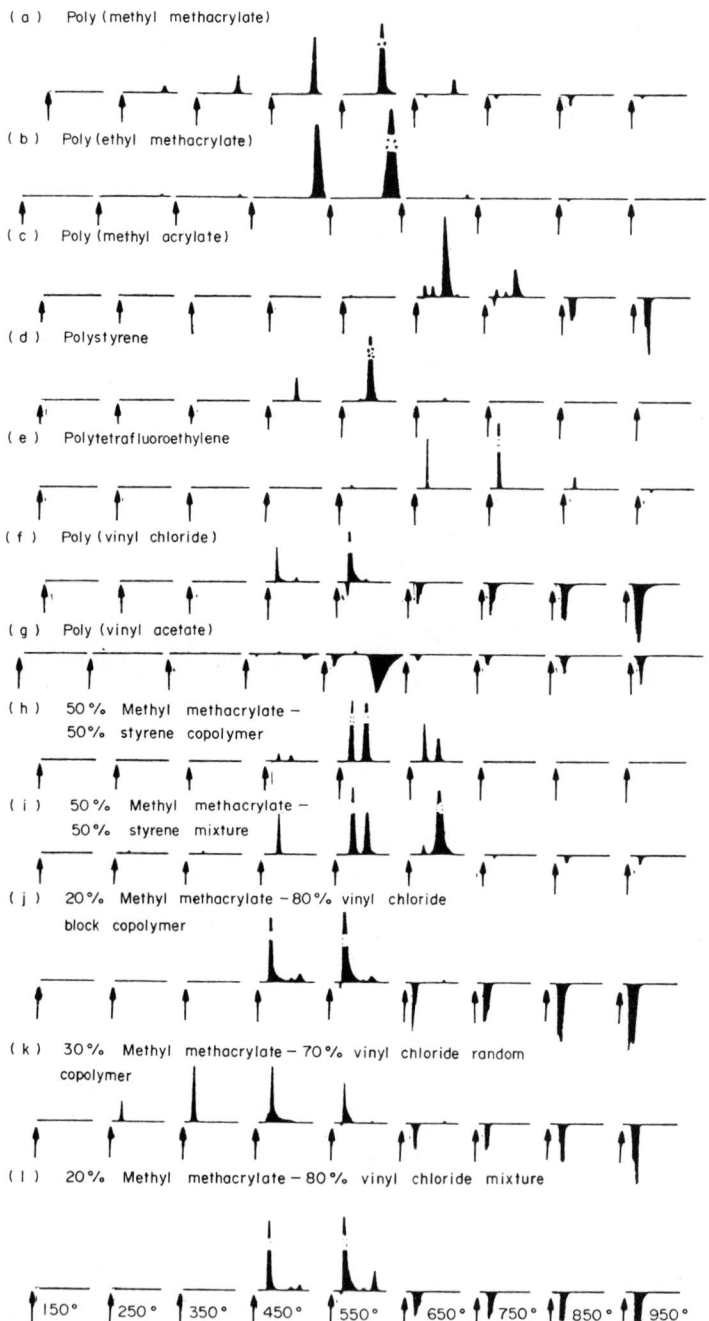

Figure 9 Sequential degradation chromatograms (temperature-stepped filament pyrolyses). This method is by far the best for characterizing samples because, in addition to product retention times, the temperature performance of the sample is used as an additional characterizing variable. The results provide an indication of the relative thermal stabilities of the samples, and may distinguish between samples which would otherwise require careful spectroscopic techniques (*e.g.* the last three sequences show that a block copolymer has been distinguished from the corresponding random copolymer) [reproduced by permission of Butterworths from *Polymer*, 1961, **2**, 39.]

This is a much preferred method of polymer identification, since even if two polymers yield degradation products of similar retention time, they are very unlikely to exhibit identical thermal stability.

27.3.2 Compositional Analysis of Polymer Mixtures and Copolymers

27.3.2.1 Mixture analysis

Once a polymer mixture has been identified, the composition may be assessed.

A characteristic GLC peak is selected for each component and a suitable pyrolysis temperature is selected to give the maximum yield of that product. Calibration curves are constructed by degrading known masses of the pure components at the selected temperature, and plots of sample mass against peak area are made. The peak areas for the characteristic peaks in the mixture can therefore be related to masses of pure components. It is important to check that the sum of the determined masses is equal to the mass of polymer mixture deposited on the filament. Any discrepancies which may arise could be due to plasticizer or filler, or to synergistic effects on the degradation processes.

27.3.2.2 Copolymer analysis

Copolymers cannot always be analyzed using the method described above. For example, methyl acrylate copolymerized with styrene produces lower yields of monomer than the homopolymer at the same temperature.[14] This is due to increased stability caused by the comonomer. In such cases, calibrations must be made using copolymers of known composition.

Copolymer composition analysis methods using PGLC, IR and chlorine estimation methods have been compared.[23] The results are shown in Table 5 and indicate that the PGLC method gives comparable results. Since the method is performed with greater facility than the alternative methods, it is now finding extensive use.

Table 5 Copolymer Composition Analysis by PGLC,
Compared with Other Methods

	% Vinyl chloride in vinyl chloride/vinyl acetate copolymers		
Copolymer	*By PGLC* *(±2%)*	*By IR analysis* *(±1%)*	*By chlorine estimation* *(mean of 2 results)*
049	55.8	54.7	60.8 ± 0.6
047	65.2	64.4	69.4 ± 0.1
075	72.2	72.3	74.1 ± 0.3
076	67.8	66.7	69.1 ± 0.9
R46/82	83.9	84.8	81.8 ± 4.4
R51/83	87.7	89.0	87.9 ± 1.0

27.3.3 The Microstructure of Copolymers

An example of such a study is the work of Shimono, Tanaka and Shono on the detailed chain structure of styrene(S)–methyl methacrylate (MMA) copolymers.[95, 96]

Very high resolution capillary chromatography was used to look at the small yields of dimers and trimers produced during the polymer degradation. Figure 10 illustrates the pyrolysis product chromatograms from the homopolymers and from the random and alternating copolymers. Temperature programmes from 120 to 250 °C were used in the GLC analysis. The results show neither dimer nor trimer peaks for the pyrolysis of poly(methyl methacrylate) whereas for polystyrene pyrolysis there is a single dimer (D_2) and a single trimer (T_4). This dimer was identified as 2,4-diphenyl-1-butene, and the trimer as 2,4,6-triphenyl-1-hexene. In the pyrolysis of the alternating copolymer an additional dimer (D_1) was characterized from its mass spectrum as the S–MMA hybrid, and the trimer T_3 was characterized as the S–MMA–S hybrid. The pyrolysis products from the random copolymer included all the products mentioned above, together with additional trimer peaks T_1 and T_2 which were were both attributed to S–S–MMA hybrid trimers.

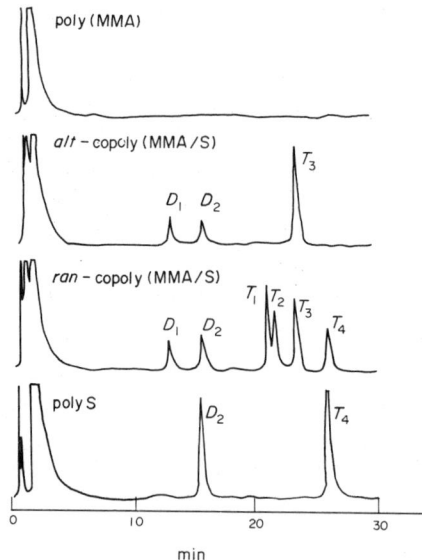

Figure 10 560 °C pyrograms of (from the top) polymers and copolymers from methyl methacrylate (MMA) and styrene (S). The significance of the peak assignments is discussed in the text [reproduced by permission of Elsevier from *J. Anal. Appl. Pyrolysis*, 1979, **1**, 79.

By assuming that the ratios of the dimer peak areas are proportional to the ratios of the corresponding diad concentrations, the authors claim that 'copolymerization theory' can be used to calculate the run number, *i.e.* the average number of monomer sequences (runs) per 100 units for each polymer. For the alternating copolymer a value of 61 was obtained, which is less than the expected value of 100. However, when the trimer areas were analyzed in terms of triads in a comparable way, a run number of 98 was obtained for the alternating copolymer, and acceptable low values were obtained for the random copolymers. The calculated run numbers appear to depend on the fitting of some adjustable parameters however.

27.3.4 Structural Analysis

Some work of Shimono *et al.*[97] exemplifies the way in which PGLC can be applied in structural analysis. 3-Methyl-1-butene was polymerized using two different catalyst systems and different polymers were obtained.

Catalyst system (a): $Al(Bu^i)_3/TiCl_3/n\text{-}C_7H_{16}/15\,°C$

Catalyst system (b): $AlCl_3/C_2H_5Cl/-78\,°C$

Two possible polymer structures were proposed (Scheme 1):

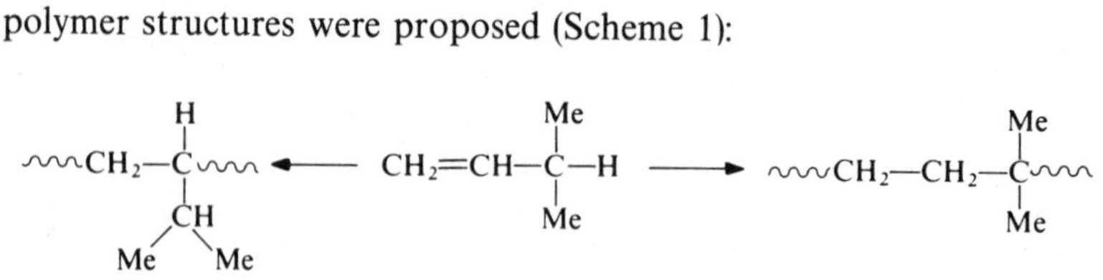

$$\text{$\sim\!\!\sim\!\!\sim$CH}_2\!-\!\underset{\underset{\displaystyle \underset{Me}{\diagdown}\!\!/Me}{\overset{|}{CH}}}{\overset{H}{C}}\!\!\sim\!\!\sim\!\!\sim \;\longleftarrow\; \text{CH}_2\!\!=\!\!\text{CH}\!-\!\underset{Me}{\overset{Me}{\underset{|}{\overset{|}{C}}}}\!-\!\text{H} \;\longrightarrow\; \sim\!\!\sim\!\!\text{CH}_2\!-\!\text{CH}_2\!-\!\underset{Me}{\overset{Me}{\underset{|}{\overset{|}{C}}}}\!\!\sim\!\!\sim\!\!\sim$$

Scheme 1

PGLC analysis of the polymer obtained using catalyst (b) indicated principally monomer and dimer peaks. However, the polymer obtained using catalyst (a) yielded smaller amounts of these materials with much larger yields of C_3 hydrocarbon. The polymer produced by system (a) was therefore assigned the structure with the C_3 side chains.

PGLC has also been used to examine the tacticity of polymers. In a study of polypropylenes,[98] in-line hydrogenation of the pyrolysis products was used to simplify the pyrograms by removing unsaturation. Such pyrograms from isotactic, syndiotactic and atactic polypropylene were examined in detail with respect to peaks corresponding to trimers, tetramers and pentamers in order

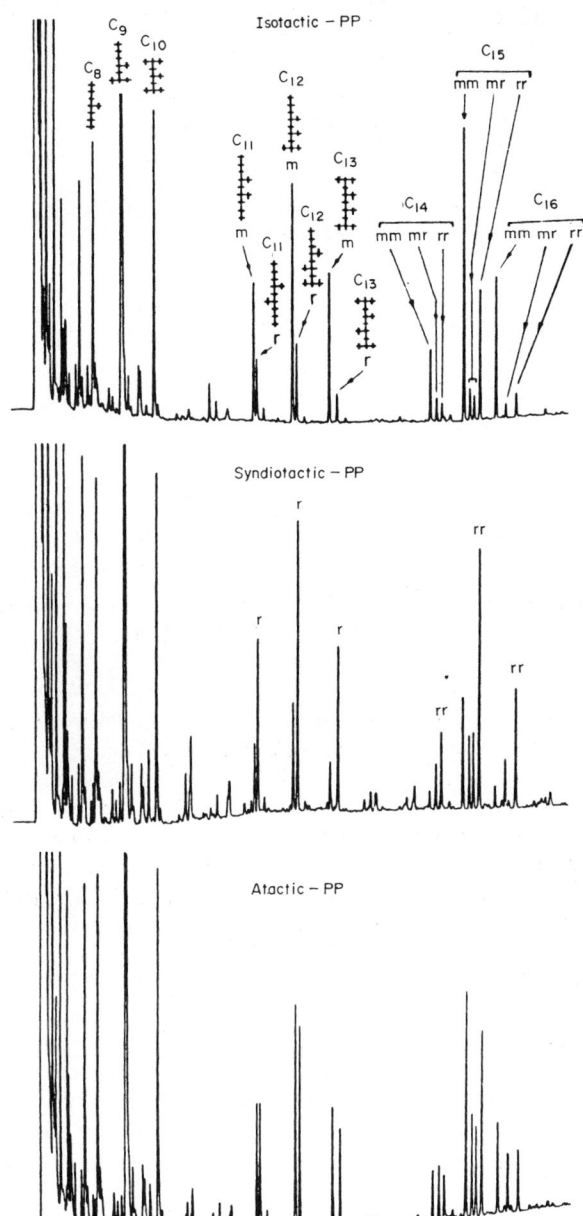

Figure 11 High resolution pyrograms of isotactic, syndiotactic and atactic polypropylenes up to pentamer regions. In-line hydrogenation of the pyrolysis products (*i.e.* removal of the unsaturation) was used to simplify the chromatograms [reproduced by permission of Butterworths from '*Analytical Pyrolysis Techniques and Applications*' (K. J. Voorhees, ed.), London, 1984.]

to assess their stereoregularity. This can be achieved because high resolution chromatography is able to separate isomeric products.

Taking the C_{12} products for example (see Figure 11), the two peaks can be attributed to (m), a *meso* form (**1**) characteristic of isotactic polypropylene, and (r), a racemic form (**2**) characteristic of syndiotactic polypropylene. Atactic polypropylene degrades to yield both isomers in roughly equal proportions.

27.3.5 Analysis of Cross-linked Networks

Pyrolyses of radiation cross-linked polybutadienes have indicated that the production of certain characteristic degradation products is dependent on the degree of cross-linking.[69] Calibration using known standards can lead to the determination of cross-link densities in uncharacterized materials.

27.3.6 Mechanistic Studies of Polymer Degradation

Many workers have used PGLC to look at polymer degradation products and, in combination with other kinetic techniques (*e.g.* DTA, TGA, TVA, *etc.*), have assigned degradation mechanisms. Some outstanding recent studies have been made on polystyrene,[77] polyamides[85] and polypropylene.[63]

PGLC, however, can be used in its own right to provide values of kinetic parameters (rate constants and activation energies) and other quantitative data which can lead to the proposal of degradation mechanisms.

For kinetic pyrolysis work, there are two important conditions which must be satisfied:

(i) Polymer samples must be very small. This is because large polymer samples restrict product diffusion and encourage secondary reactions within the polymer melt. Large samples also retard the temperature rise-time and develop large temperature gradients. Kinetic studies must therefore operate within a sample size region where the rate of polymer degradation is independent of sample size. It has been demonstrated, for example, that sample thicknesses below 250 Å for PPMA[99] and 450 Å for PS fit this criterion.[75]

(ii) Temperature rise-times must be very rapid. Since rate constants are temperature dependent it is essential that the degradation temperature is achieved virtually instantaneously.

27.3.6.1 Rate measurements using PGLC

Polymers are degraded at a specific temperature for a given duration. The resulting chromatogram gives the yield of degradation products produced in this period.

Three possible methods can be used to give the rate of formation of volatile degradation products.

(i) Method (a)

The same polymer sample is repeatedly pyrolyzed at a given temperature for a chosen duration. This gives a series of chromatograms and allows cumulative rate curves to be drawn for the individual products. A typical example of this technique is the work of Ericsson[100] on copolymer stability. This technique has several drawbacks, however. Repeated heating introduces cumulative errors in temperature and pyrolysis time. There is also the possibility of conditioning the polymer and producing a more stable variant (for example, polyacrylonitrile is stabilized by heating) which will produce anomalous kinetic results. The technique is not recommended for precise kinetic work.

(ii) Method (b)

This technique is used for polymers which degrade to a limiting yield (*e.g.* polyacrylonitrile).[19, 81] Identical masses of polymer are subjected to a single pyrolysis at a specified temperature for a series of pyrolysis times. The yields of pyrolysis products are expressed as a function of the degradation period and the limiting yields found (see Figure 12).

We define a = peak area corresponding to the volatiles evolved from the pyrolysis of chosen duration t, and a_∞ = total available yield of product ($t = \infty$). If the production of volatiles is first order with respect to the remaining polymer residue, then

$$\frac{da}{dt} = k_{obs} (a_\infty - a) \tag{1}$$

Where k_{obs} is an observed rate constant for the process. The integrated form can be derived as

$$-\ln\left(1 - \frac{a}{a_\infty}\right) = k_{obs} t \tag{2}$$

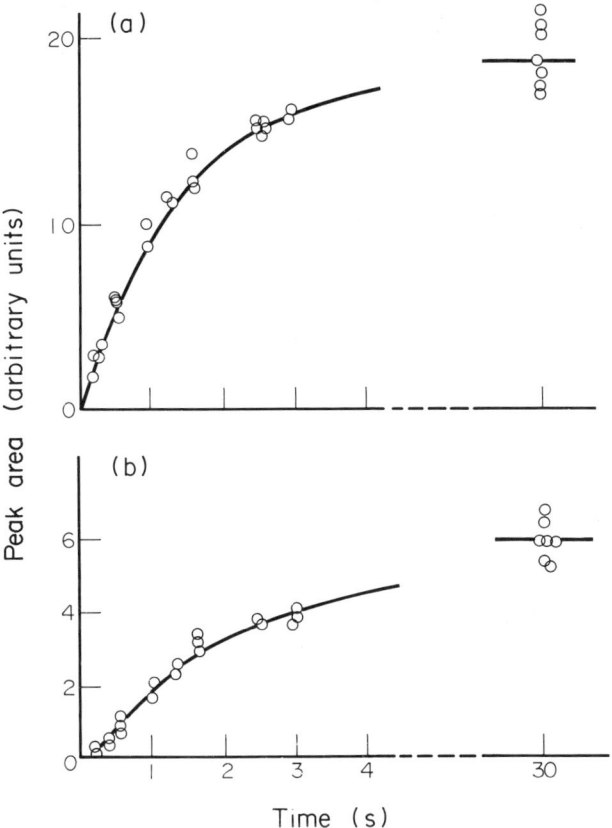

Figure 12 Conversion curves (product yield *vs.* time of pyrolysis) of products from polyacrylonitrile degrdations at 380 °C. (a) Monomer formation, (b) methacrylonitrile formation [reproduced by permission of Pergamon Press from *Eur. Polym. J.*, 1982, **18**, 452.]

If the first order relationship holds, a plot of the left-hand side of equation (2) against pyrolysis time should yield a linear plot passing through the origin. Figure 13 shows a plot obtained by Lehrle *et al.*[19] for polyacrylonitrile degradation. The gradient is the observed rate constant for the process.

(iii) Method (c)

This method can also be used for polymers which give a limiting yield of any product, but has been most often used for polymers which degrade quantitatively to monomer [for example, poly(methyl methacrylate]. An initial pyrolysis is made at the specified temperature to yield a peak of area a_1. The remaining polymer residue is then completely degraded to yield a second peak (area a_2).

The fractional conversion for the degradation can be calculated as

$$x = \frac{m}{m_o} = \frac{a_1}{a_1 + a_2} \tag{3}$$

where m = mass of monomer evolved in time t and m_0 = total mass of available monomer.

A similar first order equation can be applied where

$$\frac{dm}{dt} = k_{obs}(m_o - m) \tag{4}$$

which on integration becomes

$$-\ln(1 - x) = k_{obs} t \tag{5}$$

Figures 14(a) and 14(b) of data obtained in this way illustrate plots for poly(methyl methacrylate)[79] and polystyrene[75] respectively.

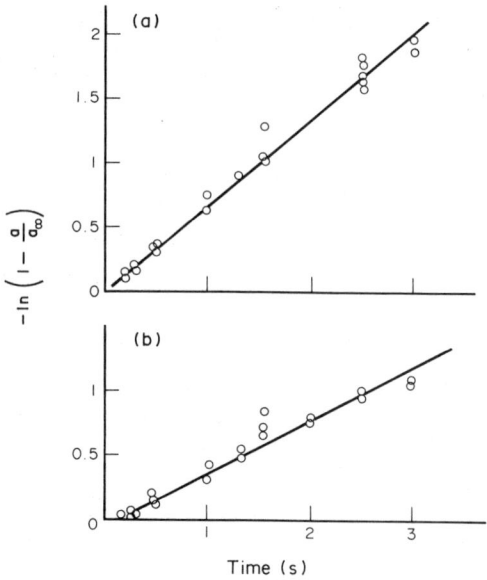

Figure 13   First order plots for products evolved from polyacrylonitrile pyrolysis at 380 °C. (a) Monomer, (b) metha-crylonitrile [reproduced by permission of Pergamon Press from *Eur. Polym. J.*, 1982, **18**, 452.]

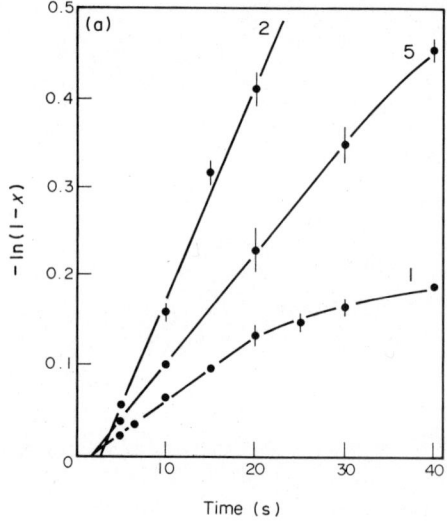

Figure 14(a)   First order plots of monomer yield from poly(MMA) pyrolyzed at three temperatures: 340 °C (plot 1), 370 °C (plot 5) and 416 °C (plot 2) [reproduced by permission of Butterworths from *Polymer*, 1967, **8**, 540.]

It can be seen that the plots deviate from linearity at high polymer conversions. This is because the first-order rate equation applies only for initial degradation rates, *i.e.* on the original polymer. At high conversions, the nature of the polymer residue has changed and the first-order equation ceases to apply. It is important that any kinetic studies are made within the linear region of this plot.

27.3.6.2   *Assignment of degradation mechanism*

In the case of polymers where monomer is the principal degradation product, the following degradation processes may be proposed:

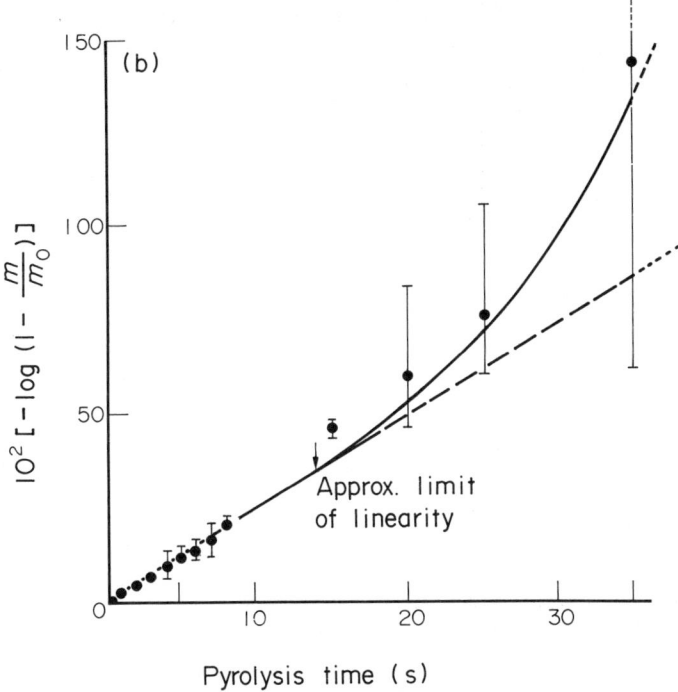

Figure 14(b) First order plot of monomer yield from polystyrene pyrolysis at 480 °C [reproduced by permission of Pergamon Press from *Eur. Polym. J.*, 1982, **18**, 521.]

(i) *Initiation* — the formation of a polymer radical. Either (a) end initiation

$$P \xrightarrow{k_i} R\cdot + X\cdot$$

or (b) scission initiation

$$P \xrightarrow{k_i'} 2R\cdot$$

can occur.

(ii) *Depropagation* — the loss of monomer units from the radical.

$$R\cdot \xrightarrow{k_d} R\cdot + M$$

When this occurs as a chain reaction, the progressive loss of monomer is often called 'unzipping'.

(iii) *Termination* — the removal of the radical from the system. Either (a) 'unzip to end of chain' — the polymer depropagates until there are no further units in the chain and the final small radical is volatilized from the system, or (b) bimolecular second-order termination,

$$R\cdot + R\cdot \xrightarrow{k_t} \text{'dead' polymer}$$

or (c) first-order termination (germinate, or certain transfer mechanisms),

$$R\cdot \xrightarrow{k_t'} \text{'dead' polymer.}$$

Kinetic schemes can be devised for these possible mechanisms[75, 79, 80] and the observed rate constant can be interpreted in terms of the rate constants for the various initiation and termination processes. Table 6 lists the appropriate expressions.

It can be seen that the way in which the observed rate constant depends on the initial degree of polymerization (D_0) is characteristic of the degradation mechanism (*i.e.* of the initiation and termination processes). Figures 15(a) and (b) illustrate this dependence for poly(methyl methacrylate) possessing laurylmercaptyl end groups.[80]

Table 6 Dependence of Observed Rate Constants on Initial Degree of Polymerization

Initiation	Termination	k_{obs}	Dependence of k_{obs} on D_o	
Chain end	Depropagation to end of chain	k_i	k_{obs} independent of D_o	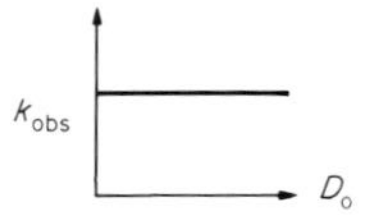
Random scission	"	$k_i' D_o$	$k_{obs} \propto D_o$	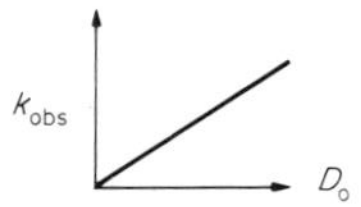
Chain end	Second order	$\left(\dfrac{k_i u}{2 k_t \rho D_o}\right)^{\frac{1}{2}} k_d$	$k_{obs} \propto \dfrac{1}{D_o^{\frac{1}{2}}}$	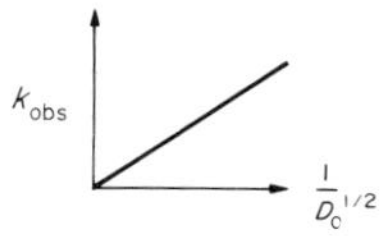
Random scission	"	$\left(\dfrac{k_i' u}{k_t \rho}\right)^{\frac{1}{2}}$	k_{obs} independent of D_o	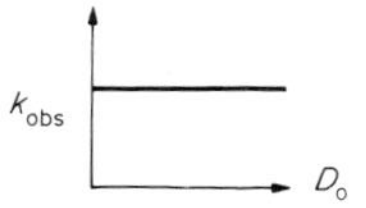

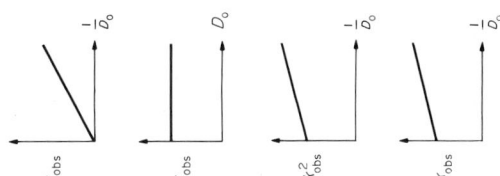

Chain end	First order	$\dfrac{k_d k_i}{k_t' D_o}$	$k_{obs} \propto \dfrac{1}{D_o}$
Random scission	"	$\dfrac{2 k_i' k_d}{k_t'}$	k_{obs} independent of D_o
Chain end and random scission	Second order	$\left(\dfrac{k_i + 2k_i'}{D_o}\right)^{\frac{1}{2}} \dfrac{k_d u^{\frac{1}{2}}}{(2k_t \rho)^{\frac{1}{2}}} = k_{obs}^2$	$k_{obs}^2 \propto \dfrac{A}{D_o} + B$
Chain end and random scission	First order	$\left(\dfrac{k_i + 2k_i'}{D_o}\right) \dfrac{k_d}{k_t'}$	$k_{obs} \propto \dfrac{X}{D_o} + Y$

u = molecular weight of monomer ρ = polymer density D_o = initial degree of polymerization

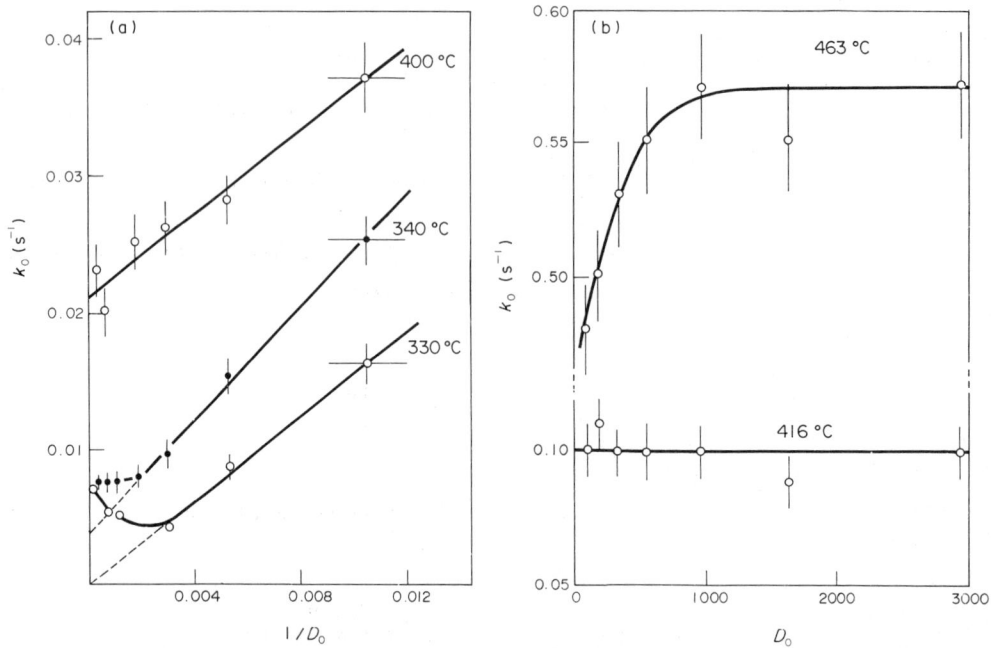

Figure 15 Monomer evolution from poly(MMA) pyrolysis: dependence of initial specific rate (k_0) on initial degree of polymerization (D_0) at five temperatures over the range 330–463 °C [reproduced by permission of Butterworths from *Polymer*, 1969, **10**, 686 and 687.]

The predominant mechanism was shown to change with temperature as outlined in the succeeding paragraphs.

At low temperatures k_{obs} is proportional to $1/D_0$; this implies end initiation and unimolecular termination.

At intermediate temperatures this situation still pertains, but the large intercept indicates that scission initiation is starting to be significant.

At high temperatures, scission initiation becomes the predominant initiation mechanism (k_{obs} is independent of D_0 at 416 °C). At higher temperatures (463 °C) there is a trend in the mechanism with molecular weight. Low molecular weight materials (k_{obs} proportional to D_0) undergo scission and then depropagate to the end of the chain. Higher molecular weight materials exhibit scission initiation with termination during unzip.

Studies on poly(methyl methacrylate) with unsaturated end groups also indicated changes in the predominant mechanisms with temperature.[79] At low and intermediate temperatures, chain-end initiation with termination during unzip is the predominant mechanism. As the temperature increases, scission initiation takes over and termination is by unzipping to end of chain.

Studies on polystyrene[75] have assigned a low temperature mechanism of principally end initiation, with increasing scission at high temperatures. A first order termination was proposed from an interpretation of overall activation energies, and this was supported by the presence of small amounts of oligomeric products amongst the degradation products.

Polyacrylonitrile degradation has been studied using the limiting yields method.[19,81] Rate constants for the production of the various degradation products were calculated and activation energies assigned from Arrhenius plots of log k_{obs} against $1/T$. Mechanisms consistent with these results were proposed.

27.4   REFERENCES

1. A. T. James and A. J. P. Martin, *The Analyst*, 1952, **77**, 915.
2. W. H. T. Davison, S. Slaney and A. L. Wragg, *Chem. Ind. (London)*, 1954, 1356.
3. E. A. Radell and H. C. Strutz, *Anal. Chem.*, 1959, **31**, 1890.
4. R. S. Lehrle and J. C. Robb, *Nature (London)*, 1959, **183**, 1671.
5. S. B. Martin, *J. Chromatogr.*, 1959, **2**, 272.
6. J. Romováček and J. Kubát, *Anal. Chem.*, 1968, **40**, 1119.

7. C. B. Honaker and A. D. Horton, *J. Gas Chromatogr.*, 1965, **3**, 396.
8. S. Tsuge and T. Takeuchi, *Anal. Chem.*, 1977, **49**, 348.
9. J. P. Schmid, P. P. Schmid and W. Simon, *Chromatographia*, 1976, **9**, 597.
10. R. L. Levy, D. L. Fanter and C. J. Wolf, *Anal. Chem.*, 1972, **44**, 38.
11. H. L. C. Meuzelaar, H. J. Ficke and H. C. Den Harink, *J. Chromatogr. Sci.*, 1975, **13**, 12.
12. R. S. Lehrle, *Lab. Pract.*, 1968, **17**, 696.
13. G. Bagby, Ph.D. Thesis, University of Birmingham, UK, 1968.
14. A. Barlow, R. S. Lehrle and J. C. Robb, 'SCI Monograph No. 17', Society of Chemical Industry, London, 1963, p. 267.
15. J. A. Cogliano, *Rev. Sci. Instrum.*, 1963, **34**, 439.
16. I. Tyden-Ericsson, *Chromatographia*, 1973, **6**, 353.
17. A. J. Martin, S. F. Sarner, O. R. Averitt, G. D. Pruder and E. J. Levy, 'Pyroprobe 100 series', Chemical Data Systems Inc., Oxford, Pennsylvania.
18. G. Wells, K. J. Voorhees and J. H. Futrell, *Anal. Chem.*, 1980, **52**, 1782.
19. R. S. Lehrle, J. C. Robb and J. R. Suggate, *Eur. Polym. J.*, 1982, **18**, 443.
20. D. L. Fanter, R. L. Levy and C. J. Wolf, *Anal. Chem.*, 1972, **44**, 43.
21. O. F. Folmer, Jr., *Anal. Chem.*, 1971, **43**, 1057.
22. C. Merrit, R. E. Sacher and B. A. Petersen, *J. Chromatogr.*, 1974, **99**, 301.
23. A. Barlow, R. S. Lehrle and J. C. Robb, *Polymer*, 1961, **2**, 27.
24. R. L. Levy, *J. Gas Chromatogr.*, 1967, **5**, 107.
25. J. Novak, *Adv. Chromatogr.*, 1974, **11**, 1.
26. D. A. Leathard, *Adv. Chromatogr.*, 1975, **13**, 265.
27. J. B. Pattison, 'A Programmed Introduction to Gas-liquid Chromatography', 2nd edn., Heyden, London, 1973.
28. T. P. Wampler and E. J. Levy, *J. Anal. Appl. Pyrolysis*, 1985, **8**, 65.
29. R. J. Lloyd, *J. Chromatogr.*, 1984, **284**, 357.
30. T. P. Wampler and E. J. Levy, *J. Anal. Appl. Pyrolysis*, 1985, **8**, 153.
31. S. A. Liebman, D. H. Ahlstrom and P. R. Griffiths, *Appl. Spectrosc.*, 1976, **30**, 335.
32. W. J. Irwin, 'Analytical Pyrolysis, A Comprehensive Guide', Dekker, New York, 1982.
33. G. Blomquist, *J. Anal. Appl. Pyrolysis*, 1979, **1**, 53.
34. E. Küllik, M. Kaljurand and M. Koel, *J. Chromatogr.*, 1976, **126**, 249.
35. K. V. Alekseeva, *J. Anal. Appl. Pyrolysis*, 1980, **2**, 19.
36. C. G. Smith, N. E. Skelly, R. A. Solomon and C. D. Chow, 'CRC Handbook of Chromatography: Polymers Volume 1', CRC Press, Boca Raton, FL, 1982.
37. B. R. Northmore, *Br. Polym. J.*, 1972, **4**, 511.
38. F. W. Willmot, *J. Chromatogr. Sci.*, 1969, **7**, 101.
39. G. Di Pasquale and T. Capaccioli, *J. Chromatogr.*, 1983, **279**, 151.
40. B. C. Cox and B. Ellis, *Anal. Chem.*, 1964, **36**, 90.
41. B. Groten, *Anal. Chem.*, 1964, **36**, 1206.
42. D. F. Nelson, J. L. Yee and P. L. Kirk, *Microchem. J.*, 1962, **6**, 225.
43. N. Iglauer and F. F. Bentley, *J. Chromatogr. Sci.*, 1974, **12**, 23.
44. J. E. Coakley and H. H. Berry, US Tech. Nat. Info. Service (15:8:1971), NAFITR-1713, p. 115.
45. R. W. May, E. F. Pearson and D. Scothern, 'Pyrolysis-gas Chromatography', Chemical Society, London, 1977.
46. E. A. Ney and A. B. Heath, *J. Inst. Rubber Ind.*, 1968, **2**, 276.
47. A. Krishen, *Anal. Chem.*, 1972, **44**, 494.
48. A. A. Foxton, D. E. Hillman and P. M. Mears, *J. Inst. Rubber Ind.*, 1969, 179.
49. R. N. Thompson, C. A. Nau and C. H. Lawrence, *Am. Ind. Hyg. Assoc. J.*, 1966, 488.
50. M. H. Cole, D. L. Petterson, V. A. Sljaka and D. S. Smith, *Rubber Chem. Technol.*, 1966, **39**, 259.
51. J. L. Wuepper, *Anal. Chem.*, 1979, **51**, 997.
52. B. B. Wheals and W. Noble, *J. Forensic Sci. Soc.*, 1974, **14**, 23.
53. W. D. Stewart, *J. Forensic Sci. Soc.*, 1974, **19**, 121.
54. G. G. Esposito, *Anal. Chem.*, 1964, **36**, 2183.
55. A. Berton, *Chim. Anal. (Paris)*, 1965, **47**, 502.
56. N. C. Jain, C. R. Foutan and P. L. Kirk, *J. Forensic Sci. Soc.*, 1965, **5**, 102.
57. J. Derminot and C. Rabourdin-Belin, *Bull. Inst. Text. Fr.*, 1971, **25**, 721.
58. J. P. Bortniak, S. E. Brown and E. H. Sild, *J. Forensic Sci. Soc.*, 1971, **16**, 380.
59. R. A. Janiak and K. A. Damerau, *J. Crim. Law, Criminol. Police Sci.*, 1968, **59**, 434.
60. N. L. Bakowski, E. C. Bender and T. O. Munson, *J. Anal. Appl. Pyrolysis*, 1985, **8**, 483.
61. S. Tsuge, Y. Sugimura and T. Nagaya, *J. Anal. Appl. Pyrolysis*, 1980, **1**, 221.
62. I. Michajlov P. Zugenmaier and H. J. Cantow, *Polymer*, 1968, **9**, 325.
63. M. T. Sousa Pessoa de Amorim, C. Comel and P. Vermande, *J. Anal. Appl. Pyrolysis*, 1982, **4**, 73.
64. D. Deur-Šiftar and V. Švob, *J. Chromatogr.*, 1970, **51**, 59.
65. J. Naveau and H. Dieu, *J. Anal. Appl. Pyrolysis*, 1980, **2**, 123.
66. M. Galin-Vacherot, H. Eustache and Pham. Quang Tho, *Eur. Polym. J.*, 1969, **5**, 211.
67. M. Galin-Vacherot, *Eur. Polym. J.*, 1971, **7**, 1455.
68. E. M. Andersson and I. Ericsson, *J. Anal. Appl. Pyrolysis*, 1979, **1**, 27.
69. K. G. Häusler, E. Schröder and B. Huster, *J. Anal. Appl. Pyrolysis*, 1980, **2**, 109.
70. M. Host and D. Deur-Sifter, *Chromatographia*, 1972, **5**, 502.
71. D. Braun and E. Čanji, *Angew. Makromol. Chem.*, 1973, **29**, 491.
72. T. Shono and K. Shinra, *Anal. Chim. Acta*, 1971, **56**, 303.
73. A. Alajbeg, P. Arpino, D. Deur-Sifter and G. Guiochon, *J. Anal. Appl. Pyrolysis*, 1980, **1**, 203.
74. R. S. Lehrle and J. C. Robb, *J. Gas Chromatogr.*, 1967, **5**, 89.
75. R. S. Lehrle, R. E. Peakman and J. C. Robb, *Eur. Polym. J.*, 1982, **18**, 517.
76. M. Tanaka, T. Shimono, Y. Yakubi and T. Shono, *J. Anal. Appl. Pyrolysis*, 1980, **2**, 207.
77. M. T. Sousa Pessoa de Amorim, C. Bouster, P. Vermonde and J. Vernon, *J. Anal. Appl. Pyrolysis*, 1981, **3**, 19.

78. A. Alajbeg and B. Stipak, *J. Anal. Appl. Pyrolysis*, 1985, **7**, 283.
79. A. Barlow, R. S. Lehrle, J. C. Robb and D. Sunderland, *Polymer*, 1967, **8**, 537.
80. G. Bagby, R. S. Lehrle and J. C. Robb, *Polymer*, 1969, **10**, 683.
81. F. A. Bell, R. S. Lehrle and J. C. Robb, *Polymer*, 1971, **12**, 579.
82. H. J. Kretzschmar, D. Gross and J. Kelm, in 'Analytical Pyrolysis', ed. C. E. R. Jones, C. A. Cramers, Elsevier, Amsterdam, 1977, p. 373.
83. S. Tsuge, T. T. Okumoto and T. Takeuchi, *Makromol. Chem.*, 1969, **123**, 123.
84. S. Tsuge, T. Okumoto and T. Takeuchi, *Macromolecules*, 1969, **2**, 200.
85. H. Ohtani, T. Nagaya, Y. Sugimura and S. Tsuge, *J. Anal. Appl. Pyrolysis*, 1982, **4**, 117.
86. H. Senoo, S. Tsuge and T. Takeuchi, *J. Chromatogr. Sci.*, 1971, **9**, 315.
87. Y. Sugimura and S. Tsuge, *J. Chromatogr. Sci.*, 1979, **17**, 269.
88. G. Montaudo, M. Przybylski and H. Ringsdorf, *Makromol. Chem.*, 1975, **176**, 1763.
89. D. O. Hummel, H. J. Dussel, H. Rosen and K. Rübenacker, *Makromol. Chem. Suppl.*, 1975, **1**, 471.
90. N. I. Zorina, T. A. Tsarfin and A. A. Karnishin, *J. Anal. Chem. USSR (Engl. Transl.)*, 1977, **32**, 936.
91. D. T. Burns, E. W. Johnson and R. F. Mills, *J. Chromatogr.*, 1975, **105**, 43.
92. J. M. Rigo, O. Riveros-Ravelo and H. Dieu, *J. Anal. Appl. Pyrolysis*, 1985, **8**, 123.
93. J. Martinex and G. Guiochon, *J. Gas Chromatogr.*, 1967, **5**, 146.
94. S. Tsuge, T. Okumoto, Y. Sugimura and T. Takeuchi, *J. Chromatogr. Sci.*, 1969, **7**, 253.
95. T. Shimono, M. Tanaka and T. Shono, *J. Anal. Appl. Pyrolysis*, 1979, **1**, 77.
96. S. Tsuge, T. Kobayashi, T. Nagaya and T. Takeuchi, *J. Anal. Appl. Pyrolysis*, 1979, **1**, 133.
97. T. Shimono, M. Tanaka and T. Shono, *J. Anal. Appl. Pyrolysis*, 1980, **1**, 189.
98. K. J. Voorhees (ed.), 'Analytical Pyrolysis — Techniques and Applications', Butterworths, London, 1984, p. 407.
99. A. Barlow, R. S. Lehrle and J. C. Robb, *Makromol. Chem.*, 1962, **54**, 230.
100. I. Ericsson; *J. Anal. Appl. Pyrolysis*, 1985, **8**, 73.

28

Crystal Structure by X-Ray Diffraction

EDWARD D. T. ATKINS
University of Bristol, UK

28.1 INTRODUCTION

Crystal structure determination from a whole crystal composed of small molecules is an established practice. The various methods, or combinations of procedures, that can be employed are documented and described in detail in many texts on the subject; see for example refs. 1–9. The availability of electronic computers and experimental techniques such as insertion of heavy atoms into the periodic structure to ease the phase problem has allowed the determination of the three-dimensional positions of the atoms in more complicated molecules. If a number of similar crystals with different heavy atoms inserted can be engineered (a technique known as isomorphous replacement), the individual atomic positions of exceedingly complex macromolecules, such as the globular proteins and enzymes, can be determined[7] exclusively by X-ray crystallography *via* Fourier synthesis. The crucial ingredient for a classical crystallographic structure determination of this type is the availability of a single (macroscopic) crystal, in which the molecule, or a chemically repeating segment of the molecule, forms a periodic lattice.

Long chain molecules like the synthetic polymers and the naturally occurring fibrous bio-polymers, even those with relatively straightforward chemically repeating sequences such as poly-ethylene and cellulose, do not form long range continuous and exact periodic structures in three dimensions. The inherent nature of these macromolecules, where the basic chemical repeat is connected covalently in a linear array which is different to the spatial bonding patterns and organization in other directions, creates a texture. In general, textures suppress the overall crystallinity in the sense that order exists only locally with amorphous regions separating the crystalline domains. The 'fringed micelle' model, illustrated in Figure 1, has been proposed for the organization of long chain polymers. Usually polymeric substances are oriented by drawing, stretching or rolling or occur naturally in an oriented form. In highly stretched systems the polymer chains tend to form long, slender crystallites, with perhaps many hundreds of repeats along the crystallite axis (often referred to as the fibre axis or chain axis), but with only a small number of repeating chains perpendicular to the crystallite axis (*e.g.* ~20 nm). In some cases the long chains fold to form lamellar-like crystallites[10,11] and in this instance the long range order extends further perpendicular to the chain direction than along the chain direction. A spectrum of textures can exist between these boundary models. In a sample of appropriate size for an X-ray diffraction experiment (*e.g.* ~20 × 10^5 nm) millions of crystallites will exist in random orientations about the fibre axis, as illustrated in Figure 2. The diffraction signals from each crystallite will, in part, overlay each other, or put another way, the X-ray spectra will be averaged in a cylindrical fashion about the fibre axis and in the recorded X-ray diffraction pattern vital information will have been lost before the processes of structure determination can even begin! In addition, misalignment of the crystallites with respect to the fibre axis produces arcing of the diffraction signals and the limited crystallite size produces broadening of the diffraction signals. Diffraction features from noncrystalline and unoriented domains and any types of disordering or paracrystallinity within the sample are superimposed. The experimental data, on which a structure analysis is to be based, are of poorer quality than those obtained in classical single crystal diffraction. The situation is even more acute for unoriented

Figure 1 The fringed micelle model for polymers representing regions or domains of order in a less well ordered matrix. If such a system is stretched, alignment of the ordered regions would be expected to occur

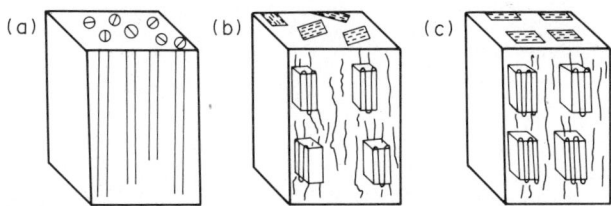

Figure 2 Simplified models for the texture of oriented polymers. (a) Uniaxially oriented crystallites in random rotational positions about the elongation or fibre direction. In reality the crystallites are not perfectly aligned. (b) Chain-folded lamellae crystals aligned with the elongation axis but in random orientation about it. (c) By rolling or squeezing, so called 'double oriented' textures can be produced. This can also happen with the crystallite model shown in (a)

samples where the diffraction information is spherically averaged by the totally random distribution of crystallites.

This chapter concentrates on structural analysis from uniaxially oriented long chain polymers exhibiting various degrees of crystalline diffraction. Some examples are shown in Figure 3. X-ray diffraction patterns of this kind are the main source of our structural knowledge of these substances. In a few cases, for polyethylene[12] and some other polymers with simple chemical repeats,[13] successful structure determinations using Fourier synthesis methods have been reported. Usually, however, the diffraction data are not sufficient to determine unequivocally the spatial coordinates of individual atoms. This problem is overcome by compensating the paucity of data with reliable stereochemical and rigid-body constraints, which can be implanted into the covalently connected polymer structure. Values of bond lengths and bond angles are obtained from single crystal structure determinations of related low molecular weight compounds and oligomers. Restrictions on rotations about bonds obtained from computer calculations on dimers and requirements on helical symmetry and helical pitch derived from the X-ray diffraction data are used to formulate a starting model (or models) that meet the basic dimensions and geometry of the chain conformation. Density measurements give valuable information on the number of chain segments in the unit cell. Calculated scattering intensities from the model(s) are compared and contrasted with measured values obtained from the X-ray diffraction patterns and refinements are performed to obtain the best fit. Fourier difference methods can be applied to locate the positions of solvent molecules or counterions in those instances where solvent or other small molecules are incorporated into the crystalline lattice. Computerized modelling plays an important role in the structure determination of linear polymer chains. Structure analyses usually proceed by successive reiteration between X-ray diffraction data and controlled model building. A flow chart outlining the procedures is shown in Figure 4.

Necessary parts of diffraction theory, crystallography and computerized modelling of polymer structures are described before proceeding to discuss some examples of structure analysis for oriented polymer samples.

28.2 SCATTERING OF X-RAYS

When a monochromatic X-ray beam is incident on a crystal, the scattered X-rays from the regularly placed atoms interfere with each other, giving strong diffraction signals in particular directions, since the interatomic distances are of the same order as the X-ray wavelength (typically ~ 0.05–0.2 nm). The directions of the diffracted beams are related to the shape and dimensions of the unit cell of the crystalline lattice, and the diffraction intensity depends on the disposition of the atoms within the unit cell.

Figure 5 shows the X-rays scattered by two electrons, one at the origin and the other specified by the position vector r. If s_0 and s are unit vectors representing the incident and scattered rays respectively, the path difference at some far away point P is given by $(s - s_0) \cdot r$, and the phase difference is $(2\pi/\lambda)(s - s_0) \cdot r$, where λ is the wavelength. It is convenient to define a scattering vector

$$S = \frac{s - s_0}{\lambda}$$

as shown in Figure 6, so that the phase difference may be written as $2\pi S \cdot r$. The dimensions of $|S|$ are reciprocal of length and S is called the reciprocal vector. If s_0 is fixed then S varies in both magnitude and direction as s rotates. The end of the vector S passes through a region called reciprocal space

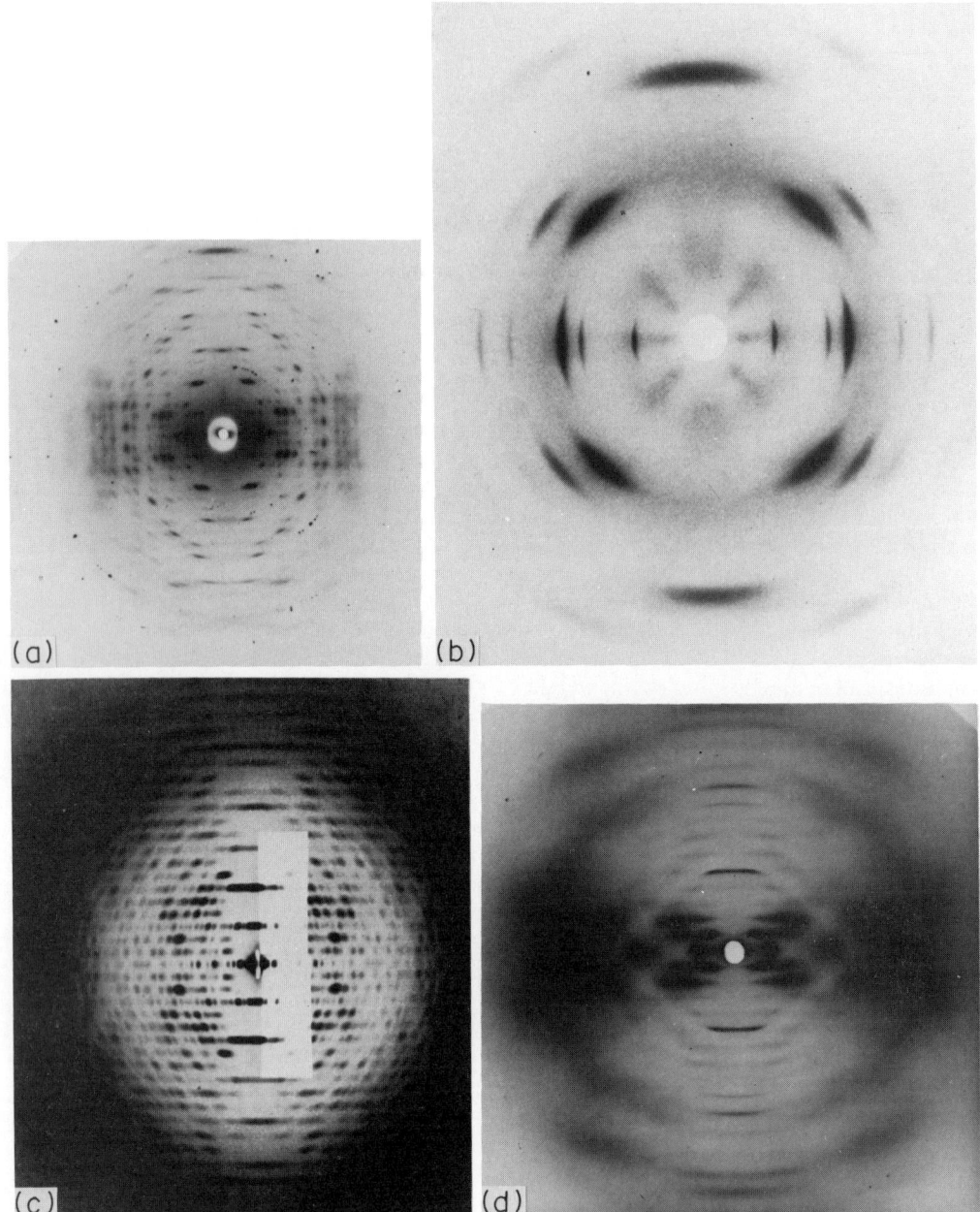

Figure 3 Examples of X-ray diffraction patterns from fibrous specimens. In each case the molecules are oriented vertically. (a) X-ray diffraction pattern from microbial polysaccharide XM6. A highly crystalline pattern with meridional diffraction signals on the 8th and 16th layer lines. The sharp dotted rings are for calibration purposes. (b) X-ray diffraction pattern of isotactic polystyrene. Note the sharpness of the equatorial diffraction signals compared with the arcs on the layer lines. This feature indicates chain-folded lamellae. (c) X-Ray diffraction pattern of oriented tobacco mosaic virus gel (reproduced by permission of the publisher from H. R. Wilson, 'Diffraction of X-rays by Proteins, Nucleic Acids and Viruses', Arnold, London, 1966). A piece of metal foil of known thickness has been placed on the film to reduce intensity of certain diffraction signals. (d) X-Ray diffraction pattern of xanthan gum; a polysaccharide based on a cellulosic backbone and decorated with side chains. Good orientation is achieved but not three-dimensional crystallinity, giving a nematic type liquid crystalline polymer pattern

and, as will be seen later, it turns out that this space is where the diffraction pattern is found. In Figure 6 the circle, or sphere in three dimensions, represents the conditions for diffraction and is known as the Ewald sphere, which has a radius of $1/\lambda$. It is an extremely useful construction for helping in the interpretation of X-ray diffraction patterns. If the diffraction angle is chosen to be 2θ

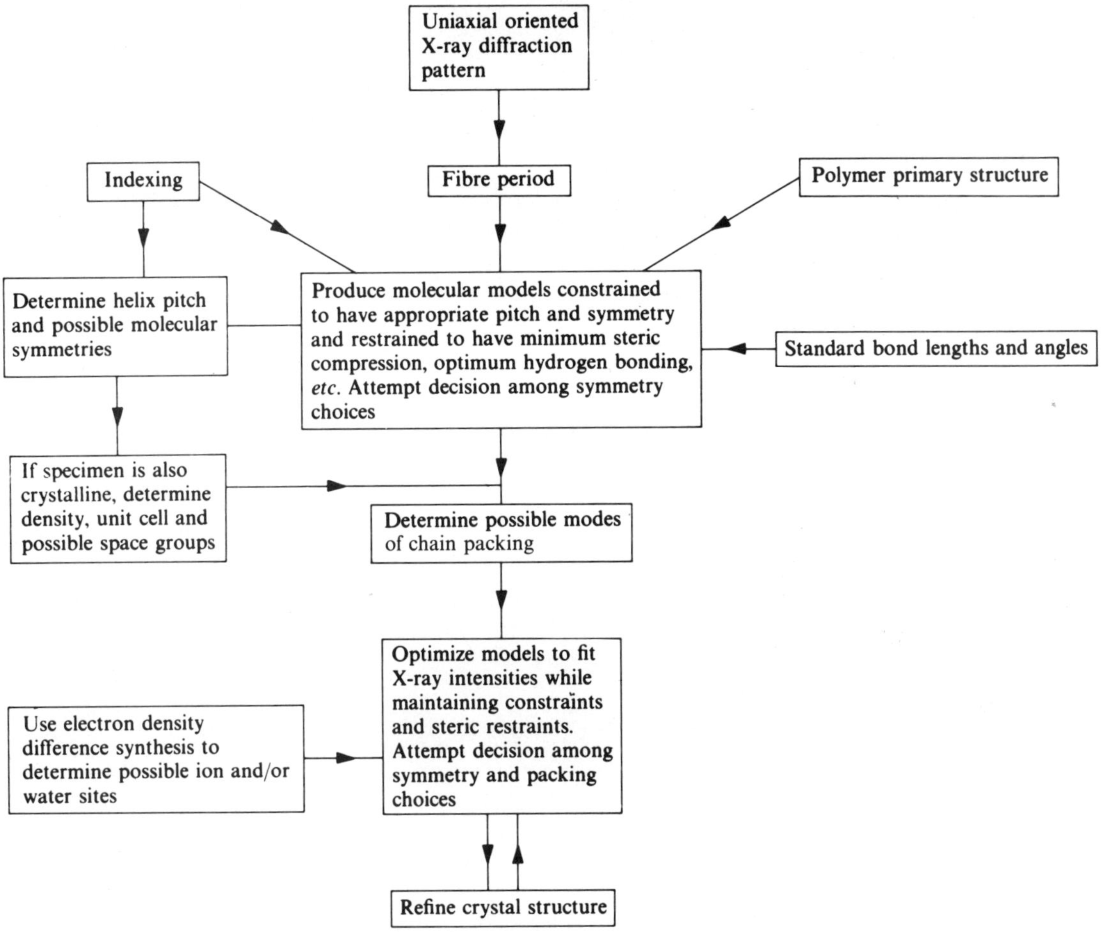

```
┌─────────────────┐
│ Uniaxial oriented│
│ X-ray diffraction│
│ pattern          │
└─────────────────┘
```

┌──────────┐   ┌──────────────┐   ┌────────────────────────┐
│ Indexing │   │ Fibre period │   │ Polymer primary structure│
└──────────┘   └──────────────┘   └────────────────────────┘

┌──────────────────┐   ┌──────────────────────────────────┐
│ Determine helix pitch│ │ Produce molecular models constrained │
│ and possible molecular│ │ to have appropriate pitch and symmetry│   ┌──────────────────────────┐
│ symmetries   │   │ and restrained to have minimum steric │◄──│ Standard bond lengths and angles │
└──────────────────┘   │ compression, optimum hydrogen bonding,│   └──────────────────────────┘
                       │ etc. Attempt decision among symmetry │
                       │ choices   │
                       └──────────────────────────────────┘

┌──────────────────┐   ┌──────────────────────┐
│ If specimen is also │ │ Determine possible modes │
│ crystalline, determine│ │ of chain packing   │
│ density, unit cell and│ └──────────────────────┘
│ possible space groups │
└──────────────────┘

┌──────────────────┐   ┌──────────────────────┐
│ Use electron density │ │ Optimize models to fit   │
│ difference synthesis to│ │ X-ray intensities while │
│ determine possible ion and/or│ │ maintaining constraints│
│ water sites   │   │ and steric restraints.   │
└──────────────────┘   │ Attempt decision among   │
                       │ symmetry and packing   │
                       │ choices   │
                       └──────────────────────┘

                       ┌──────────────────────┐
                       │ Refine crystal structure │
                       └──────────────────────┘

Figure 4 Flow chart for structure analysis of fibrous polymer structures using X-ray diffraction data and augmented with stereochemical information

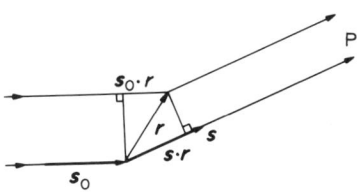

Figure 5 Phase difference between the waves scattered by two electrons situated on the ends of the vector *r*

(angle between OO′ and OP) then

$$|S| = \frac{2\sin\theta}{\lambda}$$

If we replace $|S|$ with $1/d$, where d represents the interplanar spacing of regularly arranged atoms, Bragg's Law is obtained

$$2d\sin\theta = n\lambda \qquad n = 1, 2, 3 \ldots$$

where the integer n indicates the order of diffraction. The sample is located at O in Figure 6 and O′ is the origin of reciprocal space, a distance $1/\lambda$ away in the incident X-ray beam direction.

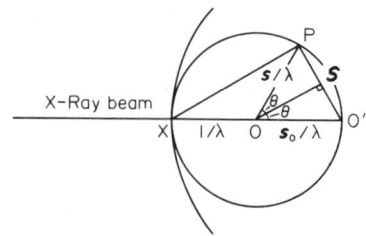

Figure 6   The Ewald sphere (circle) of diffraction, of radius $1/\lambda$, and the geometry for the diffraction condition. Part of the limiting sphere of diffraction is shown

28.2.1   Scattering Amplitude

The scattering amplitude, $F(S)$, is found by summing the exponential functions expressing the phase differences for a system of N atoms

$$F(S) \quad = \quad \sum_{j=1}^{N} f_j \exp(2\pi i S \cdot r_j) \tag{1}$$

Here f_j is the atomic scattering factor for the jth atom. If the object is continuous the set of discrete points in positions r_j can be replaced by an electron density distribution $\rho(r)$ and equation (1) can be rewritten as

$$F(S) \quad = \quad \int \rho(r) \exp(2\pi i S \cdot r) dv \tag{2}$$

where dv is the volume element of scattering matter. Equation (2) is a general equation for all types of scattering from objects of different shapes and distributions of scattering material. $F(S)$ is a complex function and $F^*(S)$ is the complex conjugate. The intensity will depend on the product $F(S)F^*(S)$ or $|F(S)|^2$.

If right-hand Cartesian coordinate systems are used with components (x,y,z) for r and (X,Y,Z) for S, equation (2) becomes

$$F(X,Y,Z) \quad = \quad \iiint_{x\,y\,z} \rho(x,y,z) \exp[2\pi i(xX \; + \; yY \; + \; zZ)] dx dy dz \tag{3}$$

If $\rho(x,y,z)$ exists only at discrete points x,y,z, then the integrals can be expressed as summations. If a simple polymer molecule is represented by a string of atoms at regular intervals nc (where n is an integer) along the z axis (therefore $x = y = 0$), the structure amplitude may be written as

$$F(Z) \quad = \quad f \sum_{n=-N}^{N} \exp(2\pi i n c Z)$$

$F(Z)$ will only have a nonzero value when $Z = 1/nc$ as shown in Figure 7.

28.2.2   The Reciprocal Lattice and Diffraction by Crystals

The crystal consists of a three-dimensional array of unit cells or a space lattice. If the primitive translations of the space lattice are represented by the vectors a, b and c then the vector $r\,(u, v, w)$ to any lattice point is given by

$$r \quad = \quad ua \; + \; vb \; + \; wc$$

where u, v and w are integers. The reciprocal lattice is defined by vectors a^*, b^* and c^* in reciprocal space where

$$a^* \quad = \quad \frac{b \; \times \; c}{V} \qquad b^* \quad = \quad \frac{c \; \times \; a}{V} \qquad c^* \quad = \quad \frac{a \; \times \; b}{V}$$

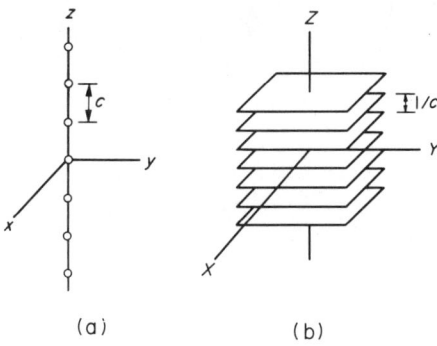

Figure 7 (a) One-dimensional lattice along z and (b) its transform in reciprocal space

The vector product $\boldsymbol{b} \times \boldsymbol{c}$ is a vector of length $|\boldsymbol{b}||\boldsymbol{c}| \sin \alpha$ and normal to the plane determined by $\boldsymbol{b}$ and $\boldsymbol{c}$. The angles α, β and γ are angles between $\boldsymbol{b}$ and $\boldsymbol{c}$, $\boldsymbol{c}$ and $\boldsymbol{a}$, and $\boldsymbol{a}$ and $\boldsymbol{b}$ respectively. A corresponding set α^*, β^* and γ^* exist for the reciprocal lattice. V is the volume of the unit cell and is equal to $\boldsymbol{a} \cdot (\boldsymbol{b} \times \boldsymbol{c}) = \boldsymbol{b} \cdot (\boldsymbol{c} \times \boldsymbol{a}) = \boldsymbol{c} \cdot (\boldsymbol{a} \times \boldsymbol{b})$. A summary of useful relationships between the real and reciprocal lattices is given in Table 1. From the definition of the reciprocal lattice it follows that

$$\boldsymbol{a} \cdot \boldsymbol{a}^* \;=\; \boldsymbol{b} \cdot \boldsymbol{b}^* \;=\; \boldsymbol{c} \cdot \boldsymbol{c}^* \;=\; 1$$

and

$$\boldsymbol{a} \cdot \boldsymbol{b}^* \;=\; \boldsymbol{a} \cdot \boldsymbol{c}^* \;=\; \boldsymbol{b} \cdot \boldsymbol{a}^* \;=\; \boldsymbol{b} \cdot \boldsymbol{c}^* \;=\; \boldsymbol{c} \cdot \boldsymbol{a}^* \;=\; \boldsymbol{c} \cdot \boldsymbol{b}^* \;=\; 0$$

Thus the a^*, b^* and c^* axes are directed perpendicular to the bc, ca and ab planes respectively. A two-dimensional example of the relationship between the real crystal and reciprocal lattice is shown in Figure 8. A set of planes in real space is represented by a point in reciprocal space. If that point touches the Ewald sphere the condition for diffraction is satisfied. Thus by allowing the reciprocal lattice to interact with the Ewald sphere the mathematical condition for diffraction can be visualized geometrically. Since the scattering vector $\boldsymbol{S}$ exists in reciprocal space it may be written as

$$\boldsymbol{S} \;=\; l\boldsymbol{a}^* \;+\; m\boldsymbol{b}^* \;+\; n\boldsymbol{c}^*$$

where l, m and n are the Miller indices of diffracting planes in the real lattice. If x_j, y_j and z_j are

Table 1 Real/Reciprocal Relationships for a Triclinic System

$$a^* = \frac{bc \sin \alpha}{V} \qquad\qquad a = \frac{b^* c^* \sin \alpha^*}{V^*}$$

$$b^* = \frac{ac \sin \beta}{V} \qquad\qquad b = \frac{a^* c^* \sin \beta^*}{V^*}$$

$$c^* = \frac{ab \sin \gamma}{V} \qquad\qquad c = \frac{a^* b^* \sin \gamma^*}{V^*}$$

$$V = \frac{1}{V^*} = abc \sqrt{1 - \cos^2\alpha - \cos^2\beta - \cos^2\gamma + 2\cos\alpha \cos\beta \cos\gamma}$$

$$V^* = \frac{1}{V} = a^* b^* c^* \sqrt{1 - \cos^2\alpha^* - \cos^2\beta^* - \cos^2\gamma^* + 2\cos\alpha^* \cos\beta^* \cos\gamma^*}$$

$$\cos\alpha^* = \frac{\cos\beta \cos\gamma - \cos\alpha}{\sin\beta \sin\gamma} \qquad\qquad \cos\alpha = \frac{\cos\beta^* \cos\gamma^* - \cos\alpha^*}{\sin\beta^* \sin\gamma^*}$$

$$\cos\beta^* = \frac{\cos\alpha \cos\gamma - \cos\beta}{\sin\alpha \sin\gamma} \qquad\qquad \cos\beta = \frac{\cos\alpha^* \cos\gamma^* - \cos\beta^*}{\sin\alpha^* \sin\gamma^*}$$

$$\cos\gamma^* = \frac{\cos\alpha \cos\beta - \cos\gamma}{\sin\alpha \sin\beta} \qquad\qquad \cos\gamma = \frac{\cos\alpha^* \cos\beta^* - \cos\gamma^*}{\sin\alpha^* \sin\beta^*}$$

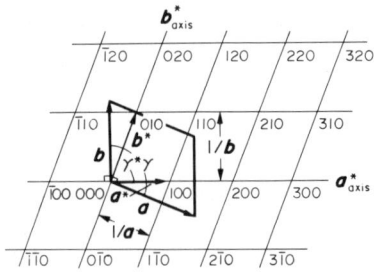

Figure 8 Relationship between the real (heavy lines) and reciprocal lattices (thin lines). This is an example of a monoclinic lattice in the first setting.[9] $a^* = 1/(a\sin\gamma)$ and $b^* = 1/(b\sin\gamma)$ (reproduced by permission of Wiley from ref. 13)

fractional coordinates of atoms in the unit cell then

$$r_j = x_j\mathbf{a} + y_j\mathbf{b} + z_j\mathbf{c}$$

and equation (1) can be rewritten as

$$F(hkl) = \sum_{j=1}^{N} f_j\exp[2\pi i(hx_j + ky_j + lz_j)] \tag{4}$$

For a known or postulated structure, $F(hkl)$ can be calculated. Separation into real and imaginary parts gives

$$F(hkl) = A(hkl) + iB(hkl)$$

where

$$A(hkl) = \sum_{j=1}^{N} f_j\cos 2\pi(hx_j + ky_j + lz_j)$$

$$B(hkl) = \sum_{j=1}^{N} f_j\sin 2\pi(hx_j + ky_j + lz_j)$$

and

$$|F(hkl)| = [A(hkl)^2 + B(hkl)^2]^{\frac{1}{2}}$$

The vector $F(hkl)$ has magnitude $|F(hkl)|$ and phase angle ϕ defined by $\tan \phi = B(hkl)/A(hkl)$. From the reciprocity property of Fourier series the expression equivalent to equations (3) and (4) for the electron density $\rho(x,y,z)$, is given by

$$\rho(x,y,z) = \sum_{-\infty}^{\infty}\sum_{-\infty}^{\infty}\sum_{-\infty}^{\infty} |F(hkl)|\exp[-2\pi i(hx + ky + lz - \phi)]$$

The summation is over all values of hkl and $\rho(x,y,z)$ varies continuously throughout the unit cell with maxima corresponding to atomic positions. If $|F(hkl)|$ and ϕ are known, the structure can be determined. Values of $|F(hkl)|$ can be obtained from the intensities on the X-ray photograph but the phase angle ϕ is lost. This is the basic problem with X-ray crystallography, known as the phase problem.

28.2.3 Systematic Absences

If the unit cell contains a number of identical groups related by certain symmetry operations, certain hkl reflections will be absent, for which there will be a systematic relationship in the h, k or l indices. These absences or space group extinctions are helpful in attempting to assign a space group symmetry to the structure. For example suppose a 2_1 screw axis passes through the origin and is parallel to the c axis. Positions (x,y,z) and $(\bar{x},\bar{y},\frac{1}{2} + z)$ are equivalent and

$$F(00l) = [1 + \exp(\pi i l)\sum_j f_j\exp(2\pi i l z_j)]$$

$F(00l)$ will only have a nonzero value when l is even, *i.e.* the $00l$ reflections are systematically absent for odd values of l.

28.2.4 Diffraction by Fibres

In the simplest case the crystallites will be uniformly distributed at all orientations about a common axis, called the fibre axis and usually denoted by c. The arrangement is analogous to a single crystal rotated with constant angular velocity about the c axis. Thus each reciprocal lattice point of a crystallite will trace out a circle centred on the c axis as shown in Figure 9(a). The condition for diffraction will be satisfied where these circles intersect the Ewald sphere, and the geometrical relationship between the distribution of reciprocal lattice points for a fibre positioned perpendicular to the incident X-ray beam and its diffraction pattern is illustrated in Figure 9(b). The layer lines allow periodicities along the fibre axis c to be separated from periodicities along a and b. The basic problem in indexing the X-ray patterns is to untangle sets of hk reflections for successive values of l. It may be noted in Figure 9 that owing to the curvature of the Ewald sphere, certain

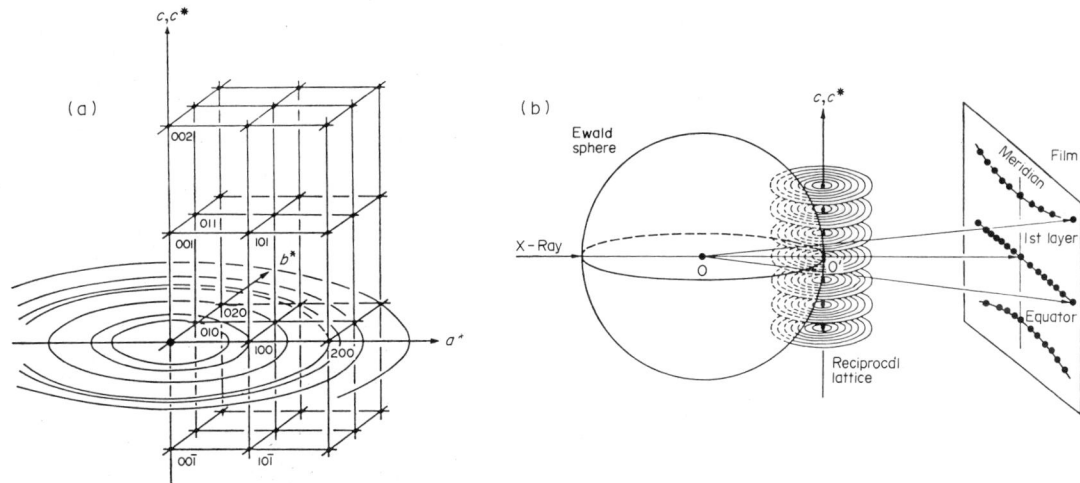

Figure 9 Diffraction pattern from a uniaxially oriented fibrous specimen. (a) Reciprocal lattice of a crystallite with concentric circular distribution of reciprocal lattice points in the equatorial ($a*b*$) plane. (b) Geometric relationship between the distribution of reciprocal lattice points and the fibre diffraction pattern (reproduced by permission of Elsevier from ref. 12)

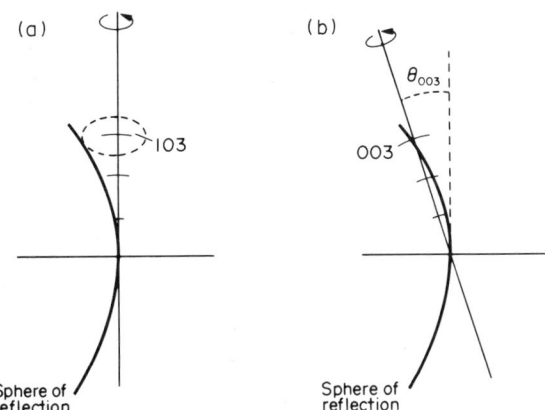

Figure 10 (a) A set of meridional ($00l$) arcs from a fibre specimen positioned perpendicular to the incident X-ray beam. Only the lower index arcs cut the Ewald sphere. (b) Effect of tilting the fibre towards the incident X-ray beam. The reciprocal lattice arcs move to intersect the Ewald sphere (in the top half of the pattern)

reciprocal lattice points, particularly those on the meridian, *i.e.* the set of 00*l* points, do not intersect the Ewald sphere. In practice, because of imperfect alignment and limited size of the crystallites within the fibre, the reciprocal lattice points are arced and broadened and some of the lower orders may extend sufficiently to intersect the Ewald sphere as shown in Figure 10(a). On higher layer lines some of the *hkl* points will trace out circles that just touch the Ewald sphere and such reciprocal lattice points will also give diffraction on the meridian (see Figures 9b and 10a). Later it will be shown that meridional diffraction signals are very useful in establishing helical symmetry and therefore it is important to decide which observed diffraction signals are meridional and which are not. The problem can be resolved by tilting the fibre, which results in tilting the reciprocal lattice, as shown in Figure 10(b), so that a particular 00*l* lattice point touches the Ewald sphere. The angle of tilt is θ, the Bragg angle for that particular reflection. To test which 00*l* reflections (or reciprocal lattice points) are absent, it is necessary to tilt the fibre for each point in turn by the appropriate Bragg angle, determined using Bragg's Law. On tilting to bring the 00*l* point into the diffracting position, its 00$\bar{l}$ counterpart in the other half of the diffraction pattern will not appear. An example of a diffraction pattern obtained with this method is shown in Figure 11 for α-poly-L-alanine. Any *hkl* reflection appearing on the meridian in the untilted fibre will split into two off-meridional reflections on tilting.

28.2.5   Unit Cell Parameters from Fibre Patterns: Indexing

To index an X-ray fibre pattern, the Bragg angles θ need to be obtained for each observable reflection. The method will depend on the particular X-ray camera used. A common arrangement is to record the diffraction pattern on a flat-film as shown in Figure 9(b). For a flat-film fibre diagram with the specimen positioned perpendicular to the beam, as illustrated in Figure 11(a), the four quadrants of the photograph are similar and the diffraction arcs are parts of circles. Figure 12 shows how the diameter 2*R* for each circle is measured and may be used to calculate the Bragg angle θ using the relation

$$\tan 2\theta \;=\; R/D$$

where *D* is the specimen to film distance. When *D* is not known accurately the photograph can be calibrated by dusting the fibre with a fine powder such as calcite, which gives a strong diffraction ring at a *d*-spacing of 0.3035 nm. From this known spacing, *D* can be calculated accurately. An internal calibration of this kind is more reliable than having a standard specimen to film distance since this can vary owing to changes in the sample position by the order of 5% in typical flat-film camera arrangements used in the laboratory.

The *d*-spacings can be found using the equation

$$d \;=\; \lambda/2\sin[\tan^{-1}(R/D)/2] \tag{5}$$

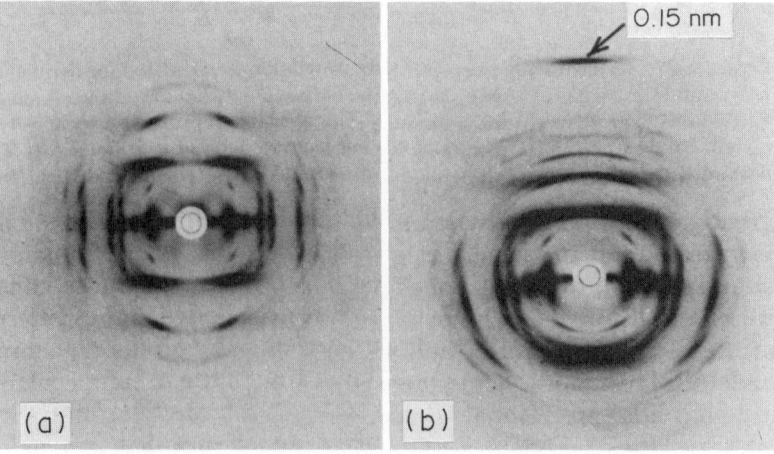

Figure 11   (a) Untilted X-ray diffraction pattern of α-poly-L-alanine. (b) Fibre axis inclined so that the meridional reflection corresponding to a spacing of 0.15 nm is recorded. Note that the equatorial line becomes distorted (reproduced by permission of Macmillan from C. H. Bamford, L. Brown, A. Elliot, W. E. Hanby and I. F. Trotter, *Nature (London)*, 1954, **173**, 27.

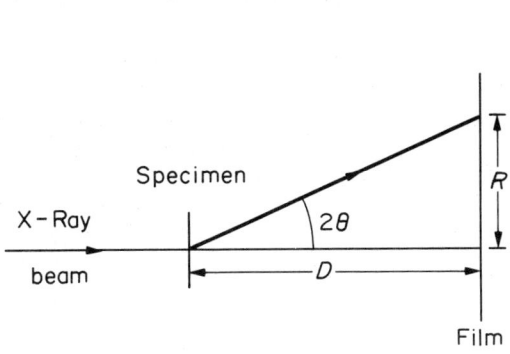
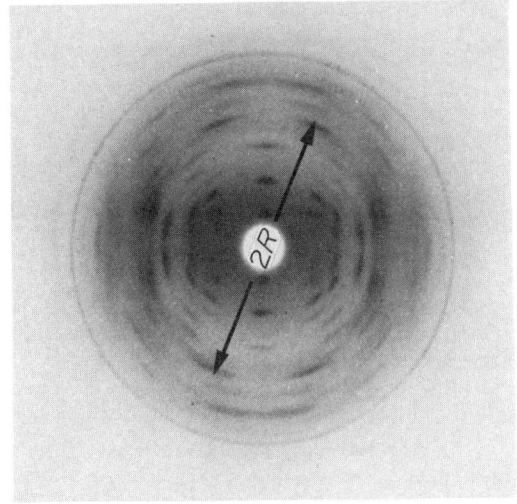

Figure 12 Method of measuring *d*-spacings from a flat-film. The values of 2*R* are measured either with a steel ruler off an original or by enlarging the photograph on to a screen. The *d*-spacings can be calculated and the specimen to film distance can be obtained accurately by calibration with, for example, fine calcite powder

It is relatively straightforward to calculate the value of *c* from the layer line spacing. In addition, it is known that all reflections on the equator have index $l = 0$, for those on the first layer line $l = 1$, and so on. It remains to determine the indices of *h* and *k* for each reflection. The measured *d*-spacings are compared with those for various sets of *hkl* planes of unit cells with different dimensions. Equations for orthorhombic, monoclinic and hexagonal unit cells are given in Table 2. The equivalent expression for the less symmetric triclinic unit cell is more complicated since $\alpha \neq \beta \neq \gamma \neq 90°$. The suspected shape of the polymer chain can be helpful. Usually planar ribbon-like structures crystallize in orthorhombic or monoclinic unit cells. The hydrogen-bonding pattern in nylon 6,6 generates a triclinic unit cell.[14] Rod-like and helical polymer chains often pack in a hexagonal or trigonal unit cell. Most indexing is done by first concentrating on matching the reflections on the equator and then extending to the other layer lines. With the aid of a computer the *d*-spacings of a number of trial unit cells can be rapidly calculated. When the agreement is adequate, say within 1%, for measured and calculated spacing, the *hkl* values are assigned to the measured data and the approximate unit cell parameters are refined using a least squares procedure. Indexing can also be aided by using or constructing various charts.[15] The Bernal chart[5] is useful for seeing how the reciprocal lattice is distorted on flat-film patterns. A Hull–Davey construction is useful for indexing orthorhombic lattices.[15] 'International Tables for Crystallography' provides useful advice on indexing.[9]

Table 2 Equations for the Interplanar Spacing d_{hkl} for Orthorhombic, Monoclinic and Hexagonal Unit Cells

Orthorhombic	$a \neq b \neq c$	$\alpha = \beta = \gamma = 90°$	$d_{hkl} = \left(\dfrac{h^2}{a^2} + \dfrac{k^2}{b^2} + \dfrac{l^2}{c^2} \right)^{-\frac{1}{2}}$
Monoclinic	$a \neq b \neq c$	$\alpha = \gamma = 90° \quad \beta \neq 90°$	$d_{hkl} = \left(\dfrac{\dfrac{h^2}{a^2} + \dfrac{l^2}{c^2} - \dfrac{2hl\cos\beta}{ac}}{\sin^2\beta} + \dfrac{k^2}{b^2} \right)^{-\frac{1}{2}}$
Hexagonal	$a = b \neq c$	$\alpha = \beta = 90° \quad \gamma = 120°$	$d_{hkl} = \dfrac{1}{\sqrt{\dfrac{4}{3a^2}\left(h^2 + k^2 + hk\right) + \dfrac{l^2}{c^2}}}$

28.2.6   Measurement of Diffraction Intensities

The crystal structure factor $F(hkl)$ is related to the observed intensity $I(hkl)$ by the expression

$$I(hkl) \;=\; pLmAK|F(hkl)|^2$$

where p is the polarization factor, L is the Lorentz factor, m is the multiplicity, A is the absorption factor and K is the scale factor.

28.2.6.1   Polarization factor p

The unpolarized incident beam from an X-ray generator will become plane polarized to a degree that is dependent on the Bragg angle θ

$$p \;=\; \frac{1}{2}(1 \;+\; \cos^2 2\theta)$$

If a monochromator is used or synchrotron X-ray radiation, this factor will change.[1,5,12] Care should be taken if using focussing X-ray cameras since different polarization factors will apply. An incorrect polarization factor can change the intensities substantially.

28.2.6.2   Lorentz factor L

This is a correction to compensate for relative lengths of time reciprocal lattice points spend in the Ewald sphere. For fibre diffraction patterns

$$L \;=\; \frac{1}{\sin 2\theta}\frac{\cos \theta}{(\cos^2\alpha \;-\; \sin^2\theta)^{1/2}}$$

where α is the angle between the reflecting plane and the fibre axis. On the equator, angle α becomes zero and the equation reduces to

$$L \;=\; \frac{1}{\sin 2\theta}$$

When $\alpha = 90°$ the equation does not operate and alternative corrections have to be applied.[13,16]

28.2.6.3   Multiplicity m

This is the number of equivalent types of crystal plane contributing to a reflection. For example, in the orthorhombic system the (100) plane is equivalent to the $(\bar{1}00)$ plane and they overlay each other in diffraction from fibres. The (210), $(\bar{2}10)$, $(2\bar{1}0)$ and $(\bar{2}\bar{1}0)$ are all equivalent. The multiplicities of these two examples are two and four respectively.

28.2.6.4   Absorption factor A

This depends on the size, shape, density and the elements within the substance and the wavelength of radiation. Usually no correction is needed for polymers with only light atoms.

28.2.6.5   Temperature and scale factors

Thermal vibrations of the atoms in a crystal suppress the ideal intensity by a factor

$$\exp[-B(\sin\theta/\lambda)^2]$$

where B is related to the mean square amplitude $\langle u^2 \rangle$ of atomic vibrations by

$$B \;=\; 8\pi^2\langle u^2 \rangle$$

Since any experimentally determined structure factor will be measured on a relative scale, it must be multiplied by a scale factor to transpose it to the absolute scale. The scale factor K and an estimate of the temperature factor B can be obtained by plotting $\ln(\bar{I}_{rel}/\Sigma f_j^2)$ against $(\sin\theta/\lambda)^2$, where $\bar{I}_{rel}$ is the average observed intensity after corrections for Lorentz and polarization factors (see ref. 1, p. 205–208).

28.2.6.6 *Oblique incidence*

In flat-film experiments the angle of incidence of the scattered beam to the film increases with the Bragg angle θ. Corrections are needed to compensate for this effect.

28.2.6.7 *Integrated intensity*

The quantity $I(hkl)$, which is a measure of the integrated intensity of the arc corresponding to the reciprocal lattice point hkl, is obtained by integrating the optical density above background over the region on the film occupied by the arc. The formal procedure is to take the product of the radial and circumferential area, obtained from microdensitometer scans after careful subtraction of background, and divide by the peak height.[17] In order to reduce the labour involved in measuring tangential traces, some workers have taken $I(hkl)$ as being proportional to the product of the area of the radial trace multiplied by a calculated arc length based on the distance of the arc from the origin and the distortion angle. Digitized data from two-dimensional microdensitometer scans are becoming more common but great care is needed in determining the background especially when so many arcs partly overlap. Measurement of intensity is the largest source of error in polymer structure determination.[18,19]

28.2.7 The *R* Factor

A measure of the agreement between the calculated $|F(hkl)|$ and the observed value is given by the reliability index R, defined as

$$R = \frac{\Sigma|\Delta F|}{\Sigma|F_o|} = \frac{\Sigma||F_o| - |F_c||}{\Sigma|F_o|}$$

where F_0 and F_c are the observed and calculated structure factor amplitudes respectively. The R factor is widely used as a guide to the goodness of fit but it is by no means a perfect parameter for establishing the correctness of a structure. It has been shown[20] that the theoretical values of R which would be obtained by using a model of the proper kind and number of atoms placed at *random* in the unit cell for an acentric structure is

$$R_{random,acentric} = 0.59$$

R values for polymer structures vary from 0.45 to 0.15. Some workers only include reflections which are observed, neglecting those that are below threshold or absent. This results in a lower value of R than is justifiable. Sometimes an R' factor is given in which the structure factors F are replaced with intensity I.

28.2.8 Density

The density of a sample is useful in determining the number of chains passing through the unit cell. In general, there must be a whole number of chain segments. The measured density is lower than the calculated crystalline density owing to a noncrystalline fraction but is usually sufficient to decide the number of chains. The density of the samples is usually measured by the flotation method. Two miscible liquids of higher and lower densities than the sample, and in which the sample will not swell or dissolve, are mixed in proportions that balance the density of the sample. An example of a polymer for which a density calculation is needed for crystal structure determination is poly(ethylene adipate). From X-ray fibre diffraction patterns a monoclinic unit cell with parameters $a=0.547$

$\pm\,0.03$ nm, $b = 0.723 \pm 0.002$ nm, c (fibre axis) $= 1.172 \pm 0.004$ nm and $\beta = 113.5°$ was established.[21] The volume of the unit cell is

$$V \;=\; abc\sin\beta \;=\; 0.4254 \text{ nm}^3 \quad \text{(see Table 1)}$$

The molecular weight M of the constitutive unit

$$\left[\!\!-(CH_2)_2\,O\overset{\displaystyle O}{\overset{\displaystyle \|}{C}}(CH_2)_4\overset{\displaystyle O}{\overset{\displaystyle \|}{C}}O\!-\!\!\right]\!\!-$$

is 172.18 and the density of the crystal is given by

$$\rho \text{ (g cm}^{-3}) \;=\; \frac{nM}{N_A V}$$

where N_A is Avogadro's number (6.022×10^{23}) and n is the number of units. If $n = 1$, $\rho = 0.67$ g cm^{-3}; if $n = 2$ then $\rho = 1.34$ g cm^{-3}. The measured value is 1.26 ± 0.04 g cm^{-3}. Thus one chain is not sufficient, two chains give a sensible fit, and three chains give a density far too high $(2.01$ g cm$^{-3})$. Thus two chains must pass through the unit cell since the fibre axis repeat of 1.172 nm is close to that in an extended polymer chain.

28.3 POLYMERS AS HELICAL STRUCTURES

A long chain molecule consists of units connected together like a string of beads. A generalized view of a polymer chain of this type is shown in Figure 13. The position of the upper unit, relative to the lower, may be described by an inclination α of the axes, the angle ψ formed by rotating O'O'' with respect to the axis OO' in the lower unit, and the angle χ turned through by the upper unit about its own axis O'O''.

Figure 14 shows the four categories of structure occurring in periodic chain molecules. The simplest case, when $\alpha = \psi = 0$, generates simple repetition, or a one-fold helix. When $\alpha = 0$, $\psi \neq 0$, a class of screw-twisted polymer chains is generated. The common type of zig-zag or two-fold helical molecules (such as polyethylene and cellulose) occurs when $\alpha \neq 0$ and $\psi = 180°$. The general case, where $\alpha \neq 0$, $\psi \neq 0$ and $\chi \neq 0$ represents the bulk of conformations of polymer chains. Almost all polymers with a regular sequence adopt a helical conformation in the crystalline state. A helix may be defined by the number of structural units in the axial identity repeat distance c. If P is the pitch of

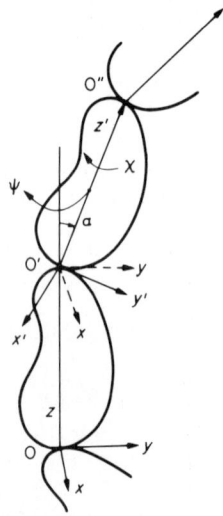

Figure 13 Schematic diagram of a polymer chain made from identical contiguous monomers (reproduced by permission of Elsevier from ref. 65)

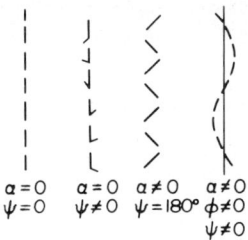

$$\begin{array}{cccc} \alpha=0 & \alpha=0 & \alpha\neq0 & \alpha\neq0 \\ \psi=0 & \psi\neq0 & \psi=180° & \phi\neq0 \\ & & & \psi\neq0 \end{array}$$

Figure 14 Diagrammatic representation of four types of helical chain structure (reproduced by permission of the publisher from B. K. Vainshtein, 'Diffraction of X-rays by Chain Molecules', Elsevier, Amsterdam, 1966)

the helix and h the projected axial advance per unit, then the number of units per turn is

$$P/h \;=\; u/t$$

where u and t represent the number of units and turns respectively in the identity period c. In structure analyses from fibre X-ray diffraction it is usual to obtain a clue to the helical nature of the polymer chains in the first instance by examining the X-ray diffraction patterns so that models can be computer-generated that are both stereochemically feasible and match the helical dimensions and symmetry indicated in the X-ray diffraction diagram. These models are the starting point for a structure determination. Thus it is important to understand the diffraction from helical structures in order to recognize the salient features in an X-ray fibre diffraction pattern.

28.3.1 Diffraction of Helical Structures

The Fourier transform of a helical structure may be developed in three stages.[22] First the Fourier transform of a continuous helical wire is derived. The process of convolution is used to obtain the Fourier transform of the helix of regularly spaced diffraction points and the basic features of the Fourier transform described. Finally the Fourier transform expression is adapted to give the diffraction pattern of a helical structure.

28.3.1.1 Fourier transform of a continuous helix

Consider a continuous helix of negligible thickness, infinite length, with radius r and pitch P as shown in Figure 15(a). The parametric equations of a helix are

$$x \;=\; r\cos(2\pi z/P) \qquad y \;=\; r\sin(2\pi z/P) \qquad z \;=\; z \tag{6}$$

The Fourier transform expression in Cartesian coordinates (see Figure 15b) may be rewritten as

$$F(X,Y,Z) \;=\; \int \exp[2\pi i(xX \;+\; yY \;+\; zZ)]\mathrm{d}v \tag{7}$$

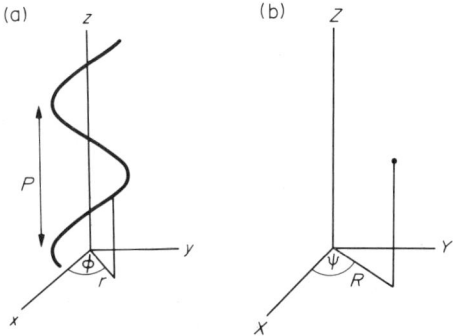

Figure 15 Real and reciprocal space coordinate systems for a continuous helix of pitch P and radius r

where X, Y, Z and x, y, z are the reciprocal space and real space coordinates respectively. For this simple helix dv is proportional to dz and substituting equation (6) into equation (7) gives

$$F(X,Y,Z) \ = \ \int_0^P \exp\{2\pi i[rX\cos(2\pi z/P) \ + \ rY\sin(2\pi z/P) \ + \ zZ]\}dz$$

Neglecting constants of proportionality and putting $R^2 = X^2 + Y^2$ and $\tan\psi = Y/X$ the expression becomes

$$F(R,\psi,Z) \ = \ \int_0^P \exp\{2\pi i[Rr\cos(2\pi z/P \ - \ \psi) \ + \ zZ]\}dz \tag{8}$$

The X-ray scattering from a continuous helix which repeats exactly after a distance P is confined to layer lines at heights $Z = n/P$ in reciprocal space (where n is an integer). Thus the integral in equation (8) has a nonzero value only when $Z = n/P$. The integral in equation (8) cannot be solved by ordinary methods of integration (since there is a trigonometric function in the argument of the exponent term) but the expression can be simplified by utilizing the Bessel identity

$$2\pi i^n J_n(X) \ \equiv \ \int_0^{2\pi} \exp(iX\cos\phi)\exp(in\phi)d\phi$$

if the substitutions $\psi = 2\pi Rr$ and $\phi = 2\pi z/P$ are made, equation (8) becomes

$$F(R,\psi,n/P) \ = \ J_n(2\pi Rr)\exp[in(\psi \ + \ \pi/2)]$$

where $J_n(X)$ is a Bessel function[23] of the first kind of order n and argument X.

28.3.1.2 Properties of Bessel functions

The shapes of Bessel functions up to order ten are illustrated in Figure 16. It may be seen that $J_0(X) = 1$ at the origin and fluctuates in a decreasing manner as X increases. All the other $J_n(X)$ are zero at the origin and the first peak is progressively attenuated and moves further from the origin as n increases.

The intensity distribution on the nth layer line is

$$I \ = \ |F(R,\psi,n/P)|^2 \ = \ |J_n(2\pi Rr)|^2$$

which is independent of ψ. Since the first maximum of each Bessel function moves further from the origin with increasing order, the centre of the diffraction pattern has a cross-shaped appearance as shown in Figure 17(a).

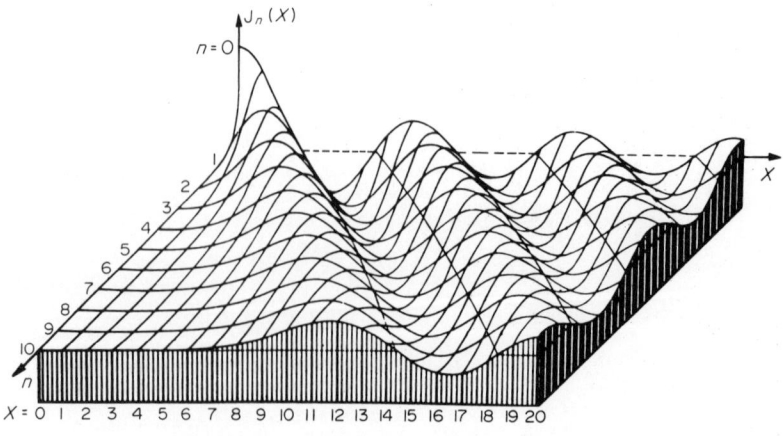

Figure 16 Plots of Bessel functions from order 0 to 10 (reproduced by permission of Dover from ref. 24)

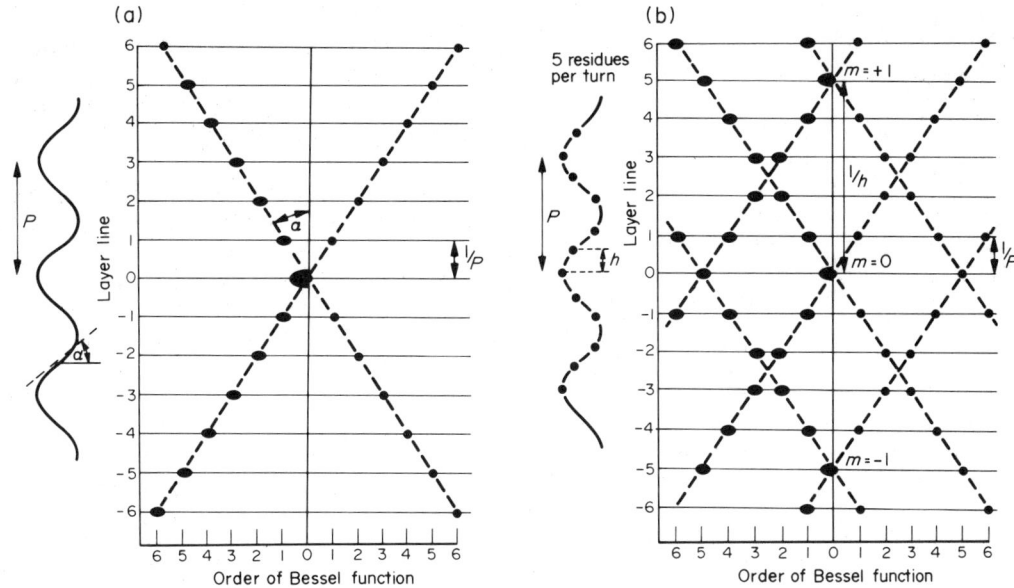

Figure 17 Diagrammatic representation of the main peaks of the Bessel functions contributing to the diffraction pattern from (a) a continuous helix of pitch P and angle α. The layer lines occur at a spacing proportional to $1/P$ and the half-angle of the cross is closely related to the pitch angle; (b) a discontinuous helix with five units per turn. The first meridional (vertical bisector) occurs on the fifth layer line at a real-lattice spacing of $1/(5/P) = P/5$ (reproduced by permission of the publisher from H. R. Wilson, 'Diffraction of X-rays by Proteins, Nucleic Acids and Viruses', Arnold, London, 1966)

28.3.1.3 Fourier transform of a discontinuous helix

A discontinuous helix is defined by a set of points occurring with a spacing, parallel to the helix axis, of h (for example), but all lying on the continuous helix. The radiation is imagined to be scattered by these points only. The axial projection of the structure repeats with periodicity h which gives rise to meridional reflection at spacings $\pm 1/h$, $\pm 2/h$ and so on. Thus the cross-shaped pattern typical of the continuous helix occurs at every point defined by $\pm m/h$ (where m is an integer) along the meridian. This process is an example of convolution. The multiplication of a continuous helix by a set of planes distance h apart in real space results in the convolution of the Fourier transform of the continuous helix with a set of points distance $1/h$ apart in reciprocal space as shown in Figure 17(b). The number of scattering points per turn of helix is given by P/h which may take any value. If the discontinuous helix repeats exactly in a distance c and if the number of scattering centres in this distance is M, then $c = Mh$, or if N is the number of turns of the helix, $c = NP$. The layer line spacing is $1/c$ and since c is the true repeat, intensity occurs only at values of $Z = l/c$ where l is the layer line index number. $l = n(c/P) + m(c/h)$ or

$$l = nN + mM \tag{9}$$

This integer equation is the selection rule governing the order n of the Bessel functions occurring on successive layer lines. For the example shown in Figure 17(b), $l = n + 5m$.

The positions of the meridional reflections and the layer lines on which they occur are valuable clues to the type of helical structure giving rise to the diffraction pattern. The first meridional diffraction signal in Figure 17(b) is on the fifth layer line and so indicates a five-fold helix. Because there is an exact number of scattering centres in one turn of the helix the layer line spacing provides a direct measure of the pitch of the helix in this case. Another example is shown in Figure 18(a) for a helix of eight units in three turns, an 8/3 helix. The values of n for the first eight layer lines are listed in Table 3. The values may be compared with those for an 8/1 helix, the corresponding diffraction pattern of which is shown in Figure 18(b). The layer line spacing and the position of the first meridional diffraction signal is the same in both cases but the general distribution of strong and weak intensities is different.

28.3.1.4 Helical structure

In practice a helical structure consists of groups of atoms lying at different radii, angles and positions along the helix. For a repeat unit of atoms with coordinates (r_j, ϕ_j, z_j) the scattering

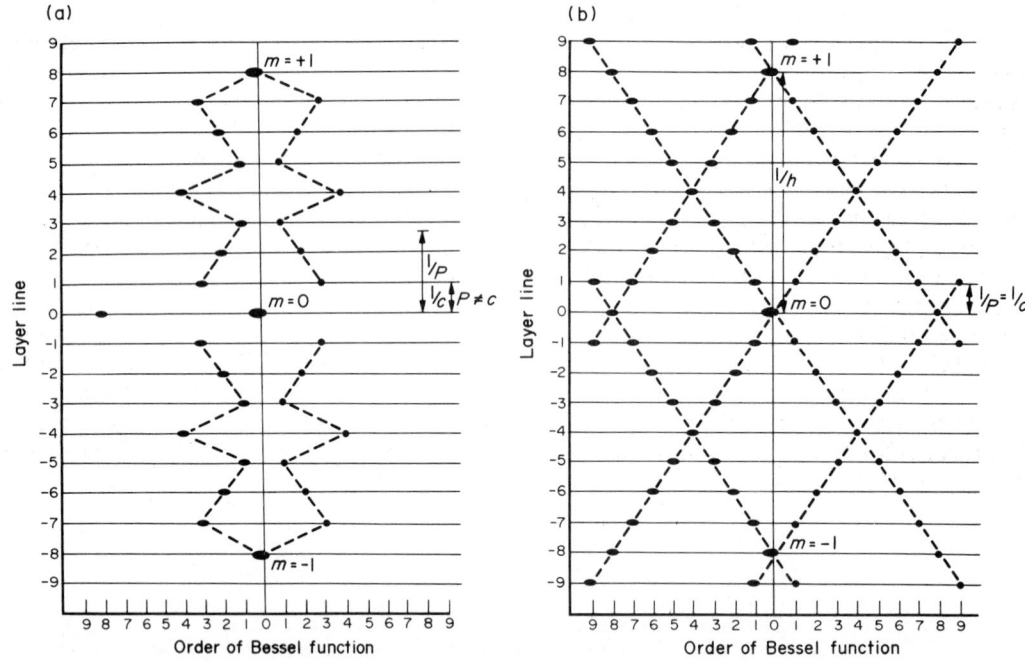

Figure 18 Diagrammatic representation of (a) an 8/3 helix, which is a structure partway between a two-fold helix and a three-fold helix (and therefore the distribution of intensities is not too dissimilar), and (b) an 8/1 helix, with eight units per turn

Table 3 Lowest Orders n of Bessel Functions Occurring on Layer Lines With Index l of 0 to 8 for 8/1 and 8/3 Helices[a,b]

8/1 Selection rule: $l = n + 8m$			8/3 Selection rule: $l = 3n + 8m$		
l	n	m	l	n	m
0	0	0	0	0	0
1	1	0	1	3	−1
2	2	0	2	−2	1
3	3	0	3	1	0
4	±4	0, 1	4	±4	−1, 2
5	−3	1	5	−1	1
6	−2	1	6	2	0
7	−1	1	7	−3	2
8	0	1	8	0	1

[a] m is an integer.
[b] The lower the value of n the stronger the diffraction. It does not matter whether n is positive or negative.

amplitude is given by

$$F(R, \psi, l/c) = \sum_n \sum_j f_j J_n(2\pi Rr) \exp\{i[n(\psi + \pi/2 - \phi_j) + 2\pi l z_j/c]\}$$

subject to to the selection rule equation (9).

28.3.1.5   Intertwining helices

There are structures where the polymer chains drape around a common axis. The first of these were discovered in biopolymers with the well known DNA double helix and the collagen triple helix. The general effect of these intertwined molecules is to modulate the intensities on the layer lines.

Structures have been discovered or proposed for polysaccharides[16,25-28] and poly(methyl methacrylate)[29] where identical intertwining chains have the same polarity and are symmetry related. For these multistranded ropes, complete layer lines can be missing and this can lead to misinterpretation of the diffraction evidence unless care is taken during the analysis. The X-ray fibre diffraction pattern illustrated in Figure 19 is obtained from a naturally occurring seaweed polysaccharide consisting of β-1,3-linked xylan, *i.e.* poly(β-(1 $\rightarrow$ 3)-D-xylose) (Figure 20). The reflections index on a hexagonal unit cell with dimenions $a = b = 1.37$ nm, c (fibre axis) $= 0.585$ nm and $\gamma = 120°$. Density considerations argue for three chains running through the unit cell.[16,30] Meridional reflections, however, occur on every even layer line, which would indicate a two-fold helix based on the helical diffraction considerations outlined earlier. The apparent inconsistency is resolved by the following model. Three identical polymer chains, each exhibiting six-fold symmetry, intertwine about a common axis. The three chains have the same polarity and are out of phase by translations of $P/3$ and $2P/3$ respectively, where $P = 1.755$ nm (3×0.585 nm) is the pitch of an individual polymer chain. This precise relationship between identical polymer chains results only in layer lines with index $l = 3m$ (m is the integer) appearing. Thus, only the $l = 0, l = 3, l = 6$, *etc.* layer lines occur, and since the first meridional diffraction signal for a six-fold helix is when $l = 6$ then the first meridional signal is found on the second *observed* layer line because of complete cancellation of layer lines $l = 1, l = 2, l = 4$

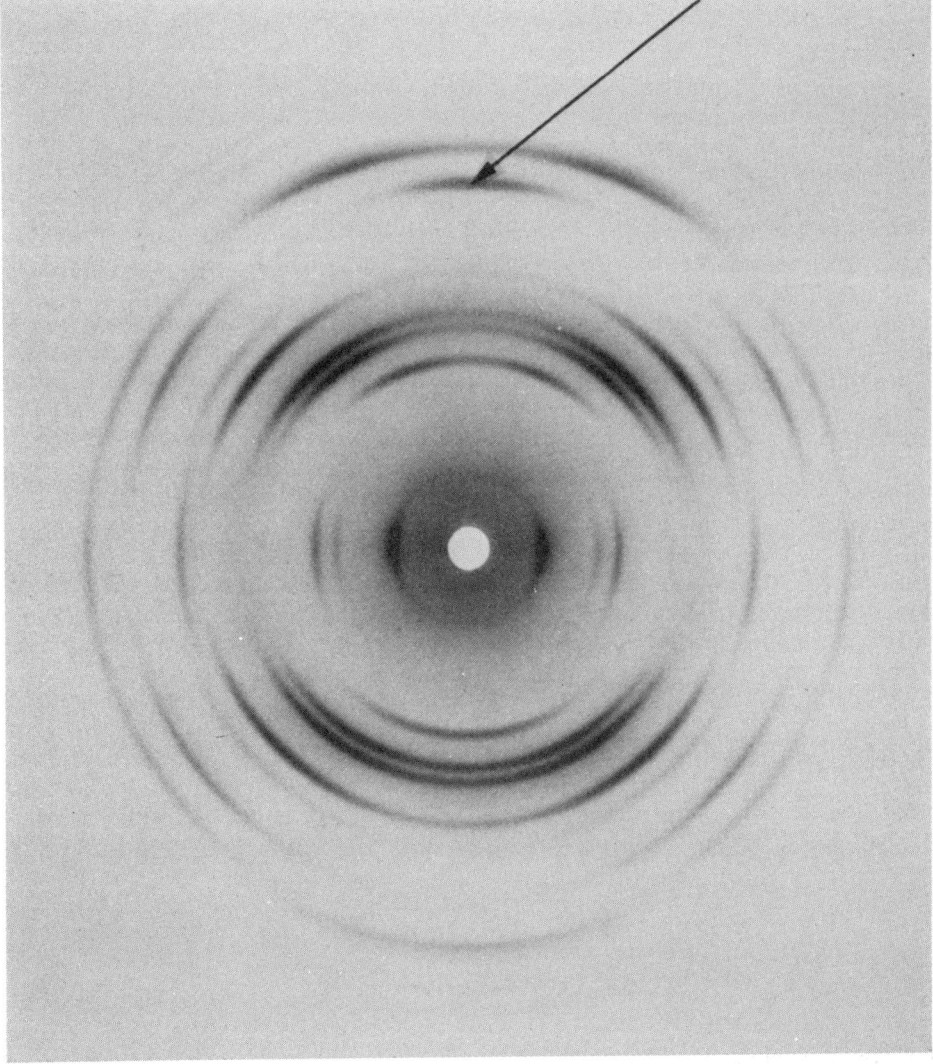

Figure 19 X-ray diffraction photograph obtained from the green seaweed polysaccharide 1,3-linked β-D-xylan. The fibre has to be tilted to bring the 002 reflection (arrowed) on to the meridian. Note the classic 100, 110, 200 triplet of reflections on the equator which index on a hexagonal unit cell. The space group is $P6_3$ (reproduced by permission of the Royal Society from ref. 30)

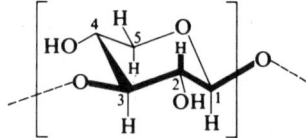

Figure 20   Repeating unit of 1,3-linked β-ᴅ-xylan

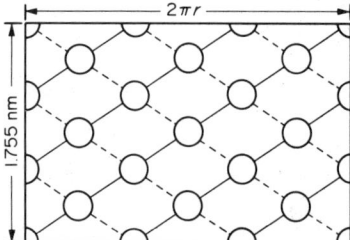

Figure 21   Radial projection of the xylan triple helix. The pitch of each helix is 1.755 nm. This is reduced by a factor of three on intertwining of the chains, each *P*/3 out of phase with each other

and *l* = 5. A radial projection of a schematic structure obeying these symmetry relationships is shown in Figure 21 and a projection of the actual structure is shown in Figure 22. An interesting experiment is to perturb the precisely defined multistrand molecular ropes in order to upset in some way the perfect symmetry relationship. When this is done the missing (or cancelled) layer lines appear.[31] This is a very reassuring test for confirming symmetry related multistrand rope models.

A double strand helical structure has been proposed[13,29] for isotactic poly(methyl methacrylate) (PMMA) based on the concept of cancellation of layer line outlined above. The X-ray diffraction pattern, Figure 23, shows layer lines of spacing 1.04 nm with the first meridional reflection on the fifth layer line. Initially five-fold single helix models were considered for the structure of isotactic PMMA.[32,33] Model-building considerations favoured helices with a pitch in the region of twice the observed layer line spacing and with about twelve monomers per turn. A model was derived where two chains, each a ten-fold helix with a pitch of 2.08 nm, were intertwined and staggered *P*/2 relative to each other so that all odd order layer lines would be completely cancelled and therefore only those as orders of 1.04 nm would appear. Unfortunately refinements of such a structure have not yet given as good a fit with the experimental intensities as perhaps would be expected for such a precisely defined structure.

The multistrand biopolymer ropes are in general acceptable in terms of current concepts of biosynthesis. If, as in the case of β-1,3-linked xylan, the three chains are polymerized together with mutual precession about a common axis, then this would seem a plausible mechanism for creation of a long chain molecular rope. How such a process occurs in the synthetic system is a very interesting problem. Some form of replica polymerization would seem appropriate.[34]

28.4   MODEL BUILDING

As pointed out in the introduction, the X-ray diffraction data from fibrous polymers do not, in general, enable positions of individual atoms to be determined directly. Thus model building is a necessary element in the structure analysis process.

28.4.1   Bond Lengths, Bond Angles and Geometry

In setting up molecular models, as much stereochemical information as possible is incorporated into the constitutive units. Standard bond lengths and bond angles can be found from many sources in the literature. Restrictions on bond rotation, such as the planarity of the amide unit, are valuable additional elements. Standard geometries for saccharide rings in the polysaccharides or aromatic rings are used in the first instance. In refinements of structures these restrictions can be relaxed if desired. If monomer or dimer (or even oligomer) structures have been crystallized and their atomic

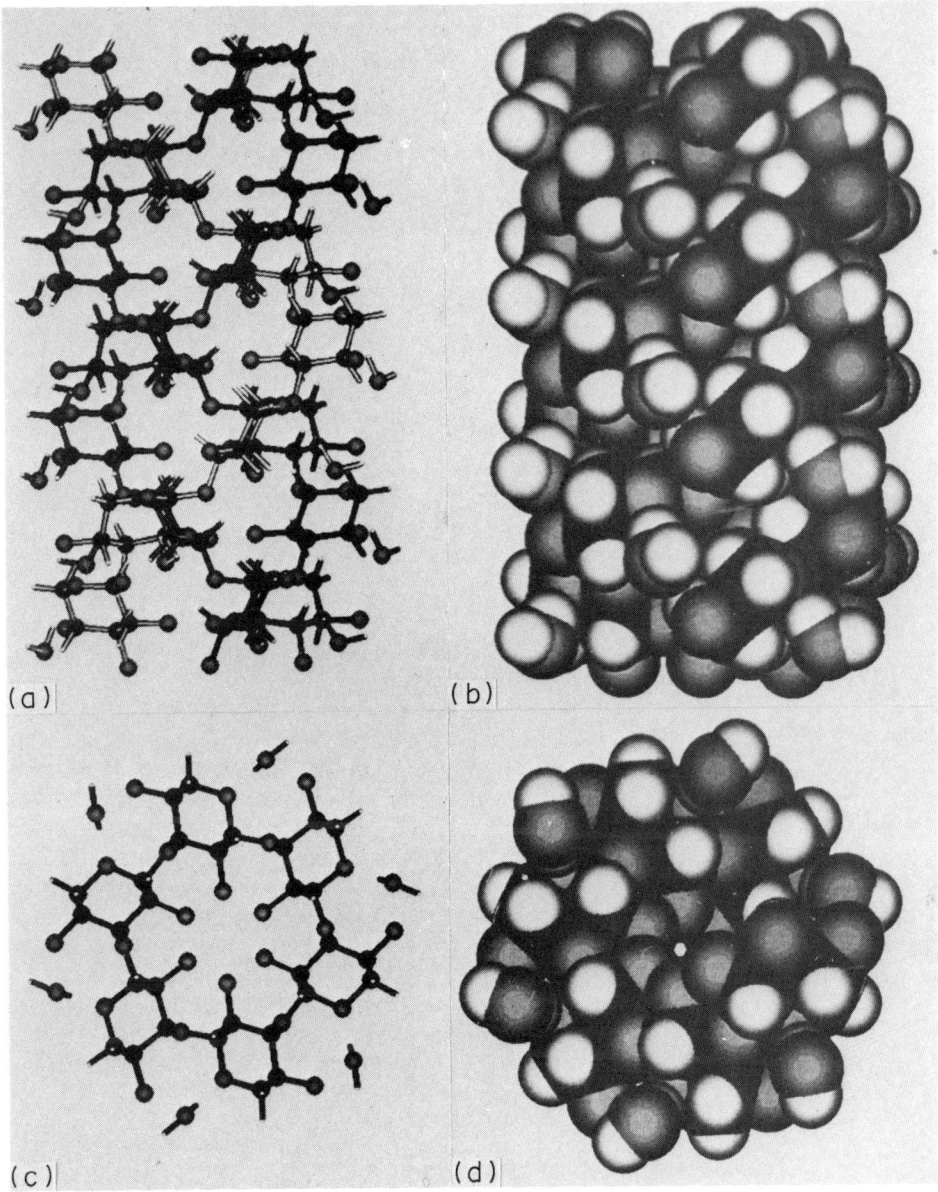

Figure 22 Triple-helix structure of 1,3-linked β-D-xylan. The three chains intertwine about a common axis. (a) and (c) are ball and stick projections perpendicular and parallel to the helix axis respectively. (b) and (d) are the corresponding space-filling projections. Hydrogen bonds in the form of a triad occur every 0.293 nm along the axis of the molecule between hydroxyl groups (reproduced by permission of Applied Science from ref. 25)

positions determined then these can be used for the basic geometry of the repeating unit in the polymer.

28.4.2 Steric Maps

The short range contacts between main-chain atoms and between side-chain and main-chain atoms can be mapped by calculating the interatomic distances as a function of the torsional angles of the bonds connecting the units together. Figure 24 shows a dipeptide unit with torsional angles about the N—C_α and C_α—C' bonds denoted by ϕ and ψ respectively. Tables are available of accepted minimum distances between atoms.[35-37] By stepping values of ϕ and ψ, a plot (called a Ramachandran plot[35]) of allowed conformations can be obtained as shown in Figure 25. This sort of

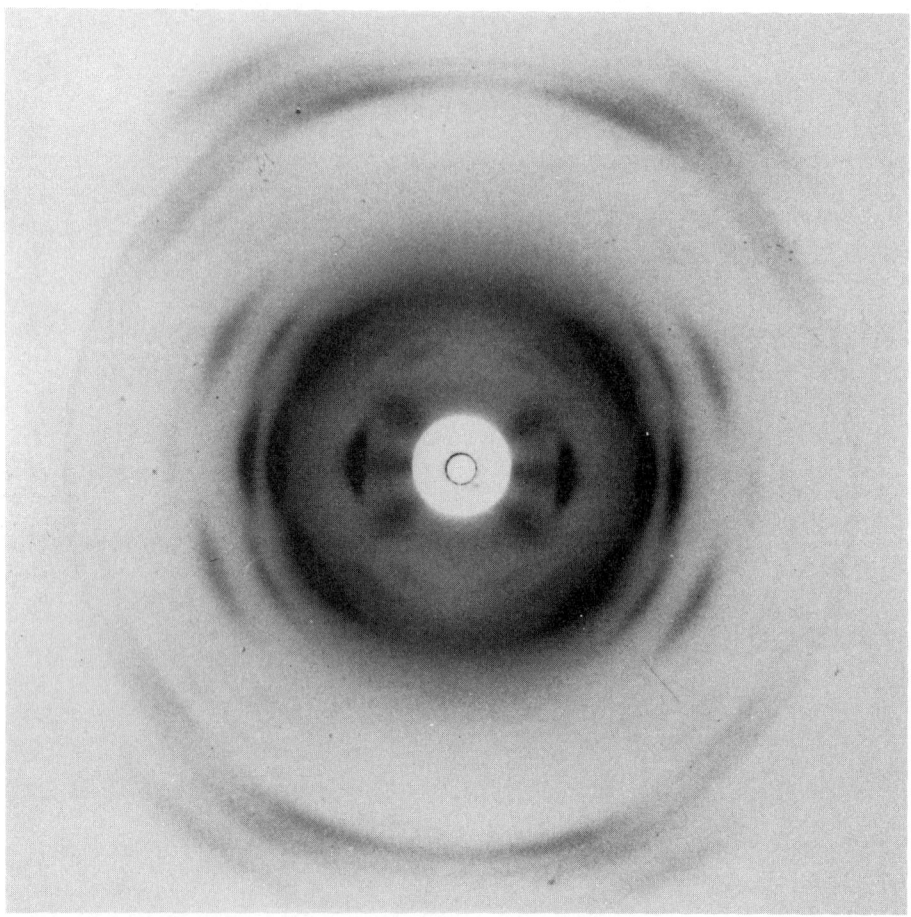

Figure 23   X-ray fibre diffraction pattern of isotactic PMMA. The fibre axis is vertical. The layer line spacing is 1.036 nm with a meridional reflection reported[13] on the fifth layer line. Note the strong halo from amorphous material

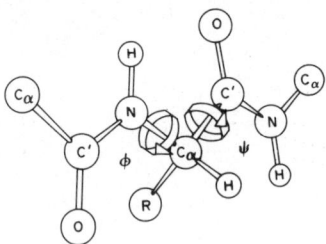

Figure 24   A dipeptide unit showing the torsional angles ϕ and ψ

plot is useful in deciding which starting positions to consider for computerized modelling of the polymer chain. If the energy of interaction between atoms as a function of ϕ and ψ is calculated this can be plotted as contours on the ϕ, ψ map. The possible sites where intraresidue hydrogen bonds occur can also be marked. Any particular set of ϕ, ψ values will, if contiguous units are linked identically, generate a helical conformation with parameters for the axial advance distance and the number of residues per turn. These parameters can be plotted to overlay the energy contours, as shown in Figure 26.

28.4.3   Potential Energy Calculations

Potential energy calculations can be used to help predict stable molecular conformations and packing of chains in the unit cell. The two common expressions for nonbonded atomic interactions

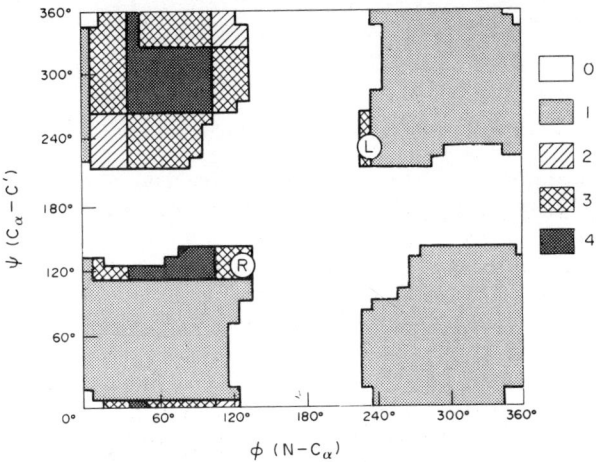

Figure 25 A Ramachandran plot or steric map for various dipeptides. In area 0 no conformations are allowed. Areas 1–4 are allowed for glycine–glycine, 2–4 for glycyl-L-alanine, and 3–4 for higher homologs. Area 4 is allowed for glycyl-L-valine and glycyl-L-isoleucine. R and L on the figure indicate positions of right- and left-handed α helices. (reproduced by permission of the American Chemical Society from G. Némethy, G. S. J. Leach and H. A. Sheraga, *J. Phys. Chem.*, 1966, **70**, 998)

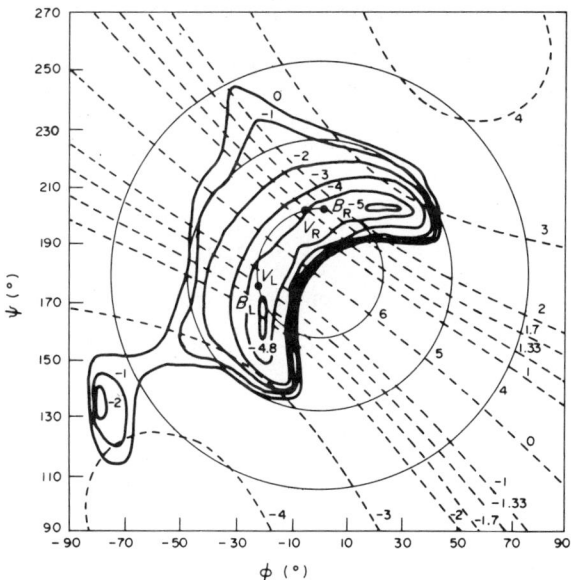

Figure 26 Conformational energy map for maltose. Heavy contours represent potential energy in kcal/maltose residue (1 kcal = 4.2 kJ); circles represent contours of fixed *h* and dotted lines indicate contours of fixed *n*. Positive values of *n* represent right-handed helices and negative values of *n* represent left-handed helices (reproduced by permission of the publishers from G. N. Ramachandran, 'Conformation in Biopolymers', Academic Press, New York, 1967)

are the Lennard-Jones '6–12' function

$$V_{ij} = \frac{a_{ij}}{r_{ij}^{12}} - \frac{b_{ij}}{r_{ij}^{6}}$$

and the Buckingham '6–exp' function

$$V_{ij} = c_{ij}\exp(-d_{ij}r_{ij}) - e_{ij}/r_{ij}^{6}$$

where V_{ij} is the nonbonded potential energy, a_{ij}, b_{ij}, c_{ij}, d_{ij} and e_{ij} are constants depending on atom type and r_{ij} is the interatomic distance.

Additional terms for the torsional potential, electrostatic, hydrogen bonds and distortion of interbond angles and variation in bond length can be incorporated if desired.[13,17,35-37]

28.4.4 Conformational Parameters for Helical Polymers

Equations have been derived[38-42] for determining the backbone conformations of a polymer chain which conforms to a given helical type. For homopolymers of the type $(-M-)_n$ three conformational parameters are defined: the distance ρ of an atom from the helix axis, the angle of rotation θ about the axis and the corresponding translation along the axis l. If r is the bond length, ϕ the bond angle and τ the torsional angle, the following equations hold[39]

$$\cos\left(\frac{\theta}{2}\right) \;=\; \cos\left(\frac{\tau}{2}\right)\sin\left(\frac{\phi}{2}\right)$$

$$l\sin\left(\frac{\theta}{2}\right) \;=\; r\sin\left(\frac{\tau}{2}\right)\sin\left(\frac{\phi}{2}\right)$$

$$2\rho^2(1 \;-\; \cos\theta) \;+\; l^2 \;=\; r^2$$

By varying τ, acceptable combinations of l and θ are sought.[43]

28.4.5 Linked Atom Least Squares Method

This procedure has been developed[44,45] to refine both the conformations of isolated helical chains and their packing in a crystal. A linked pathway is chosen along the polymer chain. Each atom is related to those already defined by a distance, a bond angle and a torsional angle, the latter being the main refinable parameter. Each linked atom structure is positioned relative to the helix axis by a further four refinable parameters. In a crystal, where intermolecular interactions are important, another four refinable parameters are provided for each independent polymer segment in the unit cell. The least squares refinement (LALS) is a constrained refinement procedure. Typical constraints are helix pitch and symmetry and continuity of the helix backbone. Model building is accomplished by varying the refinable parameters so as to minimize the function

$$\Phi \;=\; \sum k_j(\Delta P_j)^2 \;+\; \sum \lambda_h G_h \tag{10}$$

In this equation the first summation contains a term for each varied parameter. The term ΔP_j denotes the difference between the standard value for a parameter (determined by a survey of detailed crystal structures of relevant model compounds) and the value of the parameter found in the molecular model. The weighting constant for the term is represented by k_j. The first summation term in equation (10) is therefore the standard least squares expression for optimization.

The second summation term in equation (10) contains a series of Lagrangian multipliers λ_h, and constraints G_h, that must be equated to zero. This term imposes helical symmetry, pitch and chain continuity and any other constraint thought to be desirable. The same concept can be used in refining the structure against X-ray diffraction data. In this case ΔP_j is replaced with ΔF_m in the first summation term, where ΔF_m is the difference between observed and calculated structure factor amplitudes. (A more complete description of constrained least squares modelling and refinement procedure can be found in refs. 13 and 45).

28.5 STRUCTURE DETERMINATION: EXAMPLES

In the preceding sections the basic ingredients, methods and concepts involved in the usual polymer structure determination have been discussed. To illustrate how the various parts of the methodology link and operate together, examples of actual polymer structure determinations will be described. Following the procedure used in real cases is probably a good way to understand structure determination as undertaken at the present time. The first example, that of poly(3,3-diethyloxetane), is chosen because the polymer backbone conformation is a relatively straight-forward planar zig-zag or two-fold helix. The second example, a polysiloxane, is more complex and has an additional feature of interest to polymer structure analysts. Apart from the usual extended chain texture, or fibre texture, the polymer can be prepared with regular adjacent reentry chain-folded lamellar morphology as mentioned in the introduction. The results from the X-ray fibre analysis can be applied to the straight stem segments of the lamellar crystals and the chain-folding mechanism can be analyzed. In both examples the steps setting out from the polymer material to the final structure will be summarized. Thus useful information such as sample preparation, which has not been covered in the preceding sections, and the ordered steps in the analyses will be described.

28.5.1 Structure of Poly(3,3-diethyloxetane)

Poly(3,3-diethyloxetane) (PDEO) is a member of the series of polymers with the general form $[-OCH_2CR^1R^2CH_2-]_n$. The first of the series is polyoxetane, with R^1 and R^2 both being hydrogen atoms. The substitution of R^1 and R^2 with Me and Et groups introduces important changes in the properties of this family of polymers. The polyoxetanes are semicrystalline polymers and the crystal structures have been reported.[46-50] PDEO can be crystallized in two different polymorphs depending on the crystallization temperature.[51] A summary of the structure analysis[50] of the polymorph obtained when the crystallization temperature is in excess of 40 °C (called modification I) is given here.

28.5.1.1 Sample preparation

The polymer was heated in aluminum moulds to above the melting temperature (73 °C) and then allowed to crystallize slowly in a thermostatically controlled bath at 50 °C. Oriented specimens were obtained by stretching the films in an Instron dynometer at room temperature with a typical draw ratio of 5:1.

28.5.1.2 Density

The density was measured at 25 °C by flotation in water–ethanol mixtures and a value of 0.95 $\pm$ 0.01 g cm^{-3} was recorded.

28.5.1.3 X-Ray diffraction

Pinhole-collimated nickel filtered Cu radiation (Cu $K\alpha$ = 0.1542 nm) from an Elliot rotating-anode X-ray generator was used and X-ray diffraction photographs were recorded in a flat-plate camera, evacuated to reduce air scatter. The specimens were dusted with finely powdered calcite to accurately calibrate the diffraction patterns. Figure 27 shows the X-ray diffraction pattern obtained from the oriented sample. Specimens were tilted towards the incident X-ray beam at appropriate angles to register each 00l reflection. From the measured spacings of the Bragg reflections the unit cell dimensions were refined by a least squares procedure, giving a monoclinic unit cell with parameters a = 1.333(7) nm, b = 0.577(2) nm, c(fibre axis) = 0.747(2) nm and γ = 91.1(5)°. The calculated density, with two chain segments per unit, is 1.04 g cm^{-3} which matches the expected slightly lower measured value of 0.95 $\pm$ 0.01 g cm^{-3}. Meridional reflections (00l) were observed only when $l = 2n$. This selection rule favoured a two-fold helix for the chain conformation, and together with the other features of the pattern, suggest that the most probable space group is $P2_1$.

Intensities were measured with a Joyce–Loebl linear microdensitometer. Correction factors for Lorentz, polarization, multiplicity and oblique incidence on the film were applied.

28.5.1.4 Computer modelling

The first step was to generate an isolated chain molecule. Values for bond lengths and angles were assumed to be the same as those in comparable low molecular weight compounds, and thus only the chain torsion angles and orientation of the ethyl substituents remained to be determined. The atom labelling for PDEO is shown in Figure 28. The molecular conformation of the isolated chain was considered to be a planar zig-zag, giving good agreement between the measured fibre period of 0.474 nm and a calculated period of 0.485 nm, with bond lengths C—O = 0.143 nm, C—C

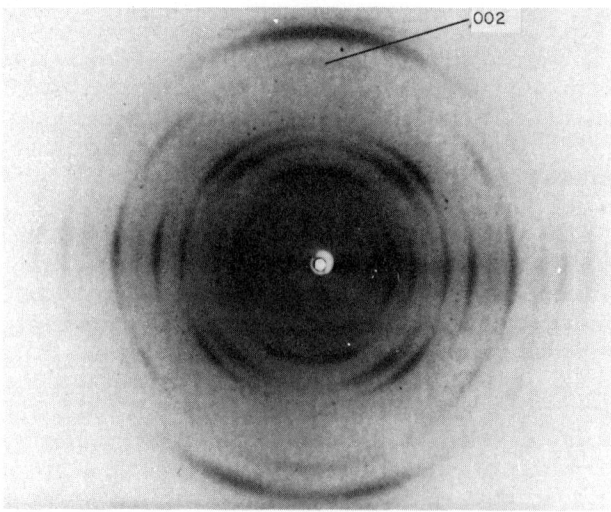

Figure 27   X-ray fibre diffraction photograph of modification I of poly(3,3-diethyloxetane). The fibre axis is tilted slightly to enhance the 002 reflection (arrowed) (reproduced by permission of Butterworths from ref. 50)

H(52)

H(51)—C(5)—H(53)

H(41)—C(4)—H(42)

H(11)   H(31)

—O—C(1)—C(2)—C(3)—

H(12)   H(32)

H(61)—C(6)—H(62)

H(71)—C(7)—H(73)

H(72)

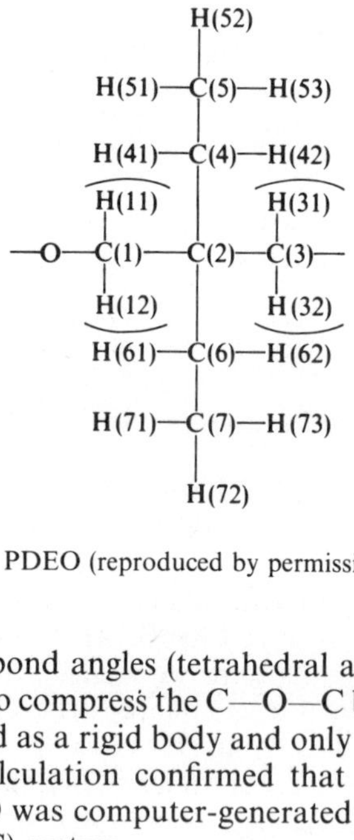

Figure 28   Atom labelling from PDEO (reproduced by permission of Butterworths from ref. 50)

= 0.154 nm, C—H = 0.109 nm and bond angles (tetrahedral angles) fixed. To match the measured fibre repeat exactly it was necessary to compress the C—O—C bond angle slightly. In trial computer models, the backbone was considered as a rigid body and only side-group orientations were treated as refinable. A minimum-energy calculation confirmed that the ethyl groups were in the *trans* conformation. The model for PDEO was computer-generated and the torsion angles refined using the linked atom least squares (LALS) system.

Two chain molecules were placed in the unit cell and packed in agreement with the space group. The space group symmetry, $P2_1$, relates one chain to the other as follows: $x_2 = 1 - x_1$, $y_2 = 1 - y_1$, $z_2 = z_1 + \frac{1}{2}$, where x_2, y_2 and z_2 are the fractional coordinates of an atom in chain 2 and x_1, y_1 and z_1 those of the equivalent atom in chain 1. Maintaining this relationship, the packed structure was refined by rotating the chains about their axes and translating them in a and b whilst allowing side-group orientations to vary. The refined, packed structure was used as the starting point for joint stereochemical and X-ray refinement. An overall temperature factor of 0.05 nm^2 was used (this was not refined). The resulting R value was 0.18. Even though refinable torsion angles and packing parameters were allowed to vary, the conformation and positions of the chains did not alter significantly during the X-ray refinement. The final coordinates are listed in Table 4. Calculated and observed structure factor amplitudes for the various reflections are tabulated in Table 5.

Projections of the crystalline structure of PDEO are shown in Figure 29. Structure analysis of PDEO allows comparison with the results for other polyoxetanes. A comparison of molecular

(a)

(b)

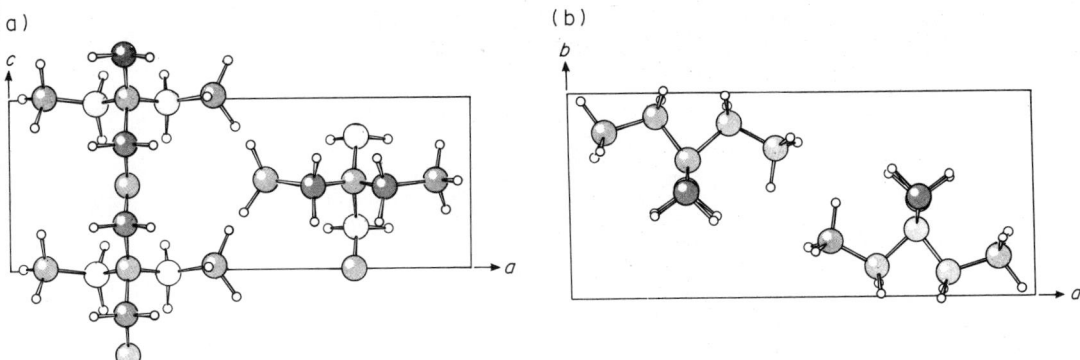

Figure 29 Projections of the structure of PDEO: (a) the *ac* projection, perpendicular to the fibre axis, and (b) the *ab* projection, along the fibre axis (reproduced by permission of Butterworths from ref. 50)

Table 4 Fractional Atomic Coordinates for Atoms in Modification I of Crystalline Poly(3,3-diethyloxetane)

Atom	$x\,(\times 10^4)$	$y\,(\times 10^4)$	$z\,(\times 10^4)$
O	2556	−3285	0
C(1)	2502	−4757	2429
C(2)	2561	−3266	5120
C(3)	2498	−4849	7733
C(4)	1882	−1156	4792
C(6)	3434	−1487	4851
C(5)	0777	−1929	5160
C(7)	4439	−2722	5261
H(11)	1809	−5840	2331
H(12)	3123	−6004	2364
H(41)	2011	−0350	2703
H(42)	2058	0121	6470
H(61)	3407	−0654	2752
H(62)	3361	−0142	6498
H(31)	1822	−5985	7579
H(32)	3060	−6230	7550
H(51)	0559	−3115	3421
H(53)	0683	−2824	7209
H(52)	0293	−0390	5115
H(71)	4767	−2219	7320
H(73)	4310	−4622	5187
H(72)	4915	−2448	3375

structures is given in Table 6. The substitution of the hydrogen atoms in the 3,3 positions for methyl and ethyl groups affects the relative stabilities of the crystal structures: in general, the structure with the greater number of *trans* bonds in the backbone is favoured.

28.5.2 Structure and Fold Conformation of Poly(tetramethyl-*p*-silylphenylenesiloxane)

The chemical structure of poly(TMPS) is

$$\left[\begin{array}{c}\text{Me}\\|\\-\text{Si}-\\|\\\text{Me}\end{array}\right.\!\!-\!\!\left\langle\rule{0pt}{2ex}\right\rangle\!\!-\!\!\left.\begin{array}{c}\text{Me}\\|\\\text{SiO}-\\|\\\text{Me}\end{array}\right]_n$$

(1)

Table 5 Calculated and Observed Structure Factor Amplitudes for Modification I of Crystalline Poly(3,3-diethyloxetane)

H	K	L	F_{calc}	F_{obs}
−2	0	0	44.7	41.6
1	1	0	61.1	55.0
2	1	0	31.1	33.4
4	0	0	29.8	27.1
4	1	0	35.2	32.4
2	2	0	21.3	24.6
3	2	0	34.6	31.2
4	2	0	18.4	13.8
−6	1	0	27.2	22.2
5	2	0	26.6	22.2
−1	0	1	20.6	41.2
2	0	1	0.6	1.5
0	1	1	26.6	24.1
−1	1	1	22.2	24.2
−2	1	1	24.5	33.8
4	0	1	6.6	1.5
0	2	1	20.0	20.3
−2	2	1	11.7	14.6
3	2	1	12.6	17.4
0	0	2	18.2	21.7
2	0	2	32.2	40.1

Table 6 Molecular Structures of Polyoxetane and its Derivatives

Polymer	System	Conformation[a]
Polyoxetane (PTO)	Mod. I (hydrate), monoclinic	Planar zigzag
	Mod. II, trigonal	$T_3GT_3\bar{G}$
	Mod. III, orthorhombic	$(T_2G_2)_2$
	Mod. IV	Planar zigzag
Poly(3,3-dimethyloxetane) (PDMO)	Mod. I	Planar zigzag
	Mod. II, monoclinic	$T_3GT_3\bar{G}$
	Mod. III, orthorhombic	$(T_2G_2)_2$
Poly(3,3-diethyloxetane) (PDEO)	Mod. I, monoclinic	Planar zigzag
	Mod. II, orthorhombic	$(T_2G_2)_2$

[a] T, G and $\bar{G}$ denote *trans*, *gauche* and minus *gauche* conformations, respectively.

The melting point and glass transition temperature of poly(TMPS) are considerably higher than those of poly(dimethylsiloxane) emphasizing the effect of replacement of alternate oxygen atoms in the chain backbone with rigid phenyl groups.[52-54] In addition to forming fibres,[55] poly(TMPS) forms spherulites and chain-folded single crystals. On raising the temperature, most chain-folded crystals thicken, but in the case of poly(TMPS) the crystals melt rather than thicken and this behaviour is likely to be associated with the details of the chain packing in the crystalline state.

28.5.2.1 *Sample preparation*

The polymer was melted, cast into a sheet approximately 1 mm thick, and quenched to room temperature and allowed to crystallize. Specimens were oriented by drawing strips of polymer cut from the sheet. The best orientation was obtained by holding the stretched strips under constant tension in a spring-loaded stretching device while annealing at 132 °C in a silicone oil bath. The density of the oriented specimen was measured as 1.06 ± 0.01 g cm^{-3} by placing samples in a density gradient column.

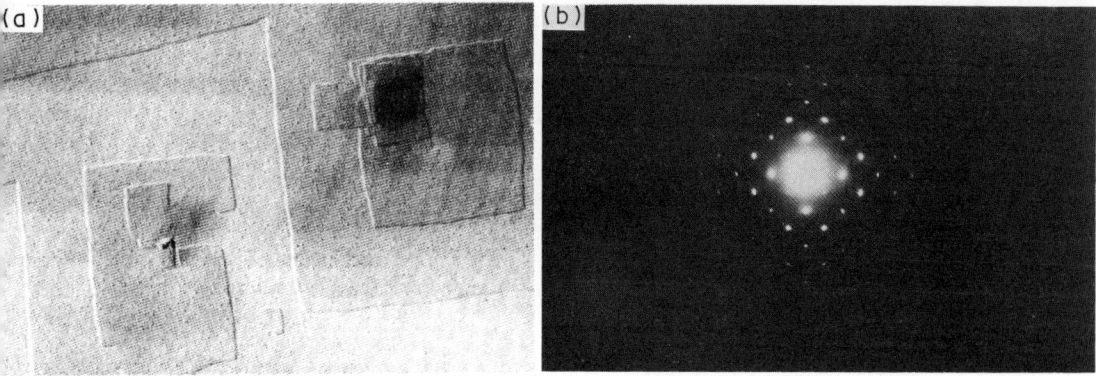

Figure 30 (a) Electron micrograph of shadowed poly(TMPS) single crystals and (b) the corresponding electron diffraction pattern from the specimen at 100 kV (reproduced by permission of Butterworths from ref. 55)

Chain-folded single crystal lamellae can be crystallized from solution and the melt.[54] A transmission electron micrograph and a corresponding electron diffraction pattern, which illustrates the tetragonal (square) symmetry of the crystals, are shown in Figure 30.

28.5.2.2 X-Ray diffraction

The X-ray fibre diffraction pattern obtained from an oriented poly(TMPS) sample is shown in Figure 31. The pattern was recorded on a flat-film using an Elliott Focussing Toroid camera and Cu *K*α radiation. Calcite was used as an internal diffraction standard. Intensities were measured with a Joyce–Loebl microdensitometer and correction factors for Lorentz, polarization, multiplicity and oblique incidence were applied.

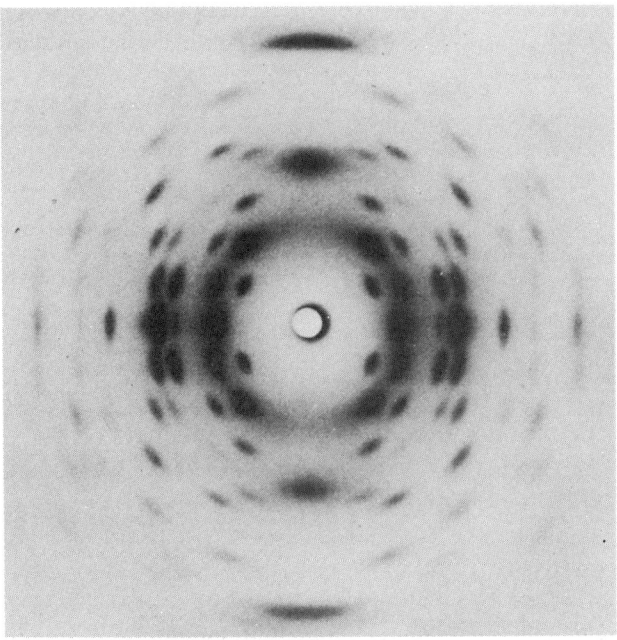

Figure 31 X-ray fibre diffraction photograph from a drawn sample of poly(TMPS). The fibre axis is vertical (reproduced by permission of Butterworths from ref. 55)

28.5.2.3 Unit cell and space group

The Bragg reflections were measured and the unit cell parameters were refined using a least squares procedure. The reflections were indexed by a tetragonal unit cell with dimensions $a = b = 0.902(3)$ nm and c(fibre axis) $= 1.543(3)$ nm. The observed meridional reflections lie on the fourth and eighth layer lines. Equatorial ($hk0$) reflections are observed only when $h = 2n$. These systematic absences suggest that the space group is either $P4_12_12$ or $P4_32_12$. (The symbols 4_1 and 4_3 represent right- and left-hand four-fold screw axes respectively.) In either case there are eight equivalent positions in the unit cell and four equivalent special positions.[9]

The calculated density based on four monomer units in the unit cell is 1.103 g cm^{-3} which is in agreement with the observed density of the drawn fibre (1.067 ± 0.01 g cm^{-3}).

28.5.2.4 Implications of the space group on model building

Space group symmetry requires the presence of eight equivalent atomic positions. The density measurements show that there should only be four monomers in the unit cell thus indicating that the crystallographic asymmetric portion of the molecule is one half the monomer unit. Since the monomer contains only one oxygen atom and one phenyl ring, the centre of mass of the phenyl ring and the oxygen must lie on crystallographic special positions, which are $(x,x,0; x,x,\frac{1}{2}; \frac{1}{2} - x, \frac{1}{2} + x, \frac{1}{4}; \frac{1}{2} + x, \frac{1}{2} - x, \frac{3}{4})$ or $(x,x,0; x,x,\frac{1}{2}; \frac{1}{2} - x, \frac{1}{2} + x, \frac{3}{4}; \frac{1}{2} + x, \frac{1}{2} - x, \frac{1}{4})$ for space groups $P4_12_12$ and $P4_32_12$ respectively. These special positions correspond to the points along the diad axes that are mutually perpendicular to each other and to the fibre axis which they intercept every $c/4$. The diad axes are oriented along the diagonals of the ab projection of the tetragonal unit cell.

In a unit cell with $P4_12_12$ or $P4_32_12$ space group symmetry the cell contents can be distributed in two ways that are consistent with the formation of continuous helices. The four residues could form a four-fold helix passing through the point $(\frac{1}{2}, 0, 0)$. In this case the choice of space group would depend on whether the helices were right- or left-handed. Alternatively, the unit cell could contain two two-fold helices passing through the corners and centre of the unit cell base with the chain at the centre of the cell having a relative translation of $+c/4(P4_12_12)$ or $-c/4(P4_32_12)$ with respect to the chain at the origin (corner). It is not possible to construct a four-fold helix meeting the special symmetry requirements placed on the positions of the oxygen atoms and the phenyl groups. Thus the structure consists of a pair of two-fold helices. Since the difference between the two space groups involves the packing of chains in the unit cell, the final choice follows the crystal structure refinement (Section 28.5.2.6).

In addition to the restrictions imposed on the molecule by the two-fold helical symmetry, the diad axis passing through the oxygen atoms requires that the torsional angle $\tau[C(1), Si, O, Si^*]$ be equal to $\tau[Si, O, Si^*, C(1)^*]$, where the atom labelling is shown in Figure 32. Even more stringent require-

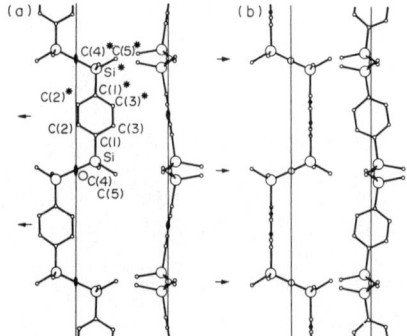

Figure 32 Projections of the two possible conformations for the poly(TMPS) molecule produced by model building. The conformations are the same except for the orientation of the phenyl groups with respect to the rest of the molecule: (a) model 1; (b) model 2. In (a) the two-fold rotation axes perpendicular to the helix axis are indicated (●) when perpendicular to the plane of the page and (→) when parallel to the plane of the page. In some of the projections the oxygen atom is obscured by the (●) symbol (reproduced by permission of Butterworths from ref. 55)

ments still are imposed in the orientation of the phenyl rings. Not only must the centre of mass of the phenyl group lie on a diad axis, but the plane of the ring must lie either perpendicular to the diad axis or contain it. These orientations place exact constraints on the allowable values for the torsion angles about the Si—C(phenyl) bonds. If the diad axis lies in the plane of the phenyl ring then $\tau[C(2), C(1), Si, O]$ must be equivalent to $\tau[C(2)^*, C(1)^*, Si^*, O]$. Alternatively, if the ring is perpendicular to the diad axis then $\tau[C(2), C(1), Si, O]$ will be equal to $\tau[C(2)^*, C(1)^*, Si^*, O] + 180°$.

28.5.2.5 *Molecular models*

Conformations for poly(TMPS) were generated using the linked atom procedure described earlier. The positions of the atoms were defined in terms of the bond lengths, bond angles and torsional angles. Bond lengths and in most cases bond angles were held constant whilst certain bond angles and torsion angles were treated as refinable parameters. These parameters were varied to allow the model chain to adopt a conformation with the appropriate helix pitch and symmetry while maintaining the additional symmetry relationships dictated by the space group. The model building was performed by varying the refinable parameters so as to minimize Φ given in equation (10) in the model-building section earlier (Section 28.4).

For the structure of poly(TMPS), covalent bond lengths, bond angles and stereochemistry of rigid groups were based on data obtained from the single crystal determinations of relevant low molecular weight compounds.[56-60] The phenyl group was fixed as a regular planar hexagon with a C—C(phenyl) distance of 1.140 nm. The silicon atoms were assumed to lie in the same place on the phenyl ring and have tetrahedral bonding geometry with a Si—C(phenyl) distance of 0.1858 nm, a Si—C(methyl) distance of 0.189 nm and a Si—O distance of 0.165 nm. Because of the flexible nature of the Si—O—Si bond, this angle was considered to be a variable parameter about a standard value of 144°.

Two-fold helical models for poly(TMPS) were computer-generated based on the experimentally determined pitch and the torsional angle constraints detailed in the previous section. The molecular variables were: (i) the Si—O—Si bond angle; (ii) the torsion angles about the Si—O bonds; and (iii) the torsion angles about the Si—C(1) bonds. For the *isolated* chain it was found that only two conformations, designated as models 1 and 2 in Figure 32, were stereochemically feasible. The nonbonded intrachain contact distances for both models are listed in Table 7. These contacts between nonbonded atoms were optimized by including an extra term[47] in equation (10).

Table 7 Nonbonded Intramolecular Distances Shorter than 0.38 nm for Models 1 and 2 of the Poly(TMPS) Chain

Model 1		Model 2	
O...C(2)	0.318 nm	O...C(2)	0.348 nm
C(3)...C(4)	0.378 nm	C(2)...C(5)	0.349 nm
C(3)...C(5)	0.337 nm	C(3)...C(4)	0.324 nm
C(4)...C(4)*	0.334 nm[a]	C(4)...C(4)*	0.334 nm[a]

[a] The contact is between C(methyl) atoms which are related by the diad axis through the oxygen atom.

Model 1 incorporates the following features: (a) the oxygen atoms reside on the helix axis; (b) the oxygen bond angle (Si—O—Si*) is 143.6°; (c) the phenyl group is displaced from the helix axis; and (d) the plane of the phenyl group lies perpendicular to the two-fold axis passing through the oxygen (see Figure 32a).

Model 2 is shown in Figure 32(b). The plane of the phenyl group is perpendicular to the diad axis through the ring.

For the sake of completeness an alternative molecular conformation which was rejected on stereochemical grounds is mentioned. This is the special case when $\tau[C(2), C(1), Si, O] = \tau[C(2)^*, C(1)^*, Si, O] = 180°$. In this case the C(1)*Si*OSiC(1)* string of atoms is planar and the centres of the phenyl groups lie on the helix axis. To match the measured fibre repeat distance it is necessary to compress the Si—O—Si bond to 122.3° which is an unacceptably low value for this bond angle and thus models incorporating this geometrical feature were rejected.

28.5.2.6　*Structure refinement*

By applying a procedure similar to that used in the model building of an isolated chain, the structural models were refined against the measured X-ray data by minimizing Φ in equation (10). In addition to the molecular parameters, the average isotropic temperature factor and the crystallographic scale factor were also refined. The resulting models were assessed in the conventional manner using the residual factors R and R'.

The chain models were placed in the unit cell at the packing positions dictated by the possible space groups $P4_12_12$ and $P4_32_12$. Least squares refinement against X-ray structure factor amplitudes were performed on the four packing models: model 1 ($P4_12_12$), model 1 ($P4_12_12$), model 2 ($P4_32_12$) and model 2 ($P4_32_12$). The same molecular parameters were varied as described in the model-building method described earlier (Section 28.4). However, the Si—O—Si bond angle was allowed to assume any value that would improve the agreement between observed (measured) and calculated structure factor amplitudes. The packing parameters of rotation about, and translation along, the helix axes were invariant and defined by the space group symmetry.

Model 1 ($P4_32_12$) gave the best agreement with observed X-ray data with values $R = 0.210$ and $R' = 0.191$ respectively. The structure has only one interchain contact of less than 0.38 nm: a C(methyl)—C(phenyl) contact of 0.352 nm. [For comparison, the structure that gave the next best agreement with the observed X-ray data was model 2 ($P4_32_12$) with $R = 0.342$ and $R' = 0.332$ respectively. In addition to the higher reliability indices this model has a short interchain contact for C(methyl)—C(phenyl) of 0.292 nm.] The conformation of the chain did not alter significantly during the X-ray refinement from the structure for model 1 shown in Figure 32(a). The final atomic coordinates of the structure are listed in Table 8. A comparison of the observed and calculated structure factor amplitudes for the refined structure is tabulated in Table 9. Projections of the crystalline structure of poly(TMPS) along the chain axis (c) and perpendicular to the chain axis (along b) are shown in Figure 33.

Considering the chain at the corner of the unit cell, the oxygen atoms lie on the helix axis at the positions $(0,0,0)$ and $(0,0,\frac{1}{2})$. The bond angle Si—O—Si* is 143.6° and is bisected by the diad axis that lies perpendicular to the chain axis. The centres of mass of the phenyl groups lie in the (110) plane on diad axes that intersect the chain axis at $z = \frac{1}{4}$ and $z = \frac{3}{4}$. The final values for the refined torsional angles are $\tau[C(1),Si,O,Si^*] = \tau[Si,O,Si^*,C(1)^*] = 107.1°$ and $\tau[C(2),C(1),Si,O] = \tau[C(2)^*,C(1)^*,Si^*,O] = -23.1°$.

The conformation of the crystalline monomer is similar to the unit in poly(TMPS). The value for the torsional angle about C—Si is $-11.5°$, compared with $-23.1°$ in the polymer. In the isolated monomer the oxygen exists as part of a hydroxyl group and the conformation is stabilized by intermolecular hydrogen bonds.[53]

Table 8　Fractional Coordinates for Atoms in the Asymmetric Unit Consisting of One Half of a Poly(TMPS) Unit[a]

Atom	$x\ (\times 10^4)$	$y\ (\times 10^4)$	$z\ (\times 10^4)$
O[b]	0	0	0
Si	−1526	0717	0415
C(1)	−1296	0947	1604
C(2)	−0258	0084	2052
C(3)	−2160	1984	2052
C(4)	2585	−1892	0098
C(5)	−0560	−3142	−0193

[a] The space group is $P4_32_12$. The remainder of the scattering material in the unit cell can be generated by the following symmetry elements: $y,x,\bar{z};\ \bar{x},\bar{y},\frac{1}{2}+z;\ \bar{y},\bar{x},\frac{1}{2}-z;\ \frac{1}{2}-y,\frac{1}{2}+x,\frac{3}{4}+z;\ \frac{1}{2}-x,\frac{1}{2}+y,\frac{3}{4}-z;\ \frac{1}{2}+y,\frac{1}{2}-x,\frac{1}{4}+z;\ \frac{1}{2}+x,\frac{1}{2}-y,\frac{1}{4}-z.$ With the first symmetry element the complementary half of the monomer unit can be produced. The second and third elements generate the second monomer of the chain at the origin, and the last four elements generate the second chain in the centre of the unit cell.
[b] This atom is at a special position and found at $0,0,\frac{1}{2};\ \frac{1}{2},\frac{1}{2},-\frac{1}{4};\ \frac{1}{2},\frac{1}{2},\frac{1}{4}.$

Table 9 Observed and Calculated[a] Structure Factor Amplitudes for Poly(TMPS)

h	k	l	F_{obs}	F_{calc}	h	k	l	F_{obs}	F_{calc}
1	0	0	sa[b]	0	3	1	2	260	150
1	1	0	1000	998	3	2	2	50	142
2	0	0	150	130					
2	1	0	940	821	0	0	3	sa	0
2	2	0	330	355	1	0	3	130	196
3	0	0	sa	0	1	1	3	60	102
3	1	0	180	113	2	0	3	40	52
3	2	0	380	321	2	1	3	310	297
					2	2	3	80	118
0	0	1	sa	0	3	0	3	140	116
1	0	1	180	202	3	1	3	180	180
1	1	1	440	583					
2	0	1	440	437	0	0	4	vs[d]	891
2	1	1	360	413	1	0	4	100	119
2	2	1	—[c]	59	1	1	4	180	209
3	0	1	—	117	2	0	4	—	16
3	1	1	170	127	2	1	4	230	182
3	2	1	250	255	2	2	4	60	152
					3	0	4	—	92
0	0	2	sa	0	3	1	4	70	157
1	0	2	470	377					
1	1	2	170	103	0	0	5	sa	0
2	0	2	160	224	1	0	5	—	29
2	1	2	230	149	1	1	5	170	199
2	2	2	160	208	2	0	5	170	111
3	0	2	—	37	2	1	5	50	113

[a] Atomic scattering factors were calculated using an analytical approximation, taken from the International Tables for Crystallography.[9] [b] Systematically absent reflections. [c] The symbol (—) indicates reflections that were not observed above the background scatter on the X-ray film. When these reflections were included in the X-ray refinement as observed reflections with unit weights and intensity of 0.0, the values of R and R' were 0.253 and 0.211, respectively, but the model remained unchanged. [d] vs = very strong.

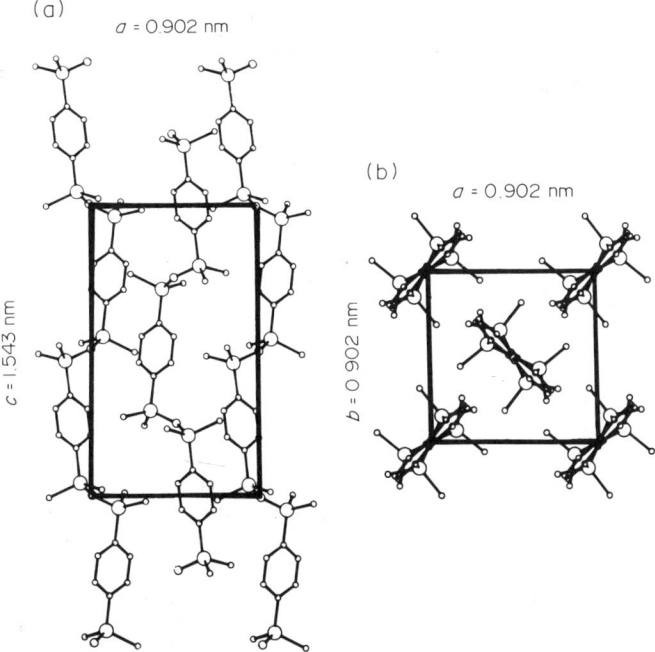

Figure 33 Crystal structure of poly(TMPS). (a) The *ac* projection is shown (perpendicular to the fibre axis) and (b) the *ab* projection is shown (looking down the fibre axis) (reproduced by permission of Butterworths from ref. 55)

28.5.2.7 *Fold conformation*

To construct a detailed model for the fold of the poly(TMPS) chain as found in the chain-folded lamellar crystals, shown in Figure 30, there are certain criteria which must be met: (i) the molecule must occupy a crystal lattice site in the structure in the (110) plane before entering and after leaving the fold; (ii) over-short nonbonded intramolecular contacts must be avoided and bond angle distortion kept to a minimum; (iii) there must be no over-short contacts between atoms in adjacent folds on a regular fold surface; and (iv) the number of residues involved in the fold should be small (because of evidence for tight or adjacent reentry folding[61]).

These criteria for an acceptable model limit the number of possible fold types. Since there is a one-quarter stagger along the *c* axis between a particular chain and its near neighbours to the left and right in the (110) fold plane (see Figure 34), it is possible for the fold to terminate at a unit in an adjacent straight stem that is staggered up or staggered down relative to the fold-initiating unit. (Owing to the two-fold helical nature of the model, there are no other new fold possibilities to be considered if another unit in the straight stem were chosen as the starting point for a fold).

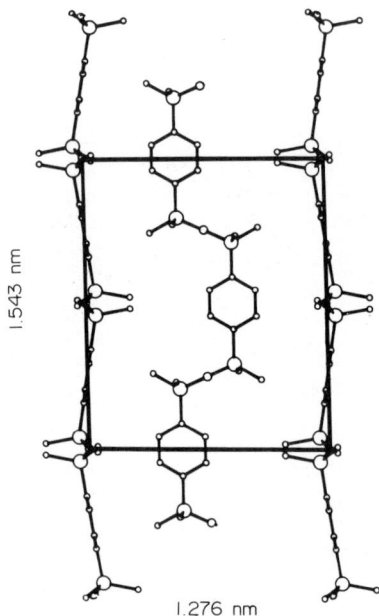

1.543 nm

1.276 nm

Figure 34 Projection of the molecules in the (110) plane. This plane corresponds to the 'fold plane' in poly(TMPS) lamellar crystals (reproduced by permission of Butterworths from ref. 55)

Models were constructed for each of these fold types, incorporating one, two and three units within the fold. The extremes of the fold were defined by having the phenyl groups of the fold-initiating and fold-terminating units in the crystal lattice (straight stems) and allowing free rotation about the Si—C(phenyl) bonds to these units.

Of the possible single unit folds only one satisfied the imposed constraints and is shown in Figure 35. The coordinates of the atoms in the fold are listed in Table 10. The model is free of nonbonded steric interferences (*i.e.* no nonbonded contacts of less than 0.32 nm).

In addition the model was constructed without distortion of the Si—O—Si bond angles from 143.6°, the value for the bond angle found in the refined crystal structure.

The manner in which a chain could be incorporated into a folded-chain lamellar crystal is shown in Figure 36. This illustrates part of a (110) plane of a poly(TMPS) chain-folded lamella. Figure 37 shows the appearance of a regular fold surface of a lamellar single crystal which incorporates the proposed fold structure.

The investigation of folds incorporating two and three units produced models which deviated minimally from the single-unit fold. Only small perturbations of the molecular parameters of the fold-initiating and fold-terminating units defined for the single-unit fold were found.

In the projection of the structure shown in Figure 36 it can be seen that the methyl groups of adjacent chains intermesh. The intermeshing of these methyl groups, when combined with the

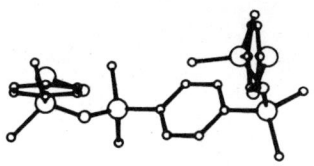

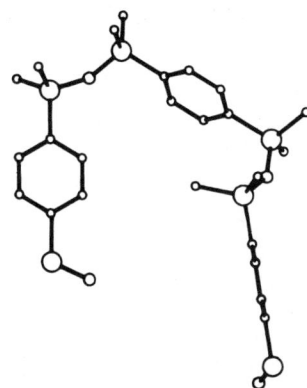

Figure 35 Conformation of poly(TMPS) fold containing single unit (reproduced by permission of Butterworths from ref. 55)

Table 10 Coordinates of the Atoms in the Single Fold of Poly(TMPS)

Atom	$x \, (\times 10^4 \, nm)$	$y \, (\times 10^4 \, nm)$	$z \, (\times 10^4 \, nm)$	Atom	$x \, (\times 10^4 \, nm)$	$y \, (\times 10^4 \, nm)$	$z \, (\times 10^4 \, nm)$
Unit one (fold-initiating unit):				Si*	4076	3099	12 419
O	0	0	0	C(1)*	2937	1972	11 480
Si	−1374	645	639	C(2)*	3451	1090	10 546
C(1)	−1166	853	2474	C(3)*	1573	2015	11 713
C(2)	−241	83	3157	C(4)*	4043	2653	14 220
C(3)	−1936	1778	3159	C(5)*	3521	4857	12 209
C(4)	2332	−1707	151				
C(5)	−505	−2834	−298	Unit three (fold-terminating unit):			
Si*	−646	1373	7055	O	5183	3745	11 388
C(1)*	−854	1650	5221	Si	5883	5156	10 913
C(2)*	−84	24	4538	C(1)	5675	5364	9078
C(3)*	−1779	1935	4538	C(2)	4750	4594	8395
C(4)*	−2174	2148	7768	C(3)	6445	6289	8395
C(5)*	814	2466	7393	C(4)	7691	5122	11 329
				C(5)	5074	6575	11 791
Unit two (fold unit):				Si*	5155	5884	4497
O	−401	−98	7751	C(1)*	5363	5676	6331
Si	91	−842	9132	C(2)*	4593	4751	7015
C(1)	1230	285	10 072	C(3)*	6288	6446	7015
C(2)	2594	242	9838	C(4)*	6807	6210	3719
C(3)	716	1167	11 006	C(5)*	4021	7313	4161
C(4)	987	−2410	8705	O	4510	4510	3858
C(5)	−1386	−1241	10 181				

inherent stiffness of the chain caused by the systematic inclusion of phenyl groups, would make self-diffusion of the molecule extremely difficult and offers an explanation for the observed melting of these crystals without prior lamellar thickening.

28.6 SUMMARY

Structure determination of polymers using X-ray diffraction is a combination of crystallography, diffraction theory (especially of helical structures) and computerized model building. The three

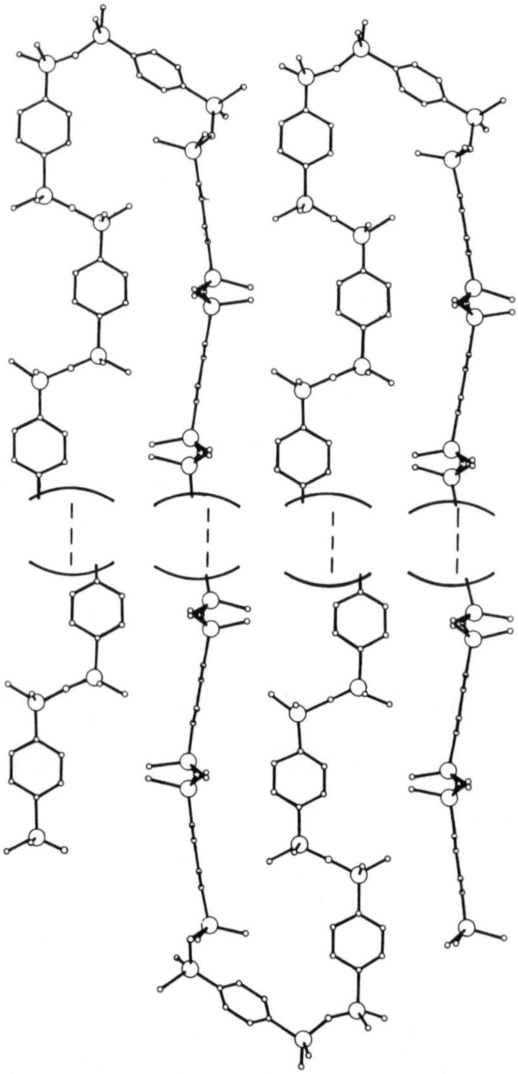

Figure 36 Projection of the (110) plane of a hypothetical chain-folded single crystal showing the way in which the folds combine with the known crystal structure to make up the complete lamellar crystal (reproduced by permission of Butterworths from ref. 55)

aspects interact constantly during the processes of structure elucidation. Care has to be exercised when intertwining helices occur, or when special symmetry relationships change the positions of the meridional diffraction signals as seen in the last example. The increasing use of computerized model-building procedures in the structure determination of fibrous polymers is providing a welcome improvement in the reproducibility and precision of determination of polymer conformation and structure and allows their presentation in a convenient form. Precision, however, is no substitute for accuracy and, in general, substantial improvements are required in the collection, processing and measurement of diffraction intensities. Three independent sets of measurements of the diffraction intensities for poly(tetramethylene terephthalate) show discrepancies of the order of 0.2 in R-factors obtained by comparison between the measured and calculated structure factor amplitudes.[18] Further improvements in the measurement and correction of intensities have been discussed[18,19,62-64] and are in the process of development. When optical density is digitized using a two-dimensional microdensitometer the data can be corrected point by point by applying the appropriate correction terms for each point in reciprocal space. Also the data can be represented on an undistorted grid.[63,64]

The methods of polymer structure determination outlined here have focussed on the diffraction from the crystalline part of the sample. Scattering from amorphous material overlays this diffraction and broad diffraction halos often obscure parts of the pattern, as may be seen in Figure 23.

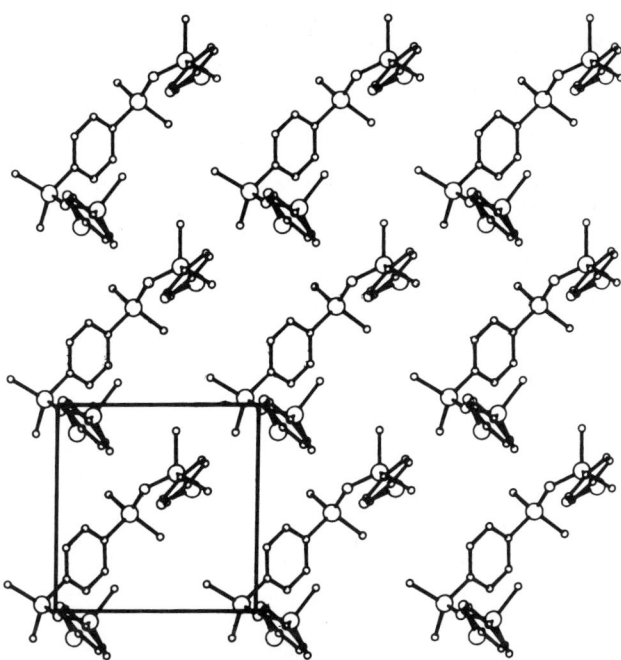

Figure 37 Topographical map of a regular chain-folded poly(TMPS) crystal. The square box indicates the position of the unit cell for the crystalline stems relative to the fold surface (reproduced by permission of Butterworths from ref. 55)

Diffraction effects emanating from distortions of the crystalline lattice occur in many polymeric substances. Detailed analyses of such distortions are reported in a number of texts[65-67] on the subject.

28.7 REFERENCES

1. G. H. Stout and L. H. Jensen, 'X-ray Structure Determination', Macmillan, London, 1968.
2. M. M. Woolfson, 'An Introduction to X-ray Crystallography', Cambridge University Press, 1970.
3. H. Lipson and W. Cochran, 'The Determination of Crystal Structures', Bell, London, 1953.
4. C. W. Bunn, 'Chemical Crystallography', 2nd edn., Clarendon Press, Oxford, 1961.
5. M. J. Bueger, 'X-ray Crystallography', Wiley, London, 1962.
6. N. F. M. Henry, H. Lipson and W. A. Wooster, 'The Interpretation of X-ray Diffraction Photographs', Macmillan, London, 1961.
7. T. L. Blundel and L. N. Johnson, 'Protein Crystallography', Academic Press, New York, 1976.
8. B. D. Cullity, 'Elements of X-ray Diffraction', 2nd edn., Addison–Wesley, London, 1978.
9. 'International Tables for X-ray Crystallography', Kynock Press, Birmingham, 1968–1972, vols. I–IV.
10. A. Keller and A. O'Connor, *Discuss. Faraday Soc.*, 1958, **25**, 114.
11. P. H. Geil, 'Polymer Single Crystals', Interscience, New York, 1963.
12. M. Kakudo and N. Kasai, 'X-ray Diffraction by Polymers', Elsevier, Amsterdam, 1972.
13. H. Tadokoro, 'Structure of Crystalline Polymers', Wiley, New York, 1979.
14. C. W. Bunn and E. V. Garner, *Proc. R. Soc. London, Ser. A*, 1947, **189**, 39.
15. L. E. Alexander, 'X-ray Diffraction Methods in Polymer Science', Wiley, New York, 1969.
16. E. D. T. Atkins and K. D. Parker, *J. Polym. Sci., Part C*, 1963, **28**, 69.
17. R. D. B. Fraser and T. P. MacRae, 'Conformation in Fibrous Proteins', Academic Press, New York, 1973.
18. I. H. Hall, in 'Fibre Diffraction Methods', ed. A. D. French and K. H. Gardner, American Chemical Society, Washington, 1980, p. 335.
19. E. D. T. Atkins, in 'Fibre Diffraction Methods', ed. A. D. French and K. H. Gardner, American Chemical Society, Washington, 1980, p. 33.
20. A. J. C. Wilson, *Acta Crystallogr.*, 1950, **3**, 397.
21. A. Turner-Jones and C. W. Bunn, *Acta Crystallogr.*, 1962, **15**, 105.
22. W. Cochran, F. H. C. Crick and V. Vand, *Acta Crystallogr.*, 1952, **5**, 581.
23. 'Bessel Functions, British Association Mathematical Tables', Cambridge University Press, 1958.
24. E. Jahnke and F. Emde, 'Tables and Functions with Formulae and Curves', Dover, New York, 1945.
25. K. Veluraja and E. D. T. Atkins, *Carbohydr. Polym.*, 1987, **7**, 133.
26. N. S. Anderson, J. W. Campbell, M. M. Harding, D. A. Rees and J. W. B. Samuel, *J. Mol. Biol.*, 1969, **45**, 85.
27. C. Upstill, E. D. T. Atkins and P. T. Attwool, *Int. J. Biol. Macromol.*, 1986, **8**, 275.
28. R. Chandrasekaran, R. P. Millane, S. Arnott and E. D. T. Atkins, *Carbohydr. Res.*, 1988, **175**, 1.
29. H. Kusanagi, H. Tadokoro and Y. Chatani, *Macromolecules*, 1976, **9**, 531.
30. E. D. T. Atkins, K. D. Parker and R. D. Preston, *Proc. R. Soc. London, Ser. B*, 1969, **173**, 209.

31. W. S. Fulton and E. D. T. Atkins, in 'Fibre Diffraction Methods', ed. A. D. French and K. H. Gardner, Americal Chemical Society, Washington, 1980, p. 385.
32. H. Tadokoro, Y. Chatani, H. Kusanagi and M. Yokoyama, *Macromolecules*, 1970, **3**, 441.
33. V. M. Coiro, P. De Santis, A. M. Liquori and L. Mazzarella, *J. Polym. Sci., Part C*, 1969, **16**, 4591.
34. R. Buter, Y. Y. Tan and G. Challa, *J. Polym. Sci., Polym. Chem. Ed.*, 1973, **11**, 2975.
35. G. N. Ramachandran and V. Sasisekharan, *Adv. Protein Chem.*, 1968, **23**, 283.
36. H. A. Scheraga, *Adv. Phys. Org. Chem.*, 1968, **6**, 103.
37. A. J. Hopfinger, 'Conformational Properties of Macromolecules', Academic Press, New York, 1973.
38. T. Shimanouchi and S. Mizushima, *J. Chem. Phys.*, 1955, **23**, 707.
39. T. Miyazawa, *J. Polym. Sci.*, 1961, **55**, 215.
40. S. Foord and E. D. T. Atkins, *Int. J. Biol. Macromol.*, 1980, **2**, 193.
41. S. Foord and E. D. T. Atkins, *Int. J. Biol. Macromol.*, 1981, **3**, 297.
42. S. Foord and E. D. T. Atkins, *Int. J. Biol. Macromol.*, 1984, **6**, 327.
43. H. Sugeta and T. Miyazawa, *Biopolymers*, 1967, **5**, 673.
44. S. Arnott and A. J. Wonacott, *Polymer*, 1966, **7**, 157.
45. P. J. Smith and S. Arnott, *Acta Crystallogr., Sect. B*, 1978, **34**, 3.
46. H. Tadokoro, Y. Takahashi, Y. Chatani and H. Kakida, *Makromol. Chem.*, 1967, **109**, 96.
47. H. Kakida, D. Makino, Y. Chatani, M. Kobayashi and H. Tadokoro, *Macromolecules*, 1970, **3**, 569.
48. Y. Takahashi, Y. Osaki and H. Tadokoro, *J. Polym. Sci., Polym. Phys. Ed.*, 1980, **18**, 1863.
49. Y. Takahashi, Y. Osaki and H. Tadokoro, *J. Polym. Sci., Polym. Phys. Ed.*, 1981, **19**, 1153.
50. M. A. Gomez, E. D. T. Atkins, C. Upstill, A. Bello and J. G. Fatou, *Polymer*, 1988, **29**, 225.
51. E. Pérez, M. A. Gómez, A. Bello and J. G. Fatou, *Colloid Polym. Sci.*, 1981, **261**, 571.
52. J. H. Magill, *J. Polym. Sci., Part A-2*, 1967, **5**, 89.
53. M. Kojima, J. H. Magill and R. L. Merker, *J. Polym. Sci., Part A-2*, 1974, **12**, 317.
54. M. N. Haller and J. H. Magill, *J. Appl. Phys.*, 1969, **40**, 4261.
55. K. H. Gardner, J. H. Magill and E. D. T. Atkins, *Polymer*, 1978, **19**, 370.
56. L. E. Alexander, M. G. Northcott and R. J. Engmann, *J. Phys. Chem.*, 1967, **71**, 4298.
57. G. S. Smith, 'Program and Abstracts, American Crystallographic Association', Bozemann, MT, July 1964.
58. G. S. Smith and L. E. Alexander, *Acta Crystallogr.*, 1963, **16**, 1015.
59. H. Steinfink, B. Post and I. Frankuchen, *Acta Crystallogr.*, 1955, **8**, 420.
60. D. Carlström and G. Falkenberg, *Acta Chem. Scand.*, 1973, **27**, 1203.
61. N. Okui and J. H. Magill, *Polymer*, 1976, **17**, 1087.
62. I. H. Hall, in 'Structure of Crystalline Polymers', ed. I. H. Hall, Elsevier, London, 1984, p. 39.
63. R. D. B. Fraser, T. P. MacRae, A. Miller and R. J. Rowlands, *J. Appl. Crystallogr.*, 1976, **9**, 81.
64. R. D. B. Fraser, E. Suzuki and T. P. MacRae, in 'Structure of Crystalline Polymers', ed. I. H. Hall, Elsevier, London, 1984, p. 1.
65. B. K. Vainshtein, 'Diffraction of X-rays by Chain Molecules', Elsevier, London, 1966.
66. R. Hosemann and S. N. Bagchi, 'Direct Analysis of Diffraction by Matter', North-Holland, Amsterdam, 1962.
67. A. Guinier, 'X-ray Diffraction in Crystals, Imperfect Crystals and Amorphous Bodies', Freeman, San Francisco, 1963.

29

Electron Diffraction from Crystalline Polymers

DOUGLAS L. DORSET

Medical Foundation of Buffalo Inc., Buffalo, NY, USA

29.1 INTRODUCTION

Soon after the Davisson–Germer experiment in 1927, the newly discovered electron diffraction was applied in both reflection and transmission modes to study the structures of linear chain molecules including polymers. Almost 60 years later, polymer physics remains the discipline in which electron diffraction techniques are most commonly used for quantitative crystal structure analysis. With the recent advent of high resolution (*ca.* 3 Å) electron microscope imaging techniques for organic specimens, diffraction studies of average crystal structure can now be correlated to local specific variations of crystal texture (including defects) as is shown in Volume 1, Chapter 34.

Unfortunately, the quantitative use of electron diffraction and microscopy for crystallographic research (hence the term 'electron crystallography') has a somewhat spotted history, largely caused by the early use of inadequate scattering models which were too reliant on X-ray diffraction theory. Misapplication of an overly simplistic theoretical framework sometimes led to structure analyses which were clearly at variance with results from, for example, X-ray crystallography. Because the literature in the field can be very confusing to a newcomer, this chapter endeavors to explain the important constraints to the use of this technique for structure analysis as well as to demonstrate its many positive features for polymer research. There also exists a vast literature on the more qualitative use of electron diffraction for the determination of unit cell dimensions and space group for many polymer crystals which cannot be covered in this review for want of space. Reference to these studies can be found in several works.[1-3]

29.2 FEATURES OF ELECTRON DIFFRACTION STRUCTURE ANALYSIS FOR POLYMER RESEARCH

Linear polymers can crystallize with varying degrees of order. An entanglement of chains with very small crystallite regions is an array that is only slightly more ordered than the amorphous state. If the material can be oriented along a fiber axis direction, then a polycrystalline fiber may be formed with rotational disorder around the orientation axis. Often polymer chains crystallize by chain folding to form thin lamellae in which the 'stem' region is similar to crystals of a monodisperse oligomer. In a least ordered expression of this, twisted lamellae can radiate from a center to produce a spherulite; in the most ordered form, a single lamellar plate can be formed which might have a surface length of several micrometers and perhaps ≤ 100 Å thickness. (Aspects of polymer crystallization have been treated in extensive detail by Geil[2] and Wunderlich.[3]) An optimal crystallographic probe for such samples is one which can obtain a useful signal from a single microcrystal.

Of the three radiations which are commonly used to resolve interatomic distances, electrons are the most efficiently scattered by matter, with a scattering cross section *ca.* a thousandfold larger than X-rays.[4] Thus, while X-ray and neutron diffraction experiments would give a radially disordered Debye–Scherrer pattern from a polycrystalline sample, electron diffraction could be used to examine each microcrystal component as a single crystal, providing a least ambiguous determinant of unit cell parameters and symmetry. (For this reason, electron diffraction patterns are sometimes used to aid the indexing of powder X-ray diffraction patterns from polymers.[5]) The wavelength of fast electrons is also much smaller than, say, Cu $K\alpha$ X-rays, *e.g.* at 100 kV the electron beam wavelength is 0.037 Å *vs.* 1.5418 Å for the X-ray source mentioned. Since the Ewald sampling sphere radius is the reciprocal of wavelength, the surface can be approximated as a plane for electron diffraction so that a whole zone of diffraction data can be collected in one exposure (Figure 1).

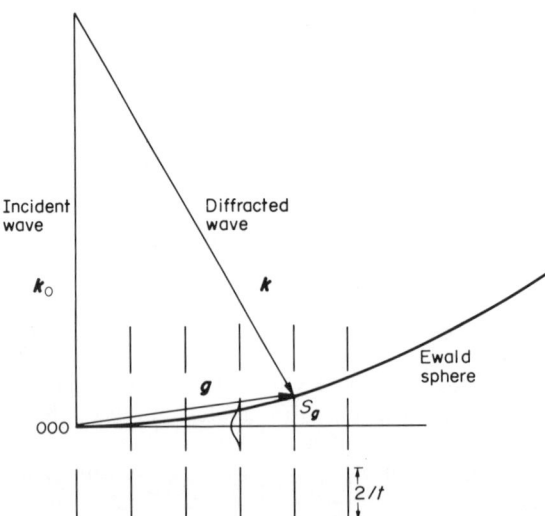

Figure 1 Reciprocal lattice net from a thin plate crystal viewed edge-on. As is clarified further in Figures 4 and 5, the diffraction spots are elongated by $2/t$ in a direction corresponding to the plate normal (where t is the plate thickness). The diffraction pattern is sampled with an Ewald sphere of radius $1/\lambda$, where λ is the electron wavelength, and k_0 and k are, respectively, incident and diffracted wave vectors. The reciprocal lattice vector g is related to these vectors by a simple vector sum. The deviation parameter s_g is the reciprocal distance of the Ewald sphere from a reflection center. When the wavelength $\lambda = 1/|k|$ is very small and the crystal is thin, it is apparent that many reflections will be sampled for any crystal orientation

Moreover, a useful electron diffraction pattern can be photographed in a matter of seconds compared to the hours required for X-ray experiments with typical laboratory sources.

The instrument normally employed for electron diffraction studies is an electron microscope for which an array of variable focus magnetic lenses can be used as an electron optical bench to set up a vast variety of diffraction modes and illumination conditions.[6-8] Given the occurrence of separate diffraction and image planes in the optical path of a lens (Figure 2), one can achieve many possible magnifications of either plane by a multiple lens sequence. The number of effective diffraction camera lengths is immense, therefore, ranging from the 0.20 to 3.6 m range often used for investigation of crystals with unit cell spacings typical of molecular crystals to, for example, > 70 m in high dispersion modes used to resolve very small reciprocal spacings (large unit cell lengths, *e.g.* ≥ 100 Å). The effective area isolated in a specimen for a diffraction experiment can be achieved either by defining an area in a magnified image with an aperture at the image plane of a lens, or by control of the incident illumination size. Specimens can be tilted and rotated on eucentric goniometer stages, they can be cooled or heated for dynamic experiments, they can be examined in the presence of solvent or a gas in environmental chambers and they can be deformed for stress–strain analysis. Many of these instrumental topics are covered in detail in standard treatises on electron microscopy for materials scientists or structural biologists.[6,7,9-13]

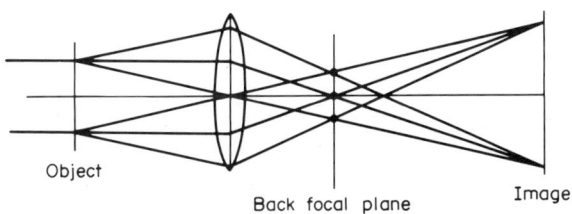

Object

Back focal plane

Image

Figure 2 Geometrical lens construction showing the spatial relationship between the irradiated object and the principal planes of the lens, *i.e.* the back focal plane where the diffraction pattern is formed and the image plane. (Although diffraction spots are drawn on the back focal plane here, the diffraction pattern from a non-periodic object is continuous as shown in Figure 4b.) If successive variable focus lenses are placed after the major objective lens, then either the back focal plane or image plane can be magnified on the screen of an electron microscope

At the accelerating voltages for typical electron microscopes, there are two basic kinds of electron diffraction experiments. Laue or transmission diffraction, the most usual kind used in structural research, considers the electron scattering from an irradiated volume element (Figure 3a) and offers the best specimen penetration. In the absence of other perturbations, the diffracted beams represent the total contents of the volume element. Illuminated areas are usually in the range 1 to 10 μm, although recently microbeam diffraction techniques have facilitated the examination of polymer samples with a beam of several hundred angstroms cross section.[14,15] Bragg or reflection diffraction (Figure 3b) is the scattering from a crystal face with a very small penetration of the sample. This diffraction mode, which is the oblique incidence geometry originally used by Davisson and Germer, is a surface technique which can be used to investigate structure at the outermost layer of the crystal (*e.g.* 20 to 40 Å for 30 to 60 kV electrons[16]) which may be less than a unit cell length.

Although both diffraction techniques have been used to investigate the structure of linear chain structures, most of the discussion in this chapter will deal with the quantitative treatment of diffraction intensity data from transmission diffraction experiments. A much more thorough discussion of practical experimental procedures also appears in a recent review on the electron diffraction structural analysis of organic materials.[17]

29.3 FUNDAMENTAL PRINCIPLES

29.3.1 Reciprocity of Structure Images and Diffraction Patterns

Crystallography is a branch of optics. Thus, any treatment of an imaging process has its corollary in terms of diffraction since, as shown in Figure 2, a diffraction pattern is formed as one proceeds along the optic axis of a lens to create an image of an object. The relation between an image and its diffraction pattern is a reciprocal one and proceeds *via* a mathematical operation known as the Fourier transform (FT). Often the analysis of image and diffraction problems can be carried out with the knowledge of merely a few typical FT pairs.

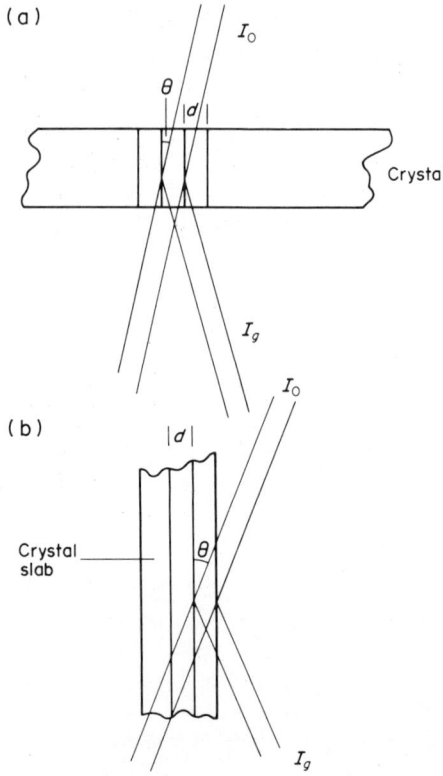

Figure 3 Geometrical representations of two diffraction geometries: (a) transmission or Laue diffraction where a volume element is irradiated and (b) reflection or Bragg diffraction where only a surface is irradiated by the incident beam. As usual, the diffraction peak positions for a periodic object are given by the Bragg equation, $n\lambda = 2d \sin\theta$

Suppose that a one-to-one mapping exists between an object mass density and its image density (an ideal case). If this object is an arbitrary shape (Figure 4a) then, as its image is formed by the perfect lens in Figure 2, a diffraction pattern will appear in the back focal plane of the lens. The distribution of wave magnitude on this plane will depend on the vectoral combination of plane wave radiation scattered by the object (amplitude + phase). (Because a recording device will measure the power of this wave front, the recorded signal is actually the intensity, which is the square of the wave amplitude.) This diffraction pattern will be related to the object structure including its symmetry. If the object is not periodic, then its diffraction pattern will be continuous. If the object has a mass distribution $f(r)$, then its FT or diffraction pattern can be expressed

$$F(s) = \int_{-\infty}^{\infty} f(r)\exp 2\pi irs \, dr = FT\, f(r) \tag{1}$$

and conversely the object wave function can be expressed

$$f(r) = \int_{-\infty}^{\infty} F(s)\exp(-2\pi irs)\, ds = FT^{-1} F(s) \tag{2}$$

These are FT pairs and can be represented also by the notation

$$f(r) \leftrightarrow F(s) \tag{3}$$

The distribution of wave amplitude in Figure 4(b) has a reciprocal relation to the image amplitude. Thus a diffraction vector magnitude $s \propto 1/r$ exists such that the signal at the intersection point for the optic axis represents zero resolution and as one proceeds away from this point the resolution of the signal improves.

When two identical objects are placed side by side (Figure 4c), interference occurs between the waves scattered by these two objects if the incident radiation is coherent (Young's fringe experiment in optics), so that its diffraction pattern (Figure 4d) contains regions where the wave fronts cancel. If

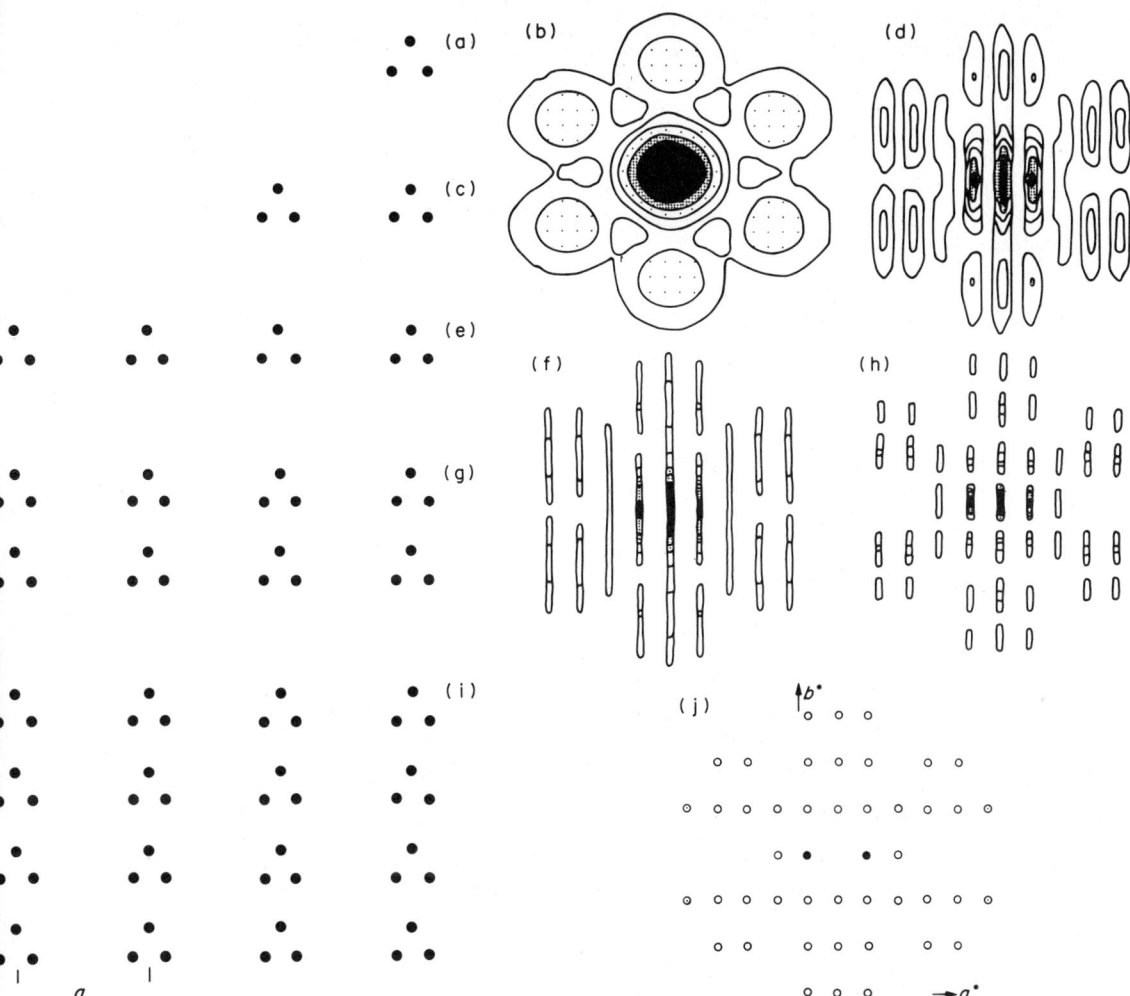

Figure 4 Diffraction from an arbitrary object — here a threefold array of Gaussian points (a). Its FT (b) is continuous but, due to Friedel symmetry where $|F_{hkl}| = |F_{\bar{h}\bar{k}\bar{l}}|$ (*i.e.* a center of inversion for diffraction amplitude), the continuous diffraction pattern has sixfold symmetry. If two such objects are placed near to each other (c), then the FT of both (d) will resemble (b) except for the interference function between them. When enough objects are placed in a periodic row with spacing a (e), then the FT (f) becomes discrete in the corresponding reciprocal direction with a spacing $a^* \propto 1/a$. However, it remains continuous in the orthogonal direction. When a second row of objects is added to the first one (g), an orthogonal interference function is seen (h) similar to the one in (d) and when the motif repeats with spacing b in this direction (i), its diffraction pattern is composed of discrete spots (j) spaced at $b^* \propto 1/b$, which sample the intensity of the continuous transform in (a) with the reciprocal lattice net $K(h,k)$. As discussed in the text, the reciprocal lattice $K(hk)$ is also the FT of the space lattice $k(na,mb)$. Thus the trigonal motif $f(r)$ is repeated in the crystal by the convolution $f(r) \otimes k(na,mb)$. Since the convolution operation is Fourier transformed to multiplication, then the reciprocal lattice is represented formally by the expression $F(s) \cdot K(h,k)$ where $F(s)$ is the transform of a single object

even more objects are added in a periodic array with repeat a (Figure 4e), then the spacing of the diffraction intensity magnitude will be proportional to $1/a$ (Figure 4f). If one repeats the structure in two dimensions then the discreteness of diffraction intensity is seen in two dimensions (Figures 4i and j); the objects can thus be said to occupy a *space lattice* and the diffraction pattern will represent a *reciprocal lattice* with inverse repeat spacings and angles dependent on the values assumed by the space lattice according to the symmetry of the array. We have, of course, created a crystalline array and the dimensional relationships between a crystal space lattice and its diffraction pattern can be found in any good crystallography text[18-20] or the International Tables for X-ray Crystallography.[21]

Let us analyze this diffraction in more detail with Fourier transforms. The diffraction pattern in Figure 4(j) is obviously related to the one in Figure 4(b) and indeed it is apparent that the wave function $F(s)$ is sampled by the reciprocal lattice function $K(s)$, at the indices h, k which represent sampling points (delta functions, as shown below), and we can represent this sampling by the multiplication $F(s) \cdot K(s)$. We also know intuitively from Figure 4 that the space lattice function $k(r)$ sampled at (na, mb), where a and b are the *unit cell* repeats and n and m are integers, is the FT of $K(s)$ so that

$$k(r) \leftrightarrow K(s) \tag{4}$$

The space lattice (compare Figures 4a and i), on the other hand, samples the average crystal structure in a different way, *i.e.* the object function $f(r)$ is repeated at intersections of the coordinates (na, mb) which define the edges of the repeating unit cell. This mathematical operation is a convolution represented as $\otimes$ so that in the image (or object) space we have an array $f(r) \otimes k(na, mb)$. We also know that

$$F(s) \cdot K(h,k) \leftrightarrow f(r) \otimes k(na, mb) \tag{5}$$

but

$$F(s) \leftrightarrow f(r) \tag{6}$$

$$K(h,k) \leftrightarrow k(na, mb) \tag{7}$$

Hence the operations multiplication and convolution are also Fourier transforms of one another. A formal definition of the convolution may be given as

$$f(r) \otimes g(r) = \int_{-\infty}^{\infty} f(r_1) g(r - r_1) \, dr_1 \tag{8}$$

A related function which will be used below is the correlation function $*$ defined by

$$f(r) * g(r) = \int_{-\infty}^{\infty} f^*(r_1) g(r_1 + r) \, dr_1 \tag{9}$$

where $f^*(r_1)$ is the complex conjugate of $f(r_1)$. In terms of Fourier transforms we can also write

$$f(r) \otimes g(r) \leftrightarrow F(s) G(s) \tag{10}$$

$$f(r) g(r) \leftrightarrow F(s) \otimes G(s) \tag{11}$$

$$f(r) * g(r) \leftrightarrow F^*(s) G(s) \tag{12}$$

$$f(r) g(r) \leftrightarrow F^*(s) * G(s) \tag{13}$$

If an object has a boundary $x = \pm na/2$, as shown in Figure 5, defined by the rectangle function $\mathrm{rect}(r)$, then its FT can be shown to be the function $\sin \pi a s / \pi s$. As a becomes very large then $\lim_{a \to \infty} \sin \pi a s / \pi s = \delta(s)$ which is the delta function. The sampling of a function by a delta function reproduces the function shifted to the center a defined by

$$\delta(x - a) = \begin{cases} 0 \text{ for } x \neq a \\ \infty \text{ for } x = a \end{cases} \tag{14}$$

Additionally

$$\int_{-\infty}^{\infty} \delta(x - a) \, dx = 1 \tag{15}$$

Thus FT $\delta(x) = 1$ but FT $\delta(x - a) = \exp 2\pi i u a$. (That is, a translation in real space corresponds to a phase shift in reciprocal space.)

If a delta function array exists with period a such that $k(x) = \sum_{n=-\infty}^{\infty} \delta(x - na)$ and it is constrained by function $s(x) = \mathrm{rect}(x)$ with width Na then we can define

$$h(x) = s(x) \cdot k(x) = s(x) \sum_{n=-\infty}^{\infty} \delta(x - na) \tag{16}$$

Its FT will be

$$H(u) = \left[\frac{\sin(\pi Nau)}{\pi Nau} Na \right] \otimes \sum_{h} \delta\left[u - \frac{h}{a} \right], \text{ where } \sum_{h} \delta\left[u - \frac{h}{a} \right] = K(u) \tag{17}$$

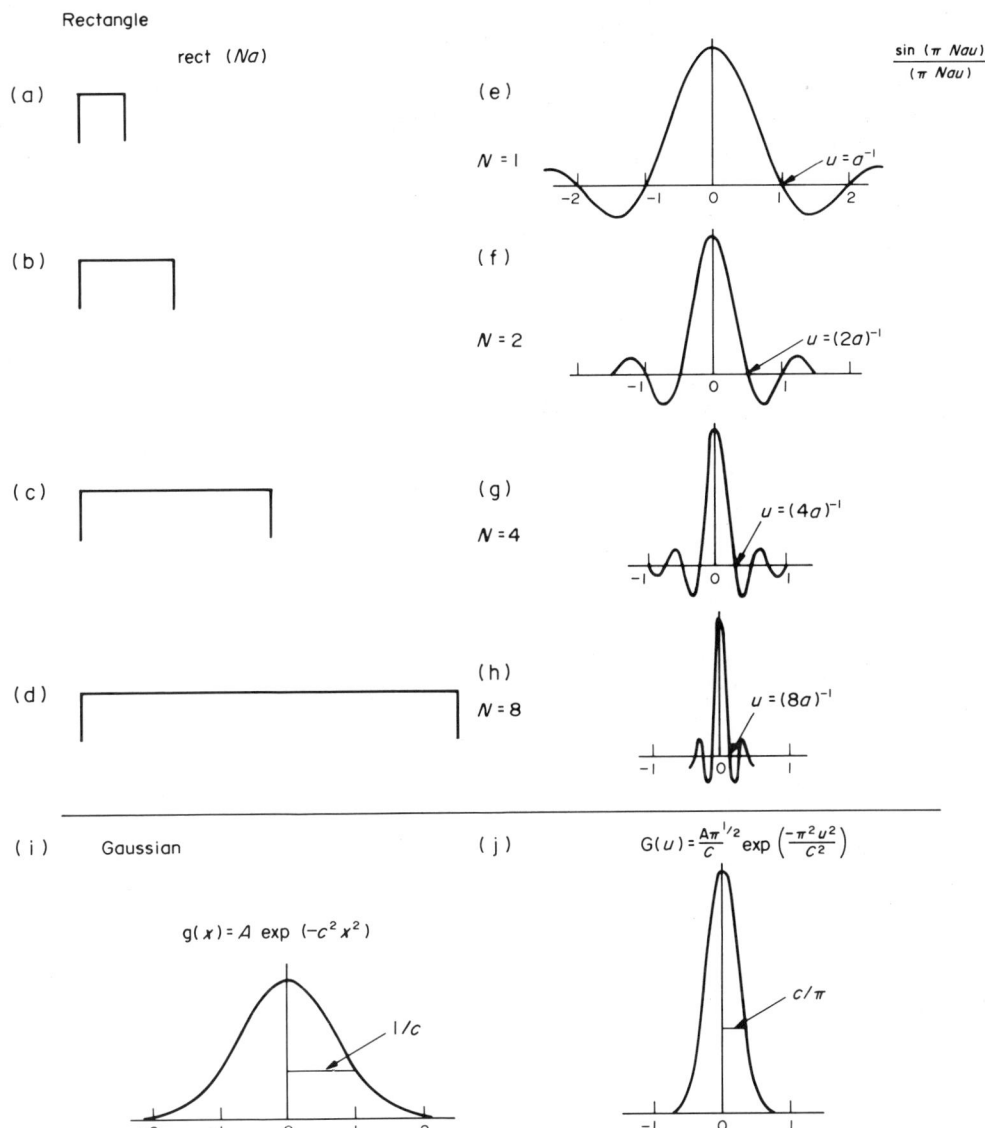

Figure 5 Basic Fourier transforms. Given a rectangle, rect(Na), then its transform is $\sin(\pi Nau)/\pi u$. (In the following, we normalize its maximum magnitude to 1.0 by the term $1/Na$.) In Figure 4, such a function could represent the finite size of a crystalline array where thickness $t = Na$ or consisting of N unit cells of lengths a. This is the *shape function* originally described by Ewald. Its transform, *i.e.* the *shape transform*, is seen to extend the maxima of diffracted spots so that the first zero is found at $1/Na$. (However, note that the transform of a single object is continuous so that the shape transform will describe an interference function for two or more objects, *i.e.* $N \geq 2$. Refer also to Figures 1 and 4.) Obviously, as the number of unit cells increases, the breadth of this shape transform will decrease. This is illustrated by (a)–(d) which give rectangle functions of increasing length which respectively transform to (e)–(h). Intuitively one can see that $\lim_{Na \to \infty} \mathrm{rect}(Na) = \mathrm{FT}\ \delta(x)$ as the crystal becomes infinitely large. Another useful function is the Gaussian $g(x)$ represented in (i). This might describe distribution of mass around an equlibrium point (*e.g.* the mass points in Figure 4 have a Gaussian distribution) or, approximately, the potential distribution in an atom. Its FT $G(u)$ is again a Gaussian, but note that the halfwidth has a reciprocal relation to the original Gaussian halfwidth. Thus, if the mass distribution described by $g(x)$ has a rather large halfwidth, then its diffraction pattern resolution will be restricted in resolution by $G(u)$. Formally stated, the crystal mass distribution in Figure 4 is convoluted by $g(r)$ such that $h(r) = f(r) \otimes k(na, mb) \otimes g(r)$ which is transformed to $H(s) = F(s) \cdot K(hk) \cdot G(s)$. Thus, $G(s)$ becomes an envelope for the diffraction pattern and demonstrates, for example, loss of diffraction resolution due to thermal motion or disorder. If the halfwidth of the atom distribution becomes very large, then the diffraction pattern will vanish. Also note

$$\lim_{c \to 0} G(u) = \delta(u)$$

This is another way of expressing the sequence of diffraction patterns in Figure 4 where

$$\sum_{n=-\infty}^{\infty} \delta(x-na) \leftrightarrow \frac{1}{a}\sum_h \delta(u-h/a) \tag{18}$$

expresses the relationship between infinite space lattices and reciprocal lattices but the *shape function* $s(x) = \text{rect}(x)$ constrains the size of the crystal so that each diffraction spot will be spread by the *shape transform* $\text{sinc}(\pi Nau) \leftrightarrow s(x)$ which has its first zero at $1/Na$. Also, since $\lim_{a\to\infty} \text{sinc}(\pi as) = \delta(s)$ the diffraction spots will be sharp points for an infinitely large crystal.

One final FT pair which is very useful involves the Gaussian $g(r)$ (Figure 5). Its FT $G(s)$ is also a Gaussian such that $G(s) \leftrightarrow g(r)$. If each object in Figure 4(i) were distributed according to $g(r)$ such that $h(r) = f(r) \otimes g(r)$ then its FT would be

$$H(s) = F(s) \cdot G(s) \tag{19}$$

which can describe the reduction of a diffraction pattern resolution by crystalline disorder or thermal motion.

The reader is referred to several excellent works[22-24] to extend his knowledge of useful FT pairs. Although the examples above have been given in one or two dimensions, the generalization to three dimensions is straightforward.

29.3.2 Kinematical Diffraction Theory

Electrons are scattered by the potential field within an atom. If $f_x(s)$ is the X-ray scattering factor of an atom with atomic number Z then the scattering factor for electrons $f_e(s)$ can be expressed by the Mott formula

$$f_e(s) = 8\pi^2 \frac{me^2}{h^2}(Z - f_x(s))/g^2$$

$$= 0.023934\,\lambda^2(Z - f_x(s))/\sin^2\theta \tag{20}$$

where $g = 4\pi\lambda^{-1}\sin\theta$. Although the scattering factor of X-rays increases linearly with atomic number, it can be readily observed[25] that the relative breadth of electron scattering factor magnitudes is much narrower than for X-rays. Like X-ray scattering factors, the electron scattering dependence on resolution is approximately Gaussian.

The propagation function (elastic scattering) for an incident electron wave through an object is represented by[24]

$$q(x, y) = \exp(-i\sigma\phi(x, y)\Delta z) \tag{21}$$

where

$$\phi(x, y) \cong \int \phi(x, y, z)\,dz \tag{22}$$

is a projected slice of the electrostatic potential for a thickness Δz. The constant $\sigma = \frac{\pi}{W\lambda} \times \frac{2}{[1 + (1 - \beta^2)^{1/2}]} = \frac{1}{\hbar c\beta}$ (where W is the accelerating voltage, λ is the electron wavelength and $\beta = v/c$ is a relativistic correction) describes the interaction between the electron and the object. Expanding the transmission function above we find

$$q(x, y) = 1 - i\sigma\phi(x, y)\Delta z + \frac{\sigma^2}{2!}\phi(x, y)^2\Delta z^2 + \dots \tag{23}$$

or in terms of its FT

$$\Phi(hk) = \text{FT } q(x, y) = \delta - i\sigma\Delta z F(hk) + \frac{\sigma^2\Delta z^2}{2!} F(hk) \otimes F(hk) + \dots \tag{24}$$

where the delta function δ represents the unscattered component of the incident electron beam.

From this expression for the diffraction pattern we find that we can measure intensities $I_{obs} \simeq \Phi(hk)^2$ which are simply related to the *kinematical* scattered wave magnitudes (*viz.* structure factor amplitudes) if only the first term of the exponential expansion is important. As usual the structure factor is related to the distribution of mass in the crystal by

$$F_{hkl} = \sum_j f_{e_j}(s) \exp(2\pi i s \cdot r) \tag{25}$$

where $s \cdot r = h\dfrac{x}{a} + k\dfrac{y}{b} + l\dfrac{z}{c}$, hkl are Miller indices of diffracted beams and x/a, y/b, z/c are fractional coordinates of atoms in the unit cell. We see that the integral in equation (1) is now approximated by a sum. The conditions for kinematical (or single) scattering of electrons are seen above to be dependent on the crystal thickness $n\Delta z$, the accelerating voltage for the electron W and also the atomic number of the unit cell contents Z. The accelerating voltage is an instrumental factor which can be controlled (generally $W \geq 100\,kV$) and, for organics, Z is appropriately small. The crystal thickness is an experimental variable. Fortunately many lamellar polymer crystals are $\leq 100\,\text{Å}$ thick, to ensure conditions approximating kinematical scattering.

29.3.3 Constraints to Interpretation of Diffraction Data

The deviation of measured structure factor magnitudes from the ideal case described by equation (25) can be caused by a number of factors as described in the following sections.

29.3.3.1 *Dynamical scattering*

In the early days of electron diffraction structure analysis it was sometimes found that measured diffraction intensity I was not proportional to the square of the structure factor magnitude. Although several beams are simultaneously excited for any crystal orientation, it was imagined that if the crystal were a mosaic, so that appropriately oriented blocks were separately responsible for individual reflections, then some form of the Blackman two beam dynamical theory (known in X-ray crystallography as 'primary extinction') could correct the observed data.[4] Although this correction had limited success, it was later found that the mosaic model was incorrect[24] since the crystals are nearly perfect within the coherence width of the incident beam. Dynamical interactions furthermore were shown to involve all the excited beams in the zone[24] as is anticipated by the expansion of equation (24) above.

The Cowley–Moodie multislice formulation[23] is often the most convenient one for computation of *n*-beam dynamical scattering, particularly if the exit wave from a series of crystals with increasing thickness is considered. In the image plane of the electron microscope the transmission function through n successive slices of a crystal is

$$\psi_n = \left[\psi_{n-1} \otimes \exp\left(\frac{ik(x^2 + y^2)}{2\Delta z} \right) \right] q_n(x, y) \tag{26}$$

where ψ_{n-1} is the transmission after the penultimate slice and $q_n(x, y)$ is that of the last slice. The exponential term is an approximation of the Ewald sphere curvature where $k = 2\pi/\lambda$. The magnitude of diffracted beams $\Phi(hk) = \text{FT}\,\psi_n$ thus involves the sequential convolution of all beams in the zone, so that these may no longer be simply related to the crystal structure. Although the variables listed above can be controlled to obtain diffracted intensity near their kinematical values, so that an *ab initio* structure analysis can be made, the most accurate description of the dynamical scattering from organic crystals has been shown experimentally to be an *n*-beam formulation[17] and not the two beam theory.

29.3.3.2 *Crystal deformation*

Diffraction contrast images (bright field, dark field) of thin organic crystals reveal that the major deformation of these is due to elastic bending over a few degrees within the coherence width of the incident beam. Also, molecular organic compounds generally crystallize from solution so that the longest unit cell axis is nearly perpendicular to the major crystal face or parallel to the incident electron beam. Thus elastic bend deformation may play a major role in altering the diffracted intensity. Following Cowley[26] the diffracted intensity from a regularly deformed crystal is expressed by

$$I(s) = \sum_i W_i(s) \exp(2\pi i r_i \cdot s) \exp(-\pi^2 c^2 s^2 z_i^2) \tag{27}$$

The autocorrelation function for the unit cell atom distribution $\phi(r)*\phi(r)$ (where $\phi(r)$ is the crystal potential) produces vector positions $w_i(r)$ which have the FT $W_i(s)$. The transform

$$\phi(x, y)*\phi(x, y) \leftrightarrow F(hk)F^*(hk) = I(hk) \tag{28}$$

is thus modulated by a Gaussian function to describe the bend deformation which depends on the position of a reflection in reciprocal space s, the amount of bending c and the z component (along the beam direction) of the Patterson peak $w_i(r)$. If z_i is large, then the measured intensity I is no longer simply related to the crystal structure.

The importance of this perturbation has been demonstrated experimentally for a number of molecular organic crystal structures.[27-29] It was also seen that this was merely a kinematical approximation and that actually an n-beam dynamical calculation should be carried out for a bend-deformed crystal lattice. Such a rigorous calculation[30] demonstrates the utility of the above simplification when n-beam scattering is not important. Fortunately, the fiber repeat of polymer chains, which is approximately parallel in the beam direction for solution grown crystals, is often not large enough to affect an *ab initio* crystal structure analysis.[31]

29.3.3.3 Multiple incoherent scattering

Since polymer crystals are often composed of stacked lamellae, a perturbation to the diffraction intensity data can result if strongly scattered beams in one layer become primary beams for successive layers, such that, instead of an intensity $I(hk)$, one measures[32] $J(hk) = I(hk) + mI'(hk)$, where

$$I'(hk) = I(hk) \otimes I(hk)$$

$$= \sum_{h_2}\sum_{k_2} I(h_2 k_2) I(h_1 - h_2, k_1 - k_2) \tag{29}$$

and m is a weighting factor. (Actually, additional terms with n-fold self convolutions of intensity should also appear in the expression for $J(hk)$.) A salient feature of this multiple incoherent scatter is that space group forbidden reflections have non-zero values, since the self-convolution involves intensities and not the phased structure factors convolved in n-beam dynamical scattering models. (On the other hand, for coherent n-beam scattering, many types of translational symmetry and thus forbidden reflections will continue to be expressed in the electron diffraction pattern if the incident beam is parallel to a zone axis.[33]) Correction for this incoherent multiple scattering is very difficult and best minimized by controlling crystal thickness.

29.3.3.4 Radiation damage

Inelastic electron scattering is also important for organic specimens and the resultant beam-induced radiation damage can be a major factor which limits the resolution of data collection, particularly in the case of direct lattice imaging. Mechanisms for beam damage are also dependent upon the nature of the specimen. For instance, materials like polyethylene will lose hydrogen and cross-link; others, such as polyoxyethylene, form stable fragments which will escape from the crystalline matrix.[34]

Radiation damage is best avoided in diffraction experiments by use of a minimal beam intensity (*e.g.* 10^{-6} A cm^{-2}) and a very fast photographic film to record the exposures (*e.g.* an X-ray film such as Kodak DEF-5 or Ceaverken CEA-15). Since diffraction concentrates information in an array of spots, high resolution information from the average crystal structure can be recorded without noticeable damage, and thus this factor is much less critical here than for direct imaging experiments, for which[35] the beam damage must be monitored by electron diffraction to assess its importance at the resolution allowed by instrumental conditions.[36]

In addition to minimal beam exposure, expedients such as specimen cooling have been utilized.[37] Although the protection of the specimen (by 'freezing in' reactive species) is probably not so favorable as proposed recently,[38] there is certainly a factor of three to fivefold to be considered.

29.3.3.5 *Solvent loss*

There are sometimes instances when solvation is important for the integrity of linear polymer crystal structures, *e.g.* polysaccharides (see Table 1). If hydration needs to be stabilized in a sample, one can either use an appropriate environmental chamber,[39] which can limit the flexible use of the

Table 1 Quantitative Electron Diffraction Structure Analyses on Linear Chains

	Compound	Phasing technique (type of data)	Ref.
1.	Polyethylene		
1.1	Oligomer polymorphs		
	1.1.1 Hexagonal	Theoretical model (2D)	a
	1.1.2 Rotationally disordered orthorhombic	Theoretical model (2D)	b
	1.1.3 Triclinic parallel ($T_\parallel$)	X-Ray crystal structure (2D)	c
	1.1.4 Orthorhombic perpendicular ($O_\perp$)	X-Ray crystal structure (2D) or direct phasing	d, e, f
	1.1.5 Hybrid orthorhombic (HS1)	Patterson function (2D) or direct phasing	f, g
	1.1.6 Hybrid orthorhombic (similar to $M_\parallel$)	Patterson function (2D)	h
1.2	*n*-Alkanes (epitaxial growth)		
	1.2.1 *n*-Hexatriacontane	Known X-Ray crystal structure (3D)	i
	1.2.2 *n*-Triatriacontane B-form	Previously suggested model (2D)	j
1.3	Polyethylene (epitaxial growth)	$O_\perp$ Subcell packing (3D)	k
2.	Poly(tetrafluoroethylene) (oligomer)	Theoretical model (2D)	l
3.	Poly(ethylene oxide) (oligomer)	X-ray structure (2D)	h
4.	Poly(ethylene sulfide)	Potential function (2D)	m, n
5.	Poly(sulfur-nitride)	Patterson map (3D — spot + fiber data)	o
6	Poly(acetylene)		
	6.1 *Cis*	Chain model (2D)	p
	6.2 *Trans*	Chain model (2D)	q
7.	Poly(diacetylene)	Patterson map (2D)	r
8.	Poly(1,11-dodecadiyne)	Potential function (2D)	s
9.	α-Poly(3,3-bis-chloromethyl oxacyclobutane)	Potential function (2D)	t
10	Poly(ε-caprolactone)		
	10.1 Solution-crystallized	X-Ray determination (2D)	u
	10.2 Epitaxial	X-Ray structure models (3D)	k
11.	Poly(trimethylene terephthalate)	Potential function (2D)	v
12.	Poly(hexamethylene terephthalate)	Potential function (2D)	w
13.	Poly(*trans*-cyclohexanediyl dimethylene succinate)	Potential function and oligomer crystal structure (3D)	x
14.	Poly(γ-methyl-L-glutamate)		
	14.1 α-Form	Theoretical model (2D)	y
	14.2 β-Form	Theoretical model (2D)	z
15.	Cellulose	Potential function (3D—texture data)	aa
16.	Cellulose triacetate	Potential function (2D)	bb
17.	Nigeran (anhydrous)	Potential function (2D)	cc
18.	Dextran		
	18.1 Low temperature form	Potential function (3D)	dd
	18.2 High temperature form	Potential function (3D)	ee

[a] D. L. Dorset, *Chem. Phys. Lipids*, 1974, **13**, 133. [b] D. L. Dorset and W. A. Pangborn, *Chem. Phys. Lipids*, 1982, **30**, 1. [c] D. L. Dorset, *Z. Naturforsch., C: Biosci.*, 1983, **38**, 511. [d] B. K. Vainshtein, A. N. Lobachev and M. M. Stasova, *Sov. Phys.-Crystallogr. (Engl. Transl.)*, 1958, **3**, 452. [e] D. L. Dorset, *Acta Crystallogr., Sect. A*, 1976, **32**, 207. [f] D. L. Dorset and H. A. Hauptman, *Ultramicroscopy*, 1976, **1**, 195. [g] D. L. Dorset, *Biochim. Biophys. Acta*, 1976, **424**, 396. [h] D. L. Dorset, *J. Colloid Interface Sci.*, 1983, **96**, 172. [i] B. Moss, D. L. Dorset, J. C. Wittmann and B. Lotz, *J. Polym. Sci., Polym. Phys. Ed.*, 1984, **22**, 1919. [j] D. L. Dorset, *J. Polym. Sci., Polym. Phys. Ed.*, 1986, **24**, 79. [k] B. Moss, D. L. Dorset, J. C. Wittmann and B. Lotz, *J. Macromol. Sci., Phys.* 1985, **24**, 99. [l] D. L. Dorset, *Chem. Phys. Lipids*, 1977, **20**, 13. [m] H. Hasegawa, W. Claffey and P. H. Geil, *J. Macromol. Sci., Phys.*, 1977, **13**, 89. [n] B. Moss and D. L. Dorset, *J. Macromol. Sci., Phys.*, 1983, **22**, 69. [o] M. Boudeulle, *Cryst. Struct. Commun.*, 1975, **4**, 9. [p] J. C. W. Chien, F. E. Karasz and K. Shimamura, *Macromolecules*, 1982, **15**, 1012. [q] K. Shimamura, F. E. Karasz, J. A. Hirsch and J. C. W. Chien, *Makromol. Chem., Rapid Commun.*, 1981, **2**, 473. [r] D. Day and J. B. Lando, *Macromolecules*, 1980, **13**, 1483. [s] M. Thakur and J. B. Lando, *Macromolecules*, 1983, **16**, 143. [t] W. Claffey, K. Gardner, J. Blackwell, J. Lando and P. H. Geil, *Philos. Mag.*, 1974, **30**, 1223. [u] F. Brisse and R. H. Marchessault, 'Fiber Diffraction Methods', Ed. A. D. French and K. H. Gardner, *ACS Symp. Ser.*, 1980, **141**, 267. [v] S. Poulin-Dandurand, S. Pérez, J. F. Revol and F. Brisse, *Polymer*, 1979, **20**, 419. [w] F. Brisse, A. Palmer, B. Moss, D. L. Dorset, W. A. Roughead and D. P. Miller, *Eur. Polym. J.*, 1984, **20**, 791. [x] F. Brisse, B. Rémillard and H. Chanzy, *Macromolecules*, 1984, **17**, 1980. [y] L. I. Tatarinova and B. K. Vainshtein, *Vysokomol. Soedin.*, 1962, **4**, 261. [z] B. K. Vainshtein, and L. I. Tatarinova, *Sov. Phys.-Crystallogr. (Engl. Transl.)*, 1966, **11**, 494. [aa] W. Claffey and J. Blackwell, *Biopolymers*, 1976, **15**, 1903. [bb] E. Roche, H. Chanzy, M. Boudeulle, R. H. Marchessault and P. Sundararajan, *Macromolecules*, 1978, **11**, 86. [cc] S. Pérez, M. Roux, J. F. Revol and R. H. Marchessault, *J. Mol. Biol.*, 1979, **129**, 113. [dd] C. Guizard, H. Chanzy and A. Sarko, *J. Mol. Biol.*, 1985, **183**, 397. [ee] C. Guizard, H. Chanzy and A. Sarko, *Macromolecules*, 1984, **17**, 100.

electron microscope, or one can use cryotechniques for which the sample is rapidly cooled in a liquid alkane (propane, butane) or liquid nitrogen and transferred in a cooled sample holder to the electron microscope vacuum. The latter technique was applied in the pioneering work of Chanzy and his co-workers[40] and the required instrumentation is now a standard accessory on most electron microscopes.

29.4 CRYSTAL STRUCTURE ANALYSIS

For the FT pair given in (1) and (2), the ease of transformation with experimental data is *not* the same for each direction. If, for example, a high resolution image is available for a crystal structure, then its transmission function is recorded on a photographic film as an intensity which can be expressed[24]

$$I(x, y) \simeq 1 - 2\sigma\phi(x, y)\Delta z \tag{30}$$

as a linear property of the lattice potential $\phi(x, y)$ assuming the weak phase object approximation holds and also that the electron microscope objective lens is in optimal defocus (*i.e.* a transfer function property similar to Zernike contrast). Thus, it is easy to calculate a diffraction pattern from this image intensity. The reverse process, *i.e.* calculation of an image from recorded diffraction intensity, is not straightforward however, since the phase relationships between diffracted beams are lost. The goal of a crystal structure analysis, therefore, is to retrieve this phase information so that a high resolution image of the molecular packing can be produced. The correctness of a proposed structure is assessed generally *via* a residual

$$R = \frac{\Sigma| |F_0| - k|F_c| |}{\Sigma|F_0|} \tag{31}$$

where calculated structure factors for a given crystal packing model are compared to the observed magnitudes and a minimum value is sought. Procedures for this phase determination are outlined in the following sections.

29.4.1 Trial and Error

If a functional group is known to dominate the scattering from the molecules in the unit cell and, moreover, a certain conformation for this group can be assumed, then a translation or rotation of this group in the unit cell can be carried out and the predicted diffraction from these trial positions can be compared to the observed data. Although this seemingly random process can be very time-consuming, it is particularly useful if relative molecular orientations can be deduced *a priori* from the dimensions and symmetry of the unit cell. A choice of orientation can also be guided by a Patterson map as described in the following section.

29.4.2 Patterson Function

The autocorrelation function of an array of near point scatterers (atoms) was shown by A. L. Patterson to be a map of interatomic vectors translated to a common origin and subject to the symmetry operations of the unit cell (*e.g.* see ref. 41). In the early days of X-ray crystallography, many organic crystal structures were determined from Patterson maps utilizing structures which were labeled with a sufficiently high Z atom so that the scattering factor would suitably dominate the total scattering from the crystal,[17] hence $\Sigma f_{heavy}^2/\Sigma f_{light}^2 \simeq 1$. In electron diffraction structure analysis the detectability of heavy atoms is less pronounced since the relative range of scattering factor magnitudes is more constrained, although Patterson maps have been used for some structures.[42] Another drawback for heavy atom derivatives is the risk of introducing significant dynamical scattering.[24]

Patterson techniques are effective for detecting characteristic groups such as rings or zigzag chains as shown in recent analyses with electron diffraction data[43] and, coupled with automated molecular orientation techniques,[17] can be used for structure analysis.

29.4.3 Continuous Diffuse Scattering

Although its presence was first described in detail for electron diffraction from molecular crystals[44] and polymers,[45] the use of continuous diffuse scatter for crystal structure analysis has been most employed in X-ray crystallography, *e.g.* by Hoppe and co-workers.[46] As shown by Amoros and Amoros,[47] if thermal motion is the origin of non-Bragg continuous diffraction with intensity

$$I_{TDS} = |F_{mol}|^2 \left\{ 1 - \exp\left(-\frac{1}{2} B|s|^2 \right) \right\}$$ (32)

then the contribution of molecular (or group) entities in the unit cell according to the magnitude of the Debye–Waller factor B is incoherent and additive. Such scattering can be used to overcome symmetry constraints to the Patterson function calculated from the discrete diffraction spots. On the other hand, other more intense contributions to diffuse scattering due to crystal defects can overwhelm thermal effects as will be shown below.

29.4.4 Direct Phasing Methods

Although so-called direct phasing methods have revolutionized small molecule X-ray crystallography, their use in electron diffraction structure analysis has been somwhat limited.[48] As formulated by Hauptman and Karle[49,50] such methods generally evaluated the probability that a linear combination of phases

$$\alpha = \alpha_n + \alpha_k + \alpha_l + \alpha_m + \ldots$$ (33)

where *h, k, l, m* are Miller indices of several reflections (often under the condition that their vector sum is zero), has a certain value, based on normalized structure factor magnitudes. Given a set of highly probable combined phase values and the freedom to choose a small number of individual phases arbitrarily to define an origin, a sufficient number of phased structure factors can often be found to find features of the molecular packing in the unit cell. The first experimental application of such techniques to electron diffraction data was for alkane chain packings.[48] Later, the effects of dynamical scattering[51] and crystal bend deformation[52] were also assessed. The major problem for their use in electron crystallography is that often a two-dimensional data set is the only one suitable for structure analysis, placing an additional constraint on the techniques.

Recently[53] it has been suggested that electron microscope (EM) crystal structure images be used to obtain lowest resolution phase information and thus serve as a basis for phase extension into the higher resolution electron diffraction data (which overcome the resolution restrictions imposed by the objective lens transfer function). Such lattice images have helped to solve a polymer crystal structure for which the number of molecules in the unit cell could not be otherwise rationalized with the space group symmetry.[54]

29.4.5 Potential Energy Minimization

Perhaps the most effective technique for phasing electron diffraction intensity data (and also powder X-ray data[55]) from complex linear polymers is to find a minimized molecular conformation packing energy with some potential function of the general type[56]

$$\varepsilon = Aa^{-n} - Ba^{-m}$$ (34)

where *A* and *B* are constants for respective repulsive and attractive contributions, $m < n$, and *a* is an interatomic distance. The starting model is often based on the complete X-ray crystal structure of an oligomer (or perhaps a sequence of these) to gain an appreciation for allowed conformational energy minima with increasing chain length. Other structure analyses have been carried out solely *via* potential function minimization on hypothetical models. An advantage of this technique is that a minimized potential energy is a figure of merit which can be used in addition to the crystallographic *R*-value for defending the probable correctness of a structural model.

29.5   REFINEMENT OF A STRUCTURE MODEL

After a likely phase assignment, which yields a physically and chemically reasonable crystal structure, efforts are made to optimize the values of valence parameters (bond lengths and angles) by some sort of refinement procedure which will also minimize the difference between calculated and observed diffraction data. A number of approaches can be used. Given the potential map derived from the phased structure factors ϕ_c, a difference synthesis[18]

$$\phi_0 - \phi_c = \frac{1}{V}\sum_h \sum_k \sum_l \Delta F_{hkl} \exp(-2\pi i r \cdot s) \tag{35}$$

can be calculated where $\Delta F = (F_0 - F_c)$ represents the difference of observed and calculated structure factors. From peaks or depressions the corresponding map will indicate directions in which individual atoms must be moved to improve the structural model. Such a technique has often been used to locate hydrogen atom positions in X-ray crystallography. Although the improved detectability of hydrogen *vs.* some atoms (*e.g.* nitrogen and oxygen) is a possibility in electron crystallography,[57] the claim of its enhanced high angle scattering contribution in the presence of carbon, when compared to X-ray diffraction, is probably not important due to the typically large hydrogen atom thermal motion for many structures.

Another technique for structure refinement is based on least-squares fit of parameters to the observed data. In this approach the following function is minimized, *i.e.*

$$\sum_{hkl} w_{hkl}[|F_0| - |kF_c(p_1, p_2, \ldots p_n)|] \frac{\partial |kF_c(p_1, \ldots p_n)|}{\partial p_j} = 0 \tag{36}$$

$$(j = 1, 2, \ldots n)$$

This is to say that various parameters of the structural model (atomic positions, thermal parameters, scale factor, occupancy, *etc.*) are adjusted to give calculated structure factors which best agree with observed data.

It was stated above that experimental conditions in electron diffraction are controlled to give observed data which are least perturbed by influences such as dynamical scattering, crystal deformation, *etc.*, so that it is possible to carry out an *ab initio* structure analysis. This does not imply, however, that such perturbations have not altered the observed data. In fact, since the corrections for dynamical scattering and crystal deformation assume fore-knowledge of the crystal structure, these must be made after the preliminary crystal model is found and will complicate the refinement procedure. The appropriateness of including such corrections in the refinement has been demonstrated by Moss[52] in her reanalyses of several polymer structures. From practical experience, difference Fourier techniques seem to be more successful for improving a structure model than least squares, since variation of refined parameters can lead to a residual minimum which no longer represents a chemically reasonable structure.

Undoubtedly part of the problem with refinement is a result of the typical use of only zonal data for crystal structure analysis.[58] Only recently have three-dimensional electron diffraction data from single crystals been used for polymer crystal structure analysis[59] and then it was shown that false *R*-value minima could be found with only the *hk0* zonal set from untilted crystals, while the correct structure was pinpointed by the more complete data set.[60]

Another aspect of this problem is the difficulty in discovering the correct structure from *R*-value minima when the number of refineable parameters *p* approaches the number of observed diffraction intensities *n*. As shown by Hamilton,[61] the correctness of either two structure models with residuals R_1 and R_0, where R_0 is the smaller, cannot be discerned until $R_1 > R_0 R$ is satisfied to allow rejection of model 1 within α, the confidence level of the test. The quantity

$$R_{p,n-p,\alpha} \equiv \left[\frac{p}{n-p} F_{p,n-p,\alpha} + 1\right]^{1/2} \tag{37}$$

is related to a distribution F expected for crystallographic residuals. There are many instances in crystallography where this point has been totally overlooked and a structure 'solution' was claimed only on the basis of a marginally lower *R*-value in a rather shallow minimum, *e.g.* the structure analysis of polyethylene from powder X-ray data.[62] In general, lowering the number of observed data reduces the precision of the structure determination.

29.6 REPRESENTATIVE CRYSTAL STRUCTURE ANALYSIS

29.6.1 Perfect Crystals

By definition all polydisperse polymer crystals are imperfect. However, the order of the stem regions in polymer lamellae is often good enough to allow a crystal structure analysis based on the fiber repeat, even though the actual unit cell which includes the (undefined) chain fold region cannot be easily addressed by diffraction techniques. In the case of solution-crystallized lamellae (by self-seeding), the disordered surfaces are separated by, for example, 100 Å and, if the crystal is bent, then, as discussed above, these features will not contribute to the diffraction intensities. Epitaxially crystallized samples, on the other hand, should give low angle data which represent the period of an average lamellar repeat. The major intensity, however, should come from the highly ordered features of the fiber repeat of the stem packing. Representative crystal structure analyses are indicated below.

29.6.1.1 Solution-crystallized lamellae

A review of quantitative crystal structure analyses, including those on solution-crystallized samples based on electron diffraction, is given in Table 1 which surveys the literature up to June 1986. A more complete compilation can be found in a recent review[17] which includes unit cell data and problems with the structure analysis. The most frequent methods used for structure analysis involve the use of Patterson maps or potential functions. One positive feature of electron diffraction studies indicated by this table is the ability to detect several polymorphic forms of the same material, as shown also by other work.[63]

29.6.1.2 Epitaxially crystallized lamellae

Epitaxial crystallization of linear chain molecules on, for example, inorganic substrates,[65-67] has often been useful for orienting chain axes to lie parallel to the best developed crystal face instead of normal to it. Some polymers can be epitaxially synthesized,[68,69] and electron diffraction data from such samples are being employed for crystal structure analysis.[70] Compounds epitaxially crystallized on organic substrates, using techniques of Wittmann *et al.*,[71-75] often produce larger area crystals than obtained with inorganics. Quantitative structure analyses on some of these are listed in Table 1. Although more success with monodisperse oligomers and other molecular crystals can be reported, it must be pointed out that at least one analysis allowed Moss *et al.*[76] to distinguish between alternative structural models which could not be discerned solely on the basis of data from solution-crystallized samples. Most recent studies by Lotz and Wittmann[77] on organic substrates reveal that the epitaxial growth is not only restricted to the polymethylene compounds for which most structural work has been done to date. A problem with crystal imperfection experienced in early studies[78] is apparently corrected *via* annealing.[108]

29.6.2 Disordered Crystals

Analyses of lattice defects have been recently assisted by the study of epitaxially crystallized oligomers. The discussions below are merely indicative of the type of studies which can be carried out.

29.6.2.1 Phase transitions

It has been long known that *n*-alkanes beyond $n\text{-}C_{38}H_{78}$, like polyethylene, have no premelt transition to a hexagonal phase. (More recently the analogy between long chain alkanes to the infinite polymer has been further established by diffraction contrast imaging to demonstrate a similarity of crystal textures.[79]) Nevertheless, as shown in early studies by Charlesby,[80] there is a lattice expansion of polyethylene, which can be followed by electron diffraction patterns on heated solution-crystallized specimens, due to a rotationally disordered chain packing found also for alkanes. Electron diffraction data from the hexagonal unit cell in alkanes were shown later[81] to be well fitted by a methylene rotor model but, unfortunately, in the projection down a long chain axis, one cannot distinguish among models which include a rotor,[82] a helix[83] or an aggregation of chain 'kink' defects.[84]

This problem was resolved recently in a study of epitaxially grown *n*-hexatriacontane crystals heated to near their hexagonal transition.[85] Features of the 0*kl* diffraction pattern dependent on the ordering of the polymethylene repeat did not change so much as did the 00*l* reflection intensities. As also found in separate X-ray and spectroscopic studies[86,87] only a chain defect model fits these data, since these defects produce chain end voids which most affect the lamellar perfection. Contributions to the diffracted intensity from the point defects themselves appear in the continuous diffuse scattering, which resembles the thermal contribution except that it is very intense and also present at low temperature. Although the diffuse scattering was analyzed with a rather primitive defect diffuse scattering model,[88] this can probably be improved with other formulations of the Laue formula[89] *viz.*

$$I(s) = C_A C_B (f_A - f_B) \left(1 + 2 \sum_1^\infty \alpha_m \cos 2\pi s \cdot x_m \right) \qquad (38)$$

which allow for correlations α_m along crystal lattice vectors x_m.

29.6.2.2   *Analysis of vacancies, defects and twins*

Localized defects are much more easily visualized in lattice images than they are in electron diffraction patterns. For example, edge dislocations were found at least 20 years ago in moiré-magnified images of multilamellar polymer crystals.[90,91] More recently, such defects have been observed directly by EM in a radiation stable polymer[92] and also in monolamellar alkane crystals.[35] Twin boundaries have also been visualized.[93]

Since diffraction is an averaging phenomenon, its use to discern defects and twins would be most important if small subdomains of the crystal lattice were produced to cause shape transform broadening of diffraction spots (see above). The random absence of structural layers in, for example, a solvated crystal could also produce similar but characteristic streaking of certain reflections in the electron diffraction pattern.[94,17]

Identification of separate structural domains could also be facilitated by microbeam diffraction techniques, for which a submicron diameter beam is used as a probe of very small domains.[14,15] The use of such techniques to study copolymers should also be mentioned[95,96] as well as its use to study microstructure in spherulites.[14,95,97] Although some of the electron diffraction patterns display rather high resolution, no quantitative use of the data has been made for structural analysis, perhaps due to the increased problem with radiation damage for such small probes.

Another aspect of polymer crystal imperfection is the case of paracrystallinity for which no regular unit cell can be described. A complete treatment of this topic can be found in the works by Guinier,[89] Vainshtein[98] and the theoretical treatise of Hosemann and Bagchi.[99]

29.6.2.3   *Binary mixtures*

Polydispersity in polymer crystals implies the capability to form solid solution beyond the limits found for monodisperse oligomers,[100] even though fractionation can be visualized for the chain folded systems.[101,102] Although solid solution formation has been often studied by powder X-ray diffraction techniques,[103] recent quantitative electron diffraction structure analyses have been made on epitaxially oriented binary single crystal alkane systems.[104,109] The symmetry criterion for chain mixing[105] was also found to be less important than relative chain length, in agreement with recent theoretical analyses.[106] This study has been recently extended to fractionated mixtures.[107]

29.7   REFERENCES

1. E. W. Fischer, in 'Newer Methods of Polymer Characterization', ed. B. Ke, Interscience, New York, 1964, p. 280.
2. P. H. Geil, 'Polymer Single Crystals', Wiley, New York, 1963.
3. B. Wunderlich, 'Macromolecular Physics', Academic Press, New York, 1973, 1976, 1980, vols. 1–3.
4. B. K. Vainshtein, 'Structure Analysis by Electron Diffraction', Pergamon Press, Oxford, 1964.
5. I. H. Hall, in 'Structure of Crystalline Polymers', ed. I. H. Hall, Elsevier, London, 1984, p. 39.
6. L. Reimer, 'Transmission Electron Microscopy', Springer, Berlin, 1984.
7. R. H. Lange and J. Blödorn, 'Das Elektronenmikroskop TEM & REM', Thieme, Stuttgart, 1981.

8. R. P. Ferrier, *Adv. Opt. Electron Microsc.*, 1969, **3**, 155.
9. R. D. Heidenreich, 'Fundamentals of Transmission Electron Microscopy', Interscience, New York, 1964.
10. P. B. Hirsch, A. Howie, R. B. Nicholson, D. W. Pashley and M. J. Whelan, 'Electron Microscopy of Thin Crystals', Butterworths, London, 1965.
11. A. W. Agar, R. H. Alderson and D. Chescoe, 'Principles and Practice of Electron Microscope Operation', North-Holland, Amsterdam, 1974.
12. J. C. H. Spence, 'Experimental High-Resolution Electron Microscopy', Clarendon Press, Oxford, 1981.
13. J. J. Hren, J. I. Goldstein and D. C. Joy, (eds.) 'Introduction to Analytical Electron Microscopy', Plenum Press, New York, 1979.
14. A. Low, D. Vesely, P. Allan and M. Bevis, *J. Mater. Sci.*, 1978, **13**, 711.
15. E. S. Sherman, W. W. Adams and E. L. Thomas, *J. Mater. Sci.*, 1981, **16**, 1.
16. Z. G. Pinsker, 'Electron Diffraction', Butterworths, London, 1953.
17. D. L. Dorset, *J. Electron Microsc. Techn.*, 1985, **2**, 89.
18. G. H. Stout and L. H. Jensen, 'X-ray Crystallography', Macmillan, New York, 1968.
19. J. W. Jeffery, 'Methods in X-ray Crystallography', Academic Press, London, 1971.
20. J. D. Dunitz, 'X-ray Analysis and the Structure of Organic Molecules', Cornell University Press, Ithaca, New York, 1979.
21. 'International Tables for X-ray Crystallography', Volumes I–IV, Kynoch Press, Birmingham, UK, 1969–1974.
22. D. C. Champeney, 'Fourier Transforms and Their Physical Applications', Academic Press, London, 1973.
23. J. D. Gaskill, 'Linear Systems, Fourier Transforms and Optics', Wiley, New York, 1978.
24. J. M. Cowley, 'Diffraction Physics', 2nd edn., North-Holland, Amsterdam, 1981.
25. P. A. Doyle and P. S. Turner, *Acta Crystallogr., Sect. A*, 1968, **24**, 390.
26. J. M. Cowley, *Acta Crystallogr.*, 1961, **14**, 920.
27. D. L. Dorset, *Z. Naturforsch., Teil A*, 1978, **33**, 964.
28. D. L. Dorset, *Acta Crystallogr., Sect. A*, 1980, **36**, 592.
29. D. L. Dorset, *Ultramicroscopy*, 1983, **12**, 19.
30. B. Moss and D. L. Dorset, *Acta Crystallogr., Sect. A.*, 1983, **39**, 609.
31. B. Moss and D. L. Dorset, *J. Polym. Sci., Polym. Phys. Ed.*, 1982, **20**, 1789.
32. J. M. Cowley, A. L. G. Rees and J. A. Spink, *Proc. Phys. Soc., London, Sect. A*, 1951, **64**, 609.
33. J. Gjønnes and A. F. Moodie, *Acta Crystallogr.*, 1965, **19**, 65.
34. D. T. Grubb, *J. Mater. Sci.*, 1974, **9**, 1715.
35. F. Zemlin, E. Reuber, E. Beckmann, E. Zeitler and D. L. Dorset, *Science*, 1985, **229**, 461.
36. D. L. Dorset and F. Zemlin, *Ultramicroscopy*, 1985, **17**, 229.
37. H. Kiho and P. Ingram, *Makromol. Chem.*, 1968, **118**, 45.
38. E. Knapek and J. Dubochet, *J. Mol. Biol.*, 1980, **141**, 147.
39. V. R. Matricardi, R. C. Moretz and D. F. Parsons, *Science*, 1972, **177**, 268.
40. K. J. Taylor, H. Chanzy and R. H. Marchessault, *J. Mol. Biol.*, 1975, **92**, 165.
41. M. J. Buerger, 'Vector Space', Wiley, New York, 1959.
42. B. K. Vainshtein, I. A. D'yakon and A. V. Ablov, *Sov. Phys.-Dokl. (Engl. Transl.)*, 1971, **15**, 645.
43. B. Moss, D. L. Dorset, J. C. Wittmann and B. Lotz, *J. Polym. Sci., Polym. Phys. Ed.*, 1984, **22**, 1919.
44. A. Charlesby, G. I. Finch and H. Wilman, *Proc. Phys. Soc. London*, 1939, **51**, 479.
45. D. G. Fischer, *Proc. Phys. Soc. London*, 1948, **60**, 99.
46. W. Hoppe, in 'Advances in Structure Research by Diffraction Methods', vol. 1, ed. R. Brill, Interscience, New York, 1964, p. 90.
47. J. L. Amoros and M. Amoros, 'Molecular Crystals. Their Transforms and Diffuse Scattering', Wiley, New York, 1968.
48. D. L. Dorset and H. A. Hauptman, *Ultramicroscopy*, 1976, **1**, 195.
49. J. Karle, in 'Advances in Structure Research by Diffraction Methods', vol. 1, ed. R. Brill, Interscience, New York, 1964, p. 55.
50. H. A. Hauptman, 'Crystal Structure Determination. The Role of the Cosine Seminvariants', Plenum, New York, 1972.
51. D. L. Dorset, B. K. Jap, M. H. Ho and R. M. Glaeser, *Acta Crystallogr., Sect. A*, 1979, **35**, 1001.
52. B. Moss and D. L. Dorset, *Acta Crystallogr., Sect. A*, 1982, **38**, 207.
53. K. Ishizuka, M. Miyazaki and N. Uyeda *Acta Crystallogr., Sect. A*, 1982, **38**, 408.
54. M. Tsuji, S. Isoda, M. Ohara, A. Kawaguchi and K. Katayama, *Polymer*, 1982, **23**, 1568.
55. H. Tadokoro, 'Structure of Crystalline Polymers', Wiley, New York, 1979.
56. E. H. Moelwyn–Hughes, 'Physical Chemistry', 2nd edn., Pergamon Press, Oxford, 1961, p. 297.
57. B. K. Vainshtein, in 'Advances in Structure Research by Diffraction Methods', vol. 1, ed. R. Brill, Interscience, New York, 1964, p. 24.
58. D. L. Dorset, *Proc.—Annu. Meet., Electron Microsc. Soc. Am., 1983*, **41**, 22.
59. F. Brisse, B. Remillard and H. Chanzy, *Macromolecules*, 1984, **17**, 1908.
60. B. Moss and F. Brisse, *Macromolecules*, 1984, **17**, 2202.
61. W. C. Hamilton, 'Statistics in Physical Science', Ronald, New York, 1964, p. 157.
62. D. L. Dorset, *Polymer*, 1986, **27**, 1349.
63. B. Lotz, F. Colonna–Cesari, F. Heitz and G. Spach, *J. Mol. Biol.*, 1976, **106**, 915.
64. F. Heitz, B. Lotz and G. Spach, *J. Mol. Biol.*, 1975, **92**, 1.
65. K. A. Mauritz, E. Baer and A. J. Hopfinger, *Macromol. Rev.*, 1978, **13**, 1.
66. S. Wellinghof, F. Rybnikar and E. Baer, *J. Macromol. Sci., Phys.* 1974, **B10**, 1.
67. Y. Ueda and M. Ashida, *J. Electron Microsc.*, 1980, **29**, 38.
68. S. E. Rickert, J. B. Lando, A. J. Hopfinger and E. Baer, *Macromolecules*, 1979, **12**, 1053.
69. H. Ishida, S. E. Rickert, A. J. Hopfinger, J. B. Lando, E. Baer and J. L. Koenig, *J. Appl. Phys.*, 1980, **51**, 5188.
70. S. E. Rickert, H. Ishida, J. B. Lando, J. L. Koenig and E. Baer, *J. Appl. Phys.*, 1980, **51**, 5194.
71. J. C. Wittmann and R. St. J. Manley, *J. Polym. Sci., Polym. Phys. Ed.*, 1977, **15**, 1089.
72. J. C. Wittmann and R. St. J. Manley, *J. Polym. Sci., Polym. Phys. Ed.*, 1978, **16**, 1891.
73. J. C. Wittmann and B. Lotz, *J. Polym. Sci., Polym. Phys. Ed.*, 1981, **19**, 1837.

74. J. C. Wittmann and B. Lotz, *J. Polym. Sci., Polym. Phys. Ed.*, 1981, **19**, 1853.
75. J. C. Wittmann, A. M. Hodge and B. Lotz, *J. Polym. Sci., Polym Phys. Ed.*, 1983, **21**, 2495.
76. B. Moss, D. L. Dorset, J. C. Wittmann and B. Lotz, *J. Macromol. Sci., Phys.*, 1985, **B24**, 99.
77. B. Lotz and J. C. Wittmann, *Makromol. Chem.*, 1984, **185**, 2043.
78. C. H. McConnell, J. R. Fryer, D. L. Dorset and F. Zemlin, *Conf. Ser.—Inst. Phys.*, 1985, **78**, 433.
79. D. L. Dorset, *J. Macromol. Sci., Phys.*, 1986, **B25**, 1.
80. A. Charlesby, *Proc. Phys. Soc. London*, 1945, **57**, 510.
81. D. L. Dorset, *Chem. Phys. Lipids*, 1974, **13**, 133.
82. J. D. Hoffmann, *J. Chem. Phys.*, 1952, **20**, 541.
83. L. D'Ilario and E. Giglio, *Acta Crystallogr., Sect. B*, 1974, **30**, 372.
84. S. Blasenbrey and W. Pechhold, *Rheol. Acta*, 1967, **6**, 174.
85. D. L. Dorset, B. Moss, J. C. Wittmann and B. Lotz, *Proc. Natl. Acad. Sci. USA*, 1984, **81**, 1913.
86. G. Strobl, B. Ewen, E. W. Fischer and W. Piesczek, *J. Chem. Phys.*, 1974, **61**, 5257.
87. M. Maroncelli, S. P. Qi, H. L. Strauss and R. G. Snyder, *J. Am. Chem. Soc.*, 1982, **104**, 6237.
88. D. L. Dorset, B. Moss and F. Zemlin, *J. Macromol. Sci., Phys.* 1985, **B24**, 87.
89. A. Guinier, 'X-ray Diffraction in Crystals, Imperfect Crystals and Amorphous Bodies', Freeman, San Francisco, 1963.
90. V. F. Holland, *J. Appl. Phys.*, 1964, **35**, 3235.
91. V. F. Holland and P. H. Lindenmeyer, *J. Appl. Phys.*, 1965, **36**, 3049.
92. S. Isoda, M. Tsuji, M. Ohara, A. Kawaguchi and K. Katayama, *Makromol. Chem., Rapid Commun.*, 1983, **4**, 141.
93. A. Kawaguchi, S. Isoda, J. Petermann and K. Katayama, *Colloid Polym. Sci.*, 1984, **262**, 429.
94. J. M. Cowley, *Acta Crystallogr., Sect. A*, 1976, **32**, 88.
95. R. M. Briber and E. L. Thomas, *J. Macromol. Sci., Phys.*, 1983, **B22**, 509.
96. R. M. Briber and E. L. Thomas, *Polymer*, 1985, **26**, 8.
97. E. J. Roche, R. S. Stein and E. L. Thomas, *J. Polym. Sci., Polym. Phys. Ed.*, 1980, **18**, 1145.
98. B. K. Vainshtein, 'Diffraction of X-rays by Chain Molecules', Elsevier, Amsterdam, 1966.
99. R. Hosemann and S. N. Bagchi, 'Direct Analysis of Diffraction by Matter', North-Holland, Amsterdam, 1962.
100. P. Smith and R. St. J. Manley, *Macromolecules*, 1979, **12**, 483.
101. F. K. Anderson, *J. Appl. Phys.*, 1964, **35**, 64.
102. R. B. Prime and B. Wunderlich, *J. Polym. Sci., Part A-2,* 1969, **7**, 2061.
103. G. I. Asbach, G. Geiger and W. Wilke, *Colloid Polym. Sci.*, 1979, **257**, 1049.
104. D. L. Dorset, *Macromolecules*, 1985, **18**, 2158.
105. A. I. Kitaigorodskii, Yu. V. Mnyukh and N. A. Nechitailo, *Sov. Phys.-Crystallogr. (Engl. Transl.)*, 1958, **3**, 303.
106. R. R. Matheson, Jr. and P. Smith, *Polymer*, 1985, **26**, 288.
107. D. L. Dorset, *Macromolecules*, 1986, **19**, 2965.
108. J. C. Wittmann, unpublished data.
109. D. L. Dorset, *Macromolecules*, 1987, **20**, 2782.

30

Small-angle X-Ray Scattering from Crystalline Polymers

IVAN H. HALL

University of Manchester Institute of Science and Technology, UK

30.1 INTRODUCTION

Scattering of X-rays at small angles (*i.e.* less than about 5°) is caused by fluctuations in electron density over lengths of about 100 Å, and has been used to measure distributions of particle size in suspension or solution, and to study morphological features of solid polymers such as void size, crazing, component segregation in copolymers and lamellar structure of crystalline polymers. This chapter will be concerned primarily with the latter and will emphasize work with oriented materials for which Figure 1 shows typical scattering patterns.

On these scattering patterns, the meridian corresponds to the orientation (or fibre) axis and so the patterns indicate a periodic alternation of electron density parallel to this axis such as would be caused by stacks in which crystalline lamellae are separated by less ordered ('amorphous') regions (in this chapter the word 'amorphous' will be used to describe all non-crystalline regions, even though they might display some degree of order.) Such a structure might be expected to produce several

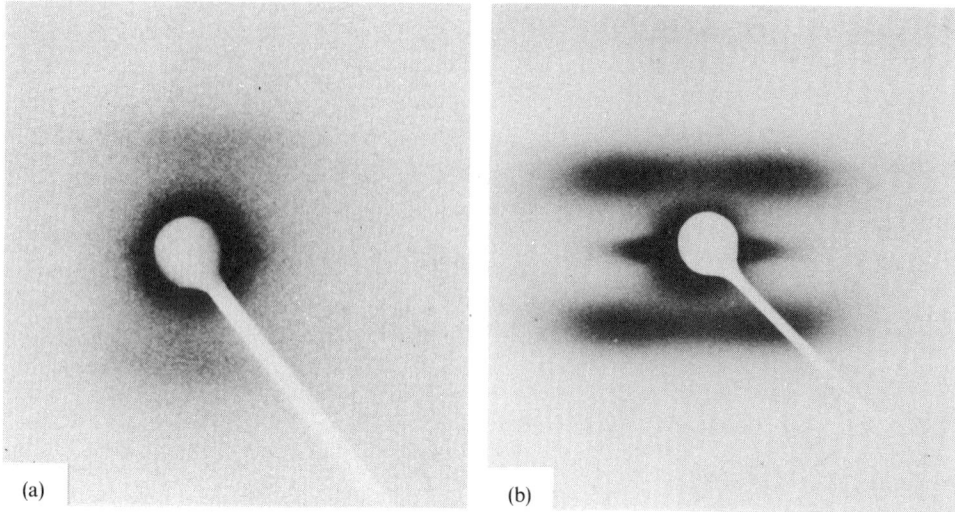

Figure 1   Typical small-angle scattering patterns of well-annealed fibres: (a) two-point pattern from poly(butylene terephthalate); and (b) four-point pattern from poly(ethylene terephthalate)

orders of diffraction; only a single broad peak is observed, which might be because of variation in
either crystal or amorphous thickness within one stack, or variation of periodicity between stacks.
The distribution of scattered intensity might also be affected by factors such as the sharpness of the
discontinuity in density at the phase boundaries, the flatness of these, and the coherence lengths of
the stacks. In this review we shall discuss the power and limitations of small-angle X-ray scattering
(SAXS) to study these features, limiting the discussion to quantitative applications. The technique
has been widely used to follow the morphological changes consequent upon various processing
treatments, using only visual inspection of the scattering pattern. Such applications are a useful
experimental tool and have produced much information, but will not be considered here.

   It will be assumed that the reader is familar with the basic ideas of scattering theory, such as
Fourier transformation, convolution, reciprocal space, and Ewald's construction. The non-specialist
should be able to follow the main arguments of the chapter without understanding these, but anyone
intending to do independent work in this subject must become familiar with these concepts which
are treated in refs. 1–4.

30.2   THEORY OF SCATTERING BY TWO-PHASE SYSTEMS

   In order to calculate scattering it is necessary to propose a model. The more detailed the model,
the more detailed will be the calculated distribution of scattering. We shall start with a very general
model, and then include further assumptions until the complete distribution of intensity can be
calculated.

30.2.1   The Isotropic Two-phase Model

30.2.1.1   *The invariant Q*

   Let a point collimated beam* be incident upon an isotropic substance comprising two uniform
phases of electron density ρ_c and ρ_a separated by sharp boundaries, the volume fractions of each
being ω_c and ω_a respectively. If $I(s)$ is the fraction of the incident intensity scattered at an angle 2θ
(the angle between the incident and scattered beam), then a quantity Q may be defined such that

$$Q = K_1 \int_0^\infty \mathbf{s}^2 I(\mathbf{s}) d\mathbf{s} = \overline{\Delta P^2} = K_2 (\rho_c - \rho_a)^2 \omega_c \omega_a \tag{1}$$

---

* Unless otherwise stated equations refer to point collimation. In many of the references cited these formulae have been
modified for slit collimation (see Section 30.3.1) and so will differ from those given here.

where $s = (2 \sin \theta)/\lambda$, λ is the wavelength of the radiation used, and $\Delta\rho$ is the difference between the local density and the mean value. K_1 and K_2 are constants which depend upon the experimental conditions and the nature of the scattering material. This is a simplified form of the equation with experimental factors gathered into constants, but which is sufficient for the present discussion. A bibliography of more extensive treatments may be found in ref. 5. $Q = \overline{\Delta\rho^2}$ is true for any scattering system, but $\overline{\Delta\rho^2}$ is only proportional to $(\rho_c - \rho_a)^2 \omega_c \omega_a$ for a two-phase system with sharp boundaries.

Whilst Q may be measured experimentally, it is related by equation (1) to three parameters, ρ_c, ρ_a and ω_c. Other information must therefore be input or assumed to deduce a value for any of them.

To determine the parameters from the scattering curve, Strobl[6] made the additional assumptions that the thickness of all amorphous layers is constant and the variation in periodicity is symmetrical about its average value. The intensity distribution is then proportional to the product of two factors, $|F(s)|^2$ and $Z(s)$. $F(s)$ (the structure factor) is the Fourier transform of the electron density profile of the amorphous layer relative to the crystalline value, and $Z(s)$ (the interference function) is the Fourier transform of a function $z'(r)$ where $z'(r)dr$ is the probability of finding a lattice point in a length dr distant r from the origin. (In this case the lattice points are the centres of the amorphous layers.) $\int Z(s)ds$ over the range Δs was shown to be unity provided $\Delta s = 1/\bar{x}$ and was centred on $s = m/\bar{x}$ where $\bar{x}$ is the average periodicity of electron density and m an integer. Thus by integrating the scattering curve over these ranges, the mean value of the structure factor at the centre of each could be obtained. Extrapolation gave $(\bar{F})^2$ at $s = 0$ which was shown[7] to be proportional to $\bar{x}^2 \omega_a^2 (\rho_c - \rho_a)^2$, where the constant of proportionality can be determined from the chemical composition of the scatterer. Together with the value of Q, this enabled ω_c, ω_a, and $\rho_c - \rho_a$ to be determined.

30.2.1.2 Phase transition zones

The theory can be extended without further assumption to show that $s^4 I(s)$ approaches a limiting value at large s and if this is called T then

$$S = \frac{2\pi T \omega_c \omega_a}{Q} \tag{2}$$

where S is the area of boundary surface between the two phases per unit volume. (Although s is large relative to small-angle scattering experiments, it is still smaller than the scattering vector from crystalline arrays of atoms.) If it is assumed that instead of sharp interfaces between phases there is a transition zone of intermediate density then the limiting value is not reached[8] and to proceed further it is necessary to make assumptions about this zone. Vonk[9] has shown that if the density falls linearly from the crystalline to the amorphous value over a length E, then the intensity distribution which would be observed if the boundaries were sharp must be multiplied by $(\sin^2 \pi Es)/(\pi Es)^2$ to give that actually observed. By truncating the series expansion of $(\sin^2 \pi Es)/(\pi Es)^2$ to two terms[10] it then follows that $s^4 I(s)$ plotted against s^2 will tend to a straight line instead of a constant value at large s.

The truncation of the series is only valid at small s and this range will not necessarily overlap the other. However, if the density is assumed to change sigmoidally in the transition zone then an analysis may be used which is exact over the whole range of s.

In this way the boundary surface per unit volume and the size of transition zones may be estimated, but if they are present[11] then the measured value of Q must also be corrected since they reduce it by an amount proportional to $E/\bar{x}$ where $\bar{x}$ is again the average periodicity of electron density.

30.2.1.3 Density fluctuations within phases

So far it has been assumed that density is uniform within a phase, but random fluctuations are likely to occur (liquid-like behaviour). This[12] causes a contribution to the scattered intensity distributed according to a power series in even powers of s and will affect determinations of parameters based on the theory so far described. Measurements at values of s at which all other contributions to the intensity have fallen to zero enable it to be estimated. It can then be extrapolated using some approximation to the power series[9,13] and subtracted. The range of s in which it is the only significant contributor to the intensity is unclear;[10] some investigators have assumed it to be constant, which is clearly invalid.

30.2.1.4   *The effect of the size of units of phases*

So far the only information derivable from the theory related to the size of the individual units of each phase is the quantity S. Ruland[13,14] has developed the theory to obtain the distribution function of the distances between phase boundaries. $I(s)$ is measured as a function of s for a point-collimated beam incident on an isotropic scatterer (the theory is also given for other arrangements) and, using the methods already described, corrected for the effects of transition zones and density fluctuations within phases. The intensity distribution is then that which would be observed from a scatterer with sharp phase boundaries, and $s^4 I(s)$ is asymptotic to a limiting value, T at large s. A function $G_1(s)$ is now defined as $G_1(s) \propto T - s^4 I(s)$ and the transform $g_1(r) = 2 \int_0^\infty G_1(s) \cos 2\pi r \cdot s \, ds$ evaluated. The interpretation of $g_1(r)$ is shown in Figure 2.

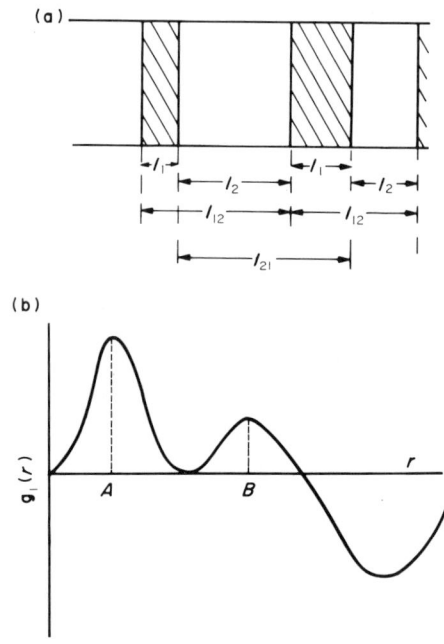

Figure 2   The interpretation of the function $g_1(r)$: (a) a sequence of crystalline (clear) and amorphous (hatched) segments; and (b) the first peak is the distribution of the lengths l_1, the second is of the lengths l_2. A and B are respectively their most probable lengths. The third peak is the sum of the distributions of l_{12} and l_{21} and any further peaks may be similarly identified

30.2.1.5   *Difficulties encountered in applying theory*

The approach described above enables morphological information to be deduced directly from the distribution of scattered intensity, using a model containing few assumptions. It has the disadvantage that the experimental measurements are difficult to perform to a high and known degree of accuracy. Many require the measurement of absolute intensity, which is error prone, and some need the integration of measured quantities over s from 0 to infinity. At small s scattering by voids increases intensity, and there is a lower limit below which measurements cannot be taken. At large s there is the problem of making proper correction for liquid scattering, and since higher powers of s are sometimes used in the integration, small errors or truncation at too low a value of s can have a large effect.

30.2.2   The One-dimensional Correlation Function

An alternative approach to the above is to calculate the scattering from a detailed proposed model and to compare this with that observed experimentally. This is exploited in the one-dimensional correlation function[15] (γ) which can be determined for a model and also from the scattering curve.

If $\Delta\rho(r_1)$ is the difference between the electron density at a point r_1 and its average value, then $\Delta\rho(r_1 - r)$ is the same difference at a point distant r from r_1. The correlation function is then defined as

$$\gamma(r) = \frac{\int_0^\infty \Delta\rho(r_1 - r)\Delta\rho(r_1)dr_1}{\int_0^\infty \Delta\rho^2(r_1)dr_1} \qquad (3)$$

γ is thus the average of the product of the electron density differences at two points a distance r apart, and is a measure of the probability of finding the density the same at these two points, showing a peak at the most probable repeat distance.

For the intensity distribution from an isotropic scatterer

$$\gamma(r) = \frac{4\pi \int_0^\infty s^2 I(s)\exp(2\pi isr)ds}{\int_0^\infty s^2 I(s)ds} \qquad (4)$$

The numerator is the Fourier transform of the intensity distribution, corrected for the fact that for an isotropic sample this is distributed over a sphere in reciprocal space. Except for an experimental constant, the denominator is Q of equation (1) which is equal to $\overline{\Delta\rho^2}$. This eliminates experimental constants, and enables relative rather than absolute intensities to be used.

The calculation of γ is not restricted to any particular model and so it can be used to distinguish between different types and investigate effects such as the presence of transition zones,[9] distortion of boundary surfaces[16] and finite lengths of crystalline and amorphous sequences. If a particular model is assumed, the volume fraction of crystalline material may be determined[15] directly from the graph of γ against r.

The same problems concerning range of integration arise as with the previous methods. The position and height of the first positive peak of the function appear to be unaffected by different extrapolations from small s to zero[15,17,18] but the depth of the first negative peak does seem to be significantly changed. Errors in the correction for liquid scatter can be important but the effect of truncation at large s and of other experimental uncertainties has received little attention in studies of crystalline polymers.

30.2.3 The Paracrystalline Lattice

30.2.3.1 *Scattering by the ideal one-dimensional paracrystal*

It is assumed that the scatterer comprises infinitely wide stacks containing alternating regions of uniform high and low electron density separated by perfectly sharp and flat boundaries perpendicular to the axis. Neighbouring stacks are parallel, but there is no coherence between them. All intensity will then be scattered along a line parallel to the stack axes. The axial alternation of electron density may be represented by the convolution of an electron density profile with a one-dimensional lattice. Since, in the small-angle region, only a few broad diffraction spots are observed, the lattice must be distorted, and such distortions are classified into two groups.[19,20] In the first group long range order is preserved, but at each lattice point there is a random displacement from the equilibrium position; in the second group long range order is destroyed, each lattice point being given by the end of a lattice vector whose other end is placed on its predecessor and whose length varies randomly about a mean value independent of the origin. This is the lattice of a one-dimensional paracrystal and is assumed to describe the locations of the crystallites in the stack.

The lattice vector is created by choosing at random amorphous and crystalline lengths from distributions which can be of different type and have different mean values. This is known as the general paracrystalline model[21] and the intensity scattered by one finite stack of N crystallites relative to that of a single electron placed at its centre is[20,22]

$$I(s) = \frac{\Delta\rho^2}{2(\pi s)^2} \text{Re}\left\{\frac{1 - H_y}{1 - H_y J_z}\left\{N(1 - J_z) + \frac{J_z(1 - H_y)[1 - (H_y J_z)^N]}{1 - H_y J_z}\right\}\right\} \qquad (5)$$

H_y and J_z are the Fourier transforms of the normalized distribution functions of the crystallite and amorphous lengths (h(y), j(z)) respectively, and (in this and subsequent equations) $\Delta\rho$ is the electron density difference between the phases. This formula is not convenient for numerical work, but since any Fourier transform may be expressed in the form $A\exp(-i\phi)$ it can be put into a form which enables the intensity distribution to be calculated for any model whose h(y) and j(z) have analytical Fourier transforms.[22] This can be compared with that observed experimentally and since each intensity distribution may be expressed as a fraction of its maximum value, absolute measurements of intensity are not necessary. A satisfactory match will then, of course, only confirm that the model predicts the correct shape of intensity distribution, not the correct magnitude.

Blundell[23] has generalized the above treatment for the case where the electron density distribution is non-uniform in the crystalline phase and can be represented by a function of the distance from the crystallite boundary. This allows the introduction of transition zones into the model.

In deriving equation (5) it was assumed that the lamellar stacks were immersed in a medium of the amorphous density,[24] whereas it is probably more realistic to assume that it is some value ρ in between those of the amorphous and crystalline phases. According to Wenig and Bramer,[25] this has the effect of adding the term

$$\frac{p\Delta\rho^2}{2(\pi s)^2}\,\mathrm{Re}\left\{[1-(H_yJ_z)^N]\left[p-\frac{1-H_y}{1-H_yJ_z}\right]\right\} \qquad (6)$$

to the right hand side of equation (5), where $p=(\rho-\rho_a)/\Delta\rho$. Matsuo et al.[26] have followed Blundell[23] and generalized the model by allowing the electron densities to follow arbitrary functions of distance from the phase boundaries. If their equations are reduced to the special case considered by Wenig and Bramer, then in agreement with them an expression is obtained which must be added to the right hand side of equation (5), but it is now

$$\frac{p\Delta\rho^2}{2(\pi s)^2}\,\mathrm{Re}\left\{[1-(H_yJ_z)^N]\left[p-\frac{(1-H_y)(1+J_z)}{1-H_yJ_z}\right]\right\} \qquad (7)$$

This differs slightly from expression (6), and without completely reworking the very complicated mathematics it is not possible to identify the source of the discrepancy. The density of the immersion medium is particularly important with small stack lengths when the intensity at the small-angle end of the experimental range is reduced.[25, 27, 28]

30.2.3.2 Scattering by three-dimensional models

The model used in the previous section would only scatter along a line parallel to the stack axis; from Figure 1 it is clear that scattering also occurs transverse to that line. Such transverse broadening can be caused by (i) finite width of lamellar stacks, (ii) by a distribution of the orientation of stack axes, or (iii) by the normals to the lamellar boundaries being inclined to these axes and these boundaries being either not all smooth or not all parallel. Case (i) smears the intensity at s along the meridian in reciprocal space into a disc whilst (ii) smears it over the cap of a sphere. Case (iii) leads to an annulus (see Figure 3) which may close to a disc if the variation in inclination of the lamellar surfaces exceeds their mean inclination to the axis. Scattered intensity will occur where this distribution in reciprocal space intersects the Ewald sphere, which may be approximated by a plane at these small angles, and so the recorded scattering pattern will be a horizontal bar perpendicular to the fibre axis for (i), an arc for (ii) and either the 'four-point' pattern of Figure (1a) or a horizontal bar for (iii). Only that fraction of total intensity intersected by the Ewald sphere will be measured and it will be necessary to multiply the observed intensity by a (Lorentz) factor which will vary with s in a way which depends on the predominant mechanism of broadening.

If intensity variation is measured along only one direction[22] and used with the theory of the preceding section to investigate the projection of electron density along stack axes, then the cause of transverse broadening must be assumed and the appropriate correction made.[22] It is better to develop theories relating the two-dimensional distribution of intensity to three-dimensional models.

For a model comprising a set of parallel stacks, all of the same width b, with lamellar faces normal to the axis, stack width is given by the Scherrer equation $b=\lambda/[\Delta(2\theta)_{0.5}]$ where $\Delta(2\theta)_{0.5}$ is the width of the diffraction spot at half peak intensity measured parallel to the equator. An indication of the shape of the stack cross-section can be obtained by combining this with an experiment in which the fibre is tilted through an angle α towards the X-ray beam, the peak intensity measured, and

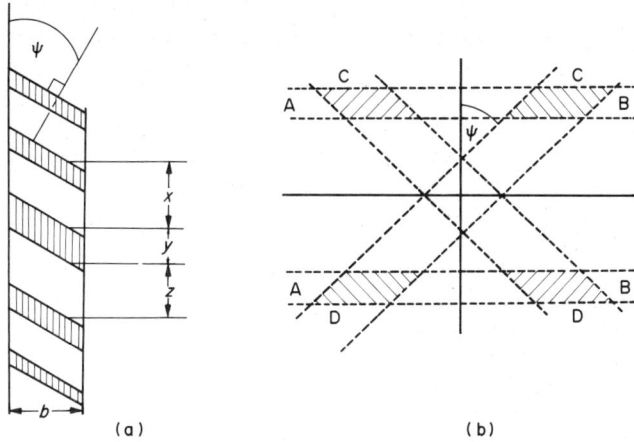

Figure 3 (a) A lamellar stack with interphase boundaries inclined to stack axis; and (b) its four-point diffraction pattern. The lattice is a series of points parallel to the axis and has a Fourier transform in the band AB. The transform of the crystallites is in the band CD and diffraction spots occur at the intersection of the transforms

$\{-\log[\mathrm{I}(\alpha)/\mathrm{I}(0)]\}^{1/2}$ plotted against $\tan\alpha$. The slope of this line is $1.21b/\bar{x}$ if the stack has square cross-section or $1.03b/\bar{x}$ if it is circular, where $\bar{x}=$ mean periodicity.[29]

If the stack widths are distributed, the Scherrer equation becomes[30]

$$\frac{\bar{b}^4}{\bar{b}^3} = \frac{1.04\lambda}{\Delta(2\theta)_1}$$

where $\Delta(2\theta)_1$ is the integral breadth of the spot, $\bar{b}^3$ the average value of b^3 and similarly for $\bar{b}^4$. This gives a quantity $k\bar{b}$ where k is of order of unity, its exact value depending on the nature of the distribution. Also

$$\mathrm{I}(\mathbf{s}_2) = K\exp\left[\frac{-\pi s_2^2}{4}\left(\frac{\bar{b}^6}{\bar{b}^4}s_2^2\right)\right]$$

where $\mathrm{I}(\mathbf{s}_2)$ is the intensity profile through the maximum and parallel to the equator, and K is a constant dependent upon experimental conditions.[30] Hence a quantity $k_1\bar{b}^2$ will be obtained where k_1 is defined similarly to k above. The two methods will only give the same result if the stacks are of uniform width.

If the lamellae surfaces are flat and parallel but with normals inclined at ψ to the stack axis, then the type of scattering pattern depends upon the parameters[31] $(b\tan\psi)/\bar{y}$ and $b/\bar{y}$ where $\bar{y}$ is the mean crystallite length. A single meridional spot is obtained for $(b\tan\psi)/\bar{y} \simeq 0.3$–$0.7$, which becomes four-point for values $\simeq 0.7$–1.0. The lobes are not parallel to the equator at larger values and become radial at about 1.7. $b/\bar{y} \simeq 1$ gives bar-type spots which become globular as this ratio increases to about four. For a similar model with infinitely long stacks of lamellae of infinite width, all perfectly oriented, but with lamellar inclination varying from stack to stack,[32] the two-dimensional intensity distribution was calculated. This confirms the qualitative description of Figure 3. The profile along the meridian depended only on the paracrystal parameters while that parallel to the equator depended on the distribution of lamellar normals. The lobes were slightly slanted to the equator.

Variation in the angle of the lamellar surface, and fluctuations in its flatness, have been introduced by considering a stack as a number of parallel fibrils with lateral coherence.[33, 34] The inclination of the phase boundary in an individual fibril is not then important; it is the inclination of the mean lamellar surface which is of importance, and this is controlled by the width of the fibril and the relative longitudinal displacement of neighbours. Individual crystallites then form a three-dimensional paracrystalline macrolattice, and the four-point diagram comprises the 001 reflections of this lattice. Calculations of intensity distribution have been performed for this type of model,[35] but the range of parameters for which these are presented are insufficient to draw any firm conclusions.

The one-dimensional model of Matsuo *et al.*[26] has been developed to include stacks with a distribution of orientations,[28, 36] all other features being retained, and Wilke and co-workers[37, 38, 39] have devised a more general method of introducing orientation distributions. This method can be applied to any structural unit with an axis about which it possesses rotational symmetry and whose axes are distributed about the fibre axis with any rotationally symmetrical function. They have applied it to investigate the effect of different orientation distributions upon a paracrystalline stack with crystalline lamellae all of same radius, thickness, and tilt angle, and to a paracrystalline macrolattice for which they have studied the influence of crystallite size, lattice parameters and distortions, and orientation. Since the actual formulae derived for these models are very complicated, with many terms needing definition, reference must be made to the original papers for more detail.

30.3 EXPERIMENTAL PROCEDURES

In order to use the theory described above to deduce information about the morphology of the scatterer it is essential that the experimental measurement of the distribution of scattered intensity be performed with sufficient resolution and accuracy. It is also important that these factors are known, otherwise false ideas may be obtained of the precision with which parameters have been determined. Regrettably, most papers give little experimental detail, and these quantities are unknown. Very great care is necessary, both in setting up apparatus and in performing the experiment, if meaningful data are to be collected and in this section we shall be concerned with the principles of instrumental design and experimental practice which lead to high and known resolution and accuracy. We shall concentrate on equipment likely to be found in most laboratories using SAXS. Some experimental difficulties may be avoidable using X-ray wavelengths longer than Cu $K\alpha$,[40] and the intense beam and tunable wavelength of synchrotron radiation[41,42] open further possibilities, but because of their restricted availability, these will not be considered further here.

30.3.1 Slit Collimation Systems

The essential requirement of apparatus to determine the distribution of scattered intensity at small angles is a fine beam of X-rays with small flare, so that the direct beam can be intercepted without blocking much of the range of scattered intensity. Since the recorded scatter is the convolution of the beam intensity profile with the actual scatter, this will also lead to high resolution. However, the finer the beam, the lower the intensity, and the simultaneous provision of adequate resolution and intensity is a major experimental difficulty. This has caused slit collimation to be widely used, the narrow width giving high resolution in a direction perpendicular to the slit length, and the length increasing the intensity. Excellent reviews of this type of camera will be found in refs. 43 and 44.

The observed scattering is the convolution of that due to the specimen with a rectangular beam profile, and is called the slit-smeared distribution. For a specimen with fibre orientation this is unimportant if the intensity distribution parallel to the fibre axis is all that is required, and the slit length is perpendicular to that axis. In practice it is difficult to achieve this setting with sufficient accuracy for quantitative work. In studies of the morphology of crystalline polymers the method has found its widest application with isotropic scatterers and the observed scattering distribution is then very different from that of the specimen.

When using methods based on the theory presented in Section 30.2.1, the usual procedure has been to convolute the theoretical distribution with the profile of a uniformly illuminated slit of infinitesimal width and infinite length, and apply the experimental data to the resulting equation. Deviations of the actual slit from the ideal that was assumed can have a significant effect,[45] and corrections have also been made for these.[46] This is to be preferred to the use of 'desmearing' procedures whereby the intensity distribution due to the specimen is recovered from the observed distribution.[47–51] These are essentially deconvolutions which are known to be error fraught, magnifying random errors, converting them to systematic ones, and introducing spurious detail,[52] particularly near the ends of the experimental range.[53] It must be realized that precisely the same smearing occurs if a rectangular entrance slit is used for the detector, whether this is a micro-densitometer or photon counter. Whilst the dangers are most marked with rectangular slits, they also occur with collimation by circular apertures. Recovery of information by deconvolution is only reliable if the beam profile is very much narrower than the width of any peaks in the recorded scattering curve. Otherwise, the calculated distribution should be convoluted with the instrumental broadening function before comparing with experimental data.

For these reasons, the use of slit collimation is not recommended in the type of experiment which is the main concern of this review, and will not be discussed further.

30.3.2 Resolution

The various factors affecting resolution have been discussed by Hall and Toy[54] who conclude that size and spacing of collimating slits and pin-holes, and the detecting system, are important, but that the thickness of the specimen and the spectral distribution of the X-rays are not, provided that the specimen thickness is not much greater than that which gives maximum intensity of scatter (about 1 mm), and β filtration or equivalent monochromatization is used.

The collimating system will cause the main beam to have an intensity profile value a' (Figure 4) at the observation plane. At small angles the scattering incident on this plane will be that due to the specimen convoluted with this profile. If some form of photon counter is used to measured the scatter, then the radiation will enter the counter through a slit and the recorded distribution will be that at the observation plane, convoluted with a 'top hat' function whose width is that of the entrance slit. Similar considerations obtain if the signal is recorded photographically and the film scanned with a microdensitometer. The entrance slit to this plays the same role as that to the counter. If the photon counter and its associated slit scan the scattered radiation at a constant angular speed with count rate being plotted against time, then the response time of the recording system will also affect the resolution. The total function by which the actual scatter is convoluted can be obtained by recording the main beam profile with the detector, using the same settings and method as those used in recording the scatter. This profile (the instrumental broadening function) must be narrow enough to meet the resolution requirements of the experiment.

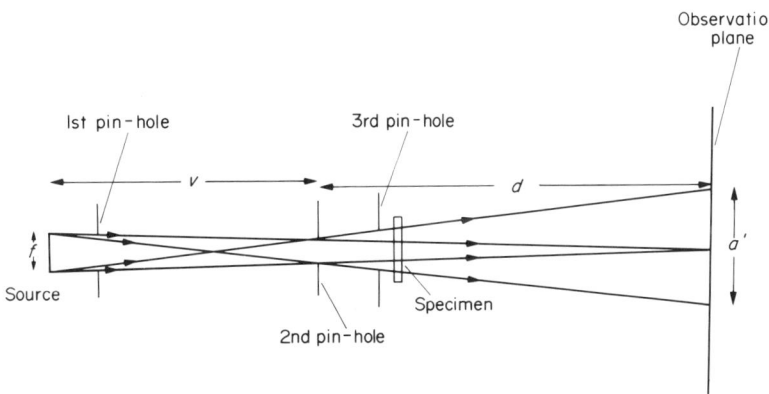

Figure 4 Schematic pin-hole collimation system

Cu $K\alpha$ X-radiation ($\lambda = 1.54$ Å) is widely used in small-angle scattering, and with this the typical distribution of scattered intensity for an oriented fibre of a crystalline polymer has a broad peak with its maximum between 0.5 and 1° and with a width at half maximum in the same range. We shall consider investigations in which recorded scattering distributions are to be matched at all points with those calculated for model structures, and for this type of experiment the instrumental broadening function must have a negligible effect. To achieve this, its width must be considerably less than that of the scattering peak, *i.e.* it should be less than about 0.1°. However, since high resolution is only obtained at the expense of beam intensity, it is undesirable to use higher resolution than is necessary for the purpose of the particular experiment; in many qualitative applications this may be much lower than that discussed here.

30.3.3 Pin-hole Collimation

The problem of maximizing intensity for a given resolution has been treated theoretically,[55, 56] showing that for maximum intensity extreme rays from the focal spot of the X-ray generator through the second aperture should intersect at the observation plane (Figure 4). However, modern high brightness X-ray generators have focal spots of about 0.1 mm width, and so a pin-hole of about 0.05 mm diameter would be required, which is not easily available. Thus intensity cannot be maximized in this way, and so it is sufficient to use ray diagrams and simple geometry to determine the dimensions for the required resolution with the available apparatus. The smallest pin-hole easily available is 0.2 mm diameter; if $f = 0.1$ mm, $v = 500$ mm and $d = 250$ mm, then the width of a' is equivalent to a divergence of 0.08° from a specimen adjacent to the second pin-hole.[22] The overall length is about the maximum that can be accommodated with commercial goniometers, the pin-holes are the smallest that are readily available, and with a high brightness source of X-rays, such as provided by a rotating-anode generator, beam intensity is adequate.[22] The resolution, which is probably the highest that can be achieved with this type of apparatus, is just within the limits set above.

30.3.4 Detectors

A very wide range of detectors are now available for use in X-ray scattering experiments,[57] but of these only proportional counters, gas-filled position sensitive detectors, and photographic film will be discussed here, since only these have found widespread application in SAXS applied to crystalline polymers. Semiconductor detectors[57] and ones using phosphors sensitive to X-rays in conjunction with high sensitivity television tubes,[57,58,59] are now becoming available and although no reports have been found of their use in studies such as these their possibilities are worthy of consideration. Gas-filled area position sensitive detectors[57,60] are becoming available, but their high cost is likely to limit their availability to major national research facilities, where specially constructed apparatus can overcome their drawbacks of bulky size, and poor spatial resolution.

30.3.4.1 *Proportional counters*

To determine the distribution of scattered intensity with angle, a proportional counter needs to be scanned through the angular range of interest. This can either be done at constant angular speed, or by step scanning, *i.e.* dwelling at each measurement point until a sufficient number of counts has accumulated. With the first method, unless the speed of scanning is very small compared with the response time of the recording system, further instrumental broadening will be introduced. The second method has the advantage that by dwelling longer in regions of weak scattering than in strong all parts can be measured with equal accuracy. However, the instrumental broadening will be the larger of the width of the entrance slit and the distance between measurement points, and so for high resolution many points will be necessary, leading to a very slow experiment. Other studies[61] have indicated that, for a given total time for the experiment, noise distortion is lower when counting takes place for the same time at each point than when it continues until the same number of counts have been included at each. The device has now been largely superseded by the linear position sensitive detector.

30.3.4.2 *The linear position sensitive detector (LPSD)*

This is a proportional counter which also defines the detection point along a line. Typically it has an active length of approximately 50 mm which is divided into about a thousand channels. Scattered photons can thus be located to 0.05 mm, but resolution is usually less than this. Depending on the model and on operating conditions, values from 0.05 to 0.25 mm have been reported.[22,57,62] Taking the higher limit, the intensity at the observation plane must be convoluted by a top hat function 0.25 mm wide, which subtends an angle of about 0.06° at a specimen 250 mm distant. In conjunction with the collimation arrangement described above, this will provide an adequate resolution for matching intensity profiles.

With these detectors there may be errors in the position assigned to photons, and sensitivity might vary with the position.[57,60,63] Of these errors, only non-uniformity of response is likely to be significant; the experimental procedure should include a regular calibration for this, and the data processing should enable correction to be made.

Should two photons arrive within a few microseconds of each other, only the first will be recorded. This 'dead time' which must elapse between successive events puts an upper limit on the usable count rate of about 10^4 counts s^{-1} for randomly occurring photons distributed uniformly over the LPSD length. At the resolution being discussed here, and with conventional X-ray generators, this is unlikely to be a problem. In any case, since statistically the number of missed events on any channel will be proportional to the number of arriving photons, it will not distort the shape of the distribution and so is unimportant unless absolute intensity is being measured.

30.3.4.3 Photographic film

Photographic film records both intensity and location of scattering and is the best form of detection for qualitative work. For quantitative work it needs to be used in conjunction with some form of microdensitometer, and until recently was less favoured than counting methods. However, reliable one- and two-dimensional scanning microdensitometers are now available, and since these scan through an aperture which can be as small as 25 μm square, resolution is as good as that which can be achieved with the LPSD and better than that with area position sensitive detectors.

Expressions are available which enable the standard deviation of the exposure to be determined[64] and these indicate that accuracy similar to that achieved with the LPSD can be obtained with similar exposure times. There are two disadvantages. Firstly, the limited dynamic range of the film,[65] requires the use of film packs when weak and strong scatter must both be measured on the same distribution, and methods are needed to ensure accurate registration of the number fields from the different films after scanning. Secondly it does not discriminate between photons of different energies. This will be important when β filtration does not provide sufficient monochromatization.

30.3.5 Parasitic Scatter

The first and third pin-hole shown in Figure 4 play no role in the collimation process but are included to reduce parasitic scatter. On the X-ray target a region of weak emission surrounds the intense focal spot. If radiation from this reaches the detector, its intensity is comparable with that which is scattered, and the purpose of the first pin-hole is to cut it out. Provided this first pin-hole is a little larger than the main beam, its size is not critical. Since the second pin-hole intersects the main beam, radiation is scattered from its edges, and is cut off by the third pin-hole, which should be very slightly larger than the beam diameter at its location.

Even with this arrangement there is a significant background (including residual noise in the detector) which needs subtracting from the measured distribution. In principle this is easily done with a position sensitive detector; the specimen is removed, the distribution is recorded for a time equal to that of the original experiment multiplied by the transmissivity of the specimen, and is then subtracted from the original distribution. In practice, it is difficult to measure the transmissivity accurately. For instrumental reasons the measurement has to be done with a much weaker beam than was used in the scattering experiment, and the attenuation inevitably changes the spectral distribution, with which transmissivity varies.[22] For accurate subtraction, parasitic scattering must be reduced as much as possible.

30.3.6 Accuracy of Intensity Measurement and Estimation of Parameter Values

When SAXS is being used quantitatively, either to determine structural models or to test proposed morphologies, it is essential to know how errors such as noise in the experimental data affect the determined quantity, or whether the differences between observed and predicted intensity are greater than likely errors of measurement. Whilst the well-known statistics of photon counting enable uncertainties to be calculated without difficulty, and similar information is available for intensity measurement from film,[64] the propagation of these errors through formulae such as equations (1–4) is not straightforward and has received scant attention in work with synthetic polymers. Analogous studies of scattering by particles in solution have investigated the effect of noise, and of limited experimental range[61,66] on the accuracy of derived parameters, and this approach needs applying to studies of solid morphology.

Similarly, in comparing experimental and predicted scattering distributions, most investigators have been content with matching by eye, and without reference to the discrepancies that would be expected, given the quality of the data. Hall *et al.*[22] provide an exception. They have used the statistical parameter χ^2, given by

$$\chi^2 = [(I_{obs} - I_{pred})/\Delta I_{obs}]^2 \tag{8}$$

where I_{obs} and I_{pred} are respectively the measured and predicted intensity, and ΔI_{obs} is the standard deviation in the measured intensity, and devised a stepwise refinement procedure which adjusted model parameters to minimize the sum of χ^2 over the experimental range. They were then able to decide which discrepancies in fit were experimentally significant, and to deduce the range of model parameters which would produce scattering distributions indistinguishable by experimental data of the accuracy available. Other best fit criteria are available,[66] and this approach needs further investigation and application.

30.4   EXPERIMENTAL INVESTIGATIONS

The theoretical and experimental procedures which have been described have been applied to study phase segregation in copolymers, and lamellar structure in homopolymers. The underlying problem in both cases is the characterization of the arrangement and the distribution of sizes of components with different electron density. However, in the first case the components are more ordered than in the second, and this leads to a scattering distribution with more orders of diffraction, and to different methods of applying the theory to the experimental results. Here we shall be concerned only with the case of homopolymers, and only with those investigations which have made quantitative comparisons between experimental data and the predictions of theoretical models. Even with these limitations there are an enormous number of reports to be considered, so, since this chapter is concerned with the way in which SAXS is used rather than the results derived from it, examples will be selected which illustrate most directly the theoretical and experimental procedures which have been discussed.

30.4.1   Determination of Density Differences Between Phases

In order to obtain a value for the invariant Q (equation 1) and so determine the density difference between phases, the integration must be carried out for all directions of scattering. This is simple for isotropic scatterers as the variation of intensity with angle need only be measured in one direction, but values are more often required for oriented materials for which the distribution must be measured in two dimensions. Area position sensitive detectors are ideal for this purpose,[67] but where they have not been available techniques have been devised to enable measurements in one dimension to be integrated. If slit collimation is used, with slit and orientation axes parallel, then integration need only be done along a line perpendicular to these axes.[11] If the sample is only slightly oriented, the axis may be placed parallel to the beam to give a radially symmetric pattern which can be integrated along one dimension (slit collimation can be used), and then multiplied by a correction factor obtained from a second experiment using pin-hole collimation with the orientation axis perpendicular to the beam. In this method the peak intensity is integrated azimuthally and the correction factor is the ratio of this integral to the value at the peak at 90° azimuth.[7] Absolute values of intensity are required to determine Q, so the incident beam intensity must be known, and it has been determined using a standard sample of known scattering power,[7,11] or absorbers of known transmissivity.[67]

With sedimented crystalline mats of polyethylene a value for $\rho_c - \rho_a$ has been reported[68] consistent with that expected from the difference between the unit cell density (ρ_c) and a quantity obtained by extrapolating measurements of the melt density to room temperature (ρ_a) confirming the model of uniform density phases with sharp boundaries and void free crystallites. Work on bulk polyethylene has yielded similar conclusions, and measurements of crystallinity and ρ_c using other techniques have enabled a value to be obtained for ρ_a.[67,69] Similar experiments with poly(ethylene terephthalate) fibres[11] yielded a lower value of $\rho_c - \rho_a$ than expected and led to the conclusion that in this material there are defects in the crystalline phase.

Vonk and Pijpers[70] combined measurements of Q with determinations of the crystallinity from the correlation function to determine the crystalline density (ρ_c). For non-linear polyethylene they obtained values greater than those obtained from wide-angle X-ray scattering. To account for this discrepancy the size of the amorphous regions between stacks would have to be large enough to be observed microscopically, which was not the case. However reasonable agreement between values of ρ_c was obtained with high density polyethylene.

By calculating Q for models with assumed values for ρ_c, ρ_a, and the sample density[71] it has been shown that changes of as little as 4% in the value assumed for ρ_c will more than double Q. This is presumably because, in these calculations, the volume fractions of the phases were obtained from the densities and the values are again very sensitive to small variations is density.[72] Similar considerations must apply when density differences are calculated from Q. In many investigations of this type wide-angle X-ray measurements of unit cell parameters have been used to determine ρ_c, and the inherent errors in such measurements are likely to cause errors in unit cell volumes[72] of about 2%. Hence the possibility of significant error cannot be discounted.

30.4.2 Evidence for Transition Zones

Vonk[9] used plots of $s^4 I(s)$ against s^2, and the correlation function, to show that transition zones of intermediate density occur between phases in many different polymers, getting good agreement between the values obtained by the two methods. Blundell[18] also found such transition zones in low density polyethylene. Evidence is, however, conflicting. Bornschlegl and Bonart[71] found no evidence of transition zones in poly(ethylene terephthalate), one of the materials studied by Vonk, nor in poly(butylene terephthalate), and neither do Adams et al.[67] with polyethylene. The reasons might lie in different methods of data treatment. The different authors used different methods to subtract the background due to density variation within phases, and it has been shown[73] that unless this is done with extreme care, very erroneous results will be obtained. Statistical counting errors in the intensity data can also cause significant error.[74]

Formulae giving the intensity distribution by the linear paracrystal have been modified to include transition zones,[23] and Blundell has calculated distributions for various model structures[75] to determine if they can be detected from the shape of this distribution. He concludes that the position and width of the diffraction peaks are unaffected, the only noticeable effect being to reduce the integrated intensity of the peak. The amount is small for the first order peak, but becomes more significant for higher orders. Since with synthetic homopolymers it is unusual to observe more than one peak, it is unlikely that direct observation of the distribution of scattered intensity will yield information on transition zones for these materials. Crist[76] arrives at a similar conclusion, which is also confirmed by work with copolymers. These have a much more regular structure, resulting in several orders of diffraction being visible, and the change in intensity between them is used to determine the thickness of transition zones.[36]

30.4.3 Investigation of the Linear Paracrystal

In order to use equation (5) to calculate the intensity distribution by a linear paracrystal, it is necessary to know the distribution of lengths of the two phases, and the number of lamellae in a stack. The calculated distribution can then be compared with that observed experimentally and parameters adjusted until the two match. The model has also been investigated by observing the ratio of the angles at which first and second order peaks occur and comparing this with calculated values.[77] Three types of length distribution have received most attention. Two of them, the top hat and Gaussian, are symmetric, the third, the Reinhold,[78] is asymmetric. Exponential functions have also been used.

Crist[76] showed that the ratio of the angles at which the peaks occur is less than two for symmetric distributions but greater for those with positive asymmetry (the tail towards long lengths). He investigated polyethylene and poly(oxymethylene) after various annealing treatments and only for quenched poly(oxymethylene) was the ratio greater than two, indicating that distributions are generally symmetric.

Blundell[18] attempted to match the experimental intensity distribution for low density polyethylene with ones calculated for symmetric (Gaussian) and asymmetric (exponential) distributions on both phases. A successful match was found only with stacks unrealistically short ($N \simeq 2$). He was led to consider a model in which the overall local crystallinity varies from one stack to another (other models assume all stacks belong to the same statistical population), and obtained a better fit for this. His work may be criticized in that his data was obtained with slit collimation and desmeared, and corrected for transition zones, both of which are likely to introduce unquantified errors, and that the effects of instrumental broadening were not considered.

Both of the above studies used isotropic material; Hall et al.[22] have used oriented poly(butylene terephthalate) (see Figure 1a). They used pin-hole collimation with an LPSD and ensured that the instrumental line broadening of the combination had negligible effect upon the recorded scattering distribution. The entrance slit to the LPSD was deep enough to admit all the scattering transverse to the meridian, so that the recorded distribution was the integral along lines normal to the meridian plotted against angle along it. They argued that the transverse broadening of the pattern required the measured intensity to be multiplied by a Lorentz factor s (see Section 30.2.3.2) before comparison with the calculated intensity. The collimation system was adjusted to minimize parasitic scatter which was measured as described in Section 30.3.5. and subtracted; no correction was made for other background scatter.

The statistical parameter $\Sigma \chi^2$ (equation 8), with ΔI_{obs} including errors in background, was used as a criterion of goodness of fit and minimized as explained in Section 30.3.6. It was found that the

Gaussian, top hat, and Reinhold distributions all predicted the correct distribution of scattering around the main peak, but only the Reinhold predicted the correct shape in the high angle tail ($> \simeq 2°$). Figure 5, the results for their best fit model, shows that a very good fit is obtained at all angles except the very smallest, where the deviation might be caused by the assumption that the medium surrounding the stacks has the same density as the amorphous regions. Provided there were more than about 20 lamellae in the stack, the number did not affect the shape of the curve. The precision with which the parameters describing the morphology could be determined by this experiment was investigated by determining the range for which $\Sigma \chi^2$ differed by an insignificant amount from the expected value of unity.

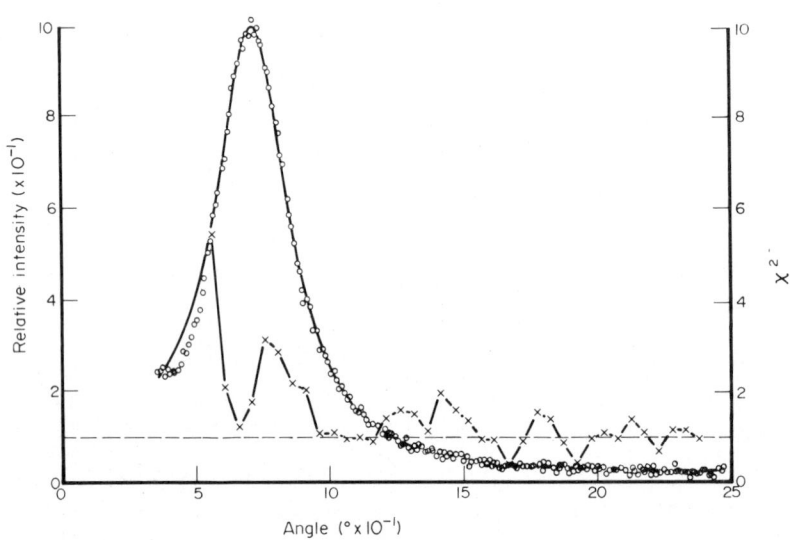

Figure 5   Comparison of calculated and measured intensity distribution for poly(butylene terephthalate).[22] The circles are experimental data, the continuous line is calculated for a model with Reinhold distributions on the lengths of both phases. The line with crosses is χ^2. The model was refined to give the best fit over the angular range for which χ^2 is plotted. At smaller angles significant deviations between the two curves was unavoidable

Investigations of polymer blends[79] have also obtained a good fit between measured and calculated scattering distributions with this type of model. However, the use by Hall *et al.*[22] of statistical tests to judge the significance of discrepancies, the good fit they have achieved over such a wide angular range, and their determination of the range of models which give an acceptable fit, appear to be unique.

30.4.4   Studies of Three-dimensional Morphology

Some of the theoretical treatments of the scattering by a three-dimensional morphological model go no further than presenting equations and sample calculations, making no reference to experimental data. These have already been discussed in Section 30.2.3.2 and will not be considered here. Others, whilst calculating two-dimensional distributions, only make quantitative comparison along one or two lines through the pattern. Thus Fronk and Wilke[38] tested their model of inclined lamellae of equal thickness by comparing the calculated intensity distribution along the meridian, the equator, and a line at 45° to each, with that observed along these directions. The scattering material was low density polyethylene oriented by various draw ratios. Reasonable agreement was obtained, but the values obtained for the morphological parameters did not agree with those from other investigations, which led them to apply their theory to a model in which the crystallites formed a paracrystalline macrolattice.[39] The predictions of this model were compared[80] with data from low density polyethylene along the meridian, along a line parallel to the equator through the point of maximum meridional intensity, and along a line about 40° to the meridian. Standard deviations of experimental data were obtained by averaging the four quadrants of the pattern and used to determine the acceptable range of morphological parameters. A good match between theory and experiment was achieved with parameters in agreement with those obtained from wide-angle scattering.

Hosemann and co-workers[33, 34, 81] use a basically similar model, but develop the theory in such a different manner that comparison with others could probably only be done numerically. They compare predicted and observed intensity distributions along the meridian and along a line parallel to the equator through the meridional point of maximum intensity, obtaining a good fit between the two sets of curves. They claim that using only a one-dimensional analysis leads to very different values for the morphological parameters.

The only work we have found in which the predictions for a three-dimensional model are compared with experimental data at all angles and azimuths is by Vonk.[82] The scattering from oriented low density polyethylene was recorded on photographic film used in conjunction with a raster-scanning microdensitometer, enabling the intensity to be measured at all points on a square lattice with 31 points in each direction. The angular separation of adjacent points was about 0.07°, and the lattice covered one asymmetric quarter of the pattern, which was either four-point or horizontal bar depending upon the annealing temperature. The two-dimensional correlation function was calculated from this data and compared with that for two different models; the first comprised finite width stacks of inclined lamellae with parallel surfaces, for which the width of the reflection perpendicular to the fibre axis was caused by the small stack width, the second was a paracrystalline macrolattice. A distinctive feature of the observed correlation function was the rows and columns of alternating positive and negative areas (Figure 6). This could not be simulated with the inclined lamellae model, and only with the paracrystalline macrolattice model if crystalline and amorphous areas in neighbouring fibrils were adjacent to each other.

The evidence from SAXS therefore favours a model for the structure of oriented crystalline polymers in which the crystallites form a three-dimensional paracrystalline macrolattice rather than one-dimensional stacks. This is, however, based on limited investigation. Now, with the ready

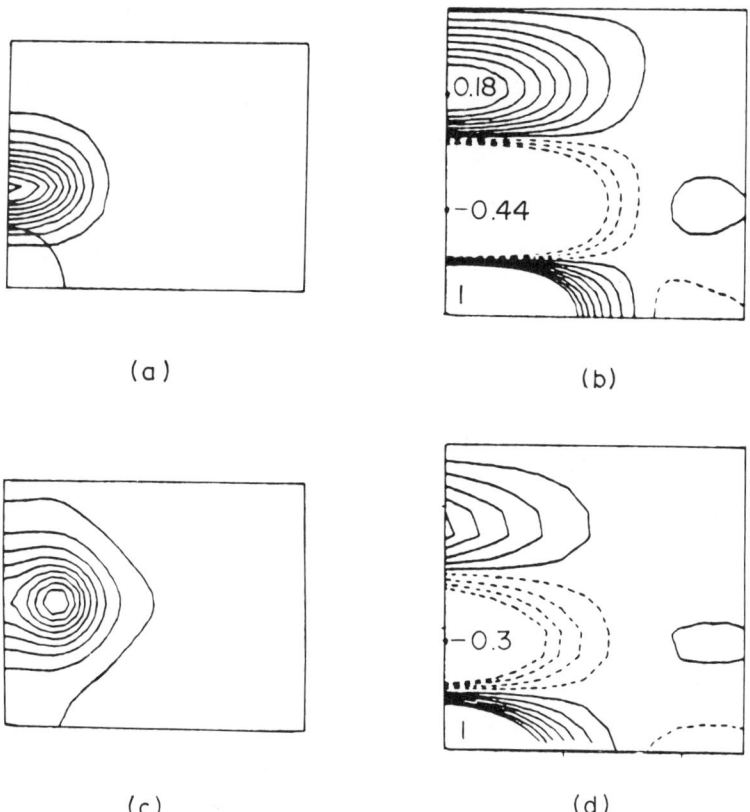

(a)   (b)

(c)   (d)

Figure 6 Contour maps showing calculated and observed scattering distributions and two-dimensional correlation functions for oriented, annealed, low density polyethylene (taken from C. G. Vonk, *Colloid Polym. Sci.*, 1979, **257**, 1021); (a) is the observed intensity distribution and (b) its correlation function, (c) is the intensity distribution calculated from (d) where (d) is the correlation function calculated from the model giving the best match to experimental data

availability of raster-scanning microdensitometers and substantial computing power, the technical difficulties of whole-pattern matching experiments have been eased. The time would seem propitious to follow up the pioneering work of Vonk.

30.5 REFERENCES

1. K. C. Holmes and D. M. Blow, *Methods Biochem. Anal.*, 1966, **13**, 113.
2. H. S. Lipson and C. A. Taylor, 'Fourier Transforms and X-ray Diffraction', Bell, London, 1958.
3. C. A. Taylor and H. S. Lipson, 'Optical Transforms', Bell, London, 1964.
4. A. Guinier, 'X-ray Diffraction in Crystals, Imperfect Crystals, and Amorphous Bodies', Freeman, San Francisco, 1963.
5. L. E. Alexander, 'X-ray Diffraction Methods in Polymer Science', Wiley, New York, 1969.
6. G. R. Strobl, *J. Appl. Crystallogr.*, 1973, **6**, 365.
7. G. R. Strobl and N. Muller, *J. Polym. Sci., Polym. Phys. Ed.*, 1973, **11**, 1219.
8. W. Ruland, *J. Appl. Crystallogr.*, 1971, **4**, 70.
9. C. G. Vonk, *J. Appl. Crystallogr.*, 1973, **6**, 81.
10. J. T. Koberstein, B. Morra and R. S. Stein, *J. Appl. Crystallogr.*, 1980, **13**, 34.
11. E. W. Fischer and S. Fakirov, *J. Mater. Sci.*, 1976, **11**, 1041.
12. W. Wiegand and W. Ruland, *Prog. Colloid Polym. Sci.*, 1979, **66**, 355.
13. W. Ruland, *Kolloid Z. Z. Polym.*, 1977, **255**, 417.
14. W. Ruland, *Kolloid Z. Z. Polym.*, 1978, **256**, 932.
15. C. G. Vonk and G. Kortleve, *Kolloid Z. Z. Polym.*, 1967, **220**, 19.
16. C. G. Vonk, *J. Appl. Crystallogr.*, 1978, **11**, 541.
17. G. Kortleve and C. G. Vonk, *Kolloid Z. Z. Polym.*, 1968, **225**, 124.
18. D. J. Blundell, *Polymer*, 1978, **19**, 1258.
19. Z. Zernicke and J. A. Prins, *Z. Phys.*, 1927, **41**, 184.
20. R. Hosemann and S. N. Bagchi, 'Direct Analysis of Diffraction by Matter', North Holland, Amsterdam, 1962.
21. R. Hosemann, *Z. Phys.*, 1950, **128**, 465.
22. I. H. Hall, E. A. Mahmoud, P. D. Carr and Y. Geng, *Colloid Polym. Sci.*, 1987, **265**, 383.
23. D. J. Blundell, *Acta Crystallogr., Sect A.*, 1970, **26**, 472.
24. C. G. Vonk, in 'Small Angle X-ray Scattering', ed. O. Glatter and O. Kratky, Academic Press, London, 1982, p. 433.
25. W. Wenig and R. Bramer, *Colloid Polym. Sci.*, 1978, **256**, 125.
26. M. Matsuo, C. Sawatari, M. Tsuji and R. St. John Manley, *J. Chem. Soc., Faraday Trans. 2*, 1983, **79**, 1593.
27. J. M. Schultz, *J. Polym. Sci., Polym. Phys. Ed.*, 1976, **14**, 2291.
28. M. Matsuo and C. Kitayama, *Polym. J.*, 1985, **17**, 479.
29. M. A. Gezalov, V. S. Kusenko and A. I. Slutsker, *Polym. Sci. USSR (Engl. Transl.)*, 1970, **12**, 2027.
30. B. Crist, Jr., *J. Appl. Crystallogr.*, 1979, **12**, 27.
31. V. I. Gerasimov and D. Ya. Tsvankin, *Polym. Sci. USSR (Engl. Transl.)*, 1969, **11**, 3013.
32. G. Blöchl and A. J. Owen, *Colloid Polym. Sci.*, 1984, **262**, 793.
33. K. Kaji, T. Mochizuki, A. Akiyama and R. Hosemann, *J. Mater. Sci.*, 1978, **13**, 972.
34. R. Hosemann, J. Loboda-Čačković and K. Kaji, *J. Appl. Crystallogr.*, 1978, **11**, 540.
35. V. I. Gerasimov, V. D. Zanegin and D. Ya. Tsvankin, *Polym. Sci. USSR (Engl. Transl.)*, 1978, **20**, 954.
36. M. Shibayama and T. Hashimoto, *Macromolecules*, 1986, **19**, 740.
37. W. Wilke and K. Göttlicher, *Colloid Polym. Sci.*, 1981, **259**, 596.
38. W. Fronk and W. Wilke, *Colloid Polym. Sci.*, 1983, **261**, 1010.
39. W. Fronk and W. Wilke, *Colloid Polym. Sci.*, 1985, **263**, 97.
40. H. K. Herglotz, in 'Structure of Crystalline Polymers', ed. I. H. Hall, Elsevier, London, 1984, p. 229.
41. C. Nave, J. R. Helliwell, P. R. Moore, A. W. Thompson, J. S. Worgan, R. J. Greenhall, A. Miller, S. K. Burley, J. Bradshaw, W. J. Pigram, W. Fuller, D. P. Siddons, M. Deutsch and R. T. Tregear, *J. Appl. Crystallogr.*, 1985, **18**, 396.
42. W. Fronk, B. Heise, B. Neppert, H.-R. Schubach and W. Wilke, *Colloid Polym. Sci.*, 1984, **262**, 99.
43. O. Kratky, in 'Small Angle X-ray Scattering', ed. O. Glatter and O. Kratky, Academic Press, London, 1982, p. 53.
44. O. Kratky and H. Stabinger, *Colloid Polym. Sci.*, 1984, **262**, 345.
45. I. R. Harrison, S. J. Kominski, W. D. Varnell and J.-I. Wang, *J. Polym. Sci., Polym. Phys. Ed.*, 1981, **19**, 487.
46. J. Goodisman, F. Delaglio and H. Brumberger, *J. Appl. Crystallogr.*, 1986, **19**, 243.
47. A. Guinier and G. Fournet, 'Small Angle Scattering of X-rays', Wiley, New York, 1955.
48. A. Dijkstra, G. Kortleve and C. G. Vonk, *Kolloid Z. Z. Polym.*, 1966, **210**, 121.
49. R. W. Hendricks and P. W. Schmidt, *Acta Phys. Austriaca*, 1967, **26**, 97.
50. C. G. Vonk, *J. Appl. Crystallogr.*, 1971, **4**, 340.
51. O. Glatter, in 'Small Angle X-ray Scattering', ed. O. Glatter and O. Kratky, Academic Press, London, 1982, p. 119.
52. M. J. Cooper, *Phys. Bull.*, 1977, **28**, 463.
53. R. W. Hendricks and L. B. Shafer, *J. Appl. Crystallogr.*, 1978, **11**, 196.
54. I. H. Hall and M. Toy, in 'Structure of Crystalline Polymers', ed. I. H. Hall, Elsevier, London, 1984, p. 181.
55. H. E. Huxley, *Acta Crystallogr.*, 1953, **6**, 457.
56. O. E. A. Bouldan and R. S. Bear, *J. Appl. Phys.*, 1949, **20**, 983.
57. U. W. Arndt, *J. Appl. Crystallogr.*, 1986, **19**, 145.
58. U. W. Arndt and D. J. Gilmore, *J. Appl. Crystallogr.*, 1979, **12**, 1.
59. G. T. Reynolds, J. R. Milch and S. M. Gruner, *Rev. Sci. Instrum.*, 1978, **49**, 1241.
60. R. W. Hendricks, *J. Appl. Crystallogr.*, 1978, **11**, 15.
61. U. Lembke and Th. Gerber, *J. Appl. Crystallogr.*, 1985, **18**, 55.
62. H. Leopold, in 'Small Angle X-ray Scattering', ed. O. Glatter and O. Kratky, Academic Press, London, 1982, p. 105.
63. T. P. Russell, R. S. Stein, M. K. Kopp, R. E. Zedler, R. W. Hendricks and J. S. Lin, *Oak Ridge Nat. Lab. Tech. Rep. ORNL/TM-6678*, 1978.

64. C. G. Vonk and A. P. Pijpers, *J. Appl. Crystallogr.*, 1981, **14**, 8.
65. U. W. Arndt, D. J. Gilmore and A. J. Wonacott, in 'The Rotation Method in Crystallography', ed. U.W. Arndt, North Holland, Amsterdam, 1977, p. 207.
66. J. J. Müller, G. Damaschun and P. W. Schmidt, *J. Appl. Crystallogr.*, 1985, **18**, 241.
67. W. W. Adams, R. M. Briber, E. S. Sherman, R. S. Porter and E. L. Thomas, *Polymer*, 1982, **23**, 1069.
68. E. W. Fischer, H. Goddar and G. F. Schmidt, *J. Polym. Sci., Polym. Lett. Ed.*, 1967, **5**, 619.
69. W. W. Adams, R. M. Briber, E. S. Sherman, R. S. Porter and E. L. Thomas, *Polymer*, 1985, **26**, 17.
70. C. G. Vonk, and A. P. Pijpers, *J. Polym. Sci., Polym. Phys. Ed.*, 1985, **23**, 2517.
71. E. Bornschlegl and R. Bonart, *Colloid Polym. Sci.*, 1980, **258**, 319.
72. I. H. Hall, in 'Structure of Crystalline Polymers', ed. I. H. Hall, Elsevier, London, 1984, p. 39.
73. U. Siemann and W. Ruland, *Colloid Polym. Sci.*, 1982, **260**, 999.
74. R.-J. Roe, *J. Appl. Crystallogr.*, 1982, **15**, 182.
75. D. J. Blundell, *Acta Crystallogr., Sect. A*, 1970, **26,** 476.
76. B. Crist, *J. Polym. Sci., Polym. Phys. Ed.*, 1973, **11**, 635.
77. B. Crist and N. Morosoff, *J. Polym. Sci., Polym. Phys. Ed.*, 1973, **11**, 1023.
78. C. Reinhold, E. W. Fischer and A. Peterlin, *J. Appl. Phys.*, 1964, **35**, 71.
79. F. P. Warner, W. J. MacKnight and R. S. Stein, *J. Polym. Sci., Polym. Phys. Ed.*, 1977, **15**, 2113.
80. W. Fronk and W. Wilke, *J. Polym. Sci., Polym. Phys. Ed.*, 1986, **24**, 839.
81. E. Ferracini, A. Ferrero, J. Loboda–Čačković, R. Hosemann and H. Čačković, *J. Macromol. Sci., Phys.*, 1974, **10**, 97.
82. C. G. Vonk, *Colloid Polym. Sci.*, 1979, **257**, 1021.

31

X-Ray Scattering from Non-crystalline and Liquid Crystalline Polymers

GEOFFREY R. MITCHELL

University of Reading, UK

31.1 INTRODUCTION

31.1.1 Scope

The absence of sharp crystalline peaks in the X-ray scattering pattern of a polymer is universally taken as a characteristic of a non-crystalline polymer. It is often thought that the diffuse peaks in

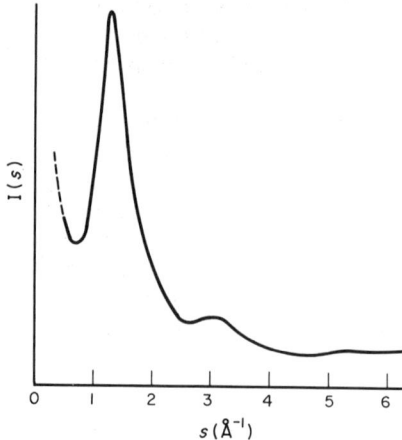

Figure 1 The X-ray scattering intensity function I(s) recorded for an isotropic sample of non-crystalline PEEK; the specimen was prepared by rapidly cooling from the melt. These data are used as an example in the initial sections of the chapter. $s = 4\pi \sin\theta/\lambda$, where the 2θ is the angle between the incident and scattered beams, λ the wavelength, in this case Cu $K\alpha$ ($\lambda = 1.54178$ Å). The data were recorded using a diffractometer with symmetrical transmission geometry equipped with an incident beam monochromator, pinhole collimation and operating in step-scan mode (see Section 31.4)

such a scattering pattern (for example Figure 1) contains no directly useful structural information. The sample simply becomes classified as 'amorphous'. This chapter is involved with techniques that allow us to look beyond such a classification and to extract a considerable quantity of structural information from the wide-angle X-ray scattering (WAXS) patterns from non-crystalline and liquid crystalline polymers. This structural information relates to chemical configurations and chain conformations, to molecular packing and to molecular orientation. This chapter is concerned with describing and illustrating the methods by which such information may be obtained. We shall be involved in the wide-angle scattering regime, that is for scattering vectors $s > 0.2$ Å^{-1} (whereby $s = 4\pi\sin\theta/\lambda$, where 2θ is the angle between the incident and scattered paths and λ is the X-ray wavelength, typically 0.79 or 1.54 Å) and hence evaluating structural features in the range 1–100 Å. Scattering at lower scattering vectors falls in the small-angle X-ray scattering regime and the reader is directed to Chapters 30 and 32 in this volume for a full description.

The reasons for the lack of structural regularity or crystallinity in synthetic polymers are many. In general, polymers based on atactic chains are 'amorphous', as are materials which have been heavily irradiated with electrons or γ rays. Molten and rapidly quenched polymer systems are also highly disordered. Crystalline polymers often have a relatively low degree of crystallinity, and hence contain a large portion of 'non-crystalline' material as may be seen from an X-ray scattering pattern (for example that of poly(tetrafluoroethylene) (PTFE) in Figure 2). Within this chapter, we shall class

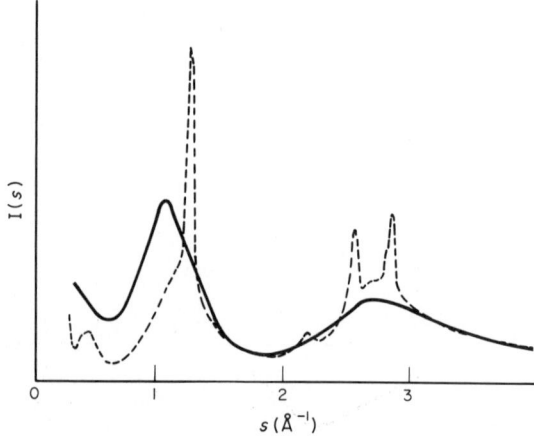

Figure 2 The X-ray scattering curves for poly(tetrafluoroethylene) in the crystal phase at 18 °C (– – –) and in the molten phase at 350 °C (——); the sharp peaks in the former indicate the crystalline structure, but note the broad components to its scattering pattern, similar to the melt pattern, which indicates a substantial non-crystalline fraction

all of these materials or components of materials as non-crystalline whatever the origin or cause of their disordered structure. The focus of the chapter will be homogeneous materials such as melts and glasses, however the concepts can be extended to heterogeneous systems such as the amorphous phase of a semi-crystalline polymer.[1]

The liquid crystalline state is also characterized by the absence of long range translational symmetry or a three dimensional lattice.[2] However, there are basic structural features, such as the long range orientational ordering of the nematic phase, or the layered structures of the smectic phase imposed on an otherwise 'liquid' like molecular organization. For certain high symmetry classes of the smectic family (for example smectic B) this extra structure approaches that of a crystalline phase[3] and its analysis falls outside the scope of this chapter. In general, similar approaches may be used in the X-ray study of non-crystalline and liquid crystalline polymers, although for the latter type the introduction of additional ordering may greatly simplify the analysis and allow additional information to be extracted.

31.1.2 Structural Information Required

It is appropriate, before embarking on the structural analysis of non-crystalline and liquid crystalline polymers, that we should lay out a structural framework which is both useful and unrestrictive for such materials. For semi-crystalline polymers (see Chapters 28–30 of this volume) we start with the crystallographic unit cell as the basic building block and then elaborate on the manner in which these are assembled together with the non-crystalline fraction. Structural descriptions of non-crystalline and liquid crystalline polymers are handicapped by the absence of the periodic lattice and we must start with the molecular chain. However, we shall not be concerned directly with chain trajectory but with local chain conformation and chemical configuration. The basic building block then becomes the chain segment, illustrated in Figure 3. The chain segment is a portion of a macromolecular chain over which the properties in terms of chain conformation and chemical configuration are broadly constant. The chain trajectory is built up of an assemblage of such structural units connected together. The molecular packing involves the interaction of such units unconnected with each other. In other words, we may partition the molecular organization into the interactions between the chain segments (interchain) and those correlations within the chain segments (intrachain).[4] The latter includes both a description of the chain conformation and its extent, while the former may involve correlations between chain segments belonging to the same macromolecular chain as indicated in Figure 3 as well as different chains. Two particular points should be borne in mind. The first, is that a structural partition of this kind does not imply or require independence of the two components as, for example, employed by the 'Random Coil' approach.[5] Clearly any such independence of chain conformation and molecular packing would be inappropriate for a liquid crystal polymer, where the two are strongly correlated.[6] The second is that chain segments should not be exclusively thought of as straight molecular cylinders, as a molecular model of an atactic poly(methyl methylacrylate) (atactic PMMA) demonstrates (Figure 4).

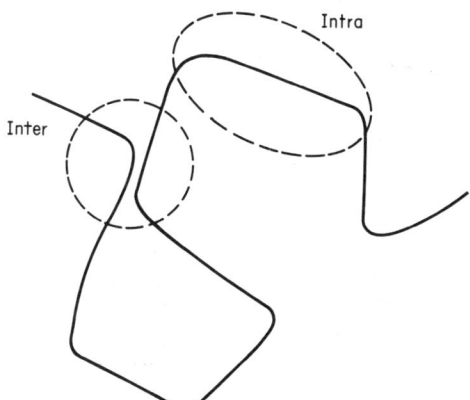

Figure 3 The solid line represents a hypothetical chain trajectory for a polymer molecule; the elliptical section represents a chain segment; and the circled components illustrate that interchain type interactions can occur between chain segments of the same molecule

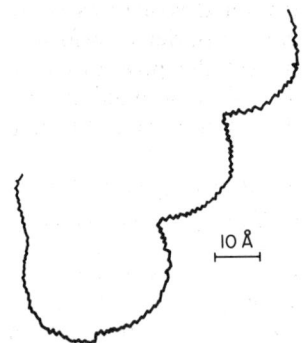

Figure 4 The projected chain trajectory for a section of an atactic poly(methyl methacrylate) molecule in which the chain conformation is based on the results of R. Lovell and A. H. Windle, *Polymer*, 1981, **22**, 175. The curved chain segment arises as a result of the unequal skeletal bond angles within each repeat unit

Thus, the basic structural building block is the chain segment, and we require quantitative information describing the nature of the segment, that is the chain conformation and chemical configuration, and interchain parameters detailing the spatial and orientational correlations of such units. X-ray scattering procedures provide a route to obtaining these structural parameters.

In many ways the situation for liquid crystal polymers is much simpler. For a nematic, we clearly need parameters describing the level of the orientational ordering,[7] and for a smectic, parameters detailing the nature of the layered structure are essential.[8] However, beneath these obvious quantities, lie the same basic structural questions we asked of the non-crystalline polymers: What is the chemical configuration? What is the chain conformation? How are the chains packed? The loose structural framework introduced above will therefore service both elements of this chapter, *i.e.* both non-crystalline and liquid crystalline polymers.

31.1.3 Approaches to the Analysis of X-Ray Scattering Patterns

The diffuse peaks in the scattering pattern of Figure 1 are related to variations in the electron density distribution within the poly(bis-1,4-phenoxy-1,4-benzoyl) sample, designated PEEK from poly(ether ether ketone). In fact the single elastically scattered X-ray intensity, $I(s)$, is the Fourier transform of the self-convolution, $Q(r)$, of the electron density, $p(r)$, for example[9]

$$I(s) = \int Q(r) \cos r.s \, dv_r \qquad (1)$$

$$\text{and} \quad Q(r) = \int p(u) \, p(u + r) \, dv_u \qquad (2)$$

$Q(r)$ is a map of interelectron vectors averaged over the structure, taking each point in space as the origin in r. In other words the X-ray scattering intensity is sensitive to spatial correlations between regions of high or low electron density. These equations have been written for the most general case, in subsequent sections reduced formulations will be presented for materials with isotropic and with anisotropic uniaxial molecular organizations. However the basic feature is that structure (*i.e* the electron distribution) is related through a Fourier transformation to the scattering. It would be most convenient if there were some universal rules which enabled the position of a diffuse peak in the scattering pattern to be converted into a real space distance corresponding to the distance between the correlated volumes of high or low electron density. Of course, if we consider crystal planes then we may use the well-established reciprocity of Bragg's Law

$$d_{hkl} = \lambda/2\sin\theta \qquad (3)$$

$$\text{or} \quad d_{hkl} = 2\pi/s \qquad (4)$$

where d_{hkl} is the interplanar spacing. Clearly for a highly disordered non-crystalline structure, such an approach is inappropriate, although examples of its use in the literature on amorphous materials abound.

Two possible models for such disordered structures would be to consider the basic unit as a cylinder or a sphere. For these cases the Fourier transformation of a random assemblage of such units can be performed analytically and the positions of the diffuse maxima arising from the correlations between the units derived. If these maxima are treated as Bragg maxima (*i.e.* using equation 3 or 4) then we find that the interunit size R is given by[10, 11]

$$R = 1.25 \, d_{Bragg} \quad \text{for spheres} \tag{5}$$

$$\text{and} \quad R = 1.11 \, d_{Bragg} \quad \text{for cylinders} \tag{6}$$

In essence we can interpret the position of diffuse maxima in the scattering patterns of disordered media, as long as we know the nature of the structural unit, and that unit conforms to one of these two options or some other simple arrangement. Table 1 lists the results of application of equations 5 and 6 to the scattering pattern of PEEK shown in Figure 1. It is perhaps reasonable to associate the largest distance to the packing of chain segments, and the other smaller distances to spatial correlations arising within the chain segments. However, the uncertainity in the precise value of the scaling constant to be used precludes any more detailed evaluation.

Table 1 Structural Distances Derived from the Scattering Pattern for PEEK shown in Figure 1

Scattering vector (Å^{-1})	Distance evaluated using		
	Equation 4 (Å)	Equation 5 (Å)	Equation 6 (Å)
1.3	4.83	6.04	5.36
2.98	2.11	2.64	2.34
5.34	1.18	1.47	1.31

One route to unravelling the complexities of the diffuse X-ray scattering of disordered polymer systems is to prepare a sample with some level of preferred global* molecular orientation.[12-14] For a liquid crystal polymer system this is particularly easy, but even for glassy polymers useful levels of orientation may be achieved by mechanical means just below the glass transition. Figure 5 shows the X-ray scattering function I(s, α) obtained from a deformed sample of PEEK with a significant level of preferred molecular orientation (see Section 31.3 for a description of the coordinate system, the definition of the angle α and of the intensity function). That there is a preferred molecular orientation

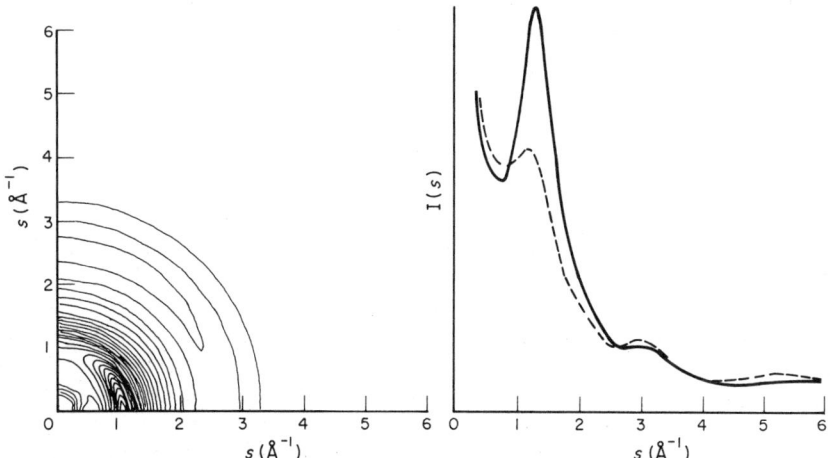

Figure 5 The scattered X-ray intensity function I(s, α) measured for a partially aligned non-crystalline sample of PEEK prepared by extrusion. The extrusion axis is vertical. The curves represent a section through this two-dimensional surface at α = 0° (– – –) *i.e.* meridional, and at α = 90° (——), *i.e.* equatorial

---

* The term *global* is used to indicate a preferred alignment about a field or other *external* axis. *Local* is used to indicate orientational correlations between chain segments in relationship to an *internal* axis. Local orientational order must exist in a liquid crystal sample, even though there is no macroscopic or global preferred alignment.

may be deduced directly from the anisotropic distribution of intensity with respect to α, the coordinate describing the angle to the mechanical deformation axis. In later sections, methods for evaluating quantitative orientational parameters from such 'arcing' will be detailed. For the present, it is sufficient to observe that the diffuse peak at $s \sim 1.3\,\text{Å}^{-1}$ is most intense at $\alpha = 90°$, while those at higher scattering vectors are more intense at $\alpha = 0°$. In other words the maxima at $s \sim 3$ and $5.5\,\text{Å}^{-1}$ intensify along a direction parallel to the direction of alignment (meridian), while the largest peak, that at $s \sim 1.3\,\text{Å}^{-1}$, intensifies in the plane normal to the mechanical axis (equator). This separation reinforces our initial surmise that the peak at $s \sim 1.3\,\text{Å}^{-1}$ represents correlations between PEEK chain segments, while those peaks at higher scattering vectors arise from correlations within the molecular chains. This identification of the structural origins of the diffuse peaks using aligned polymer samples is a powerful tool in the analysis of X-ray scattering patterns. The approach is particularly revealing when examining the scattering patterns arising from polymers with significant side chains such as poly(α-methylstyrene).[15] It is also possible to use the variation in peak position with temperature as a differentiation tool, since we would expect interchain distances to change to a greater extent with temperature than interchain distances. This approach has been applied to the analysis of polystyrene glasses.[16,17]

For polymer chains with large substituents, although we can identify the structural origins of the peaks in the scattering pattern using aligned samples, the situation for converting peak position to real space distance is unclear. We can institute an expression

$$R = k\,2\pi/s \tag{7}$$

where k is a constant dependent upon the nature of the structure and lies in the range 1 to 1.25. This type of relationship can clearly only provide a qualitative approach to the analysis of a diffuse X-ray scattering pattern. For a liquid crystalline polymer, in which an assumption that the basic units are rod-like seems much more reasonable, equation (7) has some validity. De Vries[18] and Leadbetter[8] have utilized values of k of 1.1547 and 1.117 in determining molecular packing distances, the choice relating to the level of lateral order in the material, and a value of 1.229 in calculating the molecular length of the mesogenic or structural unit. For a smectic, the maxima arising from the layered structure may be treated as Bragg maxima.[18]

The reversibility of the Fourier integral (1) enables the self-convolution $Q(r)$ of the electron density to be recovered from the intensity measurements[9]

$$Q(r) = \int I(s)\cos r.s\,\mathrm{d}v_s \tag{8}$$

However, it is not possible to invert equation (2) in order to obtain the electron density $p(r)$ from $Q(r)$. Thus in terms of real space representation we are restricted to $Q(r)$ which is a map of the interelectron vectors, averaged over the structure, taking each point in space as the origin of r. In essence, we are restricted to pair-type distribution functions. For isotropic samples, we would obtain the radial distribution function (see Section 31.2.2) and interunit distances may be read off from a plot of such a function. Table 2 lists values obtained from the radial distribution function $G(r)$ for a PEEK sample (Figure 6) derived from the scattering pattern of PEEK shown in Figure 1. The values so obtained are similar to those derived *via* equations (5) and (6) from the scattering pattern in Figure 1, apart from the presence of two peaks close to 5 Å. The radial distribution function emphasizes the very short range correlations, that is the covalently bonded distances 1.5 Å and 2.5 Å, since these are necessarily the most correlated. However, from a structural viewpoint these distances have limited value, the real interest is in distances $>2.5\,\text{Å}$, which will be sensitive to the particular chain conformation or packing mode.[19,20] Furthermore, the route to a radial distribution function is fraught with difficulties as will be discussed later (see Section 31.2.2) and thus this approach involving the inversion of the measured intensity function through equation (8), has limited

Table 2 Positions of the Maxima in the Radial Distribution
Function for PEEK shown in Figure 6

Distance from origin (Å)	1.1	2.57	4.85	6.23

The inherent resolution of the radial distribution function (see Section 31.2.2) is 0.5 Å.

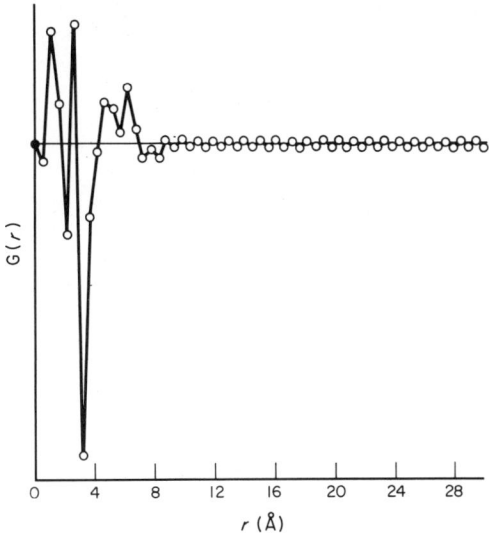

Figure 6 The atomic radial density function G(r) for PEEK derived from the scattering data of Figure 1 using the techniques detailed in Section 31.2.2 and 31.4.3; the open circles indicate the points at which the function G(r) was calculated using the sampled transform method

application for polymeric systems. The procedures, outlined above, provide interatomic distances, whereas the useful information is in molecular terms. An alternative is to start with the molecule and produce the scattering, and this will now be considered.

A particular advantage of X-ray scattering procedures, is the detailed understanding we have of the precise relationships between the positional coordinates of the atoms and the resultant scattering functions. Thus for example, an alternative representation of equation (1) is[21]

$$I(s) = \sum_{m}^{N} \sum_{n}^{N} f_n(s) f_m(s) \exp(i.s.r_{mn}) \tag{9}$$

where the summation takes place over all N atoms in the structure, and r_{mn} is the interatomic vector between the mth and nth atoms with scattering factors $f_m(s)$ and $f_n(s)$ respectively (see Section 31.2.3). Essentially, if we have a knowledge of the coordinates of the atoms, we can calculate the elastic scattering without making restrictive or unreasonable assumptions. Thus an alternative approach to understanding the diffuse scattering patterns of non-crystalline and liquid crystalline polymers is to compare the observed scattering with that calculated from molecular models using equation (9). It will be shown later that this approach is relatively straightforward because the scattering pattern can be partitioned into interchain and intrachain components, and that this considerably simplifies the molecular model 'building'.

The analysis of the diffuse X-ray scattering from disordered polymer systems is not straightforward. We have introduced three approaches which can be applied. Each makes progressively greater demands upon the analyst, but offers increasingly quantitative structural parameters in return. On the other hand a simple analysis may suffice for a particular problem. In the latter part of this chapter, examples of each of these approaches to the analysis of the X-ray scattering from non-crystalline polymers will be detailed. Firstly, a more detailed view of formulations and experimental procedures will be given.

31.2 BASIC THEORY FOR SAMPLES WITH RANDOM ORIENTATION

The mathematical relationships given in the preceding section were of the most general nature. In practice most samples have an isotropic distribution of molecular organization. That is not to say that the materials, *locally*, do not possess an anisotropic structure which, for example, liquid crystal phases by their very nature must exhibit, but that this structure when considered as a rigid body takes up all possible orientations in space. As a result the expressions relating scattered intensity with structure are considerably simplified.

31.2.1 Scattering Geometry

The basic X-ray scattering experiment involves directing an X-ray beam into a sample volume and recording the scattered intensity as a function of the scattering angle. Figure 7 defines the basic angles. K_0 and K represent the incident and scattered wave vectors with $|K_0| = 2\pi/\lambda_0$, $K = |2\pi/\lambda|$ and the scattering vector is $s = K - K_0$. Since we are only interested in the case of elastic scattering, that is scattering without a change in energy, consequently $|K_0| = |K|$ and $|s| = 4\pi \sin \theta/\lambda$. In fact for an isotropic or random orientation sample the scattering only depends on this magnitude of the scattering vector. In the next section (Section 31.3), we shall see that it is necessary to utilize polar or cartesian coordinates to describe the scattering from an aligned sample.

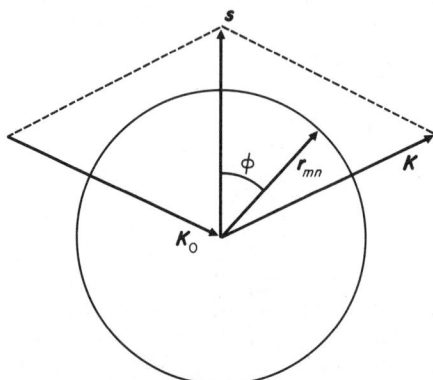

Figure 7 A schematic illustrating the definition of the scattering vector s where K_0 and K are the incident and scattered wavevectors; for an isotropic or randomly oriented structure the interatomic vector r_{mn} can take up all orientations in space

31.2.2 Scattered Intensity and Structure

For a material with no preferred orientation, the self-convolution of the electron density (equation 2) becomes the radial distribution function $W(r)$.[21]

$$W(r) = 4\pi r^2 \{\rho(r) - \rho_0\} \tag{10}$$

where $\rho(r)$ is the electron density and hence $4\pi r^2 \rho(r)\,dr$ is the average number of electrons between distances r and $r + dr$. Note that equation (13) involves the difference in density rather than total values. This is simply because the Fourier transform of the average density of the sample has the form of a δ function at zero scattering angle, of magnitude equal to the total number of electrons in the specimen, smeared out in a manner inversely related to the shape and size of the sample. Since this 'zero' angle scattering is experimentally inaccessible, the total radial density function is not attainable from just a scattering experiment, and additional measurements must be made. Moreover, it is the variations in the electron density which are of direct structural interest. Equally the correlations between electrons within the same constituent atom are of no interest in this type of study. The Fourier transform of these correlations is the so called 'independent scattering', which is equivalent to the scattering from isolated atoms and may be calculated using theoretical values for the X-ray scattering factors[22] (see Section 31.2.3). In fact it is conventional and particularly useful when dealing with non-crystalline materials to consider a reduced intensity function $i(s)$

$$i(s) = I(s) - \sum_i^M c_i f_i^2(s) \tag{11}$$

where the summation is made over the different types of atom, with a fractional composition c_i and a scattering factor $f_i(s)$. $I(s)$ is the fully corrected measured intensity function scaled in electron units (see Section 31.4.4). A plot of the s weighted reduced intensity function for PEEK is shown in Figure 8. In addition to its use in preparing the radial distribution function, the $si(s)$ scattering function displays the diffuse peaks in a more evenly weighted manner, for example the shape, position and fine detail of the broad peak at $s \sim 5.5\,\text{Å}^{-1}$ are clear in comparison to the situation in Figure 1. Furthermore, the reduced intensity function shows an additional component on the high scattering vector side to the broad interchain peak. This correlates with the presence of two distances

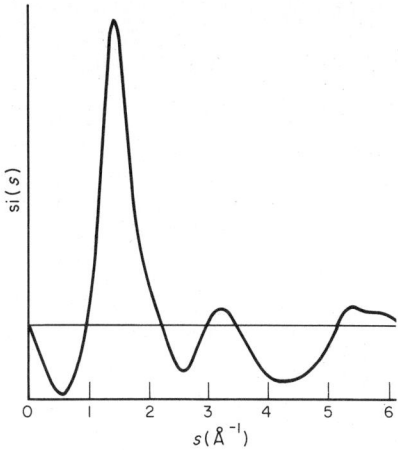

Figure 8 The reduced s-weighted X-ray scattering function si(s) for a non-crystalline sample of PEEK obtained after correcting and scaling the data in Figure 1 using the techniques detailed in Section 31.4.3; note that a shoulder at $s \sim 2\,\text{Å}^{-1}$ and the peaks at high scattering vector are much more distinctive

around 5 Å in the radial density function, and also illustrates that by obtaining high quality intensity data, attention can be given to detail. The radial density function W(r), is derived from the reduced intensity function

$$W(r) = 2/\pi \int si(s) \sin rs\, ds \qquad (12)$$

where the integration limits are, in principle, $s = 0$ and $s = \infty$. W(r) relates to correlations in electron positions and it has been used in the analysis of non-crystalline polymers.[23] However, it is possible to deconvolute from W(r) the effect of the average distribution of electrons about an atom, and thus generate a map of interatomic vectors, although still averaged with each atom at the origin. Conventionally this deconvolution is performed using Stokes' method by the division of the Fourier transform of W(r), that is i(s), and the self-convolution of the electron distribution about an atom.[21,24] The latter may be obtained using the X-ray scattering factors (Section 31.2.3) and thus the atomic density function g(r) may be obtained

$$G(r) = 4\pi r^2 \{\rho^a(r) - \rho^a\} = 2/\pi \int si(s)/g^2(s) \sin rs\, ds \qquad (13)$$

$$\text{with} \quad g^2(s) = |\Sigma\, c_i f_i(s)|^2 \qquad (14)$$

An example of this 'atomic' radial density function is shown in Figure 6. The approximations inherent in this approach are clear. For a polyatomic material it is not possible to take account of the different atomic scattering factors during this deconvolution process. The procedure has to use an 'average' atom, although for a polymer containing carbon, oxygen and hydrogen this may not be a particularly serious error. A more fundamental limitation is that although equation (13) indicates integration over all reciprocal space, in practice there must be finite limits, s_{min} and s_{max} (s_{max} must be $< 4\pi/\lambda$). s_{min} is typically in the range 0.15 to 0.25 Å^{-1} corresponding to long-range variations in electron density. Usually, little error is introduced by extrapolating i(s) smoothly to zero at $s = 0$ Å^{-1}.[25,26] However, termination at s_{max} generates two types of errors in the radial density function. The first is in the form of a loss of resolution since wavelengths $< 2\pi/s_{max}$ are lost and the second is the introduction of spurious ripples of wavelength $2\pi/s_{max}$ caused by the discontinuity in i(s) at s_{max}. These are problems which beset the use of Fourier transformations and pose real limitations to their general application. Many procedures have been suggested to 'overcome' these problems. Some involve multiplication of i(s) by some damping function so as to reduce the discontinuity at s_{max}.[27,28] Others involve the introduction of some structural information which allows the data and hence s_{max} to be extended.[29-32] The most universal approach is that of the 'sampled transform',[24] in which the transform is sampled at values of r which correspond to sine waves of wavelengths that are particular fractions of s_{max}. Different sampling procedures are possible for different termination points; if termination occurs at a maximum or minimum then

$r = (2n + 1)\pi/2s_{max}$ while at a node $r = n\pi/s_{max}$. The use of sampled transforms removes the possibility of introducing spurious ripples due to the termination step and also only calculates the radial density function at a resolution relevant to the range of scattering vectors recorded. The individual points in the plot of $G(r)$ shown in Figure 6 indicate the intervals at which the transform was calculated.

31.2.3 Scattering Factors

X-Rays are scattered by electrons, and there is a variation in the scattered intensity with angle which arises from the distribution of the electrons about an atom, as well as from the distribution of the atoms with respect to each other. It is normal to consider the first component separately by incorporating the effects into an atomic scattering factor $f(s)$. The scattering factor is essentially the Fourier transform of the electron distribution about an atom. They are derived theoretically and tabulated in 'The International Tables for Crystallography'[22] and elsewhere,[33] It is assumed that the distribution of electrons about the atom has spherical symmetry. By their very covalent nature, atoms in a polymer chain must possess an anisotropic distribution of electron density about their centres. However calculations by Stenhouse[34] show that such effects, even for carbon atoms with a large proportion of the electrons involved in bonding, on the atomic scattering factors are minimal.

31.2.4 Model Calculations

As discussed in Section 31.1.3, one possible approach to the analysis of X-ray scattering patterns is to compare the observed scattering with that calculated from models. For a sample with random orientation and hence one in which each vector r_{mn} takes all orientations (Figure 7), the average of each exponential term in equation (9) is given by allowing equal probability to all values of ϕ[21]

$$\langle \exp i s . r_{mn} \rangle = 1/4\pi r^2 \int e^{isr_{mn}\cos\phi} 2\pi r_{mn}^2 \sin\phi \, d\phi \tag{15}$$

and hence the reduced scattered intensity function $i(s)$

$$i(s) = \sum_{\substack{m \neq n}}^{N} \sum^{N} f_m(s).f_n(s) \sin sr_{mn}/sr_{mn} \tag{16}$$

This equation (16) is often called the Debye equation, the essential element of which is that the scattering only involves the magnitudes of the distances r_{mn} of each atom from every other atom. This equation is straightforward and the problem of calculating the scattering reduces to one of constructing molecular models. However, some simplification is possible. The scattering at scattering vectors $>2.5 \text{ Å}^{-1}$ arises predominantly from correlations within molecular chains[1,4] and hence comparisons in that scattering vector range need only involve the scattering calculated from single molecular chains.[4,13,15,35-38] Since these structures are statistically disordered, it is important that the intensity calculations are averaged over all configurations.[35-37]

Unless the atom positions can be expressed analytically, any molecular model that is 'built' will be of finite size and when the scattering is calculated (equation 6), there will be a small angle component which will depend upon the size and shape of the model and not on its internal atomic arrangements.[39] If comparisons with experimental data in the low angle region $s < 2 \text{ Å}^{-1}$ are required it is vital to eliminate this small angle scattering. A simple method has been developed[39] which requires very little more calculation than is involved in equation (16). The model size dependent scattering $i'_m(s)$ which should be subtracted from the calculated scattering of equation (20) is given by[39]

$$i'_m(s) = \left| \sum_i c_i f_i(s) \right|^2 v(s) \sum_{\substack{m \neq n}}^{N} \sum^{N} \sin s.r_{mn}/sr_{mn} \tag{17}$$

where $v(s)$ is the Fourier transform of a sphere of radius R, with R typically 1.8 Å[40]

$$v(s) = 9\{\sin Rs - Rs \cos Rs\}^2/(Rs)^6 \tag{18}$$

This approach is quite general and may be applied to models of arbitrary shape; the reader is referred to ref. 39 for full details.

In addition to making comparisons of the observed scattering with scattering functions calculated from molecular models, we may also compare experimental and model radial density functions.[20,41] The process of generating a radial density function is straightforward given a molecular model, it mostly being a matter of sorting interatomic vectors. However, attention has to be directed at compensating for the limited spatial resolution in the function derived from the experimental scattering.[20]

31.2.5 Orientational and Spatial Correlations

Equation (16) appears to simplify the analysis of X-ray scattering patterns considerably. However, its simplicity belies the complexity of generating models containing many chain models from which the scattering may be calculated. The algorithms required to generate models with parallel chains are straightforward.[4] However, for flexible chains, only a very few multiple chain models have been 'built' in a computer.[42-44] As indicated in the previous subsection, the generation of the molecular models may be particularly simplified by only considering the intrachain scattering. This procedure obviously only requires the generation of single molecular chains. Under certain circumstances, the procedures may be extended to take account of interchain correlations through the use of analytical functions.[4,15,35,45]

For a system of chain segments we may rewrite equation (9) by referring the intrachain segment structure to the segment centres[45]

$$i(s) = \left\langle \left| \sum_n f_n(s) \exp is.r_{cn} \right|^2 \right\rangle + \left\langle \sum_{i \neq j} \sum \exp is.r_{cij} \sum_{ni} \sum_{nj} f_{ni}(s) f_{nj}(s) f_{nj}(s) \exp is.(r_{cni} - r_{cnj}) \right\rangle \quad (19)$$

The first term represents the intrachain segment scattering, with summing over the atoms in each segment, and with r_{cn} the vector from the nth atom to the centre of the molecular segment. The second group of terms represents the interchain segment scattering, with i and j labelling different chain segments, r_{cij} is the vector between the ith and jth chain segment centres, ni labels the nth atom of the ith chain segment and thus cni labels the nth atom to segment centre for the ith segment. For spherical symmetry the first term reduces to

$$I_m(s) = \left\langle \left| \sum_n f_n(s) \sin sr_{cn}/sr_{cn} \right|^2 \right\rangle \quad (20)$$

If we assume that the relative orientations of pairs of chain segments are independent of their separation, which is perhaps a rather implausible approach, then we may separate out the statistical averaging of the second term in equation (19) into two components. In other words the averaging of $r_{cni} - r_{cnj}$ may be performed separately to that for r_{cij}. For a flexible chain system it is reasonable to assume that the distribution of the segment centres is completely isotropic and we can account for the segment–segment interactions using an analytical form of the structure factor for a random assemblage of spheres.[35,45,46] This leaves two extreme possibilities, the first is that the orientation of the ith chain segment is completely correlated with that of the jth unit, while the second is that the units are completely uncorrelated in terms of orientation. For the first possibility and taking all the segments to be parallel (although other types of correlations could be considered)[45] we obtain

$$I(s) = I_m(s) + \{H(s) - 1\}.I_m(s) \quad (21)$$

$$\text{or} \quad I(s) = I_m(s).H(s) \quad (22)$$

where $H(s)$ is the structure factor for a random assemblage of spheres; an analytical form has been given.[41] For the completely uncorrelated case equation (19) becomes

$$I(s) = I_m(s) + \{H(s) - 1\}\left\{ \left\langle \sum_n f_n(s) \sin s.r_{cn}/s.r_{cn} \right\rangle \right\}^2 \quad (23)$$

The disordered nature of the segmental packing in a non-crystalline polymer results in $H(s)$ exhibiting a constant value of ~ 1 for $s > 2.5\,\text{Å}^{-1}$, and thus at these higher scattering vectors, the scattering is dominated, in both the orientationally correlated and uncorrelated cases, by the segmental scattering $I_m(s)$. At lower scattering vectors the scattering is dominated by the interchain peak arising from the maximum in the expressions for $H(s)$. The essential difference between the correlated and uncorrelated systems in this scattering region is the s dependent multiplicative factor for the interchain segment structure factor (*cf.* equations 22 and 23). This difference manifests itself as a variation in the peak intensity for the interchain maximum, rather than changing the shape or nature of the scattering. We will see later (Section 31.5) that for structures with orientational correlations the nature of the chain segment changes, and this has a much larger impact upon the scattering pattern. Although these equations have limited application to all but the simplest of non-crystalline polymer systems, they do provide an insight into the effect of orientational correlations upon the X-ray scattering pattern.

31.2.6 Summary

In this section the basic relationships relating structure to scattering have been laid out. In Section 31.5 we shall give examples of how these expressions may be employed to extract quantitative structural parameters from samples of non-crystalline and liquid crystal polymers with a randomly oriented molecular organization.

31.3 BASIC THEORY FOR ANISOTROPIC SAMPLES

In many situations, a polymer sample will have an anisotropic structure. In other words there will be a preferred macroscopic or global molecular orientation in the specimen. This section is concerned with extending the concepts of Section 31.2 to such anisotropic materials. The presentation will be limited to those systems which exhibit uniaxial symmetry, that is, those which only contain one axis of preferred molecular alignment.

31.3.1 Choice of Coordinate System

The formulae relating to the scattering from a uniaxial system depend upon the coordinate system used. Figure 9 illustrates both cylindrical and spherical coordinates in reciprocal space. In general, conventional quantitative diffraction equipment operates with a spherical coordinate system. For that, and a variety of other reasons, many of which will transpire as this section progresses, the use of spherical coordinates is much preferred. The reader is referred elsewhere for a description of scattering from anisotropic samples using cylindrical coordinates.[9, 10, 46]

31.3.2 Orientation Distribution Functions

Macromolecular systems which have been subjected to an external field, such as stress or flow (and for liquid crystal systems electric or magnetic fields), may show a preferred alignment of the polymer chains or some other structural unit. Normally the quality of alignment is not perfect in

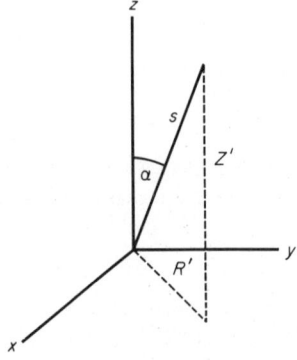

Figure 9 Spherical and cylindrical coordinates in reciprocal space

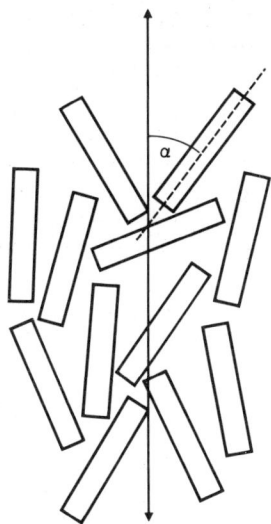

Figure 10 Structural units distributed with a preferred orientation about a field axis, such as the extension axis

relation to the field axis. In other words there is a spread of orientations about the field axis (Figure 10). For a liquid crystal system, we would expect the spread to be much smaller than for a polymer melt because of the inherent local anisotropy. We may characterize the angular distribution of the structural units with respect to the field axis by means of an orientation distribution or probability function D(α), alpha being the angle between the structural unit axis and the field axis. The symmetry of a uniaxial system results in the distribution function extending over 0 to 90°, the other three quadrants being equivalent. An example of a distribution function D(α), corresponding to the type of orientation configuration of Figure 10, is shown in Figure 11. This distribution function may be analyzed into spherical harmonic components (*cf.* Fourier Series) and

$$D(\alpha) = \sum_{n=0}^{\infty} (4n + 1) \langle P_{2n}(\cos\alpha) \rangle P_{2n}(\cos\alpha) \tag{24}$$

where the spherical harmonics, which are members of a series of Legendre polynomials, are defined in Table 3. Normally the amplitudes of the harmonics $\langle P_{2n}(\cos\alpha) \rangle$ decrease with increasing order, and when the preferred orientation is low, as for example in a deformed polymer glass, the higher order terms are negligible for $2n > 4$.[48] However, for a liquid crystal polymer system, where the degree of anisotropy is often pronounced, more than ten terms may be required to adequately describe D(α).[49]

Equation (29) may be inverted to enable the harmonic amplitudes to be obtained from D(α)

$$\langle P_{2n}(\cos\alpha) \rangle = \int_0^{\pi/2} D(\alpha) P_{2n}(\cos\alpha) \sin\alpha \, d\alpha \tag{25}$$

D(α) is normalized so that

$$\pi/2 = \int_0^{\pi/2} D(\alpha) \sin\alpha \, d\alpha \tag{26}$$

This condition allows the use of normalized Legendre polynomials, *i.e.* $P_{2n}(\cos 0) = 1$. The reader is directed elsewhere for a full account of the properties of Legendre polynomials,[50-52] but their utility should become clear as this section progresses. The component of the orientation distribution function $\langle P_2(\cos\alpha) \rangle$ is often referred to as the 'Hermans orientation function',[53] and within liquid crystal work as the 'order parameter'.

The definition of molecular orientation appears clear cut. However, there remains the problem of assigning a molecular axis to a chain segment.[54] The choice is straightforward in the case of 'molecular cylinders' (Figure 10), but for more complex molecular shapes (for example PMMA in Figure 4) the situation is confused, and any choice must be arbitrary. Later in this section it will

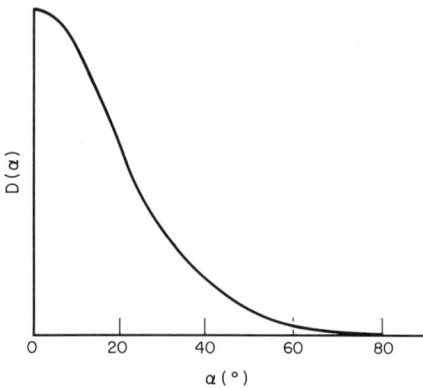

Figure 11   The orientation distribution function which might represent the level of preferred orientation in Figure 10

be shown that one major advantage of X-ray scattering in this context, is the definition of local chain segments in terms of a region which contributes to coherent X-ray scattering. In other words the working practice of X-ray scattering procedures both define the chain segment and measure its orientation.

31.3.3   Structure and Scattering Intensity

In spherical coordinates, any cylindrically symmetric function with an inversion centre can be expanded in a series of Legendre polynomials of even order (as for the orientation distribution function). Thus the cylindrical distribution function $W(r, \alpha)$, the anisotropic version of the radial distribution function may be expanded

$$W(r, \alpha) \quad = \quad \sum_0^\infty W_{2n}(r) P_{2n} \cos \alpha) \tag{27}$$

Similarly, the reduced intensity function $i(s, \alpha)$ may be expanded

$$i(s, \alpha) \quad = \quad \sum_0^\infty i_{2n}(s) P_{2n}(\cos \alpha) \tag{28}$$

$$\text{where} \quad i_{2n}(s) \quad = \quad (4n \ + \ 1) \int_0^{\pi/2} i(s, \alpha) P_{2n}(\cos \alpha) \sin \alpha \, d\alpha \tag{29}$$

The amplitudes of the spherical harmonics of the cylindrical distribution function may be related to those of the scattering function by[55, 56]

$$W_{2n}(r) \quad = \quad (-1)^n 2r/\pi \int_0^\infty s^2 i_{2n}(s) j_{2n}(rs) \, ds \tag{30}$$

where j_{2n} are spherical Bessel functions; the first few terms are shown in Table 3. For small departures from isotropy, only a few terms will be needed to construct the complete cylindrical distribution function, for example a deformed glassy sample of poly(methyl methacrylate) required four terms.[56] In such cases the first term $W_0(r)$ will dominate and this is identical with the radial distribution function of Section 31.2.2. For such systems, the 'sampled transform' method may be applied in the calculation of equation (30).[56] For a highly aligned material, such as a liquid crystal polymer system, many terms will be required[49] and it may be more useful to employ an alternative relationship[57]

$$W(r, \alpha') \quad = \quad 1/2\pi^2 \int_0^\infty \int_0^{\pi/2} s^2 i(s, \alpha) J_0(rs \sin \alpha \sin \alpha') \ \times \ \cos(sr \cos \alpha \cos \alpha') \sin \alpha \, d\alpha \, ds \tag{31}$$

where J_0 is the Bessel function of zero order, and α' and α the azimuthual angle in real and reciprocal space respectively. The use of a deconvolution route to give atomic cylindrical distribution functions[56] is identical to that described for the radial functions (Section 31.2.2).

Table 3 Legendre Polynomials $P_{2n}(x)$ and Spherical Bessel Functions $j_{2n}(x)$

$2n$	$P_{2n}(x)$	$j_{2n}(x)$
0	1	$\dfrac{\sin x}{x}$
2	$\dfrac{1}{2}(3x^2 - 1)$	$\left(\dfrac{3}{x^2} - 1\right)\dfrac{\sin x}{x} - \dfrac{3}{x^2}\cos x$
4	$\dfrac{1}{8}(35x^4 - 30x^2 + 3)$	$\left(\dfrac{105}{x^4} - \dfrac{45}{x^2} + 1\right)\dfrac{\sin x}{x} - \dfrac{5}{x^2}\left(\dfrac{21}{x^2} - 2\right)\cos x$
6	$\dfrac{1}{16}(231x^6 - 315x^4 + 105x^2 - 5)$	$\left(\dfrac{10395}{x^6} - \dfrac{4725}{x^4} + \dfrac{210}{x^2} - 1\right)\dfrac{\sin x}{x} - \dfrac{21}{x^2}\left(\dfrac{495}{x^4} - \dfrac{60}{x^2} + 1\right)\cos x$

The expressions detailed above provide a route to the calculation of the cylindrical distribution function from the observed scattering. However, we can utilize the expansions of the various functions into series of spherical harmonics in a much more powerful manner, and one which provides a direct route to both the evaluation of the orientation distribution function and the chain conformation from arcing of the measured scattering patterns.[58-60]

The scattering for an aligned polymer system is given by the convolution of the orientation distribution function $D(\alpha)$ and the scattering for a perfectly aligned system $I'(s, \alpha)$[55,58]

$$I(s, \alpha) \quad = \quad D(\alpha) * I'(s, \alpha) \tag{32}$$

Normally the convolution would involve terms coupling spatial (*i.e.* dependent upon s) and orientational (*i.e.* dependent upon α) order. Attempts have been made to deconvolute $I(s, \alpha)$ to obtain $I'(s, \alpha)$ for various glassy polymers using arbitrary orientation distribution functions.[61-63] However, for independent scattering units, we may use the properties of Legendre polynomials[52] in terms of the general Legendre addition theorem to write the convolution as[55,58,59]

$$I_{2n}(s) \quad = \quad \{2\pi/(4n \quad + \quad 1)\} D_{2n} I'_{2n}(s) \tag{33}$$

where $I_{2n}(s)$, $I'_{2n}(s)$ and D_{2n} are the spherical harmonics of the appropriate functions in equation (32). Obviously the relationship given in equation (33) is only valid for a system in which $D(\alpha)$ and $I'(s, \alpha)$ have cylindrical symmetry. Furthermore this equation is only valid over an s range in which the scattering arises solely from a single structural unit uncomplicated by interunit correlations. (A simple example of such a unit would be a crystallite, in semi-crystalline polymer. For a disordered system we would have to consider chain segments, unless some 'domain' structure existed.) Similar relationships may be established for the spherical harmonic expansion of the real space cylindrical distribution function $W(r, \alpha)$ (equation 27).[56] We may utilize equation (33) in a number of different but interrelated ways. Of course if we had a numerical knowledge of the orientation distribution function $D(\alpha)$ or its harmonics D_{2n}, then equation (33) could be inverted to deconvolute the observed scattering in order to obtain the scattering function for a perfectly aligned system. This approach would only be valid if we postulate some 'domain' type model. Such a model assumes that the spread of orientation is due to the misorientation of the domains. This type of analysis has been used to prepare a cylindrical distribution function for a liquid crystal polymer system.[49]

However, in general the above procedures are unnecessary because of a particular feature of equation (33). The 'observed' spherical harmonics, whether they are reciprocal space or real space functions, are those of a perfectly aligned system, weighted with the appropriate harmonic of the distribution function. Since the values of D_{2n} are independent of s, the only difference between $I_{2n}(s)$ and $I'_{2n}(s)$ in equation (33) is a constant. In other words, we could compare the observed $I_{2n}(s)$ functions with those calculated from perfectly aligned models, without requiring a knowledge of the orientation distribution function, as long as the comparison is restricted to peak positions, widths and shape, and relative intensities. When the two sets of curves match, not only is the structure known, but the ratio of the two gives the coefficients of the orientation distribution function! This type of analysis has been applied, in reciprocal space and real space,[57] to both non-crystalline polymers[59] and liquid crystal polymers[60,64] using individual chain segments as the structural units.

In those cases described above the structural analysis has concentrated on the intrachain scattering. In principle we could utilize the interchain scattering to provide a measure of the molecular orientation. To do so requires a model of the interchain scattering, or cylindrical distribution function. Remarkably, the model chosen by many is invariant and is a set of infinitely long parallel rods, a model almost appropriate to the liquid crystal state but certainly not for a non-crystalline polymer. Such a model exhibits scattering which is confined to the equatorial plane ($a = 90°$). It is more convenient, before proceeding further, to rewrite equation (33) in terms of the normalized amplitudes of the spherical harmonics

$$\langle P_{2n}(\cos\alpha)\rangle_D = \langle P_{2n}(\cos\alpha)\rangle / \langle P_{2n}(\cos\alpha)\rangle' \tag{34}$$

where the subscript D indicates the orientation distribution function, and for example $\langle P_{2n}(\cos\alpha)\rangle$ is given by

$$\langle P_{2n}(\cos\alpha)\rangle = \frac{\int_0^{\pi/2} I(s,\alpha) P_{2n}(\cos\alpha)\sin\alpha\, d\alpha}{\int_0^{\pi/2} I(s,\alpha)\sin\alpha\, d\alpha} \tag{35}$$

This is in essence the normalized version of equation (29). In order to operate this procedure we need values for the model harmonics $\langle P_{2n}(\cos\alpha)\rangle$ and for the simple model of a set of parallel rods the first few terms are[58]

$$P_2 = -1/2, \quad P_4 = 3/8, \quad P_6 = -5/16 \tag{36}$$

A more complete description of the application of these procedures in measuring the molecular orientation of liquid crystal polymers is given in ref. 50.

31.3.4   Model Calculations

The calculation of the scattering function for a model may be expressed in cylindrical coordinates[9] or in spherical coordinates, in which the equivalent to the Debye equation becomes[59]

$$i_{2n}^m(s) = (-1)^n(4n + 1)\sum_i\sum_j f_i(s)f_j(s)j_{2n}(sr_{ij})P_{2n}(\cos\alpha_{ij}) \tag{37}$$

where r_{ij} is the magnitude of the vector between the ith and jth atoms, and α_{ij} is the angle that the vector makes to the axis of cylindrical symmetry. This equation is analogous to the Debye equation (16) and they are equivalent when $n = 0$. This equation is used in the procedures for comparing model and experimental harmonics through equation (33).

The calculation of the harmonic components of a model cylindrical distribution function is relatively straightforward, although as for the case of the isotropic radial distribution function, care must be taken to account for the limited resolution, in terms of r, of the experimental functions. A study of polycarbonate has utilized this approach.[57]

For certain materials, invariably liquid crystal polymers, sufficiently high levels of orientation can be induced such that it is appropriate to consider the scattering along the meridian in a similar manner to that used for crystalline fibres (see Chapter 28 of this volume). In these circumstances only the structure projected onto the molecular axis needs to be considered. This naturally simplifies both the calculation and interpretation of the scattering functions. For some main chain liquid crystal copolymers it has been possible to model the meridional scattering using a relatively simple model.[65,66] The meridional scattering $Z(s)$ is given by

$$Z(s) = \text{Real}\frac{|1 + F(s)|}{|1 - F(s)|} \tag{38}$$

$$\text{with} \quad F(s) = \sum_n \cos r_n s + \sum_n \sin r_n s \tag{39}$$

where r_n are the projected atom positions in the repeat or monomer units. In fact this type of model can take account of non-randomness in the copolymer chain and the reader is directed to refs. 67 and 68 for a full description.

31.3.5   Summary

The analysis of X-ray scattering patterns of samples with a preferred molecular alignment, whether they are non-crystalline or liquid crystalline in nature, can yield parameters describing the orientation, the chain conformation and the chain packing by taking advantage of the properties of spherical harmonics. Examples of the application of these procedures are given in Section 31.6. In certain cases, for highly aligned samples the gap between non-crystalline and crystalline polymers is narrowed, at least in terms of the analysis of the scattering patterns, and the reader may find it useful to examine the methods used in the examination of crystalline polymers (see Chapter 28 in this volume).

31.4   EXPERIMENTAL PROCEDURES

The X-ray scattering from non-crystalline and liquid crystalline polymers is, by its very nature, diffuse, and hence requires more demanding experimental techniques than the collection of, for example, scattering patterns of inorganic polycrystalline powders. That is not to say that useful information cannot be obtained from simple photographic techniques, but that to fully utilize the scope of the relationships developed in the preceeding two sections, quantitative intensity data is required. It is not the intention here to provide a self-contained description of X-ray procedures, there are fine books well equipped to do so,[10,11] but rather to highlight the particular features required in the study of non-crystalline and liquid crystalline polymers.

31.4.1   Geometry

Most organic polymers of interest naturally contain a preponderance of low atomic number elements. This average low atomic number provides a material with a low absorption coefficient, for example a sample of PMMA has a linear absorption coefficient of $6\,\mathrm{cm}^{-1}$ when using Cu $K\alpha$ radiation. It is therefore relatively easy to prepare samples with the optimum thickness for transmission geometry ($t_{\mathrm{optimum}} = 1/\mu$;[10] for the example of PMMA, $t_{\mathrm{optimum}} \approx 1.5\,\mathrm{mm}$). The long path lengths inherent in such low absorption materials will result in significant multiple scattering. The probability of inelastic scattering increases as the atomic number decreases, and therefore organic polymers give a large incoherent or Compton scattering component which contains no directly useful structural information.

Due to the continous nature of the scattering, it is not possible to isolate the effects of different wavelengths in the measured scattering pattern, and it is suggested that the provision of a monochromatic beam through the means of a crystal monochromator rather than a $K\beta$ filter is essential. Although many commercial diffractometers operating in reflection geometry incorporate a scattered beam monochromator, it suffers from the fact that its acceptance window, in terms of wavelength, will exclude a varying amount of the incoherent scattering.[69] An alternative is to employ a detector with a high energy resolution operating as a spectrometer to exclude all but the incoherent signal. Such an approach has been described in its application to PMMA,[70] although such a method is cumbersome for regular use. It is instructive to examine the energy spectrum scattered by a sample of PMMA. Figure 12 shows scattered spectrum for a range of scattering vectors using Mo $K\alpha$ as the source radiation, with the first part of the scattering vector range omitted for clarity. The intensity for an energy of $\sim 17\,\mathrm{keV}$ representing the coherent signal (marked with a solid arrowhead) drops off rapidly above $s \sim 10\,\text{Å}^{-1}$. In contrast the incoherent scattering, at a slightly lower energy (marked with an open arrowhead), increases in intensity until at the highest scattering vector the incoherent scattering is almost an order of magnitude greater than the coherent signal. This large incoherent signal is the major experimental problem if intensity measurements are required at high scattering vectors since it carries no structural information. In fact, measurements at large scattering vectors are only needed to prepare a radial distribution function with a high resolution, since $\Delta r = \pi/s_{\mathrm{max}}$. As discussed in Section 31.2, radial density functions have limited application in the area of organic polymers, and hence, in general, the scattering vector range can be

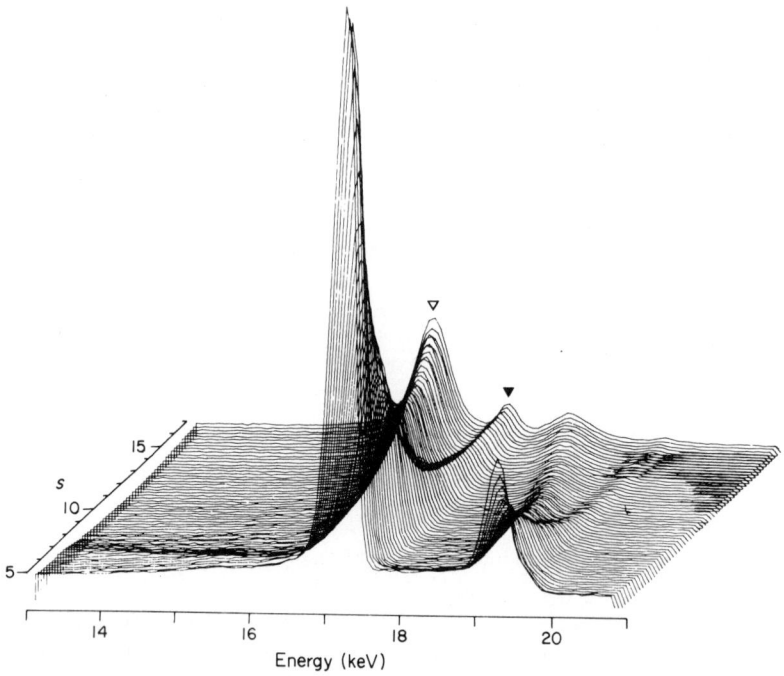

Figure 12   The energy spectra recorded at a series of scattering angles from a sample of atactic poly(methyl methacrylate) irradiated with an unfiltered beam for a Mo X-ray tube:[70] the peak at $\sim 17\,keV$ marked with a solid arrowhead corresponds to the Mo $K\alpha$ coherent X-ray scattering which contains structural information; the open arrowhead marks the incoherent or Compton scattering which contains no directly useful structural information (note that at higher scattering vectors the incoherent signal becomes more intense and moves to a lower energy as expected); the peaks at $E \sim 19\,keV$ correspond to the Mo $K\beta$ peaks

usefully limited to $s \sim 6$–$7\,\text{Å}^{-1}$. This range is easily achieved using Cu $K\alpha$ radiation, for which high power, high stability sources are readily available.

   In addition to the above problems, the use of symmetrical reflection geometry results in complex absorption and multiple scattering corrections which depend on a precise knowledge of the sample size, the slits and source geometry and any enclosing cell.[20] In contrast, for moderate scattering angles up to $\sim 100°$, the correction for absorption in transmission geometry is straightforward,[10] as is that for multiple scattering. Furthermore, the inclusion of an incident beam monochromator in transmission geometry is particularly easy, especially if used with pinhole collimation. For a disordered system, such collimation gives adequate resolution and intensity throughput. However, if higher resolution studies are required, as for example the layer ordering in smectic liquid crystal systems, a more complex focusing arrangement may be required.[71] For these various reasons, symmetrical transmission geometry is the recommended system.

   Figure 13 shows the experimental wide-angle X-ray scattering configuration used in the author's

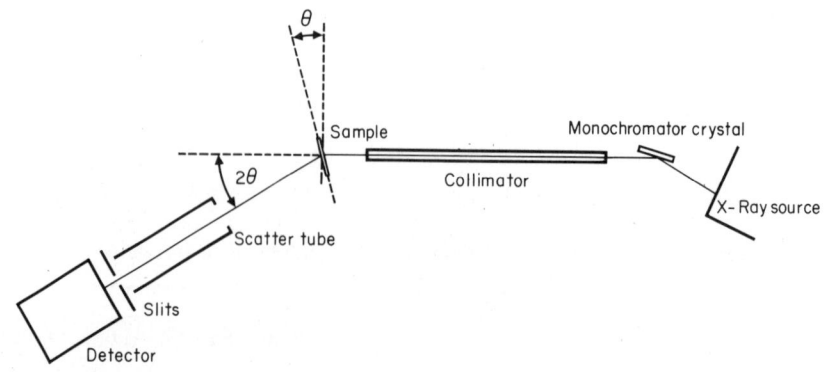

Figure 13   The schematic layout of the experimental wide-angle X-ray scattering configuration used in the author's laboratory; to study samples with a preferred molecular orientation, the sample is rotated in its own plane

laboratory, in which the factors described above have been accounted for. In order to examine samples with a preferred molecular alignment, it is necessary to rotate the sample in its plane, as shown in Figure 14. The inclusion of such a rotary stage in a transmission arrangement is relatively straightforward. For these experiments it is imperative to utilize a symmetrical arrangement (as shown) in order to record an undistorted map of the reciprocal space. The latter restriction precludes the use of film or planar electronic detector based methods for anything more than a qualitative study of an aligned polymer, if the angular range of 2θ exceeds $30°$.

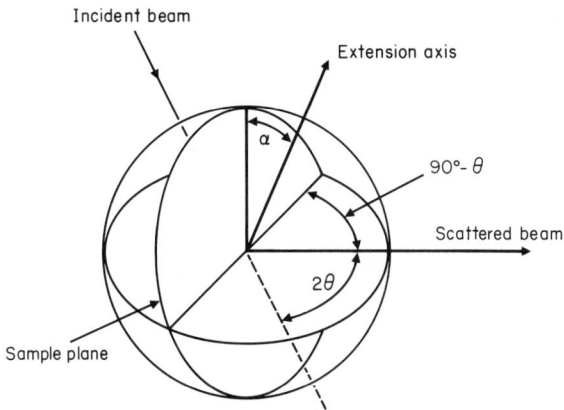

Figure 14   The scattering geometry for the examination of samples with a preferred molecular orientation

31.4.2   Data Handling

It is conventional to operate the diffraction equipment shown in Figures 13 and 14 in the so-called step-scan mode in which the scattered X-rays are counted over a fixed time at a series of defined points in reciprocal space. If the equipment operates with fixed angular intervals between points, it is useful if the increments are in Δs rather than $\Delta 2\theta$. The statistical rules for determining the number of counts to accumulate, or the count time, are well established.[10,11] Conventionally such data are 'smoothed' to minimize the effects of the statistical nature of X-ray production, scattering and detection, and we have found procedures based upon cubic splines to be most suitable.[72]

31.4.3   Intensity Corrections

The measured intensity function, either $I_{exp}(s)$ or $I_{exp}(s, \alpha)$, has to be corrected for variation with the scattering angle in polarization, absorption and multiple scattering. In addition, it may be necessary to take account of 'background' scattering, both from the air in the beam path, and from the sample stage and windows. The corrections listed below are for a symmetrical transmission geometry. Corrections for other geometries are given in the literature.[10] Normally the X-ray beam under consideration is unpolarized and the correction for polarization, $P(s)$,[73] is as follows

$$P(s) = 1 - [4m/(1 + m)]\lambda^2 S^2 + [4m/(1 + m)]\lambda^4 S^4 \qquad (40)$$

where $S = s/4\pi$, and $m = \cos^2 2\phi$ where ϕ is the Bragg angle of the crystal monochromator (if a scattered beam monochromator is also used, an additional correction factor is required). When the incident beam is polarized, as, for example, if synchrotron radiation is employed, then a more complex correction is required which involves a knowledge of the degree of polarization of the X-ray beam.[74] An alternative, particularly if only orientation measurements are required, is to use an isotropic sample as a normalization probe.[75]

The absorption correction $A(s)$ is given by[73]

$$A(s) = [1/(1 - \lambda^2 S^2)^{1/2}] \exp\{\mu t[1 - (1 - \lambda^2 S^2)^{1/2}]\} \qquad (41)$$

where t is the sample thickness and μ is the linear absorption coefficient. The introduction of cell windows *etc.*, for a liquid sample would require further correction. The absorption coefficent should take account of the differences in absorption between the coherent and incoherent signal,[11] although the effect for most samples is minimal.

Multiple scattering corrections have been developed by a number of authors,[76-79] those due to Dwiggins[80] are applicable to transmission geometry. However, it should be borne in mind that none of these procedures take account of the change in energy spectrum upon scattering. Calculations[70] suggest, at least for reflection geometry, that most multiple scattering is incoherent. For the majority of polymer samples the level of multiple scattering is $< 10\%$ at $s = 6\,\text{Å}^{-1}$ with Cu $K\alpha$ radiation.

Example plots of these three corrections for polarization, absorption and multiple scattering as a function of s, for the intensity function of PEEK shown in Figure 1, are displayed in Figure 15.

The remaining correction is the subtraction of the incoherent scattering, but unless this has been achieved experimentally,[70] the data must be first scaled to electron units, and this process is considered in the next section.

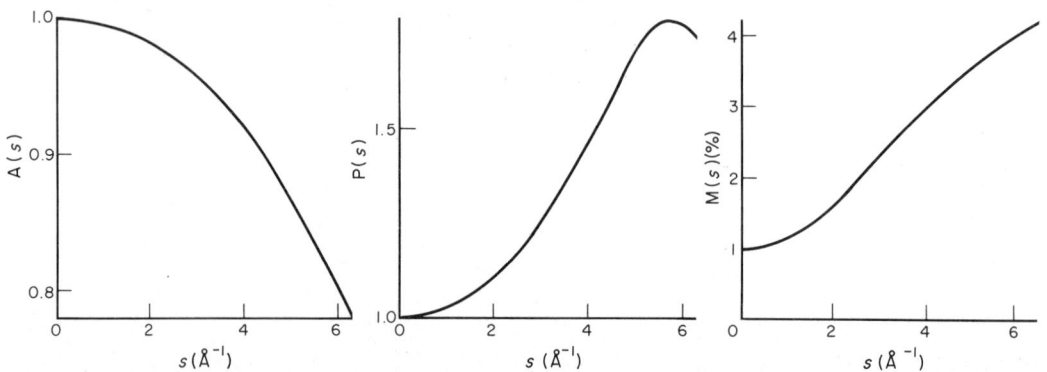

Figure 15   Examples of the corrections applied to experimental data for variations in absorption A(s), polarization P(s) and multiple scattering M(s). These curves were calculated for the experimental configuration used to obtain the data in Figure 1

31.4.4   Scaling to Electron Units

The application of the above correction procedures gives an intensity function free from experimental aberrations but still scaled in arbitrary units. The total intensity curve may be normalized using the conservation principle following a method developed by Krogh-Moe.[81] For isotropic scattering the scaling constant is given by

$$k = \left\{ \int s^2 I''(s)ds - 2\pi Z/\rho \right\} \Big/ \left\{ \int s^2 I_{corr}(s)ds \right\} \tag{42}$$

where $I''(s)$ is the total independent scattering, including both coherent and incoherent terms. The second term in the numerator of equation (42) is a correction factor (where Z is the average atomic number and ρ the mean electron density) to account for the absence of the zero-angle scattering in the experimental intensity function in the denominator. For the uniaxial case a modified form of the Krogh-Moe method may be applied[59]

$$k = \frac{\int s^2 I''(s)ds - 2\pi Z\rho}{\int \int_0^{\pi/2} s^2 I(s, \alpha)\sin\alpha\, d\alpha\, ds} \tag{43}$$

The intensity may be placed on an electron unit scale by

$$I_{eu}(s) = k.I_{corr}(s) - I_{Compton}(s) \tag{44}$$

The Compton or incoherent scattering may be calculated using tabulated values[22] or analytical expressions.[33] A similar expression may be used for the uniaxial case. The reduced intensity function is obtained using equation (11) and an example of these procedures is shown in Figure 16, using the data for PEEK presented in Figure 1. By placing the intensity functions on an electron unit scale, the

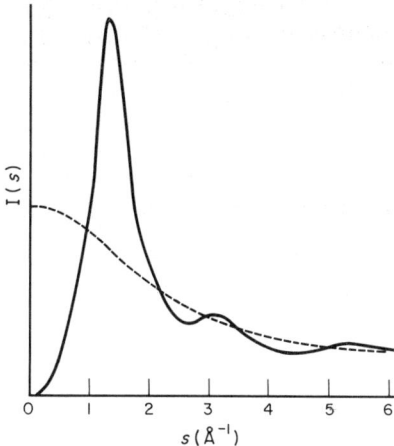

Figure 16 An illustration of the scaling procedure: the solid curve represents the fully corrected intensity function $I(s)$ obtained from the data shown in Figure 1 for a sample of PEEK using the correction factors displayed in Figure 15; the dashed line represents the 'independent' scattering involving both coherent and Compton scattering. The solid line has been multiplied by a constant k such that the second moments of the two curves are matched and hence the data $I(s)$ is scaled into electron units. The reduced intensity function of Figure 8 was obtained by subtraction of the dashed line from the solid curve and multiplying by s

experimental data may be compared directly in a quantitative manner with the functions calculated from models. In addition the radial distribution functions, which are derived from the reduced intensity function, will also be on an absolute scale.

31.4.5 Summary

In practice, all of the operations described in this section are normally performed automatically by a programmed computer system. Several such packages have been developed (for example refs. 82–84). The correction and the normalization procedures are straightforward, although time consuming, and they present no real obstacle to the generation of corrected data for analysis as described in the previous two sections. The main limitation to any serious analysis will be that of accumulating data with good statistics, and modern X-ray generation and detection apparatus have minimized the problems. The nature of the X-ray scattering geometry simplifies it's application to *in situ* measurements of the effects of, for example, temperature,[16] stress[85] and magnetic fields.[75] The continuing development of synchrotron sources and electronic area detectors opens up a new area of time-dependent structural study.[75,86]

31.5 EXAMPLES OF ANALYSIS FOR SAMPLES WITH RANDOM ORIENTATION

In the preceding sections we have assembled the background theory and the experimental tools to enable the quantitative assessment of non-crystalline and liquid crystalline polymers. This section is concerned with applying those procedures to samples without preferred orientation. Since the interest in liquid crystal systems is often centred around aligned samples, we shall not consider them further in this section. It is not the intention to provide a review of X-ray scattering studies of non-crystalline polymers or their structure; that has been adequately covered a number of times.[87-90] The purpose of this section is to use selected examples to illustrate the range of structural information available in an X-ray scattering pattern.

31.5.1 Polyethylene

Scattering studies of molten polyethylene have been a focus of interest, for nearly twenty years, as a model system for the structure of non-crystalline polymers. In fact, the earlier researchers fell into two groups, those who believed the scattering data indicated some local ordering[91−99] and those who interpreted the data as arising from a 'random coil' model.[26,100−104] For this case study, we follow more recent work, employing the quantitative procedures described in earlier sections, which shows unambigously that the molecular organization in molten polyethylene is highly disordered with an irregular chain trajectory and molecular packing.[1,4,35]

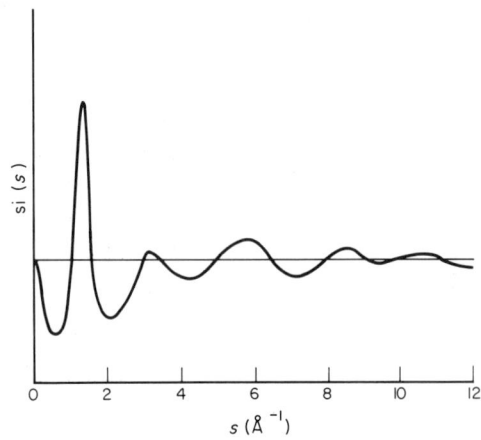

Figure 17 s weighted reduced intensity function si(s) obtained for a sample of molten polyethylene at 140 °C[35]

Figure 17 shows the reduced intensity function si(s) for a sample of molten polyethylene at 140 °C. There is a diffuse but intense peak at $s \sim 1.4\,\text{Å}^{-1}$, with a number of weaker peaks at higher scattering vectors. Figure 18 shows the radial density function, G(r), obtained by Fourier transformation of the reduced intensity function of Figure 17 employing equation (13). There are a number of sharp peaks at low r which can readily be assigned to fixed covalent distances; however, any structural analysis must focus on the less distinct peaks at larger values of r.

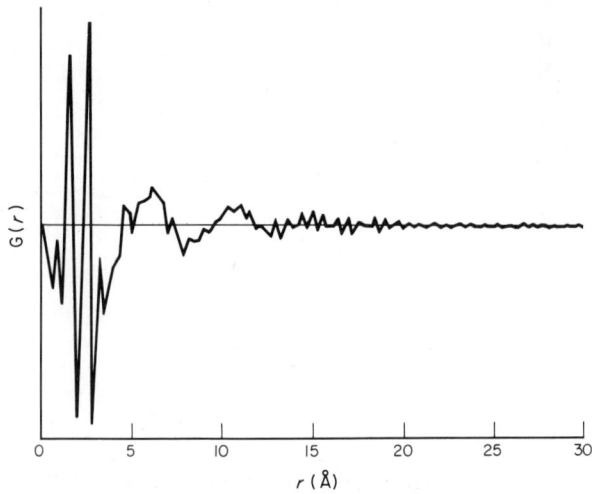

Figure 18   The atomic radial density function G(r) for a sample of molten polyethylene derived from the data of Figure 17[35]

We shall first consider the reduced intensity function si(s). Initially, as discussed in Section 31.2, the data for scattering vectors $> 2\,\text{Å}^{-1}$ will be assumed to be devoid of interchain contributions, and thus arise solely from intrachain correlations. It is well established from conformational energy calculations[35,106] that the polyethylene chain exhibits three rotational isomers at approximate values of the bond rotation angle ϕ of 0° (*trans*), 120° (*gauche*) and −120° (*−gauche*). The conformational question with polyethylene is not so much the values of these rotation angles but the

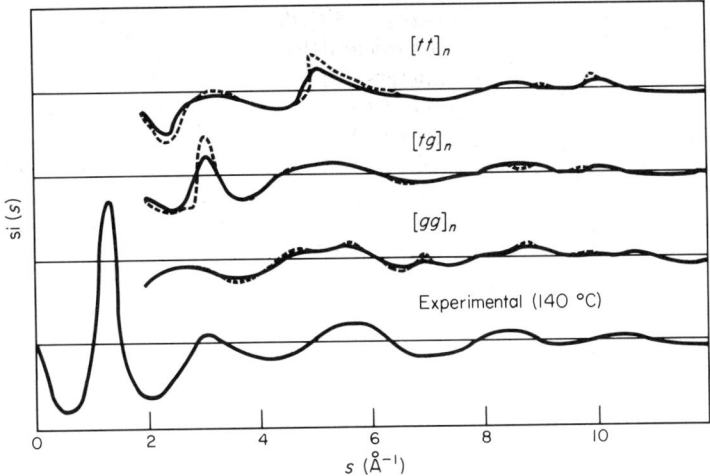

Figure 19 Model reduced intensity functions calculated for short sections of polyethylene chains placed in regular conformations of *trans* (*tt*), *trans–gauche* (*tg*) and *gauche–gauche* (*gg*), compared with the experimental function: (———) $n = 6$ and (– – –) $n = 12$

populations and distribution of the isomers along the chain. Figure 19 shows the reduced intensity functions calculated for short sections of a polyethylene chain arranged in a number of regular conformational sequences, compared with the experimental data. The absence of the peak at $s \sim 1.4 \, \text{Å}^{-1}$ in the model curves is due to the single chain nature of these models. We are not yet attempting to model the interchain structure. Clearly all of these model curves contain features much sharper than the experimental intensity data. In other words the chain conformations are too regular and extend over too great a distance. However, the all-*trans* chain does provide scattering peaks in approximately the correct positions.

Figure 20 shows the effect of introducing *gauche* isomers into an otherwise all-*trans* chain. The distribution of isomers is random, and p_t indicates the probability of the *trans* state occurring.

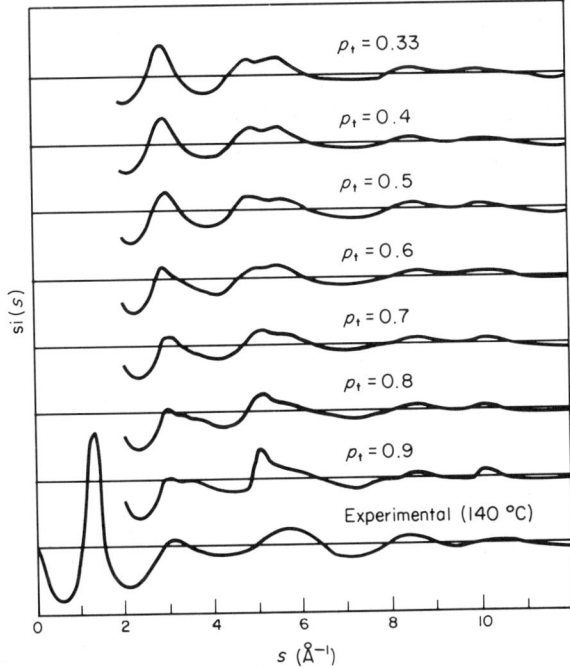

Figure 20 Reduced intensity functions calculated for random chains of polyethylene in which the probability of a *trans* isomer is as indicated in the figure, compared with the experimental data[35]

The curve for $p_t = 1$ is not drawn since that would correspond to the all-*trans* chain of Figure 19. As the *trans* state population decreases, that is the chain conformation becomes more irregular, the sharpness of the features in the model scattering curves diminish, and the diffuse peak at $s \sim 3 \,\text{Å}^{-1}$ becomes more prominent. By inspection, reasonable agreement may be obtained between the experimental si(s) curve and the model functions when p_t is in the range 0.5–0.7. In fact the agreement between the model and experiment may be enhanced by incorporating some of the detail of the conformational energy calculations, for example higher order probabilities. Figure 21 shows the scattering calculated for polyethylene chains in which the rotation states have been distributed according to the rotational isomeric scheme of Abe *et al.*,[106] for various energy differences between the *trans* and *gauche* isomers (full details are given in ref. 35). The best agreement arises when the energy dfference is 0.5–0.7 kcal mol^{-1} (1 cal = 4.19 J), which is in good agreement with experimental values obtained by spectroscopic values.[107]

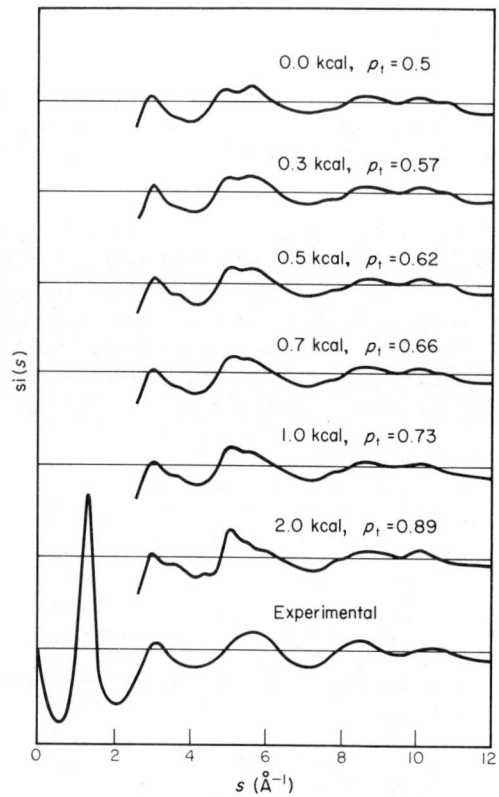

Figure 21 Reduced intensity functions si(s) calculated for random chains of polyethylene in which the distribution of rotation states is according to the work of ref. 106, but in which the energy difference between the *trans* and *gauche* states is as indicated in ref. 35 (1 cal = 4.19 J)

In essence this analysis shows without doubt that the chain conformation in molten polyethylene is irregular and 'random coil' like. In fact, one advantage of comparing the observed X-ray scattering with that calculated from models, is that any proposed model may be easily evaluated if enough structural parameters have been specified. Figure 22 shows the scattering calculated from the three models favoured in the 1970's, namely the 'Bundle Model',[91-92] the 'Meander Model'[95,96] and the random coil model developed in this analysis.[35] It is only the latter which provides any reasonable fit to the experimental data. There is clear correlation between the sharpness of features in the scattering functions and the level of regularity of the chain conformation.

The chain conformation which ensues from the above analysis is one in which typically an all-*trans* sequence of three to four bonds is followed by a *gauche* state and very ocassionally by a *gauche–gauche* sequence. The length of a chain 'segment' will thus be 4–6 Å. Since its diameter will be approximately 5–5.5 Å, the chain segment can be considered globular. As a result, the concepts of chain packing advanced in Section 31.2.5, in which the centres of distribution of the chain segments

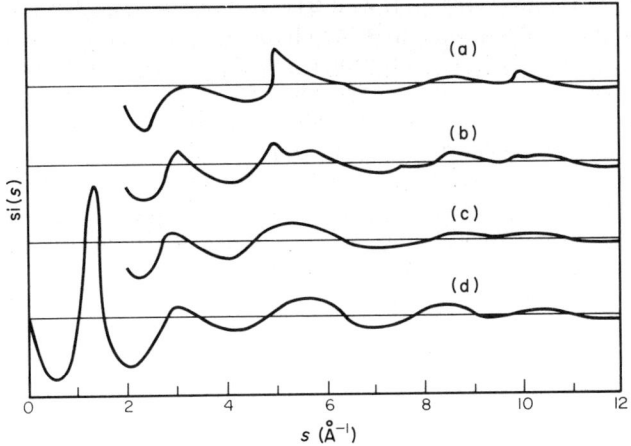

Figure 22 Reduced intensity functions si(s) calculated for three different models proposed for the molecular organization in non-crystalline polymers as relevant to molten polyethylene: (a) the bundle model;[91,92] (b) the meander model;[95,96] (c) the random coil model;[35] compared with (d) the experimental data

were represented by a set of random spheres, appears to be valid. Figure 23 shows the results of applying Section 31.2.5 to polyethylene with no orientation correlation.[35] With a disordered structure, such as polyethylene, it would be unreasonable to take a single sphere or interaction distance, since obviously the 'shape' and 'size' of the segments will vary with conformation. As a consequence a distribution of sphere sizes was used in which the breadth of the distribution varied from 0.25 to 1.0 Å. The intrachain scattering used in the calculations was essentially that from Figure 21 with an $E_{trans} - E_{gauche}$ difference of 0.5 kcal mol^{-1} (1 cal = 4.19 J). There are two important points

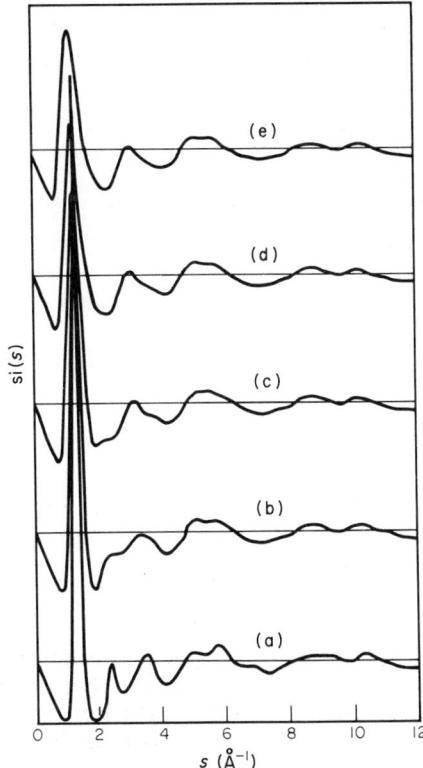

Figure 23 Scattering function si(s) calculated for a complete model of polyethylene using equations (19)–(23), for various size ranges of the interchain interactions: (a) single size interaction; (b–e) interaction range with a Gaussian distribution with standard deviation, (b) 0.25 Å, (c) 0.5 Å, (d) 0.75 Å and (e) 1.0 Å

to note from Figure 23. The first is that an acceptable match to the experimental data is obtained when the sphere diameter distribution is relatively broad. The second is that when the interchain peak is sharper in the model curve than in the experimental si(s) function, higher order interchain peaks are observed at $s > 2\,\text{Å}^{-1}$. In other words, when the spatial correlations (as measured through the breadth of the interchain peak) are in accord with the data, the interchain scattering is indeed restricted to scattering vectors $s < 2\,\text{Å}^{-1}$, justifying our original assumption in partitioning the scattering into intrachain and interchain components. This analysis has shown that the structure of the polyethylene melt is random and irregular and it is not necessary to invoke any orientational correlations to match the experimental data to that of models.

In order to illustrate the sensitivity of X-ray scattering in such structural studies, we show scattering curves for two different samples of non-crystalline polyethylene.[45] The first is a sample of polyethylene amorphotized by irradiation through the introduction of a large proportion of defects into the crystal lattice. In contrast, the second sample was held in the molten state, although it received a similar irradiation dose to the first. It would seem reasonable to expect the former to contain longer *trans* sequences and hence, perhaps, greater levels of orientational and spatial correlations, than the sample prepared from the melt. Figure 24/shows the reduced intensity functions for the two samples. The differences between these two supposedly similar 'amorphous' materials are easily seen. The intrachain scattering of the amorphotized sample contains sharper features than the melt irradiated specimen, reflecting a higher average *trans* sequence population or high p_t (*cf.* with Figures 20 and 22). The interchain scattering of the amorphotized sample is sharper, more intense and the peak occurs at a higher scattering vector, indicating closer packing of the chain segments.[45] In fact the scattering pattern for the amorphotized sample resembles that calculated for various types of the bundle model.[35] The principal points of interest are that for samples exhibiting significant levels of orientational correlation then the intrachain scattering reflects the greater anisotropy of the chain segments, and that the packing mode of those chain segments results in a sharper interchain peak at a higher scattering vector than for materials containing no orientational correlations.

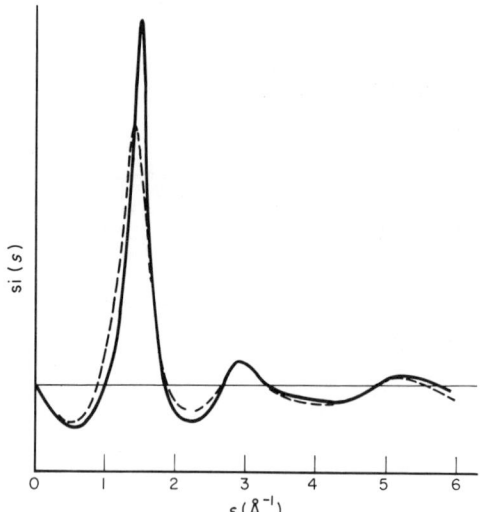

Figure 24 Experimental scattering functions si(s) for samples of polyethylene irradiated in the melt (– – –) and in the solid (——)[45]

As can be seen from the above analysis, the analysis of a scattering pattern from a non-crystalline polymer is particularly straightforward, since the scattering may be partitioned into its interchain and intrachain components. If we consider a radial density function, then it is clear the Fourier transformation process will scramble the two together. This may be seen in Figure 18, where conformationally sensitive peaks are superimposed on the broad oscillations relating to the interchain correlations. Figure 25 shows model radial density functions calculated for an all-*trans* model and a random chain model of polyethylene. The latter provides a better fit to the experimental data (Figure 18) but the comparison is not as straightforward as for the scattering analysis.

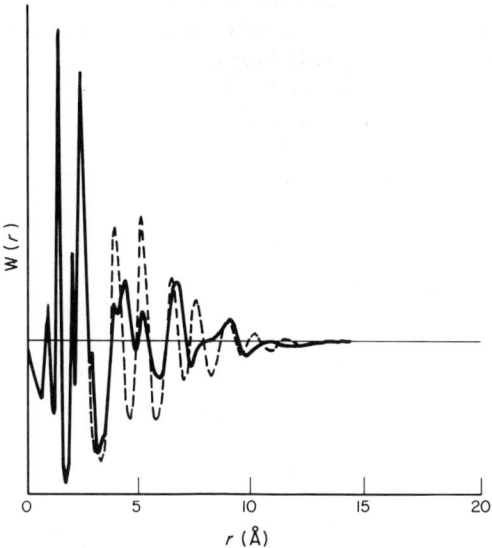

Figure 25 Comparison of model radial density functions calculated for an all-*trans* polyethylene chain (– – –) and a random polyethylene chain with $p_t = 0.6$.[35] These curves should be compared with the experimental radial density function of molten polyethylene in Figure 18

31.5.2 Natural Rubber

Natural rubber provided the first application of radial density function analysis to synthetic polymers,[108] and this was followed by other structural studies on unoriented[109,110] and on deformed samples,[85,111,112] which showed that the principal peak intensified onto the equator and hence arose from interchain correlations. Some of those authors[109,110] related 'correlation' with local parallelism of chain segments. However a recent study[15,113] has shown that the chain conformation is even more irregular than assumed by the random coil model.[114] Figure 26 shows the experimental reduced intensity function compared with the intrachain scattering calculated for the random coil model, and it is clear that there is too much 'structure' in the scattering pattern. A much better fit is obtained when the constraints of the isomeric model used by Abe *et al.*[114] are relaxed, to allow for a greater level of rotation about the single bond attached to the CH unit in the

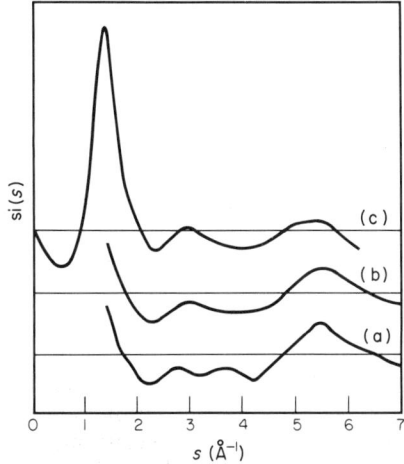

Figure 26 Calculated scattering functions for two models of polyisoprene molecules (a and b) compared with the experimental scattering function (c): (a) random chains of polyisoprene with the rotation states, their probabilities and their distribution assigned according to ref. 114, it is essentially a random coil model; (b) the random chain of (a) but with the rotation angle for the bond adjacent to the unsubstituted carbon in the isoprene unit chosen randomly from a uniform distribution between $+100°$ and $-100°$, in other words a delocalized rotation state for that bond

isoprene chain. Thus, while the experimental methods employed to probe the chain trajectory measurements are insensitive to large scale local fluctuations in intrachain order, those fluctuations have a large impact upon the X-ray scattering intensity curve. A similar result has been found for poly(dimethylsiloxane), another highly flexible polymer chain system.[36]

31.5.3   Poly(α-methylstyrene)

Natural rubber and polyethylene represent a class of materials in which the molecular chain approaches the string-like nature of most schematic views of non-crystalline polymers. There is of course a large class of polymers with bulky or lengthy side chains such as poly(α-methyl-styrene)[15,115] or the poly(n-alkyl methacrylate) family.[116-118] Poly(α-methylstyrene) will be used here as an example of this class of polymers. Of course, polymers with sufficiently long side chains usually crystallize (for example refs. 119 and 120), or form liquid crystal structures.[121-123]

Figure 27 shows the reduced intensity function for a non-crystalline sample of isotactic poly-(α-methylstyrene) obtained by rapidly cooling from the melt. The scattering pattern clearly contains more features than that for polyethylene or natural rubber. As is discussed in Section 31.2, it is useful to examine samples in which a preferred orientation has been induced by mechanical means. Figure 28 shows the reduced intensity function $si(s, \alpha)$ for an oriented sample of poly-

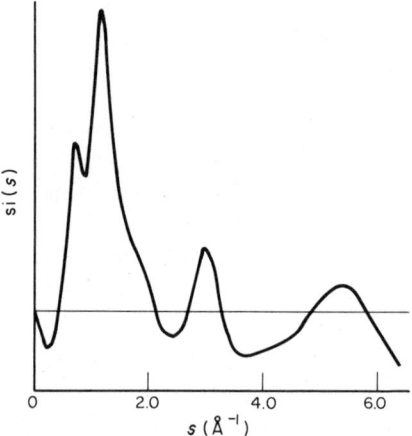

Figure 27   The experimental $si(s)$ function for a non-crystalline sample of isotactic poly(α-methylstyrene) prepared by rapidly cooling the specimen from the melt

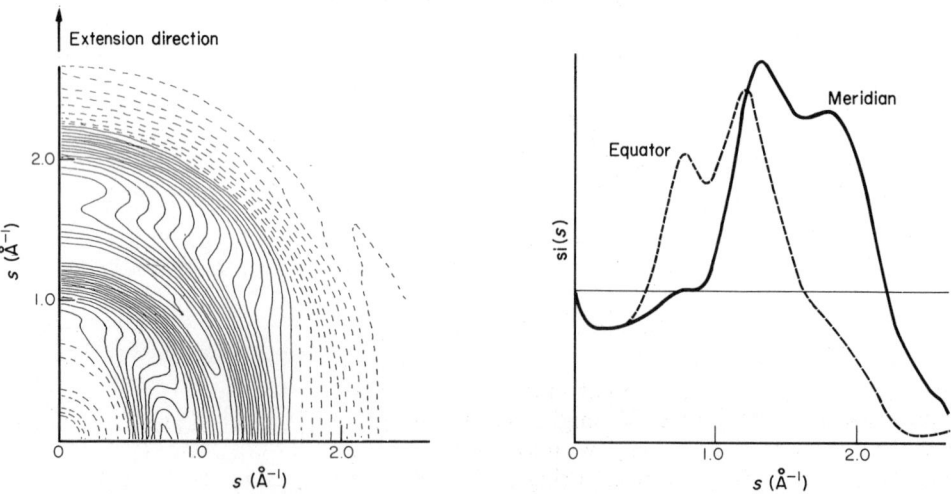

Figure 28   (a) The s weighted reduced intensity function $si(s, \alpha)$ for an oriented sample of isotactic poly(α-methylstyrene) measured at room temperature. The sample was extruded to give an extension ratio of 2. The extension axis is vertical. The dashed contours represent negative values of the intensity function. (b) The equatorial $\alpha = 90°$ section (dashed line) and the meridional $\alpha = 0°$ section (full line) of the intensity function in (a)

(α-methylstyrene). The utility of such an approach is clearly seen from the equatorial and meridional sections. The first peak is clearly equatorial, while the second contains components on both the equator and the meridian, and that at $s \sim 2 \text{Å}^{-1}$ is definitely meridional or intrachain in nature. Of course for polymers with longer side chain systems, some of the diffuse peaks which intensify onto the equator will be intraside chain in origin. However, for poly(α-methylstyrene) we can make an assignment that scattering for $s > 1.5 \text{Å}^{-1}$ is intrachain in nature. In other words there are three intrachain peaks at $s \sim 1.8, 3$ and 5.5Å^{-1}. The basic conformational parameters for poly-(α-methylstyrene) have been established by Sundararajan,[124] here we will concentrate on the values assigned to the valence angles and the conformational regularity. Steric hindrance between the side groups results in the opening out of the valence angle of the unsubstituted skeletal carbon atom, and from Sundararajan's calculations the *trans* conformation is favoured with a *trans–gauche* energy difference of 0.5 kcal mol^{-1} (1 cal = 4.19 J). Figure 29 shows the reduced intensity functions calculated for various models of isotactic poly(α-methylstyrene) chains with a random distribution of rotation states according to the study of Sundararajan.[124] Only the models with unequal skeletal bond angles provide a scattering pattern with three intrachain peaks. The best fit to the experimental si(s) function occurs when the valence angles are 110° and 128° and the bond rotation angles are 0° and 120°. Figure 30 shows similar random chains 'built' according to the scheme of ref. 124 but with varying energy differences between *trans* and *gauche*. The curves with that energy difference set to either 0 or 0.5 kcal mol^{-1} (1 cal = 4.19 J) give the best fit as long as the valence angles are 110° and 128°.[115] These two sets of curves illustrate how the conformational variables may be separated. The peak positions are strongly dependent upon the values of valence angles and of the skeletal bond rotation angles, while the shape, intensity and detail of the peaks is a function of the conformational regularity.

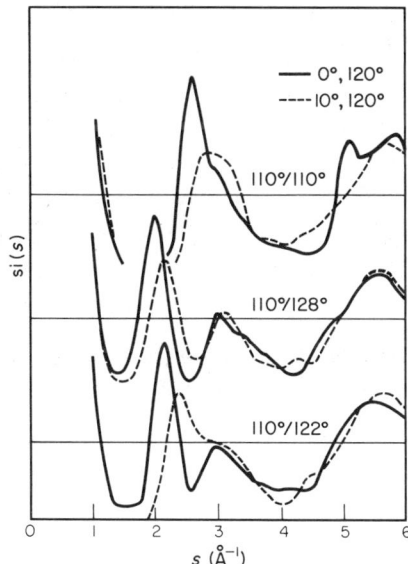

Figure 29 The s weighted reduced intensity functions si(s) calculated for random isotactic chains of poly(α-methylstyrene), in which the rotation states and their distribution are according to the scheme of ref. 124. The valence angle about the substituted skeletal carbon atom is 110°, while that about the unsubstituted carbon atom is set at 110°, 122° and 128°. The full lines represent scattering functions calculated for chains in which the *trans* state has been assigned a value of 0°, while dashed lines correspond to chains for which the *trans* state was 10°

In contrast to the scattering from polyethylene and natural rubber, poly(α-methylstyrene) shows two peaks below $s < 2 \text{Å}^{-1}$ which we might associate with interchain scattering. In fact by analogy with a detailed study of polystyrene,[16] we may assign the second of these two peaks to the inter- and intra-chain correlations between phenyl rings, this explains the apparent lack of anisotropy of this peak in the scattering from the aligned sample (Figure 28). The peak at $s \sim 0.8 \text{Å}^{-1}$ arises from supramolecular packing arrangements; the interaction distance is clearly much greater than the diameter of an individual chain. For polystyrene, it was possible to show that this supramolecular packing corresponded to a segregation of the phenyl rings into 'stacks' and the reader is directed to

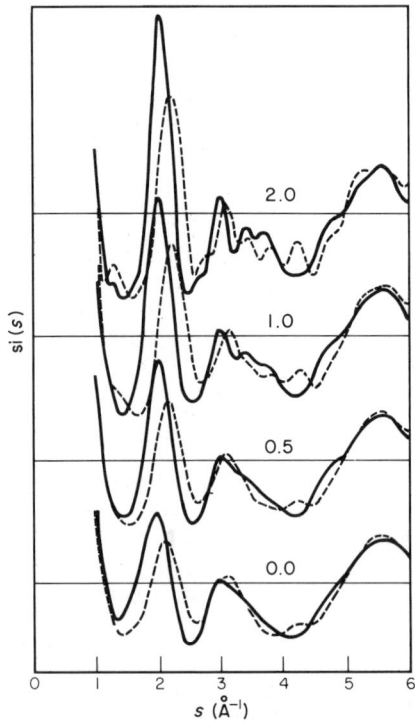

Figure 30 The model intensity functions si(s) calculated for random chains of poly(α-methylstyrene), in which the rotation states and their distribution are as assigned in Figure 29, but with the $E_{gauche} - E_{trans}$ difference as indicated in the figure, and with valence angles assigned as $110°/122°$ (dashed line) and $110°/128°$ (full line)

ref. 16 for a full description of the analysis. It is important to keep this apparent structure in perspective; from the width of the scattering peaks (the correlation length being approximately related to $2\pi/\Delta s$)[10] such a structure does not extend over much more than 2 or 3 structural units. It should not be seen as 'ordering', rather it arises as a consequence of the chemical configuration of the polymer. In other words, what other way could a polymer with bulky side groups pack densely? To describe these correlations the term 'special correlations' has been coined,[15,88] that is representing supramolecular short range correlations which arise as a result of the particular chemical configuration of the polymer. It is within this understanding that the scattering from, for example, poly-(n-alkyl methacrylates)[116-118] should be considered.

31.5.4 Other Polymers

A wide variety of polymers have been investigated in a quantitative manner. Some materials, such as poly(methyl methacrylate), have received considerable attention since the early work of Robinson *et al.*,[125] including detailed conformational analysis using both radial distribution function analysis[20,23] and reciprocal space data.[13,38] Polystyrene is another well-studied material, starting with a very early study by Katz[126] through a variety of approaches[41,4,127,128] to a detailed study in which very specific structural proposals are made with regard to the short-range stacking of phenyl groups.[16] The presence of 'special correlations' was also established between the phenyl rings of poly(phenylene sulfide) using detailed comparison of the experimental scattering data with functions derived from molecular models.[37] It has been suggested,[4] on the basis of X-ray scattering analysis, that PTFE chains exhibit some orientational correlations in the melt.

Polycarbonate was one of the first polymers to be examined in a fully quantitative manner.[129] Schubak *et al.* have made a careful study of the effect of temperature using radial density function techniques[57,128] and this has been supplemented by scattering studies which show that significant and particular changes in the molecular organization occur on annealing polycarbonate.[130-132] These structural changes were related to the embrittlement that occurs in polycarbonate upon annealing below T_g.

A number of other polymer systems have been examined[133-138] using both reciprocal and real space representations. These include a study of a blend, which showed there was significant molecular reorganization on mixing,[137] and swollen polymers.[138] However, it is important that these and any other materials and the conclusions drawn from any study of them should not be seen as a universal model for non-crystalline polymers.

31.5.5 Summary

This section has demonstrated both the versatility and the sensitivity of wide-angle X-ray scattering studies of unaligned non-crystalline polymers. We have shown that the positions, shapes and intensities of the diffuse peaks, so often dismissed as simply being due to amorphous material, do contain useful structural information relating to the levels of conformational, orientational and spatial order in the material. An important factor in the sucessful use of these procedures is to consider each polymer afresh, as the nature of the correlations will depend largely on the particular features of its chemical configuration.

31.6 EXAMPLES OF ANALYSIS FOR ANISOTROPIC SAMPLES

This section focuses on the application of the procedures described in Section 31.3 for the analysis of polymer samples with a preferred molecular alignment. As in the previous section, we will not survey the complete list of scattering studies in this area, but use selected samples to highlight particular features of such X-ray analyses. Both the areas of scattering from aligned non-crystalline polymers[139-142] and liquid crystalline polymers[50,123,143,144] have been reviewed.

31.6.1 Polypyrrole

Polypyrrole is a polymer formed from a five-membered heterocyclic unit. It is of particular interest because it forms part of a family of electrically conducting polymers which may be prepared by electrochemical procedures.[145] The electrochemical route provides simultaneously both polymerization and doping, that is the resultant film contains both the polymer and the dopant. X-Ray scattering studies of such polypyrrole films show that under certain conditions the molecular organization is anisotropic.[146,147] Careful measurements of the scattered intensity for different orientations of the scattering vector with respect to the plane of the film* yield different scattering patterns as Figure 31 shows. The intense peak in the scattering pattern for $\chi = 90°$ relates to a real

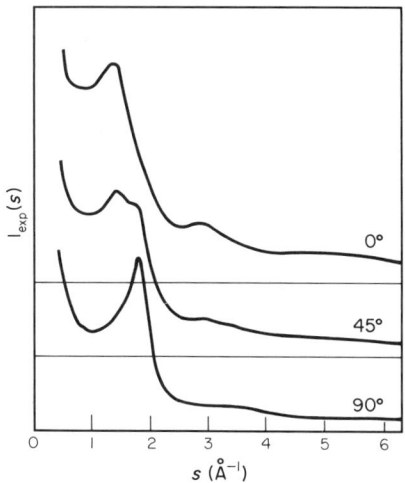

Figure 31 The experimental intensity functions I(s) recorded for a thin conducting film of polypyrrole prepared electrochemically using *para*-toluenesulfonate as the counterion.[146] The three curves correspond to data obtained with the sample orientation adjusted so that the scattering vector *s* was parallel to the plane of the film ($\chi = 0°$), or at angles of 45 or 90° to that plane. The maintenance of that geometry depends on the use of a symmetrical scattering geometry

* $\chi = 0°$ corresponds to the situation of the scattering vector *s* lying parallel to the plane of the film, while $\chi = 90°$ corresponds to *s* being normal to the film surface. The symmetrical scattering geometry ensures that these configurations are maintained over the complete scattering vector range.

space distance (using equation 7) of $\sim 3.4\,\text{Å}$ and this corresponds to the closest approach of the planes of aromatic rings. In contrast the largest peak in the scattering pattern for $\chi = 0°$ relates to the more normal distance between polymer chains of 5–$6\,\text{Å}$. From this we can conclude unambiguously that the polypyrrole chains lie preferentially with the planes of the heterocyclic units parallel to the electrode surface.[146] The level of alignment depends strongly on the preparation conditions.[147] This has obvious consequences for the electrical properties. However within the context of this chapter it is sufficent to note that anisotropy does not necessarily need to involve the long axis of the polymer chain. Moreover, it is salutary to see the amount of structural information that can be extracted from such 'amorphous' curves by making careful intensity measurements. These curves also yield conformational information and the reader is directed to refs. 146 and 147 for a complete description. Although not directly within the scope of this chapter it is useful to remember that wide-angle neutron scattering complements that of X-ray scattering, and through a combined study[148] a complete structural picture of these molecular composite films of polypyrrole has been established.

31.6.2   Poly(methyl methacrylate)

Poly(methyl methacrylate) is a widely studied material and the advantages of examining oriented samples have been thoroughly explored.[4,13,56,59,149] In this section we will apply the concepts of spherical harmonic analysis (see Section 31.3.3) to the evaluation of both the chain conformation and molecular orientation in poly(methyl methacrylate).

Figure 32 shows the reduced intensity function $si(s, \alpha)$[59] for a partially aligned non-crystalline sample of isotactic poly(methyl methacrylate), prepared by rapid cooling from the melt, followed by extrusion in a channel die at a temperature close to the glass transition. The first four spherical harmonics derived from the intensity data of Figure 32 using equation (29) are shown in Figure 33. The $si_0(s)$ term is the isotropic component. In the first anisotropic component, $si_2(s)$, negative values indicate an equatorial component, and positive values a meridional component of intrachain origin. Note that the higher orders of the harmonic functions are more or less zero, this reflecting the nature and level of the orientation distribution function. The conformational variables of particular interest for this polymer are the values of the valence angles, since steric crowding is expected to open out the valence angle around the unsubstituted skeletal carbon atom. Figures 34, 35 and 36 show sets of the first three harmonic components calculated for sections of poly(methyl methacrylate) chains placed in near all-*trans* conformations but with various valence angles. The particular utility of this spherical harmonic analysis, is that we can compare the experimental and model harmonics directly without a knowledge of the level of orientation in the sample, as long as we restrict our comparison to the positions, shapes and relative intensities of the diffuse peaks. In essence, samples with higher levels of orientation would simply give harmonics components for $2n > 0$ of greater magnitude, whilst the nature of the components would not change. In fact, as equation (33) shows, the ratio of the experimental and model curves gives the harmonics of the orientation distribution function. The $si_{2n}(s)$ functions for $n = 0$ or 1 show that valence angle about the unsubstituted backbone carbon atom is larger than $110°$, but whether the value should be $122°$ or $128°$ cannot be answered with any

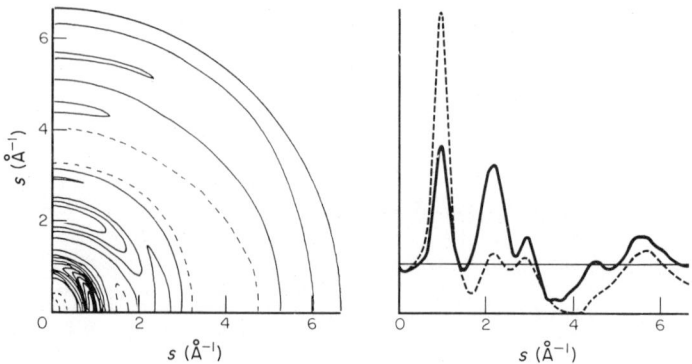

Figure 32   (a) The experimental X-ray scattering function $si(s, \alpha)$ for a partially aligned non-crystalline sample of an isotactic sample of poly(methyl methacrylate).[59] (b) Meridional $\alpha = 0°$ (——) and equatorial $\alpha = 0°$ (– – –) sections through the intensity surface of (a)

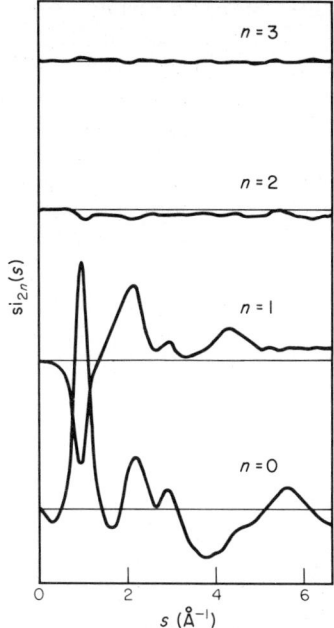

Figure 33 The first four spherical harmonics $si_{2n}(s)$ of the scattering function shown in Figure 32;[59] the vertical scale for each component is the same

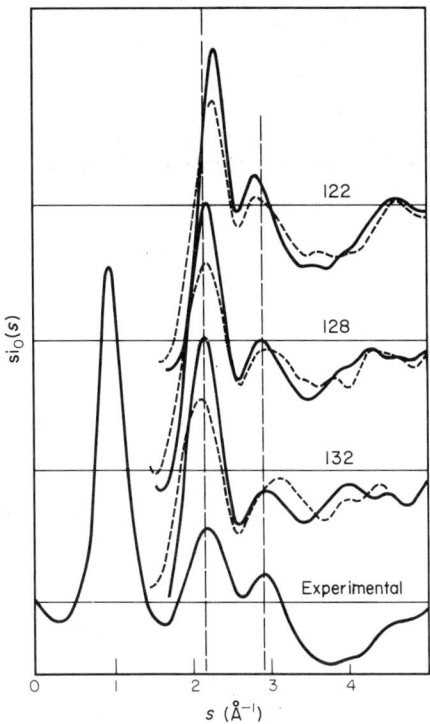

Figure 34 Calculated spherical harmonic components $si_0(s)$ of the scattering for an aligned segment of isotactic poly(methyl methacrylate) chains with a sequence of bond rotation angles fixed at near all-*trans* $(15°, 15°, 15°, 15°)_5$ but with various values for the valence angle about the unsubstituted skeletal carbon and with the side group rotation angle χ set at 0° (——) and 180° (————). The vertical dashed lines mark the positions of the peaks in the experimental intrachain scattering[59]

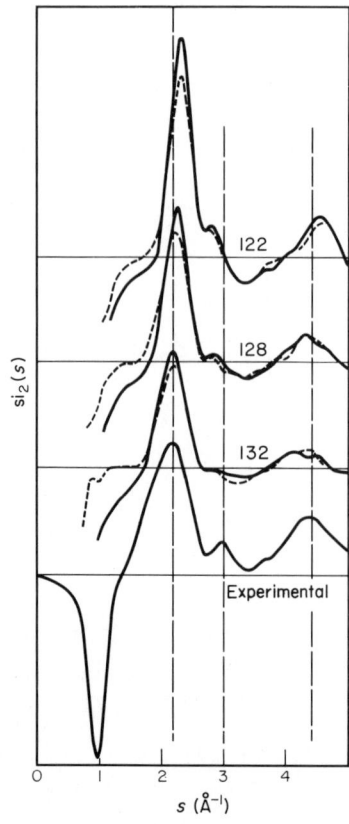

Figure 35 Calculated spherical harmonic components $si_2(s)$ for the same models as Figure 34[59]

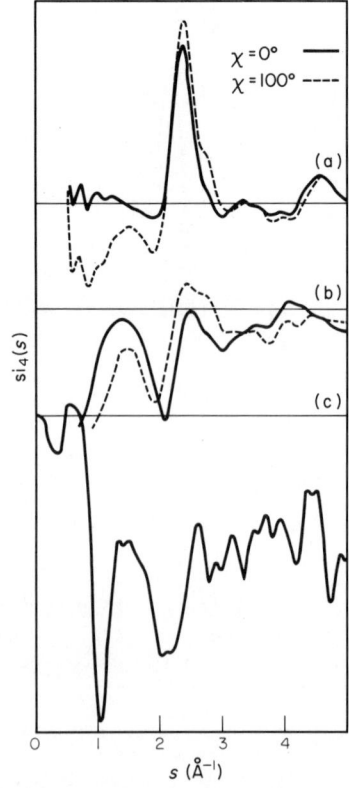

Figure 36 Calculated spherical harmonic components $si_4(s)$ for isotactic poly(methyl methacrylate) chain segments with the same conformations as Figure 34 but with: (a) skeletal valence angles $\theta_1 = 110°$ and $\theta_2 = 122°$; (b) $\theta_1 = 110°$ and $\theta_2 = 128°$; and (c) the experimental component plotted on an enlarged scale[59]

great certainty. However, if we examine the third term in the series, although the experimental function is very noisy, it clearly does not contain a large positive component. Inspection of the model curves shows that this eliminates the possibility of a 122° valence angle. Thus from the rather poorly aligned sample of poly(methyl methacrylate) (Figure 32) quite detailed conformational parameters have been obtained. Since the technique does not require a knowledge of the orientation distribution function, it could be applied to the study of conformational changes during deformation. In such a study, if there were no conformational changes other than in changes to their absolute magnitude, each set of spherical harmonic components should appear identical, and those absolute magnitudes could be used to obtain a quantitative measure of the molecular orientation.

We shall now consider an example of the use of these procedures to measure molecular orientation in atactic poly(methyl methacrylate).[48,150] Figure 37 shows the reduced intensity function for an aligned sample of atactic poly(methyl methacrylate) prepared by extrusion at 100 °C and cooled rapidly to room temperature. The experimental spherical harmonic components $si_{2n}(s)$ for $n = 0$ to 3 are shown in Figure 38. The operation of orientation measurement (equation 33 or 34) involves the division of the experimental harmonics by those for a perfectly aligned structure. In this

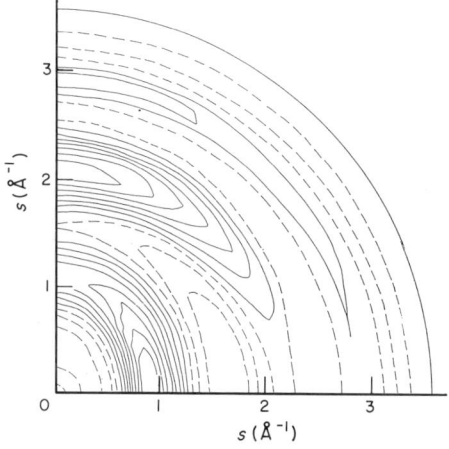

Figure 37   The experimental scattering function $si(s, \alpha)$ measured at room temperature for a partially aligned sample of atactic poly(methyl methacrylate) extruded at 100 °C to a draw ratio of 3:1 and cooled rapidly to room temperature[48]

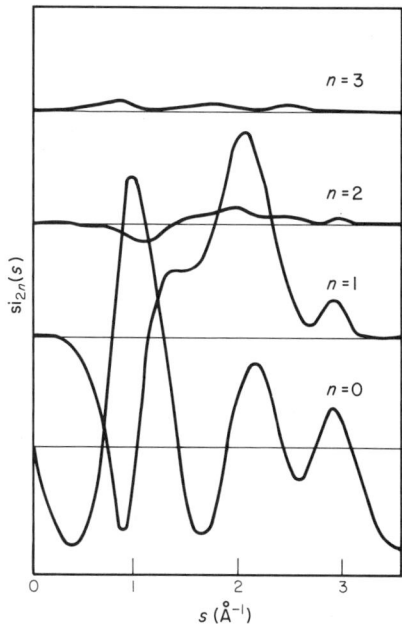

Figure 38   The first four spherical harmonic components $si_{2n}(s)$ for the scattering function shown in Figure 37

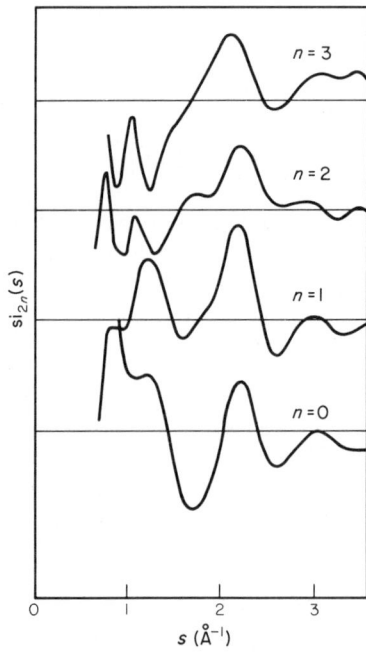

Figure 39 Calculated spherical harmonics for an aligned single chain model of a poly(methyl methacrylate) chain in an all-*trans* conformation as described in ref. 38

case we shall consider the intrachain scattering and hence we require the harmonics for an aligned chain segment. The chain conformation has been widely studied by X-ray scattering procedures, as already discussed, and the spherical harmonics for a chain segment in the confirmed conformation are shown in Figure 39.[48] The fact that these harmonics match those obtained by experiment in all but absolute magnitude indicates that the model is correct, and that the introduction of the preferred molecular alignment has not induced any significant changes in the chain conformation. We obtain the harmonics of the orientation distribution function for those chain segments by the application of equation (34), essentially by taking the ratio of the experimental to model harmonics. In principle the complete range of intrachain scattering could be used for this purpose, however in this study the most intense feature in the intrachain scattering, that is the peak at $s \sim 2 \text{ Å}^{-1}$ is used. It is useful to note that the model calculations allow the most appropriate scattering angle at which to measure the anisotropy to be determined before the measurements commence. In some situations it may be appropriate to utilize different scattering vector positions for the different parameters $\langle P_2 \rangle$, $\langle P_4 \rangle$, *etc*. The results of applying these procedures to samples of poly(methyl methacrylate), subject to different levels of uniaxial compression are shown in Figure 40. The negative values of $\langle P_2 \rangle$ indicate that the chains are preferentially aligned normal to the compression axis, as is expected. The plot of $\langle P_2 \rangle$ and $\langle P_4 \rangle$ against strain provides a useful test of deformation theories. A particular advantage of the X-ray scattering approach is that not only do we know the level of preferred molecular orientation, but perhaps more importantly we know the structural unit to which it applies, namely the chain segment. These techniques have been applied to other non-crystalline polymer systems[85,150,151] and to liquid crystalline polymers.[60,64,66] In the latter case, the choice of the chain segment is much more straightforward, because of its highly anisotropic nature, and in general the number of conformational parameters is less.

31.6.3 Side Chain Liquid Crystal Polymers

X-Ray scattering is a powerful tool in the identification of the type of liquid crystal phase present for an unknown sample. Relatively simple observation and measurement procedures allow a firm classification of nematic, smectic or cholesteric to be made. These methods are long established for low molar mass liquid crystal systems and have been described in detail elsewhere.[8,18,152,153] Their application to the study of polymeric liquid crystal systems began with the very first side chain systems.[154] Since then, examples of the use of X-ray scattering for classification abound in the

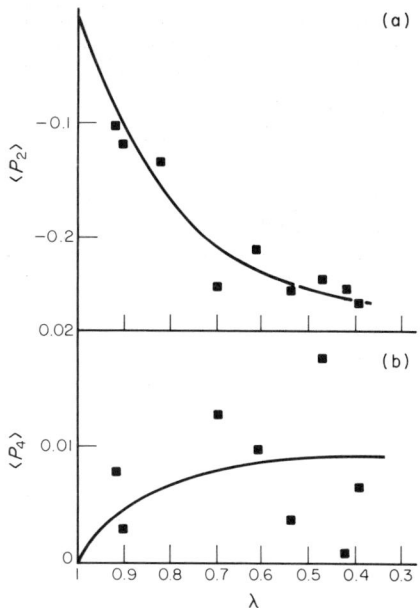

Figure 40 Plots of the orientation parameters (a) $\langle P_2(\cos\alpha)\rangle$ and (b) $\langle P_4(\cos\alpha)\rangle$ measured for samples of atactic poly(methyl methacrylate) as a function of the strain ratio.[38] The samples were uniaxially compressed at 20 °C. The full line represents a reasonable fit to the data points

literature. More quantitative studies, in terms of peak positions or the effects of temperature have concentrated on polymers exhibiting smectic phases.[155-166] A variety of novel polymeric liquid crystal compounds have been prepared, for example a side chain system with paired mesogens showed unusual layer modulation when examined using X-ray scattering techniques.[167] More quantitaive studies have been undertaken by Gudkov[168] and by Lipatov *et al.*[169] using Fourier transform techniques. A high resolution study of the nematic to smectic C transition has been reported by Nachaliel *et al.*[71] and the smectic fluctuations in the nematic phase close to the transition were found to be of the same form as displayed by low molar mass materials.

The studies outlined have concentrated on examining the basic layered structure in smectic phases; little has been directed at attempting to evaluate the local structure in such materials, other than the measurement of the level of anisotropy.[170] A study on a nematic side chain liquid crystal polymer has shown that the flexible coupling chain joining the mesogenic unit to the polymer backbone contained a significant proportion of *gauche* isomers, that is the coupling chain was not fully extended.[64] This analysis utilized the spherical harmonic analysis approach in a similar manner to the last section. A more detailed study[171] following the development of a macroscopic orientation in a magnetic field, revealed that this coupling chain showed time dependent conformational changes during reorientation, suggesting that the polymer backbone was involved in the process. This analysis was only possible through the use of the spherical harmonic procedures.

31.6.4 Main Chain Liquid Crystal Polymers

Main chain liquid crystal polymers fall into two groups: those joined by flexible coupling chains, and those comprised of 'rigid' units. There are a number of reports in the literature of the measurement of the level of molecular orientation, and these have been recently reviewed.[50] In general the study of the rigid copolymers has focused on determining the nature of the copolymerization,[65-68] while for those containing flexible coupling chains the interest is centred on the conformation of the flexible chain.[172-174] Of course X-ray scattering techniques have been widely used for classification purposes,[175-180] as discussed for the side chain liquid crystal polymers, including the study of a main chain discotic liquid crystal polymer.[181]

As a case study we will consider a main chain system of a *p*-hydroxybenzoic acid polyester prepared by acidolysis of poly(ethylene terephthalate) with *p*-acetoxybenzoic acid and polycondensation through the acetate and carboxyl groups.[49,60,65] Figure 41 shows the reduced intensity

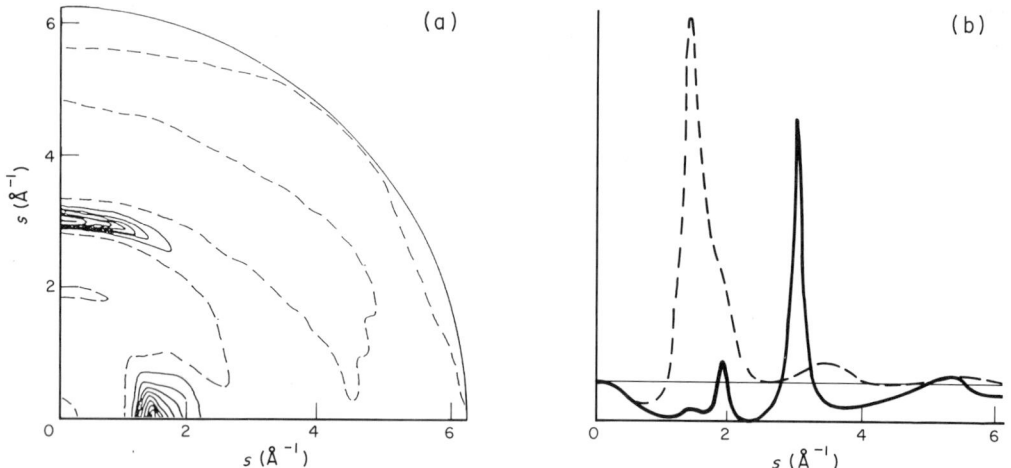

Figure 41 (a) Experimental s weighted reduced intensity function $si(s, \alpha)$ for a melt extruded pellet of random p-hydroxybenzoic acid copolyester (40 mol % polyethylene terephthalate and 60 mol % hydroxybenzoic acid) measured at room temperature.[49] The extrusion axis is vertical. (b) Meridional (——) and equatorial (– – –) sections of the intensity surface of (a).

function $si(s, \alpha)$ for an extruded pellet of such a polymer; it forms a liquid crystal phase above 190 °C but that structure can be retained by rapid cooling. The meridional section shows a number of relatively sharp peaks which are not simple orders of each other, in other words they are aperiodic. Such peaks may be related to the random copolymer nature of the chain, and in fact their positions and intensities are a sensitive measure of the randomness of the copolymer.[65–68] Their intensities may be modelled successfully using equations (38) and (39).[66] From the arcing of the interchain peak we can obtain orientation parameters $\langle P_2 \rangle = 0.58$ and $\langle P_4 \rangle = 0.37$[49] using the procedures described in equations (34)–(36). In fact the orientation distribution function plotted in Figure 11 was obtained using these and higher order terms in the series. Transformation of the scattering function of Figure 41 using equations (30) and (27) yields the cylindrical distribution function $W(r, \alpha)$ shown in Figure 42.[49] The general appearance of the function is that expected for a set of parallel rods, in that there are columns of high density situated approximately 5 Å apart. While the peaks along the meridian extend beyond the 20 Å shown, those interchain correlations extend only about 8 Å in a direction parallel to the chain direction, although a high level of lateral order is

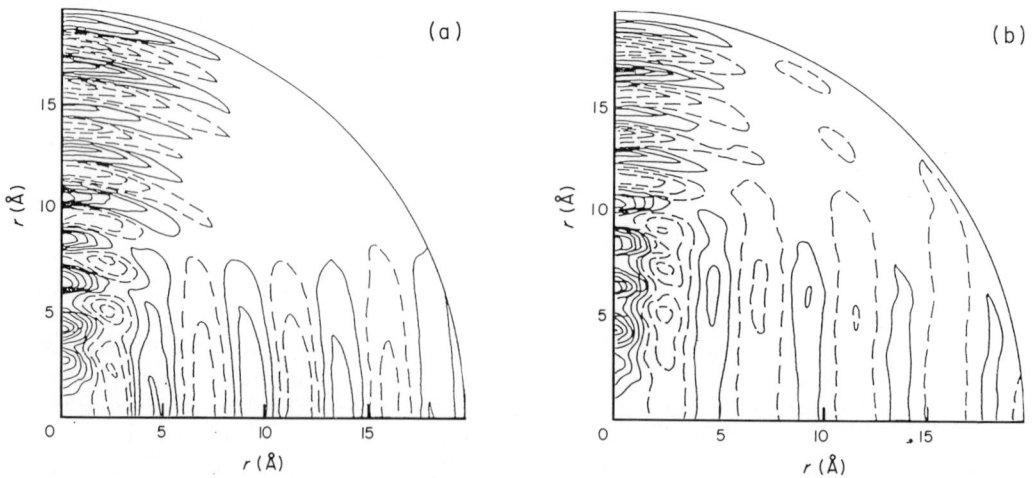

Figure 42 (a) Cylindrical distribution function derived from the scattering data of the liquid crystal copolymer shown in Figure 41. The extrusion axis is vertical. Dashed lines represent negative values of the function $W(r, \alpha)$. (b) The cylindrical distribution function resulting from the deconvolution of the $W(r, \alpha)$ function in (a) to remove the effect of global misorientation[49]

indicated. The arcing of the meridional peaks is mainly due to the range of orientations present in the sample as a whole. The fact that it is more pronounced than the arcing of the equatorial maxima is significant. It suggests that there are regions of the specimen which are better ordered laterally and are responsible for the persistent interchain correlations extending beyond 20 Å, and that in these regions the chains are better aligned with the extrusion axis than in the sample as a whole. The cylindrical distribution function shown in Figure 42(a) was deconvoluted using the orientation distribution function shown in Figure 11, to remove the effects of misorientation of the chains. As indicated in Section 31.3.3, this approach assumes a domain type structure; however, the result of the procedure shown in Figure 42(b) reveals some small maxima in the interchain peaks which are not centred on the meridional or equatorial axes. This suggests that there is some limited longitudinal correlation between adjacent chains, the correlation being limited to two or so repeat units. A more detailed analysis utilizing both the interchain and intrachain scattering[60] confirms this broad view of higher local order.

In fact this latter study utilized the intrachain scattering and a spherical harmonic analysis in a more rigorous measurement of the chain orientation. For a main chain liquid crystal polymer most of the conformational parameters are known or fixed, and thus the calculation of the spherical harmonics for a perfectly aligned system is relatively straightforward. However, there is one unknown parameter, which is the level of spatial correlation along the chain, and, therefore, in essence, the definition of the chain segment. A measure of such correlations may be obtained from the breadth of the meridional diffraction maxima, in this case the intense meridional peak at $s \sim 3 \text{ Å}^{-1}$. The breadth in terms of Δs of these meridional peaks may vary with the level of molecular alignment, as an off-meridional component of a layer line may contribute to the intensity on the meridian as a result of misorientation. Consequently it is important to match the breadths of the peaks in the calculated function to the experimental version through the use of the harmonic components $si_{2n}(s)$, since the form of the latter is unaffected by changing orientation (unless of course the correlation length is affected by different levels of orientation). Figure 43 shows a plot of half height breadth at $s = 3 \text{ Å}^{-1}$ in the calculated $si_2(s)$ against the correlation length of a statistical copolymer chain of a *p*-hydroxybenzoic acid copolyester. The essential feature is that it is not linear, as would be implied by the normal relationship $l = 2\pi/\Delta s$. This non-linearity arises from the fact that the meridional maximum under consideration is not a true Bragg peak. It emphasizes that careful investigation is required before interpreting peak breadths in anything more than a qualitative manner. By modelling the scattering we can be certain of the reliability of the correlation length so obtained. With that value we can read off the spherical harmonic parameters for a perfectly aligned chain segment of length l from the curves shown in Figure 44. These curves were prepared by calculating the scattering for various lengths of the copolymer chains using equation (37) and converting to the normalized coefficients shown *via* equation (35). The substitution of the selected harmonics for the perfectly aligned model into equation (34) yields orientational parameters, which have been obtained without making any unrealistic assumptions, and ones which relates to a specific structural scale in the sample.[60]

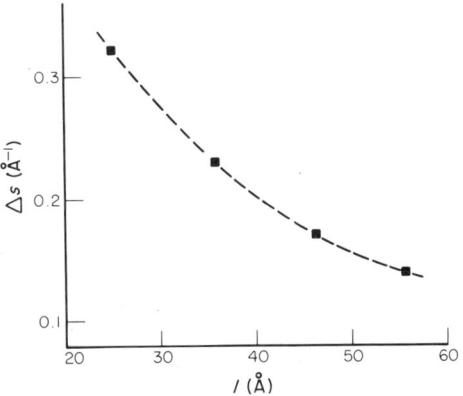

Figure 43 The half-height breadth of the peak at $s = 3 \text{ Å}^{-1}$ in the $si_2(s)$ function calculated for a model random copolymer chain, plotted as a function of the correlation length[60]

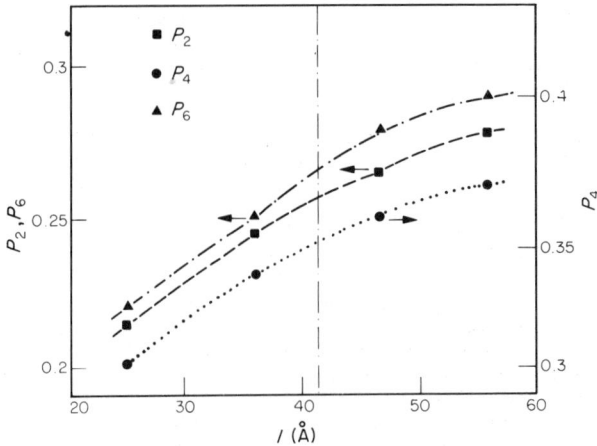

Figure 44 The normalized spherical harmonic components of the scattering calculated for a perfectly aligned random copolymer chain, plotted as a function of the correlation length. These parameters are used in equation (34) to evaluate the level of preferred molecular orientation using the intrachain scattering. The correlation length is read off from Figure 43 using the breadth of the experimental peak. The dotted and dashed vertical line indicates the correlation length measured for a typical melt extruded pellet of the liquid crystal copolymer

31.6.5 Summary

This section has demonstrated how quite detailed conformational and orientational parameters may be obtained from the scattering patterns of aligned samples. The introduction of even a modicum of preferred alignment considerably facilitates the analysis procedure. The methods of analysis are quite general and may be readily applied to a wide range of polymeric materials, whether they exhibit liquid crystal phases or not. However, the role of simple observation in the classification of anisotropy, as in the case of polypyrrole, or in the classification of liquid crystal phases must not be overlooked. There is clear potential for the exploitation of the methods described in the study of polymeric systems subjected to mechanical, magnetic and other fields including time-resolving studies, and some progress in this direction has already been made.[75,85]

31.7 SUMMARY

This chapter has demonstrated the versatility, sensitivity and suitability of wide-angle X-ray scattering for the study of the molecular arrangements in non-crystalline and liquid crystal polymers. Despite the imprecise nature of converting the positions of diffuse peaks in the scattering patterns of non-crystalline or liquid crystalline polymers to real space distances, X-ray scattering techniques provide a ready method of classifying the basic structure of such materials. However, by obtaining reliable quantitative data, coupled with some molecular modelling, useful and detailed conformational, orientational and spatial structural parameters may be obtained from randomly aligned polymer samples. In some cases the precision of the conformational analysis approaches that expected from the analysis of crystalline fibres. For aligned samples, X-ray scattering procedures provide a rigorous route to the determination of quantitative orientation parameters, particularly parameters related to a specified structural scale in the sample. By using the properties of spherical harmonics detailed chain conformational parameters may be obtained from samples with imperfect molecular alignment. Finally, X-ray scattering experimental arrangements are straightforward, and there are no unrealistic assumptions to be made in the analysis of the results. The advent of new intense sources, such as synchrotron radiation, and other developments including electronic area detectors opens up new avenues of study of the effects of external stimuli (thermal, mechanical, chemical, electrical, *etc.*) including time-resolving measurements. X-ray scattering techniques were one of the tools that unravelled the mysteries of macroscopic molecules at the beginning of this century; they remain today, and for the future, a valuable tool for the polymer scientist in the field of non-crystalline polymers.

31.8 REFERENCES

1. G. R. Mitchell, R. Lovell and A. H. Windle, *ACS Symp. Ser.*, 1980, **141**, 215.
2. S. Chandrasekhar, 'Liquid Crystals', Cambridge University Press, London, 1977.
3. G. W. Gray and J. W. G. Goodby, 'Smectic Liquid Crystals', Hill, Glasgow, 1984.
4. R. Lovell, G. R. Mitchell and A. H. Windle, *Faraday Discuss. Chem. Soc.*, 1979, **68**, 46.
5. P. J. Flory, 'Principles of Polymer Chemistry', Cornell University Press, New York, 1953.
6. P. J. Flory, *Adv. Poly. Sci.*, 1984, **59**, 1.
7. C. Zannoni, in 'The Molecular Physics of Liquid Crystals', eds. G. R. Luckhurst and G. W. Gray, Academic Press, London, 1979, p. 51.
8. A. J. Leadbetter, in 'The Molecular Physics of Liquid Crystals', eds. G. R. Luckhurst and G. W. Gray, Academic Press, London, 1979, p. 285.
9. B. K. Vainshtein, 'Diffraction of X-rays by Chain Molecules', Elsevier, Amsterdam, 1966.
10. L. E. Alexander, 'X-ray Diffraction Methods in Polymer Science', Wiley-Interscience, New York, 1969.
11. H. P. Klug and L. E. Alexander, 'X-ray Diffraction Procedures', Wiley, New York, 1974.
12. S. Krimm and A. V. Tobolsky, *Text. Res. J.*, 1951, **21**, 805.
13. R. Lovell and A. H. Windle, *Polymer*, 1981, **22**, 175.
14. S. M. Ohlberg, L. E. Alexander and E. L. Warrick, *J. Polym. Sci.*, 1958, **27**, 1.
15. G. R. Mitchell, in 'Order in the Amorphous "State" of Polymers', eds. S. E. Keinath, R. L. Miller and J. K. Rieke, Plenum Press, New York, 1985, p. 1.
16. G. R. Mitchell and A. H. Windle, *Polymer*, 1984, **25**, 906.
17. H. G. Kilian and K. Boueke, *J. Polym. Sci.*, 1962, **58**, 311.
18. A. De Vries, *Mol. Cryst. Liq. Cryst.*, 1985, **131**, 125.
19. H. H. M. Balyuzi and R. E. Burge, *Biopolymers*, 1971, **10**, 777.
20. J. R. Waring, R. Lovell, G. R. Mitchell and A. H. Windle, *J. Mater. Sci.*, 1982, **17**, 1171.
21. B. E. Warren, 'X-ray Diffraction', Addison-Wesley, Reading, 1969.
22. International Tables for X-ray Crystallography, IV, Kynoch Press, Birmingham, 1974.
23. A. Bjørnhaug, Ø. Ellefsen and B. A. Tønnesen, *J. Polym. Sci.*, 1954, **12**, 621.
24. R. Lovell, G. R. Mitchell and A. H. Windle, *Acta Crystallogr., Sect. A*, 1979, **35**, 598.
25. G. S. Cargill, *J. Appl. Crystallogr.*, 1971, **4**, 277.
26. I. Voigt-Martin and F. C. Mijlhoff, *J. Appl. Phys.*, 1976, **47**, 3942.
27. W. L. Bragg and J. West, *Philos. Mag.*, 1930, **10**, 823.
28. C. Lanczos, 'Discourse on Fourier Series', Oliver and Boyd, London, 1966, p. 65.
29. R. Kaplow, S. L. Strong and B. L. Averbach, *Phys. Rev. [Sect.] A*, 1965, **138**, 1336.
30. F. Y. Hansen, T. S. Knudsen and K. Carneiro, *J. Chem. Phys.*, 1975, **62**, 1556.
31. A. D'Anjou and F. Sanz, *J. Non-Cryst. Solids*, 1978, **28**, 319.
32. J. H. Otvos and H. Mendel, *Acta Crystallogr.*, 1962, **15**, 657.
33. F. Hadju, *Acta Crystallogr., Sect. A*, 1971, **27**, 73.
34. B. Stenhouse. P. J. Grout, N. H. March and J. Wenzel, *Philos. Mag.*, 1977, **36**, 129.
35. G. R. Mitchell, R. Lovell and A. H. Windle, *Polymer*, 1982, **23**, 1273.
36. G. R. Mitchell and A. Odajima, *Polym. J.*, 1984, **16**, 351.
37. T. P. H. Jones, G. R. Mitchell and A. H. Windle, *Colloid Polym. Sci.*, 1983, **261**, 110.
38. R. Lovell and A. H. Windle, in 'Diffraction Studies in Non-Crystalline Substances', eds. I. Hargitta and W. J. Orville-Thomas, Elsevier, Amsterdam, 1981, p. 673.
39. G. R. Mitchell, *Acta Crystallogr., Sect. A*, 1981, **37**, 487.
40. A. C. Wright, in 'Advances in Structure Research by Diffraction Methods', eds. W. Hoppe and R. Mason, Pergamon Press, Oxford, 1974.
41. S. M. Wecker, T. Davidson and J. B. Cohen, *J. Mater. Sci.*, 1972, **7**, 1249.
42. D. N. Theodorou and U. W. Suter, *Macromolecules*, 1985, **18**, 1467.
43. M. Vacatello, G. Avitabile, F. Corradini and A. Tuzi, *J. Chem. Phys.*, 1980, **73**, 548.
44. T. A. Weber and E. Helfand, *J. Chem. Phys.*, 1979, **71**, 4670.
45. G. R. Mitchell and A. Odajima, to be submitted.
46. D. W. L. Hukins, 'X-ray Diffraction by Disordered and Ordered Systems', Pergamon Press, Oxford, 1982.
47. A. C. Wright, *Faraday Discuss. Chem. Soc.*, 1970, **50**, 111.
48. G. R. Mitchell and A. H. Windle, *Polymer*. 1983, **24**, 285.
49. G. R. Mitchell and A. H. Windle, *Polymer*, 1982, **23**, 1269.
50. G. R. Mitchell and A. H. Windle, in 'Developments in Crystalline Polymers — 2', ed. D. C. Bassett, Elsevier, London, 1988, p. 115.
51. H. Kawai and S. Nomura, in 'Developments in Polymer Characterisation — 4', ed. J. V. Dawkins, Applied Science, London, 1982, p. 211.
52. E. W. Hobson, 'Theory of Spherical and Ellipsoidal Harmonics', Cambridge University Press, London, 1931.
53. P. H. Hermans and P. Platzek, *Kolloidn. Zh.*, 1939, **88**, 65.
54. F. C. Frank, *Philos. Trans. R. Soc. London, Ser. A*, 1983, **309**, 71.
55. H. D. Deas, *Acta Crystallogr.*, 1952, **5**, 54.
56. G. R. Mitchell and R. Lovell, *Acta Crystallogr., Sect. A*, 1981, **37**, 189.
57. H. R. Schubach and B. Heise, *Colloid Polym. Sci.*, 1986, **264**, 335.
58. R. Lovell and G. R. Mitchell, *Acta Crystallogr., Sect. A*, 1981, **37**, 135.
59. G. R. Mitchell and A. H. Windle, *Colloid Polym. Sci.*, 1982, **260**, 754.
60. G. R. Mitchell and A. H. Windle, *Polymer*, 1983, **24**, 1513.
61. R. Lovell and A. H. Windle, in 'The Structure of Non-Crystalline Solids', ed. P. H. Gaskell, Taylor and Francis, London, 1977, p. 27.
62. R. Lovell and A. H. Windle, *Acta Crystallogr., Sect. A*, 1977, **33**, 390.
63. W. Ruland, *Colloid Polym. Sci.*, 1977, **255**, 833.

64. G. R. Mitchell, F. J. Davis and A. S. Ashman, *Polymer*, 1987, **28**, 639.
65. G. R. Mitchell and A. H. Windle, *Colloid Polym. Sci.*, 1985, **263**, 230.
66. A. H. Windle, C. Viney, R. Golombok, A. M. Donald and G. R. Mitchell, *Faraday Discuss. Chem. Soc.*, 1985, **79**, 55.
67. R. Bonart, J. Dietrich and H. Zott, *Makromol. Chem.*, 1988, **189**, 227.
68. J. Blackwell and A. Biswas, in 'Developments in Oriented Polymers — 2', ed. I. W. Ward, Elsevier, London, 1987, p. 5.
69. W. Ruland, *Br. J. Appl. Phys.*, 1964, **15**, 1301.
70. G. R. Mitchell and A. H. Windle, *J. Appl. Cryst.*, 1980, **13**, 135.
71. E. Nachaliel, E. N. Keller, D. Davidov, H. Zimmerman and M. Deutsch, *Mol. Cryst. Liq. Cryst.*, 1987, **149**, 393.
72. M. Dixon, A. C. Wright and P. Hutchinson, *Nucl. Instrum. Methods*, 1977, **143**, 379.
73. F. Hajdu and G. Palinkas, *J. Appl. Crystallogr.*, 1972, **5**, 395.
74. J. R. Helliwell, *Rep. Prog. Phys.*, 1984, **47**, 1403.
75. G. R. Mitchell, A. S. Ashman, F. J. Davis and J. R. Ford, submitted.
76. B. E. Warren and R. L. Mozzi, *Acta Crystallogr.*, 1966, **21**, 459.
77. C. W. Dwiggins and D. A. Park, *Acta Crystallogr., Sect. A*, 1971, **27**, 264.
78. S. L. Strong and R. Kaplow, *Acta Crystallogr.*, 1967, **23**, 38.
79. G. Malet, C. Cabos, A. Escande and P. Delord, *J. Appl. Crystallogr.*, 1973, **6**, 139.
80. C. W. Dwiggins, *Acta Crystallogr., Sect. A*, 1972, **28**, 1580.
81. J. Krogh-Moe, *Acta Crystallogr.*, 1956, **9**, 951.
82. G. Johannson and M. Sandström, *Chem. Scr.*, 1973, **4**, 195.
83. P. Lecante, A. Mosset and J. Galy, *J. Appl. Crystallogr.*, 1985, **18**, 214.
84. G. R. Mitchell, 'RU-PRISM', University of Reading, UK, 1985.
85. G. R. Mitchell, *Polymer*, 1984, **25**, 1562.
86. G. Elsner, C. Riekel and H. G. Zachmann, *Adv. Polym. Sci.*, 1985, **67**, 1.
87. I. G. Voigt-Martin and J. Wendorff, in 'Encyclopedia of Polymer Science and Engineering', 2nd edn., Wiley, New York, 1985, vol. 1, p. 789.
88. A. H. Windle, *Pure Appl. Chem.*, 1985, **11**, 1627.
89. J. H. Wendorff, *Polymer*, 1982, **23**, 543.
90. J. M. O'Reilly, *CRC Crit. Rev. Solid State Mater. Sci.*, 1987, **13**, 261.
91. Yu. K. Ovchinnikov, G. S. Markova and V. A. Kargin, *Polym. Sci. USSR (Engl. Transl.)*, 1969, **11**, 369.
92. Yu. K. Ovchinnikov, Y. M. Antipov and G. S. Markova, *Polym. Sci. USSR (Engl. Transl.)*, 1975, **17**, 2081.
93. G. W. Longman, G. D. Wignall and R. P. Sheldon, *Polymer*, 1979, **20**, 1063.
94. S. Kan and T. Seto, *Rep. Prog. Polym. Phys. Jpn.*, 1976, **19**, 219.
95. W. R. Pechhold, M. E. T. Hauber and E. Liska, *Kolloidn. Zh.*, 1973, **251**, 818.
96. J. Petermann and H. Gleiter, *Philos. Mag.*, 1973, **28**, 271.
97. O. Yoda, I. Kuriyama and A. Odajima, *J. Appl. Phys.*, 1978, **49**, 5468.
98. M. Gupta and G. S. Y. Yeh, *J. Macromol. Sci., Phys.*, 1979, **16**, 225.
99. W. R. Pechhold and H. P. Grossmann, *Faraday Discuss., Chem. Soc.*, 1979, **68**, 58.
100. I. G. Voigt-Martin and F. C. Mijlhoff, *J. Appl. Phys.*, 1975, **46**, 1165.
101. A. Charlesby, *J. Polym. Sci.*, 1953, **10**, 201.
102. E. W. Fischer, G. R. Strobl, M. Dettenmaier, M. Stamm and N. Steidle, *Faraday Discuss., Chem. Soc.*, 1979, **68**, 26.
103. E. W. Fischer and M. Dettenmaier, *J. Non-Cryst. Solids*, 1978, **31**, 181.
104. M. Numakawa and A. Odajima, *Polym. J.*, 1981, **13**, 599.
105. G. R. Mitchell, R. Lovell and A. H. Windle, *Polymer*, 1980, **21**, 989.
106. A. Abe, R. L. Jernigan and P. J. Flory, *J. Am. Chem. Soc.*, 1966, **88**, 631.
107. J. R. Scherer and R. G. Synder, *J. Chem. Phys.*, 1980, **72**, 5798.
108. G. L. Simard and B. E. Warren, *J. Am. Chem. Soc.*, 1936, **58**, 507.
109. G. S. Markova, Yu. K. Ovchinnikov and E. S. Boknyan, *J. Polym. Sci., Polym. Symp.*, 1973, **42**, 671.
110. C. S. Wang and G. S. Y. Yeh, *J. Macromol. Sci., Phys.*, 1978, **B15**, 107.
111. L. E. Alexander, S. Ohlberg and G. R. Taylor, *J. Appl. Phys.*, 1955, **26**, 1068.
112. L. E. Alexander and E. R. Michalik, *Acta Crystallogr.*, 1959, **12**, 105.
113. G. R. Mitchell, to be submitted.
114. A. Abe and P. J. Flory, *Macromolecules*, 1971, **4**, 230.
115. G. R. Mitchell, submitted.
116. R. L. Miller, in 'Order in the Amorphous "State" of Polymers', eds. S. E. Keinath, R. L. Miller and J. K. Rieke, Plenum Press, New York, 1987, p. 33.
117. R. L. Miller and R. F. Boyer, *J. Polym. Sci., Polym. Phys. Ed.*, 1984, **22**, 2043.
118. R. L. Miller, R. F. Boyer and J. Heijboer, *J. Polym. Sci., Polym. Phys. Ed.*, 1984, **22**, 2021.
119. N. A. Platé and V. P. Shibaev, *Macromol. Rev.*, 1974, **9**, 117.
120. F. Andruzzi, C. Barone, D. Lupinacci and P. L. Magagnini, *Makromol. Chem. Rapid. Commun.*, 1984, **5**, 603.
121. H. Finkleman and G. Rehage, *Adv. Polym. Sci.*, 1984, **60/61**, 99.
122. V. P. Shibaev and N. A. Platé, *Adv. Polym. Sci.*, 1984, **60/61**, 175.
123. J. H. Wendorff, H. Finkleman and H. Ringsdorf, *ACS Symp. Ser.*, 1978, **74**, 12.
124. P. R. Sundararajan, *Macromolecules*, 1977, **10**, 623.
125. H. A. Robinson, R. Ruggy and E. Slantz, *J. Appl. Phys.*, 1944, **15**, 343.
126. J. R. Katz, *Trans. Faraday Soc.*, 1936, **32**, 77.
127. R. Adams, H. H. Balyuzi and R. E. Burge, *J. Mater. Sci.*, 1978, **13**, 391.
128. H. R. Schubach, E. Nagy and B. Heise, *Colloid Polym. Sci.*, 1981, **259**, 789.
129. G. D. Wignall and C. W. Longman, *J. Mater. Sci.*, 1973, **8**, 1449.
130. J. R. Saffell and A. H. Windle, *J. Polym. Sci., Polym. Lett. Ed.*, 1980, **18**, 377.
131. G. R. Mitchell and A. H. Windle, *Colloid Polym. Sci.*, 1985, **263**, 280.
132. E. Turska, J. Hurck and L. Zmudzinki, *Polymer*, 1979, **20**, 321.
133. Yu. K. Ovchinnikov, N. N. Kuz'min, Yu. A. Makhnovskii and N. F. Bakeyev, *Polym. Sci. USSR (Engl. Transl.)*, 1985, **26**, 702.

134. V. V. Shilov, N. E. Kruglyak and Yu. S. Lipatov, *J. Macromol. Sci., Phys.*, 1983, **B22**, 79.
135. V. V. Shilov, N. Ye. Kruglyak, V. V. Tsukbruk and Yu. S. Lipatov, *Polym. Sci. USSR (Engl. Transl.)*, 1980, **22**, 2654.
136. M. Numakawa and A. Odajima, *Polym. J.*, 1981, **13**, 879.
137. G. R. Mitchell and A. H. Windle, *J. Polym. Sci., Polym. Phys. Ed.*, 1985, **23**, 1967.
138. G. R. Mitchell, D. J. Brown and A. H. Windle, *Makromol Chem.*, 1983, **184**, 1937.
139. A. H. Windle, in 'Developments in Oriented Polymers — 1', ed. I. M. Ward, Applied Science, London, 1982, p. 1.
140. I. L. Hay, in 'Polymers, Part C: Physical Properties', ed. R. A. Fava, Academic Press, New York, 1980, p. 137.
141. C. R. Desper, *Crit. Rev. Macromol. Sci.*, 1973, **1**, 501.
142. G. L. Wilkes, *Adv. Poly. Sci.*, 1971, **8**, 91.
143. L. Z. Aźaroff, *Mol. Cryst. Liq. Cryst.*, 1987, **145**, 31.
144. J. H. Wendorff, in 'Liquid Crystalline Order in Polymer', ed. A. Blumstein, Academic Press, New York, 1978, p. 1.
145. R. J. Walton and J. Bargon, *Can. J. Chem.*, 1987, **64**, 76.
146. G. R. Mitchell, *Polym. Commun.*, 1986, **27**, 346.
147. G. R. Mitchell and A. Geri, *J. Phys. D*, 1987, **20**, 1346.
148. G. R. Mitchell, F. J. Davis, R. Cywinski and W. S. Howells, *J. Phys. C*, 1988, **21**, 1411.
149. A. Colebrooke and A. H. Windle, *J. Macromol. Sci., Phys.*, 1976, **B12**, 373.
150. D. J. Brown and G. R. Mitchell, *J. Polym. Sci., Lett. Ed.* 1983, **21**, 341.
151. G. R. Mitchell, *Br. Polym. J.*, 1985, **17**, 111.
152. J. Falgueirettes and P. Delord, in 'Liquid Crystals and Plastic Crystals', ed. G. W. Gray and P. A. Windsor, Ellis Horwood, Chichester, 1974, vol. 2, p. 62.
153. A. J. Leadbetter, in 'Thermotropic Liquid Crystals', ed. G. W. Gray, Wiley, Chichester, 1987, p. 1.
154. H. Finklemann and D. Day, *Makromol. Chem.*, 1979, **180**, 2269.
155. B. Hahn, J. H. Wendorff, M. Portugall and H. Ringsdorf, *Colloid Polym. Sci.*, 1981, **259**, 875.
156. S. G. Kostromin, V. V. Sinitzyn, R. V. Talroze, V. P. Shibaev and N. A. Platé, *Makromol. Chem., Rapid Commun.*, 1982, **3**, 809.
157. V. V. Tsukruk, O. A. Lokhonya, V. V. Shilov, V. A. Kuzmina and Yu. S. Lipatov, *Makromol. Chem., Rapid Commun.*, 1983, **4**, 595.
158. P. Zugenmaier and J. Mugge, *Makromol. Chem., Rapid Commun.*, 1984, **5**, 11.
159. V. V. Shilov, V. V. Tsukruk and Yu. S. Lipatov, *J. Polym. Sci., Polym. Phys. Ed.*, 1984, **22**, 41.
160. R. M. Richardson and N. J. Herring, *Mol. Cryst. Liq. Cryst.*, 1985, **123**, 143.
161. H. H. Sutherland and A. Rawas, *Mol. Cryst. Liq. Cryst.*, 1986, **138**, 179.
162. G. Decobert, J. C. Dubois, S. Esselin and C. Noël, *Liq. Cryst.*, 1986, **1**, 307.
163. R. Zentel, G. Reckert and B. Reck, *Liq. Cryst.*, 1987, **2**, 83.
164. R. Duran, D. Guillon, P. Gramain and Skoulois, *Makromol. Chem., Rapid Commun.*, 1987, **8**, 181.
165. H. H. Sutherland, S. Basu and A. Rawas, *Mol. Cryst. Liq. Cryst.*, 1987, **145**, 73.
166. P. Zugenmaier and J. Mügge, in 'Recent Advances in Liquid Crystal Polymers', ed. L. L. Chapoy, Elsevier, London, 1985, p. 267.
167. S. Diele, B. Hisgen, B. Reck and H. Ringsdorf, *Makromol. Chem., Rapid Commun.*, 1986, **7**, 267.
168. V. A. Gudkov, *Sov. Phys. — Crystallogr. (Engl. Transl.)*, 1984, **29**, 316.
169. Yu. S. Lipatov, V. V. Tsukruk, V. V. Shilov, Yu. B. Amerik and I. I. Konstantinov, in 'Advances in Liquid Crystal Research and Applications', ed. L. Bata, Pergamon Press, Oxford, 1980, p. 943.
170. F. J. Davis and G. R. Mitchell, *Polym. Commun.*, 1987, **28**, 8.
171. G. R. Mitchell and A. S. Ashman, to be submitted.
172. A. Blumstein, S. Vilasagar, S. Ponrathnam, S. B. Clough, R. B. Blumstein and G. Maret, *J. Polym. Sci., Polym. Phys. Ed.*, 1982, **20**, 877.
173. A. Blumstein, K. N. Sivaramakrishnan, R. B. Blumstein and S. B. Clough, *Polymer*, 1982, **23**, 47.
174. A. Blumstein, O. Thomas, J. Asrar, P. Makris, S. B. Clough and R. B. Blumstein, *J. Polym. Sci., Polym. Lett. Ed.*, 1984, **22**, 13.
175. A. Roviello, S. Santagata and A. Sirigu, *Makromol. Chem., Rapid. Commun.*, 1984, **5**, 209.
176. W. R. Krigbaum and J. Watanabe, *Polymer*, 1983, **24**, 1299.
177. W. R. Krigbaum, J. Watanabe and T. Ishikawa, *Macromolecules*, 1983, **16**, 1271.
178. A Roviello, S. Santagata and A. Sirigu, *Makromol. Chem., Rapid Commun.*, 1984, **5**, 141.
179. C. Noel, C. Friedrich, L. Bosio and C. Strazielle, *Polymer*, 1984, **25**, 1281.
180. J. Watanabe, H. Ono, I. Uematsu and A. Abe, *Macromolecules*, 1985, **18**, 2141.
181. O. Herrmann-Schönherr, J. H. Wendorff, W. Kreuder and H. Ringsdorff, *Makromol. Chem., Rapid Commun.*, 1986, **7**, 97.

32

Neutron Scattering from Solid Polymers

DAVID M. SADLER*
University of Bristol, UK

* *Addition in proof*: Dr. David M. Sadler died tragically at the age of 43 since the preparation of this chapter. This was his last major work to reach print, with much else uncompleted left behind. It should bear testimony to one of the varied facets of his many contributions to science, always original in subject matter, creative in approach and prodigious in output, enriching knowledge and setting an example for those who follow. A. Keller, University of Bristol.

32.1 INTRODUCTION

Neutron scattering (NS) has become the central technique for investigating polymer conformations in the solid state on a scale larger than the typical monomer–monomer distance. Since it is the long chain nature of polymers which gives them their distinctive properties (*e.g.* rubber elasticity, non-Newtonian flow, viscoelasticity) NS is now essential in relating these properties to molecular structure. The key property for molecules which are labelled is the scattering contrast between ^{1}H and ^{2}H isotopes. From a historical point of view, the emphasis early in the development of polymer science on solutions was partly because light scattering could be used to determine coil sizes. The scattering contrast is then between the different refractive indices of polymer and solvent. The fundamental results concerning Gaussian coils for the (amorphous) solid were only inferred by analogy with solutions, and these inferences were only verified much later by NS. In the case of crystalline polymers the situation was even less certain since there was no 'model system' analogous to solutions. Electron microscopy could in this case place limits on the types of conformations (they had to include chain folding) but could not, for example, determine the overall size of the molecules. Although the interest in NS has often centred on this ability to measure scattering functions for individual molecules, other uses are also important, *e.g.* concerning density distributions (as in small angle X-ray scattering) and spectroscopic information.

This chapter concerns both the technique itself and results obtained from it. It is not possible to include exhaustive references, and more complete bibliographies are often included in specific reviews referred to later. Other general reviews are given in refs. 10 to 16.

32.2 PRINCIPLES

Some aspects of scattering theory, summarizing material which is common to the large literature on this topic (*e.g.* references which include descriptions of X-ray scattering)[1-9] are dealt with in Section 32.2.4. Much of that section is not specific to neutrons, but some aspects of the theory are particularly important in connection with the labelling technique. Experimental considerations are included in Section 32.3, followed by a description of results for a wide variety of systems. In most cases an attempt will be made to identify in what way the technique has contributed to the current understanding of the subject. For this reason at least some description is given of results of other techniques and of theories.

A principal use of NS is to distinguish one molecule from its neighbours. The following sections make clear that this can be explained in terms of lack of coherence between scattering from different molecules, in an analogous way to the incoherence arising from random spin orientations. Whatever the relative concentrations of molecular species containing ^{1}H and ^{2}H, as long as the two species are chemically identical, the scattering is a superposition of that expected for the structure with ^{1}H and ^{2}H nuclei replaced with an 'average' nucleus, together with a signal related to the isotope differences. Another more common interpretation employs the analogy with solution scattering, with a minority isotopic species representing the solute. This approach does not explain so readily why concentrations higher than a few percent are permissible.

32.2.1 Scattering from Atoms

For neutrons (or X-rays) the individual scattering process from nuclei (or electrons) is sufficiently weak that the incoming wave is perturbed very little by individual atoms. The approximation due to Born is then very accurate; the amplitude of scattering is calculated as the sum of amplitudes from spherical wavelets originating from every part of the material which interacts with the incoming waves. For X-rays the scattering is related to the electron density, which in general varies smoothly in space. For neutrons most of the scattering is due to the nuclei, which act as point scatterers since their dimensions are very small compared with the neutron wavelengths. (Neutrons can also be scattered by unpaired electrons which produce net magnetic moments, but this is not as yet relevant to polymer science.) The scattering length, b, is sufficient to characterize the wavelet originating from one isolated nucleus. b^2 can be defined as the probability of an incoming neutron being scattered into one steradian per unit flux of neutrons (flux is number of neutrons per unit time per unit area of beam). The value of b depends both on the isotope and on the way the neutron spin combines with any spin the nucleus may have. Thus a chemically homogeneous system will often have a heterogeneity towards neutron scattering because of the presence of several isotopes in their natural abundances and because nuclear spins are uncorrelated except under exceptional circumstances. It is customary to quote atomic scattering characteristics so as to allow for the spin heterogeneity. The 'coherent scattering length' for one isotope only is the average b value (usually of the order of 0.5×10^{-12} cm). The incoherent scattering is zero when the nuclear spin is zero, and in other cases increases with the spin dependence of b. The coherent contribution has all the interference properties found with X-rays, for example, while for the incoherent contribution the intensities from different atoms are simply added, with no interference effects. As noted above, the theory for random mixing of nuclear spins has analogies with that for non-random mixing of isotopes.

Hydrogen is not only the most plentiful atom in most polymers, but it also has highly atypical neutron-scattering properties. Firstly ^{1}H has a very high incoherent cross section, which is very relevant to the spectroscopic use of (inelastic) NS. The structural work is based on the other characteristic of hydrogen; the very large difference between the coherent scattering length of ^{1}H (b_H is -0.374×10^{-12} cm) from that of the other hydrogen isotopes (b_D, for deuterium, is $+0.667 \times 10^{-12}$ cm). Unfortunately, the chemical effects of changing the hydrogen mass by two, although small, are not completely negligible. For example, the melting temperature of ice is increased by 4 °C on deuteration, and that of polyethylene decreased by 4 °C.

The dimensions of b^2 are area, and the scattering is often characterized by the differential cross section, which for one nucleus is simply b^2. ('Cross section' can also refer to the total scattering integrated over all solid angles, *i.e.* $4\pi b^2$ for one isolated nucleus.) Experimentally, the differential cross section can be quoted per volume of sample (the 'macroscopic' cross section $d\Sigma/d\Omega$). Other conventions exist for expressing the intensity of scattering (see below). For example, the scattering may be expressed as a multiple of that expected for one nucleus. This is similar to a convention used in X-ray scattering[5] where the intensity is often expressed as a multiple of that from one electron.

Inelastic effects are normally neglected in the limit of small q (defined in Section 32.2.2). This is reasonable since they would only be important if there were movements which were highly correlated over large distances. Inelastic scattering is discussed further in Section 32.12.

32.2.2 Scattering and the Correlation Function

The coherent scattering from neighbouring nuclei give interference effects from which structural information is available.

There is a phase difference of $(k - k_0) \cdot r_{ij}$ for the scattering between two nuclei separated in space by r_{ij}, where k_0 and k are vectors parallel to the incident and scattered neutron directions respectively. We consider here elastic scattering only, so that k and k_0 are equal in magnitude ($k = 2\pi/\lambda$). The scattering angle 2θ is commonly expressed in terms of

$$q \equiv k - k_0$$

hence

$$|q| = 4\pi \sin \theta / \lambda$$

where 2θ is the scattering angle and λ the wavelength.

In order to calculate an intensity, which can be compared to experimental values, it is necessary first to calculate the scattering for one 'assembly', which may be chosen for convenience (*e.g.* one molecule, one crystallite or even the whole sample). It is useful to choose an assembly such that the scattering from different assemblies add incoherently (*i.e.* the intensity contributions from the N assemblies per volume of sample are additive). The total amplitude of the coherent elastic scattering is given by

$$A(q) = \sum_i b_i \exp(iq \cdot r) \qquad (1)$$

where the exponential term arises from the phase differences, the origin is chosen arbitrarily, and

$$(d\Sigma/d\Omega)_{coh, e} = N|A(q)|^2 V \qquad (2)$$

where N is the number of assemblies and V the sample volume.

Most of the equations, which are relevant to structure studies, are elaborations and extensions of equation (1) whose identity as a Fourier transform ($\mathscr{F}$) is more clearly apparent when a density function $\rho(r)$ is used, which represents the total $\sum b_i$ per unit volume at the position r

$$A(q) = \int \rho(r) \exp(iq \cdot r) d^3r \qquad (3)$$

i.e.

$$A(\boldsymbol{q}) = \mathscr{F}[\rho(\boldsymbol{r})]$$

Several Fourier transform theorems are useful

$$\rho(\boldsymbol{r}) = \rho_1(\boldsymbol{r}) * \rho_2(\boldsymbol{r})$$

where the symbol * signifies a convolution, then

$$A(\boldsymbol{q}) = \mathscr{F}[\rho_1(\boldsymbol{r})] \times \mathscr{F}[\rho_2(\boldsymbol{r})]$$

For example, ρ_1 could signify the density inside one repeating 'motif', and ρ_2 a lattice of points representing the positions of an assembly. The convolution theorem also leads to

$$|A(\boldsymbol{q})|^2 = \mathscr{F}[\rho(\boldsymbol{r})] = \mathscr{F}[\gamma(\boldsymbol{r})]$$

This equation is of special interest since the 'density correlation function' $\gamma(\boldsymbol{r})$ can in principle be measured by calculating from experimental results

$$\gamma(\boldsymbol{r}) = \mathscr{F}|A(\boldsymbol{q})|^2$$

In many experiments the assemblies take up many orientations within a sample. This could be allowed for by making the requisite average of equation (3), but for an isotropic sample the averaging over orientations is best done before making the Fourier transform. The result, due to Debye, is

$$\langle |A(\boldsymbol{q})|^2 \rangle = \sum_i b_i^2 + \sum_i \sum_{\substack{j \\ i \neq j}} \frac{b_i b_j \sin(qr_{ij})}{(qr_{ij})} \qquad (4)$$

The magnitudes of all vectors $\boldsymbol{r}_{ij}$ between scattering centres is included in equation (4), and as one may expect there is no direct informaton in the intensity curve about the relative orientation of two vectors $\boldsymbol{r}_{ij}$ and $\boldsymbol{r}_{kl}$. With continuous notation, equation (4) becomes

$$\langle |A(\boldsymbol{q})|^2 \rangle = \int \gamma(r) \sin(qr)/(qr) \, \mathrm{d}r \qquad (5)$$

32.2.3 Scattering from Isotope Mixtures

For a chemically and isotopically homogeneous material, the only fluctuations in scattering power across the sample are due to spin orientations (see above); the result is a scattering signal essentially the same as for X-rays, with an incoherent background. A random mixture of isotopes (*e.g.* as a result of natural isotope abundance) would simply add to the spin incoherent scattering. Non-random mixing of isotopes, such that the isotope is the same within each molecule, is a key requirement for the application of NS to polymer science, since intra- and inter-molecular interference contributions to the scattering can then be separated.

Consider a macroscopic assembly of identical molecules which are either fully hydrogenous (Isotope 1) or fully deuterated (Isotope 2). Quite generally

$$A(\boldsymbol{q}) = \sum_i \sum_j b_i A_i A_j'$$

where
$$A_i = \exp(i\boldsymbol{R}_i \cdot \boldsymbol{q}) \quad \text{and} \quad A_j' = \exp(i\boldsymbol{r}_j \cdot \boldsymbol{q})$$

where $\boldsymbol{R}_i$ is the position of the centre of gravity of molecule i and $\boldsymbol{r}_j$ is the displacement of atom j within that molecule. The hydrogen scattering length b_i of the ith atom can be written as $\bar{b} + \delta b_i$ where $\bar{b}$ is the average over the two isotopes. Hence

$$A(\boldsymbol{q}) = \sum_i \sum_j (\bar{b} + \delta b_i) A_i A_j' + \sum_i \sum_l b_l A_i A_l''$$

where b_l represents the scattering lengths of atoms other than hydrogen and A_l'' is the corresponding scattering amplitude.

$$A(q) = \left[\sum_i \sum_j \bar{b} A_i A_j' + \sum_i \sum_l b_l A_i A_l'' \right] + \sum_i \sum_j A_i A_j' \delta b_i \qquad (6)$$

The measured intensity is obtained from the modulus square of $A(q)$, and the crux of the calculation is to identify those contributions which, over a sufficiently large number of molecules, will average to zero. The modulus square of the sum of the first and second terms in equation (6) (in square brackets) is closely related to the scattering expected from density fluctuations in a pure material. If $|A_0(q, b_k)|^2$ is the scattering for pure isotope k, then the density fluctuation term will be $|A_0(q, \bar{b})|^2$. Hence

$$A(q) = A_0(q, \bar{b}) + \sum_i \sum_j A_i A_j' \delta b_i \qquad (7)$$

A_0 is zero at small q for an incompressible homogeneous fluid in equilibrium. This can often simplify the equations for samples with very small equilibrium compressibilities. In taking the modulus squared of $A(q)$ (equation 7) the cross term averages to zero over the whole assembly if the isotopic mixing is random. The randomness must fulfil two conditions: (a) the value of $A_i A_i'$ is not correlated with whether or not molecule i and molecule i' are of the same or different isotopes; (b) there is no correlation between isotope fluctuations (in the second term of equation 6) and density fluctuations (in the first term, in square brackets).

The modulus square of the second term in equation (7) then represents the quantity that is of special interest in these experiments — the scattering function for one molecule

$$n^2 P(q) \equiv \left| \sum_j A_j' \right|^2$$

where n is the number of hydrogens per molecule. The function $P(q)$ is normalized so as to go to unity at $q = 0$. Just as in the basic equation of Debye (equation 4) there will be a series of cross terms of the type $A_i A_i^*$, $A_j' A_j'$, $\delta b_i \delta b_{i'}$, where A_i^* is a complex conjugate. In the ensemble average over all the sample, there will be many pairs of molecules which contribute the same factor $A_i A_j^*$, $A_j' A_{j'}'$. As long as the mixing is random (condition (a) above) the sum of all such terms with $i \neq i'$ will be zero, since δb_i is defined so as to sum to zero. Hence equation (6) becomes

$$|A(q)|^2 = |A_0(q, \bar{b})|^2 + \sum_k \delta b_k^2 (n^2 P(q)) N_k \qquad (8)$$

where the sum over k extends over the different isotopes present (usually 2) and N_k is the number of molecules present of isotope k.

For several years after the initial use of the technique for polymers, experiments were restricted to systems in which one isotopic species was dilute ($\simeq 1$–3% deuterated ^{2}H) compared to the other ($\simeq 99$–97% hydrogenous ^{1}H). In that case $\delta b_H = 0$, $\delta b_D \simeq b_D - b_H$ and $P(q)$ refers only to the ^{2}H species. In addition $|A_0(q, b_H)|^2$ is usually very small indeed at small q, so that equation (8) becomes

$$|A(q)|^2 = N_D n^2 P(q) (b_D - b_H)^2 \qquad (9)$$

However it has been recognized recently that equations equivalent to equation (8) enable any concentrations to be used,[17-25] the result remaining simple as long as the two isotopic species are nearly equivalent. For example, for two isotopes

$$|A(q)|^2 = |A_0(q, \bar{b})|^2 + n^2 P(q) \{ N_H (b_H - \bar{b})^2 + N_D (b_D - \bar{b})^2 \}$$
$$= |A_0(q, \bar{b})|^2 + (N_D + N_H) n^2 P(q) (\overline{b^2} - \bar{b}^2) \qquad (10)$$

This result resembles closely the result of calculating the effect of spin incoherence, which differs in form from equations (8)–(10) only in the term containing $P(q)$. For spin incoherence there are no remaining systematic interference terms for different nuclei, and equation (10) applies with $P(q) = 1$ and $n = 1$. ($P(q)$ is in this case a 'form factor' for an individual nucleus, which, for elastic scattering, is independent of angle.) In effect, the use of isotope mixtures introduces a special sort of incoherence which removes the effect of interference between molecules and makes accessible the molecular scattering factor.

Equation (8) can be rewritten to allow for different conventions for the normalization of the intensities. In the derivation of equation (9) the 'assembly' considered is a macroscopic sample, so that the result gives $V d\Sigma/d\Omega$. The term $(b^2 - \bar{b}^2)$ can be written as $c_D c_H (b_D - b_H)^2$, where the c are the volume fractions of ^2H and ^1H. Hence

$$V d\Sigma/d\Omega = |A(q)|^2 = |A_0(q, \bar{b})|^2 + N n P(q) c_D c_H (b_D - b_H)^2 \qquad (11)$$

where N is now the number of hydrogens in the sample. The quantity $nP(q)$ is a normalized intensity $I(q)$ which does not require a knowledge of n (n is often one of the less precisely known quantities in an experiment).

Equations (8)–(11) are equivalent to equations (24) and (33) of Koberstein,[23] which are more general in that two chemical species are involved. $|A_0(q, b)|^2$ is equivalent to $R_{A\langle B\rangle}(q)$ of Koberstein. Most of the term $R_{B,L}(q)$ of Koberstein comprises incoherent scattering, which is assumed to have been subtracted in the cases of equations (8)–(11) here. Although the use of $c_D \simeq c_H$ has been seen as a technical advance over the case where one of them is small, there are several restrictions which it imposes on experiments. It is important[24,25] that the ^1H and ^2H species have very similar molecular weights (which is not always straightforward and invariably very expensive). More fundamentally, these experiments require the chemical differences due to isotope to be small (see Section 32.7 on mixtures). In the case of a mixture in which the ^2H and ^1H species are sufficiently different that there is phase segregation, it could be quite difficult to detect this effect if the size of the domains of different concentration were large (see also Section 32.7). In effect two 'samples' of different c_D would be simultaneously present. Conversely, phase segregation would not necessarily prevent $P(q)$ from being measured at least in some range of q.

32.2.4 Analysis of the Molecular Scattering Factor

Given that $I(q)$ can be measured, how can structures be derived from it? In principle a Fourier transform can yield a correlation function $\gamma(r)$, which for a homopolymer can be interpreted as a histogram of the probabilities of any given separation r within the molecule. In practice this route is not often taken, but it is worth bearing in mind $\gamma(r)$, since it represents the maximum information in real space which is in principle available from the scattering. When it comes to judging the uniqueness of an interpretation it is sometimes possible to exclude some models because they would have a correlation function incompatible with the measurements. On the other hand, some models can coincidentally have remarkably similar scattering. For example, a Gaussian coil and a thin sheet both have scattering patterns of the form (constant) $\times q^{-2}$ over certain ranges of q values. Such cases as this do not represent a serious problem since a wealth of other techniques are usually available to distinguish such different models.

32.2.4.1 Guinier range of q

In the limit of the scattering as $q \to 0$ a Taylor's series expansion of the $\sin(qr_{ij})$ term of equation (4) in powers of q is possible, the usual versions being due to Guinier and Zimm respectively

$$I(q) = I(0)\exp(-R_g^2 q^2/3) \qquad (12)$$

$$I(0)/I(q) = 1 + R_g^2 q^2/3 \qquad (13)$$

where R_g^2 is the mean square radius of gyration $= \langle x^2 \rangle + \langle y^2 \rangle + \langle z^2 \rangle$ for a homopolymer, where x, y and z are coordinates with respect to the 'centre of gravity' of the molecule, and $I(q)$ replaces $\langle |A(q)|^2 \rangle$.

In the limit as $q \to 0$ these are equivalent, but in order to obtain a practicable measurement of a change in $I(q)$ it is necessary to use q values up to about R_g^{-1}, in which case the exactness of fit and the values of R_g obtained depends on which is used. For example, if intensities are calculated from equation (12) and plotted according to equation (13) a least squares fit gives a 10% difference between the R_g value used initially and derived from the plot. In effect, equations (12) and (13) are slightly model dependent, the former being a better approximation for near spherical objects (*e.g.* water soluble proteins) and the latter for Gaussian coils (for which it was originally derived). In the latter case the R_g value corresponds to a z average over the distribution of molecular sizes. Clearly, if q values larger than R_g^{-1} are used, the R_g values derived are even more model dependent. This may not however be a serious problem, since one is often more interested in trends in R_g values than in the absolute values. In some cases R_g values are obtained by fitting the scattering from a model, *e.g.* equation (19) below.

A necessary condition for the validity of plotting according to equations (12) or (13) is that there should be no disagreement between the determined value of I(0) and the value calculated from known molecular weights. If there is a disagreement this implies non-random mixing of isotopes (see below) or the existence of an additional source of intensity.

32.2.4.2 *Intermediate range of q*

As the q values are increased beyond the so-called 'Guinier range' all the interpretations have been made by calculating intensities from models and comparing with measurements. There is no well-accepted nomenclature for q ranges; 'intermediate' refers to the q range being between the Guinier range and the wide angle range associated with typical monomer–monomer distances. The intermediate range can itself be subdivided: (a) intermediate q up to about 0.1 Å^{-1} (in this range the atomic detail within the monomer unit need not be considered); (b) intermediate q with $0.1 < q < 1 \text{ Å}^{-1}$.

Several methods of calculation of I(q) can be distinguished.

(i) The simplest makes use of analytic expressions for scattering from thin sheets

$$I(q) = 2\pi n_A C_1(q)/q^2 \tag{14}$$

where n_A is the number of hydrogens per sheet, or from thin rods

$$I(q) = \pi n_L C_2(q)/q \tag{15}$$

where n_L is the number of hydrogens per length of rod. For these to be used it is not necessary for the whole molecule to be a sheet or rod; a molecule may consist of a sequence of such units joined in such a way that the interference between them is negligible over certain ranges of q. The functions C(q) are unity if the thickness of the rods or sheets is very small compared with $1/q$. The use of equations (14) and (15) can be extended to larger q by calculating the ratio C(q) of intensities for the actual structure expected (*e.g.* an extended straight sequence of chain) to that of the infinitely thin structure. If R_0 is the root mean square of the distance of hydrogens from the axis of the molecule, then for $q < R_0^{-1}$ and $q \gg$ (rod length)$^{-1}$

$$C_2(q) = \exp(-R_0^2 q^2) \tag{16}$$

similarly for the analogous case of sheets

$$C_1(q) = \exp(-D_0^2 q^2) \tag{17}$$

where D_0 is the root mean square distance of hydrogens from the sheet centre.

(ii) Debye performed an analytic calculation[26] for the scattering from a coil of N units for which the probability distribution of z links apart along the chain, being separated by a distance r, is Gaussian

$$w(r)dr = 4\pi(3/2\pi R_g^2)^{3/2} \exp(-3r^2/2R_g^2)r^2 dr \tag{18}$$

where

$$R_g^2 = za^2(1 + p)/(1 - p)$$

p is the cosine of the angle between two bonds and a is the bond length. Equation (4) is then used, the double summation extending over all pairs of units along the chain. The sum can be expressed as a geometric series, giving

$$|A(q)|^2(\text{coil}) = Nb_u^2 + (N - 1)^2 b_u^2 2\{e^{-x} - (1 - x)\}/x^2 \tag{19}$$

where $x = q^2 R_g^2$. R_g corresponds to the value for $z = N - 1$ and b_u is the scattering length for one unit. The term Nb_u^2 is normally neglected.

For solutions of polymers, excluded volume effects can make significant modifications to the Gaussian statistics (see Volume 2, Chapter 3) but for amorphous solid polymers excluded volume effects are cancelled by those of interchain packing[27] and the Debye calculation for a so-called 'phantom chain' is expected to be valid. Equation (19) simplifies to (12) in the limit of small q. For a range of q larger than R_g^{-1} the scattering is proportional to q^{-2}. For $q \simeq a^{-1}$ the details of the monomer units become significant and numerical methods can be used. In some cases 'needle scattering'[7] corresponding to equation (16) can be expected, corresponding to a distance scale where the chain can be considered as a straight rod.

(iii) The most detailed calculations, though not necessarily the most revealing, use 'Monte Carlo' methods based on equations (4) or (5). Computer programs are used to generate sets of x, y and z coordinates for each scattering unit (*e.g.* monomer). The particular advantage these methods have over analytical calculations is the ability to cope with complex models. Random number generators are used in conjunction with rules prescribed by the model. Of course, a model complete with all the spatial coordinates is necessary, and usually an average is made over a number of computer-generated structures. It is usual, in order to reduce the calculation to manageable proportions, to choose a group of monomers as the individual scattering unit, and for any given structure to select randomly only a proportion of the units in order to calculate $\gamma(r)$.

(iv) For polymers of high crystallinity, calculations can be made using the basic unit, the straight cylinders formed from the 'stem' sequences which traverse the crystals. These calculations can be further simplified by noting that within one crystallite the structure is a convolution of a single stem with a two dimensional structure describing the position of the stems. The normalized intensity ($q \gg$ (rod length)$^{-1}$) for a pair of stems separated by a distance R is

$$I(q) = (\pi n_L/(2q))(2 + J_0(q, R))C_2(q) \tag{20}$$

where $J_0 (q, R)$ is a zeroth order Bessel function. For a group of n' stems

$$I(q) = (\pi n_L/(qn'))\left\{n' + \sum_{\substack{i \ j \\ i \neq j}}^{n' \ n'} J_0(q, R_{ij})\right\}C_2(q) \tag{21}$$

The advantages of this third method are that it can cope with structures involving disorder, is fairly rapid, and involves only those physical features which have a large influence on the scattering (notably the values of the distances R_{ij}). Note that terms with n' are not neglected compared with those with n'^2 (*cf.* equation 19).

32.2.4.3 *Analysis of anisotropic scattering (Guinier region)*

In the Guinier regime of q it is not R_g which now gives the intensity (equations 12 and 13, which are derived from equation 4) but a dimension related to the projection. For example, for a homopolymer

$$R_z^2 \equiv \langle z^2 \rangle$$

where z is measured from the centre of gravity of the molecule, and is usually associated with the direction along a fibre axis (*e.g.* direction of stretch).

For fibre symmetry it is not possible to measure R_x separately from R_y. It is convenient to quote a combined dimension such as

$$R_{xy}^2 \equiv (\langle x^2 \rangle + \langle y^2 \rangle)/2$$

which would be equal to R_x if $R_x = R_y$. Equation (13) then has its counterparts in equations such as

$$I(0, 0)/I(0, q_z) = 1 + (q_z^2 R_z^2)$$
$$I(0, 0)/I(0, q_x) = 1 + (q_x^2 R_{xy}^2) \tag{22}$$

Firstly the situation when all three R values can correspond to the Guinier regime is considered. It has generally been realized that for fibre symmetry two R values are sufficient to define the scattering.[28,29] If, experimentally, values of a dimension d are obtained as a function of the angle α between the measurement direction and the symmetry axis then

$$d^2 = \langle a \rangle + \langle b \rangle \cos^2\alpha \tag{23}$$

Thus a plot of d^2 against $\cos^2 \alpha$ will give a straight line from which $\langle a \rangle$ and $\langle b \rangle$ may be determined. From there, the molecular dimensions can be calculated using

$$R_z^2 = \langle a \rangle + 2\langle b \rangle \quad \text{and} \quad R_{xy}^2 = \langle a \rangle \tag{24}$$

This method is readily extended to the case where the molecules are contained in units whose orientation is not perfect,[28, 30] *e.g.* the orientation of crystallites can be characterized independently and R values calculated which are corrected for the lack of perfect orientation.

If the molecules are highly extended (*e.g.* during deformation) they may be long enough to pose technical problems for small angle spectrometers, but this is a less severe restriction than is often realized, since the Guinier formulation implies that the xy and z dependences are separable. Equation (12) can be rewritten as

$$I(q_x, q_z) = I(0)\exp(-R_z^2 q_z^2)\exp(-R_{xy}^2 q_x^2) \tag{25}$$

where xy and z refer to the equatorial and meridional directions respectively. It is customary to measure R_z by varying q_z and keeping $q_x = 0$. It is clear however, that q_x could be kept at any other constant value, since the q_z dependence of $I(q_x, q_z)$ is the same as that of $I(0, q_z)$. With two-dimensional position-sensitive detectors $I(q_x, q_z)$ can be measured down to $q_z = 0$. The q_z dependence of $I(q_x, q_z)$ is extremely high for high R_z values, and may be 'smeared' by the combined effects of finite beam collimation and detector resolution. This has the effect of limiting the maximum R_z values which can be attained, and also of decreasing the maximum $I(q_x, q_z)$ value which is measured on the detector. This latter effect can produce some misleading results if comparisons are made with the expected 'forward scattering' $I(0, 0)$.

32.2.5 Analysis of Density Fluctuations

In contrast to Section 32.2.4, this feature of NS is directly analogous to X-ray scattering of the condensed phase (*e.g.* arising from crystal/amorphous density difference in crystalline polymers). Only rarely can units be identified which are widely enough spaced and randomly enough arranged for formulae designed for solution scattering to be applied (*e.g.* equations 12–17). If there are lamellae, for example, there is usually a 'lattice factor' (*cf.* crystals) which modulates the scattering from an individual unit and which arises from systematic interference between units.

It is very common to represent the scattering of this type as coming from a two phase structure. A special case of this arises when the boundaries between the phases are randomly enough arranged in relation to the q range used. Debye *et al.*[31] considered the case where the correlation function $\gamma(r)$ is of the type

$$\gamma(r) = e^{-r/a} \tag{26}$$

where a is a characteristic distance. In this case the scattering is of the form

$$\frac{d\Sigma}{d\Omega} = \frac{8\pi a^3 \langle \Delta\rho^2 \rangle V}{(1 + q^2 a^2)^2} \tag{27}$$

In the limit of large q this is equivalent to Porod's law,[32] which can be considered as due to isolated boundaries

$$\frac{d\Sigma}{d\Omega} = \frac{8\pi \langle \Delta\rho \rangle^2 V}{aq^4} \tag{28}$$

A more general case of continuously varying density was treated by Ornstein and Zernicke[33] for light scattering of critical opalescence. By a rather indirect route[34] they argued that

$$\gamma(r) = \text{constant} \times e^{-r/\xi}/r \tag{29}$$

where ξ is a characteristic length, which leads to

$$\frac{d\Sigma}{d\Omega} = \frac{\text{constant}}{1 + q^2 \xi^2} \tag{30}$$

Equations (26)–(30) are likely to be relevant to systems which can phase segregate (see Sections 32.6 and 32.7).

32.3 EXPERIMENTAL CONSIDERATIONS

The requirements are essentially those of any scattering experiment, and concern the generation of a beam of a suitable wavelength distribution, collimation system and a detector with a suitable angular resolution.

32.3.1 Instruments

A brief description will be given of the D11 diffractometer at the High Flux Reactor (Institut Laue Langevin, Grenoble) as an example of such instruments.[35] The initial intensity distribution with wavelength is determined by the reactor and the cold source, and has a maximum in the region 5–10 Å. A relatively generous 'slice' of this distribution is taken by a set of rotating helical slots which select neutrons according to velocity. (Normally $\Delta\lambda/\lambda$ is 8%.) The maximum specimen diameter is about 10 mm, and the divergence at the samples can be varied. An area detector contains boron in gaseous form (BF_3) so that the neutrons produce α particles whose charge makes them readily detectable. Its efficiency is approximately 50%, and varies slightly over the 64×64 array of elements, each of which is 10 mm square. The specimen to detector distance can be chosen in the range 1–40 m, and the beam divergence is normally adjusted to give a beam width at the detector position of approximately 3.5 cm (full width, half height), though sometimes less.

Pulsed neutron sources are now available.[36] No attempt is made to monochromate the beam and a wide range of wavelengths are used. The scattering patterns (intensity *vs.* angle) vary in time after the pulse and before the arrival of the next. For short times the patterns correspond to fast neutrons (small wavelength), for longer times to slow neutrons (larger wavelengths). A composite intensity *vs.* q pattern $I(q)$ must then be constructed out of the set of patterns of intensity *vs.* angle.

32.3.2 Data Reduction

The geometry of instruments such as the D11 apparatus conforms reasonably well to 'point collimation'. Methods of correcting for collimation distortions and wavelengh distributions exist (*e.g.* ref. 37); this is usually most important when sharp maxima or minima are being studied (*e.g.* Section 32.6) or for very high R_z values (see Section 32.2.4.3). In many diffractometers with one-dimensional detecting systems, the geometry corresponds to 'slit collimation', so that at least some collimation distortion is to be expected as angle is reduced.

A multidetector system requires a correction for any variations in detector efficiency which there may be. This is commonly done by using a spectrum from H_2O, which is almost entirely incoherent and hence (at small q) independent of angle.

For multidetector systems, the intensities over the counter must be averaged in some way so as to be readily analyzed.[38] For isotropic samples it is normal to average over all those cells whose centres lie on an annulus centred on the beam position. For anisotropic samples a variety of geometries can be used, for example, annuli can be divided into annular sections, or strips can be taken across the detector. In any case it is desirable to use methods which exploit the measured intensities all over the detectors which are available and not just over strips along principal directions.

32.3.3 Subtractions

For isotope labelling of molecules a measurement of $I(q)$ requires the recording of count rate *vs.* angle for: (a) the isotope mixture; (b) equivalent polymer samples of one (or both) pure isotopes; (c) the empty specimen container; (d) the container with a neutron-absorbing piece of cadmium in place of the specimen; (e) an isotropic scatterer (usually H_2O); and (f) the container only for (e). The methods of algebraic manipulation in order to extract $I(q)$ are described in detail elsewhere.[38] The parasitic intensities from the diffractometer (c and d) must be subtracted, with due allowance for that intensity contribution which is attenuated when the sample is in place. It is necessary to subtract incoherent scattering, and sometimes a coherent contribution $|A_0(q, \bar{b})|^2$ (equation 8). Subtraction of any coherent background contribution which is present is usually only a small correction. At very small q, with high ^1H it is usual to assume

$$|A_0(\boldsymbol{q}, \bar{b})|^2 \;\; \simeq \;\; |A_0(\boldsymbol{q}, b_H)|^2 \tag{31}$$

As a general rule, reliable subtractions at large q depend on taking measurements well beyond the q range used to calculate differences. It can then be verified that the 'background' contributions cancel in the way that is expected when the molecular scattering factor has decreased to a negligible degree.

For the case of density fluctuations the subtractions must allow for parasitic signals from the instrument and from incoherent scattering, and possibly a coherent diffuse signal from short range thermal motion.

32.3.4 Calibration of Intensities

A relatively simple procedure[9, 39, 40] can calibrate the macroscopic cross section $d\Sigma/d\Omega$. This is based on the true absorption being very small. Hence all the neutrons that are removed from the incident beam must be scattered. The integrated cross section is then known to be

$$\int (d\Sigma/d\Omega)d\Omega \;=\; a(1 \;-\; t)/V \tag{32}$$

where a is the sample area and t is the measured transmission. If the scattering is isotropic then its measurement at any angle must give the integrated cross section divided by 4π. Once the (differential) cross section for one specimen is calibrated in this way, cross sections for others are readily obtained by comparison. In this way the beam flux and various geometrical factors do not need to be measured directly in order to calibrate the intensity. The precision of this procedure should be only a few percent. The accuracy is not this good, however, since even for H_2O the scattering is not truly isotropic, for example because of inelastic effects which increase as wavelength decreases, and multiple scattering.[40]

32.4 AMORPHOUS POLYMERS[41]

32.4.1 Background

As soon as the chemical nature of polymers was clear, as long chain molecules which are often flexible, it was pointed out that the liquid or solvated state should correspond to the random coil.[42] The conformations could then be interpreted in terms of random walk statistics (at least over large distances), with Gaussian probabilities as a consequence. R_g is then predicted to be proportional to (molecular length)$^{1/2}$.

It was always clear that the effects of lack of complete flexibility would be significant, at least over short distances, and Kuhn introduced the length which is generally larger than the monomer unit and which, when used in random walk theory, gives the experimental coil size. Equation (18) is also intended to allow for such chemical detail. The rotational isomeric state model allows for the chemical detail in a quantitative manner,[43] by generating probability distributions from the details of the chemical bonds as characterized independently. These calculations do not require the chain conformation to be Gaussian on any distance scale, and conformations of chains which are not particularly flexible can be treated.

For solutions, which for many years were the only systems for which coil sizes could be measured experimentally, the random walk statistics are significantly modified by the finite volume occupied by the chain. Since coil segments cannot overfill space, and the coil is expanded beyond the phantom coil size by the excluded volume effects, intuition might suggest that in the amorphous solid state excluded volume would be even more important. Flory[27] predicted that in fact the interchain excluded volume effects exactly *cancel* the intrachain ones, so that the conformation is that of the phantom chain. It is easy to see that the former should indeed contract the coil, since a random coil occupies space most densely in the region around its centre of gravity, and these regions will therefore tend to repel other molecules. This argument shows in addition that the centres of gravity in the amorphous state are *not* randomly arranged in space, but tend to avoid each other (the 'correlation hole').[44]

In terms of polymer solutions, the random flight (phantom chain) conformation corresponds to a θ solvent which has an unfavourable polymer/solvent energy of contact. The polymer–polymer contacts are preferred to that degree necessary that the intrachain (self) excluded volume effects are cancelled.[27] Such a condition applies usually for only one temperature, known as the θ temperature, for any given combination of polymer and solvent (see Volume 1, Chapter 6). Flory's prediction was tested by comparing radii of gyration obtained by NS from amorphous polymers and from polymers in solution under θ conditions.

It has often been thought that the local order involved in filling space with rods, even ones which are flexible, is inconsistent in some way with the existence of random coils. There are indeed local packing effects[45] (a correlation in molecular orientation over distances of several monomers) and one must assume that for any given chain the coil is not random in the sense that the existence of a particular neighbouring molecule will bias the coil one way or another. It seems clear however, from the NS results (see below), that there is no systematic bias, other than that of excluded volume discussed above.

The meander models[46] for the amorphous state, by contrast, were based on the premise that local packing effects are dominant. Molecules were assumed to group together into bundles. Any two chain segments which are adjacent would tend to be parallel over long lengths of the chain, and the flexibility of the molecule would be far less than for one chain considered in isolation. Electron microscopy of amorphous polymers was quoted in favour of more order than would be expected from the random coil, though this has always been a controversial subject.

32.4.2   Results and Interpretation

The first measurements on amorphous poly(methyl methacrylate) (PMMA)[50] and atactic polystyrene (PS)[47-49, 51] below the glass transition temperature showed that the radii of gyration do indeed correspond within experimental error to those of the θ condition, and that R_g is proportional to $\bar{M}_w$ (Figure 1). The same result has been reported for several other liquid or glassy polymers: poly(dimethyl siloxane) (PDMS);[52] liquid polyethylene (PE);[53] quenched isotactic polystyrene (iPS),[54] polycarbonate (PC)[55] and poly(ethylene terephthlate) (PET);[56] and liquid poly(ethylene oxide) (PEO).[57]

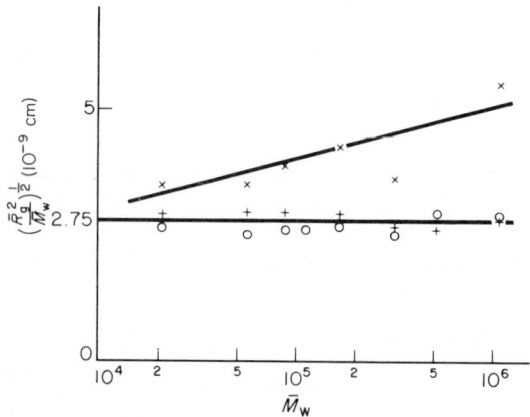

Figure 1   Radii of gyration for atactic PS: ($\bigcirc$) in the amorphous state; ($+$) θ solvent; and ($\times$) in a good solvent (carbon disulfide) (reproduced from ref. 51 by permission of the American Chemical Society)

Convincing evidence can also be obtained from data at intermediate q values. Intensities in agreement with the Debye result (equation 19) were observed for the range of the polymers listed above (see *e.g.* Figure 2). For PMMA, the bonds are not sufficiently flexible for Gaussian statistics to hold. Calculated results from the rotational isomeric state model, which needs to be applied in such cases, are in agreement with the results (see Figure 3).[58]

For most polymer scientists, any lingering doubts concerning the usefulness of the random coil idea have been dispelled by the comprehensive agreement between a substantial body of NS results and predictions based on the mainstream tradition of polymer science. It is interesting however, even if only from the point of view of a historian of scientific developments, that the meander model has also been reported to be consistent with the NS results.[59] No models can be 'proved' on the basis of scattering results only because of the loss of information inherent in measuring intensities rather than amplitudes.

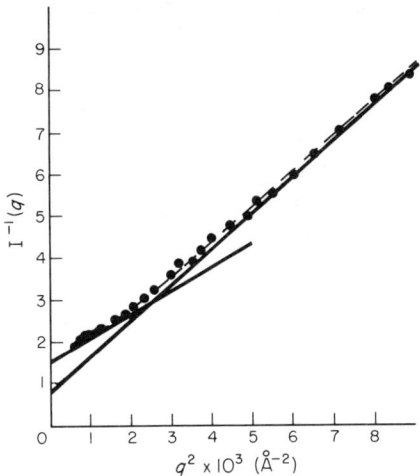

Figure 2 Comparison of intensity data for atactic PS at intermediate q values with the Debye function for a featureless random coil (equation 19): the ordinate scale is in arbitrary units, and solid lines show approximations for large and small q (reproduced from ref. 51 by permission of the American Chemical Society)

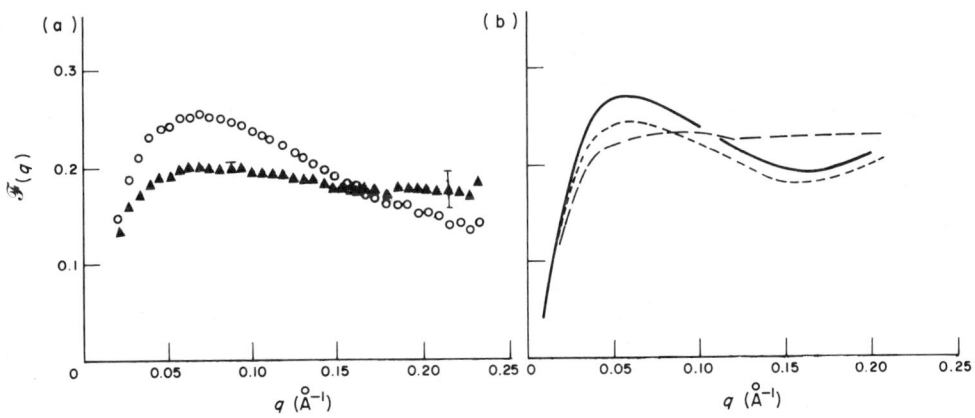

Figure 3 Comparison of intensity data at intermediate q values for PMMA. The ordinate is the Kratky function, proportional to $I(q)q^2$: (a) (○) syndiotactic and (▲) isotactic polymer; (b) calculations in the basis of the rotational isomeric state model[43] for (——) isotactic, (– – –) atactic and (— — —) syndiotactic polymer (reproduced from ref. 58 by permission of the American Chemical Society)

The results on chain conformations in liquid polymers have been seen as largely confirmatory. (It seems clear that in glassy polymers the conformations are the same as in the preexisting liquid.) The same is not true concerning the compatability of isotopic species, which will be discussed in the section on mixtures.

Experiments to test local structure on the scale of one monomer unit are carried out by means of partial labelling.[60] Intriguing results have been obtained on chains which are partially deuterated along their length. These originated in experiments on star molecules with labelled cross-links. A peak resembling a Bragg peak was found, which was also present when there were only two branches to the star[61] (*i.e.* for a linear molecule). De Gennes explained this result by noting that for a sample in which all molecules are identical, the intensity must go to zero as $q \to 0$ (because the scattering density is uniform in the limit $r \to \infty$). Since it also goes to zero at large q, a peak is inevitable. A quantitative analysis was given[44] in terms of the 'correlation hole' (see above). The effects of polydispersity can be important.[62]

NS results are seen as supporting the idea of liquid and amorphous polymers being essentially homogeneous and random in the sense explained above. Local order can however occur in the sense that some polymers turn out to have a residual crystallinity. NS on poly(vinyl chloride) (PVC) swollen with liquid have shown a scattering signal consistent with crystalline entities (perhaps lamellae).[63]

32.5 CRYSTALLINE POLYMERS[64]

32.5.1 Background

Polymer crystals are not in equilibrium. We can therefore anticipate a variety of conformations, dependent on the route by which any particular metastable state is reached. This variety has been strikingly confirmed by NS experiments.

The fact that crystals of polymers form thin platelets, which could be a monolayer, led immediately to the recognition that the chain conformation in the crystals must be folded.[65] The lack of any definitive technique to measure conformations in the bulk state gave rise to a longstanding controversy. On the one hand, the electron microscopy[66] on solution grown crystals showed unequivocally that there was *anisotropy* in the folding, whereby the fold direction is preferentially parallel to the growth face where new molecules attach to the crystals. On the other hand, even the most perfect crystal, as observed by microscopy, showed about 20% non-crystalline component as judged by the usual tests (density, heats of fusion, *etc.*). This evidence and thermodynamic arguments were invoked[67] in favour of a random type of folding which created a very disordered (amorphous-like) layer on the crystal surfaces.

The questions concerning this 'fold surface problem' can be resolved into those of: (i) chain topology (*e.g.* does the chain fold back adjacent to itself); and (ii) the detailed nature of the chain packing in the fold surface region (in particular, how liquid-like it is).

Labelling techniques, in particular using NS but also IR spectroscopy, are powerful methods for topology (the 'trajectory' of the chains) and that is what concerns most the results of this section. The second class of question still remains open to some extent, mainly because it is not very well defined (*e.g.* what exactly is 'amorphous' structure). It seems increasingly unlikely that a constrained fold surface layer only 20 Å or so thick can be genuinely liquid-like;[68] even though it may be very similar in terms of most bulk properties (*e.g.* density).[69] It has been concluded that folds are not all in the same vertical register on the basis of electron microscopy and degradation techniques (*e.g.* ref. 64 and the references therein).

Recently, the very origin of the chain-folding phenomena, considered for over 20 years to be 'nucleation', has been reassessed.[70, 71] It is likely as a result that the interpretation of the various types of folding discussed below will also have to be reassessed.

Experiments have been limited in many ways because of segregation between isotopic species. This can be particularly severe in the case of slow crystallization of polyethylene (PE),[72, 73] which has been by far the most widely studied and understood polymer for structural studies on crystals. In retrospect, this segregation can be seen as part of the more general isotopic incompatibility discussed in Section 32.7 on mixtures. The segregation signal seems to be far larger than for liquid systems. This may be because the lamellar morphology enables phases of different isotope concentration to be interleaved on a rather small scale, and also because the segregation signal from a two-phase structure increases with surface to volume ratio (equation 28). In some cases results were obtained on molten isotopic blends of PE,[53, 72] and no evidence was then found that segregation was significant prior to crystallization, at least for the molecular weights studied (but see Section 32.7). The following results refer to PE which has been crystallized sufficiently fast for segregation which occurs during crystallization to be minimized, or to polymers where the segregation effect is small.

32.5.2 Melt-quenched Crystals

R_g values have been measured as a function of molecular weight for quenched melts of PE,[73, 74] polypropylene (PP),[74-76] iPS,[77-79] hydrogenated polybutadiene (chemically similar to PE)[80] and PEO.[81] R_g is generally proportional to $\bar{M}_w^{1/2}$, where $\bar{M}_w$ is the weight average molecular weight, except for low molecular weights where R_g can be higher than expected on this basis (see Figure 4 for the case of PE and PP). In the range of $M_w^{1/2}$ where the R_g dependence of melt-quenched PE is linear, measurements on melts show the same dependence of R_g on $\bar{M}_w$. The departures from the $\bar{M}_w^{1/2}$ behaviour have been associated with the stem length.[74-76]

The conclusion seems clear enough; on a large enough scale, at least, the coil is simply 'frozen in' from the melt to the melt-quenched crystals. This may of course involve the molecule passing from one lamella to another. This was the starting point for analysis of data at intermediate q.[82-86]

A freezing in or solidification model was proposed[85, 86] which envisaged the 'freezing in' applying down to short distance scales (a few lattice spacings). This idea was supported by the fact that the scattering at intermediate q is also similar before and after melt quenching. This suggestion also fitted into the preexisting idea that the fold reentry was very random indeed.[67] Figures 5(a) and

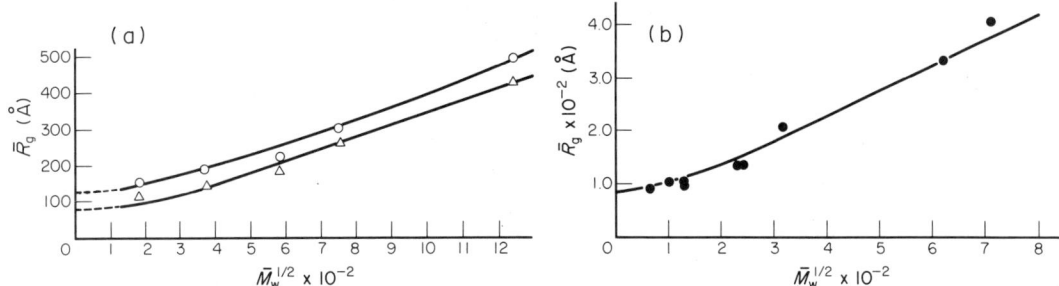

Figure 4 (a) The radius of gyration of polypropylene in the semicrystalline state produced by quenching ($\triangle$), and slow cooling ($\bigcirc$), as a function of the square root of molecular weight $\bar{M}_w$; (b) the radius of gyration of polyethylene in the semicrystallized state produced by quenching (reproduced from ref. 74 by permission of Butterworths and Co. (Publishers) Ltd.)

(b) illustrate an analytical calculation and its result which puts this idea into quantitative form; the three-dimensional Gaussian probability corresponding to a coil is projected on to a plane (a lamella) and the stem positions so defined. The result shows a dramatic disagreement with the results for PE, which are also shown in Figure 5(b). Another way of expressing this result is as follows; $\gamma(r)$ (and hence $|A(q)|^2$) are similar before and after quenching and the chains straighten locally. In order to satisfy both these conditions, the stems must on average be nearer to each other on a local scale than the freezing in model supposes, in spite of the size of the molecule remaining the same. We must therefore have subunits[87] or clusters.[88] Within each cluster the stems form into a short but fairly straight row within which the stems on average are only about 1–3 lattice spacings apart (*i.e.* folds are adjacent or 'near adjacent'). Between clusters, the folds must be very much longer (*e.g.* interlamellar ties).[88] The physical process which favours near adjacency is almost certainly the need not to overfill space in the fold surface.[89-92] For this purpose near adjacency involving stem separations of one to three nearest neighbour distances (as shown in Figure 5c) may suffice whereas the original arguments concerning the filling of space were in terms of strict adjacency. A folding pattern of 70% near adjacent and 30% random reentry (involving fully amorphous folds) was anticipated in the Monte Carlo analysis of Yoon and Flory,[93] though their model has often been interpreted in terms of 100% random folding.

32.5.3 Other Melt Grown Crystals

For PP,[75,76] iPS,[77-79] hydrogenated polybutadiene[80] and PEO segregation can be minimal. It is clear that in these cases R_g is not always the same as in the melt. R_g for PET is slightly greater after crystallization than before.[94] The inference is that for melt quenching the molecules are indeed 'trapped' for kinetic reasons, and that, given enough time, R_g can change on crystallization.

By contrast, the evidence from scattering at intermediate q is consistent with the local stem arrangements being similar for slowly and quench crystallized samples. (For PE[95] the melt was cross-linked at about one link per molecule so as to minimize segregation.) It was a common presumption at the time of the Faraday Discussion in 1979 that there is some driving force (presumably thermodynamic) in favour of adjacent folding, and that the slower the crystallization the more perfect and the more adjacently folded the crystals would be. The comparison by NS of the results of fast and slow crystallization does not support this interpretation. Nor is it likely that kinetic factors always 'trap' the molecule into randomly folded conformations.[97] This is because of more recent experiments and theories concerning growth rates. Growth rates very much higher than the ones on record in 1979 have now been observed,[98] which means that the values discussed in 1979 in connection with NS were in reality fairly modest. Secondly, it was assumed in one of the 1979 calculations[97] that only one site on an entire crystal face (of $\sim 10^4$ sites) was being added to at any instant. This would correspond to an extreme case of nucleation, whereas it now appears much more likely[70,71] that addition to most of the sites on a surface is being made simultaneously. As a consequence there is now thought to be about 10^4 more time for rearrangement than was calculated in ref. 97.

Two factors now seem likely as being more important than kinetics for determining types of folding. Current ideas suggest that the growth face can be more or less ordered according to (for example) temperature of crystallization.[64,70] Ordered (smoother) surfaces are probably more likely

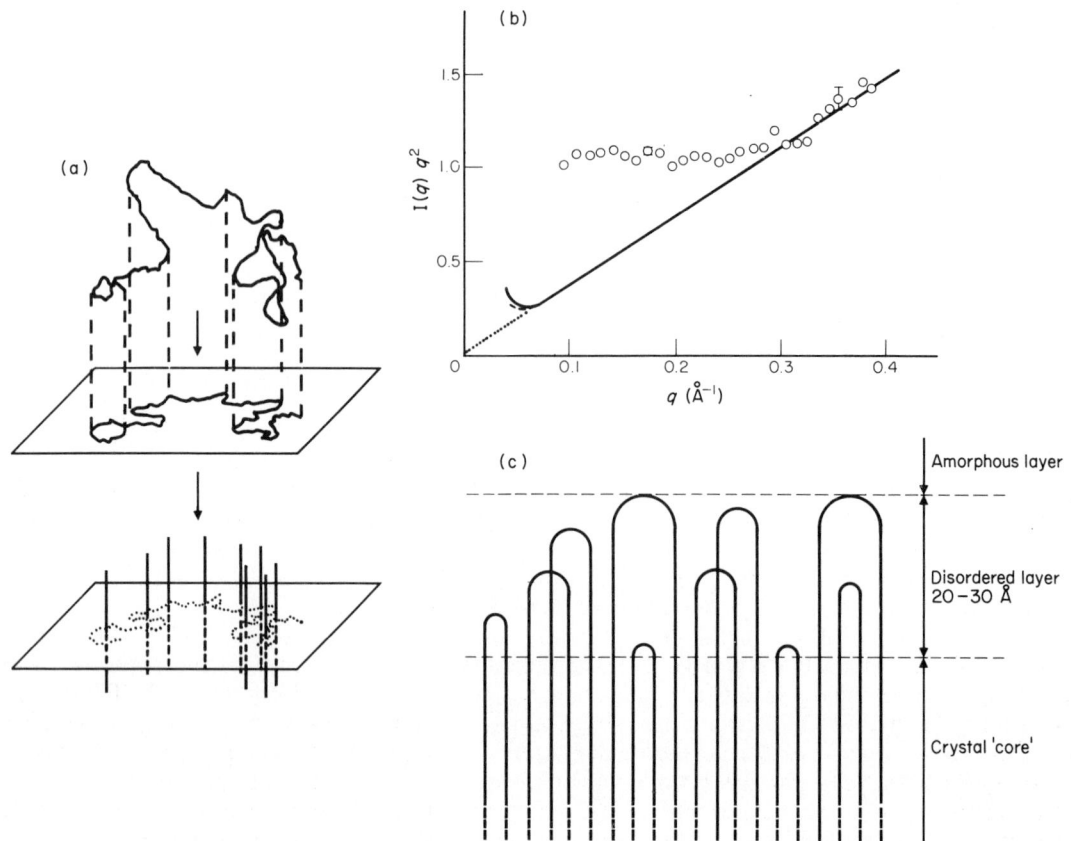

Figure 5 (a) Schematic representation of an extreme 'freezing in' model where the three-dimensional coil existing prior to crystallization is trapped into a lamella (or lamellae) with the minimum of chain movement; (b) predicted scattering (solid line) for the model in (a), and the scattering of totally uncorrelated stems (dotted line), with experimental data points included; (c) schematic model of the surface regions of lamellar crystals, showing how an intermediate disordered layer can lead to folding other than strictly adjacent yet with no surface density anomaly [(a) and (b) reproduced from ref. 87; (c) from D. M. Sadler, *Faraday Discuss. Chem. Soc.*, 1980, **69**, 91]

to lead to adjacent folding. Even more likely as a factor influencing folding is the rather obvious one of dilution in a solvent during crystallization (see below, Section 32.5.4), which is also likely to increase adjacency. This is because, if a molecule is partly attached and there are no other polymer molecules in the vicinity, there may well be a preference for the same molecule to attach next to itself to form an adjacent fold, especially if the growth surface is fairly smooth.

32.5.4 Solution Grown Crystals

For crystallization of PE at 70 °C from xylene, sedimented crystal mats give chain dimensions which increase with $\bar{M}_w$ much less steeply than with $R_g \sim \bar{M}_w^{1/2}$ (*e.g.* $R_g \sim \bar{M}_w^{0.1}$).[72] Values of molecular dimensions along (R_{xy}) and perpendicular (R_z) to the lamellar normals can be obtained from oriented mats.[28] The small anisotropy can be detected by plots according to equation (22). Figure 6 shows results[99] for crystals as grown and heat treated so as to increase the lamellar spacing in such a way as to avoid bulk melting (up to 123 °C). R_{xy} is independent of heat treatment up to 123 °C and R_z increases so that in all cases the molecules are restricted to individual lamellae. The constancy of R_{xy} implies that the region of mobility during annealing is smaller than the molecular size. For higher annealing temperatures the tendency of PE to segregate according to isotope can be exploited. (Isotope segregation is normally considered an experimental nuisance.) After annealing above the demarcation temperature of 123 °C the segregation signal is much enhanced, which shows that the molecules have been able to move and that the size of the mobile region is then probably large compared with the molecule.

The scattering at intermediate q is very different from that of melt grown crystals, being much higher at low q and lower at high q.[82, 84-86]

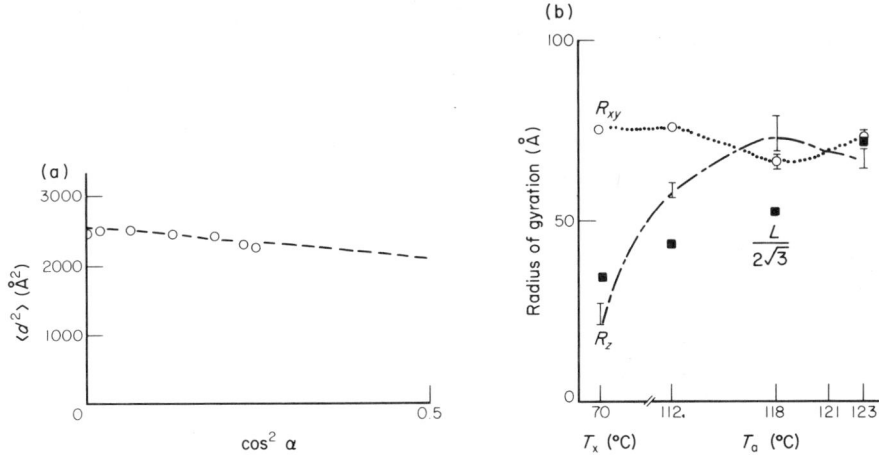

Figure 6 Radii of gyration for solution grown crystals of PE as crystallized at T_x and heat treated at T_a: (a) variation of dimension d with α, the angle between the axis of symmetry of the sample (perpendicular to lamellae) and the beam direction (see equation 23); (b) R_z and R_{xy} data from plots such as (a), with L as the lamellar repeat distance as derived by small angle X-ray periodicity (reproduced from ref. 99 by permission of Butterworth and Co. (Publishers) Ltd.)

For iPS the dependence of R_g on $\bar{M}_w$ is also different for solution grown crystals and those grown from the melt (and melt-quenched crystals).[78] However, the trend is opposite to that in PE; if R_g is expressed as $\bar{M}_w^\nu$, ν is 0.91 for iPS compared with 0.5 (the result for melts), and about 0.1 (for solution grown crystals of PE). Again, in contrast with PE, the scattering at intermediate q is similar for both solution grown and melt grown crystals.

For scattering at comparatively large q,[100-103] the details of the immediate neighbourhood of stems can be probed; for PE (see below) there is a 75% probability of adjacency along a $\{110\}$ facet. For iPS it has been shown[101] that there is no regular '330 folding' as proposed previously. The case of PE will be described in more detail (as follows) and it will be shown that a considerable amount of detail can be obtained by NS in favourable cases where there are sufficient constraints on models.

The existence of sheets has been conformed and two new features of the sheet model have been found, which were not apparent before the application of NS:

(i) R_g values do not increase with $\bar{M}_w$ faster than with $\bar{M}_w^{1/2}$ (as they would for simple sheets which were long compared with the lamellar thickness), but much slower. The superfolding model[72] explains the results in a natural way, whereby the pleated sheets, which themselves are made up of stems connected by folds, are folded back on themselves to produce stacks of sheets. This is shown schematically in Figure 7 (in projection down the chains).

(ii) The second new feature is a reduction of interference within the sheets due, for example, to the sharing of the folded sheet between two or more molecules, so that neither of them are always folding adjacently. This is apparent from the absolute values of intensity.[97,100] Within the folded sheets with the same dilution it is possible to have a range of degrees of adjacency according to the probability p_1 for adjacent reentry; for example folding with a preference for a stem separation of two lattice units ($p_1 = 0.25$)[97] would have very low adjacency. This cannot be the case[100] since a rise in intensity at a Bragg spacing corresponding to a 'double fold' is not observed (Figure 7a) and the low intensity in the very large q region leads rather directly to models where $p_1 \gtrsim 0.75$. The calibration of the absolute intensity appears to be rather accurate since very good agreement was achieved for the analogous case of alkanes (Figure 7b).[100]

It is possible to specify the model to a surprising degree of detail, such that values of p_1 and of degrees of stagger can be discussed in detail. This is largely due to the constraints put on the models by a number of techniques. Parallel 'labelling' experiments using IR[103] are fully consistent with NS and increase the degree of confidence in the details of the model.

The model is equivalent in most ways to the stacked sheet model of Stamm and co-workers.[85,86] Other work has shown that folding is not on 100 planes.[102] The inconsistency with Yoon and Flory[97] does not involve the basic sheet structure or the superfolding, but only the statistics of the folding within the rows.

The interpretation of the existence of rows of stems is straightforward in terms of a molecule attaching to a linear growth face whose molecular structure is relatively ordered (as discussed in Section 32.5.3). The dilution of stems in a sheet also seems reasonable in terms of likely attachment

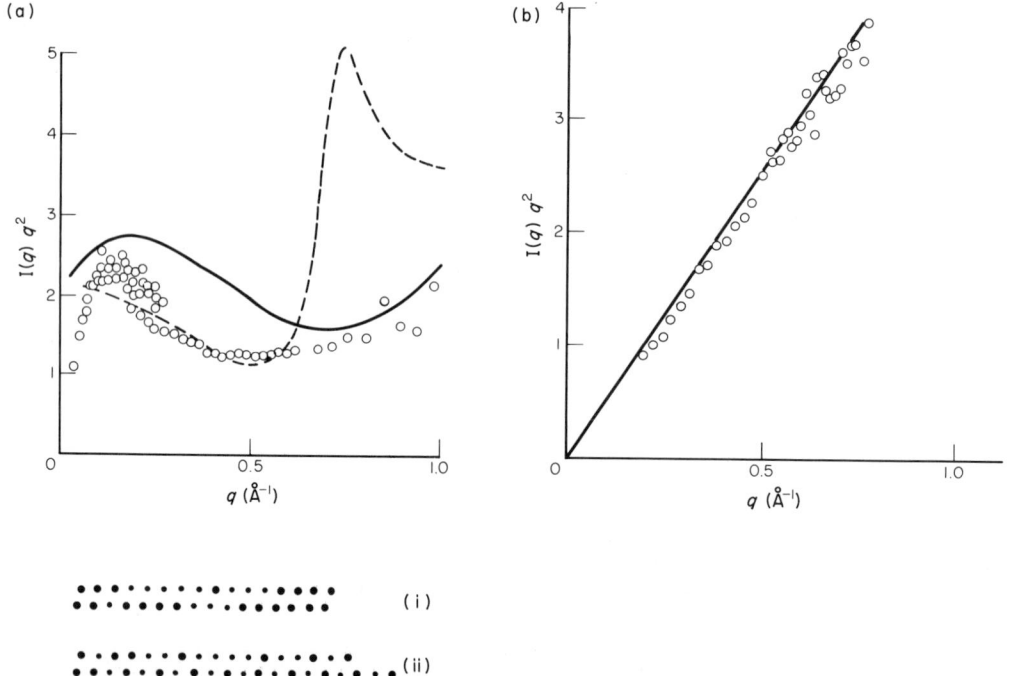

Figure 7 NS data (Kratky function) for intermediate q (values approaching the crystallographic region):[100] (a) for solution grown crystals of PE, where the solid curve shows calculations for (i) 75% adjacency and the broken curve for (ii) 25% adjacency,[97] both for a stem dilution within the sheets of 50% (the calculated intensities at $q \simeq 0.1$–0.5 for model (i) are reduced towards the experimental values by including stagger of fold heights); (b) crystalline alkanes measured and calculated (reproduced from ref. 100 by permission of Butterworth and Co. (Publishers) Ltd.)

processes. The interpretation of superfolding is not clear. It is probably associated with the high supercoolings necessary for neutron-scattering studies of PE, perhaps with the dilution in the solution phase prior to crystallization. The compactness of the conformation is very pertinent to the deformation mechanisms (Section 32.7.2).

32.5.5 Other NS Measurements

The crystal structure of PE has been refined using a powder diffractometer[104] and equivalent information has been derived by Stamm (personal communication). Inelastic scattering is described in Section 32.12.

An ingeneous method for measuring the degree of incorporation of other groups into a crystal lattice[105] is based on labelling of these groups and observing the lamellar diffraction maxima. The issue of incorporation has been highly controversial for many years. The conclusion for oxyethylene groups in poly(oxymethylene) and chloride substituent groups in PE is a substantial degree of incorporation.

32.6 BLOCK COPOLYMERS[106]

From both fundamental and technological points of view, block copolymers have analogies with crystalline polymers. A pure non-crystalline homopolymer is of limited use as a material since it is either a viscous liquid or a brittle glass. Both crystallites (crystalline polymers) and segregated domains (block copolymers) provide intermolecular links without loss of all plasticity in the mechanical properties. This explains why block copolymers generally have both glassy (hard) and rubbery (soft) sequences. Miller *et al.*[107] have made the analogy in conformational possibilities between the two classes of materials; *e.g.* how do the molecules exit and reenter segregated domains of block copolymers, and how compact are the conformations within each zone? It is readily seen how similar these are to questions discussed in Section 32.5. Another principal question is the nature of the boundary between crystalline and disordered regions on the one hand and the interphase between the hard and soft domains of a block copolymer on the other. Theoretical approaches are reviewed elsewhere;[108] they predict such properties as domain size on the basis of equilibrium, the driving forces being enthalpic (in favour of segregation) and entropic (in favour of intimate mixing, since this gives maximum freedom of conformations). The approach of Helfand and Wassermann[108] is based on the Narrow Interphase Approximation (NIA), which is in turn based on theories developed for the boundary between any unlike polymers.[109] The effects of local concentrations are taken into account in a self-consistent manner, by modifying the Gaussian statistics of the chains with an energy function which depends on the local concentration.

32.6.1 Two-phase (Interphase) Scattering

The scattering of X-rays (and neutrons where isotopic concentrations are constant within each of the chemical species) can be treated according to Section 32.2.5.[110] For domains of a given geometry (spheres, lamellae or rods) dispersed at random, the scattering should correspond to the structure factor for that domain shape. In practice the randomness in separations and presentations is rarely sufficient for interference effects between different domains never to be important. Interference is especially significant for lamellae and rods. For spheres, three q ranges can be distinguished. For the lowest q values interdomain interference is the most important and the mutual arrangement of spheres can be studied. In the intermediate range of q the interference damps out and the structure factor for individual spheres can be seen. In the largest q range the scattering is determined primarily by the interphase considered as a sheet in isolation.

Figure 8 shows data[106] corresponding to the third of the ranges for *block*-copoly(styrene/butadiene) [*block*-copoly(S/B)] samples with spherical domains, plotted according to Porod's law (equation (28) which has been modified so as to allow for the interphase thickness.[111, 106] The result is an interphase spacing of 12.8 Å, agreeing with X-ray measurements[112] and NIA theory.

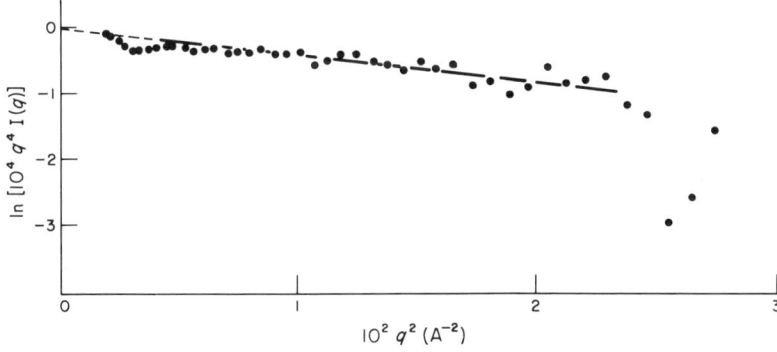

Figure 8 Plot according to Porod's law for a *block*-copoly(styrene/butadiene) with spherical domains.[106] The slope gives a width for the interface between domains of 12.8 Å (reproduced from ref. 106)

The analysis of intersphere interference is not straightforward, since the diffraction peaks are broad and strongly modulated by the form factor for spheres. Both face-centred cubic[113] and body-centred cubic[114] arrangements have been proposed for *block*-copoly(styrene/isoprene) [*block*-copoly(S/I)] and *block*-copoly(S/B) respectively. Lamellae and rods involve no such difficulties, with observations of stacked and hexagonally packed arrangements respectively. A detailed set of calculations for this problem has recently been given.[115]

An important result calculated from the NIA theory is the size of the domain as a function of the molecular weights (M) of the blocks. Clearly, one should increase with the other since the number of restrictions imposed by the links between unlike segments goes down with molecular weight. The domain size can either be measured by matching the scattered intensity to calculated form factors (see above) or by knowing both the repeat distance d of a lattice and the chemical composition. In both cases rules of the form d αM^α can be used to describe predictions and experiment. Richards[106] summarizes these rules, and concludes that there is a reasonable agreement between theoretical and measured values of α for rods and lamellae. However, for spheres the domain size is observed to increase less fast with M than predicted. Hashimoto *et al.*[112] have discussed this discrepancy in terms of a lack of equilibrium as solvent is removed during the casting of films.

32.6.2 Single Chain Measurements

NS has also been used in its more distinctive role by means of the isotope labelling technique in block copolymers. On isotropic samples with spherical domains, the radii of gyration for the 'soft' blocks were found to be the same as for the pure liquid. In one case the soft block was polyisoprene,[110] in the other polybutadiene.[116] Background subtraction difficulties were avoided by using an isotope concentration such that the average scattering length densities were the same for the two chemically different species. This means that the coherent 'background' (as treated in the preceding section) was absent (Figure 9). This is a good example of the flexibility available for experiments due to possible changes in deuteration and this approach has been adopted by most recent work. It was concluded[116] that in the case studied two effects coincidentally cancelled; a chain contraction due to chains being in domains and a chain expansion due to junctions between blocks being restricted to domain boundaries. In one sample a segregation phenomenon according to isotope, analogous to that in PE, was inferred.[116]

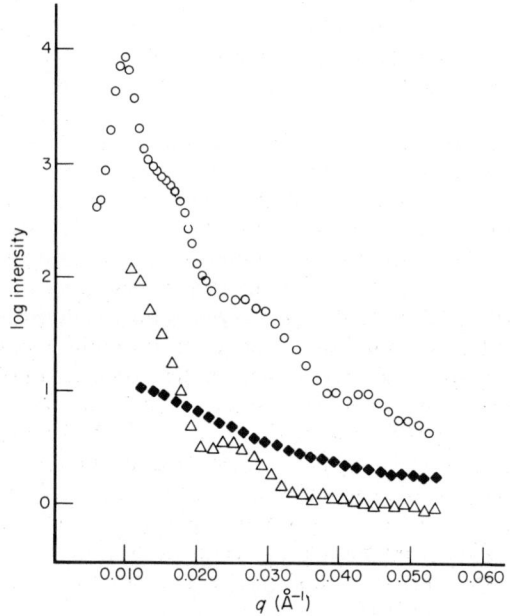

Figure 9 Scattering from *block*-copoly(styrene/butadiene) for various extents of deuteration: (○) 0.362 weight fraction, (△) 0.096 weight fraction and (◆) a mixture of the two to give no coherent scattering from the interphase structure (reproduced from ref. 116 by permission of Butterworth and Co. (Publishers) Ltd.)

For lamellar domains, the chain dimensions suggest that the molecules are extended along the lamellar normals and contracted on the planes of the lamellae compared with the dimensions expected for the free chain. Experiments on oriented lamellae of *block*-copoly (S/I)[30] separated the single chain scattering from the two-phase scattering in terms of orientation at the detector, and measured the in-plane molecular dimensions of the polystyrene (the hard segment). The contraction in in-plane dimensions compared with the free chain was less than expected for the molecule collapsing entirely on itself with no intermolecular penetration at all.

Polyurethane block copolymers with polyether and polyester components are generally lamellar and their behaviour is determined more by the crystallinity in the polyester (hard) segment than by intersegmental incompatibility. Experiments on isotropic lamellar samples of polyurethane[107] used both the method of matching the two-phase scattering to zero and the subtraction method according to Koberstein.[23] It was concluded that the former method gave more precise results, a problem with the second being lack of precise control of two-phase morphology, so that samples were not identical in this respect. It was found that the soft blocks (polyether) were larger than for the free chains, which by analogy with the experiments on oriented samples suggests a molecular extension along the lamellar normals combined with a lateral contraction. The radius of gyration of the soft segments decreased and then increased as the temperature was increased in the range 25–215 °C.[117] A systematic study by Miller *et al.*[118] also showed that the soft segment dimension decreased with temperature, and in addition that the hard segment dimension increased. Dimensions were obtained not from data in the Guinier region of q but by fitting form factors for model structures. Strong evidence was obtained that the polyether blocks were chain folded; R_g changed little on varying the block length, and good agreement was found for a fold length of three repeat units. R_g for the whole chain decreased with temperature, but increased again at very high temperatures. Detailed correlations were made with other techniques, in particular X-ray scattering.

The subject of block copolymers overlaps with mixtures (Section 32.7) since the two components of the copolymer may become more compatible with adjustment of temperature and/or chemical composition. For a copolymer comprising 1,4-polybutadiene and 1,2-polybutadiene blocks it has been possible[119] to choose a chemical composition such that increasing temperature can change the system from being (on average) significantly segregated, to being much more homogeneously mixed. In the latter case the scattering reflects the correlation hole effect discussed in Section 32.4.

32.7 POLYMER MIXTURES (BLENDS)

Blending of polymers is of considerable technological importance in order to optimize properties, and the phenomena of partial and complete demixing involves issues of fundamental interest.[120] NS, using the labelling technique, is necessarily based on mixtures (^1H and ^2H species). Even when the species involved are chemically different, NS has a role because of the possibility of labelling. Polymers which are very similar chemically have necessarily very little contrast to provide scattering signals for electromagnetic radiation. Again, NS can follow the changes in partial miscibility which occur in parts of the phase diagrams where complete segregation into two phases does not occur.

Our starting point will be the exposition of the theory of liquid mixtures by de Gennes[44] which is essentially a generalization of the Flory–Huggins theory.[27] This lattice theory enables a physical picture of the processes of critical behaviour and phase separation to be obtained. In brief, an energy parameter χ (a ratio of free energy to kT) is introduced analogously to solution theories (see Volume 2, Chapter 3) which disfavours very slightly like–unlike monomer contacts in favour of like–like. The so-called combinational entropy of mixing S_c will favour miscibility, but it is very small, of the order of k per entire molecule of N monomer (*cf.* $\sim k$ per small molecule for non-polymers — a much larger driving force). The phase behaviour can be interpreted in terms of the relative magnitude of contributions to the free energy of mixing of an energy ($\simeq N\chi kT$ per molecule) and an entropy ($TS_c \simeq kT$ per molecule). The phase diagram in terms of concentration as abscissa and temperature as ordinate has much in common with such diagrams for any other mixture, with a line which is convex upwards separating single and two-phase regions. An upper critical solution temperature (UCST) exists on this line. Complete miscibility occurs in the limit when TS_c is large compared with the energy of mixing, *i.e.* when $\chi \ll N^{-1}$, and phase separation occurs as the energy term becomes comparable or larger than TS_c ($\chi \gtrsim N^{-1}$).

In the partly miscible region ($0 < \chi \lesssim N^{-1}$) there are fluctuations whereby there is a statistical preference for like molecules to be adjacent. As a result there will be scattering from concentration fluctuations according to[44]

$$I^{-1}(q) = \frac{1}{c_A N_A P_A(q)} + \frac{1}{(1 - c_A)N_B P_B(q)} - 2\chi \qquad (33)$$

where N_A and N_B are constants related to the number of monomers for A and B, and $P_A(q)$ and $P_B(q)$ are the respective molecular form factors. For mixtures of isotopes N_A and N_B are the number of exchanged hydrogens per molecule. This can readily be seen to revert to equation (11) for $\chi = 0$. For the particular case of the $P(q)$ being described by the Guinier equation (equation 12), this becomes equivalent to the Ornstein–Zernike equation (equation 30) with characteristic length given by

$$\xi = \frac{a}{6} \{c(1 - c)(\chi_c - \chi)\}^{-1/2} \qquad (34)$$

where χ_c is the value of χ at the critical point. Clearly ξ diverges towards the critical point. This result is based on a mean field approach, *i.e.* one where the coil–coil interactions are calculated on the basis of a smoothed mean concentration. Such calculations tend to be more accurate the longer the range of interaction between molecules. For polymer blends this approach is a good approximation as long as $\xi \lesssim R_g$, in which case the fluctuation distance ξ is not larger than the range of interaction, which is of the order of R_g.

Although the results for critical scattering for non-polymers and polymers appear to be analogous, the Ornstein–Zernike treatment does not appear to be closely similar. For example, in the former case the concentration of species is allowed to vary over distances larger than the range of interaction and a term for the free energy associated with the square of the concentration gradient is essential.[121]

The criterion for at least partial miscibility is $\chi \lesssim N^{-1}$. In general χ is very much larger than this; *e.g.* for saturated hydrocarbons of different chemical types the van der Waals forces will always tend to give significant χ values.[44] From a practical point of view, this means that $\chi \sim 10^{-4}$ is often only achieved as the consequence of the near cancellation of several much larger terms. Hence very small temperature dependencies in (for example) enthalpy or non-combinational entropy can swamp the effects expected from the straightforward lattice theory. One example of this is that many pairs of polymers show a Lower Critical Solution Temperature (LCST). This means that these polymers are less miscible at higher temperatures, in contradiction to expectations from simple lattice theory, whereby randomness (miscibility) is favoured at higher temperatures.

The other principle class of theory, not utilizing a lattice, is based on an equation of state[122-124] in an analogous way to the van der Waal's equation of state theories for simple liquids. Such an approach allows non-combinational entropy terms to be included, which are often significant (see the results below) and yet do not feature in a simple lattice model. Any liquid will have an increase in entropy and energy with temperature because of the change of local structure and packing, and it is reasonable that mixing of unlike species will lead to a total entropy and energy which are not simply the result of adding terms for the pure species. It is usual to discuss the results for miscibility/phase separation in terms of the parameter χ, even though this was introduced on the basis of a strict lattice model. One can, for example, split χ into separate contributions from free volume, and enthalpic and entropic terms related to local structure.

An important part of the literature on neutron scattering of blends is concerned with derivations of scattering equations[23-25,125-127] which essentially are extensions of equations (3)–(34) for a variety of situations, a particularly important one being when only a proportion of one of the species is deuterated. Allowance can also be made for differences in segment volume and molecular weight between species and polydispersity. An analysis has been given,[127] analogous to the original derivation of the scattering law for critical opalescence by Ornstein and Zernike, which does not rely on an explicit statistical model.

As mentioned in Section 32.6, there will be an interfacial energy associated with gradients in polymer concentration, which leads to a nucleation barrier for the initiation of regions of different concentration in some (but not all) parts of the phase diagram. The study of the kinetics of nucleation, and of spinodal decomposition (in the absence of nucleation), is normally performed with light scattering, though neutron scattering can be used.[128]

32.7.1 Liquid and Amorphous Mixtures

The simple lattice theory should be most appropriate for mixtures of closely similar polymers. An example of small χ is mixtures of ^1H and ^2H polymers. The implicit assumption for the application of NS has often been that χ is then effectively zero, but in 1977 this was questioned[129] on the basis of data for poly(dimethyl siloxane).[52]

It is certain that χ cannot be identically zero, since there are certainly finite chemical differences according to isotope (melting points, refractive indices *etc.*, as well as θ temperatures and miscibility). Discussion of such differences is very common in the references given in this section. The question is whether χ is very small ($\simeq 10^{-4}$) or so extremely small that its effect in equation (33) (for example) is always negligible.

Lapp, Picot and Benoit[130] have recently confirmed that $\chi \neq 0$ ($= 1.7 \times 10^{-3}$) for ^1H and ^2H poly(dimethyl siloxane). Bates and Wignall have reported similar results for hydrogenated poly-butadiene[131,132] (similar chemically to PE but it is liquid down to the glass transition temperature) and PS,[133] and have shown data on the temperature dependence of χ consistent with the UCST behaviour expected from the lattice theory (see Figure 10). Values for χ were compared with those predicted. An important contribution to χ comes from the specific volume difference between ^1H and ^2H, a possibility discussed by Buckingham and Hentschel.[134] They separated the mixing process into: (i) compression of the two species to give the same segment volume; and then (ii) mixing at constant volume. Such volume differences in themselves contradict the premise of lattice theory but, as for non-polymers,[135] the framework of lattice theory can be retained if an adjustment (invariably positive) is made to χ. Bates and Wignall also derived a contribution to χ from polarizability differences, but Buckingham and Hentschel proposed that only the volume term should be important. However, the experimental situation is not yet clear for one of the polymers in question since Yang *et al.*[136] have reported an upper limit to χ for ^{1}H and ^{2}H PS of 10^{-6}. These authors also raise the issue of whether previous results on coil dimensions would be called into question by the existence of a finite measurable χ value. However, earlier work was restricted to small concentrations of deuterated species (c_D) where the effects of finite χ on measured R_g values are much less pronounced than for the case of the higher values of c_D employed in the recent work. The effect of finite χ is likely to be most important in the region of the critical point, *e.g.* in the one-phase region where equation (33) includes the effects of critical fluctuations. Rather paradoxically, phase separation (in the two-phase part of the phase diagram) might show only relatively subtle effects on the scattering compared with the fully miscible case, $\chi = 0$ (see also Section 32.2.3). These issues are not resolved at the time of writing.

For chemically distinct polymers van der Waals energy terms invariably disfavour mixing[44], as does the volume term mentioned in connection with the isotope mixtures. Hence it has only been possible to find a few compatible mixtures, and this has often been because of some specific energy term which favours unlike contacts. The interest in experiments on these systems is to characterize χ values for particular mixtures, and to deduce what physical processes favour mixing or demixing.

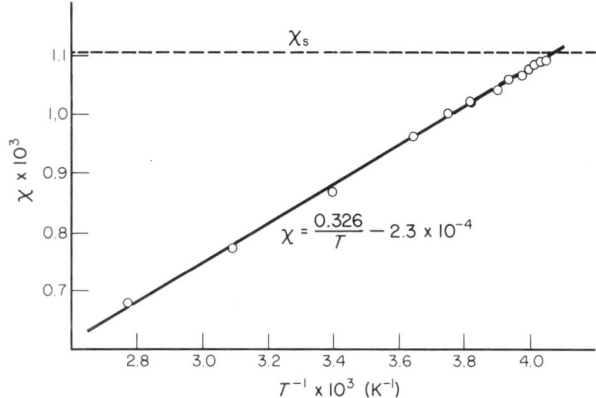

Figure 10 Temperature dependence of the segment–segment interaction parameter χ for a mixture of protonated and perdeuterated 1,4-polybutadienes (degree of polymerization 4200 and 960 respectively).[131,132] Species matched in molecular weights give scattering indicative of phase separation (reproduced from ref. 132 by permission of the American Physical Society)

Kirste and co-workers[137-140] pioneered the subject usually using one component fully deuterated and in small concentrations (up to a few percent). Several mixtures were employed, both with negative χ[140] [PS/poly(acrylonitrile) (PS/PAN), PS/PMMA, PS/poly(phenylene oxide) (PS/PPO) and PMMA/PVC] and with χ values which can become positive[139] [PS/poly(vinyl ether), (PS/PVE) and PS/poly(vinyl methyl ether), (PS/PVME)]. Figure 11(a) shows experimental values of χ from NS for deuterated PS in PVME from the work of Jelenic *et al.*[140] The critical value for χ for demixing is indicated — this of course is highly dependent on molecular weights (465 000 and 10 000 respectively for the two polymers in this case). Figure 11(b) shows the separate enthalpic and entropic contributions to χ (from the slope and intercept respectively). It is concluded that there is significantly less entropy involved in the local structure of the mixture compared with the pure components used as a reference. This entropic contribution to the free energy of mixing increases with temperature, partly because of the weighting by temperature of entropy. This can lead to demixing at higher temperatures (LCST behaviour).

Shibayama *et al.*[141] also studied PS/PVME, but with concentrations of PS up to 80 weight %. They found a significant concentration dependence of χ which was discussed in terms of changes in the fluid lattice with concentration. These authors included a detailed description of the formalism linking χ with the second virial coefficient A_2.

Poly(vinylidene fluoride)/PMMA (PVF/PMMA) was characterized as a miscible system by Hadziioannou and Stein.[142] Oligomeric systems, which allow a wider range of polymer pairs to be used and have practical advantages for experiments, have been studied by Warner *et al.*[125] and Higgins and Carter,[143] *e.g.* polyethylene glycol dimethyl ether (PEGM)/polypropylene glycol dimethyl ether (PPGM). They concluded that there is a non-combinational entropy which favours mixing, possibly due to a change to a more disordered molecular structure for PPGM. Oligomeric mixtures of PS and polybutadiene (the same pair that occurs in block copolymers) have also been

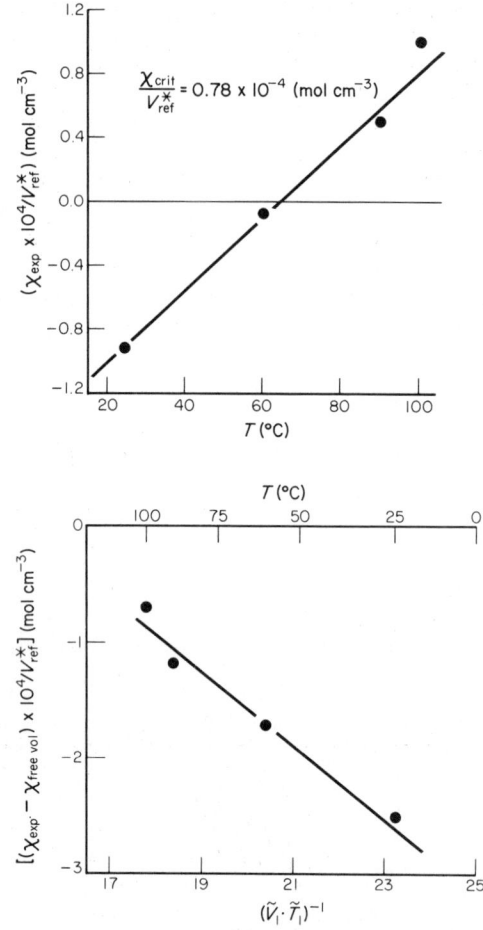

Figure 11   Values for the interaction parameter χ measured for small concentrations of deuterated PS in poly(methyl vinyl ether) $\bar{M}_w$ values 465000 and 10000 respectively: (a) the critical value of χ above which phase separation should occur, V_{ref} is an effective molar volume suitable for use in the equation of state theories; (b) shows the same data, with a correction for a free volume contribution, and plotted so that enthalpy and entropy contributions to χ correspond to slope and intercept respectively, $\tilde{V}_1$ is a reduced volume, and $\tilde{T}_1$ a reduced temperature (reproduced from ref. 140)

studied.[144] Similarly poly(dimethylphenylene oxide)/PS (PXE/PS) was found to have both entropy and enthalpy terms which favoured mixing. Compatibility can also be adjusted by the use of random copolymers, since the chemical nature of the species can then be adjusted in a continuous manner.[145, 146] In one of these cases, PXE/PS, there is clear evidence that a specific energetic interaction favours mixing.[145] As PXE is brominated the mixture becomes progressively less miscible.[147]

32.7.2   Crystalline Mixtures

The subject of isotope mixtures of otherwise similar crystalline polymers has had a complicated history. The issue of segregation first arose in the case of alkanes where it was inferred from melting point differences according to isotope that the crystal phase should be segregated.[148] However, recently it has been argued that a melting point difference does not necessarily imply segregation,[149] and calorimetric results have been interpreted in this way. It has recently been confirmed experimentally by neutron scattering that indeed there is no segregation in crystalline *n*-alkanes.[150] This leaves the situation in PE still unresolved, since sometimes there is unquestionably segregation.[73] It is possible that nominally similar PE samples are not in fact identical, since other differences (in branch content for example) could be expected to have consequences analogous to isotope differences. Almost certainly the segregation arises from kinetic effects. If there are only slight variations in crystallization rate for the two species mixed together, then the final result will be a crystal over which the isotope concentration varies. The heterogeneities will not dissipate by diffusion as in liquid mixtures. The segregation scattering signal has been analyzed according to this model of 'picture frame crystals' using equation (28)[82] and equation (27).[151-152] The distance scale for the concentration fluctuations may be fairly large in the plane of the lamellae if the lamellae are wide. For narrow crystals, and in all cases in the direction perpendicular to the lamellae, the distance scale could be expected to be similar to the lamellar spacing. This is consistent with estimates of a distance scale similar to the radius of gyration.[151-152] A different interpretation of the segregation signal has also been given.[153] It is possible that partial miscibility in the liquid prior to crystallization can have an effect (see above), since this is likely to be retained during crystallization, especially on quenching.

Experiments have also been performed on polymer mixtures where only one component is crystallizable,[155] and where both polymers crystallize but with different lattices.[154] Both cases show strong segregation after crystallization.

32.8 DEFORMATION AND RELAXATION STUDIES ON LINEAR POLYMERS

This is potentially one of the most productive applications of the technique, if only because so little information on a fundamental level is available concerning polymer structure during and after deformation. The emphasis of the discussion here will be the description of results in terms of the extent to which the deformation of the molecule is affine with the macroscopic deformation.

The limitation on the maximum R_g value measurable is often the collimation of the beam and the detected resolution rather than the size around the backstop (see Section 32.2). Figure 12 shows a contour plot of intensity from which a lower limit for $R_z \gtrsim 6000 \, \text{Å}$ can be obtained.

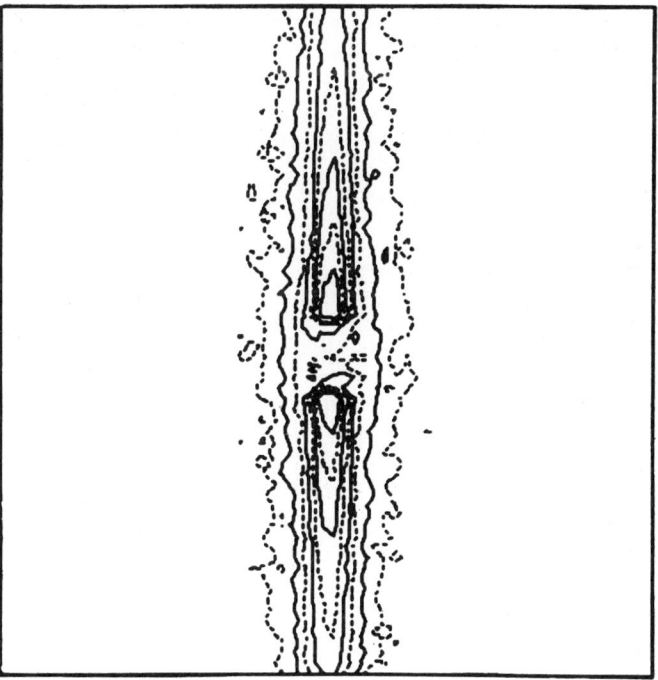

Figure 12 Contour plot of coherent neutron scattering intensity from a fibre formed from a blend of ^{1}H with ^{2}H PE, the draw direction being vertical: specimen detector distance 10.5 m; wavelenth 8 Å; detector array size 64 cm × 64 cm (reproduced from D. M. Sadler and P. J. Barham, to be published)

32.8.1 Deformation of Amorphous Polymers

PS was uniaxially stretched above its glass transition temperature T_g and quenched to below T_g before NS measurement.[156] The resulting R_g values approximated closely those expected for affine deformation, though R_z values were 5–10% lower than affine. An extrusion method was used to achieve high deformation ratios,[157] and again the molecular dimensions were affine with the samples. Large values of R_z were achieved indirectly (*cf.* equation 25). For PMMA of sufficiently high molecular weight affine deformation was also found for tensile deformation below T_g (see Figure 13).[158] For PS and PMMA the deformation was no longer affine on a small enough scale, as tested by $I(q)$ for $qR > 1$ and by R values for low molecular weight PMMA. The distance scale over which the deformation is not affine increased with temperature.

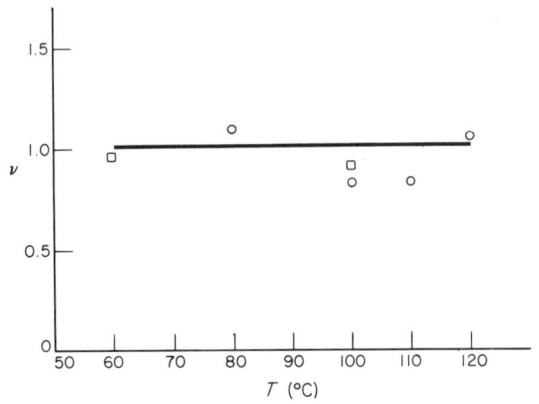

Figure 13 Ratio (v) of molecular and macroscopic extension ratio of PMMA against drawing temperature: ($\bigcirc$) $T_g = 111\,°C$, ($\square$) $T_g = 123\,°C$; the extension ratios are approximately 2; $\bar{M}_w$ of labelled species = 17 000 (reproduced from ref. 158 by permission of the American Chemical Society)

Results for deformation by compression[159, 160] of PS and PMMA below T_g were also consistent with affine deformation, though the interpretation was complicated by macroscopic heterogeneity of the samples in the beam. Recent results on amorphous PET[94] show molecular deformation substantially less than affine.

The long term aim must be the comparison of results of NS with models (*e.g.* of Ward)[161] of the mechanism of deformation.

32.8.2 Deformation of Crystalline Polymers

In this case there is extensive structural information mainly from electron microscopy[162] on changes in crystal morphology on deformation; the spherulitic texture of twisted lamellae is replaced by fibrils made up of stacks of lamellae. One approach to the subject has been by analogy with metallurgy, but a model has been proposed[163] which effectively neglects the existence of crystals and concentrates on the effects of entanglements. Another possibility, very interesting from the point of view of NS since it relates to the results in Section 32.5, is that the molecular conformation prior to deformation is the controlling factor (Barham, submitted to *Polymer*). One important and previously controversial question concerns whether the crystals melt transitorily during deformation, as suggested by the fact that the lamellar spacing after deformation corresponds to the temperature of drawing, and not to the original temperature of crystallization.[164] Over recent years a variety of studies have concluded that such melting is indeed likely.

Deformation studies on PE have been hampered in practice since it is difficult to produce material sufficiently homogeneous in isotope concentration. A favourable choice of hydrogenous PE was the basis of a set of measurements on tensile deformation.[165] It was found that for melt grown crystals drawn about ×8, R_{xy} corresponds to affine deformation, whereas R_z is less than affine by a factor of about 2. Stretching solution grown crystals by about ×30 gave R values which were changed much less than expected for affine deformation. The difference according to the initial state is almost certainly related to the difference between conformations which are compact or spread out (Section 32.5). The connection could be a direct causal one, or because of more entanglements in the latter case. Extrusion of melt grown crystals gave similar results to the tensile stretching.[166]

Drawing of polypropylene[167] has given molecular deformation which is much less than affine. A model was proposed with the molecules separated into subunits with relatively random correlations between subunits. It is not clear whether the effect of resolution on absolute intensities (Section 32.2) has been accounted for.

The isotope segregation effect in PE can be exploited in a rather interesting way, since changes in segregation imply changes in isotope concentration gradients. Such changes imply in turn that the material has been sufficiently molten for the molecules to move (see also Section 32.5.4). Studies of a PE, which was initially somewhat segregated, have been used to infer that the isotopic species desegregate on stretching.[152] Recent studies[168] have shown a very large increase in segregation if the deformation of melt grown crystals is performed at elevated temperatures.

32.8.3 Stress Relaxation[60]

Stretched samples are kept at constant length, and the ways by which isotropic molecular conformations are recovered are used to study chain dynamics (see also Section 32.12).[172-173] A number of models have been considered, the principal ones being Rouse and reptation. The former comes from considering the normal modes of a Gaussian coil and treats the molecular environment as a structureless viscous medium. The second[169,170] allows for the effects of entanglement by considering a tube formed by entanglement restrictions within which the movement would occur according to Rouse.

Both R_{xy} and R_z revert monotonically[172,173] to the isotropic values (Figure 14). This in itself is strong evidence against the reptation model, since that predicts an initial contraction within the 'tube' which would lead, for example, to an initial decrease in R_{xy} (see lines in Figure 12).

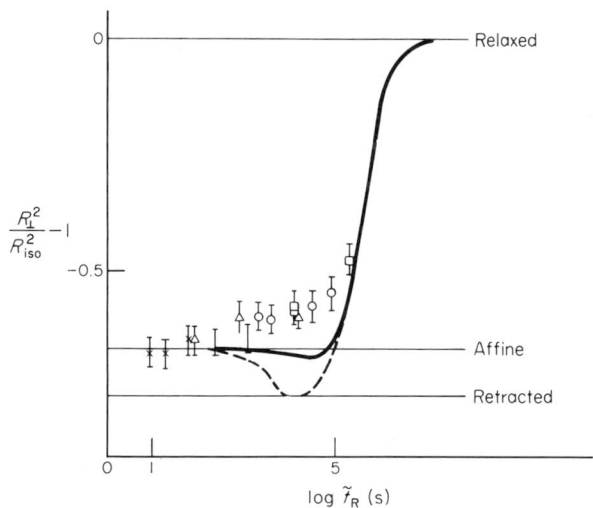

Figure 14   Variation of equatorial molecular dimensions with time of heating of a PS sample initially stretched $\times 3$; $\bar{M}_w$ of labelled PS = 760 000; temperatures are ($\times$) 117 °C, ($\triangle$) 122 °C, ($\bigcirc$) 128 °C, ($\square$) 134 °C (the times for the latter three temperatures are scaled to those at 117 °C using a time/temperature superposition); the full and dashed lines describe theoretical predictions for the reptation model; note that $R_\perp^2 = 3R_{xy}^2$ (reproduced from ref. 173)

Data at $qR_g > 1$, where R_g is the initial radius of gyration, are clearly related to local scale motion, so tests of the corresponding $P(q)$ with the Rouse model have been considered in detail.[60,172] At short time t there is an approximate, though not precise, agreement with calculations of $P(q)$ from the Rouse model. At longer times the Rouse model predicts a much more isotropic $P(q)$ than observed, which is not surprising since that model does not include effects such as entanglements. The Rouse model also predicts that the data at different times (t) should superimpose if plotted against $qt^{1/4}$. Again, agreement was found for small but not large t. Experiments are also discussed where cross-linking has inhibited all but the most local scale movements.[172]

Preliminary experiments on stress relaxation of crystalline polymers[168] show that in the temperature range where a very small proportion of material is still crystalline the molecular extension is retained. At higher temperatures and draw ratios common in crystalline polymers, the molten sample breaks under the very large stresses.

32.9 NETWORKS AND THEIR DEFORMATION[174]

Rubber elasticity has been one of the central problems in polymer science. It is striking how a characteristic unique to polymers, *i.e.* configurational disorder, can be seen to explain a restoring force which is entropic rather than energetic. NS has long been seen as offering the possibility of testing some of the important hypotheses in the subject (detailed predictions have been made by Ullman)[175] and indeed significant progress is being made.

None of the models of rubber elasticity predict a deformation of the chains between cross-links being affine with the sample dimensions, and this is borne out by the NS experiments.[176-179] The two principal models are junction affine, and the phantom chain model which allows for fluctuations in junction positions. The former predicts higher molecular deformation than the latter. Several effects could have been important which are not included in the models, *e.g.* non-Gaussian statistics of chain sequences between cross-links, trapped entanglements (sequences between cross-links cannot cross over each other during deformation) and a transmission of orientation by packing of chain segments.

The systems studied have been PS and PDMS. In the latter case the following has emerged. The chain sequences between cross-links are Gaussian, whatever the concentration v_c of the polymer in solvent during cross-linking, with chain dimensions equal to those of the free amorphous chain.[178] The behaviour on deformation[179] depends on the number-average molecular weight $\bar{M}_n$ of the network chains sequences and on the cross-link functionality f (4 to 6), (Figure 15) but rather little on v_c. For the lowest $\bar{M}_n$ the molecular deformation is the greatest and agrees with the affine junction model, whereas for the highest the deformation is slightly lower even than the phantom chain model. For $f=6$ the molecular deformation is higher than $f=4$.

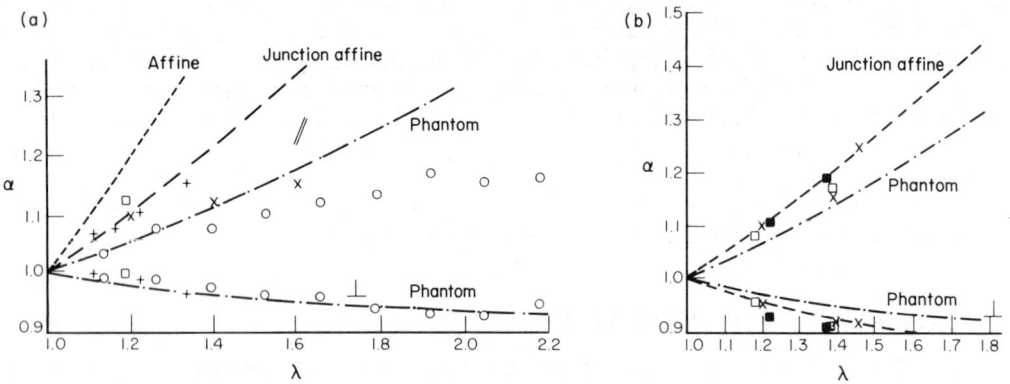

Figure 15 Molecular deformation ratios $[\alpha_\perp = R_{xy}/(R_{xy}\text{ initial})$ and $\alpha_\parallel = R_z/(R_z\text{ initial})]$ for stretched poly(DMS) networks as a function of macroscopic draw ratio λ: (a) $f=4$ and number-average molecular weight $\bar{M}_n=(\times)$ 3000, (+) 6000, (○) 25 000; (b) $f=6$ and (■) $\bar{M}_n=10 000$, $v_c=1$, (□) $\bar{M}_n=10 000$, $v_c=0.71$, (×) $\bar{M}_n=25 000$, $v_c=1$ (reproduced from ref. 179 by permission of the American Chemical Society)

These results are discussed in terms of a suggestion by Bastide that one should consider the characteristic ratio n/f of the total number of cross-links in a region to the number of cross-links which are directly connected by chain sequences. For higher n/f (promoted by smaller f and higher $\bar{M}_n$) the deformation pulls the cross-links relative to each other so as to give the highest departure from affine junction behaviour. v_c is an important experimental variable since high v_c can be expected to result in the higher number of trapped entanglements. Increasing v_c increases the mechanical modulus as would be expected. The lack of effect of v_c on molecular deformation suggests that trapped entanglements do not affect chain dimensions. Recent results on free chains mixed in with a network[180] show that the free molecules are not distorted in shape on the distance scale probed by NS, even though there is an induced orientation of segments on the scale of a monomer. This result emphasizes again the important distinction between average-segmented orientation and chain extension which may not be directly linked in a one to one manner (see also Section 32.5).

Other results on networks included an analysis of lack of randomness of cross-linking in PS networks.[181] The arrangement of cross-links in epoxy resins has also been studied and found to be uniform.[182,183]

32.10 IONOMERS

These materials are copolymers with ionic side groups distributed along non-polar chains in an irregular manner (typically 10 mol %). They can be either amorphous or semicrystalline. Segregation occurs, but in a less clearly defined manner than in block copolymers (Section 32.6). Models[184] for the structure envisage small multiplets of a few ionic groups and larger aggregates.

NS has been used to measure molecular sizes for sodium polystyrene sulfonate, which have been found to increase[185,186] as a result of the presence of the ionic groups in a way which can be correlated with a theory.[186,187] The model assumes that groups are equally spaced along the chains, that the non-polar sections obey Gaussian statistics, and that the groups participating in the aggregates are near neighbours in the hypothetical system with no aggregation.

NS has also been used to supplement information from X-ray scattering on density variations. Both techniques show a prominent peak additional to any peaks from crystalline lamellar spacings which may be present. For NS the peak is either absent (polypentenamer sulfonate)[188] or weak (nafion sulfate)[189] in the dry polymer, but increases steeply in intensity as ^2H_2O is added. It is presumed that water swells the regions of the ionic aggregates. The position of the peak is insensitive to ^2H_2O content for low contents (corresponding to a primary hydration shell), but changes to higher equivalent Bragg spacings at higher content. The geometry of the aggregates has not been defined in detail, but it has been concluded that the peak does not come from interference between aggregates and the results have been discussed in terms of a two-phase scattering model.[189]

32.11 LIQUID CRYSTALLINE POLYMERS

Recently some articles have appeared concerning comb-like polymers whose rather rigid side groups lead to the existence of nematic or smectic liquid crystalline phases.[190-191] The interest in NS is to establish how the flexible backbone fits with the liquid crystal regions. It has been found that for two polymers, which change with increasing temperature from nematic or smectic to isotropic phases,[190] the R_g values remain the same. For a nematic phase oriented in a magnetic field, R_z (along the chains) is 25% smaller than R_{xy}. Keller *et al.*[191] studied a similar system, and found a small molecular anisotropy in the nematic phase. They also oriented the smectic phase and found a very striking anisotropy (R_x, out of plane = 21 Å, R_y, in plane = 120 Å) clearly suggesting that the backbone is generally confined between the smectic layers. Wide angle diffraction peaks were also measured, indicating a doubling of the smectic layered repeat.

32.12 INELASTIC NEUTRON SCATTERING[192]

Sections 32.4–32.11 have dealt with NS which is both elastic and coherent. There is only scope here to give a very brief description of the techniques and results when changes in neutron energy occur in the scattering events. The simplest application would be as a straightforward spectroscopic tool, *e.g.* to identify maxima in intensity *vs.* the energy gained by scattered neutrons in terms of particular thermally exicted states in the sample. (NS has an advantage here that selection rules do not exclude excitations in the way that can occur in electromagnetic spectroscopy.) However, a principal feature of NS is the correlation (and hence interference) effects which are part of the scattering process. The differential cross section and scattering form factor now include a dependence on energy E ($d^2\Sigma/d\Omega\,dE$ and $S(q, \omega)$ respectively, where $\omega = Eh$). In the same way that $S(q, 0)$ is the Fourier transform of the spacial correlations $\gamma(r)$, $S(q, \omega)$ is the double Fourier transform over both spacial and time correlations as given by $\gamma(r, t)$.[1,3] $S(q, 0)$ is directly related to $|A(q)|^2$. For example, if a nucleus is moving, there will be a finite probability that the nucleus will be displaced by r in a time t, and there will be a corresponding contribution from self-correlations to $\gamma(r, t)$. The timescale is important for determining the nature of the inelastic effects, for example if the frequency of movement is very fast compared with ω then the nucleus can be considered as being 'smeared out' over a finite region of space. Such an effect will be observable even for incoherent scattering, which by definition is concerned only with the self-correlation of individual nuclei. The methods of measurement are described elsewhere,[3] and involve energy as well as angular discrimination. Energy discrimination can be achieved by measurement of wavelength (using crystal monochromators), of neutron speed (time of slight spectrometers) or of rates of spin precession (quasielastic scattering as measured by the spin echo method).

Many of the results have been for polymer solutions, but some for condensed phases are as follows.

Side group motions have been studied for polymers such as poly(propylene oxide) and PMMA, often at temperatures low enough for other motions to be frozen. Selective deuteration can be very useful in identifying which group is involved. Gabrys *et al.*[193] have recently described in detail the current status of this subject, and conclude that the motions do not correspond to a simple thermally activated angular 'hopping' (*e.g.* between sites with three-fold symmetry).

Thermal vibrations have been studied in PE, the most recent results using single crystal textures.[194] Other recent work is by Heyer *et al.*[195] Phonon dispersion curves were established and detailed information was extracted concerning the intra- and inter-molecular bonding (covalent and van der Waals respectively). See also Table V of ref. 11 which contains a comprehensive list of references to experiments of this type.

The types of molecular motion characteristic of a random coil correspond to the normal modes according to Rouse (see also Section 32.8.3). For solutions there are then three regimes of behaviour; diffusion (at the longest times), Rouse motions and local motion specific to the chemical nature of the monomer (at the shortest times). The regime of Rouse motions has been identified in a polymeric liquid.[196-201] Figure 16 shows data both from a conventional time of flight spectrometer (measuring energies of neutrons from their speeds) and from the very high resolution spin echo method. Ω in Figure 16 is an inverse time, and it is very striking how this frequency increases by over four orders of magnitude as the distance scale of the motion, as measured by $2\pi/q$, decreases from ~ 60 Å to ~ 6 Å. The straight line corresponds to $\Omega \propto q^4$ as predicted from the Rouse behaviour. Subsequently,[202,203] in the case of another polymer, poly(tetrahydrofuran), it has been possible, at a distance $\gtrsim 50$ Å, to identify effects of entanglements,[202,203] and these results have been correlated with entanglement densities as judged by viscometric data.

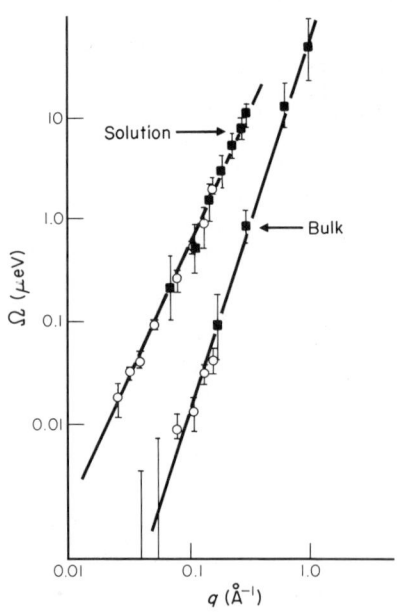

Figure 16 Inverse correlation time Ω as a function of q for individual ('labelled') 3% ^{1}H, poly(tetrahydrofuran) in an equivalent ^{2}H matrix at 110% and data for poly(THF) in CS_2 solution is also shown (reproduced from J. S. Higgins, L. K. Nicholson and J. B. Hayter, *Polymer*, 1981, **22**, 163, with permission of Butterworth and Co. (Publishers) Ltd.). *Note on Units.* Inelastic and quasi-elastic neutron scattering measures the energy difference between incident and scattered neutrons. This energy difference can be theoretically equated to $\hbar\omega$ where $\hbar = h/\pi$ and ω is an angular frequency. Consequently, the energy difference observed experimentally is equivalent to a reciprocal time. Practically, the energy difference observed may be expressed in a variety of units, *e.g.* wavenumber, frequency, electron-volts, neutron time of flight. Where small energy transfers are involved they are generally expressed in $\mu e/V$

32.13 REFERENCES

1. W. Marshall and S. W. Lovesey, 'Theory of Thermal Neutron Scattering', Clarendon Press, Oxford, 1971.
2. V. E. Turchin, 'Slow Neutrons', Israel Program for Scientific Translations, Jerusalem, 1965.
3. G. Kostorz, 'Treatise on Materials Science and Technology', vol. 15 in 'Neutron Scattering', Academic Press, New York, 1979.
4. J. E. Enderby, *Annu. Rev. Phys. Chem.*, 1983, **34**, 155.
5. A. Guinier and G. Fournet, 'Small Angle Scattering of X-rays', Wiley, New York, 1955.

6. G. E. Bacon, 'Neutron Diffraction', Clarendon Press, Oxford, 1967.
7. O. Kratky, *Pure Appl. Chem.*, 1965, **20**, 619.
8. O. Glatter and O. Kratky (eds.), 'Small Angle X-ray Scattering', Academic Press, London, 1982.
9. B. Jacrot, *Rep. Prog. Phys.*, 1976, **39**, 911.
10. J. S. Higgins and R. S. Stein, *J. Appl. Crystallogr.*, 1978, **11**, 346.
11. J. S. Higgins, in 'Treatise on Materials Science and Technology', ed. G. Kostorz, Academic Press, London, 1979, vol. 15.
12. A. Maconnachie and R. W. Richards, *Polymer*, 1978, **19**, 739.
13. D. G. H. Ballard, *Macromol. Chem. (London)*, 1980, **1**, chap. 11.
14. D. G. H. Ballard and E. Janke, *Macromol. Chem. (London)*, 1982, **2**, chap. 11.
15. R. Ullman, *Annu. Rev. Mater. Sci.*, 1980, **10**, 261.
16. J. S. Higgins, *Macromol. Chem. (London)*, 1984, **3**, chap. 9.
17. C. E. Williams, M. Nierlich, J. P. Cotton, G. Jannink, F. Boué, M. Daoud, B. Farnoux, C. Picot, P. G. de Gennes, M. Ricarde, M. Moan and C. Wolff, *J. Polym. Sci., Polym. Lett. Ed.*, 1979, **17**, 379.
18. A. Z. Akcasu, G. C. Summerfield, S. N. Jahshan, C. C. Han, C. Y. Kim and H. Yu, *J. Polym. Sci., Polym. Phys. Ed.*, 1980, **18**, 863.
19. W. Gawrisch, M. G. Brereton and E. W. Fischer, *Polym. Bull. (Berlin)*, 1981, **4**, 687.
20. G. D. Wignall, R. W. Hendricks, W. C. Koehler, J. S. Lin, M. P. Wai, E. L. Thomas and R. S. Stein, *Polymer*, 1981, **22**, 886.
21. F. Boué, M. Nierlich and L. Leibler, *Polymer*, 1982, **23**, 29.
22. S. N. Jahshan and G. C. Summerfield, *J. Polym. Sci., Polym. Phys. Ed.*, 1980, **18**, 1859.
23. J. T. Koberstein, *J. Polym. Sci., Polym. Phys. Ed.*, 1982, **20**, 593.
24. G. C. Summerfield, *J. Polym. Sci., Polym. Phys. Ed.*, 1981, **19**, 1011.
25. C. Tangari, J. A. King and G. C. Summerfield, *Macromolecules*, 1982, **15**, 132.
26. P. Debye, Rubber Reserve Co. Technical Report CR637, 1945. Reprinted in 'Light Scattering from Dilute Polymer Solutions', ed. D. McIntyre and F. Gornick, Gordon and Breach, New York, 1964.
27. P. J. Flory, 'Principles of Polymer Chemistry', Cornell University Press, Ithaca, NY, 1953.
28. D. M. Sadler, *J. Appl. Crystallogr.*, 1983, **16**, 519.
29. G. C. Summerfield and D. F. R. Mildner, *J. Appl. Crystallogr.*, 1983, **16**, 384.
30. G. Hadziioannou, C. Picot, A. Skoulios, M. L. Ionescu, A. Mathis, R. Duplessix, Y. Gallot and J. P. Lingelser, *Macromolecules*, 1982, **15**, 263.
31. P. Debye, H. R. Anderson and H. Brumberger, *J. Appl. Phys.*, 1957, **28**, 679.
32. G. Porod, *Kolloid-Z.*, 1951, **124**, 83; 1952, **125**, 51, 108.
33. L. S. Ornstein and F. Zernike, *Proc. Acad. Sci., Amsterdam*, 1914, **17**, 793.
34. R. S. Rushbrooke, in 'Physics of Simple Liquids', ed. H. N. V. Temperley, J. S. Rawlinson and R. S. Rushbrooke, North-Holland, Amsterdam, 1968.
35. W. Schmatz, T. Springer, J. Schelten and K. Ibel, *J. Appl. Crystallogr.*, 1978, **7**, 96.
36. IPNS Progress Report 198–3, Argonne National Laboratory, Argonne, Illinois, 1983.
37. O. Glatter, *J. Appl. Crystallogr.*, 1977, **10**, 415.
38. R. Ghosh, 'A Computing Guide for Small Angle Scattering Experiments, ILL report 81GH29T, 1981.
39. B. Jacrot and G. Zaccai, *Biopolymers*, 1981, **20**, 2413.
40. G. D. Wignall and F. S. Bates, *J. Appl. Crystallogr.*, 1987, **20**, 28.
41. R. W. Richards, in 'Developments in Polymer Characterisation', ed. J. V. Dawkins, Applied Science Publishers, Barking, 1978, chap. 5.
42. W. Kuhn, *Kolloid Z. Z. Polym.*, 1934, **68**, 2.
43. P. J. Flory, 'Statistical Mechanics of Chain Molecules', Wiley, New York, 1969.
44. P. -G. de Gennes, 'Scaling Concepts in Polymer Physics', Cornell University Press, New York, 1979.
45. G. D. Patterson, in 'Physical Structure of the Amorphous State', ed. G. Allen and S. E. B. Petrie, Dekker, New York, 1976.
46. W. Pechold and S. Blasenbrey, *Kolloid Z. Z. Polym.*, 1970, **241**, 955.
47. J. P. Cotton, B. Farnoux, G. Jannink, J. Mons and C. Picot, *C. R. Hebd. Seances Acad. Sci., Ser. C*, 1972, **275**, 175.
48. H. Benoit, D. Decker, J. S. Higgins, C. Picot, J. P. Cotton, B. Farnoux, G. Jannink and R. Ober, *Nature (London), Phys. Sci.*, 1973, **245**, 13.
49. D. G. H. Ballard, G. D. Wignall and J. Schelten, *Eur. Polym. J.*, 1973, **9**, 965.
50. R. G. Kirste, W. A. Kruse and J. Schelten, *Makromol. Chem.*, 1973, **162**, 299.
51. J. P. Cotton, P. Decker, H. Benoit, B. Farnoux, G. Higgins, G. Jannink, R. Ober, C. Picot and J. des Cloizeaux, *Macromolecules*, 1974, **7**, 863.
52. R. G. Kirste and B. R. Lehnen, *Makromol. Chem.*, 1976, **177**, 1137.
53. G. Lieser, E. W. Fischer and K. Ibel, *J. Polym. Sci., Polym. Lett. Ed.*, 1975, **13**, 39.
54. J. M. Guenet, C. Picot and H. Benoit, *Macromolecules*, 1979, **12**, 86.
55. D. G. H. Ballard, A. N. Burgess, P. Cheshire, E. W. Janke, A. Nevin and J. Schelten, *Polymer*, 1981, **22**, 1353.
56. K. P. McAlea, J. M. Schultz, K. H. Gardner and G. D. Wignall, *Macromolecules*, 1985, **18**, 447.
57. J. Kugler, E. W. Fischer, M. Peuscher and C. D. Eisenbach, *Makromol. Chem.*, 1983, **184**, 2325.
58. J. M. O'Reilly, D. M. Telgarden and G. D. Wignall, *Macromolecules*, 1985, **18**, 2747.
59. R. Gennant, W. Pechold and H. P. Grossmann, *Colloid Polym. Sci.*, 1977, **225**, 285.
60. F. Boué, *Adv. Polym. Sci.*, 1987, **82**, 47.
61. R. Duplessix, J. P. Cotton, H. Benoit and C. Picot, *Polymer*, 1979, **20**, 1181.
62. L. Leibler and H. Benoit, *Polymer*, 1981, **22**, 195.
63. H. R. Brown, M. Kasakevich and G. D. Wignall, *Polymer*, 1986, **27**, 1345.
64. D. M. Sadler, in 'Crystalline Polymers', ed. I. H. Hall, 1984.
65. A. Keller, *Philos. Mag.*, 1957, **2**, 1171.
66. D. C. Bassett, F. C. Frank and A. Keller, *Philos. Mag.*, 1963, **8**, 1739 and 1753.
67. P. J. Flory, *J. Am. Chem. Soc.*, 1962, **84**, 2857.
68. For example, L. Mandelkern, *Discuss. Faraday Soc.*, 1979, **68**, 310.

69. E. W. Fischer and G. F. Schmidt, *Angew. Chem.*, 1962, **74**, 551.
70. D. M. Sadler, *Polymer*, 1983, **24**, 1401.
71. D. M. Sadler and G. H. Gilmer, *Polymer*, 1984, **25**, 1446.
72. D. M. Sadler and A. Keller, *Science*, 1979, **203**, 263.
73. J. Schelten, G. D. Wignall and D. G. H. Ballard, *Polymer*, 1974, **15**, 682.
74. D. G. H. Ballard, G. W. Longman, T. L. Crowley, A. Cunningham and J. Schelten, *Polymer*, 1979, **20**, 399.
75. D. G. H. Ballard, P. Cheshire, G. W. Longman and J. Schelten, *Polymer*, 1978, **19**, 379.
76. D. G. H. Ballard, A. N. Burgess, A. Nevin, P. Cheshire, G. W. Longman and J. Schelten, *Macromolecules*, 1980, **13**, 677.
77. J. M. Guenet, *Macromolecules*, 1980, **13**, 387.
78. J. M. Guenet, *Polymer*, 1981, **22**, 313.
79. J. M. Guenet, and C. Picot, *Macromolecules*, 1983, **16**, 205.
80. B. Crist, W. W. Graessley and G. D. Wignall, *Polymer*, 1982, **23**, 1561.
81. E. W. Fischer, K. Hahn and A. R. Rennie, ILL reports, 1984.
82. D. M. Sadler and A. Keller, *Macromolecules*, 1978, **10**, 1128.
83. J. Schelten, D. G. H. Ballard, G. D. Wignall, G. W. Longman and W. Schmatz, *Polymer*, 1976, **17**, 751.
84. G. C. Summerfield, J. S. King and R. Ullman, *J. Appl. Crystallogr.*, 1978, **11**, 548.
85. M. Stamm, E. W. Fischer, M. Dettenmaier and P. Convert, *Faraday Discuss. Chem. Soc.*, 1979, **68**, 263.
86. M. Dettenmaier, E. W. Fischer and M. Stamm, *Colloid Polym. Sci.*, 1980, **258**, 343.
87. D. M. Sadler and R. Harris, *J. Polym. Sci., Polym. Phys. Ed.*, 1982, **20**, 561.
88. E. W. Fischer, K. Hahn, J. Kugler, U. Stouth, R. Born and M. Stamm, *J. Polym. Sci., Polym. Phys. Ed.*, 1984, **22**, 1491.
89. F. C. Frank, 'Growth Perfect. Cryst., Proc. Int. Conf.', ed. R. H. Doremus, B. W. Roberts and D. Turnbull, Wiley, New York, 1958, p. 529.
90. F. C. Frank, *Faraday Discuss. Chem. Soc.*, 1979, **68**, 7.
91. D. M. Sadler, *Faraday Discuss. Chem. Soc.*, 1979, **68**, 106.
92. C. M. Guttman, E. A. DiMarzio and J. D. Hoffman, *Polymer*, 1981, **22**, 597 and 1466.
93. D. Y. Yoon and P. J. Flory, *Polymer*, 1977, **18**, 509.
94. J. W. Gilmer, D. Wiswe, H. G. Zachmann, J. Kugler and E. W. Fischer, *Polymer*, 1986, **27**, 1391.
95. D. M. Sadler and S. J. Spells, *Polymer*, 1984, **25**, 1219.
96. E. A. DiMarzio, C. M. Guttman and J. D. Hoffman, *Faraday Discuss. Chem. Soc.*, 1979, **68**, 210.
97. D. Y. Yoon and P. J. Flory, *Faraday Discuss. Chem. Soc.*, 1979, **68**, 288.
98. P. J. Barham, D. A. Jarvis and A. Keller, *J. Polym. Sci., Polym. Phys. Ed.*, 1982, **20**, 1717.
99. D. M. Sadler, *Polym. Commun.*, 1985, **26**, 204.
100. S. J. Spells and D. M. Sadler, *Polymer*, 1984, **25**, 739.
101. D. M. Sadler, S. J. Spells, A. Keller and J. M. Guenet, *Polym. Commun.*, 1984, **25**, 290.
102. (a) G. D. Wignall, L. Mandelkern, C. Edwards and M. Glothin, *J. Polym. Sci., Polym. Phys. Ed.*, 1982, **20**, 245;
     (b) M. Stamm, *J. Polym. Sci., Polym. Phys. Ed.*, 1982, **20**, 235.
103. S. J. Spells, A. Keller and D. M. Sadler, *Polymer*, 1984, **25**, 749.
104. G. Avitabile, R. Napolitano, B. Pirozzi, K. D. Rouse, M. W. Thomas and B. T. M. Willis, *J. Polym. Sci., Polym. Lett. Ed.*, 1975, **13**, 351.
105. U. Kapelsky, E. W. Fischer, P. Herchenroder, J. Schelten, G. Lieser and G. Wegner, *J. Polym. Sci., Polym. Phys. Ed.*, 1979, **17**, 2117.
106. R. W. Richards, *Adv. Polym. Sci.*, 1985, **71**, 1.
107. J. A. Miller, S. L. Cooper, C. C. Han and G. Pruckmayr, *Macromolecules*, 1984, **17**; 1063.
108. E. Helfand and Z. R. Wasserman, in 'Developments in Block Copolymers', ed. I. Goodman, Applied Science, Barking, Essex, 1982, chap. 4.
109. E. Helfand, in 'Polymer Compatibility and Incompatibility', ed. K. Sole, M.M.I. Press, Midland, Michigan, 1983; E. Helfand and A. M. Sapse, *J. Chem. Phys.* 1975, **62**, 1327.
110. R. W. Richards and J. L. Thomason, *Polymer*, 1981, **22**, 581.
111. F. S. Bates, C. V. Berney and R. E. Cohen, *Macromolecules*, 1983, **16**, 1101.
112. T. Hashimoto, M. Shibayama and H. Kawai, *Macromolecules*, 1980, **13**, 1237 and 1660.
113. F. S. Bates, R. E. Cohen and C. V. Berney, *Macromolecules*, 1982, **15**, 589.
114. R. W. Richards and J. L. Thomason, *Macromolecules*, 1983, **16**, 982.
115. R. W. Richards and J. L. Thomason, *Macromolecules*, 1985, **18**, 452.
116. F. S. Bates, C. V. Berney, R. E. Cohen and G. D. Wignall, *Polymer*, 1983, **24**, 519.
117. J. A. Miller, G. Pruckmayr, E. Epperson and S. L. Cooper, *Polymer*, 1985, **26**, 1915.
118. J. A. Miller, J. M. McKenna, G. Pruckmayr, J. E. Epperson and S. L. Cooper, *Macromolecules*, 1985, **18**, 1727.
119. F. S. Bates and M. A. Hartney, *Macromolecules*, 1985, **18**, 2478.
120. D. R. Paul and S. Newman (eds.), 'Polymer Blends', Academic Press, New York, 1978.
121. E. M. Lifshitz and L. P. Pitaevskii, 'Statistical Physics', part 1, vol. 5 of 'Course of Theoretical Physics', ed. L. D. Landau and E. M. Lifshitz, Pergamon Press, Oxford, 1980.
122. P. J. Flory, R. A. Orwall and A. J. Vrij, *J. Am. Chem. Soc.*, 1964, **86**, 3507.
123. D. Patterson, S. N. Bhattacharyya and P. Picker, *Trans. Faraday Soc.*, 1968, **64**, 648.
124. R. H. Lacombe and I. C. Sanchez, *J. Phys. Chem.*, 1976, **80**, 2568.
125. M. Warner, J. S. Higgins and A. J. Carter, *Macromolecules*, 1983, **16**, 1931.
126. R. S. Stein and G. Hadziioannou, *Macromolecules*, 1984, **17**, 1059.
127. H. Benoit and M. Benmouna, *Polymer*, 1984, **25**, 1059.
128. R. G. Hill, P. E. Tomlins and J. S. Higgins, *Macromolecules*, 1985, **18**, 2555.
129. R. Koningsveld and L. A. Kleintjens, *J. Polym. Sci., Polym. Symp.*, 1977, **61**, 221.
130. A. Lapp, C. Picot and H. Benoit, *Macromolecules*, 1985, **18**, 2437.
131. F. S. Bates, S. B. Dierker and G. D. Wignall, *Macromolecules*, 1986, **19**, 1938.
132. F. S. Bates, G. D. Wignall and W. C. Koehler, *Phys. Rev. Lett.*, 1985, **55**, 2425.
133. F. S. Bates and G. D. Wignall, *Macromolecules*, 1986, **19**, 932.
134. A. D. Buckingham and H. G. E. Hentschel, *J. Polym. Sci., Polym. Phys. Ed.*, 1980, **18**, 853.

135. I. Prigogine, 'The Molecular Theory of Solutions', 1957, North-Holland, Amsterdam, chaps. XIX and XX.
136. H. Yang, R. S. Stein, C. C. Han, B. J. Bauer and E. J. Kramer, *Polym. Commun.*, 1986, **27**, 132.
137. W. A. Kruse, R. G. Kirste, J. Haas, B. J. Schmitt and D. J. Stein, *Makromol. Chem.*, 1976, **177**, 1145.
138. J. Jelenic, R. G. Kirste, B. J. Schmitt and S. Schmitt-Strecker, *Makromol. Chem.*, 1979, **180**, 2057.
139. B. J. Schmitt, R. G. Kirste and J. Jelenic, *Makromol. Chem.*, 1979, **181**, 1655.
140. J. Jelenic, R. G. Kirste, R. C. Oberthur and S. Schmitt-Strecker, *Makromol. Chem.*, 1984, **185**, 129.
141. M. Shibayama, H. Yang, R. S. Stein and C. C. Han, *Macromolecules*, 1985, **18**, 2179.
142. G. Hadziioannou and R. S. Stein, *Macromolecules*, 1984, **17**, 567.
143. J. S. Higgins and A. J. Carter, *Macromolecules*, 1984, **17**, 2197.
144. P. E. Tomlins and J. S. Higgins, *Polymer*, 1985, **26**, 1554.
145. A. Maconnachie, R. P. Kambour and R. C. Bopp, *Macromolecules*, 1984, **25**, 357.
146. R. -J. Roe and D. Rigby, *Adv. Polym. Sci.*, 1987, **82**, 103.
147. A. Maconnachie, R. P. Kambour, D. W. White, S. Rostami and D. J. Walsh, *Macromolecules*, 1984, **17**, 2645.
148. F. C. Stehling, E. Ergos and L. Mandelkern, *Macromolecules*, 1971, **4**, 672.
149. A. D. English, P. Smith and D. E. Axelson, *Polymer*, 1985, **26**, 1523.
150. D. M. Sadler, unpublished results.
151. W. Wu, *Polymer*, 1983, **24**, 43.
152. W. Wu and G. D. Wignall, *Polymer*, 1985, **26**, 661.
153. J. Schelten, G. D. Wignall, D. G. H. Ballard and G. W. Longman, *Polymer*, 1977, **18**, 1111.
154. G. D. Wignall, H. R. Child and R. J. Samuels, *Polymer*, 1982, **23**, 957.
155. J. M. Guenet and C. Picot, *Macromolecules*, 1981, **14**, 309.
156. C. Picot, R. Duplessix, D. Decker, H. Benoit, F. Boué, J. P. Cotton, M. Dauod, B. Farnoux, G. Jannink, M. Nierlich, A. J. de Vries and P. Pincus, *Macromolecules*, 1977, **10**, 436.
157. G. Hadziioannou, L. H. Wang, R. S. Stein and R. S. Porter, *Macromolecules*, 1982, **15**, 880.
158. M. Dettenmaier, A. Maconnachie, J. S. Higgins, H. H. Kausch and T. O. Nguyen, *Macromolecules*, 1986, **19**, 773.
159. J. M. Lefebvre, B. Escaig and C. Picot, *Polymer*, 1982, **23**, 1751.
160. J. M. Lefebvre, B. Escaig, G. Coulon and C. Picot, *Polymer*, 1985, **26**, 1807.
161. I. M. Ward, 'Mechanical Properties of Solid Polymers', Wiley-Interscience, New York, 1983; *Macromolecules*, 1977, **10**, 436.
162. A. Peterlin, *Adv. Chem. Ser.*, 1975, **142**, 1.
163. P. Smith, P. J. Lemstra and H. C. Boorj, *J. Polym. Sci., Polym. Phys. Ed.*, 1981, **19**, 877.
164. A. Peterlin and F. J. Balta-Calleja, *Kolloid Z. Z. Polym.*, 1970, **242**, 1093.
165. D. M. Sadler and P. J. Barham, *J. Polym. Sci., Polym. Phys. Ed.*, 1983, **21**, 309.
166. R. S. Stein *et al.*, personal communication.
167. D. G. H. Ballard, P. Cheshire, E. Janke, A. Nevin and J. Schelten, *Polymer*, 1982, **23**, 1875.
168. D. M. Sadler and P. J. Barham, ILL report, 1986.
169. S. F. Edwards, *Proc. R. Soc. London*, 1967, **9**, 92.
170. P. G. de Gennes, *J. Chem. Phys.*, 1971, **55**, 572.
171. A. Maconnachie, G. Allen and R. W. Richards, *Polymer*, 1981, **22**, 1157.
172. F. Boué, M. Nierlich and K. Osaki, *Faraday Symp. Chem. Soc.*, 1983, **18**, 83.
173. F. Boué, M. Nierlich, G. Jannink and R. C. Ball, *J. Phys. (Orsay, Fr.)*, 1982, **43**, 137.
174. R. Ullman, in 'Elastomers and Rubber Elasticity', eds. J. E. Mark and J. Lal, American Chemical Society, Washington, 1982.
175. R. Ullman, *Macromolecules*, 1982, **15**, 582 and 1395.
176. J. A. Hinkley, C. C. Han, B. Mozer and H. Yu, *Macromolecules*, 1978, **11**, 836.
177. S. B. Clough, A. Maconnachie and G. Allen, *Macromolecules*, 1980, **13**, 774.
178. M. Beltzung, C. Picot, P. Rempp and J. Herz, *Macromolecules*, 1982, **15**, 1594.
179. M. Beltzung, C. Picot and J. Herz, *Macromolecules*, 1984, **17**, 663.
180. F. Boué, J. Bastide, A. Lapp, J. Herz, C. Picot and B. Farnoux, *Europhys. Lett*, 1986, **1**, 637.
181. J. G. Weissman and L. H. Sperling, *Macromolecules*, 1985, **18**, 1720.
182. S. J. Bai, *Polymer*, 1985, **26**, 1053.
183. W. Wu and B. J. Baner, *Polymer*, 1986, **27**, 169.
184. A. Eisenberg, *Macromolecules*, 1970, **3**, 147.
185. T. R. Earnest, J. S. Higgins, D. L. Handlin and W. J. MacKnight, *Macromolecules*, 1981, **14**, 192.
186. W. C. Forsman, W. J. MacKnight and J. S. Higgins, *Macromolecules*, 1984, **17**, 490.
187. W. C. Forsman, *Macromolecules*, 1982, **15**, 1032.
188. T. R. Earnest, J. S. Higgins and W. J. MacKnight, *Macromolecules*, 1982, **15**, 1390.
189. E. J. Roche, R. S. Stein and W. J. MacKnight, *J. Polym. Sci., Polym. Phys. Ed.*, 1980, **18**, 1035.
190. R. G. Kirste and H. G. Ohm, *Makromol. Chem.*, 1985, **6**, 179.
191. P. Keller, B. Carvalko, J. P. Cotton, M. Lambert, F. Moussa and G. Pepy, *J. Phys. Lett., (Orsay, Fr.)*, 1985, **46**, L1095.
192. J. S. Higgins, in 'Developments in Polymer Characterization', ed. J. V. Dawkins 1982, Applied Science Publishing Ltd., Barking, 1982, chap. 4.
193. B. Gabrys, J. S. Higgins and D. A. Young, *Polymer*, 1985, **26**, 355.
194. J. F. Twistleton, J. W. White and P. A. Reynolds, *Polymer*, 1982, **23**, 578.
195. D. Heyer, U. Buchenau and M. Stamm, *J. Polym. Sci., Polym. Phys. Ed.*, 1984, **22**, 1515.
196. L. K. Nicholson and J. B. Hayter, *Polymer*, 1981, **22**, 137.
197. D. Richter, A. Baumgartner, K. Binder, B. Ewen and J. B. Hayter, *Phys. Rev. Lett.*, 1982, **48**, 1695.
198. F. Mezei, *J. Phys.*, 1972, **225**, 146.
199. M. R. Anderson, M. B. M. Hasegawa, D. K. Steinman, J. W. White and R. Currat, *Polymer*, 1982, **23**, 569.
200. C. Picot, *Prog. Colloid Polym. Sci.*, 1977, **75**, 83.
201. F. Boue, J. Bastide, M. Buzier, C. Colletti, A. Lapp and J. Hertz, *Prog. Colloid Polym. Sci.*, 1987, **75**, 152.
202. J. S. Higgins and J. E. Roots, *J. Chem. Soc., Faraday Trans. 2*, 1985, **81**, 757.
203. J. S. Higgins, *Br. Polym. J.*, 1987, **19**, 103.

33

Optical Microscopy

DEREK HEMSLEY
University of Technology, Loughborough, UK

33.1 INTRODUCTION

The history of the light microscope extends back some 200 years, a factor of about four times the duration that the polymer industry has been in existence. It is therefore perhaps initially surprising that such an established instrument plays an important role in investigating the microstructure of modern polymer systems. However, a closer examination of the evolution of the present day instrument and the factors behind this evolution over the past 50 years shows why this is so. The methods of examination of polymers using a light microscope are, with one notable exception, those familiar to, and regularly used by, biologists. This is hardly surprising since 'natural' and 'synthetic' polymers obviously have much in common at the microscopic level. Thus the use of phase contrast, interference and fluorescence techniques is widespread and common to both the biological and synthetic polymer fields, although developed primarily with the former in mind.

765

Although the physical and mechanical properties of most plastics and rubbers differ significantly from those of biological specimens, the primary method of specimen preparation for examination in the light microscope, thin sectioning, is nevertheless basically the same. Transmitted light methods of observation tend to dominate over the reflected light techniques traditionally used by metallographers and other materials microscopists. As discussed below, the limited amount of reflected light work carried out on polymer specimens is mainly for the examination of surfaces. These may be intentionally or unintentionally produced fracture surfaces, or surfaces arising from the manufacture of polymer products such as films, fibres or mouldings. Nevertheless some of the most useful basic literature for the polymer light microscopist is that intended for students of biology.[1, 2] The exception mentioned above is the use of transmitted polarized light methods. These are important in the observation of the crystalline structure of polymeric materials and molecular orientation effects. Both qualitative and quantitative data are available. The transmitted light polarization techniques used are 'stolen' from mineralogists, petrologists and ceramicists rather than biologists and the instruments used are standard polarizing microscopes.

It is a characteristic of polymer light microscopy that a wide variety of methods might need to be employed in the examination of a particular specimen if the maximum amount of information is to be extracted from it. This imposes demands on the flexibility of the equipment. Not all light microscopes will necessarily conveniently accommodate all the systems needed. In practice this usually dictates the use of 'research' type instruments rather than those of simpler design and construction, which are often dedicated to a more limited range of functions. Figure 1 illustrates the types of technique commonly used to examine polymer specimens.

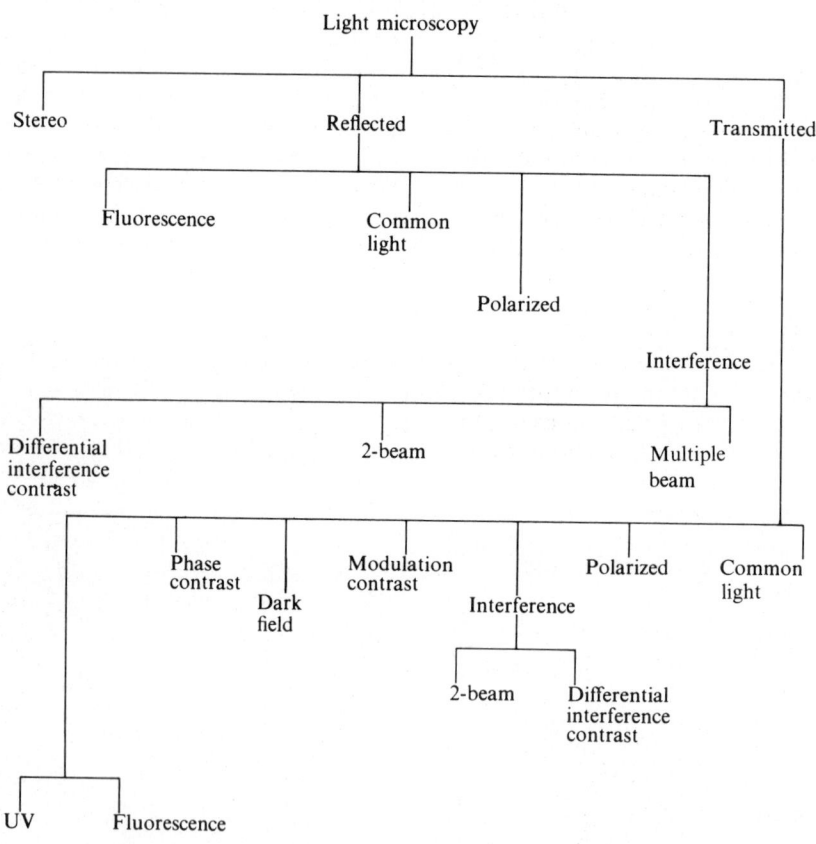

Figure 1 Light microscopy techniques used to examine polymers

33.2 THE PREPARATION OF SPECIMENS

Certain types of specimen require very little preparation for light microscopy, others need more care and the use of specialized equipment. The methods of preparation employed depend not only on the mechanical and chemical characteristics of the polymer, but also on its physical form.

Small grains, powders or large particle size latices can be examined directly once dispersed on a microscope slide and, should the optical system demand it, covered with a coverslip. It may be necessary, as in the case of a latex with a high solids content, to reduce the concentration substantially if a single layer of particles on the slide is to be achieved. Considerable care is then required to preserve the original size distribution in the final preparation. An illustration of this would be the dilution of a latex with a liquid phase which it is known will not destroy its stability. Dry powders are often best examined as a suspension in a suitable fluid. Note however that the apparent size of particles can be influenced by the relative refractive index of the particles and the liquid.[3]

Bulk specimens can be prepared by a variety of methods. The majority of synthetic polymers can be thin sectioned by microtomy for transmitted light examination. This generally yields more information on internal structure than reflected light and so, given a choice, is the preferred method.

33.2.1 Microtomy

The almost universally adopted method of producing thin sections is by microtomy. This is a well-established procedure in the biological and medical fields and the basic techniques are extensively documented.[4] Polymers can pose special problems because the spectrum of physical properties is so broad and the standard methods may need modification. It has been suggested that, for optimum results, sectioning should be carried out at a temperature just below the glass/rubber transition temperature (T_g). Certainly rigid poly(vinyl chloride), polypropylene, most polyamides, polyacetals, polystyrene and low molecular weight polyacrylics present few practical problems.

Materials with their T_g values well below ambient temperature need to be cooled to enable them to be cut satisfactorily. Failure to do this not only leads to difficulty in obtaining sections, but gross deformation can occur, resulting in modification of the microstructure. Polymers in this category include natural rubber, polybutadiene, *block*-copoly(styrene/butadiene/styrene) and silicone rubber. Cooling is carried out using solid or liquid carbon dioxide, liquid nitrogen or Peltier effect devices, although the latter are normally effective down to only modestly low temperatures.

In the case of semicrystalline polymers with a low amorphous phase T_g it appears that this dictates the sectioning temperature unless the degree of crystallinity is especially high. Thus low density polyethylene generally needs substantially greater cooling than high density polyethylene. Elevated temperatures have been used to section specimens which would otherwise be excessively brittle. High molecular weight poly(methyl methacrylate) falls into this category.

The optimum section thickness for light microscopy depends upon a number of factors, which include the size of the structural features of interest, the characteristics of the optical system being used for observation and the optical properties of the polymer. An example of where a thicker than normal section might be advantageous occurs when investigating optically anisotropic specimens in the polarizing microscope under circumstances where the intrinsic birefringence is low. Adequate image contrast may then be obtained only if the section thickness, and hence the optical path difference (OPD) displayed, is increased. Conversely, to optimize contrast in a phase contrast microscope it may be necessary to substantially reduce section thickness. The normal section thickness range is between $5\,\mu m$ and $10\,\mu m$, but under the two circumstances quoted above thicknesses of $40\,\mu m$ and $1\,\mu m$ respectively would be more appropriate.

The microtomy of polymers, particularly those with less than ideal cutting characteristics, tends to present many practical difficulties such as distortion, knife marking (due to knife edge damage) and judder marks (due to mechanical instability of the specimen or microtome produced by high cutting forces). Nevertheless microtomy continues to be the main method of preparation because these problems can be mitigated by the choice of appropriate equipment. This ranges from ultramicro-tomes furnished with a glass knife for thin small area sections, to large 'whole moulding' motor-driven microtomes with tungsten carbide tipped knives and cutting forces of up to $3\,kN$.

33.2.2 Grinding and Polishing Techniques

Some workers have successfully used grinding and polishing methods to produce thin sections in the manner normally used for petrological specimens.[5] The methods are particularly relevant when very large area sections are required, or when dealing with brittle polymers or polymer/inorganic composites. Structure modification by heating can occur if machines are used to assist with preparation of this kind so that it is essential to use suitable lubricants or coolants.

For reflected light work, specimens can be treated in similar ways to those established for metals for the production of smooth highly reflecting surfaces.[6] In the case of a metal this surface might subsequently be etched to enhance the visibility of the microstructure. A similar approach can be taken with polymers, but it is generally less popular. Ion etching[7] and chemical etching[8, 9] have both been used to reveal internal microstructure. In general these procedures need to be carried out with care and the results interpreted with caution.

33.2.3   Embedding

Small specimens, such as dry polymer powders, fibres or thin films, need to be held for sectioning by embedding them in a suitable carrier material. Waxes and epoxy or acrylic resins are widely used for this purpose. Other materials are discussed by Mason,[10] who also summarizes many of the preparative methods described above.

33.2.4   Fracture Surfaces and Replicas

The internal microstructure of a brittle polymer may be revealed for examination by fracturing. Materials which are ductile at normal temperatures and impact rates are embrittled by immersion in liquid nitrogen prior to fracturing. Although this technique has much appeal because of its simplicity, Andrews[11] has rightly pointed out that since the path of a fracture will be guided by the microstructure, then quantitative interpretation of the exposed surfaces is suspect. Rarely is a fracture surface a random plane through a polymer. As an alternative to direct examination of a surface, it is common procedure to replicate the surface and observe the replica by either reflected or transmitted light microscopy. As an example, a polyethylene surface may be replicated by thinly coating it with a solution (10% v/v) of polystyrene in toluene. After evaporation of the solvent a thin polystyrene replica can be peeled away from the polyethylene. This is then 'shadowed' by evaporating a suitable metal (*e.g.* aluminum) on to it at an angle. Variations in the thickness of the metal layer then correspond to slopes on the original surface. This coated replica may be examined by reflected light, or, after immersing it in a liquid with refractive index close to that of polystyrene, by transmitted light.

33.2.5   Melt Pressing

Thin layers of polymer for transmitted light observation can be obtained by pressing out small pieces between slide and coverslip at an elevated temperature. On cooling, such specimens are examined in the usual way but clearly they suffer from the disadvantage that the microstructure will have been modified by the thermal treatment. Also it is difficult to obtain sufficiently thin films, especially from high viscosity polymers.

33.2.6   Staining Methods

It is usually impossible to incorporate selective stains into polymers in a high enough concentration to significantly enhance contrast in thin sections. An exception is the use of osmium tetroxide to stain unsaturated rubbers. This technique was first used on rubbers to generate contrast in electron microscope images of these materials by increased electron scattering. The intense brown/black colour it imparts may however be used for contrast enhancement in light microscopy.

33.3   BRIGHT AND DARK FIELD MICROSCOPY

The absorbtion of light by pure polymer specimens is usually negligible in the visible region of the electromagnetic spectrum. The implications of this are that, in the light microscope, polymeric specimens are frequently seen as 'phase objects' needing special optical techniques to reveal structural information. The use of simple bright field transmitted light microscopy on such specimens is therefore very restricted. In reflected light, observations are confined to the study of relatively rough surfaces which, provided the specimen is capable of withstanding the hostile environment in an electron microscope, would probably be better imaged with a scanning electron microscope (SEM).

For real polymer systems containing additives such as fillers, stabilizers and pigments, there is more scope for the bright field technique, especially if the additives are particulate in nature. A major area of technological interest is quantification of the distribution and dispersion of such additives. Provided the image contrast is high enough, automatic image analysis methods can be employed. Results obtained in this way have been used to explore the effect of different mixing procedures on the distribution and dispersion of pigments.[12]

The normal bright field microscope can be modified in such a way as to exclude any undiffracted or unscattered light from the image-forming process. The instrument is then operating in the 'dark field' mode. Contrast in the image arises wherever there is a redistribution of the light passing through, or reflected from, the specimen. The background light intensity is theoretically zero and, as a result, very high image contrast is available. Thus in the case of a pigmented polymer specimen individual pigment particles appear bright against a dark background. Images of this type also lend themselves to quantitative interpretation and dark field methods have been used on a routine basis[13] to determine the distribution of carbon black in rubber samples. This method was considered more reliable and capable of working over a wider dispersion range than the traditional method of analysis based on the Cabot carbon microscopy technique, which involves very thin sectioning and highly subjective analysis.[14]

Dark field methods are also particularly useful in the reflected light mode for the investigation of subsurface features in transparent polymers. This is because specular reflection from the polymer surface takes no part in image production so that visibility of internal features is increased.

Figure 2 illustrates the character of the dark field image in comparison with bright field and includes a phase contrast image of the same specimen. This is discussed below.

33.4 POLARIZED LIGHT MICROSCOPY

Since the very earliest days of polymer science the polarizing microscope has been used to study crystalline materials and the crystalline texture of bulk polymers. The individual polymer molecule is in most cases significantly optically anisotropic and the implications of this for light microscopy are profound. Although a totally disordered polymer (such as a polymer melt) might be expected to be isotropic, any ordering of the material can give rise to marked optical anisotropy, with refractive index being a function of direction in the material. If the size of any ordered regions is within the resolution range of the microscope, then they may be observed and characterized. They will appear 'birefringent' or 'doubly refracting' when examined in the polarizing microscope.

The polarizing microscope itself is basically a standard instrument fitted with a pair of polarizing filters, one above and one below the specimen. In practice there are a number of additional

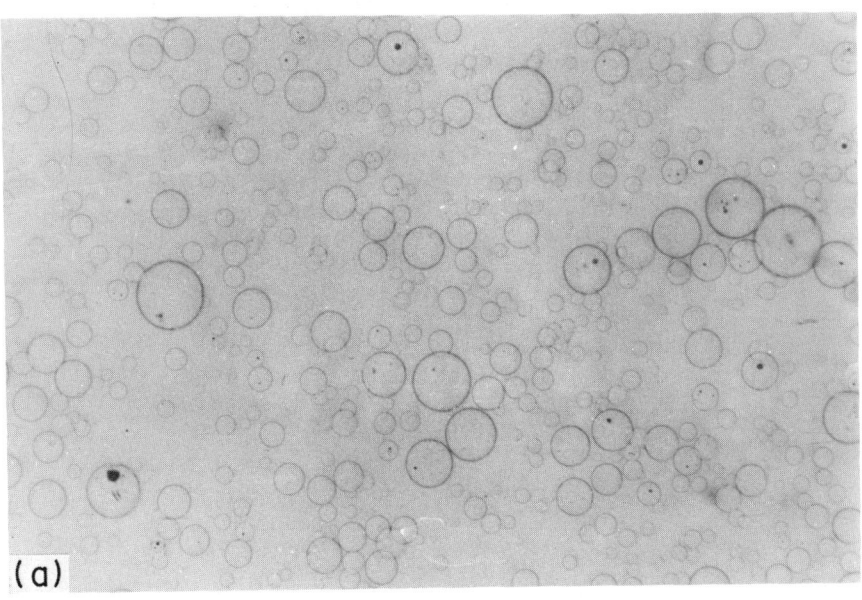

(a)

Figure 2 (a)

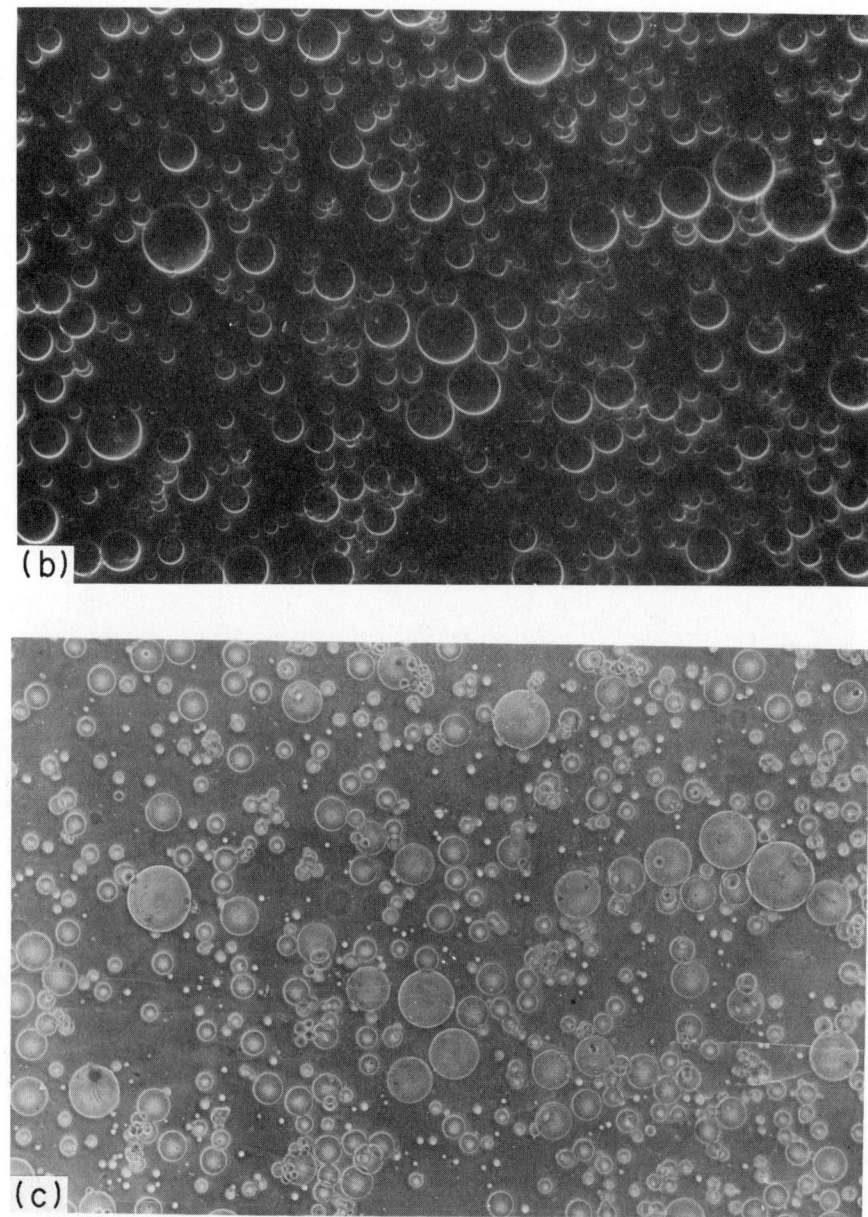

Figure 2   Spheres of poly(methyl methacrylate) imaged in three different modes: (a) bright field (common light); (b) dark field; (c) phase contrast; all × 300 magnification

requirements for the satisfactory performance and operation of the microscope and details may be found in any standard text on the subject.[15,16] These referenced texts also discuss the fundamentals of the theory of crystal optics, much of which is relevant to polymeric specimens.

If the two polarizers are in the 'crossed' position with their permitted vibration directions orthogonal, then no light will pass through the microscope in the absence of a specimen, or if the specimen is isotropic. Inserting a doubly refracting specimen gives rise to beam splitting and interference phenomena which allow light to pass through the instrument. Such specimens will then appear bright, even coloured, against a dark background.

The polarizing microscope can therefore be used in a qualitative way to image ordered regions in polymer specimens *provided the molecule is optically anisotropic*.

It is convenient to consider the applications of the polarizing instrument under a set of headings which, although rather arbitrary, do in fact reflect quite distinct areas of utilization.

33.4.1 Crystalline Texture

The most common structural feature of semicrystalline polymers that is seen in the polarizing microscope is the spherulite. This can be described as a three-dimensional assemblage of semicrystalline polymer which has grown radially from a central nucleus. These structures have been extensively studied both at the light and the electron microscope levels, the latter technique being concerned mainly with their substructure. The detailed organization of the molecular chains is still controversial but optical measurement of the sign of the birefringence of spherulites (see below) indicates a general tangential arrangement.[17] An intermediate-sized substructure directly visible in the electron microscope is the lamella.[9] Here again, their organization within the spherulite and their general nature continues to be the subject of research. In light microscopy we are concerned with coarser scale structure and need only note that these lamellae are one of the basic anisotropic units giving rise to spherulite optical anisotropy.

Figure 3 shows a typical crossed polars image of spherulites of polypropylene. On observing such an image in the microscope, various features can be noted, and in some cases quantified. The spherulite size and shape are obvious characteristics. In addition it is possible to comment on the extent to which the spherulite shows an internal 'fibrous' texture and the visibility of the dark 'maltese cross', which arises from the radial symmetry of the spherulite's structure. Some spherulites have been noted to show a concentric ring structure.[18]

Figure 3 Spherulitic structure of polypropylene; × 250 magnification, crossed polars

Certain optical characteristics of spherulites may also be measured with the microscope, but the results obtained are subject to a number of approximations and experimental errors. Keith and Padden, in a series of papers[18] concerned with the identification of four distinct types of polypropylene spherulite, showed how such characterization can be carried out. The birefringent spherulite shows two principal refractive indices, one in the radial direction and the other tangential. Their difference is a dimensionless quantity, the spherulite birefringence, which is measurable in the microscope by standard methods.[15] A spherulite is considered positive in sign if the radial refractive index is higher than the tangential. This may be checked easily in the microscope by the insertion of a birefringent accessory plate and recording whether its OPD adds to, or subtracts from, the OPD of the specimen. This is done by observing the change in polarization colours produced.

Many polymers are capable of exhibiting more than one crystal form according to the nucleation or growth conditions. These forms may be evidenced in the optical characteristics of the spherulites. This is most obvious when the sign of the spherulites changes on going from one crystal form to another, as in polypropylene (Figure 4).

The crystalline texture exhibited by products commercially manufactured from semicrystalline polymers usually differs substantially from that of laboratory-prepared specimens. In the former it is

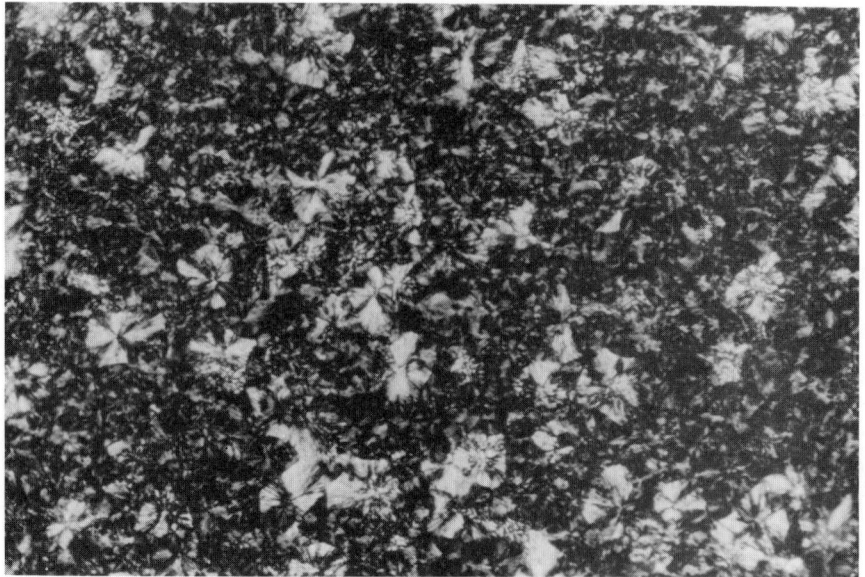

Figure 4 Spherulitic texture of injection-moulded polypropylene. Note the bright high birefringence (negative type) spherulites amongst the lower birefringence mixed forms. × 250 magnification, crossed polars

possible to relate the texture seen in thin sections to both the processing history and the likely in-service performance. Polymers vary in the ease with which this can be done. Obviously materials in which the variation in spherulite characteristics is relatively large for small differences in processing history lend themselves most effectively to this kind of analysis. For this reason optical methods have been extensively applied to polyacetals, polypropylenes, and polyamides[19] but less successfully to polyethylenes.

Since in industrial processes, such as injection moulding or extrusion, the crystallization tempera-ture will normally be a function of depth within the product, it is normal to see substantial variations in spherulitic texture between surface and interior. Indeed a section of moulding may well show a number of distinct textural layers.[20] Furthermore, many processes involve granules or pellets as feedstock. These normally show a fine scale spherulitic texture in thin section resulting from their production at high extrusion speeds followed by rapid cooling in water. Incompletely melted

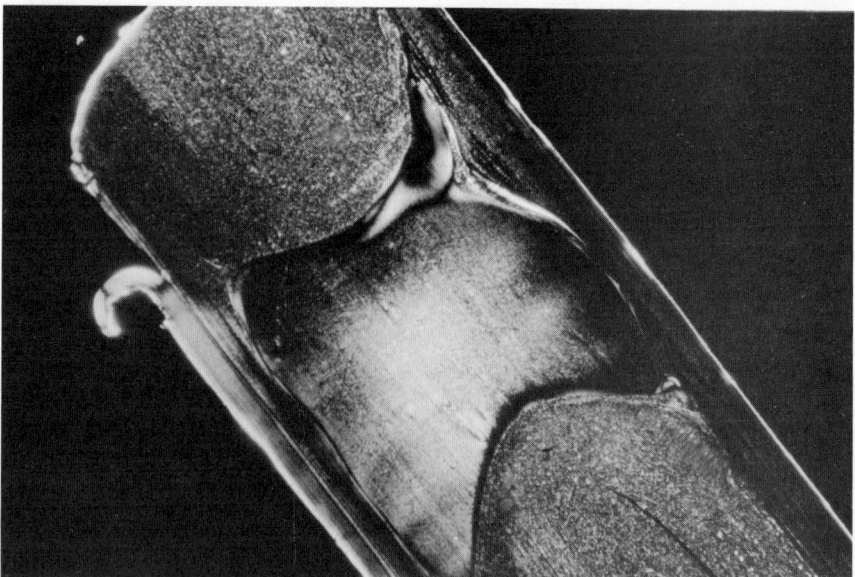

Figure 5 Incompletely melted granule in a polyethylene injection moulding; × 30 magnification, crossed polars

granules inside moulded or extruded products where the cooling conditions favour the development of large spherulitic structures can therefore be very obvious, as in Figure 5. Such features immediately cast doubt upon the machine temperature settings being employed.

The spherulitic texture of mouldings may also reveal the melt flow pattern in the mould since the rheological behaviour of the polymer can induce local homogeneous nucleation and hence modify the spherulitic texture. Similarly the meeting of flow fronts may give rise to characteristic boundaries or 'weld lines' in the moulding (Figure 6).

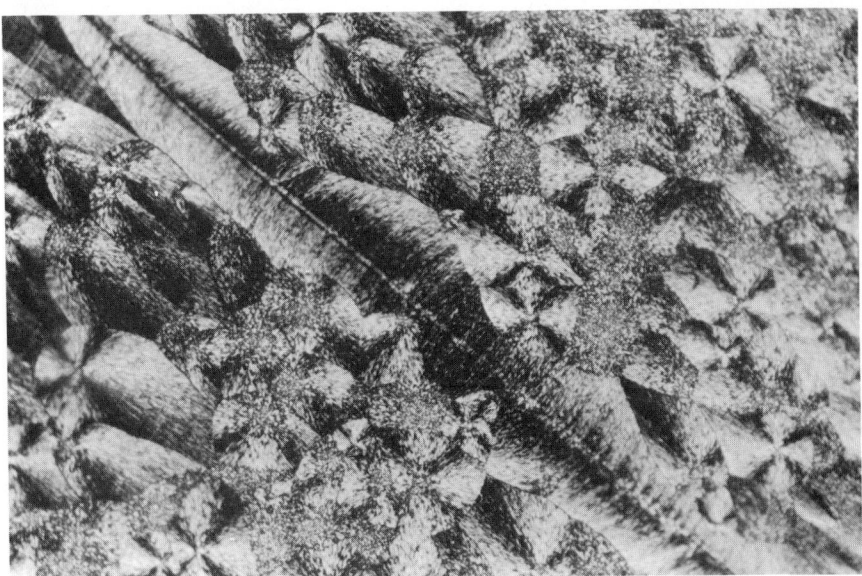

Figure 6   Internal weld line in polyacetal moulding; × 200 magnification, crossed polars

The texture of products can also be modified substantially, usually towards smaller spherulite sizes, by the incorporation of heterogeneous nucleants. In practice these may be pigments or fillers[21] as well as compounds added with the deliberate intention of reducing melt supercooling (*e.g.* benzoic acid in polypropylene). Heterogeneous nucleation may also occur at the surfaces of mouldings, as in Figure 7, and on the surface of fibres in a composite.[22]

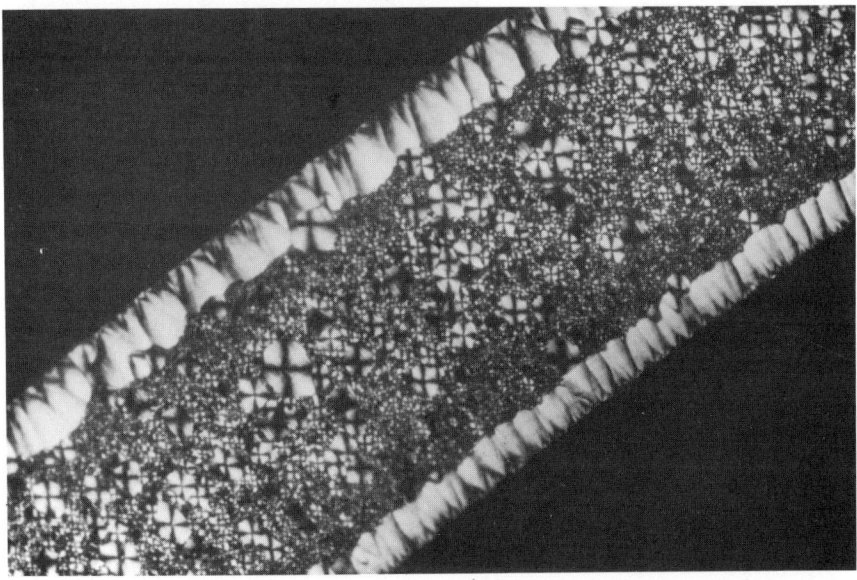

Figure 7   Surface nucleation of crystallization at the surface of extruded polypropylene sheet; cross-section, × 100, crossed polars

33.4.2 `Small Angle Light Scattering

The polarizing microscope has been used to observe the small angle light-scattering pattern (SALS) produced when polarized light passes through a specimen having spherulitic structure.[23] Details of the more usual experimental system using an optical bench, and of the interpretation of the results obtainable, have been described by Stein[24] and Samuels.[25] Using the microscope, the specimen is illuminated with near parallel light by almost closing the aperture iris in the microscope's condenser unit. Quasimonochromatic light is obtained by narrow band filtration and the system may be used with either crossed or parallel polarizers, although in practice the former is generally more easily applicable. The SALS pattern is viewed in the back focal plane of the objective lens of the microscope by means of a telescope inserted in place of the normal eyepiece, or by the use of the standard Bertrand lens often included in the tube of polarizing microscopes. A typical crossed polar SALS pattern for a polyethylene is shown in Figure 8. The separation of the diagonally opposite intensity maxima is inversely proportional to the spherulite size; thus, even in its simplest form, this technique allows spherulite size determinations. Furthermore, because of the reciprocal relationship, a small spherulite size gives a wide SALS pattern. Hence it is possible to measure sizes down to below the normal resolution limit of the light microscope. Since all the spherulites in the microscope's field of view contribute to the pattern, the measurement outlined above will give an average size.

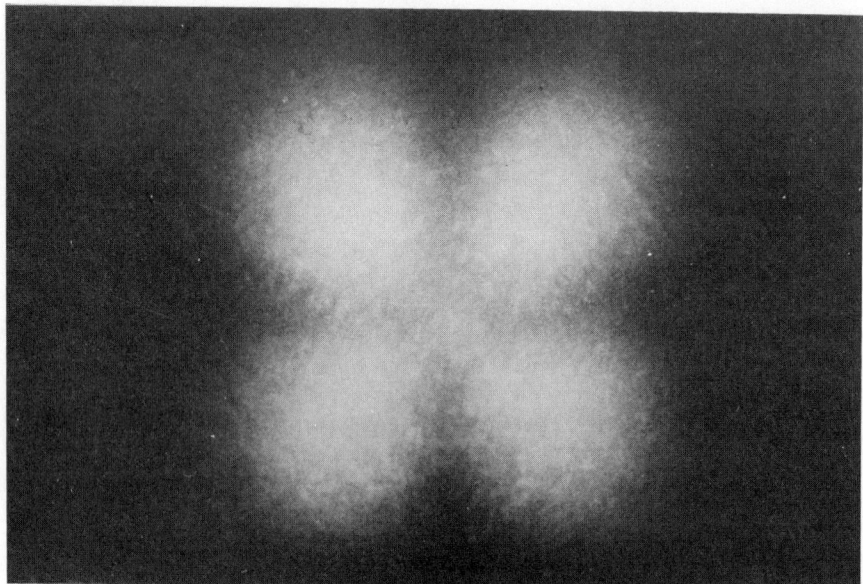

Figure 8   Polyethylene SALS pattern

33.4.3   Molecular Orientation Assessment in the Microscope

The use of birefringence measurements to assess the state of molecular orientation in polymeric materials has received considerable attention, mainly because of the relative ease with which measurements can be made. The method may be applied to both amorphous and semicrystalline polymers, although detailed interpretation of the results is usually considerably easier in the former case. Details of the theoretical background to the methods have been given by Ward,[26] Wilkes,[27] Lenk[28] and Read.[29] An interesting and rigorous review of the optical anisotropy of bulk polymers has been presented by Kawai and Nomura[30] (see Volume 2, Chapter 13 of this work) who also discuss the various techniques of orientation measurement and the information they can be expected to yield. A comparison between birefringence measurement methods, including polarized light microscopy, has been published[31] but this is limited to specimens showing high OPDs. The polarizing microscope offers the capability of highly accurate birefringence measurements on features down to a few micrometres in size. Conversely, however, large components can present practical problems. A standard polariscope is more appropriate in such cases but the accuracy of OPD measurement is reduced.

Birefringence is normally measured by first determining the OPD presented by the specimen, then applying the simple relationship $\Delta n = (\text{OPD})/t$, where t is the geometrical thickness of the specimen and Δn its birefringence. It is Δn which is directly related to molecular order. Clearly the accuracy of measurement of Δn is in part controlled by the ability to obtain an accurate value for t. For microscopical methods this is often the biggest source of error.

33.4.3.1 *Molecular orientation measurements on fibres*

The use of the polarizing microscope to characterize polymer fibres and assess orientation has a long history[32] and is well established. In many respects fibres are ideal specimens, requiring little preparation other than mounting in a liquid to remove edge diffraction effects. Generally synthetic fibres can be considered as uniaxial in optical character, with their optic axis parallel to the fibre axis, although a number of exceptions have been encountered. Three basic measurements are made between crossed polars to characterize a fibre.

(1) The extinction angle: the angle between the extinction direction and the fibre axis. This is determined by rotation of the microscope stage.

(2) The sign of birefringence: a fibre is considered positive in sign if its axial is higher than its radial refractive index. This is determined using an accessory plate (usually of 1 or 1/4 wavelength OPD), which is inserted into the microscope above or below the specimen.

(3) The magnitude of the birefringence: determined by a suitable compensation method (see below). Such measurements are usually made perpendicular to the optic axis, where the maximum birefringence for the fibre is displayed.

The results obtained can be compared with published results[33] to identify fibres in forensic work, to monitor fibre production variables or to predict other fibre properties related to molecular orientation. In the latter case the birefringence of the specimen is best compared with the maximum birefringence obtainable for a fully oriented fibre of the polymer in question.

The diameter of cylindrical fibres is measured with an eyepiece micrometer to give a figure for maximum thickness. Fibres with irregular or complex cross sections may present difficulties. The birefringence measured with the specimen mounted as described above will be the numerical difference between the axial and radial refractive indices. Using the Becke line or Van Der Kolk tests with a range of refractive index liquids[15] and a single polarizer in the microscope, the individual principal indices of fibres may be determined and the birefringence calculated by subtraction. This method does not involve fibre thickness but is intrinsically less accurate.

In the case of fibres produced from semicrystalline polymers it should be remembered that the birefringence measured will be composed of two components arising from oriented crystalline and amorphous regions in the specimen. This complication is discussed further below in relation to polymer films.

33.4.3.2 *Molecular orientation measurements on films*

Although orientation in certain cast films is all but absent, the majority of industrially produced films are highly oriented to maximize their mechanical properties. As with fibres the polarizing microscope is used to determine the nature and degree of this orientation, although the optical anisotropy is generally biaxial and thus more complicated than in fibres. Nevertheless the basic measurements of extinction angle, sign and magnitude of Δn are the same. For biaxially oriented materials the refractive properties may be represented by three indices (conventionally α, β and γ). Two of these normally reside in the plane of the film, the xy plane. The in-plane birefringence Δn_{xy} therefore involves two of the three principal axes.

The range of Δn can be large, as can the range of thicknesses. The result is that the OPDs presented to the microscope by films can cover several orders of magnitude and may be outside the measurement range of a single compensator. Out-of-the-plane birefringences, Δn_{xz} or Δn_{yz}, are needed to fully describe the refractive properties of the film, where z is the normal to the film plane. These may be obtained by tilting the specimen on a universal stage[34] or by thin sectioning in the zx or zy planes. An alternative is to measure the optic axial angle ($2V$) by using the polarizing microscope in the conoscopic mode. This is discussed in Section 33.4.3.

As with fibres the birefringences measured may incorporate contributions from crystalline and amorphous regions in the film. Separation of these components can be carried out only if additional data from supplementary techniques are available. Desper[35] and Desper and Stein[36] have proposed and used a method of separation based on combining birefringence and X-ray pole figure measurements on polyethylene film. Similar work has been done on PET films.

33.4.3.3 *Molecular orientation in bulk materials*

In principle thin-microtomed sections of bulk materials, such as extrudates or injection mouldings, could be used to explore the birefringence, and hence the orientation, at any point within a product. In practice, because the orientation may vary significantly over small distances, the procedure becomes cumbersome and handling the information is difficult. Methods to mitigate this problem have been proposed by White and Spruill.[37] However, unless simplifications can be made or symmetries assumed, the practical problems are awesome, especially for semicrystalline materials.

For optically transparent products it is tempting to carry out a birefringence investigation using a polariscope and thick specimens. An illustration of the possible folly of doing this is shown in Figure 9. This shows a cross section of a moulding with dramatic variations in birefringence through its thickness.

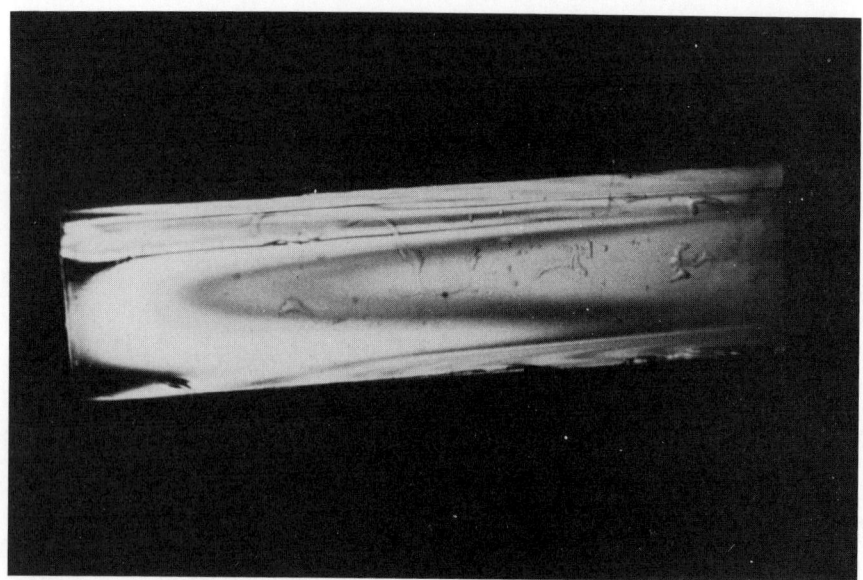

Figure 9   Cross-section of a polystyrene moulding. The intensity pattern shows considerable variation in the direction and magnitude of the molecular orientation through the cross-section; × 10 magnification, crossed polars

33.4.4 Conoscopy

The polarizing microscope can be adapted to act as a conoscope. To use this technique the specimen is first examined in the usual way between crossed polars with a high numerical aperture objective lens. The condenser unit is then adjusted to full aperture so as to illuminate the specimen with a wide-angled cone of rays. The back focal plane of the objective lens is then viewed using a telescope or Bertrand lens. The pattern of polarization colours seen is the 'conoscopic figure' or 'directions image'. Note that this differs from the method described in Section 33.4.2, where the condenser aperture is kept to a minimum.

The conoscopic technique has been applied to polymer films and more recently to polymer liquid crystal structures.[38] A typical conoscopic figure for biaxially oriented PET is shown in Figure 10. (Note that the term 'biaxial' has two distinct meanings in general use. Biaxially oriented film, that is film which has been two-way drawn, may or may not have biaxial optical characteristics.)

The conoscopic figure can give information on whether the specimen is uniaxial or biaxial in character and on the optical sign. The distance between two black 'zero-order' fringes is a measure of $2V$, the optic axial angle, which in turn can be used to determine birefringence out of the plane of the specimen. Details of the technique can be found in standard texts on crystal optics.[39]

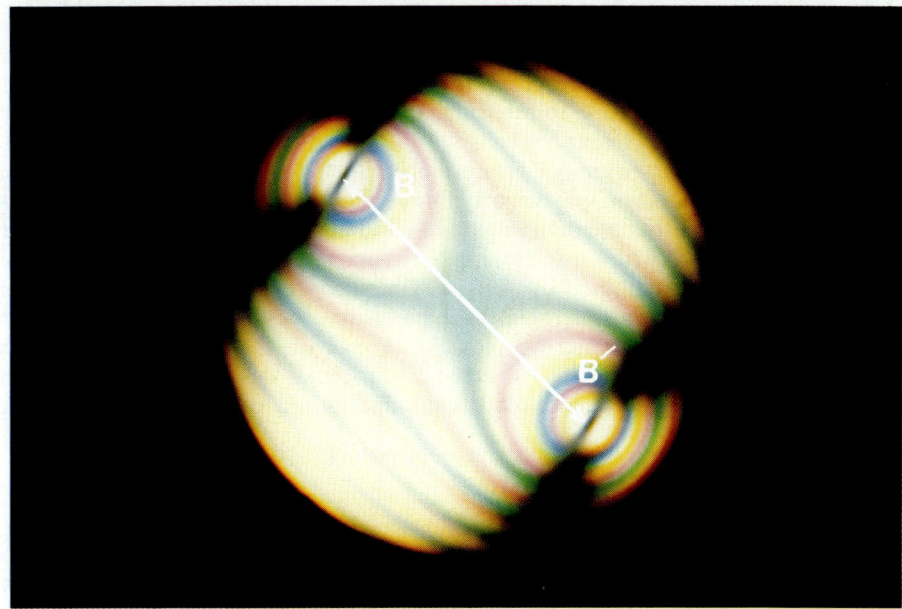

Figure 10 Conoscopic interference figure for a biaxially oriented poly(ethylene terephthalate) film

33.4.5 Types of Compensator for Polarized Light Microscopy

The OPD presented by a specimen can be estimated subjectively by comparing its polarization colour with those on standard colour charts published in most good texts on polarized light microscopy. This method breaks down if the OPD is small, because only low order greys are seen. For high OPDs the colours become less and less saturated and difficult to identify. To enable measurements to be made outside the range of the colour charts and to provide objective measurements for all OPDs, a number of compensators have been developed for the polarizing microscope. All are applicable to polymer specimens. A summary of the types available and an outline of possible applications are given in Table 1. A full description of these compensators can be found in the literature (see Born and Wolf[40] or Hartshorne and Stuart[15]).

Table 1 Types of Polarizing Microscope Compensators Used for Birefringence Measurement

Compensator type	Description	Typical applications
Senarmont	Fixed $\lambda/4$ plate in microscope. Needs graduated rotating analyzer. Range normally $0-1\lambda$	Low OPD specimens, *e.g.* most spherulites in thin sections, acrylic and acetate fibres; thin low Δn films
Elliptic (Brace Kohler)	Rotating $\lambda/10$ or $\lambda/30$ plate	Similar to Senarmont, but gives higher accuracy at very low OPDs
Berek	Simple tilting birefringent plate. Range normally $0-3\lambda$	Medium OPD specimens, *e.g.* spherulites in thick sections, thin Nylon or PET fibres, sections of oriented mouldings
Babinet Soleil	Sliding wedge type. Range normally $0-5\lambda$	Similar to Berek but gives uniform compensation over whole image plane
Ehringhaus	Multilayer tilting birefringent plate. Range up to 60λ	Very high OPD specimens, *e.g.* thick PET films and fibres; thick sections of highly oriented products

33.4.6 Other Sources of Optical Anisotropy

33.4.6.1 Stress birefringence

Unfortunately the birefringence shown by specimens does not arise only from molecular orientation. Stress can also produce effects which may be difficult to separate from those due to orientation and which may be of comparable magnitude. Photoelastic stress analysis has long been used to assess the magnitude, direction and distribution of stresses in loaded transparent or translucent components or models.[41] Assumptions made in that work are that the strains are small and that the stress optical coefficient relating stress to birefringence is independent of strain. The effect of large strains has been considered by Treloar.[43]

Stresses in microscopic specimens may have developed during product manufacture, or have been introduced by specimen preparation, particularly thin sectioning. In the latter case high strains may be involved but fortunately in the case of oriented films and fibres the problem does not usually arise, since these are often examined without sectioning. Discrimination between strain- and orientation-induced birefringence can be attempted by measuring birefringence as a function of temperature in the region of the material's T_g and relating measurements to other properties such as deformation. This can be difficult to do on the microscopic scale, even with the aid of a microhotstage. It is therefore fortunate that if polymers are subjected to processes designed to introduce molecular orientation then orientation birefringence usually dominates and stress effects can be ignored.

33.4.6.2   *Form birefringence*

Form birefringence is generated by structure within a specimen on a scale which is small in relation to the resolving power of the microscope, but large in terms of atomic or molecular dimensions. A regularity of structure consisting of phases of differing refractive indices is also required. Two well-documented model systems are alternating plates and a regular array of rods as described by Born and Wolf.[40]

The conditions for the production of form birefringence can be met by block copolymer systems in which domain structures are formed. Typical of such materials are *block*-copoly(styrene/butadiene/styrene) (SBS) in which the type and organization of domains is governed by both the relative molecular weight of the S and B blocks and the processing history. Since a system of parallel plates produces a bulk material of optically negatively uniaxial character, and a system of rods is positively uniaxial, then it is possible with the polarizing microscope to distinguish between these two structures.

The form birefringence phenomenon has been used by Canavorolo *et al.*[44] to examine rheological behaviour and transitions in SBS.

Form birefringence has also been noted as a contributor to spherulite birefringence in crystalline polymers.[45] There is however considerable disagreement as to the size of this contribution. In practice, disregarding form birefringence has not led to gross discrepancy between observed spherulite birefringences and those calculated from crystal refractive indices and morphological data, so that in some instances at least its contribution is clearly small. However, this should not exclude consideration of form birefringence effects. Certainly the known structure of spherulitically crystallized polymer provides for its occurrence.

33.4.6.3   *Liquid crystal structures*

Molecular order in polymeric liquid crystals has been examined by a variety of microscopical methods, which include polarized light. The general approach and possible interpretations of the images obtained has been covered by Hartshorne.[46] Thermotropic copolyesters have been studied in polarized light by Windle *et al.*[47] and Viney.[48] Interpretation of the data obtained is often difficult, partly because of the wide variety of structures seen, but it is clear that polarized light methods will continue to contribute in this field.

33.5   PHASE SENSITIVE METHODS

Many polymer systems are optically inhomogeneous and consist of more than one phase. The light microscopy of semicrystalline materials which can be regarded, at least to a first approximation, as consisting of separate crystalline and amorphous phases was discussed in Section 33.4.1. In this particular instance image contrast is obtained because the crystalline phase is birefringent. In multiphase systems such as polymer blends and composites, it may still be possible to contrast phases in the microscope by using crossed polars. However, this is not usually the case and other methods of contrast enhancement must be sought. The problem of imaging polymeric specimens with phases showing only low contrast, due to their having close refractive indices, can be overcome by using special phase sensitive microscope systems. These rely on some form of image processing in the back focal plane of the objective lens, or in its optical conjugate. Several systems are available, differing in sensitivity, ease of use and applicability.

33.5.1 Phase Contrast Microscopy

Using special condenser and objective lenses, the phase contrast microscope first separates light passing undeviated through the optical system from light deflected or redistributed by its interaction with features of the specimen (diffraction, refraction or scattering). The phase relationship and intensity difference between the redistributed light and the undeviated light is then modified to be more favourable to the generation of image contrast. The end result is that morphological features in the 'phase object' become considerably easier to see. Among the many accounts of the theory of phase contrast microscopy Goldstein's[49] is singularly both comprehensive and comprehensible. Figure 2 shows a dispersion of small acrylic spheres in a liquid of almost matching refractive index. Because the refractive index contrast is so small, common light microscopy shows very little contrast (Figure 2a). With the phase contrast method (Figure 2c) contrast is greatly enhanced.

Although the range of polymer refractive indices is large (roughly 1.3 to 1.7) many are close to 1.5. This means that generally mixtures of polymers will show very small refractive index differences between phases. Typical applications for the phase contrast microscope are the observation of thin sections of 'impact-modified' systems such as rubber-modified polystyrene or copoly(acrylonitrile/butadiene/styrene). Note however that the resolution of the microscope is at best unchanged by the image processing, and the usual resolving power limitation still applies.

The actual image contrast obtained is dependent upon a number of factors, some instrumental and some related to the optical properties of the specimen. The objective lens will have been designed to match particular specimen characteristics, which are unlikely to pertain exactly. Nevertheless a high contrast image can usually be obtained. Working close to the limit of resolution of the instrument it should be possible to detect features differing by only one in the fourth decimal place of refractive index from their surroundings. The system does however have a number of disadvantages: (1) features in the image are usually surrounded by a 'halo'; as a result overlapping features, or features in close proximity, become difficult to interpret; (2) large area features may be imaged non-uniformly with a contrast gradient across them; (3) there is no instrumental control over sensitivity; and (4) the system is inefficient in its use of the available light, necessitating the employment of high power light sources. Despite these shortcomings the phase contrast method is widely used to study multiphase polymers, including crystalline materials (Figure 11).

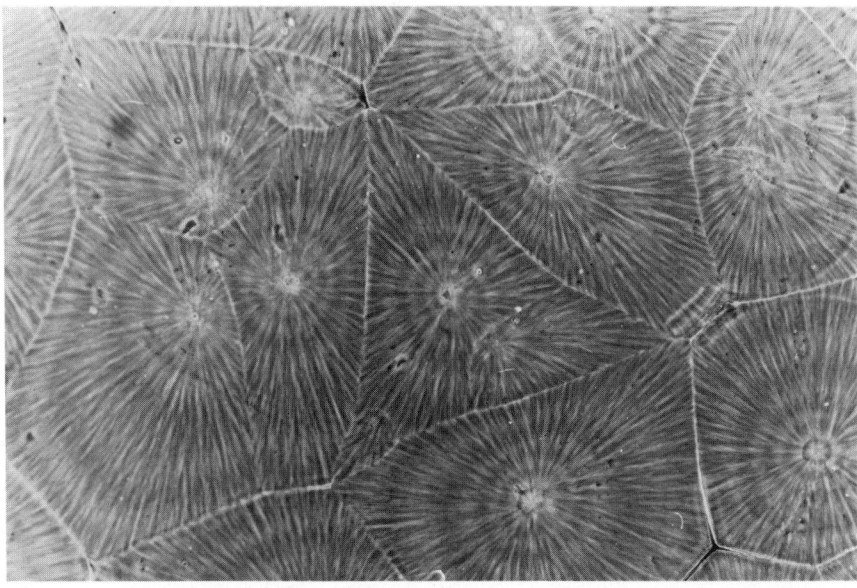

Figure 11 Spherulites of polypropylene imaged using the phase contrast technique. Note the good visibility of spherulite boundaries and the internal fibrilar structure; × 400, phase contrast

An interesting situation arises when the morphological features in the specimen are large in extent compared with thickness of the section. Variation of section thickness can then be used to 'tune' the specimen for optimum contrast in the microscope, since different thicknesses will change the phase relationship between the waves passing through the specimen.

Although the phase contrast microscope was originally available for use in a reflected as well as a transmitted light mode, the former has been superseded by other surface microscopy methods.

33.5.2 Differential Interference Contrast (DIC) Microscopy

This system, used in both a transmitted and reflected light form, generates contrast by the combination of two laterally displaced images of the specimen in such a way that optical interference takes place between them. The displacement is small compared with the resolving power of the system so that one image only is seen. Image contrast is a function of surface gradient in reflected light, or optical path length in transmitted light.

The areas of application of the transmitted light system are the same as for phase contrast. There is some dispute as to the relative sensitivity of the two methods but the advantages of avoiding the halo effect and having control over contrast are considerable. The DIC system works at full objective aperture so that lateral resolution is in no way impaired. It is based on a standard polarizing microscope and objectives, but requires a special condenser unit and access for the insertion of a Wollaston prism at the back focal plane of the objective. Note however that the lateral image shear generates contrast in one direction only and, since the beam shearing is carried out using polarization optics, the image produced by birefringent polymer specimens can be difficult to interpret. Thus the system is best confined to unoriented, amorphous specimens.

It is in the reflected light mode that DIC has made most impact on polymer microscopy. It has the capability of revealing shallow undulating surface detail, which is difficult to image with any other type of microscope, including electron microscopes. It works most effectively on smooth surfaces such as those shown by films, by high gloss mouldings and by some brittle fracture surfaces. The observation of structures only a few nanometers high has been claimed. Fracture surfaces such as that in Figure 12 show considerable detail in the 'mirror' and 'mist' regions of brittle fracture.[11] The rougher surfaces of more ductile fractures are best examined in the SEM.[50]

Figure 12 Polystyrene fracture surface imaged using reflected light interference contrast. Note the transition from the smooth textured 'mirror' region (bottom left), through the 'mist' region to the coarse texture (top right) which is associated with an accelerating fracture; × 600, differential interference contrast

33.5.3 Other Phase Sensitive Systems

A number of other phase sensitive systems have been devised. These include Hoffman modulation contrast,[51] Schlieren microscopy and such simple methods as Rheinberg illumination. Each offers its own combination of characteristics which may be particularly valuable in specific circumstances. However, the broad field of application to polymer specimens is that described above for the phase contrast microscope.

33.6 QUANTITATIVE INTERFERENCE SYSTEMS

The various types of light microscope outlined above, with the exception of the polarizing instrument, essentially yield qualitative information about the polymer specimen. There are however some instruments which are designed specifically to quantify some aspects of the image. These use the phenomenon of interference to provide data on optical path lengths in the specimen in transmitted light, and surface topography in reflected light.

33.6.1 Reflected Light Methods

These methods are well established for the examination of highly reflecting specimens such as metals. In order to use these methods on polymers their low reflectance can be overcome by evaporating metal on to them. If this is not practicable or desirable, it is possible to work with uncoated surfaces provided some degradation of the image contrast is accepted. Most microscopical interference systems generate a set of interference lines or 'fringes' across the surface of the specimen. These can be used as contour lines (with one half wavelength vertical interval) to obtain quantitative data on the surface roughness of polymer products such as films and mouldings, as shown in Figure 13. There have been successful attempts to correlate frictional characteristics with surface roughness[62] and the effectiveness of surface roughening additives. Surface interferometry has also been employed to investigate the results of chemical treatments undertaken to promote adhesion or expose internal structure.

The majority of systems are of the two-beam type which generate fringes with a $\cos^2$ intensity profile and are described in many standard texts.[52] Narrower fringes, which are therefore capable of greater vertical resolution, are produced by multiple beam or Tolansky devices.[53] These are difficult to use but have contributed to the study of film surfaces, as shown in Figure 14.

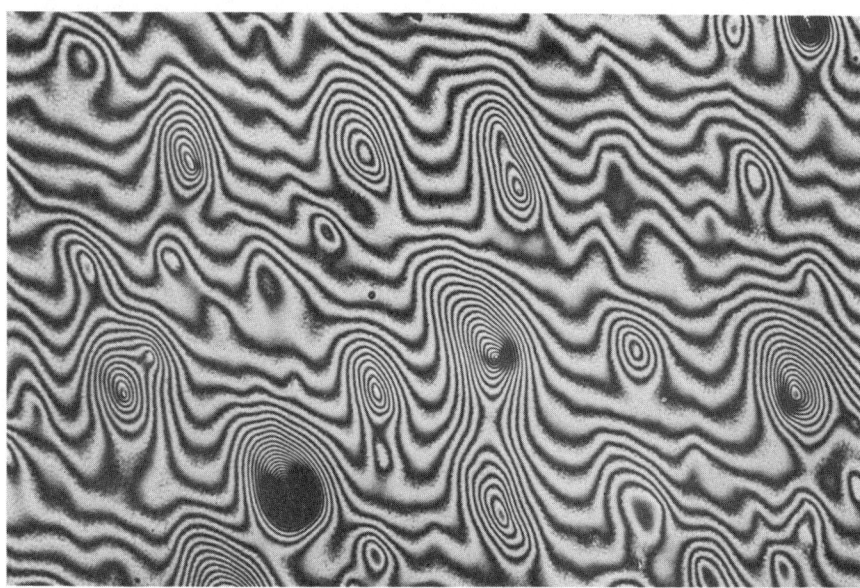

Figure 13 Surface of a polyacetal moulding showing surface pits; × 200, reflected light double beam interference

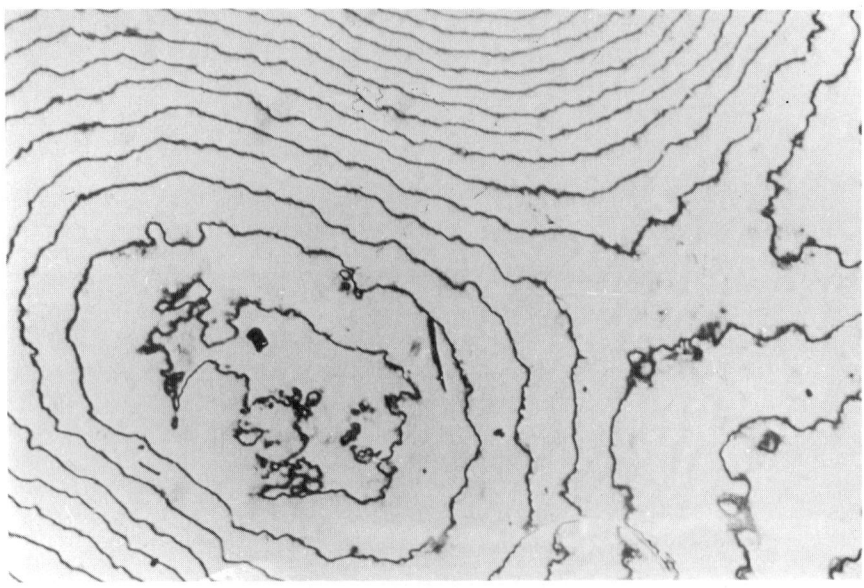

Figure 14 Surface of a polypropylene film; × 300 reflected light multiple beam interference. Note the smaller width of the interference fringes compared with those in Figure 13

33.6.2 Transmitted Light Methods

Transmitted light interference systems are diverse in the detail of their construction and operation. Many forms of interference microscope are described in the literature and a good, although now rather dated, review has been presented by Francon.[54] An almost universal feature is the separation of rays passing through the microscope into two sets. One set pass through the specimen, or chosen features of the specimen, and the other goes through a 'reference' region of the preparation. The two sets of rays are then recombined and the relative length of the optical paths compared by examining the interference pattern that results. Most systems have provision for the incorporation of compensators to allow precise measurement of the OPD down to about 3 nm.

The beam separation may be carried out by the use of birefringent components in the microscope. In such systems the beams passing through the reference and specimen regions in the object plane are plane polarized and care is required in image interpretation when working with birefringent specimens. Beam separation using, for example, a Mach–Zehnder interferometer[59] does not involve polarization optics and the constraints on the use of birefringent specimens are removed.

In almost all the systems the interference image is highest in contrast when the aperture of the microscope optics is small. This inevitably has the effect of reducing image resolution. For qualitative observation there is therefore a need for a subjective compromise between resolution and contrast. On the other hand, when the objective is quantitative measurement, resolution is sacrificed in the interests of maximum precision.

Use of these systems in the qualitative mode often reveals less about the specimen in terms of the size, shape and distribution of phases than the standard high resolution phase sensitive methods outlined in Section 33.5. Identification of the phases, if large enough to be seen, is however easier with the two-beam interference equipment, because information on the refractive indices of the phases present is immediately revealed through the interference colours displayed.

Since optical path length is the product of geometrical path length and refractive index, transmitted light methods may be used to measure either factor if the other is known. The method has been used[56] to measure the birefringence of textile fibres based on measurement of the individual refractive indices. There are significant benefits to this technique compared with the more usual birefringence measurements, in terms of the subsequent analysis of molecular orientation. Furthermore Roche and Davis[57] have usefully measured the distribution of radial birefringence in synthetic polymer fibres. Although perhaps pushing the sensitivity of interference microscopy to its limit, Katchy[58] has reported its use in detecting the small changes in density (and hence refractive index) that occur during the thermal treatment of PVC.

Given the refractive index of the specimen and the reference region of the preparation it is possible to determine geometrical thicknesses down to only 20 nm. This approach has been used in a variety

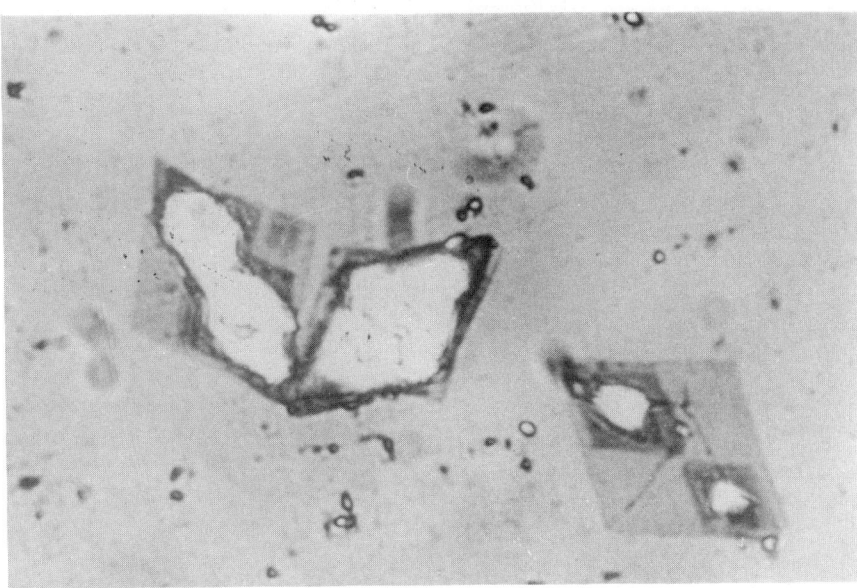

Figure 15 Single crystals of polyethylene. Measurements by microinterferometry showed these to be only 11 nm thick; × 800, transmitted light interference system

of applications, including the determination of the thickening of polymer single crystals during annealing (Figure 15), the thickness of deposited layers on polymer surfaces and the contact angles between polymers and liquid droplets.[59]

33.7 FLUORESCENCE AND ULTRAVIOLET MICROSCOPY

Microscopy at wavelengths just shorter than those in the visible spectrum has proved to be of some value in the examination of polymer systems. Imaging in the ultraviolet (UV) involves the use of special optics in the microscope, the normal lenses having too low a transmission factor at the short wavelengths. Two approaches to the use of UV have been employed. In the first an image is formed in the normal way. Image contrast is then a function of UV absorption and will give information on the distribution of absorbing species in the specimen. The image is viewed either through a UV to visible image converter or by the use of a UV sensitive TV camera. Photomicrography can be used to obtain permanent images in the usual manner.

Calvert and co-workers[60] have used the technique extensively to study the distribution of UV-absorbing additives in polyalkenes and have extended the method to make concentration measurements on local areas of the specimen and to explore the diffusion of additives.

The other approach is to irradiate the specimen with UV radiation and to image any visible light emitted by fluorescence. Both transmitted and reflected light systems are available, although the latter are more popular, at least at the higher magnifications, because of the greater efficiency offered.

The use of fluorescent dyes or stains to study polymer structure has been discussed, but little work has been carried out in this field compared with polarized light or interference methods. Nevertheless the scope for advances here seems substantial. Autofluorescence of polymers has also received little attention, although some, including poly(ethylene terephthalate) and a number of thermosets, fluoresce quite strongly. In common with a number of other polymers, PVC fluoresces when degraded and the fluorescence intensity and spectrum have been used to characterize the thermal history of PVC specimens.[61]

33.8 REFERENCES

1. E. M. Slayter, 'Optical Methods in Biology', Wiley–Interscience, New York, 1970.
2. S. Bradbury, 'The Optical Microscope in Biology', Arnold, London, 1976.

3. D. S. Skene, *J. Microsc. (Oxford)*, 1969, **89**, 63.
4. E. E. Galigher and E. N. Kozloff, 'Essentials of Practical Microtechnique', Lea and Febiger, Philadelphia, 1964.
5. P. F. Kerr, 'Optical Mineralogy', McGraw-Hill, New York, 1959.
6. 'National Metals Handbook', 8th edn., American Society of Metals, Ohio, 1973, vols. 7 and 8.
7. F. Rybnikar, *J. Appl. Polym. Sci.*, 1985, **30**, 1949.
8. R. P. Palmer and A. J. Cobbold, *Makromol. Chem.*, 1964, **74**, 174.
9. D. C. Bassett, 'Principles of Polymer Morphology', Cambridge University Press, Cambridge, 1981.
10. C. W. Mason, 'Handbook of Chemical Microscopy', 4th edn., Wiley, New York, 1983, vol. 1.
11. E. H. Andrews, 'Fracture in Polymers', Oliver and Boyd, London, 1968.
12. J. W. Ess, P. R. Hornsby, S. Y. Lin and M. J. Bevis, *Plast. Rubber, Proc. Appl.*, 1984, **4**, 17.
13. B. Mutagahywa and D. A. Hemsley, *Plast. Rubber, Proc. Appl.*, 1985, **5**, 3.
14. H. Medalia and D. F. Walker, 'Cabot Carbon Technical Report RG-124' (Revision 2), Cabot Carbon, Boston, MA, 1970.
15. N. H. Hartshorne and A. Stuart, 'Crystals and the Polarizing Microscope', 4th edn., Arnold, London, 1970.
16. A. F. Hallimond, 'The Polarizing Microscope', 3rd edn., Vickers, York, 1970.
17. A. Keller, *J. Polym. Sci.*, 1959, **39**, 151.
18. J. Padden and H. D. Keith, *J. Appl. Phys.*, 1959, **30**, 1497.
19. E. Boehme, *Plastverorbeiter*, 1982, **33**, 1464.
20. J. Bowman, N. Harris and M. Bevis, *J. Mater. Sci.*, 1975, **10**, 63.
21. D. Williams and M. Bevis, *J. Mater. Sci.*, 1982, **17**, 1915.
22. M. Folkes and R. H. Burton, *Plast. Rubber, Proc. Appl.*, 1983, **3**, 129.
23. D. A. Hemsley, 'The Light Microscopy of Synthetic Polymers', Oxford University Press, Oxford, 1984.
24. R. S. Stein and M. B. Rhodes, *J. Appl. Phys.*, 1960, **31**, 1873.
25. R. J. Samuels, *J. Polym. Sci., Part A-2*, 1971, **9**, 2165.
26. I. Ward (ed.), 'Structure and Properties of Oriented Polymers', Applied Science, London, 1975.
27. G. L. Wilkes, *Adv. Polym. Sci.*, 1971, **8**, 91.
28. R. S. Lenk, 'Polymer Rheology', Applied Science, London, 1978, p. 307.
29. B. E. Read, *Polym. Test.*, 1984, **4**, 143.
30. H. Kawai and S. Nomura, in 'Developments in Polymer Characterisation — 4', ed. J. V. Dawkins, Applied Science, London, 1983.
31. H. H. Yang, M. P. Chovinard and W. J. Lingg, *J. Polym. Sci.*, 1982, **20**, 981.
32. R. W. Moncrieff, 'Man-Made Fibres', Newnes–Butterworths, London, 1975, p. 79.
33. W. C. McCrone, 'The Particle Atlas — 5', 2nd edn., Ann Arbor Science Publishers, Ann Arbor, MI, 1979, p. 1388.
34. I. D. Muir, 'The 4-Axis Universal Stage', Microscope Publications, Chicago, 1981.
35. C. R. Desper, *J. Appl. Polym. Sci.*, 1979, **13**, 169.
36. C. R. Desper and R. S. Stein, *J. Appl. Phys.*, 1966, **37**, 3990.
37. J. L. White and J. E. Spruill, *Polym. Eng. Sci.*, 1983, **23**, 247.
38. H. J. Coles and R. Simon, in 'Recent Advances in Liquid Crystalline Polymers', ed. L. L. Chapoy, Applied Science, London, 1985, p. 331.
39. E. E. Wahlstrom, 'Optical Crystallography', 5th edn., Wiley, New York, 1979.
40. M. Born and E. Wolf, 'Principles of Optics', 2nd edn., Pergamon Press, Oxford, 1964.
41. E. G. Coker and L. N. G. Filon, 'A Treatise on Photo-Elasticity', 2nd edn., Cambridge University Press, Cambridge, 1957.
42. B. E. Read, *Polym. Test.*, 1984, **4**, 143.
43. L. R. G. Treloar, 'The Physics of Rubber Elasticity', Clarendon Press, Oxford, 1975.
44. S. Canavarolo, A. W. Birley and D. A. Hemsley, *Br. Polym. J.*, 1985, **17**, 263.
45. R. S. Stein and G. L. Wilkes, in 'Structure and Properties of Oriented Polymers', ed. I. M. Ward, Applied Science, London, 1975, p. 61.
46. N. H. Hartshorne, 'The Microscopy of Liquid Crystals', Microscope Publications, London, 1974.
47. A. H. Windle, C. Viney, R. Golombok, A. M. Donald and G. R. Mitchell, *Faraday Discuss. Chem. Soc.*, 1985, **79**, 55.
48. C. Viney, *Microscope*, 1984, **32**, 93.
49. D. J. Goldstein, *J. Microsc. (Oxford)*, 1982, **128**, 33.
50. L. Engel, H. Klingele, G. W. Ehrenstein and H. Schaper, 'An Atlas of Polymer Damage', Wolf, London, 1981.
51. R. Hoffman and L. Gross, *J. Microsc. (Oxford)*, 1977, **110**, 205.
52. E. Hecht and A. Zajac, 'Optics', Addison–Wesley, Reading, USA, 1982.
53. S. Tolansky, 'Multiple Beam Interferometry of Surfaces and Films', Oxford University Press, Oxford, 1948.
54. M. Francon, 'Progress in Microscopy', Pergamon Press, Oxford, 1961.
55. L. C. Martin, 'The Theory of the Microscope', Blackie, London, 1966, p. 387.
56. A. A. Hamza, *Text. Res. J.*, 1980, 731.
57. E. J. Roche and H. A. Davis, *Fiber Prod.*, 1984, **51**, 51.
58. E. Katchy, *J. Appl. Polym. Sci*, 1983, **28**, 1847.
59. G. W. Longman and R. P. Palmer, *J. Colloid Interface Sci.*, 1967, **24**, 185.
60. N. C. Billingham and P. Calvert, *Br. Polym. J.*, 1979, **11**, 155.
61. D. A. Hemsley, R. P. Higgs and A. Miadonye, *Polym. Commun.*, 1983, **24**, 103.
62. B. J. Burton and D. Taber, in 'Polymer Surfaces', ed. D. T. Clark and W. J. Feast, Wiley, New York, 1978, chap. 1.

34

Electron Microscopy

MASAKI TSUJI

Kyoto University, Japan

34.1 GENERAL INTRODUCTION

In 1955 Jaccodine reported thin, spirally grown, lozenge-shaped lamellar crystals, namely single crystals, of a low molecular weight linear polyethylene (PE) from dilute benzene and xylene solutions.[1] This work was extended to high molecular weights, independently, by Till,[2] Keller[3] and Fischer.[4] These authors observed the morphology of PE single crystals under a transmission electron microscope (TEM). Keller, especially, clarified that molecular chains which are much longer than the lamellar thickness ($\sim$ 10 nm) must fold back and forth at the surface of the lamella.[3] This suggestion had already been made as early as 1938 by Storks[5,6] but had passed unheeded.

This epochal concept of chain folding was squarely opposed to the traditional, so-called 'fringed micelle' model by Herrmann *et al.*,[7] which had been the working hypothesis to explain the structure of gelatin. However, since Fischer[8] and Kobayashi *et al.*[9] independently discovered that with TEM the melt-grown spherulites which had been regarded to have the fringed micelle structure were also composed of single-crystal-like lamellae, then lamellar crystals with chain folding have been considered as the basic structural constituents of crystalline polymer solids.

The solid structures of various polymers, including single crystals, have been studied with TEM. However, the application of TEM was limited to morphological investigations of polymer solids, not only because TEM did not have sufficient resolving power to resolve the individual atoms or molecules comprising the polymer solids, but also because polymer molecules are easily destroyed by electron irradiation. Consequently one observed the morphology of specimens, for example, by imaging them at low magnifications and/or investigating them by selected area electron diffraction (ED). In these cases, a small amount of irradiation ought to suffice. It is well known that TEM is designed to make a magnified image of an object or its ED pattern by slightly changing the focal length of the intermediate lens, namely by changing an electric current of the lens. That is to say, when the lenses, such as the intermediate lens and the projector lens which are below the objective lens, are focused on the image made by the objective lens, then the magnified image of the object is observed on the fluorescent screen; if their focal length is changed so that they are focused on the diffraction pattern made by the objective lens at its back focal plane, then the magnified ED pattern

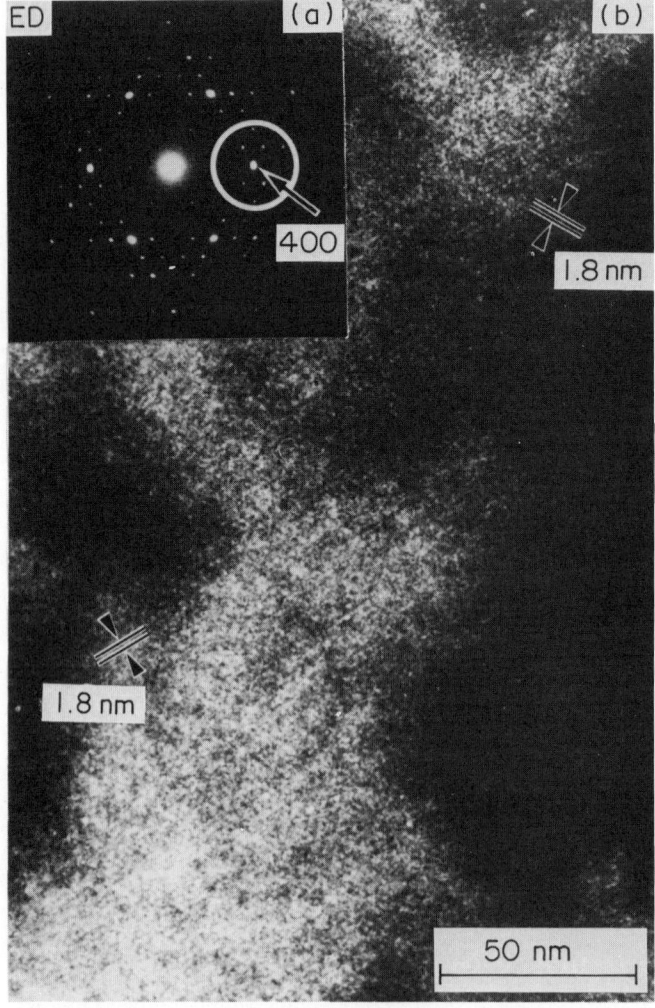

Figure 1   (a) High resolution dark field TEM image of the β-form of a PPX single crystal showing 1.8 nm 100 lattice fringes. (b) In the ED pattern, the circle shows the position and size of the objective aperture

will be observed on the screen. When an aperture, the so-called selected area aperture, is introduced in the image plane of the objective lens then the ED pattern from a corresponding domain of the specimen can be obtained. This method is called the selected area ED.

The shadow-casting method, replication method or staining method (negative or positive) can improve considerably the radiation resistivity, and therefore these are the methods used to investigate the morphology of polymer solids with high contrast, as shown in the following section. Resolution beyond a couple of nm should not be expected with these methods because of effects such as granularity of shadowing or staining materials.

The first success of high resolution electron microscopy in the field of polymer science was the dark field image (Section 34.2.2.1) showing 1.8 nm lattice fringes taken from the β-form of a poly(p-xylylene) (PPX) single crystal, which was reported in 1969.[10] Figure 1(a) shows ED from the β-form of a PPX single crystal (see Figure 3, where the circle indicates the size and position of the objective aperture). The bright area in the 400 dark field image (Figure 1b) is not uniform, and corresponds to the domain that satisfies the Bragg condition sufficiently to give a 40.0 reflection with an intensity larger than its surroundings. This demonstrates well the mosaic nature of the single crystal. Owing to the interference effect between the 40.0 reflection and its satellites, 1.8 nm lattice fringes in two directions are recognized in the same area and intersect each other at an angle of 60°. Later, there have been published several reports concerning high resolution TEM of polymers, for example lattice images of rigid polymers such as poly(p-phenylene terephthalamide) (PPTA),[11-13] the molecular image of PPX[14] and so on. Nevertheless, even now the main purpose of TEM is the observation of morphologies of polymer solids at low to medium magnifications.

Ever since Bragg[15] reported the optical analogy of X-ray diffraction patterns in 1939, the optical transform method has been much developed in the field of structure analysis using X-ray or electron diffraction.[16,17] On the basis of this principle, Klug *et al.*[18] established the so-called optical filtering method for image processing of electron micrographs, after the method already used in the field of information science.[19] On the other hand, with the advent of the space age, digital image processing for pictures which are transmitted over great distances from far ranging space probes has been rapidly advanced.[20] Such image processing techniques have become indispensable for the high resolution electron microscopy.

The author uses mainly TEM, in particular high resolution TEM, for the investigation of polymer solids. There are some recent reviews concerning the morphological observations of polymer solids by TEM and scanning electron microscopy (SEM), including sample preparation.[21-25] These are worth reading not only for beginners but also for specialists of TEM in the field of polymer science. There are also some reviews about high resolution TEM of polymers.[30,100,133,345] In this chapter, the high resolution TEM of polymers will be discussed in some detail, while other techniques will only be mentioned briefly.

34.2 TRANSMISSION ELECTRON MICROSCOPY OF POLYMERS

34.2.1 Image Formation in TEM

Provided a light microscope has an ideal objective lens in which all kinds of aberrations are corrected almost perfectly, the resolution limit of the microscope with illumination parallel to the optical axis is given by Abbe's theory[26-28] as

$$d_D = K\lambda/\sin\alpha \qquad (1)$$

where d_D is the minimum distance between two light-absorbing particles that can be recognized as separated points in the image, λ is the wavelength of the light, α is the aperture angle of the objective lens and K is a constant ($K = 0.61$ for incoherent illumination, or $K = 0.77$ for coherent illumination). Since $\alpha < \pi/2$, then $d_D > K\lambda$. Therefore the resolution of a microscope never exceeds about half the wavelength of the light used.

In electron optics the relation between aperture angle and resolving power is more complicated. Electron lenses used as an objective lens cannot be corrected spherically, though they can be stigmated almost correctly by the stigmator. Electron waves passing through the outer zones of the lens miss the Gaussian image point. The resolution limit due only to spherical aberration is given by

$$d_S = C_s\alpha^3 \qquad (2)$$

where C_s is the spherical aberration coefficient. The intensity distribution in the image plane correspondng to a point object may be considered as a Gaussian distribution whose half-breadth corresponds to the 'resolution limit'. Therefore the resolving power d of TEM may be approximately estimated from both equations (1) and (2) using the following equation

$$d^2 = d_D^2 + d_S^2 \tag{3}$$

Since α is considered sufficiently small, the optimum aperture angle α_{opt} and the resolution limit d_{opt} are given as follows (for $K = 0.61$)

$$\alpha_{opt} = 0.77\,(\lambda/C_s)^{\frac{1}{4}} \tag{4a}$$

$$d_{opt} = 0.91\,(\lambda^3\,C_s)^{\frac{1}{4}} \tag{4b}$$

For 200 keV (1 ev $= 1.6 \times 10^{-19}$ J) ($\lambda = 0.00251$ nm) and $C_s = 2.8$ mm, $d_{opt} = 0.42$ nm. This simple estimation is applicable to the amplitude contrast due to the so-called 'absorption' effect of a specimen. As is mentioned in Section 34.3, the resolution limit in high resolution TEM, owing to phase contrast, is better than the estimation here.

The contrast in TEM can be classified roughly into two types; amplitude contrast and phase contrast. Amplitude contrast is classified further into mass thickness contrast and diffraction contrast. The word 'diffraction contrast' is used for crystalline specimens, although the term 'Bragg contrast' is also used in some cases.[29] There is almost no absorption of high energy electrons when they pass through the specimen. By introducing a small opening (objective aperture) at the back focal plane of the objective lens, electrons scattered by the specimen outside of the opening will be trapped. Thus this 'absorption' effect induces amplitude contrast.

34.2.2 Amplitude Contrast

34.2.2.1 *Diffraction contrast*

There are two ways of introducing an objective aperture: a bright field (BF) mode in which an aperture is set to introduce the unscattered primary beam into the aperture; and a dark field (DF) mode in which one is set to cut the primary beam off and to make images with electrons scattered in a certain direction. Figure 2 schematically shows the effect of an objective aperture in amplitude contrast.[30] By comparing (b) with (c), it can be seen that in TEM a crystallite that gives some Bragg reflections appears as a darker area in BF and as a brighter area in DF, on the fluorescent screen. Figure 3 shows (b) DF and (c) BF images of the β-form of a PPX single crystal. The size and position of the objective aperture is indicated with circles in Figure 3(a) for both cases. Many radial stripes due to diffraction contrast are clearly seen, suggesting the existence of six (10.0) sectors in the crystal. The (40.0) DF image (Figure 3b) also shows such a sectorization and a mosaic nature of the crystal. It is considered that sectorization of a polymer single crystal comes from chain folding.[31] The shape and contrast of bright areas are slightly different from sector to sector. These may be due to the collapse of the crystal on the supporting film, though it has the tent-like or hollow pyramidal

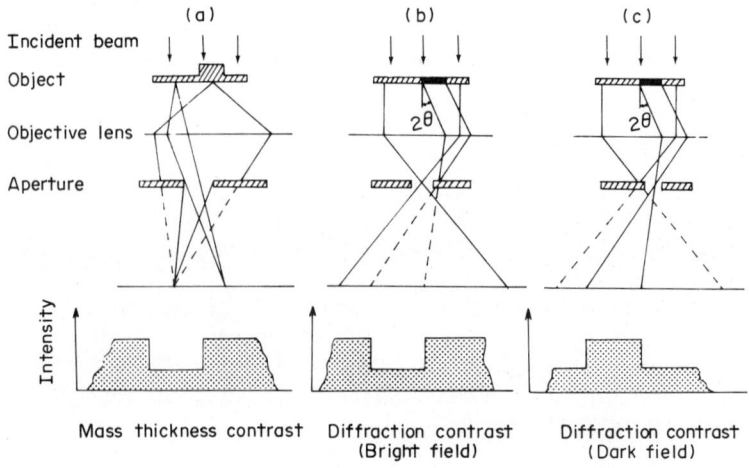

Figure 2 Effect of an objective aperture in amplitude contrast[30]

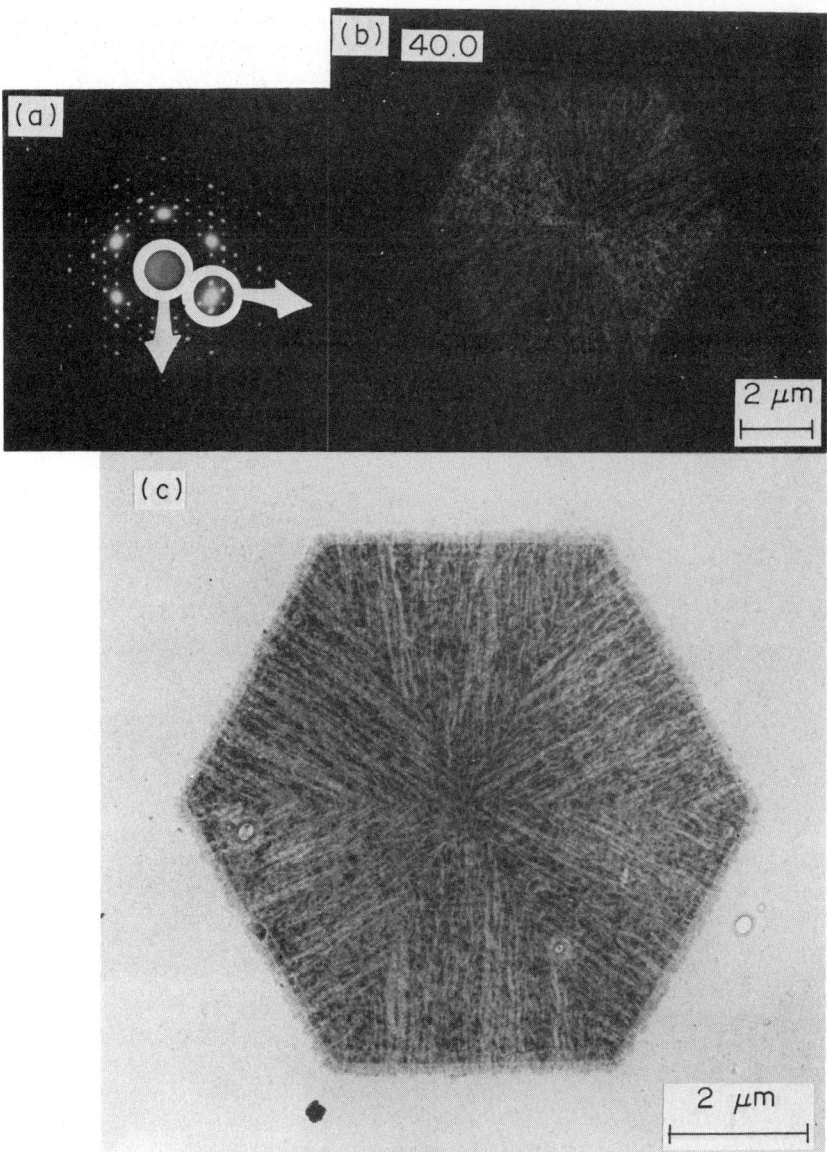

Figure 3 Dark field (b) and bright field (c) images of the β-form of a PPX single crystal. The circles in ED (a) of the crystal show the size and positions of the objective aperture used to obtain (b) and (c)

structure in solution.[32] In the DF mode, tilted illumination with an objective aperture set symmetrically on the optic axis may be better, in taking the effect of spherical and chromatic aberrations into account.[33] In present day TEMs, the DF mode using tilted illumination is naturally installed. Therefore, once the direction and angle of beam tilt are set, it is easy and simple to change the mode from BF to DF or DF to BF by conveniently pushing the mode selector. It should be noted, however, that tilted illumination changes the Bragg condition.

As a special technique, multiple dark field imaging is frequently used.[34] If the intermediate lens is strongly excited at the mode of selected area diffraction, a bright field image and many dark field images are observable at the same time, but are rather out of focus. This technique is convenient to investigate the crystallographic orientation of the specimen.

34.2.2.2 *Mass thickness contrast*

Negative staining with uranyl acetate or phosphotungstic acid (which is a familiar method in the biological research field), staining with OsO_4 or RuO_4, and shadowing with heavy metals such as a

Pt–Pd alloy will enhance mass thickness contrast, *viz.* amplitude contrast.[23] For negative staining, hydrophobicity of support film surface is harmful because it prevents uniform spreading of the specimen and/or negative stain on the support film surface, if they are in water. To suppress this effect, supporting films which are freshly made are recommended. In the literature there have been several methods reported for recovering the hydrophobicity of specimen supports, such as: adding some chemicals;[35,36] adding detergents such as Bacitracin as wetting agents;[36,37] irradiating the support surface with UV;[38] or using a glow discharge.[39-43] Figure 4 shows ribbon-like fibrils of bacterial cellulose negatively stained with uranyl sulfate[348] and clearly demonstrates the influence of hydrophobicity and the effectiveness of ion bombardment in a glow discharge.[44] In the photograph (Figure 4b and c), as a result of successful staining it can be seen that a ribbon-like fibril consists of finer microfibrils. In the case of synthetic polymers, OsO_4 is used for the staining of the copolymer, where the component with double bonds is selectively oxidized with OsO_4.[23] The Ag_2S insertion technique for the morphological observation of PPTA fibers is also considered as one of the staining techniques.[269] Kanig introduced negative staining for PE with uranyl acetate, after treatment with chlorosulfonic acid.[45] Voigt-Martin used this technique to estimate the distribution of lamellar

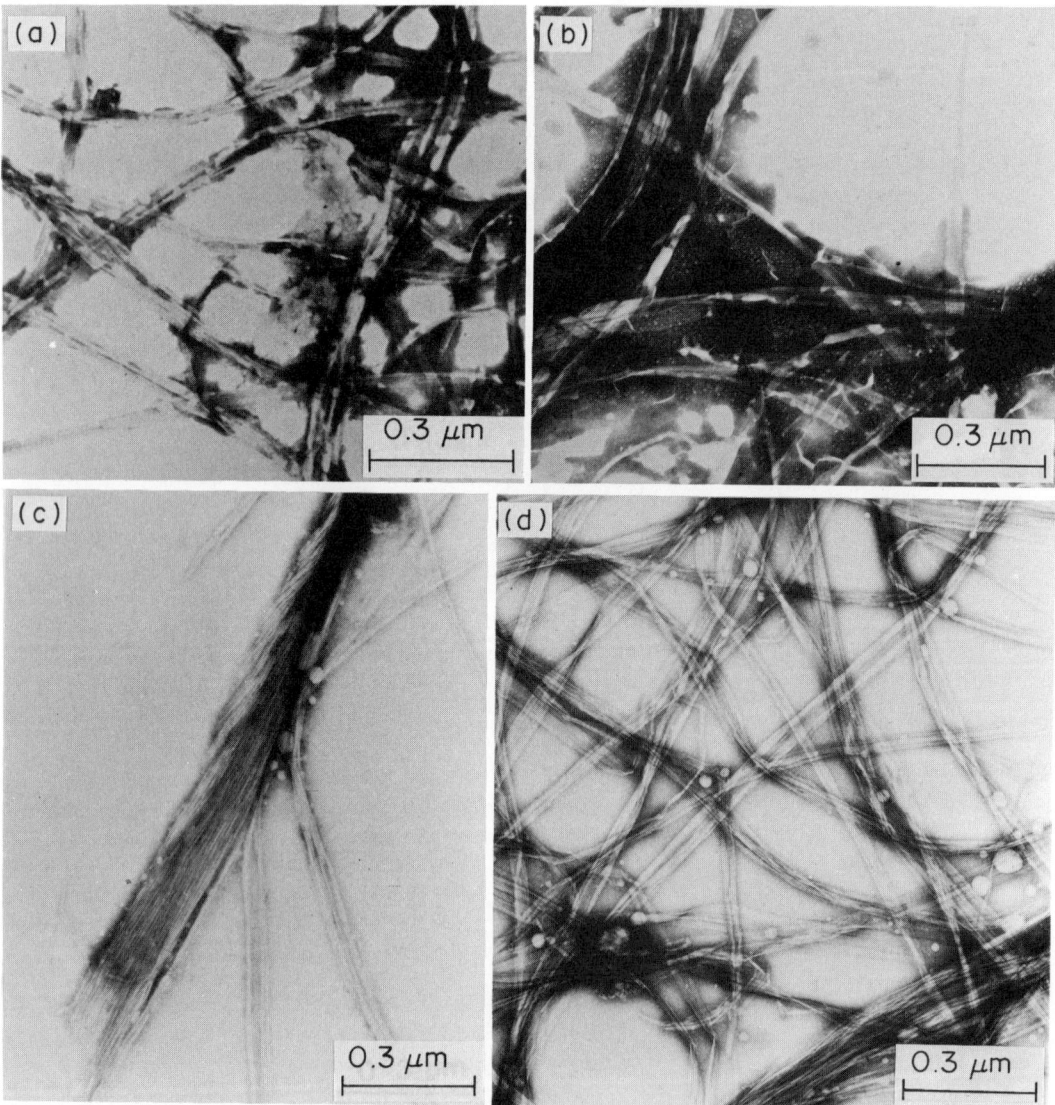

Figure 4   The effectiveness of hydrophilization of a specimen support film by ion bombardment in a grow discharge for negative staining. The specimen (ultrasonically disintegrated protofibrils of bacterial cellulose) was negatively stained with uranyl sulfate: (a, b) typical examples of uneven spreading of a negative stain due to the hydrophobicity of specimen support film (without glow discharging); and (c, d) examples of uniform spreading of a negative stain after hydrophilization by glow discharging

thicknesses in melt-crystallized PE.[348] The stacked lamellar structure of crystalline polyalkenes, like PE, is well observed with this technique, but finer structure is not recognized. Staining of synthetic polymers is extensively reviewed by Grubb.[23]

Figure 5 is a micrograph of a crystalline thin film of isotactic polystyrene (isotactic PS) shadowed with Pt–Pd at an angle of $\tan^{-1}(1/4)$.[46] It shows the existence of edge-on lamellae composed of two-dimensional and immature spherulites. Some regions, indicated by white arrows, are possibly flat-on lamellae. In the encircled area, parallel stacked lamellae have the appearance of a shish kebab. Three typical methods of metal shadowing are well known and shown schematically in Figure 6: (a) with W basket for almost all metals; (b) with W filament for some metal wires such as Au and Ag; and (c) for Pt with C. The physical properties of shadowing some metals are shown in Table 1.[47] The mass m needed for desired thickness t of evaporated layer is roughly estimated as

$$m = 4\pi r^2 \, dt / \sin \theta \tag{5}$$

where d, r and θ are the density, the distance between the source of evaporation and the specimen, and the shadowing angle, respectively.

Several metals such as Cr, Au and Ag are used for increasing image contrast and also as a reference to measure lattice spacing corresponding to Bragg reflection which appears in the ED pattern from a crystalline specimen. Al is also frequently used as a reference. For example, lattice spacings of 111 and 200 are as follows: 0.234 nm (111), 0.202 nm (200) for Al, and 0.235 nm (111), 0.204 nm (200) for Au. The best material for high resolution shadowing is W.[21] The evaporation of W, however, is not easy because of its high vaporization temperature, and thus needs a special installation with a static high voltage.[47,270] Anyhow, the shadowing method is useful for enhancing

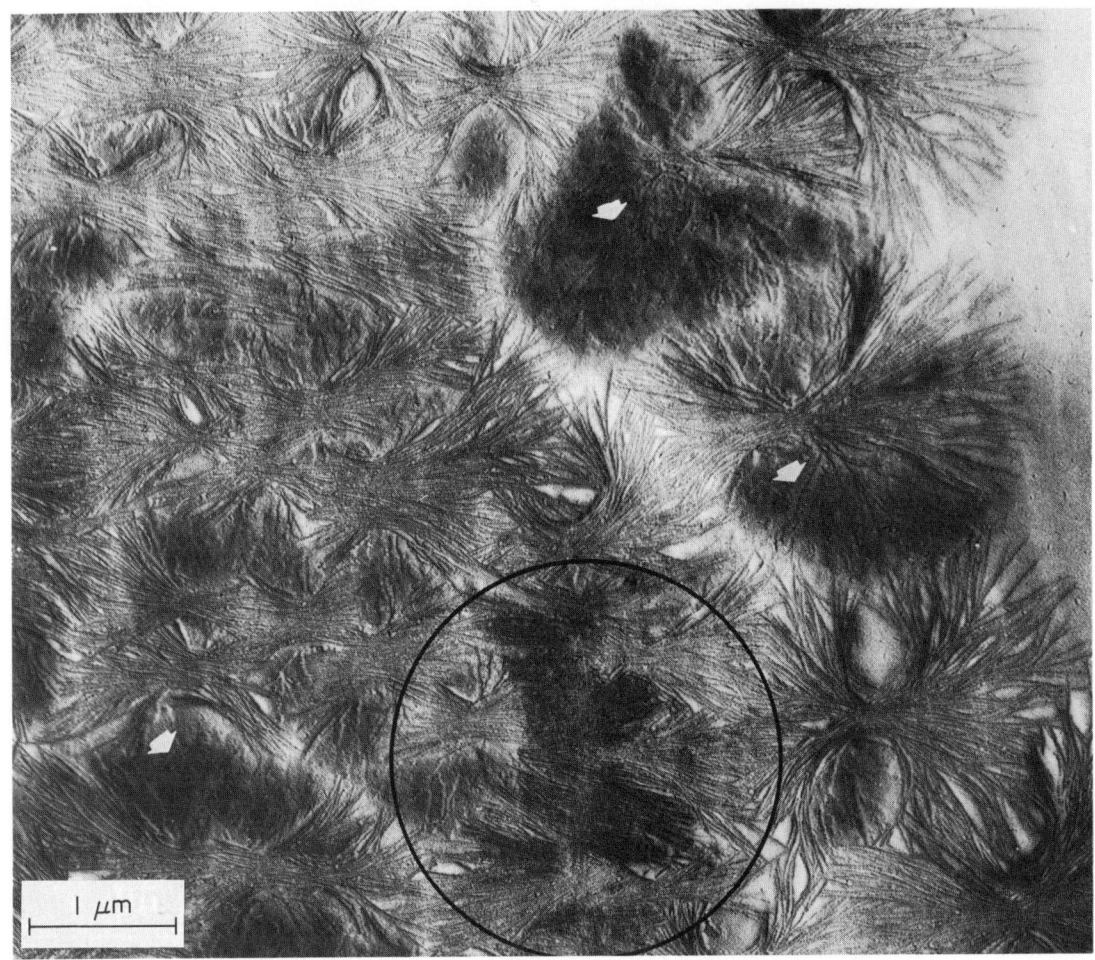

Figure 5   A thin film of isotactic PS annealed and crystallized at 165 °C for 10 min, and Pt–Pd shadowed at an angle of $\tan^{-1}(1/4)$

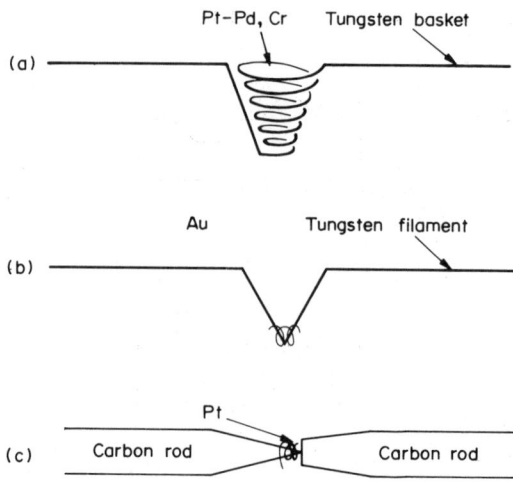

Figure 6   Three typical methods of metal shadowing

Table 1   Physical Properties of some Metals for Shadowing and of Carbon[47]

Material	Al	Au	Pt	Ag	W	Cr	Pt–Pd (80:20)	C
Density at 20 °C (g cm^{-3})	2.70	19.3	21.4	10.50	19.1	7.20	19.4	2.25
Melting temperature (°C)	660	1063	1774	961	3382	1903	—	>3500
Evaporation temperature (°C) at 10^{-5} Torr	996	1465	2090	1005	3309	1205	—	2681

the image contrast due to the topographical unevenness of the specimen surface,[48,49] but it is helpless against the inner structure of specimens with smooth surfaces.

For bulk polymers, replication combined with metal shadowing is useful to examine the surface topography.[50] If the specimen itself is soluble in a certain solvent, then single-stage replication is applicable. Two-stage replication with a thermoplastic resin-like collodion is used for a specimen which has no suitable solvent, and also for observing the so-called 'extraction replica', namely a tiny fraction of the bulk polymer. Procedures are illustrated in Figure 7 for single- and two-stage replication. Kobayashi *et al.* using the ED of an extraction replica from a PE melt-grown spherulite demonstrated that the spherulite is made up of single-crystal-like lamellae.[9]

Shadowing is applicable to the determination of the thickness of a specimen, using the shadow angle and shadow length.[21] As one of the metal-coating methods, Au decoration is sometimes used.[21] For specimens having very small steps in their surfaces, the shadowing technique is not applicable because the surface perturbations are too small. When the specimen, which is thinly coated with evaporated Au, is heated slightly, Au migrates to such steps and nucleates there.[51] Shimamura introduced this technique to investigate the fiber structure of PE, combined with the optical diffraction/filtering technique of its electron micrograph.[52]

Whilst some incident electrons are scattered elastically in all directions after passing through a specimen, others pass through without a change of direction. The scattering angle (twice the Bragg angle for a crystalline specimen) of electrons is related to the spatial frequency distribution of the inner potential of the specimen;[48,53] the larger the angle, the higher the spatial frequency, namely the finer the structure.[54] Therefore, in amplitude contrast, such a high resolution is not expected because electrons passing through a small aperture hole will make images. Figure 8 demonstrates well the effect of the size of the objective aperture in an axial BF mode, using optical filtering apparatus (see Section 34.3.4.1 and also ref. 55). If one requires an image containing information about the molecular arrangement in a crystal, then a fairly large opening must be used as the objective aperture.

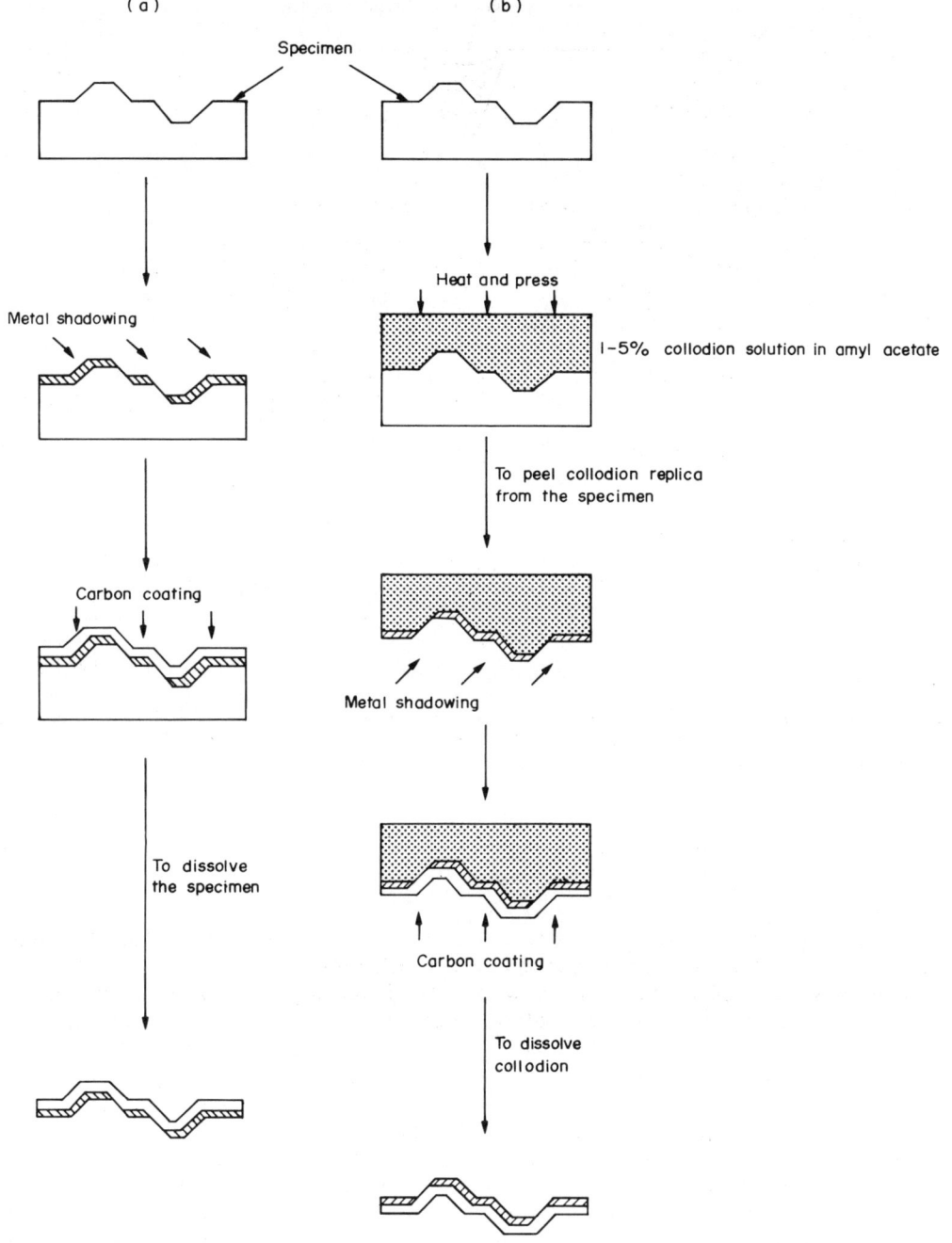

Figure 7   Procedures of two types of replication: (a) single-stage replica; and (b) two-stage replica

34.2.3   Phase Contrast

Phase contrast is related to high resolution observation at an atomic or molecular scale. The electron waves passing through a specimen are modulated in their amplitude and phase.[53] When the modulation of the amplitude is small and a large opening is used as an objective aperture, a sufficient contrast is not expected in just the focus image, namely the Gaussian image. A proper amount of defocus brings forth the optimum phase difference between the transmitted wave and the wave scattered elastically by the specimen, and gives contrast to the image, *i.e.* phase contrast.[53, 56, 49] Applications of the phase contrast technique were first introduced by Petermann *et al.* to study the morphology of polymer solids, especially their fiber structure.[57] This technique is called 'defocus

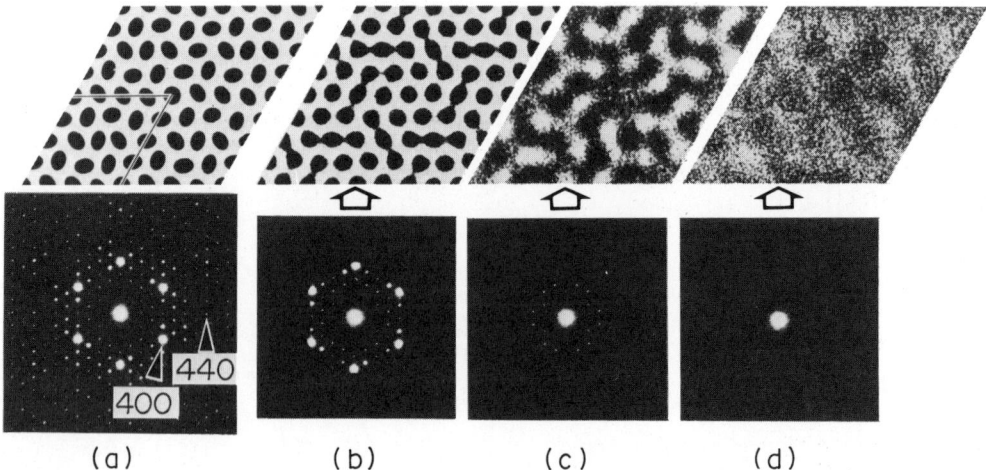

Figure 8 Optical analogy of the effect of the size of objective aperture in an axial bright field mode on image details: (a) model object was constructed on the basis of the projection of the β-form of a PPX crystal on the *ab*-plane along the chain axis (*c*-axis), with its optical diffraction pattern (see Section 34.5.1.2); (b) though reflections up to 400 were used to make the image, orientations and mutual positions of the individual ellipses are not recognized (it was confirmed that if reflections up to 440 are used, orientations and mutual positions of ellipses can be recognized in the image), (c) reflections upto 210 were used, resulting in the ellipses of the corresponding individual molecules not being resolved; and (d) using reflections up to 200

contrast'. Examples are shown in Figures 9 and 33 for an annealed thin film of isotactic PS, and in Figure 39(a) for a stretched and annealed thin film of PE. These were taken at a considerably large amount of defocus; $\sim 40\ \mu$m underfocus at 200 kV.[298,350] In Figure 9, immature spherulites with a sheaf-like appearance are seen, as in the case of shadowing (see Figure 5). Lamellae stacked in the stretching direction (fiber axis) are clearly recognized in Figure 39(a), and the averaged sequence, namely the long period, is about 30 nm. More detailed discussions for these materials will be given later (Sections 34.5.2.1 for isotactic-PS and 34.5.2.2 for PE).

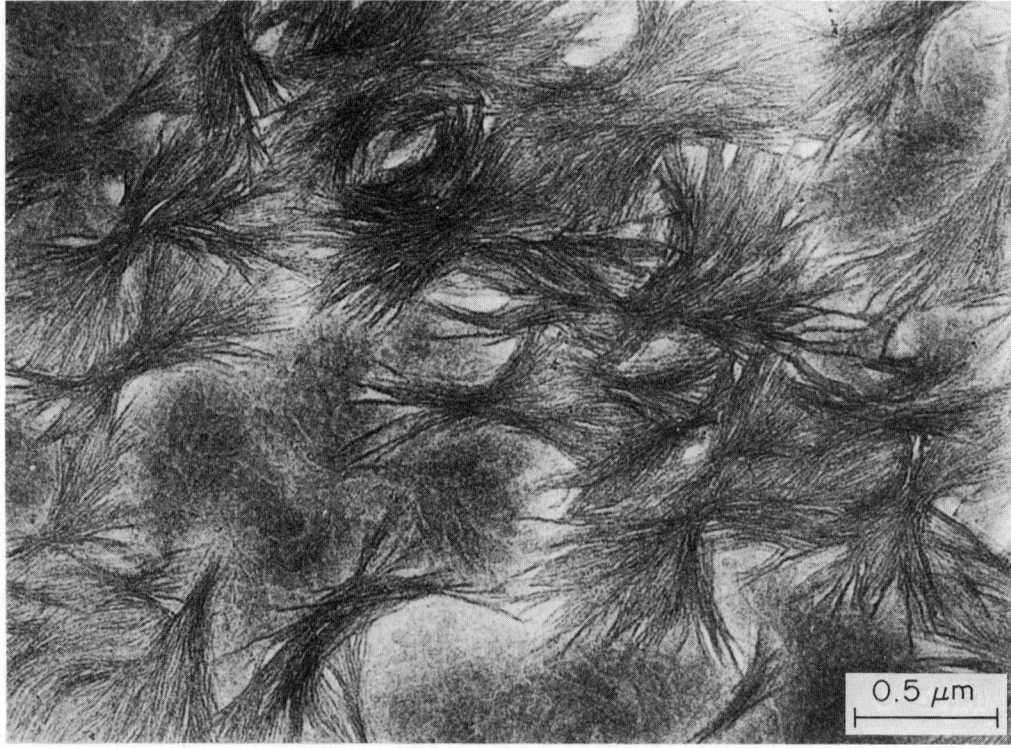

Figure 9 Defocused bright field image of a thin film of isotactic PS annealed and crystallized at about 170 °C (taken at $\sim 40\ \mu$m underfocus at 200 kV)

34.3 HIGH RESOLUTION TRANSMISSION ELECTRON MICROSCOPY

In high resolution electron microscopy, very thin specimens which can be treated as phase objects are observed using a fairly high accelerating voltage (at least 100 kV). In this case, the resolution limit should be estimated from the standpoint of the phase contrast based on Scherzer's treatment.[58] The image contrast of phase objects ought to be very weak on Gaussian image plane. Therefore, as described by Scherzer, a small amount of defocusing is beneficial to the contrast, and the amount of optimum defocus, namely 'Scherzer focus', is related to the spherical aberration coefficient, C_s. The true resolving power is inherently limited by the wavelength, maximum aperture angle, illuminating angle and defocus spread, as well as by C_s.[56,59]

In this section, the theory of image formation in high resolution TEM on an atomic or molecular level will be briefly described, together with some practical problems to obtain high resolution electron micrographs of polymer solids, especially polymer crystals.

34.3.1 Scattering of Electrons by an Object

If a periodic object (of thickness R) consists of stacked thin slices, each of which is expressed by a one-dimensional transmission function $\cos(2\pi x/d)$ with a period of d, the phase difference between the wave ψ_1 which was scattered by the first slice and propagated for a distance R, and the wave ψ_2 scattered for the first time by the last slice is given by $\pi R\lambda/d^2$.[60] This phase difference must be within $\pi/2$ according to Cowley's criterion to avoid negative interference of the waves.[60] Then, in order to resolve a distance d_{min}, the specimen thickness must not exceed approximately[61,62]

$$R_{max} = d_{min}^2/2\lambda \qquad (6)$$

This equation is derived by considering the effect of Fresnel diffraction. For $\lambda = 0.00251$ nm (200 kV electrons), equation (6) gives $R_{max} = 8.0$ nm for $d_{min} = 0.2$ nm, and $R_{max} = 50$ nm for $d_{min} = 0.5$ nm. When a specimen is sufficiently thin in relation to the TEM resolution, then the specimen is considered as a two-dimensional object according to the criterion mentioned above. Its effect on the electron beam is represented by changes of phase and amplitude which are considered to take place on a single plane. Thus the scattering from this object can be treated kinematically. Here the time-related term is neglected, and only the spatial component will be discussed in the following treatment. The spatial component of the incident wave function, which is assumed to be coherent and of unit amplitude, will be multiplied by the two-dimensional transmission function of this planar object, and can be written as

$$q(x_0) = \exp\{-i\sigma\phi(x_0) - \mu(x_0)\} \qquad (7)$$

Here σ is the interaction constant[63] as

$$\sigma = 2\pi/\lambda E\{1 + (1 - \beta^2)^{1/2}\} \qquad (8)$$

where E is the accelerating voltage. In equation (7), $\phi(x_0)$ and $\mu(x_0)$ are the projections in the beam direction of the three-dimensional potential distribution $\phi(x_0, z)$ and an effective absorption function $\mu(x_0, z)$ of the specimen, respectively, so that

$$\phi(x_0) = \int \phi(x_0, z)\,dz \qquad (9)$$

and

$$\mu(x_0) = \int \mu(x_0, z)\,dz \qquad (10)$$

For a particularly thin object and/or sufficiently high accelerating voltage with a fairly large size of objective aperture, $\mu(x_0)$ can be neglected and the specimen can be treated as a pure phase object. If the specimen is composed of light atoms, $\phi(x_0)$ is sufficiently small, and only the first order term is significant in the expansion of the exponential in equation (7)

$$q(x_0) = 1 - i\sigma\phi(x_0) \qquad (11)$$

In this case, the object is called a 'weak phase object'. The first term of equation (11) represents the unscattered incident wave and the second one the scattered wave. The Fourier transform Q, of the function q, is given by

$$Q(\pmb{u}) = \int q(\pmb{x_o}) \exp(2\pi i \pmb{u} \pmb{x_o}) d\pmb{x_o} = \delta(\pmb{u}) - i\sigma\Phi(\pmb{u}) \tag{12}$$

where

$$\Phi(\pmb{u}) = \int \phi(\pmb{x_o}) \exp(2\pi i \pmb{u} \pmb{x_o}) d\pmb{x_o} \tag{13}$$

or for periodic objects

$$\Phi(\pmb{u}) = (\lambda/\sigma) \cdot F(h, k) = (\lambda/\sigma) \sum_j f_j(h, k) \exp\{2\pi i(hx_o + ky_o)\} \tag{14}$$

(where h, k are Miller indices; $f_j(h, k)$ is the atomic scattering factor of the jth atom for electrons; and x_o, y_o are fractional coordinates.)

34.3.2   Image Formation in High Resolution TEM

The objective lens of TEM transfers the electron wave, after it has been transmitted through the object, to form its diffraction pattern (*viz.* spatial frequency distribution spectrum of the object) at the back focal plane, and then to form its inverted real image at the Gaussian image plane, as shown in textbooks of ordinary physical optics.[26, 27] The total process is interpreted in terms of two successive Fourier transforms. The amplitude transmittance of an object corresponds to the wave function of electrons transmitted through the object, as described in the previous section. With the coordinates assigned as in Figure 10, the amplitude $A(\pmb{x_f})$ (at the back focal plane of the objective lens) of the electron wave transmitted through an object will be given by the following equation[64–67]

$$A(\pmb{x_f}) = (i/\lambda f) \exp\{-ik(f + d_1 - \Delta f) - ik(x_f^2/2f)(1 - d_1/f)\} Q(\pmb{x_f}/\lambda f) \exp\{-ik \cdot \Delta f \cdot x_f^2/2f^2\} \tag{15}$$

where Δf is the amount of defocusing (being positive for underfocusing) and $k = 2\pi/\lambda$. The last factor in equation (15) is the wave aberration due to the defocussing Δf (distances d_2 and $(d_1 - \Delta f)$ in Figure 10 are fixed in TEM, where d_1 and d_2 are a pair of conjugate distances with respect to the objective lens). The diffraction intensity at the back focal plane is given by $|A(\pmb{x_f})|^2$. Then from equation (15)

$$|A(\pmb{x_f})|^2 = (1/\lambda^2 f^2) \cdot |Q(\pmb{x_f}/\lambda f)|^2 \tag{16}$$

Including the effects due to spherical aberration and defocusing, the aberration function χ is defined as follows[58]

$$\chi(\pmb{x_f}/\lambda f) \equiv \pi\lambda \cdot \Delta f \cdot (x_f^2/\lambda^2 f^2) - (\pi/2)C_s\lambda^3 \cdot (x_f^2/\lambda^2 f^2)^2 \tag{17}$$

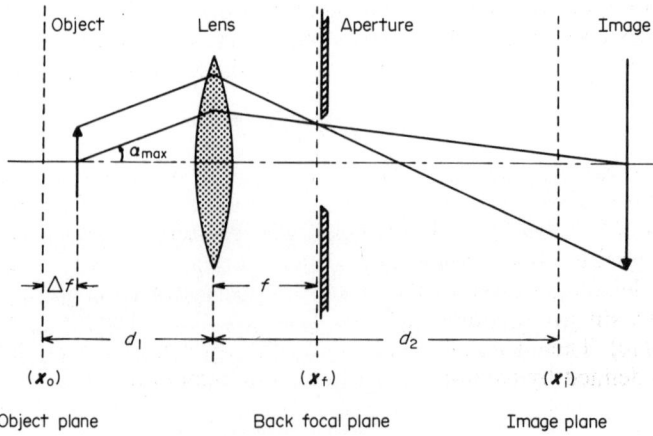

Figure 10   Image formation by an objective lens in TEM: $\alpha_{\max}$ = maximum aperture angle; f = focal length of the lens; and Δf = amount of defocus (positive for underfocus)

Thus equation (15) can be modified as

$$A(x_f) = (i/\lambda f)\exp\{-ik(f + d_1 - \Delta f) - ik(x_f^2/2f)(1 - d_1/f)\}Q(x_f/\lambda f)\exp\{-i\chi(x_f/\lambda f)\} \quad (18)$$

The amplitude $\Psi(x_i)$ at the image plane is given by the diffraction integral, namely the Fresnel diffraction, of $A(x_f)P(x_f/\lambda f)$, where P is the aperture function. Therefore

$$\Psi(x_i) = (-1/M)\exp\{-ik(f + d_1 - \Delta f) - ikx_f^2/2(d_2 - f)\} \times$$

$$\int Q(u)\exp\{-i\chi(u)\}P(u)\exp\{2\pi ix_iu/M\}\,du \quad (19)$$

where M is the magnification (cf. $M = d_2/d_1$) and $u(= x_f/\lambda f)$ is the spatial frequency. The intensity $I(x_i)$ of the image is given by

$$I(x_i) = |\Psi(x_i)|^2 = |\psi(x_i)|^2/M^2 \quad (20)$$

where

$$\psi(x_i) = \int Q(u)\exp\{-i\chi(u)\}P(u)\exp\{2\pi ix_iu/M\}\,du \quad (21)$$

If we use a circular objective aperture and an aberration function with cylindrical symmetry, both functions for $u = |u|$ are as follows

$$P(u) = \begin{matrix} 1 & \text{for } u \leqq u_{max} \\ 0 & \text{for } u > u_{max} \end{matrix} \quad (22)$$

and

$$\chi(u) = \pi\lambda \cdot \Delta fu^2 - (\pi/2)C_s\lambda^3u^4 \quad (23)$$

where $u_{max} \approx \alpha_{max}/\lambda$ and α_{max} is the maximum aperture angle of the objective lens. Noting that the potential function $\phi(x_O)$ is real, we obtain from equations (12) and (20)

$$I(x_i) \approx (1/M^2)\{1 - 2\int\sigma\Phi(u)\sin\chi(u)P(u)\exp\{2\pi ix_iu/M\}\,du\} \quad (24)$$

where $\sin\chi(u)$ is called the 'phase contrast' transfer function.[68-70] This equation indicates that the image intensity $I(x_i)$ is greatly affected by defocusing through $\sin\chi(u)$ in the case of phase contrast. If $\sin\chi(u) = 1$ (or -1) and $P(u) = 1$ for all u, then

$$I(x_i) = (1/M^2)[1 \mp 2\sigma\phi(-x_i/M)] \quad (25)$$

and we thus obtain the magnified and inverted real image which is to reflect the potential distribution ϕ, *viz.* the true structure of the object.

Under actual circumstances, $\sin\chi(u)$ oscillates and here we define u_0 as the smallest value of u (except $u = 0$) given by $\sin\chi(u) = 0$. For high resolution electron microscopy, we must operate TEM at the 'optimum defocus' which gives $|\sin\chi(u)|$ its maximum value of 1 over as wide a range of u as possible. The amount of this optimum defocus, namely the Scherzer focus, is calculated as[62]

$$\Delta f_s = [(4/3)C_s\lambda]^{1/2} \quad (26)$$

Figure 11 shows several solid curves which illustrate the defocus dependence of $\sin\chi(u)$ for $C_s = 1.06$ mm and $\lambda = 0.00142$ nm (500 keV where 1 eV = 1.6×10^{-19} J). (The effect of defocus on the image contrast will be demonstrated in Figure 27(b) in Section 34.5.1.2.ii.) The value of Δf_s for this microscope is estimated as 45 nm using equation (26). The aperture size u_{max} should be selected under the optimum defocus condition so as to cut-off all the scattering waves in the higher spatial frequency range where $\sin\chi(u)$ oscillates violently. It may be best to select u_0 at the optimum defocus as u_{max} (see Figure 11c). Then $1/u_{max}$ corresponds to the resolving power. The resolution limit at optimum defocus is defined by equation (27) for axial illumination.

$$d = 0.66\cdot(C_s\lambda^3)^{1/4} \quad (27)$$

Various criterions have been proposed to determine the optimum defocus condition, and the

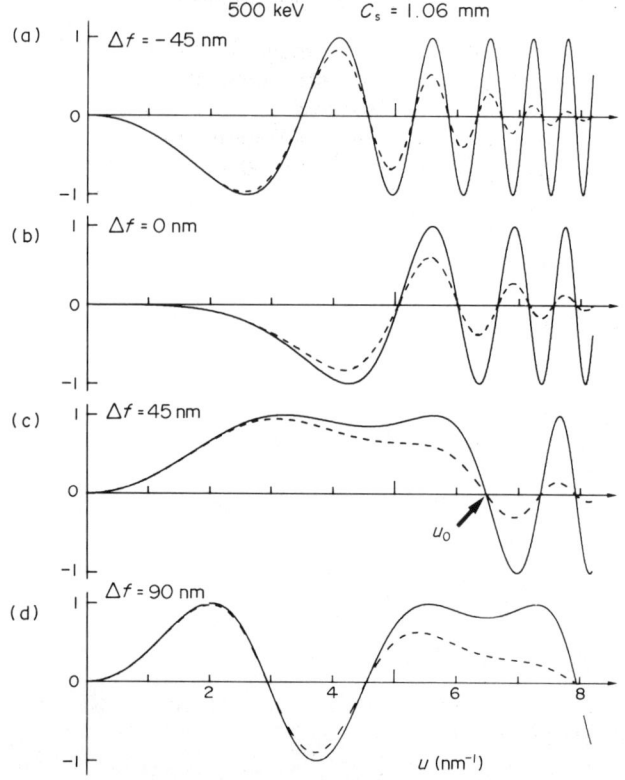

Figure 11 The defocus dependence of the phase contrast transfer functions $\sin\chi$ (solid curves) and $D_F \cdot \sin\chi$ (dashed curves) of JEOL JEM-500 ($C_s = 1.06$ mm, $\lambda = 0.00142$ nm, $\beta_i = 2 \times 10^{-4}$ rad, $\Delta = 10$ nm). In (c) $\Delta f = 45$ nm corresponds to Scherzer focus of this TEM, and see Section 34.3.3.1.ii for β_i and Δ

resulting resolution limits differ slightly.[58,71-73] The overall process of image formation in high resolution TEM is shown in Figure 12, where $D_F(u)$ means the damping factor to $\sin\chi(u)$, and this will be described in the next section. The effect of $D_F(u)$ on $\sin\chi(u)$ is shown in Figure 11, using dashed curves.

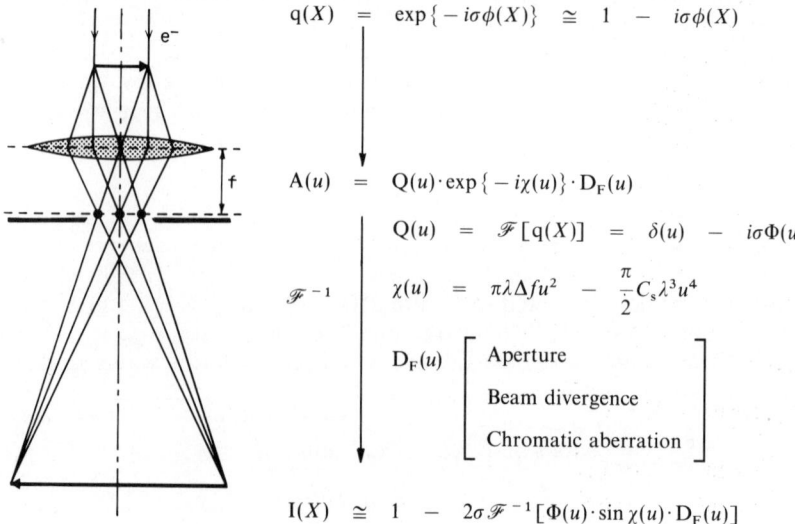

Figure 12 Overall process of image formation in high resolution TEM, where, $\mathscr{F}$ and $\mathscr{F}^{-1}$ represent the Fourier and inverse Fourier transforms, respectively

34.3.3 Some Factors Limiting Resolution in High Resolution TEM

In the previous section, the phase contrast image formation and the resolution limit in high resolution TEM were described. In this section, some problems in obtaining high resolution micrographs of polymer crystals will be discussed.

There are many factors which restrict the resolution limit of TEM.[74] The average distance between polymer chains in crystals is assumed to be 0.5 through 1 nm. To obtain information on molecular arrangement in the crystal, however, 0.2–0.3 nm resolution may be needed. Principal factors affecting the resolution of TEM imaging are summarized in Sections 34.3.3.1 and 34.3.3.2.

34.3.3.1 *Resolution-limiting factors relating to the microscope*

(i) *Spherical aberration of the objective lens*

The effect of this aberration was described in Sections 34.2.1 and 34.3.2. Electron micrographs should be taken under the optimum defocus condition (the so-called Scherzer condition), in order to obtain the image which reflects the true structure of the object, as shown in Section 34.3.2. For 200 kV TEM, the relationship between Δf_s and C_s (equation 26) and that between the resolution limit d and C_s (equation 27) suggests that, in order to clear 0.2 nm as the resolution limit, C_s needs to be smaller than 0.5 mm, and $\Delta f_s = 41$ nm underfocused for $C_s = 0.5$ mm. Several methods were proposed to estimate C_s.[74, 76]

(ii) *Defocus spread and illuminating angle*

The illuminating angle (relating to beam divergence) and the defocus spread (relating to chromatic aberration) modify the phase contrast transfer function $\sin \chi(u)$;[56] the function is multiplied by two envelope functions $S(u)$ and $E(u)$ which are expressed in terms of the illuminating angle and the parameter for defocus spread as[77]

$$S(u) = \exp\{-[(\pi\beta_i/\lambda)\cdot(C_s\lambda^3 u^2 - \Delta f\cdot\lambda)]^2\} \qquad (28)$$

and

$$E(u) = \exp\{-[(\pi\lambda u^2/2)\cdot(C_c\cdot\delta E/2E\sqrt{\ln 2})]^2\} = \exp\{-[(\pi\lambda\Delta\cdot u^2/2)]^2\} \qquad (29)$$

where β_i is the illuminating angle, Δf the defocus, C_c the chromatic aberration constant, $\delta E/E$ the fluctuation of accelerating voltage, and Δ the parameter for the defocus spread. The combined effect of these two envelope functions on $\sin \chi(u)$ is shown in Figure 11 using dashed curves. Since two envelope functions dampen $|\sin \chi(u)|$ to a considerably small value at $u > u_{max}$ [u_{max} is given from Scherzer's limit in equation (27)] as shown in Figure 11, practically we need not use the objective aperture. There have been some methods to measure β_i and/or Δ.[59, 74, 78] Table 2 shows the properties of modern electron sources at 100 kV.[25, 33, 56] To get small β_i and Δ, a field emission gun (FEG) is the best, but needs ultra high vacuum. From the view point of 'cost performance', LaB$_6$ filament is recommended.

As a damping factor to $\sin \chi(u)$, one more important function related to the granularity of the image recording medium should be introduced, which is called the modulation transfer function (MTF).[59, 74] This will be discussed later in Section 34.3.3.1.iv.

Table 2 Properties of Modern Electron Sources at 100 kV[25, 33, 56]

Source	Radiance[a] $(C s^{-1} cm^{-2} str^{-1})$	Energy width ΔE (eV)	Effective source diameter	Lifetime (h)	Vacuum required (Torr)	Emission current (μA)	Working temperature (K)
Thermionic W hairpin	5×10^5	2	30 μm	50	10^{-5}	100	2800
Thermionic W point	2×10^6	2	1–5 μm	10	10^{-5}	10	2800
Thermionic LaB$_6$	5×10^6	1	5–10 μm	500–1000	10^{-6}	50	1800
Thermal field emission W	1×10^8	0.3	5–10 nm	500–1000	10^{-9}	50–100	1800

[a] Electron flux per unit solid angle.

(iii) Astigmatism of the objective lens

In order to obtain high resolution electron micrographs with resolution better than 0.2 nm, the astigmatism of the objective lens must be compensated within about 10 nm,[79] because it introduces an extra difference in phase and/or amplitude into the diffracted electron beams according to their direction in the plane perpendicular to the optical axis of TEM. For low magnification/resolution microscopy, perforated films are used as specimens in correcting objective astigmatism. One can see the Fresnel fringe along the edge of the hole,[80] and this is not uniform if astigmatism exists. To obtain a uniform fringe on the viewing screen at a magnification of about 100 000 adjustment of a stigmator is required. For high magnification/resolution work, astigmatism will be corrected by observing phase granularity of a thin amorphous specimen, like evaporated C, on the viewing screen at a higher magnification, say at least 400 000.[56,80] Stigmating, in this case, is done so as to give no anisotropic appearance of the phase granularity pattern. This method is more accurate than the previous one. However, the resolution and contrast of the image seen on the viewing screen of a higher voltage TEM is too poor to permit compensation with much accuracy. The method shown below is useful for accurate astigmatism correction.

Astigmatism is most easily recognized from the optical diffraction pattern of a high resolution electron micrograph of a thin amorphous film.[68] As a perfectly stigmated micrograph would give a diffraction pattern consisting of concentric circular rings due to $\sin^2 \chi(u)$,[70] departure from circularity indicates the presence of astigmatism.[81,82] The usual elliptic pattern of contrasting transfer zones and gaps is shown in Figure 13(a), which is an optical diffractogram from the micrograph of a thin C film taken with a certain amount of underfocusing before the astigmatism is corrected. The transfer gaps, namely dark rings, appear at spatial frequencies u which satisfy

$$\chi(u) = n\pi \quad (n \text{ integer}) \tag{30}$$

Here the aberration function $\chi(u)$ is given by equation (23). In a high voltage TEM with small C_s, only the second term in equation (23) is significant for spatial frequencies higher than 3 nm^{-1}. When the first term dominates, the positions of the transfer gaps are given by

$$u = \{n/|\Delta f|\lambda\}^{1/2} \tag{31}$$

The magnitude of defocus is estimated using the following equation

$$|\Delta f| = L^2 n \Lambda^2 / M^2 r_n^2 \lambda \tag{32}$$

where M, Λ, r_n and L are the electron optical magnification, the He–Ne gas laser wavelength (632.8 nm), the radius of the nth gap, and the camera length of the apparatus for optical transformation, respectively. The defocus is measured for each principal astigmatism direction to determine the astigmatism difference δf between the two principal values.[68] The astigmatism is given

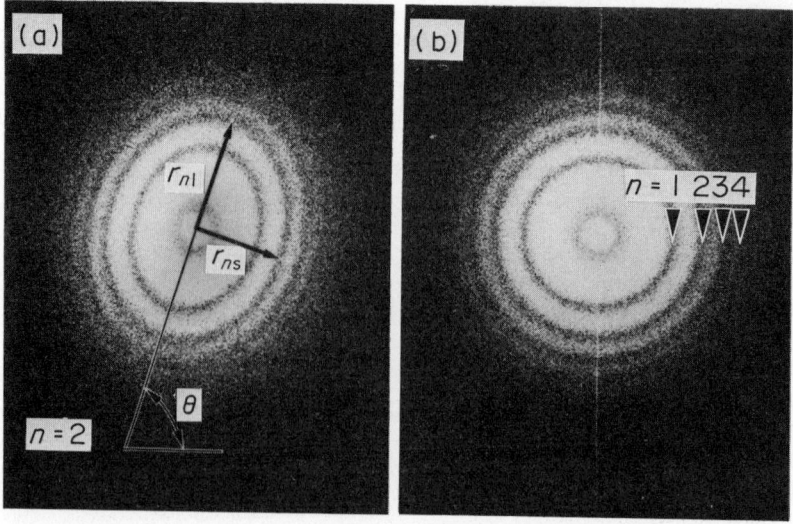

Figure 13 Optical diffractograms of images of a thin amorphous C film: (a) before and (b) after correction of astigmatism

in terms of r_{ns}, r_{nl} and θ which are observable (see Figure 13a), as

$$\delta f = (n\Lambda^2 L^2/M^2\lambda)(1/r_{ns}^2 - 1/r_{nl}^2) \tag{33}$$

When the x- and y-axis of the stigmator make an angle of $\pi/4$, the correction values of the stigmator currents are[82]

$$\Delta I_x = C\cdot\delta f\cdot\sin(|\theta - \alpha| - \pi/4)$$
$$\Delta I_y = C\cdot\delta f\cdot\sin(|\theta - \alpha + \pi/4| - \pi/4) \tag{34}$$

Here α and C are calibration constants which are to be adjusted for individual TEMs. Figure 13(b) shows the result of an astigmatism correction by the procedure mentioned above. If the image intensity is recorded with methods other than photographic emulsions, the real-time stigmating can be achieved with a scheme similar to that mentioned in this section.[83,84]

(iv) The image-recording system

In high resolution transmission electron microscopy, the best medium for image recording now available is photographic emulsion. Although various recording devices[85,86] such as an imaging plate[87,88] have been proposed for radiation sensitive specimens, the resolution of these devices is generally inferior to that of photographic emulsion. From the view point of cost performance, photographic films are recommended at present.

Though various photographic films and plates are available,[89-99] we have used Mitsubishi (MEM) and FUJI (FG) electron microscopic films. Figure 14 shows the relationship between the exposure and the optical density D of both films for 100 keV and 200 keV (1 eV = 1.6 × 10^{-19} J).[100] The developing conditions are in the figure caption. Both films show that there is quite good linearity of D against exposure, when the optical density is smaller than 1.0. Roughly speaking, MEM is about four times as sensitive as FG, but is more granular. Although some X-ray films are several times as sensitive as normal electron microscopic films such as MEM and FG, the resolution is much poorer.[56]

As shown in Figure 14, the optical density, D, is practically proportional to the electron exposure, E, up to $D = 1.0$.[89,90,99] High resolution electron micrographs should be taken at a certain exposure level to obtain the averaged optical density of about 0.5 through to 1.0. This density range is also desirable for photographic processes such as printing.

The resolution limit of photographic emulsion is represented by the 'modulation transfer function (MTF)'. For an approximate MTF, the following function $H(u)$ is proposed.[33,59,100-103]

$$H(u) = 1/\{1 + A(u/M)^2\} \tag{35}$$

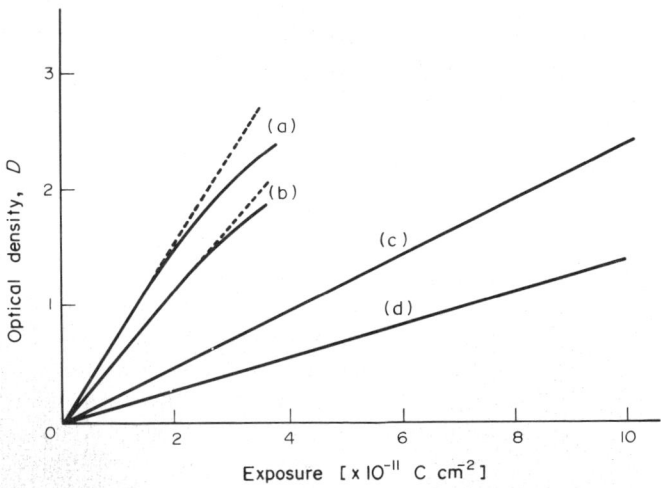

Figure 14 Relationship between electron exposure and optical density of Mitsubishi MEM (a and b) and Fuji FG (c and d) films for 100 (a and c) and 200 keV (b and d) (1 eV = 1.6 × 10^{-19} J), where MEM and FG were developed with Gekkol (Mitsubishi) in full strength and D-19 (Kodak) diluted 1:1, respectively, at 20 °C for 5 min

Here, A is a constant which may be dependent on some factors such as the kind of film, developing conditions, electron energy, exposure and so on. Also, in equation (35), u and M are the spatial frequency and electron optical magnification, respectively. When equation (35) is applicable as the MTF, the constant A is estimated from the highest spatial frequency that is recognizable in the optical diffractogram of the micrograph of an amorphous thin specimen which has been taken at a rather low magnification ($\sim 10\,000$).[59] As mentioned in Section 34.3.3.1.ii, H(u) may be introduced in $D_F(u)$ as one of the damping factors to $\sin \chi(u)$.[59,74]

34.3.3.2   Resolution-limiting factors relating to the specimens

(i) Specimen thickness

Equation (6) indicates that the specimen thickness should be smaller than R_{max} for the resolution of d_{min}, from the standpoint of the Fresnel diffraction effect.

According to the two-beam dynamical theory, the extinction distance is given approximately by[104]

$$\xi_g = \pi V_c \cos \theta / \lambda F_g$$

where V_c is the unit cell volume; θ is the Bragg angle; and F_g is the structure factor corresponding to the diffracting direction g. When the reflection indexed as g is dominant, the scattering of electrons can be treated kinematically if the thickness of crystal is much smaller than $\xi_g/2$.[105] If the specimen is rather thick, the dynamical scattering effect (multiple scattering, and so on) will be remarkable. Then TEM images do not directly reflect the crystal structure (namely, the projected potential distribution).[106-108] Using a multi-slice dynamical calculation,[106] Kawaguchi[109] demonstrated that if the PE single crystal is too thick, the structure amplitudes of 110, 200, *etc.*, greatly deviate from those for kinematical diffraction and forbidden reflections such as 010 and 100 increase in their intensity.

(ii) Specimen orientation

If individual chain stems in a polymer crystal are needed to be resolved separately in the micrographs, the electron beam must be introduced onto the crystal in the direction along its chain axis. The effect of specimen orientation is well demonstrated in Figure 25, where two kinds of high resolution micrographs of the α-form of a PPX single crystal are shown.[100] Incidence at $\langle 001 \rangle$ (namely, incidence along the molecular axis in the crystal) reveals the images of individual PPX chains in this crystal. Another example of the effect of orientation on high resolution images is demonstrated in Figure 31 for a solution-grown crystal of poly(p-phenylene sulfide) (PPS).[46,110] Detailed discussion about these specimens will be given in Section 34.5.1.1 for PPX and Section 34.5.1.2 for PPS.

It is possible to adjust the specimen orientation mechanically with a rotation/tilt or double tilt specimen holder for a side entry goniometer and with a double tilt holder for a top entry goniometer. The operation is, however, not easy for polymer specimens because of their electron irradiation damage during the operation of orientation adjustment.

(iii) Specimen drift

We should avoid the mechanical stage drift after specimen movement. If the TEM mode is changed, for example from a diffraction mode to an imaging mode, electrical hysteresis of the intermediate lens may cause focus drift. Thus, focusing and subsequent photographing should be postponed for a few minutes after changing the TEM condition to avoid the above-mentioned drifts.

We should also suppress the specimen drift due to thermal deformation and charge-up of the specimen and/or support film by applying electrons. Use of a C- and metal-coated microgrid (perforated film) is recommended.[111] A very thin C-supporting film[112] is deposited on the microgrid, as shown in Figure 15, for tiny specimens like polymer single crystals. Of course, a very wide specimen, like a stretched thin polymer film, can be deposited directly on the microgrid.[113] In such a case, but where the wide specimen has low electrical conductivity, it should be coated with a thin evaporated C layer to suppress specimen charge-up.

If the energy from the incident beam is uniformly deposited throughout the irradiated cylindrical volume, and the front and back of the thin specimen are at the same temperature, then the heat flow is considered to be purely radial. According to Isaacson,[114] the temperature rise of the very thin Au

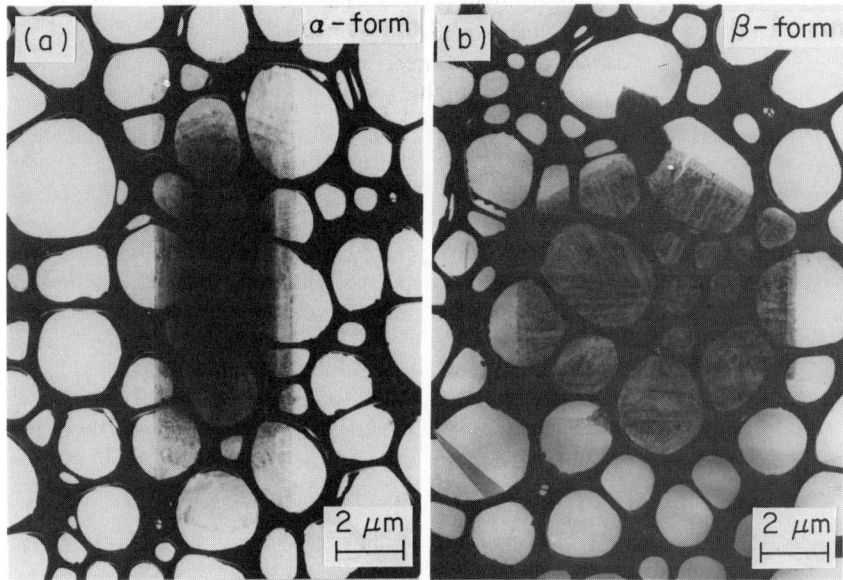

Figure 15 The α-form (a) and β-form (b) of PPX single crystals mounted on a C- and Au-coated microgrid on which a very thin C support film was deposited by indirect evaporation

support film in the center of the irradiated volume is estimated as a few degrees. Thus the temperature rise of the specimen and the radiation damage due to this temperature rise can be suppressed more or less by using an Au-coated microgrid.

We are not willing to use a carbon support film on the microgrid, even if the film is very thin, because it gives extra noise on TEM images. However, it suppresses the electron charge-up and therefore the drift of the specimen.

(iv) Radiation damage

In general, polymer crystals such as PE, poly(oxymethylene) (POM) and so on are vulnerable to electron irradiation, and the molecular chains are crosslinked to become amorphous or are decomposed in a short time on applying electrons. Both α- and β-forms of PPX crystals are, however, fairly strong, as compared with PE. Figure 16 shows the change in morphology and in the ED pattern of the β-form of a PPX single crystal due to 80 keV (1 eV = 10^{-19} J) irradiation.[115] This figure indicates that the radial stripes due to diffraction contrast disappear partially at first and then the observed area of no stripes spreads over the whole crystal, until entire stripes have disappeared. The crystalline reflections disappear from higher orders with increasing doses in the case of the α-form.[115] However, in the case of the β-form, the reflections of odd number indices such as (50.0), (41.0) and so on, disappear at first (see Figure 16c′), and the reflections having the indices of an integral multiple of four, such as (40.0), (44.0) and so on, remain (Figure 16d′). It seems from Figure 16(d′) that molecular chains are packed in the crystal with hexagonal symmetry where only one chain segment is contained in a smaller unit cell at such an irradiation dose. Then, at last, all the crystalline reflections disappear.

The disappearance of stripes due to diffraction contrast and/or of moiré fringes, as well as the disappearance of crystalline reflections, denotes the 'death' of the crystals. The effects of electron irradiation on PE crystals were discussed by Kobayashi and Sakaoku.[116,117] They illustrated that crosslinking between adjacent molecular chains in polymer crystals will take place under electron irradiation (Figure 17).[116] Kawaguchi et al. demonstrated the effect of crosslinking in a single crystal of PE due to electron irradiation on lattice distortions by using optical transforms.[118] In the case of POM, decomposition follows crosslinking and the crystal disappears.[119] As shown in Figure 16, the external shape of single crystals of PPX has not changed, suggesting that crosslinking takes place in PPX crystals under electron irradiation without decomposition. A lucid explanation of the irradiation effects on polymer crystals can be expected in the future, if we carefully observe changes in electron micrographs as well as those in ED.

Next, we will discuss quantitatively the effects of electron irradiation on polymer crystals. The changes of the lattice spacings by irradiation of 500 keV (1 eV = 1.6×10^{-19} J) are plotted in

Figure 16 Changes in morphology and in the ED pattern of the β-form of a PPX single crystal: (a')–(e') are the ED patterns corresponding to (a)–(e), respectively

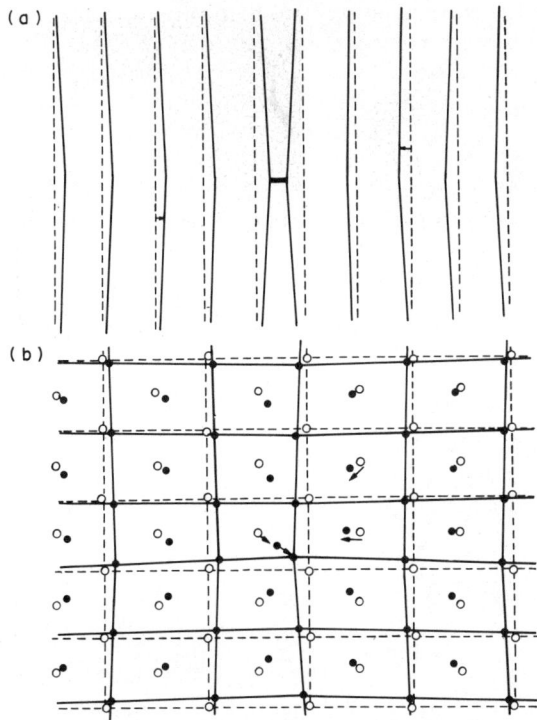

Figure 17 Distorted crystal lattice of PE with a crosslink between molecular chains due to electron irradiation:[116] (a) side view along the *b*-axis; and (b) projection along the chain axis (*c*-axis) on the *ab*-plane (dashed lines represent the original (undistorted) lattice)

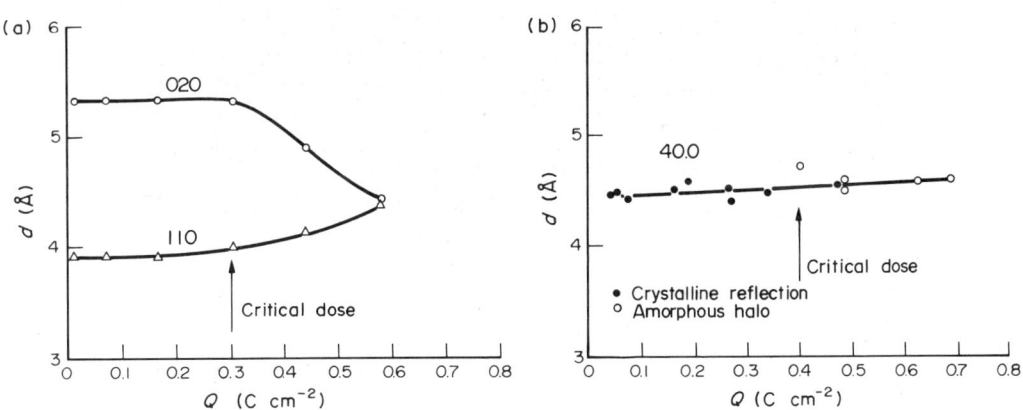

Figure 18 Variation in lattice spacing *d* of PPX (**1**) crystals ($n > 5000$) with an increasing irradiation dose Q of 500 keV (1 eV = 1.6×10^{-19} J): (a) changes of 020 and 110 spacings of the α-form single crystal; and (b) change of 400 spacing of the β-form single crystal

Figure 18 for (a) the α-form and (b) the β-form of single crystals of PPX.[14] These measurements were carried out with the single crystals mounted on an Al thin support film so as to give the diffraction rings of Al as reference. The electron beam current applied to the specimen was measured with a picoammeter connected to a Faraday cage inserted between the projector lens and the fluorescent screen. Even though such equipment is not now available, an exposure meter is installed in a present day TEM. Thus, with such a TEM, the electron beam current (*i.e.* screen current) can be measured approximately through reading the exposure meter which is connected to a viewing screen.[120,121]

The lattice spacings corresponding to 020 and 110 reflections of the α-form get closer together, starting at a dose of about 0.3 C cm^{-2} (this value may be defined as the 'critical dose'), and finally merge into each other. At a dose of about 0.5 C cm^{-2} (the 'total end point dose', TEPD), amorphous halo rings start to replace crystalline reflections. In the case of the β-form, the lattice spacing of the

40.0 reflection expands a little by irradiation, and halo rings appear at a dose of about 0.4 C cm^{-2} (this value corresponds to the 'critical dose' and also the 'TEPD'). The polymer molecules which have benzene rings in their backbone chains seem to be somewhat more resistant against electron irradiation. As for PPX (**1**) crystals, the TEPD is about 20 times larger than that of PE crystals at a given energy of electrons.[118,121] The TEPD's, namely electron doses necessary for complete damage of various polymer crystals, are summarized in Table 3, where 1 C cm^{-2} = 62 420 e nm^{-2}. The TEPD value of another accelerating voltage can be estimated from the dose–voltage relationship. The inelastic cross section or the absorption is proportional to $(m\lambda)^2$, that is $1/\beta^2$,[122–124] where m is the mass of an accelerated electron, λ the electron wavelength, and β the ratio of electron velocity to light velocity, respectively. Therefore the TEPD seems to be proportional to β^2. In the case of polymer specimens, however, it was also reported[125] that the TEPD is proportional to β^3 rather than β^2. In either case, the use of a higher accelerating voltage of electrons may be effective in reducing the radiation damage.

$$\left(\!-CH_2\!-\!\!\bigcirc\!\!-CH_2\!-\!\right)_{\!n}$$

(**1**)

Table 3   TEPD and Resolution Limit for Various Polymer Crystals at Room Temperature[46]

Polymer	TEPD, N (e nm^{-2})	Electron energy (keV)	Resolution limit (nm)		
			d_{obs}	d_{LD}	d_P
TMPS	4×10^2	200	0.32	1.00	5.0
N12	5×10^2	200	1.5	0.89	4.5
PE	6×10^2	200	0.37	0.82	4.1
iPS	2×10^3	200	0.49	0.45	2.2
PA	1×10^4	200	0.37	0.20	1.0
PPS	1×10^4	200	0.33	0.20	1.0
PPP	3×10^4	200	0.32	0.12	0.6
PPX	3×10^4	500	0.25	0.12	0.6
PPTA	5×10^4	200	0.43	0.09	0.4
(SN)$_x$	6×10^4	200	0.22	0.08	0.4

Cf. $d_{LD} = 1/0.1\sqrt{0.25\,N}$, $d_P = 5/0.1\sqrt{0.25\,N}$. Abbreviations: TMPS = poly(tetramethyl-*p*-silphenylene siloxane); PE = polyethylene; iPS = isotactic polystyrene; PS = polyacetylene; PPP = poly(*p*-phenylene); PPS = poly(*p*-phenylene sulfide); N12 = nylon 12; PPX = poly(*p*-xylene); (SN)$_x$ = polysulfurnitride; PPTA = poly(*p*-phenylene terephthalamide).

The expected resolution of such radiation sensitive polymer crystals has been calculated with the modified Rose equation[126,127]

$$d = \gamma/C\sqrt{fN} \qquad (36)$$

where C and f are the contrast and the net utilization factor, respectively. Empirically, $C = 0.1$ and $f = 0.25$ are used. In equation (36), N is the number of electrons passing through the unit area of the specimen during image recording. Here one may consider N as TEPD. The factor γ is the signal-to-noise ratio in an area of $d \times d$, and $\gamma = 5$ was deduced by visual judgment for point resolution.[128] In Table 3, TEPD, the observed resolution limit d_{obs} and the point resolution limit d_P were calculated using equation (36) with $\gamma = 5$, and are shown for various polymer crystals. Except for PPS and PPTA, d_{obs} is rather smaller than d_P. Here, d_{obs}, which was estimated by optical diffraction of the electron micrograph, should not be considered as the point resolution but as the line resolution limit in the lattice image. Therefore in order to predict the ultimate limit in the lattice resolution of a certain polymer crystal with TEPD of N, the value of γ should be re-evaluated for this case, namely for lattice imaging.

The computer simulation was thus carried out for this purpose, and visual judgment proved $\gamma = 1.5\ (\pm 0.2)$.[129] The result means that we were able to recognize and identified in the micrograph those lattice fringes for which the spacing has been predicted using equation (36) with $\gamma = 1.5$. If the optical diffraction technique is adopted to judge the resolution limit, a smaller value of γ can be expected and was actually found to be 0.85.[129] When $\gamma = 1$ is assumed, the expected resolution limit d_{LD} is calculated as shown in Table 3.

Recently 0.49 nm lattice fringes were successfully obtained in a stretched isotactic PS thin film[113] by a low dose technique with a minimum dose system (MDS)[271] (see Section 34.5.2.1.ii). This polymer does not have such a great TEPD value and then the expected point resolution limit d_p is much larger (2.2 nm) than d_{obs}. Nevertheless the expected line resolution limit d_{LD} in Table 3 is in good agreement with d_{obs}. For some polymers in Table 3, such as PPP, *etc.*, d_{obs} is much larger than d_{LD}. This is mainly due to the resolving power of the microscope itself. Furthermore, in the case of the PE crystal, 0.37 nm (200) and 0.41 nm (110) fringes were resolved with a conventional TEM at room temperature.[130-132] These values greatly exceed the d_{LD} value for PE. It should be deduced that this inconsistency is due to the choice of the values C and/or f. A study for the re-evaluation of C and f is now in progress.

For radiation sensitive specimens, MDS[271] or a low dose unit (LDU)[272] is frequently used to take high resolution images. It is noted, however, that this equipment uses an electron dose corresponding to TEPD, namely N, only for recording images. From equation (33), higher resolution is expected if N increases for a given material. There are two ways to increase N; (i) to raise the accelerating voltage; and (ii) cryo protection. The former way is not so advantageous because the image contrast and the sensitivity of photo-emulsion decreases with increasing voltage, as pointed by Katayama.[133] As regards cryo protection, ifs effectiveness has been discussed in the literature.[134,339-341] Recently we recognized the definite effect of cryo protection at 4.2 K by using a cryo electron microscope with a superconducting objective lens (JEOL JEM-2000SCM; 200 kV).[135,273] Table 4 shows the ratio $N_{4.2}/N_{300}$, where $N_{4.2}$ and N_{300} are TEPD at 4.2 K and room temperature, respectively.[46] A ratio of more than ten has been confirmed. Lattice images of a single PE crystal have been taken at an electron optical magnification of 90 000 (160 kV, 4.2 K). One

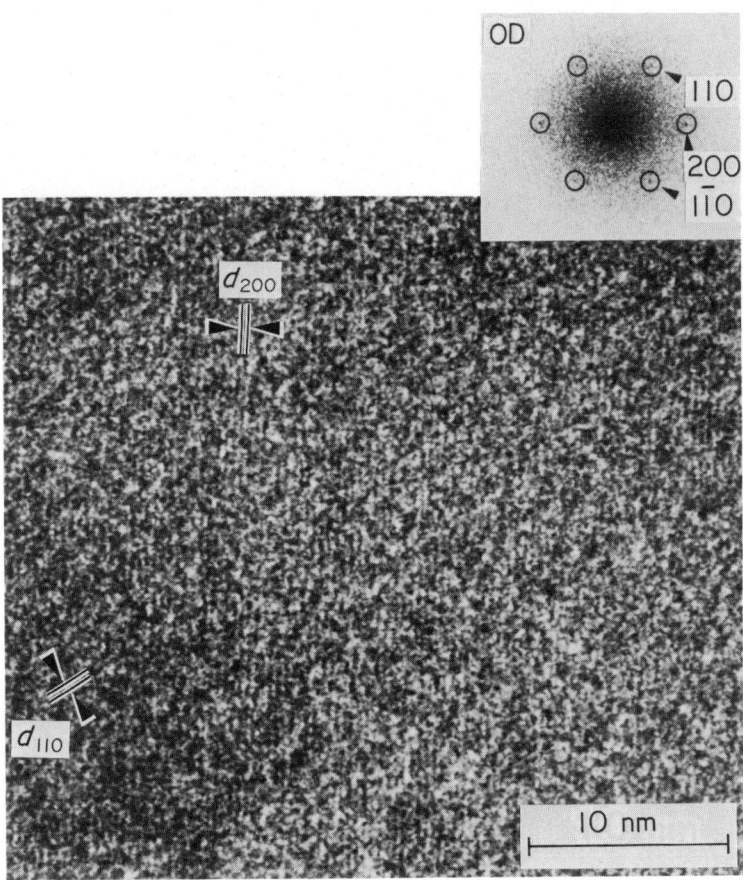

Figure 19 The lattice image of a single PE crystal taken at a direct magnification of 90 000 with a super-conducting cryo electron microscope (JEOL JEM-2000SCM) at 160 kV and 4.2 K. Inset is the correspondng optical diffractogram (OD) showing 110, 1$\bar{1}$0 and 200 reflections

Table 4 Temperature Dependence of
TEPD of Some Polymer Crystals[46]

Polymer crystals	$N_{4.2}/N_{300}$
TMPS	60
PE	15
PBPP	140
PPX	9
PLA	20

Abbreviations: TMPS = poly(tetramethyl-
p-silylphenylene   siloxane);   PE = polyethylene;   PPBP = poly(bisphenoxyphos-
phazene);   PPX = poly(*p*-xylylene);   PLA
= poly(acetic acid).

of these is shown in Figure 19, with the optical diffraction pattern (OD) from the original negative.[46] In this figure, the 200 and 110 lattice fringes are clearly seen. The 110 and/or 200 lattice fringes were also obtained from the stretched and annealed thin films of PE at $M = 75\,000$ (160 kV, 4.2 K).[132] As mentioned above, although radiation damage is one of the most important factors to restrict the resolution limit in TEM of polymer solids, cryo protection will bring us higher resolution images which will give more information about the finer structures in polymer solids.

34.3.4 Some Aspects of Image Processing in High Resolution TEM

It should be possible to improve the resolution of a TEM up to the theoretical resolving limit[26] (Section 34.3.2) to enable the direct observation of molecular and atomic images. This expectation is difficult to fulfil because of aberrations (in particular, the spherical[58] and chromatic aberrations)[136] of the objective lens in TEM and the photographic graininess[89,90] in electron microscopic films (*e.g.* granularity of photographic emulsion, the quantum noise of electrons, *etc.*). The degradation of photographed images is mainly caused by this graininess. Though various processing techniques were proposed,[137-141] including holographic filtering based on Gabor's holography,[142,143] the degradation resulting from graininess should be eliminated prior to any correction of the effect of spherical aberration or defocusing, in order to obtain high resolution images.[144]

Picture processing techniques have progressed to meet the requirements of space research, especially that of remote sensing.[20,145,146] In these techniques, pictures are converted to mathematical images (digitized), and picture processing techniques, such as geometric correction, noise removal, high frequency attenuation correction and so on, are applied.[147] These techniques have been applied to *a posteriori* image processing of electron micrographs and also to on-line processing when it is combined with a TV system directly connected to TEM.[83,84] They are now indispensable for the high-resolving electron microscopy.[55]

Image processing methods for electron microscopy fall into two broad categories: (i) analog and (ii) digital processing. Although the apparatus for analog processing (usually a purely optical method) is simple and rapid, the process has poor reproducibility and is unsatisfactory for quantitative processing. On the other hand, digital processing shows good reproducibility and enables various kinds of processing with full accuracy, if the sampling method (especially, the sampling interval) is appropriate.[147,148] This system, however, needs a large size computer for rapid handling of a great number of data.

34.3.4.1 The analog processing system (optical filtering)

The optical transform method, which was used for the first time by Bragg in 1939[15] and developed by Taylor and Lipson,[16,17] has been used for the anticipation and/or the interpretation of X-ray[149-152] or electron diffraction patterns,[109,118,153-156] and also for the analysis of electron micrographs.[68,157-164] Hosemann studied the paracrystalline structure in polymer crystals with this method.[165] In the field of high voltage and high resolution TEM, this simple method has been a *sine qua non* for accurate stigmating[81,82] of the objective lens (see Section 34.3.3.1.iii) or for the performance test of EM.[59,74] On the basis of the optical transform principle, Klug *et al.* established the so-called optical filtering method for separating superposed images in a micrograph and applied this to the analysis of large biopolymer structures.[18] This method has progressed and has been applied in the wide field of electron microscopy.[52,166-178] From the processed images, reconstruction of three-dimensional structures has also been attempted with a digital computer.[179-189] A special specimen preparation method to give a two-dimensional periodical array like a two-dimensional crystal has been proposed for image processing in biological electron microscopy.[190]

Our apparatus was constructed so as to analyze electron micrographs by optical transforms and/or remove random noises in them by optical filtering.[55] Figure 20 shows: (a) the general view, and (b) the scheme of our apparatus. This system is fixed on an iron girder (3 m long) with magnetic stands, so the system can be rearranged easily in accordance with experimental requirements. Similar systems are commercially available.[171,191]

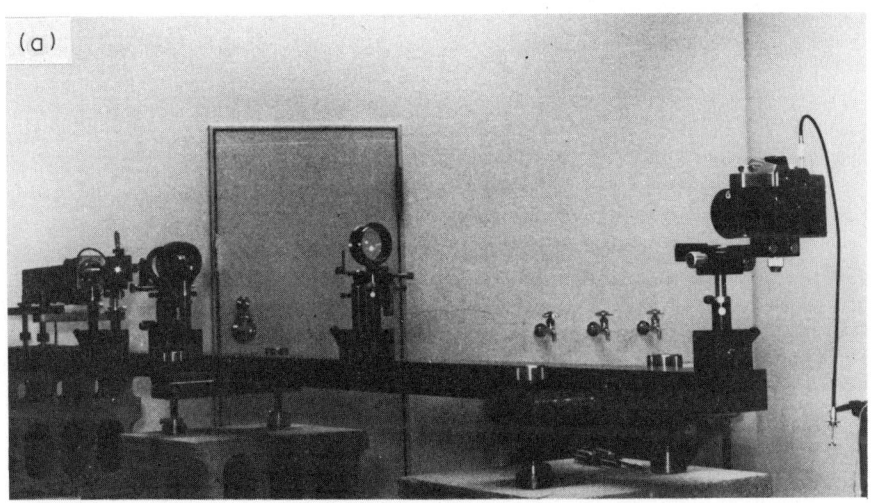

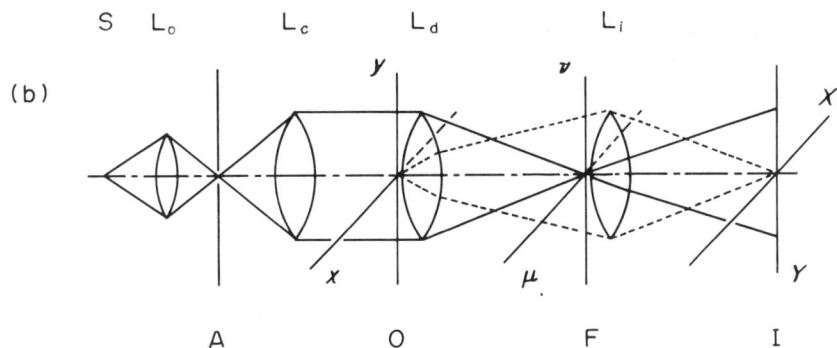

Figure 20 Apparatus for optical filtering: (a) a general view; and (b) a schematic arrangement, in which S is the light source (a He–Ne gas laser), L_o the condenser lens, L_c the collimator lens, L_d the diffraction lens, L_i the imaging lens, A the diaphragm, O the object plane, F the back focal plane of lens L_d, and I the image plane

34.3.4.2 The digital processing system

An analog processing method offers little variety in the modes of image processing. On the contrary, digital processing has a wider application, and various methods have been proposed.[20,139-141,192-200,268] Figure 21 shows: (a) a general view of our apparatus; and (b) a schematic flow of data. The facsimile receiver in this system is also used to display the simulated through-focal TEM images (see Figure 27 in Section 34.5.1.2.ii) and the results of the X-ray diffraction pattern simulation.[201]

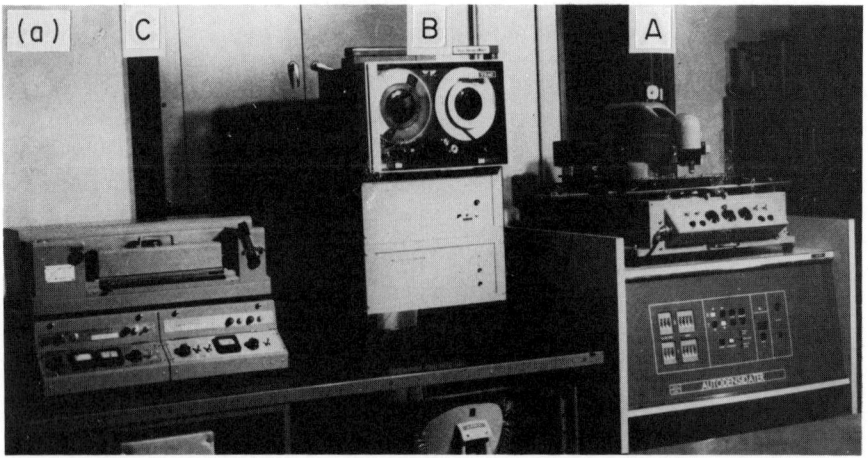

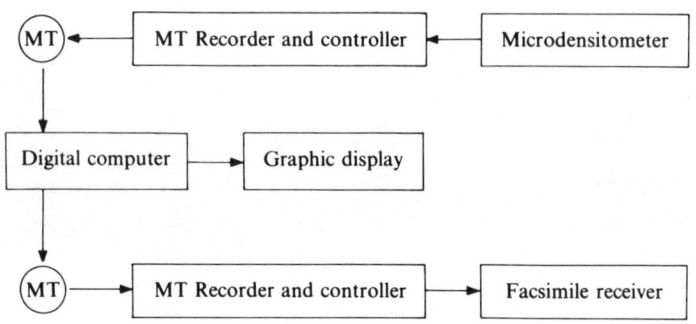

Figure 21   Apparatus for digital image processing: (a) a general view, in which A is the microdensitometer, B the magnetic tape recorder and its controller, and C the facsimile receiver; and (b) a schematic flow of data

34.3.4.3   *Theoretical treatment of image processing*

(i)   General background on spatial filtering

An optical filtering system is shown in Figure 20. Suppose that an electron micrograph with amplitude transmittance $g(x)$ is set in the object plane O in front of a lens L_d and illuminated with a collimated laser beam. Then, the amplitude distribution $G(u)$ in the back focal plane F of lens L_d can be expressed in terms of a two-dimensional Fourier transform as

$$G(u) \quad = \quad \int g(x)\exp(2\pi iux)\,dx \tag{37}$$

where the variables $u = (\mu, v)$ and $x = (x, y)$ are the spatial frequencies in Fourier space (plane F) and the position coordinates in the object plane O, respectively, and a constant phase factor is neglected. If a camera is set and the intensity $|G(u)|^2$ is recorded at this place (plane F), we can obtain the spatial frequency distribution (*viz.* an optical diffraction pattern) of the object $g(x)$. After passing through the plane F the light waves form a real image on the plane I (*i.e.* the lens L_i also produces the Fourier transform). If the coordinate axes taken in the image plane I are in the opposite direction to those in the plane O, as shown in Figure 20(b), and if an appropriate scale proportional to the magnification is defined, then the amplitude distribution $\psi(X)$ in the plane I is given by

$$\psi(X) \quad = \quad \int G(u)\exp(-2\pi iXu)\,du \tag{38}$$

where $X = (X, Y)$. When a filter with amplitude transmittance $T(u)$ is set in the plane F, the modified amplitude distribution $\psi'(X)$ on the image plane I is given by

$$\psi(X) \;=\; \int G(u)T(u)\exp(-2\pi i Xu)\,du \;=\; g(X)*t(X) \tag{39}$$

where $t(X)$ is the inverse Fourier transform of $T(u)$ and the $*$ denotes the convolution operation. The quantity $|\psi'(X)|^2$ is the intensity distribution of a filtered image. This method is generally called 'spatial frequency filtering'.[19,202] The correction of the spherical aberration will be possible with an appropriate filter $T(u)$.[197]

If random noises caused by the photographic graininess are superimposed on a periodic pattern, then $G(u)$ consists not only of a set of discrete peaks due to periodicity, but also a continuous spectrum due to noise. If we insert a filter grating $T(u)$ in the plane F which can let through the sharp peaks but block the rest of the spectrum, then the periodic structure will be enhanced with respect to the noise.[144,173]

Another method to remove random noise is called 'linear integration',[169,171,203-205] *i.e.* the shifted superposition of a periodic picture. Since the noise is regarded as random,[89,90] the $\sqrt{N}$ improvement[20] in the S/N ratio will be attained by this procedure, where N is the number of repeated superpositions. These two methods are different in practice. In a special case, however, Hashimoto *et al.*[175] and Aebi *et al.*[173] have shown that the optical image contrast formed by spatial frequency filtering using a filter grating is equivalent to the picture contrast by linear integration.

(ii) Image averaging

So-called 'image averaging' using a filter grating made up of an array of pinholes was treated mathematically by Aebi *et al.*[173] The effective averaging is determined by the spatial extent of the hole size in filter gratings.[144] Fraser and Millward discussed the effects of the size of the hole, and estimated the optimum filter hole size.[206] If both periodic and large scale irregular structures are contained in the original image to be processed, optical filtering should be carried out using a filter grating $T(u)$ which has pinholes that are rather large in size.[55,144] Such examples will be shown in Section 34.5.1.1.iii for lattice images containing edge dislocation.

Next, a typical example of a filter grating for effective averaging will be described. The filter grating, which was used in the optical filtering of the micrograph of the β-form of a PPX single crystal, is a hexagonal one (lattice constant 1 mm), having very small holes (100 μm in diameter).[14] In the experiment, only the holes which corresponded to the sharp peaks in the diffractogram were used and the others were clogged. In this case, the hole size of the filter is much smaller than the lattice constant (the hole size is 0.1 of the lattice constant), and thus rather effective averaging was successfully obtained. Tanji *et al.* reported that for effective averaging, the hole size in the filter grating should be smaller than 0.075 of the lattice constant.[207,208] Processed images were recorded on Minicopy film (copy film; Fuji Photo Film Co. Ltd.) to avoid quality degradation due to the granularity of photo-emulsion, and the film was developed at 20 °C for 6 min with Konidol Fine (Konishiroku Photo Ind. Co. Ltd.) to reduce too much high contrast of this film.[144]

A handy filter grating for optical filtering may be formed by making holes directly in the Polaroid–Land film on which the optical diffraction pattern of the micrograph is recorded. Optical filtering is very simple and useful, but several technical problems still remain. For example the multiple reflection of light between lenses causes an interference effect on the image. Lenses should be set at intervals long enough to suppress this effect.

34.4 SCANNING ELECTRON MICROSCOPY AND SCANNING TRANSMISSION ELECTRON MICROSCOPY

In the case of TEM, the specimen is illuminated rather widely with an electron beam for morphological observation. On the other hand, in the scanning electron microscope (SEM), a specimen is scanned with a focused fine beam. Using the signals from successive points in the specimen which rastered with the beam, the image is displayed on a TV monitor. In this case, the scanning raster in the TV is perfectly synchronized to the slow-scanning electron beam in the microscope; that is, the deflection coils of the cathode ray tube in the TV monitor are synchronized with the scanning coils of the incident illumination on the specimen in the microscope column.[21,23,25] The brightness at a certain point on the TV screen is modulated proportionally to the number of collected electrons which are emitted from the corresponding position of the specimen. There are two kinds of scanning electron microscopes: (i) the so-called SEM for surface observation using some of the secondary electrons emitted from the specimen surface; and (ii) scanning TEM (STEM) using electrons transmitted through thin specimens, as is the case for TEM. For both, resolution depends on the size of the electron beam as a probe rastering the specimen. Thus the field emission gun (FEG) is recommended to get a reasonable current into a fine probe (see Table 2 in Section 34.3.3.1.ii).

34.4.1 Scanning Electron Microscopy (SEM)

For TEM, the specimen must be thin enough for electrons to pass through it (Section 34.3.3.2.i). Of course, the penetrating power of the electrons increases with increasing their energy (namely, accelerating voltage), but the maximum specimen thickness seems to be at most the order of μm.[25] To observe the surface of bulk materials with TEM, the replication method is applicable (Section 34.2.2.2), but it is rather troublesome to make the replica. For these materials, SEM is easily employed.

The contrast in SEM is formed by several mechanisms. Some are briefly introduced here.[209,210] (i) Effect due to the angle between the incident beam direction and the specimen surface (surface tilt contrast); the surface normal to the incident beam direction is the darkest.
(ii) Edge effect; protuberances on the specimen surface show bright contrast (diffusion contrast).
(iii) Secondary emission rate; the image contrast depends on the amount of secondary electrons, namely, the difference of the kinds of constituent atoms in the specimens (material contrast).
(iv) Effect of metal coating; to suppress the charging of insulating specimens and/or to increase secondary emission rate, heavy metals such as Au are used to coat the specimen surface, for example, using a sputter coater. In this case, rather large fluctuations in the thickness of evaporated metals induce an additional contrast as an artifact.

An insulating specimen will be charged positively or negatively, depending on the total electron emission yield.[209] Charging will introduce artifacts in the image contrast. Detailed discussion about charging and its effect on the image contrast have been presented by Reimer.[209] As mentioned above, metal coating is frequently adopted to increase the electric conductivity of the specimen. Coating the specimen with a thin layer of evaporated metal brings the following additional benefits:[25] (i) to decrease the radiation damage, and (ii) to increase the secondary emission rate. As a coating material, an Au–Pd (60:40) alloy is recommended, as it shows finer granularity than Au.[210] Some methods are proposed to increase the conductivity,[25] including the use of a lower accelerating voltage.

In the field of polymer science, SEM is widely used. For example, Tagawa et al. showed the lamellar structure in blown films of PE, using FEG as the electron source and with a coating of an Au–Pd alloy.[211] In order to observe internal morphology of PE solids with any inherent plastic deformation, Shimamura et al. proposed a method by fracturing, combined with treatment with fuming nitric acid.[212–214] SEM combined with the energy dispersive X-ray analysis (EDXA) has been applied to the study of lignin distribution in a wood, through the use of brominated wood.[215] To improve the spatial resolution, a thin section is recommended as a specimen. This thin specimen is mounted on a collodion film deposited on a single hole C grid and then this grid is set on a holder. The holder is also used as a Faraday cage. The single hole C grid is also applicable for TEM-EDXA experiments.[216–218]

34.4.2 Scanning TEM (STEM)

The image contrast and resolution for STEM is, in principle, the same as in conventional TEM, according to the reciprocity theorem.[219] Thus, mass thickness contrast, diffraction contrast and phase contrast are expected with STEM, as in the case of conventional TEM. Of course, lattice imaging is also obtainable.[220] STEM has some advantages over conventional TEM as follows.[23]
(i) A rather thick specimen can be observed because there is almost no effect from chromatic aberration since there is no imaging lens following the specimen.
(ii) Apparent radiation damage will be reduced, because the very fine probe falls on one point of the specimen or scans a certain tiny portion. Thus the rest of the specimen is still alive. This technique is similar to the low dose technique in conventional TEM, for example MDS (Section 34.3.3.2.iv).
(iii) Microdiffraction is easily obtained. With this technique, a sequence of ED patterns from adjacent areas are observable, for example along the spherulite radius.[221]

(iv) A many beam annular dark field image is observed using an annular detector to collect one ring of the ED pattern from the specimen.

(v) Contrast is electrically adjustable as is the case for SEM. If a computer image processing system is connected directly to STEM, on-line processing is applicable.[344, 345]

Up to now, only a few works in polymer science have been reported with the use of STEM.[23] Further development is expected.

34.4.3 Scanning Tunneling Microscopy (STM)

Recently, STM has become a good tool to analyze the ultra fine structure in the surface of materials.[221] This technique is based on the tunnel effect in quantum mechanics. If a very thin and pointed probe (tip) approaches down to a certain distance from the specimen surface, electrons come out from the specimen by tunnelling through a potential barrier to the tip, *i.e.* forming a tunnel current. To control this tunnel current so that it is constant, the distance between the tip and the specimen surface is adjusted, and therefore this distance is monitored under synchronizaton of the scan of the probe along the surface. Another way, is to keep the height of the tip constant, and to monitor the tunnel current. With STM, atomic resolution is expected, but this is applicable mainly to conducting materials. Further developments are awaited.

34.5 THE APPLICATION OF ELECTRON MICROSCOPY TO THE STUDY OF POLYMER SOLIDS

Electron microscopy is widely used in the field of polymer science for morphological investigations.[21, 22, 222-224] In particular, analytical SEM is now one of the indispensable tools used by companies to analyze the constituents and structure of newly developed materials. Thus most work reported in the literature used electron microscopes in a routine way with no special techniques. In this section, our results will be described as examples of structural investigations by electron microscopy. The new techniques, except those shown in previous sections, and some important results in the structural study of polymer solids are also described.

34.5.1 Solution-grown Crystals of Polymers

The morphologies of single crystals of various polymers are reviewed extensively by Geil[225] and Wunderlich.[223] In this section, structural studies of solution-grown crystals including so-called single crystals are described. Detailed discussions, however, are mainly focused on poly(*p*-xylylene) (PPX) and poly(*p*-phenylene sulfide), with some other examples introduced briefly at the end of this section.

34.5.1.1 Poly(p-xylylene) (PPX)

The crystal structure of PPX was first reported in 1953 by Brown and Farthing[226] who discovered two crystalline modifications of α and β, corresponding to the low and high temperature modifications, respectively. In 1966, Niegsch[227] found the presence of a new transition at 270 °C through differential thermal analysis. This transition appears uniquely in the polymer prepared from di-*p*-xylylene. He also studied the morphology of both α and β single crystals of PPX grown from α-chloronaphthalene solution. Kajiura *et al.*[228] have confirmed by X-ray measurements that these two single crystals correspond to the α- and β-forms of the bulk materials observed by Brown and Farthing. Miles and Gleiter,[229] and Isoda *et al.*[230] studied the molecular mechanism of the α → β phase transition of single crystals of PPX by TEM using the heating specimen holder. Two entirely different types of single crystals (in fact, they are the so-called 'mosaic crystals') were developed from the solution (0.05%, 210 °C), showing a rectangular habit (α-modification) and a hexagonal one (β-modification).[227] The crystal thickness of the α-form was ∼ 12 nm, and that of the β-form was ∼ 8 nm.[14]

(i) α-Modification of PPX

In the case of the α-modification of single crystals of PPX, two types of diffraction patterns were obtained, corresponding to a single layer or multi-layer crystal.[14] Figure 22 shows both kinds of

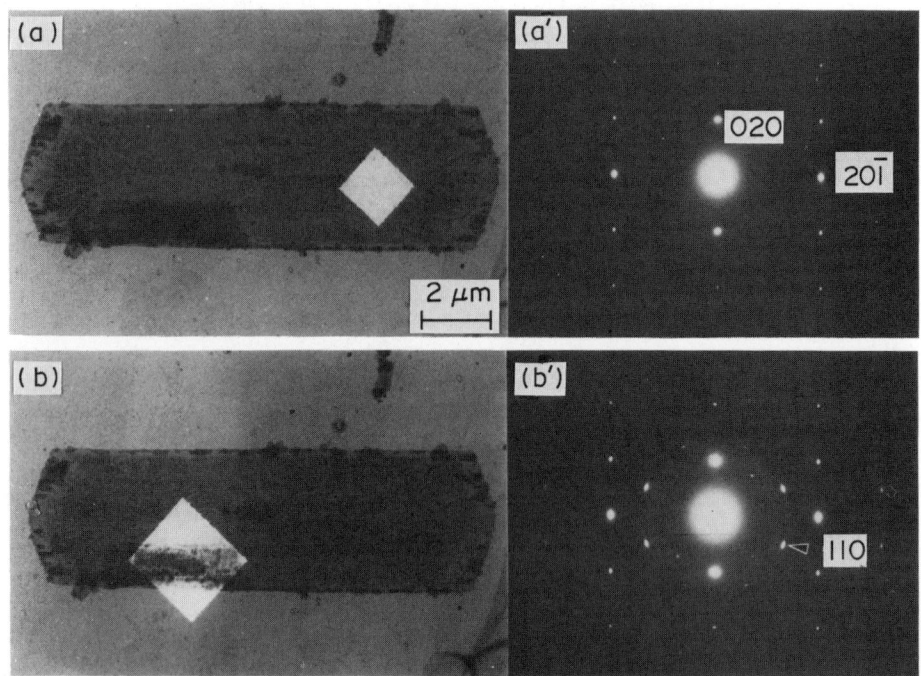

Figure 22 α-form of a PPX single crystal and the corresponding selected area ED patterns: the squares in (a) and (b) show the size and position of the selected area aperture

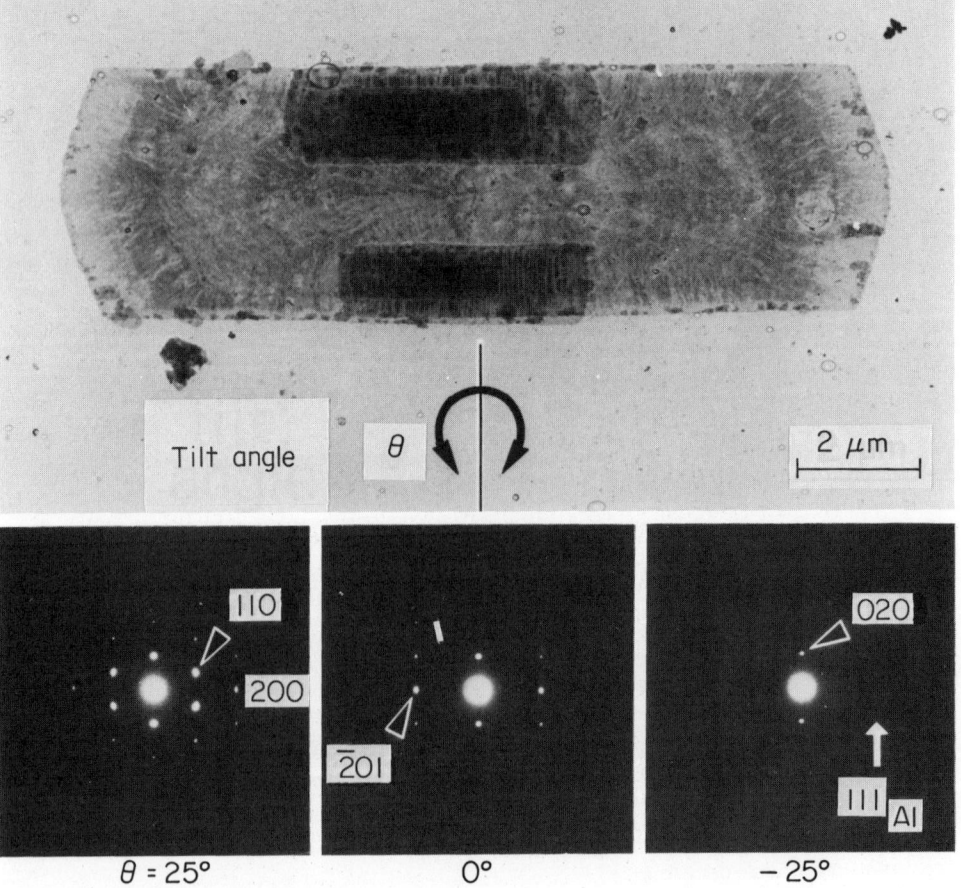

Figure 23 The change of an ED pattern of the α-form of a PPX single crystal due to specimen orientation. The specimen was tilted around the short axis of the rath crystal, using a rotation/tilt holder ($\theta = 25°$ corresponds to an $\langle 001 \rangle$ incidence)

selected area ED patterns and their double-exposed micrographs with a selected area aperture. Kubo and Wunderlich[231] reported the unit cell dimensions of the α-modification, and Iwamoto and Wunderlich[232] analyzed the crystal structure of this modification based on the result by Kubo and Wunderlich. As shown in Figure 22, a single layer crystal yields the ED pattern corresponding to an ⟨102⟩ incidence of electron beam. A multi-layer crystal yields the *hk0* reflections corresponding to an ⟨001⟩ incidence of electrons as well as the diffraction pattern of an ⟨102⟩ incidence. Accordingly it is concluded that the chain axis in the basal layer is not perpendicular to the crystal end surfaces (namely, not parallel to the direction of the incident electron beam). It also confirmed the change in the ED pattern when tilting the crystal around its *b*-axis. As demonstrated in Figure 23, tilting the crystal to a certain direction by 25° gives strong *hk0* reflections and no $\bar{2}01$ in the ED pattern. The chain axis in the layers except the basal one is, however, parallel to the incident beam direction. Figure 24 is the optical diffractogram (OD) of high resolution electron micrograph of the α-form of the single crystal.[14] It resembles Figure 22(b′), and is apparently a micrograph of a double-layer crystal. Optical filtering (Section 34.3.4) was performed to separate superposed images into two. The results are shown in Figure 25(a) for a ⟨001⟩ incidence, and in Figure 25(b) for a ⟨102⟩ incidence. Encircled reflections in the OD at the upper right corner of each figure are those used for optical filtering. The inset in each figure shows the image simulated by computer ($C_s = 1.06$ mm, $\lambda = 0.00142$ nm, $\Delta f = 40$ nm and resolution limit = 0.2 nm). Each dark ellipse in Figure 25(a) is to represent the projection of one PPX molecule viewed in the direction parallel to its molecular axis.

The parallel fringes seen in the multi-layer parts in Figure 23 are moiré image interference fringes. The radial stripes seen in the single basel layer are due to diffraction contrast, indicating that there exist six sectors, as shown by a tent-like structure,[231] prior to its deposition on the specimen support film. Moiré fringes and radial stripes can also be observed in dark field images, as long as the crystal is living under electron irradiation.

(ii) β-Modification of PPX

Most of the β-form single crystals of PPX have nearly flat and regular hexagonal shapes, as shown in Figure 3 (Section 34.2.2.1) and Figure 16 (34.3.3.2.iv). The ED pattern of the β-form single crystal

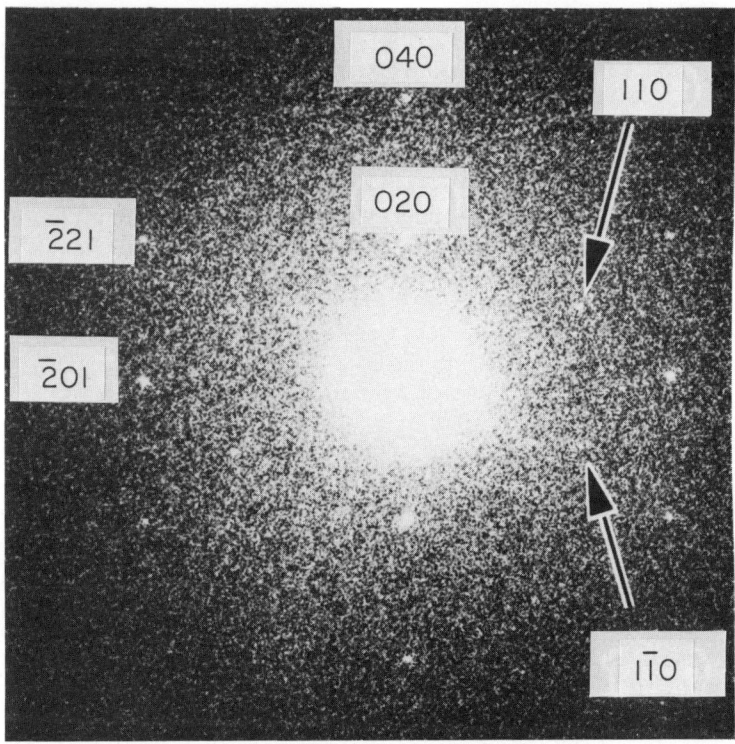

Figure 24 Optical diffractogram of the high resolution image of a multi-layered single crystal of the α-form of PPX (this pattern is similar to the ED pattern shown in Figure 22b′)

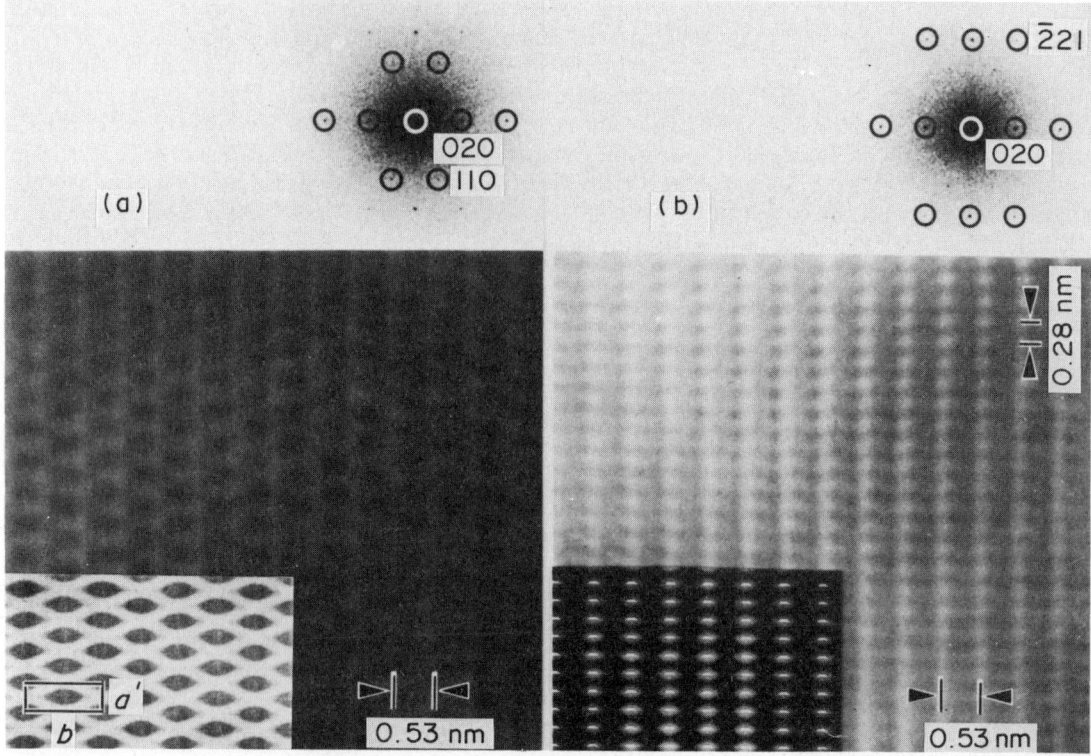

Figure 25 Optically filtered high resolution images of a multi-layered single crystal of the α-form of a PPX (a) ⟨001⟩ incidence; and (b) ⟨102⟩ incidence. The inset at the upper right corner of each figure is the optical diffractogram in which the reflections used for optical filtering are encircled. The inset at the lower left corner of each figure is the computer-simulated image on the basis of the crystallographic data of the PPX α-form. Kinematical image simulation was performed with $C_s = 1.06$ mm, $\lambda = 0.00142$ nm, $\Delta f = 40$ nm and resolution limit = 0.2 nm. In (a), $a' = a\sin\beta = 0.421$ nm and $b = 1.064$ nm

shows no systematic absences of reflections, as reported by Niegisch,[227] and has no symmetry planes (see Figure 3a or Figure 16a′). This pattern is characterized by merely six-fold rotational symmetry.[14] Spirally grown multi-layer crystals are frequently observed on the β-form.[115] The radial stripes which are seen in all the β-form crystals are due to diffraction contrast as observed in α-forms, and they indicate that there exist six (10.0) sectors in the β-form single crystal. This is also recognized in the dark field image, as shown in Figure 3(b). All the sectors yield an identical diffraction pattern. The stripes disappear under strong electron irradiation (Figure 16). Niegisch has estimated the hexagonal unit cell dimension ($a = b = 2.052$ nm, c(fiber axis) = 0.685 nm) and concluded that the molecules are all parallel to the c-axis, which is perpendicular to the single crystal end surface.[227]

The high resolution electron micrograph of the β-form single crystal was obtained using 500 kV ultra high resolution TEM (JEM-500).[14] In the case of the β-form, the electron beam is incident on the platelet crystal in its ⟨001⟩ direction and parallel to the molecular axis. Therefore individual chains comprised in the crystal are expected to be resolved. Figure 26(b) is the optical diffractogram (OD) of the micrograph and resembles ED (Figure 26a). The similarity between ED and OD patterns shows that the micrograph has sufficient information about the crystal structure of the β-form. In order to extract structural information from the noisy micrograph, optical filtering was performed using the filter grating as described in Section 34.3.4.3.ii. The processed image is shown in Figure 26(c) with a structural model in Figure 26(d).

In the case of the β-form crystal, it is concluded from the density and lattice dimensions that a unit cell must contain 16 chain segments (the length of a chain segment is the fiber period), that is to say, 16 polymer chains ought to run through a unit cell. This has made the structure analysis of the β-form crystal complicated. Prior to the structure analysis of the β-form of the PPX crystal, the polymer crystals belonging to the hexagonal or trigonal system were classified into three groups, as follows.[233]

Group 1 Each chain has a three- or six-fold axis or screw axis. The number of molecules in a unit cell is one or an integral multiple of three.

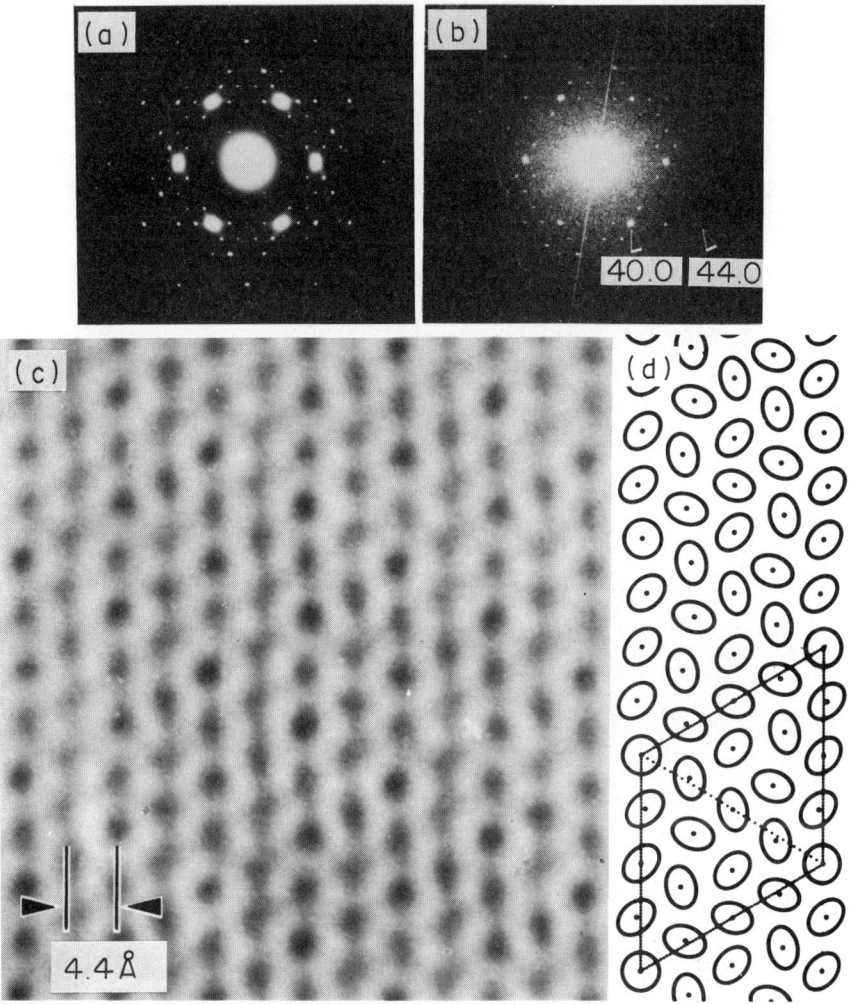

Figure 26 The β-form of a PPX single crystal: (a) the ED pattern taken at 500 kV, showing a net pattern with *hk0* reflections; (b) the optical diffractogram of the original negative taken with the ultra high resolution TEM JEM-500 operated at 500 kV. (c) the optically filtered high resolution image; and (d) the model structure analyzed by ED intensity

Group 2 Each chain has cylindrical or nearly cylindrical symmetry. In this case, a unit cell is made up of only one molecular segment and the symmetry is restricted to hexagonal or pseudo-hexagonal.

Group 3 An integral multiple of three identical chain segments constitute a unit cell, though each molecular chain has no symmetry features of Group 1 or 2.

The β-form of the PPX crystal which does not suit any of the above-mentioned prerequisites is an unusual example, and the crystal structure was not previously analyzed.

Each dark spot in Figure 26(c) represents the projection of one molecular chain on the *ab*-plane, as is the case of the α-form (Figure 25a). Some of the spots in this micrograph seem to be elliptical, showing the directions of the major axes of such ellipses. The significance of the image is its enhancement of the features of the molecular arrangement. Here the mutual positions of the molecular centers in the projection on the *ab*-plane are clearly observed, *i.e.* chains are arranged in a wave pattern of zigzag manner, for example in the 100 direction.[233] The feature of the arrangement of molecular centers seems to be represented in terms of the two-dimensional space group *P*6, which has no conditions limiting possible reflections. Indeed, the ED pattern of the β-form of a PPX single crystal has neither systematic absences of reflections nor symmetry planes but only shows six-fold rotational symmetry, as described above.

First, a two-dimensional structure model of the β-form of the PPX crystal was assumed so as to satisfy the symmetry of *P*6, taking van der Waals' radii and the high resolution image (Figure 26c) into account. Here, the orientation of the molecule at the origin of a unit cell was assigned statistically to one of three equivalent orientations to satisfy the symmetry of p6. The refinement of the structure analysis was attempted by the least squares method over the ED intensity data, which were obtained by optical densitometry of reflections which had appeared in ED patterns. A PPX 'single crystal' is a mosaic crystal which gives a spot pattern (so-called 'N-pattern') in its ED. Therefore $1/d_{hk0}$ was used as the correction factor corresponding to the Lorentz factor in X-ray diffraction, where d_{hk0} is the (hk0) spacing of the crystal.[234] The molecular conformation of the β-form of PPX was the same as that of the α-form.[232] If the molecule at the origin of a unit cell is represented with a circle and others with ellipses, this result (Figure 26d) resembles the micrograph (Figure 26c) fairly well, and some of the corresponding ellipses and spots in these figures coincide in position and in orientation with each other. Moreover, an optical mask with the same structure as this model gives an OD pattern (Figure 8(a) in Section 34.2.2.2), which is almost identical to the ED pattern of the β-form single crystal (Figures 1a, 16a′ and 26a).

Figure 27(a) schematically shows the result of crystal structure analysis of the β-form of the PPX crystal by using X-ray diffraction.[233] The crystal system and lattice constants are as follows: trigonal P3; $a = b = 2.052$ nm; $c = 0.655$ nm (molecular axis). The bold lines of the molecular backbone

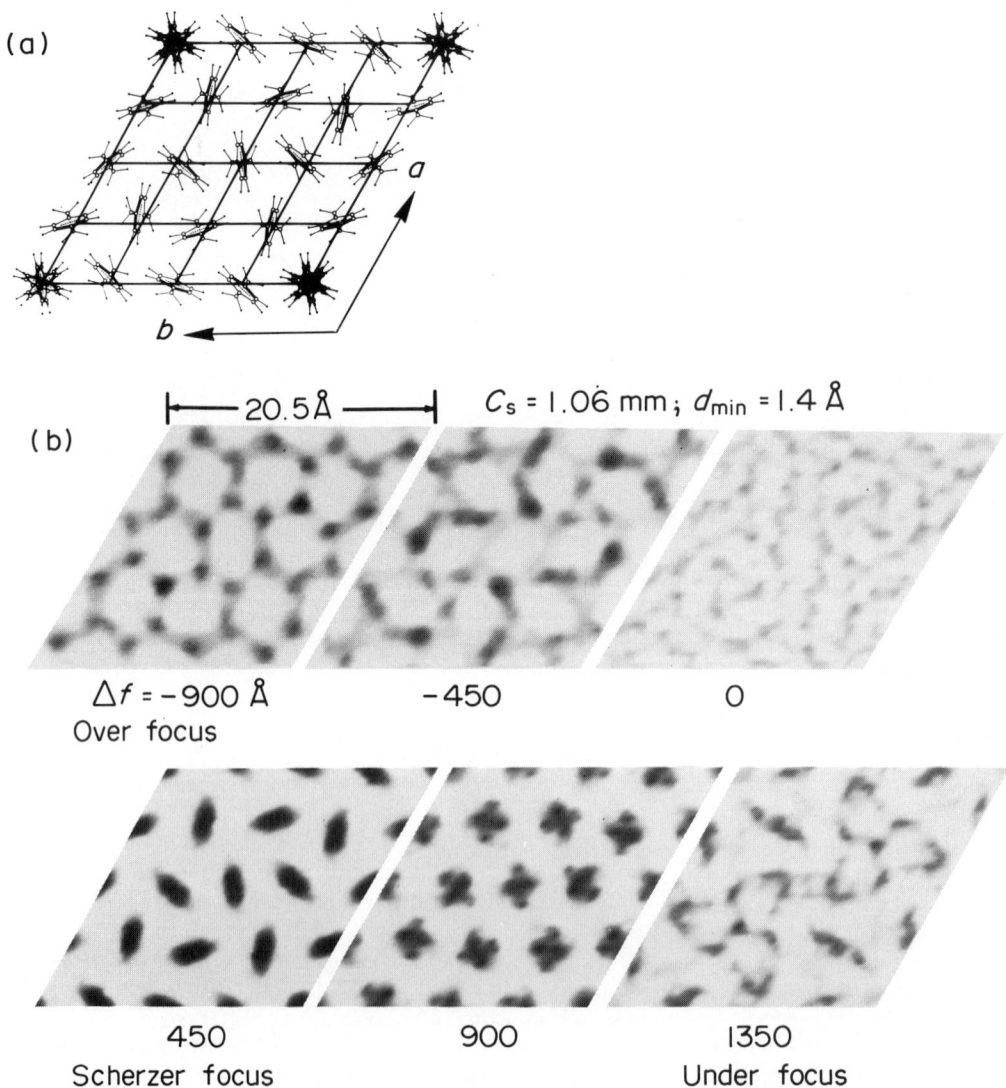

Figure 27 (a) the three-dimensional structure model of the β-form of a PPX crystal, which is shown as the ab-plane projection: trigonal P3, $a = b = 2.052$ nm, $c = 0.655$ nm, $\gamma = 120°$ [347] (the result of the final refinement should be cited from ref. 233). (b) the simulated through-focal images taken with a JEM-500, based on the structure model in (a): $\lambda = 0.00142$ nm, $C_s = 1.06$ mm, resolution limit = 0.14 nm

represent the upper part of a benzene ring in Figure 27(a). The molecule at the origin of the unit cell statistically takes one of three equivalent orientations shown in the figure.

Here we have reached the stage where the consistency of the results of X-ray analysis and the high-resolution electron micrograph should be examined. According to Kobayashi erroneous results may be obtained in the region of the focus of a microscope at its resolution limit.[235] The image intensity is greatly affected by the defocus and the spherical aberration through $\sin \chi(u)$ in phase contrast electron microscopy, as described in Section 34.3.2. Therefore it is necessary to compare the obtained electron micrograph with the simulated through-focal images, so as to confirm authenticity of the micrograph as well as the result of structure analysis. The computer simulation was performed according to the kinematical diffraction and using the following parameters: accelerating voltage $= 500$ kV, spherical aberration coefficient $C_s = 1.06$ mm, and resolving power $= 0.14$ nm). The result is shown in Figure 27(b), demonstrating the effect of defocusing. The simulated image at the Scherzer focus ($\Delta f = 45$ nm) is quite similar to the micrograph actually obtained. It is, however, not expected even with this ultra high resolution TEM to resolve individual C atoms of a PPX molecule in the *ab*-plane projection in which atoms are close together.

(iii) Direct observation of lattice defects in single crystals of PPX

Diffraction experiments (X-rays, neutrons, electrons) give only statistical data which are averaged over the whole sample. On the other hand, EM is a powerful and unique tool with which defects in the periodic structure of a crystal can be studied visually. Polymer crystals are much more imperfect than crystals of ordinary low molecular weight materials, as shown by X-ray diffraction. Based on electron micrographs with moiré image interference fringes[22, 24, 223, 236-240] (usually explained in terms of double Bragg diffraction of the electron beam) or dislocation networks[223, 240-243] (observed by virtue of the diffraction contrast) in bi-layered polymer single crystals, the imperfections in polymer crystals have been made considerably visible. Lattice defects in a PE single crystal due to electron irradiation were also investigated by the moiré method.[244] It was reported that the image intensity of the moiré fringe patterns in electron micrographs of two superimposed thin crystals is 'the square of the Patterson distribution', magnified by the factor $1/2\sin(\varepsilon/2)$ (ε is the relative rotation of the crystals).[245] This type is called a 'rotation moiré'.[223] Others are a moiré fringe from crystals of slightly different lattice spacings[223] (parallel moiré), and a tilt moiré because of an effective misorientation from the Bragg condition.[246] These three types may be partly mixed. Therefore it is not easy to analyze the structure of dislocation on a molecular or atomic scale from moiré fringes. The analysis of dislocation networks is much more difficult than that of simple moiré fringes. In the case of inorganic or organic crystals which are very resistant to electron irradiation, high-resolution images of defects and dislocations can be obtained.[247] Polymer crystals, on the contrary, are so susceptible to electron irradiation that the lattice images including the defects are difficult to take. In Figure 28 are the first successfully obtained high resolution lattice images which resolve directly the edge dislocation in the α-form and β-form of PPX single crystals.[248]

Since the constituent atoms in a polymer crystal are linked together with covalent bonds to form molecules, the direction of the backbone chain is a unique direction in the crystal. The classification of the dislocations in a PE orthorhombic crystal was first carried out by Keith and Passaglia.[249] Seto[250] classified dislocations in polymer crystals into five types on the basis of the directional relationship among the Burgers vectors $\vec{b}$, the vector of the backbone chain direction $\vec{m}$, and the vector of the dislocation line direction $\vec{l}$. Figure 29 schematically show five models of dislocation which are predicted to exist in polymer crystals. Wada and Matsui[251] reclassified the dislocations in PE in terms of these five types.

The characteristics of the dislocation types shown in Figure 29 are summarized as follows. Type (1): screw dislocation, which does not change the conformation of polymer chains. This can be first produced at the side faces of a thin polymer single crystal owing to thermal motion, and then proceeds into the crystal.[252]

Type (2): edge dislocation, which was proposed in order to account for the slip bands in the plastic deformation of nylon 66.[253] This dislocation is considered to be much less likely because of the resistance to glide motion, especially in polymers where the chemical repeat distance is very long.[249]

Type (3): edge dislocation, which is the most stable because this does not change the conformation of the polymer chains and passes through a lamellar crystal by the shortest course. This was observed through moiré fringes in lamellar crystals (such as polyoxymethylene,[238] PE[239] and so on) with TEM.

Type (4): screw dislocation, which is energetically unstable. A concentration of dislocations of this type generates a twist boundary in a crystal. Hosemann *et al.* considered, from the line profile analysis

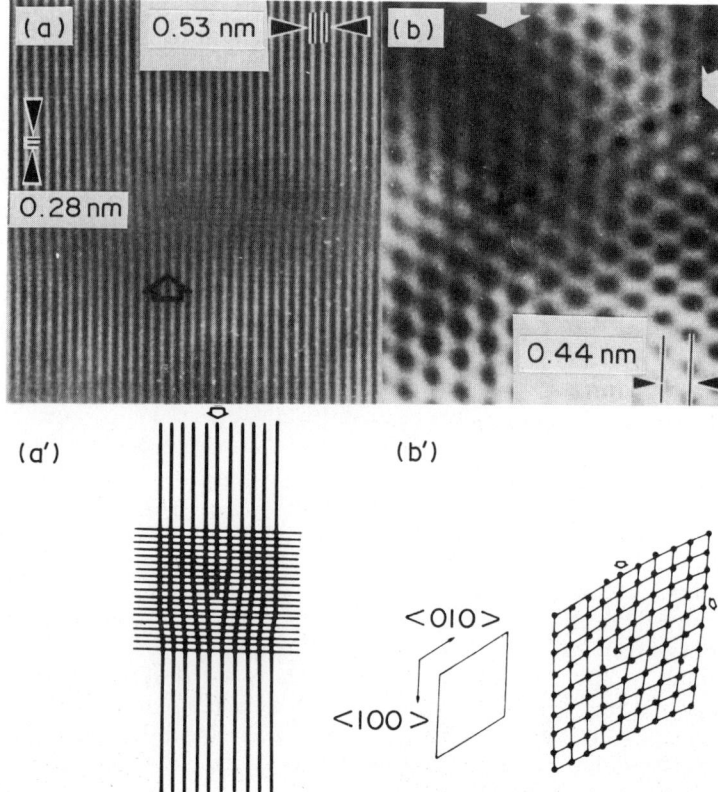

Figure 28 High resolution images of the α-form (a) and β-form (b) of PPX single crystals with lattice defects: (a) an edge type dislocation observed in 020 planes (0.53 nm in spacing), which is indicated by the arrow; (a') a schematic representation of (a), in which there is an extra half plane of 020, and no distortions observed in $\overline{2}01$ planes; (b) each dark ellipse corresponds to the projection of one molecule along the chain direction, with extra half planes of the 400 type indicated by arrows; and (b') a schematic representation of (b), in which the filled circles represent the centers of individual molecules in (b) and the lines show 400 and 040 planes

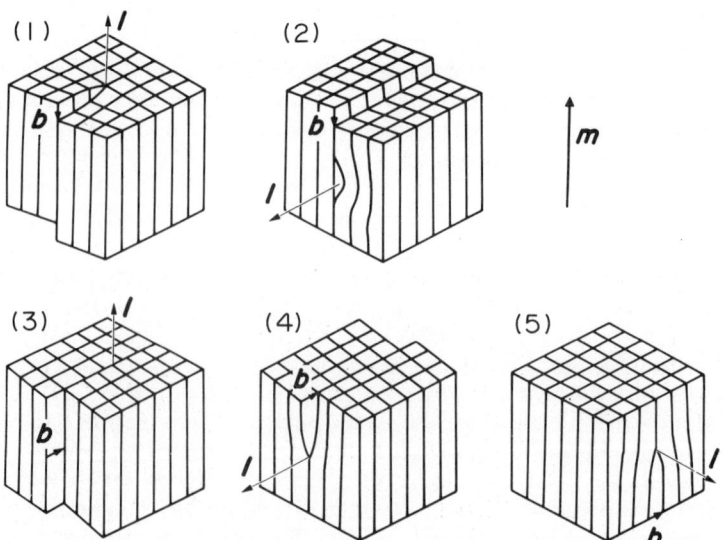

Figure 29 Models of dislocations which are predicted to exist in polymer crystals: (1), (4) have a screw dislocation; and (2), (3) and (5) an edge dislocation[250] (a square prism represents a molecular chain, $\vec{b}$ a Burgers vector, $\vec{l}$ a dislocation line vector, and $\vec{m}$ a vector which shows the chain direction)

of *hk0* X-ray reflections, that the polymer single crystals are mosaic, which they ascribed to this type of dislocation.[254]

Type (5): edge dislocation, which is generated when the ends of chains stand in a line. A tilt boundary is caused by a concentration of dislocations of this type.[255]

The screw dislocation of type (1) which has the Burgers vector parallel to the molecular axis, and the edge dislocation of type (3) whose Burgers vector is perpendicular to the molecular axis, should be very easily introduced in lamellar crystals. When the molecular axis in a thin lamellar crystal is perpendicular to the crystal's end surface, it appears to be difficult practically to observe the dislocations of types (1) and (2) because their Burgers vectors are parallel to the incident electrons. In the case of types (3), (4) and (5) dislocations can be observed using EM. The dislocation which has been observed in a PE single crystal corresponds to the partial dislocation of type (3).[240,256]

The edge dislocations in inorganic crystals have been observed on an atomic scale,[257-260] and the point defects have also been directly observed,[261-263] by high resolution electron microscopy. A high angle tilt grain boundary in a thin crystal of germanium was imaged with atomic resolution and was shown to consist of alternating columns of five- and seven-membered rings of germanium atoms.[264] The dislocations of various types are observed in the crystals of organic compounds such as hexadecachlorocopper phthalocyanine when the specimen is resistant to electron irradiation.[247]

In order to obtain high resolution images of crystal defects with TEM, the defects should extend along the direction of the incident electrons, forming columns which produce proper contrast in an image. The edge dislocation image of the β-form of a PPX single crystal in Figure 28(b) is the optically filtered image. Though the arrangement of molecules in the projection of the *ab*-plane is not very clear, the appearance of the dislocation is similar to Bragg and Nye's bubble model.[265] Two extra half planes of type 400 are designated by white arrows. The Burgers vector $\vec{b}$ of this partial dislocation is probably (1/4) [110] (see Figure 28b'). This should be type (3). Neither the mutual arrangement nor orientations of the molecular chains are defined in Figure 28(b) owing to a lack of higher resolution. Figure 28(a) is the high resolution lattice image with the lattice defect of the α-form of a PPX single crystal. Intense 020 lattice fringes (0.53 nm in spacing) and weak $\bar{2}01$ ones (0.28 nm) are observed. The upper part of this figure contains one extra half plane of 020 (indicated by an arrow). As schematically shown in Figure 28(a'), there are no distortions observed in the $\bar{2}01$ planes. Thus this dislocation is considered to be a partial one of type (3). The Burgers vector $\vec{b}$ is probably (1/2) [010] and a stacking fault should be associated with this. Details are not as yet known.

Recently, a type (2) dislocation was found in an image, with moiré fringes, of the bi-layered β-form of a PPX single crystal.[266] Read and Young[267] reported a chain end dislocation dipole in a 010 lattice image of poly[1,6-di(*N*-carbazoyl)-2,4-hexadiyne], which is one of the polydiacetylenes. Probably this corresponds to the type (5) dislocation in Figure 29, when viewed in the direction parallel to the vector $\vec{l}$.

34.5.1.2 Poly(p-phenylene sulfide) (PPS)

PPS is used in the electric and electronic field, mechanical field and so on, owing to its fairly high thermostability and small molding contraction.[274] Though the crystal structure of PPS was analyzed by Tabor *et al.*[275] and the electroconductive mechanism of doped PPS has been discussed by many workers,[276,277] the morphologies of PPS itself and doped PPS have not been studied very well. Here the electron microscopical observation of a PPS solution-grown crystal will be shown.[110,278]

Figure 30(a) is a solution-grown fibrillar crystal of PPS and its ED pattern. The thickness and the width of the crystal were measured at 11.5 nm and at 100 to 300 nm, respectively. As deduced from the micrograph of the crystal shadowed with Pt–Pd (Figure 30b), a fibrillar crystal seems to consist of finer 'microfibrils'. The ED pattern (inset of Figure 30a) shows *hk0* reflections and some other reflections such as 111 and 211, as indicated by arrows in the figure. On the basis of Ewald construction, it was concluded that the fibrillar crystal grows in a direction parallel to the *b*-axis by changing its orientation around the *b*-axis.[278] Figure 31(a) is the same ED as the inset in Figure 30(a), showing that the angle between lines drawn from the origin to the centers of the 200 and 111 reflections is 49°. In principle, the ED pattern discussed above corresponds to ⟨001⟩ incidence of electron beams onto the crystal. Occasionally the ED pattern shown in Figure 31(b) was observed. It reveals *h00* and *h11* reflections and the angle in question is 60°. In this case, the incident electron beam direction is parallel to ⟨0$\bar{1}$1⟩ of the PPS crystal (⟨011⟩ incidence is identical to ⟨0$\bar{1}$1⟩ incidence).

The total end point dose (TEPD) of PPS crystal is about 0.2 C cm^{-2} at room temperature and

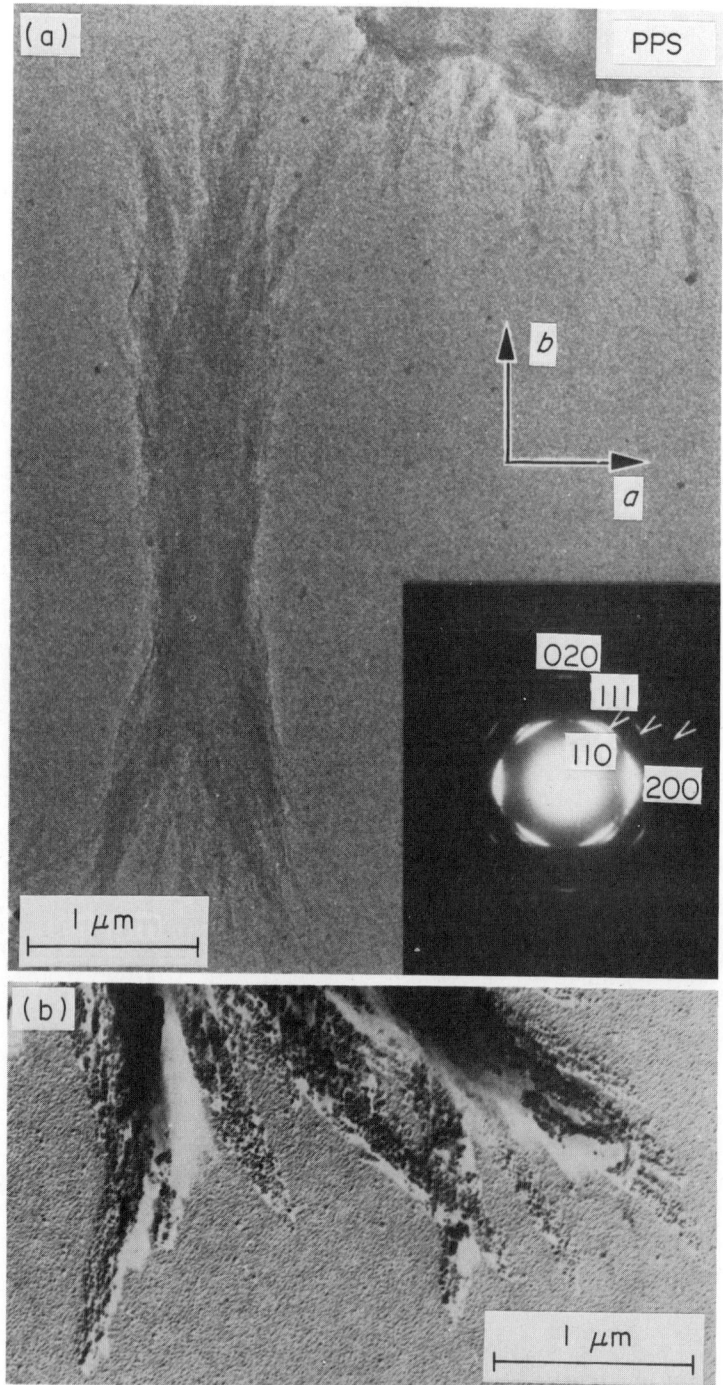

Figure 30 PPS solution-grown crystal (a) without and (b) with shadowing: the inset in (a) is a typical ED pattern of the crystal; in (b) Pt–Pd shadowing reveals that the crystal consists of several fibrillar crystals

200 kV. This means that PPS is fairly resistant to electron irradiation (see Table 3 in Section 34.3.3.2.iv). Two kinds of high resolution electron micrographs were obtained, as shown in Figures 31(a′) and (b′), which were the results of optical filtering. They are attributed to two different projections along the $\langle 001 \rangle$ and $\langle 0\bar{1}1 \rangle$ directions of the PPS crystal, respectively, judging from their optical diffractograms (OD). Figures 32(a) and (b) illustrate the projections of the crystal along $\langle 001 \rangle$ and $\langle 0\bar{1}1 \rangle$, respectively, based on the results of crystal structure analysis.[275] Comparison of Figure 31(a′) with Figure 32(a) indicates that each dark ellipse in Figure 31(a′) corresponds to a single molecular chain projected on the *ab*-plane along the chain direction. This is confirmed from

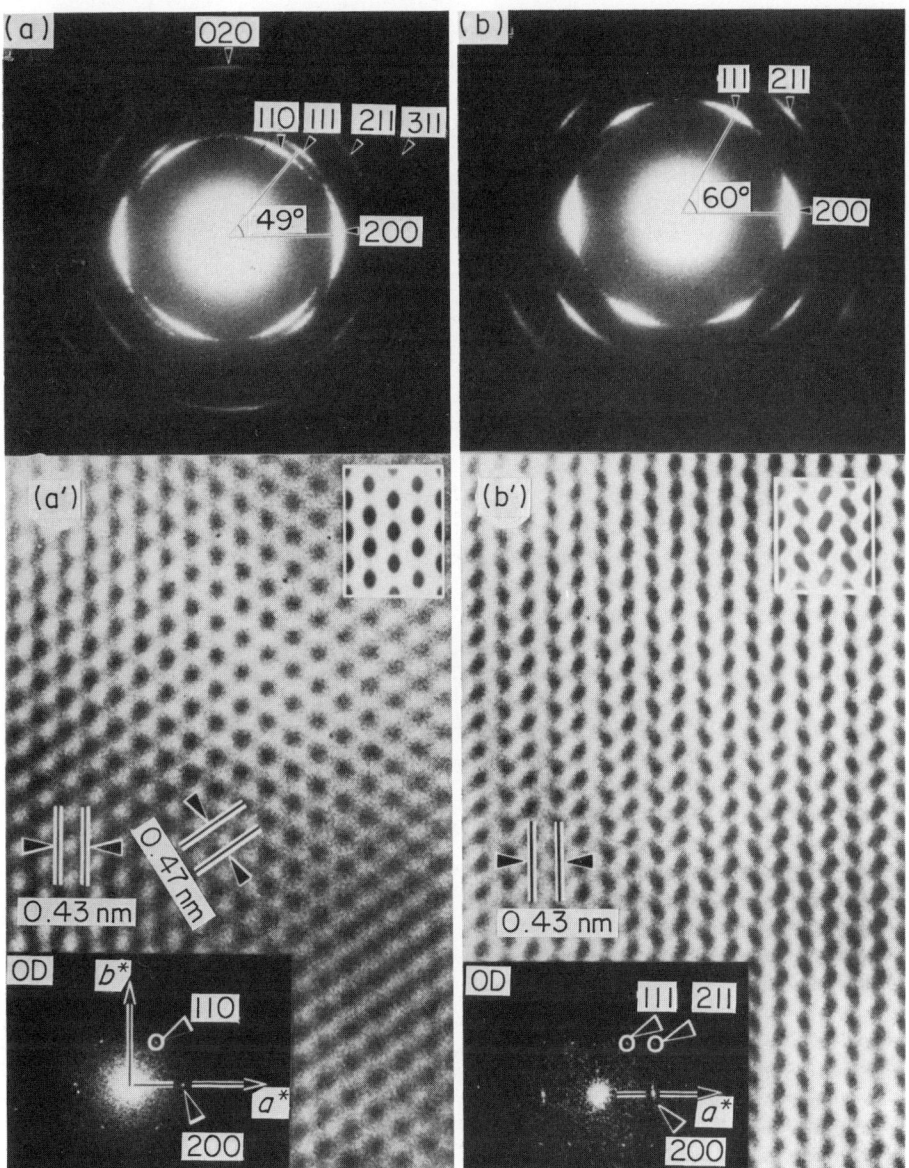

Figure 31 Two kinds of ED patterns and corresponding high resolution images of a PPS solution-grown crystal: (a) the ED usually observed, consisting basically of a *hk0* reflection; (a′) the high resolution image corresponding to ⟨001⟩ incidence (the inset at upper right corner is the simulated image: $\lambda = 0.00251$ nm, $C_s = 2.8$ mm, $\Delta f = 90$ nm); (b) the ED occasionally observed, consisting of *h00* and *h11* reflections (⟨0$\bar{1}$1⟩ or ⟨011⟩ incidence); and (b′) the high resolution image corresponding to the crystal orientation of (b) (the inset at upper right corner is the simulated image: $\lambda = 0.00251$ nm, $C_s = 0.7$ mm, $\Delta f = 48$ nm). The insets at lower left corner in (a′) and (b′) are optical diffractograms (OD) of the original negatives. In image simulation, the reflections up to 200 for (a′) and those up to 211 for (b′) were used

the simulated image which is inset at the upper right hand corner of Figure 31(a′). In the case of ⟨0$\bar{1}$1⟩ incidence, the corresponding projection (Figure 32b) is denoted as follows. In A of Figure 32(b), C atoms of an 'edge-on' phenylene group come together and S atoms are close to the phenylene. On the other hand, C atoms of a nearly 'flat-on' phenylene as in B are not so condensed and S atoms are far from the phenylene. From above, each dark ellipse in Figure 31(b′) is assigned to A and the region between ellipses to B. The simulated image (inset, upper right hand corner of Figure 31b′) proves that the above consideration is reasonable.

High resolution images of a crystal projected in two or more different crystallographic directions are very useful in determining its three-dimensional structure. In particular, they are important for the study of the three-dimensional distribution of the dopant relating to the position of the polymer chains in doped PPS.

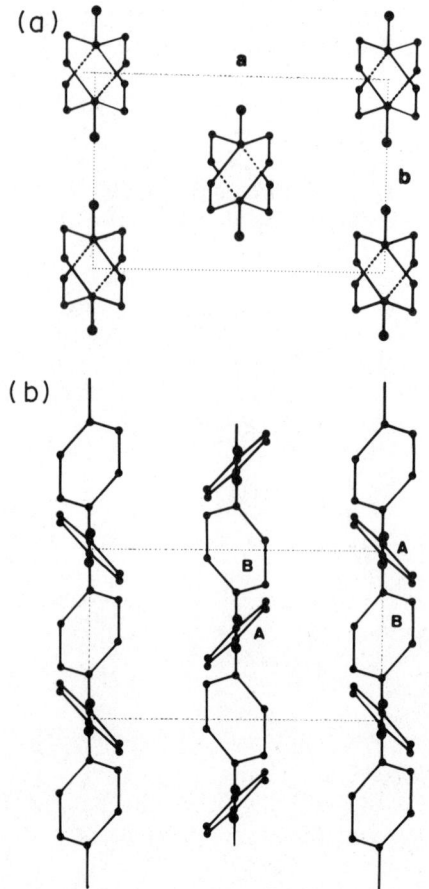

Figure 32 Projections of the crystal structure of PPS: (a) projection for the $\langle 001 \rangle$ incidence; and (b) projection for the $\langle 0\bar{1}1 \rangle$ incidence ($\langle 011 \rangle$ incidence is identical to $\langle 0\bar{1}1 \rangle$ incidence) (in (a) and (b), S atoms are indicated by larger filled circles than C atoms, and H atoms are not shown)

34.5.1.3 Other polymers

Recently the morphology of solution-grown crystals of poly(aryl ether ether ketone) (PEEK)[279] was reported, and it is similar to that of PPS. The PEEK crystal is fairly resistant to electron irradiation.[280] We have obtained lattice images of the crystal and we will be reporting on this in the near future.[281] Lattice images of poly(ether ketone) (PEK) have been obtained.[282] These materials are fairly thermostable, as is PPS.[282, 283]

Single crystals of poly(4-methyl-1-pentene) (P4M1P) are flat and square, as is well known.[284] It is not easy to take lattice images of P4M1P at room temperature because it is rather sensitive to electron bombardment. In this case, cryo protection is useful to obtain high resolution images, which show clear lattice fringes.[285] In the case of the PE single crystal, it was reported that TEPD of a very tiny crystal is much greater than that of a crystal of normal or large size.[286] Very recently, a tetragonal single crystal of poly(tetramethyl-*p*-silylphenylene siloxane) was used for high resolution observation, and the molecular arrangement in the crystal was clearly observed with a resolution of 0.32 nm.[132, 287] Preparation conditions of single crystals of various polymers were reviewed by Wunderlich[223] and Kawaguchi.[341] The procedure to make single crystals by film formation was also proposed.[342]

34.5.2 Thin Films of Polymers

There has been much work reported on the morphological observations of polymer thin films, such as poly(ethylene terephthalate) (PET),[288, 289] PE,[222] polybutene-1 (PB-1),[290] isotactic polystyrene (isotactic PS),[291, 350] poly(*p*-phenylene sulfide) (PPS),[278] poly(*p*-phenylene) (PPP),[292] poly-

(aryl ether ether ketone) (PEEK)[280] and so on. In this section, some examples of electron microscopy in the study of polymer thin films will be described.

34.5.2.1 Isotactic polystyrene (isotactic PS)

The TEPD of isotactic PS is around $0.03\,\mathrm{C\,cm^{-2}}$ ($\sim 2000\,\mathrm{e\,nm^{-2}}$) for $200\,\mathrm{keV}$ ($1\,\mathrm{eV} = 1.6 \times 10^{-19}\,\mathrm{J}$) at room temperature and about three times that of PE (Table 3 in Section 34.3.3.2.iv). The lattice spacings of i-PS crystal, in particular its *hk0* spacings, are invariant with increasing irradiation dose.[121] The largest spacing that appears in the ED pattern is 1.1 nm. It corresponds to the 110 lattice plane of isotactic PS crystal and is the most resistant to electron irradiation.[120,121] Thus lattice imaging has already been reported for isotactic PS single crystals at 120 kV.[120,121] Lattice images of unstretched and stretched thin films of isotactic PS are also obtained. In particular, the lattice imaging from stretched films is expected to help in the elucidation of the crystallization mechanism at an early stage of stress-induced crystallization.

(i) An unstretched thin film of isotactic PS[46,113]

A drop of hot solution (1–2 wt%) of *p*-xylene was spread on the surface of hot water to make a thin amorphous film of isotactic PS. The thickness of the film was around 100 nm as judged from its interference color. This is the same technique as that used to make thin films of PE and PP with a spherulitic structure.[222] The amorphous film of isotactic PS was mounted on electron microscope grids and annealed at 160–170 °C under a nitrogen atmosphere. Before annealing, unstretched films of isotactic PS were amorphous and did not give any crystalline reflections in the selected area ED pattern. Crystalline reflections, however, appeared in ED after annealing. This is the case of crystallization from the glassy state.[291,350] Specimens annealed for a short time, for example less than ten minutes, gave a sharp reflection corresponding to the 300 lattice plane, but no reflections attributed to the 110 and 220 lattice planes. By the method of defocus contrast (Section 34.2.3), a row structure, and a sheaf-like structure or immature spherulites are observable (Figure 9). Figure 33 shows the stacked lamellar structure in a isotactic PS thin film which was slightly stressed before and/or during depositing on EM grids. The change in contrast of the particle indicated by the white arrows demonstrates well the effect of defocusing. The ED pattern (Figure 33d) shows 'c-axis orientation' with a fairly strong 300 reflection. Dark striations in Figure 33(a) and bright ones in Figure 33(c) are attributed to crystalline lamellae, which is confirmed from the dark field image obtained by using the 300 reflection. Crystalline lamellae are set in the 'edge-on' position, and molecular chains of isotactic PS are set parallel to the specimen film surface. Thus it is deduced that a crystalline lamella grows in the direction normal to the 300 plane and its growing face is the 300 plane. Though the thickness of lamellae is measured as about 10 nm, for example in Figure 33(a), this value is not reliable because of a rather large amount of defocus. The distance between lamellae, namely the spacing between centers of them, however, should be constant with the change of defocus.

Figure 34 shows a high resolution image from such a specimen, which was annealed and crystallized for seven minutes at 161 °C. Latttice fringes with a spacing of 0.63 nm, corresponding to the 300 lattice plane, are observed in narrow bands whose width is about 6 nm. The band is ascribed to a lamella; strictly speaking, to the crystalline core in the lamella. Thus the thickness of the crystalline core in a lamella is about 6 nm. This value is accurate and reliable because 0.63 nm lattice fringes are used as an internal standard for magnification calibration. The average value of the center-to-center distance between successive lamellae, namely the long period, was measured as about 12 nm from images by the defocus contrast method (Figure 33), from images using shadow casting (encircled region in Figure 5), from 300 dark field images and also from small angle ED (Section 34.5.2.2).[46] If the system is assumed to be similar to a single crystal map, it may be considered that the period is equivalent to the average thickness of lamellae. Therefore it may be concluded that there exist surface layers on both sides of the lamella and their thickness is about 3 nm, half the difference between the lamellar thickness and the crystalline core thickness. The layers seem to have more or less a lack of regularity, but their detailed structure is not yet known.

The specimen was annealed for a rather long time, for example more than one hour, and showed crystalline reflections assigned to 110 and 220 on the selected area ED pattern. When a small selected area aperture is used (1 μm in diameter or less on the specimen), a single-crystal-like diffraction pattern with *hk0* reflections is observed. Figure 35(a) is a defocused micrograph of a specimen which was annealed and crystallized for about one hour at 170 °C. In the figure, the regions

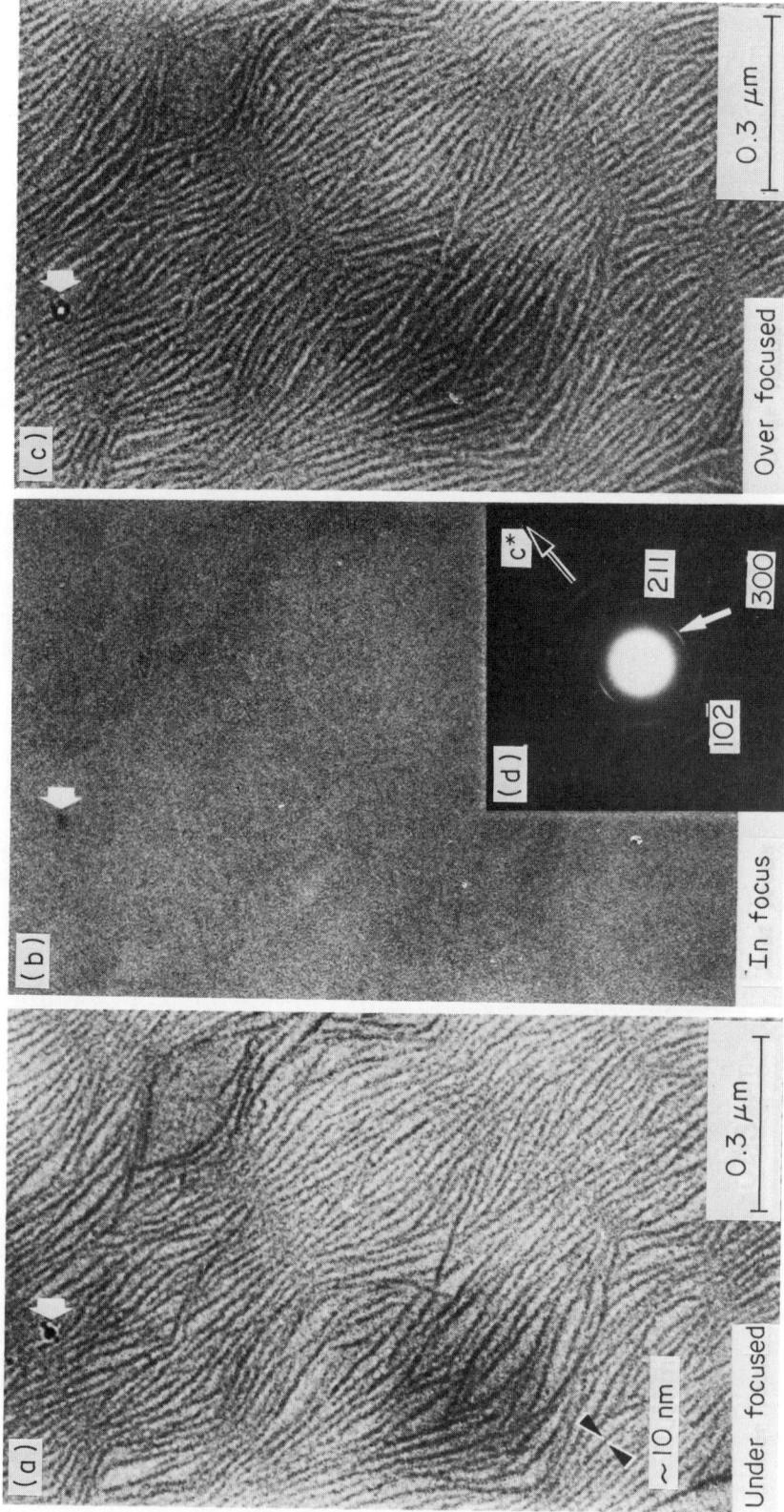

Figure 33 Bright field images (a)–(c) and the selected area ED (d) of an isotactic PS thin film annealed/crystallized at 161 °C for 7 min

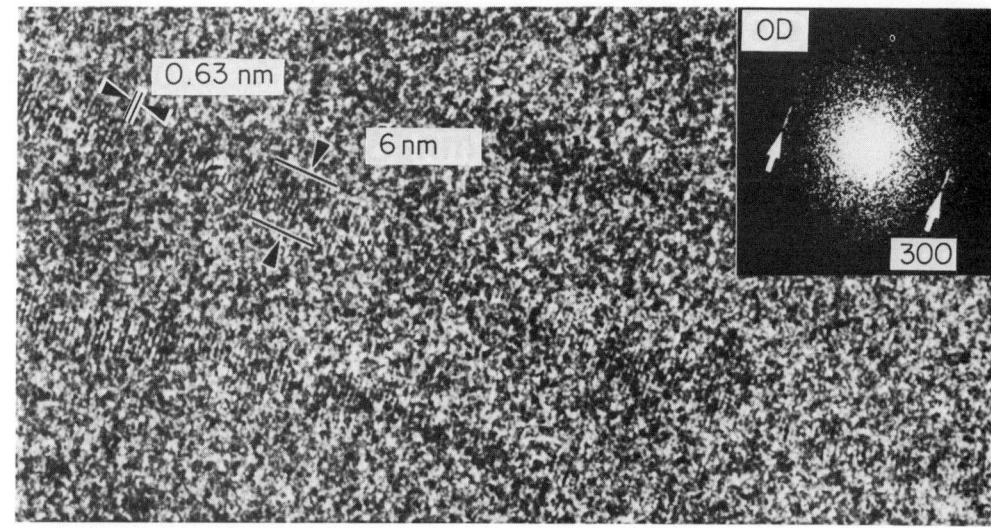

Figure 34 The high resolution image of an unstretched thin film of isotactic PS, crystallized at 161 °C for 7 min. Inset is the optical diffractogram (OD) of the image

indicated by arrows correspond to 'flat-on' lamellae because, for example, the ED pattern from the encircled area shows the single-crystal-like *hk0* N-pattern (inset in Figure 35a). This N-pattern is dominant in spite of the fact that there co-exists many edge-on lamellae within the circle. Figure 35(b) is a high resolution electron micrograph of such a specimen (annealed for about two hours at 170 °C). In the figure, 110 lattice fringes (1.1 nm) in three directions are recognized in the same area. These lattice lines intersect at an angle of 60° as is the case with a single crystal.[121] The 220 lattice fringes (0.55 nm) are also observed. The inset in Figure 35(b) is the optical diffractogram (OD) from the corresponding area in the negative. The diffractogram clearly demonstrates that there exists a single-crystal-like lamella in the 'flat-on' position.

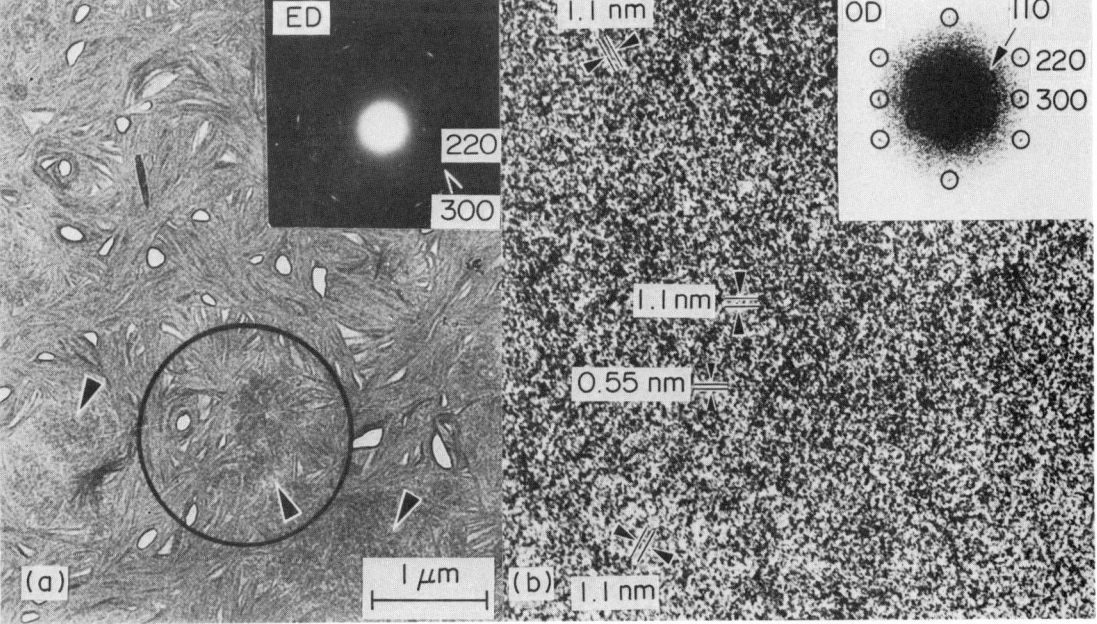

Figure 35 An unstretched thin film of isotactic PS, crystallized at 170 °C for 1–2 h: (a) the defocused bright field image, the inset being the ED pattern from the encircled area; and (b) the high resolution image, being attributed to the area indicated by an arrow in (a), with the inset being the optical diffractogram (OD) of the image

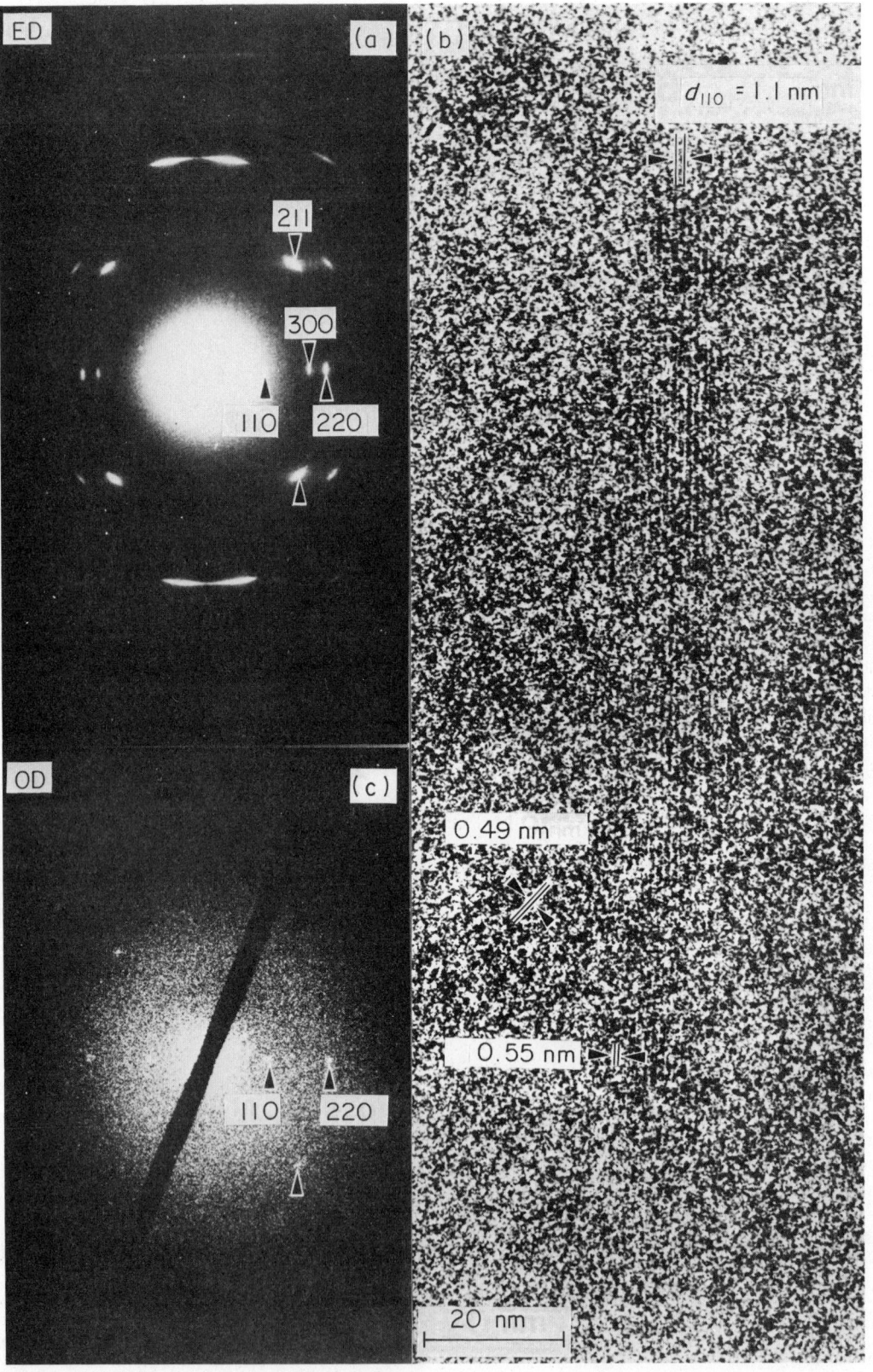

Figure 36   A stretched thin film of high molecular weight isotactic PS ($\bar{M}_w \cong 2.4 \times 10^6$); (a) the ED pattern; (b) the high resolution image; and (c) the optical diffractogram (OD) of (b), showing 110, 220 and 211 reflections (the oblique dark band is a beam stop)

(ii) A stretched thin film of isotactic PS[46,113]

Several drops of hot solution (0.3 wt %) were spread on a glass plate whose temperature was 180–190 °C. After evaporation of the solvent, the supercooled thin film of isotactic PS was vertically stretched upwards from the glass plate and then instantaneously quenched to room temperature. This procedure is essentially similar to that used by Petermann *et al.*[295] for preparing oriented thin films of various polymers. The thickness of the stretched film was about 100 nm. Figure 36(a) is the ED of a stretched thin film of isotactic PS (not annealed), which reveals a typical fiber pattern with a weak amorphous halo. Figure 36(b) shows a high resolution image of the film and Figure 36(c) is the OD of the image. It clearly indicates in the image the existence of fringes with 1.1 and 0.55 nm spacings, which are attributed to the 110 and 220 lattice planes, respectively. The 0.49 nm fringe seems to be assigned to the 211 lattice plane. In Figure 36(b), a clear 1.1 nm fringe is running in the vertical direction, namely along the fiber axis. The size of the long, slender region along the fiber axis, in which the 1.1 nm fringe is observable, is about 15 nm in width and 200 nm in length. Though 1.1 nm lattice lines are slightly curved and partly disappear, most of them pass through the region, with no lattice defects recognizable for the 110 lattice plane. A region where lattice fringes appear should directly correspond to a crystallite. It is proposed that 'shish kebabs' appear in the process of crystallization from strained melts,[296] as is the case with flow-induced crystallization from solution.[297] Taking into account the procedure for specimen preparation, the regions mentioned above seem to be ascribed to the 'shish'. Molecular chains of isotactic PS seem to be fairly extended in such a shish crystal, judging from the defect-less feature of the 110 lattice fringe in Figure 36(b).

34.5.2.2 Polyethylene (PE)

For PE thin films, in which it is very difficult to take high resolution images at room temperature due to its radiation sensitivity, dark field and strongly defocused bright field modes, as well as the surface replication method, were useful for studying its structure.[222,298] Figure 37 shows a thin film of a Marlex 6009 cast from a *p*-xylene solution, which was slightly stretched on a hot water surface, and subsequently deposited on EM grids. Defocus contrast method reveals that this film consists of lamellae whose *b*-axes more or less orient, as deduced from the ED of the encircled area B. Stretching induces changes in the molecular orientation, as demonstrated with the ED pattern from area A. This ED pattern shows that PE molecules are aligned along the stretching direction.

A PE fiber specimen was prepared by the rather strong stretching of a Sholex 6009 film case on a hot water surface and subsequently annealed at 126 °C. Figure 38(a) shows a selected area ED pattern, Figure 38(b) a dark field image taken with 110 and 200 reflections, and Figure 38(c) that taken with a 002 reflection. Figure 38(b) clearly shows that the fiber is made up of small crystallites (bright spots on a dark background) whose longitudinal thickness and lateral dimensions are about 20 nm. The bright spots align in a vertical direction, and thus the crystallites tend to align along the fiber axis, which suggests the existence of microfibrils with their lateral width of several tens of nm. Figure 39(a) is a defocused image and clearly shows a wavy stacking of lamellae as dark bands (about 20 nm thick) running in the horizontal direction (the stretching direction is vertical). Bright bands with a vertical thickness of about 10 nm correspond to the amorphous region between crystalline lamellae. The lateral dimension of these lamellae is of the order of μm and much larger than that of the crystallites shown as bright spots in Figure 38(b), whereas the thickness of the lamellae is of a magnitude similar to that of the crystallites. This fact suggests that a lamella is composed of many crystallites which do not have the same orientation, that is, the lamella has a mosaic texture. The optical transform (OD) (Figure 39b) of the image has broad but apparent intensity maxima which are indicated by arrows. It reveals the existence of a periodic texture consisting of stacked lamellae, corresponding to a low angle ED (LAED) pattern (Figure 39c). From Figures 39(b) and (c), the long period is estimated as about 30 nm. LAED patterns are observed without exciting the objective lens.[299,300,301] The specimen should be coated with evaporated C to suppress charge-up.[301]

A structure model of the PE fiber is proposed based on the information from both dark field and defocus images.[298] Recently lattice images taken at room temperature with conventional 120 kV TEM were reported by Chanzy *et al.*[131] using ultradrawn gels of high molecular weight PE. In order to directly recognize the crystallite shape and/or the orientation of individual crystallites on the enlarged micrographs, however, cryo protection is indispensable. Such micrographs have already been obtained by the authors using a cryo TEM with super-conducting objective lens,[132] as mentioned in Section 34.3.3.2.iv.

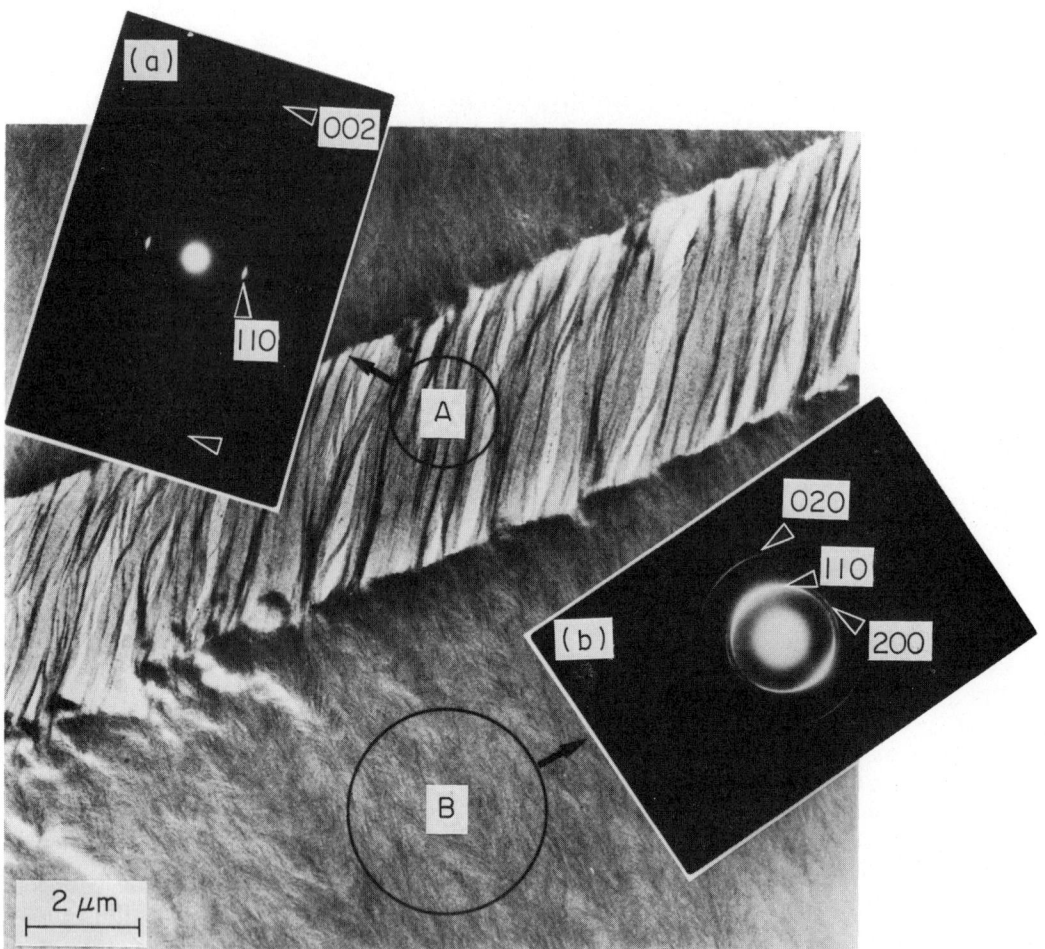

Figure 37   A PE (Marlex 6009) thin film from a *p*-xylene solution, which was cast on a hot water surface and then moderately drawn there: (a) the ED from the encircled portion A in the deformed region, showing fiber orientation, and (b) the ED from the encircled portion B in the undeformed region

34.5.2.3   Other polymers

Stretched thin films of high molecular weight nylon 12 ($\eta_r = 25$) were prepared using the same method as for stretched isotactic PS films (Section 34.5.2.1.ii), and then annealed at 170–175 °C after deposition on EM grids. The ED pattern of the film showed the fiber pattern with crystalline reflections which are indexed on the basis of the crystal structure of the γ-form of nylon 12.[302] Sometimes ED patterns showing special orientations were obtained. This pattern is characterized with a fairly strong $20\bar{2}$ reflection, suggesting that the 200 plane, namely the plane on which hydrogen bonds are set, is oriented parallel to the film surface.[303] From such specimens 020 lattice images (1.5 nm in spacing) were obtained.[303] The crystallite size in the direction parallel to the fiber axis (*b*-axis) is deduced as 8 to 15 nm because the number of lattice fringes recognized clearly within a coherent region in the micrograph is 5–10.

Morphologies of thin films with a spherulitic texture have been studied extensively by TEM for various polymers such as PPS, PE, isotactic PS and so on. The strain-induced crystallization of isotactic PS from the glassy and rubbery state was examined by Yeh and Lamber[350] using TEM. Recently Geil discussed the presence of ordered domains in amorphous polymers.[304] Such materials were PE, P4M1P, poly(vinylidene fluoride), polypivalolactone, PP, PB and PET. Thin amorphous films of these polymers were prepared by an ultraquenching technique and examined by TEM. Roche and Thomas discussed the defocus microscopy of copolymers.[305] They introduced the phase contrast transfer function, sin χ, to interpret the image contrast.

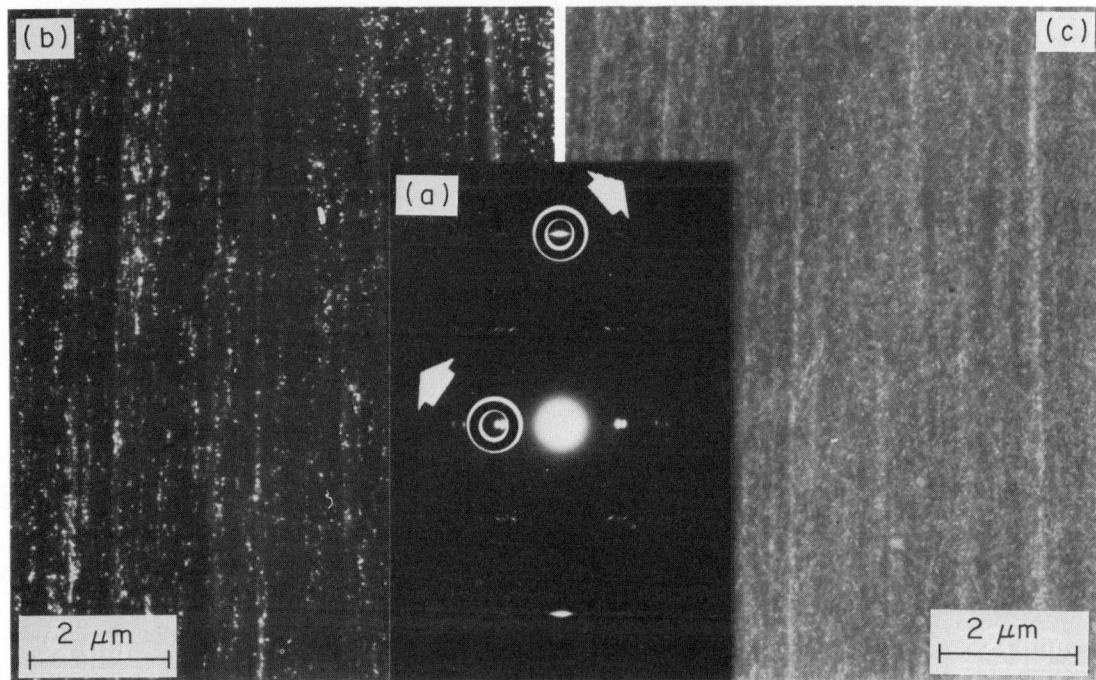

Figure 38 A drawn PE film prepared by the hot drawing of a Sholex 6009 film from a *p*-xylene solution which was cast on a hot water surface and subsequently annealed at 126 °C: (a) the ED pattern; (b) the dark field image taken with 110 and 200 reflections; and (c) the 002 dark field image of the same specimen portion as (b)

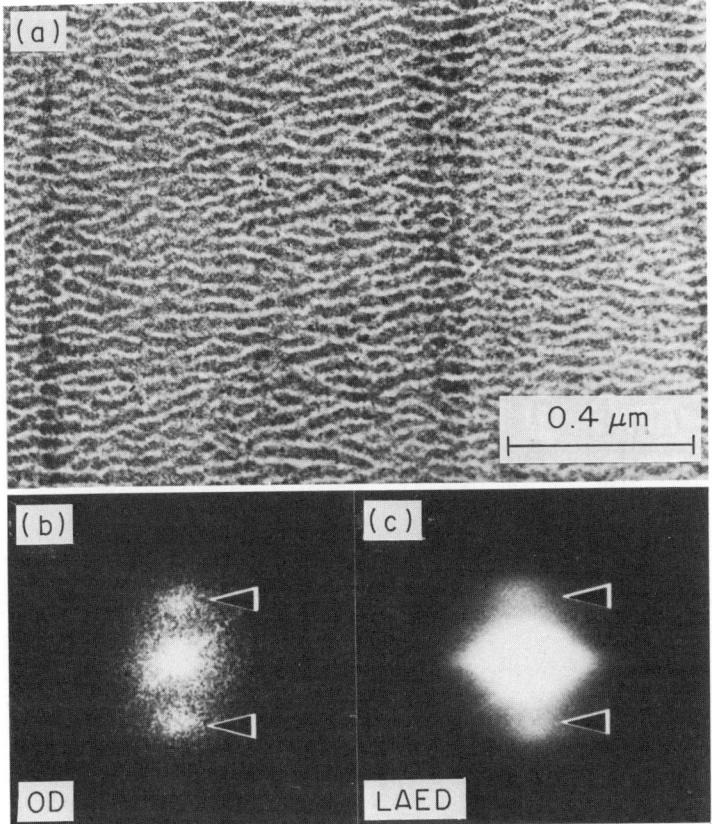

Figure 39 A drawn PE thin film which was prepared with the same material and by the same procedure as the film in Figure 38: (a) the defocused bright field image; (b) the optical diffraction (OD) pattern of (a); and (c) the low angle ED (LAED)

Kawaguchi *et al.* reported the preparation of oriented thin films of PPP and structural observations of them by TEM.[292,293] The paracrystalline nature of PPP was discussed by comparing its crystal structure with those of *p*-phenyls, in particular that of *p*-hexaphenyl. Here, *p*-phenyls such as *p*-hexaphenyl are oligomers of PPP. PPP and *p*-hexaphenyl are less radiation sensitive, and thus high resolution images were obtained for both materials. Stretched thin films of various polymers, such as PE,[306] isotactic PS,[306-308] PB-1[309] and some blends, were extensively studied using TEM by Petermann and his co-workers.[310,311] Recently, the epitaxial crystallization of PE onto uniaxially oriented PP was studied with TEM and is expected to be used for the adhesion of polymer laminates.[312]

34.5.3 Fiber Structure

Stretched or oriented thin films of isotactic PS, PE, nylon 12 and PPP showing fiber patterns in their ED were mentioned in the previous Section (34.5.2). Some of these will also be discussed here. In this section, however, other examples which are not thin films but have a fiber structure will also be described. TEM, SEM and optical microscopy of fibers have been reviewed and should be referred to in ref. 314.

34.5.3.1 Poly(p-phenylene terephthalamide) (PPTA)[298]

After annealing PPTA (Kevlar) at 400 °C, fibrillar fragments were obtained at room temperature by tearing. Figure 40(b) is a dark field image of PPTA taken with a 006 reflection. It shows a texture of alternating bright and dark bands with a period of about 500 nm. A similar texture is reported in longitudinal thin sections[314] and in fibrillar fragments.[13] Careful inspection of the figure shows that there exist microfibrils running in the direction of the fiber axis and through these bands. Although Takahashi *et al.*[315] have observed a line of bright spots in a bright band of the 006 dark field image, this is not recognizable in Figure 40(b).

Figure 40(c) is a dark field image taken with 110 and 200 reflections of the same specimen portion that was used in Figure 40(b). This figure reveals that small crystallites (bright spots in the figure) are randomly dispersed throughout the fibrillar ribbon. Moreover, bright spots seem to be placed in a row in a certain region of the ribbon. A high resolution lattice image of the PPTA fiber is shown in Figure 41. Clear 110 lattice fringes can be seen. The area where lattice fringes appeared is of the order of 10 nm × 10 nm through to 20 nm × 20 nm, and is almost of the same order as the area of a bright spot in Figure 40(c). In the case of poly(*p*-phenylene benzobisthiazole) (PBT) fiber, such an area is slightly larger than that of PPTA and 20 nm wide by 40 nm long in the fiber axis.[316] This suggests

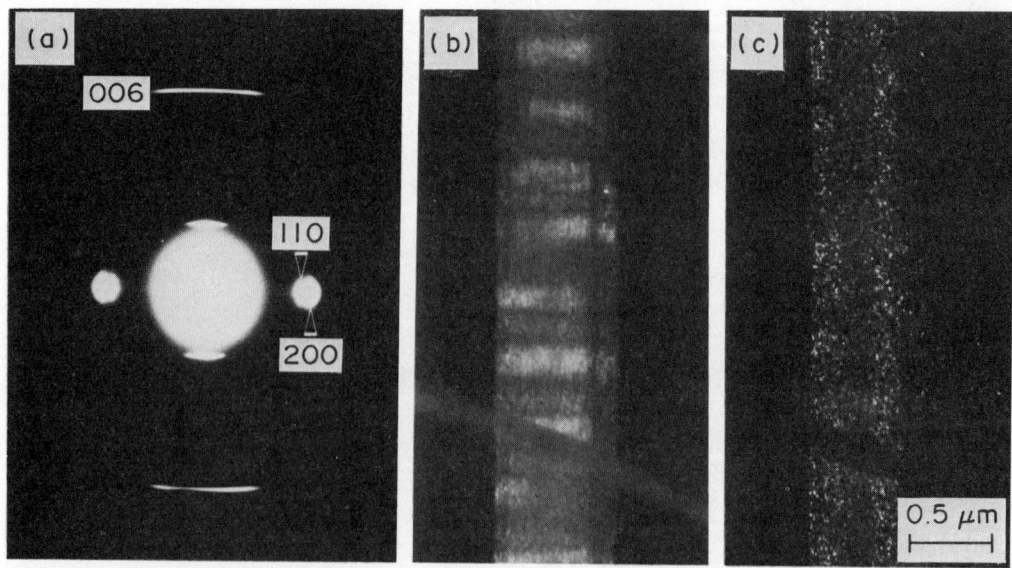

Figure 40 A PPTA fiber annealed at 400 °C: (a) the ED pattern; (b) the 006 dark field image; and (c) the 110 and 200 dark field image of the same specimen portion as (b)

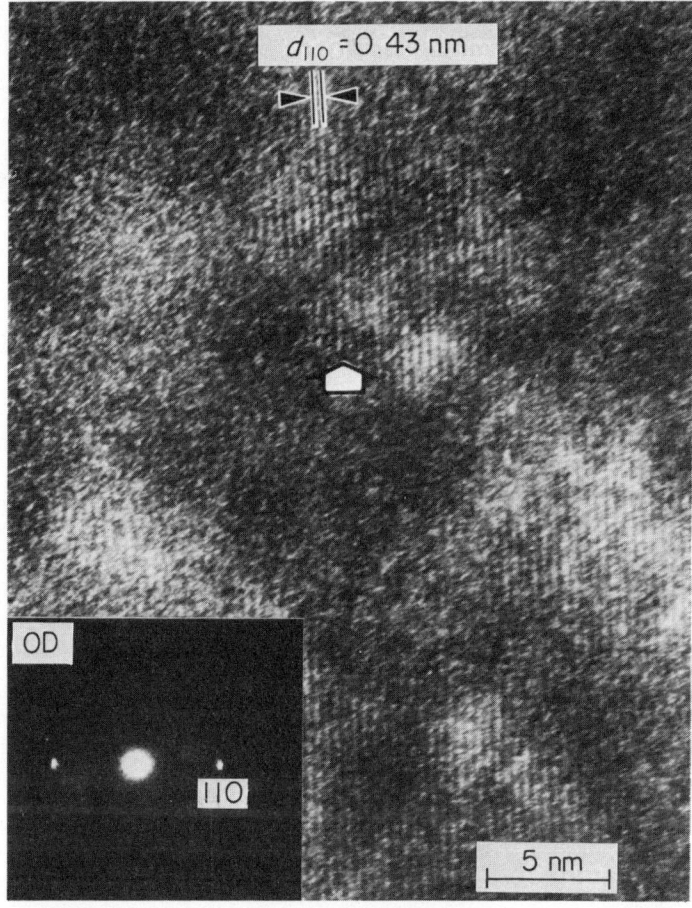

Figure 41 A high resolution image of a PPTA fiber. Inset is the optical diffractogram (OD) of the image

the reason why a PBT fiber has a greater modulus than a PPTA fiber. The direction of the 110 fringes in Figure 41, that is the c-axis of each crystallite, fluctuates slightly relative to the fiber axis in the plane of the figure. Even curved fringes can sometimes be seen.[13, 298] Examples of curved fringes are easily recognized by inspection of the region indicated by the arrow in Figure 41.

Since PPTA is made of rigid macromolecules, it is not supposed to have the two-phase structure prevailing in flexible macromolecules such as PE, but has a microfibrillar texture which is similar to the model of bundling of parallel microfibrils proposed by Peterlin.[317] Recently a SAXS with long wavelength X-rays which showed equatorial discrete maxima was reported.[318] The curved 110 fringes in the lattice images and the results of dark field imaging with a 006 reflection or 110 and 200 reflections suggested that the microfibrils are distorted by bending and/or twisting. The 006 dark field image of PPTA (Figure 40b) reveals that individual microfibrils in a fibrillar ribbon have the texture of periodic bending, and as a consequence the ribbon possesses a pleated sheet texture.[314] The 110 and 200 dark field imaging, however, reveals the existence of twisting of microfibrils in the same ribbon (Figure 40c). Therefore, it should be concluded that a microfibril in a PPTA fiber is bent along its fiber axis with a rather large periodicity (about 500 nm), and it is also twisted randomly around the axis. The Ag_2S insertion technique is also useful for morphological observations of the PPTA fiber.[269]

In the case of PBT, dark field images and lattice images showed that the crystallite size increases from several nm in width (equatorial direction) to 10–20 nm and in length (meridional direction) to 40 nm by annealing.[316] In lattice images, lattice fringes of 0.59 nm corresponding to the equatorial reflection were straight, but ones of 1.24 nm corresponding to the meridional reflection were wavy. It was deduced that this phenomenon is due to the paracrystalline nature of PBT.[319] In the diffraction pattern of PBT, many meridional and a few equatorial reflections are recognized, but the hkl ones are faint.[320, 321] Thus crystal structure analysis of this polymer is difficult. Recently determination of the idealized crystal structure of PBT has been attempted based on its high resolution images.[322]

Morphologies of polymers with rigid molecular chains including PPTA were reviewed by Takahashi.[321]

34.5.3.2 *Polymeric sulfur nitride* $[(SN)_x]$[323,324]

$(SN)_x$ is well known for its behaviour as a superconductor at low temperatures.[325] In the dark field image of a $(SN)_x$ fiber taken with 002 and 102 reflections and an image with a 020 reflection, fine striations are recognized along the chain direction (b-axis) and these correspond to the microfibrils in which the 002 or 102 lattice plane is oriented to satisfy the Bragg condition. A coherent domain (microfibril) is very long in the direction of the chain axis, but very narrow in the lateral dimension. Therefore, a microfibril in a $(SN)_x$ fiber is untwisted and distinct from that in a PPTA fiber. Such a microfibrillar texture can be also deduced from the ED pattern, where each reflection is elongated to be a streak in the direction perpendicular to the fiber axis.

The high resolution lattice images of $(SN)_x$ have been reported.[323] In the image of the $(SN)_x$ skin region of a fibrillar fragment, 002 lattice fringes (with a spacing of 0.359 nm) are observed. The lateral width of the domain where the fringes appeared is small, only about 2 nm. On the other hand, the core region gives a high resolution image of the single-crystal-like domain where the 002 and 010 fringes (with a spacing of 0.44 nm) are particularly well observed. The lateral dimension of the coherent domain is much larger than that in the skin region. This can be also deduced by comparison of the OD patterns of the images attributed to skin and core regions.

To improve the electrical conductivity, $(SN)_x$ is intercalated with halogens.[326,327] The characteristics of the ED pattern from intercalated $(SN)_x$ are that the diffraction spots from $(SN)_x$ become more streaked perpendicularly to the b-axis and that extra streaks appear between the layer lines of $(SN)_x$.[323] From the high resolution micrograph of iodinated $(SN)_x$ and ODs of the micrograph at three regions which are 50 nm in diameter, it was concluded that iodine atoms invade preferentially in the skin regions of $(SN)_x$ which are disordered in the pristine state.[323] It has also been reported that such iodine atoms are structurized there.[324]

34.5.3.3 *PE fibers and others*

For TEM, the specimen should be thin. Thus thin sections and/or thin films are prepared for TEM observation. Sometimes these specimens, however, do not represent the internal morphology of polymers in the bulk state because of plastic deformation during sectioning and the surface effect in thin films.[319] In order to expose the internal morphology of polymer solids with less deformation and to observe it by SEM, a new fracture method combined with fumic nitric acid treatment is proposed[211] (see Section 34.4.1). Recently this technique was applied to high tenacity/high modulus PE fibers and it was found that disordered regions exist but are scattered uniformly throughout the specimen.[319] Very recently lattice images of ultradrawn gels of high molecular weight PE were reported.[131] In the near future, the structural requirements for obtaining high-performance polymers will be classified. Some of the natural polymers with fibrous morphologies, for example cellulose and chitin, should be discussed here, but these will be mentioned in the following Section (34.5.4).

34.5.4   Natural Polymers

Cellulose crystals have been extensively studied by electron microscopy. The lateral crystallite size, namely the width or thickness of the microfibril of natural cellulose, is one of the important features of this polymer. Recently the effect of defocusing on the image contrast in TEM of negatively stained protofibrils of ramie cellulose was discussed on the basis of the defocus dependence of the phase contrast transfer function sin χ (see Section 34.3) using the OD technique[328] and the digital correlation method.[329] Properly defocused images suggested the existence of a certain periodical structure (about 6 nm in periodicity) along the protofibril axis. Lattice imaging of cellulose crystal was achieved by Sugiyama *et al.*[330,331] whilst Revol[332] obtained imaging for Valonia cellulose, and very recently imaging for bacterial and ramie celluloses was obtained by Kuga *et al.*[333,334] Lattice images appear to be useful in estimating the lateral crystallite size and discussing the existence of axial periodicity in cellulose.

Morphologies of natural polymers other than cellulose have also been studied by electron microscopy. If biological specimens such as the tobacco mosaic virus (TMV), T4-phage, collagen, *etc.* are included in 'natural polymers', there are too many examples to be introduced here. There are, however, not so many examples of high resolution TEM of natural polymers. Apart from cellulose, lattice images of a few materials such as β-chitin[335] and poly(hydroxybutyrate) (PHB)[336] have been reported using a 120 kV TEM. Double stranded DNA molecules were directly visualized by high resolution TEM with MDS.[337]

34.6 REFERENCES

1. R. Jaccodine, *Nature (London)*, 1955, **176**, 306.
2. P. H. Till, *J. Polym. Sci.*, 1957, **24**, 301.
3. A. Keller, *Philos. Mag.*, 1957, **2**, 1171.
4. E. W. Fischer, *Z. Naturforsch., Teil A*, 1957, **12**, 753.
5. K. H. Storks, *J. Am. Chem. Soc.*, 1938, **60**, 1753.
6. K. H. Storks, *Bell Lab. Rec.*, 1943, **21**, 390.
7. K. Herrmann, O. Gerngross and W. Abitz, *Z. Phys. Chem. (Wiesbaden)*, 1930, **B10**, 371.
8. E. W. Fischer, *Kolloidn. Z.*, 1958, **159**, 108.
9. K. Kobayashi, Y. Nishijima, S. Goto and M. Kurokawa, *Proc. 4th Int. Conf. EM, Berlin*, 1960, 728.
10. G. A. Bassett and A. Keller; cited by A. Keller, *Kolloidn. Z.*, 1969, **231**, 386.
11. M. G. Dobb, A. M. Hindeleh, D. J. Johnson and B. P. Saville, *Nature (London)*, 1975, **253**, 189.
12. S. C. Bennett, M. G. Dobb, D. J. Johnson, R. Murray and B. P. Saville, *Proc. EMAG-75, Bristol*, 1976, 329.
13. M. G. Dobb, D. J. Johnson and B. P. Saville, *J. Polym. Sci., Polym. Symp.*, 1977, **58**, 237.
14. M. Tsuji, S. Isoda, M. Ohara, A. Kawaguchi and K. Katayama, *Polymer*, 1982, **23**, 1568.
15. W. L. Bragg, *Nature (London)*, 1939, **143**, 678.
16. C. A. Taylor and H. Lipson, 'Optical Transforms', G. Bell and Sons, London, 1964.
17. H. Lipson (ed.), 'Optical Transforms', Academic Press, New York, 1972.
18. A. Klug and D. J. DeRosier, *Nature (London)*, 1966, **212**, 29.
19. E. L. O'Neill, *I.R.E. Trans. Inf. Theory*, 1956, **IT-2**, 56.
20. R. Nathan, in 'Advances in Optical and Electron Microscopy', ed. R. Barer and V. E. Cosslett, Academic Press, New York, 1971, vol. 4, p. 85.
21. R. G. Vadimsky, in 'Methods of Experimental Physics: Vol. 16, Polymers Part B: Crystal Structure and Morphology', ed. R. A. Fava, Academic Press, 1980, chap. 7.
22. D. C. Bassett, 'Principles of Polymer Morphology', Cambridge University Press, Cambridge, 1981.
23. D. T. Grubb, in 'Development in Crystalline Polymers — 1', ed. D. C. Bassett, Applied Science, Barking, 1982, chap. 1.
24. E. L. Thomas, in 'Structure of Crystalline Polymers', ed. I. H. Hall, Elsevier, Amsterdam, 1984, chap. 3.
25. 'Kobunshi no Kotaikozo II', (Solid Structure of Polymers II), in 'Experimental Methods in Polymer Science', ed. Japan Society of Polymer Science, Kyoritsu, Tokyo, 1984, vol, 17, chap. 5.
26. M. Born and E. Wolf, 'Principles of Optics', 5th edn., Pergamon Press, Oxford, 1975, chap. 8.
27. S. G. Lipson and H. Lipson, 'Optical Physics', Cambridge University Press, Cambridge, 1969, chap. 9.
28. V. E. Cosslett, 'Practical Electron Microscopy', Butterworths, London, 1951, chap. 5.
29. J. -F. Revol, *Carbohydr. Polym.*, 1982, **2**, 123.
30. K. Katayama, *Kaigai Kobunshi Kenkyu*, 1985, **31**, 192.
31. D. C. Bassett, F. C. Frank and A. Keller, *Philos. Mag.*, 1963, **8**, 1739.
32. S. Kubo and B. Wunderlich, *Makromol. Chem.*, 1972, **162**, 1.
33. L. Reimer, 'Transmission Electron Microscopy', Springer-Verlag, Berlin, 1984, chap. 4.
34. P. B. Hirsh, A. Howie, R. B. Nicholson, D. W. Pashley and M. J. Whelan, 'Electron Microscopy of Thin Crystals', Butterworths, London, 1965, chap. 13.
35. H. E. Huxley and G. Zubay, *J. Mol. Biol.*, 1960, **2**, 10.
36. D. W. Gregory and B. J. S. Pirie, *J. Microsc. (Oxford)*, 1972, **99**, 261.
37. D. W. Gregory and B. J. S. Pirie, *Proc. 5th Eur. Congr. EM, Manchester*, 1972, 234.
38. J. Trinick and A. Elliott, *J. Microsc. (Oxford)*, 1982, **126**, 151.
39. S. Fleischer, B. Fleischer and W. Stoeckenins, *J. Cell. Biol.*, 1967, **32**, 193.
40. K. Fukai, Y. Hosaka and A. Nishimura, *J. Electron Microsc.*, 1972, **21**, 331.
41. H. Akahori and T. Fukuoka, *J. Electron Microsc.*, 1975, **24**, 49.
42. J. Dubochet and M. Groom, in 'Advances in Optical and Electron Microscopy', ed. R. Barer and V. E. Cosslett, Academic Press, New York, 1982, vol. 8, p. 107.
43. E. Nemork and B. V. Johansen, *Ultramicroscopy*, 1982, **7**, 321.
44. M. Tsuji, unpublished results.
45. G. Kanig, *Kolloidn. Z*, 1973, **251**, 782.
46. M. Tsuji, A. Uemura, M. Ohara, S. Isoda, A. Kawaguchi and K. Katayama, *Koenshu-Kyoto Daigaku Nippon Kagaku Sen'i Kenkyusho*, 1987, **44**, 1.
47. K. Mihama, in 'Denshi Kembikyo' (Electron Microscopy), ed. R. Ueda, Kyoritsu, Tokyo, 1982, chap. 11.
48. R. D. Heidenreich, 'Fundamentals of Transmission Electron Microscopy', Interscience, New York, 1964, chap. 3.
49. L. Reimer, 'Transmission Electron Microscopy', Springer-Verlag, Berlin, 1984, p. 193.
50. K. Mihama, in 'Denshi Kembikyo' (Electron Microscopy), ed. R. Ueda, Kyoritsu, Tokyo, 1982, chap. 10.
51. G. A. Bassett, *Philos. Mag.*, 1958, **3**, 1042.
52. K. Shimamura, *J. Macromol. Sci., Phys.*, 1979, **B16**, 213.
53. J. M. Cowley, 'Diffraction Physics', North-Holland, Amsterdam, 1975, chap. 4.

54. B. J. Thompson, in 'Optical Transforms', ed. H. Lipson, Academic Press, New York, 1972, chap. 8.
55. M. Tsuji, S. Isoda, M. Ohara, K. Katayama and K. Kobayashi, *Bull. Inst. Chem. Res., Kyoto Univ.*, 1977, **55**, 269.
56. J. C. H. Spence, 'Experimental High-resolution Electron Microscopy', Oxford University Press, Oxford, 1981.
57. J. Petermann and H. Gleiter, *Philos. Mag.*, 1975, **31**, 929.
58. O. Scherzer, *J. Appl. Phys.*, 1949, **20**, 20.
59. M. Tsuji and R. St. J. Manley, *J. Microsc. (Oxford)*, 1983, **130**, 93.
60. J. M. Cowley, 'Diffraction Physics', North-Holland, Amsterdam, 1975, chap. 1.
61. G. R. Grinton and J. M. Cowley, *Optik (Stuttgart)*, 1971, **34**, 221.
62. J. M. Cowley, 'Diffraction Physics', North-Holland, Amsterdam, 1975, chap. 13.
63. J. M. Cowley, 'Diffraction Physics', North-Holland, Amsterdam, 1975, chap. 4.
64. J. W. Goodman, 'Introduction to Fourier Optics', McGraw-Hill, New York, 1968, chap. 5.
65. W. T. Cathey, 'Optical Information Processing and Holography', Wiley, New York, 1974, chap. 5.
66. N. Uyeda and K. Ishizuka, *J. Electron. Microsc.*, 1974, **23**, 79.
67. K. Iizuka, 'Hikari-Kogaku', Kyoritsu, Tokyo, 1977, chap. 5.
68. F. Thon, in 'Electron Microscopy in Material Science', ed. U. Valdre, Academic Press, New York, 1971, p. 562.
69. K. -J. Hanszen, in 'Advances in Optical and Electron Microscopy', ed. R. Barer and V. E. Cosslett, Academic Press, New York, 1971, vol. 4, p. 1.
70. H. P. Erickson, in 'Advances in Optical and Electron Microscopy', ed. R. Barer and V. E. Cosslett, Academic Press, New York, 1973, vol. 5, p. 163.
71. H. Hashimoto, *Denshi Kembikyo*, 1982, **17**, 41; 1983, **17**, 243; 1983, **18**, 33.
72. L. Reimer, 'Transmission Electron Microscopy', Springer-Verlag, Berlin, 1984, chap. 6.
73. K. Ishizuka, *J. Jpn. Soc. Cryst.*, 1986, **28**, 1.
74. Y. Fujiyoshi, *Denshi Kembikyo*, 1980, **15**, 72.
75. B. V. Johansen, *Micron*, 1973, **4**, 446.
76. O. L. Krivanek, *Optik (Stuttgart)*, 1976, **45**, 97.
77. R. H. Wade and J. Frank, *Optik (Stuttgart)*, 1977, **49**, 81.
78. M. Troyon, *Optik (Stuttgart)*, 1978/79, **52**, 401.
79. K. Kobayashi and N. Uyeda, *Proc. 8th Int. Congr. EM, Canberra*, 1974, **1**, 264.
80. D. Chescoe and P. J. Goodhew, 'The Operation of the Transmission Electron Microscope', Oxford University Press, Oxford, 1984.
81. O. L. Krivanek, S. Isoda and K. Kobayashi, *J. Microsc. (Oxford)*, 1977, **111**, 279.
82. S. Isoda, M. Ohara, M. Tsuji and K. Katayama, *Jpn. J. Appl. Phys.*, 1981, **20**, 2437.
83. E. D. Boyes, B. J. Muggridge and M. J. Goringe, *J. Microsc. (Oxford)*, 1982, **127**, 321.
84. Y. Kokubo, S. Moriguchi, J. Hosoi, E. Watanabe and J. Nash, *Mat. Res. Soc. Symp. Proc.*, 1984, **31**, 23.
85. E. L. Thomas and O. G. Ast, *Polymer*, 1974, **15**, 37.
86. C. J. D. Catto and K. C. A. Smith, *J. Microsc. (Oxford)*, 1975, **105**, 223.
87. N. Mori, T. Katoh, T. Oikawa, J. Miyahara and Y. Harada, *Proc. 11th Int. Congr. EM, Kyoto*, 1986, 29.
88. T. Oikawa, N. Mori, T. Katoh, Y. Harada and J. Miyahara, *Proc 11th Int. Congr. EM, Kyoto*, 1986, 439.
89. R. C. Valentine, *Lab. Invest.*, 1965, **14**, 1334.
90. R. C. Valentine, in 'Advances in Optical and Electron Microscopy', ed. R. Barer and V. E. Cosslett, Academic Press, New York, 1966, vol. 1, p. 180.
91. G. L. Jones and V. E. Cosslett, *Proc. 7th Int. Congr. EM, Grenoble*, 1970, 349.
92. H. Hashimoto, A. Kumao, S. Suzuki and H. Yotsumoto, *Proc. 7th Int. Congr. EM, Grenoble*, 1970, 351.
93. A. Fukami, M. Katoh and K. Fukushima, *Proc. 7th Int. Congr. EM, Grenoble*, 1970, 353.
94. J. A. Aznarez and F. Catalina, *Proc. 7th Int. Congr. EM, Grenoble*, 1970, 355.
95. V. Matricardi, G. Wray and D. F. Parsons, *Micron*, 1972, **3**, 526.
96. G. C. Farnell and R. B. Flint, *J. Microsc. (Oxford)*, 1973, **97**, 271.
97. M. V. King and D. F. Parsons, *Proc. 4th Int. Congr. HVEM, Toulouse*, 1975, 75.
98. V. E. Cosslett, G. L. Jones and R. A. Camps, *Proc. 3rd Int. Conf. HVEM, Oxford*, 1974, p. 147.
99. G. C. Farnell and R. B. Flint, *J. Microsc. (Oxford)*, 1975, **103**, 319.
100. M. Tsuji, *Kobunshi Kako*, 1986, **35**, 522; 1986, **35**, 574.
101. E. Zeitler and J. R. Hayes, *Lab. Invest.*, 1965, **14**, 1324.
102. D. Dorignac, M. E. C. Maclachlan and B. Jouffrey, *Ultramicroscopy*, 1976, **2**, 49.
103. K. H. Downing and D. A. Grano, *Ultramicroscopy*, 1982, **7**, 381.
104. P. B. Hirsh, A. Howie, R. B. Nicholson, D. W. Pashley and M. J. Whelan, 'Electron Microscopy of Thin Crystals', Butterworths, London, 1965, chap. 4.
105. A. Kawaguchi, Ph. D. Thesis, Kyoto University, 1979, chap. 1.
106. K. Ishizuka and N. Uyeda, *Acta. Cryatallogr., Sect. A*, 1977, **33**, 740.
107. K. Ishizuka and N. Uyeda, *Bull. Inst. Chem. Res., Kyoto Univ.*, 1977, **55**, 260.
108. M. A. O'Keefe and P. R. Buseck, *Proc. Symp. Chem. Phys. Miner., Hawaii*, 1979, p. 27.
109. A. Kawaguchi, *Polymer*, 1981, **22**, 753.
110. A. Uemura, M. Tsuji, A. Kawaguchi and K. Katayama, *J. Mater. Sci.*, in press.
111. A. Fukami, K. Adachi and M. Katoh, *Proc. 6th Int. Congr. EM, Kyoto*, 1966, **1**, 263.
112. B. V. Johansen, *Micron*, 1974, **5**, 209.
113. M. Tsuji, A. Uemura, M. Ohara, A. Kawaguchi, K. Katayama and J. Petermann, *Sen'i Gakkaishi*, 1986, **42**, T-580.
114. M. S. Isaacson, in 'Principles and Techniques of Electron Microscopy', ed. M. A. Hayat, Van Nostrand-Reinhold, New York, 1977, vol. 7, p. 1.
115. M. Tsuji, Ph. D. Thesis, Kyoto University, 1981, chap. 2.
116. K. Kobayashi, and K. Sakaoku, *Bull. Inst. Chem. Res., Kyoto Univ.*, 1964, **42**, 473.
117. K. Kobayashi and K. Sakaoku, *Lab. Invest.*, 1965, **14**, 1097.
118. A. Kawaguchi, S. Isoda, T. Haneda, M. Ohara and K. Katayama, *Bull. Inst. Chem. Res., Kyoto Univ.*, 1982, **60**, 1.
119. D. C. Bassett, *Philos. Mag.*, 1964, **10**, 595.
120. M. Tsuji, S. K. Roy and R. St. J. Manley, *Polymer*, 1984, **25**, 1573.

121. M. Tsuji, S. K. Roy and R. St. J. Manley, *J. Polym. Sci., Polym. Phys. Ed.*, 1985, **23**, 1127.
122. H. Yoshioka, *J. Phys. Soc. Jpn.*, 1957, **12**, 618.
123. L. Reimer, in 'Physical Aspects of Electron Microscopy and Microbeam Analysis', ed. B. M. Siegel and D. R. Beaman, Wiley, New York, 1975, chap. 3.
124. M. S. Isaacson, in ref. 123, chap. 14.
125. K. Kobayashi and M. Ohara, *Proc. 6th Int. Congr. EM, Kyoto*, 1966, **1**, 579.
126. R. M. Glaeser, *Proc. 3rd Int. Conf. HVEM, Oxford*, 1974, 370.
127. I. A. M. Kuo and R. M. Glaeser, *Ultramicroscopy*, 1975, **1**, 53.
128. A. Rose, *Adv. Electron.*, 1948, **1**, 131.
129. M. Tsuji, S. Moriguchi, K. J. Ihn, A. Kawaguchi and K. Katayama, *Proc 11th Int. Congr. EM, Kyoto*, 1986, 1749.
130. J. -F. Revol and R. St. J. Manley, *J. Mater. Sci. Lett.*, 1986, **5**, 249.
131. H. Chanzy, P. Smith, J. -F. Revol and R. St. J. Manley, *Polymer*, 1987, **28**, 133.
132. M. Tsuji, M. Tosaka, A. Uemura, A. Kawaguchi, K. Katayama, M. Iwatsuki and Y. Harada, *Polym. Prepr., Jpn.*, 1987, **36**, 2357.
133. K. Katayama, *Denshi Kembikyo*, 1987, **21**, 181.
134. E. Knapek, *Ultramicroscopy*, 1982, **10**, 63.
135. M. Iwatsuki, H. Kihara, K. Nakanishi and Y. Harada, *Proc. 11th Int. Congr. EM, Kyoto*, 1986, 251.
136. J. Frank, *Optik (Stuttgart)*, 1973, **38**, 519.
137. G. W. Stroke and M. Halioua, *Optik (Stuttgart)*, 1972, **35**, 50.
138. H. Okuyama, Y. Ichioka and T. Suzuki, *Jpn. J. Appl. Phys.*, 1974, **13**, 280.
139. W. O. Saxton, 'Computer Techniques for Image Processing in Electron Microscopy', Academic Press, New York, 1978.
140. A. Tonomura, T. Matsuda and J. Endo, *Jpn. J. Appl. Phys.*, 1979, **18**, 9.
141. P. W. Hawkes (ed.), 'Computer Processing of Electron Microscope Images', Springer-Verlag, Berlin, 1980.
142. D. Gabor, *Proc. R. Soc. London, Ser. A*, 1949, **197**, 454.
143. D. Gabor, *Proc. R. Soc. London, Ser. B*, 1951, **64**, 449.
144. M. Tsuji, Ph. D. Thesis, Kyoto University, 1981, chap. 4.
145. F. C. Billingsley, *Appl. Opt.*, 1970, **9**, 289.
146. R. Bernstein, *IBM J. Res. Dev.*, 1976, **20**, 40.
147. F. C. Billingsley, in 'Advances in Optical and Electron Microscopy', ed. R. Barer and V. E. Cosslett, Academic Press, New York, 1971, vol. 4, p. 127.
148. E. O. Bringham, 'The Fast Fourier Transform', Prentice-Hall, Englewood Cliffs, 1974.
149. A. R. Stokes, *Acta Crystallogr.*, 1955, **8**, 27.
150. J. Luis and M. Amorós, 'Molecular Crystals: Their Transforms and Diffuse Scattering', Wiley, New York, 1968.
151. T. R. Welberry and R. Galbraith, *J. Appl. Crystallogr.*, 1973, **6**, 87.
152. G. Border and G. Samay, *J. Polym. Sci., Polym. Symp.*, 1973, **42**, 768.
153. D. G. Fedak, T. E. Fischer and W. D. Robertson, *J. Appl. Phys.*, 1968, **39**, 5658.
154. R. Bergsten, *J. Opt. Soc. Am.*, 1974, **64**, 1309.
155. A. Kawaguchi, M. Ohara and K. Kobayashi, *J. Macromol. Sci., Phys.*, 1979, **B16**, 193.
156. N. Uyeda, *Denshi Kembikyo*, 1980, **15**, 59.
157. A. Klug and J. E. Berger, *J. Mol. Biol.*, 1964, **10**, 565.
158. W. Longley, *J. Mol. Biol.*, 1967, **30**, 323.
159. F. Thon and B. H. Siegel, *Ber. Bunsenges. Phys. Chem.*, 1970, **74**, 1116.
160. B. V. Johansen, *Micron*, 1972, **3**, 256.
161. R. W. Horne, J. M. Hobart and I. P. Ronchetti, *Micron*, 1975, **5**, 233.
162. R. Sinclair, R. Gronsky and G. Thomas, *Acta Metall.*, 1976, **24**, 789.
163. T. Tanji and H. Hashimoto, *Acta Crystallogr., Sect. A*, 1978, **34**, 453.
164. H. J. Pincus, in 'Advances in Optical and Electron Microscopy', ed. V. E. Cosslett and R. Barer, Academic Press, New York, 1978, vol. 7, p. 17.
165. R. Hosemann and S. N. Bagchi, 'Direct Analysis of Diffraction by Matter', North-Holland, Amsterdam, 1962.
166. J. H. Hitchborn and G. J. Hills, *Virology*, 1968, **35**, 50.
167. S. Boseck and H. Hager, *Optik (Stuttgart)*, 1968/69, **28**, 602.
168. D. J. DeRosier and A. Klug, *J. Mol. Biol.*, 1972, **65**, 469.
169. T. Mulvey, *J. Microsc. (Oxford)*, 1972, **98**, 232.
170. R. C. Warren and R. M. Hicks, *Micron*, 1973, **4**, 257.
171. R. W. Horne and R. Markham, 'Electron Diffraction and Optical Diffraction Techniques', in 'Practical Methods in Electron Microscopy', ed. A. M. Glauert, North-Holand, Amsterdam, 1973, vol. 1, p. 327.
172. H. P. Erickson, *J. Cell. Biol.*, 1974, **60**, 153.
173. U. Aebi, P. R. Smith, J. Dubochet, C. Henry and E. Kellenberger, *J. Supramol. Struct.*, 1974, **1**, 498.
174. A. C. Steven, U. Aebi and M. K. Showe, *J. Mol. Biol.*, 1976, **102**, 373.
175. H. Hashimoto, H. Endoh and T. Tanji, *Proc. US-Japan Seminar HVEM, Honolulu*, 1976.
176. F. P. Ottensmeyer, J. W. Andrew, D. P. Bazett-Jones, A. S. K. Chan and J. Hewitt, *J. Microsc. (Oxford)*, 1977, **109**, 259.
177. M. Osumi, M. Nagano and M. Yanagida, *J. Electron Microsc.*, 1979, **28**, 301.
178. K. Shimamura, Ph. D. Thesis, Kyoto University, 1977, chap. 3.
179. D. J. DeRosier and A. Klug, *Nature (London)*, 1968, **217**, 130.
180. R. A. Crowther, L. A. Amos, J. T. Finch, D. J. DeRosier and A. Klug, *Nature (London)*, 1970, **226**, 421.
181. P. B. Moore, H. E. Huxley and D. J. DeRosier, *J. Mol. Biol.*, 1970, **50**, 279.
182. N. A. Kieselev and A. Klug, *J. Mol. Biol.*, 1969, **40**, 155.
183. R. A. Crowther, D. J. DeRosier and A. Klug, *Proc. R. Soc. London, Ser. A*, 1970, **317**, 319.
184. A. Klug and R. A. Crowther, *Nature (London)*, 1972, **238**, 435.
185. L. A. Amos, *J. Microsc. (Oxford)*, 1974, **100**, 143.
186. P. R. Smith, U. Aebi, R. Josephs and M. Kessel, *J. Mol. Biol.*, 1976, **106**, 243.
187. B. K. Vainshtein, in 'Advances in Optical and Electron Microscopy', ed. V. E. Cosslett and R. Barer, Academic Press, New York, 1978, vol. 7, p. 282.

188. D. L. Misell, 'Image Analysis, Enhancement and Interpretation', in 'Practical Methods in Electron Microscopy', ed. A. M. Glauert, North-Holland, Amsterdam, 1978, vol. 7.
189. P. K. Luther and J. M. Sqiure, *J. Mol. Biol.*, 1980, **141**, 409.
190. R. W. Horne, *J. Microsc. (Oxford)*, 1979, **113**, 241.
191. Catalog of POLARON Equipment Ltd., England.
192. H. P. Erickson and A. Klug, *Ber. Bunsenges. Phys. Chem.*, 1970, **74**, 1129.
193. J. Frank, *Biophys. J.*, 1972, **12**, 484.
194. J. Frank and L. Al-Ali, *Nature (London)*, 1975, **256**, 376.
195. P. N. T. Unwin and R. Henderson, *J. Mol. Biol.*, 1975, **94**, 425.
196. D. L. Misell and R. E. Burge, *J. Microsc. (Oxford)*, 1975, **103**, 195.
197. N. Uyeda and K. Ishizuka, *J. Electron Microsc.*, 1975, **24**, 65.
198. R. D. B. Fraser, T. P. MacRae, E. Suzuki and C. L. Davey, *J. Microsc. (Oxford)*, 1976, **108**, 343.
199. A. N. Barrett, I. D. J. Burdett and K. A. Paton, *J. Microsc. (Oxford)*, 1978, **113**, 131.
200. L. Fang-Hua and F. Hai-Fu, *Wu Li Hsueh Pao*, 1979, **28**, 276.
201. S. Shikata, M. Tsuji and K. Katayama, unpublished results.
202. H. Thirty, *Appl. Opt.*, 1964, **3**, 39; A. Kozma and D. L. Kelly, *Appl. Opt.*, 1965, **4**, 387.
203. D. McLachlan, Jr., *Proc. Natl. Acad. Sci. USA*, 1958, **44**, 948.
204. R. Markham, J. H. Hitchborn, G. J. Hills and S. Frey, *Virology*, 1964, **22**, 342.
205. F. P. Ottensmeyer, E. E. Schmidt and A. J. Olbrecht, *Science*, 1972, **179**, 175.
206. R. D. B. Fraser and G. R. Millward, *J. Ultrastruct. Res.*, 1970, **31**, 203.
207. T. Tanji, H. Hashimoto and H. Endoh, *Proc. 5th Int. Conf. HVEM, Kyoto*, 1977, 171.
208. T. Tanji, H. Hashimoto, H. Endoh and H. Tomioka, *J. Electron Microsc.*, 1982, **31**, 1.
209. L. Reimer, 'Scanning Electron Microscopy', Springer-Verlag, Berlin, 1985.
210. Japan Society of Electron Microscopy (ed.), 'Sosadenshikenbikyo', ('Scanning Electron Microscope'), Kanto branch, Kyoritsu, Tokyo, 1976.
211. T. Tagawa and K. Ogura, *J. Polym. Sci., Polym. Phys. Ed.*, 1980, **18**, 971.
212. K. Shimamura, S. Murakami, M. Tsuji and K. Katayama, *Nippon Reoroji Gakkaishi*, 1979, **7**, 42.
213. T. Tagawa and K. Shimamura, *J. Electron Microsc.*, 1979, **28**, 314.
214. K. Shimamura, S. Murakami and K. Katayama, *Makromol. Chem., Rapid Commun.*, 1982, **3**, 199.
215. S. Saka and R. J. Thomas, *Wood Sci. Technol.*, 1982, **16**, 1.
216. S. Saka, P. Whiting, K. Fukazawa and D. A. I. Goring, *Wood Sci. Technol.*, 1982, **16**, 269.
217. S. Saka and D. A. I. Goring, *Mokuzai Gakkaishi*, 1983, **29**, 648.
218. S. -J. Kuang, S. Saka and D. A. I. Goring, *J. Wood Chem. Technol.*, 1984, **4**, 163.
219. E. Zeitler and M. G. R. Thomson, *Optik (Stuttgart)*, 1970, **31**, 258.
220. A. V. Crewe and J. Wall, *Proc. 27th Ann. EMSA*, 1969, 172.
221. N. Garcia (ed.), 'Scanning tunneling microscopy '86', North-Holland, Amsterdam, 1987.
222. K. Kobayashi, in 'Kobunshi no bussei', (Properties of polymers), ed. A. Nakajima *et al.*, Kagakudojin, Kyoto, 1969, p. 203.
223. B. Wunderlich, 'Macromolecular Physics', Academic Press, New York, 1973, vol. 1.
224. F. Khoury and E. Passaglia, in 'Treatise on Solid State Chemistry', ed. N. B. Hannay, Plenum Press, New York, 1976, vol. 3, chap. 6.
225. P. H. Geil, 'Polymer Single Crystals', Wiley Interscience, New York, 1963.
226. C. J. Brown and A. C. Farthing, *J. Chem. Soc.*, 1953, 3270.
227. W. D. Niegisch, *J. Appl. Phys.*, 1966, **37**, 4041.
228. A. Kajiura, M. Fujii, K. Kikuchi, S. Irie and H. Watase, *Kolloidn Zh.*, 1968, **224**, 124.
229. M. Miles and H. Gleiter, *J. Macromol. Sci., Phys.*, 1978, **B15**, 613.
230. S. Isoda, A. Kawaguchi and K. Katayama, *J. Polym. Sci., Polym. Phys. Ed.*, 1984, **22**, 669.
231. S. Kubo and B. Wunderlich, *Makromol. Chem.*, 1972, **162**, 1.
232. R. Iwamoto and B. Wunderlich, *J. Polym. Sci., Polym. Phys Ed.*, 1973, **11**, 2403.
233. S. Isoda, M. Tsuji, M. Ohara, A. Kawaguchi and K. Katayama, *Polymer*, 1983, **24**, 1155.
234. B. K. Vainshtein, 'Structure Analysis by Electron Diffraction', Pergamon Press, Oxford, 1964, chap. 3.
235. K. Kobayashi, *Proc. EMAG-75, Bristol*, 1976, 251.
236. A. W. Agar, F. C. Frank and A. Keller, *Phillos. Mag.*, 1959, **4**, 32.
237. V. F. Holland, *J. Appl. Phys.*, 1964, **35**, 1351.
238. D. C. Bassett, *Phillos. Mag.*, 1964, **10**, 595.
239. V. F. Holland, *J. Appl. Phys.*, 1964, **35**, 3235.
240. P. H. Lindenmeyer, *J. Polym. Sci., Part C*, 1966, **15**, 109.
241. V. F. Holland and P. H. Lindenmeyer, *Science*, 1965, **147**, 1296.
242. V. F. Holland, P. H. Lindenmeyer, R. Trivedi and S. Amelinckx, *Phys. Status Solid.*, 1965, **10**, 543.
243. V. F. Holland and P. H. Lindenmeyer, *J. Appl. Phys.*, 1965, **10**, 3049.
244. T. Nagasawa and K. Kobayashi, *J. Appl. Phys.*, 1970, **41**, 4276.
245. W. C. T. Dowell, J. L. Farrant and A. L. G. Rees, *Proc. 4th Int. Conf. EM, Berlin*, 1960, 367.
246. J. R. White, *J. Polym. Sci., Polym. Phys. Ed.*, 1974, **12**, 2375.
247. Y. Murata, T. Baird and J. R. Freyer, *Nature (London)*, 1976, **262**, 721.
248. S. Isoda, M. Tsuji, M. Ohara, A. Kawaguchi and K. Katayama, *Makromol. Chem.. Rapid Commun.*, 1983, **4**, 141.
249. H. D. Keith and E. Passaglia, *J. Res. Natl. Bur. Stand., Sect. A*, 1964, **68**, 513.
250. T. Seto, *Zairyo Kagaku*, 1967, **4**, 178.
251. Y. Wada and M. Matsui, *Kobunshi*, 1970, **19**, 658.
252. J. M. Peterson, *J. Appl. Phys.*, 1966, **37**, 4047.
253. D. A. Zaukelis, *J. Appl. Phys.*, 1962, **33**, 2797.
254. R. Hosemann, W. Wilke and F. J. Balta Calleja, *Acta Crystallogr.*, 1966, **21**, 118.
255. P. Predecki and W. O. Statton, *J. Appl. Phys.*, 1966, **37**, 4053.
256. G. A. Bassett, J. W. Menter and D. W. Pashley, *Proc. R. Soc. London, Ser. A*, 1958, **246**, 345.

257. A. Bourret, J. Desseaux and A. Renault, *J. Microsc. Spectrosc. Electron*, 1977, **2**, 467.
258. A. Bourret and J. Desseaux, *Nature (London)*, 1978, **272**, 151.
259. A. Bourret and J. Desseaux, *Philos. Mag.*, 1979, **A39**, 405.
260. A. Bourret and J. Desseaux, *Philos. Mag.*, 1979, **A39**, 419.
261. S. Iijima, *Acta Crystallogr., Sect. A*, 1973, **29**, 18.
262. S. Iijima, S. Kimura and M. Goto, *Acta Crystallogr., Sect. A*, 1973, **29**, 632.
263. S. Iijima, S. Kimura and M. Goto, *Acta Crystallogr., Sect. A*, 1974, **30**, 251.
264. O. L. Krivanek, S. Isoda and K. Kobayashi, *Philos. Mag.*, 1977, **36**, 931.
265. W. L. Bragg and J. F. Nye, *Proc. R. Soc. London, Ser. A*, 1947, **190**, 474.
266. S. Isoda and K. Katayama, Abstract of a Meeting organized by the Japan Society of Electron Microscopy, 29 January, 1983.
267. R. T. Read and R. J. Young, *J. Mater. Sci.*, 1984, **19**, 327.
268. K. Kondo, N. Nakajima, Y. Ichioka and T. Suzuki, *Oyo Butsuri*, 1978, **47**, 1140.
269. R. Hagege, M. Jarrin and M. Sotton, *J. Microsc. (Oxford)*, 1979, **115**, 65.
270. Instructions of 'Twin Hearth Electron Bombarded Source and Power Supply', Edwards High Vacuum, 1980.
271. Y. Fujiyoshi, T. Kobayashi, K. Ishizuka, N. Uyeda, Y. Ishida and Y. Harada, *Ultramicroscopy*, 1980, **5**, 459.
272. Handbook of 'Low Dose Unit for EM400', Philips, 1979.
273. M. Iwatsuki and Y. Harada, *Denshi Kembikyo*, 1987, **21**, 174.
274. Y. Watanabe, *Kobunshi*, 1984, **33**, 765.
275. B. J. Tabor, E. P. Magrè and J. Boon, *Eur. Polym. J.*, 1971, **7**, 1127.
276. K. F. Schoch, Jr., J. F. Chance and K. E. Pfeffer, *Macromolecules*, 1985, **18**, 2389.
277. T. C. Clarke, K. K. Kanazawa, V. Y. Lee, J. F. Rabolt, J. R. Reynolds and G. B. Street, *J. Polym. Sci., Polym. Phys. Ed.*, 1982, **20**, 117.
278. A. Uemura, S. Isoda, M. Tsuji, M. Ohara, A. Kawaguchi and K. Katayama, *Bull. Inst. Chem. Res., Kyoto Univ.*, 1986, **64**, 66.
279. A. J. Lovinger and D. D. Davis, *Polym. Commun.*, 1985, **26**, 322.
280. S. Kumar, D. P. Anderson and W. W. Adams, *Polymer*, 1986, **27**, 329.
281. M. Tsuji *et al.*, to be published.
282. D. H. Yan, L. H. Ge, X. M. Jin and E. Zhou, personal communication, 1987.
283. P. C. Dawson and D. J. Blundell, *Polymer*, 1980, **21**, 577.
284. G. Charlet, G. Delmas, J. -F. Revol and R. St. J. Manley, *Polymer*, 1984, **25**, 1613.
285. M. Tsuji *et al.*, to be published.
286. S. Giorgio and R. Kern, *J. Polym. Sci., Polym. Phys. Ed.*, 1984, **22**, 1931.
287. M. Tsuji, M. Ohara, A. Kawaguchi andK. Katayama, to be published.
288. G. S. Y. Yeh and P. H. Geil, *J. Macromol. Sci., Phys.*, 1967, **B1**, 235.
289. J. J. Klement and P. H. Geil, *J. Macromol. Sci., Phys.*, 1971, **B5**, 505; 1971, **B5**, 535.
290. K. Kobayashi, M. Ohara and S. Hoshino, unpublished result, (cited partly by V. F. Holland and R. L. Miller, *J. Appl. Phys.*, 1964, **35**, 3241).
291. B. C. Edwards and P. J. Phillips, *Polymer*, 1974, **15**, 351.
292. A. Kawaguchi and J. Petermann, *Mol. Cryst. Liq. Cryst.*, 1986, **133**, 189.
293. A. Kawaguchi, M. Tsuji, S. Moriguchi, A. Uemura, S. Isoda, M. Ohara, J. Petermann and K. Katayama, *Bull. Inst. Chem. Res., Kyoto Univ.*, 1986, **64**, 54.
294. M. Tsuji, unpublished results.
295. J. Petermann and R. M. Gohil, *J. Mater. Sci. Lett.*, 1979, **14**, 2260.
296. M. J. Hill and A. Keller, *J. Macromol. Sci., Phys.*, 1969, **B3**, 153.
297. A. J. Pennings, *J. Polym. Sci., Polym. Symp.*, 1977, **59**, 55.
298. K. Katayama, S. Isoda, M. Tsuji, M. Ohara and A. Kawaguchi, *Bull. Inst. Chem. Res., Kyoto Univ.*, 1984, **62**, 198.
299. G. A. Bassett and A. Keller, *Philos. Mag.*, 1964, **9**, 817.
300. R. P. Ferrier, in 'Advances in Optical and Electron Microscopy', ed. R. Barer and V. E. Cosslett, Academic Press, New York, 1969, vol. 3, p. 155.
301. 'Operation Manual of JEM-200CX Electron Microscope', JEOL Ltd.
302. K. Inoue and S. Hoshino, *J. Polym. Sci., Polym. Phys. Ed.*, 1973, **11**, 1077.
303. M. Tsuji, M. Tosaka, K. J. Ihn, A. Uemura, A. Kawaguchi and K. Katayama, Abstract of the Annual Meeting of the Society of Fiber Science Technology, Japan, 1987, p. 53.
304. P. H. Geil, in 'Order in the Amorphous 'State' of Polymers', ed. S. E. Keinath, R. L. Miller and J. K. Rieke, Plenum Press, New York, 1987, p. 83.
305. E. J. Roche and E. L. Thomas, *Polymer*, 1981, **22**, 333.
306. J. Petermann, W. Kluge and H. Gleiter, *J. Polym. Sci., Polym. Phys. Ed.*, 1979, **17**, 1043.
307. J. Petermann and R. M. Gohil, *J. Polym. Sci., Polym. Lett. Ed.*, 1980, **18**, 781.
308. H. Krug, A. Karback and J. Petermann, *Polymer*, 1984, **25**, 1687.
309. K. Wenderoth, A. Karback and J. Petermann, *Colloid Polym. Sci.*, 1985, **263**, 301.
310. R. M. Gohil and J. Petermann, *J. Macromol. Sci., Phys.*, 1980, **B18**, 217.
311. R. M. Gohil and J. Petermann, *Colloid Polym. Sci.*, 1982, **260**, 312.
312. J. Petermann, B. Broza, U. Rieke and A. Kawaguchi, *J. Mater. Sci.*, 1987, **22**, 1477.
313. H. Kawai and T. Tagawa (eds.), 'Zusetsu: Sen-i no Keitai' (Photographs of Fiber Morphologies), Asakura, Tokyo, 1982.
314. M. G. Dobb, D. J. Johnson and B. P. Saville, *J. Polym. Sci., Polym. Phys. Ed.*, 1977, **15**, 2201.
315. T. Takahashi, M. Miura and K. Sakurai, *J. Appl. Polym. Sci.*, 1983, **28**, 579.
316. K. Shimamura, J. R. Minter and E. L. Thomas, *J. Mater. Sci. Lett.*, 1983, **2**, 54.
317. A. Peterlin, *J. Macromol. Sci., Phys.*, 1973, **B8**, 83.
318. H. K. Herglotz, in 'Structure of Crystalline Polymers', ed. I. H. Hall, Elsevier, Amsterdam, 1984, chap. 6.
319. K. Shimamura and F. Yokoyama, *Sen'i Gakkaishi*, 1986, **42**, P-73.
320. E. J. Roche, T. Takahashi and E. L. Thomas, *ACS Symp. Ser.*, 1980, **141**, 303.
321. T. Takahashi, *Sen'i Gakkaishi*, 1981, **37**, P-325.

322. K. Shimamura, K. Monobe, E. L. Thomas and W. W. Adams, Abstract of the Annual Meeting of the Society of Fiber Science and Technology, Japan, 1987, p. 48.
323. A. Kawaguchi, S. Isoda, J. Petermann and K. Katayama, *Colloid Polym. Sci.*, 1984, **262**, 429.
324. S. Isoda, A. Kawaguchi, A. Uemura and K. Katayama, *Jpn. J. Appl. Phys.*, 1985, **24**, L-341.
325. R. L. Greene, G. B. Street and L. J. Suter, *Phys. Rev. Lett.*, 1975, **34**, 577.
326. G. B. Street, W. D. Gill, R. H. Geiss, R. L. Greene and J. J. Mayerle, *J. Chem. Soc., Chem. Commun.*, 1977, 407.
327. M. Akhtar, J. Kleppinger, A. G. MacDiarmid, J. Milliken, M. J. Moran, C. K. Chiang, M. J. Cohen, A. J. Heeger and D. L. Peebles, *J. Chem. Soc., Chem. Commun.*, 1977, 473.
328. M. Tsuji and R. St. J. Manley, *Colloid Polym. Sci.*, 1984, **262**, 236.
329. M. Tsuji, J. Frank and R. St. J. Manley, *Colloid Polym. Sci.*, 1986, **264**, 89.
330. J. Sugiyama, H. Harada, Y. Fujiyoshi and N. Uyeda, *Mokuzai Gakkaishi*, 1984, **30**, 98; 1985, **31**, 61.
331. J. Sugiyama, H. Harada, Y. Fujiyoshi and N. Uyeda, *Denshi Kembikyo*, 1985, **20**, 143.
332. J. -F. Revol, *J. Mater. Sci. Lett.*, 1985, **4**, 1347.
333. S. Kuga and R. M. Brown, Jr., *J. Electron Microsc. Techn.*, 1987, **6**, 349.
334. S. Kuga and R. M. Brown, Jr., *Polym. Commun.*, 1987, **28**, 311.
335. J. -F. Revol and H. Chanzy, *Biopolymers*, 1986, **25**, 1599.
336. J. -F. Revol, personal communication.
337. Y. Fujiyoshi and N. Uyeda, *Ultramicroscopy*, 1981, **7**, 189.
338. E. Knapek, G. Lefranc, H. G. Heide and I. Dietrich, *Ultramicroscopy*, 1982, **10**, 105.
339. International Experimental Study Group, *J. Microsc. (Oxford)*, 1986, **141**, 385.
340. H. Yamagishi, Y. Fujiyoshi, Y. Aoki, K. Morokawa, N. Uyeda and Y. Harada, *Denshi Kembikyo*, 1984, **19**, 32.
341. A. Kawaguchi, 'Kobunshi no Kotaikozo II' (Solid Structure of Polymers II), in 'Experimental Methods in Polymer Science', ed. Japan Society of Polymer Science, Kyoritsu, Tokyo, 1984, vol. 17, chap. 9.
342. G. N. Patel and R. D. Patel, *J. Polym. Sci., Part A-2*, 1970, **8**, 47.
343. A. V. Crewe and M. Ohtsuki, *Ultramicroscopy*, 1982, **9**, 101.
344. M. Ohtsuki, *Nippon Kessho Gakkaishi*, 1983, **25**, 38.
345. S. Isoda, *Kobunshi*, 1984, **33**, 786.
346. M. Tsuji, Ph. D. Thesis, Kyoto University, 1981, chap. 5.
347. L. F. Estis, R. H. Haschemeyer and J. S. Wall, *J. Microsc. (Oxford)*, 1981, **124**, 313.
348. I. G. Voigt-Martin and L. Mandelkern, *J. Polym. Sci., Polym. Phys. Ed.*, 1981, **19**, 1769; 1984, **22**, 1901.
349. D. J. Johnson and D. Crawford, *J. Microsc. (Oxford)*, 1973, **98**, 313.
350. G. S. Y. Yeh and S. L. Lambert, *J. Appl. Phys.*, 1971, **42**, 4614.

35

Etching and Microstructure of Crystalline Polymers

DAVID C. BASSETT
University of Reading, UK

35.1 INTRODUCTION

That the properties of substances depend upon their internal constitution is an idea so ancient that it can be taken as self-evident. As modern science has advanced, so has evidence on the constitution of matter. Some earlier hypotheses have been proved, others have fallen away. Atoms are real, so is charge, but magnetic pole is only a model, useful for calculation, with no direct correspondence to a structural unit. In this progression of knowledge, polymers have been identified as based on macromolecules, while their properties depend not only upon those of the molecules themselves but also (and most importantly, being the subject of this chapter) upon the organization of molecules within the sample. If we restrict our discussion to solids consisting of covalently bonded macromolecules, then properties such as rubber elasticity depend simply on the presence of long molecules with many conformations of equal energy. It was, however, recognized shortly after the discovery of X-rays that, for example, cellulose and eventually synthetic polyamides showed both crystalline and liquid-like organization. Such materials are stiffer, with crystallites evidently providing reinforcement. Of the nature of the crystallites one then had only the width of wide angle reflections and the existence of low angle reflections as a guide. Both were qualitatively consistent with the well-known fringed-micelle model,[1] wherein molecules passed from small crystallites, on the scale of ~ 10 nm in size, into 'amorphous' regions and *vice versa*. The important feature of this model was the inextricability of two phases, making materials composite in nature. With hindsight it can also be seen to have given undue and largely undeserved prominence to the degree of crystallinity. As we shall see, this concept is of value only for discussing relative changes within comparable textural organizations. When the nature of the organization changes, the degree of crystallinity is at best largely irrelevant and otherwise liable to mislead. Nevertheless, qualitative understanding of crystalline (strictly semi-crystalline) polymers loosely based on variable amounts of two 'phases' steadily advanced over some 30 years.

The internal organization postulated in the fringed-micelle model lay well beyond the resolution of optical microscopy. No more than unease seems to have resulted from observations such as the existence of discrete crystals of poly(oxymethylene) in 1932[2] and the presence of spherulites, first reported in 1945 in the new polymer polyethylene.[3] To be sure, it was far from obvious why spherulites should show internal radial lines whereas molecules were tangential (in non-hydrogen-bonded systems) or nearly so, but ingenious application of the fringed-micelle model was offered as an explanation. Not until 1957 did facts appear which obviously required replacement of the fringed-micelle model.

Concurrently with the rapid discovery and introduction of crystallizable macromolecules, particularly in the two decades prior to 1957, materials science had nucleated and begun to be a subject in its own right. In the study of the influence on properties of defects in and organization of crystals, a major landmark was the supposition that dislocations existed, with profound implications for mechanical and other properties. For long hypothetical, dislocations were eventually identified by ingenious use of the optical microscope. They could be detected *via* growth spirals, decoration (by silver in silver halides) and by etch pits. Then came the double-condenser electron microscope in 1955 and dislocations, their structure, interactions and dynamics could be studied in sufficiently thin metal and inorganic foils to ∼1 nm resolution. Modern materials science may very reasonably be said to have begun with the introduction of this form of electron microscope and the revelations of fine-scale microstructure which it provided.

Polymers were not exempt from this revolution in resolving power but the gains were severely circumscribed because, being organic, they were susceptible to damage by the imaging electrons to the point of destruction of crystallinity and beyond, unlike metallic and ceramic samples. What was achieved was, nevertheless, still a revolution in polymer science but one whose extent and details could not be explored very far. The new linear and stereoregular polyethylenes discovered since 1955, by using metal-alkyl catalysts, were soon shown to be crystallizable from solution as separate lamellae[4,5,6] within which molecules were chain-folded.[5] Separate crystals of macromolecules are incompatible with the fringed-micelle model. It is now known that lamellar crystallization with substantial degrees of chain-folding is characteristic of synthetic polymers crystallized from quiescent melts. Lamellae are also present, to greater or lesser extents, in many polymeric fibres. This situation has, however, taken a long while to prove. It is only very recently, following the introduction of permanganic etching,[7,8] and its development and application to numerous systems, that electron microscopic evidence, *in direct space*, has become available for representative polymeric microstructures in melt-crystallized systems, at a resolution which was available in metals 30 years ago.

We can now recognize that there is no intrinsic difference in kind between polymer lamellae crystallized from solution and from the melt. The contrary position was long held by some, partly on theoretical grounds, and then with some support from neutron-scattering data. But it was also pointed out that similar samples of the two kinds of system had not been examined in comparable ways. In particular, slowly grown melt-crystallized lamellae needed to be examined by diffraction electron microscopy, the method which had revealed most of the evidence for regular chain-folding from solution. Now that this has been done, the anticipated continuous spectrum of ordering, embracing both kinds of crystallization, has been revealed. The disappearance of this supposed dichotomy makes the conclusions on ordering of solution-grown lamellae directly relevant to those crystallized from the melt. Solution-crystallized systems are more amenable to examination, for reasons which were not always obvious. It is informative to discuss why this is so and why a technique such as permanganic etching has had to be introduced to advance knowledge of melt-crystallized polymers.

Crystalline polymers show a hierarchy of ordering from the molecular level to macroscopic dimensions. Thus molecules form lamellae typically tens of nanometres thick, while within spherulites (which may be many micrometres or even centimetres across and are the subject of Volume 2, Chapter 12) one now finds systematic placing of different types of lamellae sometimes with groupings into coarser units, traditionally called fibres, although now requiring more careful description. Such complexity was initially intractable and invited simplification for its analysis. Long ago it was found that spherulites could be dissolved from their surrounding uncrystallized matrix and even disintegrated into finer but still complex units.[9] The study of solution-grown sheaves and other embryonic objects was, however, tractable. With the advent of linear polyethylene the objects so grown were found to be lamellar, from monolayers to multilayers. They could be examined in the transmission electron microscope by simply sedimenting them on a carbon support film. Relief is accentuated by evaporating a thin film obliquely (shadowing, for example at ∼30°). External shapes and some internal structure can then be examined directly. Provided high intensities of illumination are not employed initially, detail persists in monolayers and simple objects. But in higher layers of thicker objects one observes movement which is due to mass transport. Experiment has shown that the carbon film supporting the lamellae gives them a rigidity which allows shapes and significant internal structure to be preserved.

Attempts to observe electron diffraction from lamellae are only successful at low levels of intensity and very low magnifications (usually not more than a few thousand times). Even then the effects are fleeting but the information they give has had profound consequences. In the first place they appeared to show that molecules in polyethylene were normal to monolayers only 12 nm or so thick.[5] Provided the molecular length remains much longer than the crystal thicknesses, which is the case, it must follow that molecules fold back and forth within the lamellar basal (or fold) surfaces. This conclusion remains intact even allowing for a revision of the molecular orientation (principally for polyethylene, not for most other polymers) to an inclination of $\sim 30°$ to lamellar normals.[10] Such revised information comes from diffraction, especially dark-field imaging, which reveals distinct sectors in monolayers and the orientation of the lattice within them. Sectors are usually regions bounded by a growth face and are powerful evidence, based on symmetry, for folding along growth faces. There are two effects. One, peculiar to polyethylene with its small cross-sectional area per chain, has different chain tilts (of $\sim 30°$) in different sectors which can give monolayers a hollow pyramidal habit. The other, general effect is a small lattice distortion, measurable by moire techniques, which changes systematically at sector boundaries, and is the origin of shallow dishing.[11]

This kind of evidence demonstrates that folding is preferentially along growth faces. It does not itself reveal the configuration of folds nor the probability of a molecule folding back into a lamella at the next adjacent stem position. Such information comes from neutron-scattering experiments. These have indeed confirmed that folding is along growth planes. A single molecule, however, is not confined to a series of adjacent stem positions in one plane but is placed in two or three growth planes and has an occupancy in each plane reduced to about one half.[12] The molecular shape in quenched melt-grown systems is different, revealing no planar confirmation but an average probability of adjacent re-entry of a molecule into the next site in a lamella of 0.7.[13] This figure is really a lower limit deriving from geometrical considerations of interfacial density. It is always possible to fit a molecule into a lamella by repetitive chain-folding but there are restrictions on how many may leave it without adjacent re-entry because of possible overcrowding.

The detailed electron microscopic examinations leading not only to confirmation of regularity in folding but also, *via* internal structural detail, to insight into textural processes (annealing, deformation, *etc.*) have been made on mono- and bi-layer crystals. Multilayer objects are met with more frequently as concentrations rise and introduce interlamellar connectivity. Such crystals do not lend themselves to electron microscopy; layers are frequently not in contact, while crystals remain in suspension. Severe collapse then results during preparation for electron microscopy and intrinsic features of the lamellae can become obscured. Nevertheless, all the evidence points to multilayers as consisting of chain-folded lamellae similar, though not necessarily identical to, the mono- and bi-layers formed from the most dilute solutions. We shall reach a corresponding conclusion for melt-grown systems.

Thin and presumably chain-folded lamellae were also reported in 1957 from observation of replicas of fracture surfaces.[6] For many years this technique remained the most convenient means of examining the microstructure of bulk polymers. Its application is universal provided only that fracture is brittle*, which requires the break to be made at sufficiently low temperatures (when the material is glassy). Using this technique, Geil[14] was able to reveal lamellae in an extremely wide range of polymers. Two doubts, however, remain concerning such evidence. One concerns the representativeness of the structures revealed, the other the extent and regularity of chain-folding.

Neither point is answerable without further information. On the first, a fracture surface is necessarily special because it has been followed by the propagating crack in preference to alternatives. It is perfectly possible for it to contain either special structures, or structures in special orientations. In the case of polyethylene it has now been shown that thinner crystal populations can be discriminated against to the extent of being absent from fracture surfaces and that, in general, fracture exposes {200} planes preferentially, thereby sampling the morphology of non-typical orientations and almost always excluding the view down the *b* axis (spherulitic radius).[15] This can have both advantages and disadvantages. One advantage is that one can use the preferred orientation to gain an estimate of lamellar extension along the spherulitic radius in polyethylene. One can also obtain reliable measurements of lamellar thickness, although these may refer to selected populations.[16] In general, however, it is a grave disadvantage that one does not reliably obtain a representative view of the morphology.

The essential reason for the widespread use of fracture surface replicas to study the morphology of bulk polymers is that they provide stable and detailed images whereas, because of radiation damage, polymers themselves do not.[17] If a parallel-sided thin section of a crystalline polymer is inserted

---

* It is sometimes possible to achieve a condition in which some lamellae are brittle and some not. Those which are brittle have fewer chain-folds per molecule. This could be a potentially informative technique *e.g.* in diagnosing incipient flaws relating to fracture toughness.

into a transmission electron microscope, it first appears virtually featureless because of the low variation in mass-thickness across the specimen. Soon, however, contrast appears as the imaging electrons interact with the specimen leading to either crosslinking or scission. There are differential changes in lamellar dimensions: a sideways expansion but contraction along the chain direction. What is being seen is not the original morphology but a product of it *via* a complex reaction path. When changes cease, detail in the image has become coarse and little valuable information can be gleaned as to organization at the lamellar level.

Study of little-damaged specimens has, however, allowed the determination, for example, of which crystallographic axis is along a spherulite radius.[18] By involving epitaxy it has even been possible to resolve detailed matters, notably the crystallography of twinning in α-polypropylene.[19] Such studies require rapid working at low doses, for example by performing all instrumental adjustments on one area, then switching to an adjacent one with immediate recording. For the most part they have been at low magnifications but with modern image-processing it has become possible to recover lattice fringes in suitable specimens of polyethylene[20] and other polymers.[21] This technique is most suited to revealing lattice defects, the presence of which has been related to electrical conductivity.[22] (Such defects were also identified[23] and measured[24] in various polymers long ago by moire methods at low magnification.) It is not yet a general method in the sense that any desired area can be studied, nor, since it relies on diffraction for contrast, does it reveal other than internal lamellar detail.

High resolution electron microscopy in this form does not, nor is likely to, have an impact on the problem of lamellar organization in bulk polymers, which is central to the understanding of properties.

Those techniques which have been successful in this area fall into two categories. One examines a thin section of material which has been fixed in some way to enhance contrast and, most importantly, achieve a stable image. Natural rubber and other polymers whose molecules contain double bonds have been treated, as thin films, with osmium tetroxide.[25,26] Individual lamellae with osmium at their edges can be resolved, usually edge-on, sometimes growing normal to a tensile axis as part of a filament or shish-kebab, more generally in sheaf-like or spherulitic projections. In a similar way polyethylene treated with chlorosulfonic acid[27] gains sulfur in its 'amorphous' regions. Lamellae may then be seen as white regions in a dark matrix. Visibility is optimum when lamellae are viewed sideways on, giving an image of black and white stripes. Once chlorosulfonated, polyethylene has different properties. It is much easier to section, its melting point is raised and its crystallinity reduced but its electron microscope image is both stable and detailed. This technique, which is still restricted almost entirely to polyethylene, allows the relative numbers and dimensions of lamellae to be measured, as well as their mutual disposition. The variation of the image with viewing angle away from the edge-on position can also indicate lamellar orientation (which tilting of the sample will confirm).[28] One application, for example, is to show the repetitive orientation of lamellae in banded spherulites and to test the extent to which individual lamellae, as opposed to the average orientation, are twisted (with the conclusion that they are effectively not).[29] In the context of the present chapter, chlorosulfonation is a complementary technique for polyethylene. It shows the same qualitative features as permanganic etching on the same samples, thereby attesting to their genuineness, although chlorosulfonation consistently measures smaller lamellar thicknesses than does permanganic etching, fracture surface replication and other techniques.[30]

The second alternative approach, with which this chapter is principally concerned, is that of examining detail in etched surfaces. This has the advantage over a fracture surface of giving the choice of a random (or fracture or external) surface, prepared by cutting the sample open. One needs only a satisfactory etchant to reveal genuine, fine detail and there have been many attempts, based, for example, on organic solvents, gaseous ions, activated gases and strong acids, notably fuming nitric acid for polyethylene, to provide one. The spectrum of reagents based on potassium permanganate,[7,8] which we term permanganic etchants, gives much superior detail compared with earlier methods. It has now been applied successfully to a wide range of polymers, including several polyalkenes: polyethylene (linear and branched), isotactic polystyrene, isotactic polypropylene, isotactic poly(4-methyl-1-pentene), isotactic polybutene, poly(aryl ether ether ketone) (PEEK; poly(oxy-1,4-phenyleneoxy-1,4-phenylenecarbonyl-1,4-phenylene), poly(ethylene terephthalate) and certain liquid crystalline polymers.[30] The composition of the etching reagent may need to be adjusted for each system to optimize detail. After suitable washing, etched surfaces can be examined by replication or with a scanning electron microscope. Replication (either one- or two-stage) still gives the best resolution but it is a skilled and time-consuming activity. Scanning microscopy is much quicker and the most modern instruments can resolve thin lamellae in, for example, α-polypropylene, but limited resolution and contrast effects can be potentially misleading, for example on the nature of lamellar aggregates and their orientation to a surface.

The chemistry of the reaction is oxidative but it has not been studied in detail. Empirical observations, for example that for isotactic polystyrene of a particular colour, works best and that aqueous reagents are generally more aggressive, suggest that some subtlety is involved. The suitability for electron microscopy is due in part to the fact that although interlamellar regions are etched more deeply, no great depth is involved. Treatment with fuming nitric acid, by comparison, will render polyethylene friable by eating away interlamellar regions almost entirely.[31] Such friable materials are not easy to replicate; in consequence the main applications of nitric acid etching have been to drawn fibres,[32] where penetrability is decreased, or to residual lamellar fragments.[33] These differences between etchants have been discussed elsewhere.[34]

Expectations of an etchant are that it will remove material selectively according to penetrability and chemical potential. This is the case for permanganic etching, but it is not the whole story as has recently become evident. Physical factors are also involved, specifically differential damage to different populations in a sample.[35] Thus, in spherulites of linear polyethylene the first-forming or dominant lamellae often stand proud after etching because later-forming, infilling lamellae have been more susceptible to the etchant (Figure 1). It has been shown that this is not an effect which is intrinsic to the different populations but arises from the different mechanical response of the populations to cutting the sample open prior to etching. Suitable annealing and recrystallization treatments will remove the difference in etchability: fully for linear polyethylene[35] but only partly for branched material.[36]

The way in which permanganic etching achieves its effects is thus somewhat complex. But the reality is that it can selectively remove material from any surface of a suitable polymer exposing crisp detail to lamellar resolutions and beyond. Such information must have a major impact on our understanding of crystallization processes and resulting properties. In the remainder of this chapter examples will be cited of the range of applications with the intention of illustrating both the achievements and potential of this technique. For the most part this will concern only shapes. While it is desirable and possible[7] to combine this with diffraction, applications of such complementary use are still few.

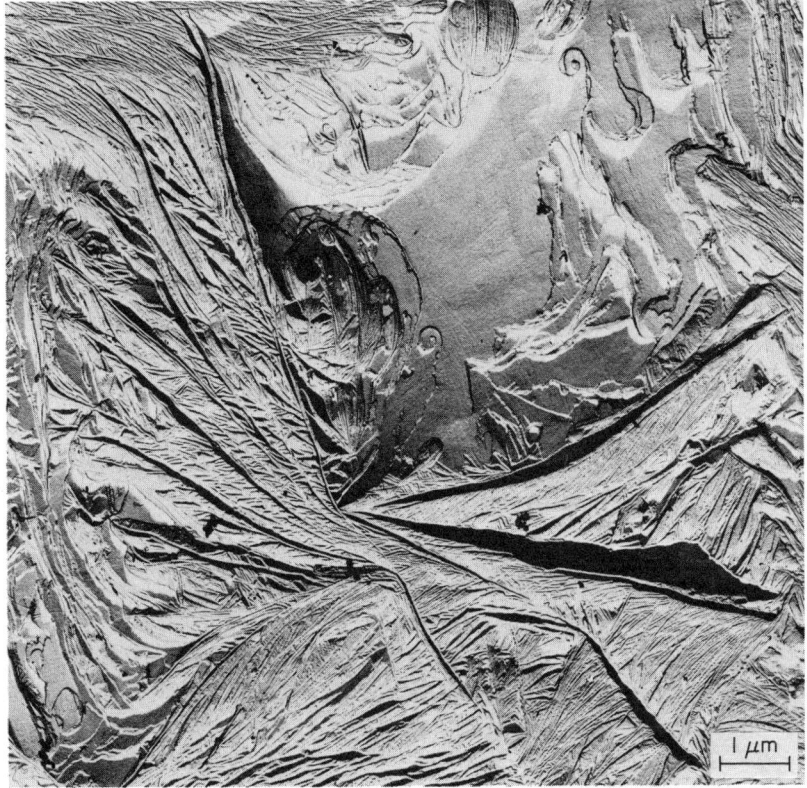

Figure 1 Linear polyethylene ($\bar{M}_w = 5.4 \times 10^4$, $\bar{M}_n = 1.7 \times 10^4$) crystallized at 128 °C for 3 h, then quenched. This texture shows radiating dominant lamellae, both edge-on and flat-on standing proud in a structured matrix of lamellae also grown at 128 °C but, from their often ridged habit, indicating isothermal fractional crystallization. Lamellae crystallized on quenching are evident at the top left hand corner

35.1.1   A Basic Technique

Among the various reagents used, the one which is routinely used in this laboratory for polyethylene and isotactic polypropylene is a freshly prepared 0.7% w/v solution of potassium permanganate in a 2:1 mixture of concentrated sulfuric:orthophosphoric acids.[8] A suitable sample of polymer, which might be a 2 mm cube, is immersed below the surface of 5 cm^3 of this reagent in a stoppered tube, possibly with agitation at room temperature for, say, 1 h. (The extent of etching is monitored, if need be, by the appearance of the sample under Nomarski Differential Interference Contrast Optics). Etching is stopped by decanting the reagent and washing first with a mixture of hydrogen peroxide and aqueous sulfuric acid (precooled almost to its freezing point with solid carbon dioxide), then with distilled water. The hydrogen peroxide is added in the proportion of 1 cm^3 (100 volumes strength) to 25 cm^3 of 2:7 sulfuric acid, *i.e.* 2 volumes sulfuric acid to 7 volumes of water. Related procedures are used for other permanganic etchants.

This recipe is a more delicate etchant than that first introduced[7] which, nevertheless, was itself very satisfactory for an extensive study of lamellae in polyethylene.[15,37] The phosphoric acid slows the etching rate to *ca.* 3 Å s^{-1}, *i.e. ca.* 1 μm h^{-1}, for the lateral surfaces of polyethylene lamellae. The method produces very crisp detail, thereby facilitating the study of α-polypropylene, and avoids the production of spurious detail or 'artefacts' encountered with polyethylene etched with solutions of sulfuric acid alone.[8] There is another important advantage. The thickness removed in such treatments is about 1 μm, *i.e.* less than typical lamellar widths, which are some microns in polyethylene. It therefore follows that the great majority of lamellae seen in an etched surface will also have been present in the surface exposed before etching. This allows study of cut, fracture and free surfaces, which typically have different characteristic morphologies. Mild etching, such as that described, allows these to be distinguished and characterized, whereas when, say, *ca.* 10 μm is removed by etching, this distinction would probably be lost.

35.2   ETCHED MICROSTRUCTURES

35.2.1   Component Polymers

Study of structures in polymeric systems begins, at the lowest resolution, with identification of components in a mixture. Such applications embrace blends where one is likely to be interested in the intimacy or scale of mixing and effects of interfaces. Another is in composites, for example of carbon fibres in a polymer matrix, where the contact between the two components is of crucial importance. Permanganic etching can contribute at varying levels to these matters. One would expect, *a priori*, that different polymers would etch at different rates, so that differences of depth within defined boundaries would result. An example is shown in Figure 2. If, however, there is intimate mixing as, for example, in certain blends of polyethylene and ethylene–propylene rubbers, then this, too, can be ascertained on inspection by the lack of clear internal boundaries. Still at low resolutions, one can observe whether fibres have all been wetted by the matrix or whether there are some gaps present between the two. At higher resolution one can observe lamellar structure which may not only identify each polymer by its appearance but also allow inferences to be drawn as to the influence of the interface. Thus in Figure 2 the internal banding immediately identifies the polyethylene. In this instance (but not universally) one can see that the centre of the circular banding is away from the boundary. At first sight this may seem to show that nucleation was away from the interpolymer boundary but one must always be aware of the third dimension and the possibility that the true centre of the spherulite might be located on the interphase junction out of plane. The radial orientation of lamellae in the banded spherulite is then relevant. If they lie in the plane of the specimen then so will the centre of the spherulite and one can draw a firm conclusion. It is possible to obtain a measure of the number of nucleation centres and the relative importance of interfacial nucleation, which affects crystallization kinetics and properties. When there is profuse nucleation on interfaces, growth will normally be away from them, giving a preferred orientation to the resulting microstructure.

35.2.2   Lamellar Morphologies

This is the area where permanganic etching is making original and fundamental contributions to polymer science, namely the revelation of lamellar morphologies. In this term we include such features as the number, shape, dimensions, orientation and substructure of lamellae as functions of

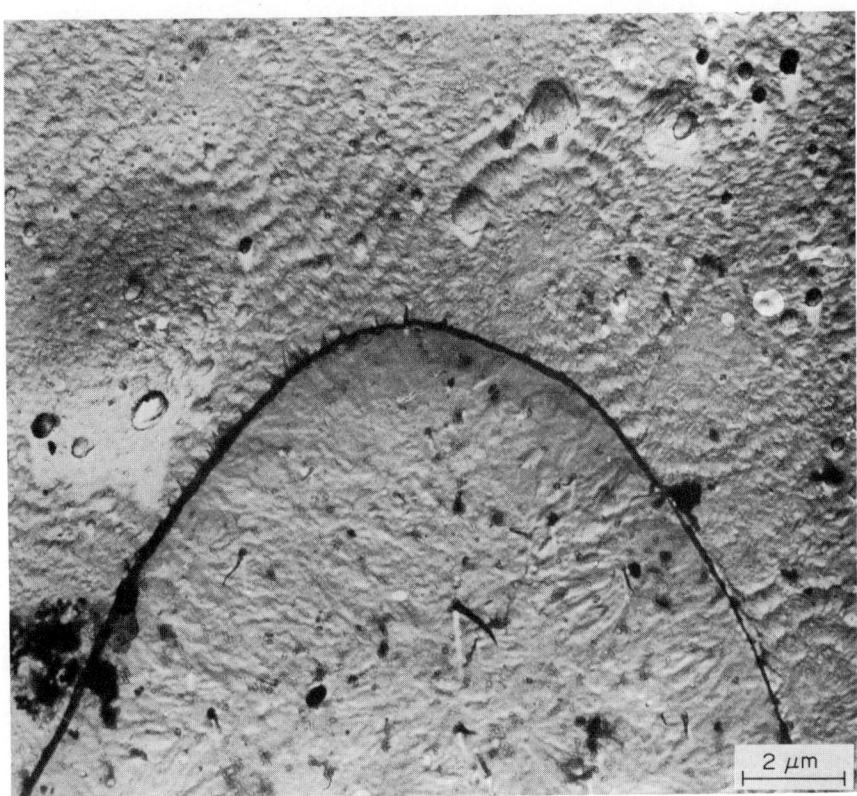

Figure 2 A blend of linear polyethylene (above) with isotactic polypropylene (below). The polypropylene is etched more deeply and each polymer displays its own characteristic texture. In this instance there is no clear evidence for nucleation on the interphase boundary

position, sample history, molecular composition and constitution. It is the richness of this record which makes polymer morphology so useful.[11] This record of past history remains in a crystalline polymer essentially because once laid down the lengths of the molecules make it difficult for them to reorganize substantially. A lamella will retain a thickness characteristic of its growth conditions, especially temperature, while the pattern and relative placement of lamellae usually reflect the molecular composition of a sample, notably the length, tacticity, branching ratio and distributions thereof. Annealing and deformation leave their own marks, which can be read and often quantified.[11]

Permanganic etching does indeed confirm in detail that crystalline polymers are profusely lamellar. There are many parallels with solution-crystallized lamellae, not least evidence for distinct sectorization (which will be discussed presently). We shall find that there is a continuity between solution and melt-grown crystals. This is a position which, with hindsight, may appear obvious but powerful voices have argued for a dichotomy.[38, 39] Expectation, both intuitive and theoretical, has played its part in this. It may appear reasonable, *a priori*, that individual lamellae formed from solution are different in kind from those formed under crowded growth conditions in the melt. Indeed molecular conformations measured by neutron scattering showed major differences between samples of polyethylene prepared in the two ways.[12] But like was not being compared with like: the extremes of the spectrum of crystallization were being compared.[40] What we shall now attempt is to develop a picture based, for the first time, on evidence in real space, of the nature of organization in melt-crystallized polymers to the lamellar level and without preconceptions.

If one starts with an extreme, that of linear polyethylene crystallized very slowly for days at 130 °C, to high crystallinity, one finds (Figure 3) that the sample is effectively full of lamellae of the same high thickness. On the other hand the crystallinity achievable by branched (linear low density) polyethylene is less, even for prolonged crystallization, and here one finds well-defined lamellae within an apparently unstructured matrix (Figure 4). In fact there generally is structure in the matrix which higher magnification will reveal. The quantification and identification of this will be discussed presently. First we point out that the properties of a sample with such a complex texture will vary

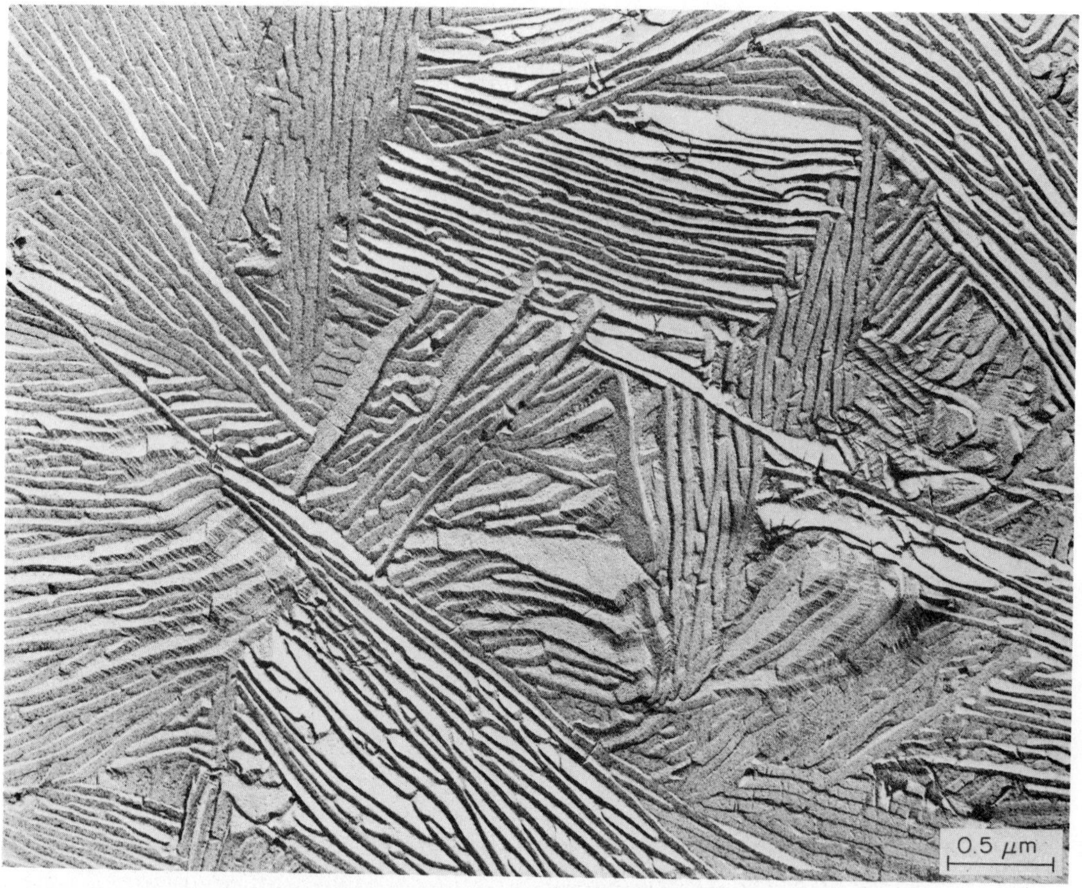

Figure 3 The same linear polyethylene as in Figure 1 but now crystallized at 130 °C. The different widths of lamellae show that this, too, developed by a framework of individual dominant lamellae. The inclined chain orientation is apparent, *e.g.* at lower left

Figure 4 In a linear low density polyethylene (butyl branched) crystallized at 118 °C, dominant lamellae are well separated and often curved

from point to point. Furthermore, mechanical and other properties will depend on precisely how these textural elements are arranged with respect to each other: the simple concept of degree of crystallinity is not adequate to describe situations in which different crystallites have different environments.[41]

Quantification of the morphology can proceed with complementary thermal analysis measurements. Elementary thermodynamics shows that a lamellar crystal of thickness l of a solid which should melt at T_m will show a melting point depression equal to $T_m[(2\sigma_e)/\Delta h \cdot l]$ where Δh is the enthalpy of fusion per unit volume and σ_e the surface Gibbs function of the lamellar surfaces.[11] In a melting endotherm, which is conveniently measured by differential scanning calorimetry (DSC) of a small sample, we have, therefore, a record of the different lamellar thicknesses and their proportions within a specimen. As an illustration (Figure 5) consider a linear low density (butyl-branched) polyethylene sample showing four well-defined endothermic peaks or shoulders in its DSC curve and four sets of lamellae of corresponding thickness in its electron micrograph. These lamellae are, moreover, not randomly placed but are positioned systematically. Any quantitative theory, for example of mechanical response, would require knowledge of both the amounts and siting of each component type of lamella as basic data.

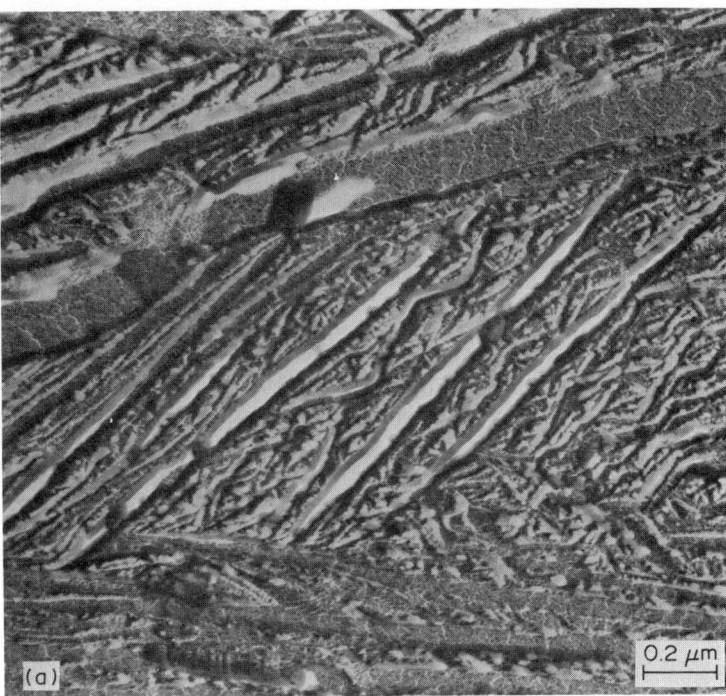

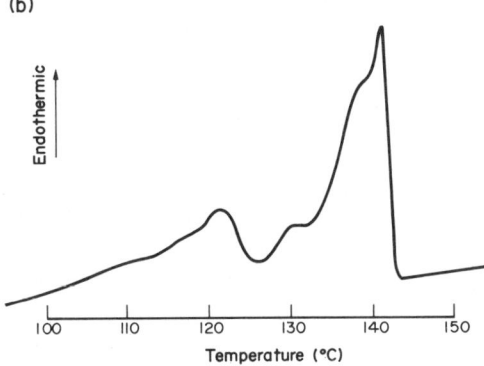

Figure 5 (a) A butyl-branched linear low density polyethylene crystallized at 0.5 GPa to a texture with a comparable pattern, but thicker lamellae, to that of Figure 1. The four principal lamellar thicknesses are evident in the DSC melting endotherm of (b)

One can obtain such data, in general and not merely for special cases, by complementary use of thermal analysis and permanganic etching plus electron microscopy. Consider the DSC melting endotherms of Figure 6, which relate to annealing a linear low density (ethyl-branched) polyethylene at progressively higher temperatures within the original melting endotherm.[11] They reveal the effect of melting part of the sample at the selected annealing temperature T_A (*i.e.* that part of the sample which melts at or below T_A) and recrystallizing it. In this instance recrystallization was for 15 min at T_A followed by quenching, but in general one has a choice within extremes of immediate quenching and prolonged annealing at T_A. Comparison of the endotherm of an annealed sample with that of the original indicates how much material has had its condition changed. Permanganic etching can locate this in the texture by the altered appearance. The results of many experiments show that, in general, in a crystallized sample melting occurs in the reverse order to crystallization. In other words the least stable material is that which crystallized last. In deformed samples, similarly, the locations of the lowest melting regions help identify details of the deformation processes. However, especially for melt-crystallized specimens, one must address the question of why different regions have differing melting points.

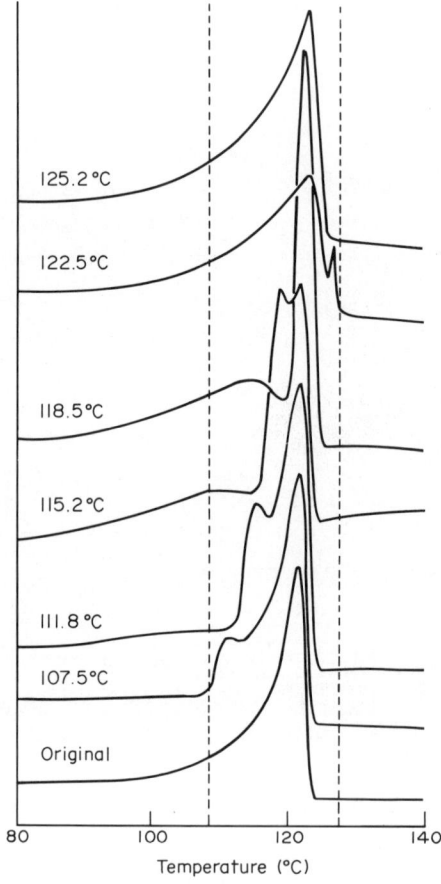

Figure 6 DSC melting endotherms of an ethyl-branched linear low density polyethylene after annealing for 15 min at the stated temperatures, then quenching. Material melted by the annealing temperature may recrystallize either isothermally to a higher melting point or to a lower one on quenching. The latter course is adopted more by shorter and more-branched molecules

One answer which has already been given is that they possess different lamellar thicknesses but this simply transfers the question to how this comes about. Other factors which may also be involved include, for example, different crystal structures, possibly originating in deformation, strained regions, changes in surface free energy or any cause which increases the chemical potential of a lamella. Such factors are also those which should enhance etching so that one might expect to discern relative features in an etched surface although one must remember that secondary effects, notably penetrability and differential deformation during sample preparation, can also be highly significant and noticeable.[35,36]

Regions of thinner lamellae were found in the first published observations of permanganic etching (Figure 7). They comprised some 15% of a sample of linear polyethylene of $\bar{M}_w = 26\,000$ and polydispersity $\bar{M}_w/\bar{M}_n = 1.3$ which had been crystallized at 130 °C for 3 weeks, then quenched.[15] These thinner lamellae had crystallized in pockets on quenching and were formed from shorter molecules. Such fractional crystallization was anticipated and was confirmed by selective extraction in which the lower-melting population is removed with a solvent and its molecular properties measured.[42] Similar experiments on growth at low supercoolings with linear low density polyethylene also show fractionation partly by molecular length but also by branch content with later-crystallizing molecules being more branched.[30] The segregation of shorter molecules is potentially serious for mechanical properties because cracks are better able to propagate through such samples. But one must consider the degree to which fractionation persists under more typical commercial crystallization conditions of high supercooling. One would not expect much fractional crystallization under these circumstances but permanganic etching has shown differences of etchability.[37] At low supercoolings the pockets of thin lamellae were usually etched more deeply than the wider and thicker ones formed before them. A similar effect is observable in rapidly crystallized polyethylene blends, but to a lesser extent (Figure 8). Evidently there is a difference in etchability related to the time of crystallization. However, with the demonstration that secondary deformation on cutting a sample open can lead to such variation there is no necessity to relate such differences in etchability to fractional crystallization.[35]

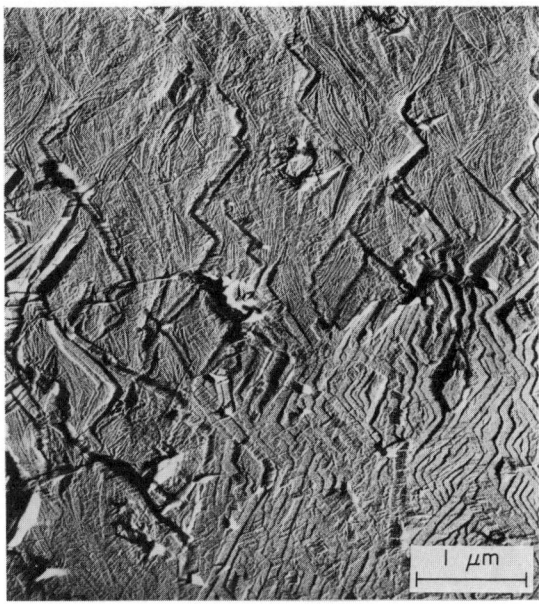

Figure 7 The concentration of shorter molecules at the edge of a spherulite of linear polyethylene. They have crystallized as thin lamellae on quenching (above) in contrast to the thicker, ridged lamellae (below) formed isothermally at 130 °C

An alternative approach to the origin of different lamellar populations is selectively to extract those of lower melting point, then to recrystallize the residue and examine whether or not similar populations and properties recur. An early example was that of linear polyethylene crystallized at a pressure of 0.5 GPa.[43] Such so-called chain-extended samples generally show one or two small peaks ($< ca.$ 130 °C) in their DSC melting endotherms in addition to their preponderant very high melting peaks (~ 142 °C) due to lamellae of thicknesses approaching 1 μm. The shortest molecules give rise to the low peak(s); if they are extracted the same peaks do not recur. Without extraction the short molecules can concentrate into pockets and induce brittleness;[44] in their absence this segregation and related brittleness also disappear.

Such strong fractional crystallization and segregation appears to be especially pronounced in polyethylene. Measurements of the molecular weight of isotactic polystyrene showed a reduction by a factor of only one quarter in $\bar{M}_w$ between a population formed at very high crystallization temperature and that crystallized on quenching when isothermal growth was interrupted.[45] (A factor in excess of two is typically found in polyethylene.[35]) This represents a maximum for isotactic polystyrene and suggests that any effects of fractional crystallization in this polymer are

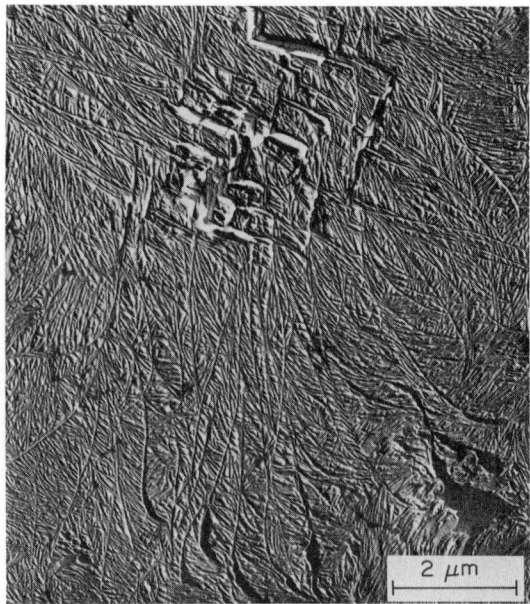

Figure 8   In a 1:1 blend of linear and branched polyethylenes, the low density polymer has crystallized rapidly on quenching but its dominant lamellae stand proud in an etched cut surface

small compared to those found in polyethylene. A similar conclusion has been reached for isotactic polypropylene based on the reproduction of its typical broad-melting endotherm after extraction of the lower molecular weight portion and recrystallization: it appears that all molecules share in the different lamellar populations present in the texture. A detailed study of PEEK has also inferred that textural differences related to spherulitic crystallization do not derive from variations in molecular constitution.[46]

This polymer typically and interestingly shows two peaks in its melting endotherm, a phenomenon shared with other low crystallinity materials such as isotactic polystyrene and poly(ethylene terephthalate). This endotherm is very broad, extending from the crystallization (T_c) or annealing (T_A) temperatures to 340–350 °C. In a sample crystallized by heating from the glass to, say, 200 °C the breadth of the endotherm is well over 100 K. A lower, usually smaller, peak is superposed on this broad endotherm and is found some 10–20 K above T_A or T_c as appropriate. The first interpretation of these peaks[47] was that the lower represented the vestiges of the originally crystallized polymer, which had mostly transformed during the measurement into higher-melting material. This is the familiar and well-attested behaviour found in polyethylene and certain other polymers. Three sets of authors have, however, independently found that it does not apply to PEEK.[46,48,49] At its simplest this is because the upper broad-melting endotherm forms before the lower-melting peak and so cannot derive from conversion of it. Moreover, the location of the lower-melting material has been shown, by procedures outlined above using permanganic etching (Figure 9), to be in the later-crystallizing lamellae, infilling the framework of a spherulite. The evidence is against this separation being a consequence of fractional crystallization: the population appears to have a textural origin deriving from its position in a spherulite. If there were substantial fractionation involved, then it should be revealed in melting endotherms obtained by procedures similar to those used to obtain the results of Figure 6, as a peak below the annealing temperature decreasing with time but increasing with temperature. This is due to molecules which are unable to crystallize under the selected conditions, but do so on cooling. The effect is almost absent in PEEK so that explanations other than fractionation need to be sought for the second population. It has been suggested as arising from increasingly severe restrictions on crystallization as growth proceeds.

Identification of which lamellae are present in a sample, their location and molecular constitution is also highly relevant to the processes by which polymeric texture is established. It can be supplemented by a study of the interfaces of developing objects. Optimum conditions are when the temperature can be reduced sufficiently quickly that two stages of growth, before and after cooling, are easily identified. Because growth rates of polymers increase strongly with super-cooling, this is most easily accomplished by examining growth at low supercoolings, where the

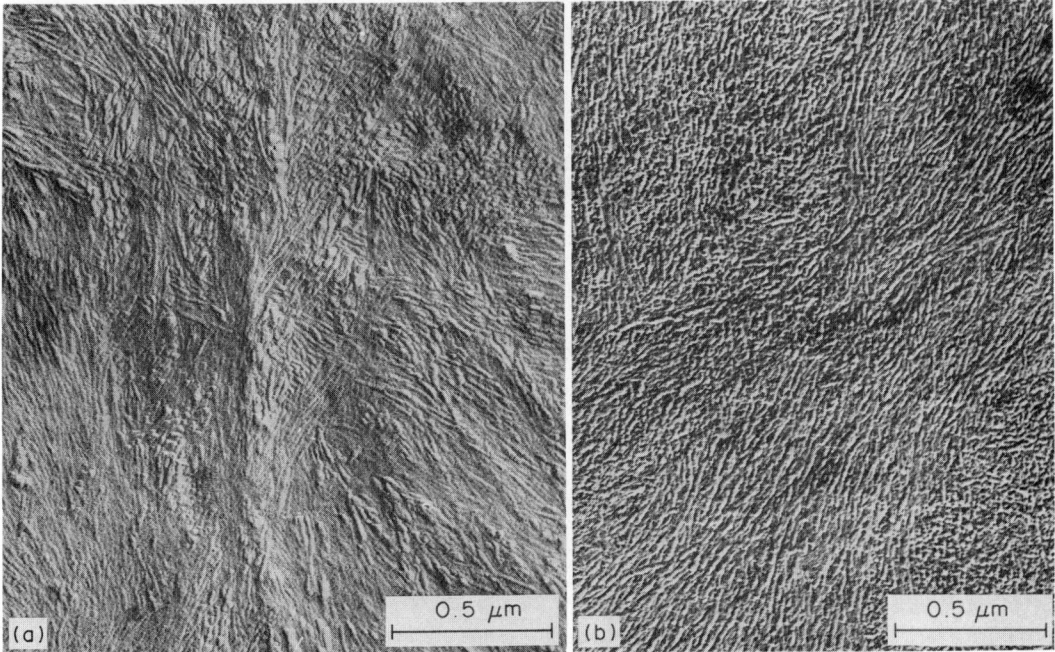

Figure 9 (a) An etched cut surface of PEEK crystallized at 330 °C so that it has two nearly equal peaks in its DSC melting endotherm. (b) The same sample but after material giving the lower peak had been melted, then quenched. The textures in (a) and (b) have the same separation of dominant lamellae but the intervening regions have been etched more deeply in (b) providing confirmation that the lower-melting material is located between dominant lamellae

objects produced are also the simplest. Much work has been done in this area, especially in relation to the development of spherulites. This is the subject of a separate chapter but certain points are worth emphasizing here as examples of how permanganic etching and electron microscopy can provide a firm basis of factual information.

It became apparent following the discovery of chain-folded lamellar crystallization that, saving arguments concerning chain-folding, polymer spherulites contain polycrystalline arrays of lamellae growing outwards with a common radial direction.[50] This is not a unique habit to organic polymers but the polycrystalline development stands in contrast to single-crystalline forms such as dendrites found in other circumstances. Texture was at and below the limit of optical resolution but could be coarsened by the addition of low mass or atactic species together called 'impurities'.[51] In these instances groups of lamellae formed which appeared, in projection, to be fibres. Fibres of multiple lamellae separated by impurities were considered as an intermediate level of order between lamellae and spherulites.[51] The branching and splaying of fibres would lead geometrically to an eventual spherical envelope. Permanganic etching has now revealed the actual situation in several polymers. Not surprisingly it is more complex than simple models had supposed and the concept of fibrous groups of lamellae as essential components of spherulites has not been substantiated.[41]

As an illustration of how permanganic etching provides new information, consider Figure 10. This was one of the first taken of a randomly chosen interior view of a melt-crystallized polymer. It appears to show fibres based on ridged crystals and was unexpectedly complex compared to fracture surfaces of the same sample. These simply showed lamellae radiating outwards (Figure 11) because a fracture surface is preferentially {200}, *i.e.* along ridges based on {201} facets. The view in a fracture surface is partial not only in orientation but also in its under-representation of thin lamellar populations, which are often obscured or avoided.[16] We may confirm this by etching a fracture surface when we find (Figure 12) that ridges lie parallel to it and thin lamellae become evident. The most informative view is down the growth direction, which can often be located relative to the centre of an off-diametral section of a spherulite or, as in Figure 7, by inspection. When one does this, it is evident, in the samples described here, that the impression of fibrous aggregates is an illusion. These spherulites grow as a framework of individual lamellae, with a ridged habit, leaving interstices of still-molten polymer. Solidification of the polymers in these interstices fills them with lamellae, of essentially the same thickness if grown isothermally, or of thinner ones in the regions where segregated species had failed to crystallize prior to quenching, *i.e.* towards the outer regions of

Figure 10   At the crystallization temperature (130 °C) this sample of linear polyethylene consisted of individual dominant ridged lamellae enclosing columns of melt

Figure 11   The appearance of a fracture surface of a linear polyethylene crystallized at 130 °C is of linear traces of radiating lamellae. Ridges as in Figure 10 are present but are not revealed because fracture is along their length

spherulites. The objects which branch and splay to establish spherulitic envelopes are not groups of lamellae but individuals, termed dominants. This qualitative difference is coupled with a substantial reduction in concentration gradients of segregated species when only one lamella rather than a group crystallizes in a comparatively large volume. Every system examined has shown this basic feature: branching of individual dominants, frequently identified with giant screw dislocations, and

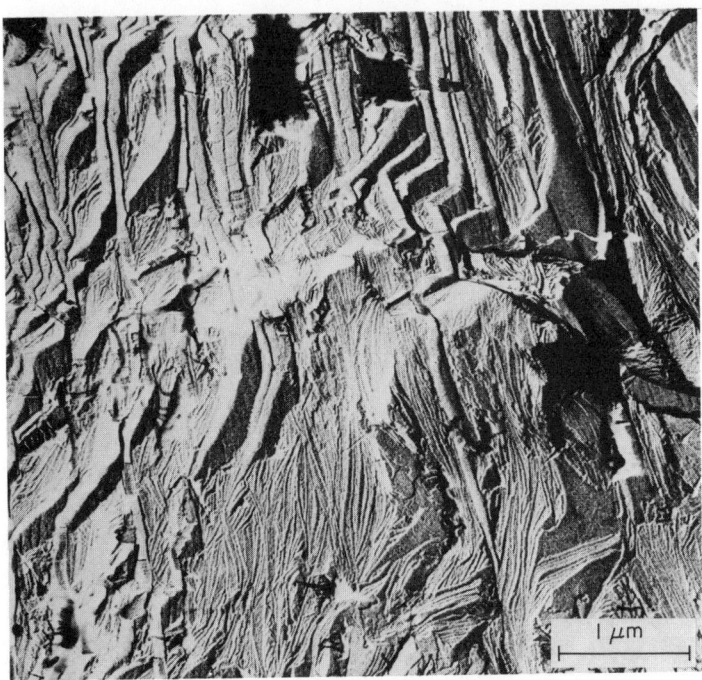

Figure 12 For linear polyethylene crystallized at 130 °C, confirmation of preferential fracture along the length of ridges is provided by etching a fracture surface similar to that in Figure 11. The result reveals the ridges lying parallel to the fracture surface

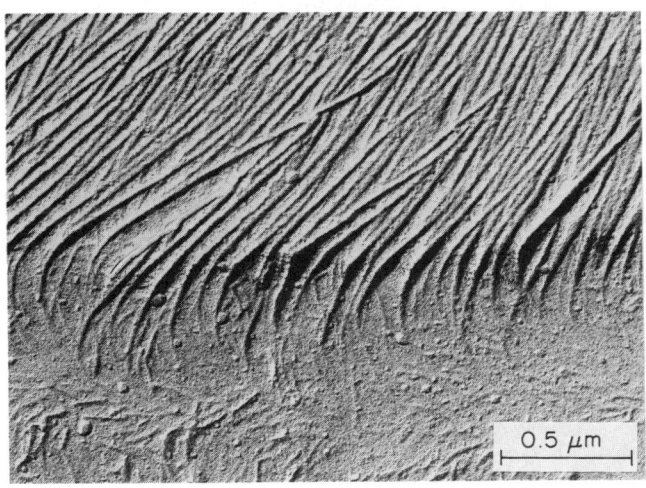

Figure 13 The crystal–liquid interface of crystallizing 1:1 blend of linear:branched polyethylene consists of separated individual dominant lamellae

splaying to give divergent orientation (Figure 13).[37,45,46,52,53] Nor should the term splaying be interpreted here too rigidly as giving only two-dimensional divergence: lamellae normally diverge in three dimensions, individual lamellae frequently bending and sometimes twisting in the process. Phenomenologically individual dominants often appear as if they repel each other[45,52] and it is of considerable interest to identify the origin of the forces involved. Two principal candidates are the modified conformations of cilia, *i.e.* uncrystallized portions of partly crystallized molecules[45] and the forces due to the drag of crystallizing molecules.[37] The flexibility of thin lamellae seems, however, to be a principle cause of their divergent orientations in spherulites. Of course if one adds uncrystallizable species or has a whole polymer with a considerable proportion of short molecules in its distribution, then mass balance will become important and it is no surprise that under these circumstances, but not otherwise, one begins to see aggregations of lamellae (Figure 14).[41]

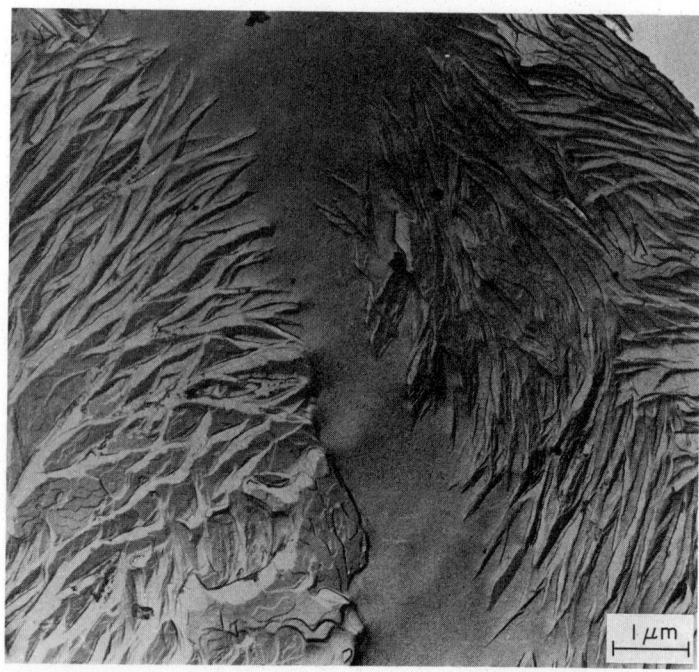

Figure 14 The grouping of lamellae so that they project as 'fibres' in the optical microscope only occurs for isotactic polystyrene at high levels of doping, *e.g.* a 1:1 blend of isotactic:atactic polymer crystallized at 190 °C

Illustrations of different etched polymers show the architecture of melt-grown polymers being established by branching and splaying of dominant lamellae as well as relevant individual detail. Thus for isotactic polystyrene (Figure 15) we observe that nucleation in a high molecular weight polymer $\bar{M}_w \approx 10^6$ occurs along threads, presumably of molecules stretched between entanglements,

Figure 15 Nucleation occurs along lines, sometimes visible as threads, in this sample of high molecular weight isotactic polystyrene. Similar effects occur in polyethylene

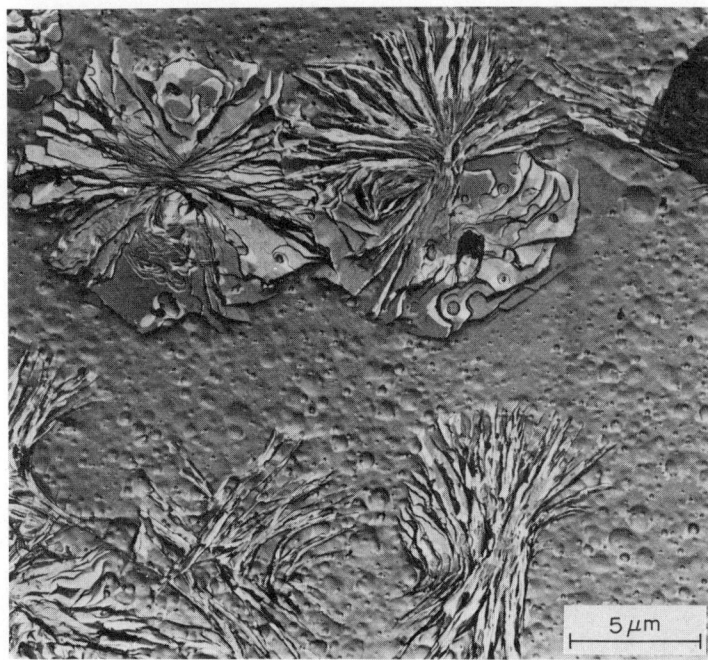

Figure 16 High symmetry polymers such as isotactic polystyrene develop radiating growth in three dimensions around individual dominants with equiaxed lamellae. Radial extension can be achieved with the involvement of screw dislocations

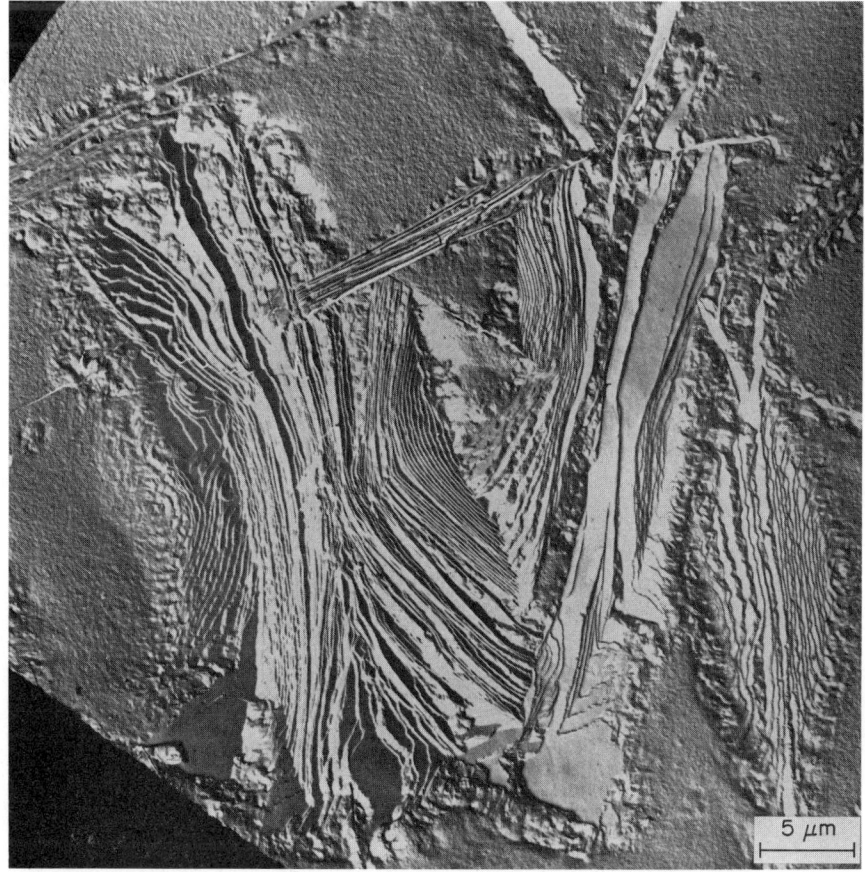

Figure 17 Developing growth of the tetragonal form of isotactic poly(4-methylpentene-1). Note the separated dominants, with splaying around an embryonic screw dislocation in the middle right, and the dishing of outer layers

but even here lamellae splay and bend apart. Individual lamellae show internal sector boundaries similar to those identified in solution-grown polymers.[30] When polystyrene lamellae develop into axialites and spherulites (Figure 16) they retain the basic hexagonal habit and extend radially with the involvement of giant screw dislocations. Neither in this hexagonal polymer nor in the tetragonal isotactic poly(4-methylpentene-1) has any degeneration and elongation of habit due to 'cellulation' been observed: indeed no explicit evidence for this phenomenon, supposed to be how fibres in spherulites develop,[51,54] has ever been produced. In both polymers splaying of dominants can be observed (Figure 17) and in isotactic polystyrene the separation of adjacent dominants, termed λ, have been measured as functions of relevant crystallization variables.[55] This has shown, *inter alia*, that only with heavy doping, for example with 50% atactic polymer, are groups of lamellae found which project optically as separate 'fibres' (*cf.* Figure 14).

Isotactic polypropylene provides further exemplification of these points. This polymer occurs in two crystal forms, monoclinic and hexagonal, which are immediately distinguishable in their appearance after etching. The hexagonal form (Figure 18) has extensive lamellae, and micrographs often show hexagonal etch pits round screw dislocations, whereas the monoclinic form is famous for its characteristic twinning known as crosshatching (Figure 19). At the very highest growth temperatures, $>155\,°C$, crosshatching is absent and growth proceeds by the splaying of lath-like lamellae.[52,56] At 155 °C and below twinning occurs with a common b axis and an 80° 40′ twin angle

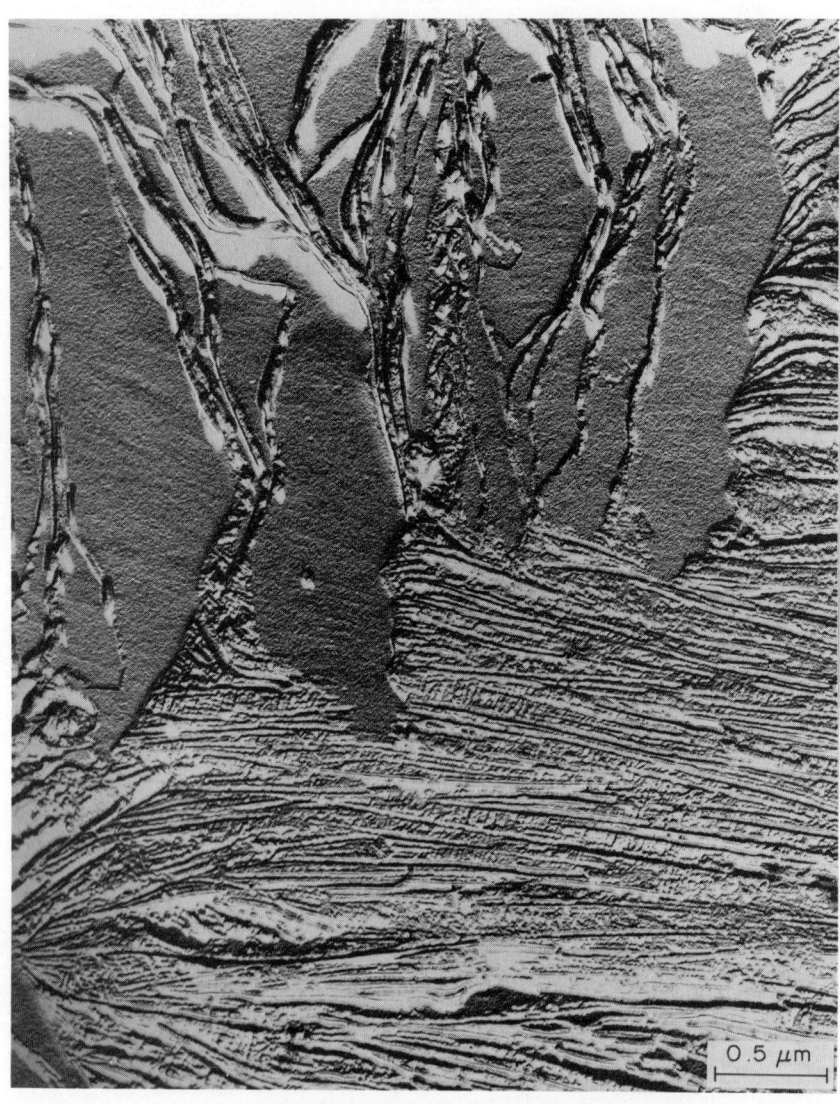

Figure 18   Two crystal forms of isotactic polypropylene. The upper layers are hexagonal, as their edges show, while those at the bottom are monoclinic. In this view we see radial dominants and subsidiary crosshatching in the outer regions of a spherulite

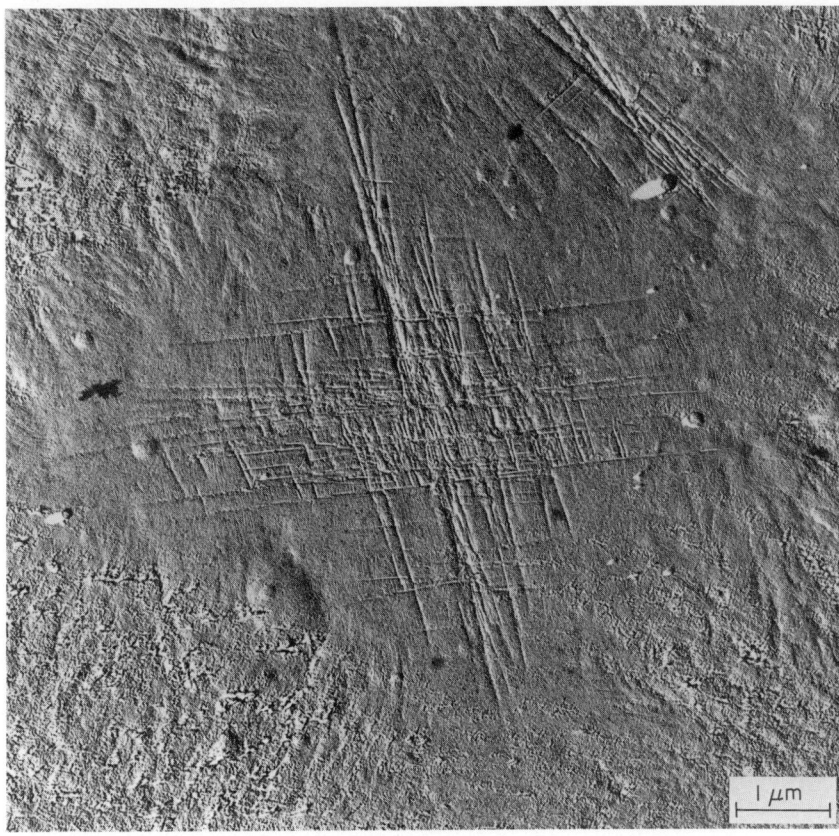

Figure 19 The central 'quadrite' of an isotactic polypropylene spherulite grown at 155 °C. The structure is constructed from individual lamellae with the twin axis (b) normal to the page

in the a^*c plane to give quadrites. These then develop as two superposed sets of laths, with splaying, bending and some twisting around their long axes to develop fully three-dimensional growth with a^* radial. Such objects have been grown to very large sizes (> 0.2 mm) and as such have shown that so-called axialites differ from spherulites primarily in their relatively slow randomization around the radial growth direction.[57]

35.2.3 Melt-crystallized Lamellae

The conventional picture of a polymer crystal lamella is that of a more or less uniform platelet with the molecular chain axis normal to it. Permanganic etching has shown that this is by no means always true. Polyethylene has been shown to occur in a structure in which repetition of several {201} facets forms a ridged sheet but this is found only within a very restricted range when the ratio of lamellar thickness:molecular length is approximately three.[15, 37] (Pockets of such ridged habit, due to segregated polymer, can often be identified in rapidly crystallized polyethylene homopolymer). Longer molecular lengths at low supercoolings give planar {201} lamellae, *i.e.* with molecules inclined at ~35° to normals, while at higher supercoolings dominant lamellae seen down the *b* axis are S- or C-shaped.[37] As first measured the S-shaped lamellae were based on near {201} surfaces separated by what must be strained surface regions where molecules are normal to lamellae. This index varied little if at all with crystallization temperature but greater curvature was found (Figure 20) in linear low density (butyl-branched) polyethylene. Evidently S- or C-profiled lamellae are strained. This can arise either from internal stresses (for which fold interactions have been suggested to be responsible)[58] or from external stresses of crystallization.[37] It is notable that only the dominants are curved: their associated subsidiaries appear most often to have planar {201} fold surfaces.[37] Moreover curvature can be induced by rapid quenching into normally planar lamellae, such as those of poly(4-methylpentene-1) (Figure 21). In either case, the shapes are novel and unsuspected evidence of details of growth processes, which permanganic etching has now made available for study.

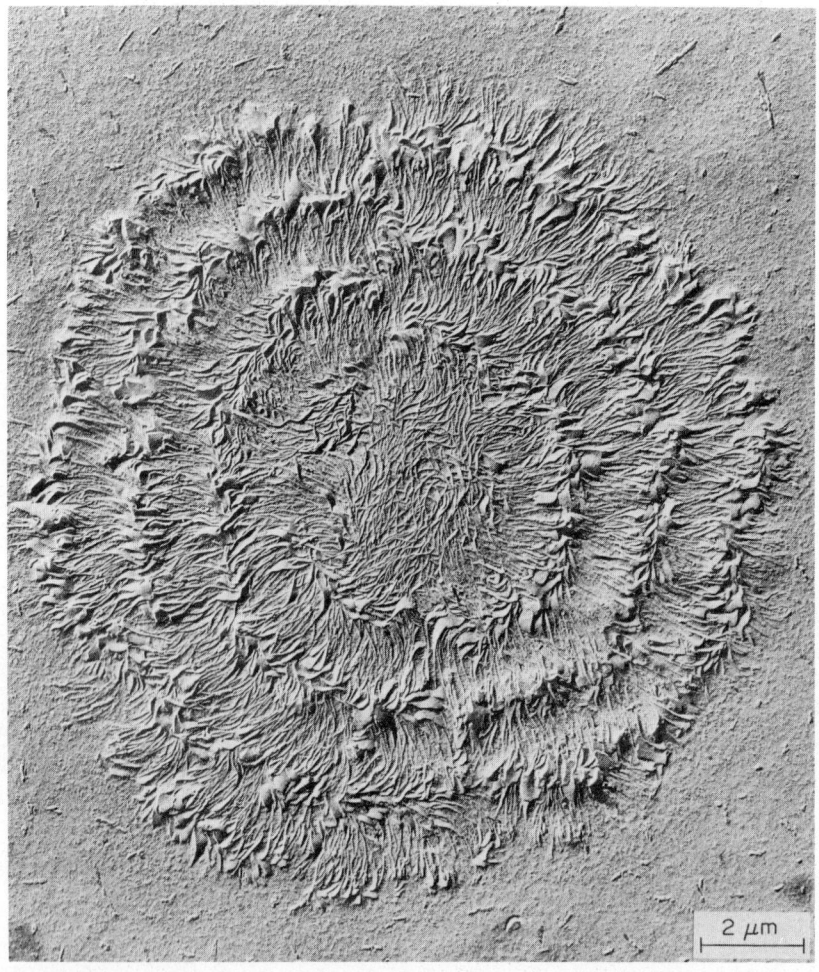

Figure 20   A banded spherulite. Linear low density polyethylene (butyl branched) crystallized at 113 °C. Note the curved lamellae

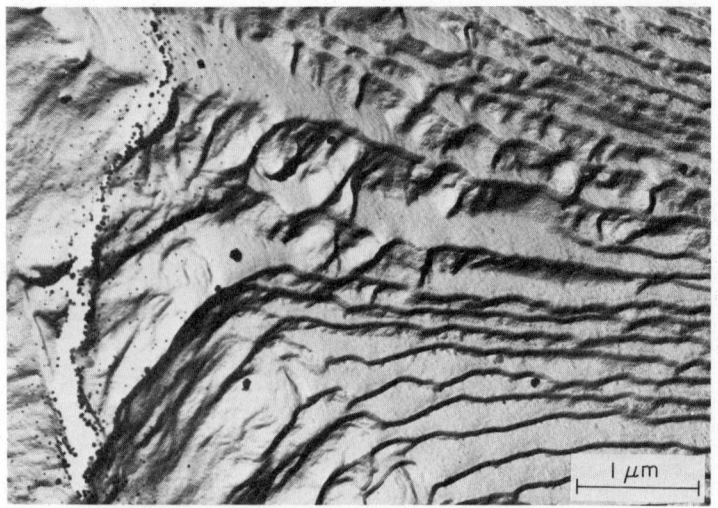

Figure 21   Separated, planar dominant lamellae have become curved on quenching in this sample of isotactic poly-(4-methylpentene-1)

All such observations add to our knowledge of what constitutes a lamella. It has recently been possible to go further and to isolate and study individual melt-crystallized lamellae in polyethylene.[59] This began with a study of so-called early objects, *i.e.* those which form first. Figure 22 shows one such revealed by permanganic etching. Its habit is elliptic, *i.e.* continuously curved, a matter of some interest in current theories of polymeric crystallization since this crystal grew well within regime I, *i.e.* with supposedly smooth growth surfaces.[60,61] We can also see something of the three-dimensional development around a screw dislocation, noting especially that successive layers are separated. Related experiments with isolactic poly(4-methylpentene-1) have shown a gradual change from smooth facets and a square habit to rough facets and circular lamellae with increasing crystallization temperature (Figure 23). Continuously curved lamellae have now been extracted from

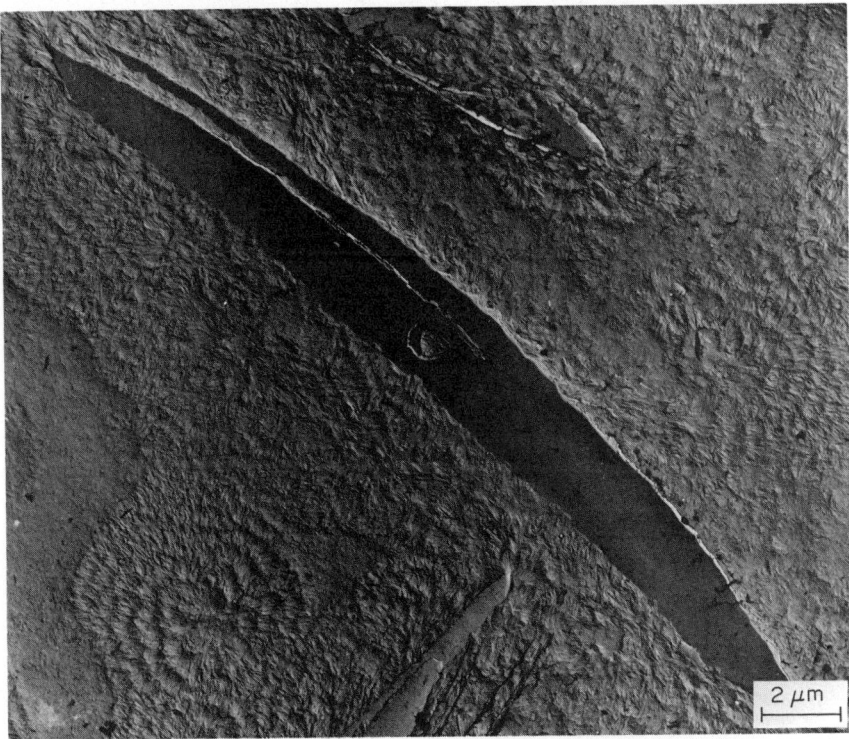

Figure 22 The beginnings of growth at 130 °C for the linear polyethylene of Figure 1. Successive layers of the spiral terrace around the central screw dislocation are not in contact

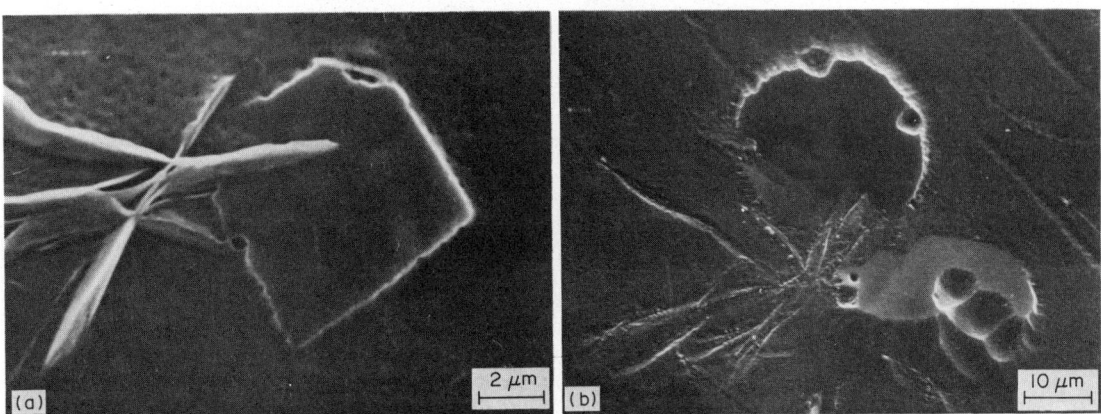

Figure 23 The changing crystal habit of isotactic-poly(4-methylpentene-1) with crystallization temperature (a) 210 °C, square, (b) 242.8 °C, circular. Scanning electron micrographs of etched surfaces

a quenched polyethylene by selective dissolution and sedimented on a carbon substrate in the manner familiar for solution-grown crystals. They then appear as in Figure 24 and can now be studied by electron diffraction and dark-field electron microscopy. The latter has revealed distinct sectors (Figure 25) indicating a degree of regularity in chain-folding and illustrating continuity with solution-grown lamellae. Of special interest is the asymmetric development of upper layers. These give the multilayered crystal an arrow-shaped appearance which is not present in the orthorhombic subcell structure. This is consistently associated with a particular geometrical condition relating successive layers, in these electron diffraction confirms that the average chain orientation is at *ca.* 35° to the lamella surface, *i.e.* the fold surfaces are close to {201}. Any chain axis will thus make

Figure 24   Crystals of linear polyethylene crystallized at 130 °C, as in Figure 22 but after dissolution from their matrix and sedimentation on a substrate. Note the asymmetric growth of the darker layers which are above and below the central lamella

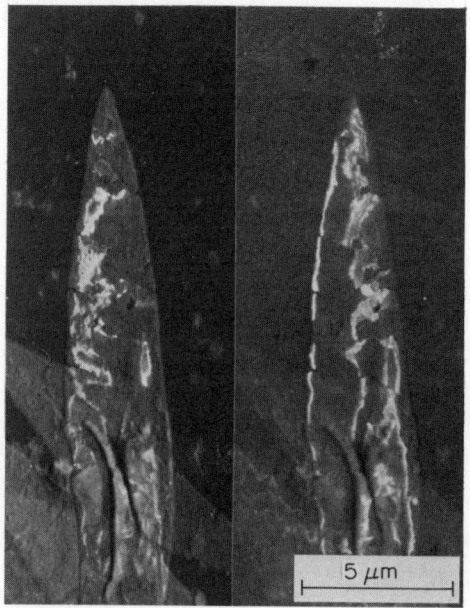

Figure 25   Dark field micrographs, taken through alternate (200) reflections, of a lamella of linear polyethylene prepared as described in Figure 24

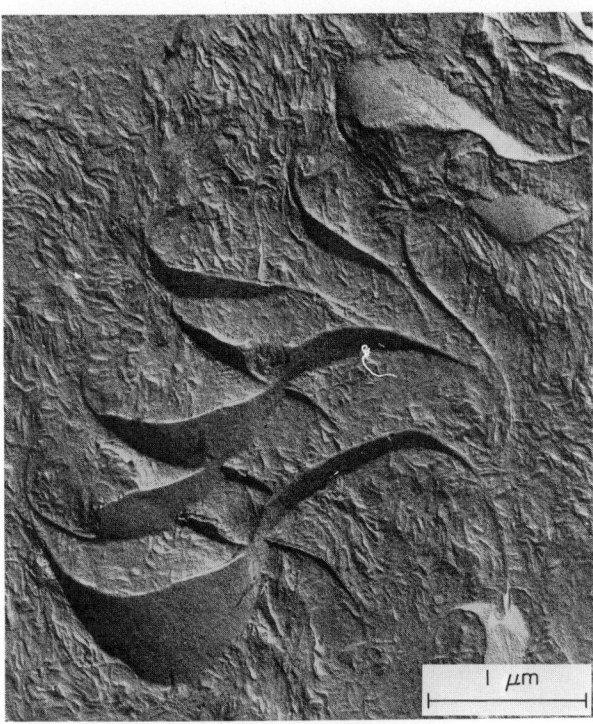

Figure 26 The beginnings of a banded spherulite of polyethylene developing around embryonic screw dislocations, each associated with differently oriented lamellae

supplementary acute and obtuse angles with the lamella surface. It is found that the inner straight edges of the dark terraces (see Figure 24), which is where growth has stopped, all have the acute angle between their growth face and the layer over which they are growing. In other words growth appears to have come to a halt because of some kind of generalized impingement which has limited the number of layers in the spiral terrace and also selected a consistent hand for the screw dislocation related to the sense of chain tilt. But this constraint must be released at some point because of the observed separation of lamellae. In this way an array of screw dislocations of the same hand is generated. Moreover each is associated with an increment of twist of the upper and lower layers of the spiral terrace, as is observed (Figure 26), because of 'repulsion' from the parent layer. It has been suggested[59] that this may be relevant to the way in which banded spherulites are formed.

Banded spherulites are eye-catching objects, the banding of which is a consequence of the mean chain orientation in the section being examined spiralling around the radius.[11] No convincing explanation of this phenomenon has yet appeared. Indeed there is still disagreement as to whether individual lamellae twist with the mean orientation or not. It is known that banded spherulites in polyethylene are based on S-profiled dominant lamellae,[29] as may be seen for a projection down the radius in the centre of a spherulite, although this is not a general phenomenon as lamellae in banded spherulites of α-poly(vinylidene fluoride) are planar.[62] Complementary chlorosulfonation has been used to show that an individual S is not substantially twisted around the radius.[29] Rather, as permanganic etching reveals, increments of twist occur at what, topologically, are screw dislocations of limited development. These features can be seen in Figures 20 and 26. In other words banded spherulites in polyethylene are constructed from a framework of S- or C-profiled dominant lamellae with the twist propagating non-uniformly around radial arrays of screw dislocations of consistent sign. The sign is the same as that reported above. It has, therefore, been suggested that a similar growth mechanism to that seen in Figure 24, and selection of screw dislocations of one hand, may be responsible for the growth of banded spherulites in polyethylene.[59]

35.2.4 Aspects of Deformed Morphologies

Polymers find many applications in oriented form and there is much interest in details of the various deformation processes involved. Permanganic etching can be used successfully in many of these areas although conditions for etching must change for high draw ratios. A major factor is the decrease in penetrability within a sample, which will hinder differential attack, although with nitric acid this so reduces the resulting friability that this reagent finds its major use as an etchant in drawn samples.[32] This changed penetrability with increasing draw is illustrated in Figures 27 and 28 for a sample containing banded spherulites drawn through a neck. Before the neck (Figure 27) good lamellar resolution is achieved; one can study the developing inhomogeneity of draw and observe that the original S-shaped profiles become planar. Beyond the neck or with increasing draw ratio resolution is impaired (Figure 28), although the legacy of the original bands is readily seen. Few attempts have as yet, been made to modify the etchant for drawn fibres. Even so, present formulations have already allowed detailed investigation of the fabrication of drawn polypropylene film.[63]

Detailed study has also been made of the uniaxial extension of pressure-crystallized samples to draw ratios as high as 12. These experiments were designed as models, partly to follow what happened to the thick original lamellae (whose thickness is not reproducible at atmospheric pressure) and partly to reveal the extent to which new lamellae were created. What was found was that a proportion, decreasing with draw ratio, of thick lamellae remained but became increasingly sheared and rotated towards the tensile axis. At the same time a transformed matrix of new lamellae developed, which could be revealed particularly clearly by selective heat treatment in the manner described earlier.[64]

There is thus already much scope for application of permanganic etching to reveal differences in morphology with deformation. In future they will need increasingly to be combined with additional evidence, specifically from diffraction studies, to relate molecular structure to lamellar and other textural elements, since the original relationship prior to deformation, often evident on inspection from symmetry, can no longer be relied upon. The involvement of diffraction is an area where there

2 μm

Figure 27 A banded spherulite of linear polyethylene after extension by a factor of 1.11. The profile of the lamellae has become more planar

Figure 28 At higher draw ratios less detail is revealed in banded spherulites when treated with the standard etchant although the residues of bands can still be discerned

have been successful preliminary experiments[7] but where further development of techniques is still required. For example an etched surface can be coupled with detachment replication[65] to remove the outer layer for diffraction study. Alternatively it is sometimes possible to remove some lamellar fragments from an etched surface in normal replication.[66] This can be useful in assessing orientation but one may well have no control over which area is being selected and not be able to obtain diffraction from a desired area. In principle one can thin down a sample to 100 nm or less, perhaps beginning with a 5 μm thick section, with the etching reagent to obtain surface morphology and diffraction information on the same sample. Alternatively it could be advantageous to combine diffraction data from thin sections with morphology from permanganic etching on equivalent, but not identical, portions of the same sample.

It is in such directions that the range of application of permanganic etching may well develop beyond what is now a large body of proven capability in previously inaccessible regions of real space. What has already been achieved confirms the substantial gains which are available when this new tool and a morphological approach are applied across the science and technology of crystalline polymers.

35.3 REFERENCES

1. O. Gerngross, K. Herrmann and W. Abitz, *Z. Phys. Chem., Abt. B.* 1930, **10**, 371.
2. E. Sauter, *Z. Phys. Chem., Abt. B*, 1932, **18**, 417.
3. C. W. Bunn and T. C. Alcock, *Trans. Faraday Soc.*, 1945, **41**, 317.
4. P. H. Till, Jr., *J. Polym. Sci.*, 1957, **24**, 301.
5. A. Keller, *Philos. Mag.*, 1957, **2**, 1171.
6. E. W. Fischer, *Z. Naturforsch., Teil A*, 1957, **12**, 753.
7. R. H. Olley, A. M. Hodge and D. C. Bassett, *J. Polym. Sci., Polym. Phys. Ed.*, 1979, **17**, 627.
8. R. H. Olley and D. C. Bassett, *Polymer*, 1982, **23**, 1707.
9. A. Keller, *Nature (London)*, 1952, **169**, 913.
10. D. C. Bassett, F. C. Frank and A. Keller, *Philos. Mag.*, 1963, **8**, 1739, 1753.
11. D. C. Bassett, 'Principles of Polymer Morphology', Cambridge University Press, Cambridge, 1981.
12. S. J. Spells and D. M. Sadler, *Polymer*, 1984, **25**, 739.
13. J. D. Hoffman, C. M. Guttman and E. A. DiMarzio, *Faraday Discuss. Chem. Soc.*, 1979, **68**, 177.
14. P. H. Geil, 'Polymer Single Crystals', Wiley, New York, 1963.
15. D. C. Bassett and A. M. Hodge, *Proc. R. Soc. London, Ser. A*, 1978, **359**, 121.
16. D. C. Bassett, B. A. Khalifa and R. H. Olley, *J. Polym. Sci., Polym. Phys. Ed.*, 1977, **15**, 995.

17. D. T. Grubb, *J. Mater. Sci.*, 1974, **9**, 1715.
18. A. J. Lovinger and D. D. Davis, *J. Appl. Phys.*, 1985, **58**, 2843.
19. A. J. Lovinger, *J. Polym. Sci., Polym. Phys. Ed.*, 1983, **21**, 97.
20. J. -F. Revol and R. St. J. Manley, *J. Mater. Sci. Lett.*, 1986, **5**, 249.
21. H. D. Chanzy, P. Smith and J. -F. Revol, *J. Polym. Sci., Polym. Lett. Ed.*, 1986, **24**, 557.
22. P. H. J. Yeung and R. J. Young, *Polymer*, 1986, **27**, 202.
23. A. W. Agar, F. C. Frank and A. Keller, *Philos. Mag.*, 1959, **4**, 32.
24. D. C. Bassett, *Philos. Mag.*, 1964, **10**, 595.
25. E. H. Andrews, *Proc. R. Soc. London, Ser. A*, 1964, **277**, 562.
26. B. C. Edwards and P. J. Phillips, *Polymer*, 1974, **15**, 491.
27. G. Kanig, *Kunstoffe*, 1974, **64**, 470.
28. A. M. Hodge and D. C. Bassett, *J. Mater. Sci.*, 1977, **12**, 2065.
29. D. C. Bassett and A. M. Hodge, *Polymer*, 1978, **19**, 469.
30. D. C. Bassett, in 'Developments in Crystalline Polymers', vol. 2, ed. D. C. Bassett, Applied Science, London, 1988, p. 67.
31. R. P. Palmer and A. J. Cobbold, *Makromol. Chem*, 1964, **74**, 174.
32. K. Sakaoku and A. Peterlin, *J. Macromol. Sci., Phys.*, 1967, **1**, 103.
33. A. Keller and S. Sawada, *Makromol. Chem*, 1964, **74**, 190.
34. R. H. Olley, *Sci. Prog. (Oxford)*, 1986, **70**, 17.
35. A. M. Freedman, D. C. Bassett, A. S. Vaughan and R. H. Olley, *Polymer*, 1986, **27**, 1163.
36. A. M. Freedman, D. C. Bassett and R. H. Olley, *J. Macromol. Sci., Phys.*, 1988, **27**, 319.
37. D. C. Bassett and A. M. Hodge, *Proc. R. Soc. London, Ser. A*, 1981, **377**, 25, 61; D. C. Bassett, A. M. Hodge and R. H. Olley, *Proc. R. Soc. London, Ser. A*, 1981, **377**, 39.
38. P. J. Flory, *J. Am. Chem. Soc.*, 1962, **84**, 2857.
39. D. Y. Yoon and P. J. Flory, *Faraday Discuss. Chem. Soc.*, 1979, **68**, 177.
40. D. C. Bassett, A. M. Hodge and R. H. Olley, *Faraday Discuss. Chem. Soc.*, 1979, **68**, 218.
41. D. C. Bassett, *CRC Crit. Rev. Solid State Mater. Sci.*, 1984, **12**, 97.
42. A. Mehta and B. Wunderlich, *Colloid Polym. Sci.*, 1975, **253**, 193.
43. D. C. Bassett and D. R. Carder, *Philos. Mag.*, 1973, **28**, 513.
44. Ref. 11, Figures 9.1 and 9.2.
45. D. C. Bassett and A. S. Vaughan, *Polymer*, 1985, **26**, 717.
46. D. C. Bassett, R. H. Olley and I. A. M. al Raheil, 1988, submitted to *Polymer*.
47. D. J. Blundell and B. N. Osborn, *Polymer*, 1983, **24**, 953.
48. S. Z. D. Cheng, M. -Y. Cao and B. Wunderlich, *Macromolecules*, 1986, **19**, 1868.
49. P. Cebe and S. -D. Hong, *Polymer*, 1986, **27**, 1183.
50. A. Keller, in 'Growth and Perfection of Crystals', ed. R. H. Doremus, B. W. Roberts and D. Turnbull, Wiley, New York, 1958, p. 499.
51. H. D. Keith and F. J. Padden, *J. Appl. Phys.*, 1963, **34**, 2409.
52. D. C. Bassett and R. H. Olley, *Polymer*, 1984, **25**, 935.
53. D. Patel and D. C. Bassett, unpublished results.
54. H. D. Keith, *J. Polym. Sci., Part A*, 1964, **2**, 4339.
55. A. S. Vaughan, D. C. Bassett and R. H. Olley, in 'Morphology of Polymers', ed. B. Sedlacek, de Gruyter, Berlin, 1986, p. 387.
56. F. L. Binsbergen and B. G. M. de Lange, *Polymer*, 1968, **9**, 23.
57. R. H. Olley and D. C. Bassett 1988, submitted to *Polymer*.
58. H. D. Keith and F. J. Padden, *Polymer*, 1984, **25**, 28.
59. D. C. Bassett, R. H. Olley and I. A. M. al Raheil, 1988, *Polymer*, 1988, **29**, 1539.
60. J. D. Hoffman, G. S. Frolen, G. S. Ross and J. I. Lauritzen, *J. Res. Nat. Bur. Stand., Sect. A*, 1975, **79**, 671.
61. D. M. Sadler, *Nature (London)*, 1987, **326**, 174.
62. D. C. Bassett, in 'Morphology of Polymers', ed. B. Sedlacek, de Gruyter, Berlin, 1986, p. 47.
63. R. H. Olley and D. C. Bassett, unpublished results.
64. Ref. 11, Figure 9.15.
65. D. C. Bassett, *Phillos. Mag.*, 1961, **6**, 1053.
66. D. R. Norton and A. Keller, *Polymer*, 1985, **26**, 704.

36

Thermal Analysis

MICHAEL J. RICHARDSON

National Physical Laboratory, Teddington, UK

36.1 INTRODUCTION

Thermal analysis is defined[1] by ICTA (the International Confederation for Thermal Analysis) as a 'term covering a group of techniques in which a physical property of a substance and/or its reaction product(s) is measured as a function of temperature'. A generous interpretation of this encompasses a huge range of techniques and this chapter is confined to the specific areas of differential thermal analysis (DTA) and its more quantitative development, differential scanning calorimetry (DSC, this abbreviation is also used for the calorimeter). These, together with thermogravimetry (Volume 1, Chapter 37) are currently the most widely used methods of thermal analysis in polymer science, although dynamic mechanical methods are rapidly increasing in popularity.[2]

DTA/DSC curves reflect changes in the energy of the system under investigation — changes that may be chemical or physical in origin. The technique is therefore particularly useful for polymers because polymerization or structural changes are almost invariably accompanied by energetic effects so that crystallization and melting, curing and other reactions, and the glass transition all show characteristic DSC curves. Small samples (a few mg) and rapid experimentation (heating, $q(+)$, or cooling, $q(-)$, rates of up to 50–100 K min^{-1} are common) mean that thermal analysis finds applications in both research laboratories and routine quality control.

DTA has long been used for the characterization of rocks and minerals. The sample (denoted by subscript s) and an inert reference (subscript r) material are heated at a given rate and the signal is the differential temperature ($\Delta T = T_s - T_r$) between them as recorded by thermocouples embedded in the materials (Figure 1a). The initial offset (S, Figure 2) is a summation of the several material and instrumental factors (conductivity, heat capacity, material and packing densities) that govern the rate of heat transfer between the source and both sensors. At a phase change the sample takes in heat of transition and the temperature rise is temporarily halted with a consequent rise in the magnitude of ΔT, which eventually returns to a different 'steady state' line concomitant with the changed physical properties of the new phase (Figure 2). The location and magnitude of such events provides a useful 'fingerprint' for many materials and may, under favourable conditions, yield semiquantitative information. The results discussed in one of the first reviews of DTA devoted to polymers[3] were obtained with this type of instrument, often home-made. A major problem, however, is the

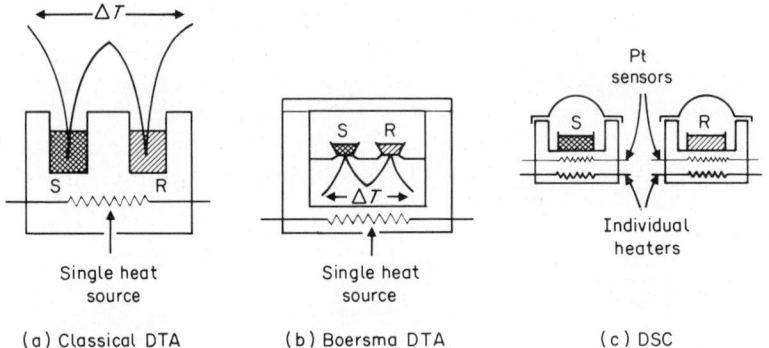

Figure 1 The three types of DTA: (a) conventional DTA; (b) quantitative (Boersma) DTA/heat flux DSC; and (c) power compensation DSC (reproduced by courtesy of Perkin Elmer from Thermal Analysis Newsletter No. 9, 1970)

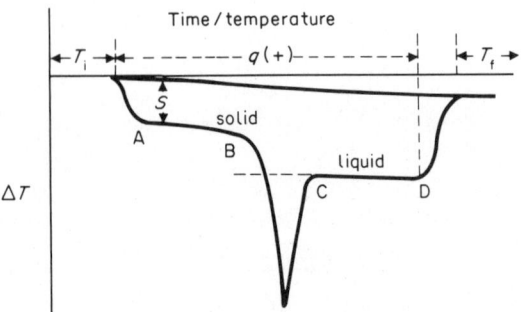

Figure 2 The melting curve by conventional DTA or heat flux DSC. Quasisteady state regions are AB, CD. T_i and T_f are isothermal regions shown with exaggerated offset to illustrate the baseline used to define the signal S

difficulty in obtaining a reproducible thermal path between heat source and sensors: powders are troublesome enough to pack consistently and the task is impossible for irregular lumps. In addition there may be a discontinuity at the melting point if the sample pulls away from the thermocouple, giving a meaningless signal — alternatively it may adhere strongly, with obvious implications for subsequent cleaning operations.

The present generation of DTA instruments dates back to the work of Boersma[4] who discussed the design parameters required when the sample is removed from the heater–sensor path and there is a controlled heat leak between sample and reference cells (Figure 1b). Later advances in signal amplification and linear temperature programming were such that it became practicable to use milligram-sized samples although the sophisticated electronics required meant that commercial instrumentation became increasingly common and this is very much the current situation.

The expression 'differential scanning calorimeter' was originally applied to an instrument of different design in which the signal was a differential power ($\Delta P = P_s - P_r$), *i.e.* that required to keep both sample and reference at the same programmed temperature.[5,6] Separate heaters must be provided (Figure 1c) but a very advantageous feature is that the purely electrical nature of the measurements means that the conversion (or calibration) factors to heat capacity or enthalpy (Section 36.3.2.2) are independent of temperature. The corresponding quantities for a Boersma-type DTA show considerable temperature dependence although this handicap is now generally overcome by some form of electronic linearization. It is now customary to refer to all potentially quantitative instruments as differential scanning calorimeters although the distinction can be made between 'heat flux' (ΔT) and 'power compensation' (ΔP) DSC. Subsequent sections of this chapter will not generally distinguish between the two types; references to DTA will imply that applications are concerned more with the temperatures at which events occur than with their magnitudes. Potential confusion arises from the opposite signs of ΔT and ΔP for similar events. At T_m, for example, $T_s < T_r$ and $\Delta T < 0$ whereas, because latent heat must be supplied, $P_s > P_r$ and $\Delta P > 0$. Standardized conditions have been recommended by ICTA but, as either signal is related to a heat capacity (Section 36.2), it is unfortunate that sight is lost of its physical significance.

36.2 THEORY

The detailed theory of DSC[4-8] is mainly of interest to instrument makers although there are important implications for those concerned with the melting of low molecular weight (MW) materials. For these, the observed shape of the melting curve must be corrected because heat of fusion has to be supplied over a very narrow range of temperature and there are gross perturbations to the 'steady state' conditions of the solid and liquid states (AB and CD in Figure 2). The melting of polymers is a more gradual process and a simpler theory is adequate although this still serves to indicate the factors that influence the observed signal.

Starting from an isothermal baseline the signal on heating (S, Figure 2) is related to the overall difference in the 'thermal capacity' of the sample and reference cells. The 'thermal capacity' may be thought of as being made up of two components, one of which (ΔW) is related to such material constants as mass and heat capacity and the other of which (I) is a very general term that groups together the various expressions for heat exchange between the two cells and their surroundings (Figure 3). Both ΔW and I are temperature dependent and, most important, should be reproducible from one run to another. With this proviso

$$KS = \Delta W + I \tag{1}$$

where K is a factor that is further discussed below. If three experiments are performed: (i) empty

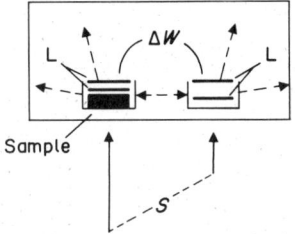

Figure 3 DSC cells showing the differential heat capacity $\Delta W(= \Delta W_e + mC_p)$ and potential routes ($-- \to$) for heat loss (I). The lids L help to keep I reproducible from one run to another. The signal S is related to $\Delta W + I$

calorimeter (subscript e); (ii) calorimeter + sample (subscript s); and (iii) calorimeter + calibrant (subscript c), and m and C_p are mass and specific heat respectively then

$$KS_e = \Delta W_e + I \tag{2}$$

$$KS_s = \Delta W_e + m_s C_{ps} + I \tag{3}$$

$$KS_c = \Delta W_e + m_c C_{pc} + I \tag{4}$$

$$C_{ps} = \frac{(S_s - S_e)m_c}{(S_c - S_e)m_s}C_{pc} = \frac{K(S_s - S_e)}{m_s} \tag{5}$$

Equation (5) is obtained by subtracting (2) from (3) and from (4) and dividing $(3) - (2)$ by $(4) - (2)$. The second form of (5) shows that K is the ordinate-to-C_p calibration factor. The three operations leading to (5) are not normally found in DSC work on polymers, nor are they essential, but they are deliberately introduced here to emphasize important aspects of the technique. Firstly there is the purely practical feature that conditions should be kept as well defined as possible (reproduciblity of $\Delta W_e + I$) otherwise K, or its area-to-enthalpy analogue (Section 36.3.2.2) will vary. Secondly, since thermal events in polymers are relatively sluggish, they can be thought of as regions of unusual C_p behaviour, integration of which gives the overall enthalpy change for the event. This is the key to the successful application of DSC in the study of polymer glasses, crystallinity, crystallization and melting, and cure. In whichever way DSC curves are analyzed, some form of baseline is essential. This is represented by S_e in equation (5), but a separate curve is not always needed. The broken curve of Figure 2 shows an internal reference, the extrapolated liquid, and other possibilities are considered later in the chapter.

36.3 INSTRUMENTATION AND EXPERIMENTAL ASPECTS

Figure 1(a)–(c) shows the transition from classical DTA to power compensation DSC. Between these limits are several modified sensor systems[9] (Figure 4), which are often associated with the

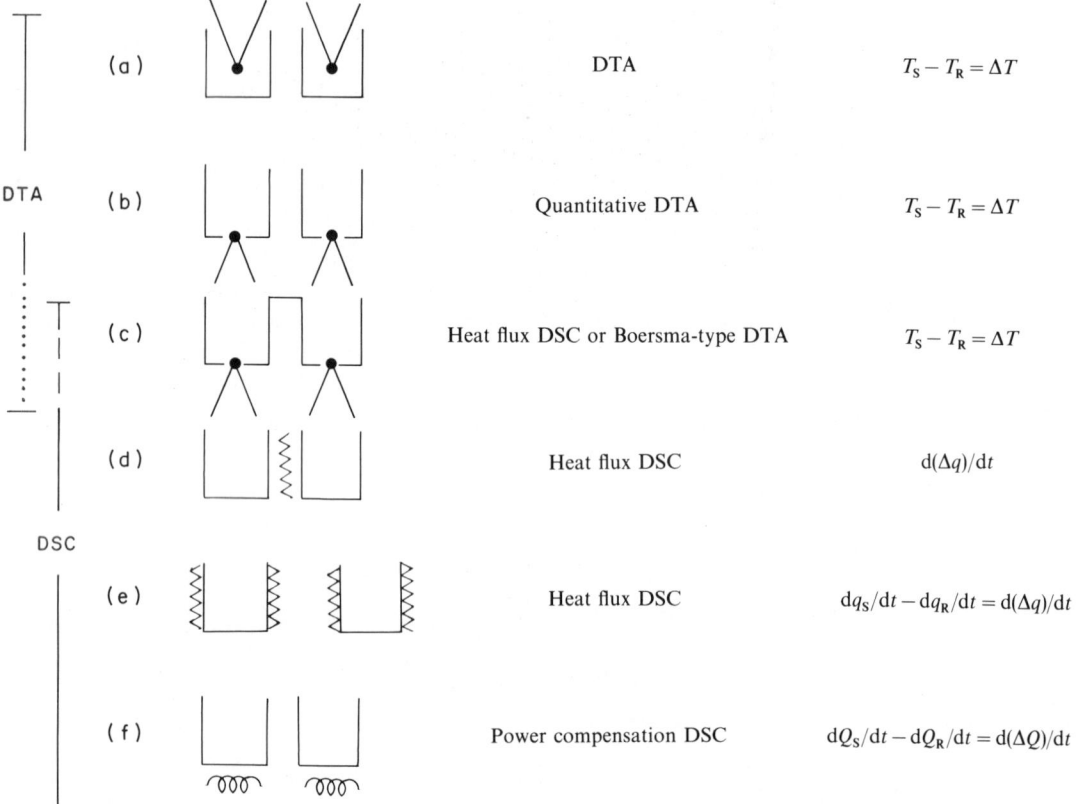

Figure 4 Sensor configuration showing the transition from conventional DTA to power compensation DSC (reproduced by courtesy of the Royal Society of Chemistry from ref. 9)

names of specific manufacturers. Du Pont, Stanton Redcroft and Netzsch produce Boersma-type calorimeters (Figure 4c), a five-junction thermopile (Figure 4d) is used in Mettler instruments, the Setaram series is based on the Calvet principle (Figure 4e) and power compensation DSC (Figure 4f) is always associated with the Perkin Elmer Company. Thermocouple instruments are also made by Shinku–Riku and Shimadzu, with another Japanese company, Seiko, using a thermopile detector in their calorimeters. While there are naturally differences of detail between individual instruments, certain features are fairly standard: operating temperatures from 100–150 K to 700–1000 K (a region of purely thermogravimetric interest so far as most polymers are concerned); heating and cooling rates of from a few tenths of a degree to many tens of degrees per minute, the extremes are normally used to impose some thermal history on the sample (annealing or quenching), 10–20 K min^{-1} are the most commonly used rates for routine work; and a signal-to-noise ratio that gives satisfactory results with sub-milligram quantities of polymer but can still cope with 20–40 mg samples, as size decreases the selection of a truly representative sample becomes increasingly difficult.

36.3.1 Sample Preparation

Polymers are typically contained in aluminum pans that hold up to about 50 mm^3. Pans should be lidded so that each presents an approximately constant external environment to the cell. Without this precaution the term I of equations (1)–(4) and Figure 3 will be a function of the contents with extreme differences coming at high temperatures with, for example, carbon black and titania-filled polymers. Polymer–solvent systems may be studied using sealable pans that will withstand pressures of a few atmospheres; because of the small quantities involved it is always advisable to check the composition of such systems at the end of any series of experiments by piercing the pan, evaporating the solvent and reweighing. In principle any form of sample (film, powder, granular, or liquid) is acceptable but irregular chunks will obviously give increased thermal lag (Section 36.3.2.1) and when possible such material (*e.g.* pellets) should be given a flat base by cutting with a scalpel followed by light rubbing on a fine grade of abrasive paper. Such precautions minimize the distorted signal that often occurs on first taking a polymer above the glass temperature (T_g) or melting point (T_m), when flow first becomes possible. This is especially relevant for drawn material which retracts on heating, sometimes giving gross artefacts. Procedures have been described to overcome such effects: if retraction is to be discouraged, the film can be clamped[10] and fibres can be wrapped around a spool with the ends tied;[11] if the relaxed form is sought, material can be subdivided or encapsulated with metal powder[10] or silicone oil[12] as heat transfer media. For convenience of handling it is tempting to compact freeze dried polymer by compression in a small die; this, however, must be done with great caution because it is easy to store energy in polymers by this procedure (Section 36.5.4). If 'as received' material is not of specific interest, sample preparation is much simplified by preheating in the DSC to the maximum temperature expected in subsequent runs, when any flow or relaxation can take place before measurements are started.

36.3.2 Calibration

Most generally the abscissa of a DSC curve is a time, but except for isothermal work it is most usefully shown as a temperature using the relevant heating or cooling rate. Equation (5) shows how the ordinate is related to heat capacity (C_p) which on integration over a range of temperature gives an enthalpy change. Quantitative operation of a DSC therefore requires calibration with respect to temperature, heat capacity and/or enthalpy. These are discussed below from a somewhat different viewpoint to that usually recommended in the manufacturers' manuals. Particular emphasis is placed on simple thermodynamic aspects (common to most forms of calorimetry), which are often neglected and have led to the publication of many 'heats of fusion', for example, that are devoid of any physical significance.

A recent IUPAC publication[13] has useful discussions of reference materials that are suitable for the calibration of DTA and DSC equipment.

36.3.2.1 Temperature

This is most usefully considered as being made up of two terms;[14] an isothermal correction (δT_i) to the indicated temperature (T_{in}) and a thermal lag (δT). δT_i is simply obtained from the stepwise

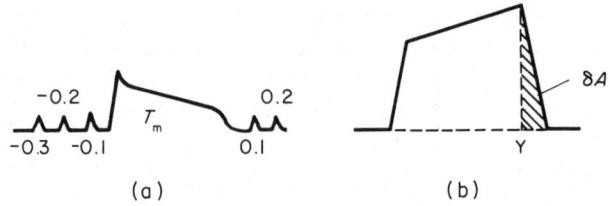

Figure 5 Temperature calibration: (a) isothermal; and (b) scanning, showing the 'enthalpy lag' δA at the end of a run

melting[15] of a material of known T_m (Figure 5a); pure metals melt over less than 0.1 K and $\delta T_i = T_m - T_{in}$ can be defined with this precision. Because thermal lag depends on the heating or cooling rate, sample size and geometry, and physical properties, an individual determination is desirable. This can be achieved by using the 'enthalpy lag'[14] at the end of a run (Figure 5b). In the absence of lag, there would be an immediate return to the isothermal baseline at Y and thermal lag is given by $m_x C_{px} \delta T = F \delta A$, where F is the area-to-enthalpy calibration factor (Section 36.3.2.2) and δA is defined in Figure 5b. The determination of δT under a variety of conditions gives a useful insight into heat transfer to polymeric materials. Figure 6 shows δT as polished 0.6×6 mm discs of medium and high MW polystyrene (PS) are thermally cycled through T_g. 20 000 MW PS flows above T_g and makes good sample/pan contact; cooling initially retains this contact (bracketed, Figure 6a) but on further cooling the glassy sample eventually parts from its pan (an event that is often revealed by noise in the signal) to give an increased δT. Very high MW PS does not flow on the timescale of an average DSC experiment, merely relaxing above T_g to give a slightly uneven surface (and increased δT), which persists throughout subsequent thermal cycling although thermal contact is still better in the molten (rubbery) state. The variation in δT shown in Figure 6 is much greater than the uncertainty in δT_i and is ignored in the conventional temperature calibration, which uses the extrapolated onset of melting of a pure material (preferably a metal). Such a calibration actually corresponds[14] to δT in the limit of a vanishingly small sample; it is also only valid for heating experiments because nucleation problems mean that crystallization temperatures are poorly reproducible in a DSC. Care is clearly needed when claiming precise temperatures in a scanning experiment. Additional problems are posed by superheating effects, which are specific to certain structures, and these are considered in Section 36.6.2.

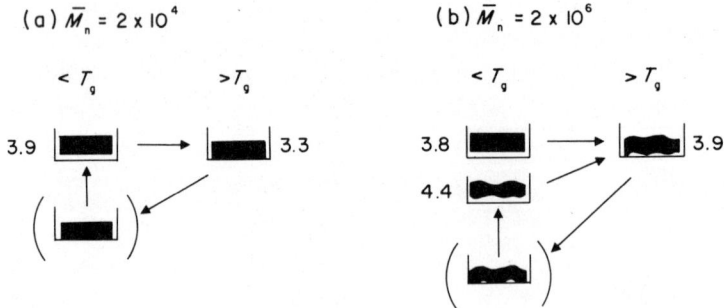

Figure 6 Thermal lag at 20 K min⁻¹ of polystyrene discs of MW (a) 20 000 and (b) 2 000 000 on cycling in the DSC; the lag (K) is shown

36.3.2.2 Heat capacity and enthalpy

Detailed descriptions are available for the experimental procedures and data treatment required when DSC is being used to measure heat capacities.[16-18] The main computational problem is to allow for variations in the slope of the isothermal baselines, which may be random or progressive; the latter implies long term instrumental instability, or drift, or the use of samples which may be losing some volatile component that condenses on to the ambient surroundings, especially if these are cooled. The calibration factor K is not usually calculated as a specific component of equation (5) but its behaviour as a function of temperature and from day to day gives a good indication of instrumental performance. It should, for example, be independent of temperature for both power compensation (a theoretical requirement) and heat flux (if successfully linearized) DSC. Some deviations from 'constant' conditions can be accepted because the final quantity C_p (equation 5) is a ratio of signals so that systematic errors will tend to cancel. A good criterion of instrumental performance, which takes in both ordinate and abscissa calibrations, is the degree to which C_p can be reproduced from both heating and cooling experiments. These should be made in thermally inert regions (where hysteresis effects are absent) and molten polymers or glasses well below T_g make ideal test materials.

Calibration with respect to enthalpy requires an area that corresponds to a well defined enthalpy change — a heat of fusion $\Delta H(T_m)$ is commonly used, that of indium being especially popular. A calorimeter, however, can only respond to the overall energy demands of its contents and an observed melting curve will almost certainly contain heat capacity contributions as well as $\Delta H(T_m)$. Resolution into the appropriate components is often difficult, especially for polymers, but the correct procedures were fully described in early work concerned with adiabatic calorimetry.[19] Three constructions which define areas appropriate for calibration purposes are shown in Figure 7. Both Figure 7(a) and 7(b) require additional data for the empty pan, and this is only avoided in Figure 7(c) by using the supercooled liquid as the baseline (equation 19). It is much safer to generate the curve for a supercooled liquid experimentally, as shown, because it may be distorted by inherent instrumental curvature (which is normally compensated by equivalent curvature for the empty pan).

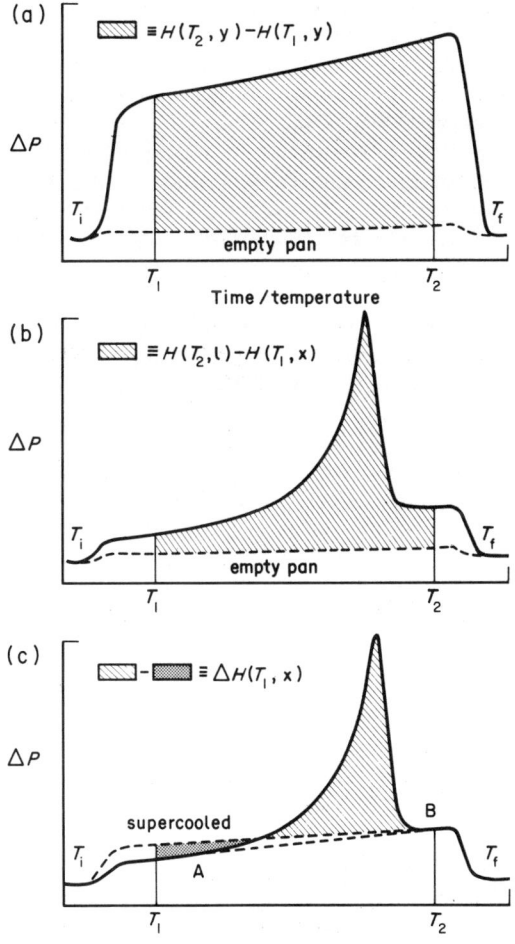

Figure 7 Area (enthalpy) calibration: (a) heat capacity only; (b) heat capacity + fusion; and (c) fusion using internal calibration (supercooled liquid); y represents the phase: x (crystalline), g (glass); or l (liquid). Division of the relevant enthalpy change by the corresponding area (shaded) gives the calibration factor F

The enthalpy change represented in Figure 7(a) is of the form $C_p\Delta T$ only, while in 7(b) the fusion term is predominant (the ΔP scale may be taken to be $\times 10$–100 that of 7a). Subtraction of curve 7(a) from 7(b) gives 7(c), a true heat of fusion, but only at some arbitrary temperature T_1, which is preferably far enough below T_m to avoid the premelting region. The correction of $\Delta H(T_1)$ from T_1 to T_m depends on $\Delta C_p = C_{pl} - C_{ps}$ and may easily amount to a few per cent. The baselines normally used are simple geometrical constructions (*e.g.* AB, Figure 7c) having no particular meaning and which make this 'calibration' process very questionable. There is no need for such ambiguity, as thermodynamically well defined materials that are suitable for all types of calibration suggested by Figure 7 are available.[13] A true calorimeter should give identical F values (within experimental error) whatever calibration scheme is used. If the chosen scheme is based on unsound foundations, the apparent F may still reproduce satisfactorily on a day-to-day basis, showing instrumental stability, but will disagree with values found by other procedures. Modern instrumentation is such that F and K should be reproducible to $\pm 1\%$ on a fairly routine basis.

36.4   DSC AND POLYMERS: GENERAL COMMENTS

Specific applications of DSC to polymers will be discussed in the following sections. The present section presents a general picture that emphasizes the calorimetric nature of the technique. Quantitative aspects have already been highlighted because the thermodynamic analysis which usually follows any calorimetric experiment demands well founded data. By contrast, many reported applications of DSC use the temperature scale alone, *i.e.* when an event occurs and how it changes in response to physical or chemical treatments. This is perfectly acceptable for quality control purposes or for establishing broad generalizations but it does not realize the full potential of the technique and there is no reason why this should be so with modern instrumentation. Microprocessors are standard items and remove the drudgery of the many routine calculations that are required to produce heat capacity curves. These, in turn, may be transformed to the enthalpy curves that are the foundations of so many quantitative applications of DSC.

Figure 8 shows DSC curves for a quenched and a partially crystalline polymer. The former is a familiar sight in most manufacturers' brochures illustrating, as it does, the principle applications of DSC to polymers. Main uses are confined to the temperature region between T_g and T_m. The glass transition region appears as a step of variable magnitude (very often with attendant fine structure); above T_g recrystallization or annealing are indicated by exothermic processes and these are followed by melting peaks, the location and magnitude of which contain information on crystal size and quality. Reactions, usually exothermic, are qualitatively similar to recrystallization although they generally extend over several decades of temperature. There is an extensive literature on most of these events and this will be referred to *via* reviews and the more recent papers. The most comprehensive and useful book, containing a wealth of information for a variety of topics, is that edited by Turi.[20] Thermal analysis is reviewed biennially in the Annual Review issues of Analytical Chemistry (even years/volume numbers) with applications to polymers, rubbers and coatings being considered in the alternating odd years. The proceedings of the various national, regional and international thermal analysis societies are often available as extended abstracts. The highest

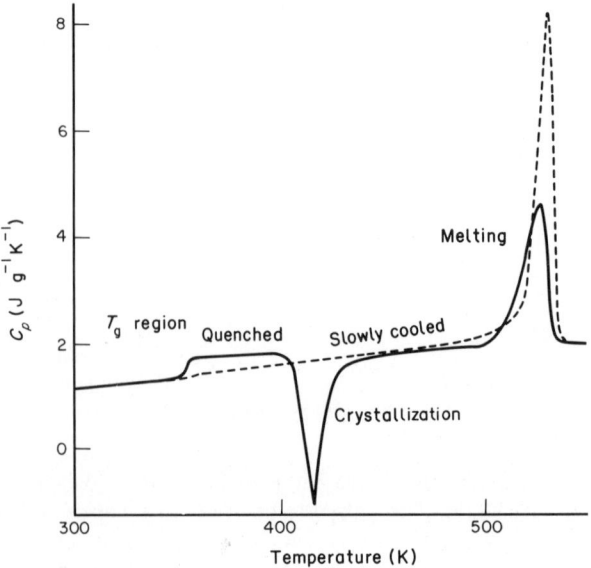

Figure 8   DSC curves for quenched (amorphous) and slowly cooled partly crystalline poly(ethylene terephthalate) (PET)

incidence of polymer-related papers is in the NATAS (North Americal Thermal Analysis Society) Proceedings; general meetings in Europe tend to concentrate more on inorganic materials.

36.5 THE GLASS TRANSITION

There is a fairly abrupt increase (ΔC_p) in C_p at the glass transition temperature T_g with the detailed geometry of the transition region depending very much on experimental conditions. Typical cooling and subsequent heating curves are shown in Figure 9. Cooling always gives the simple curve of Figure 9(a) and similar heating curves (*i.e.* free of peaks) are found whenever a '100%' amorphous polymer is perturbed in any way, for example by crystallinity, diluents or blending. A variety of 'definitions' of T_g may be found and are shown in Figure 9. The three commonest are T_o, $T(\frac{1}{2}\Delta C_p)$ and T_{inf}. The latter two of these tend to merge when fine structure is absent and under these conditions they are a reasonably accurate representation of T_g. The breadth of the transition region, $T_a - T_o$, is useful when discussing compatibility in mixed systems.

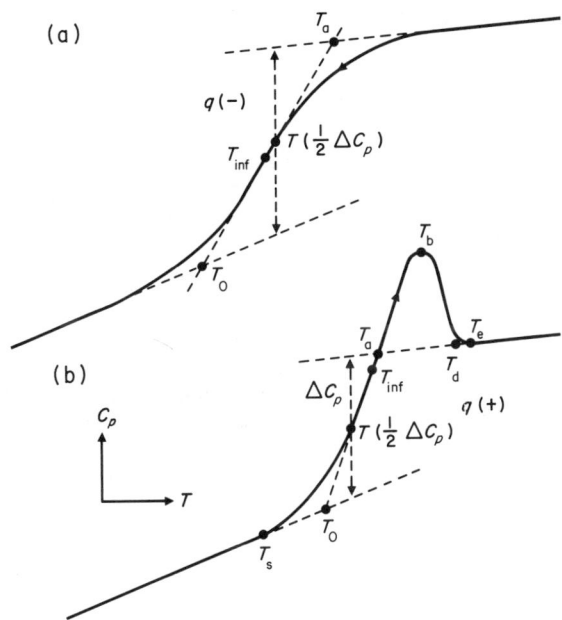

Figure 9 DSC curves of (a) cooling, and (b) heating in the glass transition region showing points that have been used to define T_g. T_o is the extrapolated 'onset' temperature (T_a is the equivalent value on the liquid side of the transition), $T(\frac{1}{2}\Delta C_p)$ is the temperature where half the heat capacity increment (ΔC_p) has been reached and T_{inf} is the point of inflection; other points correspond to the start s, peak b, end e, and extrapolated end d, temperatures. A more rigorous definition of T_g is shown in Figure 12

If DSC is to have any basic value in characterizing a glass, the same result should emerge from both cooling and subsequent heating curves (they both refer to the same glass). This is clearly not the case for the several points on the schematic curves of Figures 9(a) and (b) which summarize common observations. A first step must therefore be the derivation of a consistent glass temperature that defines a particular glass rather than the conditions under which it is measured.

36.5.1 Derivation of T_g

The step-like increase in heat capacity at T_g represents a complex summation of conformational and vibrational effects.[21] The important feature insofar as DSC measurements are concerned is the rapid decrease in the relaxation time from years in the glass to fractions of a second in the liquid. Rate effects may therefore be expected to play an important role as the timescales for experimentation and molecular motion become comparable; this is illustrated in Figure 10, which shows that the T_g region for the same polystyrene glass shifts to progressively higher temperature with increase in heating rate. The various points of Figure 9 (onset, inflection, peak *etc.*) may be

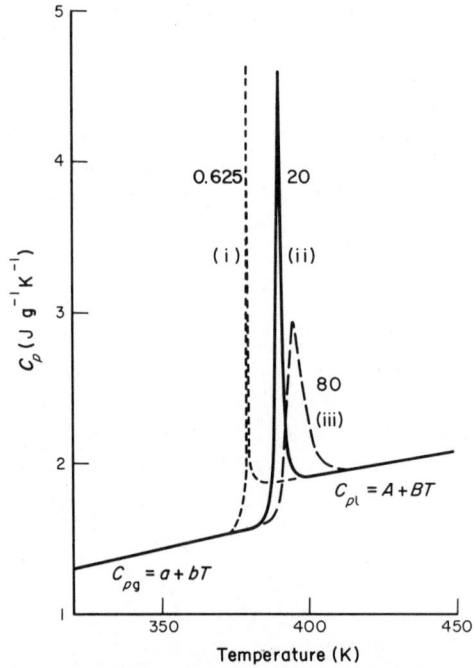

Figure 10   Heat capacity curves for a well annealed PS glass $(q(-) = 2.47 \text{ K d}^{-1})$ for the heating rates (K min^{-1}) shown; (i) does not show the considerable noise that is associated with very slow heating rates

arbitrarily taken as T_g but, since the several curves do not overlap (the extreme conditions of Figure 10 were chosen to emphasize this point), no unique value emerges. There is no reason — other than the subjective one of experimental convenience — to prefer one heating rate to another so that 'T_g' appears to be a function of the technique and correlation with other procedures should not be expected. Some form of extrapolation to zero heating rate may be possible[22] but this, even if justified, loses the rapidity of operation that is one of the main attractions of DSC. Fortunately there is a method that makes it possible to retain this feature but still obtain a value of T_g that is unique to a given glass,[15,23,24] *i.e.* there is no heating rate dependence. The procedure is based on the definition of T_g as the point of intersection of enthalpy curves (Figure 11) for the glassy and liquid states, *i.e.* $H_g(T_g) = H_1(T_g)$ where H_g and H_1 are extrapolated from low and high temperatures respectively. The only effect that heating rate has in this procedure is a local perturbation in the T_g region, which has no effect on the extrapolation. The procedure is the exact analogue of the well known dilatometric method of determining T_g[25] with enthalpy replacing specific volume (V). In V–T operations it is never suggested that the step in the expansion curve ($\alpha = \partial V/\partial T$) as $\alpha_g \rightarrow \alpha_1$ should be

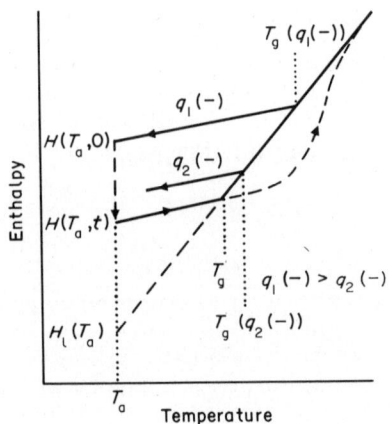

Figure 11   Schematic enthalpy curves in the T_g region showing the effect of isothermal annealing at T_a

used to define T_g even though kinetic effects are less important because of the very low heating rates that are normal in dilatometry.

The relative slopes of V–T and H–T curves above and below T_g are very different. For polystyrene,[26] for example, $\alpha_l/\alpha_g = 2.5$ and $C_{pl}/C_{pg} = 1.2$ so that it is very difficult to obtain an accurate value of T_g by visual inspection of enthalpy curves. A simple calculation overcomes this problem. Most amorphous polymers have extensive temperature ranges (many tens of degrees) below and above T_g where C_p is a linear function of temperature; the respective equations (Figure 10) may be integrated to give

$$H_g(T) \;=\; aT \;+\; \frac{1}{2}bT^2 \;+\; P \tag{6}$$

$$H_l(T) \;=\; AT \;+\; \frac{1}{2}BT^2 \;+\; Q \tag{7}$$

where P and Q are integration constants. Q–P is obtained by substituting T_1 in equation (6) and T_2 in equation (7) and substracting to give $H_l(T_2) - H_g(T_1)$, an experimental quantity, the shaded area of Figure 12. T_1 and T_2 are convenient temperatures in the glassy and liquid states, respectively, well away from the transition region. T_g then emerges as a solution of the quadratic equation $H_g(T_g) = H_l(T_g)$ that will henceforth be referred to as $T_g(H)$. The simple computational procedures required (least squares solutions giving a, b, and A, B, integration from $T_1 \rightarrow T_2$, etc.) are well within the capabilities of the microprocessors that are standard with most modern instruments and $T_g(H)$ values are routinely available with reproducibilities of ± 1 K. They have a clear thermodynamic meaning and there is no justification for using sophisticated programming to define some arbitrary point (onset, *etc.*) as 'T_g'. The only geometrical construction that has validity[27] is a graphical representation of the above ideas (Figure 12) where $T_g(H)$ describes a discrete step in the C_p–T curve at a location which is determined by the requirement that area $(A + C) =$ area B. $T_g(H)$ characterizes a glass formed in a particular way, *i.e.* it is independent of heating, but not cooling, rate. Variation of cooling rate is an important route to a variety of glasses (Section 36.5.2). The absence of rate effects is clearly shown by the data of Figure 13 where $T_g(-20$ K min$^{-1}) = 421.4$ K and $T_g(-20$, $+20$ K min$^{-1}) = 421.7$ K; data in brackets refer to thermal treatment through T_g.

Curves similar to those of Figure 13 use the empty pan to give the $C_p = 0$ baseline and remove any instrumental curvature. It is only valid to use a single 'sample' DSC curve for a construction similar to that of Figure 12 (equalization of areas) if independent experiments have shown that curvature is absent — otherwise extrapolation of the internal baselines for the glass and liquid may be seriously in error.

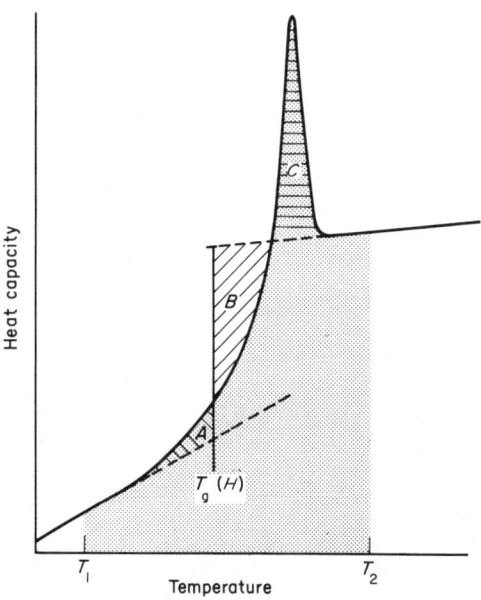

Figure 12 Graphical construction showing how $T_g(H)$ is defined. Area $(A + C) =$ Area B

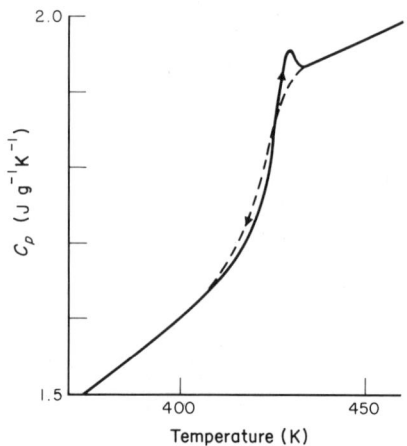

Figure 13   Cooling and subsequent heating curves (all 20 K min^{-1}) for a polycarbonate with $\bar{M}_{\rm w}/\bar{M}_{\rm n} = 38\,100/14\,400$

36.5.2 The Effect of Thermal History

The intent of most reports in the literature that use DSC to determine $T_{\rm g}$ is to obtain a simple one-parameter ($T_{\rm g}$) characterization of a given polymer. The associated fine structure that commonly occurs is a complication that is generally ignored but which was shown in Section 36.5.1 to be essential [in its contribution to $H_1(T_2) - H_{\rm g}(T_1)$] in deriving $T_{\rm g}(H)$, the only satisfactory parameter for a given glass. 'Given glass' should be emphasized because, as shown in Figure 11, the enthalpy, and hence $T_{\rm g}(H)$, may be varied by cooling through $T_{\rm g}$ at different rates; quenching gives a high free volume state whereas slow cooling allows relaxation to a relatively more stable condition. Clearly the full characterization of a polymer glass should give not only $T_{\rm g}(H)$ but also the glass formation conditions. This section will consider the magnitude of 'thermal history' effects and their implication for the routine characterization of polymers.

Figure 14 shows C_p curves for the same heating rate for polystyrene glasses formed by cooling through $T_{\rm g}$ at three different rates; the calculated values of $T_{\rm g}(H)$ are also shown. There is no direct

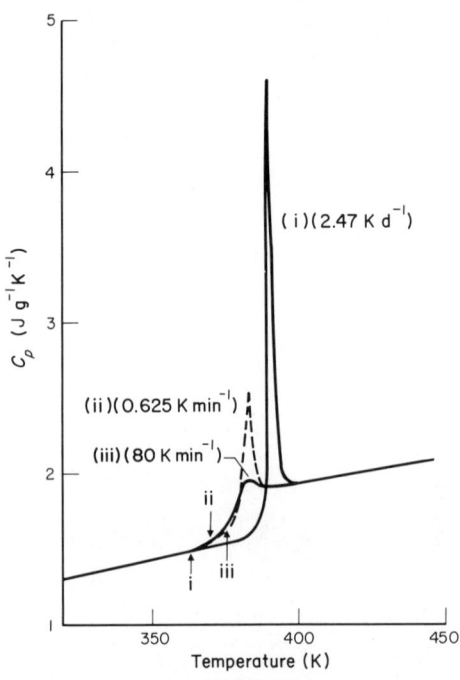

Figure 14   Heat capacity curves for an anionic PS ($\bar{M}_{\rm n} = 37\,000$) cooled at the rates shown (K min^{-1}); $q(+) = 20$ K min^{-1}.
The arrows show calculated values of $T_{\rm g}(H)$

correlation between $T_g(H)$ and the location of the transition region; in fact their relative order is reversed if any of the commonly given 'definitions' of T_g is used. The difference reflects the importance of kinetic factors in the T_g region and an analysis of the observed curves (Section 36.5.3) forms an important part of any discussion of the nature of the transition.[28,29] Taken at face value, the curves of Figure 14 give an absurd result with the transition region shifting to higher temperatures with decrease in $q(-)$ although it is clear from Figure 11 that any form of annealing must be paralleled by a reduction in T_g — the behaviour shown by $T_g(H)$ and plotted in Figure 15. Most polymers show comparable trends[30] so that for the limited window provided by the majority of polymer processing procedures $(q(-) \approx 50\text{-}500 \text{ K min}^{-1})$ the spread of T_g will be small — perhaps 3-4 K and measurements on 'as received' material will tend to give consistent results. In addition, when $q(-) > q(+)$ the peaks characteristic of kinetic overshoot are small or non-existent (Figure 14) and the point of inflection is a good approximation to $T_g(H)$ (Figure 9) so that this commonly used definition of T_g becomes acceptable. The apparent consistency of T_g for materials with $q(-) > q(+)$ can be very misleading when other situations are encountered. The example in Figure 14 in which $q(-) = 2.47 \text{ K d}^{-1}$ has already been mentioned and, although this particular history is unlikely to be met with in routine operations, analogous effects may result from other physical processes such as isothermal annealing, mechanical stress and strain, and solvent or vapour treatment. Figure 16

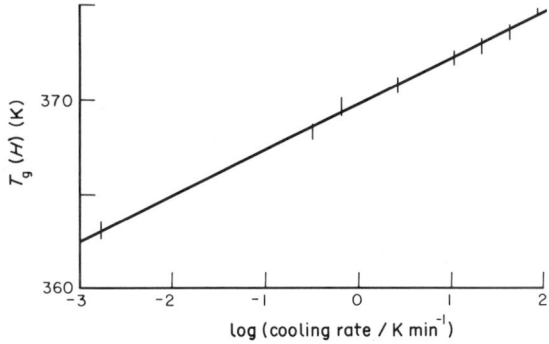

Figure 15 The influence of cooling rate on $T_g(H)$. Anionic PS ($\bar{M}_n = 37\,000$). Bars show the range of experimental data

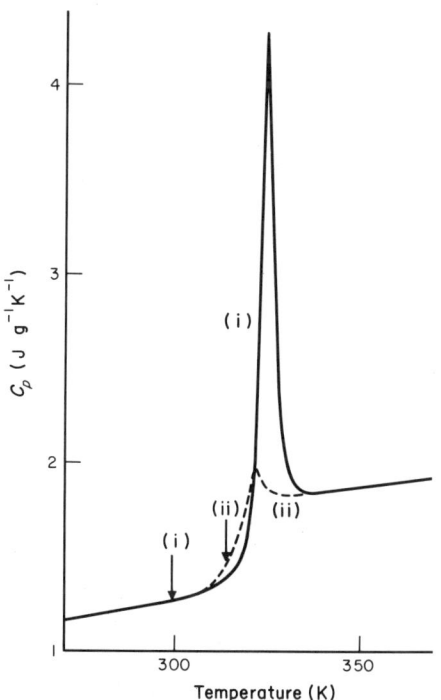

Figure 16 The effect of prolonged isothermal annealing (four years at room temperature) on the C_p curve (i) and the glass temperature (arrowed) of poly(vinyl acetate) ($\bar{M}_w/\bar{M}_n = 840\,000/103\,000$). The rerun (ii) shows that the polymer is physically unchanged

shows C_p curves for a poly(vinyl acetate) glass that has effectively annealed almost to the equilibrium state on storage at room temperature $T_g(H) = 298 \text{ K} \approx T(\text{ambient})$.

36.5.3 Enthalpy Relaxation

The physical ageing exemplified by the full curve of Figure 16 has important practical consequences and DSC has been widely used in recent years to monitor the processes involved.[31] Overall enthalpy changes are quite small and the importance of calorimetric operation cannot be too strongly emphasized.[32] Relaxation studies follow the decrease in enthalpy from some standard state to the extrapolated equilibrium liquid[33] (Figure 11) (although the validity of this limit has been queried[34]). The standard state should be attained after cooling at a modest rate[31] because quenching gives an ill-defined history, whether brought about by immersion in a refrigerant or by rapid cooling in the DSC itself. In the latter case it is impossible to be sure that 'steady state' conditions have been attained before passing through T_g. The same criticism could be applied to the converse process, reheating, after annealing for a time t at T_a (which cannot be too far below T_g, otherwise the process is too slow), to the equilibrium liquid at T_2 so that the shape of the $C_p - T$ curve in the transition region is lost. It is therefore tempting, after the appropriate time t, to quench to T_1, well below T_g, so that full details can be observed on subsequent heating. This procedure is hazardous because of the wide distribution of relaxation times[35] and it is much better to rely on the calorimetric performance of the DSC to give the enthalpy change $H_1(T_2) - H_g(T_a, t)$; *i.e.* the point where steady state conditions are actually attained is unimportant. The enthalpy decrease can be measured for a succession of annealing times by direct subtraction of the standard reference curve $(t = 0)$ from subsequent curves to give $H(T_a, 0) - H(T_a, t)$ but $H(T_a, 0)$ is a fairly arbitrary state [determined by $q(-)$] and a more useful quantity is $H(T_a, t) - H_1(T_a)$, which is a direct measure of the departure from equilibrium (Figure 11). Although an extrapolation is required to obtain $H_1(T_a)$, through the quantity $H_1(T_2) - H_1(T_a)$, this is generally no problem because the condition $C_{pl} = A + BT$ holds for most polymers. There are occasional problems, as with poly(vinyl chloride) (PVC), for example, when crystallinity prevents an unambiguous definition of C_{pl} and $H_1(T_a)$ must be treated as an additional parameter to be determined.[36] It is generally found that $H(T_a, t) - H_1(T_a)$ is linear with $\ln t$[33,37,38] and if the slope of this curve is $-d$ the relaxation time $\tau(t)$ is given by[37]

$$\tau(t) = [H(T_a, t) - H_1(T_a)]t/d \qquad (8)$$

The complex behaviour of $\tau(t)$ was recognized by Soviet workers some years before DSC was widely available and an excellent summary has been given.[39]

The effects of cooling and heating cycles on glasses in general,[29] and on polymers[40,41] in particular, are very conveniently studied by *in situ* DSC experiments. The 'structural' state of a glass is characterized by the fictive temperature T_f, which for calorimetric measurements relates the observed enthalpy at T to the equilibrium enthalpy at T_f. At high temperatures equilibrium is rapid and $T_f = T$ but well below the transition region the structure is frozen and the limiting low temperature value $T_f' = T_g(H)$. The derivative

$$dT_f/dT = (C_p - C_{pg})_T/(C_{pl} - C_{pg})_{T_f} \qquad (9)$$

reflects this with limiting values of zero and one at low and high temperatures respectively. The temperature dependence of T_f is given by[42]

$$T_{f,m} = T_0 + \sum_{j=1}^{m} \Delta T_j \left[1 - \exp\left(- \sum_{k=j}^{m} \Delta T_k / q_k \tau_{0k} \right)^{\beta} \right] \qquad (10)$$

where $T_{f,m}$ is the fictive temperature after m temperature steps (*ca.* 1 K is sufficiently small). The summation starts from an initial temperature T_0 and q is the heating or cooling rate. Retardation times are given by

$$\tau_{0k} = A \exp\left[\frac{x \Delta h}{RT_k} + \frac{(1 - x)\Delta h}{RT_{f,k-1}} \right] \qquad (11)$$

where A is a time factor, Δh is an activation enthalpy, x is a partitioning factor $(0 \leqslant x \leqslant 1)$ between temperature and structure (T_f) and β is a measure of the width of the relaxation spectrum $(0 < \beta \leqslant 1)$.

The activation enthalpy is obtained directly from the change of T_f' with cooling rate

$$\Delta h = -R\partial \ln q(-)/\partial(1/T_f') \tag{12}$$

and the remaining parameters A, x and β are obtained by matching experimental and theoretical dT_f/dT curves. This procedure has been used to model the behaviour of several polymers in the T_g region with reasonable success,[43-45] although it is not possible to find a unique set of parameters that fits both quenched and highly relaxed glasses.[46,47] Of the four parameters, only Δh may be obtained directly, the remainder only emerging after complex curve-fitting operations and an alternative, more experimental, approach that gives x fairly rapidly has been suggested.[48]

36.5.4 Sub-T_g Peaks — the Effect of Stress

Recent theories[28,49,50] suggest that extended low temperature annealing of quenched glasses should give sub-T_g peaks on subsequent heating through the transition region. The effect is not common, PVC[51] for long being the only clear example (Figure 17) but, with its hierarchy of structural regimes, PVC is hardly a typical polymer and Figure 17 may reflect an unrelated phenomenon. More convincing experimental support has recently been forthcoming *via* PET,[52] PS[53] and PS/PMMA blends.[54]

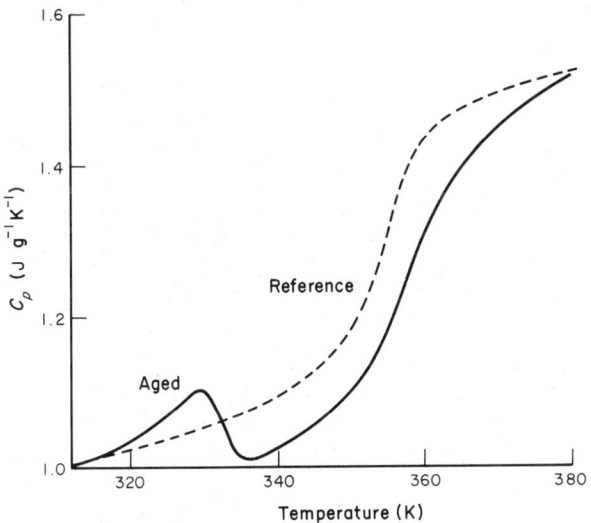

Figure 17 The sub-T_g peak that develops on isothermal ageing of PVC at room temperature; $q(-) = 40$ K min^{-1} for the reference curve, $q(+) = 20$ K min^{-1}

It is more usual to find sub-T_g peaks after programmes which feature some form of stress, especially after cooling through T_g under pressure.[55-59] Typical curves are shown in Figure 18[60] where it is seen that the low temperature peak increases in magnitude and shifts to lower temperatures with increase in pressure. When a direct comparison of the behaviour of the same normal one atmosphere glasses is possible, $T_g(V) \approx T_g(H)$ [the comparison is surprisingly difficult to make because the differing timescales for dilatometry and DSC mean that, in general, $q(-)(\text{DSC}) > q(-)(\text{dilatometry})$] but the two diverge on densification. Pressure can only decrease the specific volume and with it $T_g(V)$ but enthalpy first decreases and then increases[57,61] so that for the more densified glasses $T_g(H) > T_g(V)$.

The earlier model[42] of relaxation in the T_g region has been widened to include the effects of stress in general by assuming that the effect is to increase τ_0, a procedure that accounts for the disparate effects of hydrostatic and tensile stresses and vapour annealing.[62]

The densified polymers of Figure 18 are formed under conditions that overlap with those used in injection moulding and it is worthwhile examining the potential of DSC for the characterization of this complex process. A comparison of curves (4) and (5) with (1) and (2) of Figure 18 shows obvious similarities, with DSC having sufficient discrimination to distinguish the skin/core structure arising

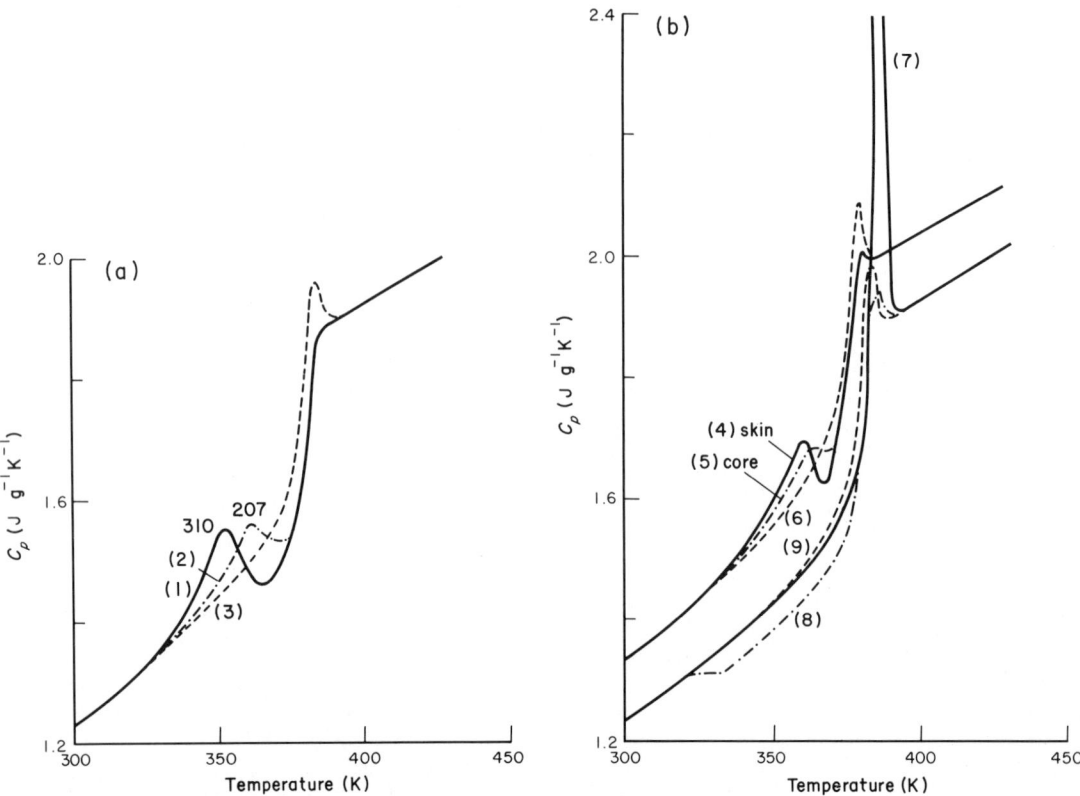

Figure 18 Heat capacity for PS glasses formed under stress: (a) (1), (2): cooled through T_g(1 K min^{-1}) under the pressures shown (MN m^{-2}); (b) (4), (5): sections from an injection-moulded bar (samples obtained by the layer removal technique, G. J. Sandilands and J. R. White, *Polymer*, 1980, **21**, 338); (7), (8): freeze-dried powder, as received (7) and after densification at room temperature (8); (3), (6), (9) are the corresponding reference curves run after cooling (20 K min^{-1}) in the DSC. All $q(\underset{\sim}{+}) = 20$ K min^{-1}. (4)–(6) are displaced by $+0.1$ J g^{-1} K^{-1}

from the different solidification conditions that these two regions experience. The resultant internal stresses can lead to distortion of poorly designed or badly moulded articles and there are obvious implications for the use of DSC as a quality control or fault-finding technique.

The internal stress in a polymer densified at room temperature (perhaps accidentally, as by compaction in a press, see Section 36.3.1) soon relaxes on heating,[63] giving curves that may be very misleading with respect to the original structure [compare curves (7) and (8), Figure 18].

36.5.5 Molecular Weight and Crosslinking

The glass temperature of a polymer increases with MW to a limiting value $T_{g\infty}$ that is usually attained in the MW range 10^4–10^5. One of the simplest relationships[64] expresses the change as

$$T_g = T_{g\infty} - J/M_n \qquad (13)$$

Low MW polymers show marked deviations[65] and better agreement is found for[66]

$$\frac{1}{T_g} = \frac{1}{T_{g\infty}} + \frac{J'}{M_n} \qquad (14)$$

which may be transformed to a modified form[67] of equation (13) in which M_n is replaced by $M_n + Q$, where Q, J and J' are constants for a given polymer. Equations (13) and (14) are both based on free volume concepts: statistical mechanical[68,69] and thermodynamic[70] theories have also been published (although the assumed continuity of entropy of mixing at T_g, on which the last is based, has been disputed[71] and defended[72]) and the most appropriate should emerge from a comparison of theory and experiment. In practice experimental data are scattered; trends are clear but it is not possible to state unambiguously that results fall on a single curve or are better represented by a family of linear regions.[73] Statistical mechanical theories contain molecular parameters that are normally allowed to vary to give 'best' fits so that there is a certain arbitrariness in their choice and this becomes apparent when different methods of calculation are used.[74-76]

Scatter is inevitable when T_g data are based on a variety of experimental techniques and when thermal history is almost always neglected. For example, even assuming that $T_g(V) = T_g(H)$ *for the same glass* (Section 36.5.4), experimental conditions are such that cooling rates in dilatometry are slower by about two orders of magnitude with respect to DSC; under their respective natural cooling conditions (and, by implication, ignoring thermal history) the two techniques should therefore give glasses with T_g values that differ by 4–5 K (Figure 15). As the parameter J of equation (13) is about 10^5 for PS[65] (a fairly representative polymer), T_g rises by some 10 K as M_n increases from $10^4 \rightarrow \infty$. With uncertainties of 4–5 K in this figure due to thermal history effects alone (and totally neglecting random experimental errors) it is clear that misleading T_g–MW relationships can easily occur.

The convenience of operation and small sample size required (materials may be obtained using GPC) should make DSC an ideal technique for the study of T_g–MW behaviour and, through this, gain an insight into the appropriate theories. Data have been reported for PS,[77,78] PP,[79] poly(methyl methacrylate) (PMMA)[80] and poly(vinylpyrrolidone)[81] although even here thermal histories are often not clearly defined. Samples may be described as 'quenched', *i.e.* cooled at the maximum rate possible for a given instrument. This has the benefit that kinetic effects are minimized so that $T(\frac{1}{2}\Delta C_p)$ is a good approximation to $T_g(H)$. However, since $q(-)$ follows from what is essentially a natural cooling curve, it will pass through a maximum at some intermediate temperature but will be low in the initial stages and also towards the end as ambient temperature is approached. Attention to the details of thermal history greatly reduces experimental scatter (Figure 19) so that the process may be reversed and the observed curve used to measure M_n. This is a very simple procedure that is most useful in the oligomeric region where conventional methods are at their limits.

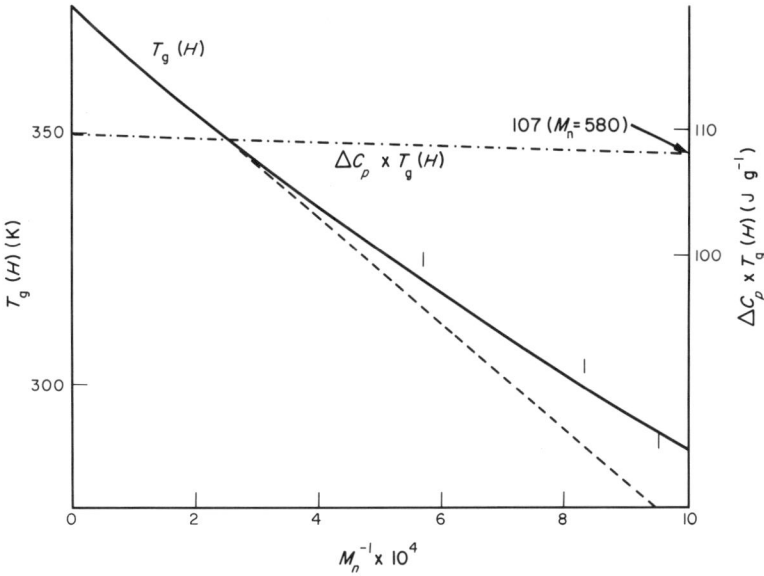

Figure 19 The influence of MW on T_g and the product $\Delta C_p T_g$. Anionic PS glasses formed by cooling through T_g at 5 K min^{-1}. All points, other than the three shown, fall within ± 2 K of the full line. The broken line corresponds to $J = 1.05 \times 10^5$ (equation 13)

It has been suggested[82] that for linear amorphous polymers $\Delta C_p T_g \approx$ constant (ΔC_p is the heat capacity increment at T_g). Exceptions and trends due to specific features (chain stiffness, hydrogen bonding) have been discussed but a homologous series provides a good test covering a wide range of temperatures with consistent molecular parameters. Data for PS show that the product is remarkably constant for $M_n \geqslant 580$ (Figure 19).

Introduction of crosslinks gives an increase in T_g[83-85] although ΔC_p decreases more rapidly than the '$\Delta C_p T_g \approx$ constant' relationship would suggest and the T_g region eventually becomes undetectable by DSC. The elevation of T_g can be regarded as a summation of effects due to crosslinking and copolymerization. The former inevitably enhances T_g; the latter may be of either sign, and an experimental distinction has been made.[86]

The increase in T_g on crosslinking (Figure 20) is the basis of a procedure that is widely used (from printed circuit boards to pipeline coatings) for the characterization of the degree of cure of a resin. An initial determination of T_g on 'as received' material (modified only by the imposition of a known thermal history) is followed by a suitable period at a temperature high enough to allow any residual cure to take place; cooling to give the same thermal history as before is followed by a second determination of T_g. Any increase in T_g due to unreacted material in the original sample is immediately obvious from a comparison of the two T_g curves (a decrease would imply degradation caused by too high a cure temperature) and a detailed product specification will state what change in T_g is acceptable.

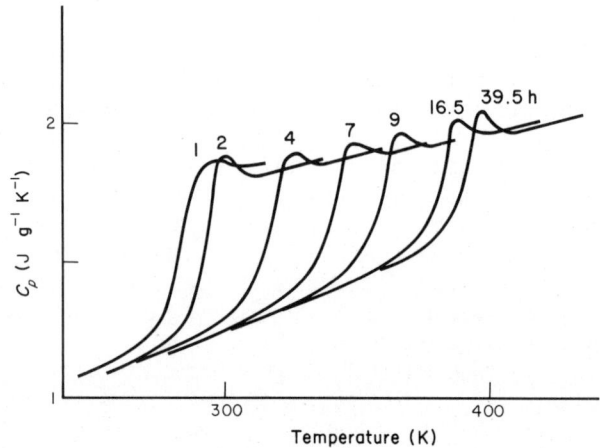

Figure 20 The increase in T_g of an epoxy resin with crosslinking. Cure times (h at 398 K) are shown

An extent of reaction, P(t), can be defined by

$$P(t) = \frac{T_{gt} - T_{go}}{T_{g\infty} - T_{go}} \tag{15}$$

where t is the cure time and subscripts o and ∞ refer to the original homogeneous mixture and the fully cured material respectively. By using P(t), the cure of two epoxy system was shown to be first order in the initial stages with rate constants similar to those for branched ether crosslink formation.[87] The width of the T_g region did not change with P(t) but this is a function of the system studied and a change in width may correlate with some other parameter giving valuable supplementary information.[88]

36.5.6 Copolymers

The glass temperature of a random copolymer is generally a monotonic function of composition but examples showing extremes at intermediate compositions are well documented[89] and must be explained by any theory. Extensive discussions have been given,[89-91] an essential point being that DSC measurements must be supplemented by details of the sequence distribution along the chain. Using this approach and treating PMMAs of varying tacticity as copolymers[92] (triad compositions being determined by NMR) gives basic T_g data for 100% isotactic and syndiotactic polymer (315 and 397 K respectively) that are little influenced by the origin (entropic[93] or free volume[89]) of the copolymer equation.

Random copolymers give only a single T_g but the situation is much more complex for diblock (AB) and triblock (ABA) copolymers with incompatible blocks. Two transition regions, roughly corresponding to those of homopolymer A and B, may occur although they are often at lower temperatures than would be expected on the basis of MW effects alone.[94,95] An intermediate T_g originating in another phase is sometimes seen[96] but a detailed analysis is difficult with several unanswered questions concerning compatibility and detectability of T_g in domains that may be of molecular dimensions, and the nature of any interfacial regions. Future progress in this complex area must utilize the full potential of DSC with particular attention being paid to ΔC_p, the width of the transition, and thermal history.[97]

36.5.7 Multicomponent Systems of High and Low Molecular Weight

The thermal behaviour of polymer blends has been comprehensively reviewed.[98,99] A discussion of the many T_g–composition relationships for compatible blends indicates that the more recent treatments should also be applicable to polymer/low MW diluent systems[99] which cover a very wide range of materials and applications.[100] Problems common to DSC work on amorphous multi-component systems (the most widely used) are discussed in this section. Compatibility or incompatibility is generally considered to be indicated by the presence of one or two T_g regions respectively (for binary systems). The latter certainly implies some degree of incompatibility but the location, magnitude (ΔC_p) and width of both transitions must all be considered, as shifts away from pure polymer behaviour may suggest partial solubility of one or both components in the other. The presence of a single T_g region strongly suggests compatibility but once again the detailed geometry of the curve must be carefully examined, especially if the T_g values of the original polymers are close together. One useful procedure is to anneal by slow cooling[101] or isothermal ageing[102] to encourage the appearance of kinetic, or hysteresis, peaks as in Figure 14 (although the magnitude of these appears to decrease whenever conditions deviate from '100% amorphous homopolymer'[101]). Using this approach a single broad transition was resolved into two, both of which were perturbed from homopolymer values showing the mutual partial solubility discussed above.[102]

It remains unclear what is the minimum 'probe' size that can be resolved by DSC and this must always be borne in mind, especially for systems with low concentrations of one component. A related problem is the effect of surface energy and interfacial composition on T_g when domain sizes are small; the only discussion[103] suggests an enhancement of T_g by perhaps 5 K for a 6 nm sphere.

Although the enthalpy of mixing, ΔH_{mix}, is the dominant term in the phase behaviour of polymer blends, its direct determination presents insuperable difficulties because of the mechanical work that must be put into the formation of any blend. If, however, the system has a lower critical solution temperature in an experimentally accessible region (where degradation can be avoided), it is in principle possible to observe the reverse process, demixing, in direct heating experiments and initial results have been reported for poly(ethylene acrylate)/poly(vinylidene fluoride) blends[104] and some model electron donor–acceptor complexes.[105] The temperature dependence of ΔH_{mix} is related to the excess heat capacity of the mixture which can also be obtained by careful DSC work; heat capacities up to 4% greater than simple additivity have been found in the molten state but there is little effect in the glass.[106]

Low MW diluents are widely used as plasticizers and DSC has long been used to study the variation of T_g with plasticizer content.[107] Attention has usually concentrated on T_g itself but the magnitude of ΔC_p and breadth of the transition region (which generally pass through minima and maxima, respectively) give valuable supplementary information that is available with little additional experimental effort.[100]

Plasticizers are intended to remain compatible over the widest range of temperature and composition but there is now considerable interest in gels, another type of polymer/diluent system in which phase separation plays a key role. Gels have historically featured a crystallizable polymer and some form of crystallite is thought to form the 'crosslinking' unit[108] but the formation of gels from atactic PS poses more difficult conceptual problems. DSC curves show an exotherm on cooling followed by a glass transition with the processes reversed on heating. A straightforward explanation is that the cooling exotherm is due to demixing and the subsequent T_g region provides the (glassy) crosslinks,[109] although it has been suggested that some type of molecular aggregate is formed.[110]

Although diluents are generally introduced deliberately, the fortuitous effects of water should never be forgotten. Any sample that is heated in a non-hermetically sealed pan should always be reweighed at the end of any series of experiments. Losses of one or two per cent are common and, if the missing mass is regained after overnight exposure to the atmosphere, moisture is the most likely explanation. Although only a small percentage, the associated heat of vaporization is not negligible and the first DSC curve of a set can show artefacts in the 350–400 K range; there is often slow drift if an isothermal region falls within this range. Obviously the rate of loss depends on sample mass and geometry. Special care is needed with crosslinked systems where the plasticizer effect increases with degree of crosslinking.[111]

A review[112] of the effects of particulate fillers on T_g shows conflicting results with a cautionary note regarding variable thermal histories. Consideration of the stress field around a filler particle suggests the T_g should increase[113] but a variety of techniques shows a reduction for iron/epoxy composites containing up to 25 volume per cent of iron.[114]

36.6 MELTING AND CRYSTALLIZATION

The most common applications of DSC are to the melting process which, in principle, contains information on both the quality (temperature) and quantity (peak area) of crystallinity in a polymer. The majority of reports discuss only peak temperatures which, taken in isolation, can be very misleading because it is rare to find a polymer that does not anneal on heating at a rate that may be slow relative to its previous thermal history — quenched in a mould, for example. Nevertheless the presence or absence of a particular peak may be sufficient to characterize a specific grade for a particular application, thus small quantities of polyethylene in polypropylene, which are difficult to detect by other methods, are readily observed.

The annealing effects referred to above can be discouraged or emphasized depending on whether the primary concern is with the original material or the morphological changes *en route* to a more stable structure. Interest in 'as received' polymer (*e.g.* for the correlation of crystallinity with mechanical properties) implies rapid heating to minimize the potential for change or some form of pretreatment, such as etching or irradiation, that attempts to stabilize the original structure. The following sections will discuss the effects of material and instrumental variables on temperatures and heats of fusion and show how basic thermodynamic data can be obtained. More detailed examples are reviewed in the three volumes[115-117] by Wunderlich (the final one is especially useful in this respect); other valuable sources are refs. 118 and 119, with the latter emphasizing quantitative aspects.

36.6.1 Melting Points — the Influence of Experimental Variables

36.6.1.1 Sample mass and geometry

DSC curves for the melting of different masses (discs of constant diameter but variable thickness) of polyethylene are shown in Figure 21.[17] Reduction to the common specific heat capacity scale demonstrates that only the peak region is affected — C_{pc} and C_{pl} in thermally 'inert' regions are unchanged, as they should be, being materials constants that serve as useful internal references (C_{pc} depends on previous history but should be reproducible for a given sample). This behaviour is

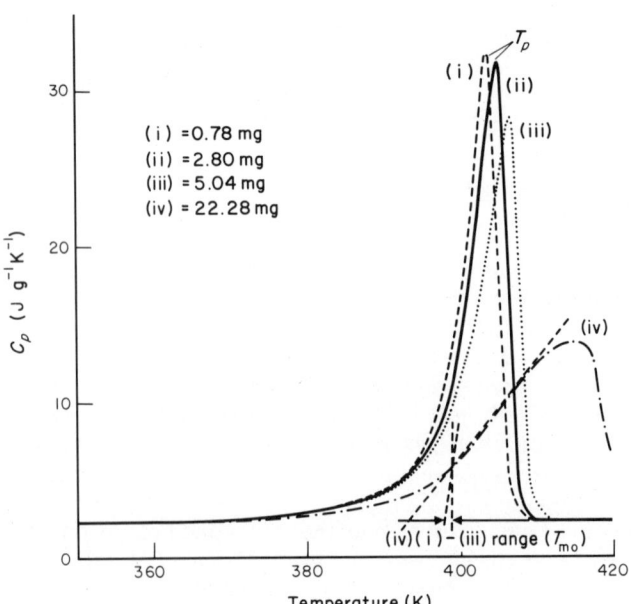

Figure 21   The effect of sample mass (thickness) on the DSC melting curves of polyethylene

qualitatively similar to that of any pure low MW material although polymers, having a distribution of crystal sizes and perfections, melt over a broader range of temperature. The shift of the melting peak temperature, T_p, with increasing mass reflects the finite time needed for heat transfer in any material and in accurate work on low MW materials the melting point is taken as the extrapolated onset of melting T_{mo}.[6] An equivalent construction is possible for the rather idealized polymer of Figure 21 where it is seen that T_{mo} values cover only a narrow range of temperature if the most massive sample is neglected. The description 'idealized' is used because the initial structure in this sample is fairly uniform and the melting region is correspondingly narrow; for many polymers it is impossible to locate T_{mo} so that T_p is the only well defined point on the melting curve.

Figure 21 shows how T_p is influenced by mass for a sample of constant geometry. When the mass is held constant but the geometry changed, to decrease the surface area to volume ratio, the resulting poorer thermal contact leads to broad, shallow curves, with T_p displaced to higher temperatures.[17]

36.6.1.2 Heating rates

For a low MW material, T_p shifts to higher temperatures with increase in heating rate, $q(+)$. The peak width $\Delta T = T_p - T_{mo}$ is proportional to $q(+)^{\frac{1}{2}}$ so that it is possible to obtain T_{mo} from a graph of T_p against $q(+)^{\frac{1}{2}}$. In principle the same procedure may be used for a polymer[120] provided each experiment in a given set refers to the same size and geometry of sample. Positive deviations from linearity at low $q(+)$ are a sign of annealing *en route* to T_m; they are also associated with higher values of ΔH.[121] Many examples are known when 'positive deviations' are due to the appearance of a totally new crystal form at slow heating rates. This is clearly illustrated in Figure 22, which shows that the highest melting component completely vanishes as $q(+)$ is raised from $10 \rightarrow 40$ K min^{-1}. For a quenched polymer it may never be possible to suppress reorganization[120] by increasing $q(+)$ and the metastable structure must be stabilized by crosslinking or etching[117] — processes which inevitably introduce their own uncertainties.

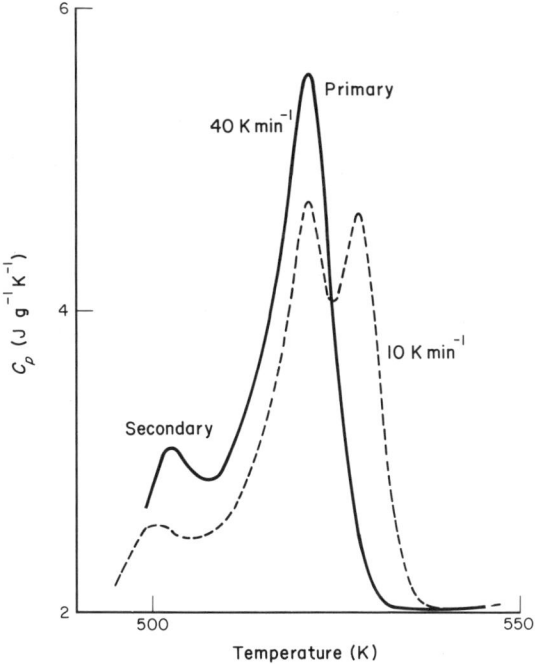

Figure 22 The influence of $q(+)$ on the melting of PET isothermally crystallized at 490 K for 30 min. The highest melting component only develops on slow heating $q(+) = 10$ K min^{-1}

36.6.2 Melting Points — Idealized Crystals and Superheating

The T_m described above normally refers to lamellar crystals of thickness L. It is related to $T_{m\infty}$, the equilibrium melting temperature of a crystal of infinite thickness by

$$T_m = T_{m\infty}[1 - (2\sigma_e/L\Delta H_f')] \tag{16}$$

where σ_e is the surface energy and $\Delta H'_f$ the heat of fusion per unit volume. $T_{m\infty}$ is of fundamental importance in the thermodynamic theory of polymer crystallization[122] and it is important to know it with some accuracy. Indirect methods are normally used,[117] plotting T_m against $1/L$ (equation 16) or T_m against T_x (the crystallization temperature) and extrapolating to $T_m = T_x$. In some cases, however, much larger crystals are available (especially for polyethylene); they may be in the original 'as-polymerized' material[123] or may have been crystallized under stress.[124-127] Most show unusual melting behaviour with peaks appearing perhaps 20 K above the expected $T_{m\infty}$ value. Rate effects are very important and equilibrium is only attained if the material is held at the appropriate temperature for times of the order of days.[125] The 'appropriate temperature' is progressively increased to define the melting point through the disappearance of enthalpy of fusion. There are clearly no problems in attaining thermal equilibrium on the timescales indicated and the super-heating effect that is observed is thought to represent the transition to a partially ordered melt[128] — the entropy of fusion $[\Delta S(x, T_m)]$ in $T_m = \Delta H(x, T_m)/\Delta S(x, T_m)$ (x is the degree of crystallinity) is low because the conformations of tie molecules are restricted.

36.6.3 Enthalpy Curves and the Heat of Fusion

36.6.3.1 Background thermodynamic requirements

Although DSC is a very simple technique it should be clear from the discussion of T_g and T_m that the correct interpretation of results, although straightforward, requires rather more effort than is usually invested in this stage. This is nowhere more true than in the determination of the enthalpy of fusion or transition. The basic procedures were developed in careful work using adiabatic calorimetry[19] but have generally been totally neglected in DSC work so that the majority of 'heats of fusion', for example, have no thermodynamic significance whatever. It is understandable that various geometrical constructions should have been used before microcomputers became widely available but there is no reason why these should now be employed only to generate erroneous data more rapidly.

The basic principles required[17,129,130] were introduced in Section 36.3.2.2 in the discussion of enthalpy calibration. Integration of a DSC-derived 'C_p'–T curve gives an enthalpy change $H_y(T_2) - H_x(T_1)$ where T_1 and $T_2(T_2 > T_1)$ are the integration limits and x and y represent phases. Putting $y = 1$ (liquid) and letting $x =$ degree of crystallinity, the heat of fusion is defined by

$$\Delta H(x, T_m) \;=\; H_1(T_m) \;-\; H_x(T_m) \tag{17}$$

This can be written in terms of three measurable quantities, A, B, C (Figure 23)

$$\Delta H(x, T_m) \;=\; \underbrace{[H_1(T_2) \;-\; H_x(T_1)]}_{A} \;-\; \underbrace{[H_1(T_2) \;-\; H_1(T_m)]}_{B} \;-\; \underbrace{[H_x(T_m) \;-\; H_x(T_1)]}_{C} \tag{18}$$

Examination of equation (18) and Figure 23 shows that the conditions that must be met are: (i) (a) T_2 must be far enough above T_m for the curve to have returned to the liquid region at this temperature, (b) T_1 must be below the onset of melting i.e. $x = x_c$ ($x_c =$ constant) for all $T \leqslant T_c$ where $T_c \geqslant T_1$ (Figure 23); and (ii) the extrapolations implied by (a) B and (b) C must be feasible.

Conditions (ia) and (iia) are related and are easily met because C_{pl} is generally a linear function of temperature (Figures 13, 14, 16) over many decades that are experimentally accessible by direct cooling/supercooling or by quenching to the amorphous state when the region above T_g (and prior to crystallization) represents highly supercooled C_{pl}. Conditions (ib) and (iib) are more difficult to meet. There are no immediately obvious criteria that will decide the onset of melting although it is tempting to take that point when C_{px} becomes a non-linear function of temperature inasmuch as the extrapolation required (C) becomes somewhat uncertain — but this is hardly a justification for associating non-linear behaviour with melting. Even if conditions (ib) and (iib) could be met with total confidence the quantity $\Delta H(x, T_m)$ is of doubtful value although it is the implicit aim of most 'area-measuring' techniques. The enthalpic crystallinity, x_H, is defined as $\Delta H(x, T_m)/\Delta H(x = 1, T_m)$ but the denominator usually refers to $T_{m\infty}$ or room temperature (Section 36.6.5) rather than T_m, an arbitrary temperature that is itself a function of x (more perfect crystals melt at higher temperatures). A much simpler and more useful procedure is to eliminate the uncertainties both of measuring T_m (and $T_{m\infty}$) and of making the extrapolation C (equation 18) by defining $\Delta H(x, T)$ as

$$\Delta H(x, T) \;=\; \underbrace{[H_1(T_2) \;-\; H_x(T)]}_{A} \;-\; \underbrace{[H_1(T_2) \;-\; H_1(T)]}_{B} \tag{19}$$

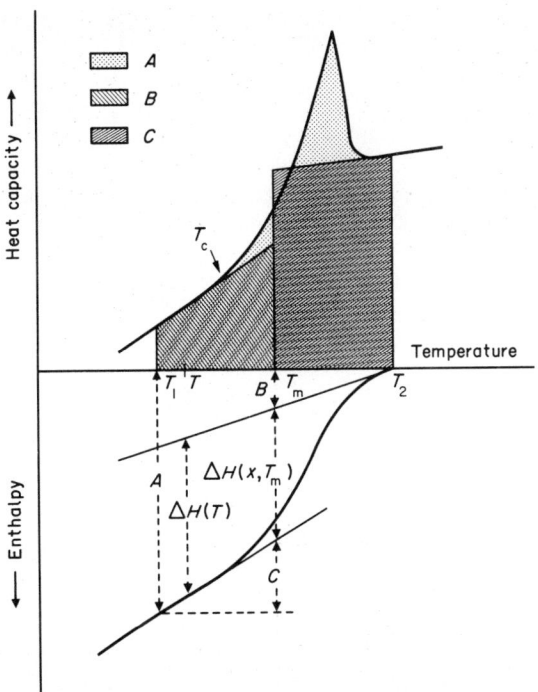

Figure 23 Schematic DSC melting curve (a) and integrated enthalpy changes (b) with respect to $H_1(T_2) = 0$. A, B and C are defined in equation (18). Crystallinity is constant below T_c.

where $T < T_m$. Crystallinity need not be in the x_c region and in fact $\Delta H(x, T)$ may now be used to define the onset of melting: the relative crystallinity of two samples is given by $\Delta H(x_1, T)/\Delta H(x_2, T)$ and if both are at temperatures in their x_c regions $x_{1c}/x_{2c} = $ constant, a decrease from this condition (assuming $x_1 < x_2$) implies that $x_1 < x_{1c}$ and melting has started. The crystallinities could, of course, both decrease in parallel but this is unlikely, especially if very different routes are·used in the preparation of samples.

36.6.3.2 DSC procedures

Equations (18) and (19) and Figure 23 show how ΔH can be obtained from well defined components of a DSC curve or curves — a single run may require a subsidiary experiment to better define the baseline, C_{pl}, see below. In the simplest case (equation 19), Figure 23(a) (with area $C = 0$) shows that $\Delta H(x, T)$ is represented by the difference between the areas below the experimental curve (A) and the extrapolated liquid (B). The latter defines the 'baseline' which is often experimentally accessible. since polymers readily supercool (polyethylene is the notable exception). The baseline cannot be drawn between two cursors on the melting curve, *i.e.* the microprocessor equivalent of the incorrect construction that is normally used to define 'ΔH'. As already discussed, $\Delta H(x, T_m)$ requires an additional extrapolation (term C in equation 18) from lower temperature to give a baseline that is discontinuous at T_m. There is little advantage in constructing more complex baselines (*e.g.* sigmoidal) to represent the changing amounts of crystalline and amorphous phases in the main melting region; any curve produced in this way is so grossly affected by material and experimental variables that it has only tenuous relevance to the original structure (dotted region, Figure 24).

The two terms of equation (19) are shown in Figure 24(a) for poly(ethylene terephthalate) (PET). The resultant heats of fusion (Figure 24b) demonstrate the need for the thermodynamic background of Section 36.6.3.1 in the definition of $\Delta H(T)$; in general $C_{pc} < C_{pl}$ and so $\Delta H(T)$ is an increasing function of temperature (when $x = x_c$) until the decrease in x becomes the dominant effect. Curves similar to those of Figure 24 can be drawn for the PE results of Figure 21 or the PET results of Figure 22. For the former, the curves superimpose at all points other than in the melting region, for the latter they merge only at the start and in the molten state because structural changes occur soon after heating is started. The overall enthalpy changes in both cases are independent of sample mass (Figure 21) or heating rate (Figure 22) showing that a DSC will behave as a true calorimeter: there is

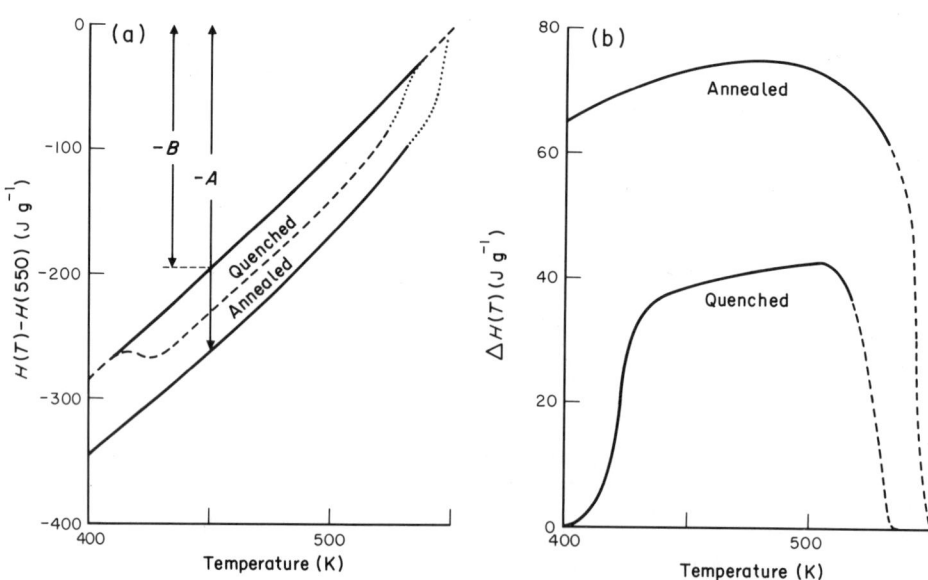

Figure 24   Enthalpy changes (a) and heats of fusion (b) of quenched and partially crystalline PET. The dotted regions are strongly influenced by experimental parameters

less ambiguity about using it for its enthalpy-measuring capability than as a thermometric device. The latter must be hedged about with numerous qualifications if absolute, rather than relative (peak locations), temperatures are to be obtained.

36.6.4   Crystallization

36.6.4.1   *Non-isothermal crystallization*

The final stages of most polymer fabrication processes involve rapid cooling below a glass transition or crystallization temperature. In the latter case it might be thought that DSC experiments at fairly high $q(-)$ would be widely used to simulate these forming procedures but this is not the case. Crystallization work is still relatively neglected and the majority of reports still infer low temperature structural properties of polymers from melting curves even though these may be perturbed by phase transitions and annealing phenomena whereas crystallization generally goes directly to the relevant structure.

Figure 25 shows crystallization and subsequent melting curves for a high density polyethylene (HDPE) sample. Agreement of heating and cooling curves in thermally inert regions is a stern test of quantitative performance. 'Thermally inert' should be emphasized: in Figure 25(a), for example, the

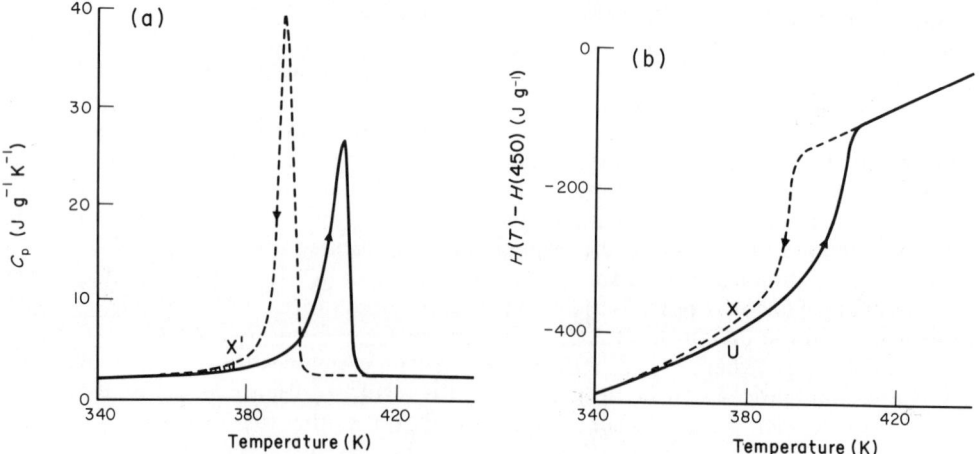

Figure 25   The crystallization and melting of polypropylene: (a) heat capacity; and (b) enthalpy changes. Secondary crystallization is represented by the shaded area in (a) or by XU in (b)

extensive region of secondary crystallization (shaded) only slightly affects the C_p curve but if cooling were stopped at X', equilibrated, and the heating curve then started, sufficient isothermal crystallization could occur, giving the imbalance XU (Figure 25b). Overall accuracies of C_p are $\pm 1\%$ for DSC work on stable materials and integration to give the enthalpy curves of Figure 25(b) generally finds the cycle polymer(liquid, T_2) $\to$ (solid, T_1) $\to$ (liquid, T_2) balancing with a residual of perhaps 3–4 J g^{-1} in an overall change of 400–500 J g^{-1}. Subtraction of the (liquid) baseline contribution generally gives a final $\Delta H(T)$ of 40–200 J g^{-1} with an uncertainty of some 3–4% (errors in the two terms of equation (19) cancel each other to a limited extent).

The cooling curve of Figure 25(a) gives information about the crystallization process (onset and peak temperatures, peak height and half-width *etc.*) as well as $\Delta H(T)$, but these data are essentially of relative value [apart from $\Delta H(T)$] and of most use in comparative experiments as, for example, in studying the effects of nucleating agents, when both the location and shape of the event may be changed.[131-133]

Although non-isothermal crystallization is a good approximation to practical conditions, little progress has been made towards a fundamental description of the crystallization curve itself in terms of basic mechanisms. An integrated Avrami expression has been used[134] but the Avrami approach has limitations even under isothermal conditions and uncertainties in the parameters reduce the operation to a curve-fitting exercise.

36.6.4.2 *Isothermal crystallization*

Strict adherence to the definition of thermal analysis given in Section 36.1 should exclude isothermal events but these have always fallen within the scope of DSC and this convention is followed here.

The evolution with time of the eventual crystallinity is given by $\Delta H_t/\Delta H_\infty$ where ΔH_t represents the partial area after time t (Figure 26) and ΔH_∞ is the final heat of crystallization. Although isothermal crystallization experiments are simple in principle there are a number of experimental difficulties. A schematic crystallization curve showing a quench from the stable liquid to the desired crystallization temperature T_x is given in Figure 26 and emphasizes the problems that will be encountered. Unless crystallization is very slow (high T_x) the approach to isothermal equilibrium at T_x will merge with the onset of the process and the definition of $t = 0$ is uncertain (dilatometry has the same problem); dummy experiments to temperatures where only the liquid polymer is stable allow a background correction to be made.[135] At the other, long-time, extreme, slow drift characterizes secondary crystallization processes and the true location of the isothermal baseline may be concealed. The limit may be found from preliminary work (which is also checked at the end) in which the polymer is crystallized, without particular initial care, followed by annealing at $T > T_x$ so that on cooling to T_x a stable baseline is rapidly attained. Figure 27 shows curves obtained for PET crystallizing at 480, 490 and 500 K; the large variation of rate with temperature and the crystallization 'tails' are characteristic and obvious. The former emphasizes the need for good stability and reproducibility of temperature and this has been shown to vary by less than 0.02 K from results for a polyethylene crystallized at 396.9, 397.0 and 397.1 K.[135] Secondary crystallization introduces uncertainty into the value of ΔH; for PET, for example, there are peaks associated with both primary and secondary processes[136] (Figure 22) and the latter merges into background noise and drift even though subsequent melting curves show clear changes with time of crystallization.[136]

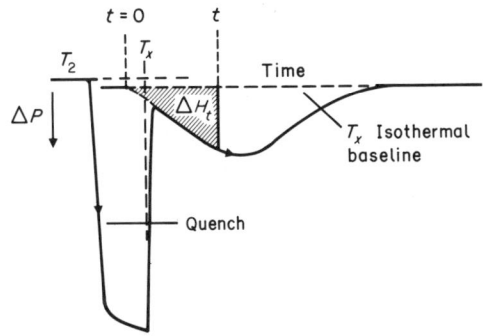

Figure 26 Schematic curve for isothermal crystallization showing the approach to T_x from $T_2(> T_m)$

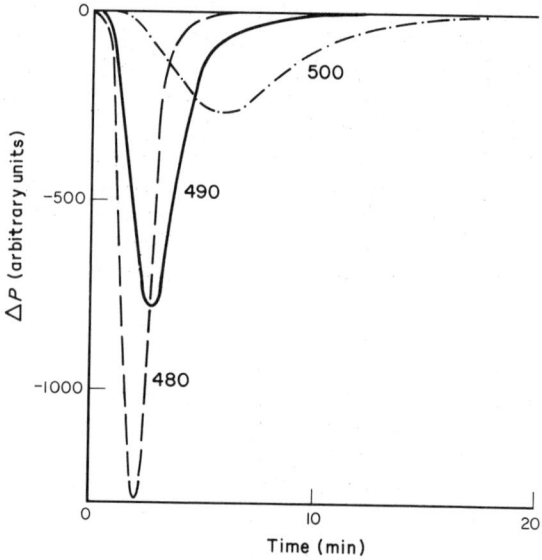

Figure 27 Isothermal crystallization of PET at the temperatures indicated

Crystallization kinetics of polymers in bulk are almost invariably treated using the Avrami approach which considers the evolution of the volume fraction (x_t) crystallized

$$\ln(1 - x_t) = -Zt^n \qquad (20)$$

where Z is the rate constant, t the time and the exponent n is characteristic of the crystallization mechanism. A double logarithmic plot of equation (20) gives n, which is almost invariably found to be a fractional value (often between three and four) and the basic Avrami assumptions must be heavily qualified for crystallization from a viscous melt. A very useful discussion has been given[137] of the effect of uncertainties in the experimental quantities (t, ΔH_t, ΔH_∞) on the Avrami 'constants'. Problems are not specific to DSC, similar results being obtained using dilatometry,[138] and improvements are required in the theory rather than the technique; many of the original Avrami assumptions are not valid for crystallizing polymers.[137]

36.6.5 Crystallinity

It was shown in Section 36.6.3.1 how the heat of fusion could be referred to any value of T. Room temperature (RT) is particularly important because it is to this that most other measurements of crystallinity refer. The x_H scale is normally calibrated for $\Delta H(x = 1, RT)$ using the inverted relationship

$$\Delta H(x = 1, RT) = \Delta H(x, RT)/x_x \qquad (21)$$

where x_x is crystallinity measured by another, non-enthalpic procedure. The implicit assumption $x_x = x_H$ has been made but, if incorrect, the source of error is well defined and, with improved knowledge, a modified relationship (see later) may be obtainable. Unfortunately the situation is generally confused by using not $\Delta H(x, RT)$ but $\Delta H(x?, T?)$, where the queries indicate that the measured value of ΔH refers to an unknown crystallinity (but almost certainly $< x_{RT}$) and temperature. With soundly based data it is possible to proceed further and analyze the data with respect to yet another implicit assumption, i.e. that a simple two-phase, crystal + amorphous, model holds. Recent spectroscopic work has shown that interfacial regions are significant and can be quantified[139] but it is still rare to find that they are considered even though the procedures required were available early in the development of DSC.[140]

It has been suggested that DSC measurements alone can give crystallinity without recourse to external calibration.[141,142] The method is based on two different responses to x — the heat of fusion and the ΔC_p step at T_g. The latter roughly follows the relationship $\Delta C_p T_g \approx$ constant (Section 36.5.5) when $x = 0$ and is assumed to vanish when $x = 1$. Measurements on two different samples are needed and appear to give good results[141] for poly(dimethylsiloxane) and polybutadiene but the method fails for PET[143] for which the fall-off in ΔC_p at T_g is much more rapid than the decrease in crystallinity, a common observation for many polymers.[144]

36.6.6 Multicomponent Systems

DSC is potentially a simple and useful technique for measuring the interaction parameter χ between polymers. The melting point T_m of a crystallizable polymer (2) in a compatible blend with an amorphous component (1) is given by[145]

$$\frac{1}{T_m} - \frac{1}{T_m^\circ} = -\frac{R V_{2u}}{\Delta H_{2u} V_{1u}} \cdot \chi V_1^2 \tag{22}$$

where T_m° is the melting point of the pure polymer, V_u and ΔH_u the molar volume and heat of fusion of the repeat unit respectively, V the volume fraction, R the gas constant, and MWs of both components are $> 10^4$.

Equation (22) implies a simple relationship between T_m^{-1} and composition but the difficulties in determining T_m using DSC have already been discussed (Section 36.6.1–2). These must be overcome if equation (22) is to be used successfully because it is based on purely thermodynamic considerations and T_m must refer to the equilibrium value. This calls for low $q(+)$ or, preferably, thermal pretreatment to facilitate lamellar reorganization to the equilibrium state. Equation (22) makes no allowance for changes in crystal perfection, morphology and lamellar thickness, and even crystal structure due to the changed melt environment. Extensive preliminary work is required to define conditions that simulate those of the crystallizing pure component with regard to the degree of supercooling,[146] but even then changing conditions during growth, as the concentration of crystallizable units decreases and can only be replenished by transport through (if $T_{g1} > T_{g2}$) an increasingly viscous medium, play havoc with the kinetics and make it impossible to maintain correspondence with pure component 2. It is possible to allow for morphological changes[147] to obtain self-consistent χ values but the overall effect of environment on the melting point of a polymer is a summation of several contributions, some quite subtle. These have been investigated in a series of experiments using model compounds, *i.e.* polymer crystals in glassy and rubbery matrices of differing compatibility.[148-150] The model work has shown that even for incompatible components, interfacial effects or a rigid matrix may lead to a small depression in T_m. There is some evidence that the rigid matrix restricts lamellar reorganization and under the condition $T_g(\text{matrix}) > T_{2m}$ the melting point is directly observable.

DSC has been used for the construction of phase diagrams for polymer/diluent systems. These have mainly formed simple eutectics[151,152] although schematic DTA curves have been given for a more complex situation.[153] The eutectic curves clearly illustrate both the potential and the problems of experimental work on polymer systems. The diluent branch can be calculated using χ as a variable. There is good agreement between theory and experiment using a reasonable value for χ but only a very qualitative approximation to the liquidus curve of the polymeric component is possible because of the metastable nature of the polymer crystals formed.

When the diluent is water, it is common to relate any deficit in the observed heat of fusion to a 'bound water' fraction[154,155] and a very comprehensive treatment of this has been given.[156] It is essential to use as many different approaches as possible (*e.g.* DSC measurements of C_p as a function of composition both below and above the melting point of water) to ensure that the enthalpy deficit refers to a genuine structural state as opposed to a slow approach to equilibrium.

Somewhat different types of polymer/diluent systems are represented by the gels that form by physical, rather than chemical, crosslinks. These may be glassy (Section 36.5.7) or, more commonly, crystalline regions[157-159] and it is the behaviour of these that is of interest. Gels may be examined at all stages from the initial one, when the signal is relatively weak (typical polymer concentrations are less than 10%), through to the dry product which is usually obtained by exchange of the original solvent by a more volatile liquid. The detection of concentration/dissolution/melting/crystallization events is a powerful tool for the study of these increasingly important systems.

36.7 POLYMERIZATION

Any chemical reaction is accompanied by enthalpy changes which can be followed by calorimetric methods and DSC is widely used to study polymerizing systems, especially those based on epoxides.[160-162] Both rate and extent of reaction can be monitored using either isothermal or scanning modes of operation. It was originally hoped that the latter would provide equivalent information to that obtained from a family of isothermal curves, and in a fraction of the time, but, as with thermogravimetry (Volume 1, Chapter 37) this has not proved possible for reasons that relate more to the materials studied than to the technique of DSC.

For any reacting system the fractional conversion α is defined as $\Delta h_t / \Delta h_r$, where Δh_t is the partial heat of reaction from $t = 0 \to t$ and Δh_r is the total heat of reaction ($t = 0 \to \infty$). The rate of conversion is given by

$$d\alpha/dt \;=\; (dh/dt)/\Delta h_r \;=\; kf(\alpha) \tag{23}$$

where k is the rate constant, which is generally assumed to have the usual Arrhenius form $A\exp(-E/RT)$. A is the preexponential factor, E the activation energy and $f(\alpha)$ is some function of α — for an nth order reaction it is $(1-\alpha)^n$. In this case

$$\ln(d\alpha/dt) \;=\; \ln A \;+\; n\ln(1-\alpha) \;-\; E/RT \tag{24}$$

so that for an isothermal reaction n is available from a graph of $\ln(\text{rate})$ against $\ln(1-\alpha)$ and, in a scanning situation, from a multiple regression analysis. Whatever method of analysis is chosen, Δh_r is the fundamental quantity on which any extent of reaction is based and its derivation is considered in the following section.

36.7.1 Heats of Reaction

Figure 28 shows the DSC curve for the cure of a commercial coating material; it includes the glass transition region (A) of the reactants (R), the exothermic cure region (B) and the region of fully cured products (P, shown at C). The rerun curve shows only T_g for the products (D) before merging with the original curve at C. For the particular example of Figure 28, the large peak at A (due to isothermal annealing at room temperature, Section 36.5.2) shows that the sample has been stored for some time — how long could be decided from a calibration curve. The reaction represented by Figure 28 is physically simple in the sense that there is no loss of mass, even in unsealed pans (although a loss of one or two per cent of moisture is common if material is not stored in a dessicator), and also in that it proceeds to completion as judged by the linear $C_p - T$ behaviour at the highest temperatures (region C), which agrees with that for the rerun. If volatile products are formed, the reaction must be carried out in sealed pans or under pressure[163] to conserve mass and avoid an uncertain correction for vaporization or to pressure-shift simultaneous vaporization events. It is often difficult to ensure that the reaction goes to completion in isothermal work; the final stages may be lost in instrumental noise and/or insensitivity and in this sense scanning experiments give a better approach to Δh_r. Even here there may be problems if an upper temperature limit is set

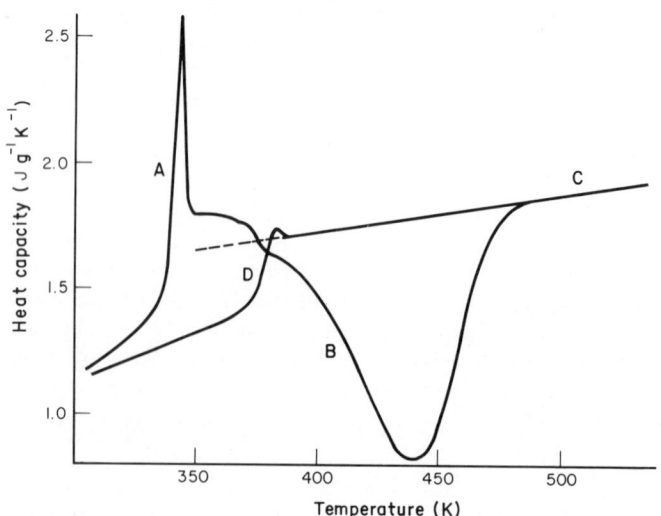

Figure 28 The cure of a powder coating compound. Lettered regions are identified in the text

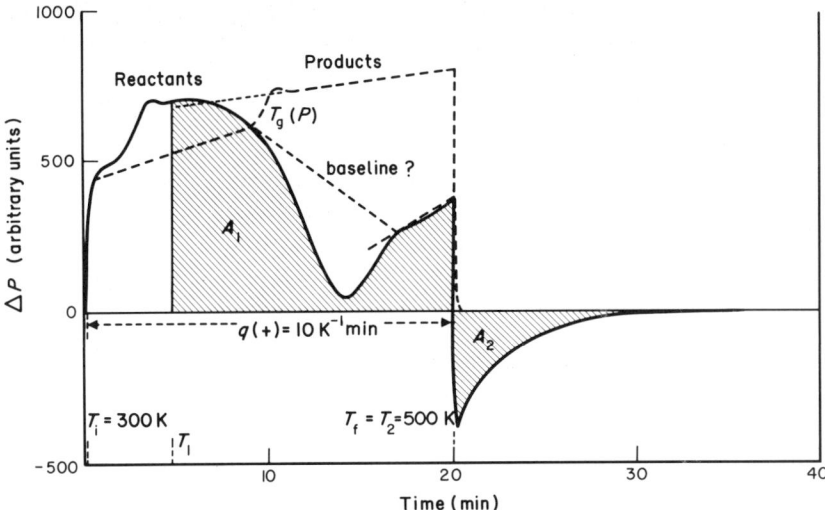

Figure 29 Cure of a printed circuit board material, using both scanning and isothermal modes of operation to avoid competition with degradation at higher tempertures. See text for explanation of symbols

by the start of degradation (Figure 29) when the definition of Δh_r by any kind of conventional baseline becomes meaningless (see below).

Since cure perturbs what would be a simple heat capacity curve, it is normal to construct a baseline that simulates the latter (or its change with composition as reaction progresses) and assume that the area defined by this baseline and the cure curve defines Δh_r. In Figure 28 a simple construction is to join the fairly obvious start and end of reaction — although this implicitly assumes that $\Delta C_p = C_p(\text{P}) - C_p(\text{R}) \approx 0$. There is no analogous construction for Figure 29, which is a composite full-minus-empty curve (see Figure 7) for a slow-to-cure printed circuit board material together with the equivalent curve for the product. Scanning was stopped at $T_2 = 500$ K as a compromise between thermal degradation at higher temperatures and a reasonably short time (<20 min) for the completion of cure isothermally at T_2. Any treatment that considers only the dynamic portion of Figure 29 is seriously flawed since it can take no account of the isothermal contribution (A_2, a negative quantity) to the overall enthalpy change. In fact no simple baseline is appropriate but this is not serious because a thermodynamically correct value of Δh_r is easily obtained.[164] The shaded area ($A_1 + A_2$) of Figure 29 represents $H(\text{P,r},T_2) - H(\text{R,l},T_1)$, the corresponding quantity for the product alone is $H(\text{P,r},T_2) - H(\text{P,r},T_1)$ and subtraction of the two gives $\Delta h_r(T_1)$ $= H(\text{P, r, } T_1) - H(\text{R,l},T_1)$, where r represents the rubbery crosslinked state; a short extrapolation below $T_g(\text{P})$ is required if, as shown, $T_1 < T_g(\text{P})$. Sufficient information is available in Figure 29 to convert $\Delta h_r(T_1)$ to $\Delta h_r(T)$ using $\Delta C_p = C_p(\text{P}) - C_p(\text{R})$ and $\Delta T = T - T_1$. The appropriate 'baselines' for Figure 29 are therefore the curve for $C_p(\text{P})$ and the 'no signal' isothermal baseline. Anything else, *e.g.* as sketched, will neglect an appreciable fraction of Δh_r even when the isothermal contribution is small. Tests should always be made to check the validity of derived values of Δh_r by curing under different time/temperature regimes which should give consistent values unless the reaction mechanism varies with temperature — which is valuable information in itself.

Although scanning experiments can clearly show the completion of cure, the kinetic interpretation of results is simplified for isothermal conditions — but in attaining this state any sample requires a finite time to equilibrate at the chosen reaction temperature T_r. The most widely used procedure[165] is to equilibrate the instrument at T_r, insert the sample and, when the reaction is judged to be complete, to repeat the procedure with the cured product. It is hoped that any difference between the two curves in the initial approach to the isothermal region is due to early cure (the maximum rate for an *n*th order reaction is at $t = 0$) which can thus be estimated. However the physical placement of a sample in the DSC introduces an unquantifiable effect and a better approach for modern calorimeters is to program rapidly (>150 K min^{-1}) to T_r, recording data at all times and, on completion, cool and repeat; the experimental configuration remains unchanged throughout the several operations. If there is any uncertainty regarding instrumental overshoot at the end of the period of rapid heating, a dummy experiment can be carried out using a standard melting point material such as indium: the instrument is set to isotherm at $T_m - \Delta T$ and when equilibrium has been attained a check is made to see if the sample has melted for a given value of ΔT (if melting has occurred the material will remain molten because supercooling of at least a few degrees is required to induce recrystallization). Trial and error soon demonstrates what value of ΔT is characteristic of a given combination of instrument and heating rate; for a power compensation instrument $\Delta T < 0.1$ K for $q(+) = 160$ K min^{-1}.

36.7.2   General Comments on Kinetics

Polymerizing systems are far removed from the gas phase reactions of classical chemical kinetics and concepts that are valid for the latter will at best be gross approximations in highly viscous media. Each polymer system must therefore be treated on its own merits. On a fundamental level DSC may only provide confirmation of a detailed reaction mechanism that is based on other techniques. On a more pragmatic level, extents of reaction are obtainable from Δh_r and isothermal or simple scanning curves (Figure 28). For the more complex case of Figure 29, residual cure can be estimated for a series of samples and a kinetic curve formed in this way. No attempt is made here to survey the contentious topic of kinetic parameters and DSC;[166] full reviews of the many procedures that have been used for the treatment of isothermal and dynamic DSC curves are readily available[160, 161, 167-169] together with useful additional experimental details.[170]

Although isothermal experiments are intrinsically simpler to interpret, many polymerizations are extremely complex with several simultaneous processes that have different activation energies. As a result, the reaction path changes with temperature and a dynamic curve gives useful empirical information that summarizes behaviour over the whole temperature region. Derived parameters represent some 'average reaction'; they have no fundamental meaning but allow reasonable estimates to be made of degrees of conversion at other heating rates or under isothermal conditions. It is with this type of application in mind that the peak temperature method[171, 172] [the shift of reaction peak position with $q(+)$] for E and A has been adapted to form the basis of ASTM Standard E698-79. This rightly emphasizes the potential hazards of uncritical use and recommends a trial calculation of the half-life for isothermal cure at some convenient temperature, which is then checked experimentally. If theory and experiment agree other predictions may be made with some confidence. Many of the concepts of crystallization, polymerization kinetics and degradation have similarities and a useful summary has been given.[173]

36.7.3   Other Systems — Photopolymerization

Although previous remarks have been mainly concerned with epoxides, DSC has been applied to reacting systems as diverse as rubbers[162] and template polymerization.[174] There is much current interest in photopolymerization, which is rapid and may be carried out in solvent-free conditions.[175-178] Several manufacturers now offer 'bolt-on' equipment, which allows the ready conversion of existing DSC instrumentation so that a sample can be exposed to a known dose of UV light.

36.8   POLYMER STABILITY

Stability is more commonly studied using thermogravimetric methods (Volume 1, Chapter 37) but there are occasion when DSC proves helpful.[179] A reduction in MW due to thermal degradation implies a decrease in T_g and this is conveniently monitored using DSC.[180] An important practical application is in the determination of the oxidative induction time of stabilized polymers[100] and rubbers,[181] which forms the basis of ASTM D3895-80. Samples are programmed to the required temperature with the usual inert carrier gas in use; this is then changed to oxygen and the time to the start of the exothermic oxidation reaction noted. Special precautions are needed in sample preparation and to ensure ready access of oxygen, but the result is a very useful measurement. In general isothermal oxidation conditions are used, but some advantages have been claimed for scanning procedures when higher temperatures replace longer times as indicators of improved stability.[182]

36.9   OTHER TRANSITIONS

Previous sections have emphasized the three main areas of DSC studies: the glass transition, crystallization and melting, and heats of reaction. A variety of other applications, mainly to phase changes, can only be briefly referred to here. DSC is very widely used in the study of polymer liquid crystals[183-186,207] and of the complexes that are formed between polyethers and metal ions[187] and other low MW components.[188] The many types of comb-like polymers[189-194,207] are interesting in that some have the unusual property that $T_g > T_m$, since T_g refers to the main chain whereas crystallinity is due to relatively short side chains, *e.g. n*-alkyl groups.

36.9.1 Low Temperature Transitions

Some comb-like polymers have another unusual property — the appearance of a sub-T_g transition at temperatures considerably below those discussed in Section 36.5.4. In general, the low temperature transitions that are often seen in DMA/DMTA work do not appear in DSC traces. However it seems clear that certain poly[N-(n-alkyl)maleimides][190,191] and esters of poly(itaconic acid)[192,193] show low temperature (< 200 K) events that are related to a local side chain glass transition. The magnitudes of the changes in C_p associated with these events are not large and careful quantitative work is needed to demonstrate their presence.

36.9.2 Liquid/Liquid Transitions

A variety of evidence[195] suggests that there may be a transition, T_{ll}, in molten polymers at temperatures given very approximately by $T_{ll} \approx 1.2\,T_g$. The evidence, which includes DSC data for poly(methyl methacrylate) (PMMA)[195] and polystyrene (PS),[196] is controversial and contrary views have been expressed.[197,198] The relevant DSC curves show increased slopes at T_{ll} but all relate to direct recorder traces and it would be more appropriate to have more quantitative data in the form of heat capacity curves. By using these, all instrumental curvature is eliminated by subtraction of the contribution due to the empty curve (equation 5). The existing evidence for PMMA[199] and PS[200] suggests that C_{pl} is an excellent linear function of temperature which conflicts with the concept of T_{ll}. It should be noted that work related to T_{ll} must of necessity extend to fairly high temperatures and samples should be checked for stability by allowing isothermal periods at the upper temperatures. Any drift implies degradation (for a correctly operating DSC), which could also perturb the dynamic curve — low MW volatile compounds would have the same effect but these would be expected to have vaporized earlier.

36.10 ANALYTICAL APPLICATIONS

The quantitative capability of DSC can be developed as a useful analytical tool, the basis for which is the dilution of the thermal response of the host polymer (P) by an additional component (A), which may be another polymer, a reinforcing fibre, or a low MW additive or filler. The most direct application features the heat capacity in a morphologically insensitive region, below T_g (where C_p is almost independent of morphology[201,202]) or above T_m, and assumes additivity of the contributions due to $C_p(P)$ and $C_p(A)$ (the assumption should be checked by direct measurements on model mixtures). The accuracy obviously increases with the difference between $C_p(P)$ and $C_p(A)$ and this is usually at a maximum when P is molten. Some typical C_p values (J g^{-1} K^{-1}) for specific systems for which the procedure has been used (especially for monitoring product homogeneity), are poly(ether ether ketone)/carbon fibre (650 K) 2.18/1.46, polypropylene/limestone (400 K) 2.52(supercooled)/ 0.93 and epoxy/silica (420 K) 2.00/0.91. Crystalline regions should be avoided because component A may affect crystallization, giving gross deviations from additivity.

A natural extension is to more specific effects such as the magnitude of the ΔC_p step at T_g (*e.g.* for determination of the dry rubber content of a latex[203]) although care is always needed to ensure that additivity prevails, or, if not (Section 36.5.6–7), to generate a suitable calibration curve.

36.11 CONCLUDING REMARKS

This chapter has naturally concentrated on DSC, but it would be misleading to give the impression that it is a universal panacea. The more techniques to which a problem can be exposed, the greater the chance of a solution.[204] The complementary nature of DSC and DMA/DMTA was touched upon in Section 36.9.1 and the growth of simultaneous methods, already available for

DSC/TG (plus analytical procedures for evolved gases) and for DSC/optical measurements, can be expected; DSC/FTIR[205] and DSC/SAXS[206] have already been reported.

As a technique DSC is very simple and useful results (approximate temperatures and magnitudes of events) can readily be obtained by newcomers after only a brief perusal of the instrument manual. Full quantitative operation is not difficult but requires rather greater effort, especially in the treatment of data, which should be routine with the widespread availability of microcomputers. Unfortunately these still tend to be used to perform calculations, based on the manipulation of chart recorder traces, which do not give thermodynamically meaningful data. For this reason the chapter has emphasized the quantitative capability of DSC, a feature that certainly justifies the calorimetric aspect of the name and brings the power of thermodynamics to bear on a wide range of problems of polymer science and technology.

36.12 REFERENCES

1. R. C. Mackenzie, *Pure Appl. Chem.*, 1985, **57**, 1737.
2. R. E. Wetton, in 'Developments in Polymer Characterisation', ed. J. V. Dawkins, Elsevier, London, 1986, vol. 5, p. 179.
3. B. Ke, in 'Newer Methods of Polymer Characterization', ed. B. Ke, Wiley, New York, 1964.
4. S. L. Boersma, *J. Am. Ceram. Soc.*, 1955, **38**, 281.
5. M. J. O'Neill, *Anal. Chem.*, 1964, **36**, 1238.
6. A. P. Gray, in 'Analytical Calorimetry', ed. R. S. Porter and J. F. Johnson, Plenum Press, New York, vol. 1, p. 209.
7. R. A. Baxter, in 'Thermal Analysis', Academic Press, New York, 1969, vol. 1, p. 65.
8. S. C. Mraw, *Rev. Sci. Instrum.*, 1982, **53**, 228.
9. R. C. Mackenzie, *Proc. Anal. Div. Chem. Soc.*, 1980, **17**, 217.
10. K. Latrous, J. P. Cavrot and F. Rietsch, *Eur. Polym. J.*, 1981, **17**, 1205.
11. R. E. Lyon, R. J. Farris and W. J. MacKnight, *J. Polym. Sci., Polym. Lett. Ed.*, 1983, **21**, 323.
12. R. B. Sanderson and J. B. Badenhorst, in 'Order in the Amorphous State of Polymers', ed. S. E. Keinath, R. L. Miller and J. K. Rieke, Plenum Press, New York, 1987, p. 427.
13. K. N. Marsh (ed.), 'Materials for the Realization of Physicochemical Properties', Blackwell, Oxford, 1987.
14. M. J. Richardson and N. G. Savill, *Thermochim. Acta*, 1975, **12**, 213.
15. J. H. Flynn, *Thermochim. Acta*, 1974, **8**, 69.
16. U. Gaur, A. Mehta and B. Wunderlich, *J. Therm. Anal.*, 1978, **13**, 71.
17. M. J. Richardson, in 'Developments in Polymer Characterisation', ed. J. V. Dawkins, Applied Science, London, 1978, vol. 1, p. 205.
18. M. J. Richardson, in 'Compendium of Thermophysical Property Measurement Methods', ed. K. D. Maglic, A. Cezairliyan and V. E. Peletsky, Plenum Press, New York, 1984, vol. 1, p. 669.
19. M. Dole, *Fortschr. Hochpolym. -Forsch.*, 1960, **2**, 221.
20. E. A. Turi (ed.), 'Thermal Characterization of Polymeric Materials', Academic Press, New York, 1981.
21. R. -J. Roe and A. E. Tonelli, *Macromolecules*, 1978, **11**, 114.
22. S. Strella and P. F. Erhardt, *J. Appl. Polym. Sci.*, 1969, **13**, 1373.
23. M. J. Richardson and N. G. Savill, *Polymer*, 1975, **16**, 753.
24. P. Peyser and W. D. Bascom, *J. Macromol. Sci., Phys.*, 1977, **13**, 597.
25. A. J. Kovacs, *Fortschr. Hochpolym. -Forsch.*, 1963, **3**, 394.
26. M. J. Richardson and N. G. Savill, *Polymer*, 1977, **18**, 413.
27. S. M. Ellerstein, *Appl. Polym. Symp.*, 1966, **2**, 111.
28. A. R. Ramos, J. M. Hutchinson and A. J. Kovacs, *J. Polym. Sci., Polym. Phys. Ed.*, 1984, **22**, 1655.
29. C. T. Moynihan, A. J. Easteal, J. Wilder and J. Tucker, *J. Phys. Chem.*, 1974, **78**, 2673.
30. M. J. Richardson and N. G. Savill, *Br. Polym. J.*, 1979, **11**, 123.
31. S. E. B. Petrie, *J. Macromol. Sci., Phys.*, 1976, **12**, 225.
32. R. R. Lagasse, *J. Polym. Sci., Polym. Phys. Ed.*, 1982, **20**, 279.
33. A. S. Marshall and S. E. B. Petrie, *J. Appl. Phys.*, 1975, **46**, 4223.
34. R. -J. Roe, *Macromolecules*, 1981, **14**, 1586.
35. R. E. Robertson, *J. Appl. Phys.*, 1978, **49**, 5048.
36. J. L. G. Ribelles, R. Diaz-Calleja, R. Ferguson and J. M. G. Cowie, *Polymer*, 1987, **28**, 2262.
37. H. -J. Ott, *Colloid Polym. Sci.*, 1979, **257**, 486.
38. H. Yoshida and Y. Kobayashi, *J. Macromol. Sci., Phys.*, 1982, **21**, 565.
39. Yu. A. Sharanov and M. V. Volkenshtein, in 'The Structure of Glass', ed. E. A. Porai-Koshits, Consultants Bureau, New York, 1966, vol. 6, p. 62.
40. H. Sasabe and C. T. Moynihan, *J. Polym. Sci., Polym. Phys. Ed.*, 1978, **16**, 1447.
41. J. M. O'Reilly, *J. Appl. Phys.*, 1979, **50**, 6083.
42. M. A. DeBolt, A. J. Easteal, P. B. Macedo and C. T. Moynihan, *J. Am. Ceram. Soc.*, 1976, **59**, 16.
43. I. M. Hodge and G. S. Huvard, *Macromolecules*, 1983, **16**, 371.
44. I. M. Hodge, *Macromolecules*, 1983, **16**, 898.
45. V. P. Privalko, S. S. Demchenko and Y. S. Lipatov, *Macromolecules*, 1986, **19**, 901.
46. G. C. Stevens and M. J. Richardson, *Polym. Commun.*, 1985, **26**, 77.
47. J. J. Tribone, J. M. O'Reilly and J. Greener, *Macromolecules*, 1986, **19**, 1732.
48. I. Avramov, E. Grantscharova and I. Gutzow, *J. Non-Cryst. Solids*, 1987, **91**, 386.
49. J. M. Hutchinson, in 'Molecular Dynamics and Relaxation Phenomena in Glasses', ed. Th. Dorfmüller and G. Williams, Springer, Heidelberg, 1987, p. 172.
50. I. M. Hodge and A. R. Berens, *Macromolecules*, 1982, **15**, 762.
51. K. -H. Illers, *Makromol. Chem.*, 1969, **127**, 1.

52. M. R. Tant and G. L. Wilkes, *J. Appl. Polym. Sci.*, 1981, **26**, 2813.
53. H. S. Chen and T. T. Wang, *J. Appl. Phys.*, 1981, **52**, 5898.
54. A. R. Schultz and A. L. Young, *Macromolecules*, 1980, **13**, 663.
55. R. M. Kimmel and D. R. Uhlmann, *J. Appl. Phys.*, 1971, **42**, 4917.
56. W. C. Dale and C. E. Rogers, *J. Appl. Polym. Sci.*, 1972, **16**, 21.
57. A. Weitz and B. Wunderlich, *J. Polym. Sci., Polym. Phys. Ed.*, 1974, **12**, 2473.
58. J. B. Yourtee and S. L. Cooper, *J. Appl. Polym. Sci.*, 1974, **18**, 897.
59. R. E. Wetton and H. G. Moneypenny, *Br. Polym. J.*, 1975, **7**, 51.
60. I. G. Brown, R. E. Wetton, M. J. Richardson and N. G. Savill, *Polymer*, 1978, **19**, 659.
61. C. Price, *Polymer*, 1975, **16**, 585.
62. I. M. Hodge and A. R. Berens, *Macromolecules*, 1985, **18**, 1980.
63. V. A. Bershtein, V. M. Yegorov, L. G. Razgulyayeva and V. A. Stepanov, *Polym. Sci. USSR (Engl. Transl.)*, 1978, **20**, 2560.
64. T. G. Fox and P. J. Flory, *J. Appl. Phys.*, 1950, **21**, 581.
65. R. F. Boyer, *Macromolecules*, 1974, **7**, 142.
66. K. Ueberreiter and G. Kanig, *J. Colloid Sci.*, 1952, **7**, 569.
67. R. F. Fedors, *Polymer*, 1979, **20**, 518.
68. J. H. Gibbs, *J. Chem. Phys.*, 1956, **25**, 185.
69. J. H. Gibbs and E. A. DiMarzio, *J. Chem. Phys.*, 1958, **28**, 373.
70. P. R. Couchman, *J. Mater. Sci.*, 1980, **15**, 1680.
71. M. Goldstein, *Macromolecules*, 1985, **18**, 277.
72. P. R. Couchman, *Macromolecules*, 1987, **20**, 1712.
73. J. M. G. Cowie, *Eur. Polym. J.*, 1975, **11**, 297.
74. R. P. Kusy and A. R. Greenberg, *Polymer*, 1984, **25**, 600.
75. A. R. Greenberg and R. P. Kusy, *Polymer*, 1984, **25**, 927.
76. I. Havlicek and L. Nicolais, *Polymer*, 1986, **27**, 921.
77. A. Rudin and D. Burgin, *Polymer*, 1975, **16**, 291.
78. P. Claudy, J. M. Letoffe, Y. Camberlain and J. P. Pascault, *Polym. Bull. (Berlin)*, 1983, **9**, 208.
79. J. M. G. Cowie, *Eur. Polym. J.*, 1973, **9**, 1041.
80. R. P. Kusy, W. F. Simmons and A. R. Greenberg, *Polymer*, 1981, **22**, 268.
81. D. T. Turner and A. Schwartz, *Polymer*, 1985, **26**, 757.
82. R. F. Boyer, *J. Macromol. Sci., Phys.*, 1973, **7**, 487.
83. F. Rietsch, D. Daveloose and D. Froelich, *Polymer*, 1976, **17**, 859.
84. J. H. Glans and D. T. Turner, *Polymer*, 1981, **22**, 1540.
85. L. H. Judovits, R. C. Bopp, U. Gaur and B. Wunderlich, *J. Polym. Sci., Polym. Phys. Ed.*, 1986, **24**, 2725.
86. A. R. Greenberg and R. P. Kusy, *J. Appl. Polym. Sci.*, 1980, **25**, 1785.
87. G. C. Stevens and M. J. Richardson, *Polymer*, 1983, **24**, 851.
88. C. Feger and W. J. MacKnight, *Macromolecules*, 1985, **18**, 280.
89. N. W. Johnston, *J. Macromol. Sci., Rev. Macromol. Chem.*, 1976, **14**, 215.
90. P. R. Couchman, *Macromolecules*, 1982, **15**, 770.
91. J. S. Roman, E. L. Madruga and J. Guzman, *Polym. Commun.*, 1984, **25**, 373.
92. J. Biros, T. Larina, J. Trekoval and J. Pouchly, *Colloid Polym. Sci.*, 1982, **260**, 27.
93. J. M. Barton, *J. Polym. Sci., Part C*, 1970, **30**, 573.
94. S. Krause, Z. Lu and M. Iskandar, *Macromolecules*, 1982, **15**, 1076.
95. S. Bywater, *Polym. Eng. Sci.*, 1984, **24**, 104.
96. G. C. Meyer and J. M. Widmaier, *J. Polym. Sci., Polym. Phys. Ed.*, 1982, **20**, 389.
97. U. Gaur and B. Wunderlich, *Macromolecules*, 1980, **13**, 1618.
98. W. J. MacKnight, F. E. Karasz and J. R. Fried, in 'Polymer Blends', ed. D. R. Paul and S. Newman, Academic Press, New York, 1978, vol. 1, p. 186.
99. J. R. Fried, in 'Developments in Polymer Characterisation', ed. J. V. Dawkins, Applied Science, London, 1983, vol. 4, p. 39.
100. H. E. Bair, in 'Thermal Characterization of Polymeric Materials' ed. E. A. Turi, Academic Press. New York, 1981, p. 845.
101. S. Lau, J. Pathak and B. Wunderlich, *Macromolecules*, 1982, **15**, 1278.
102. J. L. Feijoo, A. J. Muller and J. R. Acosta, *J. Mater. Sci., Lett.*, 1986, **5**, 1193.
103. P. R. Couchman and F. E. Karasz, *J. Polym. Sci., Polym. Phys. Ed.*, 1977, **15**, 1037.
104. M. Ebert, R. W. Garbella and J. H. Wendorff, *Makromol. Chem., Rapid Commun.*, 1986, **7**, 65.
105. J. M. Rodriguez-Parada and V. Percec, *J. Polym. Sci., Polym. Chem. Ed.*, 1986, **24**, 579.
106. R. S. Barnum, S. H. Goh, J. W. Barlow and D. R. Paul, *J. Polym. Sci., Polym. Lett. Ed.*, 1985, **23**, 395.
107. L. H. Dunlap, C. R. Foltz and A. G. Mitchell, *J. Polym. Sci., Polym. Phys. Ed.*, 1972, **10**, 2223.
108. S. Wellinghoff, J. Shaw and E. Baer, *Macromolecules*, 1979, **12**, 932.
109. J. Arnauts and H. Berghmans, *Polym. Commun.*, 1987, **28**, 66.
110. J. Francois, J. Y. S. Gan and J. -M. Guenet, *Macromolecules*, 1986, **19**, 2755.
111. T. S. Ellis, F. E. Karasz and G. ten Brinke, *J. Appl. Polym. Sci.*, 1983, **28**, 23.
112. P. Peyser, *Polym. -Plast. Technol. Eng.*, 1978, **10**, 117.
113. E. Alfthan, A. De Ruvo and M. Rigdahl, *Int. J. Polym. Mater.*, 1979, **7**, 163.
114. P. S. Theocaris, G. C. Papanicolaou and E. P. Sideridis, *J. Reinf. Plast. Compos.*, 1982, **1**, 92.
115. B. Wunderlich, 'Macromolecular Physics', Academic Press, New York, 1973, vol. 1.
116. B. Wunderlich, 'Macromolecular Physics', Academic Press, New York, 1976, vol. 2.
117. B. Wunderlich, 'Macromolecular Physics', Academic Press, New York, 1980, vol. 3.
118. S. W. Shalaby, in 'Thermal Characterization of Polymer Materials' ed. E. A. Turi, Academic Press, New York, 1981, p. 235.
119. V. B. F. Mathot, *Polymer*, 1984, **25**, 579.
120. K. -H. Illers, *Eur. Polym. J.*, 1974, **10**, 911.

121. V. I. Selikhova, Yu. A. Zubov, N. F. Bakeyev and G. P. Belov, *Polym. Sci. USSR (Engl. Transl.)*, 1977, **19**, 879.
122. L. Mandelkern, 'Crystallization of Polymers', McGraw-Hill, New York, 1964.
123. H. D. Chanzy, E. Bonjour and R. H. Marchessault, *Colloid Polym. Sci.*, 1974, **252**, 8.
124. E. Hellmuth and B. Wunderlich, *J. Appl. Phys.*, 1965, **36**, 3039.
125. A. M. Rijke and L. Mandelkern, *J. Polym. Sci., Part A-2*, 1970, **8**, 225.
126. D. J. Blundell, F. N. Cogswell, P. J. Holdsworth and F. M. Willmouth, *Polymer*, 1977, **18**, 204.
127. A. J. Pennings and A. Zwijnenburg, *J. Polym. Sci., Polym. Phys. Ed.*, 1979, **17**, 1011.
128. H. G. Zachmann, *Kolloid. Z.*, 1967, **216–217**, 180.
129. M. J. Richardson, *Polym. Testing*, 1984, **4**, 101.
130. V. B. F. Mathot and M. F. J. Pijpers, *J. Therm. Anal.*, 1983, **28**, 349.
131. H. N. Beck, *J. Appl. Polym. Sci.*, 1975, **19**, 371.
132. R. Legras, C. Bailly, M. Daumerie, J. M. Dekoninck, J. P. Mercier, V. Zichy and E. Nield, *Polymer*, 1984, **25**, 835.
133. D. Garcia, *J. Polym. Sci., Polym. Phys. Ed.*, 1984, **22**, 2063.
134. C. N. Velisaris and J. C. Seferis, *Polym. Eng. Sci.*, 1986, **26**, 1574.
135. J. N. Hay and P. J. Mills, *Polymer*, 1982, **23**, 1380.
136. M. Yagpharov, *J. Therm. Anal.*, 1986, **31**, 1073.
137. D. Grenier and R. E. Prud'homme, *J. Polym. Sci., Polym. Phys. Ed.*, 1980, **18**, 1655.
138. J. N. Hay, *Br. Polym. J.*, 1979, **11**, 137.
139. M. Glotin and L. Mandelkern, *Colloid Polym. Sci.*, 1982, **260**, 182.
140. A. Peterlin and G. Meinel, *J. Appl. Phys.*, 1965, **36**, 3028.
141. M. Sh. Yagfarov, *Polym. Sci. USSR (Engl. Transl.)*, 1969, **11**, 1355.
142. M. Sh. Yagfarov, *Polym. Sci. USSR (Engl. Transl.)*, 1980, **21**, 2631.
143. S. Y. Hobbs and G. I. Mankin, *J. Polym. Sci., Part A-2*, 1971, **9**, 1907.
144. J. Menczel and B. Wunderlich, *Polym. Prepr., Am. Chem. Soc., Div. Polym. Sci.*, 1986, **27**(1), 255.
145. T. Nishi and T. T. Wang, *Macromolecules*, 1975, **8**, 909.
146. E. Roerdink and G. Challa, *Polymer*, 1978, **19**, 173.
147. T. K. Kwei and H. L. Frisch, *Macromolecules*, 1978, **11**, 1267.
148. I. R. Harrison and J. Runt, *J. Polym. Sci., Polym. Phys. Ed.*, 1980, **18**, 2257.
149. J. P. Runt, *Macromolecules*, 1981, **14**, 420.
150. P. B. Rim and J. P. Runt, *Macromolecules*, 1984, **17**, 1520.
151. P. Smith and A. J. Pennings, *Polymer*, 1974, **15**, 413.
152. S. L. Hager and T. B. Macrury, *J. Appl. Polym. Sci.*, 1980, **25**, 1559.
153. B. Wunderlich, in 'Thermal Characterization of Polymeric Materials', ed. E. A. Turi, Academic Press, New York, 1981, p. 91.
154. B. Bogdanov and M. Mihailov, *J. Therm. Anal.*, 1985, **30**, 1027.
155. T. Hatakeyama, K. Nakamura, H. Yoshida and H. Hatakeyama, *Thermochim. Acta*, 1985, **88**, 223.
156. J. Pouchly, J. Biros and S. Benes, *Makromol. Chem.*, 1979, **180**, 745.
157. E. D. T. Atkins, M. J. Hill, D. A. Jarvis, A. Keller, E. Sarhene and J. S. Shapiro, *Colloid Polym. Sci.*, 1984, **262**, 22.
158. R. C. Domszy, R. Alamo, C. O. Edwards and L. Mandelkern, *Macromolecules*, 1986, **19**, 310.
159. J. -M. Guenet, *Macromolecules*, 1986, **19**, 1961.
160. R. B. Prime in 'Thermal Characterization of Polymeric Materials', ed. E. A. Turi, Academic Press, New York, 1981, p. 435.
161. J. M. Barton, *Adv. Polym. Sci.*, 1985, **72**, 111.
162. D. W. Brazier, in 'Developments in Polymer Degradation', ed. N. Grassie, Applied Science, London, 1981, vol. 3, p. 27.
163. P. F. Levy, G. Nieuweboer and L. C. Semanski, *Thermochim. Acta*, 1970, **1**, 429.
164. S.-S. Chang, *J. Therm. Anal.*, 1988, **34**, 135.
165. J. M. Barton, *Thermochim. Acta*, 1983, **71**, 337.
166. J. Sestak, in 'Comprehensive Analytical Chemistry', ed. G. Svehla, Elsevier, Amsterdam, 1984, vol. XII, part D.
167. J. H. Flynn, *J. Therm. Anal.*, 1988, **34**, 367.
168. T. Provder, R. M. Holsworth, T. H. Grentzer and S. A. Kline, *Adv. Chem. Ser.*, 1983, **203**, 233.
169. J. Galy, A. Sabra and J. -P. Pascault, *Polym. Eng. Sci.*, 1986, **26**, 1514.
170. N. S. Schneider, J. F. Sprouse, G. L. Hagnauer and J. K. Gillham, *Polym. Eng. Sci.*, 1979, **19**, 304.
171. H. E. Kissinger, *Anal. Chem.*, 1957, **29**, 1702.
172. T. Ozawa, *J. Therm. Anal.*, 1970, **2**, 301.
173. Y. P. Khanna and T. J. Taylor, *Polym. Eng. Sci.*, 1987, **27**, 764.
174. G. O. R. Alberda van Ekenstein and Y. Y. Tan, *Polym. Commun.*, 1984, **25**, 105.
175. F. R. Wight and G. W. Hicks, *Polym. Eng. Sci.*, 1978, **18**, 378.
176. G. R. Tryson and A. R. Schultz, *J. Polym. Sci., Polym. Phys. Ed.*, 1979, **17**, 2059.
177. H. Eckhardt, T. Prusik and R. R. Chance, *Macromolecules*, 1983, **16**, 732.
178. J. G. Kloosterboer, G. M. M. van de Hei, R. G. Gossink and G. C. M. Dortant, *Polym. Commun.*, 1984, **25**, 322.
179. N. C. Billingham, D. C. Bott and A. S. Manke, in 'Developments in Polymer Degradation', ed. N. Grassie, Applied Science, London, 1981, vol. 3, p. 63.
180. H. K. Toh and B. L. Funt, *J. Appl. Polym. Sci.*, 1982, **27**, 4171.
181. B. Stenberg and F. Bjork, *J. Appl. Polym. Sci.*, 1986, **31**, 487.
182. L. Koski and K. Saarela, *J. Therm. Anal.*, 1982, **25**, 167.
183. H. Finkelmann, in 'Polymer Liquid Crystals', ed. A. Ciferri, A. R. Krigbaum and R. B. Meyer, Academic Press, New York, 1982, p. 35.
184. C. K. Ober, J. -I. Jin and R. W. Lenz, *Adv. Polym. Sci.*, 1984, **59**, 103.
185. B. Wunderlich and J. Grebowicz, *Adv. Polym. Sci.*, 1984, **60/61**, 1.
186. C. Noel, in 'Recent Advances in Liquid Crystalline Polymers', ed. L. L. Chapoy, Elsevier, London, 1985, p. 135.
187. Y. L. Lee and B. Crist, *J. Appl. Phys.*, 1986, **60**, 2683.
188. K. Suehiro and Y. Nagano, *Makromol. Chem.*, 1983, **184**, 669.
189. Ref. 117, p. 325.

190. J. I. Gonzalez de la Campa, J. M. Barrales-Rienda and J. Gonzalez Ramos, *J. Polym. Sci., Polym. Phys. Ed.*, 1980, **18**, 2197.
191. J. M. Barrales-Rienda, F. Fernandez-Martin and C. R. Galicia, *J. Macromol. Sci., Phys.*, 1984, **23**, 93.
192. J. M. G. Cowie, R. Ferguson and I. J. McEwen, *Polymer*, 1982, **23**, 605.
193. J. M. G. Cowie, I. J. McEwen and M. Y. Pedram, *Macromolecules*, 1983, **16**, 1151.
194. P. B. Rim, *J. Macromol. Sci., Phys.*, 1984, **23**, 549.
195. L. R. Denny, R. F. Boyer and H. -G. Elias, *J. Macromol. Sci., Phys.*, 1986, **25**, 227.
196. S. J. Stadnicki, J. K. Gilham and R. F. Boyer, *J. Appl. Polym. Sci.*, 1976, **20**, 1245.
197. G. D. Paterson, H. E. Bair and A. E. Tonelli, *J. Polym. Sci., Polym. Symp.*, 1976, **54**, 249.
198. J. Chen, C. Kow, L. J. Fetters and D. J. Plazek, *J. Polym. Sci., Polym. Phys. Ed.*, 1982, **20**, 1565.
199. U. Gaur, S. F. Lau, B. B. Wunderlich and B. Wunderlich, *J. Phys. Chem. Ref. Data*, 1982, **11**, 1065.
200. U. Gaur and B. Wunderlich, *J. Phys. Chem. Ref. Data*, 1982, **11**, 313.
201. S. S. Chang and A. B. Bestul, *J. Chem. Phys.*, 1972, **56**, 503.
202. B. Wunderlich and H. Bauer, *Adv. Polym. Sci.*, 1970, **7**, 151.
203. D. R. Burfield, *Polym. Commun.*, 1983, **24**, 178.
204. T. Masuko, R. L. Simeone, J. H. Magill and D. J. Plazek, *Macromolecules*, 1984, **17**, 2857.
205. J. T. Koberstein, I. Gancarz and T. C. Clarke, *J. Polym. Sci., Polym. Phys. Ed.*, 1986, **24**, 2487.
206. J. T. Koberstein and T. P. Russell, *Macromolecules*, 1986, **19**, 714.
207. N. A. Platé and V. P. Shibaev, 'Comb-Shaped Polymers and Liquid Crystals', Plenum Press, New York, 1987.

37

Thermogravimetric Analysis

JAMES R. MacCALLUM
University of St Andrews, UK

37.1 INTRODUCTION

The major techniques of thermal analysis can readily be grouped into three headings, differential scanning calorimetry, DSC, differential thermal analysis, DTA, and thermogravimetry, TG. The latter technique is used widely as a means for assessing the thermal stability of polymeric materials and, as such, it can provide valuable technical information. It must be borne in mind that the conclusions drawn with regard to stability are only relevant within the context of loss-of-weight of the sample. There are reactions which can take place on heating a polymer and which alter drastically the physical properties of the material without an observed change in weight. The side chain condensation of polyacrylonitrile provides a good illustration of this type of reaction.

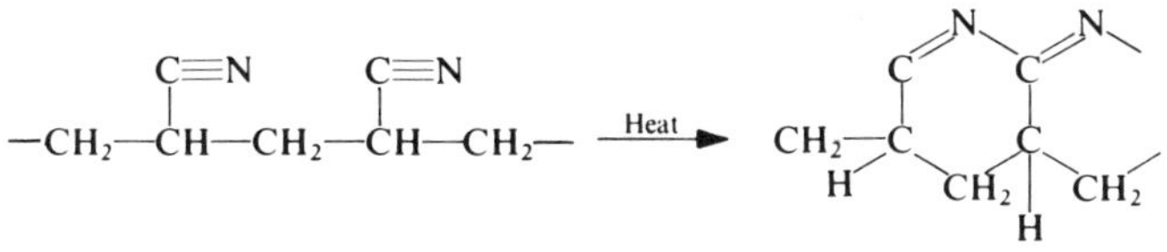

Scheme 1

Some years ago the demand for polymers which could be used in high temperature applications stimulated investigations into finding relationships between thermal stability and chemical structure. A number of qualitative trends were recognized[1] and although a number of attempts were made to quantify observed behaviour, in general it was concluded that the systems were so complex that little progress could be made along the lines, for example, developed for studies in the gas phase.

The interest in producing non-metallic materials with high temperature stability is growing again and it is important that the difficulties and drawbacks of TG be appreciated by those who apply the technique. It is worth noting that the techniques of DTA and TG can be combined to measure in more detail some of the chemical changes which take place when a polymer is heated beyond its point of decomposition.

The process of TG can be summarized as shown in Scheme 2.

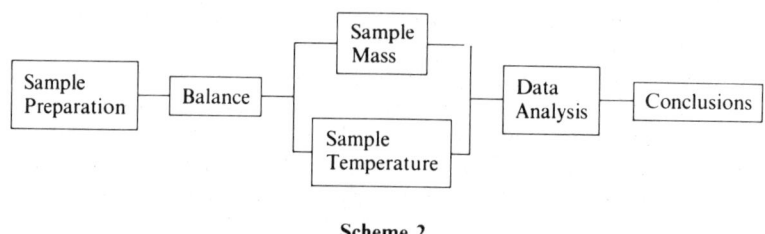

Scheme 2

37.2 EXPERIMENTAL PROCEDURES

37.2.1 Sample Preparation

Obviously it is important that whatever data are obtained on thermogravimetry of a polymer they must represent the thermal decomposition behaviour of the pure polymer. Thus, catalyst residues and additives must be removed before testing the material. The presence of the former impurity presents something of a problem for non-soluble polymers, as the process of residue removal can be complex and on occasions almost impossible. Since most organic polymers break down by some form of chain mechanism the occurrence of a minor amount of initiation by impurities can have a significant effect on conversion to volatile products. This factor is almost certainly the cause of discrepancy for results reported for the same polymer type prepared by different synthetic processes.

37.2.2 Sample Size

Two transport phenomena can be significant in the experimental procedures; they are heat conduction, and product diffusion.

37.2.3 Heat Conduction

Polymers are not efficient conductors of heat and consequently it is possible to achieve a condition of fairly high thermal gradients within a sample, with subsequent effects on reproducibility and analysis of measured data. Since samples are supported in a holder of some design and material, the optimum condition is to use thin films of test polymer. This means very small samples should be examined. It is also necessary to check that the material comprising the holder does not have any interaction with the polymer.

37.2.4 Product Diffusion

Slow diffusion of the primary products of thermal decomposition out of the sample holder can result in secondary breakdown of these substances resulting in further reaction with the parent polymer. Build-up of concentration of products within the sample may lead to the observed rate of decomposition being controlled by the physical process of effusion of products rather than the required condition of chemical control of rate of decomposition.

37.3 BALANCE

A vast range of equipment has been designed to enable investigators to perform thermogravimetric experiments. The early designs were produced for research into the thermal decomposition of inorganic salts. The field has been comprehensively surveyed by Escoubes and co-workers and details of construction and operation are given in that review.[2]

It is worth commenting on two aspects of experimental procedure which require the attention of those who construct their own apparatus. An important feature in attaining good temperature control of the sample is the rate of transfer of heat from the surrounding furnace to the sample and its holder. This is particularly relevant for the condition in which thermogravimetry is performed on a continuously pumped sample. Under these circumstances the occurrence of primary product breakdown referred to previously is minimized.

It is tempting to reduce the gap between sample holder and furnace wall to a minimum, consistent with unrestricted balance operation. However, although heat transfer may be facilitated, the pumping away of volatile products will be restricted and quite large pressure differentials can result between the volumes above and below the sample.[3] The outcome of this pressure gradient is recorded by the balance as an apparent mass *increase* resulting in a serious underestimate of the true situation. The higher the pressure gradient, that is the faster the true rate of weight loss, the more significant the error.

The second aspect of achieving good temperature control, which requires careful experimental design, is the measurement of sample temperature. The actual measurement is performed using a thermocouple and the key factor is that placement of the thermocouple should be such that the weighing mechanism should not be impeded. A compromise must be struck between accurate determination and control of sample temperature and retention of freedom of balance movement.

Many designs of apparatus allow control of the surrounding atmosphere in the heated volume of the apparatus. Changing from an efficiently evacuated system to one involving a flow of inert gas such as nitrogen will result in slowing down to some extent of product evolution. However, when air or pure oxygen is used then the whole pattern of decomposition will change. Indeed there is a good case for pursuing this experimental approach for those concerned with investigating the combustion of polymers. However, while such studies are of great technological importance they are not readily related to the direct results of thermogravimetry of polymers.

Once the experimental procedures are functioning satisfactorily then the investigator collects data in the form of the mass of the sample as a function of time of heating or temperature of sample. The problem to be faced is how to analyze the data in such a way that meaningful conclusions can be drawn.

37.4 DATA ANALYSIS

37.4.1 General Background

Historically the approach to data interpretation in thermogravimetry of polymers developed from the background established for gas and liquid phase kinetics. An extra facet developed, however, in that the equipment available made possible use of variable temperature experiments. The vast majority of gas phase measurements had been performed using isothermal conditions resulting in the necessity for doing several experiments to obtain the basic parameters — activation energy and pre-exponential factor, assuming a simple mechanism.

The great hope for programmed TG was that from only one experimental measurement it would be possible to observe the activation energy and pre-exponential factor, again assuming a simple mechanism and consequently using an uncomplicated rate expression. The theories and concepts of gas phase kinetics, which had been developed from isothermal measurements, were translated to the analyses of data collected during experiments carried out with varying temperature. A large amount of data was analyzed yielding orders of reaction, activation energies and pre-exponential factors for a range of polymers. Critical examination of these 'kinetic parameters' makes it quite clear that they cannot have mechanistic significance. Orders greater than three, activation energies significantly higher than carbon–carbon bond strengths and pre-exponential factors of the order of 10^{20} s^{-1} indicate that while numbers can be collected, the application of these parameters in any kinetic relationship to mechanism is untenable.

Further complications arose when the results of isothermal experiments were compared, using the same conceptual kinetic analysis, with programmed experimental results, yielding the conclusion that the two techniques did not produce the same kinetic parameters.[4] A possible explanation was proposed based on the premise that the Arrhenius equation required modification when used in the analysis of data collected during temperature programmed experiments.[5] There resulted a long and complicated debate on the validity of this proposal, the principle point of contention being around whether or not time can be a state function.[6, 7] An alternative approach has not been disproved.[8] The significant common factor which emerged from this debate is the conclusion that more than one run would be required for any thermogravimetric investigation, thereby removing the advantage which was originally attributed to programmed TG.

Since control of temperature and rate of decomposition can be more readily achieved using isothermal experiments, that is the condition recommended and analyzed in this review. The concensus opinion is that this procedure is more satisfactory than the alternative of temperature variation with time of reaction.

37.4.2   Kinetic Expressions

The general expression used to describe the thermal decomposition of a polymer is

$$-\frac{d(1-\alpha)}{dt} = k(1-\alpha)^n \tag{1}$$

α is the extent of conversion and is given by

$$\alpha = W_e/W_0 \tag{2}$$

W_e is the mass of polymer evolved as volatile fragments and W_0 is the initial mass. If W_r represents the mass of undegraded polymer then

$$1-\alpha = W_r/W_0 \tag{3}$$

k is described as a rate constant and is assumed to obey the Arrhenius relationship

$$k = A\exp(-E/RT) \tag{4}$$

in which A is the pre-exponential factor and E the activation energy for the process of weight loss; n can be defined as the order of the reaction. It should be noted that mass changes are being observed, with no correlation between chemical processes and mass change.

The form of equation (1) and the terms used in the expression are analogous to the approach applied in gas and liquid phase kinetics. It must be stressed that A, E and n do not generally have any real significance other than that their use provides a convenient means for summarizing the experimental data. Thus α is evaluated on the basis of fractional mass loss with no regard for the actual chemical nature of the volatile fragments or the residue. Conventional usage in chemical kinetics expresses change by determining moles of reactant gone or moles of product produced accepting knowledge of the stoicheometric relationship between the two.

Provided A, E and n are constant with time, equation (1) can be applied in its integrated forms. For $n=1$,

$$\ln(1-\alpha) = kt \tag{5}$$

and for $n \neq 1$,

$$\frac{1}{(1-\alpha)^n} - 1 = kt \tag{6}$$

In both equations (5) and (6) the apparent rate constant, k, has the dimensions of $(\text{time})^{-1}$ which in conventional kinetics would make it a first order rate coefficient.

For a set of experimental data collected isothermally the computer fitting of α and t will yield n and k. Repeating the procedure at different temperatures will allow evaluation of A and E.

Assumption of the form of equation (1) is somewhat restrictive and since, following traditional methodology, the aim of the investigator is to solve for E and A, the following approach can be used.

$$-\frac{d(1-\alpha)}{dt} = kf(1-\alpha) \tag{7}$$

$$F(1-\alpha) = kt \tag{8}$$

$F(1-\alpha)$ is the integrated expression for $f(1-\alpha)$ and no form is proposed. However it can be assumed that for isothermal experiments performed over a range of temperatures $F(1-\alpha)$ has the same value for a given conversion α, and equation (8) is rewritten as equation (9) which rearranges to equation (10).

$$\ln[F(1-\alpha)] = \ln A - E/RT + \ln t \tag{9}$$

$$E/RT + \ln[F(1-\alpha)] - \ln A = \ln t \tag{10}$$

For a range of temperatures the logarithm of the time taken to reach a fixed conversion α plotted against the reciprocal of the temperature of the experiment will yield E the activation energy for TG within the range chosen. A plot of the intercepts obtained from these analyses (given by equation 11) *vs.* α should allow extrapolation to α equal to zero, thereby producing A.

$$\text{Intercept} = \ln[F(1-\alpha)] - \ln A \qquad (11)$$

Use of this procedure is illustrated in Figure 1 and the derived activation energies shown in Table 1.[9] This set of results indicates that as the reaction progresses the apparent activation energy increases, implying involvement of further reactions at advanced decomposition. Use of equations (5) or (6) would not have indicated this trend in E. However, it should be noted that no conclusions whatsoever may be drawn with regard to the mechanism of reaction.

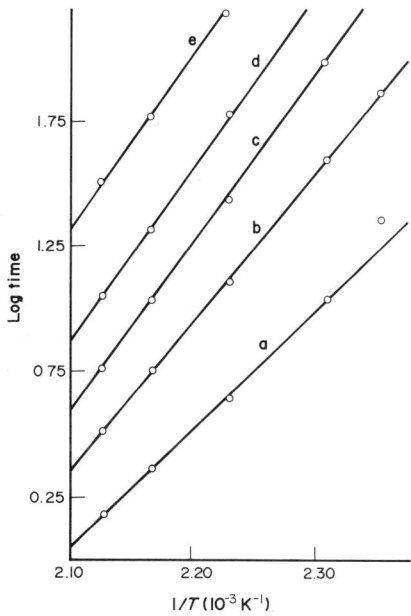

Figure 1 Plot of ln(time to fixed conversion) *vs.* $1/T$ according to equation (10) for head-to-head poly(α-methylstyrene). Percentage decomposition as follows: (a) 12.5; (b) 25.0; (c) 37.5; (d) 50.0; (e) 62.5 (Reproduced by permission of Pergamon Press from *Eur. Polym. J.*, 1972, **8**, 809)

Table 1 Energies of Activation Corresponding to the Different Extents of Decomposition for Head-to-head Poly(α-methylstyrene) shown in Figure 1

Curve	Percentage decomposition	Energy of activation (kJ mol^{-1})
a	12.5	85.0
b	25.0	106.4
c	37.5	111.9
d	50.0	117.0
e	62.5	121.0

A number of instruments incorporate a facility for electronic differentiation of the mass *vs.* time curve allowing direct determination of $-\mathrm{d}(1-\alpha)/\mathrm{d}t$. It is also possible to read the mass/time data directly to a computer and use a mathematical routine to obtain the rate of change. For this procedure, data analysis can be performed from equation (7) which can be rewritten as shown in equation (12) in which r replaces $-\mathrm{d}(1-\alpha)/\mathrm{d}t$. Equation (12) rearranges to equation (13).

$$r = A \exp(-E/RT) f(1-\alpha) \qquad (12)$$

$$\ln r = \ln A + \ln[f(1-\alpha)] - E/RT \qquad (13)$$

For a fixed conversion α and a set of data collected from isothermal experiments a plot of logarithm of the rate *vs.* $1/T$ for that fixed conversion will allow evaluation of E, and a plot of the intercepts (using equation 14) *vs.* α will by extrapolation yield A.

$$\text{Intercept} = \ln A + \ln[f(1-\alpha)] \qquad (14)$$

It is relevant to state that both electronic and mathematical differentiation of conversion/time curves can lead in certain circumstances to systematic errors and use of the integrated expression, equation (10), is strongly recommended.

The analyses described above imply a relatively simple rate expression can be used to describe the process of thermogravimetry. It does not take into account the possibility of two independent concurrent reactions taking place for which a more complex statement would apply. In this situation the process could be described by the expression

$$-\frac{d(1-\alpha)}{dt} = k_1(1-\alpha_1) + k_2(1-\alpha_2) \qquad (15)$$

α_1 and α_2, the fractional conversions for reactions 1 and 2 having rate constants k_1 and k_2 can only be obtained by product analysis and thus use of equation (15) is beyond thermogravimetry.

37.5 CONCLUSIONS

The question which must inevitably be asked is 'What is the value of the parameters derived from thermogravimetry?' A number of negative aspects can be advanced. The 'kinetic' data derived as n, A and E have no mechanistic significance and in a number of ways it is unfortunate that this terminology has been transfered from homogeneous gas and liquid phase kinetics.

A summary of values published for the activation energy of the thermal decomposition of polystyrene is given in Table 2. It can be seen that the magnitude of this parameter varies significantly from investigator to investigator and it can be concluded that either the method of analysis is suspect or the sample varied from laboratory to laboratory. Whatever the explanation it is clear that care must be exercised in interpreting the significance of E.

Table 2 Energies of Activation Reported for the Thermal Decomposition of Polystyrene

Activation energy (kJ mol^{-1})	*Ref.*
250	10
241	11
235	12
226	13
224	14
164	15

On heating a polymer, decomposition takes place in a highly viscous medium and the possibility of physical rather than chemical control of the degradation reactions must always be considered. However, in spite of these difficulties it is clear that high values of E are associated with thermally stable materials, and certain structural features such as ladder polymers or intramolecular side group condensation produce stable polymers. It is important that investigators using TG do more than simply report n, A and E. The use of the technique can be justified only if an attempt is made to relate trends in values to the structures of materials.

While thermogravimetry of polymers has not yielded the depth of information originally hoped for, nevertheless it provides a very useful technique for assessing thermal stability of polymers. It may be that in the future, with increasing numbers of n, A and E data becoming available, it will be possible to predict stability from structure using these parameters.

37.6 REFERENCES

1. R. H. Still, in 'Developments in Polymer Degradation', ed. N. Grassie, Applied Science, London, 1977, vol. 1, p. 1.
2. M. Escoubes, E. Eyraud and E. Robens, *Thermochim. Acta*, 1984, **82**, 15 and 23.

3. J. R. MacCallum and W. W. Wright, *Polym. Degrad. Stab.*, 1981, **3**, 397.
4. J. R. MacCallum and J. Tanner, *Eur. Polym. J.*, 1970, **6**, 907.
5. J. R. MacCallum and J. Tanner, *Nature (London)*, 1970, **225**, 1127.
6. P. Holba and J. Šesták, *Z. Phys. Chem.*, Neue Folge, 1972, **80**, 1.
7. R. M. Felder and E. P. Stahel, *Nature (London)*, 1970, **228**, 1085.
8. J. R. MacCallum, *Thermochim. Acta*, 1982, **53**, 375.
9. J. Atkinson and J. R. MacCallum, *Eur. Polym. J.*, 1972, **8**, 809.
10. E. S. Freeman and B. Carroll, *J. Phys. Chem.*, 1958, **62**, 394.
11. C. D. Doyle, *J. Appl. Polym. Sci.*, 1961, **5**, 285.
12. J. R. MacCallum and J. Tanner, *Eur. Polym. J.*, 1970, **6**, 1033.
13. T. Ozawa, *Bull. Chem. Soc. Jpn.*, 1965, **38**, 1881.
14. A. W. Coats and J. P. Redfern, *Nature (London)*, 1964, **201**, 68.
15. J. Boon and G. Challa, *Makromol. Chem.*, 1965, **84**, 25.

7

Photon Correlation Spectroscopy: Technique and Scope

TERENCE A. KING
University of Manchester, UK

7.1 INTRODUCTION

The light-scattering technique of photon correlation spectroscopy is now firmly established as a valuable tool for the study of the structure and dynamics of macromolecules. Over the last 25 years there has been steady development of the technique and a growing, and now extensive, range of applications to polymer systems. The determination of translational and rotational diffusion coefficients and internal molecular flexing provides a means of polymer characterization for molecular weight and molecular weight distributions, giving size and structural information. Study of polymer dynamics in isolated or weakly interacting molecules in dilute solutions, and strongly interacting molecules in semidilute and concentrated solutions, has provided experimental data concurrent with the extensive development of theoretical models of polymers using renormalization group and scaling theories. Photon correlation spectroscopy is also an important technique in the study and characterization of crosslinked and physical gels, solid and melt bulk polymers, colloids, liquid crystals, aspects of polymer phase transitions and polyelectrolytes. As well as the application to synthetic polymers the technique has found broad application to biopolymers; these applications are not addressed here.

Molecular scattering of light has been investigated over the last 100 years, and scattering at the incident wavelength, Rayleigh and Rayleigh–Debye scattering, has been used extensively in more recent times for characterization of molecular weight, size, shape and interactions. The treatment of total intensity light scattering and its application to static molecular and particle characterization is well documented.[1-5] Photon correlation spectroscopy is a more recently developed inelastic light-scattering process, involving frequency broadening of the central Rayleigh line about the incident light wave frequency. The earlier discoveries of the inelastic light-scattering processes of Raman and Brillouin scattering, in which there is a frequency shift between the incident and scattered light, have also been applied extensively to polymer systems. Combined Rayleigh–Brillouin scattering has proven fruitful in the study of bulk systems. We concentrate here only on the technique and application of photon correlation spectroscopy.

As well as the name 'photon correlation spectroscopy' many other names have been used in the literature to describe the method, in which broadening of the central Rayleigh line is measured, dependent on the way in which the scattered signal is detected or processed. These names include: light-beating spectroscopy, self-beat spectroscopy, quasielastic light scattering, Rayleigh linewidth spectroscopy, laser Doppler spectroscopy, intensity fluctuation spectroscopy and optical mixing spectroscopy. For standardization we adopt here the general name of photon correlation spectroscopy (PCS). This name has recently come into more general use since processing of the scattered signal is usually done by digital time autocorrelation of detected photons.

Light incident on a medium induces an oscillatory polarization of the medium which in turn acts as a secondary source of radiation in the form of scattering. Fluctuations in the dielectric constant of the medium induce net scattering of radiation. The intensity, angular distribution, polarization and frequency shift of the scattered light are determined by the size, shape and molecular interactions of the scattering centres. With this process information on the structure, interactions and molecular dynamics within the scattering medium can be derived.

The frequency distribution of light scattered from macromolecules was investigated by Pecora (1964 and subsequently)[6-8] who showed that the macromolecular translational diffusion coefficient, the rotational diffusion coefficient and internal motion dynamics may be derived. Also in 1964, the previously demonstrated[9] technique of optical mixing was used[10] to measure the small frequency broadening of the Rayleigh-scattered peak from dilute suspensions of polystyrene latex particles. The subsequent growth of photon correlation spectroscopy and its application to polymers and biopolymers is described in a number of books and reviews.[11-30] For early and general introductions see the books of Berne and Pecora[17] and Chu[12] and to two review proceedings.[13,18,20-22] Attention is also drawn to Chu's account of the application of PCS to polymer solutions in Chapter 8 of this volume.

7.2 PHOTON CORRELATION SPECTROSCOPY TECHNIQUES

7.2.1 Summary of Light Scattering

Light is scattered by a molecule in solution if the molecule has a polarizability different from its surroundings.[31-33] The molecular polarizability difference gives a spatial inhomogeneity to the dielectric constant of the medium or equivalently a refractive index difference. The classical mechanism of light scattering involves the electric field of the incident light inducing an oscillating dipole moment in the molecule which re-radiates to form the scattered radiation. Inhomogeneities of the medium may arise from spontaneous thermal fluctuations or concentration fluctuations, for example for a polymer solution or colloidal dispersion.

The intensity of the scattered light depends on the intensity of the incident light, the scattering angle, the solution parameters and the light polarization. Figure 1(a) illustrates a form of the scattering geometry. A linearly polarized, monochromatic plane wave of wavelength λ is incident on the scattering medium and an optical detector is at the point P. For Rayleigh scattering the molecules have dimensions $\ll \lambda$, so that the molecules sense a uniform field E_0 and the scattering is not sensitive to the shape of the molecule.

Conservation of momentum in the scattering process leads to a wavevector diagram as shown in Figure 1(b). The quantity $q = k_s - k_i$ is the difference between the wavevectors of the scattered and incident waves. Since in Rayleigh scattering the wavelength of the scattered light is very near to that of the incident light, $k_s \simeq k_i = 2\pi/\lambda_i$. Then

$$q \;=\; 2k_i \sin\left[\frac{\theta}{2}\right] \;=\; \frac{4\pi n_0}{\lambda_i}\sin\left[\frac{\theta}{2}\right] \tag{1}$$

where θ is the angle of scattering, n_0 is the medium refractive index and λ_i the vacuum wavelength of the light. If a is the largest dimension of the scattering molecules, when $qa \ll 1$ they behave as point scatterers.

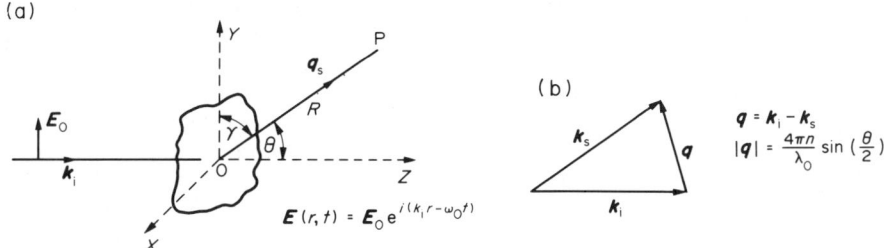

Figure 1 (a) Light-scattering geometry; and (b) wavevectors and conservation of momentum diagram with definition of wavevector *q*

The electric field of the scattered light is in general

$$E_s(R,t) = \frac{k_s \wedge (k_s \wedge E_0)\alpha}{4\pi\varepsilon R} e^{i(\omega_0 t - k_s \cdot R)} e^{iq \cdot r} \tag{2}$$

for a molecule of polarizability α and where ε is the relative dielectric constant of the medium surrounding the molecule.

For molecules of size $\ll \lambda$ (Rayleigh scattering) the polarized scattered light intensity I_s at position distance R is

$$\frac{I_s}{I_0} = \left[\frac{k_i}{n_0}\right]^4 V^2 \frac{\sin^2 \phi}{16\pi^2 R^2} [n^2 - n_0^2]^2 \tag{3}$$

where V is the molecular volume and $n_0 (= \sqrt{\varepsilon})$ is the refractive index of the surrounding medium. When the polarizability of the molecule α is isotropic such that the molecule α is homogeneous with refractive index n, then $\alpha = (n^2 - n_0^2)V$. The scattered light intensity is seen to be proportional to k_0^4 (i.e. $1/\lambda_i^4$), to I_0 and inversely proportional to R^2.

The total intensity of scattered light contains information on the static properties of the scattering medium, *i.e.* the size and shape of the scattering molecules and the thermodynamic quantities. The other basic quantity which can be measured by light scattering is the frequency distribution of the scattered light; this carries information on dynamical quantities, such as diffusion coefficients, internal motion and molecular velocity. There light is scattered from refractive index fluctuations, which for a polymer solution arise from the polarizability difference between the solute and solvent.

7.2.2 Dynamic Light Scattering

The phase ψ_s of the scattered field E_s is made up of the phase of the incident field at position r, $\omega_0 t - k_i r$ plus a phase shift due to propagation of the scattered field from position r to position R, $-k_s(R - r)$, so that $\psi_s = (\omega_0 t - k_s \cdot R) + (k_s - k_i) \cdot r$. The scattered electric field from a single molecule is

$$E_{s,j}(t) = A_j(t) e^{-i\omega_0 t} e^{i\psi_s(t)} \tag{4}$$

where $A_j(t)$ is the amplitude of the scattered light from molecule j and $\psi_j(t)$ is the phase difference from the optical path of light scattered from position r_j compared to that at the origin, $\psi_j(t) = q \cdot r(t)$, *i.e.* the phase of the scattered field depends on the position r of the scattering molecule.

The total scattered field is

$$E_s(q,t) = \sum_j^N A_j(t) e^{i\omega_0 t} e^{iq \cdot r_j(t)} \tag{5}$$

Since the molecules are moving, r is a function of time and ψ_s has a dependence determined by the molecular dynamics. The scattered field varies in time due to translational diffusional motion or to changes in $A_j(t)$ induced by rotational or internal motion. A less common case is for smaller values of molecular occupational number N, where fluctuations in N lead to scattered light fluctuations;[34] polymer solutions at typical concentrations provide many molecules in the scattering volume. For example a laser beam focused into a polymer solution illuminates a sample volume defined by a $100 \, \mu m$ focal spot diameter, with detection optics collecting from that volume; the sample volume is $\sim 10^{-6} \, cm^3$. A polymer of molecular weight $M \sim 10^5$ at a concentration of $1 \, mg \, cm^{-3}$ is equivalent to about 10^{10} molecules in the sample volume.

When the scattering molecule is undergoing Brownian motion, r is a random variable and E_s has a randomly modulated phase. The scattered light is broadened in frequency with an optical frequency distribution $S(\omega)$ as illustrated in Figure 2(a). Since the particle motion contains no preferred direction, the spectrum contains a continuous distribution of frequencies centred around ω_0. The correlation function of the electric field $G^{(1)}(\tau)$ is also a measure of the frequency distribution and contains information on the molecular motion.[35-40] It is the Fourier transform of the power spectrum $S(\omega)$

$$G^{(1)}(\tau) = \langle E_s^*(t) \, E_s(t + \tau) \rangle \tag{6}$$

where $\langle \; \rangle$ denotes a time or ensemble average and τ is the correlation time

$$S(\omega) = \frac{1}{2\pi} \int_0^\infty G^{(1)}(\tau) \, e^{i\omega\tau} \, d\tau \tag{7}$$

Discussions of the nature of correlation functions and molecular dynamics can be found in refs. 41 and 42.

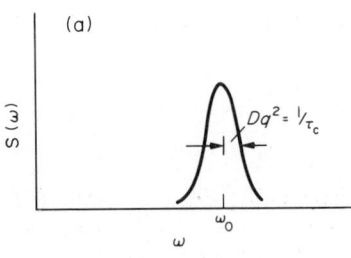

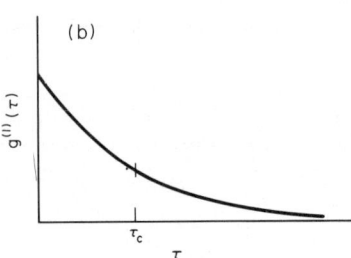

Figure 2   (a) Illustration of an optical spectrum of scattered light; and (b) electric field correlation function

An illustration of a scattered optical spectrum and its normalized field correlation function is shown in Figure 2. The broadening of the Rayleigh-scattered light spectrum contains information on the motion of the scattering molecules. We are concerned here only with the broadening of the central Rayleigh component of the scattered spectrum; the other types of inelastic scattering, the Brillouin doublet and Stokes and anti-Stokes Raman scattering, both occur at much greater frequency shifts.

The normalized electric field correlation illustrated in Figure 2(b) is

$$g^{(1)}(\tau) = \frac{G^{(1)}(\tau)}{G^{(1)}(0)} \tag{8}$$

In terms of amplitude and phase time dependences

$$g^{(1)} = e^{(i\omega_0 t)} \frac{\langle A^*(0)A(\tau) \rangle}{\langle |A(0)|^2 \rangle} \left\langle e\{-i\boldsymbol{q}\cdot[r(\tau)-r(0)]\} \right\rangle = e^{(i\omega_0 t)} \, C_A(\tau) \, C_\psi(\tau) \tag{9}$$

where $A(\tau)$ is the scattering amplitude per molecule and $C_A(\tau)$ and $C_\psi(\tau)$ are the amplitude and phase correlation functions. For small molecules $(a \ll q^{-1})$ or spherical molecules the amplitude part of the autocorrelation function becomes

$$C_A(\tau) = \frac{\langle A^*(0) A(\tau) \rangle}{\langle |A(0)|^2 \rangle} = 1 \tag{10}$$

Then $g^{(1)}(\tau)$ carries information on the translational diffusion coefficient D_T through $C_\psi(\tau)$. This is related to the intermediate structure factor $G_s(r,\tau)$, which is the probability of finding a particle at position r at time τ if it was at the origin at $\tau = 0$.

$$C_\psi(\tau) = \langle e^{\{-i\mathbf{q}\cdot[r(\tau)-r(0)]\}} \rangle = \int G_s(r, \tau) e^{(-i\mathbf{q}\cdot r)} \mathrm{d}^3 r \tag{11}$$

Thus for spherical identical scatterers undergoing Brownian motion in solution

$$g^{(1)}(\tau) = e^{-D_T q^2 \tau} e^{-i\omega_0 \tau} \tag{12}$$

The associated optical spectrum is

$$S(\omega) = \frac{\langle I_s \rangle D_T q^2 / \pi}{(\omega - \omega_0)^2 + (D_T q^2)^2} \tag{13}$$

which is a Lorentzian function centred at ω_0 with a halfwidth $D_T q^2$.

7.2.2.1 Translational diffusion

An estimate of the degree of broadening and correlation decay time, assuming a system of polystyrene with $\bar{M}_w = 670\,000$ in cyclohexane with $D_T = 1.66 \times 10^{-7}\,\mathrm{cm^2\,s^{-1}}$ and observed at $90°$, leads to a linewidth of about $2 \times 10^4\,\mathrm{Hz}$ and a characteristic decay time of the correlation function $\tau_c = 1/D_T q^2 = 5 \times 10^{-5}\,\mathrm{s}$.

The diffusion coefficient D may be related to the molecular friction factor f through the Einstein relation

$$D_T = \frac{kT}{f} \tag{14}$$

For a spherical molecule of radius a, $f = 6\pi\eta a$, where η is the dynamic viscosity of the solvent, then

$$D_T = \frac{kT}{6\pi\eta a} = \frac{kT}{6\pi\eta R_h} \tag{15}$$

where R_h is the hydrodynamic radius of the molecule.

7.2.2.2 Larger particles

When the molecular size is not much less than q^{-1}, the scattering intensity is reduced by intramolecular interference.

For molecules with size $\sim 1/q$, the scattered intensity I_s is given as

$$I_s = I_0 \left[\frac{q}{n_0}\right]^4 \frac{V^2 \sin^2 \phi}{16\pi^2 R^2} [n^2 - n_0^2] P(\theta) \tag{16}$$

where $P(\theta)$ is the molecular interference form factor. The most general form of $P(\theta)$ is described by Mie scattering theory.[2-5] The Rayleigh–Debye approximation is valid when $(n - n_0)a \ll \lambda/4\pi$. Then for an isotropic molecule $P(\theta) = |1/V \int_V e^{i\mathbf{q}\cdot r} \mathrm{d}v|^2$ where V is the volume of the scattering molecule.

Values of $P(\theta)$ have been calculated for many scattering particle shapes.[4] For any shape of particle at small scattering angles, $P(\theta) = 1 - qR_g^2/3$ where R_g is the radius of gyration. If the molecule is not small and is not spherical or optically isotropic then rotational diffusion will contribute a time-dependent scattering amplitude; also for flexible molecules intramolecular dynamics gives a similar contribution.

The form of the correlation function $G^{(1)}$ can be calculated for model systems. Here we consider rod-like molecules and flexible coils. For a rod-like molecule of length L, the dynamics of the molecule will contain contributions from translation and rotation. The form of $G^{(1)}(\tau)$ for a molecule having a rotational diffusion coefficient D_R is[7,43,44]

$$|G^{(1)}| \quad = \quad I_s e^{-q^2 D_T \tau} [\, S_0 + S_1 e^{-6 D_R \tau} \quad + \quad \ldots \,] \tag{17}$$

where $S_0, S_1 \ldots$ are weighting factors as shown in Figure 3(b). For $qL \leq 3$, $S_0 = 1$ and $S_1 = 0$, while for $qL = 6$, $S_1 = 0.1$ and is significant compared with S_0.

In a similar way the form of $G^{(1)}$ for a flexible coil is[8]

$$G^{(1)}(\tau) \quad = \quad I_s e^{-q^2 D_T t} [P_0 \quad + \quad P_1 e^{-2\tau/\tau_1} \quad + \quad \ldots] \tag{18}$$

where $P_0, P_1 \ldots$ are weighting factors for the translational and internal modes of motion (Figure 3a) and τ_1 is the characteristic time for the longest intramolecular relaxation.

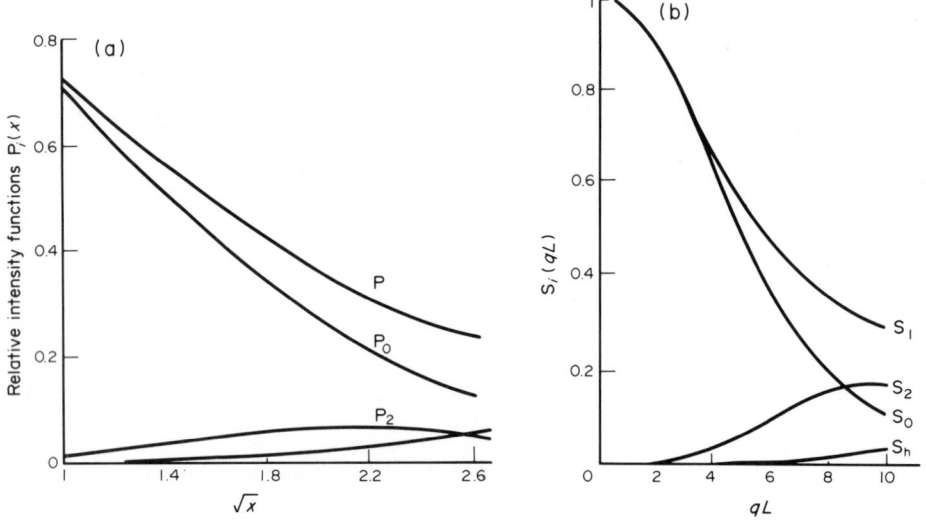

Figure 3 (a) Relative intensity functions $P(x)$ for a flexible polymer, $x = \frac{1}{6} q^2 \langle l^2 \rangle_e$ with $\langle l^2 \rangle_e$ = equilibrium mean square end-to-end distance of the polymer.[7] (b) Relative intensity functions for an optically rigid rod of length L. S_0: pure translation, S_1: translation–rotation, S_h: sum of all higher terms.[8]

7.2.2.3 *Depolarized scattering*

For the incident field E_i shown in Figure 1, linearly polarized with E_i perpendicular to the scattering plane, optically isotropic point scatterers give a scattered field intensity I_{VV}, also linearly polarized perpendicular to the scattering plane. Scatterers which are optically anisotropic or geometrically anisotropic give a second depolarized scattered component,[44-49] intensity I_{HV}, polarized in the scattering plane. The depolarized scattered electric field has its amplitude modulated by rotation of the scatterer.

For a rigid particle with rotational symmetry the correlation function for the depolarized field $E_{s,d}$ is

$$|G_d^{(1)}(\tau)| \quad = \quad I_{s,d} e^{-[D_T q^2 + 6 D_R]\tau} \tag{19}$$

At small angle $Dq^2 \to 0$ and then the correlation function is determined by the rotational diffusion coefficient D_R. D_R has values typically within the range $D_R \sim 10^2$–10^6 s^{-1} corresponding respectively to large rod-like molecules and small spherical molecules.

7.2.3 Linewidth Measurement

The scattered light intensity is proportional to the square of the time average of the electric field

$$\text{Scattered intensity} \quad = \quad \langle I_s \rangle \propto \langle |E_s| \rangle^2 \tag{20}$$

where $\langle\ \rangle$ denotes the time average.

In order to measure the very small optical linewidths indicated in Section 7.2.1 optical mixing techniques are employed since dispersing optical instruments, such as diffraction gratings and Fabry–Perot interferometers, do not have sufficient resolving power. There is a large body of work on optical mixing, from which PCS has emerged, in the field of optical coherence, the quantum theory of light and intensity fluctuation interferometry.[50-52] Since photon correlation instruments generally have greater efficiency than real time spectrum analyzers these are usually preferred over power spectrum measurements in PCS, while power spectrum analysis is used in laser velocimetry. There are two basic forms of optical mixing: heterodyne and intensity fluctuation (self-beat). A note of clarification is essential here since the literature contains conflicting nomenclature.[16] By heterodyne mixing we refer to mixing the scattered light with a reference light wave (local oscillator) unshifted or shifted in frequency from the incident light beam—this is usually and most conveniently derived from the laser source providing the incident beam.* In self-beat optical mixing the scattered wave is not mixed with a reference signal but is directly detected. The nomenclature adopted here is taken as a consensus of that appearing in the literature.

The fluctuating scattered light intensity is illustrated together with related quantities in Figure 4. The normalized electric field autocorrelation function is

$$g^{(1)}(\tau) \quad = \quad \frac{\langle E_s^*(t) E_s(t+\tau) \rangle}{\langle I_s \rangle} \tag{21}$$

This quantity can be measured by heterodyne detection. With a reference beam E_r, the total field at the detector is

$$E_{\text{tot}}(t) \quad = \quad E_s(t) \quad + \quad E_r(t) \tag{22}$$

whose intensity is

$$I_{\text{tot}} \quad = \quad |E_{\text{tot}}|^2 \quad = \quad I_s(t) \quad + \quad I_r(t) \quad + \quad 2\text{Re}\{E_s(t) E_r^*(t)\} \tag{23}$$

where Re indicates the real part of the expression in brackets. Then the total intensity correlation function is given by

$$G_{\text{tot}}^{(2)} \quad = \quad G_s^{(2)} \quad + \quad G_r^{(2)} \quad + \quad 2\langle I_s \rangle \langle I_r \rangle \quad + \quad 2\text{Re}\{G_s^{(1)} G_r^{(1)}\} \tag{24}$$

with

$$E_s(t) \quad = \quad E_{0s}(t)\, e^{i[\omega_{0s}t + \phi_s(t)]} \tag{25}$$

and

$$E_r(t) \quad = \quad E_{0r}(t)\, e^{i[\omega_{0r}t + \phi_r(t)]} \tag{26}$$

Then

$$G_s^{(1)}(\tau) \quad = \quad \langle I_s \rangle\, e^{i\omega_{0s}\tau}\, g^{(1)}(\tau) \tag{27}$$

$$G_s^{(2)}(\tau) \quad = \quad \langle I_s \rangle^2 [1 \quad + \quad g^{(2)}(\tau)] \tag{28}$$

From which

$$G_{\text{tot}}^{(2)} \quad = \quad \langle I_s \rangle^2 [1 \quad + \quad g^{(2)}(\tau)] \quad + \quad I_r^2 \quad + \quad 2I_r \langle I_s \rangle \{1 \quad + \quad \cos(\omega_{0s} - \omega_{0r})\tau\, g^{(1)}(\tau)\} \tag{29}$$

For $I_r \gg \langle I_s \rangle$, the first term may be neglected and $G_{\text{tot}}^{(2)} \propto \text{Re}\{G^{(1)}(\tau)\}$ shifted to be around the frequency ω_{0r}. As described later heterodyne mixing and detection has some valuable advantages in polymer studies with PCS.

---

* When mixing the scattered light with a reference signal which is unshifted in frequency, this is termed homodyne detection in other areas.

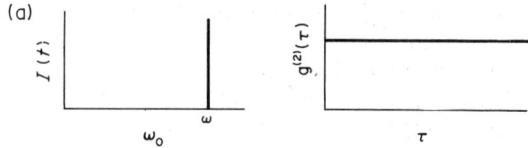

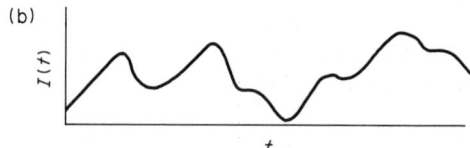

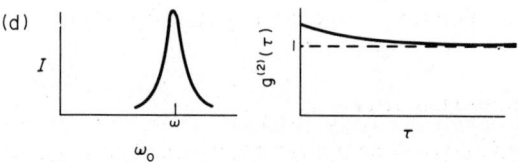

Figure 4   (a) Spectrum and correlation function of ideal laser source; (b) fluctuating scattered light intensity; (c) standardized detected photon counts; and (d) optical spectrum and $g^{(2)}(\tau)$ of scattered light under heterodyne detection

In self-beat detection the intensity autocorrelation function is determined as

$$G^{(2)}(\tau) \ = \ \lim_{T \to \infty} \ \frac{1}{2T} \ \int_{-T}^{T} \ I_s(t)\,I_s(t \ + \ \tau)\,\mathrm{d}t \tag{30}$$

It is the Fourier transform of the power spectrum and is readily measured by digital techniques. The normalized form of $G^{(2)}(\tau)$ is

$$g^{(2)}(\tau) \ = \ \frac{\langle E_s^*(\tau)\,E_s(t)\,E_s^*(t \ + \ \tau)\,E_s(t \ + \ \tau)\rangle}{\langle I_s \rangle^2} \tag{31}$$

With some restrictions (such that the scattered field is a Gaussian random process), the correlation functions $g^{(1)}(\tau)$ and $g^{(2)}(\tau)$ are connected through the Siegert relation

$$g^{(2)}(\tau) \ = \ 1 \ + \ |g^{(1)}(\tau)|^2 \tag{32}$$

Experimentally in self-beat PCS the intensity autocorrelation function is measured as

$$C(\tau) \ = \ g^{(2)}(\tau) \ = \ B[1 + b|g^{(1)}(\tau)|^2] \tag{33}$$

Here B is a constant and b is a geometric factor dependent on the detector area.[53] For $g^{(1)}(\tau) = e^{i\omega_0 t}\,e^{-D_T q^2 \tau}$ we find $g^{(2)}(\tau) = B[1 + be^{-2D_T q^2 \tau}]$.

The detector has an average photocurrent $\langle i \rangle$ which is proportional to the average light intensity: $\langle i \rangle \propto \langle I_s \rangle$. Since the scattered light is normally at low level and in the form of discrete photon pulses, the scattered signal and hence correlation function is most usefully recorded using digital photon detection. In terms of photon counts

$$g^{(2)}(\tau) \ = \ \frac{1}{\langle n^2 \rangle} \ \sum_{i=1}^{N_c} n_i\, n_{i+p} \tag{34}$$

where $\tau = p\,\Delta T$, ΔT = channel width, N_c = number of correlation channels and $\langle n \rangle$ = average number of protons counted in time ΔT.

Heterodyne optical mixing has several features which are valuable in application to polymer systems.[54-58] With heterodyne detection the autocorrelation function is linearly proportional to $g^{(1)}(\tau)$, the first order scattered electric field correlation function, and leads to useful simplification in multiexponential data analysis. A double exponential first order correlation function of the form

$$g^{(1)}(\tau) \;=\; A\,e^{-at} \;+\; B\,e^{-bt} \tag{35}$$

with constants A and B and decay frequencies a and b converts by use of the relation $g^{(2)}(\tau) = 1 + |g^{(1)}(\tau)|^2$ to a current autocorrelation function in self-beat optical mixing

$$i(\tau) \propto 1 \;+\; A^2\,e^{2a\tau} \;+\; B^2\,e^{-2b\tau} \;+\; 2AB\,e^{-(a+b)\tau} \tag{36}$$

In heterodyne mixing the equation for analysis is simpler, being

$$i(\tau) \;\propto\; 1 \;+\; A\,e^{-at} \;+\; B\,e^{-bt} \tag{37}$$

This is valuable in polydispersity analysis and the derivation of molecular weight distributions using multiexponential algorithms. With self-beat detection (i) the presence of the background term A needs to be derived before the function can be simplified by taking the square root, and (ii) application of self-beat detection is restricted to applications giving scattered fields with Gaussian statistics. Practical considerations are that heterodyne mixing provides (i) improved signal-to-noise under conditions of equivalent measurement, and (ii) inadvertent partial heterodyning, by stray light acting as a reference when operating in the direct detection configuration, can be avoided.

7.2.4 Photon Correlation Spectrometers

The main components and layout of a photon correlation spectrometer in the intensity fluctuation (self-beat) configuration to measure the angular distribution of scattered light are shown in Figure 5. We describe here the usual form of PCS using[57-60] digital autocorrelation. Light scattered by the sample, held in a special cell, is detected normally as a function of angle θ, but also sometimes at a fixed angle, and analyzed by a time autocorrelator or spectrum analyzer. The light source is usually a laser, although non-laser sources can be used under special circumstances,[61] and the detector is a

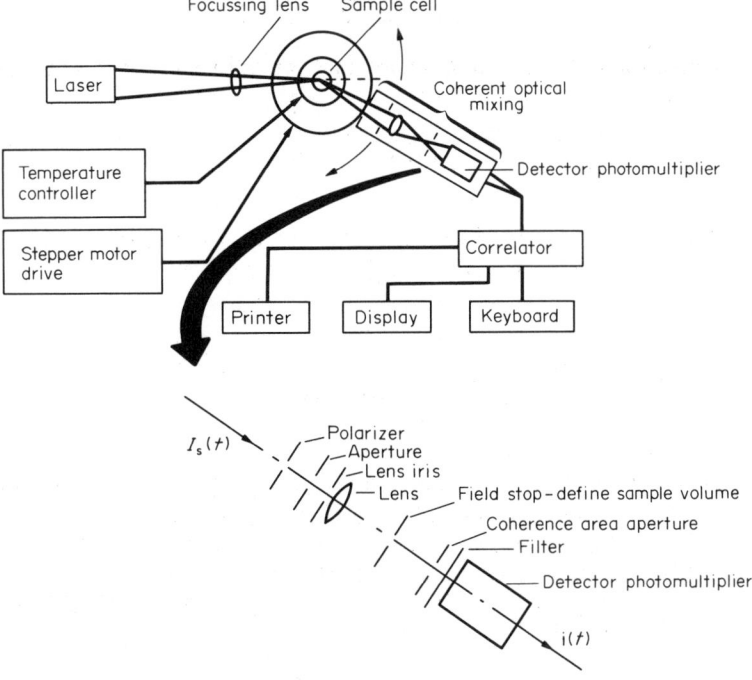

Figure 5 Schematic of self-beat photon correlation spectrometer and detection optics

photomultiplier or sometimes a sensitive photodiode. The signal from the correlator is stored and analyzed by a computer.

The laser provides continuous high intensity in a single radiation mode which can be focused to a small volume with a dimension $\simeq 2F\lambda$, where F is the f-number of the focusing lens. The most common lasers used are He–Ne ($\lambda = 632.8$ nm or 543.2 nm), Ar ion ($\lambda = 488.0$ nm or 514.5 nm) and He–Cd ($\lambda = 441.6$ nm). Sample fluorescence or stray light is removed by a line filter in front of the detector.

For strongly scattering samples the He–Ne laser is used at a typical power level of 3–10 mW. The argon ion laser provides higher stable powers up to 1 W for weakly scattering systems but requires to be used within the power limitations of the sample and avoiding sample heating which induces convection. The laser is required to operate with minimal intensity fluctuations with an instability of $< 1\%$; for the heterodyne operation a single longitudinal mode of the laser is usually selected by the use of an etalon in the laser cavity.

Detection photomultipliers must have the required characteristics of sensitivity, wavelength response, speed and low noise and are used with a high speed pulse amplifier–discriminator to convert the photon signal into a form suitable for the correlator. The photomultiplier can also be obtained with low noise and free of self-correlation, induced by after-pulsing in the detector process, by testing and selecting the photomultipliers; detector self-correlation would distort the measured correlation function at short correlation times.

Several modern and complete commercial spectrometers are available, satisfying the requirements for a general purpose instrument. Since PCS can be used to carry out a wide range of different types of measurement or may be applied to particularly difficult systems the need may arise for a purpose-made instrument. Vibrations of the spectrometer are required to be avoided by the use of stable optical construction and a vibration isolation system to prevent coupling of vibrations to the instrument. Relative motion of the laser with the sample should be avoided by mounting laser source, scattering chamber and detector on a common baseplate.

Efficient operation requires as much light as possible to be scattered into a single coherence area at the detector. The input focusing lens is selected to condense light into the scattering volume but without adding an unacceptably large spread of input angles leading to error definition of wavevector $\boldsymbol{q}$. The radius of the laser beam at the focus of a lens of focal length f is $f\alpha$, where α is the divergence of the incident beam and is typically 0.5 mrad. The preferred small focal length of the lens conflicts with the error induced in q of $\Delta q \simeq (2\pi/d)\cos(\theta/2)$ where d is the diameter of the scattering volume.

In optical mixing to study fluctuations in the scattered field a collection aperture is used to limit the detection area near to one coherence area.[38] The coherence area A_c is $A_c = \lambda^2 R^2/wd \sin\theta$ where w is the laser beam diameter in the scattering volume of dimension d and R is the scattering volume to detector distance. For large A_c the laser beam diameter w and scattering volume size should be small.

The sample is normally held in a round or square cell similar to a spectrometer cuvette and in a temperature-controlled index-matched enclosure. Careful temperature control is required to avoid convective effects. Measurements can be made up to high temperatures; in the author's laboratory a light-scattering cell for operation up to 1000 °C for use with inorganic glass samples has been constructed. Sample volumes can be very small (*e.g.* ~ 0.1 cm^3) or larger dependent on requirements. For accurate measurement great care is required on sample preparation to avoid contamination from dust. Sample cleaning by micropore filtering and centrifugation is generally used and solvent distillation may also be employed. Centrifugation of the sample in specially constructed combined centrifuge–sample cells is effective in avoiding contamination. Special signal processing has also been proposed to counter contamination.[62, 63]

The autocorrelator receives a continuous signal and records the correlation function $G(\tau) = \lim\limits_{T \to \infty} \dfrac{1}{2T} \displaystyle\int_{-T}^{T} I(t)\, I(t - \tau)\, dt$. With digital photon detection and the time range divided into finite channels of time width T, the correlator computes the function $G(kT) = \displaystyle\sum_{i=0}^{N-1} n_k n_{i+k}$ where k is the channel number and n_i is the number of counts in channel i. A correlation function and associated residuals from data analysis are shown in Figure 6. A full multibit correlator performs the product operation with the number of bits in each channel corresponding to the sample counts. To speed up computation the signal can be converted to single bit by limiting with a clipping level n_c such that $n_i = 0$ if $n_i < n_c$ and $n_i = 1$ for $n_i \geq n_c$. The clipped correlation function is a good approximation to the true correlation function if the signal obeys Gaussian random statistics, *i.e.* the signal is from a large number of independent scatterers undergoing random Brownian motion. If the

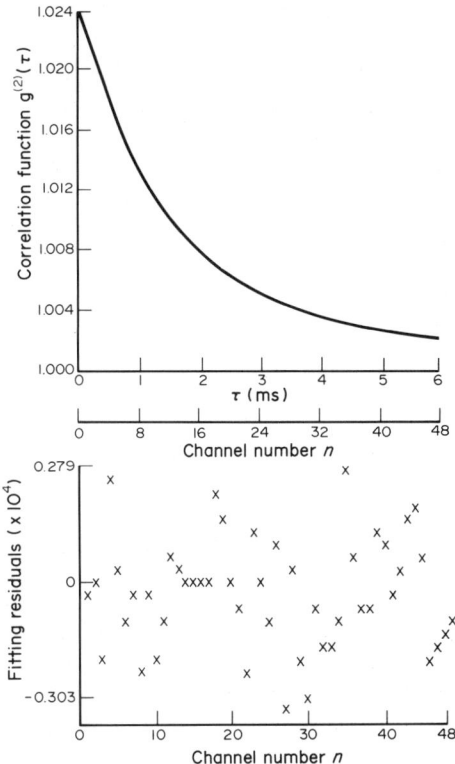

Figure 6 (a) Autocorrelation function for two-component polystyrene, $\bar{M}_w = (1.1 + 0.498) \times 10^6$ in benzene at 25 °C. Sampling time 200 μs, heterodyne mode. (b) Data residuals from S-exponential analysis; see Section 7.2.5

sample contains only a few scatterers or the scatterers have dynamically coupled scattering then the clipped correlation function will be distorted.

A number of correlation systems have been developed.[57-61] A wide range of correlator sample times can be provided typically from 10^{-7} s up to minutes; sample times less than 100 ns can be provided often with single bit operation or in burst-mode operation. An alternative method of measuring the correlation for fast times has been explored using pulse interval timing between pairs of scattered photon pulses.[64] The correlator channels may be distributed in time in various ways as well as with a linear time range. Delaying a group of channels at the end of the time range by a long delay enables the baseline of the correlation function to be obtained more accurately by measurement rather than by computer analysis. A more recent and very valuable development[58,65] for polymer applications is the use of channels distributed exponentially in time (logarithmic correlator) or with several linear groups of channels with linear time ranges and distributed to cover different time regions. These latter instruments are able to record the correlation function simultaneously over a very broad range of time.

In many polymer systems the correlation function is characterized by a distribution of relaxation times, for example in polydisperse systems or in highly concentrated or bulk samples.[66] The PCS correlation function can then be highly non-exponential and great care is required in the collection and analysis of the data. In these cases collection of the data is required over a wide range of delay times to derive the relaxation times adequately. The use of exponential sampling provides a wide range of effective correlation times.

The limiting precision of PCS is set by sample considerations and noise error sources. The signal-to-noise (S-N-R) ratio is influenced by light intensity, data collection time and stray light. The concentration of the sample affects the scattered intensity and for a polymer–solvent system has a minimum value set by the ratio of signal photons from the solute compared to stray photons from the solvent. For a correlation function of the form $G(\tau) = 1 + a e^{-\Gamma \tau}$, the number of decay times in a measurement run time of T_e is ΓT_e, leading to a S-N-R of $(\Gamma T_e)^{\frac{1}{2}}$. The S-N-R can be improved by increasing T_e, the duration of the measurement, or by increasing Γ by scattering at longer angles, since $\Gamma \propto D q^2$. Stray light from contamination and dust scattering, flare off cell walls and optical components and impurity scattering leads to distortion of the correlation function and may cause mixed self-beat–heterodyne detection to occur.

7.2.5 Photon Correlation Spectroscopy Data Analysis

Fitting of the correlation function $C(\tau) \cong g(\tau)$ by a sum of exponential components is often required. In many cases there may be no prior knowledge of the number or weighting of the exponentials. In general the time correlation function can be written as

$$C(\tau) \;=\; \sum_{i=1}^{p} A_i \, e^{-\tau/\tau_i} \tag{38}$$

where p is the required number of exponentials. One of the most common methods employed is that of non-linear least squares fitting, such as the Marquardt routine.[67] These routines are not reliable in fitting more than two exponentials with background.

In the case of polydisperse systems the general form of the time correlation function is the Laplace transform of a distribution linewidth function $G(\Gamma)$

$$C(\tau) \;=\; g^{(1)}(\tau) \;=\; \int_{0}^{\infty} G(\Gamma)\, e^{-\Gamma\tau}\, d\Gamma \tag{39}$$

In principle, knowledge of $G(\Gamma)$ allows the particle size distribution to be derived, when the scattering intensity as a function of particle size is known. In recent years a number of methods have been developed[68-84] to extract $G(\Gamma)$ from $C(\tau)$.

For spherical monodisperse particles the normalized self-beat photon correlation function is

$$g^{(2)}(\tau) \;=\; 1 + C|g^{(1)}(\tau)|^2 \;=\; 1 \;+\; C e^{-2D_T q^2 \tau} \tag{40}$$

where C is a constant. In the heterodyne mode the correlation function may be reduced to

$$g^{(2)}(\tau) \;=\; 1 \;+\; C g^{(1)}(\tau) \;=\; 1 \;+\; C e^{-D_T q^2 \tau} \tag{41}$$

With a polydisperse system, the distribution of molecular sizes gives a distribution in D_T and the correlation function is made up of a distribution of exponentials. Then from equation (39)

$$g^{(1)}(\tau) \;=\; \int_{0}^{\infty} e^{-\Gamma\tau} C(\Gamma)\, p(\Gamma)\, d\Gamma \tag{42}$$

with $G(\Gamma) = C(\Gamma)p(\Gamma)$ in which $p(\Gamma)$ takes account of the variation of the scattering power with particle size. Equation (42) is of the form of a Laplace transform and to derive the linewidth distribution requires Laplace inversion of the data. This presents substantial difficulty as in the inversion small changes in the data can give rise to large errors in the inverted information.

A general purpose and flexible method of inverting PCS data has been developed[74] under the name CONTIN. This has been widely applied in PCS studies with excellent results for PCS data having low noise. The program contains safeguarding constraints to avoid the ill-posed nature of the inversion.

An early method of analysis was based on a cumulant expansion[85,86] of the correlation function

$$\ln|g^{(1)}(\tau)| \;=\; -\bar{\Gamma}\tau \;+\; \frac{1}{2!}\mu_2\tau^2 \;-\; \frac{1}{3!}\mu_3\tau^3 \;+\; \frac{1}{4!}[\mu_4 - 3\mu_2^2]\,\tau^4 \;+\; \cdots \;=\; \sum_{k=1}^{\infty} k_m(\Gamma)\frac{(-\tau)^m}{m!} \tag{43}$$

where k_m is the cumulant and $\mu_i = \int_{0}^{\infty} (\Gamma - \bar{\Gamma})^i\, G(\Gamma)\, d\Gamma$. Equation (43) may be fitted by a least squares routine to the correlation function and values for $\mu_2, \mu_3, \ldots$ obtained. The average width $\bar{\Gamma} = \int_{0}^{\infty} \Gamma G(\Gamma)\, d\Gamma$ = mean relaxation time. The variance is $\mu_2/\bar{\Gamma}^2$ with $\mu_2 = \int_{0}^{\infty} (\Gamma - \bar{\Gamma})^2\, G(\Gamma)\, d\Gamma$. For low q, $qR < 1$ and $\bar{\Gamma} = \bar{D}_T q^2$ where $\bar{D}_T$ is the z-average D_T. Practically it is difficult to obtain accurate values of the moments greater than the second and the method is restricted to relatively low unimodal distributions. When the moments have been obtained $G(\Gamma)$ may be reconstructed using Pearson's method.[87] Cumulant analysis is discussed further in Chapter 8.

An eigenvalue expansion multiexponential method has been developed[69,71] of particular utility in application to polymer systems. The linewidth distribution is represented as a series of exponentially spaced linewidths such that

$$G(\Gamma)d\Gamma = G^*(\ln\Gamma)d(\ln\Gamma) \tag{44}$$

$$G^*(\ln\Gamma) = \sum_n a_n \delta(\ln\Gamma - \ln\Gamma_n) \tag{45}$$

where n is the number of functions whose upper value is limited by the noise level. The closest points that can be resolved are given by

$$\frac{\Gamma_n}{\Gamma_{n-1}} = \exp\left[\frac{\pi}{\omega_{max}}\right] \tag{46}$$

$$\omega_{max} = \left[-\frac{2}{\pi}\right]\ln\left[\frac{\text{noise level}}{\sqrt{\pi}}\right] \tag{47}$$

This defines a resolution limit such that two linewidths Γ_n and Γ_{n-1} cannot be resolved unless they are separated by at least the factor $\exp(\pi/\omega_{max})$ which is dependent on the noise level of the correlation function. The theoretical linewidth resolution when noise-limited from equation (47) is shown in Figure 7(a), with the maximum noise level permitting resolution of two decay components shown in Figure 7(b).

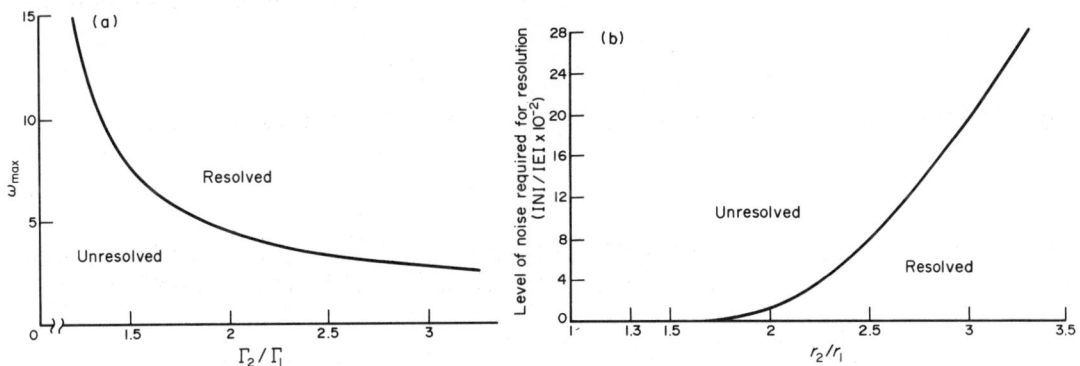

Figure 7 (a) Noise-limited theoretical linewidth resolution $\omega_{max} = (-2/\pi)\ln(\text{noise level}/\sqrt{\pi})$; and (b) maximum noise level for resolution of two decay components

Using the eigenvalue analysis it has been shown[65,72,76,78] that the Laplace inversion can be reduced to a linear fitting problem using the defined linewidths. When correlation function data points are placed with exponentially spaced delay times τ, $\tau e^{\Delta\tau}$, $\tau e^{2\Delta\tau} \ldots \tau n \ldots$ interpolation between these points enables the correlation function to be reconstructed for all values of τ from the values at τ_n, within the limits set by the noise bandwidth.

A method of histograms has been used[88-90] in which $G(\bar{\Gamma}_j)$ is assumed to be constant over a range $[\bar{\Gamma}_j - \Delta\Gamma/2]$ to $[\bar{\Gamma}_j + \Delta\Gamma/2]$ and one histogram element is approximated as a single exponential. Then

$$g^{(1)}(\tau) = \sum_{j=1}^{n} G[\bar{\Gamma}_j] \int_{\bar{\Gamma}_j - \frac{\Delta\Gamma}{2}}^{\bar{\Gamma}_j + \frac{\Delta\Gamma}{2}} \exp[-\bar{\Gamma}_j\tau]d\Gamma \tag{48}$$

The analysis reduces to the fitting of exponentials to the data and is an approximation to the eigenvalue expansion method. This is related to a more general method of splines.

A method of linear programming (SIMPLEX) or non-negative least squares has been assessed[91] for PCS. It has been shown that even finely structured $G(\tau)$ functions can be recovered from the Laplace transform, avoiding convergence problems and negative amplitudes. The program incorporates a technique of sequences of residuals defining the choice of the maximum retrievable resolution.

Analysis of PCS decay curves by a new S-exponential sum method has been demonstrated.[92] This has been applied to the extraction of mean linewidths and moments of unimodal polymer distributions, mixtures of monodisperse polymers, and bimodal polydisperse systems; an example is shown in Figure 6.[93-95]

The use of a profiled function incorporating the measured mean and polydispersity index has been shown[81-83] to improve the inversion and resolution of closely spaced peaks; this also provides a rapid method for computation.

The progress made in recent times in photon correlation instrumentation and in methods of data analysis now enables several exponential components to be reliably extracted from PCS data having low noise content. The interactive operation of the measurement with the data analysis,[96,97] aided by an on-line computer, enables the best parameters for each measurement to be set up.

7.2.6 Related Techniques

There are a number of promising developments in photon correlation spectroscopy techniques which add greater versatility to application of the method to polymers. It is valuable to arrange the spectrometer such that both the PCS linewidth-scattered light (for dynamic information) and also the total intensity angular distribution (for static information) can be measured. We have seen that depolarized I_{HV} scattering can provide additional discrimination where more than one dynamical process contributes to the linewidth.[44-49] New compact spectrometers can be designed using fibre-optic links;[98] this is now routinely used in the application of PCS to biomedical blood flow measurements.[99,100]

The technique of fluorescence correlation spectroscopy[101-105] enables weakly scattering systems to be investigated and with the use of attached chromophores allows specific features of molecular dynamics to be emphasized. Similarly the use of resonance-enhanced scattering techniques in Rayleigh scattering,[106,107] for example by using tuned lasers, may be used to increase the scattering sensitivity. Combination of the PCS technique with a microscope attachment for detection enables very small samples to be studied and has been successfully applied to biological systems.[108] Development of the PCS method for use in surface scattering[109] and with fibre-optic evanescent wave methods[110] would lead to novel configurations. Several other PCS configurations have been investigated which have potential application to polymers; these include forced Rayleigh scattering,[111] a differential scattering technique to observe small conformational changes in macromolecules,[112] the use of multidetectors[113] and anticorrelations,[114] and application of oscillatory electric fields to the scattering system.[115-118] Finally, future technical developments on correlation methods with microprocessor and transputer-based systems may provide faster and more compact instruments.

7.3 APPLICATIONS OF PHOTON CORRELATION SPECTROSCOPY TO POLYMERS*

7.3.1 Scope

PCS was applied to the study of polymer characterization and polymer dynamics from the inception of the technique. Early studies were on the diffusion and sizing of colloidal particles and on the translational diffusion of polymer coils in solution under variation of conditions, molecular weight, solvent and concentration. This work was extended to the study of a variety of polymer systems which included gels, liquid crystals, bulk polymers, biopolymers, micelles, polymer adsorption and polyelectrolytes. More recently PCS has been applied to a systematic study of polymer dynamics in association with new theoretical studies using scaling and renormalization techniques and to a wide range of polymer systems, particularly semidilute and concentrated systems, gels and bulk polymers to provide information on collective dynamics. Generally the study of polymer dynamics by PCS has provided a complementary technique to investigation of static properties by small angle X-ray scattering and neutron diffraction.

---

* See Chapter 8 for the application of PCS to the characterization of polymers in solution.

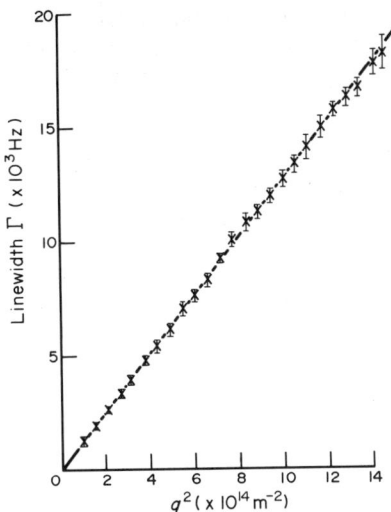

Figure 8 Linewidth variation with q^2 for polystyrene in benzene $\lambda_i = 488$ nm, $T = 25\,°C$, heterodyne detection. $\bar{M}_w = 1.1 \times 10^6$, $c = 4.16 \times 10^{-4}\,\text{g cm}^{-3}$. The solid line is calculated[93] from $(\bar{D}_0)_z = 2.18 \times 10^{-4}\,\bar{M}_z^{-0.55}\,\text{cm}^2\,\text{s}^{-1}$

7.3.2 Polymer and Colloid Characterization

The use of PCS for the determination of translational diffusion coefficients of polymers in solution is well established. Examples covering a wide range of systems are contained in refs. 119–138. For monodisperse polymer solutions which can be well cleaned, values of D_T can be obtained to an accuracy of better than 1%. An example of the q^2 plot of the linewidth is shown in Figure 8 for a sample of polystyrene in benzene. For a polydisperse polymer solution, each contributing narrow molecular weight range in the distribution has an associated diffusion coefficient D_i. The PCS autocorrelation function is then a sum of exponentials with relaxation time $(q^2 D_i)^{-1}$, each weighted by the scattering power. Fitting the correlation function to a single exponential gives only an ill-defined average value. Where the correlation times are distinct and well separated multiexponential fitting can provide the separate values up to a number of about five, dependent on experimental conditions. When the exponential components are close together only recently developed data analysis as described in Section 7.2.5 enables meaningful data to be extracted.

The cumulant analysis* in which the logarithm of the correlation function is fitted to a power series, gives as the first linear cumulant $k_1 = q^2 \bar{D}_z$ providing an apparent z-average of the diffusion coefficient, defined as $\bar{D}_z = \Sigma D_i^z M_i C_i P_i(q) / \Sigma M_i C_i P_i(q)$. This quantity differs from the true z-average diffusion coefficient unless the scattering factors $P_i(q) = 1$, i.e. unless the molecules are small or $\theta \to 0°$. The $\bar{D}_z$ has been combined with the weight average sedimentation coefficient to give, by the Svedberg equation, the weight average molecular weight.[139] The second cumulant k_2 gives the variance in D and is a measure of the polymer polydispersity. The cumulant analysis has been applied to the characterization of many polymers (for example refs. 140–161), and is often used as a rapid assay of the polydispersity, although it is restricted to unimodal distributions of relatively narrow polydispersity. When the molecules are sufficiently large for form factors to be important, the angular dependence of k_1 and k_2 provides additional information on the moments of the size distribution. The cumulant analysis can also be used to obtain $\bar{M}_n$ and $\bar{M}_w$ for polymers for which D as a function of M is known. Study of D_T has been made to investigate polymers in many situations such as temperature variation,[150] at large $\bar{M}_w$,[151] for chain stiffness and statistical lengths,[152] and for comparison of good and poor solvent conditions.[153]

7.3.3 Polymer Dynamics

Photon correlation spectroscopy has been applied to the study of polymer dynamics since the early 1960s. It was soon found that accurate values of the self-diffusion coefficient could be obtained

* See Chapter 8.

under a wide variation of polymer type, size, concentration and solution conditions. Of course care was needed in the sample preparation and in the measurement technique. It is, in a way, unfortunately too easy in PCS to make a measurement and obtain a result, but it requires some diligence in sample preparation and measurement to make such a result a reliable one.

PCS has been used extensively to extract rotational and intramolecular dynamics in the presence of translational diffusion. The use of depolarized scattering reviewed in Section 7.2.2 has been applied[162-165] to polymers, biopolymers and colloids.

In recent years the understanding of polymer chain dynamics has advanced significantly in dilute, semi-dilute and concentrated ranges. PCS is a valuable technique for measurement in each of those regions. The high level of interest in the application of light scattering to polymer dynamics is heightened by the advances in understanding from new theoretical approaches of scaling and renormalization group, leading to new insights and novel predictions.[166-177] The application of PCS also complements the techniques of small angle X-ray scattering and neutron diffraction, which are sensitive to spatial dimensions of polymers up to 10 nm, whereas the q values relevant to visible light scattering probe over dimensions of $30 < q^{-1} < 3000$ nm. Thus PCS provides a probe of polymer dynamics on a scale matching overall polymer dimensions.

In PCS, information from the intensity correlation function $\langle I(q,t)I(q,0)\rangle$ provides the intermediate scattering function $S(q,t)$

$$\langle I(q,t)\,I(q,0)\rangle \quad - \quad \langle I^2\rangle \quad = \quad \beta\,S^2(q,t) \tag{49}$$

$$S(q,t) \quad = \quad \frac{1}{V}\sum \exp\{i\mathbf{q}[\mathbf{r}_i(t) \quad - \quad \mathbf{r}_j(0)]\} \tag{50}$$

where V is the scattering volume, $r_{i,j}(t)$ is the position of the i,j monomers at time t and β is a geometrical factor. Polymer models in exact forms have been developed for $S(q,t)$, in order to interpret scattering data; these include models for isolated chains under theta conditions and for chains with hydrodynamic interaction through the preaveraged Oseen tensor. Calculations of $S(q,t)$ for a flexible polymer have been made by Pecora[6-8] using a Green function solution of the Fokker–Planck equation for the Rouse–Zimm bead-and-spring model.

All the dynamical regions shown in Figure 9 are accessible by PCS. Here the dimension R is a characteristic polymer size, R_h or R_g. For $qR \ll 1$, PCS is sensitive to fluctuations greater than the size of the polymer, the correlation linewidth is $\Gamma = D_T q^2$, and the translation diffusion coefficient D_T is determined. For $qR \sim 1$ both translational diffusion and internal polymer dynamics are sensed and as the condition is emphasized, $qR \gg 1$, internal dynamics dominate.

With $qR \ll 1$ the recent work in renormalization group theory and the 'blob' model has predicted polymer dynamical dependences and has been thoroughly investigated by PCS polymer translation diffusion. Under theta conditions $D(c \to 0) \sim M^{-\frac{1}{2}}$ for molecular weight M. For good solvent conditions $D(c \to 0) \propto M^{-a}$ with $0.5 < a < 0.6$; for smaller values of M smaller values of a

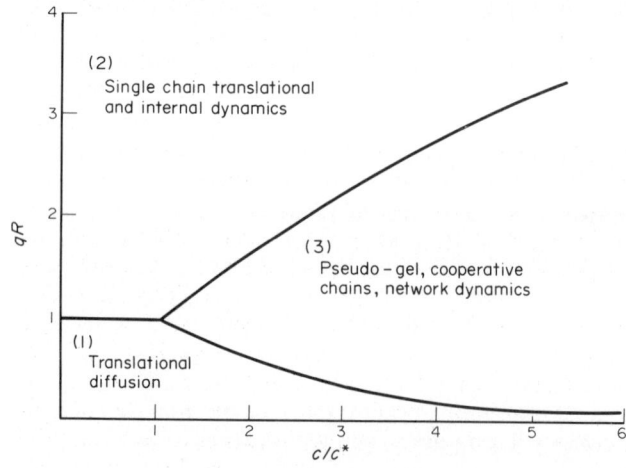

Figure 9 Polymer dynamical regimes accessible by PCS for variation of polymer radius R or concentration c. The overlap concentration $c^* \simeq 3N/r\pi R_g^3$, where N = degree of polymerization

than predicted may be explained by the hydrodynamic radius R_h not reaching the asymptotic limit.[178]

Under the condition $qR \gg 1$ the PCS correlation function enables the fundamental mode of internal dynamics to be measured[178-189] and to be related to polymer models[190-199] such as the simple or extended bead-and-spring model and dynamic scaling exponents. This mode is controlled by the balance between the elastic and viscous forces on the chain. Within the asymptotic limit, predictions of the large scale intramolecular relaxation rate have given $\Gamma \sim (kT/\eta)q^3$ within the Zimm model with hydrodynamic interaction, and $\Gamma \sim (kT/\eta)q^4$ for the Rouse model with no hydrodynamic interaction.

The concentration dependence of the translational diffusion coefficient has been investigated for several systems. For dilute solutions the diffusion coefficient can be written

$$D_T = \frac{kT}{f} \left[1 - \frac{N\bar{v}c}{M} \right] [1 + 2A_2 Mc + 3A_3 Mc^2 + \ldots] \tag{51}$$

where f is the friction coefficient of the polymer in solution, A_2 and A_3 are the second and third virial coefficients of osmotic pressure and $\bar{v}$ is the partial specific volume. At the theta temperature D_T depends on concentration only by the concentration dependence of f as $f = f_0 (1 + k_f c + k'_f c^2 + \ldots)$ and the concentration dependence of D for dilute solutions is first order in c

$$D(c) = D_0(1 + k_D c) \tag{52}$$

where $D_0 = kT/f_0$ is the translational diffusion coefficient at infinite dilution. The thermodynamic and hydrodynamic dependence of k_D is given as

$$k_D = 2A_2 M - k_f - \frac{N\bar{v}}{M} \tag{53}$$

This relation has been used to determine the second virial coefficient A_2 by PCS.[124]

In the semidilute region $c \geq c^*$ the polymer coils overlap while the polymer volume fraction is still relatively low. This region has been of considerable interest recently in the application of the new theoretical methods mainly based on scaling techniques.[200-204] PCS provides a measure of the cooperative diffusion coefficient D_c which is associated with a hydrodynamic correlation length ξ_h

$$D_c = \frac{kT}{6\pi\eta_0\xi_h} \tag{54}$$

where η_0 is the viscosity of the solvent. The concentration dependence of D_c and ξ_h depends on the quality of the solvent. For good solvents ξ_h is identified with the average distance between nearest chain contacts ξ_ρ and scales with monomer concentration ρ as $\xi_h \sim \rho^{-\frac{3}{4}}$. For theta solvents the pair interaction is absent and the hydrodynamic correlation length is associated with ternary contacts, i.e. $\xi_h \sim \rho^{-1}$. A detailed study of the semidilute region by PCS has been made[205-215] and also of the cross-over between good and theta solvents.[206-210] PCS studies have shown that the concept of a sharp cross-over between good and theta conditions to be untenable.

Following de Gennes,[200] polymers in the semidilute region have physical entanglements with lifetime T_r and a mean distance between entanglements ξ. For frequencies $\omega > 1/T_r$ the entanglements do not relax and the solution behaves as a network with permanent crosslinks, termed a pseudo-gel. The following sequence of scaling laws has been proposed: good solvent: $\xi_h \sim \rho^{-\frac{3}{4}}$, marginal solvent: $\xi_h \sim \rho^{-\frac{1}{2}}$ and theta solvent: $\xi_h \sim \rho^{-1}$. A body of data has been drawn up supporting this analysis. In PCS the autocorrelation function is found to be strongly non-exponential in the semidilute region with evidence of slow decay modes.[216] The concept of physical entanglements is unresolved and requires further investigation. Recent studies[217,218] have been made by PCS on high molecular weight polystyrene in an ethanol/ethyl acetate mixed solvent system in the semidilute regions. Using multiexponential analysis two components were resolved: a fast gel mode and a slow hydrodynamic mode. PCS has been shown to be a valuable probe of the semidilute region and further study aiming for a detailed understanding of the general area of condensed polymer systems is merited.

7.3.4 Gels

The random collective dynamics of a network is describable by a diffusive process; crosslinked polymer gels have been shown to be amenable to study by PCS.[219-231] Since the wavelength of light is much larger than the average distance between crosslinks, in one of the first analyses[219, 225, 229] a continuum model has been developed for the process. Fluctuations of the polymer network around its equilibrium position are driven by an osmotic force tending to equalize the concentrations and an elastic force maintaining the position of the network; the fluctuations are damped by the frictional force between the polymer network and solvent. Concentration fluctuations in the network may be described by a displacement vector $u(r, t)$ giving the displacement of a point r at time t from its average position. Small deformations of the network follow the equation

$$\rho \frac{\partial^2 u}{\partial t^2} = \nabla \cdot \tilde{\sigma} - \rho \frac{\partial u}{\partial t} \tag{55}$$

where ρ is the average density and $\tilde{\sigma}$ is a stress tensor. Solution of the equation of motion leads to the normal modes in the gel following a diffusion equation with diffusion coefficient D

$$D = \frac{\text{elastic modulus}}{\text{friction coefficient}} = \frac{E}{f} \tag{56}$$

$$\text{longitudinal mode: } E = L = K + \frac{4}{3}\mu; \quad \text{transverse mode: } E = \mu$$

where L, K and μ refer to the longitudinal, bulk and shear moduli. From PCS measurements the quantities L/f and μ/f can be obtained and in principle light scattering enables the viscoelastic parameters of the gel to be determined. Scaling arguments may be used to relate K, f and D to the network concentration or to the number of polymer units between crosslinks N, *i.e.* $K \sim \phi^{1/4} \sim M^{1/5}$, $f \sim \phi^{-1/2} \sim N^{2/5}$, $D \sim \phi^{3/4} \sim N^{-3/5}$.

Polyacrylamide–water gels,[219] crosslinked networks[220, 221] and pseudo-gels[222] have been investigated, including PS–benzene, PS–ethyl acetate and PDMS–toluene. It has been shown[221, 223, 229] that the dynamical behaviour of gels is analogous to that of semidilute solutions, although experimental data fit better to scaling laws for semidilute solutions than for gels, probably due to the effects of pendant chains in networks of variable density. In PCS experiments stray light scattered by inhomogeneities in gels or semidilute solutions gives a reference signal unshifted in frequency and inducing heterodyne optical mixing. Then the correlation function is $C(\tau) = A = e^{-Dq^2\tau} + B$ where A is an amplitude and B a background. As for semidilute solutions, recent developments in exponential sampling in PCS measurements and multiexponential analysis should provide a powerful method to identify relaxation processes and separate informative signals from spurious scattering.

7.3.5 Bulk Polymers

Light scattering in bulk polymers is due to fluctuations in the local dielectric constant. Theoretical approaches predict frequency-shifted Brillouin peaks and a dynamic Rayleigh peak, centred around the incident frequency and associated with relaxation of the moduli, and slowly relaxing adiabatic and isothermal density fluctuations. Theoretical treatments have been made by Mountain[232] and Rytov[233] and more recently by Wang and co-workers.[234] Near the glass transition temperature T_g relaxation of the longitudinal modulus is slow enough to be observed by PCS, the density fluctuations are isothermal and the scattering is weak. In polymers there may also be a strong central peak due to anisotropy fluctuations.

Considerable progress has been made in the application of PCS to bulk polymers, particularly in the work of Patterson,[235-237] Wang[238-240] and co-workers. A pure sample of poly(ethyl methacrylate) (PEMA) has been prepared[241] by thermal polymerization from the monomer, providing a weakly scattering anisotropic polymer near T_g (65 °C). The scattering from this sample was attributed to thermal density fluctuations. The observed average relaxation time changed by a factor of ten times for a 5 °C change near T_g. This is characteristic of the primary glass transition.

A distribution of relaxation times has been determined[242] for poly(methyl methacrylate) (PMMA) with two peaks in the distribution being observed near T_g; this is a general phenomenon in the poly(alkyl methacrylate)s. PCS has been applied[238] to melt samples of poly(vinyl acetate) which indicates that the dynamical properties of melts can be explored by this technique. Use of the PCS technique with exponential sampling enables a very wide range of relaxation times to be simultaneously recorded and heterodyne detection may be beneficial in this study.

7.3.6 Liquid Crystals

Nematic liquid crystals are strong scatterers of light resulting from thermal fluctuations associated with the liquid crystal director. The dynamic fluctuations may be ascribed to three elastic and six viscotic constants. Lorentzian broadening with linewidths Γ_1 and Γ_2 results from two overdamped modes within the nematic liquid crystal due to splay–bend and twist–bend combinations. PCS has been used by several groups[243–246] to determine the ratio of elastic to viscotic constants in nematic liquid crystals. Self-beat and heterodyne optical mixing have been used to measure the viscoelastic ratios $[k_{22}/\gamma_1]$, where k_{22} is a Franck elastic constant and γ_1 the twist viscosity. The scattered intensity relates to ratios of the elastic constants and PCS gives elastic/viscotic constant ratios. As for scattering from other bulk media, stray elastically scattered light is unavoidable and operation with partial heterodyning can be circumvented by purposely working in the heterodyne mode. Partial heterodyning is undesirable since it can lead to multiple exponential decay constants and make data analysis intractable.

Application of electric or magnetic fields to the nematic liquid crystal enables the elastic/viscotic constants to be separated into individual elastic and viscotic constants.[247–250] PCS has been used[251] to measure k_{22}/γ_1 for the polymer–monomer liquid crystalline mixture polysiloxane–pentylcyanobiphenyl and with an applied electric field to determine k_{22} and γ_1.

7.3.7 Polymer Adsorption

Several studies have been made of polymer adsorption to determine particle size and adsorbed polymer layer thickness.[252–256] The Stokes–Einstein equation $D_T = kT/6\pi\eta R_h$ is used to derive the hydrodynamic radius R_h from the translational diffusion coefficient D_T. Accurate values of the adsorbed layer thickness can be obtained if attention is paid to experimental conditions with averaging over a full angular distribution. Accuracy in the measurement of layer thickness is improved by making the measurements as a function of particle concentration, particularly for aqueous dispersions at low ionic strength, to eliminate structural effects from interparticle interactions.

Where the adsorbed layer thickness is to be derived from the difference in hydrodynamic radii of the coated and uncoated particles then accurate measurement of the radii is essential since the adsorbed layer is usually only a small fraction of the particle radii. PCS studies have been made on the system poly(ethylene oxide)/polystyrene latex/water for narrow fractions of PEO on surfactant free latices.[257] The hydrodynamic thickness of the adsorbed layer δ_h increases more rapidly with $\bar{M}_w$ than does the free coil radius R_h. For constant polymer concentration it is predicted that $\delta_h \sim M^{0.8}$; this is in good agreement with experimental data, an example[258] of which is shown in Figure 10.

To make this application more suitable for PCS a differential scattering technique is required. A spectrometer designed for differential measurements has been described.[112]

7.3.8 Phase Transitions

In recent years there has been great interest in the coil–globule transition of a flexible polymer from the random coil state at the theta temperature to a collapsed globule at reduced temperatures.[259,260] The transition may be induced by reducing the solvency as well as the temperature over the region between the theta temperature and the coexistence curve at which phase separation takes place. With reduction in temperature from the theta temperature, the hydrodynamic radius R_h decreases from its theta value, $R_h \sim N^{1/2}$, to the collapsed value, $R_h \sim N^{1/3}$, where N is the degree of polymerization. Within the theories of classical mean field and renormalization group, in the temperature region in which the polymer is collapsing, $R_h \sim N^{1/3}|\tau|^{-1/3}$ with τ being the reduced temperature $\tau = (T - \theta)/T$.

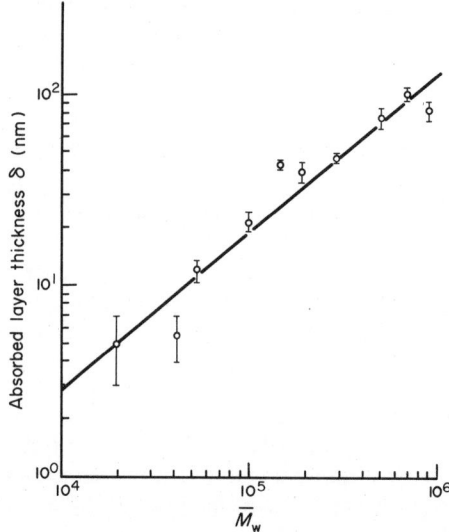

Figure 10 Adsorbed layer thickness δ determined by PCS.[258] Measurements of δ for poly(ethylene oxide) fractions $(M_w/M_n = 1.03$ to $1.85)$ at different molecular weights adsorbed at $T = 25\,°C$ onto surfactant free polystyrene latex $(R_h = 116\,nm)$. Gradient $= 0.85 \pm 0.05$

Various experimental studies of the coil–globule transition have been made by small angle neutron scattering, viscosity, elastic light scattering and PCS. There are stringent experimental requirements to be satisfied to observe the transition. Very high dilution is required in order to observe isolated chain collapse without interchain overlap aggregation. Large molecular weight polymers with narrow molecular weight distributions are also required.

Measurements[261,262] have been made on polystyrene of $\bar{M}_w = 26 \times 10^6$ in cyclohexane. Figure 11 shows the collapse transition for a highly diluted sample of polystyrene $(\bar{M}_w = 27 \times 10^6)$ in cyclohexane as the temperature is reduced below the theta temperature.[262] It has been confirmed[260] that the molecular weight dependence is $R_h \sim \bar{M}_w^{1/3}\,|\tau|^{-1/3}$, with a smooth crossover between the theta and collapse regimes, though studies made on a very pure high molecular weight sample $\bar{M}_w = 41 \times 10^6$ with observation of the transition.

Some dramatic phase transition effects have been observed by PCS by Tanaka (reviewed in ref. 263) in polyacrylamide/acetone/water gel caused by changes in solvent composition and temperature. The gel undergoes collapse when the acetone concentration is increased or the temperature is lowered. Ionization of the network has been shown to play a role in the transition. A non-ionized gel undergoes a continuous change in equilibrium volume when the acetone concentration or temperature is changed. When a small number of the polyacrylamide chain units are hydrolyzed, to form ionizable acid groups, a sharp reversible volume collapse occurs, with up to a 500-fold volume reduction. This volume collapse can also be induced by change in the pH of the gel fluid. The large

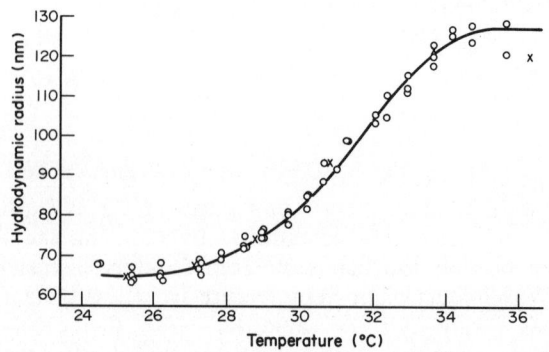

Figure 11 Hydrodynamic radius for high molecular weight polystyrene $\bar{M}_w = 27 \times 10^6$ in cyclohexane,[262] showing collapse of the polymer for reduction in temperature below the theta temperature $(T \simeq 35\,°C)$

volume collapse has been attributed to the change in osmotic pressure of hydrogen ions dissociated from the ionizable groups.

7.3.9 Electrophoretic Scattering and Application to Polyelectrolytes

Macroions in solution in an electric field acquire a terminal velocity

$$v = \mu E \tag{57}$$

where μ is termed the electrophoretic mobility. Light scattered from the moving ions undergoes a Doppler shift with frequency change

$$v_E = \frac{2\mu E}{\lambda} \sin\left[\frac{\theta}{2}\right] \cos \alpha \tag{58}$$

By heterodyne mixing with unscattered laser light the frequency shift v_E is derived and may be analyzed with a spectrum analyzer or autocorrelator giving

$$g(\tau) = A + Be^{-D_T q^2 \tau} \cos(2\pi v_E \tau) \tag{59}$$

The light-scattering cell in electrophoretic light scattering requires careful design to provide a homogeneous electric field and to ensure that extraneous effects such as bubble formation at the electrodes, electro-osmosis and Joule heating induced convection are avoided.

A few studies of polyelectrolytes by PCS have been initiated. In a study of Sodium–polystyrene-sulfonate of $M = 10^5$ in water, PCS was used to obtain information on the conformation of the polyions.[264] The cooperative diffusion coefficient in the semidilute region ($c > c^*$) was found to follow a power law, $D_{eff} \sim c^{0.52 \pm 0.02}$; at high concentration a constant D_{eff} was observed. These results indicated that the conformation of the polyion was rod-like at low concentration and stayed extended between entanglements in the semidilute region. PCS has been used to demonstrate a conformational change of a weak acid polyelectrolyte upon ionization,[265] and a few other studies on charged polymers can be cited.[266-269] Theories of light scattering for polyelectrolyte solutions predict a non-exponential autocorrelation function in which there is a coupling between ionic relaxation and hydrodynamic self-diffusion. These aspects are relevant to several topics of techno-logical interest, such as polyelectrolyte steric stabilization, and justify further exploration.

7.4 REFERENCES

1. H. C. Van de Hulst, 'Light Scattering by Small Particles', Wiley, New York, 1962.
2. D. McIntyre and F. Gornick (eds.), 'Light Scattering from Dilute Polymer Solutions', Gordon and Breach, New York, 1964.
3. I. L. Fabelinskii, 'Molecular Scattering of Light', Plenum Press, New York, 1967.
4. M. Kerker, 'The Scattering of Light and Other Electromagnetic Radiation', Academic Press, New York, 1969.
5. M. B. Huglin, 'Light Scattering from Polymer Solutions', Academic Press, New York, 1972.
6. R. Pecora, *J. Chem. Phys.*, 1964, **40**, 1604.
7. R. Pecora, *J. Chem. Phys.*, 1965, **43**, 1562.
8. R. Pecora, *J. Chem. Phys.*, 1968, **49**, 1032.
9. A. T Forrester, R. A. Gudmunsden and P. O. Johnson, *Phys. Rev.*, 1955, **99**, 1691.
10. H. Z. Cummins, N. Knable and Y. Yeh, *Phys. Rev. Lett.*, 1964, **12**, 150.
11. H. Z. Cummins and L. H. Swinney, *Prog. Opt.*, 1970, **8**, 133.
12. B. Chu, 'Laser Light Scattering', Academic Press, New York, 1974.
13. H. Z. Cummins and E. R. Pike (eds.), 'Photon Correlation and Light Beating Spectroscopy', Plenum Press, New York, 1974.
14. S. H. Chen and S. Yip (eds.) 'Spectroscopy in Biology and Chemistry — Neutrons, X-Rays and Lasers', Academic Press, New York, 1974.
15. B. Crosignani, P. Di Porto and M. Bertolotti, 'Statistical Properties of Scattered Light', Academic Press, New York, 1975.
16. P. N. Pusey and J. M. Vaughan, *Dielectr. Relat. Mol. Processes*, 1975, **2**, 48.
17. B. J. Berne and R. Pecora, 'Dynamic Light Scattering with Applications to Chemistry, Biology and Physics', Wiley, New York, 1976.
18. H. Z. Cummins and E. R. Pike (eds.), 'Photon Correlation Spectroscopy and Velocimetry', Plenum Press, New York, 1977.

19. A. D. Buckingham, *Philos. Trans. R. Soc. London, Ser. A*, 1979, **293**, 209.
20. V. Degiorgio, M. Corti and M. Giglio (eds.), 'Light Scattering in Liquids and Macromolecular Solutions', Plenum Press, New York, 1980.
21. S. H. Chen, B. Chu and R. Nossal (eds.), 'Scattering Techniques Applied to Supramolecular and Nonequilibrium Systems', Plenum Press, New York, 1981.
22. D. B. Satelle, W. I. Lee and B. R. Ware (eds.), 'Biomedical Applications of Laser Light Scattering', Elsevier, New York, 1982.
23. H. Z. Cummins and A. P. Levanyuk (eds.), 'Light Scattering near Phase Transitions', North-Holland, Amsterdam, 1983.
24. B. E. Dahneke (ed.), 'Measurement of Suspended Particles by Quasi-Elastic Light Scattering', Wiley–Interscience, New York, 1983.
25. J. C. Earnshaw and M. W. Steer, 'The Application of Laser Light Scattering to the Study of Biological Motion', Plenum Press, New York, 1983.
26. E. D. Schulz–DuBois (ed.), 'Photon Correlation Techniques in Fluid Mechanics', Springer-Verlag, New York, 1983.
27. W. Hess and R. Klein, *Adv. Phys.*, 1983, **32**, 173.
28. B. R. Ware, 'Light Scattering' in 'Optical Techniques in Biological Research', ed. D. L. Rousseau, Academic Press, New York, 1984.
29. W. Burchard, *Chimia*, 1985, **39**, 10.
30. R. Pecora (ed.), 'Dynamic Light Scattering; Applications of Photon Correlation Spectroscopy', Plenum Press, New York, 1985.
31. E. R. Pike, in ref. 13, p. 9.
32. V. Degiorgio, in ref. 18, p. 142.
33. V. Degiorgio, in ref. 25, p. 9.
34. R. F. Voss and I. Clarke, *J. Phys. A: Math. Gen.*, 1976, **9**, 561.
35. V. Degiorgio and J. B. Lastovka, *Phys. Rev. A*, 1971, **4**, 2033.
36. M. Bertolotti, in ref. 13, p. 41.
37. H. Z. Cummins, in ref. 13, p. 225.
38. E. Jakeman, in ref. 13, p. 75.
39. E. R. Pike, in ref. 18, p. 3.
40. B. R. Ware, in 'Optical Techniques in Biological Research', ed. D. L. Rousseau, Academic Press, New York, 1984, p. 1.
41. G. Williams, *Dielectr. Relat. Mol. Processes*, 1979, 4.
42. B. J. Berne, in 'Physical Chemistry: An Advanced Treatise', ed. H. Eyring, D. Henderson and W. Jost, Academic Press, Press, New York, vol. 8B, p. 539.
43. H. Z. Cummins, F. D. Carlson, T. J. Herbert and G. Woods, *Biophys. J.*, 1969, **9**, 518.
44. D. R. Bauer, J. I. Brauman and R. Pecora, *Macromolecules*, 1975, **8**, 443.
45. T. Norisuye and H. Yu, *J. Chem. Phys.*, 1978, **68**, 4038.
46. K. Moro and R. Pecora, *J. Chem. Phys.*, 1978, **69**, 3254.
47. B. Herpigny and J. P. Boon, in ref. 20, p. 91.
48. R. Pecora, in ref. 21, p. 173.
49. K. Zero and R. Pecora, in ref. 30, p. 59.
50. R. J. Glauber (ed.), 'Quantum Optics', Academic Press, New York, 1969.
51. S. M. Kay and A. Maitland (eds.), 'Quantum Optics', Academic Press, New York, 1970.
52. R. Loudon, 'The Quantum Theory of Light', Oxford University Press, Oxford, 1983.
53. E. Jakeman, C. J. Oliver and E. Pike, *J. Phys. A: Gen. Phys.*, 1970, **3**, L45.
54. E. Jakeman, *J. Phys. A: Gen. Phys.*, 1972, **5**, 1972.
55. H. M. Fijnaut and F. C. van Rijswijk, in ref. 18, p. 465.
56. P. J. Nash and T. A. King, *J. Phys. E*, 1985, **18**, 319.
57. C. J. Oliver, in ref. 13, p. 151.
58. C. J. Oliver, in ref. 21, p. 87.
59. C. J. Oliver, in ref. 21, p. 121.
60. N. C. Ford, in ref. 30, p. 7.
61. E. Jakeman, P. N. Pusey and J. M. Vaughan, *Opt. Commun.*, 1976, **17**, 305.
62. Y. Alon and A. Hochberg, *Rev. Sci. Instrum.*, 1975, **46**, 388.
63. R. C. O'Driscoll and D. N. Pinder, *J. Phys. E*, 1980, **13**, 192.
64. H. C. Kelly, *J. Phys. A, Math. Gen.*, 1972, **5**, 104.
65. N. Ostrowsky, D. Sornette, P. Parker and E. R. Pike, *Opt. Acta*, 1981, **28**, 1059.
66. H. Lee, A. M. Jamieson and R. Simha, *Macromolecules*, 1979, **12**, 329.
67. D. W. Marquardt, *SIAM J. Soc. Ind. Appl. Maths.*, 1963, **11**, 431.
68. E. R. Pike, in ref. 21, p. 179.
69. J. G. McWhirter and E. R. Pike, *J. Phys. A: Math. Gen.*, 1978, **11**, 1729.
70. S. W. Provencher, *Makromol. Chem.*, 1979, **180**, 201.
71. J. G. McWhirter, *Opt. Acta*, 1980, **27**, 83.
72. D. Sornette and N. Ostrowsky, in ref. 21, 1981, p. 755.
73. S. W. Provencher, *Comp. Phys. Commun.*, 1982, **27**, 213.
74. S. W. Provencher, *Comp. Phys. Commun.*, 1982, **27**, 229.
75. S. W. Provencher, in ref. 26, p. 322.
76. M. Bertero and E. R. Pike, in ref. 26, p. 298.
77. G. R. Danovich and I. N. Serdyuk, in ref. 26, p. 315.
78. N. Ostrowsky and D. Sornette, in ref. 26, p. 286.
79. E. R. Pike, D. Watson and F. McNeil Watson, in ref. 24, p. 107.
80. B. Chu, in ref. 25, p. 53.
81. M. Bertero, P. Brianzi and E. R. Pike, *Inverse Probl. I*, 1985, **1**.
82. M. Bertero, P. Brianzi and E. R. Pike, *Proc. R. Soc. London, Ser. A*, 1982, **383**, 15.
83. M. Bertero, P. Brianzi, E. R. Pike, G. de Villiers, K. H. Lan and N. Ostrowsky, *J. Chem. Phys.*, 1985, **82**, 1551.

84. M. Onelin and J. R. Ford, *J. Phys. Chem.*, 1984, **88**, 6566.
85. D. E. Koppel, *J. Chem. Phys.*, 1972, **57**, 4814.
86. S. P. Lee and B. Chu, *Appl. Phys. Lett.*, 1974, **24**, 575.
87. K. Pearson, *Philos. Trans. R. Soc.*, 1894, **185**, 71.
88. B. Chu, E. Gulari and E. Gulari, *Phys. Sci.*, 1979, **19**, 476.
89. E. Gulari, Y. Tsunashimo and B. Chu, *J. Chem. Phys.*, 1979, **70**, 3965.
90. G. C. Fletcher and D. J. Ramsey, *Opt. Acta*, 1983, **30**, 1183.
91. K. Zimmermann, M. Delaye and P. Licinio, *J. Chem. Phys.*, 1985, **82**, 2228.
92. P. J. Nash and T. A. King, *J. Chem. Soc., Faraday Trans. 2*, 1983; **79**, 989.
93. P. J. Nash and T. A. King, *J. Chem. Soc., Faraday Trans. 2*, 1985, **81**, 881.
94. P. J. Nash and T. A. King, *J. Chem. Soc., Faraday Trans. 2*, 1985, **81**, 897.
95. P. J. Nash and T. A. King, *J. Chem. Soc., Faraday Trans. 2*, 1985, **81**, 913.
96. J. R. Ford and B. Chu, in ref. 26, p. 303.
97. B. Chu, J. R. Ford and H. S. Dhadwal, *Methods Enzymol.*, 1985, **117**, 256.
98. R. G. W. Brown and A. P. Jackson, *J. Phys. E*, 1987, **20**, 1312, 1503.
99. R. F. Bonner, T. R. Clem, P. D. Bowen and R. C. Bowman, in ref. 21, p. 685.
100. R. J. Gush and T. A. King, *Med. Biol. Eng. Comp.*, 1987, **25**, 391.
101. H. E. Lessing, in ref. 18, p. 526.
102. H. Geerts, in ref. 25, p. 143.
103. P. Kask, T. Kändler, P. Piksarv, M. Pooga and E. Lippmon, in ref. 26, p. 393.
104. E. L. Elson, *Ann. Rev. Phys. Chem.*, 1985, **36**, 379.
105. J. Schneider, J. Ricka and T. Binkert, *Rev. Sci. Instrum.*, 1988, **59**, 588.
106. D. R. Bauer, B. Hudson and R. Pecora, *J. Chem. Phys.*, 1975, **63**, 588.
107. R. Chiarello and L. Reinisch, *J. Chem. Phys.*, 1988, **88**, 1253.
108. R. P. C. Johnson, in ref. 25, p. 147.
109. J. G. H. Joosten and H. M. Fijnaut, in ref. 20, p. 157.
110. K. H. Lan, N. Ostrowsky and D. Sornette, *Phys. Rev. Lett.*, 1986, **57**, 17.
111. F. Rondelez, in ref. 20, p. 243.
112. D. S. Cannell, *Rev. Sci. Instrum.*, 1975, **46**, 706.
113. W. G. Griffin, M. C. A. Griffin and F. Boué, *Macromolecules*, 1987, **20**, 2187.
114. W. G. Griffin and P. N. Pusey, *Phys. Rev. Lett.*, 1979, **43**, 110.
115. A. J. Bennett and E. E. Uzgiris, *Phys. Rev. A*, 1973, **8**, 2662.
116. K. S. Schmitz, *Chem. Phys. Lett.*, 1976, **42**, 137.
117. K. S. Schmitz, *Chem. Phys. Lett.*, 1979, **63**, 259.
118. T. Fujikado, R. Hayakawa and Y. Wada, *Biopolymers*, 1979, **18**, 2303.
119. H. Z. Cummins and P. N. Pusey, in ref. 18, p. 164.
120. B. Chu, in ref. 21, p. 231.
121. W. Hess, in ref. 20, p. 31.
122. W. Burchard, *Adv. Polym. Sci.*, 1983, **48**, 1.
123. T. A. King, A. Knox, W. J. Lee and J. D. G. McAdam, *Polymer*, 1973, **14**, 151.
124. T. A. King, A. Knox and J. D. G. McAdam, *Polymer*, 1973, **14**, 293.
125. M. Adam and M. Delsanti, *J. Phys. (Orsay, Fr.)*, 1976, **37**, 1045.
126. E. Gulari, E. Gulari, Y. Tsunashima and B. Chu, *Polymer*, 1979, **20**, 347.
127. B. Chu and T. Nose, *Macromolecules*, 1979, **12**, 590.
128. B. Chu and T. Nose, *Macromolecules*, 1979, **12**, 599.
129. M. Adam and M. Delsanti, *J. Phys., (Orsay, Fr.)*, 1980, **41**, 713.
130. J. Roots and B. Nystrom, *Macromolecules*, 1980, **13**, 1595.
131. J. E. Martin, *Macromolecules*, 1984, **17**, 1279.
132. K. Huber, S. Bantle, P. Lutz and W. Burchard, *Macromolecules*, 1985, **18**, 1461.
133. A. Barooah and S. H. Chen, *J. Polym. Sci., Polym. Phys. Ed.*, 1985, **23**, 2451.
134. A. Barooah, C. K. Sun and S. H. Chen, *J. Polym. Sci., Polym. Phys. Ed.*, 1986, **24**, 817.
135. K. Venkataswamy, A. M. Jamieson and R. G. Petschek, *Macromolecules*, 1986, **19**, 124.
136. M. Adam and M. Delsanti, *Macromolecules*, 1977, **10**, 1229.
137. Y. F. Maa and S. H. Chen, *Macromolecules*, 1987, **20**, 138.
138. P. M. Cotts, R. D. Miller, P. T. Trefonas III, R. West and G. N. Fickes, *Macromolecules*, 1987, **20**, 1046.
139. R. J. Goldberg, *J. Phys. Chem.*, 1953, **57**, 194.
140. P. N. Pusey, in 'Industrial Polymers: Characterization by Molecular Weight', ed. J. H. S. Green and R. Dietz, Transcripta Books, London, 1973, p. 26.
141. J. C. Selser and Y. Yeh, *Biophys. J.*, 1976, **16**, 847.
142. T. A. King and M. F. Treadaway, *J. Chem. Soc., Faraday Trans. 2*, 1977, **73**, 1616.
143. E. Gulari, E. Gulari, Y. Tsunashima and B. Chu, *J. Chem. Phys.*, 1979, **70**, 3965.
144. J. C. Selser, *Macromolecules*, 1979, **12**, 909.
145. B. Chu and E. Gulari, *Macromolecules*, 1979, **12**, 445.
146. E. Gulari, E. Gulari, Y. Tsunashima and B. Chu, *J. Chem. Phys.*, 1979, **70**, 3965.
147. E. R. Pike, in ref. 21, p. 179.
148. A. Z. Akcasu, *Macromolecules*, 1982, **15**, 1321.
149. P. J. Nash and T. A. King, *J. Soc. Photo-Opt. Instrum. Eng.*, 1983, **369**, 622.
150. G. Allen, P. Vasudevan, E. Y. Hawkins and T. A. King, *J. Chem. Soc., Faraday Trans. 2*, 1977, **73**, 449.
151. K. Ohbayashi, M. Minoda and H. Utiyama, in ref. 21, p. 749.
152. G. V. Laivins and D. G. Gray, *Macromolecules*, 1985, **18**, 1746.
153. S. Luzzati, M. Adam and M. Delsanti, *Polymer*, 1986, **27**, 834.
154. J. W. Pope and B. Chu, *Macromolecules*, 1984, **17**, 2633.
155. M. Naoki, I. H. Park, S. L. Wunder and B. Chu, *J. Polym. Sci., Polym. Phys. Ed.*, 1985, **23**, 2567.

156. B. Chu, Q. Ying, C. Wu, J. R. Ford and H. S. Dhadwal, *Polymer*, 1985, **26**, 1408.
157. Q.-C. Ying, B. Chu, R. Qian, J. Bao, J. Zhang and C. Xu, *Polymer*, 1985, **26**, 1401.
158. B. Chu, Q.-C. Ying, C. Wu, J. R. Ford and H. S. Dhadwal, *Polymer*, 1985, **26**, 1408.
159. B. Chu and C. Wu, *Macromolecules*, 1987, **20**, 93.
160. C. Wu, W. Buck and B. Chu, *Macromolecules*, 1987, **20**, 98.
161. B. Chu, C. Wu and J. Zuo, *Macromolecules*, 1987, **20**, 700.
162. T. A. King, A. Knox and J. D. G. McAdam, *Biopolymers*, 1973, **12**, 1917.
163. D. R. Bauer, J. I. Brauman and R. Pecora, *Ann. Rev. Phys. Chem*, 1976, **27**, 443.
164. D. R. Bauer, S. J. Opella, D. J. Nelson and R. Pecora, *J. Am. Chem. Soc.*, 1975, **97**, 258.
165. S. R. Aragon, *Macromolecules*, 1987, **20**, 370.
166. F. Brochard and P. G. de Gennes, *Macromolecules*, 1977, **10**, 1157.
167. M. Benmouna and A. Z. Akcasu, *Macromolecules*, 1978, **11**, 1187.
168. J. des Cloizeaux, *J. Phys. Lett., (Orsay, Fr.)*, 1978, **39**, 151.
169. G. Weill and J. des Cloizeaux, *J. Phys. Lett., (Orsay, Fr.)*, 1979, **40**, 99.
170. P. G. de Gennes, 'Scaling Concepts in Polymer Physics', Cornell University Press, Ithaca, NY, 1979.
171. A. Z. Akcasu and C. C. Han, *Macromolecules*, 1979, **12**, 276.
172. A. Z. Akcasu, M. Benmouna and C. C. Han, *Polymer*, 1980, **21**, 866.
173. J. Francois, T. Schwartz and G. Weill, *Macromolecules*, 1980, **13**, 564.
174. R. Ullman, *Macromolecules*, 1981, **14**, 746.
175. C. C. Han and A. Z. Akcasu, *Polymer*, 1981, **22**, 1019.
176. P. Vidakovic and F. Rondelez, *Macromolecules*, 1983, **16**, 253.
177. D. W. Schaefer and C. C. Han, in ref. 30, p. 181.
178. A. Z. Akcasu and C. M. Gutteman, *Macromolecules*, 1985, **18**, 938.
179. J. D. G. McAdam and T. A. King, *Chem. Phys.*, 1974, **6**, 109.
180. J. M. Schurr, *Q. Rev. Biophys.*, 1976, **9**, 109.
181. T. A. King and M. F. Treadaway, *J. Chem. Soc., Faraday Trans. 2*, 1976, **72**, 1473.
182. M. Adam and M. Delsanti, *J. Phys. Lett., (Orsay, Fr.)*, 1977, **38**, 271.
183. W. Burchard, *Polymer*, 1979, **20**, 577.
184. R. Pecora, in ref. 21, p. 161.
185. C. C. Han and A. Z. Akcasu, *Macromolecules*, 1981, **14**, 1080.
186. Y. Tsunashima, N. Nemoto and M. Kurata, *Macromolecules*, 1983, **16**, 584.
187. M. Eisele and W. Burchard, *Macromolecules*, 1984, **17**, 1636.
188. P. J. Nash and T. A. King, *Polymer*, 1985, **26**, 1003.
189. J. T. Koberstein, C. Picot and H. Benoit, *Polymer*, 1985, **26**, 673.
190. P. G. de Gennes and E. Dubois-Violette, *Physics*, 1967, **3**, 37, 181.
191. M. Daoud and G. Jannink, *J. Phys. (Orsay, Fr.)*, 1976, **37**, 973.
192. A. Perico, P. Piaggio and C. Cuniberti, *J. Chem. Phys.*, 1975, **62**, 4911.
193. A. Perico, P. Piaggio and C. Cuniberti, *J. Chem. Phys.*, 1975, **62**, 2690.
194. R. Kapral, D. Ng and S. G. Whittington, *J. Chem. Phys.*, 1976, **64**, 539.
195. S. R. Aragon and R. Pecora, *J. Chem. Phys.*, 1977, **66**, 2506.
196. M. Bixon and R. Zwanzig, *J. Chem. Phys.*, 1978, **68**, 1890.
197. P. G. de Gennes, *Nature (London)*, 1979, **282**, 367.
198. M. Benmouna and A. Z. Akcasu, *Macromolecules*, 1980, **13**, 409.
199. W. Hess, W. Jilge and R. Klein, *J. Polym. Sci., Polym. Phys. Ed.*, 1981, **19**, 849.
200. P. G. de Gennes, *Macromolecules*, 1976, **9**, 587.
201. P. G. de Gennes, *Macromolecules*, 1976, **9**, 594.
202. A. Z. Akcasu and M. Benmouna, *Macramolecules*, 1978, **11**, 1193.
203. M. Daoud and G. Jannink, *J. Phys. Lett., (Orsay, Fr.)*, 1980, **41**, 217.
204. P. Mathiez, C. Mouttet and G. Weisbuch, *J. Phys. (Orsay, Fr.)*, 1980, **41**, 519.
205. E. Geissler and A. M. Hecht, *J. Chem. Phys.*, 1976, **65**, 103.
206. D. Bailey, T. A. King and D. N. Pinder, *Chem. Phys.*, 1976, **12**, 161.
207. J. P. Munch, S. Candau, J. Herz and G. Hild, *J. Phys. (Paris)*, 1977, **38**, 971.
208. M. Adam and M. Delsanti, *Macromolecules*, 1977, **10**, 1229.
209. J. Roots and B. Nystrom, *Macromolecules*, 1980, **13**, 1595.
210. J. P. Munch, J. Herz, S. Boileau and S. Candau, *Macromolecules*, 1981, **14**, 1370.
211. S. J. Candau, I. Butler and T. A. King, *Polymer*, 1983, **24**, 1601.
212. E. J. Amis and C. C. Han, *Polymer*, 1982, **23**, 1403.
213. E. J. Amis, C. C. Han and Y. Matsushita, *Polymer*, 1984, **25**, 650.
214. J. E. Martin, *Macromolecules*, 1986, **19**, 1278.
215. R. Borsali, M. Duval, H. Benoit and M. Benmouna, *Macromolecules*, 1987, **20**, 1112.
216. W. Brown, *Macromolecules*, 1985, **18**, 1713.
217. W. Brown, *Macromolecules*, 1986, **19**, 1083.
218. W. Brown and R. Johnsen, *Macromolecules*, 1986, **19**, 2002.
219. T. Tanaka, L. Hocker and G. Benedek, *J. Chem. Phys.*, 1973, **59**, 5151.
220. T. A. King, A. Knox and J. D. G. McAdam, *J. Polym. Sci., Polym. Phys. Ed.*, 1974, **44**, 195.
221. J. P. Munch, S. Candau and G. Hild, *J. Polym. Sci., Polym. Phys. Ed.*, 1977, **15**, 11.
222. J. P. Munch, P. Lemarechal, S. Candau and J. Herz, *J. Phys., (Orsay, Fr.)*, 1977, **38**, 1499.
223. M. Adam and M. Delsanti, *Macromolecules*, 1977, **10**, 1229.
224. A. M. Hecht and E. Geissler, *J. Phys. (Paris)*, 1978, **39**, 631.
225. T. Tanaka and D. J. Fillmore, *J. Chem. Phys.*, 1979, **70**, 1214.
226. R. Nossal, in ref. 21, p. 301.
227. S. Candau, J. Bastide and M. Delsanti, *Adv. Polym. Sci.*, 1982, **44**, 27.
228. H. M. Tan, A. Hillner, E. Moet and E. Baer, *Macromolecules*, 1983, **16**, 28.

229. T. Tanaka, in ref. 30, p. 347.
230. J. Francois, Y. S. Gau and J. M. Guenet, *Macromolecules*, 1986, **19**, 2755.
231. S. J. Candau, Y. Dormoy, P. H. Mutin, F. Debeauvais and J. M. Guenet, *Polymer*, 1987, **28**, 1334.
232. R. D. Mountain, *J. Res. Natl. Bur. Stand., Sect. A*, 1966, **70**, 207.
233. S. M. Rytov, *Sov. Phys. -JETP, (Engl. Transl.)*, 1970, **31**, 1163.
234. C. H. Wang and E. W. Fischer, *J. Chem. Phys.*, 1985, **82**, 632.
235. G. D. Patterson, *Adv. Polym. Sci.*, 1983, **48**, 125.
236. G. D. Patterson, in ref. 30, p. 245.
237. G. D. Patterson and A. Munoz-Rojas, *Ann. Rev. Phys. Chem.*, 1987, **38**, 191.
238. C. H. Wang, G. Fytas and E. W. Fischer, *J. Chem. Phys.*, 1985, **82**, 4332.
239. G. Fytas, C. H. Wang, E. W. Fischer and K. Mehler, *J. Polym. Sci., Polym. Phys. Ed.*, 1986, **24**, 1859.
240. G. Meier, J.-U. Hagenah, C. H. Wang, G. Fytas and E. W. Fischer, *Polymer*, 1987, **28**, 1640.
241. G. Patterson, J. R. Stevens and C. P. Lindsey, *J. Macromol. Sci., Phys.*, 1980, **18**, 641.
242. G. D. Patterson, P. J. Carroll and J. R. Stevens, *J. Polym. Sci., Polym. Phys. Ed.*, 1983, **21**, 613.
243. J. D. Litster, in ref. 13, p. 475.
244. Orsay group, *Phys. Rev. Lett.*, 1969, **22**, 1361.
245. D. C. Van Eck and W. Westera, *Mol. Cryst. Liq. Cryst.*, 1977, **38**, 319.
246. J. P. Van der Meulen and R. J. J. Zijlstra, *J. Phys. (Orsay, Fr.)*, 1982, **43**, 411.
247. Orsay group, *J. Chem. Phys.*, 1969, **51**, 816.
248. J. L. Martinand and G. Durand, *Solid State Commun.*, 1972, **10**, 815.
249. M. S. Sefton, A. R. Bowdler and H. J. Coles, *Mol. Cryst. Liq. Cryst.*, 1985, **1**, 151.
250. F. M. Leslie and C. M. Waters, *Mol. Cryst. Liq. Cryst.*, 1985, **123**, 101.
251. M. S. Sefton and H. J. Coles, *Polymer*, 1985, **26**, 1319.
252. M. J. Garvey, Th. F. Tadros and B. Vincent, *J. Colloid Interface Sci.*, 1976, **55**, 440.
253. Th. Van den Boomgaard, T. A. King, Th. F. Tadros, H. Tang and B. Vincent, *J. Colloid Interface Sci.*, 1978, **66**, 68.
254. K. G. Barnett, T. L. Crowley, T. Cosgrove, B. Vincent, A. Burgess, T. A. King, J. D. Turner, J. Schelten and Th. F. Tadros, *Polymer*, 1981, **22**, 283.
255. T. Kato, N. Nakamura, M. Kawaguehi and A. Takahashi, *Polym. J.*, 1981, **13**, 1037.
256. M. A. Cohen Stuart, F. H. W. H. Waajen, T. Cosgrove, B. Vincent and T. L. Crowley, *Macromolecules*, 1984, **17**, 1825.
257. T. Cosgrove, T. L. Crowley, M. A. Cohen Stuart and B. Vincent, *ACS Symp. Ser.*, 1984, **240**, 147.
258. J. D. Turner and T. A. King, 1988, in press.
259. C. Williams, F. Brochard and H. Frisch, *Ann. Rev. Phys. Chem.*, 1981, **32**, 433.
260. P. Vidakovic and F. Rondelez, *Macromolecules*, 1984, **17**, 418.
261. S. T. Sun, I. Nishio, G. Swislow and T. Tanaka, *J. Chem. Phys.*, 1980, **73**, 5971.
262. R. J. Gush and T. A. King, 1988, in press.
263. T. Tanaka, in ref. 30, p. 347.
264. F. Grüner, W. P. Lehmann, H. Fahlbusch and R. Weber, in ref. 26, p. 348.
265. J. P. Meullenet, A. Schmitt and S. Candau, *Chem. Phys. Lett.*, 1978, **55**, 523.
266. C. W. Lantman, W. J. MacKnight, D. G. Peitler, S. K. Sinha and R. D. Lundberg, *Macromolecules*, 1987, **20**, 1096.
267. P. N. Pusey, in ref. 26, p. 335.
268. P. N. Pusey and R. J. A. Tough, in ref. 30, p. 85.
269. J. M. Schurr and K. S. Schmitz, *Ann. Rev. Phys. Chem.*, 1986, **37**, 271.

Subject Index

937